INFERENCE:

Confidence interval for parameter = *statistic* ± *critical value* × *SD*(*statistic*)

Test statistic $= \dfrac{statistic - parameter}{SD(statistic)}$ Use the *SE* whenever the *SD* is unknown.

Parameter	Statistic	*SD*(statistic)	*SE*(statistic)
p	$\hat{p}$	$\sqrt{\frac{pq}{n}}$	$\sqrt{\frac{\hat{p}\hat{q}}{n}}$
$p_1 - p_2$	$\hat{p}_1 - \hat{p}_2$	$\sqrt{\frac{p_1q_1}{n_1} + \frac{p_2q_2}{n_2}}$	$\sqrt{\frac{\hat{p}_1\hat{q}_1}{n_1} + \frac{\hat{p}_2\hat{q}_2}{n_2}}$
μ	$\bar{y}$	$\frac{\sigma}{\sqrt{n}}$	$\frac{s}{\sqrt{n}}$
$\mu_1 - \mu_2$	$\bar{y}_1 - \bar{y}_2$	$\sqrt{\frac{\sigma_1^2}{n_1} + \frac{\sigma_2^2}{n_2}}$	$\sqrt{\frac{s_1^2}{n_1} + \frac{s_2^2}{n_2}}$
μ_d	$\bar{d}$	$\frac{\sigma_d}{\sqrt{n}}$	$\frac{s_d}{\sqrt{n}}$
σ_ε	$s_e = \sqrt{\frac{\sum(y - \hat{y})^2}{n - 2}}$		
β_1	b_1		$\frac{s_e}{s_x\sqrt{n - 1}}$
μ_ν	$\hat{y}_\nu$		$\sqrt{SE^2(b_1) \times (x_\nu - \bar{x})^2 + \frac{s_e^2}{n}}$
y_ν	$\hat{y}_\nu$		$\sqrt{SE^2(b_1) \times (x_\nu - \bar{x})^2 + \frac{s_e^2}{n} + s_e^2}$

Pooling: For testing difference between proportions: $\hat{p}_{pooled} = \dfrac{y_1 + y_2}{n_1 + n_2}$

For testing difference between means: $s_p = \sqrt{\dfrac{(n_1 - 1)s_1^2 + (n_2 - 1)s_2^2}{n_1 + n_2 - 2}}$

Substitute these pooled estimates in the respective SE formulas for both groups when assumptions and conditions are met.

Chi-square: $\chi^2 = \sum \dfrac{(Obs - Exp)^2}{Exp}$

SECOND CANADIAN EDITION

STATS

DATA AND MODELS

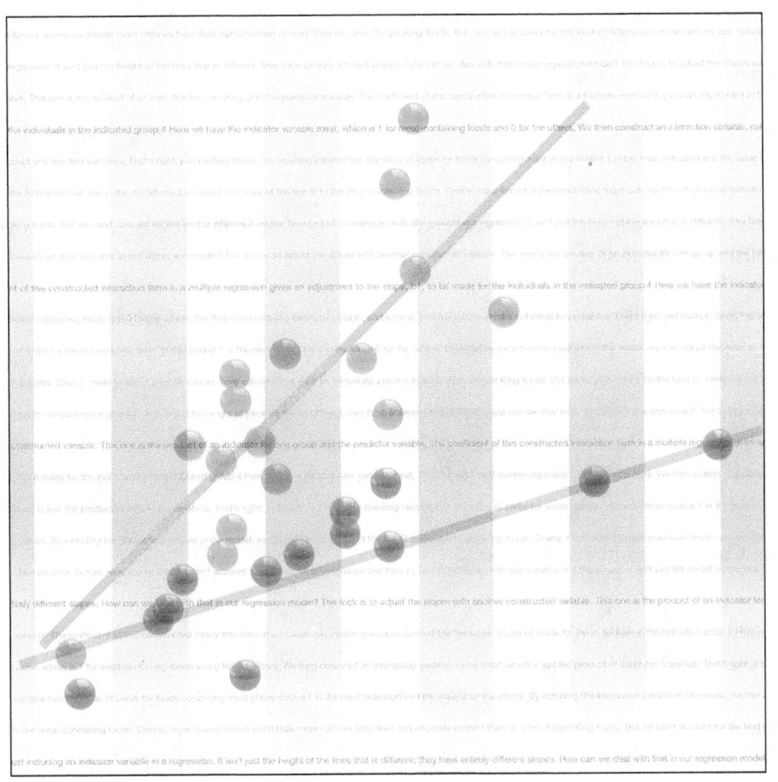

RICHARD D. DE VEAUX
WILLIAMS COLLEGE

PAUL F. VELLEMAN
CORNELL UNIVERSITY

DAVID E. BOCK
CORNELL UNIVERSITY

AUGUSTIN M. VUKOV
UNIVERSITY OF TORONTO

AUGUSTINE C.M. WONG
YORK UNIVERSITY

With contributions from Craig Burkett, University of Toronto

PEARSON

Toronto

Acquisitions Editor: *David S. Le Gallais*
Marketing Manager: *Michelle Bish*
Program Manager: *Patricia Ciardullo*
Project Manager: *Marissa Lok*
Senior Developmental Editor: *John Polanszky*
Media Editor: *Ben Zaporozan*
Media Producer: *Kelli Cadet*
Production Services: *Kayci Wyatt, Electronic Publishing Services*
Permissions Project Manager: *Erica Mojzes*
Photo Permissions Research: *Divya Narayanan, Lumina Datamatics*
Text Permissions Research: *Haydee Hidalgo, Electronic Publishing Services*
Art Director: *Zena Denchik*
Cover Designer: *Anthony Leung*
Interior Designer: *Electronic Publishing Services*
Cover Image: *Based on figure 28.5 from page 848*

Credits and acknowledgments for material borrowed from other sources and reproduced, with permission, in this textbook appear on the appropriate page within the text.

Library and Archives Canada Cataloguing in Publication

De Veaux, Richard D., author
Stats : data and models/Richard D. De Veaux, Williams College, Paul F. Velleman, Cornell University, David E. Bock, Cornell University, Augustin M. Vukov, University of Toronto, Augustine C.M. Wong, York University.—Second Canadian edition.
Includes index.
Revision of: Stats : data and models / Richard D. De Veaux . . . [et al.].— Canadian ed.—Toronto : Pearson Canada, © 2012.
ISBN 978-0-321-82842-2 (bound)
1. Statistics—Textbooks. 2. Mathematical statistics—Textbooks. I. Velleman, Paul F., 1949–, author II. Bock, David E., author III. Vukov, Augustin M., author IV. Wong, Augustine C. M., author V. Title.

QA276.12.S74 2014 519.5 C2014-905255-3

1 2 3 4 5 6 7 8 9 10

ISBN 978-0-321-8284-2

*To Sylvia, who has helped me in more ways than she'll ever know,
and to Nicholas, Scyrine, Frederick, and Alexandra,
who make me so proud in everything that they are and do*

—Dick

*To my sons, David and Zev, from whom I've learned so much,
and to my wife, Sue, for taking a chance on me*

—Paul

*To Greg and Becca, great fun as kids and great friends as adults,
and especially to my wife and best friend, Joanna, for her
understanding, encouragement, and love*

—Dave

*To my adopted country, Canada, my vibrantly multicultural
city of Toronto, and the University of Toronto without which this
dedication would never have happened.*

—Gus

*To anyone I may ever teach, that you may gain as much joy
from learning Statistics as I have.*

—Craig

Meet the Authors

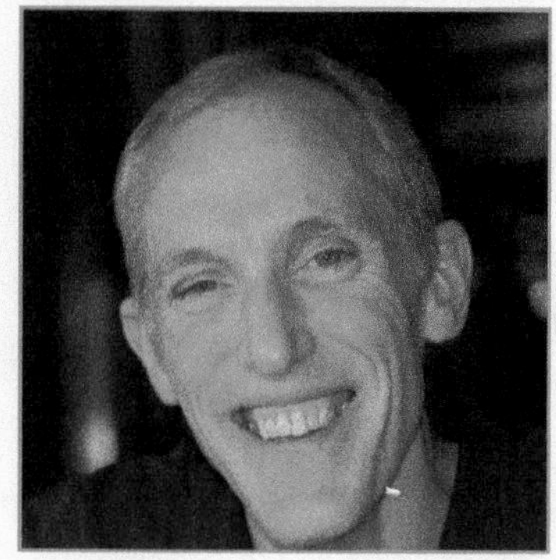

Richard D. De Veaux is an internationally known educator and consultant. He has taught at the Wharton School and the Princeton University School of Engineering, where he won a "Lifetime Award for Dedication and Excellence in Teaching." Since 1994, he has been Professor of Statistics at Williams College. Dick has won both the Wilcoxon and Shewell awards from the American Society for Quality. He is a fellow of the American Statistical Association (ASA). In 2008, he was named Statistician of the Year by the Boston Chapter of the ASA. Dick is also well known in industry, where for more than 25 years he has consulted for such Fortune 500 companies as American Express, Hewlett-Packard, Alcoa, DuPont, Pillsbury, General Electric, and Chemical Bank. Because he consulted with Mickey Hart on his book *Planet Drum*, he has also sometimes been called the "Official Statistician for the Grateful Dead." His real-world experiences and anecdotes illustrate many of this book's chapters.

Dick holds degrees from Princeton University in Civil Engineering (B.S.E.) and Mathematics (A.B.) and from Stanford University in Dance Education (M.A.) and Statistics (Ph.D.), where he studied dance with Inga Weiss and Statistics with Persi Diaconis. His research focuses on the analysis of large data sets and data mining in science and industry.

In his spare time, he is an avid cyclist and swimmer. He also is the founder of the "Diminished Faculty," an a cappella Doo-Wop quartet at Williams College and sings bass in the college concert choir. Dick is the father of four children.

Paul F. Velleman has an international reputation for innovative Statistics education. He is the author and designer of the multimedia Statistics program *ActivStats*, for which he was awarded the EDUCOM Medal for innovative uses of computers in teaching statistics, and the ICTCM Award for Innovation in Using Technology in College Mathematics. He also developed the award-winning statistics program, *Data Desk*, and the Internet site Data and Story Library (DASL) (lib.stat.cmu.edu/DASL/), which provides data sets for teaching Statistics. Paul's understanding of using and teaching with technology informs much of this book's approach.

Paul has taught Statistics at Cornell University since 1975. He holds an A.B. from Dartmouth College in Mathematics and Social Science, and M.S. and Ph.D. degrees in Statistics from Princeton University, where he studied with John Tukey. His research often deals with statistical graphics and data analysis methods. Paul co-authored (with David Hoaglin) *ABCs of Exploratory Data Analysis.* Paul is a Fellow of the American Statistical Association and of the American Association for the Advancement of Science. Paul is the father of two boys.

David E. Bock taught mathematics at Ithaca High School for 35 years. He has taught Statistics at Ithaca High School, Tompkins-Cortland Community College, Ithaca College, and Cornell University. Dave has won numerous teaching awards, including the MAA's Edyth May Sliffe Award for Distinguished High School Mathematics Teaching (twice), Cornell University's Outstanding Educator Award (three times), and has been a finalist for New York State Teacher of the Year.

Dave holds degrees from the University at Albany in Mathematics (B.A.) and Statistics/Education (M.S.). Dave has been a reader and table leader for the AP Statistics exam, serves as a Statistics consultant to the College Board, and leads workshops and institutes for AP Statistics teachers. He has recently served as K–12 Education and Outreach Coordinator and a senior lecturer for the Mathematics Department at Cornell University. His understanding of how students learn informs much of this book's approach.

Dave and his wife relax by biking or hiking, spending much of their free time in Canada, the Rockies, or the Blue Ridge Mountains. They have a son, a daughter, and four grandchildren.

Augustin M. Vukov enjoys his daily walk to work at the University of Toronto, where he has studied and taught for several decades. During that time, one of his major responsibilities has been lecturing and coordinating for the large multi-section introductory service course—The Practice of Statistics—which serves the needs of nearly 1000 students per year, who come from a great variety of disciplines. Having taught so many students their required Stats course, he was not surprised to hear his new family doctor, his new ophthalmologist and a prospective new dentist say "Your name looks familiar. Ah yes, I used to sit in your Statistics lecture!". When not equipping future doctors and dentists with important tools, he enjoys travel and music, both preferably Brazilian, and summer softball, though he no longer gets to chase fly balls and smack home runs for the department, since the team has folded up shop. Oh well, more time to write!

Augustine C.M. Wong is a professor of Statistics at York University. He completed his Ph.D. at the University of Toronto in 1990 and taught at the University of Waterloo and University of Alberta before coming to York in 1993. His research interests include asymptotic inference, computational methods in statistics, and likelihood-based methods. He is an author or co-author of over 60 research articles and 2 book chapters. At York, he teaches various statistics courses at both the undergraduate and graduate levels.

Craig Burkett is a former aerospace engineer and high school teacher who now teaches Statistics at the University of Toronto. He also works as a statistical consultant, providing advice and expertise to anyone who will listen to him. Having worked with engineers, doctors, marketing firms, airlines, educators, pharmaceutical companies, and a variety of other researchers, Craig really enjoys the variety statistical consulting provides. And the money.

When not crunching numbers or preparing classes, Craig can be found riding his bike, playing rugby or classical piano (though not at the same time) or trying to renovate his condo by himself. When not doing any of those things, he can be found trying to undo the damage he has done to his condo.

Contents

Part III Gathering Data

Part IV Randomness and Probability

Part V From the Data at Hand to the World at Large

*Optional section.

*Optional section.

Part VII Inference When Variables Are Related

Part VIII Inference By Other Means

*Optional section.

Preface

We are often asked why we write Statistics texts. After all, it takes a lot of work to find new and better examples, to keep datasets current, and to make a book an enjoyable and effective learning tool. So we thought we'd address that question first.

We do it because it's fun.

Of course, we care about teaching students to think statistically; we are teachers and professional statisticians. But Statistics can be a particularly challenging subject to teach. The student encounters many new concepts, many new methods, and many new terms. And we want to change the way students think about the world. From the start, our challenge has been to write a book that students would read, learn from, and enjoy. And we return to that goal with each new edition.

The book you hold is shorter, quicker to the point, and, we hope, even more readable than the first Canadian edition. Of course, we've kept our conversational style and background anecdotes.[1] But we've tightened discussions and adjusted the order of some topics to move the story we tell about Statistics even more quickly to interesting real-world questions. We've focused even more on statistical thinking.

More and more instructors are using examples from Statistics to provide intuitive examples of how a little bit of math can help us say a lot about the world. So students expect Statistics to be about real-world insights. This second Canadian edition of *Stats: Data and Models* keeps your students engaged and interested because we show Statistics in action right from the start. Students will be solving problems of the kind they're likely to encounter in real life sooner. In Chapter 4, they will be comparing groups and in Chapter 6, they'll see relationships between two quantitative variables—and, of course, always with real, modern data.

There are few things more fun and useful for students than being empowered to discover something new about the world. And few things more fun for authors than helping students make those discoveries.

So, What's New in the Second Canadian Edition?

We've rewritten sections throughout the book to make them clearer and more interesting, along with many new up-to-the-minute motivating examples.

We've added some new features, each with the goal of making it even easier for students to put the concepts of Statistics together into a coherent whole.

1. New and improved pedagogical tools: A new section head list at the beginning of each chapter provides a road map. Section heads within each chapter are reorganized, numbered, and reworded to be clear and specific. Chapter study materials now include Learning Objectives as well as Terms. Students who understand the Objectives and know the Terms are well on their way to being ready for any tests.

2. Streamlined design: Our goal has always been an accessible text. This edition sports a new design that clarifies the purpose of each text element. Essential supporting material is clearly boxed and shaded, so students know where to focus their study efforts. Enriching—and often entertaining—side material is boxed, but not shaded.

3. Streamlined content: Our reorganization has shortened the book from 33 to 30 chapters. Each chapter is still a focused discussion, and most can be taught in one lesson. We've reduced time spent on secondary topics. The result is a more readable text.

[1] and our footnotes.

4. Content changes: Here's how we've reorganized or changed the content:

   a. Chapter 1 now gets down to business immediately rather than just providing an introduction to the book's features.

   b. The discussion of re-expression or transformation of data is no longer a separate chapter, but instead is dispersed throughout other chapters.

   c. The discussions of probability and random variables are tighter—and a chapter shorter—but also with a clearer introduction to continuous distributions. The geometric distribution has been moved to the exercises.

   d. We've added an optional section on Logistic Regression to the Inference for Regression chapter, an optional section "How ANOVA Works—the Gory Details" to our chapter on multi-factor ANOVA, and the test for a set of betas to the first chapter on multiple regression.

   e. We've removed the 10% Condition when inference is about means.

5. Exercises: We've updated most exercises that use real-world data, retired some that were getting old, and added new exercises. As in the previous edition, many of the exercises feature Canadian content and data with up-to-date references to Aboriginal peoples, sports, health care, education, the environment, and other social and political issues. And we continue to add exercises requiring computer simulations.

6. Major sections have new numbers to help with navigation and reading assignments.

7. On the Computer sections now include instructions for R and StatCrunch.

Our Approach

Statistics is practiced with technology. We think a modern statistics text should recognize that fact from the start. And so do our students. You won't find tedious calculations worked by hand. But you will find equation forms that favour intuition over calculation. You'll find extensive use of real data—even large data sets. And, most important, you'll find a focus on statistical thinking rather than calculation. The question that motivates each of our hundreds of examples is not "how do you *find* the answer?" but "how do you *think* about the answer?"

We've been guided in the choice and order of topics by several fundamental principles. First, we have tried to ensure that each new topic fits into the growing structure of understanding that we hope students will build. We know that learning a new subject like Statistics requires putting together many new ideas, and we have tried to make that process as natural as possible. That goal has led us to cover some topics in a different order than is found in other texts.

As one example, we introduce inference by looking at a confidence interval for a proportion rather than the more traditional approach that starts with the one-sample mean. Everyone has seen opinion polls. Most people understand that pollsters use a sample of voters to try to predict the preferences of the entire population and that the result is an estimate with a margin of error. Showing how to construct and interpret a confidence interval in this context introduces key concepts of inference in a familiar setting. We next examine hypothesis tests for a proportion. The mechanics use formulas and a test statistic that is now familiar, so it is easier to focus on the logic and structure of a formal hypothesis test. After hypothesis tests we examine inference about the difference between two proportions. We build on the understanding of confidence intervals and hypothesis tests as we introduce the added complexity of comparing two groups. When we then turn our attention to the more complicated topic of small-sample inference for means, the only new issue is the t-distribution. However, should you wish to discuss the z-test and confidence interval for means (known sigma or large n) earlier, in tandem with the z-procedures for proportions, optional sections are included in those chapters.

Textbooks can be defined more by what they choose not to cover than by what they do cover. We've structured this text so that each new topic fits into a student's growing understanding. Several topic orders can support this goal. While this text contains material sufficient for two semesters, it offers good choices for those teaching a one-semester course, for which we recommend coverage of Chapters 1–22, followed by (time permitting) one or more of Chapters 23, 24, 25, and 29. Some topics in Chapters 9–11 may be de-emphasized or omitted entirely depending on the needs of the particular audience. Likewise, the material on probability and random variables in chapters 12–14 is considerable and portions may be de-emphasized. Optional sections, indicated by asterisks, are indeed just that and may be omitted without harm.

GAISE Guidelines

The Guidelines for Assessment and Instruction in Statistics Education (GAISE) report adopted by the American Statistical Association urges that Statistics education should

1. emphasize Statistical literacy and develop Statistical thinking,
2. use real data,
3. stress conceptual understanding rather than mere knowledge of procedures,
4. foster active learning,
5. use technology for developing concepts and analyzing data, and
6. make assessment a part of the learning process.

We've designed our text and supporting materials to support this approach to the introductory course. We urge you to think about these guidelines with each class meeting.

Our Goal: Read This Book!

The best text in the world is of little value if students don't read it. Here are some of the ways we have made this edition even more approachable:

- ***Readability.*** This book doesn't read like other Statistics texts. Our style is both colloquial and informative, engaging students to actually read the book to see what it says. We've tightened the discussions and eliminated digressions.
- ***Humour.*** We know that humour is the best way to promote learning. You will find quips and wry comments throughout the narrative, in margin notes, and in footnotes.
- ***Informality.*** Our informal diction doesn't mean that we treat the subject matter lightly or informally. We try to be precise and, wherever possible, we offer deeper explanations and justifications than those found in most introductory texts.
- ***Focused lessons.*** The chapters are shorter than in most other texts so instructors and students can focus on one topic at a time.
- ***Consistency.*** We try to avoid the "do what we say, not what we do" trap. Having taught the importance of plotting data and checking assumptions and conditions, we model that behavior right through the rest of the book. (Check the exercises in Chapter 24. You'll see that we still require and demonstrate the plots and checks that were introduced in the early chapters.) This consistency helps reinforce these fundamental principles.
- ***The need to read.*** Students who plan just to skim the book may find our presentation frustrating. The important concepts, definitions, and sample solutions don't sit in little boxes. Statistics is a consistent story about how to understand the world when we have data. The story can't be told piecemeal. This is a book that needs to be read, so we've tried to make the reading experience enjoyable.

Mathematics

Mathematics can

1. provide a concise, clear statement of important concepts.
2. describe calculations to be performed with data.
3. embody proofs of fundamental results.

Of these, we emphasize the first. Mathematics can make discussions of Statistics concepts, probability, and inference clear and concise. We don't shy away from using math where it can clarify without intimidating. But we know that some students are discouraged by equations, so we always provide a verbal description and a numerical example as well. Some theorems about Statistics are quite interesting, and many are important. Often, though, their proofs are not enlightening to introductory Statistics students and can distract the audience from the concepts we want them to understand. So we avoid them here.

Nor do we slide in the opposite direction and concentrate on calculation. Although statistics calculations are generally straightforward, they are also usually tedious. And, more to the point, they are often unnecessary. Today, virtually all statistics are calculated with technology, so there is little need for students to spend time summing squared deviations by hand. We have selected the equations that do appear for their focus on illuminating concepts and methods. Although these equations may be the best way to understand the concepts, they may not be optimal for hand calculation. When that happens, we give an alternative formula, better suited for hand calculation, for those who find following the process a better way to learn about the result.

Technology and Data

To experience the real world of Statistics, use modern technology to explore real data sets.

Technology. We assume that you are using some form of technology—a statistics package, a calculator, a spreadsheet, or some combination of these—in your Statistics course. We also assume that you'll put little emphasis on calculating answers by hand, even though we often show how. However, this is not a technology-heavy book. The role of technology in this book is to get the calculations out of the way so we can focus on statistical thinking. We discuss generic computer output, but we don't adopt any particular statistics software. We do offer guidance to help students get started on eight common software platforms: Excel®, Minitab®, JMP®, SPSS®, TI-83/84 Plus graphing calculators, StatCrunch®, and R®. The **On the Computer** section at the end of most chapters is specific to the methods learned in that chapter.

Data. Because we use technology for computing, we don't limit ourselves to small, artificial data sets. You'll find some small data sets, but we also base examples and exercises on real data with a moderate number of cases—usually more than you would want to enter by hand into a program or calculator. Machine-readable versions of the data are included on the book's website, **www.pearsoncanada.ca/deveaux**.

Continuing Features

Enhancing Understanding

Where Are We Going? Each chapter starts with a paragraph that points out the kinds of questions students will learn how to answer in the chapter. A new chapter outline helps organize major topics for the students.

Each chapter ends with a **What Have We Learned?** summary, which includes new learning objectives and definitions of terms introduced in the chapter. Students can think of these as study guides.

In each chapter, our innovative **What Can Go Wrong?** sections highlight the most common errors that people make and the misconceptions they have about Statistics. One of our goals is to arm students with the tools to detect statistical errors and to offer practice in debunking misuses of statistics, whether intentional or not.

Margin and in-text boxed notes. Throughout each chapter, boxed margin and in-text notes enhance and enrich the text. Boxes with essential or practical information are screened. Conversational notes that enhance the text and entertain the reader are unscreened.

Math Boxes. In many chapters we present the mathematical underpinnings of the statistical methods and concepts. By setting these proofs, derivations, and justifications apart from the narrative, we allow the student to continue to follow the logical development of the topic at hand, yet also refer to the underlying mathematics for greater depth. Depending on the nature of the course, instructors can decide whether or not to assign these boxes to their students.

By Hand. Even though we encourage the use of technology to calculate statistical quantities, we realize the pedagogical benefits of occasionally doing a calculation by hand. The By Hand boxes break apart the calculation of many simpler formulas to help the student through the calculation of a worked example.

Reality Check. We regularly remind students that Statistics is about understanding the world with data. Results that make no sense are probably wrong, no matter how carefully we think we did the calculations. Mistakes are often easy to spot with a little thought, so we ask students to stop for a reality check before interpreting their result.

Notation Alert. Throughout this book, we emphasize the importance of clear communication, and proper notation is part of the vocabulary of Statistics. We've found that it helps students when we are clear about the letters and symbols statisticians use to mean very specific things, so we've included Notation Alerts whenever we introduce a special notation that students will see again.

Connections. Each chapter has a Connections feature to link key terms and concepts with previous discussions and to point out the continuing themes. Connections help students fit newly learned concepts into a growing understanding of Statistics.

Learning by Example

For Example. As we introduce each important concept, we provide a focused example applying it—usually with real up-to-the-minute data. Many For Examples carry the discussion through the chapter, picking up the story and moving it forward as students learn more about the topic.

Step-by-Step Examples: Think, Show, Tell. Step-by-Step examples repeat the mantra of Think, Show, and Tell in every chapter. These longer, worked examples guide students through the process of analyzing the problem with the general explanation on the left and the worked-out problem on the right. They emphasize the importance of thinking about a Statistics question (What do we know? What do we hope to learn? Are the assumptions and conditions satisfied?) and reporting our findings (the Tell step). The Show step contains the mechanics of calculating results and conveys our belief that it is only one part of the process. The result is a better understanding of the concept, not just number crunching.

Testing Understanding

Just Checking. Just Checking questions are quick checks throughout the chapter; most involve very little calculation. These questions encourage students to pause and think about what they've just read. The Just Checking answers are at the end of the exercise sets in each chapter so students can easily check themselves.

Exercises. Exercises have been updated with the most recent data. Many come from news stories; some from recent research articles. Whenever possible, the data are available on MyStatLab and the Companion Website for the book so students can explore them further.

Technology

***ActivStats* Pointers.** MyStatLab for this book includes *ActivStats*, so we've included occasional pointers to the *ActivStats* activities when they parallel discussions in the book. Many students choose to look at these first, before reading the chapter or attending a class on each subject.

Data Sources. Most of the data used in examples and exercises are from real-world sources, and whenever we can, we include references to the Internet data sources used, often in the form of URLs. The data we use are usually on MyStatLab and the Companion Website. If you seek the data—or an updated version of the data—on the Internet, we try to direct you to a good starting point.

On the Computer. In the real world, Statistics is practiced with computers. We prefer not to choose a particular Statistics program. Instead, at the end of most chapters, we summarize what students can find in the most common packages, often with annotated output. We then offer specific guidance for several of the most common packages (Excel®, JMP®, Minitab®, R®, SPSS®, StatCrunch®, and TI-83/84 Plus[2]) to help students get started with the software of their choice.

[2]For brevity, we will write TI-83/84 Plus for the TI-83 Plus and/or TI-84 Plus. Keystrokes and output remain the same for the TI-83 Plus and the TI-84 Plus, so instructions and examples serve for both calculators.

Supplements

For Instructors

Instructor resources are password protected and available for download via the Pearson online catalogue at **http://catalogue.pearsoned.ca.**

Instructor's Solutions Manual. This resource provides complete, detailed, worked-out solutions for all the exercises in the textbook and is available through the online catalogue in both PDF and Word formats.

TestGen and Test Item File. For your convenience, our testbank is available in two formats. TestGen is a computerized testbank containing a broad variety of multiple-choice, short answer, and more complex problem questions. Questions can be searched and identified by question type, level of difficulty, and skill type (computational or conceptual). Each question has been checked for accuracy and is available in the latest version of TestGen software. This software package allows instructors to custom design, save, and generate classroom tests. The test program permits instructors to edit, add, or delete questions from the test bank; edit existing graphics and create new ones; analyze test results; and organize a database of tests and student results. This software allows for greater flexibility and ease of use. It provides many options for organizing and displaying tests, along with search and sort features. The same questions can also be found in a Test Item File available in Word format. The TestGen testbank and Test Item File can be downloaded from the online catalogue.

PowerPoint® Presentations. These PowerPoint® lecture slides provide an outline to use in a lecture setting, presenting definitions, key concepts, and figures from the textbook. These lecture slides can be downloaded from the online catalogue.

Clicker Questions. Clicker questions provide in-the-moment opportunities to engage and to assess students' understanding during lecture. Clicker Questions in PowerPoint® format can be downloaded from the online catalogue.

Image Library. The Image Library provides access to many of the images, figures, and tables in the textbook and is available to instructors on the online catalogue.

Pearson Custom Library

For enrollments of at least 25 students, you can create your own textbook by choosing the chapters that best suit your own course needs. *To begin building your custom text, visit www.pearsoncustomlibrary.com.* You may also work with a dedicated Pearson Custom Editor to create your ideal text—publishing your own original content or mixing and matching Pearson content. *Contact your local Pearson representative to get started.*

For Students

Student Solutions Manual. This solutions manual provides complete worked-out solutions to all of the odd-numbered exercises in the book, expanding on the answers provided in the Appendix at the back of the text. (ISBN-13: 978-0-321-99162-1).

CourseSmart for Students. CourseSmart goes beyond traditional expectations—providing instant, online access to the textbooks and course materials you need at an average savings of 60%. With instant access from any computer and the ability to search your text, you'll find the content you need quickly, no matter where you are. And with online tools like highlighting and note-taking, you can save time and study efficiently. See all the benefits at **www.coursesmart.com/students.**

Companion Web Site. The Companion Web Site provides additional resources (and data files) for instructors and students: **www.pearsoncanada.ca/deveaux.**

Technology Resources

MyStatLab™ Online Course (access code required)

MyStatLab is a course management system that delivers **proven results** in helping individual students succeed.

- MyStatLab can be successfully implemented in any environment—lab-based, hybrid, fully online, traditional—and demonstrates the quantifiable difference that integrated usage has on student retention, subsequent success, and overall achievement.
- MyStatLab's comprehensive online gradebook automatically tracks students' results on tests, quizzes, homework, and in the study plan. Instructors can use the gradebook to provide positive feedback or intervene if students have trouble. Gradebook data can be easily exported to a variety of spreadsheet programs, such as Microsoft Excel. You can determine which points of data you want to export, and then analyze the results to determine success.

MyStatLab provides **engaging experiences** that personalize, stimulate, and measure learning for each student. In addition to the following resources, each course includes a full interactive online version of the accompanying textbook.

- **Tutorial Exercises with Multimedia Learning Aids:** The homework and practice exercises in MyStatLab align with the exercises in the textbook, and they regenerate algorithmically to give students unlimited opportunity for practice and mastery. Exercises offer immediate helpful feedback, guided solutions, sample problems, animations, videos, and eText clips for extra help at point-of-use.
- **StatTalk Videos:** *24 Conceptual Videos to Help You Actually Understand Statistics.* Fun-loving statistician Andrew Vickers takes to the streets of Brooklyn, New York, to demonstrate important statistical concepts through

interesting stories and real-life events. These fun and engaging videos will help students actually understand statistical concepts. Available with an instructor's user guide and assessment questions.

- **Getting Ready for Statistics:** A library of questions now appears within each MyStatLab to offer the developmental math topics students need for the course. These can be assigned as a prerequisite to other assignments, if desired.
- **Conceptual Question Library:** In addition to algorithmically regenerated questions that are aligned with your textbook, there is a library of 1000 Conceptual Questions available in the assessment manager that requires students to apply their statistical understanding.
- **StatCrunch®:** MyStatLab integrates the web-based statistical software, StatCrunch, within the online assessment platform so that students can easily analyze data sets from exercises and the text. In addition, MyStatLab includes access to **www.StatCrunch.com,** a website where users can access more than 15,000 shared data sets, conduct online surveys, perform complex analyses using the powerful statistical software, and generate compelling reports.
- **Statistical Software Support:** Knowing that students often use external statistical software, we make it easy to copy our data sets, both from the ebook and the MyStatLab questions, into software such as StatCrunch, Minitab, Excel, and more. Students have access to a variety of support tools—Technology Tutorial Videos, Technology Study Cards, and Technology Manuals for select titles—to learn how to effectively use statistical software.

And, MyStatLab comes from a **trusted partner** with educational expertise and an eye on the future.

- Knowing that you are using a Pearson product means knowing that you are using quality content. That means that our eTexts are accurate and our assessment tools work. Whether you are just getting started with MyStatLab, or have a question along the way, we're here to help you learn about our technologies and how to incorporate them into your course.

To learn more about how MyStatLab combines proven learning applications with powerful assessment, visit **www.mystatlab.com** or contact your Pearson representative.

MathXL® for Statistics Online Course (access code required)

MathXL® is the homework and assessment engine that runs MyStatLab. (MyStatLab is MathXL plus a learning management system.)

With MathXL for Statistics, instructors can:

- Create, edit, and assign online homework and tests using algorithmically generated exercises correlated at the objective level to the textbook.
- Create and assign their own online exercises and import TestGen tests for added flexibility.
- Maintain records of all student work, tracked in MathXL's online gradebook.

With MathXL for Statistics, students can:

- Take chapter tests in MathXL and receive personalized study plans and/or personalized homework assignments based on their test results.
- Use the study plan and/or the homework to link directly to tutorial exercises for the objectives they need to study.
- Students can also access supplemental animations and video clips directly from selected exercises.
- Knowing that students often use external statistical software, we make it easy to copy our data sets, both from the eText and the MyStatLab questions, into software like StatCrunch™, Minitab, Excel, and more.

MathXL for Statistics is available to qualified adopters. For more information, visit our website at **www.mathxl.com**, or contact your Pearson representative.

StatCrunch®

StatCrunch is powerful web-based statistical software that allows users to perform complex analyses, share data sets, and generate compelling reports of their data. The vibrant online community offers more than 15,000 data sets for students to analyze.

- **Collect.** Users can upload their own data to StatCrunch or search a large library of publicly shared data sets, spanning almost any topic of interest. Also, an online survey tool allows users to quickly collect data via web-based surveys.
- **Crunch.** A full range of numerical and graphical methods allow users to analyze and gain insights from any data set. Interactive graphics help users understand statistical concepts, and are available for export to enrich reports with visual representations of data.
- **Communicate.** Reporting options help users create a wide variety of visually-appealing representations of their data.

Full access to StatCrunch is available with MyStatLab, and StatCrunch is available by itself to qualified adopters. StatCrunch Mobile is now available to access from your mobile device. For more information, visit our website at **www.StatCrunch.com**, or contact your Pearson representative.

Acknowledgments

A very big thanks goes to Shivon Sue-Chee for her hard work on the exercise sections, data sets, and appendix solutions. Her hard work and conscientiousness were a marvel. And what a huge contribution was made by Craig Burkett. Going beyond the call of duty far too often, his substantial contributions to the text made this edition possible. Also, thanks to Aaron Springford and Andy Leung for their help in updating the Instructor's Manual. They often went above and beyond in their efforts to help us correct not only the manual, but the text itself.

Thank you also to the team at Pearson and EPS who supported the revision of this text throughout the writing and production process including John Polanszky, Senior Developmental Editor; Marissa Lok, Project Manager; Patricia Cardullo, Program Manager; Michelle Bish, Marketing Manager; Anthony Leung, Designer; Kayci Wyatt, Content Project Manager, Electronic Publishing Services; Janice Dyer, copy editor; Rachelle Redford, proofreader; and Eric Smith and Jingjing Wu, tech checkers.

Finally, thank you to the reviewers who provided feedback on our work and helped guide our plans for this new edition, including:

Julius Bankole, *College of New Caledonia*
Ken Butler, *University of Toronto at Scarborough*
Henry Kolacz, *University of Alberta*
Dot Miners, *Brock University*
Allyson Rozell, *Kwantlen Polytechnic University*
Gary Sneddon, *Mount Saint Vincent University*
Asokan Mulayath Variyath, *Memorial University of Newfoundland*

Index of Application

Note: (E) = Exercise, (FE) = For Example, (JC) = Just Checking, (IE) = in-text example (SBS) = Step-by-Step

Entertainment

Environment

Finance

Food and Drinks

Games

Genetics

Health

Law

Media

Miscellaneous

People

Politics

Psychology

Science

Social Issues

Sports

Statistics

Surveys

Technology

Telecommunication

Transport

Web

chapter

1 Stats Starts Here

Where are we going?

Statistics gets no respect. People say things like, "you can prove anything with Statistics." People will write off a claim based on data as "just a statistical trick." And Statistics courses don't have the reputation of being students' first choice for a fun elective.

But Statistics can be fun, though you might not guess it from some of the book covers popping up now and then at bookstores, with titles like *Statistics for the Terrified* and *Statistics without Tears*. Well, we certainly hope to calm your fears and hold back the tears while helping you learn to think more clearly, critically, and statistically with data. This is a book about understanding the world by using data. So we'd better start by understanding data. There's more to that than you might have thought.

"But where shall I begin?" asked the White Rabbit.
"Begin at the beginning," the King said gravely,
"and go on till you come to the end: then stop."

—Lewis Carroll, Alice's Adventures in Wonderland

1.1 What Is Statistics?

People around the world have one thing in common—they all want to figure out what's going on. You'd think with the amount of information available today this would be an easy task, but actually, as the amount of information grows, so does our need to understand what it can tell us.

At the base of all this information, on the Internet and all around us, are data. We'll talk about data in more detail in the next section, but for now, think of **data** as any collection of numbers, characters, images, or other items that provide information about something. What sense can we make of all this data? You certainly can't make a coherent picture from random pieces of information. Whenever there are data and a need for understanding the world, you'll find Statistics.

This book will help you develop the skills you need to understand and communicate the knowledge that can be learned from data. By thinking clearly about the question you're trying to answer and learning the statistical tools to show what the data are saying, you'll acquire the skills to tell clearly what it all means. Our job is to help you make sense of the concepts and methods of Statistics and to turn them into a powerful, effective approach to understanding the world through data.

Data vary. Ask different people the same question and you'll get a variety of answers. Statistics helps us to make sense of the world by seeing past the underlying variation to find patterns and relationships. This book will teach you skills to help with this task and ways of thinking about variation that are the foundation of sound reasoning about data.

Consider the following:

- If you have a Facebook account, you have probably noticed that the ads you see online tend to match your interests and activities. Coincidence? Hardly. According to the *Wall Street Journal*[1], much of your personal information has probably been sold to marketing or tracking companies. Why would Facebook give you a free account and let you upload as much as you want to its site? Because your data are valuable! Using your Facebook profile, a company might build a profile of your interests and activities: what movies and sports you like; your age, sex, education level, and hobbies; where you live; and, of course, who your friends are and what *they* like. From Facebook's point of view, your data are a potential gold mine. Gold ore in the ground is neither very useful nor pretty. But with skill, it can be turned into something both beautiful and valuable. What we're going to talk about in this book is how you can mine your own data and learn valuable insights about the world.

[1] blogs.wsj.com/digits/2010/10/18/referers-how-facebook-apps-leak-user-ids/

Q: What is Statistics?
A: Statistics is a way of reasoning, along with a collection of tools and methods, designed to help us understand the world.
Q: What are statistics?
A: Statistics (plural) are particular calculations made from data.
Q: So what is data?
A: You mean, "What are data?" Data is the plural form. The singular is datum.
Q: OK, OK, so what are data?
A: Data are values along with their context.

The ads say, "Don't drink and drive; you don't want to be a statistic." But you can't be a statistic.
We say: "Don't be a datum."

Statistics is about *variation*.
Data vary because we don't see everything, and because even what we do see and measure, we measure imperfectly.
But Statistics helps us understand and model this variation so that we can see through it to the underlying truths and patterns.[4]

- By January 2013, Canada's 27 million wireless subscribers were sending over 270 million text (SMS) messages on average per day (96.5 billion in total in 2012, almost tripling the 2009 total of 35.3 billion).[2] Some of these messages were sent or read while the sender or the receiver was driving. How dangerous is texting while driving?

How can we study the effect of texting while driving? One way is to measure reaction times of drivers faced with an unexpected event while driving and texting. Researchers at the University of Utah tested drivers on simulators that could present emergency situations. They compared reaction times of sober drivers, drunk drivers, and texting drivers.[3] The results were striking. The texting drivers actually responded more slowly and were more dangerous than those who were above the legal limit for alcohol.

In this book, you'll learn how to design and analyze experiments like this. You'll learn how to interpret data and to communicate the message you see to others. You'll also learn how to spot deficiencies and weaknesses in conclusions drawn by others that you see in newspapers and on the Internet every day. Statistics can help you become a more informed citizen by giving you the tools to understand, question, and interpret data.

How do we assess the risk of genetically engineered foods being considered by the Canadian Food Inspection Agency or the safety and effectiveness of new drugs submitted to Health Canada for approval? How can we predict the number of new cases of AIDS by regions of the country or the number of customers likely to respond to a sale at the mall? Or determine whether enriched early education affects later performance of school children, and whether vitamin C really prevents illness? Statistics is all about answering real-world questions such as these.

Statistics in a Word

It can be fun and sometimes useful to summarize a discipline in only a few words. So,

Economics is about . . . *Money (and why it is good).*

Psychology: *Why we think what we think (we think).*

Biology: *Life.*

Anthropology: *Who?*

History: *What, where, and when?*

Philosophy: *Why?*

Engineering: *How?*

Accounting: *How much?*

In such a caricature, Statistics is about . . . ***Variation.***

[2] www.txt.ca

[3] "Text Messaging During Simulated Driving," Drews, F. A. et al., Human Factors: hfs.sagepub.com/content/51/5/762

[4] Seeing the forest *through* the trees

1.2 Data

©Dorling Kindersley

Amazon.com opened for business in July 1995, billing itself as "Earth's Biggest Bookstore." By 1997, Amazon had a catalogue of more than 2.5 million book titles and had sold books to more than 1.5 million customers in 150 countries. In 2010, the company's sales reached $34.2 billion (a nearly 40% increase from the previous year). Amazon has sold a wide variety of merchandise, including a $400,000 necklace, yak cheese from Tibet, and the largest book in the world (see photo at left). Amazon.ca launched in June 2002, following up in April 2011 with a new shipping facility built in Mississauga, Ontario. How did Amazon become so successful and how can it keep track of so many customers and such a wide variety of products? The answer to both questions is *data.*

But what are data? Think about it for a minute. What exactly *do* we mean by "data"? Do data have to be numbers? The amount of your last purchase in dollars is numerical data. But your name and address in Amazon's database are also data, even though they are not numerical.

Let's look at some hypothetical data values that Amazon.ca might collect:

105-2686834-3759466	Quebec	Nashville	Kansas	10.99	819	N	B00000I5Y6	Katherine H.
105-9318443-4200264	Nova Scotia	Orange County	Boston	16.99	902	Y	B000002BK9	Samuel P.
105-1872500-0198646	Alberta	Bad Blood	Chicago	15.98	403	N	B000068ZVQ	Chris G.
103-2628345-9238664	Ontario	Let Go	Mammals	11.99	416	N	B0000010AA	Monique D.
002-1663369-6638649	Quebec	Best of Kansas	Kansas	10.99	819	N	B002MXA7Q0	Katherine H.

Try to guess what they represent. Why is that hard? Because there is no *context*. If we don't know what attributes or characteristics are being measured and what is measured about them, the values displayed are meaningless. We can make the meaning clear if we organize the values into a **data table** such as this one:

Order Number	Name	Province	Price	Area Code	Previous Album Download	Gift?	ASIN	New Purchase Artist
105-2686834-3759466	Katherine H.	Quebec	10.99	819	Nashville	N	B0000015Y6	Kansas
105-9318443-4200264	Samuel R	Nova Scotia	16.99	902	Orange County	Y	B000002BK9	Boston
105-1372500-0198646	Chris G.	Alberta	15.98	403	Bad Blood	N	B000068ZVQ	Chicago
103-2628345-9238664	Monique D.	Ontario	11.99	416	Let Go	N	B0000010AA	Mammals
002-1663369-6638649	Katherine H.	Quebec	10.99	819	Best of Kansas	N	B002MXA7Q0	Kansas

Now we can see that these are purchase records for album download orders from Amazon. The column titles tell *what* has been recorded. Each row is about a particular purchase.

What information would provide a **context**? Newspaper journalists know that the lead paragraph of a good story should establish the "Five W's": *who, what, when, where,* and (if possible) *why*. Often, we add *how* to the list as well. The answers to the first two questions are essential. If we don't know *what* attributes or characteristics are measured and *who* they are measured on, the values displayed are meaningless.

Who and What

In general, the rows of a data table correspond to individual **cases** about whom (or about which, if they're not people) we record some characteristics. Those characteristics (called variables—see Section 1.2) constitute the "*what*" of the study. The "*who*" or the cases go by different names, depending on the situation.

Data trumps intuition Amazon monitors and evolves its Web site to better serve customers and maximize sales. To decide which changes to make, analysts experiment with new designs, offers, recommendations, and links. Statisticians want to know how long you'll spend browsing the site and whether you'll follow the links or purchase the suggested items. As Ronny Kohavi, former director of Data Mining and Personalization for Amazon, said, "Data trumps intuition. Instead of using our intuition, we experiment on the live site and let our customers tell us what works for them . . . Data is king at Amazon."

Individuals who answer a survey are called **respondents**. People on whom we experiment are **subjects** or (in an attempt to acknowledge the importance of their role in the experiment) **participants**. Animals, plants, Web sites, and other inanimate subjects are often called **experimental units**. Often we simply call *cases* what they are: for example, *customers, economic quarters*, or *companies*. In a database, rows are called **records**—in this example, purchase records. Perhaps the most generic term is cases, but in any event the rows represent the *who* of the data.

Look at all the columns to see exactly what each row refers to. Here the cases are different purchase records. You might have thought that each customer was a case, but notice that, for example, Katherine H. appears twice, both in the first and the last row. Each row of data is information about a single case, and that case is often identified in the very first column. In this example, it's the order number. If you collect the data yourself, you'll know what the cases are. But, often you'll be looking at data that someone else collected and you'll have to ask or figure that out yourself.

Often the cases are a **sample** from some larger **population** that we'd like to understand. Amazon doesn't care about just these customers; it wants to understand the buying patterns of *all* its customers, and, generalizing further, it wants to know how to attract other Internet users who may not have made a purchase from Amazon's site. To be able to generalize from the sample of cases to the larger population, we'll want the sample to be *representative* of that population—a kind of snapshot image of the larger world.

We must know *who* and *what* to analyze data. Without knowing these two, we don't have enough information to start. Of course, we'd always like to know more. The more we know about the data, the more we'll understand about the world. If possible, we'd like to know the *when* and *where* of data as well. Values recorded in 1803 may mean something different than similar values recorded last year. Values measured in Tanzania may differ in meaning from similar measurements made in Mexico. And knowing *why* the data were collected may also tell us much about its reliability and quality.

How the Data Are Collected

A S *Activity:* **Collect data in an experiment on yourself.** With the computer, you can experiment on yourself and then save the data. Go on to the subsequent related activities to check your understanding.

How the data are collected can make the difference between insight and nonsense. As we'll see later, data that come from a voluntary survey on the Internet are almost always worthless. One primary concern of Statistics, to be discussed in Part III, is the design of sound methods for collecting data. Throughout this book, whenever we introduce data, we'll provide a margin note listing the W's (and H) of the data. Identifying the W's is a habit that we recommend you adopt.

The first step of any data analysis is to know what you are trying to accomplish and what you want to know (*why?*). To help you use Statistics to understand the world and make decisions, we'll lead you through the entire process of *thinking* about the problem, *showing* what you've found, and *telling* others what you've learned. Every guided example in this book is broken into these three steps: *Think*, *Show*, and *Tell*. Identifying the problem and the *who* and *what* of the data is a key part of the *Think* step of any analysis. Make sure you know these before you proceed to *Show* or *Tell* anything about the data.

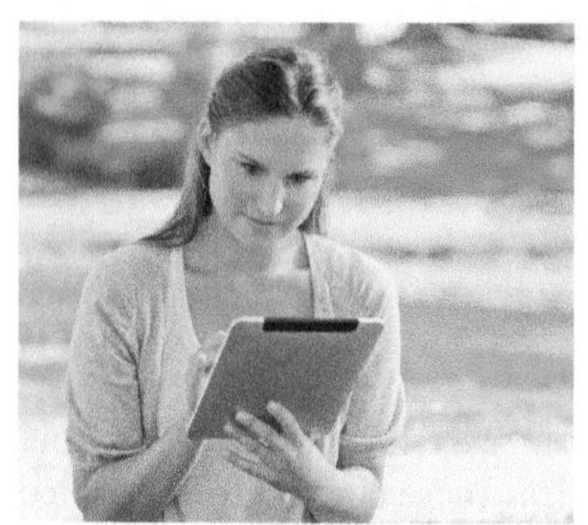

Wavebreakmedia Ltd./Shutterstock

For Example IDENTIFYING THE "WHO"

In December 2011, *Consumer Reports* published an evaluation of 25 tablets from a variety of manufacturers.

QUESTION: Describe the population of interest, the sample, and the *Who* of the study.

ANSWER: The magazine is interested in the performance of tablets currently offered for sale. It tested a sample of 25 tablets, which are the "Who" for these data. Each tablet selected represents all similar tablets offered by that manufacturer.

1.3 Variables

The characteristics (the *"what"*) recorded about each individual are called **variables**. They are usually found as the columns of a data table with a name in the header that identifies what has been recorded. In the Amazon data table we find the variables *Order Number, Name, Province, Price,* and so on.

Categorical Variables

Some variables just tell us what group or category each individual belongs to. Are you male or female? Pierced or not? We call variables like these **categorical**, or **qualitative, variables**. (You may also see them called **nominal variables** because they name categories.) Some variables are clearly categorical, like the variable *Province*. Its values are text and those values tell us what category the particular case falls into. But numerals are often used to label categories, so categorical variable values can also be numerals. For example, Amazon collects telephone area codes that *categorize* each phone number into a geographical region. So area code is considered a categorical variable even though it has numeric values. (But see the story in the following box.)

Tuja66/Fotolia

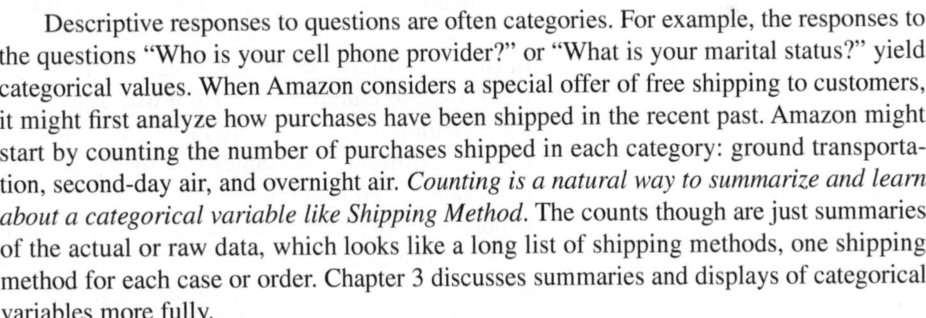

Area codes—numbers or categories? The *What* and *Why* of area codes are not as simple as they may first seem. When area codes were first introduced, AT&T was still the source of all telephone equipment, and phones had dials.

To reduce wear and tear on the dials, the area codes with the lowest digits (for which the dial would have to spin least) were assigned to the most populous regions—those with the most phone numbers and thus the area codes most likely to be dialed. New York City was assigned 212, Chicago 312, and Los Angeles 213, while Halifax was 902, Quebec City 418, and Vancouver 604. For that reason, at one time the numerical value of an area code could be used to guess something about the population of its region. Since the advent of push-button phones, area codes have now become just categories.

Descriptive responses to questions are often categories. For example, the responses to the questions "Who is your cell phone provider?" or "What is your marital status?" yield categorical values. When Amazon considers a special offer of free shipping to customers, it might first analyze how purchases have been shipped in the recent past. Amazon might start by counting the number of purchases shipped in each category: ground transportation, second-day air, and overnight air. *Counting is a natural way to summarize and learn about a categorical variable like Shipping Method.* The counts though are just summaries of the actual or raw data, which looks like a long list of shipping methods, one shipping method for each case or order. Chapter 3 discusses summaries and displays of categorical variables more fully.

"Far too many scientists have only a shaky grasp of the statistical techniques they are using. They employ them as an amateur chef employs a cookbook, believing the recipes will work without understanding why. A more *cordon bleu* attitude . . . might lead to fewer statistical soufflés failing to rise."

—*The Economist*, **June 3, 2004, "Sloppy stats shame science"**

Quantitative Variables

When a variable contains measured numerical values with measurement *units*, we call it a **quantitative variable**. Quantitative variables typically record an amount or degree of something. For quantitative variables, its measurement **units** provide a meaning for the numbers. Even more important, units such as yen, cubits, carats, angstroms, nanoseconds, kilometres per hour, or degrees Celsius tell us the *scale* of measurement, so we know how far apart two values are. Without units, the values of a measured variable have no meaning. It does little good to be promised a raise of 5000 a year if you don't know whether it will be paid in Euros, dollars, pennies, yen, or Estonian krooni.

Aside from the process of measurement, quantitative variables may also arise from the process of counting. Counts also measure the amounts of things. How many songs are on your iPod? How many siblings do you have? To measure these quantities, we would naturally count, instead of measure, and we'd consider the units to be "number of . . ." or just "counts." Amazon might be interested in the number of teenage customers visiting its site each month, in order to track customer growth and forecast sales. So, Who = months and What = # teens (whereas when Amazon counts the number of ground, second-day air, and overnight shipments during the last month, these counts are being used to analyze and summarize the categorical variable Shipping Method, where Who = purchases, What = method). Chapter 3 discusses quantitative variables. We'll see how to display and summarize them there.

Sometimes a variable with numeric values can be treated as either categorical or quantitative depending on what we want to know from it. Amazon could record your *Age* in years. That seems quantitative, and it would be if the company wanted to know the average age of those customers who visit their site after 3 a.m. But suppose Amazon wants to decide which album to feature on its site when you visit. Then placing your age into one of the categories *Child, Teen, Adult,* or *Senior* might be more useful. So, sometimes whether a variable is treated as categorical or quantitative is more about the question we want to ask rather than an intrinsic property of the variable itself. You can always convert a quantitative variable to a categorical one, by simply breaking up the range of values into several intervals. There is some loss of information, but if this facilitates a particular method of analysis, it may still be worthwhile doing.

Identifiers

For a categorical variable like *Sex*, each individual is assigned one of two possible values, say *M* or *F*. But for a variable with ID numbers, such as *Student ID*, each individual receives a unique value. A variable like this, used solely to uniquely identify each of the cases, is called an **identifier variable**. Identifiers are useful, but not typically for analysis.

Amazon wants to know who you are when you sign in again and doesn't want to confuse you with some other customer. So it assigns you a unique identifier. Amazon also wants to send you the right product, so it assigns a unique Amazon Standard Identification Number (ASIN) to each item it carries. You'll want to recognize when a variable is playing the role of an identifier so you aren't tempted to analyze it.

Identifier variables themselves don't tell us anything useful about their categories because we know there is exactly one individual in each. However, they are crucial in this era of large data sets because by uniquely identifying the cases, they make it possible to combine data from different sources, protect (or violate) privacy, and provide unique labels. Many large databases are *relational* databases. In a relational database, different data tables link to one another by matching identifiers. In the Amazon example, *Transaction Number* is clearly an identifier. What about a CD's ASIN? It depends on the purpose of the study. If we want to determine the best-selling CD, then its ASIN is a categorical variable of interest and we would count how often each ASIN appears in a row or transaction. But if we are examining the range of prices for CDs, the ASIN could be used as an identifier variable. The IP (Internet protocol) address of your computer is another identifier, needed so that the electronic messages sent to you can find you.

Privacy and the Internet You have many identifiers: a social insurance number, a student ID number, a passport number, a health insurance number, and probably a Facebook account name. Privacy experts are worried that Internet thieves may match your identity in these different areas of your life, allowing, for example, your health, education, and financial records to be merged. Even online companies, such as Facebook and Google, are able to link your online behaviour to some of these identifiers, which carries with it both advantages and dangers. The National Strategy for Trusted Identities in Cyberspace (www.wired.com/images_blogs/threatlevel/2011/04/NSTICstrategy_041511.pdf) proposes ways that we may address this challenge in the near future.

Ordinal Variables

A typical course evaluation survey asks, "How valuable do you think this course will be to you?" 1 = Worthless; 2 = Slightly; 3 = Middling; 4 = Reasonably; 5 = Invaluable. Is *Educational Value* categorical or quantitative? Often the best way to tell is to look to the *why* of the study. A teacher might just count the number of students who gave each response for her course, treating *Educational Value* as a categorical variable. When she wants to see whether the course is improving, she might treat the responses as the *amount* of perceived value—in effect, treating the variable as quantitative.

But what are the units? There is certainly an *order* of perceived worth: Higher numbers indicate higher perceived worth. A course that averages 4.5 seems more valuable than one that averages 2, but we should be careful about treating *Educational Value* as quantitative. To treat it as quantitative, the teacher will have to imagine that it has "educational value units" or some similar arbitrary construct. Because there are no natural units, she should be extremely cautious. Variables that report order without natural units are often called **ordinal variables**. But saying "that's an ordinal variable" doesn't get you off the hook. You must still look to the *why* of your study and understand what you want to learn from the variable to decide whether to treat it as categorical or quantitative. Ordinal variables are often viewed or classified as a subclass of categorical variables, along with nominal variables, since nominal and ordinal variables both arise through a process of categorizing your cases (though sometimes they may be treated as if quantitative).

For Example IDENTIFYING "WHAT" AND "WHY" OF TABLETS

RECAP: A *Consumer Reports* article about 25 tablets lists each tablet's manufacturer, cost, battery life (hrs.), operating system (iOS/Android/Blackberry), and overall performance score (0–100).

QUESTION: Are these variables categorical or quantitative? Include units where appropriate, and describe the "Why" of this investigation.

ANSWER: The variables are

- manufacturer (categorical)
- cost (quantitative, $)
- battery life (quantitative, hrs.)
- operating system (categorical)
- performance score (quantitative, no units)

The magazine hopes to provide consumers with the information to choose a good tablet.

Just Checking

In the 2004 Tour de France, Lance Armstrong made history by winning the race for an unprecedented sixth time. In 2005, he became the only seven-time winner and once again set a new record for the fastest average speed. And on January 17, 2013, on the Oprah Winfrey show, he made history yet again, confessing to a long history of extensive doping! You can find data on all the Tour de France races online. Here are the first three and last 11 lines of our data set, which goes up to 2012. Keep in mind that the entire data set has nearly 100 entries. For now, the Armstrong entries remain in our data file, but they, like his medals, will vanish by the time we publish our third edition!

1. List as many of the W's as you can for this data set.
2. Classify each variable as categorical or quantitative; if quantitative, identify the units.

Year	Winner	Country of Origin	Total Time (h/min/s)	Avg. Speed (km/h)	Stages	Total Distance Ridden (km)	Starting Riders	Finishing Riders
1903	Maurice Garin	France	94.33.00	25.3	6	2428	60	21
1904	Henri Cornet	France	96.05.00	24.3	6	2388	88	23
1905	Louis Trousseller	France	112.18.09	27.3	11	2975	60	24
⋮								
2002	Lance Armstrong	USA	82.05.12	39.93	20	3278	189	153
2003	Lance Armstrong	USA	83.41.12	40.94	20	3427	189	147
2004	Lance Armstrong	USA	83.36.02	40.53	20	3391	188	147
2005	Lance Armstrong	USA	86.15.02	41.65	21	3608	189	155
2006	Óscar Periero	Spain	89.40.27	40.78	20	3657	176	139
2007	Alberto Contador	Spain	91.00.26	38.97	20	3547	189	141
2008	Carlos Sastre	Spain	87.52.52	40.50	21	3559	199	145
2009	Alberto Contador	Spain	85.48.35	40.32	21	3460	180	156
2010	Andy Schleck	Luxembourg	91.59.27	39.590	20	3642	180	170
2011	Cadel Evans	Australia	86.12.22	39.788	21	3430	198	167
2012	Bradley Wiggins	Great Britain	87.34.47	39.83	20	3488	198	153

There's a world of data on the Internet These days, one of the richest sources of data is the Internet. With a bit of practice, you can learn to find data on almost any subject. Many of the data sets we use in this book were found in this way. The Internet has both advantages and disadvantages as a source of data. Among the advantages are that often you'll be able to find even more current data than those we present. The disadvantage is that references to Internet addresses can "break" as sites evolve, move, and die.

Our solution to these challenges is to offer the best advice we can to help you search for the data, wherever they may be residing. We usually point you to a Web site. We'll sometimes suggest search terms and offer other guidance.

Some words of caution, though: Data found on Internet sites may not be formatted in the best way for use in statistics software. Although you may see a data table in standard form, an attempt to copy the data may leave you with a single column of values. You may have to work in your favourite statistics or spreadsheet program to reformat the data into variables. You will also probably want to remove commas from large numbers and extra symbols such as money indicators ($, ¥, £); few statistics packages can handle these.

WHAT CAN GO WRONG?

- **Don't label a variable as categorical or quantitative without thinking about the data and what they represent.** The same variable can sometimes take on different roles.
- **Don't assume that a variable is quantitative just because its values are numbers.** Categories are often given numerical labels. Don't let that fool you into thinking they have quantitative meaning. Look at the context.
- **Always be skeptical.** One reason to analyze data is to discover the truth. Even when you are told a context for the data, it may turn out that the truth is a bit (or even a lot) different. The context colours our interpretation of the data, so those who want to influence what you think may slant the context. A survey that seems to be about all students may in fact report just the opinions of those who visited a fan Web site. The question that respondents answered may be posed in a way that influences responses.

What Have We Learned?

Learning Objectives

Understand that data are values, whether numerical or labels, together with their context.

- *Who, what, why, where, when* (and *how?*)—the W's—help nail down the context of the data.
- We must know *who, what,* and *why* to be able to say anything useful based on the data. The *who* are the cases. The *what* are the variables. A variable gives information about each of the cases. The *why* helps us decide which way to treat the variables.
- Stop and identify the W's whenever you have data, and be sure you can identify the cases and the variables.

Consider the source of your data and the reasons the data were collected. That can help you understand what you might be able to learn from the data.

Identify whether a variable is being used as categorical (sometimes called qualitative) or quantitative.

- Categorical variables identify a category for each case. Usually we think about the counts of cases that fall in each category. (An exception is an identifier variable that just names each case.)
- Quantitative variables record measurements or amounts of something; they must have units.
- Sometimes we may treat the same variable as categorical or quantitative depending on what we want to learn from it, which means some variables can't be pigeonholed as one type or the other.

Review of Terms

The key terms are in chapter order so you can use this list to review the material in the chapter.

Data Recorded values whether numbers or labels, together with their context (p. 1).

Data table An arrangement of data in which each row represents a case and each column represents a variable (p. 3).

Context The context ideally tells *who* was measured, *what* was measured, *how* the data were collected, *where* the data were collected, and *when* and *why* the study was performed (p. 3).

Case An individual about whom or which we have data (p. 3).

Respondent Someone who answers, or responds to, a survey (p. 4).

Subject A human experimental unit. Also called a participant (p. 4).

Participant A human experimental unit. Also called a subject (p. 4).

Experimental unit An individual in a study for which or for whom data values are recorded. Human experimental units are usually called subjects or participants (p. 4).

Record Information about an individual in a database (p. 4).

Sample A subset of a population, examined in hope of learning about the population (p. 4).

Population The entire group of individuals or instances about whom we hope to learn (p. 4).

Variable A variable holds information about the same characteristic for many cases (p. 5).

Categorical (or qualitative) variable A variable that names categories with words or numerals (p. 5).

Nominal variable A variable whose values are used only to name categories (p. 5).

Quantitative variable A variable in which the numbers are values of measured quantities with units (p. 5).

Unit A quantity or amount adopted as a standard of measurement, such as dollars, hours, or grams (p. 5).

Identifier variable A categorical variable that records a unique value for each case, used to name or identify it (p. 6).

Ordinal variable A variable whose categorical values possess some kind of order (p. 7).

On the Computer DATA

Most often we find statistics on a computer using a program, or package, designed for that purpose. There are many different statistics packages, but they all do essentially the same things. If you understand what the computer needs to know to do what you want and what it needs to show you in return, you can figure out the specific details of most packages pretty easily.

For example, to get your data into a computer statistics package, you need to tell the computer:

- Where to find the data. This usually means directing the computer to a file stored on your computer's disk or to data on a database. Or it might just mean that you have copied the data from a spreadsheet program or Internet site and it is currently on your computer's clipboard. Usually the data is in the form of a data table. Most computer statistics packages prefer the delimiter that marks the division between elements of a data table to be a tab character and the delimiter that marks the end of a case to be a return character.
- Where to put the data. (Usually this is handled automatically.)
- What to call the variables. Some data tables have variable names as the first row of the data, and often statistics packages can take the variable names from the first row automatically.

Exercises

For the description of data in each of the exercises below, identify the W's (and How, if known), name the variables, specify for each variable whether its use indicates it should be treated as categorical or quantitative, and, for any quantitative variable, identify the units in which it was measured (or note that they were not provided).

1. **The news** Find a newspaper or magazine article in which some data are reported. For the data discussed in the article, answer the questions above. Include a copy of the article with your report.

2. **The Internet** Find an Internet source that reports on a study and describes the data. Print out the description and answer the questions above.

3. **Orientation** A study conducted by a team of American and Canadian researchers found that during ovulation, a woman can tell whether a man is gay or straight by looking at his face. To explore the subject, the authors conducted three investigations, the first of which involved 40 undergraduate women who were asked to guess the sexual orientation of 80 men based on photos of their face. Half of the men were gay, and the other half were straight. All held similar expressions in the photos or were deemed to be equally attractive. None of the women were using any contraceptive drugs at the time of the test. The result: the closer a woman was to her peak ovulation, the more accurate her guess. (http://consumer.healthday.com/mental-health-information-25/

behavior-health-news-56/does-ovulation-boost-a-woman-s-gaydar-654279.html)

4. **Blindness** A study begun in 2011 examines the use of stem cells in treating two forms of blindness, Stargardt's disease and dry age-related macular degeneration. Each of the 24 patients entered one of two separate trials in which embryonic stem cells were to be used to treat the condition. (www.redorbit.com/news/health/2066060/stem_cell_trials_to_begin_for_blindness_treatment/)

5. **Investments** Which mutual funds to choose? An investor with some RRSP funds to invest was considering investing in China or India funds. She went to www.globeinvestor.com, did a search, and found 48 China/India/Chindia funds, where she viewed the one-month, one-year, and five-year returns for each of these funds.

6. **Biomass of trees** Because of the difficulty of determining the biomass (or volume) of trees in the wild, Canadian forestry ecology researchers chose a sample of 25 trees in a BC forest, from each of three different species, then measured each tree's diameter (at chest height), and estimated its height via a special viewing instrument (and basic Pythagorean math for triangles). Later these trees were chopped down and the biomass of each was determined by a precise method.

7. **Air travel** Transport Canada and the Canadian Transportation Agency monitor airlines for safety and customer service. For each flight, carriers must report the type of aircraft, the number of passengers, whether or not the flights departed and arrived on schedule, and any mechanical problems.

8. **Tracking sales** A start-up company is building a database of customers and sales information. For each customer, it records name, ID number, region of the country (1 = East, 2 = Central, 3 = Prairies, 4 = West), date of last purchase, amount of purchase, and item purchased.

9. **Cars** A survey of automobiles parked in student and staff lots at a large university recorded the make, country of origin, type of vehicle (car, van, SUV, etc.), and age.

10. **Stats students** An online survey of students in a large Statistics class asked them to report their height, shoe size, sex, which degree program they were in, and their birth order (1 = only child or first born). The data were used for classroom illustrations.

11. **Honesty** Coffee stations in offices often just ask users to leave money in a tray to pay for their coffee, but many people cheat. Researchers at Newcastle University alternately taped two posters over the coffee station. During one week, it was a picture of flowers; during the other, it was a pair of staring eyes. They found that the average contribution was significantly higher when the eyes poster was up than when the flowers were there. Apparently, the mere feeling of being watched—even by eyes that were not real—was enough to encourage people to behave more honestly. (Source: *NY Times,* Dec. 10, 2006)

12. **Molten iron** The Cleveland Casting Plant is a large, highly automated producer of grey and nodular iron automotive castings for Ford Motor Company. According to an article in *Quarterly Engineering* (7[1995]), Cleveland Casting is interested in keeping the pouring temperature of the molten iron (in degrees Fahrenheit) close to the specified value of 2550 degrees. Cleveland Casting measured the pouring temperature for 10 randomly selected crankshafts.

13. **Weighing bears** Because of the difficulty of weighing a bear in the woods, researchers caught and measured 54 bears, recording their weight, neck size, length, and sex. They hoped to find a way to estimate weight from the other, more easily determined quantities.

14. **Schools** The provincial or territorial ministry of education requires local school districts to keep the following records on all students: age, race or ethnicity, days absent, current grade level, standardized test scores in reading and mathematics, and any disabilities or special educational needs.

15. **Tim Hortons doughnuts** At the Tim Hortons Web site (www.timhortons.com), there is a link called "Nutritional Information" that gives nutrition facts for various categories of food that it sells, such as doughnuts, sandwiches, etc. Under doughnuts, for example, we see a listing of all the doughnut types, and for each, information is provided about the number of calories and the amounts of trans fat, total fat, sodium, sugar, protein, and "% daily value" of iron and of calcium.

16. **Trudeaumania?** An April 2013 Harris/Decima poll indicated that 57% of Canadians held a favourable view of new Liberal leader Justin Trudeau, while only 30% held an unfavourable impression. These survey results were based on 1006 telephone respondents, via the company's national telephone omnibus survey. For each respondent, gender, province, party preference, and age group (18–34, 35–54, 55+) was recorded, along with the answer to the favourability question.

17. **Babies** Medical researchers at a large city hospital investigating the impact of prenatal care on newborn health collected data from 882 births between 1998 and 2000. They kept track of the mother's age, the number of weeks the pregnancy lasted, the type of birth (Caesarean, induced, natural), the level of prenatal care the mother had had (none, minimal, adequate), the birth weight and sex of the baby, and whether the baby exhibited health problems (none, minor, major).

18. **Flowers** In a study appearing in the journal *Science*, a research team reported that plants in southern England are flowering earlier in the spring. Records of the first

flowering dates for 385 species over a period of 47 years show that flowering has advanced an average of 15 days per decade, an indication of climate warming, according to the authors.

19. **Herbal medicine** Scientists at a major pharmaceutical firm conducted an experiment to study the effectiveness of an herbal compound to treat the common cold. They exposed each patient to a cold virus, then gave them either the herbal compound or a sugar solution known to have no effect on colds. Several days later they assessed each patient's condition using a cold severity scale ranging from 0–5. They found no evidence of the benefits of the compound.

20. **Vineyards** Business analysts hoping to provide information helpful to grape growers compiled the following data about vineyards: size (acres), number of years in existence, province, varieties of grapes grown, average case price, gross sales, and percent profit.

21. **Streams** As research for an ecology class, students at Laurentian University collect data on streams each year. They record a number of biological, chemical, and physical variables, including the stream name, the substrate of the stream (limestone, shale, or mixed), the acidity of the water (pH), the temperature (°C), and the BCI (a numerical measure of biological diversity).

22. **Fuel economy** Transport Canada and Natural Resources Canada track the fuel economy of automobiles. Among the data they collect are the manufacturer (Ford, Toyota, etc.); the vehicle type (car, SUV, etc.); and the weight, horsepower, and gas mileage (L/100 km) for city and highway driving.

23. **Refrigerators** A *Consumer Reports* article evaluating refrigerators listed 41 models, giving the brand, cost, size (cu. ft.), type (such as top freezer), estimated annual energy cost, overall rating (good, excellent, etc.), and the repair history for that brand (percentage requiring repairs over the past five years).

24. **Walking in circles** People who get lost in the desert, mountains, or woods often seem to wander in circles rather than walk in straight lines. To see whether people naturally walk in circles in the absence of visual clues, researcher Andrea Axtell tested 32 people on a football field. One at a time, they stood at the centre of one goal line, were blindfolded, and then tried to walk to the other goal line. She recorded each individual's sex, height, handedness, the number of yards each was able to walk before going out of bounds, and whether each wandered off course to the left or the right. No one made it all the way to the far end of the field without crossing one of the sidelines.[5]

25. **Kentucky Derby 2012** The Kentucky Derby is a horse race that has been run every year since 1875 at Churchill Downs, Louisville, Kentucky. The race started as a 1.5-mile race, but in 1896, it was shortened to 1.25 miles because experts felt that 3-year-old horses shouldn't run such a long race that early in the season. (It has been run in May every year but one—1901—when it took place on April 29). Here are the data for the first four and several recent races.

2012	I'll Have Another	M. Gutierrez	D. O'Neill	Reddam Racing	2:01.83
2011	Animal Kingdom	J. Velazquez	H. G. Motion	Team Valor	2:02.04
2010	Super Saver	C. Borel	T. Pletcher	WinStar Farm	2:04.45
2009	Mine That Bird	C. Borel	B. Woolley	Double Eagle Ranch	2:02.66
2008	Big Brown	K. Desormeaux	R. Dutrow	IEAH Stables, Pompa et al	2:01.82
2007	Street Sense	C. Borel	C. Nafzger	James Tafel	2:02.17
⋮					
1878	Day Star	J. Carter	L. Paul	T. J. Nichols	2:37.25
1877	Baden Baden	W. Walker	E. Brown	Daniel Swigert	2:38
1876	Vagrant	R. Swim	J. Williams	William Astor	2:38.25
1875	Aristides	O. Lewis	A. Williams	H. P. McGrath	2:37.75

Source: http://horsehats.com/KentuckyDerbyWinners.html

26. **Stanley Cup 2012** Lord Stanley's Cup is the oldest professional sports trophy in North America and has been awarded to the championship-winning hockey team every year since 1893. It was awarded to amateur champions in the early years, before the professional leagues had been formed. Below are some data for recent and not-so-recent Stanley Cup finals. The current best-of-seven championship format was not the case until 1939, as you can see from the earlier championship series listed (some were best of five, some best of three).

[5]*Source: STATS* No. 39, Winter 2004.

Year	Winning Team	Series (W-L)	Finalist	Time of Winning Goal	Winning Goal Scored By
1926	Montreal Maroons	3–1	Victoria Cougars	N/A	N/A
1927	Ottawa Senators	2–0	Boston Bruins	27:30	Denneny
1928	New York Rangers	3–2	Montreal Canadiens	43:35	Boucher
1929	Boston Bruins	2–0	New York Rangers	58:02	Carson
1930	Montreal Canadiens	2–0	Boston Bruins	21:00	Morenz
1931	Montreal Canadiens	3–2	Chicago Blackhawks	29:59	Gagnon
1932	Toronto Maple Leafs	3–0	New York Rangers	55:07	Bailey
1933	New York Rangers	3–1	Toronto Maple Leafs	67:34	Cook
1934	Chicago Blackhawks	3–1	Detroit Red Wings	90:05	March
1935	Montreal Maroons	3–0	Toronto Maple Leafs	36:18	Northcott
1936	Detroit Red Wings	3–1	Toronto Maple Leafs	49:45	Kely
1937	Detroit Red Wings	3–2	New York Rangers	19:22	Barry
1938	Chicago Blackhawks	3–1	Toronto Maple Leafs	36:45	Voss
1939	Boston Bruins	4–1	Toronto Maple Leafs	37:54	Conacher
1940	New York Rangers	4–2	Toronto Maple Leafs	62:07	Hextall
⋮					
2010	Chicago Blackhawks	4–2	Philadelphia Flyers	64:06	Kane
2011	Boston Bruins	4–3	Vancouver Canucks	14:37	Bergeron
2012	Los Angeles Kings	4–2	New Jersey Devils	112:45	Carter

(Note: above, 29:59 means 9:59 into the second period, since regular time periods are 20 minutes)

Just Checking ANSWERS

1. Who—Tour de France races; What—year, winner, country of origin, total time, average speed, stages, total distance ridden, starting riders, finishing riders; How—official statistics at race; Where—France (for the most part); When—1903 to 2009; Why—to see progress in speeds of cycling racing

2.

Variable: Year	Type—Quantitative or Categorical	Units—Years
Variable: Winner	Type—Categorical	
Variable: Country of Origin	Type—Categorical	
Variable: Total Time	Type—Quantitative	Units—Hours/minutes/seconds
Variable: Average Speed	Type—Quantitative	Units—Kilometres per hour
Variable: Stages	Type—Quantitative	Units—Counts (stages)
Variable: Total Distance	Type—Quantitative	Units—Kilometres
Variable: Starting Riders	Type—Quantitative	Units—Counts (riders)
Variable: Finishing Riders	Type—Quantitative	Units—Counts (riders)

chapter

2

Displaying and Describing Categorical Data

2.1 One Categorical Variable

2.2 Exploring the Relationship Between Two Categorical Variables

Where are we going?

What is your field of study: Humanities, Social Sciences, or Sciences? What is your blood type? Which candidate do you plan to vote for?

Data are not just numbers—the answers to all these questions are categories. We'll study categorical variables like these, but things get even more interesting when we look at how categorical variables work together. Are Grade 5 students more likely to experience bullying than students in Grade 9? Are homosexuals more likely to be victims of crime than heterosexuals?

You often see categorical data displayed in pie charts or bar charts, or summarized in tables, sometimes in confusing ways. With a little skill, however, it's not hard to do it right.

A single death is a tragedy; a million is a statistic.

—Joseph Stalin

Moviestore Collection/Alamy

What happened on the *Titanic* at 11:40 on the night of April 14, 1912, is well known. Frederick Fleet's cry of "Iceberg, right ahead" and the three accompanying pulls of the crow's nest bell signalled the beginning of a nightmare that has become legend. By 2:15 a.m., the *Titanic*, thought by many to be unsinkable, had sunk, leaving almost 1500 passengers and crewmembers on board to meet their icy fate.

Here are some data about the passengers and crew aboard the *Titanic*. Each case (row) of the data table represents a person on board the ship. The variables are the person's *Survival* status (*Dead* or *Alive*), *Age* (*Adult* or *Child*), *Sex* (*Male* or *Female*), and ticket *Class* (*First*, *Second*, *Third*, or *Crew*).

Who	People on the *Titanic*
What	Survival status, age, sex, ticket class
When	April 14, 1912
Where	North Atlantic
How	A variety of sources and Internet sites
Why	Historical interest

Table 2.1

Part of a data table showing four variables for nine people aboard the *Titanic*.

Survival	Age	Sex	Class
Dead	Adult	Male	Third
Dead	Adult	Male	Crew
Dead	Adult	Male	Third
Dead	Adult	Male	Crew
Dead	Adult	Male	Crew
Dead	Adult	Male	Crew
Alive	Adult	Female	First
Dead	Adult	Male	Third
Dead	Adult	Male	Crew

A S ***Video:* The Incident** tells the story of the *Titanic* and includes rare film footage.

The problem with a data table like this—and in fact with all data tables—is that you can't see what's going on. And seeing is just what we want to do. We need ways to show the data so that we can see patterns, relationships, trends, and exceptions.

2.1 One Categorical Variable

Florence Nightingale Museum, London

So, what should we do with data like these? There are three things you should always do first with data:

The Three Rules of Data Analysis

1. **Make a picture.** A display of your data will reveal things you are not likely to see in a table of numbers and will help you to *Think* clearly about the patterns and relationships that may be hiding in your data.
2. **Make a picture.** A well-designed display will *Show* the important features and patterns in your data. A picture will also show you the things you did not expect to see: the extraordinary (possibly wrong) data values or unexpected patterns.
3. **Make a picture.** The best way to *Tell* others about your data is with a well-chosen picture.

These are the three rules of data analysis. There are pictures of data throughout the book, and new kinds keep showing up. These days, technology makes it easy to draw pictures of data, so there is no reason not to follow the three rules.

Figure 2.1

A Picture to Tell a Story. Florence Nightingale (1820–1910), a founder of modern nursing, was also a pioneer in health management statistics and epidemiology. She was the first female member of the British Statistical Society and was granted honourary membership in the newly formed American Statistical Association.

To argue forcefully for better hospital conditions for soldiers, she and her colleague, Dr. William Farr, invented this display, which showed that in the Crimean War, far more soldiers died of illness and infection than of battle wounds. Her campaign succeeded in improving hospital conditions and nursing for soldiers.

Florence Nightingale went on to apply statistical methods to a variety of important health issues and published more than 200 books, reports, and pamphlets during her long and illustrious career.

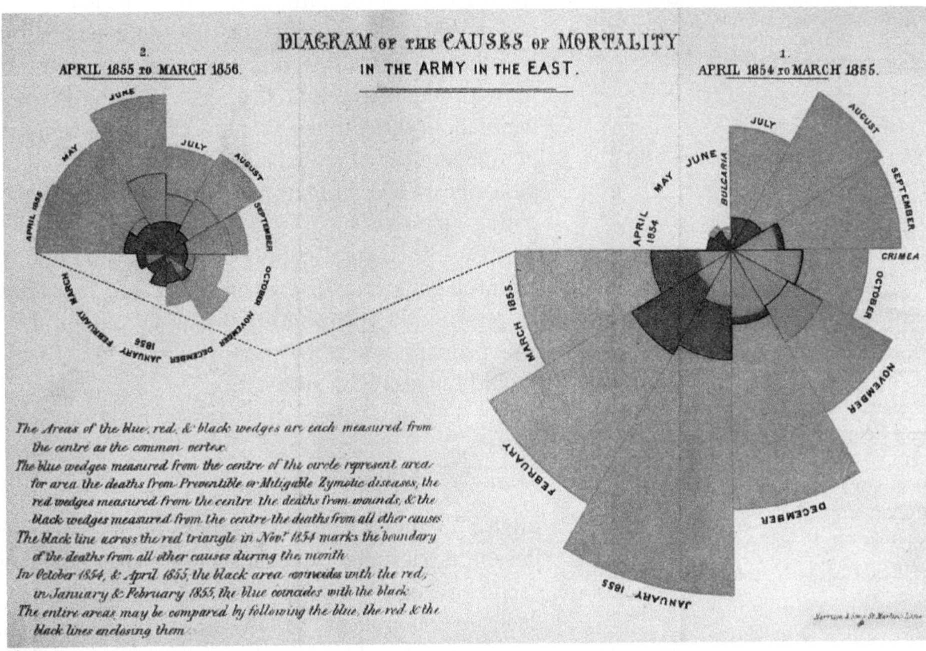

Statistical chart invented by Florence Nightingale to show the predominance of disease as a cause of mortality in the British army during the Crimean War, April 1854 to March 1855 and April 1855 to March 1956 (litho), English School, (19th century) / Florence Nightingale Museum, London, UK / The Bridgeman Art Library

A S ***Activity:* Make and examine a table of counts.** Even data on something as simple as hair colour can reveal surprises when you organize them in a data table.

Frequency Tables

Looking at the list of 2201 passengers and observing each one's class is not very informative. We need to organize and summarize the data more effectively. Even when we have thousands of cases, a variable like ticket *Class*, with only a few categories, is

Class	Count
First	325
Second	285
Third	706
Crew	885

Table 2.2
A frequency table of the *Titanic* passengers and crew.

Class	%
First	14.77
Second	12.95
Third	32.08
Crew	40.21

Table 2.3
A relative frequency table for the same data.

easy to organize. We just count the number of cases corresponding to each category and put them in a **frequency table**. For ticket Class, these are "First," "Second," "Third," and "Crew."

For a variable with dozens or hundreds of categories, it will be much harder to read a frequency table. You might want to combine categories into larger headings. For example, instead of counting the number of students from each state in the U.S. who attend McGill University, you might group the states into regions like "Northeast," "South," "Midwest," "Mountain States," and "West." If the number of cases in several categories is relatively small, you can put them together into one category labelled "Other."

Counts are useful, but sometimes we want to know the fraction or proportion of the data in each category, so we divide the counts by the total number of cases. Often we multiply by 100 to express these proportions as percentages. A **relative frequency table** displays the *proportions* or *percentages*, rather than the counts, of the values in each category. Both types of tables show how the cases are distributed across the categories. In this way, they describe the **distribution** of a categorical variable because they name the possible categories and tell how frequently each occurs.

The Area Principle

Now that we have the frequency table, we're ready to follow the three rules of data analysis and make a picture of the data. But a bad picture can distort our understanding rather than help it. Figure 2.2 is a colourful graph of the *Titanic* data. What impression do you get about who was aboard the ship?

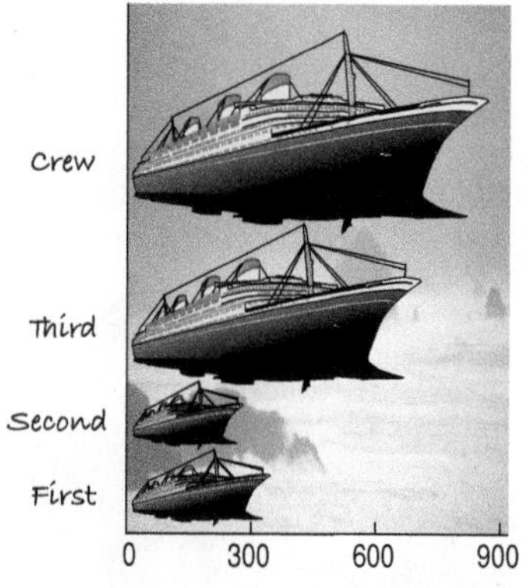

Figure 2.2
How many people were in each class on the *Titanic*? From this display it looks as though the service must have been great, since most aboard were crewmembers. Although the length of each ship here corresponds to the correct number, the impression is all wrong. In fact, only about 40% were crew.

It looks like most of the people on the *Titanic* were crewmembers, with a few passengers along for the ride. That doesn't seem right. What's wrong? The lengths of the ships do match the number of people in each ticket class category. However, experience and psychological tests show that our eyes tend to be more impressed by the area than by other aspects of each ship image. So, even though the *length* of each ship matches up with one of the totals, it's the associated *area* in the image that we notice. There were about three times as many crew as second-class passengers, and the ship depicting the number of crew is about three times longer than the ship depicting second-class passengers. The problem is that it occupies about nine times the area.[1] That just isn't a correct impression.

The best data displays observe a fundamental principle of graphing data called the area principle. The **area principle** says that the area occupied by a part of the graph should correspond to the magnitude of the value it represents. Violations of the area principle are a common way to lie (or, since most mistakes are unintentional, we should say err) with statistics.

Bar Charts

Figure 2.3 shows a chart that obeys the area principle. It's not as visually entertaining as the ships, but it does give an accurate visual impression of the distribution. The height of each bar shows the count for its category. The bars are the same width, so their heights determine their areas, and the areas are proportional to the counts in each class. Now it's easy to see that the majority of people on board were not crew, as the ships picture led us to believe. We can also see that there were about three times as many crew as second-class passengers. And there were more than twice as many third-class passengers as either first- or second-class passengers, something you may have missed in the frequency table. Bar charts make these kinds of comparisons easy and natural.

[1]Edward Tufte would say the Lie Factor here is 3.0, since the impression or displayed effect is three times bigger than it should be (*The Visual Display of Quantitative Information*).

A **bar chart** may be used to display the distribution of a categorical variable, showing the counts for each category next to each other for easy comparison. Bar charts should have spaces between the bars to indicate that these are freestanding bars that could be rearranged into any order. The bars are lined up along a common base.

Usually they stick up like this 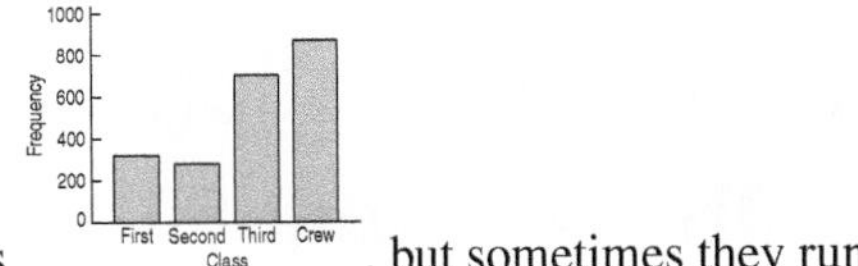, but sometimes they run

sideways like this 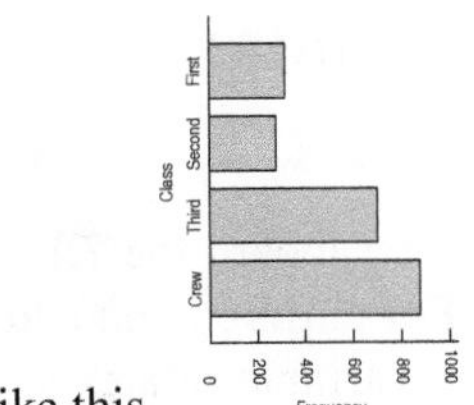.

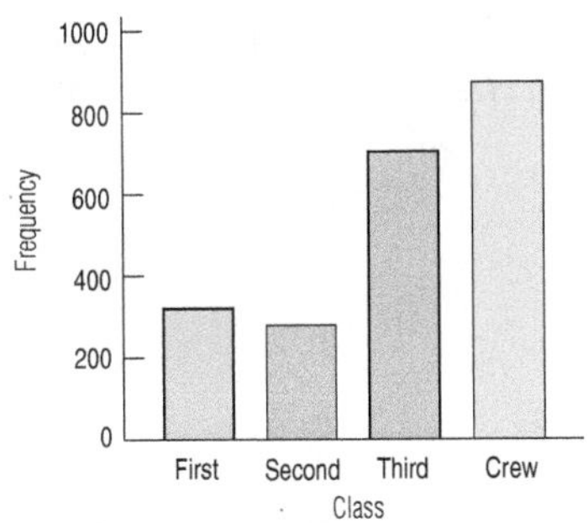

Figure 2.3

People on the* Titanic *by ticket class. With the area principle satisfied, we can see the true distribution more clearly.

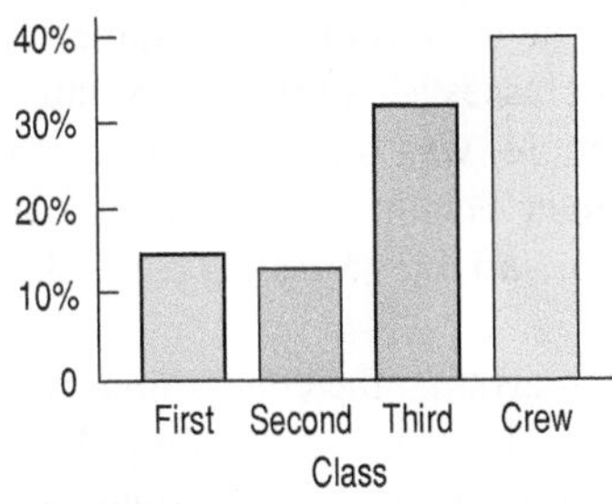

Figure 2.4

The relative frequency bar chart looks the same as the bar chart (Figure 2.3), but shows the proportion of people in each category rather than the counts. The vertical axis may also read 0, 0.10, 0.20, 0.30, 0.40.

A S ***Activity:* Bar Charts.** Watch bar charts grow from data, then use your statistics package to create some bar charts for yourself.

If we really want to draw attention to the relative proportion of passengers falling into each of these classes, we could replace the counts with percentages (or proportions) and use a **relative frequency bar chart** (Figure 2.4).

Pie Charts

Another common display that shows how a whole group breaks into several categories is a pie chart. **Pie charts** show the whole group of cases as a circle. They slice the circle into pieces whose size is proportional to the fraction of the whole in each category.

Pie charts give a quick impression of how a whole group is partitioned into smaller groups. Because we're used to cutting up pies into 2, 4, or 8 pieces, pie charts are good for seeing relative frequencies near 1/2, 1/4, or 1/8. For example, you may be able to tell that the pink slice, representing the second-class passengers, is very close to 1/8 of the total (Figure 2.5). It's harder to see that there were about twice as many third-class as first-class passengers. Which category had the most passengers? Were there more crew or more third-class passengers? Comparisons such as these are easier in a bar chart. No matter what display you choose for your variable, make sure that the data are counts or percentages of cases. For example, a pie chart showing the price of each ticket class wouldn't make sense. Price is quantitative. So before making a bar chart or a pie chart, always check the **Categorical Data Condition** that the data are counts or percentages of individuals in categories.

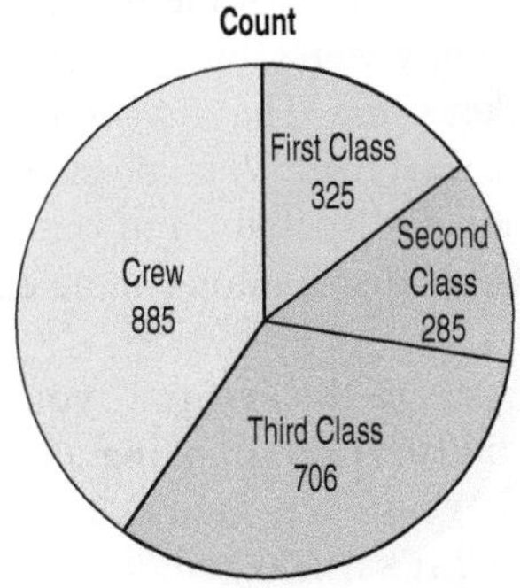

Figure 2.5

The number of *Titanic* crew and passengers in each class.

THINK BEFORE YOU DRAW

Our first rule of data analysis is *Make a picture*. But what kind of picture? We don't have a lot of options—yet. There's more to Statistics than pie charts and bar charts, and knowing when to use each type of graph is a critical first step in data analysis. That decision depends in part on what type of data we have.

If you want to make a relative frequency bar chart or a pie chart, you'll need to also make sure that the categories don't overlap, so no individual is counted twice. If the categories do overlap, you can still make a bar chart, but the percentages won't add up to 100%. For the *Titanic* data, either kind of display is appropriate because the categories don't overlap.

For any statistical analysis, a crucial decision is the choice of a proper method. Be sure to think about the situation at hand, and make sure to check that the type of analysis you plan is appropriate. The Categorical Data Condition is just the first of many such checks.

2.2 Exploring the Relationship Between Two Categorical Variables

Contingency Tables: Children and First-Class Ticket Holders First?

A S ***Activity:* Children at Risk.** This activity looks at the fates of children aboard the *Titanic*; the subsequent activity shows how to make such tables on the computer.

Only 32% of all those aboard the *Titanic* survived. Was the survival rate the same for men and women? For children and adults? For all ticket classes? It's often more interesting to ask if one variable relates to another. For example, was there a relationship between the kind of ticket a passenger held and the passenger's chances of surviving? To answer this question, we need to look at the two categorical variables *Survival* and ticket *Class* together.

To look at two categorical variables together, we often arrange the counts in a two-way table. Table 2.4 shows each person aboard the *Titanic* classified according to both their ticket *Class* and their *Survival*. Because the table shows how the individuals are distributed along each variable, contingent on the value of the other variable, such a table is called a **contingency table**.

Table 2.4

***Contingency table of ticket* class *and* survival.** The bottom line of "Total" is the same as the previous frequency table.

	Class				
	First	Second	Third	Crew	Total
Alive	203	118	178	212	**711**
Dead	122	167	528	673	**1490**
Total	**325**	**285**	**706**	**885**	**2201**

Each **cell** of the table gives the count for a combination of values of the two variables. The margins of the table, on the right and at the bottom, give totals. The bottom line of the table just shows the frequency distribution of ticket *Class*. The right column of the table shows the frequency distribution of the variable *Survival*. When presented like this, in the margins of a contingency table, the frequency distribution of one of the variables is called its **marginal distribution**. The marginal distribution can be expressed either as counts or percentages.

If you look down the column for second-class passengers to the first row, you'll find the cell containing the 118 second-class passengers who survived. Looking at the cell to its right we see that *more* third-class passengers (178) survived. But, does that mean that third-class passengers were more *likely* to survive? It's true that more third-class passengers survived, but there were many more third-class passengers on board the ship. To compare the two numbers fairly, we need to express them as percentages—but as a percentage of what?

For any cell, there are three choices of percentage. We could express the 118 second-class survivors as 5.4% of all 2201 passengers on the *Titanic* (the *overall percent*), as 16.6% of all 711 survivors (the *row percent*), or as 41.4% of all 285 second-class passengers (the *column percent*). Each of these percentages is potentially interesting.

Statistics programs offer all three. Unfortunately, they often put them all together in each cell of the table. The resulting table holds lots of information, but it can be hard to understand:

Table 2.5

***Another contingency table of ticket* Class.** This time we see not only the counts for each combination of *Class* and *Survival* (in bold), but also the percentages these counts represent. For each count, there are three choices for the percentage: by row, by column, and by table total. There's probably too much information here for this table to be useful.

Survival		Class				
		First	Second	Third	Crew	Total
Alive	Count	**203**	**118**	**178**	**212**	**711**
	% of Row	28.6	16.6	25.0	29.8	100
	% of Column	62.5	41.4	25.2	24.0	32.3
	% of Overall Total	9.2	5.4	8.1	9.6	32.3
Dead	Count	**122**	**167**	**528**	**673**	**1490**
	% of Row	8.2	11.2	35.4	45.2	100
	% of Column	37.5	58.6	74.8	76.0	67.7
	% of Overall Total	5.6	7.6	24.0	30.6	67.7
Total	Count	**325**	**285**	**706**	**885**	**2201**
	% of Row	14.8	12.9	32.1	40.2	100
	% of Column	100	100	100	100	100
	% of Overall Total	14.8	12.9	32.1	40.2	100

To simplify the table, let's first pull out the overall percent values:

Table 2.6

A contingency table of class by survival with only the table percentages

Survival	Class				
	First	Second	Third	Crew	Total
Alive	9.2%	5.4%	8.1%	9.6%	32.3%
Dead	5.6%	7.6%	24.0%	30.6%	67.7%
Total	14.8%	12.9%	32.1%	40.2%	100%

These percentages tell us what percent of all passengers belong to each combination of column and row category (called the **joint distribution** of the two variables). For example, we see that although 8.1% of the people aboard the *Titanic* were surviving third-class ticket holders, only 5.4% were surviving second-class ticket holders. Comparing these percentages, you might think that the chances of surviving were better in third class than in second. But be careful. There were many more third-class than second-class passengers on the *Titanic*, so we should naturally expect there to be more third-class survivors. Comparing the overall percentages, 5.4% and 8.1%, doesn't answer the question about whether second-class passengers were more or less likely to survive than third-class passengers. Overall percentages can be useful, but they don't answer questions like this, involving comparisons of percentages of one variable across levels of the other.

PERCENT OF WHAT?

The English language can be tricky when we talk about percentages. If you're asked, "What percent *of the survivors* were in second class?" it's pretty clear that we're interested only in survivors. So we should restrict our attention only to the survivors and look at the number of second-class passengers among all the survivors—in other words, the row percent.

But if you're asked, "What percent were second-class passengers who survived?" you have a different question. Be careful; now we need to consider everyone on board, so 2201 should be the denominator, and the answer is the overall percent (5.4%).

And if you're asked, "What percent of the second-class passengers survived?" you have yet another question. Now the denominator is the 285 second-class passengers, and the answer is the column percent (41.4%).

Always be sure to ask "percent of what?" That will help you to know *Who* we're talking about and whether we want *row*, *column*, or *overall* percentages.

For Example FINDING MARGINAL DISTRIBUTIONS

QUESTION: A recent Gallup poll asked 1008 U.S. residents aged 18 and over whether they planned to watch the upcoming Super Bowl. The pollster also asked those who planned to watch whether they were looking forward more to seeing the football game or the commercials. The results are summarized in the table:

Response	Male	Female	Total
	Sex		
Game	279	200	**479**
Commercials	81	156	**237**
Won't watch	132	160	**292**
Total	**492**	**516**	**1008**

QUESTION: What's the marginal distribution of *Response*?

ANSWER: To determine the percentages for the three responses, divide the count for each response by the total number of people polled:

$$\frac{479}{1008} = 47.5\%, \quad \frac{237}{1008} = 23.5\%, \quad \frac{292}{1008} = 29.0\%$$

According to the poll, 47.5% of U.S. adults were looking forward to watching the Super Bowl game, 23.5% were looking forward to watching the commercials, and 29% didn't plan to watch at all.

Conditional Distributions

Rather than looking at the overall percentages, it's more interesting to ask whether the chance of survival *depended* on ticket class. We can look at this question in two ways. First, we could ask how the distribution of ticket *Class* changes between survivors and nonsurvivors. To do that, we look at the *row percentages*:

Table 2.7

The conditional distribution of ticket *Class* conditioned on each value of survival: alive and dead.

Survival		First	Second	Third	Crew	Total
		Class				
Alive		203	118	178	212	711
		28.6%	16.6%	25.0%	29.8%	100%
Dead		122	167	528	673	1490
		8.2%	11.2%	35.4%	45.2%	100%

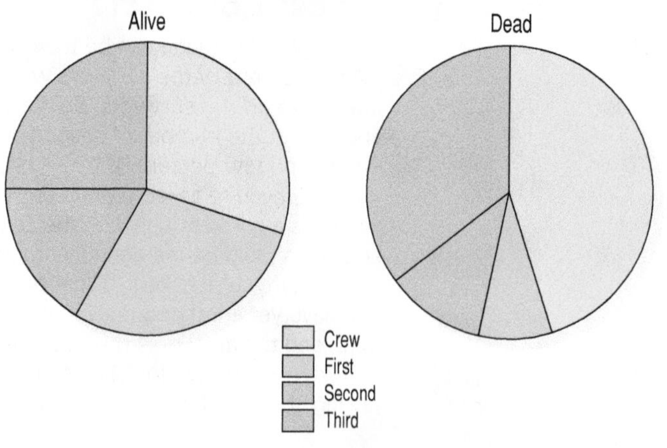

Figure 2.6

Pie charts of the conditional distributions of ticket class for survivors and non-survivors, separately. Do the distributions appear to be the same? We're primarily concerned with percentages here, so pie charts are a reasonable choice.

By focusing on each row separately, we see the distribution of *Class* for each condition of surviving or not. The sum of the percentages in each row is 100%, and we divide that up by ticket class. First, we restrict our attention to survivors, and make a pie chart for them. Then we focus on the nonsurvivors and make their pie chart. These pie charts show the distribution of ticket classes *for each row* of the table: survivors and nonsurvivors. The distributions we create this way are called **conditional distributions** because they show the distribution of one variable for just those cases that satisfy a condition on another variable.

For Example FINDING CONDITIONAL DISTRIBUTIONS

RECAP: The table shows results of a poll asking adults whether they were looking forward to the Super Bowl game, looking forward to the commercials, or didn't plan to watch.

Response	Sex: Male	Female	Total
Game	279	200	**479**
Commercials	81	156	**237**
Won't watch	132	160	**292**
Total	**492**	**516**	**1008**

QUESTION: How does interest primarily in the commercials differ for men and women?

ANSWER: Find the percent of males who responded "commercials" and repeat for females:

$$\frac{81}{492} = 16.5\%, \quad \frac{156}{516} = 30.2\%$$

Almost twice as many women (30%) as men (17%) look forward to seeing Super Bowl commercials more than the game itself.

But we still haven't answered our question. Instead, we can turn the conditioning around and look at the distribution of *Survival* for each category of ticket *Class*. The *column percentages* show us whether the chance of surviving was roughly the same *for each of the four classes*. Now the percentages in each column add to 100%, because we've restricted our attention, in turn, to each of the four ticket classes:

Table 2.8

A contingency table of class by survival with only counts and column percentages. Each column represents the conditional distribution of *survival* for a given category of ticket *Class*.

Survival		Class: First	Second	Third	Crew	Total
Alive	Count	203	118	178	212	711
	% of Column	62.5%	41.4%	25.2%	24.0%	32.3%
Dead	Count	122	167	528	673	1490
	% of Column	37.5%	58.6%	74.8%	76.0%	67.7%
Total	Count	325	285	706	885	2201
		100%	100%	100%	100%	100%

Looking at how the percentages change across each row, it sure looks like ticket class made a difference in terms of whether a passenger survived. To make it more vivid, we could display the percentages for surviving and not surviving for each *Class* in a side-by-side bar chart (Figure 2.7):

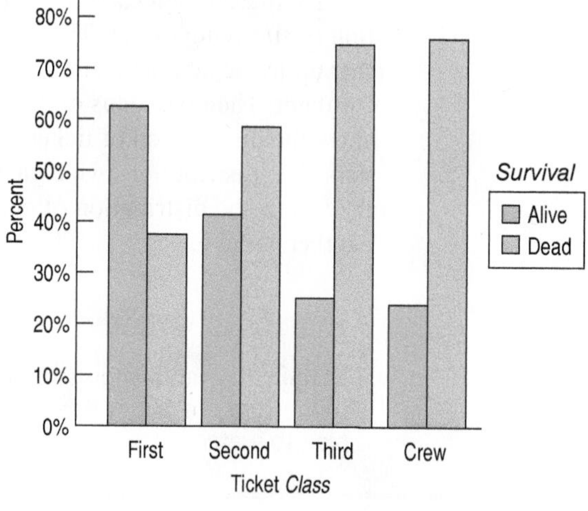

Figure 2.7

Side-by-side bar charts showing the conditional distribution of *Survival* for each category of ticket *Class*. The corresponding pie charts would have only two categories in each of four pies, so bar charts seem the better alternative.

These bar charts are simple because for the variable *Survival*, we have only two alternatives: Alive and Dead. When we have only two categories, we really only need to know the percentage of one of them. Knowing the percentage that survived tells us the percentage that died. Hence, we can simplify the display even more by dropping one category. Here are the percentages of dying across the classes displayed in one chart:

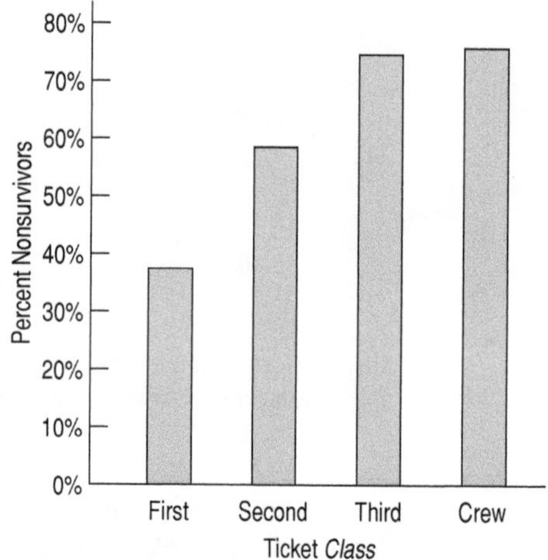

Figure 2.8

Bar chart showing just non-survivor percentages for each value of ticket *Class*. Because we have only two values, the second bar doesn't add any information.

Now it's easy to compare the risks. Among first-class passengers, 37.5% perished, compared to 58.6% for second-class ticket holders, 74.8% for those in third class, and 76.0% for crewmembers.

If the risk had been about the same across the ticket classes, we would have said that survival was *independent* of class. But it's not. The differences we see among these conditional distributions suggest that survival may have depended on ticket class. You may find it useful to look at the conditional distributions of each variable in a contingency table in order to explore the dependence between them.

It is interesting to know that *Class* and *Survival* are associated. That's an important part of the *Titanic* story. And we know how important this is because the margins show us the actual numbers of people involved.

Variables can be associated in many ways and to different degrees. The best way to tell whether two variables are associated is to ask whether they are not.[2] In a contingency table, when the distribution of *one* variable is the same for all categories of another, we say that the variables are **independent**. That tells us there's no association between these variables. We'll see a way to formally check for independence later in the book. For now, we'll just compare the distributions.

[2]This kind of "backward" reasoning shows up surprisingly often in science—and in Statistics. We'll see it again later on when we discuss tests of hypotheses.

For Example LOOKING FOR ASSOCIATIONS BETWEEN VARIABLES

RECAP: The table shows results of a poll asking adults whether they were looking forward to the Super Bowl game, looking forward to the commercials, or didn't plan to watch.

Response	Sex: Male	Female	Total
Game	279	200	**479**
Commercials	81	156	**237**
Won't watch	132	160	**292**
Total	**492**	**516**	**1008**

QUESTION: Does it seem that there's an association between interest in Super Bowl TV coverage and a person's sex?

ANSWER: First find the distribution of the three responses for the men (the column percentages):

$$\frac{279}{492} = 56.7\%$$

$$\frac{81}{492} = 16.5\%$$

$$\frac{132}{492} = 26.8\%$$

Then, do the same for the women who were polled, and display the two distributions with a side-by-side bar chart:

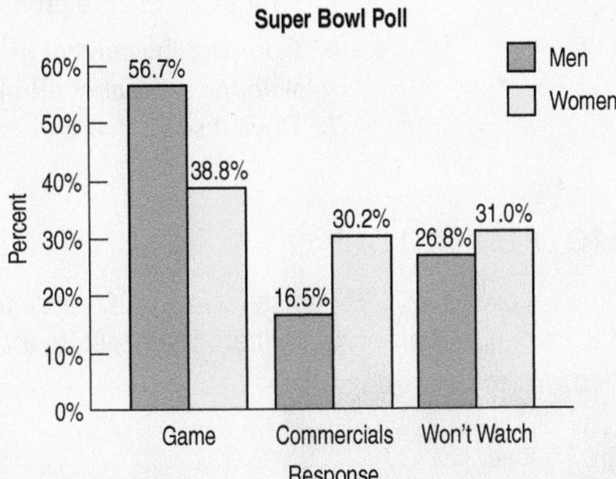

Based on this poll, it appears that women are only slightly less interested than men in watching the Super Bowl telecast: 31% of the women said they didn't plan to watch, compared to just under 27% of men. Among those who planned to watch, however, there appears to be an association between the viewer's sex and what the viewer is most looking forward to. While more women are interested in the game (39%) than the commercials (30%), the margin among men is much wider: 56.7% of men said they were looking forward to seeing the game, compared to only 16.5% who cited the commercials. (We could also mimic Figure 2.7 and show the conditional distribution of *Response* for males, in blue, on one side, and the conditional distribution of *Response* for females, in yellow, on the other side. Your software may well construct it this way).

Figure 2.9

***A segmented bar chart for* Class *by* Survival.** Notice that although the totals for survivors and nonsurvivors are quite different, the bars are the same height because we have converted the numbers to percentages. Compare this display with the side-by-side pie charts of the same data in Figure 2.6.

Segmented Bar Charts

We could display the *Titanic* information by dividing up bars rather than circles. The resulting **segmented bar chart** treats each bar as the "whole" and divides it proportionally into segments corresponding to the percentage in each group. We can clearly see (Figure 2.9) that the distributions of ticket *Class* are different, indicating again that survival was not independent of ticket *Class*.

Just Checking

Below are data from the Ontario Health Supplement survey of 4888 individuals (over age 15 with no history of physical or sexual abuse) about the effect of spanking (or slapping) on future antisocial behaviours during teen or adult years.[3] Externalizing of problems—including illicit drugs and antisocial behaviour—were assessed by an interviewer-administered questionnaire.

	Never spanked	Rarely spanked	Sometimes/ often spanked	Total
Externalizing problems	75	309	240	624
No externalizing problems	921	2146	1197	4264
Total	996	2455	1437	4888

1. What percent of those rarely spanked had externalizing problems?
2. What percent of those with externalizing problems were rarely spanked?
3. What percent of those in the study were rarely spanked and had externalizing problems?
4. Determine the (marginal) distribution of *Spanking (Never, Rarely, Sometimes)*
5. Determine the conditional distribution of the variable *Spanking* for those with externalizing problems.
6. Compare the percent of the never spanked who had externalizing problems with the percent of all individuals who had externalizing problems.
7. Does it seem that *Externalizing* and *Spanking* are independent? Explain.

Step-by-Step Example EXAMINING CONTINGENCY TABLES

Medical researchers followed 6272 Swedish men for 30 years to see if there was any association between the amount of fish in their diet and prostate cancer. Their results are summarized in this table.

We asked for a picture of a man eating fish. This is what we got.
Pieter De Pauw/Fotolia

		Prostate Cancer	
		No	Yes
Fish Consumption	Never/seldom	110	14
	Small part of diet	2420	201
	Moderate part	2769	209
	Large part	507	42

Table 2.9

Question: Is there an association between fish consumption and prostate cancer?

[3]H. L. MacMillan, M. H. Boyle, M. Y. Wong, E. K. Duku, J. E. Fleming, and C. A. Walsh, 1999, "Slapping and spanking in childhood and its association with lifetime prevalence of psychiatric disorders in a general population sample." *CMAJ*, 161(7): 805–809.

THINK ➡ Plan State what the problem is about.

I want to know if there is an association between fish consumption and prostate cancer.

Variables Identify the variables and report the W's.

The individuals are 6272 Swedish men followed by medical researchers for 30 years. The variables record their fish consumption and whether or not they were diagnosed with prostate cancer.

Be sure to check the appropriate condition.

✓ **Categorical Data Condition:** I have counts for both fish consumption and cancer diagnosis. The categories of diet do not overlap, and the diagnoses do not overlap. It's okay to draw pie charts or bar charts.

SHOW ➡ Mechanics Check the marginal distributions first before looking at the two variables together.

Fish Consumption	Prostate Cancer: No	Prostate Cancer: Yes	Total
Never/Seldom	110	14	**124 (2.0%)**
Small part of diet	2420	201	**2621 (41.8%)**
Moderate part	2769	209	**2978 (47.5%)**
Large part	507	42	**549 (8.8%)**
Total	**5806 (92.6%)**	**466 (7.4%)**	**6272 (100%)**

Two categories of the diet are quite small, with only 2.0% Never/Seldom eating fish and 8.8% in the "Large part" category. Overall, 7.4% of the men in this study had prostate cancer.

Then, make appropriate displays to see whether there is a difference in the relative proportions. These pie charts compare fish consumption for men who have prostate cancer to fish consumption for men who don't.

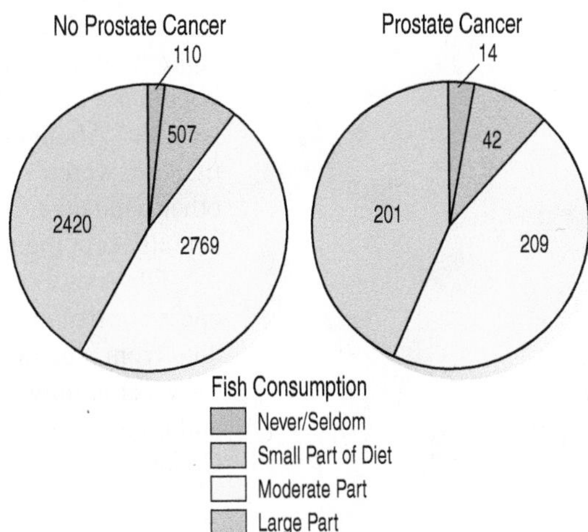

It's hard to see much difference in the pie charts. So, I made a display of the row percentages. Because there are only two alternatives, I chose to display the risk of prostate cancer for each group:

Both pie charts and bar charts can be used to compare conditional distributions. Here we compare prostate cancer rates based on differences in fish consumption.

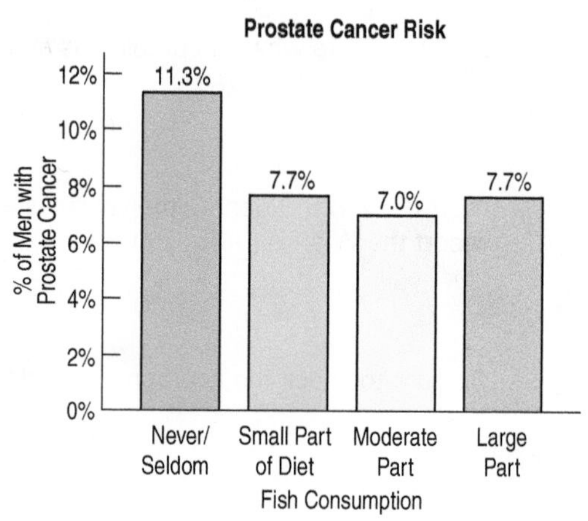

TELL ➡ Conclusion Interpret the patterns in the table and displays in context. If you can, discuss possible real-world consequences. Be careful not to overstate what you see. The results may not generalize to other situations.

Overall, there is a 7.4% rate of prostate cancer among men in this study. Most of the men (89.3%) ate fish either as a moderate or small part of their diet. From the pie charts, it's hard to see a difference in diet between the cancer and non-cancer groups. But in the bar chart, it looks like the cancer rate for those who never/seldom ate fish may be somewhat higher.

However, only 124 of the 6272 men in the study fell into this category, and only 14 of them developed prostate cancer. More study would probably be needed before we would recommend that men change their diets.[4]

This study is an example of looking at a sample of data to learn something about a larger population, one of the main goals of this book. We care about more than these particular 6272 Swedish men. We hope that learning about their experiences will tell us something about the value of eating fish in general. That raises several questions. What population do we think this sample might represent? Do we hope to learn about all Swedish men? About all men? How do we know that other factors besides that amount of fish they ate weren't associated with prostate cancer? Perhaps men who eat fish often have other habits that distinguish them from the others, and maybe those other habits are what actually kept their cancer rates lower.

Observational studies, like this one, often lead to contradictory results because we can't control all the other factors. In fact, a later paper, published in 2011, based on data from a cancer prevention trial on 3400 men from 1994 to 2003, showed that some fatty acids may actually increase the risk of prostate cancer.[5] We'll discuss the pros and cons of observational studies and experiments where we can control the factors in Chapter 11.

[4] "Fatty fish consumption and risk of prostate cancer," *Lancet*, June 2001. The original study actually used pairs of twins, which enabled the researchers to discern that the risk of cancer for those who never ate fish actually was substantially greater. Using pairs is a special way of gathering data. We'll discuss such study design issues and how to analyze the data in later chapters.

[5] "Serum phospholipid fatty acids and prostate cancer risk: Results from the Prostate Cancer Prevention Trial," *American Journal of Epidemiology*, 2011.

WHAT CAN GO WRONG?

- **Don't violate the area principle.** This is probably the most common mistake in a graphical display. It is often made in the cause of artistic presentation. Here, for example, are two displays of the pie chart of the *Titanic* passengers by class:

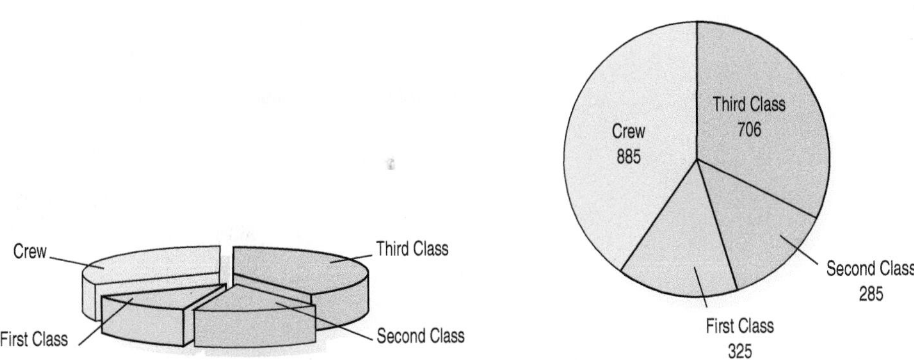

  The one on the left looks pretty, doesn't it? But showing the pie on a slant violates the area principle and makes it much more difficult to compare fractions of the whole made up of each class—the principal feature that a pie chart ought to show.

- **Keep it honest.** Here's a pie chart that displays data on the percentage of high school students who engage in specified dangerous behaviours as reported by the Centers for Disease Control. What's wrong with this plot?

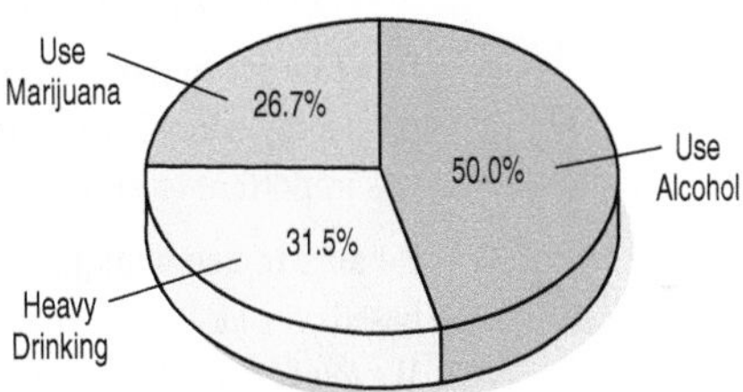

  Try adding up the percentages. Or look at the 50% slice. Does it look right? Then think: What are these percentages of? Is there a "whole" that has been sliced up? In a pie chart, the proportions shown by each slice of the pie must add up to 100% and each individual must fall into *only one category*. Of course, showing the pie on a slant makes it even harder to detect the error.

  Here's another one. The following chart shows the average number of texts in various time periods by American cell phone customers in the period 2006 to 2011.

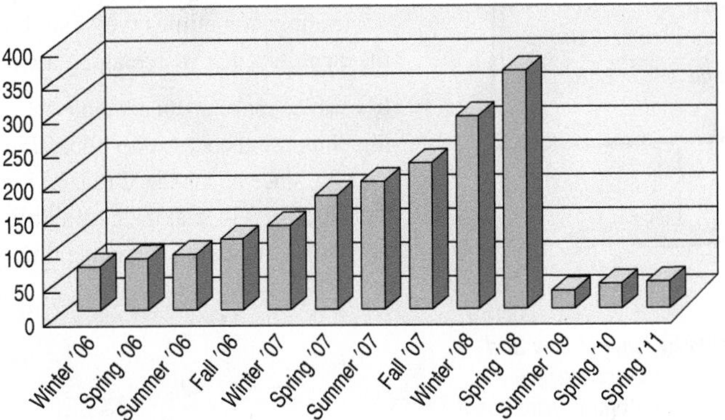

  It may look as though text messaging decreased suddenly sometime around 2010, which probably doesn't match your experience. In fact, this chart has several problems. First, it's not a bar chart. Bar charts display counts of categories. This bar chart is a plot of a quantitative variable (average number of texts) against time—although to make it worse, some of the time periods are missing. Even though these flaws are already fatal, the worst mistake is one that can't be seen from the plot. In 2010, the company reporting the data switched from reporting the average number of texts per year (reported each quarter) to the average number of texts per month. So, the numbers in the last three quarters should be multiplied by 12 to make them comparable to the rest.

Looks like things didn't change much in the final years of the twentieth century—until you read the bar labels and see that the last three bars represent single years, while all the others are for pairs of years. Of course, the false depth makes it harder to see the problem.

- **Don't confuse similar-sounding percentages.** These percentages sound similar but are different:

		Class				
		First	Second	Third	Crew	Total
Survival	Alive	203	118	178	212	**711**
	Dead	122	167	528	673	**1490**
	Total	**325**	**285**	**706**	**885**	**2201**

- The percentage of the passengers who were both in first class and survived: This would be 203/2201, or 9.2%.
- The percentage of the first-class passengers who survived: This is 203/325, or 62.5%.
- The percentage of the survivors who were in first class: This is 203/711, or 28.6%.

In each instance, pay attention to the *who* implicitly defined by the phrase. Often there is a restriction to a smaller group (all aboard the *Titanic*, those in first class, and those who survived, respectively) before a percentage is found. Your discussion of results must make these differences clear.

- **Don't forget to look at the variables separately, too.** When you make a contingency table or display a conditional distribution, be sure you also examine the marginal distributions. It's important to know how many cases are in each category.
- **Be sure to use enough individuals.** When you consider percentages, take care that they are based on a large enough number of individuals. Take care not to make a report such as this one: *We found that 66.67% of the rats improved their performance with training. The other rat died.*
- **Don't overstate your case.** Independence is an important concept, but it is rare for two variables to be entirely independent. We can't conclude that one variable has no effect whatsoever on another. Usually all we know is that little effect was observed in our study. Other studies of other groups under other circumstances could find different results.

Entering Centreville	
Established	1793
Population	7943
Elevation	710
Average	3482

Simpson's Paradox

- **Be wary of looking only at overall percentages or averages.** Sometimes averages can be misleading; sometimes they just don't make sense at all. Be careful when averaging that the quantities you're averaging are comparable. The Centreville sign says it all.

It's easy to make up an example showing that looking at *overall* percentages can be misleading. Suppose there are two pilots, Pierre and Usha. Pierre argues that he's the better pilot of the two, since he managed to land 83% of his last 120 flights on time compared with Usha's 78%. But let's look at the data a little more closely. Here are the results for each of their last 120 flights, broken down by the time of day they flew:

Table 2.10

***On-time flights by* Time of Day *and* Pilot.** Look at the percentages within each *Time of Day* category. Who has a better on-time record during the day? At night? Who is better overall?

		Time of Day		
		Day	Night	Overall
Pilot	Pierre	90 out of 100 90%	10 out of 20 50%	100 out of 120 83%
	Usha	19 out of 20 95%	75 out of 100 75%	94 out of 120 78%

Look at the day- and night-time flights separately. For day flights, Usha had a 95% on-time rate, and Pierre only a 90% rate. At night, Usha was on time 75% of the time, and Pierre only 50%. So Pierre is better "overall," but Usha is better both during the day and at night. How can this be?

What's going on here is a problem known as **Simpson's paradox**, named for the statistician who first described it in the 1960s. It comes up rarely in real life, but there have been several well-publicized cases of it. As we can see from the pilot example, the problem is combining percentages over different groups. Usha has mostly night flights, which are more difficult, so her *overall average* is heavily influenced by her night-time average. Pierre, on the other hand, benefits from flying mostly during the day, with its higher on-time percentage. With their very different patterns of flying conditions, taking an overall percentage is misleading. It's not a fair comparison.

The moral of Simpson's paradox is to be careful when you combine across different levels of a second variable. It's always better to compare percentages *within* each level of that other variable (*controlling* for that variable, we often say). The overall percentage may be misleading. From a practical standpoint, the issue often is whether you are unknowingly averaging over an important unidentified variable buried in your data (lurking). Think about and try to identify variables that have the potential to alter your understanding of a relationship. They may not be so easy to catch as "time of day" or "graduate school" in the previous and next examples.

Simpson's Paradox

One famous example of Simpson's paradox arose during an investigation of admission rates for men and women at the University of California at Berkeley's graduate schools. As reported in an article in *Science*[6], about 45% of male applicants were admitted, but only about 30% of female applicants got in. It looked like a clear case of discrimination. However, when the data were broken down by school (Engineering, Law, Medicine, etc.), it turned out that within each school, the women were admitted at nearly the same or, in some cases, much higher rates than the men. How could this be? Women applied in large numbers to schools with very low admission rates (Law and Medicine, for example, admitted fewer than 10% of all applicants). Men tended to apply to Engineering and Science. Those schools have admission rates above 50%. When the average was taken, the women had a much lower overall rate, but the average didn't really make sense.

All of the methods of this chapter work with *categorical variables*. Focusing on the total, row, or column percentages in a contingency table changes who you are interested in. Restricting the *who* of the study in different ways answers different questions.

What Have We Learned?

Learning Objectives

Make and interpret a frequency table for a categorical variable.

- We can summarize categorical data by counting the number of cases in each category, sometimes expressing the resulting distribution as percentages.

[6] P. J. Bickel, E. A. Hammel, and J. W. O'Connell (1975). "Sex Bias in Graduate Admissions: Data from Berkeley." *Science*, 187(4175): 398–404. doi:10.1126/science.187.4175.398.

Make and interpret a bar chart or pie chart.

- We display categorical data using the area principle in either a **bar chart** or a **pie chart**.

Make and interpret a contingency table.

- When we want to see how two categorical variables are related, we put the counts (and/or percentages) in a two-way table called a **contingency table**.

Make and interpret bar charts and pie charts of marginal distributions.

- We look at the **marginal distribution** of each variable (found in the margins of the table). We also look at the **conditional distribution** of a variable within each category of the other variable.
- Comparing conditional distributions of one variable across categories of another tells us about the association between variables. If the conditional distributions of one variable are the same for every category of the other, the variables are **independent**.

Review of Terms

The key terms are in chapter order so you can use this list to review the material in the chapter.

Frequency table A table that lists the categories in a categorical variable and gives the count (or percentage) of observations for each category (p. 15).

Relative frequency table A table that displays the proportions or percentages, rather than the counts, of the values in each category (p. 16).

Distribution The distribution of a variable gives

- the possible values of the variable, and
- the relative frequency of each value (p. 16).

Area principle In a statistical display, each data value should be represented by the same amount of area (p. 16).

Bar chart A chart that shows bars whose heights represent the count (or percentage) of observations for each category of a categorical variable (p. 17).

Relative frequency bar chart A chart that shows bars whose heights represent the percentages, instead of the counts, of observations for each category of a categorical variable (p. 17).

Pie chart A chart that shows how a "whole" divides into categories by showing a wedge of a circle whose area corresponds to the proportion in each category (p. 17).

Categorical data condition The methods in this chapter are appropriate for displaying and describing categorical data. Be careful not to use them with quantitative data (p. 17).

Contingency table A table that displays counts and, sometimes, percentages of individuals falling into named categories on two or more variables. The table categorizes the individuals on all variables at once to reveal possible patterns in one variable that may be contingent on the category of the other (p. 18).

Marginal distribution In a contingency table, the distribution of either variable alone. The counts or percentages are the totals found in the margins (last row or column) of the table (p. 18).

Joint distribution The distribution of both variables in a contingency table, expressed as a percentage or a count. (p. 19).

Conditional distribution The distribution of a variable restricting the *who* to consider only a smaller group of individuals (p. 21).

Independent Variables are said to be independent if the conditional distribution of one variable is the same for each category of the other. We'll show how to check for independence in a later chapter (p. 22).

Segmented bar chart A chart that displays the conditional distribution of a categorical variable within each category of another variable (p. 23).

Simpson's paradox Relationships among proportions taken within different groups or subsets can appear to contradict relationships among the overall proportions (p. 29).

On the Computer DISPLAYING CATEGORICAL DATA

Although every package makes a slightly different bar chart, they all have similar features:

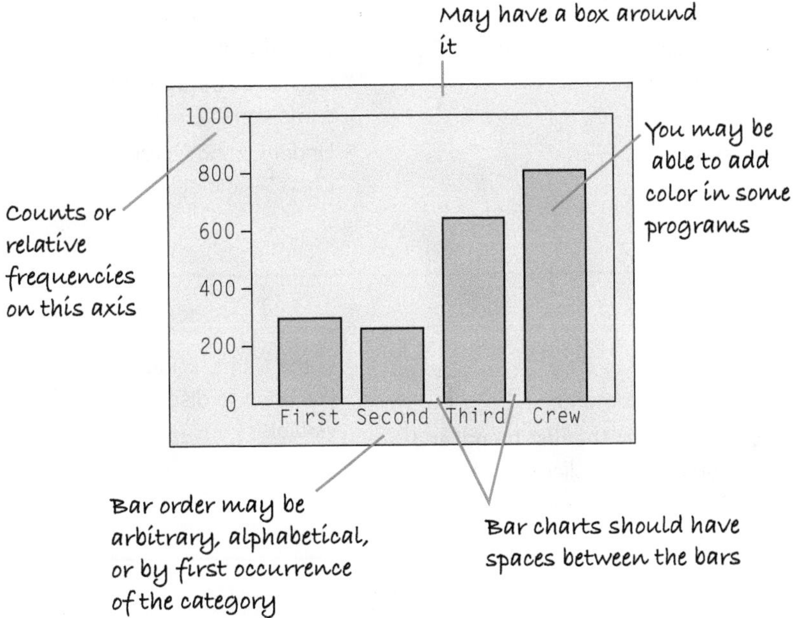
May have a box around it
1000
800
600
400
200
0
First
Second
Third
Crew
Counts or relative frequencies on this axis
You may be able to add color in some programs
Bar order may be arbitrary, alphabetical, or by first occurrence of the category
Bar charts should have spaces between the bars

Sometimes the count or a percentage is printed above or on top of each bar to give some additional information. You may find that your statistics package sorts category names in annoying orders by default. For example, many packages sort categories alphabetically or by the order the categories are seen in the data set. Often, neither of these is the best choice.

EXCEL

- First make a Pivot Table (Excel's name for a frequency table). From the **Data** menu choose **Pivot Table** and **Pivot Chart Report**.
- When you reach the Layout window, drag your variable to the row area and drag your variable again to the data area. This tells Excel to count the occurrences of each category. Once you have an Excel Pivot Table, you can construct bar charts and pie charts.
- Click inside the Pivot Table.
- Click the Pivot Table Chart Wizard button. Excel creates a bar chart.
- A longer path leads to a pie chart; see your Excel documentation.

COMMENTS

Excel uses the Pivot Table to specify the category names and find counts within each category. If you already have that information, you can proceed directly to the Chart Wizard.

EXCEL 2007

To make a bar chart:

- Select the variable in Excel you want to work with.
- Choose the **Column** command from the Insert tab in the Ribbon.
- Select the appropriate chart from the drop-down dialog.

To change the bar chart into a pie chart:

- Right-click the chart and select **Change Chart Type...** from the menu. The Chart type dialog opens.
- Select a pie chart type.
- Click the **OK** button. Excel changes your bar chart into a pie chart.

JMP

JMP makes a bar chart and frequency table together.

- From the **Analyze** menu, choose **Distribution**.
- In the Distribution dialog, drag the name of the variable into the empty variable window beside the label "Y, Columns"; click **OK**.

To make a pie chart:

- Choose **Chart** from the **Graph** menu.
- In the Chart dialog, select the variable name from the Columns list.
- Click on the button labelled "Statistics," and select "N" from the drop-down menu.
- Click the "**Categories, X, Levels**" button to assign the same variable name to the x-axis.
- Under Options, click on the second button—labelled "Bar Chart"—and select "Pie" from the drop-down menu.

MINITAB

To make a bar chart:

- Choose **Bar Chart** from the **Graph** menu.
- Select "Counts of unique values" in the first menu, and select "Simple" for the type of graph. Click **OK**.
- In the Chart dialog, enter the name of the variable that you wish to display in the box labelled "Categorical variables."
- Click **OK**.

R

To make a bar chart or pie chart in **R**, first create the frequency table for the desired variable:

- **table(X)** will give a frequency table for a single variable X.
- **barplot(table(X))** will give a bar chart for X.
- **pie(table(X))** will give a pie chart.

COMMENTS

Stacked bar charts of two variables, X and Y, can be made using **barplot(xtabs(~X+Y))** or directly from a two-way table of counts or percentages. Legends and other options are available for all charts using various functions.

SPSS

To make a bar chart:

- Open the **Chart Builder** from the **Graphs** menu.
- Click the **Gallery** tab.
- Choose **Bar Chart** from the list of chart types.
- Drag the appropriate bar chart onto the canvas.
- Drag a categorical variable onto the x-axis drop zone.
- Click **OK**.

COMMENT

A similar path makes a pie chart by choosing **Pie chart** from the list of chart types.

STATCRUNCH

To make a bar chart or pie chart:

- Click on **Graphics**.
- Choose the type of plot ≫ **with data** or ≫ **with summary**.
- Choose the variable name from the list of **Columns**; if using summaries, also choose the counts.
- Click on **Next**.
- Choose **Frequency/Counts** or (usually) **Relative frequency/Percents**. Note that you may elect to group categories under a specified percentage as "Other."
- Click on **Create Graph**.

TI-83/84 PLUS

The TI-83 won't do displays for categorical variables.

Exercises

1. **Graphs in the news** Find a bar chart of categorical data from a newspaper, a magazine, or the Internet.
   a) Is the graph clearly labelled?
   b) Does it violate the area principle?
   c) Does the accompanying article tell the W's of the variable?
   d) Do you think the article correctly interprets the data? Explain.

2. **Graphs in the news II** Find a pie chart of categorical data from a newspaper, a magazine, or the Internet.
   a) Is the graph clearly labelled?
   b) Does it violate the area principle?
   c) Does the accompanying table tell the W's of the variable?
   d) Do you think the article correctly interprets the data? Explain.

3. **Tables in the news** Find a frequency table of categorical data from a newspaper, a magazine, or the Internet.
   a) Is it clearly labelled?
   b) Does it display percentages or counts?
   c) Does the accompanying article tell the W's of the variable?
   d) Do you think the article correctly interprets the data? Explain.

4. **Tables in the news II** Find a contingency table of categorical data from a newspaper, a magazine, or the Internet.
   a) Is it clearly labelled?
   b) Does it display percentages or counts?
   c) Does the accompanying article tell the W's of the variable?
   d) Do you think the article correctly interprets the data? Explain.

5. **Forest fires 2010** Every year, thousands of forest fires burn throughout Canada. According to the National Forestry Database,[7] in 2010, a fewer-than-average 6986 forest fires were started, affecting 3 048 926 hectares of land. Of these, 3279 fires were caused by lightning, affecting 2 633 036 hectares of land; 3608 were caused by human activities, affecting 413 664 hectares of land; while the remainder were of unknown cause. Determine the relative frequency distribution of 2010 forest fire causes and describe it in a sentence or two.

6. **Forest fires 2010 by region** Of those forest fires discussed in Exercise 5, 737 were in Quebec, 931 in Ontario, 1845 in Alberta, and 1673 in BC. Determine the relative frequency distribution of forest fire locations within Canada, and describe it in a sentence or two.

7. **Teen smokers** The organization Monitoring the Future (www.monitoringthefuture.org) asked 2048 U.S. Grade 8 students who said they smoked cigarettes what brands they preferred. The table below shows brand preferences for two regions of the country. Write a few sentences describing the similarities and differences in brand preferences among Grade 8 students in the two regions listed.

Brand Preference	South	West
Marlboro	58.4%	58.0%
Newport	22.5%	10.1%
Camel	3.3%	9.5%
Other (over 20 brands)	9.1%	9.5%
No Usual Brand	6.7%	12.9%

[7]Source: National Forestry Database http://nfdp.ccfm.org/compendium/fires/tables-_index_e.php

8. Bad countries According to the January 2008 Canada's World Poll (conducted by Environics, principal sponsor The Simon Foundation[8]), when asked to name countries that stand out as a negative force in the world today, half (52%) of Canadians pointed to their nearest neighbour, the U.S. Other countries making the list were Iran (21%), Iraq (19%), China (13%), Afghanistan (11%), Pakistan (9%), North Korea (8%), and Russia (6%). Another 4% identified Israel or the Palestinian territories. Four percent named Canada. (To see who the good guys are, you'll have to read the report.)

a) Display this information in an appropriate display. Explain clearly which displays are appropriate and which are not and why.

b) Some could not name any country as a negative force. Can you determine what percent this was from the given information? Explain. Could it be possibly be over 30%?

c) Is it true that 49% (21 + 19 + 9) named Iran or Iraq or Pakistan as negative forces? Explain.

9. Oil spills as of 2010 Data from the International Tanker Owners Pollution Federation Limited give the cause of spillage for 460 oil tanker accidents resulting in spills of more than 700 tons of oil from 1970–2010 (www.itopf.com/information-services/data-and-statistics/statistics). Following are some displays.

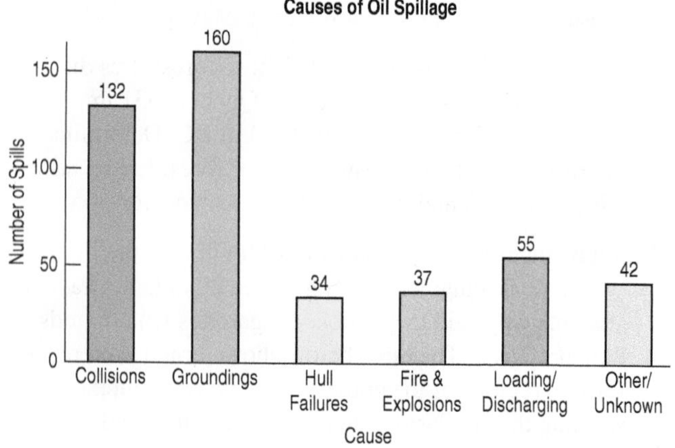

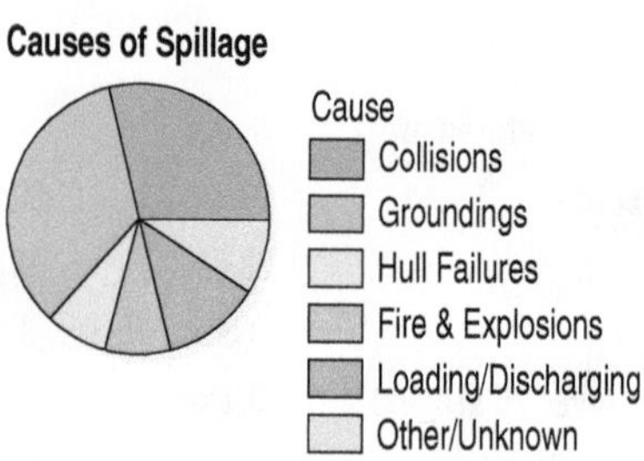

a) Write a brief report interpreting what the displays show.

b) Is a pie chart an appropriate display for these data? Explain.

10. Winter Olympics 2010 Twenty-six countries won medals in the 2010 Winter Olympics in Vancouver. The table lists them, along with the total number of medals each won:

Country	Medals	Country	Medals
United States	37	Poland	6
Germany	30	Italy	5
Canada	26	Japan	5
Norway	23	Finland	5
Austria	16	Australia	3
Russia	15	Belarus	3
South Korea	14	Slovakia	3
China	11	Croatia	3
Sweden	11	Slovenia	3
France	11	Latvia	2
Switzerland	9	Great Britain	1
Netherlands	8	Estonia	1
Czech Republic	6	Kazakhstan	1

a) Try to make a display of these data. What problems did you encounter?

b) Can you find a way to organize the data so that the graph is more successful?

11. Global warming The Pew Research Center for the People and the Press (http://people-press.org) asked a representative sample of U.S. adults about global warming, repeating the question over time. In October 2010, the responses reflected an increased belief that global warming is real and due to human activity. Here's a display of the percentages of respondents choosing each of the major alternatives offered:

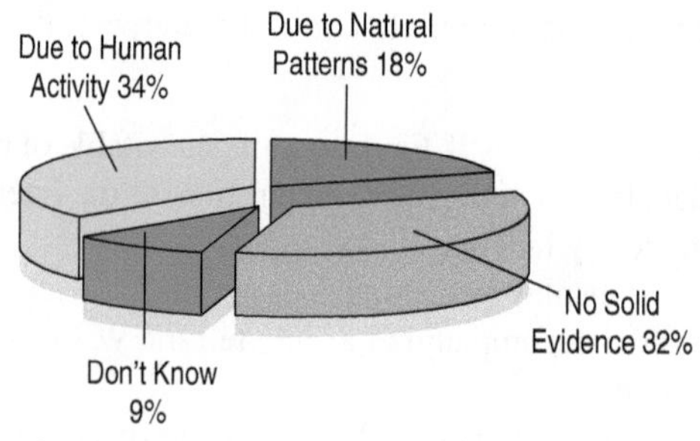

List the errors in this display.

[8] Canada's World Poll's principal sponsor is the Simmons Foundation, and it is also sponsored by the Environics Institute, the Canadian Broadcasting Corporation (CBC), the *Globe and Mail*, and *Le Devoir*.

12. Modalities A survey of athletic trainers[9] asked what modalities (treatment methods, such as ice, whirlpool, ultrasound, or exercise) they commonly use to treat injuries. Respondents were each asked to list three modalities. The article included the figure here.

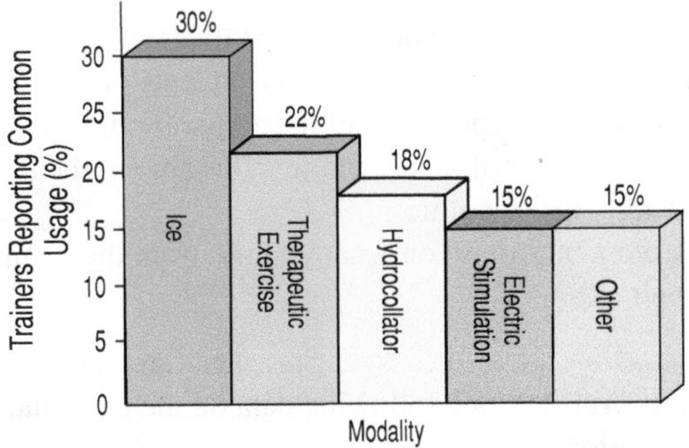

a) What problems do you see with the graph?
b) Consider the percentages for the named modalities. Do you see anything odd about them?

13. Complications The survey of athletic trainers reported in Exercise 12 summarizes the complications most commonly associated with cryotherapy (ice). The trainers were asked to report the most common side effects. Of those identifying cryotherapy, 86 respondents reported allergic reactions, 23 reported burns, 16 reported pain intolerance, and 6 reported frostbite.
a) Make an appropriate display of these data.
b) Specify the *Who* for these data. Would the data provide the most useful information about the risks of cryotherapy?

T **14. Aboriginal languages 2011** Many Aboriginal languages in Canada are endangered, and extinction for some may be only a generation away. In the 2011 Census, a vast majority of Canadians of Aboriginal identity reported English as their mother tongue. Following are Census counts for Aboriginal people who reported an Aboriginal language as their mother tongue (excluding multiple responses, e.g., English and Aboriginal language).
a) Find the relative frequency distribution of Aboriginal mother tongue.
b) Portray it with two different types of graphs. Point out the most interesting or salient characteristic(s) of the distribution. Which type of graph do you prefer? Explain.
c) The first seven in the list are all *Algonquian* languages. Repeat part b) after combining the first seven languages into a category named "Algonquian languages." Included in the "Other" category are 2670 people with other Algonquian languages as their mother tongue, so adjust for this, too.

Aboriginal mother tongue	Total = 200 700
Cree	78 200
Ojibway	17 625
Oji-Cree	9 835
Innu/Montagnais	10 785
Mi'kmaq	7 635
Atikamekw	5 820
Blackfoot	2 860
Inuktitut	33 500
Dene	11 215
Dakota/Stoney/Siouan	4 200
Other Aboriginal languages	19 025

Source: Statistics Canada, 2011 Census of Population, Statistics Canada catalogue no. 98-314-XCB2011016 (Canada, Code01)

15. Spatial distribution We divided a tract of land into four equal quadrants, in the east–west direction. We randomly selected 100 longleaf pine trees from the tract of land and counted the number of trees in each quadrant, yielding the counts below:

Quadrant 1	Quadrant 2	Quadrant 3	Quadrant 4
18	21	22	39

a) Is it possible that the longleaf pine trees are randomly dispersed over the region, which would result in a uniform or even distribution over quadrants? Find the relative frequency distribution of quadrant location. Do you think the proportions are similar enough to support this theory? (We will learn how to formally test such theories or *hypotheses* in later chapters.)
b) Counts for these same quadrants were also obtained for a random sample of 50 trees of a different species selected from this tract of land:

Quadrant 1	Quadrant 2	Quadrant 3	Quadrant 4
6	12	14	18

Find the distribution of quadrant location for this species, and compare it with the distribution in part a). Are they fairly similar? Do you think the differences are within the bounds of chance or random variation in

[9]*Source*: Scott F. Nadler, Michael Prybicien, Gerard A. Malanga, and Dan Sicher, 2003, "Complications from therapeutic modalities: results of a national survey of athletic trainers," *Archives of Physical Medical Rehabilitation*, 84.

the samples? (We'll learn how to answer this question more precisely when we discuss tests of hypotheses.)

16. Politics Students in an Intro Stats course were asked to describe their politics as "Liberal," "Moderate," or "Conservative." Here are the results:

		Politics			
		L	M	C	Total
Sex	Female	35	36	6	77
	Male	50	44	21	115
	Total	**85**	**80**	**27**	**192**

a) What percent of the class is male?
b) What percent of the class considers themselves to be "Conservative"?
c) What percent of the males in the class consider themselves to be "Conservative"?
d) What percent of all students in the class are males who consider themselves to be "Conservative"?

17. Politics revisited Look again at the table of political views for the Intro Stats students in Exercise 16.
a) Find the conditional distributions (percentages) of political views for the females.
b) Find the conditional distributions (percentages) of political views for the males.
c) Make a graphical display that compares the two distributions.
d) Do the variables *Politics* and *Sex* appear to be independent? Explain.

18. More politics Look once more at the table summarizing the political views of Intro Stats students in Exercise 16.
a) Produce a graphical display comparing the conditional distributions of males and females among the three categories of politics.
b) Comment briefly on what you see from the display in part a).

T **19. Canadian languages 2011** Statistics Canada provides, on its Web site, the following data on the Canadian population.
a) What percent of Canadian citizens speak only English?
b) What percent of Canadian citizens speak French?
c) What percent of Quebec residents speak French?
d) What percent of French-speaking Canadians live in Quebec?
e) Do you think that language knowledge and province of residence are independent? Explain.

		Knowledge of official languages[1]				
	Geographic name	**English only**	**French only**	**Both**	**Neither**	**Total**
	Canada	22 564 665	4 165 015	5 795 575	595 920	33 121 175
Province	Newfoundland and Labrador	485 740	135	23 450	625	509 950
	Prince Edward Island	120 590	130	17 005	715	138 435
	Nova Scotia	814 670	875	93 435	1635	910 615
	New Brunswick	426 675	66 380	245 885	955	739 900
	Quebec	363 860	4 047 175	3 328 725	76 195	7 815 955
	Ontario	10 984 360	42 980	1 395 805	298 920	12 722 065
	Manitoba	1 074 335	1 490	103 145	14 135	1 193 095
	Saskatchewan	965 925	430	46 570	5 395	1 018 310
	Alberta	3 321 815	3 205	235 565	49 600	3 610 185
	British Columbia	3 912 955	2 050	296 645	144 560	4 356 205
	Yukon	29 050	90	4 420	95	33 655
	Northwest Territories	37 045	45	3 720	235	41 040
	Nunavut	27 665	35	1 200	2 860	31 765

[1]Refers to the ability of the individual to conduct a conversation in English only, in French only, in both English and French, or in neither English nor French at the time of the census (May 10, 2011).

Source: Adapted from Statistics Canada, Language Highlight Tables, 2011 Census (Catalogue no.: 98-314-XWE2011002)

T **20. Tattoos** A study by the University of Texas Southwestern Medical Center examined 626 people to see if there was an increased risk of contracting hepatitis C associated with having a tattoo. If the subject had a tattoo, researchers asked whether it had been done in a commercial tattoo parlour or elsewhere. Write a brief description of the association between tattooing and hepatitis C, including an appropriate graphical display.

	Tattoo done in commercial parlour	Tattoo done elsewhere	No tattoo
Has hepatitis C	17	8	18
No hepatitis C	35	53	495

T **21. Weather forecasts** Just how accurate are the weather forecasts we hear every day? The following table compares the daily forecast with a city's actual weather for a year.

		Actual Weather	
		Rain	No rain
Forecast	Rain	27	63
	No rain	7	268

a) On what percent of days did it actually rain?
b) On what percent of days was rain predicted?
c) What percent of the time was the forecast correct?
d) Do you see evidence of an association between the type of weather and the ability of forecasters to make an accurate prediction? Write a brief explanation, including an appropriate graph.

22. Attraction, repulsion In a certain geographical region, 200 one-metre-square quadrants were randomly selected. In each quadrant, it was noted whether there were any specimens of grass species *Agropyron* and whether there were any specimens of grass species *Agrostis*. We are often interested in whether species tend to attract or repel each other. The following summary counts were recorded:

	Agrostis present	Agrostis absent
Agropyron present	10	70
Agropyron absent	40	80

a) How often is *Agrostis* present in plots where *Agropyron* is present? On the other hand, how often is *Agrostis* present when *Agropyron* is absent?
b) Find the conditional distribution of *Agropyron* presence/absence when *Agrostis* is present, and repeat for when *Agrostis* is absent. Plot these two distributions and compare. Describe the nature of the association between these two species. Do they attract or repel? (In later chapters, we will learn how to do a precise test of the association hypothesis.)

T **23. Drivers' licenses 2011** The following table shows the number of licensed American drivers (in millions) by age and by sex (www.dot.gov).

Age	Male Drivers (millions)	Female Drivers (millions)	Total
19 and Under	5.1	4.9	10.0
20–24	8.7	8.6	17.3
25–29	9.2	9.2	18.4
30–34	8.9	8.9	17.8
35–39	9.7	9.6	19.3
40–44	9.9	9.8	19.7
45–49	10.6	10.7	21.3
50–54	10.1	10.2	20.3
55–59	8.7	8.9	17.6
60–64	7.2	7.3	14.5
65–69	5.3	5.4	10.7
70–74	3.8	4.0	7.8
75–79	2.9	3.2	6.1
80–84	2.0	2.4	4.4
85 and Over	1.4	1.7	3.1
Total	103.5	104.8	208.3

a) What percent of total drivers are under 20?
b) What percent of total drivers are male?
c) Write a few sentences comparing the number of male and female licensed drivers in each age group.
d) Do a driver's age and sex appear to be independent? Explain.

24. Fat and fatter Following are estimated population counts (in thousands), using data from the National Population Health Survey, on changes from one body mass index (BMI) category to another between 1995 and 2005 for the Canadian household population aged 18 to 56 (who reported their height and weight in the survey). BMI is defined as body weight in kilograms divided by height (in metres) squared.

	2005			
1995	Underweight	Normal	Overweight	Obese
Underweight	91	241	?	?
Normal	110	4850	2348	265
Overweight	?	547	3188	1383
Obese	?	66	270	1448

Source: Adapted from the Statistics Canada CANSIM database http://cansim2.statcan.gc.cak table 104-7030, May 1, 2008.

The "?" indicates a value too small to be estimated reliably. You may plug in zero for the "?", if needed, to answer any questions below.
a) What percent of Canadians were normal in weight during both study years?

b) What percent of Canadians were overweight or obese in 2005?
c) What percent of normal weight Canadians in 1995 became overweight or obese by 2005?
d) What percent of normal weight Canadians in 2005 were overweight or obese in 1995?
e) What percent of overweight or obese Canadians in 1995 got their weight down to normal by 2005?

25. Anorexia Hearing anecdotal reports that some patients undergoing treatment for the eating disorder anorexia seemed to be responding positively to the antidepressant Prozac, medical researchers conducted an experiment to investigate. They found 93 women being treated for anorexia who volunteered to participate. For one year, 49 randomly selected patients were treated with Prozac and the other 44 were given an inert substance called a placebo. At the end of the year, patients were diagnosed as healthy or relapsed, as summarized in the table:

	Prozac	Placebo	Total
Healthy	35	32	67
Relapse	14	12	26
Total	**49**	**44**	**93**

Do these results provide evidence that Prozac might be helpful in treating anorexia? Explain.

T 26. Neighbourhood density How do people distribute themselves in our cities? From the 2001 Canadian census, we know what types of neighbourhoods people live in, for example, whether the neighbourhoods are high, medium, or low density. Low density means that 66.6% or more of the housing stock is composed of single family dwellings, semi-detached dwellings, or mobile homes, while less than 33.3% means high density, and in between is medium-density. Following is a table with neighbourhood density data for some larger cities.

a) Are these percentages column percentages, row percentages, or table percentages?

	Density		
	High	**Medium**	**Low**
Quebec City	30%	24%	46%
Montreal	47%	19%	34%
Ottawa	22%	37%	40%
Toronto	23%	31%	47%
Winnipeg	10%	26%	64%
Calgary	6%	27%	67%
Edmonton	12%	30%	58%
Vancouver	25%	38%	37%

Source: Adapted from Statistics Canada Canadian Social Trends, 11-008-XIE2008001, vol. 85, June 2008. www.statcan.gc.ca/bsolc/olc-cel/olc-cel?catno=11-008-x&lang=eng.

b) Use segmented bar charts to show and compare the different percentages of city dwellers' high-, medium-, and low-density neighbourhoods for each of the cities (i.e., the conditional distribution of density, by city). Order the cities according to percentage of high density. Do you prefer this plot or some other type?
c) Can you construct a table showing what percent of all the high-density neighbourhood dwellers live in each city, what percent of all the medium-density dwellers live in each city, and what percent of all the low-density dwellers live in each city? If not, what additional information would you need? Would row total counts suffice? Would the total sample size and marginal distribution of city suffice?
d) Do you think that these data prove that the Latin or French temperament causes people to live closer to each other?

T 27. Smoking gene? In a study of Icelandic smokers, the association of smoking and a genetic variant is shown below.

a) Find the marginal distribution of genotype for these individuals.
b) Find the conditional distribution of genotype for each category of smoker.

Genotype Status and SQ Level of 13 945 Icelandic Smokers

Parameter	Genotype of rs1051730			Total *n* (frequency)	Frequency of T allele
	GG	GT	TT		
Cigarettes per day (SQ level)					
1 to 10 (0)	1 743	1 558	326	3 627 (0.260)	0.305
11 to 20 (1)	2 727	2 865	810	6 402 (0.459)	0.350
21 to 30 (2)	1 145	1 416	427	2 988 (0.214)	0.380
31 and more (3)	341	448	139	928 (0.067)	0.391
All levels (frequency)	5956 (0.427)	6 287 (0.451)	1 702 (0.122)	13 945 (1.000)	0.347

Source: Stephen J. Chanock and David J. Hunter. "Genomics: When the smoke clears." *Nature*, 452, 537–538 (3 April 2008). Reprinted by permission from Macmillan Publishers Ltd. www.nature.com/nature

c) Plot the conditional distributions and compare. Does the allele T seem to be related to smoking? If so, do you think that this proves that presence of T increases susceptibility to nicotine addiction?

28. Pet ownership The U.S. Census Bureau reports on the number of households owning various types of pets. Specifically, they keep track of dogs, cats, birds, and horses.

Income Distribution of Households Owning Pets (Percent)

Income	Pet: Dog	Cat	Bird	Horse
Under $12 500	14	15	16	9
$12 500 to $24 999	20	20	21	21
$25 000 to $39 999	24	23	24	25
$40 000 to $59 999	22	22	21	22
$60 000 and over	20	20	18	23
Total	100	100	100	100

a) Do you think the income distributions of the households who own these different animals should be roughly the same? What is your guess? Explain.

b) The table shows column percentages. Are these the percentages of income levels for each type of animal owned or the percentages of type of pet owned for each income class?

c) Do the data support your initial guess? Explain.

29. Antidepressants and bone fractures. For a period of five years, physicians at McGill University Health Centre followed more than 5000 adults over the age of 50. The researchers were investigating whether people taking a certain class of antidepressants (SSRIs) might be at greater risk of bone fractures. Their observations are summarized in the table.

	Taking SSRI	No SSRI	Total
Experienced fractures	14	244	258
No fractures	123	4627	4750
Total	137	4871	5008

Do these results suggest there's an association between taking SSRI antidepressants and experiencing bone fractures? Explain.

T **30. Blood proteins** Blood samples were obtained from a random sample of 100 individuals, and blood type (A or B) along with presence or absence of a certain blood protein were determined for each individual. The following counts were obtained:

Type A, protein	Type A, no protein	Type B, protein	Type B, no protein
5	35	20	40

Examine the association between the presence of the protein and blood type by reorganizing the data into an appropriate two-way table. Give relevant conditional distributions and describe any observed association.

31. Cell phones A researcher wanted to determine the possible association between using a cell phone and car accidents. He surveyed 180 drivers and found that 78 regularly used a cell phone while driving, of which 20 had had crashes in the past year, whereas 102 drivers did not use cell phones while driving, of which 10 had had accidents in the past year.

a) Create a two-by-two table of crash versus non-crash, cell phone versus non-cell phone for these data. Find the relevant conditional distributions (condition on cell phone ownership), and comment on any observed association.

b) On the basis of this study, would you conclude that the use of a cell phone increases the risk of a car accident? Explain your answer.

T **32. Twins** In 2000, the *Journal of the American Medical Association (JAMA)* published a study that examined pregnancies that resulted in the birth of twins.[10] Births were classified as preterm with intervention (induced labour or Caesarean), preterm without procedures, or term/post-term. Researchers also classified the pregnancies by the level of prenatal medical care the mother received (inadequate, adequate, or intensive). The data, from the years 1995–1997, are summarized in the following table. Figures are in thousands of births.

Twin Births 1995–1997 (in thousands)

Level of Prenatal Care	Preterm (induced or Caesarean)	Preterm (without procedures)	Term or post-term	Total
Intensive	18	15	28	61
Adequate	46	43	65	154
Inadequate	12	13	38	63
Total	76	71	131	278

[10]From "Trends in Twin Birth Outcomes and Prenatal Care Utilization in the United States, 1981–1997" by Dr. Michael D. Kogan et al. JAMA 284[2000]: 335–341. Reprinted with permission.

a) What percent of these mothers did not receive adequate medical care during their pregnancies?
b) What percent of all twin births were preterm?
c) Among the mothers who did not receive adequate medical care, what percent of the twin births were preterm?
d) Create an appropriate graph comparing the outcomes of these pregnancies by the level of medical care the mother received.
e) Write a few sentences describing the association between these two variables.

33. Blood pressure A company held a blood pressure screening clinic for its employees. The results are summarized in the table below by age group and blood pressure level.

		Age		
		Under 30	30–49	Over 50
Blood Pressure	Low	27	37	31
	Normal	48	91	93
	High	23	51	73

a) Find the marginal distribution of blood pressure level.
b) Find the conditional distribution of blood pressure level within each age group.
c) Compare these distributions with a segmented bar graph.
d) Write a brief description of the association between age and blood pressure among these employees.
e) Do you think that this proves that people's blood pressure increases as they age? Explain.

34. Self-reported BMI Self-reported measures are often unreliable. For example, people tend to understate their weight and overstate their height, resulting in understatement of one's body mass index or BMI (weight divided by height squared). Analyses based on understated BMIs may overstate health risks of overweight status. The following table shows estimated counts in the entire Canadian population (based on a sample) where the true height, weight, and BMI were accurately measured and compared with self-reported measures for each individual. The counts are in thousands, but just ignore the three extra zeroes in any calculations. (Due to rounding to nearest 1000, column entries may not add perfectly to totals)
a) Give the conditional distribution of self-reported BMI for each of the three measured overweight classes.

	Measured BMI category (kg/m^2)									
	Underweight (less than 18.5)		Normal weight (18.5 to 24.9)		Overweight (25.0 to 29.9)		Obese class I (30.0 to 34.9)		Obese class II/III (35 or more)	
	'000	%	'000	%	'000	%	'000	%	'000	%
Self-reported BMI category (kg/m^2)										
Total	**402**	**100**	**10 859**	**100**	**8 746**	**100**	**4 288**	**100**	**1 562**	**100**
Underweight (less than 18.5)	271	67	308	3	1	0	6	0	0	0
Normal weight (18.5 to 24.9)	131	33	10 163	94	2 651	30	120	3	4	0
Overweight (25.0 to 29.9)	0	0	388	4	5 851	67	1 894	44	134	9
Obese class I (30.0 to 34.9)	0	0	0	0	244	3	2 247	52	603	39
Obese class II/III (35.0 or more)	0	0	0	0	0	0	22	1	822	53

Source: Adapted from Statistics Canada, Estimates of obesity based on self-report versus direct measures, *Health Reports*, 82-003XIE2008002, vol 19 no. 2, May, 2008. www.statcan.gc.ca/bsolc/olc-cel/olc-cel?catno=82-003-XWE&lang=eng.

b) Give the conditional distribution of measured BMI for each of the three self-reported overweight classes.
c) Why do you know in advance that the two categorical variables in this study will not be independent? How is this clear by visual inspection of the table?
d) What have you learned about self-reported measures of overweight?
e) Name another (categorical) variable that it might be wise to incorporate into such an analysis, for example, breaking up the data according to values of such a variable prior to analysis.
f) This analysis shows a general underestimation of overweight. How does this affect our estimates of the health risk (for diabetes, etc.) to Canadians of being

overweight when self-reported data are used? Will risk of overweight to health be over- or underestimated?

35. Aboriginal identity 2011 Following are 2011 estimated counts of the Aboriginal identity population for Canada, provinces, and territories.

a) Aside from the placement of marginal totals at the top and left rather than bottom and right, the table is not exactly the standard contingency table. Why not?

b) If you re-labelled and changed the numbers in the second column of numbers, you would produce a standard contingency table. How would you recalculate and rename this column? (But don't bother to do it.)

c) What percent of Canadians are Inuit from Nunavut?

d) What percent of Canadian Aboriginals are Inuit from Nunavut?

e) What percent of Aboriginals are Inuit?

f) What percent of Aboriginals are from Nunavut?

Geographic name	Total population	Aboriginal population[11]	North American Indian	Métis	Inuit	Non-Aboriginal population
Canada	**32 852 320**	**1 400 685**	**851 560**	**451 795**	**59 440**	**31 451 640**
Newfoundland and Labrador	507 270	35 800	19 315	7 660	6 265	471 470
Prince Edward Island	137 375	2 230	1 520	410	55	135 145
Nova Scotia	906 175	33 845	21 895	10 050	695	872 325
New Brunswick	735 835	22 620	16 120	4 850	485	713 215
Quebec	7 732 520	141 915	82 420	40 960	12 570	7 590 610
Ontario	12 651 790	301 430	201 105	86 015	3 360	12 350 365
Manitoba	1 174 345	195 895	114 225	78 835	575	978 450
Saskatchewan	1 008 760	157 740	103 210	52 450	290	851 020
Alberta	3 567 980	220 700	116 670	96 870	1 985	3 347 280
British Columbia	4 324 455	232 285	155 020	69 475	1 570	4 092 170
Yukon Territory	33 320	7 705	6 590	840	175	25 610
Northwest Territories	40 795	21 160	13 350	3 245	4 340	19 640
Nunavut	31 700	27 360	130	130	27 070	4 335

Source: Statistics Canada, 2011 National Household Survey, Statistics Canada Catalogue no. 99-011-X2011027

g) What percent of the people of Nunavut are Inuit?

h) What percent of Nunavut Aboriginals are Inuit?

i) What percent of Inuits live in Nunavut?

j) What percent of Ontario Aboriginals could not be simply classified as Inuit, Métis, or N.A. Indian?

k) It is no surprise that the distribution of Aboriginal identity group (Inuit, etc.) differs according to region in Canada. Find the percentage of total provincial population for each Aboriginal identity group (Inuit, Métis, N.A. Indian) for Newfoundland, Ontario, Saskatchewan, and Alberta; graph suitably, and compare and comment. (Note that since we are omitting the "others" in each province, these are not proper conditional distributions. But if we include "others," the graph will be harder to read, since this will be such a large and dominating percentage.)

[11]Includes the Aboriginal groups (North American Indian, Métis, and Inuit), multiple Aboriginal responses, and Aboriginal responses not included elsewhere.

36. Dim sum 2011 When in China, one of the authors would sometimes mention a favourite Chinese treat, called dim sum—and some types of dim sum like *har gow, cheong fun, siu mei*—to people in various parts of the country. But they didn't understand! These expressions are in the Cantonese dialect and are not usually understood by Mandarin speakers. Of the many languages spoken in Canada, Chinese has grown rapidly, becoming the third most common mother tongue after English and French due to large numbers of immigrants. Following are "language spoken most often at home" Census counts for Mandarin and Cantonese (the two major dialects, but not the only dialects) in several cities:

a) How many categorical variables do we have here? Describe each in a few words.

b) Find the conditional distribution of spoken Chinese dialect for each year in Toronto. Repeat for Calgary and again for Vancouver.

c) Can you estimate the proportion of Chinese-speaking Calgarians whose dialect was Mandarin in 2006?

Calgary:

	Cantonese	Mandarin
2011	16 920	9 900
2006	12 785	5 345

Toronto:

	Cantonese	Mandarin
2011	156 425	91 670
2006	129 925	44 990

Vancouver:

	Cantonese	Mandarin
2011	113 610	83 825
2006	94 760	51 465

Source: Statistics Canada, Census of Population, 2011 & 2006 Census of Population, Statistics Canada catalogue no. 97-555-XCB2006039

d) Would you prefer to use pie charts or bar charts to display all the information captured in your part b) calculations? Explain. Create your display and tell what you learned.

e) Collapse (add) the counts across the variable "city," give the resulting two-way table by year and dialect, and find the conditional distribution of dialect for each year. What did you learn? People from Hong Kong generally speak Cantonese, while those from other parts of China more often speak Mandarin. What is likely happening in terms of immigration patterns?

f) Using just 2011 data, find the marginal distribution of dialect and the conditional distribution of dialect for each city. Which conditional distribution differs most from the marginal distribution?

37. Hospitals Most patients who undergo surgery make routine recoveries and are discharged as planned. Others suffer excessive bleeding, infection, or other postsurgical complications and have their discharges from the hospital delayed. Suppose your city has a large hospital and a small hospital, each performing major and minor surgeries. You collect data to see how many surgical patients have their discharges delayed by postsurgical complications, and find the results shown in the following table.

		Discharge Delayed	
		Large hospital	**Small hospital**
Procedure	**Major surgery**	120 of 800	10 of 50
	Minor surgery	10 of 200	20 of 250

a) Overall, for what percent of patients was discharge delayed?

b) Were the percentages different for major and minor surgery?

c) Overall, what were the discharge delay rates at each hospital?

d) What were the delay rates at each hospital for each kind of surgery?

e) The small hospital claims that it has a lower rate of postsurgical complications. Do you agree?

f) Explain, in your own words, why this confusion occurs.

38. Delivery service A company must decide which of two delivery services they will contract with. During a recent trial period they shipped numerous packages with each service and kept track of how often deliveries did not arrive on time. Here are the data:

Delivery Service	Type of Service	Number of Deliveries	Number of Late Packages
Pack Rats	Regular	400	12
	Overnight	100	16
Boxes R Us	Regular	100	2
	Overnight	400	28

a) Compare the two services' overall percentage of late deliveries.

b) Based on the results in part a), the company has decided to hire Pack Rats. Do you agree that they deliver on time more often? Why or why not? Be specific.

c) The results here are an instance of what phenomenon?

39. Graduate admissions A 1975 article in the magazine *Science* examined the graduate admissions process at Berkeley for evidence of sex discrimination. The following table shows the number of applicants accepted to each of four graduate programs.

		Males accepted (of applicants)	Females accepted (of applicants)
Program	1	511 of 825	89 of 108
	2	352 of 560	17 of 25
	3	137 of 407	132 of 375
	4	22 of 373	24 of 341
	Total	**1022 of 2165**	**262 of 849**

a) What percent of total applicants were admitted?

b) Overall, were a higher percentage of males or females admitted?

c) Compare the percentage of males and females admitted in each program.

d) Which of the comparisons you made do you consider to be the most valid? Why?

40. Be a Simpson! Can you design a Simpson's paradox? Two companies are vying for a city's "Best Local Employer" award, to be given to the company most committed to hiring local residents. While both employers hired 300 new people in the past year, Company A brags that it deserves the award because 70% of its new jobs went to local residents, compared to only 60% for Company B. Company B concedes that those percentages are correct, but points out that most of its new jobs were full-time, while most of Company A's were part-time. Not only that, says Company B, a higher percentage of its full-time jobs went to local residents than did Company A's, and the same was true for part-time jobs. Thus, Company B argues, it's a better local employer than Company A. Show how it's possible for Company B to fill a higher percentage of both full-time and part-time jobs with local residents, even though Company A hired more local residents overall. You can make up some numbers for illustrative purposes.

Just Checking ANSWERS

1. 12.6%
2. 49.5%
3. 6.3%
4. 20.4% Never, 50.2% Rarely, 29.4% Sometimes
5. 12.0% Never, 49.5% Rarely, 38.5% Sometimes
6. 7.5% of the never spanked had externalizing problems compared to 12.8% of all subjects.
7. Since those never spanked appear less likely to have problems, they may not be independent.

MathXL

MyStatLab

Go to MathXL at www.mathxl.com or MyStatLab at www.mystatlab.com. You can practise exercises for this chapter as often as you want. The guided solutions will help you find answers step by step. You'll find a personalized study plan available to you too!

chapter

3

Displaying and Summarizing Quantitative Data

3.1 Displaying Quantitative Variables with Graphs

3.2 Describing Quantitative Variables with Numbers

Numerical quantities focus on expected values, graphical summaries on unexpected values.

—*John Tukey*

Where are we going?

If someone asked you to summarize a variable, what would you say? You might start by making a picture. For quantitative data, that first picture would probably be a histogram. We've all looked at histograms, but what should we look *for*? We'll describe histograms, and we'll often do more—such as report numerical summaries of the centre (like the mean and median) and the spread (like the standard deviation). The latter are less commonly reported, but in Statistics, they can be even more important.

Toru Hanai/Reuters

On March 11, 2011, the most powerful earthquake ever recorded in Japan created a wall of water that devastated the northeast coast of Japan and left 20 000 people dead or missing. Tsunamis like this are most often caused by earthquakes beneath the sea that shift the earth's crust, displacing a large mass of water. The 2011 tsunami in Japan was caused by a 9.0 magnitude earthquake. It was particularly noted for the damage it caused to the Fukushima Daiichi nuclear power plant, bringing it perilously close to a complete meltdown and causing international concern.

As disastrous as it was, the Japan tsunami was not nearly as deadly as the tsunami of December 26, 2004, that occurred off the west coast of Sumatra. This disaster killed an estimated 297 248 people, making it the most lethal tsunami on record. The earthquake that caused it was a magnitude 9.1 earthquake, more than 25% more powerful than the Japanese earthquake.[1] Were these earthquakes truly extraordinary, or did they just happen at unlucky times and places? The U.S. National Geophysical Data Center (NGDC)[2] has data on more than 5000 earthquakes dating back to 2150 B.C.E. We have estimates of the magnitude of the underlying earthquake for the 1318 earthquakes that were known to cause tsunamis. What can we learn from these data?

[1]Earthquake magnitudes are measured on a logarithmic scale.

[2]www.ngdc.noaa.gov.

3.1 Displaying Quantitative Variables with Graphs

Let's start with a picture. For categorical variables, it is easy to draw the distribution because it's natural to count the cases in each category. But for quantitative variables, there are no categories. Instead, we usually slice up all the possible values into bins and then count the number of cases that fall into each bin. The bins, together with these counts, give the **distribution** of the quantitative variable and provide the building blocks for one common type of display of the distribution, called a **histogram**. By representing the counts as bars and plotting them against the bin values, the histogram displays the distribution at a glance.

Histograms

Here is a histogram of the *Magnitudes* (on the Richter scale) of the 1318 earthquakes in the NGDC data:

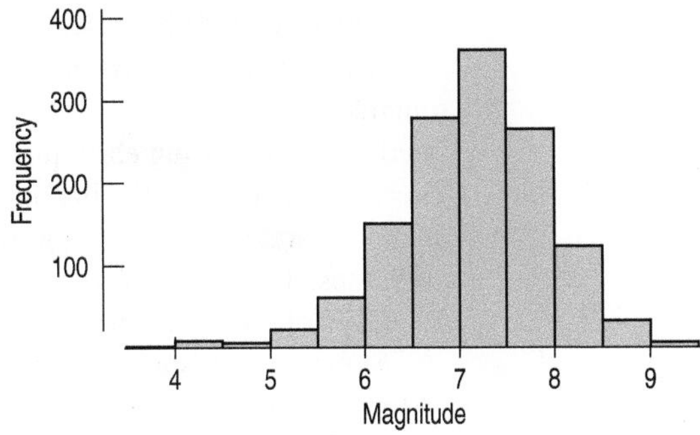

Figure 3.1
A histogram shows the distribution of magnitudes (in Richter scale units) of tsunami-generating earthquakes.

Who	1318 earthquakes known to have caused tsunamis for which we have data or good estimates
What	Magnitude (Richter scale[3]), depth (m), date, location, and other variables
When	From 2000 B.C.E. to the present
Where	All over the earth

The height of each bar of the histogram shows the number of cases that fall in its bin. In this histogram of earthquake magnitudes, each bin has a width of 0.5 units, so, for example, the height of the tallest bar says that there were about 350 earthquakes with magnitudes between 7.0 and 7.5.

DESIGNING YOUR HISTOGRAM

Different features of the distribution may appear more obvious at different bin width choices. When you use technology, it's usually easy to vary the bin width interactively so you can make sure that a feature you think you see isn't a consequence of a certain bin width choice.

Looking at the histogram (Figure 3.1), you can see the entire distribution of earthquake magnitudes. Does this distribution look as you expected? It is often a good idea to imagine what the distribution might look like before you make the display. That way you'll be less likely to be fooled by errors in the data or when you accidentally graph the wrong variable. We can see that these earthquakes typically have magnitudes around 7. Most are between 5.5 and 8.5, but one is less than 4 and a few are 9 or bigger. Now we can answer the question about the Sumatra and Japan tsunamis. With values of 9.1 and 9.0, it's clear that these earthquakes were extraordinarily powerful—among the largest on record.

In a histogram, the bins slice up all the values of the quantitative variable, so that every value falls in a bin. Choosing the bin width used to be an important exercise in introductory statistics courses. Today, technology will choose a default width for you, but you can usually vary the bin width interactively. You'll probably want to have between 5 and 30 bins (unless your data set is very large), and you'll want the bin boundaries to be aesthetically pleasing (such as 7.0 and 7.5 instead of 7.098 and 7.464). You may find that

[3]Technically, Richter scale values are in units of log dyne-cm. But the Richter scale is so common now that usually the units are assumed. The U.S. Geological Survey gives the background details of Richter scale measurements on its website, www.usgs.gov.

some features of the distribution appear or disappear as you change the bin width. Here's a histogram with a smaller bin width of 0.2 for the tsunami data:

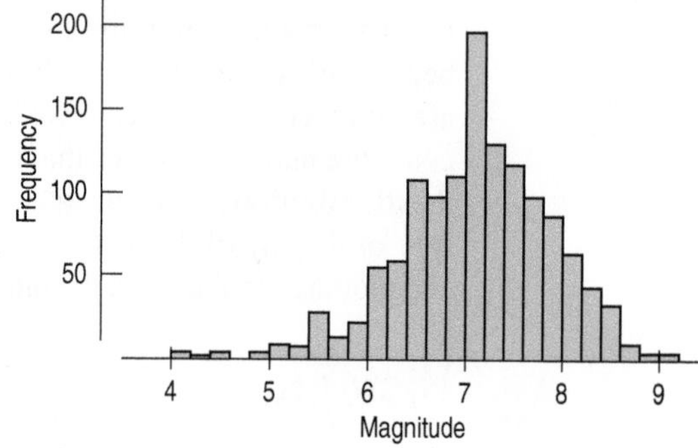

Figure 3.2
A histogram of earthquake magnitudes with a smaller bin size highlights the surprising number of earthquakes with values at 7.0.

The overall distribution appears similar, as it should, but we're struck by two features that we didn't see before. There is a spike around magnitude 7.0. The 1318 earthquake magnitudes include historical values, some more than 2000 years old. Many of the values were estimated by experts and not measured by modern seismographs. Perhaps the experts rounded some of the values to 7, reasoning that this was a typical value for a moderately sized tsunami-causing earthquake. That might explain the overabundance of values of 7.0. There is also a gap in the histogram between 4.6 and 4.8. Unlike a bar chart that uses gaps just to separate categories, a **gap** in a histogram indicates that there is a bin (or bins) with no values. There were no earthquakes among the 1318 that had values between 4.6 and 4.8. Gaps often reveal interesting features about the distribution.

Sometimes it is useful to make a **relative frequency histogram**, replacing the counts on the vertical axis with the percentage of the total number of cases falling in each bin. Of course, the shape of the histogram is exactly the same; only the vertical scale is different.

Figure 3.3
A relative frequency histogram looks just like a frequency histogram except for the labels on the *y*-axis, which now show the percentage of earthquakes in each bin.

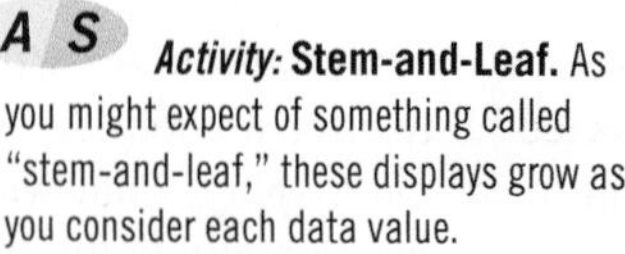

***Activity:* Stem-and-Leaf.** As you might expect of something called "stem-and-leaf," these displays grow as you consider each data value.

Stem-and-Leaf Displays

Histograms provide an easy-to-understand summary of the distribution of a quantitative variable, but they don't show the data values themselves. Here's a histogram of the pulse rates of 24 women, taken by a researcher at a health clinic:

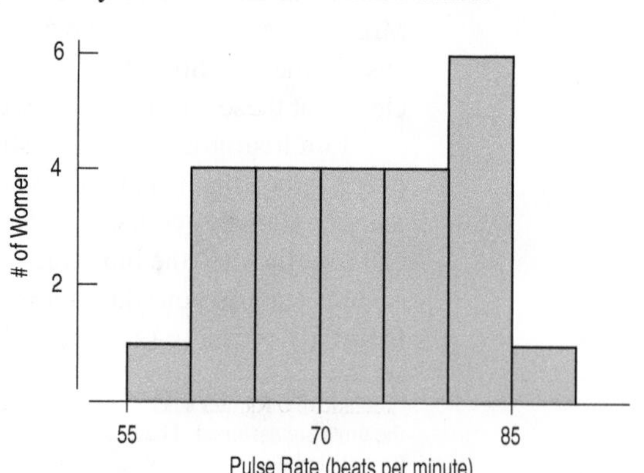

Figure 3.4
The pulse rates of 24 women at a health clinic.

The stem-and-leaf display was devised by John W. Tukey, one of the greatest statisticians of the twentieth century. It is called a "stemplot" in some texts and computer programs, but we prefer Tukey's original name for it.

The story seems pretty clear. We can see the entire span of the data and can easily see what a typical pulse rate might be. But is that all there is to these data?

A **stem-and-leaf display** is like a histogram, but it shows the individual values. It's also easier to make by hand. Here's a stem-and-leaf display of the same data:

```
5 | 6
6 | 0444
6 | 8888
7 | 2222
7 | 6666
8 | 000044
8 | 8
```

Pulse Rate
(5|6 means 56 beats/min)

Turn the stem-and-leaf on its side (or turn your head to the right) and squint at it. It should look roughly like the histogram of the same data. Does it?[4]

What does the line at the top of the display that says 5 | 6 mean? It stands for a pulse of 56 beats per minute (bpm). We've taken the tens place of the number and made that the "stem." Then we sliced off the ones place and made it a "leaf." The next line down is 6 | 0444. The four leaves tell us that there are four pulse rates in this row. Attach each leaf to the common stem of 6, and you get one pulse rate of 60 and three of 64 bpm.

Stem-and-leaf displays are especially useful when you make them by hand for batches of fewer than a few hundred data values. They are a quick way to display—and even to record—numbers. Because the leaves show the individual values, we can sometimes see even more in the data than the distribution's shape. Take another look at all the leaves of the pulse data. See anything unusual? At a glance, they are all even. With a bit more thought you can see that they are all multiples of four—something you couldn't possibly determine from a histogram. How do you think the nurse took the pulses? Counting beats for a full minute or counting for only 15 seconds and multiplying by four?

BY HAND MAKING A STEM-AND-LEAF

Stem-and-leaf displays work like histograms, but they show more information. They use part of the number itself (called the stem) to name the bins. To make the "bars," they use the next digit of the number. For example, if we had a test score of 76, we could write it 7 | 6 where 7 serves as the stem and 6 as the leaf. Then, to display the scores 83, 76, and 88 together, we would write

```
7 | 6
8 | 38
```

For the pulse data, we have

```
5 | 6
6 | 04448888
7 | 22226666
8 | 0000448
```

Pulse Rate
(5|6 means 56 beats/min)

(continued)

[4]You could make the stem-and-leaf with the higher values on the top. Putting the lower value at the top matches the histogram, when turned; putting the higher value at the top matches the way a vertical axis works in other displays.

This display is OK, but a little crowded. A histogram might split each line into two bars. With a stem-and-leaf, we can do the same by putting the leaves 0–4 on one line and 5–9 on another, as we saw earlier:

```
5 | 6
6 | 0444
6 | 8888
7 | 2222
7 | 6666
8 | 000044
8 | 8
```

Pulse Rate
(5|6 means 56 beats/min)

We could have stretched out the scale even more, using five lines for each distinct stem, assigning leaves 0 and 1 to the first line, 2 and 3 to the second line, etc. We need an equal number of possible leaves per stem, to maintain the integrity of our display. Here's what that would look like just for the pulses from 60 to 69:

```
6 | 0
6 |
6 | 444
6 |
6 | 8888
```

Pulse Rate
(6|4 means 64 beats/min)

That's probably too spread out to be useful. Which resolution you choose is up to you. For numbers with three or more digits, you'll need to decide on the "resolution" as well. Usually you'll truncate (or round) the number to two digits, using the first digit as the stem and the second as the leaf. So, if you had 432, 540, 571, and 638, you might display them as

```
6 | 3
5 | 47
4 | 3
```

with an indication that 6|3 means 630 (though actually 6|3 could be any value from 630 to 639, due to the truncation process). Some stats packages will simply say "leaf unit = 10," telling you that a leaf of 3 is 3 tens, or 30. Could you also use two-digit leaves? Yes, if doing it by hand. You would have to separate them by commas or spaces—for example, 5|40,71—but this is tricky stuff for computer software, and frankly that extra digit adds little useful information to the display, while leaving you little space on each row for very many leaves.

But what if the data were 520, 541, 542, 547, 530 . . .? In this case, you would use two-digit stems (52, 53, 54), one digit leaves, and perhaps do some splitting to get a more useful picture.

When you make a stem-and-leaf by hand, make sure to give each leaf the same width in order to preserve the area principle. (That can lead to some fat 1s and thin 8s, but it makes the display honest.)

Dotplots

A S ***Activity:* Dotplots.** Click on points to see their values and even drag them around.

A **dotplot** is a simple display. It just places a dot along an axis for each case in the data. It's like a stem-and-leaf display but with dots instead of digits for all the leaves. Dotplots are a great way to display a small data set (especially if you forget how to write the digits

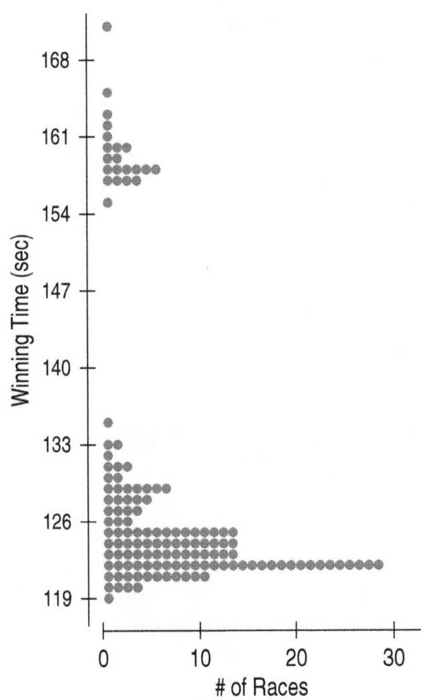

Figure 3.5
A dotplot of Kentucky Derby winning times plots each race as its own dot. We can see two distinct groups corresponding to the two different race distances.

from 0 to 9). Here's a dotplot (Figure 3.5) of the time (in seconds) that the winning horse took to win the Kentucky Derby in each race between the first derby in 1875 and the 2011 derby.

Dotplots show basic facts about distribution. We can find the slowest and quickest races by finding times for the topmost and bottommost dots. It's also clear that there are two clusters of points, one just below 160 seconds and the other at about 122 seconds. Something strange happened to the derby times. Once we know to look for it, we can find out that in 1896 the distance of the race was changed from 1.5 miles (2.4 km) to the current 1.25 miles (2.0 km). That explains the two clusters of winning times.

Some dotplots stretch out horizontally, with the counts on the vertical axis, like a histogram. Others, such as the one shown here, run vertically, like a stem-and-leaf display. Some dotplots place points next to each other when they would otherwise overlap. Others just place them on top of one another. Newspapers sometimes offer dotplots using little pictures, called pictographs, instead of dots.

Think Before You Draw

Suddenly, we face a lot more options when it's time to invoke our first rule of data analysis and "make a picture." You'll need to *Think* carefully to decide which type of graph to make. In the previous chapter, you learned to check the Categorical Data Condition before constructing a pie or bar chart. Now, before making a stem-and-leaf display, a histogram, or a dotplot, you need to check the

> **Quantitative Data Condition**; that is, that the data are values of a quantitative variable whose units are known.

Although a bar chart and histogram may look somewhat similar, they're not the same display. You can't display categorical data in a histogram or quantitative data in a bar chart. Always check the condition that confirms what type of data you have before proceeding with your display.

Step back from a histogram or stem-and-leaf display. What can you say about the distribution? Every histogram can tell its own story, but you need to develop a vocabulary to help you explain it to others. When you describe a distribution, you should always discuss its **shape**, **centre**, and **spread**, as well as any unusual features or departures from the overall pattern.

Shape

1. *Does the histogram have a single, central hump or several separated humps?* These humps are called **modes**.[5] The earthquake magnitudes have a single mode at about 7. A histogram with one peak, such as the earthquake magnitudes, is dubbed **unimodal**; histograms with two peaks are *bimodal*, and those with three or more are called *multimodal*.[6] For example, Figure 3.6 below shows a bimodal histogram.

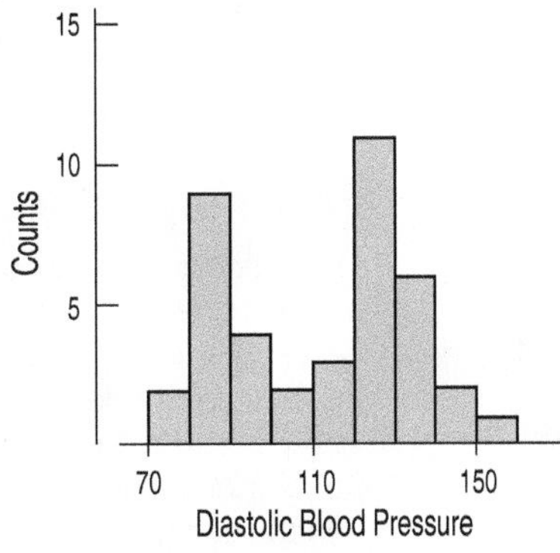

Figure 3.6
A bimodal histogram has two apparent peaks. Always ask: Are there two distinguishable groups present?

[5]Well, technically, the value on the horizontal axis of the histogram is the mode, but anyone asked to point to the mode would point to the hump.

[6]Apparently, statisticians don't like to count past two.

A histogram (Figure 3.7) that doesn't appear to have any mode and in which all the bars are approximately the same height is called **uniform**.

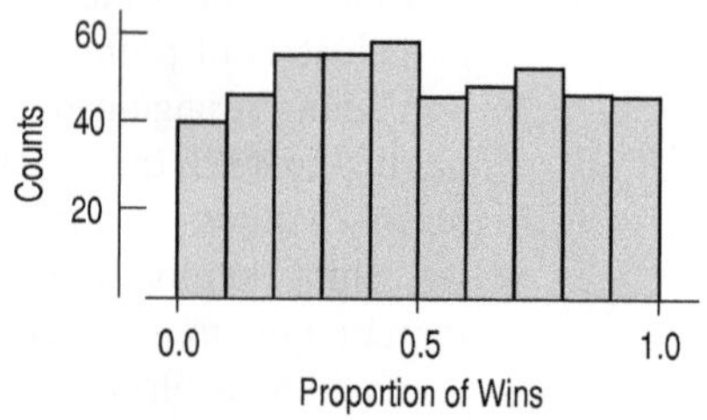

Figure 3.7

In this histogram, the bars are all about the same height. The histogram doesn't appear to have a mode and is called uniform.

2. *Is the histogram* **symmetric**? Can you fold it (Figure 3.8) along a vertical line through the middle and have the edges match pretty closely, or are more of the values on one side?

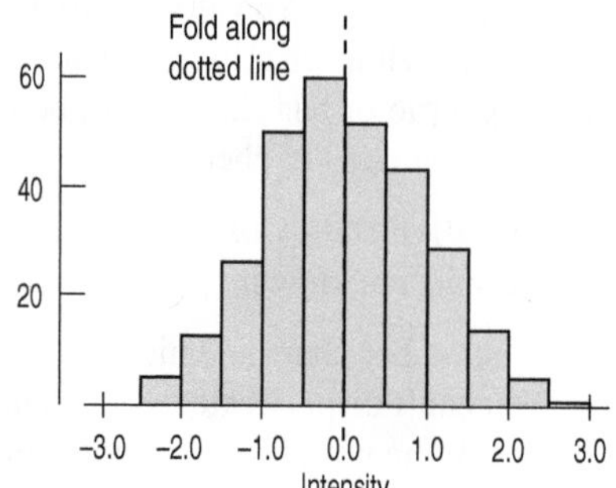

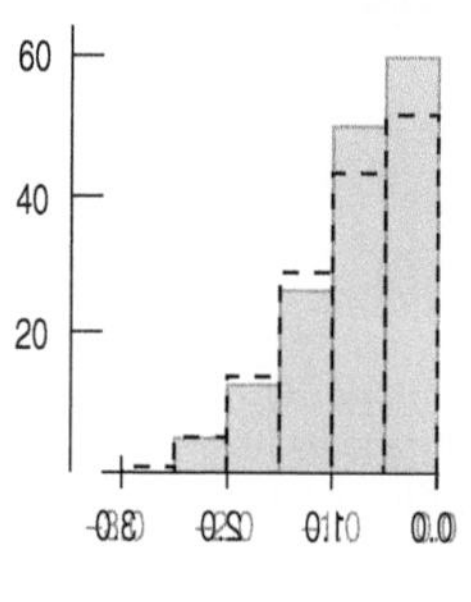

Figure 3.8

A symmetric histogram can fold in the middle so that the two sides almost match.

A S ***Activity:*** **Attributes of Distribution Shape.** This activity and the others on this page show off aspects of distribution shape through animation and example, then let you make and interpret histograms with your statistics package.

The (usually) thinner ends of a distribution are called the **tails**. If one tail stretches out farther than the other (Figure 3.9), the histogram is said to be **skewed** to the side of the longer tail.

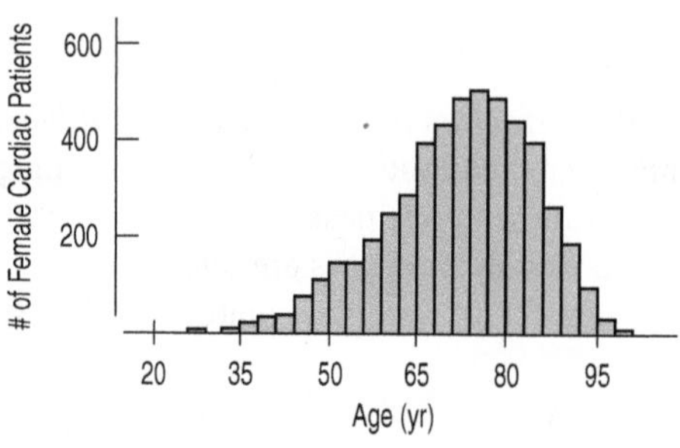

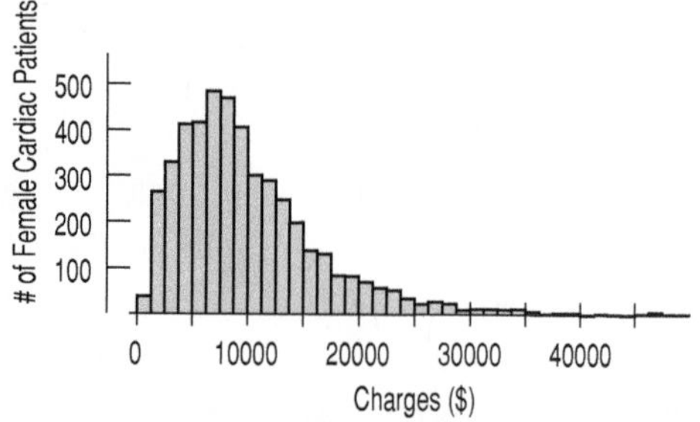

Figure 3.9

Two skewed histograms showing data on two variables for all female heart attack patients in New York State in one year. The blue one (age in years) is skewed to the left (or negatively skewed). The purple one (charges in $) is skewed to the right (or positively skewed).

Pie à la mode? You've heard of pie à la mode. Is there a connection between pie and the mode of a distribution? Actually, there is! The mode of a distribution is a *popular* value near which a lot of the data values gather. And "à la mode" means "in style"—*not* "with ice cream." That just happened to be a *popular* way to have pie in Paris around 1900.

WHAT IS THE MODE?

The *mode* is sometimes defined as the single value that appears most often. That definition is fine for categorical variables because all we need to do is count the number of cases for each category. For quantitative variables, the mode is more ambiguous. What is the mode of the Kentucky Derby times? Well, seven races were timed at 122.2 seconds—more than any other race time. Should that be the mode? Probably not. For quantitative data, it makes more sense to use the term "mode" in the more general sense of the peak of the histogram rather than as a single summary value. In this sense, the important feature of the Kentucky Derby races is that there are two distinct modes, representing the two different versions of the race and warning us to consider those two versions separately.

©Corbis Sygma

3. *Do any unusual features stick out?* Often such features tell us something interesting or exciting about the data. You should always mention any stragglers, or **outliers,** that stand off away from the body of the distribution. For example, if you're collecting data on nose lengths and Pinocchio is in the group, you'd probably notice him, and you'd certainly want to mention it.

   Outliers can affect almost every method we discuss in this course, so we'll always be on the lookout for them. An outlier can be the most informative part of your data. Or it might just be an error (find it and fix it if you can). But don't throw it away without comment. Treat it specially and discuss it when you tell about your data. Sometimes we point it out, try to explain it, and set it aside, rather than have it distort our analysis and understanding of the true story present in the rest of the data. Be sure to look for outliers. Always. Figure 3.10 is a histogram with some clear outliers.

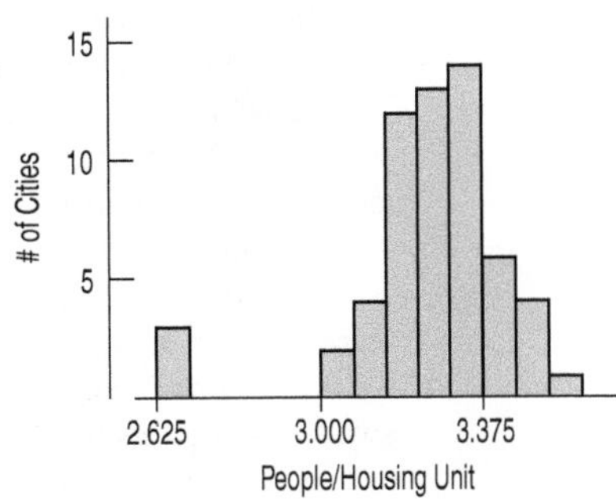

Figure 3.10
A histogram with outliers. There are three cities in the leftmost bar.

A bit later in this chapter you'll learn a handy rule of thumb for deciding when a point might be considered an outlier.

Are there any gaps in the distribution? The Kentucky Derby data we saw in the dotplot on page 49 has a large gap between two groups of times, one near 120 seconds and one near 160. Gaps help us see multiple modes and encourage us to notice when the data may come from different sources or contain more than one group.

For Example DESCRIBING HISTOGRAMS

A credit card company wants to see how much customers in a particular segment of their market use their credit cards. They have provided you with data[7] on the amount spent by 500 selected customers during a three-month period and have asked you to summarize the expenditures. Of course, you begin by making a histogram.

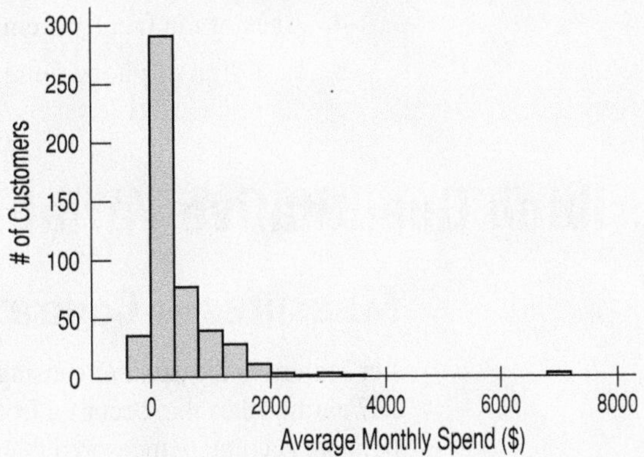

QUESTION: Describe the shape of this distribution.

ANSWER: The distribution of expenditures is unimodal and skewed to the high end. There is an extraordinarily large value at about $7000.

[7] These data are real, but cannot be further identified for privacy reasons.

Why do I get the feeling that we're not in math class anymore?
How you characterize a distribution is often a judgment call. Is that point way out on the right really an outlier, or is it just an indication of a long tail to the distribution? Generally, we start by looking at the main body of the data. If a point is separated from it by only a small gap, then it's not that unusual and probably not an outlier. If the main body seems roughly symmetric, then more distant stragglers are best regarded as outliers. If the main body of the data is skewed, then the long tail that continues that skewness is part of the overall pattern, so points would need to be farther away to be called outliers.

When we summarize data, our goal is usually more than just developing detailed knowledge of the data we have at hand. We want to know what the data say about the world, so we'd like to know whether the patterns we see in histograms and summary statistics generalize to other individuals and situations. Scientists generally don't care about the particular guinea pigs in their experiment, but rather about what their reaction to different treatments says about how other animals (and, perhaps, humans) might respond.

Because we want to see broad patterns rather than focus on the details of the data set we're looking at, many of the most important concepts in Statistics are not defined as precisely as most concepts in Mathematics. Whether a histogram is symmetric or skewed, whether it has one or more modes, whether a case is far enough from the rest of the data to be considered an outlier—these are all somewhat vague concepts. They all require judgment.

You may be used to finding a single correct and precise answer, but in Statistics there may be more than one interpretation. That may make you a little uncomfortable at first, but soon you'll see that leaving room for judgment brings you both power and responsibility. It means that your own knowledge about the world and your judgment matter. You'll use them, along with the statistical evidence, to draw conclusions and make decisions about the world.

Just Checking

It's often a good idea to think about what the distribution of a data set might look like before we collect the data. What do you think the distribution of the following data sets will look like? Be sure to discuss shape. Where do you think the centre might be? How spread out do you think the values will be?

1. Number of kilometres run by Saturday morning joggers at a park.
2. Hours spent by Canadian adults watching CBC's *Hockey Night in Canada*.[8]
3. Amount of winnings of all people playing Lotto 6/49.
4. Ages of the faculty members at your school.
5. Last digit of phone numbers on your campus.

3.2 Describing Quantitative Variables with Numbers

Measures of Centre: The Mean

Let's return to the tsunami-causing earthquakes. But this time, let's look at just 25 years of data: 207 earthquakes that occurred from 1987 through 2011 (see Figure 3.11 below). These should be more accurately measured than prehistoric quakes because seismographs were in wide use during that time period. If you tried to attempt the impossible by pointing to one number to describe the magnitude of all these earthquakes, where would you point? When a histogram is unimodal and symmetric, most people would point to the *centre* of the distribution, where the histogram peaks. The typical tsunami-causing earthquake has a magnitude of about 7.0.

[8]Saturday night hockey is broadcast on CBC television. It is one of the highest-rated Canadian television programs, and is the world's oldest television sports program still on the air.

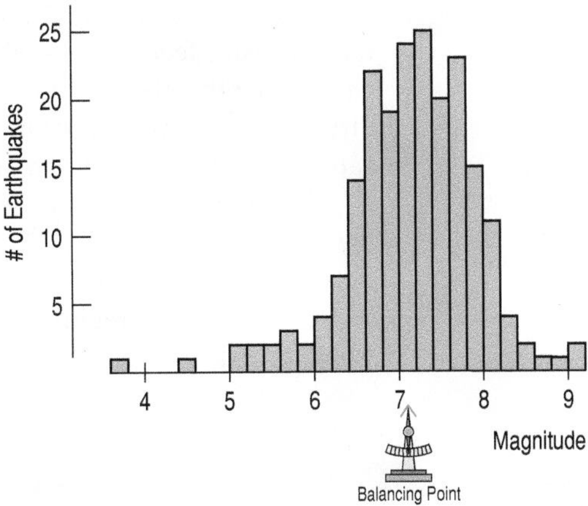

Figure 3.11
Magnitudes of tsunami-causing earthquakes during 1987–2011. The mean is located at the balancing point of the histogram (if it were a solid object).

NOTATION ALERT

In Algebra, you used letters to represent values in a problem, but it didn't matter what letter you picked. You could call the width of a rectangle *X* or you could call it *w* (or *Fred*, for that matter). But in Statistics, the notation is part of the vocabulary. For example, *n* is always the number of data values. Always.

We will point out such special notation conventions. Think of them as part of the language you need to learn in this course.

Here's another one: Whenever we put a bar over a variable, it means "find the mean."

In everyday language, sometimes "average" *does* mean what we want it to mean. We don't talk about your grade point mean or a baseball player's batting mean or the Dow Jones Industrial mean. So we'll continue to say "average" when that seems most natural. When we do, though, you may assume that what we mean is the mean.

If we want to *calculate* a number, we can *average* the data. The average tsunami-causing earthquake magnitude is 7.08, about what we might expect from the histogram. You already know how to average values, but this is a good place to introduce notation that we'll use throughout the book. We use the Greek capital letter sigma, Σ, to mean "sum" (sigma is "S" in Greek), and we write

$$\bar{y} = \frac{Total}{n} = \frac{\sum y}{n}$$

The formula says to add up all the values of the variable and divide that sum by the number of data values, *n*—just as you've always done.[9]

Once you have averaged the data, you'd expect the result to be called the *average*, but that would be too easy. Informally, we speak of the "average person," but we don't add up people and divide by the number of people. To make this distinction, the value we calculated is called the **mean**, $\bar{y}$, and pronounced "*y*-bar."

The **mean** feels like the centre because it is the point at which the histogram balances:

The centre of balance makes sense when the data are fairly symmetric. But data are not always this well behaved. If the distribution is skewed or has outliers, the centre is not so well defined and the mean may not be what we want.

Measures of Centre and Skewed Distributions

The mean salary for the Toronto Maple Leaf players in the 2012–2013 season was $1 996 143. Is this a typical player's salary? Figure 3.12 is a histogram of their salaries. Let's balance the histogram:

Figure 3.12
Toronto Maple Leaf players salaries, 2012–2013.

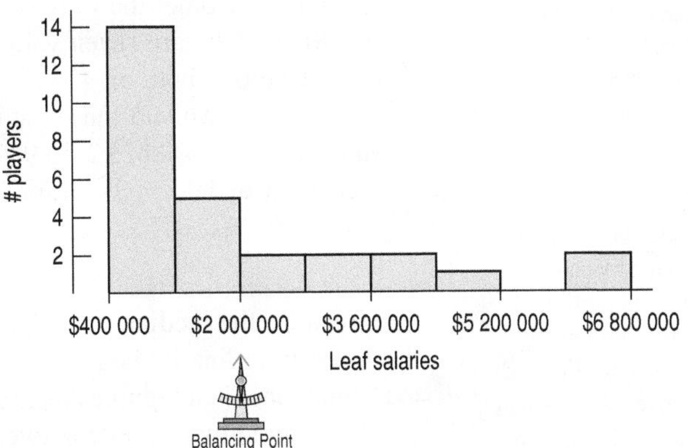

[9]You may also see the variable called *x* and the equation written $\bar{x} = \frac{Total}{n} = \frac{\Sigma x}{n}$. Don't let that throw you. You are free to name the variable anything you want, but we'll generally use *y* for variables like this that we want to summarize, model, or predict. (Later we'll talk about variables that are used to explain, model, or predict *y*. We'll call them *x*.)

The mean is about \$2 million, but 19 of 28 players had salaries below that, so the mean doesn't feel like a good overall summary. Why is the balancing point so high? It's easy to see that the very big salaries on the right are pulling the balancing point in that direction. That may make the mean an inappropriate summary of the centre. Should a summary of the centre be so influenced by some super-star salaries?

There are other ways to describe the centre. How about the value right in the middle—the one that splits the data in half?[10] We call that value the **median**. The median salary is only \$1 150 000, which seems like a more typical summary. The median splits the histogram into two regions of equal[11] area (blue and pink in Figure 3.13), regardless of where the bars are.

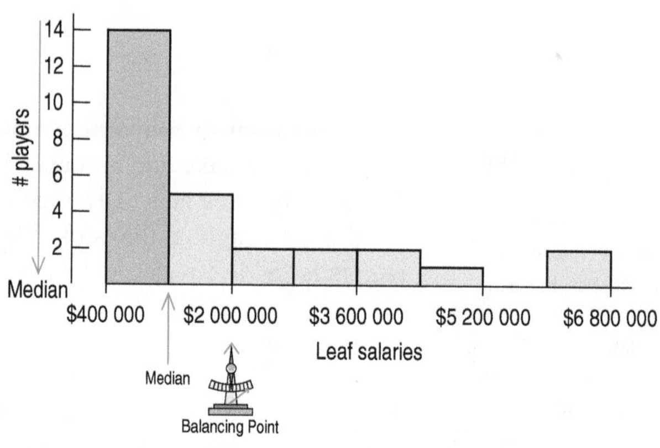

Figure 3.13
The median splits the area of the histogram in half at \$1 150 000, at the upper edge of the blue rectangle. Because the distribution is skewed to the right, the mean is higher than the median. The high salaries at the right have pulled the mean toward them, away from the median.

BY HAND FINDING THE MEDIAN

Finding the median of a batch of n numbers is easy as long as you remember to order the values first. If n is odd, the median is the middle value. Counting in from the ends, we find this value in the $\frac{n+1}{2}$ position.

When n is even, there are two middle values. So in this case, the median is the average of the two values in positions $\frac{n}{2}$ and $\frac{n}{2} + 1$.

Here are two examples:

Suppose the batch has the values: 14.1, 3.2, 25.3, 2.8, −17.5, 13.9, 45.8.

First, we order the values: −17.5, 2.8, 3.2, 13.9, 14.1, 25.3, 45.8.

Since there are seven values, the median is the $(7 + 1)/2 = 4$th value counting from the top or bottom: 13.9.

Suppose we had the same batch with another value at 35.7. Then the ordered values are: −17.5, 2.8, 3.2, 13.9, 14.1, 25.3, 35.7, 45.8. The median is the average of the 8/2, or 4th, and the $(8/2) + 1$, or 5th, values. So the median is $(13.9 + 14.1)/2 = 14.0$.

A S ***Activity:* The Centre of a Distribution.** Compare measures of centre by dragging points up and down and seeing the consequences. Another activity shows how to find summaries with your statistics package.

A **resistant measure** is one that is not affected too much by changes, big or small, to a small portion of the data. Such a measure will pay little attention to outliers. If recording errors are discovered in a small subset of your data, resistant measures may not even need to be recalculated, as they will barely move. The median is one example of a resistant measure.

Because the median considers only the order of the values, it is **resistant** to values that are extraordinarily large or small; it simply notes that they are one of the "big ones" or the "small ones" and ignores their distance from the centre.

How should we choose between the mean and the median? In the salary data, it certainly seems more sensible to use the median salary, or salary of the player who stands

[10] When the number of values is odd, we can find the middle value; when it's even, we have to average the two middle ones. Here there are 28 players, so after putting their salaries in order, we average the 14th and 15th biggest values.

[11] Approximately equal, since the precise median depends on where the data points actually lie in the middle bin(s).

right in the middle. The distribution is skewed to the right, and even one very big salary will pull the mean away from the centre. Like a child who sits way out on a see-saw, one extreme value can pull the balancing point toward it. For a distribution like this, the median is usually a better descriptor.

For the 207 recent tsunami-causing earthquakes, it doesn't seem to make much difference—the mean is 7.10 and the median is 7.20. When the data are symmetric, the mean and median will be close, but when the data are skewed, the median is likely to be a better choice. So why not just use the median? Well, for one, the median can go overboard. It's not just resistant to occasional outliers, but can be unaffected by changes in most of the data values. By contrast, the mean includes input from each data value and gives each one equal weight. It's also easier to work with, so when the distribution is unimodal and symmetric, we'll use the mean.

Of course, to choose between mean and median as a descriptor of centre, we'll start by looking at the data. If the histogram is roughly symmetric and there are no outliers, we'll prefer the mean. However, if the histogram is skewed or has outliers, we're usually better off with the median. If you're not sure, report both and discuss why they might differ. If symmetric, except for one or two clear outliers, you might also consider reporting the outliers separately, and the mean for the rest of the data.

For Example DESCRIBING CENTRE

RECAP: You want to summarize the expenditures of 500 credit card company customers, and have looked at a histogram.

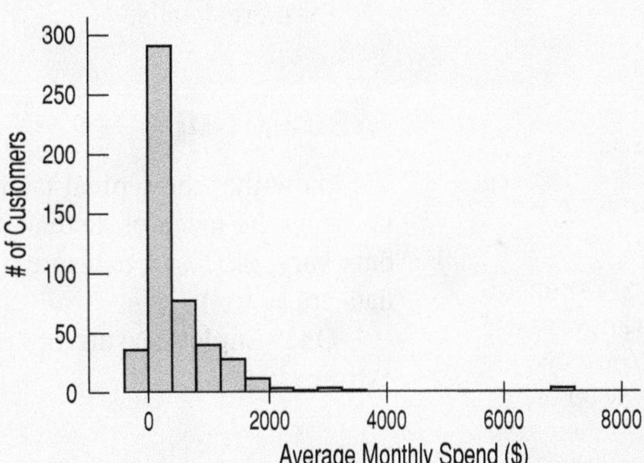

QUESTION: You have found the mean expenditure to be $478.19 and the median to be $216.28. Which is the more appropriate measure of centre, and why?

ANSWER: Because the distribution of expenditures is skewed, the median is a very appropriate measure of centre. The median is the middle value, with half the expenditures above and half below it, and, unlike the mean, it's not affected by the large outlying value or by the skewness of the distribution.

WHEN TO EXPECT SKEWNESS

Even without making a histogram, we can expect some variables to be skewed. When values of a quantitative variable are bounded on one side but not the other, the distribution may be skewed. For example, incomes and waiting or survival times can't be less than zero, so they are often skewed to the right. Amounts of things (e.g., dollars, employees) are often skewed to the right for the same reason. If a test is too easy, the distribution will be skewed to the left, because many scores will bump against 100%. And combinations of things are often skewed. In the case of a hockey team's salaries, often a small group of superstars will have very inflated . . . er, impressive salaries . . . so including them along with the more typical players and the minimum salary players can lead to a very skewed (or even bimodal) distribution.

But there are also times when the purpose or intended use of the measure is key, rather than just trying to provide a useful descriptor of the data. For example, the players on a hockey team might be keenly interested in the median salary, as this is what the "average" player is making, while the team owner would be more interested in the average or mean salary, since this is how much is being paid out per player and directly related to total payroll (mean × number of players = total payroll). Many sports teams have highly right-skewed salary distributions due to the presence of a few star or superstar players, hence a much bigger mean than median salary.

Or if you are interested in your potential long-term returns from some type of investment, the mean of past returns would interest you more than the median, since the longer you hold the investment, the closer your average return will be to the long-term mean (assuming future returns will be similar to past returns). This is a consequence of the Law of Large Numbers—to be discussed in later chapters.

Some other possible measures of centre include:

- *Mode(s) or modal bin(s).* See the note on page 50 about these peaks or high-frequency values in a distribution. The modes of a bimodal distribution, for example, would be far more informative than a mean or median. And if I could win a lottery by guessing the approximate price of the next resale condo sold in Toronto, please tell me the mode—not the mean or median—of the recent distribution of sale prices, since more houses sell around the modal price than any other price.
- *Trimmed mean.* We can make the ordinary mean more *resistant* by trimming off a small percent (like 5% or 10%) of the biggest and smallest values before averaging, creating what is called a trimmed mean. See exercise #35 at the end of this chapter for more details.

Measuring Spread: The Standard Deviation

WHAT WE DON'T KNOW

Statistics pays close attention to what we *don't* know as well as what we do know. Understanding how spread out the data are is a first step in understanding what a summary *cannot* tell us about the data. It's the beginning of telling us what we don't know.

We know that the typical tsunami-creating earthquake has a magnitude around 7, but knowing the mean or median doesn't tell us about the entire distribution. The more the data vary, the less a measure of centre can tell us. We need to know how spread out the data are as well.

One simple measure of spread is the **range**, defined as the difference between the extremes:

$$\text{Range} = \max - \min$$

For the tsunami-causing earthquake magnitudes, the minimum is 3.7 and the maximum is 9.1, so the range is $9.1 - 3.7 = 5.4$. Notice that the range is *a single number*, not an interval of values, as you might think from its use in common speech. Of course, if there are any outliers in the data, the max, the min, or both will be outliers themselves, so the range is very sensitive to outliers.

Because the range uses only two values, it ignores much of the information about how individual values vary. A more powerful approach uses the **standard deviation**, which takes into account how far *each* value is from the mean. Like the mean, the standard deviation is appropriate only for reasonably symmetrical data.

NOTATION ALERT

s^2 always means the variance, and s always denotes the standard deviation.

One way to think about spread is to examine how far each data value is from the mean. This difference is called a *deviation*. For example, consider the two samples of data in Figure 3.14.

Both samples have the same central value (5.0) and range (10.0), but in (orange) sample (a), the data tend to cluster more tightly around the centre. This is captured in the size of the deviations or distances from the centre. The third smallest data point is 1.5 units below the centre in (blue) sample (b) whereas the third smallest data point is 3 units below the centre in sample (a) Similarly, if you run through all the corresponding data points, left to right, in each graph, you will see that each data point is closer to the middle in sample (b), except for the maximum (10.0), the minimum (0.0), and the mean (5.0), which have similar deviations of +5, –5, and 0 respectively, in each case.

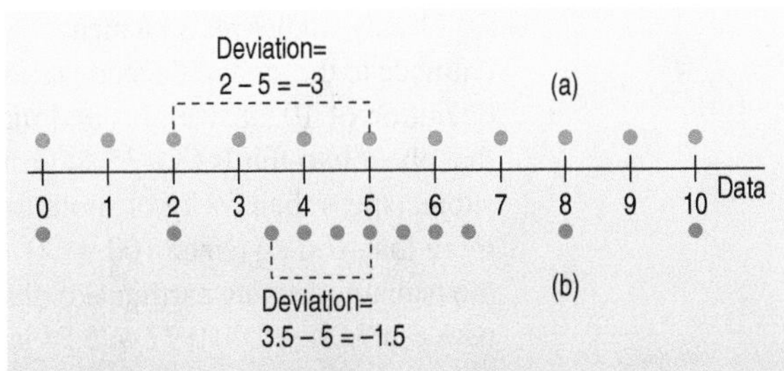

Figure 3.14
Deviations reflect spread. The sample on the top has bigger deviations, on average, than the sample on the bottom, though the ranges are identical.

One single deviation or even a few of them won't be very useful. We want to use all the information about data spread—use all the deviations together—combining them all into one useful summary number. Our first thought might be to just average the deviations, but the positive and negative differences exactly cancel each other out! The average deviation is always zero—not very helpful.

To keep them from cancelling out, one reasonable approach would be to take absolute values of the deviations, but instead we will elect to square them. Statisticians rarely see a deviation not worth squaring! There are very good reasons for squaring, having to do with things like the geometry of right triangles—all of which is beyond the level of mathematical sophistication in this course. Squaring always gives a positive or zero value, so the sum cannot be negative. (Can it be zero? When?) That's great. Squaring also emphasizes larger differences (for example, a deviation of 3 is thrice a deviation of 1, but after squaring, the former becomes $3^2 = 9$ times bigger than $1^2 = 1$)—a feature that turns out to be both good and bad.

When we add up these squared deviations and find their average (almost), we call the result the **variance**:

$$s^2 = \frac{\sum(y - \bar{y})^2}{n - 1}$$

Why almost? It *would* be a mean if we divided the sum by n. Instead, we divide by $n - 1$. Why? There are good technical reasons, but we'll put them off until later.

The variance will play an important role later in this book, but it has a problem as a measure of spread. Whatever the units of the original data are, the variance is in *squared* units. We want measures of spread to have the same units as the data. And we probably don't want to talk about squared dollars, or mpg^2. So to get back to the original units, we take the square root of s^2. The result, s, is the **standard deviation**.

Putting it all together, the standard deviation of the data is found by the following formula:

$$\sqrt{\frac{\sum(y - \bar{y})^2}{n - 1}}$$

You will almost always rely on a calculator or computer to do the calculating. For the recent 207 tsunami-causing earthquakes, we found the standard deviation of the magnitudes to be 0.77 (in Richter scale units).

Does this number "0.77" have any simple or intuitive interpretation? Is it an average or mean absolute deviation (MAD)? Not quite, since squaring, averaging, and square-rooting absolute deviations does not somehow cancel the first and last operation to produce a simple average. But it is in the same ballpark, usually a bit larger than the MAD, so it is close to being an average (absolute) deviation from the mean. And for distributions that are roughly symmetrical and mound- or bell-shaped in appearance, with no extreme outliers, we can say more, based on the *Empirical Rule* first discovered some 300 years ago:

i. About 68% of the data will lie within 1 standard deviation of the mean.
ii. About 95% of the data will lie within 2 standard deviations of the mean.
iii. Virtually all of the data will lie within 3 standard deviations of the mean.

Many studies report a mean and standard deviation for various subject characteristics upon entrance to the study. If a study reported the mean mass of 50 subjects was 70 kg with standard deviation of 10 kg (weights and mass generally are somewhat bell-shaped), we would know that about two-thirds (30–35 subjects) had a mass of between 60 kg and 80 kg, with just a few subjects less than 50 kg or more than 90 kg. Likely no one had a mass of less than 40 kg or more than 100 kg (since $100 = 70 + 3 \times 10$, and $40 = 70 - 3 \times 10$). Now, check this out for the tsunami-causing earthquake data set. Determine how many earthquakes have magnitudes between $7.08 - 2 \times 0.77 = 5.54$ and $7.08 + 2 \times 0.77 = 8.62$. Locate these two points on the histogram, and you can see that about 95% of tsunami-causing earthquake magnitudes fall between these numbers.

The empirical rule gives reasonable approximations even for mildly skewed mound-shaped distributions.[12] Keep in mind that it is rather common to find data in the vicinity of 1 standard deviation from the mean, but rather uncommon to find data more than 2 standard deviations from the mean for such distributions. One particularly important bell-shaped distribution is called the Normal model. And it just so happens that the mean and standard deviation together provide not just a good but a complete description of the Normal model. But we'll have to wait until Chapter 5 to ring this bell.

The standard deviation (like its partner in crime, the mean) is not very resistant to changes to small portions of the data set, and a single outlier can hugely inflate its value—and remember that an outlier's big deviation also gets squared! With the tsunami-causing earthquake data, does the extremely small value of 3.7 greatly inflate the standard deviation? It could, but the data set is so big that the contribution of one squared deviation to the total sum of squared deviations is swamped by a couple of hundred other terms, so there is little effect here. Change the total number of observations to, say, 30, and one very small value would have a much bigger effect on the mean and standard deviation. Or pull that small value much farther off to the left, and its effect on the calculation might well be felt.

BY HAND FINDING THE STANDARD DEVIATION

To find the standard deviation, start with the mean, $\bar{y}$. Then find the *deviations* by taking $\bar{y}$ from each value: $(y - \bar{y})$. Square each deviation: $(y - \bar{y})^2$.

Now just add up the squared deviations and divide by $n - 1$. That gives you the variance, s^2. To find the standard deviation, s, take the square root.

Suppose the batch of values is 4, 3, 10, 12, 8, 9, and 3.

The mean is $\bar{y} = 7$. So find the deviations by subtracting 7 from each value:

Original Values	Deviations	Squared Deviations
4	4 − 7 = −3	(−3)² = 9
3	3 − 7 = −4	(−4)² = 16
10	10 − 7 = 3	9
12	12 − 7 = 5	25
8	8 − 7 = 1	1
9	9 − 7 = 2	4
3	3 − 7 = −4	16

Add up the squared deviations: $9 + 16 + 9 + 25 + 1 + 4 + 16 = 80$.

Now, divide by $n - 1$: $80/6 = 13.33$.

Finally, take the square root: $s = \sqrt{13.33} = 3.65$

[12] To make similar types of statements using means and standard deviations about an arbitrary distribution of any shape, we would have to invoke Chebyshev's Rule, which, due to its complete generality, has more theoretical than practical value, so we won't discuss it here (but see The Worse Case margin note on page 127 in Chapter 5).

Step-by-Step Example SUMMARIZING A DISTRIBUTION

One of the authors owned a Nissan Maxima for 8 years. Being a statistician, he recorded the car's fuel efficiency (in mpg[13]) each time he filled the tank. He wanted to know what fuel efficiency to expect as "ordinary" for his car. (Hey, he's a statistician. What would you expect?[14]) Knowing this, he was able to predict when he'd need to fill the tank again and to notice if the fuel efficiency suddenly got worse, which could be a sign of trouble.

Question: How would you describe the distribution of *fuel efficiency* for this car?

THINK ➡ Plan · State what you want to find out.

I want to summarize the distribution of Nissan Maxima fuel efficiency.

Variable Identify the variable and report the *W*'s.

The data are the fuel efficiency values in miles per gallon for the first 100 fill-ups of a 1989 Nissan Maxima between 1989 and 1992.

Be sure to check the appropriate condition.

✓ **Quantitative Data Condition:** The fuel efficiencies are quantitative with units of miles per gallon. Histograms are appropriate displays for displaying the distribution. Numerical summaries are appropriate as well.

SHOW ➡ Mechanics Make a histogram and boxplot. Based on the shape, choose appropriate numerical summaries.

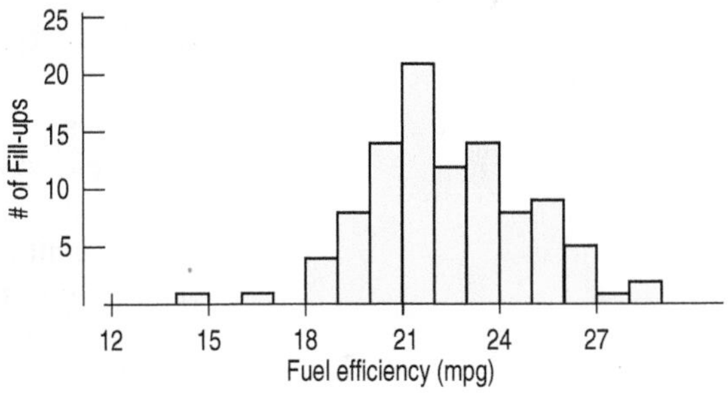

REALITY CHECK A value of 22 mpg seems reasonable for such a car. The spread is reasonable, although the range looks a bit large.

A histogram of the data shows a fairly symmetric distribution with a low outlier.

Count	100
Mean	22.4 mpg
Std. Dev.	2.45
Median	22.0

The mean and median are close, so the outlier doesn't seem to be a problem. I can use the mean and standard deviation.

TELL ➡ Conclusion Summarize and interpret your findings in context. Be sure to discuss the distribution's shape, centre, spread, and unusual features (if any).

The distribution of mileage is unimodal and roughly symmetric with a mean of 22.4 mpg. There is a low outlier that should be investigated, but it does not influence the mean very much. The standard deviation suggests that from tankful to tankful, I can expect the car's fuel economy to differ from the mean on average by about 2.45 mpg.

[13]Data collected by one of the authors in the U.S., where they use miles per gallon to measure fuel efficiency

[14]He also recorded the time of day, temperature, price of gas, and phase of the moon. (OK, maybe not phase of the moon.)

While the standard deviation is the most important and commonly reported measure of spread, it may not be particularly useful as a descriptor for skewed data. In the next section, we will learn more about summarizing skewed distributions.

Measuring Spread: Quartiles and the Interquartile Range

Another way to describe the spread of a variable might be to ignore the extremes and concentrate on the middle of the data. We could, for example, find the range of just the middle half of the data. What do we mean by the middle half? Divide the data in half at the median. Now divide both halves in half again, cutting the data into four quarters. We call these new dividing points **quartiles**. One quarter of the data lies below the **lower quartile** (denoted Q_1), and one quarter of the data lies above the **upper quartile** (denoted Q_3), so half the data lies between them. The quartiles border the middle half of the data.

For any percentage of the data, there is a corresponding **percentile**—the value that leaves that percentage of the data below it. The lower and upper quartiles are also known as the 25th and 75th percentiles of the data, respectively, since 25% of the data fall below the lower quartile and 75% of the data fall below the upper quartile. The median is the 50th percentile.

BY HAND FINDING QUARTILES

A simple way to find the quartiles is to start by splitting the batch into two halves at the median. (When n is odd, some statisticians include the median in both halves; others omit it. We will omit it.) The lower quartile is the median of the lower half, and the upper quartile is the median of the upper half.

Here are our two previous examples again.

The ordered values of the first batch were −17.5, 2.8, 3.2, 13.9, 14.1, 25.3, and 45.8, with a median of 13.9. Notice that 7 is odd; we exclude the median in both halves to get −17.5, 2.8, 3.2 and 14.1, 25.3, 45.8.

Each half has 3 values, so the median of each is the 2nd value. So, the lower quartile is 2.8 and the upper quartile is 25.3.

The second batch of data had the ordered values −17.5, 2.8, 3.2, 13.9, 14.1, 25.3, 35.7, and 45.8.

Here n is even, so the two halves of 4 values are −17.5, 2.8, 3.2, 13.9 and 14.1, 25.3, 35.7, 45.8.

Now the lower quartile is $(2.8 + 3.2)/2 = 3.0$ and the upper quartile is $(25.3 + 35.7)/2 = 30.5$.

The difference between the quartiles tells us how much territory the middle half of the data covers and is called the **interquartile range**. It's commonly abbreviated IQR (and pronounced "eye-cue-are"):

$$IQR = upper\ quartile - lower\ quartile$$

Why the quartiles?
Could we use other percentiles besides the quartiles to measure the spread? Sure, we could, but the IQR is the most commonly used percentile difference and the one you're most likely to see in practice.

For the earthquakes, there are 103 values below the median and 103 values above the median (excluding the median in each half). The midpoint of the lower half is the 52nd value in the ordered data; that turns out to be 6.7. In the upper half, we take the 155th value, finding a magnitude of 7.6 as the third quartile. The *difference* between the quartiles gives the IQR:

$$\begin{aligned} IQR &= 7.6 - 6.7 \\ &= 0.9 \end{aligned}$$

Now we know that the middle half of the earthquake magnitudes extends across a (interquartile) range of 0.9 Richter scale units. This seems like a reasonable summary of the spread of the distribution, as we can see from Figure 3.15:

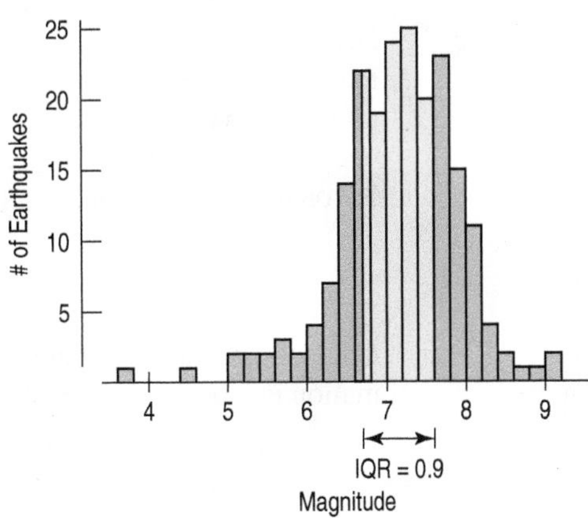

Figure 3.15
The quartiles bound the middle 50% of the values of the distribution. This gives a visual indication of the spread of the data. Here we see that the IQR is 0.9 Richter scale units.

***Activity:* Displaying Spread.** What does the IQR look like on a histogram? How about the standard deviation?

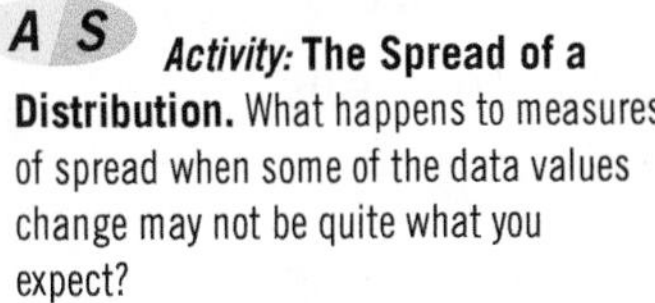

***Activity:* The Spread of a Distribution.** What happens to measures of spread when some of the data values change may not be quite what you expect?

Even if the distribution is skewed or has some outliers, the IQR should provide useful information. The upper and lower quartiles should also be reported along with the IQR, for skewed data, so that we can see the *difference in spread* on both sides of the median. And we will see in the next chapter how useful the IQR can be for comparing the spread of two or more distributions (with side-by-side boxplots). The one exception is when the data are strongly bimodal. For example, remember the dotplot of winning times in the Kentucky Derby (page 49)? Because the race distance was changed, we have data on two different races, and they really shouldn't be summarized together.

SO, WHAT IS A QUARTILE ANYWAY?

Finding the quartiles sounds easy, but surprisingly, the quartiles are not well-defined. It's not always clear how to find a value such that exactly one quarter of the data lies above or below that value. We offered a simple rule for finding quartiles in the box above: Find the median of each half of the data split by the median. When *n* is odd, we exclude the median with each of the halves. Some other texts include the median in each half before finding the quartiles. Both methods are commonly used. If you are willing to do a bit more calculating, there are several other methods that locate a quartile somewhere between adjacent data values. We know of at least six different rules for finding quartiles. Remarkably, each one is in use in some software package or calculator.

So don't worry too much about getting the "exact" value for a quartile. All of the methods agree pretty closely when the data set is large. When the data set is small, different rules will disagree more, but in that case there's little need to summarize the data anyway.

Remember, Statistics is about understanding the world, not about calculating the right number. The "answer" to a statistical question is a sentence about the issue raised in the question.

NOTATION ALERT

We always use Q1 to label the lower (25%) quartile and Q3 to label the upper (75%) quartile. We skip the number 2 because the median would, by this system, naturally be labelled Q2—but we don't usually call it that.

5-Number Summary

The **5-number summary** of a distribution reports its median, quartiles, and extremes (maximum and minimum). The 5-number summary for the recent tsunami-causing earthquake *Magnitudes* looks like this:

Max	9.1
Q3	7.6
Median	7.2
Q1	6.7
Min	3.7

It's a good idea to report the number of data values and the identity of the cases (the *Who*). Here there are 207 earthquakes.

The 5-number summary provides a good overview of the distribution of magnitudes of these tsunami-causing earthquakes. For a start, we can see that the median magnitude is 7.2. Because the IQR is only 7.6 − 6.7 = 0.9, we see that many quakes are close to the median magnitude. Indeed, the quartiles show us that the middle half of these earthquakes had magnitudes between 6.7 and 7.6. One quarter of the earthquakes had magnitudes above 7.6 and one quarter were below 6.7, although one tsunami was caused by a quake measuring only 3.7 on the Richter scale.

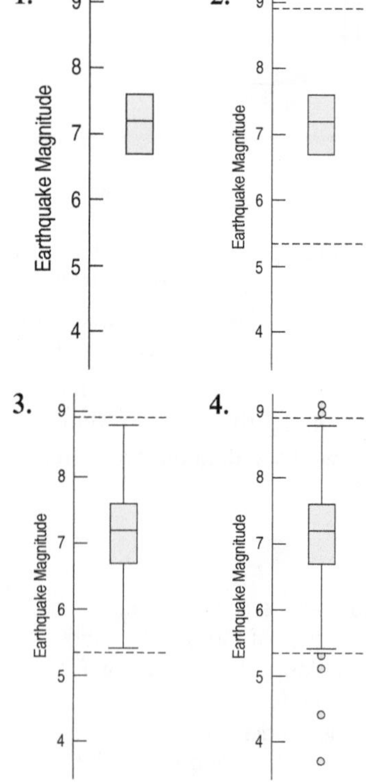

Boxplots

Once we have a 5-number summary of a (quantitative) variable, we can display that information in a **boxplot**. To make a boxplot of the earthquake magnitudes, follow these steps:

1. Draw a single vertical axis spanning the extent of the data.[15] Draw short horizontal lines at the lower and upper quartiles and at the median. Then connect them with vertical lines to form a box. The box can have any width that looks OK.[16]
2. To construct the boxplot, erect "fences" around the main part of the data. Place the upper fence 1.5 IQRs above the upper quartile and the lower fence 1.5 IQRs below the lower quartile. For the earthquake magnitude data, we compute

$$\textit{Upper fence} = Q3 + 1.5\ \text{IQR} = 7.6 + 1.5 \times 0.9 = 8.95$$

and

$$\textit{Lower fence} = Q1 - 1.5\ \text{IQR} = 6.7 - 1.5 \times 0.9 = 5.35$$

The fences are just for construction and are not part of the display. We show them here with dotted lines for illustration. You should never include them in your boxplot.
3. We use the fences to grow "whiskers." Draw lines from the ends of the box up and down to the *most extreme data values found within the fences*. If a data value falls outside one of the fences, we do *not* connect it with a whisker.
4. Finally, we display any data values lying beyond the fences with special symbols. These are worthy of consideration as possible outliers. Let's call them *suspect outliers*.

A S **Boxplots.** Watch a boxplot under construction

A boxplot highlights several features of the distribution. The central box shows the middle half of the data, between the quartiles. The height of the box is equal to the IQR. If the median is roughly centred between the quartiles, then the middle half of the data is roughly symmetric. If the median is not centred, the distribution is skewed. The whiskers show skewness as well if

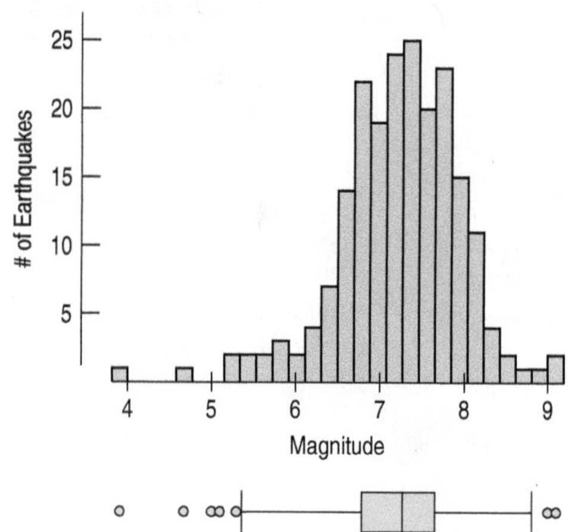

Figure 3.16

By turning the boxplot and putting it on the same scale as the histogram, we can compare both displays of the earthquake magnitudes and see how each represents the distribution.

Why 1.5 IQRs? One of the authors asked the prominent statistician, John W. Tukey, the originator of the boxplot, why the outlier nomination rule cut at 1.5 IQRs beyond each quartile. He answered that the reason was that 1 IQR would be too small and 2 IQRs would be too large. That works for us.

[15] The axis could also run horizontally.

[16] Some computer programs draw wider boxes for larger data sets. This can be useful when comparing groups.

they are not roughly the same length. Any outliers are displayed individually, both to keep them out of the way for judging skewness and to encourage you to give them special attention. They may be mistakes, or they may be the most interesting cases in your data.

For the recent tsunami-causing earthquake data, the central box contains all the earthquakes whose magnitudes are between 6.7 and 7.6 on the Richter scale. From the shape of the box, it looks like the central part of the distribution of earthquakes is roughly symmetrical, but the longer lower whisker indicates that the distribution stretches out slightly at the lower end. We also see a few very small magnitude earthquakes and the two large earthquakes, which we've discussed. Boxplots are particularly good at pointing out possible outliers.

Step-by-Step Example SHAPE, CENTRE, AND SPREAD: FLIGHT CANCELLATIONS

MO_SES Premium/Shutterstock

The U.S. Bureau of Transportation Statistics (www.bts.gov) reports data on airline flights. Let's look at data giving the percentage of flights cancelled each month between 1995 and 2011.

Question: How often are flights cancelled?

Who	Months
What	Percentage of flights cancelled at U.S. airports
When	1995–2011
Where	United States

THINK ➡ Variable Identify the *variable*, and decide how you wish to display it.

To identify a variable, report the *W*'s.

Select an appropriate display based on the nature of the data and what you want to know.

I want to learn about the monthly percentage of flight cancellations at U.S airports.

I have data from the U.S. Bureau of Transportation Statistics giving the percentage of flights cancelled at U.S. airports each month between 1995 and 2011.

✓ **Quantitative Data Condition:** Percentages are quantitative. A histogram and numerical summaries would be appropriate.

SHOW ➡ Mechanics We usually make histograms with a computer or graphing calculator.

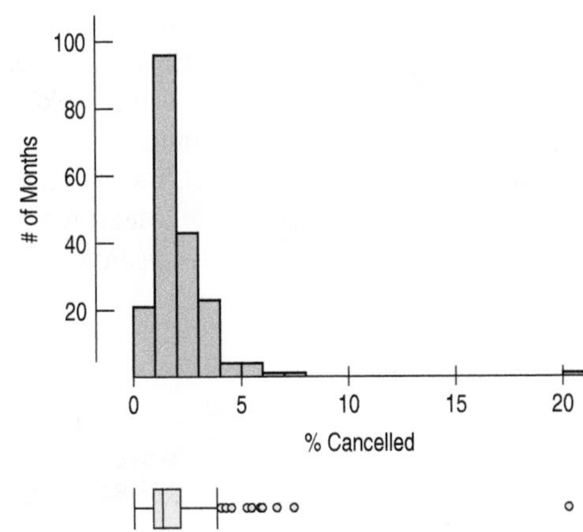

The histogram shows a distribution skewed to the high end and one extreme outlier, a month in which more than 20% of flights were cancelled.

REALITY CHECK It's always a good idea to think about what you expect to see so that you can check whether the histogram looks like what you expected.

In most months, fewer than 5% of flights are cancelled and usually only about 2% or 3%. That seems reasonable.

With 201 cases, we probably have more data than you'd choose to work with by hand. The summary values given here are from technology.

The 5-number summary should be appropriate for these data:

Count	201
Max	20.24
Q3	2.54
Median	1.730
Q1	1.312
Min	0.540
IQR	1.227

TELL ➡ Interpretation Describe the shape, centre, and spread of the distribution. Report on the symmetry, number of modes, and any gaps or outliers. You should also mention any concerns you may have about the data.

The distribution of cancellations is skewed to the right, and this makes sense: The values can't fall below 0%, but can increase almost arbitrarily due to bad weather or other events.

The median is 1.73% and the IQR is 1.23%. The low IQR indicates that in most months the cancellation rate is close to the median. In fact, it's between 1.31% and 2.54% in the middle 50% of all months, and in only 1/4 of the months were more than 2.54% of flights cancelled.

There is one extraordinary value: 20.2%. Looking it up, I find that the extraordinary month was September 2001. The attacks of September 11 shut down air travel for several days, accounting for this outlier.

Thinking About Variation

Why do banks favour a single line that feeds several teller windows rather than separate lines for each teller? The average waiting time is the same, but the time you can expect to wait is less variable when there is a single line, and people prefer consistency.

Statistics is about variation, so spread is an important fundamental concept. Measures of spread help us to be precise about what we don't know. If many data values are scattered far from the centre, the IQR and the standard deviation will be large. If the data values are close to the centre, then these measures of spread will be small. If all our data values were exactly the same, we'd have no question about summarizing the centre, and all measures of spread would be zero—and we wouldn't need statistics. You might think this would be a big plus, but it would make for a boring world. Fortunately (at least for statistics), data do vary.

Measures of spread tell how well other summaries describe the data. That's why we always (always!) report a spread along with any summary of the centre.

For Example DESCRIBING SPREAD

RECAP: The histogram showed you that the distribution of credit card expenditures is skewed, and you have used the median to describe the centre. The quartiles are \$73.84 and \$624.80.

QUESTION: What is the IQR, and why is it a suitable measure of spread?

ANSWER: For these data, the interquartile range (IQR) is \$624.80 – \$73.84 = \$550.96. Like the median, the IQR is not affected by the outlying value or by the skewness of the distribution, so it is an appropriate measure of spread for the given expenditures.

Just Checking

6. Statistics Canada reports the median family income in its summary of census data. Why do you suppose they use the median instead of the mean? What might be the disadvantages of reporting the mean?
7. You've just bought a new car that claims to get a highway fuel efficiency of 9.8 litres per 100 kilometres. Of course, your mileage will "vary." If you had to guess, would you expect the IQR of gas mileage attained by all cars like yours to be 10, 1, or 0.1 L/100 km? Why?
8. A company selling a new MP3 player advertises that the player has a mean lifetime of five years. If you were in charge of quality control at the factory, would you prefer that the standard deviation of lifespans of the players you produce be two years or two months? Why?

What to *Tell* About a Quantitative Variable—A Summary

A S *Activity:* **Playing with Summaries.** Here's a statistics game about summaries that even some experienced statisticians find . . . well, challenging. Your intuition may be better. Give it a try!

HOW "ACCURATE" SHOULD WE BE?

Don't think you should report means and standard deviations to a zillion decimal places; such implied accuracy is really meaningless. Although there is no ironclad rule, statisticians commonly report summary statistics to one or two decimal places more than the original data.

What should you *Tell* about a quantitative variable?

- Start by making a histogram or stem-and-leaf display and discuss the shape of the distribution.
- Next, discuss the centre *and* spread.
  - If the shape is skewed, report the 5-number summary and be sure to discuss the median and IQR. You may want to include the mean and standard deviation as well, but you should point out why the mean and median differ.
  - If the shape is symmetric, report the mean and standard deviation and possibly the median and IQR as well. For unimodal symmetric data, the IQR is usually a bit larger than the standard deviation. If that's not true for your data set, look again to make sure the distribution isn't skewed and there are no outliers.
  - Always pair the median with the IQR (both implicit in the 5-number summary) and the mean with the standard deviation. It's not useful to report one without the other. Reporting a centre without a spread is dangerous. You may think you know more than you do about the distribution. Reporting only the spread leaves us wondering where we are.
- Also discuss any unusual features.
  - If there are multiple modes, try to understand why. If you can identify a reason for separate modes (for example, women and men typically have heart attacks at different ages), it may be a good idea to split the data into separate groups.
  - If there are any clear outliers, you should point them out. If you are reporting the mean and standard deviation, report them with the outliers present and with the outliers omitted. The differences may be revealing. (Of course, the median and IQR won't be affected very much by the outliers.)

For Example CHOOSING SUMMARY STATISTICS

RECAP: You have provided the credit-card company's board of directors with a histogram of customer expenditures, and you have summarized the centre and spread with the median and IQR. Knowing a little about statistics, the directors now insist on having the mean and standard deviation as summaries of the spending data.

QUESTION: Although you know that the mean is $478.19 and the standard deviation is $741.87, you need to explain to them why these are not suitable summary statistics for the expenditures data. What would you give as reasons?

ANSWER: The high outlier at $7000 pulls the mean up substantially and inflates the standard deviation. Locating the mean value on the histogram shows that it is not a typical value at all, and the standard deviation suggests that expenditures vary much more than they do. The median and IQR are more resistant to the presence of skewness and outliers, giving more realistic descriptions of centre and spread.

WHAT CAN GO WRONG?

A data display should tell a story about the data. To do that it must speak in a clear language, making plain what variable is displayed, what any axis shows, and what the values of the data are. And it must be consistent in those decisions.

A display of quantitative data can go wrong in many ways. The most common failures arise from only a few basic errors:

- **Don't make a histogram of a categorical variable.** Just because the variable contains numbers doesn't mean it's quantitative. Here's a histogram of the insurance policy numbers of some workers. It's not very informative because the policy numbers are just labels. A histogram or stem-and-leaf display of a categorical variable makes no sense. A bar or pie chart would be more appropriate.
- **Don't look for shape, centre, and spread of a bar chart.** A bar chart showing the sizes of the piles displays the distribution of a categorical variable, but the bars could be arranged in any order from left to right. Concepts like symmetry, centre, and spread make sense only for quantitative variables.

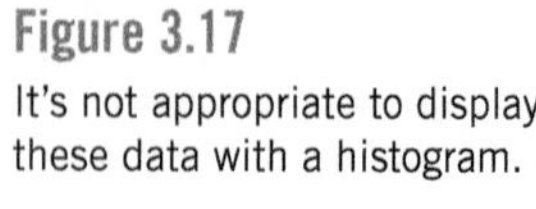

Figure 3.17
It's not appropriate to display these data with a histogram.

- **Don't use bars in every display—save them for histograms and bar charts.** In a bar chart, the bars indicate how many cases of a categorical variable are piled in each category. Bars in a histogram indicate the number of cases piled in each interval of a quantitative variable. In both bar charts and histograms, the bars represent counts of data values. Some people create other displays that use bars to represent individual data values. Beware: Such graphs are neither bar charts nor histograms. For example, a student was asked to make a histogram from data showing the number of juvenile bald eagles seen during each of the 13 weeks in the winter of 2003–2004 at a site in Rock Island, IL. Instead, he made this plot:

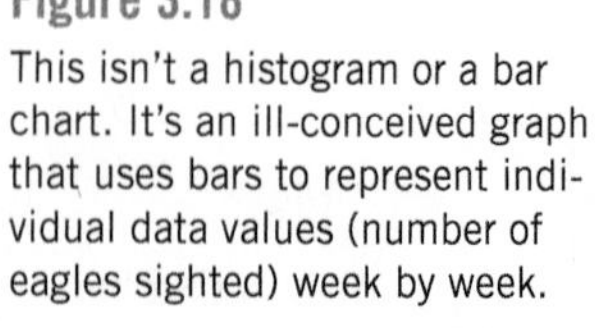

Figure 3.18
This isn't a histogram or a bar chart. It's an ill-conceived graph that uses bars to represent individual data values (number of eagles sighted) week by week.

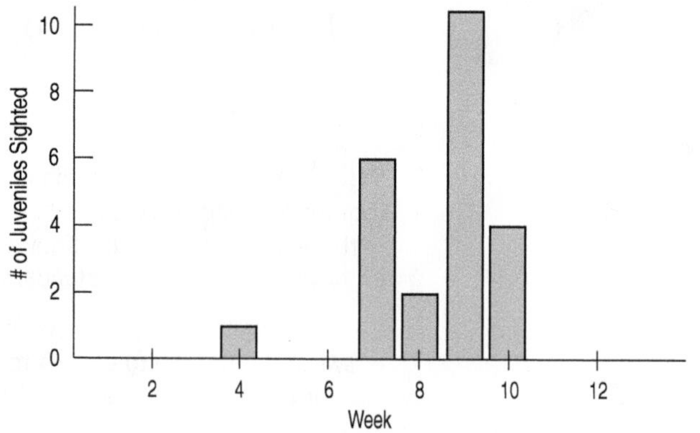

Look carefully. That's not a histogram. A histogram shows *What* we've measured along the horizontal axis and counts of the associated *Who* represented as bar heights. This student has it backwards. He used bars to show counts of birds for each week.[17] We need counts of weeks. A correct histogram should have a tall bar at "0" to show there were many weeks when no eagles were seen, like this:

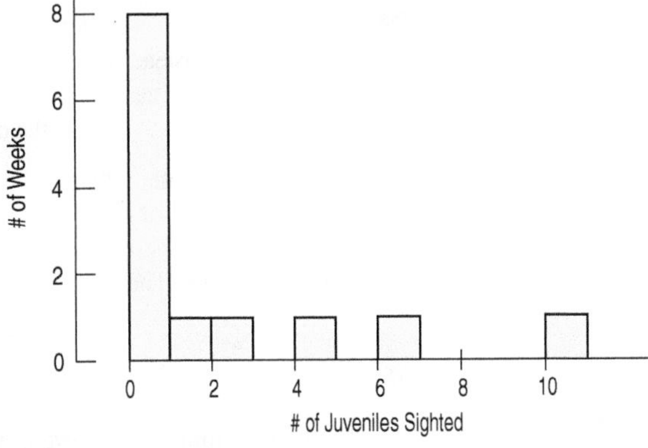

Figure 3.19

A histogram of the eagle sighting data shows the number of weeks in which different counts of eagles occurred. This display shows the distribution of juvenile eagle sightings.

- **Choose a bin width appropriate to the data.** Computer programs usually do a pretty good job of choosing histogram bin widths. Often, there's an easy way to adjust the width, sometimes interactively. Here are the tsunami-causing earthquakes with two (rather extreme) choices for the bin size:

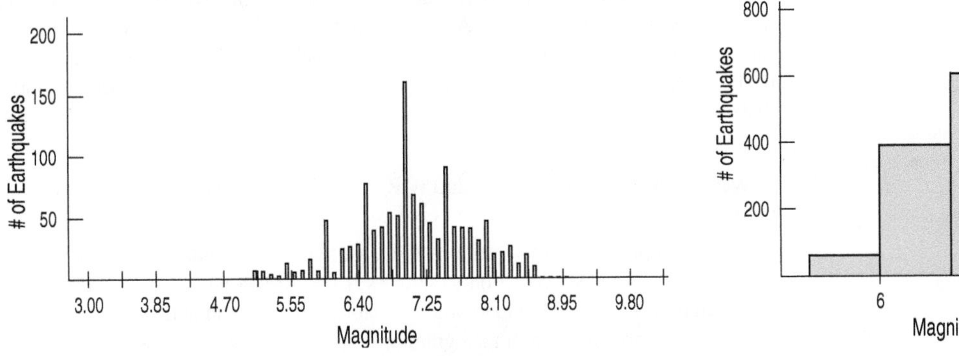

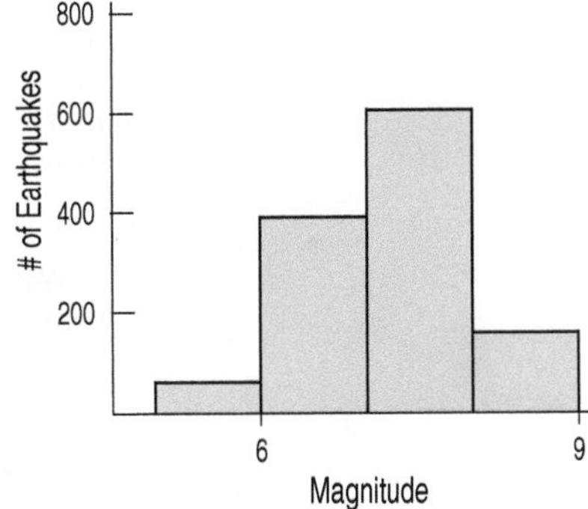

The task of summarizing a quantitative variable is relatively simple, and there is a simple path to follow. However, you need to watch out for certain features of the data that make summarizing them with a number dangerous. Here's some advice:

- **Don't forget to do a reality check.** Don't let the computer or calculator do your thinking for you. Make sure the calculated summaries make sense. For example, does the mean look like it is in the centre of the histogram? Think about the spread: An IQR of 50 L/100 km would clearly be wrong for gas mileages. And no measure of spread can be negative. The standard deviation can take the value 0, but only in the very unusual case that all the data values equal the same number. If you see the IQR or standard deviation equal to 0, it's probably a sign that something's wrong with the data.
- **Don't forget to sort the values before finding the median or percentiles.** It seems obvious, but when you work by hand, it's easy to forget to sort the data first before counting in to find medians, quartiles, or other percentiles. Don't report that the median of the five values 194, 5, 1, 17, and 893 is 1 just because 1 is the middle number.
- **Don't worry about small differences when using different methods.** Finding the 10th percentile or the lower quartile in a data set sounds easy enough, but it turns out that the

(continued)

[17]Edward Tufte, in his book *The Visual Display of Quantitative Information*, proposes that graphs should have a high data-to-ink ratio. That is, we shouldn't waste a lot of ink to display a single number when a dot would do the job. Actually, replace the bars by dots, and we get a timeplot (discussed in the next chapter).

definitions are not exactly clear. If you compare different statistics packages or calculators, you may find that they give slightly different answers for the same data. These differences, however, are unlikely to be important in interpreting the data, the quartiles, or the IQR, so don't let them worry you.

Gold Card Customers—Regions National Bank

Month	April 2007	May 2007
Average U.S. Postal Code	45 034.34	38 743.34

- **Don't compute numerical summaries of a categorical variable.** The mean (5-digit) U.S. postal code or the standard deviation of Canadian social insurance numbers is not meaningful. If the variable is categorical, you should instead report summaries such as percentages of individuals in each category. It is easy to make this mistake when using technology to do the summaries for you. After all, the computer doesn't care what the numbers mean.
- **Don't report too many decimal places.** Statistical programs and calculators often report a ridiculous number of digits. A general rule for numerical summaries when you report them is to use one or two more digits than the data. For example, earlier we saw a dotplot of Kentucky Derby race times. The mean and standard deviation of those times could be reported as

$$\bar{y} = 130.63401639344262 \text{ sec}$$
$$s = 13.66448201942662 \text{ sec}$$

but the race times are reported only to the nearest quarter-second, so the extra digits are meaningless.

- **Don't round in the middle of a calculation.** Don't *report* too many decimal places, but it's best not to do *any* rounding until the end of calculation. Even though you might report the mean of the earthquakes as 7.08, it's really 7.08339. Use the more precise number in your calculations if you're finding the standard deviation by hand—or be prepared to see small differences in your final result.
- **Watch out for multiple modes.** The summaries of the Kentucky Derby times are meaningless for another reason. As we saw in the dotplot, the derby was initially a longer race. It would make much more sense to report that the old 1.5 mile derby had a mean time of 159.6 seconds, while the current derby has a mean time of 124.6 seconds. If the distribution has multiple modes, consider separating the data into different groups and summarizing each group separately.
- **Beware of outliers.** The median and IQR are resistant to outliers, but the mean and standard deviation are not. To help spot outliers:
  - **Don't forget to make a picture (make a picture, make a picture).** The sensitivity of the mean and standard deviation to outliers is one reason why you should always make a picture of the data. Summarizing a variable with its mean and standard deviation when you have not looked at a histogram or dotplot to check for outliers or skewness invites disaster. You may find yourself making absurd or dangerously wrong conclusions about the data. And, of course, you should demand no less of others. Don't accept a mean and standard deviation blindly without some evidence that the variable they summarize is unimodal, reasonably symmetric, and free of outliers.

CONNECTIONS

Distributions of quantitative variables, like those for categorical variables, show the possible values and their relative frequencies. A histogram shows the distribution of values in a quantitative variable with adjacent bars. Don't confuse them with bar charts, which display categorical variables. For categorical data, the mode is the category with the biggest count. For quantitative data, modes are peaks in the histogram.

The shape of the distribution of a quantitative variable is an important concept in most of the subsequent chapters. We will be especially interested in distributions that are unimodal and symmetric.

In addition to their shape, we summarize distributions with centre and spread, usually pairing a measure of centre with a measure of spread: median with IQR and mean with standard deviation. We favour the mean and standard deviation when the shape is unimodal and symmetric, but choose the median and IQR (along with the 5-number summary) for skewed distributions or when there are outliers we can't otherwise set aside.

What Have We Learned?

Learning Objectives

Make and interpret displays of the distribution of a variable.

- Display the distribution of a quantitative variable with a histogram, stem-and-leaf display, or dotplot.
- Understand distributions in terms of their shape, centre, and spread.

Describe the shape of a distribution.

- A **symmetric** distribution has roughly the same shape reflected around the centre.
- A **skewed** distribution extends farther on one side than on the other.
- A **unimodal** distribution has a single major hump or mode; a bimodal distribution has two; multimodal distributions have more.
- **Outliers** are values that lie far from the rest of the data.
- Report any other **unusual feature** of the distribution, such as gaps.

Describe the centre of a distribution by computing the mean and median, and know when it is best to use each.

- The **mean** is the sum of the values divided by the count. It is a suitable measure of centre for unimodal, symmetric distributions.
- The **median** is the middle value; half the values are above and half are below the median. It is generally a better measure of centre when the distribution is skewed or has outliers.

Compute the standard deviation and the 5-number summary, with IQR, and know when it is best to use each to summarize the spread.

- The **standard deviation** is roughly the square root of the average squared difference between each data value and the mean. It is the summary of choice for the spread of unimodal, symmetric variables.
- A **5-number summary** consists of the median, the quartiles, and the extremes of the data. The quartiles, along with the extremes, show the difference in spread on each side of the median. The difference between the quartiles is the IQR. This summary often provides a better summary of spread for skewed distributions or data with outliers.
- We'll report the 5-number summary and IQR, when the distribution is skewed. If it's symmetric, we'll summarize the distribution with the mean and standard deviation (and possibly the 5-number summary as well). Always pair the median with the 5-number summary & IQR and the mean with the standard deviation.

Recognize that the methods of this chapter assume that the data are quantitative.

- The **Quantitative Data Condition** serves as a reminder to check that the data are quantitative. A good way to be sure is to report the measurement units.

Use the **5-number summary** to make a boxplot. Use the boxplot's outlier nomination rule to identify cases that may deserve special attention.

- A **boxplot** shows the quartiles as the upper and lower ends of a central box, the median as a line across the box, and "whiskers" that extend to the most extreme values that are not nominated as outliers.
- Boxplots display separately any case that is more than 1.5 IQRs beyond each quartile. These cases should be considered as possible outliers.

Review of Terms

Distribution Slices up all the possible values of a quantitative variable into equal width bins and gives the number of values (or counts) falling into each bin (p. 45).

Histogram (relative frequency histogram) Uses adjacent bars to show the distribution of a quantitative variable. Each bar represents the frequency (or relative frequency) of values falling in each bin (p. 45).

Gap A region of the distribution where there are no values (p. 46).

Stem-and-leaf display Shows quantitative data values in a way that sketches the distribution of the data. It's best described in detail by example (p. 47).

Dotplot Graphs a dot for each case against a single axis (p. 48).

Quantitative Data Condition The data are values of a quantitative variable whose units are known (p. 49).

Shape To describe the shape of a distribution, look for

- single versus multiple modes
- symmetry versus skewness (p. 49).

Centre The place in the distribution of a variable that you'd point to if you wanted to attempt the impossible by summarizing the entire distribution with a single number. Measures of centre include the mean and median (p. 49).

Spread A numerical summary of how tightly the values are clustered around the "centre." Measures of spread include the standard deviation and IQR (p. 49).

Mode A hump or local high point in the shape of the distribution of a variable. The apparent location of modes can change as the scale of a histogram is changed (p. 49).

Unimodal Having one mode. This is a useful term for describing the shape of a histogram when it's generally mound-shaped. Distributions with two modes are called **bimodal**. Those with more than two are **multimodal** (p. 49).

Uniform A distribution that's roughly flat is said to be uniform (p. 50).

Symmetric A distribution is symmetric if the two halves on either side of the centre look approximately like mirror images of each other (p. 50).

Tails The parts of a distribution that typically trail off on either side. Distributions can be characterized as having long tails (if they straggle off for some distance) or short tails (if they don't) (p. 50).

Skewed A distribution is skewed if it's not symmetric and one tail stretches out farther than the other. Distributions are said to be **skewed left** (or negatively) when the longer tail stretches to the left (or in the negative direction), and **skewed right** (or positively) when it goes to the right (or in the positive direction) (p. 50).

Outliers Extreme values that don't appear to belong with the rest of the data. They may be unusual values that deserve further investigation or just mistakes; there's no obvious way to tell. Don't delete outliers automatically—you have to think about them. Outliers can affect many statistical analyses, so you should always be alert for them (p. 51).

Mean Found by summing all the data values and dividing by the count:

$$\bar{y} = \frac{Total}{n} = \frac{\sum y}{n}$$

It is usually paired with the standard deviation (p. 53).

Median The middle value with half of the data above and half below it. If n is even, it is the average of the two middle values. It is usually paired with the IQR (p. 54).

Resistant A calculated summary measure is said to be resistant if it is affected only a limited amount by any small portion of the data, such as outliers (p. 54).

Range The difference between the lowest and highest values in a data set. *Range* = *max* − *min* (p. 55).

Variance The sum of squared deviations from the mean, divided by the count minus one.

$$s^2 = \frac{\sum(y - \bar{y})^2}{n - 1}$$

It is useful in calculations later in the book (p. 57).

Standard deviation The square root of the variance.

$$s = \sqrt{\frac{\sum(y - \bar{y})^2}{n - 1}}$$

It is usually reported along with the mean (p. 56).

Percentile The *i*th percentile is the number that falls above *i* percent of the data (p. 60).

Quartile The lower quartile (Q1) is the value with a quarter of the data below it. The upper quartile (Q3) has three quarters of the data below it. The median and quartiles divide data into four parts of equal numbers of data values (p. 60).

Interquartile range (IQR) The difference between the first and third quartiles. $IQR = Q3 - Q1$. It is usually reported along with the median and 5-number summary (p. 60).

5-number summary A useful summary of data consisting of minimum, Q1, median, Q3, and maximum (p. 61).

Boxplot Displays the 5-number summary as a central box with whiskers that extend to the non-outlying data values. Boxplots are particularly effective for comparing groups and for displaying possible outliers (p. 62).

On the Computer DISPLAYING AND SUMMARIZING QUANTITATIVE VARIABLES

Almost any program that displays data can make a histogram, but some will do a better job of determining where the bars should start and how they should partition the span of the data.

The vertical scale may be counts or proportions. Sometimes it isn't clear which. But the shape of the histogram is the same either way.

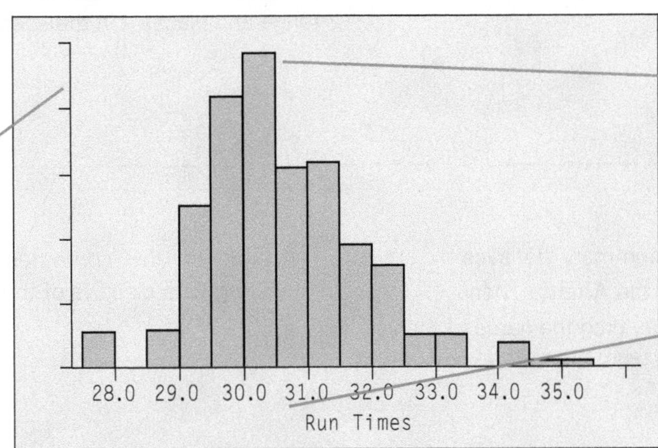

Most packages choose the number of bars for you automatically. Often you can adjust that choice.

The axis should be clearly labelled so you can tell what "pile" each bar represents. You should be able to tell the lower and upper bounds of each bar.

Many statistics packages offer a pre-packaged collection of summary measures. The result might look like this:

```
Variable: Weight
N = 234
Mean = 143.3          Median = 139
St. Dev. = 11.1       Q1 = 134.0
                      Q3 = 148.0
```

Alternatively, a package might make a table for several variables and summary measures.

Variable	N	mean	median	stdev	Q1	Q3
Weight	234	143.3	139	11.1	134.0	148.0
Height	234	68.3	68.1	4.3	66.0	71.0
Score	234	86	88	9	86	91

A S ***Case Study:* Describing Distribution Shapes.** Who's safer in a crash—passengers or the driver? Investigate with your statistics package.

It is usually easy to read the results and identify each computed summary. You should be able to read the summary statistics produced by any computer package.

Packages often provide many more summary statistics than you need. Of course, some of these may not be appropriate when the data are skewed or have outliers. It is your responsibility to check a histogram or stem-and-leaf display and decide which summary statistics to use.

It is common for packages to report summary statistics to many decimal places of "accuracy." Of course, it is rare that data have such accuracy in the original measurements. Just because a package calculates to six or seven digits beyond the decimal point doesn't mean that those digits have any meaning. Generally, it's a good idea to round these values, allowing perhaps one more digit of precision than was given in the original data.

Displays and summaries of quantitative variables are among the simplest things you can do in most statistics packages.

EXCEL

Excel cannot make histograms or dotplots without a third party add-in.

To calculate summaries, for example, the mean:

- Click on an empty cell.
- Go to the Formulas tab in the Ribbon. Click on the drop down arrow next to "AutoSum" and choose "**Average**."
- Enter the data range in the formula displayed in the empty box you selected earlier.
- Press **Enter**. This computes the mean for the values in that range.

To compute the standard deviation:

- Click on an empty cell.
- Go to the Formulas tab in the Ribbon and click the drop down arrow next to "AutoSum" and select "**More functions . . .**"
- In the dialog window that opens, select "**STDEV**" from the list of functions and click **OK**. A new dialog window opens. Enter a range of fields into the text fields and click **OK**.

Excel computes the standard deviation for the values in that range and places it in the specified cell of the spreadsheet.

JMP

To make a histogram and find summary statistics:

- Choose **Distribution** from the **Analyze** menu.
- In the **Distribution** dialogue, drag the name of the variable that you wish to analyze into the empty window beside the label **Y, Columns**.
- Click **OK**. JMP computes standard summary statistics along with displays of the variables.

MINITAB

To make a histogram:

- Choose **Histogram** from the **Graph** menu.
- Select "Simple" for the type of graph and click **OK**.
- Enter the name of the quantitative variable you wish to display in the box labelled "Graph variables." Click **OK**.

To calculate summary statistics:

- Choose **Basic statistics** from the **Stat** menu. From the **Basic Statistics** submenu, choose **Display Descriptive Statistics**.
- Assign variables from the variable list box to the Variables box. MINITAB makes a Descriptive Statistics table.

R

For a quantitative variable X:

- **summary(X)** gives a 5-number summary and the mean.
- **mean(X)** gives the mean and **sd(X)** gives the standard deviation.
- **hist(X)** produces a histogram and **boxplot(X)** makes a boxplot.

COMMENTS

Many other summaries are available, including **min( )**, **max ( )**, **quantile(X,prob=p)** (where p is a probability between 0 and 1), and **median( )**. Both hist and boxplot have many options.

SPSS

To make a histogram in SPSS open the Chart Builder from the Graphs menu:

- Click the **Gallery** tab.
- Choose **Histogram** from the list of chart types.
- Drag the histogram onto the canvas.
- Drag a scale variable to the *y*-axis drop zone.
- Click **OK**.

To calculate summary statistics:

- Choose **Explore** from the **Descriptive Statistics** submenu of the **Analyze** menu. In the Explore dialog, assign one or more variables from the source list to the Dependent List and click the **OK** button.

STATCRUNCH

To make a histogram, dotplot, or stem-and-leaf plot:

- Click on **Graphics**.
- Choose the type of plot.
- Choose the variable name from the list of **Columns**.
- Click on **Next**.
- (For a histogram) Choose **Frequency** or (usually) **Relative frequency**, and (if desired) set the axis scale by entering the **Start** value and **Bin width**.
- Click on **Create Graph**.

To calculate summaries:

- Click on **Stat**.
- Choose **Summary Stats » Columns**.
- Choose the variable name from the list of **Columns**.
- Click on **Calculate**.

COMMENTS

- You may need to hold down the Ctrl or command key to choose more than one variable to summarize.
- Before calculating, click on **Next** to choose additional summary statistics.

TI-83/84 PLUS

To make a histogram:

- Turn a **STATPLOT** on.
- Choose the **histogram icon** and specify the List where the data are stored.
- **ZoomStat**, then adjust the **WINDOW** appropriately.

To calculate summary statistics:

- Choose **1-VarStats** from the **STAT CALC** menu and specify the List where the data are stored. You must scroll down to see the 5-number summary.
- To make a boxplot, set up a **STAT PLOT** using the boxplot icon.

COMMENTS

If the data are stored as a frequency table (say, with data values in L1 and frequencies in L2), set up the Plot with Xlist: L1 and Freq: L2.

Exercises

1. **Histogram** Find a histogram that shows the distribution of a variable in a newspaper, a magazine, or on the Internet.
   a) Does the article identify the *W*'s?
   b) Discuss whether the display is appropriate for the data.
   c) Discuss what the display reveals about the variable and its distribution.
   d) Does the article accurately describe and interpret the data? Explain.

2. **Not a histogram** Find a graph other than a histogram that shows the distribution of a quantitative variable in a newspaper, a magazine, or on the Internet.
   a) Does the article identify the *W*'s?
   b) Discuss whether the display is appropriate for the data.
   c) Discuss what the display reveals about the variable and its distribution.
   d) Does the article accurately describe and interpret the data? Explain.

3. **In the news** Find an article in a newspaper, a magazine, or on the Internet that discusses an "average."
   a) Does the article discuss the *W*'s for the data?
   b) What are the units for the variable?
   c) Is the average used the median or the mean? How can you tell?
   d) Is the choice of median or mean appropriate for the situation? Explain.

4. **In the news II** Find an article in a newspaper, a magazine, or on the Internet that discusses a measure of spread.
   a) Does the article discuss the *W*'s for the data?
   b) What are the units for the variable?
   c) Does the article use the range, IQR, or standard deviation?
   d) Is the choice of measure of spread appropriate for the situation? Explain.

5. **Thinking about shape** Would you expect distributions of these variables to be uniform, unimodal, or bimodal? Symmetric or skewed? Explain.
   a) The number of speeding tickets each student in their final year of university has ever had
   b) Players' scores (number of strokes) at the RBC Canadian Open golf tournament in a given year
   c) Weights of female babies born in a particular hospital over the course of a year
   d) The length of the average hair on the heads of students in a large class

6. **More shapes** Would you expect distributions of these variables to be uniform, unimodal, or bimodal? Symmetric or skewed? Explain.
   a) Ages of people at a pee-wee hockey game
   b) Number of siblings of people in your class
   c) Pulse rates of college- or university-age males
   d) Number of times each face of a die shows in 100 tosses

T 7. **Sugar in cereals** The histogram displays the sugar content (as a percent of weight) of 49 brands of breakfast cereals.

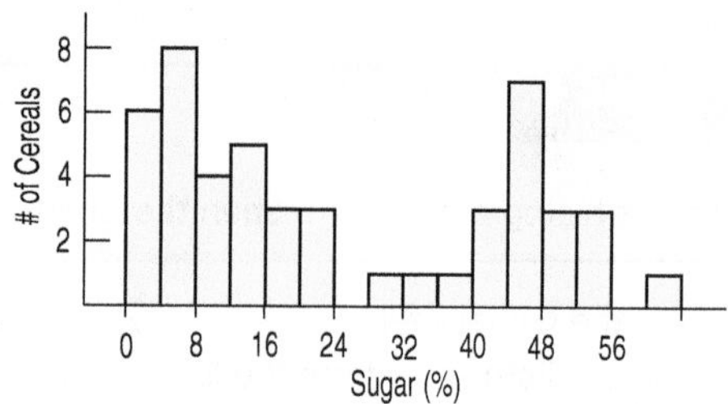

   a) Describe this distribution.
   b) What do you think might account for this shape?

8. **Chorus heights** The display shows the heights of some of the singers in a chorus, collected so that the singers could be positioned on stage with shorter ones in front and taller ones in back.

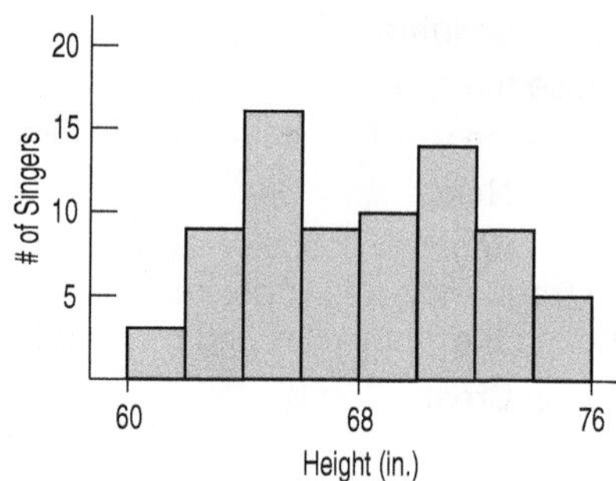

   a) Describe the distribution.
   b) Can you account for the features you see here?

9. **Test scores** Below is a histogram of 110 students' test scores on a one-hour test in STA220 at the University of Toronto during summer 2007. The test was out of 50, not 100.

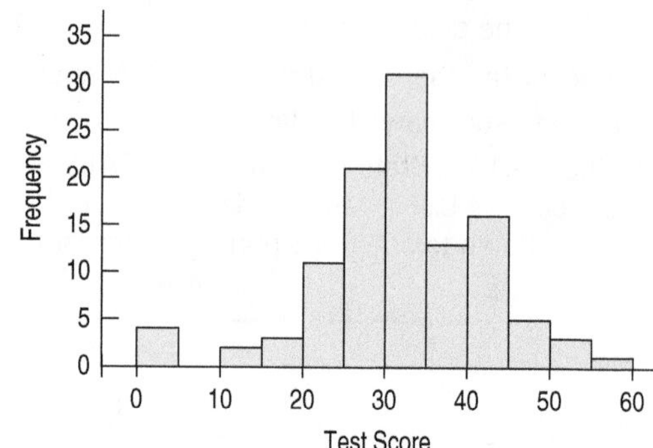

a) Approximately what percentage of students got A grades? An A was 40 or higher, since the test was out of 50 (some scored above 50 due to bonus marks).
b) What percent got C or D grades? (i.e., between 50% and 69%, or 25 and 35 in the graph since any scores of 35 went into the higher bin)
c) Write a brief description of this distribution (shape, centre, spread, and unusual features). Can you account for any of the features you see here?

10. Run times One of the authors collected the times (in minutes) it took him to run four miles on various courses during a 10-year period. Here is a histogram of the times.

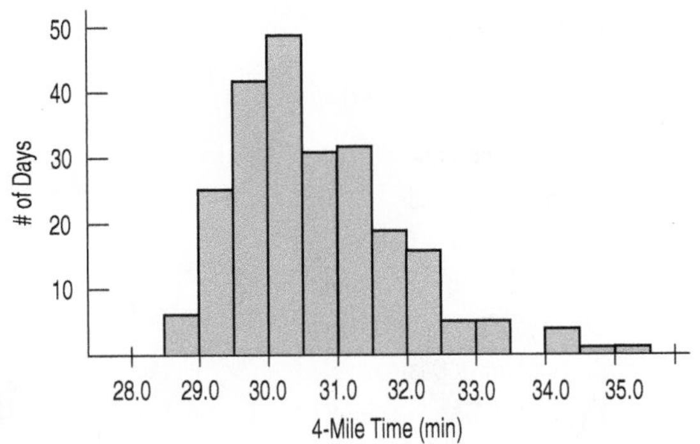

Describe the distribution and summarize the important features. What is it about running that might account for the shape you see?

11. Election 2011 Below is a histogram of percentages of rejected ballots for each of the 308 federal electoral districts in the 2011 Canadian federal election.

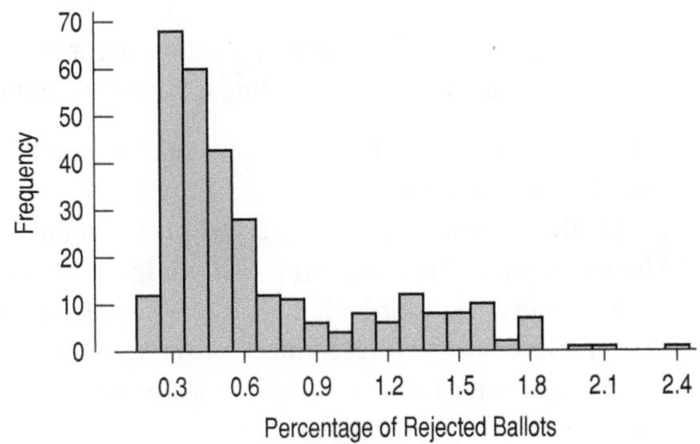

a) From the histogram, would you expect the mean or median to be larger? Explain. What can you say about the actual numerical value of the median from the graph?
b) Write a few sentences describing this distribution (shape, centre, spread, unusual features). Precise numerical calculations are not necessary (though some numbers are). Is there anything of possible practical importance to be learned from this graph? Any further investigation you'd suggest?

12. E-mails A university professor saved every e-mail received from students in a large Introductory Statistics class during an entire term. He then counted, for each student who had sent him at least one e-mail, how many e-mails each student had sent.

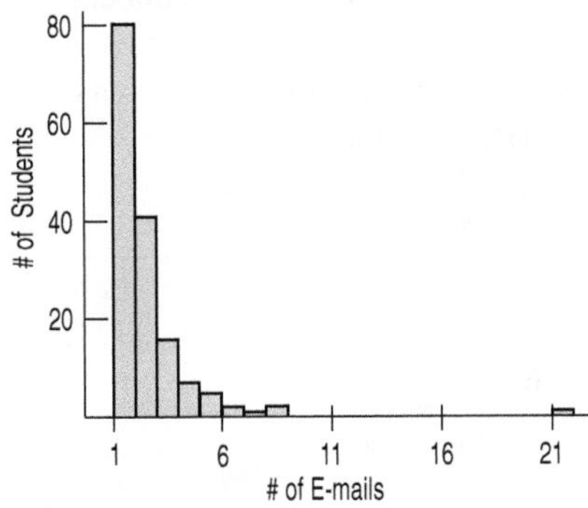

a) From the histogram, would you expect the mean or the median to be larger? Explain. What can you say about the actual value of the median from this graph?
b) Write a few sentences describing this distribution (shape, centre, spread, and unusual features).

13. Summaries Here are costs of 10 electric smoothtop ranges rated very good or excellent by *Consumer Reports* on their website www.consumerreports.org.

$850 900 1400 1200 1050 1000 750 1250 1050 565

Find these statistics *by hand* (no calculator!):
a) Mean
b) Median and quartiles
c) Range and IQR

14. Tornadoes 2011 Here are the annual numbers of deaths from tornadoes in the United States from 1998 through 2011 (*Source:* NOAA):

130 94 40 40 555 54 35 38 67 81 125 21 45 544

Find these statistics *by hand* (no calculator!):
a) Mean
b) Median and quartiles
c) Range and IQR

15. Mistake A clerk entering salary data into a company spreadsheet accidentally put an extra "0" in the boss's salary, listing it as $2 000 000 instead of $200 000. Explain how this error will affect these summary statistics for the company payroll:
a) Measures of centre: median and mean
b) Measures of spread: range, IQR, and standard deviation

16. Sick days During contract negotiations, a company seeks to change the number of sick days employees may take, saying that the annual "average" is seven days of absence per employee. The union negotiators counter that the "average" employee misses only three days of work each year. Explain how both sides might be correct, identifying the measure of centre you think each side is using and why the difference might exist.

17. Standard deviation I For each lettered part, a) through c), examine the two given sets of numbers. Without doing any calculations, decide which set has the larger standard deviation and explain why. Then check by finding the standard deviations *by hand.*

	Set 1	Set 2
a)	3, 5, 6, 7, 9	2, 4, 6, 8, 10
b)	10, 14, 15, 16, 20	10, 11, 15, 19, 20
c)	2, 6, 6, 9, 11, 14	82, 86, 86, 89, 91, 94

18. Standard deviation II For each lettered part, a) through c), examine the two given sets of numbers. Without doing any calculations, decide which set has the larger standard deviation and explain why. Then check by finding the standard deviations *by hand.*

	Set 1	Set 2
a)	4, 7, 7, 7, 10	4, 6, 7, 8, 10
b)	100, 140, 150, 160, 200	10, 50, 60, 70, 110
c)	10, 16, 18, 20, 22, 28	48, 56, 58, 60, 62, 70

19. Payroll A small warehouse employs a supervisor at \$1200 a week, an inventory manager at \$700 a week, six stock boys at \$400 a week each, and four drivers at \$500 a week each.

a) Find the mean and median wage.
b) How many employees earn more than the mean wage?
c) Which measure of centre best describes a typical wage at this company, the mean or the median?
d) Which measure of spread would best describe the payroll, the range, the IQR, or the standard deviation? Why?

T **20. Singers** The frequency table shows the heights (in inches) of 130 members of a choir.

Height	Count	Height	Count
60	2	69	5
61	6	70	11
62	9	71	8
63	7	72	9
64	5	73	4
65	20	74	2
66	18	75	4
67	7	76	1
68	12		

a) Find the 5-number summary and IQR.
b) Find the mean and standard deviation.
c) Display these data with a histogram.
d) Write a few sentences describing the distribution of heights.

T **21. Alberta casinos 2013** Below are the numbers of gaming machines in each of Alberta's 24 casinos:

Apex (Gold Dust) Casino	330
Baccarat Casino	330
Boomtown Casino	415
Camrose Casino	200
Cash Casino – Calgary	600
Cash Casino – Red Deer	299
Casino by Vanshaw	230
Casino Calgary	862
Casino Dene	170
Casino Edmonton	860
Casino Lethbridge	437
Casino Yellowhead	783
Century Casino – Calgary	525
Century Casino – Edmonton	758
Cowboys (Stampede) Casino	400
Deerfoot Inn & Casino	767
Eagle River Casino & Travel Plaza	250
Elbow River Casino	600
Great Northern Casino	419
Grey Eagle Casino & Bingo	600
Jackpot Casino	204
Palace Casino	704
River Cree Resort & Casino	1000
Stoney Nakoda Resort & Casino	299

Source: www.abgamblinginstitute.ualberta.ca/LibraryResources/ReferenceSources/AlbertaCasinos

Construct a stem-and-leaf plot and a dotplot for these data. Describe the distribution (including a 5-number summary).

T **22. He shoots, he scores** During his 20 seasons in the NHL, the "Great One", Wayne Gretzky, scored 50% more points than anyone who ever played professional hockey. He accomplished this amazing feat while playing in 280 fewer games than Gordie Howe, the previous record holder. Here are the total number of points scored by Gretzky during each season, in order, from the 1979–80 season to the 1998–99 season:

137 164 212 196 205 208 215 183 149 168 142 163 121 65 130 48 102 97 90 62

a) Create a stem-and-leaf display for these data with 7 to 10 bins (rows). You will need to use split stems, and to either truncate (i.e., chop off last digit) or round the values. Another alternative is to carry two-digit leaves separated by commas.
b) Describe the shape and any unusual features of the distribution. What might explain it?
c) Give the most appropriate and useful numerical summary measures for this distribution, and explain your choice.

23. How tall? Students in a large class were asked to estimate the teacher's height in centimetres. Here's a histogram of their estimates:

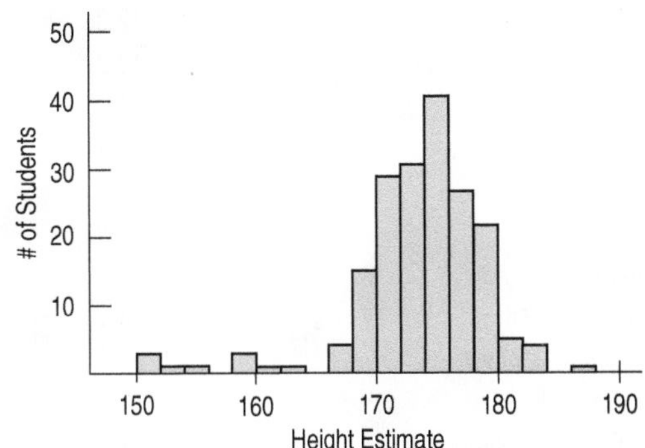

You want a good guess of the teacher's true height (or at least a good guess of how tall the students think the teacher is). How would you summarize these data? Why? What might explain guesses as low as 150 centimetres?

24. He shoots, he scores again In Exercise 22, you examined the total number of points scored by hockey great Wayne Gretzky during his 20-year career in the NHL.

a) Would you use the median or the mean to describe the centre of this distribution? Why?

b) Find the median.

c) Without actually finding the mean, would you expect it to be higher or lower than the median? Explain.

25. World Series champs On Oct 23, 1993, Joe Carter hit the 9th inning game-winning home run (with the Jays trailing 6–5) to make the Toronto Blue Jays World Series champions for a second straight year. Here is a stem-and-leaf display of the number of home runs hit by Joe Carter during the 1983–1998 seasons. (*Source:* www.baseballreference.com/players/c/cartejo01.shtml). Describe the distribution, mentioning its shape and any unusual features.

```
0 | 0
0 |
1 | 3
1 | 58
2 | 14
2 | 5779
3 | 02334
3 | 5
(3|2 means 32)
```

26. Bird species 2010 The Cornell Lab of Ornithology holds an annual Christmas Bird Count (www.birdsource.org), in which bird watchers at various locations around the country see how many different species of birds they can spot. Here are some of the counts reported from sites in Texas during the 2010 event:

150	216	177	150
166	156	159	160
164	169	158	231
175	150	178	183
199	154	164	203

a) Create a stem-and-leaf display of these data.

b) Write a brief description of the distribution. Be sure to discuss the overall shape as well as any unusual features.

27. Hurricanes 2010 The following data give the number of hurricanes classified as major hurricanes in the Atlantic Ocean each year from 1944 through 2010, as reported by NOAA (*Source:* www.nhc.noaa.gov):

3, 3, 1, 2, 4, 3, 8, 5, 3, 4, 2, 6, 2, 2, 4, 2, 2, 7, 1, 2, 6, 1, 3, 1, 0, 5, 2, 1, 0, 1, 2, 3, 2, 1, 2, 2, 2, 3, 1, 1, 1, 3, 0, 1, 3, 2, 1, 2, 1, 1, 0, 5, 6, 1, 3, 5, 3, 4, 2, 3, 6, 7, 2, 2, 5, 2, 5

a) Create a dotplot of these data.

b) Describe the distribution.

28. Horsepower Create a stem-and-leaf display for these horsepowers of autos reviewed by *Consumer Reports* one year, and describe the distribution:

155	103	130	80	65
142	125	129	71	69
125	115	138	68	78
150	133	135	90	97
68	105	88	115	110
95	85	109	115	71
97	110	65	90	
75	120	80	70	

29. World Series Champs again Students were asked to make a histogram of the number of home runs hit by Joe Carter from 1983–1998 (see Exercise 25). One student submitted the following display (he noted that Carter had hit no home runs in 1983 while playing in only 23 games):

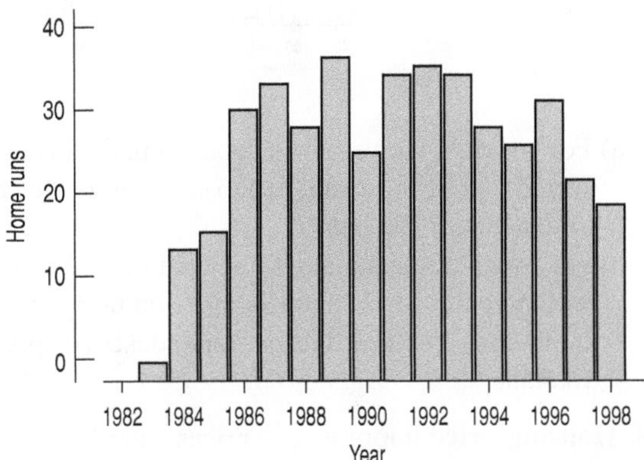

a) Comment on this graph.

b) Create your own histogram of the data.

30. Return of the birds 2010 Students were given the assignment to make a histogram of the data on bird counts reported in Exercise 26. One student submitted the following display:

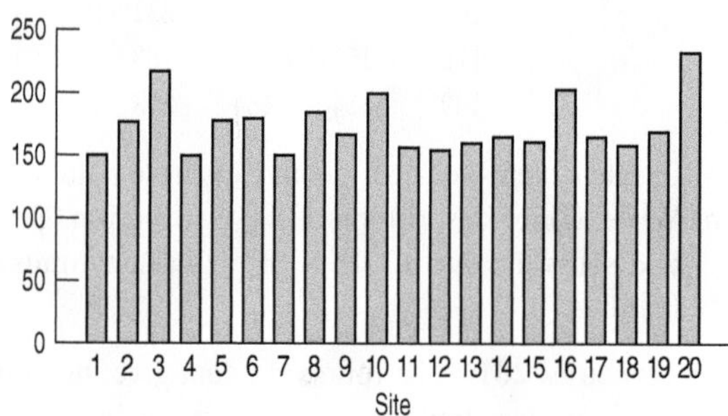

a) Comment on this graph.
b) Create your own histogram of the data.

31. Acid rain Two researchers measured the pH (a scale on which a value of 7 is neutral and values below 7 are acidic) of water collected from rain and snow over a six-month period in Allegheny County, Pennsylvania. Describe their data with a graph and a few sentences.

4.57 5.62 4.12 5.29 4.64 4.31 4.30 4.39 4.45
5.67 4.39 4.52 4.26 4.26 4.40 5.78 4.73 4.56
5.08 4.41 4.12 5.51 4.82 4.63 4.29 4.60

32. Sip size Researchers in Cornell University's Food Sciences department study how people experience foods. One study considered how much liquid people typically take into their mouths in one "sip" (in millilitres). The researchers also recorded the height (metres) and weight (kilograms) of the participants. Here are histograms of three of the variables from that study:

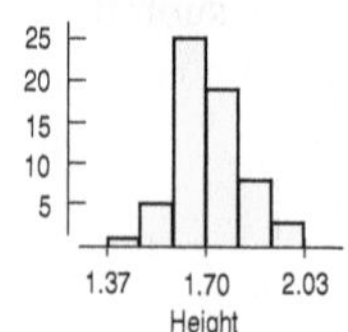

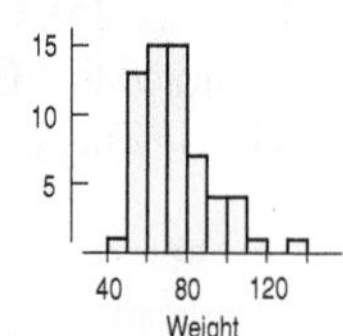

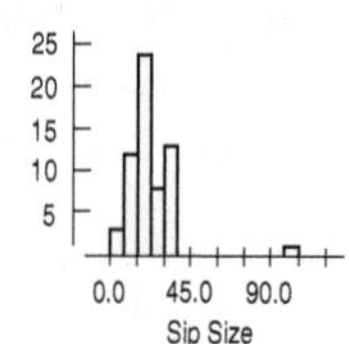

a) For which of the variables depicted in the histograms would you be most satisfied to summarize the centre with a mean? Explain.
b) For which of the variables depicted in the histograms would you most strenuously insist on using an IQR rather than a standard deviation to describe spread? Explain.

33. Housing price boom House prices were booming in some Canadian cities during 2007–2008. Below are percentage increases in new home prices in major cities:

Metropolitan Area	Percentage increase in new housing prices between March 2007 and March 2008
St. John's	12.0
Halifax	12.8
Charlottetown	1.4
Saint John, Fredericton and Moncton	2.4
Quebec	3.9
Montreal	4.5
Ottawa–Gatineau	3.1
Toronto and Oshawa	4.5
Hamilton	3.9
St. Catharines–Niagara	4.3
Kitchener	3.4
London	4.0
Windsor	−0.6
Greater Sudbury and Thunder Bay	6.3
Winnipeg	15.0
Regina	27.8
Saskatoon	46.2
Calgary	5.3
Edmonton	13.5
Vancouver	6.1
Victoria	1.2

Source: Adapted from Statistics Canada, "New Housing Price Index," *The* Daily, May 12, 2008. www.statcan.gc.ca/daily-quotidien/080512/dq080512a-eng.htm

a) Give a stem-and-leaf plot for these data, after rounding (not truncating) the values to the nearest 1.0. Split your stems, writing each one twice, once for the five lowest-value leaves and once for the five highest-value leaves.
b) Calculate the 5-number summary using the stem-and-leaf plot.
c) Find the mean. Did you expect it to be larger or smaller than the median? Why?
d) Give a useful brief report about annual metropolitan area house price increases (including any relevant summary measures).
e) Data taken over time or space can show a dependence on time or location. We can easily plot data versus time (see next chapter), but showing dependence on location is trickier. Do you think that the data above show any apparent or identifiable spatial (geographical) patterns? If so, explain. (Optional: Find a map of Canada and put bars at each city listed above, with height equal to new housing percentage increase.)

34. Getting high 2011 The European School Survey Project on Alcohol and other Drugs issued in 2011 its report: *Substance Use Among Students in 36 European Countries,* (www.espad.org). Among other issues, the survey investigated the percentages of 15–16 year olds who had used alcohol together with pills to get high. Here are the results for the 36 European countries. Create an appropriate graph of these data, and describe the distribution.

Country	%Booze+Pills	Country	%Booze+Pills
Albania	2	Latvia	6
Belgium	2	Liechtenstein	7
Bosnia & Herz.	1	Lithuania	4
Bulgaria	4	Malta	8
Croatia	10	Moldava	1
Cyprus	4	Monaco	5
Czech Republic	16	Montenegro	2
Denmark	3	Norway	2
Estonia	4	Poland	5
Faroe Islands	3	Portugal	3
Finland	10	Romania	3
France	7	Russian Fed.	3
Germany	8	Serbia	2
Greece	4	Slovak Republic	8
Hungary	10	Slovenia	4
Iceland	2	Sweden	4
Ireland	5	Ukraine	2
Italy	3	United Kingdom	5

35. Trimmed mean We can make the ordinary mean a more resistant measure by trimming off a certain percent of the biggest values and the smallest values. Let TrMean(x%) represent the mean of the data after placing the data in order from smallest to biggest and deleting x% at each end. So the 10% trimmed mean, for example, is the average of the middle 80% of the data. If you can't trim off exactly x%, trim off the percentage that is closest to x% (a computer would interpolate).

a) Find the mean, median and 10% trimmed mean of the following data: 6, 8, 8, 8, 1, 5, 7, 7, 6, 8.

b) Find the mean, median and 5% trimmed mean of the data from Exercise 34.

c) Explain the relative order of the three measures in each of the above data sets. Note that the ordinary mean is just a 0% trimmed mean and the median is essentially a 50% trimmed mean. A 5% or 10% trimmed mean can be a nice compromise between these two measures, more resistant than the mean, while using more of the data than the median.

T **36. Raptors 2011** Below are the lengths of all the Toronto Raptor loss streaks in the 2011–2012 season:

3 2 8 1 2 3 4 1 1 1 2 1 3 2 1 4 4

a) Construct a stem-and-leaf plot for these data.

b) Calculate the median, the 5% trimmed mean (see previous exercise), and the ordinary mean. Place these three measures in order from smallest to largest, and explain why you would expect them to be in this order for data such as these.

c) What is the mode of these data (highest frequency value)?

d) Can you think of another variable (categorical) that perhaps we should take into account in order to improve an analysis of loss streaks (rather than just ignore such a variable and let it lurk)?

37. Zip codes Holes-R-Us, an Internet company that sells piercing jewellery, keeps transaction records on its sales. At a recent sales meeting, one of the staff presented a histogram and summary statistics for the U.S. zip codes of the last 500 U.S. customers so that they might understand where sales are coming from. Comment on the usefulness and appropriateness of the display and summary statistics.

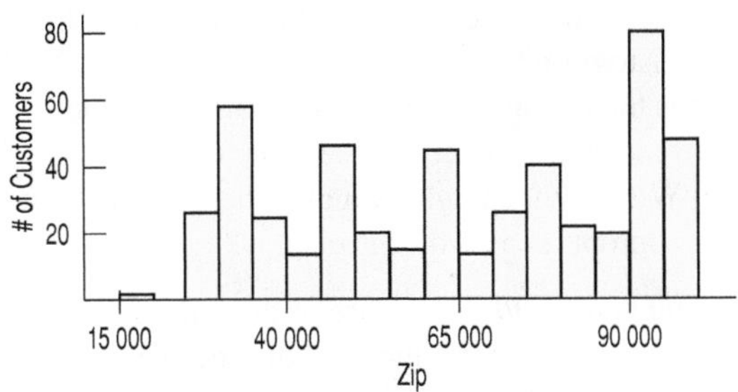

Count	500
Mean	64 970.0
StdDev	23 523.0
Median	64 871
IQR	44 183
Q1	46 050
Q3	90 233

38. Industry codes An investment analyst has been assigned to analyze the companies that make up the Fortune 800, a collection of the largest U.S. companies. He codes each company by the type of industry to which it belongs. Here is a table showing some of the industry types and the corresponding codes he uses.

Industry	Industry Code
Financial services	1
Food/drink/tobacco	2
Health	3
Insurance	4
Retailing	6
Forest products	9
Aerospace/defense	11
Energy	12
Capital goods	14
Computers/communications	16
Entertainment/information	17
Consumer nondurables	18
Electric utilities	19

The analyst produces the following histogram of the industry codes for the companies:

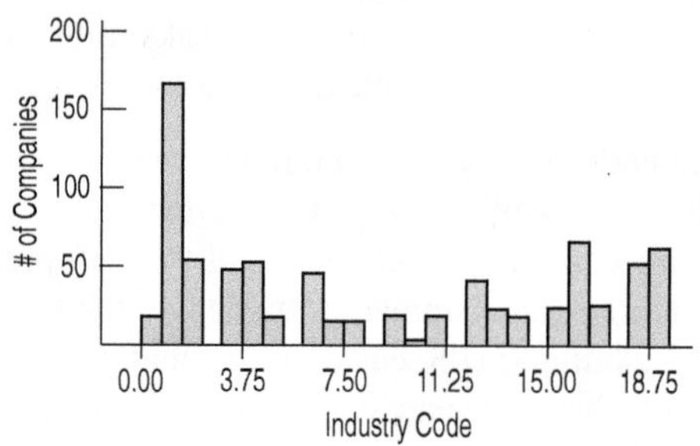

a) What might account for the gaps seen in the histogram?
b) Is the histogram unimodal? While the mode is 1, the median is 6. Interpret these numbers.
c) What advice might you give the analyst about the appropriateness of this display?

39. Golf courses One measure of the difficulty of a golf course is its length: the total distance (in yards) from tee to hole for all 18 holes. Here are the histogram and summary statistics for the lengths of all the golf courses in Vermont.

Count	45
Mean	5 892.91 yd
StdDev	386.59
Min	5 185
Q1	5 585.75
Median	5 928
Q3	6 131
Max	6 796

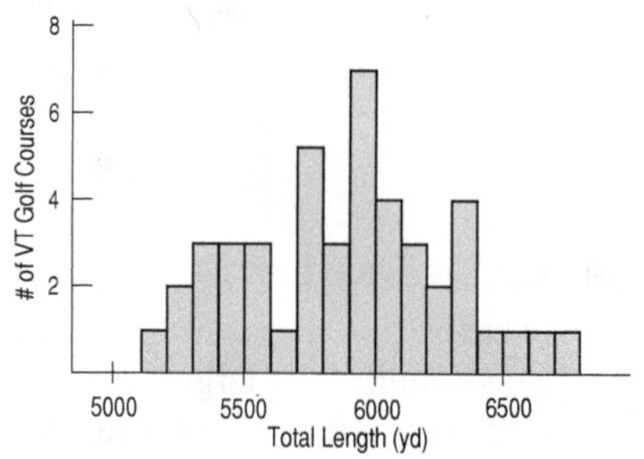

a) By hand, construct a boxplot directly below the histogram, using the same x-axis.
b) Between what lengths do the central 50% of these courses lie?
c) What summary statistics would you use to describe these data?
d) Write a brief description of these data (shape, centre, and spread).
e) Compute the intervals $(\bar{y} \pm s)$ and $(\bar{y} \pm 2\,s)$. Count, approximately, using the histogram, the number of golf courses with lengths within each interval, and convert to percents. Are the percents close to what you expected?

40. Women's Olympic alpine 2010 The women's super combined alpine event consists of a downhill and a slalom. Here are the total times (adding the two race times) for the 2010 Vancouver Olympics.

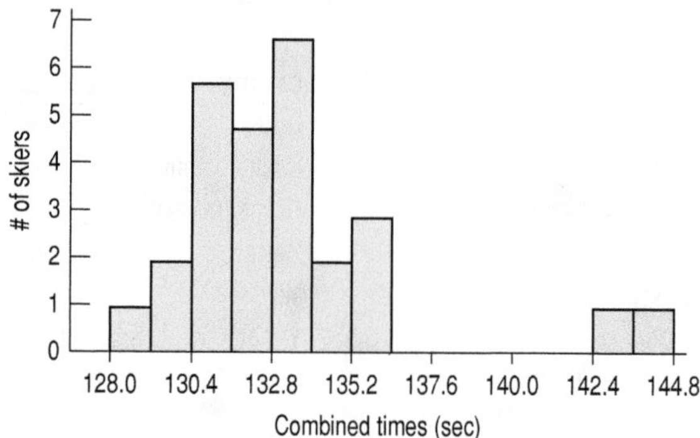

Count	28
Mean	133.43
StdDev	3.40
Min	129.14
Q1	131.09
Median	132.87
Q3	134.11
Max	144.15

a) By hand, construct a boxplot directly below the histogram, using the same x-axis (assume data points fall at midpoints of bins when helpful).
b) Between what lengths do the central 50% of these times lie?
c) Using appropriate summary statistics, write a brief description of these data (shape, centre, and spread).

T **41. Math scores 2009** The Programme for International Student Assessment (PISA) reported 2009 average mathematics performance scores for 15-year-olds in 34 Organisation for Economic Co-operation and Development (OECD) nations. Korea led the group, with an average score of 546, while Mexico had the lowest average of 419. (Non-OECD China topped everyone, with a 600 average). The average scores for each nation are given below:[18]

Country	Score	Country	Score
Australia	514	Czech Republic	493
Austria	496	Denmark	503
Belgium	515	Estonia	512
Canada	527	Finland	541
Chile	421	France	497

[18]Based on data from nces.ed.gov/nationsreportcard/.

Country	Score	Country	Score
Germany	513	New Zealand	519
Greece	466	Norway	498
Hungary	490	Poland	495
Iceland	507	Portugal	487
Ireland	487	Slovakia	497
Israel	447	Slovenia	501
Italy	483	Spain	483
Japan	529	Sweden	494
Korea	546	Switzerland	534
Luxembourg	489	Turkey	445
Mexico	419	United Kingdom	492
Netherlands	526	United States	487

a) Find the 5-number summary and IQR, the mean, and the standard deviation of these national averages. Can you explain why the mean is a bit smaller than the median?

b) Which summary of the data would you report? Why?

c) Write a brief summary of the performance of 15-year-olds in OECD countries. Be sure to comment on the performance of Canada and the United States.

d) In Canada, the standard deviation of the individual students' scores was 88 and the distribution was fairly bell-shaped. About 68% of students scored between _______ and _______ (fill in the blanks). Only about 5% scored less than _______ or more than _______ (fill in the blanks). Only a real math genius could have scored above _______ (fill in the blank).

42. Incarceration rates 2013 Below is a portion of the data set you can find online at http://www.pearsoncanada.ca/deveaux showing incarceration rates (prisoners per 100 000 of nation's population) for 221 nations of the world, according to the World Prison Brief at the International Centre for Prison Studies:

Country	Rate
United States of America	716
St. Kitts and Nevis	649
Seychelles	641
Virgin Islands (USA)	539
Rwanda	527
Cuba	510
Russian Federation	490
: :	
: :	
: :	
Central African Republic	19
Comoros	19
Faroe Islands	17

a) Graph these data.

b) Calculate appropriate summary statistics.

c) Write a few sentences about these data. (Include shape, centre, spread, unusual features.) Discuss where Canada and the U.S. stand respectively in this distribution.

43. First Nations 2010 Below are the first few entries from a data file you can find online at www.pearsoncanada.ca/deveaux showing the sizes of First Nations registry groups in Ontario. Registry groups are usually the same as bands, except that occasionally a band may consist of more than one registry group, like the Six Nations of Grand River in Ontario, which consists of 13 registry groups denoted in the complete data file. Hence in Ontario, there are 126 bands but 138 registry groups. (*Source:* www.aadnc-aandc.gc.ca/DAM/DAM-INTER-HQ/STAGING/texte-text/ai_rs_pubs_sts_ni_rip_rip10_rip10_1309289046808_eng.pdf)

Registry Group	Size
Aundeck-Omni-Kaning	752
Batchewana_First_Nation	2460
Dokis	988
Garden_River_First_Nation	2396
Henvey_Inlet_First_Nation	654
Magnetawan	233
M'Chigeeng_First_Nation	2329
Mississauga	1104
Nipissing_First_Nation	2245
...	
...	

a) Using appropriate graphical displays and summary statistics, write a report on the distribution of registry group size in Ontario.

b) Suppose we were to analyze bands rather than registry groups. Explain how each of the following would change if you analyzed bands rather than registry groups: mean, median, standard deviation, IQR, histogram. Use your logic here, not new calculations.

44. Election 2011 Maritimes Below are results from the 2011 Canadian federal election for the Maritime provinces, showing the riding name, winning candidate, and percentage of rejected ballots:

Electoral district	Percentage of Ballots Rejected	Elected candidate
Newfoundland and Labrador		
Avalon	0.5	Andrews, Scott Liberal
Bonavista—Gander—Grand Falls—Windsor	0.5	Simms, Scott Liberal
Humber—St. Barbe—Baie Verte	0.3	Byrne, Gerry Liberal
Labrador	0.5	Penashue, Peter Conservative
Random—Burin—St. George's	0.5	Foote, Judy Liberal
St. John's East	0.3	Harris, Jack NDP
St. John's South—Mount Pearl	0.3	Cleary, Ryan NDP
Prince Edward Island		
Cardigan	0.5	MacAulay, Lawrence Liberal
Charlottetown	0.6	Casey, Sean Liberal
Egmont	0.8	Shea, Gail Conservative
Malpeque	0.4	Easter, Wayne Liberal
Nova Scotia		
Cape Breton—Canso	0.9	Cuzner, Rodger Liberal
Central Nova	0.6	MacKay, Peter G. Conservative
Cumberland—Colchester—Musquodoboit Valley	0.6	Armstrong, Scott Conservative
Dartmouth—Cole Harbour	0.6	Chisholm, Robert NDP
Halifax	0.5	Leslie, Megan NDP
Halifax West	0.5	Regan, Geoff Liberal
Kings—Hants	0.5	Brison, Scott Liberal
Sackville—Eastern Shore	0.6	Stoffer, Peter NDP
South Shore—St. Margaret's	0.7	Keddy, Gerald Conservative
Sydney—Victoria	0.7	Eyking, Mark Liberal
West Nova	0.8	Kerr, Greg Conservative
New Brunswick		
Acadie—Bathurst	1.3	Godin, Yvon NDP
Beauséjour	1.2	LeBlanc, Dominic Liberal
Fredericton	0.5	Ashfield, Keith Conservative
Fundy Royal	0.6	Moore, Rob Conservative
Madawaska—Restigouche	1.6	Valcourt, Bernard Conservative
Miramichi	1.1	O'Neill Gordon, Tilly Conservative
Moncton—Riverview—Dieppe	0.7	Goguen, Robert Conservative
New Brunswick Southwest	0.6	Williamson, John Conservative
Saint John	0.5	Weston, Rodney Conservative
Tobique—Mactaquac	0.8	Allen, Mike Conservative

Source: Elections Canada

a) Make a suitable display of the percentages of rejected ballots.
b) Find the mean and standard deviation.
c) Report the 5-number summary.
d) Why do the mean and median differ here?
e) Which of b) and c) above does a better job of summarizing the distribution of percentage rejected ballots? Why?
f) Suppose the true percentage for Egmont was 1.8% and not 0.8%. How would you expect the mean, median, standard deviation, and IQR to change? Explain your expectations for each (no computations, please).
g) Write a brief report about rejected ballots in the Maritimes.

T **45. Bi-lingual Ni-lingual 2011** Canada is officially bilingual, according equal treatment of English and French languages, but individual Canadians who are fluent in English and French are uncommon in many regions of the country. And how many Canadians speak neither English nor French? Below are data from the 2011 Census for major metropolitan areas in Canada. Compute the proportion of the total bilingual population in each city, and the proportion knowledgeable in neither language.

a) Examine the distribution of the proportion of city residents who are bilingual, using suitable graphical and numerical methods. Summarize and report your findings.

Population by knowledge of official language, by census metropolitan area (2011 Census)

City	Total number	English only	French only	Both English and French	Neither English nor French
Abbotsford-Mission (BC)	167 815	151 740	45	7 830	8 200
Barrie (ON)	185 345	172 680	150	11 775	740
Brantford (ON)	133 915	127 665	35	5 620	600
Calgary (AB)	1 205 175	1 088 785	1 040	89 345	26 000
Edmonton (AB)	1 146 600	1 045 145	1 310	83 640	16 500
Greater Sudbury/Grand Sudbury (ON)	159 200	95 390	1 665	61 795	350
Guelph (ON)	140 410	127 285	55	11 620	1 450
Halifax (NS)	386 440	338 470	380	46 405	1 190
Hamilton (ON)	712 580	656 790	480	45 060	10 250
Kelowna (BC)	177 615	165 370	65	11 515	660
Kingston (ON)	155 405	135 645	315	18 860	580
Kitchener-Cambridge-Waterloo (ON)	472 095	434 105	220	30 275	7 490
London (ON)	469 010	432 600	240	31 095	5 065
Moncton (NB)	136 145	68 060	4 650	63 245	195
Montréal (QC)	3 785 915	280 785	1 401 455	2 039 035	64 635
Oshawa (ON)	353 460	326 965	245	24 640	1 605
Ottawa-Gatineau (ON)	1 222 760	556 405	105 280	547 625	13 450
Peterborough (ON)	117 595	109 420	30	7 895	250
Québec (QC)	756 405	1 550	479 895	273 335	1 620
Regina (SK)	208 085	194 785	140	11 700	1 455
Saguenay (QC)	156 585	150	125 335	31 050	45
Saint John (NB)	126 280	107 145	160	18 695	280
Saskatoon (SK)	256 900	239 330	125	15 350	2 105
Sherbrooke (QC)	199 190	3 060	111 455	83 855	815
St. Catharines-Niagara (ON)	386 525	353 445	490	29 910	2 680
St. John's (NL)	194 935	180 240	40	14 345	305
Thunder Bay (ON)	119 925	110 485	135	8 755	550
Toronto (ON)	5 541 880	4 872 010	5 000	424 265	240 600
Trois-Rivières (QC)	149 710	190	106 775	42 535	210
Vancouver (BC)	2 292 115	1 997 605	1 265	164 785	128 460
Victoria (BC)	339 725	303 850	185	33 060	2 635
Windsor (ON)	316 515	279 865	410	31 650	4 600
Winnipeg (MB)	721 115	635 575	1 050	75 715	8 775

Source: Statistics Canada. 2012. Language Highlight Tables. 2011 Census. Statistics Canada Catalogue no. 98-314-XWE2011002. Ottawa. released October 24, 2012

b) Make changes to your graph above (by hand is fine) so that the cities in Quebec or New Brunswick have different symbols (in shape or colour) than the other cities. Does that group of cities appear to differ from the rest in bilingualism?

c) Examine the distribution of the proportion of city residents who speak neither English nor French, using suitable graphical and numerical methods. Summarize and report your findings.

T **46. First Nations 2010 on reserves** Open up the data file First Nations 2010 available online at http://www.pearsoncanada.ca/deveaux. For each band, you see both the band size as well as the number living on reserve. Let's analyze the latter variable.

a) Construct a histogram for band counts of those living on reserve. Describe the distribution as best you can without doing any computations. How does it differ from the distribution of band sizes?

b) Instead of the actual count of those living on reserve, we might be interested in the proportion of the band that lives on reserve. Do you think its distribution should be similar to the distribution of on-reserve counts? Compute this proportion for each band by dividing the reserve count by the band size. Examine the distribution graphically and describe it as best you can without doing precise calculations. From your graph, guess the mean and standard deviation (this will certainly be an easier task after you have read Chapter 5).

c) Compute the mean and standard deviation for the on-reserve proportions. Are they close to your guesses?

d) For the on-reserve proportions, do you expect the mean and median to be similar? Why? For the living on reserve counts, do you expect the mean and median to be similar? Why?

e) Are the bands lying far from the middle of the distribution of proportions likely to be larger or smaller bands? Why?

f) Is the mean of the proportions the same as the overall proportion of the (band registered) First Nations population living on reserve?

T **47. Election 2011 again** Below are a graph and some calculations describing the distribution of voter turnout percentage for all federal ridings in the 2011 federal election in Canada.

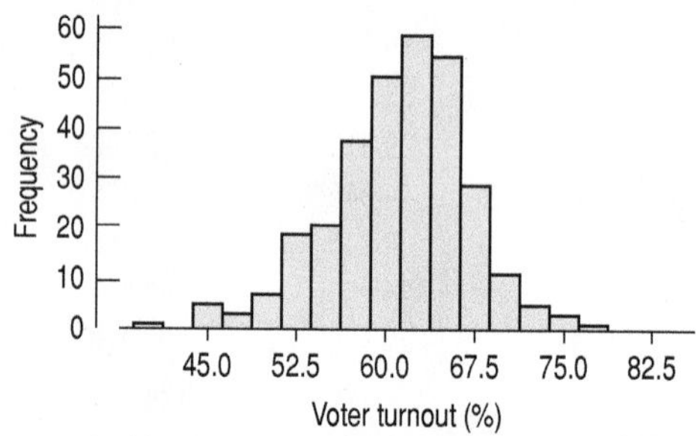

Descriptive Statistics: Voter turnout percentage

N:	308	Q1:	57.700
Mean:	61.068	Median:	61.700
St.Dev.:	5.772	Q3:	65.075
Minimum:	40.300	Maximum:	77.000

a) What does the relationship between the mean and median suggest?

b) Give a brief report about percentage of voter turnout in the 2011 election.

c) Work out the two numbers: "$\bar{y} + 2s$" and "$\bar{y} - 2s$" and count the number of observations that lie in between (it is easier to count the number that lie outside, then subtract). Is it close to 95%? Did you expect it to be? To count, either use the actual data that are available online, or estimate from the graph above, assuming that data are evenly spaced or spread out across any bin.

d) Since the mean of the percentages is 61.07, we can assume that the percentage of eligible voters in all these ridings combined who went to vote was 61.07%—true or not? Explain.

48. Family income 2011 According to Statistics Canada, Income Statistics Division, the mean after-tax family income in 2011 was \$63 000, while the median was \$50 700. In constant 2011 dollars, the mean and median in 1980 were \$54 000 and \$49 500 respectively.

a) Why did you expect to find a higher mean than median in both years? Draw a very rough plot of what the histogram for 2011 incomes might look like.

b) How much higher was the mean compared with the median in 2011, in percentage terms? How much higher in 1980? What does this suggest about how the income distribution has changed in Canada? Draw two very rough histograms to represent what the two distributions might look like in comparison to each other. Are the rich getting richer, and the poor poorer? Explain your answer.

c) Suppose that in 2030, the mean and median in actual 2030 dollars are \$80 000 and \$60 000 respectively. Have the rich gotten richer, and the poor poorer? Say what you can about what appears to be going on with incomes, based on this limited information.

49. Run times continued For the times provided in Exercise 10, explain how you would expect the mean, median, standard deviation, range, and IQR to change if we were to:

a) Change all times above 32.0 to 32.0.

b) Add 50 new run times, all between 30.0 and 31.0.

c) Add 30 new run times around 32.7 and another 30 times around 28.7 (the median and mean are 30.5 and 30.7 respectively).

d) Subtract 1 minute from each time.

e) Add 5 times around 29.5 and another 5 times around 35.0.

f) Change 10 times of 32.0 to 32.5 and 10 times of 29.4 to 28.9.

50. Test scores continued For the data provided in Exercise 9, explain how you would expect the mean, median, standard deviation, range, and IQR to change (with brief explanation), if you were to:

a) Delete all the test scores between 0 and 5.

b) Change all the test scores above 50 to scores of exactly 50.

c) Correct some data entry errors: One score recorded as 35 was actually 3.5 and another score recorded as 45 was actually 4.5.

d) Delete half the test scores between 30 and 35.

e) Add 30 additional graded tests with scores all between 30 and 35.

f) Add 20 additional test papers, where 10 scored 50 and the other 10 scored 15.
g) Adjust all the test scores upward by +5.
h) Double all the raw test scores so that you get percentage scores out of 100.

51. Grouped data Suppose that the data on heights provided in Exercise 20 had been organized into only six bins, as follows:

Height	Count
60–62	17
63–65	32
66–68	37
69–71	24
72–74	15
75–77	5

a) If you no longer had the actual data, but just this grouped data, could you still calculate the mean and standard deviation? Try the following: Pretend that each value in a bin lies exactly at the midpoint (i.e., 17 members stand 61 inches, 32 stand 64 inches, etc.). Now proceed exactly as you did before for the ungrouped data, and find the mean and standard deviation. Did you get good approximations to the actual values? Why do you think this method works as well as it does?
b) Write down a formula for the mean as you calculated it in part a), using "m_i" for midpoint of the *i*th bin and "f_i" for the frequency (count) of the ith bin. Do likewise for the variance.

52. Grouped data II Look back at the points scored per season by Wayne Gretzky in Exercise 22.
a) Compute his mean points per season.
b) Group the data into bins: 40–60, 60–80, etc., and find the counts for each bin. Multiply each bin midpoint (50, 70, . . .) by its respective bin count. Add all of these products. Divide by 20, the number of seasons. This is your 'grouped mean' calculation. We are pretending that each data point lies right at the midpoint, for this calculation (obviously they don't, but errors high and low may just balance out nicely in certain calculations). How close is this approximation to the actual mean you calculated in part (a)?

53. Unequal bin widths Look back at the histogram of times in Exercise 10.
a) Construct by hand another histogram, for these data (estimate counts as best you can), where the bins are 28.5–29.0, 29.0–29.5, 29.5–30.0, 30.0–30.5, 30.5–31.0, 31.0–31.5, 31.5–32.0, 32.0–32.5, 32.5–35.5. Plot as usual frequency for the height of each bin's rectangle. Any problems? If so, explain.
b) Redo your histogram but for the last bin, plot the frequency divided by 6. Do you prefer this histogram or the one in part a)? How does the area of the rectangle over 32.5–35.5 compare with the area of all the rectangles starting from 32.5 in the original histogram? What basic principle (mentioned in Chapter 2) was violated in part a) and corrected in part b)? (The vertical axis is now a measure of "frequency density")

54. Unequal bin widths II We emphasize using equal widths for the bins in a histogram. What's the problem with using unequal bin widths?
a) Draw a histogram for data containing five measurements between 10 and 11 cm, four measurements between 11 and 12 cm, three measurements between 12 and 13 cm, two measurements between 13 and 14 cm, and one measurement between 14 and 15 cm. Plot frequency (count) on the *y*-axis, as usual.
b) Now combine the last two bins into one, so that the final bin is 13 to 15 cm, which hence contains three measurements. Draw a histogram now, the usual way. What's wrong? Was any basic principle (mentioned in Chapter 2) violated?
c) Correct the histogram by plotting frequency divided by bin width on the *y*-axis. Why does this correct the problem?

55. Rock concert accidents Crowd Management Strategies (www.crowdsafe.com) monitors accidents at rock concerts. In their database, they list the names and other variables of victims whose deaths were attributed to "crowd crush" at rock concerts. Here are the histogram and boxplot of the victims' ages for data from a recent one-year period:

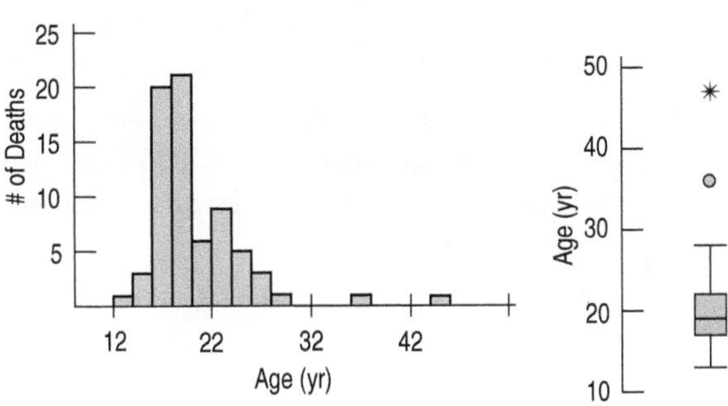

a) What features of the distribution can you see in both the histogram and the boxplot?
b) What features of the distribution can you see in the histogram that you could not see in the boxplot?
c) What summary statistic would you choose to summarize the centre of this distribution? Why?
d) What summary statistic would you choose to summarize the spread of this distribution? Why?

56. Slalom times 2010 The Men's Giant Slalom skiing event consists of two runs whose times are added together for a final score. Two displays of the giant slalom times in the 2010 Winter Olympics are shown below.

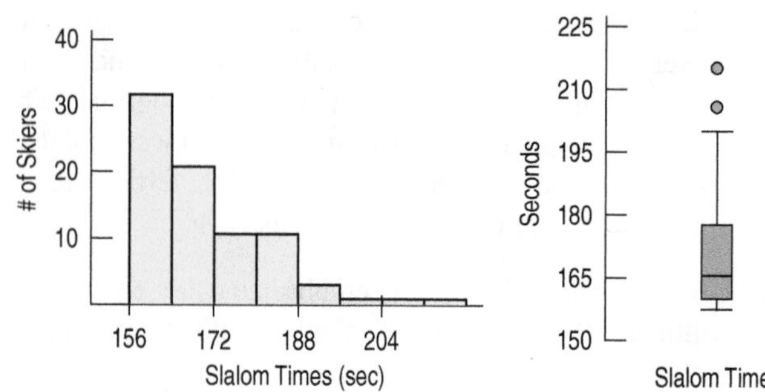

a) What features of the distribution can you see in both the histogram and the boxplot?
b) What summary statistic would you choose to summarize the centre of this distribution? Why?
c) What summary statistic would you choose to summarize the spread of this distribution? Why?

Just Checking ANSWERS

(Thoughts will vary.)

1. Roughly symmetric, slightly skewed to the right. Centre around 5 kilometres, few over 15 kilometres.
2. Bimodal or trimodal, with peaks around zero, three, and maybe six hours, since many people watch no hockey, and others watch most of one or two games. Spread from zero hours to about six hours.
3. Strongly skewed to the right, with almost everyone at \$0; a few small prizes, with the winner an outlier.
4. Fairly symmetric, somewhat uniform, perhaps slightly skewed to the right. Centre in the 40s? Few ages below 25 or above 70.
5. Uniform, symmetric. Centre near 5. Roughly equal counts for each digit 0–9.
6. Incomes are probably skewed to the right and not symmetric, making the median the more appropriate measure of centre. The mean will be influenced by the high end of family incomes and not reflect the "typical" family income as well as the median would. It will give the impression that the typical income is higher than it is.
7. An IQR of 10 L/km would mean that only 50% of the cars get gas mileages in an interval 10 L/km wide. Fuel economy doesn't vary that much. 1 L/km is reasonable. It seems plausible that 50% of the cars will be within about 1 L/km of each other. An IQR of 0.1 L/km would mean that the gas mileage of half the cars varies little from the estimate. It's unlikely that cars, drivers, and driving conditions are that consistent.
8. We'd prefer a standard deviation of two months. Making a consistent product is important for quality. Customers want to be able to count on the MP3 player lasting somewhere close to five years, and a standard deviation of two years would mean that lifespans were highly variable.

chapter

4 Understanding and Comparing Distributions

Where are we going?

Do people who drink green tea live longer? Does the presence of some gene allele affect one's health? Are tsunami-creating earthquakes bigger in certain geographic regions? Have they gotten bigger in recent times? These are the kinds of questions for which Statistics can really help. Some simple graphical displays and summaries can start us thinking about patterns, trends, and models—something we'll do throughout the rest of this book.

Who	Days during 2011
What	Average daily wind speed (miles per hour), average barometric pressure (mb), average daily temperature (degrees Celsius)
When	2011
Where	Hopkins Forest, Western Massachusetts
Why	Long-term observations to study ecology and climate

I'm not an outlier; I just haven't found my distribution yet!

—*Ronan M. Conroy*

The Hopkins Memorial Forest is a 2500-acre preserve in Massachusetts, New York, and Vermont managed by the Williams College Center for Environmental Studies (CES). As part of their mission, CES monitors forest resources and conditions over the long term.

One of the variables measured in the forest is wind speed. Each day, remote sensors record the minimum, maximum, and average wind speed (in miles per hour). Wind is caused as air flows from areas of high pressure to areas of low pressure. Centres of low pressure often accompany storms, so both high winds and low pressure are associated with some of the fiercest storms. Wind speeds can vary greatly during a day and from day to day, but if we step back a bit farther, we can see patterns. By modelling these patterns, we can understand things about *Average Wind Speed* and the insights it provides about weather that we may not have known.

In Chapter 2, we looked at the association between two categorical variables using contingency tables and displays. Here we'll look at different ways to examine the relationship between two variables when one is quantitative and the other indicates groups to compare. We are given wind speed averages for each day of the year. But we can gather the days together into groups and compare the wind speeds among them. If we partition *Time* in this way, we'll gain enormous flexibility. When we change the size of the groups, from the entire year, to seasons, to months, and finally to days, different patterns emerge that together increase our understanding of the entire pattern.

Let's start with the "big picture." Here's a histogram and 5-number summary of the daily *Average Wind Speed* for every day in 2011.

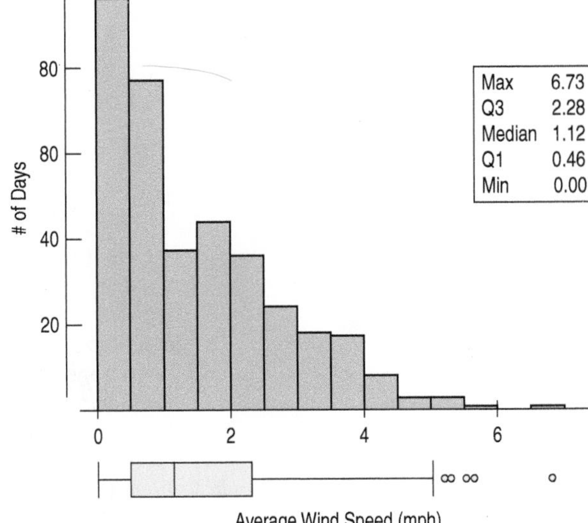

Figure 4.1

A histogram of daily *Average Wind Speed* for 2011. It is unimodal and skewed to the right. The boxplot below the histogram suggests some possible outliers that may warrant our attention.

We can see that the distribution is unimodal and skewed to the right. Median daily wind speed is about 1.12 mph. On half of the days, the average wind speed was between 0.46 and 2.28 mph, because the quartiles frame the middle half of the data. We also see some possible outliers, including one rather windy 6.73 mph day. Were these unusual weather events or just the windiest days of the year? To answer that, we'll need to work with the summaries a bit more.

4.1 Comparing Groups

It is almost always more interesting to compare groups. Is it windier in the winter or the summer? Are any months particularly windy? Are weekends a special problem? Let's split the year into two groups: April through September (Summer) and October through March (Winter). To compare the groups, we create two histograms, being careful to use the same scale. Figure 4.2 displays average daily wind speeds for Summer (on the left) and for Winter (on the right):

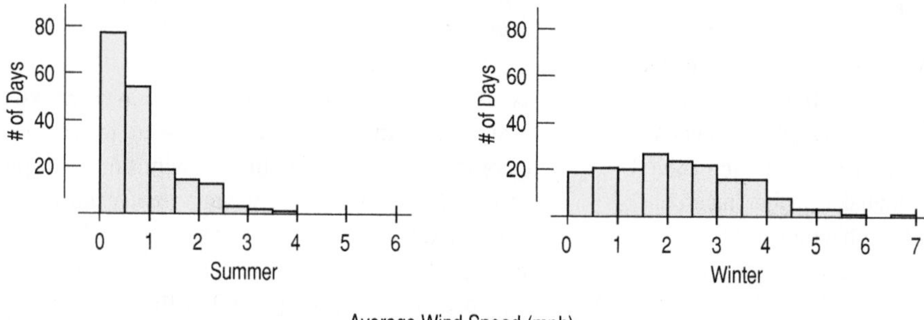

Figure 4.2
Histograms of *Average Wind Speed* for days in Summer (left) and Winter (right) show very different patterns.

The distribution of wind speeds for the summer months is unimodal and skewed to the right. By contrast, the distribution for the winter months is less strongly skewed and more nearly uniform. During the summer, a typical day has a mean wind speed of less than 1 mph, and few have average speeds above 3 mph. In winter, the typical wind speed is higher, days with wind speeds above 3 mph are not unusual, and there is at least one noticeably high value. There is also more variability in the winter (IQR = 1.91 mph) than in summer (IQR = 0.79 mph)

Summaries for *Average Wind Speed* by Season

Group	Mean	StdDev	Median	IQR
Winter	2.17	1.33	2.07	1.91
Summer	0.85	0.74	0.62	0.79

For Example COMPARING GROUPS WITH STEM-AND-LEAF

The Nest Egg Index, devised by the investment firm of A.G. Edwards, is a measure of saving and investment performance for each of the 50 states. It is based on 12 economic factors, including participation in retirement savings plans, personal debt levels, and home ownership. The average index is 100 and the numbers indicate the percentage above or below the average. There are only 50 values, so a *back-to-back stem-and-leaf plot* is an effective display. Here's one comparing the nest egg index in the Northeast and Midwest states to those in the South and West. In this display, the stems run down the middle of the plot, with the leaves for the two regions to the left or right. Be careful when you read the values on the left: 5|8 means nest egg index of 85% of the southern or western states.

South and West		Northeast and Midwest
5778	8	8
12344	9	03
6667778899	9	67
02334	10	012233334
56	10	6779
	11	122444

(4|9|3 means 94% for a South/West state and 93% for a Northeast/Midwest state)

QUESTION: How do nest egg indices compare for these regions?

ANSWER: Nest egg indices were generally higher for the northeastern or midwestern states with most having indices above 100, while most states in the south or west had indices below 100. There were nine northeastern or midwestern states with higher indices than any states in the south or west.

4.2 Comparing Boxplots

A S ***Video:* Can Diet Prolong Life?** Here's a subject that's been in the news: Can you live longer by eating less? (Or would it just seem longer?) Look at the data in subsequent activities, and you'll find that you can learn a lot by comparing two groups with boxplots.

Are some months windier than others? Even residents may not have a good idea which parts of the year are the most windy. (Do you know for your hometown?) And what about the spreads? Are wind speeds equally variable from month to month, or do some months show more variation?

Histograms or stem-and-leaf plots are a fine way to look at one distribution or two, but it would be hard to see patterns by comparing 12 histograms. Boxplots offer an ideal balance of information and simplicity, hiding the details while displaying the overall summary information. We often plot them side by side for groups or categories we wish to compare.

By placing boxplots side by side, we can easily see which groups have higher medians, which have the greater IQRs, where the central 50% of the data is located in each group, and which have the greater overall range. And, when the boxes are in some natural order, we can get a general idea of patterns in both the centres and the spreads. Equally important, we can see past any outliers in making these comparisons because boxplots display them separately.

Figure 4.3 shows boxplots of the *Average Daily Wind Speed* by month:

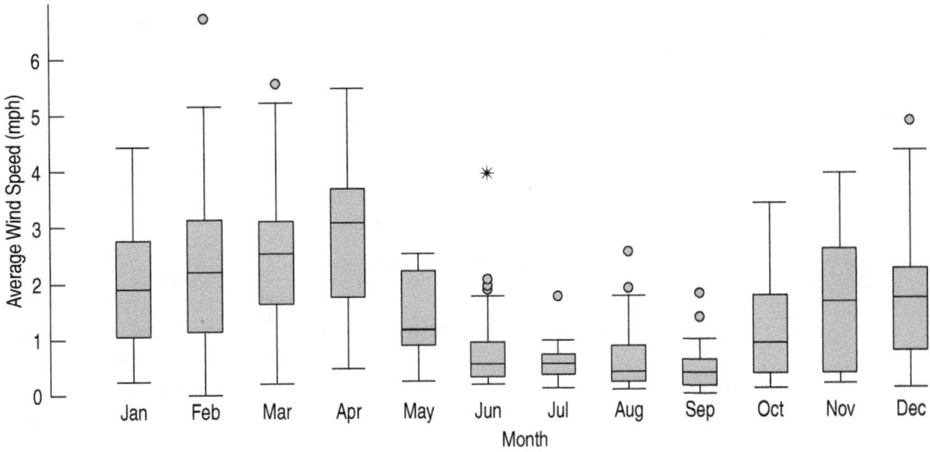

Figure 4.3

Boxplots of the *Average Daily Wind Speed* plotted for each *Month* show seasonal patterns in both the centres and spreads. New outliers appear because they are now judged relative to the *Month* in which they occurred.

Here we see that wind speeds tend to decrease in the summer. The months in which the winds are both strongest and most variable are November through April.

When we looked at a boxplot for wind speeds of the entire year (Figure 4.1), there were only five suspect outliers. But the monthly boxplots show different outliers than before because some days that seemed ordinary when placed against the entire year's data look like outliers for the month that they're in. Outliers are context dependent. That windiest day in August (the outlier at about 2.5 mph) certainly wouldn't stand out in November or December, but for August it was a remarkable day—as we'll soon see.

nra/Shutterstock

For Example COMPARING DISTRIBUTIONS

Roller coaster[1] riders want a coaster that goes fast. There are two main types of roller coasters: those with wooden tracks and those with steel tracks. Do they typically run at different speeds? Here are their boxplots:

QUESTION: Compare the speeds of wood and steel roller coasters.

ANSWER: Overall, wooden-track roller coasters are slower than steel-track coasters. In fact, the fastest half of the steel coasters are faster than three quarters of the wooden coasters. Although the IQRs of the two groups are similar, the range of speeds among steel coasters is larger than the range for wooden coasters. The distribution of speeds of wooden coasters appears to be roughly symmetric, but the speeds of the steel coasters are somewhat skewed to the right with a high outlier at 120 mph. We should look into why that steel coaster is so fast.

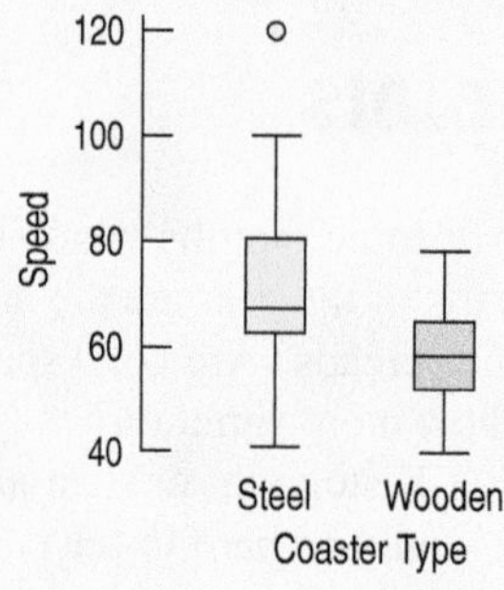

Step-by-Step Example COMPARING GROUPS

Of course, we can compare groups even when they are not in any particular order. Most scientific studies compare two or more groups. It is almost always a good idea to start an analysis of data from such studies by comparing boxplots for the groups. Here's an example.

For her class project, a student compared the efficiency of various coffee mugs. She compared four different mugs, testing each of them eight different times. Each time, she heated water to 180°F[2], poured it into a mug, and sealed it. (We'll learn the details of how to set up experiments in Chapter 11.) After 30 minutes, she measured the temperature again and recorded the difference in temperature. Because these are temperature differences, smaller differences mean that the liquid stayed hot—just what we would want in a coffee mug.

Question: What can we say about the effectiveness of these four mugs?

THINK ➡ Plan State what you want to find out. **Variables** Identify the *variables* and report the *W*'s.	I want to compare the effectiveness of the four different mugs in maintaining temperature. I have eight measurements of Temperature Change for each of the mugs.
Be sure to check the appropriate condition.	✓ **Quantitative Data Condition:** The temperature changes are quantitative, with units of °F. Boxplots are appropriate displays for comparing the groups. Numerical summaries of each group are appropriate as well.

[1]See the roller coaster database at www.rcdb.com.

[2]This experiment was conducted in the U.S., where, like in the Bahamas, Belize, Palau, Cayman Islands—and likely on your oven's temperature readings—the Fahrenheit scale still prevails. To convert to Celsius, the formula is °C = (°F − 32)/1.8. So, 180°F = 82°C (and this was much more pleasant for your Canadian authors than converting the data and all text references to °C!)

SHOW ➡ **Mechanics** Report the 5-number summaries of the four groups. Including the IQR is a good idea.

	Min	Q1	Median	Q3	Max	IQR
CUPPS	6°F	6	8.25	14.25	18.50	8.25
Nissan	0	1	2	4.50	7	3.50
SIGG	9	11.50	14.25	21.75	24.50	10.25
Starbucks	6	6.50	8.50	14.25	17.50	7.75

Make a picture. Because we want to compare the distributions for four groups, boxplots are an appropriate choice.

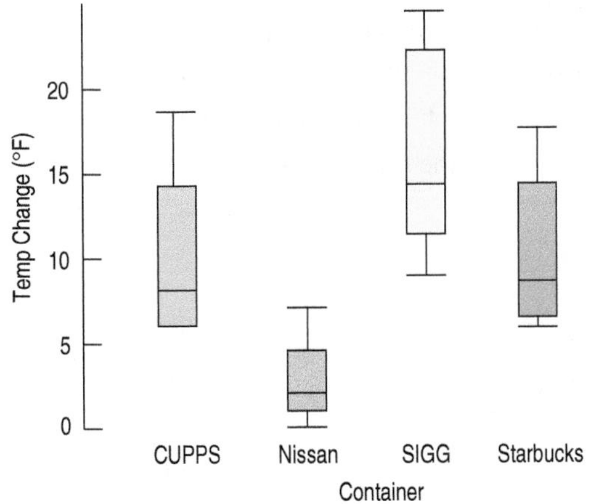

TELL ➡ **Conclusion** Interpret what the boxplots and summaries say about the ability of these mugs to maintain heat. Compare the shapes, centres, and spreads, and note any outliers.

The individual distributions are all slightly skewed to the high end. The Nissan cup does the best job of keeping liquids hot, with a median loss of only 2 °F, and the SIGG cup does the worst, typically losing 14 °F. The difference is large enough to be important; a coffee drinker would be likely to notice a 14° drop in temperature. And the mugs are clearly different; 75% of the Nissan tests showed less heat loss than any of the other mugs in the study. The IQR of results for the Nissan cup is also the smallest of these test cups, indicating that it is a consistent performer.

Just Checking

Spoiled ballots are a real threat to democracy. Below are displays of data from Elections Canada showing percentage of spoiled ballots (number of spoiled ballots divided by total number of ballots cast) for all 308 electoral ridings in the 2011

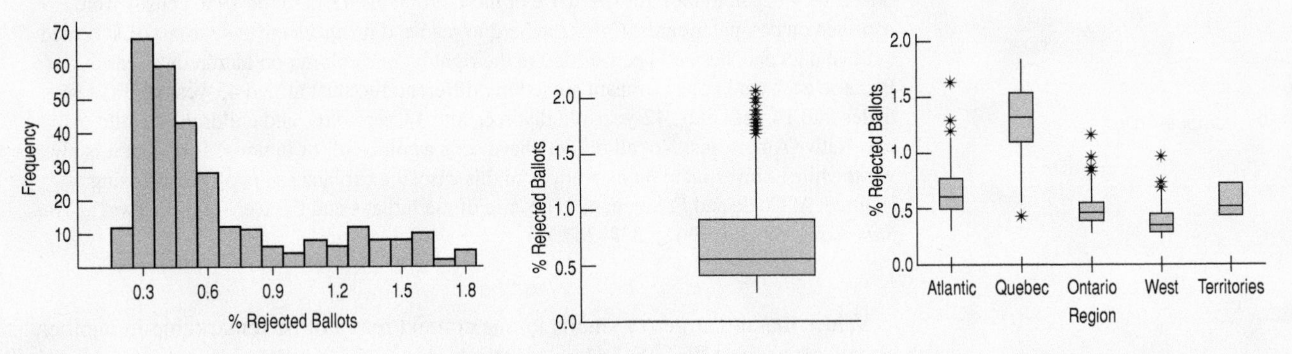

Canadian federal election, and another display comparing results by region. (See Exercise 26 for some rather disheartening results from the 2000 federal election.)

1. Describe what the histogram says about spoiled ballots.
2. What does the histogram suggest that you can't see in the boxplot?
3. Are there many outliers present in the nationwide data? Explain.
4. Describe the regional variation in spoiled ballots.
5. Why is one of the regional boxplots missing whiskers?
6. Could it possibly be true that three quarters of results in one region were worse than all the individual riding results in two other regions? Explain.
7. Which region had the worst result? The best result?
8. How does the last display above clarify what is seen in the histogram?

4.3 Outliers

In the boxplots for the daily *Average Wind Speed by Month*, several days are nominated by the boxplots as possible outliers. Cases that stand out from the rest of the data almost always deserve attention. An outlier is a value that doesn't fit with the rest of the data, but exactly how different it should be to receive special treatment is a judgment call. Boxplots provide a rule of thumb to highlight these unusual cases, but that rule is only a rough guide, and it doesn't tell you what to do with them. So, what *should* you do with outliers?

Outliers arise for many reasons. They *may* be the most important values in the data set, pointing out an exceptional case or illuminating a pattern by being the exception to the rule. They may be values that just happen to lie above the limits *suggested* by the box plot rule. Or they may be errors. A decimal point may have been misplaced, digits transposed, or digits repeated or omitted. Some outliers are obviously wrong. For example, if a class survey includes a student who claims to be 170 inches tall (about 14 feet, or 4.3 metres), you can be sure that's an error. But, maybe the units were wrong. (Did the student mean 170 centimetres, or about 65 inches?). There are many ways errors can creep into data sets and such errors occur remarkably often. If you can identify the correct value, then you should certainly use it. One important reason to look into outliers is to correct errors in your data.

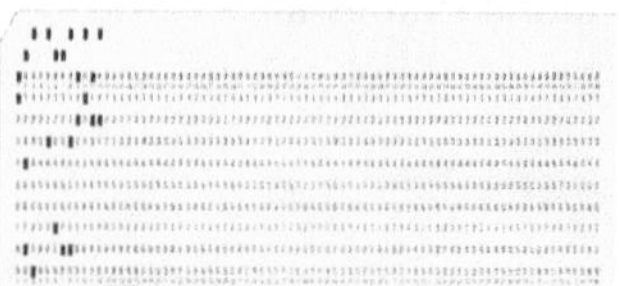
1950s, computer card

14-year-old widowers? Two researchers, Ansley Coale and Fred Stephan, looking at data from the 1950 census, noticed that the number of widowed 14-year-old boys had increased from 85 in 1940 to a whopping 1600 in 1950. The number of divorced 14-year-old boys had increased, too, from 85 to 1240. Oddly, the number of teenaged widowers and divorcés decreased for every age group after 14, from ages 15 to 19. When Coale and Stephan also noticed a large increase in the number of young Native Americans in the Northeast United States, they began to look for the cause of these problems. Data in the 1950 census were recorded on computer cards. Cards are hard to read and mistakes are easy to make. It turned out that data punches had been shifted to the right by one column on hundreds of cards. Because each card column meant something different, the shift turned 43-year-old widowed males into 14-year-olds, 42-year-old divorcés into 14-year-olds, and children of white parents into Native Americans. Not all outliers have such a colourful (or famous) story, but it is always worthwhile to investigate them. And, as in this case, the explanation is often surprising. (Source: A. Coale and F. Stephan, "The case of the Indians and the teen–age widows," *J. Am. Stat. Assoc.* 57 [Jun 1962]: 338–347.)

Values that are large (or small) in one context may not be remarkable in another. The boxplots of *Average Wind Speed by Month* show possible outliers in several of the months.

The windiest day in February was an outlier not only in that windy month, but for the entire year as well. The windiest day in June was a "far" outlier—lying more than 3 IQRs from the upper quartile for that month, but it wouldn't have been unusual for a winter day. And, the windiest day in August seemed much windier than the days around it, even though it was relatively calm.

Many outliers are not wrong; they're just different. And most repay the effort to understand them. You can sometimes learn more from the extraordinary cases than from summaries of the entire data set.

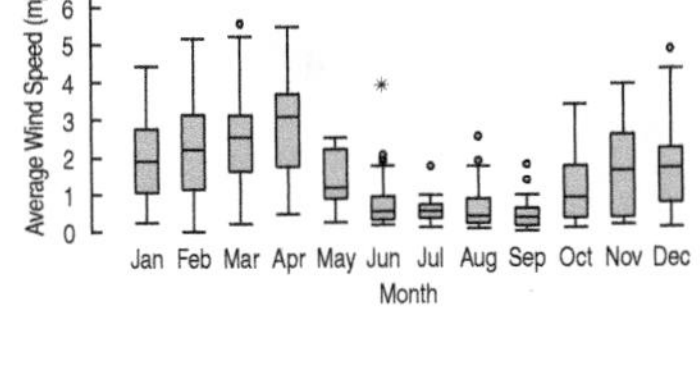

What about those outliers in the Hopkins wind data? That blustery day in February turned out to be a blast that brought four days of subzero temperatures (−17°F) to the region. John Quinlan of the National Weather Service predicted at the time: "The Berkshires have definitely had some of the highest gusts with this wind event. This is definitely the worst we'll be seeing."

What about that June day? It was June 2. A search for weather events on the Internet for June 2, 2011, finds a rare tornado in Western Massachusetts. *The Berkshire Eagle* reported: "It probably never occurred to most Berkshire residents horrified by the photos and footage of recent tornado damage in Alabama and Missouri that tornadoes would soon strike close to home. They did, of course, on Wednesday, sowing destruction and in some cases costing lives in Springfield, West Springfield, Monson and other towns near enough to us that the destruction had a sobering effect."

Finally, let's look at the more extreme outlier in August. It wouldn't be remarkable in most other months. It turns out that that was Hurricane Irene, whose eye passed right over the Hopkins Forest. According to *The New York Times*, "it was probably the greatest number of people ever threatened by a single storm in the United States." In fact, the average wind speed that day, 2.53 mph, was much lower than the predictions for Irene had warned.

Not all outliers are as dramatic as Hurricane Irene, but all deserve attention. If you can correct an error, you should do so (and note the correction). If you can't correct it, or if you confirm that it is correct, you can simply note its existence and leave it in the data set. But the safest path is to report summaries and analyses with *and* without the outlier so that readers can judge the influence of the outlier for themselves.

There are two things you should *never* do with outliers. You should not leave an outlier in place and proceed as if nothing were unusual; analyses of data with outliers are very likely to be wrong. The other thing you should never do is omit an outlier from the analysis without comment. If you want to exclude an outlier, you must announce your decision and, to the extent you can, justify it. But a case lying just over the fence suggested by the boxplot may just be the largest (or smallest) value at the end of a stretched-out tail. A histogram is often a better way to see more detail about how the outlier fits in (or doesn't) with the rest of the data, and how large the gap is between it and the rest of the data.

A S ***Case Study:* Are passengers or drivers safer in a crash?** Practise the skills of this chapter by comparing these two groups.

For Example CHECKING OUT THE OUTLIERS

RECAP: We've looked at the speeds of roller coasters and found a difference between steel- and wooden-track coasters. We also noticed an extraordinary value.

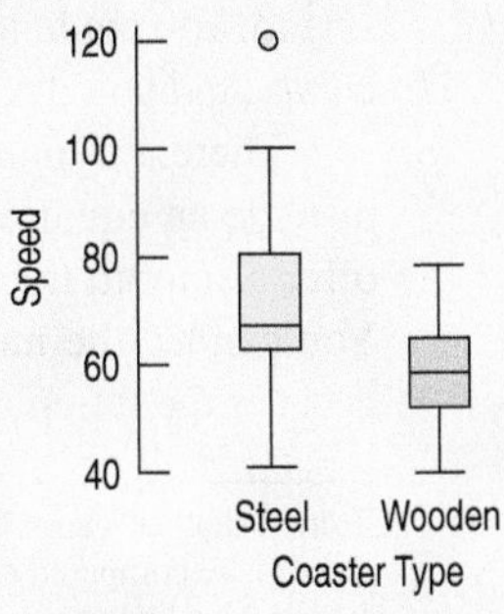

QUESTION: The fastest coaster in this collection turns out to be the "Top Thrill Dragster" at Cedar Point amusement park. What might make this roller coaster unusual? You'll have to do some research, but that's often what happens with outliers.

ANSWER: The Top Thrill Dragster is easy to find in an Internet search. We learn that it is a "hydraulic launch" coaster. That is, it doesn't get its remarkable speed just from gravity, but rather from a kick-start by a hydraulic piston. That could make it different from the other roller coasters.

(You might also discover that it is no longer the fastest roller coaster in the world.)

4.4 Timeplots

The Hopkins Forest wind speeds are reported as daily averages.[3] Previously, we grouped the days into months or seasons, but we could look at the wind speed values day by day. Whenever we have data measured over time, it is a good idea to look for patterns by plotting the data in time order. Figure 4.4 plots the daily average wind speeds over time:

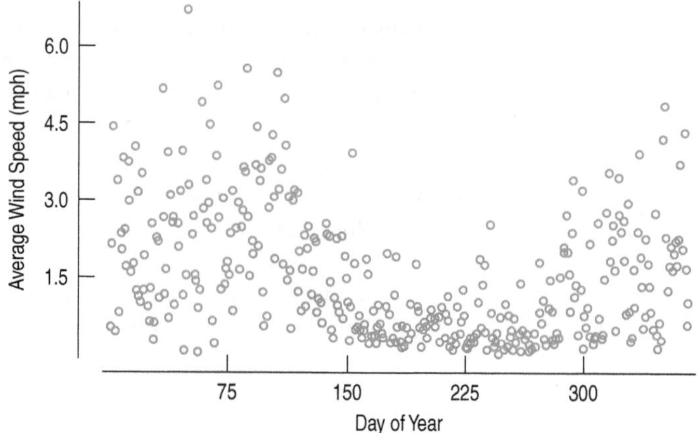

Figure 4.4
A timeplot of *Average Wind Speed* shows the overall pattern and changes in variation.

A display of values against time is called a *time series plot* or more simply, **timeplot**. This timeplot reflects the pattern that we saw when we plotted the wind speeds by month. But without the arbitrary divisions between months, we can see a calm period during the summer, starting around day 150 (the beginning of June), when the wind is relatively mild and doesn't vary greatly from day to day. We can also see that the wind becomes both more variable and stronger during the late fall and winter months.

*Smoothing Timeplots

Timeplots often show a great deal of point-to-point variation, as the one in Figure 4.4 does. We usually want to see past this variation to understand any underlying smooth trends, and then also think about how the values vary around that trend—the timeplot version of centre and spread. You could take your pencil, squint at the plot a bit to blur things, and sketch a smooth curve. And you'd probably do a pretty good job. But we can also ask a computer to do it for us.

There are many ways for computers to run a smooth trace through a timeplot. Some provide an equation that gives a typical value for any given time point, while others just offer a smooth trace. For our purposes here, it doesn't matter all that much which is used. You can try the methods offered by your favourite program and pick the one that looks best for the timeplot you are looking at and the questions you'd like to answer.

[3]Here's another warning that the word "average" is used to mean different things. Many "daily average wind speeds" are computed as the midrange—the average of the highest and lowest speeds seen during the day. But the Hopkins Forest values are actually measured continuously all day and averaged over all those observations. When you see the word "average," it is wise to check for the details.

You'll often see timeplots drawn with all the points connected, especially in financial publications. This trace through the data falls at one extreme, following every bump. For the wind speeds, the daily fluctuations are so large that connecting the points (Figure 4.5) doesn't help us a great deal:

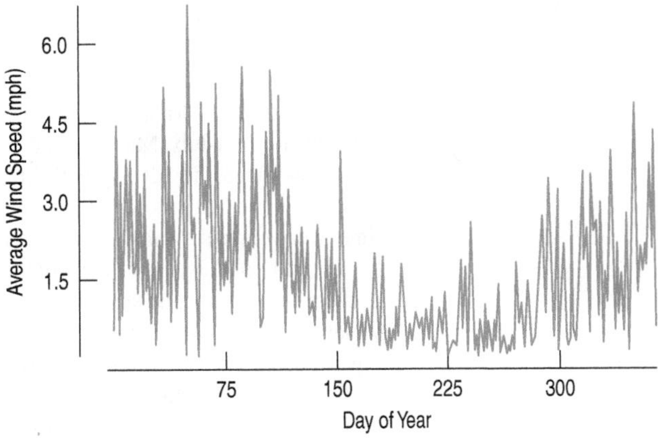

Figure 4.5
The *Average Wind Speeds* of Figure 4.4, drawn connecting all the points. Sometimes this can help you see the underlying pattern, but here there is too much daily variation for this to be very useful.

A smooth trace can highlight long-term patterns and help you see them through the more local variation. Figure 4.6 shows the daily average wind speed values with a smooth trace found by a method called *lowess*, available in many statistics programs.

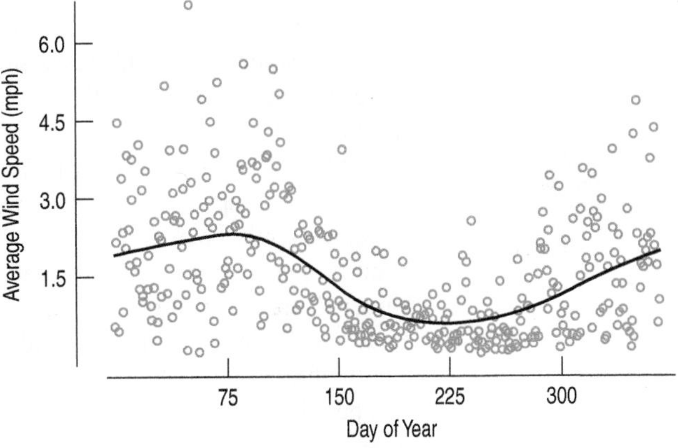

Figure 4.6
The *Average Wind Speeds* of Figure 4.4, with a smooth trace added to help your eye see the long-term pattern.

With the smooth trace, it's a bit easier to see a pattern. The trace helps your eye follow the main trend and alerts you to points that don't fit the overall pattern.

RECAP: We have looked at the current speeds of the world's fastest roller coasters. Have coasters been getting faster? Our data include coasters going back to 1909. Here's a timeplot with a *lowess* smooth trace:

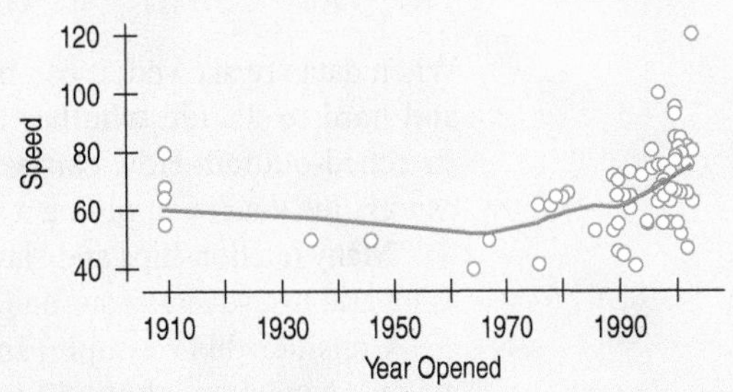

QUESTION: What does the timeplot say about the trend in roller-coaster speeds?

ANSWER: Roller coasters do seem to have been getting faster. Speeds were stagnant, or even declining, until 1970, but have increased (on average) since then. The increase seems to have been more rapid since 1990, pushed along by a few outliers.

How do smoothers work? If you planned to go camping in the Hopkins Forest around July 15 and wanted to know how windy it was likely to be, you'd probably look at a typical value for mid-July. Maybe you'd consider all the wind speeds for July, or maybe only those from July 10th to July 20th. That's just the kind of thing we do when we smooth a timeplot.

One simple way to smooth is with a **moving average**. To find a smooth value for a particular time, we average the values around that point in an interval called the "window." To find the value for the next point, we *move* the window by one point in time and take the new *average*. The size of the window you choose affects how smooth the resulting trace will be. For the Hopkins Forest winds, we might use a five-day moving average. Stock analysts often use a 50- or 200-day moving average to help them (attempt) to see the underlying pattern in stock price movements.

Can we use smoothing to predict the future? We have only the values of a stock in the past, but (unfortunately) none of the future values. We could use the recent past as our window and take the simple average of those values. A more sophisticated method, **exponential smoothing**, gives more weight to the recent past values and less and less weight to values as they recede into the past.

Looking into the Future

People have always wanted to peer into the future, and we predict that they will continue to do so! Sometimes that makes sense. Most likely, the Hopkins Forest climate follows regular seasonal patterns. It's probably safe to predict a less windy June next year and a windier December. But we certainly wouldn't predict another hurricane on August 28.

Other patterns are riskier to extend into the future. If a stock has been rising, will it continue to go up? No stock has ever increased in value indefinitely, and no stock analyst has consistently been able to forecast when a stock's value will turn around. Stock prices, unemployment rates, and other economic, social, or psychological concepts are much harder to predict than physical quantities. The path a ball will follow when thrown from a certain height at a given speed and direction is well understood. Even some dogs can predict the flight of a frisbee well enough to run to where it will be within reach and catch it. The path interest rates will take is much less clear. Unless you have strong (nonstatistical) reasons for doing otherwise, you should resist the temptation to think that any trend you see will continue, even into the near future.

Statistical models often tempt those who use them to think beyond the data. We'll pay close attention later in this book to understanding when, how, and how much we can justify doing that.

4.5 Re-expressing Data

Re-expressing to Improve Symmetry

When data are skewed, it can be hard to summarize them simply with a centre and spread, and hard to decide whether the most extreme values are outliers or just part of the stretched-out tail. How can we say anything useful about such data? The secret is to *re-express* the data by applying a simple function to each value.

Many relationships and "laws" in the sciences and social sciences include functions such as logarithms, square roots, and reciprocals. Similar relationships often show up in data.

Consider this example: In 1980, large companies' chief executive officers (CEOs) made, on average, about 42 times what workers earned. In the next two decades, CEO compensation soared compared to the average worker. By 2008, that multiple had jumped[4] to 344. What does the distribution of the compensation of Fortune 500 companies' CEOs look like? Here's a histogram and boxplot for 2010 compensation:

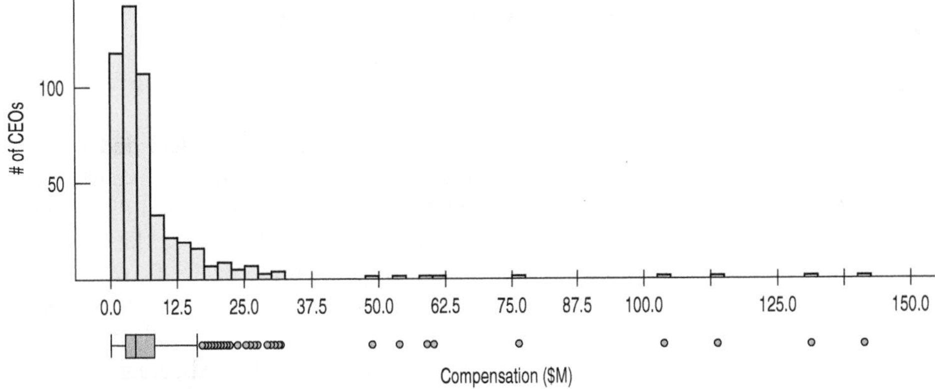

Figure 4.7
CEOs' compensation for the Fortune 500 companies in 2010.

We have 500 CEOs and about 57 possible histogram bins, most of which are empty—but don't miss the tiny bars straggling out to the right. The boxplot indicates that some CEOs received extraordinarily high pay, while the majority received relatively little. But look at the values of the bins. The first bin, with about a quarter of the CEOs, covers compensations of \$0 to \$2 500 000. Imagine receiving a salary survey with these categories:

What is your income?
a. \$0 to \$2 500 000
b. \$2 500 001 to \$5 000 000
c. \$5 000 001 to \$7 500 000
d. More than \$7 500 000

The reason that the histogram seems to leave so much of the area blank is that the salaries are spread all along the axis from about \$35 000 000 to \$150 000 000. After \$50 000 000 there are so few for each bin that it's very hard to see the tiny bars. What we *can* see from this histogram and boxplot is that this distribution is highly skewed to the right.

It can be hard to decide what we mean by the "centre" of a skewed distribution, so it's hard to pick a typical value to summarize the distribution. What would you say was a typical CEO total compensation? The mean value is \$8 035 770, while the median is "only" \$4 780 000. Each tells us something different about the data.

One approach is to **re-express**, or **transform**, the data by applying a simple function to make the skewed distribution more symmetric. For example, we could take the square root or logarithm of each pay value. Taking logs works pretty well for the CEO compensations, as you can see:[5]

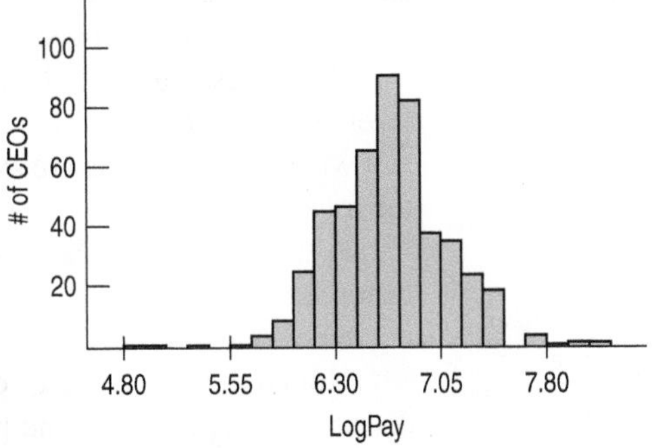

Figure 4.8
The logarithms of 2010 CEO compensations are much more nearly symmetric.

[4]www.faireconomy.org/files/executive_excess_2008.pdf

[5]One CEO—Steve Jobs of Apple—accepted no pay. When we took logs, his pay dropped out because we couldn't take log(0).

The histogram of the logs of the total CEO compensations is much more nearly symmetric, so we can see that log compensation is generally between 5, which corresponds to $100 000, and 7.5, corresponding to $31 600 000. And it's easier to talk about a typical value for the logs. The mean log compensation is 6.68, while the median is 6.67. (That's $4 786 301 and $4 677 351, respectively, but who's counting?)

Against the background of a generally symmetric main body of data, it's easier to decide whether the largest or smallest pay values are outliers. In fact, the three most highly compensated CEOs are identified as outliers by the boxplot rule of thumb even after this re-expression. It's impressive to be an outlier CEO in annual compensation. It's even more impressive to be an outlier in the log scale!

Variables that are skewed to the right often benefit from a re-expression by square roots, logs, or reciprocals. Those skewed to the left may benefit from squaring the data (or using the exponential function). Because computers and calculators can do the calculating, re-expressing data is quite easy. Consider re-expression as a helpful tool whenever you have skewed data.

DEALING WITH LOGARITHMS

You have probably learned about logs in math courses and seen them in psychology or science classes. In this book, we use them only for making data behave better. Base 10 logs are the easiest to understand, but natural logs are often used as well. (Either one is fine.) You can think of base 10 logs as roughly one less than the number of digits you need to write the number. So 100, which is the smallest number to require three digits, has a $\log_{10}$ of 2. And 1000 has a $\log_{10}$ of 3. The $\log_{10}$ of 500 is between 2 and 3, but you'd need a calculator to find that it's approximately 2.7. All salaries of "six figures" have $\log_{10}$ between 5 and 6. Logs are incredibly useful for making skewed data more symmetric. But don't worry—nobody does logs without technology and neither should you. Often, remaking a histogram or other display of the data is as easy as pushing another button.

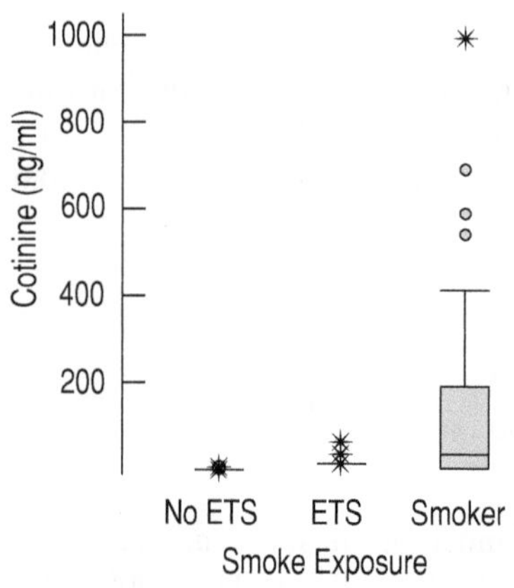

Figure 4.9

Cotinine levels (nanograms per millilitre) for three groups with different exposures to tobacco smoke. Can you compare the ETS (exposed to smoke) and No ETS groups?

Re-expressing to Equalize Spread Across Groups

Researchers measured the concentration (nanograms per millilitre) of cotinine in the blood of three groups of people: nonsmokers who have not been exposed to smoke, nonsmokers who have been exposed to smoke (ETS), and smokers. Cotinine is left in the blood when the body metabolizes nicotine, so this measure gives a direct measurement of the effect of passive smoke exposure. The boxplots of the cotinine levels of the three groups (Figure 4.9) tell us that the smokers have higher cotinine levels. However, if we want to compare the levels of the passive smokers (the ETS group) to those of the nonsmokers, we're in trouble, because on this scale, the cotinine levels for both nonsmoking groups are too low to be seen.

Re-expressing can help alleviate the problem of comparing groups that have very different spreads. For measurements like the cotinine data, whose values can't be negative and whose distributions are skewed to the high end, a good first guess at a re-expression is the logarithm.

After taking logs (Figure 4.10), we can compare the groups and see that the nonsmokers exposed to environmental smoke (ETS) do show increased levels of (log) cotinine, although not the high levels found in the blood of smokers.

Notice that the same re-expression has also improved the symmetry of the cotinine distribution for smokers and pulled in most of the apparent outliers in all of the groups. It is not unusual for a re-expression that improves one aspect of data to improve others as well. We'll talk about other ways to re-express data throughout the book.

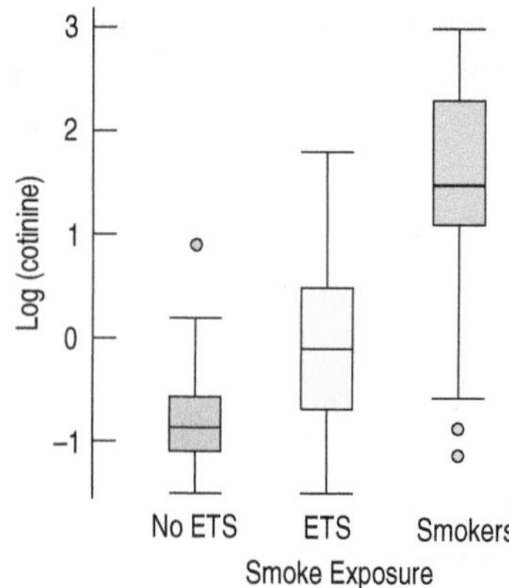

Figure 4.10

Blood cotinine levels after taking logs. What a difference a log makes!

WHAT CAN GO WRONG?

- **Avoid inconsistent scales.** Parts of displays should be mutually consistent—no fair changing scales in the middle or plotting two variables on different scales but on the same display. When comparing two groups, be sure to compare them on the same scale.
- **Label clearly.** Variables should be identified clearly and axes labelled so a reader knows what the plot displays.

  Here's a remarkable example of a plot gone wrong. It illustrated a news story about rising college costs. It uses timeplots, but it gives a misleading impression. First think about the story you're being told by this display. Then try to figure out what has gone wrong.

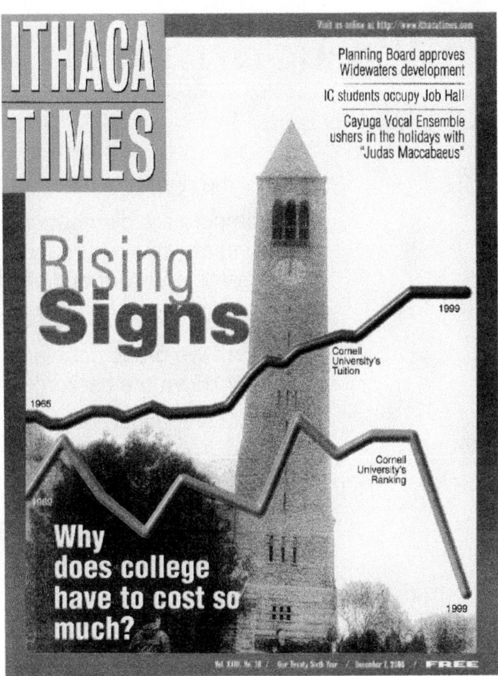

Ithaca Times

What's wrong? Just about everything.

- The horizontal scales are inconsistent. Both lines show trends over time, but exactly for what years? The tuition sequence starts in 1965, but rankings are graphed from 1989. Plotting them on the same (invisible) scale makes it seem that they're for the same years.
- The vertical axis isn't labelled. That hides the fact that there are actually two distinct vertical axes. It should show dollar units on one vertical axis, say the left side, and ranking units on the right-side vertical axis.

This display violates three of the rules. And it's even worse than that: It violates a rule that we didn't even bother to mention.

- The two inconsistent scales for the vertical axis don't point in the same direction! The line for Cornell's rank shows that it has "plummeted" from 15th place to 6th place in academic rank. Most of us think that's an *improvement*, but that's not the message of this graph.

- **Beware of outliers.** If the data have outliers and you can correct them, you should do so. If they are clearly wrong or impossible, you should remove them and report on them. Otherwise, consider summarizing the data both with and without the outliers.
- **Be careful when comparing groups that have very different spreads.** As the cotinine data showed, a simple re-expression can often make comparisons across groups with very different spreads much easier.

We discussed the value of summarizing a distribution with shape, centre, and spread in Chapter 3, and we developed several ways to measure these attributes. Now we've seen the value of comparing distributions for different groups and of looking at patterns in a quantitative variable measured over time. Although it can be interesting to summarize a single variable for a single group, it is almost always more interesting to compare groups and look for patterns across several groups or over time. We'll continue to make comparisons like these throughout the rest of our work.

What Have We Learned?

Learning Objectives

Choose the right tool for comparing distributions.

- Compare the distributions of two or three groups with histograms.
- Compare several groups with boxplots, which make it easy to compare centres and spreads and spot outliers, but hide much of the detail of distribution shape.

Treat outliers with attention and care.

- When we group data in different ways, different cases may emerge as outliers.
- Track down the background for outliers—it may be informative.

Timeplots graph individual values over time.

Re-express data to make it easier to work with.

- Re-expression can make skewed distributions more nearly symmetric.
- Re-expression can make the spreads of different groups more nearly comparable.

Review of Terms

Comparing distributions When comparing the distributions of several groups using histograms or stem-and-leaf displays, consider their:

- Shape
- Centre
- Spread (p. 89)

Comparing boxplots When comparing groups with boxplots:

- Compare the shapes. Do the boxes look symmetric or skewed? Are there differences between groups?
- Compare the medians. Which group has the higher centre? Is there any pattern to the medians?
- Compare the IQRs. Which group is more spread out? Is there any pattern to how the IQRs change?
- Using the IQRs as a background measure of variation, do the medians seem to be different, or do they just vary in the way that you'd expect from the overall variation?
- Check for possible outliers. Identify them if you can and discuss why they might be unusual. Of course, correct them if you find that they are errors (p. 89).

Timeplot A timeplot (often called a *time series plot*) displays data that change over time. Often, successive values are connected with lines to show trends more clearly. Sometimes a smooth curve is added to the plot to help show long-term patterns and trends (p. 94).

On the Computer COMPARING DISTRIBUTIONS

Most programs for displaying and analyzing data can display plots to compare the distributions of different groups. Typically, these are boxplots displayed side by side.

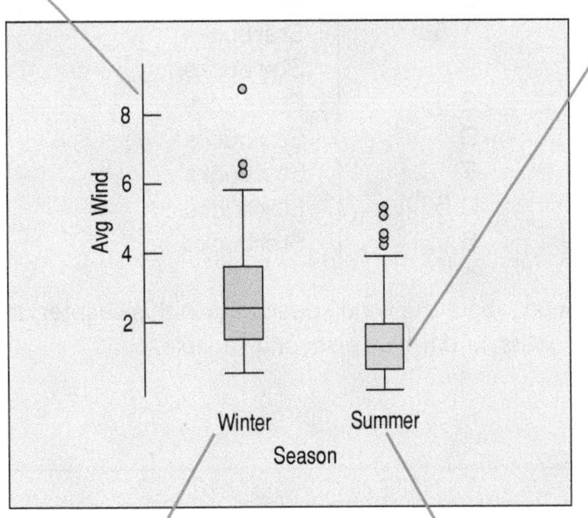
Side-by-side boxplots should be on the same y-axis scale so they can be compared.
Some programs offer a graphical way to assess how much the medians differ by drawing a band around the median or by "notching" the boxes.
Avg Wind
8
6
4
2
Winter
Summer
Season
Boxes are typically labeled with a group name. Often they are placed in alphabetical order by group name—not the most useful order.

There are two ways to organize data when we want to compare groups. Each group can be in its own variable (or list, on a calculator). In this form, the experiment comparing cups would have four variables, one for each type of cup:

CUPPS	Nissan	SIGG	Starbucks
6	12	2	13
6	16	1.5	7
6	9	2	7
18.5	23	3	17.5
10	11	0	10
17.5	20.5	7	15.5
11	12.5	0.5	6
6.5	24.5	6	6

But there's another way to think about and organize the data. What is the variable of interest (the *What*) in this experiment? It's the number of degrees lost by the water in each cup. And the *Who* is each time she tested a cup. We could gather all of the temperature values into one variable and put the names of the cups in a second variable listing the individual results, one on each row. Now, the *Who* is clearer—it's an experimental run, one row of the table. Most statistics packages prefer data on groups organized in this way.

Mug	Temperature Difference	Mug	Temperature Difference
CUPPS	6	SIGG	12
CUPPS	6	SIGG	16
CUPPS	6	SIGG	9
CUPPS	18.5	SIGG	23
CUPPS	10	SIGG	11
CUPPS	17.5	SIGG	20.5
CUPPS	11	SIGG	12.5
CUPPS	6.5	SIGG	24.5
Nissan	2	Starbucks	13
Nissan	1.5	Starbucks	7
Nissan	2	Starbucks	7
Nissan	3	Starbucks	17.5
Nissan	0	Starbucks	10
Nissan	7	Starbucks	15.5
Nissan	0.5	Starbucks	6
Nissan	6	Starbucks	6

That's actually the way we've thought about the wind speed data in this chapter, treating wind speeds as one variable and the groups (whether seasons, months, or days) as a second variable.

EXCEL

Excel cannot make boxplots.

JMP

- Choose **Fit y by x**.
- Assign a continuous response variable to **Y, Response** and a nominal group variable holding the group names to **X**, **Factor**, and click **OK**. JMP will offer (among other things) dotplots of the data.
- Click the red triangle and, under **Display Options**, select Boxplots.

Note: If the variables are of the wrong type, the display options might not offer boxplots.

MINITAB

- Choose **Boxplot** . . . from the **Graph** menu.

If your data are in the form of one quantitative variable and one group variable:

- Choose **One Y** and **with Groups.**

If your data are in separate columns of the worksheet:

- Choose **Multiple Ys.**

R

If the data for each group are in separate variables, X1, X2, X3 . . . then:

- **boxplot(X1,X2,X3. . .)** will produce side-by-side boxplots.

If the quantitative values are in variable X and the grouping values are in categorical variable Y:

- **boxplot(X~Y)** will produce side-by-side boxplots—one for each level of Y.

SPSS

To make a boxplot in SPSS, open the **Chart Builder** from the Graphs menu:

- Click the **Gallery** tab.
- Choose **Boxplot** from the list of chart types.
- Drag a single or 2-D (side-by-side) boxplot onto the canvas.
- Drag a scale variable to the *y*-axis drop zone.
- To make side-by-side boxplots, drag a categorical variable to the *x*-axis drop zone.
- Click **OK**.

STATCRUNCH

To make a boxplot:

- Click on **Graphics**.
- Choose **Boxplot**.
- Choose the variable name from the list of **Columns**.
- Click on **Next**.

TI-83/84 PLUS

To make a boxplot, set up a **STAT PLOT** by using the boxplot icon.

To compare groups with boxplots, enter the data in lists, for example in L1 and L2.

Set up STATPLOT's Plot1 to make a boxplot of the L1 data:

- Turn the plot On.
- Choose the first boxplot icon (so the plot will indicate outliers).
- Specify Xlist:L1 and Freq:1, and select the Mark you want the calculator to use for displaying any outliers.
- Use ZoomStat to display the boxplot for L1. You can now TRACE to see the statistics in the 5-number summary.
- Then set up Plot2 to display the L2 data. This time when you use ZoomStat with both plots turned on, the display shows the boxplots in parallel.

Exercises

1. **In the news** Find an article in a newspaper or a magazine or online that compares two or more groups of quantitative data.
   a) Does the article discuss the *W*'s?
   b) Is the choice of graphics appropriate for the situation? Explain.
   c) Discuss what the display reveals about the groups.
   d) Does the article accurately describe and interpret the data? Explain.

2. **In the news again** Find an article in a newspaper or a magazine or online that shows a timeplot.
   a) Does the article discuss the *W*'s?
   b) Is the timeplot appropriate for the data? Explain.
   c) Discuss what the timeplot reveals about the variable.
   d) Does the article accurately describe and interpret the data? Explain.

3. **Graduation?** A survey of major universities asked what percentage of incoming freshmen usually graduate "on time" in four years. Use the summary statistics given to answer these questions.

	% on Time
Count	48
Mean	68.35
Median	69.90
StdDev	10.20
Min	43.20
Max	87.40
Range	44.20
25th %tile	59.15
75th %tile	74.75

a) Would you describe this distribution as symmetric or skewed? Explain.
b) Are there any outliers? Explain.
c) Create a boxplot of these data.
d) Write a few sentences about the graduation rates.

4. Vineyards Here are summary statistics for the sizes (in acres) of Finger Lakes vineyards.

Count	36
Mean	46.50 acres
StdDev	47.76
Median	33.50
IQR	36.50
Min	6
Q1	18.50
Q3	55
Max	250

a) Would you describe this distribution as symmetric or skewed? Explain.
b) Are there any outliers? Explain.
c) Create a boxplot of these data.
d) Write a few sentences about the sizes of the vineyards.

5. Dots to boxes For the blood pressure measurements: 110 113 116 118 120 122 124 126 132 134 144 164, create a dotplot (by hand or by computer). Take the dotplot and superimpose upon it a boxplot, following the steps below:
i. Divide the data into two halves, spot the median, and draw a bar there.
ii. Divide the upper portion of the data into two halves, spot the upper quartile, and draw a bar there.
iii. Divide the lower portion of the data into two halves, spot the lower quartile, and draw a bar there.
iv. Finish drawing the box.
v. Draw a whisker on each side extending out from the box through the data points, one data point at a time. Stop when extending to the next data point would be too far to travel, i.e., further than 150% of the width of the box.
vi. Draw an asterisk or star on any data point(s) beyond the whiskers that you drew.

6. Dots to boxes again For the blood pressure measurements: 102 115 124 128 129 130 132 133 135 137 142 160, create a dotplot (by hand or by computer). Take the dotplot and superimpose upon it a boxplot, following the steps below:
i. Divide the data into two halves, spot the median, and draw a bar there.
ii. Divide the upper portion of the data into two halves, spot the upper quartile, and draw a bar there.
iii. Divide the lower portion of the data into two halves, spot the lower quartile, and draw a bar there.
iv. Finish drawing the box.
v. Draw a whisker on each side extending out from the box through the data points, one data point at a time. Stop when extending to the next data point would be too far to travel, i.e., further than 150% of the width of the box.
vi. Draw an asterisk or star on any data point(s) beyond the whiskers that you drew.

7. Hospital stays The U.S. National Center for Health Statistics complliles data on the length of stay by patients in short-term hospitals and publishes its finding in *Vital and Health Statistics*. Data from a sample of 39 male patients and 35 female patients on length of stay (in days) are displayed in the histograms below.

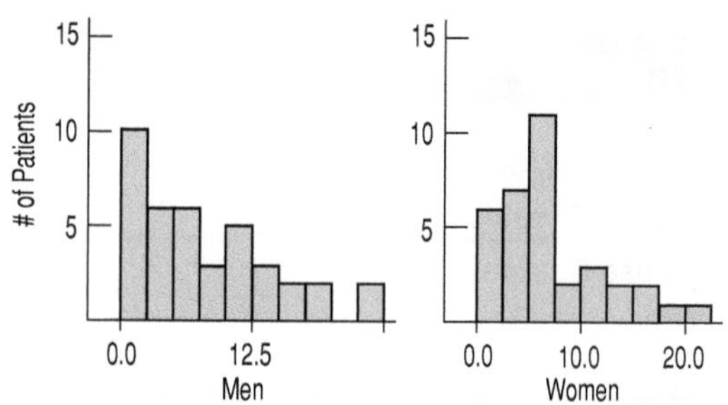

a) What would you suggest be changed about these histograms to make them easier to compare?
b) Describe these distributions by writing a few sentences comparing the duration of hospitalization for men and women.
c) Can you suggest a reason for the peak in women's length of stay?

8. Deaths 2009 A National Vital Statistics Report (www.cdc.gov/nchs/) indicated that nearly 290 000 black Americans died in 2009, compared with just over 2 million white Americans. Below are histograms displaying the distributions of their ages at death.

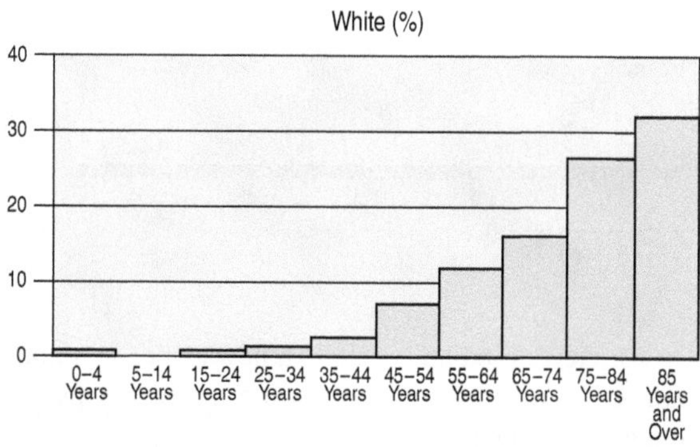

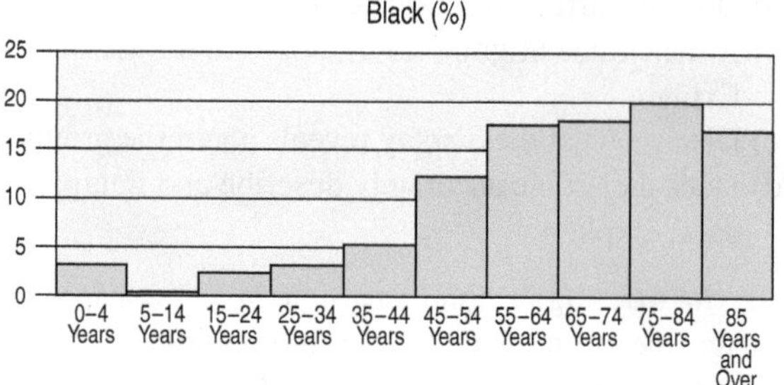

a) Describe the overall shapes of these distributions.
b) How do the distributions differ?
c) Look carefully at the bar definitions. Where do these plots violate the rules for statistical graphs?

9. Women's basketball Here are boxplots of the points scored during the first 10 games of the season for both Bini and Yan.

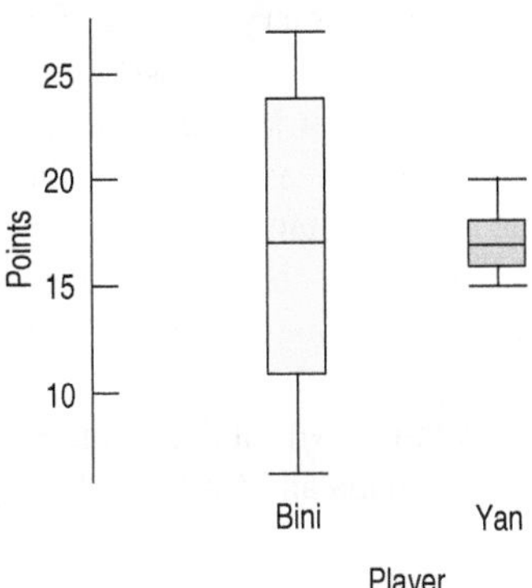

a) Summarize the similarities and differences in their performance so far.
b) The coach can take only one player to the provincial championship. Which one should she take? Why?

10. Wines The boxplots display case prices (in dollars) of varieties of wines produced by vineyards along three of the Finger Lakes in New York State.

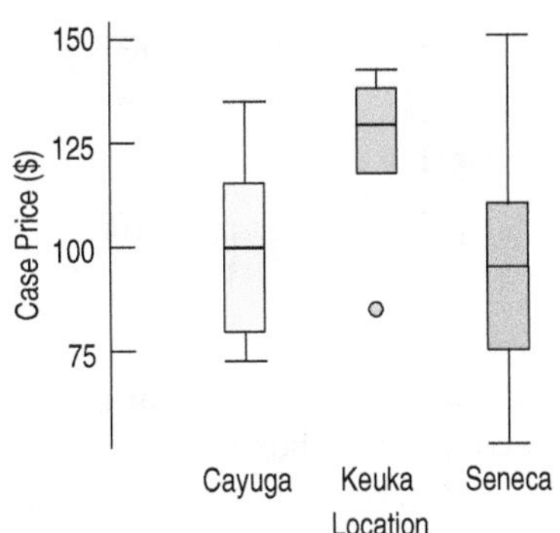

a) Which lake region produces the most expensive wine?
b) Which lake region produces the cheapest wine?
c) In which region are the wines generally more expensive?
d) Write a few sentences describing these wine prices.

11. Test scores Three Statistics classes all took the same test. Histograms and boxplots of the scores for each class are shown below. Match each class with the corresponding boxplot.

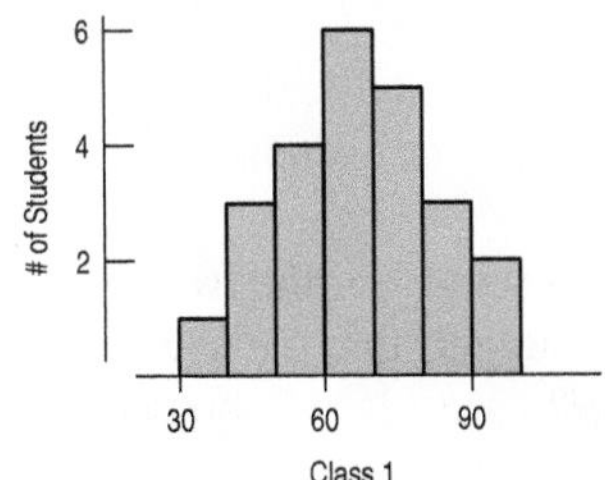

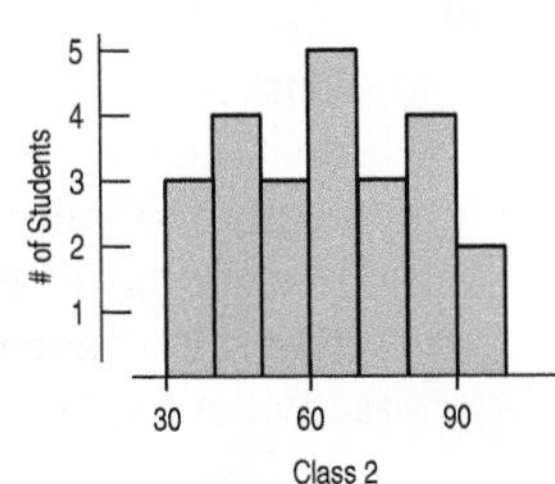

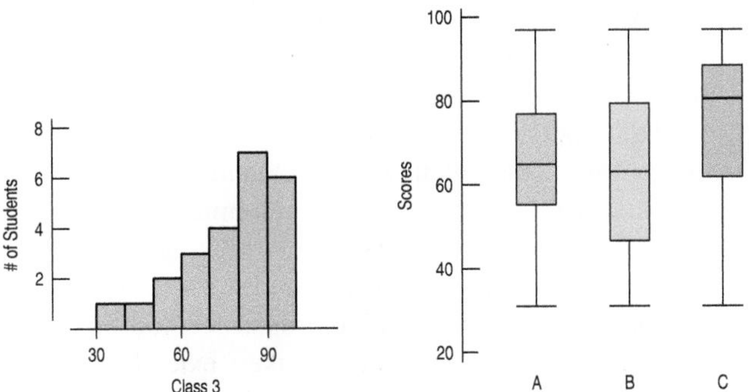

12. Cereals Sugar is a major ingredient in many breakfast cereals. The histogram displays the sugar content as a percentage of weight for 49 brands of cereal. The boxplots compare sugar content for adult and children's cereals.

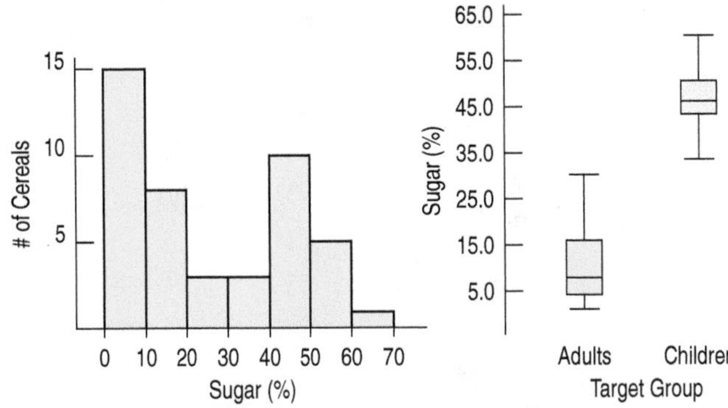

a) What is the range of the sugar contents of these cereals?
b) Describe the shape of the distribution.
c) What aspect of breakfast cereals might account for this shape?
d) Are all children's cereals higher in sugar than adult cereals?
e) Which group of cereals varies more in sugar content? Explain.

13. Caffeine A student study of the effects of caffeine asked volunteers to take a memory test two hours after drinking soda. Some drank caffeine-free cola, some drank regular cola (with caffeine), and others drank a mixture of the two (getting a half-dose of caffeine). Here are the 5-number summaries for each group's scores (number of items recalled correctly) on the memory test:

	n	Min	Q1	Median	Q3	Max
No caffeine	15	16	20	21	24	26
Low caffeine	15	16	18	21	24	27
High caffeine	15	12	17	19	22	24

a) Describe the *W*'s for these data.
b) Name the variables and classify each as categorical or quantitative.
c) Create parallel boxplots to display these results as best you can with this information.
d) Write a few sentences comparing the performances of the three groups.

14. PISA math scores 2009 Below are some summary statistics available at www.pisa.oecd.org showing mean, standard deviation, 5th, 10th, 25th, 75th, 90th, and 95th percentiles for 2009 PISA math scores of 15-year-old students in various Organisation for Economic Co-operation and Development (OECD) countries, and also China.

a) Since you do not have sufficient information to construct standard boxplots, construct modified boxplots instead, showing whiskers extending to 5th and 95th percentiles on each side of the box, with the box constructed as usual, but marking off the mean rather than the median inside the box. Use such plots to compare the countries and then summarize the noticeable differences among countries.

b) Do the distributions appear to be close to symmetric or very skewed? Explain. Are the mean and standard deviation useful summary numbers for these data?

c) If the distributions are mound-shaped and not very skewed, then the empirical rule (see chapter 3) should hold. Calculate the mean + two standard deviations and the mean – two standard deviations for Canada and the U.S. Can you check if more than 90% of scores lie between these two numbers for each country? Explain. Can you confirm that between 50% and 80% of the data fall within the mean ± one standard deviation for Canada and for the U.S.? Explain.

Country	Mean	StDev	5th	10th	25th	75th	90th	95th
Australia	514	94	357	392	451	580	634	665
Canada	527	88	379	413	468	588	638	665
Finland	541	82	399	431	487	599	644	669
France	497	101	321	361	429	570	622	652
Germany	513	98	347	380	443	585	638	666
Japan	529	94	370	407	468	595	648	677
Korea	546	89	397	430	486	609	659	689
Mexico	419	79	289	318	366	472	520	547
UK	492	87	348	380	434	552	606	635
USA	487	91	337	368	425	551	607	637
China	600	103	421	462	531	674	726	757

Source: Based on data from OECD PISA 2009 online database http://pisa2009.acer.edu.au, accessed on Feb 22, 2013.

15. Population growth 2010 Following is a "back-to-back" stem-and-leaf display that shows two data sets at once—one going to the left, one to the right. The display compares the percent change in population for two regions of the United States (based on census figures for 2000 and 2010). The fastest growing state was Nevada at 35%. To show the distributions better, this display breaks each stem into two lines, putting leaves 0–4 on one stem and leaves 5–9 on the other.

NE/MW States		S/W States
4433332200	0	134
987777666555	0	7799
2	1	0023344
5	1	57889
	2	1
	2	5
	3	4
	3	5

Population Growth rate
(2|1|0 means 12% for a NE/NW state and 10% for a S/W state)

a) Use the data displayed in the stem-and-leaf display to construct comparative boxplots.

b) Write a few sentences describing the difference in growth rates for the two regions of the United States.

16. Reading An educator believes that new reading activities for elementary schoolchildren will improve reading comprehension scores. She randomly assigns Grade 3 students to an eight-week program in which some will use these activities and others will experience traditional teaching methods. At the end of the experiment, both groups take a reading comprehension exam. Here are their scores:

Control: 17 28 37 48 43 42 42 55 62 85 60 55 10 53 42 42 42 41 33 37 20 26

New Activities: 49 54 57 58 43 24 33 71 67 62 52 53 56 59 43 43 46 61

a) Construct a back-to-back stem-and-leaf display, similar to the one in the previous exercise (no need to split stems).

b) Describe the difference in scores for the two groups. Do these results suggest that the new activities might be better?

17. Derby speeds 2011 How fast do horses run? Kentucky Derby winners top 30 miles per hour, as shown in this graph. The graph shows the percentage of Derby winners that have run *slower* than each given speed. Note that few have won running less than 33 miles per hour, but about 86% of the winning horses have run less than 37 miles per hour. (A cumulative frequency graph like this is called an "ogive.")

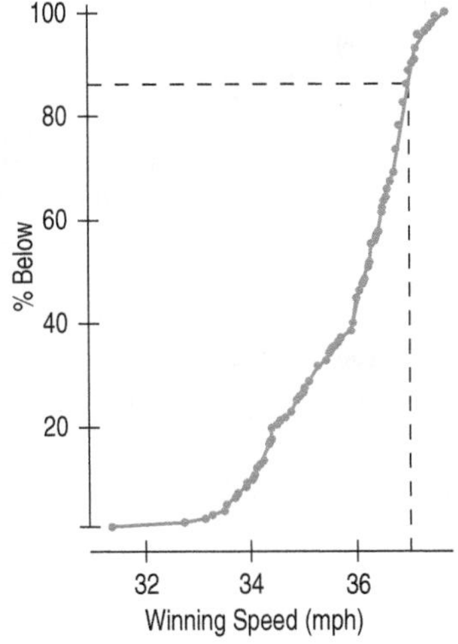

a) Estimate the median winning speed.
b) Estimate the quartiles.
c) Estimate the range and the IQR.
d) Create a boxplot of these speeds.
e) Write a few sentences about the speeds of the Kentucky Derby winners.

18. Framingham The Framingham Heart Study recorded the cholesterol levels of more than 1400 men. Here is an ogive of the distribution of these cholesterol measures. (An ogive shows the percentage of cases at or below a certain value.) Create an approximate boxplot for these data and write a few sentences describing the distribution.

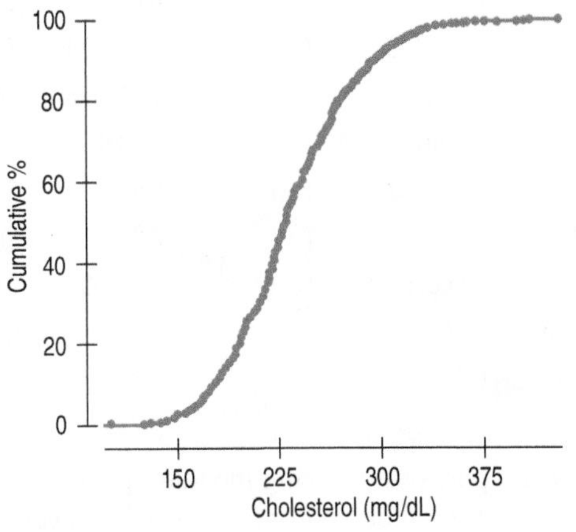

19. Reading scores A class of Grade 4 students took a diagnostic reading test and the scores were reported by reading grade-level. The 5-number summaries for the 14 boys and 11 girls are shown:

	Min	Q1	Median	Q3	Max
Boys:	2.0	3.9	4.3	4.9	6.0
Girls:	2.8	3.8	4.5	5.2	5.9

a) Which group had the highest score?
b) Which group had the greatest range?
c) Which group had the greatest interquartile range?
d) Which group's scores appear to be more skewed? Explain.
e) Which group generally did better on the test? Explain.
f) If the mean reading level for boys was 4.2 and for girls was 4.6, what is the overall mean for the class?

20. Cloud seeding In an experiment to determine whether seeding clouds with silver iodide increases rainfall, 52 clouds were randomly assigned to be seeded or not. The amount of rain they generated was then measured (in acre-feet).

	n	Mean	Median	SD	IQR	Q1	Q3
Unseeded	26	164.59	44.20	278.43	138.60	24.40	163
Seeded	26	441.98	221.60	650.79	337.60	92.40	430

a) Which of the summary statistics are most appropriate for describing these distributions? Why?
b) Do you see any evidence that seeding clouds may be effective? Explain.

T **21. PISA Math by gender 2009** At the PISA Web site, www.pisa.oecd.org, there are summary statistics for math scores of 15-year-old males and females separately. For each OECD country below, we show the mean math scores for males, the mean math scores for females, and also the difference of the two means, followed by graphs of the data.

Country	Males	Females	Difference
Australia	519	509	10
Austria	506	486	19
Belgium	526	504	22
Canada	533	521	12
Chile	431	410	21
Czech Republic	495	490	5
Denmark	511	495	16
Estonia	516	508	9
Finland	542	539	3
France	505	489	16
Germany	520	505	16
Greece	473	459	14
Hungary	496	484	12
Iceland	508	505	3
Ireland	491	483	8
Israel	451	443	8
Italy	490	475	15

Japan	534	524	9
Korea	548	544	3
Luxembourg	499	479	19
Mexico	425	412	14
Netherlands	534	517	17
New Zealand	523	515	8
Norway	500	495	5
Poland	497	493	3
Portugal	493	481	12
Slovakia	498	495	3
Slovenia	502	501	1
Spain	493	474	19
Sweden	493	495	−2
Switzerland	544	524	20
Turkey	451	440	11
United Kingdom	503	482	20
United States	497	477	20

Source: Based on data from OECD PISA 2009 online database http://pisa2009.acer.edu.au, accessed on Feb 22, 2013.

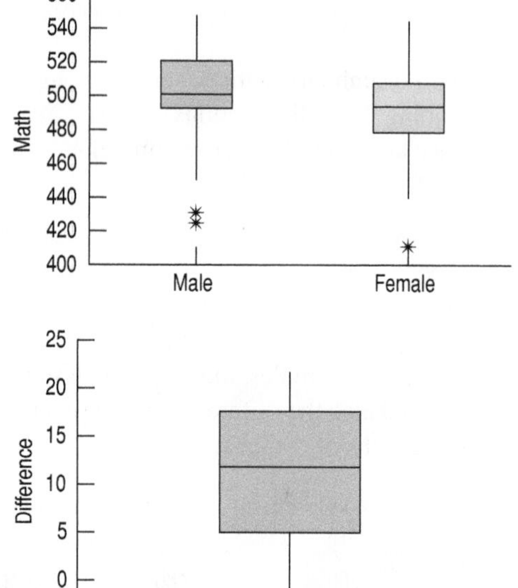

a) Use the graph(s) to compare performance of male and female students.

b) Locate Canada in all plots above and place an "x" to mark the spot. Repeat for the U.S. and for Mexico (using different symbols). For each plot, indicate which countries are outliers and what makes them unusual.

c) Give the 5-number summaries for male scores, for female scores, and for differences. Confirm that Mexico is a suspect outlier by the 1.5 IQR rule in the distribution of female scores.

d) What do you learn from the top graph that is not apparent in the other one? What do you learn from the bottom graph that is not apparent in the other one? Which graph is more useful for learning about gender differences in OECD countries? Why?

T **22. Cholesterol** A study examining the health risks of smoking measured the cholesterol levels of people who had smoked for at least 25 years and people of similar ages who had smoked for no more than five years and then stopped. Create appropriate graphical displays for both groups and write a brief report comparing their cholesterol levels.

Smokers				Ex-Smokers		
225	211	209	284	250	134	300
258	216	196	288	249	213	310
250	200	209	280	175	174	328
225	256	243	200	160	188	321
213	246	225	237	213	257	292
232	267	232	216	200	271	227
216	243	200	155	238	163	263
216	271	230	309	192	242	249
183	280	217	305	242	267	243
287	217	246	351	217	267	218
200	280	209		217	183	228

T **23. Industrial experiment** Engineers at a computer production plant tested two methods for accuracy in drilling holes into a PC board. They tested how fast they could set the drilling machine by running 10 boards at each of two different speeds. To assess the results, they measured the distance (in inches) from the centre of a target on the board to the centre of the hole. The data and summary statistics are shown in the table:

	Distance (in.)	Speed	Distance (in.)	Speed
	0.000101	Fast	0.000098	Slow
	0.000102	Fast	0.000096	Slow
	0.000100	Fast	0.000097	Slow
	0.000102	Fast	0.000095	Slow
	0.000101	Fast	0.000094	Slow
	0.000103	Fast	0.000098	Slow
	0.000104	Fast	0.000096	Slow
	0.000102	Fast	0.975600	Slow
	0.000102	Fast	0.000097	Slow
	0.000100	Fast	0.000096	Slow
Mean	0.000102		**Mean**	0.097647
StdDev	0.000001		**StdDev**	0.308481

Write a report summarizing the findings of the experiment. Include appropriate visual and verbal displays of the distributions, and make a recommendation to the engineers if they are most interested in the accuracy of the method.

24. Women researchers 2009 Examine the display of data from Eurostat showing the percentage of female researchers in each of several employment categories for 35 countries, mostly EU members, but not exclusively (e.g., Japan is included).

a) Graph the data with four side-by-side boxplots (one for each employment category). Point out any outliers, indicating the country and its particular distinction.

b) Describe the distribution of these nations' percentages of women researchers, across all sectors, with words and numbers.

c) Compare the distributions for the three different sectors (shape, centre, spread, and unusual features).

d) Which countries are in the upper quarter for women researchers in the business enterprise sector?

e) In higher education, 75% of countries have at least ___% women researchers (fill in the blank).

Country	All sectors	Govt.	Higher Education	Business Enterprise
Belgium	32.7	32.5	39	24
Bulgaria	47.6	54	43.2	43.4
Czech Republic	28.9	37.2	35.4	15.7
Denmark	31.7	36.2	41	23.9
Germany	24.9	32.4	34.7	12.7
Estonia	42.5	61.4	46	27.5
Ireland	33.3	38.9	38.7	25.8
Spain	38.1	48.5	39.8	28.8
France	26.9	35.1	34.4	19.9
Italy	33.8	43.8	37.8	20.7
Cyprus	35.6	46.3	36.5	27.3
Latvia	52.4	53.3	52.1	53.4
Lithuania	51	54.1	53.3	31.2
Luxembourg	21.2	35.5	35.8	11.4
Hungary	32.1	40	36.1	21
Malta	29.4	44	29.5	26.6
Netherlands	25.9	30.4	36.9	14.2
Austria	28.4	43.1	37.8	16.3
Poland	39.5	41.2	42.1	22.1
Portugal	45.8	60.4	49.6	30.2
Romania	44.7	49.5	45.6	37.6
Slovenia	35.7	45.5	40.7	23.4
Slovakia	42.5	44.6	44.6	21.8
Finland	31.4	42.4	46.6	17
Sweden	35.7	38.9	44.5	25.5
United Kingdom	37.9	35.3	43.7	19.1
Iceland	41	46.8	43.8	31.8
Norway	35.2	42	44.1	22.1
Switzerland	30.2	33.5	34.4	18.7
Croatia	46.4	51.8	45.4	40.7
Macedonia	51.3	50.6	49.5	80
Turkey	36.3	29.2	40.6	23.6
Russia	41.9	45.2	40.7	39.9
Japan	13.6	14.6	23.9	7.6
South Korea	15.8	18.4	25	11.4

Source: http://epp.eurostat.ec.europa.eu. © European Union, 1995–2012

25. Crime comparisons Below are crime rates per 100 000 residents for 18 of the biggest Canadian and U.S. cities. Rates are shown for six classes of criminal activity.

a) For each type of crime, graphically compare Canadian cities with U.S. cities. Point out crimes where rates tend to be higher in the U.S. or higher in Canada or similar in both countries.

City	Homicide	Aggravated Assault	Robbery	Break and Enter	Motor Vehicle Theft	Theft	Country
Toronto	1.7	157	107	553	365	1 692	Canada
Montreal	2.1	152	173	1 195	800	2 068	Canada
Vancouver	2	164	187	1 430	1 058	4 415	Canada
Calgary	1.7	132	105	814	580	2 616	Canada
Edmonton	2	180	134	986	539	2 559	Canada
Ottawa	1	123	96	690	558	1 835	Canada
Quebec	1.7	76	70	925	230	1 771	Canada
Winnipeg	2.5	276	251	1 228	1 425	2 779	Canada
Hamilton	1.3	144	86	815	698	1 832	Canada
LA	10.6	607	298	636	674	1 726	USA
NYC	7.8	474	372	453	428	1 785	USA
Philadelphia	8.1	350	270	507	492	2 199	USA
Washington	7.4	265	171	452	484	2 223	USA
Detroit	10.6	472	229	735	919	2 280	USA
Houston	7.7	433	242	960	645	2 724	USA
Boston	2.1	N/A	106	408	426	1 640	USA
Dallas	8.5	377	252	1 081	747	3 146	USA
Phoenix	7.2	354	170	1 111	1 010	3 254	USA

Source: Adapted from Statistics Canada, Juristat. 85-002-XIE2001011 Vol. 21 no. 11. Released December 18, 2001.

b) Which Canadian city could pass for American in aggravated assault? Which U.S. city could pass for Canadian in homicide rate? Which Canadian city is an outlier in one of the last three listed crime categories?

c) What extraneous (i.e., lurking, or not included) quantitative variable might be important and worth taking into account, in some appropriate fashion, in order to make these Canada–U.S. comparisons as honest as possible?

T **26. Election 2000 and spoiled ballots** Open the Election 2000 data set available online at www.pearsoncanada.ca/deveaux, a small part of which is shown below.

District & Elected Member	Province	% Rejected Ballots
Abitibi—Baie-James—Nunavik [Guy St-Julien (Lib.)]	Quebec	2.28
Acadie—Bathurst [Yvon Godin (N.D.P.)]	NB	1.22
Ahuntsic [Eleni Bakopanos (Lib.)]	Quebec	2.37
Algoma—Manitoulin [Brent St. Denis (Lib.)]	Ontario	0.43
:	:	:

a) Use boxplots to compare the percent of rejected ballots by province (ignore territories) and point out the most noticeable differences among the provinces. Comments?

b) Are there any outliers for the western provinces? What would you suggest or recommend to Canadian officials?

c) Are there any outliers for the province of Quebec? If so, which party had the winning candidate for each outlier? Which outlier(s) are a particularly grievous smirch upon the democratic process?

d) Is the mean percentage of rejected ballots per riding bigger or smaller than the median for Manitoba? For B.C.? Explain without using any calculations. From the boxplots, determine which province is most likely to have a smaller mean than median.

T **27. Election 2011** Open the 2011 Election data set available online at www.pearsoncanada.ca/deveaux, a small part of which is shown below.

Electoral district	Province	Voter turnout (%)	Elected candidate	Party
Avalon	NL	56.8	Andrews, Scott	Liberal
Bonavista—Gander—Grand Falls—Windsor	NL	44.8	Simms, Scott	Liberal
Humber—St. Barbe—Baie Verte	NL	50.9	Byrne, Gerry	Liberal
Labrador	NL	52.9	Penashue, Peter	Conservative
⋮	⋮	⋮	⋮	⋮

a) Use boxplots to compare the percentage of voter turnout by province, and point out the more (voting) motivated provinces and the less motivated provinces.
b) Do you consider the ridings designated as suspect outliers in the boxplots to be genuine outliers? Explain.
c) For Canadian ridings, overall, are there any suspect outliers (by the boxplot rule)? If so, can you explain the cause or nature of any of the suspect outliers?
d) Is the mean or median smaller for Manitoba? For B.C.? No calculations, but explain your answer.

28. Election 2006 Open up the Election 2006 data set available online at www.pearsoncanada.ca/deveaux, a small part of which is displayed below.

Prov. Voter	Electoral District	Region	% Rejected Ballots	% Turnout
NL	Avalon	Atlantic	1.65	59.6
NL	Bonavista—Gander—Grand Falls—Windsor	Atlantic	0.50	54.2
NL	Humber—St. Barbe—Baie Verte	Atlantic	0.58	54.8
NL	Labrador	Atlantic	0.45	58
⋮	⋮	⋮	⋮	⋮

a) Use boxplots to compare the percentage of voter turnout among the regions (ignore the territories), and point out the differences among regions. Be sure to comment on more (voting) motivated and less motivated regions.
b) Which ridings are considered to be suspect outliers within their respective regions by the 1.5 IQR rule? Are those ridings also suspect outliers for Canada as a whole?
c) Which regional distribution is least symmetric (ignore territories and the Atlantic region due to small numbers)? Use any appropriate graphs.
d) Find the mean and standard deviation for turnout percentages in all Quebec ridings. Compute "$\bar{y} + (2 \times s)$" and "$\bar{y} - (2 \times s)$". Predict the percent of ridings whose turnout percentages will fall between these two numbers (see Empirical Rule, chapter 3). Go through the Quebec data and count how many (and what percent) actually do fall between these two numbers.
e) If you averaged the turnout percentages for all ridings in Ontario, would you get the percentage turnout for the province?

29. Homicides 2011 Examine the displayed data on homicides from Statistics Canada. The rate shown is the average homicide rate per 100 000 population over the 11 years from 2001 to 2011. Cities were classified into three size categories. We also classified them by geographic region below.

a) Use graphs to examine the relationship between crime rate and city size (category). Describe any patterns you see.
b) What type of variable is city size, if determined precisely by a count of residents? What type is it in our analysis above? Is this change a good thing or a bad thing? If you wanted to use the actual city sizes, can you think of another type of graph to use to explore the relationship between crime rate and city size? If so, produce this graph. (More on this in Chapter 6.)
c) Use graphs to examine the relationship between crime rate and geographical region. Describe any geographical patterns you see.
d) We have used the average homicide rate over 10 years. Is there another relationship we should examine to more fully understand homicide patterns in Canada? If so, explain what type of data plot(s) you would like to request from StatsCan.

City	Size	Population	Region	Homicide Rate
Toronto	500 000+	5 841 126	Ontario	1.78
Montreal	500 000+	3 917 875	Quebec	1.53
Vancouver	500 000+	2 426 200	West	2.43
Calgary	500 000+	1 268 704	West	1.98
Edmonton	500 000+	1 196 704	West	3.07
Ottawa	500 000+	945 489	Ontario	1.20
Winnipeg	500 000+	765 770	West	3.51
Quebec	500 000+	760 961	Quebec	0.62
Hamilton	500 000+	749 722	Ontario	1.43
Kitchener	250 000–500 000	498 876	Ontario	0.90
London	250 000–500 000	496 156	Ontario	1.30
Halifax	250 000–500 000	409 662	East	1.92
St. Cath-Niagara	250 000–500 000	405 216	Ontario	1.42
Oshawa	250 000–500 000	370 448	Ontario	1.18
Victoria	250 000–500 000	361 685	West	1.19
Windsor	250 000–500 000	332 539	Ontario	1.48
Gatineau	250 000–500 000	310 449	Quebec	1.15
Saskatoon	250 000–500 000	272 771	West	2.88
Regina	100 000–250 000	219 286	West	3.67
Sherbrooke	100 000–250 000	200 695	Quebec	0.36
St. John's	100 000–250 000	197 524	East	0.67
Abbotsford-Mission	100 000–250 000	177 064	West	2.98
Sudbury	100 000–250 000	164 196	Ontario	1.60
Kingston	100 000–250 000	164 089	Ontario	1.67
Saguenay	100 000–250 000	151 394	Quebec	0.74
Trois-Rivieres	100 000–250 000	147 657	Quebec	0.82
Saint John	100 000–250 000	128 534	East	0.92
Thunder Bay	100 000–250 000	127 049	Ontario	1.86

Source: Adapted from 'Homicide in Canada' - Statistics Canada, Catalogue no. 85-002X, Juristat, ISSN 1209-6393

30. Eye and hair colour A survey of 1021 school-age children was conducted by randomly selecting children from several large urban elementary schools. Two of the questions concerned eye and hair colour. In the survey, the following codes were used:

Hair Colour	Eye Colour
1 = Blond	1 = Blue
2 = Brown	2 = Green
3 = Black	3 = Brown
4 = Red	4 = Grey
5 = Other	5 = Other

The Statistics students analyzing the data were asked to study the relationship between eye and hair colour. They produced this plot:

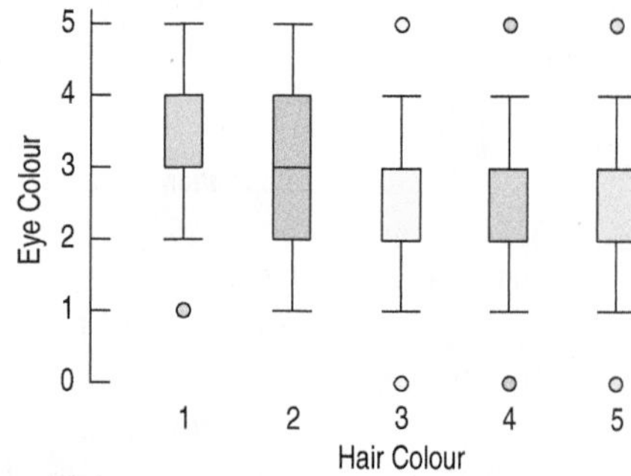

Is their graph appropriate? If so, summarize the findings. If not, explain why not.

T **31. He shoots, he scores the hat-trick** Look once more at the data on total points scored each season by Wayne Gretzky, seen in Exercise 22, Chapter 3.
a) Find the 5-number summary.
b) Construct a boxplot.
c) Explain why the season when Gretzky scored only 48 points would not be deemed a suspect outlier in a boxplot.
d) Plot total points versus year. What do you learn from this plot that is masked in the boxplot?

T **32. Drunk driving 2008** Accidents involving drunk drivers in the USA account for about 40% of all deaths on the nation's highways. The table tracks the number of alcohol-related fatalities for 26 years. (*Source:* www.alcoholalert.com)

Year	Deaths (thousands)	Year	Deaths (thousands)
1982	26.2	1996	17.7
1983	24.6	1997	16.7
1984	24.8	1998	16.7
1985	23.2	1999	16.6
1986	25.0	2000	17.4
1987	24.1	2001	17.4
1988	23.8	2002	17.5
1989	22.4	2003	17.1
1990	22.6	2004	16.9
1991	20.2	2005	16.9
1992	18.3	2006	15.8
1993	17.9	2007	15.4
1994	17.3	2008	13.8
1995	17.7		

a) Create a stem-and-leaf display or a histogram of these data.
b) Create a timeplot.
c) Using features apparent in the stem-and-leaf display (or histogram) and the timeplot, write a few sentences about deaths caused by drunk driving.

T **33. Hurricanes 2010 again** Here are the number of Atlantic hurricanes classified as major for the years 1944 to 2010:

3, 3, 1, 2, 4, 3, 8, 5, 3, 4, 2, 6, 2, 2, 4, 2, 2, 7, 1, 2, 6, 1, 3, 1, 0, 5, 2, 1, 0, 1, 2, 3, 2, 1, 2, 2, 2, 3, 1, 1, 1, 3, 0, 1, 3, 2, 1, 2, 1, 1, 0, 5, 6, 1, 3, 5, 3, 4, 2, 3, 6, 7, 2, 2, 5, 2, 5

You examined these data in the previous chapter's exercises. Now create a timeplot of the data and tell if there is anything new that you have learned. (Use a smoother if your software has one available.)

34. World Series champs again What type of plot was the graph of Joe Carter's home runs shown in Exercise 29 in Chapter 3? Summarize what you see.

T **35. Tsunamis 2013** Look again at the magnitudes of tsunami generating earthquakes (updated in this data file to early 2013). Delete all data from 1950 and earlier, so that only relatively recent data, from 1951 to the present, will be examined. Use computer software to answer the following questions about these more recent tsunami-creating earthquakes:
a) Examine graphically and describe the distribution of tsunami-creating earthquake magnitudes since 1951.
b) Plot magnitude versus year and apply a smoother to the plot. Do you learn anything more from this plot that is not observable from your plot(s) in part a)?
c) Now examine how magnitude depends on country with side-by-side boxplots. Any problems? Any suggestions?
d) Delete data for all countries that have suffered 10 or fewer tsunami-creating earthquakes. Give side-by-side boxplots again. Where did quakes tend to be biggest? Smallest? Which three countries had the biggest quakes (and how big)? Where and when was the smallest quake?

36. Assets Here is a histogram of the assets (in millions of dollars) of 79 companies chosen from the *Forbes* list of top corporations.

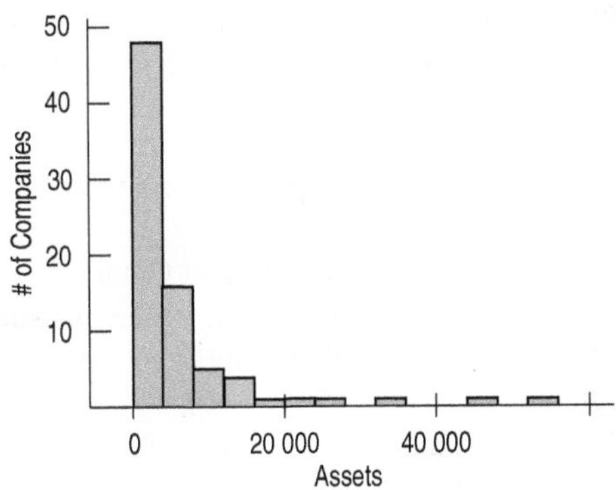

a) What aspect of this distribution makes it difficult to summarize or discuss centre and spread?
b) What would you suggest doing with these data if we want to understand them better?

37. Assets again Here are the same data you saw in Exercise 36 after re-expressions as the square root of assets and the logarithm of assets.

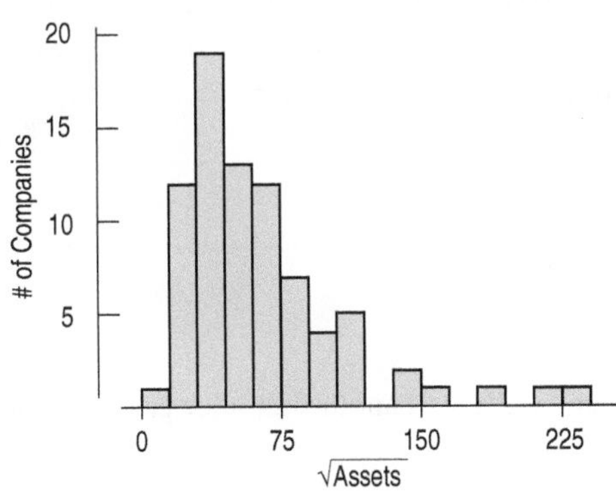

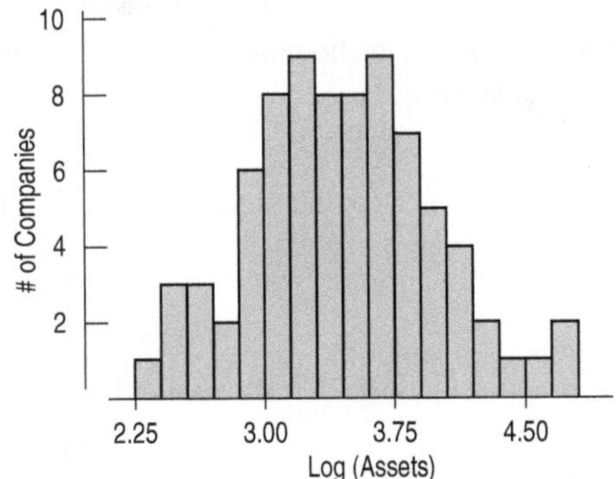

a) Which re-expression do you prefer? Why?
b) In the square root re-expression, what does the value 50 actually indicate about the company's assets?
c) In the logarithm re-expression, what does the value 3 actually indicate about the company's assets?

38. Rainmakers The table lists the amount of rainfall (in acre-feet) from the 26 clouds seeded with silver iodide discussed in Exercise 20:

2745	703	302	242	119	40	7
1697	489	274	200	118	32	4
1656	430	274	198	115	31	
978	334	255	129	92	17	

a) Why is acre-feet a good way to measure the amount of precipitation produced by cloud seeding?
b) Plot these data, and describe the distribution.
c) Create a re-expression of these data that produces a more advantageous distribution.
d) Explain what your re-expressed scale means.

39. Stereograms Stereograms appear to be composed entirely of random dots. However, they contain separate images that a viewer can "fuse" into a three-dimensional (3D) image by staring at the dots while defocusing the eyes. An experiment was performed to determine whether knowledge of the embedded image affected the time required for subjects to fuse the images. One group of subjects (group NV) received no information or just verbal information about the shape of the embedded object. A second group (group VV) received both verbal information and visual information (specifically, a drawing of the object). The experimenters measured how many seconds it took for the subject to report that he or she saw the 3D image.
a) What two variables are discussed in this description?
b) For each variable, is it quantitative or categorical? If quantitative, what are the units?

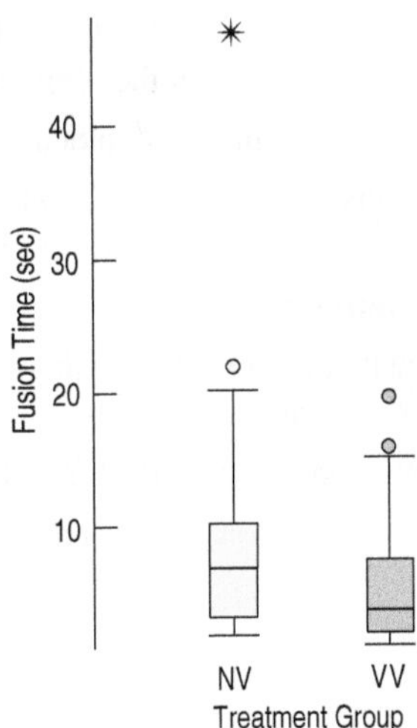

c) The boxplots above compare the fusion times for the two treatment groups. Write a few sentences comparing these distributions. What does the experiment show?

T **40. Stereograms revisited** Because of the skewness of the distributions of fusion times, we might consider a re-expression. Here are the boxplots of the log of fusion times. Is it better to analyze the original fusion times or the log fusion times? Explain.

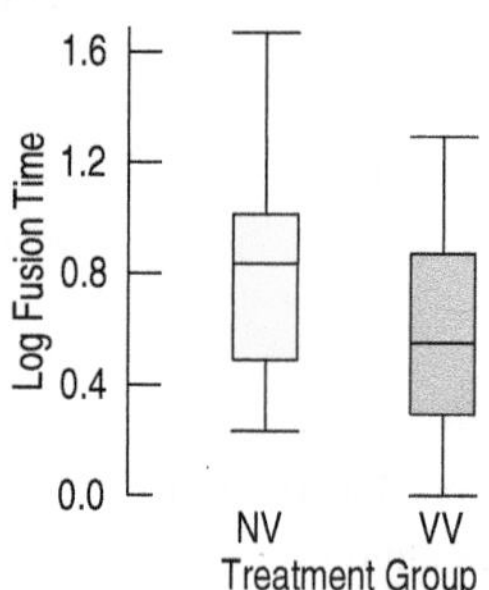

T **41. First Nations 2010 by province** Let's compare sizes of First Nations bands in Ontario and Saskatchewan.

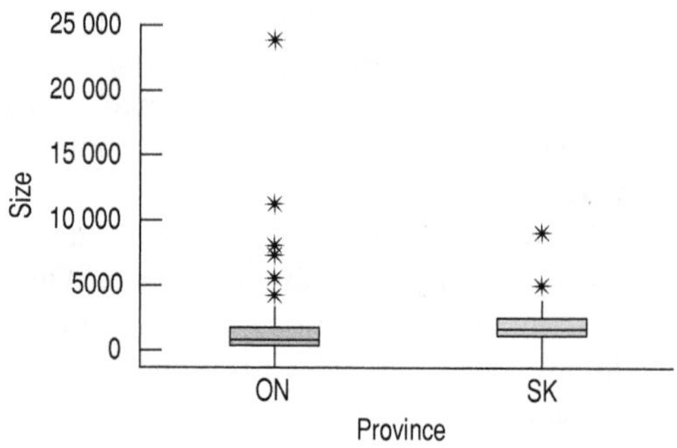

a) Using the boxplots, what is the approximate size of the largest band in Ontario? Do band sizes tend to be bigger in one province than in the other? From the plot, what can you say about the difference in median band size? In mean band size?

b) How could we improve readability of the boxplots above?

c) Describe the shapes of the two distributions. Do you consider most of the suspect outliers to be genuine outliers? Explain.

d) If we were to exclude the biggest Ontario band, the Six Nations of the Grand River, which measures of centre and spread would hardly need the trouble of recalculation, and why?

e) Why is the mean band size bigger than the median band size in both provinces?

f) Seventy-five percent of Saskatchewan bands are at least what size? Half of Saskatchewan bands are between ___ and ___ in size (fill in the blanks). Just estimate from the boxplots as best you can.

g) Calculate the logarithms (any base) of all band sizes in both provinces, and produce side-by-side boxplots. Describe the major effects of this transformation.

h) After this transformation, does the Six Nations still appear to be such an extreme outlier? Are there new outliers (if so, which band is the farthest outlying)?

Just Checking Answers

1. The distribution is somewhat bimodal with one big group centred around 0.3% and another smaller group extending from about 1% to 2%. Each group shows some right skewness.
2. The boxplot doesn't show the bimodality well.
3. The suspect outliers in the boxplot are not really outliers, but rather members of what appears to be a distinct subgroup.
4. Quebec results are poorest, with 3/4 of ridings exceeding 1% spoiled ballots, and also showing the largest variability in results. Western provinces did best at limiting spoiled ballots, with most ridings around 0.3%.
5. Only three territories for the far-right boxplot.
6. Yes, we see that the lower quartile of Quebec's boxplot (lower edge of the box) exceeds the highest result from the West, and the territories.
7. Worst result is in Quebec. Best is in the West.
8. We see that the unusually poor results in Quebec created the bimodality in the histogram.

chapter

5

The Standard Deviation as a Ruler and the Normal Model

Where are we going?

A graduate school admissions officer is looking at the files of two candidates from overseas, one with a paper-based TOEFL test score of 520, another with an Internet-based TOEFL test score of 75. Which candidate scored better? How do we compare things when they're measured on different scales?

To answer a question like this, we need to *standardize* the results. To do that, first we need a base value for comparison—for that, we'll often use the mean. Next, we need to know how far away we are from the mean. We need some sort of ruler. Fortunately, the standard deviation is just what we need.

The idea of measuring distances from means by counting standard deviations shows up throughout Statistics, and it starts right here.

All models are wrong—but some are useful.

—George Box, famous statistician

The women's heptathlon in the Olympics consists of seven track-and-field events: 200 m and 800 m runs, 100 m high hurdles, shot put, javelin, high jump, and long jump. In the 2008 Olympics, Nataliya Dobrynska of the Ukraine posted a long jump of 6.63 metres—about half a metre farther than the mean distance for all contestants. Hyleas Fountain of the United States won the 200 m run with a time of 23.21 seconds—about a second and a half faster than the average. Which performance deserves more points? It's not clear how to compare them. They aren't measured in the same units, or even in the same *direction* (longer jumps are better but shorter times are better.)

We want to compare individual values, not whole groups as we did earlier. To see which value is more extraordinary, we need a way to judge them against the background of data they come from. The tool we'll use is one you've already seen: the standard deviation. We'll judge how unusual a value is by how far, in standard deviation units, it lies from the mean. Statisticians use the standard deviation as a ruler throughout Statistics, and we'll do that for the rest of the book.

We'll start with stem-and-leaf displays (Figure 5.1) because they show individual values, and because it is easy to orient them either high-to-low or low-to-high so the best performances can be at the top of both displays.

Which of the two winning scores is the better one? Dobrynska's 6.63 m long jump is 0.52 m longer than the mean jump of 6.11 m. How many standard deviations better than the mean is that? The standard deviation for this event was 0.238 m, so her jump was $(6.63 - 6.11)/0.238 = 2.18$ standard deviations better than the mean. Fountain's winning

Figure 5.1

Stem-and-leaf displays for the 200 m race and the long jump in the 2008 Olympic Heptathlon. Hyleas Fountain (green scores) won the 200 m, and Nataliya Dobrynska (red scores) won the long jump. The stems for the 200 m race run from faster to slower and the stems for the long jump from longer to shorter so that the best scores are at the top of each display.

200 m Race Stem	Leaf	Long Jump Stem	Leaf
23	233	66	3
23	699	65	3
24	22333334	64	0578
24	5566677999	63	368
25	000234444	62	1
25	599	61	11235668
26	1	60	24689
		59	267778
		58	8

23|3=23.3 seconds   66|3=6.63 meters

	Event	
	Long Jump	200 m
Mean (all contestants)	6.11 m	24.71 s
SD	0.238 m	0.700 s
n	36	38
Dobrynska	6.63 m	24.39 s
Fountain	6.38 m	23.21 s

200 m run was 1.50 seconds faster than the mean, and that's only (23.21 − 24.71)/0.700 = −2.14, or 2.14 standard deviations faster than the mean. That's a winning performance, but just a bit less impressive than Dobrynska's long jump. To combine all seven events—each with its own scale—into a single score, Olympic judges use tables based on similar calculations. In the final standings (including all seven events), Dobrynska took the gold medal and Fountain the silver.

5.1 Standardizing with z-Scores

Expressing a distance from the mean in standard deviations *standardizes* the performances. To **standardize** a value, we subtract the mean and then divide this difference by the standard deviation:

$$z = \frac{y - \bar{y}}{s}$$

NOTATION ALERT

We always use the letter z to denote values that have been standardized with the mean and standard deviation.

The values are called **standardized values**, and are commonly denoted with the letter z. Usually we just call them **z-scores**.

z-scores measure the distance of a value from the mean in standard deviations. A z-score of 2 says that a data value is two standard deviations above the mean. It doesn't matter whether the original variable was measured in fathoms, dollars, or carats; those units don't apply to z-scores. Changing the original variable's units (like from metres to feet) will not affect a z-score. Data values below the mean have negative z-scores, so a z-score of −1.6 means that the data value was 1.6 standard deviations below the mean. Of course, regardless of the direction, the farther a data value is from the mean, the more unusual it is, so a z-score of −1.3 is more extraordinary than a z-score of 1.2. Looking at the z-scores, we can see that Dobrynska's long jump with a z-score of 2.18 is slightly more impressive than Fountain's 200 m race with a z-score of −2.14.

		Event	
		Long Jump	200 m Run
	Mean SD	6.11 0.238	24.71 0.700
Dobrynska	Performance z-score	6.63 (6.63 − 6.11)/0.238 = 2.18	24.39 s (24.39 − 24.71)/0.700 = −0.46
Fountain	Performance z-score	6.38 (6.38 − 6.11)/0.238 = 1.13	23.21 (23.21 − 24.71)/0.700 = −2.14

By using the standard deviation as a ruler to measure statistical distance from the mean, we can compare values from the same data set, or even two values measured on different variables or with different scales. However, the comparison must still make sense. The z-score simply tells us how far a value is from its mean in terms of standard deviations.

For Example STANDARDIZING SKIING TIMES

The men's super-combined skiing event debuted in the 2010 Winter Olympics in Vancouver. It consists of two races: a downhill and a slalom. Times for the two events are added together, and the skier with the lowest total time wins. At Vancouver, the mean slalom time was 52.67 seconds with a standard deviation of 1.614 seconds. The mean downhill time was 116.26 seconds with a standard deviation of 1.914 seconds. Bode Miller of the United States, who won the gold medal with a combined time of 164.92 seconds, skied the slalom in 51.01 seconds and the downhill in 113.91 seconds.

QUESTION: On which race did he do better compared to the competition?

ANSWER: Standardize each of his times by subtracting the mean and dividing by the standard deviation, as follows:

$$z_{Slalom} = \frac{51.01 - 52.67}{1.614} = -1.03$$

$$z_{Downhill} = \frac{113.91 - 116.26}{1.914} = -1.23$$

Keeping in mind that faster times are *below* the mean, Miller's downhill time of 1.23 SDs below the mean is even more remarkable than his slalom time, which was 1.03 SDs below the mean.

Just Checking

1. Your Statistics teacher has announced that the lower of your two tests will be dropped. You got a 90 on test 1 and an 80 on test 2. You're all set to drop the 80 until she announces that she grades "on a curve." She standardized the scores in order to decide which is the lower one. If the mean on the first test was 88 with a standard deviation of 4 and the mean on the second was 75 with a standard deviation of 5
   a) Which one will be dropped?
   b) Does this seem "fair"?

5.2 Shifting and Scaling

Who	80 male participants of the NHANES survey between the ages of 19 and 24 who measured between 68 and 70 inches tall
What	Their weights
Unit	Kilograms
When	2001–2002
Where	United States
Why	To study nutrition and health issues and trends
How	National survey

There are two steps to finding a z-score. First, the data are *shifted* by subtracting the mean. Then, they are *rescaled* by dividing by the standard deviation. How do these operations work?

Shifting to Adjust the Centre

Since the 1960s, the Centers for Disease Control and Prevention's National Center for Health Statistics has been collecting health and nutritional information on people of all ages and backgrounds. The most recent survey, the National Health Interview Survey (NHIS), interviewed nearly 75 000 children about their health. The previous survey, the National Health And Nutrition Examination Survey (NHANES) 2001–2002,[1] measured a wide variety of variables, including body measurements, cardiovascular fitness, blood chemistry, and demographic information on more than 11 000 individuals.

[1]www.cdc.gov/nchs/nhanes.htm

Included in this group were 80 men between 19 and 24 years old of average height (between 5′8″ and 5′10″ tall). Figure 5.2 displays a histogram and boxplot of their weights:

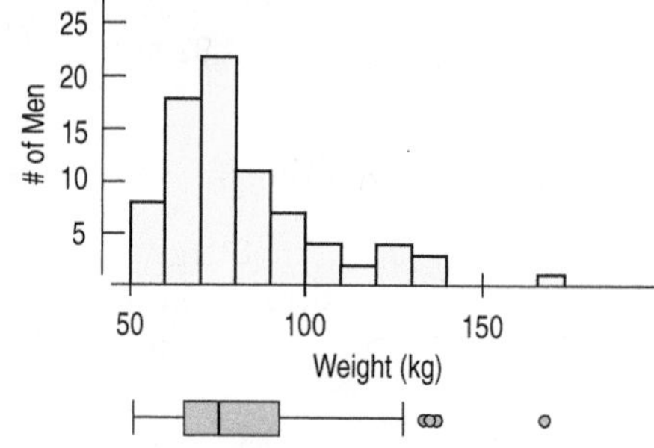

Figure 5.2
Histogram and boxplot for the men's weights. The shape is skewed to the right with several high outliers.

Their mean weight is 82.36 kg. For this age and height group, the National Institutes of Health recommends a maximum healthy weight of 74 kg, but we can see that some of the men are heavier than the recommended weight. To compare their weights to the recommended maximum, we could subtract 74 kg from each of their weights. What would that do to the centre, shape, and spread of the histogram? Figure 5.3 gives us the new picture:

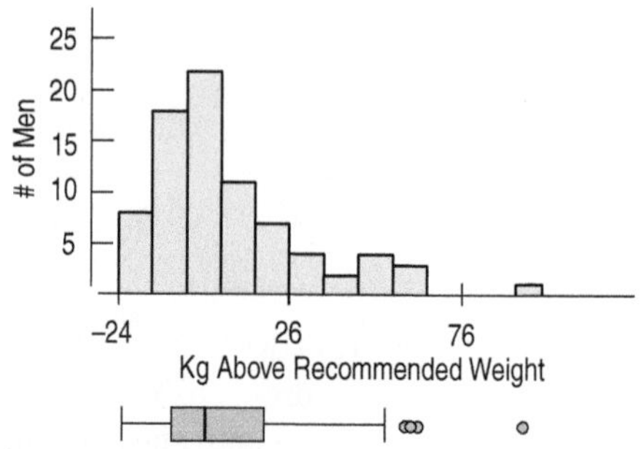

Figure 5.3
Subtracting 74 kilograms shifts the entire histogram down, but leaves the spread and the shape exactly the same.

A S *Activity:* **Changing the Baseline.** What happens when we shift data? Do measures of centre and spread change?

On average, they weigh 82.36 kg, so on average they're 8.36 kg overweight. And, after subtracting 74 from each weight, the mean of the new distribution is $82.36 - 74 = 8.36$ kg. In fact, when we **shift** the data by adding (or subtracting) a constant to each value, all measures of position (centre, percentiles, min, max) will increase (or decrease) by the same constant.

What about the spread? What does adding or subtracting a constant value do to the spread of the distribution? Look at the two histograms again. Adding or subtracting a constant changes each data value equally, so the entire distribution just shifts. Its shape doesn't change and neither does the spread. None of the measures of spread we've discussed—not the range, not the IQR, not the standard deviation—changes.

Adding (or subtracting) a constant to every data value adds (or subtracts) the same constant to measures of position, but leaves measures of spread unchanged.

Rescaling to Adjust the Scale

Not everyone thinks naturally in metric units. Suppose we want to look at the weights in pounds instead. We'd have to rescale the data. Because there are about 2.2 pounds in every kilogram, we'd convert the weights by multiplying each value by 2.2. Multiplying or dividing each value by a constant—**rescaling** the data—changes the measurement units. Figure 5.4 shows histograms of the two weight distributions, plotted on the same scale, so you can see the effect of multiplying:

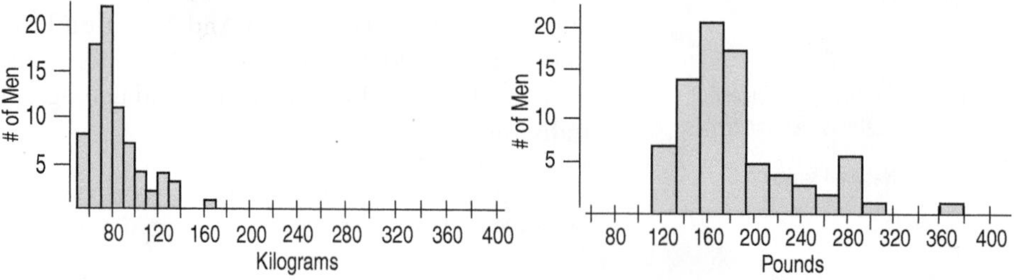

Figure 5.4
Men's weights in both kilograms and pounds. How do the distributions and numerical summaries change?

A S *Simulation:* **Changing the Units.** Change the centre and spread values for a distribution and watch the summaries change (or not, as the case may be).

What happens to the shape of the distribution? Although the histograms don't look exactly alike, we see that the shape really hasn't changed: Both are unimodal and skewed to the right.

What happens to the mean? Not too surprisingly, it gets multiplied by 2.2 as well. The men weigh 82.36 kg on average, which is 181.19 pounds. As the boxplots and 5-number summaries below show, all measures of position act the same way. They all get multiplied by this same constant.

What happens to the spread? Take a look at the boxplots (Figure 5.5). The spread in pounds (on the right) is larger. How much larger? If you guessed 2.2 times, you've figured out how measures of spread get rescaled.

Figure 5.5

The boxplots (drawn on the same scale) show the weights measured in kilograms (on the left) and pounds (on the right). Because 1 kg is 2.2 lb, all the points in the right box are 2.2 times larger than the corresponding points in the left box. So each measure of position and spread is 2.2 times as large when measured in pounds rather than kilograms.

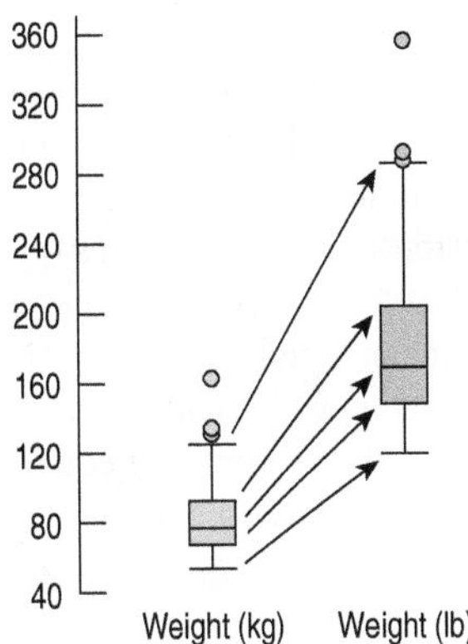

	Weight (kg)	Weight (lb)
Min	54.3	119.46
Q1	67.3	148.06
Median	76.85	169.07
Q3	92.3	203.06
Max	161.5	355.30
IQR	25	55
SD	22.27	48.99

When we multiply (or divide) all the data values by any constant, all measures of position (such as the mean, median, and percentiles) and measures of spread (such as the range, the IQR, and the standard deviation) are multiplied (or divided) by that same constant.

Mark van Manen/MCT/Newscom

For Example RESCALING THE MEN'S COMBINED TIMES

RECAP: The times in the men's combined event at the winter Olympics are reported in minutes and seconds. The mean and standard deviation of the 34 final super combined times at the 2010 Olympics were 168.93 seconds and 2.90 seconds, respectively.

QUESTION: Suppose instead that we had reported the times in minutes—that is, that each individual time was divided by 60. What would the resulting mean and standard deviation be?

Dividing all the times by 60 would divide both the mean and the standard deviation by 60:

Mean = 168.93/60 = 2.816 minutes SD = 2.90/60 = 0.048 minutes.

Just Checking

2. SAT test scores are required of applicants to many U.S. universities. In 1995, the Educational Testing Service (ETS) adjusted the scores of SAT tests. Before ETS recentred the SAT Verbal test, the mean of all test scores was 450.

a) How would adding 50 points to each score affect the mean?

b) The standard deviation was 100 points. What would the standard deviation be after adding 50 points?

c) Suppose we drew boxplots of test takers' scores a year before and a year after the '+50' recentring. How would the boxplots of the two years differ?

3. A company manufactures wheels for in-line skates. The diameter of the wheels has a mean of 3 inches and a standard deviation of 0.1 inches. Because so many of its customers use the metric system, the company decides to report their production statistics in millimetres (1 inch = 25.4 mm). They report that the standard deviation is now 2.54 mm. A corporate executive is worried about this increase in variation. Should he be concerned? Explain.

Non-linear versus linear transformations

Shifting and rescaling are examples of linear transformations of data. Sometimes we use *non-linear transformations*, such as log, square root, etc., to re-express data in a useful manner—perhaps to change the shape to something close to Normal. Linear transformations do not change the shape of the distribution, and measures of centre and spread change in an easily predictable manner. On the other hand, non-linear transformations will change the shape of the distribution, and measures of centre and spread do not change in a predictable manner. You will have to re-compute them after transformation. So, for example, in the earlier example, the mean weight of 80 NHANES participants was 82.36 kg. If we take the logs of all 80 weights, the new distribution of the log-weights will be less skewed to the right, and the mean of this distribution will *not* equal the log of 82.36. Re-compute the mean and standard deviation or other relevant summary measures using the log-transformed data.

A S *Activity:* **Standardizing.** What if we both shift and rescale? The result is so nice that we give it a name.

Back to *z*-scores

Standardizing data into *z*-scores is just shifting them by the mean and rescaling them by the standard deviation. Now we can see how standardizing affects the distribution. When we subtract the mean of the data from every data value, we shift the mean to zero. As we have seen, such a shift doesn't change the standard deviation.

When we *divide* each of these shifted values by s, however, the standard deviation should be divided by s as well. Since the standard deviation was s to start with, the new standard deviation becomes 1.

How, then, does standardizing affect the distribution of a variable? Let's consider the three aspects of a distribution: the shape, centre, and spread.

z-scores have mean 0 and standard deviation 1.

- Standardizing into *z*-scores does not change the *shape* of the distribution of a variable.
- Standardizing into *z*-scores changes the *centre* by making the mean 0.
- Standardizing into *z*-scores changes the *spread* by making the standard deviation 1.

Step-by-Step Example WORKING WITH STANDARDIZED VARIABLES

The School of Graduate Studies (SGS) at the University of Toronto requires non-Canadian applicants to demonstrate facility in the English language if English is not the applicant's primary language nor language of instruction. TOEFL test scores may be used as proof. SGS requires a minimum score of 580 on the paper-based TOEFL test (PBT). In a recent year, the mean and standard deviation for all paper-based TOEFL writers, who were graduate school applicants, were 548 and 57, respectively. An alternative test is the Internet-based TOEFL test (iBT), which had a mean of 82 and standard deviation of 23 for graduate school writers.[2] If SGS wanted to require a similar standing on the Internet-based test for applicants, what minimal score would you recommend that they accept?

THINK ➡ **Plan** State what you want to find out. **Variables** Identify the variables and report the *W*'s (if known).	I want to know what iBT score corresponds approximately to the required PBT score, based on having knowledge only of the mean and standard deviation for both tests for graduate-level test writers.
Check the appropriate conditions.	✓ **Quantitative Data Condition:** Scores for both tests are quantitative where the units are points.

[2]Test statistics from Educational Testing Service (www.ets.org).

SHOW ➡ **Mechanics** Standardize the variables.	The PBT minimal score is 580, which standardized gives a z-score of $$z = \frac{580 - 548}{57} = 0.56$$
The raw score is z standard deviations above the mean.	So a PBT score of 580 is 0.56 standard deviations above the mean of all test takers. For the iBT, 0.56 standard deviations above the mean is $82 + 0.56\,(23) = 94.9$.
TELL ➡ **Conclusion** Interpret your results in context.	I'd recommend accepting an IBT score of 95 in place of the PBT score of 580. [Caution: Identical z-scores does not necessarily mean identical percentile standing—unless the two distributions have similar shapes.]

5.3 Density Curves and the Normal Model

A z-score gives us an indication of how unusual a value is because it tells us how far it is from the mean. If the data value sits right at the mean, it's not very far at all and its z-score is 0. A z-score of 1 tells us the data value is 1 standard deviation above the mean, while a z-score of -1 tells us that the value is 1 standard deviation below the mean. How far from 0 does a z-score have to be to be interesting or unusual? There is no universal standard, but the larger the score (negative or positive), the more unusual it is. We know that 50% of the data lie between the quartiles. For symmetric data, the standard deviation is usually a bit smaller than the IQR, and it's not uncommon for at least half of the data to have z-scores between -1 and 1. But no matter what the shape of the distribution, a z-score of 3 (plus or minus) or more is rare, and a z-score of 6 or 7 shouts out for attention.

To say more about how big we expect a data value or its z-score to be, we need to *model* the data's distribution. A model will let us say much more precisely how often we'd be likely to see data values of different sizes. Of course, like all models of the real world, the model will be wrong—wrong in the sense that it can't match reality exactly. But it can still be useful. Like a physical model, it's something we can look at and manipulate in order to learn more about the real world.

Models help our understanding in many ways. Just as a model of an airplane in a wind tunnel can give insights even though it doesn't show every rivet,[3] models of data give us summaries that we can learn from and use, even though they don't fit each data value exactly. It's important to remember that they're only models of reality and not reality itself. But without models, what we can learn about the world at large is limited to only what we can say about the data we have at hand.

What do we mean by a statistical model? To model the frequency distribution of a quantitative variable, start by imagining a histogram for a variable based on lots of data; for example, think about the weights, in grams, of at-term newborns. If we observed, say, 5000 births, we might get a histogram resembling the one in Figure 5.6a.

We can easily imagine fitting a curve going through the tops of the rectangles. But what if we observed 5000 more births? The heights of the bars would roughly double, changing the fitted curve. We don't want our model to depend on the number of observations. What we are really interested in modelling is the relative frequencies or proportions

[3]In fact, the model is useful *because* it doesn't have every rivet. It is because models offer a simpler view of reality that they are so useful as we try to understand reality.

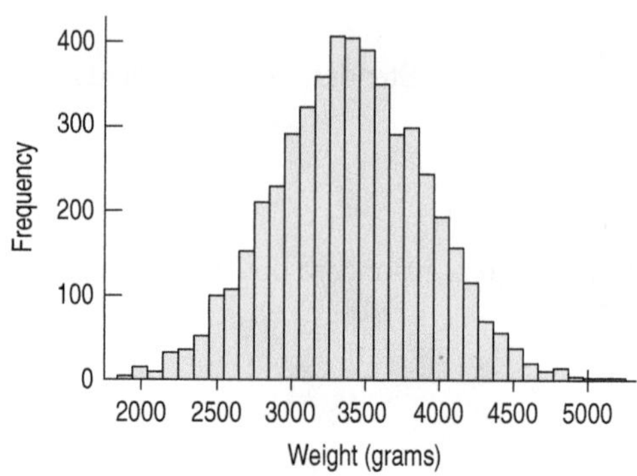

Figure 5.6a
Simulated weights of 5000 newborns

of births in certain ranges. So let's switch to relative frequencies, which removes the dependence on the amount of data. But there is still a problem: If we halve the bin width, the relative frequencies all decrease by about 50%, changing the picture again. How can we stabilize our picture of the distribution? What we'll do is set the *area of a rectangle* equal to the *proportion* or *relative frequency* of individuals falling into that bin:

Area of Rectangle = *Width* × *Height* = *Relative Frequency*

Now solve the above for *Height,* and we get:

Height = *Relative Frequency of Bin* ÷ *Bin Width* (also known as *frequency density)*

The histogram above becomes:

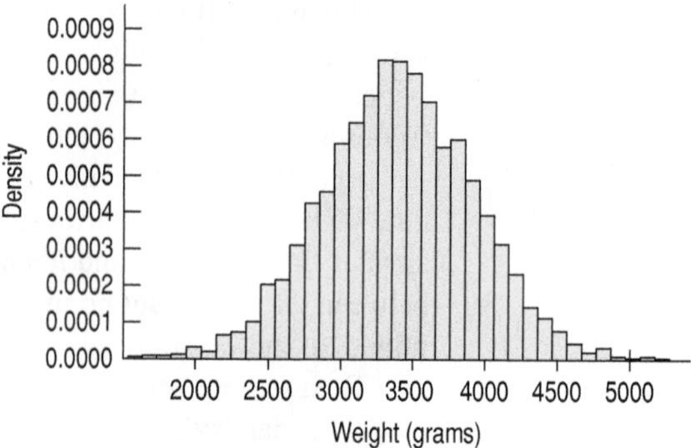

Figure 5.6b
Histogram of newborn weights, with Area = Relative Frequency or Proportion

So area equals relative frequency, and if you add up the relative frequencies or areas of all the rectangles (Figure 5.6b) you have to get 1.00. Decrease or increase the bin width and things change a bit as you trace a path across the tops of the rectangles, but the overall pattern stays about the same and the total area stays put at 1.00. What would happen if you could keep adding data (while gradually decreasing the bin width)? You would get a smoother and smoother shape as you get closer and closer to a true unclouded picture of the underlying distribution.

Now you have to decide what type of smooth curve to fit to the data. Such curves are called **density curves**. In Figure 5.6c, we fitted a *Normal* density curve, which will be discussed in more detail shortly.

Of interest are the proportions of data in specified ranges. If we want to estimate the proportion of at-term newborns weighing between 2450 and 2750 grams, we could add up the areas of the three corresponding rectangles, shown in red in Figure 5.6c, or approximate it by calculating the area under the fitted curve between 2450 and 2750 (Figure 5.6c–second diagram). If we are interested in the proportion weighing between 2500 and 2700, we take the area under the curve between 2500 and 2700, even though they are not cutpoints, or boundaries, of bins. Once we use the smooth curve model, there no longer are any bins or cutpoints; we can cut anywhere on the weight-axis. Finding areas under a curve actually requires the use of calculus, but for the most common models, we can use tables or computer software.

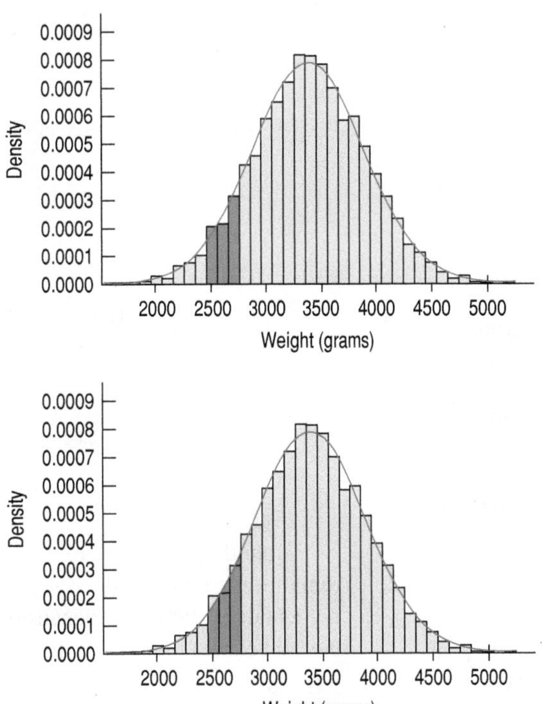

Figure 5.6c
Matching up areas under the histogram and under the density curve

So, if we believe that we can adequately model a distribution with some particular density curve, we use areas under that density curve to approximate percentages of the data falling in ranges of interest. The area under the density curve between any two numbers a and b will represent the proportion of the data that lie between those same two numbers, as shown below (Fig. 5.6d):

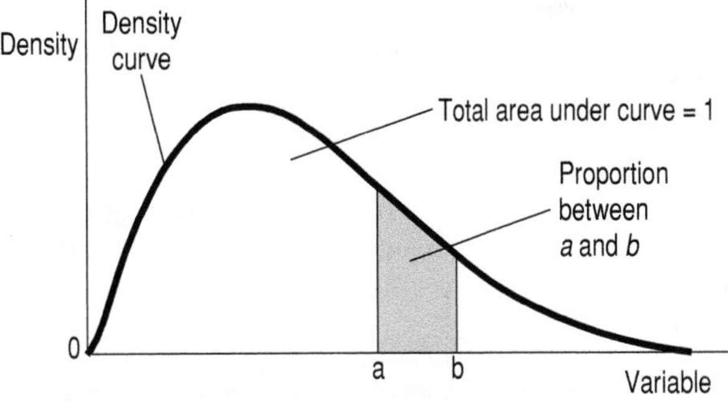

Figure 5.6d
Area under the density curve between *a* and *b* equals the proportion of the data between *a* and *b*.

There are many possible curves that might serve as useful models, but any density curve has to satisfy the following conditions:

1. Always positive or zero (or we'd have negative percentages)
2. Total area under the curve above the x-axis equal to 1.00 (or we would not cover exactly 100% of the observations)

One density curve stands out in particular, due to its central role in statistical theory (which we will delve into further in Chapter 15): the Normal curve. This is a bell-shaped curve (not the only bell-shaped curve used in Statistics), and you could draw it—if you had to—using the following formula:

$$y = \frac{1}{\sigma\sqrt{2\pi}}e^{-\frac{1}{2}\left(\frac{x-\mu}{\sigma}\right)^2} \text{ for } -\infty < x < +\infty$$

You cannot draw this curve until you choose values for the two *parameters* appearing in the equation, μ and σ, which control the centre and dispersion of the distribution, respectively. For example, Figure 5.7 shows two different Normal curves for different choices of these parameters, one modelling a set of rather average-looking blood pressure measurements, the other modelling another set of blood pressure measurements with a higher centre and greater dispersion:

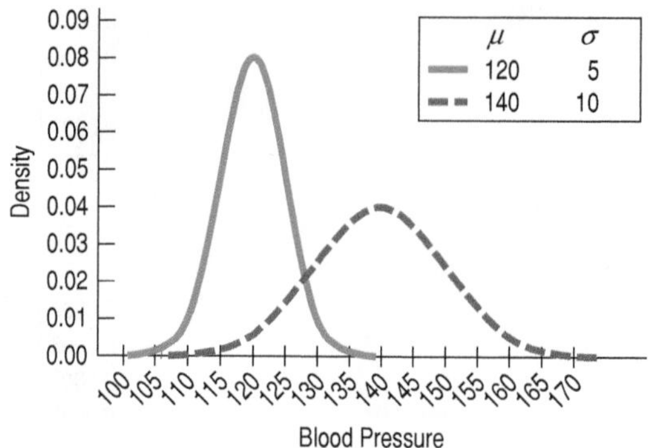

Figure 5.7

Two Normal curves with different choices for parameters.

Is the Normal normal?

Don't be misled. The name "Normal" doesn't mean that these are the *usual* shapes for histograms. The name follows a tradition of positive thinking in Mathematics and Statistics in which functions, equations, and relationships that are easy to work with or have other nice properties are called "normal," "common," "regular," "natural," or similar terms. It's as if by calling them ordinary, we could make them actually occur more often and simplify our lives.

Note that the curve extends indefinitely in both directions, though after a while, the remaining area is close enough to zero to just be ignored. The curve with the bigger σ is not as tall as the other, because the total area has to be maintained at 1.00. In a sense, all Normal curves are essentially the same. For example, just re-label the x-axis with numbers $-4, -3, -2, -1, 0, 1, 2, 3, 4, \ldots$, and re-label the y-axis with numbers 0.00, 0.05, 0.10, . . . 0.45, and you get the Normal curve with $\mu = 0$, $\sigma = 1$, on the left.

The parameters μ and σ in the Normal curve equation are indeed just the mean and standard deviation of an idealized Normal frequency distribution. We write $N(\mu, \sigma)$ to represent a Normal model with a mean of μ and standard deviation of σ. Why the Greek? Well, *this* mean and standard deviation are not numerical summaries of data. They are part of the model—numbers that we choose to help specify the model. Such numbers are called **parameters** of the model.

NOTATION ALERT

$N(\mu, \sigma)$ always denotes a Normal model. The μ, pronounced "mew," is the Greek letter for "m," and always represents the mean in a model. The σ, sigma, is the lowercase Greek letter for "s," and always represents the standard deviation in a model.

We don't want to confuse the model parameters with summaries of the data such as $\bar{y}$ and s, so we use special symbols. In Statistics, we almost always use Greek letters for parameters. By contrast, summaries of data are called **statistics** and are usually written with Latin letters. If you're thinking that statistics calculated from data might be used to estimate (unknown) model parameters—you're quite right! But this big and important topic—called *statistical inference*—will be deferred to later chapters in this text.

If we model data with a Normal model and standardize the values using the corresponding μ and σ, we still call the standardized value a ***z*-score**, and we write

$$z = \frac{y - \mu}{\sigma}$$

Usually it's easier to standardize data first (using its mean and standard deviation). Then we need only the model $N(0, 1)$. The Normal model with mean 0 and standard deviation 1 is called the **standard Normal model** (or the **standard Normal distribution**).

A S ***Activity:* The Normal Model.** Learn more about the Normal model and see what data drawn at random from a Normal model might look like.

But be careful. You shouldn't use a Normal model for just any data set. Remember that standardizing won't change the shape of the distribution. If the distribution is not unimodal and symmetric to begin with, standardizing won't make it Normal.

When we use the Normal model, we assume that the distribution of the data is, well, Normal. Practically speaking, there's no way to check whether this **Normality Assumption** is true. In fact, it is almost certainly not perfectly true. Real data don't behave like mathematical models. Models are idealized; real data are real. The good news, however, is that to use a Normal model, it's usually sufficient to check the following condition:

IS THE STANDARD NORMAL A STANDARD?

Yes. We call it the "standard Normal" because it models standardized values. It is also a "standard" because this is the particular Normal model that we almost always work with.

Nearly Normal Condition. The shape of the data's distribution is unimodal and fairly symmetric. Check this by making a histogram (or a Normal probability plot, which we'll explain later).

Don't model data with a Normal model without checking whether this condition is satisfied. All models make ***assumptions***. Whenever we model—and we'll do that often—we'll be careful to point out the assumptions that we're making. And, what's

> One in a million. These magic 68, 95, 99.7 values come from the Normal model. As a model, it can give us corresponding values for any z-score. For example, it tells us that fewer than 1 out of a million values have z-scores smaller than −5.0 or larger than +5.0. So if someone tells you you're "one in a million," they must really admire your z-score.

even more important, we'll check the associated ***conditions*** in the data to make sure that those assumptions are reasonable.

The 68–95–99.7 Rule for Normal Models

Normal models give us an idea of how extreme a value is by telling us how likely it is to find one that far from the mean. We'll soon show how to find these numbers precisely—but one simple rule is often all we need.

It turns out that in a Normal model, approximately 68% of the values fall within one standard deviation of the mean, approximately 95% of the values fall within two standard deviations of the mean, and approximately 99.7%—almost all—of the values fall within three standard deviations of the mean. These facts are summarized in a rule that we call (let's see . . .) the **68-95-99.7 Rule.**[4]

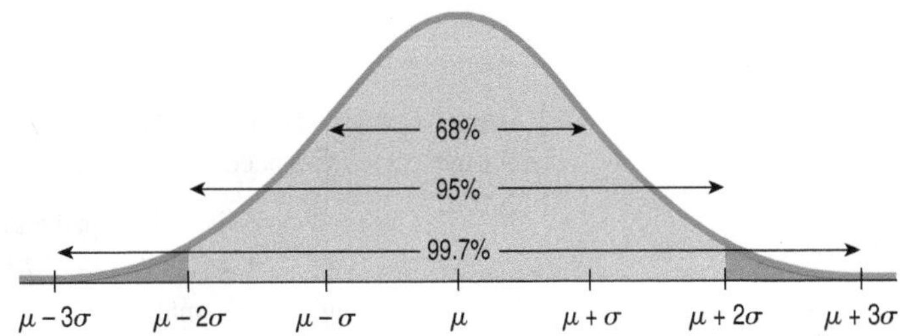

Figure 5.8
Reaching out one, two, and three standard deviations on a Normal model gives the 68–95–99.7 Rule.

For Example THE 68–95–99.7 RULE AND CLARA HUGHES

QUESTION: In the Vancouver 2010 Olympic Games, once again Canada's female speed skaters excelled, winning four medals in the six events (only the Germans won as many). Perhaps Canada's most amazing female athlete, Clara Hughes (two cycling Olympic medals in 1996 summer games and four speed skating medals over three winter games) had a time of 6:55.73 minutes in the 5000 metre race, 4.81 seconds behind the gold medalist Martina Sablikova. Her time was slightly faster than one standard deviation below the mean. Where do you think she finished, out of 15 competitors? Though there were only 15 race times, their distribution was in fact nearly Normal.

ANSWER: From the 68–95–99.7 Rule, we expect 68% of the skaters to be within one standard deviation of the mean. Of the remaining 32%, we expect half on the high end and half on the low end. 16% of 15 is 2.4, so we'd expect about two or possibly three racers to have skated faster than Hughes. Did she win the bronze medal? (Yes!)

Just Checking

4. As a group, the Dutch are among the tallest people in the world. The average Dutch man is 184 cm tall (and the average Dutch woman is 170.8 cm tall). If a Normal model is appropriate and the standard deviation for men is about 8 cm, what percentage of all Dutch men will be over 2 metres tall?
5. Suppose it takes you 20 minutes, on average, to drive to school, with a standard deviation of 2 minutes. Suppose a Normal model is appropriate for the distributions of driving times.
   a) How often will you arrive at school in less than 22 minutes?
   b) How often will it take you more than 24 minutes?
   c) Do you think the distribution of your driving times is unimodal and symmetric?
   d) What does this say about the accuracy of your predictions? Explain.

[4] This rule is just the observation-based "empirical rule" of Chapter 3, made exact for the Normal model. The empirical rule was first published by Abraham De Moivre in 1733, 75 years before the Normal model was discovered.

The First Three Rules for Working with Normal Models

1. Make a picture.
2. Make a picture.
3. Make a picture.

A S *Activity:* **Working with Normal Models.** Well, actually playing with them. This interactive tool lets you do what this chapter's figures can't do: move them!

Although we're thinking about models, not histograms of data, the three rules don't change. To help you think clearly, a simple hand-drawn sketch is all you need. Even experienced statisticians sketch pictures to help them think about Normal models. You should too.

Of course, when we have data, we'll also need to make a histogram to check the Nearly Normal Condition to be sure we can use the Normal model to model the data's distribution. Other times, we may be told that a Normal model is appropriate based on prior knowledge of the situation or on theoretical considerations.

How to sketch a Normal curve that looks Normal To sketch a good Normal curve, you need to remember only three things:

- The Normal curve is bell-shaped and symmetric around its mean. Start at the middle, and sketch to the right and left from there.
- Even though the Normal model extends forever on either side, you need to draw it only for 3 standard deviations. After that, there's so little left that it isn't worth sketching.
- The place where the bell shape changes from curving downward to curving back up—the *inflection point*—is located exactly one standard deviation away from the mean.

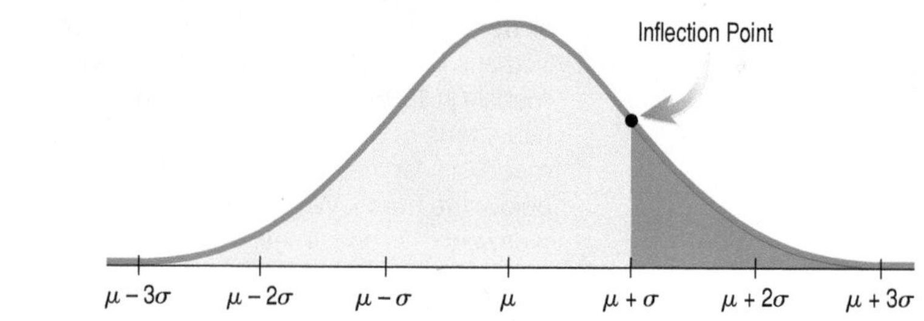

Step-by-Step Example WORKING WITH THE 68–95–99.7 RULE

TOEFL scores had a distribution that was roughly unimodal and symmetric with an overall mean of about 540 and a standard deviation of 60 for all test takers, including graduate students, undergraduates, and professionals.

Question: Suppose you earned a 600 on this TOEFL exam. Where do you stand among all students who took that test?

You could calculate your z-score and find out that it is $z = (600 - 540)/60 = 1.0$, but what does that tell you about your percentile? You'll need the Normal model and the 68–95–99.7 Rule to answer that question.

THINK ➡ Plan State what you want to know.	I want to see how my score compares with all other students. To do that, I'll need to model the distribution.
Variables Identify the variable and report the *W*'s.	Let y = my TOEFL score. Scores are quantitative but have no meaningful units other than points.
Be sure to check the appropriate conditions.	✓ **Nearly Normal Condition:** If I had data, I would check the histogram. I have no data, but I am told that the scores are roughly unimodal and symmetric.
Specify the parameters of your model.	I will model TOEFL score with a $N(540,60)$ model.

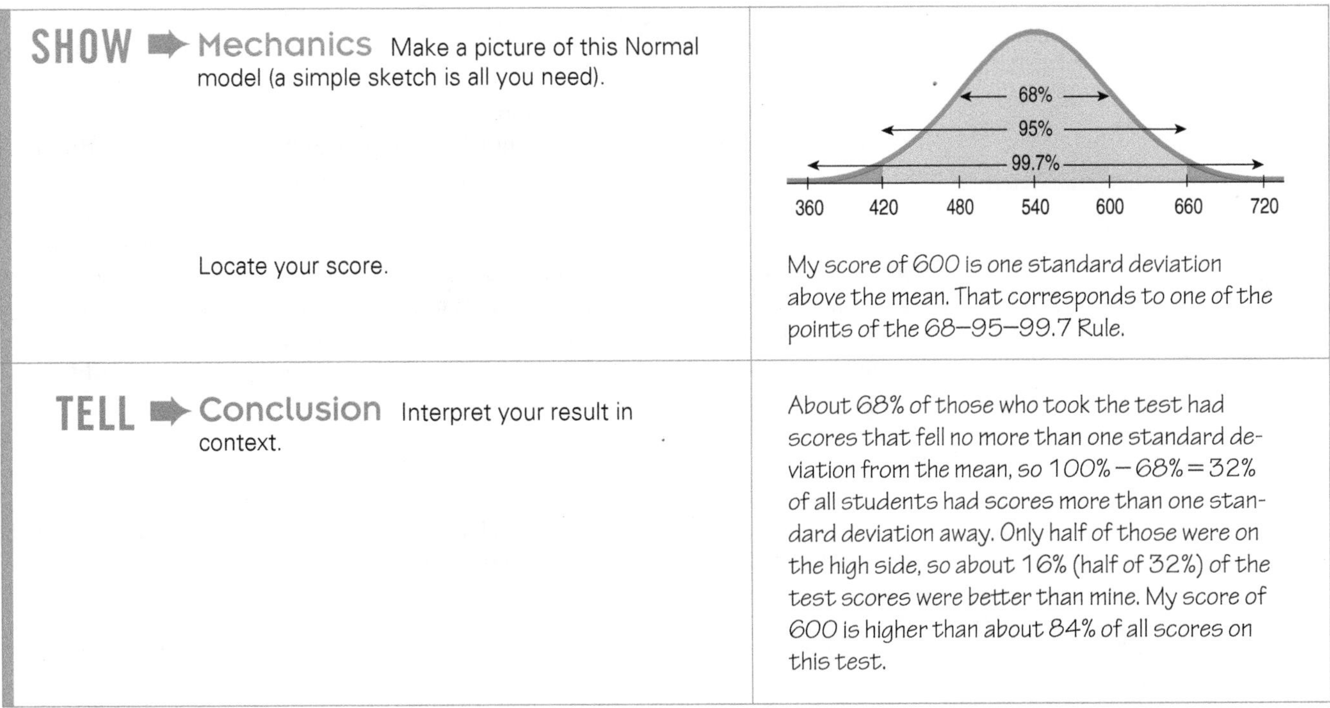

SHOW ➡ Mechanics Make a picture of this Normal model (a simple sketch is all you need).

Locate your score.

My score of 600 is one standard deviation above the mean. That corresponds to one of the points of the 68–95–99.7 Rule.

TELL ➡ Conclusion Interpret your result in context.

About 68% of those who took the test had scores that fell no more than one standard deviation from the mean, so 100% − 68% = 32% of all students had scores more than one standard deviation away. Only half of those were on the high side, so about 16% (half of 32%) of the test scores were better than mine. My score of 600 is higher than about 84% of all scores on this test.

The upper bound of TOEFL scoring was actually 677, a score that sits between two and three standard deviations above the mean of 540. Hence the upper tail of the distribution is slightly compressed as compared with a perfect Normal. If we extended the scoring a bit, we'd only get more information about less than (100 − 95)/2 = 2.5% of students—students who clearly have superlative English skills, so it would hardly pay.

5.4 Finding Normal Percentiles

A TOEFL score of 600 is easy to assess, because we can think of it as one standard deviation above the mean. If your score was 648, though, where do you stand among the rest of the people tested? Your z-score is 1.80, so you're somewhere between one and two standard deviations above the mean. We figured out that no more than 16% of people scored better than 600. By the same logic, no more than 2.5% of people scored better than 660. Can we be more specific than "between 16% and 2.5%"?

When the value doesn't fall exactly one, two, or three standard deviations from the mean, we can look it up in a table of **Normal percentiles** or use technology.[5] We convert our data to z-scores before using the table. Your TOEFL score of 648 has a z-score of (648 − 540)/60 = 1.80.

> **The Worst Case**
>
> Our 68–95–99.7 Rule applies to Normal models. What if we're dealing with a distribution that's strongly skewed, or bimodal? *Chebyshev's Rule says: In any distribution, at least* $1 - 1/k^2$ *of the values must lie within k standard deviations of the mean.* For example, if $k = 2$, $1 - 1/2^2 = 3/4$; no matter what the shape of the distribution, *at least 75%* of the values must be within 2 standard deviations of the mean. For $k = 3$, $1 - 1/3^2 = 8/9$, so at *least 89%* of the values must lie within 3 standard deviations of the mean. However, it's rare in practice for data to come closer to the Chebyshev extremes than to the 68–95–99.7 percents.

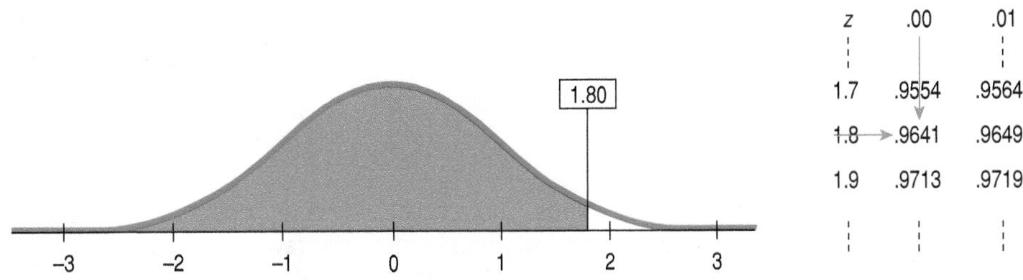

Figure 5.9

A table of Normal percentiles (Table Z in Appendix C) lets us find the percentage of individuals in a standard Normal distribution falling below any specified z-score value.

[5]See Table Z in Appendix C. Many calculators and statistics computer packages do this, too.

A S *Activity:* **The Normal Table.** Table Z just sits there, but this version of the Normal table changes so it always *makes a picture* that fits.

A S *Activity:* **Your Pulse *z*-Score.** Is your pulse rate high or low? Find its *z*-score with the Normal Model Tool.

A S *Activity:* **Normal Models.** Normal models have several interesting properties—see them here.

In the piece of the table shown (Figure 5.9), you can find the *z*-score by looking down the left column for the first two digits, 1.8, and across the top row for the third digit, 0. The table gives the percentile as 0.9641. That means that 96.4% of the *z*-scores are less than 1.80. Only 3.6% of people, then, scored better than 648 on this TOEFL.

These days, finding percentiles from a Normal probability table is a "desert island" method—something we might do if we desperately needed a Normal percentile and were stranded miles from the mainland with only a Normal probability table. (Of course, you might feel just that way during a Statistics exam, so it's a good idea to know how to do it.) Fortunately, most of the time, we can use a calculator, a computer, a smart phone or tablet app, or the Internet. Most technology tools that find Normal percentiles offer to draw the picture as well. With modern technology tools, it's easy to answer a question like what percentage of test takers scored between 550 and 650 on the paper-based TOEFL, once you specify the precise Normal model.

Other Models

Of course, the Normal is not the only model for data. There are models for skewed data and we'll see models for variables that can take on only a few values (Binomial and Poisson, to name two). But the Normal will return later on in an important and surprising way.

Step-by-Step Example PRACTICE—WORKING WITH NORMAL MODELS PART I

The Normal model is our first model for data. It's the first in a series of modelling situations where we step away from the data at hand to make more general statements about the world. We'll become more practised in thinking about and learning the details of models as we progress through the book. To give you some practice in thinking about the Normal model, here are several problems that ask you to find percentiles in detail.

Question: What proportion of the TOEFL scores fell between 510 and 600?

THINK ➡ Plan State the problem.

I want to know the proportion of TOEFL scores between 510 and 600.

Variables Name the variable.

Let y = TOEFL score.

Check the appropriate conditions and specify which Normal model to use.

✓ **Nearly Normal Condition:** We are told that TOEFL scores are nearly Normal.

I'll model TOEFL scores with a $N(540,60)$ model, using the mean and standard deviation specified for them.

SHOW ➡ Mechanics Make a picture of this Normal model. Locate the desired values and shade the region of interest.

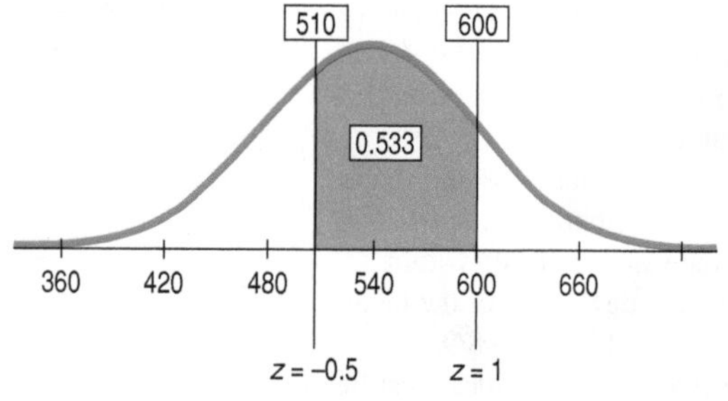

Locate the desired values. The easiest way is to use a technology that allows you to input the raw data values directly, after specifying the model and parameters.

Standardizing the two scores, I find

$$z = \frac{(y - \mu)}{\sigma} = \frac{(600 - 540)}{60} = 1.00$$

and

$$z = \frac{(510 - 540)}{60} = -0.50$$

Alternatively, you can convert 510 and 600 to *z*-scores before using technology or a table to find the two areas. Then you need to subtract to find the area *between* the two values.	The area under the N(540, 60) curve between 510 and 600 is the same as the area under the N(0, 1) curve between these two *z*-scores. From Table Z, the *Area* $(z < 1.0) = 0.8413$ and *Area* $(z < -0.5) = 0.3085$, so the proportion of *z*-scores *between* them is $0.8413 - 0.3085 = 0.5328$, or 53.28%.
TELL ➡ **Conclusion** Interpret your result in context.	The Normal model estimates that about 53.3% of TOEFL scores fell between 510 and 600.

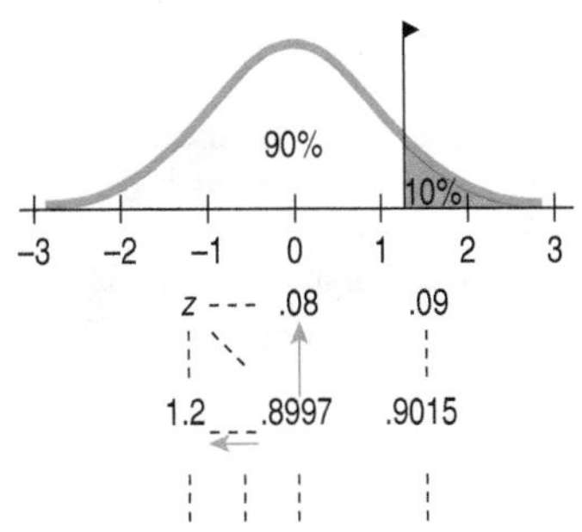

From Percentiles to Scores

Finding areas from *z*-scores is the simplest way to work with the Normal model. But sometimes we start with areas and are asked to work backward to find the corresponding *z*-score or even the original data value. For instance, what *z*-score cuts off the top 10% in a Normal model?

Make a picture like the one shown, shading the rightmost 10% of the area. Notice that this is the 90th percentile. Look in Table Z for an area of 0.900. The exact area is not there, but 0.8997 is pretty close. That shows up in the table with 1.2 in the left margin and .08 in the top margin. The *z*-score for the 90th percentile, then, is approximately $z = 1.28$.

Of course, computers and calculators will determine the cut point more precisely (and more easily).

Step-by-Step Example WORKING WITH NORMAL MODELS PART II

Question: Shuang is hoping to score among the top 10% on the TOEFL. How high a score does she need?

THINK ➡ **Plan** State the problem.	How high a TOEFL score does Shuang need to be in the top 10% of all test takers?
Variable Define the variable.	Let y = Shuang's score.
Check to see if a Normal model is appropriate, and specify which Normal model to use.	✓ **Nearly Normal Condition:** I am told that TOEFL scores are nearly Normal. I'll model them with *N*(540,60).
SHOW ➡ **Mechanics** Make a picture of this Normal model. Locate the desired percentile approximately by shading the rightmost 10% of the area. The cutoff score is the 90th percentile in order for Shuang to make the top 10%. Find the corresponding *z*-score using Table Z as shown earlier.	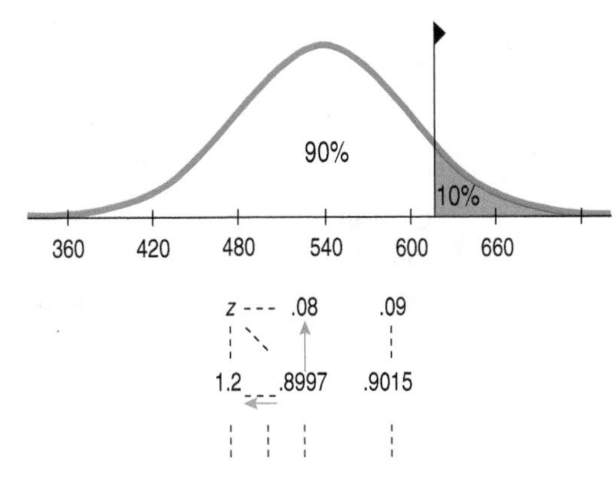 The cut point, when standardized, is $z = 1.28$.

Convert the *z*-score back to the original units.	A *z*-score of 1.28 is 1.28 standard deviations above the mean. Since the SD is 60, that's 76.8 points. The cutoff is 77 points above the mean of 540, or 617.
TELL ➡ **Conclusion** Interpret your results in the proper context.	To place in the top 10%, Shuang needs a score of 617.

Step-by-Step Example MORE WORKING WITH NORMAL MODELS

Working with Normal percentiles can be a little tricky, depending on how the problem is stated. Here are a few more worked examples of the kind you're likely to see.

A cereal manufacturer has a machine that fills the boxes. Boxes are labelled "16 ounces," so the company wants to have that much cereal in each box, but since no packaging process is perfect, there will be minor variations. If the machine is set at exactly 16 ounces and the Normal model applies (or at least the distribution is roughly symmetric), then about half of the boxes will be underweight, making consumers unhappy and exposing the company to bad publicity and possible lawsuits. To prevent underweight boxes, the manufacturer has to set the mean a little higher than 16.0 ounces.

Based on their experience with the packaging machine, the company believes that the amount of cereal in the boxes fits a Normal model with a standard deviation of 0.2 ounces. The manufacturer decides to set the machine to put an average of 16.3 ounces in each box. Let's use that model to answer a series of questions about these cereal boxes.

Question 1: What fraction of the boxes will be underweight?

THINK ➡ **Plan** State the problem.	What proportion of boxes weigh less than 16 ounces?
Variable Name the variable.	Let y = weight of cereal in a box.
Check to see if a Normal model is appropriate.	✓ **Nearly Normal Condition:** I have no data, so I cannot make a histogram, but I am told that the company believes the distribution of weights from the machine is Normal.
Specify which Normal model to use.	I'll use a $N(16.3, 0.2)$ model.
SHOW ➡ **Mechanics** Make a picture of this Normal model. Locate the value you're interested in on the picture, label it, and shade the appropriate region. **REALITY CHECK** Estimate from the picture the percentage of boxes that are underweight. (This will be useful later to check that your answer makes sense.) It looks like a low percentage. Less than 20%, for sure.	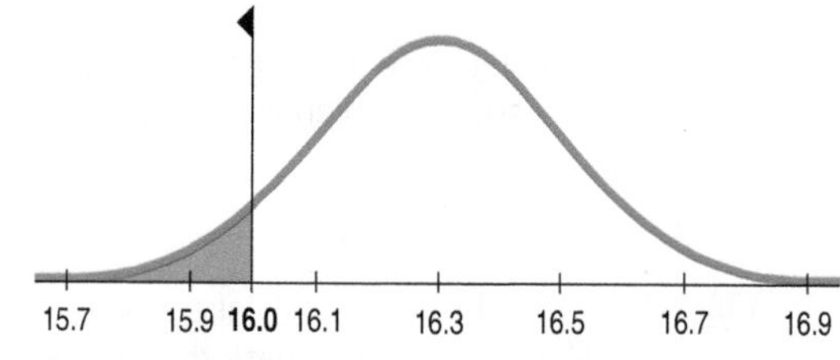 I want to know what fraction of the boxes will weigh less than 16 ounces.
Convert your cutoff value into a *z*-score if using the table of standard Normal areas.	$z = \frac{y - \mu}{\sigma} = \frac{16 - 16.3}{0.2} = -1.50$
Find the area with your software or use the Normal table.	Area $(y < 16)$ = Area $(z < -1.50) = 0.0668$
TELL ➡ **Conclusion** State your conclusion, and check that it's consistent with your earlier guess. It's below 20%—seems okay.	I estimate that approximately 6.7% of the boxes will contain less than 16 ounces of cereal.

Question 2: The company's lawyers say that 6.7% is too high. They insist that no more than 4% of the boxes can be underweight. So the company needs to set the machine to put a little more cereal in each box. What mean setting do they need?

THINK ➡ Plan State the problem.

What mean weight will reduce the proportion of underweight boxes to 4%?

Variable Name the variable.

Let y = weight of cereal in a box.

Check to see if a Normal model is appropriate.

✓ **Nearly Normal Condition:** *I am told that a Normal model applies.*

Specify which Normal model to use. This time you are not given a value for the mean!

I don't know μ, the mean amount of cereal. The standard deviation for this machine is 0.2 ounces. The model is $N(\mu, 0.2)$.

REALITY CHECK We found out earlier that setting the machine to $\mu = 16.3$ ounces made 6.7% of the boxes too light. We'll need to raise the mean a bit to reduce this fraction.

No more than 4% of the boxes can be below 16 ounces.

SHOW ➡ Mechanics Make a picture of this Normal model. Centre is at μ (since you don't know the mean). Shade the region below 16 ounces.

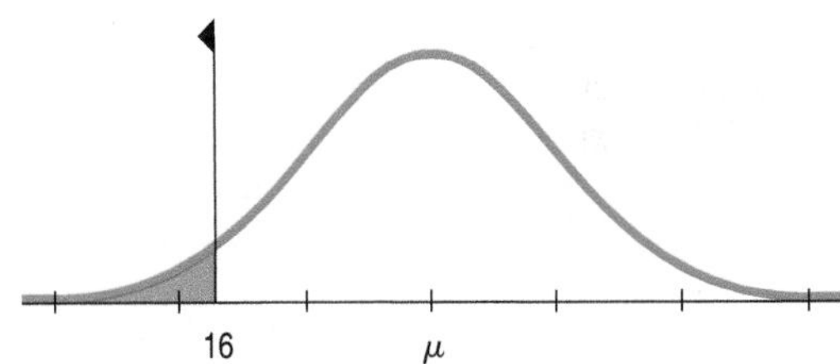

Using software or the Normal table, find the z-score that cuts off the lowest 4%.

The z-score that has a 0.04 area to the left of it is $z = -1.75$.

Use this information to find μ. It's located 1.75 standard deviations to the right of 16. Since σ is 0.2, that's 1.75×0.2, or 0.35 ounces more than 16. (Or set $(16 - \mu)/0.2 = -1.75$ and solve)

For 16 to be 1.75 standard deviations below the mean, the mean must be

$$16 + 1.75(0.2) = 16.35 \text{ ounces.}$$

TELL ➡ Conclusion Interpret your result in context. This makes sense; we knew it would have to be just a bit higher than 16.3.

The company must set the machine to average 16.35 ounces of cereal per box.

Question 3: The company president vetoes that plan, saying the company should give away less free cereal, not more. Her goal is to set the machine no higher than 16.2 ounces and still have only 4% underweight boxes. The only way to accomplish this is to reduce the standard deviation. What standard deviation must the company achieve, and what does that mean about the machine?

THINK ➡ Plan State the problem.

What standard deviation will allow the mean to be 16.2 ounces and still have only 4% of boxes underweight?

Variable Name the variable.

Let y = weight of cereal in a box.

Check conditions to be sure that a Normal model is appropriate.

✓ **Nearly Normal Condition:** *The company believes that the weights are described by a Normal model.*

Specify which Normal model to use. This time you don't know σ.

I know the mean, but not the standard deviation, so my model is $N(16.2, \sigma)$.

REALITY CHECK We know the new standard deviation must be less than 0.2 ounces.

SHOW ➡ **Mechanics** Make a picture of this Normal model. Centre it at 16.2, and shade the area you're interested in. We want 4% of the area to the left of 16 ounces.

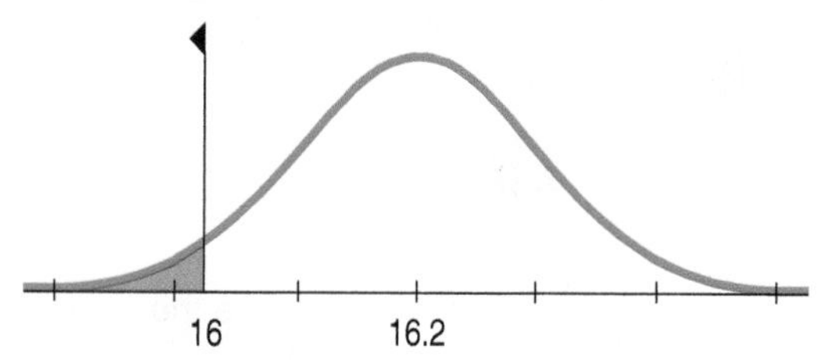

Find the z-score that cuts off the lowest 4%.

I know that the z-score with 4% below it is $z = -1.75$.

We need 16 standardized to equal −1.75. Write this down and solve for σ.

$$z = \frac{y - \mu}{\sigma}$$

$$-1.75 = \frac{16 - 16.2}{\sigma}$$

$$1.75\sigma = 0.2$$

$$\sigma = 0.114$$

TELL ➡ **Conclusion** Interpret your result in context.

As we expected, the standard deviation is lower than before—actually, quite a bit lower.

The company must get the machine to box cereal with a standard deviation of only 0.114 ounces. This means the machine must be more consistent (by nearly a factor of two) in filling the boxes.

5.5 Normal Probability Plots

In the examples we've worked through, we've assumed that the underlying data distribution was roughly unimodal and symmetric, so using a Normal model makes sense. When you have data, you must *check* to see whether a Normal model is reasonable. How? Make a picture, of course! Drawing a histogram of the data and looking at the shape is one good way to see if a Normal model might work.

There's a more specialized graphical display that can help you to decide whether the Normal model is appropriate: the **Normal probability plot**.[6] If the distribution of the data is roughly Normal, the plot is roughly a diagonal straight line. Systematic deviations from a straight line indicate that the distribution is not Normal. This plot is usually able to show deviations from Normality more clearly than the corresponding histogram, but it's usually easier to understand how a distribution fails to be Normal by looking at its histogram.

Some data on a car's fuel efficiency provide an example of data that are nearly Normal. The overall pattern of the Normal probability plot (Figure 5.10 on the next page) is straight. The two trailing low values correspond to the values in the histogram that trail off the low end. They're not quite in line with the rest of the data set. The Normal probability plot shows us that they're a bit lower than we'd expect of the lowest two values in a Normal model.

By contrast, the Normal probability plot of the men's *weights* from the NHANES Study (Figure 5.11) is far from straight. The weights are skewed to the high end, and the plot is curved. We'd conclude from these pictures that approximations using the 68-95-99.7 Rule for these data would not be very accurate.

[6] Also called a *Normal quantile plot.*

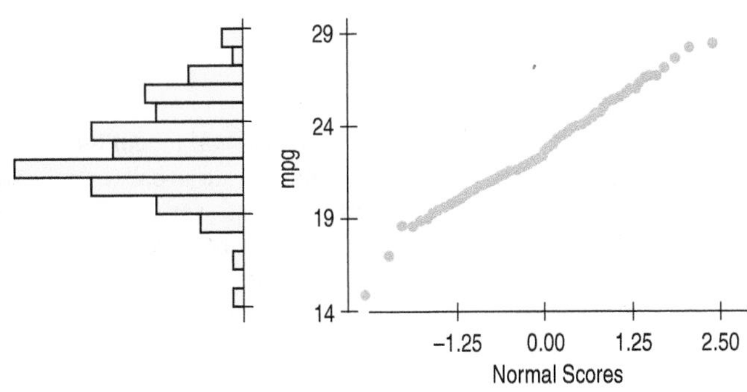

Figure 5.10

Histogram and Normal probability plot for gas mileage (mpg or miles per gallon) recorded by one of the authors over the eight years he owned a Nissan Maxima. The vertical axes are the same, so each dot on the probability plot would fall into the bar on the histogram immediately to its left. On the other hand, if you project the dots straight down onto the horizontal axis, and make a histogram, it will look perfectly Normal.

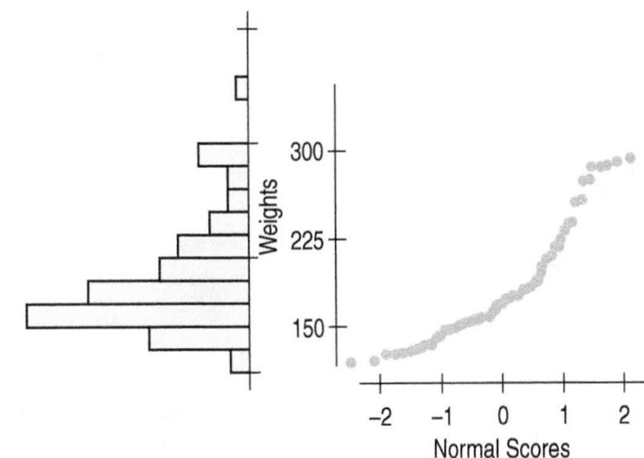

Figure 5.11

Histogram and Normal probability plot for men's weights. Note how a skewed distribution corresponds to a bent probability plot.

How Does a Normal Probability Plot Work?

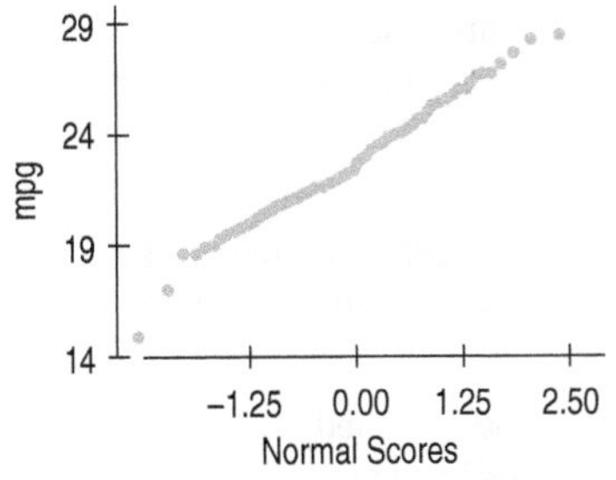

Why does the Normal probability plot work like that? We looked at 100 fuel efficiency measures for the author's Nissan car. The smallest of these has a z-score of -3.16. The Normal model can tell us what value to expect for the smallest z-score in a batch of 100 if a Normal model were appropriate. That turns out to be -2.58. So our first data value is smaller than we would expect from the Normal.

We can continue this and ask a similar question for each value. For example, the fourteenth smallest fuel efficiency has a z-score of almost exactly -1, and that's just what we should expect (well, -1.08 to be exact). A Normal probability plot takes each data value and plots it against the z-score you'd expect that point to have if the distribution were perfectly Normal.[7]

A S ***Activity:* Assessing Normality.** This activity guides you through the process of checking the nearly Normal condition using your statistics package.

When the values match up well, the line is straight. If one or two points are surprising from the Normal's point of view, they don't line up. When the entire distribution is skewed or different from the Normal in some other way, the values don't match up very well at all and the plot will bend or show some type of curvature.

It turns out to be tricky to find the values we expect. They're called *Normal scores,* but you can't easily look them up in the tables. That's why probability plots are best made with technology and not by hand.

The best advice on using Normal probability plots is to see whether they are essentially straight, aside from very random-looking squiggles. If so, then your data look like data from a Normal model. If not—and particularly if you see clear curving patterns or points way out of line with the rest—make a histogram to understand how they differ from the model.

---

[7]A Normal probability plot is also often drawn with the two axes switched, putting the data on the x-axis and the standard Normal scores on the y-axis.

WHAT CAN GO WRONG?

- **Don't use a Normal model when the distribution is not unimodal and fairly symmetric.** Normal models are so easy and useful that it is tempting to use them even when they don't describe the data very well. That can lead to wrong conclusions. Don't use a Normal model without first checking the **Nearly Normal Condition.** Look at a picture of the data to check that it is unimodal and symmetric. A histogram, or a Normal probability plot, can help you tell whether a Normal model is appropriate.

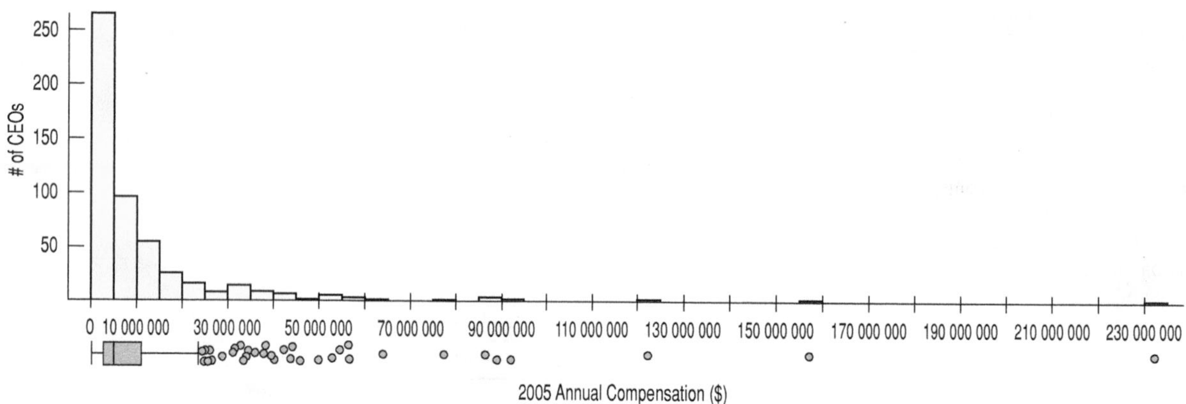

Annual compensations for Fortune 500 CEOs in 2005 are displayed in the histogram above. In that year, they had a mean total compensation of $10 307 311.87 with a standard deviation of $17 964 615.16. Using the Normal model rule, we should expect about 68% of the CEOs to have had compensations between −$7 657 303.29 and $28 271 927.03. In fact, more than 90% of the CEOs had annual compensations in this range. What went wrong? The distribution is extremely skewed, not symmetric. Using the 68–95–99.7 Rule for data like these will lead to silly results.

- **Don't use the mean and standard deviation when outliers are present.** Both means and standard deviations can be distorted by outliers, and no model based on distorted values will do a good job. A z-score calculated from a distribution with outliers may be misleading. It's always a good idea to check for outliers. How? Make a picture.

- **Don't round your results in the middle of a calculation.** We *reported* the mean of the heptathletes' 200 m run as 24.71 seconds. More precisely, it was 24.70894736842105 seconds.

  You should use all the precision available in the data for all the intermediate steps of a calculation. Using the more precise value for the mean (and also carrying 15 digits for the SD), the z-score calculation for Fountain's run comes out to

$$z = \frac{23.21 - 24.70894736842105}{0.7002346690571983} = -2.140635753495674$$

  We'd report that as −2.141, as opposed to the rougher rounded-off value of 2.14 we got earlier from the table. So, . . .

- **Do what we say, not what we do.** When we showed the z-score calculations for Fountain, we rounded the mean to 24.7 seconds and the SD to 0.70 seconds. Then to make the story clearer we used *those values* in the displayed calculation.

  We'll continue to show simplified calculations in the book to make the story simpler. When you calculate with full precision, your results may differ slightly from ours. So, we also advise . . .

- **Don't worry about minor differences in results.** Because various calculators and programs may carry different precision in calculations, your answers may differ slightly from those we show in the text and Step-By-Steps, or even from the values given in the answers in the back of the book. Those differences aren't anything to worry about. They're not the main story Statistics tries to tell.

CONNECTIONS

Changing the centre and spread of a variable is similar to changing its *units*. All other aspects of the context do not depend on the choice or modification of measurement units, which can make the numbers easier to work with, but does not alter the meaning.

Standardizing shifts and rescales to produce a mean of 0 and standard deviation of 1, but does not affect the *shape* of a distribution. A histogram or boxplot of standardized values looks just the same as the histogram or boxplot of the original values, except, perhaps, for the numbers on the axes.

When we summarized *shape, centre,* and *spread* for histograms, we often compared them to unimodal, symmetric shapes. You couldn't ask for a nicer example than the Normal model. And if the shape *is* like a Normal, we can use the mean and standard deviation to standardize the values and reduce any Normal model to the standard Normal.

What Have We Learned?

Learning Objectives

Understand how *z*-scores facilitate comparisons by standardizing variables to have zero mean and unit standard deviation. Recognize normally distributed data by making a histogram and checking whether it is unimodal, symmetric, and bell-shaped, or by making a Normal probability plot using technology and checking whether the plot is roughly a straight line.

- The Normal model is a distribution that will be important for much of the rest of this course.
- Before using a Normal model, check that the data are plausibly from a normally distributed population.
- A Normal probability plot provides evidence that the data are Normally distributed if it is roughly linear.

Understand how to use the Normal model to judge whether a value is extreme.

- Standardize values to make *z*-scores and obtain a standard scale. Then refer to a standard Normal distribution.
- Use the 68-95-99.7 Rule as a rule-of-thumb to judge whether a value is extreme.

Know how to refer to tables or technology to find the probability of a value randomly selected from a Normal model falling in any interval.

- Know how to perform calculations about Normally distributed values and probabilities.

Review of Terms

Standardizing We standardize to eliminate units. Standardized values can be compared and combined even if the original variables had different units and magnitudes (p. 116).

Standardized value A value found by subtracting the mean and dividing by the standard deviation (p. 116).

***z*-score** A *z*-score tells how many standard deviations a value is from the mean; *z*-scores have a mean of zero and a standard deviation of one. When working with data, use the statistics $\bar{y}$ and s:

$$z = \frac{y - \bar{y}}{s}$$

When working with models, use the parameters μ and σ:

$$z = \frac{y - \mu}{\sigma} \text{ (p. 116).}$$

Shifting Adding a constant to each data value adds the same constant to the mean, the median, and the quartiles, but does not change the standard deviation or IQR (p. 118).

Rescaling Multiplying each data value by a constant multiplies both the measures of position (mean, median, and quartiles) and the measures of spread (standard deviation and IQR) by that constant (p. 118).

Density curve A model for the frequency distribution of data using areas under the curve to represent relative frequencies (p. 122).

Parameter A numerically valued attribute of a model. For example, the values of μ and σ in a $N(\mu, \sigma)$ model are parameters (p. 124).

Statistic A value calculated from data to summarize aspects of the data. For example, the mean, ($\bar{y}$), and standard deviation, s, are statistics (p. 124).

Normal model A useful family of models for unimodal, symmetric distributions (p. 124).

Standard Normal model A Normal model, $N(\mu,\sigma)$, with mean $\mu = 0$ and standard deviation $\sigma = 1$. Also called the **standard Normal distribution** (p. 124).

68–95–99.7 Rule In a Normal model, approximately 68% of values fall within one standard deviation of the mean, approximately 95% fall within two standard deviations of the mean, and approximately 99.7% fall within three standard deviations of the mean (p. 125).

Nearly Normal Condition A distribution is nearly Normal if it is unimodal and fairly symmetric. Check by looking at a histogram or a Normal probability plot (p. 124).

Normal percentile The Normal percentile corresponding to a *z*-score gives the percentage of values in a standard Normal distribution found at that *z*-score or below (p. 127).

Normal probability plot A display to help assess whether a distribution of data is approximately Normal. If the plot is nearly straight, the data satisfy the **Nearly Normal Condition** (p. 132).

On the Computer NORMAL PROBABILITY PLOTS

The best way to tell whether your data can be modelled well by a Normal model is to make a picture or two. We've already talked about making histograms. Normal probability plots are almost never made by hand because the values of the Normal scores are tricky to find. But most statistics software make Normal plots, though various packages call the same plot by different names and array the information differently.

EXCEL

Excel offers a "Normal probability plot" as part of the Regression command in the Data Analysis extension, but (as of this writing) it is not a correct Normal probability plot and should not be used.

JMP

To make a "Normal Quantile Plot" in JMP:

- Make a histogram using **Distributions** from the **Analyze** menu.
- Click on the drop-down menu next to the variable name.
- Choose **Normal Quantile Plot** from the drop-down menu.
- JMP opens the plot next to the histogram.

COMMENTS

JMP places the ordered data on the vertical axis and the Normal scores on the horizontal axis. The vertical axis aligns with the histogram's axis, a useful feature.

MINITAB

To make a "Normal Probability Plot" in MINITAB:

- Choose **Probability Plot** from the **Graph** menu.
- Select "Single" for the type of plot. Click **OK**.
- Enter the name of the variable in the "Graph variables" box. Click **OK**.

COMMENTS

MINITAB places the ordered data on the horizontal axis and the Normal scores on the vertical axis.

R

To make a Normal probability (Q-Q) plot for x:

- **qqnorm(X)** will produce the plot.

To standardize a variable *x*:

- z=(X–mean (X))/sd(X) will create a standardized variable *Z*.

COMMENTS

By default, R places the ordered data on the vertical axis and the Normal scores on the horizontal axis, but that can be reversed by setting **datax = TRUE** inside qqnorm.

SPSS

To make a Normal "P-P plot" in SPSS:

- Choose **P-P** from the **Graphs** menu.
- Select the variable to be displayed in the source list.
- Click the arrow button to move the variable into the target list.
- Click the **OK** button.

COMMENTS

SPSS places the ordered data on the horizontal axis and the Normal scores on the vertical axis. You may safely ignore the options in the P-P dialog.

STATCRUNCH

To make a Normal probability plot:

- Click on **Graphics**.
- Choose **QQ Plot.**
- Choose the variable name from the list of **Columns**.
- Click on **Create Graph**.

To work with Normal percentiles:

- Click on **Stat**.
- Choose **Calculators » Normal**.
- Choose a lower tail (≤) or upper tail (≥) region.
- Enter the *z*-score cutoff, and then click on **Compute** to find the probability.

  OR

  Enter the desired probability, and then click on **Compute** to find the *z*-score cutoff.

TI-83/84 PLUS

To create a "Normal Percentile Plot" on the TI-83:

- Set up a **STAT PLOT** using the last of the Types.
- Specify your datalist and the axis you choose to represent the data.
- Although most people wouldn't open a statistics package just to find a Normal model value they could find in a table, you *would* use a calculator for that function.
- So, to find what percent of a Normal model lies between two *z*-scores, choose **normalcdf** from the **DISTRibutions** menu and enter the command **normalcdf(zLeft, zRight).**
- To find the *z*-score that corresponds to a given percentile in a Normal model, choose **invNorm** from the **DISTRibutions** menu and enter the command **invNorm(percentile).**

COMMENTS

We often want to find Normal percentages from a certain *z*-score to infinity. On the calculator, indicate "infinity" as a very large *z*-score, say, 99. For example, the percentage of a Normal model over 2 standard deviations above the mean can be evaluated with **normalcdf(2, 99).**

To make a Normal Probability plot:

- Turn a STATPLOT On.
- Tell it to make a Normal probability plot by choosing the last of the icons.
- Specify your datalist and which axis you want the data on. (Use Y to make the plot look like those here.)
- Specify the Mark you want the plot to use.
- Now ZoomStat does the rest.

Exercises

1. **Payroll** Here are the summary statistics for the weekly payroll of a small company: lowest salary = \$300, mean salary = \$700, median = \$500, range = \$1200, IQR = \$600, first quartile = \$350, standard deviation = \$400.
   a) Do you think the distribution of salaries is symmetric, skewed to the left, or skewed to the right? Explain why.
   b) Between what two values are the middle 50% of the salaries found?
   c) Suppose business has been good and the company gives every employee a \$50 raise. Tell the new value of each of the summary statistics.
   d) Instead, suppose the company gives each employee a 10% raise. Tell the new value of each of the summary statistics.

2. **Hams** A specialty foods company sells gourmet hams by mail order. The hams vary in size from 4.15 to 7.45 pounds, with a mean weight of 6 pounds and standard deviation of 0.65 pounds. The quartiles and median weights are 5.6, 6.2, and 6.55 pounds.
   a) Find the range and the IQR of the weights.
   b) Do you think the distribution of the weights is symmetric or skewed? If skewed, which way? Why?
   c) If these weights were expressed in kilograms (1 kg = 2.2 pounds) what would the mean, standard deviation, quartiles, median, IQR, and range be?
   d) When the company ships these hams, the box and packing materials add 900 grams. What are the mean, standard deviation, quartiles, median, IQR, and range of weights of boxes shipped (in kg)?
   e) One customer made a special order of a 5 kg ham. Which of the summary statistics of part d might *not* change if that data value were added to the distribution?

3. **SAT or ACT?** Each year thousands of high-school students in the U.S. take either the SAT or the ACT, standardized tests used in the college admissions process. Combined SAT Math and Verbal scores go as high as 1600, while the maximum ACT composite score is 36. Since the two exams use very different scales, comparisons of performance are difficult. A convenient rule of thumb is $SAT = 40 \times ACT + 150$; that is, multiply an ACT score by 40 and add 150 points to estimate the equivalent SAT score. An admissions officer reported the following statistics about the ACT scores of 2355 students who applied to her college one year. Find the summaries of equivalent SAT scores.

Lowest score = 19	Mean = 27	Standard deviation = 3
Q3 = 30	Median = 28	IQR = 6

4. **Cold U?** A high school senior in the U.S. uses the Internet to get information on December temperatures in the Canadian town where he'll be going to college. He finds a Web site with some statistics, but they are given in degrees Celsius. The conversion formula is $°F = \frac{9}{5}°C + 32$. Determine the Fahrenheit equivalents for the summary information below.

Maximum temperature = 11°C	Range = 33°	Mean = 1°
Standard deviation = 7°	Median = 2°	IQR = 16°

5. **Temperatures** A town's January high temperatures average 2°C with a standard deviation of 6°, while in July the mean high temperature is 24° and the standard deviation is 5°. In which month is it more unusual to have a day with a high temperature of 13°? Explain.

6. **Placement exams** An incoming freshman took her college's placement exams in French and Mathematics. In French, she scored 82, and in Math, 86. The overall results on the French exam had a mean of 72 and a standard deviation of 8, while the mean Math score was 68, with a standard deviation of 12. On which exam did she do better compared with the other freshmen?

7. **Final exams** Anna, a language major, took final exams in both French and Spanish and scored 83 on both. Her roommate Megan, also taking both courses, scored 77 on the French exam and 95 on the Spanish exam. Overall, student scores on the French exam had a mean of 81 and a standard deviation of 5, and the Spanish scores had a mean of 74 and a standard deviation of 15.
   a) To qualify for language honours, a major must maintain at least an 85 average for all language courses taken. So far, which student qualifies?
   b) Which student's overall performance was better?

8. **MP3s** Two companies market new batteries targeted at owners of personal music players. DuraTunes claims a mean battery life of 11 hours, while RockReady advertises 12 hours.
   a) Explain why you would also like to know the standard deviations of the battery lifespans before deciding which brand to buy.
   b) Suppose those standard deviations are 2 hours for DuraTunes and 1.5 hours for RockReady. You are headed for 8 hours at the beach. Which battery is most likely to last all day? Explain.
   c) If your beach trip is all weekend, and you probably will have the music on for 16 hours, which battery is most likely to last? Explain.

9. **Cattle** The Virginia Cooperative Extension reports that the mean weight of yearling Angus steers is 1152 pounds. Suppose that weights of all such animals can be described by a Normal model with a standard deviation of 84 pounds.
   a) How many standard deviations from the mean would a steer weighing 1000 pounds be?
   b) Which would be more unusual, a steer weighing 1000 pounds, or one weighing 1250 pounds?

T 10. **Car speeds** John Beale of Stanford, CA, recorded the speeds of cars driving past his house, where the speed limit read 20 mph (miles per hour). The mean of 100 readings was 23.84 mph, with a standard deviation of 3.56 mph. (He actually recorded every car for a two-month period. These are 100 representative readings.)
   a) How many standard deviations from the mean would a car going under the speed limit be?
   b) Which would be more unusual, a car travelling 34 mph or one going 10 mph?

11. **More cattle** Recall that the beef cattle described in Exercise 9 had a mean weight of 1152 pounds, with a standard deviation of 84 pounds.
   a) Cattle buyers hope that yearling Angus steers will weigh at least 1000 pounds. To see how much over (or under) that goal the cattle are, we could subtract 1000 pounds from all the weights. What would the new mean and standard deviation be?
   b) Suppose such cattle sell at auction for 40 cents a pound. Find the mean and standard deviation of the sale prices (in dollars) for all the steers.

T 12. **Car speeds again** For the car speed data in Exercise 10, recall that the mean speed recorded was 23.84 mph, with a standard deviation of 3.56 mph. To see how many cars are speeding, John subtracts 20 mph from all speeds.
   a) What is the mean speed now? What is the new standard deviation?
   b) His friend in Inuvik, NWT, wants to study the speeds, so John converts all the original miles per hour readings to kilometres per hour by multiplying all speeds by 1.609 (km per mile). What is the mean now? What is the new standard deviation?

13. **Cattle, part III** Suppose the auctioneer in Exercise 11 sold a herd of cattle whose minimum weight was 980 pounds, median was 1140 pounds, standard deviation was 84 pounds, and IQR was 102 pounds. They sold for 40 cents a pound, and the auctioneer took a $20 commission on each animal. Then, for example, a steer weighing 1100 pounds would net the owner $0.40(1100) - 20 =$ $420. Find the minimum, median, standard deviation, and IQR of the net sale prices.

14. **Caught speeding** Suppose police set up radar surveillance on the Stanford street described in Exercise 10. They handed out a large number of tickets to drivers going a mean of 28 mph, with a standard deviation of 2.4 mph, a maximum of 33 mph, and an IQR of 3.2 mph. Local law prescribes fines of $100 plus $10 per mile per hour over the 20 mph speed limit. For example, a driver convicted of going 25 mph would be fined $100 + 10(5) =$ $150. Find the mean, maximum, standard deviation, and IQR of all the potential fines.

15. **Professors** A friend tells you about a recent study dealing with the number of years of teaching experience among current college professors. He remembers the mean but can't recall whether the standard deviation was 6 months, 6 years, or 16 years. Tell him which one it must have been, and why.

16. **Rock concerts** A popular band on tour played a series of concerts in large venues. They always drew a large crowd, averaging 21 359 fans. While the band did not announce (and probably never calculated) the standard deviation, which of these values do you think is most likely to be correct: 20, 200, 2000, or 20 000 fans? Explain your choice.

17. **Trees** A forester measured 27 of the trees in a large wooded area that is up for sale. He found a mean diameter of 26 cm and a standard deviation of 12 cm. Suppose these trees provide an accurate description of the whole forest and that a Normal model applies.
   a) Draw the Normal model for tree diameters.
   b) In what size range would you expect the central 95% of all trees to be?
   c) About what percent of the trees should be less than 2 cm in diameter?
   d) About what percent of the trees should be between 14 and 26 cm in diameter?
   e) About what percent of the trees should be over 38 cm in diameter?

18. **Rivets** A company that manufactures rivets believes the shear strength (in kg) is modelled by $N(400, 25)$.
   a) Draw and label the Normal model.
   b) Would it be safe to use these rivets in a situation requiring a shear strength of 375 kg? Explain.
   c) About what percent of these rivets would you expect to fall below 450 kg?
   d) Rivets are used in a variety of applications with varying shear strength requirements. What is the maximum shear strength for which you would feel comfortable approving this company's rivets? Explain your reasoning.

19. **Trees, part II** Later on, the forester in Exercise 17 shows you a histogram of the tree diameters he used in analyzing the wooded area that was for sale. Do you

think he was justified in using a Normal model? Explain, citing some specific concerns.

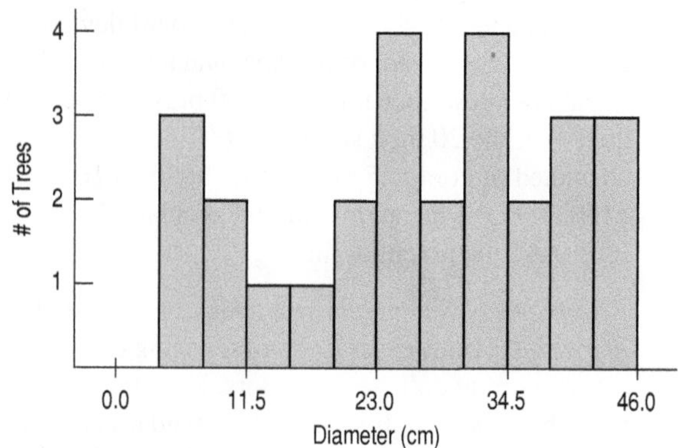

T **20. Car speeds, the picture** Here is the histogram, boxplot, and Normal probability plot of the 100 readings for the car speed data from Exercise 10. Do you think it is appropriate to apply a Normal model here? Explain.

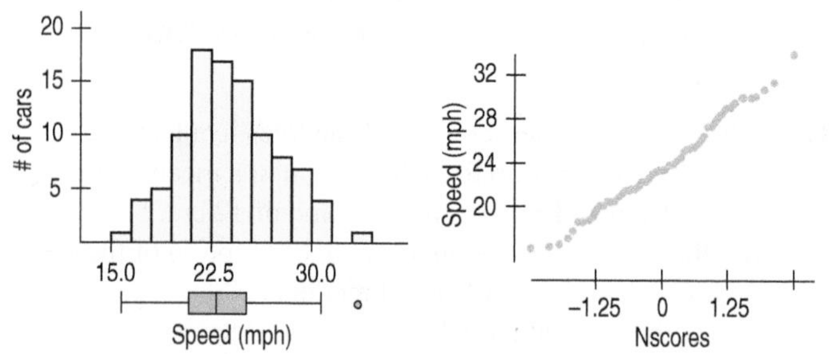

T **21. Winter Olympics 2010 downhill** Fifty-nine men completed the men's alpine downhill race in Vancouver. The gold medal winner finished in 114.3 seconds. Here are the times (in seconds) for all competitors (*Source:* espn.go.com/olympics/winter/2010/results/_/sport/1/event/2):

114.3	115.0	115.7	116.7	118.6	119.8
114.4	115.2	115.8	116.7	118.7	120.0
114.4	115.2	116.0	117.0	118.8	120.1
114.5	115.3	116.1	117.2	118.9	120.6
114.6	115.3	116.2	117.2	119.2	121.7
114.7	115.4	116.2	117.4	119.5	121.7
114.8	115.4	116.3	117.7	119.6	122.6
114.8	115.5	116.4	117.9	119.7	123.4
114.9	115.6	116.6	118.1	119.8	124.4
114.9	115.7	116.7	118.4	119.8	

a) The mean time was 117.34 seconds, with a standard deviation of 2.465 seconds. If the Normal model is appropriate, what percent of times will be less than 114.875 seconds?

b) What is the actual percent of times less than 114.875 seconds?

c) Why do you think the two percentages don't agree?

d) Make a histogram of these times. What do you see?

22. Check the model Recall that the mean of the 100 car speeds in Exercises 10 and 20 was 23.84 mph, with a standard deviation of 3.56 mph.

a) Using a Normal model, what values should border the middle 95% of all car speeds?

b) Here are some summary statistics.

Percentile		Speed
100%	Max	34.060
97.5%		30.976
90.0%		28.978
75.0%	Q3	25.785
50.0%	Median	23.525
25.0%	Q1	21.547
10.0%		19.163
2.5%		16.638
0.0%	Min	16.270

From your answer in part a), how well does the model do in predicting those percentiles? Are you surprised? Explain.

T **23. TV watching** A survey of 200 university students showed the following distribution of the number of hours of television watched per week.

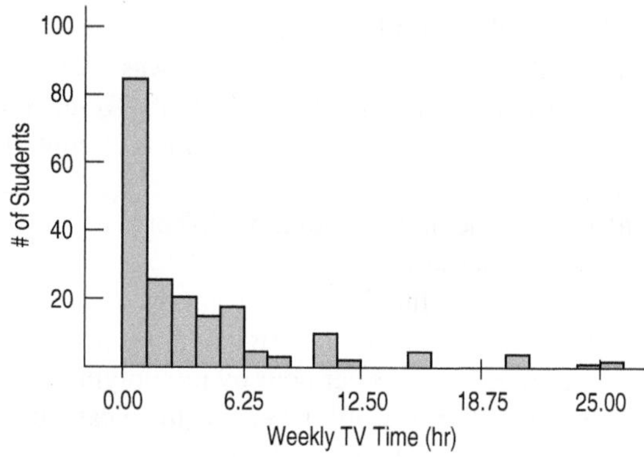

The mean is 3.66 hours, with a standard deviation of 4.93 hours.

a) According to the Normal model, what percent of students will watch fewer than one standard deviation below the mean number of hours?

b) For these data, what does that mean? Explain.

c) Explain the problem in using the Normal model for these data.

T **24. Receivers 2010** NFL data from the 2010 football season reported the number of yards gained by each of the league's 191 wide receivers:

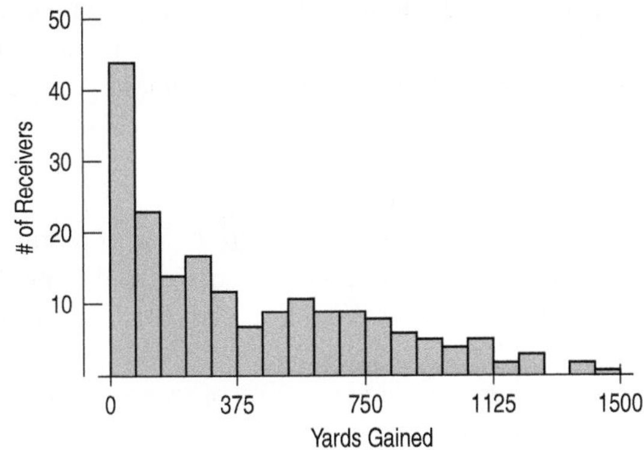

The mean is 397.15 yards, with a standard deviation of 362.4 yards.

a) According to the Normal model, what percent of receivers would you expect to gain more yards than 2 standard deviations above the mean number of yards?
b) For these data, what does that mean?
c) Explain the problem in using a Normal model here.

25. Normal models What percent of a standard Normal model is found in each region? Be sure to draw a picture first.

a) $z > 1.5$
b) $z < 2.25$
c) $-1 < z < 1.15$
d) $|z| < 0.5$

26. Normal models, again What percent of a standard Normal model is found in each region? Draw a picture first.

a) $z > -2.05$
b) $z < -0.33$
c) $1.2 < z < 1.8$
d) $|z| < 1.28$

27. More Normal models In a standard Normal model, what value(s) of z cut(s) off the region described? Don't forget to draw a picture.

a) the highest 20%
b) the highest 75%
c) the lowest 3%
d) the middle 90%

28. Yet another Normal model In a standard Normal model, what value(s) of z cut(s) off the region described? Remember to draw a picture first.

a) the lowest 12%
b) the highest 30%
c) the highest 7%
d) the middle 50%

29. Normal cattle Using $N(1152, 84)$, the Normal model for weights, in pounds, of Angus steers from Exercise 9, what percent of steers weigh

a) over 1250 pounds?
b) under 1200 pounds?
c) between 1000 and 1100 pounds?

30. More cattle Based on the model $N(1152, 84)$ for steer weights, what are the cutoff values for

a) the highest 10% of the weights?
b) the lowest 20% of the weights?
c) the middle 40% of the weights?

31. Cattle, finis Consider the Angus weights model $N(1152, 84)$ one last time.

a) What weight represents the 40th percentile?
b) What weight represents the 99th percentile?
c) What's the IQR of the weights of these Angus steers?

32. TOEFL scores, Internet test The mean and standard deviation for all 2007 writers of the Internet-based TOEFL were 78 and 24, respectively.[8] The test results are a bit left or negatively skewed.

a) The maximum score possible is 120. How does this suggest that a Normal model might not be a very accurate approximation?
b) The middle 68% of scores were between 52 and 104. If you used a Normal model, what proportion would you expect to find between these two scores?
c) The middle 95% of scores were between 29 and 115. If you used a Normal model, what proportion would you expect to find between these two scores? What scores does the Normal model predict for an interval to capture 95% of the data?

33. Cholesterol Assume the cholesterol levels of adult women can be described by a Normal model with a mean of 188 mg/dL and a standard deviation of 24.

a) Draw and label the Normal model.
b) What percent of adult women do you expect to have cholesterol levels over 200 mg/dL?
c) What percent of adult women do you expect to have cholesterol levels between 150 and 170 mg/dL?
d) Estimate the interquartile range of the cholesterol levels.
e) Above what value are the highest 15% of women's cholesterol levels?

34. Tires A tire manufacturer believes that the treadlife of its snow tires can be described by a Normal model with a mean of 52 000 km and standard deviation of 2500 km.

a) If you buy a set of these tires, would it be reasonable for you to hope they'll last 60 000 km? Explain.
b) Approximately what fraction of these tires can be expected to last less than 50 000 km?

[8] www.ets.org/Media/Research/pdf/71943_web.pdf

c) Approximately what fraction of these tires can be expected to last between 50 000 and 55 000 km?
d) Estimate the IQR of the treadlifes.
e) In planning a marketing strategy, a local tire dealer wants to offer a refund to any customer whose tires fail to last a certain number of kilometres. However, the dealer does not want to take too big a risk. If the dealer is willing to give refunds to no more than 1 of every 25 customers, for how many kilometres can he guarantee these tires to last?

35. Kindergarten Companies who design furniture for elementary school classrooms produce a variety of sizes for kids of different ages. Suppose the heights of kindergarten children can be described by a Normal model with a mean of 97.0 cm and standard deviation of 4.9 cm.
a) What fraction of kindergarten kids should the company expect to be less than 91 cm tall?
b) In what height interval should the company expect to find the middle 80% of kindergarteners?
c) At least how tall are the biggest 10% of kindergarteners?

36. Body temperatures Most people think that the "Normal" adult body temperature is 37.0°C. That figure, based on a nineteenth-century study, has recently been challenged. In a 1992 article in the *Journal of the American Medical Association,* researchers reported that a more accurate figure may be 36.8°C. Furthermore, the standard deviation appeared to be around 0.4°C. Assume that a Normal model is appropriate.
a) In what interval would you expect most people's body temperatures to be? Explain.
b) What fraction of people would be expected to have body temperatures above 37.0°C?
c) Below what body temperature are the coolest 20% of all people?

37. Undercover? We learned in this chapter that the average Dutch man is 184 cm tall. The standard deviation of Caucasian adult male heights is about 7 cm. The average Greek 18-year-old in Athens is 167.8 cm tall. How easily could the average Dutch man hide in Athens? (Let's assume he dyes his hair, if necessary.) That is, would his height make him sufficiently extraordinary that he'd stand out easily? Assume heights are nearly Normal.

38. Big mouth! A Cornell University researcher measured the mouth volumes of 31 men and 30 women. She found a mean of 66 cc for men (SD = 17 cc) and a mean of 54 cc for women (SD = 14.5 cc). The man with the largest mouth had a mouth volume of 111.2 cc. The woman with the largest mouth had a mouth volume of 95.8 cc.
a) Which had the more extraordinarily large mouth?
b) If the distribution of mouth volumes is nearly Normal, what percentage of men and of women should have even larger mouths than these?

39. Helmet sizes Adult women's head circumferences are approximately Normally distributed with a mean of 56.0 cm and standard deviation of 1.8 cm. You are manufacturing a new type of helmet for female recruits in the Canadian Armed Forces, and are planning on producing three sizes: small, medium, large. Each size will fit a range of head circumferences. Assume female recruits have similar head circumferences to the general population.
a) If we wanted to adequately fit 99.7% of recruits, the small size should fit what head size at the smallest, and the large size should fit what head size at the biggest?
b) If medium is to fit about 60% of recruits, it should fit what head circumferences?
c) If small fits those with head circumferences between 51 cm and 54 cm, what percent of recruits will need this size?
d) If the large size is to fit the biggest 15% of heads, what is the smallest head circumference suitable for a large helmet?
e) If you decide to make a special production run of a few extra small helmets, suitable for the smallest 2% of recruit circumferences, what circumference range will it fit?

40. Newborns At-term newborns in Canada vary in weight according to, approximately, a Normal distribution, with mean of 3500 grams and standard deviation of 500 grams.
a) Heavy birth weight (HBW) babies are those weighing over 4500 grams. Approximately how many at-term newborns among the next 10 000 will be HBW babies?
b) Low birth weight (LBW) babies are those weighing less than 2500 grams. Approximately how many at-term newborns among the next 10 000 will be LBW babies? How can you answer this using what you learned in part a), with no new calculation?
c) Approximately how many at-term newborns among the next 10 000 will be babies weighing between 3300 and 4300 grams?
d) A very low birth weight (VLBW) is sometimes defined as less than 1500 grams. If we wanted to set a VLBW limit at the 0.1^{th} percentile of the Canadian distribution, i.e., at a weight such that only 1 in 1000 babies weigh less, what would this VLBW limit equal?
e) One in ten at-term babies weighs more than _________. Fill in the blank.

41. TOEFL scores again TOEFL test scores are commonly used to assess English language ability. The paper-based test scores are close to Normally distributed, though the top scores are slightly more tightly compressed than the bottom scores, perhaps due to bumping up against a maximum score of 677.

The mean and standard deviation for all writers of the paper-based test in 2007 were approximately 540 and 60, respectively.[9]

a) Some schools require applicants for admission to obtain a TOEFL score of at least 600. Approximately what proportion of those writing the test actually obtained such a score?
b) The 10th and 90th percentiles were reported as 460 and 620, respectively. See if the Normal model gives you very similar values.
c) The IQR was actually 84. What would you expect using the Normal model?
d) That maximum of 677 seems to cheat the top students; i.e., if a true Normal distribution applied, with mean and standard deviation as indicated above, what percent of test-takers get their grade reduced to 677 from something potentially higher? (Find the probability of scoring higher than 677.)

42. Blackjack If you play the card game of Blackjack using the well-known *basic strategy*, you can expect to lose an average 3 cents for every 10-dollar bet, in the long run (learn to count cards and you can do even better). Suppose that thousands of people played Blackjack using the basic strategy, each playing 10 000 times and betting $10 each time. A histogram of the final net gains (or losses) for those thousands of players would closely resemble a Normal distribution with a mean of approximately −$30.00 and a standard deviation of approximately $1140.00.[10] How can we make such a prediction without actually observing thousands of players? Such predictability follows from the laws of probability, which we will study later in this book.

a) What percent of the players lost money (have a negative outcome)?
b) What percent of the players won over $1000.00?
c) The luckiest 10% of players won over _________ dollars. (Fill in the blank.)
d) The unluckiest 5% of players lost more than _________ dollars. (Fill in the blank.)

43. Eggs Hens usually begin laying eggs when they are about six months old. Young hens tend to lay smaller eggs, often weighing less than the desired minimum weight of 54 grams.

a) The average weight of the eggs produced by the young hens is 50.9 grams, and only 28% of their eggs exceed the desired minimum weight. If a Normal model is appropriate, what would the standard deviation of the egg weights be?
b) By the time these hens have reached the age of one year, the eggs they produce average 67.1 grams, and 98% of them are above the minimum weight. What is the standard deviation for the appropriate Normal model for these older hens?
c) Are egg sizes more consistent for the younger hens or the older ones? Explain.
d) A certain farmer finds that 8% of his eggs are underweight and that 12% weigh over 70 grams. Estimate the mean and standard deviation of his eggs (Challenging).

44. Tomatoes Agricultural scientists are working on developing an improved variety of Roma tomatoes. Marketing research indicates that customers are likely to bypass Romas that weigh less than 70 grams. The current variety of Roma plants produces fruit that average 74 grams, but 11% of the tomatoes are too small. It is reasonable to assume that a Normal model applies.

a) What is the standard deviation of the weights of Romas now being grown?
b) Scientists hope to reduce the frequency of undersized tomatoes to no more than 4%. One way to accomplish this is to raise the average size of the fruit. If the standard deviation remains the same, what target mean should they have as a goal?
c) The researchers produce a new variety with a mean weight of 75 grams, which meets the 4% goal. What is the standard deviation of the weights of these new Romas?
d) Based on their standard deviations, compare the tomatoes produced by the two varieties.

45. Music library Corey has 4929 songs in his computer's music library. The lengths of the songs have a mean of 242.4 seconds and standard deviation of 114.51 seconds. A Normal probability plot of the song lengths looks like this:

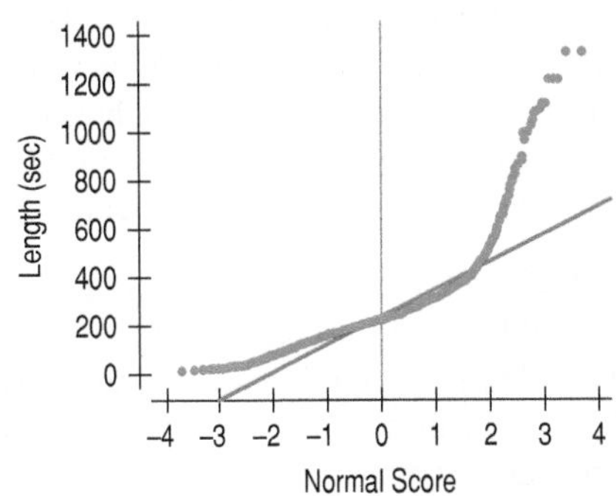

a) Do you think the distribution is Normal? Explain.
b) If it isn't Normal, how does it differ from a Normal model?

T **46. Wisconsin ACT math** The histogram shows the distribution of mean ACT mathematics scores for all Wisconsin public schools in 2011. The three vertical lines show the mean and one standard deviation above

[9] www.ets.org/Media/Research/pdf/71943_web.pdf
[10] Courtesy of http://wizardofodds.com/blackjack

and below the mean. 78.8% of the data points are between the two outermost vertical lines.

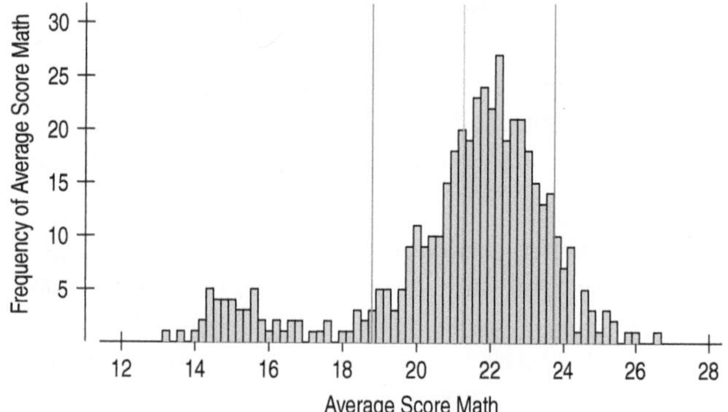

a) Give two reasons that a Normal model is not appropriate for these data.

b) The Normal probability plot on the left shows the distribution of these scores. The plot on the right shows the same data with the Milwaukee area schools (mostly near the lower mode) removed. What do these plots tell you about the shape of the distributions?

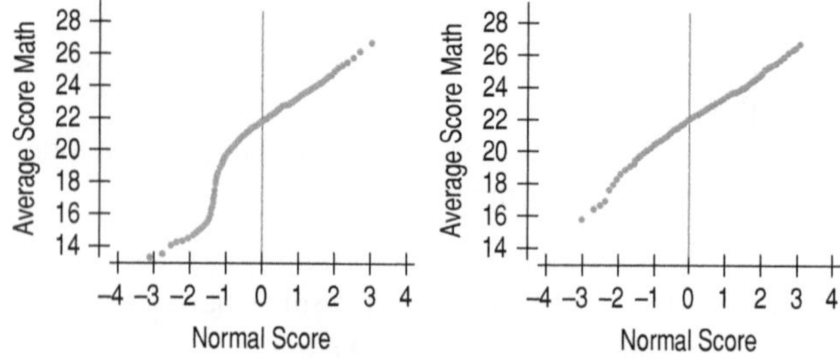

T **47. Tsunamis 2013 again** We discussed tsumani-causing earthquakes in the previous chapter. Plot the magnitudes of these earthquakes (updated to early 2013 in our data file) in a histogram, boxplot, and NPP. Does the distribution appear roughly Normal? If not, how does it appear to depart from Normality, and how do you see possible Normality or lack thereof in each graph?

T **48. First Nations bands 2010 again** Plot the data on sizes of the 138 First Nations registry groups in Ontario from Exercise 43 in Chapter 3 in a histogram, boxplot, and NPP. Does the distribution appear roughly Normal? If not, how does it appear to depart from Normality, and how do you see possible Normality or lack thereof in each graph?

49. Simulations Use your statistical software to draw a random sample of 30 observations from a standard Normal distribution (look for menu items like *Random* and *Normal* and plug in *0* and *1* for the parameters). Plot the data in a histogram and in a Normal probability plot.

a) Compare the two plots. Is it easier to judge *bell-shapedness* or straightness?

b) Repeat four more times, and answer the same question. Is it easier to be fooled by the purely random squiggles in the data when viewing the NPPs or the histograms? Or put another way, in which type of plot is it easier to see the essential pattern through the noise or random jitter?

50. More simulations Use your statistical software to draw a random sample of 30 observations from a *uniform distribution* extending from 0 to 1 (look for menu items like *Random* and *Uniform* and plug in *0* and *1* for the parameters). Plot the data in a histogram and in a Normal probability plot.

a) Compare the two plots. Is the shape of the histogram roughly uniform or flat? What shape do you see in the NPP?

b) Repeat four more times and answer the same questions. Try to explain why the NPP has such a shape here. In which type of plot is it easier to see through the noise or random squiggles to the underlying pattern?

51. Some more simulations Use your statistical software to draw a random sample of 100 observations from a standard Normal distribution (look for menu items like *Random* and *Normal* and plug in *0* and *1* for the parameters). Plot the data in a histogram and in a Normal probability plot.

a) Is the NPP nearly perfectly straight? Is the histogram nearly perfectly bell-shaped? Double the original number of bins in the histogram. Can the choice of number of bins change one's impression of the bell-shapedness? Is it easier to judge bell-shapedness of the histogram or straightness of the NPP?

b) Repeat four more times, and answer the same questions. Is it more difficult to see through the random variation or noise in your data to the underlying pattern with a histogram or NPP?

52. And more simulations Use your statistical software to draw a random sample of 100 observations from a uniform distribution extending from 0 to 1 (look for menu items like *Random* and *Uniform* and plug in *0* and *1* for the parameters). Plot the data in a histogram and in a Normal probability plot.

a) Is the histogram nearly perfectly flat? Change the bin cut points or endpoints to 0, 0.1, 0.2 . . . 0.9, 1.0. Does your impression of the flatness change at all? What is the shape of the NPP? Try to explain why the NPP has such a shape here.

b) Repeat four more times, and answer the same questions. Is it more difficult to see through the random variation or noise in your data to the underlying pattern with a histogram or NPP?

Just Checking ANSWERS

1. **a)** On the first test, the mean is 88 and the SD is 4, so $z = (90 - 88)/4 = 0.5$. On the second test, the mean is 75 and the SD is 5, so $z = (80 - 75)/5 = 1.0$. The first test has the lower z-score, so it is the one that will be dropped.
   b) The second test is one standard deviation above the mean, farther away than the first test, so it's the better score relative to the class.
2. **a)** The mean would increase to 500.
   b) The standard deviation is still 100 points.
   c) The two boxplots would look nearly identical (the shape of the distribution would remain the same), but the later one would be shifted 50 points higher.
3. The standard deviation is now 2.54 millimetres, which is the same as 0.1 inches. Nothing has changed. The standard deviation has "increased" only because we're reporting it in millimetres now, not inches.
4. The mean is 184 cm, with a standard deviation of 8 cm. Two metres is 200 cm, which is two standard deviations above the mean. We expect 5% of the men to be more than two standard deviations below or above the mean, so half of those, 2.5%, are likely to be above 2 metres.
5. **a)** We know that 68% of the time, we'll be within one standard deviation (two minimum) of 20. So 32% of the time we'll arrive in less than 18 or more than 22 minutes. Half of those times (16%) will be greater than 22 minutes, so 84% will be less than 22 minutes.
   b) 24 minutes is two standard deviations above the mean. Because of the 95% rule, we know 2.5% of the times will be more than 24 minutes.
   c) Traffic incidents may occasionally increase the time it takes to get to school, so the driving times may be skewed to the right, and there may be outliers.
   d) If so, the Normal model would not be appropriate and the percentages we predict would not be accurate.

MathXL
MyStatLab

Go to MathXL at www.mathxl.com or MyStatLab at www.mystatlab.com. You can practise exercises for this chapter as often as you want. The guided solutions will help you find answers step by step. You'll find a personalized study plan available to you too!

part I

Review

Exploring and Understanding Data

Quick Review

It's time to put it all together. Real data don't come tagged with instructions for use. So let's step back and look at how the key concepts and skills we've seen work together. This brief list and the review exercises that follow should help you check your understanding of Statistics so far.

- We treat data two ways: categorical and quantitative.
- To describe categorical data:
  - Make a picture. Bar graphs work well for comparing counts in categories.
  - Summarize the distribution with a table of counts or relative frequencies (percents) in each category.
  - Pie charts and segmented bar charts display divisions of a whole.
  - Compare distributions with plots side by side.
  - Look for associations between variables by comparing conditional distributions.
- To describe quantitative data:
  - Make a picture. Use histograms, boxplots, stem-and-leaf displays, or dotplots. Stem-and-leafs are great when working by hand and good for small data sets. Histograms are a good way to see the distribution. Boxplots are best for comparing several distributions.
  - Describe distributions in terms of their shape, centre, and spread, and note any unusual features, such as gaps or outliers.
  - The shape of most distributions you'll see will likely be uniform, unimodal, or bimodal. It may even be multimodal. If it is unimodal, then it may be symmetric or skewed.
  - A 5-number summary provides a good numerical description of a distribution: minimum, Q1, median, Q3, and maximum. It gives information about shape, centre, and spread, and is particularly appropriate for describing skewed distributions. The median and IQR ($Q_3 - Q_1$) are *resistant* measures of centre and spread.
  - A distribution that is severely skewed may benefit from re-expressing the data. If it is skewed to the high end, taking logs often works well.
  - If the distribution is unimodal and symmetric, describe its centre and spread with the mean and standard deviation.
  - Use the standard deviation as a ruler to tell how unusual an observed value may be, or to compare or combine measurements made on different scales.
  - Shifting a distribution by adding or subtracting a constant affects measures of position but not measures of spread. Rescaling by multiplying or dividing by a constant affects both.
  - When a distribution is roughly unimodal and symmetric, a Normal model may be useful. For Normal models, the 68–95–99.7 Rule is a good rule of thumb.
  - If the Normal model fits well (check a histogram or Normal probability plot), then Normal percentile tables or functions found in most statistics technology can provide more detailed values.

Need more help with some of this? It never hurts to reread sections of the chapters! And in the following pages we offer you more opportunities[1] to review these concepts and skills.

The exercises that follow use the concepts and skills you've learned in the first six chapters. To be more realistic and more useful for your review, they don't tell you which of the concepts or methods you need. But neither will your exam.

[1]If you doubted that we are teachers, this should convince you. Only a teacher would call additional homework exercises "opportunities."

Review Exercises

T **1. Homicide across Canada 2011** Below are annual homicide rates (per 100 000 population) in Canadian cities bigger than 100 000 in population, averaged over the period 2001–2011.

City	Homicide Rate
Regina	3.67
Winnipeg	3.51
Edmonton	3.07
Abbotsford	2.98
Saskatoon	2.88
Vancouver	2.43
Kelowna	2.06
Calgary	1.98
Halifax	1.92
Thunder Bay	1.86
Toronto	1.78
Kingston	1.67
Greater Sudbury	1.60
Montreal	1.53
Windsor	1.48
Hamilton	1.43
St. Catharines–Niagara	1.42
London	1.30
Ottawa	1.20
Moncton	1.20
Victoria	1.19
Oshawa	1.18
Brantford	1.16
Gatineau	1.15
Barrie	1.03
Saint John	0.92
Kitchener	0.90
Peterborough	0.83
Trois-Rivières	0.82
Saguenay	0.74
St. John's	0.67
Guelph	0.66
Quebec City	0.62
Sherbrooke	0.36

Source: Statistics Canada, Canadian Centre for Justice Statistics, Homicide Survey.

a) Display the data with an appropriate graph, and describe the distribution as best you can. Include any appropriate summary statistics.

b) The graph in a) misses many important factors, such as changes over time and over space (geography). We could plot homicide rates versus year to assess the effect of time. Instead, let's look at geographical effects. Group cities by region—for example Atlantic Canada, Quebec, Ontario, Prairies, Western Canada—and assess if there are any geographical trends.

2. Prenatal care Results of a 1996 American Medical Association report about the infant mortality rate for twins carried for the full term of a normal pregnancy are shown below, broken down by the level of prenatal care the mother had received.

Full-Term Pregnancies, Level of Prenatal Care	Infant Mortality Rate Among Twins (deaths per thousand live births)
Intensive	5.4
Adequate	3.9
Inadequate	6.1
Overall	5.1

a) Is the overall rate the average of the other three rates? Should it be? Explain.

b) Do these results indicate that adequate prenatal care is important for pregnant women? Explain.

c) Do these results suggest that a woman pregnant with twins should be wary of seeking too much medical care? Explain.

3. Singers The boxplots shown display the heights (in inches) of 130 members of a choir.

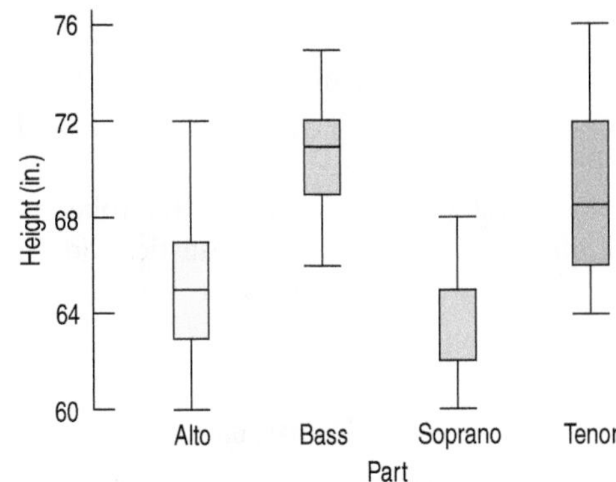

a) It appears that the median height for sopranos is missing, but actually the median and the upper quartile are equal. How could that happen?

b) Write a few sentences describing what you see.

4. **Dialysis** In a study of dialysis, researchers found that "of the three patients who were currently on dialysis, 67% had developed blindness and 33% had their toes amputated." What kind of display might be appropriate for these data? Explain.

5. **Beanstalks** Beanstalk Clubs are social clubs for very tall people. To join, a man must be over 6′2″ tall, and a woman over 5′10″. Heights of adults are approximately Normally distributed, with mean heights of 69.1″ for men and 64.0″ for women. The respective standard deviations are 2.8″ and 2.5″.
   a) You are probably not surprised to learn that men are generally taller than women, but what does the greater standard deviation for men's heights indicate?
   b) Who are more likely to qualify for Beanstalk membership, men or women? Explain.

6. **Bread** Tobermory Bakery is trying to predict how many loaves to bake. In the past 100 days, they have sold between 95 and 140 loaves per day. Here is a histogram of the number of loaves they sold for the past 100 days.

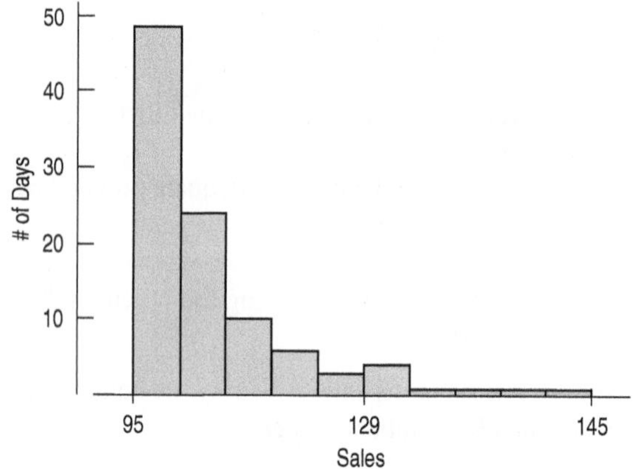

   a) Describe the distribution.
   b) Which should be larger, the mean number of sales or the median? Explain.
   c) Here are the summary statistics for Tobermory Bakery's bread sales. Use these statistics and the histogram above to create a boxplot. You may approximate the values of any outliers.

Summary of Sales	
Median	100
Min	95
Max	140
25th %tile	97
75th %tile	105.5

   d) For these data, the mean was 103 loaves sold per day, with a standard deviation of 9 loaves. Do these statistics suggest that Tobermory Bakery should expect to sell between 94 and 112 loaves on about 68% of days? Explain.

7. **University survey** Public relations staff at a Canadian university phoned 850 local residents. After identifying themselves, the callers asked the survey participants their ages, whether they had attended college or university, and whether they had a favourable opinion of the university. The official report to the university's directors claimed that, in general, people had very favourable opinions about their alma mater.
   a) Identify the *W*'s of these data.
   b) Identify the variables, classify each as categorical or quantitative, and specify units if relevant.
   c) Are you confident about the report's conclusion? Explain.

8. **Acid rain** Based on long-term investigation, researchers have suggested that the acidity (pH) of rainfall in Shenandoah Mountains storms can be described by the Normal model $N(4.9, 0.6)$.
   a) Draw and carefully label the model.
   b) What percent of storms produce rainfall with pH over 6?
   c) What percent of storms produce rainfall with pH under 4?
   d) The lower the pH, the more acidic the rain. What is the pH level for the most acidic 20% of all storms?
   e) What is the pH level for the least acidic 5% of all storms?
   f) What is the IQR for the pH of rainfall?

9. **Fraud detection** A credit card bank is investigating the incidence of fraudulent card use. The bank suspects that the type of product bought may provide clues to the fraud. To examine this situation, the bank looks at the Standard Industrial Code (SIC) of the business related to the transaction. This is a code that was used by Statistics Canada and the U.S. Census Bureau to identify the type of every registered business in North America.[2] For example, 1011 designates Meat and Meat Products (except Poultry), 1012 is Poultry Products, 1021 is Fish Products, 1031 is Canned and Preserved Fruits and Vegetables, and 1032 is Frozen Fruits and Vegetables.

[2] Since 1997, the SIC has been replaced by the NAICS, a code of six letters.

A company intern produces the following histogram of the SIC codes for 1536 transactions:

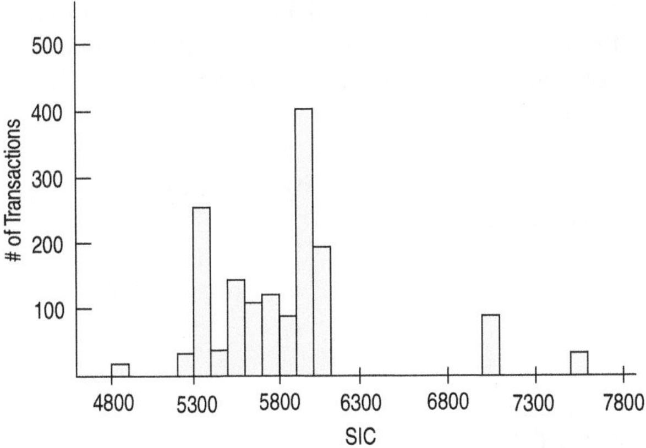

He also reports that the mean SIC is 5823.13 with a standard deviation of 488.17.

a) Comment on any problems you see with the use of the mean and standard deviation as summary statistics.

b) How well do you think the Normal model will work on these data? Explain.

10. Streams As part of the course work, a class at a northern BC college collects data on streams each year. Students record a number of biological, chemical, and physical variables, including the stream name, the substrate of the stream (*limestone*, *shale*, or *mixed*), the pH, the temperature (°C), and the BCI, a measure of biological diversity.

Group	Count	%
Limestone	77	44.8
Mixed	26	15.1
Shale	69	40.1

a) Name each variable, indicate whether it is categorical or quantitative, and give the units if available.

b) These streams have been classified according to their substrate—the composition of soil and rock over which they flow—as summarized in the table. What kind of graph might be used to display these data? .

11. Cramming One Thursday, researchers gave students enrolled in a section of basic French a set of 50 new vocabulary words to memorize. On Friday, the students took a vocabulary test. When they returned to class the following Monday, they were retested—without advance warning. Both test scores (side by side) for the 25 students are shown in the table.

Fri	Mon	Fri	Mon
42	36	50	47
44	44	34	34
45	46	38	31
48	38	43	40
44	40	39	41
43	38	46	32
41	37	37	36
35	31	40	31
43	32	41	32
48	37	48	39
43	41	37	31
45	32	36	41
47	44		

a) Create a graphical display to compare the two distributions of scores.

b) Write a few sentences about the scores reported on Friday and Monday.

c) Create a graphical display showing the distribution of the *changes* in student scores.

d) Describe the distribution of changes.

12. Band sizes 2011 Below is a chart of First Nations registry groups (similar to bands) in Canada by size, on December 31, 2010[3]:

Size of band	Number of bands
<100	20
100–249	75
250–499	112
500–999	167
1 000–1 999	139
2 000+	118

a) Which types of graphs are appropriate for displaying these data? Name them all.

b) Present the data in an appropriate graph.

c) The size of a band is actually what type of variable? How is it that we can use a pie chart for such a variable?

[3]*Registered Indian Population by Sex and Residence 2010.* Ottawa: Indian and Northern Affairs Canada, 2008. http://www.aadnc-aandc.gc.ca/DAM/DAM-INTER-HQ/STAGING/texte-text/ai_rs_pubs_sts_ni_rip_rip10_rip10_1309289046808_eng.pdf. Reproduced with the permission of the Minster of Public Works and Government Services Canada, 2010.

d) Is it possible to convert a categorical variable to a quantitative one? Or a quantitative variable to a categorical one? If so, what are the possible advantages and disadvantages?

13. **Off or on reserve 2011** Below is a table with information about age and residence status of the registered First Nations population of Canada as of December 31, 2010[4]:

Age Group	0–24	25–34	35–64	65+
Residence Status				
On reserve	49.5%	14.6%	30.2%	5.7%
Off reserve	38.7%	15.1%	38.4%	7.7%

a) Compare the two conditional age distributions, using pie charts. Describe the difference.
b) The total living on-reserve was 459 159, and the total living off-reserve was 365 182. Compare the four conditional residence distributions, using an appropriate bar chart(s), and describe the relationship between residence status and age.
c) Are the variables of interest here quantitative or categorical? Explain clearly.
d) Would it be possible for the average age of First Nations males to be less than the average age of First Nations females on reserve as well as off reserve, but for the average age of all First Nations males to be greater than the average age of all First Nations females? Explain as best you can (you can make up some hypothetical numbers).

T 14. **Accidents** Progressive Insurance asked customers who had been involved in auto accidents how far they were from home when the accident happened. The data are summarized in the table.

Miles from Home	% of Accidents
Less than 1	23
1 to 5	29
6 to 10	17
11 to 15	8
16 to 20	6
Over 20	17

a) Create an appropriate graph of these data.
b) Do these data indicate that driving near home is particularly dangerous? Explain.

[4]Ibid.

T 15. **Hard water** In an investigation of environmental causes of disease, data were collected on the annual mortality rate (deaths per 100 000) for males in 61 large towns in England and Wales. In addition, the water hardness was recorded as the calcium concentration (parts per million, ppm) in the drinking water.
a) What are the variables in this study? Indicate whether each variable is quantitative or categorical and what the units are.
b) Here are histograms of calcium concentration and mortality. Describe the distributions of the two variables.

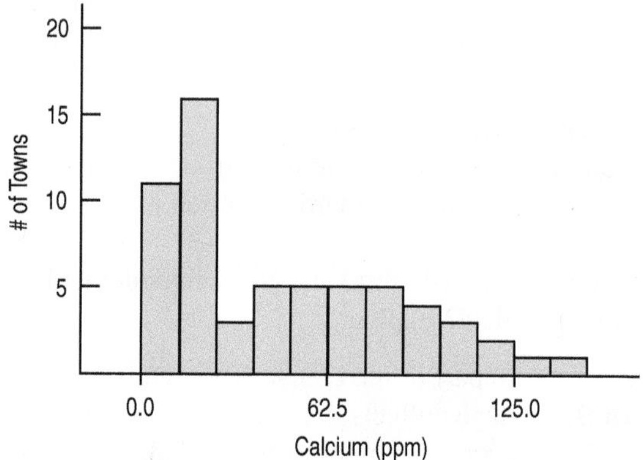

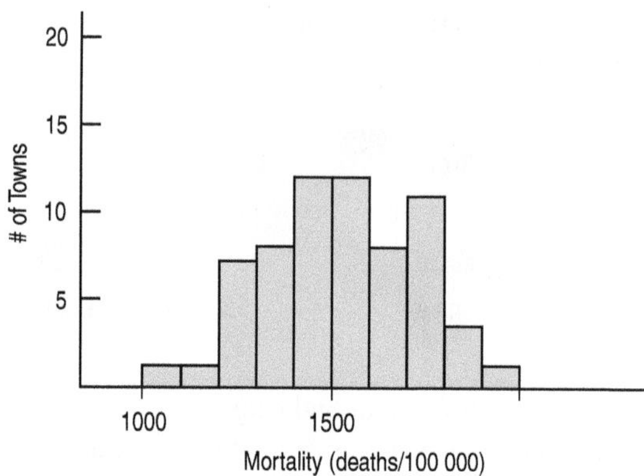

T 16. **Hard water II** The data set from England and Wales also notes for each town whether it was south or north of Derby. Here are some summary statistics and a comparative boxplot for the two regions.

Summary of Mortality

Group	Count	Mean	Median	StdDev
North	34	1631.59	1631	138.470
South	27	1388.85	1369	151.114

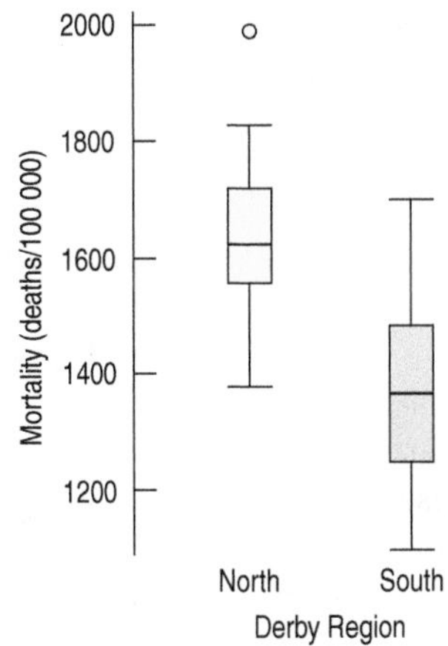

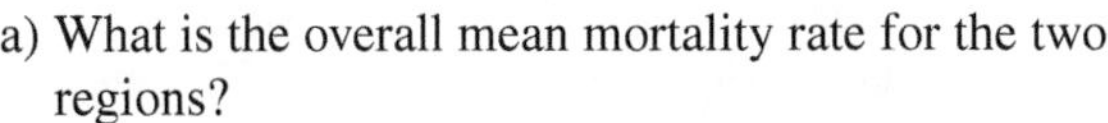

a) What is the overall mean mortality rate for the two regions?

b) Do you see evidence of a difference in mortality rates? Explain.

17. Seasons The histograms below show the average daily temperatures in January and in July for 60 large U.S. cities.

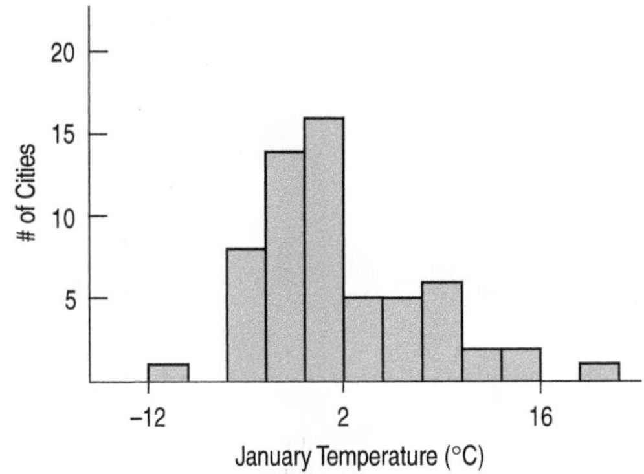

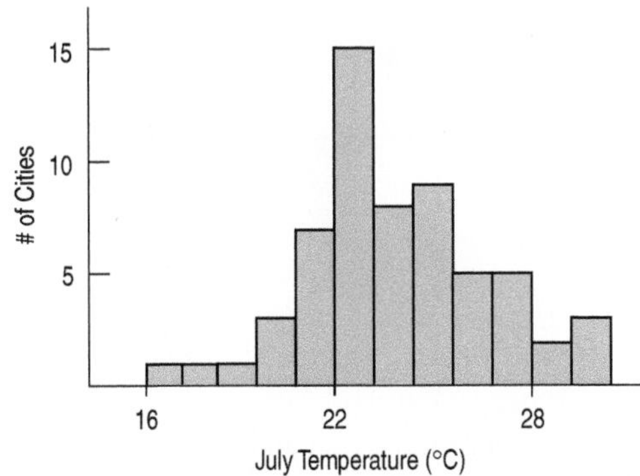

a) What aspect of these histograms makes it difficult to compare the distributions?

b) What differences do you see between the distributions of January and July average temperatures?

c) Differences in temperatures (July–January) for each of the cities are displayed in the boxplot below. Write a few sentences describing what you see.

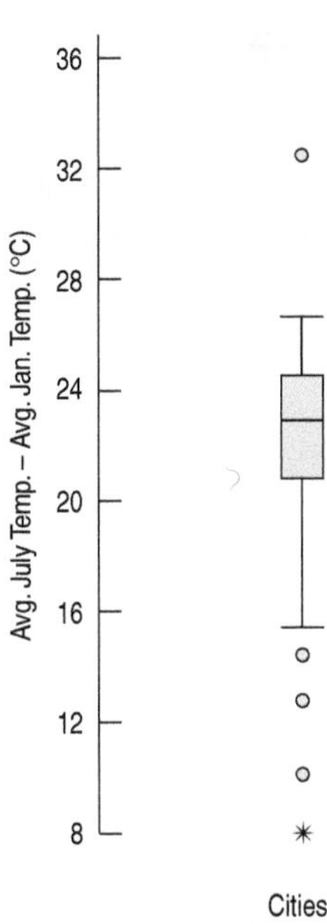

T 18. Old Faithful It is a common belief that Yellowstone's most famous geyser erupts once an hour at very predictable intervals. The histogram below shows the time gaps (in minutes) between 222 successive eruptions. Describe this distribution.

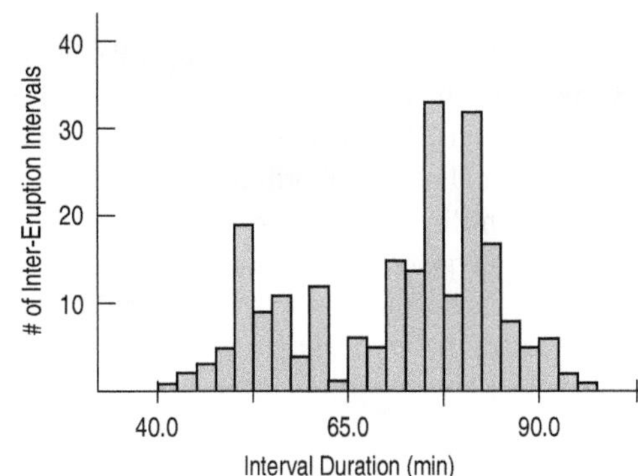

T 19. Old Faithful? Does the duration of an eruption have an effect on the length of time that elapses before the next eruption?

a) The histogram below shows the duration (in minutes) of those 222 eruptions. Describe this distribution.

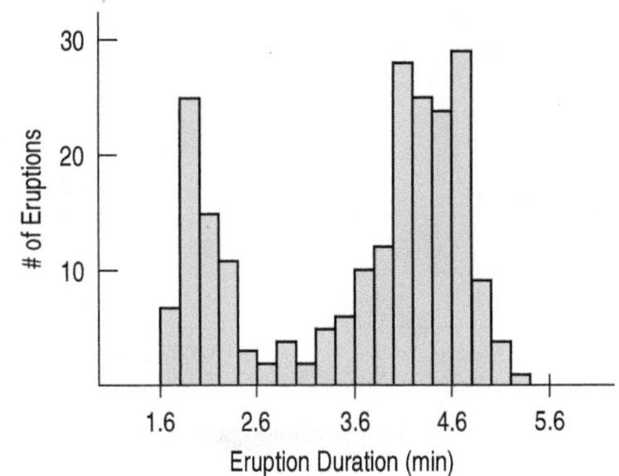

b) Which summary statistics would you choose to describe this distribution? Explain your choices.
c) Let's classify the eruptions as "long" or "short," depending on whether they last at least three minutes. Describe what you see in the comparative boxplots.

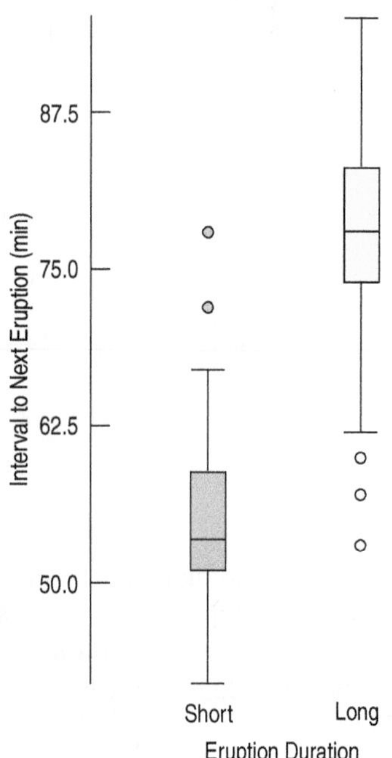

20. Very, very rich According to the World Wealth Report 2008[5], we have the following distributions of High Net Worth Individuals (HNWIs; \$1 million or more in assets) and Ultra High Net Worth Individuals (Ultra-HNWIs; \$30 million or more in assets):

a) What is this type of display called? To make it look similar to displays in this text, what would you have to change about the vertical axis in this graph?

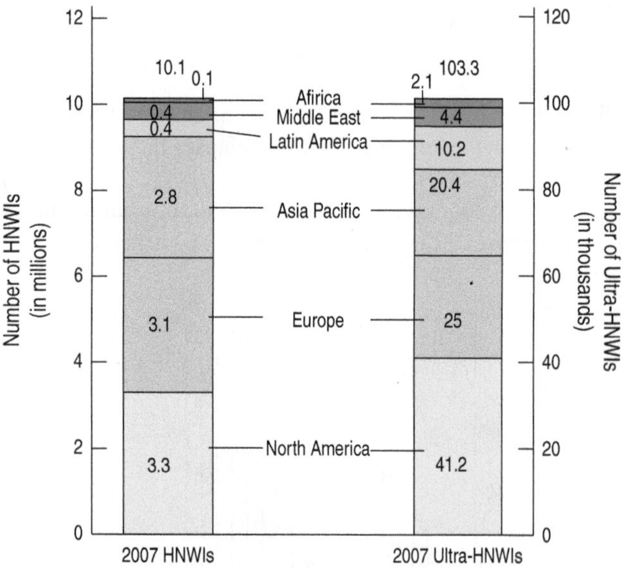

[5]Capgemini/Merril Lynch, "World Wealth Report 2008," www.ml.com/media/100502.pdf.

b) Answer the following or explain why you can't:
   (i) What percent of all Ultra-HNWIs are in Asia-Pacific?
   (ii) What percent of Latin Americans are Ultra-HNWIs?
   (iii) What percent of HNWIs in North America are Ultra-HNWIs?
c) Explain the major differences between the two distributions displayed.
d) If you entered the numbers above into a 2 × 6 table of counts, would you then have a proper contingency table? If not, how would you convert it to one?

21. Liberty's nose Is the Statue of Liberty's nose too long? Her nose measures 4′6″ (4.5 feet), but she is a large statue, after all. Her arm is 42 feet long. That means her arm is 42/4.5 = 9.3 times as long as her nose. Is that a reasonable ratio? Shown in the table are arm and nose lengths of 18 women in a Statistics class, and the ratio of arm-to-nose length for each.

Arm (cm)	Nose (cm)	Arm/Nose Ratio
73.8	5.0	14.8
74.0	4.5	16.4
69.5	4.5	15.4
62.5	4.7	13.3
68.6	4.4	15.6
64.5	4.8	13.4
68.2	4.8	14.2
63.5	4.4	14.4
63.5	5.4	11.8
67.0	4.6	14.6
67.4	4.4	15.3
70.7	4.3	16.4
69.4	4.1	16.9
71.7	4.5	15.9
69.0	4.4	15.7
69.8	4.5	15.5
71.0	4.8	14.8
71.3	4.7	15.2

a) Make an appropriate plot and describe the distribution of the ratios.
b) Summarize the ratios numerically, choosing appropriate measures of centre and spread.
c) Is the ratio of 9.3 for the Statue of Liberty unrealistically low? Explain.

22. Winter Olympics 2010 speed skating The top 34 women's 500-m speed skating times are listed in the following table.

a) The mean finishing time was 40.72 seconds, with a standard deviation of 9.82 seconds. If the Normal model is appropriate, what percent of the times should be within 0.5 second of 40.72?
b) What percent of the times actually fall within this interval?
c) Explain the discrepancy between parts a) and b).

Nation	Athlete	Result
Korea	Sang-Hwa Lee	38.249
Germany	Jenny Wolf	38.307
China	Beixing Wang	38.487
Netherlands	Margot Boer	38.511
China	Shuang Zhang	38.530
Japan	Sayuri Yoshii	38.566
Russian Federation	Yulia Nemaya	38.594
China	Peiyu Jin	38.686
United States	Heather Richardson	38.698
Germany	Monique Angermuller	38.761
China	Aihua Xing	38.792
Japan	Nao Kodaira	38.835
Canada	Christine Nesbitt	38.881
Netherlands	Thijsje Oenema	38.892
DPR Korea	Hyon-Suk Ko	38.893
Japan	Shihomi Shinya	38.964
Japan	Tomomi Okazaki	38.971
United States	Elli Ochowicz	39.002
Kazakhstan	Yekaterina Aydova	39.024
United States	Jennifer Rodriguez	39.182
Netherlands	Laurine van Riessen	39.302
Canada	Shannon Rempel	39.351
Germany	Judith Hesse	39.357
Russian Federation	Olga Fatkulina	39.359
Czech Republic	Karolina Erbanova	39.365
Korea	Bo-Ra Lee	39.396
Russian Federation	Svetlana Kaykan	39.422
Italy	Chiara Simionato	39.480
United States	Lauren Cholewinski	39.514
Korea	Jee-Min Ahn	39.595
Australia	Sophie Muir	39.649
Russian Federation	Yekaterina Malysheva	39.782
Korea	Min-Jee Oh	39.816
Canada	Anastasia Bucsis	39.879
Belarus	Svetlana Radkevich	39.899
Netherlands	Annette Gerritsen	97.952

23. Simpson's paradox A study in South Africa focusing on the impact of health insurance identified 1590 children at birth and then sought to conduct follow-up health studies five years later. Only 416 of the original group participated in the five-year follow-up study. This made researchers concerned that the follow-up group might not accurately resemble the total group in terms of health insurance. The table summarizes the two groups by race and by presence of medical insurance when the child was born. Carefully explain how this study demonstrates Simpson's paradox.[6]

		Number (%) Insured	
		Follow-up	**Not traced**
Race	**Black**	36 of 404 (8.9%)	91 of 1048 (8.7%)
	White	10 of 12 (83.3%)	104 of 126 (82.5%)
	Overall	46 of 416 (11.1%)	195 of 1174 (16.6%)

24. Sluggers Babe Ruth was the first great "slugger" in baseball. His record of 60 home runs in one season held for 34 years until Roger Maris hit 61 in 1961. Mark McGwire, with the aid of steroids, shocked the baseball world with 70 home runs in 1998. Listed below are the home run totals for each season McGwire played. Also listed are Babe Ruth's home run totals.

McGwire: 3*, 49, 32, 33, 39, 22, 42, 9*
9*, 39, 52, 58, 70, 65, 32*, 29*

Ruth: 54, 59, 35, 41, 46, 25, 47, 60, 54, 46, 49, 46, 41, 34, 22

a) Find the 5-number summary for McGwire's career.
b) Do any of his seasons appear to be outliers? Explain.
c) McGwire played in only 18 games at the end of his first big league season, and missed major portions of some other seasons because of injuries to his back and knees. Those seasons are marked with asterisks in the list above. Omit these values and make parallel boxplots comparing McGwire's career to Babe Ruth's.
d) Write a few sentences comparing the two sluggers.
e) Create a back-to-back stem-and-leaf display comparing the careers of the two players (Omit those injured seasons).
f) What aspects of the distributions are apparent in the stem-and-leaf displays that the boxplots did not clearly show?

25. Be quick! Avoiding an accident when driving can depend on reaction time. That time, measured from the moment the driver first sees the danger until he or she

[6]*Birth to Ten Study*, Medical Research Council, South Africa.

gets a foot on the brake pedal, is thought to follow a Normal model with a mean of 1.5 seconds and a standard deviation of 0.18 seconds.

a) Use the 68–95–99.7 Rule to draw the Normal model.
b) Write a few sentences describing driver reaction times.
c) What percent of drivers have a reaction time less than 1.25 seconds?
d) What percent of drivers have reaction times between 1.6 and 1.8 seconds?
e) What is the interquartile range of reaction times?
f) Describe the reaction times of the slowest 1/3 of all drivers.

26. Music and memory Is it a good idea to listen to music when studying for a big test? In a study conducted by some Statistics students, 62 people were randomly assigned to listen to rap music, Mozart, or no music while attempting to memorize objects pictured on a page. They were then asked to list all the objects they could remember. Here are the 5-number summaries for each group:

	n	Min	Q1	Median	Q3	Max
Rap	29	5	8	10	12	25
Mozart	20	4	7	10	12	27
None	13	8	9.5	13	17	24

a) Describe the *W*'s for these data: *Who, What, Where, Why, When, How.*
b) Name the variables and classify each as categorical or quantitative.
c) Create parallel boxplots as best you can from the summary statistics to display these results.
d) Write a few sentences comparing the performances of the three groups.

27. Mail Here are the number of pieces of mail received at a school office for 36 days:

123	70	90
80	78	72
52	103	138
112	92	93
118	118	106
95	131	59
151	115	97
100	128	130
66	135	76
143	100	88
110	75	60
115	105	85

a) Plot these data.
b) Find appropriate summary statistics.
c) Write a brief description of the school's mail deliveries.
d) What percent of the days actually lie within one standard deviation of the mean? Comment.

T **28. Birth order** Is your birth order related to your choice of program of study? A Statistics professor at a large university polled his students to find out what their specialist programs were and what position they held in the family birth order. The results are summarized in the table.

a) What percent of these students are oldest or only children?
b) What percent of Humanities specialists are oldest children?
c) What percent of oldest children are Humanities students?
d) What percent of the students are oldest children specializing in the Humanities?

		Birth Order*				
		1	2	3	4+	Total
Field	Math/Science	34	14	6	3	**57**
	Agriculture	52	27	5	9	**93**
	Humanities	15	17	8	3	**43**
	Other	12	11	1	6	**30**
	Total	**113**	**69**	**20**	**21**	**223**

* 1 = oldest or only child

29. Herbal medicine Researchers for the Herbal Medicine Council collected information on people's experiences with a new herbal remedy for colds. They went to a store that sold natural health products. There they asked 100 customers whether they had taken the cold remedy and, if so, to rate its effectiveness (on a scale from 1 to 10) in curing their symptoms. The Council concluded that this product was highly effective in treating the common cold.

a) Identify the *W*'s of these data.
b) Identify the variables, classify each as categorical or quantitative, and specify units, if relevant.
c) Are you confident about the Council's conclusion? Explain.

T **30. Birth order revisited** Consider again the data on birth order and university programs of study in Exercise 28.

a) What is the marginal distribution of field?
b) What is the conditional distribution of field for the oldest children?
c) What is the conditional distribution of field for the children born second?
d) Do you think that program of study appears to be independent of birth order? Explain.

31. Engines One measure of the size of an automobile engine is its "displacement," the total volume (in litres or cubic inches) of its cylinders. Summary statistics for several models of new cars are shown below. These displacements were measured in cubic inches.

Summary of Displacement	
Count	38
Mean	177.29
Median	148.5
StdDev	88.88
Range	275
25th %tile	105
75th %tile	231

a) How many cars were measured?
b) Why might the mean be so much larger than the median?
c) Describe the centre and spread of this distribution with appropriate statistics.
d) Your neighbour is bragging about the 227-cubic-inch engine he bought in his new car. Is that engine unusually large? Explain.
e) Are there any engines in this data set that you would consider to be outliers? Explain.
f) Is it reasonable to expect that about 68% of car engines measure between 88 and 266 cubic inches? (That's 177.29 ± 88.88.) Explain.
g) We can convert all the data from cubic inches to cubic centimetres (cc) by multiplying by 16.4. For example, a 200-cubic-inch engine has a displacement of 3280 cc. How would such a conversion affect each of the summary statistics?

32. Engines, again Horsepower is another measure commonly used to describe auto engines. Below are the summary statistics and histogram displaying horsepowers of the same group of 38 cars.

Summary of Horsepower	
Count	38
Mean	101.7
Median	100
StdDev	26.4
Range	90
25th %ile	78
75th %ile	125

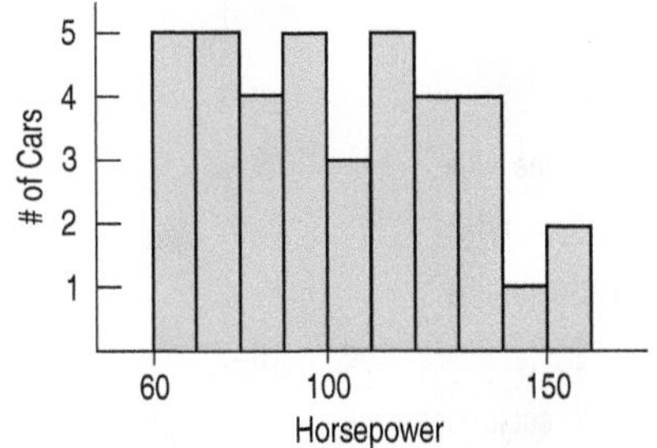

a) Describe the shape, centre, and spread of this distribution.
b) What is the interquartile range?
c) Are any of these engines outliers in terms of horsepower? Explain.
d) Do you think the 68–95–99.7 Rule applies to the horsepower of auto engines? Explain.
e) From the histogram, make a rough estimate of the percentage of these engines whose horsepower is within one standard deviation of the mean.
f) A fuel additive boasts in its advertising that it can "add 10 horsepower to any car." Assuming that is true, what would happen to each of these summary statistics if this additive were used in all the cars?

33. Toronto students Below is information about region of birth of Toronto public school students (2005). Make a pie chart with two categories: Canada and Outside of Canada. Next, make both a pie chart and a bar chart showing the breakdown for those born outside of Canada. In the pie chart, do some logical grouping, since there are so many categories (10 categories at most).

Region of Birth	Number of Students
Canada	192956
US	2297
English-speaking Caribbean/reginon	4853
Non-English speaking Caribbean	260
Central & South America & Mexico	4767
Central Africa	376
Eastern Africa	3240
Northern Africa	727
Southern Africa	199
Western Africa	1098
UK & Republic of Ireland	752
South & Western Europe	2700
Eastern Europe	9163
Central Asia	400
Eastern Asia	19335
Southeast Asia	3002
South Asia	22927
Western Asia	9605
Oceania	253

34. Pay According to the U.S. Bureau of Labor Statistics, the mean hourly wage for chief executives in 2009 was \$80.43 and the median hourly wage was \$77.27. By contrast, for general and operations managers, the mean hourly wage was \$53.15 and the median was \$44.55. Are these wage distributions likely to be symmetric, skewed left, or skewed right? Explain.

35. Toronto teams Below are salaries for the 2008 season for two popular Toronto sports teams.

Maple Leaf Salaries in $	Blue Jay Salaries in $
2 150 000	12 000 000
2 500 000	11 625 000
4 500 000	10 000 000
525 000	6 750 000
1 400 000	4 835 000
600 000	3 750 000
3 500 000	3 687 500
850 000	2 250 000
3 000 000	1 900 000
525 000	1 550 000
700 000	1 125 000
4 250 000	700 000
5 000 000	640 000
1 400 000	410 000
900 000	404 100
2 240 000	402 500
1 750 000	400 300
1 700 000	395 800
4 000 000	395 500
3 350 000	392 200
850 000	13 200 000
500 000	

Source: Data Courtesy Sports-Reference.com

a) Using appropriate methods, compare the two distributions with graphs and numbers.

b) Which team would you rather play for if you are a true superstar? A new guy starting out? An average player? (Just think financially, please.)

c) If you were the owner, which salary structure would you prefer to see, and why?

36. Men with brooms 2013 Canadian men will do the sweeping, as long as it is on ice. Canada excels not only in hockey, but also in the sport of curling. Below (at the bottom of this page) is a portion of the complete data set (available online at http://www.pearsoncanada.ca/deveaux) showing scores in games between Canada and the U.S. in the men's world championships since 1968. The tournaments were played in Canada in the following years: 1968, 1973, 1978, 1980, 1981, 1983, 1986, 1987, 1991, 1995, 1996, 1998, 1999, 2003, 2005, 2007, 2009, 2011, and 2013.

a) Clearly, Canada usually wins. Let's look at the score differential. Calculate the difference between the Canadian and the American score for each game. Examine the distribution with an appropriate plot. Describe the distribution as best you can (shape, centre, spread).

b) Check for possible Normality of the score differentials via a Normal probability plot, and comment on what the shape of the plot suggests.

c) Maybe the analysis above is missing some important factors, such as time or location. Examine possible dependence on time, graphically, and comment on what you see. How does this plot help you better understand the distribution examined in parts a) and b)?

d) Examine the possible home country effect (Does playing in Canada improve the point differential?) with appropriate graphs, and comment.

37. Some assembly required A company that markets build-it-yourself furniture sells a computer desk that is advertised with the claim "less than an hour to assemble." However, through post-purchase surveys the company has learned that only 25% of its customers succeeded in building the desk in under an hour; 5% said it took them over two hours. The company assumes that consumer assembly time follows a Normal model (exercise continues on the next page.

Year	Round	Canada	U.S.	Score
2013	8	Brad Jacobs	Brady Clark	7–2
2012	5	Glenn Howard	Heath McCormick	8–7
2011	8	Jeff Stoughton	Pete Fenson	5–3
2010	2	Kevin Koe	Pete Fenson	6–3
2009	13	Kevin Martin	John Shuster	9–6
⋮				
1970	6	Don Duguid	Art Tallackson	10–8
1969	F	Ron Northcott	Raymond "Bud" Somerville	9–6
1969	5	Ron Northcott	Raymond "Bud" Somerville	10–12
1968	SF	Ron Northcott	Raymond "Bud" Somerville	12–2
1968	2	Ron Northcott	Raymond "Bud" Somerville	10–6

Source: World Curling Federation (WCF), http://results.worldcurling.org/Associations.aspx.

a) Find the mean and standard deviation of the assembly time model. (Challenging)
b) One way the company could solve this problem would be to change the advertising claim. What assembly time should the company quote so that 60% of customers succeed in finishing the desk in that time?
c) Wishing to maintain the "less than an hour" claim, the company hopes that revising the instructions and labelling the parts more clearly can improve the 1-hour success rate to 60%. If the standard deviation stays the same, what new lower mean time does the company need to achieve?
d) Months later, another post-purchase survey shows that new instructions and part labelling did lower the mean assembly time, but only to 55 minutes. Nonetheless, the company did achieve the 60%-in-an-hour goal, too. How was that possible?

T **38. Profits** The following stem-and-leaf display shows profits as a percent of sales for 29 of the Forbes 500 largest U.S. corporations. The stems are split; each stem represents a span of 5%, from a loss of 9% to a profit of 25%.

a) Find the 5-number summary.
b) Draw a boxplot for these data.
c) Find the mean and standard deviation.
d) Describe the distribution of profits for these corporations.

chapter

6 Scatterplots, Association, and Correlation

Where are we going?

Is the price of running shoes related to how long they last? Is your alertness in class related to how much (or little) sleep you got the night before?

In this chapter, we'll look at relationships between two quantitative variables. We'll start by looking at scatterplots and describing the essence of what we see—the direction, form, and strength of the association. Then, as we did for histograms, we'll find a quantitative summary of what we learned from the display. We'll use the correlation to measure the strength of the association we see in the scatterplot.

"The cause of lightning" Alice said very decidedly, for she felt quite sure about this, "is the thunder—no, no" she hastily corrected herself, "I meant it the other way."

"It's too late to correct it," said the Red Queen, "When you've once said a thing, that fixes it, and you must take the consequences."

—Lewis Carroll

Humans have fairly big and heavy brains. Most other species have comparatively smaller brains. What affects the brain size for a particular species? Why are ours so large? For bigger brains, perhaps more time is needed in the womb, which is called the gestation period. Let's explore this relationship for a number of placental mammal species. Below, we plot one point for each of 23 different species. On the x-axis, we show the average gestation period, in days, for that species, and on the y-axis, we show the average neonatal brain weight (grams). When we plot one quantitative variable versus another quantitative variable, the resultant graph is called a **scatterplot**.

Who	Placental mammal species
What	Gestation period and neonatal brain weight
Units	Days; Grams
When	1974
Where	All over the Earth
How	Published in *The American Naturalist*
Why	To learn about vertebrate growth

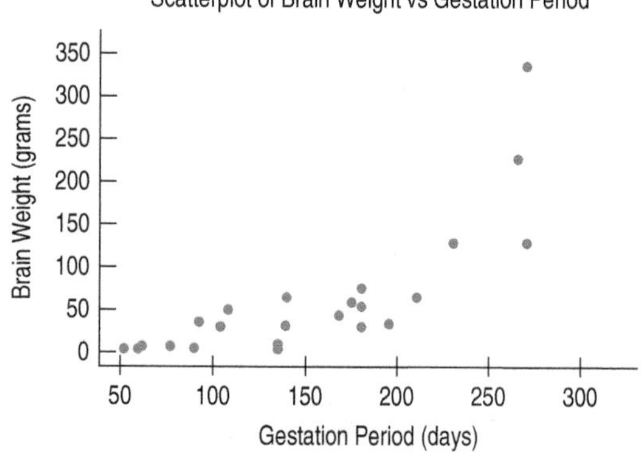

Figure 6.1

Scatterplot of neonatal brain weight versus gestation period for a number of placental mammal species. Is the general pattern roughly straight or curved? Are any species somewhat unusual compared with the rest? Are there distinct groups or clusters? (Do you recognize any of the plotted species?)

So what do you see? Brain weight clearly increases, in general, as gestation time increases, which we call a **positive association**. The general pattern though does not closely resemble a straight line but instead curves up. For example, if you look at greater gestation periods—say, over 200 days—the brain weight "dividend," or increase, for every extra day in the womb seems to be greater than when gestation periods are shorter (like an acceleration in brain-growth rate).

If it were a straight-line type pattern, this would not be the case, and every extra day of gestation would produce a roughly similar brain weight dividend anywhere in the plot. There may be other reasonable ways to measure brain size—for example, taking the logarithm of each brain weight—that might produce a more straight-line like or linear pattern, but let's postpone the topic of data re-expression until later in this chapter .

Are any species unusual? It is hard to say with such a curved pattern, but when you get to around 270 days in the plot, one species' brain weight seems unusually high, at about 340 g. And what about the species with a brain weight of 130 g? Does this appear unusually low to you? Hey, did you notice that one of those species is you guys (i.e., humans[1])?

Are there distinct groups or clusters present? Look very closely at the eight species to the left. Could this be a separate group whose pattern is curving upward very sharply? This is not at all clear from the plot, so let's find out more about the actual species included. It turns out that seven of those eight species on the left are carnivores, while all the remaining species are primates. Figure 6.2 is another plot of the same data, distinguishing the carnivores from the primates:

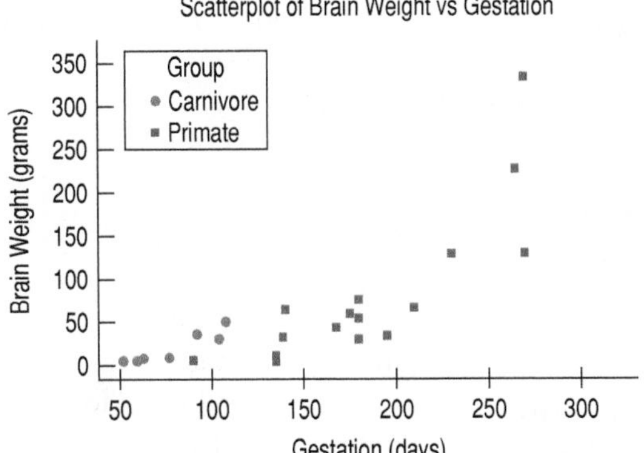

Figure 6.2
Introducing a categorical variable into the scatterplot of brain weight versus gestation, in order to separate the primates from the carnivores

The answer is pretty clear now. There are two distinct subgroups, with different relationships between brain weight and gestation for each group. The relationship between gestation and brain weight now appears to curve upward even more sharply than before, particularly for carnivores. It might be best to separate the data according to subgroup and analyze primates and carnivores separately. We have discovered the existence of an important lurking variable in our initial analysis, namely carnivore/primate status. Lurking variables are hidden variables that lurk or stand behind an analysis but may have an important influence on the relationship being studied.

Sometimes we can see separate subgroups by careful inspection of the scatterplot, but in other cases we have to use our knowledge of categorical variables that may make a difference, introduce the categorical variable into the plot, and see if the relationship between the two quantitative variables is similar or not for the different subgroups.

A S *Activity:* **Heights of Husbands and Wives.** Husbands are usually taller than their wives. Or are they?

Scatterplots may be the most common displays for data. Just looking at them, you can see patterns, trends, relationships, and even the occasional extraordinary value sitting apart from the others. As the great philosopher Yogi Berra[2] once said, "You can observe a lot by watching."[3] Scatterplots are the best way to start observing the relationship between two *quantitative* variables.

Relationships between variables are often at the heart of what we'd like to learn from data:

- Are grades higher now than they used to be?
- Do people tend to reach puberty at a younger age than in previous generations?
- Does applying magnets to parts of the body relieve pain? If so, are stronger magnets more effective?
- Do students learn better when they get more sleep?

Questions such as these relate two quantitative variables and ask whether there is an **association** between them. Scatterplots are the ideal way to *picture* such associations.

[1]I confess, I'm Klingon.
[2]Hall of Fame catcher and manager of the New York Mets and Yankees.
[3]But then he also said, "I really didn't say everything I said." So we can't really be sure.

You've seen one scatterplot earlier in this book—the plot of average wind speed versus day of year—in Chapter 4. When the x-variable is time, the scatterplot becomes a timeplot. Timeplots can show quite complex patterns, such as cyclical and seasonal patterns, often requiring what are called *time series methods of analysis*, which we won't be discussing in this text.

6.1 Scatterplots

A S **Activity: Making and Understanding Scatterplots.** See the best way to make scatterplots—using a computer.

Look for **Direction**: What's my sign—positive, negative, or neither?

Look for **Form**: straight, curved, something exotic, or no pattern?

Look for **Strength**: How much scatter?

Look for **Unusual Features:** Are there outliers or subgroups?

Everyone has seen a scatterplot. But if asked, many people would find it hard to say what to look for in a scatterplot. What are the main features or aspects of the plot that we should focus on and discuss in order to accurately report the story being told by our scatterplot?

The **direction** of the association is important. A pattern that runs from the upper left to the lower right is said to be *negative*. A pattern running the other way is called *positive*. In our example, there was a positive association between brain weight and gestation period.

The second thing to look for in a scatterplot is its **form**. If there is a straight-line relationship, it will appear as a cloud or swarm of points stretched out in a generally consistent, straight form. In this case, we describe the association as **linear**.

If the relationship isn't straight but curves gently, while still increasing or decreasing steadily like in the chapter opening example , we can often find ways to make it more nearly straight. But if it curves sharply—up and then down , we will need to resort to the multiple regression methods of Chapter 27.

The third feature to look for in a scatterplot is the **strength** of the relationship. At one extreme, do the points appear tightly clustered in a single stream (whether straight, curved, or bending all over the place)? Or, at the other extreme, does the swarm of points seem to form a vague cloud through which we can barely discern any trend or pattern? For example, the brain weight and gestation plot shows a moderate degree of scatter, indicating a moderately strong relationship between a species' gestation period and its neonatal brain weight.

Finally, always look for the unexpected. Often the most interesting thing to see in a scatterplot is something you never thought to look for. One example of such a surprise is an **outlier** standing away from the overall pattern of the scatterplot. Such a point is almost always interesting and always deserves special attention. *Clusters or subgroups* that stand away from the rest of the plot or show a trend in a different direction than the rest of the plot should raise questions about why they are different. They may be a clue that you should split the data into subgroups (like primates and carnivores in our example) instead of looking at all the cases together.

For Example COMPARING PRICES WORLDWIDE

If you travel overseas, you know that what's really important is not the amount in your wallet but the amount it can buy. UBS (one of the largest banks in the world) prepared a report comparing prices, wages, and other economic conditions in cities around the world for their international clients. Some of the variables they measured in 73 cities are *Cost of Living, Food Costs, Average Hourly Wage,* average number of *Working Hours* per year, average number of *Vacation Days,* hours of work (at the average wage) needed to buy an *iPod,* minutes of work needed to buy a *Big Mac,* and *Women's Clothing Costs.*[4] For your burger fix, you might want to live in Chicago, Toronto, or Tokyo where it takes only about 12 minutes of work to afford a Big Mac. In Mexico City, Jakarta, and Nairobi, you'd have to work more than two hours.

Of course, these variables are associated, but do they consistently reflect costs of living? Plotting pairs of variables can reveal how and even if they are associated. The variety of these associations illustrates different directions and kinds of association patterns you might see in other scatterplots.

QUESTION: Describe the patterns shown by each of these plots.

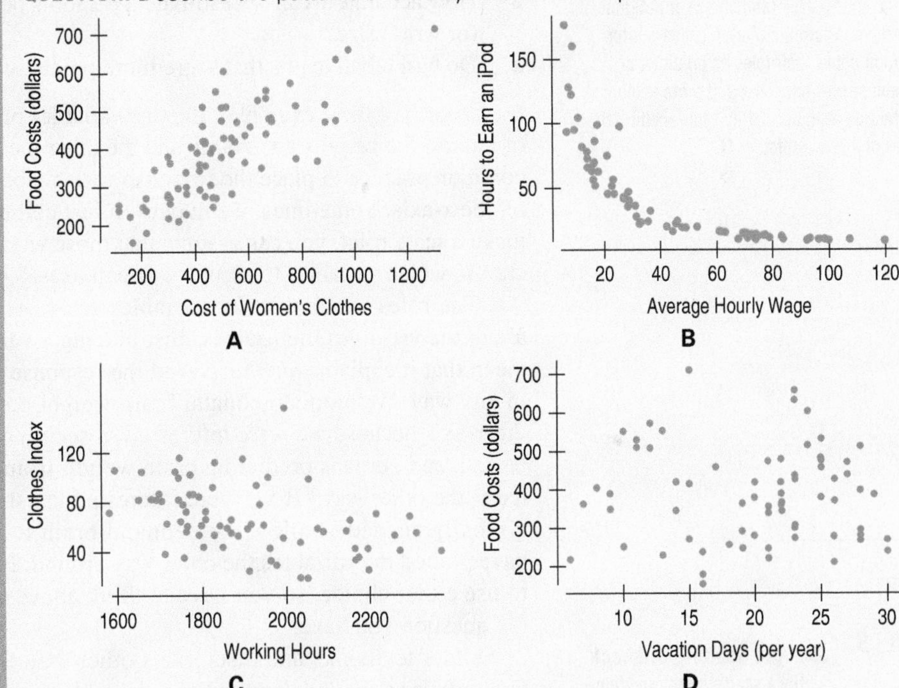

ANSWER: In Plot A, the association between *Food Costs* and *Cost of Women's Clothes* is positive and straight. In Plot B, the association between *Hours to Earn an iPod* and *Average Hourly Wage* has a negative direction, but the form is not straight. In Plot C, the association between the *Clothes Index* and *Working Hours* has a negative direction and is straight. There may be a high outlier. In Plot D, there does not appear to be any association between *Food Costs* and *Vacation Days.*

A figure from Newton's *Enumeration of lines of the third order*[5] labels the axes in the modern way.

Scatterplot history The idea of using two axes at right angles to define a field on which to display values can be traced back to the philosopher, scientist, and mathematician René Descartes (1596–1650). The playing field he defined in this way is formally called a Cartesian plane in his honour. Sir Isaac Newton (1642–1727) may have been the first to use the now-standard convention of denoting the vertical axis *y* and the horizontal axis *x*.

---

[4]Detail of the methodology can be found in the report *Prices and Earning: A comparison of purchasing power around the globe/2009 edition.* http://www.qfc.com.qa/files/Prices_and_Earnings.pdf

[5]Sir Isaac Newton's *Enumeration of lines of the third order, generation of curves by shadows, organic description of curves, and construction of equations by curves.* p. 86. Translated from the Latin. With notes and examples, by C.R.M. Talbot. Available at quod.lib.umich.edu/u/umhistmath/ABQ9451.0001.001/81?rgn=full+text;view=pdf

Both Descartes and Newton used the Cartesian plane to plot functions, as you have probably done yourself. Plotting individual data values works in a similar way. Each case is displayed as a dot or symbol on a scatterplot at a position that corresponds to its values on two variables—one measured along the x-axis and the other along the y-axis. When you work with data, you should label the axes with the names of the variables rather than just using y and x.

Roles for Variables

Which variable should go on the x-axis and which on the y-axis? What we want to know about the relationship can tell us how to make the plot. We often have questions:

- Do older houses sell for less than newer ones of comparable size and quality?
- Do people who drink more coffee tend to have higher blood pressure?
- How accurately can we estimate a person's percent body fat by measuring their waist (or wrist) size?
- Do basketball teams that score more points sell more tickets to their games?

NOTATION ALERT

In Statistics, the assignment of variables to the x- and y-axes (and choice of notation for them in formulas) often conveys information about their roles as predictor or response variable. So x and y are reserved letters as well, but not just for labelling the axes of a scatterplot.

In each of these examples, the two variables play different roles. We'll call our variable of interest the **response variable** and the other the **explanatory** or **predictor variable**. It's common practice to place the response variable on the y-axis and the explanatory variable on the x-axis. Sometimes, we may even just call them the **x-** and **y-variables**. When you make a scatterplot, you can assume that those who view it will think this way, so carefully choose which variables to assign to which axes.

The roles we choose for variables are more about how we *think* about them than about the variables themselves. Just placing a variable on the x-axis doesn't necessarily mean that it explains *anything*. And the response variable may not actually *respond* to it in any way. We plotted neonatal brain weight on the y-axis against gestation period on the x-axis because we were interested in seeing how brain weight depends on gestation period, and perhaps predicting brain weight from gestation period. Could we have plotted it the other way? If we were interested in estimating how much time in gestation is typically needed to allow the neonatal brain to reach some particular size, we might have plotted the variables the other way around. For some scatterplots, it can make sense to use either choice, so you have to think about how the choice of role helps to answer the question you have.

A S *Self-Test:* **Scatterplot Check.** Can you identify a scatterplot's direction, form, and strength?

Older textbooks and disciplines other than Statistics sometimes refer to the x- and y-variables as the *independent* and *dependent* variables, respectively. The idea was that the y-variable depended on the x-variable and the x-variable acted independently to make y respond. These names, however, conflict with other uses of the same terms in Statistics.

6.2 Correlation

Data collected from students in a Statistics class included their *Height* (in inches) and *Weight* (in pounds), and are displayed in Figure 6.3. It's no great surprise that there is a positive association between the two. As you might suspect, taller students tend to weigh more. (If we had reversed the roles and chosen weight as the explanatory variable, we might say that heavier students tend to be taller.)[6] And the form of the scatterplot is fairly straight as well, although there appears to be a high outlier.

[6]The young son of one of the authors, when told (as he often was) that he was tall for his age, used to point out that, actually, he was young for his height.

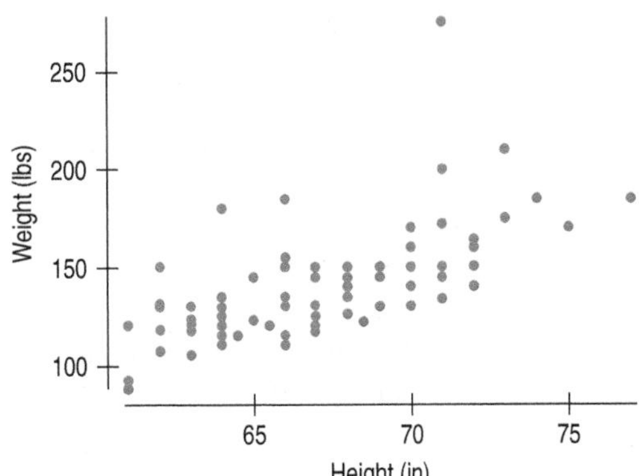

Figure 6.3

Weight (lbs) versus *Height (in)* from a Statistics class.

Who	Students
What	Height (inches), Weight (pounds)
Where	Ithaca, NY
Why	Data for class
How	Survey

The pattern in the scatterplot looks straight, and the positive association is clear, but how strong is that association? If you had to put a number (say, between 0 and 1) on the strength, what would you say? The strength of the association shouldn't depend on the units we choose for x or y. For example, if we had measured weight in *kg* instead of *lbs* and height in *cm* instead of *inches,* the plot (Figure 6.4) would look the same.

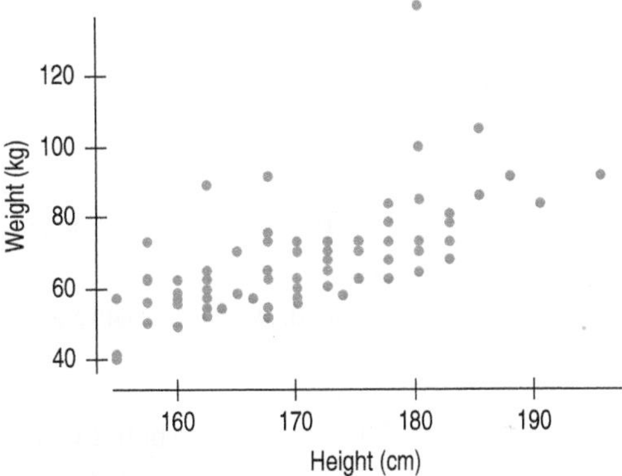

Figure 6.4

Changing the units (from pounds to kilograms for *Weight* and from inches to centimetre's for *Height*) did not change the direction, form, or strength of the association between the variables.

If we subtract the mean from each variable, that shouldn't change the strength of the association either. We would just move the means of both variables to zero. This makes it easier to see what each point contributes to the strength of the association because we can tell which points are above and which are below the means of each variable. Figure 6.5 colours them to make this even clearer. The green points in the upper right (first quadrant) and lower left (third quadrant) are consistent with a positive association, while the red points in the other two quadrants are consistent with a negative association. Points lying on the x- or y-axis don't really add any information; they are coloured blue.

A S ***Activity:*** Correlation. Here's a good example of how correlation works to summarize the strength of a linear relationship and disregard scaling.

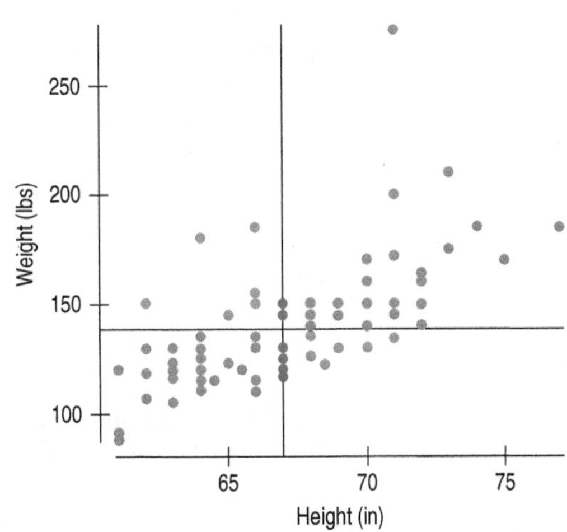

Figure 6.5

In this scatterplot, points are coloured according to how they affect the association: green for positive, red for negative, and blue for neutral.

We said that the units shouldn't affect our measure of strength. Now that we've subtracted means, a natural way to remove the units is to standardize each variable and work instead with the z-scores. Recall that to find z-scores for any variable, we subtract its mean and divide by its standard deviation. So instead of plotting (x, y), we plot

$$(z_x, z_y) = \left(\frac{x - \bar{x}}{s_x}, \frac{y - \bar{y}}{s_y}\right)$$

z-scores do just what we want. They centre the plotted points around the origin and they remove the original units, replacing them with standard deviations. As we've seen, measuring differences in standard deviations is a fundamental idea in Statistics.

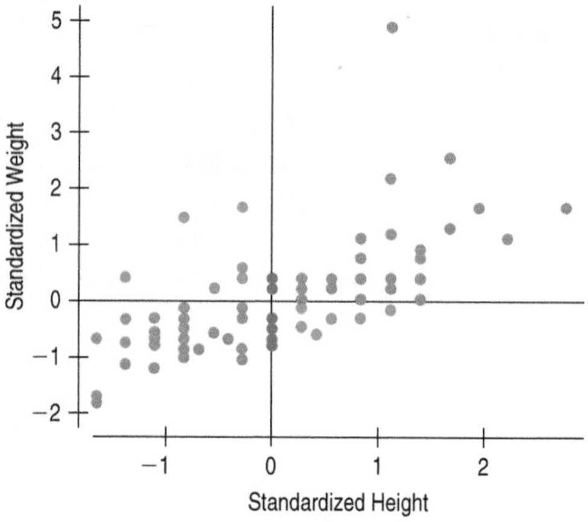

Figure 6.6

Standardized *Weight* and *Height*. We use the standard deviation to judge the distance of each point from the center of the plot and the influence it should have.

For the green points in Figure 6.6, both z-scores are positive, or both are negative. In either case, their product, $z_x z_y$, is positive. But the red points have opposite signs, so their products are negative. We can measure the strength of the association by adding up these products:

$$\sum z_x z_y$$

This summarizes both the direction and strength of the association. Points farther from the origin have larger z-scores, so they'll contribute more to the sum. But the sum can keep growing as we consider more points. To keep that from happening, the natural (for Statisticians, anyway) thing to do is to divide the sum by $n - 1$.[7]

The result is the famous **correlation coefficient**:

$$r = \frac{\sum z_x z_y}{n - 1}$$

Dividing the sum by $n - 1$ serves two purposes. It adjusts the strength for the number of points and it makes the correlation lie between values of -1 and $+1$. For the students' heights and weights, the correlation is 0.644. Because it is based on z-scores, which have no units, the correlation has no units either. It will stay the same if you change from inches to centimetres, fathoms, or Ångstroms.

There is an alternative formula for the correlation coefficient, using x and y in their original units, that is also faster to compute:

$$r = \frac{s_{xy}}{s_x \times s_y}$$

where

$$s_{xy} = \frac{\sum (x - \bar{x})(y - \bar{y})}{(n - 1)}$$

is called the covariance of x and y. Plug in x for y in the covariance formula, and you will see that the covariance of x and x is just the variance of x. Covariance generalizes

[7] Yes, the same $n - 1$ as in the standard deviation calculation. And we offer the same promise to explain it better later.

■ **NOTATION ALERT**

The letter r is always used for correlation, so you can't use it for anything else in Statistics. Whenever you see an r, it's safe to assume it's a correlation.

the concept of variance. And the covariance of x standardized and y standardized produces the correlation. For a faster computational formula, just plug in the formulas for the two standard deviations, and the covariance, and cancel out all the "$n - 1$" terms.[8]

BY HAND FINDING THE CORRELATION COEFFICIENT

Start with the summary statistics for both variables: $\bar{x}, \bar{y}, s_x$, and s_y. Then find the deviations as we did for the standard deviation, but now in both x and y: $(x - \bar{x})$ and $(y - \bar{y})$. Form the products for all data pairs, and add up all the products. Divide the sum by $(n - 1)$ to get the covariance, then divide again by the product of the two standard deviations $s_x \times s_y$, to get the correlation coefficient.

Here we go:

Suppose the data pairs are:

x	6	10	14	19	21
y	5	3	7	8	12

We calculate $\bar{x} = 14, \bar{y} = 7, s_x = 6.20$, and $s_y = 3.39$.

Deviations in x	Deviations in y	Product
$6 - 14 = -8$	$5 - 7 = -2$	$-8 \times -2 = 16$
$10 - 14 = -4$	$3 - 7 = -4$	16
$14 - 14 = 0$	$7 - 7 = 0$	0
$19 - 14 = 5$	$8 - 7 = 1$	5
$21 - 14 = 7$	$12 - 7 = 5$	35

Add up the products: $16 + 16 + 0 + 5 + 35 = 72$.
Dividing the above value by $(n - 1) = (5 - 1)$ gives a covariance of 18.00.
Divide by $s_x \times s_y = 6.20 \times 3.39 = 21.02$ to get the correlation coefficient:

$$r = 18/21.02 = 0.856$$

Assumptions and Conditions for Correlation

Correlation measures the strength of *linear* association between two *quantitative* variables. To interpret a correlation, we must assume that there is a true underlying linear relationship. Of course, in general, we can't know that. But we *can* check whether that assumption is plausible by looking at the data we do have. To distinguish between what we *assume* about variables from what we can *check* by looking at the data, we'll call the things we check *conditions*. Checking the conditions almost always means either thinking about how the data were collected or looking at a graph—but if you follow the first three rules, you'll have already made the plot, so that takes no extra effort.

There are three conditions to check before you use a correlation:

- **Quantitative Variables Condition** Don't make the common error of calling an association involving a categorical variable a correlation. Correlation is only about quantitative variables.
- **Straight Enough Condition** The best check for the assumption that the variables are truly linearly related is to look at the scatterplot to see whether it looks reasonably straight. That's a judgment call, but not a difficult one.

[8]Which gives $r = \dfrac{\sum(x - \bar{x})(y - \bar{y})}{\sqrt{\sum(x - \bar{x})^2 \sum(y - \bar{y})^2}}$

- **No Outliers Condition** Outliers can distort the correlation dramatically, making a weak association look strong or a strong one look weak. Outliers can even change the sign of the correlation. But it's easy to see outliers in the scatterplot, so to check this condition, just look.

Correlations are often reported without supporting data or plots. Nevertheless, you can still think about the conditions. Be cautious in interpreting (or accepting others' interpretations of) a correlation when you can't check the conditions.

A S *Activity:* **Correlation and Relationship Strength.** What does a correlation of 0.8 look like? How about 0.3?

For Example CORRELATIONS FOR SCATTERPLOT PATTERNS

Look back at the scatterplots of the economic variables in cities around the world (p. 161). The correlations for those plots are 0.774, −0.783, −0.576, and −0.022, respectively.

QUESTION: Check the conditions for using correlation. If you feel they are satisfied, interpret the correlation.

ANSWER: All of the variables examined are quantitative and none of the plots shows a severe outlier. However, the relationship between *Hours to Earn an iPod* and *Average Wage* is not straight, so the correlation coefficient isn't an appropriate summary. For the others:

A correlation of 0.774 between *Food Costs* and *Women's Clothing Costs* indicates a fairly strong positive association.

A correlation of −0.576 between *Clothes Index* and *Working Hours* indicates a moderate negative association.

The small correlation value of −0.022 between *Food Costs* and *Vacation Days* suggests that there may be no linear association between them.

Just Checking

Your Statistics teacher tells you that the correlation between the scores (points out of 50) on Exam 1 and Exam 2 was 0.75.

1. Before answering any questions about the correlation, what would you like to see? Why?
2. If she adds 10 points to each Exam 1 score, how will this change the correlation?
3. If she standardizes scores on each exam, how will this affect the correlation?
4. In general, if someone did poorly on Exam 1, is the student likely to have done poorly or well on Exam 2? Explain.
5. If someone did poorly on Exam 1, can you be sure that the student did poorly on Exam 2 as well? Explain.

Step-by-Step Example LOOKING AT ASSOCIATION

When your blood pressure is measured, it is reported as two values, systolic blood pressure and diastolic blood pressure. How are these variables related to each other? Do they tend to be both high or both low? Let's examine their relationship with a scatterplot.

THINK ➡ Plan State what you are trying to investigate.	I'll examine the relationship between two measures of blood pressure.

Variables Identify the two quantitative variables whose relationship we wish to examine. Report the *W*'s and be sure both variables are recorded for the same individuals.	The variables are systolic and diastolic blood pressure (SBP and DBP) recorded in millimetres of mercury (mm Hg) for each of 1406 participants in the Framingham Heart Study, a long-running health study in Framingham, MA.[9]
Plot Make the scatterplot. Use a computer program or graphing calculator if you can.	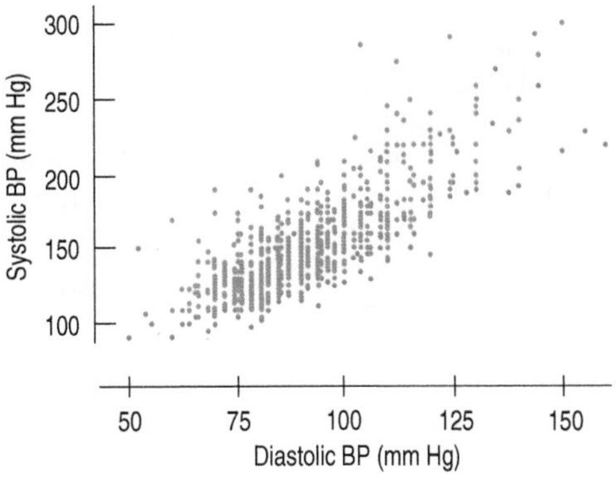
Check the conditions.	✓ **Quantitative Variables Condition:** Both SBP and DBP are quantitative with units of mm Hg. ✓ **Straight Enough Condition:** The scatterplot is quite straight. ✓ **Outliers?** There are a few cases with moderately high SBP (compared with their DBP), but not extreme enough to matter, considering the large number of cases.
REALITY CHECK Looks like a strong positive linear association. We shouldn't be surprised if the correlation coefficient is positive and fairly large.	I have two quantitative variables that satisfy the conditions, so correlation is a suitable measure of association.
SHOW ➡ **Mechanics** We usually calculate correlations with technology. Here we have 1406 cases, so we'd never try it by hand.	The correlation coefficient is $r = 0.792$.
TELL ➡ **Conclusion** Describe the direction, form, and strength you see in the plot, along with any unusual points or features. Be sure to state your interpretations in the proper context.	The scatterplot shows a positive association, with higher SBP going with higher DBP. The plot is generally straight with a moderate amount of scatter. The correlation of 0.792 is consistent with what I saw in the scatterplot.

Correlation Properties

Here's a useful list of facts about the correlation coefficient:

- The sign of a correlation coefficient gives the direction of the association.
- Correlation is always between −1 and +1. Correlation *can* be exactly equal to −1.0 or +1.0, but these values are unusual in real data because they mean that all the data points fall *exactly* on a single straight line.

A S *Activity:* **Construct Scatterplots with a Given Correlation.** Try to make a scatterplot that has a given correlation. How close can you get?

[9]www.nhlbi.nih.gov/about/framingham

HEIGHT AND WEIGHT, AGAIN

We could have measured the students' weights in stones. A stone is a measure in the pre-metric UK system of measures equal to 14 pounds. And we could have measured heights in hands. Hands are still commonly used to measure the heights of horses. A hand is 4 inches. But no matter what units we use to measure the two variables, *the correlation* stays the same.

- Correlation treats x and y symmetrically. The correlation of x with y is the same as the correlation of y with x.
- Correlation has no units. This fact can be especially appropriate when the data's units are somewhat vague to begin with (IQ score, personality index, socialization, and so on). Correlation is sometimes given as a percentage, but we discourage that because it suggests a percentage of *something*—and correlation, lacking units, has no "something" of which to be a percentage.
- Correlation is not affected by changes in the centre or scale of either variable. Changing the units or baseline of either variable has no effect on the correlation coefficient. Correlation depends only on the z-scores, and they are unaffected by changes in centre or scale.
- Correlation measures the strength of the *linear* association between the two variables. Variables can be strongly associated but still have a small correlation if the association isn't linear.
- Correlation is sensitive to outliers. A single outlying value can make a small correlation large or make a large one small.

How strong is strong? You'll often see correlations characterized as "weak," "moderate," or "strong," but be careful. There's no agreement on what those terms mean. The same numerical correlation might be strong in one context and weak in another. You might be thrilled to discover a correlation of 0.7 between the new summary of the economy you've come up with and stock market prices, but you'd consider it a design failure if you found a correlation of "only" 0.7 between two tests intended to measure the same skill. Deliberately vague terms like "weak," "moderate," or "strong" that describe a linear association can be useful additions to the numerical summary that correlation provides. But be sure to include the correlation and show a scatterplot so others can judge for themselves.

For Example CHANGING SCALES

RECAP: Several measures of prices and wages in cities around the world show a variety of relationships, some of which we can summarize with correlations.

QUESTION: Suppose that, instead of measuring prices in dollars and recording work time in hours, we had used Euros and minutes. How would those changes affect the conditions, the values of correlation, or our interpretation of the relationships involving those variables?

ANSWER: Not at all! Correlation has no units, so the conditions, value of r, and interpretation are all unaffected by changes to the units of the individual variables.

Correlation Tables

It is common in some fields to compute the correlations between every pair of variables in a collection of variables and arrange these correlations in a table (see Table 6.1). The rows and columns of the table name the variables, and the cells hold the correlations.

Correlation tables are compact and give a lot of summary information at a glance. They can be an efficient way to start to look at a large data set, but a dangerous one. By presenting all of these correlations without any checks for linearity and outliers, the correlation table risks showing truly small correlations that have been inflated by outliers, truly large correlations that are hidden by outliers, and correlations of any size that may be meaningless because the underlying form is not linear.

The diagonal cells of a correlation table always show correlations of exactly 1. (Can you see why?) Correlation tables are commonly offered by computer statistics packages. These same packages often offer simple ways to make all the scatterplots that go with these correlations.[10]

[10]A table of scatterplots arranged just like a correlation table is sometimes called a *scatterplot matrix*.

Table 6.1

A correlation table of data reported by *Forbes* magazine on several financial measures for large companies. From this table, can you be sure that the variables are linearly associated and free from outliers?

	Assets	Sales	Market Value	Profits	Cash Flow	Employees
Assets	1.000					
Sales	0.746	1.000				
Market Value	0.682	0.879	1.000			
Profits	0.602	0.814	0.968	1.000		
Cash Flow	0.641	0.855	0.970	0.989	1.000	
Employees	0.594	0.924	0.818	0.762	0.787	1.000

6.3 Warning: Correlation ≠ Causation

Whenever we have a strong correlation, it's tempting to try to explain it by imagining that the predictor variable has *caused* the response to change. Humans are like that; we tend to see causes and effects in everything.

Sometimes this tendency can be amusing. A scatterplot (Figure 6.7) of the human population (y) of Oldenburg, Germany, in the beginning of the 1930s plotted against the number of storks nesting in the town (x) shows a tempting pattern.

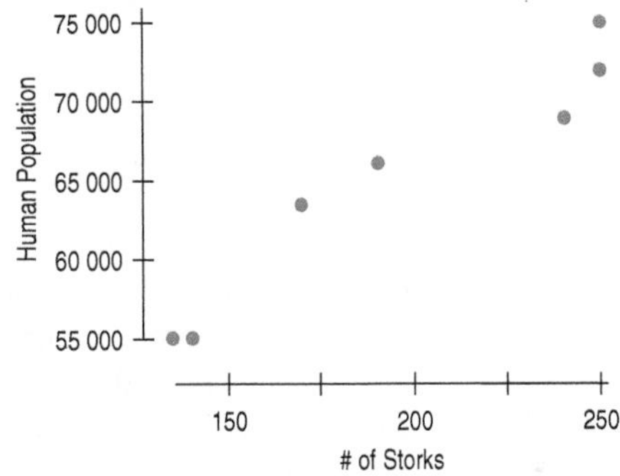

Figure 6.7

A scatterplot of the number of storks in Oldenburg, Germany, plotted against the population of the town for seven years in the 1930s. The association is clear. How about the causation?

(*Source: Ornithologishe Monatsberichte*, 44, no. 2)

Anyone who has seen the beginning of the movie *Dumbo* remembers Mrs. Jumbo anxiously waiting for the stork to bring her new baby. Even though you know it's silly, you can't help but think for a minute that this plot shows that storks are the culprits. The two variables are obviously related to each other (the correlation is 0.97!), but that doesn't prove that storks bring babies.

It turns out that storks nest on house chimneys. More people means more houses, more nesting sites, and so more storks. The causation is actually in the *opposite* direction, but you can't tell from the scatterplot or correlation. You need additional information—not just the data—to determine the real mechanism.

Does cancer cause smoking? Even if the correlation of two variables is due to a causal relationship, the correlation itself cannot tell us what causes what.

Sir Ronald Aylmer Fisher (1890–1962) was one of the greatest statisticians of the twentieth century. Fisher testified in court (paid by the tobacco companies) that a causal relationship might underlie the correlation of smoking and cancer:

> "Is it possible, then, that lung cancer . . . is one of the causes of smoking cigarettes? I don't think it can be excluded . . . the pre-cancerous condition is one involving a certain amount of slight chronic inflammation . . .
>
> A slight cause of irritation . . . is commonly accompanied by pulling out a cigarette, and getting a little compensation for life's minor ills in that way. And . . . is not unlikely to be associated with smoking more frequently."

Ironically, the proof that smoking indeed is the cause of many cancers came from experiments conducted following the principles of experiment design and analysis that Fisher himself developed—and that we'll see in Chapter 11.

Scatterplots and correlation coefficients *never* prove causation. That's one reason it took so long for the U.S. Surgeon General to get warning labels on cigarettes. Although there was plenty of evidence that increased smoking was *associated* with increased levels of lung cancer, it took years to provide evidence that smoking actually *causes* lung cancer.

A scatterplot of the damage (in dollars) caused to a house by fire would show a strong correlation with the number of firefighters at the scene. Surely the financial damage doesn't cause firefighters. And firefighters do seem to cause damage, spraying water all around and chopping holes. Does that mean we shouldn't call the fire department? Of course not. There is an underlying variable that leads to both more damage and more firefighters: the size of the blaze. We say that damage and firefighters are exhibiting a **common response** to the size of the blaze.

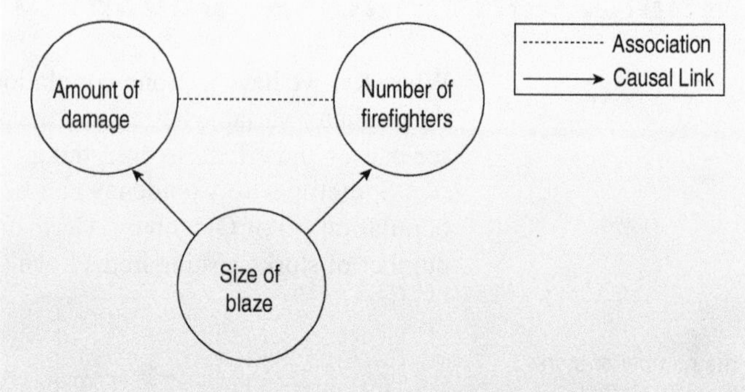

Figure 6.8
Amount of damage and number of firefighters exhibit a common response to the size of the blaze.

A hidden variable that stands behind a relationship but that may affect or change our understanding of the relationship is called a **lurking variable**. You can often debunk cause-and-effect claims made about data by finding a variable lurking behind the scenes.

In the example of cancer and smoking, while it is now well established that smoking causes cancer, what might have been some other possible explanations for the observed association? Maybe cancer causes increased smoking (as suggested by Fisher). Maybe both cancer and smoking are a common response to some gene(s), making one genetically predisposed to both developing cancer and nicotine addiction. Or—one more theory—maybe smokers differ from nonsmokers in certain lifestyle characteristics (alcohol, diet, exercise, etc.) and perhaps it is those lifestyle differences that cause the cancer, not the smoking. See Figure 6.9 below showing a *confounding* of effects.

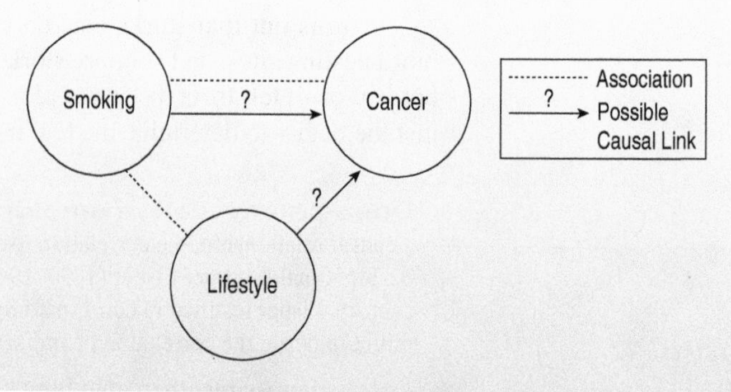

Figure 6.9
The effect of smoking on cancer is confounded with the effect of lifestyle difference.

Lifestyle and smoking effects are **confounded**, making it difficult to determine which (if any) of the two confounded variables is the causal culprit. Maybe both have a causal link to cancer, maybe neither. Maybe some other undiscovered lurking variable is the real culprit, its hidden effects confounded with smoking habits. Such confounding and common response explanations for the increased incidence of cancer among smokers were eventually ruled out, through great effort over many years, via many different types of corroborative investigations. Proving causation in the absence of controlled experiments is exceedingly difficult.

6.4 Straightening Scatterplots

Correlation is a suitable measure of strength for straight relationships only. When a scatterplot bends but shows a steadily increasing pattern or a steadily decreasing pattern, we can often straighten the form of the plot by re-expressing one or both variables.

Some camera lenses have an adjustable aperture, the hole that lets the light in. The size of the aperture is expressed by a mysterious number called the f/stop. Each increase of one f/stop number corresponds to a halving of the light that is allowed to come through. The f/stops of one digital camera are

f/stop:	2.8	4	5.6	8	11	16	22	32

When you halve the shutter speed, you cut down the light, so you have to open the aperture one notch. We could experiment to find the best f/stop value for each shutter speed. A table of recommended shutter speeds and f/stops for a camera lists the relationship like this:

Shutter speed:	1/1000	1/500	1/250	1/125	1/60	1/30	1/15	1/8
f/stop:	2.8	4	5.6	8	11	16	22	32

The correlation of these shutter speeds and f/stops is 0.979. That sounds pretty high. But a high correlation doesn't necessarily mean a *linear* relationship. When we check the scatterplot (we *always* check the scatterplot), it shows that something is not quite right:

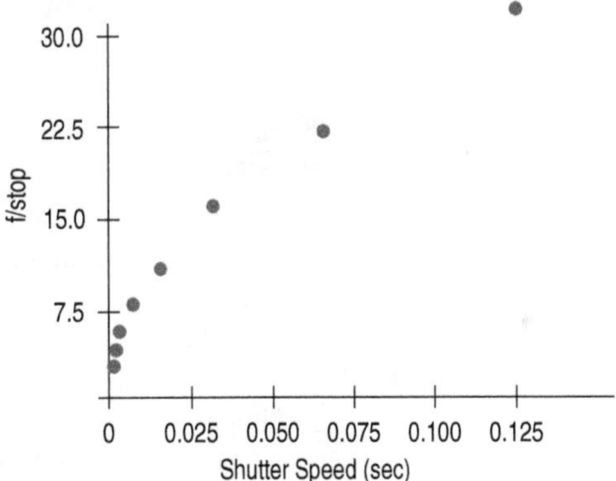

Figure 6.10

A scatterplot of *f/stop* vs. *Shutter Speed* shows a bent relationship.

We can see that the f/stop is not *linearly* related to the shutter speed. Can we find a transformation of *f/stop* that straightens out the line?

In Chapter 4, we saw that **re-expressing** data by raising each value to a power or taking the logarithm could make the shape of the histogram of the data more nearly symmetric and could make the spreads of different groups more nearly equal. Now we hope to make a curved relationship straighter. Remarkably, these three goals are closely related, so we can use the same methods to achieve all of them.

The Ladder of Powers

The secret is to select a re-expression from a simple collection of functions—the powers and the logarithm. We can raise each data value in a quantitative variable to the same power. It is important to recall a few facts about powers:

- The ½ power is the same as taking the square root.
- The –1 power is the reciprocal: $y^{-1} = 1/y$
- Putting those together, $y^{-1/2} = 1/\sqrt{y}$

In addition, we'll use the logarithm in place of the "0" power. Although it doesn't matter for this purpose, the base 10 logarithm is usually easier to think about.[11]

Re-expressions in scientific laws

Scientific laws often include simple re-expressions. For example, in Psychology, Fechner's Law states that sensation increases as the logarithm of stimulus intensity ($S = k \log R$).

[11] You may have learned facts about calculating with logarithms. Don't worry; you won't need them here.

Where to start? It turns out that certain kinds of data are more likely to be helped by particular re-expressions. Knowing that gives you a good place to start your search for a re-expression.

Power	Name	Comment
2	The square of the data values, y^2.	Try this for unimodal distributions that are skewed to the left or to re-express *y* in a scatterplot that bends downward.
1	The raw data—no change at all. This is "home base." The farther you step from here up or down the ladder, the greater the effect.	Data that can take on both positive and negative values with no bounds are less likely to benefit from re-expression.
1/2	The square root of the data values, $\sqrt{y}$.	Counts often benefit from a square root re-expression. For counted data, start here.
"0"	Although mathematicians define the "0-th" power differently,[12] for us the place is held by the logarithm. You may feel uneasy about logarithms. Don't worry; the computer or calculator does the work.[13]	Measurements that cannot be negative, and especially values that grow by percentage increases such as salaries or population sizes, often benefit from a log re-expression. When in doubt, start here. If your data have zeros, try adding a small constant to each value before finding the logs.
−1/2	The reciprocal square root, $1/\sqrt{y}$.	This is an uncommon re-expression, but sometimes useful. Change the sign and take the *negative* of the reciprocal square root if you want to preserve the direction of relationships.
−1	The reciprocal, $1/y$.	Ratios of two quantities (kilometres per hour, for example) often benefit from a reciprocal. (You have about a 50–50 chance that the original ratio was taken in the "wrong" order for simple statistical analysis and would benefit from re-expression.) Often, the reciprocal will have simple units (hours per km). Put a *negative* sign in front if you want to preserve the direction of relationships. If your data have zeros, try adding a small constant to all values before finding the reciprocal.

The powers and logarithm have an important advantage as re-expressions: They are *invertible*. That is, you can get back to the original values easily. For example, if we re-express *y* by the square root, we can get back to the original values by squaring the re-expressed values. To undo a logarithm, raise 10 to the power of each value.[14]

The positive powers and the logarithm preserve the order of values—the larger of two values is still the larger one after re-expression. Negative powers reverse the order ($4 > 2$ but $\frac{1}{4} < \frac{1}{2}$) so we sometimes negate the values ($-\frac{1}{4} > -\frac{1}{2}$) to preserve the order.

Most important, the *effect* that any of these re-expressions has on straightening a scatterplot is ordered. Choose a power. If it doesn't straighten the plot by enough, then try the next power. If it overshoots so the plot bends the other way, then step back. That insight makes it easy to find an effective re-expression, and it leads to referring to these functions as the **ladder of powers**.

F/Stops Again

Following the advice in the table, we can try taking the logarithm of the f/stop values.

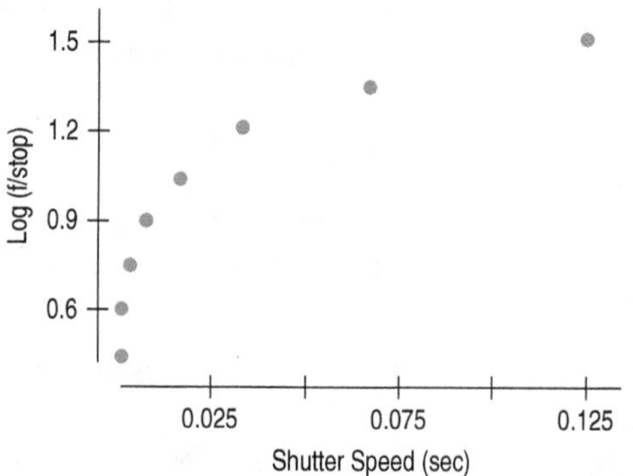

Figure 6.11
Log(f/stop) values vs. *Shutter Speed*.

[12]You may remember that for any nonzero number *y*, $y^0 = 1$ This is not a very exciting re-expression for data; every data value would be the same. We use the logarithm in its place.

[13]Your calculator or software package probably gives you a choice between "base 10" logarithms and "natural (base *e*)" logarithms. Don't worry about that. It doesn't matter at all which you use; they have exactly the same effect on the data. If you want to choose, base 10 logarithms can be a bit easier to interpret.

[14]Your computer or calculator knows how.

But, as Figure 6.11 shows, the logarithm of f/stop values is even *more* curved vs. *Shutter Speed*. The Ladder of Powers says that we'd do better if we move in the opposite direction. That would take us to the square.

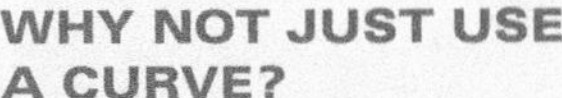
Figure 6.12
(f/stop)^2 vs. *Shutter Speed.*

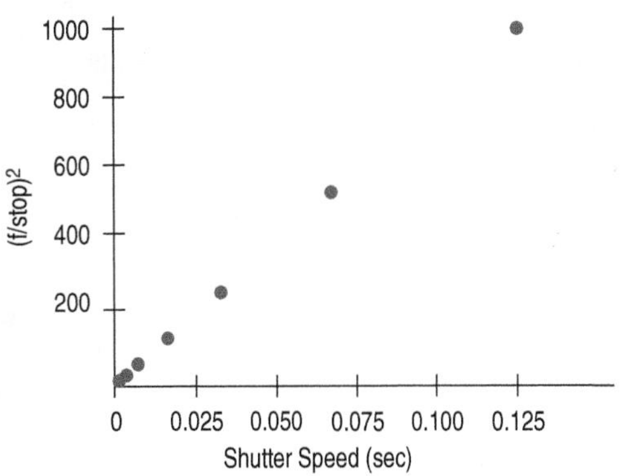

WHY NOT JUST USE A CURVE?

While it is possible to fit various curves to data using our familiar "least-squares" criterion, the mathematics and calculations are considerably more difficult. Straight lines are easier to fit and to understand. We know how to think about the slope and the *y*-intercept, for example. Powerful and simple statistical methods are available for analyzing linear associations (discussed in later chapters). So, it's usually better to linearize.

That scatterplot looks much straighter.[15] It's appropriate to summarize the strength of this relationship with a correlation, which turns out to be 0.998.[16] (If the square was in the right direction but not strong enough, you might next try the stronger exponential function, though not in our ladder.)

Could we also try re-expressing the *x*-variable? Sure, or even both variables. In fact here, taking the square root of *Shutter Speed* works just as well!

But watch out for scatterplots that turn around. Re-expression can straighten many bent relationships, but not those that go up and then down or down and then up. You should refuse to analyze such data with methods that require a roughly linear form.

WHAT CAN GO WRONG?

- **Don't say "correlation" when you mean "association."** How often have you heard the word "correlation"? Chances are pretty good that when you've heard the term, it's been misused. When people want to sound scientific, they often say "correlation" when talking about the relationship between two variables. It's one of the most widely misused Statistics terms, and given how often statistics are misused, that's saying a lot. One of the problems is that many people use the specific term correlation when they really mean the more general term *association.* "Association" is a deliberately vague term describing the relationship between two variables.

  "Correlation" is a precise term that measures the strength and direction of the linear relationship between quantitative variables.

- **Don't correlate categorical variables.** Did you know that there's a strong correlation between playing an instrument and drinking coffee? No? One reason might be that the statement doesn't make sense. People who misuse the term "correlation" to mean "association" often fail to notice whether the variables are quantitative. But correlation is valid only for *quantitative* variables. Be sure to check this condition.

- **Don't confuse correlation with causation.** One of the most common mistakes people make in interpreting statistics occurs when they observe a high correlation between two variables and jump to the perhaps tempting conclusion that one thing must be causing the other. Scatterplots and correlations *never* demonstrate causation. At best, these statistical tools can only reveal an association between variables, and that's a far cry from establishing cause and effect.

---

[15]Sometimes, we can do a "reality check" on our choice of re-expression. In this case, a bit of research reveals that f/stops are related to the diameter of the open shutter. Since the amount of light that enters is determined by the *area* of the open shutter, which is related to the diameter by squaring, the square re-expression seems reasonable. Not all re-expressions have such nice explanations, but it's a good idea to think about them.

[16]That's a high correlation, but it does not confirm that we have the best re-expression. After all, the correlation of the original relationship was 0.979—high enough to satisfy almost anyone. To judge a re-expression, look at the scatterplot. Your eye is a better judge of straightness than any statistic.

- **Make sure the association is linear.** Not all associations between quantitative variables are linear. Correlation can miss even a strong nonlinear association. A student project evaluating the quality of brownies baked at different temperatures reports a correlation of −0.05 between judges' scores and baking temperature. That seems to say there is no relationship until we look at the scatterplot:

The relationship between brownie taste *Score and Baking Temperature is* strong, but not at all linear.

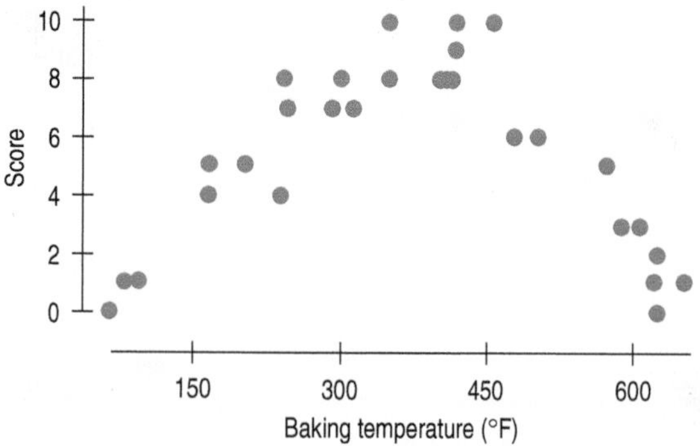

There is a strong association, but the relationship is not linear. Don't forget to check the **Straight Enough Condition**.

- **Don't assume the relationship is linear just because the correlation coefficient is high.** The correlation of f/stops and shutter speeds is 0.979 and yet the relationship is clearly not straight. A high correlation is no guarantee of straightness. Nor is it safe to use just the correlation to judge the best re-expression. It's always important to look at the scatterplot.

A scatterplot of *f/stop* vs. *Shutter Speed* shows a bent relationship even though the correlation is $r = 0.979$.

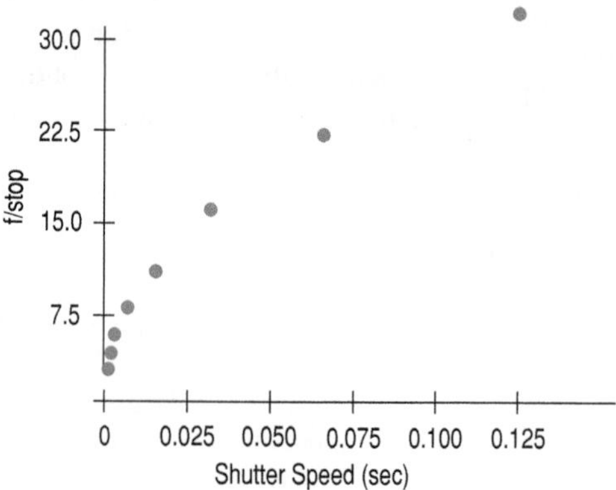

- **Beware of outliers.** You can't interpret a correlation coefficient safely without a background check for outliers (and subgroups). Here's a silly example:

  The relationship between IQ and shoe size among comedians shows a surprisingly strong positive correlation of 0.50. To check assumptions, we look at the scatterplot.

AP Images

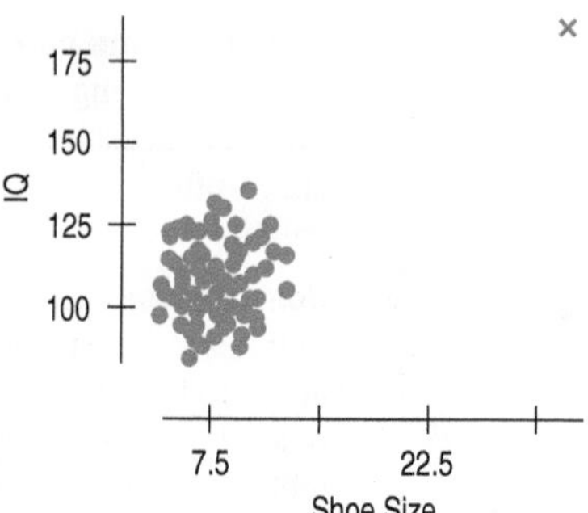

What is the relationship? The correlation is 0.50, but it is all due to the outlier (the green *x*) in the upper right corner. Who does that point represent?

The outlier is Bozo the Clown, known for his large shoes and widely acknowledged to be a comic "genius." Without Bozo the correlation is near zero.

Even a single outlier can dominate the correlation value. That's why you need to check the scatterplot for outliers.

Scatterplots are the basic tool for examining the relationship between two quantitative variables. We start with a picture when we want to understand the distribution of a single variable, and we always make a scatterplot to begin to understand the relationship between two quantitative variables.

We used z-scores as a way to measure the statistical distance of data values from their means. Now we've seen the z-scores of x and y working together to build the correlation coefficient. Correlation is a summary statistic like the mean and standard deviation—only it summarizes the strength of a linear relationship.

What Have We Learned?

Learning Objectives

Make a scatterplot to display the relationship between two quantitative variables.

- Look at the direction, form, and strength of the relationship, and for any outliers that stand away from the overall pattern. Look for distinct subgroups.

Provided the form of the relationship is straight, summarize its strength with a correlation, r.

- The sign of the correlation gives the direction of the relationship.
- $-1 \le r \le 1$; A correlation of 1 or -1 is a perfect linear relationship. A correlation of 0 indicates that there is no linear relationship.
- Correlation has no units, so shifting or scaling the data, standardizing, or even swapping the variables has no effect on the numerical value.

A S *Simulation:* **Correlation, Centre, and Scale.** If you have any lingering doubts that shifting and rescaling the data won't change the correlation, watch nothing happen right before your eyes!

A large correlation is not a sign of a causal relationship.

Recognize nonlinearity, and straighten plots when possible.

- If a scatterplot of y vs. x isn't straight, correlation isn't appropriate.
- Re-expressing one or both variables can often improve the straightness of the relationship.
- The powers, roots, and the logarithm provide an ordered collection of re-expressions so you can search up and down the *ladder of powers* to find an appropriate one.

Review of Terms

Scatterplots Shows the relationship between two quantitative variables measured on the same cases (p. 158).

Association

- **Direction:** A positive direction or association means that, in general, as one variable increases, so does the other. When increases in one variable generally correspond to decreases in the other, the association is negative (p. 160).
- **Form:** The simplest form is straight, but you should certainly describe other patterns you see in scatterplots (p. 160).
- **Strength:** A scatterplot is said to show a strong association if there is little scatter around the underlying relationship (p. 160).

Outlier A point that does not fit the overall pattern seen in the scatterplot (p. 160).

Response variable, explanatory variable, *y*-variable, *x*-variable In a scatterplot, you must choose a role for each variable. Assign the response to the y-axis the response variable that you hope to predict or explain. Assign to the x-axis the explanatory or predictor variable that accounts for, explains, predicts, or is otherwise responsible for the y-variable (p. 162).

Correlation coefficient A numerical measure of the direction and strength of a linear association (p. 164).

$$r = \frac{\sum z_x z_y}{n - 1}$$

Common response Changes in both x and y are caused by a lurking variable (p. 170).

Lurking variable A variable not present in our analysis that may influence our understanding of the relationship between x and y (p. 170).

Confounded variables Variables whose effects on the response variable, y, are entangled and difficult to distinguish (p. 170).

Re-expression We re-express data by taking the logarithm, the square root, the reciprocal, or some other mathematical operation of all values of a variable (p. 171).

Ladder of Powers Places in order the magnitude of effects that many re-expressions have on the data (p. 172).

On the Computer SCATTERPLOTS AND CORRELATION

Statistics packages generally make it easy to look at a scatterplot to check whether the correlation is appropriate. Some packages make this easier than others.

Many packages allow you to modify or enhance a scatterplot, altering the axis labels, the axis numbering, the plot symbols, or the colours used. Some options, such as colour and symbol choice, can be used to display additional information on the scatterplot.

EXCEL

To make a scatterplot in Excel:

- Select the columns of data to use in the scatterplot. You can select more than one column by holding down the control key while clicking.
- In the Insert tab, click on the **Scatter** button and select the **Scatter with only Markers** chart from the menu.

Unfortunately, the plot this creates is often statistically useless. To make the plot useful, we need to change the display:

- With the chart selected, click on the **Gridlines** button in the Layout tab to cause the Chart Tools tab to appear.
- Within Primary Horizontal Gridlines, select **None**. This will remove the gridlines from the scatterplot.
- To change the axis scaling, click on the numbers of each axis of the chart, and click on the **Format Selection** button in the Layout tab.
- Select the **Fixed** option instead of the Auto option, and type a value more suited for the scatterplot. You can use the pop-up dialog window as a straightedge to approximate the appropriate values.

Excel automatically places the leftmost of the two columns you select on the *x*-axis, and the rightmost one on the *y*-axis. If that's not what you'd prefer for your plot, you'll want to switch them.

To switch the X- and Y-variables:

- Click the chart to access the **Chart Tools** tabs.
- Click on the **Select Data** button in the Design tab.
- In the pop-up window's Legend Entries box, click on **Edit**.
- Highlight and delete everything in the Series *x* Values line, and select new data from the spreadsheet. (Note that selecting the column would inadvertently select the title of the column, which would not work well here.)
- Do the same with the Series *y* Values line.
- Press **OK**, then press **OK** again.

To find the correlation in Excel:

- Click on a blank cell in the spreadsheet.
- Go to the **Formulas** tab in the Ribbon and click **More Functions —> Statistical.**
- Choose the **CORREL** function from the drop-down menu of functions.
- In the dialogue that pops up, enter the range of one of the variables in the space provided.
- Enter the range of the other variable in the space provided.
- Click OK

COMMENTS

The correlation is computed in the selected cell. Correlations computed this way will update if any of the data values are changed.

JMP

To make a scatterplot and compute correlation:

- Choose **Fit Y by X** from the **Analyze** menu.
- In the Fit Y by X dialogue, drag the *y* variable into the **"Y, Response"** box, and drag the *x* variable into the **"X, Factor"** box.
- Click the **OK** button.

Once JMP has made the scatterplot, click on the red triangle next to the plot title to reveal a menu of options:

- Select **Density Ellipse** and select 0.95. JMP draws an ellipse around the data and reveals the **Correlation** tab.
- Click the blue triangle next to Correlation to reveal a table containing the correlation coefficient.

MINITAB

To make a scatterplot:

- Choose **Scatterplot** from the **Graph** menu.
- Choose "Simple" for the type of graph. Click **OK**.
- Enter variable names for the Y variable and X variable into the table. Click **OK**.

To compute a correlation coefficient:

- Choose **Basic Statistics** from the **Stat** menu.
- From the Basic Statistics submenu, choose **Correlation.** Specify the names of at least two quantitative variables in the "Variables" box.
- Click **OK** to compute the correlation table.

R

To make a scatterplot of two variables, X and Y:

- **plot(X,Y)** or equivalently **plot(Y~X)** will produce the plot.
- **cor(X,Y)** finds the correlation.

COMMENTS

Your variables X and Y may be variables in a data frame. If so, and DATA is the name of the data frame, then you will need to attach the data frame, or use

with(DATA,plot(Y~X)) or plot (Y~X,data=DATA).

SPSS

To make a scatterplot in SPSS:

- Open the Chart Builder from the **Graphs** menu.
- Click the **Gallery** tab.
- Choose Scatterplot from the list of chart types.
- Drag the scatterplot onto the canvas.
- Drag a scale variable you want as the response variable to the *y*-axis drop zone.
- Drag a scale variable you want as the factor or predictor to the *x*-axis drop zone.
- Click **OK**.

To compute a correlation coefficient:

- Choose **Correlate** from the **Analyze** menu.
- From the Correlate submenu, choose **Bivariate**.
- In the Bivariate Correlations dialogue, use the arrow button to move variables between the source and target lists.

Make sure the **Pearson** option is selected in the Correlation Coefficients field.

STATCRUNCH

To make a scatterplot:

- Click on **Graphics**.
- Choose **Scatter Plot**.
- Choose **X** and **Y** variable names from the list of **Columns**.
- Click on **Create Graph**.

To find a correlation:

- Click on **Stat**.
- Choose **Summary Stats » Correlation**.
- Choose two variable names from the list of **Columns**. (You may need to hold down the ctrl or command key to choose the second one.)
- Click on **Calculate**.

TI-83/84 PLUS

To create a scatterplot:

- Set up the **STAT PLOT** by choosing the **scatterplot** icon (the first option). Specify the lists where the data are stored as Xlist and Ylist.
- Set the graphing **WINDOW** to the appropriate scale and **GRAPH** (or take the easy way out and just ZoomStat!).

To find the correlation:

- Go to **STAT CALC** menu and select **8: LinReg(a + bx)**.
- Then specify the lists where the data are stored. The final command you will enter should look like **LinReg(a + bx) L1, L2.**

COMMENTS

Notice that if you **TRACE** the scatterplot, the calculator will tell you the *x*- and *y*-value at each point.

If the calculator does not tell you the correlation after you enter a LinReg command, try this: Hit **2nd CATALOG.** You now see a list of everything the calculator knows how to do. Scroll down until you find **DiagnosticOn.** Hit **ENTER** twice. (It should say Done.) Now and forevermore (or until you change batteries), you can find a correlation by using your calculator.

Exercises

1. Association Suppose you were to collect data for each pair of variables given below. You want to make a scatterplot. Which variable would you use as the explanatory variable and which as the response variable? Why? What would you expect to see in the scatterplot? Discuss the likely direction, form, and strength.

a) Apples: weight in grams, weight in ounces
b) Apples: circumference (cm), weight (g)
c) University freshmen: shoe size, grade point average
d) Gasoline: number of kilometres you drove since filling up, litres remaining in your tank

2. Association II Suppose you were to collect data for each pair of variables given below. You want to make a scatterplot. Which variable would you use as the explanatory variable and which as the response variable? Why? What would you expect to see in the scatterplot? Discuss the likely direction, form, and strength.

a) T-shirts at a store: price each, number sold
b) Scuba diving: depth, water pressure
c) Scuba diving: depth, visibility
d) All elementary-school students: weight, score on a reading test

3. Association III Suppose you were to collect data for each pair of variables given below. You want to make a scatterplot. Which variable would you use as the explanatory variable and which as the response variable? Why? What would you expect to see in the scatterplot? Discuss the likely direction, form, and strength.

a) When climbing a mountain: altitude, temperature
b) For each week: ice-cream cone sales, air conditioner sales
c) People: age, grip strength
d) Drivers: blood alcohol level, reaction time

4. Association IV Suppose you were to collect data for each pair of variables given below. You want to make a scatterplot. Which variable would you use as the explanatory variable and which as the response variable? Why? What would you expect to see in the scatterplot? Discuss the likely direction, form, and strength.

a) Long-distance calls: time (minutes), cost
b) Lightning strikes: distance from lightning, time delay of the thunder
c) A streetlight: its apparent brightness, your distance from it
d) Cars: weight of car, age of owner

5. Scatterplots Which of the scatterplots below show

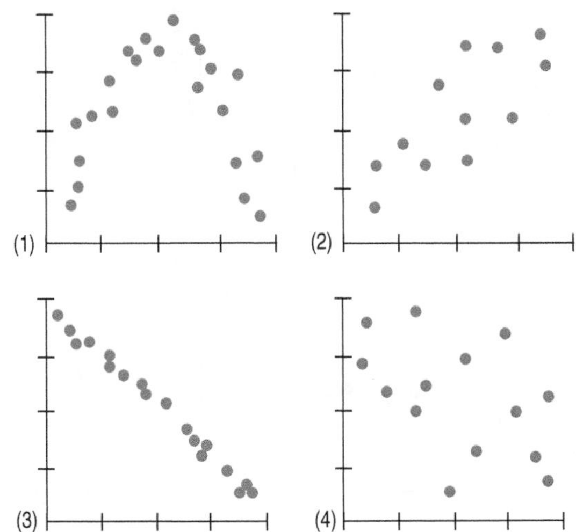

a) little or no association?
b) a negative association?
c) a linear association?
d) a moderately strong association?
e) a very strong association?

6. Scatterplots II Which of the scatterplots below show

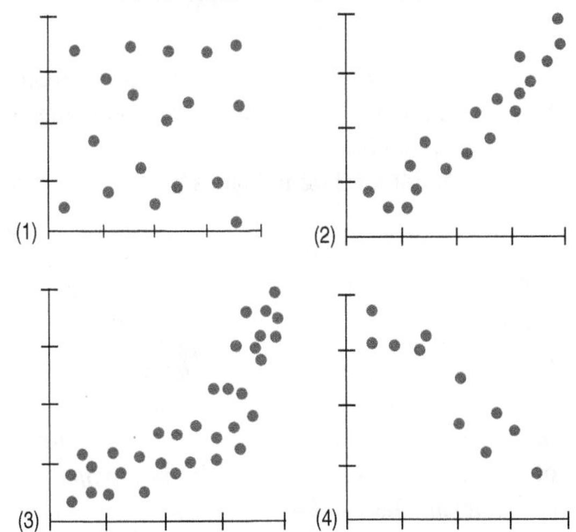

a) little or no association?
b) a negative association?
c) a linear association?
d) a moderately strong association?
e) a very strong association?

7. Performance IQ scores versus brain size A study examined brain size (measured as pixels counted in a digitized magnetic resonance image [MRI] of a cross-section of the brain) and IQ (four performance scales of the Weschler IQ test) for university students. The scatterplot shows the performance IQ scores versus brain size. Comment on the association between brain size and IQ as seen in this scatterplot.

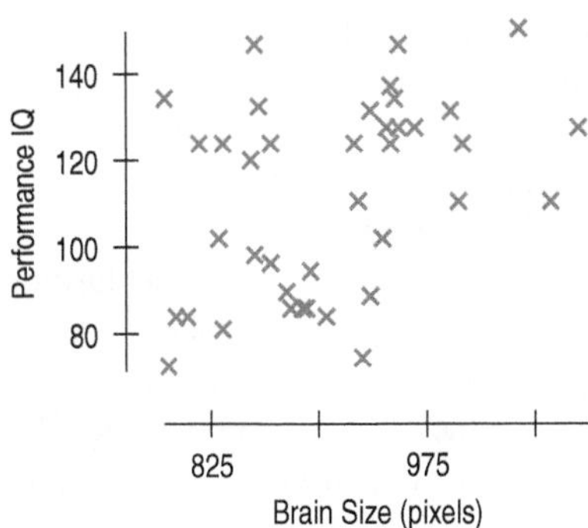

8. Kentucky Derby 2011 The fastest horse in Kentucky Derby history was Secretariat in 1973. The scatterplot shows speed (in miles per hour) of the winning horses each year.

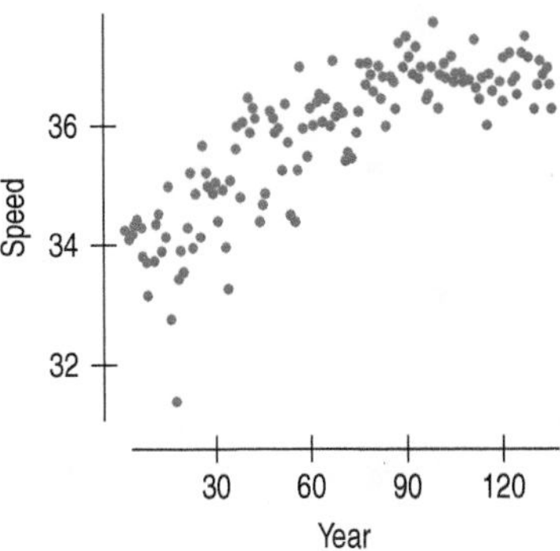

What do you see? In most sporting events, performances have improved and continue to improve, so surely we anticipate a positive direction. But what of the form? Has the performance increased at the same rate throughout the past 130 years?

9. Firing pottery A ceramics factory can fire eight large batches of pottery a day. Sometimes in the process a few of the pieces break. To better understand the problem, the factory records the number of broken pieces in each batch for three days and then creates the scatterplot shown.

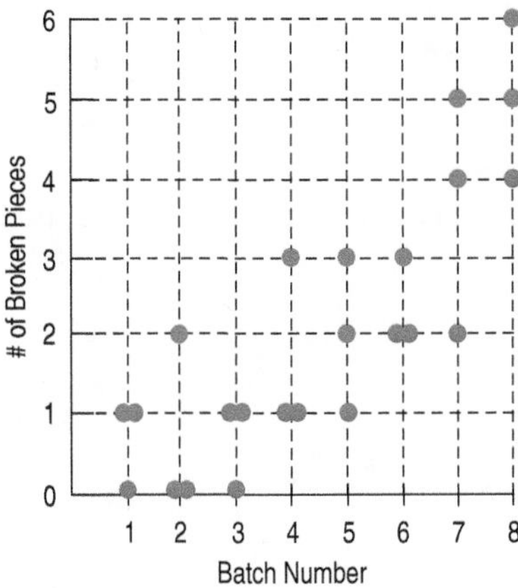

a) Make a histogram showing the distribution of the number of broken pieces in the 24 batches of pottery examined.

b) Describe the distribution as shown in the histogram. What feature of the problem is more apparent in the histogram than in the scatterplot?
c) What aspect of the company's problem is more apparent in the scatterplot?
d) In the histogram, "batch" was a "lurking variable," lurking behind the data display, and not visible. Plotting the number of broken pieces against batch helped to explain the variation in breakage. Can you think of any other lurking variables to investigate here?

10. Coffee sales Owners of a new coffee shop tracked sales for the first 20 days and displayed the data in a scatterplot (by day):

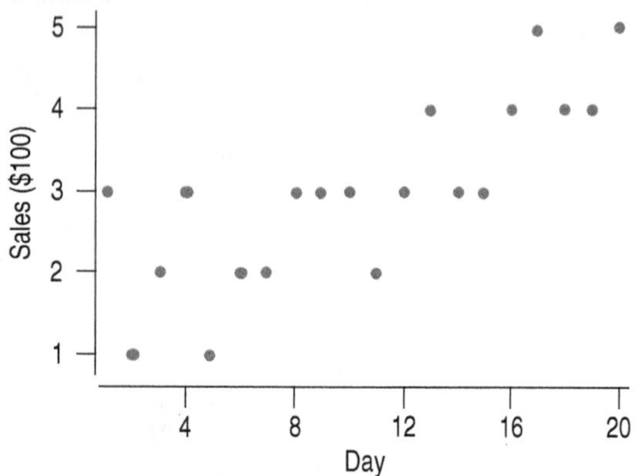

a) Make a histogram of the daily sales since the shop has been in business.
b) State one fact that is obvious from the scatterplot but not from the histogram.
c) State one fact that is obvious from the histogram but not from the scatterplot.

11. Matching Here are several scatterplots. The calculated correlations are −0.923, −0.487, 0.006, and 0.777. Which is which?

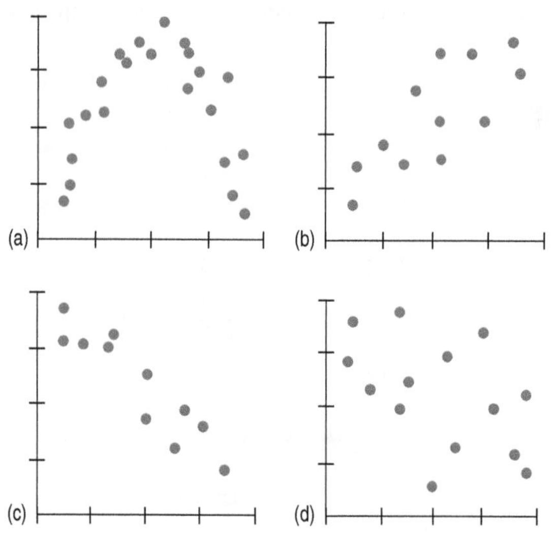

12. Matching II Here are several scatterplots. The calculated correlations are −0.977, −0.021, 0.736, and 0.951. Which is which?

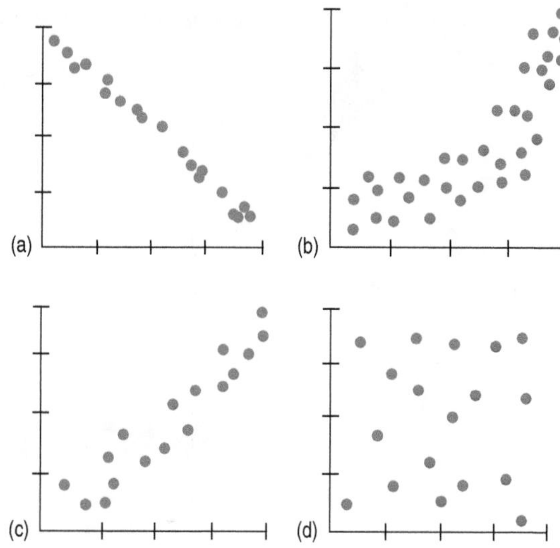

13. Correlation facts If we assume that the conditions for correlation are met, which of the following are true? If false, explain briefly.
a) A correlation of −0.98 indicates a strong, negative association.
b) Multiplying every value of x by 2 will double the correlation.
c) The units of the correlation are the same as the units of y.

14. Correlation facts II If we assume that the conditions for correlation are met, which of the following are true? If false, explain briefly.
a) A correlation of 0.02 indicates a strong positive association.
b) Standardizing the variables will make the correlation 0.
c) Adding an outlier can dramatically change the correlation.

T **15. Roller coasters** Roller coasters get all their speed by dropping down a steep initial incline, so it makes sense that the height of that drop might be related to the speed of the coaster. Here's a scatterplot of top *Speed* (miles per hour) and largest *Drop* (feet) for 75 roller coasters around the world.

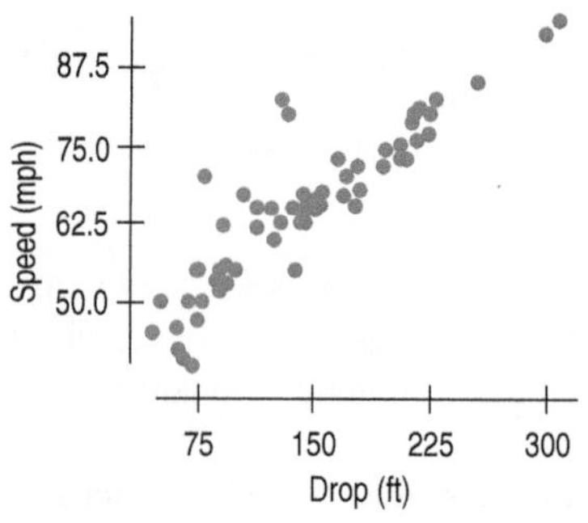

a) Does the scatterplot indicate that it is appropriate to calculate the correlation? Explain.
b) In fact, the correlation of *Speed* and *Drop* is 0.91. Describe the association.

16. Antidepressants A study compared the effectiveness of several antidepressants by examining the experiments in which they had passed the FDA requirements. Each of those experiments compared the active drug to a placebo, an inert pill given to some of the subjects. In each experiment, some patients treated with the placebo had improved, a phenomenon called the *placebo effect*. Patients' depression levels were evaluated on the Hamilton Depression Rating Scale, where larger numbers indicate greater improvement. The scatterplot shown compares mean improvement levels for the antidepressants and placebos for several experiments.

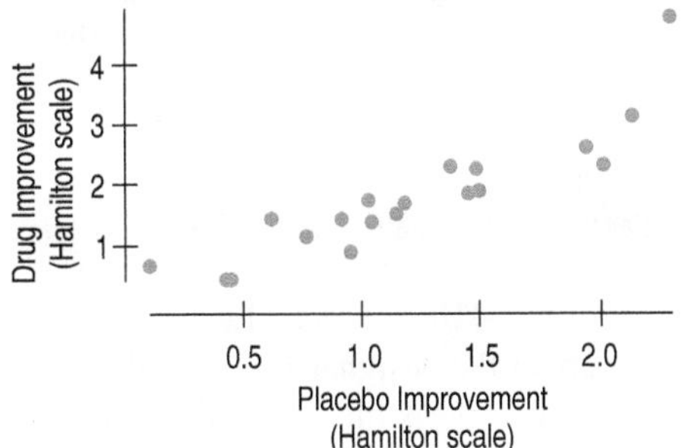

a) Is it appropriate to calculate the correlation? Explain.
b) The correlation is 0.898. Explain what you have learned about the results of these experiments.

17. Lunchtime Does how long children remain at the lunch table help predict how much they eat? The table gives data on 20 toddlers observed over several months at a nursery school. "Time" is the average number of minutes a child spent at the table when lunch was served. "Calories" is the average number of calories the child consumed during lunch, calculated from careful observation of what the child ate each day.

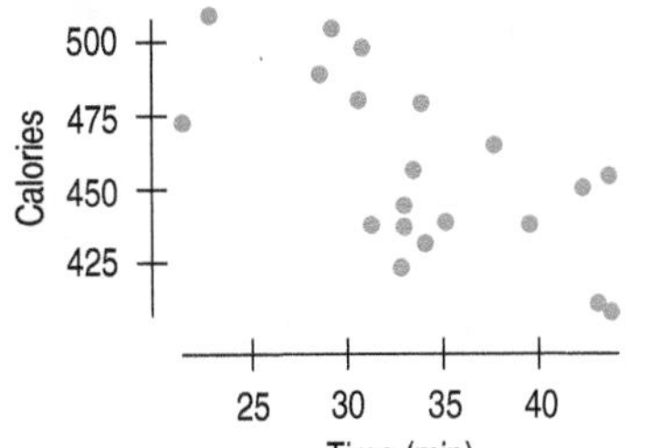

Calories	Time
472	21.4
498	30.8
465	37.7
456	33.5
423	32.8
437	39.5
508	22.8
431	34.1
479	33.9
454	43.8
450	42.4
410	43.1
504	29.2
437	31.3
489	28.6
436	32.9
480	30.6
439	35.1
444	33.0
408	43.7

a) Find the correlation for these data.
b) Suppose we were to record time at the table in hours rather than in minutes. How would the correlation change? Why?
c) Write a sentence or two explaining what this correlation means for these data. Remember to write about food consumption by toddlers rather than about correlation coefficients.
d) One analyst concluded, "It is clear from this correlation that toddlers who spend more time at the table eat less. Evidently something about being at the table causes them to lose their appetites." Explain why this explanation is not an appropriate conclusion from what we know about the data.

18. Vehicle weights The Minnesota Department of Transportation hoped that they could measure the weights of big trucks without actually stopping the vehicles by using a newly developed "weight-in-motion" scale. To see if the new device was accurate, they conducted a calibration test. They weighed several trucks when stopped (*Static Weight*), assuming that this weight was correct. Then they weighed the trucks again while they were moving to see how well the new scale could estimate the actual weight. Their data are given in the table.

Weight of a Truck (thousands of pounds)	
Weight-in-Motion	**Static Weight**
26.0	27.9
29.9	29.1
39.5	38.0
25.1	27.0
31.6	30.3
36.2	34.5
25.1	27.8
31.0	29.6
35.6	33.1
40.2	35.5

a) Make a scatterplot for these data.
b) Describe the direction, form, and strength of the plot.
c) Write a few sentences telling what the plot says about the data. (*Note*: The sentences should be about weighing trucks, not about scatterplots.)
d) Find the correlation.
e) If the trucks were weighed in kilograms, how would this change the correlation? (1 kilogram = 2.2 pounds)
f) Do any points deviate from the overall pattern? What does the plot say about a possible recalibration of the weight-in-motion scale?

19. Fuel economy 2010 Here are advertised horsepower ratings and expected gas mileage, in miles per U.S. gallon, for several 2010 vehicles. (www.kbb.com)

Car	hp	mpg
Audi A4	211	30
BMW 3 series	230	28
Buick LaCrosse	182	30
Chevy Cobalt	155	37
Chevy Suburban	320	21
Ford Expedition	310	20
GMC Yukon	320	21
Honda Civic	140	34
Honda Accord	177	31
Hyundai Elantra	138	35
Lexus IS 350	306	25
Lincoln Navigator	310	20
Mazda Tribute	171	28
Toyota Camry	169	33
Volkswagen Beetle	150	28

a) Make a scatterplot for these data.
b) Describe the direction, form, and strength of the association.
c) Find the correlation between horsepower and miles per gallon.
d) Write a few sentences telling what the plot says about fuel economy.
e) Convert the mileage to litres per 100 km (= 235.2 ÷ mpg) and repeat pars a) to d). Did the correlation change? Which scale for fuel consumption do you prefer? Why?

20. Drug abuse A survey was conducted in the United States and 10 countries of Western Europe to determine the percentage of teenagers who had used marijuana and other drugs. The results are summarized in the table.

	Percent Who Have Used	
Country	**Marijuana**	**Other Drugs**
Czech Rep.	22	4
Denmark	17	3
England	40	21
Finland	5	1
Ireland	37	16
Italy	19	8
N. Ireland	23	14
Norway	6	3
Portugal	7	3
Scotland	53	31
U.S.	34	24

a) Create a scatterplot.
b) What is the correlation between the percent of teens who have used marijuana and the percent who have used other drugs?
c) Write a brief description of the association.
d) Do these results confirm that marijuana is a "gateway drug"—that is, that marijuana use leads to the use of other drugs? Explain.

21. Burgers Fast food is often unhealthy because much of it is high in both fat and sodium. But are the two related? Here are the fat and sodium contents of several brands of burgers. Analyze the association between fat content and sodium.

Fat (g)	19	31	34	35	39	39	43
Sodium (mg)	920	1500	1310	860	1180	940	1260

22. Burgers II In the previous exercise, you analyzed the association between the amounts of fat and sodium in fast food hamburgers. What about fat and calories? Here are data for the same burgers. Analyze the association using correlation and scatterplots.

Fat (g)	19	31	34	35	39	39	43
Calories	410	580	590	570	640	680	660

23. Attendance 2010 American League baseball games are played under the designated hitter rule, meaning that pitchers, often weak hitters, do not come to bat. Baseball owners believe that the designated hitter rule means more runs scored, which in turn means higher attendance. Is there evidence that more fans attend games if the teams score more runs? Data collected from American League games during the 2010 season indicate a correlation of 0.667 between runs scored and the number of people at the game. (www.mlb.com)

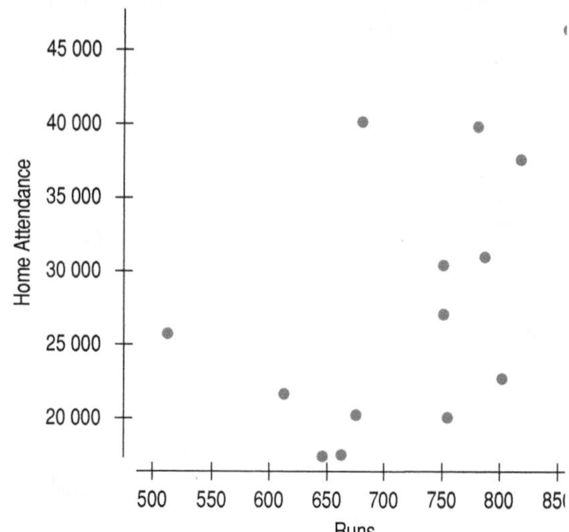

a) Does the scatterplot indicate that it's appropriate to calculate a correlation? Explain.
b) Describe the association between attendance and runs scored.
c) Does this association prove that the owners are right that more fans will come to games if the teams score more runs?

24. Second inning 2010 Perhaps fans are just more interested in teams that win. The correlation matrix and scatterplot below are based on American League teams for the 2010 season (espn.go.com). Are the teams that win necessarily those which score the most runs?

	Wins	Runs	Attendance
Wins	1.00		
Runs	0.919	1.00	
Attendance	0.533	0.538	1.00

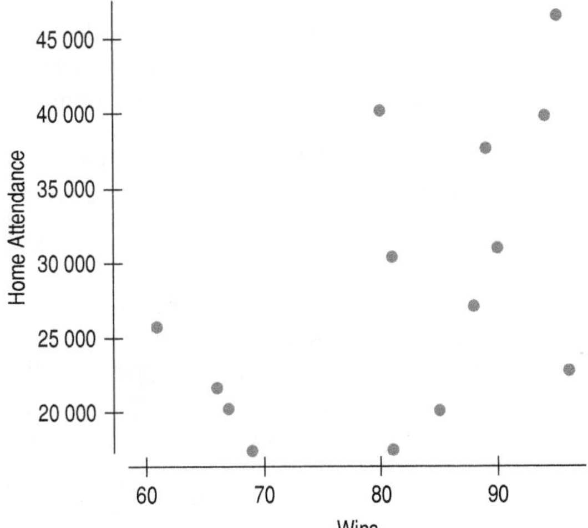

a) Do winning teams generally enjoy greater attendance at their home games? Describe the association.
b) Is attendance more strongly associated with winning or scoring runs? Explain.
c) How strongly is scoring more runs associated with winning more games?

25. **Politics** A candidate for office claims that "there is a correlation between television watching and crime." Criticize this statement in statistical terms.

26. **Association V** A researcher investigating the association between two variables collected some data and was surprised when he calculated the correlation. He had expected to find a fairly strong association, yet the correlation was near 0. Discouraged, he didn't bother making a scatterplot. Explain to him how the scatterplot could still reveal the strong association he anticipated.

27. **Height and reading** A researcher studies children in elementary school and finds a strong positive linear association between height and reading scores.
a) Does this mean that taller children are generally better readers?
b) What might explain the strong correlation?

28. **Smart phones and life expectancy** A survey of the world's nations in 2010 shows a strong positive correlation between percentage of the country using smart phones and life expectancy in years at birth.
a) Does this mean that smart phones are good for your health?
b) What might explain the strong correlation?

T 29. **Hard water** In a study of streams in the Adirondack Mountains, the following relationship was found between the pH of the water and the water's hardness (measured in grains):

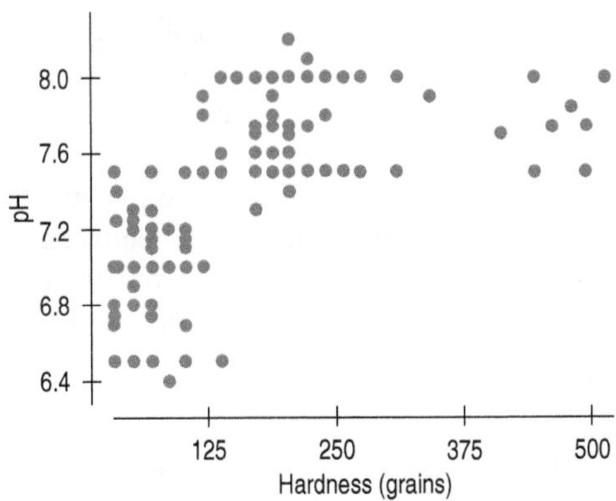

Is it appropriate to summarize the strength of association with a correlation? Explain.

30. **Traffic headaches** A study of traffic delays in 68 U.S. cities found the following relationship between total delays (in total hours lost) and mean highway speed:

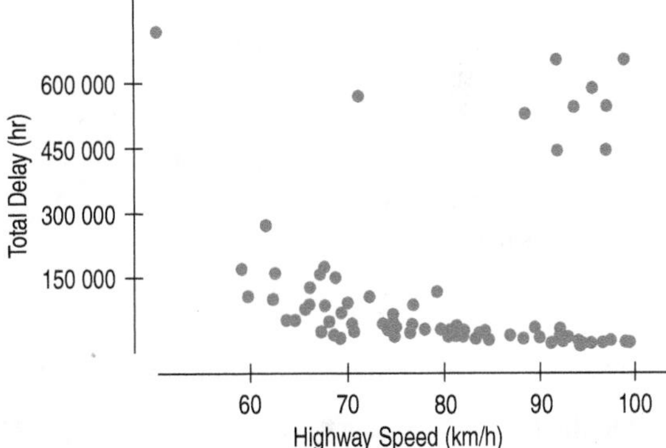

Is it appropriate to summarize the strength of association with a correlation? Explain.

31. **Correlation errors** Your Economics instructor asks the class to investigate factors associated with the gross domestic product (GDP) of nations. Each student examines a different factor (life expectancy, literacy rate, etc.) for a few countries and reports to the class. Apparently, some of your classmates do not understand statistics very well because you know several of their conclusions are incorrect. Explain the mistakes in their statements below.
a) "My correlation of −0.772 shows that there is almost no association between GDP and infant mortality rate."
b) "There was a correlation of 0.44 between GDP and continent."

32. **More correlation errors** Students in the Economics class discussed in Exercise 31 also wrote these conclusions. Explain the mistakes they made.
a) "There was a very strong correlation of 1.22 between life expectancy and GDP."
b) "The correlation between literacy rate and GDP was 0.83. This shows that countries wanting to increase their standard of living should invest heavily in education."

33. Baldness and heart disease Medical researchers followed 1435 middle-aged men for a period of five years, measuring the amount of baldness present (none = 1, little = 2, some = 3, much = 4, extreme = 5) and presence of heart disease (no = 0, yes = 1). They found a correlation of 0.089 between the two variables. Comment on their conclusion that this shows that baldness is not a possible cause of heart disease.

34. Sample survey A California polling organization is checking its database to see if the three data sources they used sampled the same zip codes (five-digit U.S. postal codes). The variable *Datasource* = 1 if the data source is MetroMedia, 2 if the data source is DataQwest, and 3 if it's RollingPoll. The organization finds that the correlation between *Zip Code* and *Datasource* is –0.0229. It concludes that the correlation is low enough to state that there is no dependency between zip code and the source of data. Comment.

T **35. Thrills 2011** Since 1994, the Best Roller Coaster Poll (www.ushsho.com/bestrollercoasterpoll.htm) has been ranking the world's best roller coasters. In 2011, Bizarro earned the top steel coaster rank for the sixth straight year. Here are data on the top 10 steel coasters from this poll:

Rank	Roller Coaster	Park	Location	Initial Drop (ft.)	Duration (sec)	Height (ft.)	Max Speed (mph)	Max Vert Angle (degrees)	Length (ft.)
1	Bizarro	Six Flags New England	MA	221	155	208	77	72	5400
2	Expedition GeForce	Holiday Park	DE	184	75	188	74.6	82	4003
3	Intimidator 305	Kings Dominion	VA	300	98	305	93	85	5100
4	Kawasemi	Tobu Zoo	JP		60	108	54	67.4	2454
5	Nemesis	Alton Towers	UK	104	80	43	50	40	2349
6	Piraten	Djurs Sommerland	DK	100	61	105	56	70	2477
7	Goliath	Walibi World	NL	152	92	155	66.9	70	3984
8	Millennium Force	Cedar Point	OH	300	120	310	93	80	6595
9	Katun	Mirabilandia	IT	148	142	169	65		3937
10	iSpeed	Mirabilandia	IT	181	60	180	74.6	90	3281

What do these data indicate about the *Length* of the track and the *Duration* of the ride you can expect?

T **36. Thrills II** For the roller coaster data in Exercise 35:

a) Examine the relationship between *Initial Drop* and *Max Speed.*

b) Examine the relationship between *Initial Drop* and *Height.*

c) What conclusions can you safely draw about the initial drop of a roller coaster? Is *Initial Drop* strongly correlated with other variables as well?

T **37. Thrills III** For the roller coaster data in Exercise 35:

a) Explain why in looking for a variable that explains rank, you will be hoping for a negative correlation.

b) Do any of the provided variables provide a strong predictor for roller coaster rank?

c) What other (unaccounted for) variables might help explain the rank?

T **38. Ontario migration 2012** Below are data showing the net interprovincial migration each year for Ontario (number arriving minus number leaving):

Year	Ontario Net Migration	Year	Ontario Net Migration	Year	Ontario Net Migration
1979	−4 325	1991	−11 627	2003	637
1980	−22 362	1992	−11 045	2004	−6 935
1981	−33 247	1993	−14 189	2005	−11 172
1982	−5 665	1994	−9 420	2006	−17 501
1983	23 585	1995	−2 841	2007	−20 047
1984	36 400	1996	−2 822	2008	−14 750
1985	33 885	1997	1 977	2009	−15 601
1986	33 562	1998	9 231	2010	−4 662
1987	42 601	1999	16 706	2011	−4 007
1988	35 215	2000	22 369	2012	−8 091
1989	9 739	2001	18 623		
1990	−5 961	2002	5 354		

Source: Adapted from Statistics Canada. CANSIM table 051-0004, March 31, 2013

a) Calculate the correlation between Ontario net migration and year.
b) A reporter claimed that the low correlation between year and migration shows that Ontario migration has remained fairly steady over this period of time. Do you agree with this interpretation? Explain.

39. Bigger and bigger According to the UK National Sizing Survey, women's average body-size measurements have increased over time as follows:

Year	Bust size (in)	Waist size (in)	Hips size (in)
1920	32	20	32
1940	33	21	33
1960	34	24	33
1980	35	26	37
2000	36	28	38

Source: P. Treleaven, (September 2007), "How to fit into your clothes," *Significance* 4(3), 113–117.

a) Calculate all possible correlations among the four variables.
b) Plot the body-size measurement sizes against each other and against time. Describe the relationships you see. Which relationship is the strongest?
c) Offer some plausible explanation(s) for these high correlations.

40. Obvious correlations Looking at the largest 152 urban areas in Canada, the correlation between the number of renter-occupied dwellings and number of owner-occupied dwellings per urban area is 0.955, while the correlation between the percentage of renter-occupied dwellings and percentage of owner-occupied dwellings is −1.0. Explain why these two correlation values are just what you would expect.

41. Census at school The International *CensusAtSchool* Project collects data on primary and secondary school students from various countries, including Canada. We selected a random sample of 111 Canadian secondary school students age 14 and over. Below are the first four rows of the 111-row worksheet of data, available at http://www.pearsoncanada.ca/deveaux. All measurements are in centimetres and were taken in the classroom. Middlefinger is distance from wrist to tip of middle finger. Wristbone is distance from wrist bone to elbow. Armspan is distance from tip of left hand to tip of right hand with arms spread wide. Foot is length of right foot.

Gender	Age	Height	Armspan	Wristbone	Middlefinger	Foot
Boy	14	166	169	25	19	26
Boy	14	160	170	27	19	28
Boy	14	167	163	26	20	26
Girl	16	165	165	24	17	22
:	:	:	:	:	:	:

Source: The Royal Statistical Society Centre for Statistical Education, CensusAtSchool project. http://rds.censusatschool.org.uk

a) Find all the pair-wise correlations. Which two variables possess the strongest linear correlation?
b) Suppose we are interested in the relationship between armspan and height. Below is a plot of armspan versus height:

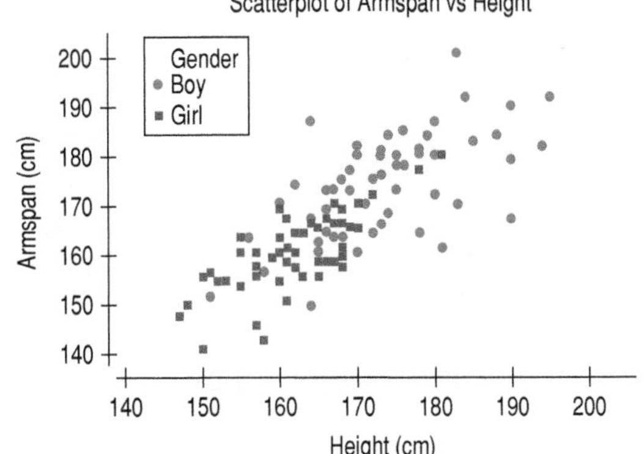

Describe the nature of the relationship between armspan and height.
c) In the plot above, we also included information on another variable, gender. Does gender appear to be an important variable in understanding the relationship between armspan and height? If we omitted the information about gender in the plot above, it would become a __________ variable. Fill in the blank.

42. Interprovincial migration 2012 Below are data showing the net interprovincial migration each year for Ontario and Alberta (the number arriving minus the number leaving). Note that each year actually stands for a non-calendar year period ending in that particular year.

Year	Ontario	Alberta
1978/1979	−4 325	33 426
1979/1980	−22 362	41 435
1980/1981	−33 247	44 250
1981/1982	−5 665	36 562
1982/1983	23 585	−11 650
1983/1984	36 400	−31 986
1984/1985	33 885	−20 771
1985/1986	33 562	−3 831
1986/1987	42 601	−29 998
1987/1988	35 215	−23 223
1988/1989	9 739	−1 528
1989/1990	−5 961	5 593
1990/1991	−11 627	8 983
1991/1992	−11 045	2 983
1992/1993	−14 189	−1 181
1993/1994	−9 420	−1 630
1994/1995	−2 841	−556
1995/1996	−2 822	7 656
1996/1997	1 977	26 282
1997/1998	9 231	43 089
1998/1999	16 706	25 191

1999/2000	22 369	22 674
2000/2001	18 623	20 457
2001/2002	5 354	26 235
2002/2003	637	11 903
2003/2004	−6 935	10 606
2004/2005	−11 172	34 423
2005/2006	−17 501	45 795
2006/2007	−20 047	33 809
2007/2008	−14 750	15 317
2008/2009	−15 601	13 184
2009/2010	−4 662	−3 271
2010/2011	−4 007	8 443
2011/2012	−8 091	28 170

Source: Adapted from Statistics Canada CANSIM table 051-0004, March 31, 2013.

a) Plot Alberta net migration versus Ontario net migration for each year.
b) Describe the nature of the association.
c) Calculate the correlation.
d) Why would you expect a negative association? Is it possible to have a positive correlation here? Why (or why not)? If there were only two provinces in Canada, what would the correlation have to equal, and why?

43. Planets (more or less) On August 24, 2006, the International Astronomical Union voted that Pluto is not a planet. Some members of the public have been reluctant to accept that decision. Let's look at some of the data. Is there any pattern to the locations of the planets? The table shows the average distance of each of the traditional nine planets from the Sun.

Planet	Position Number	Distance from Sun (million miles)
Mercury	1	36
Venus	2	67
Earth	3	93
Mars	4	142
Jupiter	5	484
Saturn	6	887
Uranus	7	1784
Neptune	8	2796
Pluto	9	3666

a) Make a scatterplot and describe the association. (Remember: direction, form, and strength!)
b) Why would you not want to talk about the correlation between a planet's *Position Number* and *Distance* from the sun?
c) Make a scatterplot showing the logarithm of *Distance* vs. *Position Number*. What is better about this scatterplot?

44. Human Development Index 2010 The United Nations Development Programme (UNDP) uses the Human Development Index (HDI) in an attempt to summarize in one number the progress in health, education, and economics of a country. In 2010, the HDI was as high as 0.938 for Norway and as low as 0.14 for Zimbabwe. The gross domestic product per capita (GDPPC) is used to summarize the overall economic strength of a country. Here is a plot of GDPPC against HDI for 169 countries throughout the world:

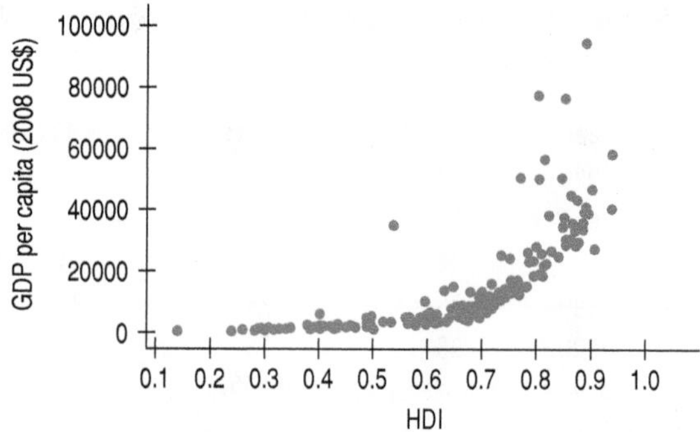

a) GDPPC is measured in dollars. Incomes and other economic measures tend to be highly right-skewed. Taking logs often makes the distribution more unimodal and symmetric. Compare the histogram of GDPPC to the histogram of log(GDPPC).
b) Use the log re-expression to make a scatterplot of log(GDPPC) against HDI. Comment on the effects of the re-expression.
c) The HDR report classifies countries into high, medium, and low development based in part on their HDI. Look at the scatterplot of log(GDPPC) against HDI for only the 127 medium- and high-development countries (currently viewed as the highest 75%, so HDI > 0.48 here). Does this relationship appear stronger or weaker than when we include all 169 countries?
d) Find the correlation of log(GDPPC) with HDI for all 169 countries. Find the correlation for only the 127 medium- and high-development countries. Which is higher? (Restricting the range of a variable usually reduces the correlation, but not always.)

45. Mandarin or Cantonese 2011 Below are data on Canadian cities showing mother-tongue counts from the 2011 census:

Urban Centre	Cantonese	Mandarin
Abbotsford (BC)	310	400
Barrie (ON)	215	295
Brantford (ON)	125	90
Calgary (AB)	20 175	11 380
Edmonton (AB)	13 595	7 295
Guelph (ON)	770	650
Halifax (NS)	460	805
Hamilton (ON)	1 975	2 215
Kelowna (BC)	235	210
Kingston (ON)	335	395
Kitchener-Cambridge-Waterloo (ON)	1 390	3 060

London (ON)	1 165	1 415
Moncton (NB)	70	75
Montréal (QC)	10 515	10 180
Oshawa (ON)	730	645
Ottawa-Gatineau (ON-QC)	6 570	7 275
Peterborough (ON)	90	110
Québec (QC)	130	270
Regina (SK)	560	505
Saguenay (QC)	0	15
Saint John (NB)	45	125
Saskatoon (SK)	865	790
Sherbrooke (QC)	20	65
St. Catharines-Niagara (ON)	640	555
St. John's (NL)	105	165
Sudbury (ON)	140	90
Thunder Bay (ON)	75	75
Toronto (ON)	170 485	100 045
Trois-Rivières (QC)	10	20
Vancouver (BC)	128 110	90 190
Victoria (BC)	2 930	1 945
Windsor (ON)	1 200	1 085
Winnipeg (MB)	3 220	2 080

Source: Statistics Canada, 2011 Census of Population and Statistics Canada, catalogue no. 98-314-XCB (last modified: 2013-02-13)

a) Find the correlation between these two mother-tongue counts. Are you surprised that the correlation is so high? Explain.
b) Plot Cantonese speaker count versus Mandarin speaker count. Describe the relationship. Is a correlation a useful and appropriate summary measure here?
c) Take the log (base 10) of both columns of numbers using your computer or calculator. Replot the data. Have things improved in any way? Calculate the correlation. Is the correlation more appropriate now?
d) What is one obvious reason (for example, a lurking variable) for such a high correlation between the two log counts here? Suggest another useful way of defining the two variables that would produce a smaller correlation. (In real-life studies, there are always issues about how to best define and measure the relevant variables.)

46. HDI and phones 2010 In Exercise 44, we examined the relationship between GDPPC and the Human Development Index for a large number of countries. The number of phone subscribers per 100 people is also associated with economic progress in a country. Here's a scatterplot of phone subscribers per 100 people against HDI:

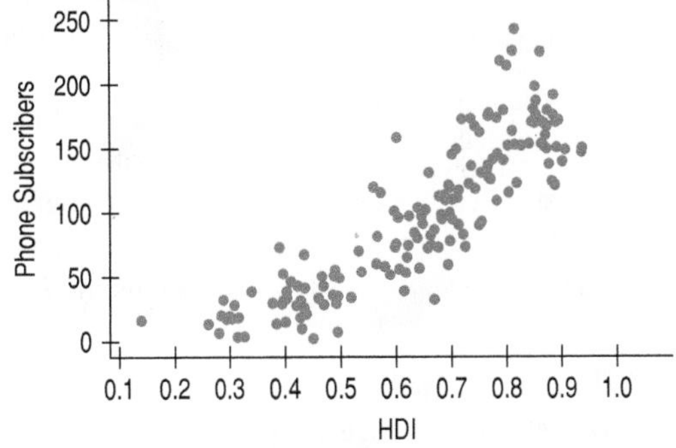

a) The square root transformation is often useful for counts. Examine the histogram of *Phone Subscribers* and the histogram of square-rooted *Phone Subscribers*. Comment.
b) Use the re-expression in part a) for the scatterplot against HDI. Comment.

Just Checking ANSWERS

1. We know the scores are quantitative. We should check to see if the Straight Enough Condition and the No Outliers Condition are satisfied by looking at a scatterplot of the two scores.
2. It won't change.
3. It won't change.
4. They are likely to have done poorly. The positive correlation means that low scores on Exam 1 are associated with low scores on Exam 2 (and similarly for high scores).
5. No. The general association is positive, but the correlation is not 1.0, so individual performances may vary considerably.

MathXL

MyStatLab

Go to MathXL at www.mathxl.com or MyStatLab at www.mystatlab.com. You can practise exercises for this chapter as often as you want. The guided solutions will help you find answers step by step. You'll find a personalized study plan available to you too!

chapter 7

Linear Regression

Statisticians, like artists, have the bad habit of falling in love with their models.

—George Box, famous statistician

Where are we going?

We know that the sales price of a downtown condo is related to its size, but how much should you expect to pay for a 750-square foot condo? Hubble discovered that galaxies farther away from us are moving away faster than nearer galaxies, but what velocity would you predict for a galaxy that is 5 million light-years away? The correlation alone won't tell us the answer.

We need a model to be able to use one variable to predict another. Using linear regression, we can better understand the relationship between two quantitative variables and make predictions.

The Whopper™ has been Burger King's signature sandwich since 1957. One Triple Whopper with cheese provides 71 grams of protein—all the protein you need in a day. It also supplies 1230 calories and 82 grams of fat. The Daily Value (based on a 2000-calorie diet) for fat is 65 grams, so after a Triple Whopper you'll want the rest of your calories that day to be fat-free.[1]

Of course, the Whopper isn't the only item Burger King (BK) sells. How are fat and protein related on the entire BK menu? The Figure 7.1 scatterplot of the *Fat* (in grams) versus the *Protein* (in grams) for foods sold at Burger King shows a positive, moderately strong linear relationship.

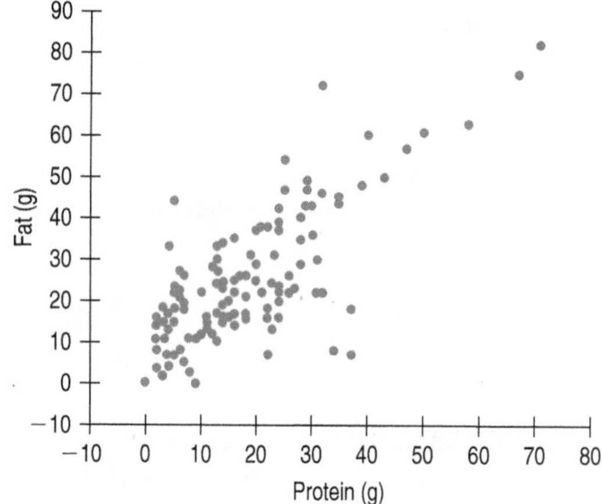

Figure 7.1

Total *Fat* versus *Protein* for 122 items on the BK menu. The Triple Whopper is in the upper right corner. It's extreme, but is it out of line?

7.1 Least Squares: The Line of "Best Fit"

If there are 25 grams of protein in your lunch, how much fat should you expect to consume at Burger King? The correlation between *Fat* and *Protein* is 0.76, which tells us that the linear association seen in the scatterplot is fairly strong. But *strength* of the relationship is only part of the picture. The correlation says, "The linear association between these two variables is fairly strong," but it doesn't tell us *how to predict one variable from the other.*

Of course, the points in the scatterplot don't all line up perfectly. But anyone looking at the plot can see that the points move from the lower left to the upper right in a

A S **Video: Manatees and Motorboats.** Are motorboats killing more manatees in Florida? Here's the story on video.

[1]Sorry about the fries.

Who	Items on the Burger King menu
What	Protein content and total fat content
Unit	Grams of protein, Grams of fat
How	Supplied by BK on request or at their Web site

A S ***Activity:* Linear Equations.** For a quick review of linear equations, view this activity and play with the interactive tool.

A S ***Activity:* Residuals.** Residuals are the basis for fitting lines to scatterplots. See how they work.

consistent way. You could probably sketch a straight line that summarizes that relationship pretty well. We'll take that one step further and find a **linear model** that gives an equation of a straight line through the data. We want to predict the fat content of any Burger King food given its protein, so fat content is our response variable and protein is the explanatory variable or predictor. This is an important distinction and will affect the line that we fit to the data.

No straight line can go through all the points, but a linear model can summarize the general pattern with only a couple of parameters. Like all **models** of the real world, the line will be wrong—wrong in the sense that it can't match reality *exactly*. But it can help us understand how the variables are associated. Without models, our understanding of the world is limited to only the data we have at hand.

Not only can't we draw a line through all the points, the best line might not even hit any of the points. Then how can it be the "best" line? We want to find the line that somehow comes closer to all the points than any other line. Some of the points will be above the line and some below. For example, the line might suggest that a BK Tendercrisp® chicken sandwich (without mayonnaise) with 31 grams of protein[2] should have 36.7 grams of fat when, in fact, it actually has only 22. We call the estimate made from a model the **predicted value**, and write it as $\hat{y}$ (called *y-hat*) to distinguish it from the observed response, y. The difference between the observed response and its associated predicted value is called the **residual**. The residual value tells us how far off the model's prediction is for that case or point. The BK Tendercrisp chicken residual would be $y - \hat{y} = 22 - 36.6 = -14.6$ g of fat.

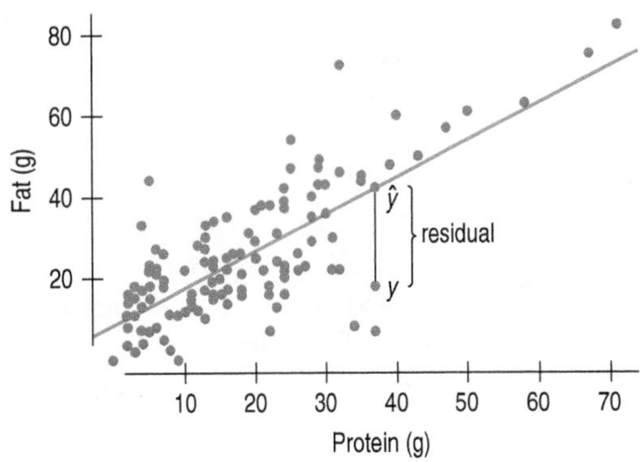

Figure 7.2
The residual is the difference between the actual fat (our response variable) and the predicted fat. So it is a difference in the vertical direction in our plot.

To find the residuals, we always subtract the predicted value from the observed one.

$$\text{Residual} = \text{Observed response} - \text{Predicted response}$$

The negative residual tells us that the actual fat content of the BK Tendercrisp chicken is about 14.6 grams *less* than the model predicts for a typical Burger King menu item with 31 grams of protein. We often use residuals in this way. The line tells us the average fat content we'd expect for a given amount of protein. Items with negative residuals have less fat than we'd expect, while those with positive residuals have more.

RESIDUALS

A positive residual means the actual value is larger than the predicted value—an underestimate. A negative residual shows that the model made an overestimate.

A S ***Activity:* The Least Squares Criterion.** Does your sense of "best fit" look like the least squares line?

Our challenge now is how to find the right line.

The size of the residuals tells us how well our line fits the data. A line that fits well will have small residuals. But we can't just add up all the residuals to see how well we've done, because some of the residuals are negative. We faced the same issue when we calculated a standard deviation to measure spread. And we deal with it the same way here: by squaring the residuals. Now that they're all positive, adding them up makes sense. Squaring also emphasizes the large residuals. After all, points near the line are consistent with the model, but we're more concerned about points far from the line. When we add all the squared residuals together, that sum indicates how well the line we drew fits the

[2]The sandwich comes with mayo unless you ask for it without. That adds an extra 24 grams of fat, which is more than the original sandwich contained.

data—the smaller the sum, the better the fit. A different line will produce a different sum, maybe bigger, maybe smaller. The **line of best fit** is the line for which the sum of the squared residuals is smallest, hence called the **least squares** line.

You might think that finding this line would be pretty hard. Surprisingly, it's not, although it was an exciting mathematical discovery when Legendre published it in 1805.

Who's on first In 1805, the French mathematician Adrien-Marie Legendre was the first to publish the *least squares* solution to the problem of fitting a line to data when the points don't all fall exactly on the line. The main challenge was how to distribute the errors "fairly." After considerable thought, he decided to minimize the sum of the squares of what we now call the residuals. When Legendre published his paper, however, the German mathematician Carl Friedrich Gauss claimed he had been using the method since 1795. Gauss later referred to the *least squares* solution as "*our* method" (*principium nostrum*), which certainly didn't help his relationship with Legendre.

7.2 The Linear Model

NOTATION ALERT:

"Putting a hat on it" is standard Statistics notation to indicate that something has been predicted by a model. Whenever you see a hat over a variable name or symbol, you can assume it is the predicted version of that variable or symbol (and look around for the model). In a linear model, we use b_1 for the slope and b_0 for the y-intercept.

You may remember from Algebra that a straight line can be written as

$$y = mx + b$$

where m is the slope and b is the y-intercept. We'll use this form for our linear model, but in Statistics we use slightly different notation:

$$\hat{y} = b_0 + b_1 x$$

We write $\hat{y}$ (y-hat) to emphasize that the points that satisfy this equation are just our *predicted* values, not the actual data values (which scatter around the line). If the model is a good one, the data values will scatter closely around it.

We write b_0 and b_1 for the intercept and slope of the line. The bs are also called the *coefficients* of the linear model. The coefficient b_1 is the **slope**, which tells how rapidly $\hat{y}$ changes with respect to x. The coefficient b_0 is the **intercept**, which tells where the line hits (intercepts) the y-axis.[3]

For the Burger King menu items, the line of best fit is

$$\widehat{Fat} = 8.4 + 0.91\ Protein.$$

A S *Simulation:* **Interpreting Equations.** This demonstrates how to use and interpret linear equations.

What does this mean? The **slope**, 0.91, says that a Burger King item with one more gram of protein can be expected, on average, to have 0.91 more grams of fat. Less formally, we might say that, on average, Burger King foods pack about 0.91 grams of fat per gram of protein. Slopes are always expressed in y-units per x-unit. They tell how the y-variable changes (in its units) for a one-unit change in the x-variable. When you see a phrase like "students per teacher" or "kilobytes per second," think slope.

What about the **intercept**, 8.4? Algebraically, that's the value the line takes when x is zero. Here, our model predicts that even a BK item with no protein would have, on average, about 8.4 grams of fat. Is that reasonable? Well, there are two items on the menu with 0 grams of protein (apple fries and applesauce). Neither has any fat either (they are essentially pure carbohydrates), but we could imagine a protein-less item with 8.4 grams of fat. But often 0 is not a plausible value for x (the predicted price of a 0 square foot condo?). Then the intercept serves only as a starting value for our predictions, and we don't interpret it as a meaningful predicted value.

[3]We change from $mx + b$ to $b_0 + b_1x$ for a reason: Eventually we'll want to add more x-variables (predictors) to the model to make it more realistic, and we don't want to use up the entire alphabet. What would we use after m? The next letter is n and that one's already taken. o? See our point? Sometimes subscripts are the best approach.

For Example A LINEAR MODEL FOR HURRICANES

The barometric pressure at the centre of a hurricane is often used to measure the strength of the hurricane because it can predict the maximum wind speed of the storm. A scatterplot shows that the relationship is straight, strong, and negative. It has a correlation of −0.879.

Using technology to fit the straight line, we find

$$\widehat{MaxWindSpeed} = 955.27 - 0.897\ CentralPressure$$

Jeff Schmaltz, MODIS Rapid Response, NASA Goddard Space Flight Center/NASA

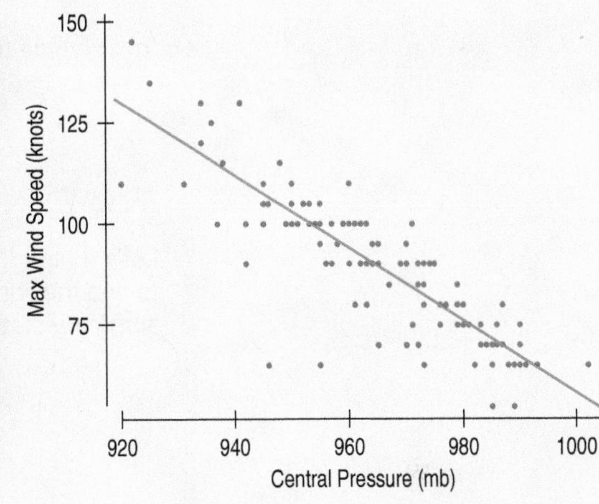

QUESTION: Interpret this model. What does the slope mean in this context? Does the intercept have a meaningful interpretation?

ANSWER: The negative slope says that as *CentralPressure* falls, *MaxWindSpeed* increases. This is consistent with the way hurricanes work: Low central pressure pulls in moist air, driving the rotation and the resulting destructive winds. The slope's value says that, on average, the maximum wind speed is about 0.897 knots faster for every 1-millibar drop in central pressure.

It's not meaningful, however, to interpret the intercept as the wind speed predicted for a central pressure of 0—that would be a vacuum. Instead, it is merely a starting value for the model.

7.3 Finding the Least Squares Line

SLOPE

$$b_1 = r\frac{s_y}{s_x}$$

UNITS OF *y* PER UNIT OF *x*

Get into the habit of identifying the units by writing down "*y*-units per *x*-unit," with the unit names put in place. You'll find it'll really help you to *Tell* about the line in context.

How do we find the slope and intercept of the least squares line? The formulas are simple. The model is built from the summary statistics we've used over and over: the correlation (to tell us the strength of the linear association), the standard deviations (to give us the units), and the means (to tell us where to put the line). The slope[4] of the line is just

$$b_1 = r\frac{s_y}{s_x}$$

We've already seen that the correlation tells us the *sign* as well as the strength of the relationship, so it's no surprise to see that the slope inherits this sign as well.

Correlations don't have units, but slopes do. How x and y are measured—what units they have—doesn't affect their correlation, but can change their slope. If children grow an average of 3 inches per year, that's the same as 0.21 millimetres per day. For the slope, it

[4]The slope may also be written: $b_1 = \frac{s_{xy}}{s_x^2} = \frac{covariance\ of\ x\ and\ y}{variance\ of\ x}$

INTERCEPT
$$b_0 = \bar{y} - b_1\bar{x}$$

matters whether you express age in days or years and whether you measure height in inches or millimetres. Changing the units of x and y affects their standard deviations directly. And that's how the slope gets its units—from the ratio of the two standard deviations. So the units of the slope are a ratio, too. The units of the slope are always the units of y *per* unit of x.

What about the intercept? If you had to predict the y-value for a data point whose x-value was the average, $\bar{x}$, you'd probably pick $\bar{y}$, and that's just what the least squares line does. Putting that into our equation, we find

$$\bar{y} = b_0 + b_1\bar{x}$$

Or, rearranging the terms, we get an equation for the intercept:

$$b_0 = \bar{y} - b_1\bar{x}$$

For Example FINDING THE REGRESSION EQUATION

Let's try out the slope and intercept formulas on the Burger King data and see how to find the line. We checked the conditions when we calculated the correlation. The summary statistics are as follows:

Protein	Fat
$\bar{x} = 18.0$ g	$\bar{y} = 24.8$ g
$s_x = 13.5$ g	$s_y = 16.2$ g
$r = 0.76$	

So, the slope is

$$b_1 = r\frac{s_y}{s_x} = 0.76 \times \frac{16.2 \text{ g fat}}{13.5 \text{ g protein}} = 0.91 \text{ grams of fat per gram of protein.}$$

The intercept is

$$b_0 = \bar{y} - b_1\bar{x} = 24.8 \text{ g fat} - 0.91 \text{ g fat/g protein} \times 18.0 \text{ g protein} = 8.4 \text{ g fat}$$

Putting these results back into the equation gives

$$\widehat{Fat} = 8.4 \text{ g fat} + 0.91 \text{ g fat/g protein}$$

or more simply,

$$\widehat{Fat} = 8.4 + 0.91\ Protein.$$

In summary, the predicted fat of a Burger King food item in grams is 8.4 + 0.91 × the protein content (in grams).

With an estimated linear model, it's easy to predict fat content for any menu item we want. For example, for the BK Tendercrisp chicken sandwich with 31 grams of protein, we can plug in 31 grams for the amount of protein and see that the predicted fat content is $8.4 + 0.91(31) = 36.6$ grams of fat. (Actually, rounding has caught up with us. Using full precision not shown here, the predicted value to one decimal place should be 36.7. Always use full precision until you report your results). Because the BK Tendercrisp chicken sandwich actually has 22 grams of fat, its residual is

$$Fat - \widehat{Fat} = 22 - 36.7 = -14.7 \text{ g}$$

Least squares lines are commonly called regression lines. In a few pages, we'll see that this name is an accident of history. For now, you just need to know that *regression* almost always means *the model fit by least squares*.

To use a regression model, we should check the same **Conditions** for regression as we did for correlation, making sure that the variables are **Quantitative**, that the relationship is **Straight Enough**, and that there are **No Outliers**.

A S ***Activity:* Find a Regression Equation.** Now that we've done it by hand, try it with technology using the statistics package paired with your version of *MyStatLab*.

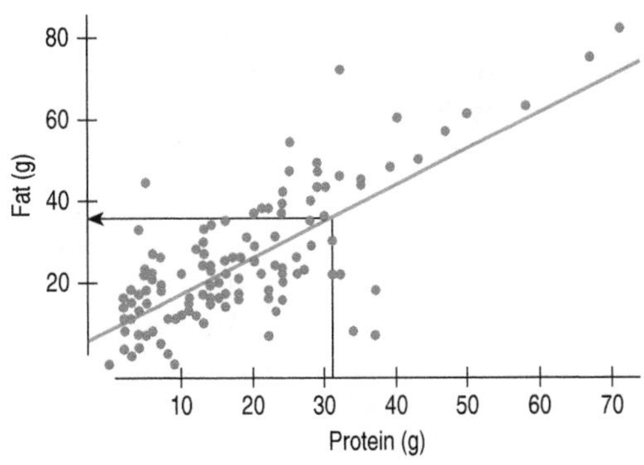

Figure 7.3
Burger King menu items with the regression line. The regression model lets us make predictions. The predicted fat value for an item with 31 grams of protein is 36.6 g.

Step-by-Step Example CALCULATING A REGRESSION EQUATION

Zuma Wire Service/Alamy

During the evening rush hour of August 1, 2007, an eight-lane steel truss bridge spanning the Mississippi River in Minneapolis, Minnesota, collapsed without warning, sending cars plummeting into the river, killing 13 and injuring 145. Although similar events had brought attention to North America's aging infrastructure, this disaster put the spotlight on the problem and raised the awareness of the general public.

How can we tell which bridges are safe?

Most states conduct regular safety checks, giving a bridge a structural deficiency score on various scales. The New York State Department of Transportation uses a scale that runs from 1 to 7, with a score of 5 or less indicating "deficient." Many factors contribute to the deterioration of a bridge, including amount of traffic, material used in the bridge, weather, and bridge design.

New York has more than 17 000 bridges. We have available data on the 194 bridges of Tompkins County.

One natural concern is the age of a bridge. A model that relates age to safety score might help the DOT to focus inspectors' efforts where they are most needed.

Question: Is there a relationship between the age of a bridge and its safety rating?

THINK ➡ Plan State the problem.

I want to know whether there is a relationship between the age of a bridge in Tompkins County, NY, and its safety rating.

Variables Identify the variables and report the *W*'s.

I have data giving the Safety Score and Age for 194 bridges constructed since 1865.

Just as we did for correlation, check the conditions for a regression by making a picture. Never fit a regression without looking at the scatterplot first.

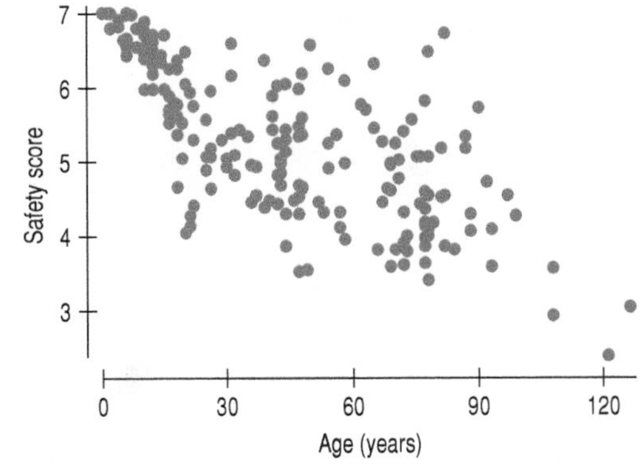

Conditions:

✓ **Quantitative Variables:** We'll assume the Safety Scores are determined on a reasonably precise quantitative scale. Age is in years.

✓ **Straight Enough:** Scatterplot looks straight.

✓ **No Outliers:** None are evident.

It is okay to use a linear regression to model this relationship.

SHOW ➡ Mechanics Find the equation of the regression line. Summary statistics give the building blocks of the calculation.	**Age** $\bar{x} = 44.9$ $s_x = 30.75$
(We generally report summary statistics to one more digit of accuracy than the data. We do the same for intercept and predicted values, but for slopes we usually report an additional digit. Remember, though, not to round until you finish computing an answer.)[5]	**Safety Score** $\bar{y} = 5.2779$ $s_y = 1.0297$ **Correlation** $r = -0.681$
Find the slope, b_1.	$b_1 = r\frac{s_y}{s_x} = -0.681\frac{1.0297}{30.75}$ $= -0.0228$ *Safety Score points per year*
Find the intercept, b_0.	$b_0 = \bar{y} - b_1\bar{x} = 5.2779 - (-0.0228)44.9$ $= 6.30$
Write the equation of the model, using meaningful variable names.	The least squares line is $\hat{y} = 6.30 - 0.0228x$ or $\widehat{\text{Safety Score}} = 6.30 - 0.0228\ \text{Age}$
TELL ➡ Conclusion Interpret what you have found in the context of the question. Discuss in terms of the variables and their units.	The condition of the bridges in Tompkins County, NY, decreases with the age of the bridges at the time of inspection. Bridge condition declines by about 0.023 points on the 1 to 7 scale for each year of age. The model uses a base or intercept of 6.3, which is quite reasonable because a new bridge (0 years of age) should have a safety score near the maximum of 7. Because I have only data from one county, I can't tell from these data whether this model would apply to bridges in other counties of New York or in other locations.

7.4 Regression to the Mean

Suppose you were told that a new male student was about to join the class, and you were asked to guess his height in inches. What would be your guess? Since you have no other information, a good guess would be the mean height of all male students. (If you thought about this in *z*-scores, you'd say $\hat{z}_{Height} = 0$). Now suppose you are also told that this student has a grade point average (*GPA*) of 3.9—about two SDs above the mean *GPA*. Would that change your guess? Probably not. The correlation between *GPA* and *Height* is near 0, so knowing something about *GPA* wouldn't tell you anything about *Height*.

[5] We warned you that we'll round in the intermediate steps of a calculation to show the steps more clearly, and we've done that here. If you repeat these calculations yourself on a calculator or statistics program, you may get somewhat different results. When calculated with more precision, the intercept is 6.40495 and the slope is −0.02565.

On the other hand, if you were told that, measured in centimetres, the student's *Height* was two SDs above the mean, what could you say about his *Height* in inches? There's a perfect correlation between *Height* in inches and *Height* in centimetres, so you'd predict that he's two SDs above the mean *Height* in inches as well. Or, writing this as z-scores, you could say: $\hat{z}_{Htin} = z_{Htcm}$.

What if you're told that the student is two SDs above the mean in *Shoe Size*? Would you still guess that he's of average *Height*? You might guess that he's taller than average, since there's a positive correlation between *Height* and *Shoe Size*. But would you guess that he's two SDs above the mean? When there was no correlation, we didn't move away from the mean at all. With a perfect correlation, we moved our guess the full two SDs. Any correlation between these extremes should lead us to guess somewhere between zero and two SDs above the mean. As we'll see in a minute, if we think in z-scores, the relationship is remarkably simple. When the correlation between x and y is r, and x has a z-score of z_x, the predicted value of y is just $\hat{z}_y = rz_x$. For example, if our student was two SDs above the mean *Shoe Size*, the formula tells us to guess $r \times 2$ standard deviations above the mean for his *Height*.

Why is correlation "r"? In his original paper, Galton standardized the variables and so the slope was the same as the correlation. He used r to stand for this coefficient from the (standardized) regression.

How did we get that? We know from the least squares line that $\hat{y} = b_0 + b_1x$. We also know that $b_0 = \bar{y} - b_1\bar{x}$. If we standardize x and y, both their means will be 0, and so for standardized variables, $b_0 = 0$. And the slope $b_1 = r\frac{s_y}{s_x}$. For standardized variables, both s_y and s_x are 1, so $b_1 = r$. In other words, $\hat{z}_y = rz_x$.

Even more important than deriving the formula is understanding what it says about regression. While we rarely standardize variables to do a regression, we can think about what the z-score regression means. It says that when the correlation is r, and an x-value lies k standard deviations from its mean, our prediction for y is $r \times k$ standard deviations from its mean.

The z-score equation has another profound consequence. It says that for linear relationships, you can never predict that y will be *farther* from its mean than x was from its mean. That's because r is always between -1 and 1. So, each predicted y tends to be closer to its mean (in standard deviations) than its corresponding x was. Sir Francis Galton discovered this property when trying to predict heights of offspring from their parents. What he found was that even though tall parents tend to have tall children, their children are closer to the mean of all offspring than their parents were. He called this phenomenon "regression to mediocrity," because he had hoped that tall parents would have children whose heights were even more extraordinary. But that's not what he found. Often when people relate essentially the same variable in two different groups, or at two different times, they see this same phenomenon—the tendency of the response variable to be closer to the mean than the explanatory variable. Unfortunately, people try to interpret this by thinking that the performance of those far from the mean is deteriorating, but it's just a mathematical fact about the correlation. So, today we try to be less judgmental about this phenomenon and we call it **regression to the mean**. We managed to get rid of the term "mediocrity," but the name regression stuck as a name for the whole least squares fitting procedure—and that's where we get the term **regression line**.[6]

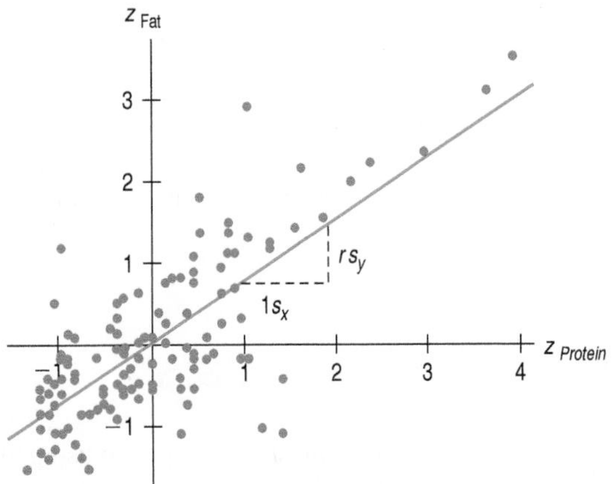

Figure 7.4
Standardized Fat vs. *Standardized Protein* with the regression line. Each one-standard-deviation change in *Protein* results in a predicted change of r standard deviations in *Fat*.

[6]Just think. If children were more extraordinary than their parents, rather than less, we might be fitting linear progressions.

Just Checking

A scatterplot of house price (in dollars) versus house size (in square feet) for houses sold recently in your small Maritime city shows a relationship that is straight, with only moderate scatter and no outliers. The correlation between house price and house size is 0.85 and the equation of the regression model is

$$\widehat{Price} = 9564 + 122.74\, Size$$

1. What does the slope of 122.74 mean?
2. What are the units of the slope?
3. Your house is 2000 sq. ft. bigger than your neighbour's house. How much more do you expect it to be worth?
4. You read an ad about a house that is 2 standard deviations above the mean in size. What can you say about its price?
5. The mean price of all these houses is \$238 000, with standard deviation of \$63 790. What is the predicted price for a house that is 2 SDs above the mean in size?

MATH BOX

To get from $\hat{z}_y = rz_x$ back to our equation $\hat{y} = b_0 + b_1x$, first remember how we standardize. For the x's: $z_x = \frac{x - \bar{x}}{s_x}$ and for the y's: $z_y = \frac{y - \bar{y}}{s_y}$.

So, $\hat{z}_y = rz_x$ becomes

$$\frac{\hat{y} - \bar{y}}{s_y} = r\frac{x - \bar{x}}{s_x}$$

or

$$\hat{y} - \bar{y} = rs_y\frac{x - \bar{x}}{s_x} = r\frac{s_y}{s_x}(x - \bar{x})$$

Now, we're nearly there. If we rearrange terms, we'll have

$$\hat{y} = \left(\bar{y} - r\frac{s_y}{s_x}\bar{x}\right) + \left(r\frac{s_y}{s_x}\right)x$$

So, if we let $r\frac{s_y}{s_x} = b_1$ and $\bar{y} - r\frac{s_y}{s_x}\bar{x} = \bar{y} - b_1\bar{x} = b_0$, we can write

$$\hat{y} = b_0 + b_1x$$

And we get the equations for the slope and intercept for free.

7.5 Examining the Residuals

The linear model we are using assumes that the relationship between the two variables is a straight line. The residuals are the part of the data that *hasn't* been modelled. Conceptually, we think

$$Data = Model + Residual$$

NOTATION ALERT

Galton used *r* for correlation, so we use *e* for "error." But it's not a mistake. Statisticians often refer to variability not explained by a model as error.

or, equivalently,

$$Residual = Data - Model$$

More precisely, the actual individual residuals are defined as:

Residual = Observed response – Predicted response, or, in symbols,

$$e = y - \hat{y}$$

When we want to know how well the model fits, we can ask instead what the model missed. No model is perfect, so it's important to know how and where it fails. To see that, we look at the residuals.

For Example KATRINA'S RESIDUAL

RECAP: The linear model relating hurricanes' wind speeds to their central pressures was

$$\widehat{MaxWindSpeed} = 955.27 - 0.897\ CentralPressure$$

Let's use this model to make predictions and see how those predictions do.

QUESTION: Hurricane Katrina had a central pressure measured at 920 millibars. What does our regression model predict for her maximum wind speed? How good is that prediction, given that Katrina's actual wind speed was measured at 110 knots?

ANSWER: Substituting 920 for the central pressure in the regression model equation gives

$$\widehat{MaxWindSpeed} = 955.27 - 0.897(920) = 130.03$$

The regression model predicts a maximum wind speed of 130 knots for Hurricane Katrina.

The residual for this prediction is the observed value minus the predicted value:

$$110 - 130 = -20\ kts$$

In the case of Hurricane Katrina, the model predicts a wind speed 20 knots higher than was actually observed.

Residuals help us to see whether the model makes sense. When a regression model is appropriate, it should model the underlying relationship. Nothing interesting should be left behind. So after we fit a regression model, we usually plot the residuals in the hope of finding . . . nothing interesting, as in Figure 7.5 below.

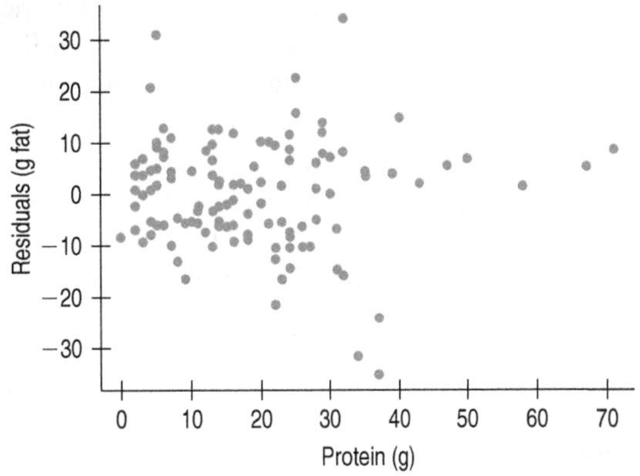

Figure 7.5

The residuals for the BK menu regression look appropriately boring. There are no obvious patterns, although there are a few points with large residuals. The negative ones turn out to be grilled chicken items, which are among the lowest fat (per protein) items on the menu. The two high outliers contain hash browns, which are high fat per protein items.

A scatterplot of the residuals versus the *x*-values should be the most boring scatterplot you've ever seen. It shouldn't have any interesting features, like a direction or shape. It should stretch horizontally, with about the same amount of scatter throughout. It should show no bends and it should have no outliers. If you see any of these features, find out what the regression model missed. For example, in Figure 7.6 below, we see a curving scatterplot and the corresponding bent residual plot (note how the bend is accentuated in the residual plot):

Figure 7.6
A case of the bends.

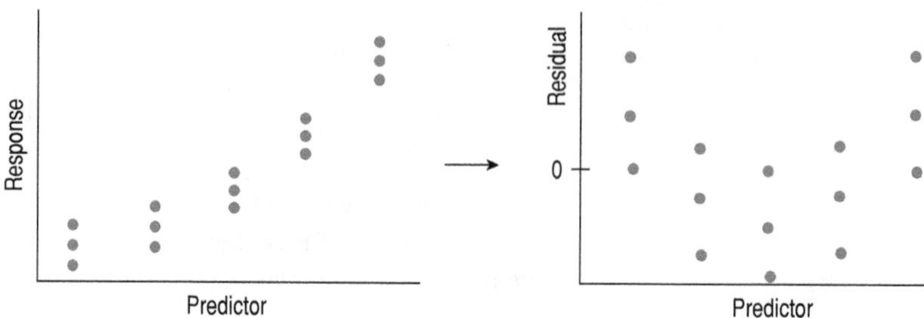

Most computer statistics packages plot the residuals against the predicted values $\hat{y}$, rather than against x.[7] When the slope is positive, they're virtually identical except for the axis labels. When the slope is negative, the two versions are mirror images. Since all we care about is the patterns (or, better, lack of patterns) in the plot, it really doesn't matter which way we plot the residuals.

If the residuals show no interesting pattern when we plot them against x, we can look at how big they are. Since their mean is always zero, though, it's only sensible to look at how much they vary (from zero). The standard deviation of the residuals, s_e, will give us a measure of how much the points spread around the regression line, but for this summary to make sense, the residuals should all share the same underlying spread. That's why we check to make sure that the residual plot has about the same amount of scatter throughout the plot.

This gives us a new assumption: the **Equal Variance Assumption**. The associated condition to check is the **Does the Plot Thicken? Condition**. We check to make sure that the spread is about the same throughout. We can check that either in the original scatterplot of y against x or in the scatterplot of residuals.

For example, the variability of measurements on the response variable often increases as the measurements get bigger, such as illustrated below:

Figure 7.7
A steady trend in dispersion. The plot thickens (as we go from left to right).

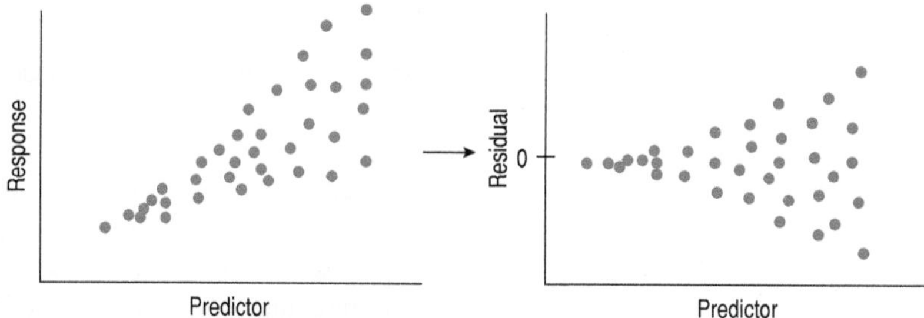

This is not the only way the problem of uneven dispersion can arise, but it is the easiest to deal with. Often a non-linear transformation, like a logarithm, will fix it (sometimes even fixing other problems as well, like non-linearity). This problem may also arise when we mix together different subgroups in our analysis—again with a simple solution—separate the groups. But at other times, there may be no simple or easy fix for big differences in response variability across the scatterplot.

The Residual Standard Deviation

We estimate the standard deviation of the residuals in almost the way you'd expect:

$$s_e = \sqrt{\frac{\Sigma e^2}{n-2}}$$

Why $n - 2$ rather than $n - 1$? We used $n - 1$ for s when we estimated the mean. Now we're estimating both a slope and an intercept. Looks like a pattern—and it is. We subtract one more for each parameter we estimate.

We don't need to subtract the mean because the mean of the residuals $\bar{e} = 0$. The $n - 2$, like the $n - 1$ in the denominator of the variance, is an adjustment we'll discuss in more detail in a later chapter.

[7] They have a good reason for this choice. When regression models use more than one x-variable, there can be many x's to plot against, but there's still only one $\hat{y}$.

For the Burger King foods, the standard deviation of the residuals is 10.6 grams of fat. That looks about right in the scatterplot of residuals, as a fairly typical distance from the mean of zero. The residual for the BK Tendercrisp chicken was −14.7 grams, just under 1.5 standard deviations in size.

It's a good idea to make a histogram of the residuals (Figure 7.8). If we see a unimodal, symmetric histogram, then we can apply the 68–95–99.7 Rule to see how well the regression model describes the data. In particular, we would know that about 95% of the residuals should be no bigger in size than $2s_e$. The Burger King residuals look like:

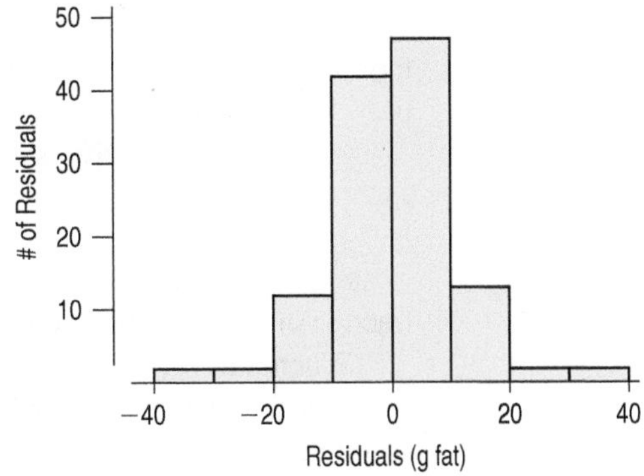

Figure 7.8
The histogram of residuals is symmetric and unimodal, centred at 0, with a standard deviation of 10.6. Only a few values lie outside of 2 standard deviations. The low ones are the chicken items mentioned before. The high ones contain hash browns.

Sure enough, most are less than 2 × 10.6, or 21.2 g of fat in size. In other words, *about 95% of the fat measurements lie within 21.2 grams of their predicted values.*

7.6 R^2—The Variation Accounted For by the Model

The variation in the residuals is the key to assessing how well the model fits the data. Let's compare the variation of the response variable with the variation of the residuals. The total fat has a standard deviation of 16.2 grams. The standard deviation of the residuals is 10.6 grams. If the correlation were 1.0 and the model predicted the fat values perfectly, the residuals would all be zero and have no variation. We couldn't possibly do any better than that.

On the other hand, if the correlation were zero, the model would simply predict 24.8 grams of fat (the mean) for all menu items (Figure 7.9). The residuals from that prediction would just be the observed fat values minus their mean. These residuals would have the same variability as the original data because, as we know, just subtracting the mean doesn't change the spread.

How well does the BK regression model do? Look at the boxplots in Figure 7.9. The variation in the residuals (on the right) is smaller than in the data, but certainly bigger than

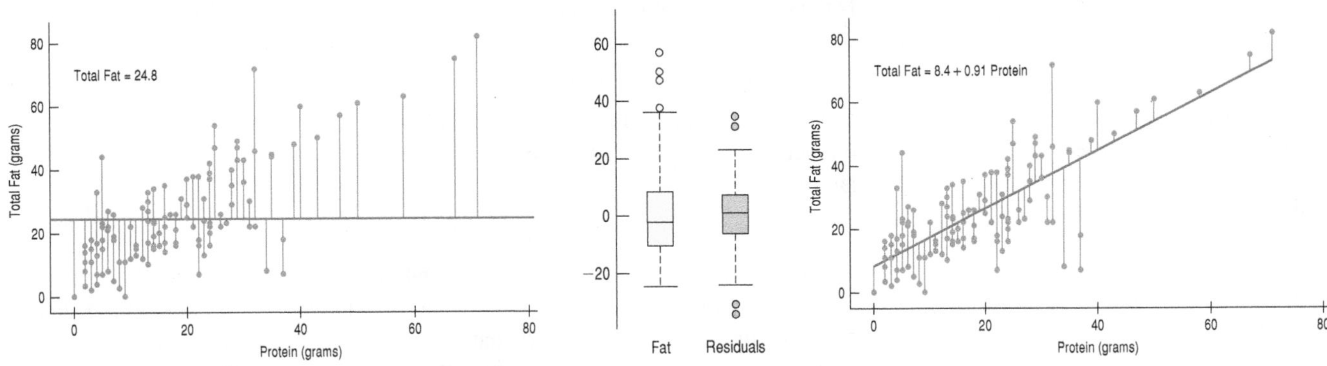

Figure 7.9
Compare the variability of total *Fat* with the residuals from the regression. In the boxplots, we compare the deviations of the individual *Fat* values from their mean of 24.8 (shown in the plot on the left), with the deviations or residuals from the line of best fit (on the right). The variation left in the residuals is unaccounted for by the model, but it's less than the variation in the original data.

TWICE AS STRONG?

Is a correlation of 0.80 twice as strong as a correlation of 0.40? Not if you think in terms of R^2. A correlation of 0.80 means an R^2 of $0.80^2 =$ 64%. A correlation of 0.40 means an R^2 of $0.40^2 =$ 16%—only a quarter as much of the variability accounted for. A correlation of 0.80 gives an R^2 *four* times as strong as a correlation of 0.40 and accounts for four times as much of the variability.

zero. That's nice to know, but how much of the variation is still left in the residuals? If you had to put a number between 0% and 100% on the fraction of the variation left in the residuals, what would you say?

All regression models fall somewhere between the two extremes of zero correlation and perfect correlation. We'd like to gauge where our model falls. Can we use the correlation to do that? Well, a regression model with correlation −0.5 is doing as well as one with correlation +0.5. They just have different directions. If we *square* the correlation coefficient, we'll get a value between 0 and 1, and the direction won't matter. The squared correlation, r^2, gives the fraction of the data's variation accounted for by the model, and $1 - r^2$ is the fraction of the original variation left in the residuals. For the Burger King model, $r^2 = 0.76^2 = 0.58$, and $1 - r^2$ is 0.42, so 42% of the variability in total *Fat* has been left in the residuals. How close was that to your guess?

All regression analyses include this statistic, although by tradition, it is written with a capital letter, R^2, pronounced "*R*-squared," and also may be called the *coefficient of determination*. An R^2 of 0 means that none of the variance in the data is in the model; all of it is still in the residuals. It would be hard to imagine using that model for anything. Because R^2 is a fraction of a whole, it is often given as a percentage.[8] For the Burger King data, R^2 is 58%.

When interpreting a regression model, you need to *Tell* what R^2 means. According to our linear model, 58% of the variability in the fat content of Burger King sandwiches is accounted for by variation in the protein content.

How can we see that R^2 is really the fraction of variance accounted for by the model? It's a simple calculation. The variance of the fat content of the Burger King foods is $16.2^2 = 262.44$. If we treat the residuals as data, the variance of the residuals is 110.53.[9] As a fraction, that's 110.53/262.44 = 0.42 *or* 42%. That's the fraction of the variance that is not accounted for by the model. The fraction that is accounted for is 100% − 42% = 58%, just the value we got for R^2.

Just Checking

Back to our regression of house price (in dollars) on house size (in square feet):

$$\widehat{Price} = 9564 + 122.74\ Size$$

The R^2 value is reported as 71.4%, and the standard deviation of the residuals is 63 790.

6. What does the R^2 value mean about the relationship of price and size?

7. Is the correlation of price and size positive or negative? How do you know?

8. If we measured the size in square metres instead of square feet, would the R^2 value change? How about the slope?

9. You find that your house is worth $80 000 more than the regression model predicts. You are undoubtedly pleased, but is this actually a surprisingly large value?

Some Extreme Tales

One major company developed a method to differentiate between proteins. To do so, they had to distinguish between regressions with R^2 of 99.9% and 99.8%. For this application, 99.8% was not high enough. On the other hand, the head of a hedge fund reports that although his regressions give R^2 below 2%, they are highly successful because those used by his competition are even lower.

R^2 is always between 0% and 100%. But what's a "good" R^2 value? The answer depends on the kind of data you are analyzing and on what you want to do with it. Just as with correlation, there is no value for R^2 that automatically determines that the regression is "good." Data from scientific experiments often have R^2 in the 80% to 90% range and even higher. Data from observational studies and surveys, though, often show relatively

[8] By contrast, we usually report correlation coefficients as decimal values between −1.0 and 1.0.

[9] This isn't quite the same as squaring the s_e that we discussed a couple of pages ago (due to different denominators, $n - 1$ vs $n - 2$), but it's very close. We'll deal with the distinction in Chapter 24.

weak associations because it's so difficult to measure responses reliably. An R^2 of 50% to 30% or even lower might be taken as evidence of a useful regression. The standard deviation of the residuals can give us more information about the usefulness of the regression by telling us how much scatter there is around the line.

As we've seen, an R^2 of 100% is a perfect fit, with no scatter around the line. The s_e would be zero. All of the variance is accounted for by the model and none is left in the residuals at all. This sounds great, but it's too good to be true for real data.[10]

Along with the slope and intercept for a regression, you should always report R^2 so that readers can judge for themselves how successful the regression is at fitting the data. Statistics is about variation, and R^2 measures the success of the regression model in terms of the fraction of the variation of the response variable accounted for by the regression. R^2 is the first part of a regression that many people look at because, along with the scatterplot, it tells whether the regression model is even worth thinking about.

For Example INTERPRETING R^2

Recap: Our regression model that predicts maximum wind speed in hurricanes based on the storm's central pressure has $R^2 = 77.3\%$.

QUESTION: What does that say about our regression model?

ANSWER: An R^2 of 77.3% indicates that 77.3% of the variation in maximum wind speed can be accounted for by the hurricane's central pressure. Other factors, such as temperature and whether the storm is over water or land, may account for some of the remaining variation.

A Tale of Two Regressions

Regression slopes may not behave exactly the way you'd expect at first. Our regression model for the Burger King sandwiches was $\widehat{Fat} = 8.4 + 0.91\ Protein$. That equation allowed us to estimate that a sandwich with 31 grams of protein would have 36.7 grams of fat. Suppose, though, that we knew the fat content and wanted to predict the amount of protein. It might seem natural to think that by solving our equation for *Protein* we'd get a model for predicting *Protein* from *Fat*. But that doesn't work.

To see why it doesn't work, let's *make a picture* (see Figure 7.10). On the left side is a scatterplot with simulated data for two variables *x* and *y*, showing the least-squares fit for predicting *y* from *x*. Note how reasonable the line seems: i.e., at any *x*, the data points are roughly equally above and below the line, and the line nicely slices through the data (in a vertical sense) as you move from low *x* to high *x*, or from left to right.

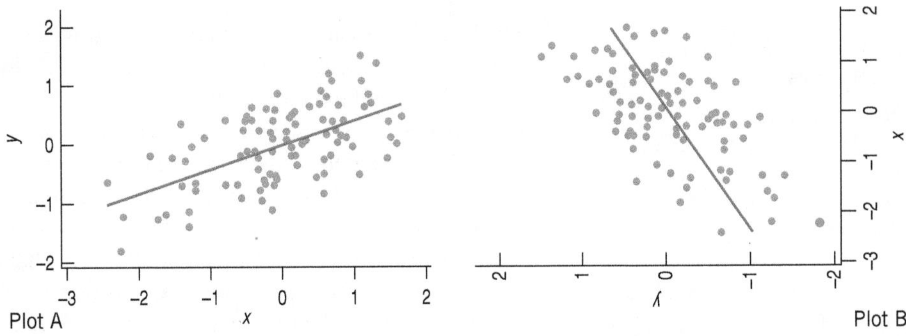

Figure 7.10

Plot A is a scatterplot with best fit line to predict *y* from *x*.
Plot B is the same graph but rotated 90 degrees, so *x* is the vertical axis and *y* is the horizontal axis. Is the same line good for predicting *x* from *y*?

Now examine the plot on the right in Figure 7.10. We see a very poorly fitted straight line if predicting the vertical axis variable from the horizontal axis variable.

[10]If you see an R^2 of 100%, it's a good idea to figure out what happened. You may have discovered a new law of physics, but it's much more likely that you accidentally regressed two variables that measure the same thing.

For example, at values on the right side of the horizontal axis, the predicted values on the line fall way below the data points, so predictions here are very low, while on the left side of the graph the line shoots high above the actual data. For a value of 1 on the horizontal axis, the prediction is off the graph! Yet the plot on the right is exactly the same as the plot on the left, except for rotating the plot by 90 degrees (which is why the symbol y is upside down and the positive direction on the horizontal axis is to the left, not to the right[11]). Clearly, the line's tilt is too extreme, and a good model would have a less severe slope. To find the best fit, we have to reverse the roles of the two variables and refit, rather than solving the prediction equation from the graph on the left for x in terms of y, which would just give us the hideous prediction line displayed on the right side.

Our original model was $\hat{y} = b_0 + b_1x$, but now we want to find an $\hat{x}$ based on a value of y. We don't have y in our original model, only $\hat{y}$, and that makes all the difference. Our model doesn't fit the data values perfectly, and the least squares criterion focuses on the *vertical* (y) errors the model makes in using x to predict y—not on *horizontal* errors related to prediction of x.

A quick look at the equations reveals why. Simply solving our original equation for x would give a new line whose slope is the reciprocal of ours. To model y in terms of x, our slope is $b_1 = r\frac{s_y}{s_x}$. To model x in terms of y, we'd need to use the slope $b_1 = r\frac{s_x}{s_y}$. That's *not* the reciprocal of ours.

Sure, if the correlation, r, were 1.0 or –1.0, the slopes *would* be reciprocals, but that would happen only if we had a perfect fit. Real data don't follow perfect straight lines, so in the real world y and $\hat{y}$ aren't the same, r is a fraction, and the slopes of the two models are not simple reciprocals of one another.

Protein	Fat
$\bar{x} = 18.0\text{g}$	$\bar{y} = 24.8\text{g}$
$s_x = 13.5\text{g}$	$s_y = 16.2\text{g}$
$r = 0.76$	

If we want to predict x from y, we need to create that particular model from the data. For example, to predict *Protein* from *Fat* we find the slope, $b_1 = 0.76\frac{13.5}{16.2} = 0.63$ grams of protein per gram of fat. The regression model is then $\widehat{Protein} = 2.29 + 0.63\ Fat$, so we'd predict that a sandwich with 36.7 grams of fat should have 25.4 grams of protein—not the 31 grams that we used in the first equation (when we began this discussion).

Moral of the story: *Think.* Decide which variable you want to use (the explanatory variable) to predict values for the other (the response variable). Then find the model that does that. If, later, it turns out that you were mistaken and actually want to make predictions in the other direction, you'll need to start over and fit the correct model from scratch.

7.7 Regression Assumptions and Conditions

MAKE A PICTURE (OR TWO)

You can't check the conditions just by checking boxes. You need to examine both the original scatterplot of y against x before you fit the model, and the plot of residuals afterward. These plots can save you from making embarrassing errors and losing points on the exam.

The linear regression model may be the most widely used model in all of Statistics. It has everything we could want in a model: two easily estimated parameters, a meaningful measure of how well the model fits the data, and the ability to predict new values. It even provides an easy way to see violations of conditions in plots of the residuals.

Like all models, though, linear models are only appropriate if some assumptions are true. We can't confirm assumptions, but we often can check related conditions.

First, be sure that both variables are quantitative. It makes no sense to perform a regression on categorical variables. After all, what could the slope possibly mean? Always check the **Quantitative Variables Condition**.

The linear model only makes sense if the relationship is linear. It is easy to check the associated **Straight Enough Condition**. Just look at the scatterplot of y vs. x. You don't need a *perfectly* straight plot, but it must be straight enough for the linear model to make sense. If you try to model a curved relationship with a straight line, you'll usually get just

[11]We could flip it over like a page in a book to correct the direction of values on the horizontal axis, but that wouldn't change the story (just make it easier to read).

what you deserve. If the scatterplot is not straight enough, stop here. You can't use a linear model for *any* two variables, even if they are related. They must have a *linear* association, or the model won't mean a thing.

The response variable needs to exhibit a roughly similar amount of dispersion or spread regardless of the value of x. Fitting the model and using it for inference both depend on this assumption. If the scatterplot of y vs. x looks equally spread out everywhere and (often more vividly) if the *residual plot* of residuals *vs*. predicted values also has a consistent spread, then the assumption is reasonable. The most common violation of that equal variance assumption is for the data or residuals to spread out more for *larger* values of x, so a good mnemonic for this check is the **Does the Plot Thicken? Condition**.

Outlying points can dramatically change a regression model. They can even change the sign of the slope, which could give a very misleading impression of the relationship between the variables. So check the **Outlier Condition**. Check the scatterplot of y against x and the residual plots to be sure there are no outliers. Residual plots often show violations more clearly and may reveal other unexpected patterns or interesting quirks in the data. Of course, any outliers are likely to be interesting and informative, so be sure to look into why they are unusual.

To summarize:

Before starting, be sure to check the

- **Quantitative Variable Condition** If either y or x is categorical, you can't make a scatterplot and you can't perform a regression. Stop.

From the scatterplot of y against x, check the

- **Straight Enough Condition** Is the relationship between y and x straight enough to proceed with a linear regression model?
- **Outlier Condition** Are there any outliers that might dramatically influence the fit of the least squares line?
- **Does the Plot Thicken? Condition** Does the spread of the data around the generally straight relationship seem to be consistent (aside from some natural random variation) for all values of x?

After fitting the regression model, make a plot of residuals against the predicted values (or x) and look for

- Any bends that would violate the **Straight Enough Condition**
- Any outliers that weren't clear before
- Any change in the spread of the residuals from one part of the plot to another

Step-by-Step Example REGRESSION

If you plan to hit the fast food joints for lunch, you should have a good breakfast. Nutritionists, concerned about "empty calories" in breakfast cereals recorded facts about 77 breakfast cereals, including the *Calories* and *Sugar* content (in grams) per serving.

Question: How are calories and sugar content related in breakfast cereals?

THINK ➡ Plan State the problem.	I am interested in the relationship between sugar content and calories in cereals.
Variables Name the variables and report the *W*'s.	I have two quantitative variables, *Calories* and *Sugar* content, measured on 77 breakfast cereals. The units of measurement are calories and grams of sugar, respectively.

Check the conditions for a regression by making a picture. Never fit a regression without looking at the scatterplot first.

✓ **Quantitative Variables:** Both variables are quantitative.

I'll check the remaining conditions from the scatterplot (and later, the residual plot):

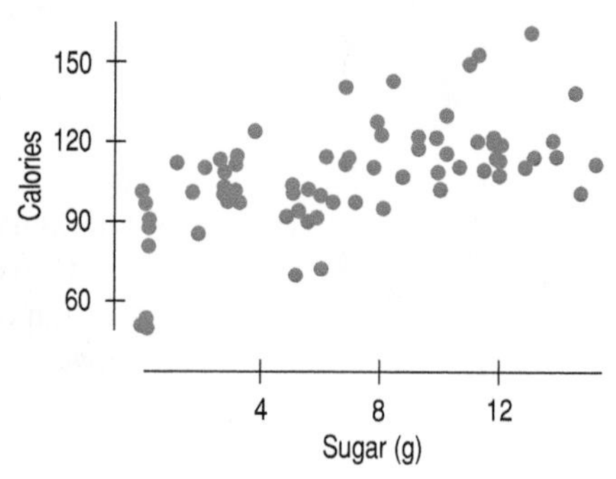

✓ **Outlier Condition:** There are no obvious outliers or groups.

✓ **Straight Enough:** The scatterplot looks straight to me.

✓ **Does the Plot Thicken?** The spread around the line looks about the same throughout, but I'll check this and the previous condition again in a residual plot, after fitting the model.

So far so good, so I will proceed to fit a regression model to these data.

SHOW ➡ Mechanics If there are no clear violations of the conditions, fit a straight line model of the form $\hat{y} = b_0 + b_1x$ to the data. Summary statistics give the building blocks of the calculation.

Calories

$\bar{y} = 107.0$ calories

$s_y = 19.5$ calories

Sugar

$\bar{x} = 7.0$ grams

$s_x = 4.4$ grams

Correlation

$r = 0.564$

Find the slope.

$$b_1 = r\frac{s_y}{s_x} = 0.564\frac{19.5}{4.4}$$

$= 2.50$ calories per gram of sugar

Find the intercept.

$b_0 = \bar{y} - b_1\bar{x} = 107 - 2.50(7) = 89.5$ calories

So the least squares line is

$\hat{y} = 89.5 + 2.50x$,

Write the equation using meaningful variable names.

or $\widehat{Calories} = 89.5 + 2.50$ Sugar.

Squaring the correlation gives

State the value of R^2.

$R^2 = 0.564^2 = 0.318$ or 31.8%.

TELL ➡ Conclusion Describe what the model says in words and numbers. Be sure to use the names of the variables and their units.

The key to interpreting a regression model is to start with the phrase "b_1 y-units per x-unit," substituting the estimated value of the slope for b_1 and the names of the respective units. The intercept is then a starting or base value. It may (as in this example) be meaningful, or (when $x = 0$ is not realistic) it may just be a starting value.

R^2 gives the fraction of the variability of y accounted for by the linear regression model.

Find the standard deviation of the residuals, s_e, and compare it to the original s_y.

The scatterplot shows a positive linear relationship and no outliers. The least squares regression line fit through these data has the equation

$$\widehat{Calories} = 89.5 + 2.50\ Sugar$$

Cereals have about 2.50 calories more per gram of sugar.

And sugar-free cereals would average about 89.5 calories.[12]

The R^2 says that 31.8% of the variability in *Calories* is accounted for by variation in *Sugar* content.

$s_e = 16.2$ calories. That's smaller than the original cereal calorie count SD of 19.5, but still fairly large.

THINK AGAIN ➡ Check Again Even though we looked at the scatterplot before fitting a regression model, a plot of the residuals is essential to any regression analysis because it is the best check for additional patterns and interesting quirks in the data.

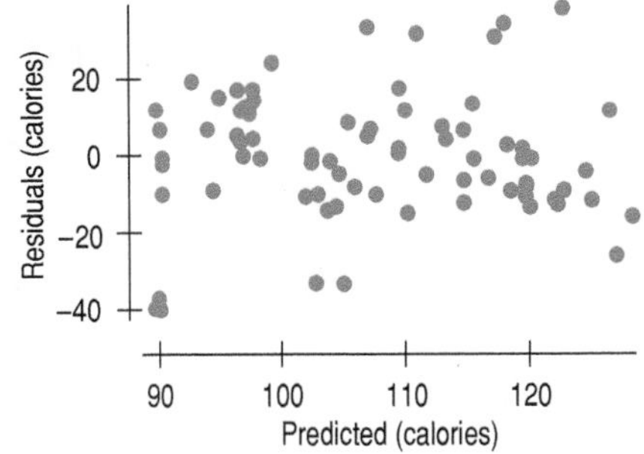

The residuals show a horizontal direction, a shapeless form, and roughly equal scatter for all predicted values. The linear model appears to be appropriate.

> **Regression: adjective, noun, or verb?**
> You may see the term *regression* used in different ways. There are many ways to fit a line to data, but the term *regression line* or *regression* without any other qualifiers always means *least squares*. People also use *regression* as a verb when they speak of *regressing* a y-variable on an x-variable to mean fitting a linear model.

Reality Check: Is the Regression Reasonable?

Statistics don't come out of nowhere. They are based on data. The results of a statistical analysis should reinforce your common sense, not fly in its face. If the results are surprising, then either you've learned something new about the world or your analysis is wrong.

Whenever you perform a regression, think about the coefficients and ask whether they make sense. Is a slope of 2.5 calories per gram of sugar reasonable? That's hard to say right off. We know from the summary statistics that a typical cereal has about 100 calories and 7 grams of sugar. A gram of sugar contributes some calories (actually, 4, but you don't need to know that), so calories should go up with increasing sugar. The direction of the slope seems right.

To see if the *size* of the slope is reasonable, a useful trick is to consider its order of magnitude. We'll start by asking if deflating the slope by a factor of 10 seems reasonable. Is 0.25 calories per gram of sugar enough? The 7 grams of sugar found in the average cereal would contribute less than 2 calories. That seems too small.

Now let's try inflating the slope by a factor of 10. Is 25 calories per gram reasonable? Then the average cereal would have 175 calories from sugar alone. The average cereal has only 100 calories per serving, though, so that slope seems too big.

We have tried inflating the slope by a factor of 10 and deflating it by 10 and found both to be unreasonable. So, like Goldilocks, we're left with the value in the middle that's just right. And an increase of 2.5 calories per gram of sugar is certainly *plausible.*

[12]Predictions at $x = 0$ only makes sense if $x = 0$ is within the range of the observed data.

The small effort of asking yourself whether the regression equation is plausible is repaid whenever you catch errors or avoid saying something silly or absurd about the data. It's too easy to take something that comes out of a computer at face value and assume that it makes sense.

Always be skeptical and ask yourself if the answer is reasonable.

WHAT CAN GO WRONG?

There are many ways in which data that appear at first to be good candidates for regression analysis may be unsuitable. And there are ways that people use regression that can lead them astray. Here's an overview of the most common problems.

- **Don't fit a straight line to a nonlinear relationship.** Linear regression is suited only to relationships that are, well, *linear.* Fortunately, we can often easily improve the linearity by using re-expression (as discussed in Chapter 6).
- **Don't ignore outliers.** Outliers can have a serious impact on the fitted model. You should identify them and think about why they are extraordinary. If they turn out not to be obvious errors, read the next chapter for advice.
- **Don't invert the regression.** The BK regression model was $\widehat{Fat} = 8.4 + 0.91\ Protein$. Knowing protein content, we can predict the amount of fat. But that doesn't let us switch the regression around. We can't use this model to predict protein values from fat values. To swap the predictor–response roles of the variables in a regression, we must fit a new regression equation (it turns out to be $\widehat{Protein} = 2.29 + 0.63\ Fat$).
- **Don't choose a model based on R^2 alone.** Although R^2 measures the strength of the linear association, a high R^2 does not demonstrate the *appropriateness* of the regression. The relationship may still be clearly curved or have very uneven spread about the line. A single outlier or data that separate into two groups may hugely inflate or deflate the value of R^2. Always look at the scatterplot!

CONNECTIONS

The linear model is one of the most important models in Statistics. Chapter 6 talked about the usual assignment of variables to the y- and x-axes. That didn't matter to correlation, but it does matter to regression because the response variable y is predicted by x in the regression model.

The connection of R^2 to correlation is obvious, although it may not be immediately clear that just by squaring the correlation we can learn the fraction of the variability of y accounted for by a regression on x. We'll return to this in (much) later chapters on the topic of Regression.

We made a big fuss about knowing the units of your quantitative variables. We didn't need units for correlation, but without the units we can't define the slope of a regression. A regression makes no sense if you don't know the *Who*, the *What*, and the *Units* of both your variables.

We've summed squared deviations before when we computed the standard deviation and variance. That's not coincidental. They are closely connected to regression.

What Have We Learned?

Learning Objectives

Model linear relationships with a model that shows the response y as a linear (straight-line) function of the explanatory or x-variable.

- Examine residuals to assess how well a model fits the data.

The best fit model is the one that minimizes the sum of the squared residuals—the Least Squares Model.

Find the coefficients of the least squares line from the correlation and the summary statistics of each variable: $b_1 = r\frac{s_y}{s_x}$

$$b_0 = \bar{y} - b_1\bar{x}$$

Understand the correlation coefficient as the number of standard deviations by which one variable is expected to change for a one standard deviation change in the other. Since r is always less than 1 in magnitude, as a consequence we find a natural (though sometimes surprising) *regression to the mean.*

Always examine the residuals to check for violations of assumptions and conditions and to identify any outliers.

Always report and interpret R^2, which reports the fraction of the variation of y accounted for by the linear model.

- R^2 is the square of the correlation coefficient.

Be sure to check the conditions for regression before reporting or interpreting a regression model. Check that:

- The relationship is Straight Enough.
- A scatterplot does not "thicken" in any region.
- There are no outliers.

Interpret a regression slope as y-units per x-unit.

Review of Terms

Linear model An equation of the form

$$\hat{y} = b_0 + b_1x$$

where the x-variable is being used as an *explanatory variable* to help predict the *response variable* y. To interpret a linear model, we need to know the variables (along with their *W*'s) and their units (p. 189).

Model An equation or formula that simplifies and represents reality (p. 189).

Predicted (fitted) value The value of $\hat{y}$ found for a given x-value in the data. A predicted value is found by substituting the x-value in the regression equation. The predicted values are the values on the fitted line; the points $(x, \hat{y})$ all lie exactly on the fitted line (p. 189).
The predicted values are found from the linear model that we fit:

$$\hat{y} = b_0 + b_1x.$$

Residuals The differences between the observed values of the response variable y and the corresponding values predicted by the regression model—or, more generally, values predicted by any model ($\hat{y}$) (p. 189).

$$\text{Residual} = \text{Observed } y\text{-value} - \text{Predicted } y\text{-value} = y - \hat{y}$$

Regression line (Line of best fit) The particular linear equation

$$\hat{y} = b_0 + b_1x$$

that satisfies the least squares criterion is called the least squares regression line. Casually, we often just call it the regression line, or the line of best fit (p. 190).

Least squares The least squares criterion specifies the unique line that minimizes the variance of the residuals or, equivalently, the sum of the squared residuals (p. 190).

Slope The slope, b_1, gives a value in "y-units *per* x-unit." Changes of one unit in x are associated with changes of b_1 units in predicted values of y (p. 190).

The slope can be found by

$$b_1 = r\frac{s_y}{s_x}$$

Intercept The intercept, b_0, gives a starting value in *y*-units. It's the $\hat{y}$-value when *x* is 0. You can find the intercept from $b_0 = \bar{y} - b_1\bar{x}$ (p. 190).

Regression to the mean Because the correlation is always less than 1.0 in magnitude, each predicted $\hat{y}$ tends to be fewer standard deviations from its mean than its corresponding *x* was from its mean (p. 195).

Standard deviation of the residuals (s_e) The standard deviation of the residuals is found by $s_e = \sqrt{\frac{\sum e^2}{n-2}}$. When the residuals are roughly Normally distributed (check their histogram), their sizes can be well described by using this standard deviation and the 68–95–99.7 Rule (p. 198).

Coefficient of determination R^2 The square of the correlation between *y* and *x* (p. 200).

- R^2 gives the fraction of the variability of *y* accounted for by the least squares linear regression on *x*.
- R^2 is an overall measure of how successful the regression is in linearly relating *y* to *x*.

On the Computer REGRESSION

All statistics packages make a table of results for a regression. These tables may differ slightly from one package to another, but all are essentially the same—and all include much more than we need to know for now. Every computer regression table includes a section that looks something like this:

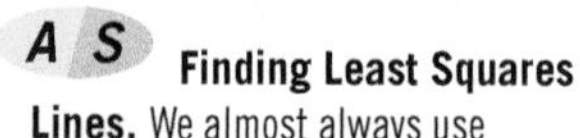

Finding Least Squares Lines. We almost always use technology to find regressions. Practice now—just in time for the exercises.

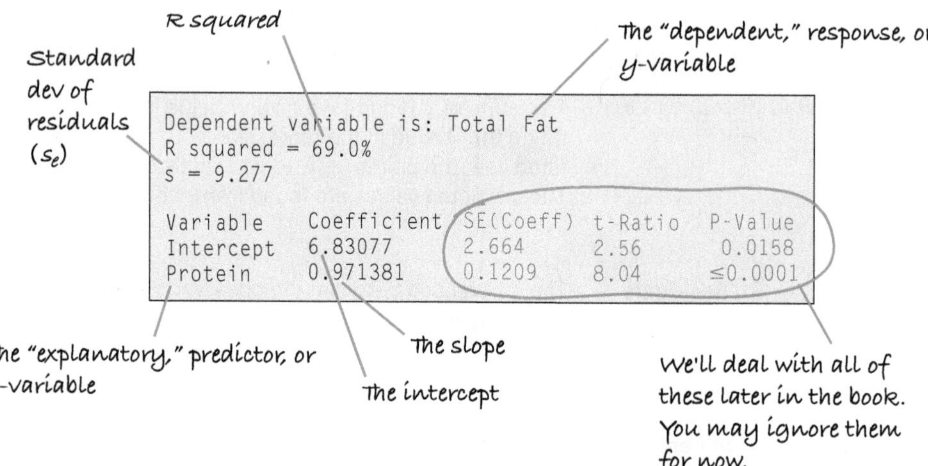

The slope and intercept coefficient are given in a table such as this one. Usually the slope is labelled with the name of the *x*-variable, and the intercept is labelled "Intercept" or "Constant." So the regression equation shown here (for data different from the current chapter example) is

$$\widehat{Fat} = 6.83077 + 0.97138\ Protein.$$

It is not unusual for statistics packages to give many more digits of the estimated slope and intercept than could possibly be estimated from the data. (The original data were reported to the nearest gram.) Ordinarily, you should round most of the reported numbers to one digit more than the precision of the data, and the slope to two. We will learn about the other numbers in the regression table later in the book. For now, all you need to be able to do is find the coefficients, the s_e, and the R^2 value.

EXCEL

Make a scatterplot of the data. You will now see in the menu, under Chart Tools > Design, a choice of various Chart Layouts. Choose Layout 9, which show the fitted line and its equation, and R^2.

COMMENTS

The computer section for Chapter 6 shows how to make a scatterplot.

JMP

- Choose **FitY by X** from the **Analyze** menu.
- Specify the *y*-variable in the Select Columns box and click the *y*, **Response** button.
- Specify the *x*-variable and click the **X, Factor** button.
- Click **OK** to make a scatterplot.
- In the scatterplot window, click on the red triangle beside the heading labelled "**Bivariate Fit . . .**" and choose **Fit Line**. JMP draws the least squares regression line on the scatterplot and displays the results of the regression in tables below the plot.

MINITAB

- Choose **Regression** from the **Stat** menu.
- From the Regression submenu, choose **Fitted Line Plot**.
- In the Fitted Line Plot dialog, click in the **ResponseY** box, and assign the *y*-variable from the variable list.
- Click in the **Predictor X** box, and assign the *x*-variable from the Variable list. Make sure that the Type of Regression Model is set to Linear. Click the **OK** button.

COMMENTS

Alternatively, **Stat** > **Regression** > **Regression** may be used—no graph but more output options.

R

- **lm(Y ~ X)** produces the linear model.
- **summary(lm(Y ~ X))** produces more information about the model.

COMMENTS

Typically, your variables X and Y will be in a data frame. If DATA is the name of the data frame, then

- **lm(Y ~ X,data = DATA)** is the preferred syntax.

SPSS

- Choose **Interactive** from the **Graphs** menu.
- From the interactive Graphs submenu, choose **Scatterplot**.
- In the Create Scatterplot dialog, drag the *y*-variable into the ***y*-axis target**, and the *x*-variable into the ***x*-axis target**.
- Click on the **Fit** tab.
- Choose **Regression** from the **Method** popup menu. Click the **OK** button.

STATCRUNCH

- Click on **Stat**.
- Choose **Regression » Simple Linear**.
- Choose X and Y variable names from the list of columns.
- Click on **Next** (twice) to **Plot the fitted line** on the scatterplot.
- Click on **Calculate** to see the regression analysis.
- Click on **Next** to see the scatterplot.

COMMENTS

Remember to check the scatterplot to be sure a linear model is appropriate.

Note that before you **Calculate**, clicking on **Next** also allows you to:

- enter an X-value for which you want to find the predicted Y-value
- save all the fitted values
- save the residuals
- ask for a residuals plot

TI-83/84 PLUS

To find the equation of the regression line (add the line to a scatterplot), choose **LinReg(a+bx)**, tell it the list names, and then add a comma to specify a function name (from **VARS Y-Vars 1:Function**). The final command looks like

LinReg (a+bx)L1, L2,Y1

- To make a residuals plot, set up a **STATPLOT** as a scatterplot.
- Specify your explanatory data list as Xlist.
- For Ylist, import the name RESID from the **LIST NAMES** menu. **ZoomStat** will now create the residuals plot.

COMMENTS

Each time you execute a **LinReg** command, the calculator automatically computes the residuals and stores them in a data list named RESID. If you want to see them, go to **STAT EDIT**. Space through the names of the lists until you find a blank. Import RESID from the LIST NAMES menu. Now every time you have the calculator compute a regression analysis, it will show you the residuals.

Exercises

1. Regression equations Fill in the missing information in the table below.

	$\bar{x}$	s_x	$\bar{y}$	s_y	r	$\hat{y} = b_0 + b_1x$
a)	10	2	20	3	0.5	
b)	2	0.06	7.2	1.2	−0.4	
c)	12	6			−0.8	$\hat{y} = 200 - 4x$
d)	2.5	1.2		100		$\hat{y} = -100 + 50x$

2. More regression equations Fill in the missing information in the table below.

	$\bar{x}$	s_x	$\bar{y}$	s_y	r	$\hat{y} = b_0 + b_1x$
a)	30	4	18	6	−0.2	
b)	100	18	60	10	0.9	
c)		0.8	50	15		$\hat{y} = -10 + 15x$
d)			18	4	−0.6	$\hat{y} = 30 - 2x$

3. Residuals Tell what each of the residual plots below indicates about the appropriateness of the linear model that was fit to the data.

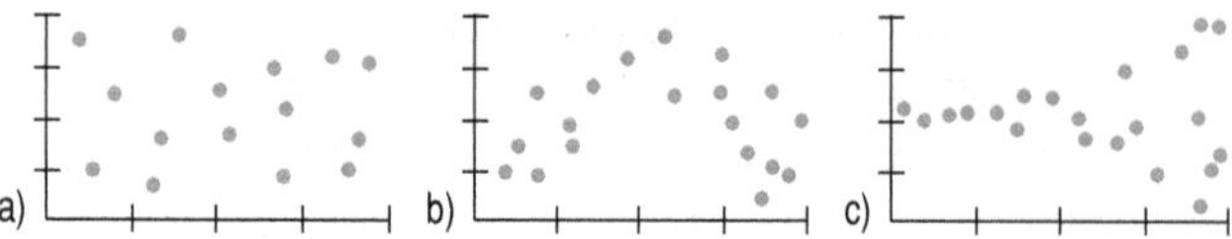

4. Residuals II Tell what each of the residual plots below indicates about the appropriateness of the linear model that was fit to the data.

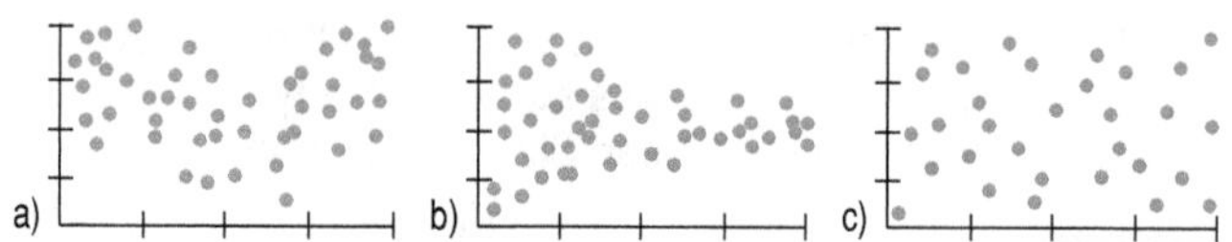

5. Least squares Consider the four points (10, 10), (20, 50), (40, 20), and (50, 80). The least squares line is $\hat{y} = 7.0 + 1.1x$. Explain (or show) what *least squares* means using these data as a specific example.

6. Least squares II Consider the four points (200, 1950), (400, 1650), (600, 1800), and (800, 1600). The least squares line is $\hat{y} = 1975 - 0.45x$. Explain (or show) what *least squares* means using these data as a specific example.

7. Real estate The Re/Max August/Sept 2008 Market Report lists the asking price and size, in square feet, of 38 condos under 1500 square feet in downtown Toronto. A regression to predict the asking price, in thousands of dollars, from size has an R^2 of 70.2%. The plot of residuals indicates that a linear model is appropriate.

a) What are the variables and units in this regression?
b) What units does the slope have?
c) Do you think the slope is positive or negative? Explain.

8. Money and degrees The 2001 Canadian Census provides data on a number of variables for each census tract (neighbourhood). For example, we can find the average income and percent with a Bachelor's degree for each of the 237 census tracts in Ottawa-Gatineau. A regression to predict average income in dollars from percent with degree has an R^2 of 54.2%. The plot of residuals versus predictor indicates that a linear model is appropriate.

a) What are the variables and units in this regression?
b) What units does the slope have?
c) Do you think the slope is positive or negative? Explain.

9. Real estate again The regression of price on size of condos in Toronto, as described in Exercise 7, had $R^2 = 70.2\%$. Write a sentence (in context, of course) summarizing what the R^2 value says about this regression.

10. Money again The regression of average income on percent with degree for Ottawa-Gatineau census tracts, as described in Exercise 8, had $R^2 = 54.2\%$. Write a sentence (in context, of course) summarizing what the R^2 value says about this regression.

11. Real estate redux The regression of price on size of homes in Toronto, as described in Exercise 7, had $R^2 = 70.2\%$.

a) What is the correlation between size and price?
b) What would you predict about the price of a home one standard deviation above average in size?
c) What would you predict about the price of a home two standard deviations below average in size?

12. More money The regression of average income on percent with degree for census tracts in Ottawa-Gatineau, as described in Exercise 8, had $R^2 = 54.2\%$.

a) What is the correlation between average income and percent with degree?
b) What would you predict about the average income of a census tract one standard deviation above average in percent with degree?
c) What would you predict about the average income of a census tract two standard deviations below average in percent with degree?

13. More real estate Consider the Toronto downtown condo data from Exercise 7 again. The regression analysis gives the model $\widehat{Price} = 49.30 + 0.37\ Size$.

a) Explain what the slope of the line says about condo prices and condo size.
b) What asking price would you predict for a 1000-square-foot condo in this market?
c) A real estate agent shows a potential buyer a 1200-square-foot condo, saying that the asking price is \$6000 less than what one would expect for a condo of this size. What is the asking price, and what is this \$6000 called?

14. Lots of money Consider the Ottawa-Gatineau census data from Exercise 8 again. The regression analysis gives the model $\widehat{Average\ income} = \$23\,129 + 527.4\ Degree$.

a) Explain what the slope of the line says about average income and percent with degree.
b) What average income would you predict for a census tract with 50% degree holders?
c) One census tract with 60% degree holders had an average income of \$81 000. Is this more or less than you'd expect? By how much? What's that called?
d) Another census tract with 42% degree holders had an average income of \$25 000. Is this more or less than you'd expect? By how much? What's that called?
e) Make up a rough plot of what the data and fitted line might look like here, and indicate where the census tracts in parts c) and d) would lie. Note that these two tracts were rather unusual in their average incomes, as the standard deviation of the residuals was about 7000.

T **15. Cigarettes** Is the nicotine content of a cigarette related to the "tars"? A collection of data (in milligrams) on 29 cigarettes produced the scatterplot, residual plot, and regression analysis shown:

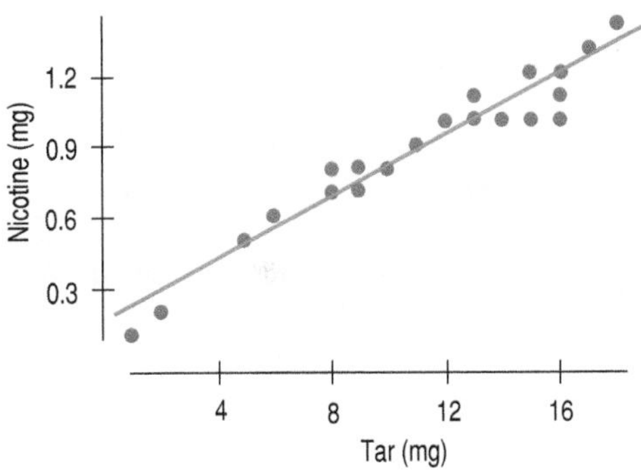

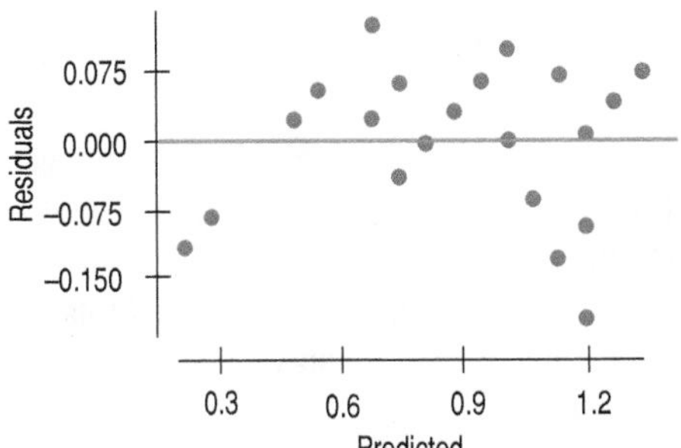

Dependent variable is: nicotine

R squared = 92.4%

Variable	Coefficient
Constant	0.154030
Tar	0.065052

a) Do you think a linear model is appropriate here? Explain.
b) Explain the meaning of R^2 in this context.

16. Attendance 2010 In the previous chapter, you looked at the relationship between the number of wins by American League baseball teams and the average attendance at their home games for the 2010 season. Here are the scatterplot, the residuals plot, and part of the regression analysis:

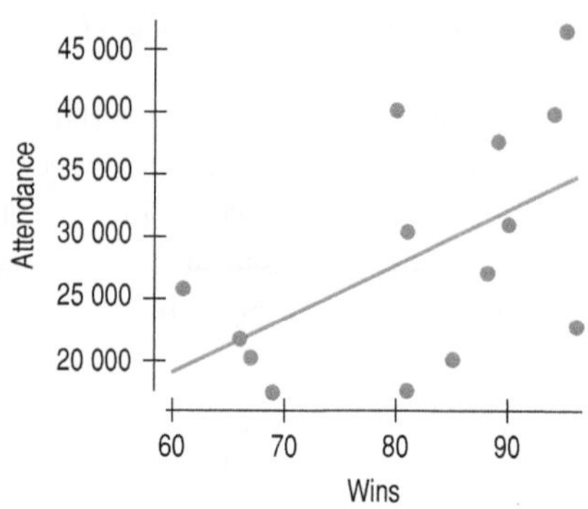

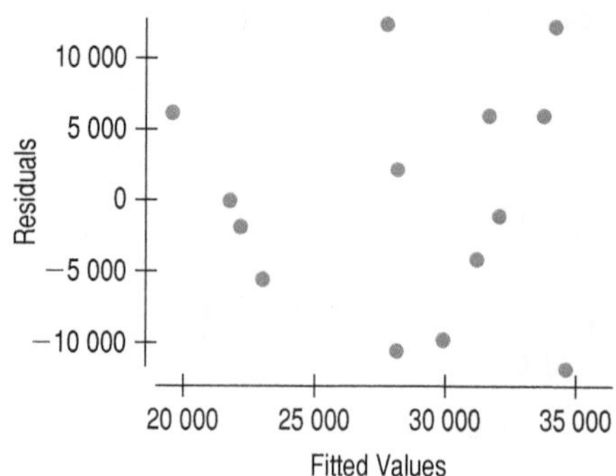

Dependent variable is Home Attendance
R-squared = 28.4%

Variable	Coefficient
Constant	−6760.5
Wins	431.22

a) Do you think a linear model is appropriate here? Explain.
b) Interpret the meaning of R^2 in this context.
c) Do the residuals show any pattern worth remarking on?
d) The point in the upper right of the plots is the New York Yankees. What can you say about the residual for the Yankees?

17. Another cigarette Consider again the regression of nicotine content on tar (both in milligrams) for the cigarettes examined in Exercise 15.
a) What is the correlation between tar and nicotine?
b) What would you predict about the average nicotine content of cigarettes that are two standard deviations below average in tar content?
c) If a cigarette is one standard deviation above average in nicotine content, what do you suspect is true about its tar content?

18. Second inning 2010 Consider again the regression of *Average Attendance* on *Wins* for the baseball teams examined in Exercise 16.
a) What is the correlation between *Wins* and *Average Attendance*?
b) What would you predict about the *Average Attendance* for a team that is 2 standard deviations above average in *Wins*?
c) If a team is 1 standard deviation below average in attendance, what would you predict about the number of games the team has won?

19. Last cigarette Take another look at the regression analysis of tar and nicotine content of the cigarettes in Exercise 15.
a) Write the equation of the regression line.
b) Estimate the nicotine content of cigarettes with 4 milligrams of tar.
c) Interpret the meaning of the slope of the regression line in this context.
d) What does the y-intercept mean?
e) If a new brand of cigarette contains 7 milligrams of tar with a residual of −0.5 mg, what is the nicotine content?

20. Last inning 2010 Refer again to the regression analysis for average attendance and games won by American League baseball teams, seen in Exercise 16.
a) Write the equation of the regression line.
b) Estimate the *Average Attendance* for a team with 50 *Wins*.
c) Interpret the meaning of the slope of the regression line in this context.
d) In general, what would a negative residual mean in this context?
e) The San Francisco Giants, the 2010 World Champions, are not included in these data because they are a National League team. During the 2010 regular season, the Giants won 92 games and averaged 41 736 fans at their home games. Calculate the residual for this team, and explain what it means.

21. What slope? If you create a regression model for predicting the weight of a car (in pounds) from its length (in feet), is the slope most likely to be 3, 30, 300, or 3000? Explain.

22. What slope again? If you create a regression model for estimating the height of a pine tree (in centimetres) based on the circumference of its trunk (in centimetres), is the slope most likely to be 0.1, 1, 10, or 100? Explain.

23. Misinterpretations A Biology student who created a regression model to use a bird's height when perched for predicting its wingspan made these two statements. Assuming the calculations were done correctly, explain what is wrong with each interpretation.
a) My R^2 of 93% shows that this linear model is appropriate.
b) A bird 30 cm tall will have a wingspan of 34 cm.

24. More misinterpretations A Sociology student investigated the association between a country's literacy rate and life expectancy, then drew the conclusions listed below. Explain why each statement is incorrect. (Assume that all the calculations were done properly.)
a) The literacy rate determines 64% of the life expectancy for a country.
b) The slope of the line shows that an increase of 5% in literacy rate will produce a two-year improvement in life expectancy.

25. ESP People who claim to "have ESP" participate in a screening test in which they have to guess which of several images someone is thinking of. You and a friend both took the test. You scored two standard deviations above the mean, and your friend scored one standard deviation below the mean. The researchers offer everyone the opportunity to take a retest.
a) Should you choose to take this retest? Explain.
b) Now explain to your friend what his decision should be and why.

26. SI jinx Players in any sport who are having great seasons—turning in performances that are much better than anyone might have anticipated—often are pictured on the cover of *Sports Illustrated.* Frequently, their performances then falter somewhat, leading some athletes to believe in a "*Sports Illustrated* jinx." Similarly, it is common for phenomenal rookies to have less stellar second seasons—the so-called "sophomore slump." While fans, athletes, and analysts have proposed many theories about what leads to such declines, a statistician might offer a simpler (statistical) explanation. Explain.

T **27. SAT scores** The SAT is a test often used in the U.S. as part of an application to college. SAT scores are between 200 and 800. Tests are given in both math and verbal areas. Doing the SAT-Math problems also involves the ability to read and understand the questions, but can a person's verbal score be used to predict the math score? Verbal and math SAT scores of a high-school graduating class are displayed in the scatterplot, with the regression line added.

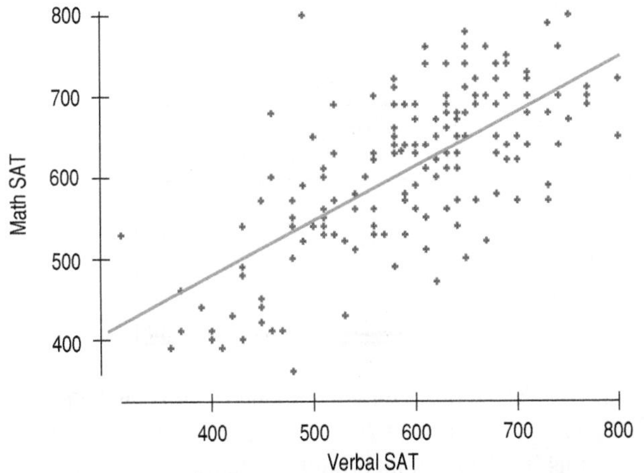

a) Describe the relationship.
b) Are there any students whose scores do not seem to fit the overall pattern?
c) For these data, $r = 0.685$. Interpret this statistic.
d) These verbal scores averaged 596.3 with a standard deviation of 99.5, and the math scores averaged 612.2 with a standard deviation of 96.1. Write the equation of the regression line.
e) Interpret the slope of this line.
f) Predict the math score of a student with a verbal score of 500.
g) Every year some student scores a perfect 1600. Based on this model, what would that student's residual be for their math score?

28. Success in college Many U.S. colleges use SAT scores in the admissions process because they believe these scores provide some insight into how a high-school student will perform at the college level. Suppose the entering freshmen at a certain college have mean combined SAT score of 1833, with a standard deviation of 123. In the first semester these students attained a mean GPA of 2.66, with a standard deviation of 0.56. A scatterplot showed the association to be reasonably linear, and the correlation between SAT score and GPA was 0.47.
a) Write the equation of the regression line.
b) Explain what the y-intercept of the regression line indicates.
c) Interpret the slope of the regression line.
d) Predict the GPA of a freshman who scored a combined 1400.
e) Based upon these statistics, how effective do you think SAT scores would be in predicting academic success during the first term of the freshman year at this college? Explain.
f) As a student, would you rather have a positive or a negative residual in this context? Explain.

T **29. SAT, take 2** Suppose we wanted to use SAT math scores to estimate verbal scores based on the information in Exercise 27.
a) What is the correlation?
b) Write the equation of the line of regression predicting verbal scores from math scores.
c) In general, what would a positive residual mean in this context?
d) A person tells you her math score was 500. Predict her verbal score.
e) Using that predicted verbal score and the equation you created in Exercise 27, predict her math score.
f) Why doesn't the result in part e) come out to 500?

30. Success, part 2 Based on the statistics for college freshmen given in Exercise 28, what SAT score might be expected among freshmen who attained a first-term GPA of 3.0?

T **31. Used cars 2013** Classified ads at www.auto123.com (on July 22, 2013) offered used Toyota Corollas for sale in central Ontario. Listed below are ages and advertised prices for some Corolla CEs with automatic transmission:

Age (years)	Price (dollars)
1	13 900
1	14 610
2	13 998
2	16 887
3	10 900
3	12 995
3	14 995
4	10 999
4	11 988
4	11 995
4	12 495
5	8 988
5	9 488
5	10 488
5	10 989
7	11 400
8	6 995
9	6 988
10	10 000
12	2 700

a) Make a scatterplot for these data.
b) Describe the association between age and price of a used Corolla.
c) Do you think a linear model is appropriate?
d) Computer software says that $R^2 = 74.8\%$. What is the correlation between age and price?
e) Explain the meaning of R^2 in this context.
f) Why doesn't this model explain 100% of the variability in the price of a used Corolla?

T **32. Drug abuse** In the exercises of the last chapter you examined results of a survey conducted in the United States and 10 countries of Western Europe to determine the percentage of teenagers who had used marijuana and other drugs. Below is the scatterplot. Summary statistics showed that the mean percent that had used marijuana was 23.9, with a standard deviation of 15.6%. An average of 11.6% of teens had used other drugs, with a standard deviation of 10.2%.

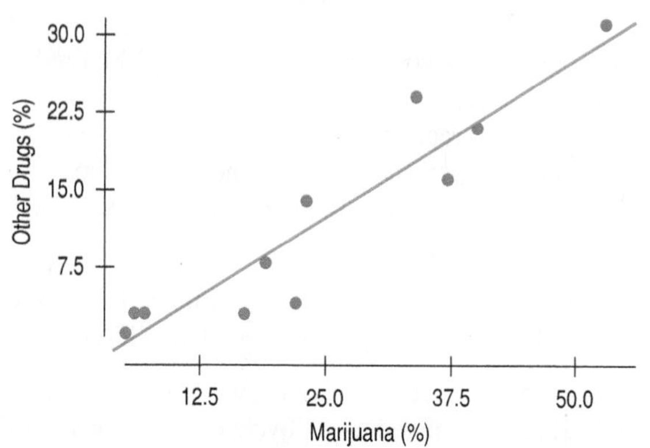

a) Do you think a linear model is appropriate? Explain.
b) For this regression, R^2 is 87.3%. Interpret this statistic in this context.
c) Write the equation you would use to estimate the percentage of teens who use other drugs from the percentage who have used marijuana.
d) Explain in context what the slope of this line means.
e) Do these results confirm that marijuana is a "gateway drug"—that is, that marijuana use leads to the use of other drugs?

33. More used cars 2013 Use the advertised prices for Toyota Corollas given in Exercise 31 to create a linear model for the relationship between a car's age and its price.

a) Find the equation of the regression line.
b) Explain the meaning of the slope of the line.
c) Explain the meaning of the *y*-intercept of the line.
d) If you want to sell a 6-year-old Corolla, what price seems appropriate?
e) You have a chance to buy one of two cars. They are about the same age and appear to be in equally good condition. Would you rather buy the one with a positive residual or a negative residual? Explain.
f) You see a "For Sale" sign on a 5-year-old Corolla stating the asking price as $9000. What is the residual?
g) Would this regression model be useful in establishing a fair price for a 17-year-old car? Explain.

34. Veggie burgers Recently Burger King introduced a meat-free burger. The nutrition label is shown here.

Nutrition Facts

Calories	330
Fat	10g*
Sodium	760mg
Sugars	5g
Protein	14g
Carbohydrates	43g
Dietary Fibre	4g
Cholesterol	0

* (2 grams of saturated fat)

RECOMMENDED DAILY VALUES (based on a 2,000-calorie/day diet)

Iron	20%
Vitamin A	10%
Vitamin C	10%
Calcium	6%

a) Use the regression model created in this chapter, $\widehat{Fat} = 8.4 + 0.91\,Protein$, to predict the fat content of this burger from its protein content.

b) What is its residual? How would you explain the residual?
c) Write a brief report about the fat and protein content of this new menu item. Be sure to talk about the variables by name and in the correct units.

35. Burgers In the last chapter's exercises, you examined the association between the amounts of fat and calories in fast-food hamburgers. Here are the data:

Fat (g)	19	31	34	35	39	39	43
Calories	410	580	590	570	640	680	660

a) Create a scatterplot of *Calories vs. Fat*.
b) Interpret the value of R^2 in this context.
c) Write the equation of the line of regression.
d) Use the residuals plot to explain whether your linear model is appropriate.
e) Explain the meaning of the y-intercept of the line.
f) Explain the meaning of the slope of the line.
g) A new burger containing 28 grams of fat is introduced. According to this model, its residual for *Calories* is 133. How many calories does the burger have?

36. Chicken Chicken sandwiches are often advertised as a healthier alternative to beef because many are lower in fat. Tests on 11 brands of fast food chicken sandwiches produced the following summary statistics and scatterplot from a graphing calculator:

	Fat (g)	Calories
Mean	20.6	472.7
St. Dev.	9.8	144.2
Correlation	0.947	

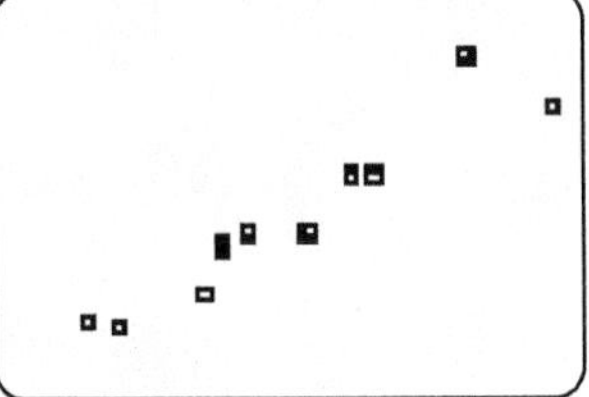

a) Do you think a linear model is appropriate in this situation?
b) Describe the strength of this association.
c) Write the equation of the regression line.
d) Explain the meaning of the slope.
e) Explain the meaning of the y-intercept.
f) What does it mean if a certain sandwich has a negative residual?
g) If a chicken sandwich and a burger each advertised 35 grams of fat, which would you expect to have more calories? (See Exercise 35.)
h) McDonald's Filet-O-Fish sandwich has 26 grams of fat and 470 calories. Does the fat/calorie relationship in this sandwich appear to be very different from that found in chicken sandwiches or in burgers (see Exercise 35)? Explain.

37. A second helping of burgers In Exercise 35, you created a model that can estimate the number of calories in a burger when the fat content is known.

a) Explain why you cannot use that model to estimate the fat content of a burger with 600 calories.
b) Using an appropriate model, estimate the fat content of a burger with 600 calories.

38. Cost of living 2008 The *Worldwide Cost of Living Survey City Rankings* determine the cost of living in the 25 most expensive cities in the world.[13] These rankings scale New York City as 100, and express the cost of living in other cities as a percentage of the New York cost. For example, the table indicates that in Tokyo the cost of living was 22.1% higher than New York in 2007, and increased to 27.0% higher in 2008.

City	2007	2008	City	2007	2008
Moscow	134.4	142.4	Tel Aviv	97.7	105.0
Tokyo	122.1	127.0	Sydney	94.9	104.1
London	126.3	125.0	Dublin	99.6	103.9
Oslo	105.8	118.3	Rome	97.6	103.9
Seoul	122.4	117.7	St. Petersburg	103.0	103.1
Hong Kong	119.4	117.6	Vienna	96.9	102.3
Copenhagen	110.2	117.2	Beijing	95.9	101.9
Geneva	109.8	115.8	Helsinki	93.3	101.1
Zurich	107.6	112.7	New York City	100.0	100.0
Milan	104.4	111.3	Istanbul	87.7	99.4
Osaka	108.4	110.0	Shanghai	92.1	98.3
Paris	101.4	109.4	Amsterdam	92.2	97.0
Singapore	100.4	109.1			

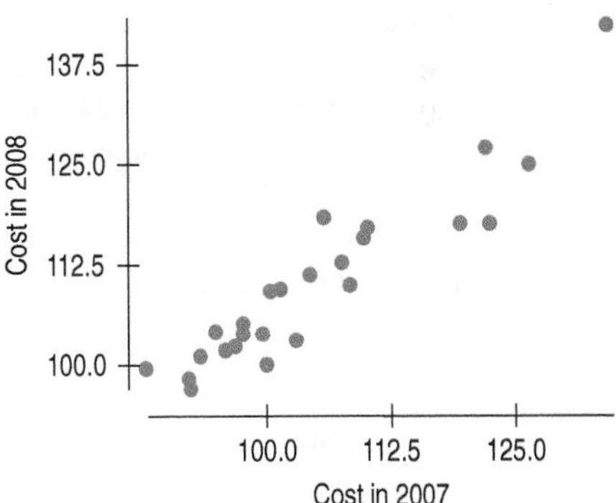

a) Above is a scatterplot. Describe the association between costs of living in 2007 and 2008.
b) The correlation is 0.938. Find and interpret the value of R^2.
c) The regression equation predicting the 2008 cost of living from the 2007 figure is $\widehat{Cost08} = 21.75 + 0.84Cost07$. Use this equation to find the residual for Oslo.
d) Explain what the residual means.

39. New York bridges We saw in this chapter that in Tompkins County, NY, older bridges were in worse condition than newer ones. Tompkins is a rural area. Is this relationship true in New York City as well? Here are data on the

[13] www.finfacts.com/costofliving.htm

Safety Score (as measured by the state Department of Transportation) and *Age at Inspection* for bridges in New York City.

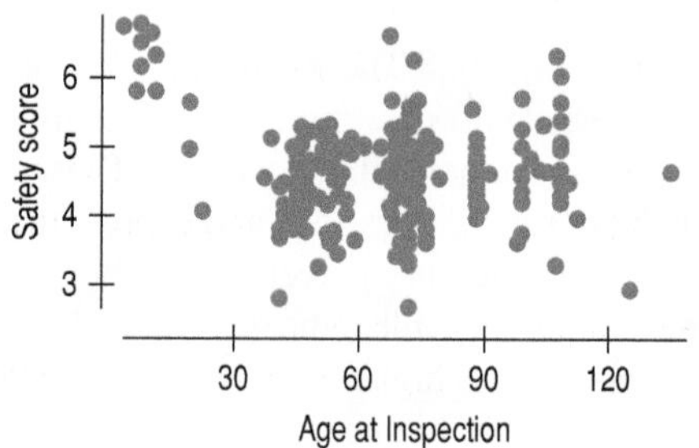

Dependent variable is: Safety Score
R-squared = 2.6%
s = 0.6708

Variable	Coefficient
Intercept	4.95147
Age at Inspection	−0.00481

a) New York State defines any bridge with a safety score less than 5 as *deficient.* What does this model predict for the safety scores of New York City bridges?
b) Our earlier model found that the safety scores of bridges in Tompkins County was decreasing at about 0.023 per year. What does this model say about New York City bridges?
c) How much faith would you place in this model? Explain.

40. Candy 2009 The table shows the increase in Halloween candy sales over a 6-year period as reported by the National Confectioners Association (www.candyusa.com). Using these data, estimate the amount of candy sold in 2009. Discuss the appropriateness of your model and your faith in the estimate. Then comment on the fact that NCA reported 2009 sales of \$2.207 million. (Enter *Year* as 3, 4, . . .)

2003	1.993
2004	2.041
2005	2.088
2006	2.146
2007	2.202
2008	2.209

41. Climate change 2011 The earth's climate is getting warmer due to the increase in atmospheric levels of carbon dioxide (CO_2), a greenhouse gas. Here is a scatterplot showing the mean annual CO_2 concentration in the atmosphere, measured in parts per million (ppm) at the top of Mauna Loa in Hawaii, and the mean annual air temperature over both land and sea across the globe, in degrees Celsius (°C) for the years 1959 to 2011.

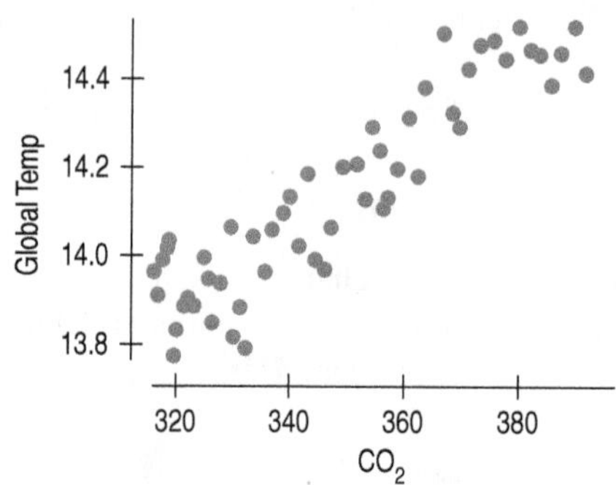

A regression predicting *Temperature* from CO_2 produces the following output table (in part):

Dependent variable is Global Temperature (°C)
R-squared = 84.0%

Variable	Coefficient
Intercept	11.0276
CO_2 (ppm)	0.0089

a) What is the correlation between CO_2 and *Temperature*?
b) Explain the meaning of *R*-squared in this context.
c) Give the regression equation.
d) What is the meaning of the slope in this equation?
e) What is the meaning of the *y*-intercept of this equation?
f) Here is a scatterplot of the residuals vs. CO_2 Does this plot show evidence of the violation of any assumptions behind the regression? If so, which ones?

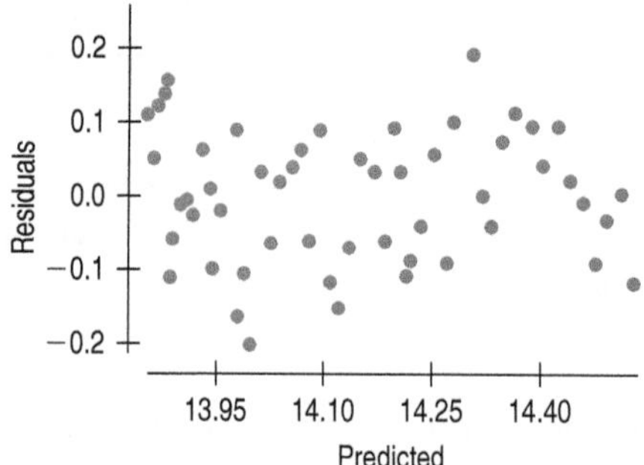

g) CO_2 levels will probably reach 400 ppm by 2020. What mean *Temperature* does the regression predict for that concentration of CO_2?

42. Birthrates 2009 The table shows the number of live births per 1000 women aged 15–44 years in the United States, starting in 1965. (National Center for Health Statistics, www.cdc.gov/nchs/)

Year	1965	1970	1975	1980	1985	1990	1995	2000	2005	2009
Rate	19.4	18.4	14.8	15.9	15.6	16.4	14.8	14.4	14.0	13.5

a) Make a scatterplot and describe the general trend in *Birthrates*. (Enter *Year* as years since 1900: 65, 70, 75, etc.)
b) Find the equation of the regression line.
c) Check to see if the line is an appropriate model. Explain.
d) Interpret the slope of the line.

e) The table gives rates only at 5-year intervals. Estimate what the rate was in 1978.

f) In 1978, the birthrate was actually 15.0. How close did your model come?

g) Predict what the *Birthrate* will be in 2010. Comment on your faith in this prediction.

h) Predict the *Birthrate* for 2025. Comment on your faith in this prediction.

T **43. Body fat** It is difficult to accurately determine a person's body fat percentage without immersing him or her in water. Researchers hoping to find ways to make a good estimate immersed 20 male subjects, then measured their waists and recorded their weights.

a) Create a model to predict *% Body Fat* from *Weight*.

b) Do you think a linear model is appropriate? Explain.

c) Interpret the slope of your model.

d) Is your model likely to make reliable estimates? Explain.

e) What is the residual for a person who weighs 190 pounds and has 21% body fat?

Waist (in.)	Weight (lb)	Body Fat (%)	Waist (in.)	Weight (lb)	Body Fat (%)
32	175	6	33	188	10
36	181	21	40	240	20
38	200	15	36	175	22
33	159	6	32	168	9
39	196	22	44	246	38
40	192	31	33	160	10
41	205	32	41	215	27
35	173	21	34	159	12
38	187	25	34	146	10
38	188	30	44	219	28

T **44. Body fat, again** Would a model that uses the person's *waist* size be able to predict the *% Body Fat* more accurately than one that uses a person's weight? Using the data in Exercise 43, create and analyze that model.

T **45. Heptathlon 2012** We discussed the women's 2008 Olympic heptathlon in Chapter 5. Here are the results from the high jump (metres), 800-metre run (seconds), and long jump (metres) for the women who successfully completed all three events in the 2012 Olympics:

Athlete	Country	Points	High Jump	Long Jump	800 m
Jessica Ennis	GBR	6955	1.86	6.48	128.65
Lilli Schwarzkopf	GER	6649	1.83	6.30	130.50
Tatyana Chernova	RUS	6628	1.80	6.54	129.56
Lyudmyla Yosypenko	UKR	6618	1.83	6.31	133.28
Austra Skujyté	LTU	6599	1.92	6.25	140.59
Antoinette Nana Djimou Ida	FRA	6576	1.80	6.13	133.62
Jessica Zelinka	CAN	6480	1.68	5.91	129.15
Kristina Savitskaya	RUS	6452	1.83	6.21	132.27
Laura Ikauniece	LAT	6414	1.83	6.13	132.13
Hanna Melnychenko	UKR	6392	1.80	6.40	132.90
Brianne Theisen	CAN	6383	1.83	6.01	129.27
Dafne Schippers	NED	6324	1.80	6.28	135.52
Nadine Broersen	NED	6319	1.86	5.94	136.98
Jessica Samuelsson	SWE	6300	1.77	6.18	131.31
Katarina Johnson-Thompson	GBR	6267	1.89	6.19	130.76
Sharon Day	USA	6232	1.77	5.85	131.31
Yana Maksimava	BLR	6198	1.89	5.99	133.37
Eliška Klučinová	CZE	6109	1.80	6.13	136.08
Ellen Sprunger	SUI	6107	1.71	5.88	137.54
Olga Kurban	RUS	6084	1.80	5.83	139.82
Marisa De Aniceto	FRA	6030	1.71	5.76	136.20
Györgyi Farkas	HUN	6013	1.80	6.07	137.83
Grit Šadeiko	EST	6013	1.74	6.11	143.01
Sofia Ifadidou	GRE	5947	1.68	5.81	142.03
Ivona Dadic	AUT	5935	1.80	6.00	135.90
Sarah Cowley	NZL	5873	1.80	6.00	139.01
Louise Hazel	GBR	5856	1.59	5.77	138.78
Ida Marcussen	NOR	5846	1.68	5.82	133.62
Chantae McMillan	USA	5688	1.68	5.37	160.55

Source: www.espn.com

Let's examine the association among these events. Perform a regression to predict high jump performance from the 800-metre results. Set aside Ms. McMillan in your analysis, in view of her very atypical 800 m time.

a) What is the regression equation? What does the slope mean?

b) What is the R^2 value?

c) Do good high jumpers tend to be fast runners? (Be careful—low times are good for running events and high distances are good for jumps.)

d) What does the residual plot reveal about the model?

e) Do you think this is a useful model? Would you use it to predict high jump performance? (Compare the residual standard deviation to the standard deviation of the high jumps.)

T **46. Heptathlon 2012 again** We saw the data for the women's 2012 Olympic heptathlon in Exercise 45. Are the two jumping events associated? Perform a regression of the long jump results on the high jump results. Set aside in your analysis the last place (points) finisher, an apparent outlier.

a) What is the regression equation? What does the slope mean?

b) What is the R^2 value?

c) Do good high jumpers tend to be good long jumpers?

d) What does the residual plot reveal about the model?

e) Do you think this is a useful model? Would you use it to predict long jump performance? (Compare the residual standard deviation to the standard deviation of the long jumps.)

47. Hard water In an investigation of environmental causes of disease, data were collected on the annual mortality rate (deaths per 100 000) for males in 61 large towns in England and Wales. In addition, the water hardness was recorded as the calcium concentration (parts per million, ppm) in the drinking water. The following display shows the relationship between mortality and calcium concentration for these towns:

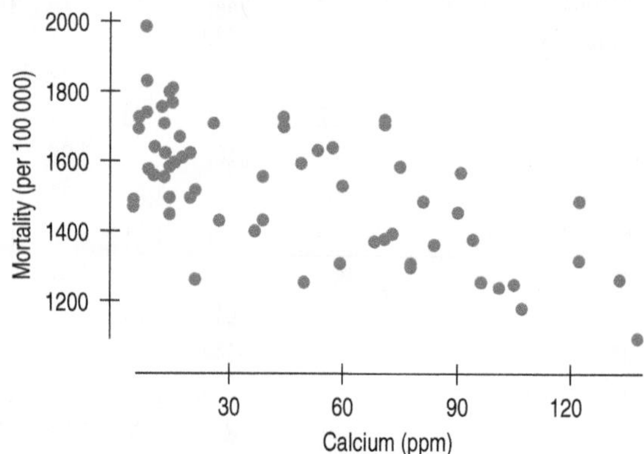

a) Describe what you see in this scatterplot, in context.

b) Here is the regression analysis of mortality and calcium concentration. What is the regression equation?

Dependent variable is: Mortality

$R^2 = 43\%$

s = 143.0

Variable	Coefficient
Intercept	1676
Calcium	−3.23

c) Interpret the slope and y-intercept of the line, in context.

d) The largest residual, with a value of −348.6, is for the town of Exeter. Explain what this value means.

e) The hardness of Derby's municipal water is about 100 ppm of calcium. Use this equation to predict the mortality rate in Derby.

f) Explain the meaning of R^2 in this situation.

48. Gators Wildlife researchers monitor many wildlife populations by taking aerial photographs. Can they estimate the weights of alligators accurately from the air? Here is a regression analysis of the weight of alligators (in pounds) and their length (in inches) based on data collected from captured alligators.

Dependent variable is: Weight

R-squared = 83.6%

s = 54.01

Variable	Coefficient
Intercept	−393
Length	5.9

a) Did they choose the correct variables to use as the response variable and the predictor? Explain.

b) What is the correlation between an alligator's length and weight?

c) Write the regression equation.

d) Interpret the slope of the equation in this context.

e) Do you think this equation will allow the scientists to make accurate predictions about alligators? What part of the regression analysis indicates this? What additional concerns do you have?

49. Brakes The table below shows stopping distances in metres for a car tested three times at each of five speeds. We hope to create a model that predicts stopping distance from the speed of the car.

Speed (km/h)	Stopping Distances (m)
32	19.5, 18.9, 18.0
48	34.8, 36.0, 32.0
64	46.7, 52.1, 50.3
80	70.4, 61.9, 72.6
95	96.6, 97.9, 84.1

a) Explain why a linear model is not appropriate. Be sure to examine the residual plot.

b) Re-express the data to straighten the scatterplot, by using the square roots of the stopping distances.

c) Fit an appropriate model. Is the residual plot any better after re-expression?

d) Estimate the stopping distance for a car travelling 90 km/h.

50. Ontario interprovincial migration 2012 Below are data showing the net interprovincial migration each year for Ontario (number arriving minus number leaving). The data are not for calendar years, but for a one-year period ending in the displayed year.

Year	Ontario Net Migration	Year	Ontario Net Migration
1979	−4 325	1996	−2 822
1980	−22 362	1997	1 977
1981	−33 247	1998	9 231
1982	−5 665	1999	16 706
1983	23 585	2000	22 369
1984	36 400	2001	18 623
1985	33 885	2002	5 354
1986	33 562	2003	637
1987	42 601	2004	−6 935
1988	35 215	2005	−11 172
1989	9 739	2006	−17 501
1990	−5 961	2007	−20 047
1991	−11 627	2008	−14 750
1992	−11 045	2009	−15 601
1993	−14 189	2010	−4 662
1994	−9 420	2011	−4 007
1995	−2 841	2012	−8 091

Source: Adapted from Statistics Canada CANSIM table 051-0004, March 31, 2013.

a) Fit a least squares line to predict Ontario interprovincial migration from year.
b) Use your equation to predict Ontario's interprovincial migration in 2014.
c) Do you think your prediction is reasonable? Do you think your regression line is an appropriate model?

T **51. Immigrant commuters** Recent immigrants are more likely than Canadian-born to use public transit. Below are data for selected Census Metropolitan Areas (CMA) on the percentage of individuals who commute to work by public transit, for recent immigrants (<10 years since arrival) and for Canadian-born:

CMA	Canadian Born	Recent Immigrant
Montreal	20.1	47.8
Toronto	20.0	35.4
Ottawa–Gatineau	17.8	33.5
Calgary	12.7	24.6
Winnipeg	13.6	24.3
Vancouver	10.9	20.1
Edmonton	8.8	19.3
Victoria	10.3	17.2
Hamilton	8.3	16.4
London	6.7	15.1
Windsor	3.3	10.0
Kitchener	4.4	8.8
Abbotsford	2.1	1.2

Source: Adapted from Statistics Canada trends and conditions in Census Metropolitan Areas, immigrants in Canada's Census Metropolitan Areas, 89-613-MIE2004003, www.statcan.gc.ca/bsolc/olc-cel/olc-cel?catno=89-613-MIE&lang=eng.

a) Plot the immigrant percentage versus the Canadian-born percentage. Describe the plot.
b) Fit a line to predict the immigrant percent from the Canadian-born percent and superimpose this line on the data scatterplot.
c) For every additional 1% ridership by Canadian-born commuters, how much additional ridership (as a percent) by recent immigrants do we tend to find, on average?
d) What is unusual about Montreal? Use some numbers (via the regression) to explain more precisely this unusualness.
e) Plot residuals from the regression versus the fitted (predicted) values or versus the Canadian-born percentage. Describe the pattern you see and what it suggests. Do you think Montreal has anything to do with the pattern?

T **52. Math and gender 2009** Below are mean PISA (Programme for International Student Assessment) math scores for samples of 15-year-old male and female students from a number of randomly selected schools in each of various OECD and other industrialized nations (4000–10 000 per country).

Country	Mean Male Score	Mean Female Score
OECD		
Australia	519	509
Austria	506	486
Belgium	526	504
Canada	533	521
Chile	431	410
Czech Republic	495	490
Denmark	511	495
Estonia	516	508
Finland	542	539
France	505	489
Germany	520	505
Greece	473	459
Hungary	496	484
Iceland	508	505
Ireland	491	483
Israel	451	443
Italy	490	475
Japan	534	524
Korea	548	544
Luxembourg	499	479
Mexico	425	412
Netherlands	534	517
New Zealand	523	515
Norway	500	495
Poland	497	493
Portugal	493	481
Slovak Republic	498	495
Slovenia	502	501
Spain	493	474
Sweden	493	495
Switzerland	544	524
Turkey	451	440
United Kingdom	503	482
United States	497	477
Partners		
Albania	372	383
Argentina	394	383
Azerbaijan	435	427
Brazil	394	379
Bulgaria	426	430
Colombia	398	366
Croatia	465	454
Dubai (UAE)	454	451
Hong Kong-China	561	547
Indonesia	371	372
Jordan	386	387
Kazakhstan	405	405
Kyrgyzstan	328	334
Latvia	483	481
Liechtenstein	547	523

Lithuania	474	480
Macao-China	531	520
Montenegro	408	396
Panama	362	357
Peru	374	356
Qatar	366	371
Romania	429	425
Russian Federation	469	467
Serbia	448	437
Shanghai-China	599	601
Singapore	565	559
Chinese Taipei	546	541
Thailand	421	417
Trinidad and Tobago	410	418
Tunisia	378	366
Uruguay	433	421

Source: Based on data from OECD PISA 2009 online database, http://pisa2009.acer.edu.au, accessed on Feb 22, 2013.

a) Plot female mean score versus male mean score. Describe the relationship.

b) Fit a straight line for predicting female means from male means, and find and interpret the R^2 value.

c) Plot residuals versus male means or versus predicted (fitted) values. What do you learn?

d) In which country or countries do the female students perform unusually poorly in relation to the male students, according to the regression model? Give the residual(s).

53. Regressing Test scores over the years in an introductory statistics course have shown a 0.6 correlation between term test and final exam. Suppose test and exam both have a mean of 70 with similar standard deviations.

a) Predict the exam score of someone who gets 85 on the test.

b) Predict the test score of someone who gets 79 on the exam.

c) Is something wrong? Explain.

Just Checking ANSWERS

1. An increase in home size of one square foot is associated with an increase of \$122.74 , on average, in price.
2. Units are dollars per square foot.
3. About \$245 480, on average.
4. You would guess the price to be about $2 \times 0.85 = 1.7$ standard deviations above the average price.
5. 1.7 SDs above the mean price is $238\,000 + 1.7 \times 63\,790 = \$346\,443$
6. Differences in the size of houses account for about 71.4% of the variation in house prices.
7. It's positive. The correlation and the slope have the same sign.
8. The correlation won't change, so neither will R^2. The slope will change because the units have changed.
9. You shouldn't be surprised since the value of your house is only slightly more than one standard deviation above the regression prediction. If it were more than two standard deviations above the predicted value, you might well be pleasantly surprised.

MathXL

MyStatLab

Go to MathXL at www.mathxl.com or MyStatLab at www.mystatlab.com. You can practise exercises for this chapter as often as you want. The guided solutions will help you find answers step by step. You'll find a personalized study plan available to you too!

chapter

8 Regression Wisdom

Regression is a powerful tool for forecasting. Economists using it successfully predicted ten out of the last two recessions.

—Hershey M. Friedman

Where are we going?

What happens when we fit a regression model to data that aren't straight or that have outliers? How bad will the predictions be? How can we tell if the model is appropriate or not? Questions like these can be much more important than knowing how to tell your software to execute a least-square fit to some data. In this chapter, we'll see how to tell whether a regression model is sensible and what to do if it isn't.

Regression is used every day throughout the world to predict customer loyalty, numbers of admissions at hospitals, sales of automobiles, and many other things. Because regression is so widely used, it's also widely abused and misinterpreted. This chapter presents examples of regressions in which things are not quite as simple as they seem at first, and shows how you can still use regression to discover what the data have to say.

8.1 Examining Residuals

No regression analysis is complete without a display of the residuals to check that the linear model is reasonable. Because the residuals are what is "left over" after the model describes a relationship, they often reveal subtleties that were not clear from a plot of the original data. Sometimes these are additional details that help confirm or refine our understanding. Sometimes they reveal violations of the regression conditions that require our attention.

The fundamental assumption in working with a linear model is that the relationship you are modelling is, in fact, linear. That sounds obvious, but you can't take it for granted. It may be hard to detect nonlinearity from the scatterplot you looked at before you fit the regression model. Sometimes you can see important features such as nonlinearity more readily when you plot the residuals.

A S *Activity:* **Construct a Plot with a Given Slope.** How's your feel for regression lines? Can you make a scatterplot that has a specified slope?

STRAIGHT ENOUGH CONDITION

We can't *know* whether the **Linearity Assumption** is true, but we can see if it's *plausible* by checking the **Straight Enough Condition**.

Getting the "Bends": When the Residuals Aren't Straight

Jessica Meir and Paul Ponganis study emperor penguins at the Scripps Institution of Oceanography's Center for Marine Biotechnology and Biomedicine at the University of California at San Diego. Says Jessica:[1]

> *Emperor penguins are the most accomplished divers among birds, making routine dives of 5–12 minutes, with the longest recorded dive over 27 minutes. These birds can also dive to depths of over 500 metres! Since air-breathing animals like penguins must hold their breath while submerged, the duration of any given dive depends on how much oxygen is in the bird's body at the beginning of the dive, how quickly that oxygen gets used, and the lowest level of oxygen the bird can tolerate. The rate of oxygen depletion is primarily determined by the penguin's heart rate. Consequently, studies of heart rates during dives can help us understand how these animals regulate their oxygen consumption in order to make such impressive dives.*

[1]Pearson Education

The researchers equip emperor penguins with devices that record their heart rates during dives. Figure 8.1 is a scatterplot of the *Dive Heart Rate* (beats per minute) and the *Duration* (minutes) of dives by these high-tech penguins:

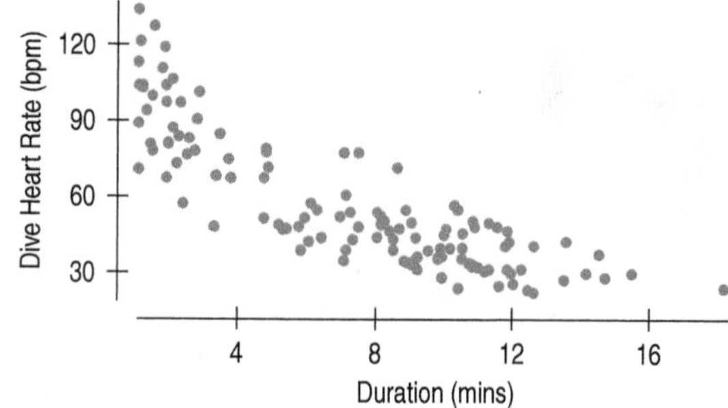

Figure 8.1
The scatterplot of *Dive Heart Rate* in beats per minute (bpm) versus *Duration* (minutes) shows a strong, roughly linear, negative association.

The scatterplot looks fairly linear with a moderately strong negative association ($R^2 = 71.5\%$). The linear regression equation

$$\widehat{DiveHeartRate} = 96.9 - 5.47\ Duration$$

says that for longer dives, the average dive heart rate decreases by about 5.47 beats per dive minute, starting from a heart rate of 96.9 beats per minute, on average.

The scatterplot of the residuals against *Duration* in Figure 8.2 makes things much clearer. The **Linearity Assumption** says we should not see a pattern, but instead there's a strong bend, starting high on the left, dropping down in the middle of the plot, and rising again at the right. Graphs of residuals often reveal patterns such as this that were not readily apparent in the original scatterplot.

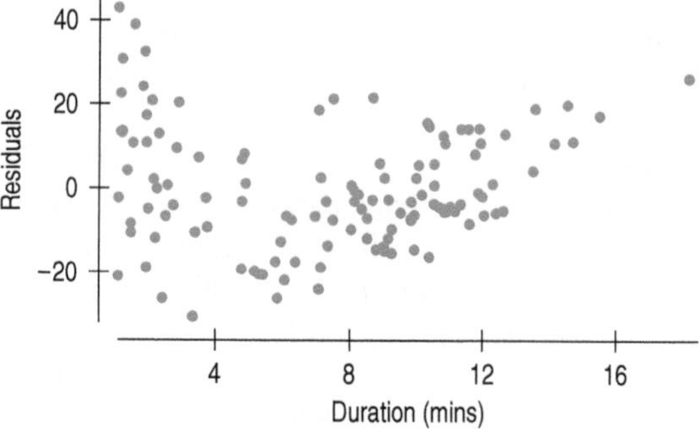

Figure 8.2
Plotting the residuals against *Duration* reveals a bend. It was also in the original scatterplot, but here it's easier to see.

Now looking back at the original scatterplot, we can see that the scatter of points isn't really straight. There's a slight bend to that plot, but the bend is much easier to see in the residuals. Even though it means rechecking the **Straight Enough Condition** *after* you run the regression, it's always a good idea to check your scatterplot of the residuals for bends that you might have overlooked in the original scatterplot. And you might also wonder—is it possible to unbend the bends? A re-expression (logarithms for example) of one or both of the variables might well help straighten out the relationship (see Section 6.4).

Sifting the Data for Groups

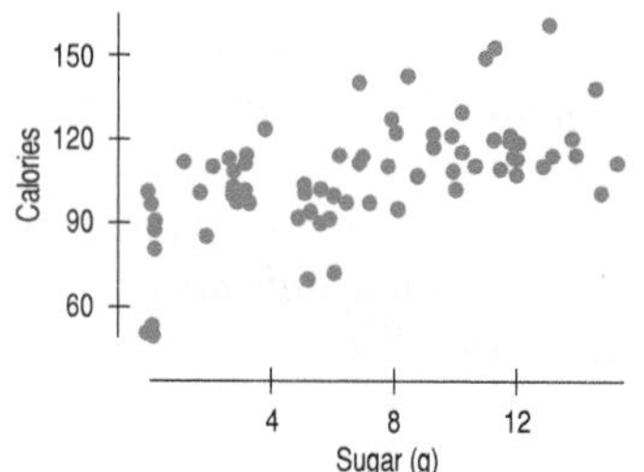

In the step-by-step analysis in Chapter 7 to predict *Calories* from *Sugar* content in breakfast cereals, we examined scatterplots of the data and of the residuals. Our first impression was that they had no particular structure. But look again carefully. Are there cereals that stand apart from the rest? Are there any distinct groups of data points (cereals)?

Look at Figure 8.3, a histogram of the residuals. How would you describe its shape? It looks like there might be small modes on both sides of the central body of the data. One group of cereals seems to stand out (on the left) as having large negative residuals, with fewer calories than our model predicts. Another stands out (on the right) with large positive residuals. Whenever we suspect multiple modes, we ask whether there might be different subgroups in the data.

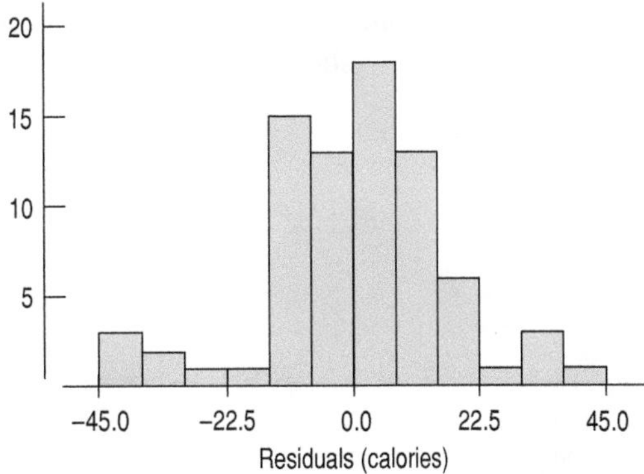

Figure 8.3
A histogram of the regression residuals shows small modes both above and below the central large mode. These may be worth a second look.

Figure 8.4 shows the residual scatterplot, with the cereals in those modes highlighted. We can see that those two groups stand away from the central pattern in this plot. If you look at the data scatterplot, there too you can readily spot these same two cereal clusters standing apart from the central pattern. The high-residual cereals are Just Right Fruit & Nut; Muesli Raisins, Dates & Almonds; Muesli Peaches & Pecans; Mueslix Crispy Blend; and Nutri-Grain Almond Raisin. Do these cereals seem to have something in common? They all present themselves as "healthy." This might be surprising, but in fact "healthy" cereals often contain more fat, and therefore more calories, than we might expect from looking at their sugar content alone.

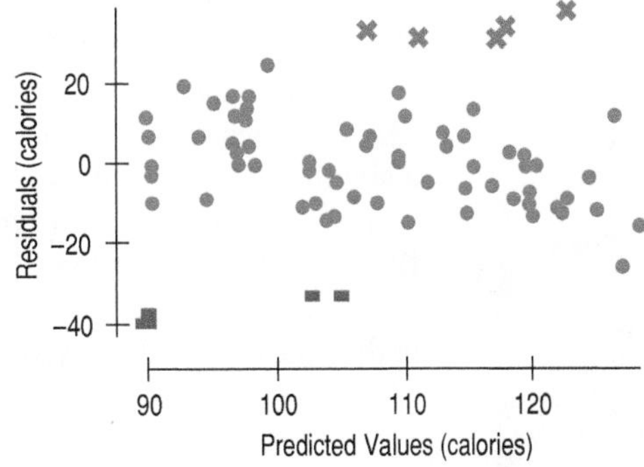

Figure 8.4
A scatterplot of the residuals versus predicted values for the cereal regression, highlighting some apparent clusters. The green "x" points are cereals whose calorie content is much higher than the linear model predicts. The red "–" points show cereals with many fewer calories than the model predicts. Is there something special about these cereals?

The low-residual cereals are Puffed Rice, Puffed Wheat, three bran cereals, and Golden Crisps. You might not have grouped these cereals together before. What they have in common is a low calorie count *relative to their sugar content*—even though their sugar contents are quite different.

These observations may not lead us to question the overall linear model, but they do help us understand that other factors may be part of the story. A careful examination of the data and residuals often leads us to discover groups of observations that are different from the rest.

When we discover that there is more than one group in a regression, we may decide to analyze the groups separately, using a different model for each group. Or we can stick with the original model and simply note that there are groups that are a little different. Either way, the model will be wrong, but useful, so it will improve our understanding of the data.

Subsets

Sometimes it's not careful data inspection, but rather our special knowledge about the data that suggests the possibility of distinct subgroups or subsets. Cereal manufacturers aim cereals at different segments of the market. Supermarkets and cereal manufacturers try to attract different customers by placing different types of cereals on certain shelves. Cereals for kids tend to be on the "kid's shelf," at their eye level. Toddlers wouldn't be likely to grab a box from this shelf and beg, "Mom, can we please get this All-Bran with Extra Fibre?"

Here's an important unstated condition for fitting models: **All the data must come from the same population.**

Should we take this extra information into account in our analysis? Figure 8.5 shows a scatterplot of *Calories* and *Sugar,* coloured according to the shelf on which the cereals were found and with a separate regression line fit for each. The top shelf is clearly different. We might want to report two regressions, one for the top shelf and one for the bottom two shelves.[2]

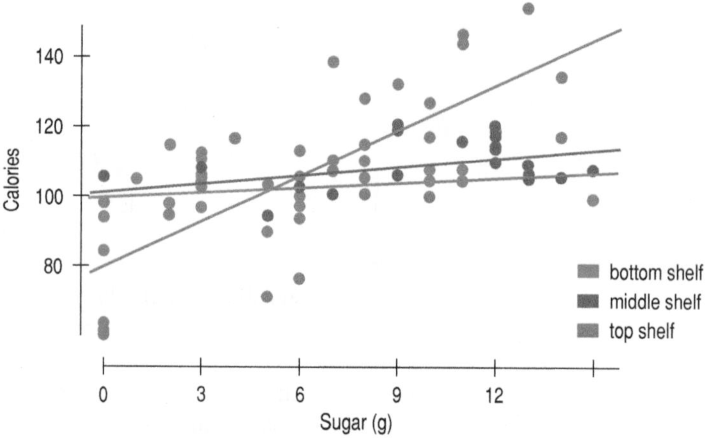

Figure 8.5
Calories and Sugar, coloured according to the shelf on which the cereal was found in a supermarket, with regression lines fit for each shelf individually. Do these data appear homogeneous? That is, do all the cereals seem to be from the same population of cereals? Or are there different kinds of cereals that we might want to consider separately?

8.2 Outliers, Leverage, and Influence

St Petersburg Times/ZUMAPRESS/Newscom

The outcome of the 2000 U.S. presidential election was determined in Florida amid much controversy. Even years later, historians continue to debate who really won the election. The main race was between George W. Bush and Al Gore, but two minor candidates played a significant role. To the political right of the main party candidates was Pat Buchanan, while to the political left was Ralph Nader. Generally, Nader earned more votes than Buchanan throughout the state. We would expect counties with larger vote totals to give more votes to each candidate. Here's a regression relating *Buchanan* vote totals by county in the state of Florida to *Nader* vote totals:

Dependent variable is: Buchanan
R-squared = 42.8%

Variable	**Coefficient**
Intercept	50.3
Nader	0.14

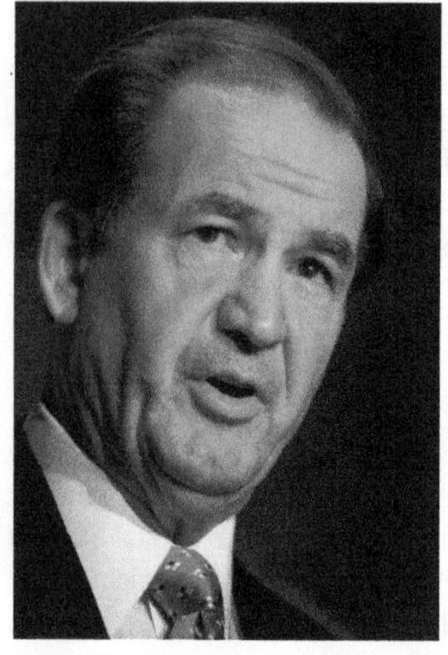

Jim Cole/AP Images

The regression model,

$$\widehat{Buchanan} = 50.3 + 0.14\,Nader,$$

says that, in each county, Buchanan received about 0.14 times (or 14% of) the vote Nader received, starting from a base of 50.3 votes.

This seems like a reasonable regression, with an R^2 of almost 43%. But we've violated all three Rules of Data Analysis by going straight to the regression summary without making a picture.

Here's a scatterplot that shows the votes for Buchanan in each county of Florida plotted against votes for Nader. The striking outlier is Palm Beach County.

The so-called "butterfly ballot," used only in Palm Beach County, was a source of controversy. It has been claimed that the format of this ballot confused voters so that some who intended to vote for Democrat Al Gore punched the wrong hole next to his name and as a result voted for Buchanan.

[2]More complex models can take into account both sugar content and shelf information. This kind of *multiple regression* model is a natural extension of the model we're using here. We'll see such models in Chapters 27 and 28.

Figure 8.6

Votes received by Buchanan against votes for Nader in all Florida counties in the presidential election of 2000. The red "x" point is Palm Beach County, home of the "butterfly ballot."

Buchanan (votes)
3000
2250
1500
750
2500
7500
Nader (votes)

The scatterplot shows a strong, positive, linear association, and one striking point. With Palm Beach removed from the regression, the R^2 jumps from 42.8% to 82.1% and the slope of the line changes to 0.1 (the blue line in Figure 8.7), suggesting that Buchanan received only about 10% of the votes that Nader received. Palm Beach County now stands out, not as a Buchanan stronghold, but rather as a clear violation of the model that begs for explanation.

Figure 8.7

The red line shows the effect that one unusual point can have on a regression.

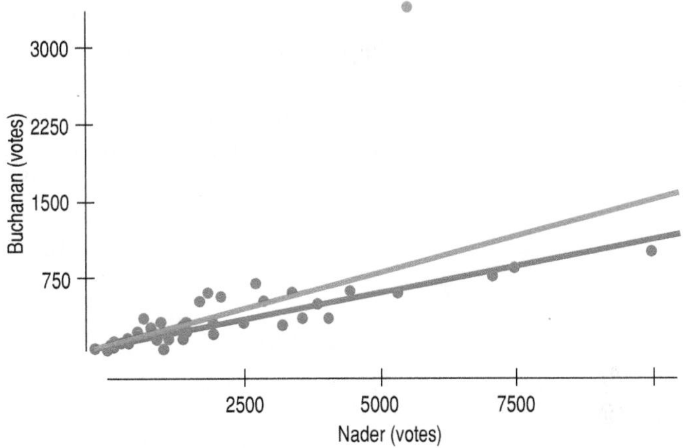

Alamy Limited

"Give me a place to stand and I will move the Earth."

—*Archimedes (287–211 BCE)*

A S ***Activity:* Leverage.** You may be surprised to see how sensitive a regression line is to a single influential point.

"For whoever knows the ways of Nature will more easily notice her deviations; and, on the other hand, whoever knows her deviations will more accurately describe her ways."

—*Francis Bacon (1561–1626)*

One of the great values of models is that they help us to see when and how data values are unusual. In regression, a point can stand out in two different ways. First, a data value can have a large residual, as Palm Beach County does in this example. Because they seem to be different from the other cases, points whose residuals are large always deserve special attention.

A data point can also be unusual if its x-value is far from the mean of the x-values. Such a point is said to have high **leverage**. The physical image of a lever is exactly right. We know the line must pass through $(\bar{x}, \bar{y})$, so you can picture that point as the fulcrum of the lever. Just as sitting farther from the hinge on a see-saw gives you more leverage to pull it your way, points with values far from $\bar{x}$ pull more strongly on the regression line.

A point with high leverage has the potential to change the regression line, but it doesn't always use that potential. If the point lines up with the pattern of the other points, then including it doesn't change our estimate of the line. How can you tell if a high-leverage point actually changes the model? Just fit the linear model twice, both with and without the point in question. We say that a point is **influential** if omitting it from the analysis changes the model enough to make a meaningful difference.[3]

Influence depends on both leverage and residual; a case with high leverage whose y-value sits right on the line fit to the rest of the data is not influential. Removing that case won't change the slope, even if it does affect R^2. A case with modest leverage but a very

[3]How big a change would make a difference depends on the circumstances and is a judgment call. Note that some textbooks use the term *influential point* for any observation that influences the slope, intercept, or R^2. We'll reserve the term for points that influence the slope.

large residual (such as Palm Beach County) can be influential. Of course, if a point has enough leverage, it can pull the line right to it. Then it's highly influential, but its residual is small. The only way to be sure is to fit both regressions.

Unusual points in a regression often tell us a great deal about the data and the model. We face a challenge: The best way to identify unusual cases is against the background of a model, but a model dominated by a single case is unlikely to be useful for identifying unusual cases. (That insight's at least 400 years old. See the Francis Bacon quote in the margin on p. 225.) You can set aside cases and discuss what the model looks like with and without them, but arbitrarily deleting cases can give a false sense of how well the model fits the data. Your goal should be understanding the data, not making R^2 as big as you can.

In 2000, George W. Bush won Florida (and thus the presidency) by only a few hundred votes, so Palm Beach County's residual was big enough to be meaningful. It's the rare unusual case that determines a presidency, but all are worth examining and trying to understand.

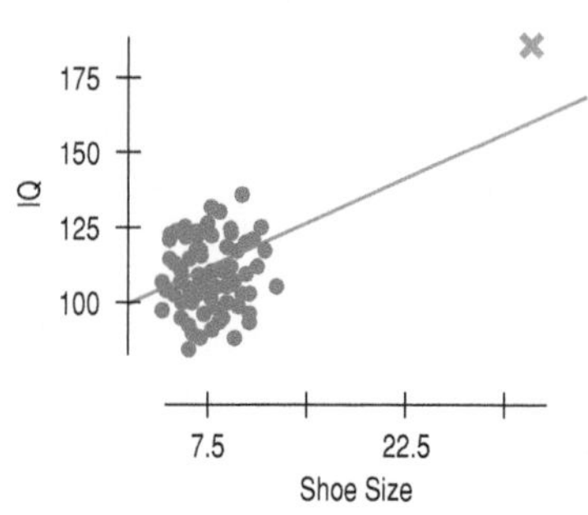

Figure 8.8

Bozo's extraordinarily large shoes give his data point high leverage in the regression. Wherever Bozo's IQ falls, the regression line will follow.

A point with so much influence that it pulls the regression line close to it can make its residual deceptively small. Influential points like that can have a shocking effect on the regression. Figure 8.8 is a plot of *IQ* against *Shoe Size*, with fitted regression line, again from the fanciful study of intelligence and foot size in comedians we saw in Chapter 6. What if Bozo's IQ had instead been very low—say, 50—as shown in Figure 8.9? The slope of the line drops from approximately +2 IQ points per shoe size to about –1 IQ point per shoe size.

No matter where Bozo's IQ is, the line tends to follow it because his shoe size, being so far from the mean shoe size, makes this a high-leverage point.

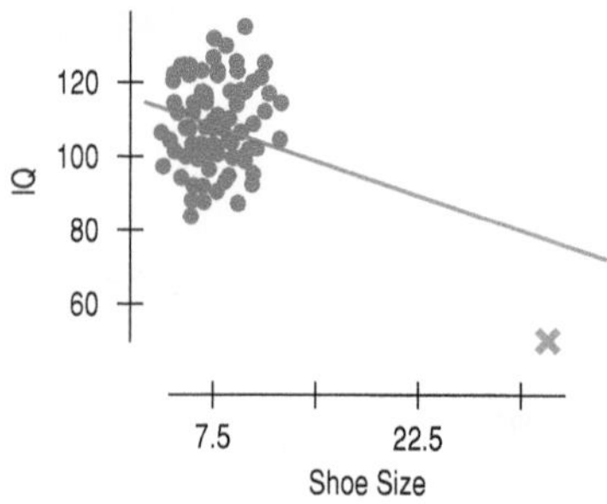

Figure 8.9

If Bozo's IQ were low, the regression slope would change from positive to negative. A single influential point can change a regression model drastically.

Without Bozo, there is little correlation between shoe size and IQ, and $R^2 = 0.7\%$, whereas including Bozo with the high IQ yields an $R^2 = 24.8\%$, so almost all of the variance accounted for is due to one case, namely Bozo. Even though this example is far-fetched, similar situations occur all the time in real life. For example, a regression of sales against floor space for hardware stores that looked primarily at small-town businesses could be dominated in a similar way if a Home Depot were included.

WARNING

Influential points can hide in plots of residuals. Points with high leverage pull the line close to them, so they often have small residuals. You'll spot influential points more easily in scatterplots of the original data or by fitting a regression model with and without the points.

Just Checking

Each of these scatterplots shows an unusual point. For each, tell whether the point is a high-leverage point, would have a large residual, or is influential.

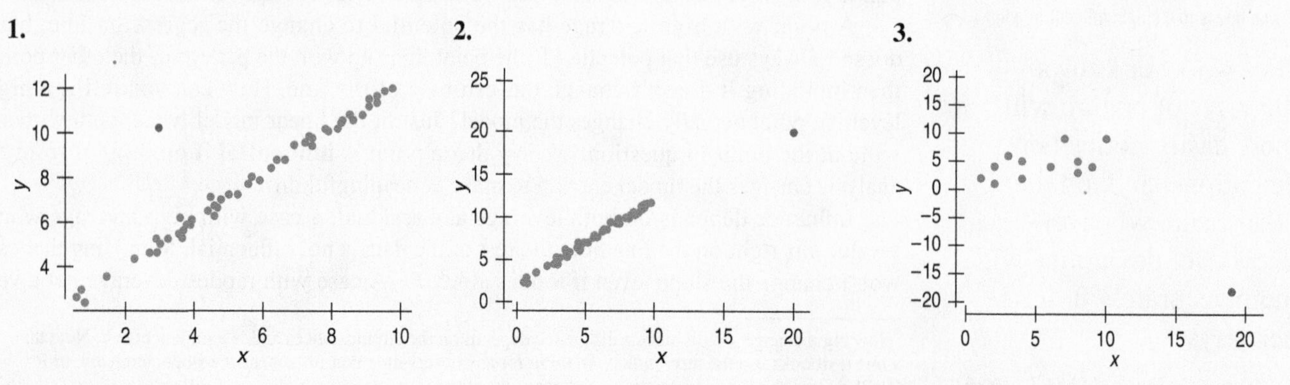

8.3 Extrapolation: Reaching Beyond the Data

A S *Case Study:* **Predicting Manatee Kills.** Can we use regression to predict the number of manatees that will be killed by power boats this year?

"Prediction is difficult, especially about the future."

—*Niels Bohr, Danish physicist*

Linear models give a predicted value for each case in the data. Put a new x-value into the equation, and it gives a predicted value, $\hat{y}$, to go with it. But when the new x-value lies far from the data we used to build the regression, how trustworthy is the prediction?

The simple answer is that the farther the new x-value is from $\bar{x}$, the less trust we should place in the predicted value. Once we venture into new x territory, such a prediction is called an **extrapolation**. Extrapolations are dubious because they require the very questionable assumption that nothing about the relationship between x and y changes even at extreme values of x and beyond.

Extrapolations require caution and can easily get us into deep trouble. When the x-variable is *Time*, extrapolation becomes an attempt to peer into the future. People have always wanted to see into the future, and it doesn't take a crystal ball to foresee that they always will.

Bill Amend/Universal Uclick

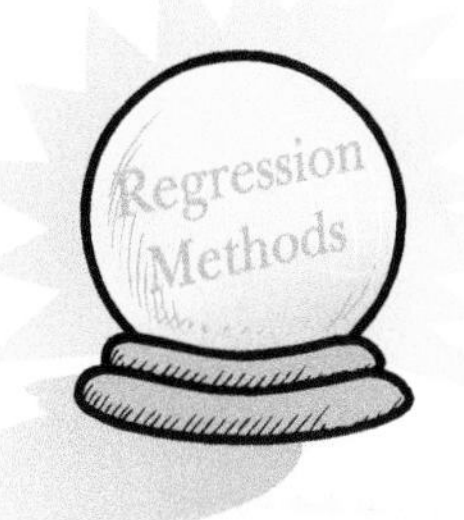

Those with a more scientific outlook may use a linear model as their digital crystal ball. Linear models are based on the x-values of the data at hand and cannot be trusted beyond that span. Some physical phenomena do exhibit a kind of "inertia" that allows us to guess that current systematic behaviour will continue, but be careful in counting on that sort of regularity in phenomena, such as stock prices, sales figures, hurricane tracks, or public opinion.

Extrapolating from current trends is so tempting that even professional forecasters sometimes expect too much from their models—and sometimes the errors are striking. In the mid-1970s, oil prices surged and long lines at gas stations were common. In 1970, oil cost about US\$17 a barrel (in 2005 dollars)—about what it had cost for 20 years or so. But then, within just a few years, the price surged to over \$40. In 1975, a survey of 15 top econometric forecasting models (built by groups that included Nobel prize–winning economists) found predictions for 1985 oil prices that ranged from \$300 to over \$700 a barrel (in 2005 dollars). How close were these forecasts?

WHEN THE DATA ARE YEARS . . .

. . . we usually don't enter them as four-digit numbers. Here we used 0 for 1970, 10 for 1980, and so on. Or we may simply enter two digits, using 82 for 1982, for instance. Rescaling years like this often makes calculations easier and equations simpler. We recommend you do it, too. But be careful: If 1982 is 82, then 2004 is 104 (not 4), right?

Figure 8.10 on the next page shows a scatterplot of oil prices from 1971 to 1981 (in 2005 dollars).

The regression model

$$\widehat{Price} = 3.08 + 6.90\ \textit{Years Since 1970}$$

says that prices had been going up on average \$6.90 per year, or \$69 over 10 years. If you assume that they would keep going up, it's not hard to imagine almost any price you want.

So how did the forecasters do? Well, Figure 8.11 shows the period from 1981 to 1998, in which oil prices didn't exactly continue that steady increase. In fact, they went down so much that by 1998, prices (adjusted for inflation) were the lowest they'd been since before World War II:

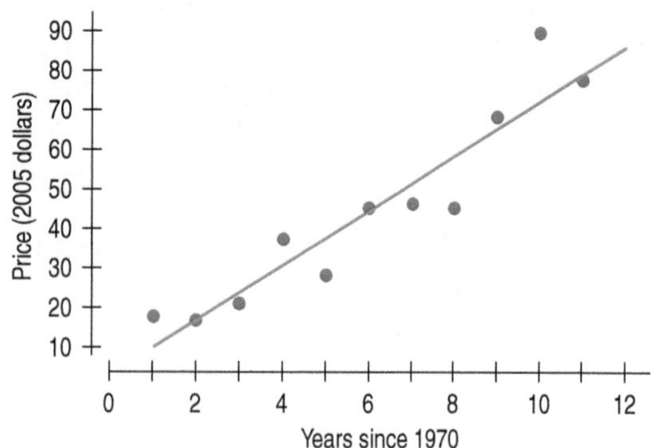

Figure 8.10

The scatterplot shows that a barrel of oil increased by an average of about $7 per year from 1971 to 1981.

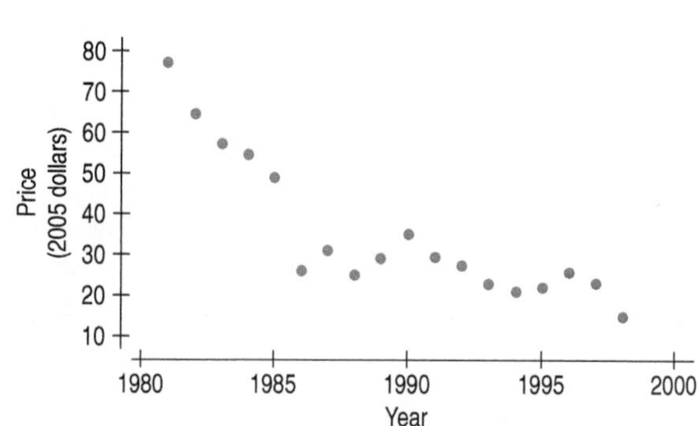

Figure 8.11

This scatterplot of oil prices from 1981 to 1998 shows a fairly constant decrease of about US$3 per barrel per year.

Not one of the experts' models predicted that.

Of course, these decreases clearly couldn't continue, or oil would be free by now. The Energy Information Administration offered two *different* 20-year forecasts for oil prices after 1998 (AE098 and AE099), and both called for relatively modest increases in oil prices. So how accurate have *these* forecasts been? Here's a timeplot of the EIA's predictions and the actual prices (in 2005 dollars):

Figure 8.12

Here are the EIA forecasts with the actual prices from 1981 to 2012. Neither forecast predicted the sharp run-up in the past few years.

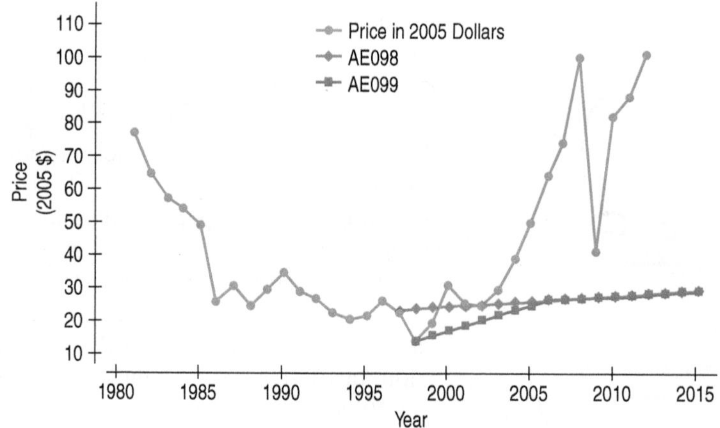

Oops! They seemed to have missed the sharp run-up in oil prices between 2004 and 2008. And they also missed the sharp drop in prices at the beginning of 2009 back to about $40 per barrel and the bounce back up to $100 after that (and the subsequent drop, and . . .).

Where do you think oil prices will go in the next decade? *Your* guess may be as good as anyone's!

Of course, knowing that extrapolation requires careful thought and caution doesn't stop people from doing it. The temptation to see into the future is hard to resist. So our more realistic advice is this:

If you extrapolate into the future, at least don't believe blindly that the prediction will come true.

For Example EXTRAPOLATION: REACHING BEYOND THE DATA

The U.S. Census Bureau (www.census.gov) reports the median age at first marriage for men and women. Here's a regression of median *Age* (at first marriage) for men against *Year* (since 1890) at every census from 1890 to 1940:

R-squared = 92.6%
s = 0.2417

Variable	Coefficient
Intercept	25.7
Year	−0.04

The regression equation is

$$\widehat{Age} = 25.7 - 0.04\ Year.$$

QUESTION: What would this model predict as the age at first marriage for men in the year 2000?

ANSWER: When *Year* counts from 0 in 1890, the year 2000 is "110." Substituting 110 for *Year*, we find that the model predicts a first marriage *Age* of 25.7 − 0.04 × 110 = 21.3 years old.

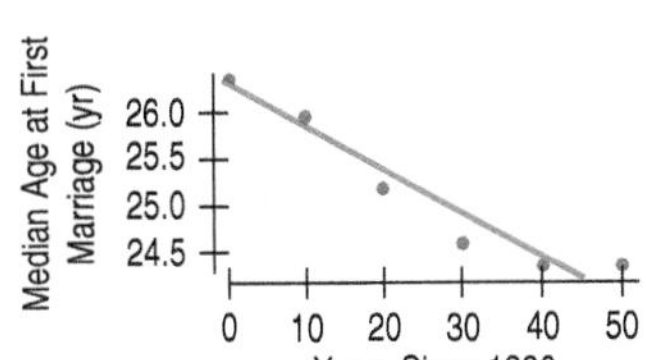

The median age at which men first married fell at the rate of about a year every 25 years from 1890 to 1940.

QUESTION: In the year 2010, the median *Age* at first marriage for men was 28.2 years. What went wrong?

ANSWER: It is never safe to extrapolate very far beyond the data. The regression was fit for years up to 1940. To see how absurd a prediction from that period can be when extrapolated into the present, look at a scatterplot of the median *Age* at first marriage for men for all the data from 1890 to 2010:

Median *Age* at first marriage (years of age) for men in the United States vs. *Year*. The regression model above was fit only to the first 50 years of the data (shown in blue), which looked nicely linear. But the linear pattern could not have continued, and in fact it changed in direction, steepness, and strength.

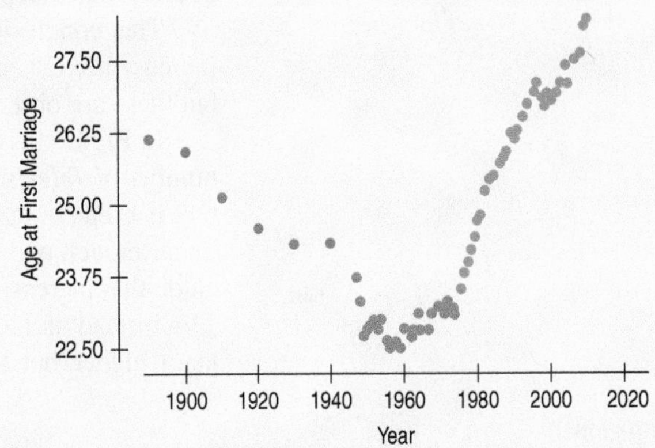

Now we can see why the extrapolation failed. Although the trend in *Age* at first marriage was linear and negative for the first part of the century, after World War II, it levelled off for about 30 years. Since 1980 or so, it has risen steadily. To characterize age and first marriage, we should probably treat these three time periods separately.

8.4 Cautions

Lurking Variables and Causation

In Chapter 6, we tried to make it clear that no matter how strong the correlation is between two variables, there's no simple way to show that one variable causes the other. Putting a regression line through a cloud of points just increases the temptation to think and to say that the x-variable *causes* the y-variable. Let's repeat the point again: No matter how strong the association, no matter how large the R^2 value, no matter how straight the line, you cannot conclude from a regression alone that one variable causes the other.

CAUSAL INTERPRETATIONS

One common way to interpret a regression slope is to say that "a change of 1 unit in *x* results in a change of b_1 units in *y*." This way of saying things encourages causal thinking. Beware.

There's always the possibility that some third variable is driving both of the variables you have observed. With observational data, as opposed to data from a designed experiment, there is no way to be sure that a **lurking variable** is not the cause of any apparent association.

Here's an example: Figure 8.13 shows the *Life Expectancy* (average of men and women, in years) for each of 40 countries of the world, plotted against the square root of the number of *Doctors per person* in the country. (The square root is used here to make the relationship satisfy the Straight Enough Condition.)

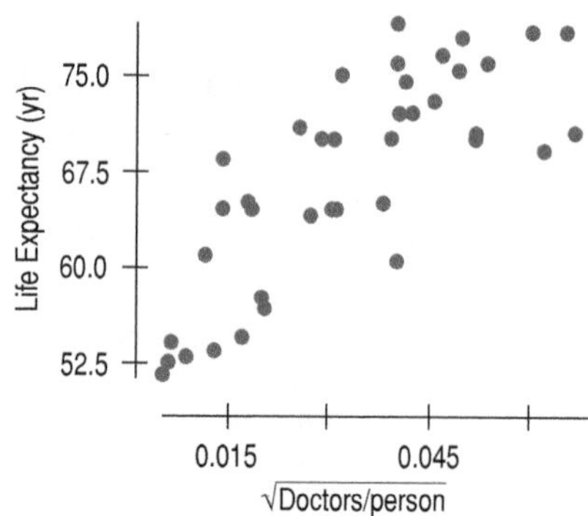

Figure 8.13
The relationship between life expectancy (years) and availability of doctors (measured as $\sqrt{\text{doctors per person}}$) for countries of the world is strong, positive, and linear.

The strong positive association ($R^2 = 62.4\%$) seems to confirm our expectation that more doctors per person improves healthcare, leading to longer lifetimes and a greater life expectancy. The strength of the association would *seem* to argue that we should send more doctors to developing countries to increase life expectancy.

That conclusion is about the consequences of a change. Would sending more doctors increase life expectancy? Specifically, do doctors *cause* greater life expectancy? Perhaps, but these are observed data, so there may be another explanation for the association.

In Figure 8.14, the similar-looking scatterplot's *x*-variable is the square root of the number of *Televisions per person* in each country. The positive association in this scatterplot is even *stronger* than the association in the previous plot ($R^2 = 72.5\%$). We can fit the linear model, and use the number of TVs to help predict life expectancy. Should we conclude that increasing the number of TVs actually extends lifetimes? If so, we should send TVs instead of doctors to developing countries. Not only is the correlation with life expectancy higher, but TVs are much cheaper than doctors and possibly more fun.

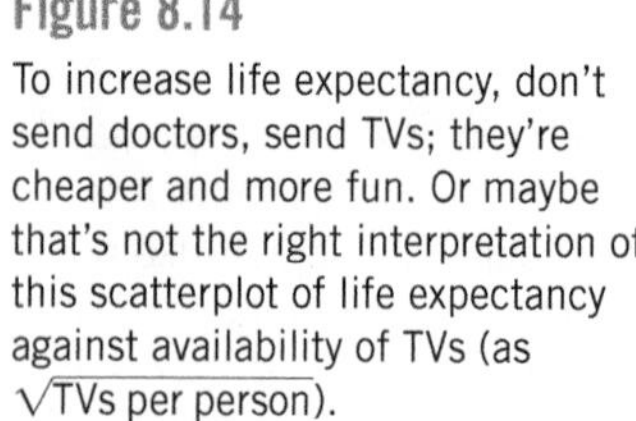

Figure 8.14
To increase life expectancy, don't send doctors, send TVs; they're cheaper and more fun. Or maybe that's not the right interpretation of this scatterplot of life expectancy against availability of TVs (as $\sqrt{\text{TVs per person}}$).

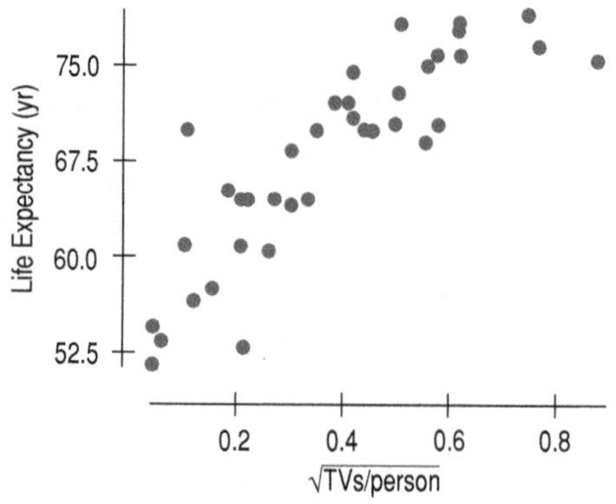

What's wrong with this reasoning? Maybe we were a bit hasty earlier when we concluded that doctors *cause* longer lives. Maybe there's a lurking variable here. Countries with higher standards of living have both longer life expectancies *and* more doctors (and more TVs). Could higher living standards cause changes in the other variables? If so, then improving living standards might be expected to prolong lives, increase the number of doctors, and increase the number of TVs, as illustrated in Figure 8.15 below:

Figure 8.15
Common response to living standards.

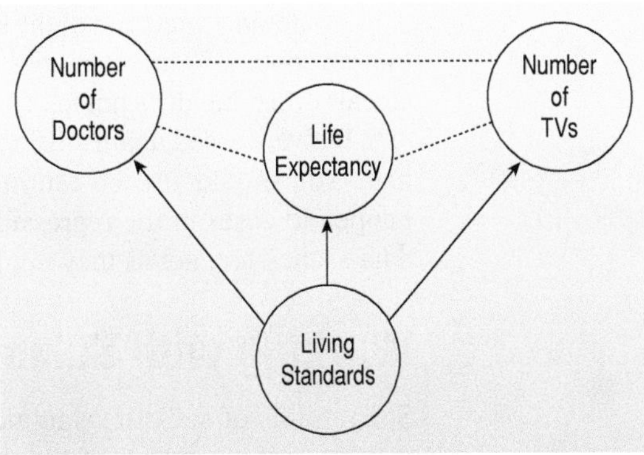

To think about lurking variables you must think "outside the box." What variables are not in your data that ought to be?[4]

From this example, you can see how easy it is to fall into the trap of mistakenly inferring causality from a regression. For all we know, doctors (or TVs!) *do* increase life expectancy. But we can't tell that from data like these, no matter how much we'd like to. Resist the temptation to conclude that x causes y from a regression, no matter how obvious that conclusion seems to you.

For Example CAUSATION AND REGRESSION

Our regression model in the previous chapter predicting hurricane wind speeds from the central pressure was reasonably successful. The negative slope indicates that in general, storms with lower central pressures have stronger winds.

QUESTION: Can we conclude that lower central barometric pressure causes the higher wind speeds in hurricanes?

ANSWER: No. While it may be true that lower pressure causes higher winds, a regression model for observed data like these cannot demonstrate causation. Perhaps higher wind speeds reduce the barometric pressure, or perhaps pressure and wind speed are both driven by some other variable we have not observed.

(As it happens, in hurricanes it is reasonable to say that the low central pressure at the eye is responsible for the high winds because it draws moist, warm air into the centre of the storm, where it swirls around, generating the winds. But as is often the case, things aren't quite that simple. The winds themselves contribute to lowering the pressure at the centre of the storm as it becomes organized into a hurricane. The lesson is that to understand causation in hurricanes, we must do more than just model the relationship of two variables; we must study the mechanism itself.)

Predicting Changes in the Response Variable

Not only is it incorrect and dangerous to interpret association with causation, but when using regression there's a more subtle danger. Never interpret a regression slope coefficient as predicting how y is likely to change if its x value in the data were changed. Here's an example: In Chapter 7, we found a regression model relating calories in breakfast cereals to their sugar content as

$$\widehat{Calories} = 89.5 + 2.50\,Sugar.$$

It might be tempting to interpret this slope as implying that adding 1 gram of sugar is expected to *lead to* a change of 2.50 calories. Predicting a change of b_1 in y for a 1 unit change in x might be true in some instances, but it isn't a consequence of the regression. In fact, adding a gram of sugar to a cereal increases its calorie content by 3.90 calories—that's the calorie content of a gram of sugar.

[4]If you wonder about the graphic: this is one of the boxes you should think outside of. There are many others. Stay alert for them.

A safer interpretation of the slope is that cereals that have a gram more sugar in them tend to have about 2.50 more calories per serving. That is, the regression model describes how the cereals differ, but does not tell us how they might change if circumstances were different.

We've warned against using a regression model to predict far from the x-values in the data. You should also be cautious about using a regression model to predict what might happen to cases in the regression if they were changed. Regression models describe the data as they are, not as they might be under other circumstances.

Working with Summary Values

Scatterplots of statistics summarized over groups tend to show less variability than we would see if we measured the same variable on individuals. This is because the summary statistics themselves vary less than the data on the individuals do—a fact we will make more specific in coming chapters.

In Chapter 6, we looked at the heights and weights of individual students, portrayed again in Figure 8.16 below. We saw a correlation of 0.644, so R^2 is 41.5%.

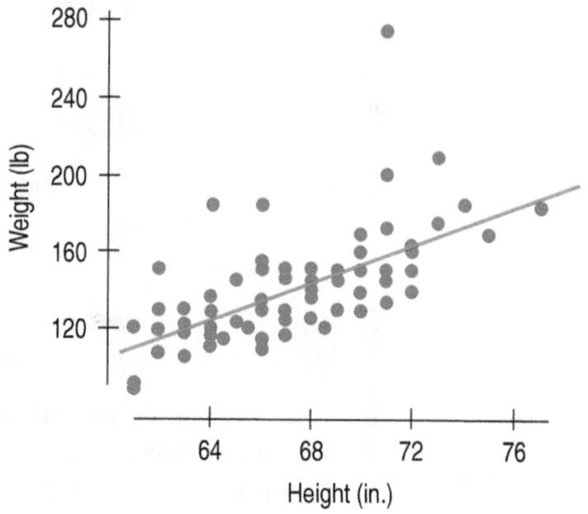

Figure 8.16
Weight (lb) against Height (in.) for a sample of men. There's a strong, positive, linear association.

Suppose, instead of data on individuals, we knew only the mean weight for each height value. The scatterplot of mean weight by height in Figure 8.17 shows less scatter. And the R^2 would increase to 80.1%.

Scatterplots of summary statistics show less scatter than the baseline data on individuals and can give a false impression of how well a line summarizes the relationship. There's no simple correction for this phenomenon. Once we're given summary data, there's no simple way to get the original values back.

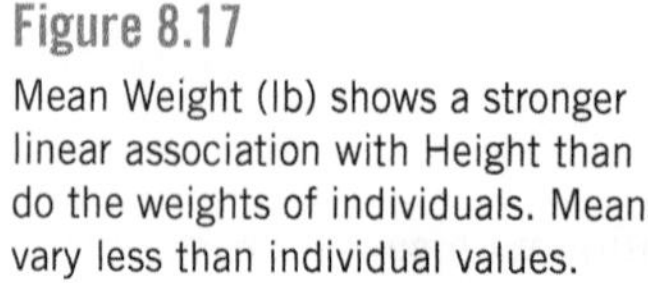

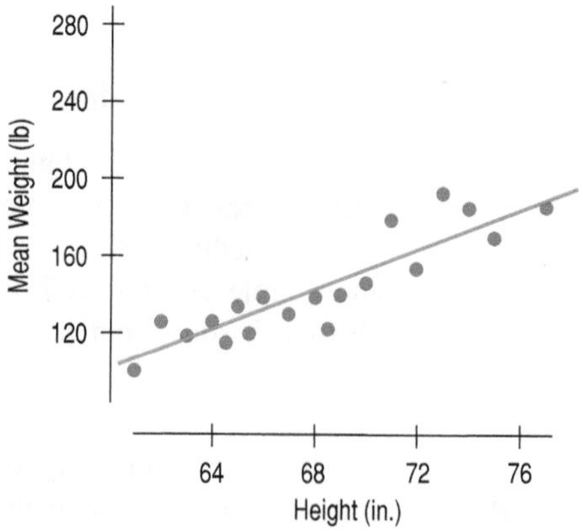

Figure 8.17
Mean Weight (lb) shows a stronger linear association with Height than do the weights of individuals. Means vary less than individual values.

In the life expectancy and TVs example, we have no good measure of exposure to doctors or to TVs on an individual basis. But if we did, we should expect the scatterplot to show more variability and the corresponding R^2 to be smaller. The bottom line is that you should be a bit suspicious of conclusions based on regressions of summary data. They may look better than they really are.

Restricted Range

In the scatterplot of individuals' weights and heights above, block out all those shorter than 72 inches. What do you think is the correlation now? Or block out all those taller than 64 inches. Guess the correlation. Block out everyone not between 64 and 68 inches. Guess the correlation again. In each case, the correlation appears to be close to zero, and certainly much less than the 0.644 correlation based on data over the full range of heights.

Correlations reported from a study may be smaller than expected if the range of observations on one or both variables has been restricted in some way. For example, we would expect the correlation between students' high school average and GPA for McGill University students to be less than it would be if all high school graduates (not just the best) were able to enrol and study at this fine university. If you just study the relationship for McGill students, you might be fooled into thinking high school averages don't say very much about potential success at university.

moodboard/Fotolia LLC

For Example USING SEVERAL OF THESE METHODS TOGETHER

Motorcycles designed to run off-road, often known as dirt bikes, are specialized vehicles.

We have data on 104 dirt bikes that were available for sale in 2005. Some cost as little as $3000, while others were substantially more expensive. Let's investigate how the size and type of engine contribute to the cost of a dirt bike. As always, we start with a scatterplot.

Here's a scatterplot of the manufacturer's suggested retail price (MSRP) in dollars against the engine displacement, along with a regression analysis:

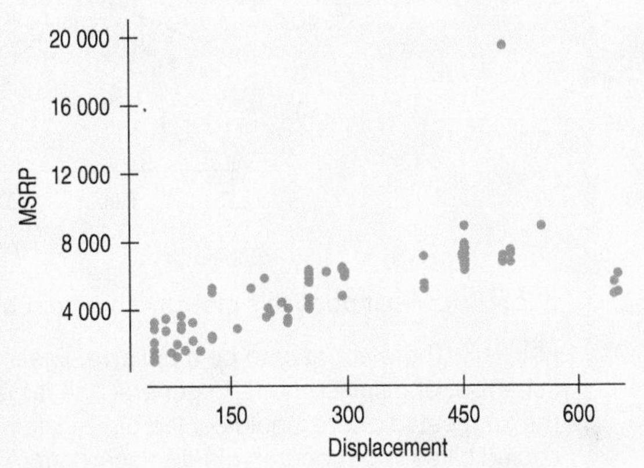

Dependent variable is: MSRP
R-squared = 49.9% s = 1737

Variable	Coefficient
Intercept	2273.67
Displacement	10.0297

QUESTION: What do you see in the scatterplot?

ANSWER: There is a strong positive association between the engine displacement of dirt bikes and the manufacturer's suggested retail price. One of the dirt bikes is an outlier; its price is more than double that of any other bike.

The outlier is the Husqvarna TE 510 Centennial. Most of its components are handmade exclusively for this model, including extensive use of carbon fibre throughout. That may explain its $19 500 price tag! Clearly, the TE 510 is not like the other bikes. We'll set it aside for now and look at the data for the remaining dirt bikes.

QUESTION: What effect will removing this outlier have on the regression? Describe how the slope, R^2, and s_e will change.

ANSWER: The TE 510 was an influential point, tilting the regression line upward. With that point removed, the regression slope will get smaller. With that dirt bike omitted, the pattern becomes more consistent, so the value of R^2 should get larger and the standard deviation of the residuals, s_e, should get smaller.

With the outlier omitted, here's the new regression and a scatterplot of the residuals:

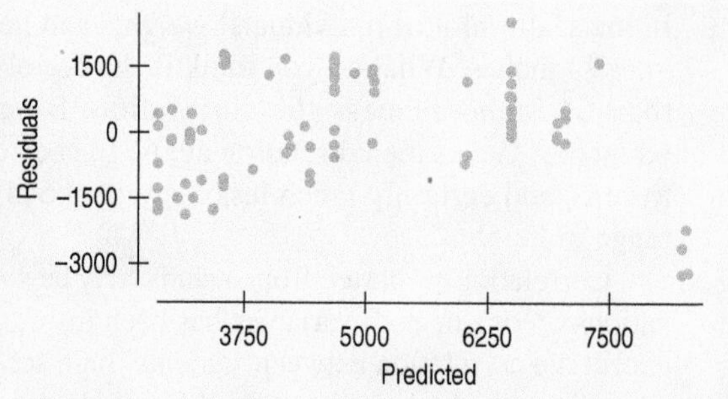

Dependent variable is: MSRP

R-squared = 61.3%   s = 1237

Variable	Coefficient
Intercept	2411.02
Displacement	9.05450

QUESTION: What do you see in the residuals plot?

ANSWER: The points at the far right don't fit well with the other dirt bikes. Overall, there appears to be a bend in the relationship, so a linear model may not be appropriate.

Let's try a re-expression. Here's a scatterplot showing MSRP against the cube root of displacement to make the relationship closer to straight. (Since displacement is measured in cubic centimetres, its cube root has the simple units of centimetres.) In addition, we've coloured the plot according to the cooling method used in the bike's engine: liquid or air. Each group is shown with its own regression line, as we did for the cereals on different shelves

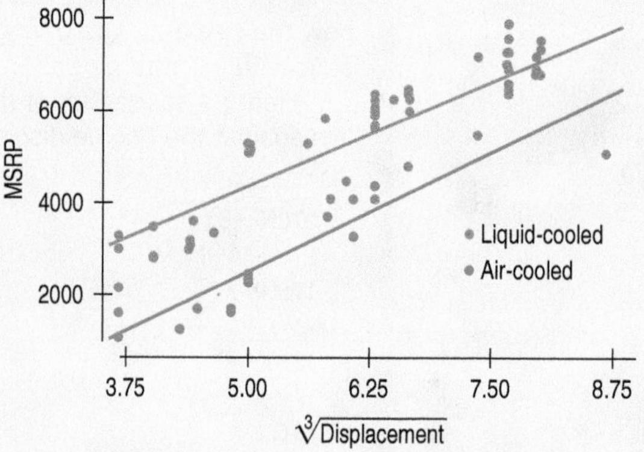

QUESTION: What does this plot say about dirt bikes?

ANSWER: There appears to be a positive, linear relationship between MSRP and the cube root of displacement. In general, the larger the engine a bike has, the higher the suggested price. Liquid-cooled dirt bikes, however, typically cost more than air-cooled bikes with comparable displacement. A few liquid-cooled bikes appear to be much less expensive than we might expect, given their engine displacements (but without separating the groups we might have missed that because they look more like air-cooled bikes).[5]

WHAT CAN GO WRONG?

This entire chapter has included warnings about things that can go wrong in a regression analysis. So let's just recap. When you make a linear model,

- **Make sure the relationship is straight.** Check the Straight Enough Condition. Always examine the residuals for evidence that the Linearity Assumption has failed. It's often

[5]Jiang Lu, Joseph B. Kadane, and Peter Boatwright, "The dirt on bikes: An illustration of CART models for brand differentiation," provided data on 2005-model bikes, lib.stat.cmu.edu/datasets/dirtbike_aug.csv.

easier to see deviations from a straight line in the residuals plot than in the scatterplot of the original data.

- **Be on guard for different groups in your regression.** Check for evidence that the data consist of separate subsets. If you find subsets that behave differently, consider fitting a different linear model to each subset.
- **Beware of extrapolating.** Beware of extrapolation beyond the x-values that were used to fit the model. Although it's common to use linear models to extrapolate, the practice is dangerous.
- **Beware especially of extrapolating into the future!** Be especially cautious about extrapolating into the future with linear models. To predict the future, you must assume that future changes will continue at the same rate you've observed in the past. Predicting the future is particularly tempting and particularly dangerous.
- **Look for unusual points.** Unusual points always deserve attention, and may well reveal more about your data than the rest of the points combined. Always look for them and try to understand why they stand apart. A scatterplot of the data is a good way to see high-leverage and possibly influential points. A scatterplot of the residuals against the predicted values is a good tool for finding points with large residuals (as is a histogram).
- **Beware of high-leverage points, and especially of those that are influential.** Influential points can alter the regression model a great deal. The resulting model may say more about one or two points than about the overall relationship.
- **Consider comparing two regressions.** To see the impact of outliers on a regression, it's often wise to run two regressions, one with and one without the extraordinary points, and then to discuss the differences.
- **Treat unusual points honestly.** If you remove enough carefully selected points, you will eventually get a regression with a high R^2, but it won't give you much understanding. Some variables are not related in a way that's simple enough for a linear model to fit very well. When that happens, report the failure and stop.
- **Beware of lurking variables.** Think about lurking variables before interpreting a linear model. It's particularly tempting to explain a strong regression by thinking that the x-variable *causes* the y-variable. A linear model alone can never demonstrate such causation, in part because it cannot eliminate the chance that a lurking variable has caused the variation in y.
- **Watch out when dealing with data that are summaries.** Be cautious in working with data values that are themselves summaries, such as means or medians. Such statistics are less variable than the data on which they are based, so they tend to inflate the impression of the strength of a relationship.
- **Watch out when dealing with range-restricted data.** Not including data over a full range of the variables will often deflate the impression of the strength of a relationship.

CONNECTIONS

We should always be alert to things that could go wrong if we were to use statistics without thinking carefully. Regression opens new vistas of potential problems, but each one relates to issues we've thought about before.

It is always important that our data be from a single homogeneous group and not made up of disparate groups. We looked for multiple modes in single variables. Now we check scatterplots for evidence of subgroups in our data. As with modes, it's often best to split the data and analyze the groups separately.

Our concern with unusual points and their potential influence also harks back to our earlier concern with outliers in histograms and boxplots—and for many of the same reasons. As we've seen here, regression offers such points new scope for mischief.

The risks of interpreting linear models as causal arose in Chapters 6 and 7. And they're important enough to mention again in later chapters.

What Have We Learned?

Learning Objectives

Be skeptical of regression models. Always plot and examine the residuals for unexpected behaviour. Be alert to a variety of possible violations of the standard regression assumptions and know what to do when you find them.

Be alert for subgroups in the data.

- Often these will turn out to be separate groups that should not be analyzed together in a single analysis.
- Often identifying subgroups can help us understand what is going on in the data.

Be especially cautious about extrapolating beyond the data.

Look out for unusual and extraordinary observations.

- Cases that are extreme in *x* have high leverage and can affect a regression model strongly.
- Cases that are extreme in *y* have large residuals and may also influence a regression model.
- Running two regressions, one with and one without the unusual points, can help you assess whether their influence on the model is appropriate or desirable. If undesirable, consider removing and reporting these cases separately.

Interpret regression models appropriately. Don't infer causation from a regression model.

Be careful when you interpret regressions based on summarized data or with restricted data ranges. Regressions with summaries tend to look stronger than the regression based on all the individual cases, whereas restricting the ranges of variables usually makes regressions weaker.

Diagnose and treat nonlinearity.

- If a scatterplot of *y* vs. *x* isn't straight, a linear regression model isn't appropriate.
- Re-expressing one or both variables can often improve the straightness of the relationship.
- The powers, roots, and the logarithm provide an ordered collection of re-expressions so you can search up and down the "ladder of powers" (see Section 6.4) to find an appropriate one.

Review of Terms

Outlier Any data point that stands away from the others can be called an outlier. In regression, cases can be extraordinary in two ways: by standing far apart in the y-direction and having a large residual or by standing far apart in the x-direction and having high leverage (p. 225).

Leverage Data points whose *x*-values are far from the mean of *x* are said to exert leverage on a linear model. High-leverage points pull the line close to them, and so they can have a large effect on the line, sometimes completely determining the slope and intercept. With high enough leverage, their residuals can be deceptively small (p. 225).

Influential point A point that, if omitted from the data, results in a very different regression model (p. 225).

Extrapolation Although linear models provide an easy way to predict values of *y* for a given value of *x*, it is unsafe to predict for values of *x* far from the ones used to find the linear model equation. At times, such extrapolation may pretend to see into the future, but the predictions should not be trusted (p. 227).

Lurking variable A variable that is not explicitly part of a model but affects the way the variables in the model appear to be related. Because we can never be certain that observational data are not hiding a lurking variable that influences the relationship between *x* and *y*, it is never safe to conclude that a linear model demonstrates a causal relationship, no matter how strong the linear association (p. 230).

On the Computer REGRESSION DIAGNOSTICS

Most statistics technology offers simple ways to check whether your data satisfy the conditions for regression. We have already seen that these programs can make a simple scatterplot. They can also help us check the conditions by plotting residuals.

EXCEL

The Data Analysis add-in for Excel includes a Regression command. The dialog box it shows offers to make plots of residuals.

COMMENTS

Do not use the Normal probability plot offered in the regression dialog. It is not what it claims to be and is wrong.

JMP

- From the **Analyze** menu, choose **FitY by X**. Select **Fit Line**.
- Under Linear Fit, select **Plot Residuals.** You can also choose to **Save Residuals**.
- Then, from the **Distribution** menu, choose **Normal quantile plot** or **histogram** for the residuals.

MINITAB

- From the **Stat** menu, choose **Regression**.
- From the **Regression** submenu, select **Regression** again.
- In the Regression dialogue, enter the response variable name in the "Response" box and the predictor variable name in the "Predictor" box.
- To specify saved results, in the Regression dialog, click **Storage**.
- Check "Residuals" and "Fits." Click **OK**.
- To specify displays, in the Regression dialog, click **Graphs**.
- Under "Residual Plots," select "Individual plots" and check "Residuals versus fits" (or select "Four in One" for a set of four very useful residual plots).
- Click **OK**. Now back in the Regression dialog, click **OK**. Minitab computes the regression and the requested saved values and graphs.

R

Save the regression model object by giving it a name, such as "myreg":

- **myreg = lm(Y~X) or myreg = lm(Y~X, data = DATA)** where DATA is the name of the data frame.
- **plot(residuals(myreg)~predict(myreg))** plots the residuals against the predicted values.
- **qqnorm(residuals(myreg))** gives a normal probability plot of the residuals.
- **plot(myreg)** gives similar plots (but not exactly the same).

SPSS

- From the **Analyze** menu, choose **Regression**.
- From the Regression submenu, choose **Linear**.
- After assigning variables to their roles in the regression, click the "**Plots . . .**" button.

In the Plots dialog, you can specify a Normal probability plot of residuals and scatterplots of various versions of standardized residuals and predicted values.

COMMENTS

A plot of ***ZRESID** against ***PRED** will look most like the residual plots we've discussed. SPSS standardizes the residuals by dividing by their standard deviation. (There's no need to subtract their mean; it must be zero.) The standardization doesn't affect the scatterplot.

STATCRUNCH

To create a residuals plot:

- Click on **Stat.**
- Choose **Regression » Simple Linear** and choose X and Y.
- Click on **Next** and click on **Next** again.
- Indicate which type of residuals plot you want.
- Click on **Calculate.**

COMMENTS

Note that before you click on **Next** for the second time you may indicate that you want to save the values of the residuals. Residuals becomes a new column, and you may use that variable to create a histogram or residuals plot.

TI-83/84 PLUS

- To make a residuals plot, set up a **STATPLOT** as a scatterplot.
- Specify your explanatory data list as **Xlist.**
- For **Ylist,** import the name **RESID** from the **LIST NAMES** menu. **ZoomStat** will now create the residuals plot.

COMMENTS

Each time you execute a **LinReg** command, the calculator automatically computes the residuals and stores them in a data list named **RESID**. If you want to see them, go to **STAT EDIT**. Space through the names of the lists until you find a blank. Import **RESID** from the **LIST NAMES** menu. Now every time you have the calculator compute a regression analysis, it will show you the residuals.

Exercises

T **1. Marriage age 2010** Is there evidence that the age at which women get married has changed over the past 100 years? The scatterplot shows the trend in age at first marriage for American women (www.census.gov).

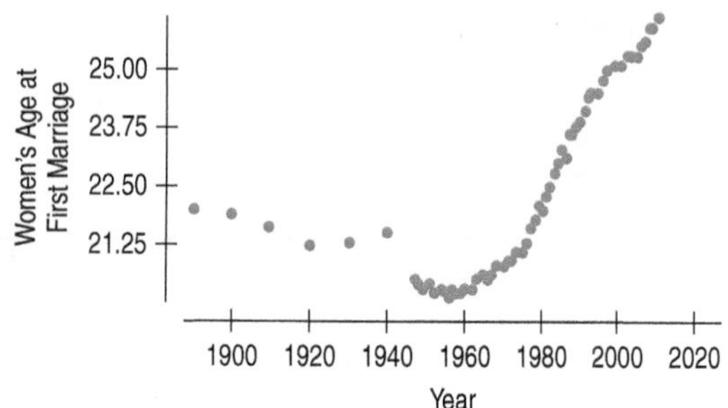

a) Is there a clear pattern? Describe the trend.
b) Is the association strong?
c) Is the correlation high? Explain.
d) Is a linear model appropriate? Explain.

T **2. Smoking 2009** The Centers for Disease Control and Prevention track cigarette smoking in the United States. How has the percentage of people who smoke changed since the danger became clear during the last half of the 20th century? The scatterplot shows percentages of smokers among men 18–24 years of age, as estimated by surveys, from 1965 through 2009 (www.cdc.gov/nchs/).

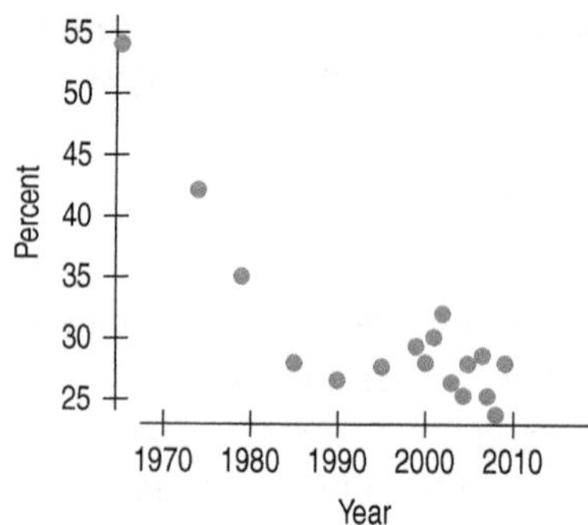

a) Is there a clear pattern? Describe the trend.
b) Is the association strong?
c) Is a linear model appropriate? Explain.

3. Gas mileage One of the important factors determining a car's fuel efficiency is its weight. Let's examine this relationship for 11 cars. In this exercise, we are using the U.S. units for fuel efficiency, miles per (U.S.) gallon.

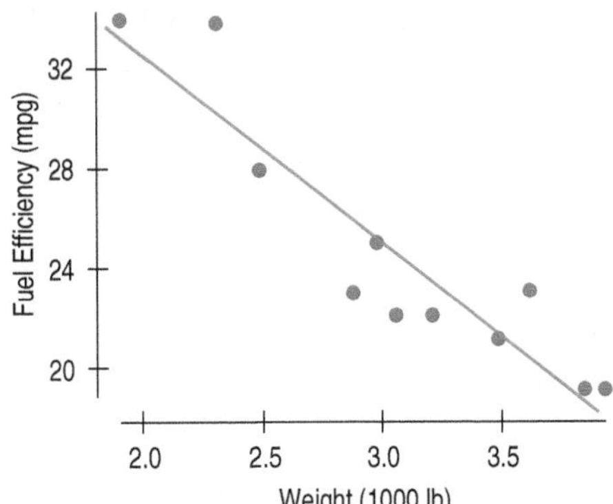

a) Describe the association between these variables shown in the scatterplot.

b) Here is the regression analysis for the linear model. What does the slope of the line say about this relationship?

Dependent variable is: Fuel Efficiency
R-squared = 85.9%

Variable	Coefficient
Intercept	47.9636
Weight	−7.65184

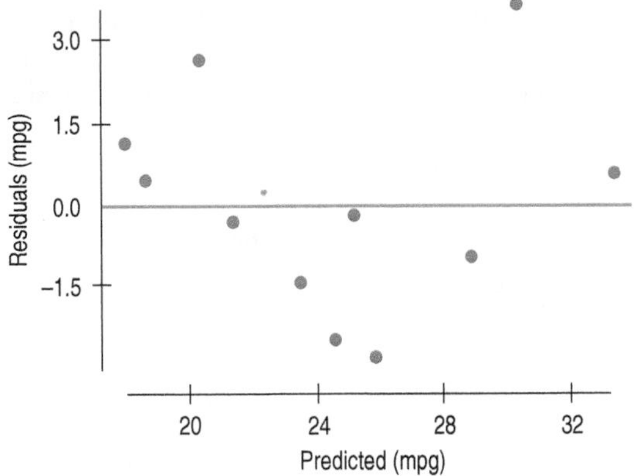

c) Do you think this linear model is appropriate? Use the residual plot to explain your decision.

4. Gas mileage revisited Let's try the re-expressed variable fuel consumption (gal/100 mi) to examine the fuel efficiency of the 11 cars in Exercise 3. Here are the revised regression analysis and residual plot:

Dependent variable is: Fuel Consumption
R-squared = 89.2%

Variable	Coefficient
Intercept	0.624932
Weight	1.17791

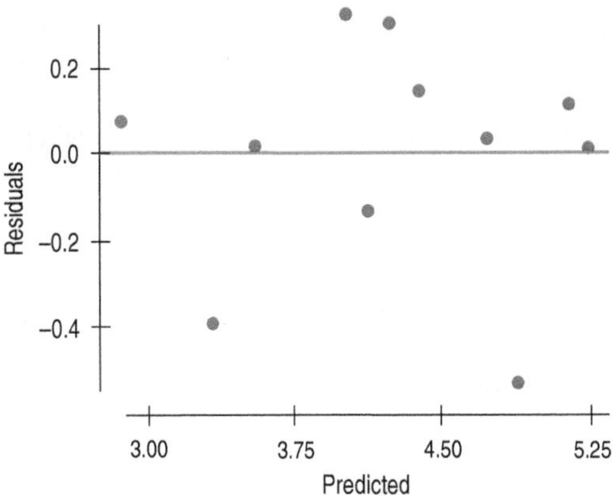

a) Explain why this model appears to be better than the model used in Exercise 3.

b) Using the regression analysis above, write an equation of the model.

c) Interpret the slope of the line.

d) Based on this model, how many miles per gallon would you expect a 3500-pound car to get?

e) Canadians might wish instead to know the number of litres per 100 km expected for a 3500-pound car. What would you estimate? (1 US gallon = 3.7854 litres; 1 mile = 1.61 km)

5. Good model? In justifying his choice of a model, a student wrote, "I know this is the correct model because $R^2 = 99.4\%$."

a) Is this reasoning correct? Explain.

b) Does this model allow the student to make accurate predictions? Explain.

6. Bad model? A student who has created a linear model is disappointed to find that her R^2 value is a very low 13%.

a) Does this mean that a linear model is not appropriate? Explain.

b) Does this model allow the student to make accurate predictions? Explain.

T **7. Movie dramas** Here's a scatterplot of the production budgets (in millions of dollars) versus the running time (in minutes) for major release movies in 2005. Dramas are plotted as red x's and all other genres are plotted as blue dots. (The re-make of *King Kong* is plotted as a black "–". At the time, it was the most expensive movie ever made, and not typical of any genre.) A separate least squares regression line has been fitted to each group. For the following questions, examine the plot:

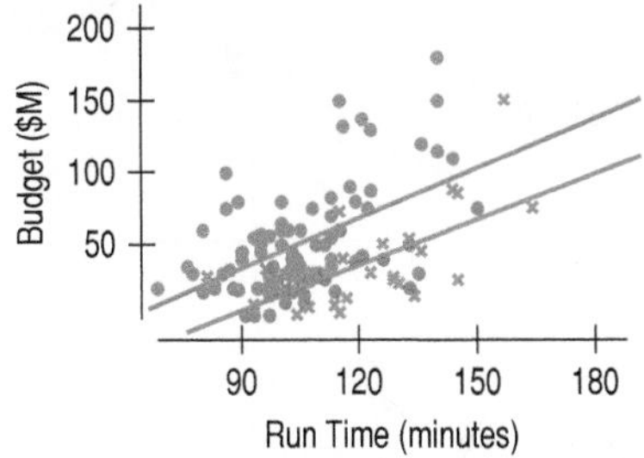

a) What are the units for the slopes of these lines?
b) In what way are dramas and other movies similar with respect to this relationship?
c) In what way are dramas different from other genres of movies with respect to this relationship?

8. Smoking 2009, women and men In Exercise 2, we examined the percentage of men aged 18–24 who smoked from 1965 to 2009 according to the Centers for Disease Control and Prevention. How about women? Here's a scatterplot showing the corresponding percentages for both men and women:

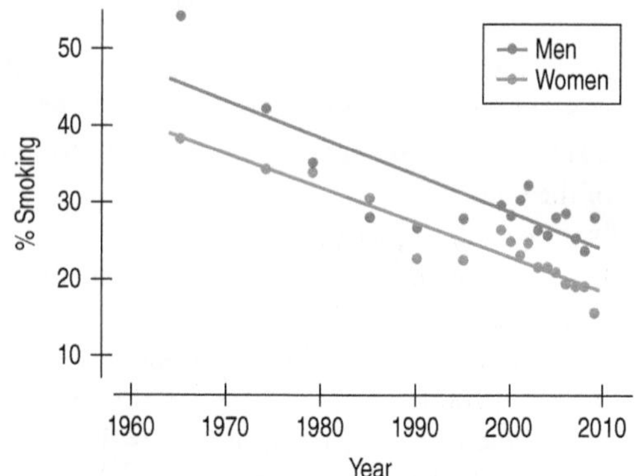

a) In what ways are the trends in smoking behaviour similar for men and women?
b) How do the smoking rates for women differ from those for men?
c) Viewed alone, the trend for men may have seemed to violate the Linearity Condition. How about the trend for women? Does the consistency of the two patterns encourage you to think that a linear model for the trend in men might be appropriate? (Note: there is no correct answer to this question; it is raised for you to think about.)

9. Oakland passengers The scatterplot below shows the number of passengers departing from Oakland (CA) airport month by month since the start of 1997 (up to early 2006).[6] Time is shown as years since 1990, with fractional years used to represent each month. (Thus June of 1997 is 7.5—halfway through the seventh year after 1990.)

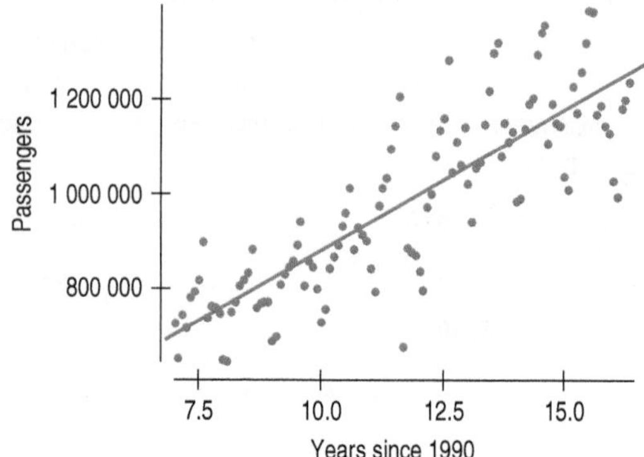

Here's a regression and the residuals plot:

Dependent variable is: Passengers
R-squared = 71.1% s = 104330

Variable	Coefficient
Constant	282584
Year-1990	59704.4

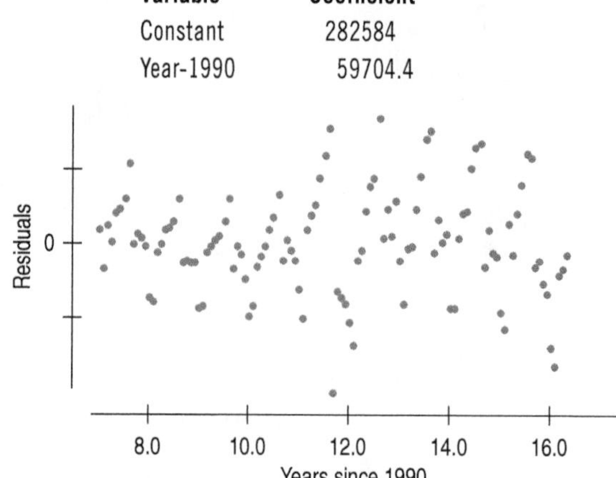

a) Interpret the slope and intercept of the regression model.
b) What does the value of R^2 say about how successful the model is?
c) Interpret s_e in this context.
d) Would you use this model to predict the numbers of passengers in 2010 (*YearsSince1990* = 20)? Explain.
e) There's a point near the middle of this timespan with a large negative residual. Can you explain this outlier?

10. Tracking hurricanes 2010 Below are data on the average tracking error, per year, (in nautical miles) made by the National Hurricane Center in predicting the path of hurricanes. The following scatterplot shows the trend in the 24-hour tracking errors since 1970 (www.nhc.noaa.gov).

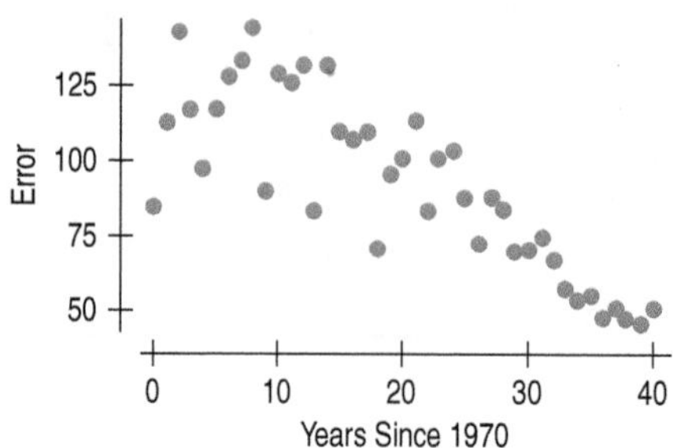

Dependent variable is: Error
R-squared = 68.7% s = 16.44

Variable	Coefficient
Intercept	132.301
Years-1970	−2.00662

a) Interpret the slope and intercept of the model.
b) Interpret s_e in this context.
c) The Center would like to achieve an average tracking error of 45 nautical miles by 2015. Will they make it? Defend your response.
d) What if their goal were an average tracking error of 25 nautical miles?
e) What cautions would you state about your conclusion?

[6] www.oaklandairport.com/

11. **Unusual points** Each of these four scatterplots shows a cluster of points and one "stray" point. For each, answer these questions:
   a) In what way is the point unusual? Does it have high leverage, a large residual, or both?
   b) Do you think that point is an influential point?
   c) If that point were removed, would the correlation become stronger or weaker? Explain.
   d) If that point were removed, would the slope of the regression line increase or decrease? Explain.

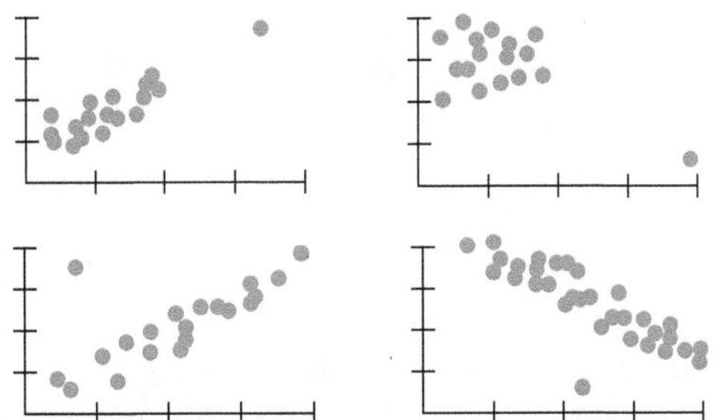

12. **More unusual points** Each of the following scatterplots shows a cluster of points and one "stray" point. For each, answer these questions:
   a) In what way is the point unusual? Does it have high leverage, a large residual, or both?
   b) Do you think that point is an influential point?
   c) If that point were removed, would the correlation become stronger or weaker? Explain.
   d) If that point were removed, would the slope of the regression line increase or decrease? Explain.

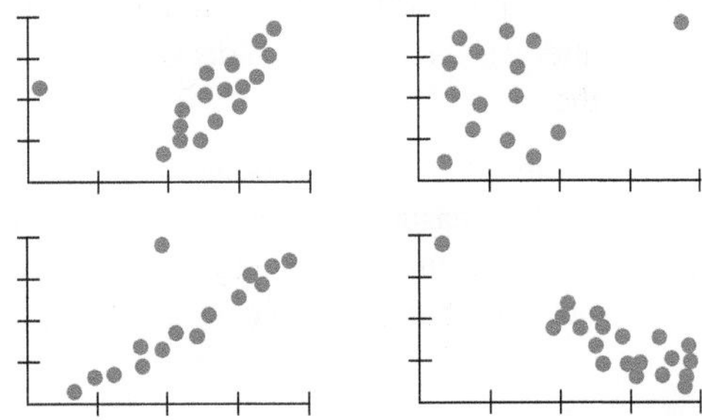

13. **The extra point** The scatterplot shows five blue data points at the left. Not surprisingly, the correlation for these points is $r = 0$. Suppose *one* additional data point is added at one of the five positions suggested below in green. Match each point (a–e) with the correct new correlation from the list given.
   a) −0.90 b) −0.40 c) 0.00 d) 0.05 e) 0.75

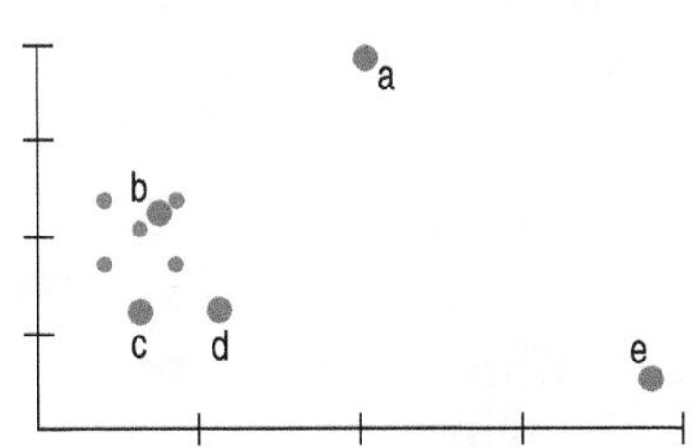

14. **The extra point revisited** The original five points in Exercise 13 produce a regression line with slope 0. Match each of the green points (a–e) with the slope of the line after that one point is added:
   a) −0.45 b) −0.30 c) 0.00 d) 0.05 e) 0.85

15. **What's the cause?** Suppose a researcher studying health issues measures blood pressure and the percentage of body fat for several adult males and finds a strong positive association. Describe three different possible cause-and-effect relationships that might be present.

16. **What's the effect?** A researcher studying violent behaviour in elementary-school children asks the children's parents how much time each child spends playing computer games and has their teachers rate each child on the level of aggressiveness they display while playing with other children. Suppose that the researcher finds a moderately strong positive correlation. Describe three different possible cause-and-effect explanations for this relationship.

17. **Reading** To measure progress in reading ability, students at an elementary school take a reading comprehension test every year. Scores are measured in "grade-level" units; that is, a score of 4.2 means that a student is reading at slightly above the expected level for a Grade 4 student. The school principal prepares a report to parents that includes a graph showing the mean reading score for each grade. In his comments, he points out that the strong positive trend demonstrates the success of the school's reading program.

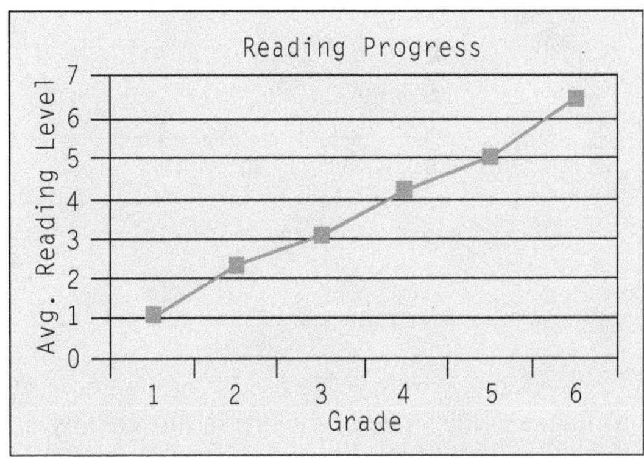

   a) Does this graph indicate that students are making satisfactory progress in reading? Explain.
   b) What would you estimate the correlation between *Grade* and *Average Reading Level* to be?
   c) If, instead of this plot showing average reading levels, the principal had produced a scatterplot of the reading levels of all the individual students, would you expect the correlation to be the same, higher, or lower? Explain.
   d) Although the principal did not do a regression analysis, someone as statistically astute as you might. (But don't bother.) What value of the slope of that line would you view as demonstrating acceptable progress in reading comprehension? Explain.

18. Grades A U.S. college admissions officer, defending the college's use of SAT scores in the admissions process, produced the graph below. It shows the mean GPAs for last year's freshmen, grouped by SAT scores. How strong is the evidence that *SAT Score* is a good predictor of *GPA*? What concerns you about the graph—the statistical methodology or the conclusions reached?

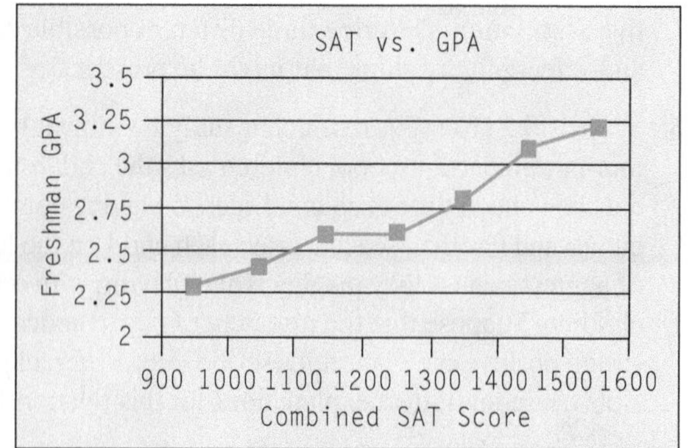

19. Heating After keeping track of his heating expenses for several winters, a homeowner in the U.S. believes he can estimate the monthly cost from the average daily Fahrenheit temperature by using the model $\widehat{Cost} = 133 - 2.13\ Temp$. The residuals plot for his data is shown.

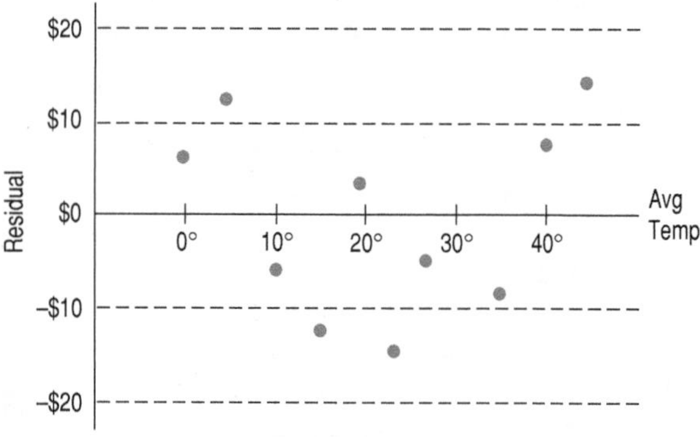

a) Interpret the slope of the line in this context.
b) Interpret the y-intercept of the line in this context.
c) During months when the temperature stays around freezing, would you expect cost predictions based on this model to be accurate, too low, or too high? Explain.
d) What heating cost does the model predict for a month that averages 10°F?
e) During one of the months on which the model was based, the temperature did average 10°F. What were the actual heating costs for that month?
f) Do you think the homeowner should use this model? Explain.
g) Would this model be more successful if the temperature were expressed in degrees Celsius? Explain.

20. Speed How does the speed at which you drive affect your fuel economy? To find out, researchers in the United States drove a compact car for 200 miles at speeds ranging from 35 to 75 miles per hour, and measured fuel efficiency in miles per gallon (mpg). From their data, they created the model $\widehat{Fuel\ Efficiency} = 32 - 0.1\ Speed$ and created this residual plot:

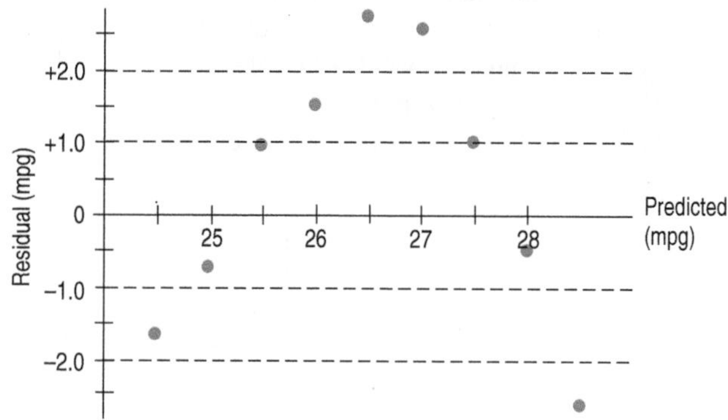

a) Interpret the slope of this line in context.
b) Explain why it's silly to attach any meaning to the y-intercept.
c) When this model predicts high *Fuel Efficiency,* what can you say about those predictions?
d) What *Fuel Efficiency* does the model predict when the car is driven at 50 mph?
e) What was the actual *Fuel Efficiency* when the car was driven at 45 mph?
f) Do you think there appears to be a strong association between *Speed* and *Fuel Efficiency*? Explain.
g) Do you think this is the appropriate model for that association? Explain.
h) If we use km/h and litres/100 km in our model instead, the scatterplot and residual plots should look essentially the same. True or false? Explain.

21. Interest rates Here's a plot showing the federal rate on 3-month U.S. Treasury bills from 1950 to 1980, and a regression model fit to the relationship between the *Rate* (in %) and *Years Since 1950* (www.gpoaccess.gov/eop/).

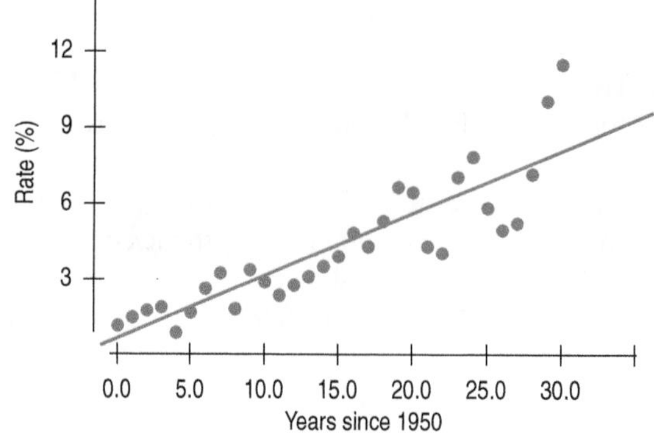

Dependent variable is: Rate
R-squared = 77.4%   s = 1.239

Variable	Coefficient
Intercept	0.640282
Year-1950	0.247637

a) What is the correlation between *Rate* and *Year*?
b) Interpret the slope and intercept.
c) What does this model predict for the interest rate in the year 2000?
d) Would you expect this prediction to have been accurate? Explain.

22. Age of couples 2010 The graph shows the ages of both men and women at first marriage (www.census.gov).

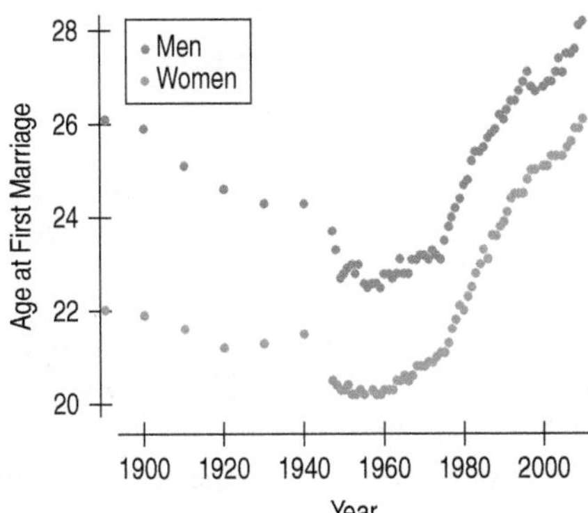

Clearly, the patterns for men and women are similar. But are the two lines getting closer together?

Here's a timeplot showing the *difference* in average age (men's age – women's age) at first marriage, the regression analysis, and the associated residuals plot.

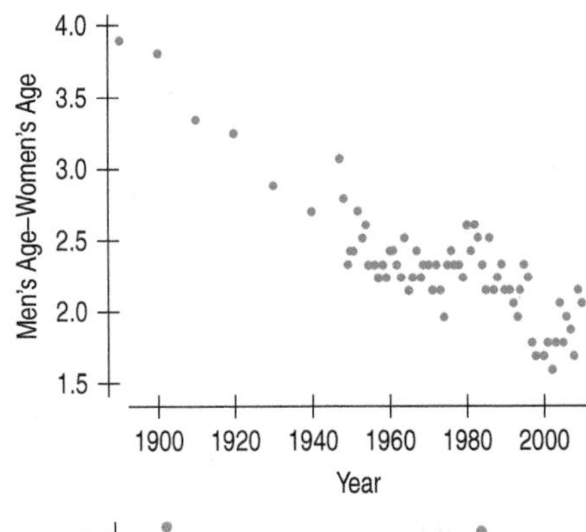

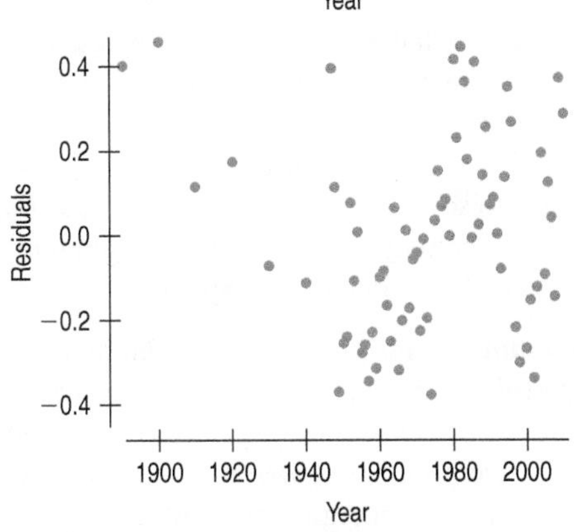

Dependent variable is: Age Difference
R-squared = 75.5%   s = 0.2319

Variable	Coefficient
Intercept	33.396
Year	−0.01571

a) What is the correlation between *Age Difference* and *Year*?
b) Interpret the slope of this line.
c) Predict the average age difference in 2015.
d) Describe reasons why you might not place much faith in that prediction.

23. Interest rates revisited In Exercise 21, you investigated the federal rate on 3-month U.S. Treasury bills between 1950 and 1980. The scatterplot below shows that the trend changed dramatically after 1980, so we computed a new regression model just for the years 1980 to 2005.

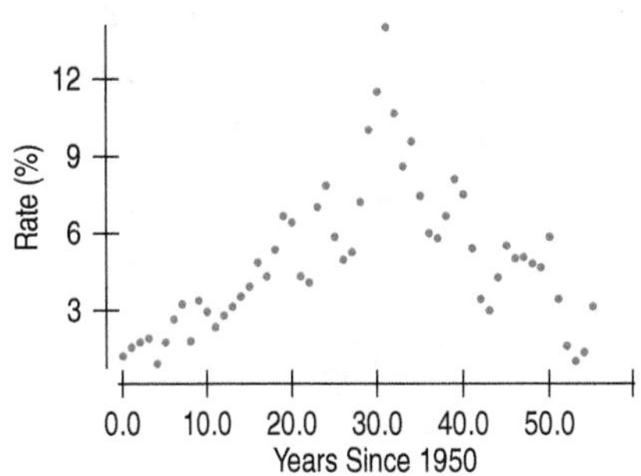

Dependent variable is: Rate
R-squared = 74.5%   s = 1.630

Variable	Coefficient
Intercept	21.0688
Year-1950	−0.356578

a) How does this model compare to the one in Exercise 21?
b) What does this model estimate the interest rate was in 2000? How does this compare to the rate you predicted in Exercise 21?
c) Do you trust this newer predicted value? Explain.
d) Given these two models, what would you predict the interest rate on 3-month Treasury bills will be in 2020?

24. Ages of couples again Has the trend of decreasing difference in age at first marriage seen in Exercise 22 gotten stronger recently? The scatterplot and residual plot for the data from 1980 through 2010, along with a regression for just those years, are below.

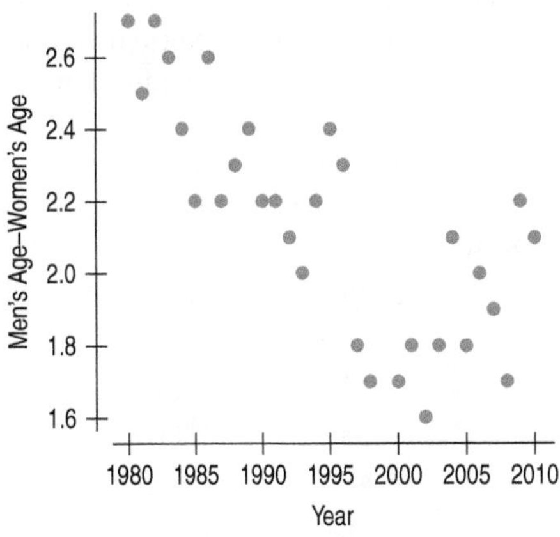

Dependent variable is: Men – Women
R-squared = 56.5% s = 0.212

Variable	Coefficient
Intercept	53.512
Year	−0.0258

a) Is this linear model appropriate for the post-1980 data? Explain.
b) What does the slope say about marriage ages?
c) Explain why it's not reasonable to interpret the y-intercept.

25. Gestation For humans, pregnancy lasts about nine months. In other species of animals, the length of time from conception to birth varies. Is there any evidence that the gestation period is related to the animal's lifespan? The first scatterplot shows *Gestation Period* (in days) vs. *Life Expectancy* (in years) for 18 species of mammals. The highlighted point at the far right represents humans.

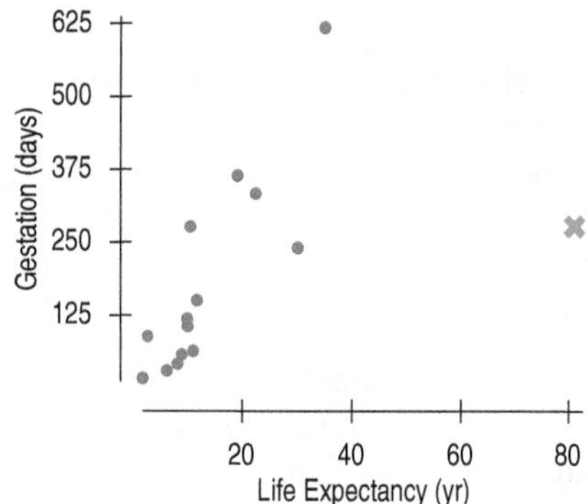

a) For these data, $r = 0.54$, not a very strong relationship. Do you think the association would be stronger or weaker if humans were removed? Explain.
b) Is there reasonable justification for removing humans from the data set? Explain.
c) Here are the scatterplot and regression analysis for the 17 nonhuman species. Comment on the strength of the association.

Dependent variable is: Gestation
R-squared = 72.2%

Variable	Coefficient
Constant	−39.5172
LifExp	15.4980

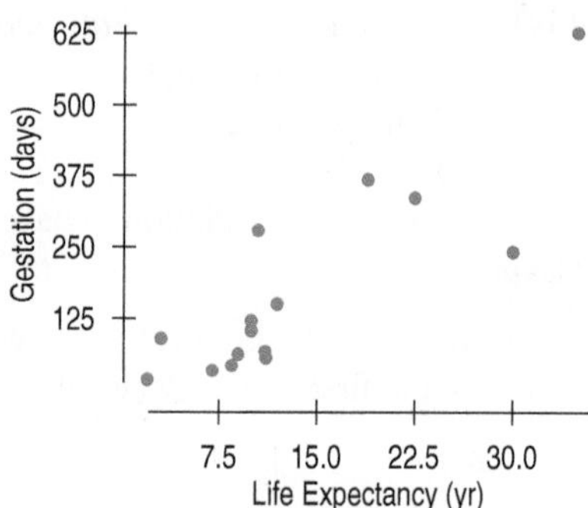

d) Interpret the slope of the line.
e) Some species of monkeys have a life expectancy of about 20 years. Estimate the expected gestation period of one of these monkeys.

26. Swim the lake 2010 People swam across Lake Ontario 48 times between 1974 and 2010 (www.soloswims.com). We might be interested in whether they are getting any faster or slower. Here is the regression of the crossing *Times* (minutes) against the *Year* of the crossing and the residuals plot:

Dependent variable is: Time
R-squared = 2.0% s = 449.9

Variable	Coefficient
Intercept	−9943.083
Year	5.64227

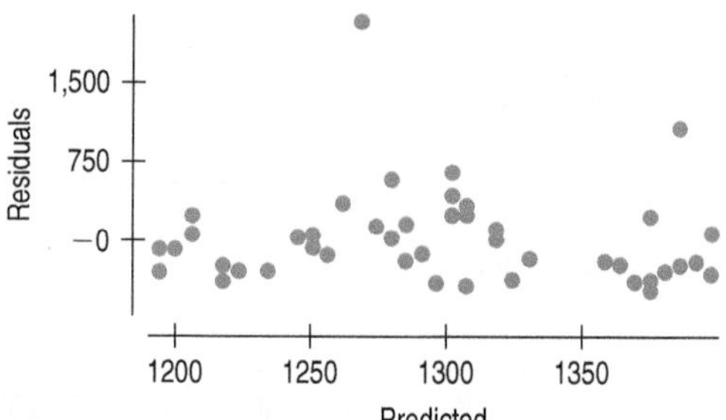

a) What does the R^2 mean for this regression?
b) Are the swimmers getting faster or slower? Explain.
c) The outlier seen in the residuals plot is a crossing by Vicki Keith in 1987 in which she swam a round trip, north to south, and then back again. Clearly, this swim doesn't belong with the others. Would removing it change the model a lot? Explain.

27. Elephants and hippos We removed humans from the scatterplot in Exercise 25 because our species was an outlier in life expectancy. The resulting scatterplot shows two points that now may be of concern. The point in the upper right corner of this scatterplot is for elephants, and the other point at the far right is for hippos.

a) By removing one of these points, we could make the association appear to be stronger. Which point? Explain.

b) Would the slope of the line increase or decrease?
c) Should we just keep removing animals to increase the strength of the model? Explain.
d) If we remove elephants from the scatterplot, the slope of the regression line becomes 11.6 days per year. Do you think elephants were an influential point? Explain.

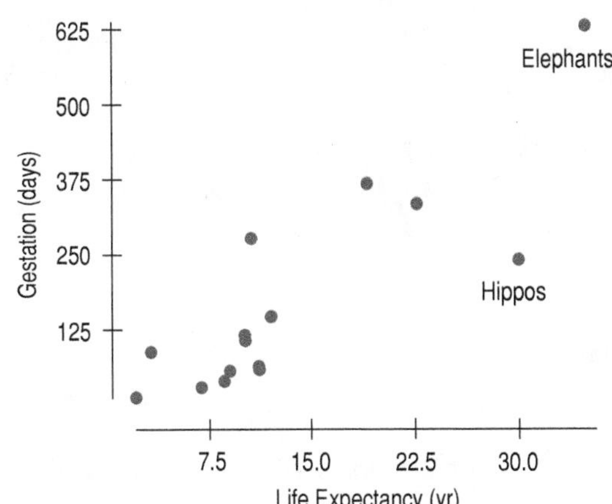

28. Another swim 2010 In Exercise 26, we saw that Vicki Keith's round-trip swim of Lake Ontario was an obvious outlier among the other one-way times. Here is the new regression after this unusual point is removed:

Dependent variable is: Time
R-Squared = 6.0% s = 326.1

Variable	Coefficient
Intercept	−13123.2
Year	7.21677

a) In this new model, the value of s_e is smaller. Explain what that means in this context.
b) Now would you be willing to say that the Lake Ontario swimmers are getting faster (or slower)?

29. Marriage age 2010 revisited Suppose you wanted to predict the trend in marriage age for American women into the early part of this century.
a) How could you use the data graphed in Exercise 1 to get a good prediction? Marriage ages in selected years starting in 1900 are listed below. Use all or part of these data to create an appropriate model for predicting the average age at which women will first marry in 2020.

1900–1950 (10-yr intervals):	21.9, 21.6, 21.2, 21.3, 21.5, 20.3
1955–2010 (5-yr intervals):	20.3, 20.3, 20.6, 20.8, 21.1, 22.0, 23.3, 23.9, 24.5, 25.1, 25.3, 26.1

b) How much faith do you place in this prediction? Explain.

c) Do you think your model would produce an accurate prediction about your grandchildren, say, 50 years from now? Explain.

30. South American Immigrants Below are data on immigration to Canada from South America compared with total immigrants, by year:

Year	Total immigrants	South America
1986	99 343	6 699
1987	152 031	10 798
1988	161 534	7 237
1989	191 516	8 663
1990	216 424	8 986
1991	232 776	10 674
1992	254 856	10 471
1993	256 754	9 616
1994	224 395	7 936
1995	212 875	7 545
1996	226 061	6 114
1997	216 034	5 689
1998	174 184	4 971
1999	189 971	5 610
2000	227 429	6 782
2001	250 638	8 510
2002	229 049	8 825
2003	221 349	11 070
2004	235 824	12 292
2005	262 240	13 720
2006	251 644	13 423
2007	236 754	12 899

Source: Adapted from Statistics Canada, CANSIM table 051-0006, June 30, 2013.

a) Fit a model to the data to predict the annual percentage of total immigrants who are from South America in future years. You may ignore any number of earlier years, and start your fit at the year of your choice. (First you need to compute percents from the last two columns shown).
b) Predict the percent of South American immigrants in 2012 using your model. The percent was in fact 3.8. What was the error of your prediction? Are you surprised?
c) In fact, the percents in 2008–2012 were 5.1, 5.2, 4.4, 4.1, 3.8, respectively. Add these points to your data plot, and now draw a (straight or curving) line out from 2007 or earlier to 2018 indicating where you think things are headed. What do you predict for 2018? Can you see the future? (Good luck! Check back at www.statcan.gc.ca or see our 3rd Canadian edition to see how well you did.)

31. Life expectancy 2010 Data for 26 Western Hemisphere countries can be used to examine the association between life expectancy and the birth rate (number of births per 1000 population).

Country	Birth Rate (births/ 1000 population)	Life Expectancy
Argentina	18	77
Bahamas, The	16	70
Barbados	13	74
Belize	27	68
Bolivia	26	67
Canada	10	81
Chile	15	77
Colombia	18	74
Costa Rica	17	78
Dominican Republic	22	74
Ecuador	21	75
El Salvador	25	72
Guatemala	28	70
Honduras	26	70
Jamaica	20	74
Mexico	20	76
Nicaragua	23	72
Panama	20	77
Paraguay	28	76
Puerto Rico	12	79
United States	14	78
Uruguay	14	76
Venezuela	21	74
Virgin Islands	12	79

a) Create a scatterplot relating these two variables and describe the association.
b) Find the equation of the regression line.
c) Interpret the value of R^2.
d) Make a plot of the residuals. Are any countries unusual?
e) Is the line an appropriate model?
f) If you conclude that there is an outlier, set it aside and re-compute the regression.
g) If government leaders want to increase life expectancy, should they encourage women to have fewer children? Explain.

Ⓣ **32. Tour de France 2012** We met the Tour de France data set in Chapter 1 (in Just Checking). One hundred years ago, the fastest rider finished the course at an average speed of about 25.3 kph. In 2005, a doped-up Lance Armstrong averaged 41.65 kph for the fastest average speed in history.
a) Make a scatterplot of *Avg Speed* against *Year*. Describe the relationship of *Avg Speed* by *Year*, being careful to point out any unusual features in the plot.
b) Find the regression equation of *Avg Speed* on *Year*.
c) Are the conditions for regression met? Comment.

Ⓣ **33. Rich getting richer 2011** Below are annual percentages of Canadian families that earned over $150 000 (in constant 2011 dollars) from 1986–2011. This percentage has gone up, but alas, we're not all getting richer! During the same period, the percentage earning under $40 000 increased from 41.3% in 1986 to 41.9% in 2011, partly attributable to an increase in part-time workers.

Year	over $150 000 (%)	Year	over $150 000 (%)
1986	3.7	1999	4.5
1987	4.1	2000	5
1988	4.5	2001	5.1
1989	4.8	2002	5.3
1990	4.5	2003	5.4
1991	4.2	2004	5.6
1992	4.1	2005	5.9
1993	3.4	2006	6.4
1994	3.7	2007	6.8
1995	3.7	2008	7.1
1996	3.9	2009	7.3
1997	4	2010	7.2
1998	4.5	2011	7.7

Source: Adapted from Statistics Canada, CANSIM table 202-0408 (accessed June 30, 2013).

a) Plot the data appropriately and describe the relationship.
b) Fit a straight line to the data. What percent of the variation in the percentage earning over $150 000 is explained by the straight line relationship?
c) For the year 1997, what is the predicted percent, using the linear model, and what is the residual for that observation? Does this model predict too high or too low for years around 1997?
d) Now use your best judgment to create a model from these data that you would consider using for prediction into some future years (you may ignore some earlier years and start your fit where you choose). What is R^2 for your model? Use it to predict the percentage earning over $150 000 in 2013 and in 2040, and comment on the reliability of your two predictions.

Ⓣ **34. Second stage 2012** Look once more at the data from the Tour de France. In Exercise 32, we looked at the whole history of the race, but now let's consider just the post-World War II era.
a) Find the regression of *Avg Speed* by *Year* only for years from 1947 to the present. Are the conditions for regression met?
b) Interpret the slope.
c) In 1979, Bernard Hinault averaged 39.8 kph, while in 2005 Lance Armstrong averaged 41.65 kph. Armstrong's drug use may or may not have been truly exceptional, but which of these two *racing speeds* would you consider to be the more exceptional? And why?

Ⓣ **35. Gender gap 2011** Below is the difference in median income, for full-time, full-year workers, between males and females in Canada, by year (in constant 2011 dollars):

Year	Difference in Median Earnings (Male – Female)
1976	20 500
1977	20 600
1978	20 400
1979	19 500
1980	19 900
1981	18 800
1982	18 800
1983	17 500
1984	17 800
1985	18 300
1986	17 000
1987	17 600
1988	18 500
1989	17 600
1990	16 100
1991	15 200
1992	14 500
1993	14 000
1994	14 800
1995	13 400
1996	13 200
1997	13 500
1998	13 300
1999	15 100
2000	14 500
2001	14 000
2002	13 600
2003	14 100
2004	13 700
2005	13 500
2006	12 800
2007	13 700
2008	12 800
2009	11 700
2010	12 200
2011	12 600

Source: Adapted from Statistics Canada, CANSIM table 202-0101, June 30, 2013.

a) Plot the difference in median earnings versus year. Describe the relationship.
b) Fit a straight line and assess your model by examining both R^2 and a plot of residuals versus year.
c) Predict the difference in the year 2070 using your fitted straight-line model from part b). Do you think this is likely to be high or low?
d) Do you think a quadratic curve might be a better model? Below is a scatterplot of residuals after fitting a quadratic curve by the same least squares principle:

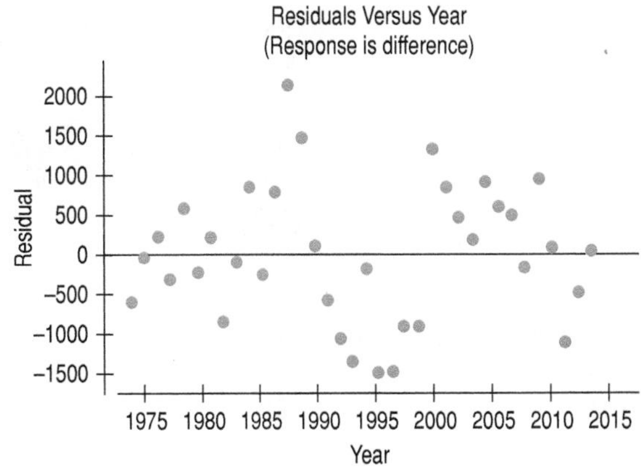

Compare this residual plot with your plot from part b). What does it suggest about appropriate choice of model?

T **36. Interprovincial migration 2012** In Chapter 6 exercises, we examined data showing the net interprovincial migration each year for Ontario and Alberta (number arriving minus number leaving). Below are the first few and last few lines of the data file available at www.pearsoncanada.ca/deveaux.

Year	Ontario	Alberta
1978/1979	–4 325	33 426
1979/1980	–22 362	41 435
1980/1981	–33 247	44 250
:	:	:
:	:	:
2009/2010	–4 662	–3 271
2010/2011	–4 007	8 443
2011/2012	–8 091	–8 170

Source: Adapted from Statistics Canada. CANSIM table 051-0004, March 31, 2013.

a) Plot Alberta net migration versus Ontario net migration for each year.
b) Fit a straight line model to the data for predicting Alberta migration from Ontario migration. What percent of the variation in Alberta migration numbers is explained by Ontario migration?
c) Examine the residuals. Are they approximately Normal? Is the plot of residuals versus predicted values (or the explanatory variable) fairly random-looking?
d) Plot residuals versus year. Is there a pattern? What do you learn from this plot? Does the pattern you see have anything to do with the Alberta economy and its dependence on oil prices? Do you think oil prices might be a better predictor of Alberta migration numbers?

T **37. Math and gender 2009** Below are mean PISA (Programme for International Student Assessment) math scores for samples of 15-year-old male and female students from a number of randomly selected schools in each of various OECD and other industrialized nations (4000–10 000 per country).

Country	Male mean	Female mean
OECD		
Australia	519	509
Austria	506	486
Belgium	526	504
Canada	533	521
Chile	431	410
Czech Republic	495	490
Denmark	511	495
Estonia	516	508
Finland	542	539
France	505	489
Germany	520	505
Greece	473	459
Hungary	496	484
Iceland	508	505
Ireland	491	483
Israel	451	443
Italy	490	475
Japan	534	524
Korea	548	544
Luxembourg	499	479
Mexico	425	412
Netherlands	534	517
New Zealand	523	515
Norway	500	495
Poland	497	493
Portugal	493	481
Slovak Republic	498	495
Slovenia	502	501
Spain	493	474
Sweden	493	495
Switzerland	544	524
Turkey	451	440
United Kingdom	503	482
United States	497	477
OECD total	496	481
OECD average	501	490
Partners		
Albania	372	383
Argentina	394	383
Azerbaijan	435	427
Brazil	394	379
Bulgaria	426	430
Colombia	398	366
Croatia	465	454
Dubai (UAE)	454	451
Hong Kong-China	561	547
Indonesia	371	372
Jordan	386	387
Kazakhstan	405	405
Kyrgyzstan	328	334
Latvia	483	481
Liechtenstein	547	523
Lithuania	474	480
Macao-China	531	520
Montenegro	408	396
Panama	362	357
Peru	374	356
Qatar	366	371
Romania	429	425
Russian Federation	469	467
Serbia	448	437
Shanghai-China	599	601
Singapore	565	559
Chinese Taipei	546	541
Thailand	421	417
Trinidad and Tobago	410	418
Tunisia	378	366
Uruguay	433	421

Source: Based on data from OECD PISA 2009 online database, http://pisa2009.acer.edu.au, accessed on Feb 22, 2013.

a) Plot male mean score versus female mean score, and superimpose the least-squares regression line on the plot (for prediction of male means from female means).
b) Plot residuals from the regression versus female means or versus predicted (fitted) values. What do you learn?
c) Plot a histogram and Normal probability plot of the residuals. Are they approximately Normally distributed?
d) Plot residuals versus row number. Are the OECD countries less variable than the non-OECD countries?
e) Which two countries have the greatest leverage on the fit? Explain in simple words what this means.
f) The slope does not differ much from 1.0. What would a slope of 1.0 indicate about the nature of the relationship? If we fitted a model with the slope fixed at 1.0, what prediction equation would you expect to get? (Hint: Find the mean scores for males and females.)
g) If we examined only those countries with means over 500 for both sexes, what would happen to the correlation? Why?

T **38. Mandarin or Cantonese 2011** Below are data on Canadian cities, showing mother-tongue counts from the 2011 Census.

Urban Centre	Cantonese	Mandarin
Abbotsford (BC)	310	400
Barrie (ON)	215	295
Brantford (ON)	125	90
Calgary (AB)	20 175	11 380
Edmonton (AB)	13 595	7 295
Guelph (ON)	770	650
Halifax (NS)	460	805
Hamilton (ON)	1 975	2 215

Kelowna (BC)	235	210
Kingston (ON)	335	395
Kitchener-Cambridge-Waterloo (ON)	1 390	3 060
London (ON)	1 165	1 415
Moncton (NB)	70	75
Montréal (QC)	10 515	10 180
Oshawa (ON)	730	645
Ottawa-Gatineau (ON-QC)	6 570	7 275
Peterborough (ON)	90	110
Québec (QC)	130	270
Regina (SK)	560	505
Saguenay (QC)	0	15
Saint John (NB)	45	125
Saskatoon (SK)	865	790
Sherbrooke (QC)	20	65
St. Catharines-Niagara (ON)	640	555
St. John's (NL)	105	165
Sudbury (ON)	140	90
Thunder Bay (ON)	75	75
Toronto (ON)	170 485	100 045
Trois-Rivières (QC)	10	20
Vancouver (BC)	128 110	90 190
Victoria (BC)	2 930	1 945
Windsor (ON)	1 200	1 085
Winnipeg (MB)	3 220	2 080

Source: Statistics Canada, 2011 Census of Population and Statistics Canada, catalogue no. 98-314-XCB. Last modified: 2013-02-13.

a) Plot Cantonese speaker count versus Mandarin speaker count. Would you fit a regression line for predicting Cantonese speakers from Mandarin speakers? Explain.

b) Take the log (base 10) of both columns of numbers using your computer or calculator (why do you lose one city?). Now would you use regression for prediction purposes? What percentage of the variation in Cantonese counts is explained by the linear association with Mandarin counts?
And stick to this log-log model for the remaining parts of this exercise.

c) Predict the Cantonese log count in a city where the Mandarin log count is 4.0. What would you predict for the actual Cantonese count?

d) What is the residual for the city of Vancouver?

e) Which city has the most surprising number of Cantonese speakers (considering the number of resident Mandarin speakers)? Give the residual.

f) Which cities have the greatest leverage on the fit?

39. Bigger and bigger According to the UK National Sizing Survey, women's average body size measurements have increased over time as follows:

Year	Bust size (in)	Waist size (in)	Hips size (in)
1920	32	20	32
1940	33	21	33
1960	34	24	33
1980	35	26	37
2000	36	28	38

Source: P. Treleaven, 2007, "How to Fit Into Your Clothes," *Significance 4*(3), 113–117.

a) Compute all correlations among all pairs of the four variables.

b) Why are the correlations so high?

c) Predict the average waist size for UK women in 1950 and in 2080. Comment on the reliability of your predictions.

d) Predict average waist size for UK women if their average bust size increases to 44. Comment on the reliability of your prediction.

40. More Mandarin 2011 Mandarin speakers in Canada are increasing rapidly. Here are 2001 and 2011 Census figures on Mandarin mother-tongue residents in some urban areas:

Urban Area	2001 Count	2011 Count
Abbotsford (BC)	105	400
Calgary (AB)	3 235	11 380
Edmonton (AB)	2 230	7 295
Guelph (ON)	290	650
Halifax (NS)	170	805
Hamilton (ON)	740	2 215
Kingston (ON)	190	395
Kitchener-Cambridge-Waterloo (ON)	645	3 060
London (ON)	330	1 415
Montréal (QC)	3 935	10 180
Oshawa (ON)	95	645
Ottawa-Gatineau (ON-QC)	3 245	7 275
Regina (SK)	130	505
Saskatoon (SK)	250	790
St. Catharines - Niagara (ON)	160	555
Toronto (ON)	35 315	100 045
Vancouver (BC)	47 165	90 190
Victoria (BC)	890	1 945
Windsor (ON)	425	1 085
Winnipeg (MB)	685	2 080

Source: Statistics Canada, 2011 Census of Population and Statistics Canada, catalogue no. 98-314-XCB. Last modified: 2013-02-13.

a) Plot the 2011 count against the 2001 count. Would you fit a regression line for predicting 2011 counts from 2001 counts? Explain.

b) Take the log (base 10) of both columns of numbers using your computer or calculator. Re-plot the data. Now would you use regression for prediction purposes? Stick to this log-log model for the remaining parts of this exercise.
c) Predict the 2011 log-count in a city where the log of the 2001 count is 4.0. What would you predict for the actual 2011 count?
d) Which city's 2011 count is the most surprising, and why? Give the residual for this city. How would the fitted line change if you removed this city?
e) Which cities have the greatest leverage on the fit?
f) Take anti-logs of both sides of your fitted equation, i.e., rewrite it without any logs in the equation. If the slope of the fitted log-log model happened to be 1.00, what simple interpretation would you now get of this prediction equation?
g) What is the correlation between the two log-counts? What is one obvious reason (for example, a lurking variable) for such a high correlation? Suggest another useful way of defining the two variables that would produce a smaller correlation. (In real life studies, there are always issues about how to best define and measure the relevant variables.)

41. Census at school The International *CensusAtSchool* Project collects data on primary and secondary school students from various countries, including Canada. We selected a random sample of 111 Canadian secondary school students, age 14 and over, from 2006–2007. Below are the first four rows of the 111 row worksheet of data. See Chapter 6, Question 41, for more information. All measurements are in centimetres and were taken in the classroom. Middlefinger is distance from wrist to tip of middle finger. Wristbone is distance from wrist bone to elbow. Armspan is distance from tip of left hand to tip of right hand with arms spread wide. Foot is length of right foot.

Gender	Age	Height	Armspan	Wrist-bone	Middle-finger	Foot
Boy	14	166	169	25	19	26
Boy	14	160	170	27	19	28
Boy	14	167	163	26	20	26
Girl	16	165	165	24	17	22
⋮						

Source: The Royal Statistical Society Centre for Statistical Education, CensusAtSchool project.

Below is a table of all the pairwise correlations:

	Age	Wrist-bone	Middle-finger	Foot	Armspan
Wristbone	−0.008				
Middlefinger	0.059	0.360			
Foot	0.188	0.436	0.461		
Armspan	0.219	0.449	0.565	0.548	
Height	0.272	0.397	0.560	0.542	0.781

a) Which variable appears to be the single best predictor of middlefinger? What percentage of the variation in middlefinger measurements can be explained by a linear association with this predictor variable?
b) Which variable appears to be the single best predictor of armspan? What percentage of the variation in armspan measurements can be explained by a linear association with this predictor variable?

Below is a plot of armspan versus height, including information about gender.

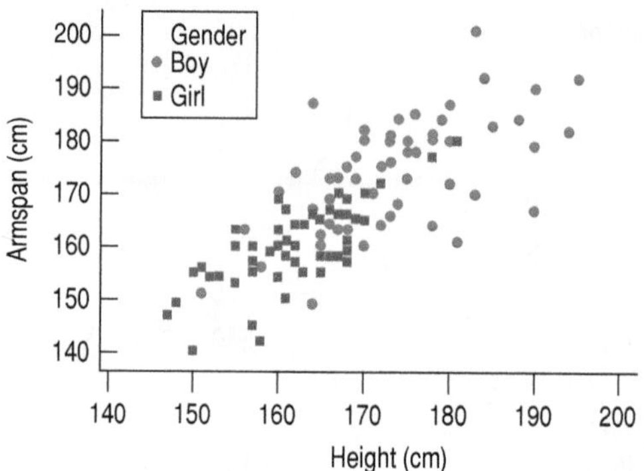

c) Regress armspan on height and give the prediction equation. Interpret the slope.
d) Examine a histogram and Normal probability plot of the residuals from the regression above and describe the distribution of the residuals.
e) Plot the residuals against height or the fitted (predicted) values. What do you learn from this plot?
f) Which child has the most unusually long or short armspan considering his or her height? (Give the row number and gender.) How much longer or shorter than the (appropriate) average is it?
g) It is not clear from the plot above whether gender is an important variable in understanding the relationship between armspan and height. Plot the residuals from the regression above versus gender. Is there any pattern? For either gender, is there a tendency for the model to over-predict or under-predict armspan? If so, what would be your modelling recommendation?

42. Census at school, boys versus girls Separate the *CensusAtSchool* data in the previous exercise according to gender, and run the two regressions of armspan on height, once for each gender.
a) Describe the differences suggested by the two fitted models.
b) Estimate the difference in armspan between a boy and girl if each of them is 160 cm tall. What if each of them is 170 cm tall?
c) For the girls, which observation appears to have the most leverage? Do you think it is influential? Check to confirm your suspicion.
d) Which girl has the most unusually long or short armspan, considering her height? How much longer or shorter than the (appropriate) average is it?

e) Predict the armspan of a boy who is 130 cm tall. Predict the armspan of a boy who is 185 cm tall. Is either prediction more reliable than the other? Why?

43. Planet distances and years 2006 At a meeting of the International Astronomical Union (IAU) in Prague in 2006, Pluto was determined not to be a planet, but rather the largest member of the Kuiper belt of icy objects. Let's examine some facts. The table shows the nine sun-orbiting objects formerly known as planets:

Planet	Position Number	Distance from Sun (million miles)	Length of Year (Earth years)
Mercury	1	36	0.24
Venus	2	67	0.61
Earth	3	93	1.00
Mars	4	142	1.88
Jupiter	5	484	11.86
Saturn	6	887	29.46
Uranus	7	1784	84.07
Neptune	8	2796	164.82
Pluto	9	3707	247.68

a) Plot the length of the year against the distance from the sun. Describe the shape of your plot.
b) Re-express one or both variables to straighten the plot. (square root, reciprocal, or log). Use the re-expressed data to create a model describing the length of a planet's year based on its distance from the sun.
c) Comment on how well your model fits the data.

44. Asian immigrants 2012 Below are data on immigration to Canada from countries in Asia:

Year	Total immigrants	Total Asian immigrants	Percentage from Asia
1963	93 151	3 912	4.2
1964	112 606	6 526	5.8
1965	146 758	11 684	8
1966	194 743	14 238	7.3
1967	222 876	21 247	9.5
1968	183 974	22 076	12
1969	161 531	23 706	14.7
1970	147 713	21 451	14.5
1971	121 900	22 369	18.4
1972	122 006	23 325	19.1
1973	184 200	43 193	23.4
1974	218 465	50 566	23.1
1975	187 881	47 382	25.2
1976	149 429	44 091	29.5
1977	114 914	31 380	27.3
1978	86 313	24 029	27.8
1979	112 096	46 871	41.8
1980	143 498	71 793	50
1981	128 794	48 898	38
1982	121 331	41 749	34.4
1983	89 377	36 998	41.4
1984	88 599	42 091	47.5
1985	84 339	38 614	45.8
1986	99 343	41 650	41.9
1987	152 031	67 306	44.3
1988	161 534	80 938	50.1
1989	191 516	93 026	48.6
1990	216 424	112 902	52.2
1991	232 776	121 000	52
1992	254 856	140 385	55.1
1993	256 754	147 847	57.6
1994	224 395	141 916	63.2
1995	212 875	129 435	60.8
1996	226 061	144 397	63.9
1997	216 034	138 319	64
1998	174 184	101 492	58.3
1999	189 971	113 086	59.5
2000	227 429	140 178	61.6
2001	250 571	155 682	62.1
2002	229 071	141 358	61.7
2003	221 352	132 851	60
2004	235 824	135 658	57.5
2005	262 239	159 968	61
2006	251 652	149 523	59.4
2007	236 762	133 383	56.3
2008	247247	140961	57.0
2009	252177	141196	56.0
2010	280682	164408	58.6
2011	248746	146951	59.1
2012	257730	149235	57.9

Source: Adapted from Statistics Canada, CANSIM database, http://cansim2.statcan.gc.ca, table 051-0006, July 2, 2013.

a) We want to predict the percentage of total immigrants who are from Asia based on the year. You may use two straight line models (which we call *piecewise linear*) depending on the time period. Draw your fitted models on the scatterplot.
b) Predict the percent of Asian immigrants in 1955 using your model. The percent was in fact 3.3. What was the error of your prediction? Are you surprised?
c) Use your model to predict Asian immigration for 2027. On your graph from part a), draw a (straight or not straight) line out from 2007 to 2027 indicating where you think things are headed. Can you see the future? (Good luck! Check www.statcan.gc.ca/ or our 6th Canadian edition to see how well you did.)

45. Placental mammals In the introductory example in Chapter 6, we plotted (average) neonatal brain weight versus (average) gestation period for a number of placental mammal species. We are interested in finding a good model for predicting the neonatal brain weight of a species from its gestation period. Below are the data for a sample of 23 species.

Brain weight (g)	Gestation (days)	Brain weight (g)	Gestation (days)
8.78	135	129.00	270
4.00	90	128.00	230
4.00	135	227.00	265
30.80	139	335.00	270
29.00	180	6.80	63
64.00	140	3.82	52
58.00	175	7.73	77
75.00	180	3.34	60
53.00	180	35.30	92
33.50	195	49.50	108
43.00	168	30.00	104
65.00	210		

Source: J.A. Witmer, 1992, *Data Analysis: An Introduction.* Table 4-1 "Brain Weight and Gestation Period of Various Mammals" p. 50, © 1992 Prentice-Hall, Inc. Reproduced by permission of Pearson Education, Inc.

a) Plot neonatal brain weight versus gestation. Fit a straight line, show it on the data scatterplot, and report R^2. Plot the residuals versus the predicted values. What do you learn?

b) Compute the logarithms of the neonatal brain weights, and plot this variable versus gestation. Fit a straight line, show it on the scatterplot, and report R^2. Plot residuals versus predicted values. Compare this model with the one in part a). Which model do you prefer and why? From your plots, do you see any possible problem(s) with the log-transformed model?

c) Take the square roots of brain weights, plot them versus gestation, and fit a straight line. Compare your results with the log transformation of part b). Which transformation do you think works best and why?

d) Using the log-transformed model of part b), predict the neonatal brain weight of an individual species whose gestation period is 200 days (start by predicting the log brain weight).

e) Using the square root model of part c), predict the neonatal brain weight of an individual species whose gestation period is 200 days.

46. Placental mammals again Continuing with the previous exercise, the species listed in the last 7 rows of the table were all carnivores, while primates occupy the first 16 rows. Enter into a new column in your data worksheet—"p" or "c" for each species, to denote whether primate or carnivore.

a) Plot brain weight versus gestation, using different symbols for the primates and carnivores. Repeat using log of brain weight. What do you learn from these plots?

b) Use the logarithm of brain weight. Fit a model to predict brain weight for the primates only. Fit another model using only the carnivore data. Does fitting separate models make sense from a statistical standpoint? From a practical standpoint? Draw by hand the two fitted straight lines on the scatterplot with all the (log-transformed) data. How do carnivores differ from primates?

c) Are humans (row 16) an outlier?

Just Checking ANSWERS

1. Not high leverage, not influential, large residual
2. High leverage, not influential, small residual
3. High leverage, influential, not large residual

MathXL

MyStatLab

Go to MathXL at www.mathxl.com or MyStatLab at www.mystatlab.com. You can practise exercises for this chapter as often as you want. The guided solutions will help you find answers step by step. You'll find a personalized study plan available to you too!

part II Review

Exploring Relationships Between Variables

Quick Review

You have now survived your second major unit of Statistics. Here's a brief summary of the key concepts and skills:

- We treat data two ways: as categorical and as quantitative.
- To explore relationships in categorical data, check out Chapter 3.
- To explore relationships in quantitative data:
  - Make a picture. Use a scatterplot. Put the explanatory variable on the x-axis and the response variable on the y-axis. (If there is a categorical variable present, forming distinct groups, show this in your scatterplot using different symbols for each group.)
  - Describe the association between two quantitative variables in terms of direction, form, and strength.
  - The amount of scatter determines the strength of the association.
  - If, as one variable increases so does the other, the association is positive. If one increases as the other decreases, it's negative.
  - If the form of the association is linear, calculate a correlation to measure its strength numerically, and possibly do a regression analysis to model it.
  - Correlations closer to -1 or $+1$ indicate stronger linear associations. Correlations near 0 indicate weak linear relationships, but other forms of association may still be present.
  - The line of best fit is also called the least squares regression line because it minimizes the sum of the squared residuals.
  - The regression line predicts values of the response variable from values of the explanatory variable.
  - A residual is the difference between the true value of the response variable and the value predicted by the regression model.
  - The slope of the line is a rate of change, best described in "y-units" per "x-unit."
  - R^2 gives the fraction of the variation in the response variable that is accounted for by the model.
  - The standard deviation of the residuals measures the amount of scatter around the line.
  - Outliers and influential points can distort any of our models.
  - If you see a pattern (curve) in the plot of residuals versus explanatory variable, your chosen model is not appropriate; try a different model. You may, for example, be able to straighten the relationship by re-expressing one of the variables.
  - To straighten bent relationships or to ameliorate trends in dispersion, re-express the data using logarithms or a power (squares, square roots, reciprocals, etc.).
  - Always remember that an association is not necessarily an indication that one of the variables causes the other.

Need more help with some of this? Try rereading some sections of Chapters 6 through 8. And go on to the next page for more opportunities to review these concepts and skills.

"One must learn by doing the thing; though you think you know it, you have no certainty until you try."

—Sophocles (495–406 B.C.E.)

Review Exercises

1. **College** Every year, *US News and World Report* publishes a special issue on many U.S. colleges and universities. The scatterplots below have *Student/Faculty Ratio* (number of students per faculty member) for the colleges and universities on the y-axes plotted against 4 other variables. The correct correlations for these scatterplots appear in the list of six possible correlations below. Match them.

−0.98 −0.71 −0.51 0.09 0.23 0.69

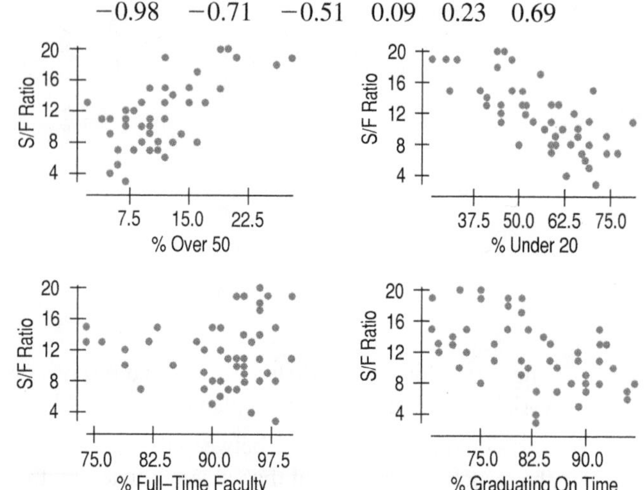

2. **Togetherness** Are good grades in high school associated with family togetherness? A random sample of 142 high school students was asked how many meals per week their families ate together. Their responses produced a mean of 3.78 meals per week, with a standard deviation of 2.2. Researchers then matched these responses against the students' grade point averages (GPAs). The scatterplot appeared to be reasonably linear, so they created a line of regression. No apparent pattern emerged in the residuals plot. The equation of the line was $\widehat{GPA} = 2.73 + 0.11\ Meals$.
   a) Interpret the y-intercept in this context.
   b) Interpret the slope in this context.
   c) What was the mean GPA for the students in this study?
   d) If a student in this study had a negative residual, what did that mean?
   e) Upon hearing of this study, a counsellor recommended that parents who want to improve their children's grades should get the family to eat together more often. Do you agree with this interpretation? Explain.

T 3. **Vineyards** Here are the scatterplot and regression analysis for *Case Prices* of 36 wines from vineyards in the Finger Lakes region of New York State and the *Ages* of the vineyards.

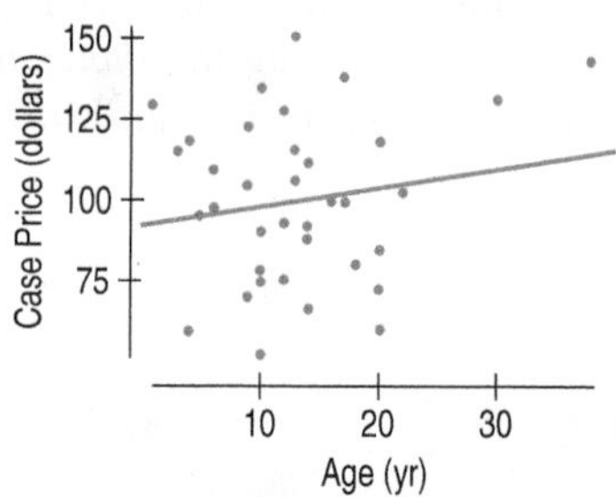

Dependent variable is: Case Price
R-squared = 2.7%

Variable	Coefficient
Constant	92.7650
Age	0.567284

   a) Does it appear that vineyards in business longer get higher prices for their wines? Explain.
   b) What does this analysis tell us about vineyards in the rest of the world?
   c) Write the regression equation.
   d) Explain why that equation is essentially useless.

T 4. **Vineyards again** Instead of *Age*, perhaps the *Size* of the vineyard (in acres) is associated with the price of the wines. Look at the scatterplot:

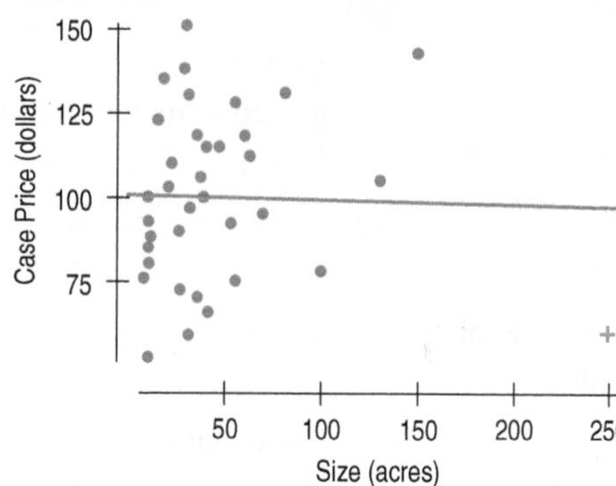

   a) Do you see any evidence of an association?
   b) What concern do you have about this scatterplot?
   c) If the red "+" data point is removed, would you expect the correlation to become stronger or weaker? Explain.
   d) If the red "+" data point is removed, would you expect the slope of the line to increase or decrease? Explain.

5. **Lurking variables** Each of the following scatterplots contains two distinct subgroups (+'s for males, o's for females). In each case, estimate the correlation between y and x:
   a) ignoring gender
   b) taking gender into account, i.e., separately by gender

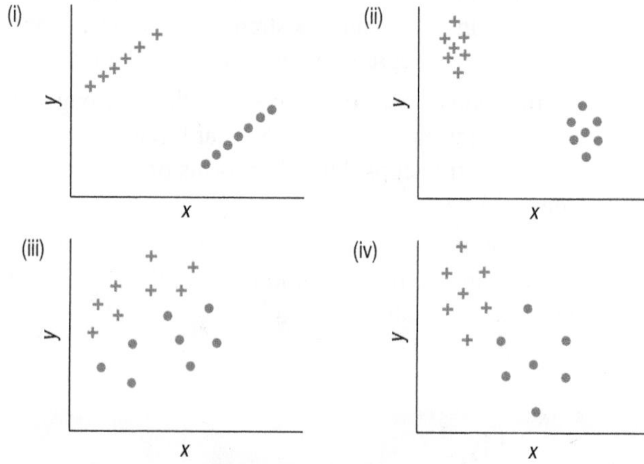

6. **Residuals** Below is a plot of primary school enrolment percentage of females (in the population in the correct age group) versus gross national income per capita for a number of countries. Also shown is the least squares straight-line fit.

a) Show and describe by drawing on the plot and with words what it is that makes the drawn line the "least squares" fit.
b) If we construct the usual graph of residuals versus predictor, what would it look like? Draw it, roughly, but put some numbers on each axis. If we were instead to construct the graph of residuals versus predicted values, how would it compare with the first residual plot?
c) What is the approximate numerical value of the biggest (in magnitude) residual?
d) Would this model predict too high or too low for countries with incomes per capita in the \$10 000 to \$30 000 range? Do you get positive or negative residuals for these countries?

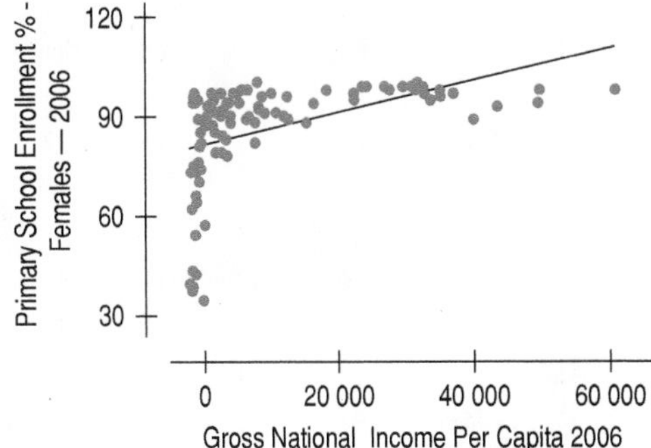

7. Acid rain Biologists studying the effects of acid rain on wildlife collected data from 163 streams in the Adirondack Mountains. They recorded the *pH* (acidity) of the water and the *BCI*, a measure of biological diversity, and they calculated $R^2 = 27\%$. Here's a scatterplot of *BCI* against *pH*:

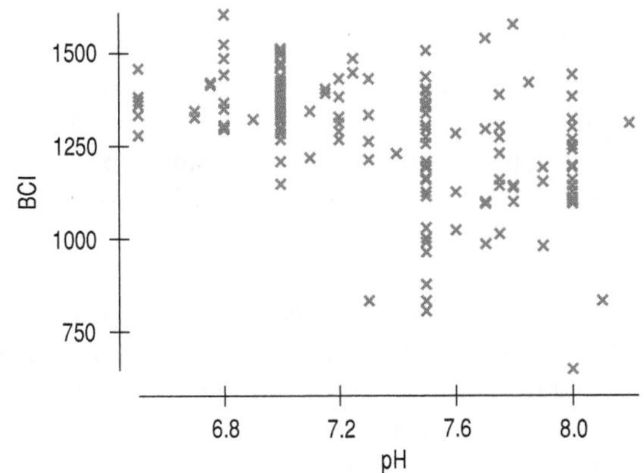

a) What is the correlation between *pH* and *BCI*?
b) Describe the association between these two variables.
c) If a stream has average *pH*, what would you predict about the *BCI*?
d) In a stream where the *pH* is 3 standard deviations above average, what would you predict about the *BCI*?

8. Manatees 2010 Marine biologists warn that the growing number of powerboats registered in Florida threatens the existence of manatees. The data below come from the Florida Fish and Wildlife Conservation Commission (myfwc.com/research/manatee/) and the National Marine Manufacturers Association (www.nmma.org/).

Year	Manatees Killed	Powerboat Registrations (in 1000s)
1982	13	447
1983	21	460
1984	24	481
1985	16	498
1986	24	513
1987	20	512
1988	15	527
1989	34	559
1990	33	585
1992	33	614
1993	39	646
1994	43	675
1995	50	711
1996	47	719
1997	53	716
1998	38	716
1999	35	716
2000	49	735
2001	81	860
2002	95	923
2003	73	940
2004	69	946
2005	79	974
2006	92	988
2007	73	992
2008	90	932
2009	97	949
2010	83	914

a) In this context, which is the explanatory variable?
b) Make a scatterplot of these data and describe the association you see.
c) Find the correlation between *Boat Registrations* and *Manatee Deaths*.
d) Interpret the value of R^2.
e) Does your analysis prove that powerboats are killing manatees?

9. A manatee model 2010 Continue your analysis of the manatee situation from the previous exercise.
a) Create a linear model of the association between *Manatee Deaths* and *Powerboat Registrations*.
b) Interpret the slope of your model.
c) Interpret the *y*-intercept of your model.
d) How accurately did your model predict the high number of manatee deaths in 2010?
e) Which is better for the manatees, positive residuals or negative residuals? Explain.
f) What does your model suggest about the future for the manatee?

10. Grades A Statistics instructor created a linear regression equation to predict students' final exam scores from their midterm exam scores. The regression equation was $\widehat{Fin} = 10 + 0.9\ Mid$.

a) If Susan scored a 70 on the midterm, what did the instructor predict for her score on the final?

b) Susan got an 80 on the final. How big is her residual?

c) Suppose that the standard deviation of the final was 12 points and the standard deviation of the midterm was 10 points. What is the correlation between the two tests?

d) How many points would someone need to score on the midterm to have a predicted final score of 100?

e) Suppose someone scored 100 on the final. Explain why you can't estimate this student's midterm score from the information given.

f) One of the students in the class scored 100 on the midterm but got overconfident, slacked off, and scored only 15 on the final exam. What is the residual for this student?

g) No other student in the class "achieved" such a dramatic turnaround. If the instructor decides not to include this student's scores when constructing a new regression model, will the R^2 value of the regression increase, decrease, or remain the same? Explain briefly.

h) Will the slope of the new line increase or decrease?

11. Traffic Highway planners investigated the relationship between traffic density (number of automobiles per mile) and the average speed (in miles per hour) of the traffic on a moderately large city thoroughfare. The data were collected at the same location at 10 different times over a span of three months. They found a mean traffic density of 68.6 cars per mile (cpm) with a standard deviation of 27.07 cpm. Overall, the cars' average speed was 26.38 mph, with a standard deviation of 9.68 mph. The researchers found the regression line for these data to be $\widehat{Speed} = 50.55 - 0.352\ Density$.

a) What is the value of the correlation coefficient between speed and density?

b) What percent of the variation in average speed is explained by traffic density?

c) Predict the average speed of traffic on the thoroughfare when the traffic density is 50 cpm.

d) What is the value of the residual for a traffic density of 56 cpm with an observed speed of 32.5 mph?

e) The data set initially included the point (density = 125 cpm, speed = 55 mph). This point was considered an outlier and was not included in the analysis. Will the slope increase, decrease, or remain the same if we redo the analysis and include this point?

f) Will the correlation become stronger, weaker, or remain the same if we redo the analysis and include this point (125, 55)?

g) A Canadian member of the research team measured the speed of the cars in kilometres per hour (1 km $\approx$ 0.62 miles) and the traffic density in cars per kilometre. Find the value of his calculated correlation between speed and density.

T **12. Cramming** One Thursday, researchers gave students enrolled in a section of basic Spanish a set of 50 new vocabulary words to memorize. On Friday, the students took a vocabulary test. When they returned to class the following Monday, they were retested—without advance warning. Here are the test scores for the 25 students (where each row gives us paired scores for the same student).

Fri.	Mon.	Fri.	Mon.	Fri.	Mon.
42	36	48	37	39	41
44	44	43	41	46	32
45	46	45	32	37	36
48	38	47	44	40	31
44	40	50	47	41	32
43	38	34	34	48	39
41	37	38	31	37	31
35	31	43	40	36	41
43	32				

a) What is the correlation between *Friday* and *Monday* scores?

b) What does a scatterplot show about the association between the scores?

c) What does it mean for a student to have a positive residual?

d) What would you predict about a student whose *Friday* score was one standard deviation below average?

e) Write the equation of the regression line.

f) Predict the *Monday* score of a student who earned a 40 on Friday.

T **13. Aboriginals rising 2011** The data below (next page) show Aboriginal populations in a sample of Canadian cities.

a) Plot the 2011 percents versus the 2001 percents. Guess the correlation, and confirm your guess.

b) Fit a regression model to predict the 2011 percent from the 2001 percent.

c) Interpret the slope, assuming the intercept is close enough to zero to be considered zero.

d) What distinction is held by Prince Albert? If you set it aside, are there still problems? (Check the residual scatterplot. Does the plot thicken?)

e) Including Prince Albert, take the log (base 10) of both variables and refit the model. Did taking logs help?

f) If a census area had an Aboriginal percent of 10% in 2001, what would you predict for the percent in 2011?

CMA/CA	2001 Number of Aboriginal Persons	2001 Percentage of Population	2011 Number of Aboriginal Persons	2011 Percentage of Population
Prince Albert	11 640	29.2	15 780	38.5
Sault Ste. Marie	5 610	7.2	8 070	10.3
Prince George	7 980	9.4	9 930	12.0
Greater Sudbury	7 385	4.8	13 410	8.5
Saskatoon	20 275	9.1	23 890	9.3
Winnipeg	55 755	8.4	78 415	11.0
Regina	15 685	8.3	19 785	9.5
Kamloops	5 470	6.4	8 265	8.6
Thunder Bay	8 200	6.8	11 675	9.8
Edmonton	40 930	4.4	61 770	5.4
Victoria	8 695	2.8	14 200	4.2
Calgary	21 915	2.3	33 375	2.8
Hamilton	7 270	1.1	11 980	1.7
Vancouver	36 860	1.9	52 375	2.3
London	5 640	1.3	8 475	1.8
Ottawa-Hull	13 485	1.3	30 570	2.5
Toronto	20 300	0.4	36 990	0.7
Montreal	11 085	0.3	26 285	0.7

Source: Statistics Canada: Aboriginal peoples of Canada: A demographic profile, 2003, and Statistics Canada, 2011 National Household Survey, Statistics Canada Catalogue no. 99-011-X2011028.

14. Quebec abortions Below are annual rates of abortion performed on females under the age of 20, per 1000 Quebec females (aged 15–49) over the period 1974–2005 (data not available for 1991).

Year	Rate	Year	Rate	Year	Rate
1974	2.4	1985	9.9	1996	18.7
1975	2.9	1986	11.1	1997	18.7
1976	3.7	1987	10.7	1998	21.0
1977	3.9	1988	12.2	1999	21.8
1978	6.1	1989	14.1	2000	21.8
1979	7.6	1990	16.3	2001	20.9
1980	8.8	1991	*	2002	20.4
1981	8.7	1992	17.3	2003	19.4
1982	9.4	1993	17.3	2004	19.7
1983	8.4	1994	19.1	2005	18.6
1984	8.8	1995	16.3		

Source: Adapted from Statistics Canada CANSIM database, table 106-9002, (accessed: 2013-07-05).

a) Plot the rate versus the year. Describe the trend.
b) Calculate the correlation.
c) Fit a straight line model, and use it to predict the abortion rate in 2007.
d) Examine the plot of residuals versus year. What does this plot tell you? Are you likely to over- or underpredict the rate in 2007?
e) Is there another approach you might take to constructing a model for the relationship between abortion rates and year in Quebec (say, fitting two different models, depending on the time period)? Try it and give your best prediction for the year 2007.

15. Cars with horses Can we predict the horsepower of the engine that manufacturers will put in a car by knowing the weight of the car? Here are the regression analysis and residual plot:

Dependent variable is: Horsepower
R-squared = 84.1%

Variable	Coefficient
Intercept	3.49834
Weight	34.3144

Residuals (horsepower): 12.5, 0.0, –12.5
Predicted (horsepower): 80, 100, 120, 140

a) Write the equation of the regression line.
b) Do you think the car's weight is measured in pounds or thousands of pounds? Explain.
c) Do you think this linear model is appropriate? Explain.

d) The highest point in the residuals plot, representing a residual of 22.5 horsepower, is for a Chevy weighing 2595 pounds. How much horsepower does this car have?

16. Colourblind Although some women are colourblind, this condition is found primarily in men. Why is it wrong to say there's a strong correlation between sex and colourblindness?

T **17. Old Faithful** There is evidence that eruptions of Old Faithful can best be predicted by knowing the duration of the previous eruption.

a) Describe what you see in the scatterplot of *Intervals* between eruptions vs. *Duration* of the previous eruption.

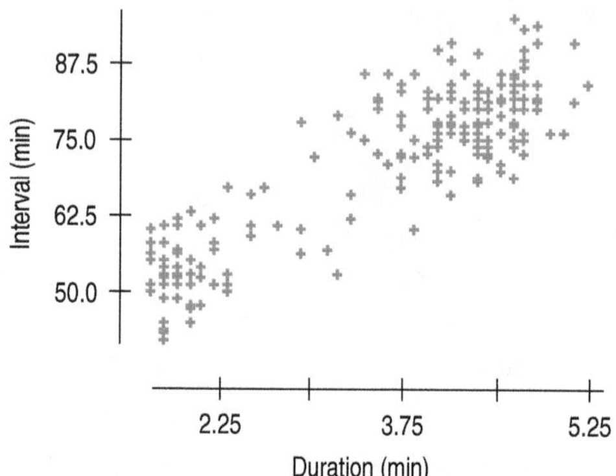

b) Write the equation of the line of best fit. Here's the regression analysis:

Dependent variable is: Interval
R-squared = 77.0% s = 6.16 min

Variable	Coefficient
Intercept	33.9668
Duration	10.3582

c) Carefully explain what the slope of the line means in this context.

d) How accurate do you expect predictions based on this model to be? Cite statistical evidence.

e) If you just witnessed an eruption that lasted 4 minutes, how long do you predict you'll have to wait to see the next eruption?

f) So you waited, and the next eruption came in 79 minutes. Use this as an example to define a residual.

T **18. Crocodile lengths** The ranges inhabited by the Indian gharial crocodile and the Australian saltwater crocodile overlap in Bangladesh. Suppose a very large crocodile skeleton is found there, and we wish to determine the species of the animal. Wildlife scientists have measured the lengths of the heads and the complete bodies of several crocs (in centimetres) of each species, creating the following regression analyses:

Indian Crocodile
Dependent variable is: IBody
R-squared = 97.2%

Variable	Coefficient
Intercept	−69.3693
IHead	7.40004

Australian Crocodile
Dependent variable is: ABody
R-squared = 98.0%

Variable	Coefficient
Intercept	−20.2245
AHead	7.71726

a) Do the associations between the sizes of the heads and bodies of the two species appear to be strong? Explain.

b) In what ways are the two relationships similar? Explain.

c) What is different about the two models? What does that mean?

d) The crocodile skeleton found had a head length of 62 cm and a body length of 380 cm. Which species do you think it was? Explain why.

T **19. How old is that tree?** One can determine how old a tree is by counting its rings, but that requires cutting the tree down. Can we estimate the tree's age simply from its diameter? A forester measured 27 trees of the same species that had been cut down, and counted the rings to determine the ages of the trees.

Diameter (inches)	Age (yr)	Diameter (inches)	Age (yr)
1.8	4	10.3	23
1.8	5	14.3	25
2.2	8	13.2	28
4.4	8	9.9	29
6.6	8	13.2	30
4.4	10	15.4	30
7.7	10	17.6	33
10.8	12	14.3	34
7.7	13	15.4	35
5.5	14	11.0	38
9.9	16	15.4	38
10.1	18	16.5	40
12.1	20	16.5	42
12.8	22		

a) Find the correlation between *Diameter* and *Age*. Does this suggest that a linear model may be appropriate? Explain.

b) Create a scatterplot and describe the association.

c) Create the linear model.

d) Check the residuals. Explain why a linear model is probably not appropriate.

e) If you used this model, would it generally overestimate or underestimate the ages of very large trees? Explain.

T **20. Improving trees** In the last exercise, you saw that the linear model had some deficiencies. Let's create a better model.

a) Perhaps the cross-sectional area of a tree would be a better predictor of its age. Since area is measured in square units, try re-expressing the data by squaring the diameters. Does the scatterplot look better?

b) Create a model that predicts *Age* from the square of *Diameter*.
c) Check the residuals plot for this new model. Is this model more appropriate? Why?
d) Estimate the age of a tree 18 inches in diameter.

21. Big screen An electronics website collects data on the size of new HD flat panel televisions (measuring the diagonal of the screen in inches) to predict the cost (in hundreds of dollars). Which of these is most likely to be the slope of the regression line: 0.03, 0.3, 3, 30? Explain.

T 22. Smoking and pregnancy 2006 The Child Trends Data Bank monitors issues related to children. The table shows a 50-state average of the percent of expectant mothers who smoked cigarettes during their pregnancies.

Year	% Smoking While Pregnant	Year	% Smoking While Pregnant
1990	19.2	2000	14.0
1991	18.7	2001	13.8
1992	17.9	2002	13.3
1993	16.8	2003	12.7
1994	16.0	2004	10.9
1995	15.4	2005	10.1
1996	15.3	2006	10.0
1997	14.9		
1998	14.8		
1999	14.1		

a) Create a scatterplot and describe the trend you see.
b) Find the correlation.
c) How is the value of the correlation affected by the fact that the data are averages rather than percentages for each of the 50 states?
d) Write a linear model and interpret the slope in this context.

T 23. Down the drain Most water tanks have a drain plug so that the tank may be emptied when it's to be moved or repaired. How long it takes a certain size of tank to drain depends on the size of the plug, as shown in the table. Create a model. (Hint: Replace *Drain Time* by $1/\sqrt{\text{Drain Time}}$)

Diameter of Plug (in.)	Drain Time (min.)	Diameter of Plug (in.)	Drain Time (min.)
$\frac{3}{8}$	140	$1\frac{1}{4}$	13
$\frac{1}{2}$	80	$1\frac{1}{2}$	10
$\frac{3}{4}$	35	2	5
1	20		

24. Tips It's commonly believed that people use tips to reward good service. A researcher for the hospitality industry examined tips and ratings of service quality from 2645 dining parties at 21 different restaurants. The correlation between ratings of service and tip percentages was 0.11. (Source: M. Lynn and M. McCall, "Gratitude and Gratuity." *Journal of Socio-Economics* 29: 203–214)
a) Describe the relationship between *Quality of Service* and *Tip Size*.
b) Find and interpret the value of R^2 in this context.

25. U.S. Cities Data from 50 large U.S. cities show the mean *January Temperature* (in Fahrenheit degrees) and *Latitude*. Describe what you see in the scatterplot.

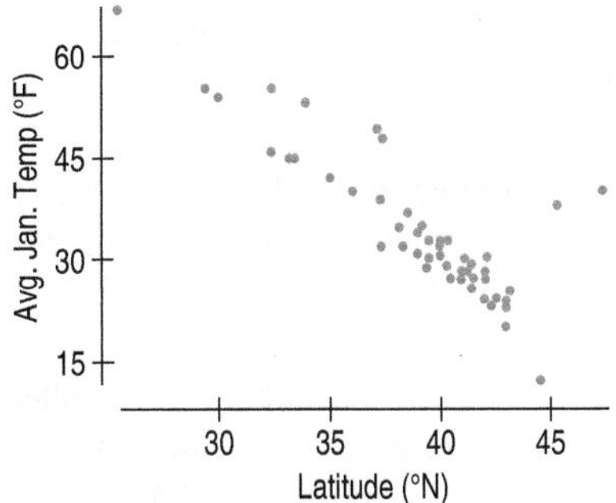

26. Correlations The study of U.S. cities in Exercise 25 found the mean *January Temperature* (degrees Fahrenheit), *Altitude* (feet above sea level), and *Latitude* (degrees north of the equator) for 55 cities. Here's the correlation matrix:

	Jan. Temp.	Latitude	Altitude
Jan. Temp.	1.000		
Latitude	−0.848	1.000	
Altitude	−0.369	0.184	1.000

a) Which seems to be more useful in predicting *January Temperature*: *Altitude* or *Latitude*? Explain.
b) If *Temperature* were measured in degrees Celsius, what would be the correlation between *Temperature* and *Latitude*?
c) If *Temperature* were measured in degrees Celsius and *Altitude* in meters, what would be the correlation? Explain.
d) What would you predict about the *January Temperature* in a city whose *Altitude* is two standard deviations higher than the average *Altitude*?

27. Winter in the city Summary statistics for the data relating the latitude and average January temperature (degrees Fahrenheit) for 55 large U.S. cities are given below.

Variable	Mean	StdDev
Latitude	39.02	5.42
JanTemp	26.44	13.49

Correlation = −0.848

a) What percent of the variation in *January Temperature* can be explained by variation in *Latitude*?
b) What is indicated by the fact that the correlation is negative?

c) Write the equation of the line of regression for predicting *January Temperature* from *Latitude*.
d) Explain what the slope of the line means.
e) Do you think the y-intercept of the line is meaningful? Explain.
f) The latitude of Denver is 40° N. Predict the mean January temperature (°F) there.
g) If the residual for a city is positive, what does that mean?
h) How do your answers to parts a), c), d), and f) change if temperature is in degrees Celsius?

28. Depression The September 1998 issue of the *American Psychologist* published an article by Kraut et al. that reported on an experiment examining "the social and psychological impact of the Internet on 169 people in 73 households during their first 1 to 2 years online." In the experiment, 73 households were offered free Internet access for 1 or 2 years in return for allowing their time and activity online to be tracked. The members of the households who participated in the study were also given a battery of tests at the beginning and again at the end of the study. The conclusion of the study made news headlines: Those who spent more time online tended to be more depressed at the end of the experiment. Although the paper reports a more complex model, the basic result can be summarized in the following regression of *Depression* (at the end of the study, in "depression scale units") vs. *Internet Use* (in mean hours per week):

Dependent variable is: Depression
R-squared = 4.6% s = 0.4563

Variable	Coefficient
Intercept	0.5655
Internet use	0.0199

The news reports about this study clearly concluded that using the Internet causes depression. Discuss whether such a conclusion can be drawn from this regression. If so, discuss the supporting evidence. If not, say why not.

T **29. Jumps 2008** How are Olympic performances in various events related? The plot shows winning long-jump and high-jump distances, in metres, for the Summer Olympics from 1912 through 2008.

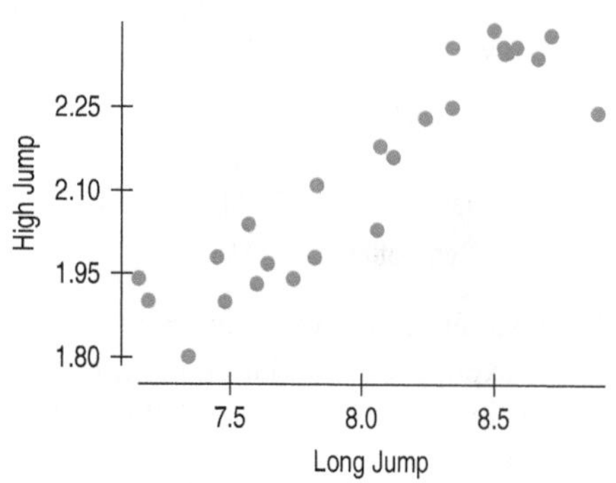

a) Describe the association.
b) Do long-jump performances somehow influence the high-jumpers? How do you account for the relationship you see?
c) The correlation for the given scatterplot is 0.920. If we converted the jump lengths to centimetres by multiplying by 100, would that make the actual correlation higher or lower?
d) What would you predict about the long jump in a year when the high-jumper jumped one standard deviation better than the average high jump?

T **30. Modelling jumps 2008** Here are the summary statistics for the Olympic long jumps and high jumps displayed in the scatterplot above:

Event	Mean	StdDev
High Jump	2.13880	0.191884
Long Jump	8.03960	0.521380

Correlation = 0.920

a) Write the equation of the line of regression for estimating *High Jump* from *Long Jump*.
b) Interpret the slope of the line.
c) In a year when the long jump is 8.9 m, what high jump would you predict?
d) Why can't you use this line to estimate the *Long Jump* for a year when you know the *High Jump* was 2.25 m?
e) Write the equation of the line you need to make that prediction.

31. French Consider the association between a student's score on a French vocabulary test and the weight of the student. What direction and strength of correlation would you expect in each of the following situations? Explain.
a) The students are all in Grade 3.
b) The students are in Grades 3 through 11 in the same Chicago school district.
c) The students are in Grade 10 in Quebec.
d) The students are in Grades 3 through 11 in Quebec.

32. Dollars on education Suppose you want to investigate the relationship between standardized test scores and dollars spent for schools in your province. For each school, an average test score (for the same grade) and the average dollars spent per student in 2013 were determined. Someone suggests using the ratio of these two—dollars per test point—to compare schools for efficiency. The schools with lowest ratios will be deemed the best or most efficient. Ignoring the issue of whether standardized test scores are a good measure of education quality, comment on the appropriateness of such a metric (measure) of efficiency, and suggest a better approach, using your knowledge of regression and residuals. Draw a hypothetical data plot pointing out the best and worst schools.

33. Lunchtime Does how long children remain at the lunch table help predict how much they eat? Twenty toddlers were observed over several months at a nursery school. The table and graph show the average number of minutes the kids stayed at the table and the average number of calories they consumed during lunch, calculated from careful observation of what the children ate each day. Create and interpret a model for the toddlers' lunchtime data.

Calories	Time	Calories	Time
472	21.4	450	42.4
498	30.8	410	43.1
465	37.7	504	29.2
456	33.5	437	31.3
423	32.8	489	28.6
437	39.5	436	32.9
508	22.8	480	30.6
431	34.1	439	35.1
479	33.9	444	33.0
454	43.8	408	43.7

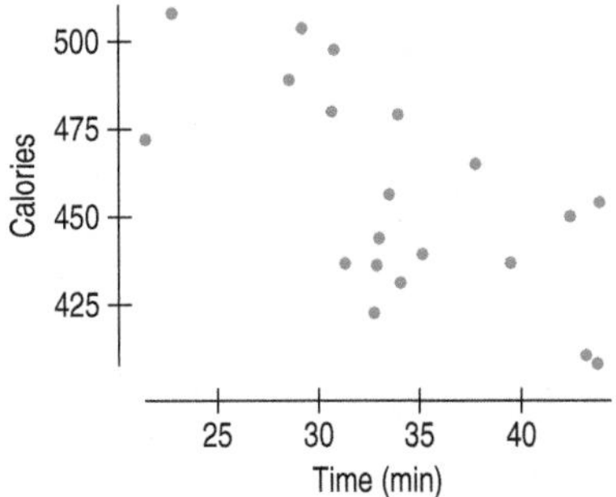

34. Chips A start-up company has developed an improved electronic chip for use in laboratory equipment. The company needs to project the manufacturing cost, so it develops a spreadsheet model that takes into account the purchase of production equipment, overhead, raw materials, depreciation, maintenance, and other business costs. The spreadsheet estimates the cost of producing 10 000 to 200 000 chips per year, as seen in the table. Develop a regression model to predict *Cost* based on the *Level* of production. (Hint: Take logs of both variables)

Chips Produced (1000s)	Cost per Chip ($)	Chips Produced (1000s)	Cost per Chip ($)
10	146.10	90	47.22
20	105.80	100	44.31
30	85.75	120	42.88
40	77.02	140	39.05
50	66.10	160	37.47
60	63.92	180	35.09
70	58.80	200	34.04
80	50.91		

35. Tobacco and alcohol Are people who use tobacco products more likely to consume alcohol? Here are data on household spending (in pounds) taken by the British government on 11 regions in Great Britain. Do tobacco and alcohol spending appear to be related? What questions do you have about these data? What conclusions can you draw?

Region	Alcohol	Tobacco
North	6.47	4.03
Yorkshire	6.13	3.76
Northeast	6.19	3.77
East Midlands	4.89	3.34
West Midlands	5.63	3.47
East Anglia	4.52	2.92
Southeast	5.89	3.20
Southwest	4.79	2.71
Wales	5.27	3.53
Scotland	6.08	4.51
Northern Ireland	4.02	4.56

36. Models Find the predicted value of y, using each fitted regression model below for $x = 10$.

a) $\hat{y} = 2 + 0.8 \ln x$

b) $\widehat{\log y} = 5 - 0.23x$

c) $\widehat{1/\sqrt{y}} = 17.1 - 1.66x$

37. Vehicle weights The Minnesota Department of Transportation hoped that they could measure the weights of big trucks without actually stopping the vehicles by using a newly developed "weigh-in-motion" scale. After installation of the scale, a study was conducted to find out whether the scale's readings correspond to the true weights of the trucks being monitored. In Exercise 18 of Chapter 6, you examined the scatterplot for the data they collected, finding the association to be approximately linear with $R^2 = 93\%$. Their regression equation is $\widehat{Wt} = 10.85 + 0.64\,Scale$, where both the scale reading and the predicted weight of the truck are measured in thousands of pounds.

a) Estimate the weight of a truck if this scale read 31 200 pounds.

b) If that truck actually weighed 32 120 pounds, what was the residual?

c) If the scale reads 35 590 pounds, and the truck has a residual of −2440 pounds, how much does it actually weigh?

d) In general, do you expect estimates made using this equation to be reasonably accurate? Explain.

e) If the police plan to use this scale to issue tickets to trucks that appear to be overloaded, will negative or positive residuals be a greater problem? Explain.

38. Profit How are a company's profits related to its sales? Let's examine data from 71 large U.S. corporations. All amounts are in millions of dollars.

a) Histograms of *Profits* and *Sales* and histograms of the logarithms of *Profits* and *Sales* are seen below. Why are the re-expressed data better for regression?

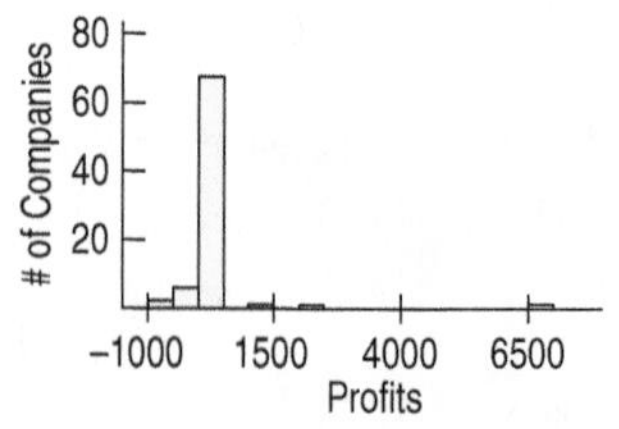

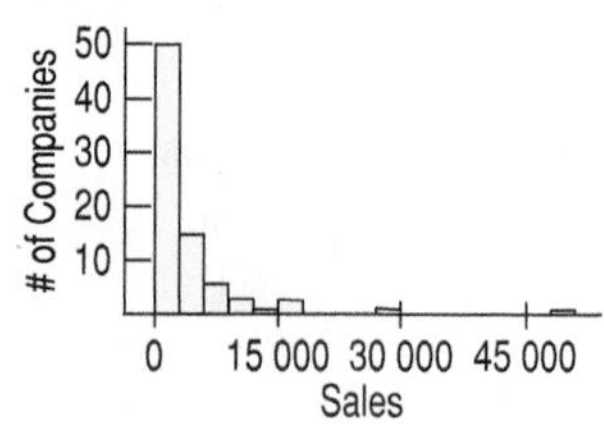

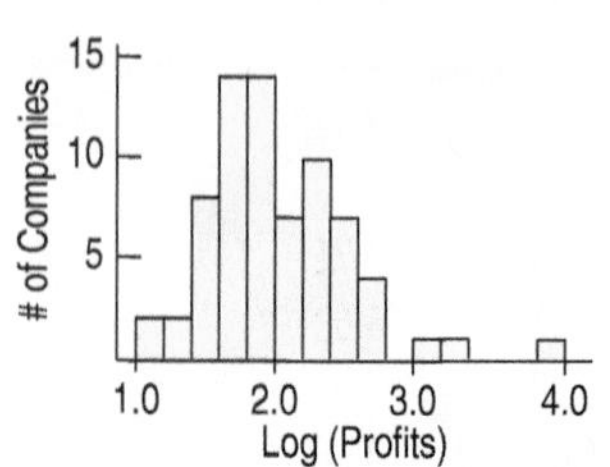

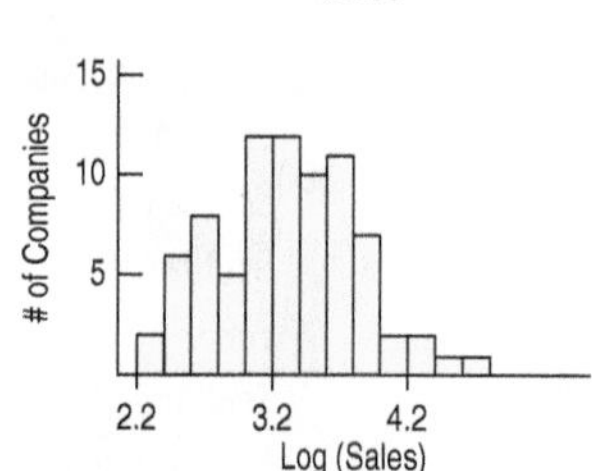

b) Here are the scatterplot and residual plot for the regression of logarithm of *Profits* vs. log of *Sales*. Do you think this model is appropriate? Explain.

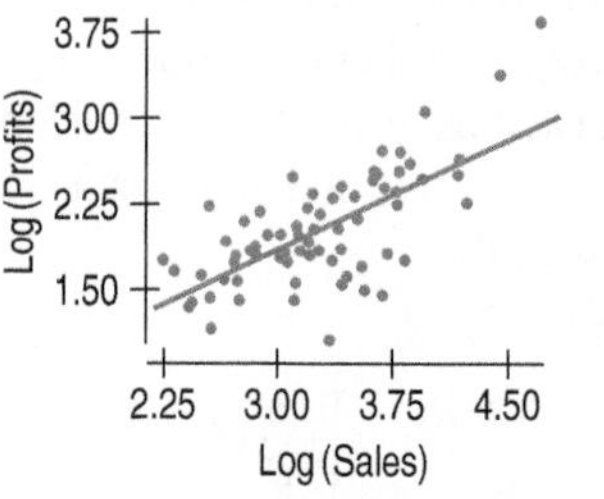

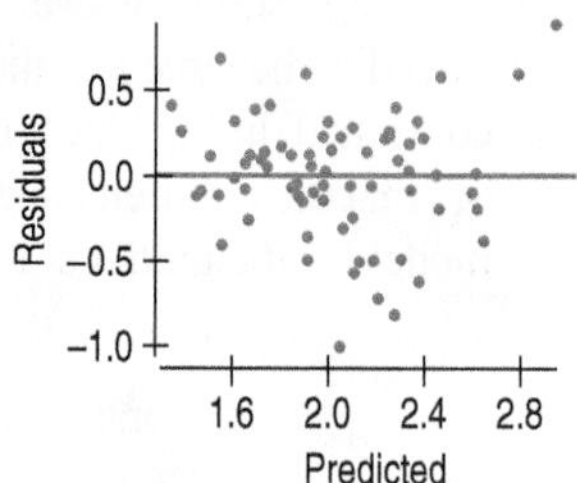

c) Here's the regression analysis. Write the equation.

Dependent variable is: Log Profit
R-squared = 48.1%

Variable	Coefficient
Intercept	−0.106259
LogSales	0.647798

d) Use your equation to estimate profits earned by a company with sales of $2.5 billion (that's 2500 million.)

chapter

9 Understanding Randomness

The generation of random numbers is too important to be left to chance.

—Robert R. Coveyou, Oak Ridge National Laboratory

Where are we going?

Few things in life are certain. But that doesn't mean that we all understand randomness. We'll be formal about it in a few chapters, but for now we'll simulate what we want to understand to see the patterns that emerge.

"The most decisive conceptual event of twentieth century physics has been the discovery that the world is not deterministic. . . . A space was cleared for chance."

— *Ian Hocking*
The Taming of Chance

We all know what it means for something to be random. Or do we? Many children's games rely on chance outcomes. Rolling dice, spinning spinners, and shuffling cards all select outcomes at random. Adult games use randomness as well, from card games to lotteries and bingo. What's the most important aspect of the randomness in these games? *It must be fair.*

What is it about random selection that makes it seem fair? It's really two things. First, nobody can guess the outcome before it happens. Second, when we want things to be fair, usually some underlying set of outcomes will be equally likely (although in many games, some combinations of outcomes are more likely than others).

Randomness is not always what we might think of as "at random." **Random** outcomes have a lot of structure, especially when viewed in the long run. You can't predict how a fair coin will land on any single toss, but you're pretty confident that if you flipped it thousands of times you'd see about 50% heads. As we will see, randomness is an essential tool of statistics; statisticians don't think of randomness as the annoying tendency of things to be unpredictable or haphazard. In fact, without deliberately applying randomness, we couldn't do most of statistics, and this book would stop right about here.

But truly random values are surprisingly hard to get. Just to see how fair humans are at selecting, pick a number at random from below. Go ahead. Quickly look at the numbers below, and pick a number at random.

1 2 3 4

9.1 What Is Randomness?

Did you pick 3? If so, you've got company. Almost 75% of all people pick the number 3. About 20% pick either 2 or 4. If you picked 1, well, consider yourself a little different. Only about 5% choose 1. Psychologists have proposed reasons for this phenomenon, but for us, it simply serves as a lesson that we've got to find a better way to choose things at random.

So how should we **generate random numbers**? It's surprisingly difficult to get random values even when they're equally likely. Computers (and smartphones) have become a popular way to generate them. Even though they often do much better than humans, computers can't generate truly random numbers either. Computers follow programs. Start a program from the same place, and it will always follow exactly the same path. So numbers generated by a program are not truly random. Technically, "random" numbers generated

A S *Activity:* **Random Behaviour.** *MyStatLab's* Random Experiment Tool lets you experiment with truly random outcomes. We'll use it a lot in the coming chapters.

A S *Activity:* **Truly Random Values on the Internet.** This activity will take you to a Web site (www.random.org) that generates all the truly random numbers you could want.

this way are *pseudorandom* numbers. Pseudo-random number generators (PRNGs) are algorithms that can automatically create long runs of numbers with good random properties but eventually the sequence repeats (or the memory usage grows without bound). Fortunately, pseudorandom values are virtually indistinguishable from truly random numbers even though they aren't really random. For most purposes, that's good enough.

DILBERT: © Scott Adams/Distributed by United Feature Syndicate, Inc. © 2001 Scott Adams. Distributed by Universal Uclick. Reprinted with permission. All rights reserved

There *are* ways to generate random numbers so that they are both equally likely and truly random. In the past, entire books of carefully generated random numbers were published. The books never made the best-seller lists and probably didn't make for great reading, but they were quite valuable to those who needed truly random values.[1] And, along with almost anything else, you can find truly random numbers on the Internet. Several sites are available that use methods like timing the decay of a radioactive element to generate truly random digits[2]. A string of random digits might look like this:

2217726304387410092537086270581997622725849795907032825001108963
3217535822643800292254644943760642389043766557204107354186024508
8906427308645681412198226653885873285801699027843110380420067664
8740522639824530519902027044464984322000946238678577902639002954
8887003319933147508331265192321413908608674496383528968974910533
6944182713168919406022181281304751019321546303870481407676636740
6070204916508913632855351361361043794293428486909462881431793360
7706356513310563210508993624272872250535395513645991015328128202

Soon we'll discuss ways to use such random digits to apply randomness to real situations. The best ways we know to generate data that give a fair and accurate picture of the world rely on randomness, and how we draw conclusions from those data depends on understanding randomness, too.

An ordinary deck of playing cards, like the ones used in bridge and many other card games, consists of 52 cards. There are numbered cards (2 through 10), face cards (Jack, Queen, King), and the Ace, all of whose value depends on the game you are playing. Each card is also marked by one of four suits (clubs, diamonds, hearts, or spades) whose significance is also game-specific.

Aren't you done shuffling yet? Even something as common as card shuffling may not be as random as you might think. If you shuffle cards by the usual method, in which you split the deck in half and try to let cards fall roughly alternately from each half, you're doing a "riffle shuffle."

How many times should you shuffle cards to make the deck random? A surprising fact was discovered by statisticians Persi Diaconis, Ronald Graham, and W. M. Kantor. It takes seven riffle shuffles. Fewer than seven leaves order in the deck, but after that, more shuffling does little good. However, most people don't shuffle that many times.

When computers were first used to generate hands in bridge tournaments, some professional bridge players complained that the computer was making too many "weird" hands—hands with 10 cards of one suit, for example.[3] Suddenly these hands were appearing more often than players were used to when cards were shuffled by hand. The players assumed that the computer was doing something wrong. But it turns out that it's humans who hadn't been shuffling enough to make the decks really random and have those "weird" hands appear as often as they should.

[1] You'll find a table of random digits of this kind in the back of this book.
[2] For example, www.random.org or www.randomnumbers.info.
[3] See more at www.acbl.org/learn/computerHands.html.

ALEXANDER NEMENOV/ AFP/Getty Images/ Newscom

Action Plus Sports Images/Alamy

ZUMA Press, Inc./Alamy

Practical Randomness

Suppose a cereal manufacturer puts pictures of famous athletes on cards in boxes of cereal in the hope of boosting sales. The manufacturer announces that 20% of the boxes contain a picture of Sidney Crosby, 30% have a picture of Christine Sinclair, and the rest include a picture of Clara Hughes. You want all three pictures. How many boxes of cereal do you expect to have to buy to get the complete set?

How can we answer questions like this? Well, one way is to buy hundreds of boxes of cereal to see what might happen. But let's not. Instead, we'll consider using a random model. Why random? When we pick a box of cereal off the shelf, we don't know what picture is inside. We'll assume that the pictures are randomly placed in the boxes and that the boxes are distributed randomly to stores around the country. Why a model? Because we won't actually buy the cereal boxes. We can't afford all those boxes and we don't want to waste food, so we need an imitation of the real process that we can manipulate and control. In short, we're going to *simulate* reality. A **simulation** mimics reality by using random numbers to represent the outcomes of real events. Just as pilots use flight simulators to learn about and practice real situations, we can learn a great deal about the real events by carefully modeling the randomness and analyzing the simulation results.

9.2 Simulating Randomness

Modern physics has shown that randomness is not just a mathematical game; it is fundamentally the way the universe works.

Regardless of improvements in data collection or in computer power, the best we can ever do, according to quantum mechanics . . . is predict the probability that an electron, or a proton, or a neutron, or any other of nature's constituents, will be found here or there. Probability reigns supreme in the microcosmos.

—Brian Greene, *The Fabric of the Cosmos: Space, Time, and the Texture of Reality* (p. 91)

The question we've asked is how many boxes you would expect to buy to get a complete card collection. But we can't answer our question by completing a card collection just once. We want to understand the *typical* number of boxes to open, how that number varies, and, often, the shape of the distribution. So we'll have to do this over and over. We call each time we obtain a simulated answer to our question a **trial**.

For the sports cards, a trial's outcome is the number of boxes. We'll need at least three boxes to get one of each card, but with really bad luck, you could empty the shelves of several supermarkets before finding the last card you need to get all three. So the possible outcomes of a trial are 3, 4, 5, or lots more. But we can't simply pick one of those numbers at random, because they're not equally likely. We'd be surprised if we only needed three boxes to get all the cards, but we'd probably be even more surprised to find that it took exactly 7359 boxes. In fact, the reason we're doing the simulation is that it's hard to guess how many boxes we'd expect to open.

Building a Simulation

We know how to find equally likely random digits. How can we get from there to simulating the trial outcomes? The digits from 0 to 9 are equally likely to occur. Because 20% of the boxes contain Sidney Crosby's picture, we'll use two of the 10 digits to represent that outcome. Three of the 10 digits can model 30% of the boxes with Christine Sinclair cards, and the remaining five digits can represent the 50% of boxes with Clara Hughes. So we can interpret the digits 0 and 1 as finding Crosby; 2, 3, and 4 as finding Sinclair; and 5 through 9 as finding Hughes to simulate opening one box. Opening one box is the basic building block, called a **component** of our simulation. But the component's outcome isn't the result we want. We need to observe a sequence of components until our card collection is complete. The *trial's* outcome is called the **response variable**; for this simulation that's the *number* of boxes (components) in the sequence.

Let's look at the steps for making a simulation:

Specify how to model a component outcome using equally likely random digits:

1. **Identify the component to be repeated.** In this case, our component is the opening of a box of cereal.
2. **Explain how you will model the component's outcome.** We'll assign outcomes to the equally likely random digits as follows:

0, 1 Crosby   2, 3, 4 Sinclair   5, 6, 7, 8, 9 Hughes

Specify how to simulate trials:

3. **State clearly what the response variable is.** What are we interested in? We want to find out the number of boxes it might take to get all three pictures.
4. **Explain how you will combine the components into a trial to model the response variable.** We pretend to open boxes (repeat components) by looking at each random digit and indicating what picture it represents. Continue until we've found all three. The trial's outcome is the number of boxes (components).

Put it all together to run the simulation:

5. **Run several trials.** For example, consider the third line of random digits shown earlier (p. 264):

   8906427308645681412198226653885873285801699027843110380420067664.

   Let's see what happened.

The first random digit, 8, means you get Hughes' picture. So the first component's outcome is Hughes. The second digit, 9, means Hughes' picture is also in the next box. Continuing to interpret the random digits, we get Crosby's picture (0) in the third, Hughes' (6) again in the fourth, and finally Sinclair (4) on the fifth box. Since we've now found all three pictures, we've finished one trial of our simulation. This trial's outcome is five boxes.

Now we keep going, running more trials by looking at the rest of our line of random digits:

89064 2730 8645681 41219 822665388587328580 169902 78431 1038 042006 7664.

It's best to create a chart to keep track of what happens:

Trial number	Component Outcomes	Trial Outcomes: y = Number of boxes
1	89064 = Hughes, Hughes, Crosby, Hughes, Sinclair	5
2	2730 = Sinclair, Hughes, Sinclair, Crosby	4
3	8645681 = Hughes, Hughes, Sinclair,. . . ,Crosby	7
4	41219 = Sinclair, Crosby, Sinclair, Crosby, Hughes	5
5	822665388587328580 = Hughes, Sinclair, . . . , Crosby	18
6	169902 = Crosby, Hughes, Hughes, Hughes, Crosby, Sinclair	6
7	78431 = Hughes, Hughes, Sinclair, Sinclair, Crosby	5
8	1038 = Crosby, Crosby, Sinclair, Hughes	4
9	042006 = Crosby, Sinclair, Sinclair, Crosby, Crosby, Hughes	6
10	7664 . . . = Hughes,Hughes, Hughes, Sinclair . . .	?

Analyze the response variable:

6. **Collect and summarize the results of all the trials.** You know how to summarize and display a quantitative response variable. You'll certainly want to report the shape, centre, and spread, and depending on the question asked, you may want to include more.
7. **State your conclusion,** as always, in the context of the question you wanted to answer. Based on this simulation, we estimate that customers hoping to complete their card collection will need to open a median of 5 boxes, but it could take a lot more.

A S *Activity:* **Bigger Samples Are Better.** The random simulation tool can generate lots of outcomes with a single click, so you can see more of the long run with less effort.

If you fear that these may not be accurate estimates because we ran only nine trials, you are absolutely correct. The more trials the better, and nine is woefully inadequate. Twenty trials is probably a reasonable minimum if you are doing this by hand. Even better, use a computer and run a few thousand trials.

For Example SIMULATING A DICE GAME

A version of the game of 21 can be played with an ordinary six-sided die. Competitors each roll the die repeatedly, trying to get the highest total less than or equal to 21. If your total exceeds 21, you lose.

Suppose your opponent has rolled an 18. Your task is to try to beat him by getting more than 18 points without going over 21. How many rolls do you expect to make, and what are your chances of winning?

QUESTION: How will you simulate the components?

ANSWER: A component is one roll of the die. Simulate each roll by looking at a random digit from a table or Web site. The digits 1 through 6 will represent the results on the die; ignore digits 7–9 and 0.

QUESTION: What's the response variable? How will you combine components to obtain a response?

ANSWER: There are two response variables. Count the number of times you roll the die, and keep track of whether you win or lose. Add components until the total is greater than 18, counting the number of rolls. If the total is greater than 21, it is a loss; if not, it is a win.

QUESTION: How would you use these random digits to run trials?

91129 58757 69274 92380 82464 33089

Show your method clearly for two trials.

ANSWER: Mark the discarded digits in colour.

Trial #1:	9	**1**	**1**	**2**	9	**5**	8	7	**5**	7	**6**				
Total:		1	2	4		9			14		20				Outcomes: 6 rolls, won
Trial #2:	9	**2**	7	**4**	9	**2**	**3**	8	0	8	**2**	**4**	**6**		
Total:		2		6		8	11				13	17	23		Outcomes: 7 rolls, lost

QUESTION: Suppose you run 30 trials, getting the outcomes tallied here. What is your conclusion?

Number of rolls		Result	
4	///	Won	𝍸 𝍸 𝍸 𝍸 /
5	𝍸 𝍸	Lost	𝍸 ////
6	𝍸 𝍸 /		
7	𝍸		
8	/		

ANSWER: Based on the simulation, when competing against an opponent who has a score of 18, you expect a typical turn to last five or six rolls, and you should win about 70% of the time.

Just Checking

The Stanley Cup Finals consists of up to seven games. The first team to win four games wins the series. The first two are played at one team's home arena, the next two at the other team's arena, and the final three (if needed) are played alternately between the two arenas. Records over the past century show that there is a home-ice advantage; the home team has about a 55% chance of winning. Does the current system of alternating arenas even out the home-ice advantage? How often will the team that begins at home win the series?

Let's set up the simulation:

1. What is the component to be repeated?
2. How will you model each component from equally likely random digits?
3. What is the response variable?
4. How will you model a trial by combining components? What rule will you use for stopping the collection of components? How will you use components to generate the outcome of each trial?
5. How will you analyze the response variable?

Step-by-Step Example SIMULATION

Fifty-seven students participated in a lottery for a particularly desirable dorm room—a triple with a fireplace and private bath in the tower. Twenty of the participants were members of the same varsity team. When all three winners were members of the team, the other students cried foul.

Question: Could an all-team outcome reasonably be expected to happen if everyone had a fair shot at the room?

THINK ➡ Plan State the problem. Identify the important parts of your simulation.

I'll use a simulation to investigate whether it's unlikely that three varsity athletes would get the great room in the dorm if the lottery were fair.

Components Identify the components.

A component is the selection of a student.

Outcomes State how you will model each component using equally likely random digits.

I'll look at two-digit random numbers.

Let 00–19 represent the 20 varsity applicants.

Let 20–56 represent the other 37 applicants.

There are 20 and 37 students in the two groups. This time you must use *pairs* of random digits (and ignore some of them) to represent the 57 students.

Skip 57–99. If I get a number in this range, I'll throw it away and go back for another two-digit random number.

Response Variable Define your response variable.

The response variable is whether or not all three selected students are on the varsity team.

Trial Define a trial. Explain how you will combine the components to simulate each trial's outcome. In each of these trials, you can't choose the same student twice, so you'll need to ignore a random number if it comes up a second or third time. Be sure to mention this in describing your simulation.

Each trial consists of identifying pairs of digits as V (varsity) or N (non-varsity) until three people are chosen, ignoring out-of-range or repeated numbers (X)—I can't put the same person in the room twice.

SHOW ➡ Mechanics Run several trials. Carefully record the random numbers, indicating:

1. the corresponding component outcomes (in this case, varsity, non-varsity, or ignored number) and
2. the resulting trial outcome.

Trial Number	Component Outcomes	Trial Outcome: All Varsity?
1	74 02 94 39 02 77 55 X V X N X X N	No
2	18 63 33 25 V X N N	No
3	05 45 88 91 56 V N X X N	No
4	39 09 07 N V V	No
5	65 39 45 95 43 X N N X N	No
6	98 95 11 68 77 12 17 X X V X X V V	Yes
7	26 19 89 93 77 27 N V X X X N	No
8	23 52 37 N N N	No
9	16 50 83 44 V N X N	No
10	74 17 46 85 09 X V N X V	No

Analyze Summarize the results across all trials to answer the initial question.

"All varsity" occurred once, or 10% of the time.

TELL ➡ Conclusion Describe what the simulation shows, and interpret your results in the context of the real world.	In my simulation of "fair" room draws, the three people chosen were all varsity team members only 10% of the time. While this result could happen by chance, it is not particularly likely. I'm suspicious, but I'd need many more trials and a smaller frequency of the all-varsity outcome before I would make an accusation of unfairness.

WHAT CAN GO WRONG?

- **Don't overstate your case.** Let's face it: In some sense, a simulation is *always* wrong. After all, it's not the real thing. We didn't buy any cereal or run a room draw. So beware of confusing what *really* happens with what a simulation suggests *might* happen. Never forget that future results will not match your simulated results exactly.
- **Model outcome chances accurately.** A common mistake in constructing a simulation is to adopt a strategy that may appear to produce the right kind of results, but that does not accurately model the situation. For example, in our room draw, we could have gotten 0, 1, 2, or 3 team members. Why not just see how often these digits occur in random digits from 0 to 9, ignoring the digits 4 and up?

  3 2 1 7 9 0 0 5 9 7 3 7 9 2 5 2 4 1 3 8
  3 2 1 x x 0 0 x x x 3 x x 2 x 2 x 1 3 x

  This "simulation" makes it seem fairly likely that three team members would be chosen. There's a big problem with this approach, though: The digits 0, 1, 2, and 3 occur with equal frequency among random digits, making each outcome appear to happen 25% of the time. In fact, the selection of 0, 1, 2, or all 3 team members are not all equally likely outcomes. In our correct simulation, we estimated that all three would be chosen only about 10% of the time. If your simulation overlooks important aspects of the real situation, your model will not be accurate.
- **Run enough trials.** Simulation is cheap and fairly easy to do. Don't try to draw conclusions based on 5 or 10 trials (even though we did for illustration purposes here). We'll make precise how many trials to use in later chapters. For now, err on the side of large numbers of trials.

Activity: **Estimating Summaries from Random Outcomes.** See how well you can estimate something you can't know just by generating random outcomes.

CONNECTIONS

Simulations often generate many outcomes of a response variable, and we are often interested in the distribution of these responses. The tools we use to display and summarize the distribution of any real variable are appropriate for displaying and summarizing randomly generated responses as well.

Make histograms, boxplots, and normal probability plots of the response variables from simulations, and summarize them with measures of centre and spread. Be especially careful to report the variation of your response variable.

Don't forget to think about your analyses. Simulations can hide subtle errors. A careful analysis of the responses can save you from erroneous conclusions based on a faulty simulation.

You may be less likely to find an outlier in simulated responses, but if you find one, you should certainly determine how it happened.

What Have We Learned?

Learning Objectives

Know how to obtain random digits from computer programs or the Internet.

Be able to design and execute, by hand, a simulation of a random phenomenon, specifying the design of the simulation to properly reflect the relative frequencies of random events.

Understand that although individual random outcomes cannot be predicted, in the long run, simulated values tend to settle down. The goal of simulations is to obtain and understand those values.

Review of Terms

Random An outcome is random if we know the possible values it can have, but not which particular value it takes (p. 263).

Generating random numbers Random numbers are hard to generate. Nevertheless, several Web sites offer an unlimited supply of equally likely random values (p. 263).

Simulation A simulation models a real-world situation by using random-digit outcomes to mimic the uncertainty of a response variable of interest (p. 265).

Trial The sequence of several components representing events that we are pretending will take place (p. 265).

Simulation component A component that uses equally likely random digits to model simple random occurrences whose outcomes may not be equally likely (p. 265).

Response variable Values of the response variable record the results of each trial with respect to what we were interested in (p. 265).

On the Computer SIMULATION

Simulations are best done with the help of technology simply because the more trials, the better the simulation—and computers are fast. Special computer programs designed for simulation are available, but most statistics packages can generate random numbers and support a simulation.

All technology-generated random numbers are *pseudorandom*. The random numbers available on the Internet may technically be better, but the differences won't matter for any simulation of modest size. Pseudorandom numbers generate the next random value from the previous one by a specified algorithm. But they have to start somewhere. This starting point is called the "seed." Most programs let you set the seed. There's usually little reason to do this, but if you wish to, go ahead. If you reset the seed to the same value, the programs will generate the same sequence of "random" numbers.

A S *Activity:* **Creating Random Values.** Learn to use your statistics package to generate random outcomes.

EXCEL

The RAND function generates a random value between 0 and 1. Multiply to scale it up to any range you like. Use the INT function to turn the result into an integer.

COMMENTS

Published tests of Excel's random-number generation have declared it to be inadequate. However, for simple simulations, it should be okay. Don't trust it for important large simulations.

JMP

- In a new column, in the **Cols** menu choose **Column Info . . .**
- In the dialogue, click the **New Property** button, and choose **Formula** from the drop-down menu.
- Click the **Edit Formula** button.
- In the **Functions (grouped)** window, click on **Random. Random Integer** 10, for example, will generate a random integer between 1 and 10.

MINITAB

- In the **Calc** menu, choose **Random Data . . .**
- In the **Random Data** submenu, choose **Uniform . . .**

A dialogue guides you in specifying details of range and number of columns to generate.

R

R has many built-in random number generators.
To generate *k* random integers between 1 and *n*,

- **sample(1:n,k,replace =T)** where *k* and *n* are numbers.
- **sample(1:n,k)** will ensure that there are no repeats. Here $k \le n$.
- **sample(c(0,1),k,replace =T)** will simulate Bernoulli. Trials with p = 0.5.

COMMENTS

To generate *k* random digits, use **sample(0:9,k,replace = T)**.
To simulate Bernoulli trials with p = 0.1 instead of 0.5, use:

sample(c(0,0,0,0,0,0,0,0,0,1),k, replace = T)

SPSS

The **RV.UNIFORM(min, max)** function returns a random value that is equally likely between the min and max limits.

STATCRUNCH

To generate a list of random numbers:

- Click on **Data**.
- Choose **Simulate data » Uniform**.

Enter the number of rows and columns of random numbers you want. (Often you'll specify the desired number of random values as **Rows** and just 1 **Column**.)

- Enter the interval of possible values for the random numbers (**a ≤ x < b**).
- Click on **Simulate.** The random numbers will appear in the data table.

COMMENTS

To get random *integers* from 0 to 99, set **a** = 0 and **b** = 100, and then simply ignore the decimal places in the numbers generated.

OR

- Under **Data** choose **Computer expression**.
- As Y, choose the **Column** of random numbers.
- Choose **Floor(Y)** from the functions.
- Click on **Set Expression**.
- Click on **Compute**.

TI-83/84 PLUS

To generate random numbers, use **5:RandInt** from the **Math PRB** menu. This command will produce any number of random integers in a specified range.

COMMENTS

Some examples:
RandInt(0,1) randomly chooses a 0 or a 1. This is an effective simulation of a coin toss.
RandInt(1,6,2) randomly returns two integers between 1 and 6. This is a good way to simulate rolling two dice. To run several trials, just hit ENTER repeatedly.
RandInt(0,57,3) produces three random integers between 0 and 57—a nice way to simulate the chapter's dorm room lottery.

Exercises

1. Coin toss Is a coin flip random? Why or why not, in your opinion?

2. Casino A casino claims that its electronic video roulette machine is truly random. What should that claim mean?

3. Random outcomes For each of the following scenarios, decide if the outcome is random.
a) Flip a coin to decide who takes out the trash. Is who takes out the trash random?
b) A friend asks you to quickly name a professional sports team. Is the sports team named random?
c) Names are selected out of a hat to decide roommates in a dormitory. Is your roommate for the year random?

4. More random outcomes For each of the following scenarios, decide if the outcome is random.
a) You enter a contest in which the winning ticket is selected from a large drum of entries. Was the winner of the contest random?
b) When playing a board game, the number of spaces you move is decided by rolling a six-sided die. Is the number of spaces you move random?
c) Before flipping a coin, your friend asks you to "call it." Is your choice (heads or tails) random?

5. Components Playing a board game, you are stuck in jail until you roll "doubles" (same numbers) on a pair of dice. You will simulate using random digits to estimate how many turns you expect to remain in jail. Identify the component if you were to run a simulation.

6. More components Your local burger joint gives away a game piece with each purchase. Game pieces are labelled *Burger*, *Fries*, and *Shake* and are equally likely to be received. You will simulate to estimate how many purchases you will need to collect the entire set to win a free meal. Identify the component if you were to run a simulation.

7. Response variable For the board game described in Exercise 5, how will you combine components to model a trial? What is the response variable?

8. Response variable, take two For the burger joint contest described in Exercise 6, how will you combine components to model a trial? What is the response variable?

9. The lottery Many provinces run lotteries, giving away millions of dollars if you match a certain set of winning numbers. How are those numbers determined? Do you think this method guarantees randomness? Explain.

10. Games Many kinds of games rely on randomness. Cite three different methods commonly used in the attempt to achieve this randomness, and discuss the effectiveness of each.

11. Birth defects The Society of Obstetricians and Gynecologists of Canada reports that out of every 100 babies born in Canada, three are born with a serious congenital anomaly (or birth defect). How would you assign random numbers to conduct a simulation on this statistic?[4]

12. Colour blind By some estimates, about 10% of all males have some colour perception defect, most commonly red–green colour blindness. How would you assign random numbers to conduct a simulation based on this proportion?

13. Geography An elementary school teacher with 20 students plans to have each student make a poster about two different Canadian provinces and territories. The teacher first numbers the provinces and territories in

[4] www.phac-aspc.gc.ca/publicat/cac-acc02/index-eng.php

alphabetical order and then creates sets of random digits to assign to each student his or her two provinces and territories:

01 – Alberta	02 – British Columbia
03 – Manitoba	04 – New Brunswick
05 – Newfoundland and Labrador	06 – Northwest Territories
07 – Nova Scotia	08 – Nunavut
09 – Ontario	10 – Prince Edward Island
11 – Quebec	12 – Saskatchewan
13 –Yukon	

Here are the random digits:

88063 56513 31056 32105 08993

a) Which two provinces and territories does the first student get?
b) Which two provinces and territories does the second student get?

14. Get rich Your province's BigBucks Lottery prize has reached $100 000 000, and you decide to play. You have to pick five numbers between 1 and 60, and you'll win if your numbers match those drawn by the lotto. You decide to pick your "lucky" numbers using a random number table. Which numbers do you play, based on these random digits?

43680 98750 13092 76561 58712

15. Play the lottery Some people play lotteries by always playing the same favourite "lucky" number. Assuming that the lottery is truly random, is this strategy better, worse, or the same as choosing different numbers for each play? Explain.

16. Play it again, Sam In Exercise 14, you imagined playing the lottery by using random digits to decide what numbers to play. Is this a particularly good or bad strategy? Explain.

17. Bad simulations Explain why each of the following simulations fails to model the real situation properly:
a) Use a random integer from 0 through 9 to represent the number of heads that appear when nine coins are tossed.
b) A basketball player takes a foul shot. Look at a random digit, using an odd digit to represent a good shot and an even digit to represent a miss.
c) Use five random numbers from 1 through 13 to represent the denominations of the cards in a poker hand.

18. More bad simulations Explain why each of the following simulations fails to model the real situation properly:
a) Use random numbers 2 through 12 to represent the sum of the faces when two dice are rolled.
b) Use a random integer from 0 through 5 to represent the number of boys in a family of five children.
c) Simulate a baseball player's performance at bat by letting 0 = an out, 1 = a single, 2 = a double, 3 = a triple, and 4 = a home run.

19. Wrong conclusion A Statistics student properly simulated the length of checkout lines in a grocery store and then reported, "The average length of the line will be 3.2 people." What's wrong with this conclusion?

20. Another wrong conclusion After simulating the spread of a disease, a researcher wrote, "Twenty-four percent of the people contracted the disease." What should the correct conclusion be?

21. Election You're pretty sure that your candidate for class president has about 55% of the votes in the entire school. But you're worried that only 100 students will show up to vote. How often will the underdog (the one with 45% support) win? To find out, you set up a simulation.
a) Describe how you will simulate a component and its outcomes.
b) Describe how you will simulate a trial.
c) Describe the response variable.

22. Two pairs or three of a kind? When drawing five cards randomly from a deck, which is more likely, two pairs or three of a kind? A pair is exactly two of the same denomination. Three of a kind is exactly three of the same denomination. (Don't count three 8s as a pair—that's three of a kind. And don't count four of the same kind as two pair—that's four of a kind, a very special hand.) How could you simulate five-card hands? Be careful: Once you've picked the 8 of spades for a hand, you can't get it again until the next hand.
a) Describe how you will simulate a component and its outcomes.
b) Describe how you will simulate a trial.
c) Describe the response variable.

23. Cereal In the chapter's example, 20% of the cereal boxes contained a picture of Sidney Crosby, 30% Christine Sinclair, and the rest Clara Hughes. Suppose you buy five boxes of cereal. Estimate the probability that you end up with a complete set of the pictures. Your simulation should have at least 20 runs.

24. Cereal, again Suppose you really want the Sidney Crosby picture. How many boxes of cereal do you need to buy to be pretty sure of getting at least one? Your simulation should use at least 10 runs.

25. Multiple choice You take a quiz with six multiple choice questions. After you studied, you estimated that you would have about an 80% chance of getting any individual question right. What are your chances of getting them all right? Your simulation should use at least 20 runs.

26. Lucky guessing? A friend of yours who took the multiple choice quiz in Exercise 25 got all six questions right, but now claims to have guessed blindly on every question. If each question offered four possible answers, do you believe her? Explain, basing your argument on a simulation involving at least 10 runs.

27. **Beat the lottery** Many provinces run lotteries to raise money. A Web site advertises that it knows "how to increase YOUR chances of winning the lottery." They offer several systems and criticize others as foolish. One system is called "*Lucky Numbers*." People who play the "*Lucky Numbers*" system just pick a "lucky" number to play, but maybe some numbers are luckier than others. Let's use a simulation to see how well this system works.

To make the situation manageable, simulate a simple lottery in which a single digit from 0 to 9 is selected as the winning number. Pick a single value to bet, such as 1, and keep playing it over and over. You'll want to run at least 100 trials. (If you can program the simulations on a computer or programmable calculator, run several hundred. Or generalize the questions to a lottery that chooses two- or three-digit numbers—for which you'll need thousands of trials.)

a) What proportion of the time do you expect to win?
b) Would you expect better results if you picked a "luckier" number, such as seven? (Try it if you don't know.) Explain.

28. **Random is as random does** The "beat the lottery" Web site discussed in Exercise 27 suggests that because lottery numbers are random, it is better to select your bet randomly. For the same simple lottery, generate each bet by choosing a separate random value between 0 and 9. Play many games. What proportion of the time do you win?

29. **It evens out in the end** The "beat the lottery" Web site in Exercise 27 notes that in the long run we expect each value to turn up about the same number of times. That leads to their recommended betting strategy. First, watch the lottery for a while, recording all the winners. Then bet the value that has turned up the least, because we expect it will need to turn up more often to even things out. If there is more than one "rarest" value, just take the lowest one (since it doesn't matter). Simulating the simplified lottery described in Exercise 27, play many games with this system. What proportion of the time do you win?

30. **Play the winner?** Another strategy for beating the lottery is the reverse of the system described in Exercise 29. Simulate the simplified lottery described in that exercise. Each time, bet the number that just turned up. The Web site suggests that this method should do worse. Does it? Play many games and see.

31. **Driving test** You are about to take the road test for your driver's license. You hear that only 34% of candidates pass the test the first time, but the percentage rises to 72% on subsequent retests. Estimate the average number of tests drivers take in order to get a license. Your simulation should use at least 20 runs.

32. **Still learning?** As in Exercise 31, assume that your chance of passing the driver's test is 34% the first time and 72% for subsequent retests. Estimate the percentage of those tested who still do not have a driver's license after two attempts.

33. **Basketball strategy** Late in a basketball game, the team that is behind often fouls someone in an attempt to get the ball back. Usually the opposing player will get to shoot foul shots "one and one," meaning he gets a shot, and then a second shot only if he makes the first one. Suppose the opposing player has made 72% of his foul shots this season. Estimate the number of points he will score in a one-and-one situation.

34. **Blood donors** A person with type-O-positive blood can receive blood only from other type-O donors. About 44% of the population has type-O blood. At a blood drive, how many potential donors do you expect to examine in order to get three units of type-O blood?

35. **Free groceries** To attract shoppers, a supermarket runs a weekly contest that involves "scratch-off" cards. With each purchase, customers get a card with a black spot obscuring a message. When the spot is scratched away, most of the cards simply say, "Sorry—please try again." But during the week, 100 customers will get cards that make them eligible for a drawing for free groceries. Ten of the cards are worth \$200, 10 others say \$100, 20 are worth \$50, and the rest are worth \$20. To register the cards, customers write their names on them and put them in a barrel at the front of the store. At the end of the week, the store manager draws cards at random, awarding the lucky customers free groceries in the amount specified on their card. The drawings continue until the store has given away more than \$500 in free groceries. Estimate the average number of winners each week.

36. **Find the ace** A new electronics store holds a contest to attract shoppers. Once an hour someone in the store is chosen at random to play the Music Game. Here's how it works: An ace and four other cards are shuffled and placed face down on a table. The customer gets to turn cards over one at a time, looking for the ace. The person wins \$100 worth of free CDs or DVDs if the ace is the first card, \$50 if it is the second card, and \$20, \$10, or \$5 if it is the third, fourth, or fifth card chosen. What is the average dollar amount of music the store will give away?

37. **The family** Many couples want to have both a boy and a girl. If they decide to continue to have children until they have one child of each sex, what would the average family size be? Assume that boys and girls are equally likely.

38. **A bigger family** Suppose a couple will continue having children until they have at least two children of each sex (two boys *and* two girls). How many children might they expect to have?

39. Dice game You are playing a children's game in which the number of spaces you get to move is determined by the rolling of a die. You must land exactly on the final space in order to win. If a roll makes you overshoot the final space, you have to stay at your current position and then make another roll for your next turn. If you are 10 spaces away, how many turns might it take you to win?

40. The hot hand A basketball player with a 65% shooting percentage has just made six shots in a row. The announcer says this player "is hot tonight! She's in the zone!" Assume the player takes about 20 shots per game. Is it unusual for her to make six or more shots in a row during a game?

41. The Stanley Cup final The Stanley Cup final ends when a team wins four games. Suppose that sports analysts consider one team a bit stronger, with a 55% chance to win any individual game. Estimate the likelihood that the underdog wins the series.

42. Teammates Four couples at a dinner party play a board game after the meal. They decide to play as teams of two and to select the teams randomly. All eight people write their names on slips of paper. The slips are thoroughly mixed, then drawn two at a time. How likely is it that every person will be teamed with someone other than the person he or she came to the party with?

43. Second team Suppose the couples in Exercise 42 choose the teams by having one member of each couple write their names on the cards and the other people each pick a card at random. How likely is it that every person will be teamed with someone other than the person he or she came with?

44. Job discrimination? A company with a large sales staff announces openings for three positions as regional managers. Twenty-two of the current salespersons apply: 12 men and 10 women. After the interviews, when the company announces the newly appointed managers, all three positions go to women. The men complain of job discrimination. Do they have a case? Simulate a random selection of three people from the applicant pool, and make a decision about the likelihood that a fair process would result in hiring all women.

45. Smartphones A proud legislator claims that your province's new law banning texting and hand-held phones while driving has reduced cell phone use to less than 12% of all drivers. While waiting for your bus the next morning, you notice that 4 of the 10 people who drive by are using their cell phones. Does this cast doubt on the legislator's figure of 12%? Use a simulation to estimate the likelihood of seeing at least 4 of 10 randomly selected drivers using their cell phones if the actual rate of usage is 12%. Explain your conclusion clearly.

46. Rainfall According to Environment Canada, the yearly rainfall amounts in Vancouver can be described by a normal model with a mean of 1055.4 mm and a standard deviation of 172 mm. A weather forecaster in Vancouver examined the existing records at random to look for two years with yearly rainfall over 1000 mm. How many years do you think he will need to examine?

47. Tires A tire manufacturer believes that the tread life of its snow tires can be described by a normal model with a mean of 32 000 km and a standard deviation of 2500 km. You buy four of these tires, hoping to drive them at least 30 000 km. Estimate the chances that all four last at least that long.

Just Checking ANSWERS

1. The component is one game.
2. Generate random numbers and assign numbers from 00 to 54 to the home team's winning and from 55 to 99 to the visitors' winning.
3. The response is who wins the series.
4. Generate components until one team wins four games. Record which team wins the series.
5. Analyze the response variable by counting up the proportion of wins by the team that starts at home.

MathXL

MyStatLab

Go to MathXL at www.mathxl.com or MyStatLab at www.mystatlab.com. You can practise exercises for this chapter as often as you want. The guided solutions will help you find answers step by step. You'll find a personalized study plan available to you too!

chapter

10

Sample Surveys

The student remained unimpressed, saying, "You mean that just a sample of a few thousand can tell us exactly what over 250 MILLION people are doing?" Finally, the professor, somewhat disgruntled with the skepticism, replied, "Well, the next time you go to the campus clinic and they want to do a blood test . . . tell them that's not good enough . . . tell them to TAKE IT ALL!!"

Where are we going?

We see surveys all the time. How can asking just a thousand people tell us much about a national election? And how *are* the respondents selected? It turns out that there are many ways to select a good sample, but there are just three main ideas to understand.

In June 2008, an Ipsos Reid poll conducted a survey on behalf of the Dominion Institute to compare Canadians' knowledge of American history and of Canadian history. They asked a random sample of 1024 Canadian adults 20 questions—10 questions about the history of each country. The respondents had a higher average percentage of correct scores on questions about American history (47%) than they did on Canadian history (42%). Ipsos Reid went on to report that "it appears Canadians know more about the history and politics of their neighbours to the south than they do about their own country."[1] That step from a small sample to the entire population is impossible without understanding Statistics. To make business decisions, to do science, to choose wise investments, or to understand how voters think they'll vote in the next election, we need to stretch beyond the data at hand to the world at large.

10.1 The Three Big Ideas of Sampling

To make that stretch, we need three ideas. You'll find the first one natural. The second may be more surprising. The third is one of the strange but true facts that often confuse those who don't know statistics.

Idea 1: Examine a Part of the Whole

The first idea is to draw a sample. We'd like to know about an entire group of individuals—a **population**—but examining all of them is usually impractical, if not impossible. So we settle for examining a smaller group of individuals—a **sample**—selected from the population.

A S *Activity:* **Populations and Samples.** Explore the differences between populations and samples.

THE *W*'s AND SAMPLING

The population we are interested in is usually determined by the *Why* of our study. The sample we draw will be the *Who*. *When* and *How* we draw the sample may depend on what is practical.

You do this every day. For example, suppose you wonder if your friends will like the vegetable soup you're cooking for dinner tonight. To decide whether it meets your standards, you only need to try a small amount. You might taste just a spoonful or two. You certainly don't have to consume the whole pot. You trust that the taste will *represent* the flavour of the entire pot. The idea behind your tasting is that a small sample, if selected properly, can represent the entire population.

It's hard to go a day without hearing about the latest opinion poll. These polls are examples of **sample surveys**, designed to ask questions of a small group of people in the hope of learning something about the entire population. Most likely, you've never been selected to be part of one of these national opinion polls. That's true of most people. So how can the pollsters claim that a sample is representative of the entire population? The answer is that professional pollsters work quite hard to ensure that the "taste"—the sample that they take—represents the population.

[1] www.historicacanada.ca/content/polls

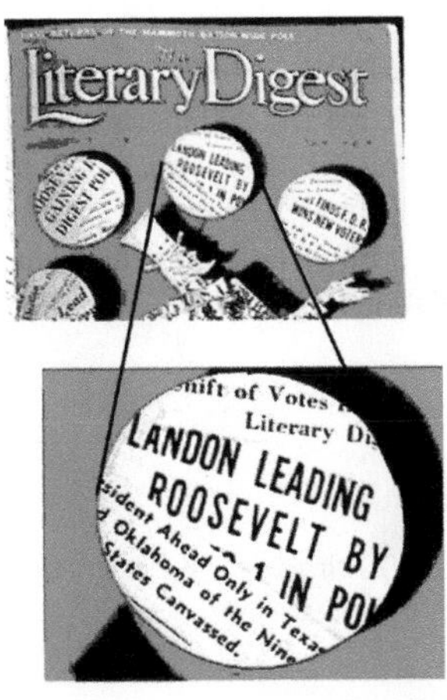

The New York Public Library/Art Resource

Keystone Pictures USA/Alamy

In 1936, a young pollster named George Gallup used a subsample of only 3000 of the 2.4 million responses that the *Literary Digest* received to reproduce the wrong prediction of Landon's victory over Roosevelt. He then used an entirely different sample of 50 000 and predicted that Roosevelt would get 56% of the vote to Landon's 44%. His sample was apparently much more *representative* of the actual voting populace. The Gallup Organization went on to become one of the leading polling companies.

Bias

Selecting a sample to represent the population fairly is more difficult than it sounds. Polls or surveys most often fail because they use a sampling method that tends to over- or underrepresent parts of the population. The method may overlook subgroups that are harder to find (such as the homeless or those who use only cell phones) or favour others (such as Internet users who like to respond to online surveys). Sampling methods that, by their nature, tend to over- or underemphasize some characteristics of the population are said to be **biased**. Bias is the bane of sampling—the one thing above all to avoid. Conclusions based on samples drawn with biased methods are inherently flawed. There is usually no way to fix bias after the sample is drawn and no way to salvage useful information from it.

Here's a famous example of a really dismal failure. By the beginning of the twentieth century, it was common for newspapers to ask readers to return "straw" ballots on a variety of topics. (Today's Internet surveys are the same idea, gone electronic.)

During the period from 1916 to 1936, the magazine *Literary Digest* regularly surveyed public opinion and forecasted election results correctly. During the 1936 U.S. presidential campaign between Alf Landon and Franklin Delano Roosevelt, it mailed more than 10 million ballots and got back an astonishing 2.4 million. (Polls were still a relatively novel idea, and many people thought it was important to send back their opinions.) The results were clear: Alf Landon would be the next president by a landslide, 57% to 43%. In fact, Roosevelt won, 62% to 37%, and, perhaps coincidentally, the magazine went bankrupt soon afterward.

What went wrong? One problem was that *Literary Digest*'s sample wasn't representative. Where would *you* find 10 million names and addresses to sample? The *Digest* used the phone book, as modern surveys do.[2] But in 1936, at the height of the Great Depression, telephones were a real luxury, so they under-sampled poor voters. The campaign of 1936 focused on the economy, and those who were less well off were more likely to vote for Roosevelt (a Democrat). So the magazine's sample was hopelessly biased.

How do modern polls get their samples to *represent* the entire population? You might think that they'd handpick individuals to sample with care and precision. But in fact, they do something quite different: They select individuals to sample *at random*.

Idea 2: Randomize

Think back to the soup sample. Suppose you add some salt to the pot. If you sample it from the top before stirring, you'll get the misleading idea that the whole pot is salty. If you sample from the bottom, you'll get an equally misleading idea that the whole pot is bland. By stirring, you *randomize* the amount of salt throughout the pot, making each taste more typical of the whole pot.

Not only does randomization protect you against factors that you know are in the data, it can also help protect against factors you don't even know are there. Suppose, while you weren't looking, a friend added a handful of peas to the soup. If they're down at the bottom of the pot, and you don't randomize the soup by stirring, your test spoonful won't have any peas. By stirring in the salt, you *also* randomize the peas throughout the pot, making your sample taste more typical of the overall pot *even though you didn't know the peas were there.* So randomizing protects us even in this case.

How do we "stir" people in a survey? We select them at random. **Randomizing** protects us from the influences of *all* the features of our population by making sure that, *on average,* the sample looks like the rest of the population. The importance of deliberately using randomization is one of the great insights of Statistics.

A S *Video:* **The *Literary Digest* Poll and the Election of 1936.** Hear the story of one of the most famous polling failures in history.

[2]Today phone numbers are computer-generated to make sure that unlisted numbers are included. But special methods must be used for the growing part of the population who do not have "landline" access.

A S *Activity:* **Sampling from Some Real Populations.** Draw random samples to see how closely they resemble each other and the population.

Why not match the sample to the population? Rather than randomizing, we could try to design our sample so that the people we choose are typical in terms of every characteristic we can think of. We might want the income levels of those we sample to match the population. How about age? Political affiliation? Marital status? Having children? Living in the suburbs? We can't possibly think of all the things that might be important. Even if we could, we wouldn't be able to match our sample to the population for all these characteristics.

For Example IS A RANDOM SAMPLE REPRESENTATIVE?

Here are summary statistics comparing two samples of 8000 drawn at random from a company's database of 3.5 million customers:

Age (yr)	White (%)	Female (%)	# of Children	Income Bracket (1–7)	Wealth Bracket (1–9)	Homeowner? (% Yes)
61.4	85.12	56.2	1.54	3.91	5.29	71.36
61.2	84.44	56.4	1.51	3.88	5.33	72.30

QUESTION: Do you think these samples are representative of the population? Explain.

ANSWER: The two samples look very similar with respect to these seven variables. It appears that randomizing has automatically matched them pretty closely. We can reasonably assume that since the two samples don't differ too much from each other, they don't differ much from the rest of the population either.

Idea 3: It's the Sample Size

A S *Activity:* **Does the Population Size Matter?** Here's the narrated version of this important idea about sampling.

How large a random sample do we need for the sample to be reasonably representative of the population? Most people think that we need a large percentage, or *fraction*, of the population, but it turns out that what matters is the *number* of individuals in the sample, not the size of the population. A random sample of 100 students in a university represents the student body just about as well as a random sample of 100 voters represents the entire electorate of Canada. This is the *third* idea and probably the most surprising one in designing surveys.

How can it be that only the size of the sample, and not the population, matters? Well, let's return one last time to that pot of soup. If you're cooking for a banquet rather than just for a few people, your pot will be bigger, but do you need a bigger spoon to decide how the soup tastes? Of course not. The same-size spoonful is probably enough to make a decision about the entire pot, no matter how large the pot. The *fraction* of the population that you've sampled doesn't matter.[3] It's the **sample size** itself that's important.

How big a sample do you need? That depends on what you're estimating. To get an idea of what's really in the soup, you'll need a large enough taste to get a *representative* sample from the pot. For a survey that tries to find the proportion of the population falling into a category, you'll usually need several hundred respondents to say anything precise enough to be useful.[4]

A friend who knows that you are taking Statistics asks your advice on her study. What can you possibly say that will be helpful? Just say, "If you could just get a larger sample, it would probably improve your study." Even though a larger sample might not be worth the cost, it will almost always make the results more precise.

What do the pollsters do? How do professional polling agencies do their work? The most common polling method today is to contact respondents by telephone. Computers generate random telephone numbers, so pollsters can even call some people with unlisted phone numbers. The interviewer may then ask to speak with the adult who is home who had the most recent birthday or use some other essentially random way to select among those available. In phrasing questions, pollsters often list alternative responses

[3]Well, that's not exactly true. If the population is small enough and the sample is more than 10% of the whole population, it *can* matter. It *doesn't* matter when our sample is a very small fraction of the population.
[4]Chapter 16 gives the details behind this statement and shows how to decide on a sample size for a survey.

A S **Video: Frito-Lay Sampling for Quality.** How does a potato chip manufacturer make sure to cook only the best potatoes?

(such as candidates' names) in different orders, to avoid biases that might favour the first name on the list.

Do these methods work? The Pew Research Center for the People and the Press, reporting on one survey, said that

> *Across five days of interviewing, surveys today are able to make some kind of contact with the vast majority of households (76%). But because of busy schedules, skepticism and outright refusals, interviews were completed in just 38% of households that were reached using standard polling procedures.*

Nevertheless, studies indicate that those actually sampled can give a good snapshot of larger populations from which the surveyed households were drawn.

Does a Census Make Sense?

Why bother determining the right sample size? Wouldn't it be better just to include everyone and "sample" the entire population? Such a special sample is called a **census**. Although a census would appear to provide the best possible information about a population, there are a number of reasons why it might not.

First, it can be difficult to complete a census. Some individuals in a population will be hard (and expensive) to locate. Or a census might just be impractical. If you were a taste-tester for the Hostess™ Company, you probably wouldn't want to census *all* the Twinkies on the production line. Not only might this be life-endangering, but you wouldn't have any left to sell.

Second, populations rarely stand still. In populations of people, babies are born and folks die or leave the country. In opinion surveys, events may cause a shift in opinion during the survey. A census takes longer to complete and the population changes while you work. A sample surveyed in just a few days may give more accurate information.

Third, taking a census can be more complex than sampling. For example, the Canadian Census may record too many university students. Some could be counted once with their families and then counted a second time in a report filed by their schools.

10.2 Populations and Parameters

Any quantity that we calculate from data could be called a "statistic." But in practice, we usually use a statistic to estimate a population parameter.

A S **Activity: Statistics and Parameters.** Explore the difference between statistics and parameters.

Remember: Population model parameters are not just unknown—usually they are *unknowable.* The best we can do is to estimate them with sample statistics.

A study found that Canadians under 25 years old are less likely to "buckle up" and fasten their seatbelts when in cars. In 2011, Transportation Canada reported that 7.0% of Canadians under 25 never or rarely wear seatbelts.[5] We're sure they didn't take a census, so what does the 7.0% mean? We can't know what percentage of Canadians under 25 wear seatbelts. Reality is just too complex. But we can simplify the question by building a model.

Models use mathematics to represent reality. Parameters are the key numbers in those models. A parameter used in a model for a population is sometimes called (redundantly) a **population parameter**.

But let's not forget about the data. We use summaries of the data to estimate the population parameters. As we know, any summary found from the data is a **statistic**. Sometimes you'll see the (also redundant) term **sample statistic**.[6]

We've already met two parameters in Chapter 5: the mean, μ, and the standard deviation, σ. We'll try to keep denoting population model parameters with Greek letters and the corresponding statistics with Latin letters. Usually, but not always, the letter used for the statistic and the parameter correspond in a natural way. So the standard deviation of the data is s, and the corresponding parameter is σ. In Chapter 6, we used r to denote the sample correlation. The corresponding correlation in a model for the population would be called ρ (rho). In Chapter 7, b_1 represented the slope of a linear regression estimated from the data. But when we think about a (linear) *model* for the population, we denote the slope parameter β_1 (beta).

[5] http://www.tc.gc.ca/media/documents/roadsafety/tp2436e-rs201101.pdf
[6] Where else besides a sample *could* a statistic come from?

NOTATION ALERT

This entire table is a notation alert.

Get the pattern? Good. Now it breaks down. We denote the mean of a population model with μ. We denote the sample mean by putting a bar over anything when we average it, so we write $\bar{y}$ (rather than the Latin equivalent m). What about proportions? Suppose we want to talk about the proportion of Canadians under 25 years old who don't wear seatbelts. We'll use p for the population model parameter (since Greek π is already reserved for the number 3.1415926 . . .) and $\hat{p}$ for the proportion from the data (since, like $\hat{y}$ in regression, it's an estimated value).

Here's a table summarizing the notation:

Name	Statistic	Parameter
Mean	$\bar{y}$	μ (mu, pronounced "meeoo," not "moo")
Standard deviation	s	σ (sigma)
Correlation	r	ρ (rho)
Regression coefficient	b	β(beta, pronounced "baytah")
Proportion	$\hat{p}$	p (pronounced "pee"[7])

We draw samples because we can't work with the entire population. When the statistics we compute from the sample accurately reflect the corresponding parameters, then the sample is said to be **representative**. A biased sampling methodology tends to over- or underestimate the parameter(s) of interest.

Just Checking

1. Various claims are often made for surveys. Why are each of the following claims not correct?
   a) It is always better to take a census than to draw a sample.
   b) Stopping students on their way out of the cafeteria is a good way to sample if we want to know about the quality of the food there.
   c) We drew a sample of 100 from the 3000 students in a school. To get the same level of precision for a town of 30 000 residents, we'll need a sample of 1000.
   d) A poll taken at a Statistics support Web site garnered 12 357 responses. The majority said they enjoy doing statistics homework. With a sample size that large, we can be pretty sure that most Statistics students feel this way, too.
   e) The true percentage of all Statistics students who enjoy the homework is called a "population statistic."

10.3 Simple Random Samples

How would you select a representative sample? Most people would say that every individual in the population should have an equal chance to be selected, and certainly that seems fair. But it's not sufficient. There are many ways to give everyone an equal chance that still wouldn't give a representative sample. Consider, for example, a school that has equal numbers of males and females. We could sample like this: Flip a coin. If it comes up heads, select 100 female students at random. If it comes up tails, select 100 males at random. Everyone has an equal chance of selection, but every sample is of only a single sex—hardly representative.

We need to do better. Suppose we insist that every possible *sample* of the size we plan to draw has an equal chance to be selected. This ensures that situations like the one just described are not likely to occur and still guarantees that each person has an equal chance of being selected. What's different is that with this method, each *combination* of people has an equal chance of being selected as well. A sample drawn in this way is called a **Simple Random Sample**, usually abbreviated **SRS**. An SRS is the standard against which we measure other sampling methods, and the sampling method on which the basic theory of working with sampled data is based.

[7]Just in case you weren't sure.

To select a sample at random, we first need to define where the sample will come from. The **sampling frame** is a list of individuals from which the sample is drawn. For example, to draw a random sample of students at a university, we might obtain a list of all registered full-time students and sample from that list. In defining the sampling frame, we must deal with the details of defining the population. Are part-time students included? How about those who are attending school elsewhere and transferring credits back to the university?

Once we have a sampling frame, the easiest way to choose an SRS is to assign a random number to each individual in the sampling frame. We then select only those whose random numbers satisfy some rule.[8] Let's look at some ways to do this.

For Example USING RANDOM NUMBERS TO GET AN SRS

There are 80 students enrolled in an introductory Statistics class; you are to select a sample of five.

QUESTION: How can you select an SRS of five students using these random digits found on the Internet?

05166 29305 77482

ANSWER: First I'll number the students from 00 to 79. Taking the random numbers two digits at a time gives me 05, 16, 62, 93, 05, 77, and 48. I'll ignore 93 because the students were numbered only up to 79. And so I don't pick the same person twice, I'll skip the repeated number 05. My simple random sample consists of students with the numbers 05, 16, 62, 77, and 48.

ERROR OKAY, BIAS NO WAY!

Sampling variability is sometimes referred to as *sampling error,* making it sound as if it's some kind of mistake. It's not. We understand that samples will vary, so "sampling error" is to be expected. It's *bias* we must strive to avoid. Bias means our sampling method distorts our view of the population, and that will surely lead to mistakes.

- We can be more efficient when we're choosing a larger sample from a sampling frame stored in a data file. First we obtain random numbers with several digits (say, from 0 to 10 000) and assign one to each individual. Then we arrange the random numbers in numerical order, keeping the names along with them. Choosing the first n names from this list will give us a random sample of that size.
- Often the sampling frame is so large that it would be too tedious to number everyone consecutively. If our intended sample size is approximately 10% of the sampling frame, we can assign each individual a single random digit 0 to 9. Then we select only those with a specific random digit, say, 5.

Samples drawn at random generally differ one from another. Each draw of random numbers selects *different* people for our sample. These differences lead to different values for the variables we measure. We call these sample-to-sample differences **sampling variability**. Surprisingly, sampling variability isn't a problem; it's an opportunity. In future chapters we'll investigate what the variation in a sample can tell us about its population.

For Example DESIGNING A SAMPLE

What do the frosh at your school think of the food served on campus? Food Services has asked you to design a survey. The registrar is willing to make available an alphabetical list of all frosh as a computer file.

QUESTION: How would you draw a simple random sample?

ANSWER: I would list the names in one column of a spreadsheet (or a variable in a statistics program). Suppose there are N names on the list.

In the next column (or variable), I would generate N random numbers between 1 and 10 times N, at the least (so there will be few duplicate numbers). I could get these from a suitable Web site or from software.

Then I would sort the random numbers into numerical order, "carrying along" the names to randomize the order of the names.

My sample would take names starting at the top of the sorted list.

---

[8]Chapter 9 presented ways of finding and working with random numbers.

10.4 Other Sampling Designs

Simple random sampling is not the only fair or unbiased way to sample. More complicated designs may save time or money or help avoid sampling problems. All statistical sampling designs have in common the idea that chance, rather than human choice, is used to select the sample.

Stratified Sampling

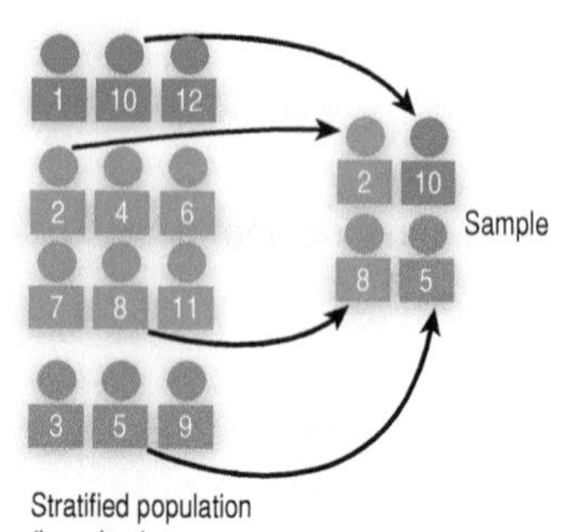

Stratified population (by colour)

Reprinted with permission of Dan Kernler, Associate Professor of Mathematics, Elgin Community College, Elgin, IL

Designs that are used to sample from large populations—especially populations residing across large areas—are often more complicated than simple random samples. Sometimes the population is first sliced into groups, called **strata**, before the sample is selected. We usually try to create these strata so that they are relatively homogeneous—i.e., so that individuals within the same stratum are somewhat more similar to each other than to individuals in other strata (or that strata differ substantially from each other), and if we succeed, significant benefits will accrue, as discussed below. However, there are also times when their formation is dictated by various practical sampling issues. Once the strata are defined, random sampling is used within each stratum, after which all the results from the various strata are combined to produce population estimates. This common sampling technique is called **stratified random sampling**.

Why would we want to complicate things? Here's an example. Suppose we want to learn how students feel about funding for the soccer team at a large university. The campus is 60% men and 40% women, and we suspect that men and women have different views on the funding. If we use simple random sampling to select 100 people for the survey, we could end up with 70 men and 30 women or 35 men and 65 women. Our resulting estimates of the level of support for the soccer funding could vary widely. To help reduce this sampling variability, we can decide to force a representative balance, selecting 60 men at random and 40 women at random. This would guarantee that the proportions of men and women within our sample match the proportions in the population, and that should make such samples more accurate in representing population opinion.

You can imagine the importance of stratifying by race, income, age, and other characteristics, depending on the questions in the survey. Samples taken within a stratum vary less, so our estimates can be more precise. This reduced sampling variability is the most important benefit of effective stratifying.

There is great flexibility in this method. Since we sample independently from the various strata, we are free to use different sampling methods or sampling rates best suited to each stratum (e.g., we may choose a smaller sampling fraction in strata where the data are more difficult or expensive to collect). Another advantage is that aside from obtaining overall estimates, we will also have estimates for each stratum. For example, we might split Canada into five or six geographic regions (the strata) and have separate teams run surveys separately within each region, varying sampling procedures as needed. We could later combine the data from all regions to produce national estimates, and, if desired, we can easily produce the regional estimates, too.

Stratified sampling can also help us notice important differences among groups. But be careful. As we saw in Chapter 2, if we unthinkingly combine group data, we risk reaching the wrong conclusion, becoming victims of Simpson's paradox.

For Example STRATIFYING THE SAMPLE

RECAP: You're trying to find out what frosh think of the food served on campus. Food Services believes that men and women typically have different opinions about the importance of the salad bar.

QUESTION: How should you adjust your sampling strategy to allow for this difference?

ANSWER: I will stratify my sample by drawing an SRS of men and a separate SRS of women—assuming that the data from the registrar include information about each person's sex.

Cluster and Multistage Sampling

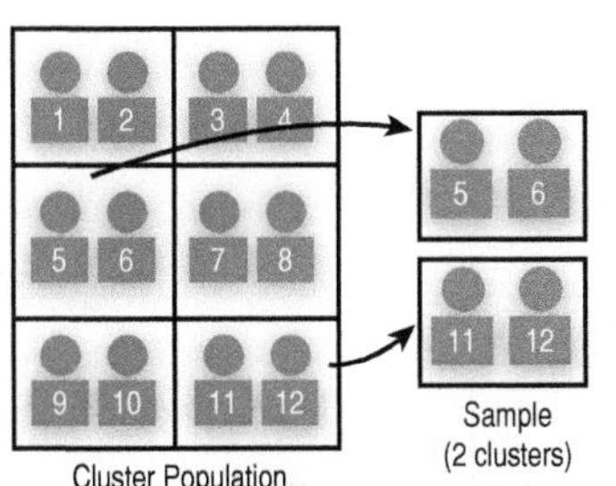

Reprinted with permission of Dan Kernler, Associate Professor of Mathematics, Elgin Community College, Elgin, IL

Suppose we wanted to assess the reading level of this textbook based on the length of the sentences. Simple random sampling could be awkward; we'd have to number each sentence, then find, for example, the 576th sentence or the 2482nd sentence, and so on. Doesn't sound like much fun, does it?

It would be much easier to pick a number of *pages* at random and count the lengths of the sentences on those pages. Splitting the population into **clusters** (each page is a cluster of sentences) can make sampling more practical. Then we could simply select a number of clusters at random and perform a census within each of them (measure all sentences on the page). This sampling design is called simple (or one-stage) **cluster sampling**. If instead of a complete census of individuals in the cluster, we randomly select only a portion of the individuals in each selected cluster (measure only some of the sentences on each selected page), this introduces a second stage of sampling, and voilà, we have *two-stage (cluster) sampling.* With proper random selection of clusters (and within clusters, if there are multiple stages of sampling), cluster sampling will give us an unbiased sample.

For Example CLUSTER SAMPLING

RECAP: In trying to find out what frosh think about the food served on campus, you've considered both an SRS and a stratified sample. Now you have run into a problem: It's simply too difficult and time-consuming to track down the individuals whose names were chosen for your sample. Fortunately, frosh at this school are all housed in 12 frosh residences.

QUESTIONS: How could you use this fact to draw a cluster sample? How might that alleviate the problem? What concerns do you have?

ANSWER: To draw a cluster sample, I would select several residences at random and then try to contact everyone in each selected residence. I could save time by simply knocking on doors on a given evening and interviewing people. I hope to achieve a high response rate on that evening, or, if not, that the people I'm able to contact are representative of the residence (the usual non-response issue). A major concern would be the differences among residences. If I think that students' opinions about food service will not depend much on the particular residence, this sampling approach should work well. However, if I expect big differences among residences, the sampling variation resulting from this design could be great (if you choose different residences, the results change significantly), and it might be better to treat the residences as strata and to draw a random sample from each of the 12 residences—if feasible.

Stratification vs. Clustering

It appears that we could split a population into strata or clusters. What is the difference? We create strata by dividing the entire population into subpopulations, which we try to define wisely (differing along one or more important dimensions). Then we draw a separate random sample from each and every stratum. We sample *from* or *within* strata,

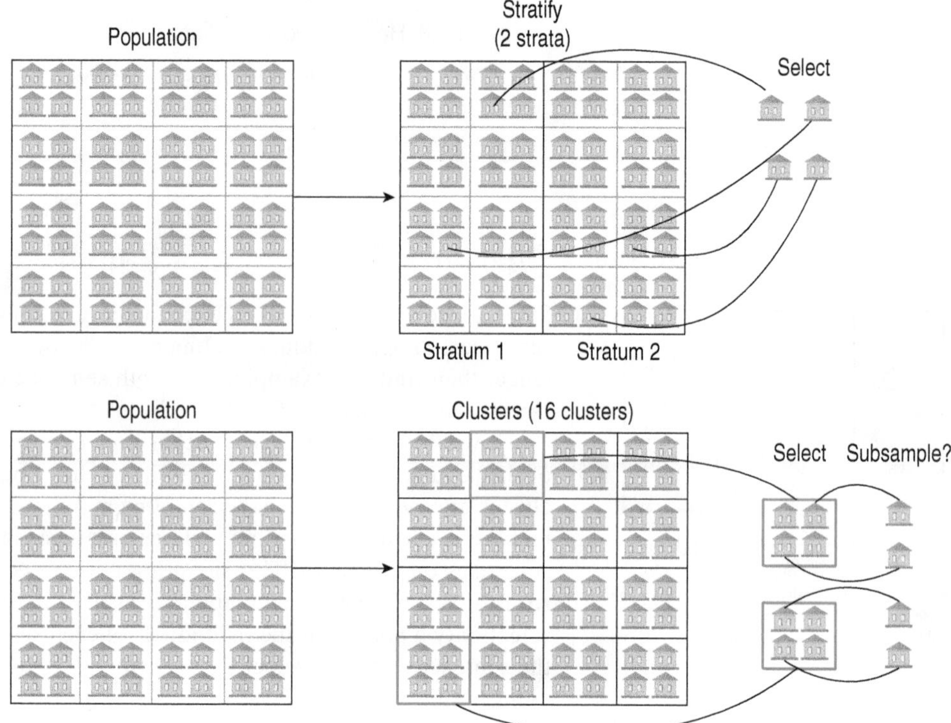

Figure 10.1

Stratified and Cluster Sampling. Above, we stratify geographically (east, west) before drawing the sample. Below, we view the city blocks as clusters, then draw two of them, perhaps following up with a subsample of only half of each block (second stage of sampling).

whereas clusters are *sampling units*—to be picked at random from a list. Clusters are randomly selected; strata are not. We need to learn something about each and every stratum, so we randomly select some individuals from each one, but we do not randomly select the strata themselves (instead, we *form* strata). Clusters are chosen at random from a list (or frame) of clusters, and we do not study all of them; only some lucky clusters make it to our final sample. Strata are often few and very large (e.g., geographic regions or size categories of businesses: large, medium, small), therefore we cannot skip over any of them. In contrast, clusters tend to be smaller and more numerous (such as blocks in a city or enumeration areas in a large region), so we don't need to study all of them, just enough to ensure representativeness. Figure 10.1 compares the two approaches. Imagine 16 city blocks with four households in each. We wish to sample four households. Which scheme gives more precise estimates (with less sampling variability)? Which costs more? How could you combine the two approaches?[9]

Why sample individuals in clusters, rather than directly? Generally, for practical reasons. It may be too difficult to list all population individuals, such as households in a city, but we may be able to obtain a map showing blocks, on which we can number the blocks and then select some blocks at random. From each selected block, we can now sample some or all of its households. It is easier to get at the final sampled households indirectly by first sampling clusters of them, because the needed frame is readily available. As a bonus, we only need to list households in the selected blocks, and not the entire city. And then there is the improved cost efficiency—in many cases, continued sampling from the same cluster can add extra data very cheaply. Once you arrive at a selected city block, it is fast and easy to go from household to household on the same block. Or suppose we want to sample students. Students are clustered in schools—it is easier to list schools than students, and more convenient to sample many students from a few randomly chosen schools, than to sample students spread out all over the city. Of course, a simple random

[9]The cluster sample would be more convenient and cheaper than the stratified sample, but less precise. We could combine methods by stratifying geographically and then selecting a cluster (block) or two at random from each stratum (region).

sample of students in the city is more precise than a sample using only some selected schools, if the samples are of the same size—but that is not the issue. What we care about is *maximizing the precision obtained for some total expenditure*, not for some fixed sample size, and the same expenditure can yield a much bigger cluster sample than a simple random sample.

If clusters are big, we often *subsample* selected ones. We choose at random some, but not all, of the individuals in each selected cluster, since typically there are diminishing returns from continued observation within a cluster when the clusters are somewhat homogeneous, as is all too often the case with naturally existing clusters (for example, people in the same city block or enumeration area tend to be somewhat similar in income, ethnicity, etc.). As a general rule, the more similar the individuals in a cluster are, the fewer individuals we should bother to study, whereas we should obtain data on many or all of the individuals in clusters that are very heterogeneous, since we are getting good additional information on the cheap. When we subsample and study just a portion of each selected cluster, we are introducing a *second stage* of sampling.

Often, individuals exist in a hierarchy of clusters—for example, adults residing in some areas of a city are clustered together in households, which in turn are clustered together in blocks, which are clustered in enumeration areas. To sample adults, you could start with a list of enumeration areas and randomly select some. Then, from each selected area, get a map and select some blocks, go to each selected block and select some households, and, finally, select one adult from each household. You would work your way down, in successive stages of sampling (four stages, in this case) until you finally get to the population elements. This is **multistage sampling**.

Sometimes we use a variety of sampling methods together. In trying to assess the reading level of this book, we might worry that it starts out easy and then gets harder as the concepts become more difficult. If so, we'd want to avoid samples that select heavily from early or from late chapters. To guarantee a fair mix, we could randomly choose one chapter from each of the eight parts of the book and then randomly select a few pages from each of those chapters. If that made too many sentences in total, we might select a few sentences at random from each of the chosen pages. In other words, we

i. stratify by the part of the book, so that each part is a stratum that we must sample from;
ii. randomly choose a chapter (first-stage cluster or sampling unit) from each part;
iii. within each selected chapter, choose some pages (second-stage clusters or sampling units) at random; and
iv. perhaps finally, select an SRS of sentences from each selected page.

Most surveys conducted by professional polling organizations use some combination of stratified and multistage cluster sampling as well as simple random samples or systematic samples (discussed later in the chapter). In the example of sampling adults in a city, it would add to our precision if we used readily available information about enumeration areas to organize them into several strata, such as high income, moderate income, and low income (or urban/suburban/rural), before sampling, and then selecting, say, two enumeration areas from each of the three strata, rather than selecting an SRS of six enumeration areas from the entire list.

For Example COMBINING METHODS TO IMPROVE PRECISION

RECAP: Having learned that frosh are housed in separate residences allowed you to sample their attitudes about the campus food by going to residences chosen at random, but you're still concerned about possible differences in opinions between men and women. It turns out that these frosh residences house the sexes on alternate floors.

QUESTION: How can you design a sampling plan that uses this fact to your advantage?

ANSWER: Select floors at random, in two stages, and stratify where convenient. Choose four residences at random and then select two of the male residence floors and two of the female residence floors at random in each of the chosen residences. Then try to interview everyone on each selected floor, or, if that's just too many, choose an SRS of rooms on each floor. (Or, better yet, read on about systematic sampling, and consider a systematic sample along the chosen floor, taking, say, every fifth room).

In this example, we combined stratification—based on our knowledge about possible sex differences—with two-stage cluster sampling, with residences as the first-stage sampling units or clusters and floors the second-stage sampling units or clusters.

Systematic Samples

Some samples select individuals systematically. For example, you might survey every tenth person on an alphabetical list of students. To make it random, you must start the systematic selection from a randomly selected individual. When the order of the list is not associated in any way with the responses sought, **systematic sampling** is similar to drawing an SRS, while it can be much less expensive. It will give more accurate (less variable) results than an SRS when the list order is associated to some extent with the response variable, as long as the systematic sampling rule doesn't coincide with some "cycle" in the sampling frame (for example, if every tenth house in our frame were a corner house, we'd have a cycle).

Think about the reading-level sampling example again. Suppose we have chosen a chapter of the book at random, then three pages at random from that chapter, and now we want to select a sample of 10 sentences from the 73 sentences found on those pages. Instead of numbering each sentence so we can pick a simple random sample, it would be easier to sample systematically. A quick calculation shows $73/10 = 7.3$, so we can get our sample by just picking every seventh sentence on the page. But where should you start? At random, of course. We've accounted for $10 \times 7 = 70$ of the sentences, so we'll throw the extra three into the starting group and choose a sentence at random from the first 10. Then we pick every seventh sentence after that and record its length.

Just Checking

2. We need to survey a random sample of the 300 passengers on a flight from Vancouver to Tokyo. Name each sampling method described below.
   - **a)** Pick every tenth passenger as people board the plane.
   - **b)** From the boarding list, randomly choose 5 people flying first class and 25 of the other passengers.
   - **c)** Randomly generate 30 seat numbers and survey the passengers who sit there.
   - **d)** Randomly select a seat position (right window, right centre, right aisle, etc.), and survey all the passengers sitting in those seats.

Step-by-Step Example DESIGNING A SURVEY

The assignment says, "Conduct your own sample survey to find out how many hours per week students at your school spend watching television during the school year." Let's see how we might do this step by step. (Remember, though—actually collecting the data from your sample can be difficult and time-consuming.)

Question: How would you design this survey?

THINK ➡ Plan State what you want to know.

I want to design a study to find out how many hours of television students at my school watch.

Population and Parameter Identify the *W*'s of the study. The *Why* determines the population and the associated sampling frame. The *What* identifies the parameter of interest and the variables measured. The *Who* is the sample we actually draw. The *How, When,* and *Where* are given by the sampling plan.

Often, thinking about the *Why* will help you see whether the sampling frame and plan are adequate to learn about the population.

The population studied was students at our school. I obtained a list of all students currently enroled and used it as the sampling frame. The parameter of interest was the number of television hours watched per week during the school year, which I attempted to measure by asking students how much television they had watched the previous week.

Sampling Plan Specify the sampling method and the sample size, *n*. Specify how the sample was actually drawn. What is the sampling frame?

The description should, if possible, be complete enough to allow someone to replicate the procedure, drawing another sample from the same population in the same manner. A good description of the procedure is essential, even if it could never practically be repeated.

Terms like "simple random sample," "cluster sample," and "stratified sample" often appear in this part of the discussion. Be sure to describe how the randomization was performed.

I decided against stratifying by class or sex because I didn't think television watching would differ much between males and females or across classes. I selected a simple random sample of students from the list. I obtained a list of random digits, matched it to an alphabetically arranged list of students, and selected all students who were assigned a "4." This method generated a sample of 212 students from the population of 2133 students.

SHOW ➡ Sampling Practice Specify *When, Where,* and *How* the sampling was performed. Specify any other details of your survey, such as how respondents were contacted, what incentives were offered to encourage them to respond, how nonrespondents were treated, and so on.

The survey was taken over the period October 15 to October 25. I sent surveys to selected students by e-mail, with the request that they respond by e-mail as well. Students who could not be reached by e-mail were handed the survey in person.

TELL ➡ Summary and Conclusion This report should include a discussion of all the elements. In addition, it's good practice to discuss any special circumstances. Professional polling organizations report the *When* of their samples but will also note, for example, any important news that might have changed respondents' opinions during the sampling process. In this survey, perhaps, a major news story or sporting event might change students' television viewing behaviour.

During the period October 15 to October 25, I randomly selected 212 students using a simple random sample from a list of all students currently enroled. The survey they received asked the following question: "How many hours did you spend watching television last week?"

Of the 212 students surveyed, 110 responded. It's possible that the nonrespondents differed in the number of television hours watched from those who responded, but I was unable to follow

The question you ask also matters. It's better to be specific ("How many hours did you watch television last week?") than to ask a general question ("How many hours of television do you usually watch in a week?").

The report should show a display of the data, provide and interpret the statistics from the sample, and state the conclusions that you reached about the population.

up on them due to limited time and funds. The 110 respondents reported an average 3.62 hours of television watching per week. The median was only two hours per week. A histogram of the data shows that the distribution is highly right-skewed, indicating that the median might be a more appropriate summary of the typical television watching of the students.

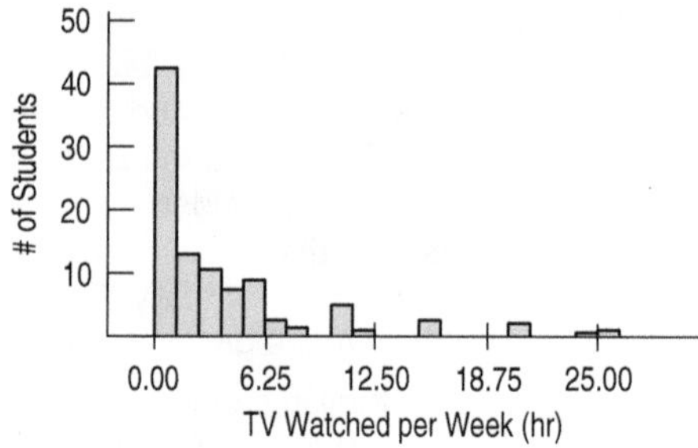

Most of the students (90%) watched between 0 and 10 hours per week, while 30% reported watching less than one hour per week. A few watched much more. About 3% reported watching more than 20 hours per week.

10.5 Defining the "Who": You Can't Always Get What You Want

The population is determined by the *Why* of the study. Unfortunately, the sample is just those we can reach to obtain responses—the *Who* of the study. This difference could undermine even a well-designed study.

Before you start a survey, think first about the population you want to study. You may find that it's not the well-defined group you thought it was. Who, exactly, is a student, for example? Even if the population seems well-defined, it may not be a practical group from which to draw a sample. For example, election polls want to sample from all those who will vote in the next election—a population that is impossible to identify before election day.

Next, you must specify the sampling frame—the individuals who are actually available to be sampled. (Do you have a list of students to sample from? How about a list of registered voters?) Usually, the sampling frame is not the group you *really* want to know about. (All those registered to vote are not equally likely to show up.) The sampling frame limits what your survey can find out.

Then there's your target sample. These are the individuals for whom you *intend* to measure responses. You're not likely to get responses from all of them. Nonresponse is a problem in many surveys.

Finally, there's your sample—the actual respondents. These are the individuals about whom you *do* get data and can draw conclusions. Unfortunately, they might not be representative of the sampling frame or the population.

At each step, the group we can study may be constrained further. The *Who* keeps changing, and each constraint can introduce biases. A careful study should address the question of how well each group matches the population of interest. One of the main benefits of simple random sampling is that it never loses its sense of who's *Who*. The *Who* in an SRS is the population of interest from which we've drawn a representative sample. That's not always true for other kinds of samples.

10.6 The Valid Survey

It isn't sufficient to just draw a sample and start asking questions. We'll want our survey to be *valid*. A valid survey yields the information we are seeking about the population we are interested in. Before setting out to survey, ask yourself:

- What do I want to know?
- Am I asking the right respondents?
- Am I asking the right questions?
- What would I do with the answers if I had them; would they address the things I want to know?

These questions may sound obvious, but there are a number of pitfalls to avoid.

Know what you want to know. Before considering a survey, understand what you hope to learn and about whom you hope to learn it. If you don't know that, you can't even judge whether you have a valid survey. Far too often, people decide to perform a survey without any clear idea of what they hope to learn.

Use the right frame. A valid survey obtains responses from the appropriate respondents. Be sure you have a suitable *sampling frame*. Have you identified the population of interest and sampled from it appropriately? A company might survey customers who returned warranty registration cards—after all, that's a readily available sampling frame. But if the company wants to make their product more attractive, they may be missing the most important population—the customers who rejected their product in favour of one from a competitor. Those might be the folks who can tell the company what to improve.

Tune your instrument. The survey instrument—the questionnaire itself—can be a source of errors. It is often tempting to ask questions you don't really need, since you're sending out the survey anyway. (Who knows? It might be interesting to find out . . . .) But beware—longer questionnaires yield fewer responses and thus a greater chance of nonresponse bias. So, for each question in your survey instrument, ask yourself, "What would I do with the responses if I had them?" If you don't have a good use for the answer, don't ask the question.

Ask specific rather than general questions. People are not very good at estimating their typical behaviour. It is better to ask "How many hours did you sleep last night?" than "How much do you usually sleep?" Sure, some responses will include some unusual events (My dog was sick; I was up all night.) but overall, you'll get better data.

Ask for quantitative results when possible. "How many magazines did you read last week?" is better than "How much do you read: a lot, a moderate amount, a little, or none at all?"

Be careful in phrasing questions. A respondent may not understand the question—or may understand the question differently than the researcher intended it. ("Does anyone in your family belong to a union?" Do you mean just me, my spouse, and my children? Or does "family" include my father, my siblings, and my second cousin once removed? What about my grandfather, who is staying with us? I think he once belonged to the Auto Workers Union.) Respondents are unlikely (or may not have the opportunity) to ask for clarification. A question like "Do you approve of the recent actions of the prime minister?" is likely not to measure what you want if many respondents don't know who the prime minister is or what actions he or she recently made.

Respondents may even lie or shade their responses if they feel embarrassed by the question ("Did you have too much to drink last night?"), are intimidated or insulted by the question ("Could you understand our new *Instructions for Dummies* manual, or was it too difficult for you?"), or if they want to avoid offending the interviewer ("Would you hire a man with a tattoo?" asked by a tattooed interviewer). Also, be careful to avoid phrases that have double or regional meanings. "Is that restaurant cool?" might elicit a response about the climate from your grandmother.

Even subtle differences in phrasing can make a difference. In January 2006, the *New York Times* asked half of the 1229 U.S. adults in their sample the following question:

> *After 9/11, President Bush authorized government wiretaps on some phone calls in the U.S. without getting court warrants, saying this was necessary to reduce the threat of terrorism. Do you approve or disapprove of this?*

They found that 53% of respondents approved. But when they asked the other half of their sample a question with only slightly different phrasing,

> *After 9/11, George W. Bush authorized government wiretaps on some phone calls in the U.S. without getting court warrants. Do you approve or disapprove of this?*

only 46% approved.

Be careful in phrasing answers. It's often a good idea to offer choices rather than inviting a free response. Open-ended answers can be difficult to analyze. "How did you like the movie?" may start an interesting debate, but it may be better to give a range of possible responses. Be sure to phrase them in a neutral way. When asking "Do you support higher school taxes?" positive responses could be worded "Yes," "Yes, it is important for our children," or "Yes, our future depends on it." But those are not equivalent answers.

By permission of John L. Hart FLP and Creators Syndicate, Inc.

The best way to protect a survey from such unanticipated measurement errors is to perform a pilot survey. A **pilot** is a trial run of the survey you eventually plan to give to a larger group, using a draft of your survey questions administered to a small sample drawn from the same sampling frame you intend to use. By analyzing the results from this smaller survey, you can often discover ways to improve your instrument.

10.7 Common Sampling Mistakes, or How to Sample Badly

Bad sample designs yield worthless data. Many of the most convenient forms of sampling can be seriously biased. And there is no way to correct for the bias from a bad sample. So it's wise to pay attention to sample design—and to beware of reports based on poor samples.

Mistake 1: Sample Badly with Volunteers

One of the most common dangerous sampling methods is a voluntary response sample. In a voluntary response sample, a large group of individuals is invited to respond, and all who do respond are counted. This method is used by call-in shows, 900 numbers, Internet polls, and letters written to Members of Parliament. Voluntary response samples are almost always biased, and so conclusions drawn from them are almost always wrong.

It's often hard to define the sampling frame of a voluntary response study. Practically, the frames are groups, such as Internet users who frequent a particular Web site, or those who happen to be watching a particular TV show at the moment. But those sampling frames don't correspond to interesting populations.

Even within the sampling frame, voluntary response samples are often biased toward those with strong opinions or those who are strongly motivated. People with very negative opinions tend to respond more often than those with equally strong positive opinions. The sample is not representative, even though every individual in the population may have been offered the chance to respond. The resulting **voluntary response bias** invalidates the survey.

How often do people write to their Member of Parliament when they're pretty happy about how things are going? People respond when they feel strongly. Which survey would *you* be more likely to respond to—one that asked, "Should the minimum age to drive a car be raised to 25?" or one that asked, "Should shoe sizes be adjusted so women's and men's sizes correspond?"

A S *Activity:* **Sources of Sampling Bias.** Here's a narrated exploration of sampling bias.

If you had it to do over again, would you have children? Ann Landers, the advice columnist, asked parents this question. The overwhelming majority—70% of the more than 10 000 people who wrote in—said no, kids weren't worth it. A more carefully designed survey later showed that about 90% of parents are actually happy with their decision to have children. What accounts for the striking difference in these two results? What kind of parents do you think are most likely to respond to the original question?

For Example BIAS IN SAMPLING

RECAP: You're trying to find out what frosh think of the food served on campus, and have thought of a variety of sampling methods, all time-consuming. A friend suggests that you set up a "Tell Us What You Think" Web site and invite frosh to visit the site to complete a questionnaire.

QUESTION: What's wrong with this idea?

ANSWER: Letting each frosh decide whether to participate makes this a voluntary response survey. Students who are dissatisfied might be more likely to go to the Web site to record their complaints, and this could give me a biased view of the opinions of all frosh.

Mistake 2: Sample Badly, but Conveniently

Internet convenience surveys are worthless. As voluntary response surveys, they have no well-defined sampling frame (all those who use the Internet and visit their site?) and thus report no useful information. Do not believe them.

Another sampling method that doesn't work is convenience sampling. As the name suggests, in **convenience sampling** we simply include the individuals who are convenient for us to sample. Unfortunately, this group may not be representative of the population. In 2006, a trade group for liquor retailers put out a press release with the headline: "Millions of Kids Buy Internet Alcohol, Landmark Survey reveals." In fact, they surveyed 1001 teenagers via the Internet and found 2.1% reported that they had bought alcohol online. This was a convenient way to collect data and surely easier than drawing a simple random sample, but perhaps the trade group shouldn't have drawn the conclusion from this study that "millions of kids buy Internet alcohol." Why not? Many teenagers might refuse to participate in the study, some might not be telling the truth, and many may have just been missed![10]

Many surveys conducted at shopping malls suffer from the same problem. People in shopping malls are not necessarily representative of the population of interest. Mall shoppers tend to be more affluent and include a larger percentage of teenagers and retirees than the population at large. To make matters worse, survey interviewers tend to select individuals who look "safe," or easy to interview.

For Example BIAS IN SAMPLING

RECAP: To try to gauge frosh opinion about the food served on campus, Food Services suggests that you just stand outside a school cafeteria at lunchtime and stop people to ask them questions.

QUESTIONS: What's wrong with this sampling strategy?

ANSWER: This would be a convenience sample, and it's likely to be biased. I would miss people who use the cafeteria for dinner, but not for lunch, and I'd never hear from anyone who hates the food so much that they have stopped coming to the school cafeterias.

Mistake 3: Sample from a Bad Sampling Frame

An SRS from an incomplete sampling frame introduces bias because the individuals included may differ from the ones not in the frame. People in prison, homeless people, students, and long-term travellers are all likely to be missed. In telephone surveys, people who have only cell phones or who use VOIP Internet phones are missing from the sampling frame.

Mistake 4: Undercoverage

Many survey designs suffer from **undercoverage**, in which some portion of the population is not sampled at all or has a smaller representation in the sample than it has in the population. Undercoverage can arise for a number of reasons, but it's always a potential source of bias.

Telephone surveys are usually conducted when you are likely to be home, interrupting your dinner. If you eat out often, you may be less likely to be surveyed, a possible source of undercoverage.

WHAT CAN GO WRONG?

- **Watch out for nonrespondents.** A common and serious potential source of bias for most surveys is **nonresponse bias**. No survey succeeds in getting responses from everyone. The problem is that those who don't respond may differ from those who do. And they may differ on just the variables we care about. The lack of response will bias the results. Rather than sending out a large number of surveys for which the response rate will be low, it is often better to design a smaller randomized survey for which you have the resources to ensure a high response rate. One of the problems with nonresponse bias is that it's usually impossible to tell what the nonrespondents might have said.

Do people like filling out questionnaires? A business school student wanted to study how people feel about filling out questionnaires. To do this, he designed—you guessed it—a questionnaire and gave it to a sample of all the other students at his business school. He was surprised and happy to learn that nearly all the questionnaires he got back indicated a positive attitude toward answering surveys. From this, he concluded that people in general must not mind filling out questionnaires.

[10]Agresti & Franklin 2009, 174–175.

- **Don't bore respondents with surveys that go on and on and on and on . . .** Surveys that are too long are more likely to be refused, reducing the response rate and biasing *all* the results. For each question you consider including in the survey, ask yourself, "What would I do if I knew the answer to this question?" If you don't have a use for the answer, then don't ask the question.
- **Work hard to avoid influencing responses.** Response bias refers to anything in the survey design that influences the responses.[11] Response biases include the tendency of respondents to tailor their responses to try to please the interviewer; the ways in which the wording of the questions can change responses; and the natural unwillingness of respondents to reveal personal facts, or admit to illegal or unapproved behaviour.

A S *Video:* **Biased Question Wording.** Watch a hapless interviewer make every mistake in the book.

A SHORT SURVEY

Given the fact that those who understand Statistics are smarter and better looking than those who don't, don't you think it is important to take a course in Statistics?

- **Make sure the question wording is neutral.** Many surveys, especially those conducted by special-interest groups, present one side of an issue before the question itself. For example, asking a question like

  Given that the threat of nuclear war is higher now than it has ever been in human history, and the fact that a nuclear war poses a threat to the very existence of the human race, would you favour an all-out nuclear test ban?

  will probably result in a higher percentage of people in favour of the ban than the simpler

  Are you in favour of or opposed to a nuclear test ban?

A researcher distributed a survey to an organization before some economizing changes were made. She asked how people felt about a proposed cutback in secretarial and administrative support on a seven-point scale from Very Happy to Very Unhappy. But virtually all respondents were very unhappy about the cutbacks, so the results weren't particularly useful. If she had pretested the question, she might have chosen a scale that ran from Unhappy to Outraged.

Remember the Literary Digest Survey? It turns out that they were wrong on two counts. First, their list of 10 million people was not representative. There was a selection bias in their sampling frame. There was also a nonresponse bias. We know this because the magazine also surveyed a systematic sample in Chicago, sending the same question used in the larger survey to every third registered voter. They still got a result in favour of Landon, even though Chicago voted overwhelmingly for Roosevelt in the election. This suggests that the Roosevelt supporters were less likely to respond to the survey. There's a modern version of this problem: It's been suggested that those who screen their calls with caller ID or an answering machine, and so might not talk to a pollster, may differ in wealth or political views from those who just answer the phone.

How to Think About Biases

- **Look for biases in any survey you encounter.** If you design one of your own, ask someone else to help look for biases that may not be obvious to you. And do this *before* you collect your data. There's no way to recover from a biased sample or a survey that asks biased questions. Sorry, it just can't be done.

  A bigger sample size for a biased study just gives you a bigger useless study. A really big sample gives you a really big useless study. (Think of the 2.4 million *Literary Digest* responses.)
- **Spend your time and resources reducing biases.** No other use of resources is as worthwhile as reducing the biases.
- **If you can, pilot your survey.** Administer the survey in the exact form that you intend to use it to a small sample drawn from the population you intend to sample. Look for misunderstandings, misinterpretation, confusion, or other possible biases. Then refine your survey instrument.
- **Always report your sampling methods in detail.** Others may be able to detect biases where you did not expect to find them.

CONNECTIONS

With this chapter, we take our first formal steps to relate our sample data to a larger population. Some of these ideas have been lurking in the background as we sought patterns and summaries for data. Even when we only worked with the data at hand, we often thought about implications for a larger population of individuals.

[11]Response bias is not the opposite of nonresponse bias. (We don't make these terms up; we just try to explain them.)

Notice the ongoing central importance of models. We've seen different kinds of models in previous chapters. Here we recognize the value of a model for a population. The parameters of such a model are values we will often want to estimate using statistics such as those we've been calculating. The connections to summary statistics for centre, spread, correlation, and slope are obvious.

We now have a specific application for random numbers. The idea of applying randomness deliberately showed up in Chapter 9 for simulation. Now we need randomization to get good-quality data from the real world.

What Have We Learned?

Learning Objectives

Know the three ideas of sampling:

- Examine a part of the whole: A sample can give information about the population.
- Randomize to make the sample representative.
- The sample size is what matters. It's the size of the sample—and not its fraction of the larger population—that determines the precision of the statistics it yields.

Be able to draw a Simple Random Sample (SRS) using a table of random digits or a list of random numbers from technology or an Internet site:

- In a **simple random sample** (SRS), every possible group of *n* individuals has an equal chance of being our sample.

Understand the advantages and disadvantages of other sampling methods:

- **Stratified samples** can reduce sampling variability by identifying homogeneous subgroups and then randomly sampling within each.
- **Cluster samples** randomly select among conveniently available subgroups, making sampling tasks more manageable.
- **Systematic samples** can work in some situations and are often the least expensive method of sampling. But we still want to start them randomly.
- **Multistage samples** involves multiple stages of random sampling, working down through successively smaller clusters of population elements. When clusters are somewhat homogeneous, this sub-sampling of the clusters is much more efficient than sampling entire clusters.

Identify and avoid causes of bias:

- **Voluntary response** samples are almost always biased and should be avoided and distrusted.
- **Convenience samples** are likely to be flawed for similar reasons.
- Bad sampling frames can lead to samples that don't represent the population of interest.
- **Undercoverage** occurs when individuals from a subgroup of the population are selected less often than they should be.
- **Nonresponse bias** can arise when sampled individuals will not or cannot respond.
- **Response bias** arises when respondents' answers might be affected by external influences, such as question wording or interviewer behaviour.
- Use best practices when designing a sample survey to improve the chance that your results will be valid.

Review of Terms

Population The entire group of individuals or instances about whom we hope to learn (p. 276).

Sample A (representative) subset of a population, examined in hope of learning about the population (p. 276).

Sample survey A study that asks questions of a sample drawn from some population in the hope of learning something about the entire population. Polls taken to assess voter preferences are common sample surveys (p. 276).

Bias Any systematic failure of a sampling method to represent its population. It is almost impossible to recover from bias, so efforts to avoid it are well spent. Common errors include

- relying on voluntary response
- undercoverage of the population
- nonresponse bias
- response bias (p. 277)

Randomization The best defence against bias is randomization, in which each individual is given a fair, random chance of selection (p. 277).

Sample size The number of individuals in a sample. The sample size determines how well the sample represents the population, not the fraction of the population sampled (p. 278).

Census A sample that consists of the entire population (p. 279).

Population parameter A numerically valued attribute of a model for a population. We rarely expect to know the true value of a population parameter, but we do hope to estimate it from sampled data. For example, the mean income of all employed people in the country is a population parameter (p. 279).

Statistic, sample statistic Values calculated for sampled data. Those that correspond to, and thus estimate, a population parameter are of particular interest. For example, the mean income of all employed people in a representative sample can provide a good estimate of the corresponding population parameter. The term "sample statistic" is sometimes used, usually to parallel the corresponding term "population parameter" (p. 279).

Representative A sample is said to be representative if the statistics computed from it accurately reflect the corresponding population parameters (p. 280).

Simple random sample (SRS) A simple random sample of sample size n is a sample in which each set of n elements in the population has an equal chance of selection (p. 280).

Sampling frame A list of individuals from whom the sample is drawn. Individuals who may be in the population of interest, but who are not in the sampling frame, cannot be included in any sample (p. 281).

Sampling variability The natural tendency of randomly drawn samples to differ, one from another. Sometimes—unfortunately—called *sampling error*, sampling variability is no error at all, but just the natural result of random sampling (p. 281).

Stratified random sample A sampling design in which the population is divided into several subpopulations, or **strata,** and random samples are then drawn from each stratum. If the strata are homogeneous, but are different from each other, a stratified sample may yield more consistent results (p. 282).

Cluster sample A sampling design in which entire groups, or **clusters,** are chosen at random. Cluster sampling is usually selected as a matter of convenience, practicality, or cost (p. 283).

Multistage sample A sampling scheme that involves multiple stages of random sampling, where at each successive stage, we sample from lists of ever smaller clusters (hierarchical in nature). For example, a government agency concerned with gas leaks in city homes might select an SRS of some of the current enumeration areas in the city, followed by an SRS of blocks in each selected area. They might then visit an SRS, or perhaps a systematic sample, of homes in each selected block (p. 285).

Systematic sample A sample drawn by selecting individuals systematically from a sampling frame. When there is no relationship between the order of the sampling frame and the variables of interest, a systematic sample can be representative (p. 286).

Pilot A small trial run of a survey to check whether questions are clear. A pilot study can reduce errors due to ambiguous questions (p. 290).

Voluntary response bias Bias introduced to a sample when individuals can choose on their own whether to participate in the sample. Samples based on voluntary response are always invalid and cannot be recovered, no matter how large the sample size (p. 291).

Convenience sample A sample of individuals who are conveniently available. Convenience samples often fail to be representative because every individual in the population is not equally convenient to sample (p. 291).

Undercoverage A sampling scheme that biases the sample in a way that gives a part of the population less representation than it has in the population suffers from undercoverage (p. 292).

Nonresponse bias Bias introduced when a large fraction of those sampled fails to respond. Those who do respond are likely to not represent the entire population. Voluntary response bias is a form of nonresponse bias, but nonresponse may occur for other reasons. For example, those who are at work during the day won't respond to a telephone survey conducted only during working hours (p. 292).

Response bias Anything in a survey design that influences responses. One typical response bias arises from the wording of questions, which may suggest a favoured response. Voters, for example, are more likely to express support of "the prime minister" than support of the particular person holding that office at the moment (p. 293).

On the Computer SAMPLING

Computer-generated pseudorandom numbers are usually good enough for drawing random samples, but there is little reason not to use the truly random values available on the Internet.

Here's a convenient way to draw an SRS of a specified size using a computer-based sampling frame. The sampling frame can be a list of names or of identification numbers arrayed, for example, as a column in a spreadsheet, statistics program, or database:

1. Generate random numbers of enough digits so that each exceeds the size of the sampling frame list by several digits. This makes duplication unlikely.
2. Assign the random numbers arbitrarily to individuals in the sampling frame list. For example, put them in an adjacent column.
3. Sort the list of random numbers, *carrying* along the sampling frame list.
4. Now the first *n* values in the sorted sampling frame column are an SRS of *n* values from the entire sampling frame.

Exercises

1. **Roper** Through their *Roper Reports Worldwide*, GfK Roper conducts a global consumer survey to help multinational companies understand different consumer attitudes throughout the world. The researchers interview 1000 people aged 13–65 from each of 30 countries. Their samples are designed so that they get 500 males and 500 females in each country.[11]
   a) Are they using a simple random sample? How do you know?
   b) What kind of design do you think they are using?

2. **Student centre survey** For their class project, a group of Statistics students decide to survey the student body to assess opinions about the proposed new student centre. Their sample of 200 includes 50 first-year students, 50 sophomores, 50 juniors, and 50 seniors.
   a) Do you think the group was using an SRS? Why?
   b) What kind of sampling design do you think they used?

3. **Emoticons** As their question of the day for Web site visitors, www.gamefaqs.com asked, *"Do you ever use emoticons when you type online?"* Of the 87 262 respondents, 27% said that they did not use emoticons.
   a) What kind of sample was this?
   b) How much confidence would you place in using 27% as an estimate of the fraction of people who use emoticons?

[11] www.gfkamerica.com

4. Drug tests Major League Baseball tests players to see whether they are using performance-enhancing drugs. Officials select a team at random, and a drug-testing crew shows up unannounced to test all 40 players on the team. Each testing day can be considered a study of drug use in Major League Baseball.
a) What kind of sample is this?
b) Is that choice appropriate?

5. Gallup At its Web site (www.galluppoll.com), the Gallup Poll publishes results of a new survey each day. Scroll down to the end and you'll find a statement that includes words such as these:

Results are based on telephone interviews with 1008 national adults, aged 18 and older, conducted April 2–5, 2010. . . . In addition to sampling error, question wording and practical difficulties in conducting surveys can introduce error or bias into the findings of public opinion polls.

a) For this survey, identify the population of interest.
b) Gallup performs its surveys by phoning numbers generated at random by a computer program. What is the sampling frame?
c) What problems, if any, would you be concerned about in matching the sampling frame with the population?

6. Gallup World The Gallup World Poll reports results on its Web site of surveys conducted in various places around the world. At the end of one of these reports, they describe their methods, including explanations, such as the following:

Results are based on face-to-face interviews with randomly selected national samples of approximately 1000 adults, aged 15 and older, who live permanently in each of the 21 sub-Saharan African nations surveyed. Those countries include Angola (areas where land mines might be expected were excluded), Benin, Botswana, Burkina Faso, Cameroon, Ethiopia, Ghana, Kenya, Madagascar (areas where interviewers had to walk more than 20 kilometres from a road were excluded), Mali, Mozambique, Niger, Nigeria, Senegal, Sierra Leone, South Africa, Tanzania, Togo, Uganda (the area of activity of the Lord's Resistance Army was excluded from the survey), Zambia, and Zimbabwe . . . . In all countries except Angola, Madagascar, and Uganda, the sample is representative of the entire population.

a) Gallup is interested in sub-Saharan Africa. What kind of survey design are they using?
b) Some of the countries surveyed have large populations. (Nigeria is estimated to have about 130 million people.) Some are quite small. (Togo's population is estimated at 5.4 million.) Nonetheless, Gallup sampled 1000 adults in each country. How does this affect the precision of its estimates for these countries?

7–16. What did they do? For each of the following reports about statistical studies, identify the following items (if possible). If you can't tell, then say so—this often happens when we read about a survey.
a) The population
b) The population parameter of interest
c) The sampling frame
d) The sample
e) The sampling method, including whether randomization was employed
f) Any potential sources of bias you can detect and any problems you see in generalizing to the population of interest

7. A business magazine mailed a questionnaire to the human resource directors of all *Fortune 500* companies and received responses from 23% of them. Those responding reported that they did not find that such surveys intruded significantly on their work day.

8. A question posted on the Lycos Web site on June 18, 2000, asked visitors to say whether they thought that marijuana should be legally available for medicinal purposes.[12]

9. Consumers Union asked all subscribers whether they had used alternative medical treatments and, if so, whether they had benefitted from them. For almost all of the treatments, approximately 20% of those responding reported cures or substantial improvement in their condition.

10. The Gallup Poll interviewed 1007 randomly selected U.S. adults aged 18 and older from March 23 to 25, 2007. Gallup reports that when asked when (if ever) the effects of global warming will begin to happen, 60% of respondents said the effects had already begun. Only 11% thought they would never happen.

11. Researchers waited outside a bar they had randomly selected from a list of such establishments. They stopped every tenth person who came out of the bar and asked whether he or she thought drinking and driving was a serious problem.

12. Hoping to learn what issues may resonate with voters in the coming election, the campaign director for a mayoral candidate selects one block from each of the city's election districts. Staff members go there and interview all the residents they can find.

13. The Environmental Protection Agency took soil samples at 16 locations near a former industrial waste dump and checked each for evidence of toxic chemicals. They found no elevated levels of any harmful substances.

14. Provincial police set up roadblocks to estimate the percentage of trucks with up-to-date registration, insurance, and safety inspection stickers. They usually find problems with about 10% of the trucks they stop.

---

[12] www.lycos.com

15. A company packaging snack foods maintains quality control by randomly selecting 10 cases from each day's production and weighing the bags. Then they open one bag from each case and inspect the contents.

16. Dairy inspectors visit farms unannounced and take samples of the milk to test for contamination. If the milk is found to contain dirt, antibiotics, or other foreign matter, the milk will be destroyed and the farm reinspected until purity is restored.

17. **Mistaken poll** A local television station conducted a "PulsePoll" to predict the winner in the upcoming mayoral election. Evening news viewers were invited to phone in their votes, with the results to be announced on the late-night news. Based on the phone calls, the station predicted that Amabo would win the election with 52% of the vote. They were wrong: Amabo lost, getting only 46% of the vote. Do you think the station's faulty prediction is more likely to be a result of bias or sampling error? Explain.

18. **Another mistaken poll** Prior to the mayoral election discussed in Exercise 17, the newspaper also conducted a poll. The paper surveyed a random sample of registered voters stratified by political party, age, sex, and area of residence. This poll predicted that Amabo would win the election with 52% of the vote. The newspaper was wrong: Amabo lost, getting only 46% of the vote. Do you think the newspaper's faulty prediction is more likely to be a result of bias or sampling error? Explain.

19. **Parent opinion, part 1** In a large city school system with 20 elementary schools, the school board is considering the adoption of a new policy that would require elementary students to pass a test in order to be promoted to the next grade. The PTA wants to find out whether parents agree with this plan. Listed below are some of the ideas proposed for gathering data. For each, indicate what kind of sampling strategy is involved and what (if any) biases might result.

a) Put a big ad in the newspaper asking people to log their opinions on the PTA Web site.
b) Randomly select one of the elementary schools and contact every parent by phone.
c) Send a survey home with every student, and ask parents to fill it out and return it the next day.
d) Randomly select 20 parents from each elementary school. Send them a survey, and follow up with a phone call if they do not return the survey within a week.

20. **Parent opinion, part 2** Let's revisit the school system described in Exercise 19. Four new sampling strategies have been proposed to help the PTA determine whether parents favour requiring elementary students to pass a test in order to be promoted to the next grade. For each, indicate what kind of sampling strategy is involved and what (if any) biases might result.

a) Run a poll on the local television news, asking people to dial one of two phone numbers to indicate whether they favour or oppose the plan.
b) Hold a PTA meeting at each of the 20 elementary schools, and tally the opinions expressed by those who attend the meetings.
c) Randomly select one class at each elementary school and contact each of those parents.
d) Go through the district's enrollment records, selecting every 40th parent. PTA volunteers will go to those homes to interview the people chosen.

21. **Roller coasters** Canada's Wonderland has opened a new roller coaster. It is so popular that people are waiting for up to three hours for a two-minute ride. Concerned about how patrons (who paid a large amount to enter the park and ride on the rides) feel about this, park staff survey every 10th person in the line for the roller coaster, starting from a randomly selected individual.

a) What kind of sample is this?
b) Is it likely to be representative?
c) What is the sampling frame?

22. **Playground** Some people have been complaining that the children's playground at a municipal park is too small and is in need of repair. Managers of the park decide to survey city residents to see if they believe the playground should be rebuilt. They hand out questionnaires to parents who bring children to the park. Describe possible biases in this sample.

23. **Social life** A question posted on the gamefaqs.com Web site on August 1, 2011, asked visitors to the site, "Do you have an active social life outside the Internet?" 22% of the 55 581 respondents said "No" or "Not really, most of my personal contact is online."

a) What kind of sample was this?
b) How much confidence would you place in using 22% as an estimate of the fraction of people who have no active social life outside the Internet. Explain.

24. **Churches** For your political science class, you'd like to take a survey from a sample of all the Catholic Church members in your city. A list of churches shows 17 Catholic churches within the city limits. Rather than try to obtain a list of all members of all these churches, you decide to pick three churches at random. For those churches, you'll ask to get a list of all current members and contact 100 members at random.

a) What kind of design have you used?
b) What could go wrong with your design?

25. **Wording the survey** Two members of the PTA committee in Exercises 19 and 20 have proposed different questions to ask in seeking parents' opinions.

Question 1: Should elementary school-age children have to pass high-stakes tests in order to remain with their classmates?

Question 2: Should schools and students be held accountable for meeting yearly learning goals by testing students before they advance to the next grade?

a) Do you think responses to these two questions might differ? How? What kind of bias is this?
b) Propose a question with more neutral wording that might better assess parental opinion.

26. **Banning Ephedra** An online poll at a popular Web site asked:

A nationwide ban of the diet supplement Ephedra went into effect recently. The herbal stimulant has been linked to 155 deaths and many more heart attacks and strokes. Ephedra manufacturer NVE Pharmaceuticals, claiming that the FDA lacked proof that Ephedra is dangerous if used as directed, was denied a temporary restraining order on the ban yesterday by a federal judge. Do you think that Ephedra should continue to be banned nationwide? Sixty-five percent of 17 303 respondents said "yes."

Comment on each of the following statements about this poll:

a) With a sample size that large, we can be pretty certain we know the true proportion of Americans who think Ephedra should be banned.
b) The wording of the question is clearly very biased.
c) The sampling frame is all Internet users.
d) This is a voluntary response survey, so the results can't be reliably generalized to any population of interest.

27. **Another ride** The survey of patrons waiting in line for the roller coaster in Exercise 21 asks whether they think it is worthwhile to wait a long time for the ride and whether they'd like the amusement park to install more roller coasters. What biases might cause a problem for this survey?

28. **Play it again** The survey described in Exercise 22 asked,

Many people believe this playground is too small and in need of repair. Do you think the playground should be repaired and expanded even if that means raising the entrance fee to the park?

Describe two ways this question may lead to response bias.

29. **Survey questions** Examine each of the following questions for possible bias. If you think the question is biased, indicate how and propose a better question.

a) Should companies that pollute the environment be compelled to pay the costs of cleanup?
b) Given that 18-year-olds are old enough to vote and to serve in the military, is it fair to set the drinking age at 21?

30. **More survey questions** Examine each of the following questions for possible bias. If you think the question is biased, indicate how and propose a better question.

a) Do you think high school students should be required to wear uniforms?
b) Given humanity's great tradition of exploration, do you favour continued funding for space flights?

31. **Phone surveys** Anytime we conduct a survey, we must take care to avoid undercoverage. Suppose we plan to select 500 names from the city phone book, call their homes between noon and 4 p.m., and interview whoever answers, anticipating contact with at least 200 people.

a) Why is it difficult to use a simple random sample here?
b) Describe a more convenient, but still random, sampling strategy.
c) What kinds of households are likely to be included in the eventual sample of opinion? Who will be excluded?
d) Suppose, instead, that we continue calling each number, perhaps in the morning or evening, until an adult is contacted and interviewed. How does this improve the sampling design?
e) Random-digit dialling machines can generate the phone calls for us. How would this improve our design? Is anyone still excluded?

32. **Cell phone survey** What about drawing a random sample only from cell phone exchanges? Discuss the advantages and disadvantages of such a sampling method compared with surveying randomly generated telephone numbers from non-cell phone exchanges. Do you think these advantages and disadvantages have changed over time? How do you expect they'll change in the future?

33. **Arm length** How long is your arm compared with your hand size? Put your right thumb at your left shoulder bone, stretch your hand open wide, and extend your hand down your arm. Put your thumb at the place where your little finger is, and extend down the arm again. Repeat this a third time. Now your little finger will probably have reached the back of your left hand. If the fourth hand width goes past the end of your middle finger, turn your hand sideways and count finger widths to get there.[13]

a) How many hand and finger widths is your arm?
b) Suppose you repeat your measurement 10 times and average your results. What parameter would this average estimate? What is the population?
c) Suppose you now collect arm lengths measured in this way from nine friends and average these 10 measurements. What is the population now? What parameter would this average estimate?
d) Do you think these 10 arm lengths are likely to be representative of the population of arm lengths in your community? In the country? Why or why not?

34. **Fuel economy** Occasionally, when I fill my car with gas, I figure out how many litres/100 kilometres my car got. I wrote down those results after six fill-ups in the past few months. Overall, it appears my car gets 8.7 litres per 100 kilometres.

a) What statistic have I calculated?
b) What is the parameter I'm trying to estimate?
c) How might my results be biased?
d) When the Office of Energy Efficiency (of Natural Resources Canada) checks a car like mine to predict its fuel economy, what parameter is it trying to estimate?

35. **Accounting** Between quarterly audits, a company likes to check on its accounting procedures to address any problems before they become serious. The accounting staff processes payments on about 120 orders each day. The next day, the supervisor rechecks 10 of the transactions to be sure they were processed properly.

a) Propose a sampling strategy for the supervisor.
b) How would you modify that strategy if the company makes both wholesale and retail sales, requiring different bookkeeping procedures?

36. **Happy workers?** A manufacturing company employs 14 project managers, 48 foremen, and 377 labourers. In an effort to keep informed about any possible sources of employee discontent, management wants to conduct job satisfaction interviews with a sample of employees every month.

a) Do you see any danger of bias in the company's plan? Explain.
b) Propose a sampling strategy that uses a simple random sample.
c) Why do you think a simple random sample might not provide the representative opinion the company seeks?
d) Propose a better sampling strategy.
e) Listed below are the last names of the project managers. Use random numbers to select two people to be interviewed. Be sure to explain your method carefully.

Barrett	Bowman	Chen
DeLara	DeRoos	Grigorov
Maceli	Mulvaney	Pagliarulo
Rosica	Smithson	Tadros
Williams	Yamamoto	

37. **Quality control** Sammy's Salsa, a small local company, produces 20 cases of salsa a day. Each case contains 12 jars and is imprinted with a code indicating the date and batch number. To help maintain consistency, at the end of each day, Sammy selects three bottles of salsa, weighs the contents, and tastes the product. Help Sammy select the sample jars. Today's cases are coded 07N61 through 07N80.

a) Carefully explain your sampling strategy.
b) Show how to use random numbers to pick the three jars for testing.
c) Did you use a simple random sample? Explain.

38. **A fish story** Concerned about reports of discoloured scales on fish caught downstream from a newly sited chemical plant, scientists set up a field station in a shoreline public park. For one week they asked fishers there to bring any fish they caught to the field station for a brief inspection. At the end of the week, the scientists said that 18% of the 234 fish that were submitted for inspection displayed the discolouration. From this information, can the researchers estimate what proportion of fish in the river have discoloured scales? Explain.

39. **Sampling methods** Consider each of these situations. Do you think the proposed sampling method is appropriate? Explain.

a) We want to know what percentage of local family doctors are accepting new patients. We call the offices of 50 doctors randomly selected from local Yellow Page listings.
b) We want to know what percentage of local businesses anticipate hiring additional employees in the upcoming month. We randomly select a page in the Yellow Pages and call every business listed there.

40. **More sampling methods** Consider each of these situations. Do you think the proposed sampling method is appropriate? Explain.

a) We want to know if there is neighbourhood support to turn a vacant lot into a playground. We spend a Saturday afternoon going door-to-door in the neighbourhood, asking people to sign a petition.
b) We want to know if students at our university are satisfied with the selection of food available on campus. We go to the largest cafeteria and interview every 10th person in line.

[13] According to Da Vinci, the distance from your elbow to the tip of your hand should be 2.5 hand lengths.

Just Checking ANSWERS

1. **a)** It can be hard to reach all members of a population, and it can take so long that circumstances change, affecting the responses. A well-designed sample is often a better choice.
   b) This sample is probably biased—students who didn't like the food at the cafeteria might not choose to eat there.
   c) No, only the sample size matters, not the fraction of the overall population.
   d) Students who frequent this Web site might be more enthusiastic about Statistics than the overall population of Statistics students. A large sample cannot compensate for bias.
   e) It's the population "parameter." "Statistics" describe samples.

2. **a)** systematic
   b) stratified
   c) simple
   d) cluster

MathXL

MyStatLab

Go to MathXL at www.mathxl.com or MyStatLab at www.mystatlab.com. You can practise exercises for this chapter as often as you want. The guided solutions will help you find answers step by step. You'll find a personalized study plan available to you too!

chapter

11 Experiments and Observational Studies

Where are we going?

Experiments are the "gold standard" of data collection. No drug comes to market without at least one Health Canada-approved experiment to demonstrate its safety and effectiveness. Much of what we know in science and social science comes from carefully designed experiments.

The four principles of experimental design (control what you can, randomize for the rest, replicate the trials, and, when appropriate, block to remove identifiable variation) describe what makes a sound experiment and how to understand the results.

You should treat as many patients as possible with the new drugs while they still have the power to heal.

—nineteenth-century physician Armand Trusseau

Who gets good grades? And, more importantly, why? Is there something schools and parents could do to help weaker students improve their grades? Some people think they have an answer: music! No, not your iPod, but an instrument. In a study conducted at Mission Viejo High School in California, researchers compared the scholastic performance of music students with that of nonmusic students. Guess what? The music students had a much higher overall grade point average than the nonmusic students, 3.59 to 2.91. Not only that: A whopping 16% of the music students had all As, compared with only 5% of the nonmusic students.[1]

As a result of this study and others, many parent groups and educators pressed for expanded music programs in North American schools. They argued that the work ethic, discipline, and feeling of accomplishment fostered by learning to play an instrument also enhance a person's ability to succeed in school. They thought that involving more students in music would raise academic performance. What do you think? Does this study provide solid evidence? Or are there other possible explanations for the difference in grades? Is there any way to really prove such a conjecture?

11.1 Observational Studies

This research tried to show an association between music education and grades. But it wasn't a survey. Nor did it assign students to get music education. Instead, it simply observed students "in the wild," recording the choices they made and the outcome. Such studies are called **observational studies**. In observational studies, researchers don't *assign* choices; they simply observe them. In addition, this was a **retrospective study**, because researchers first identified subjects who studied music and then collected data on their past grades.

What's wrong with concluding that music education causes good grades? One high school during one academic year may not be representative of a whole country. That's true, but the real problem is that the claim that music study *caused* higher grades depends on there being *no other differences* between the groups that could account for the differences in grades, and studying music was not the *only* difference between these two groups of students.

We can think of plenty of lurking variables that might cause the groups to perform differently. Students who study music may have better work habits to start with, and this makes them successful in both music and course work. Music students may have more

[1] There are similar studies in Canada and the same results can be found on http://conditionformusiced.ca/html/sec5-research.articles.php.

RETROSPECTIVE STUDIES CAN GIVE VALUABLE CLUES

For rare illnesses, it's not practical to draw a large enough sample to see many ill respondents, so the only option remaining is to use retrospective data. For example, researchers can interview those who have become ill. The likely causes of both Legionnaires' disease and HIV were initially identified from retrospective studies of the small populations that were initially infected. But to confirm the causes, researchers needed laboratory-based experiments.

parental support (someone had to pay for all those lessons), and that support may have enhanced their academic performance, too. Maybe they came from wealthier homes and had other advantages. Or it could be that smarter kids just like to play musical instruments.

Observational studies are valuable for discovering trends and possible relationships. They are used widely in public health and marketing. Observational studies that try to discover variables related to rare outcomes, such as specific diseases, are often retrospective. They first identify people with the disease and then look into their history and heritage in search of things that may be related to their condition. But retrospective studies have a restricted view of the world because they are usually restricted to a small part of the entire population. And because retrospective records are based on historical data, they can have errors. (Do you recall *exactly* what you ate even yesterday? How about last Wednesday?)

A somewhat better approach is to observe individuals over time, recording the variables of interest and ultimately seeing how things turn out. For example, we might start by selecting young students who have not begun music lessons. We could then track their academic performance over several years, comparing those who later choose to study music with those who do not. Identifying subjects in advance and collecting data as events unfold would make this a **prospective study**.

Although an observational study may identify important variables related to the outcome we are interested in, there is no guarantee that we have found the most important related variables. Students who choose to study an instrument might still differ from the others in some important way that we failed to observe. It may be this difference—whether we know what it is or not—rather than music itself that leads to better grades. It's just not possible for observational studies, whether prospective or retrospective, to demonstrate a causal relationship.

shimmo/istockphoto

For Example DESIGNING AN OBSERVATIONAL STUDY

In early 2007, a larger-than-usual number of cats and dogs developed kidney failure; many died. Initially, researchers didn't know why, so they used an observational study to investigate.

QUESTION: Suppose you were called on to plan a study seeking the cause of this problem. Would your design be retrospective or prospective? Explain why.

ANSWER: I would use a retrospective observational study. Even though the incidence of disease was higher than usual, it was still rare. Surveying all pets would have been impractical. Instead, it makes sense to locate some who were sick, and ask about their diets, exposure to toxins, and other possible causes.

11.2 Randomized, Comparative Experiments

Is it *ever* possible to prove a cause-and-effect relationship? Well, yes it is, but we would have to take a different approach. We could take a group of Grade 3 students, randomly assign half to take music lessons, and forbid the other half to do so. Then we could compare their grades several years later. This kind of study design is called an **experiment**.

An experiment requires a **random assignment** of subjects to treatments. Only an experiment can justify a claim like "Music lessons cause higher grades." Questions such as "Does taking vitamin C reduce the chance of getting a cold?" and "Does working with computers improve performance in Statistics class?" and "Is this drug a safe and effective treatment for that disease?" require a designed experiment to establish cause and effect.

Experiments study the relationship between two or more variables. An experimenter must identify at least one explanatory variable, called a **factor**, to manipulate, and at least one **response variable** to measure. What distinguishes an experiment from other types of investigation is that the experimenter actively and deliberately manipulates the factors to control the details of the possible treatments, and assigns the subjects to those treatments *at random*. The experimenter then observes the response variable and *compares* responses for different groups of subjects who have been treated differently. For example, we might design an experiment to see whether the amount of sleep and exercise you get affects your performance.

The individuals on whom or which we experiment are referred to as **experimental units**; however, humans who are experimented on are commonly called *subjects* or *participants*. When we recruit subjects for our sleep deprivation experiment by advertising in Statistics class, we'll probably have better luck if we invite them to be participants than if we advertise that we need experimental units!

The specific values that the experimenter chooses for a factor are called the **levels** of the factor. We might assign our participants to sleep for four, six, or eight hours. Often there are several factors at a variety of levels. (Our subjects will also be assigned to a treadmill for 0 or 30 minutes.) The combination of specific levels from all the factors that an experimental unit receives is known as its **treatment**. (Our subjects could have any one of six different treatments—three sleep levels, each at two exercise levels.)

AN EXPERIMENT:

Manipulates the factor levels to create treatments.
Randomly assigns subjects to these treatment levels.
Compares the responses of the subject groups across treatment levels.

How should we assign our participants to these treatments? Some students prefer four hours of sleep, while others need eight. Some exercise regularly; others are couch potatoes. Should we let the students choose the treatments they'd prefer? No. That would not be a good idea. To have any hope of drawing a fair conclusion, we must assign our participants to their treatments at *random*.

The need for random assignment is a lesson that was hard for some to accept. For example, physicians might naturally prefer to assign patients to the therapy that they think best rather than have a random element, such as a coin flip, determine the treatment. But we've known for more than a century that for the results of an experiment to be valid, there is no way to avoid deliberate randomization.

The Women's Health Initiative is a major 15-year research program funded by the U.S. National Institutes of Health to address the most common causes of death, disability, and poor quality of life in older women. It consists of both an observational study with more than 93 000 participants and several randomized comparative experiments. The goals of this study include

- giving reliable estimates of the extent to which known risk factors predict heart disease, cancers, and fractures;
- identifying "new" risk factors for these and other diseases in women;
- comparing risk factors, presence of disease at the start of the study, and new occurrences of disease during the study across all study components; and
- creating a future resource to identify biological indicators of disease, especially substances and factors found in blood.

In other words, the study seeks to identify possible risk factors and assess how serious they might be. It seeks to build up data that might be checked retrospectively as the women in the study continue to be followed. There would be no way to find out these things with an experiment because the task includes identifying new risk factors. If we don't know those risk factors, we could never control them as factors in an experiment.

By contrast, one of the clinical trials (randomized experiments) that received much press attention randomly assigned postmenopausal women to take either hormone replacement therapy or an inactive pill. The results, published in 2002 and 2004, concluded that hormone replacement with estrogen carried increased risks of stroke.

No drug can be sold in Canada without first showing, in a suitably designed experiment approved by Health Canada, that it's safe and effective. The small print on the booklet that comes with many prescription drugs usually describes the outcomes of that experiment.

For Example DETERMINING THE TREATMENTS AND RESPONSE VARIABLE

RECAP: In 2007, deaths of a large number of pet dogs and cats were ultimately traced to contamination of some brands of pet food. The manufacturer now claims that the food is safe, but before it can be released, it must be tested.

A S *Video:* **An Industrial Experiment.** Manufacturers often use designed experiments to help them perfect new products. Watch this video about one such experiment.

QUESTION: In an experiment to test whether the food is now safe for dogs to eat,[2] what would be the treatments and what would be the response variable?

ANSWER: The treatments would be ordinary size portions of two dog foods: the new one from the company (the test food) and one that I was certain was safe (perhaps prepared in my kitchen or laboratory). The response would be a veterinarian's assessment of the health of the test animals.

11.3 The Four Principles of Experimental Design

A methodology for designing experiments was proposed by Sir R.A. Fisher in his innovative book *The Design of Experiments* (1935). His idea of experimental design can be summarized by the following four principles.

1. **Control**. We control sources of variation other than the factors we are testing by making conditions as similar as possible for all treatment groups. For human subjects, we try to treat them alike. However, there is always a question of degree and practicality. Controlling extraneous sources of variation reduces the variability of the responses, making it easier to detect differences among the treatment groups.

   Making generalizations from the experiment to other levels of the controlled factor can be risky. For example, suppose we test two laundry detergents and carefully control the water temperature at 40°C. This would reduce the variation in our results due to water temperature, but what could we say about the detergents' performance in cold water? Not much. It would be hard to justify extrapolating the results to other temperatures.

   Although we control both experimental factors and other sources of variation, we think of them very differently. We control a factor by assigning subjects to different factor levels because we want to see how the response will change at those different levels. We control other sources of variation to *prevent* them from changing and affecting the response variable.
2. **Randomize**. As in sample surveys, *randomization* allows us to equalize the effects of unknown or uncontrollable sources of variation. It does not eliminate the effects of these sources, but it spreads them out across the treatment groups so that we can see past them. If experimental units were not assigned to treatments at random, we would not be able to use statistics to draw conclusions from an experiment. Assigning subjects to treatments at random reduces bias due to uncontrolled sources of variation. Randomization protects us even from effects we didn't know about. There's an adage that says "*control what you can, and randomize the rest.*"
3. **Replicate**. Two kinds of replication show up in comparative experiments. First, we should apply each treatment to a number of subjects. Only with replication can we

Scott Adams/Universal Uclick

DILBERT: © Scott Adams/Distributed by United Feature Syndicate, Inc.

[2] It may disturb you (as it does us) to think of deliberately putting dogs at risk in this experiment, but in fact that is what is done. The risk is borne by a small number of dogs so that the far larger population of dogs can be kept safe.

A S *Activity:* **The Rules of Experimental Design.** Watch an animated discussion of the design of experiments.

A S *Activity:* **Perform an Experiment.** How well can you read pie charts and bar charts? Find out as you serve as the subject in your own experiment.

estimate the variability of responses. If we have not assessed the variation, the experiment is not complete. The outcome of an experiment on a single subject is an anecdote, not data.

A second kind of replication shows up when the experimental units are not a representative sample from the population of interest. We may believe that what is true of the students in Psych 101 who volunteered for the sleep experiment is true of all humans, but we'll feel more confident if our results for the experiment are *replicated* in another part of the country, with people of different ages, and at different times of the year. *Replication* of an entire experiment with the controlled sources of variation at different levels is an essential step in science.

4. **Blocking**. Blocking is not required in an experimental design, but it is a compromise between randomization and control. The ability of randomizing to equalize variation across treatment groups works best in the long run. For example, if we're allocating players to two six-player soccer teams from a pool of 12 children, we might do so at random to equalize the talent. But what if there were two 12-year-olds and ten 6-year-olds in the group? Randomizing may place both 12-year-olds on the same team. In the long run, if we did this over and over, it would all equalize. But wouldn't it be better to assign one 12-year-old to each group (at random) and five 6-year-olds to each team (at random)? By doing this, we would improve fairness in the short run. This approach makes the division more fair by recognizing the variation in *age* and allocating the players at random *within* each age level. When we do this, we call the variable *age* a *blocking variable*. The levels of *age* are called blocks.

Sometimes, attributes of the experimental units that we are not studying and that we can't control may nevertheless affect the outcomes of an experiment. If we group similar individuals together and then randomize within each of these *blocks*, we can remove much of the variability due to the difference among the blocks. Blocking in experiments has much in common with stratification in surveys.

For Example CONTROL, RANDOMIZE, AND REPLICATE

RECAP: We're planning an experiment to see whether the new pet food is safe for dogs to eat. We'll feed some animals the new food and others a food known to be safe, comparing their health after a period of time.

QUESTIONS: In this experiment, how will you implement the principles of control, randomization, and replication?

ANSWER: I'd control the portion sizes eaten by the dogs. To reduce possible variability from factors other than the food, I'd standardize other aspects of their environments—housing the dogs in similar pens and ensuring that each got the same amount of water, exercise, play, and sleep time, for example. I might restrict the experiment to a single breed of dog and to adult dogs to further minimize variation.

To equalize traits, pre-existing conditions, and other unknown influences, I would randomly assign dogs to the two feed treatments.

I would replicate by assigning more than one dog to each treatment to allow for variability among individual dogs. If I had the time and funding, I might replicate the entire experiment using, for example, a different breed of dog.

If experiments can determine cause and observational studies can't, then why don't we just run experiments all the time? Well, for one, they are a lot more expensive and time-consuming. A properly controlled experiment may involve hundreds of people, last for many years, and cost tens of millions of dollars. Observational studies (particularly retrospective ones) can be very cheap once you have the data, and can even be carried out by a single person! There are also a limited number of subjects available for experimentation, and we want to make sure we use them in the best way possible.

Diagrams

An experiment is carried out over time with specific actions occurring in a specified order. A diagram of the procedure can help in thinking about experiments.[3]

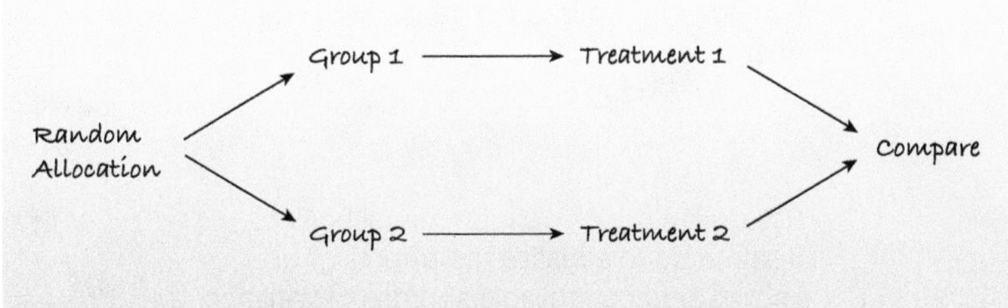

A completely randomized experiment is the ideal simple design, just as a *simple random sample* is the ideal simple sample—and for many of the same reasons.

The diagram emphasizes the random allocation of subjects to treatment groups, the separate treatments applied to these groups, and the ultimate comparison of results. It's best to specify the responses that will be compared. A good way to start comparing results for the treatment groups is with boxplots.

Step-by-Step Example DESIGNING AN EXPERIMENT

Vibe Images/Fotolia

An ad for OptiGro plant fertilizer claims that you will grow "juicier, tastier" tomatoes with this product. You'd like to test this claim, and wonder whether you might be able to get by with half the specified dose. How can you set up an experiment to check out the claim?

Of course, you'll have to get some tomatoes, try growing some plants with the product and some without, and see what happens. But you'll need a clearer plan than that. How should you design your experiment?

Let's work through the design step by step. We'll design the simplest kind of experiment, a **completely randomized experiment in one factor**. Since this is a *design* for an experiment, most of the steps are part of the *Think* stage. The statements in the right column are the kinds of things you would need to say in *proposing* an experiment. You'd need to include them in the "methods" section of a report once the experiment is run.

Questions: How would you design an experiment to test OptiGro fertilizer?

THINK ➡ Plan State what you want to know.	I want to know whether tomato plants grown with OptiGro yield juicier, tastier tomatoes than plants raised in otherwise similar circumstances but without the fertilizer.
Response Specify the response variable.	I'll evaluate the juiciness and taste of the tomatoes by asking a panel of judges to rate them on a scale from 1 to 7 in juiciness and in taste.
Treatment Specify the factor levels and the treatments.	The factor is fertilizer, specifically OptiGro fertilizer. I'll grow tomatoes at three different factor levels: some with no fertilizer, some with half the specified amount of OptiGro, and some with the full dose of OptiGro. These are the three treatments.

[3] Diagrams of this sort were introduced by David Moore in his textbooks and are widely used.

Experimental Units Specify the experimental units.

I'll obtain 24 tomato plants of the same variety from a local garden store.

Experimental Design Observe the principles of design:

Control any sources of variability you know of and can control.

I'll locate the farm plots near each other so that the plants get similar amounts of sun and rain and experience similar temperatures. I will weed the plots equally and otherwise treat the plants alike.

Randomly assign experimental units to treatments to equalize the effects of unknown or uncontrollable sources of variation.

Specify how the random numbers needed for randomization will be obtained.

I'll randomly divide the plants into three groups. I will use random numbers from a table to determine the assignment.

Replicate results by placing more than one plant in each treatment group.

There are eight plants in each treatment group.

Make a Picture A diagram of your design can help you think about it clearly.

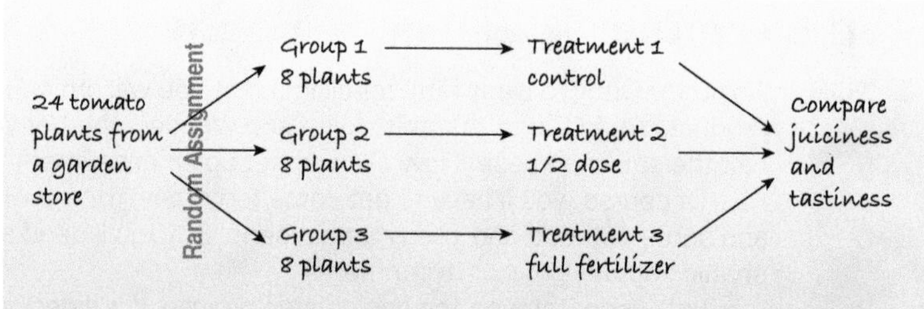

Specify any other experiment details. You must give enough details so that another experimenter could exactly replicate your experiment. It's generally better to include details that might seem irrelevant than to leave out matters that could turn out to make a difference.

Specify how to measure the response.

I will grow the plants until the tomatoes are mature, as judged by reaching a standard colour.

I'll harvest the tomatoes when ripe and store them for evaluation.

I'll set up a numerical scale of juiciness and one of tastiness for the taste testers. Several people will taste slices of tomato and rate them.

SHOW ➡ Once you collect the data, you'll need to display them and compare the results for the three treatment groups.

I will display the results with side-by-side boxplots to compare the three treatment groups.

I will compare the means of the groups.

TELL ➡ To answer the initial question, we ask whether the differences we observe in the means of the three groups are meaningful.

Because this is a randomized experiment, we can attribute significant differences (as defined on the next page) to the treatments. To do this properly, we'll need methods from what is called "statistical inference," the subject of the rest of this book.

If the differences in taste and juiciness among the groups are greater than I would expect by knowing the usual variation among tomatoes, I may be able to conclude that these differences can be attributed to treatment with the fertilizer.

Does the Difference Make a Difference?

A S *Activity:* **Graph the Data.** Do you think there's a significant difference in your perception of pie charts and bar charts? Explore the data from your plot perception experiment.

If the differences among the treatment groups are big enough, we'll attribute the differences to the treatments—but how can we decide whether the differences are big enough?

Would we expect the group means to be identical? Not really. Even if the treatment made no difference whatsoever, there would still be some variation. We assigned the plants to treatments at random. But a different random assignment would have led to different results. Even a repeat of the *same* treatment on a different randomly assigned set of plants would lead to a different mean. The real question is whether the differences we observed are about as big as we'd get just from the randomization alone, or whether they're bigger than that. If we decide that they're bigger, we'll attribute the differences to the treatments. In that case we say the differences are **statistically significant**.

How will we decide if something is different enough to be considered statistically significant? Later chapters will offer methods to help answer that question, but to get some intuition, think about deciding whether a coin toss is fair. If we flip a fair coin 100 times, we expect, *on average*, to get 50 heads. Suppose we get 54 heads out of 100. That doesn't seem very surprising. It's well within the bounds of ordinary random fluctuations. What if we'd seen 94 heads? That's clearly outside the bounds. We'd be pretty sure that the coin flips were not random. But what about 74 heads? Is that far enough from 50% to arouse our suspicions? That's the sort of question we need to ask of our experiment results.

In statistics terminology, 94 heads would be a statistically significant difference from 50, and 54 heads would not. Whether 74 is *statistically significant* would depend on the chance of getting a result that is much higher than 50 (i.e., getting 74 or more heads in 100 flips of a fair coin) and on our tolerance for believing that rare events can happen.

Back at the tomato stand, we ask whether the differences we see among the treatment groups are the kind of differences we'd expect from randomization. A good way to get a feeling for that is to look at how much our results vary among plants that get the *same* treatment. Boxplots of our results by treatment group can give us a general idea.

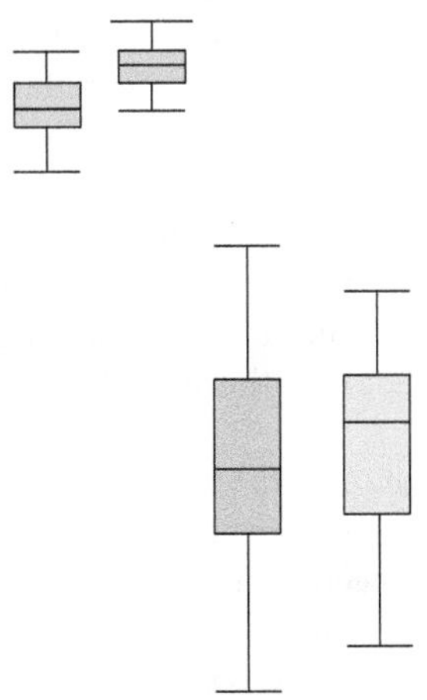

Figure 11.1

The boxplots in both pairs have centres the same distance apart, but when the spreads are large, it is more likely that the observed difference may be just from random fluctuation. (And what about the sample sizes? Do you think that might matter too?)

For example, Figure 11.1 shows two pairs of boxplots whose centres differ by exactly the same amount. In the upper set, that difference appears to be larger than we'd expect just by chance. Why? Because the variation is quite small *within* treatment groups, so the larger difference between the groups is unlikely to be just from the randomization. In the bottom pair, that same difference *between* the centres looks less impressive. There the variation *within* each group swamps the difference *between* the two means. It is quite possible that the difference is statistically significant in the upper pair and not statistically significant in the lower pair.

Later we'll see statistical tests that quantify this intuition. For now, the important point is that a difference is statistically significant if we don't believe that it's likely to have occurred only by chance.

Just Checking

1. At one time, a method called "gastric freezing" was used to treat people with peptic ulcers. An inflatable bladder was inserted down the esophagus and into the stomach, and then a cold liquid was pumped into the bladder. Now you can find the following notice on the Internet site of a major insurance company:

[Our company] does not cover gastric freezing (intragastric hypothermia) for chronic peptic ulcer disease . . . . Gastric freezing for chronic peptic ulcer disease is a non-surgical treatment which was popular about 20 years ago but

now is seldom performed. It has been abandoned due to a high complication rate, only temporary improvement experienced by patients, and a lack of effectiveness when tested by double-blind, controlled clinical trials.

What did that "controlled clinical trial" (experiment) probably look like? (Don't worry about "double-blind"; we'll get to that soon.)

a) What was the factor in this experiment?
b) What was the response variable?
c) What were the treatments?
d) How did researchers decide which subjects received which treatment?
e) Were the results statistically significant?

Experiments and Samples

Experiments are rarely performed on random samples from a population. Don't describe the subjects in an experiment as a random sample unless they really are. More likely, the randomization was in assigning subjects to treatments.

Both experiments and sample surveys use randomization to get unbiased data. But they do so in different ways and for different purposes. Sample surveys try to estimate population parameters, so the sample needs to be as representative of the population as possible. By contrast, experiments try to assess the effects of treatments. *Experimental units are rarely drawn randomly from a population.* For example, a medical experiment may deal only with patients who have the disease under study. The randomization is in the assignment of their therapy. We want a sample to exhibit the diversity and variability of the population, but for an experiment the more homogeneous the subjects, the more easily we'll spot differences in the effects of the treatments.

Unless the experimental units are chosen from the population at random, you should *be cautious about generalizing experiment results to larger populations* until the experiment has been repeated under different circumstances. Results become more persuasive if they remain the same in completely different settings, such as in a different season, in a different country, or for a different species, to name a few. Hence, the *Who* of an experiment always needs to be described carefully.

Even without choosing experimental units from a population at random, experiments can draw stronger conclusions than surveys. By looking only at the differences across treatment groups, experiments cancel out many sources of bias. For example, the entire pool of subjects may be a biased representation of the population. (University students may need more sleep, on average, than the general population.) When we assign subjects randomly to treatment groups, all the groups are still biased, but *in the same way*. When we consider the differences in their responses, these biases cancel out, allowing us to see the *differences* due to treatment effects more clearly.

11.4 Control Treatments

A S *Activity:* **Control Groups in Experiments** Is a control group really necessary?

Suppose you wanted to test a $300 piece of software designed to shorten download times. You could just try it on several files and record the download times, but you probably want to *compare* the speed with what would happen *without* the software installed. Such a baseline measurement is called a control treatment, and the experimental units to whom it is applied are called a **control group**.

This is a use of the word "control" in an entirely different context. Previously, we controlled extraneous sources of variation by keeping them constant. Here, we use a control treatment as another *level* of the factor in order to compare the treatment results to a situation in which "nothing happens." That's what we did in the tomato experiment when we used no fertilizer on the eight tomatoes in Group 1.

Blinding

Humans are notoriously susceptible to errors in judgment.[4] All of us. When we know what treatment was assigned, it's difficult not to let that knowledge influence our assessment of the response, even when we try to be careful.

Suppose you were trying to advise your school on which brand of cola to stock in the school's vending machines. You set up an experiment to see which of the three competing brands students prefer (or whether they can tell the difference at all). But people have brand loyalties. You probably prefer one brand already. So if you knew which brand you were tasting, it might influence your rating. To avoid this bias, it would be better to disguise the brands as much as possible. This strategy is called **blinding** the participants to the treatment.[5]

But it isn't just the subjects who should be blind. Experimenters themselves often subconsciously behave in ways that favour what they believe. Even technicians may treat plants or test animals differently if, for example, they expect them to die. An animal that starts doing a little better than others by showing an increased appetite may get fed a bit more than the experimental protocol specifies.

People are so good at picking up subtle cues about treatments that the best (in fact, the *only*) defense against such biases in experiments on human subjects is to keep *anyone* who could affect the outcome or the measurement of the response from knowing which subjects have been assigned to which treatments. So, not only should your cola-tasting subjects be blinded, but also *you*, as the experimenter, shouldn't know which drink is which, either—at least until you're ready to analyze the results.

BLINDING BY MISLEADING

Social science experiments can sometimes blind subjects by misleading them about the purpose of a study. One of the authors participated as an undergraduate volunteer in a (now infamous) psychology experiment using such a blinding method. The subjects were told that the experiment was about three-dimensional spatial perception and were assigned to draw a model of a horse. While they were busy drawing, a loud noise and then groaning were heard coming from the room next door. The *real* purpose of the experiment was to see how people reacted to the apparent disaster. The experimenters wanted to see whether the social pressure of being in groups made people react to the disaster differently. Subjects had been randomly assigned to draw either in groups or alone; that was the treatment. The experimenter had no interest in how well the subjects could draw the horse, but the subjects were blinded to the treatment because they were misled.

There are two main classes of individuals who can affect the outcome of an experiment:

- those who could influence the results (the subjects, treatment administrators, or technicians)
- those who evaluate the results (judges, treating physicians, etc.)

When all the individuals in one of these classes are blinded, an experiment is said to be **single-blind**. When everyone in *both* classes is blinded, we call the experiment **double-blind**. If only some of the individuals in a class are blinded—for example, if subjects are not told of their treatment but the administering technician is not blinded—there is a substantial risk that subjects can discern their treatment from subtle cues in the technician's behaviour or that the technician might inadvertently treat subjects differently. Such experiments cannot be considered truly blind.

In our tomato experiment, we certainly don't want the people judging the taste to know which tomatoes got the fertilizer. That makes the experiment single-blind. We might

[4] For example, here we are in Chapter 11 and you're still reading the footnotes.

[5] C. S. Peirce, in the same 1885 work in which he introduced randomization, also recommended blinding.

also not want the people caring for the tomatoes to know which ones were being fertilized, in case they might treat them differently in other ways, too. We can accomplish this double-blinding by having some fake fertilizer for them to put on the other plants.

THE LANARKSHIRE MILK STUDY

In 1930, a now famous study[6] examined the impact of providing milk to school-children. The main objective was to examine weight gain; it was thought that children who were provided milk would gain more weight. Participants were randomized into control groups (who received nothing) and treatment groups, but when the results came back, it was actually the control groups who had larger weights! How was this possible? One of the problems with the study was that the teachers were not blind to the treatment. Some of the students were quite poor, and the teachers compassionately moved them into the treatment group. Since they were poor, they didn't eat as much in general, which explains the lower weights.

This study is a textbook[7] example of the need for double-blinding.

For Example BLINDING

RECAP: In our experiment to see if the new pet food is now safe, we're feeding one group of dogs the new food, and another group a food we know to be safe. Our response variable is the health of the animals as assessed by a veterinarian.

QUESTIONS: Should the vet be blinded? Why or why not? How would you do this? (Extra credit: Can this experiment be double-blind? Would that mean that the test animals wouldn't know what they were eating?)

ANSWER: Whenever the response variable involves judgment, it is a good idea to blind the evaluator to the treatments. The veterinarian should not be told which dogs ate which foods.

EXTRA CREDIT: There is a need for double-blinding. In this case, the workers who care for and feed the animals should not be aware of which dogs are receiving which food. We'll need to make the "safe" food look as much like the "test" food as possible.

Placebos

A S *Activity:* **Blinded Experiments.** This narrated account of blinding isn't a placebo!

Often, simply applying *any* treatment can induce an improvement. Every parent knows the medicinal value of a kiss to make a toddler's scrape or bump stop hurting. Some of the improvement seen with a treatment—even an effective treatment—can be due simply to the act of treating. To separate these two effects, we can use a control treatment that mimics the treatment itself.

A "fake" treatment that looks just like the treatments being tested is called a sham treatment, or **placebo**. Placebos are the best way to blind subjects from knowing whether they are receiving the treatment or not. One common version of a placebo in drug testing is a "sugar pill." Especially when psychological attitude can affect the results, control group subjects treated with a placebo may show an improvement.

The placebo effect is stronger when placebo treatments are administered with authority or by a figure who appears to be an authority. "Doctors" in white coats generate a stronger effect than salespeople in polyester suits. But the placebo effect is not reduced much even when subjects know that the effect exists. People often suspect that they've gotten the placebo if nothing at all happens. So, drug manufacturers have recently gone so far in making placebos realistic that they cause the same side effects as the drug being tested! Such "active placebos" usually induce a stronger placebo effect. When those side effects include loss of appetite or hair, the practice may raise ethical questions.

[6] See: Student, *Biometrika* Vol. 23, No. 3/4 (Dec., 1931), pp. 398–406 for a critique of this experiment.
[7] Pun intended.

The fact is that subjects treated with a placebo sometimes improve. It's not unusual for 20% or more of subjects given a placebo to report reduction in pain, improved movement, or greater alertness, or even to demonstrate improved health or performance. This **placebo effect** highlights both the importance of effective blinding and the importance of comparing treatments with a control. Placebo controls are so effective that you should use them as an essential tool for blinding whenever possible.

The best experiments are usually

- randomized
- double-blind
- comparative
- placebo-controlled

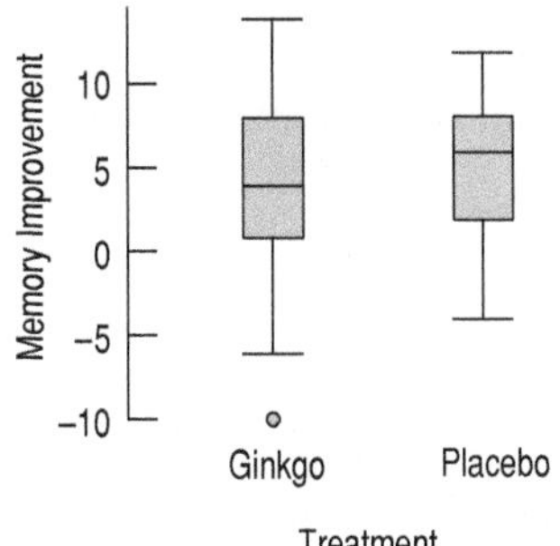

Does *Ginkgo biloba* improve memory? Researchers investigated the purported memory-enhancing effect of *Ginkgo biloba* tree extract.[8] In a randomized, comparative, double-blind, placebo-controlled study, they administered treatments to 230 elderly community members. One group received Ginkoba™ according to the manufacturer's instructions. The other received a similar-looking placebo. Thirteen different tests of memory were administered before and after treatment. The placebo group showed greater improvement on seven of the tests; the treatment group on the other six. None showed any significant differences. In the margin are boxplots of one measure.

By permission of John L. Hart FLP and Creators Syndicate, Inc.

11.5 Blocking

We wanted to use 18 tomato plants of the same variety for our experiment, but suppose the garden store had only 12 plants left. We drove to the nursery and bought six more plants of that variety. We worry that the tomato plants from the two stores are different somehow, and, in fact, they don't really look the same.

How can we design the experiment so that the differences between the stores don't mess up our attempts to see differences among fertilizer levels? We can't measure the effect of a store the same way as we can the fertilizer because we can't assign it as we would a factor in the experiment. You can't tell a tomato what store to come from.

Because stores may vary in the care they give plants or in the sources of their seeds, the plants from either store are likely to be more like each other than they are like the plants from the other store. When groups of experimental units are similar, it's often a good idea to gather them together into blocks. By **blocking**, we isolate the variability attributable to the differences between the blocks so that we can more clearly see the differences caused by the treatments. Here, we would define the plants from each store to be a block. The randomization is introduced when we randomly assign treatments within each block.

[8] P. R. Solomon, F. Adams, A. Silver, J. Zimmer, R. De Veaux, 2002, "Ginkgo for memory enhancement. A randomized controlled trial." *JAMA* 288: 835–840.

In a **completely randomized design**, each of the 18 plants would have an equal chance to land in each of the three treatment groups. But we realize that the store may have an effect. To isolate the store effect, we block on store by assigning the plants from each store to treatments at random. So we now have six treatment groups, three for each block. Within each block, we'll randomly assign the same number of plants to each of the three treatments. The experiment is still fair because each treatment is still applied (at random) to the same number of plants and to the same proportion from each store: four from store A and two from store B. Because the randomization occurs only within the blocks (plants from one store cannot be assigned to treatment groups for the other), we call this a **randomized block design**.

In effect, we conduct two parallel experiments, one for tomatoes from each store, and then combine the results. The picture tells the story:

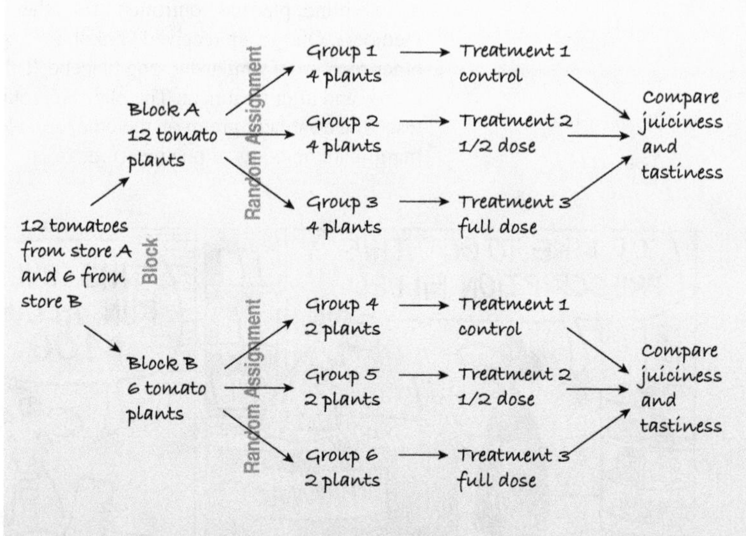

In a retrospective or prospective study, subjects are sometimes paired because they are similar in ways *not* under study. **Matching** subjects in this way can reduce variation in much the same way as blocking. For example, a retrospective study of music education and grades might match each student who studies an instrument with someone of the same sex who is similar in family income but doesn't study an instrument. When we compare grades of music students with those of nonmusic students, the matching would reduce the effects of variation in income and gender.

Blocking is the same idea for experiments as stratifying is for sampling. Both methods group together subjects that are similar and randomize within those groups as a way to remove unwanted variation. (But be careful to keep the terms straight. Don't say that we "stratify" an experiment or "block" a sample.) We use blocks to reduce variability so we can see the effects of the factors; we're not usually interested in studying the effects of the blocks themselves.

For Example BLOCKING

RECAP: In 2007, pet food contamination put cats at risk, as well as dogs. Our experiment should probably test the safety of the new food on both animals.

QUESTIONS: Why shouldn't we randomly assign a mix of cats and dogs to the two treatment groups? What would you recommend instead?

ANSWER: Dogs and cats might respond differently to the foods, and that variability could obscure my results. Blocking by species can remove that superfluous variation. I'd randomize cats to the two treatments (test food and safe food) separately from the dogs. I'd measure their responses separately and look at the results afterward.

Just Checking

2. Recall the experiment about gastric freezing, an old method for treating peptic ulcers that you read about in the first Just Checking. Doctors would insert an inflatable bladder down the patient's esophagus and into the stomach and then pump in a cold liquid. A major insurance company now states that it doesn't cover this treatment because "double-blind, controlled clinical trials" failed to demonstrate that gastric freezing was effective.
   a) What does it mean that the experiment was double-blind?
   b) Why would you recommend a placebo control?
   c) Suppose researchers suspected that the effectiveness of the gastric freezing treatment might depend on whether a patient had recently developed the peptic ulcer or had been suffering from the condition for a long time. How might the researchers have designed the experiment?

11.6 Adding More Factors

Some gardeners water frequently, making sure that the plants are never dry. Others let Mother Nature take her course and leave the watering to her. The makers of OptiGro want to ensure that their product will work under a wide variety of watering conditions. Maybe we should include the amount of watering as part of our experiment. Can we study a second factor at the same time and still learn as much about fertilizer?

We now have two factors (fertilizer at three levels and irrigation at two levels). We combine them in all possible ways to yield six treatments:

	No Fertilizer	Half Fertilizer	Full Fertilizer
No Added Water	1	2	3
Daily Watering	4	5	6

If we allocate the original 24 plants, the experiment now assigns four plants to each of these six treatments at random. This experiment is a **completely randomized two-factor experiment** because any plant could end up assigned at random to any of the six treatments (and we have two factors).

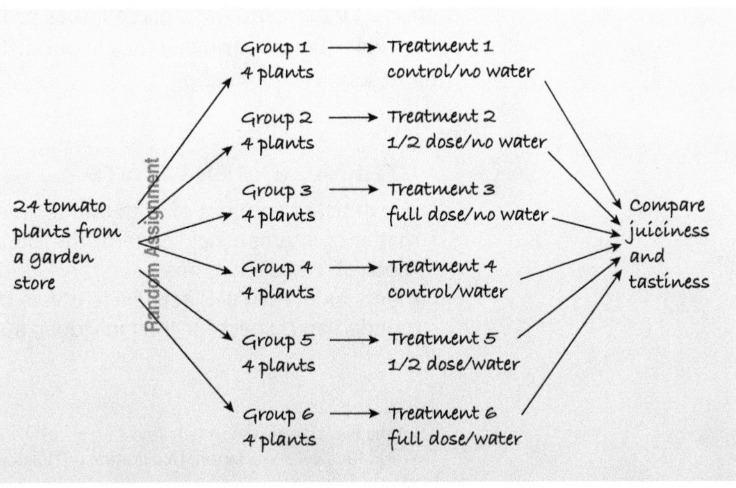

THINK LIKE A STATISTICIAN

With two factors, we can account for more of the variation. That lets us see the underlying patterns more clearly.

It's often important to include several factors in the same experiment in order to see what happens when the factor levels are applied in different *combinations*. A common misconception is that applying several factors at once makes it difficult to separate the effects of the individual factors. You may hear people say that experiments should always be run "one factor at a time." In fact, just the opposite is true: Experiments with more than one factor are both more efficient and provide more information than one-at-a-time experiments. There are many ways to design efficient multifactor experiments.

11.7 Confounding

Professor Stephen Ceci of Cornell University performed an experiment to investigate the effect of a teacher's classroom style on student evaluations. He taught a class in developmental psychology during two successive terms to a total of 472 students in two very similar classes. He kept everything about his teaching identical (same text, same syllabus, same office hours, etc.) and modified only his style in class. During the fall term, he maintained a subdued demeanour. During the spring term, he used expansive gestures and lectured with more enthusiasm, varying his vocal pitch and using more hand gestures. He administered a standard student evaluation form at the end of each term.

The students in the fall term class rated him only an average teacher. Those in the spring term class rated him an excellent teacher, praising his knowledge and accessibility, and even the quality of the textbook. On the question "How much did you learn in the course?" the average response changed from 2.93 to 4.05 on a five-point scale.[9]

How much of the difference he observed was due to his difference in manner, and how much might have been due to the season of the year? Fall term in Ithaca, New York (home of Cornell University), starts out colourful and pleasantly warm, but ends cold and bleak. Spring term starts out bitter and snowy and ends with blooming flowers and singing birds. Might students' overall happiness have been affected by the season and reflected in their evaluations?

Unfortunately, there's no way to tell. Nothing in the data enables us to tease apart these two effects, because all the students who experienced the subdued manner did so during the fall term and all who experienced the expansive manner did so during the spring. When the levels of one factor are associated with the levels of another factor, we say that these two factors are **confounded**.

In some experiments, such as this one, it's just not possible to avoid some confounding. Professor Ceci could have randomly assigned students to one of two classes during the same term, but then we might question whether mornings or afternoons were better, or whether he really delivered the same class the second time (after practising on the first class). Or he could have had another professor deliver the second class, but that would have raised more serious issues about differences in the two professors and concern over more serious confounding.

ETHICAL EXPERIMENTS

The ethical treatment of both human and animal subjects in experiments has been a matter of international concern. The *Declaration of Helsinki*,[10] developed by the World Medical Association, has been the foundation for most ethical policies for experiments on human subjects since it was originally proposed in 1964. The Declaration is founded on respect for the individual, subjects' rights to self-determination, and the

[9] But the two classes performed almost identically well on the final exam.

[10] World Medical Association, Declaration of Helsinki, www.wma.net/en/30publications/10policies/b3/index.html.

right to make informed decisions about participating in the research. The subject's welfare must always take precedence over the interests of science and safety, and ethical considerations always take precedence over laws and regulations, according to the Declaration. It has been revised six times; the 1975 revision introduced the importance of oversight by an "independent committee," which led to the development of Institutional Review Boards (IRBs), found today at all research universities and centres in Canada.

Later revisions dealt with (among other issues) the use of placebo controls in medical trials. Is it ethical to administer an inert placebo to some subjects if a proven effective alternative exists that could be used as a control treatment? Knowledge of the principles of experiment design can help you understand this question.

The 2008 revision states that:[11]

The benefits, risks, burdens and effectiveness of a new intervention must be tested against those of the best current proven intervention, except in the following circumstances:

- *The use of placebo, no treatment, is acceptable in studies where no current proven intervention exists; or*
- *Where for compelling and scientifically sound methodological reasons the use of placebo is necessary to determine the efficacy or safety of an intervention and the patients who receive placebo or no treatment will not be subject to any risk of serious or irreversible harm. Extreme care must be taken to avoid abuse of this option.*

In April 2008, the U.S. Food and Drug Administration (FDA) declared that clinical trials conducted outside of the U.S. did not need to comply with the Declaration of Helsinki. Agency officials said that the Declaration favours available treatments over placebos, which the FDA prefers in some circumstances, and that the periodic amendments to the Declaration could create confusion among drug companies. FDA spokesperson Rachel Behrman said, "We're saying that our regulations should not depend on any documents outside our control."[12] But others (see, for example, an editorial in the *British Medical Journal*[13]) claim that the decision was a response to pressures from drug companies that wished to be free to conduct tests in poor countries using inert placebos rather than more expensive alternative therapies, and not be required to provide effective drugs to all participants in a study at its conclusion, as the Declaration demands.

This is an ongoing debate with substantial consequences. You can follow much of it online with search terms, such as "FDA rules" and "Helsinki."

For Example CONFOUNDING

RECAP: After many dogs and cats suffered health problems caused by contaminated foods, we're trying to find out whether a newly formulated pet food is safe. Our experiment will feed some animals the new food and others a food known to be safe, and a veterinarian will check the response.

QUESTION: Why would it be a bad design to feed the test food to some dogs and the safe food to cats?

ANSWER: This would create confounding. We would not be able to tell whether any differences in animals' health were attributable to the food they had eaten or to differences in how the two species responded.

[11]World Medical Association, Declaration of Helsinki, www.wma.net/en/30publications/10policies/b3/index.html.
[12]*Science*, 9 May 2008: Vol. 320: 5877, p. 731.
[13]*British Medical Journal*, 2009: 338:b1559.

A two-factor example. Confounding can also arise from a badly designed multifactor experiment. Here's a classic. A credit card bank wanted to test the sensitivity of the market to two factors: the annual fee charged for a card and the annual percentage rate charged. Not wanting to scrimp on sample size, the bank selected 100 000 people at random from a mailing list. It sent out 50 000 offers with a low rate and no fee, and 50 000 offers with a higher rate and a $50 annual fee. Guess what happened? That's right—people preferred the low-rate, no-fee card. No surprise. In fact, they signed up for that card at over twice the rate as the other offer. And because of the large sample size, the bank was able to estimate the difference precisely. But the question the bank really wanted to answer was, "How much of the change was due to the rate, and how much was due to the fee?" Unfortunately, there's simply no way to separate out the two effects. If the bank had sent out all four possible different treatments—low rate with no fee, low rate with $50 fee, high rate with no fee, and high rate with $50 fee—each to 25 000 people, it could have learned about both factors and could have also seen what happens when the two factors occur in combination.

Lurking or Confounding?

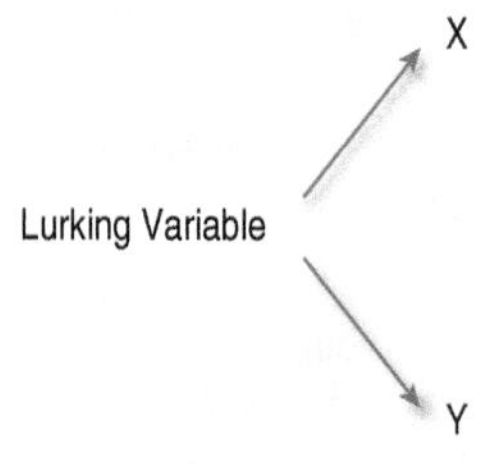

Our discussion of confounding may remind you of the problem of lurking variables we discussed in Chapters 6 and 8, where we explained how lurking variables can operate via common response or confounding to make causal connections difficult to establish. Confounding is a general concept and is not limited to just lurking variables. Effects of explanatory variables may also become confounded. In either case, confounding interferes with our ability to interpret our analyses simply. Effects of different variables become difficult or impossible to disentangle.

Confounding variables and lurking variables are alike in that they interfere with our ability to interpret our analyses simply. Each can mislead us, but there are important differences in both how and where the confusion may arise.

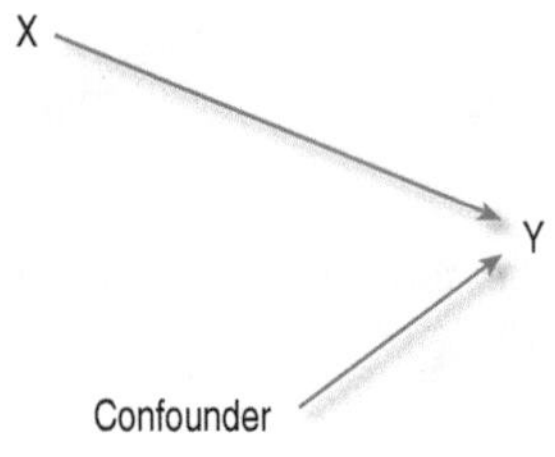

A lurking variable creates an association between two other variables that tempts us to think that one may cause the other. This can happen in a regression analysis or an observational study when a lurking variable influences both the explanatory and response variables. Recall that countries with more TV sets per capita tend to have longer life expectancies. We shouldn't conclude it's the TVs "causing" longer life. We suspect instead that a generally higher standard of living may mean that people can afford more TVs and get better health care, too. Our data revealed an association between TVs and life expectancy, but economic conditions were a likely lurking variable. A lurking variable, then, is usually thought of as a variable associated with both y and x that makes it appear that x may be causing y.

Confounding can arise in experiments when some other variable becomes associated with a factor and may also have an effect on the response variable. In a designed experiment, the experimenter *assigns* treatments at random to subjects rather than just observing them, in an attempt to neutralize the effects of all possible lurking variables. But problems can arise that ruin our exquisite plan. For example, because Professor Ceci used one teaching style in the fall and the other in spring, he was unable to tell how much of his students' reactions was attributable to his teaching and how much to the weather. A confounding variable, then, is a variable associated with the experimental factor that may also have some effect on the response. Because of the confounding, we find that we can't tell whether any effect we see was caused by our factor or by the confounding variable—or even by both working together.

It's worth noting that the role of blinding in an experiment is to combat a possible source of confounding. There's a risk that knowledge about the treatments could lead the subjects or those interacting with them to behave differently or could influence judgments made by the people evaluating the responses. That means we won't know whether the treatments really do produce different results or if we're being fooled by these confounding influences.

WHAT CAN GO WRONG?

- **Don't give up just because you can't run an experiment.** Sometimes we can't run an experiment because we can't identify or control the factors. Sometimes it would simply be unethical to run the experiment. (Consider randomly assigning students to take—and be graded in—a Statistics course deliberately taught to be boring and difficult or one that had an unlimited budget to use multimedia, real-world examples, and field trips to make the subject more interesting.) If we can't perform an experiment, often an observational study is a good choice.
- **Beware of confounding.** Use randomization whenever possible to ensure that the factors not in your experiment are not confounded with your treatment levels. Be alert to confounding that cannot be avoided, and report it along with your results.
- **Bad things can happen even to good experiments.** Protect yourself by recording additional information. For example, an experiment in which the air conditioning failed for two weeks, which affected the results, was saved by recording the temperature (although that was not originally one of the factors) and estimating the effect the higher temperature had on the response.[14]

  It's generally good practice to collect as much information as possible about your experimental units and the circumstances of the experiment. For example, in the tomato experiment, it would be wise to record details of the weather (temperature, rainfall, sunlight) that might affect the plants and any facts available about their growing situation. (Is one side of the field in shade sooner than the other as the day proceeds? Is one area lower and a bit wetter?) Sometimes we can use this extra information during the analysis to reduce biases.
- **Don't spend your entire budget on the first run.** Just as it's a good idea to pretest a survey, it's always wise to try a small pilot experiment before running the full-scale experiment. You may learn, for example, how to choose factor levels more effectively, about effects you forgot to control, and about unanticipated confounding.

CONNECTIONS

The fundamental role of randomization in experiments clearly points back to our discussions of randomization, to our experiments with simulations, and to our use of randomization in sampling. The similarities and differences between experiments and samples are important to keep in mind and can make each concept clearer.

If you think that blocking in an experiment resembles stratifying in a sample, you're quite right. Both are ways of removing the variation we can identify to help us see past the variation in the data.

Experiments compare groups of subjects that have been treated differently. Graphics such as boxplots that help us compare groups are closely related to these ideas. Think about what we look for in a boxplot to tell whether two groups look really different, and you'll be thinking about the same issues as experiment designers.

Generally, we're going to consider how different the mean responses are for different treatment groups. And we're going to judge whether those differences are large by using standard deviations as rulers. (That's why we need to replicate results for each treatment; we need to be able to estimate those standard deviations.) The discussion in Chapter 5 introduced this fundamental statistical thought, and it's going to keep coming back over and over again. Statistics is about variation.

We'll see a number of ways to analyze results from experiments in subsequent chapters.

[14] R. D. DeVeaux and M. Szelewski, 1989, "Optimizing automatic splitless injection parameters for gas chromatographic environmental analysis." *Journal of Chromatographic Science* 27:9, 513–518.

What Have We Learned?

Learning Objectives

Recognize observational studies:

- A retrospective study looks at an outcome in the present and looks for facts in the past that relate to it.
- A prospective study selects subjects and follows them as events unfold.

Know the elements of a designed randomized experiment:

- *Experimental units* (sometimes called *subjects or participants*) are assigned at random to *treatments*.
- The experimenter manipulates *factors*, setting them to specified *levels* to establish the treatments.
- A quantitative *response variable* is measured or observed for each experimental unit.
- We can attribute statistically significant differences in the response to the differences among the treatments.

State and apply the Four Principles of Experimental Design:

- *Control* sources of variation other than the factors being tested. Make the conditions as similar as possible for all treatment groups except for differences among the treatments.
- *Randomize* the assignment of participants to treatments.
- *Replicate* by applying each treatment to more than one participant.
- *Block* the experiment by grouping together participant who are similar in important ways that you cannot control.

Work with *blinding* and *control groups*:

- A *single-blind* study is one in which *either* all those who can affect the results *or* all those who evaluate the results are kept ignorant of which subjects receive which treatments.
- A *double-blind* study is one in which *both* all who can affect the results *and* all who evaluate the results are ignorant of which subjects receive which treatments.
- A *control group* is assigned to a null treatment or to the best available alternative treatment.
- Control participants are often administered a *placebo* or null treatment that mimics the treatment being studied but is known to be inactive. This is one way to blind participants.

Understand the differences between experiments and surveys.

- Surveys try to estimate facts (parameter) about a population, so they require a representative random sample from that population.
- Experiments try to estimate the differences in the effects of treatments. They randomize a group of experimental units to treatments, but there is no need for the experimental units to be a representative sample from the population.
- Be alert for possible confounding due to a variable that is not under control affecting the responses differentially.

Terms

Observational study A study based on data in which no manipulation of factors has been employed (p. 302).

Retrospective study An observational study in which subjects are selected and then their previous conditions or behaviours are determined. Because retrospective studies are not based on random samples, they usually focus on estimating differences between groups or associations between variables (p. 302).

Prospective study An observational study in which subjects are followed to observe future outcomes. Because no treatments are deliberately applied, a prospective study is not an experiment. Nevertheless, prospective studies typically focus on estimating differences among groups that might appear as the groups are followed during the course of the study (p. 303).

Experiment An experiment manipulates factor levels to create treatments, randomly assigns subjects to these treatment levels, and then compares the responses of the subject groups across treatment levels (p. 303).

Random assignment To be valid, an experiment must assign experimental units to treatment groups at random. This is called random assignment (p. 303).

Factor A variable whose levels are controlled by the experimenter. Experiments attempt to discover the effects that differences in factor levels may have on the responses of the experimental units (p. 303).

Response variable A variable whose values are compared across different treatments. In a randomized experiment, large response differences can be attributed to the effect of differences in treatment level (p. 303).

Experimental units Individuals on whom an experiment is performed. Usually called *subjects* or *participants* when they are human (p. 304).

Level The specific values that the experimenter chooses for a factor are called the levels of the factor (p. 304).

Treatment The process, intervention, or other controlled circumstance applied to randomly assigned experimental units. Treatments are the different levels of a single factor or are made up of combinations of levels of two or more factors (p. 304).

Principles of experimental design

- **Control** aspects of the experiment that we know may have an effect on the response, but that are not the factors being studied (p. 305).
- **Randomize** subjects to treatments to even out effects that we cannot control (p. 305).
- **Replicate** over as many subjects as possible. Results for a single subject are just anecdotes. If, as often happens, the subjects of the experiment are not a representative sample from the population of interest, replicate the entire study with a different group of subjects, preferably from a different part of the population (p. 305).
- **Block** to reduce the effects of identifiable attributes of the subjects that cannot be controlled (p. 306).

Statistically significant When an observed difference is too large for us to believe that it is likely to have occurred naturally, we consider the difference to be statistically significant. Subsequent chapters will show specific calculations and give rules, but the principle remains the same (p. 309).

Control group The experimental units assigned to a baseline treatment level, typically either the default treatment, which is well understood, or a null, placebo treatment. Their responses provide a basis for comparison (p. 310).

Blinding Any individual associated with an experiment who is not aware of how subjects have been allocated to treatment groups is said to be blinded (p. 311).

Single-blind/Double-blind There are two main classes of individuals who can affect the outcome of an experiment:

- those who could influence the results (the subjects, treatment administrators, or technicians)
- those who evaluate the results (judges, treating physicians, etc.)

When every individual in either of these classes is blinded, an experiment is said to be single-blind. When everyone in both classes is blinded, we call the experiment double-blind (p. 311).

Placebo A treatment known to have no effect, administered so that all groups experience the same conditions. Many subjects respond to such a treatment (a response known as the placebo effect). Only by comparing with a placebo can we be sure that the observed effect of a treatment is not due simply to the placebo effect (p. 312).

Placebo effect The tendency of many human subjects (often 20% or more of experiment subjects) to show a response when administered a placebo (p. 313).

Blocking When groups of experimental units are similar, it is often a good idea to gather them together into blocks. By blocking, we isolate the variability attributable to the differences between the blocks so that we can see the differences caused by the treatments more clearly (p. 313).

Designs In a **completely randomized design**, all experimental units have an equal chance of receiving any treatment (p. 314).
In a **randomized block design**, the randomization occurs only within blocks (p. 314).

Matching In a retrospective or prospective study, subjects who are similar in ways not under study may be matched and then compared with each other on the variables of interest. Matching, like blocking, reduces unwanted variation (p. 314).

Completely randomized two-factor experiment Experimental units assigned at random to any of the treatments when two factors are involved (p. 315).

Confounding When the levels of one factor are associated with the levels of another factor in such a way that their effects cannot be separated (p. 316).

On the Computer EXPERIMENTS

Most experiments are analyzed with a Statistics package. You should almost always display the results of a comparative experiment with side-by-side boxplots. You may also want to display the means and standard deviations of the treatment groups in a table.

The analyses offered by statistics packages for comparative randomized experiments fall under the general heading of Analysis of Variance, usually abbreviated ANOVA. These analyses are beyond the scope of this chapter.

Exercises

1. **Tips** A pizza delivery driver, always trying to increase tips, runs an experiment on his next 40 deliveries. He flips a coin to decide whether or not to call a customer from his mobile phone when he is five minutes away, hoping this slight bump in customer service will lead to a slight bump in tips. After 40 deliveries, he will compare the average tip percentage between the customers he called and those he did not. What are the experimental units and how did he randomize treatments?

2. **Tomatoes** You want to compare the tastiness and juiciness of tomatoes grown with three amounts of a new fertilizer: none, half the recommended amount, and the full recommended amount. You allocate 6 tomato plants to receive each amount of fertilizer, assigning them at random. What are the experimental units? What is the response variable?

3. **Tips II** For the experiment described in Exercise 1, list the factor, the levels, and the response variable.

4. **Tomatoes II** For the experiment described in Exercise 2, name the factor and its levels. How might the response be measured?

5. **Tips again** For the experiment in Exercise 1, name some variables the driver did or should have controlled. Was the experiment randomized and replicated?

6. **Tomatoes again** For the experiment in Exercise 2, discuss variables that could be controlled or that could not be controlled. Is the experiment randomized and replicated?

7. **More tips** Is the experiment in Exercise 1 blinded? Can it be double-blinded? Explain.

8. **More tomatoes** If the tomato taster in Exercise 2 doesn't know how the tomatoes have been treated, is the experiment single or double-blind? How might the blinding be improved further?

9. **Block that tip** The driver in Exercise 1 wants to know about tipping in general. So he recruits several other

drivers to participate in the experiment. Each driver randomly decides whether to phone customers before delivery and records the tip percentage. Is this experiment blocked? Is that a good idea?

10. **Blocking tomatoes** To obtain enough plants for the tomato experiment in Exercise 2, experimenters have to purchase plants from two different garden centres. They then randomly assign the plants from each garden centre to all three fertilizer treatments. Is the experiment blocked? Is that a good idea?

11. **Confounded tips** For the experiment in Exercise 1, name some confounding variables that might influence the experiment's results.

12. **Tomatoes finis** What factors might confound the results of the experiment in Exercise 2?

13. **Facebook usage and grades** For her Statistics class experiment, researcher K. Morrison decided to study how usage of the Web site Facebook affected Grade 12 students' performance in school. She proposed to collect information from a random sample of Grade 12 students and examine the relationship between the average amount of time spent on Facebook each day and the average graduating grade.
a) Is this an experiment? If not, what kind of study is it?
b) If there is a relationship between the average amount of time spent on Facebook each day and the average graduating grade, why can't we conclude that the differences in average graduating grades are caused by differences in average amount of time spent on Facebook each day?

14. **Heart attacks and height** Researchers who examined the health records of thousands of males found that men who died of myocardial infarction (heart attack) tended to be shorter than men who did not.
a) Is this an experiment? If not, what kind of study is it?
b) Is it correct to conclude that shorter men are at higher risk for heart attack? Explain.

15. **MS and vitamin D** Multiple sclerosis (MS) is an autoimmune disease that strikes more often the farther people live from the equator. Could vitamin D—which most people get from the sun's ultraviolet rays—be a factor? Researchers compared vitamin D levels in blood samples from 150 U.S. military personnel who have developed MS with blood samples of nearly 300 who have not. The samples were taken, on average, five years before the disease was diagnosed. Those with the highest blood vitamin D levels had a 62% lower risk of MS than those with the lowest levels. (The link was only in whites, not in blacks or Hispanics.)
a) What kind of study was this?
b) Is that an appropriate choice for investigating this problem? Explain.
c) Who were the subjects?
d) What were the variables?

16. **Super Bowl commercials** When spending large amounts to purchase advertising time, companies want to know what audience they'll reach. In January 2013, a poll asked 1008 American adults whether they planned to watch the upcoming Super Bowl. Men and women were asked separately whether they were looking forward more to the football game or to watching the commercials. Among the men, 16% were planning to watch and were looking forward primarily to the commercials. Among women, 30% were looking forward primarily to the commercials.
a) Was this a stratified sample or a blocked experiment? Explain.
b) Was the design of the study appropriate for the advertisers' question?

17. **Menopause** Researchers studied the herb black cohosh as a treatment for hot flashes caused by menopause. They randomly assigned 351 women aged 45 to 55 who had reported at least two hot flashes a day to one of five groups: (1) black cohosh, (2) a multiherb supplement with black cohosh, (3) the multiherb supplement plus advice to consume more soy foods, (4) estrogen replacement therapy, or (5) a placebo. After a year, only the women given estrogen replacement therapy had symptoms different from those of the placebo group.[15]
a) What kind of study was this?
b) Is that an appropriate choice for this problem?
c) Who were the subjects?
d) Identify the treatment and response variables.

18. **Honesty** Coffee stations in offices often ask users to leave money in a tray to pay for their coffee, but many people cheat. Researchers at Newcastle University replaced the picture of flowers on the wall behind the coffee station with a picture of staring eyes. They found that the average contribution increased significantly above the well-established standard when people felt they were being watched, even though the eyes were patently not real.[16]
a) Was this a survey, an observational study, or an experiment? How can you tell?
b) Identify the variables.
c) What does "increased significantly" mean in a statistical sense?

19–38. **What's the design?** *Read each brief report of statistical research, and identify*
a) whether it was an observational study or an experiment.
If it was an observational study, identify (if possible)
b) whether it was retrospective or prospective.

---

[15] *Annals of Internal Medicine* 145:12, 869–897.
[16] *New York Times*, December 10, 2006.

c) the subjects studied and how they were selected.
d) the response variable of interest.
e) the nature and scope of the conclusion the study can reach.

If it was an experiment, identify (if possible)
b) the subjects studied.
c) the factor(s) in the experiment and the number of levels for each.
d) the number of treatments.
e) the response variable measured.
f) the design (completely randomized, blocked, or matched).
g) whether it was blind (or double-blind).
h) the nature and scope of the conclusion the experiment can reach.

19. Over a four-month period, among 30 people with bipolar disorder, patients who were given a high dose (10 g/day) of omega-3 fats from fish oil improved more than those given a placebo.[17]

20. The leg muscles of men aged 60 to 75 were 50% to 80% stronger after they had participated in a 16-week, high-intensity resistance-training program twice a week.[18]

21. In a test of roughly 200 men and women, those with moderately high blood pressure (averaging 164/89 mm Hg) did worse on tests of memory and reaction time than those with normal blood pressure.[19]

22. Among a group of disabled women aged 65 and older who were tracked for several years, those who had a vitamin B12 deficiency were twice as likely to suffer severe depression as those who did not.[20]

23. An examination of the medical records of more than 360 000 Swedish men showed that those who were overweight or who had high blood pressure had a higher risk of kidney cancer.[21]

24. To research the effects of "dietary patterns" on blood pressure in 459 subjects, subjects were randomly assigned to three groups and had their meals prepared by dieticians. Those who were fed a diet low in fat and cholesterol and high in fruits, vegetables, and low-fat dairy foods (known as the DASH diet) lowered their systolic blood pressure by an average of 6.7 points compared with subjects fed a control diet.

25. After menopause, some women take supplemental estrogen. There is some concern that if these women also drink alcohol, their estrogen levels will rise too high. Twelve volunteers who were receiving supplemental estrogen were randomly divided into two groups, as were 12 other volunteers not on estrogen. In each case, one group drank an alcoholic beverage, the other a nonalcoholic beverage. An hour later, everyone's estrogen level was checked. Only those on supplemental estrogen who drank alcohol showed a marked increase.

26. Is diet or exercise effective in combating insomnia? Some believe that cutting out desserts can help alleviate the problem, while others recommend exercise. Forty volunteers suffering from insomnia agreed to participate in a month-long test. Half were randomly assigned to a special no-desserts diet; the others continued desserts as usual. Half of the people in each of these groups were randomly assigned to an exercise program, while the others did not exercise. Those who ate no desserts and engaged in exercise showed the most improvement.

27. Some gardeners prefer to use nonchemical methods to control insect pests in their gardens. Researchers have designed two kinds of traps and want to know which design will be more effective. They randomly choose 10 locations in a large garden and place one of each kind of trap at each location. After a week, they will count the number of bugs in each trap.

28. Researchers have linked an increase in the incidence of breast cancer in Italy to dioxin released by an industrial accident in 1976. The study identified 981 women who lived near the site of the industrial explosion and were under age 40 at the time. Fifteen of the women had developed breast cancer at an unusually young average age of 45. Medical records showed that these women had heightened concentrations of dioxin in their blood and that each tenfold increase in dioxin level was associated with a doubling of the risk of breast cancer.[22]

29. In 2002, the journal *Science* reported that a study of women in Finland indicated that having sons shortened the lifespans of mothers by about 34 weeks per son, but that daughters helped to lengthen the mothers' lives. The data came from church records from the period 1640 to 1870.

30. In 2001, a report in the *Journal of the American Cancer Institute* indicated that women who work nights have a 60% greater risk of developing breast cancer. Researchers based these findings on the work histories of 763 women with breast cancer and 741 women without the disease.

31. The May 4, 2000, issue of *Science News* reported that, contrary to popular belief, depressed individuals cry no more often in response to sad situations than nondepressed people. Researchers studied 23 men and 48 women with major depression, and 9 men and 24 women with no depression. They showed the subjects a sad film about a boy whose father had died, noting

---

[17] *Archives of General Psychiatry* 56 [1999], 407.
[18] *Journal of Gerontology* 55A [2000], B336
[19] *Hypertension* 36 [2000], 1079.
[20] *American Journal of Psychiatry* 157 [2000], 715.
[21] *New England Journal of Medicine* 3434 [2000], 1305.
[22] *Science News*, Aug. 3, 2002.

whether or not the subjects cried. Women cried more often than men, but there were no significant differences between the depressed and nondepressed groups.

32. Scientists at a major pharmaceutical firm investigated the effectiveness of an herbal compound to treat the common cold. They exposed each subject to a cold virus, then gave him or her either the herbal compound or a sugar solution known to have no effect on colds. Several days later they assessed the patients' conditions, using a cold severity scale ranging from 0 to 5. They found no evidence of benefits associated with the compound.

33. Scientists examined the glycogen content of rats' brains at the rats' normal bedtimes and after they had been kept awake for an extra 6, 12, or 24 hours. The scientists found that glycogen was 38% lower among rats that had been sleep-deprived for 12 hours or more, and that the levels recovered during subsequent sleep. These researchers speculate that we may need to sleep in order to restore the brain's energy fuel.[23]

34. Some people who race greyhounds give the dogs large doses of vitamin C in the belief that the dogs will run faster. Investigators at the University of Florida tried three different diets in random order on each of five racing greyhounds. They were surprised to find that when the dogs ate high amounts of vitamin C they ran more slowly.[24]

35. Some people claim they can get relief from migraine headache pain by drinking a large glass of ice water. Researchers plan to enlist several people who suffer from migraines in a test. When a participant experiences a migraine headache, he or she will take a pill that may be a standard pain reliever or a placebo. Half of each group will also drink ice water. Participants will then report the level of pain relief they experience.

36. Weight is an issue for both humans and their pets. A dog food company wants to compare a new lower-calorie food with their standard dog food to see if it's effective in helping inactive dogs maintain a healthy weight. They have found several dog owners willing to participate in the trial. The dogs have been classified as small, medium, or large breeds, and the company will supply some owners of each size of dog with one of the two foods. The owners have agreed not to feed their dogs anything else for a period of 6 months, after which the dogs' weights will be checked.

37. Athletes who had suffered hamstring injuries were randomly assigned to one of two exercise programs. Those who engaged in static stretching returned to sports activity in a mean of 37.4 days ($SD = 27.6$ days). Those assigned to a program of agility and trunk stabilization exercises returned to sports in a mean of 22.2 days ($SD = 8.3$ days).[25]

38. Pew Research compared respondents to an ordinary five-day telephone survey with respondents to a four-month-long rigorous survey designed to generate the highest possible response rate. They were especially interested in identifying any variables for which those who responded to the ordinary survey were different from those who could be reached only by the rigorous survey.

39. **Omega-3** Exercise 19 described an experiment that showed that high doses of omega-3 fats might be of benefit to people with bipolar disorder. The experiment involved a control group of subjects who received a placebo. Why didn't the experimenters just give everyone the omega-3 fats to see if they improved?

40. **Insomnia** Exercise 26 described an experiment that showed that getting exercise helped people sleep better. The experiment involved other groups of subjects who didn't exercise. Why didn't the experimenters just have everyone exercise and see if their ability to sleep improved?

41. **Omega-3 revisited** Exercises 19 and 39 described an experiment investigating a dietary approach to treating bipolar disorder. Researchers randomly assigned 30 subjects to two treatment groups, one group taking a high dose of omega-3 fats and the other a placebo.
a) Why was it important to randomize in assigning the subjects to the two groups?
b) What would be the advantages and disadvantages of using 100 subjects instead of 30?

42. **Insomnia again** Exercises 26 and 40 described an experiment investigating the effectiveness of exercise in combating insomnia. Researchers randomly assigned half of the 40 volunteers to an exercise program.
a) Why was it important to randomize in deciding who would exercise?
b) What would be the advantages and disadvantages of using 100 subjects instead of 40?

43. **Omega-3, finis** Exercises 19, 39, and 41 described an experiment investigating the effectiveness of omega-3 fats in treating bipolar disorder. Suppose some of the 30 subjects were very active people who walked a lot or got vigorous exercise several times a week, while others tended to be more sedentary, working office jobs and watching a lot of television. Why might researchers choose to block the subjects by activity level before randomly assigning them to the omega-3 and placebo groups?

44. **Insomnia, at last** Exercises 26, 40, and 42 described an experiment investigating the effectiveness of exercise in

[23] *Science News*, July 20, 2002.
[24] *Science News*, July 20, 2002.
[25] *Journal of Orthopaedic & Sports Physical Therapy* 34 [March 2004], 3.

combating insomnia. Suppose some of the 40 subjects had maintained a healthy weight, but others were quite overweight. Why might researchers choose to block the subjects by weight level before randomly assigning some of each group to the exercise program?

45. Tomatoes, next season Describe a strategy to randomly split the 24 tomato plants into the three groups for the chapter's completely randomized single factor test of OptiGro fertilizer.

46. Tomatoes for dessert The chapter also described a completely randomized two-factor experiment testing OptiGro fertilizer in conjunction with two different routines for watering the plants. Describe a strategy to randomly assign the 24 tomato plants to the six treatments.

47. Shoes A running-shoe manufacturer wanted to test the effect of its new sprinting shoe on 100-metre dash times. The company sponsored five athletes who were running the 100-metre dash in the 2012 Summer Olympic Games. To test the shoe, it had all five runners run the race with a competitor's shoe and then again with its new shoe. The company used the difference in times as the response variable.

a) Suggest some improvements to the design.

b) Why might the shoe manufacturer not be able to generalize the results they found to all runners?

48. Swimsuits A swimsuit manufacturer wants to test the speed of its newly designed suit. The company designs an experiment by having six randomly selected Olympic swimmers swim as fast as they can with their old swimsuit first and then swim the same event again with the new, expensive swimsuit. The company will use the difference in times as the response variable. Criticize the experiment and point out some of the problems with generalizing the results.

49. Hamstrings Exercise 37 discussed an experiment to see if the time it took athletes with hamstring injuries to be able to return to sports was different depending on which of two exercise programs they engaged in.

a) Explain why it was important to *randomly* assign the athletes to the two different treatments.

b) There was no control group consisting of athletes who did not participate in a special exercise program. Explain the advantage of including such a group in this experiment.

c) How might blinding have been used in this experiment?

d) One group returned to sports activity in a mean of 37.4 days ($SD = 27.6$ days) and the other in a mean of 22.2 days ($SD = 8.3$ days). Do you think this difference is statistically significant? Explain.

50. Diet and blood pressure Exercise 24 reported on an experiment that showed that subjects fed the DASH diet were able to lower their blood pressure by an average of 6.7 points compared to a group fed a "control diet." All meals were prepared by dieticians.

a) Why were the subjects randomly assigned to the diets instead of letting people pick what they wanted to eat?

b) Why were the meals prepared by dieticians?

c) Why did the researchers need the control group? If the DASH diet group's blood pressure was lower at the end of the experiment than at the beginning, wouldn't that prove the effectiveness of that diet?

d) What additional information would you want to know to decide whether an average reduction in blood pressure of 6.7 points is statistically significant?

51. Mozart Will listening to a Mozart piano sonata make you smarter? In a 1995 study published in the journal *Psychological Science*, Rauscher, Shaw, and Ky reported that when students were given a spatial reasoning section of a standard IQ test, those who listened to Mozart for 10 minutes improved their scores more than those who simply sat quietly.

a) These researchers said the differences were statistically significant. Explain what that means in this context.

b) Steele, Bass, and Crook tried to replicate the original study. In their study, also published in *Psychological Science* (1999), the subjects were 125 college students who participated in the experiment for course credit. Subjects first took the test. Then they were assigned to one of three groups: listening to a Mozart piano sonata, listening to music by Philip Glass, and sitting for 10 minutes in silence. Three days after the treatments, they were retested. Draw a diagram displaying the design of this experiment.

c) These boxplots show the differences in score before and after treatment for the three groups. Did the Mozart group show improvement?

d) Do you think the results prove that listening to Mozart is beneficial? Explain.

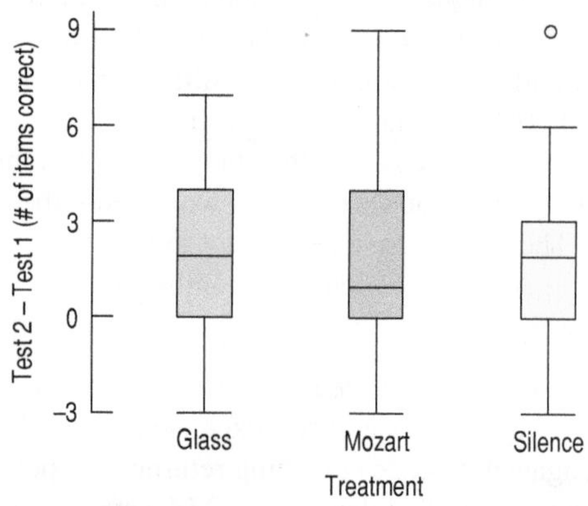

52. Mozart, 2nd movement The *New York Times* (Sept. 9, 1991) reported that "A sake brewer [near] Tokyo is marketing a Mozart brand of sake, asserting that Mozart is regularly played at the brewery." "It seems that the malt grows well especially with Mozart music in the background," the company director said. Suppose, just for the sake (as it were) of discussion, you wished to design an experiment to test whether this is true. Assume you have the full cooperation of the sake brewery. Specify how you would design the experiment. Indicate factors and response and how they would be measured, controlled, or randomized.

53. Frumpies The makers of Frumpies, "the breakfast of rug rats," want to improve their marketing, so they consult you.

a) They first want to know what fraction of children ages 10 to 13 like their celery-flavoured cereal. What kind of study should they perform?

b) They are thinking of introducing a new flavour, maple-marshmallow Frumpies, and want to know whether children will prefer the new flavour to the old one. Design a completely randomized experiment to investigate this question.

c) They suspect that children who regularly watch the Saturday morning cartoon show starring Frump, the flying teenage warrior rabbit who eats Frumpies in every episode, may respond differently to the new flavour. How would you take that into account in your design?

54. Contrast Bath treatments use the immersion of an injured limb alternately in water of two contrasting temperatures. Those who use the method claim that it can reduce swelling. Researchers compared three treatments: contrast baths and exercise (1), contrast baths alone (2), and exercise alone (3).[26] They reported the following boxplots comparing the change in hand volume after treatment:

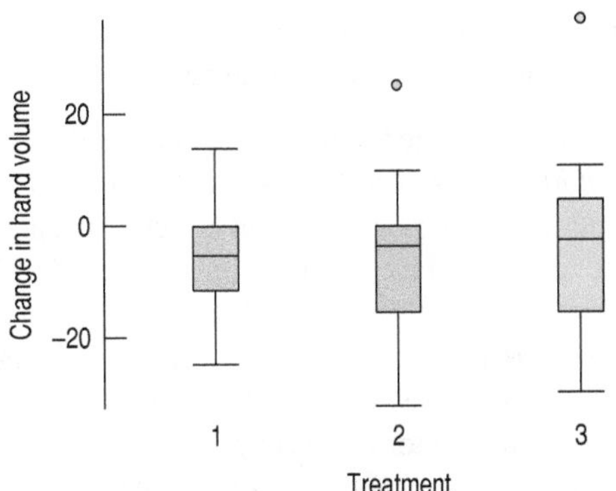

a) The researchers concluded that the differences were not statistically significant. Explain what that means in context.

b) The title says that the study was randomized and controlled. Explain what that probably means for this study.

[26]R. G. Janssen, D. A. Schwartz, and P.F. Velleman, 2009, "A randomized controlled study of contrast baths on patients with carpal tunnel syndrome." *Journal of Hand Therapy*.

c) The study did not use a placebo treatment. What was done instead? Do you think that was an appropriate choice? Explain.

55. Wine A 2001 Danish study published in the *Archives of Internal Medicine* casts significant doubt on suggestions that adults who drink wine have higher levels of "good" cholesterol and fewer heart attacks. These researchers followed a group of individuals born at a Copenhagen hospital between 1959 and 1961 for 40 years. Their study found that in this group the adults who drank wine were richer and better educated than those who did not.

a) What kind of study was this?

b) It is generally true that people with high levels of education and high socio-economic status are healthier than others. How does this call into question the supposed health benefits of wine?

c) Can studies such as these prove causation (that wine helps prevent heart attacks, that drinking wine makes one richer, that being rich helps prevent heart attacks, etc.)? Explain.

56. Swimming Recently, a group of adults who swim regularly for exercise were evaluated for depression. It turned out that these swimmers were less likely to be depressed than the general population. The researchers said the difference was statistically significant.

a) What does "statistically significant" mean in this context?

b) Is this an experiment or an observational study? Explain.

c) News reports claimed this study proved that swimming can prevent depression. Explain why this conclusion is not justified by the study. Include an example of a possible confounding variable.

d) But perhaps it is true. We wonder if exercise can ward off depression, and whether anaerobic exercise (like weight training) is as effective as aerobic exercise (like swimming). We find 120 volunteers not currently engaged in a regular program of exercise. Design an appropriate experiment.

57. Dowsing A water dowser claims to be able to sense the presence of water using a forked stick. Suppose we wish to set up an experiment to test his ability. We get 20 identical containers, fill some with water, and ask the dowser to tell which ones are full and which empty.

a) How will we randomize this procedure?

b) The dowser correctly identifies the contents of 12 out of 20 containers. Do you think this level of success is statistically significant? Explain.

c) How many correct identifications (out of 20) would the dowser have to make to convince you that the forked-stick trick works? Explain.

58. Healing A medical researcher suspects that giving post-surgical patients large doses of vitamin E will speed their recovery times by helping their incisions heal more quickly. Design an experiment to test this conjecture. Be sure to identify the factors, levels, treatments, response variable, and the role of randomization.

59. Reading Some schools teach reading using phonics (the sounds made by letters) and others using whole language (word recognition). Suppose a school district wants to know which method works better. Suggest a design for an appropriate experiment.

60. Gas mileage Do cars get better gas mileage with premium instead of regular unleaded gasoline? While it might be possible to test some engines in a laboratory setting, we'd rather use real cars and real drivers in real day-to-day driving, so we get 20 volunteers. Design the experiment.

61. Weekend deaths A study published in the *New England Journal of Medicine* (August 2001) suggests that it's dangerous to enter a hospital on a weekend. During a 10-year period, researchers tracked over 4 million emergency admissions to hospitals in Ontario, Canada. Their findings revealed that patients admitted on weekends had a much higher risk of death than those who went to the emergency room on weekdays.
a) The researchers said the difference in death rates was "statistically significant." Explain what that means in this context.
b) What kind of study was this? Explain.
c) If you think you're quite ill on a Saturday, should you wait until Monday to seek medical help? Explain.
d) Suggest some possible explanations for this troubling finding.

62. Shingles A research doctor has discovered a new ointment that she believes will be more effective than the current medication in the treatment of shingles (a painful skin rash). Eight patients have volunteered to participate in the initial trials of this ointment. You are the statistician hired as a consultant to help design a completely randomized experiment.
a) Describe how you will conduct this experiment.
b) Suppose the eight patients' last names start with the letters A to H. Using the random numbers listed below, show which patients you will assign to each treatment. Explain your randomization procedure clearly.

41098 18329 78458 31685 55259

c) Can you make this experiment double-blind? If so, explain how.
d) The initial experiment revealed that males and females may respond differently to the ointment. Further testing of the drug's effectiveness is now planned, and many patients have volunteered. What changes in your first design, if any, would you make for this second stage of testing?

63. Beetles Hoping to learn how to control crop damage by a certain species of beetle, a researcher plans to test two different pesticides in small plots of corn. A few days after application of the chemicals, he'll check the number of beetle larvae found on each plant. The researcher wants to know whether either pesticide works and whether there is a significant difference in effectiveness between them. Design an appropriate experiment.

64. LSAT prep Can test prep courses actually help raise LSAT scores? One organization says that the 30 students they tutored achieved an average gain of 15 points when they retook the test.
a) Explain why this does not necessarily prove that the test prep course caused the scores to go up.
b) Propose a design for an experiment that could test the effectiveness of the test prep course.
c) Suppose you suspect that the test prep course might be more helpful for students whose initial scores were particularly low. How would this affect your proposed design?

65. Safety switch An industrial machine requires an emergency shutoff switch that must be designed so that it can be easily operated with either hand. Design an experiment to find out whether workers will be able to deactivate the machine as quickly with their left hands as with their right hands. Be sure to explain the role of randomization in your design.

66. Washing clothes A consumer group wants to test the effectiveness of a new "organic" laundry detergent and make recommendations to customers about how to best use the product. They intentionally get grass stains on 30 white T-shirts to see how well the detergent will clean them. They want to try the detergent in cold water and in hot water on both the "regular" and "delicates" wash cycles. Design an appropriate experiment, indicating the number of factors, levels, and treatments. Explain the role of randomization in your experiment.

67. Skydiving, anyone? A humour piece published in the *British Medical Journal* notes that we can't tell for sure whether parachutes are safe and effective because there has never been a properly randomized, double-blind, placebo-controlled study of parachute effectiveness in skydiving.[27] (Yes, this is the sort of thing statisticians find funny . . .) Suppose you were designing such a study:
a) What is the factor in this experiment?
b) What experimental units would you propose?[28]
c) Explain what would serve as a placebo for this study.
d) What would the treatments be?
e) What would be the response variable for such a study?
f) What sources of variability would you control?
g) How would you randomize this "experiment"?
h) How would you make the experiment double-blind?

---

[27] Gordon Smith and Jill Pell, "Parachute use to prevent death and major trauma related to gravitational challenge: Systematic review of randomized control trials," *BMJ*, 2003:327.
[28] Don't include your Statistics instructor!

Just Checking ANSWERS

1. a) The factor was type of treatment for peptic ulcer.
   b) The response variable could be a measure of relief from gastric ulcer pain or an evaluation by a physician of the state of the disease.
   c) Treatments would be gastric freezing and some alternative control treatment.
   d) Treatments should be assigned randomly.
   e) No. The Web site reports "lack of effectiveness," indicating that no large differences in patient healing were noted.
2. a) Neither the patients who received the treatment nor the doctor who evaluated them to see if they had improved knew what treatment they had received.
   b) The placebo is needed to accomplish blinding. The best alternative would be using body-temperature liquid rather than the freezing liquid.
   c) The researchers should block the subjects by the length of time they had had the ulcer, then randomly assign subjects in each block to the freezing and placebo groups.

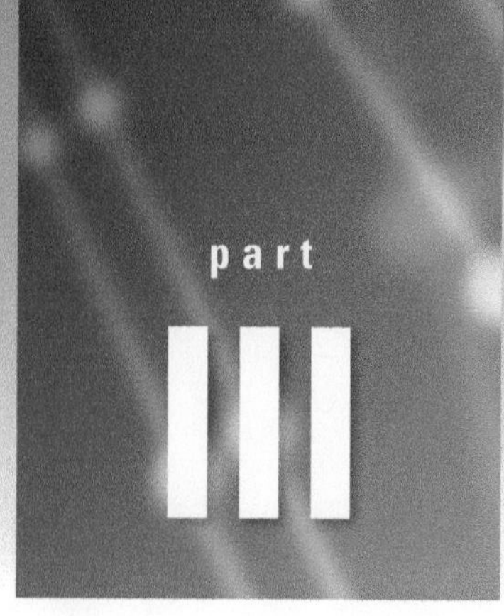

Review

Gathering Data

Quick Review

Before you can make a boxplot, calculate a mean, describe a distribution, or fit a line, you must have meaningful data to work with. Getting good data is essential to any investigation. No amount of clever analysis can make up for badly collected data. Here's a brief summary of the key concepts and skills:

- The way you gather data depends both on what you want to discover and on what is practical.
- To get some insight into what might happen in a real situation, model it with a **simulation** using random numbers.
- To answer questions about a target population, collect information from a sample with a **survey** or poll.
  - Choose the sample randomly. Random sampling designs include simple, stratified, systematic, cluster, and multistage.
  - A simple random sample draws without restriction from the entire target population.
  - When there are subgroups within the population that may respond differently, use a stratified sample.
  - Avoid bias, a systematic distortion of the results. Sample designs that allow undercoverage or response bias and designs such as voluntary response or convenience samples don't faithfully represent the population.
  - Samples will naturally vary one from another. This sample-to-sample variation is called sampling error. Each sample only approximates the target population.
- **Observational studies** collect information from a sample drawn from a target population.
  - Retrospective studies examine existing data. Prospective studies identify subjects in advance, then follow them to collect data as the data are created, perhaps over many years.
  - Observational studies can spot associations between variables but cannot establish cause and effect. It's impossible to eliminate the possibility of lurking or confounding variables.
- To see how different treatments influence a response variable, design an **experiment**.
  - Assign subjects to treatments randomly. If you don't assign treatments randomly, your experiment is not likely to yield valid results.
  - Control known sources of variation as much as possible. Reduce variation that cannot be controlled by using blocking, if possible.
  - Replicate the experiment, assigning several subjects to each treatment level.
  - If possible, replicate the entire experiment with an entirely different collection of subjects.
  - A well-designed experiment can provide evidence that changes in the factors cause changes in the response variable.

Now for more opportunities to review these concepts and skills . . .

Review Exercises

1–18. What design? *Analyze the design of each research example reported. Is it a sample, an observational study, or an experiment? If a sample, what is the population, the parameter of interest, and the sampling procedure? If an observational study, was it retrospective or prospective? If an experiment, describe the factors, treatments, randomization, response variable, and any blocking, matching, or blinding that may be present. In each, what kind of conclusions can be reached?*

1. Researchers identified 242 children in the Cleveland area who had been born prematurely (at about 29 weeks). They examined these children at age 8 and again at age 20, comparing them to another group of 233 children not born prematurely. According to their report, published in the *New England Journal of Medicine*, the "preemies" engaged in significantly less risky behaviour than the others. Differences between the groups showed up in the use of alcohol and marijuana, conviction of crimes, and teenage pregnancy.

2. The journal *Circulation* reported that among 1900 people who had heart attacks, those who drank an average of 19 cups of tea a week were 44% more likely than nondrinkers to survive at least three years after the attack.

3. Researchers at the Purina Pet Institute studied Labrador retrievers for evidence of a relationship between diet and longevity. At eight weeks of age, puppies of the same sex and weight were randomly assigned to one of two groups—a total of 48 dogs in all. One group was allowed to eat all they wanted, while the other group was fed a low-calorie diet (about 75% as much as the others). The median lifespan of dogs fed the restricted diet was 22 months longer than that of other dogs. (*Science News* 161, no. 19)

4. The radioactive gas radon, found in some homes, poses a health risk to residents. To assess the level of contamination in their area, a municipal health department wants to test a few homes. If the risk seems high, they will publicize the results to emphasize the need for home testing. Officials plan to use the local property tax list to randomly choose 25 homes from various areas of the municipality.

5. Data were collected over a decade from 1021 men and women with a recent history of precancerous colon polyps. Participants were randomly assigned to receive folic acid (a B vitamin) or a placebo. The study concluded that those receiving the folic acid may actually increase their risk of developing additional precancerous growths. Previous studies suggested that taking folic acid may help to prevent colorectal cancer. (Source: *JAMA*, 2007, 297)

6. In a study appearing in the journal *Science*, a research team reported that plants in southern England are flowering earlier in the spring. Records of the first flowering dates for 385 species over a period of 47 years indicate that flowering has advanced an average of 15 days per decade, an indication of climate warming, according to the authors.

7. Fireworks manufacturers face a dilemma. They must be sure that the rockets work properly, but test firing a rocket essentially destroys it. On the other hand, not testing the product leaves open the danger that they sell a bunch of duds, leading to unhappy customers and loss of future sales. The solution, of course, is to test a few of the rockets produced each day, assuming that if those tested work properly, the others are ready for sale.

8. People who read the last page of a mystery novel first generally like stories better. Researchers recruited 819 college students to read short stories. For one story, they were given a spoiler paragraph beforehand. On the second and third story, the spoiler was incorporated as the opening paragraph or not given at all. Overall, participants liked the stories best after first reading spoilers. (Source: *Psychological Science*, August 12, 2011)

9. Does keeping a child's lunch in an insulated bag, even with ice packs, protect the food from warming to temperatures where germs can proliferate? Researchers used an electric temperature gun on 235 lunches at preschools 90 minutes before they were to be eaten. Of the lunches with ice packs, over 90% of them were at unsafe temperatures. The study was of particular interest because preschoolers develop up to four times as many foodborne infections as do adults. (Source: *Science News*, August 9, 2011)

10. Some doctors have expressed concern that men who have vasectomies seem more likely to develop prostate cancer. Medical researchers used a national cancer registry to identify 923 men who had had prostate cancer and 1224 men of similar ages who had not. Roughly one quarter of the men in each group had undergone a vasectomy, many more than 25 years before the study. The study's authors concluded that there is strong evidence that having the operation presents no long-term risk for developing prostate cancer. (*Science News*, July 20, 2002)

11. Widely used antidepressants may reduce ominous brain plaques associated with Alzheimer's disease. In the study, mice genetically engineered to have large amounts of brain plaque were given a class of antidepressants that boost serotonin in the brain. After a single dose, the plaque levels dropped. After four months, the mice had about half the brain plaques as the mice that didn't take the drug. (Source: *Proceedings of the National Academy of Sciences*, August 22, 2011)

12. An artisan wants to create pottery that has the appearance of age. He prepares several samples of clay with four different glazes and test fires them in a kiln at three different temperature settings.

13. Tests of gene therapy on laboratory rats have raised hopes of stopping the degeneration of tissue that characterizes chronic heart failure. Researchers at the University of California, San Diego, used hamsters with cardiac disease, randomly assigning 30 to receive the gene therapy and leaving the other 28 untreated. Five weeks after treatment, the gene therapy group's heart muscles stabilized, while those of the untreated hamsters continued to weaken. (*Science News*, July 27, 2002)

14. People aged 50 to 71 were initially contacted in the mid-1990s to participate in a study about smoking and bladder cancer. Data were collected from more than 280 000 men and 186 000 women from eight U.S. states who answered questions about their health, smoking history, alcohol intake, diet, physical activity, and other lifestyle factors. When the study ended in 2006, about half the

bladder cancer cases in adults age 50 and older were traceable to smoking. (Source: *Journal of the American Medical Association*, August 17, 2011)

15. An orange-juice processing plant will accept a shipment of fruit only after several hundred oranges selected from various locations within the truck are carefully inspected. If too many of those checked show signs of unsuitability for juice (bruised, rotten, unripe, etc.) the whole truckload is rejected.

16. A soft-drink manufacturer must be sure the bottle caps on the soft drinks are fully sealed and will not come off easily. Inspectors pull a few bottles off the production line at regular intervals and test the caps. If they detect any problems, they will stop the bottling process to adjust or repair the machine that caps the bottles.

17. Physically fit people are less likely to die of cancer. A report in the May 2002 issue of *Medicine and Science in Sports and Exercise* followed 25 892 men aged 30 to 87 for 10 years. The most physically fit men had a 55% lower risk of death from cancer than the least fit group.

18. Does the use of computer software in Introductory Statistics classes lead to better understanding of the concepts? A professor teaching two sections of Statistics decides to investigate. She teaches both sections using the same lectures and assignments, but gives one class statistics software to help them with their homework. The classes take the same final exam, and graders do not know which students used computers during the semester. The professor is also concerned that students who have had calculus may perform differently from those who have not, so she plans to compare software versus no-software scores separately for these two groups of students.

19. **Point spread** When taking bets on sporting events, bookmakers often include a "point spread" that awards the weaker team extra points. In theory this makes the outcome of the bet a toss-up. Suppose a gambler places a \$10 bet and picks the winners of five games. If he's right about fewer than three of the games, he loses. If he gets three, four, or all five correct, he's paid \$10, \$20, and \$50, respectively. Estimate the amount such a bettor might expect to lose over many weeks of gambling.

20. **The lottery** Many people spend a lot of money trying to win huge jackpots in provincial lotteries. Let's play a simplified version using only the numbers from 1 to 20. You bet on three numbers. The province picks five winning numbers. If your three are all among the winners, you are rich!
    a) Simulate repeated plays. How long did it take you to win?
    b) In real lotteries, there are many more choices (often 54) and you must match all five winning numbers. Explain how these changes affect your chances of hitting the jackpot.

21. **Everyday randomness** Aside from casinos, lotteries, and games, there are other situations you encounter in which something is described as "random" in some way. Give three different examples. Describe how randomness is (or is not) achieved in each.

22. **Cell phone risks** Researchers at the Washington University School of Medicine randomly placed 480 rats into one of three chambers containing radio antennas. One group was exposed to digital cell phone radio waves, the second to analog cell phone waves, and the third group to no radio waves. Two years later the rats were examined for signs of brain tumors. In June 2002 the scientists said that differences among the three groups were not statistically significant.
    a) Is this a study or an experiment? Explain.
    b) Explain what "statistically significant" means in this context.
    c) Comment on the fact that this research was supported by funding from Motorola, a manufacturer of cell phones.

23. **Tips** In restaurants, servers rely on tips as a major source of income. Does serving candy after the meal produce larger tips? To find out, two waiters determined randomly whether or not to give candy to 92 dining parties. They recorded the sizes of the tips and reported that guests getting candy tipped an average of 17.8% of the bill, compared with an average tip of only 15.1% from those who got no candy.[28]
    a) Was this an experiment or an observational study? Explain.
    b) Is it reasonable to conclude that the candy caused guests to tip more? Explain.
    c) The researchers said the difference was statistically significant. Explain what that means in this context.

24. **Tips, second course** In another experiment to see if getting candy after a meal would induce customers to leave a bigger tip, a waitress randomly decided what to do with 80 dining parties. Some parties received no candy, some just one piece, and some two pieces. Others initially got just one piece of candy, and then the waitress suggested that they take another piece. She recorded the tips received, finding that, in general, the more candy, the higher the tip, but the highest tips (23%) came from the parties who got one piece and then were offered more.[29]
    a) Diagram this experiment.
    b) How many factors are there? How many levels?
    c) How many treatments are there?
    d) What is the response variable?

---

[28] "Sweetening the till: The use of candy to increase restaurant tipping." *Journal of Applied Social Psychology* 32, no. 2 [2002]: 300–309.
[29] "Sweetening the till: The use of candy to increase restaurant tipping." *Journal of Applied Social Psychology* 32, no. 2 [2002]: 300–309.

e) Did this experiment involve blinding? Double blinding?
f) In what way might the waitress, perhaps unintentionally, have biased the results?

25. Timing A Léger Marketing online poll conducted in June of 2013 showed Justin Trudeau of the Liberal party with a four-point lead over current Prime Minister Stephen Harper. Considering the survey design, can you think of any reasons why Stephen Harper might still win the election, were it to be held immediately?

26. Laundry An experiment to test a new laundry detergent, SparkleKleen, is being conducted by a consumer advocate group. They would like to compare its performance with that of a laboratory standard detergent they have used in previous experiments. They will stain 16 swatches of cloth with 2 tsp of a common staining compound and then use a well-calibrated optical scanner to detect the amount of the stain that is left after washing with detergent. To save time in the experiment, several suggestions have been made. Comment on the possible merits and drawbacks of each one.
a) Since data for the laboratory standard detergent are already available from previous experiments, for this experiment wash all 16 swatches with SparkleKleen, and compare the results with the previous data.
b) Use both detergents with eight separate runs each, but to save time, use only a 10-second wash time with very hot water.
c) To ease bookkeeping, successively run all of the standard detergent washes on eight swatches, then run all of the SparkleKleen washes on the other eight swatches.
d) Rather than run the experiment, use data from the company that produced SparkleKleen, and compare them with past data from the standard detergent.

27. When to stop? You play a game that involves rolling a die. You can roll as many times as you want, and your score is the total for all the rolls. But ... if you roll a six your score is zero and your turn is over. What might be a good strategy for a game like this?
a) One of your opponents decides to roll four times, then stop (hoping not to get the dreaded six before then). Use a simulation to estimate his average score.
b) Another opponent decides to roll until she gets at least 12 points, then stop. Use a simulation to estimate her average score.
c) Propose another strategy that you would use to play this game. Using your strategy, simulate several turns. Do you think you would beat the two opponents?

28. Rivets A company that manufactures rivets believes the shear strength of the rivets they manufacture follows a Normal model with a mean breaking strength of 950 pounds and a standard deviation of 40 pounds.
a) What percentage of rivets selected at random will break when tested under a 900-pound load?
b) You're trying to improve the rivets and want to examine some that fail. Use a simulation to estimate how many rivets you might need to test to find three that fail at 900 pounds (or below).

29. Homecoming A university Statistics class conducted a survey concerning community attitudes about the school's large homecoming celebration. That survey drew its sample in the following manner: Telephone numbers were generated at random by selecting one of the local telephone exchanges (first three digits) at random and then generating a random four-digit number to follow the exchange. If a person answered the phone and the call was to a residence, then that person was taken to be the subject for interview. (Undergraduate students and those under voting age were excluded, as was anyone who could not speak English.) Calls were placed until a sample of 200 eligible respondents had been reached.
a) Did every telephone number that could occur in that community have an equal chance of being generated?
b) Did this method of generating telephone numbers result in a simple random sample (SRS) of local residences? Explain.
c) Did this method generate an SRS of local voters? Explain.
d) Did this method generate an unbiased sample of households? Explain.

30. Youthful appearance *Readers' Digest* reported results of several surveys that asked graduate students to examine photographs of men and women and try to guess their ages. Researchers compared these guesses with the number of times the people in the pictures reported having sexual intercourse. It turned out that those who had been more sexually active were judged as looking younger, and that the difference was described as "statistically significant." Psychologist David Weeks, who compiled the research, speculated that lovemaking boosts hormones that "reduce fatty tissue and increase lean muscle, giving a more youthful appearance."
a) What does "statistically significant" mean in this context?
b) Explain in statistical terms why you might be skeptical about Dr. Weeks's conclusion. Propose an alternative explanation for these results.

31. Smoking and Alzheimer's Medical studies indicate that smokers are less likely to develop Alzheimer's disease than people who never smoked.
a) Does this prove that smoking may offer some protection against Alzheimer's? Explain.
b) Offer an alternative explanation for this association.
c) How would you conduct a study to investigate this issue?

32. Antacids A researcher wants to compare the performance of three types of antacid in volunteers suffering from acid reflux disease. Because men and women may react differently to this medication, the subjects are split into two groups, by sex. Subjects in each group are randomly assigned to take one of the antacids or to take a sugar pill made to look the same. The subjects will rate their level of discomfort 30 minutes after eating.

a) What kind of design is this?
b) The experiment uses volunteers rather than a random sample of all people suffering from acid reflux disease. Does this make the results invalid? Explain.
c) How may the use of the placebo confound this experiment? Explain.

33. **Sex and violence** Does the content of a television program affect viewers' memory of the products advertised in commercials? Design an experiment to compare the ability of viewers to recall brand names of items featured in commercials during programs with violent content, sexual content, or neutral content.

34. **Pubs** In England, a Leeds University researcher said that the local watering hole's welcoming atmosphere helps men get rid of the stresses of modern life and is vital for their psychological well-being. Author of the report, Dr. Colin Gill, said rather than complain, women should encourage men to "pop out for a swift half." "Pub-time allows men to bond with friends and colleagues," he said. "Men need break-out time as much as women and are mentally healthier for it." Gill added that men might feel unfulfilled or empty if they had not been to the pub for a week. The report, commissioned by alcohol-free beer brand Kaliber, surveyed 900 men on their reasons for going to the pub. More than 40% said they went for the conversation, with relaxation and a friendly atmosphere being the other most common reasons. Only 1 in 10 listed alcohol as the overriding reason.
Let's examine this news story from a statistical perspective.

a) What are the Ws: *who, what, when, where, why?*
b) What population does the researcher think the study applies to?
c) What is the most important thing about the selection process that the article does *not* tell us?
d) How do *you* think the 900 respondents were selected? (Name a method of drawing a sample that is likely to have been used.)
e) Do you think the report that only 10% of respondents listed alcohol as an important reason for going to the pub might be a biased result? Why?

35. **Age and party 2008** The Pew Research Center collected data from national exits polls conducted by NBC News after the 2008 U.S. presidential election. The following table shows information regarding voter age and party preference:

Age	Republican	Democrat	Other	Total
18–29	260	390	351	1001
30–44	320	379	300	999
45–64	329	369	300	998
65 +	361	392	251	1004
Total	1270	1530	1202	4002

a) What sampling strategy do you think the pollsters used? Explain.
b) What percentage of the people surveyed were Democrats?
c) Do you think this is a good estimate of the percentage of voters in the United States who are registered Democrats? Why or why not?
d) In creating this sample design, what question do you think the pollsters were trying to answer?

36. **Bias?** Political analyst Michael Barone has written that "conservatives are more likely than others to refuse to respond to polls, particularly those polls taken by media outlets that conservatives consider biased" (*The Weekly Standard,* March 10, 1997). The Pew Research Foundation tested this assertion by asking the same questions in a national survey run by standard methods and in a more rigorous survey that was a true SRS with careful follow-up to encourage participation. The response rate in the "standard survey" was 42%. The response rate in the "rigorous survey" was 71%.

a) What kind of bias does Barone claim may exist in polls?
b) What is the population for these surveys?
c) On the question of political position, the Pew researchers report the following table:

	Standard Survey	Rigorous Survey
Conservative	37%	35%
Moderate	40%	41%
Liberal	19%	20%

What makes you think these results are incomplete?

d) The Pew researchers report that differences between opinions expressed on the two surveys were not statistically significant. Explain what "statistically significant" means in this context.

37. **Save the grapes** Vineyard owners have problems with birds that like to eat the ripening grapes. Grapes damaged by birds cannot be used for winemaking (or much of anything else). Some vineyards use scarecrows to try to keep birds away. Others use netting that covers the plants. Owners really would like to know if either method works and, if so, which one is better. One owner has offered to let you use his vineyard this year for an experiment. Propose a design. Carefully indicate how you would set up the experiment, specifying the factor(s) and response variable.

38. Bats It's generally believed that baseball players can hit the ball farther with aluminum bats than with the traditional wooden ones. Is that true? And, if so, how much farther? Players on your local high school baseball team have agreed to help you find out. Design an appropriate experiment.

39. Acupuncture. Research reported in 2008 brings to light the effectiveness of treating chronic lower back pain with different methods. One-third of nearly 1200 volunteers were administered conventional treatment (drugs, physical therapy, and exercise). The remaining patients got 30-minute acupuncture sessions. Half of these patients were punctured at sites suspected of being useful and half received needles at other spots on their bodies. Comparable shares of each acupuncture group, roughly 45%, reported decreased back pain for at least six months after their sessions ended. This was almost twice as high as those receiving the conventional therapy, leading the researchers to conclude that results were statistically significant.

a) Why did the researchers feel it was necessary to have some of the patients undergo a "fake" acupuncture?
b) Because patients had to consent to participate in this experiment, the subjects were essentially self-selected—a kind of voluntary response group. Explain why that does not invalidate the findings of the experiment.
c) What does "statistically significant" mean in the context of this experiment?

40. NBA draft lottery Professional basketball teams hold a draft each year in which they get to pick the best available college and high school players. In an effort to promote competition, teams with the worst records get to pick first, theoretically allowing them to add better players. To combat the fear that teams with no chance of making the playoffs might try to get better draft picks by intentionally losing late-season games, the NBA's Board of Governors adopted a weighted lottery system in 1990. Under this system, the 11 teams that did not make the playoffs were eligible for the lottery. The NBA prepared 66 cards, each naming one of the teams. The team with the worst win–loss record was named on 11 of the cards, the second-worst team on 10 cards, and so on, with the team having the best record among the nonplayoff clubs getting only one chance at having the first pick. The cards were mixed, then drawn randomly to determine the order in which the teams could draft players. Suppose there are two exceptional players available in this year's draft and your favourite team had the third-worst record. Use a simulation to find out how likely it is that your team gets to pick first or second. Describe your simulation carefully.

41. Security Twenty first-class passengers and 120 coach passengers are scheduled on a flight. In addition to the usual security screening, 10% of the passengers will be subjected to a more complete search.

a) Describe a sampling strategy to randomly select those to be searched.
b) Here is the first-class passenger list and a set of random digits. Select two passengers to be searched, carefully demonstrating your process.

65436 71127 04879 41516 20451 02227 94769 23593

Bergman	Cox	Fontana	Perl
Bowman	DeLara	Forester	Rabkin
Burkhauser	Delli-Bovi	Frongillo	Roufaiel
Castillo	Dugan	Furnas	Swafford
Clancy	Febo	LePage	Testut

c) Explain how you would use a random number table to select the coach passengers to be searched.

42. Profiling? Among the 20 first-class passengers on the flight described in Exercise 41, there were four businessmen from the Middle East. Two of them were the two passengers selected to be searched. They complained of profiling, but the airline claims that the selection was random. What do you think? Support your conclusion with a simulation.

43. Par 4 In theory, a golfer playing a par-4 hole tees off, hitting the ball in the fairway, then hits an approach shot onto the green. The first putt (usually long) probably won't go in, but the second putt (usually much shorter) should. Sounds simple enough, but how many strokes might it really take? Use a simulation to estimate a pretty good golfer's score based on these assumptions:

a) The tee shot hits the fairway 70% of the time.
b) A first approach shot lands on the green 80% of the time from the fairway, but only 40% of the time otherwise.
c) Subsequent approach shots land on the green 90% of the time.
d) The first putt goes in 20% of the time, and subsequent putts go in 90% of the time.

44. The back nine Use simulations to estimate more golf scores, similar to the procedure in Exercise 43.

a) On a par 3, the golfer hopes the tee shot lands on the green. Assume that the tee shot behaves like the first approach shot described in Exercise 43.
b) On a par 5, the second shot will reach the green 10% of the time and hit the fairway 60% of the time. If it does not hit the green, the golfer must play an approach shot as described in Exercise 43.
c) Create a list of assumptions that describe your golfing ability, and then simulate your score on a few holes. Explain your simulation clearly.

chapter

12 From Randomness to Probability

. . . Nature almost surely operates by combining chance with necessity, randomness with determinism . . .

—*Eric Chaisson, Epic of Evolution: Seven Ages of the Cosmos*

Where are We Going?

Flip a coin. Can you predict the outcome? It's hard to guess the outcome of just one flip because the outcome is random. If it's a fair coin, though, you can predict the *proportion* of heads you're likely to see in the long run.

It's this long-term predictability of randomness that we'll use throughout the rest of the book. To do that, we'll need to talk about the probability of different outcomes and learn some rules for dealing with them.

Early humans saw a world filled with random events. To help them make sense of the chaos around them, they sought out seers, consulted oracles, and read tea leaves. As science developed, we learned to recognize some events as predictable. We can now forecast the change of seasons, tell precisely when eclipses will occur, and even make a reasonably good guess at how warm it will be tomorrow. But many other events are still essentially random. Will the coffee shop have your favourite doughnut in stock? Will the headache remedy you take actually cure your pounding headache? Will you be able to get tickets for the U2 concert? Will the Toronto Maple Leafs make it to the playoffs this year? Will the stock market go up or down?

People assess randomness every day. The weatherperson forecasts a 40% chance of rain today. The newspaper reports that 17% of voters are undecided. Susie thinks she has a 95% chance of earning an A in her Statistics course. The odds of winning the Lotto 6/49 jackpot are about 1 in 14 million. These numbers give us a sense of how likely it is for a random event to actually happen. Where do these numbers come from? How does the weatherperson make her forecast? She does it *empirically* by looking at past weather maps that are similar to today's, and finding the percent of those that were rainy days. How do we know the odds of a 6/49 jackpot win? The number-crunchers work it out *theoretically*. How does Susie arrive at her chance of earning an A? She does it *subjectively*: she thinks about the course material and her study habits, and she may even add a touch of optimism.

12.1 Random Phenomena

Let's say that every day you drive through the intersection at Portage and Main. Even though it may seem that the light is never green when you get there, you know this can't actually be. In fact, if you really try, you might recall sailing through a green light once in a while.

What's random here? The light itself is governed by a timer. Its pattern isn't haphazard. In fact, the light may even be red at precisely the same times each day. It's the pattern of *your driving* that is random. Now, we're certainly not insinuating that you can't keep the car on the road. At the precision level of the 30 seconds or so when the light is red or green, the time you arrive at the light *is random*. Even if you try to leave your house at exactly the same time every day, whether the light is red or green as you reach the intersection is a **random phenomenon**.[1]

Is the colour of the light completely unpredictable? When you stop to think about it, it's clear that you expect some kind of *regularity* in your long-run experience. Some

[1]If you somehow managed to leave your house at *precisely* the same time every day and there was *no* variation in the time it took you to get to the light, then there wouldn't be any randomness, but that's not very realistic.

fraction of the time, the light will be green as you arrive at the intersection. How can we determine what that fraction is?

You might record what happens at the intersection each day and graph the *accumulated percentage* of green lights:

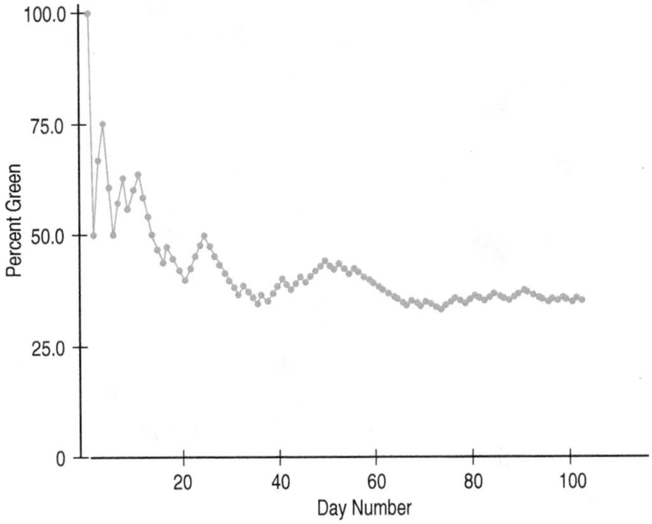

Figure 12.1

The overall percentage of times the light is green settles down as we see more outcomes.

Day	Light	% Green
1	Green	100
2	Red	50
3	Green	66.7
4	Green	75
5	Red	60
6	Red	50
⋮	⋮	⋮

The first day you recorded the light, it was green. Then for the next five days, it was red, then green again, then green, red, and red. If we plot the percentage of green lights against days, the graph would start at 100% (the first time, the light was green, so 1 out of 1 equals 100%). The next day it was red, so the accumulated percentage dropped to 50% (1 out of 2 green). The third day it was green again (2 out of 3, or 67% green), then green (3 out of 4, or 75% green), then red twice in a row (3 out of 5, for 60% green, and then 3 out of 6, for 50% green), and so on. As you collect a new observation for each day, each new outcome becomes a smaller and smaller fraction of the accumulated experience. As a result, in the long run, the graph settles down. As it settles down, we can see that in fact the light is green about 35% of the time.

A **phenomenon** consists of **trials**. Each **trial** has an outcome. **Outcomes** combine to make **events**.

When talking about random phenomena, it helps to define our terms. We aren't interested in the traffic light *all* the time. You pull up to the intersection only once a day, so you care about the colour of the light only at this particular time.[2] In general, each occasion when we observe a random phenomenon is called a **trial**. At each trial, we note the value of the random phenomenon, and call that the trial's **outcome**. (If this language reminds you of Chapter 9, that's *not* unintentional).

NOTATION ALERT

We often use capital letters—and usually from the beginning of the alphabet—to denote events.

For the traffic light, there are really three possible outcomes: red, yellow, or green. Often we're more interested in a combination of outcomes rather than in the individual ones. When you see the light turn yellow, what do *you* do? If you race through the intersection, then you treat the yellow more like a green light. If you step on the brakes, you treat it more like a red light. Either way, you might want to group the yellow with one or the other. When we combine outcomes like that, the resulting combination is an **event**. If you always run yellow lights, then you might want to combine green lights and yellow lights into the event "going through the intersection." We can use the letter **A** to represent this event. Individual outcomes that are not combined with any other outcomes are also events. If you always run yellow lights, then you might want to think of a red light as the event "can't go through the intersection."

We sometimes talk about the collection of *all possible outcomes* and call that event the **sample space**.[3] We'll denote the sample space **S**. The sample space is just a set that contains all the possible outcomes. For the traffic light, **S** = {red, green, yellow}.

[2]Even though the randomness here comes from the uncertainty in your arrival time, we can think of the light itself as showing a colour at random.

[3]Mathematicians like to use the term "space" as a fancy name for a set. Sort of like referring to that closet they gave you in residence as your "living space." But remember that it's really just the set of all outcomes.

The Law of Large Numbers

INTERFOTO/Personalities/INTERFOTO/Alamy

"For even the most stupid of men . . . is convinced that the more observations have been made, the less danger there is of wandering from one's goal."

—*Jacob Bernoulli, 1713, discoverer of the LLN*

EMPIRICAL PROBABILITY

For any event A,

$$P(A) = \frac{\text{\# times A occurs}}{\text{total \# of trials}}$$

in the long run.

A S ***Activity:* What Is Probability?** The best way to get a feel for probabilities is to experiment with them. We'll use this random-outcomes tool many more times.

The law of averages in everyday life "Dear Abby: My husband and I just had our eighth child. Another girl, and I am really one disappointed woman. I suppose I should thank God she was healthy, but, Abby, this one was supposed to have been a boy. Even the doctor told me that the law of averages was in our favour 100 to one." (Abigail Van Buren, 1974. Quoted in Karl Smith, *The Nature of Mathematics*. 6th ed. Pacific Grove, CA: Brooks/Cole, 1991, p. 589. Reprinted by permission.)

What's the *probability* of a green light at Portage and Main? Based on the graph, it looks like the relative frequency of green lights settles down to about 35%. Since you encounter a green light about 35% of the time, saying that the probability is about 0.35 seems like a reasonable answer. But do random phenomena always behave well enough for this to make sense? Perhaps the relative frequency of an event can bounce back and forth between two values forever, never settling on just one number.

Fortunately, a principle called the **Law of Large Numbers** (LLN) gives us the guarantee we need. It simplifies things if we assume that the events are **independent**. Informally, this means that the outcome of one trial doesn't affect the outcomes of the others. (Formally, we'll see a definition for independent events in the next chapter.) For independent trials, one form of the LLN says that as the number of trials increases, the long-run *relative frequency* of repeated events gets closer and closer to a single value.

Although the LLN wasn't proven until the eighteenth century, everyone expects the kind of long-run regularity that the law describes from everyday experience. In fact, the first person to prove the LLN, Jacob Bernoulli, thought it was pretty obvious, too, as his remark quoted in the margin shows.[4]

Because the LLN guarantees that relative frequencies settle down in the long run, we can now officially give a name to the value that they approach. We call it the **probability** of the event. If the relative frequency of green lights at that intersection settles down to 35% in the long run, we say that the probability of encountering a green light is 0.35, and write $P(\text{green}) = 0.35$. Because it is based on repeatedly observing the event's outcome, this definition of probability is often called **empirical probability**.

The Nonexistent Law of Averages

Even though the LLN seems natural, it is often misunderstood because the idea of the *long run* is hard to grasp. Many people believe, for example, that an outcome of a random event that hasn't occurred in many trials is "due" to occur. Many gamblers bet on numbers that haven't been seen for a while, mistakenly believing that they're likely to come up sooner. A common term for this is the "Law of Averages." After all, we know that in the long run the relative frequency will settle down to the probability of that outcome, so now we have some "catching up" to do, right?

Wrong. The Law of Large Numbers says nothing about short-run behaviour. Relative frequencies even out only *in the long run*. And, according to the LLN, the long run is *really* long (*infinitely* long, in fact).

The so-called Law of Averages doesn't exist at all. But you'll hear people talk about it as if it does. A good hockey team has lost their last six games, are they *due* for a win in their next game? If you've been doing particularly well in weekly quizzes in Statistics class, are you *due* for a bad grade? No. This isn't the way random phenomena work. There is *no* Law of Averages for short runs.

You've just flipped a fair coin and seen six heads in a row. Does the coin "owe" you some tails? Suppose you spend that coin and your friend gets it in change. When she starts flipping the coin, should we expect a run of tails? Of course not. Each flip is a new event. The coin can't "remember" what it did in the past, so it can't "owe" any particular outcomes in the future.

To see how this works in practice, the authors ran a simulation of 100 000 flips of a fair coin. We collected 100 000 random numbers, letting the numbers 0 to 4 represent heads and the numbers 5 to 9 represent tails. In our 100 000 "flips," there were 2981 streaks of at least five heads. The "Law of Averages" suggests that the next flip after a run

[4]In case you were wondering, Jacob's reputation was that he was every bit as nasty as this quotation suggests. He and his brother, who was also a mathematician, fought publicly over who had accomplished the most.

Don't let yourself think that there's a Law of Averages that promises short-term compensation for recent deviations from expected behaviour. A belief in such a "Law" can lead to money lost in gambling and to poor business decisions.

of five heads should be tails to even things out. Actually, the next flip was heads more often than tails: 1550 times to 1431 times. That's 51.9% heads. You can perform a similar simulation easily on a computer. Try it!

The lesson of the LLN is that sequences of random events don't compensate in the *short* run and don't need to do so to get back to the right long-run probability. If the probability of an outcome doesn't change and the events are independent, the probability of any outcome in another trial is *always* what it was, no matter what has happened in other trials.

Just Checking

1. One common proposal for beating the lottery is to note which numbers have come up lately, eliminate those from consideration, and bet on numbers that have not come up for a long time. Proponents of this method argue that in the long run, every number should be selected equally often, so those that haven't come up are due. Explain why this is faulty reasoning.

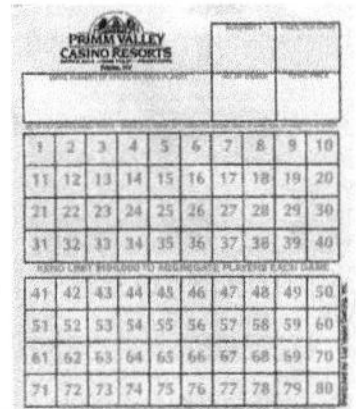

Keno and the Law of Averages Of course, sometimes an apparent drift from what we expect means that the probabilities are, in fact, *not* what we thought. If you get 10 heads in a row, maybe the coin has heads on both sides!

Keno is a simple casino game in which numbers from 1 to 80 are chosen. The numbers, as in most lottery games, are supposed to be equally likely. Payoffs are made depending on how many of those numbers you match on your card. A group of graduate students from a Statistics department decided to take a field trip to Reno, Nevada.[5] They (*very* discreetly) wrote down the outcomes of the games for a couple of days, then drove back to test whether the numbers were, in fact, equally likely. It turned out that some numbers were *more likely* to come up than others. Rather than bet on the Law of Averages and put their money on the numbers that were "due," the students put their faith in the LLN—and all their (and their friends') money on the numbers that had come up before. After they had pocketed more than $50 000, they were escorted off the premises and invited never to show their faces in that casino again.

12.2 Modelling Probability

A S *Activity:* **Multiple Discrete Outcomes.** The world isn't all heads or tails. Experiment with an event with four random alternative outcomes.

OJO Images Ltd/Alamy

Probability was first studied extensively by a group of French mathematicians who were interested in games of chance.[6] Rather than *experiment* with the games (and risk losing their money), they developed mathematical models of **theoretical probability**. To make things simple (as we usually do when we build models), they started by looking at games in which the different outcomes were equally likely. Fortunately, many games of chance are like that. Any of 52 cards is equally likely to be the next one dealt from a well-shuffled deck. Each face of a die is equally likely to land up (or at least it *should* be).

When outcomes are equally likely, their probability is easy to compute—it's just 1 divided by the number of possible outcomes. So the probability of rolling a three with a fair die is 1/6. The probability of picking the ace of spades from the top of a well-shuffled deck is 1/52.

It's almost as simple to find probabilities for events that are made up of several *equally likely* outcomes. We simply count all the outcomes that the event contains. The probability

[5]Although Nevada legalized gambling in 1931, it was not until Raymond Smith, the father of Nevada's gambling industry, opened Harold's Club in Reno that Reno's gambling industry became a tourist attraction. Las Vegas was up and coming but relatively small and quiet back then.

[6]Ok, gambling.

■ **NOTATION ALERT**

We often use capital letters—and usually from the beginning of the alphabet—to denote events. We *always* use P to denote probability. So,

$$P(\mathbf{A}) = 0.35$$

means "the probability of the event **A** is 0.35."

When being formal, use decimals (or fractions) for the probability values, but sometimes, especially when speaking more informally, it's easier to use percentages.

of the event is the number of outcomes in the event divided by the total number of possible outcomes. We can write

$$P(\mathbf{A}) = \frac{\text{\# outcomes in A}}{\text{\# of possible outcomes}}$$

For example, the probability of drawing a face card (Jack, Queen, King) from a deck is

$$P(\text{face card}) = \frac{\text{\# face cards}}{\text{\# cards}} = \frac{12}{52} = \frac{3}{13}$$

Is that all there is to it? Finding the probability of any event when the outcomes are equally likely is straightforward, but not necessarily easy. It gets hard when the number of outcomes in the event (and in the sample space) gets big. Think about flipping two coins. The sample space is $\mathbf{S} = \{HH, HT, TH, TT\}$ and each outcome is equally likely. So what's the probability of getting *exactly* one head and one tail? Let's call that event **A**. Well, there are two outcomes in the event $\mathbf{A} = \{HT, TH\}$ out of the four possible equally likely ones in **S**, so $P(\mathbf{A}) = \frac{2}{4}$, or $\frac{1}{2}$.

Okay, now flip 100 coins. What's the probability of exactly 67 heads? Well, first, how many outcomes are in the sample space? $\mathbf{S} = \{HHHHHHH \ldots HHH, HHHHHHH \ldots HHT, \ldots\}$ Hmm. A lot. In fact, there are 1 267 650 600 228 229 401 496 703 205 376 different possible outcomes when flipping 100 coins. To answer the question, we'd still have to figure out how many ways there are to get 67 heads. We'll see how in Chapter 14; stay tuned!

Don't get trapped into thinking that random events are always equally likely. The chance of winning the Lotto 6/49 jackpot is close to 1 in 14 million[7]—that's a really small chance. Regardless, people continue to buy tickets. In an attempt to understand why, an interviewer asked someone who had just purchased a lottery ticket, "What do you think your chances are of winning the lottery?" The reply was, "Oh, about 50–50." The shocked interviewer asked, "How do you get that?" to which the response was, "Well, the way I figure it, either I win or I don't!" The moral of the story is that events are *not* always equally likely.

Oh no, this complicates things. How can we get theoretical probabilities for events when outcomes are not equally likely? Keep reading this chapter, and wait for the next couple of chapters, for some useful ways to handle this. Sometimes these calculations can be messy and involve what is called 'combinatorics'; we'll avoid dealing with the messier ones in this course.

Personal Probability

What's the probability that you will earn a mark of A in this Statistics course? You may be able to come up with a number that seems reasonable. Of course, no matter how confident or depressed you feel about your chances for success, your probability should be between 0 and 1. How did you come up with this probability? Is it an empirical probability? Not unless you plan on taking the course over and over (and over . . .) and calculating the proportion of times you get an A.[8] Is it a theoretical probability? Unless you assume the outcomes (grades A, B, C, and so on) are equally likely, it will be hard to find the theoretical probability. But we use probability in a third sense as well.

We use the language of probability in everyday speech to express a degree of uncertainty *without* basing it on long-run relative frequencies or mathematical models. Your personal assessment of your chances of getting an A expresses your uncertainty about the outcome. That uncertainty may be based on how comfortable you're feeling in the course or about your midterm grade, but it can't be based on long-run behaviour. We call this third kind of probability **personal** or ***subjective* probability**.

Although subjective probabilities may be based on experience, they're not based on either long-run relative frequencies or equally likely events. So they don't display the kind

[7]There are actually 13 983 816 different ways you can select your numbers. There are equally likely outcomes, but only one of them wins the top prize.

[8]And unless you *really* forget what you learned over the summer, your chances of getting an A won't stay the same, so your course grades probably won't be independent either (as required by the LLN)!

of consistency that we'll need probabilities to have. For that reason, we'll stick to formally defined probabilities. You should be alert to the difference.

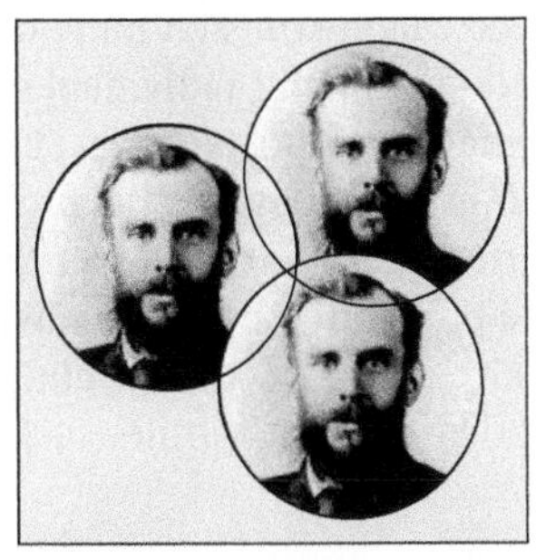

The Royal Society

John Venn (1834–1923) created the Venn diagram. His book on probability, The Logic of Chance, *was "strikingly original and considerably influenced the development of the theory of Statistics,"*[9] *according to John Maynard Keynes, one of the founders of Economics.*

Which kind of probability? The line between personal probability and the other two probabilities can be a little fuzzy. When a weather forecaster predicts a 40% probability of rain, is this a personal probability or a relative frequency probability? The claim may be that 40% of the time, when the weather radar map appears a certain way, it has rained (over some period of time). Or the forecaster may be stating a personal opinion based on years of experience that reflects a sense of what has happened in similar situations in the past. When you hear a probability stated, it's good to try to ascertain what kind of probability is intended.

The First Three Rules for Working with Probability

1. Make a picture.
2. Make a picture.
3. Make a picture.

We're dealing with probabilities now, not data, but the three rules don't change. The most common kind of picture to make is called a Venn diagram. We'll use Venn diagrams throughout the rest of this chapter. Even experienced statisticians make Venn diagrams to help them think about probabilities of compound or overlapping events. You should, too.

12.3 Formal Probability Rules

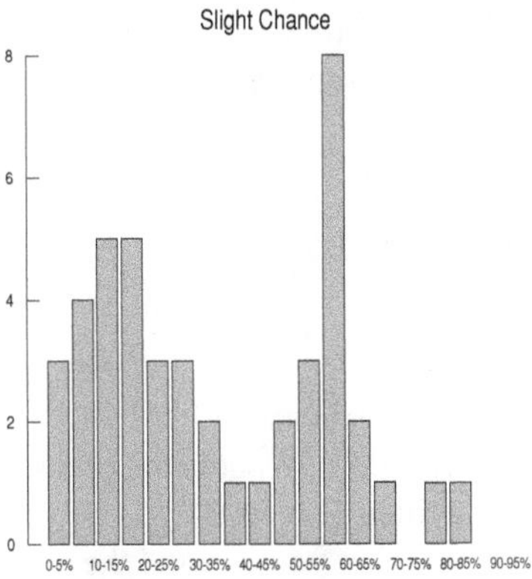

For some people, the phrase "50/50" means something vague like "I don't know" or "whatever." But when we discuss probabilities of outcomes, it takes on the precise meaning of *equally likely*. Speaking vaguely about probabilities will get us into trouble, so whenever we talk about probabilities, we'll need to be precise.[10] And to do that, we'll need to develop some formal rules about how probability works.

One of the authors ran an experiment while teaching a communication class at UBC, Vancouver. The students were given several subjective terms (for example, "slight chance," "doubtful," "pretty sure") and asked to quantify what the terms meant to them, to the nearest 5%. As you can see (in the margin), there's a *slight chance* the students didn't agree on what this phrase meant!

SURPRISING PROBABILITIES

We've been careful to discuss probabilities only for situations in which the outcomes were finite, or even countably infinite. But if the outcomes can take on *any* numerical value at all (we say they are *continuous*), things can get surprising. For example, what is the probability that a randomly selected child will be *exactly* 1 metre tall? Well, if we mean 1.00000 . . . metres, the answer is zero. Any randomly selected child—even one whose height would be recorded as 1 metre, won't be *exactly* 1 metre tall (to an infinite number of decimal places). But, if you've grown taller than 1 metre, there must have been a time in your life when you actually *were* exactly 1 metre tall, even if only for a fraction of a second. So this is an outcome with probability 0 that not only has happened—it has happened to *you.*

We've seen another example of this already in Chapter 5 when we worked with the Normal model. We said that the probability of any *specific* value—say, $z = 0.5$—is zero. The model gives a probability for any *interval* of values, such as $0.49 < z < 0.51$. The probability is smaller if we ask for $0.499 < z < 0.501$, and smaller still for $0.49999999 < z < 0.50000001$. Well, you get the idea. Continuous probabilities are useful for the mathematics behind much of what we'll do, but it's easier to deal with probabilities for countable outcomes.

[9] http://mathforum.org/library/drmath/view/52427.html

[10] And to be precise, we will be talking only about sample spaces where we can enumerate (that means we can begin to make a list, but may need to write something like 1, 2, 3, . . .) all the outcomes. Mathematics calls this a countable number of outcomes.

The simplest way to deal with probabilities is to base the calculations on individual outcomes. But any interesting analysis is likely to come from combining individual outcomes into events. In a database containing customer information on postal codes, each individual postal code is an outcome. Often we're interested in questions like the probability that the customer is from Manitoba or from the Maritime provinces, rather than from a particular postal code. We could always compute the probability of each individual outcome separately and add them up, but as we've already seen, this might get tedious. Instead, we'll want to develop rules[11] for the probability of events—rules such as the following:

1. If the probability is 0, the event can't occur. Likewise, if the probability is 1, the event *always* occurs. Even if you think an event is very unlikely, its probability can't be negative, and even if you're sure it will happen, its probability can't be greater than 1. Making this more formal gives the **probability assignment rule**.

   A probability is a number between 0 and 1.

   For any event **A**, $0 \leq P(\mathbf{A}) \leq 1$.

2. If a random phenomenon has only one possible outcome, it's not very interesting (and it's certainly not very random). So we need to distribute the probabilities among all the outcomes a trial can have. How can we do that so that it makes sense? For example, consider what you're doing as you read this book. The possible outcomes might be:

   A. You read to the end of this chapter before stopping.
   B. You finish this section but stop reading before the end of the chapter.
   C. You bail out before the end of this section.

   When we assign probabilities to these outcomes, the first thing to be sure of is that we distribute all of the available probability. Something in the sample space always occurs, so the probability of the sample space is 1.

   Making this more formal gives the **total probability rule**.

   The set of all possible outcomes of a trial must have probability 1.

   $P(\mathbf{S}) = 1$

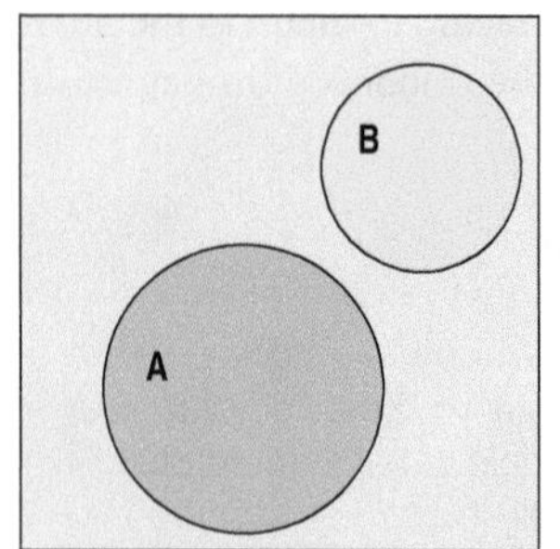

Two disjoint events, **A** and **B**.

3. Suppose the probability that (**A**) a randomly selected student is in second year is 0.20, and the probability that (**B**) he or she is in third year is 0.30. What is the probability that the student is in *either* second or third year, written *P*(**A** *or* **B**)? If you guessed 0.50, you've deduced the *addition rule*, which says that you can add the probabilities of events that are disjoint.[12] To see whether two events are disjoint, we break them apart into their component outcomes and check whether they have any outcomes in common. **Disjoint** (or **mutually exclusive**) events have no outcomes in common. Making this more formal gives the **addition rule for disjoint events:**

   For two disjoint events **A** and **B**, the probability that one or the other occurs is the sum of the probabilities of the two events.

   $P(\mathbf{A} \text{ or } \mathbf{B}) = P(\mathbf{A}) + P(\mathbf{B})$, provided that **A** and **B** are disjoint.

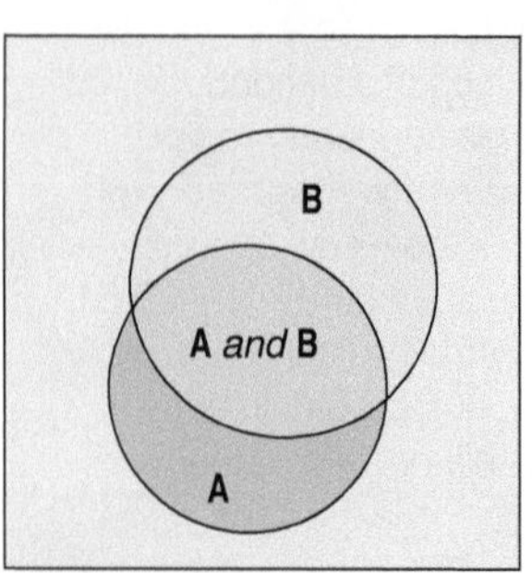

Two events **A** and **B** that are not disjoint. The event (**A** *and* **B**) is their intersection.

We can always add the probabilities of two events that each consist of only a single outcome. Because they have no outcomes to share, they must be disjoint. This gives us an easy way to check whether the probabilities we've assigned to the possible outcomes are **legitimate**.

[11] Actually, in mathematical terms, these three rules are called the axioms of probability. An axiom is a universally accepted truth. The axioms of probability are useful to derive other rules for probability.

[12] You may see *P*(**A** *or* **B**) written as $P(\mathbf{A} \cup \mathbf{B})$. The symbol $\cup$ means "union," representing the outcomes in event A *or* event B (or both). You may see *P*(**A** *and* **B**) written as $P(\mathbf{A} \cap \mathbf{B})$. The symbol $\cap$ means "intersection," representing outcomes that are in both event **A** *and* event **B**.

A S **Addition Rule for Disjoint Events.** Experiment with disjoint events to explore the addition rule.

From the total probability rule, the sum of the probabilities of all possible outcomes must be exactly one. No more, no less. And because individual outcomes are disjoint, we can add their probabilities to check that the sum is exactly 1. For example, if we were told the probabilities of selecting at random a first year, second year, third year, or fourth year student from all the undergraduates at a school were 0.25, 0.23, 0.22, and 0.20, respectively, we would know that something was wrong. These "probabilities" sum to only 0.90, so this is not a legitimate probability assignment. Either a value is wrong or we just missed some possible outcomes, like the "special student," "pre-first-year," or "more than four years" categories that soak up the remaining 0.10. Similarly, a claim that the probabilities were 0.26, 0.27, 0.29, and 0.30 would be wrong because these "probabilities" sum to 1.12, which is more than 1.

But be careful. The addition rule for disjoint events doesn't work for events that aren't disjoint. If the probability of owning an iPod is 0.70 and the probability of owning a computer is 0.90, the probability of owning either an iPod or a computer may be pretty high, but it is *not* 1.60! Why can't you add probabilities like this? Because these events are not disjoint. You can own both. Shortly, we'll see how to add probabilities for events like these, but we'll need another rule.

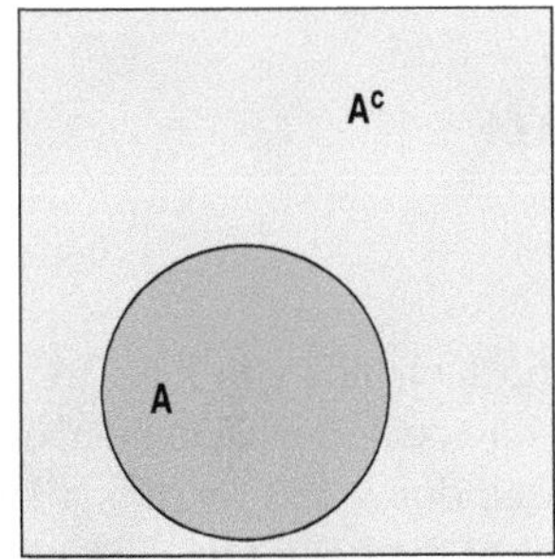

The set **A** and its complement $\mathbf{A^C}$. Together, they make up the entire sample space **S**. So the complement rule is really just an application of the total probability rule, $P(\mathbf{A}) + P(\mathbf{A^C}) = P(\mathbf{S}) = 1$.

Here's another rule. Suppose the probability that you get to class on time is 0.8. What's the probability that you don't get to class on time? Yes, it's 0.2. The set of outcomes that are *not* in the event **A** is called the **complement** of **A**, and is denoted $\mathbf{A^C}$. This leads to the **complement rule**:

The probability of an event occurring is 1 minus the probability that it doesn't occur.

$$P(\mathbf{A}) = 1 - P(\mathbf{A^C})$$

Putting the Rules to Work

When you get to the light at Portage and Main, it's either red, green, or yellow. We know that *P*(green) is 0.35. What would be a legitimate probability assignment for all three colours? Suppose we find out that *P*(yellow) is 0.04.

We can find the probability that the light is green *or* yellow from the addition rule for disjoint events, because these are disjoint events—the light can't be both green and yellow at the same time.[13]

$$P(\text{green or yellow}) = 0.35 + 0.04 = 0.39$$

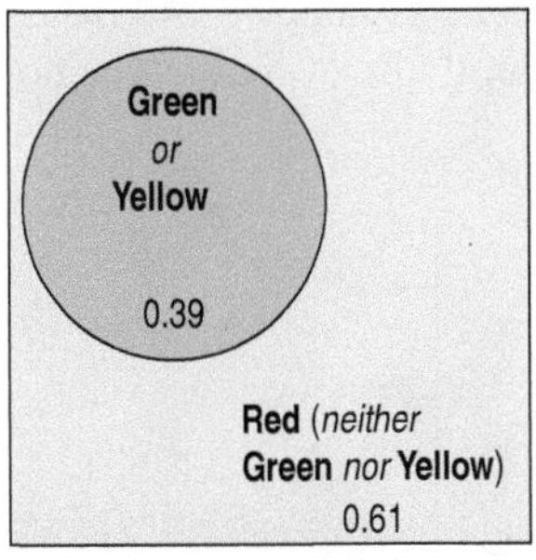

Figure 12.2

Red light. "Green or yellow" and "red" are complementary events. Their probabilities add to one.

So what's left for red? That's the only remaining alternative, so the complement rule tells us that

$$P(\text{red}) = P(\text{not green or yellow}) = 1 - P(\text{green or yellow}) = 1 - 0.39 = 0.61$$

This is a typical application of the probability rules. In most situations where we want to find a probability, we often combine rules.

Imagine Ali is a very daring driver—almost reckless—and whenever he sees a yellow light he puts his foot down and races through it. Belinda is the exact opposite; she always stops for yellow lights. Say Ali and Belinda are driving down the same street and arrive at the intersection of Portage and Main at the same time. What is the probability that Ali keeps driving through the intersection? That's the same as asking if the light is green or yellow. And we know that this event has a probability of 0.39. What is the probability that Ali has to stop? That's the same as asking if the light is red. And we know that this event has a probability of 0.61.

$$P(\text{Ali drives through}) = 0.39$$

$$P(\text{Ali stops}) = 0.61$$

What about Belinda? The probability that she drives through the intersection is the probability that the light is green, which is 0.35. The probability that she has to stop can be

[13]At least that's the assumption for Portage and Main. There are some places, especially in Europe, where traffic lights are sometimes simultaneously green and yellow, but we'll ignore that here.

found using the complement rule or the addition rule for disjoint events. Using the complement rule, the probability that she stops is 1 minus the probability she drives through, or $1 - 0.35 = 0.65$. Using the addition rule for disjoint events, the probability that she has to stop is the probability that the light is red or yellow, or $0.61 + 0.04 = 0.65$ (Whew—glad to see these came to the same answer. Something would be wrong if they didn't.)

$$P(\text{Belinda drives through}) = 0.35$$
$$P(\text{Belinda stops}) = 0.65$$

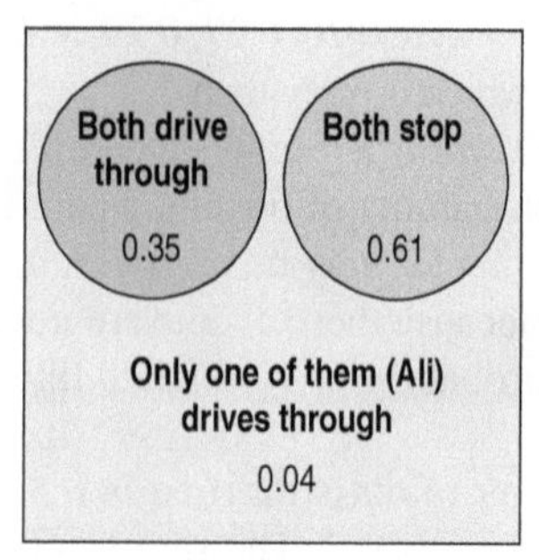

Figure 12.3

Three disjoint events. Their probabilities add to one

Everybody drives through green lights; nobody drives through red lights. Ali drives through a yellow light but Belinda doesn't. So the only way they can both drive through the intersection is if the light is green. And the only way they both stop is if the light is red.

$$P(\text{Ali and Belinda both drive through}) = 0.35$$
$$P(\text{Ali and Belinda both stop}) = 0.61$$

The General Addition Rule

When you get to the light at Portage and Main, there is a 65% chance you have to stop because the light is red. But if a bus is blocking the intersection you still have to stop, even if the light is green. Suppose the empirical research has been done, and there is a 16% chance you have to stop behind a bus regardless of the colour of the light. Only 3% of the time will you have a double reason to stop: The light is red and a bus is blocking your way.

Now what is the probability you have to stop at the intersection? What is the probability of **A** = {stopping for a red light} *or* **B** = {stopping for a bus}? We know $P(\mathbf{A}) = 0.35$ and $P(\mathbf{B}) = 0.16$. But $P(\mathbf{A} \text{ or } \mathbf{B})$ is not simply the sum $P(\mathbf{A}) + P(\mathbf{B})$, because the events **A** and **B** are not disjoint. Sometimes the light is red and a bus is blocking the intersection. So what can we do? We'll need a new probability rule.

As Figure 12.4 shows, we can't use the addition rule and add the two probabilities because the events are not disjoint; they overlap. There's an outcome in the *intersection* of **A** and **B**. Figure 12.4 represents the sample space. In reality, of course, there could be other reasons for stopping aside from a red light or bus, but let's assume we will only stop for these two reasons. Sitting outside the circle, then, is simply the event that we do not stop.

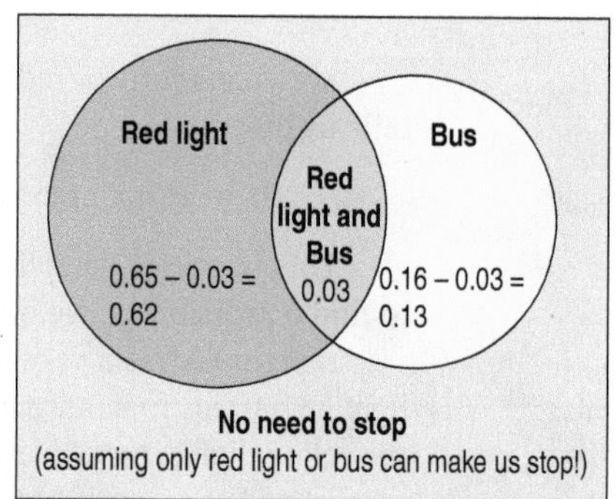

Figure 12.4

Red light or bus: non-disjoint events. A red light and a bus obstruction can both happen. Some probabilities also correspond to specific disjoint regions. Notice that we had to break up the probability of a red light, 0.65, into two parts of the two disjoint regions composing that event (and likewise for the bus).

The reason we can't simply add the probabilities of **A** and **B** is that we'd count the probability of stopping for a red light *and* stopping for a bus twice. If we add the two probabilities, we have to compensate by subtracting out the probability of *stopping* for a red light *and* stopping for a bus. So,

$$\begin{aligned} P(\text{stop}) &= P(\text{stopping for a red light or stopping for a bus}) \\ &= P(\text{stopping for a right light}) + P(\text{stopping for a bus}) - \\ &\quad P(\text{stopping for a red light } \textit{and} \text{ stopping for a bus}) \\ &= 0.65 + 0.16 - 0.03 \\ &= 0.78 \end{aligned}$$

This method works in general. We add the probabilities of two events and then subtract out the probability of their intersection. This approach gives us the **general addition rule**, which does not require disjoint events:

$$P(\mathbf{A}\ or\ \mathbf{B}) = P(\mathbf{A}) + P(\mathbf{B}) - P(\mathbf{A}\ and\ \mathbf{B}).$$

The addition rule for disjoint events is not unrelated to the general addition rule. In fact, the general addition rule automatically becomes the addition rule for disjoint events. When events **A** and **B** are disjoint, it's impossible for them both to happen at the same time. So when **A** and **B** are disjoint, $P(\mathbf{A}\ and\ \mathbf{B}) = 0$, and the last term in the general addition rule disappears, leaving the addition rule for disjoint events.

"Would you like dessert or coffee?" We've all heard this question and we know we can say, "Yes to coffee, no to dessert" or "No to coffee, yes to dessert" or "Yes to both" or "No to both." But technically the question is ambiguous. Are you expected to decide between them, or can you have both? If you simply say, "Yes," what does that mean? Suppose you'd been asked a different question: "Do you want to go to the movies or go out to dinner?" We know we are expected to decide between the two choices, and so we respond, "Let's go to the movies" or "Let's go to dinner." Like the previous question, this one is also technically ambiguous. What happens if you simply say, "Yes"? Suppose you'd been asked yet another question: "Do you want to go out or stay in?" This question isn't ambiguous. You either decide to stay in or to go out.

Why are the first two questions ambiguous? Sometimes when we say *or*, we mean *one or the other but not both*, and sometimes we mean one or the other or possibly both. If we mean *one or the other but not both*, we are using the *exclusive* "or." If we mean *one or the other or possibly both*, we are using the *inclusive* "or." Probability uses the *inclusive* "or." When we ask for the probability that **A** *or* **B** occurs, we mean **A** or **B** or both. And we know *that* probability is $P(\mathbf{A}) + P(\mathbf{B}) - P(\mathbf{A}\ and\ \mathbf{B})$. The general addition rule subtracts the probability of the outcomes in **A** *and* **B** because we've counted those outcomes *twice*. But they're still there.

Why is the third question not ambiguous? Because we can't stay in and go out at the same time, these events are disjoint (or mutually exclusive).

If we really mean **A** or **B**, but *not* both, we have to get rid of the outcomes in {**A** *and* **B**}. So $P(\mathbf{A}\ or\ \mathbf{B}$, but *not both*$) = P(\mathbf{A}\ or\ \mathbf{B}) - P(\mathbf{A}\ and\ \mathbf{B}) = P(\mathbf{A}) + P(\mathbf{B}) - 2 \times P(\mathbf{A}\ and\ \mathbf{B})$. Now we've subtracted $P(\mathbf{A}\ and\ \mathbf{B})$ twice—once because we don't want to double-count these events and a second time because we really didn't want to count them at all. At this point, it would be a good idea to make a picture to be sure you see how this works!

Just Checking

2. Earlier in the book, we suggested that you sample some pages of this book at random to see whether they include a graph or other data display. We actually did just that. We drew a representative sample and found the following:
   - 48% of pages had some kind of data display,
   - 27% of pages had an equation, and
   - 7% of pages had both a data display and an equation.

   a) Display these results in a Venn diagram.

   b) What is the probability that a randomly selected sample page had neither a data display nor an equation?

   c) What is the probability that a randomly selected sample page had a data display but no equation?

Step-by-Step Example USING THE GENERAL ADDITION RULE

Police report that 78% of drivers stopped on suspicion of impaired driving are given a breath test, 36% are given a blood test, and 22% receive both tests. What is the probability that a randomly selected impaired driving suspect is given

1. a test?
2. a breath test or a blood test, but not both?
3. neither test?

THINK ➡ Plan Define the events we're interested in. There are no conditions to check; the general addition rule works for any event!

The blue region represents **A** but not **B**. The green intersection region represents **A** *and* **B**. Note that since $P(\mathbf{A}) = 0.78$ and $P(\mathbf{A} \textit{ and } \mathbf{B}) = 0.22$, the probability of **A** but not **B** must be $0.78 - 0.22 = 0.56$.

Plot Make a picture, and use the given probabilities to find the probability for each region.

The yellow region is **B** but not **A**.

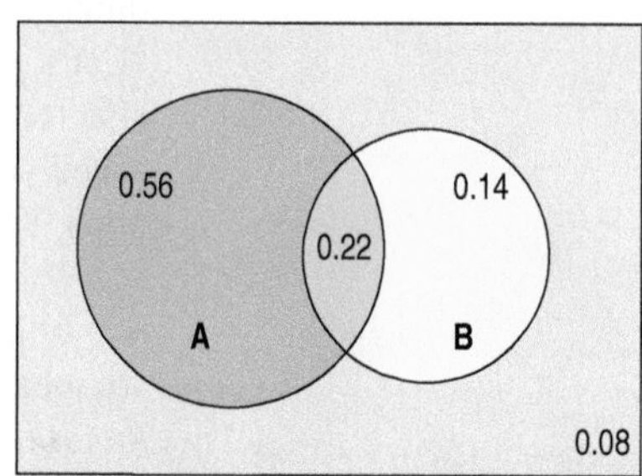

The grey region outside both circles represents the outcome neither **A** nor **B**. All the probabilities must total 1, so you can determine the probability of that region by subtraction.

Then figure out what you want to know. The probabilities can come from the diagram or from a formula. Sometimes translating the words to equations is the trickiest step.

Let **A** = {suspect is given a breath test}.

Let **B** = {suspect is given a blood test}.

I know that

$P(\mathbf{A}) = 0.78$

$P(\mathbf{B}) = 0.36$

$P(\mathbf{A} \textit{ and } \mathbf{B}) = 0.22$

So

$P(\mathbf{A} \textit{ and } \mathbf{B}^C) = 0.78 - 0.22 = 0.56$

$P(\mathbf{B} \textit{ and } \mathbf{A}^C) = 0.36 - 0.22 = 0.14$

$P(\mathbf{A}^C \textit{ and } \mathbf{B}^C) = 1 - (0.56 + 0.22 + 0.14)$

$= 0.08$

Question 1. What is the probability that the suspect is given a test?

SHOW ➡ **Mechanics** The probability the suspect is given a test is $P(\mathbf{A} \text{ or } \mathbf{B})$. We can use the general addition rule, or we can add the probabilities seen in the diagram.	$P(\mathbf{A} \text{ or } \mathbf{B}) = P(\mathbf{A}) + P(\mathbf{B}) - P(\mathbf{A} \text{ and } \mathbf{B})$ $= 0.78 + 0.36 - 0.22$ $= 0.92$ OR $P(\mathbf{A} \text{ or } \mathbf{B}) = 0.56 + 0.22 + 0.14 = 0.92$
TELL ➡ **Conclusion** Don't forget to interpret your result in context.	92% of all suspects are given a test.

Question 2. What is the probability that the suspect gets either a blood test or a breath test but NOT both?

SHOW ➡ **Mechanics** We can use the rule, or just add the appropriate probabilities seen in the Venn diagram.	$P(\mathbf{A} \text{ or } \mathbf{B} \text{ but NOT both}) = P(\mathbf{A} \text{ or } \mathbf{B}) - P(\mathbf{A} \text{ and } \mathbf{B})$ $= 0.92 - 0.22 = 0.70$ OR $P(\mathbf{A} \text{ or } \mathbf{B} \text{ but NOT both}) = P(\mathbf{A} \text{ and } \mathbf{B}^C) + P(\mathbf{B} \text{ and } \mathbf{A}^C)$ $= 0.56 + 0.14 = 0.70$
TELL ➡ **Conclusion** Interpret your result in context.	70% of the suspects get exactly one of the tests.

Question 3. What is the probability that the suspect gets neither test?

SHOW ➡ **Mechanics** Getting neither test is the complement of getting one or the other. Use the complement rule or notice that "neither test" is represented by the region outside both circles.	$P(\text{neither test}) = 1 - P(\text{either test})$ $= 1 - P(\mathbf{A} \text{ or } \mathbf{B})$ $= 1 - 0.92 = 0.08$ OR $P(\mathbf{A}^C \text{ and } \mathbf{B}^C) = 0.08$
TELL ➡ **Conclusion** Interpret your result in context.	Only 8% of the suspects get no test.

WHAT CAN GO WRONG?

- **Beware of probabilities that don't add up to 1.** To be a legitimate probability assignment, the sum of the probabilities for all possible outcomes must total 1. If the sum is less than 1, you may need to add another category ("other") and assign the remaining probability to that outcome. If the sum is more than 1, check that the outcomes are disjoint. If they're not, then you can't assign probabilities by just counting relative frequencies.

- **Don't use the addition rule for disjoint events when you should use the general addition rule.** Events must be disjoint to use the addition rule for disjoint events. The probability of being under 80 *or* a female is not the probability of being under 80 *plus* the probability of being a female; this sum may be more than 1. To find the probability of being under 80 *or* a female, first add the two individual probabilities and then subtract the probability of being under 80 *and* a female.

CONNECTIONS

We saw in the previous three chapters that randomness plays a critical role in gathering data. That fact alone makes it important that we understand how random events behave. The rules and concepts of probability give us a language to talk and think about random phenomena. From here on, randomness will be fundamental to how we think about data, and probabilities will show up in every chapter.

Our interest in probability extends back to the start of the book. We've talked about "relative frequencies" often. But—let's be honest—they are just the same as probabilities once we talk about selecting one individual at random. For example, you can now rephrase the 68–95–99.7 Rule to talk about the *probability* that a value randomly selected from a Normal model will fall within one, two, or three standard deviations of the mean.

Why not just say "probability" from the start? Well, we didn't need any of the formal rules in this chapter (or the next one), so there was no point weighing down the discussion. And "relative frequency" is the right intuitive way to think about probability in this course, so you've been thinking right all along.

Keep it up.

What Have We Learned?

Learning Objectives

We've learned that probability is usually based on long-run relative frequencies. We've thought about the Law of Large Numbers and noted that it speaks only of long-run behaviour. Because the long run is a very long time, we need to be careful not to misinterpret the Law of Large Numbers. Even when we've observed a string of heads, we shouldn't expect extra tails in subsequent coin flips.

We've also learned some basic rules for combining probabilities of outcomes to find probabilities of more complex events. These include

- the probability assignment rule
- the total probability rule
- the addition rule for disjoint events
- the complement rule
- the general addition rule

Terms

Random phenomenon A phenomenon is random if we know what outcomes could happen, but not which particular values will happen (p. 336).

Trial A single attempt or realization of a random phenomenon (p. 337).

Outcome The value measured, observed, or reported for an individual instance of a trial (p. 337).

Event A collection of outcomes. Usually, we identify events so that we can attach probabilities to them. We denote events with bold capital letters, such as **A**, **B**, or **C** (p. 337).

Sample Space The collection of all possible outcome values. The sample space has a probability of 1 (p. 337).

Law of Large Numbers The *relative frequency* of an event in repeated independent trials gets closer and closer to its *true* or long-run relative frequency as the number of trials increases (p. 338).

Independence (informally) Two events are *independent* if learning that one event occurs does not change the probability that the other event occurs (p. 338).

Probability The probability of an event is a number between 0 and 1 that reports the likelihood of an event's occurrence. We write $P(A)$ for the probability of the event A (p. 338).

Empirical probability The long-run relative frequency of an event's occurrence (p. 338). The Law of Large Numbers assures us of its existence, when we can perform repeated independent trials.

Theoretical probability The probability that comes from a model (such as equally likely outcomes) (p. 339).

Personal probability A probability that is subjective and represents your personal degree of belief (p. 340).

Probability assignment rule A probability is a number between 0 and 1. For any event A, $0 \leq P(A) \leq 1$ (p. 342).

Total probability rule Since the probability of any event is between 0 and 1, inclusive, the entire sample space must be 1. $P(S) = 1$ (p. 342).

Disjoint or mutually exclusive Two events are disjoint if they share no outcomes in common. If A and B are disjoint, then knowing that A occurs tells us that B cannot occur. Disjoint events are also called mutually exclusive events (p. 342).

Addition rule for disjoint events If A and B are disjoint events, then the probability of A *or* B is $P(A \text{ or } B) = P(A) + P(B)$ (p. 342).

Legitimate probability assignment An assignment of probabilities to outcomes is legitimate if

- each probability is between 0 and 1 (inclusive)
- the sum of the probabilities is 1 (p. 342)

Complement rule The probability of an event occurring is 1 minus the probability that it doesn't occur. $P(A) = 1 - P(A^C)$ (p. 343).

General addition rule For any two events, A and B, the probability of A *or* B is $P(A \text{ or } B) = P(A) + P(B) - P(A \text{ and } B)$ (p. 345).

Exercises

1. **Sample spaces** For each of the following, list the sample space and tell whether you think the events are equally likely.
   a) Toss two coins; record the order of heads and tails.
   b) A family has three children; record the number of boys.
   c) Flip a coin until you get a head or three consecutive tails; record the order of head and tails.
   d) Roll two dice; record the larger number.

2. **Sample spaces II** For each of the following, list the sample space and tell whether you think the events are equally likely.
   a) Roll two dice; record the sum of the numbers.
   b) A family has three children; record each child's gender in order of birth.
   c) Toss four coins; record the number of tails.
   d) Toss a coin 10 times; record the longest run of heads.

3. **Roulette** A casino claims that its roulette wheel is truly random. What should that claim mean?

4. **Rain** The weather reporter on television makes predictions such as a 25% chance of rain. What do you think such a phrase means?

5. **Winter** Comment on the following quotation:

   "What I think is our best determination is it will be a colder than normal winter," said Pamela Naber Knox, a Wisconsin state climatologist. "I'm basing that on a couple of different things. First, in looking at the past few winters, there has been a lack of really cold weather. Even though we are not supposed to use the law of averages, we are due."[14]

6. **Snow** After an unusually dry autumn, a radio announcer is heard to say, "Watch out! We'll pay for these sunny days later on this winter." Explain what he's trying to say, and comment on the validity of his reasoning.

[14]Associated Press, Fall 1992, quoted by Schaeffer et al.

7. **Cold streak** A batter who failed to get a hit in seven consecutive times at bat hits a game-winning home run. When speaking with reporters after the game, he says he was very confident on his last time up at bat because he knew he was "due for a hit." Comment on the batter's reasoning.

8. **Crash** Commercial airplanes have an excellent safety record. Nevertheless, there are crashes occasionally, with the loss of many lives. In the weeks following a crash, airlines often report a drop in the number of passengers, probably because people are afraid to risk flying.
   a) A travel agent suggests that, since the law of averages makes it highly unlikely to have two plane crashes within a few weeks of each other, flying soon after a crash is the safest time. What do you think?
   b) If the airline industry proudly announces that it has set a new record for the longest period of safe flights, would you be reluctant to fly? Are the airlines due to have a crash?

9. **Fire insurance** Insurance companies collect annual payments from homeowners in exchange for paying to rebuild houses that burn down.
   a) Why should you be reluctant to accept a $300 payment from your neighbour to replace his house should it burn down during the coming year?
   b) Why can the insurance company make that offer?

10. **Jackpot** On February 11, 2009, the AP news wire released the following story:

    (*LAS VEGAS, Nev.*)—A man in town to watch the NCAA basketball tournament hit a $38.7 million jackpot on Friday, the biggest slot machine payout ever. The 25-year-old software engineer from Los Angeles, whose name was not released at his request, won after putting three $1 coins in a machine at the Excalibur hotel-casino, said Rick Sorensen, a spokesman for slot machine maker International Game Technology.[15]

    a) How can the Excalibur afford to give away millions of dollars on a $3 bet?
    b) Why was the maker willing to make a statement? Wouldn't most businesses want to keep such a huge loss quiet?

11. **Spinner** The plastic arrow on a spinner for a child's game stops rotating to point at a colour that will determine what happens next. Which of the following probability assignments are possible?

	Probabilities of . . .			
	Red	Yellow	Green	Blue
a)	0.25	0.25	0.25	0.25
b)	0.10	0.20	0.30	0.40
c)	0.20	0.30	0.40	0.50
d)	0	0	1.00	0
e)	0.10	0.20	1.20	−0.50

[15]Reprinted by permission from AP news wire, February 11, 2009 © 2005 The Associated Press. All rights reserved.

12. **Scratch off** Many stores run "secret sales": Shoppers receive cards that determine how large a discount they get, but the percentage is revealed by scratching off that black stuff (what *is* that?) only after the purchase has been totalled at the cash register. The store is required to reveal (in the fine print) the distribution of discounts available. Which of these probability assignments are plausible?

	Probabilities of . . .			
	10% off	20% off	30% off	50% off
a)	0.20	0.20	0.20	0.20
b)	0.50	0.30	0.20	0.10
c)	0.80	0.10	0.05	0.05
d)	0.75	0.25	0.25	−0.25
e)	1.00	0	0	0

13. **Car repairs** A consumer organization estimates that over a one-year period, 17% of cars will need to be repaired once, 7% will need repairs twice, and 4% will require three or more repairs. What is the probability that a car chosen at random will need
    a) No repairs?
    b) No more than one repair?
    c) Some repairs?

14. **Stats projects** In a large Introductory Statistics lecture hall, the professor reports that 55% of the students enrolled have never taken a Calculus course, 32% have taken only one semester of Calculus, and the rest have taken two or more semesters of Calculus. The professor randomly assigns students to groups of three to work on a project. What is the probability that the first groupmate you meet has studied
    a) Two or more semesters of Calculus?
    b) Some Calculus?
    c) No more than one semester of Calculus?

15. **Ethanol** A Nanos Poll in August 2008 asked 500 adult Ontarians if they thought Ontario should produce more, produce less, or maintain current production of ethanol. Ethanol is made from corn, and can be used as a gasoline additive to increase fuel efficiency.
    Here's how they responded:

Response	Number
More ethanol production	160
Less ethanol production	136
Maintain current production level	155
No opinion	49
Total	**500**

If we select a person at random from this sample of 500 adults,
   a) What is the probability that the person responded, "More ethanol production"?
   b) What is the probability that the person responded, "Maintain current production level" or had no opinion?

16. Optimism A Nanos Poll conducted in May 2012 asked 1000 Canadians whether they think the next generation will have a standard of living that is higher, the same, or lower than what Canadians have today. Here's how they responded:

Response	Number
Higher	255
Same	263
Lower	373
Unsure	110
Total	**1000**

If we select a person at random from this sample of 1000 adults,

a) What is the probability that the person responded "Higher"?

b) What is the probability that the person responded "Higher" or "Same"?

17. Census Canada conducts a census every five years. Federal, provincial, and local governments use census data to shape public policy. In 2006, Canada tried something no other country had ever tried—census questionnaires that could be completed online. Sixty-six percent of households completed their census questionnaire by the May 16 deadline. (Households that didn't respond by the deadline received follow-up letters and phone calls.) Nineteen percent of households responded online; sixteen percent of households responded on time and online.

a) What percent of households filed on time or filed online?

b) What percent of households did not file on time but used the online service when they eventually responded?

c) What percent of households filed on time but did not file online?

d) What percent of households neither filed on time nor online?

18. Census returns In 2006, 80% of households received a short census form with only eight questions. The remainder received a long census form with 53 questions. As previously mentioned in Exercise 17, census questionnaires could be completed online. Fifteen percent of households received the short form and completed their census online; 4% of households received the long form and completed their census online.

a) What percent of households filed online?

b) What percent of households received the short form and filed online?

c) What percent of households received the short form or filed online?

d) What percent of households received the long form and did not file online?

19. Homes Real estate ads suggest that 64% of homes for sale have garages, 21% have swimming pools, and 17% have both features. What is the probability that a home for sale has

a) A pool or a garage?

b) Neither a pool nor a garage?

c) A pool but no garage?

20. Immigration In 2007, 52% of all immigrants to Canada were females, 26% were under 18 years old, and 12% were females under 18 years old. Find the probability that a randomly selected person who immigrated to Canada in 2007 was

a) Female and at least 18 years old.

b) Either female or under 18 years old.

c) Male and at least 18 years old.

21. Emigration In 2007, 51% of all emigrants from Canada were males, 82% were at least 18 years old, and 42% were males of at least 18 years of age. Find the probability that a randomly selected person who emigrated from Canada in 2007 was

a) Male and under 18 years old.

b) Either male or at least 18 years of age.

c) Female and under 18 years old.

22. Workers Employment data at a large company reveal that 72% of the workers are married, 44% are university graduates, and half of the university graduates are married. What is the probability that a randomly chosen worker

a) Is neither married nor a college graduate?

b) Is married but not a college graduate?

c) Is married or a college graduate?

23. Movies A video store investigates the types of movies their customers rent. They record the age of the renter and the movie genre of 240 recent rentals. Suppose the data breaks down as follows:

	Type of Movie			
Age of Renter	**Documentary**	**Comedy**	**Mystery**	**Total**
12–20	14	9	8	31
21–29	15	14	9	38
30–39	9	21	39	69
40–49	7	22	17	46
50 and over	6	38	12	56
Total	**51**	**104**	**85**	**240**

If we select a person at random from this sample,

a) What is the probability that the person rented a documentary?

b) What is the probability that the person is younger than 30 years old?

c) What is the probability that the person is younger than 30 *and* rented a documentary?

d) What is the probability that the person is younger than 30 *or* rented a documentary?

24. Birth order A survey of students in a large Introductory Statistics class asked about their birth order (1 = oldest or only child) and which college of the university they were enrolled in. Here are the data:

Birth Order	1 or only	2 or more	Total
Arts & Sciences	34	23	57
Agriculture	52	41	93
Human Ecology	15	28	43
Other	12	18	30
Total	113	110	223

Suppose we select a student at random from this class.

a) What is the probability we select a Human Ecology student?

b) What is the probability that we select a first-born student?

c) What is the probability that the person is first-born *and* a Human Ecology student?

d) What is the probability that the person is first-born *or* a Human Ecology student?

25. Gambling The game of Craps starts with a player tossing a pair of dice. If the dice add up to 2 or 3 or 12, the player's turn is up and they "crap out." What is the probability of crapping out?

26. Red cards You shuffle a deck of cards, and then start turning them over one at a time. The first one is red. So is the second. And the third. In fact, you are surprised to get 10 red cards in a row. You start thinking, "The next one is due to be black!"

a) Are you correct in thinking that there's a higher probability that the next card will be black than red? Explain.

b) Is this an example of the Law of Large Numbers? Explain.

27. Poker A poker hand consists of five cards. You are dealt all four aces and the king of spades. Your buddy is dealt 10 ♥, 9 ♠, 7 ♣, 4 ♦, 2 ♥. Does your hand have a smaller probability than your buddy's? Explain.

28. Lottery Lilith never selects 1, 2, 3, 4, 5, and 6 when she gets a Lotto 6/49 ticket. After all, everyone knows this will never win. Next time she buys a ticket, however, she selects 4, 9, 14, 22, 36, and 41, believing her choices are better than 1, 2, 3, 4, 5, and 6. Is Lilith's reasoning correct? Explain.

29. Simulations Probabilities at times can be hard to determine mathematically. By building a simulation to model the trials, we can estimate probabilities using computer power instead. Use simulations to estimate the probability that at least two people have the same birthday in a random group of 40 people. Assume there are 365 equally likely birthdays for any individual in the group. Use software to draw randomly from the integers 1–365. Repeat your simulations at least 40 times. Combine your results with classmates if possible.

30. Simulations again Suppose two hockey teams are equally matched, and play a series of 10 games. Use simulations to estimate the probability that at least one of the two teams will have a win streak of at least four games at some time in the series. Use software to draw randomly from the digits 0 and 1. Repeat your simulation at least 50 times. Combine results with classmates if possible. Are you making any assumptions? (If a team wins four straight, is it a hot streak or might it just be chance?)

Just Checking ANSWERS

1. The Law of Large Numbers works only in the long run, not in the short run. The random methods for selecting lottery numbers have no memory of previous picks, so there is no change in the probability that a certain number will come up.

2. a)

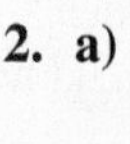

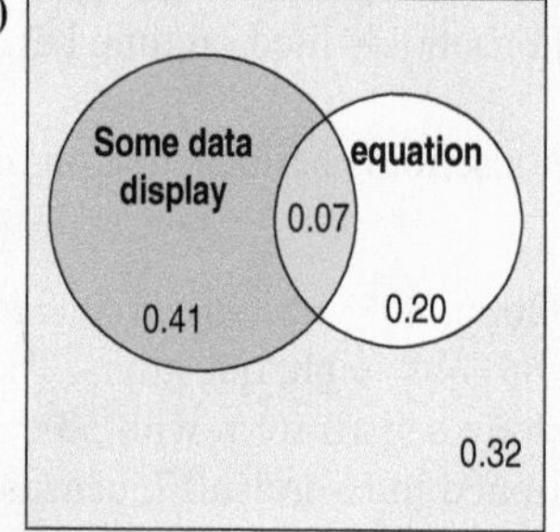

b) 0.32

c) 0.41

chapter

13 Probability Rules!

Where are we going?

Is the probability of a car accident the same for all age groups? Insurance companies want to know so they can set their rates. Is everyone at equal risk for getting the flu next winter? Medical researchers use factors like age, sex, lifestyle, and family history to estimate different flu probabilities for different individuals. These are examples of *conditional* probabilities. We'll see how the probability of an event can change when we learn more about the situation at hand. Sometimes the results are surprising.

Natural selection is a mechanism for generating an exceedingly high degree of improbability.

—Ronald Fisher

13.1 Probability on Condition

Imagine that while touring England, you spend a day exploring the ancient monument of Stonehenge. As you wander the site, wondering about its purpose, you hear a voice from home. You introduce yourself and are surprised to find out that the person you've just met is from the same province as you. Should you be surprised by this coincidence? Probably not, but it depends on your home province. If you live in Ontario, you shouldn't be at all surprised. But if you're an Inuktituk-speaking Nunavummiuq, meeting another Inuktituk-speaking Nunavummiuq, you should be totally surprised!

A lot of people live in Ontario (about 12 million). Every day, thousands of Ontarians are on holiday, many of them in England. There are probably hundreds of Ontarians at Stonehenge every single day. So if you're from Ontario, meeting another Ontarian shouldn't be at all surprising. By comparison, very few people live in Nunavut (about 33 000). Not many of these people will be exploring England. There likely will be days when there are no Inuktituk-speaking Nunavummiut in England, let alone at Stonehenge.

If you ask a travelling Canadian the name of their home province, it could be said that you are conducting a trial. The sample space would be all the possible outcomes for this trial, **S** = {Alberta, British Columbia, Manitoba, New Brunswick, Newfoundland and Labrador, Northwest Territories, Nova Scotia, Nunavut, Ontario, Prince Edward Island, Quebec, Saskatchewan, Yukon}. The probability of meeting an Ontarian is higher than the probability of meeting a Nunavummiuq, as there are simply more travelling Ontarians to be met. The 13 different outcomes in the sample space are not *equally likely*.

We can combine the outcomes in different ways to make different events. For example, **A** = {New Brunswick, Newfoundland and Labrador, Nova Scotia, Prince Edward Island} defines the event that the person is from the Atlantic provinces, and **B** = {New Brunswick, Northwest Territories, Nunavut, Quebec, Yukon} defines the event that French is an official language of the person's home province. (Did you know that there are 11 official languages in the Northwest Territories? English, French, and nine Aboriginal languages.)

What about the probabilities of events **A** and **B**? The outcomes in the sample space are not equally likely, so even though event **A** contains four outcomes, it will not have probability 4/13. Similarly, event **B** will not have probability 5/13.

Conditional Probability—It Depends . . .

Suppose a keen Statistics student polled all Canadians visiting Stonehenge on Canada Day. He asked the Canadians which official language they prefer to use, French or English, and their province of residence. (Sadly, he met no Nunavummiut that day.)

	English	French	Total
Atlantic	15	9	24
Quebec	5	50	55
Ontario	89	4	93
West	70	6	76
Total	179	69	248

Table 13.1
The distribution of official language most commonly used by region.

Table 13.1 is a *contingency table* giving counts of the visitors by their language preference and region of residence.

Let's focus on the data and make the sample space just the set of these 248 people. If we select a person at random from this group, the probability that we select a French speaker is just the corresponding relative frequency (since we're equally likely to select any of the 248 people). There are 69 French speakers out of a total of 248, giving a probability of

$$P(\text{French}) = 69/248 = 0.2782$$

The same method works for more complicated events, like intersections. For example, what's the probability of selecting a French speaker from Atlantic Canada? Well, nine French speakers of the 248 sampled live in Atlantic Canada, so the probability is

$$P(\text{French and Atlantic}) = 9/248 = 0.0363$$

The probability of selecting a Quebecer is

$$P(\text{Quebecer}) = 55/248 = 0.2218$$

What if we are given the information that the selected person is from Quebec? Would that change the probability that the person is a French speaker? You bet it would! The table shows that Quebecers are much more likely to speak French than people from the rest of Canada. When we restrict our focus to Quebecers, we look only at the Quebecer's row of the table, which gives the conditional distribution of language preference given "Quebec." Of the 55 Quebecers in the sample, a whopping 50 of them prefer to use French.

We write the probability that a selected person prefers French *given that we have selected a Quebecer* as

$$P(\text{French} \mid \text{Quebec}) = 50/55 = 0.9091$$

For Atlantic Canada, we look at the conditional distribution of language given "Atlantic" shown in the top row of the table. There, of the 24 people from Atlantic Canada in the sample, nine said French was their preferred language. So, $P(\text{French} \mid \text{Atlantic}) = 9/24 = 0.375$.

In general, when we want the probability of an event from a *conditional* distribution, we write $P(\mathbf{B} \mid \mathbf{A})$ and say "the probability of **B** *given* **A**." A probability that takes into account a given *condition* such as this is called a **conditional probability**. When we work with conditional probabilities, we are *certain* that the event after the vertical line has occurred.

Let's look at what we did. We worked with the counts, but we could work with the probabilities just as well. There were 50 French speakers from Quebec, and there were 55 Quebecers. So we calculate the probability to be 50/55, or 0.909.

To find the probability of the event **B** *given* the event **A**, we restrict our attention to the outcomes in **A**. We then find in what fraction of *those* outcomes **B** also occurred. Formally, we write:

NOTATION ALERT
$P(\mathbf{B} \mid \mathbf{A})$ is the conditional probability of **B** *given* **A**.

$$P(\mathbf{B} \mid \mathbf{A}) = \frac{P(\mathbf{A} \textit{ and } \mathbf{B})}{P(\mathbf{A})}.$$

Thinking this through, we can see that it's just what we've been doing, but with probabilities rather than with counts. Look back at the French speakers who were from Quebec. How did we calculate the probability of selecting a French speaker given that we know we've selected a Quebecer, $P(\text{French} \mid \text{Quebec})$?

The rule says to use probabilities. It says to find $P(\mathbf{A} \textit{ and } \mathbf{B})/P(\mathbf{A})$. We know that we've selected a person from Quebec, so let **A** be the event that the person is from Quebec. We want to find the probability that the person (from Quebec) speaks French, so let **B** be the event that the person is a French speaker.

A S **Birth weights and Smoking.** Does smoking increase the chance of having a baby with a low birth weight? Check the conditional probabilities.

$$P(\mathbf{B} \mid \mathbf{A}) = \frac{P(\mathbf{A} \textit{ and } \mathbf{B})}{P(\mathbf{A})} = \frac{P(\text{Quebec} \textit{ and } \text{French})}{P(\text{Quebec})} = \frac{50/248}{55/248} = \frac{50}{55} = 0.9091$$

Applying the conditional probability rule gives the same probabilities we arrived at when we used counts. It doesn't matter which we use, because the total number in the sample cancels out.

A S **Conditional Probability.** Simulation is great for seeing conditional probabilities at work. Here's a hands-on way to get a feel for them.

Let's take our rule out for another spin. What's the probability that we have selected a Westerner *given* that the person speaks English? We know that we've selected a person who speaks English, so let **A** be the event that the person speaks English. We want to find the probability that the person (who speaks English) is from the West, so let **B** be the event that the person is a Westerner.

$$P(\mathbf{B}\,|\,\mathbf{A}) = \frac{P(\mathbf{A}\ \mathit{and}\ \mathbf{B})}{P(\mathbf{A})} = \frac{P(\text{English } \mathit{and} \text{ West})}{P(\text{English})} = \frac{70/248}{179/248} = \frac{70}{179} = 0.3911$$

To use the formula for conditional probability, we have to insist on one restriction. The formula doesn't work if $P(\mathbf{A})$ is 0. But this restriction is fine, since $P(\mathbf{A}) = 0$ means that **A** can't possibly occur. Of course, this means that thinking about the probability that **B** occurs given that **A** has occurred would be nonsense anyway (unless we are dealing with continuous sample spaces, in which case zero probability events do occur!—but the math changes as well).

13.2 Independence

What exactly do we mean when we say events are *independent*? We've already used the word and said that events are independent if the outcome of one event *does not influence* the probability of the other. Conditional probability is the driving force behind the concept of independence.

- **It's raining, better grab the umbrella.** You don't like getting rained on, so you usually take an umbrella with you when you go out in the rain. But sometimes you don't bother with the umbrella if it's only light rain or if you're in a hurry. Other times, it starts to rain after you've gone out without an umbrella and you end up soaked. Sometimes the forecast says rain is likely, so you take your umbrella, only to find the forecaster got it wrong and no rain appears. Of course, when the weather is fine, you don't bother to tote an unnecessary umbrella around.[1] Let's define some events. Let **B** be the event "you have an umbrella with you." Let **A** be the event "it's raining." Suppose you're out and it starts to rain. What's the probability that you have an umbrella with you? In other words, what is $P(\mathbf{B}\,|\,\mathbf{A})$? Pretty high, since you don't like being rained on. Suppose you're out and it isn't raining. What's the probability that you have an umbrella with you? In other words, what is $P(\mathbf{B}\,|\,\mathbf{A}^c)$? Pretty low, since you don't like carrying an unnecessary umbrella. The probability that you have an umbrella with you when it's raining is not the same as when it's not raining. This is an example of **dependent events**. The probability of **B** is not the same conditional on **A** as it is conditional on $\mathbf{A}^c$. For dependent events, $P(\mathbf{B}\,|\,\mathbf{A}) \neq P(\mathbf{B}\,|\,\mathbf{A}^c)$.
- **Flip a coin, roll a die.**[2] If you flip a coin, it comes up heads or tails. When you roll a die, it comes up 1, 2, 3, 4, 5, or 6. Let's define some events. Let **B** be the event "die comes up 2." Let **A** be the event "the coin lands tails up." Suppose you flip a coin and it lands tails up. What's the probability that you get a 2 when you roll the die? In other words, what is $P(\mathbf{B}\,|\,\mathbf{A})$? Does the die say, "Oh my, the coin came up tails so I'd better redistribute my weight so that I have a high chance of landing 2 up"? Of course not. The die completely ignores the outcome of the coin toss and $P(\mathbf{B}\,|\,\mathbf{A})$ remains $1/6$, as expected. Now suppose you flip the coin and it lands heads up. What's the probability you get a 2 when you roll the die? In other words, what is $P(\mathbf{B}\,|\,\mathbf{A}^c)$? Again, the die ignores the outcome of the coin toss and $P(\mathbf{B}\,|\,\mathbf{A}^c)$ remains $1/6$, as expected. Regardless of which face comes up on the coin flip, the probability that the die will land 2 up is $1/6$. These are an example of **independent events**. The probability of **B** is exactly the same conditional on **A** as it is conditional on $\mathbf{A}^c$. For independent events, $P(\mathbf{B}\,|\,\mathbf{A}) = P(\mathbf{B}\,|\,\mathbf{A}^c) = P(\mathbf{B})$. In fact, if any two of these are equal, they all must be equal.

A S **Probabilities of Compound Events.** The Random Tool lets you experiment with compound random events to see if they are independent.

Events **A** and **B** are **independent** if (and only if) the probability of **B** is the same when we are given that **A** has occurred. That is:

$$P(\mathbf{B}|\mathbf{A}) = P(\mathbf{B}).$$

[1]Unless you live in Vancouver, in which case an umbrella is standard fare year-round.
[2]Die is the singular of dice. A bit of Latin for you.

A S **Independence.** Are smoking and low birth weight independent? The general multiplication rule helps us find out. See p. 354.

Let's go back to the sample space of the 248 Canadians polled at Stonehenge. Are speaking French and living in the West independent? Let **A** be the event "speaks French" and let **B** be the event "from the West." Does $P(\mathbf{B} \mid \mathbf{A}) = P(\mathbf{B} \mid \mathbf{A}^c) = P(\mathbf{B})$? (Compare any two)

$$P(\mathbf{B} \mid \mathbf{A}) = \frac{P(\mathbf{A} \text{ and } \mathbf{B})}{P(\mathbf{A})} = \frac{P(\text{French } and \text{ West})}{P(\text{French})} = \frac{6/248}{69/248} = \frac{6}{69} = 0.087$$

$$P(\mathbf{B} \mid \mathbf{A}^c) = \frac{P(\mathbf{A}^c \text{ and } \mathbf{B})}{P(\mathbf{A}^c)} = \frac{P(\text{English } and \text{ West})}{P(\text{English})} = \frac{70/248}{179/248} = \frac{70}{179} = 0.3911$$

Since $P(\mathbf{B} \mid \mathbf{A}) \neq P(\mathbf{B} \mid \mathbf{A}^c)$, we conclude that speaking French and being a Westerner are dependent events. Alternatively, we could have noted that $P(\mathbf{B} \mid \mathbf{A}) \neq P(\mathbf{B}) = P(\text{West}) = 76/248 = 0.306$.

Are you concerned about which event to label **A** and which one to label **B**? If so, relax. It doesn't make any difference. The numeric probabilities you calculate will probably differ, but your conclusion (whether the events are dependent or independent) will be exactly the same regardless of how you assign the labels.

Suppose you conduct a survey asking respondents if they support the Green Party. Consider two respondents: Hector and Seema. Will their answers be independent events? Let **A** be the event "Seema answers yes" and let **B** be the event "Hector answers yes." Does $P(\mathbf{B} \mid \mathbf{A}) = P(\mathbf{B} \mid \mathbf{A}^c) = P(\mathbf{B})$? Probably, but only if we assume that Hector and Seema don't know each other and don't talk to each other during the survey. But if Hector and Seema talk or are a couple, the probabilities for Seema's responses might be influenced by Hector's opinions (or vice versa).

Many statistics methods require an **Independence Assumption**, but *assuming* independence doesn't make it true. Always *think* about whether an assumption is reasonable.

Independence is the most important idea in this chapter. Spend some time getting your head around its definition and meaning. We'll be using independence in all remaining chapters of this book.

The General Multiplication Rule

We've *already* written this rule. We just need to rearrange the equation a bit. The equation in the definition for conditional probability contains the probability of **A** *and* **B**. Rearranging the equation gives

$$P(\mathbf{A} \text{ and } \mathbf{B}) = P(\mathbf{A}) \times P(\mathbf{B} \mid \mathbf{A}).$$

A S **The General Multiplication Rule.** The best way to understand the general multiplication rule is with an experiment. See the rule at work here.

This is the **general multiplication rule**. The probability that two events, **A** and **B**, *both* occur is the probability that event **A** occurs multiplied by the probability that event **B** occurs *given* that event **A** occurs.

Of course, there's nothing special about which set we call **A** and which one we call **B**. So we should be able to state this the other way around, and indeed we can. It is equally true that

$$P(\mathbf{A} \text{ and } \mathbf{B}) = P(\mathbf{B}) \times P(\mathbf{A} \mid \mathbf{B}).$$

Remember the 248 Canadians polled at Stonehenge? What is the probability a person randomly selected from this sample space speaks French and lives in the West? Let **A** be the event "speaks French" and let **B** be the event "from the West."

$$P(\mathbf{A} \text{ and } \mathbf{B}) = P(\mathbf{A}) \times P(\mathbf{B} \mid \mathbf{A}) = \frac{69}{248} \times \frac{6}{69} = \frac{6}{248} = 0.0242$$

Yes, we could have found this directly from the contingency table. And yes, we could have found it using $P(\mathbf{A} \text{ and } \mathbf{B}) = P(\mathbf{B}) \times P(\mathbf{A} \mid \mathbf{B})$.

The Multiplication Rule for Independent Events

We can always use the general multiplication rule: $P(\mathbf{A} \text{ and } \mathbf{B}) = P(\mathbf{A}) \times P(\mathbf{B} \mid \mathbf{A})$. But when events **A** and **B** are independent, we can write $P(\mathbf{B})$ for $P(\mathbf{B} \mid \mathbf{A})$ because the probability of **B** does not depend on **A**, and we can simplify the rule to

$$P(\mathbf{A} \text{ and } \mathbf{B}) = P(\mathbf{A}) \times P(\mathbf{B}).$$

The multiplication rule for independent events says that for independent events, to find the probability that both events occur, we simply multiply the probabilities together. Formally:

For two independent events **A** and **B**, the probability that both **A** and **B** occur is the product of the probabilities of the two events.

$P(\mathbf{A}\ \textit{and}\ \mathbf{B}) = P(\mathbf{A}) \times P(\mathbf{B})$, provided that **A** and **B** are independent.

A S **Multiplication Rule for Independent Events.** Experiment with independent random events to explore the Multiplication Rule for Independent Events.

Putting the Multiplication Rule for Independent Events to Work

When you get to the light at Portage and Main, it's either red, green, or yellow. The probability of it being green is 0.35, the probability of it being yellow is 0.04, and the probability of it being red is 0.61. Let **A** be the event "the light is green."

So, what is the probability that the first green light of your commuting week doesn't show up until Wednesday? For that to happen, you'd have to see red or yellow on Monday, red or yellow on Tuesday, and then green on Wednesday. Combining all those probabilities could get messy. But wait. We can simplify this by saying we want *not* green on Monday and Tuesday and then green on Wednesday. These events are independent. (Can you explain why?) And the multiplication rule extends in the obvious manner to any number of independent events. Hence:

$$P(\text{not green Monday AND not green Tuesday AND green Wednesday})$$
$$= P(\mathbf{A}^c) \times P(\mathbf{A}^c) \times P(\mathbf{A})$$
$$= 0.65 \times 0.65 \times 0.35 = 0.147875$$

What is the probability that we have to stop *at least once* during the week? Having to stop at least once means that we have to stop for the light either 1, 2, 3, 4, or 5 times during the week. We could calculate each of those probabilities and add them up, but that would be a lot of work. There's an easier way to find the answer by using the probability rules. The phrase "at least" is often a tip-off to think about the *complement.* Something that happens *at least once* does happen. Happening *at least once* is the complement of *not happening*. Let's put this together:

$$P(\text{having to stop at the light at least once in 5 days})$$
$$= 1 - P(\text{not having to stop in any of the 5 days})$$
$$= 1 - P(\text{green 5 days in a row})$$
$$= 1 - P(\mathbf{A}) \times P(\mathbf{A}) \times P(\mathbf{A}) \times P(\mathbf{A}) \times P(\mathbf{A})$$
$$= 1 - P(\mathbf{A})^5$$
$$= 1 - (0.35)^5$$
$$= 1 - 0.00525$$
$$= 0.99475$$

In informal English, you may see "some" used to mean "at least one." "What is the probability that some of the eggs in that carton are broken?" means at least one.

So be prepared to stop.

Just Checking

1. Opinion polling organizations contact their respondents by telephone, calling the households using randomly generated telephone numbers. These surveys used to work quite well: In 1981, interviewers obtained responses from 72% of randomly generated telephone numbers. By 2003, the response rate had dropped to 48%, probably because most people had installed call display and simply stopped answering these pesky calls (Curtin, Presser, & Singer, 2005).

We can reasonably assume each household's response to be independent of the others.

a) What is the probability that an interviewer successfully contacts the next household on her list?

b) What is the probability that the interviewer successfully contacts both of the next two households on her list?

c) What is the probability that the interviewer's first successful contact is the third household on the list?

d) What is the probability the interviewer makes at least one successful contact among the next five households on the list?

Step-by-Step Example MULTIPLICATION RULE

In 2001, Masterfoods, the manufacturer of M&M's® milk chocolate candies, decided to add another colour to the standard colour lineup of brown, yellow, red, orange, blue, and green. To decide which colour to add, they surveyed kids in nearly every country of the world and asked them to vote for purple, pink, or teal. The global winner was purple! In Japan, the percentages of responses were 38% pink, 36% teal, and only 16% purple.

1. What is the probability that a Japanese M&M's survey respondent selected at random preferred either pink or teal?
2. If two respondents were picked at random, what is the probability that they *both* selected purple?
3. If three respondents were picked at random, what is the probability that *at least one* preferred purple?

THINK ➡ The probability of an event is its long-term relative frequency. It can be determined in several ways: by looking at many replications of an event, by deducing it from equally likely events, or by using some other information. Here, we are told the relative frequencies of the three responses.

The M&M's Web site reports the proportions of Japanese votes by colour. These give the probability of selecting a voter who preferred each of the colours:

$P(\text{pink}) = 0.38$

$P(\text{teal}) = 0.36$

$P(\text{purple}) = 0.16$

Make sure the probabilities are legitimate. Here, they're not. Either there was a mistake or the other voters must have chosen a colour other than the three given. A check of the reports from other countries shows a similar deficit, so probably we're seeing those who had no preference or who wrote in another colour.

Each is between 0 and 1, but they don't add up to 1. The remaining 10% of the voters must not have expressed a preference or must have written in another colour. I'll put them together into "no preference" and add: $P(\text{no preference}) = 0.10$.

With this addition, I have a legitimate assignment of probabilities.

The global winner was purple, but only 16% of Japanese people voted for purple. Does this make sense?

The vote was held worldwide. If a large percent of people from all over the world, except Japan, all voted for purple, then purple could easily be the overall winner.

Question 1. What is the probability that a Japanese M&M's survey respondent selected at random prefers either pink or teal?

THINK ➡ Plan Decide which rules to use and check the conditions they require.

SHOW ➡ **Mechanics** Show your work.	The events "pink" and "teal" are individual outcomes (a respondent can't choose both colours), so they are disjoint. I can apply the addition rule. $P(\text{pink or teal}) = P(\text{pink}) + P(\text{teal})$ $= 0.38 + 0.36 = 0.74$
TELL ➡ **Conclusion** Interpret your results in the proper context.	The probability that the respondent selected pink or teal is 0.74.
Question 2. If we pick two respondents at random, what is the probability that they both said purple?	
THINK ➡ **Plan** The word "both" suggests we want *P*(A *and* B), which calls for the multiplication rule. Think about the assumption.	For both respondents to pick purple, each one has to pick purple. **Independence Assumption:** It's unlikely that the choice made by one respondent affected the choice of the other, so the events seem to be independent. I can use the multiplication rule.
SHOW ➡ **Mechanics** Show your work.	$P(\text{both purple})$ $= P(\text{first respondent picks purple and second respondent picks purple})$ $= P(\text{first respondent picks purple}) \times P(\text{second respondent picks purple})$ $= 0.16 \times 0.16 = 0.0256$
TELL ➡ **Conclusion** Interpret your results in the proper context.	The probability that both respondents pick purple is 0.0256.
Question 3. If we pick three respondents at random, what is the probability that at least one preferred purple?	
THINK ➡ **Plan** The phrase "at least . . ." often flags a question best answered by looking at the complement, and that's the best approach here. The complement of "at least one preferred purple" is "none of them preferred purple." Think about the assumption.	$P(\text{at least one picked purple})$ $= P(\{\text{none picked purple}\}^C)$ $= 1 - P(\text{none picked purple}).$ $P(\text{none picked purple}) = P(\text{not purple and not purple and not purple})$ **Independence Assumption:** These are independent events because they are choices by three randomly selected respondents. I can use the multiplication rule for independent events.

SHOW ➡ **Mechanics** We calculate P(none purple) using the multiplication rule. Then we can use the complement rule to get the probability we want.	P(none picked purple) = P(first not purple) × P(second not purple) × P(third not purple) = $[P(\text{not purple})]^3$. P(not purple) = 1 − P(purple) = 1 − 0.16 = 0.84. So P(none picked purple) = $(0.84)^3$ = 0.5927. P(at least 1 picked purple) = 1 − P(none picked purple) = 1 − 0.5927 = 0.4073.
TELL ➡ **Conclusion** Interpret your results in the proper context.	There is about a 40.7% chance that at least one of the respondents picked purple.

Independent ≠ Disjoint

Are disjoint events independent? Both concepts seem to have similar ideas of separation and distinctness about them, but in fact disjoint events *cannot* be independent.[3] Let's see why.

Consider the two disjoint events {you get an A in this course} and {you get a B in this course}. They're disjoint because they have no outcomes in common. Suppose you learn that you *did* get an A in the course. Now what is the probability that you got a B? You can't get both grades, so it must be 0.

Think about what that means. Knowing that the first event (getting an A) occurred changes your probability for the second event (down to 0). So these events aren't independent.

Mutually exclusive events can't be independent. They have no outcomes in common, so if one occurs, the other doesn't. A common error is to treat disjoint events as if they were independent, and apply the multiplication rule for independent events. Don't make that mistake.

Because events ***A*** and ***B*** are mutually exclusive, learning that ***A*** happened tells us that ***B*** didn't. The probability of ***B*** has changed from whatever it was to zero. So disjoint events ***A*** and ***B*** are not independent.

Just Checking

2. The American Association for Public Opinion Research (AAPOR) is an association of about 1600 individuals who share an interest in public opinion and survey research. They report that typically as few as 10% of random phone calls result in a completed interview. Reasons are varied, but some of the most common include no answer, refusal to cooperate, and failure to complete the call.

[3]Well, technically, two disjoint events can be independent, but only if the probability of one of the events is 0. For practical purposes, though, we can ignore this case. After all, as statisticians we don't anticipate having data about things that don't happen.

Which of the following events are independent, which are disjoint, and which are neither independent nor disjoint?

a) **A** = Your telephone number is randomly selected. **B** = You're not at home at dinner time when they call.

b) **A** = As a selected subject, you complete the interview. **B** = As a selected subject, you refuse to cooperate.

c) **A** = You are not at home when they call Monday at 11 a.m. **B** = You are employed full time.

Dependence ≠ Causality

Is there a causal relationship between dependent events? The word "dependent" might suggest that there is, but in fact there need not be anything causal between the two events. Let's see why. Consider the two events **B** = {geese fly south} and **A** = {the leaves change colour}. The probability that geese fly south given that the leaves change colour is quite high; after all, both events occur in the fall. The probability that geese fly south given that the leaves don't change colour is quite low; the leaves stay green in the summer months but very few geese fly south during the summer. A and B are dependent events, $P(\mathbf{B} \mid \mathbf{A}) \neq P(\mathbf{B} \mid \mathbf{A}^c)$. Did the geese leaving cause the leaves to change colour? Did the leaves changing colour cause the geese to fly south? The answer is no to both questions. A common error is to assume a causal relationship between dependent events. Don't make that mistake.

Depending on Independence

It may seem easier to think about independent events than to deal with conditional probabilities. Someone may estimate the probability of a compound event by multiplying the probabilities of its component events together without asking seriously whether those probabilities are independent.

For example, experts have assured us that the probability of a major commercial nuclear plant failure is so small that we should not expect such a failure to occur even in a span of hundreds of years. After only a few decades of commercial nuclear power, however, the world has seen four major failures (Chernobyl, Three Mile Island, Fukushima Daiichi, and Kyshtym). How could the estimates have been so wrong?

A S **Is There a Hot Hand in Basketball?** Even the experts can be fooled about independent events. Most coaches and fans believe that basketball players sometimes get "hot" and make more of their shots. What do the conditional probabilities say?

One simple part of the failure calculation is to test a particular valve and determine that valves such as this one fail only once in, say, 100 years of normal use. For a coolant failure to occur, several valves must fail. So we need the compound probability, *P*(valve 1 fails *and* valve 2 fails *and* . . .). A simple risk assessment might multiply the small probability of one valve failure together as many times as needed.

But if the valves all came from the same manufacturer, a flaw in one might be found in the others. And maybe when the first fails, it puts additional pressure on the next one in line. Or maybe the common power supply fails (called a *common-cause failure*). In either case, the events aren't independent, so we can't simply multiply the probabilities together.

Whenever you see probabilities multiplied together, stop and ask whether you think they are really independent.

13.3 Tables and Conditional Probability

One of the easiest ways to think about conditional probabilities is with contingency tables. We did that earlier in the chapter when we began our discussion. But sometimes we're given probabilities without a table. You can often construct a simple table to correspond to the probabilities.

For example, police report that 78% of drivers stopped on suspicion of impaired driving get a breath test, 36% a blood test, and 22% both. That's enough information. Translating percentages to probabilities, what we know looks like this:

		Breath Test		
		Yes	No	Total
Blood Test	Yes	0.22		**0.36**
	No			
	Total	**0.78**		**1.00**

Notice that the 0.78 and 0.36 are *marginal* probabilities and so they go into the *margins*. The 0.22 is the probability of getting both tests—a breath test *and* a blood test, so that's a *joint* probability. Those belong in the interior of the table.

Because the cells of the table show disjoint events, the probabilities always add to the marginal totals going across rows or down columns. So, filling in the rest of the table is quick:

		Breath Test		
		Yes	No	Total
Blood Test	Yes	0.22	0.14	**0.36**
	No	0.56	0.08	**0.64**
	Total	**0.78**	**0.22**	**1.00**

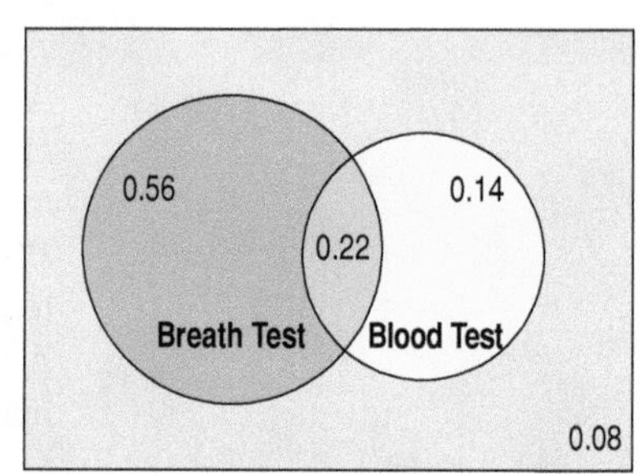

Compare this with the Venn diagram. Notice which entries in the table match up with the sets in the diagram. Whether a Venn diagram or a table is better to use will depend on what you are given and the questions you're being asked. Try both.

Step-by-Step Example ARE THE EVENTS DISJOINT? INDEPENDENT?

Police report that 78% of drivers are given a breath test, 36% a blood test, and 22% both tests.

Questions:

1. Are giving an impaired driving suspect a blood test and a breath test mutually exclusive?
2. Are giving the two tests independent?

THINK ➡ Plan Define the events we're interested in.	Let A = {suspect is given a breath test}. Let B = {suspect is given a blood test}.
State the given probabilities.	I know that $P(\mathbf{A}) = 0.78$ $P(\mathbf{B}) = 0.36$ $P(\mathbf{A} \text{ and } \mathbf{B}) = 0.22$

Question 1. Are giving an impaired driving suspect a blood test and a breath test mutually exclusive?

SHOW ➡ Mechanics Disjoint events cannot both happen at the same time, so check to see if $P(\mathrm{A} \textit{ and } \mathrm{B}) = 0$.	$P(\mathbf{A} \textit{ and } \mathbf{B}) = 0.22$. Since some suspects are given both tests, $P(\mathbf{A} \textit{ and } \mathbf{B}) \neq 0$. The events are not mutually exclusive.
TELL ➡ Conclusion State your conclusion in context.	22% of all suspects get both tests, so a breath test and a blood test are not mutually exclusive.

Question 2. Are the two tests independent?

THINK ➡ Plan Make a table.

		Breath Test		
		Yes	No	Total
Blood Test	Yes	0.22	0.14	**0.36**
	No	0.56	0.08	**0.64**
	Total	**0.78**	**0.22**	1.00

SHOW ➡ Mechanics Does getting a breath test change the probability of getting a blood test? That is, does $P(B|A) = P(B)$?

Because the two probabilities are not the same, the events are not independent.

$$P(B|A) = \frac{P(A \text{ and } B)}{P(A)} = \frac{0.22}{0.78} \approx 0.28$$

$$P(B) = 0.36$$

$$P(B|A) \neq P(B)$$

TELL ➡ Conclusion Interpret your results in context.

Overall, 36% of the drivers get blood tests, but only 28% of those who get a breath test also get a blood test. Since suspects who get a breath test are less likely to have a blood test, the two events are not independent.

Just Checking

3. Canadian farmers grow a large quantity of genetically modified corn.

19% of corn is genetically modified to be resistant to insects.

70% of corn is genetically modified to be resistant to herbicides.

13.3% of corn is genetically modified to be resistant to both.

a) Make a contingency table for the variables *resistant to insects* and *resistant to herbicides.*

b) If you randomly select an ear of corn, what is the probability it has been genetically modified to be resistant to either insects or herbicides?

c) If you randomly select an ear of corn that you know has been modified to be resistant to insects, what is the probability it has also been modified to be resistant to herbicides?

d) Are the genetic modifications of resistance to insects and resistance to herbicides independent?

"Why," said the Dodo, "the best way to explain it is to do it."

—*Lewis Carroll*

Tree Diagrams

For men, binge drinking is defined as having five or more drinks in a row, and for women it is defined as having four or more drinks in a row. (The difference is because of the average difference in weight.) According to a study by the Harvard School of Public Health,[4] 44% of university students engage in binge drinking, 37% drink moderately, and 19% abstain entirely. Another study, published in the *American Journal of Health Behavior,* finds that among binge drinkers aged 21 to 34, 17% have been involved in an alcohol-related automobile accident, while among non-bingers of the same age, only 9% have been involved in such accidents.

[4]H. Wechsler, G. W. Dowdall, A. Davenport, & W. DeJong, "Binge drinking on campus: Results of a national study."

What is the probability that a randomly selected university student will be a binge drinker who has had an alcohol-related car accident?

To start, we see that the probability of selecting a binge drinker is about 44%. To find the probability of selecting someone who is both a binge drinker and a driver with an alcohol-related accident, we would need to pull out the general multiplication rule and multiply the probability of one of the events by the conditional probability of the other given the first.

Or we *could* make a picture. Which would you prefer?

We thought so.

Venn diagrams can't handle conditional probabilities, and tables can get confusing. The kind of picture that helps us think through this kind of reasoning is called a **tree diagram**, because it shows sequences of events as paths that look like branches of a tree. It is a good idea to make a tree diagram almost any time you plan to use the general multiplication rule. The number of different paths we can take can get large, so we usually draw the tree starting from the left and growing vine-like across the page, although sometimes you'll see them drawn from the bottom up or top down.

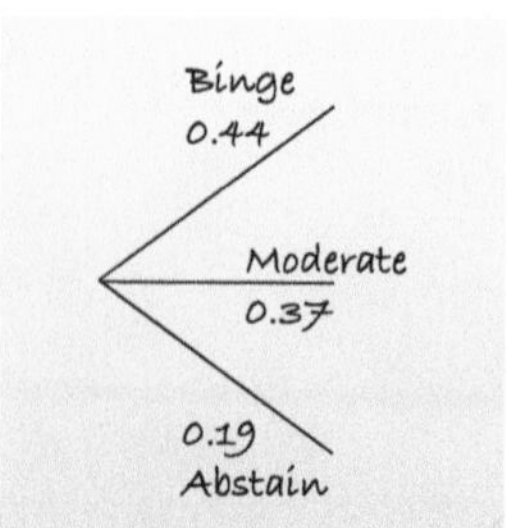

Figure 13.1

We can diagram the three outcomes of drinking and indicate their respective probabilities with a simple tree diagram.

The first branch of our tree separates students according to their drinking habits. We label each branch of the tree with a possible outcome and its corresponding probability (see Figure 13.1).

Notice that we cover all possible outcomes with the branches. The probabilities add up to one. But we're also interested in car accidents. The probability of having an alcohol-related accident *depends* on one's drinking behaviour. Because the probabilities are *conditional,* we draw the alternatives separately on each branch of the tree:

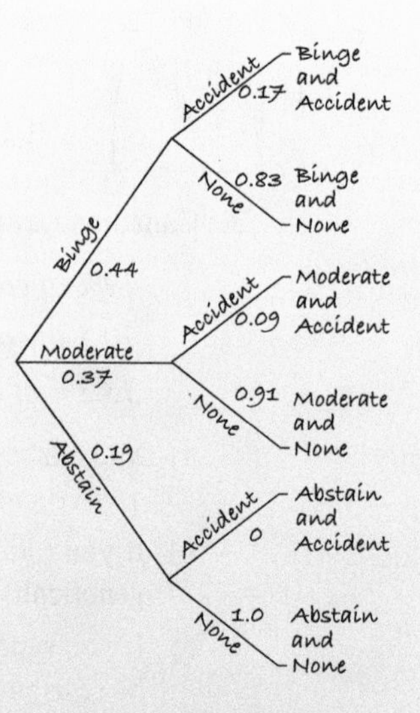

Figure 13.2

Extending the tree diagram, we can show both drinking and accident outcomes. The accident probabilities are conditional on the drinking outcomes, and they change depending on which branch we follow. Because we are concerned only with alcohol-related accidents, the conditional probability $P(\text{Accident} \mid \text{Abstinence})$ must be 0.

On each of the second set of branches, we write the possible outcomes associated with having an alcohol-related car accident (having an accident or not) and the associated probability. These probabilities are different because they are *conditional* depending on the student's drinking behaviour. (It shouldn't be too surprising that those who binge drink have a higher probability of alcohol-related accidents.) The probabilities add up to one, because given the outcome on the first branch, these outcomes cover all the possibilities. Looking back at the general multiplication rule, we can see how the tree depicts the calculation. To find the probability that a randomly selected student will be a binge drinker who has had an alcohol-related car accident, we follow the top branches. The probability of selecting a binger is 0.44. The conditional probability of an accident *given* binge drinking is 0.17. The general multiplication rule tells us that to find the *joint* probability of being a binge drinker and having an accident, we multiply these two probabilities together:

$$P(\text{Binge } \textit{and} \text{ Accident}) = P(\text{Binge}) \times P(\text{Accident} \mid \text{Binge})$$
$$= 0.44 \times 0.17 = 0.075$$

And we can do the same for each combination of outcomes:

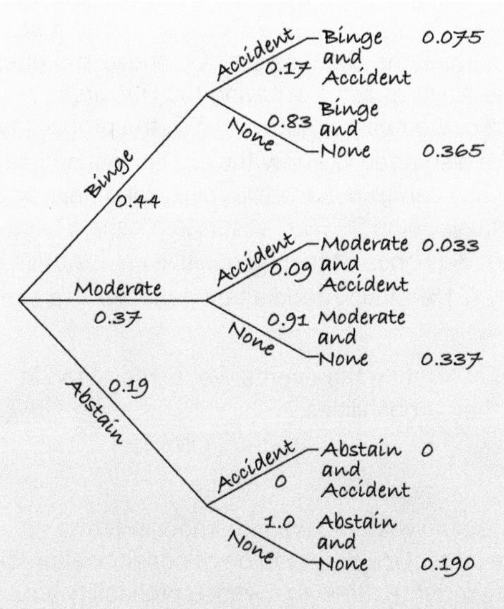

Figure 13.3
We can find the probabilities of compound events by multiplying the probabilities along the branch of the tree that leads to the event, just the way the general multiplication rule specifies. The probability of abstaining and having an alcohol-related accident is, of course, zero.

All the outcomes at the far right are disjoint because at each branch of the tree, we chose between disjoint alternatives. And they are *all* the possibilities, so the probabilities on the far right must add up to one.

Because the final outcomes are disjoint, we can add up their probabilities to get probabilities for compound events. For example, what's the probability that a selected student has had an alcohol-related car accident? We simply find *all* the outcomes on the far right in which an accident has happened. There are three and we can add their probabilities: $0.075 + 0.033 + 0 = 0.108$—almost an 11% chance.

13.4 Reversing the Conditioning and Bayes' Rule

If we know a student has had an alcohol-related accident, what's the probability that the student is a binge drinker? That's an interesting question, but we can't just read it from the tree. The tree gives us $P(\text{Accident} \mid \text{Binge})$, but we want $P(\text{Binge} \mid \text{Accident})$—conditioning in the other direction. The two probabilities are definitely *not* the same. We have reversed the conditioning.

We may not have the conditional probability we want, but we do know everything we need to know to find it. To find a conditional probability, we need the probability that both events happen divided by the probability that the given event occurs. We have already found the probability of an alcohol-related accident: $0.075 + 0.033 + 0 = 0.108$.

The joint probability that a student is both a binge drinker and someone who's had an alcohol-related accident is found at the top branch: 0.075. We've restricted the *who* of the problem to the students with alcohol-related accidents, so we divide the latter by the former to find the conditional probability:

$$P(\text{Binge} \mid \text{Accident}) = \frac{P(\text{Binge } \textit{and} \text{ Accident})}{P(\text{Accident})}$$

$$= \frac{0.075}{0.108} = 0.694$$

The chance that a student who has an alcohol-related car accident is a binge drinker is more than 69%! As we said, reversing the conditioning is rarely intuitive, but tree diagrams help us keep track of the calculation when there aren't too many alternatives to consider.

Step-by-Step Example REVERSING THE CONDITIONING

If your HIV test is positive and you are a man at low risk of infection, what are the chances that you actually have the virus? The nurse told you that the test was 99.99% effective, so it seems that the chances must be pretty high that you have HIV. How high do you think the chances are? Go ahead and guess. Guess low.

We'll call **HIV** the event of actually having the HIV virus, '+' the event of testing positive, and '−' the event of testing negative. To start a tree, we need to know $P(\textbf{HIV})$, the probability of having the HIV virus. For men at low risk for infection, this is 1 in 10 000. We also need to know the conditional probabilities $P(+ \mid \textbf{HIV})$ and $P(+ \mid \textbf{HIV}^C)$. Diagnostic tests can make two kinds of errors. They can give a positive result for a healthy person (a *false positive*) or a negative result for a sick person (a *false negative*). Being 99.99% accurate means a false-positive rate of 0.01%. That is, someone who doesn't have the disease has a 0.01% chance of testing positive anyway. We can write $P(+ \mid \textbf{HIV}^C) = 0.0001$. HIV tests have a 0.1% false-negative rate, so only 0.1% of sick people test negative. We can write $P(- \mid \textbf{HIV}) = 0.001$.

THINK ➡ Plan Define the events we're interested in and their probabilities.

Let **HIV** = {Having HIV} and $\textbf{HIV}^C$ = {No HIV}

+ = {*Testing positive*} *and*

− = {*Testing negative*}

Figure out what you want to know in terms of the events. Use the notation of conditional probability to write the event whose probability you want to find.

I know that $P(+ \mid \textbf{HIV}^C) = 0.0001$ and $P(- \mid \textbf{HIV}) = 0.001$. I also know that $P(\textbf{HIV}) = 0.0001$.

I'm interested in the probability of having HIV given a positive test result: $P(\textbf{HIV} \mid +)$.

SHOW ➡ Plot Draw the tree diagram. When probabilities are very small like these are, be careful to keep all the significant digits.

To finish the tree, we need $P(\textbf{HIV}^C)$, $P(- \mid \textbf{HIV}^C)$, and $P(+ \mid \textbf{HIV})$. We can find each of these from the Complement Rule:

$$P(\textbf{HIV}^C) = 1 - P(\textbf{HIV}) = 0.9999$$
$$P(- \mid \textbf{HIV}^C) = 1 - P(+ \mid \textbf{HIV}^C) = 1 - 0.0001 = 0.9999$$
$$P(+ \mid \textbf{HIV}) = 1 - P(- \mid \textbf{HIV}) = 1 - 0.001 = 0.999$$

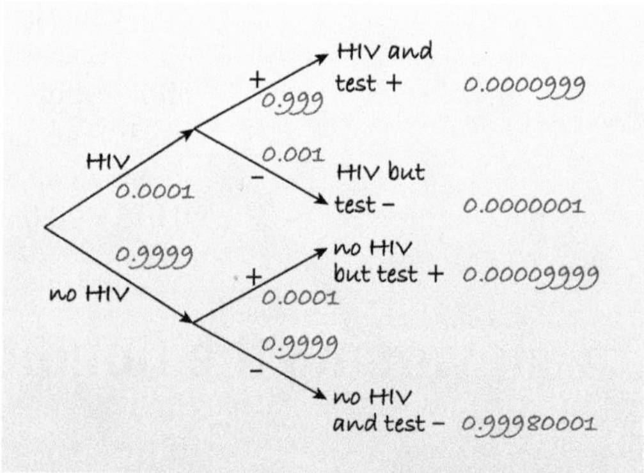

Check: 0.0000999 + 0.0000001 + 0.00009999 + 0.99980001 = 1)

Mechanics Multiply along the branches to find the probabilities of the four possible outcomes. It's a good idea to check your work by seeing if they total 1.

Add up the probabilities corresponding to the condition of interest—in this case, testing positive. We can add the probabilities from the tree twigs that correspond to testing positive because the tree shows disjoint events.

$$P(+) = P(\textbf{HIV} \text{ and } +) + P(\textbf{HIV}^C \text{ and } +)$$
$$P(+) = 0.0000999 + 0.00009999 = 0.00019989$$

Divide the probability of both events occurring (here, having HIV given a positive test) by the probability of satisfying the condition (testing positive).

$$P(\textbf{HIV} \mid +) = \frac{P(\textbf{HIV} \text{ and } +)}{P(+)} = \frac{0.0000999}{0.00019989} = 0.4998$$

TELL ➡ Conclusion Interpret your result in context.

The chance of having HIV after you test positive is less than 50%.

With events of low probability, the result can be surprising. That's the reason patients who test positive for HIV, for example, are always told to seek medical counselling. They may have only a small chance of actually being infected. Universal drug or disease testing can have unexpected consequences if people interpret testing positive as being positive.

Pearson Education

The Reverend Thomas Bayes is credited posthumously with the rule that is the foundation of Bayesian Statistics.

Bayes' Rule

When we have $P(\mathbf{A} \mid \mathbf{B})$ but want the *reverse* probability $P(\mathbf{B} \mid \mathbf{A})$, we need to find $P(\mathbf{A}$ *and* $\mathbf{B})$ and $P(\mathbf{A})$. A tree is often a convenient way of finding these probabilities. It can work even when we have more than two possible events, as we saw in the binge-drinking example. Instead of using the tree, we *could* write the calculation algebraically, showing exactly how we found the quantities that we needed: $P(\mathbf{A}$ *and* $\mathbf{B})$ and $P(\mathbf{A})$. The result is a formula known as Bayes' Rule, after the Reverend Thomas Bayes (c. 1702–1761), who was credited with the rule after his death, when he could no longer defend himself. Bayes' Rule is quite important in Statistics and is the foundation of an approach to statistical analysis known as Bayesian Statistics. Although the simple rule deals with two alternative outcomes, it can be extended to the situation in which there are more than two branches to the first split of the tree. The principle remains the same (although the math gets more difficult). Bayes' Rule is just a formula for reversing the probability from the conditional probability that you're originally given. Bayes' Rule for two events says that $P(\mathbf{B} \mid \mathbf{A}) = \dfrac{P(\mathbf{A} \mid \mathbf{B})P(\mathbf{B})}{P(\mathbf{A} \mid \mathbf{B})P(\mathbf{B}) + P(\mathbf{A} \mid \mathbf{B}^C)P(\mathbf{B}^C)}$. Try it with the HIV testing probabilities. (It's easier to just draw the tree, isn't it?)

For Example REVERSING THE CONDITIONING

A recent Maryland highway safety study found that in 77% of all accidents, the driver was wearing a seatbelt. Accident reports indicated that 92% of those drivers escaped serious injury (defined as hospitalization or death), but only 63% of the non-belted drivers were so fortunate.

QUESTION: What's the probability that a driver who was seriously injured wasn't wearing a seatbelt?

ANSWER: Let B = the driver was wearing a seatbelt, and NB = no belt.

Let I = serious injury or death, and OK = not seriously injured.

I know $P(\text{B}) = 0.77$ so $P(\text{NB}) = 1 - 0.77 = 0.23$.

Also, $P(\text{OK}|\text{B}) = 0.92$, so $P(\text{I}|\text{B}) = 0.08$

and $P(\text{OK}|\text{NB}) = 0.63$, so $P(\text{I}|\text{NB}) = 0.37$

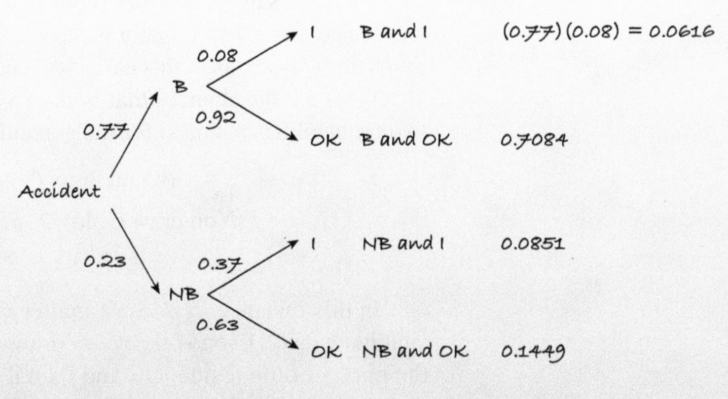

$$P(\text{NB} \mid \text{I}) = \frac{P(\text{NB } \textit{and } \text{I})}{P(\text{I})} = \frac{0.0851}{0.0616 + 0.0851} = 0.58$$

Even though only 23% of drivers weren't wearing seatbelts, they accounted for 58% of all the deaths and serious injuries.

Just some advice from your friends, the authors: *Please buckle up!* (We want you to finish this course.)

Photos.com/Getty Images

Nicholas Saunderson (1682–1739) was a blind English mathematician who invented a tactile board to help other blind people do mathematics. And he may have been the true originator of "Bayes' Rule."

Who discovered Bayes' Rule? Stigler's "Law of Eponymy" states that discoveries named for someone (eponyms) are usually named for the wrong person. Steven Stigler, who admits he didn't originate the law, is an expert on the history of Statistics, and he suspected that the law might apply to Bayes' Rule. He looked at the possibility that another candidate—one Nicholas Saunderson—was the real discoverer, not the Reverend Bayes. He assembled historical evidence and compared probabilities that the historical events would have happened *given* that Bayes was the discoverer of the rule, with the corresponding probabilities *given* that Saunderson was the discoverer. Of course, what he really wanted to know were the probabilities that Bayes or Saunderson was the discoverer *given* the historical events. How did he *reverse* the conditional probabilities? He used Bayes' Rule and concluded that, actually, it's more likely that Saunderson is the real originator of the rule.

But that doesn't change our tradition of naming the rule for Bayes and calling the branch of Statistics arising from this approach Bayesian Statistics.

Sampling Without Replacement

Room draw is a process for assigning residence rooms to students who live on campus. Sometimes, when students have equal priority, they are randomly assigned to the currently available rooms. When it's time for you and your friend to draw, there are 12 rooms left. Three are in Gold Hall, a very desirable residence with spacious, wood-panelled rooms. Four are in Silver Hall, centrally located, but not quite as desirable. And five are in Wood Hall, a new residence with cramped rooms located over a kilometre from the centre of campus at the edge of the woods.

You get to draw first, and then your friend will draw. Naturally, you would both like to score rooms in Gold. What are your chances? In particular, what is the chance that you will *both* get rooms in Gold?

When you go first, the chance that *you* will draw one of the Gold rooms is 3/12. Suppose you do. Now, with you clutching your prized room assignment, what chance does your friend have? At this point there are only 11 rooms left and just 2 left in Gold, so your friend's chance is now 2/11.

Using our notation, we write

$$P(\text{Friend draws Gold} \mid \text{You draw Gold}) = 2/11.$$

The reason the denominator changes is that we draw these rooms *without replacement.* That is, once one is drawn, it doesn't go back into the pool.

We often sample without replacement. When we draw from a very large population, the change in the denominator is too small to worry about. But when there's a small population to draw from, as in this case, we need to take note and adjust the probabilities.

What are the chances that *both* of you will luck out? Well, now we've calculated the two probabilities we need for the general multiplication rule, so we can write:

$$\begin{aligned} &P(\text{You draw Gold } \textit{and} \text{ Friend draws Gold}) \\ &= P(\text{You draw Gold}) \times P(\text{Friend draws Gold} \mid \text{You draw Gold}) \\ &= 3/12 \times 2/11 = 1/22 = 0.045 \end{aligned}$$

In this instance, it doesn't matter who went first, or even if the rooms were drawn simultaneously. Even if the room draw was accomplished by shuffling cards containing the names of the residences and then dealing them out to 12 applicants (rather than by each student drawing a room in turn), we can still *think* of the calculation as having taken place in two steps:

$$\xrightarrow{3/12} \text{Gold} \xrightarrow{2/11} \text{Gold} \mid \text{Gold}$$

That is, one of you has a probability of 3/12 of drawing a Gold room and the other then has a probability of 2/11 of also drawing a Gold room. It doesn't matter whose draw we think of first. The probability changes for the second person nonetheless. The diagram shows this ordering of our thoughts.

Sampling without replacement involves dependent events. Let's define some events. Let **B** be the event "you draw Gold." Let **A** be the event "friend draws Gold." Suppose your friend draws first and gets Gold. What is the probability that you also draw Gold? In other words, what is $P(\mathbf{B} \mid \mathbf{A})$? $P(\mathbf{B} \mid \mathbf{A}) = 2/11$. Suppose your friend draws first and doesn't get Gold. What is the probability that you draw Gold? In other words, what is $P(\mathbf{B} \mid \mathbf{A}^c)$? $P(\mathbf{B} \mid \mathbf{A}^c) = 3/11$. Since $P(\mathbf{B} \mid \mathbf{A}) \neq P(\mathbf{B} \mid \mathbf{A}^c)$ these are dependent events.

WHAT CAN GO WRONG?

- **Don't use a simple (special case) probability rule where a general rule is appropriate.** Don't assume independence without reason to believe it. Don't assume that outcomes are disjoint without checking that they are. Remember that the general rules always apply, even when outcomes are in fact independent or disjoint.
- **Don't find probabilities for samples drawn without replacement as if they had been drawn with replacement.** Remember to adjust the denominator of your probabilities. This warning applies only when we draw from small populations or draw a large fraction of a finite population. When the population is very large relative to the sample size, the adjustments make very little difference, and we ignore them.
- **Don't reverse conditioning naively.** As we have seen, the probability of **A** *given* **B** may not, and, in general does not, resemble the probability of **B** *given* **A**. The true probability may be counterintuitive.
- **Don't confuse "disjoint" with "independent."** Disjoint events *cannot* happen at the same time. When one happens, you know the other does not, so $P(\mathbf{B} \mid \mathbf{A}) = 0$. Independent events *must* be able to happen at the same time. When one happens, you know it has no effect on the other, so $P(\mathbf{B} \mid \mathbf{A}) = P(\mathbf{B})$.
- **Don't multiply probabilities of events if they're not independent.** The probability of selecting a student at random who is over 2 metres tall *and* on the basketball team is *not* the probability the student is over 2 metres tall *times* the probability he's on the basketball team. Knowing that the student is over 2 metres tall changes the probability of his being on the basketball team. You can't multiply these probabilities. The multiplication of probabilities of events that are not independent is one of the most common errors people make in dealing with probabilities. The only way you can multiply probabilities is if you know the appropriate conditional probabilities.
- **Don't confuse addition and multiplication when finding probabilities.** To find the probability that events **A** and **B** have both occurred, use the general multiplication rule, or use the multiplication rule if they are independent. To find the probability that either event **A** *or* **B** (or both of them) have occurred, use the general additive rule, or use the additive rule if they are disjoint.

CONNECTIONS

This chapter shows the unintuitive side of probability. If you've been thinking, "My mind doesn't work this way," you're probably right. Humans don't seem to find conditional and compound probabilities natural and often have trouble with them. Even statisticians make mistakes with conditional probability.

We began thinking about independence in Chapter 2 when we looked at contingency tables and asked whether the distribution of one variable was the same for each category of another. Then in Chapter 10, we saw that independence was fundamental to drawing a simple random sample. For computing compound probabilities, we again ask about independence. And we'll continue to think about independence throughout the rest of the book.

Our guiding principle is that statistics is about understanding the world. The events discussed in this chapter are close to the kinds of real-world situations in which understanding probabilities matters. The methods and concepts presented are the tools you need to understand the part of the real world that deals with the outcomes of complex, uncertain events.

What Have We Learned?

Learning Objectives

Know the general rules of probability and how to apply them.

- The General Multiplication Rule says that $P(\text{A } and \text{ B}) = P(\text{A}) \times P(\text{B} \mid \text{A})$

Know that the conditional probability of an event B given the event A is

$$P(\text{B} \mid \text{A}) = \frac{P(\text{A } and \text{ B})}{P(\text{A})}$$

Know how to define and use independence.

- Events A and B are independent if $P(\text{A} \mid \text{B}) = P(\text{A})$

Know how to make and use tables, Venn diagrams, and tree diagrams to solve problems involving probability.

Know how to use tree diagrams and Bayes' Rule to reverse the conditioning.

Review of Terms

Conditional probability

$$P(\text{B} \mid \text{A}) = \frac{P(\text{A } and \text{ B})}{P(\text{A})}$$

$P(\text{B} \mid \text{A})$ is read "the probability of B *given* A." (p. 354)

Independence Events A and B are independent when $P(\text{B} \mid \text{A}) = P(\text{B})$ (p. 355).

Independence Assumption We often require events to be independent. (So you should think about whether this assumption is reasonable.) (p. 356)

General multiplication rule For any two events, A and B, the probability of A *and* B is $P(\text{A } and \text{ B}) = P(\text{A}) \times P(\text{B} \mid \text{A})$ (p. 356).

Multiplication rule If A and B are independent events, then the probability of A *and* B is $P(\text{A } and \text{ B}) = P(\text{A}) \times P(\text{B})$. This is easily extended to any number of independent events. (p. 357).

Tree diagram A display of conditional events or probabilities that is helpful in thinking through conditioning (p. 364).

Exercises

1. Pet ownership Suppose that 25% of people have a dog, 29% of people have a cat, and 12% of people own both. What is the probability that someone owns a dog or a cat?

2. Sports What is the probability that a person liked to watch football, given that she also likes to watch basketball?

	Football	No Football
Basketball	27	13
No Basketball	38	22

3. Sports again From Exercise 2, if someone doesn't like to watch basketball, what is the probability that she will be a football fan?

4. Field goals A nervous kicker usually makes 70% of his first field goal attempts. If he makes his first attempt, his success rate rises to 90%. What is the probability that he makes his first two kicks?

5. Titanic On the *Titanic*, the probability of survival was 0.323. Among first class passengers, it was 0.625. Were *survival* and *first class ticket* independent events? Explain.

6. Facebook Facebook reports that 70% of its users are from outside the United States and that 50% of its users log on to Facebook every day. Suppose that 20% of its users are U.S. users who log on every day. Make a probability table. Why is a table better than a tree here?

7. Ethanol A Nanos Poll in August 2008 asked 500 adult Ontarians if they thought Ontario should produce more, produce less, or maintain current production of ethanol. Ethanol is made from corn, and can be used as a gasoline additive to increase fuel efficiency.

Here's how they responded:

Response	Number
More ethanol production	160
Less ethanol production	136
Maintain current production level	155
No opinion	49
Total	**500**

Suppose we select three people at random from this sample.

a) What is the probability that all three responded, "Less ethanol production"?
b) What is the probability that none responded, "Maintain current production level"?
c) What assumption did you make in computing these probabilities?
d) Explain why you think that assumption is reasonable.

8. Optimistic A Nanos Poll in November 2008 asked 1000 Canadian adults to compare Canada's international reputation in 2008 with its international reputation in 2007. Here's how they responded:

Response	Number
Not improved	251
No change	313
Somewhat improved	288
Improved	120
No opinion	28
Total	**1000**

Let's call someone who responded "Somewhat improved" or "Improved" an "optimist." If we select two people at random from this sample,

a) What is the probability that both are optimists?
b) What is the probability that neither is an optimist?
c) What is the probability that one is an optimist and one isn't?
d) What assumption did you make in computing these probabilities?
e) Explain why you think that assumption is reasonable.

9. M&Ms The Masterfoods company says that before the introduction of purple, yellow candies made up 20% of their plain M&Ms; red another 20%; and orange, blue, and green each made up 10%. The rest were brown. If you pick three M&Ms in a row, what is the probability that

a) They are all brown?
b) The third one is the first red one?
c) None are yellow?
d) At least one is green?

10. Blood Canadian Blood Services says that about 46% of the Canadian population have Type O blood, 42% have Type A, 9% have Type B, and the rest have Type AB. Among four potential donors, what is the probability that

a) All are Type O?
b) No one is Type AB?
c) They are not all Type A?
d) At least one person is Type B?

11. Disjoint or independent? In Exercise 9, you calculated probabilities of getting various M&Ms. Some of your answers depended on the assumption that the outcomes described were *disjoint*; that is, they could not both happen at the same time. Other answers depended on the assumption that the events were *independent*; that is, the occurrence of one of them doesn't affect the probability of the other. Do you understand the difference between disjoint and independent?

a) If you draw one M&M, are the events of getting a red one and getting an orange one disjoint or independent or neither?
b) If you draw two M&Ms one after the other, are the events of getting a red on the first and a red on the second disjoint or independent or neither?
c) Can disjoint events ever be independent? Explain.

12. Disjoint or independent, again In Exercise 10, you calculated probabilities involving various blood types. Some of your answers depended on the assumption that the outcomes described were *disjoint*; that is, they could not both happen at the same time. Other answers depended on the assumption that the events were *independent*; that is, the occurrence of one of them doesn't affect the probability of the other. Do you understand the difference between disjoint and independent?

a) If you examine one person, are the events that the person is Type A and that the person is Type B disjoint or independent or neither?
b) If you examine two people, are the events that the first is Type A and the second is Type B disjoint or independent or neither?
c) Can disjoint events ever be independent? Explain.

13. Dice You roll a fair die three times. What is the probability that

a) You roll all 6s?
b) You roll all odd numbers?
c) None of your rolls gets a number divisible by 3?
d) You roll at least one 5?
e) The numbers you roll are not all 5s?

14. Slot machine A slot machine has three wheels that spin independently. Each has 10 equally likely symbols: four bars, three lemons, two cherries, and a bell. If you play, what is the probability that

a) You get three lemons?
b) You get no fruit symbols?
c) You get three bells (the jackpot)?
d) You get no bells?
e) You get at least one bar (an automatic loser)?

15. Champion bowler A certain bowler can bowl a strike 70% of the time. What is the probability that she
a) Goes three consecutive frames without a strike?
b) Makes her first strike in the third frame?
c) Has at least one strike in the first three frames?
d) Bowls a perfect game (12 consecutive strikes)?

16. The train To get to work, a commuter must cross train tracks. The time the train arrives varies slightly from day to day, but the commuter estimates he'll get stopped on about 15% of work days. During a certain five-day work week, what is the probability that he
a) Gets stopped on Monday and again on Tuesday?
b) Gets stopped for the first time on Thursday?
c) Gets stopped every day?
d) Gets stopped at least once during the week?

17. Ottawa Seventy-six percent of Ottawans speak English only when they're at home, 10% speak only French, 11% speak a non-official language, and the remainder speak more than one language. You are conducting a poll by calling Ottawans at random. In your first three calls, what is the probability you talk to
a) Only people who speak only English at home?
b) No people who speak only French at home?
c) At least one person who speaks a non-official language at home?

18. Ottawa revisited Sixty percent of Ottawans drive their own vehicles to work, 7% ride as passengers in someone else's vehicle, 22% take public transit, 10% walk or cycle, and the remainder get to work some other way. You are conducting a poll by calling Ottawans at random. In your first four calls, what is the probability you talk to
a) Only people who drive their own vehicles to work?
b) No one who takes public transit to work?
c) At least one person who walks or cycles to work?

19. Tires You bought a new set of four tires from a manufacturer who just announced a recall because 2% of those tires are defective. What is the probability that at least one of yours is defective?

20. Pepsi For a sales promotion, the manufacturer places winning symbols under the caps of 10% of all Pepsi bottles. You buy a six-pack. What is the probability that you win something?

21. Lottery tickets The Ontario Lottery Association claims that your odds of winning a prize on an Instant Win Crossword game are 1 in 3.09. This means that any ticket has probability $1/3.09 = 0.324$ of winning a prize. Every Friday you buy one Crossword game.
a) What is the probability you don't win a prize next Friday?
b) What is the probability you don't win a prize six Fridays in a row?
c) If you haven't won a prize for the past six Fridays, what is the probability you will win a prize on your next game?
d) What is the probability you will win on two of your next three games?

22. Craps The game of Craps starts with a player tossing a pair of dice. If dice add up to 2 or 3 or 12, the player's turn is up and they "crap out." You're determined to win at this game.
a) What is the probability that you crap out on your next turn?
b) What is the probability that you crap out on your next four turns?
c) If you've crapped out on your last three turns, what is the probability that you don't crap out on your next turn?
d) What is the probability that you crap out on two of your next three turns?

23. Cards You draw a card at random from a standard deck of 52 cards. Find each of the following conditional probabilities:
a) The card is a heart, given that it is red.
b) The card is red, given that it is a heart.
c) The card is an ace, given that it is red.
d) The card is a queen, given that it is a face card.

24. Pets In its monthly report, a local animal shelter states that it currently has 24 dogs and 18 cats available for adoption. Eight of the dogs and six of the cats are male. Find each of the following conditional probabilities if an animal is selected at random:
a) The pet is male, given that it is a cat.
b) The pet is a cat, given that it is female.
c) The pet is female, given that it is a dog.

25. Men's health The table shows the approximate probabilities that an adult Canadian male has high blood pressure and/or high cholesterol.

		Blood Pressure	
		High	OK
Cholesterol	High	0.11	0.21
	OK	0.16	0.52

a) What is the probability that a man has both conditions?
b) What is the probability that he has high blood pressure?
c) What is the probability that a man with high blood pressure has high cholesterol?
d) What is the probability that a man has high blood pressure if it's known that he has high cholesterol?

26. Death penalty The table shows the political affiliation of American voters and their positions on the death penalty.

		Death Penalty	
		Favour	Oppose
Party	Republican	0.26	0.04
	Democrat	0.12	0.24
	Other	0.24	0.10

a) What is the probability that a randomly chosen voter favours the death penalty?
b) What is the probability that a Republican favours the death penalty?

c) What is the probability that a voter who favours the death penalty is a Democrat?
d) A candidate thinks she has a good chance of gaining the votes of anyone who is a Republican or in favour of the death penalty. What portion of the voters is that?

27. Movies A video store investigates the types of movies their customers rent. They record the customer's age and the movie genre of 240 recent rentals. Suppose the data break down as follows:

	Type of Movie			
Age	**Documentary**	**Comedy**	**Mystery**	**Total**
12–20	14	9	8	31
21–29	15	14	9	38
30–39	9	21	39	69
40–49	7	22	17	46
50 and over	6	38	12	56
Total	**51**	**104**	**85**	**240**

a) Among 21 to 29-year-olds, what is the probability that a person rented a documentary?
b) Among people who rented documentaries, what is the probability that the renter was 21 to 29 years old?
c) If a renter was over 30 years old, what is the probability that the person selected a comedy?
d) If the renter didn't select a comedy, what is the probability the person was over 30?

28. Birth order A survey of students in a large Introductory Statistics class asked about their birth order (1 = oldest or only child) and which faculty of the university they were studying under. Here are the data:

	Birth Order		
Faculty	**1 or only**	**2 or more**	**Total**
Arts & Sciences	34	23	57
Agriculture	52	41	93
Human Ecology	15	28	43
Other	12	18	30
Total	**113**	**110**	**223**

a) If we select a student at random, what is the probability that the person is an Arts and Sciences student who is a second child (or more)?
b) Among the Arts and Sciences students, what is the probability that a student is a second child (or more)?
c) Among second children (or more), what is the probability that the student is enrolled in Arts and Sciences?
d) What is the probability that a first or only child is enrolled in the Agriculture College?
e) What is the probability that an Agriculture student is a first or only child?

29. Sick kids Seventy percent of kids who visit a doctor have a fever, and 30% of kids with a fever have sore throats. What is the probability that a kid who goes to the doctor has a fever and a sore throat? What did you assume to answer this question?

30. Sick cars Twenty percent of cars that are inspected have faulty pollution control systems. The cost of repairing a pollution control system exceeds $100 about 40% of the time. When a driver takes her car in for inspection, what is the probability that she will end up paying more than $100 to repair the pollution control system? What did you assume to answer this question?

31. Cards again You are dealt a hand of three cards, one at a time. Find the probability of each of the following.
a) The first heart you get is the third card dealt.
b) Your cards are all red (that is, all diamonds or hearts).
c) You get no spades.
d) You have at least one ace.

32. Another hand You pick three cards at random from a deck. Find the probability of each event described below.
a) You get no aces.
b) You get all hearts.
c) The third card is your first red card.
d) You have at least one diamond.

33. Batteries A junk box in your room contains a dozen old batteries, five of which are totally dead. You start picking batteries one at a time and testing them. Find the probability of each outcome.
a) The first two you choose are both good.
b) At least one of the first three works.
c) The first four you pick all work.
d) You have to pick five batteries to find one that works.

34. Shirts The soccer team's shirts have arrived in a big box, and people just start grabbing them, looking for the right size. The box contains 4 medium, 10 large, and 6 extra-large shirts. You want a medium for you and one for your sister. Find the probability of each event described.
a) The first two you grab are the wrong sizes.
b) The first medium shirt you find is the third one you check.
c) The first four shirts you pick are all extra-large.
d) At least one of the first four shirts you check is a medium.

35. Cards III If you draw a card at random from a well-shuffled deck, is getting an ace independent of the suit? Explain.

36. Pets again The local animal shelter in Exercise 24 reported that it currently has 24 dogs and 18 cats available for adoption; eight of the dogs and six of the cats are male. Are the species and sex of the animals independent? (For example, are the events *dog* and *male* independent events?) Explain.

37. Movies, final showing. In Exercise 27, we looked at results of a survey about movie rentals.
a) Are being under 30 and being over 50 disjoint? Explain.
b) Are being under 30 and being over 50 independent? Explain.
c) Are renting a comedy and being under 30 disjoint? Explain.
d) Are renting a comedy and being under 30 independent? Explain.

38. Birth order, finis In Exercise 28, we looked at the birth orders and faculty choices of some Intro Stats students.
a) Are enrolling in Agriculture and Human Ecology disjoint? Explain.
b) Are enrolling in Agriculture and Human Ecology independent? Explain.
c) Are being first-born and enrolling in Human Ecology disjoint? Explain.
d) Are being first-born and enrolling in Human Ecology independent? Explain.

39. Men's health, again Given the table of probabilities from Exercise 25, are high blood pressure and high cholesterol independent? Explain.

Cholesterol		Blood Pressure	
		High	OK
	High	0.11	0.21
	OK	0.16	0.52

40. Death penalty, reborn Given the table of probabilities from Exercise 26, are being a Republican and favouring the death penalty independent events? Explain.

Party		Death Penalty	
		Favour	Oppose
	Republican	0.26	0.04
	Democrat	0.12	0.24
	Other	0.24	0.10

41. Television According to Statistics Canada, about 46% of the television we watch is Canadian programming on Canadian channels. (If this seems high to you, just think of news programs and hockey games.) Close to 15% of what we watch is foreign programming on foreign channels, and we don't watch any Canadian programming on foreign channels.
a) What proportion of our television watching is foreign programming on Canadian channels?
b) Are watching Canadian programming and watching a Canadian channel independent? Explain.

42. Snoring After surveying 995 adults, 81.5% of whom were over 30, the National Sleep Foundation reported that 36.8% of all the adults snored. Thirty-two percent of the respondents were snorers over the age of 30.
a) What percent of the respondents were under 30 and did not snore?
b) Is snoring independent of age? Explain.

43. Luggage Leah is flying from Moncton to Vancouver with a connection in Montreal. The probability that her first flight leaves on time is 0.15. If the flight is on time, the probability that her luggage will make the connecting flight in Montreal is 0.95, but if the first flight is delayed, the probability that the luggage will make it is only 0.65.
a) Are the first flight leaving on time and the luggage making the connection independent events? Explain.
b) What is the probability that Leah's luggage arrives in Vancouver with her?

44. Postsecondary A private high school report contains these statistics:

70% of incoming Grade 9 students attended public primary schools.

75% of the students who came from public primary school go on to university after completing high school.

90% of the students who did not come from public primary schools go on to university after completing high school.

a) Is there any evidence that a private-school student's chances of going to university may depend upon what kind of primary school the student attended? Explain.
b) What percent of this private high school's students go on to university?

45. Late luggage Remember Leah from Exercise 43? Suppose you pick her up at the Vancouver airport and her luggage is not there. What is the probability that Leah's first flight was delayed?

46. Further postsecondary What percent of students who go on to university from the private high school in Exercise 44 attended a public primary school?

47. Absenteeism A company's records indicate that on any given day, about 1% of their day shift employees and 2% of the night shift employees will miss work. Sixty percent of the employees work the day shift.
a) Is absenteeism independent of shift worked? Explain.
b) What percent of employees are absent on any given day?

48. Lungs and smoke Suppose that 23% of adults smoke cigarettes. It's known that 57% of smokers and 13% of nonsmokers develop a certain lung condition by age 60.
a) Explain how these statistics indicate that lung condition and smoking are not independent.
b) What is the probability that a randomly selected 60-year-old has this lung condition?

49. Absenteeism, part II At the company described in Exercise 47, what percent of the absent employees are on the night shift?

50. Lungs and smoke, second puff Based on the statistics in Exercise 48, what is the probability that someone with the lung condition was a smoker?

51. Drunks Police often set up impaired driving checkpoints—roadblocks where drivers are asked a few brief questions to allow the officer to judge whether the person may have been drinking. If the officer does not suspect a problem, drivers are released to go on their way.

Otherwise, drivers are detained for a Breathalyzer test that will determine whether they will be arrested. The police say that based on the brief initial stop, trained officers can make the right decision 80% of the time. Suppose the police operate a sobriety checkpoint after 9:00 p.m. on a Saturday night, a time when national traffic safety experts suspect that about 12% of drivers have been drinking.

a) You are stopped at the checkpoint and, of course, have not been drinking. What is the probability that you are detained for further testing?
b) What is the probability that any given driver will be detained?
c) What is the probability that a driver who is detained has actually been drinking?
d) What is the probability that a driver who was released had actually been drinking?

52. **Polygraphs** Lie detectors are controversial instruments, barred from use as evidence in Canadian courts. Nonetheless, some employers use lie detector screening as part of their hiring process in the hope that they can avoid hiring people who might be dishonest. There has been some research, but no agreement, about the reliability of polygraph tests. Based on this research, suppose that a polygraph can detect 65% of lies, but incorrectly identifies 15% of true statements as lies.

A certain company believes that 95% of its job applicants are trustworthy. The company gives everyone a polygraph test, asking, "Have you ever stolen anything from your place of work?" Naturally, all the applicants answer, "No," but the polygraph identifies some of those answers as lies, making the person ineligible for a job. What is the probability that a job applicant rejected under suspicion of dishonesty was actually trustworthy?

53. **Dishwashers** Digby's Diner employs three dishwashers. Alma washes 40% of the dishes and breaks only 1% of those she handles. Filip and Kai each wash 30% of the dishes. Filip breaks only 1% of the dishes he washes, but Kai breaks 3% of the dishes he washes. (He, of course, will need a new job soon. . . .) You go to Digby's for dinner one night and hear a dish break at the sink. What is the probability that Kai is on the job?

54. **Parts** A company manufacturing electronic components for home entertainment systems buys electrical connectors from three suppliers. The company prefers to use supplier A because only 1% of those connectors prove to be defective, but supplier A can deliver only 70% of the connectors needed. The company also must purchase connectors from two other suppliers: 20% from supplier B and the rest from supplier C. The rates of defective connectors from B and C are 2% and 4%, respectively. You buy one of these components, and when you try to use it you find that the connector is defective. What is the probability that your component came from supplier A?

Just Checking ANSWERS

1. a) 0.48
b) $(0.48)(0.48) = 0.2304$
c) $(1 - 0.48)^2(0.48) = 0.129792$
d) $1 - (1 - 0.48)^5 = 0.96198$

2. a) Independent
b) Disjoint
c) Neither

3. a)

		Resistant to Insects		
		Yes	No	
Resistant to	Yes	0.133	0.567	0.7
Herbicides	No	0.057	0.243	0.3
		0.19	0.81	1

b) P(resistant to herbicide OR resistant to insects) = P(resistant to herbicide) + P(resistant to insects) − P(resistant to herbicide AND resistant to insects) $= 0.70 + 0.19 - 0.133 = 0.757$

c) P(resistant to herbicides|resistant to insects) = P(resistant to herbicide AND resistant to insects)/P(resistant to insects) $= 0.133/0.19 = 0.70$

d) Yes they are.
P(resistant to herbicides|resistant to insects) = P(resistant to herbicide) = 0.70

chapter

14 Random Variables

A statistics major was completely hungover the day of his final exam. It was a true/false test, so he decided to flip a coin for the answers. The professor watched the student the entire time as he was flipping the coin . . . writing the answer . . . flipping the coin . . . writing the answer. At the end of two hours, everyone else had left except this student. The professor walked over and said, 'Listen, I see that you didn't study, you didn't even look at the questions. If you are just flipping a coin for your answers, what is taking so long?' Still flipping the coin, the student bitterly replied, 'Shhh! I am checking my answers!'

—Friedman, Friedman, and Amoo (2002)

Where are we going?

How long do products last? How can you reduce your risk for developing hepatitis? Businesses, medical researchers, and other scientists use probability concepts and methods to help answer questions like these. To do that, they model the probability of outcomes using a special kind of variable—a *random* variable. We'll learn how to talk about and predict random behaviour using a few of the most common and important probability models.

NOTATION ALERT

The most common letters for random variables are *X*, *Y*, and *Z*. But be cautious. If you see any capital letter, it just might denote a random variable.

Insurance companies make bets. They bet that you're going to live a long life. You bet that you're going to die sooner. Both you and the insurance company want the company to stay in business, so it's important to find a "fair price" for your bet. Of course, the right price for *you* depends on many factors, and nobody can predict exactly how long you'll live. But when the company averages over enough customers, it can make reasonably accurate estimates of the amount it can expect to collect on a policy before it has to pay its benefit.

WHAT IS AN ACTUARY?

Actuaries are the daring people who put a price on risk, estimating the likelihood and costs of rare events so that they can be insured. That takes financial, statistical, and business skills. It also makes them invaluable to many businesses. Actuaries are rather rare themselves; only about 19 000 work in North America. Perhaps because of this, they are well paid. If you enjoy this course and are good at math, you may want to look into a career as an actuary. Contact the Society of Actuaries or the Canadian Institute of Actuaries (who, despite what you may think, did not pay for this blurb).

Here's a simple example. An insurance company offers a "death and disability" policy that pays \$10 000 when you die or \$5000 if you are permanently disabled. It charges a premium of only \$50 a year for this benefit. Is the company likely to make a profit selling such a plan? To answer this question, the company needs to know the *probability* that its clients will die or be disabled in any year. From actuarial information like this, the company can calculate an average or mean outcome for such a policy.

14.1 Expected Value or Mean

A S *Activity:* **Random Variables.** Learn more about random variables from this animated tour.

We'll want to build a probability model in order to answer the questions about the insurance company's risk. First we need to define a few terms. The amount the company pays out on an individual policy is called a **random variable** because its numeric value is based on the outcome of a random event. We use a capital letter, like X, to denote a random variable. We'll denote a particular value that it can have by the corresponding lower-case letter, in this case x. For the insurance company, x can be \$10 000 (if you die that year), \$5000 (if you are disabled), or \$0 (if neither occurs). Because we can list all the outcomes, we might formally call this random variable a **discrete random variable**. Otherwise, we'd call it a **continuous random variable**. The collection of all the possible values and the probabilities that they occur is called the **probability model or probability distribution** for the random variable.

Suppose, for example, that the death rate in any year is 1 out of every 1000 people, and that another 2 out of 1000 suffer some kind of disability. Then we can display the probability model for this insurance policy in a table like:

Table 14.1

The probability model shows all the possible values of the random variable and their associated probabilities.

Policyholder Outcome	Payout x	Probability $P(X = x)$
Death	10 000	$\frac{1}{1000}$
Disability	5 000	$\frac{2}{1000}$
Neither	0	$\frac{997}{1000}$

Given these probabilities, what should the insurance company expect to pay out? They can't know exactly what will happen in any particular year, but they can calculate what to expect, at least on average. In a probability model, we call this the **expected value** and denote it in two ways: $E(X)$ or μ. $E(X)$ is short hand for expected value of X. The symbol μ makes sense since it is a parameter of a model, and not an average calculated from observed data. To understand the expected value calculation, it's useful to imagine some convenient number of clients. In this case, 1000 works because the probabilities all have 1000 as the denominator. If the outcomes in a year matched the probabilities exactly, then out of 1000 clients, 1 died, 2 were disabled, and 997 were safe. The average payout per client would be:

NOTATION ALERT

The expected value (or mean) of a random variable is written $E(X)$ or μ.

$$\mu = E(X) = \frac{10\,000(1) + 5000(2) + 0(997)}{1000}$$

So our expected payout comes to \$20 000, or \$20 per policy.

We can rewrite it as separate terms each divided by 1000.

$$\begin{aligned}\mu &= E(X)\\ &= \$10\,000\left(\frac{1}{1000}\right) + 5000\left(\frac{2}{1000}\right) + 0\left(\frac{997}{1000}\right)\\ &= \$20.\end{aligned}$$

If they sell policies for \$50 each, that's an expected profit of \$30 per customer—not bad!

This is how we calculate an **expected value** in general. Given a (discrete) probability model like the one in Table 14.1, we just multiply each possible outcome by the probability that it occurs, and find the sum of those products

$$\mu = E(X) = \sum xP(x)$$

where $P(x)$ is short for $P(X = x)$. Be sure that every possible outcome is included in the sum. And verify that you have a valid probability model to start with—the probabilities should each be between 0 and 1 and should sum to one.

For Example LOVE AND EXPECTED VALUES

On Valentine's Day, the Quiet Nook restaurant offers a Lucky Lovers Special that could save couples money on their romantic dinners. When the waiter brings the bill, he also brings the four aces from a deck of cards. He shuffles them and lays them

out face down on the table. The couple then gets to turn one card over. If it's a black ace, they owe the full amount; but if it's the ace of hearts, the waiter gives them a \$20 Lucky Lovers discount. If they first turn over the ace of diamonds (hey—at least it's red!), they then get to turn over one of the remaining cards, earning a \$10 discount for finding the ace of hearts this time (but no discount for a black ace).

QUESTION: Based on a probability model for the size of the Lucky Lovers discounts the restaurant will award, what's the expected discount for a couple?

Let X = the Lucky Lovers discount. The probabilities of the three outcomes are:

$$P(X = 20) = P(A\heartsuit) = \frac{1}{4}$$

$$P(X = 10) = P(A\diamondsuit, \text{then } A\heartsuit) = P(A\diamondsuit) \cdot P(A\heartsuit \mid A\diamondsuit) = \frac{1}{4} \cdot \frac{1}{3} = \frac{1}{12}$$

$$P(X = 0) = P(X \neq 20 \text{ or } 10) = 1 - \left(\frac{1}{4} + \frac{1}{12}\right) = \frac{2}{3}$$

My probability model is:

Outcome	A♥	A♦, then A♥	Black Ace Appears
X	20	10	0
$P(X = x)$	$\frac{1}{4}$	$\frac{1}{12}$	$\frac{2}{3}$

$$E(X) = 20 \cdot \frac{1}{4} + 10 \cdot \frac{1}{12} + 0 \cdot \frac{2}{3} = \frac{70}{12} \approx 5.83$$

The Quiet Nook can expect to give couples an average discount of \$5.83.

The Law of Large Numbers Revisited

Can you recall the Law of Large Numbers (LLN) from Chapter 12? It tells us that the relative frequency of an event converges—gets closer and closer—to a number called its probability when we repeat the random phenomenon over and over in an identical independent fashion. Well, we have a similar result for sample means. As we take more and more independent observations on a random variable X, the sample average of all those observations also must converge in the long run to a number! Voilà—its expected value! This gives us a tremendously useful and practical interpretation of an expected value—as the long run average for X under repeated observations.

This result actually follows easily from our earlier version of the LLN. Consider the example above. The expected value of \$5.83 is certainly not what we expect to see from one customer—in fact, it's not even a possible outcome. But suppose we observe many customers and average their discounts—what will this sample average converge to in the long run? For example, suppose for the next 100 customers, we find that 30 get the \$20 discount, 8 get the \$10 discount, and 62 get no discount. For this sample of data, the sample mean $\bar{x}$ is:

$$\frac{20 + 20 + \cdots + 10 + 10 + \cdots + 0 + 0 + \cdots}{100} = \frac{20(30) + 10(8) + 0(62)}{100}$$

$$= 20 \cdot \frac{30}{100} + 10 \cdot \frac{8}{100} + 0 \cdot \frac{62}{100}$$

Where are those relative frequencies headed? Of course, toward the probabilities, so if we increase the sample size from 100, this sample mean calculation will look more and more like:

$$20 \cdot \frac{1}{4} + 10 \cdot \frac{1}{12} + 0 \cdot \frac{2}{3}$$

which of course is just the expected value of X!

Just Checking

1. One of the authors took his minivan in for repair recently because the air conditioner was cutting out intermittently. The mechanic identified the problem as dirt in a control unit. He said that in about 75% of such cases, drawing down and then recharging the coolant a couple of times cleans up the problem—and costs only \$60. If that fails, then the control unit must be replaced at an additional cost of \$100 for parts and \$40 for labour.

a) Define the random variable and construct the probability model.

b) What is the expected value of the cost of this repair?

c) What does that mean in this context?

Oh—in case you were wondering—the \$60 fix worked!

14.2 Variance and Standard Deviation

NOTATION ALERT

The standard deviation of a random variable X is written as $SD(X)$ or σ.

Of course, this expected value (or mean) is not what actually happens to any *particular* policyholder. No individual policy actually costs the company \$20. We are dealing with random events, so some policyholders receive big payouts, others nothing. Because the insurance company must anticipate this variability around the average outcome, it also needs to know the *standard deviation* of the random variable. We denote it as $SD(X)$ or σ.

For data, we calculated the **standard deviation** by first computing each observation's deviation from the mean and squaring it. We do that with (discrete) random variables as well. First, we find the deviation of each payout from the mean (expected value):

VARIANCE AND STANDARD DEVIATION

$$\sigma = Var(X)$$
$$= \sum (x - \mu)^2 P(x)$$
$$\sigma = SD(X) = \sqrt{Var(X)}$$

Policyholder Outcome	Payout x	Probability $P(X = x)$	Deviation $(x - \mu)$
Death	10 000	$\frac{1}{1000}$	$(10\,000 - 20) = 9980$
Disability	5 000	$\frac{2}{1000}$	$(5000 - 20) = 4980$
Neither	0	$\frac{997}{1000}$	$(0 - 20) = -20$

Next, we square each deviation. The **variance** of X is just the average of those squared deviations. Taking an expected value is the way we average in a probability model, so we multiply each squared deviation by the corresponding probability and sum those products:

$$Var(X) = 9980^2\left(\frac{1}{1000}\right) + 4980^2\left(\frac{2}{1000}\right) + (-20)^2\left(\frac{997}{1000}\right) = 149\,600$$

Finally, we take the square root to get the standard deviation:

$$SD(X) = \sqrt{149\,600} \approx \$386.78$$

The insurance company can expect an average payout of \$20 per policy, with a standard deviation of \$386.78.

Think about that. The company charges \$50 for each policy and expects to pay out \$20 per policy. Sounds like an easy way to make \$30. In fact, most of the time (probability 997/1000) the company pockets the entire \$50. But would you consider selling your roommate such a policy? The problem is that occasionally the seller of the policy loses big.

With probability 1/1000, it will pay out \$10 000, and with probability 2/1000, it will pay out \$5000. That may be more risk than you would be willing to take on for such a small profit. (But remember that the insurance company sells thousands of such policies—what would be the expected value and standard deviation then? See Section 14.3 and Exercise #37 and #38.)

Here are the explicit formulas for the calculations shown above. You can also write the variance and standard deviation as σ^2 and σ respectively.

$$\sigma^2 = Var(X) = \sum (x - \mu)^2 P(x)$$
$$\sigma = SD(X) = \sqrt{Var(X)}$$

There is an alternative formula for the variance that can make computations quicker: $\sigma^2 = \sum x^2 P(x) - \mu^2$. If there are other random variables in the study, add subscripts for clarity, like $\mu_X, \mu_Y, \sigma_X, \sigma_Y$. And just as with sample statistics, the variance is in squared units, while the standard deviation is in the same units as the observations (as is the expected value).

For Example FINDING THE STANDARD DEVIATION

RECAP: The probability model for the Lucky Lovers restaurant discount is

Outcome	A♥	A♦, then A♥	Black Ace Appears
x	20	10	0
$P(X = x)$	$\frac{1}{4}$	$\frac{1}{12}$	$\frac{2}{3}$

We found that the restaurant grants an average discount of $\mu = \$5.83$.

QUESTION: What's the standard deviation of the restaurant's discount?

First find the variance:

$$Var(X) = \sum (x - \mu)^2 \cdot P(x)$$
$$= (20 - 5.83)^2 \cdot \frac{1}{4} + (10 - 5.83)^2 \cdot \frac{1}{12} + (0 - 5.83)^2 \cdot \frac{2}{3}$$
$$\approx 74.306$$

So, $SD(X) = \sqrt{74.306} \approx \$8.62$

Couples can expect the Lucky Lovers discounts to average \$5.83, with a standard deviation of \$8.62.

Step-by-Step Example EXPECTED VALUES AND STANDARD DEVIATIONS FOR DISCRETE RANDOM VARIABLES

As the head of inventory for Knowway Computer Company, you were thrilled that you had managed to ship two computers to your biggest client the day the order arrived. You are horrified, though, to find out that someone had restocked refurbished computers in with the new computers in your storeroom. The shipped computers were selected randomly from the 15 computers in stock, but four of those were actually refurbished.

If your client gets two new computers, things are fine. If the client gets one refurbished computer, it will be sent back at your expense—\$100—and you can replace it. However, if both computers are refurbished, the client will cancel the order this month and you'll lose a total of \$1000.

Question: What's the expected value and the standard deviation of the company's loss?

THINK ➡ **Plan** State the problem.	I want to find the company's expected loss for shipping refurbished computers and the standard deviation.
Variable Define the random variable. **Plot** Make a picture. This is another job for tree diagrams.	

If you prefer calculation to drawing, find $P(\mathbf{NN})$ and $P(\mathbf{RR})$, then use the complement rule to find $P(\mathbf{NR}\ or\ \mathbf{RN})$.

Let X = amount of loss.

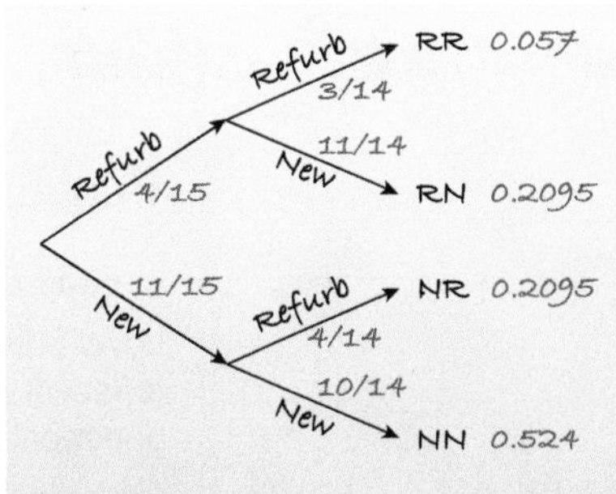

Model List the possible values of the random variable, and determine the probability model.

Outcome	x	$P(X = x)$
Two refurbs	1000	$P(\mathbf{RR}) = 0.057$
One refurb	100	$P(\mathbf{NR}\ or\ \mathbf{RN}) = 0.2095 + 0.2095 = 0.419$
New/new	0	$P(\mathbf{NN}) = 0.524$

SHOW ➡ Mechanics Find the expected value.

$$E(X) = 0(0.524) + 100(0.419) + 1000(0.057) = \$98.90$$

Find the variance.

$$Var(X) = (0 - 98.90)^2(0.524) + (100 - 98.90)^2(0.419) + (1000 - 98.90)^2(0.057) = 51\ 408.79$$

Find the standard deviation.

$$SD(X) = \sqrt{51408.79} = \$226.735$$

TELL ➡ Conclusion Interpret your results in context.

The cost to the firm has an expected value of \$98.90, with a standard deviation of \$226.74 reflecting the large range of possible losses.

REALITY CHECK Both numbers seem reasonable. The expected value of \$98.90 is between the extremes of \$0 and \$1000, and there's great variability in the outcome values.

In fact, neither the expected cost nor the standard deviation are particularly informative here since this is a one-time only situation, with just three possible outcomes. The expected value is more usefully interpreted as a *long-term average outcome.* (If the distribution were closer to bell-shaped, these parameters would be much more useful as a summary.)

14.3 Combining Random Variables

Our insurance company expects to pay out an average of \$20 per policy, with a standard deviation of about \$387. If we take the \$50 premium into account, we see the company makes a profit of 50 – 20 = \$30 per policy. Suppose the company lowers the premium by \$5 to \$45. It's pretty clear that the expected profit also drops an average of \$5 per policy, to 45 − 20 = \$25.

What about the standard deviation? We know that adding or subtracting a constant from data shifts the mean but doesn't change the variance or standard deviation. The same is true of random variables.[1]

$$E(X \pm c) = E(X) \pm c \qquad Var(X \pm c) = Var(X) \qquad SD(X \pm c) = SD(X)$$

For Example ADDING A CONSTANT

RECAP: We've determined that couples dining at the Quiet Nook can expect Lucky Lovers discounts averaging \$5.83 with a standard deviation of \$8.62. Suppose that for several weeks the restaurant has also been distributing coupons worth \$5 off any one meal (one discount per table).

QUESTION: If a couple dining there on Valentine's Day brings a coupon, what will be the mean and standard deviation of the total discounts they'll receive?

Let D = total discount (Lucky Lovers plus the coupon); then $D = X + 5$.

$$E(D) = E(X + 5) = E(X) + 5 = 5.83 + 5 = \$10.83$$
$$Var(D) = Var(X + 5) = Var(X) = 8.62^2$$
$$SD(D) = \sqrt{Var(X)} = \$8.62$$

The restaurant can expect that the discounts to couples with the coupon will average \$10.83. The standard deviation is still \$8.62.

Back to insurance. . . . What if the company decides to double all the payouts—that is, pay \$20 000 for death and \$10 000 for disability? This would double the average payout per policy and also increase the variability in payouts. We have seen that multiplying or dividing all data values by a constant changes both the mean and the standard deviation by the same factor. Variance, being the square of standard deviation, changes by the square of the constant. The same is true of random variables. In general, multiplying each value of a random variable by a constant multiplies the mean by that constant and the variance by the *square* of the constant.

$$E(aX) = aE(X) \qquad Var(aX) = a^2Var(X) \qquad SD(aX) = |a|SD(X)$$

For Example DOUBLE THE LOVE

RECAP: On Valentine's Day at the Quiet Nook, couples may get a Lucky Lovers discount averaging \$5.83 with a standard deviation of \$8.62. When two couples dine together on a single bill, the restaurant doubles the discount offer—\$40 for the ace of hearts on the first card and \$20 on the second.

QUESTION: What are the mean and standard deviation of discounts for such foursomes?

$$E(2X) = 2E(X) = 2(5.83) = \$11.66$$
$$Var(2X) = 2^2 Var(X) = 2^2 \cdot 8.62^2 = 297.2176, \text{ so}$$
$$SD(2X) = 297.2176 = \$17.24$$

If the restaurant doubles the discount offer, two couples dining together will average \$11.66 in savings with a standard deviation of \$17.24.

An insurance company sells policies to more than just one person. How can the company find the total expected value (and standard deviation) of policies taken over all

[1]The rules in this section are true for both discrete *and* continuous random variables.

policyholders? Consider a simple case: just two customers, Mr. Ecks and Ms. Wye. With an expected payout of \$20 on each policy, we might predict a total of \$20 + \$20 = \$40 to be paid out on the two policies. Nothing surprising there. The expected value of the sum is the sum of the expected values.

$$E(X + Y) = E(X) + E(Y)$$

What about the variability? Is the risk of insuring two people the same as the risk of insuring one person for twice as much? We wouldn't expect both clients to die or become disabled in the same year. Because we've spread the risk, the standard deviation should be smaller. Indeed, this is the fundamental principle behind insurance. By spreading the risk among many policies, a company can keep the standard deviation quite small and predict costs more accurately.

But how much smaller is the standard deviation of the sum? It turns out that, if the random variables are independent, there is a simple **Addition Rule for Variances**: *The variance of the sum of two independent random variables is the sum of their individual variances.*

For Mr. Ecks and Ms. Wye, the insurance company can expect their outcomes to be independent, so (using X for Mr. Ecks's payout and Y for Ms. Wye's)

$$\begin{aligned} Var(X + Y) &= Var(X) + Var(Y) \\ &= 149\,600 + 149\,600 \\ &= 299\,200 \end{aligned}$$

If they had insured only Mr. Ecks for twice as much, there would only be one outcome rather than two *independent* outcomes, so the variance would have been

$$Var(2X) = 2^2 Var(X) = 4 \times 149\,600 = 598\,400, \text{ or}$$

twice as big as with two independent policies.

Of course, variances are in squared units. The company would prefer to know standard deviations, which are in dollars. The standard deviation of the payout for two independent policies is $\sqrt{299\,200} = \$546.99$. But the standard deviation of the payout for a single policy of twice the size is $\sqrt{598\,400} = \$773.56$, or about 40% more.

So, if the company has only two customers, it will have an expected annual total payout of \$40 with a standard deviation of about \$547.

For Example ADDING THE DISCOUNTS

RECAP: The Valentine's Day Lucky Lovers discount for couples averages \$5.83 with a standard deviation of \$8.62. We've seen that if the restaurant doubles the discount offer for two couples dining together on a single bill, they can expect to save \$11.66 with a standard deviation of \$17.24. Some couples decide instead to get separate bills and pool their two discounts.

QUESTION: You and your amour go to this restaurant with another couple and agree to share any benefit from this promotion. Does it matter whether you pay separately or together?

Let X_1 and X_2 represent the two separate discounts, and T the total; then $T = X_1 + X_2$.

$$E(T) = E(X_1 + X_2) = E(X_1) + E(X_2) = 5.83 + 5.83 = \$11.66,$$

so the expected saving is the same either way.

The cards are reshuffled for each couple's turn, so the discounts couples receive are independent. It's okay to add the variances:

$$Var(T) = Var(X_1 + X_2) = Var(X_1) + Var(X_2) = 8.62^2 + 8.62^2 = 148.6088$$

$$SD(T) = \sqrt{148.6088} = \$12.19$$

When two couples get separate bills, there's less variation in their total discount. The standard deviation is \$12.19, compared to \$17.24 for couples who play for the double discount on a single bill.

PYTHAGOREAN THEOREM OF STATISTICS

We often use the standard deviation to measure variability, but when we add independent random variables, we use their variances. Think of the Pythagorean theorem. In a right triangle (only), the *square* of the length of the hypotenuse is the sum of the *squares* of the lengths of the other two sides:

$$c^2 = a^2 + b^2.$$

For independent random variables (only), the square of the standard deviation of their sum is the sum of the squares of their standard deviations:

$SD^2(X + Y) =$
$SD^2(X) + SD^2(Y).$

It's simpler to write this with *variances:*
For independent random variables, X and Y,
$Var(X + Y) = Var(X) + Var(Y).$

In general,

- *The mean of the sum of two random variables is the sum of the means.*
- *The mean of the difference of two random variables is the difference of the means.*
- *If the random variables are independent, the variance of their sum or difference is the sum of the variances.*

$$E(X \pm Y) = E(X) \pm E(Y)$$
$$Var(X \pm Y) = Var(X) + Var(Y) \text{ when the variables are independent}$$

Wait a minute! Is that third part correct? Do we always *add* variances? Yes. Think about the two insurance policies. Suppose we want to know the mean and standard deviation of the *difference* in payouts to the two clients. Since each policy has an expected payout of \$20, the expected difference is 20 − 20 = \$0. If we also subtract variances, we get \$0 too, and that surely doesn't make sense. Note that if the outcomes for the two clients are independent, the difference in payouts could range from \$10 000 − \$0 = \$10 000 to \$0 − \$10 000 = −\$10 000, a spread of \$20 000. The variability in differences increases as much as the variability in sums. If the company has two customers, the difference in payouts has a mean of \$0 and a standard deviation of about \$547 (again).

For Example WORKING WITH DIFFERENCES

RECAP: The Lucky Lovers discount at the Quiet Nook averages \$5.83 with a standard deviation of \$8.62. Just up the street, the Wise Fool restaurant has a competing Lottery of Love promotion. There a couple can select a specially prepared chocolate from a large bowl and unwrap it to learn the size of their discount. The restaurant's manager says the discounts vary with an average of \$10.00 and a standard deviation of \$15.00.

QUESTION: How much more can you expect to save at the Wise Fool? With what standard deviation?

ANSWER: Let W = discount at the Wise Fool, X = the discount at the Quiet Nook, and D = the difference: $D = W - X$. These are different promotions at separate restaurants, so the outcomes are independent.

$$E(W - X) = E(W) - E(X) = 10.00 - 5.83 = \$4.17$$
$$\begin{aligned} SD(W - X) &= \sqrt{Var(W - X)} \\ &= \sqrt{Var(W) + Var(X)} \\ &= \sqrt{15^2 + 8.62^2} \\ &\approx \$17.30 \end{aligned}$$

Discounts at the Wise Fool will average \$4.17 more than at the Quiet Nook, with a standard deviation for the difference of \$17.30.

For random variables, does $X + X + X = 3X$? Be careful. Insuring one person for \$30 000 is not the same risk as insuring three people for \$10 000 each (as we saw earlier in the case of two people). When each instance represents a different outcome for the same random variable, it's easy to fall into the trap of writing all of them with the same symbol. Don't make this common mistake. Make sure you write each instance as a *different* random variable. Just because each random variable describes a similar situation doesn't mean that each random outcome will be the same.

These are *random* variables, not the variables you saw in Algebra. Being random, they take on different values each time they're evaluated. So what you really mean is $X_1 + X_2 + X_3$. Written this way, it's clear that the sum shouldn't necessarily equal three times *anything*.

For Example SUMMING A SERIES OF RANDOM OUTCOMES

RECAP: The Quiet Nook's Lucky Lovers promotion offers couples discounts averaging $5.83 with a standard deviation of $8.62. The restaurant owner is planning on serving 40 couples on Valentine's Day.

QUESTION: What's the expected total of the discounts the owner will give? With what standard deviation?

ANSWER: Let $X_1, X_2, X_3, \ldots X_{40}$ represent the discounts to the 40 couples, and T the total of all the discounts. Then:

$$\begin{aligned} T &= X_1 + X_2 + X_3 + \ldots + X_{40} \\ E(T) &= E(X_1 + X_2 + X_3 + \ldots + X_{40}) \\ &= E(X_1) + E(X_2) + E(X_3) + \ldots + E(X_{40}) \\ &= 5.83 + 5.83 + 5.83 + \ldots + 5.83 \\ &= \$233.20 \end{aligned}$$

Reshuffling cards between couples makes the discounts independent, so:

$$\begin{aligned} SD(T) &= \sqrt{Var(X_1 + X_2 + X_3 + \ldots + X_{40})} \\ &= \sqrt{Var(X_1) + Var(X_2) + Var(X_3) + \ldots + Var(X_{40})} \\ &= \sqrt{8.62^2 + 8.62^2 + 8.62^2 + \ldots + 8.62^2} \\ &\approx 54.52 \end{aligned}$$

The restaurant owner can expect the 40 couples to win discounts totalling $233.20, with a standard deviation of $54.52.

Just Checking

2. Suppose the time it takes a customer to get and pay for seats at the ticket window of a baseball park is a random variable with a mean of 100 seconds and a standard deviation of 50 seconds. When you get there, you find only two people in line in front of you.
   a) How long is your expected wait for your turn to get tickets?
   b) What's the standard deviation of your wait time?
   c) What assumption did you make about the two customers in finding the standard deviation?

What about the Actual Probability Model?

Although we know how to find the expected value and the variance for the sum of two independent random variables, that's not all we'd like to know. It would be nice if we could go directly from models of each random variable to a model for their sum. But the probability model for the sum of two random variables is *not* necessarily the same as the model we started with *even when the variables are independent*.

That's easy to see for discrete random variables because we can just think about their possible values. When we looked at insuring one person, either we paid out \$0, \$5000, or \$10 000. Suppose we insure three people. The possibilities are not just \$0, \$15 000, or \$30 000. We might have to pay out in total \$5000, or \$20 000, or other possible combinations of the three outcomes. Even though the expected values add, the probability model is different. The probability model for $3X$ is not the same as the model for $X_1 + X_2 + X_3$. That's another reason to use the subscripted form rather than write $X + X + X$.

Fortunately, there are some cases where we readily know the model for a sum or difference, such as for Normal random variables (Section 14.6). As well as some shocking (and very useful) news about what happens to the model when you add a large number of random variables together (Central Limit Theorem in Chapter 15).

*Correlation and Covariance

In Chapter 6, we saw that the association of two variables could be measured with their correlation. What about random variables? We can talk about the correlation between random variables, too. But it's easier to start with a related concept called covariance.

If X is a random variable with expected value $E(X) = \mu$ and Y is a random variable with expected value $E(Y) = v$, then the covariance of X and Y is defined as

$$Cov(X, Y) = E[(X - \mu)(Y - v)]$$

The covariance, like the correlation, measures how X and Y vary together (*co* = *together*). Think about a scatterplot of two highly positively correlated variables. When X is above its mean, Y tends to be above its mean. And when X is below its mean, Y tends to be below its mean. So, on average, the product in the covariance formula is positive.

Covariance has a few properties worth remembering:

- $Cov(X, Y) = Cov(Y, X)$
- $Cov(X, X) = Var(X)$
- $Cov(cX, dY) = cdCov(X, Y)$, for any constants c and d
- If X and Y are independent, then $Cov(X, Y) = 0$, but be careful not to assume that the converse is true; it is possible for two variables to have zero covariance yet not be independent.

The covariance gives us the extra information we need to find the variance of the sum or difference of two random variables when they are not independent:

$$Var(X \pm Y) = Var(X) + Var(Y) \pm 2Cov(X, Y)$$

When X and Y are independent, their covariance is zero, so we have the Pythagorean Theorem of statistics, as we saw earlier.[2]

Covariance can take any number between $-\infty$ and $+\infty$. If X and Y have large values, the covariance will be large as well. And that's its problem. If we know the covariance between X and Y is –64.5, it doesn't give us a real sense of how negatively related they are. To fix the "problem," we can divide the covariance by each of the standard deviations to get the **correlation**:

$$Corr(X, Y) = \frac{Cov(X, Y)}{\sigma_X \sigma_y}$$

Correlation of random variables is usually denoted as ρ. Correlation of random variables has many of the properties we saw in Chapter 6 for the sample correlation coefficient, r.

- The sign of a correlation coefficient gives the direction of the association between random variables.

[2]If you want to follow the geometry one step further, the more general form for correlated variables is related to the Law of Cosines.

- Correlation is always between -1 and $+1$.
- Correlation treats X and Y symmetrically. The correlation of X with Y is the same as the correlation of Y with X.
- Unlike covariance, correlation has no units.
- Correlation is not affected by changes in the centre or scale of either variable.

14.4 The Binomial Model

ALEXANDER NEMENOV/AFP/Getty Images/ Newscom

Suppose a cereal manufacturer puts pictures of famous athletes on cards in boxes of cereal, in the hope of increasing sales. The manufacturer announces that 20% of the boxes contain a picture of Sidney Crosby, 30% a picture of Christine Sinclair, and the rest a picture of Clara Hughes.

Sound familiar? In Chapter 9, we simulated finding the number of boxes we'd need to open to get one of each card. This isn't an easy problem to solve mathematically with probability calculations, but fortunately the most common and important questions in situations like this can be answered using the Binomial Model.

You're a huge Sidney Crosby fan. You don't care about completing the whole sports card collection, but you really hope to find some Sidney Crosby pictures. If you open, say, five boxes, how likely is it that you will find one Crosby card? Two Crosby cards? No Crosby cards (aw shucks . . .)? We'll see that the Binomial Model is just what we need to answer these questions. But first we need to define Bernoulli trials (named after Jacob Bernoulli, whom you can recall from chapter 12), which are the building blocks for this model.

We'll keep the assumption that pictures are distributed at random and we'll trust the manufacturer's claim that 20% of the cards are Crosby. So when you open a box, the probability that you succeed in finding Crosby is 0.20. Now we'll call the act of opening *each* box a trial, and note that:

- There are only two possible outcomes (called *success* and *failure*) on each trial. Either you get Crosby's picture (success), or you don't (failure).
- The probability of success, denoted p, is the same on every trial. Here $p = 0.20$.
- The trials are independent. Finding Crosby in the first box does not change what might happen when you reach for the next box.

Situations like this occur often, and are called **Bernoulli trials**. Common examples of Bernoulli trials include tossing a coin, looking for defective products rolling off an assembly line, or even shooting free throws in a basketball game. Just as we found equally likely random digits to be the building blocks for our simulation, we can use Bernoulli trials to build a wide variety of useful probability models.

The Critical Assumption of Independence

One of the important requirements for Bernoulli trials is that the trials be independent. Sometimes that's a reasonable assumption. Is it true for our example? We said that whether we find a Sidney Crosby card in one box has no effect on the probabilities in other boxes. This is *almost* true. Technically, if exactly 20% of the boxes have Sidney Crosby cards, then when you find one, you've reduced the number of remaining Sidney Crosby cards. With a few million boxes of cereal, though, the difference is hardly worth mentioning.

A S *Activity:* **Bernoulli Trials.** Guess what! We've been generating Bernoulli trials all along. Look at the Random Simulation Tool in a new way.

NOTATION ALERT

Now we have two more reserved letters. Whenever we deal with Bernoulli trials, p represents the probability of success, and q the probability of failure. (Of course, $q = 1 - p$)

But if you knew there were two Sidney Crosby cards hiding in the 10 boxes of cereal on the grocery store shelf, then finding one in the first box you try would clearly change your chances of finding Crosby in the next box.

If we had an infinite number of boxes, there wouldn't be a problem. It's selecting from a finite population that causes the probabilities to change, making the trials not independent. Obviously, taking two out of 10 boxes changes the probability. Taking even a few hundred out of millions, though, makes very little difference. Fortunately, we have a rule of thumb for the in-between cases. It turns out that if we randomly sample less than 10% of the population, we can pretend that the trials are independent and still calculate probabilities that are sufficiently accurate.

The 10% Condition: Bernoulli trials must be independent. When you randomly select more than 10% of a population, the independence assumption fails and you cannot use the Binomial model.[3]

Dependence among trials can arise in other ways as well, such as if there is a learning or practice effect from trial to trial (in say a memory training experiment). So, also check that there is little "carry-over" effect from one trial to the next.

The Binomial Model

Suppose you buy five boxes of cereal. What is the probability you get *exactly* two pictures of Sidney Crosby? We want to find the probability of getting two successes among five Bernoulli trials.

A S *Activity:* **The Binomial Distribution.** It's more interesting to combine Bernoulli trials. Simulate this with the Random Tool to get a sense of how Binomial models behave.

We're interested in the *number of successes* in the five trials, so we'll call it X = number of successes. We want to find $P(X = 2)$. This is an example of a **Binomial probability**. It takes two parameters to define this **Binomial model**: the number of trials, n, and the probability of success, p. We denote this model Binom(n, p). Here, $n = 5$ trials, and $p = 0.2$, the probability of finding a Sidney Crosby card in any trial.

Exactly two successes in five trials means two successes and three failures. It seems logical that the probability should be $(0.2)^2(0.8)^3$. Too bad! It's not that easy. That calculation would give you the probability of finding Crosby in the first two boxes and not in the next three—*in that order*. But you could find Crosby in the third and fifth boxes and still have two successes. The probability of those outcomes in that particular order is (0.8)(0.8) (0.2)(0.8)(0.2). That's also $(0.2)^2(0.8)^3$. In fact, the probability will always be the same, no matter what order the successes and failures occur in. Anytime we get two successes in five trials, no matter what the order, the probability will be $(0.2)^2(0.8)^3$. We just need to take into account all of the possible orders in which the outcomes can occur.

Fortunately, these possible orders are *disjoint*. (For example, if your two successes came on the first two trials, they couldn't come on the last two.) So we can use the addition rule and add up the probabilities for all the possible orderings. Since the probabilities are all the same, we only need to know how many orders are possible. For small numbers, we can just make a tree diagram and count the branches. For larger numbers this isn't practical, so we let the computer or calculator do the work.

Each different order in which we can have k successes in n trials may be called a *combination*. The total number of ways that can happen is written $\binom{n}{k}$ or ${}_nC_k$ and pronounced "n choose k." We evaluate this with the following formula:

$${}_nC_k = \frac{n!}{k!(n-k)!} \text{ where } n! = n \times (n-1) \times \cdots \times 1$$

For two successes in five trials,

$${}_5C_2 = \frac{5!}{2!(5-2)!} = \frac{5 \times 4 \times 3 \times 2 \times 1}{2 \times 1 \times 3 \times 2 \times 1} = \frac{5 \times 4}{2 \times 1} = 10.$$

So there are 10 ways to get two Crosby pictures in five boxes, and the probability of each is $(0.2)^2(0.8)^3$. Now we can find what we wanted:

$$P(X = 2, \text{ or exactly 2 successes in 5 trials}) = 10(0.2)^2(0.8)^3 = 0.2048$$

In general, the probability of exactly k successes in n trials is ${}_nC_k p^k q^{n-k}$ since $p^k q^{n-k}$ is the probability of any one sequence of k successes (and hence $n-k$ failures), while ${}_nC_k$ is the number of such sequences.

[3]In this case, we would need to use the Hypergeometric model, which is even more complicated than the Binomial.

Using this formula, we could find the expected value by adding up $kP(X = k)$ for all values, but it would be a long, hard way to get an answer that you already know intuitively. What's the expected value? If we have five boxes and Crosby's picture is in 20% of boxes, then we would expect on average to have $5(0.2) = 1$ success. If we had 100 trials with probability of success 0.2, how many successes would you expect? Can you think of any reason not to say 20? It seems so simple that most people wouldn't even stop to think about it. Multiply the probability of success by n. In other words, $E(X) = np$.[4]

The standard deviation is less obvious; you can't just rely on your intuition. Fortunately, the formula for the standard deviation also boils down to something simple: $SD(X) = \sqrt{npq}$. In 100 boxes of cereal, we expect to find 20 Sidney Crosby cards, with a standard deviation of $\sqrt{100 \times 0.8 \times 0.2} = 4$ pictures.

Time to summarize. A Binomial probability model describes the probability distribution for the number of successes in a specified number of trials. It takes two parameters to specify this model: the number of trials n and the probability of success p.

Binomial probability model for bernoulli trials: Binom(n, p)

n = number of trials

p = probability of success (and $q = 1 - p$ = probability of failure)

X = number of successes in n trials

$$P(X = x) = {}_nC_x p^x q^{n-x}, \text{ where } {}_nC_x = \frac{n!}{x!(n-x)!}$$

Mean: Expected number of successes in n trials $= \mu = np$

Standard Deviation: $\sigma = \sqrt{npq}$

Binom(n, p) will be our shorthand for saying "Binomial Model with parameters n and p." But be sure to plug in the actual values for n and p that match your actual study. The assumptions and conditions you need to check are those for Bernoulli trials.

MATH BOX

To derive the formulas for the mean and standard deviation of a Binomial model, we start with the most basic situation.

Consider a single Bernoulli trial with probability of success p. Let's find the mean and variance of the number of successes.

Here's the probability model for the number of successes:

x	0	1
$P(X = x)$	q	p

Find the expected value:
$$E(X) = 0q + 1p$$
$$E(X) = p$$

And now the variance:
$$\begin{aligned} Var(X) &= (0 - p)^2 q + (1 - p)^2 p \\ &= p^2 q + q^2 p \\ &= pq(p + q) \\ &= pq(1) \\ Var(X) &= pq \end{aligned}$$

[4]It's an amazing fact that this does work out the long way. That is, $\sum_{x=0}^{n} x \cdot {}_nC_x p^x q^{n-x} = np$, something that is not obvious. In the Math Box, we work it out another way.

What happens when there is more than one trial, though? A Binomial model simply counts the number of successes in a series of n independent Bernoulli trials. That makes it easy to find the mean and standard deviation of a Binomial random variable, Y.

Let $Y = X_1 + X_2 + X_3 + \cdots X_n$ where $X_i = 1$ if we have a success in the ith trial, and $= 0$ if we have a failure. Then

$$\begin{aligned} E(Y) &= E(X_1 + X_2 + X_3 + \cdots + X_n) \\ &= E(X_1) + E(X_2) + E(X_3) + \cdots + E(X_n) \\ &= p + p + p + \cdots + p \text{ (There are } n \text{ terms.)} \end{aligned}$$

So, as we thought, the mean is $E(Y) = np$.

And since the trials are independent, the variances add:

$$\begin{aligned} Var(Y) &= Var(X_1 + X_2 + X_3 + \cdots + X_n) \\ &= Var(X_1) + Var(X_2) + Var(X_3) + \cdots + Var(X_n) \\ &= pq + pq + pq + \cdots + pq \text{ (Again, there are } n \text{ terms.)} \\ Var(Y) &= npq \end{aligned}$$

Voilà! The standard deviation is $SD(Y) = \sqrt{npq}$

For Example SPAM AND THE BINOMIAL MODEL

RECAP: The communications monitoring company Postini has reported that 91% of e-mail messages are spam. Your inbox contains 25 messages.

QUESTIONS: What are the mean and standard deviation of the number of real messages you'll find in your inbox? What is the probability that you'll find only one or two real messages?

ANSWER: Since spam comes from many different sources, I will assume that messages arrive independently and at random, with the probability of success (a real message) $p = 1 - 0.91 = 0.09$. Let $X =$ the number of real messages among 25. I can use the model Binom(25, 0.09).

$$E(X) = np = 25(0.09) = 2.25$$

$$SD(X) = \sqrt{npq} = \sqrt{25(0.09)(0.91)} = 1.43$$

$$\begin{aligned} P(X = 1 \text{ or } 2) &= P(X = 1) + P(X = 2) \\ &= {}_{25}C_1\,(0.09)^1(0.91)^{24} + {}_{25}C_2(0.09)^2(0.91)^{23} \\ &= 0.2340 + 0.2777 \\ &= 0.5117 \end{aligned}$$

Among 25 e-mail messages, I expect an average of 2.25 that aren't spam, with a standard deviation of 1.43 messages. There's just over a 50% chance that one or two of my 25 e-mails will be real messages.

Step-by-Step Example WORKING WITH A BINOMIAL MODEL

Canadian Blood Services is a not-for-profit organization that collects nearly one million units of blood per year. About 6% of people have O-negative blood. They are called *universal donors* because O-negative blood can be given to anyone, regardless of the recipient's blood type. Suppose that 20 blood donors arrive at one of this organization's donation centres.

Questions:

1. What are the mean and standard deviation of the number of universal donors among them?
2. What is the probability that there are two or three universal donors?

THINK ➡ **Plan** State the questions.	I want to know the mean and standard deviation of the number of universal donors among 20 people, and the probability that there are two or three of them.
Check to see that these are Bernoulli trials.	✓ There are two outcomes: success = O-negative failure = other blood types ✓ $p = 0.06$ because volunteers tend to show up more or less at random (at least with respect to blood type). ✓ **10% Condition:** Trials are not independent because the population is finite, but fewer than 10% of all possible donors are lined up.
Variable Define the random variable.	Let X = number of O-negative donors among $n = 20$ people.
Model Specify the model.	I can model X with Binom(20, 0.06).
SHOW ➡ **Mechanics** Find the expected value and standard deviation.	$E(X) = np = 20(0.06) = 1.2$ $SD(X) = \sqrt{npq} = \sqrt{20(0.06)(0.94)} \approx 1.06$ $P(X = 2 \text{ or } 3) = P(X = 2) + P(X = 3)$ $= {}_{20}C_2(0.06)^2(0.94)^{18}$ $+ {}_{20}C_3(0.06)^3(0.94)^{17}$ $\approx 0.2246 + 0.0860$ $= 0.3106$
TELL ➡ **Conclusion** Interpret your results in context.	In groups of 20 randomly selected blood donors, I expect to find on average 1.2 universal donors, with a standard deviation of 1.06. About 31% of the time, I'd find two or three universal donors among the 20 people.

*14.5 The Poisson Model

In the early 1990s, a leukemia cluster was identified in the Massachusetts town of Woburn. Many more cases of leukemia, a cancer that originates in a cell in the marrow of bone, appeared in this small town than would be predicted. Was it evidence of a problem in the town, or was it chance? That question led to a famous trial in which the families of eight leukemia victims sued, and became grist for the book and movie *A Civil Action*.

When rare events occur together or in clusters, people often want to know if that happened just by chance or whether something else is going on. If we assume that the events occur independently, we can use a Binomial model to find the probability that a cluster of events like this might occur. For rare events, p will be quite small, and when n is large, it may be difficult to compute the exact probability that a certain size cluster might occur.

To see why, let's try to compute the probability that a cluster of cases of size x occurs in Woburn. We'll use the U.S. *national average* of leukemia incidence to get a value for p, and the population of Woburn as our value for n. In the United States in the early 1990s, there were about 30 800 new cases of leukemia each year and about 280 000 000 people,

NOTATION ALERT

We almost always use λ (the Greek letter lambda) for the mean of a Poisson model.

giving a value for p of about 0.00011. The population of Woburn was about $n = 35\ 000$. We'd expect $np = 3.85$ new cases of leukemia in Woburn. How unlikely would eight or more new cases be? To answer that, we'll need to calculate the complement, adding the probabilities of no cases, exactly one case, etc., up to seven cases. Each of those probabilities would have the form ${}_{35000}C_x p^x q^{35000-x}$. To find ${}_{35000}C_x$, we'll need 35 000!—a daunting number even for computers. In the eighteenth century, when people first were interested in computing such probabilities, it was impossible. We can easily go beyond the capabilities of even today's computers by making n large enough and p small enough.

Simeon Denis Poisson[5] was a French mathematician interested in events with very small probability. He originally derived his model to approximate the Binomial model when the probability of a success, p, is very small and the number of trials, n, is very large. Poisson's contribution was providing a simple approximation to find that probability. When you see the formula, however, you won't necessarily see the connection to the Binomial. The Poisson's parameter is often denoted by λ, the mean of the distribution. To use the **Poisson model** to approximate the Binomial, we'll make their means match, so we set $\lambda = np$.

Trinity Mirror/Mirrorpix/Alamy

Poisson probability model for successes: Poisson (λ)

λ = mean number of successes

X = number of successes

$$P(X = x) = \frac{e^{-\lambda}\lambda^x}{x!} \qquad \text{(recall that } e = 2.71828\ldots\text{)}$$

Expected value: $E(X) = \lambda$

Standard deviation: $SD(X) = \sqrt{\lambda}$

The Poisson model is a reasonably good approximation of the Binomial when $n \geq 20$ and $p \leq 0.05$ with $np \leq 10$.

Using Poisson's model, we can easily find the probabilities of a given size cluster of leukemia cases. Using $\lambda = np = 35\ 000 \times 0.00011 = 3.85$ new cases a year, we find that the probability of seeing exactly x cases in a year is $P(X = x) = \frac{e^{-3.85}3.85^x}{x!}$. By adding up these probabilities for $x = 0, 1, \ldots 7$, we'd find that the probability of eight or more cases with $\lambda = 3.85$ is about 0.043. That's small but not terribly unusual.

In spite of its origins as an approximation to the Binomial, the Poisson model is also used directly to model the probability of the occurrence of events for a variety of phenomena. It's a good model to consider whenever your data consist of counts of occurrences (over time, or space). It requires only that the events be independent and that the mean number of occurrences stays constant for the duration of the data collection (and beyond, if we hope to make predictions).

One nice feature of the Poisson model is that it scales according to the sample size.[6] For example, if we know that the average number of occurrences in a town the size of Woburn, with 35 000 people, is 3.85, we know that the average number of occurrences in a town of only 3500 residents is 0.385. We can use that new value of λ to calculate the probabilities for the smaller town.

The Poisson distribution, with some help from Professor Jeff Rosenthal of the University of Toronto (or maybe it was the other way around), determined that the 200 major Lotto 6/49 prizes won by retail lottery sellers during the period of 1999–2006 had a probability of 0.00000000013 of occurring just by chance (as reported on CBC's *The Fifth Estate*). Clearly, so many prizes were not won purely by luck, and fraud investigations—and reforms—followed. Probability or statistical analysis can often uncover serious problems,

POISSON IN HISTORY

In Thomas Pynchon's book *Gravity's Rainbow,* Roger Mexico is a statistician using the Poisson model for the bombings over London during World War II. "The rockets *are* distributing about London just as Poisson's equation in the textbook predicts. As the data keep coming in, Roger looks more and more like a prophet . . . all I'm doing is plugging numbers into a well-known equation, you can look it up in the book and do it yourself. . . ."

[5] *Poisson* is a French name (meaning "fish"), properly pronounced "pwa sohn" and not with an "oy" sound as in poison.

[6] Because *Poisson* means "fish" in French, you can remember this fact by the bilingual pun "*Poisson* scales!"

though the underlying explanation of a rare phenomenon may be difficult or impossible to discover (for another example, see the butterfly ballot box example in Chapter 8).

One of the properties of the Poisson model is that, as long as the mean rate of occurrences stays constant, the occurrence of past events doesn't change the probability of future events. That is, even though events that occur according to the Poisson model may *appear* to cluster, the probability of *another* event occurring is still the same (just as with the Binomial model). This is counterintuitive and may be one reason why so many people believe in the form of the "Law of Averages" that we argued against in Chapter 12. When the Poisson model was used to model bombs dropping over London in World War II, the model said that even though several bombs had hit a particular sector, the probability of *another* hit was still the same. So there was no point in moving people from one sector to another, even after several hits. You can imagine how difficult it must have been trying to convince people of that! This same phenomenon leads many people to think they see patterns in which an athlete, a slot machine, or a financial market gets "hot." But the occurrence of several successes in a row is not unusual and does not constitute evidence that the mean has changed. Careful studies have been made of "hot streaks," but none has ever convincingly demonstrated that they actually exist.

Just Checking

The Pew Research Center (www.pewresearch.org) reports that they are actually able to contact only 76% of the randomly selected households drawn for a telephone survey.

3. Explain why these phone calls can be considered Bernoulli trials.

4. How would you model the number of successful contacts from a list of 1000 sampled households? Explain.

5. Suppose that in the course of a year, legitimate survey organizations (not folks pretending to take a survey but actually trying to sell you something or change your vote) sample 70 000 of the approximately 107 000 000 households in the United States. You wonder if people from your school are represented in these polls. Counting all the living alumni from your school, there are 10 000 households. How would you model the number of these alumni households you'd expect will be contacted next year by a legitimate poll?

14.6 Continuous Models

Up to now, we have dealt with discrete random variable, variables whose outcomes can be listed. We then assign probabilities one at a time to each outcome. But suppose instead the random variable is to represent the survival time of a mouse given some treatment for a deadly disease. We might measure it in weeks, or more accurately in days, or perhaps in hours, or minutes, or seconds. Unless we measure it very coarsely, there are a huge number of possible outcomes. Instead, imagine the actual outcome as any real number from zero on up (perhaps zero to infinity, since we can always assign negligibly small probabilities beyond some point), i.e., a portion of the real line, and choose any unit you like, say hours. Now, what is the probability the mouse survives exactly 40 hours? Exactly 40 hours means 40.0000000 . . . hours—with zeros out to infinity. How about making that a zero probability? Seems reasonable enough, when you think of this outcome as one infinitesimally small point on the real line. The essential probabilities that we will need are the probabilities of intervals, such as the probability of surviving at most 40 hours (0 to 40). The probability of dieing after 40 hours, when measuring in hours or rounding to the nearest hour, is also just an interval: from 39.5 to

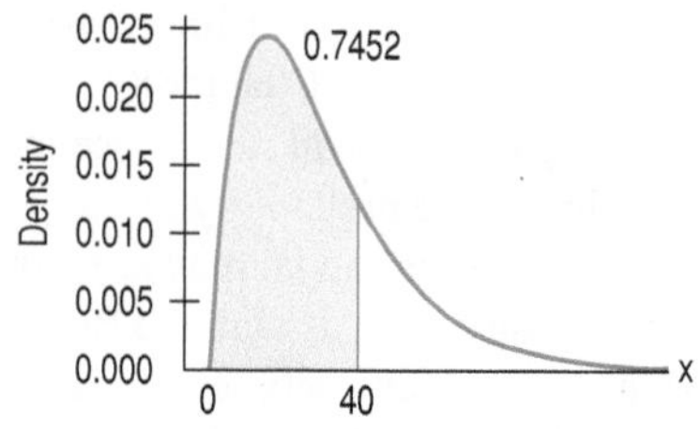

Figure 14.1
Hypothetical density curve with $P(0 < X < 40)$ = area under curve between 0 and 40 = 0.7452.

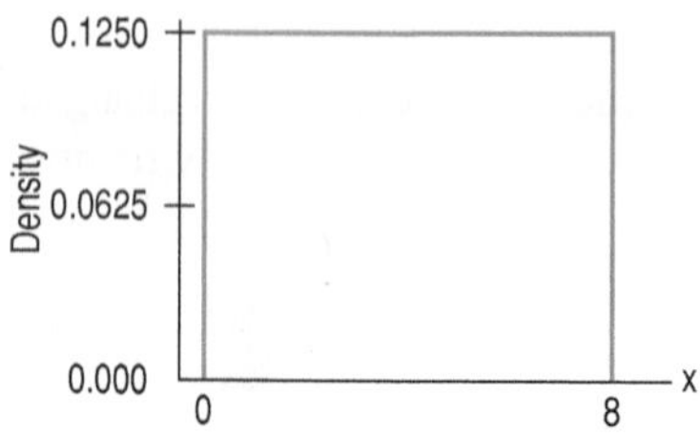

Figure 14.2
Uniform model over the interval (0, 8).

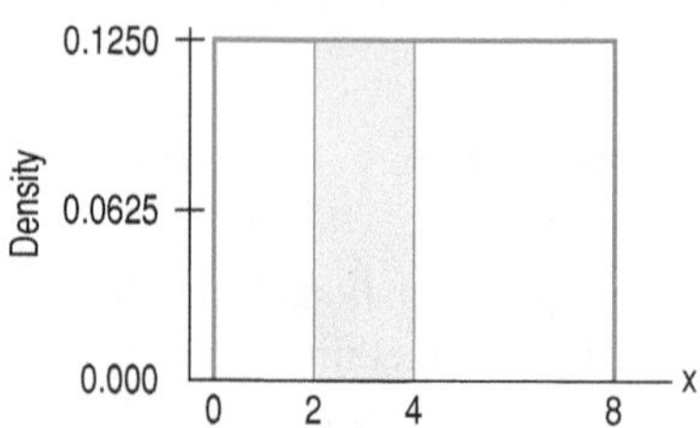

Figure 14.3
Find $P(2 < X < 4)$ by using the correct area under the curve.

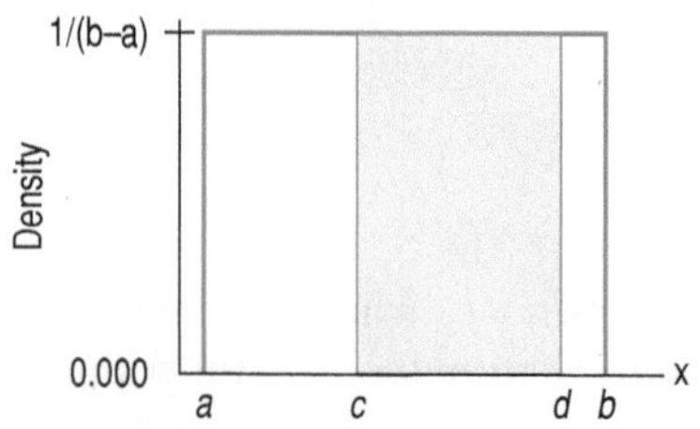

Figure 14.4
Uniform model on (a, b) showing $P(c < X < d)$.

40.5. To assign probabilities to intervals rather than to individual outcomes, we need to use a *density curve*. Does that sound familiar?

Recall the most famous density curve, the Normal curve, discussed in Chapter 5. There, we used it to model relative frequencies (or proportions) of numerical outcomes. Well, probabilities are just relative frequencies, in the long run. So, no surprise, we can also use density curves to model probabilities. The area under the curve will now represent probability instead of relative frequency (see Figure 14.1). A density curve used as a probability model for a continuous random variable is often referred to as a *probability density function*. Finding areas under a curve in general requires the use of calculus (integration), but for important models, tables or software should be available.

Some continuous random variables have Normal density curve models; others may need to be modelled with skewed or, uniform, or other-shaped curves. Regardless of shape, all continuous random variables have means (which we also call *expected values*) and variances. In this book we won't worry about how to calculate them, but we can still work with models for continuous random variables when we're given these parameters. Let's start with one simple example of a continuous model, the Uniform model, and then we'll revisit the Normal model.

The Uniform Model

One of the authors often takes the Bloor subway line westbound from the Bay Street station in Toronto when going to the airport. His wait time for the next train is a random variable. Subway trains tend to run at regular intervals; let's say it is off-peak hours and the trains are only running every 8 minutes. We might model the wait time for the next subway with a uniform model on the interval 0 to 8 minutes, since any wait time is as likely as any other wait time, given my rather random arrival at the station. We'll ignore the small probability that the wait might be longer than 8 minutes (models only have to be useful, not perfect). In other words, the density curve should not clump up anywhere, and should look like Figure 14.2.

What is the height of the curve? Recall that *total probability must equal 1*. So the total area under the curve must equal one. Since the area of a rectangle is the height multiplied by the width, the height of this flat curve must equal $1/8$. The probability that the wait will be between 2 and 4 minutes is the area under the curve between these two numbers (Figure 14.3), which is easily seen to be $(4 - 2) \times 1/8$ or 0.25.

In general, we say that X has a uniform distribution (or follows a Uniform model) on the interval (a, b) if the density curve is flat above the interval (a, b) as shown in Figure 14.4.

Clearly $P(c < X < d) = \frac{d - c}{b - a}$ by working out the area of the proper rectangle. Even without calculus, you should be able to guess that the mean of this distribution, or $E(X)$, must equal $(a + b)/2$. The variance is $(b - a)^2/12$, but this requires calculus to confirm.[7]

If only all continuous models were so easy to work with! But flat density curve models are rare. Most are not flat, but clump up somewhere, with either a symmetrical spread around the centre or some skewness. There are some important skewed density curves that we will discuss much later in this text (like the chi-squared and F distributions), but for now, the key model we need to understand and work with is the Normal model.

The Normal Probability Model

In Chapter 5, we introduced the Normal curve and learned how to calculate areas under the curve, as well as how to find percentile points. Recall Fig 14.5.

Shifting or scaling a Normal random variable—that is, adding a constant to X or multiplying X by a constant—produces another Normal random variable, but with a new mean

[7]For those of you who have studied some calculus, the formulas for the mean and variance are μ_x or $E(X) = \int xf(x)dx$ and σ_x^2 or $Var(X) = \int (x - \mu_x)^2 f(x)dx$, where $f(x)$ is the density curve for X.

Figure 14.5
The Normal Model

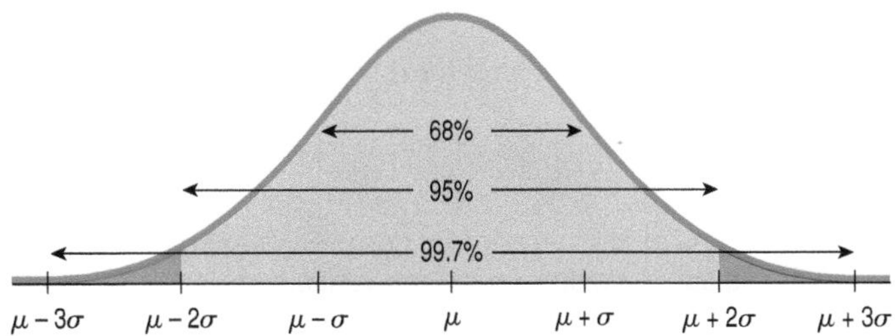

A S ***Activity:* Numeric Outcomes.** You've seen how to simulate discrete random outcomes. There's a tool for simulating continuous outcomes, too.

A S ***Activity:* Means of Random Variables.** Experiment with continuous random variables to learn how their expected values behave.

and standard deviation (whereas similar statements are not true for Binomial or Poisson random variables).

Here's an example. A company manufactures small stereo systems. At the end of the production line, the stereos are packaged and prepared for shipping. Stage 1 of this process is called "packing." Workers must collect all the system components (a main unit, two speakers, a power cord, an antenna, and some wires), put each in plastic bags, and then place everything inside a protective Styrofoam form. The packed form then moves on to Stage 2, called "boxing." There, workers place the form and a packet of instructions in a cardboard box, close it, then seal and label the box for shipping.

The company says that times required for the packing stage can be described by a Normal model with a mean of 9 minutes and standard deviation of 1.5 minutes. The times for the boxing stage can also be modelled as Normal, with a mean of 6 minutes and standard deviation of 1 minute.

We know that when we add two random variables together, the sum generally has a different probability model than the one we started with. Amazingly, independent Normal random variables behave better.

The probability model for the sum of any number of independent Normal random variables is also Normal.

And we already know what the mean and variance of that sum will be because those both add. Using these facts, let's continue the example above:

Step-by-Step Example PACKAGING STEREOS

©PhotoDisc

Consider the company that manufactures and ships small stereo systems that we just discussed.

Recall that times required to pack the stereos can be described by a Normal model with a mean of 9 minutes and standard deviation of 1.5 minutes. The times for the boxing stage can also be modelled as Normal, with a mean of 6 minutes and standard deviation of 1 minute.

Questions:

1. What is the probability that packing two consecutive systems takes over 20 minutes?
2. What percentage of the stereo systems take longer to pack than to box?

Question 1: What is the probability that packing two consecutive systems takes over 20 minutes?

THINK ➡ **Plan** State the problem.	I want to estimate the probability that packing two consecutive systems takes over 20 minutes.
Variables Define your random variables.	Let P_1 = time for packing the first system P_2 = time for packing the second T = total time to pack two systems
Write an appropriate equation.	$T = P_1 + P_2$
Think about the assumptions. Sums of independent Normal random variables follow a Normal model. Such simplicity isn't true in general.	✓ **Normal Model Assumption:** We are told that both random variables follow Normal models. ✓ **Independence Assumption:** We can reasonably assume that the two packing times are independent.

SHOW ➡ Mechanics Find the expected value.

$$E(T) = E(P_1 + P_2)$$
$$= E(P_1) + E(P_2)$$
$$= 9 + 9 = 18 \text{ minutes}$$

For sums of independent random variables, variances add. (We don't need the variables to be Normal for this to be true—just independent.)

Since the times are independent,

$$Var(T) = Var(P_1 + P_2)$$
$$= Var(P_1) + Var(P_2)$$
$$= 1.5^2 + 1.5^2$$
$$Var(T) = 4.50$$

Find the standard deviation.

$$SD(T) = \sqrt{4.50} \approx 2.12 \text{ minutes}$$

Now we use the fact that both random variables follow Normal models to say that their sum is also Normal.

I'll model *T* with $N(18, 2.12)$.

Sketch a picture of the Normal model for the total time, shading the region representing over 20 minutes.

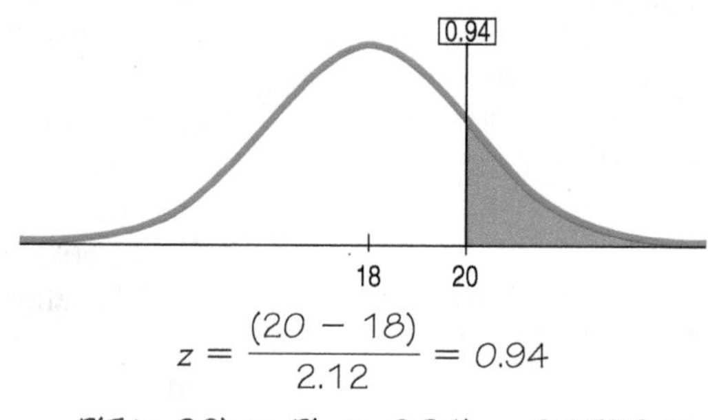

Find the *z*-score for 20 minutes.

$$z = \frac{(20 - 18)}{2.12} = 0.94$$

Use technology or Table Z to find the probability.

$$P(T > 20) = P(z > 0.94) = 0.1736$$

TELL ➡ Conclusion Interpret your results in context.

There's a little more than a 17% chance that it will take a total of over 20 minutes to pack two consecutive stereo systems.

Question 2: What percentage of the stereo systems take longer to pack than to box?

THINK ➡ Plan State the question.

I want to estimate the percentage of the stereo systems that take longer to pack than to box.

Variables Define your random variables.

Let P = time for packing a system
B = time for boxing a system
D = difference in times to pack and box a system

Write an appropriate equation.

$D = P - B$

What are we trying to find? Notice that we can tell which of two quantities is greater by subtracting and asking whether the difference is positive or negative.

The probability that it takes longer to pack than to box a system is the probability that the difference $P - B$ is greater than zero.

Don't forget to think about the assumptions.

Is D Normal? A difference is also a sum; D is the sum of the independent Normals P and −B, hence Normal.

✓ **Normal Model Assumption:** We are told that both random variables follow Normal models.

✓ **Independence Assumption:** We can assume that the times it takes to pack and to box a system are independent.

SHOW ➡ Mechanics Find the expected value.

$$E(D) = E(P - B)$$
$$= E(P) - E(B)$$
$$= 9 - 6 = 3 \text{ minutes}$$

For the difference of independent random variables, variances add.	Since the times are independent, $Var(D) = Var(P - B)$ $= Var(P) + Var(B)$ $= 1.5^2 + 1^2$ $Var(D) = 3.25$
Find the standard deviation.	$SD(D) = \sqrt{3.25} \approx 1.80$ minutes
State what model you will use.	I'll model D with $N(3, 1.80)$
Sketch a picture of the Normal model for the difference in times, and shade the region representing a difference greater than zero.	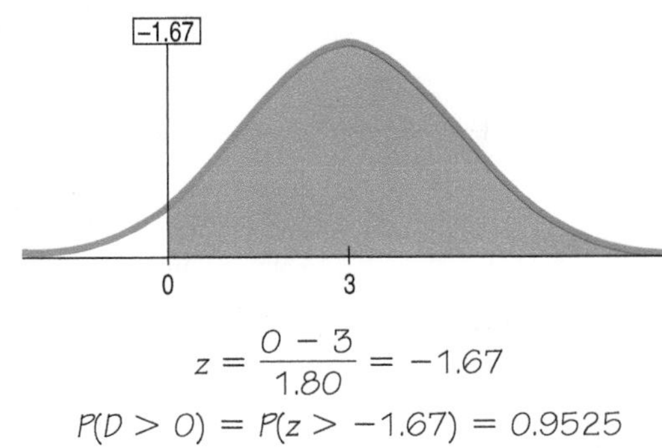
Find the *z*-score for 0 minutes, then use Table *Z* or technology to find the probability.	$z = \frac{0 - 3}{1.80} = -1.67$ $P(D > 0) = P(z > -1.67) = 0.9525$
TELL ➡ **Conclusion** Interpret your result in context.	About 95% of all the stereo systems will require more time for packing than for boxing.

14.7 Approximating the Binomial with a Normal Model

A S *Activity:* **Normal Approximation.** Binomial probabilities can be hard to calculate. With the Simulation Tool you'll see how well the normal model can approximate the Binomial—a much easier method.

Suppose Canadian Blood Services anticipates the need for at least 1850 units of O-negative blood this year. It estimates that it will collect blood from 32 000 donors. How great is the risk that they will fall short of meeting their need? We've just learned how to calculate such probabilities. We can use the Binomial model with $n = 32\,000$ and $p = 0.06$. The probability of getting *exactly* 1850 units of O-negative blood from 32 000 donors is ${}_{32000}C_{1850} \times 0.06^{1850} \times 0.94^{30150}$. No calculator on earth can calculate that first term (it has more than 100 000 digits).[8] And that's just the beginning. The problem said *at least* 1850, so we have to do it again for 1851, for 1852, and all the way up to 32 000. No thanks.

When we're dealing with a large number of trials like this, making direct calculations of the probabilities becomes tedious (or outright impossible). Here the Normal model comes to the rescue.

The Binomial model has a mean of $np = 1920$ and a standard deviation of $\sqrt{npq} \approx 42.48$. We could try approximating its distribution with a Normal model, using the same mean and standard deviation. Remarkably enough, that turns out to be a very good approximation. (We'll see why in the next chapter.) With that approximation, we can find the *probability*:

$$P(X < 1850) = P\left(z < \frac{1850 - 1920}{42.28}\right) \approx P(z < -1.65) \approx 0.05$$

There seems to be about a 5% chance that Canadian Blood Services will run short of O-negative blood.

Can we always use a Normal model to make estimates of Binomial probabilities? No. Consider the Sidney Crosby situation—pictures in 20% of the cereal boxes. If we buy five

[8]If your calculator *can* find Binom(32000,0.06), then it's smart enough to use an approximation. Read on to see how you can, too.

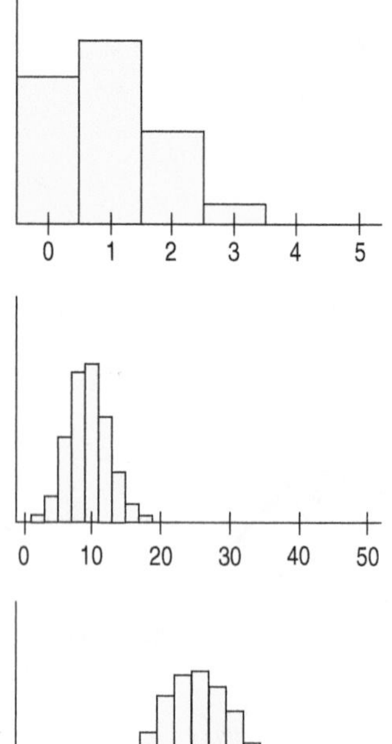

boxes, the actual Binomial probabilities that we get zero, one, two, three, four, or five pictures of Crosby are 33%, 41%, 20%, 5%, 1%, and 0.03%, respectively. The first histogram shows that this probability model is skewed. That makes it clear that we should not try to estimate these probabilities by using a Normal model.

Now suppose we open 50 boxes of this cereal and count the number of Sidney Crosby pictures we find. The second histogram shows this probability model. It is centred at $np = 50(0.2) = 10$ pictures, as expected, and it appears to be fairly symmetric around that centre. Let's have a closer look.

The third histogram again shows Binom(50, 0.2), this time magnified somewhat and centred at the expected value of 10 pictures of Crosby. It looks close to Normal, for sure. With this larger sample size, it appears that a Normal model might be a useful approximation.

A Normal model, then, is a close enough approximation only for a large enough number of trials. And what we mean by "large enough" depends on the probability of success. We'd need a larger sample if the probability of success were very low (or very high). It turns out that a Normal model works pretty well if we expect to see at least 10 successes and 10 failures. That is, we check the **Success/Failure Condition.**

The Success/Failure Condition: A Binomial model is approximately Normal if we expect at least 10 successes and 10 failures:

$$np \geq 10 \text{ and } nq \geq 10.$$

*Why 10? (Well, Actually 9)

It's easy to see where the magic number 10 comes from. You just need to remember how Normal models work. The problem is that a Normal model extends infinitely in both directions. But a Binomial model must have between 0 and n successes, so if we use a Normal to approximate a Binomial, we have to cut off its tails. That's not very important if the centre of the Normal model is so far from 0 and n that the lost tails have only a negligible area. More than three standard deviations should do it, because a Normal model has little probability past that.

So the mean needs to be at least three standard deviations from 0 and at least three standard deviations from n. Let's look at the 0 end.

We require:	$\mu - 3\sigma > 0$
Or, in other words:	$\mu > 3\sigma$
For a Binomial, that's:	$np > 3\sqrt{npq}$
Squaring yields:	$n^2p^2 > 9npq$
Now simplify:	$np > 9q$
Since $q \leq 1$, we can require:	$np > 9$

For simplicity, we usually require that np (and nq for the other tail) be at least 10 to use the Normal approximation, the Success/Failure Condition.[9]

For Example SPAM AND THE NORMAL APPROXIMATION TO THE BINOMIAL

RECAP: The communications monitoring company Postini has reported that 91% of e-mail messages are spam. Recently, you installed a spam filter. You observe that over the past week it okayed only 151 of 1422 e-mails you received, classifying the rest as junk. Should you worry that the filtering is too aggressive?

[9]Looking at the final step, we see that we need $np > 9$ in the worst case, when q (or p) is near 1, making the Binomial model quite skewed. When q and p are near 0.5—say between 0.4 and 0.6—the Binomial model is nearly symmetric and $np > 5$ ought to be safe enough. Although we'll always check for 10 expected successes and failures, keep in mind that for values of p near 0.5, we can be somewhat more forgiving.

QUESTION: What is the probability that no more than 151 of 1422 e-mails are real messages?

ANSWER: I assume that messages arrive randomly and independently, with a probability of success (a real message) $p = 0.09$. The model Binom(1422, 0.09) applies, but will be hard to work with. Checking conditions for the Normal approximation, I see that:

✓ These messages represent less than 10% of all e-mail traffic.

✓ I expect $np = (1422)(0.09) = 127.98$ real messages and $nq = (1422)(0.91) = 1294.02$ spam messages, both far greater than 10.

It's okay to approximate this Binomial probability by using a Normal model.

$$\mu = np = 1422(0.09) = 127.98$$

$$\sigma = \sqrt{npq} = \sqrt{1422(0.09)(0.91)} \approx 10.79$$

$$P(x \le 151) = P\left(z \le \frac{151 - 127.98}{10.79}\right)$$
$$= P(z \le 2.13)$$
$$= 0.9834$$

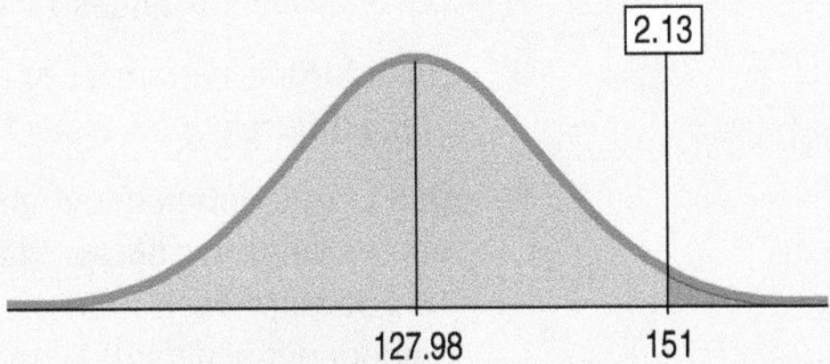

Among my 1422 e-mails, there's over a 98% chance that no more than 151 of them were real messages, so the filter may be working properly.

But the Normal is continuous, the Binomial discrete

There's a small problem with approximating a Binomial model with a Normal model. The Binomial is *discrete*, giving probabilities for specific counts, but the Normal models a *continuous* random variable. It is not possible to list all the (uncountably infinite) outcomes and their probabilities for a continuous random variable, as we could for discrete random variables.[10]

As we saw in the previous section, models for continuous random variables only assign probabilities for *intervals* of outcomes. So, when we use the Normal model, we no longer calculate the probability that the random variable equals a *particular* value (well you can, but you just get zero), but only that it lies *between* two values. We won't calculate the probability of getting exactly 1850 units of blood, but we have no problem approximating the probability of getting 1850 *or more*, which was, after all, what we really wanted.[11]

WHAT CAN GO WRONG?

- **Probability models are still just models.** Models can be useful, but they are not reality. Think about the assumptions behind your models. Are your dice really perfectly fair? (They are probably pretty close.) But when you hear that the probability of a nuclear accident is 1/10 000 000 per year, is that likely to be a precise value? Question probabilities as you would data.

[10]In fact, some people use an adjustment called the "continuity correction" to help with this problem. It's related to the suggestion we make in the next footnote.

[11]If we really had been interested in a single value, we might have approximated it by finding the probability of getting between 1849.5 and 1850.5 units of blood. And likewise we could improve our Normal approximation for $P(X \ge 1850)$ by changing the 1850 to 1849.5 before standardizing.

- **If the model is wrong, so is everything else.** Before you try to find the mean or standard deviation of a random variable, check to make sure the probability model is reasonable. As a start, the probabilities in your model should add up to 1. If not, you may have calculated a probability incorrectly or left out a value of the random variable. For instance, in the insurance example, the description mentions only death and disability. Good health is by far the most likely outcome, not to mention the best for both you and the insurance company (who gets to keep your money). Don't overlook that.

To find the expected value of the sum or difference of random variables, we simply add or subtract means. Centre is easy; spread is trickier. Watch out for some common traps such as the first four items below:

- **Watch out for variables that aren't independent.** You can add expected values of any two random variables, but you can only add variances of independent random variables. Suppose a survey includes questions about the number of hours of sleep people get each night and also the number of hours they are awake each day. From their answers, we find the mean and standard deviation of hours asleep and hours awake. The expected total must be 24 hours; after all, people are either asleep or awake.[12] The means still add just fine. Since all the totals are exactly 24 hours, however, the standard deviation of the total will be 0. We can't add variances here because the number of hours you're awake depends on the number of hours you're asleep. Be sure to check for independence before adding variances.
- **Don't forget: Variances of independent random variables add. Standard deviations don't.**
- **Don't forget: Variances of independent random variables add, even when you're looking at the difference between them.**
- **Don't write independent instances of a random variable with notation that looks like they are the same variables.** Make sure you write each instance as a different random variable. Just because each random variable describes a similar situation doesn't mean that each random outcome will be the same. These are *random* variables, not the variables you saw in Algebra. Write $X_1 + X_2 + X_3$ rather than $X + X + X$.
- **Don't assume you have the Bernoulli trials required for the Binomial model without checking.** Having two possible outcomes isn't enough. Be sure to check the requirements: two possible outcomes per trial ("success"and "failure", though a failure can simply be all the outcomes other than the one deemed a success), a constant probability of success, and independence. Remember to check the 10% Condition when sampling without replacement.
- **Don't use the Normal approximation to the Binomial with small *n*.** To use a Normal approximation in place of a Binomial model, there must be at least 10 expected successes and 10 failures.
- **Don't assume everything's Normal.** Just because a random variable is continuous or you happen to know a mean and standard deviation doesn't mean that a Normal model will be useful. You must *Think* about whether the **Normality Assumption** is justified. Using a Normal model when it is a poor representation of the random behaviour will lead to wrong answers and misleading conclusions.

CONNECTIONS

We've seen means, variances, and standard deviations of data. We know that they estimate parameters of models for these data. Now we're looking at the probability models directly. We have only parameters because there are no data to summarize.

It should be no surprise that expected values and standard deviations adjust to shifts and changes of units in the same way as the corresponding data summaries. The fact that we can add variances of independent random quantities is fundamental and will explain why a number of statistical methods work the way they do.

We've encountered an important new discrete model, the Binomial model, for counts and revisited the Normal model as applied to continuous random variables. And we'll see in the next chapter how the facts that we've learned about the Binomial model can help explain how proportions behave (since a proportion is just a count divided by the sample size).

[12] Although some students do manage to attain a state of consciousness somewhere between sleeping and wakefulness during Statistics class.

What Have We Learned?

Learning Objectives

Understand how probability models relate probabilities to outcomes.

- For discrete random variables, probability models assign a probability to each possible outcome.
- For continuous random variables, areas under density curves assign probabilities to intervals of outcomes.

Know how to find the mean, or expected value, of a discrete probability model from $\mu = \sum xP(x)$ and the standard deviation from $\sigma = \sqrt{\sum (x - \mu)^2 P(x)}$.

Foresee the consequences of shifting and scaling random variables, specifically

$$E(X \pm c) = E(X) \pm c \qquad E(aX) = aE(X)$$
$$Var(X \pm c) = Var(X) \qquad Var(aX) = a^2 Var(X)$$
$$SD(X \pm c) = SD(X) \qquad SD(aX) = |a| SD(X)$$

Understand that when adding or subtracting random variables, the expected values add or subtract as well: $E(X \pm Y) = E(X) \pm E(Y)$. However, when adding or subtracting independent random variables, the variances *add:*

$$Var(X \pm Y) = Var(X) + Var(Y)$$

Know that Bernoulli trials show up in lots of places. We use a Binomial model when we're interested in the number of successes in a certain number of Bernoulli trials.

- If n is large and we expect at least 10 successes and 10 failures, we can use a Normal model to approximate the Binomial model.
- If n is large but p is so small that we expect fewer than 10 successes, we can use a Poisson model to approximate the Binomial model.

Review of Terms

Random variable Assumes any of several different values as a result of some random event. Random variables are denoted by a capital letter, such as X (p. 376).

Discrete random variable A random variable that can take one of a finite number[13] of distinct outcomes (p. 376).

Continuous random variable A random variable that can take on any of an uncountably infinite number of outcomes, typically, an interval of values on the real line (p. 376).

Probability model or probability distribution A function that associates a probability P with each value of a discrete random variable X, denoted $P(X = x)$ or $P(x)$, or with any interval of values of a continuous random variable, using a density curve (p. 376).

Expected value A random variable's theoretical long-run average value, the centre of its model. Denoted μ or $E(X)$, it is found (if the random variable is discrete) by summing the products of variable values and their respective probabilities (p. 376):

$$\mu = E(X) = \sum xP(x)$$

Standard deviation of a random variable Describes the spread in the model and is the square root of the variance, denoted $SD(X)$ or σ (p. 379).

Variance The expected value of the squared deviations from the mean. For discrete random variables, it can be calculated as (p. 379).

$$\sigma^2 = Var(X) = \sum (x - \mu)^2 P(x).$$

[13]Actually, there could be an infinite number of outcomes as long as they're *countable*. Essentially, that means we can imagine listing them all in order, like the counting numbers 1, 2, 3, 4, 5, ...

Changing a random variable by a constant $E(X \pm c) = E(X) \pm c$ $\quad Var(X \pm c) = Var(X)$ $\quad SD(X \pm c) = SD(X)$
$E(aX) = aE(X)$ $\quad Var(aX) = a^2 Var(X)$ $\quad SD(aX) = |a| SD(X)$ (p. 382).

Addition Rule for Expected Values of Random Variables $E(X \pm Y) = E(X) \pm E(Y)$ (p. 383).

Addition Rule for Variances of Random Variables (Pythagorean Theorem of Statistics) (p. 384)

If X and Y are independent: $Var(X \pm Y) = Var(X) + Var(Y)$,

and $SD(X \pm Y) = \sqrt{Var(X) + Var(Y)}$.

***Convariance** The convariance of random variables X and X is $Cov(X, Y) = E[(X - \mu)(Y - \nu)]$ where $\mu = E(X)$ and $\nu = E(Y)$. In general (no need to assume independence)

$$Var(X \pm Y) = Var(X) + Var(Y) \pm 2Cov(X, Y) \text{ (p. 386).}$$

Bernoulli trials A sequence of trials where (p. 387)

1. There are exactly two possible outcomes (usually denoted *success* and *failure*).
2. The probability of success is constant.
3. The trials are independent.

Binomial probability distribution A Binomial distribution is appropriate for a random variable that counts the number of successes in n Bernoulli trials (p. 388).

Binomial Model $P(X = x) = {}_nC_x p^x q^{n-x}$, where ${}_nC_x = \frac{n!}{x!(n - x)!}$ (p. 388).

***Poisson model** A discrete model used to model counts for a wide variety of phenomena, where we are counting events occurring over time or space, such as the number of arrivals of customers arriving in a queue or calls arriving into a call centre. Also sometimes used to approximate Binomial probabilites for rare events. (p. 392)

On the Computer RANDOM VARIABLES

Statistics packages deal with data, not with random variables. Nevertheless, the calculations needed to find means and standard deviations of random variables are little more than weighted means. Most packages can manage that, but then they are just being overblown calculators. For technological assistance with these calculations, we recommend you pull out your calculator.

Most statistics packages offer functions that compute probabilities for various probability models. Some technology solutions automatically use the Normal approximation for the Binomial when the exact calculations become unmanageable. The only important differences among these functions are how they are named and the order of their arguments. In these functions, pdf stands for "probability density function"—what we've been calling a probability model. The letters cdf stand for "cumulative distribution function," the technical term when we want to accumulate probabilities over a range of values. These technical terms show up in many of the function names.

For example, in Excel, **Binomdist(*x*, *n*, prob, cumulative)** computes Binomial probabilities. If cumulative is set to false, the calculation is only for one value of *x*. That is **Binomdist(2,5,.2,false)** would be the probability of finding Sidney Crosby's picture twice in five boxes with probability of success = 0.2. **Binomdist(2,5,.2,true)** would be the probability of finding 0, 1, or 2 (up to and including 2) Crosby pictures out of 5.

EXCEL

Binomdist(*x, n, prob, cumulative*)

COMMENTS

Set cumulative = *true* for cdf, *false* for pdf.
Excel's function fails when *x* or *n* is large.
It may not use the Normal approximation.

JMP

Binomial Probability (*prob, n, x*) (pdf)
Binomial Distribution (*prob, n, x*) (cdf)
Poisson Probability (*mean, k*) (pdf)
Poisson Distribution (*mean, k*) (cdf)

MINITAB

- Choose **Probability Distributions** from the **Calc** menu.
- Choose **Binomial** from the Probability Distributions submenu.
- To calculate the probability of getting *x* successes in *n* trials, choose **Probability.**
- To calculate the probability of getting *x* or fewer successes among *n* trials, choose **Cumulative Probability.**
- For Poisson, choose **Poisson** from the Probability Distribution submenu.

R

The library stats contains the pdfs and cdfs of several common distributions

- **library(stats)**
- **dbinom(x,n,p)** # Gives $P(X = x)$ for binomial with n and p.
- **pbinom(x,n,p)** # Gives the cumulative distribution function $P(X \geq x)$ for the binomial.
- **dpois(x,mean)** # Gives $P(X = x)$ for the Poisson.
- **ppois(x,mean)** # Gives the cumulative distribution function $P(X \geq x)$ for the Poisson.

SPSS

PDF.Poisson(*x, mean*)
CDF.Poisson(*x, mean*)
PDF.BINOM(*x, n, prob*)
CDF.BINOM(*x, n, prob*)

STATCRUNCH

To calculate binomial probabilities:

- Click on **Stat**.
- Choose **Calculators » Binomial.**
- Enter the parameters, **n** and **p**.
- Choose a specific outcome (=) or a lower tail (≤ or <), or upper tail (≥ or >) sum.
- Enter the number of successes **x**.
- Click on **Compute**.

To calculate Poisson probabilities:

- Click on **Stat**.
- Choose **Calculators » Poisson**.
- Enter the **mean**.
- Choose a specific outcome (=) or a lower tail (≤ or <) or upper tail (≥ or >) sum.
- Enter the number of successes **x**.
- Click on **Compute**.

TI-83/84 PLUS

To calculate the mean and standard deviation of a discrete random variable, enter the probability model in two lists:

- In one list (say, L1), enter the *x*-values of the variable.
- In a second list (say, L2), enter the associated probabilities $P(X = x)$.
- From the **STAT CALC** menu, select 1-VarStats and specify the two lists. You'll see the mean and standard deviation.

binompdf(*n, prob, x*)
binomcdf(*n, prob, x*)
poissonpdf(*mean, x*)
poissoncdf(*mean, x*)

COMMENTS

You can enter the probabilities as fractions; the calculator will change them to decimals for you.
Notice that the calculator knows enough to call the standard deviation σ, but mistakenly says $\bar{x}$ when it should say μ. Make sure you don't make that mistake!
Find the distribution commands in the **2nd DISTR** menu (the calculator refers to models as "distributions").

Exercises

1. **Expected value** Find the expected value of each random variable:

a)

x	10	20	30
$P(X = x)$	0.3	0.5	0.2

b)

x	2	4	6	8
$P(X = x)$	0.3	0.4	0.2	0.1

2. **Expected value II** Find the expected value of each random variable:

a)

x	0	1	2
$P(X = x)$	0.2	0.4	0.4

b)

x	100	200	300	400
$P(X = x)$	0.1	0.2	0.5	0.2

3. **Spinning the wheel** A wheel comes up green 50% of the time and red 50% of the time. If it comes up green, you win \$100. If it comes up red, you win nothing.
   a) Intuitively, how much do you expect to win per spin of the wheel, on average?
   b) Calculate the expected value of the game.

4. **Stock market** A day trader buys an option on a stock that will return \$100 profit if the stock goes up today and lose \$200 if it goes down. If the trader thinks there is a 75% chance that the stock will go up, what is his expected value of the option?

5. **Pick a card, any card** You draw a card from a deck. If you get a red card, you win nothing. If you get a spade, you win \$5. For any club, you win \$10 plus an extra \$20 for the ace of clubs.
   a) Create a probability model for the amount you win at this game.
   b) Find the expected amount you'll win.
   c) How much would you be willing to pay to play this game?

6. **You bet!** You roll a die. If it comes up a six, you win \$100. If not, you get to roll again. If you get a six the second time, you win \$50. If not, you lose.
   a) Create a probability model for the amount you win at this game.
   b) Find the expected amount you'll win.
   c) How much would you be willing to pay to play this game?

7. **Kids** A couple plans to have children until they get a girl, but they agree that they will not have more than three children even if all are boys. (Assume boys and girls are equally likely.)
   a) Create a probability model for the number of children they'll have.
   b) Find the expected number of children.
   c) Find the expected number of boys they'll have.

8. **Carnival** A carnival game offers a \$100 cash prize for anyone who can break a balloon by throwing a dart at it. It costs \$5 to play, and you're willing to spend up to \$20 trying to win that cash prize. You estimate that you have about a 10% chance of hitting the balloon on any throw.
   a) Create a probability model for the possible outcomes when you play this carnival game.
   b) Find the expected number of darts you'll throw.
   c) Find your expected winnings.

9. **Software** A small software company bids on two contracts. It anticipates a profit of \$50 000 if it gets the

larger contract and a profit of \$20 000 on the smaller contract. The company estimates there is a 30% chance it will get the larger contract and a 60% chance it will get the smaller contract. Assuming the contracts will be awarded independently, what is the expected profit?

10. **Racehorse** A man buys a racehorse for \$20 000 and enters it in two races. He plans to sell the horse afterward, hoping to make a profit. If the horse wins both races, its value will jump to \$100 000. If it wins one of the races, it will be worth \$50 000. If it loses both races, it will be worth only \$10 000. The man believes there's a 20% chance that the horse will win the first race and a 30% chance it will win the second one. Assuming that the two races are independent events, find the man's expected profit.

11. **Variation 1** Find the standard deviations of the random variables in Exercise 1.

12. **Variation 2** Find the standard deviations of the random variables in Exercise 2.

13. **Pick another card** Find the standard deviation of the amount you might win drawing a card in Exercise 5.

14. **The die** Find the standard deviation of the amount you might win rolling a die in Exercise 6.

15. **Kids again** Find the standard deviation of the number of children the couple in Exercise 7 may have.

16. **Darts** Find the standard deviation of your winnings throwing darts in Exercise 8.

17. **Repairs** The probability model below describes the number of repair calls that an appliance repair shop may receive during an hour.

Repair Calls	0	1	2	3
Probability	0.1	0.3	0.4	0.2

a) How many calls should the shop expect per hour, on average?
b) What is the standard deviation?

18. **Red lights** A commuter must pass through five traffic lights on her way to work and will have to stop at each one that is red. She estimates the probability model for the number of red lights she hits, as shown below.

X = # of red	0	1	2	3	4	5
$P(X = x)$	0.05	0.25	0.35	0.15	0.15	0.05

a) How many red lights should she expect to hit each day, on average?
b) What is the standard deviation?

19. **Defects** A consumer organization inspecting new cars found that many had appearance defects (dents, scratches, paint chips, etc.). While none had more than three of these defects, 7% had three, 11% had two, and 21% had one defect. Find the expected number of appearance defects in a new car, and the standard deviation.

20. **Insurance** An insurance policy costs \$100 and will pay policyholders \$10 000 if they suffer a major injury (resulting in hospitalization) or \$3000 if they suffer a minor injury (resulting in lost time from work). The company estimates that each year, 1 in every 2000 policyholders may have a major injury, and 1 in 500 may have a minor injury.
a) Create a probability model for the profit on a policy.
b) What is the company's expected profit on this policy?
c) What is the standard deviation?

21. **Cancelled flights** Mary is deciding whether to book the cheaper flight home from university after her final exams, but she's unsure when her last exam will be. She thinks there is only a 20% chance that the exam will be scheduled after the last day she can get a seat on the cheaper flight. If it is and she has to cancel the flight, she will lose \$150. If she can take the cheaper flight, she will save \$100.
a) If she books the cheaper flight, what can she expect to gain, on average?
b) What is the standard deviation?

22. **Day trading again** An option to buy a stock is priced at \$200. If the stock closes above 30 on May 15, the option will be worth \$1000. If it closes below 20, the option will be worth nothing. If it closes between 20 and 30 (inclusively), the option will be worth \$200. A trader thinks there is a 50% chance that the stock will close in the 20–30 range, a 20% chance that it will close above 30, and a 30% chance that it will fall below 20 on May 15.
a) How much is her expected gain?
b) What is the standard deviation of her gain?
c) Should she buy the stock option?

23. **Contest** You play two games against the same opponent. The probability that you win the first game is 0.4. If you win the first game, the probability that you also win the second is 0.2. If you lose the first game, the probability that you win the second is 0.3.
a) Are the two games independent? Explain your answer.
b) What is the probability that you lose both games?
c) What is the probability that you win both games?
d) Let random variable X be the number of games you win. Find the probability model for X.
e) What are the expected value and standard deviation of X?

24. **Contracts** Your company bids for two contracts. You believe the probability that you will get contract 1 is 0.8. If you get contract 1, the probability that you will also get contract 2 will be 0.2. If you do not get contract 1, the probability that you will get contract 2 will be 0.3.
a) Are the two contracts independent? Explain.
b) Find the probability that you get both contracts.

c) Find the probability that you get no contract.
d) Let X be the number of contracts you get. Find the probability model for X.
e) Find the expected value and standard deviation of X.

25. Batteries In a group of 10 batteries, three are dead. You choose two batteries at random.
a) Create a probability model for the number of good batteries you get.
b) What is your expected number of good ones?
c) What is the standard deviation?

26. Kittens In a litter of seven kittens, three are female. You pick two kittens at random.
a) Create a probability model for the number of male kittens you get.
b) What is the expected number of males?
c) What is the standard deviation?

27. Random variables Given independent random variables X and Y, with means and standard deviations as shown, find the mean and standard deviation of each of the random variables below. Also, X_1 and X_2 are independent variables with the same distribution as X.
a) $3X$
b) $Y + 6$
c) $X + Y$
d) $X - Y$
e) $X_1 + X_2$

	Mean	SD
X	10	2
Y	20	5

28. Random variables II Given independent random variables X and Y, with means and standard deviations as shown, find the mean and standard deviation of each of the variables below. Also, Y_1 and Y_2 are independent variables with the same distribution as Y.
a) $X - 20$
b) $0.5Y$
c) $X + Y$
d) $X - Y$
e) $Y_1 + Y_2$

	Mean	SD
X	80	12
Y	12	3

29. Random variables III Given independent random variables X and Y, with means and standard deviations as shown, find the mean and standard deviation of each of the variables below. Also, Y_1, Y_2, Y_3, and Y_4 are independent variables with the same distribution as Y.
a) $0.8Y$
b) $2X - 100$
c) $X + 2Y$
d) $3X - Y$
e) $Y_1 + Y_2 + Y_3 + Y_4$

	Mean	SD
X	120	12
Y	300	16

30. Random variables fini Given independent random variables X and Y, with means and standard deviations as shown, find the mean and standard deviation of each of the variables below. X_1, X_2, and X_3 are independent variables with the same distribution as X.
a) $2Y + 20$
b) $3X$
c) $0.25X + Y$
d) $X - 5Y$
e) $X_1 + X_2 + X_3$

	Mean	SD
X	80	12
Y	12	3

31. Eggs A grocery supplier believes that in a dozen eggs, the mean number of broken ones is 0.6 with a standard deviation of 0.5 eggs. You buy three dozen eggs without checking them.
a) What is your expected number of broken eggs?
b) What is the standard deviation?
c) What assumptions did you have to make about the eggs to answer this question? Do you think that assumption is warranted? Explain.

32. Garden A company selling vegetable seeds in packets of 20 estimates that the mean number of seeds that will actually grow is 18, with a standard deviation of 1.2 seeds. You buy five different seed packets.
a) What is your expected number of bad (non-growing) seeds?
b) What is the standard deviation of the number of bad seeds?
c) What assumption did you make about the seeds? Do you think that assumption is warranted? Explain.

33. Repair calls Find the mean and standard deviation of the number of repair calls the appliance shop in Exercise 17 will receive during an 8-hour day. State any assumptions you are making.

34. Stop! Find the mean and standard deviation of the number of red lights the commuter in Exercise 18 will hit on her way to work during a five-day work week. State any assumptions you are making.

35. Tickets A delivery company's trucks occasionally get parking tickets. Based on past experience, the company expects that the trucks will average 1.3 tickets a month, with a standard deviation of 0.7 tickets.
a) If they have 18 trucks, what are the mean and standard deviation of the total number of parking tickets the company will have to pay this month?
b) What assumption did you make in answering this question?
c) What would you consider to be an unusually bad month for the company?

36. Donations Organizers of a televised fundraiser know from past experience that most people donate small amounts ($10–$25), some donate larger amounts ($50–$100), and a few people make very generous donations of $250, $500, or more. Historically, pledges average about $32 with a standard deviation of $54.
a) If 120 people call in pledges, what are the mean and standard deviation of the total amount raised?
b) What assumption did you make in answering this question?

c) The organizers are worried that people are pledging less than normal. How low would the 120-person total need to be to convince you that this is the case?

37. Fire! An insurance company estimates that it should make an expected annual profit of $150 on each homeowner's policy written, with a standard deviation of $6000.
a) Why is the standard deviation so large?
b) If it writes only two of these policies, what are the mean and standard deviation of the annual profit?
c) If it writes 10 000 of these policies, what are the mean and standard deviation of the annual profit?
d) Do you think the company is likely to be profitable? Explain.
e) What assumptions underlie your analysis? Can you think of circumstances under which those assumptions might be violated? Explain.

38. Casino A casino knows that people play the slot machines in hopes of hitting the jackpot, but that most of them lose their dollar. Suppose a certain machine pays out an average of $0.92, with a standard deviation of $120.
a) Why is the standard deviation so large?
b) If you play five times, what are the mean and standard deviation of the casino's profit?
c) If gamblers play this machine 1000 times in a day, what are the mean and standard deviation of the casino's profit?
d) Do you think the casino is likely to be profitable? Explain.

39. Bernoulli Can we use probability models based on Bernoulli trials to investigate the following situations? Explain.
a) We roll 50 dice to find the distribution of the number of spots on the faces.
b) How likely is it that in a group of 120 people the majority will have Type A blood, given that Type A is found in 43% of the population?
c) We deal five cards from a deck and get all hearts. How likely is that?
d) To predict the outcome of a vote on the school budget, we poll 500 of the 3000 likely voters to see how many favour the proposed budget.
e) A company realizes that about 10% of its packages are not being sealed properly. In a case of 24, is it likely that more than three are unsealed?

40. Bernoulli 2 Can we use probability models based on Bernoulli trials to investigate the following situations? Explain.
a) You are rolling five dice and need to get at least two 6s to win the game.
b) We record the eye colours found in a group of 500 people.
c) A manufacturer recalls a doll because about 3% have buttons that are not properly attached. Customers return 37 of these dolls to the local toy store. Is the manufacturer likely to find any dangerous buttons?
d) A city council of 11 Conservatives and 8 Liberals picks a committee of four at random. What's the probability they choose all Liberals?
e) A 2002 Rutgers University study found that 74% of high-school students have cheated on a test at least once. Your local high-school principal conducts a survey in homerooms and 322 of the 481 students admit to cheating.

41. Crosby again Let's take one last look at the Sidney Crosby picture search. You know his picture is in 20% of the cereal boxes. You buy five boxes to see how many pictures of Crosby you might get.
a) Describe how you would simulate the number of pictures of Crosby you might find in five boxes of cereal.
b) Run at least 30 trials.
c) Based on your simulation, estimate the probabilities that you get no pictures of Crosby, one picture, two pictures, etc.
d) Calculate the actual probability model.
e) Compare the distribution of outcomes in your simulation to the probability model.

42. Seatbelts Suppose 75% of all drivers always wear their seatbelts. Let's investigate how many of the drivers might be belted among five cars waiting at a traffic light.
a) Describe how you would simulate the number of seatbelt-wearing drivers among the five cars.
b) Run at least 30 trials.
c) Based on your simulation, estimate the probabilities there are no belted drivers, exactly one, two, etc.
d) Calculate the actual probability model.
e) Compare the distribution of outcomes in your simulation to the probability model.

43. On time A Transport Canada report about air travel found that, nationwide, 76% of all flights are on time. Suppose you are at the airport and your flight is one of 50 scheduled to take off in the next two hours. Can you consider these departures to be Bernoulli trials? Explain.

44. Lost luggage A Transport Canada report about air travel found that airlines misplace about five bags per 1000 passengers. Suppose you are travelling with a group of people who have checked 22 pieces of luggage on your flight. Can you consider the fate of these bags to be Bernoulli trials? Explain.

45. Coins and intuition You flip a fair coin 100 times.
a) Intuitively, how many heads do you expect?
b) Use the formula for expected value to verify your intuition.

46. Roulette and intuition An American roulette wheel has 38 slots, of which 18 are red, 18 are black, and 2 are green (0 and 00). You spin the wheel 38 times.
a) Intuitively, how many times would you expect the ball to wind up in a green slot?

b) Use the formula for expected value to verify your intuition.

47. Lefties Assume that 13% of people are left-handed. We select 12 people at random.
a) Find the mean and standard deviation of the number of right-handers in the group.
b) What is the probability that they're not all right-handed?
c) What is the probability that there are no more than 10 righties?
d) What is the probability that there is exactly six of each?
e) What is the probability that the majority is right-handed?

48. Arrows An Olympic archer is able to hit the bull's-eye 80% of the time. Assume each shot is independent of the others. She shoots 10 arrows.
a) Find the mean and standard deviation of the number of bull's-eyes she may get.
b) What is the probability that she never misses?
c) What is the probability that there are no more than eight bull's-eyes?
d) What is the probability that there are exactly eight bull's-eyes?
e) What's the probability that she hits the bull's-eye more often than she misses?

49. Vision It is generally believed that nearsightedness affects about 12% of all children. A school district tests the vision of 169 incoming kindergarten children. How many would you expect to be nearsighted, on average? With what standard deviation?

50. International students At a certain university, 6% of all students come from outside Canada. Incoming students are assigned at random to freshman dorms, where they live in residential clusters of 40 freshmen sharing a common lounge area. How many international students would you expect to find in a typical cluster, on average? With what standard deviation?

51. Tennis, anyone? A particular tennis player makes a successful first serve 70% of the time. Assume that each serve is independent of the others. If she serves six times, what is the probability that she gets
a) all six serves in?
b) exactly four serves in?
c) at least four serves in?
d) no more than four serves in?

52. Frogs A wildlife biologist examines frogs for a genetic trait he suspects may be linked to sensitivity to industrial toxins in the environment. Previous research had established that this trait is usually found in one of every eight frogs. He collects and examines a dozen frogs. If the frequency of the trait has not changed, what is the probability that he finds the trait in
a) none of the 12 frogs?
b) at least two frogs?
c) three or four frogs?
d) no more than four frogs?

53. And more tennis Suppose the tennis player in Exercise 51 serves 80 times in a match.
a) What are the mean and standard deviation of the number of good first serves expected?
b) Verify that you can use a Normal model to approximate the distribution of the number of good first serves.
c) Use the 68–95–99.7 Rule to describe this distribution.
d) What is the probability that she makes at least 65 first serves?

54. More arrows The archer in Exercise 48 will be shooting 200 arrows in a large competition.
a) What are the mean and standard deviation of the number of bull's-eyes she might get?
b) Is a Normal model an appropriate approximation here? Explain.
c) Use the 68–95–99.7 Rule to describe the distribution of the number of bull's-eyes she may get.
d) Would you be surprised if she made only 140 bull's-eyes? Explain.

55. Apples An orchard owner knows that he'll have to use about 6% of the apples he harvests for cider because they will have bruises or blemishes. He expects a tree to produce about 300 apples.
a) Describe an appropriate model for the number of cider apples that may come from that tree. Justify your model.
b) Find the probability there will be no more than a dozen cider apples.
c) Is it likely there will be more than 50 cider apples? Explain.

56. Frogs, part II Based on concerns raised by his preliminary research, the biologist in Exercise 52 decides to collect and examine 150 frogs.
a) Assuming the frequency of the trait is still one in eight, determine the mean and standard deviation of the number of frogs with the trait that he will find in his sample.
b) Verify that he can use a Normal model to approximate the distribution of the number of frogs with the trait.
c) He found the trait in 22 of his frogs. Do you think this proves that the trait has become more common? Explain.

57. Lefties again A lecture hall has 200 seats with folding arm tablets, 30 of which are designed for left-handers. The average size of classes that meet there is 188, and we can assume that about 13% of students are left-handed. What is the probability that a right-handed student in one of these classes is forced to use a lefty arm tablet?

58. No-shows An airline, believing that 5% of passengers fail to show up for flights, overbooks (sells more tickets than there are seats). Suppose a plane will hold 265 passengers, and the airline sells 275 tickets. What is the probability that the airline will not have enough seats, so someone gets bumped?

59. Annoying phone calls A newly hired telemarketer is told he will probably make a sale on about 12% of his phone calls. The first week he called 200 people, but only made 10 sales. Should he suspect he was misled about the true success rate? Explain.

60. The euro Shortly after the introduction of the euro coin in Belgium, newspapers around the world published articles claiming the coin is biased. The stories were based on reports that someone had spun the coin 250 times and gotten 140 heads—that's 56% heads. Do you think this is evidence that spinning a euro is unfair? Explain.

***61. Hurricanes 2010, redux** We first looked at the occurrences of hurricanes in Chapter 3 exercises and found that they arrive with a mean of 2.70 per year. Suppose the number of hurricanes can be modelled by a Poisson distribution with this mean.

a) What is the probability of no hurricanes next year?
b) What is the probability that during the next two years, there will be exactly one hurricane?

***62. Bank tellers** You are the only bank teller on duty at your local bank. You need to run out for 10 minutes, but you don't want to miss any customers. Suppose the arrival of customers can be modelled by a Poisson distribution with a mean of two customers per hour.

a) What is the probability that no one will arrive in the next 10 minutes?
b) What is the probability that two or more people arrive in the next 10 minutes?
c) You've just served two customers who came in one after the other. Is this a better time to run out?

***63. HIV again** In Chapter 13, we saw that the probability of contracting HIV is small, with p about 0.0005 for a new case in a given year. In a town of 8000 people,

a) What is the expected number of new cases?
b) Use the Poisson model to approximate the probability that there will be at least one new case of HIV next year.

***64. Lotto 6/49 scandal** Earlier in the chapter, we mentioned Professor Jeff Rosenthal's work on the Lotto 6/49 scandal. He estimated—conservatively—that for the 5713 major-prize lottery wins over the period 1996–2006, the expected number of wins by retailers was 123.2. In fact, retailers won 200 times. Using the Poisson distribution, Rosenthal calculated the chances of their winning this many times or more. The calculation is a bit messy, so let's do a similar but simpler calculation. Assume that conditions (purchasing patterns, etc.) have not changed since Rosenthal's study was conducted.

a) What is the expected number of retailer wins among the next 40 major-prize wins in Ontario if just chance is at work? (Adjust the expected wins in 5713 prizes to get the expected wins in 40 prizes.)
b) Use the expected value from part a) in the Poisson model to approximate the probability that retailers will win two or more prizes among the next 40 major wins.

65. Seatbelts II Police estimate that 80% of drivers now wear their seatbelts. They set up a safety roadblock, stopping cars to check for seatbelt use.

a) What is the probability that the first 10 drivers are all wearing their seatbelts?
b) If police stop 30 cars during the first hour, find the mean and standard deviation of the number of drivers who will be wearing seatbelts.
c) If they stop 120 cars during this safety check, what's the probability they find at least 20 drivers not wearing their seatbelts?

66. Rickets Vitamin D is essential for strong, healthy bones. Our bodies produce vitamin D naturally when the skin is exposed to sunlight, or it can be taken as a dietary supplement. Although the bone disease rickets was largely eliminated in England during the 1950s, some people there are concerned that this generation of children is at increased risk because they are more likely to watch television or play computer games than spend time outdoors. Recent research indicated that about 20% of British children are deficient in vitamin D. Suppose doctors test a group of elementary school children.

a) What is the probability that the first 10 children tested are all okay?
b) They test 50 Grade 3 students. Find the mean and standard deviation of the number who are deficient in vitamin D.
c) If they test 320 children at this school, what is the probability that no more than 50 of them have the vitamin deficiency?

67. ESP Scientists wish to test the mind-reading ability of a person who claims to "have ESP." They use five cards with different and distinctive symbols (square, circle, triangle, line, squiggle). Someone picks a card at random and thinks about the symbol. The "mind reader" must correctly identify which symbol was on the card. If the test consists of 100 trials, how many would this person need to get right to convince you that ESP may actually exist? Explain.

68. True–False A true–false test consists of 50 questions. How many does a student have to get right to convince you that he is not merely guessing? Explain.

69. Hot hand A basketball player who ordinarily makes about 55% of his free throw shots has made four in a row. Is this evidence that he has a "hot hand" tonight? That is, is this streak so unusual that it means the probability he makes a shot must have changed? Explain.

70. New bow Our archer in Exercise 48 purchases a new bow, hoping that it will improve her success rate to more than 80% bull's-eyes. She is delighted when she first tests her new bow and hits six consecutive bull's-eyes. Do you think this is compelling evidence that the new bow is better? In other words, is a streak like this unusual for her? Explain.

71. Hotter hand Our basketball player in Exercise 69 has new sneakers, which he thinks improve his game. Over his past 40 shots, he's made 32—much better than the 55% he usually shoots. Do you think his chances of making a shot really increased? In other words, is making at least 32 of 40 shots really unusual for him? (Do you think it's his sneakers?)

72. New bow, again The archer in Exercise 70 continues shooting arrows, ending up with 45 bull's-eyes in 50 shots. Now are you convinced that the new bow is better? Explain.

73. Continuous uniform model Suppose that X is continuous and uniformly distributed over the outcomes 0 to 360. (For example, you might be spinning a bottle and observing the final angle.)

a) Draw a picture of the flat density curve to model this random variable. What is the correct height for the curve (recall that total probability is always equal to 1)?
b) Find $P(20 < X < 110)$ using the correct area under the density curve. Shade in this area in your picture of the density curve.

74. Continuous models Suppose that X is a continuous random variable with the following density curve model:

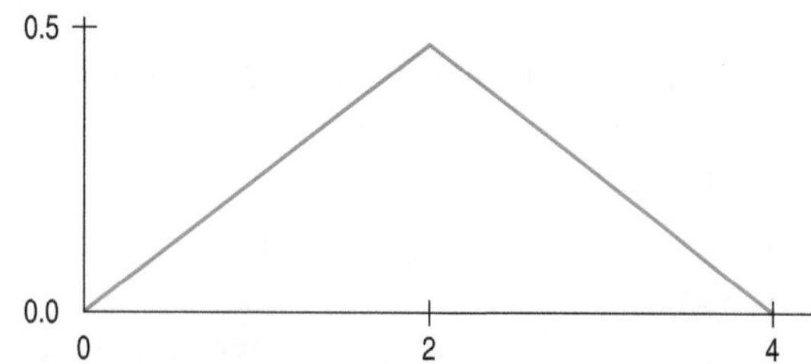

a) Find $P(0 < X < 1)$ and shade in the relevant area.
b) Find $P(1 \leq X \leq 3)$ and shade in the relevant area.

75. Cereal The amount of cereal that can be poured into a small bowl varies with a mean of 1.5 ounces and a standard deviation of 0.3 ounces. A large bowl holds a mean of 2.5 ounces with a standard deviation of 0.4 ounces. You open a new box of cereal and pour one large and one small bowl.

a) How much more cereal do you expect there to be in the large bowl than in the small bowl?
b) What's the standard deviation of this difference?
c) If the difference follows a Normal model, what is the probability that the small bowl contains more cereal than the large one?
d) What are the mean and standard deviation of the total amount of cereal in the two bowls?
e) If the total follows a Normal model, what is the probability you poured out more than 4.5 ounces of cereal in the two bowls together?
f) The amount of cereal the manufacturer puts in the boxes is a random variable with a mean of 16.3 ounces and a standard deviation of 0.2 ounces. Find the expected amount of cereal left in the box, and the standard deviation.

76. Pets The Canadian Veterinary Medical Association claims that the annual cost of medical care for dogs averages \$100 with a standard deviation of \$30, and for cats averages \$120 with a standard deviation of \$35.

a) What is the expected difference in the cost of medical care for dogs and cats?
b) What is the standard deviation of that difference?
c) If the difference in costs can be described by a Normal model, what is the probability that medical expenses are higher for someone's dog than for her cat?
d) Do you have any concerns regarding your answers above?

77. More cereal In Exercise 75, we poured a large and a small bowl of cereal from a box. Suppose the amount of cereal that the manufacturer puts in the boxes is a random variable with mean 16.2 ounces, and standard deviation 0.1 ounces.

a) Find the expected amount of cereal left in the box.
b) What is the standard deviation?
c) If the weight of the remaining cereal can be described by a Normal model, what is the probability that the box still contains more than 13 ounces?

78. More pets You're thinking about getting two dogs and a cat. Assume that annual veterinary expenses are independent and have a Normal model with the means and standard deviations described in Exercise 76.

a) Define appropriate variables and express the total annual veterinary costs you might accrue.
b) Describe the model for this total cost. Be sure to specify the model's name, expected value, and standard deviation.
c) What is the probability that your total expenses will exceed \$400?
d) About how much would a typical year cost you in expenses (give a range)? How much money should you set aside if you wanted to be fairly sure that your annual bill will be covered?

79. Medley In the 4×100 medley relay event, four swimmers swim 100 yards, each using a different stroke. A college team preparing for the national championship looks at the times their swimmers have posted and creates a model based on the following assumptions:

- The swimmers' performances are independent.
- Each swimmer's times follow a Normal model.
- The means and standard deviations of the times (in seconds) are as shown:

Swimmer	Mean	SD
1 (backstroke)	50.72	0.24
2 (breaststroke)	55.51	0.22
3 (butterfly)	49.43	0.25
4 (freestyle)	44.91	0.21

a) What are the mean and standard deviation for the relay team's total time in this event?

b) The team's best time so far this season was 3:19.48. (That's 199.48 seconds.) Do you think the team is likely to swim faster than this at the championship? Explain.

80. Bikes Bicycles arrive at a bike shop in boxes. Before they can be sold, they must be unpacked, assembled, and tuned (lubricated, adjusted, etc.). Based on past experience, the shop manager makes the following assumptions about how long this may take:

- The times for each setup phase are independent.
- The times for each phase follow a Normal model.
- The means and standard deviations of the times (in minutes) are as shown:

Phase	Mean	SD
Unpacking	3.5	0.7
Assembly	21.8	2.4
Tuning	12.3	2.7

a) What are the mean and standard deviation for the total bicycle set-up time?

b) A customer decides to buy a bike like one of the display models but wants a different colour. The shop has one, still in the box. The manager says they can have it ready in half an hour. Do you think the bike will be set up and ready to go as promised? Explain.

c) Most days, the bike shop completes this process on three bikes. The manager schedules 90 minutes per day in total for store employees to complete this task. Are the workers going to be able to complete this task in 90 minutes (of their combined work time)?

81. Farmers' market A farmer has 100 lb of apples and 50 lb of potatoes for sale. The market price for apples (per pound) each day is a random variable with a mean of 0.5 dollars and a standard deviation of 0.2 dollars. Similarly, for a pound of potatoes, the mean price is 0.3 dollars and the standard deviation is 0.1 dollars. It also costs him $2 to bring all the apples and potatoes to the market. The market is busy with eager shoppers, so we can assume that he'll be able to sell all of each type of produce at that day's price.

a) Define your random variables, and use them to express the farmer's net income.

b) Find the mean.

c) Find the standard deviation of the net income.

d) Do you need to make any assumptions in calculating the mean? How about the standard deviation?

82. Bike sale The bicycle shop in Exercise 80 will be offering two specially priced children's models at a sidewalk sale. The basic model will sell for $120 and the deluxe model for $150. Past experience indicates that sales of the basic model will have a mean of 5.4 bikes with a standard deviation of 1.2, and sales of the deluxe model will have a mean of 3.2 bikes with a standard deviation of 0.8 bikes. The cost of setting up for the sidewalk sale is $200.

a) Define random variables and use them to express the bicycle shop's net income.

b) What is the mean of the net income?

c) What is the standard deviation of the net income?

d) Do you need to make any assumptions in calculating the mean? How about the standard deviation?

83. Geometric model Suppose we keep flipping a loonie until we see a head. Let X = the flip on which we observe that first head. For example, if we get a tail, then a head, $X = 2$. Assume this is a typical loonie, flipped high and randomly.

a) Are these Bernoulli trials?

b) Can the Binomial model be applied here? Explain.

c) Find the probability that the first head occurs on the fifth flip.

d) The actual probability model for X is called the Geometric model, with parameter p, where here $p = 0.5$, the probability of success. The general formula is $p(x) = p\,q^{x-1}$ for $x = 1, 2, \ldots$ Use this formula to find the probability that the first head occurs on the third flip. (It is not so easy to prove, but $E(X) = 1/p$.)

e) Do you think this is a fairly symmetric distribution or a very skewed distribution? Why?

84. Geometric model again Suppose we inspect objects produced by a process in which 10% on average are defective. We will keep sampling and testing objects one at a time until we find a defective one.

a) Are these Bernoulli trials?

b) Can the Binomial model be applied here? Explain.

c) Find the probability that we find the first defective object upon inspecting the fourth article.

d) The actual probability model for X is called the Geometric model, with parameter p, where here $p = 0.1$, the probability of success. The general formula is $p(x) = p\,q^{x-1}$ for $x = 1, 2, \ldots$ Use this formula to find the probability that the first defective is found on the tenth inspection. (Note that even though the number of possible outcomes for X is infinite, X is still a discrete *RV*, since we can list and eventually reach any outcome via that list. We say that such a set of outcomes is *countably infinite*, whereas values on the real line are *uncountably infinite.*)

e) It can be shown that the mean of this distribution is $1/p$. What is the mode?

***85. Correlated RVs** Investors often purchase negatively correlated investment vehicles to minimize risk. Suppose you purchase two investments. Based on past experience, you believe that each has a mean gain of 5% with standard deviation of 2% (over some time frame). Let X = the percentage return from investment A, and let Y = the percentage return from investment B. If you invest equally, your percentage return will be $W = (X + Y)/2$. Find the variance of W if:

a) The returns are independent.
b) The returns have a correlation of -0.5 (covariance of -2).
c) The returns have a correlation of 0.5 (covariance of 2).
d) The returns have a correlation of 1.0 (covariance of 4).
e) Which of the above situations is least risky (the mean return is the same in each case)?

***86. Correlated RVs again** Suppose you purchase two investments. Based on past experience, you believe that each has a mean gain of 10% with standard deviation of 4% (over some time frame). Let X = the percentage return from investment A, and let Y = the percentage return from investment B. If you invest 70% in A and 30% in B, your percentage return will be $W = .7X + .3Y$. Find the variance of W if:

a) The returns are independent.
b) The returns have a correlation of -0.8 (covariance of -12.8).
c) The returns have a correlation of 0.2 (covariance of 3.2).
d) The returns have a correlation of -1.0 (covariance of -16).
e) Which of the above situations is least risky (the mean return is the same in each case)?

Just Checking ANSWERS

1. a)

Outcome	X = cost of repair	Probability
Recharging works	\$ 60	0.75
Replace control unit	\$200	0.25

b) $60(0.75) + 200(0.25) = \$95$
c) Car owners with this problem will spend \$95 on average to get it fixed.

2. a) $100 + 100 = 200$ seconds
b) $\sqrt{50^2 + 50^2} = 70.7$ seconds
c) The times for the two customers are independent.

3. There are two outcomes (contact, no contact). The probability of contact is 0.76, and random calls should be independent.

4. Binomial, with $n = 1000$ and $p = 0.76$. For actual calculations, we could approximate using a Normal model with $\mu = np = 1000(0.76) = 760$ and $\sigma = \sqrt{npq} = \sqrt{1000(0.76)(0.24)} \approx 13.5$.

5. Poisson, with $l = np = 10\,000\left(\frac{70\,000}{107\,000\,000}\right) = 6.54$.

MathXL

MyStatLab

Go to MathXL at www.mathxl.com or MyStatLab at www.mystatlab.com. You can practise exercises for this chapter as often as you want. The guided solutions will help you find answers step by step. You'll find a personalized study plan available to you too!

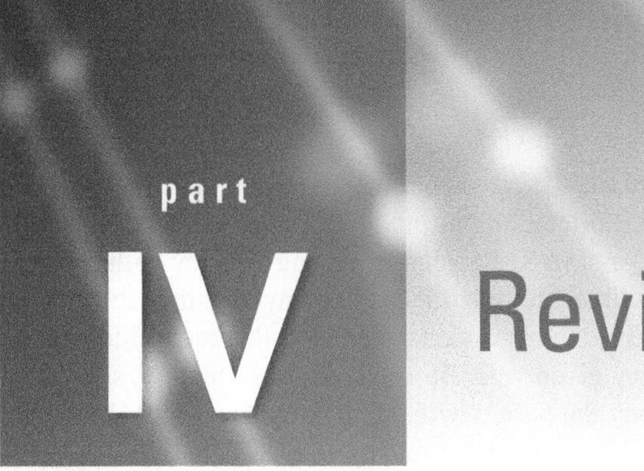

Review

Randomness and Probability

Quick Review

Here's a brief summary of the key concepts and skills in probability and probability modelling:

- The Law of Large Numbers says that the more times we try something, the closer the results will come to theoretical prediction.
  - Don't mistakenly misinterpret the Law of Large Numbers as the "Law of Averages." There's no such thing.
- Basic rules of probability can handle most situations.
  - To find the probability that an event OR another event happens, add their probabilities and subtract the probability that both happen.
  - To find the probability that an event AND another independent event both happen, multiply the probabilities (the second probability being conditional on the first event).
  - Conditional probabilities tell you how likely one event is to happen, knowing that another event has happened.
  - Mutually exclusive events (also called "disjoint") cannot both happen at the same time.
  - Two events are independent if the occurrence of one doesn't change the probability that the other happens.
- A probability model for a random variable describes the theoretical distribution of outcomes.
  - The mean of the theoretical distribution of a random variable is also called the expected value of the random variable.
  - For sums or differences of independent random variables, variances add.
  - To estimate probabilities involving quantitative variables, you may be able to use a Normal model—but only if the distribution of the variable is unimodal and symmetric.
  - To estimate the probability that you'll get a certain number of successes in a specified number of independent trials, use a Binomial model.
  - To estimate the probability that you'll get a certain number of occurrences of a relatively rare phenomenon or a certain number of occurrences of a phenomenon occurring over time or space, consider using a Poisson model.

Ready? Here are some opportunities to check your understanding of these ideas.

Review Exercises

1. **Quality control** A consumer organization estimates that 29% of new cars have a cosmetic defect, such as a scratch or a dent, when they are delivered to car dealers. This same organization believes that 7% have a functional defect—something that does not work properly—and that 2% of new cars have both kinds of problems.
   a) If you buy a new car, what's the probability that it has some kind of defect?
   b) What's the probability it has a cosmetic defect but no functional defect?
   c) If you notice a dent on a new car, what's the probability it has a functional defect?
   d) Are the two kinds of defects disjoint events? Explain.
   e) Do you think the two kinds of defects are independent events? Explain.

2. **Workers** A company's human resources officer reports a breakdown of employees by job type and sex as shown in the table.

		Sex	
		Male	Female
Job Type	Management	7	6
	Supervision	8	12
	Production	45	72

   a) What's the probability that a worker selected at random is
      i. female?
      ii. female or a production worker?
      iii. female, if the person works in production?
      iv. a production worker, if the person is female?

b) Do these data suggest that job type is independent of being male or female? Explain.

3. **Airfares** Each year a company must send three officials to a meeting in China and five officials to a meeting in France. Airline ticket prices vary from time to time, but the company purchases all tickets for a country at the same price. Past experience has shown that tickets to China have a mean price of $1200, with a standard deviation of $150, while the mean airfare to France is $900, with a standard deviation of $100.
   a) Define random variables and use them to express the total amount the company will have to spend to send these officials to the two meetings.
   b) Find the mean and standard deviation of this total cost.
   c) Find the mean and standard deviation of the difference in price of a ticket to China and a ticket to France.
   d) Do you need to make any assumptions in calculating these means? How about the standard deviations?

4. **Autism** Psychiatrists estimate that about 1 in 100 adults has autism. What's the probability that in a city of 20 000, there are more than 300 people with this condition? Be sure to verify that a Normal model can be used here.

5. **A game** To play a game, you must pay $5 for each play. There is a 10% chance you will win $5, a 40% chance you will win $7, and a 50% chance you will win only $3.
   a) What are the mean and standard deviation of your net winnings?
   b) You play twice. Assuming the plays are independent events, what are the mean and standard deviation of your total winnings?

6. **Emergency switch** Safety engineers must determine whether industrial workers can operate a machine's emergency shutoff device. Among a group of test subjects, 66% were successful with their left hands, 82% with their right hands, and 51% with both hands.
   a) What percent of these workers could not operate the switch with either hand?
   b) Are success with right and left hands independent events? Explain.
   c) Are success with right and left hands mutually exclusive? Explain.

7. **Facebook** According to Pew Research, 50% of adults and 75% of teenagers were using a social networking site in early 2012. Most of that activity was on Facebook, so let's assume these probabilities apply strictly to Facebook. Among a group of 10 people, what's the probability that
   a) at least one of the people was not on Facebook if they were all adults?
   b) at least one of the people was not on Facebook if they were all teenagers?
   c) at least one of the people was not on Facebook if half were teenagers?

8. **Deductible** A car owner may buy insurance that will pay the full price of repairing the car after an at-fault accident, or save $12 a year by getting a policy with a $500 deductible. Her insurance company says that about 0.5% of drivers in her area have an at-fault auto accident during any given year. Based on this information, should she buy the policy with the deductible or not? How does the value of her car influence this decision?

9. **More Facebook** Using the percentages from Exercise 7, suppose there is a group of five teenagers. What's the probability that
   a) all will be on Facebook?
   b) exactly one will be on Facebook?
   c) at least three will be on Facebook?

10. **At fault** The car insurance company in Exercise 8 believes that about 0.5% of drivers have an at-fault accident during a given year. Suppose the company insures 1355 drivers in that city.
    a) What are the mean and standard deviation of the number who may have at-fault accidents?
    b) Can you describe the distribution of these accidents with a Normal model? Explain.

11. **Friend me?** One hundred fifty-eight teenagers are standing in line for a big movie premier night. Refer to Exercise 7.
    a) What are the mean and standard deviation of the number of Facebook users we might expect?
    b) Can we use a Normal model in this situation?
    c) What the probability that no more than 110 of the teenagers are Facebook users?

12. **Child's play** In a board game, you determine the number of spaces you may move by spinning a spinner and rolling a die. The spinner has three regions: Half of the spinner is marked "5," and the other half is equally divided between "10"and "20." The six faces of the die show 0, 0, 1, 2, 3, and 4 spots. When it's your turn, you spin and roll, adding the numbers together to determine how far you may move.
    a) Create a probability model for the outcome on the spinner.
    b) Find the mean and standard deviation of the spinner results.
    c) Create a probability model for the outcome on the die.
    d) Find the mean and standard deviation of the die results.
    e) Find the mean and standard deviation of the number of spaces you get to move.

13. **Language** Neurological research has shown that in about 80% of people, language abilities reside in the brain's left side. Another 10% display right-brain language centres, and the remaining 10% have two-sided language control. (The latter two groups are mainly left-handers.)[1]

[1]*Science News*, 161 no. 24 [2002].

a) Assume that a first-year composition class contains 25 randomly selected people. What's the probability that no more than 15 of them have left-brain language control?
b) In a randomly assigned group of five of these students, what's the probability that no one has two-sided language control?
c) In the entire first-year class of 1200 students, how many would you expect to find of each type?
d) What are the mean and standard deviation of the number of these students who will be right-brained in language abilities?
e) If an assumption of Normality is justified, use the 68–95–99.7 Rule to describe how many students in the first-year class will have right-brain language control.

14. **Play again** If you land in a "penalty zone" on the game board described in Exercise 12, your move will be determined by subtracting the roll of the die from the result on the spinner. Now what are the mean and standard deviation of the number of spots you may move?

15. **Beanstalks** In some cities, tall people who want to meet and socialize with other tall people can join Beanstalk Clubs. To qualify, a man must be over 6′2″ tall and a woman over 5′10.″ According to the National Health Survey, heights of adults may have a Normal model with mean heights of 69.1″ for men and 64.0″ for women. The respective standard deviations are 2.8″ and 2.5″.
a) You're probably not surprised to learn that men are generally taller than women, but what does the greater standard deviation for men's heights indicate?
b) Are men or women more likely to qualify for Beanstalk membership?
c) Beanstalk members believe that height is an important factor when people select their spouses. To investigate, we select at random a married man and, independently, a married woman. Define two random variables, and use them to express how many inches taller the man is than the woman.
d) What's the mean of this difference?
e) What's the standard deviation of this difference?
f) What's the probability that the man is taller than the woman (that the difference in heights is greater than 0)?
g) Suppose a survey of married couples reveals that 92% of the husbands were taller than their wives. Based on your answer to part f), do you believe that people choose spouses independent of height? Explain.

16. **Stocks** Since the stock market began in 1872, stock prices have risen in about 73% of the years. Assuming that market performance is independent from year to year, what's the probability that
a) the market will rise for three consecutive years?
b) the market will rise three out of the next five years?
c) the market will fall during at least one of the next five years?
d) the market will rise during a majority of years over the next decade?

17. **Multiple choice** A multiple choice test has 50 questions, with four answer choices each. You must get at least 30 correct to pass the test, and the questions are very difficult.
a) Are you likely to be able to pass by guessing on every question? Explain.
b) Suppose, after studying for a while, you believe you have raised your chances of getting each question right to 70%. How likely are you to pass now?

18. **Stock strategy** Many investment advisors argue that after stocks have declined in value for two consecutive years, people should invest heavily because the market rarely declines three years in a row.
a) Since the stock market began in 1872, there have been two consecutive losing years eight times. In six of those cases, the market rose during the following year. Does this confirm the advice?
b) Overall, stocks have risen in value during 95 of the 130 years since the market began in 1872. How is this fact relevant in assessing the statistical reasoning of the advisors?

19. **Insurance** A 65-year-old woman takes out a $100 000 term life insurance policy. The company charges an annual premium of $520. Estimate the company's expected profit on such policies if mortality tables indicate that only 2.6% of women age 65 die within a year.

20. **Teen smoking** The Centers for Disease Control say that about 23% of high-school students smoke tobacco (down from a high of 38% in 1997). Suppose you randomly select high school students to survey them on their attitudes toward scenes of smoking in the movies. What's the probability that
a) none of the first four students you interview is a smoker?
b) there are no more than two smokers among 10 people you choose?

21. **Passing stats** Molly's university offers two sections of Statistics 101. From what she has heard about the two professors listed, Molly estimates that her chances of passing the course are 0.80 if she gets Professor Scedastic and 0.60 if she gets Professor Kurtosis. The registrar uses a lottery to randomly assign the 120 enrolled students based on the number of available seats in each class. There are 70 seats in Professor Scedastic's class and 50 in Professor Kurtosis's class.
a) What's the probability that Molly will pass Statistics?
b) At the end of the semester, we find out that Molly failed. What's the probability that she had Professor Kurtosis?

22. **Teen smoking II** Suppose that, as reported by the Centers for Disease Control, about 23% of high school students smoke tobacco. You randomly select 120 high school students to survey them on their attitudes toward scenes of smoking in the movies.
a) What's the expected number of smokers?

b) What's the standard deviation of the number of smokers?
c) The number of smokers among 120 randomly selected students will vary from group to group. Explain why that number can be described with a Normal model.
d) Using the 68–95–99.7 Rule, create and interpret a model for the number of smokers among your group of 120 students.

23. Random variables Given independent random variables with means and standard deviations as shown, find the mean and standard deviation of each of the variables below. The variables X_1 and X_2 are independent with the same distribution as X.

a) $X + 50$
b) $10Y$
c) $X + 0.5Y$
d) $X - Y$
e) $X_1 + X_2$

	Mean	SD
X	50	8
Y	100	6

24. Merger Explain why the facts you know about variances of independent random variables might encourage two small insurance companies to merge. (*Hint:* Think about the expected amount and potential variability in payouts for the separate and the merged companies.)

25. Youth survey According to a recent Gallup survey, 93% of teens use the Internet, but there are differences in how teen boys and girls say they use computers. The telephone poll found that 77% of boys had played computer games in the past week, compared with 65% of girls. On the other hand, 76% of girls said they had e-mailed friends in the past week, compared with only 65% of boys.

a) For boys, the cited percentages are 77% playing computer games and 65% using e-mail. That total is 142%, so there is obviously a mistake in the report. No? Explain.
b) Based on these results, do you think playing games and using e-mail are mutually exclusive? Explain.
c) Do you think whether a teen e-mails friends is independent of being a boy or a girl? Explain.

26. Meals A university student on a seven-day meal plan reports that the amount of money he spends daily on food varies with a mean of \$13.50 and a standard deviation of \$7.

a) What are the mean and standard deviation of the amount he might spend in two consecutive days?
b) What assumption did you make in order to find that standard deviation? Are there any reasons you might question that assumption?
c) Estimate his average weekly food costs and the standard deviation.
d) Do you think it likely he might spend less than \$50 in a week? Explain, including any assumptions you make in your analysis.

27. Travel to Kyrgyzstan Your pocket copy of *Kyrgyzstan on 4237 ± 360 Soms a Day* claims that you can expect to spend about 4237 soms each day with a standard deviation of 360 soms. How well can you estimate your expenses for the trip?

a) Your budget allows you to spend 90 000 soms. To the nearest day, how long can you afford to stay in Kyrgyzstan, on average?
b) What's the standard deviation of your expenses for a trip of that duration?
c) You doubt that your total expenses will exceed your expectations by more than two standard deviations. How much extra money should you bring? On average, how much of a "cushion" will you have per day?

28. Picking melons Two stores sell watermelons. At the first store, the melons weigh an average of 22 pounds, with a standard deviation of 2.5 pounds. At the second store, the melons are smaller, with a mean of 18 pounds and a standard deviation of 2 pounds. You select a melon at random at each store.

a) What's the mean difference in weights of the melons?
b) What's the standard deviation of the difference in weights?
c) If a Normal model can be used to describe the difference in weights, what's the probability that the melon you got at the first store is heavier?

29. Home sweet home 2011 According to Statistics Canada, 67% of Canadian households own the home they live in. A mayoral candidate conducts a survey of 820 randomly selected homes in your city and finds only 523 owned by the current residents. The candidate then attacks the incumbent mayor, saying that there is an unusually low level of home ownership in the city. Do you agree? Explain.

30. Buying melons The first store in Exercise 28 sells watermelons for 32 cents a pound. The second store is having a sale on watermelons—only 25 cents a pound. Find the mean and standard deviation of the difference in the price you may pay for melons randomly selected at each store.

31. Who's the boss 2011? According to Industry Canada, 17% of all small businesses were owned by women at the end of 2011. You call some small business firms doing business locally, assuming that the national percentage is true in your area.

a) What's the probability that the first three you call are all owned by women?
b) What's the probability that none of your first four calls finds a firm that is owned by a woman?
c) Suppose none of your first five calls found a firm owned by a woman. What's the probability that your next call does?

32. Jerseys A Statistics professor comes home to find that all four of his children got white team shirts from soccer camp this year. He concludes that this year, unlike other years, the camp must not be using a variety of colours. But then he finds out that in each child's age group there are four teams, only one of which wears white shirts. Each child just happened to get on the white team at random.

a) Why was he so surprised? If each age group uses the same four colours, what's the probability that all four kids would get the same colour shirt?
b) What's the probability that all four would get white shirts?
c) We lied. Actually, in the oldest child's group there are six teams instead of the four teams in each of the other three groups. How does this change the probability you calculated in part b)?

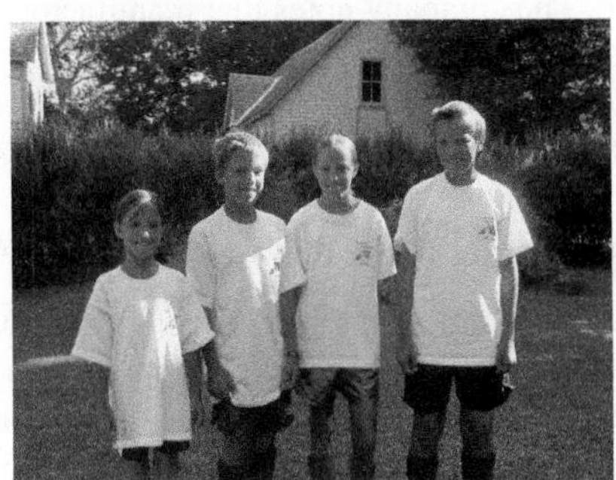

Courtesy of Richard De Veaux

33. **When to stop?** In Exercise 27 of the Review Exercises for Part III, we posed this question:

You play a game that involves rolling a die. You can roll as many times as you want, and your score is the total for all the rolls. But . . . if you roll a 6, your score is 0 and your turn is over. What might be a good strategy for a game like this?

You attempted to devise a good strategy by simulating several plays to see what might happen. Let's try calculating a strategy.

a) It can be shown that on average, a 6 would appear for the first time on the 6th role. So, roll *one time less* than that. Assuming all those rolls were not 6s, what's your expected score?
b) What's the probability that you can roll that many times without getting a 6?

34. **Plan B** Here's another attempt at developing a good strategy for the dice game in Exercise 33. Instead of stopping after a certain number of rolls, you could decide to stop when your score reaches a certain number of points.

a) How many points would you expect a roll to *add* to your score?
b) In terms of your current score, how many points would you expect a roll to *subtract* from your score?
c) Based on your answers in parts a) and b), at what score will another roll "break even"?
d) Describe the strategy this result suggests.

35. **Technology on campus 2005** Every five years, the Conference Board of the Mathematical Sciences surveys college math departments. In 2005, the board reported that 51% of all undergraduates taking Calculus I were in classes that used graphing calculators and 21% were in classes that used computer assignments. Suppose that 10% used both calculators and computers.

a) What percent used neither kind of technology?
b) What percent used calculators but not computers?
c) What percent of the calculator users had computer assignments?
d) Based on this survey, do calculator and computer use appear to be independent events? Explain.

36. **Dogs** A census by the municipal dog control officer found that 18% of homes kept one dog as a pet, 4% had two dogs, and 1% had three or more. If a salesman visits two homes selected at random, what's the probability he encounters

a) no dogs?
b) some dogs?
c) dogs in each home?
d) more than one dog in each home?

***37.** **O-rings** Failures of O-rings on the space shuttle are fairly rare, but often disastrous, events. If we are testing O-rings, suppose that the probability of a failure of any one O-ring is 0.01. Let X be the number of failures in the next 10 O-rings tested.

a) Clearly the Binomial model can be applied here. What other model might you use to model X?
b) What is the mean number of failures in the next 10 O-rings?
c) What is the probability that there is exactly one failure in the next 10 O-rings? Use the model from your answer to part a).
d) What is the probability that there is at least one failure in the next 10 O-rings? Use the model from your answer to part a).

***38.** **Volcanoes** Almost every year, there is some incidence of volcanic activity on the island of Japan. In 2005, there were five volcanic episodes, defined as either eruptions or sizable seismic activity. Suppose the mean number of episodes is 2.4 per year. Let X be the number of episodes in the two-year period 2013–2014.

a) What model might you use to model X?
b) What is the mean number of episodes in this period?
c) What is the probability that there will be no episodes in this period?
d) What is the probability that there are more than three episodes in this period?
e) How much activity would convince you that there is a significant increase in volcanic activity in the region?

39. **Socks** In your sock drawer you have four blue socks, five grey socks, and three black ones. Half asleep one

morning, you grab two socks at random and put them on. Find the probability you end up wearing
a) two blue socks.
b) no grey socks.
c) at least one black sock.
d) a green sock.
e) matching socks.

40. Coins A coin is tossed 36 times.
a) What are the mean and standard deviation of the number of heads?
b) Suppose the resulting number of heads is unusual, two standard deviations above the mean. How many "extra" heads were observed?
c) If the coin were tossed 100 times, would you still consider the same number of extra heads unusual? Explain.
d) In the 100 tosses, how many extra heads would you need to observe in order to say the results were unusual?
e) Explain how these results refute the Law of Averages but confirm the Law of Large Numbers.

41. The Drake equation Not the latest hit by the Toronto rapper, but an equation developed in 1961 by astronomer Frank Drake. He was trying to estimate the number of extraterrestrial civilizations in our galaxy that might be able to communicate with us via radio transmissions. Now largely accepted by the scientific community, the Drake equation has helped spur efforts by radio astronomers to search for extraterrestrial intelligence. It has been so Successful, that Drake says: "You Can Thank Me Later."

Here is the equation:

$$N_c = N \cdot f_p \cdot n_e \cdot f_l \cdot f_i \cdot f_c \cdot f_L$$

Okay, it looks a little messy, but here's what it means:

Factor	What It Represents	Possible Value
N	Number of stars in the Milky Way Galaxy	200–400 billion
f_p	Probability that a star has planets	20%–50%
n_e	Number of planets in a solar system capable of sustaining earth-type life	1? 2?
f_l	Probability that life develops on a planet with a suitable environment	1%–100%
f_i	Probability that life evolves intelligence	50%?
f_c	Probability that intelligent life develops radio communication	10%–20%
f_L	Fraction of the planet's life for which the civilization survives	$\frac{1}{1\,000\,000}$?
N_c	Number of extraterrestrial civilizations in our galaxy with which we could communicate	?

So, how many ETs are out there? That depends; values chosen for the many factors in the equation depend on ever-evolving scientific knowledge and one's personal guesses. But now, some questions.
a) What quantity is calculated by the first product $N \cdot f_p$?
b) What quantity is calculated by the product $N \cdot f_p \cdot n_e \cdot f_l \cdot f_i$?
c) What probability is calculated by the product $f_l \cdot f_i$?
d) Which of the factors in the formula are conditional probabilities? Restate each in a way that makes the condition clear.

Note: A quick Internet search will find you a site where you can Find Your Love with the Drake equation. Make Me Proud!

42. Recalls In a car rental company's fleet, 70% of the cars are American brands, 20% are Japanese, and the rest are German. The company notes that manufacturers' recalls seem to affect 2% of the American cars, but only 1% of the others.
a) What's the probability that a randomly chosen car is recalled?
b) What's the probability that a recalled car is American?

43. Pregnant? Suppose that 70% of the women who suspect they may be pregnant and purchase an in-home pregnancy test are actually pregnant. Further suppose that the test is 98% accurate. What's the probability that a woman whose test indicates that she is pregnant actually is?

44. Door prize You are among 100 people attending a charity fundraiser at which a large-screen television will be given away as a door prize. To determine who wins, 99 white balls and one red ball have been placed in a box and thoroughly mixed. The guests will line up and, one at a time, pick a ball from the box. Whoever gets the red ball wins the television, but if the ball is white, it is returned to the box. If none of the 100 guests gets the red ball, the television will be auctioned off for additional benefit to the charity.
a) What's the probability that the first person in line wins the television?
b) You are the third person in line. What's the probability that you win the television?
c) What's the probability that the charity gets to sell the television because no one wins?
d) Suppose you get to pick your spot in line. Where would you want to be to maximize your chances of winning?
e) After hearing some protest about the plan, the organizers decide to award the prize by not returning the white balls to the box, thus ensuring that one of the 100 people will draw the red ball and win the television. Now what position in line would you choose to maximize your chances?

chapter 15

Sampling Distribution Models

The n's justify the means.

—Apocryphal statistical saying

Where are we going?

A 2008 poll of 1000 randomly sampled Canadians reported that 91% believe that Canada's healthcare system is better than that of the United States.[1] But how much can we trust this statistic? After all, only 1000 people were polled, leaving out of the study some 22 million or so other Canadian adults. Maybe the true proportion of all Canadians who feel this way is 95% or 85%, or could it be as little as 75%? Is almost anything possible? Just how reliable are proportions based on random samples? We'll see in this chapter that we can be surprisingly precise about how much we expect proportions from random samples to vary. This will enable us to start generalizing from samples we have at hand to the population at large.

The Canadian military mission in Afghanistan was a controversial one. In late April 2007, after a rash of troop fatalities earlier that month, an Angus Reid poll of 1009 randomly chosen Canadians reported that 52% were in favour of an early withdrawal of troops (before the actual mission mandate end). At nearly the same time, a Strategic Counsel poll of 1000 Canadians reported 64% in favour of an early withdrawal. Why the difference? Should we be surprised to find that we could get proportions this different from (presumably) properly selected random samples drawn from the same population? You're probably used to seeing that observations vary, but how much variability among polls should we expect to see?

Why do sample proportions vary at all? How can surveys conducted at essentially the same time by organizations asking the same questions get different results? The answer is at the heart of Statistics. The proportions vary from sample to sample because the samples are composed of different people.

It's actually pretty easy to predict how much a proportion will vary under circumstances like this. Understanding the variability of our estimates will let us actually use that variability to better understand the world.

15.1 Sampling Distribution of a Proportion

We've talked about *Think*, *Show*, and *Tell*. Now we have to add *Imagine*. To understand the Strategic Counsel poll, we want to imagine the results from all the random samples of size 1000 that Strategic Counsel didn't take. What would the histogram of all the sample proportions look like?

For Canadians' support for early withdrawal, where do you expect the centre of that histogram to be? Of course, we don't *know* the answer to that (and probably never will). But we know that it will be at the true proportion in the population, and we can call that p. (See the Notation Alert on the next page.) For the sake of discussion here, let's suppose that *60% of all Canadian adults* supported early withdrawal, so we'll use $p = 0.60$.

How about the *shape* of the histogram? We don't have to just imagine. We can simulate. We want to simulate a bunch of random samples that we didn't really draw. To do that, we'll simulate many independent random samples of size 1000, keeping the same probability of success. We chose 0.60 for p as a reasonable value. Here's a histogram of the sample proportions (denoted $\hat{p}$) supporting early withdrawal for 2000 independent samples of 1000 adults when the true proportion is $p = 0.60$.

It should come as no surprise that we don't get the same proportion for each sample we draw, even though the underlying true value is the same for the population. Each $\hat{p}$ comes from a different simulated sample. The histogram in Figure 15.1 is a simulation of

A S *Activity:* **Sampling Distribution of a Proportion.** You don't have to imagine—you can simulate.

[1] In the same June 2008 study, 1000 Americans were polled and 45% of them thought Canada's healthcare system was superior.

Figure 15.1
A histogram of sample proportions for 2000 simulated samples of 1000 adults drawn from a population with $p = 0.60$. The sample proportions vary, but their distribution is centred at the true proportion, p.

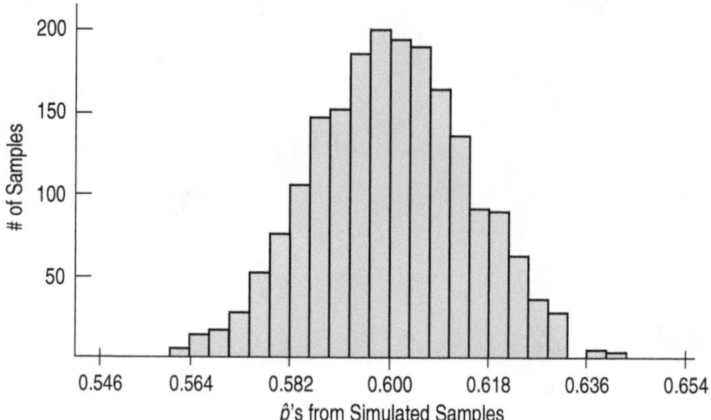

Who	Canadian adults
What	When to bring the troops home
When	April 2007
Where	Canada
Why	Public attitudes

IMAGINE

We see only the sample that we actually drew, but by simulating or modelling, we can *Imagine* what we might have seen had we drawn other possible random samples.

NOTATION ALERT

We use p as our symbol for the true (generally unknown) proportion of successes in the population.[2] This violates our "Greek letters for parameters" rule, but if we stuck to that, our natural choice would be π, which we prefer to reserve for the universal constant 3.1415926 . . . Recall that in Chapter 7 we introduced $\hat{y}$ as the predicted value for y. The "hat" here plays a similar role. It indicates that the observed proportion in our data is our *estimate* of the parameter p. Likewise, the sample proportion of failures estimates the true failure proportion and is $\hat{q}$. The "hat" signifies an estimate of what is underneath.

what we'd get if we could see *all the proportions from all possible samples*. That distribution has a special name. It is called the **sampling distribution** of the proportion.

Does it surprise you that the histogram is unimodal? Symmetric? That it is centred at p? You probably don't find any of this shocking. Does the shape remind you of any model we've already discussed? It's an amazing and fortunate fact that a Normal model is just the right one for the histogram of sample proportions.

This fact was proved in 1718 by the French mathematician Abraham de Moivre, though there is no reason you should guess that the Normal model would be the one we need here.[3]

Modelling how sample proportions vary from sample to sample is one of the most powerful ideas we'll see in this course. A **sampling distribution model** for how a sample proportion varies from sample to sample allows us to quantify that variation and to talk about how likely it is that we'd observe a sample proportion in any particular interval.

To use a Normal model, we need to specify two parameters: its mean and standard deviation. The centre of the histogram is naturally at p, so we'll put μ, the mean of the Normal, at p.

What about the standard deviation? Usually, the mean gives us no information about the standard deviation. Suppose we told you that a batch of bike helmets had a mean diameter of 26 centimetres and asked what the standard deviation was. If you said, "I have no idea," you'd be exactly right. There's no information about σ from knowing the value of μ.

But there's a special fact about proportions. With proportions we get something for free. Once we know the mean, p, we automatically also know the standard deviation for $\hat{p}$, the proportion of successes:

$$\sigma(\hat{p}) = SD(\hat{p}) = \sqrt{\frac{p(1-p)}{n}} = \sqrt{\frac{pq}{n}}$$

When we draw simple random samples of n individuals, the proportions we find will vary from sample to sample. As long as n is reasonably large,[4] we can model the distribution of these sample proportions with a probability model that is

$$N\left(p, \sqrt{\frac{pq}{n}}\right)$$

Although we'll never know the true proportion of adults who favour early withdrawal, we're supposing it for the moment to be 60%. Once we put the centre at $p = 0.60$, the standard deviation for the Strategic Counsel poll is

$$SD(\hat{p}) = \sqrt{\frac{pq}{n}} = \sqrt{\frac{(0.60)(0.40)}{1000}} = 0.0155, \text{ or } 1.55\%$$

North Wind Picture Archives/Alamy

Abraham de Moivre (1667–1754)

[2]But be careful. We've already used capital P for a general probability. And we'll soon see another use of P in Chapter 17. There are a lot of ps in this course; you'll need to think clearly about the context to keep them straight.

[3]Well, the fact that we spent most of Chapter 5 on the Normal model might have been a hint.

[4]For smaller n, we can just use a Binomial model, which we saw in Chapter 14.

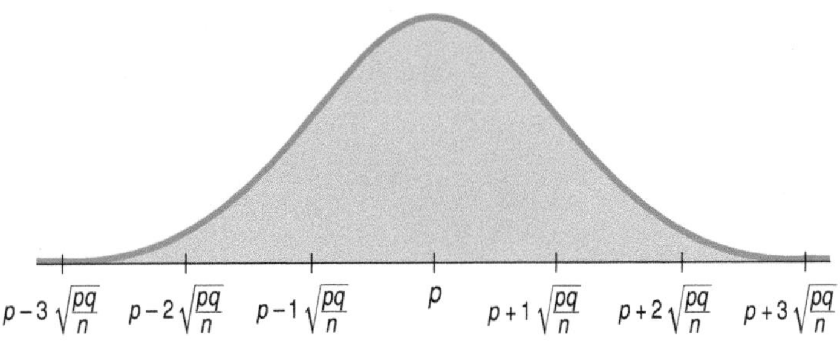

Figure 15.2

A Normal model centred at p with a standard deviation of $\sqrt{\frac{pq}{n}}$ is a good model for a collection of proportions found for many random samples of size n from a population with success probability p.

Here's a picture of the Normal model for our simulation histogram:

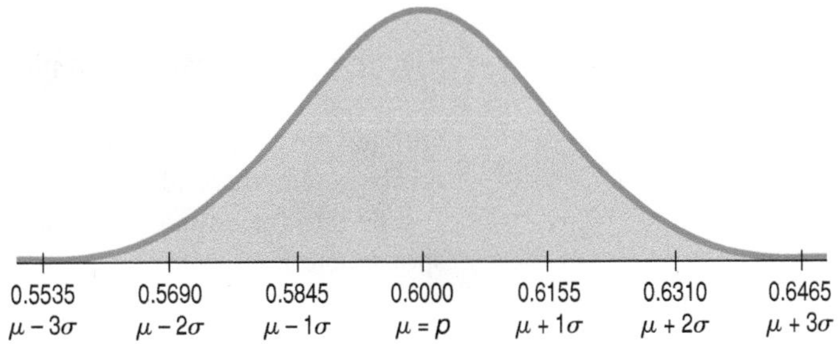

Figure 15.3

Using 0.60 for p gives this Normal model for Figure 15.1's histogram of the sample proportions of adults favouring early withdrawal (n = 1000).

A S *Simulation:* **Simulating Sampling Distributions.** Watch the Normal model appear from random proportions.

A S *Simulation:* **The Standard Deviation of a Proportion.** Do you believe this formula for standard deviation? Don't just take our word for it—convince yourself with an experiment.

Because we have a Normal model, we can use the 68–95–99.7 Rule or look up other probabilities using a table or technology. For example, we know that 95% of Normally distributed values are within two standard deviations of the mean, and 99.7% are within three standard deviations, so we should expect about 19 out of 20 such polls to give results within $2 \times 1.55\% = 3.1\%$[5] of the true percentage, and nearly all to give results within $3 \times 1.55\% = 4.65\%$ of the truth. So, if we assume for the moment that the true percentage is 60%, the Strategic Counsel result of 64% seems possible, if unlikely (falling between two and three standard deviations from the centre), but the other poll result of 52% would be essentially impossible,[6] as it reaches out more than five standard deviations from the centre!

This very natural and expected variability of sample proportions about the true population proportion as we imagine going from one random sample to another is called **sampling error**, though a better name for it might be **sampling variability**. Even considering sampling error, our guess of 60% isn't looking so good; neither poll seems likely if the truth is 60% (and the difference between them is too big to just be sampling error)—but is there just one truth here?[7]

MATH BOX

How do we know that sample proportions have a mean of p and a standard deviation of $\sqrt{\frac{pq}{n}}$?

Remember from Chapter 14 that if we have a sequence of independent trials where we record a success or failure on each trial (Bernoulli trials), that the total number of successes, Y, out of n trials has a *mean* of np and a *standard deviation* of $\sqrt{npq}$.

[5]The standard deviation always has the same units as the data. Here, our units are percentage points. But that can be confusing, because the standard deviation is not 1.55% of anything; it is 1.55 percentage points. If that's confusing, write the units as "percentage points" instead of percent.

[6]We ignore the small difference in sample size, 1000 versus 1009, since the effect on the standard deviation is negligible.

[7]While the polls were only a couple of days apart, there were fine differences in the wording of the questions, possible slight differences in the sampled populations, and a high "not sure" rate of 14% in the Angus Reid poll. Eliminating the "not sure" makes it 60.5% for early withdrawal. The lesson here is that in addition to *sampling error*, there is always *nonsampling error* affecting results. In this chapter, though, we focus on understanding sampling errors—which can be modelled precisely (what a relief!).

The sample proportion $\hat{p}$ is found by taking the total number of successes Y and dividing by the sample size:

$$\hat{p} = \frac{Y}{n}$$

So,

$$E(\hat{p}) = E\left(\frac{Y}{n}\right) = \frac{E(Y)}{n} = \frac{np}{n} = p$$

the true proportion of successes in the population.

When the expected value of a sample statistic equals a population parameter, we say that the statistic is an *unbiased estimator* of the parameter. So we have just proven that the sample proportion is an unbiased estimator of the true population proportion for simple random samples. Sample proportions will generally miss the target (the true proportion), of course, but there will be no systematic tendency to over- or undershoot the target (though bias can still be introduced via nonsampling errors such as undercoverage, nonresponse, etc.).

And

$$SD(\hat{p}) = SD\left(\frac{Y}{n}\right) = \frac{SD(Y)}{n} = \frac{\sqrt{npq}}{\sqrt{n^2}} = \sqrt{\frac{npq}{n^2}} = \sqrt{\frac{pq}{n}}$$

If we know the mean, p, of the sample proportion, then we know the standard deviation as well. This is a special property of sample proportions.

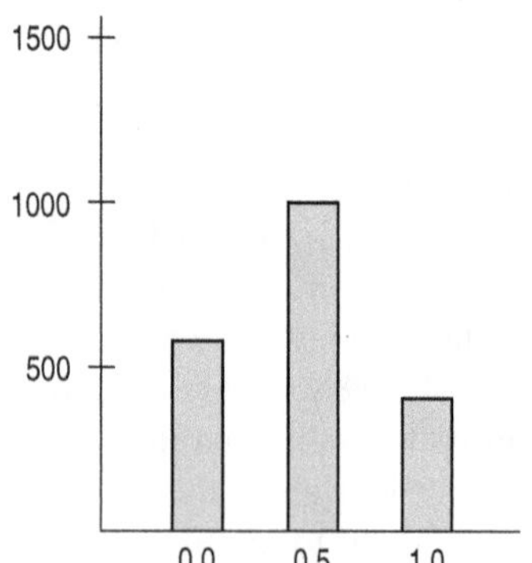

Figure 15.4
Proportions from samples of size two can take on only three possible values. A Normal model does not work well.

How Good Is the Normal Model?

Stop and think for a minute about what we've just said. It's a remarkable claim. We've said that if we draw repeated random samples of the same size, n, from some population and measure the proportion of successes, $\hat{p}$, we see in each sample, then the collection of these proportions will pile up around the underlying population proportion, p, and that a histogram of the sample proportions can be modelled well by a Normal model.

There must be a catch. Suppose the samples were of size two, for example. Then the only possible proportion values would be 0, 0.5, and 1. There's no way the histogram, such as in Figure 15.4, could ever look like a Normal model with only three possible values for the variable.

Well, there *is* a catch. The claim is only approximately true. (But that's okay. After all, models are only supposed to be approximately true.) And the model becomes a better and better representation of the distribution of the sample proportions as the sample size gets bigger.[8] Samples of size one or two just aren't going to work very well. But the distributions of proportions from many larger samples do have histograms that are remarkably close to a Normal model as long as p isn't too close to 0 or 1.

Populations with a true proportion, p, close to 0 or 1 can be a problem. Suppose a basketball coach surveys students to see how many male high school seniors are over 2.0 metres. What will the proportions in samples of size 1000 look like? If the true proportion of students that tall is 0.001, then the coach is likely to get only a few seniors over 2.0 m in any random sample of 1000. Most samples will have proportions of $\{0/1000, 1/1000, 2/1000, 3/1000, 4/1000, \ldots\}$ with very few samples having a higher proportion. Figure 15.5 shows results from a simulation of 2000 surveys of size 1000 with $p = 0.001$.

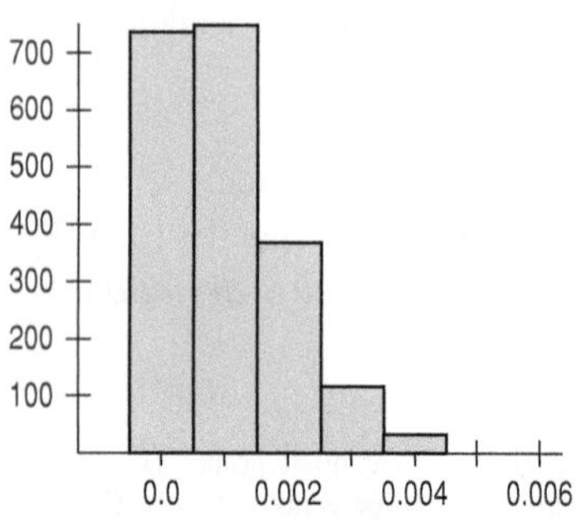

Figure 15.5
The distribution of sample proportions for 2000 samples of size 1000 with $p = 0.001$. Because the true proportion is so small, the sampling distribution is skewed to the right and the Normal model won't work well.

The problem is that the distribution is skewed to the right because p is so close to 0. (Had p been very close to 1, it would have been skewed to the left). So, even though n is large, p is too small, and so the Normal model still won't work well.

[8]Formally, we say the claim is true in the limit as n grows.

15.2 Assumptions and Conditions

To use a model, we usually must make some assumptions. To use the Normal sampling distribution model for a sample proportion, we need two assumptions:

The Independence Assumption: The sampled values must be *independent* random draws from the population under study.

The Sample Size Assumption: The sample size, n, must be large enough.

Of course, assumptions are hard—often impossible—to check. That's why we *assume* them. But, as we saw in Chapter 7, we should check to see whether the assumptions are reasonable. To think about the Independence Assumption, we often wonder whether there is any reason to think that the data values might affect each other. Fortunately, we can often check *conditions* that provide information about the assumptions. Check these conditions before using the Normal model to model the distribution of sample proportions:

Randomization Condition: If your data come from an experiment, subjects should have been randomly assigned to treatments. If you have a survey, your sample should be a *simple random sample* of the population. In practice, other more practical sampling designs are often used (e.g., multistage samples), and while the sampling distribution will still tend to be Normal for large enough n, our formula for the SD of the proportion falls apart and often gives an underestimate of the true SD. Statisticians then have to derive the correct formula mathematically, or they may decide to "bootstrap" it (a computer-intensive approach, discussed in Chapter 30). Statistics Canada bootstraps nearly all of their SDs due to complex sample designs.

If not a proper random sample, try to be sure that the sampling method was not biased and that the data values were drawn in an independent fashion and may reasonably be thought of as representative of some relevant population.

10% Condition: The sample size, n, must be no larger than 10% of the population in order to validate our formula for the SD. For national polls, the total population is usually very large, so the sample is a small fraction of the population.

Success/Failure Condition: The sample size has to be big enough so that we expect at least 10 successes and at least 10 failures. When np and nq are at least 10, we have enough data for sound conclusions. For the Strategic Counsel survey, a "success" might be favouring early withdrawal. If $p = 0.60$, we expect $1000 \times 0.60 = 600$ successes and $1000 \times 0.40 = 400$ failures. Both are at least 10, so we certainly expect enough successes and enough failures for the condition to be satisfied.

The terms "success" and "failure" for the outcomes that have probability p and q are common in statistics. But they are completely arbitrary labels. When we say that a disease occurs with probability p, we certainly don't mean that getting sick is a "success" in the ordinary sense of the word.

These last two conditions seem to conflict with each other. The **Success/Failure Condition** wants sufficient data. How much depends on p. If p is near 0.5, we need a sample of only 20 or so. If p is only 0.01, however, we'd need a sample of 1000. But the **10% Condition** says that a sample should be no larger than 10% of the population. If the sample were more than 10% of the population, this is certainly not a bad thing, but we would need to make adjustments to our method for analyzing the data.[9] However, this is rare in practice. Often, as in polls that sample from all Canadian adults or industrial samples from a day's production, the populations are much larger than 10 times the sample size.

A Sampling Distribution Model for a Proportion

We've simulated repeated samples and looked at a histogram of the sample proportions. We modelled that histogram with a Normal model. Why do we bother to model it? Because this model will give us insight into how much the sample proportion will vary from sample to

[9]See optional section 15.5: Finite Population Correction.

TERMINOLOGY ALERT

A word of caution. In earlier chapters, we would plot and examine the *distribution of the sample*, a display of the actual data that were collected in one sample. But now we've plotted a *sampling distribution*—a display of summary statistics ($\hat{p}$s, for example) for many different samples. "Sample distribution" and "sampling distribution" sound a lot alike, but they refer to very different things. (Sorry about that—we didn't make up the terms.) And the distinction is critical. Whenever you read or write something about one of these, think very carefully about what the words signify.

sample. We've simulated many of the other random samples we might have gotten. The model is an attempt to show the distribution from *all* possible random samples. But how do we know that a Normal model will really work? Is this just an observation based on some simulations that *might* be approximately true some of the time?

It turns out that this model can be justified theoretically and that the larger the sample size, the better the model works. That's the result de Moivre proved. We won't bother you with the math because, in this instance, it really wouldn't help your understanding.[10] Nevertheless, the fact that we can think of proportions from random samples as random quantities and then say something this specific about their distribution is a fundamental insight—one that we will use in each of the next four chapters.

We have changed our point of view in a very important way. No longer is a proportion just something we compute for a set of data. We now see it as a random quantity that has a (probability) distribution, and thanks to de Moivre we have a model for that distribution. We call that the **sampling distribution model for the proportion**, and we'll make good use of it. When the assumptions and conditions mentioned earlier are met, you can use the Normal model to model the distribution of the sample proportion. In other words:

We have now answered the question raised at the start of the chapter. To know how variable a sample proportion is, we need to know the proportion and the size of the sample. That's all.

THE SAMPLING DISTRIBUTION MODEL FOR A PROPORTION

Provided that the sampled values are independent and the sample size is large enough, the sampling distribution of $\hat{p}$ is modelled by a Normal model with mean p and standard deviation $SD(\hat{p}) = \sqrt{\frac{pq}{n}}$.

Without the sampling distribution model, the rest of Statistics just wouldn't exist. Sampling models are what make Statistics work. They inform us about the amount of variation we should expect when we sample. Suppose we spin a coin 100 times to decide whether it's fair. If we get 52 heads, we're probably not surprised. Although we'd expect 50 heads, 52 doesn't seem particularly unusual for a fair coin. But we would be surprised to see 90 heads; that might really make us doubt that the coin is fair. How about 64 heads? Harder to say. That's a case where we need the sampling distribution model. The sampling model quantifies the variability, telling us how surprising any sample proportion is. And it enables us to make informed decisions about how precise our estimate of the true value of a parameter might be. That's exactly what we'll be doing for the rest of this book.

These sampling distributions are very abstract in nature. Even though we depend on them, we never actually get to see them. We only imagine (or simulate) them. They're important because they enable us to say something about the population when all we have is data from a sample. This is the huge leap of Statistics. Rather than thinking about the sample proportion as a fixed quantity calculated from our data, we now think of it as a random variable—our value is just one of many we might have seen had we chosen a different random sample. By imagining what *might* happen if we were to draw many, many samples from the same population, we can learn a lot about how close the statistics computed from our one particular sample may be to the corresponding population parameters they estimate. That's the path to the *margin of error* you hear about in polls and surveys. We'll see how to determine that in the next chapter.

[10]The proof is pretty technical. We're not sure it helps *our* understanding all that much either.

[11]BMI = mass in kg/(height in m)^2

For Example USING THE SAMPLING DISTRIBUTION MODEL FOR PROPORTIONS

The U.S. Centers for Disease Control and Prevention report that 22% of 18-year-old women in the United States have a body mass index (BMI)[11] of 25 or more—a value considered by the U.S. National Heart Lung and Blood Institute to be associated with increased health risk.

As part of a routine health check at a large university, the physical education department usually requires students to come in to be measured and weighed. This year, the department decided to try out a self-report system. It asked 200 randomly selected female students to report their heights and weights (from which their BMIs could be calculated). Only 31 of these students had BMIs greater than 25.

QUESTION: Is this proportion of high-BMI students unusually small?

ANSWER: First, check the conditions:

✓ **Randomization Condition:** The department drew a random sample, so the respondents should be independent and randomly selected from the population.

✓ **10% Condition:** 200 respondents is less than 10% of all the female students at a "large university."

✓ **Success/Failure Condition:** The department expected $np = 200(0.22) = 44$ "successes" and $nq = 200(0.78) = 156$ "failures," both at least 10.

It's okay to use a Normal model to describe the sampling distribution of the proportion of respondents with BMIs above 25.

The phys ed department observed $\hat{p} = \frac{31}{200} = 0.155$.

The department expected $E(\hat{p}) = p = 0.22$, with $SD(\hat{p}) = \sqrt{\frac{pq}{n}} = \sqrt{\frac{(0.22)(0.78)}{200}}$ $= 0.029$, so $z = \frac{\hat{p} - p}{SD(\hat{p})} = \frac{0.155 - 0.22}{0.029} = -2.24$.

By the 68–95–99.7 Rule, I know that values more than two standard deviations below the mean of a Normal model show up less than 2.5% of the time. Perhaps women at this university differ from the general population, or self-reporting may not provide accurate heights and weights.

Just Checking

1. You want to poll a random sample of 100 students on campus to see if they are in favour of the proposed location for the new student centre. Of course, you'll get just one number, your sample proportion, $\hat{p}$. But if you imagined all the possible samples of 100 students you could draw and imagined the histogram of all the sample proportions from these samples, what shape would it have?
2. Where would the centre of that histogram be?
3. If you think that about half the students are in favour of the plan, what would the standard deviation of the sample proportions be?

Step-by-Step Example WORKING WITH SAMPLING DISTRIBUTION MODELS FOR PROPORTIONS

Suppose that about 13% of the population is left-handed.[12] A 200-seat school auditorium has been built with 15 "lefty seats," seats that have the built-in desk on the left rather than the right arm of the chair. (For the right-handed readers among you, have you ever tried to take notes in a chair with the desk on the left side?)

Question: In a class of 90 students, what's the probability that there will not be enough seats for the left-handed students?

THINK ➡ Plan State what we want to know.

I want to find the probability that in a group of 90 students, more than 15 will be left-handed. Since 15 out of 90 is 16.7%, I need the probability of finding more than 16.7% left-handed students out of a sample of 90 if the proportion of lefties is 13%.

Model Think about the assumptions and check the conditions.

You might be able to think of cases where the **Independence Assumption** is not plausible—for example, if the students are all related, or if they were selected for being left- or right-handed. But for students in a class, the assumption of independence with respect to handedness seems reasonable.

✓ **Independence Assumption:** The 90 students in the class are not a random sample, but it is reasonable to assume that the probability that one student is left-handed is not changed by the fact that another student is right- or left-handed. It is also reasonable to assume that the students in the class are representative of students in general with respect to handedness.

✓ **10% Condition:** Ninety is surely less than 10% of all students so there is no need for special formulas. (Even if the school itself is small, I'm thinking of the population of all possible students who could have gone to the school.)

✓ **Success/Failure Condition:**

$$np = 90(0.13) = 11.7 \geq 10$$
$$nq = 90(0.87) = 78.3 \geq 10$$

State the parameters and the sampling distribution model.

[We could also work with the count instead of the proportion and use the Binomial model with Normal approximation, as in Chapter 14]

The population proportion is $p = 0.13$. The conditions are satisfied, so I'll model the sampling distribution of $\hat{p}$ with a Normal model with mean 0.13 and a standard deviation of

$$SD(\hat{p}) = \sqrt{\frac{pq}{n}} = \sqrt{\frac{(0.13)(0.87)}{90}} \approx 0.035$$

My model for $\hat{p}$ is N(0.13, 0.035).

SHOW ➡ Plot Make a picture. Sketch the model and shade the area we're interested in, in this case the area to the right of 16.7%.

Mechanics Use the standard deviation as a ruler to find the z-score of the cutoff proportion. We see that 16.7% lefties would be just over one standard deviation above the mean.

Find the resulting probability from a table of Normal probabilities, a computer program, or a calculator.

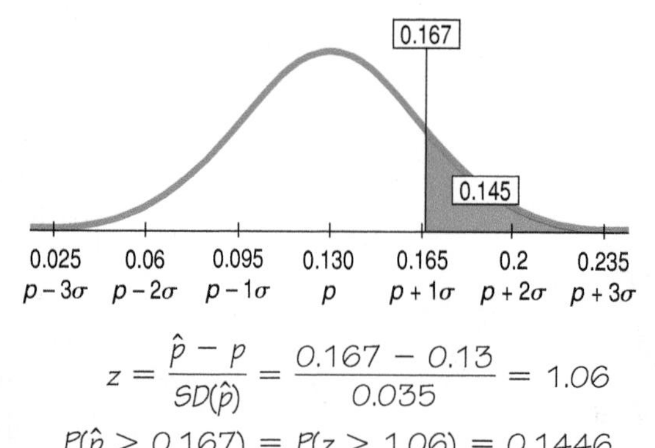

$$z = \frac{\hat{p} - p}{SD(\hat{p})} = \frac{0.167 - 0.13}{0.035} = 1.06$$

$$P(\hat{p} > 0.167) = P(z > 1.06) = 0.1446$$

[12] Actually, it's quite difficult to get an accurate estimate of the proportion of lefties in the population. Estimates range from 8% to 15%.

TELL ➡ Conclusion Interpret the probability in the context of the question.	There is about a 14.5% chance that there will not be enough seats for the left-handed students in the class.

15.3 The Sampling Distribution of Other Statistics

The sampling distribution for the proportion is especially useful because the Normal model provides such a good approximation. But, it might be useful to know the sampling distribution for *any* statistic that we can calculate, not just the sample proportion. Is the Normal model a good model for all statistics? Would you expect the sampling distribution of the minimum or the maximum or the variance of a sample to be Normally distributed? What about the median? Here's an example to help us find out.

For Example SIMULATING THE SAMPLING DISTRIBUTIONS OF OTHER STATISTICS

In Chapter 5, we looked at a study of body measurements of 250 men. The median of all 250 men's weights is 176.125 lbs and the variance is 730.9 lbs^2. We can treat these 250 men as a population, draw repeated random samples of 10, and compute the median, the variance, and the minimum for each sample.

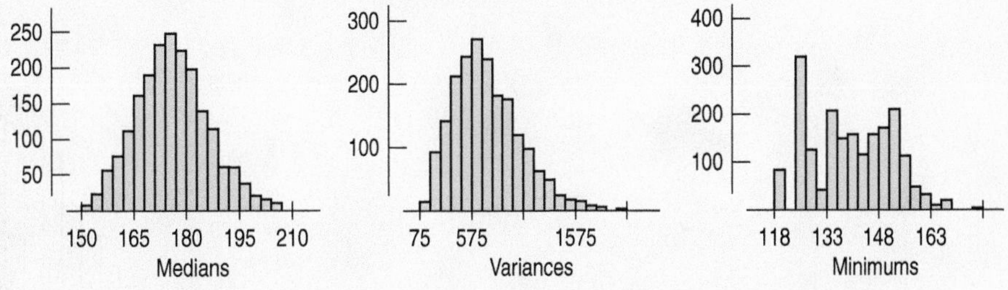

Each of these histograms depicts the estimated sampling distribution of its respective statistic. And it is easy to see that they aren't all the same. The sampling distribution of the medians is unimodal and symmetric. The sampling distribution of the variances is skewed to the right. And the sampling distribution of the minimums is, well, messy.

We can simulate to find the sampling distribution of *any* statistic we like: the maximum, the IQR, the 37th percentile, anything. But, both the proportion and the mean (as we'll see in the next section) have sampling distributions that are well approximated by a Normal model. And that observation hints at a fundamental fact about the sampling distribution models for the two summary statistics that we use most often.

Simulating the Sampling Distribution of a Mean

Here's a simple simulation. Let's start with one fair die. If we toss this die 10 000 times, what should the histogram of the numbers on the face of the die look like? Here are the results of a simulated 10 000 tosses:

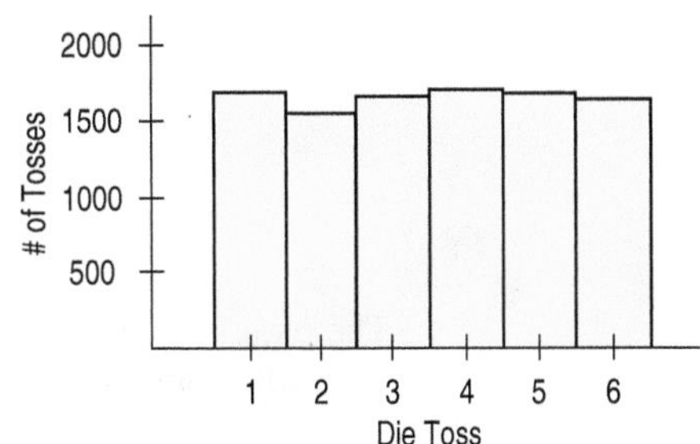

Now let's toss a *pair* of dice and record the average of the two. If we repeat this (or at least simulate repeating it) 10 000 times, recording the average of each pair, what will the histogram of these 10 000 averages look like? Before you look, think about it for a minute. Is getting an average of 1 on *two* dice as likely as getting an average of 3 or 3.5?

Let's see:

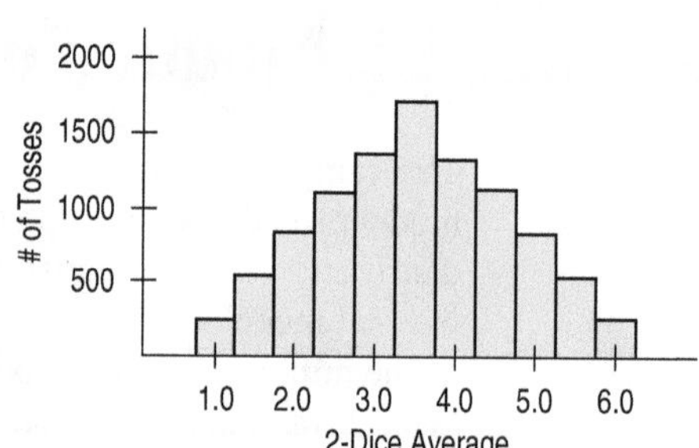

We're much more likely to get an average near 3.5 than we are to get one near 1 or 6. Without calculating those probabilities exactly, it's fairly easy to see that the *only* way to get an average of 1 is to get two ones. To get a total of seven (for an average of 3.5), though, there are many more possibilities. This distribution even has a name: the *triangular* distribution.

What if we average three dice? We'll simulate 10 000 tosses of three dice and take their average:

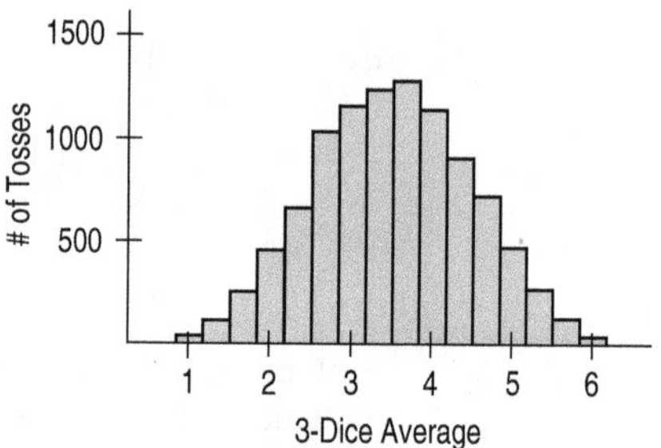

What's happening? First notice that it's getting harder to have averages near the ends. Getting an average of 1 or 6 with three dice requires all three to come up 1 or 6, respectively. That's less likely than for two dice to come up both 1 or both 6. The distribution is being pushed toward the middle. But what's happening to the shape? (This distribution doesn't have a name, as far as we know.)

Let's continue this simulation to see what happens with larger samples. Here's a histogram of the averages for 10 000 tosses of five dice:

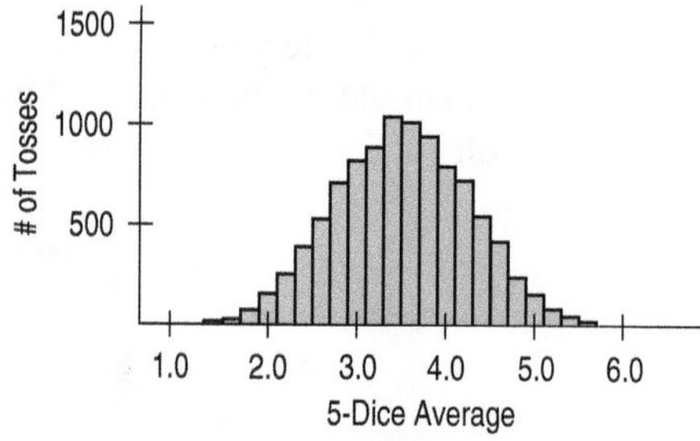

The pattern is becoming clearer. Two things continue to happen. The first fact we knew already from the Law of Large Numbers. It says that as the sample size (number of dice) gets larger, each sample average is more likely to be closer to the population mean. So we see the shape continuing to tighten around 3.5. But the shape of the distribution is the surprising part. It's becoming bell-shaped. And not just bell-shaped; it's approaching the Normal model.

Are you convinced? Let's skip ahead and try 20 dice. The histogram of averages for 10 000 throws of 20 dice looks like this:

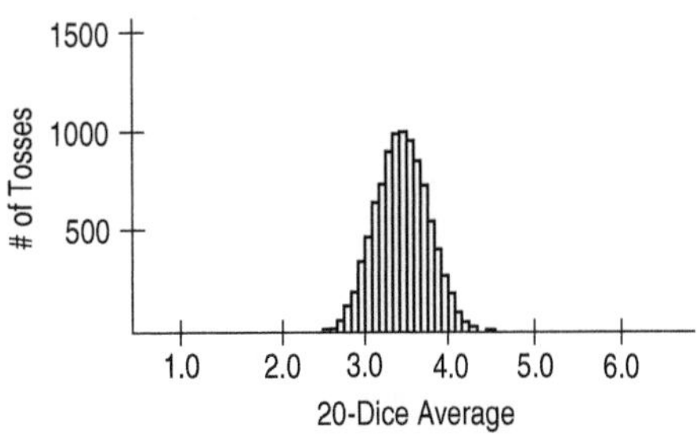

A S ***Activity:* The Sampling Distribution Model for Means.** Don't just sit there reading about the simulation—do it yourself.

Now we see the Normal shape again (and notice how much smaller the spread is). But can we count on this happening for situations other than dice throws? What kinds of sample means have sampling distributions that we can model with a Normal model? It turns out that Normal models work well amazingly often.

15.4 The Central Limit Theorem: The Fundamental Theorem of Statistics

The dice simulation may look like a special situation. But it turns out that what we saw with dice is true for means of repeated samples for almost every situation. When we looked at the sampling distribution of a proportion, we had to check only a few conditions. For means, the result is even more remarkable. *There are almost no conditions at all.*

Let's say that again: The sampling distribution of *any* mean becomes more nearly Normal as the sample size grows. All we need is for the observations to be independent and collected with randomization. We don't even care about the shape of the population distribution![13] This surprising fact is the result of a famous French mathematician named Pierre-Simon Laplace who proved a fairly general form in 1810. At the time, Laplace's theorem caused quite a stir (at least in mathematics circles) because it is so unintuitive. Laplace's result is called the **Central Limit Theorem** [14] (CLT).

Georgios Kollidas/Shutterstock

"The theory of probabilities is at bottom nothing but common sense reduced to calculus."

—*Laplace, in Théorie analytique des probabilités, 1812*

Why should the Normal model show up again for the sampling distribution of means as well as proportions? We're not going to try to persuade you that it is obvious, clear, simple, or straightforward. In fact, the CLT is surprising and a bit weird. Not only does the distribution of means of many random samples get closer and closer to a Normal model as the sample size grows, *this is true regardless of the shape of the population distribution*! Even if we sample from a skewed or bimodal population, the Central Limit Theorem tells us that means of repeated random samples will tend to follow a Normal model as the sample size grows. Of course, you won't be surprised to learn that it works better and faster the closer the population distribution is to a Normal model. And it works better for larger samples. If the data come from a population that's exactly Normal to start with, then

[13]Okay, one technical condition. The data must come from a population with a finite variance. You probably can't imagine a population with an infinite variance, but statisticians can construct such things, so we have to discuss them in footnotes like this. It really makes no difference in how you think about the important stuff, so you can just forget we mentioned it.

[14]The word "central" in the name of the theorem means "fundamental." It doesn't refer to the centre of a distribution.

the observations themselves are Normal. If we take samples of size 1, their "means" are just the observations—so, of course, they have a Normal sampling distribution.

> **Pierre-Simon Laplace** Laplace was one of the greatest scientists and mathematicians of his time. In addition to his contributions to probability and statistics, he published many new results in mathematics, physics, and astronomy (where his nebular theory was one of the first to describe the formation of the solar system in much the way it is understood today). He also played a leading role in establishing the metric system of measurement.
>
> His brilliance, though, sometimes got him into trouble. A visitior to the Academie des Sciences in Paris reported that Laplace let it be widely known that he considered himself the best mathematician in France. The effect of this on his colleagues was not eased by the fact that Laplace was right.

But what if we start with a distribution that's strongly skewed, like the CEO compensations we saw back in Chapter 4?

Figure 15.6
The distribution of CEO compensations is highly right skewed.

If we sample from this distribution, there's a good chance that we'll get an exceptionally large value. So some samples will have sample means much larger than others. Here is the sampling distribution of the means from 1000 samples of size 10:

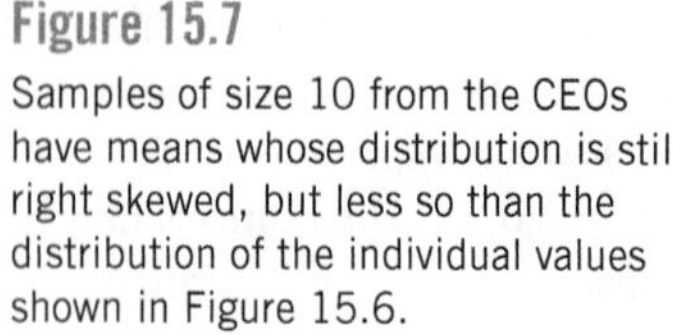

Figure 15.7
Samples of size 10 from the CEOs have means whose distribution is still right skewed, but less so than the distribution of the individual values shown in Figure 15.6.

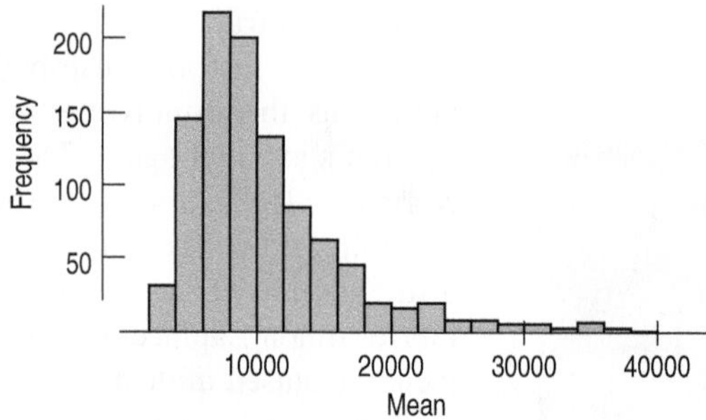

This distribution is not *as* skewed as the original distribution, but still strongly right skewed. We wouldn't ordinarily find a mean of such a skewed distribution, but bear with us for a few steps. The mean of these values is around 10 000 (in $1000s), which actually matches the mean of all 500 CEOs in our population. What happens if we take a larger sample? Here is the sampling distribution of means from samples of size 50:

Figure 15.8
Means of samples of size 50 have a distribution that is only moderately right skewed.

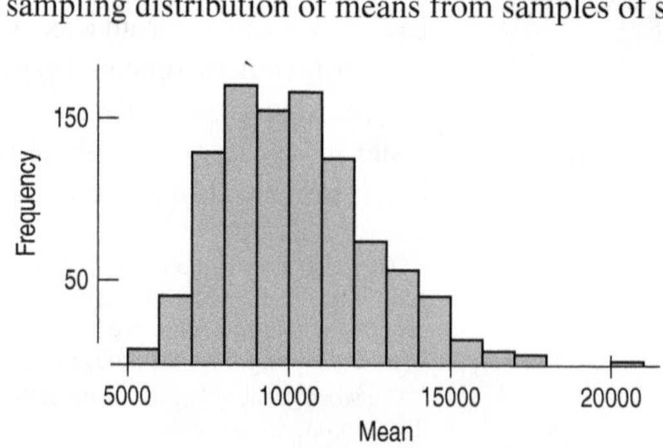

This distribution is less skewed than the corresponding distribution from smaller samples and its mean is again near 10 000. Will this continue as we increase the sample size? Let's try samples of size 100:

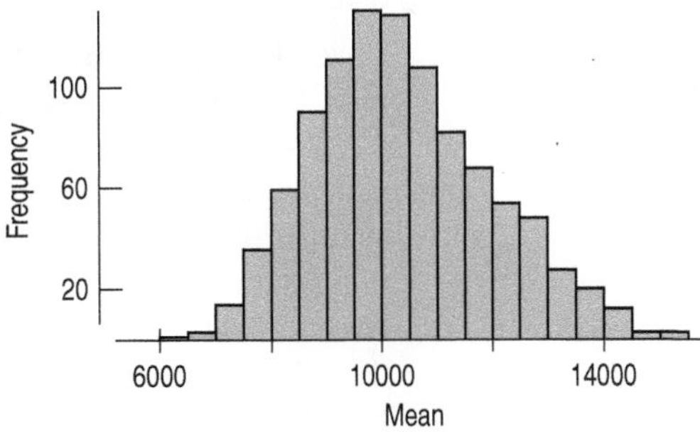

Figure 15.9
The means of samples of size 100 are nearly symmetric.

As we take larger samples, the distribution of means becomes more and more symmetric. By the time we get to samples of size 200, the distribution is quite symmetrical and, of course, has a mean quite close to 10 000.

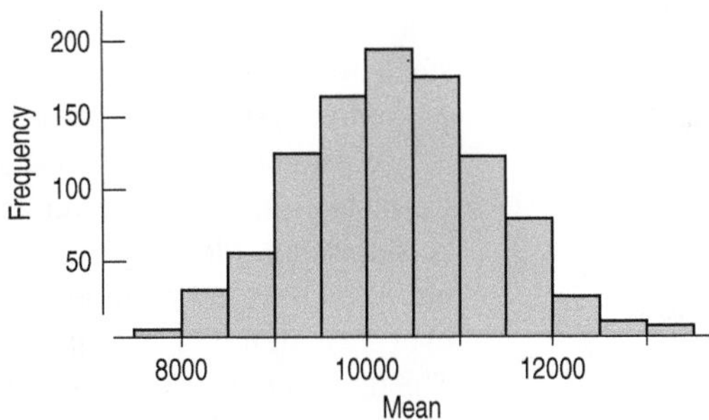

Figure 15.10
When the sample size is 200, even though we started with a highly skewed distribution, the sampling distribution of means is now almost perfectly symmetrical.

For another example, think about a really bimodal population, one that consists of only zeros and ones. The CLT says that even means of samples from this population will follow a Normal sampling distribution model. But wait. Suppose we have a categorical variable and we assign a 1 to each individual in the category and a 0 to each individual not in the category. And then we find the mean of these zeros and ones. That's the same as counting the number of individuals who are in the category and dividing by n. That mean will be . . . the *sample proportion*, p, of individuals who are in the category (a "success"). So maybe it wasn't so surprising after all that proportions, like means, have Normal sampling distribution models; they are actually just a special case of Laplace's remarkable theorem. Of course, for such an extremely bimodal population, we'll need a reasonably large sample size—and that's where the special conditions for proportions come in.

THE CENTRAL LIMIT THEOREM (CLT)

The mean[15] of a random sample is a random variable whose sampling distribution can be approximated by a Normal model if the sample size is big enough. The larger the sample, the better the approximation will be.

Assumptions and Conditions

A S *Activity:* **The Central Limit Theorem.** Does it really work for samples from non-Normal populations?

The CLT requires essentially the same assumptions as we saw for modelling proportions:

Independence Assumption: The sampled values must be *independent* random draws from the population under study.

Sample Size Assumption: The sample size must be sufficiently large.

[15]And hence the sample total, too, since the mean and the total only differ by the multiplicative constant n.

We can't check these directly, but we can think about whether the **Independence Assumption** is plausible. We can also check some related conditions:

Randomization Condition: Proper random sampling can help ensure independence as well as representativeness of the study population.

Large Enough Sample Condition: Although the CLT tells us that a Normal model is useful in thinking about the behaviour of sample means when the sample size is large enough, it doesn't tell us how large a sample we need. The truth is, it depends; there's no one-size-fits-all rule. If the population is unimodal and symmetric, even a fairly small sample (like five or ten) is okay. If the population is strongly skewed, like the compensation for CEOs, it can take a pretty large sample (like 50, 100, or even more) for a Normal model to adequately describe the distribution of sample means. You'll just need to think about your sample size in the context of what you know about the population, and then tell whether you believe the **Large Enough Sample Condition** has been met.

But Which Normal?

A S *Activity:* **The Standard Deviation of Means.** Experiment to see how the variability of the mean changes with the sample size.

A S *Activity:* **The Sampling Distribution of the Mean.** The CLT tells us what to expect. In this activity you can work with the CLT or simulate it if you prefer.

The CLT says that the sampling distribution of any mean or proportion is approximately Normal. But which Normal model? We know that any Normal is specified by its mean and standard deviation. For proportions, the sampling distribution is centred at the population proportion. For means, it's centred at the population mean. What else would we expect?

What about the standard deviations, though? We noticed in our dice simulation that the histograms got narrower as we averaged more and more dice together. This shouldn't be surprising. Means vary less than the individual observations. Think about it for a minute. Which would be more surprising, having *one* person in your Statistics class who is over 205 cm tall or having the *mean* of 100 students taking the course be over 205 cm? The first event is fairly rare.[16] You may have seen somebody this tall in one of your classes sometime. But finding a class of 100 whose mean height is over 205 cm tall just won't happen. Why? Because *means have smaller standard deviations than individuals.*

How much smaller? Well, we have good news and bad news. The good news is that the standard deviation of y falls as the sample size grows. The bad news is that it doesn't drop as fast as we might like. It only goes down by the *square root* of the sample size. Why? The Math Box will show you that the Normal model for the sampling distribution of the mean has a standard deviation equal to

$$SD(\bar{y}) = \frac{\sigma}{\sqrt{n}}$$

where σ is the standard deviation of the population. To emphasize that this is a standard deviation *parameter* of the sampling distribution model for the sample mean, $\bar{y}$, we write $SD(\bar{y})$ or $\sigma(\bar{y})$.

On rare occasion, we may sample more than 10% of the population. In this case, the SD formula above overestimates the true SD (but the formula can be corrected—see the Finite Population Correction section in this chapter).

THE SAMPLING DISTRIBUTION MODEL FOR A MEAN

When a simple random sample is drawn from any (much bigger) population with mean μ and standard deviation σ, its sample mean, $\bar{y}$, has a sampling distribution with the same *mean* μ but whose *standard deviation* is $\frac{\sigma}{\sqrt{n}}$ (and we write $\sigma(\bar{y}) = SD(\bar{y}) = \frac{\sigma}{\sqrt{n}}$). No matter what population the random sample comes from, the *shape* of the sampling distribution is approximately Normal as long as the sample size is large enough. The larger the sample used, the more closely the Normal model approximates the sampling distribution for the mean.

[16]If students are a random sample of adults, fewer than 1 out of 10 000 should be taller than 205 cm. Why might university students not really be a random sample with respect to height? Even if they're not a perfectly random sample, a university student over 205 cm tall is still rare.

MATH BOX

Why is $SD(\bar{y}) = \frac{\sigma}{\sqrt{n}}$? We know that $\bar{y}$ is a sum divided by *n:*

$$\bar{y} = \frac{y_1 + y_2 + y_3 + \ldots + y_n}{n}$$

As we saw in Chapter 14, when a random variable is divided by a constant its variance is divided by the *square* of the constant:

$$Var(\bar{y}) = \frac{Var(y_1 + y_2 + y_3 + \ldots + y_n)}{n^2}$$

We draw the y's randomly from a big population, ensuring they are independent. For independent random variables, variances add:

$$Var(\bar{y}) = \frac{Var(y_1) + Var(y_2) + Var(y_3) + \ldots + Var(y_n)}{n^2}$$

All n of the y's were drawn from our population, so they all have the same variance, σ^2:

$$Var(\bar{y}) = \frac{\sigma^2 + \sigma^2 + \sigma^2 + \ldots + \sigma^2}{n^2} = \frac{n\sigma^2}{n^2} = \frac{\sigma^2}{n}$$

The standard deviation of $\bar{y}$ is the square root of this variance:

$$SD(\bar{y}) = \sqrt{\frac{\sigma^2}{n}} = \frac{\sigma}{\sqrt{n}}$$

In a very similar manner, we can also prove that the expected value of the sample mean, $E(\bar{y})$ is equal to μ, which shows that the mean of the measurements, $\bar{y}$, in a simple random sample is an *unbiased estimator* of the population mean, μ (in the absence of non-sampling sources of bias).

We now have two closely related sampling distribution models that we can use *when the appropriate assumptions and conditions are met.* Which one we use depends on which kind of data we have:

- When we have categorical data, we calculate a sample proportion, $\hat{p}$; the sampling distribution of this random variable has a Normal model with a mean at the true proportion p, and a standard deviation of $SD(\hat{p}) = \sqrt{\frac{pq}{n}} = \frac{\sqrt{pq}}{\sqrt{n}}$. We'll use this model in Chapters 16 through 19.
- When we have quantitative data, we calculate a sample mean, $\bar{y}$; the sampling distribution of this random variable has a Normal model with a mean at the true mean, μ, and a standard deviation of $SD(\bar{y}) = \frac{\sigma}{\sqrt{n}}$. We'll use this model in Chapters 20, 21, and 22.

The means of these models are easy to remember, so all you need to be careful about are the standard deviations. Remember that these are standard deviations of the *statistics* $\hat{p}$ and $\bar{y}$. They both have a square root of n in the denominator. That tells us that the larger the sample, the less either statistic will vary. The only difference is in the numerator. If you just start by writing $SD(\bar{y})$ for quantitative data and $SD(\hat{p})$ for categorical data, you'll be able to remember which formula to use.

For Example USING THE CLT FOR MEANS

RECAP: A university physical education department asked a random sample of 200 female students to self-report their heights and weights, but the percentage of students with body mass indexes over 25 seemed suspiciously low. One possible explanation may be that the respondents "shaded" their weights down a bit. The CDC reports that the mean weight of 18-year-old women is 65.4 kg, with a standard deviation of 23.45 kg, but these 200 randomly selected women reported a mean weight of only 63.7 kg.

QUESTION: Based on the Central Limit Theorem and the 68–95–99.7 Rule, does the mean weight in this sample seem exceptionally low, or might this just be random sample-to-sample variation?

ANSWER: The conditions check out okay:

✓ **Randomization Condition:** The women were a random sample and their weights can be assumed to be independent.
✓ **10% Condition:** They sampled fewer than 10% of all women at the college.
✓ **Large Enough Sample Condition:** The distribution of university women's weights is likely to be unimodal and reasonably symmetric, so the CLT applies to means of even small samples; 200 values is plenty.

The sampling model for sample means is approximately Normal with $E(\bar{y}) = 65.4$ and $SD(\bar{y}) = \frac{\sigma}{\sqrt{n}} = \frac{23.45}{\sqrt{200}} = 1.66$. The expected distribution of sample means is

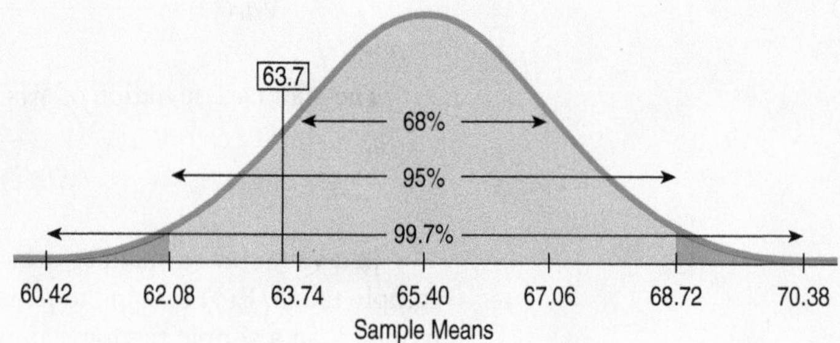

The 68–95–99.7 Rule suggests that although the reported mean weight of 63.7 kg is somewhat lower than expected, it does not appear to be unusual. Such variability is not all that extraordinary for samples of this size.

Step-by-Step Example WORKING WITH THE SAMPLING DISTRIBUTION MODEL FOR THE MEAN

The mean weight of adult men in Canada is 82 kg with a standard deviation of 16 kg.[17]

Question: An elevator in our building has a weight limit of 1000 kg. What's the probability that if 10 men get on the elevator, they will overload its weight limit?

THINK ➡ **Plan** State what we want to know.	Asking the probability that the total weight of a sample of 10 men exceeds 1000 kg is equivalent to asking the probability that their mean weight is greater than 100 kg.
Model Think about the assumptions and check the conditions.	✓ **Independence Assumption:** We don't have a random sample, but it's reasonable to think that the weights of 10 men entering the elevator will be independent of each other and reasonably representative of the general population. (But there could be exceptions—for

[17]Author's estimate, using Canadian Community Health Survey data and other sources.

example, if they were all from the same family or if the elevator were in a building with a diet clinic!)

Note that if the sample were larger, we'd be less concerned about the shape of the distribution of all weights.

✓ **Large Enough Sample Condition:** I believe the distribution of population weights is only mildly skewed, so my sample of 10 men seems large enough to use the CLT.

State the parameters and the sampling model.

The mean for all weights is $\mu = 82$ kg and the standard deviation is $\sigma = 16$ kg. Since the conditions are satisfied, the CLT says that the sampling distribution of $\bar{y}$ has a Normal model with mean 82 and standard deviation

$$SD(\bar{y}) = \frac{\sigma}{\sqrt{n}} = \frac{16}{\sqrt{10}} \approx 5.06$$

SHOW ➡ **Plot** Make a picture. Sketch the model and shade the area we're interested in. Here the mean weight of 100 kg appears to be far out on the right tail of the curve.

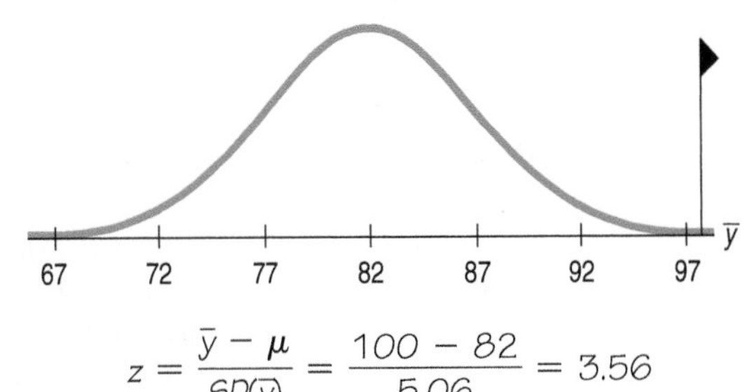

Mechanics Use the standard deviation as a ruler to find the *z*-score of the cutoff mean weight. We see that an average of 100 kg is more than three standard deviations above the mean.

$$z = \frac{\bar{y} - \mu}{SD(\bar{y})} = \frac{100 - 82}{5.06} = 3.56$$

Find the resulting probability from a table of Normal probabilities such as Table Z, a computer program, or a calculator.

$$P(\bar{y} > 100) = P(z > 3.56) = 1 - 0.9998 = 0.0002$$

TELL ➡ **Conclusion** Interpret your result in the proper context, being careful to relate it to the original question.

The chance that a random collection of 10 men will exceed the elevator's weight limit is only 0.0002. So, if they are a random sample, it is quite unlikely that 10 people will exceed the total weight allowed on the elevator.

About Variation

Means vary less than individual data values. That makes sense. If the same test is given to many sections of a large course and the class average is, say, 80%, some students may score 95% because individual scores vary a lot. But we'd be shocked (and pleased!) if the *average* score of the students in any section was 95%. Averages are much less variable. Not only do group averages vary less than individual values, but common sense suggests that averages should be more consistent for larger groups. The fact that $SD(\bar{y}) = \frac{\sigma}{\sqrt{n}}$ has n in the denominator shows that the variability of sample

means decreases as the sample size increases. There's a catch, though. The standard deviation of the sampling distribution declines only with the square root of the sample size and not, for example, with $1/n$. The mean of a random sample of four has half $\left(\frac{1}{\sqrt{4}} = \frac{1}{2}\right)$ the standard deviation of an individual data value. To cut the standard deviation in half again, we'd need a sample of 16, and a sample of 64 to halve it once more.

If only we had a much larger sample, we could get the standard deviation of the sampling distribution *really* under control so that the sample mean could tell us still more about the unknown population mean, but larger samples cost more and take longer to survey. And while we're gathering all that extra data, the population itself may change, or a news story may alter opinions. There are practical limits to most sample sizes. As we shall see, that nasty square root limits how much we can make a sample tell about the population. This is an example of something that's known as the Law of Diminishing Returns.

A Billion Dollar Misunderstanding? In the late 1990s, the Bill and Melinda Gates Foundation began funding an effort to encourage the breakup of large schools into smaller schools. Why? It had been noticed that smaller schools were more common among the best-performing schools than one would expect. In time, the Annenberg Foundation, the Carnegie Corporation, the Center for Collaborative Education, the Center for School Change, Harvard's Change Leadership Group, the Open Society Institute, Pew Charitable Trusts, and the U.S. Department of Education's Smaller Learning Communities Program all supported the effort. Well over a billion dollars was spent to make schools smaller.

But was it all based on a misunderstanding of sampling distributions? Statisticians Howard Wainer and Harris Zwerling[18] looked at the mean test scores of schools in Pennsylvania. They found that indeed 12% of the top-scoring 50 schools were from the smallest 3% of Pennsylvania schools—substantially more than the 3% we'd naively expect. But then they looked at the bottom 50. There they found that 18% were small schools! The explanation? Mean test scores are, well, means. We are looking at a rough real-world simulation in which each school is a trial. Even if all Pennsylvania schools were equivalent, we'd expect their mean scores to vary. How much? The CLT tells us that means of test scores vary according to $\frac{\sigma}{\sqrt{n}}$. Smaller schools have (by definition) smaller n's, so the sampling distributions of their mean scores naturally have larger standard deviations. It's natural, then, that small schools have both higher and lower mean scores.

On October 26, 2005, the *Seattle Times* reported:[19]

> *The Gates Foundation announced last week it is moving away from its emphasis on converting large high schools into smaller ones and instead giving grants to specially selected school districts with a track record of academic improvement and effective leadership. Education leaders at the Foundation said they concluded that improving classroom instruction and mobilizing the resources of an entire district were more important first steps to improving high schools than breaking down the size.*

The Real World and the Model World

Be careful. We have been slipping smoothly between the real world, in which we draw random samples of data, and a magical mathematical model world, in which we describe how the sample means and proportions we observe in the real world might behave if we could see the results from every random sample that we might have drawn. Now we have *two* distributions to deal with. The first is the real world distribution of the sample, which we might display with a histogram (for quantitative data) or with a bar chart or table (for categorical data). The second is the math world *sampling distribution* of the statistic, which we may be able to model with a Normal model based on the Central Limit Theorem. Don't confuse the two.

[18]Wainer, H. and Zwerling, H., 2006, "Legal and empirical evidence that smaller schools do not improve student achievement." *The Phi Delta Kappan* 87:300–303. Discussed in Howard Wainer, 2007, "The most dangerous equation." *American Scientist* pp. 249–256; also at www.americanscientist.org.

[19]Copyright 2005, Seattle Times Company. Used with permission.

For example, don't mistakenly think the CLT says that the *data* are Normally distributed as long as the sample is large enough. In fact, as samples get larger, we expect the distribution of the data to look more and more like the population from which they are drawn—skewed, bimodal, whatever—but not necessarily Normal. You can collect a sample of CEO salaries for the next 1000 years,[20] but the histogram will never look Normal. It will be skewed to the right. The Central Limit Theorem doesn't talk about the distribution of the data from the sample. It talks about the sample *means* and sample *proportions* of many different random samples drawn from the same population. Of course, the CLT does require that the sample be big enough when the population shape is not unimodal and symmetric, but the fact that, even then, a Normal model is useful is still a very surprising and powerful result.

The CLT in Nature

The CLT we have given here is adequate for our purposes, assuming independent observations (as is common in statistical studies), but actually there have been many generalizations of this theorem over the years. It turns out that even if the individual random variables exhibit some dependence (though not too great), or do not all come from the same distribution (as long as each has but a small effect on the total), Normality will still arise if enough variables are averaged (or summed). Perhaps this more general CLT effect helps explain all those variables in nature that possess an approximately Normal frequency distribution. Are heights in a genetically homogenous population nearly Normal because the final result comes from summing (or averaging) many small genetic and environmental influences? Possibly.[21] Was your height a sum of such random influences, somewhat like summing the random values in a sample?

Many other variables are formed by multiplying many small influences together rather than adding them. But guess what? Take the log (converting multiplication to addition), and you may just get Normality!

15.5 Sampling Distribution Models

A S *Simulation:* **The CLT for Real Data.** Why settle for a picture when you can see it in action?

Let's summarize what we've learned about sampling distributions. At the heart is the idea that *the statistic itself is a random quantity*. We can't know what our statistic will be because it comes from a random sample. It's just one instance of something that happened for our particular random sample. A different random sample would have given a different result. This sample-to-sample variability is what generates the sampling distribution. The sampling distribution shows us the distribution of possible values that the statistic could have had.

We could simulate that distribution by pretending to take lots of samples. While a sampling distribution can take on pretty much any form depending on the particular statistic and study design, fortunately for the mean or the proportion, the CLT tells us that we can usually model the sampling distribution with a Normal model.

Two basic truths about sampling distributions are:

1. Sampling distributions arise because samples vary. Each random sample will contain different cases, and therefore a different value of the statistic.
2. Although we can simulate a sampling distribution, the Central Limit Theorem saves us the trouble for means and proportions.

---

[20] Don't forget to adjust for inflation.

[21] But add gender to the mix of influences on height, and if the average height difference between males and females is very big, we can lose the Normality, possibly ending up with bi-modality. What went wrong? One variable had a big effect on the final result.

Here's a picture showing the process going into the sampling distribution model:

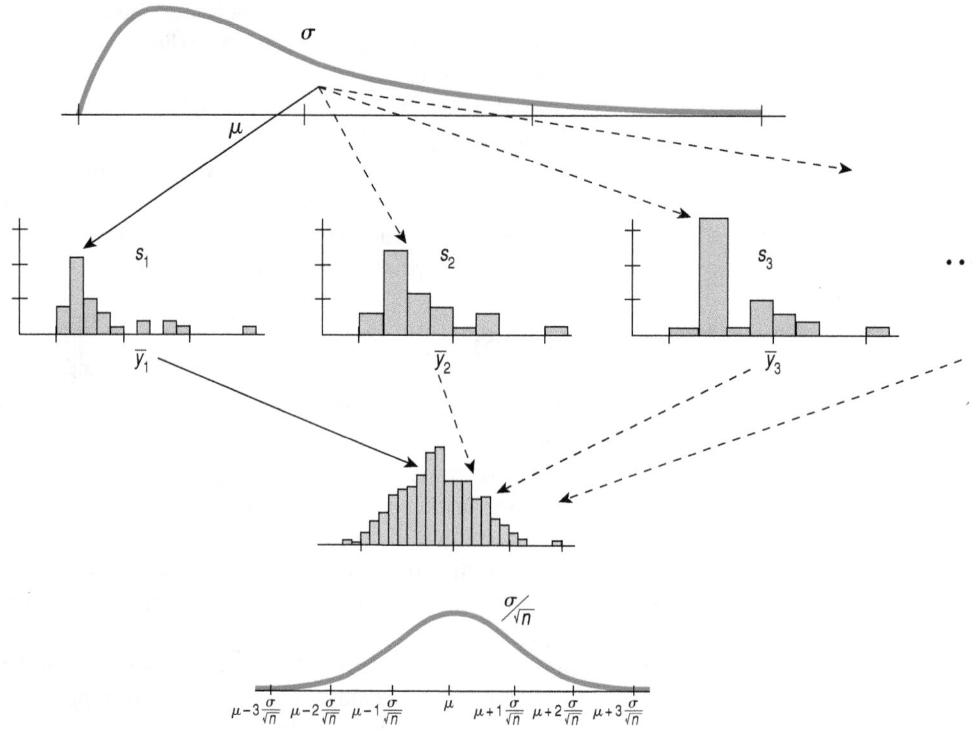

Figure 15.11
We start with a population model, which can have any shape. It can even be bimodal or skewed (as this one is). We label the mean of this model μ and its standard deviation, σ.

We draw one real sample (solid line) of size n and show its histogram and summary statistics. We imagine (or simulate) drawing many other samples (dotted lines), which have their own histograms and summary statistics.

We (imagine) gathering all the means into a histogram.

The CLT tells us we can model the shape of this histogram with a Normal model. The mean of this Normal is μ, and the standard deviation is $SD(\bar{y}) = \frac{\sigma}{\sqrt{n}}$.

Just Checking

4. Human gestation times have a mean of about 266 days, with a standard deviation of about 16 days. If we record the gestation times of a sample of 100 women, do we know that a histogram of the times will be well modelled by a Normal model?
5. Suppose we look at the *average* gestation times for a sample of 100 women. If we imagined all the possible random samples of 100 women we could take and looked at the histogram of all the sample means, what shape would it have?
6. Where would the centre of that histogram be?
7. What would be the standard deviation of that histogram?

The View from Here . . .

We've seen that when it comes to understanding the data's story, our biggest nemesis is variability. Whether we're looking at information collected in a survey or the results of an experiment, we have only a narrow glimpse of the larger world we wish we knew about. That glimpse is subject to error, the variability that naturally arises from sample to sample and group to group. What we've now learned is that this unavoidable variability is predictable and quantifiable. The Central Limit Theorem quantifies sampling error.

In this chapter, our discussion has been purely theoretical. We've imagined that we know the population parameters—proportions, means, and standard deviations—and we've learned to describe the behaviour of statistics observed in many, many samples. But that's not the way the real world works. In practice, we'll never know the population parameters, and we'll have data from just a single study. We poll 1000 voters, but we want to know whom the entire electorate will choose. We've seen a new medicine lower blood pressure more than a placebo, but we need to know whether that difference is statistically significant.[22] We now must combine our ability to design studies and summarize the data with the CLT's insights about sampling variability in order to use the sample statistics we observe to make inferences about the population parameters we wish we knew.

That's the path that lies ahead. In Chapters 16 to 19, we'll learn how to make inferences about proportions. In Chapters 20 to 22, we'll turn our attention to means. The Central Limit Theorem will be with us all the way.

*Finite Population Correction

When drawing a simple random sample from a finite population of size N, the actual variance of the sample proportion and sample mean are given by the formulas:

$$SD(\hat{p}) = \sqrt{\frac{pq}{n}} \times \sqrt{\frac{N-n}{N-1}} \quad and \quad SD(\bar{y}) = \frac{\sigma}{\sqrt{n}} \times \sqrt{\frac{N-n}{N-1}}$$

The latter multiplicative part, $\sqrt{\frac{N-n}{N-1}}$, appended to our earlier formulas, is called the *finite population correction*. This factor is very close to one if N is much bigger than n, in which case these formulas reduce to our earlier formulas for SD's. Also note that they give you the correct result of $SD = 0$ if you sample the entire population, since then the sample mean (or proportion) must equal the true mean (or proportion), and hence has no variability at all.

The limitation of our earlier formulas and derivations is that we assumed independent observations. But when you sample (without replacement) from the population, you change it. Imagine sampling five cards from a deck of 52. Every time you draw, say, a face card, you reduce the number of face cards left and the chance that the next card will be a face card. Suppose instead that you draw five cards from 10 decks shuffled together. Every time you draw a card now, the remaining deck composition changes little, and the draws are nearly independent.

Our earlier formulas are good enough most of the time, but what if we fail to correct for finiteness when we should (if we violate that *10% Condition*)? The result is that we overestimate the variability of our sample, and our results will actually be more precise or accurate than suggested by our calculations.[23] Not a bad way to err; we call this a *conservative error* (sorry, but there is no place for *liberal* thinking in Statistics). But what is this worth if we may lose the Normality? Fear not—more general versions of the CLT (relaxing the Independence Assumption) assure us of Normality even if we take a very big bite out of the population.

WHAT CAN GO WRONG?

- **Don't confuse the sampling distribution with the distribution of the sample.** When you take a sample, you always look at the distribution of the values, usually with a histogram, and you may calculate summary statistics. Examining the distribution of the sample like this is wise. But that's not the sampling distribution. The sampling distribution is an imaginary collection of the values that a statistic *might* have taken for all the random samples—the one you got and the ones that you didn't get. We use the sampling distribution model to make statements about how the statistic varies.

[22]We first encountered "statistically significant" in Chapter 11. The term refers to results so unusual that they're probably not merely sample-to-sample variation. You'll be seeing—and saying—those words a lot from here on.

[23]And the same goes for two sample procedures to be discussed in later chapters.

- **Beware of observations that are not independent.** The CLT depends crucially on the assumption of independence. If our elevator riders are related, are all from the same school (for example, an elementary school), or in some other way aren't similar enough to a random sample, then the statements we try to make about the mean are going to be wrong. Unfortunately, this isn't something you can check in your data. You have to think about how the data were gathered. Good sampling practice and well-designed randomized experiments ensure independence.
- **Watch out for small samples from skewed populations.** The CLT assures us that the sampling distribution model is Normal if n is large enough. If the population is nearly Normal, even small samples (like our 10 elevator riders) work. If the population is very skewed, then n will have to be large before the Normal model will work well. If we sampled 15 or even 20 CEOs and used $\bar{y}$ to make a statement about the mean of all CEOs' compensation, we'd likely get into trouble because the underlying data distribution is so skewed. Unfortunately, there's no good rule of thumb.[24] It just depends on how skewed the data distribution is. Always plot the data to check.

CONNECTIONS

The concept of a sampling distribution connects to almost everything we have done. The fundamental connection is to the deliberate application of randomness in random sampling and randomized comparative experiments. The distribution of statistic values arises directly because different random samples and randomized experiments would generate different statistic values in a predictable manner.

The connection to the Normal distribution is obvious. We first introduced the Normal model before because it was "nice." As a unimodal, symmetric distribution with 99.7% of its area within three standard deviations of the mean, the Normal model is easy to work with. Now we see that the Normal holds a special place among distributions because we can use it to model the sampling distributions of the mean and the proportion.

We used simulation to understand sampling distributions. In fact, some important sampling distributions were discovered first by simulation.

What Have We Learned?

Learning Objectives

Model the variation in statistics from sample to sample with a sampling distribution.

- The Central Limit Theorem tells us that the sampling distribution of both the sample proportion and the sample mean are Normal, for sufficiently large samples.

Understand that, usually, the mean of a sampling distribution is the value of the parameter estimated.

- For the sampling distribution of $\hat{p}$ the mean is p.
- For the sampling distribution of $\bar{y}$, the mean is μ.

Interpret the standard deviation of a sampling distribution.

- The standard deviation of a sampling model is the most important information about it.
- The standard deviation of the sampling distribution of a proportion is $\sqrt{\frac{pq}{n}}$, where $q = 1 - p$
- The standard deviation of the sampling distribution of a mean is $\frac{\sigma}{\sqrt{n}}$, where σ is the population standard deviation.

[24]For proportions, of course, there is a rule: the **Success/Failure Condition.** That works for proportions because the standard deviation of a proportion is linked to its mean.

Understand that the Central Limit Theorem is a *limit* theorem.

- The sampling distribution of the mean is Normal, *no matter what the underlying distribution of the data is, however . . .*
- The CLT says that this happens in the limit, as the sample size grows. The Normal model applies sooner when sampling from a unimodal, symmetric population and more gradually when the population is very non-Normal.

Review of Terms

Sampling distribution model Different random samples give different values for a statistic. The sampling distribution model shows the behaviour of the statistic over all the possible samples for the same size n (p. 420).

Sampling variability (sampling error) The variability we expect to see from one random sample to another. It is sometimes called sampling error, but sampling variability is the better term (p. 421).

Sampling distribution model for a proportion If assumptions of independence and random sampling are met, and we expect at least 10 successes and 10 failures, then the sampling distribution of a proportion is modelled by a Normal model with a mean equal to the true proportion value, p, and a standard deviation equal to $\sqrt{\frac{pq}{n}}$ (p. 424).

Sampling distribution model for a mean If assumptions of independence and random sampling are met, and the sample size is large enough, the sampling distribution of the sample mean is modelled by a Normal model with a mean equal to the population mean, μ, and a standard deviation equal to $\frac{\sigma}{\sqrt{n}}$ (p. 432).

Central Limit Theorem (CLT) States that the sampling distribution model of the sample mean (and proportion) from a random sample is approximately Normal for large n, regardless of the distribution of the population, as long as the observations are independent (p. 429).

Exercises

1. **Web site** An investment Web site can tell what devices are used to access the site. The site managers wonder whether they should enhance the facilities for trading via smartphones so they want to estimate the proportion of users who access the site that way (even if they also use their computers sometimes). They draw a random sample of 200 investors from their customers. Suppose that the true proportion of smartphone users is 36%.
   a) What would you expect the shape of the sampling distribution for the sample proportion to be?
   b) What would be the mean of this sampling distribution?
   c) What would be the standard deviation of the sampling distribution?

2. **Marketing** The proportion of adult women in Canada is approximately 51%. A marketing survey company telephones 400 people at random.
   a) What proportion of the sample of 400 would you expect to be women?
   b) What would the standard deviation of the sampling distribution be?
   c) How many women, on average, would you expect to find in a sample of that size?

3. **Sample maximum** The distribution of scores on a Statistics test for a particular class (of size 5) is skewed to the left. The professor wants to predict the maximum score and so wants to understand the distribution of the sample maximum. She simulates the distribution of the maximum of the test for 30 different tests (with $n = 5$). The histogram below shows a simulated sampling distribution of the sample maximum from these tests.

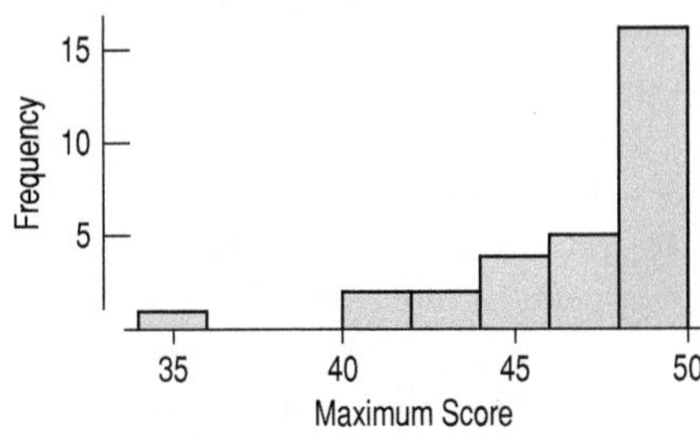

   a) Would a Normal model be a useful model for this sampling distribution? Explain.
   b) The mean of this distribution is 46.3 and the SD is 3.5. Would you expect about 95% of the samples to have their maximums within 7 of 46.3? Why or why not?

4. **Soup** A machine is supposed to fill cans with 16 oz of soup. Of course there, will be some variation in the amount actually dispensed, and measurement errors are often approximately normally distributed. The manager would like to understand the variability of the variances of the samples, so he collects information from the last 250 batches of size 10 and plots a histogram of the variances:

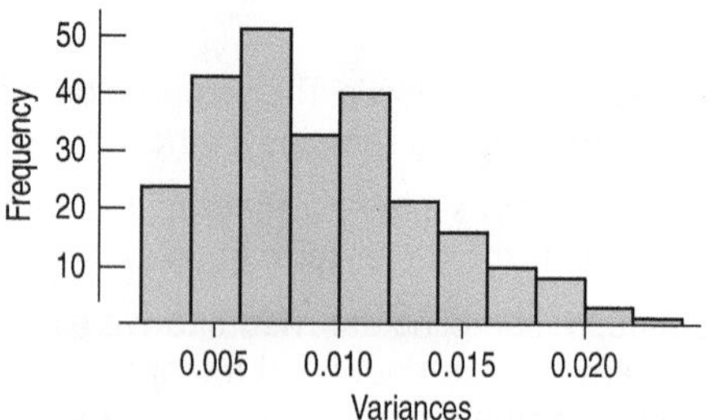

a) Would a Normal model be a useful model for this sampling distribution? Explain.

b) The mean of this distribution is 0.009 and the SD is 0.004. Would you expect about 95% of the samples to have their variances within 0.008 of 0.009? Why or why not?

5. **Market research** A marketing researcher for a phone company surveys 100 people and finds that the proportion of customers who are likely to switch providers when their contract expires is 15%.

a) What is the standard deviation of the sampling distribution of the proportion?

b) If she wants to reduce the standard deviation by half, how large a sample would she need?

6. **Market research II** A market researcher for a provider of iPod accessories wants to know the proportion of customers who own cars to assess the market for a new iPod car charger. A survey of 500 customers indicates that 76% own cars.

a) What is the standard deviation of the sampling distribution of the proportion?

b) How large would the standard deviation have been if he had surveyed only 125 customers (assuming the proportion is about the same)?

7. **Send money** When it sends out its fundraising letter, a philanthropic organization typically gets a return from about 5% of the people on their mailing list. To see what the response rate might be for future appeals, it did a simulation using samples of size 20, 50, 100, and 200. For each sample size, it simulated 1000 mailings with success rate $p = 0.05$, and constructed the histogram of the 1000 sample proportions, shown below. Explain how these histograms demonstrate what the Central Limit Theorem says about the sampling distribution model for sample proportions. Be sure to talk about shape, centre, and spread.

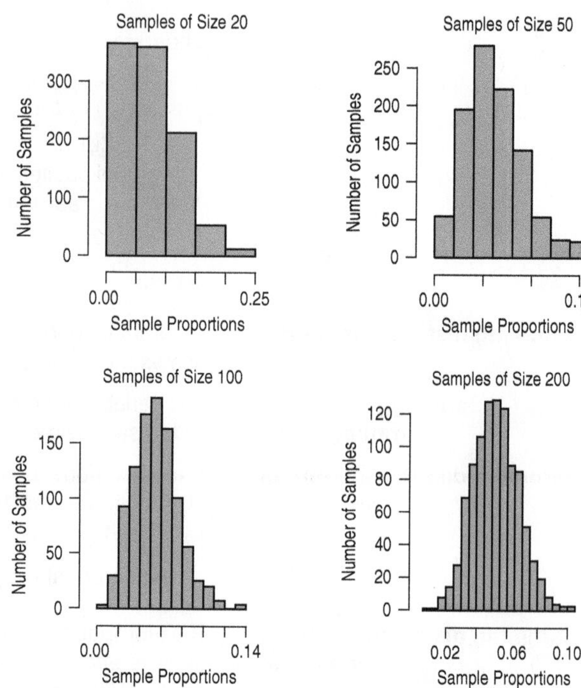

8. **Character recognition** An automatic character recognition device can successfully read about 85% of handwritten credit card applications. To estimate what might happen when this device reads a stack of applications, the company did a simulation using samples of size 20, 50, 75, and 100. For each sample size, they simulated 1000 samples with success rate $p = 0.85$, and constructed the histogram of the 1000 sample proportions, shown here. Explain how these histograms demonstrate what the Central Limit Theorem says about the sampling distribution model for sample proportions. Be sure to talk about shape, centre, and spread.

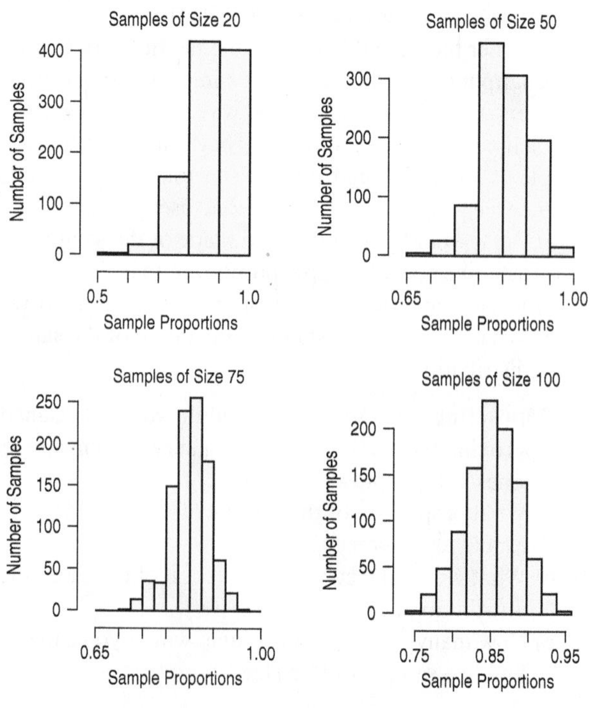

9. Send more money The philanthropic organization in Exercise 7 expects about a 5% success rate when it sends fundraising letters to the people on their mailing list. In Exercise 7, you looked at the histograms showing distributions of sample proportions from 1000 simulated mailings for samples of size 20, 50, 100, and 200. The sample statistics from each simulation were as follows:

n	mean	st. dev.
20	0.0497	0.0479
50	0.0516	0.0309
100	0.0497	0.0215
200	0.0501	0.0152

a) According to the Central Limit Theorem, what should the theoretical mean and standard deviations be for these sample sizes?
b) How close are those theoretical values to what was observed in these simulations?
c) Looking at the histograms in Exercise 7, at what sample size would you be comfortable using the Normal model as an approximation for the sampling distribution?
d) What does the Success/Failure Condition say about the choice you made in part c)?

10. Character recognition, again The automatic character recognition device discussed in Exercise 8 successfully reads about 85% of handwritten credit card applications. In Exercise 8, you looked at the histograms showing distributions of sample proportions from 1000 simulated samples of size 20, 50, 75, and 100. The sample statistics from each simulation were as follows:

n	mean	st. dev.
20	0.8481	0.0803
50	0.8507	0.0509
75	0.8481	0.0406
100	0.8488	0.0354

a) According to the Central Limit Theorem, what should the theoretical mean and standard deviations be for these sample sizes?
b) How close are those theoretical values to what was observed in these simulations?
c) Looking at the histograms in Exercise 8, at what sample size would you be comfortable using the Normal model as an approximation for the sampling distribution?
d) What does the Success/Failure Condition say about the choice you made in part c)?

11. Coin tosses In a large class of introductory Statistics students, the professor has each person toss a coin 16 times and calculate the proportion of his or her tosses that were heads. The students then report their results, and the professor plots a histogram of these proportions.
a) What shape would you expect this histogram to be? Why?
b) Where do you expect the histogram to be centred?
c) How much variability would you expect among these proportions?
d) Explain why a Normal model should not be used here.

12. M&Ms The candy company claims that 10% of the M&Ms it produces are green. Suppose that the candies are packaged at random in small bags containing about 50 M&Ms. A class of elementary school students learning about percents opens several bags, counts the various colours of the candies, and calculates the proportion that are green.
a) If we plot a histogram showing the proportions of green candies in the various bags, what shape would you expect it to have?
b) Can that histogram be approximated by a Normal model? Explain.
c) Where should the centre of the histogram be?
d) What should the standard deviation of the proportion be?

13. More coins Suppose the class in Exercise 11 repeats the coin-tossing experiment.
a) The students toss the coins 25 times each. Use the 68–95–99.7 Rule to describe the sampling distribution model.
b) Confirm that you can use a Normal model here.
c) They increase the number of tosses to 64 each. Draw and label the appropriate sampling distribution model. Check the appropriate conditions to justify your model.
d) Explain how the sampling distribution model changes as the number of tosses increases.

14. Bigger bag Suppose the class in Exercise 12 buys bigger bags of candy, with 200 M&Ms each. Again the students calculate the proportion of green candies they find.
a) Explain why it's appropriate to use a Normal model to describe the distribution of the proportion of green M&Ms they might expect.
b) Use the 68–95–99.7 Rule to describe how this proportion might vary from bag to bag.
c) How would this model change if the bags contained even more candies?

15. Just (un)lucky? One of the students in the introductory Statistics class in Exercise 13 claims to have tossed her coin 200 times and found only 42% heads. What do you think of this claim? Explain.

16. Too many green ones? In a really large bag of M&Ms, the students in Exercise 14 found 500 candies, and 12% of them were green. Is this an unusually large proportion of green M&Ms? Explain.

17. Speeding Provincial police believe that 70% of the drivers travelling on a particular section of the Trans-Canada

Highway exceed the speed limit. They plan to set up a radar trap and check the speeds of 80 cars.

a) Using the 68–95–99.7 Rule, draw and label the distribution of the proportion of these cars the police will observe speeding.

b) Do you think the appropriate conditions necessary for your analysis are met? Explain.

18. Smoking pot The Canadian Drug Use and Monitoring Survey estimates that 33% of 15- to 24-year-olds used cannabis in 2008. Using the 68–95–99.7 Rule, describe the sampling distribution model for the proportion of cannabis users among a randomly selected group of fifty 15- to 24-year-olds. Be sure to discuss your assumptions and conditions.

19. Vision It is generally believed that nearsightedness affects about 12% of all children. A school district has registered 170 incoming kindergarten children.

a) Can you apply the Central Limit Theorem to describe the sampling distribution model for the sample proportion of children who are nearsighted? Check the conditions and discuss any assumptions you need to make.

b) Sketch and clearly label the sampling model, based on the 68–95–99.7 Rule.

c) How many of the incoming students might the school expect to be nearsighted? Explain.

20. Mortgages In early 2009, rising unemployment and the house-price slump in Alberta led to a mortgage delinquency rate of 0.54% of all big bank mortgages (rising for the nineteenth straight month), the highest delinquency rate in Canada. Suppose a large bank holds 1731 mortgages.

a) Can you apply the Central Limit Theorem to describe the sampling distribution model for the sample proportion of delinquencies? Check the conditions and discuss any assumptions you need to make.

b) Sketch and clearly label the sampling model, based on the 68–95–99.7 Rule.

c) How many of these homeowners might the bank expect to be delinquent on their mortgages? Explain.

21. Loans Based on past experience, a bank believes that 7% of the people who receive loans will not make payments on time. The bank has recently approved 200 loans.

a) What are the mean and standard deviation of the proportion of clients in this group who may not make timely payments?

b) What assumptions underlie your model? Are the conditions met? Explain.

c) What is the probability that over 10% of these clients will not make timely payments?

22. Contacts Assume that 30% of students at a university wear contact lenses.

a) We randomly pick 100 students. Let $\hat{p}$ represent the proportion of students in this sample who wear contacts. What is the appropriate model for the distribution of $\hat{p}$? Specify the name of the distribution, the mean, and the standard deviation. Be sure to verify that the conditions are met.

b) What is the approximate probability that more than one third of this sample wear contacts?

23. Big families According to Statistics Canada, 8.3% of Canadian families have five or more members. If we take a random sample of 400 families, use the 68–95–99.7 Rule to describe the sampling distribution model for the percentage of those 400 families that will be of size five or bigger. Are the appropriate conditions met?

24. Binge drinking A national study found that 44% of university students engage in binge drinking (five drinks at a sitting for men, four for women). Use the 68–95–99.7 Rule to describe the sampling distribution model for the proportion of students in a randomly selected group of 200 who engage in binge drinking. Do you think the appropriate conditions are met?

25. Big families, again Exercise 23 mentioned that 8.3% of Canadian families have five or more members. Suppose we take a random sample of 625 rural farm families and find that 16% of those families have five or more members. Is this a very unusual result? Would you conclude that farm families differ from other Canadian families? Why?

26. Binge sample After hearing of the national result that 44% of students engage in binge drinking, a professor surveyed a random sample of 244 students at his university and found that 96 of them admitted to binge drinking in the past week. Should he be surprised at this result? Explain.

27. Polling Just before a Quebec referendum on sovereignty, a local newspaper polls 400 voters in an attempt to predict whether it would succeed. Suppose that sovereignty has the support of 48% of the voters. What is the probability that the newspaper's sample will lead them to predict victory? Be sure to verify that the assumptions and conditions necessary for your analysis are met.

28. Seeds Information on a packet of seeds claims that the germination rate is 92%. What is the probability that more than 95% of the 160 seeds in the packet will germinate? Be sure to discuss your assumptions and check the conditions that support your model.

29. Apples When a truckload of apples arrives at a packing plant, a random sample of 150 is selected and examined for bruises, discolouration, and other defects. The whole truckload will be rejected if more than 5% of the sample is unsatisfactory. Suppose that 8% of the apples on the truck do not meet the desired standard. What is the probability that the shipment will be accepted anyway?

30. Airline no-shows About 12% of ticketed passengers do not show up for their flights, so airlines like to overbook for flights. If an airplane has sold tickets to 400 people for the next flight, how many seats, at the least, would you hope to be available on the plane to be very sure of having enough seats for all who show up? Comment on the assumptions and conditions that support your model, and explain what "very sure" means to you.

31. Nonsmokers You are planning to open a new "Peking Duck" specialty restaurant in Beijing, and need to plan the seating arrangement. Unlike in Canada, you'll need to plan for smokers as well as nonsmokers. While some nonsmokers do not mind being seated in the smoking section of a restaurant, you estimate that about 60% of the customers will demand a smoke-free area. Your restaurant will have 120 seats. How many seats should be in the nonsmoking area to be very sure of having enough seating there? Comment on the assumptions and conditions that support your model, and explain what "very sure" means to you.

32. Meals A restauranteur anticipates serving about 180 people on a Friday evening, and believes that about 20% of the patrons will order the chef's steak special. How many of those meals should he plan on serving to be pretty sure of having enough steaks on hand to meet customer demand? Justify your answer, including an explanation of what "pretty sure" means to you.

33. Sampling A sample is chosen randomly from a population that can be described by a Normal model.

a) What is the sampling distribution model for the sample mean? Describe shape, centre, and spread.

b) If we choose a larger sample, what is the effect on this sampling distribution model?

34. Sampling, part II A sample is chosen randomly from a population that was strongly skewed to the left.

a) Describe the sampling distribution model for the sample mean if the sample size is small.

b) If we make the sample larger, what happens to the sampling distribution model's shape, centre, and spread?

c) As we make the sample larger, what happens to the expected distribution of the data in the sample?

35. Waist size A study measured the waist size of 250 men, finding a mean of 36.33 inches and a standard deviation of 4.02 inches. Here is a histogram of these measurements

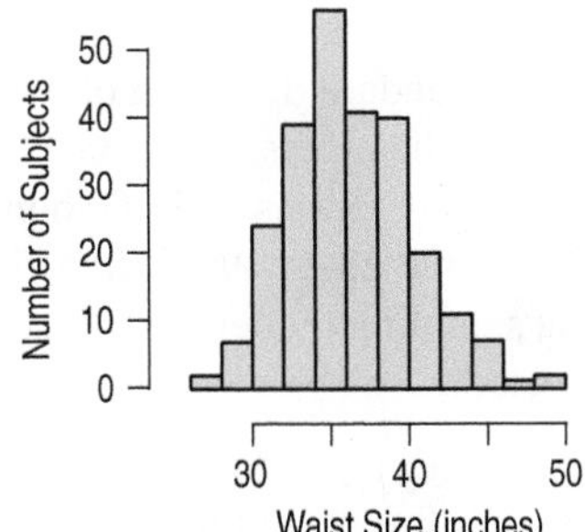

a) Describe the histogram of waist size.

b) To explore how the mean might vary from sample to sample, the researchers simulated by drawing many samples of size 2, 5, 10, and 20, with replacement, from the 250 measurements. Here are histograms of the sample means for each simulation. Explain how these histograms demonstrate what the Central Limit Theorem says about the sampling distribution model for sample means.

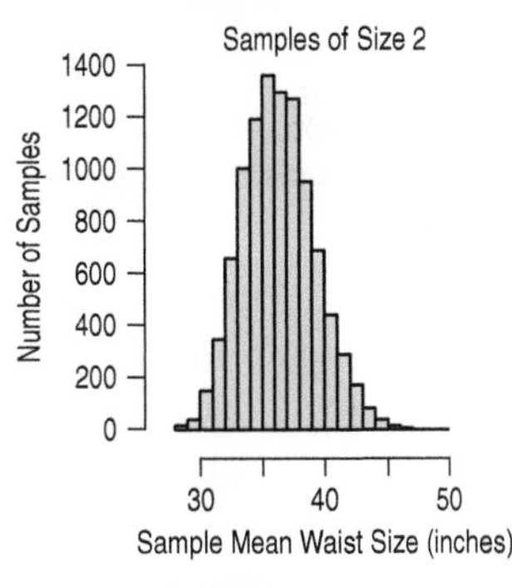

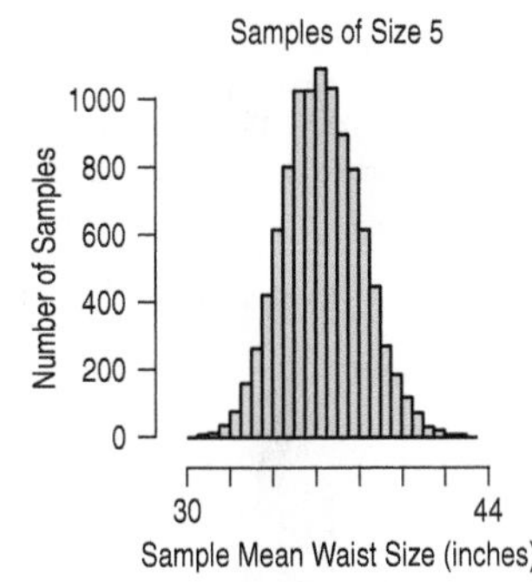

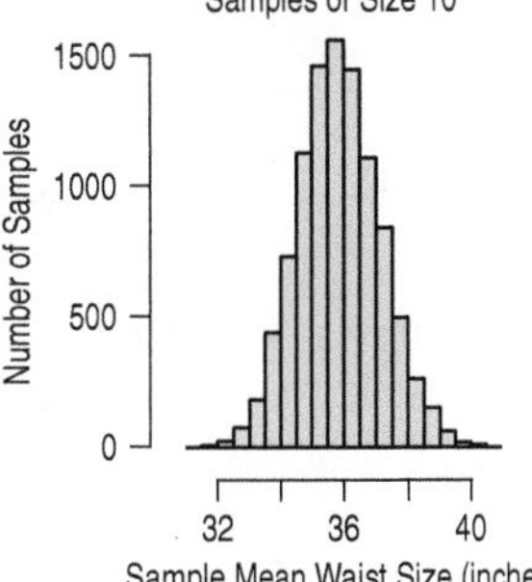

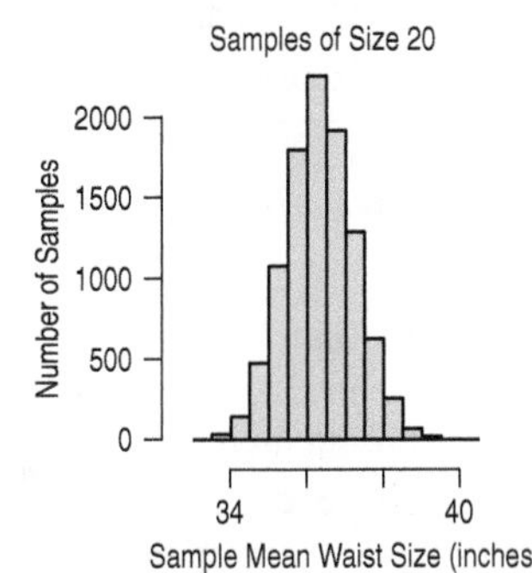

36. CEO compensation In Chapter 3, we saw the distribution of the total compensation of the chief executive officers (CEOs) of the 800 largest U.S. companies (the *Fortune* 800). The average compensation (in thousands of dollars) is 10 307.31 and the standard deviation is 17 964.62. Here is a histogram of their annual compensations:

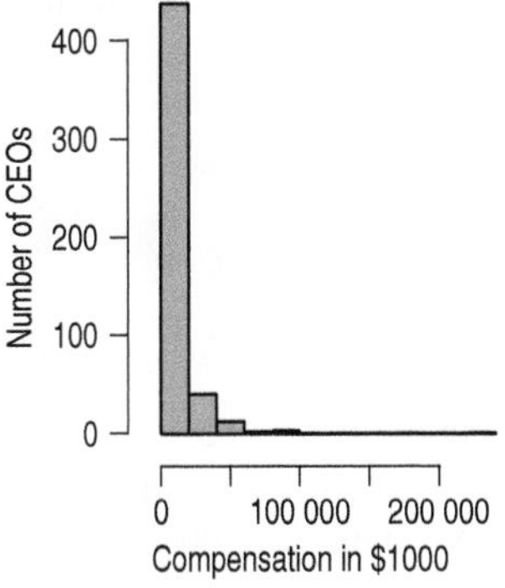

a) Describe the histogram of total compensation. A research organization simulated sample means by drawing samples of 30, 50, 100, and 200, with replacement, from the 800 CEOs. The histograms show the distributions of means for many samples of each size.

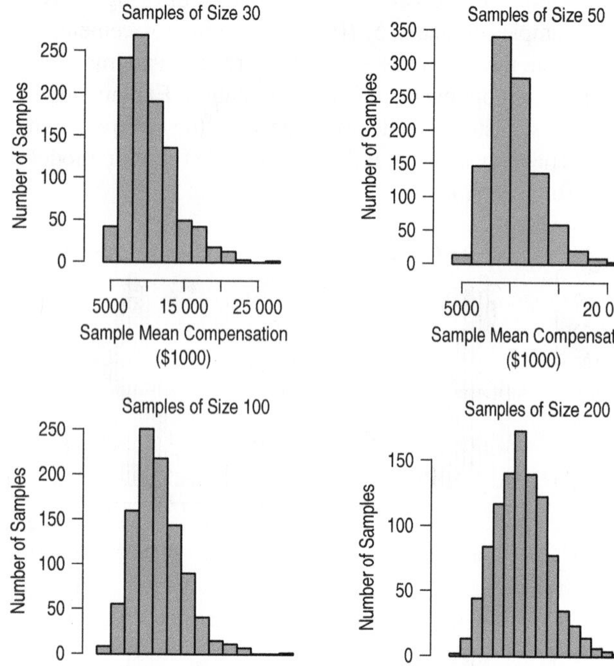

b) Explain how these histograms demonstrate what the Central Limit Theorem says about the sampling distribution model for sample means. Be sure to talk about shape, centre, and spread.

c) Comment on the "rule of thumb" that "With a sample size of at least 30, the *sampling* distribution of the mean is nearly Normal"[25]?

37. Waist size revisited Researchers measured the waist sizes of 250 men in a study on body fat. The true mean and standard deviation of the waist sizes for the 250 men are 36.33 inches and 4.019 inches, respectively. In Exercise 35, you looked at the histograms of simulations that drew samples of sizes 2, 5, 10, and 20 (with replacement). The summary statistics for these simulations were as follows:

n	mean	st. dev.
2	36.314	2.855
5	36.314	1.805
10	36.341	1.276
20	36.339	0.895

a) According to the Central Limit Theorem, what should the theoretical mean and standard deviation be for each of these sample sizes?

b) How close are the theoretical values to what was observed in the simulation?

c) Looking at the histograms in Exercise 35, at what sample size would you be comfortable using the Normal model as an approximation for the sampling distribution?

d) What about the shape of the distribution of waist size explains your choice of sample size in part c)?

38. CEOs revisited In Exercise 36, you looked at the annual compensation for 800 CEOs, for which the true mean and standard deviation were (in thousands of dollars) 10 307.31 and 17 964.62, respectively. A simulation drew samples of sizes 30, 50, 100, and 200 (with replacement) from the total annual compensations of the *Fortune* 800 CEOs. The summary statistics for these simulations were as follows:

n	mean	st. dev.
30	10 251.73	3359.64
50	10 343.93	2483.84
100	10 329.94	1779.18
200	10 340.37	1230.79

a) According to the Central Limit Theorem, what should the theoretical mean and standard deviation be for each of these sample sizes?

b) How close are the theoretical values to what was observed from the simulation?

c) Looking at the histograms in Exercise 36, at what sample size would you be comfortable using the Normal model as an approximation for the sampling distribution?

d) What about the shape of the distribution of total compensation explains your answer in part c)?

39. GPAs A university's data about incoming high school graduates indicates that the mean of their high school averages was 73, with a standard deviation of 7; the distribution was roughly mound-shaped and only slightly skewed. The students are randomly assigned to first-year writing seminars in groups of 25. What might the mean high school average of one of these seminar groups be? Describe the appropriate sampling distribution model—shape, centre, and spread—with attention to assumptions and conditions. Make a sketch using the 68–95–99.7 Rule.

40. Home values Assessment records indicate that the value of homes in a small city is skewed right, with a mean of \$280 000 and standard deviation of \$90 000. To check the accuracy of the assessment data, officials plan to conduct a detailed appraisal of 100 homes selected at random. Using the 68–95–99.7 Rule, draw and label an appropriate sampling model for the mean value of the homes selected.

41. The trial of the pyx In 1150, it was recognized in England that coins should have a standard weight of

[25] Actually, this rule of thumb is generally qualified with the condition that the population not be severely skewed or have extreme outliers.

precious metal as the basis for their value. A guinea, for example, was supposed to contain 128 grains of gold. (There are 12.7 grains in a gram.) In the "trial of the pyx," coins minted under contract to the Crown were weighed and compared to standard coins (which were kept in a wooden box called the pyx). Coins were allowed to deviate by no more than 0.28 grains—roughly equivalent to specifying that the standard deviation should be no greater than 0.09 grains (although they didn't know what a standard deviation was in 1150). In fact, the trial was performed by weighing 100 coins at a time and requiring the *sum* to deviate by no more than 100 × 0.28 or 28 grains—equivalent to the sum having a standard deviation of about nine grains.

a) In effect, the trial of the pyx required that the mean weight of the sample of 100 coins have a standard deviation of 0.09 grains. Explain what was wrong with performing the trial in this manner.
b) What should the limit have been on the standard deviation of the mean?

Note: Because of this error, the Crown was exposed to being cheated by private mints that could mint coins with greater variation and then, after their coins passed the trial, select out the heaviest ones and recast them at the proper weight, retaining the excess gold for themselves. The error persisted for over 600 years, until sampling distributions became better understood.

42. Safe cities Allstate Insurance Company identified the 10 safest and 10 least-safe cities from among the 200 largest cities in the United States, based on the mean number of years that drivers went between automobile accidents. The cities on both lists were all smaller than the 10 largest cities. Using facts about the sampling distribution model of the mean, explain why this is not surprising.

Exercises 43–60 require the use of Normal tables or technology.

43. Pregnancy Assume that the duration of human pregnancies can be described by a Normal model with mean 266 days and standard deviation 16 days.

a) What percentage of pregnancies should last between 270 and 280 days?
b) At least how many days should the longest 25% of all pregnancies last?
c) Suppose a certain obstetrician is currently providing prenatal care to 60 pregnant women. Let $\bar{y}$ represent the mean length of their pregnancies. According to the Central Limit Theorem, what is the distribution of this sample mean, $\bar{y}$? Specify the model, mean, and standard deviation.
d) What is the probability that the mean duration of these patients' pregnancies will be less than 260 days?

44. Calgary in January The average January temperature in Calgary varies from year to year. These average January temperatures themselves have an average, over many past years, of –7.9 degrees Celsius, with a standard deviation of 5.0 degrees. Assume a Normal model applies.

a) During what percentage of years does Calgary have a January average above 0 degrees?
b) In the 20% coldest Januaries, the average January temperature reaches no higher than how many degrees?
c) Suppose you will be in Calgary for the next four years. Describe the sampling distribution model for the average of the next four January average temperatures. Assume climatic conditions are not changing.
d) What is the probability that the average of the next four January average temperatures will exceed −4 degrees? If this happens, would you consider this evidence of climatic warming (at least locally, in Calgary), or just Normal variation?

45. Pregnant again The duration of human pregnancies may not actually follow a Normal model, as described in Exercise 43.

a) Explain why it may be somewhat skewed to the left.
b) If the correct model is in fact skewed, does that change your answers to parts a), b), and c) of Exercise 43? Explain why or why not for each.

46. At work Some business analysts estimate that the length of time people work at a job has a mean of 6.2 years and a standard deviation of 4.5 years.

a) Explain why you suspect this distribution may be skewed to the right.
b) Explain why you could estimate the probability that 100 people selected at random had worked for their employers an average of 10 years or more, but you could not estimate the probability that an individual had done so.

47. Dice and dollars You roll a die. You win nothing if the number of spots is odd. You win $1 for a two or a four, and $10 for a six.

a) Find the expected value and standard deviation of your prospective winnings.
b) You play twice. Find the mean and standard deviation of your total winnings.
c) You play 40 times. What is the probability that you win at least $100?

48. New game You pay $10 and roll a die. If you get a six, you win $50. If not, you get to roll again. If you get a six this time, you get your $10 back.

a) Create a probability model for this game.
b) Find the expected value and standard deviation of your prospective winnings.
c) You play this game five times. Find the expected value and standard deviation of your average winnings.
d) 100 people play this game. What is the probability that the person running the game makes a profit?

49. Canadian families In the 2011 census, Statistics Canada reported the following distribution for family size in Canada.

Size	Percent of Families
6 or more	2.9
5	5.4
4	14.3
3	15.6
2	34.1
1	27.6

a) Find the mean and standard deviation of family size (treat "6 or more" as "6").
b) If we select a random sample of 40 families, would you expect their sizes to follow a Normal model? Explain.
c) Consider the mean sizes of random samples of 40 families. Describe the sampling model for these means (shape, centre, and spread).

50. Museum membership A museum offers several levels of membership, as shown in the table.

Member Category	Amount of Donation ($)	Percent of Members
Individual	50	41
Family	100	37
Sponsor	250	14
Patron	500	7
Benefactor	1000	1

a) Find the mean and standard deviation of the donations.
b) During the museum's annual membership drive, it hopes to sign up 50 new members each day. Would you expect the distribution of the donations for a day to follow a Normal model? Explain.
c) Consider the mean donation of the 50 new members each day. Describe the sampling model for these means (shape, centre, and spread).

51. Canadian families, again In a certain rural area, there are 36 farm families whose average size is 3.0. What is the probability that the average family size will be at least 3.0 if we were to draw a random sample of 36 families from the entire Canadian population, as described in Exercise 49? Does this calculation tell you anything about farm families?

52. Joining the museum One of the museum's phone volunteers sets a personal goal of getting an average donation of at least $100 from the new members she enrols during the membership drive from Exercise 50. If she gets 80 new members and they can be considered a random sample of all the museum's members, what is the probability that she can achieve her goal?

53. Pollution Carbon monoxide (CO) emissions for a certain kind of truck vary with mean 2.9 L/km and standard deviation 0.4 L/km. A company has 80 of these trucks in its fleet. Let $\bar{y}$ represent the mean CO level for the company's fleet.
a) What is the approximate model for the distribution of $\bar{y}$? Explain.
b) Estimate the probability that $\bar{y}$ is between 3.0 and 3.1 L/km.
c) There is only a 5% chance that the fleet's mean CO level is greater than what value?

54. Potato chips The weight of potato chips in a medium-size bag is stated to be 300 g. The amount that the packaging machine puts in these bags is believed to have a Normal model with mean 306 g and standard deviation 3.6 g.
a) What fraction of all bags sold are underweight?
b) Some of the chips are sold in "bargain packs" of three bags. What is the probability that none of the three is underweight?
c) What is the probability that the mean weight of the three bags is below the stated amount?
d) What is the probability that the mean weight per bag of a 24-bag case of potato chips is below 300 g?

55. Tips A waiter believes the distribution of his tips has a model that is slightly skewed to the right, with a mean of $9.60 and a standard deviation of $5.40.
a) Explain why you cannot determine the probability that a given party will tip him at least $20.
b) Can you estimate the probability that the next four parties will tip an average of at least $15? Explain.
c) Is it likely that his 10 parties today will tip an average of at least $15? Explain.

56. Groceries Grocery store receipts show that customer purchases have a skewed distribution with a mean of $32 and a standard deviation of $20.
a) Explain why you cannot determine the probability that the next customer will spend at least $40.
b) Can you estimate the probability that the next 10 customers will spend an average of at least $40? Explain.
c) Is it likely that the next 50 customers will spend an average of at least $40? Explain.

57. More tips The waiter in Exercise 55 usually waits on about 40 parties over a weekend of work.
a) Estimate the probability that he will earn at least $500 in tips.
b) How much does he earn on the best 10% of such weekends?

58. More groceries Suppose the store in Exercise 56 had 312 customers today.

a) Estimate the probability that the store's revenues were at least \$10 000.

b) If, in a typical day, the store serves 312 customers, how much does the store take in on the worst 10% of such days?

59. Big stats course STA220H is a large introductory Statistics course at the University of Toronto, with 720 students enrolled. On the term test, the average score was 72, with a standard deviation of 16. There are 24 tutorials, with 30 students in each. Xiu Xiu teaches one, and Sabri teaches another. Assume that tutorial groups are formed quite randomly.

a) We select a student at random from Sabri's tutorial. Find the probability that this student's test score is at least 80. Assume that scores are Normally distributed.

b) We select one student at random from each tutorial. Find the probability that Sabri's student outperformed Xiu Xiu's student by at least seven marks on the test. (Hint: Work with the difference between the two scores.) Assume that scores are Normally distributed.

c) What is the probability that the average for Sabri's tutorial is at least 80? Do you need to assume that scores are Normally distributed?

d) What is the probability that the average test score in Sabri's tutorial is at least seven marks higher than the average test score for Xiu Xiu's tutorial? (Hint: Work with the difference between the two tutorial averages.) Do you need to assume that scores are Normally distributed?

60. Milk Although most of us buy milk by the litre, farmers measure daily production in kilograms or pounds. Ayrshire cows average 47 pounds of milk a day, with a standard deviation of six pounds. For Jersey cows, the mean daily production is 43 pounds, with a standard deviation of 5 pounds. Assume that Normal models describe milk production for these breeds.

a) We select an Ayrshire at random. What is the probability that she produces more than 50 pounds of milk in one day?

b) What is the probability that a randomly selected Ayrshire gives more milk than a randomly selected Jersey on one day? (Hint: Work with the difference in milk production between the two cows.)

c) A farmer has 20 Jerseys. What is the probability that the average production for this small herd exceeds 45 pounds of milk on one day?

d) A neighbouring farmer has 10 Ayrshires. What is the probability that his herd average is at least five pounds higher than the average for the Jersey herd on one day? (Hint: Work with the difference of the two means.)

For the exercises below, go to www.onlinestatbook.com/stat_sim/sampling_dist/index.html where there is a Java applet illustrating sampling distributions. Click on the ***Begin*** *button. According to the questions below, select first the desired population shape, at the top. Next, where you see* ***Mean*** *and N= **5**, change these according to the question (click on the small arrows). Below you can also change* ***None*** *and **N=5** if the question asks you to compare two different statistics or two different sample sizes. After these are set, you are ready to sample and learn. Always start by choosing* ***Sample: Animated****, which produces a single sample and calculates the chosen statistic. Then click on* ***Sample: 1000*** *several times to add more samples, until the picture of the sampling distribution stabilizes nicely. Note that measures of centre and spread appear to the side of all distributions.*

61. Mean or median? Choose the (default) Normal population at the top. What is the centre of this population? Choose **Mean** and **N=25,** and below choose **Median** and **N=25.** Draw samples and examine the two sampling distributions. Which of these sampling distributions are centred at the population centre (unbiased)? Which statistic would you prefer to use as an estimator of the true population centre—if unknown to you—for this type of population? Explain.

62. Effect of sample size Choose the (default) Normal population. Choose **Mean** at both lower levels, but choose **N=5** at one level and **N=25** at the other. Predict the mean and standard deviation of each of the two sampling distributions you generate. Now generate the sampling distributions and confirm your predictions.

63. Central Limit Theorem Choose the skewed population at the top, and **N=5** and **N=25** at the two levels below, along with the **Mean** in both cases. Predict the shapes, means, and standard deviations of the two sampling distributions you generate. Generate the sampling distributions and compare with your predictions.

64. Central Limit Theorem again Use your computer mouse to create an odd-looking population at the top. Choose **Mean** and **N=25** at the first level below. Generate the sampling distribution. What phenomenon does its shape reflect? Try again to create another odd population that you think might lead to a very different sampling distribution below. Did you succeed?

65. Central Limit Theorem, finis Compare the sampling distribution of the sample mean when sampling from the uniform population versus sampling from the skewed population (both provided at the top). Start with a small sample size and gradually increase it, first for the uniform population, then for the skewed population. About how big does the sample size have to be to produce a nearly Normal sampling distribution for the mean, for each type of population? Why the difference?

66. **Another sampling distribution** Use the Normal population at the top. Every statistic has its own sampling distribution. Let's examine the sampling distribution of the sample variance. Choose **N = 5** and **Variance** at one of the levels below. The sample variance calculated uses the denominator n rather than the $n - 1$ we are accustomed to using. Is the centre (mean) of the sampling distribution of this sample variance equal to the actual population variance (printed out at the top next to the population)? If so, we say that this statistic is an unbiased estimator of the population variance. Try changing the statistic to **Var (U).** This statistic uses $n - 1$ as its denominator. Which version of the sample variance do you prefer? Explain. (We'll learn more about this denominator, which we call the *degrees of freedom*, in Chapter 20.)

Just Checking ANSWERS

1. A Normal model (approximately).
2. At the actual proportion of all students who are in favour.
3. $SD(\hat{p}) = \sqrt{\frac{(0.5)(0.5)}{100}} = 0.05$
4. No, this is a histogram of individuals. It may or may not be Normal, but we can't tell from the information provided.
5. A Normal model (approximately).
6. 266 days.
7. $\frac{16}{\sqrt{100}} = 1.6$ days.

chapter

16 Confidence Intervals for Proportions

Where are we going?

On June 30, 2009, the annual Angus Reid Strategies Canada Day poll surveyed 1000 adult Canadians about their sources of national pride.[1] They reported that, overall, 42% indicated that the monarchy made them feel ("very or moderately") proud, but with a "3.1% margin of error" for this result. Among younger respondents—those aged 18 to 34 years—the percentage dropped to 34%, but with a bigger margin of error. What does this mean? Are younger Canadians less reliable (as well as less impressed with the monarchy)? What is a margin of error? How is it found, and how should it be interpreted? We'll answer those questions in this chapter.

Far better an approximate answer to the right question . . . than an exact answer to the wrong question.

—John Tukey

Coral reef communities are home to one quarter of all marine plants and animals worldwide. These reefs support large fisheries by providing breeding grounds and safe havens for young fish of many species. Coral reefs are seawalls that protect shorelines against tides, storm surges, and hurricanes, and are sand "factories" that produce the limestone and sand that make beaches. Beyond the beach, these reefs are major tourist attractions for snorkelers and divers, driving a tourist industry worth tens of billions of dollars.

But marine scientists say that 10% of the world's reef systems have been destroyed in recent times. At current rates of loss, 70% of the reefs could be gone in 40 years. Pollution, global warming, outright destruction of reefs, and increasing acidification of the oceans are all likely factors in this loss.

Dr. Drew Harvell's lab studies corals and the diseases that affect them. They sampled sea fans at 19 randomly selected reefs along the Yucatan peninsula and diagnosed whether the animals were affected by the disease *aspergillosis*.[2] In specimens collected at a depth of 12 metres (40 feet) from the Las Redes Reef in Akumal, Mexico, these scientists found that 54 of 104 sea fans[3] sampled were infected with that disease.

Of course, we care about much more than these particular 104 sea fans. We care about the health of coral reef communities throughout the Caribbean. What can this study tell us about the prevalence of the disease among sea fans?

We have a sample proportion, which we write as $\hat{p}$, of $54/104$, or 51.9%. Our first guess might be that this observed proportion is close to the population proportion, p. But we also know that because of natural sampling variability, if the researchers had drawn a second sample of 104 sea fans at roughly the same time, the proportion infected from that sample probably wouldn't have been exactly 51.9%.

What *can* we say about the population proportion, p? To begin answering this question, think about how different the sample proportion might have been if we'd taken another random sample from the same population. But wait. Remember—we aren't actually going to take more samples. We just want to *imagine* how the sample proportions might vary from sample to sample. In other words, we want to know about the *sampling distribution* of the sample proportion of infected sea fans.

[1] The top sources of national pride were the flag (86%), the Armed Forces (84%), and hockey (72%).

[2] K. M. Mullen, C. D. Harvell, A. P. Alker, D. Dube, E. Jordán-Dahlgren, J. R. Ward, and L. E. Petes, 2006, "Host range and resistance to aspergillosis in three sea fan species from the Yucatan." *Marine Biology*, Springer-Verlag.

[3] That's a sea fan in the picture on the next page. Although they look like broccoli, they are actually colonies of genetically identical animals.

16.1 A Confidence Interval

Who	Sea fans
What	Percent infected
When	June 2000
Where	Las Redes Reef, Akumal, Mexico (12 metres or 40 feet deep)
Why	Research

Gennadiy Pozniyakov/Fotolia

Sea fans

A S ***Activity:*** **Confidence Intervals and Sampling Distributions.** Simulate the sampling distribution, and see how it gives a confidence interval.

Let's look at our model for the sampling distribution. What do we know about it? We know it's approximately Normal (under certain assumptions, which we should be careful to check) and that its mean is the proportion of all infected sea fans on the Las Redes Reef. Is the infected proportion of *all* sea fans 51.9%? No, that's just $\hat{p}$, our estimate. We don't know the proportion, p, of all the infected sea fans; that's what we're trying to find out. We do know, though, that the sampling distribution model of $\hat{p}$ is centred at p, and we know that the standard deviation of the sampling distribution is $\sqrt{\frac{pq}{n}}$.

Now we have a problem: Since we don't know p, we can't find the true standard deviation of the sampling distribution model. We do know the observed proportion, $\hat{p}$, so of course we just use what we know, and we estimate. That may not seem like a big deal, but it gets a special name. Whenever we estimate the standard deviation of a sampling distribution, we will call it a **standard error** (SE).[4] For a sample proportion, $\hat{p}$, the standard error is

$$SE(\hat{p}) = \sqrt{\frac{\hat{p}\hat{q}}{n}}.$$

For the sea fans, then:

$$SE(\hat{p}) = \sqrt{\frac{\hat{p}\hat{q}}{n}} = \sqrt{\frac{(0.519)(0.481)}{104}} = 0.049 = 4.9\%$$

Now we know that the sampling model for $\hat{p}$ should look like this:

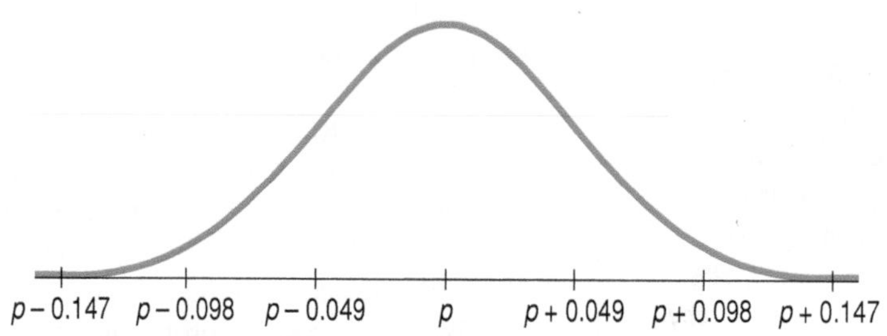

Figure 16.1
The sampling distribution model for $\hat{p}$ is Normal with a mean of p and a standard deviation we estimate to be 0.049.

NOTATION ALERT

Remember that $\hat{p}$ is our sample-based estimate of the true proportion p. Recall also that q is just shorthand for $1 - p$ and $\hat{q} = 1 - \hat{p}$.

When we use $\hat{p}$ to estimate the standard deviation of the sampling distribution model, we call that the **standard error**, and write $SE(\hat{p}) = \sqrt{\frac{\hat{p}\hat{q}}{n}}$.

Great. What does that tell us? Well, because it's Normal, it says that about 68% of all samples of 104 sea fans will have $\hat{p}$'s within 1 *SE*, 0.049, of p. And about 95% of all these samples will be within $p \pm 2$ *SEs*. But where is *our* sample proportion in this picture? And what value does p have? We still don't know!

We do know that for 95% of random samples, $\hat{p}$ will be no more than two *SE*s away from p. So let's look at this from $\hat{p}$'s point of view. If I'm $\hat{p}$, there's a 95% chance that p is no more than two *SE*s away from me. If I reach out two *SE*s, or 2×0.049 away from me on both sides, I'm 95% sure that p will be within my grasp. Now, I've got him! Probably. Of course, even if my interval does catch p, I still don't know its true value. The best I can do is an interval, and even then I can't be positive it contains p.

[4] This isn't such a great name because it isn't standard and nobody made an error. But it's much shorter and more convenient than saying, "the estimated standard deviation of the sampling distribution of the sample statistic." Some prefer to call the standard deviation of a sampling distribution the standard error, and its data-based estimate the estimated standard error.

ACME p-trap: Guaranteed*
to capture p.
*with 95% confidence
p̂ − 2 SE
p̂
p̂ + 2 SE

Figure 16.2
Reaching out two SEs on either side of $\hat{p}$ makes us 95% confident that we'll trap the true proportion, p.

So what can we really say about p? Here's a list of things we'd like to be able to say, in order of strongest to weakest, and the reasons we can't say most of them:

A S *Activity:* **Can We Estimate a Parameter?** Consider these four interpretations of a confidence interval by simulating to see whether they could be right.

1. **"51.9% of *all* sea fans on the Las Redes Reef are infected."** It would be nice to be able to make absolute statements about population values with certainty, but we just don't have enough information to do that. There's no way to be sure that the population proportion is the same as the sample proportion; in fact, it almost certainly isn't. Observations vary. Another sample would yield a different sample proportion.
2. **"It is *probably* true that 51.9% of all sea fans on the Las Redes Reef are infected."** No. In fact, we can be pretty sure that whatever the true proportion is, it's not exactly 51.900%. So the statement is not true.
3. **"We don't know exactly what proportion of sea fans on the Las Redes Reef is infected but we *know* that it's within the interval 51.9% ± 2 × 4.9%. That is, it's between 42.1% and 61.7%."** This is getting closer, but we still can't be certain. We can't know *for sure* that the true proportion is in this interval—or in any particular interval.
4. **"We don't know exactly what proportion of sea fans on the Las Redes Reef are infected, but the interval from 42.1% to 61.7% *probably* contains the true proportion."** We've now fudged twice—first by giving an interval and second by admitting that we only think the interval "probably" contains the true value. And this statement is true.

That last statement may be true, but it's a bit wishy-washy. We can tighten it up by quantifying what we mean by "probably." We saw that 95% of the time when we reach out two *SE*s from $\hat{p}$ we capture p, *so we can be 95% confident that this is one of those times*. After putting a number on the probability that this interval covers the true proportion, we've given our best guess of where the parameter is and how certain we are that it's within some range.

5. **"We are 95% confident that between 42.1% and 61.7% of Las Redes sea fans are infected."** Statements like these are called **confidence intervals**. They're the best we can do.

Each confidence interval discussed in the book has a name. You'll see many different kinds of confidence intervals in the following chapters. Some will be about more than one sample, some will be about statistics other than proportions, and some will use models other than the Normal. The interval calculated and interpreted here is sometimes called a **one-proportion *z*-interval.**[5]

[5] In fact, this confidence interval is so standard for a single proportion that you may see it simply called a "confidence interval for the proportion."

Just Checking

The Angus Reid Strategies Canada Day poll asked Canadians about various sources of national pride. Fifty-eight percent of respondents said they were ("very or moderately") proud of the Canadian healthcare system. We can estimate a 95% confidence interval to be 0.58 ± 0.03, or between 55% and 61%. Are the following statements correct? Explain.

1. In the Angus Reid sample, somewhere between 55% and 61% of respondents reported that they were proud of Canadian healthcare.
2. We can be 95% confident that 58% of Canadians are proud of their healthcare system.
3. We are 95% confident that between 55% and 61% of all Canadians are proud of their healthcare system.
4. We know that between 55% and 61% of all Canadians are proud of their healthcare system.
5. Ninety-five percent of all Canadians are proud of their healthcare system.

16.2 Interpreting Confidence Intervals: What Does "95% Confidence" Really Mean?

A S *Activity:* **Confidence Intervals for Proportions.** This new interactive tool makes it easy to construct and experiment with confidence intervals. We'll use this tool for the rest of the course—sure beats calculating by hand!

What do we mean when we say we have 95% confidence that our interval contains the true proportion? Formally, what we mean is that "95% of samples of this size will produce confidence intervals that capture the true proportion." This is correct, but a little long winded, so we sometimes say, "we are 95% confident that the true proportion lies in our interval." Our uncertainty is about whether the particular sample we have at hand is one of the successful ones or one of the 5% that fail to produce an interval that captures the true value.

In Chapter 15, we saw that proportions vary from sample to sample. If other researchers select their own samples of sea fans, they'll also find some infected by the disease, but each person's sample proportion will almost certainly differ from ours. When they each try to estimate the true rate of infection in the entire population, they'll centre *their* confidence intervals at the proportions they observed in their own samples. Each of us will end up with a different interval.

Our interval guessed the true proportion of infected sea fans to be between about 42% and 62%. Another researcher whose sample contained more infected fans than ours did might guess between 46% and 66%. Still another who happened to collect fewer infected fans might estimate the true proportion to be between 23% and 43%. And so on. Every possible sample would produce yet another confidence interval. Although wide intervals like these can't pin down the actual rate of infection very precisely, we expect that most of them should be winners, capturing the true value. Nonetheless, some will be duds, missing the population proportion entirely.

The figure below shows the confidence intervals produced by simulating 20 different random samples. The red dots are the proportions of infected fans in each sample, and the blue segments show the confidence intervals found for each. The green line represents the true rate of infection in the population, so you can see that most of the intervals caught

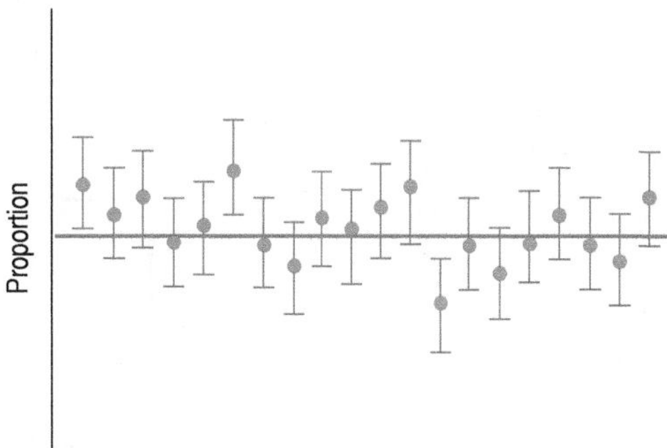

The horizontal green line shows the true percentage of all sea fans that are infected. Most of the 20 simulated samples produced confidence intervals that captured the true value, but a few missed.

it—but a few missed. (And notice again that it is the *intervals* that vary from sample to sample; the green line doesn't move.)

Of course, there's a huge number of possible samples that *could* be drawn, each with its own sample proportion. These are just some of them. Each sample proportion can be used to make a confidence interval. That's a large pile of possible confidence intervals, and ours is just one of those in the pile. Did *our* confidence interval "work"? We can never be sure, because we'll never know the true proportion of all the sea fans that are infected. However, the Central Limit Theorem assures us that 95% of the intervals in the pile are winners, covering the true value, and only 5% are duds. *That's* why we're 95% confident that our interval is a winner!

For Example POLLS AND MARGIN OF ERROR

In June 2008, a survey of 1000 Canadians conducted by the Strategic Counsel for CTV and the *Globe and Mail* found that 68% of Canadians support gay marriage. The researchers reported their margin of error to be ±3.1%.

QUESTION: It is standard among pollsters to use a 95% confidence level unless otherwise stated. Given that, what do they mean by claiming a margin of error of ±3.1% in this context?

ANSWER: If this polling were done repeatedly, 95% of all random samples would yield estimates that come within 3.1 percentage points (0.031) of the true percentage (proportion) of all Canadians who support gay marriage.

16.3 Margin of Error: Certainty versus Precision

We've just claimed that with a certain confidence we've captured the true proportion of all infected sea fans. Our confidence interval had the form

$$\hat{p} \pm 2\,SE(\hat{p})$$

The extent of the interval on either side of $\hat{p}$ is called the **margin of error** (*ME*). We'll want to use the same approach for many other situations besides estimating proportions. Often, confidence intervals look like this:

$$Estimate \pm ME$$

The margin of error for our 95% confidence interval was two *SE*s or $2 \times SE$. What if we wanted to be more confident? To be more confident, we'll need to capture p more often, and to do that we'll need to make the interval wider. For example, if we want to be 99.7% confident, the margin of error will have to be $3 \times SE$.

Figure 16.3
Reaching out three SEs on either side of $\hat{p}$ makes us 99.7% confident we'll trap the true proportion, p. Compare with Figure 16.2.

The more confident we want to be, the larger the margin of error must be. We can be 100% confident that the proportion of infected sea fans is between 0% and 100%, but this isn't likely to be very useful. On the other hand, we could give a confidence interval from 51.8% to 52.0%, but we can't be very confident about a precise statement like this. Every confidence interval is a balance between certainty and precision.

A S *Activity:* **Balancing Precision and Certainty.** What percent of parents expect their kids to pay for university with a student loan? Investigate the balance between the precision and the certainty of a confidence interval.

The tension between certainty and precision is always there. Fortunately, in most cases we can be both sufficiently certain and sufficiently precise to make useful statements. There is no simple answer to the conflict. You must choose a confidence level yourself. The data can't do it for you. The choice of confidence level is somewhat arbitrary. The most commonly chosen confidence levels are 90%, 95%, and 99%, but any percentage can be used. (In practice, though, using something like 92.9% or 97.2% is likely to make people think you're up to something.)

Paws, Inc./Universal Uclick

For Example FINDING THE MARGIN OF ERROR (TAKE 1)

RECAP: A June 2008 Strategic Counsel poll of 1000 Canadians reported a margin of error of ±3.1 percentage points. It is a convention among pollsters to use a 95% confidence level and to report the "worst case" margin of error, based on $p = 0.5$.

QUESTION: How did Strategic Counsel calculate their margin of error?

ANSWERS: Assuming $p = 0.5$, for random samples of size $n = 1000$,

$$SD(\hat{p}) = \sqrt{\frac{pq}{n}} = \sqrt{\frac{(0.5)(0.5)}{1000}} = 0.0158$$

For a 95% confidence level, $ME = 2(0.0158) = 0.032$.[6]

The margin of error is just a bit over ±3%.

---

[6]Using 1.96 *SE*s instead of 2 *SE*s would produce their reported *ME* of ± 3.1%. See "Critical Values" below for an explanation of this small difference.

Critical Values

■ **NOTATION ALERT**

We'll put an asterisk on a letter to indicate a critical value, so z^* is always a critical value from a standard Normal model. Critical values are just percentile points of the model, chosen to match the probability requirements.

In our sea fans example, we used 2 SEs to give us a 95% confidence interval. To change the confidence level, we'd need to change the *number* of SEs so that the size of the margin of error corresponds to the new level. This number of SEs is called the **critical value**. Here, it's based on the standard Normal model, so we denote it z^*. For any confidence level, we can find the corresponding critical value from a computer, a calculator, or a Normal probability table, such as Table Z.

For a 95% confidence interval, you'll find the precise critical value is $z^* = 1.96$. That is, 95% of a Normal model is found within ± 1.96 standard deviations of the mean. We've been using $z^* = 2$ from the 68–95–99.7 Rule because it's easy to remember. What about 90% confidence? See Figure 16.4.

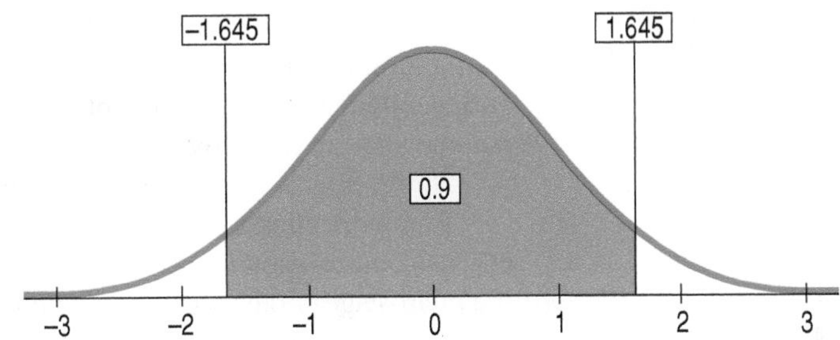

Figure 16.4
For a 90% confidence interval, the critical value is 1.645 because, for a Normal model, 90% of the values are within 1.645 standard deviations from the mean.

For Example FINDING THE MARGIN OF ERROR (TAKE 2)

RECAP: In June 2008, a Strategic Counsel poll of 1000 Canadians estimated that 68% of Canadians support gay marriage. They reported a 95% level of confidence with a margin of error of ± 3.1.

QUESTIONS: Using the critical value of z and the standard error based on the observed proportion, what would be the margin of error for a 90% level of confidence? What's good and bad about this change in the level of confidence?

ANSWER: With $n = 1000$ and $\hat{p} = 0.68$,

$$SE(\hat{p}) = \sqrt{\frac{\hat{p}\hat{q}}{n}} = \sqrt{\frac{(0.68)(0.32)}{1000}} = 0.0148$$

For a 90% confidence level, $z^* = 1.645$, so

$$ME = 1.645(0.0148) = 0.024$$

Now the margin of error is closer to $\pm 2\%$, producing a narrower interval. That makes for a more precise estimate of voter belief, but provides less certainty that the interval actually contains the true proportion of voters supporting gay marriage.

16.4 Assumptions and Conditions

We've just made some pretty sweeping statements about sea fans. Those statements were possible because we used a Normal model for the sampling distribution. But is that model appropriate?

We've also said that the same basic ideas will work for other statistics. One difference for those statistics is that some will have sampling distribution models that are different than the Normal. But the background theory (which we won't bother with) is so similar for all of those models that they share the same basic assumptions and conditions about independence and sample size. Then for each statistic, we usually tack on a special-case assumption or two. We'll deal with those as we get to them in later chapters.

We saw the assumptions and conditions for using the Normal to model the sampling distribution for a proportion in the last chapter. Because they are so crucial to making sure our confidence interval is useful, we'll repeat them here.

We can never be certain that an assumption is true, but we can decide intelligently whether it is reasonable. When we have data, we can often decide whether an assumption is plausible by checking a related condition. However, we want to make a statement about the world at large, not just about the data we collected. So the assumptions we make are not just about how our data look, but about how representative they are.

Independence Assumption

A S *Activity:* **Assumptions and Conditions.** Here's an animated review of the assumptions and conditions.

Independence Assumption: We first need to *Think* about whether the independence assumption is plausible. We often look for reasons to suspect that it fails. We wonder whether there is any reason to believe that the data values somehow affect each other. (For example, might the disease in sea fans be contagious?) Whether you decide that the **Independence Assumption** is plausible depends on your knowledge of the situation. It's not one you can check by looking at the data.

However, now that we have data, there are two conditions that we can check:

Randomization Condition: Were the data sampled at random or generated from a properly randomized experiment? Proper randomization can help ensure independence. However, if a multi-stage or other complex sampling design was used, observations may be dependent and our simple SE formula will no longer be correct, often understating the true SE.

10% Condition:[7] Samples are almost always drawn without replacement. Of course, we'd like to have a large sample, but when the population itself is small, the sample may be a big chunk of the population, and then the probability of success may be very different for the last few individuals we draw than it was for the first few. For example, if most of the women from a population have already been sampled, the chance of drawing a woman from the remaining population is lower. If less than 10% of the population is sampled, probabilities change little, and the effect on independence is negligible. However, if you sample more than 10% of the population, our SE formula falls apart, producing an overestimate of the true SE. In that case, if you report the Confidence Interval formula given in this chapter, be sure to note that the reported confidence level is *conservative* (lower than the true level). (Or apply the "finite population correction" as discussed in the optional section in Chapter 15.)

Sample Size Assumption

The model we use for inference is based on the Central Limit Theorem. The **Sample Size Assumption** addresses the question of whether the sample is large enough to make the sampling model for the sample proportions approximately Normal (and to allow us to estimate its SD with the SE). It turns out that we need more data as the proportion gets closer and closer to either extreme (0 or 1). We can check this assumption with the:

Success/Failure Condition: We must expect at least 10 "successes" and at least 10 "failures". Recall that, by tradition, we arbitrarily label one alternative (usually the outcome being counted) as a "success" even if it's something bad (like a sick sea fan). The other alternative is, of course, then a "failure."

A S *Activity:* **A Confidence Interval for *p*.** View the video story of pollution in Chesapeake Bay and use the interactive tool to make a confidence interval for the analysis.

ONE-PROPORTION *Z*-INTERVAL

When the conditions are met, we are ready to find a level C confidence interval for the population proportion, p. The confidence interval is $\hat{p} \pm z^* \times SE(\hat{p})$, where the standard deviation of the proportion is estimated by $SE(\hat{p}) = \sqrt{\frac{\hat{p}\hat{q}}{n}}$ and the critical value, z^*, specifies the number of SEs needed for C% of random samples to yield confidence intervals that capture the true parameter value.

[7] To be clear, sampling more than 10% of the population is definitely a good thing. But you need to recognize that the SE formula presented here overestimates the true SE. We will gladly sacrifice independence for precision any day!

Step-by-Step Example A CONFIDENCE INTERVAL FOR A PROPORTION

Who Adults in the United States

What Response to a question about the death penalty

When October 2010

Where United States

How 510 adults were randomly sampled and asked by the Gallup Poll

Why Public opinion research

The last execution in Canada for the crime of murder took place in 1962. There is little support in Canada for bringing back capital punishment, unlike in the United States where capital punishment has considerable public support (there were 3125 inmates on death row as of January 1, 2013).

In October 2010, the Gallup Poll[8] asked 510 randomly sampled U.S. adults the question, "Generally speaking, do you believe the death penalty is applied fairly or unfairly in this country today?" Of these, 58% answered "fairly," 36% said "unfairly," and 7% said they didn't know.

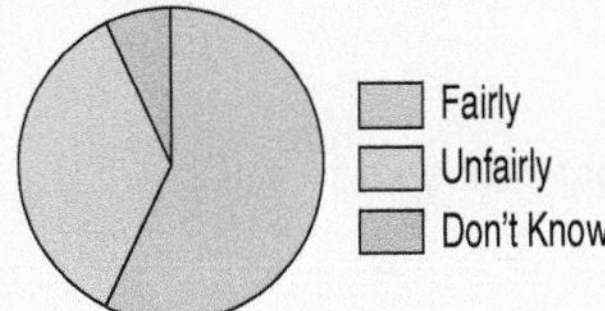

Question: From this survey, what can we conclude about the opinions of *all* adults?

To answer this question, we'll build a confidence interval for the proportion of all U.S. adults who believe the death penalty is applied fairly. There are four steps to building a confidence interval for proportions: plan, model, mechanics, and conclusion.

THINK ➡ Plan State the problem and the *W*s.

Identify the *parameter* you wish to estimate.

Identify the *population* about which you wish to make statements.

Choose and state a confidence level.

I want to find an interval that is likely with 95% confidence to contain the true proportion, p, of U.S. adults who think the death penalty is applied fairly. I have a random sample of 510 U.S. adults.

Model Think about the assumptions and check the conditions.

✓ **Independence Assumption:** Gallup phoned a random sample of U.S. adults. It is very unlikely that any of their respondents influenced each other.

✓ **Randomization Condition:** Gallup drew a random sample from all U.S. adults. I don't have details of their randomization but assume that I can trust it, and that it is similar to an SRS.

✓ **10% Condition:** Although sampling was necessarily without replacement, there are many more U.S. adults than were sampled. The sample is certainly less than 10% of the population.

✓ **Success/Failure Condition**

$n\hat{p} = 510(58\%) = 296 \geq 10$ and

$n\hat{q} = 510(42\%) = 214 \geq 10,$

so the sample appears to be large enough to use the Normal model.

State the sampling distribution model for the statistic.

Choose your method.

The conditions are satisfied, so I can use a Normal model to find a **one-proportion z-interval**.

[8] www.galluppoll.com

SHOW ➡ Mechanics Construct the confidence interval.	$n = 510, \hat{p} = 0.58$
First find the standard error. (Remember: It's called the "standard error" because we don't know p and have to use $\hat{p}$ instead.)	$SE(\hat{p}) = \sqrt{\frac{\hat{p}\hat{q}}{n}} = \sqrt{\frac{(0.58)(0.42)}{510}} = 0.022$
Next find the margin of error. We could informally use 2 for our critical value, but 1.96 is more accurate.	Because the sampling model is Normal, for a 95% confidence interval, the critical value $z^* = 1.96$. The margin of error is
Write the confidence interval (CI).	$ME = z^* \times SE(\hat{p}) = 1.96(0.022) = 0.043$
REALITY CHECK The CI is centred at the sample proportion and is about as wide as we might expect for a sample of 510.	So the 95% confidence interval is 0.58 ± 0.043 or $(0.537, 0.623)$
TELL ➡ Conclusion Interpret the confidence interval in the proper context. We're 95% confident that our interval captured the true proportion.	I am 95% confident that between 53.7% and 62.3% of all U.S. adults think that the death penalty is applied fairly.

Just Checking

Think some more about the 95% confidence interval we just created for the proportion of U.S. adults who think the death penalty is applied fairly.

6. If we wanted to be 98% confident, would our confidence interval need to be wider or narrower?
7. Our margin of error was about $\pm 4\%$. If we wanted to reduce it to $\pm 3\%$, would our level of confidence be higher or lower?
8. If the Gallup organization had polled more people, would the interval's margin of error have been larger or smaller?

Choosing Your Sample Size

The question of how large a sample to take is an important step in planning any study. We weren't ready to make that calculation when we first looked at study design in Chapter 10, but now we can—and we always should.

Suppose a candidate is planning a poll and wants to estimate voter support within 3 percentage points with 95% confidence. How large a sample does she need?

Let's look at the margin of error:

$$ME = z^* \sqrt{\frac{\hat{p}\hat{q}}{n}}$$

And now plug in to get: $0.03 = 1.96\sqrt{\frac{\hat{p}\hat{q}}{n}}$

We want to find n, the sample size. To find n we need a value for $\hat{p}$. We don't know $\hat{p}$ because we don't have a sample yet, but we can probably guess a value. The worst case—the

WHAT DO I USE INSTEAD OF $\hat{p}$?

Often we have an estimate of the population proportion based on experience or perhaps a previous study. If so, use that value as $\hat{p}$ in calculating what size sample you need. If not, the cautious approach is to use $p = 0.5$ in the sample size calculation; that will determine the largest sample necessary regardless of the true proportion.

value that makes $\hat{p}\hat{q}$ (and therefore n) largest—is 0.50, so if we use that value for $\hat{p}$, we'll certainly be safe. Our candidate probably expects to be near 50% voter support anyway.

Our equation, then, is

$$0.03 = 1.96\sqrt{\frac{(0.5)(0.5)}{n}}$$

To solve for n, we first multiply both sides of the equation by $\sqrt{n}$ and then divide by 0.03:

$$0.03\sqrt{n} = 1.96\sqrt{(0.5)(0.5)}$$

$$\sqrt{n} = \frac{1.96\sqrt{(0.5)(0.5)}}{0.03} \approx 32.667$$

Notice that evaluating this expression tells us the *square root* of the sample size. We need to square that result to find n:

$$n \approx (32.667)^2 \approx 1067.1$$

To be safe, we round up and conclude that we need at least 1068 respondents to keep the margin of error as small as 3 percentage points with a confidence level of 95%.

For Example CHOOSING A SAMPLE SIZE

RECAP: The Strategic Counsel poll that estimated that 68% of all Canadians support gay marriage had a margin of error of ±3.1%. Suppose another organization planning a follow-up survey of opinions on gay marriage wants to determine a 95% confidence interval with a margin of error of no more than ±2.0%.

QUESTION: How large a sample do they need? Use the Strategic Counsel estimate as the basis for your calculation.

ANSWER:

$$ME = z^*\sqrt{\frac{\hat{p}\hat{q}}{n}}$$

$$0.02 = 1.96\sqrt{\frac{(0.68)(0.32)}{n}}$$

$$\sqrt{n} = \frac{1.96\sqrt{(0.68)(0.32)}}{0.02} \approx 45.715$$

$$n = 45.715^2 = 2089.8$$

This organization's survey will need about 2090 respondents.

Popular opinion polls

Public opinion polls often sample 1000 people, which gives an ME of 3 percentage points when $p = 0.5$. But businesses and nonprofit organizations typically use much larger samples to estimate the proportion of people who will accept a direct mail offer. Why? Because that proportion is very low—often far below 5%. An ME of 3% wouldn't be precise enough. An ME like 0.1% would be more useful, and that requires a very large sample size.

Unfortunately, bigger samples cost more money and require more effort. Because the standard error declines only with the *square root* of the sample size, to cut the standard error (and thus the ME) in half, we must *quadruple* the sample size.

Often a margin of error of 5 percentage points or less is acceptable, but different circumstances call for different standards. For a pilot study, a margin of error of 10% may be fine, so a sample of 100 will do quite well. In a close election, a polling organization might want to get the margin of error down to 2%. Drawing a large sample to get a smaller ME, however, can run into trouble. It takes time to survey 2400 people, and a survey that extends over a week or more may end up hitting a target that moves during the time of the survey. An important event can change public opinion in the middle of the survey process. Professional polling organizations often report both an instantaneous estimate and a rolling average of several recent polls in order to increase the sample size.

Keep in mind that the sample size for a survey is the number of respondents, not the number of people to whom questionnaires were sent or whose phone numbers were dialled. And keep in mind that *a low response rate turns any study essentially into a voluntary response study*, which is of little value for inferring population values. It's almost always better to spend resources on increasing the response rate than on surveying a larger group. A full or nearly full response by a modest-size sample can yield useful results.

Surveys are not the only place where proportions pop up. Banks sample huge mailing lists to estimate what proportion of people will accept a credit card offer. Even pilot studies may mail offers to over 50 000 customers. Most don't respond; that doesn't make the sample smaller—they simply said, "No, thanks." Those who do respond want the card. To the bank, the response rate[9] is $\hat{p}$. With a typical success rate around 0.5%, the bank needs a very small margin of error—often as low as 0.1%—to make a sound business decision. That calls for a large sample, and the bank must take care in estimating the size needed. For our election poll calculation we used $p = 0.5$, both because it's safe and because we honestly believed p to be near 0.5. If the bank used 0.5, they'd get an absurd answer. Instead, they base their calculation on a proportion closer to the one they expect to find.

For Example SAMPLE SIZE REVISITED

A credit card company is about to send out a mailing to test the market for a new credit card. From that sample, they want to estimate the true proportion of people who will sign up for the card nationwide. To be within a tenth of a percentage point (0.001) of the true rate with 95% confidence, how big does the test mailing have to be? A pilot study suggests that about 0.5% of the people receiving the offer will accept it. Using that estimate the company finds:

$$ME = 0.001 = z^*\sqrt{\frac{\hat{p}\hat{q}}{n}} = 1.96\sqrt{\frac{(0.005)(0.995)}{n}}$$

$$(0.001)^2 = 1.96^2\frac{(0.005)(0.995)}{n} \Rightarrow n = \frac{1.96^2(0.005)(0.995)}{(0.001)^2}$$

$$= 19\,111.96 \text{ or } 19\,112$$

That's a lot, but it's actually a reasonable size for a trial mailing such as this. But if they had assumed a conservative 0.50 for the value of p, they would have found

$$ME = 0.001 = z^*\sqrt{\frac{pq}{n}} = 1.96\sqrt{\frac{(0.5)(0.5)}{n}}$$

$$(0.001)^2 = 1.96^2\frac{(0.5)(0.5)}{n} \Rightarrow n = \frac{1.96^2(0.5)(0.5)}{(0.001)^2} = 960\,400$$

Quite a different (and expensive) result.

Try to be conservative with your guesses when planning the sample size. Figure 16.5 graphs pq versus p and shows that pq peaks at $(0.5)(0.5) = 0.25$, so the closer p is to 0.5, the bigger pq is. If, for example, you think p should be somewhere between 0.7 and 0.9, let $pq = (0.7)(0.3) = 0.21$. In other words, choose the value for p from your range guess that lies closest to 0.5. Even if you guess badly, your confidence interval will still be valid, just wider than hoped.

[9]In marketing studies, every mailing yields a response—"yes" or "no"—and "response rate" means the proportion of customers who accept an offer. That's not the way we use the term for survey response.

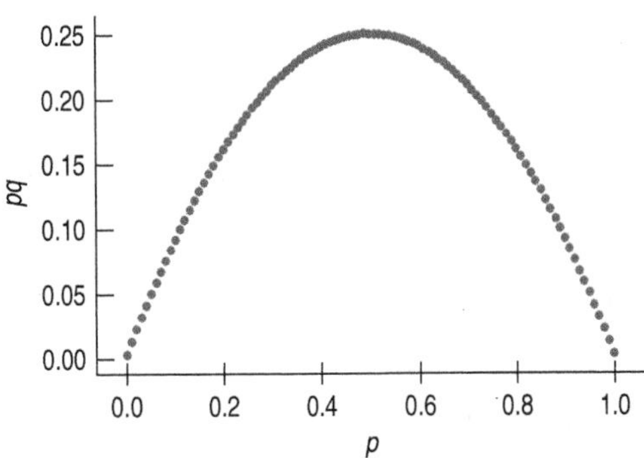

Figure 16.5

Graph of pq versus p. Choose the p from its plausible range that maximizes pq.

*16.5 The Plus Four Confidence Interval for Small Samples

When the **Success/Failure Condition** fails, all is not lost. The Normal sampling model fails, but we know the exact sampling distribution model for the sample count—the Binomial distribution. We can use this underlying distribution to derive what are called *exact Binomial confidence intervals* for the parameter p. Many statistics software programs will compute these exact confidence intervals, but there is another approach, involving a simple adjustment to the Normal-based confidence interval, which seems to work even better (though that may seem strange—how something *approximate* can be better than something said to be *exact*!). It's called the *Plus Four CI* for p.

All we do is add four *phony* observations to our sample—two successes and two failures. So instead of the proportion $\hat{p} = \frac{y}{n}$, we use the adjusted proportion $\tilde{p} = \frac{y+2}{n+4}$ and, for convenience, we write the phony sample size as $\tilde{n} = n + 4$. We modify the interval by using these adjusted values for both the centre of the interval and the margin of error. Now the adjusted interval is

$$\tilde{p} \pm z^*\sqrt{\frac{\tilde{p}(1-\tilde{p})}{\tilde{n}}}$$

This method gives better performance overall,[10] and especially for proportions near 0 or 1. It has the additional advantage that we no longer need to check the Success/Failure Condition that (the two counts) $n\hat{p}$ and $n\hat{q}$ are both greater than 10.

For Example A PLUS FOUR CONFIDENCE INTERVAL

Surgeons examined their results to compare two methods for a surgical procedure used to alleviate pain on the outside of the wrist. A new method was compared with the traditional "freehand" method for the procedure. Of 45 operations using the "freehand" method, three were unsuccessful, for a failure rate of 6.7%. With only three failures, the data don't satisfy the Success/Failure Condition, so we can't use the regular Normal-based confidence interval.

QUESTION: What is the confidence interval using the Plus Four method?

ANSWER: There were 42 successes and three failures. Adding two "pseudo-successes" and two "pseudo-failures," we find

[10]By "better performance," we mean that a 95% confidence interval has more nearly a 95% chance of covering the true population proportion. Simulation studies have shown that our original unadjusted confidence interval is in fact less likely than 95% to cover the true population proportion when the sample size is small or the proportion is very close to 0 or 1 while 'exact Binomial' confidence intervals have slightly more than 95% confidence. The original idea for this method can be attributed to E. B. Wilson. The approach discussed here was proposed by Agresti and Coull (A. Agresti and B. A. Coull 1998, "Approximate is better than 'exact' for interval estimation of binomial proportions." *The American Statistician*, 52: 119–129).

$$\tilde{p} = \frac{3 + 2}{45 + 4} = 0.102$$

A 95% confidence interval for the failure rate is then

$$0.102 \pm 1.96\sqrt{\frac{0.102(1 - 0.102)}{49}} = 0.102 \pm 0.085$$

or (0.017, 0.187).

Notice that although the observed failure rate of 0.067 is contained in the interval, it is not at the centre of the interval—something we haven't seen with any of the other confidence intervals we've considered.

*16.6 Large Sample Confidence Intervals Based on Normally Distributed Estimators

Looking back at the derivation of the confidence interval for a proportion suggests that this method may be broadly applicable and easily adapted for other parameters. The key features appear to be the following:

I. We can construct an estimator of the parameter that has an approximately Normal sampling distribution centred at the parameter of interest (so Figure 16.1 applies); and
II. We can derive a formula for the standard deviation of this sampling distribution and evaluate it with sufficient accuracy, perhaps using the sample data, to give us the relevant SE, from which we can then calculate the ME $= z^* \times$ SE.

Now, what do large samples have to do with this? Well, many estimators (such as means or proportions or differences thereof) will be approximately Normal if the sample size is big, regardless of the population distribution, due to that marvel of probability theory, the Central Limit Theorem. And for a big enough sample, we can also use the sample information to estimate the standard deviation of the sampling distribution with sufficient accuracy. Non-Normality or excessive variability in the SE estimate will make this approach fall apart (in which case, you may need small sample methods, like the Plus Four method discussed in the previous section or the methods for means in Chapters 20–22).

If we are estimating a parameter θ, using an estimator $\hat{\theta}$ whose sampling distribution is Normal with mean equal to θ and standard deviation denoted $SD(\hat{\theta})$, the confidence interval takes the form:

$$\text{Confidence Interval for } \theta\text{: } Estimate \pm ME = \hat{\theta} \pm z^*SD(\hat{\theta})$$

However, this equation can only be used if you can evaluate the $SD(\hat{\theta})$. Usually you will not know the true value of $SD(\hat{\theta})$, but if you can estimate it accurately enough via an appropriate formula and data from a large enough sample, you can then plug in this estimate, denoted $SE(\hat{\theta})$, to get the:

$$\text{Large Sample Confidence Interval for } \theta\text{: } Estimate \pm ME = \hat{\theta} \pm z^*SE(\hat{\theta})$$

Now let's apply the above approach to the estimation of the mean of a population of numbers.

Confidence Interval for the Mean of a Quantitative Population When the Sample Is Big

To estimate the mean μ of a quantitative population, assume we draw a simple random sample. From the sampled items, we take measurements on the variable of interest: $y_1, y_2, \ldots, y_n$, and compute their mean, $\bar{y}$, which will be used to estimate the population mean. What do we know about $\bar{y}$,? By the Central Limit Theorem, for size n, the sampling distribution of $\bar{y}$ is approximately Normal. In Chapter 15, we derived a formula for the standard deviation of this sampling distribution:

$SD(\bar{y}) = \frac{\sigma}{\sqrt{n}}$, which we can estimate from the data by $\frac{s}{\sqrt{n}}$ which we can write more simply as $SE(\bar{y})$.

Now plug into the generic confidence interval formula based on an unbiased Normally distributed estimator:

$$\text{Confidence Interval for } \mu\text{: } Estimate \pm ME = \hat{\theta} \pm z^*SD(\hat{\theta})$$

$$= \hat{\mu} \pm z^*SD(\hat{\mu}) = \bar{y} \pm z^*SD(\bar{y}) = \bar{y} \pm z^*\frac{\sigma}{\sqrt{n}}$$

And the latter, we can estimate by $\bar{y} \pm z^*\frac{s}{\sqrt{n}}$ for *big enough n*.[11]

How big should n be in order to achieve approximate Normality and an accurate estimate of the SD? Let's say that *n should be at least 60* (other texts may differ in their advice, ranging from $n = 30$ to $n = 100+$, simply due to the question of how close we need to be to the specified level of confidence). For smaller n, use the methods described in Chapter 20. And even if $n > 60$, *inspect the data* for severe outliers or extreme skewness. (How often have we warned you about that?) If extremely skewed, a bigger n might be needed, or perhaps you should think about a transformation of the data—say, with a logarithm—before doing any analysis (or maybe you should focus your inference on the median rather than the mean).

And—of course, of course, of course—just as with proportions, make sure that the observations are independent (of each other) and that you are positioned to generalize beyond your sample to a larger population.

Consider an example. A team of researchers tested 150 farm-raised salmon (collected from various sources) for organic contaminants, in particular the level of the carcinogenic insecticide mirex.[12] We expect some variation in the individual salmon measurements, but we may be primarily interested in the mean concentration for salmon raised in this type of environment. The researchers found the mean mirex concentration in this sample to be 0.0913 ppm with a standard deviation of 0.0495 ppm. The histogram of the data showed no extreme skewness or outliers. Assuming this to be a simple random sample of farm-raised salmon, a 95% confidence interval for the true mean level of mirex in farm-raised salmon would be (for convenience, we'll use $z^* = 2$ instead of $z^* = 1.96$):

$$\bar{y} \pm z^*\frac{s}{\sqrt{n}} = 0.0913 \pm 2 \times \frac{0.0495}{\sqrt{150}}$$

$$= 0.0913 \pm 2 \times 0.00404 = 0.0913 \pm 0.00808$$

$$= (0.0832, 0.0994)$$

So we can say with 95% confidence that the true mean concentration for salmon raised in this type of environment is between 0.0832 ppm and 0.0994 ppm.

---

[11] Why not use the true population standard deviation, σ, instead of its estimate, s? Since σ is a parameter of the population, you need the entire population in order to calculate it. The best we can do is a sample estimate—the sample standard deviation, s. Like the sample mean, it will not exactly equal the corresponding population characteristic due to sampling variation. However, if you find yourself in the rare situation where you actually believe you know the true population σ (perhaps from prior experience with the process), then go ahead and use it. The real thing is always better than an estimate.

[12] Ronald A. Hites, Jeffrey A. Foran, David O. Carpenter, M. Coreen Hamilton, Barbara A. Knuth, and Steven J. Schwager, 2004, "Global assessment of organic contaminants in farmed salmon," *Science* Vol. 303, no. 5655, pp. 226–229.

Choosing the Sample Size

Just as with proportions, we should think carefully about the sample size before running a study. To set the margin of error, proceed just as you did with proportions, setting the formula for the margin of error equal to the requirement and then solving for n.

Set:

$$z^* \frac{\sigma}{\sqrt{n}} = \textit{required margin or error}$$

and solve for n. But what should we do about that σ? Make an educated guess, perhaps based on some past experience, a pilot study, or similar studies known to you. It may be easier to guess the likely range of the data, call it R, and then use R/4 as a (very rough) ballpark estimate of the standard deviation of the measurements. Remember, guessing badly will not affect the validity of your conclusions, just the actual margin of error attained, but try to err on the conservative side, if possible, by guessing high rather than low.

As a final comment, we note that not all confidence intervals are formed in this way, nor are they all symmetric about the estimate, but, fortunately, they may all be interpreted exactly the same way. For example, we know that if the procedure were repeated over and over, 95 times in 100, the true value of the parameter being estimated would be caught between the limits of the confidence interval!

WHAT CAN GO WRONG?

Confidence intervals are powerful tools. Not only do they tell what we know about the parameter value, but—more important—they also tell what we *don't* know. To use confidence intervals effectively, you must be clear about what you say about them.

WHAT *CAN* I SAY?

Confidence intervals are based on random samples, so the interval is random, too. The Central Limit Theorem tells us that 95% of random samples will yield intervals that capture the true value. That's what we mean by being 95% confident.

Technically, we should say, "I am 95% confident that the interval from 42.1% to 61.7% captures the true proportion of infected sea fans." That formal phrasing emphasizes that *our confidence (and our uncertainty) is about the interval, not the true proportion.* But you may choose a more casual phrasing, like "I am 95% confident that between 42.1% and 61.7% of the Las Redes fans are infected." Because you've made it clear that the uncertainty is yours and you didn't suggest that the randomness is in the true proportion, this is okay. Keep in mind that it's the interval that is random and is the focus of both our confidence and our doubt.

Don't Misstate What the Interval Means

- **Don't suggest that the parameter varies.** A statement like, "There is a 95% chance that the true proportion is between 42.7% and 51.3%" sounds as though you think the population proportion wanders around and sometimes happens to fall between 42.7% and 51.3%. When you interpret a confidence interval, make it clear that *you* know that the population parameter is fixed and that it is the interval that varies from sample to sample.
- **Don't claim that other samples will agree with yours.** Keep in mind that the confidence interval makes a statement about the true population proportion. An interpretation such as, "In 95% of samples of Canadian adults, the proportion who think marijuana should be decriminalized will be between 42.7% and 51.3%" is just wrong. The interval isn't about sample proportions but about the population proportion.
- **Don't be certain about the parameter.** Saying, "Between 42.1% and 61.7% of sea fans are infected" asserts that the population proportion cannot be outside that interval. Of course, we can't be absolutely certain of that. Just pretty sure.

- **Don't forget: It's about the parameter.** Don't say, "I'm 95% confident that $\hat{p}$ is between 42.1% and 61.7%." Of course you are—in fact, we calculated that $\hat{p} = 51.9\%$ of the fans in our sample were infected. So we already *know* the sample proportion. The confidence interval is about the (unknown) population parameter, *p*.
- **Don't claim to know too much.** Don't say, "I'm 95% confident that between 42.1% and 61.7% of all the sea fans in the world are infected." You didn't sample from all 500 species of sea fans found in coral reefs around the world. Just those of this type on the Las Redes Reef.
- **Do take responsibility.** Confidence intervals are about *un*certainty. *You* are the one who is uncertain, not the parameter. You have to accept the responsibility and consequences of the fact that not all the intervals you compute will capture the true value. In fact, about 5% of the 95% confidence intervals you find will fail to capture the true value of the parameter. You *can* say, "I am 95% confident that between 42.1% and 61.7% of the sea fans on the Las Redes Reef are infected."[13]
- **Do treat the whole interval equally.** Although a confidence interval is a set of plausible values for the parameter, don't think that the values in the middle of a confidence interval are somehow "much more plausible" than values nearer the edges.

Margin of Error Too Large to Be Useful

We know we can't be exact, but how precise do we need to be? A confidence interval that says that the percentage of infected sea fans is between 10% and 90% wouldn't be of much use. Most likely, you have some sense of how large a margin of error you can tolerate. What can you do?

One way to make the margin of error smaller is to reduce your level of confidence. But that may not be a useful solution. It's a rare study that reports confidence levels lower than 80%. Levels of 95% or 99% are more common.

The time to think about whether your margin of error is small enough to be useful is when you design your study. Don't wait until you compute your confidence interval. To get a narrower interval without giving up confidence, you need to have less variability in your sample proportion. How can you do that? Choose a larger sample.

Violations of Assumptions

Confidence intervals and margins of error are often reported along with poll results and other analyses. But it's easy to misuse them and wise to be aware of the ways things can go wrong.

- **Watch out for biased sampling.** Don't forget about the potential sources of bias in surveys we discussed in Chapter 10. Just because we have more statistical machinery now doesn't mean we can forget what we've already learned. A questionnaire that finds that most people enjoy filling out surveys suffers from nonresponse bias if only 10% of the surveys are returned. Even though now we're able to put confidence intervals around this (biased) estimate, the nonresponse has made it useless.
- **Think about independence.** The assumption that the values in our sample are mutually independent is one that we usually cannot check. It always pays to think about it, though. For example, the disease affecting the sea fans might be contagious, so that fans growing near a diseased fan are more likely themselves to be diseased. Such contagion would violate the Independence Assumption and could severely affect our sample proportion. It could be that the proportion of infected sea fans on the entire reef is actually quite small, and the researchers just happened to find an infected area. To avoid this, the researchers should be careful to sample sites far enough apart to make contagion unlikely and improve independence. This is always an issue with any sort of convenience sample.

CONNECTIONS

Now we can see a practical application of sampling distributions. To find a confidence interval, we lay out an interval measured in standard deviations. We're using the standard deviation as a ruler again. But now the standard deviation we need is the standard deviation of the sampling distribution. That's the one that tells how much the proportion varies. (And when we estimate it from the data, we call it a standard error.)

[13] When we are being very careful, we say, "95% of samples of this size will produce confidence intervals that capture the true proportion of infected sea fans on the Las Redes Reef."

What Have We Learned?

Learning Objectives

Construct a confidence interval for a proportion, p, as the statistic, $\hat{p}$, plus or minus a **margin of error**.

- The margin of error consists of a **critical value** based on the sampling model times a **standard error** based on the sample.
- The critical value is found from the standard Normal model.
- The standard error of a sample proportion is calculated as $\sqrt{\frac{\hat{p}\hat{q}}{n}}$.

Interpret a confidence interval correctly.

- You can claim to have the specified level of confidence that the interval you have computed actually covers the true value.

Understand the relationship of the sample size, n, to both the certainty (confidence level) and precision (margin of error).

- For the same sample size and true population proportion, more certainty means less precision (wider interval) and more precision (narrower interval) implies less certainty.

Know and check the assumptions and conditions for finding and interpreting confidence intervals.

- Independence Assumption and Randomization Condition
- 10% Condition
- Success/Failure Condition

Be able to invert the calculation of the margin of error to find the sample size required, given a guess at the proportion, a confidence level, and a desired margin of error.

Review of Terms

Standard error When we estimate the standard deviation of a sampling distribution using statistics found from the data, the estimate is called a standard error (p. 452).

$$SE(\hat{p}) = \sqrt{\frac{\hat{p}\hat{q}}{n}}$$

Confidence interval A level C confidence interval for a model parameter is an interval of values often of the form

$$\textit{estimate} \pm \textit{margin of error}$$

found from data in such a way that C% of all random samples will yield intervals that capture the true parameter value (p. 453).

One-proportion *z*-interval A confidence interval for the true value of a proportion based on the Normal approximation. The confidence interval is

$$\hat{p} \pm z^* SE(\hat{p}),$$

where z^* is a critical value from the standard Normal model corresponding to the specified confidence level (p. 453).

Margin of error In a confidence interval, the extent of the interval on either side of the observed statistic value is called the margin of error. A margin of error is typically the product of a critical value from the sampling distribution and a standard error from the data. A small margin of error corresponds to a confidence interval that pins down the parameter precisely. A large margin of error corresponds to a confidence interval that gives relatively little information about the estimated parameter. For a proportion, (p. 455)

$$ME = z^* \sqrt{\frac{\hat{p}\hat{q}}{n}}$$

Critical value The number of standard errors to move away from the mean of the sampling distribution to correspond to the specified level of confidence. The critical value, denoted z^*, is usually found from a table or with technology (p. 457).

On the Computer CONFIDENCE INTERVALS FOR PROPORTIONS

Confidence intervals for proportions are so easy and natural that many statistics packages don't offer special commands for them. Statistics programs can use the "raw data" for computations. For proportions, the raw data are the "success" and "failure" status for each case. Usually, these are given as 1 or 0, but they might be category names like "yes" and "no." Often we just know the proportion of successes, $\hat{p}$, and the total count, n. Computer packages usually deal easily with summary data like this, and the statistics routines found on many graphing calculators also allow you to create confidence intervals from summaries of the data—usually all you need to enter are the number of successes and the sample size.

In some programs, you can reconstruct variables of 0s and 1s with the given proportions. But even when you have (or can reconstruct) the raw data values, you may not get exactly the same margin of error from a computer package as you would find working by hand. The reason is that some packages make approximations or use other methods. The result is very close but not exactly the same. Fortunately, Statistics means never having to say you're certain, so the approximate result is good enough.

EXCEL

Inference methods for proportions are not part of the standard Excel tool set.

For summarized data, you can type the calculation into any cell and evaluate it.

JMP

For a **categorical** variable that holds category labels, the **Distribution** platform includes tests and intervals for proportions.

For summarized data, put the category names in one variable and the frequencies in an adjacent variable. Designate the frequency column to have the **role of frequency**. Then use the **Distribution** platform.

COMMENTS

JMP uses slightly different methods for proportion inferences than those discussed in this text. Your answers are likely to be slightly different, especially for small samples.

MINITAB

Choose **Basic Statistics** from the **Stat** menu.

- Choose **1Proportion** from the Basic Statistics submenu.
- If the data are category names in a variable, assign the variable from the variable list box to the **Samples in columns** box. If you have summarized data, click the **Summarized Data** button and fill in the number of trials and the number of successes.
- Click the **Options** button and specify the remaining details.
- If you have a large sample, check **Use test and interval based on Normal distribution.**

Click the **OK** button.

COMMENTS

When working from a variable that names categories, MINITAB treats the last category as the "success" category. You can also specify how the categories should be ordered.

R

To create a confidence interval for *k* successes in *n* trials, use prop.test(k,n).

SPSS

SPSS does not find confidence intervals for proportions.

STATCRUNCH

To create a confidence interval for a proportion using summaries:

- Click on Stat.
- Choose Proportions » One sample » with summary.
- Enter the Number of successes (*x*) and Number of observations (*n*).
- Click on Next.
- Indicate Confidence Interval (Standard-Wald), and then enter the Level of confidence.
- Click on Calculate.

To create a confidence interval for a proportion using data:

- Click on Stat.
- Choose Proportions » One sample » with data.
- Choose the variable Column listing the Outcomes.
- Enter the outcome to be considered a Success.
- Click on Next.
- Indicate Confidence Interval, and then enter the Level of confidence.
- Click on Calculate.

TI-83/84 PLUS

To calculate a confidence interval for a population proportion:

- Go to the **STATS TESTS** menu and select **A:1-PropZInt**.
- Enter the number of successes observed and the sample size.
- Specify a confidence level.
- Calculate the interval.

COMMENTS

Beware: When you enter the value of *x*, you need the count, not the percentage. The count must be a whole number. If the number of successes is given as a percentage, you must first form the product *np* and then round the result.

Exercises

1. **Lying about age** Pew Research, in November 2011, polled a random sample of 799 teens about Internet use. 49% of those teens reported that they had misrepresented their age online to gain access to Web sites and online services.
   a) Explain the meaning of $\hat{p} = 0.49$ in the context of this situation.
   b) Calculate the standard error of $\hat{p}$.
   c) Explain what this standard error means in the context of this situation.

2. **Single moms** In a 2010 Pew Research study on trends in marriage and family, 4% of randomly selected respondents said that the trend of more single women having children is "a good thing." The 95% confidence interval is from 2.9% to 5.1% ($n = 1229$).
   a) Interpret this interval in this context.
   b) Explain the meaning of "95% confident" in this context.

3. **Lying about age again** The 95% confidence interval for the number of teens in Exercise 1 who reported that they had misrepresented their age online is from 45.6% to 52.5%.
   a) Interpret the interval in this context.
   b) Explain the meaning of "95% confident" in this context.

4. **More single moms** In Exercise 2, we saw that 4% of people in the sample think the trend of more single women having children is a "good thing."
   a) Are the conditions for constructing a confidence interval met?
   b) How many people would need to be surveyed for a 95% confidence interval to ensure the margin of error would be less than 3%? (Hint: Don't use $p = 0.5$. Be more realistic and use the data we already have.)

5. **Margin of error** A television newscaster reports the results of a poll of voters, and then says, "The margin of error is plus or minus 4%." Carefully explain what that means.

6. **Margin of error II** A medical researcher estimates the percentage of children who are exposed to lead-based paint, adding that he believes his estimate has a margin of error of about 3%. Explain what the margin of error means.

7. **Conditions** Consider each situation described below. Identify the population and the sample, explain what p and $\hat{p}$ represent, and tell whether the methods of this chapter can be used to create a confidence interval.
   a) Police set up an auto checkpoint at which drivers are stopped and their cars inspected for safety problems. They find that 14 of the 134 cars stopped have at least one safety violation. They want to estimate the percentage of all cars that may be unsafe.
   b) A television talk show asks viewers to register their opinions on prayer in schools by logging on to a Web site. Of the 602 people who voted, 488 favoured prayer in schools. They want to estimate the level of support among the general public.
   c) A school is considering requiring students to wear uniforms. It surveys parent opinion by sending a questionnaire home with all 1245 students. Three hundred and eighty surveys are returned, with 228 families in favour of the change.
   d) A college admits 1632 new first-year students, and four years later 1388 of them graduate on time. The college wants to estimate the percentage of all their first year enrollees who graduate on time.

8. **More conditions** Consider each situation described. Identify the population and the sample, explain what p and $\hat{p}$ represent, and tell whether the methods of this chapter can be used to create a confidence interval.
   a) A consumer group hoping to assess customer experiences with auto dealers surveys 167 people who recently bought new cars. Three percent of them expressed dissatisfaction with the salesperson.
   b) What percent of university students have cell phones? Researchers questioned 2883 students as they entered a football stadium, and 243 indicated that they had phones with them.
   c) Researchers randomly check 240 potato plants in a field in P.E.I. and only seven show signs of blight. How severe is the blight problem for the Canadian potato industry?
   d) Twelve of the 309 employees of a small company suffered an injury on the job last year. What can the company expect in future years?

9. **Conclusions** A catalogue sales company promises to deliver orders placed on the Internet within three days. Follow-up calls to a few randomly selected customers show that a 95% confidence interval for the proportion of all orders that arrive on time is 88% ±6%. What does this mean? Are these conclusions correct? Explain.
   a) Between 82% and 94% of all orders arrive on time.
   b) 95% of all random samples of customers will show that 88% of orders arrive on time.
   c) 95% of all random samples of customers will show that 82% to 94% of orders arrive on time.
   d) We are 95% sure that between 82% and 94% of the orders placed by the customers in this sample arrived on time.
   e) On 95% of the days, between 82% and 94% of the orders will arrive on time.

10. **More conclusions** In January 2002, two students made worldwide headlines by spinning a Belgian euro 250

times and getting 140 heads—that's 56%. That makes the 90% confidence interval (51%, 61%). What does this mean? Are these conclusions correct? Explain.

a) Between 51% and 61% of all Belgian euros are unfair.
b) We are 90% sure that in this experiment this euro landed heads on between 51% and 61% of the spins.
c) We are 90% sure that spun Belgian euros will land heads between 51% and 61% of the time.
d) If you spin a Belgian euro many times, you can be 90% sure of getting between 51% and 61% heads.
e) 90% of all spun Belgian euros will land heads between 51% and 61% of the time.

11. Confidence intervals Several factors are involved in the creation of a confidence interval. Among them are the sample size, the level of confidence, and the margin of error. Which statements are true?

a) For a given sample size, higher confidence means a smaller margin of error.
b) For a specified confidence level, larger samples provide smaller margins of error.
c) For a fixed margin of error, larger samples provide greater confidence.
d) For a given confidence level, halving the margin of error requires a sample twice as large.

12. Confidence intervals, again Several factors are involved in the creation of a confidence interval. Among them are the sample size, the level of confidence, and the margin of error. Which statements are true?

a) For a given sample size, reducing the margin of error will mean lower confidence.
b) For a certain confidence level, you can get a smaller margin of error by selecting a bigger sample.
c) For a fixed margin of error, smaller samples will mean lower confidence.
d) For a given confidence level, a sample nine times as large will make a margin of error one third as big.

13. Cars What fraction of cars is made in Japan? The computer output below summarizes the results of a random sample of 50 autos. Explain carefully what it tells you.

z-Interval for proportion

With 90.00% confidence,

0.29938661 < p(Japan) < 0.46984416

14. Parking in Toronto In a June 2009 *Toronto Star* study of 100 downtown Toronto parkers, researchers counted how many parkers, out of 100 observed, bumped or mounted the curb while parking.[14] Below is the computer output for these 100 observations. Assuming these parkers are representative of all those who might be attempting to park in this area at the same time of day, what can you conclude?

z-Interval for proportion

With 95.00% confidence,

0.237840 < p(bump or mount) < 0.422160

[14]Diana Zlomislic, *Toronto Star*, June 6, 2009.

15. Mislabelled seafood In December 2011, *Consumer Reports* published their study of labelling of seafood sold in New York, New Jersey, and Connecticut. They purchased 190 pieces of seafood from various kinds of food stores and restaurants in the three states and genetically compared the pieces to standard gene fragments that can identify the species. Laboratory results indicated that 22% of these packages of seafood were mislabelled, incompletely labelled, or misidentified by store or restaurant employees.

a) Construct a 95% confidence interval for the proportion of all seafood packages in those three states that are mislabelled or misidentified.
b) Explain what your confidence interval says about seafood sold in these three states.
c) A 2009 report by the Government Accountability Board says that the Food and Drug Administration has spent very little time recently looking for seafood fraud. Suppose an official said, "That's only 190 packages out of the billions of pieces of seafood sold in a year. With the small number tested, I don't know that one would want to change one's buying habits." (An official was quoted similarly in a different but similar context). Is this argument valid? Explain.

16. Mislabelled seafood, second course The December 2011 *Consumer Reports* study described in Exercise 15 also found that 12 of the 22 "red snapper" packages tested were a different kind of fish.

a) Are the conditions for creating a confidence interval satisfied? Explain.
b) Construct a 95% confidence interval.
c) Explain what your confidence interval says about "red snapper" sold in these three states.

17. Canadian pride In the 2009 Canada Day poll of 1000 Canadians by Angus Reid Strategies, 58% said they were proud of Canadian health care. A year earlier, 50% of a similar-size sample reported being proud of Canadian health care.

a) Find the margin of error for the 2009 poll if we want 90% confidence in our estimate of the percentage of national adults who are proud of the Canadian heath care.
b) Explain what that margin of error means.
c) If we wanted to be 99% confident, would the margin of error be larger or smaller? Explain.
d) Find that margin of error.
e) In general, if all other aspects of the situation remain the same, will smaller margins of error produce greater or less confidence in the interval?
f) Do you think there's been a change from 2008 to 2009 in the real proportion of Canadian adults who are proud of the Canadian health care system? Explain.

18. **Canada's birthday 2009** A June 2009 Harris-Decima poll found that only 21% of a random sample of 1000 adults knew that Canada was turning 142 years old.
   a) Find the margin of error for this poll if we want 95% confidence in our estimate of the percent of Canadian adults who know Canada's age.
   b) Explain what that margin of error means.
   c) If we only need to be 90% confident, will the margin of error be larger or smaller? Explain.
   d) Find that margin of error.
   e) In general, if all other aspects of the situation remain the same, would smaller samples produce smaller or larger margins of error?

19. **Contributions, please** Philanthropic organizations, like the MS Society, rely on contributions. They send free mailing labels and greeting cards to potential donors on their list and ask for a voluntary contribution. To test a new campaign, one philanthropic organization sent letters to a random sample of 100 000 potential donors and received 4781 donations.
   a) Give a 95% confidence interval for the true proportion of those from their entire mailing list who may donate.
   b) A staff member thinks that the true rate is 5%. Given the confidence interval you found, do you find that percentage plausible?

20. **Take the offer** A major credit card company is planning a new offer for their current cardholders. The offer will give double airline miles on purchases for the next six months if the cardholder goes online and registers for the offer. To test the effectiveness of the campaign, the company sent out offers to a random sample of 50 000 cardholders. Of those, 1184 registered.
   a) Give a 95% confidence interval for the true proportion of cardholders who will register for the offer.
   b) If the acceptance rate is only 2% or less, the campaign won't be worth the expense. Given the confidence interval you found, what would you say?

21. **Teenage drivers** An insurance company checks police records on 582 accidents selected at random and notes that teenagers were at the wheel in 91 of them.
   a) Create a 95% confidence interval for the percentage of all auto accidents that involve teenage drivers.
   b) Explain what your interval means.
   c) Explain what "95% confidence" means.
   d) A politician urging tighter restrictions on drivers' licenses issued to teens says, "In one of every five auto accidents, a teenager is behind the wheel." Does your confidence interval support or contradict this statement? Explain.

22. **Junk mail** Direct mail advertisers send solicitations (a.k.a. "junk mail") to thousands of potential customers in the hope that some will buy the company's product. The acceptance rate is usually quite low. Suppose a company wants to test the response to a new flyer, and sends it to 1000 people randomly selected from their mailing list of over 200 000 people. They get orders from 123 of the recipients.
   a) Create a 90% confidence interval for the percentage of people the company contacts who may buy something.
   b) Explain what this interval means.
   c) Explain what "90% confidence" means.
   d) The company must decide whether or not to do a mass mailing. The mailing won't be cost-effective unless it produces at least a 5% return. What does your confidence interval suggest? Explain.

23. **Safe food** Some food retailers propose subjecting food to a low level of radiation to improve safety, but sale of such "irradiated" food is opposed by many people. Suppose a grocer wants to find out what his customers think. He has cashiers distribute surveys at checkouts and ask customers to fill them out and drop them in a box near the front door. He gets responses from 122 customers, of whom 78 oppose the radiation treatments. What can the grocer conclude about the opinions of all his customers?

24. **Local news** The mayor of a small city suggests that the federal government locate a new prison there, arguing that the construction project and resulting jobs will be good for the local economy. A total of 183 residents show up for a public hearing on the proposal, and a show of hands finds only 31 in favour of the prison project. What can the city council conclude about public support for the mayor's initiative?

25. **Death penalty, again** In the survey on the death penalty you read about in the chapter, the Gallup Poll actually split the sample at random, asking 510 respondents the question quoted earlier: "Generally speaking, do you believe the death penalty is applied fairly or unfairly in this country today?" The other 510 were asked, "Generally speaking, do you believe the death penalty is applied unfairly or fairly in this country today?" Seems like the same question, but sometimes the order of the choices matters. Suppose that for the second way of phrasing it, only 48% said they thought the death penalty was fairly applied.
   a) What kind of bias may be present here?
   b) If we combine them, considering the overall group to be one larger random sample, what is a 95% confidence interval for the proportion of the general public that thinks the death penalty is being fairly applied?
   c) How does the margin of error based on this pooled sample compare with the margins of error from the separate groups? Why?

26. **Gambling** A city ballot includes an initiative that would permit casino gambling. The issue is hotly contested, and two groups decide to conduct polls to predict the outcome. The local newspaper finds that 53% of 1200 randomly selected voters plan to vote "yes," while a university Statistics class finds 54% of 450 randomly

selected voters in support. Both groups will create 95% confidence intervals.

a) Without finding the confidence intervals, explain which one will have the larger margin of error.
b) Find both confidence intervals.
c) Which group concludes that the outcome is too close to call? Why?

27. Rickets Vitamin D, whether ingested as a dietary supplement or produced naturally when the skin is exposed to sunlight, is essential for strong, healthy bones. The bone disease rickets was largely eliminated in England during the 1950s, but now there is concern that a generation of children more likely to watch television or play computer games than spend time outdoors is at increased risk. A recent study of 2700 children randomly selected from all parts of England found 20% of them deficient in vitamin D.

a) Find a 98% confidence interval.
b) Explain carefully what your interval means.
c) Explain what "98% confidence" means.

28. Pregnancy In 1998, a San Diego reproductive clinic reported 49 live births to 207 women under the age of 40 who had previously been unable to conceive.

a) Find a 90% confidence interval for the success rate at this clinic.
b) Interpret your interval in this context.
c) Explain what "90% confidence" means.
d) Do these data refute the clinic's claim of a 25% success rate? Explain.

29. Studying poverty To study poverty in the city of Halifax, researchers will randomly select 30 blocks from the city, then randomly select 10 households from each of those 30 blocks to be interviewed, producing a total of 300 interviews. Based on the interviews, the families will be classified as living below the poverty line or not. Suppose that we find that 45 families out of the 300 meet the criterion for living in poverty. Give a 95% confidence interval for the true proportion of families living in poverty in Halifax, with the usual approach discussed in this chapter. What is wrong with this approach here? Do you think you have over or underestimated the true SE of the sample proportion? Explain.

30. Legal music A random sample of 168 students were asked how many songs were in their digital music library and what fraction of them were legally purchased. Overall, they reported having a total of 117 079 songs, of which 23.1% were legal. The music industry would like a good estimate of the fraction of songs in students' digital music libraries that are legal.

a) Think carefully. What is the parameter being estimated? What is the population? What is the sample size?
b) Construct a 95% confidence interval for the fraction of legal digital music, if you can. If not, explain the problem.

31. Deer ticks Wildlife biologists inspect 153 deer taken by hunters and find 32 of them carrying ticks that test positive for Lyme disease.

a) Create a 90% confidence interval for the percentage of deer that may carry such ticks.
b) If the scientists want to cut the margin of error in half, how many deer must they inspect?
c) What concerns do you have about this sample?

32. Pregnancy II The San Diego reproductive clinic in Exercise 28 wants to publish updated information on its success rate.

a) The clinic wants to cut the stated margin of error in half. How many patients' results must be used?
b) Do you have any concerns about this sample? Explain.

33. Graduation It's believed that as many as 25% of adults over 50 never graduated from high school. We wish to see if this percentage is the same among the 25 to 30 age group.

a) How many of this younger age group must we survey to estimate the proportion of nongraduates to within 6 percentage points, with 90% confidence?
b) Suppose we want to cut the margin of error to 4 percentage points. What's the necessary sample size?
c) What sample size would produce a margin of error of 3 percentage points?

34. Hiring In preparing a report on the economy, we need to estimate the percentage of businesses that plan to hire additional employees in the next 60 days.

a) How many randomly selected employers must we contact to create an estimate in which we are 98% confident with a margin of error of 5%?
b) Suppose we want to reduce the margin of error to 3%. What sample size will suffice?
c) Why might it not be worth the effort to try to get an interval with a margin of error of only 1%?

35. Graduation, again As in Exercise 33, we hope to estimate the percentage of adults aged 25 to 30 who never graduated from high school. What sample size would allow us to increase our confidence level to 95% while reducing the margin of error to only 2 percentage points?

36. Better hiring info Editors of the business report in Exercise 34 are willing to accept a margin of error of 4% but want 99% confidence. How many randomly selected employers will they need to contact?

37. Pilot study A province's environmental ministry worries that many cars may be violating clean air emissions standards. The agency hopes to check a sample of vehicles in order to estimate that percentage with a margin of error of 3% and 90% confidence. To gauge the size of the problem, the agency first picks 60 cars and finds nine with faulty emissions systems. How should be sampled for a full investigation?

38. Another pilot study During routine screening, a doctor notices that 22% of her adult patients show higher-than-normal levels of glucose in their blood—a possible warning signal for diabetes. Hearing this, some medical researchers decide to conduct a large-scale study, hoping to estimate the proportion to within 4 percentage points with 98% confidence. How many randomly selected adults must they test?

39. Approval rating A newspaper reports that the premier's approval rating stands at 65%. The article adds that the poll is based on a random sample of 972 adults and has a margin of error of 2.5%. What level of confidence did the pollsters use?

40. Amendment A television news reporter says that a proposed constitutional amendment is likely to win approval in an upcoming referendum because a poll of 1505 likely voters indicated that 52% would vote in favour. The reporter goes on to say that the margin of error for this poll was 3.0%.

a) Explain why the poll is actually inconclusive.
b) What confidence level did the pollsters use?

41. Bad countries In the Canada's World poll, conducted by Environics in January 2008 (pre-Obama), 52% of respondents pointed to the United States as a negative force in the world (4% named Canada as a negative force). Among young people (aged 15–24), the percentage rose to 63%. A total of 2001 Canadians were randomly sampled for the poll. Assume it was a simple random sample. Answer the following questions at the 99% confidence level.

a) What is the margin of error for the estimate of the proportion pointing to the United States as a negative force? Give a 99% confidence interval for the true population proportion.
b) The figure for British Columbia was the highest, at 58%. Can we safely take the margin of error to be the same as in part a)? Explain.
c) If young people constituted, say, 25% of the sample, what would be the margin of error for their reported 63% figure?
d) How big would the sample have to be to reduce by half the margin of error for the overall 52% estimate?

42. More Canadian pride In the Angus Reid 2009 Canada Day poll of 1000 adult Canadians, hockey was a source of pride for 78% of males and 66% of females. Recall that the reported margin of error for the total sample is 3.1% (you may round this to 3%) at the 95% confidence level.

a) Can we safely say with 95% confidence that between 75% and 81% of all adult Canadian males feel pride in the sport of hockey? Explain.
b) Determine a 99% confidence interval for the true population proportion of Canadian females who feel pride in our national game of hockey. You may make any reasonable assumptions, but state them.
c) To reduce the margin of error in part b) by three-quarters, or 75%, how big should the entire sample of Canadians have been?

43. Simulations Use your computer software to generate a sample of size 100 from a Bernoulli distribution. This distribution has only two outcomes, 0 and 1, with complementary probabilities. Set $p(1) = 0.8$. Consider "1" to be a success and "0" to be a failure.

a) From the sample, calculate a 90% confidence interval for the true population proportion of successes. Does it contain the true population proportion of 0.8?
b) Repeat part a) 99 more times, obtaining 99 additional samples and confidence intervals. How many confidence intervals contained the true proportion? What percentage of confidence intervals would you expect to contain the true proportion if you repeated these simulations many times? For 100 simulated samples, if X = number of confidence intervals that contain 0.8, what is the probability distribution of X?

44. More simulations Use your computer software to generate a sample of size 50 from a Bernoulli distribution. This distribution has only two outcomes, 0 and 1, with complementary probabilities. Set $p(1) = 0.3$. Consider "1" to be a success and "0" to be a failure.

a) From the sample, calculate an 80% confidence interval for the true population proportion of successes. Does it contain the true population proportion of 0.3?
b) Repeat part a) 49 more times, obtaining 49 additional samples and confidence intervals. How many confidence intervals contained the true proportion? What percentage of confidence intervals would you expect to contain the true proportion if you repeated these simulations many times? For 50 simulated samples, if X = number of confidence intervals that contain 0.3, what is the probability distribution of X?

45. Dogs Canine hip dysplasia is a degenerative disease that causes pain in many dogs. Sometimes advanced warning signs appear in puppies as young as six months. A veterinarian checked 42 puppies and found five with early hip dysplasia. She considers this group to be a random sample of all puppies.

a) Explain why we cannot construct the customary Normal approximation based confidence interval for the rate of occurrence of early hip dysplasia among all puppies.
b) Construct a "Plus Four" confidence interval and interpret it in context.

46. Fans A survey of 81 randomly selected people standing in line to enter a football game found that 73 of them were home-team fans.

a) Explain why we cannot construct the customary Normal approximation based confidence interval for the proportion of all people at the game who are fans of the home team.
b) Construct a "Plus Four" confidence interval and interpret it in context.

T *47. **Departures 2011** What are the chances your flight will leave on time? The U.S. Bureau of Transportation Statistics of the Department of Transportation publishes information about airline performance. Here are a histogram and summary statistics for the percentage of flights departing on time each month from 1995 through 2011.[15]

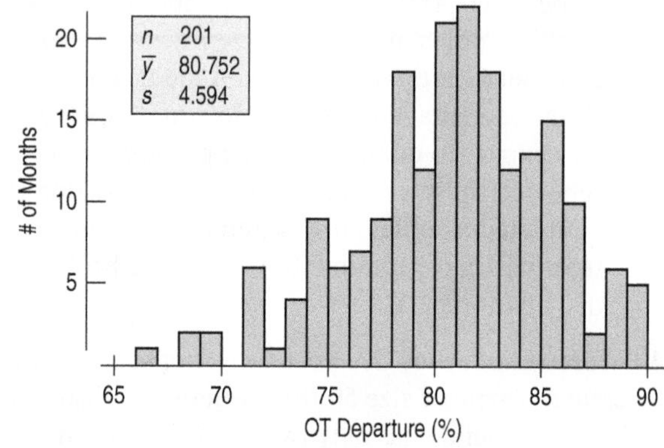

There is no evidence of a trend over time.
a) Check the assumptions and conditions for inference.
b) Find a 90% confidence interval for the true mean monthly on-time departure rate.
c) Interpret this interval for a traveller planning to fly.

T *48. **Late arrivals 2011** Will your flight get you to your destination on time? The U.S. Bureau of Transportation Statistics reported the percentage of flights that were late each month from 1995 through 2011. Here is a histogram, along with some summary statistics:

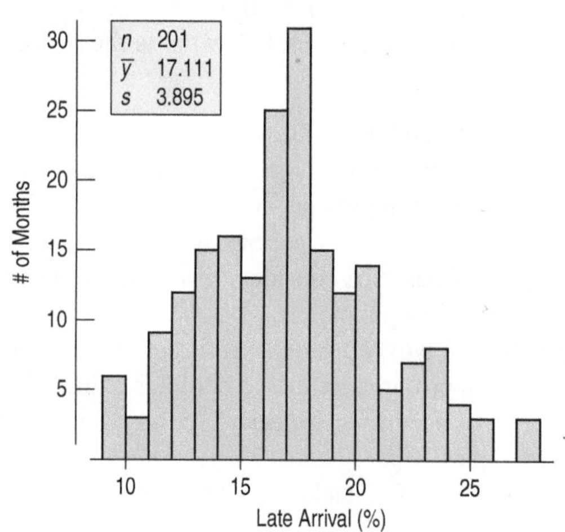

[15]www.transtats.bts.gov/HomeDrillChart.asp

We can consider these data to be a representative sample of all months. There is no evidence of a time trend ($r = 0.07$).
a) Check the assumptions and conditions for inference about the mean.
b) Find a 99% confidence interval for the true mean monthly late-arrival rate.
c) Interpret this interval for a traveller planning to fly.

*49. **Speed of light** In 1897, Michelson conducted an experiment to measure the speed of light (which is known to be a physical constant throughout the universe). He reported the results of 100 independent trials. The measurements are in kilometres per second and were coded by subtracting 299 000 from each. The average of the 100 coded measurements was 852.4, with a standard deviation of 79.0.
a) Estimate the standard deviation of the mean speed of light measurement.
b) Report a 99% confidence interval for the true speed of light.
c) State in words what this interval means.
d) In order to trust your reported level of confidence, is there anything else you might like to see or check?
e) The true speed of light is now known to be 299 710.5 km/sec, corresponding to a coded value of 710.5. What does this indicate about Michelson's experiment?

*50. **Hot dogs** A nutrition lab tested 60 "reduced sodium" hot dogs and found the mean sodium content to be 318 mg, with a standard deviation of 32 mg.
a) Estimate the standard deviation of the mean sodium content of the sample.
b) Find and interpret a 90% confidence interval for the true mean sodium content for this type of hot dog.
c) In order to trust your reported level of confidence, is there anything more you need to do or know?

*51. **Hot dog margin of error** To reduce the margin of error by 50% in the example of "reduced sodium" hot dogs from Exercise 50, how much bigger would the sample need to be? How big would it need to be to produce a margin of error equal to 4.0 mg?

*52. **Speed of light margin of error** To reduce the margin of error in Exercise 49 by 2/3 (to 1/3 of what was actually obtained), how much larger would the sample need to be? How much larger would it have to be to produce a margin of error equal to 10.0 km/sec?

Just Checking ANSWERS

1. No. We know that in the sample, exactly 58% said they were proud; there's no need for a margin of error.
2. No, we are 95% confident that the percentage falls in some interval, not exactly on a particular value.
3. Yes. That's what the confidence interval means.
4. No. We don't know for sure that's true; we are only 95% confident.
5. No. That's our level of confidence, not the proportion who are proud.
6. Wider.
7. Lower.
8. Smaller.

MathXL
MyStatLab

Go to MathXL at www.mathxl.com or MyStatLab at www.mystatlab.com. You can practise exercises for this chapter as often as you want. The guided solutions will help you find answers step by step. You'll find a personalized study plan available to you too!

chapter

17 Testing Hypotheses About Proportions

Where are we going?

Do people aged 18 to 24 really prefer Pepsi to Coke? Does this new allergy medication really reduce symptoms more than a placebo? There are times when we want to make a decision. In order to do that we'll propose a model for the situation at hand and test a hypothesis about that model. The result will help us answer the real-world question.

"Half the money I spend on advertising is wasted; the trouble is I don't know which half."

—John Wanamaker

Ingots are huge pieces of metal, sometimes weighing more than 9000 kilograms (20 000 pounds), made in a giant mould. They must be cast in one large piece for use in fabricating large structural parts for cars and planes. If they crack while being made, the crack can propagate into the zone required for the part, compromising its integrity. Airplane manufacturers insist that metal for their planes be defect-free, so the ingot must be remade if any cracking is detected.

Even though the metal from the cracked ingot is recycled, the scrap cost runs into the tens of thousands of dollars. Metal manufacturers would like to avoid cracking if at all possible, but the casting process is complicated and not everything is completely under control. In one plant, only about 80% of the ingots have been free of cracks. In an attempt to reduce the cracking proportion, the plant engineers and chemists recently implemented some changes in the casting process. Since then, 400 ingots have been cast and only 17% of them have cracked. Should management declare victory? Has the cracking rate really decreased, or was the 17% just the result of luck?

We can treat the 400 ingots cast with the new method as a random sample. We know that each random sample will have a somewhat different proportion of cracked ingots. Is the 17% we observe merely a result of natural sampling variability, or is this lower cracking rate strong enough evidence to assure management that the true cracking rate now is really below 20%?

People want answers to questions like this all the time. Has the Prime Minister's approval rating changed since last month? Has teenage smoking decreased in the past five years? Is the global temperature increasing? Did the Super Bowl ad we bought actually increase sales? To answer such questions, we test *hypotheses* about models.

17.1 Hypotheses

A S *Activity:* **Testing a Claim.** Can we really draw a reasonable conclusion from a random sample? Run this simulation before you read the chapter, and you'll gain a solid sense of what we're doing here.

We want to know if the changes made by the engineers have lowered the cracking rate from 20%. Humans are natural skeptics, so to test whether the changes have worked, we'll assume that they haven't. We'll make the **hypothesis** that the cracking rate is still 20% and see if the data can convince us otherwise. Hypotheses are models that we adopt temporarily—until we can test them once we have data. This starting hypothesis to be tested is called the **null hypothesis**—null because it assumes that nothing has changed. We denote it H_0. It specifies a parameter value—here that the cracking rate is 20%— which we usually write in the form H_0: *parameter* = *hypothesized value*.

Which value to take is often obvious from the *Who* and *What* of the data. But sometimes it takes a bit of thinking to translate the question we hope to answer into a hypothesis about a parameter. For the ingots, we can write H_0: $p = 0.20$.

Hypothesis n.;
pl. {Hypotheses}.
A supposition; a proposition or principle which is supposed or taken for granted, in order to draw a conclusion or inference for proof of the point in question; something not proved, but assumed for the purpose of argument.
—*Webster's Unabridged Dictionary, 1913*

NOTATION ALERT

Capital H is the standard letter for hypotheses, and H_0 always labels the null hypothesis.

To remind us that the parameter value comes from the null hypothesis, p is sometimes written as p_0, and the standard deviation is

$$SD(\hat{p}) = \sqrt{\frac{p_0 q_0}{n}}$$

The **alternative hypothesis**, which we denote H_A, contains the values of the parameter that we consider plausible if we reject the null hypothesis. Our null hypothesis is that $p = 0.20$. What's the alternative? Management is interested in *reducing* the cracking rate, so their alternative is H_A: $p < 0.20$.

What would convince you that the cracking rate had actually gone down? If you observed a cracking rate *much lower* than 20% in your sample, you'd likely be convinced. If only four out of the next 400 ingots crack (for a rate of 1%), most folks would conclude that the changes helped. But if the sample cracking rate is 19.8%, you should be skeptical. After all, observed proportions do vary, so we wouldn't be surprised to see some difference. How much smaller must the sample cracking rate be before we *are* convinced that things have changed? That's the crucial question in a hypothesis test. As usual in Statistics, when we think about how big the change has to be, we naturally think of using the standard deviation as the ruler to measure that change. So let's start by finding the standard deviation of the sample cracking rate.

Since the company changed the process, 400 new ingots have been cast. The sample size of 400 is big enough to satisfy the **Success/Failure Condition**. (We expect $0.20 \times 400 = 80$ ingots to crack under H_0.) Although not a random sample, the engineers think that whether an ingot cracks should be independent from one ingot to the next, so the Normal sampling distribution model should work well. If the null hypothesis is true, the standard deviation of the sampling model is

$$SD(\hat{p}) = \sqrt{\frac{pq}{n}} = \sqrt{\frac{(0.20)(0.80)}{400}} = 0.02$$

NOTE THAT ABOVE WE WROTE SD($\hat{p}$) AND NOT SE($\hat{p}$).

In the previous chapter, we *estimated* the unknown standard deviation of the sampling distribution of $\hat{p}$ by plugging in $\hat{p}$ for an unknown p in the formula above, and we named it the *standard error*[1] of $\hat{p}$. Here, though, there is no need to use a sample estimate for p in this formula, since we know the true value of p (well, not really, but under our null hypothesis, we assume we know it). When constructing a confidence interval for p, there is no null hypothesis p—no assumed value for p—so we have no choice but to estimate the SD($\hat{p}$), turning it into the SE($\hat{p}$).

Now we know both parameters of the Normal sampling distribution model—$p = 0.20$ and $SD(\hat{p}) = 0.02$—so we can find out how likely it would be to see the observed rate of 17%. Since we are using a Normal model, we find the z-score:

$$z = \frac{0.17 - 0.20}{0.02} = -1.5$$

Then we ask, "How likely is it to observe a value at least 1.5 standard deviations below the mean of a Normal model?" The answer (from a calculator, computer program, or the Normal table) is about 0.067, as shown in Figure 17.1. This is the probability of observing a cracking rate of 17% or less in a sample of 400 if the null hypothesis is true and the true cracking rate is still 20%. Management now must decide whether an event that would happen 6.7% of the time by chance is strong enough evidence to conclude that the true cracking proportion has decreased.

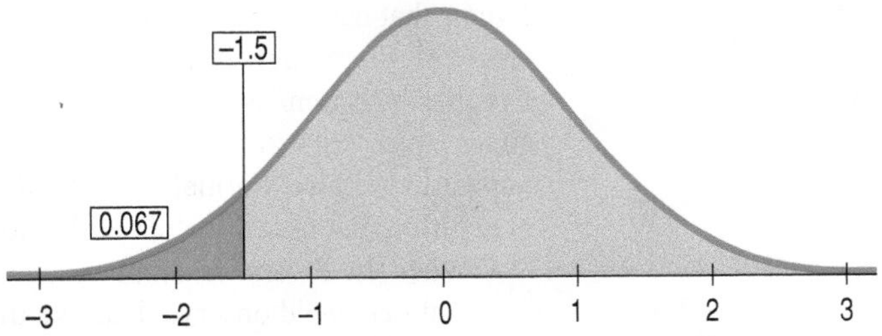

Figure 17.1
How likely is a z-score of -1.5 (or lower)? This is what it looks like. The red area is 0.067, or 6.7%, of the total area under the curve.

[1]Not all books use the same terminology. The standard deviation (of $\hat{p}$) and standard error (of $\hat{p}$) are sometimes referred to as the standard error (of $\hat{p}$) and estimated standard error (of $\hat{p}$), respectively, which may help avoid confusion with population and sample standard deviations (which measure variation among individuals, not the possible values of a statistic). In either case, we are simply making a distinction between the true standard deviation of the sampling distribution of a statistic and an estimate thereof. And if the statistic is estimating a parameter unbiasedly, a *deviation* from the parameter is just an *error* in the estimate.

A Trial as a Hypothesis Test

Does the reasoning of hypothesis tests seem backwards? That could be because we usually prefer to think about getting things right rather than getting them wrong. You have seen this reasoning before in a different context. This is the logic of jury trials.

Let's suppose a defendant has been accused of robbery. In British common law and systems derived from it, the null hypothesis is that the defendant is innocent. Instructions to juries are quite explicit about this.

The evidence takes the form of facts that seem to contradict the presumption of innocence. For us, this means collecting data. In the trial, the prosecutor presents evidence. ("If the defendant were innocent, wouldn't it be remarkable that the police found him at the scene of the crime with a bag full of money in his hand, a mask on his face, and a getaway car parked outside?")

The next step is to judge the evidence. Evaluating the evidence is the responsibility of the jury in a trial, but it falls on your shoulders in hypothesis testing. The jury considers the evidence in light of the *presumption* of innocence and judges whether the evidence against the defendant would be plausible *if the defendant were in fact innocent.*

Like the jury, we ask, "Could these data plausibly have happened by chance if the null hypothesis were true?" If they are very unlikely to have occurred, then the evidence raises serious doubt about the null hypothesis.

A S *Activity:* **The Reasoning of Hypothesis Testing.** Our reasoning is based on a rule of logic that dates back to ancient scholars. Here's a modern discussion of it.

Ultimately, you must make a decision. The standard of "beyond a reasonable doubt" is wonderfully ambiguous because it leaves the jury to decide the degree to which the evidence contradicts the hypothesis of innocence. Juries don't explicitly use probability to help them decide whether to reject that hypothesis. But when you ask the same question of your null hypothesis, you have the advantage of being able to quantify exactly how surprising the evidence would be were the null hypothesis true.

How unlikely is unlikely? Some people set rigid standards, like 1 time out of 20 (0.05) or 1 time out of 100 (0.01). But if *you* have to make the decision, you must judge for yourself in each situation whether the probability of observing your data is small enough to leave you with little "reasonable doubt."

17.2 P-Values

Beyond a Reasonable Doubt

We ask whether the data were unlikely beyond a reasonable doubt. We've just calculated that probability. The probability that the observed statistic value (or an even more extreme value) could occur if the null model were true—in this case, 0.067—is the P-value.

The fundamental step in our reasoning is the question: "Are the data surprising, given the null hypothesis?" And the key calculation is to determine exactly how likely the data we observed would be were the null hypothesis a true model of the world. So we need a *probability*. Specifically, we want to find the probability of seeing data like these (or something even less likely) *given* that the null hypothesis is true. Statisticians are so thrilled with their ability to measure precisely how surprised they are that they give this probability a special name. It's called a **P-value**.

When the P-value is high, we haven't seen anything unlikely or surprising at all. Events that have a high probability of happening happen all the time. The data are consistent with the model from the null hypothesis, and we have no reason to reject the null hypothesis. But many other similar hypotheses could also account for the data we've seen, so *we haven't proven that the null hypothesis is true*. The most we can say is that it doesn't appear to be false. Formally, we "fail to reject" the null hypothesis. That's a pretty weak conclusion, but it's all we're entitled to.

NOTATION ALERT

We have many P's to keep straight. We use an uppercase P for probabilities, as in *P*(A), and for the special probability we care about in hypothesis testing, the P-value.

We use lowercase *p* to denote our model's underlying proportion parameter, and $\hat{p}$ to denote our observed proportion statistic.

When the P-value *is* small enough, it says that we are very surprised. It means that it's very unlikely we'd observe data like these if our null hypothesis were true. The model we started with (the null hypothesis) and data we collected are at odds with each other, so we have to make a choice. Either the null hypothesis is correct and we've just seen something remarkable, or the null hypothesis is wrong, and we were wrong to use it as the basis for computing our P-value. If you believe in data more than in assumptions, then, given that choice, you should reject the null hypothesis.

"You must find the accused not guilty of the offence unless the Crown proves beyond a reasonable doubt that he/she is guilty of it."

—*Canadian Judicial Council*

DON'T THINK THE NULL HYPOTHESIS IS PROVEN TRUE

Every child knows that he (or she) is at the "centre of the universe," so it's natural to suppose that the sun revolves around the Earth. The fact that the sun appears to rise in the east every morning and set in the west every evening is *consistent* with this hypothesis and *seems* to lend support to it, but it certainly doesn't prove it, as we all eventually come to understand.

What to Do with an "Innocent" Defendant

If the evidence is not strong enough to reject the defendant's presumption of innocence, what verdict does the jury return? They say "not guilty." Notice that they do not say that the defendant is innocent. All they say is that they have not seen sufficient evidence to convict, to reject innocence. The defendant may, in fact, be innocent, but the jury has no way to be sure.

Said statistically, the jury's null hypothesis is H_0: innocent defendant. If the evidence is too unlikely given this assumption, the jury rejects the null hypothesis and finds the defendant guilty. But—and this is an important distinction—if there is *insufficient evidence* to convict the defendant, the jury does not decide that H_0 is true and declare the defendant innocent. Juries can only *fail to reject* the null hypothesis and declare the defendant "not guilty."

In the same way, if the data are not particularly unlikely under the assumption that the null hypothesis is true, then the most we can do is to "fail to reject" our null hypothesis. We never declare the null hypothesis to be true (or "accept the null"), because we simply do not know whether it's true or not. (After all, more evidence may come along later.)

In the trial, the burden of proof is on the prosecution. In a hypothesis test, the burden of proof is on the unusual claim. The null hypothesis is the ordinary state of affairs, so it's the alternative to the null hypothesis that we consider unusual and for which we must marshal evidence.

Imagine a clinical trial testing the effectiveness of a new headache remedy. In Chapter 11 we saw the value of comparing such treatments to a placebo. We hope to prove that the new treatment is more effective than placebo. To do this, we set up its negation as our null hypothesis—"the new treatment is no more effective than placebo"—and then do our best to destroy its credibility. How we choose the null is important, because results are variable and some patients will improve even when administered the placebo treatment. If we use only six people to test the drug, the results are likely *not to be clear* and we'll be unable to discredit the null hypothesis. Does this mean the drug doesn't work? Of course not. It simply means that we don't have enough evidence to reject our (null) assumption. That's why we don't start by assuming that the drug *is more effective*. If we were to do that, then we could test just a few people, find that the results aren't clear, and claim that since we've been unable to reject our assumption of drug effectiveness, it must be so! Health Canada is unlikely to be impressed by that argument.

Just Checking

1. A research team wants to know if aspirin helps to thin blood. The null hypothesis says that it doesn't. They test 12 patients, observe the proportion with thinner blood, and get a P-value of 0.32. They proclaim that aspirin doesn't work. What would you say?
2. An allergy drug has been tested and found to give relief to 75% of the patients in a large clinical trial. Now the scientists want to see if the new, improved version works even better. What would the null hypothesis be?
3. The new drug is tested and the P-value is 0.0001. What would you conclude about the new drug?

17.3 The Reasoning of Hypothesis Testing

Hypothesis tests follow a carefully structured path. To avoid getting lost as we navigate down it, we divide that path into four distinct sections.

"The null hypothesis is never proved or established, but is possibly disproved, in the course of experimentation. Every experiment may be said to exist only in order to give the facts a chance of disproving the null hypothesis."

—*Sir Ronald Fisher, The Design of Experiments*

NOTATION ALERT

H_A always labels the alternative hypothesis. Sometimes, it is written H_1.

1. Hypotheses

First, we state the null hypothesis. That's usually the skeptical claim that nothing's different. Are we considering a (New! Improved!) possibly better method? The null hypothesis says, "Oh yeah? Convince me!" To convert a skeptic, we must pile up enough evidence against the null hypothesis that we can reasonably reject it.

In statistical hypothesis testing, hypotheses are almost always about model parameters. To assess how unlikely our data may be, we need a null model. The null hypothesis specifies a particular parameter value to use in our model. In the usual shorthand, we write H_0: *parameter* = *hypothesized value*. The **alternative hypothesis**, H_A, contains the values of the parameter we consider plausible when we reject the null.

For Example WRITING HYPOTHESES

The provincial Ministry of Transportation claims that 80% of candidates pass driving tests, but a newspaper reporter's survey of 90 randomly selected local teens who have taken the test found only 61 who passed.

QUESTION: Does this finding suggest that the passing rate for teenagers is lower than the Ministry's reported overall rate? Write appropriate hypotheses.

ANSWER: I'll assume that the passing rate for teenagers is the same as the Ministry's overall rate of 80%, unless there's strong evidence that it's lower.

$$H_0: p = 0.80$$
$$H_A: p < 0.80$$

2. Model

WHEN THE CONDITIONS FAIL . . .

You might proceed with caution, explicitly stating your concerns. Or you may need to do the analysis with and without an outlier, or on different subgroups, or after re-expressing the response variable. Or you may not be able to proceed at all.

To plan a statistical hypothesis test, specify the *model* you will use to test the null hypothesis and the parameter of interest. Of course, all models require assumptions, so you will need to state them and check any corresponding conditions.

Your Model step should end with a statement such as

Because the conditions are satisfied, I can model the sampling distribution of the proportion with a Normal model.

Watch out, though. Your Model step could end with

Because the conditions are not satisfied, I can't proceed with the test.

If that's the case, stop and reconsider.

Each test in the book has a name that you should include in your report. We'll see many tests in the chapters that follow. Some will be about more than one sample, some will involve statistics other than proportions, and some will use models other than the Normal (and so will not use *z*-scores). The test about proportions that we discuss here is called a **one-proportion *z*-test.**[2]

ONE-PROPORTION *Z*-TEST

The conditions for the one-proportion *z*-test are the same as for the one-proportion *z*-interval. We test the hypothesis H_0: $p = p_0$ using the statistic $z = \frac{(\hat{p} - p_0)}{SD(\hat{p})}$. We use the hypothesized proportion to find the standard deviation, $SD(\hat{p}) = \sqrt{\frac{p_0 q_0}{n}}$.

When the conditions are met and the null hypothesis is true, this statistic follows the standard Normal model, so we can use that model to obtain a P-value.

A S ***Activity:* Was the Observed Outcome Unlikely?** Complete the test you started in the first activity for this chapter. The narration explains the steps of the hypothesis test.

[2] It's also called the "one-sample test for a proportion."

For Example CHECKING THE CONDITIONS

RECAP: The provincial Ministry of Transportation claims that 80% of candidates pass driving tests. A reporter has results from a survey of 90 randomly selected local teens who had taken the test.

QUESTION: Are the conditions for inference satisfied?

✓ The 90 teens surveyed were a random sample of local teenage driving candidates.
✓ 90 is fewer than 10% of the teenagers who take driving tests in a province.
✓ We expect $np_0 = 90(0.80) = 72$ successes and $nq_0 = 90(0.20) = 18$ failures. Both are at least 10.

ANSWER: The conditions are satisfied, so it's okay to use a Normal model and perform a one-proportion *z*-test.

3. Mechanics

CONDITIONAL PROBABILITY

Did you notice that a P-value is a conditional probability? It's the probability that the observed results could have happened *if the null hypothesis is true.*

Under "Mechanics," we place the actual calculation of our test statistic from the data. Different tests we encounter will have different formulas and different test statistics. Usually, the mechanics are handled by a statistics program or calculator, but it's good to have the formulas recorded for reference and to know what's being computed. The ultimate goal of the calculation is to obtain a P-value—the probability that the observed statistic value (or an even more contradictory value) will occur if the null model is correct. If the P-value is small enough, we have evidence against the null hypothesis. And draw yourself a picture. You can usually catch a silly mistake in the calculations simply by sketching the Normal distribution and shading the area on either side of your *z*-score. Don't say we didn't warn you!

For Example FINDING A P-VALUE

RECAP: The Ministry of Transportation claims that 80% of candidates pass driving tests, but a survey of 90 randomly selected local teens who have taken the test found only 61 who passed.

QUESTION: What's the P-value for the one-proportion *z*-test?

ANSWER: I have $n = 90$, $x = 61$, and a hypothesized $p = 0.80$.

$$\hat{p} = \frac{61}{90} \approx 0.678$$

$$SD(\hat{p}) = \sqrt{\frac{p_0 q_0}{n}} = \sqrt{\frac{(0.8)(0.2)}{90}} \approx 0.042$$

$$z = \frac{\hat{p} - p_0}{SD(\hat{p})} = \frac{0.678 - 0.800}{0.042} \approx -2.90$$

$$P\text{-value} = P(z < -2.90) = 0.002$$

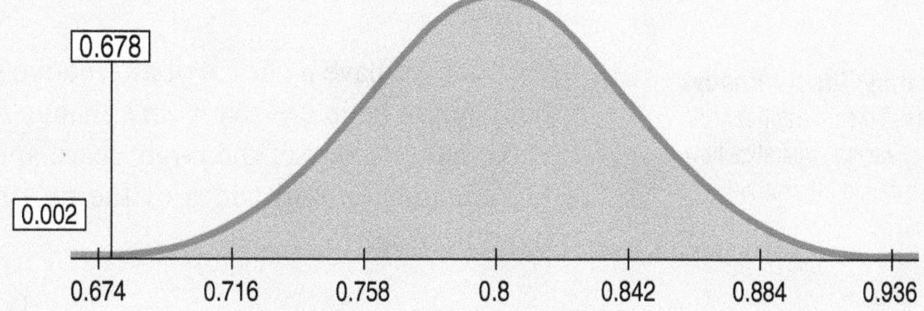

4. Conclusion

The conclusion in a hypothesis test is always a statement about the null hypothesis. The conclusion must state how strong the evidence is for rejection of the null hypothesis, as well as any decision we make based on the evidence. And, as always, the conclusion should be stated in context.

For Example STATING THE CONCLUSION

RECAP: The Ministry of Transportation claims that 80% of candidates pass driving tests. Data from a reporter's survey of randomly selected local teens who had taken the test produced a P-value of 0.002.

QUESTIONS: What can the reporter conclude? And how might the reporter explain what the P-value means for the newspaper story?

ANSWER: Because the P-value of 0.002 is very low, there is good evidence to reject the null hypothesis. These survey data provide strong evidence that the passing rate for teenagers taking the driving test is lower than 80%.

If the passing rate for teenage driving candidates were actually 80%, we'd expect to see success rates this low in only about 1 in 500 samples (0.2%). This seems quite unlikely, casting doubt that the Ministry's stated success rate applies to teens.

Your conclusion about the null hypothesis should never be the end of a testing procedure. Often, there are actions to take or policies to change. In our ingot example, management must decide whether to continue the changes proposed by the engineers. The decision always includes the practical consideration of whether the new method is worth the cost. Suppose management concludes that the cracking percentage has been reduced. They must still evaluate how much the cracking rate will be reduced and how much it will cost to accomplish the reduction. The *size of the effect* is always a concern when we test hypotheses. *A good way to look at the* **effect size** *is to examine a confidence interval.*

How much does it cost? Formal tests of a null hypothesis base the decision of whether to reject the null hypothesis solely on the size of the P-value. But in real life, we want to evaluate the costs of our decisions as well. How much would you be willing to pay for a faster computer? Shouldn't your decision depend on how much faster? And on how much more it costs? Costs are not just monetary, either. Would you use the same standard of proof for testing the safety of an airplane as for the speed of your new computer?

17.4 Alternative Alternatives

Tests on the ingot data can be viewed in two different ways. We know the old cracking rate is 20%, so the null hypothesis is

$$H_0: p = 0.20$$

A S *Activity:* **The Alternative Hypotheses.** This interactive tool provides easy ways to visualize how one- and two-tailed alternative hypotheses work.

But we have a choice of alternative hypotheses. A metallurgist working for the company might be interested in *any* change in the cracking rate due to the new process. Even if the rate got worse, she might learn something useful from it. She's interested in possible changes on both sides of the null hypothesis. So she would write her alternative hypothesis as

$$H_A: p \neq 0.20$$

An alternative hypothesis such as this is known as a **two-sided alternative,**[3] because we are equally interested in deviations on either side of the null hypothesis value. For

[3]It is also called a *non-directional alternative*.

two-sided alternatives, the P-value is the probability of deviating at least as much as our actual sample in *either* direction from the null hypothesis value.

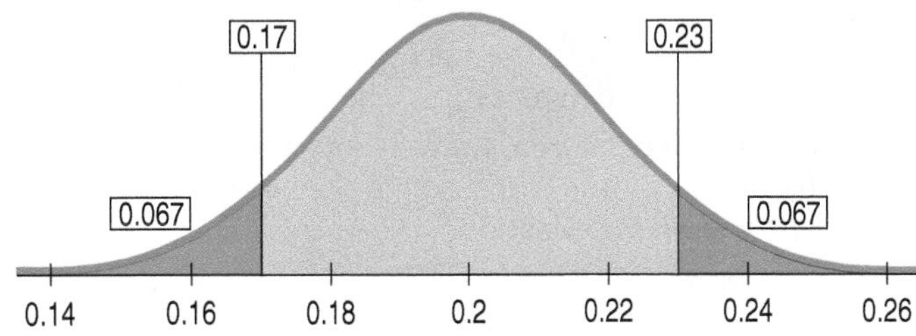

But management is really interested only in *lowering* the cracking rate below 20%. The scientific value of knowing how to *increase* the cracking rate may not appeal to them. The only alternative of interest to them is that the cracking rate *decreases*. They would write their alternative hypothesis as

$$H_A: p < 0.20$$

An alternative hypothesis that focuses on deviations from the null hypothesis value in only one direction is called a **one-sided alternative.**[4]

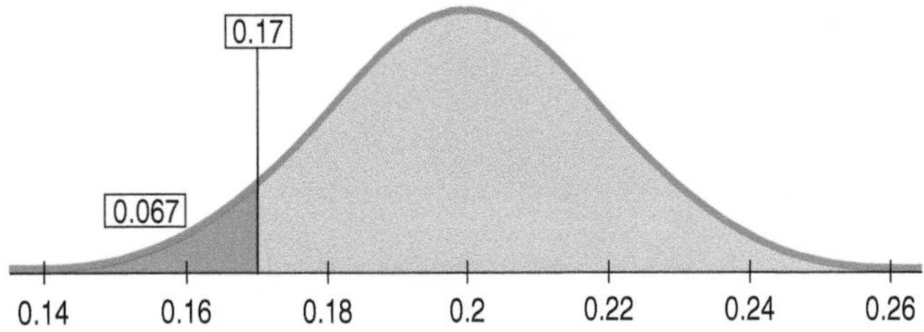

For a hypothesis test with a one-sided alternative, the P-value is the probability of deviating *only in the direction of the alternative* away from the null hypothesis value. For the same data, the one-sided P-value is half the two-sided P-value. So, a one-sided test will reject the null hypothesis more often. *If you aren't sure which to use, a two-sided test is always more conservative.* Be sure you can justify the choice of a one-sided test from the *Why* of the situation (and not from the data).

Some statisticians believe that you should never do one-sided tests, as they consider them an artificial way to "buy" yourself more power, or more significant results. We'll leave that discussion to a more philosophical course, but one thing you should certainly never do is decide on the alternative hypothesis *after* looking at your observed test statistic.[5] If a one-sided alternative is interesting to you, it should be interesting before you see the direction of the data!

Step-by-Step Example TESTING A HYPOTHESIS

Advances in medical care, such as prenatal ultrasound examination, now make it possible to determine a child's sex early in a pregnancy. There is a fear that in some cultures some parents may use this technology to select the sex of their children. A study from one hospital in Punjab, India,[6] reports that, in 1993, 56.9% of the 550 live births that year were boys. It's a medical fact that male babies are slightly more common than female babies. The study's authors report a baseline for this region of 51.7% male live births.

Question: Is there evidence that the proportion of male births has changed?

[4] Also called a directional alternative.

[5] Besides, you'll never see the test statistic go in both directions!

[6] E. Booth, M. Verma, and R. S. Beri 1994, "Fetal sex determination in infants in Punjab, India: Correlations and implications," *BMJ* 309: 1259–1261.

THINK ➡ **Plan** State what we want to know.

Define the variables and discuss the *W*'s.

I want to know whether the proportion of male births at a hospital has changed from the established baseline of 51.7%. The data are the recorded sexes of the 550 live births from a hospital in Punjab, India, in 1993, collected for a study on fetal sex determination. The parameter of interest, p, is the proportion of male births:

Hypotheses The null hypothesis makes the claim of no difference from the baseline.

Before seeing the data, we were interested in any change in male births, so the alternative hypothesis is two-sided.[7]

$$H_0: p = 0.517$$
$$H_A: p \neq 0.517$$

Model Think about the assumptions and check the appropriate conditions.

For testing proportions, the conditions are the same ones we had for making confidence intervals, except that we check the **Success/Failure Condition** with the *hypothesized* proportions rather than with the *observed* proportions.

✓ **Independence Assumption:** There is no reason to think that the sex of one baby can affect the sex of other babies, so births can reasonably be assumed to be independent with regard to the sex of the child.

✓ **Randomization Condition:** The 550 live births are not a random sample, so I must be cautious about any general conclusions. I hope that this is a representative year, and I think that the births at this hospital may be typical of this area of India.

✓ **10% Condition:** I would like to be able to make statements about births at this and similar hospitals in this region around this time. These 550 births are fewer than 10% of all of those births.

✓ **Success/Failure Condition:** Both $np_0 = 550(0.517) = 284.35$ and $nq_0 = 550(0.483) = 265.65$ are greater than 10; I expect the births of at least 10 boys and at least 10 girls, so the sample is large enough.

Specify the sampling distribution model.

Tell what test you plan to use.

The conditions are satisfied, so I can use a Normal model and perform a *one-proportion z-test.*

SHOW ➡ **Mechanics** The null model gives us the mean, and (because we are working with proportions) the mean gives us the standard deviation.

The null model is a Normal distribution with a mean of 0.517 and a standard deviation of

$$SD(\hat{p}) = \sqrt{\frac{p_0 q_0}{n}} = \sqrt{\frac{(0.517)(1 - 0.517)}{550}}$$
$$= 0.0213$$

We find the *z*-score for the observed proportion to find out how many standard deviations it is from the hypothesized proportion.

The observed proportion, $\hat{p}$, is 0.569, so

$$z = \frac{\hat{p} - p_0}{SD(\hat{p})} = \frac{0.569 - 0.517}{0.0213} = 2.44$$

Make a picture. Sketch a Normal model centred at $p_0 = 0.517$. Shade the region to the right of the observed proportion, and because this is a two-tail test, also shade the corresponding region in the other tail.

The sample proportion lies 2.44 standard deviations above the mean.

[7]If we had a clear theory or concern *a priori* (before seeing the data) about a possible excess of male births in this region, a one-sided alternative could be justified. The P-value, calculated at the end of this example, would then be only 0.0073.

From the z-score, we can find the P-value, which tells us the probability of observing a value that extreme (or more). Use technology or a table (see below).

Because this is a two-tail test, the P-value is the probability of observing an outcome more than 2.44 standard deviations from the mean of a Normal model *in either direction.* We must therefore *double* the probability we find in the upper tail.

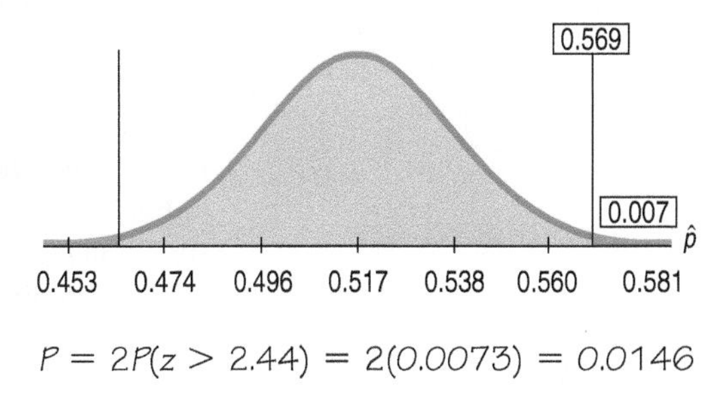

$P = 2P(z > 2.44) = 2(0.0073) = 0.0146$

TELL ➡ Conclusion State your conclusion in context.

This P-value is roughly 1 time in 70. That's clearly significant, but don't jump to other conclusions. We can't be sure how this deviation came about. For instance, we don't know whether this hospital or time period is typical.

The P-value of 0.0146 says that if the true proportion of male babies were still at 51.7%, then an observed proportion as different as 56.9% male babies would occur at random only about 15 times in 1000. With a P-value this small, we have strong evidence that the birth ratio of boys to girls is not equal to its natural level. It appears that the proportion of boys increased in this region around this time.[8]

Here's a portion of a Normal table that gives the probability we needed for the hypothesis test. At $z = 2.44$, the table gives the percentile (shaded blue) as 0.9927. The upper-tail probability (shaded red) is, therefore, $1 - 0.9927 = 0.0073$; so, for our two-sided test, the P-value is $2(0.0073) = 0.0146$ (Figure 17.2).

Figure 17.2

A portion of a Normal table with the look-up for $z = 2.44$ indicated. Since $1 - 0.9927 = 0.0073$, the P-value we want is $2(0.0073) = 0.0146$.

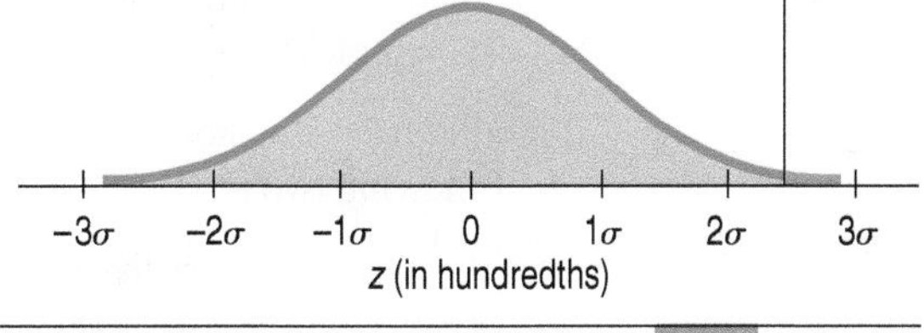

z	0.00	0.01	0.02	0.03	0.04	0.05
1.9	0.9713	0.9719	0.9726	0.9732	0.9738	0.9744
2.0	0.9772	0.9778	0.9783	0.9788	0.9793	0.9798
2.1	0.9821	0.9826	0.9830	0.9834	0.9838	0.9842
2.2	0.9861	0.9864	0.9868	0.9871	0.9875	0.9878
2.3	0.9893	0.9896	0.9898	0.9901	0.9904	0.9906
2.4	0.9918	0.9920	0.9922	0.9925	0.9927	0.9929
2.5	0.9938	0.9940	0.9941	0.9943	0.9945	0.9946
2.6	0.9953	0.9955	0.9956	0.9957	0.9959	0.9960

A S ***Activity:* Practice with Testing Hypotheses About Proportions.** Here's an interactive tool that makes it easy to see what's going on in a hypothesis test.

17.5 P-Values and Decisions: What to Tell About a Hypothesis Test

Hypothesis tests are particularly useful when we must make a decision. Is the defendant guilty or not? Should we choose print advertising or television? Questions like these cannot always be answered with the margins of error of confidence intervals. The absolute nature of the hypothesis test decision, however, makes some people (including the authors) uneasy. If possible, it's usually a good idea to report a confidence interval for the parameter of interest as well.

[8] So there was an increase. Just how big was it? Use a confidence interval.

How small should the P-value be for you to reject the null hypothesis? A jury needs enough evidence to show the defendant as guilty "beyond a reasonable doubt." How does that translate to P-values? The answer is that it's highly context-dependent. When we're screening for a disease and want to be sure we treat all those who are sick, we may be willing to reject the null hypothesis of no disease with a P-value as large as 0.10. We would rather treat the occasional healthy person than fail to treat someone who was really sick. But a long-standing hypothesis, believed by many to be true, needs stronger evidence (and a correspondingly small P-value) to reject it.

See if you require the same P-value to reject each of the following null hypotheses:

- A renowned musicologist claims that she can distinguish between the works of Mozart and Haydn simply by hearing a randomly selected 20 seconds of music from any work by either composer. What's the null hypothesis? If she's just guessing, she'll get 50% of the pieces correct, on average. So our null hypothesis is that p is 50%. If she's for real, she'll get more than 50% correct. We present her with 10 pieces of Mozart or Haydn chosen at random. She gets 9 out of 10 correct. It turns out that the P-value associated with that result is 0.011. (In other words, if you tried to just guess, you'd get at least 9 out of 10 correct only about 1% of the time.) What would *you* conclude? Most people would probably reject the null hypothesis and be convinced that the musicologist has some ability to do as she claims. Why? Because the P-value is small and we don't have any particular reason to doubt the alternative.

"Extraordinary claims require extraordinary proof."

—*Carl Sagan*

- On the other hand, imagine a student who bets that he can make a flipped coin land the way he wants just by thinking hard. To test him, we flip a fair coin 10 times. Suppose he gets 9 out of 10 right. This also has a P-value of 0.011. Are you willing now to reject this null hypothesis? Are you convinced that he's not just lucky? What amount of evidence *would* convince you? We require more evidence if rejecting the null hypothesis would contradict long-standing beliefs or other scientific results. Of course, with sufficient evidence we should revise our opinions (and scientific theories). That's how science makes progress.

Another factor in choosing a P-value is the importance of the issue being tested. Consider the following two tests:

- A researcher claims that the proportion of university students who hold part-time jobs now is higher than the proportion known to hold such jobs a decade ago. You might be willing to believe the claim (and reject the null hypothesis of no change) with a P-value of 0.10.
- An engineer claims that the proportion of rivets holding the wing on an airplane that are likely to fail is below the proportion at which the wing would fall off. What P-value would be small enough to get you to fly on that plane?

A S *Activity:* **Hypothesis Tests for Proportions.** You've probably noticed that the tools for confidence intervals and for hypothesis tests are similar. See how tests and intervals for proportions are related—and an important way in which they differ.

Your conclusion about any null hypothesis should always be accompanied by the P-value of the test. Don't just declare the null hypothesis rejected or not rejected. The P-value show the strength of evidence against the null hypothesis and will help readers use their own criteria for making a decision.

To complete your analysis (particularly if the null is rejected), follow up your test with a confidence interval to help provide a good idea of the size of the effect, and hence its practical importance.

Just Checking

4. A bank is testing a new method for getting delinquent customers to pay their past-due credit card bills. The standard way was to send a letter (costing about \$0.40) asking the customer to pay. That worked 30% of the time. They want to test a new method that involves sending a DVD to customers encouraging them

to contact the bank and set up a payment plan. Developing and sending the video costs about $10.00 per customer. What is the parameter of interest? What are the null and alternative hypotheses?

5. The bank sets up an experiment to test the effectiveness of the DVD. They mail it out to several randomly selected delinquent customers and keep track of how many actually do contact the bank to arrange payments. The bank's statistician calculates a P-value of 0.003. What does this P-value suggest about the DVD?

6. The statistician tells the bank's management that the results are clear and that they should switch to the DVD method. Do you agree? What else might you want to know?

Step-by-Step Example AFTER A SIGNIFICANT TEST, REPORT THE CONFIDENCE INTERVAL

In the previous Step-by-Step Example, we discovered evidence of an increased proportion of male babies in Punjab, India, in 1993. This should not be the end of our analysis. The next logical question should be:

Question: How big an increase are we talking about? Let's find a confidence interval for the proportion of male births.

THINK AGAIN ➡ Model Check the conditions.

The conditions are identical to those for the hypothesis test, with one difference. Now we are not given a hypothesized proportion, p_0, so we must instead work with the observed proportion $\hat{p}$.

Specify the sampling distribution model.

Tell what method you plan to use.

✓ **Success/Failure Condition:** Both $n\hat{p} = 550(0.569) = 313$and $n\hat{q} = 237$ are at least 10.

The conditions are satisfied, so I can model the sampling distribution of the proportion with a Normal model and find a *one-proportion z-interval.*

SHOW MORE ➡ Mechanics We can't find the sampling model standard deviation from the null model proportion. (In fact, we've just rejected it.) Instead, we find the standard error of $\hat{p}$ from the *observed* proportion. Other than that substitution, the calculation looks the same as for the hypothesis test.

With this large a sample size the difference is negligible, but in smaller samples it could make a bigger difference.

$$SE(\hat{p}) = \sqrt{\frac{\hat{p}\hat{q}}{n}} = \sqrt{\frac{(0.569)(1-0.569)}{550}}$$
$$= 0.0211$$

The sampling model is Normal, so for a 95% confidence interval, the critical value $z^* = 1.96$.

The margin of error is

$$ME = z^* \times SE(\hat{p}) = 1.96(0.0211) = 0.041$$

So the 95% confidence interval is

$$0.569 \pm 0.041 \text{ or } (0.528, 0.610)$$

TELL ALL ➡ Conclusion Confidence intervals help us think about the size of the effect. Comment on the possible magnitude of the change from the baseline.

We are 95% confident that the true proportion of male births is between 52.8% and 61.0%. The change from the baseline of 51.7% male births might be quite substantial.

For Example A CONFIDENCE INTERVAL FOR THE TEEN DRIVERS

RECAP: In a large city, only 61 of 90 randomly selected teenagers passed their driving tests, leading the investigating newspaper reporter to reject the Ministry's claim of an 80% passing rate.

QUESTION: Based on this sample, what might the reporter say about the true passing rate for teenage driving candidates?

ANSWER: With $x = 61$ and $n = 90$, $\hat{p} = 0.678$. I'll find a 95% confidence interval for p:

$$\hat{p} \pm z^{*}SE(\hat{p}) = 0.678 \pm 1.96\sqrt{\frac{(0.678)(0.322)}{90}}$$
$$= 0.678 \pm 0.097 = (0.581, 0.775)$$

Based on this investigation, it's unlikely that the true passing rate for teenagers in this city is any higher than 77.5%, and it may be as low as 58%.

Follow Your Plan A study published in the *Canadian Medical Association Journal* (CMAJ) revealed that many published studies reported hypothesis tests that were not part of the original research plan. The researchers compared the original protocols for randomized clinical trials approved by the Canadian Institutes of Health Research (CIHR) between 1990 and 1998 with the subsequent reports published in journals. They found that 40% of the publications focused on research outcomes that were different from the variable the researchers originally set out to analyze. The authors suggested that this was done when the data failed to fit the original hypothesis, declaring, "This is an ethical and scientific issue." CMAJ editor Dr. John Hoey says, "Once you change the objective of the trial so it fits with the data, it's no longer a randomized trial, it's a fishing expedition."[9]

Ideally, hypotheses should be stated even before the data are collected, but in reality, you often have the data before you formulate your hypotheses. What you may *not* do is analyze the data first, and *then* state your hypotheses to fit what you find.

That's "drawing the target around the arrow." And it's wrong.

17.6 *Large Sample Tests of Hypotheses Based on Normally Distributed Estimators

Just as with confidence intervals, the procedure discussed in this chapter may be easily extended to test hypotheses about other parameters, such as means, as long as we have a sufficiently large sample (for small sample tests of means, use the methods described in Chapters 20–22).

Suppose we wish to test a hypothesis about a parameter θ, and are able to construct an estimator $\hat{\theta}$ whose distribution is approximately Normal, centred at θ, with a standard deviation denoted SD($\hat{\theta}$). The test statistic is just the estimate of the parameter, $\hat{\theta}$, standardized, using the value for θ specified by our null hypothesis, denoted θ_0:

$$z = \frac{(\hat{\theta} - \theta_0)}{SD(\hat{\theta})} \text{ or in words: } \frac{\text{Estimate} - \text{Hypothesized Value}}{\text{Standard Deviation of the Estimator}}$$

But just as with confidence intervals, we can only use this method if we can accurately evaluate all parts of it, and the denominator presents an issue. First, we need a formula for this standard deviation, and then we must be able to plug in the values for all of its parts. However, if the sample is big enough, the good news is that we can just plug in sample estimates for any unknowns. We use the sample data to calculate the numerical value of the z-statistic, and then, as illustrated in Figure 17.1, find the P-value to base our test on.

[9]http://news.utoronto.ca/_bulletin/2004-10-12.pdf

Testing for the Mean of a Quantitative Population When the Sample Is Big

Suppose we wish to test a hypothesis about the mean of a quantitative population, denoted μ, where the null hypothesized value is μ_0. We may use the sample mean $\bar{y}$ to estimate μ without bias, and, for a large sample, we know that $\bar{y}$ will be nearly Normal due to the Central Limit Theorem. The standard deviation of the sampling distribution of $\bar{y}$ has the formula $SD(\bar{y}) = \frac{\sigma}{\sqrt{n}}$, which may be estimated from the sample data by $\frac{s}{\sqrt{n}}$, which we call the standard error of $\bar{y}$, or $SE(\bar{y})$.

Our test statistic then becomes $z = \frac{\bar{y} - \mu_0}{SE(\bar{y})}$, where the SE is $\frac{s}{\sqrt{n}}$. Keep in mind that substituting this estimate for the true standard deviation in the denominator should have some effect on the distribution of this statistic, but the effect is negligible if the sample is sufficiently large—say, greater than 60. If we introduce too much variability into the denominator, the distribution actually changes into something other than a standard Normal. In fact, it becomes the "*t*" distribution, a sampling model that will be discussed in great detail in Chapter 20, where we will get around to dealing with the small sample problem. As always, be sure to inspect the data for extreme outliers or severe skewness. The latter might force you to transform the data or focus your inference on the median rather than the mean (in which case, the methods in Chapters 29 and 30 might help).

And just as with proportions, make sure that the observations are mutually independent and that your sampling procedure will allow you to generalize beyond your sample to a bigger population—do you have a valid random sample or a reasonably representative sample?

Continuing the example from Chapter 16, a team of researchers had tested 150 farm-raised salmon for organic contaminants. They found a mean concentration of the carcinogenic insecticide mirex to be 0.0913 ppm with a standard deviation of 0.0495 ppm. A histogram of the data showed no extreme skewness or outliers. A level higher than 0.08 ppm is considered dangerous. Is there *clear evidence* of excessively high levels of mirex? This requires a test of hypothesis. After all, even if the true mean level were 0.08 ppm, half of the time we would get a sample mean bigger than 0.08 purely by chance. Is the observed mean level too high to just be chance variation?

$\bar{y} = 0.0913$ ppm, with an estimated standard deviation $= \frac{s}{\sqrt{n}} = \frac{0.0495}{\sqrt{150}} = 0.00404$, so our test statistic is $z = \frac{\bar{y} - \mu_0}{SE(\bar{y})} = \frac{0.0913 - 0.08}{0.00404} = 2.80$, from which it follows that the P-value $= P(Z > 2.80) = (1 - 0.9974) = 0.0026$, which is small enough to provide very convincing evidence of an excessively high mean level of the mirex contaminant.

WHAT CAN GO WRONG?

Hypothesis tests are so widely used—and so widely misused—that we've devoted all of the next chapter to discussing the pitfalls involved, but there are a few issues that we can talk about already.

- **Don't base your null hypotheses on what you see in the data.** You are not allowed to look at the data first and then adjust your null hypothesis so that it will be rejected. When your sample value turns out to be $\hat{p} = 51.8\%$, with a standard deviation of 1%, don't form a null hypothesis like H_0: $p = 49.8\%$, knowing that you can reject it. You should always *Think* about the situation you are investigating and make your null hypothesis describe the "nothing interesting" or "nothing has changed" scenario. No peeking at the data!

- **Don't base your alternative hypothesis on the data, either.** Again, you need to *Think* about the situation. Are you interested only in knowing whether something has *increased*? Then write a one-sided (upper-tail) alternative. Or would you be equally interested in a change in either direction? Then you want a two-sided alternative. You should decide whether to do a one- or two-sided test based on what results would be of interest to you, not what you see in the data.
- **Don't make your null hypothesis what you want to show to be true.** Remember, the null hypothesis is the status quo, the nothing-is-strange-here position a skeptic would take. You wonder whether the data cast doubt on that. You can reject the null hypothesis, but you can never "accept" or "prove" the null.
- **Don't forget to check the conditions.** The reasoning of inference depends on randomization. No amount of care in calculating a test result can recover from a biased sample. The probabilities we compute depend on the Independence Assumption. And our sample must be large enough to justify our use of a Normal model.
- **Don't accept the null hypothesis.** You may not have found enough evidence to reject it, but you surely have not proven it's true!

CONNECTIONS

Hypothesis tests and confidence intervals share many of the same concepts. Both rely on sampling distribution models, and because the models are the same and require the same assumptions, both check the same conditions. They also calculate many of the same statistics. Like confidence intervals, hypothesis tests use the standard deviation of the sampling distribution as a ruler, as we first saw in Chapter 5.

For testing, we find ourselves looking once again at z-scores, and we compute the P-value by finding the distance of our test statistic from the centre of the null model. P-values are conditional probabilities. They give the probability of observing the result we have seen (or one even more extreme) *given* that the null hypothesis is true.

The standard Normal model is here again as our connection between z-score values and probabilities.

What Have We Learned?

Learning Objectives

Know how to formulate a null and an alternative hypothesis for a question of interest.

- The null hypothesis specifies a parameter and a (null) value for that parameter.
- The alternative hypothesis specifies a range of plausible values should we reject the null.

Be able to perform a hypothesis test for a proportion.

- The null hypothesis has the form H_0: $p = p_0$.
- We find the standard deviation of the sampling distribution of the sample proportion by assuming that the null hypothesis is true:

$$SD(\hat{p}) = \sqrt{\frac{p_0 q_0}{n}}$$

- We refer the statistic $z = \frac{\hat{p} - p_0}{SD(\hat{p})}$ to the standard Normal model.

Understand P-values.

- A P-value is the estimated probability of observing a statistic value at least as far from the (null) hypothesized parameter as the one we have actually observed, assuming the null hypothesis to actually be true.
- A small P-value indicates that the statistic we have observed would be unlikely were the null hypothesis true. That leads us to doubt the null.
- A large P-value just tells us that we have insufficient evidence to doubt the null hypothesis. In particular, it does not prove the null to be true.

Know the reasoning of hypothesis testing.

- State the **hypotheses**.
- Determine (and check assumptions for) the sampling distribution **model**.
- Calculate the test statistic—the **mechanics**.
- State your **conclusions and decisions**.

Be able to decide on a one-sided or two-sided alternative hypothesis, and justify your decision.

Know that confidence intervals and hypothesis tests go hand in hand in helping us think about models.

- A hypothesis test assesses the plausibility of the value of a parameter value, by determining the degree to which the sample provides contradictory evidence (the *P*-value).
- A confidence interval shows the range of plausible values for the parameter.

Review of Terms

Null hypothesis The claim being assessed in a hypothesis test. Usually, the null hypothesis is a statement of "no change from the traditional value," "no effect," "no difference," or "no relationship." For a claim to be a testable null hypothesis, it must specify a value for some population parameter that can form the basis for assuming a sampling distribution for a test statistic (p. 478).

Alternative hypothesis Proposes what we should conclude if we find the null hypothesis to not be plausible (p. 479).

P-value The probability of observing a value for a test statistic at least as far from the hypothesized parameter as the statistic value actually observed if the null hypothesis is true. A small P-value indicates either that the observation is improbable or that the probability calculation was based on incorrect assumptions. The assumed truth of the null hypothesis is the assumption under suspicion (p. 480).

One-proportion z-test A test of the null hypothesis that the proportion parameter for a single population equals a specified value (H_0: $p = p_0$) by referring the statistic $z = \frac{\hat{p} - p_0}{SD(\hat{p})}$ to a standard Normal model (p. 482).

Effect size The difference between the null hypothesis value and the true value of a model parameter (p. 484).

Two-sided alternative (Non-directional alternative) An alternative hypothesis is two-sided (H_A: $p \neq p_0$) when we are interested in deviations in either direction away from the null hypothesized parameter value (p. 484).

One-sided alternative (Directional alternative) An alternative hypothesis is one-sided (H_A: $p > p_0$ or H_A: $p < p_0$) when we are interested in deviations in only one direction away from the hypothesized parameter value (p. 485).

On the Computer HYPOTHESIS TESTS FOR PROPORTIONS

Hypothesis tests for proportions are so easy and natural that many statistics packages don't offer special commands for them. Most statistics programs want to know the "success" and "failure" status for each case. Usually these are given as 1 or 0, but they might be category names like "yes" and "no." Often, we just know the proportion of successes, $\hat{p}$ and the total count, n. Computer packages don't usually deal naturally with summary data like this, but the statistics routines found on many graphing calculators do. These calculators allow you to test hypotheses from summaries of the data—usually, all you need to enter are the number of successes and the sample size.

In some programs, you can reconstruct the original values. But even when you have reconstructed (or can reconstruct) the raw data values, often you won't get exactly the same test statistic from a computer package as you would find working by hand. The reason is that when the packages treat the proportion as a mean, they make some approximations. The result is very close, but not exactly the same.

EXCEL

Inference methods for proportions are not part of the standard Excel tool set.

COMMENTS

For summarized data, type the calculation into any cell and evaluate it.

JMP

For a **categorical** variable that holds category labels, the **Distribution** platform includes tests and intervals of proportions. For summarized data, put the category names in one variable and the frequencies in an adjacent variable. Designate the frequency column to have the **role** of **frequency**. Then use the **Distribution** platform.

COMMENTS

JMP uses slightly different methods for proportion inferences than those discussed in this text. Your answers are likely to be slightly different.

MINITAB

- Choose **Basic Statistics** from the **Stat** menu.
- Choose **1Proportion** from the Basic Statistics submenu.
- If the data are category names in a variable, assign the variable from the variable list box to the **Samples in columns** box.
- If you have summarized data, click the **Summarized Data** button and fill in the number of trials and the number of successes.
- Click the **Options** button and specify the remaining details.
- If you have a large sample, check **Use test and interval based on Normal distribution**.
- Click the **OK** button.

COMMENTS

When working from a variable that names categories, MINITAB treats the last category as the "success" category. You can specify how the categories should be ordered.

R

In library(stats):

- **prop.test(X, n, p = NULL, alternative = c("two. sided," "less," "greater"), conf.level = 0.95, correct = FALSE)**

will test the hypothesis that $p = p_0$ against various alternatives with $\alpha = 0.05$. For example, with 260 successes out of 500, at $\alpha = 0.05$, to test that $p = 0.5$ vs. $p \neq 0.5$, use: **prop.test(260,500,0.5,"two.sided," correct = FALSE)**

SPSS

SPSS does not find hypothesis tests for proportions.

STATCRUNCH

To test a hypothesis for a proportion using summaries:

- Click on **Stat**.
- Choose **Proportions** » **One sample** » **with summary**.
- Enter the **Number of successes** (x) and **Number of observations** (n).
- Click on **Next**.
- Indicate **Hypothesis Test**, then enter the hypothesized **Null** proportion, and choose the **Alternative** hypothesis.
- Click on **Calculate**.

To test a hypothesis for a proportion using data:

- Click on **Stat**.
- Choose **Proportions** » **One sample** » **with data**.
- Choose the variable **Column** listing the **Outcomes**.
- Enter the outcome to be considered a **Success**.
- Click on **Next**.
- Indicate **Hypothesis Test**, then enter the hypothesized **Null proportion**, and choose the **Alternative** hypothesis.
- Click on **Calculate**.

TI-83/84 PLUS

To do the mechanics of a hypothesis test for a proportion:

- Select **5:1-PropZTest** from the **STAT TESTS** menu.
- Specify the hypothesized proportion.
- Enter the observed value of *x*.
- Specify the sample size.
- Indicate what kind of test you want: one-tail lower tail, two-tail, or one-tail upper tail.
- Calculate the result.

COMMENTS

Beware: When you enter the value of *x*, you need the *count*, not the percentage. The count must be a whole number. If the number of successes is given as a percent, you must first multiply *np* and round the result to obtain *x*.

Exercises

1. **Better than aspirin?** A very large study showed that aspirin reduced the rate of first heart attacks by 44%. A pharmaceutical company thinks they have a drug that will be more effective than aspirin, and plans to do a randomized clinical trial to test the new drug.
   a) What is the null hypothesis the company will use?
   b) What is their alternative hypothesis?

2. **Psychic** A friend of yours claims to be psychic. You are skeptical. To test this you take a stack of 100 playing cards and have your friend try to identify the suit (hearts, diamonds, clubs, or spades), without looking, of course!
   a) State the null hypothesis for your experiment.
   b) State the alternative hypothesis.

3. **Better than aspirin II?** A clinical trial compares the new drug described in Exercise 1 to aspirin. The group using the new drug had somewhat fewer heart attacks than those in the aspirin group.
   a) The P-value from the hypothesis test was 0.28. What do you conclude?
   b) What would you have concluded if the P-value had been 0.004?

4. **Psychic after all?** The "psychic" friend from Exercise 2 correctly identified more than 25% of the cards.
   a) A hypothesis test gave a P-value of 0.014. What do you conclude?
   b) What would you conclude if the P-value had been 0.245?

5. **Psychic again (you should have seen this coming)** If you were to do a hypothesis test on your experiment with your "psychic" friend from Exercise 2, would your alternative hypothesis be one- or two-sided? Explain why.

6. **Expensive medicine** Developing a new drug can be an expensive process, resulting in high costs to patients. A pharmaceutical company has developed a new drug to reduce cholesterol, and it will conduct a clinical trial to compare the effectiveness to the most widely used current treatment. The results will be analyzed using a hypothesis test.
   a) If the test yields a low P-value and the researcher rejects the null hypothesis that the new drug is not more effective, but it actually is not better, what are the consequences of such an error?
   b) If the test yields a high P-value and the researcher fails to reject the null hypothesis, but the new drug *is* more effective, what are the consequences of such an error?

7. **Hypotheses** Write the null and alternative hypotheses you would use to test each of the following situations:
   a) The premier of the province is concerned about his "negatives"—the percentage of provincial residents who express disapproval of his job performance. His political committee pays for a series of television ads, hoping that they can keep the negatives below 30%. They will use follow-up polling to assess the ads' effectiveness.
   b) Is a coin fair?
   c) Only about 20% of people who try to quit smoking succeed. Sellers of a motivational tape claim that listening to the recorded messages can help people quit.

8. **More hypotheses** Write the null and alternative hypotheses you would use to test each of the following situations.
   a) In the 1950s, only about 40% of high school graduates went on to college or university. Has the percentage changed?
   b) Twenty percent of cars of a certain model needed costly transmission work after being driven between 50 000 and 100 000 miles. The manufacturer hopes that a redesign of a transmission component will solve this problem.
   c) We field-test a new flavour of soft drink, planning to market it only if we are sure that over 60% of the people tested like the flavour.

9. **Negatives** After the political ad campaign described in Exercise 7a), pollsters check the premier's negatives. They test the hypothesis that the ads produced no change against the alternative that the negatives are now below 30% and find a P-value of 0.22. Which conclusion is appropriate? Explain.
   a) There's a 22% chance that the ads worked.
   b) There's a 78% chance that the ads worked.
   c) There's a 22% chance that the poll they conducted is correct.
   d) There's a 22% chance that natural sampling variation could produce poll results like these if there's really no change in public opinion.

10. **Dice** The seller of a loaded die claims that it will favour the outcome 6. We don't believe that claim, and roll the die 200 times to test an appropriate hypothesis. Our P-value turns out to be 0.03. Which conclusion is appropriate? Explain.
    a) There's a 3% chance that the die is fair.
    b) There's a 97% chance that the die is fair.
    c) There's a 3% chance that a loaded die could randomly produce the results we observed, so it's reasonable to conclude that the die is fair.
    d) There's a 3% chance that a fair die could randomly produce the results we observed, so it's reasonable to conclude that the die is loaded.

11. **Relief** A company's old antacid formula provided relief for 70% of the people who used it. The company tests a new formula to see if it is better and gets a P-value of 0.27. Is it reasonable to conclude that the new formula and the old one are equally effective? Explain.

12. **Cars** A survey investigating whether the proportion of today's Grade 12 students who own their own cars is higher than it was a decade ago finds a P-value of 0.017. Is it reasonable to conclude that more Grade 12 students have cars? Explain.

13. **He cheats!** A friend of yours claims that when he tosses a coin he can control the outcome. You are skeptical and want him to prove it. He tosses the coin, and you call heads; it's tails. You try again and lose again.
    a) Do two losses in a row convince you that he really can control the toss? Explain.
    b) You try a third time, and again you lose. What's the probability of losing three tosses in a row if the process is fair?

c) Would three losses in a row convince you that your friend cheats? Explain.
d) How many times in a row would you have to lose to be pretty sure that this friend really can control the toss? Justify your answer by calculating a probability and explaining what it means.

14. Candy Someone hands you a box of a dozen chocolate-covered candies, telling you that half are vanilla creams and the other half are peanut butter. You pick candies at random and discover that the first three you eat are all vanilla.
a) If there really were six vanilla and six peanut butter candies in the box, what is the probability that you would have picked three vanillas in a row?
b) Do you think there really might have been six of each? Explain.
c) Would you continue to believe that half are vanilla if the fourth one you try is also vanilla? Explain.

15. Smartphones Many people have trouble setting up all the features of their smartphones, so a company has developed what it hopes will be easier instructions. The goal is to have at least 96% of customers succeed. The company tests the new system on 200 people, of whom 188 were successful. Is this strong evidence that the new system fails to meet the company's goal? A student's test of this hypothesis is shown. How many mistakes can you find?

$H_0: \hat{p} = 0.96$

$H_A: \hat{p} \neq 0.96$

SRS, $0.96(200) > 10$

$\frac{188}{200} = 0.94; SD(\hat{p}) = \sqrt{\frac{(0.94)(0.06)}{200}} = 0.017$

$z = \frac{0.96 - 0.94}{0.017} = 1.18$

$P = P(z > 1.18) = 0.12$

There is strong evidence that the new instructions do not work.

16. Got milk? In November 2001, the *Ag Globe Trotter* newsletter reported that 90% of adults drink milk. A regional farmers' organization planning a new marketing campaign across its multicounty area polls a random sample of 750 adults living there. In this sample, 657 people said that they drink milk. Do these responses provide strong evidence that the 90% figure is not accurate for this region? Correct the mistakes you find in a student's attempt to test an appropriate hypothesis.

$H_0: \hat{p} = 0.9$

$H_A: \hat{p} < 0.9$

SRS, $750 > 10$

$\frac{657}{750} = 0.876; SD(\hat{p}) = \sqrt{\frac{(0.88)(0.12)}{750}} = 0.012$

$z = \frac{0.876 - 0.90}{0.012} = -2$

$P = P(z > -2) = 0.977$

There is more than a 97% chance that the stated percentage is correct for this region.

17. Dowsing In a rural area, only about 30% of the wells that are drilled find adequate water at 100 metres deep or less. A local man claims to be able to find water by "dowsing"—using a forked stick to indicate where the well should be drilled. You check with 80 of his customers and find that 27 have wells less than 100 metres deep. What do you conclude about his claim?
a) Write appropriate hypotheses.
b) Check the necessary assumptions.
c) Perform the mechanics of the test. What is the P-value?
d) Explain carefully what the P-value means in this context.
e) What is your conclusion?

18. Abnormalities In the 1980s, it was generally believed that congenital abnormalities affect about 5% of the nation's children. Some people believe that the increase in the number of chemicals in the environment in recent years has led to an increase in the incidence of abnormalities. A recent study examined 384 children and found that 46 of them showed signs of an abnormality. Is this strong evidence that the risk has increased?
a) Write appropriate hypotheses.
b) Check the necessary assumptions.
c) Perform the mechanics of the test. What is the P-value?
d) Explain carefully what the P-value means in this context.
e) What is your conclusion?
f) Do environmental chemicals cause congenital abnormalities?

19. Educated Statistics Canada monitors many aspects of secondary and post-secondary education in the regular census. The 2011 National Household Survey (NHS) released in June 2013 showed that 64.1% of the total population aged 25–64 had a post-secondary degree. In comparison, the 2006 census showed that 60.7% held a similar qualification.

Is it appropriate to conduct a hypothesis test to determine if there is a change in the proportion of degrees awarded between 2006 and 2011? Discuss why or why not.

20. More educated To test the figures quoted in the previous question, you run an SRS of 4000 people aged 25–64 to see if the proportion with a post-secondary degree in 2013 has changed since 2011. You find that 2630 have a degree.
a) Write appropriate hypotheses.
b) Check the assumptions and conditions.

c) Perform the test and find the P-value.
d) State your conclusion.
e) Do you think this difference is meaningful? Explain.

21. Contributions, please, part II In Exercise 19 of Chapter 16, you learned that charitable organizations like the MS Society send free mailing labels and greeting cards to potential donors on their list and ask for a voluntary contribution. Suppose they test a new campaign by sending letters to a random sample of 100 000 potential donors and receive 4781 donations. A staff member worries that the contribution rate will be lower than their past average rate of 5% if they run this campaign as currently designed.
a) What are the hypotheses?
b) Are the assumptions and conditions for inference met?
c) Do you think the rate would drop? Explain.

22. Take the offer, part II In Exercise 20 of Chapter 16, you learned that a major credit card company is planning a new offer for their current cardholders. They will give double airline miles on purchases for the next six months if the cardholder goes online and registers for this offer. To test the effectiveness of this campaign, the company recently sent out offers to a random sample of 50 000 cardholders. Of those, 1184 registered. A staff member suspects that the success rate for the full campaign will be comparable to the standard 2% rate that they are used to seeing in similar campaigns. What do you predict?
a) What are the hypotheses?
b) Are the assumptions and conditions for inference met?
c) Do you think the rate would change if they use this fundraising campaign? Explain.

23. Law school According to the Ontario Universities' Application Centre, in the year 2012, 29.6% of law school applicants were accepted to an Ontario school.[10] A training program claims that 83 of the 240 students trained in 2012 were admitted to law school. You can safely consider these trainees to be representative of the population of law school applicants. Has this program demonstrated a real improvement over the provincial average?
a) What are the hypotheses?
b) Check the conditions and find the P-value.
c) Would you recommend this program based on what you see here? Explain.

24. Rural med school applicants In 2002, 2246 students from Ontario applied to the province's medical schools; in 2003, 2702 applied. Of the two-year total (4948), there were 4588 applicants from urban areas and 360 from rural areas.[11] GPAs and MCAT scores were similar for the urban and rural applicants, who had similar acceptance rates (about 20% of applicants). Since Ontario's population is 13% rural, is there reason for concern about a lack of rural applicants for medical school? Are rural students underrepresented among the applicants, or are we just seeing some natural random variation in the results for this two-year period?
a) What are the hypotheses?
b) Verify that the conditions are satisfied for a test, and find the P-value.
c) Should we be concerned about a lack of rural applicants to medical school?

25. Pollution A company with a fleet of 150 cars found that the emissions systems of seven out of the 22 they tested failed to meet pollution control guidelines. Is this strong evidence that more than 20% of the fleet might be out of compliance? Test an appropriate hypothesis and state your conclusion. Be sure the appropriate assumptions and conditions are satisfied before you proceed.

26. Scratch and dent An appliance manufacturer stockpiles washers and dryers in a large warehouse for shipment to retail stores. Sometimes, in handling, the appliances get damaged. Even though the damage may be minor, the company must sell those machines at drastically reduced prices. The company goal is to keep the level of damaged machines below 2%. One day an inspector randomly checks 60 washers and finds that five of them have scratches or dents. Is this strong evidence that the warehouse is failing to meet the company goal? Test an appropriate hypothesis and state your conclusion. Be sure the appropriate assumptions and conditions are satisfied before you proceed.

27. Twins In 2001, a U.S. national vital statistics report indicated that about 3% of all births produced twins. Is the rate of twin births the same among very young mothers? Data from a large city hospital found that only seven sets of twins were born to 469 teenage girls. Test an appropriate hypothesis and state your conclusion. Be sure the appropriate assumptions and conditions are satisfied before you proceed.

28. Hockey 2013 During the 2013 National Hockey League playoffs, the home team won 59 of the 86 games. Is this strong evidence of a home-ice advantage in professional hockey? Test an appropriate hypothesis and state your conclusion. Be sure the appropriate assumptions and conditions are satisfied before you proceed.

29. WebZine A magazine is considering the launch of an online edition. It plans to go ahead only if it's convinced that more than 25% of current readers would subscribe. The magazine contacted a simple random sample of 500 current subscribers, and 137 of those surveyed expressed interest. What should the company do? Test an appropriate hypothesis and state your conclusion. Be sure the appropriate assumptions and conditions are satisfied before you proceed.

---

[10] www.ouac.on.ca/statistics/law-school-application-statistics/
[11] www.pubmedcentral.nih.gov/articlerender.fcgi?artid=1479467

30. Seeds Garden centre staff want to store leftover packets of vegetable seeds for sale the following spring, but are concerned that the seeds may not germinate at the same rate a year later. The manager finds a packet of last year's green bean seeds and plants them as a test. Although the packet claims a germination rate of 92%, only 171 of 200 test seeds sprout. Is this evidence that the seeds have lost viability during a year in storage? Test an appropriate hypothesis and state your conclusion. Be sure the appropriate assumptions and conditions are satisfied before you proceed.

31. Female executives A company is criticized because only 13 of 43 people in executive-level positions are women. The company explains that although this proportion is lower than it might wish, it's not surprising given that only 40% of all its employees are women. What do you think? Test an appropriate hypothesis and state your conclusion. Be sure the appropriate assumptions and conditions are satisfied before you proceed.

32. Jury Census data for a certain municipality show that 19% of the adult residents are South Asian. Suppose 72 people are called for jury duty and only nine of them are South Asian. Does this apparent underrepresentation of South Asians call into question the fairness of the jury selection system? Explain.

33. Hockey injuries and relative age Is there a relationship between (relative) age and injury prevalence in Canadian youth hockey? Below are summary data from the Canadian Hospitals Injury Reporting and Prevention Program (for boys 10–15 years of age, 1995–2002). Quartile 1 included those with January, February, and March birth dates and quartile 4 included those with October, November, and December birth dates; hence those born in quartile 1 were relatively older than their peers born in quartile 4. The other two quartiles do not affect any of the conclusions that may be drawn from this study, and hence are omitted to simplify things.

Number Injured by Age Category

	Atom (10–11y)	Peewee (12–13y)	Bantam (14–15y)
1st Quartile	221	506	646
4th Quartile	156	340	433

a) Test, at the Atom level, whether an injured youth is equally likely to be from either quartile. If not, then there may be a "relative age" effect on likelihood of injury. "Equally likely" means that half of injured players in the two quartiles (377 in total) should be found in quartile 4, on average.

b) Repeat part a) for the Bantam level injured players.

c) The question of interest here is whether relative age influences chance of injury, but researchers have also found that there is some selection bias for relatively older players. How does this affect conclusions drawn from these data about the question of interest? What additional data would you try to obtain to more accurately assess the increased risk for older players?

34. Acid rain A study of the effects of acid rain on trees in the Hopkins Forest shows that 25 of 100 trees sampled exhibit some sort of damage from acid rain. This rate seemed to be higher than the 15% quoted in a recent *Environmetrics* article on the average proportion of damaged trees in the Northeast U.S. Does the sample suggest that trees in the Hopkins Forest are more susceptible than trees from the rest of the region? Comment, and write up your own conclusions based on an appropriate confidence interval as well as a hypothesis test. Include any assumptions you made about the data.

35. Lost luggage An airline's public relations department says that the airline rarely loses passengers' luggage. It further claims that on those occasions when luggage is lost, 90% is recovered and delivered to its owner within 24 hours. A consumer group that surveyed a large number of air travellers found that only 103 of 122 people who lost luggage on that airline were reunited with the missing items by the next day. Does this cast doubt on the airline's claim? Explain.

36. TV ads A start-up company is about to market a new computer printer. It decides to gamble by running commercials during the Super Bowl. The company hopes that name recognition will be worth the high cost of the ads. The goal of the company is that over 40% of the public recognize its brand name and associate it with computer equipment. The day after the game, a pollster contacts 420 randomly chosen adults and finds that 181 of them know that this company manufactures printers. Would you recommend that the company continue to advertise during Super Bowls? Explain.

37. John Wayne Like many others, the famous actor John Wayne died of cancer. But is there more to this story? In 1955, Wayne was in Utah shooting the film *The Conqueror*. Across the state line, in Nevada, the United States military was testing atomic bombs. Radioactive fallout from those tests drifted across the filming location. A total of 46 of the 220 people working on the film eventually died of cancer. Cancer experts estimate that one would expect only about 30 cancer deaths in a group this size.

a) Is the death rate observed in the movie crew unusually high?

b) Does this prove that exposure to radiation increases the risk of cancer?

38. TOEFL scores Of those undergraduate-level students who took the 2008 paper-based TOEFL exam, 47% earned scores of 540 or higher. One teacher wondered if the performance of her students was different. Thirty of her 54 students achieved scores of 540 or better. Does she have cause to brag? Explain.

39. Obama or Harper? A poll by the Strategic Counsel, conducted June 12–22, 2008 (before Obama was elected president and while Stephen Harper was prime minister) surveyed random samples of 1000 Canadians and 1000 Americans.[12] When asked about which political figure they admired most, the top two responses from Canadians were:

Barack Obama: 26%

Stephen Harper: 21%

Would it be fair to say that more Canadians would choose Obama as their most admired political figure than would choose Harper (at the time of this survey)? Do a (two-sided) test, after throwing out the 53% (round to 530) who chose someone else.

40. Your health care system or mine? In the same poll mentioned in Exercise 39, it was reported that 45% of Americans felt Canada had the superior healthcare system, while 42% thought the United States should stick with its own. Would it be fair to say that more Americans preferred the Canadian healthcare system over their own (at the time of this survey)? Do a (two-sided) test, after throwing out the 13% (round to 130) who were undecided. (Considering the misinformation in some U.S. media about Canadian health care, the 45% figure is surprisingly high!)

***41. Body temperature** A researcher measured the body temperature of a random sample of 64 adults and found the mean and standard deviation to be 36.83°C and 0.38°C, respectively. A temperature of 37.0°C is commonly assumed to be "normal." Do these data suggest otherwise?

a) Provide an answer using a test of hypothesis.
b) Provide an answer using a confidence interval.

***42. Hot dogs** In Exercise 50 of Chapter 16, laboratory tests of "sodium reduced" hot dogs produced a mean sodium content of 318 mg with a standard deviation of 32 mg for a sample of 60 hot dogs of this type. To qualify for "reduced sodium" labelling, regulation require 30% less sodium content than for regular hot dogs. Here, this works out to be 325 mg—in other words, if the sodium content per hot dogs of a certain brand averages to less than 325 mg, they qualify. Based on the data, should this particular brand be labelled as "reduced sodium"?

a) Provide an answer using a proper test of hypothesis. Is there good evidence that sodium is adequately reduced to qualify this brand as "reduced sodium"?
b) Provide an answer using the confidence interval you calculated in Exercise 50 of Chapter 16.

***43. Speed of light** In Exercise 49 of Chapter 16, we looked at Michelson's 100 measurements of the speed of light. The mean and standard deviation he obtained were 299 852.4 km/sec and 79.0 km/sec, respectively. Use his data to test the null hypothesis that the true speed of light is equal to 299 710.5 km/sec. What does this indicate about Michelson's experiment, given current knowledge that the true constant speed of light is 299 710.5 km/sec?

***44. Wind power** Should you generate electricity with your own personal wind turbine? That depends on whether you have enough wind on your site. To produce enough energy, your site should have an annual average wind speed above 8 miles per hour (13 kilometres per hour), according to the Wind Energy Association. One candidate site was monitored for a year, with wind speeds recorded every six hours. A total of 1114 readings of wind speed averaged 8.019 mph with a standard deviation of 3.813 mph. You've been asked to make a statistical report to help the landowner decide whether to place a wind turbine at this site.

a) Below are some plots of the data. Can we analyze the data using inferential methods discussed in this chapter (optional section)? Explain.

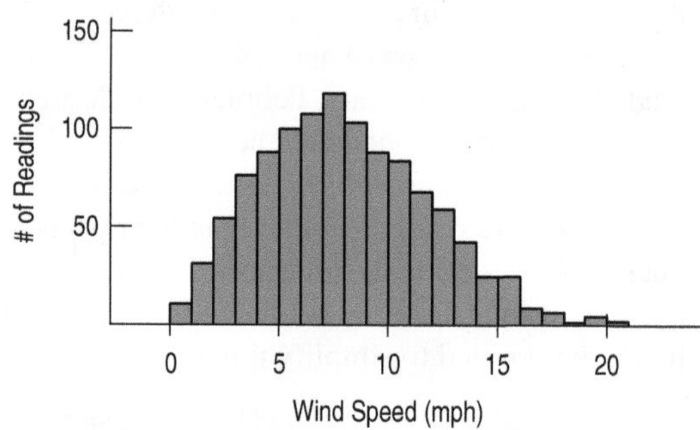

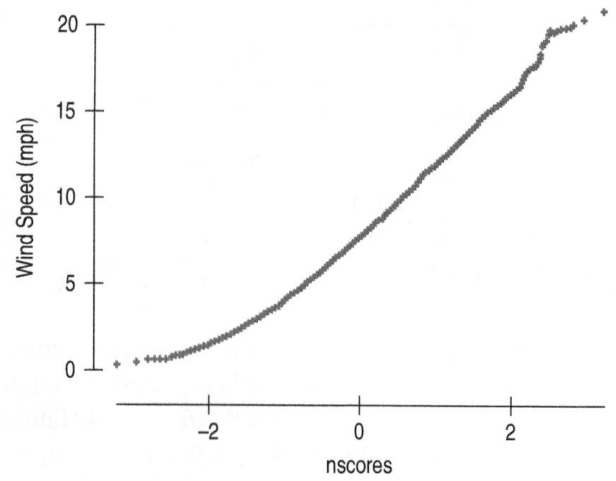

[12]www.ctv.ca/servlet/ArticleNews/story/CTVNews/20080629/poll_us_canada_080629/20080629?hub=Politics

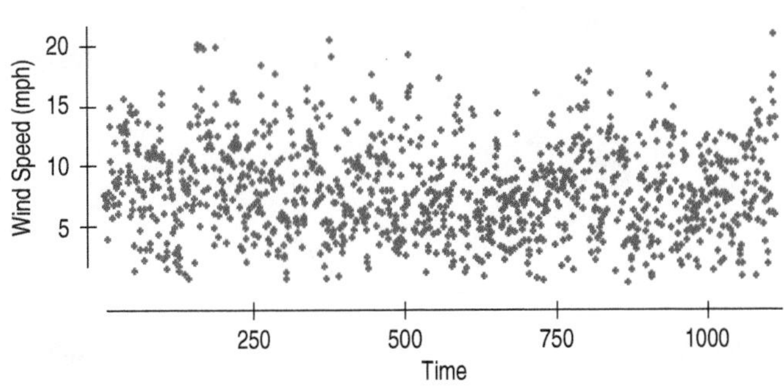

b) Is there strong evidence that the average wind speed is in excess of 8 mph? Do a proper test.

Just Checking ANSWERS

1. You can't conclude that the null hypothesis is true. You can conclude only that the experiment was unable to disprove the null hypothesis. They were unable, on the basis of 12 patients, to show that aspirin was effective.
2. The null hypothesis is H_0: $p = 0.75$.
3. A P-value of 0.0001 strongly contradicts the assumed null hypothesis, that is, there is strong evidence that the improved version of the drug gives relief to a higher proportion of patients.
4. The parameter of interest is the proportion, p, of all delinquent customers who will pay their bills.

$$H_0: p = 0.30 \text{ and } H_A: p > 0.30.$$

5. The very low P-value leads us to reject the null hypothesis. There is strong evidence that the DVD is more effective in getting people to start paying their debts than just sending a letter had been.
6. All we know is that there is strong evidence to suggest that $p > 0.30$. We don't know how much higher than 30% the new proportion is. We'd like to see a confidence interval to see if the new method is worth the cost.

chapter

18 More About Tests

... They make things admirably plain, But one hard question will remain: If one hypothesis you lose, Another in its place you choose ...

—*James Russell Lowell, Credidimus Jovem Regnare*

Where are we going?

A news headline reports a "statistically significant" increase in global temperatures. Another says that studies have found no "statistically significant" benefits of taking vitamin C to prevent colds. What does *significance* really mean? Knowing how the researchers arrived at their conclusions can help you decide whether you agree with their findings. We'll look at hypothesis testing in more depth in this chapter.

Who	Florida motorcycle riders aged 20 and younger involved in motorcycle accidents
What	Percentage wearing helmets
When	2001–2003
Where	Florida
Why	Assessment of injury rates commissioned by the National Highway Traffic Safety Administration (NHTSA)

In 2000, Florida changed its motorcycle helmet law. No longer are riders 21 and older required to wear helmets. Under the new law, those under 21 still must wear helmets, but a report by the Preusser Group (www.preussergroup.com) suggests that helmet use may have declined in this group too.

It isn't practical to survey young motorcycle riders. (For example, how could you construct a sampling frame? If you contacted licensed riders, would they admit to riding illegally without a helmet?) The researchers adopted a different strategy. Police reports of motorcycle accidents record whether the rider wore a helmet and give the rider's age. Before the change in the helmet law, 60% of youths involved in motorcycle accidents had been wearing their helmets. The Preusser study looked at accident reports during 2001–2003, the three years following the law change. They observed 781 young riders who were involved in accidents. Of these, 396 (or 50.7%) were wearing helmets. Is this evidence of a decline in helmet-wearing, or just the natural fluctuation of such statistics?

18.1 Choosing the Hypotheses

How do we choose the null hypothesis? The appropriate null arises directly from the context of the problem. It is dictated not by the data, but by the situation. One good way to identify both the null and alternative hypotheses is to think about why the study is being done and what we hope to learn from the test. In this case we are interested in seeing whether the proportion using a helmet has decreased, but our null hypothesis is what we want to disprove, namely that the proportion is unchanged at 60%. The alternative is what we hope to prove—that the proportion has decreased.

A pharmaceutical company wanting to develop and market a new drug needs to show that the new drug is effective. A typical null hypothesis might state that the proportion of patients who recover after receiving a new drug is the same as the proportion who recover among those who receive a placebo. The alternative hypotheses would be that the new drug cures a higher proportion of patients. Similarly, if we wanted to test athletes' strength training with new equipment, we would be interested in whether the new equipment increases their strength. In that case, the null hypothesis would be that their mean strength hasn't changed. The alternative would be that the mean has increased.

To write a null hypothesis, identify a parameter and choose a null value that relates to the question at hand. Even though the null usually means no difference or no change, you can't automatically interpret "null" to mean zero. A claim that "nobody" wears a motorcycle helmet would be absurd. The null hypothesis for the Florida study is that the true rate of helmet use remained unchanged at $p = 0.60$ among young riders after the law changed. You need to find the value for the parameter in the null hypothesis from the context of the problem.

There is a temptation to state your *claim* as the null hypothesis. As we have seen, however, you cannot prove a null hypothesis true any more than you can prove a defendant innocent. So, it makes more sense to use what you want to show as the *alternative*. This way, if the data support your claim, the P-value will be small. When you then reject the null, you will be left with what you want to show.

For Example WRITING HYPOTHESES

The diabetes drug Avandia® was approved to treat Type 2 diabetes in 1999. But in 2007 an article in the *New England Journal of Medicine* (NEJM)[1] raised concerns that the drug might carry an increased risk of heart attack. This study combined results from a number of other separate studies to obtain an overall sample of 4485 diabetes patients taking Avandia®. People with Type 2 diabetes are known to have about a 20.2% chance of suffering a heart attack within a seven-year period. According to the article's author, Dr. Steven E. Nissen,[2] the risk found in the NEJM study was equivalent to a 28.9% chance of heart attack over seven years. The FDA is the U.S. government agency responsible for relabelling Avandia® to warn of the risk if it is judged to be unsafe. Although the statistical methods they use are more sophisticated, we can get an idea of their reasoning with the tools we have learned.

QUESTION: What null hypothesis and alternative hypothesis about seven-year heart attack risk would you test? Explain.

$$H_0: p = 0.202$$
$$H_A: p > 0.202$$

ANSWER: The parameter of interest is the proportion of diabetes patients suffering a heart attack in seven years. The FDA is concerned only with whether Avandia® *increases* the seven-year risk of heart attacks above the baseline value of 20.2%, so a one-sided upper-tail test is appropriate.

One-sided or two? In the 1930s, a series of experiments was performed at Duke University in an attempt to see whether humans were capable of extrasensory perception, or ESP. Psychologist Karl Zener designed a set of cards with five symbols, later made infamous in the movie *Ghostbusters*:

In the experiment, the "sender" selects one of the five cards at random from a deck and then concentrates on it. The "receiver" tries to determine which card it is. If we let p be the proportion of correct responses, what's the null hypothesis? The null hypothesis is that ESP makes no difference. Without ESP, the receiver would just be guessing, and since there are five possible responses, there would be a 20% chance of guessing each card correctly. So, H_0 is $p = 0.20$. What's the alternative? It seems that it should be $p > 0.20$, a one-sided alternative. But some ESP researchers have expressed the claim that if the proportion guessed were much lower than expected, that would show an "interference" and should be considered evidence for ESP as well. So they argue for a two-sided alternative.

---

[1]Steven E. Nissen and Kathy Wolski, 2007, "Effect of rosiglitazone on the risk of myocardial infarction and death from cardiovascular causes," *NEJM*: 356.

[2]Interview reported in the *New York Times* (May 26, 2007).

18.2 How to Think About P-Values

WHICH CONDITIONAL?

Suppose that as a political science major you are offered the chance to be an intern at the Office of the Prime Minister. There would be a very high probability that next summer you'd be in Ottawa. That is, *P*(Ottawa | Intern) would be high. But if we find a student in Ottawa, is it likely that he's an intern at the office of the prime minister? Almost surely not; *P*(Intern | Ottawa) is low. You can't switch around conditional probabilities. The P-value is *P*(data | H_0). We might wish we could report *P*(H_0 | data), but these two quantities are NOT the same.

A P-value is actually a conditional probability. It tells us the probability of getting results at least as unusual as the observed statistic, *given* that the null hypothesis is true. We can write P-value = *P*(observed statistic value [or even more extreme] | H_0).

Writing the P-value this way helps to make clear that the P-value is *not* the probability that the null hypothesis is true. It is a probability about the data. Let's say that again:

The P-value is not the probability that the null hypothesis is true.

The P-value is not even the conditional probability that the null hypothesis is true given the data. We would write that probability as *P*(H_0 | observed statistic value). This is the conditional probability for the P-value, in reverse. It would be nice to know this, but it's impossible to calculate without making additional assumptions. As we saw in Chapter 13, reversing the order in a conditional probability is difficult, and the results can be counterintuitive.

We can find the P-value, *P*(observed statistic value | H_0), because H_0 gives the parameter values that we need to find the required probability. But there's no direct way to find *P*(H_0 | observed statistic value).[3] As tempting as it may be to say that a P-value of 0.03 means there's a 3% chance that the null hypothesis is true, that just isn't right. All we can say is that, given the null hypothesis, there's a 3% chance of observing the statistic value that we have actually observed (or one even more unlike the null value).

What to Do with a Small P-Value

"The wise man proportions his belief to the evidence."

—*David Hume, Enquiry Concerning Human Understanding, 1748*

We know that a small P-value means that the result we just observed is unlikely to occur if the null hypothesis is true. So we have evidence against the null hypothesis. An even smaller P-value implies stronger evidence against the null hypothesis, but it doesn't mean that the null hypothesis is "less true" (see "How Guilty Is the Suspect" on page 506).

How small the P-value has to be for you to reject the null hypothesis depends on a lot of things, not all of which can be precisely quantified. Your belief in the null hypothesis will influence your decision. Your trust in the data, in the experimental method if the data come from a planned experiment, or in the survey protocol if the data come from a designed survey all influence your decision. The P-value should serve as a measure of the strength of the evidence against the null hypothesis, but should never serve as a hard and fast rule for decisions. You have to take that responsibility on yourself.

As a review, let's look at the helmet law example from the chapter opener. Did helmet wearing among young riders decrease after the law allowed older riders to ride without helmets? What is the evidence?

Step-by-Step Example ANOTHER ONE-PROPORTION *Z*-TEST

Question: Has helmet use in Florida declined among riders under the age of 21 subsequent to the change in the helmet laws?

THINK ➡ Plan State the problem and discuss the variables and the *W*'s.	I want to know whether the rate of helmet wearing among Florida's motorcycle riders under the age of 21 decreased after the law changed to allow older riders to go without helmets. I have data from accident records showing 396 of 781 young riders were wearing helmets. I am only looking at riders

[3]The approach to statistical inference known as Bayesian Statistics addresses the question in just this way, but it requires more advanced mathematics and more assumptions. See p. 367 for more about the founding father of this approach.

	involved in accidents, so it is unclear to what extent my conclusions apply to the entire population of young riders.
Hypotheses The null hypothesis is established by the rate set before the change in the law. The study was concerned with safety, so they'll want to know of any decline in helmet use. Hence a one-sided alternative is appropriate.	The percentage before the law was passed was 60% so I'll use that as my null hypothesis value. The alternative is one-sided because I'm only interested in seeing if the rate decreased (to something less than 60%). $H_0: p = 0.60$ $H_A: p < 0.60$
SHOW ➡ **Model** Check the conditions.	✓ **Independence Assumption:** The data are for riders involved in accidents during a three-year period. Individuals are independent of one another. ✓ **Randomization Condition:** No randomization was applied, but we may consider these riders to be a representative sample of future riders (well, those who might be in an accident anyway). ✓ **Success/Failure Condition:** We'd expect $np = 781(0.6) = 468.6$ helmeted riders and $nq = 781(0.4) = 312.4$ non-helmeted riders. Both are at least 10.
Specify the sampling distribution model and name the test.	The conditions are reasonably satisfied, so I will use the Normal model and perform **a one-proportion z-test.**
SHOW ➡ **Mechanics** Find the standard deviation of the sampling model using the hypothesized proportion.	There were 396 helmet wearers among the 781 accident victims. $\hat{p} = \frac{396}{781} = 0.507$
Find the *z*-score for the observed proportion.	$SD(\hat{p}) = \sqrt{\frac{p_0 q_0}{n}} = \sqrt{\frac{(0.60)(0.40)}{781}} = 0.0175$
Make a picture. Sketch a Normal model centred at the hypothesized helmet rate of 60%. This is a lower-tail test, so shade the region to the left of the observed rate.	$z = \frac{\hat{p} - p_0}{SD(\hat{p})} = \frac{0.507 - 0.60}{0.0175} = -5.31$
Given this *z*-score, the P-value is obviously very low.	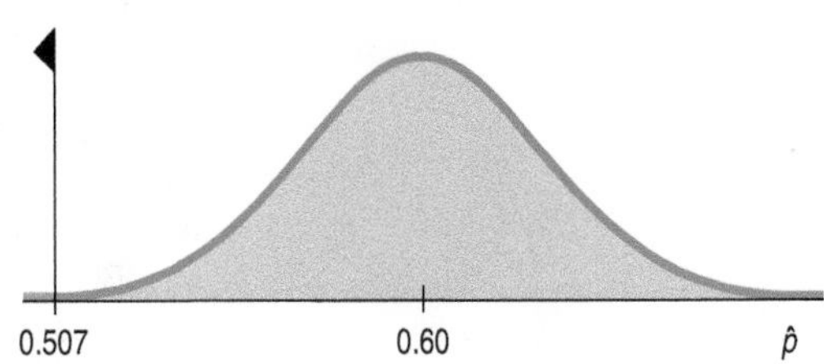 The observed helmet rate is 5.31 standard deviations below the former rate. The corresponding P-value is less than 0.001.

TELL ➡ Conclusion Link the P-value to your decision about the null hypothesis, and then state your conclusion in context.	The P-value says that if the true rate of helmet-wearing among riders under 21 were still 60%, the probability of observing a rate no higher than 50.7% in a sample like this is less than 1 chance in 1000, so I reject the null hypothesis. There is strong evidence that there has been a decline in helmet use among riders under 21.

The P-value in the helmet example is quite small—less than 0.001. That's strong evidence to suggest that the rate has decreased since the law was changed. But it doesn't say that it was "a lot lower." To answer that question, you'd need to construct a confidence interval:

$$\hat{p} \pm z^*\sqrt{\frac{\hat{p}\hat{q}}{n}} = 0.507 \pm 1.96(0.0179) = (0.472, 0.542)$$

(using 95% confidence).

There is strong evidence that the rate is no longer 60%, but the small P-value by itself says nothing about how much lower the rate might be. The confidence interval provides that information; the rate seems to be closer to 50% now. Whether a change from 60% to 50% makes an important difference in safety is a judgment that depends on the situation, but not on the P-value. Not coincidentally, on July 1, 2008, Florida required a motorcycle "endorsement" for all motorcycle riders. For riders under 21, that requires a motorcycle safety course. Although only about 70% of motorcycle riders are endorsed, the percentage of unendorsed riders involved in crashes dropped considerably after 2008.[4]

HOW GUILTY IS THE SUSPECT?

We might like to know $P(H_0 \mid \text{data})$, but when you think about it, we can't talk about the probability that the null hypothesis is true. The null is not a random event, so either it is true or it isn't. The data, however, are random in the sense that if we were to repeat a randomized experiment or draw another random sample, we'd get different data and expect to find a different statistic value. So we can talk about the probability of the data given the null hypothesis, and that's the P-value.

But it does make sense that the smaller the P-value, the more confident we can be in declaring that we doubt the null hypothesis. Think again about the jury trial. Our null hypothesis is that the defendant is innocent. Then the evidence starts rolling in. A car the same colour as his was parked in front of the bank. Well, there are lots of cars that colour. The probability of that happening (given his innocence) is pretty high, so we're not persuaded that he's guilty. The bank's security camera showed the robber was male and about the defendant's height and weight. Hmm. Could that be a coincidence? If he's innocent, then it's a little less likely that the car and description would both match, so our P-value goes down. We're starting to question his innocence a little. Witnesses said the robber wore a blue jacket just like the one the police found in a garbage can behind the defendant's house. Well, if he's innocent, then that doesn't seem very likely, does it? If he's really innocent, the probability that all of these could have happened is getting pretty low. Now our P-value may be small enough to be called "beyond a reasonable doubt" and lead to a conviction. Each new piece of evidence strains our skepticism a bit more. The more compelling the evidence—the more *unlikely* it would be were he innocent—the more convinced we become that he's guilty.

But even though it may make us more confident in declaring him guilty, additional evidence does not make him any guiltier. Either he robbed the bank or he didn't. Additional evidence (like the teller picking him out of a police lineup) just makes us more confident that we did the right thing when we convicted him. The lower the P-value, the more comfortable we feel about our decision to reject the null hypothesis, but the null hypothesis doesn't get any more false.

[4]www.ridesmartflorida.com

For Example THINKING ABOUT THE P-VALUE

RECAP: A *New England Journal of Medicine* paper reported that the seven-year risk of heart attack in diabetes patients taking the drug Avandia® was increased from the baseline of 20.2% to an estimated risk of 28.9% and said the P-value was 0.03.

QUESTION: How should the P-value be interpreted in this context?

ANSWER: The P-value $= P(\hat{p} \geq 28.9\% \mid p = 20.2\%)$. That is, it's the probability of seeing such a high heart attack rate among the people studied if, in fact, taking Avandia® really didn't increase the risk at all.

What to Do with a High P-Value

A S *Video:* **Is There Evidence for Therapeutic Touch?** This video shows the experiment and tells the story.

A S *Activity:* **Testing Therapeutic Touch.** Perform the one-proportion *z*-test using *MyStatLab* technology. The test in *MyStatLab* is two-sided. Do you think this is the appropriate choice?

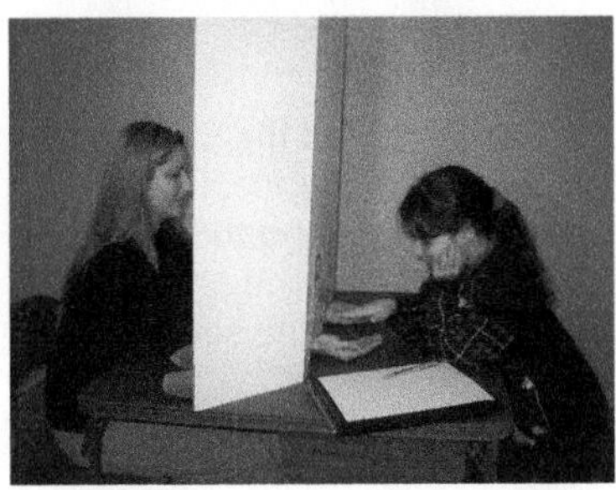
Paul Velleman

Therapeutic touch (TT), taught in some schools of nursing, is a therapy in which the practitioner moves her hands near, but does not touch, a patient in an attempt to manipulate a "human energy field." Therapeutic touch practitioners believe that by adjusting this field they can promote healing. However, no instrument has ever detected a human energy field, and no experiment has ever shown that TT practitioners can detect such a field.

In 1998, the *Journal of the American Medical Association* published a paper reporting work by a then nine-year-old girl.[5] She had performed a simple experiment in which she challenged 15 TT practitioners to detect whether her unseen hand was hovering over their left or right hand (selected by the flip of a coin).

The practitioners "warmed up" with a period during which they could see the experimenter's hand, and each said that they could detect the girl's human energy field. Then a screen was placed so that the practitioners could not see the girl's hand, and they attempted 10 trials each. Overall, of 150 trials, the TT practitioners were successful only 70 times—a success proportion of 46.7%.

The null hypothesis here is that the TT practitioners were just guessing. If that were the case, since the hand was chosen using a coin flip, the practitioners would guess correctly 50% of the time. So the null hypothesis is that $p = 0.5$ and the alternative that they could actually detect a human energy field is (one-sided) $p > 0.5$.

What would constitute evidence that they weren't guessing? Certainly, a very high proportion of correct guesses out of 150 would convince most people. Exactly how high the proportion of correct guesses has to be for you to reject the null hypothesis depends on how small a P-value you need to be convinced (which, in turn, depends on how often you're willing to make mistakes—a topic we'll discuss later in the chapter).

But let's look again at the TT practitioners' proportion. Does it provide any evidence that they weren't guessing? The proportion of correct guesses is 46.7%—that's *less* than the hypothesized value, not greater! When we find $SD(\hat{p}) = 0.041$ (or 4.1%), we can see that 46.7% is almost 1 SD *below* the hypothesized proportion:

$$SD(\hat{p}) = \sqrt{\frac{p_0 q_0}{n}} = \sqrt{\frac{(0.5)(0.5)}{150}} \approx 0.041$$

The observed proportion, $\hat{p}$ is 0.467.

$$z = \frac{\hat{p} - p_0}{SD(\hat{p})} = \frac{0.467 - 0.5}{0.041} = -0.805$$

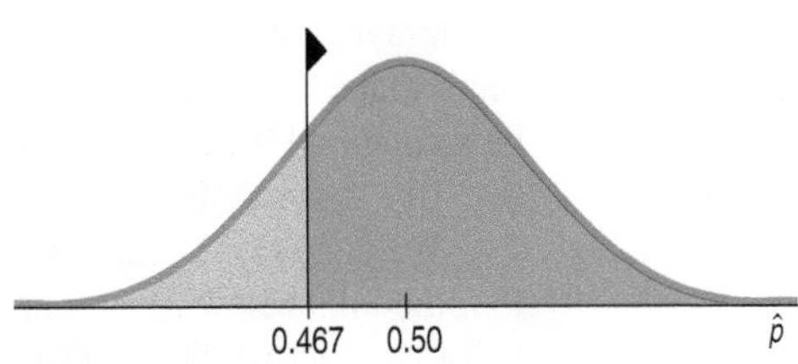

[5]L. Rosa, E. Rosa, L. Sarner, and S. Barrett, "A Close Look at Therapeutic Touch," *JAMA* 279(13) [1 April 1998]: 1005–1010.

The observed success rate is 0.805 standard deviations below the hypothesized mean.

$$P = P(z > -0.805) = 0.790$$

If the practitioners had been highly successful, we would have seen a low P-value. In that case, we would then have concluded that they could actually detect a human energy field.

But that's not what happened. What we observed was a $\hat{p} = 0.467$ success rate. The P-value for this proportion is greater than 0.5 because the observed value is on the "wrong" side of the null hypothesis value. Wrong, that is, if you wanted to find evidence against the null hypothesis. To convince us, the practitioners should be doing better than guessing, not worse!

Obviously, we won't be rejecting the null hypothesis; for us to reject it, the P-value would have to be quite small. But a P-value of 0.790 seems so big it is almost strange. With a success rate even lower than chance, we could have concluded right away that we have no evidence for rejecting H_0. In fact, for a one-sided test, any time our test statistic is on the "wrong" side of the hypothesis, we know immediately that the P-value is at least 0.50 and we know that we have no evidence to reject the null hypothesis.

Big P-values just mean that what we've observed isn't surprising. That is, the results are in line with our assumption that the null hypothesis models the world, so we have no reason to reject it. A big P-value doesn't prove that the null hypothesis is true, but it certainly offers no evidence that it's *not* true. When we see a large P-value, all we can say is that we "don't reject the null hypothesis."

For Example MORE ABOUT P-VALUES

RECAP: The question of whether the diabetes drug Avandia® increased the risk of heart attack was raised by a study in the *New England Journal of Medicine.* This study estimated the seven-year risk of heart attack to be 28.9% and reported a P-value of 0.03 for a test of whether this risk was higher than the baseline seven-year risk of 20.2%. An earlier study (the ADOPT study) had estimated the seven-year risk to be 26.9% and reported a P-value of 0.27.

QUESTION: Why did the researchers in the ADOPT study not express alarm about the increased risk they had seen?

ANSWER: A P-value of 0.27 means that a heart attack rate at least as high as the one they observed could be expected in 27% of similar experiments even if, in fact, there were no increased risk from taking Avandia®. That's not remarkable enough to reject the null hypothesis. In other words, the ADOPT study wasn't convincing.

18.3 Alpha Levels and Significance

A S *Activity:* **Rejecting the Null Hypothesis.** See alpha levels at work in the animated hypothesis-testing tool.

Sometimes we need to make a firm decision about whether to reject the null hypothesis. A jury must *decide* whether the evidence reaches the level of "beyond a reasonable doubt." A business must *select* a Web design. You need to decide which section of Statistics to enrol in.

When the P-value is small, it tells us that our data are rare, *given the null hypothesis.* As humans, we are suspicious of rare events. If the data are "rare enough," we just don't think that could have happened due to chance. Since the data *did* happen, something must be wrong. All we can do now is to reject the null hypothesis.

But how rare is "rare"?

We can define "rare event" arbitrarily by setting a threshold for our P-value. If our P-value falls below that point, we'll reject the null hypothesis. We call such results **statistically significant**. The threshold is called an **alpha level**. Not surprisingly, it's labelled with the Greek letter α. Common α levels are 0.10, 0.05, and 0.01. You have the option—almost the *obligation*—to consider your alpha level carefully and choose an appropriate one for the situation. If you're assessing the safety of air bags, you'll want a low alpha level; even 0.01 might not be low enough. If you're just wondering whether folks prefer their pizza with or without pepperoni, you might be happy with $\alpha = 0.10$. It can be hard to justify your choice of α, though, so often we arbitrarily choose 0.05. Note, however, that you must select the alpha level *before* you look at the data. Otherwise, you can be accused of cheating by tuning your alpha level to suit the data.

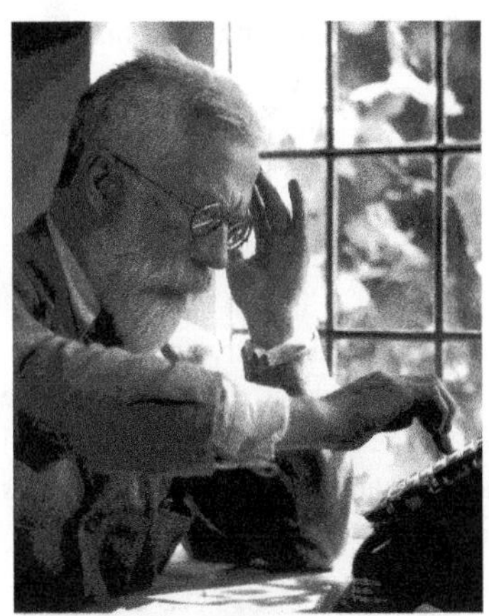

A. Barrington Brown/Science Source

Sir Ronald Fisher (1890–1962) was one of the founders of modern Statistics.

Where did the value 0.05 come from? In 1931, in a famous book called *The Design of Experiments*, Sir Ronald Fisher discussed the amount of evidence needed to reject a null hypothesis. He said that it was *situation dependent*, but remarked, somewhat casually, that for many scientific applications, 1 out of 20 *might be* a reasonable value. Since then, some people—indeed some entire disciplines—have treated the number 0.05 as sacrosanct.

NOTATION ALERT

The first Greek letter, α, is used in statistics for the threshold value of a hypothesis test. You'll hear it referred to as the alpha level. Common values are 0.10, 0.05, 0.01, and 0.001.

The alpha level of the test is also called its **significance level**. When we reject the null hypothesis, we say that the test is "significant at that level." For example, we might say that we reject the null hypothesis "at the 5% level of significance."

What can you say if the P-value does not fall below α?

When you have not found sufficient evidence to reject the null according to the standard you have established, you should say that "the data have failed to provide sufficient evidence to reject the null hypothesis." Don't say that you "accept the null hypothesis." You certainly haven't proven or established it; it was assumed to begin with. Say that you've failed to reject it.

Think again about the therapeutic touch example. The P-value was 0.788. This is so much larger than any reasonable alpha level that we can't reject H_0. For this test, we'd conclude, "We fail to reject the null hypothesis. There is insufficient evidence to conclude that the practitioners are performing better than they would if they were just guessing."

The automatic nature of the reject/fail-to-reject decision when we use an alpha level may make you uncomfortable. If your P-value falls just slightly above your alpha level, you're not allowed to reject the null. Yet a P-value just barely below the alpha level leads to rejection. If this bothers you, you're in good company. Many statisticians think it better to report the P-value than to base a decision on an arbitrary alpha level. When you do decide to declare a verdict, it's always a good idea to report the P-value as an indication of the strength of the evidence.

It could happen to you! Of course, if the null hypothesis *is* true, no matter what alpha level you choose, you still have a probability α of rejecting a true null hypothesis by mistake. This is the rare event we want to protect ourselves against. When we do reject the null hypothesis, no one ever thinks that *this* is one of those rare times. As statistician Stu Hunter notes, "*The statistician says 'rare events do happen—but not to me!'*"

IT'S IN THE STARS

Some disciplines carry the idea further and code P-values by their size. In this scheme, a P-value between 0.05 and 0.01 gets highlighted by *. A P-value between 0.01 and 0.001 gets **, and a P-value less than 0.001 gets ***. This can be a convenient summary of the weight of evidence against the null hypothesis if it's not taken too literally. But we warn you against making a black-and-white decision near the boundaries. The boundaries are a matter of tradition, not science; there is nothing special about 0.05. A P-value of 0.051 should be looked at very seriously and not casually thrown away just because it's larger than 0.05, and one that's 0.009 is not very different from one that's 0.011.

Sometimes it's best to report that the conclusion is not yet clear and to suggest that more data be gathered. (In a trial, a jury may "hang" and be unable to return a verdict.) In these cases, the P-value is the best summary we have of what the data say or fail to say about the null hypothesis.

Practical vs. Statistical Significance

What do we mean when we say that a test is statistically significant? All we mean is that the test statistic had a P-value lower than our alpha level. Don't be lulled into thinking that statistical significance carries with it any sense of practical importance or impact.

For large samples, even small, unimportant ("insignificant") deviations from the null hypothesis can be statistically significant. On the other hand, if the sample is not large

Statistically significant? Yes, but is it important? A large insurance company mined its data and found a statistically significant ($P = 0.04$) difference between the mean value of policies sold in 2001 and 2002. The difference in the mean values was \$9.83. Even though it was statistically significant, management did not see this as an important difference when a typical policy sold for more than \$1000. On the other hand, even a clinically important improvement of 10% in the cure rate with a new treatment is not likely to be statistically significant in a study of fewer than 225 patients. A small clinical trial would probably not be conclusive.

enough, even large financially or scientifically "significant" differences may not be statistically significant.

In addition to the P-value on which we base statistical significance, it's good practice to report a confidence interval, to indicate the range of plausible values for the parameter.

*Large Sample Test for the Mean

In Chapter 17 (optional section), we discussed tests of means of quantitative populations, assuming the sample is *big enough* to produce a Normally distributed sample mean (using the Central Limit Theorem) and to allow us to estimate the SD of the Normal sampling model with sufficient accuracy. Typically, a sample size over 60 is adequate, but one should *always* check the data for very extreme skewness or outliers. Here is another example.

For Example ARE THEY SPEEDING?

Motor vehicle crashes are the leading cause of death for people between 4 and 33 years old. In the past 10 years, motor vehicle accidents have claimed the lives of over 40 000 people a year in the United States. This means that, on average, motor vehicle crashes resulted in about 110 deaths each day, or 1 death every 13 minutes. Speeding is a contributing factor in 31% of all fatal accidents, according to the National Highway Traffic Safety Administration.

College Terrace is a neighbourhood of Palo Alto, California on the edge of Stanford University. The town has implemented a series of traffic-calming measures over the past decade including building speed tables, centre median islands, and traffic circles, and reducing the number of through streets. The posted speed limit in College Terrace is 25 mph (40 kph). The town reports that speeds and traffic volumes in the neighbourhood have been "noticeably reduced" so they are seeking approval to make the changes permanent.

A Palo Alto resident, convinced that cars were still speeding by his house, measured speeds of every car going by his house by using two infrared photoelectric gates connected to a microcontroller. We can examine 250 of the speeds randomly selected from more than 2000 cars that he recorded.

QUESTION: Is the mean speed of cars greater than 25 mph?

ANSWER:

$$H_0: \mu = 25 \; mph$$
$$H_A: \mu > 25 \; mph$$

Independence Assumption: The data were a census of cars over several days. We have drawn a random sample of about 10% of the data. That random selection makes it reasonable to assume that the car speeds in our sample are independent, although cars driving near each other may not have been.

Sample size condition: The histogram of the data is unimodal and fairly symmetric, with no outliers. The sample size of 250 is more than big enough to ensure a Normal sampling model for the sample mean and to provide a good enough estimate of its standard deviation.

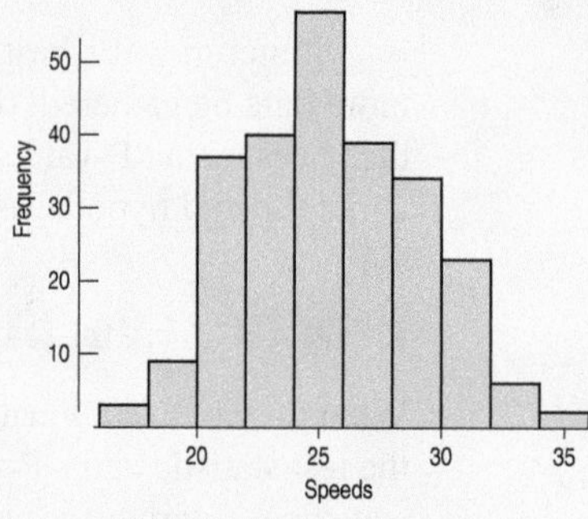

$$n = 250;$$
$$\bar{y} = 25.55; \; s = 3.618$$

Using s in place of σ, we get: $SE(\bar{y}) = \frac{3.618}{\sqrt{250}} = 0.2288$

$$z = \frac{25.55 - 25}{0.2288} = 2.404$$

$$P(z > 2.404) = 0.0081$$

With a P-value of 0.0081, I reject the null hypothesis and conclude that the cars are, on average, going faster than 25 mph. A confidence interval shows that a plausible set of values for the true mean speed is 25.55 ± 1.96 (0.2288) or between 25.1 and 26.0 mph.

In this example, the hypothesis test leads us to conclude that the true mean speed is greater than 25 mph. But, would the town of Palo Alto want to act on that conclusion and try further traffic calming measures? As the confidence interval shows, cars are not on average going very *much* faster than 25 mph. In fact, the true mean speed is probably no more than 26 mph. The difference between the speed limit of 25 mph and the sample mean of 25.55 mph is "statistically significant," but it doesn't seem meaningful. For example, would it be a large enough difference to convince the town of Palo Alto to spend more money on traffic control? Probably not. It's always a good idea when you test a hypothesis to check the confidence interval and think about the likely values for the mean.

18.4 Critical Values for Hypothesis Tests

When we make a confidence interval, we find a **critical value**, z^*, to correspond to our selected confidence level. Critical values can also be used as a shortcut for hypothesis tests. Before computers and calculators were common, P-values were hard to find. It was easier to select a few common alpha levels (0.05, 0.01, 0.001, for example) and learn the corresponding critical values for the Normal model (values that produce a P-value exactly equal to alpha). Rather than looking up the probability corresponding to your z-score in the table, you'd just check your z-score directly against these z^* values. Any z-score larger in magnitude (that is, more extreme) than a particular critical value has to be less likely, so it will have a P-value smaller than the corresponding alpha. If we are willing to settle for a flat reject/fail-to-reject decision, comparing an observed z-score with the critical value for a specified alpha level gives a shortcut path to that decision.

For the motorcycle helmet example, if we choose $\alpha = 0.05$, we then need to find the critical z-value that cuts off an area of 0.05 to the left under a standard Normal curve. Using the Normal table, we get $z^* = -1.64$. To reject H_0, our z-score has to be less than the critical value. Our calculated z-score is -5.31, so clearly we can reject the null hypothesis at the 5% level. This is perfectly correct and does give us a yes/no decision, but it gives us less information about the hypothesis because we don't have the P-value to think about. With technology, P-values are easy to find. And since they give more information about the strength of the evidence, you should always report them. In the motorcycle helmet example, the very tiny P-value that we calculated tells us that the evidence of a change is not just adequate at the 5% level, but is in fact overwhelming!

Here are the traditional critical values from the standard Normal model:[6]

α	1-sided	2-sided
0.05	1.645	1.96
0.01	2.33	2.576
0.001	3.09	3.29

1.96 or 2 If you want to make a decision on the fly without technology, remember "2"—our old friend from the 68–95–99.7 Rule. It's roughly the critical value for testing a hypothesis against a two-sided alternative at $\alpha = 0.05$.

A more exact critical value for z is 1.96. Although 2 is good enough for most decisions, 1.96 is so well known in statistics (and, apparently, not a special value in any other field) that it is almost a badge of the Statistics student. Want to know if someone has studied Statistics? Say "1.96" to them and see if there's any flash of recognition.

[6] In a sense, these are the flip side of the 68–95–99.7 Rule. When we studied that rule, we chose simple statistical distances from the mean and recorded the areas between the tails. Now we select convenient tail areas (0.05, 0.01, and 0.001, either on one side or adding both together) and record the corresponding statistical distances.

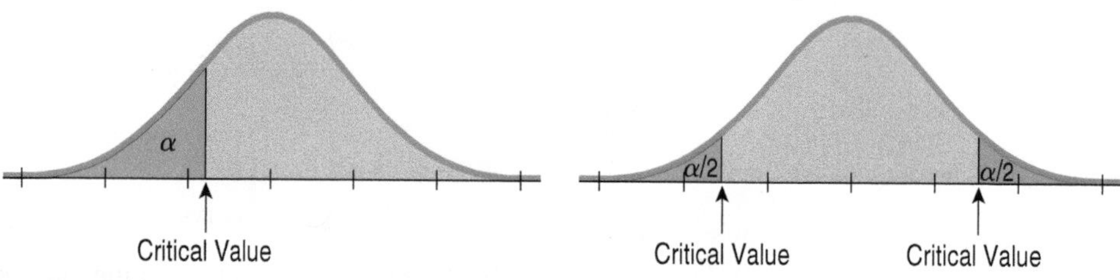

Figure 18.1
When the alternative is one-sided, the critical value puts all of α on one side.

Figure 18.2
When the alternative is two-sided, the critical value splits α equally into two tails.

Confidence Intervals and Tests

For the motorcycle helmet example, a 95% confidence interval would give $0.507 \pm 1.96 \times 0.0179 = (0.473, 0.541)$, or 47.3% to 54.1%. If the previous rate of helmet compliance had been, say, 50%, we would not have been able to reject the null hypothesis because 50% is in the interval, so it's a plausible value. Indeed, *any* hypothesized value for the true proportion of helmet wearers in this interval is consistent with the data. Any value outside the confidence interval would make a null hypothesis that we would reject, but we'd feel more strongly about values far outside the interval.

Confidence intervals and hypothesis tests are built from the same calculations.[7] They have the same assumptions and conditions. As we have just seen, you can approximate a hypothesis test by examining the confidence interval. Just ask whether the null hypothesis value is consistent with a confidence interval for the parameter at the corresponding confidence level. Because confidence intervals are naturally two-sided, they correspond to two-sided tests. For example, a 95% confidence interval corresponds to a two-sided hypothesis test at $\alpha = 5\%$. In general, a confidence interval with a confidence level of $C\%$ corresponds to a two-sided hypothesis test with an α level of $(100 - C)\%$.

The relationship between confidence intervals and one-sided hypothesis tests is a little more complicated. One-sided tests are often accompanied by *one-sided confidence intervals*. These intervals take the form of $(-\infty, a)$ or (b, ∞). The former says that the parameter is no bigger than a, while the latter interval says that the parameter is at least as big as b. If you accept the alternative hypothesis that $p < 0.6$, for example, you know p is smaller than 0.6, but how much smaller? A one-sided confidence interval from the data might tell you that p is *no bigger than 0.53*. Just as with two-sided tests and confidence intervals, a precise relationship exists, such that values lying in, for example, the 95% one-sided confidence interval are precisely those values for the null hypothesis that we could not reject in a one-sided 5% level test. We will not explore this topic any further here. In practice, for one-sided tests, you may see either type of confidence interval reported. However, in this text we will stick to the more traditional two-sided confidence intervals regardless of the nature of the test alternative.

If you want to use a two-sided confidence interval to perform a one-sided test, how might you proceed? If we have a 95% confidence interval (more generally, a $C\%$ confidence interval), and the entire interval lies within the alternative (and not on the other or wrong side), you may then reject the null at not just the 5% level, but at the more impressive 2.5% level (more generally, at the $\frac{1}{2}(100 - C)\%$ level). Because you specified a direction from the null *a priori*, you earn some benefit.

[7] As we saw in Chapter 17, this is not *exactly* true for proportions. For a confidence interval, we estimate the standard deviation of $\hat{p}$ from $\hat{p}$ itself. For the corresponding hypothesis test, we use the model's standard deviation for $\hat{p}$, based on the null hypothesis value p_0. These calculations usually give similar results. When the SE and the null model's SD differ considerably, you're likely to reject H_0 (because the observed proportion is far from your hypothesized value) and in that case don't use the confidence interval to perform a test.

MATH BOX

Suppose we are testing a null hypothesis against a two-sided alternative hypothesis:

$$H_0: p = p_0 \text{ vs. } H_A: p \neq p_0$$

and we use $\alpha = 0.05$.

Our test statistic would be $z = \frac{\hat{p} - p_0}{SD(\hat{p})}$. We would perform the test by checking whether $\frac{\hat{p} - p_0}{SD(\hat{p})} < -1.96$ or $\frac{\hat{p} - p_0}{SD(\hat{p})} > 1.96$. If either were true, we'd reject H_0.

We can rearrange the inequalities to solve for p_0:

$$p_0 > \hat{p} + 1.96SD(\hat{p}) \text{ or } p_0 < \hat{p} - 1.96SD(\hat{p})$$

These inequalities are symmetric, so we can combine them and say that we reject H_0 if the null value p_0 falls outside $\hat{p} \pm 1.96SD(\hat{p})$.

But that looks remarkably like a 95% confidence interval for p:

$$\hat{p} \pm 1.96SE(\hat{p})$$

Except for the fact that in the confidence interval we use $SE(\hat{p})$ instead of $SD(\hat{p})$, they are the same statement. That is, *we reject* H_0 *when the confidence interval fails to cover the hypothesized value* p_0. In most cases, the difference between the *SE* and the *SD* is small. We said that a confidence interval holds the *plausible* values of p, so we shouldn't be surprised to find that we would reject a hypothesized p_0 if it isn't a plausible value, and fail to reject it if it is plausible.

For Example MAKING A DECISION BASED ON A CONFIDENCE INTERVAL

RECAP: The baseline seven-year risk of heart attacks for diabetics is 20.2%. In 2007, a *NEJM* study reported a 95% confidence interval equivalent to 20.8% to 40.0% for the risk among patients taking the diabetes drug Avandia®.

QUESTION: What did this confidence interval suggest to the FDA about the safety of the drug?

ANSWER: The *FDA* could be 95% confident that the interval from 20.8% to 40.0% included the true risk of heart attack for diabetes patients taking Avandia®. Because the lower limit of this interval was higher than the baseline risk of 20.2%, there was evidence of an increased risk.

Just Checking

1. An experiment to test the fairness of a roulette wheel gives a z-score of 0.62. What would you conclude?
2. In the last chapter, we encountered a bank that wondered if it could get more customers to make payments on delinquent balances by sending them a DVD urging them to set up a payment plan. Well, the bank just got back the results of their test of this strategy and reported that a 90% confidence interval for the success rate is (0.29, 0.45). Their old send-a-letter method had worked 30% of the time. Can you reject the null hypothesis that the proportion is still 30%? Explain.
3. Given the confidence interval the bank found in their trial of DVDs, what would you recommend that they do? Should they scrap the DVD strategy?

18.5 Decision Errors

A S *Activity:* **Type I and Type II Errors.** View an animated exploration of Type I and Type II errors—a good backup for the reading in this section.

Nobody's perfect. Even with lots of evidence, we can still make the wrong decision. In fact, when we perform a hypothesis test, we can make mistakes in *two* ways:

I. The null hypothesis is true, but we mistakenly reject it.
II. The null hypothesis is false, but we fail to reject it.

These two types of errors are known as **Type I** and **Type II errors**. One way to keep the names straight is to remember that we start by assuming the null hypothesis is true, so a Type I error is the first kind of error we could make.

False positive consequences. Some false-positive results mean no more than an unnecessary chest X-ray. But for a drug test or a disease like AIDS, a false-positive result that is not kept confidential could have serious consequences.

In medical disease testing, the null hypothesis is usually the assumption that a person is healthy. The alternative is that he or she has the disease we're testing for. So a Type I error is a *false positive*: A healthy person is diagnosed with the disease. A Type II error, in which an infected person is diagnosed as disease-free, is a *false negative*. These errors have other names, depending on the particular discipline and context.

Which type of error is more serious depends on the situation. In the jury trial, a Type I error occurs if the jury convicts an innocent person. A Type II error occurs if the jury fails to convict a guilty person. Which seems more serious? In medical diagnosis, a false negative could mean that a sick patient goes untreated. A false positive might mean that the person must undergo further tests. But for a drug test or a disease like AIDS, a false-positive result that is not kept confidential could have serious consequences. If you are testing a new treatment for epilepsy, a Type I error will lead to future patients getting a useless treatment; a Type II error means a useful treatment will remain undiscovered. Which of these errors seems more serious? It depends on the situation, the cost, and your point of view.

Here's an illustration of the situations:

A S *Activity:* **Hypothesis Tests Are Random.** Simulate hypothesis tests and watch Type I errors occur. When you conduct real hypothesis tests you'll never know, but simulation can tell you when you've made an error.

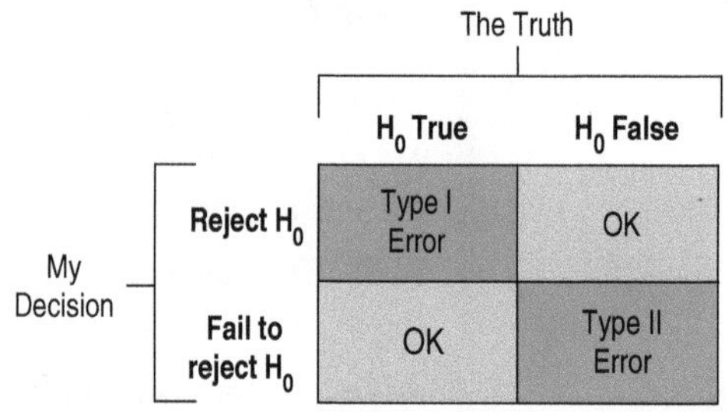

How often will a Type I error occur? It happens when the null hypothesis is true but we've had the bad luck to draw an unusual sample. To reject H_0, the P-value must fall below α. When H_0 is true, that happens *exactly* with probability α. So when you choose level α, you're setting the probability of a Type I error to α.

What if H_0 is not true? Then we can't possibly make a Type I error. You can't get a false positive from a sick person. A Type I error can happen only when H_0 is true.

NOTATION ALERT

In Statistics, α is almost always saved for the alpha level. But β's will also be used for the parameters of a linear model. Fortunately, it's usually clear whether we're talking about a Type II error probability or the slope or intercept of a regression model.

When H_0 is false and we *fail* to reject it, we have made a Type II error. We assign the letter β to the probability of this mistake. What's the value of β? That's harder to assess than α because we don't know what the value of the parameter really is. When H_0 is true, it specifies a single parameter value. But when H_0 is false, we don't know the parameter value and there are many possible values. We can compute the probability β for any parameter value in H_A.

So β is not one single number but is actually a function of the parameter values contained in H_A. Not surprisingly, the value of β gets smaller as we move away from the null hypothesized value—if a coin produces heads 90% of the time on average, it is less likely that any data produced will support a null hypothesis of 50–50 balance, than if the coin produces heads 51% of the time. β when $p = 0.9$ is smaller than β when $p = 0.51$. The further the truth (true value of p) lies from the null, the easier it is to make the right decision, the harder to make the wrong decision. A small β is desirable, but we can't make β small for all values in H_A, and fortunately we never need to. Some values in H_A will be of more practical importance than others—more about this later in our Effect Size discussion.

FINDING β

The null hypothesis specifies a single value for the parameter, so it's easy to calculate the probability of a Type I error. But the alternative gives a whole range of possible values, and we may want to find a β for several or for specially chosen ones.

We could reduce β for *all* alternative parameter values by increasing α. By making it easier to reject the null, we'd be more likely to reject it whether it's true or not. So we'd reduce β—the chance that we fail to reject a false null—but we'd make more Type I errors. This tension between Type I and Type II errors is inevitable. In the political arena, consider the ongoing debate between those who favour provisions to reduce Type I errors in the courts (supporting rights of accused persons, providing legal representation for those who can't afford it, etc.) and those who advocate changes to reduce Type II errors (admitting into evidence confessions made when no lawyer is present, eavesdropping on conferences with lawyers, restricting paths of appeal, etc.).

The only way to reduce *both* types of error is to collect more evidence or, in statistical terms, to collect more data. Too often, studies fail because their sample sizes are too small to detect the changes or effects the researchers are looking for.

Of course, what we really want is to detect a false null hypothesis. When H_0 is false and we reject it, we have done the right thing. A test's ability to detect a false null hypothesis is called the **power** of the test. In a jury trial, power is the ability of the criminal justice system to convict people who are guilty—a good thing!

For Example THINKING ABOUT ERRORS

RECAP: A published study found the risk of heart attack to be increased in patients taking the diabetes drug Avandia®. The issue of the *New England Journal of Medicine* (NEJM) in which that study appeared also included an editorial that said, in part, "A few events either way might have changed the findings for myocardial infarction[8] or for death from cardiovascular causes. In this setting, the possibility that the findings were due to chance cannot be excluded."

QUESTION: What kind of error would the researchers have made if, in fact, their findings were due to chance? What could be the consequences of this error?

ANSWER: The null hypothesis said the risk didn't change, but the researchers rejected that model and claimed evidence of a higher risk. If these findings were just due to chance, they rejected a true null hypothesis—a Type I error.

If, in fact, Avandia® carried no extra risk, then patients might be deprived of its benefits for no good reason.

Fisher and $\alpha = 0.05$ Why did Sir Ronald Fisher suggest 0.05 as a criterion for testing hypotheses? It turns out that he had in mind small initial studies. Small studies have relatively little power. Fisher was concerned that they might make too many Type II errors—failing to discover an important effect—if too strict a criterion were used.

The increased risk of Type I errors arising from a generous criterion didn't concern him as much because he expected a significant result to be followed with a replication or a larger study. The probability that two independent studies tested at $\alpha = 0.05$ would both make Type I errors is $0.05 \times 0.05 = 0.0025$, so Fisher was confident that Type I errors in initial studies were not a major concern.

The widespread use of the relatively generous 0.05 criterion even in large studies is most likely not what Fisher had in mind.

18.6 Power and Sample Size

When we failed to reject the null hypothesis about TT practitioners, did we prove that they were just guessing? No, it could be that they actually *can* discern a human energy field, but we just couldn't tell. For example, suppose they really have the ability to get 53% of the trials right but just happened to get only 47% in our particular experiment. As shown earlier, with these data we wouldn't have rejected the null. And if we retained the null even though the true proportion was actually greater than 50%, we would have made a Type II error because we failed to detect their ability.

[8] Doctorese for "heart attack."

Remember, we can never prove a null hypothesis true. We can only fail to reject it. But when we fail to reject a null hypothesis, it's natural to wonder whether we looked hard enough. Might the null hypothesis actually be false and our test too weak to tell?

When the null hypothesis actually *is* false, we hope our test is strong enough to reject it. We'd like to know how likely we are to succeed. The power of the test gives us a way to think about that. The **power** of a test is the probability that it correctly rejects a false null hypothesis. When the power is high, we can be confident that we've looked hard enough. We know that β is the probability that a test *fails* to reject a false null hypothesis, so the power of the test is the probability that it *succeeds* in rejecting the false null hypothesis: $1 - \beta$. We might have just written $1 - \beta$, but power is such an important concept that it gets its own name.

Whenever a study fails to reject its null hypothesis, the test's power comes into question. Was the sample size big enough to detect an effect had there been one? Might we have missed an effect large enough to be interesting just because we failed to gather sufficient data or because there was too much variability in the data we could gather? The therapeutic touch experiment failed to reject the null hypothesis that the TT practitioners were just guessing. Might the problem be that the experiment simply lacked adequate power to detect their ability?

For Example ERRORS AND POWER

RECAP: The study of Avandia® published in the NEJM combined results from 47 different trials—a method called meta-analysis.[9] The drug's manufacturer, GlaxoSmithKline (GSK), issued a statement that pointed out, "Each study is designed differently and looks at unique questions: For example, individual studies vary in size and length, in the type of patients who participated, and in the outcomes they investigate." Nevertheless, by combining data from many studies, meta-analyses can achieve a much larger effective sample size.

QUESTION: How could this larger sample size help?

ANSWER: If Avandia® really did increase the seven-year heart attack rate, doctors needed to know. To overlook that would be a Type II error (failing to detect a false null hypothesis), resulting in patients being put at greater risk. Increasing the sample size could increase the power of the analysis, making it more likely that researchers will detect the danger if there is one.

Sensitivity and specificity The terms *sensitivity* and *specificity* often appear in medical studies. In these studies, the null hypothesis is that the person is healthy and the alternative is that the person is sick. The *specificity* of a test measures its ability to correctly identify only the healthy (the ones who should test negative). It's defined as

$$\frac{\text{number of true negatives}}{(\text{number of true negatives} + \text{number of false positives})}$$

So, *specificity* $= 1 - \alpha$.

Specificity gives you the probability that the test will correctly identify you as healthy when you are healthy.

The *sensitivity* is the ability to detect the disease. It's defined as:

$$\frac{\text{number of true positives}}{(\text{number of true positives} + \text{number of false negatives})}$$

So *sensitivity* $= 1 - \beta =$ power of the test. Sensitivity gives you the probability that the test will correctly identify you as sick when you are sick.

[9]One of the challenges of a good meta-analysis is digging out the studies that may have been *filed away* and not published, possibly due to lack of statistical significance (the "file drawer" problem). Yet, without those studies, conclusions of a meta-analysis can be seriously flawed and biased.

Effect Size

When we think about power, we imagine that the null hypothesis is false. The value of the power depends on how far the truth lies from the value we hypothesize. We call the distance between the null hypothesis value, p_0, and the truth, p, the **effect size**.

The effect size is unknown, of course, since it involves the true p, but it is central to how we think about the power of a hypothesis test. A larger effect is easier to see and results in larger power. Small effects are naturally more difficult to detect. They'll result in more Type II errors and therefore lower power. When we design a study we imagine possible effect sizes and look at their consequences. How can we decide what effect sizes to look at?

One way to think about the effect size is to ask, "*How big a difference would matter?*" The answer depends on who is asking it and why. For example, if therapeutic touch practitioners could detect a human energy field about 53% of the time, it might not be a sufficient improvement over chance for health insurers to conclude it was worth covering. But *any* real ability to detect a previously unknown human energy field would be of great interest to scientists.[10]

The power of the test depends on both the effect size and the standard deviation of our Normal sampling model (for $\hat{p}$) which, in turn, depends on the square root of the sample size, n. Once we decide what effect size matters, practically speaking, we can estimate the sample size we'll need. This is a common calculation to perform when designing a study so you can be sure to gather a large enough sample (and, speaking practically, to gather enough funds to pay for that large a sample). We made a similar calculation in Chapter 16, based on how large a margin of error we wanted—and this is essentially the same idea.

Effect size and power are also an important part of our conclusions whenever we fail to reject the null hypothesis. The natural question to ask in that event is whether we tried hard enough—whether we had sufficient power to discern a difference. After all, if we based our test on only three observations, we might have easily missed even a large effect.

Whenever a hypothesis test fails to reject the null, the question of power can arise. In the therapeutic touch experiment, if the researchers considered 75% a reasonably interesting effect size (keeping in mind that 50% is the level of guessing), they could determine that the TT experiment would have been able to detect such an ability with a power of 99.99%[11] (assuming an α-level of 0.01). So there is only a very small chance that the study would have failed to detect a practitioner's ability at that level, had it existed. If, on the other hand, they thought that even an increase to 51% was interesting, then they would need a sample size of over 20 000 trials to have a 90% chance of detecting it!

Just Checking

4. Remember the bank that is sending out DVDs to try to get customers to make payments on delinquent loans? It is looking for evidence that the costlier DVD strategy produces a higher success rate than the letters it has been sending. Explain what a Type I error is in this context and what the consequences would be to the bank.
5. What would be a Type II error in the bank experiment context, and what would the consequences be?
6. For the bank, which situation would have higher power: a strategy that works really well, actually getting 60% of people to pay off their balances, or a strategy that barely increases the payoff rate to 32%? Explain briefly.

---

[10] And probably lead to a Nobel Prize.

[11] See upcoming Math Box for exact calculation.

A Picture Worth $\frac{1}{P(z > 3.09)}$ Words

It makes intuitive sense that the larger the effect size, the easier it should be to see it. Obtaining a larger sample size decreases the probability of a Type II error, so it increases the power. It also makes sense that the more we're willing to accept a Type I error, the less likely we will be to make a Type II error.

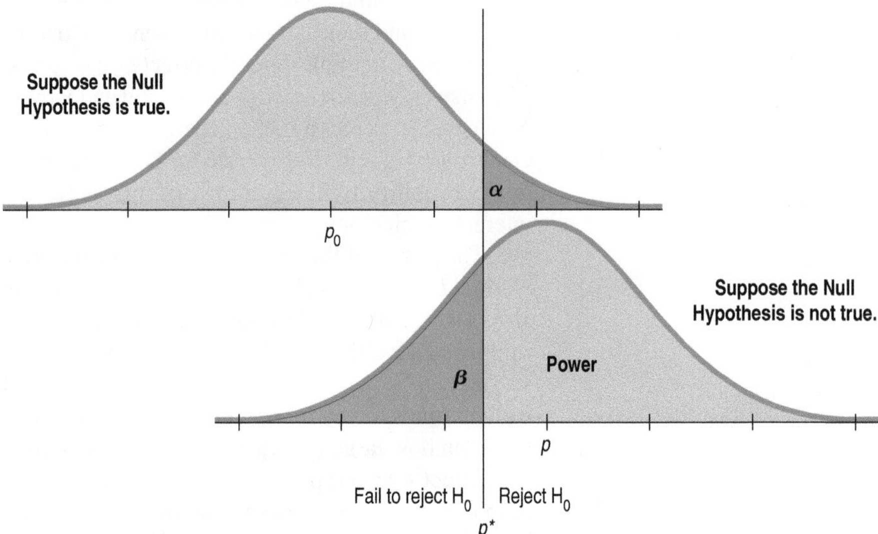

Figure 18.3

The power of a test is the probability that it rejects a false null hypothesis. The upper figure shows the null hypothesis model. We'd reject the null in a one-sided test if we observed a value of $\hat{p}$ in the red region to the right of the critical value, p^*. The lower figure shows the assumed true model. If the true value of p is greater than p_0, then we're more likely to observe a value that exceeds the critical value and make the correct decision to reject the null hypothesis. The power of the test is the purple region on the right in the lower figure. Of course, even when we draw samples whose observed proportions are distributed around p, we'll sometimes get a value in the red region on the left and make a Type II error of failing to reject the null.

NOTATION ALERT

We've attached symbols to many of the p's. Let's keep them straight. p is a true proportion parameter. p_0 is a hypothesized value of p. $\hat{p}$ is an observed proportion. p^* is a critical value of a proportion corresponding to a specified α.

A S *Activity:* **The Power of a Test.** Power is a concept that's much easier to understand when you can visualize what's happening.

Figure 18.3 above shows a good way to visualize the relationships among these concepts. Suppose we are testing H_0: $p = p_0$ against the alternative H_A: $p > p_0$. We'll reject the null if the observed proportion, $\hat{p}$, is big enough. By big enough, we mean $\hat{p} > p^*$ for some critical value p^*, which is determined from our α-level (shown as the red region in the right tail of the upper curve). For example, suppose that p^* were 0.65; then we would be willing to believe the ability of therapeutic touch practitioners if they were successful in at least 65% of our trials. This is what the upper model in Figure 18.3 shows. It's a picture of the sampling distribution model for the proportion when the null hypothesis is true. If the null were true, then this would be a picture of that truth. We'd make a Type I error whenever the sample gave us $\hat{p} > p^*$, because we would reject the (true) null hypothesis. And unusual samples like that would happen only with probability α.

In reality, though, the null hypothesis is rarely *exactly* true. The lower probability model in Figure 18.3 supposes that H_0 is not true. In particular, it supposes that the true value is p, not p_0. (Perhaps the TT practitioner really can detect the human energy field 75% of the time.) It shows a distribution of possible observed $\hat{p}$ values around this true value. Because of sampling variability, sometimes $\hat{p} < p^*$ and we fail to reject the (false) null hypothesis. Suppose a TT practitioner with a true ability level of 75% is actually successful on fewer than 65% of our tests. Then we'd make a Type II error. The area under the curve to the left of p^* in the bottom model represents how often this happens. The probability is β. In this picture, β is less than half, so most of the time we *do* make the right decision. The *power* of the test—the probability that we make the right decision—is shown as the region to the right of p^*. It's $1 - \beta$.

MATH BOX

Calculating the Power of the Test

Power calculations are always relevant, but they can be quite difficult to perform for many tests. In this case, however, it's not so hard. Let's give it a try, using the TT example on page 517, where the null hypothesis is H_0: $p = 0.5$, with $\alpha = 0.01$, and we are interested in the power of the test when the true $p = 0.75$, for $n = 150$ trials.

First, restate the criterion for rejection of the null hypothesis, in terms of a critical value p^*:

We reject H_0 at the 1% level if

$$z = \frac{\hat{p} - p_0}{SD(\hat{p})} = \frac{\hat{p} - 0.5}{0.0408} > z^* = 2.33 \text{ (from tables), since}$$

$$SD(\hat{p}) = \sqrt{\frac{(0.5)(0.5)}{150}} = 0.0408, \text{ under the null.}$$

Or, rewriting the above, if $\hat{p} > 0.595$ (the critical value for $\hat{p}$, denoted p^* in text diagram).

Power = Probability of rejecting H_0 when H_A is true, and in particular, when p equals 0.75

$$= P(\hat{p} > 0.595 \text{ when } p = 0.75)$$

or standardized:

$$= P\left(\frac{\hat{p} - p}{SD(\hat{p})} > \frac{0.595 - p}{SD(\hat{p})}\right)$$

Since we are assuming that $p = 0.75$, plug this in for p in the top and also in the SD formula in the bottom, which gives $SD(\hat{p}) = \sqrt{\frac{(0.75)(0.25)}{150}} = 0.0354$.

$$\text{Power} = P\left(\frac{\hat{p} - p}{SD(\hat{p})} > \frac{0.595 - 0.75}{0.0354}\right) = P(z > -4.38) = 0.99999$$

(to 5 decimal places, using software).

So, if we have 150 trials, and the true success rate is 75%, we are essentially certain of making the correct decision!

We calculate p^* based on the upper model because p^* depends only on the null model and the alpha level. No matter what the true proportion, no matter whether the practitioners can detect a human energy field 90%, 53%, or 2% of the time, p^* doesn't change. After all, we don't *know* the truth, so we can't use it to determine the critical value. But we always reject H_0 when $\hat{p} > p^*$.

How often we are able to reject H_0 when it's *false* depends on the effect size. We can see from Figure 18.3 that if the effect size were larger (the true proportion were farther from the hypothesized value), the bottom curve would shift to the right, making the power greater.

We can see several important relationships from this figure:

- Power $= 1 - \beta$.
- Reducing α to lower the chance of committing a Type I error will move the critical value, p^*, to the right (in this example). This will have the effect of increasing β, the probability of a Type II error, and correspondingly reducing the power.
- The larger the real difference between the hypothesized value p_0 and the true population value p, the smaller the chance of making a Type II error and the greater the power of the test. If the two proportions are very far apart, the two models will barely overlap, and we will not be likely to make any Type II errors at all—but then, we are unlikely to really need a formal hypothesis-testing procedure to see such an obvious difference. If the TT practitioners were successful almost all the time, we'd be able to see their ability with even a small experiment.

Determining the Sample Size

Figure 18.3 shows that if we reduce the chances of a Type II error (β), we increase our risk of a Type I error (α). That would be true if nothing else changes. But in fact there is a way to reduce β, while maintaining our desired α. Can you think of it?

If we can make both curves shown in Figure 18.4 narrower, with the size of the red α area fixed, the red β area will decrease and the power of the test will increase.

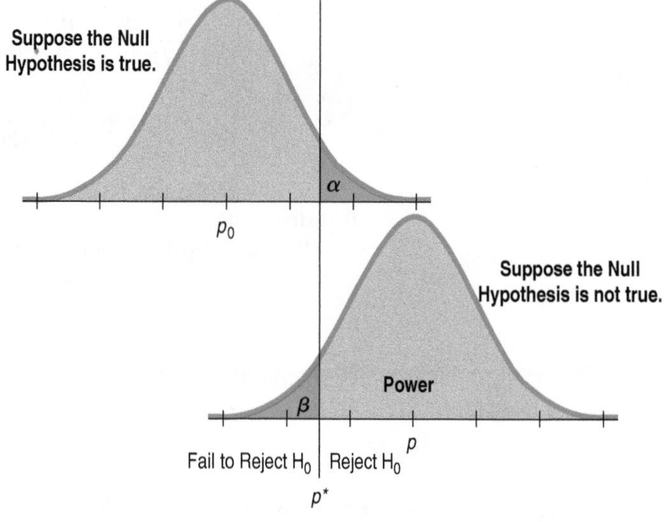

Figure 18.4
Making the standard deviations of these curves smaller will increase the power without changing the alpha level. The bar at p^* will slide to the left as the curves get narrower, increasing the power. The two proportions are just as far apart as in Figure 18.3, but the Type II error rate is reduced.

A S *Activity:* **Power and Sample Size.** Investigate how the power of a test changes with the sample size. The interactive tool is really the only way you can see this easily.

How can we accomplish that? The only way is to reduce the standard deviations by increasing the sample size. (Remember, these are pictures of sampling distribution models, not of data.) Increasing the sample size works regardless of the true population parameters. But recall the curse of diminishing returns. The standard deviation of the sampling distribution model decreases only as the *square root* of the sample size, so to *halve* the standard deviations we must *quadruple* the sample size.

So sample size is at issue once again! Just as with confidence intervals, at the design stage of a study we have to give careful thought to choosing an appropriate sample size. But instead of controlling the margin of error, now we want to control the Type II error rate or, equivalently, the power. How should we proceed?

First, we set α appropriately depending on the consequences of a Type I error. Then we determine a sample size big enough to keep the Type II error rate (β) low or the power ($1 - \beta$) high for alternatives to the null (*effect sizes*) of practical importance. Before beginning the TT study, we should ask what departures from $p = 0.5$ are so big and important that we cannot allow them to slip by undiscovered. Suppose we decide that a success rate of 75%[12] is remarkable and we (or health insurers) need to know about it. We should choose n so that the power is high (maybe 0.8 or 0.9) when $p = 0.75$. While it is possible to develop a precise formula for this simple test, we will emphasize the use of software here (as well as for later tests). One hundred and fifty trials were conducted in this particular study, which produced a power of essentially 1.0 (see Math Box), but if researchers decided that a power of, say, 0.9 was adequate, then from the output below, we can see that 31 trials would have been sufficient.

POWER AND SAMPLE SIZE

Test for One Proportion

Testing proportion = 0.5 (versus > 0.5), Alpha = 0.05

Alternative Proportion	Sample Size	Target Power	Actual Power
0.75	31	0.9	0.905785

On the other hand, if a scientist suggests that even a value as small as $p = 0.51$ is hugely important, constituting the discovery of a previously unknown energy field, how many trials might now be needed? Likely many more in order to discover such a small effect. Below is the answer—over 21 400 trials!

[12]Technically, our power calculation is accurate only if we use one practitioner for all trials or a different practitioner for each trial. If we have multiple trials for each of multiple practioners, there is a clustering effect or dependence for data from any one practitioner, invalidating our independence assumption.

POWER AND SAMPLE SIZE

Test for One Proportion

Testing proportion = 0.5 (versus > 0.5), Alpha = 0.05

Alternative Proportion	Sample Size	Target Power	Actual Power
0.51	21 406	0.9	0.900002

For Example SAMPLE SIZE, ERRORS, AND POWER

RECAP: The meta-analysis of the risks of heart attacks in patients taking the diabetes drug Avandia® combined results from 47 smaller studies. As GlaxoSmith-Kline (GSK), the drug's manufacturer, pointed out in their rebuttal, "Data from the ADOPT clinical trial did show a small increase in reports of myocardial infarction among the Avandia®-treated group . . . however, the number of events is too small to reach a reliable conclusion about the role any of the medicines may have played in this finding."

QUESTION: Why would this smaller study have been less likely to detect the difference in risk? What are the appropriate statistical concepts that underlie combining the smaller studies? What are the possible drawbacks of a very large sample?

ANSWER: Smaller studies are subject to greater sampling variability; that is, the sampling distributions they estimate have a larger standard deviation for the sample proportion. That gives small studies less power: They'd be less able to discern whether an apparently higher risk was merely the result of chance variation or evidence of real danger. Health Canada doesn't want to restrict the use of a drug that's safe and effective (Type I error), nor do they want patients to continue taking a medication that puts them at risk (Type II error). Larger sample sizes can reduce the risk of both kinds of error. Greater power (the probability of rejecting a false null hypothesis) means a better chance of spotting a genuinely higher risk of heart attacks. Of course, larger studies are more expensive. Their larger power also makes them able to detect differences that may not be clinically or financially significant. This tension among the power, cost, and sample size lies at the heart of most clinical trial designs.

On September 23, 2010, "The U.S. Food and Drug Administration announced that it will significantly restrict the use of the diabetes drug Avandia (rosiglitazone) to patients with Type 2 diabetes who cannot control their diabetes on other medications. These new restrictions are in response to data that suggest an elevated risk of cardiovascular events, such as heart attack and stroke, in patients treated with Avandia." (www.fda.gov/drugs/DrugSafety/PostmarketDrugSafetyInformationfor PatientsandProviders/ucm226956.htm)

WHAT CAN GO WRONG?

- **Don't interpret the P-value as the probability that H_0 is true.** The P-value is about the data, not the hypothesis. It's the probability of observing data this unusual, *given* that H_0 is true, not the other way around.
- **Don't believe too strongly in arbitrary alpha levels.** There's not really much difference between a P-value of 0.051 and a P-value of 0.049, but sometimes it's regarded as the difference between night (having to retain H_0) and day (being able to shout to the world that your results are "statistically significant"). It may just be better to report the P-value and a confidence interval and let the world decide along with you.
- **Don't confuse practical and statistical significance.** A large sample size can make it easy to discern even a trivial change from the null hypothesis value. On the other hand, an important difference can be missed if your test lacks sufficient power.
- **Don't forget that in spite of all your care, you might make a wrong decision.** We can never reduce the probability of a Type I error (α) or of a Type II error (β) to zero (but increasing the sample size helps).
- **If you fail to reject the null hypothesis, don't think that a bigger sample would be more likely to lead to rejection.** If the results you looked at were "almost" significant, it's

enticing to think that because you would have rejected the null had similar observations come from a larger sample, then a larger sample would surely lead to rejection. Don't be misled. Remember, each sample is different, and a larger sample won't necessarily duplicate your current observations. Maybe they'd be statistically significant, but maybe not. On the other hand, if a small study (small n) yields weak evidence ($P < 0.10$, for example) for the hoped-for effect, proceeding to a bigger future study can make sense. Examining the confidence interval can also help. If none of the plausible parameter values in the interval would matter to you (for example, because none would be *practically* significant), then even a larger study is unlikely to be worthwhile.

CONNECTIONS

All of the hypothesis tests we'll see boil down to the same question: "Is the difference between two quantities large enough to draw conclusions that go beyond this particular sample?" We always measure "how large" by finding a ratio of this difference to the standard deviation of the sampling distribution of the statistic. Using the standard deviation as our ruler for inference is one of the core ideas of statistical thinking.

We've discussed the close relationship between hypothesis tests and confidence intervals. They are often two sides of the same coin.

This chapter also has natural links to the discussion of probability, to the Normal model, and to the two previous chapters on inference.

What Have We Learned?

Learning Objectives

Review: Know the reasoning of hypothesis testing.

- State the **hypotheses**.
- Determine (and check assumptions for) the sampling distribution **model**.
- Calculate the test statistic—the **mechanics**.
- State your **conclusions and decisions**.

Know how to formulate a null and alternative hypothesis for a question of interest.

- The null hypothesis specifies a parameter and a (null) value for that parameter.
- The alternative hypothesis specifies a range of plausible values should we fail to reject the null.
- Be able to decide on a two-sided or one-sided alternative hypothesis, and justify your decision.

Understand P-values.

- A P-value is the estimated probability of observing a statistic value at least as far from the (null) hypothesized value as the one we have actually observed.
- A small P-value indicates that the statistic we have observed would be unlikely were the null hypothesis true. That leads us to doubt the null.
- A large P-value just tells us that we have insufficient evidence to doubt the null hypothesis. In particular, it does not prove the null to be true.

Know that α is the probability of making a Type I error—rejecting the null hypothesis when it is, in fact, true.

Know that statistical significance simply implies strong evidence that the null hypothesis is false, not that the difference is important.

Know that sometimes P-values are compared to a pre-determined α-level, called a critical value, to decide whether to reject the null hypothesis.

Know the value of reporting the estimated effect size.

- A test may be statistically significant, but practically meaningless if the estimated effect is of trivial importance.

Be aware of the risks of making errors when testing hypotheses.

- A Type I error can occur when rejecting a null hypothesis if that hypothesis is, in fact, true. The probability of this is α.
- A Type II error can occur when failing to reject a null hypothesis if that hypothesis is, in fact, false. The probability of this is β (not one number, but a function of the alternatives to the null).

Understand the concept of the power of a test.

- We are particularly concerned with power when we fail to reject a null hypothesis.
- The power of a test reports, for some specified effect size, the probability that the test would correctly reject the false null hypothesis.
- Remember that increasing the sample size will generally improve the power of any test.

Review of Terms

Statistically significant When the P-value falls below the alpha level, we say that the test is "statistically significant" at that alpha level (p. 508).

Alpha level (α) The threshold P-value that determines when we reject a null hypothesis. If we observe a statistic whose P-value based on the null hypothesis is less than α, we reject that null hypothesis (p. 508).

Significance level The alpha level is also called the significance level of the test, most often in a phrase such as a conclusion that a particular test is "significant at the 5% significance level" (p. 509).

Critical value The value in the sampling distribution model of the statistic whose P-value is equal to the alpha level. The critical value is often denoted with an asterisk, as z^*, for example (p. 511).

Type I error The error of rejecting a null hypothesis when in fact it is true (called a "false positive" in diagnostic tests). The probability of a Type I error is α (p. 512).

Type II error The error of failing to reject a null hypothesis when in fact it is false (called a "false negative" in diagnostic tests) (p. 512).

β The probability of a Type II error. β depends on the specified alternative value for the parameter (or effect size) (p. 512).

Power The probability that a hypothesis test will correctly reject a false null hypothesis is the power of the test. To find power, we must specify a particular alternative parameter value as the "true" value. For any specific value in the alternative, the power is $1 - \beta$ (p. 513).

Effect size The difference between the null hypothesis value and true value of a model parameter (p. 516).

On the Computer HYPOTHESIS TESTS

Reports about hypothesis tests generated by technologies don't follow a standard form. Most will name the test and provide the test statistic value, its standard deviation, and the P-value. But these elements may not be labelled clearly. For example, the expression "Prob > $|z|$" means the probability (the "Prob") of observing a test statistic whose magnitude (the absolute value tells us this) is larger than that of the one (the "z") found in the data (which, because it is written as "z," we know follows a Normal model). That is a fancy (and not very clear) way of saying P-value. In some packages, you can specify that the test be one-sided. Others might report three P-values, covering the ground for both one-sided tests and the two-sided test.

Sometimes a confidence interval and hypothesis test are automatically given together. The confidence interval ought to be for the corresponding confidence level: $1 - \alpha$.

Often, the standard deviation of the statistic is called the "standard error," and usually that's appropriate because we've had to estimate its value from the data. That's not the case for proportions, however: We get the standard deviation for a proportion from the null hypothesis value. Nevertheless, you may see the standard deviation called a "standard error," even for tests with proportions.

It's common for statistics packages and calculators to report more digits of "precision" than could possibly have been found from the data. You can safely ignore them. Round values such as the standard deviation to one digit more than the number of digits reported in your data.

Here are the kind of results you might see. This is not from any program or calculator we know of, but it shows some of the things you might see in typical computer output.

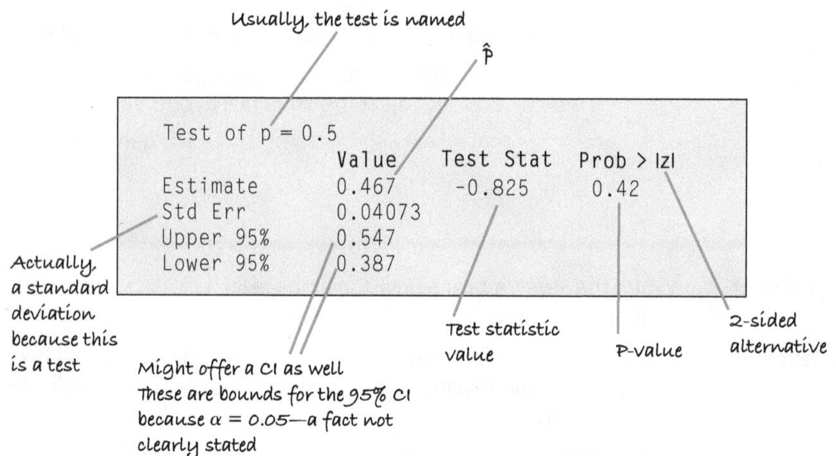

For information on hypothesis testing with particular statistics packages, see the On the Computer section in the previous chapter.

CALCULATING POWER AND SAMPLE SIZE

Good statistical packages will readily compute the power of a test for a given sample size, or the sample size needed to achieve a certain power, for any specified effect size. Common test procedures should be available and listed in the menu for you to choose from.

EXCEL

Power calculations are not in Excel's standard data analysis toolkit, but you can try to imitate the individual steps shown in the Math Box titled Calculating the Power of the Test.

JMP

- Choose **Power and Sample Size** from the **DOE** menu.
- Choose **One Sample Proportion** from the submenu.
- Indicate the **Null Proportion** and alternative **Proportion** that you are testing, select the **Alpha** value, and choose a one- or two-sided alternative.
- Fill in your desired **Sample size** or **Power**. Whichever you leave blank will be calculated.
- Click **Continue**.

COMMENTS

Specify the **alternative** Proportion that represents a departure from the null hypothesis that is large enough to be of practical importance. If the sample size is fixed and you need to know the power, leave **Power** blank, but if you are planning a study and need to know the sample size required to achieve desired power, fill in the **Power** sought and leave **Sample size** blank. If you leave both **Sample size** and **Power** blank, you will get a power curve, showing power as a function of sample size.

MINITAB

- Choose **Power** *and* **Sample Size** from the **Stat** menu.
- Choose **1 Proportion** from the submenu.
- Fill in the **Hypothesized** *p* that you are testing.
- Fill in any two of: **Sample sizes, Alternative values of** *p*, **Power values**.
- Use **Options** if you want a one-sided alternative or a significance level different from 0.05.
- Use **Graphs** if you want to produce an entire power curve of power values as a function of *p*.
- Click **OK**.

COMMENTS

Specify the **alternative value of** *p* that represents a departure from the null big enough to be of practical importance. If the sample size is fixed and you need to know the power, leave **Power values** blank, but if you are planning a study and need to know the sample size required to achieve desired power, fill in the **Power values** sought (multiple entries are fine) and leave **Sample sizes** blank.

R

R has a package {pwr} (not installed by default) that does various power calculations.

SPSS

The IBM SPSS SamplePower add-on makes power calculations simple. Otherwise, you can try to imitate the individual steps shown in the Math Box titled Calculating the Power of the Test.

STATCRUNCH

Select **Stat**>**Proportion Statistics**>**One Sample**>**Power/Sample size**. At the top of the *one sample p calculator*, click on **Hypothesis Test Power.** This allows you to find the effect size, power, or sample size given any two of the values as inputs. Proceed as follows:

- Enter **Alpha**, the **Target** (null) proportion, and the type of **Alternative** hypothesis.
- Enter two of the following three values: **True proportion**, **Power**, and **Sample size**.
- Click **Compute** to calculate the value left blank while updating the power curve.
- Click and drag the red dot in the power curve to interactively investigate the power/difference relationship. Note how the values below for difference and power change accordingly.

T1-83/84 PLUS

Power calculations are not automatic, but you can imitate the individual calculation steps shown in the Math Box titled Calculating the Power of the Test.

Exercises

1. **One-sided or two?** In each of the following situations, is the alternative hypothesis one-sided or two-sided? What are the hypotheses?

   a) A business student conducts a taste test to see whether students prefer Diet Coke or Diet Pepsi.

b) PepsiCo recently reformulated Diet Pepsi in an attempt to appeal to teenagers. They run a taste test to see if the new formula appeals to teenagers more than the standard formula.
c) A budget override in a small town requires a two-thirds majority to pass. A local newspaper conducts a poll to see if there's evidence it will pass.
d) One financial theory states that the stock market will go up or down with equal probability. A student collects data over several years to test the theory.

2. **Which alternative?** In each of the following situations, is the alternative hypothesis one-sided or two-sided? What are the hypotheses?
a) A university dining service wants to know if students prefer plastic or metal cutlery. They conduct a survey to find out.
b) In recent years, 10% of third-year students have applied for study abroad. The dean's office conducts a survey to see if the percentage changed this year.
c) A pharmaceutical company conducts a clinical trial to see if more patients who take a new drug experience headache relief than the (previously established rate of) 22% who claim relief after taking a placebo.
d) At a small computer peripherals company, only 60% of the hard drives produced passed all their performance tests the first time. Management recently invested a lot of resources into the production system and now conducts a test to see if it helped.

3. **P-value** A medical researcher has tested a new treatment for poison ivy against the traditional ointment. He concludes that the new treatment, with a P-value of 0.047, is more effective. Explain what the P-value means in this context.

4. **Another P-value** Have harsher penalties and ad campaigns increased seatbelt use among drivers and passengers? Observations of commuter traffic failed to find evidence of a significant change compared with three years ago. Explain what the study's P-value of 0.17 means in this context.

5. **Alpha** A researcher developing scanners to search for hidden weapons at airports has concluded that a new device is significantly better than the current scanner. He made this decision based on a test using $\alpha = 0.05$. Would he have made the same decision at $\alpha = 0.10$? How about $\alpha = 0.01$? Explain.

6. **Alpha again** Environmentalists concerned about the impact of high-frequency radio transmissions on birds found that there was no evidence of a higher mortality rate among hatchlings in nests near cell towers. They based this conclusion on a test using $\alpha = 0.05$. Would they have made the same decision at $\alpha = 0.10$? How about $\alpha = 0.01$? Explain.

7. **Significant?** Public health officials believe that 90% of children have been vaccinated against measles. A random survey of medical records at many schools across the country found that, among more than 13 000 children, only 89.4% had been vaccinated. A statistician would reject the 90% hypothesis with a P-value of $P = 0.011$.
a) Explain what the P-value means in this context.
b) The result is statistically significant, but is it important? Comment.

8. **Significant again?** A new reading program may reduce the number of elementary school students who read below grade level. The company that developed this program supplied materials and teacher training for a large-scale test involving nearly 8500 children in several school districts. Statistical analysis of the results showed that the percentage of students who did not attain the grade-level standard was reduced from 15.9% to 15.1%. The hypothesis that the new reading program produced no improvement was rejected with a P-value of 0.023.
a) Explain what the P-value means in this context.
b) Even though this reading method has been shown to be significantly better, why might you not recommend that your local school adopt it?

9. **Groceries** In January 2011, Yahoo surveyed 2400 U.S. men. A total of 1224 of the men identified themselves as the primary grocery shopper in their household.
a) Estimate the percentage of all U.S. males who identify themselves as the primary grocery shopper. Use a 98% confidence interval. Check the conditions first.
b) A grocery store owner had been told that 45% of men are the primary grocery shopper for their family, and targets his advertising accordingly. He wishes to conduct a hypothesis test to see if the fraction is in fact different from 45%. What does your confidence interval indicate? Explain.
c) What is the level of significance of this test? Explain.

10. **Is the euro fair?** Soon after the euro was introduced as currency in Europe, it was widely reported that someone had spun a euro coin 250 times and gotten heads 140 times. We wish to test a hypothesis about the fairness of spinning the coin.
a) Estimate the true proportion of heads. Use a 95% confidence interval. Don't forget to check the conditions first.
b) Does your confidence interval provide evidence that the coin is unfair when spun? Explain.
c) What is the significance level of this test? Explain.

11. **Rural med school applicants, again** In Exercise 24 in Chapter 17, we examined medical school applications from the rural Ontario population. In 2002, 2246 students from Ontario applied to the province's medical schools; in 2003, 2702 applied. Of the two-year total

(4948), there were 4588 applicants from urban areas and 360 from rural areas.[13] Of course, there are many sources of random variation in the results from any one-year or two-year period.

a) Give a 95% confidence interval for the percentage of applications generally coming from rural students around this time period.
b) Since the population of Ontario is 13% rural, is there reason for concern about a lack of rural applicants for medical school? Use your 95% confidence interval to test the null hypothesis that the general percentage of applications coming from rural students is no different than the rural population percentage of 13%.

12. Hard times In June 2010, a random poll of 800 working men found that 9% had taken on a second job to help pay the bills. (www.careerbuilder.com)

a) Estimate the true percentage of men that are taking on second jobs by constructing a 95% confidence interval.
b) A pundit on a TV news show claimed that only 6% of working men had a second job. Use your confidence interval to test whether his claim is plausible given the poll data.

13. Loans Before lending someone money, banks must decide whether they believe the applicant will repay the loan. One strategy used is a point system. Loan officers assess information about the applicant, totalling points they award for the person's income level, credit history, current debt burden, and so on. The higher the point total, the more convinced the bank is that it's safe to make the loan. Any applicant with a lower point total than a certain cutoff score is denied a loan.

We can think of this decision as a hypothesis test. Since the bank makes its profit from the interest collected on repaid loans, their null hypothesis is that the applicant will repay the loan and therefore should get the money. Only if the person's score falls below the minimum cutoff will the bank reject the null and deny the loan. This system is reasonably reliable, but, of course, sometimes there are mistakes.

a) When a person defaults on a loan, which type of error did the bank make?
b) Which kind of error is it when the bank misses an opportunity to make a loan to someone who would have repaid it?
c) Suppose the bank decides to lower the cutoff score from 250 points to 200. Is that analogous to choosing a higher or lower value of α for a hypothesis test? Explain.
d) What impact does this change in the cutoff value have on the chance of each type of error?

14. Spam Spam filters try to sort your e-mails, deciding which are real messages and which are unwanted. One method used is a point system. The filter reads each incoming e-mail and assigns points to the sender, the subject, key words in the message, and so on. The higher the point total, the more likely it is that the message is unwanted. The filter has a cutoff value for the point total; any message rated lower than that cutoff passes through to your inbox, and the rest, suspected to be spam, are diverted to the junk mailbox.

We can think of the filter's decision as a hypothesis test. The null hypothesis is that the e-mail is a real message and should go to your inbox. A higher point total provides evidence that the message may be spam; when there's sufficient evidence, the filter rejects the null, classifying the message as junk. This usually works pretty well, but, of course, sometimes the filter makes a mistake.

a) When the filter allows spam to slip through into your inbox, which kind of error is that?
b) Which kind of error is it when a real message gets classified as junk?
c) Some filters allow the user (that's you) to adjust the cutoff. Suppose your filter has a default cutoff of 50 points, but you reset it to 60. Is that analogous to choosing a higher or lower value of α for a hypothesis test? Explain.
d) What impact does this change in the cutoff value have on the chance of each type of error?

15. Second loan Exercise 13 described the loan score method a bank uses to decide which applicants to lend money to. Only if the total points awarded for various aspects of an applicant's financial condition fail to add up to a minimum cutoff score set by the bank will the loan be denied.

a) In this context, explain what is meant by the power of the test.
b) What could the bank do to increase the power?
c) What is the disadvantage of doing that?

16. More spam Consider again the points-based spam filter described in Exercise 14. When the points assigned to various components of an e-mail exceed the cutoff value you've set, the filter rejects its null hypothesis (that the message is real) and diverts that e-mail to a junk mailbox.

a) In this context, explain what is meant by the power of the test.
b) What could you do to increase the filter's power?
c) What is the disadvantage of doing that?

17. Obesity in First Nations youth According to the March 2007 "Healthy Weights for Healthy Kids" report of the Standing Committee on Health, 55% of First Nations children living on reserves are ether overweight or obese, which is far above the (already too high) national average.[14] Suppose the federal and provincial governments

[13] www.ncbi.nlm.nih.gov/pmc/articles/PMC1479467/

[14] www.parl.gc.ca/HousePublications/Publication.aspx?DocId=2795145&Language=E&Mode=1&Parl=39&Ses=1

decide to allocate funds for a new program targeted specifically at First Nations youth to help reduce obesity in this community. They decide to make it a trial program, to be evaluated after four years when data will be available. They will continue with the program only if there is strong evidence that the obesity rate in First Nations youth has decreased.

a) In words, what will their hypotheses be?
b) What would a Type I error be?
c) What would a Type II error be?
d) For each type of error, tell who would be harmed.
e) What would the power of the test represent in this context?

18. Alzheimer's Testing for Alzheimer's disease can be a long process, consisting of lengthy tests and medical diagnosis. Recently, a group of researchers (Solomon et al., 1998) devised a seven-minute test to serve as a quick screen for the disease for use in the general population of senior citizens. A patient who tested positive would then go through the more expensive battery of tests and medical diagnosis. The authors reported a false positive rate of 4% and a false negative rate of 8%.

a) Put this in the context of a hypothesis test. What are the null and alternative hypotheses?
b) What would a Type I error mean?
c) What would a Type II error mean?
d) Which is worse here, a Type I or Type II error? Explain.
e) What is the power of this test?

19. Testing cars A clean air standard requires that vehicle exhaust emissions not exceed specified limits for various pollutants. Suppose government regulators double-check a random sample of cars that a suspect repair shop has certified as okay. They will revoke the shop's license if they find significant evidence that the shop is certifying vehicles that do not meet standards.

a) In this context, what is a Type I error?
b) In this context, what is a Type II error?
c) Which type of error would the shop's owner consider more serious?
d) Which type of error might environmentalists consider more serious?

20. Quality control Production managers on an assembly line must monitor the output to be sure that the level of defective products remains small. They periodically inspect a random sample of the items produced. If they find a significant increase in the proportion of items that must be rejected, they will halt the assembly process until the problem can be identified and repaired.

a) In this context, what is a Type I error?
b) In this context, what is a Type II error?
c) Which type of error would the factory owner consider more serious?
d) Which type of error might customers consider more serious?

21. Cars again As in Exercise 19, government regulators are checking up on repair shops to see if they are certifying vehicles that do not meet pollution standards.

a) In this context, what is meant by the power of the test the regulators are conducting?
b) Will the power be greater if they test 20 or 40 cars? Why?
c) Will the power be greater if they use a 5% or a 10% level of significance? Why?
d) Will the power be greater if the repair shop's inspectors are only a little out of compliance or a lot? Why?

22. Production Consider again the task of the quality control inspectors in Exercise 20.

a) In this context, what is meant by the power of the test the inspectors conduct?
b) They are currently testing five items each hour. Someone has proposed that they test 10 each hour instead. What are the advantages and disadvantages of such a change?
c) Their test currently uses a 5% level of significance. What are the advantages and disadvantages of changing to an alpha level of 1%?
d) Suppose that, as a day passes, one of the machines on the assembly line produces more and more items that are defective. How will this affect the power of the test?

23. Equal opportunity? A company is sued for job discrimination because only 19% of the newly hired candidates were minorities when 27% of all applicants were minorities. Is this strong evidence that the company's hiring practices are (prima facie) discriminatory?

a) Is this a one-tailed or a two-tailed test? Why?
b) In this context, what would a Type I error be?
c) In this context, what would a Type II error be?
d) In this context, describe what is meant by the power of the test.
e) If the hypothesis is tested at the 5% level of significance instead of 1%, how will this affect the power of the test?
f) The lawsuit is based on the hiring of 37 employees. Is the power of the test higher than, lower than, or the same as it would be if it were based on 87 hires?

24. Stop signs Highway safety engineers test new road signs, hoping that increased reflectivity will make them more visible to drivers. Volunteers drive through a test course with several of the new- and old-style signs and rate which kind shows up the best.

a) Is this a one-tailed or a two-tailed test? Why?
b) In this context, what would a Type I error be?
c) In this context, what would a Type II error be?
d) In this context, describe what is meant by the power of the test.

e) If the hypothesis is tested at the 1% level of significance instead of 5%, how will this affect the power of the test?

f) The engineers hoped to base their decision on the reactions of 50 drivers, but time and budget constraints may force them to cut back to 20. How would this affect the power of the test? Explain.

25. **Dropouts** A Statistics professor has observed that for several years about 13% of the students who initially enrol in his Introductory Statistics course withdraw before the end of the semester. A salesman suggests that he try a statistics software package that gets students more involved with computers, predicting that it will cut the dropout rate. The software is expensive, and the salesman offers to let the professor use it for a semester to see if the dropout rate goes down significantly. The professor will have to pay for the software only if he chooses to continue using it.
a) Is this a one-tailed or two-tailed test? Explain.
b) Write the null and alternative hypotheses.
c) In this context, explain what would happen if the professor makes a Type I error.
d) In this context, explain what would happen if the professor makes a Type II error.
e) What is meant by the power of this test?

26. **Ads** A company is willing to renew its advertising contract with a local radio station only if the station can prove that more than 20% of the residents of the city have heard the ad and recognize the company's product. The radio station conducts a random phone survey of 400 people.
a) What are the hypotheses?
b) The station plans to conduct this test using a 10% level of significance, but the company wants the significance level lowered to 5%. Why?
c) What is meant by the power of this test?
d) For which level of significance will the power of this test be higher? Why?
e) They finally agree to use $\alpha = 0.05$, but the company proposes that the station call 600 people instead of the 400 initially proposed. Will that make the risk of Type II error higher or lower? Explain.

*27. **Wind power again** In Chapter 17 Exercise 44, you were asked to make a statistical report to help a land owner decide whether to place a wind turbine at his site, by testing if the mean wind speed for a candidate site is greater than 8 miles per hours. If so, there is enough electricity so that it is financially feasible to use one's own personal wind turbine. Suppose you set $\alpha = 0.05$.
a) Explain in this context the consequences of committing a Type I error.
b) Explain in this context the consequences of committing a Type II error.
c) If you are worried about being sued for giving bad advice to purchase the turbine, should you increase or decrease the value of α? Explain.
d) What is meant by the power of the test that you are conducting?
e) Suppose you repeat this test next year, when the wind speeds are generally greater. Will the power of the test increase, decrease, or remain the same?
f) What could you do to improve the power of the test this year?
g) Suppose we increase the sample size, and get a P-value = 0.001. The greater the wind speed, the greater the savings. Does that tiny P-value mean that the land owner can count on very big savings? What else might be useful?

*28. **More hot dogs** In Chapter 17, Exercise 42, a lab had to decide if a certain brand of hot dogs qualify for "reduced sodium" labelling (30% less sodium than regular hot dogs, or $\mu < 325$ mg). Suppose the lab has to make a decision and sets $\alpha = 0.05$.
a) Explain in this context the consequences of committing a Type I error.
b) Explain in this context the consequences of committing a Type II error.
c) Suppose you are very worried about the consequences of erroneously labelling the brand as reduced sodium and hence choose a very small significance level of $\alpha = 0.001$. How does this affect the probability of a Type II error and what are the practical consequences?
d) What is meant by the power of the test that the lab is conducting?
e) Suppose the lab repeats this test next year, when the average sodium content has actually increased a bit for this brand. Will the power of the test increase, decrease, or remain the same?
f) What could you do to improve the power of the test?
g) If you get a very tiny P-value, like 0.001, that means we can count on a sodium reduction very much more than the requisite 30%. Do you agree?

29. **Dropouts, part II** Initially 203 students signed up for the Stats course in Exercise 25. They used the software suggested by the salesman, and only 11 dropped out of the course.
a) Should the professor spend the money for this software? Support your recommendation with an appropriate test.
b) Explain carefully what your P-value means in this context.

30. **Testing the ads** The company in Exercise 26 contacts 600 people selected at random, and only 133 remember the ad.
a) Should the company renew the contract? Support your recommendation with an appropriate test.

b) Explain carefully what your P-value means in this context.

31. Two coins Two coins are in a drawer. They look the same, but one coin produces heads 90% of the time when spun, while the other one produces heads only 30% of the time. You select one of the coins. You are allowed to spin it *once* and then must decide whether the coin is the 90%- or the 30%-head coin. Your null hypothesis is that your coin produces 90% heads.
a) What is the alternative hypothesis?
b) Given that the outcome of your spin is tails, what would you decide? What if it were heads?
c) How large is α in this case?
d) How large is the power of this test? (*Hint:* How many possibilities are in the alternative hypothesis?)
e) How could you lower the probability of a Type I error and increase the power of the test at the same time?

32. Faulty or not? You are in charge of shipping computers to customers. You learn that a faulty disk drive was put into some of the machines. There's a simple test you can perform, but it's not perfect. All but 4% of the time, a good disk drive passes the test, but unfortunately, 35% of the bad disk drives pass the test, too. You have to decide on the basis of one test whether the disk drive is good or bad. Make this a hypothesis test.
a) What are the null and alternative hypotheses?
b) Given that a computer fails the test, what would you decide? What if it passes the test?
c) How large is α for this test?
d) What is the power of this test? (*Hint:* How many possibilities are in the alternative hypothesis?)

33. Hoops A basketball player with a poor foul-shot record practises intensively during the off-season. He tells the coach that he has raised his proficiency from 60% to 80%. Dubious, the coach asks him to take 10 shots, and is surprised when the player hits 9 out of 10. Did the player prove that he has improved?
a) Suppose the player really is no better than before—still a 60% shooter. What is the probability that he can hit at least 9 of 10 shots anyway? (*Hint:* Use a Binomial model.)
b) If that is what happened, now the coach thinks the player has improved when he has not. Which type of error is that?
c) If the player really can hit 80% now, and it takes at least 9 out of 10 successful shots to convince the coach, what's the power of the test?
d) List two ways the coach and player could increase the power to detect any improvement.

34. Pottery An artist experimenting with clay to create pottery with a special texture has been experiencing difficulty with these pieces. About 40% break in the kiln during firing. Hoping to solve this problem, she buys some more expensive clay from another supplier. She plans to make and fire 10 pieces and will decide to use the new clay if at most one of them breaks.
a) Suppose the new, expensive clay really is no better than her usual clay. What is the probability that this test convinces her to use it anyway? (*Hint:* Use a Binomial model.)
b) If she decides to switch to the new clay and it is no better, what kind of error did she commit?
c) If the new clay really can reduce breakage to only 20%, what is the probability that her test will not detect the improvement?
d) How can she improve the power of her test? Offer at least two suggestions.

35. Hoops II In Exercise 33, the basketball player was hoping to improve his foul-shot percentage. Suppose the coach asks him to take 80 shots, and we test the null hypothesis that his percentage is still 60% against the alternative that he has improved, at the 5% significance level.
a) What percentage of shots does he have to hit for us to reject the null hypothesis?
b) Calculate the power of the test if he now can hit on 75% of shots. Imitate the calculation in this chapter's second Math Box (pages 518–519).
c) If your software includes power calculations, confirm your calculation from part b). Also find the number of shots necessary to raise the power of the test to 0.99 when his true foul-shot percent-age has risen to 75%.

36. Pottery II In Exercise 34, the artist was hoping to reduce the breakage by switching to a new type of clay. Suppose she fires up 60 pieces, and tests the null hypothesis that the breakage rate is still 40% against the alternative of reduced breakage using the new clay, at the 5% significance level.
a) What percentage of the 60 tested pieces have to break for her to reject the null hypothesis?
b) Calculate the power of the test if breakage in fact is only 30% for the new clay. Imitate the calculation in this chapter's second Math Box (pages 518–519).
c) If your software includes power calculations, confirm your calculation in part b). Also find the number of pieces needed in the study to raise the power to 0.80 when the new clay's true breakage rate is 30%.

37. Simulations Use your computer software to generate a sample of size 75 from a Bernoulli distribution. This distribution has only two outcomes, 0 and 1, with complementary probabilities. Set the probability of the outcome "1" equal to 0.8. Consider "1" to be a success and "0" to be a failure.
a) For your sample, test the null hypothesis that $p = 0.8$, at the 5% significance level, with a two-sided alternative. What do you conclude? Did you make the right decision?

b) Draw 99 additional samples and repeat part a) for each, performing 99 additional 5% level tests. How often did you make the correct decision? The wrong decision? What percent of tests would you expect to produce a correct decision if you repeated these simulations many times? For 100 simulated samples, if X = number wrong decisions, what is the probability distribution (model) of X?

38. More simulations Use your computer software to generate a sample of size 60 from a Bernoulli distribution. This distribution has only two outcomes, 0 and 1, with complementary probabilities. Set the probability of "1" equal to 0.3. Consider "1" to be a success and "0" to be a failure.

a) For your sample, test the null hypothesis that $p = 0.3$, at the 10% significance level, with a two-sided alternative. What do you conclude? Did you make the right decision?

b) Draw 49 additional samples and repeat part a) for each, performing 49 additional 10% level tests. How often did you make the correct decision? The wrong decision? What percent of tests would you expect to produce a correct decision if you repeated these simulations many times? For 50 simulated samples, if X = number wrong decisions, what is the probability distribution (model) of X?

39. Still more simulations Use your computer software to generate a sample of size 75 from a Bernoulli distribution. This distribution has only two outcomes, 0 and 1, with complementary probabilities. Set the probability of "1" equal to 0.8. Consider "1" to be a success and "0" to be a failure.

a) For your sample, test the null hypothesis that $p = 0.7$, at the 5% significance level, for a two-sided alternative. What do you conclude? Did you make the correct decision?

b) Draw 99 additional samples and repeat part a) for each, performing 99 additional 5% level tests. How often did you make the correct decision? The wrong decision? Based on your simulations, estimate the Type II error probability and the power of the test when the true $p = 0.8$. Combine with results of other students in your class to get a better estimate of the power.

c) Optional: Use software or follow the Math Box calculation on page 513 to find the exact power of the test that you estimated by simulations in part b).

40. Yet more simulations Use your computer software to generate a sample of size 60 from a Bernoulli distribution. This distribution has only two outcomes, 0 and 1, with complementary probabilities. Set the probability of "1" equal to 0.3. Consider "1" to be a success and "0" to be a failure.

a) For your sample, test the null hypothesis that $p = 0.5$, at the 10% significance level, for two-sided alternative. What do you conclude? Did you make the correct decision?

b) Draw 49 additional samples and repeat part a) for each, performing 49 additional 10% level tests. How often did you make the correct decision? The wrong decision? Based on your simulations, estimate the Type II error probability and the power of the test when the true $p = 0.3$. Combine with results of other students in your class to get a better estimate of the power.

c) Optional: Use software or follow the Math Box calculation on page 513 to find the exact power of the test that you estimated by simulations in part b).

41. Multiple tests Suppose that 10 different research centres are investigating the same null hypothesis, but with different groups of subjects. Each tests the same null hypothesis at the 5% level. If the null is true, what is the chance of one or more centres producing a significant study (at 5% level)? What is the distribution of the number of centres that produce a (5% level) significant study? If (exactly) one centre gets a P-value < 0.05, what should you conclude? (What we really need is a *meta-analysis*, which is a methodology for combining the evidence from many studies properly.)

42. Lots of tests Suppose you run a study to compare a new drug with a placebo, and run a new test at the 5% level every day on the evolving data. The null hypothesis says there is no difference. On the 20th day, you find a significance difference at the 5% level. Should you celebrate the drug's success? Explain.

Just Checking ANSWERS

1. With a z-score of 0.62, you can't reject the null hypothesis. The experiment shows no evidence that the wheel is not fair.
2. At $\alpha = 0.10$ for a two-sided alternative (or at $\alpha = 0.05$ for a one-sided alternative), you can't reject the null hypothesis because 0.30 is contained in the 90% confidence interval—it's plausible that sending the DVDs is no more effective than just sending letters.
3. The confidence interval is from 29% to 45%. The DVD strategy is more expensive and may not be worth it. We can't distinguish the success rate from 30% given the results of this experiment, but 45% would represent a large improvement. The bank should consider another trial, increasing their sample size to get a narrower confidence interval.
4. A Type I error would mean deciding that the DVD success rate is higher than 30% when it really isn't. They would adopt a more expensive method for collecting payments that would be no better than the less expensive strategy.
5. A Type II error would mean deciding that there's not enough evidence to say that the DVD strategy works when in fact it does. The bank would fail to discover an effective method for increasing their revenue from delinquent accounts.
6. 60%; the larger the effect size, the greater the power. It's easier to detect an improvement to a 60% success rate than to a 32% rate.

MathXL
MyStatLab

Go to MathXL at www.mathxl.com or MyStatLab at www.mystatlab.com. You can practise exercises for this chapter as often as you want. The guided solutions will help you find answers step by step. You'll find a personalized study plan available to you too!

chapter 19

Comparing Two Proportions

Statistics is not a discipline like physics, chemistry, or biology where we study a subject to solve problems in the same subject. We study statistics with the main aim of solving problems in other disciplines.

—C. R. Rao

Where are we going?

Is the proportion of men who like our new Web page the same as the proportion of women who like it? Do people really feel better about the economy this month compared to last month, or was that increase just a result of sampling variation? It's much more common to compare two proportions than to test whether one is equal to a given number. Comparing two proportions is very much like testing one. The standard error is different, but the concepts are essentially the same.

Who	6971 male drivers
What	Seatbelt use
Why	Highway safety
When	2007
Where	Massachusetts

Do men take more risks than women? Psychologists have documented that, in many situations, men choose riskier behaviour than women do. But what is the effect of having a woman by their side? A recent seatbelt observation study in Massachusetts[1] found that, not surprisingly, male drivers wear seatbelts less often than women do. The study also noted that men's belt-wearing jumped more than 16 percentage points when they had a female passenger. Seatbelt use was recorded at 161 locations in Massachusetts, using random-sampling methods developed by the U.S. National Highway Traffic Safety Administration (NHTSA). Female drivers wore belts more than 70% of the time, regardless of the sex of their passengers. Of 4208 male drivers with female passengers, 2777 (66%) were belted. But among 2763 male drivers with male passengers only, 1363 (49.3%) wore seatbelts. This was only a random sample, but it suggests there may be a shift in men's risk-taking behaviour when women are present. What would we estimate the true size of that gap to be?

Comparisons between two percentages are much more common than questions about isolated percentages. And they are more interesting. We often want to know how two groups differ, whether a treatment is better than a placebo control, or whether this year's results are better than last year's.

19.1 The Standard Deviation of the Difference Between Two Proportions

We know the difference between the proportions of men wearing seatbelts seen in the *sample*. It's 16.7%. But what's the *true* difference for all men? We know that our estimate probably isn't exactly right. To say more, we need a way to judge how large the sample difference is. That means we'll need to estimate its standard deviation.

How do we find this estimate? We know how to estimate the standard deviation of the proportion in each group, but how about their difference? The answer comes from realizing that we have to start with the variances and the Pythagorean Theorem of Statistics from Chapter 14:

The variance of the sum or difference of two independent random variables is the sum of their variances.

This is such an important (and powerful) idea in Statistics that it's worth pausing a moment to review the reasoning. Here's some intuition about why variation increases even when we subtract two random quantities.

[1]Massachusetts Traffic Safety Research Program (June 2007).

For independent random variables, **variances add.**

A S *Activity:* **Compare Two Proportions.** Does a preschool program help disadvantaged children later in life?

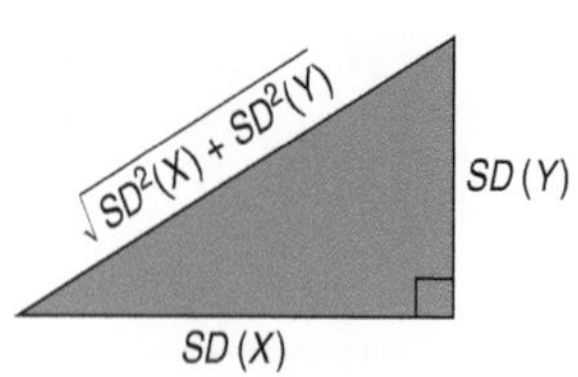

Let's say everyone in your classroom has a full box of cereal that's labelled 16 ounces (454 grams). We know that's not exact: there's some variation from box to box. Now everyone pours a two-ounce bowl of cereal. There'll also be some variation in how much each person pours. How much variation will there be in the amount of cereal that is left in each box? Even though the amount left in the box is the *difference* between the original amount and the amount that was poured, there will be even more variation than there was in either the original amount or the amount poured. This is because the amount left depends on *both* the variation in how much was originally in the box and the variation in the amount poured.

According to our rule, the variance of the amount of cereal left in the box would now be the *sum* of the two *variances.*

We want a standard deviation, not a variance, but that's just a square root away. We can write symbolically what we've just said:

$$Var(X - Y) = Var(X) + Var(Y), \text{ so}$$

$$SD(X - Y) = \sqrt{SD^2(X) + SD^2(Y)} = \sqrt{Var(X) + Var(Y)}.$$

Be careful, though—this simple formula applies only when X and Y are independent. Just as the Pythagorean Theorem works only for right triangles, our formula works only for independent random variables. Always check for independence before using it.

Fortunately, proportions observed in independent random samples *are* independent, so we can put the two proportions in for X and Y and add their variances. We just need to use careful notation to keep things straight.

Combining independent random quantities always *increases* the overall variation, so even for *differences* of independent random variables, **variances add.**

When we have two samples, each can have a different size and proportion value, so we keep them straight with subscripts. Often we choose subscripts that remind us of the groups. For our example, we might use "M" and "F," but generically we'll just use "1" and "2." We will represent the two sample proportions as $\hat{p}_1$ and $\hat{p}_2$, and the two sample sizes as n_1 and n_2.

The standard deviations of the sample proportions are $SD(\hat{p}_1) = \sqrt{\frac{p_1q_1}{n_1}}$ and $SD(\hat{p}_2) = \sqrt{\frac{p_2q_2}{n_2}}$, so the variance of the difference in the proportions is

$$Var(\hat{p}_1 - \hat{p}_2) = \left(\sqrt{\frac{p_1q_1}{n_1}}\right)^2 + \left(\sqrt{\frac{p_2q_2}{n_2}}\right)^2 = \frac{p_1q_1}{n_1} + \frac{p_2q_2}{n_2}$$

The standard deviation is the square root of that variance:

$$SD(\hat{p}_1 - \hat{p}_2) = \sqrt{\frac{p_1q_1}{n_1} + \frac{p_2q_2}{n_2}}$$

But we don't know the true values of p_1 and p_2. When we have the sample proportions in hand from the data, we use them to estimate the true proportions in the SD formula, giving us the standard error of the difference:

$$SE(\hat{p}_1 - \hat{p}_2) = \sqrt{\frac{\hat{p}_1\hat{q}_1}{n_1} + \frac{\hat{p}_2\hat{q}_2}{n_2}}$$

For Example FINDING THE STANDARD ERROR OF A DIFFERENCE IN PROPORTIONS

A recent survey of 886 randomly selected teenagers (aged 12–17) found that more than half of them had online profiles.[2] Some researchers and privacy advocates are concerned about the possible access to personal information about teens in public places on the Internet. There appear to be differences between boys and girls in their online behaviour. Among teens aged 15–17, 57% of the 248 boys had posted profiles, compared to 70% of the 256 girls. Let's start the process of estimating how large the true gender gap might be.

[2]Princeton Survey Research Associates International for the Pew Internet & American Life Project.

QUESTION: What is the standard error of the difference in sample proportions?

ANSWER: Because the boys and girls were selected at random, it's reasonable to assume their behaviours are independent, so it's okay to use the Pythagorean Theorem of Statistics:

$$SE(\hat{p}_{boys}) = \sqrt{\frac{0.57 \times 0.43}{248}} = 0.0314$$

$$SE(\hat{p}_{girls}) = \sqrt{\frac{0.70 \times 0.30}{256}} = 0.0286$$

$$SE(\hat{p}_{girls} - \hat{p}_{boys}) = \sqrt{0.0314^2 + 0.0286^2} = 0.0425$$

19.2 Assumptions and Conditions When Comparing Proportions

Before we complete our example, we need to check assumptions and conditions.

Independence Assumptions

Independence Assumption: Within each group, the data should be based on results for independent individuals. We can't check that for certain, but we *can* check the following:

Randomization Condition: The data in each group should ideally be drawn independently and at random[3] from the target population or generated by a randomized comparative experiment. If not, you need to argue that the cases may be considered reasonably representative of some larger populations to which your conclusions may apply.

The 10% Condition: If the data are sampled without replacement and the sample is bigger than 10% of the population, the individual observations will not be independent and our SE formula requires an adjustment (but you may still proceed using our SE formula as long as you note that your P-value and confidence level will be conservative since the calculated SE is an overestimate).

Because we are comparing two groups in this way, we need an additional Independence Assumption. In fact, this may be the most important of these assumptions. If it is violated, these methods just won't work.

Independent Groups Assumption: The two groups we're comparing must also be independent of each other. Usually, the independence of the groups from each other is evident from the way the data were collected.

Why is the independent groups assumption so important? If we compare husbands with their wives, or a group of subjects before and after some treatment, we can't just add the variances. Subjects' performance before a treatment might very well be related to their performance after the treatment. So the proportions are not independent and the Pythagorean-style variance formula does not hold. A different method of analysis would be required if the subjects or cases are linked between the two samples in some fashion (a method we won't be discussing in this book, except for the case of quantitative data, in Chapter 22).

Sample Size Condition

Each of the groups must be big enough. As with individual proportions, we need larger groups to estimate proportions that are near 0% or 100%. We usually check the Success/Failure Condition for each group.

Success/Failure Condition: Both groups are big enough that at least 10 successes and at least 10 failures have been observed in each.

---

[3]Our SE formula assumes simple random samples. It should be a good approximation or a bit conservative for stratified or systematic samples, but could be a severe underestimate for cluster or multistage samples.

For Example CHECKING ASSUMPTIONS AND CONDITIONS

RECAP: Among randomly sampled teens aged 15–17, 57% of the 248 boys had posted online profiles, compared with 70% of the 256 girls.

QUESTION: Can we use these results to make inferences about all 15–17-year-olds?

ANSWER:

- **Randomization Condition:** The sample of boys and the sample of girls were both chosen randomly.
- **10% Condition:** 248 boys and 256 girls are each less than 10% of all teenage boys and girls.
- **Independent Groups Assumption:** Because the samples were selected at random, it's reasonable to believe the boys' online behaviours are independent of the girls' online behaviours.
- **Success/Failure Condition:** Among the boys, 248(0.57) = 141 had online profiles and the other 248(0.43) = 107 did not. For the girls, 256(0.70) = 179 successes and 256(0.30) = 77 failures. All counts are at least 10.

Because all the assumptions and conditions are satisfied, it's okay to proceed with inference.

(Note that when we find the observed counts of successes and failures, we round off to whole numbers. We're using the reported percentages to recover the actual counts.)

19.3 A Confidence Interval for the Difference Between Two Proportions

WHY NORMAL?

In Chapter 14, we learned that sums and differences of independent Normal random variables follow a Normal model. That's the reason we use a Normal model for the difference of two independent proportions.

We're almost there. We just need one more fact about proportions. We already know that for large enough samples, each of our proportions has an approximately Normal sampling distribution. The same is true of their difference.

THE SAMPLING DISTRIBUTION MODEL FOR A DIFFERENCE BETWEEN TWO INDEPENDENT PROPORTIONS

Provided that the sampled values are independent, the samples are independent, and the sample sizes are large enough, the sampling distribution of $\hat{p}_1 - \hat{p}_2$ is modelled by a Normal model with mean $\mu = p_1 - p_2$ and standard deviation

$$SD(\hat{p}_1 - \hat{p}_2) = \sqrt{\frac{p_1 q_1}{n_1} + \frac{p_2 q_2}{n_2}}.$$

The sampling distribution model and the standard deviation give us all we need to find a margin of error for the difference in proportions—or at least they would if we knew the true proportions, p_1 and p_2. However, we don't know the true values, so we'll work with the observed proportions, $\hat{p}_1$ and $\hat{p}_2$, and use $SE(\hat{p}_1 - \hat{p}_2)$ to estimate the standard deviation. The rest is just like a one-proportion z-interval.

A TWO-PROPORTION Z-INTERVAL

When the conditions are met,[4] we are ready to find the confidence interval for the difference of two proportions, $p_1 - p_2$. The confidence interval is

$$(\hat{p}_1 - \hat{p}_2) \pm z^* \times SE(\hat{p}_1 - \hat{p}_2)$$

[4]What if the sample size condition is not met? Just as in the case of one proportion, there is a "plus four" adjustment that works surprisingly well. Add one phony success and one phony failure to each of the samples, and apply the confidence interval formula given in this box to your artificially enhanced samples (altering the $\hat{p}_i$, $\hat{q}_i$, and n_i values).

where we find the standard error of the difference,

$$SE(\hat{p}_1 - \hat{p}_2) = \sqrt{\frac{\hat{p}_1\hat{q}_1}{n_1} + \frac{\hat{p}_2\hat{q}_2}{n_2}}.$$

from the observed proportions.

The critical value z^* depends on the particular confidence level, C, that we specify.

For Example FINDING A TWO-PROPORTION Z-INTERVAL

RECAP: Among randomly sampled teens aged 15–17, 57% of the 248 boys had posted online profiles, compared to 70% of the 256 girls. We calculated the standard error for the difference in sample proportions to be $SE(\hat{p}_{girls} - \hat{p}_{boys}) = 0.0425$ and found that the assumptions and conditions required for inference checked out okay.

QUESTION: What does a confidence interval say about the difference in online behaviour?

ANSWER: A 95% confidence interval for $p_{girls} - p_{boys}$ is

$$(\hat{p}_{girls} - \hat{p}_{boys}) \pm z^* SE(\hat{p}_{girls} - \hat{p}_{boys})$$
$$(0.70 - 0.57) \pm 1.96(0.0425)$$
$$0.13 \pm 0.083$$
$$(4.7\%, 21.3\%)$$

We can be 95% confident that among teens aged 15–17, the proportion of girls who post online profiles is between 4.7 and 21.3 percentage points higher than the proportion of boys who do. It seems clear that teen girls are more likely to post profiles than are boys the same age (since the confidence interval lies entirely above zero).

Step-by-Step Example A TWO-PROPORTION Z-INTERVAL

Now we are ready to be more precise about the passenger-based gap in male drivers' seatbelt use. We'll estimate the difference with a confidence interval using a **two-proportion *z*-interval**.

Question: How much difference is there in the proportion of male drivers who wear seatbelts when sitting next to a male passenger and the proportion who wear seatbelts when sitting next to a female passenger?

THINK ➡ Plan State what you want to know. Discuss the variables and the *W*'s. Identify the parameter you wish to estimate. (It usually doesn't matter in which direction we subtract, but it is much easier for people to understand the meaning of the interval when we choose the direction with a positive difference.)	I want to know the true difference in the population proportion, p_M, of male drivers who wear seatbelts when sitting next to a man, and p_F, the proportion who wear seatbelts when sitting next to a woman. The data are from a random sample of drivers in Massachusetts in 2007, observed according to procedures developed by the NHTSA. The parameter of interest is the difference $p_F - p_M$.
Choose and state a confidence level.	I will find a 95% confidence interval for this parameter.
Model Think about the assumptions and check the conditions.	✓ **Randomization Condition:** The NHTSA methods are more complex than a simple random sample but should result in a suitable random sample allowing us to generalize beyond the sample, and to assume that driver behaviour is independent from car to car. Lacking more knowledge of their actual sampling design, we'll assume the resultant precision is similar to that of an SRS, justifying our SE formula. (However, if cluster sampling was used, our SE estimate may be too small.)

✓ **10% Condition:** The samples include far fewer than 10% of all male drivers accompanied by male or by female passengers.

✓ **Independent Groups Assumption:** It's reasonable to believe that seatbelt use among drivers with male passengers and those with female passengers are independent.

The Success/Failure Condition must hold for each group.

✓ **Success/Failure Condition:** Among male drivers with female passengers, 2777 wore seatbelts and 1431 did not; of those driving with male passengers, 1363 wore seatbelts and 1400 did not. Each group contained far more than 10 successes and 10 failures.

State the sampling distribution model for the statistic.

Choose your method.

Under these conditions, the sampling distribution of the difference between the sample proportions is approximately Normal, so I'll find a **two-proportion z-interval.**

SHOW ➡ Mechanics Construct the confidence interval.

I know

$n_F = 4208,\ n_M = 2763$

The observed sample proportions are

As often happens, the key step in finding the confidence interval is estimating the standard deviation of the sampling distribution model of the statistic. Here the statistic is the difference in the proportions of men who wear seatbelts when they had a female passenger and those who had a male passenger. Substitute the data values into the formula.

$$\hat{p}_F = \frac{2777}{4208} = 0.660,\ \hat{p}_M = \frac{1363}{2763} = 0.493$$

I'll estimate the SD of the difference with

$$SE(\hat{p}_F - \hat{p}_M) = \sqrt{\frac{\hat{p}_F\hat{q}_F}{n_F} + \frac{\hat{p}_M\hat{q}_M}{n_M}}$$

$$= \sqrt{\frac{(0.660)(0.340)}{4208} + \frac{(0.493)(0.507)}{2763}}$$

$$= 0.012$$

The sampling distribution is Normal, so the critical value for a 95% confidence interval, z^*, is 1.96. The margin of error is the critical value times the SE.

$$ME = z^* \times SE(\hat{p}_F - \hat{p}_M)$$

$$= 1.96(0.012) = 0.024$$

The confidence interval is the *statistic* ±*ME*.

The observed difference in proportions is $\hat{p}_F - \hat{p}_M = 0.660 - 0.493 = 0.167$, so the 95% confidence interval is

$$0.167 \pm 0.024$$

or 14.3% to 19.1%

TELL ➡ Conclusion Interpret your confidence interval in the proper context. (Remember: We're 95% confident that our interval captured the true difference.)

I am 95% confident that the proportion of male drivers who wear seatbelts when driving next to a female passenger is between 14.3 and 19.1 percentage points higher than the proportion who wear seatbelts when driving next to a male passenger.[5]

[5] And since the interval lies *entirely above zero*, we have convincing evidence of a passenger gender effect on male driver seatbelt use.

This is an interesting result—but be careful not to try to say too much! In Massachusetts, overall seatbelt use is relatively low, so we can't be certain that these results generalize to other regions. And these were two different groups of men, so we can't say that, individually, men are more likely to buckle up when they have a woman passenger. You can probably think of several alternative explanations; we'll suggest just a couple. Perhaps age is a lurking variable: Maybe older men are more likely to wear seatbelts and also more likely to be driving with their wives. Or maybe men who don't wear seatbelts have trouble attracting women!

Just Checking

A listener-sponsored radio station plans to launch a special appeal for additional contributions from current members. Unsure of the most effective way to contact people, they run an experiment. They randomly select two groups of current members. They send the same request for donations to everyone, but it goes to one group by e-mail and to the other group by regular mail. The station was successful in getting contributions from 26% of the members they e-mailed but only from 15% of those who received the request by regular mail. A 90% confidence interval estimated the difference in donation rates to be 11% ± 7%.

1. Interpret the confidence interval in this context.
2. Based on this confidence interval, what conclusion would we reach if we tested the hypothesis that there's no difference in the response rates to the two methods of fundraising? Explain.

19.4 Testing for a Difference Between Proportions

Prejudice in the Penitentiary?

Who	Women serving time in federal penitentiaries
What	Security risk classification
When	2003
Where	Federal penitentiaries in Canada
Why	To compare the treatment of Aboriginal and non-Aboriginal women inmates

Correctional Services Canada uses a rating system to classify prison inmates according to security risk level. A higher security risk level classification has consequences such as withdrawal of certain privileges and increased likelihood of serving out a longer sentence.

In a study published in the *Canadian Journal of Criminology and Criminal Justice*, University of Toronto criminologists Anthony Doob and Cheryl Webster contend that Aboriginal women are often classified at a higher security risk level compared with other female prisoners. In 2003, they found that 60% (41) of 68 Aboriginal women inmates were classified as medium security risk, while 42% (112) of 266 non-Aboriginal women inmates were similarly classified, though fewer infractions had actually been committed by the Aboriginal women while incarcerated. (31% of the Aboriginal women inmates had committed infractions, compared with 53% of the non-Aboriginal women inmates).[6]

How should we interpret the observed 18% difference? Is it indicative of a real difference in general medium risk classification rates between Aboriginal and non-Aboriginal women—or might this difference be the result of some random variation that naturally occurs from sample to sample (due to numerous chance factors at work in any given year)? After all, if you toss two loonies 100 times each, would you expect to see equal counts of heads in the two samples? I hope not. So we have to rule out random sampling variation before we can conclude that a real difference exists.

The question calls for a hypothesis test. The parameter of interest is the true *difference* between the medium security risk classification rates of the two groups.

[6]www.magazine.utoronto.ca/leading-edge/prejudice-in-the-penitentiary

What is the appropriate null hypothesis? That's easy: We hypothesize that there is no difference in the proportions. This is such a natural null hypothesis that we rarely consider any other. But instead of writing H_0: $p_1 = p_2$, we usually express it in a slightly different way. To make it relate directly to the *difference*, we hypothesize that the difference in proportions is zero:

$$H_0: p_1 - p_2 = 0$$

We know that the standard error of the difference in proportions is

$$SE(\hat{p}_1 - \hat{p}_2) = \sqrt{\frac{\hat{p}_1\hat{q}_1}{n_1} + \frac{\hat{p}_2\hat{q}_2}{n_2}}$$

and we could just plug in the numbers, but we can do even better. The secret is that proportions and their standard deviations are linked. There are two proportions in the standard error formula—but look at the null hypothesis. It says that these proportions are equal. To do a hypothesis test, we *assume* that the null hypothesis is true. So there should be just a single value of $\hat{p}$ in the SE formula (and, of course, $\hat{q}$ is just $1 - \hat{p}$).

How would we do this for this risk classification example? If the null hypothesis is true, then among all female inmates, the two groups have the same medium-risk proportion. Overall, we saw $41 + 112 = 153$ medium risk inmates out of a total of $68 + 266 = 334$ inmates. The overall proportion of medium-risk inmates was $153/334 = 0.4581$.

Combining the counts like this to get an overall proportion is called **pooling**. Whenever we have data from different sources or different groups that we believe came from the same underlying population, we can pool them to get better estimates.

When we have counts for each group, we can find the pooled proportion as

$$\hat{p}_{\text{pooled}} = \frac{Success_1 + Success_2}{n_1 + n_2}$$

where $Success_1$ is the number of successes in group 1 and $Success_2$ is the number of successes in group 2. That's the overall proportion of success.

When we have only proportions and not the counts, as in the inmate risk classification example, we have to reconstruct the number of successes by multiplying the sample sizes by the proportions:[7]

$$Success_1 = n_1\hat{p}_1 \text{ and } Success_2 = n_2\hat{p}_2$$

If these calculations don't come out to whole numbers, round them first. There must have been a whole number of successes, after all. (This is the *only* time you should round values in the middle of a calculation.)

We then put this pooled value into the formula, substituting it for *both* sample proportions in the standard error formula:

$$SE_{\text{pooled}}(\hat{p}_1 - \hat{p}_2) = \sqrt{\frac{\hat{p}_{\text{pooled}}\hat{q}_{\text{pooled}}}{n_1} + \frac{\hat{p}_{\text{pooled}}\hat{q}_{\text{pooled}}}{n_2}}$$
$$= \sqrt{\frac{0.4581 \times (1 - 0.4581)}{68} + \frac{0.4581 \times (1 - 0.4581)}{266}}$$

This works out to 0.068. (You may factor out the product of the two proportions if you wish.)

Naturally, we'll reject our null hypothesis if we see a large enough difference in the two proportions. How can we decide whether the difference we see, $\hat{p}_1 - \hat{p}_2$ is large? The answer is the same as always: We just compare it to its standard deviation.

Unlike the one-sample test, the null hypothesis here doesn't provide a standard deviation, so we'll use a standard error (here, pooled). Since the sampling distribution is Normal, divide the observed difference by its standard error to get a *z*-score. Then find the corresponding P-value so you can assess the evidence. We call this procedure a **two-proportion *z*-test**.

---

[7]Hence we may also calculate the pooled p using $\hat{p}_{\text{pooled}} = \frac{n_1\hat{p}_1 + n_2\hat{p}_2}{n_1 + n_2}$, so our pooled estimate is just a weighted average of the two sample proportions.

We didn't pool proportions for the confidence interval, because we weren't assuming the two proportions were equal. Combining them doesn't usually make a big practical difference, but we should take advantage of it for the hypothesis test.

A TWO-PROPORTION Z-TEST

The conditions for the two-proportion z-test are the same as for the two-proportion z-interval. We are testing the hypothesis

$$H_0: p_1 - p_2 = 0$$

Because we hypothesize that the proportions are equal, we pool the groups to find

$$\hat{p}_{pooled} = \frac{Success_1 + Success_2}{n_1 + n_2}$$

and use that pooled value to estimate the standard error:

$$SE_{pooled}(\hat{p}_1 - \hat{p}_2) = \sqrt{\frac{\hat{p}_{pooled}\hat{q}_{pooled}}{n_1} + \frac{\hat{p}_{pooled}\hat{q}_{pooled}}{n_2}}$$

Now we find the test statistic,

$$z = \frac{(\hat{p}_1 - \hat{p}_2) - 0}{SE_{pooled}(\hat{p}_1 - \hat{p}_2)}$$

When the conditions are met and the null hypothesis is true, this statistic follows the standard Normal model, so we can use that model to obtain a P-value.

Step-by-Step Example A TWO-PROPORTION Z-TEST

Question: Are the medium security classification rates of the Aboriginal and non-Aboriginal women inmates really different?

THINK ➡ Plan State what you want to know. Discuss the variables and the *W*'s.

I want to know whether medium risk classification rates differ for Aboriginal and non-Aboriginal women inmates. The data are for female federal penitentiary inmates in Canada in 2003.

Hypotheses Did the researchers have a clear *a priori* hypothesis of unfavourable treatment of Aboriginal women and set out to prove this? If so, a one-sided alternative could be justified. But the details are not so clear, so we will take the conservative approach of a two-sided alternative, reflecting our uncertainty about what to expect and an interest in detecting any difference in classification practices.

H_0: There is no difference in medium risk classification rates in the two groups:

$$p_{Aboriginal} - p_{Other} = 0$$

H_A: The rates are different: $p_{Aboriginal} - p_{Other} \neq 0$

Model Think about the assumptions and check the conditions.

✓ **Independence Assumption:** Classifications of different inmates should be more or less independent.

✓ **Randomization Condition:** The inmates were not randomly selected, but we will view them as representative of the larger collection of recent past, present, and possible future inmates incarcerated under similar conditions.

✓ **10% Condition:** The number of possible inmates is large.

✓ **Independent Groups Assumption:** We have two unrelated groups of women (no linking or pairing of women in one group with specific women in the other group).

	✓ **Success/Failure Condition:** For Aboriginal women, the counts were 41 and 68 − 41 = 27. For non-Aboriginal women, the counts were 112 and 266 − 112 = 154. The observed numbers of both successes and failures are much more than 10 for both groups.[8]
State the null model. Choose your method.	Because the conditions are satisfied, I'll use a Normal model and perform a **two-proportion *z*-test.**
SHOW ➡ Mechanics The hypothesis is that the proportions are equal, so pool the sample data. Use the pooled SE to estimate $SD(\hat{p}_{Aboriginal} - \hat{p}_{Other})$	$n_{Aboriginal} = 68,\ y_{Aboriginal} = 41,\ \hat{p}_{Aboriginal} = 0.603$ $n_{Other} = 266,\ y_{Other} = 112,\ \hat{p}_{Other} = 0.421$ $\hat{p}_{pooled} = \frac{y_{Aboriginal} + y_{Other}}{n_{Aboriginal} + n_{Other}} = \frac{41 + 112}{68 + 266} = 0.4581$ $SE_{pooled}(\hat{p}_{Aboriginal} - \hat{p}_{Other})$ $= \sqrt{\frac{\hat{p}_{pooled}\hat{q}_{pooled}}{n_{Aboriginal}} + \frac{\hat{p}_{pooled}\hat{q}_{pooled}}{n_{Other}}}$ $= \sqrt{\frac{(0.4581)(0.5419)}{68} + \frac{(0.4581)(0.5419)}{266}}$ ≈ 0.0677 The observed difference in sample proportions is $\hat{p}_{Aboriginal} - \hat{p}_{Other} = 0.603 - 0.421 = 0.182$
Make a picture. Sketch a Normal model centred at the hypothesized difference of 0. Shade the region to the right of the observed difference, and because this is a two-tailed test, also shade the corresponding region in the other tail. The observed difference in proportions is 0.182. Find its SE and standardize. Find the P-value using Table Z or technology. Because this is a two-tailed test, we must *double* the probability we find in the upper tail.	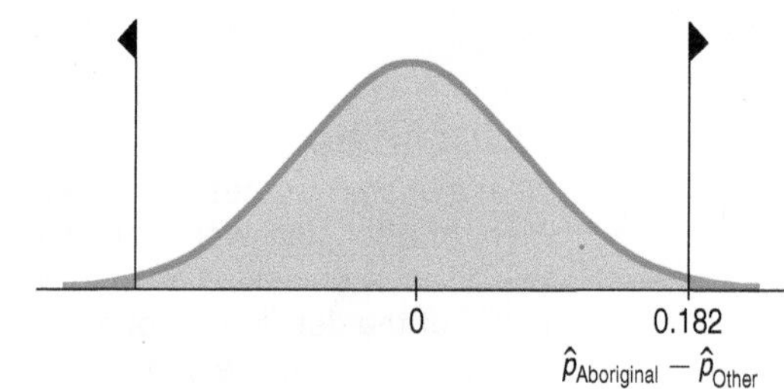 $z = \frac{(\hat{p}_{Aboriginal} - \hat{p}_{Other}) - 0}{SE_{pooled}(\hat{p}_{Aboriginal} - \hat{p}_{Other})} = \frac{0.182 - 0}{0.0677} = 2.69$ $P = 2P(z \geq 2.69) = 0.0072$
TELL ➡ Conclusion Link the P-value to your decision about the null hypothesis, and state your conclusion in context.	The P-value of 0.0072 says that if there really were no genuine difference in security risk between the two groups, then the difference observed in this study would happen only 7 times in 1000 by chance. This is so small that I reject the null hypothesis of no difference and conclude that there is a difference in the rate of medium-security risk classification, with Aboriginal women more likely to be classified as a medium security risk. It appears that a *prima facie* case exists for the possibility of systemic bias, though other lurking variables may explain this statistically significant difference.

[8]This is one of those situations in which the traditional term "success" seems a bit weird. A success here is an inmate deemed a security risk. "Success" and "failure" are arbitrary labels left over from studies of gambling games.

Improving the Success/Failure Condition

The vaccine Gardasil® was introduced to prevent the strains of human papillomavirus (HPV) that are responsible for almost all cases of cervical cancer. In randomized placebo-controlled clinical trials,[9] only one case of HPV was diagnosed among 7897 women who had received the vaccine, compared with 91 cases diagnosed among 7899 women who had received a placebo. The one observed HPV case ("success") doesn't appear to meet the at-least-10-successes criterion mentioned earlier for confidence intervals.

However, for the two-proportion z-test, the *proper* success/failure test stems from the null hypothesis and hence uses the *expected* frequencies, which we can find from the pooled proportion. In this case,

$$\hat{p}_{\text{pooled}} = \frac{91 + 1}{7899 + 7897} = 0.0058$$

$$n_1\hat{p}_{\text{pooled}} = 7899(0.0058) = 46$$

$$n_2\hat{p}_{\text{pooled}} = 7897(0.0058) = 46$$

so we can proceed with the hypothesis test.

Often it is easier just to check the observed numbers of successes and failures. If they are both greater than 10, you don't need to look further. But keep in mind that the correct condition actually uses the *expected* frequencies rather than the *observed* ones.

Just Checking

3. The Canadian Drug Use and Monitoring Survey (CADMUS) 2011 survey estimated that 2.6% of Canadians aged 15–24 had used the drug ecstasy in the previous year, based on its sample of 671 Canadians in this age group. A similar study in 2004 estimated that 4.4% of Canadians in this age group had used ecstasy in the previous year, based on its sample of 2085 Canadians in this age group. To see if the percentage using ecstasy has decreased over time, why can't we just use a one-proportion z-test of H_0: $p = 0.044$ and see what the P-value of $\hat{p} = 0.026$ is?
4. For surveys like this, which has more variability: the percentage of respondents who say they have used ecstasy in either year or the difference in the percentages between the two years?

For Example A TWO-PROPORTION Z-TEST

RECAP: Another concern of the study on teens' online profiles was safety and privacy. In the random sample, girls were less likely than boys to say that they are easy to find online from their profiles. Only 19% (62 girls) of 325 teen girls with profiles say that they are easy to find, while 28% (75 boys) of the 268 boys with profiles say the same.

QUESTION: Are these results evidence of a real difference between boys and girls? Perform a two-proportion z-test and discuss what you find.

ANSWER:

$$H_0: p_{\text{boys}} - p_{\text{girls}} = 0$$
$$H_A: p_{\text{boys}} - p_{\text{girls}} \neq 0$$

✓ **Randomization Condition:** The sample of boys and the sample of girls were both chosen randomly, ensuring independence of the observations and generalizability.

[9] *Quadrivalent Human Papillomavirus Vaccine: Recommendations of the Advisory Committee on Immunization Practices (ACIP)*, National Center for HIV/AIDS, Viral Hepatitis, STD and TB Prevention [May 2007].

✓ **10% Condition:** 268 boys and 325 girls are each less than 10% of all teenage boys and girls with online profiles.

✓ **Independent Groups Assumption:** Because the samples were selected at random, it's reasonable to believe the boys' perceptions are independent of the girls'.

✓ **Success/Failure Condition:** Among the girls, there were 62 "successes" and 263 failures, and among boys, 75 successes and 193 failures. These counts are at least 10 for each group.

Because all the assumptions and conditions are satisfied, it's okay to do a **two-proportion *z*-test**:

$$\hat{p}_{pooled} = \frac{75 + 62}{268 + 325} = 0.231$$

$$SE_{pooled}(\hat{p}_{boys} - \hat{p}_{girls}) = \sqrt{\frac{0.231 \times 0.769}{268} + \frac{0.231 \times 0.769}{325}} = 0.0348$$

$$z = \frac{(0.28 - 0.19) - 0}{0.0348} = 2.59$$

$$P(z > 2.59) = 0.0048$$

This is a two-tailed test, so the P-value = 2(0.0048) = 0.0096. Because this P-value is very small, I reject the null hypothesis. This study provides strong evidence that there really is a difference in the proportions of teen girls and boys who say they are easy to find online.[10]

Sample Size Determination

When designing a study, we need to carefully plan the sample size, or we may end up with rather noninformative results, lacking either in precision (with a big margin of error in the confidence interval) or in power (so we have little ability to detect differences that matter).

To control the margin of error for the difference of the sample proportions, follow the basic approach discussed in earlier chapters: Set the formula for the margin of error equal to the desired margin of error and solve for the sample size after guessing at unknowns in the formula. Here you will have two proportions to guess, but you can always go with the conservative approach and plug in 0.5 for any proportion rather than guess.

Suppose that in the example of teen online profile privacy, we want the width of the confidence interval to be no more than 6 percentage points, or, equivalently, we want a margin of error equal to 0.03 or less, with 95% confidence. How large would the samples have to be? Assume equal sample sizes and denote the common sample size by n. Using the worst case scenario with all proportions equal to 0.5, we set:

$$\textbf{\textit{Desired margin of error}} = 0.03 = 2\sqrt{\frac{p_1q_1}{n_1} + \frac{p_2q_2}{n_2}} = 2\sqrt{\frac{(.5)(.5)}{n} + \frac{(.5)(.5)}{n}}$$

Then solve for n to get $n = 2222$ (2222 girls and 2222 boys).

To control the power of the test for equal proportions, we recommend the use of software. Unfortunately, this calculation depends not just on the difference between the two proportions, but also on the specific values of the two proportions, so we have no choice but to do some guesswork. Make a guess at one proportion,[11] then think about how big a difference from that proportion would represent a clinically or practically important change or difference in order to get your second proportion.

[10]How big is the difference? A confidence interval would be a good way to wrap things up.

[11]If one treatment is a control group, a standard treatment, or a "before" treatment, making a reasonable guess may not be so hard. Using a value of 0.5 would produce the most conservative calculation, just as with margin of error calculations.

Consider the teen online profile privacy example again. How big should the samples be if we consider a difference of 5 percentage points to be important and we want a 90% chance of discovering a difference of this magnitude when it actually exists?

Using your software, specify all of the following:

- *The nature of the alternative (one-sided or two-sided).* Choose a two-sided alternative here.
- *The alpha level of the test.* Let's go with $\alpha = 0.05$.
- *The values of both proportions.* This is the hard part. Let's guess that about a quarter of teens have easy-to-find profiles. Set one proportion at 0.25 and the other 5 percentage points lower, at 0.20.
- *The desired power of the test.* Let's target the power at 0.9.

We see from the computer output below that a sample of 1464 boys and 1464 girls would be required to achieve the desired power for a 5% difference in the true proportions.

POWER AND SAMPLE SIZE

Test for Two Proportions

Testing proportion 1 = proportion 2 (versus not =)
Calculating power for proportion 2 = 0.25
Alpha = 0.05

Proportion 1	Sample Size	Target Power	Actual Power
0.2	1464	0.9	0.900057

The sample size is for each group.

WHAT CAN GO WRONG?

- **Don't use two-sample proportion methods when the samples aren't independent.** These methods give wrong answers when this assumption of independence is violated. Good random sampling is usually the best insurance of independent groups. Make sure there is no relationship between the two groups. For example, you can't compare the proportion in your sample who favour the Liberals with the proportion *in the same sample* who favour the Conservatives! The responses are not independent because you've asked the same people!

  Randomized block designs, where we pair subjects before administering treatments (Chapter 11), arise often and are important, but they can't be analyzed with the methods of this chapter. The pairing of subjects creates a dependence between observations in the two treatment groups.

- **Interpret inference methods with caution where there was no randomization.** If the data do not come from random samples or from a properly randomized experiment, then inference about the differences in proportions must be interpreted cautiously. Statistical significance means that something more than chance is at work. A genuine difference has been demonstrated, but in just what populations? How far beyond the data at hand can you reasonably generalize?

- **Don't interpret a significant difference in proportions causally unless the study is a randomized experiment.** In an example in this chapter, we discovered that Aboriginal women in prison are more likely to be classified as high risk than non-Aboriginal women. The data, though, are based on an observational study, so the observed difference here is only suggestive of possible systemic bias based on race; there may be other differences between Aboriginal women inmates and other female inmates affecting risk classification. In an observational study, there is always the danger that other lurking variables not accounted for are the real reason for an observed difference. They may be difficult to guess, but that doesn't mean they don't exist, or can't be found with further careful investigation. Be careful not to jump to conclusions about causality.

CONNECTIONS

In Chapter 2, we looked at contingency tables for two categorical variables. Differences in proportions are just 2 × 2 contingency tables. You'll often see data presented in this way. For example, we could show the seatbelt use data as

	Female Passenger	Male Passenger	Total
Seatbelt	2777	1363	4140
No seatbelt	1431	1400	2831
Total	4208	2763	6971

We tested whether the column percentages of belted drivers were the same for the two different passenger-gender groups being compared in the study.

The inference methods we are learning follow a consistent pattern. Although we have a different standard error, the step-by-step procedures are almost identical. For a test, we divide the statistic by its standard error to get a z-score and P-value.

What Have We Learned?

Learning Objectives

Know how to construct and interpret confidence intervals for the difference between proportions of two independent groups.

- Check the Assumptions and Conditions before making inferences.

Be able to perform and interpret a two-sample z-test of whether proportions in two independent groups are equal.

- Because we are hypothesizing that the proportions are equal, we pool the counts to find a *pooled standard error* estimate of the standard deviation.

For all comparisons of two groups, keep in mind that we require both independence of observations within each group and independence between the groups themselves.

- The methods discussed in this chapter are not appropriate for paired or matched data.

Review of Terms

Variances of independent random variables add The variance of a sum or difference of independent random variables is the sum of the variances of those variables (p. 534).

Sampling distribution of the difference between two proportions The sampling distribution $\hat{p}_1 - \hat{p}_2$ is, under appropriate assumptions, modelled by a Normal model with mean $\mu = p_1 - p_2$ and standard deviation $SD(\hat{p}_1 - \hat{p}_2) = \sqrt{\frac{p_1 q_1}{n_1} + \frac{p_2 q_2}{n_2}}$ (p. 536).

Two-proportion z-interval Gives a confidence interval for the true difference in proportions, $p_1 - p_2$, in two independent groups.

The confidence interval is $(\hat{p}_1 - \hat{p}_2) \pm z^* \times SE(\hat{p}_1 - \hat{p}_2)$, where z^* is a critical value from the standard Normal model corresponding to the specified confidence level (p. 537).

Pooling Data from two populations may sometimes be combined, or *pooled*, to estimate a parameter when the parameter is assumed to be the same in both populations. If our null hypothesis states that two proportions are equal, we pool the data to provide an estimate of the common proportion, and then use that pooled value in SE calculations to make them more precise (p. 540).

Two-proportion z-test Test the null hypothesis H_0: $p_1 - p_2 = 0$ by referring the statistic

$$z = \frac{\hat{p}_1 - \hat{p}_2}{SE_{pooled}(\hat{p}_1 - \hat{p}_2)}$$

to a standard Normal model (p. 540).

On the Computer TWO SAMPLE METHODS FOR PROPORTIONS

It is so common to test against the null hypothesis of no difference between the two true proportions that most statistics programs simply assume this null hypothesis. And most will automatically use the pooled SE. If you wish to test a different null (say, that the true difference is 0.3), you may have to search for a way to do it.

As with inference for single proportions, most statistics programs want the "success" and "failure" status for each case, usually specified by category names like "yes" and "no." Often we just know the proportions of successes, $\hat{p}_1$ and $\hat{p}_2$, and the sample sizes, n_1 and n_2.

EXCEL

Inference methods for proportions are not part of the standard Excel tool set, nor are power and sample size calculations.

COMMENTS

For summarized data, type the calculation into any cell and evaluate it.

JMP

For a **categorical variable** that holds category labels, the **Distribution** platform includes tests and intervals of proportions.

- For summarized data, put the category names in one variable and the frequencies in an adjacent variable.
- Designate the frequency column to have the **role** of **frequency**. Then use the **Distribution** platform.

For power and sample size calculations, proceed as follows:

- Choose **Power and Sample Size** from the **DOE** menu.
- Choose **Two Sample Proportions** from the submenu.
- Choose your **Alpha** value and select a one- or two-sided alternative.
- Indicate the null difference between the **Proportions** (likely zero).
- Next, choose two reasonable values for **Proportion 1** and **Proportion 2**, which are separated enough to be of practical importance.
- Fill in either your desired **Sample sizes** or **Power**. What you leave blank will be calculated.
- Click **Continue**.

COMMENTS

JMP uses slightly different methods for proportion inferences than those discussed in this text. Your answers are likely to be slightly different.

MINITAB

To find hypothesis tests and confidence intervals for proportions:

- Choose **Basic Statistics** from the **Stat** menu.
- Choose 2 **Proportions** . . . from the Basic Statistics submenu. If the data are organized as category names in one column and data values in another, assign the variables from the variable list box to the **Samples in one column** box. "Subscripts" are the category names.
- If the data are organized as two separate columns of responses, click on **Samples in different columns:** and assign the variables from the variable list box. If you have summarized data, click the **Summarized Data** button and fill in the number of trials and the number of successes for each group.
- Click the **Options** button and specify the remaining details. Remember to click the **Use pooled estimate of *p* for test** box when testing the null hypothesis of no difference between proportions.
- Click the **OK** button.

To compute the power of the two independent proportions test for a given sample size, or to find the sample size needed to achieve the target power:

- Choose **Power and Sample Size** from the **Stat** menu.
- Choose 2 **Proportions** from the submenu.
- The null hypothesis is zero difference between the proportions. Fill in your guess or estimate for one of the proportions labelled **Proportion 2**.

- Fill in a value for **Proportion 1 value** reflecting a difference from **Proportion 2** that matters.
- Fill in either **Sample size(s)** or **Power value(s)**. The sample size is assumed to be the same for each group.
- Use **Options . . .** if you want a one-sided alternative, or significance level different from 0.05.
- Use **Graph. . .** if you want to produce an entire power curve of power values as a function of **Proportion 1** (Release 16 or later).
- Click **OK**.

COMMENTS

When working from a variable that names categories, MINITAB treats the last category as the "success" category. You can specify how the categories should be ordered.

For power calculations, fill **Proportion 1** and **Proportion 2** to reflect the type of difference you want to detect. Make reasonable guesses. If one group is a control group or standard treatment, estimate **Proportion 2** for that treatment based on prior knowledge, and then think how big a difference would be important to get **Proportion 1**. If the sample size is fixed and you need to know the power, leave **Power value** blank. If you are planning a study and need to know the sample size required to achieve desired power, fill in the **Power value(s)** sought (multiple entries are fine), and leave **Sample size(s)** blank.

R

- Create a **vector** of **successes** = **c(X1,X2)** where X1 is the number of successes in group 1 and X2 is the number of successes in group 2.
- Create a vector of **sample.sizes** = **c(n1,n2)** where n1 and n2 are the sample sizes for the two groups, respectively.
- To test the equality of the two proportions:
  - **prop.test(successes,sample.sizes,alternative = "two.sided",conf.level = 0.95,correct = FALSE)**

will test the null hypothesis of equal proportions against the two-sided alternative and give a 95% confidence level.

The package {pwr} does various power calculations, but they are not straightforward in the case of proportions, due to a definition of effect size that uses the arcsine transformation (which stabilizes variances of proportions).

COMMENTS

R uses the χ^2 statistic instead of the *z*-statistic for testing. To find the corresponding *z*-statistic, take the square root of the X-squared (χ^2) statistic in the output. The P-values are the same. Other possibilities for alternative are "greater" and "less."

SPSS

SPSS does not provide hypothesis tests for proportions. However, power and sample size calculations can be performed with the IBM SPSS SamplePower add-on.

STATCRUNCH

To do inference for the difference between two proportions using summaries:

- Click on **Stat**.
- Choose **Proportions » Two sample » with summary.**
- Enter the **Number of successes** (x) and **Number of observations** (n) in each group.
- Click on **Next**.
- Indicate **Hypothesis Test**, then enter the hypothesized **Null proportion** difference (usually 0), and choose the **Alternative** hypothesis.

  OR

  Indicate **Confidence Interval**, and then enter the **Level** of confidence.
- Click on **Calculate**.

To do inference for the difference between two proportions using data:

- Click on **Stat**.
- Choose **Proportions » Two sample » with data.**
- For each group, choose the variable **Column** listing the **Outcomes**, and enter the **outcome** to be considered a **Success**.
- Click on **Next**.

- Indicate **Hypothesis Test**, then enter the hypothesized **Null** proportion difference (usually 0), and choose the **Alternative** hypothesis.

OR

Indicate **Confidence Interval**, and then enter the **Level** of confidence.

- Click on **Calculate**.

Power calculations may be done using **Stat>Proportion Statistics>Two sample>Power/Sample size.** Click on **Hypothesis Test Power**. Fill in all the boxes except for the one that you want to determine, either **Power** or **Sample Size**

TI-83/84 PLUS

To calculate a confidence interval for the difference between two population proportions:

- Select **B:2-PropZInt** from the **STAT TESTS** menu.
- Enter the observed counts and the sample sizes for both samples.
- Specify a confidence level.
- Calculate the interval.

To do the mechanics of a hypothesis test for equality of population proportions:

- Select **6:2-PropZTest** from the **STAT TESTS** menu.
- Enter the observed counts and sample sizes.
- Indicate what kind of test you want: one-tail upper tail, lower tail, or two-tail.
- Calculate the result.

Power and sample size calculations are not available.

COMMENTS

Beware: When you enter the value of x, you need the count, not the percentage. The count must be a whole number.

Exercises

1. **Online social networking** In April 2011, the Pew Internet and American Life Project surveyed 799 teenagers aged 12 to 17. Among the study's findings, 93% of teen social media users have a Facebook account, while only 24% have a MySpace account. What does it mean to say that the difference in proportions is "significant"?

2. **Science news** At the end of 2010, the Pew Project for Excellence in Journalism investigated where people are getting their news. In the study, 34% of responders said they read news online as opposed to 31% who favoured newspapers. What does it mean to say that the difference is not significant?

3. **Name recognition** A political candidate runs a week-long series of television ads designed to attract public attention to his campaign. Polls taken before and after the ad campaign show some increase in the proportion of voters who now recognize this candidate's name, with a P-value of 0.033. Is it reasonable to believe the ads may be effective?

4. **Origins** In a 1993 Gallup poll, 47% of U.S. respondents agreed with the statement: "*God created human beings pretty much in their present form at one time within the last 10 000 years or so.*" When Gallup asked the same question in 2001, only 45% of respondents agreed. Is it reasonable to conclude that there was a change in public opinion given that the P-value for the difference is 0.37? Explain.

5. **Revealing information** A random sample of 886 teens were asked which of several personal items of information they thought were okay to share with someone they had just met. Forty-four percent said it was okay to share their e-mail addresses, but only 29% said they would give out their cell phone numbers. A researcher claims that a two-proportion z-test could tell whether there was a real difference among all teens. Explain why that test would not be appropriate for these data.

6. **Regulating access** When a random sample of 935 parents were asked about rules in their homes, 77% said they had rules about the kinds of television shows their children could watch. Among the 790 parents whose teenage children had Internet access, 85% had rules about the kinds of Internet sites their teens could visit. That looks like a difference, but can we tell? Explain why a two-sample z-test would not be appropriate here.

7. **Gender gap** A local member of parliament fears he has a problem with women voters. His campaign staff plans to run a poll to assess the situation. They'll randomly sample 300 men and 300 women, asking if they have a favourable impression of the candidate. Obviously the staff can't know this, but suppose the candidate has a positive image with 59% of males but only with 53% of females.
   a) What kind of sampling design is his staff planning to use?
   b) What difference would you expect the poll to show?

c) Sampling error means the poll won't reflect the difference perfectly. What is the standard deviation for the difference in the proportions?
d) Sketch a sampling model for the size difference in proportions of men and women with favourable impressions of the candidate that might appear in a poll like this.
e) Could the campaign staff be misled by the poll, concluding in error that there really is no gender gap? Explain.

8. Buy it again? A consumer magazine plans to poll car owners to see if they are happy enough with their vehicles that they would purchase the same model again. They'll randomly select 450 owners of North American-made cars and 450 owners of Japanese models. Obviously the actual opinions of the entire population can't be known, but suppose 76% of owners of North American cars and 78% of owners of Japanese cars would purchase another.
a) What kind of sampling design is the magazine planning to use?
b) What difference would you expect their poll to show?
c) Sampling error means the poll won't reflect the difference perfectly. What is the standard deviation for the difference in the proportions?
d) Sketch a sampling model for the difference in proportions that might appear in a poll like this.
e) Could the magazine be misled by the poll, concluding that owners of North American cars are much happier with their vehicles than are owners of Japanese cars? Explain.

9. Arthritis The U.S. Centers for Disease Control and Prevention reported the results of a survey of randomly selected Americans aged 65 and older, which found that 411 of 1012 men and 535 of 1062 women suffered from some form of arthritis.
a) Are the assumptions and conditions necessary for inference satisfied? Explain.
b) Create a 95% confidence interval for the difference in the proportions of senior men and women who have this disease.
c) Interpret your interval in this context.
d) Does this confidence interval suggest that arthritis is more likely to afflict women than men? Explain.

10. Prostate cancer There has been debate among doctors over whether surgery can prolong life among men suffering from prostate cancer, a type of cancer that typically develops and spreads very slowly. Recently, the *New England Journal of Medicine* published results of some Scandinavian research. Men diagnosed with prostate cancer were randomly assigned to either undergo surgery or not. Among the 347 men who had surgery, 16 eventually died of prostate cancer, compared with 31 of the 348 men who did not have surgery.
a) Was this an experiment or an observational study? Explain.
b) Are the assumptions and conditions necessary for inference satisfied? Explain.
c) Create a 95% confidence interval for the difference in rates of death for the two groups of men.
d) Based on your confidence interval, is there evidence that surgery may be effective in preventing death from prostate cancer? Explain.

11. Pets Researchers at the U.S. National Cancer Institute released the results of a study that investigated the effect of weed-killing herbicides on house pets. They examined 827 dogs from homes where an herbicide was used on a regular basis, diagnosing malignant lymphoma in 473 of them. Of the 130 dogs from homes where no herbicides were used, only 19 were found to have lymphoma.
a) What is the standard error of the difference in the two proportions?
b) Construct a 95% confidence interval for this difference.
c) State an appropriate conclusion.

12. Carpal tunnel The painful wrist condition called carpal tunnel syndrome can be treated with surgery or less invasive wrist splints. Recently, *Time* magazine reported on a study of 176 patients. Among the half that had surgery, 80% showed improvement after three months, but only 48% of those who used the wrist splints improved.
a) What is the standard error of the difference in the two proportions?
b) Construct a 95% confidence interval for this difference.
c) State an appropriate conclusion.

13. Ear infections A new vaccine was recently tested to see if it could prevent the painful and recurrent ear infections that many infants suffer from. *The Lancet*, a medical journal, reported a study in which babies of about a year old were randomly divided into two groups. One group received vaccinations; the other did not. During the following year, only 333 of 2455 vaccinated children had ear infections, compared to 499 of 2452 unvaccinated children in the control group.
a) Are the conditions for inference satisfied?
b) Find a 95% confidence interval for the difference in rates of ear infection.
c) Use your confidence interval to explain whether you think the vaccine is effective.

14. **Anorexia** The *Journal of the American Medical Association* reported on an experiment intended to see if the drug Prozac® could be used as a treatment for the eating disorder anorexia nervosa. The subjects, women being treated for anorexia, were randomly divided into two groups. Of the 49 who received Prozac®, 35 were deemed healthy a year later, compared to 32 of the 44 who got the placebo.
a) Are the conditions for inference satisfied?
b) Find a 95% confidence interval for the difference in outcomes.
c) Use your confidence interval to explain whether you think Prozac® is effective.

15. **Another ear infection** In Exercise 13 you used a confidence interval to examine the effectiveness of a vaccine against ear infections in babies. Suppose that instead you had conducted a hypothesis test. (Answer these questions *without* actually doing the test.)
a) What hypotheses would you test?
b) State a conclusion based on your confidence interval.
c) If that conclusion is wrong, which type of error did you make?
d) What would be the consequences of such an error?

16. **Anorexia again** In Exercise 14 you used a confidence interval to examine the effectiveness of Prozac® in treating anorexia nervosa. Suppose that instead you had conducted a hypothesis test. (Answer these questions *without* actually doing the test.)
a) What hypotheses would you test?
b) State a conclusion based on your confidence interval.
c) If that conclusion is wrong, which type of error did you make?
d) What would be the consequences of such an error?

17. **Teen smoking, part I** A Vermont study published by the American Academy of Pediatrics examined parental influence on teenagers' decisions to smoke. A group of students who had never smoked were questioned about their parents' attitudes toward smoking. These students were questioned again two years later to see if they had started smoking. The researchers found that, among the 284 students who had indicated that their parents disapproved of kids smoking, 54 had become established smokers. Among the 41 students who had initially said their parents were lenient about smoking, 11 became smokers. Do these data provide strong evidence that parental attitude influences teenagers' decisions about smoking?
a) What kind of design did the researchers use?
b) Write appropriate hypotheses.
c) Are the assumptions and conditions necessary for inference satisfied?
d) Test the hypothesis and state your conclusion.
e) Explain what your P-value means in this context.
f) If it is later found that parental attitudes actually do influence teens decisions to smoke, which type of error did you commit?

18. **Depression** A study published in the *Archives of General Psychiatry* examined the impact of depression on a patient's ability to survive cardiac disease. Researchers identified 450 people with cardiac disease, evaluated them for depression, and followed the group for four years. Of the 361 patients with no depression, 67 died. Of the 89 patients with minor or major depression, 26 died. Among people who suffer from cardiac disease, are depressed patients more likely to die than nondepressed ones?
a) What kind of design was used to collect these data?
b) Write appropriate hypotheses.
c) Are the assumptions and conditions necessary for inference satisfied?
d) Test the hypothesis and state your conclusion.
e) Explain what your P-value means in this context.
f) If your conclusion is actually incorrect, which type of error did you commit?

19. **Teen smoking, part II** Consider again the Vermont study discussed in Exercise 17.
a) Create a 95% confidence interval for the difference in the proportion of teens who may smoke and have approving parents and those who may smoke and have disapproving parents.
b) Interpret your interval in this context.
c) Carefully explain what "95% confidence" means.

20. **Depression revisited** Consider again the study of the association between depression and cardiac disease survivability in Exercise 18.
a) Create a 95% confidence interval for the difference in survival rates.
b) Interpret your interval in this context.
c) Carefully explain what "95% confidence" means.

21. **Pregnancy** A reproductive clinic reported 42 live births to 157 women under the age of 38, but only 7 live births for 89 clients aged 38 and older. Is this strong evidence of a difference in the effectiveness of the clinic's methods between younger and older women?
a) Was this an experiment? Explain.
b) Test an appropriate hypothesis and state your conclusion in context.
c) If you concluded that there was a difference, estimate that difference with a confidence interval and interpret your interval in context.

22. **Birthweight** The *Journal of the American Medical Association* reported a study examining the possible impact of air pollution caused by the 9/11 attack on

New York's World Trade Center on the weight of babies. Researchers found that 8% of 182 babies born to mothers who were exposed to heavy doses of soot and ash on September 11 were classified as having low birth weight. Only 4% of 2300 babies born in another New York City hospital whose mothers had not been near the site of the disaster were similarly classified. Does this indicate a possibility that air pollution might be linked to a significantly higher proportion of low-weight babies?

a) Was this an experiment? Explain.
b) Test an appropriate hypothesis, and state your conclusion in context.
c) If you concluded there is a difference, estimate that difference with a confidence interval and interpret that interval in context.

23. Politics and sex One month before an election, a poll of 630 randomly selected voters showed 54% planning to vote for a certain candidate. A week later it became known that the candidate was involved in a scandal, and a new poll showed only 51% of 1010 voters supporting him. Do these results indicate a decrease in voter support for his candidacy?

a) Test an appropriate hypothesis, and state your conclusion.
b) If your conclusion turns out to be wrong, would you be making a Type I or a Type II error?
c) If you concluded there was a difference, estimate that difference with a confidence interval and interpret your interval in context.

24. Mammograms Regular mammogram screening may detect breast cancer early, resulting in fewer deaths from that disease. One study that investigated this issue over a period of 18 years was published during the 1970s. Among 30 565 women who had never had mammograms, 196 died of breast cancer, while only 153 of 30 131 who had undergone screening died of breast cancer.

a) Do these results support that mammograms may be an effective screening tool to reduce breast cancer deaths?
b) If your conclusion is incorrect, what type of error have you committed?
c) If you found a difference, estimate it with a confidence interval and interpret your interval in context.

25. Pain Researchers comparing the effectiveness of two pain medications randomly selected a group of patients who had been complaining of a certain kind of joint pain. They randomly divided these people into two groups, then administered the pain killers. Of the 112 people in the group who received medication A, 84 said this pain reliever was effective. Of the 108 people in the other group, 66 reported that pain reliever B was effective.

a) Write a 95% confidence interval for the percentage of people who may get relief from this kind of joint pain by using medication A. Interpret your interval.
b) Write a 95% confidence interval for the percentage of people who may get relief by using medication B. Interpret your interval.
c) Do the intervals for A and B overlap? What do you think this means about the comparative effectiveness of these medications?
d) Find a 95% confidence interval for the difference in the proportions of people who may find these medications effective. Interpret your interval.
e) Does this interval contain zero? What does that mean?
f) Why do the results in parts c) and e) seem contradictory? If we want to compare the effectiveness of these two pain relievers, which is the correct approach? Why?

26. Gender gap Candidates for political office realize that different levels of support among men and women may be a crucial factor in determining the outcome of an election. One candidate found that 52% of 473 men polled said they will vote for him, but only 45% of the 522 women in the poll expressed support.

a) Write a 95% confidence interval for the percentage of male voters who may vote for this candidate. Interpret your interval.
b) Write a 95% confidence interval for the percentage of female voters who may vote for him. Interpret your interval.
c) Do the intervals for males and females overlap? What do you think this means about the gender gap?
d) Find a 95% confidence interval for the difference in the proportions of males and females who will vote for this candidate. Interpret your interval.
e) Does this interval contain zero? What does that mean?
f) Why do the results in parts c) and e) seem contradictory? If we want to see if there is a gender gap among voters with respect to this candidate, which is the correct approach? Why?

27. Sensitive men A *Time* magazine article about survey of men's attitudes noted that "Young men are more comfortable than older men talking about their problems." The survey reported that 80 of 129 men between 18–24 years of age and 98 of 184 men between 25–34 years of age said they were comfortable talking about their problems. What do you think? Is *Time's* interpretation justified by these numbers?

28. Mammograms redux In 2001, the conclusion of the study outlined in Exercise 24 was questioned. A new nine-year study was conducted in Sweden, comparing 21 088 women who had mammograms with 21 195 who did not. Of the women who underwent screening, 63 died of breast cancer, compared to 66 deaths among the control group. (Source: *The New York Times,* Dec. 9, 2001.) Do these results support the effectiveness of regular mammograms in preventing deaths from breast cancer?

29. Online activity checks Are more parents checking up on their teens' online activities? A Pew survey in 2004 found that 33% of 868 randomly sampled teens said that their parents checked to see what Web sites they visited. In 2006, the same question posed to 811 teens found 41% reporting such checks. Do these results provide evidence that more parents are checking?

30. Computer gaming Who plays online or electronic games? A survey in 2006 found that 69% of 223 boys aged 12–14 said they "played computer or console games like Xbox or PlayStation . . . or games online." Of 248 boys aged 15–17, only 62% played these games. Is this evidence of a real age-based difference?

31. Food preference GfK Roper Consulting gathers information on consumer preferences around the world to help companies monitor attitudes about health, food, and healthcare products. They asked people in many different cultures how they felt about the following statement:

I have a strong preference for regional or traditional products and dishes from where I come from.

In a random sample of 800 respondents, 417 of 646 people who live in urban environments agreed (either completely or somewhat) with that statement, compared to 78 out of 154 people who live in rural areas. Based on this sample, is there evidence that the percentage of people agreeing with the statement about regional preferences differs between all urban and rural dwellers?

32. Fast food The global survey we learned about in Exercise 31 also asked respondents how they felt about the statement, "I try to avoid eating fast foods." The random sample of 800 included 411 people 35 years old or younger, and of those, 197 agreed (completely or somewhat) with the statement. Of the 389 people over 35 years old, 246 people agreed with the statement. Is there evidence that the percentage of people avoiding fast food is different in the two age groups?

33. Lake Ontario fish According to Environmental Defense's "Up to the Gills" 2009 update on pollution in Great Lakes fish species, mercury contamination was the cause of four fish advisories issued for 44 various fish-size combinations in Lake Ontario in 2009, whereas there were 11 mercury-caused fish advisories for the same 44 species-size combinations in 2003.[12] Test for a significant difference between years, if you can. If you cannot, explain precisely why not.

34. First openly gay premier In a February 2013 Harris-Decima poll, it was reported that 31% of a random sample of Canadians felt that Kathleen Wynne's swearing in as the first female Ontario premier and the first openly gay premier in Canada was a very significant breakthrough for women in Canadian politics, while 11% felt that it was not at all significant (a majority felt it was somewhat or not that significant). At the end of the report, it was stated that 1000 randomly selected Canadians were interviewed and that a sample of this size has a margin of error of 3.1%, 19 times in 20. (www.harris-decima.ca/news/releases)

a) Give a 95% confidence interval for the difference between the percentage of Canadians who felt this event was a very significant breakthrough and the percentage who felt it was not significant at all, if you can. If you cannot, explain clearly why not.

b) In the same poll, it was reported that the 40% of those aged 18–24 felt it was a very significant breakthrough, while 28% of those over the age of 50 felt it was very significant. Give a 95% confidence interval for the true population difference in this opinion between 18–24 year olds and those over 50, if you can. If you cannot, explain clearly why not.

35. Sex and reading preference For a class project, a University of Toronto student studied the difference in reading preferences between males and females observed in the Toronto subway system.

Reading Material	Male	Female
Novels	23	30
School Notes	8	4
Newspaper	18	10
Magazine	3	5
Other	4	1

a) Test for a difference in the proportion of males and females who read newspapers on the subway.

b) Test for a difference in the proportion of males and females who read novels on the subway.

c) Is there a danger in comparing one category at a time via a series of 5% level tests? Explain. What would happen if there were not five but 40 categories of reading material to compare in this way?

[12]www.environmentaldefence.ca/reports/pdf/UpToTheGills_2009%20Final.pdf

36. Stabbed in London, shot in Toronto In London, England, during 2006–07, there were 167 homicides: 68 by stabbing and 29 by shooting. In the same period in Toronto, there were 154 homicides: 47 by stabbing and 72 by shooting.[13]

a) Test for a significant difference in the proportions of murders committed by shooting between the two cities. If significant, report and interpret the appropriate 95% confidence interval.

b) Test for a significant difference in the proportions of murders committed by stabbing between the two cities. If significant, report and interpret the appropriate 95% confidence interval.

c) We could also test for significant differences in the proportions of murders committed by several other methods. If we did this for several methods, say four different ones, each at the 5% level, what would happen, roughly, to the chance of making one or more Type I errors while performing the four tests? In other words, if there are no differences in any of these average rates between the two cities, would there be a 5% chance of a falsely significant test result? Explain.

37. Homicide victims Cases of female victims of male-perpetrated homicide are well publicized, but the majority of victims of male violence—at the homicide level—are in fact male. In Canada in 2007, 90% of those accused of homicide were male and 73% of victims were male. Gender clearly matters. What about age? Below are some data on the victims in 2007 by gender and age:

	Male victim	Female victim
0–11 years	14	11
12–29 years	202	45
30 or more years	216	106
Total	432	162

Source: Adapted from Statistics Canada CANSIM database, http://cansim2.statcan.gc.ca, table 253-0003, October 23, 2008.

a) Compute and compare the (conditional) victim age distributions for males and females.

b) Test for a significant difference in the proportions of victims aged 30 or more.

c) Test for a significant difference in the proportions of victims aged 11 or less.

d) Test for a significant difference in the proportions of victims aged 12–29.

e) Describe your findings from the tests above—in other words, describe the nature of the association between gender and age for the victims of homicide. Instead of doing all these tests, an alternative would be the chi-square test of association, to be discussed in Chapter 23. What is one key issue of concern when examining a question via multiple tests of hypothesis (hunting for significance)?

f) Considering the young children murdered (age 0–11), is there a significant gender effect? Are such victims equally divided between the genders? Do a test.

38. Hockey injuries and relative age Is there a relationship between (relative) age and injury prevalence in Canadian youth hockey? Below are summary data from the Canadian Hospitals Injury Reporting and Prevention Program (boys 10–15 years of age; 1995–2002). Quartile 1 included those with January, February, and March birth dates and quartile 4 included those with October, November, and December birth dates; hence those born in quartile 1 were relatively older than their peers born in quartile 4. The other two quartiles do not affect any of the conclusions that may be drawn from this study and are omitted to simplify things.

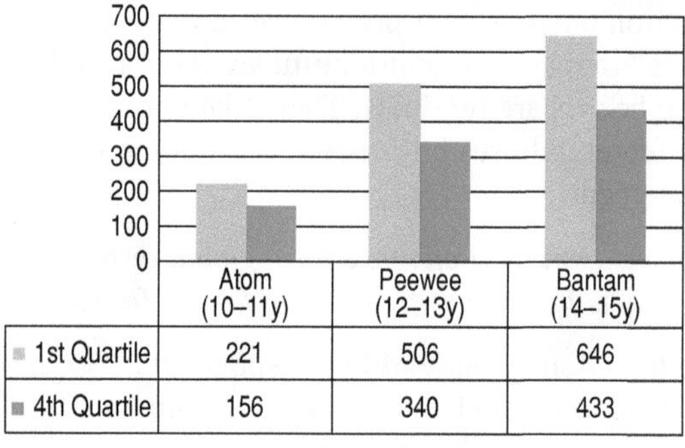

a) Test whether the relative age effect is the same at the Atom level and the Bantam level; that is, whether the likelihood that the injured player (given he is from one of the two age groups shown) is relatively older is the same at both levels.

b) Repeat part a), but compare Peewee and Bantam.

c) Repeat part a), but compare Atom and Peewee.

d) We are investigating the relationship between level of play and the relative age effect on injury. Comment on any possible problems inherent in examining a single question via multiple tests.

39. Carpal tunnel again In Exercise 12, two methods for treating carpal tunnel syndrome (wrist splints and surgery) were compared.

a) Instead of 176 patients, about how many patients in total would have been needed to reduce the margin of error to 5 percentage points (with 95% confidence)?

[13]D. Spiegelhalter and A. Barnett, "London murders: A predictable pattern?" *Significance*, March 2009.

b) We think that about half, or 50%, of patients improve using wrist splints. An improvement of 10% (from say 50% to 60%) from use of surgery would be big enough to matter clinically. For a test of hypothesis (two-sided), how many patients should be included if we wanted an 80% chance of detecting an improvement of this magnitude (power = 0.80) when using surgery? (You need software for this.)

40. More pain In Exercise 25, we compared the effectiveness of two pain medications.

a) About how many patients in each group (assuming equal sizes) would have been needed to bring down the margin of error to 10 percentage points (with 95% confidence)?

b) We think that a difference of about 20% in effectiveness would be a clinically important difference between the medications. If we think one will be effective in about 50% of patients, how many subjects would be needed to have a 90% probability of detecting a genuine difference of 20% in effectiveness (power = 0.90) in a two-sided test of hypothesis? (You need software for this.)

Just Checking ANSWERS

1. We're 90% confident that if members are contacted by e-mail, the donation rate will be between 4 and 18 percentage points higher than if they received regular mail.
2. Since a difference of 0 is not in the confidence interval, we'd reject the null hypothesis. There is evidence that more members will donate if contacted by e-mail.
3. The proportion from the sample in 2004 has variability, too. If we do a one-proportion z-test, we won't take that variability into account and our P-value will be incorrect.
4. The difference in the proportions between the two years has more variability than either individual proportion. The variance of the difference is the sum of the two variances.

Review

From the Data at Hand to the World at Large

Quick Review

What do samples really tell us about the populations from which they are drawn? Are the results of an experiment meaningful, or are they just sampling error? Statistical inference based on our understanding of sampling models can help answer these questions. Here's a brief summary of the key concepts and skills:

- Sampling models describe the variability of sample statistics and are often Normal due to a remarkable result called the Central Limit Theorem.
  - When the number of trials is sufficiently large, proportions found in different samples vary according to an approximately Normal model.
  - When samples are sufficiently large, the means of different samples vary, with an approximately Normal model.
  - The variability of sample statistics decreases as sample size increases.
  - Statistical inference procedures are often based on the Central Limit Theorem.
  - No inference procedure is valid unless the underlying assumptions are true. Always check the conditions before proceeding.
- A confidence interval uses a sample statistic (such as a proportion) to estimate a range of plausible values for the parameter of a population model.
  - All confidence intervals involve an estimate of the parameter, a margin of error, and a level of confidence.
  - For confidence intervals based on a given sample, the higher the confidence level, the bigger the margin of error.
  - At a given level of confidence, the larger the sample, the smaller the margin of error.
- A hypothesis test proposes a model for the population, then examines an observed statistic to see if that model is plausible.
  - A null hypothesis suggests a parameter value for the population model. Assuming the null hypothesis to be true, there should be nothing unusual, or surprisingly different about the sample results.
  - The alternative hypothesis states what we will believe if the sample results turn out to be inconsistent with our null model.
  - We compare the difference between the statistic and the hypothesized value with the standard deviation of the statistic. It's the sampling distribution of this ratio that gives us a P-value.
  - The P-value of the test is the probability that the null model could produce results at least as extreme as those observed in the sample or the experiment just as a result of sampling error.
  - A low P-value indicates evidence against the null model. If it is sufficiently low, we have evidence to reject the null model.
  - A high P-value indicates that the sample results are not inconsistent with the null model, so we cannot reject it. However, this does not prove the null model is true.
  - Sometimes we will mistakenly reject the null hypothesis even though it's actually true—that's called a Type I error. If we fail to reject a false null hypothesis, we commit a Type II error.
  - The power of a test measures its ability to detect a false null hypothesis.
  - You can lower the risk of a Type I error by requiring a higher standard of proof (lower P-value) before rejecting the null hypothesis. But this will raise the risk of a Type II error and decrease the power of the test.
  - The best way to increase the power of a test is to design a study based on a larger sample.

And now for some opportunities to review these concepts and skills . . .

Review Exercises

1. **Herbal cancer** A report in the *New England Journal of Medicine* noted growing evidence that the herb *Aristolochia fangchi* can cause urinary tract cancer in those who take it. Suppose you are asked to design an experiment to study this claim. Imagine that you have data on urinary tract cancers in subjects who have used this herb and similar subjects who have not used it, and that you can measure incidences of cancer and precancerous lesions in these subjects. State the null and alternative hypotheses you would use in your study.

2. **Colour-blind** Medical literature says that about 8% of males are colour-blind. A university's introductory Psychology course is taught in a large lecture hall. Among the students, there are 325 males. Each semester when the professor discusses visual perception, he shows the class a test for colour-blindness. The percentage of males who are colour-blind varies from semester to semester.
   a) Is the sampling distribution model for the sample proportion likely to be Normal? Explain.
   b) What are the mean and standard deviation of this sampling distribution model?
   c) Sketch the sampling model, using the 68–95–99.7 Rule.
   d) Write a few sentences explaining what the model says about this professor's class.

3. **Birth days** During a two-month period in 2002, 72 babies were born at the Tompkins Community Hospital in upstate New York. The table shows how many babies were born on each day of the week.

Day	Births
Mon.	7
Tues.	17
Wed.	8
Thurs.	12
Fri.	9
Sat.	10
Sun.	9

   a) If births are uniformly distributed across all days of the week, how many would you expect on each day?
   b) Only seven births occurred on a Monday. Does this indicate that women might be less likely to give birth on a Monday? Explain.
   c) Are the 17 births on Tuesdays unusually high? Explain.
   d) Can you think of any reasons why births may not occur completely at random?

4. **Third parties** Support for the New Democratic Party[1] has come in at around 15% in federal polls and elections for decades. Assuming the party's support remains at 15.0% in the future, and that 1000 individuals are randomly sampled in each poll, answer the following:
   a) How low will NDP support drop in 2.5% of future polls?
   b) How high will NDP support rise in about four of the next 25 polls?
   c) The Liberals tend to get 35% or so of popular support. In future polls, would the Liberal percent tend to vary more, less, or about the same as the NDP percent? Explain.

5. **Leaky gas tanks** It is estimated that 40% of service stations have gas tanks that leak to some extent. A new program in California is designed to lessen the prevalence of these leaks. We want to assess the effectiveness of the program by seeing if the percentage of service stations whose tanks leak has decreased. To do this, we randomly sample 27 service stations in California and determine whether there is any evidence of leakage. In our sample, only seven of the stations exhibit any leakage. Is there evidence that the new program is effective?
   a) What are the null and alternative hypotheses?
   b) Check the assumptions necessary for inference.
   c) Test the null hypothesis.
   d) What do you conclude (in plain English)?
   e) If the program actually works, have you made an error? What kind?
   f) What two things could you do to decrease the probability of making this kind of error?
   g) What are the advantages and disadvantages of taking those two courses of action?

6. **Surgery and germs** Joseph Lister (for whom Listerine is named!) was a British physician who was interested in the role of bacteria in human infections. He suspected that germs were involved in transmitting infection, so he tried using carbolic acid as an operating room disinfectant. In 75 amputations, he used carbolic acid 40 times. Of the 40 amputations using carbolic acid, 34 of the patients lived. Of the 35 amputations without carbolic acid, 19 patients lived. The question of interest is whether carbolic acid is effective in increasing the chances of surviving an amputation.
   a) What kind of a study is this?
   b) What do you conclude? Support your conclusion by testing an appropriate hypothesis.
   c) What reservations do you have about the design of the study?

[1] Third parties can make important contributions. The NDP's predecessor, the CCF, fought through a doctor's strike in Saskatchewan in July 1962 to implement North America's first universal health insurance system, leading soon thereafter to similar plans across Canada.

7. **Scrabble** Using a computer to play many simulated games of Scrabble, researcher Charles Robinove found that the letter "A" occurred in 54% of the hands. This study had a margin of error of $\pm 10\%$.[2]
   a) Explain what the margin of error means in this context.
   b) Why might the margin of error be so large?
   c) Probability theory predicts that the letter "A" should appear in 63% of the hands. Does this make you concerned that the simulation might be faulty? Explain.

8. **Dice** When one die is rolled, the number of spots showing has a mean of 3.5 and a standard deviation of 1.7. Suppose you roll 10 dice. What's the approximate probability that your total is between 30 and 40 (that is, the average for the 10 dice is between 3 and 4)? Specify the model you use and the assumptions and conditions that justify your approach.

9. **Net-newsers** In June 2008, the Pew Research Foundation sampled 3615 U.S. adults and asked about their choice of news sources. They identified 13% as "Net-newsers" who regularly get their news from online sources rather than TV or newspapers.
   a) Pew reports a margin of error of $\pm 2\%$ for this result. Explain what the margin of error means.
   b) Pew's survey included 2802 respondents contacted by landline and 813 contacted by cell phone. If the percentage of Net-newsers is the same in both groups and Pew estimated those percentages separately, which group would have the larger margin of error? Explain.
   c) Pew reports that 82% of the 470 Net-newsers in their survey get news during the course of the day, far more than other respondents. Find a 95% confidence interval for this statistic.
   d) How does the margin of error for your confidence interval compare with the values in parts a) and b)? Explain why.

10. **Rally tally** "Rally tally was really 52 800, experts' averaged figures show" reported a Toronto newspaper one day about the size of a rally at Queen's Park in Toronto. Reporters used the average of six experts' assessments, all of whom were given the same photos of the crowd that day. With reference to appropriate statistical concepts (like bias and variability), answer the following:
    a) Why do you think they averaged the six estimates?
    b) If they could average not just six, but hundreds of experts' assessments, could they expect this average to stabilize around some value as they average more and more assessments? Explain.
    c) Could they expect to hit on the true rally size (at the time the photos were taken) if enough assessments are averaged? Explain.

11. **Bimodal** We are sampling randomly from a distribution known to be bimodal.
    a) As our sample size increases, what's the expected shape of the sample's distribution?
    b) What's the expected value of our sample's mean? Does the size of the sample matter?
    c) How is the variability of this sample's mean related to the standard deviation of the population? Does the size of the sample matter?
    d) How is the shape of the sampling distribution model affected by the sample size?

12. **Vitamin D 2012** In 2012, the *American Journal of Clinical Nutrition* reported that 31% of Australian adults over age 25 have a vitamin D deficiency. The data came from the AusDiab study of 11 218 Australians.
    a) Do these data meet the assumptions necessary for inference? What would you like to know that you don't?
    b) Create a 95% confidence interval.
    c) Interpret the interval in this context.
    d) Explain what "95% confidence" means in this context.

13. **Twins** There is some indication in medical literature that doctors may have become more aggressive in inducing labour or doing preterm Caesarean sections when a woman is carrying twins. Records at a large hospital show that, of the 43 sets of twins born in 1990, 20 were delivered before the 37th week of pregnancy. In 2000, 26 of 48 sets of twins were born preterm. Does this indicate an increase in the incidence of early births of twins? Test an appropriate hypothesis and state your conclusion.

14. **Eclampsia** It's estimated that 50 000 pregnant women worldwide die each year of eclampsia, a condition involving elevated blood pressure and seizures. A research team from 175 hospitals in 33 countries investigated the effectiveness of magnesium sulfate in preventing the occurrence of eclampsia in at-risk patients. Results are summarized below.[3]

	Total Subjects	Reported side effects	Developed eclampsia	Deaths
Magnesium sulfate	4999	1201	40	11
Placebo Treatment	4993	228	96	20

    a) Write a 95% confidence interval for the increase in the proportion of women who may develop side effects from this treatment. Interpret your interval.
    b) Is there evidence that the treatment may be effective in preventing the development of eclampsia? Test an appropriate hypothesis and state your conclusion.

15. **Eclampsia II** Refer again to the research summarized in Exercise 14. Is there any evidence that when eclampsia

[2]*Chance*, 15, no. 1 [2002].

[3]*Lancet*, June 1, 2002.

does occur, the magnesium sulfate treatment may help prevent the woman's death?

a) Write an appropriate hypothesis.
b) Check the assumptions and conditions.
c) Find the P-value of the test.
d) What do you conclude about the magnesium sulfate treatment?
e) If your conclusion is wrong, which type of error have you made?
f) Name two things you could do to increase the power of this test.
g) What are the advantages and disadvantages of those two options?

16. Eggs The ISA Babcock Company supplies poultry farmers with hens, advertising that a mature B300 Layer produces eggs with a mean weight of 60.7 grams. Suppose that egg weights follow a Normal model with standard deviation 3.1 grams.

a) What fraction of the eggs produced by these hens weigh more than 62 grams?
b) What's the probability that a dozen randomly selected eggs average more than 62 grams?
c) Using the 68–95–99.7 Rule, sketch a model of the total weights of a dozen eggs.

17. No appetite for privatization In a June 2009 Harris-Decima poll,[4] 82% of Canadians indicated that they felt the Canadian health care system is superior to the American health care system, while 8% felt that the American system is superior. Twelve percent felt that there should be more privatization (fee-based services) of the Canadian health care system, while 55% wanted more public elements. However, in Quebec, 19% felt the American system is superior, and 16% wanted more fee-based services. It was reported that the poll was based on a random national sample of 1000 Canadians, and that a sample of the same size has a margin of error of 3.1%, 19 times in 20.

a) Is the actual margin of error for the 12% figure (for more privatization) equal to 3.1%? If not, what is it?
b) Can you determine the margin of error for the Quebec figure of 16%? If not, what do you need to know? Would it help if we told you that the population of Quebec is 23% of the Canadian population?
c) To reduce the margin of error for Quebec by 75%, how large would the total sample have to be? (Of course, it would be cheaper to over-sample from just Quebec.)

18. Enough eggs? One of the important issues for poultry farmers is the production rate—the percentage of days on which a given hen actually lays an egg. Ideally, that would be 100% (an egg every day), but realistically, hens tend to lay eggs on about three of every four days. ISA Babcock wants to advertise the production rate for the B300 Layer (see Exercise 16) as a 95% confidence interval with a margin of error of $\pm 2\%$. How many hens must they collect data on?

19. Teen deaths 2005 Traffic accidents are the leading cause of death among people aged 16 to 20. In May 2005, the National Highway Traffic Safety Administration reported that even though only 6.3% of licensed drivers are between 16 and 20 years old, they were involved in 12.6% of all fatal crashes. Insurance companies have long known that teenage boys were high risks, but what about teenage girls? One insurance company found that the driver was a teenage girl in 44 of the 388 fatal accidents they investigated. Is this strong evidence that the accident rate is lower for girls than for teens in general?

a) Test an appropriate hypothesis and state your conclusion.
b) Explain what your P-value means in this context.

20. Perfect pitch A recent study of perfect pitch tested 2700 students in American music conservatories. It found that 7% of non-Asian and 32% of Asian students have perfect pitch. A test of the difference in proportions resulted in a P-value of <0.0001.

a) What are the researchers' null and alternative hypotheses?
b) State your conclusion.
c) Explain in this context what the P-value means.
d) The researchers claimed that the data prove that genetic differences between the two populations cause a difference in the frequency of occurrence of perfect pitch. Do you agree? Why or why not?

21. Largemouth bass Organizers of a fishing tournament believe that the lake holds a sizable population of largemouth bass. They assume that the weights of these fish have a model that is skewed to the right with a mean of 3.5 pounds and a standard deviation of 2.2 pounds.

a) Explain why a skewed model makes sense here.
b) Explain why you cannot determine the probability that a largemouth bass randomly selected ("caught") from the lake weighs over three pounds.
c) Each fisherman in the contest catches five fish each day. Can you determine the probability that someone's catch averages over three pounds? Explain.
d) The 12 fishermen competing each caught the limit of five fish. What's the probability that the total catch of 60 fish averaged more than three pounds per fish?

22. Cheating A Rutgers University study found that many high school students cheat on tests. The researchers surveyed a random sample of 4500 high school students across the U.S.; 74% of them said they had cheated at least once.

a) Create a 90% confidence interval for the level of cheating among high school students. Don't forget to check the appropriate conditions.
b) Interpret your interval.
c) Explain what "90% confidence" means.
d) Would a 95% confidence interval be wider or narrower? Explain without actually calculating the interval.

---
[4]www.harrisdecima.ca/news/releases

23. Language Neurological research has shown that in about 80% of people, language abilities reside in the brain's left side. Another 10% display right-brain language centres, and the remaining 10% have two-sided language control. (The latter two groups are mainly left-handers.)[5]

a) We select 60 people at random. Is it reasonable to use a Normal model to describe the possible distribution of the proportion of the group that has left-brain language control? Explain.
b) What's the probability that our group has at least 75% left-brainers?
c) If the group had consisted of 100 people, would that probability be higher, lower, or about the same? Explain why, without actually calculating the probability.
d) How large a group would almost certainly guarantee at least 75% left-brainers? Explain.

24. Following RIM It seems hard to believe now, but back in 2009, the price of Research In Motion (now renamed Blackberry) shares on the TSE was about $80 in late April, and was at $80 again in late July. Sometimes it appears that there may be *autocorrelation* in stock price daily movements, which means that what happens on one day may affect or correlate with what happens on the next day. Good days may have follow-through or carry-over effects to the next day or two (or, maybe after a good day, we tend to "correct" to the downside the next day, which would be negative autocorrelation). Tracking RIM's up and down days, there were 16 times it followed an up day with an up day, and 13 times it followed an up day with a down day. There were 15 times it followed a down day by a down day and 13 times it followed a down day by an up day. If you want to play the short-term trends, should you, say, buy on an up day, hoping for positive follow-through the next day? Test the null hypothesis that there is no relationship between up days and price direction on the next day using these data.

25. Crohn's disease In 2002, the medical journal *The Lancet* reported that 335 of 573 patients suffering from Crohn's disease responded positively to injections of the arthritis-fighting drug infliximab.

a) Create a 95% confidence interval for the effectiveness of this drug.
b) Interpret your interval in context.
c) Explain carefully what "95% confidence" means in this context.

26. Narrower or wider? A marketing researcher is running a study to estimate the proportion of adults in Canada who spent more than $1000 on electronics in the past six months. He obtained a simple random sample of 4000 Canadian adults and calculated a 95% confidence interval estimate of this proportion. Estimates were also obtained, from the same sample, for the proportion of men who spent more than $1000 on electronics, the proportion of women who spent more than $1000 on electronics, and the proportion of Ontarians who spent more than $1000 on electronics. A similar study was conducted in the U.S. that selected 16 000 adults at random. Assume that there are equal numbers of males and females in each population, that Ontario's population is 25% of the Canadian population, and that the U.S. population is nine times bigger than the Canadian population. Answer the following questions, assuming that the true proportion of the population who spent more than $1000 is actually fairly similar for each population studied.

a) Since we selected a bigger percentage of the Canadian population than the U.S. population, the precision should be greater, resulting in a narrower confidence interval for the Canadian estimate. Do you agree? Explain why or why not.
b) The margin of error for the U.S. estimate will be about one-fourth of the margin of error for the Ontario estimate. Do you agree? Explain why or why not.
c) The margin of error for the estimate of the proportion of Canadian women who spent more than $1000 will be about half as big as the margin of error for the proportion of Ontarians who spent more than $1000. Do you agree? Explain why or why not.

27. Alcohol abuse Growing concern about binge drinking among university students has prompted one large school to conduct a survey to assess the size of the problem on its campus. The university plans to randomly select students and ask how many have been drunk during the past week. If the school hopes to estimate the true proportion among all its students with 90% confidence and a margin of error of $\pm 4\%$, how many students must be surveyed?

28. Errors An auto parts company advertises that its special oil additive will make the engine "run smoother, cleaner, longer, with fewer repairs." An independent laboratory decides to test part of this claim. It arranges to use a taxicab company's fleet of cars. The cars are randomly divided into two groups. The company's mechanics will use the additive in one group of cars but not in the other. At the end of a year, the laboratory will compare the percentage of cars in each group that required engine repairs.

a) What kind of a study is this?
b) Will they do a one-tailed or a two-tailed test?
c) Explain what a Type I error would be in this context.
d) Explain what a Type II error would be in this context.
e) Which type of error would the additive manufacturer consider more serious?
f) If the cabs with the additive do indeed run significantly better, can the company conclude it is an effect of the additive? Can they generalize this result and recommend the additive for all cars? Explain.

[5] *Science News*, 161, no. 24 (2002)

29. Preemies Among 242 Cleveland-area children born prematurely at low birth weights between 1977 and 1979, only 74% graduated from high school. Among a comparison group of 233 children of normal birth weight, 83% were high school graduates.[6]

a) Create a 95% confidence interval for the difference in graduation rates between children of normal birth weights and children of very low birth weights. Be sure to check the appropriate assumptions and conditions.

b) Does this provide evidence that premature birth may be a risk factor for not finishing high school? Use your confidence interval to test an appropriate hypothesis.

c) Suppose your conclusion is incorrect. Which type of error did you make?

30. Safety Observers in Texas watched children at play in eight communities. Of the 814 children seen biking, roller skating, or skateboarding, only 14% wore a helmet.

a) Create and interpret a 95% confidence interval.

b) What concerns do you have about this study that might make your confidence interval unreliable?

c) Suppose we want to do this study again, picking various communities and locations at random, and hope to end up with a 98% confidence interval having a margin of error of $\pm 4\%$. How many children must we observe?

31. Fried PCs A computer company recently experienced a disastrous fire that ruined some of its inventory. Unfortunately, during the panic of the fire, some of the damaged computers were sent to another warehouse, where they were mixed with undamaged computers. The engineer responsible for quality control would like to check out each computer to decide whether it's undamaged or damaged. Each computer undergoes a series of 100 tests. The number of tests it fails will be used to make the decision. If it fails more than a certain number, it will be classified as damaged and then scrapped. From past history, the distribution of the number of tests failed is known for both undamaged and damaged computers. The probabilities associated with each outcome are listed in the table below:

Number of tests failed	0	1	2	3	4	5	>5
Undamaged (%)	80	13	2	4	1	0	0
Damaged (%)	0	10	70	5	4	1	10

The table indicates, for example, that 80% of the undamaged computers have no failures, while 70% of the damaged computers have two failures.

a) To the engineers, this is a hypothesis-testing situation. State the null and alternative hypotheses.

b) Someone suggests classifying a computer as damaged if it fails any of the tests. Discuss the advantages and disadvantages of this test plan.

c) What number of tests would a computer have to fail to be classified as damaged if the engineers want to have the probability of a Type I error equal to 5%?

d) What's the power of the test plan in part c)?

e) A colleague points out that increasing α by just 2% would increase the power substantially. Explain.

32. Power We are replicating an experiment. How will each of the following changes affect the power of our test? Indicate whether it will increase, decrease, or remain the same, assuming that all other aspects of the situation remain unchanged.

a) We increase the number of subjects from 40 to 100.

b) We require a higher standard of proof, changing from $\alpha = 0.05$ to $\alpha = 0.01$.

33. Name recognition An advertising agency won't sign an athlete to do product endorsements unless it is sure the person is known to more than 25% of its target audience. The agency always conducts a poll of 500 people to investigate the athlete's name recognition before offering a contract. Then it tests H_0: $p = 0.25$ against H_A: $p > 0.25$ at a 5% level of significance.

a) Why does the company use upper-tail tests in this situation?

b) Explain what Type I and Type II errors would represent in this context, and describe the risk that each error poses to the company.

c) The company is thinking of changing its test to use a 10% level of significance. How would this change the company's exposure to each type of risk?

34. Name recognition, part II The advertising company described in Exercise 33 is thinking about signing an Olympic athlete to an endorsement deal. In its poll, 27% of the respondents could identify her.

a) Fans who never took Statistics can't understand why the company did not offer this Olympic athlete an endorsement contract even though the 27% recognition rate in the poll is above the 25% threshold. Explain it to them.

b) Suppose that further polling reveals that this Olympic athlete really is known to about 30% of the target audience. Did the company initially commit a Type I or Type II error in not signing her?

c) Would the power of the company's test have been higher or lower if the player were more famous? Explain.

35. NIMBY In March 2007, the Gallup Poll split a sample of 1003 randomly selected U.S. adults into two groups at random. Half ($n = 502$) of the respondents were asked,

"Overall, do you strongly favor, somewhat favor, somewhat oppose, or strongly oppose the use of nuclear energy as one of the ways to provide electricity for the U.S.?"

---

[6] "Outcomes in young adulthood for very-low-birth-weight infants," *New England Journal of Medicine*, 346, no. 3 (2002).

They found that 53% were either "somewhat" or "strongly" in favour. The other half (n = 501) were asked,

"Overall, would you strongly favor, somewhat favor, somewhat oppose, or strongly oppose the construction of a nuclear energy plant in your area as one of the ways to provide electricity for the U.S.?"

Only 40% were somewhat or strongly in favour. This difference is an example of the *NIMBY* (Not In My Back-Yard) phenomenon and is a serious concern to policy-makers and planners. How large is the difference between the proportion of American adults who think nuclear energy is a good idea and the proportion who would be willing to have a nuclear plant in their area? Construct and interpret an appropriate confidence interval.

36. Dropouts One study comparing various treatments for the eating disorder anorexia nervosa initially enlisted 198 subjects, but found overall that 105 failed to complete their assigned treatment programs. Construct and interpret an appropriate confidence interval. Discuss any reservations you have about this inference.

37. Home ice There may be some home ice advantage in hockey, and it may depend to some degree on the particular team. In 2008–09, the Vancouver Canucks won 24 of 41 home games and 21 of 41 away games. The Calgary Flames won 27 of 41 home games and 19 of 41 away games.

a) Is the difference in home-winning percentages between the two teams statistically significant?
b) Is the difference between home-and-away winning percentages statistically significant for the Flames? For the Canucks?
c) We looked through the teams' records[7] to find the team with the biggest relative home ice advantage, and discovered that this distinction belonged to the last place Islanders who won 17 home games and 9 away, the biggest home-to-away ratio of nearly 2 to 1. How impressive is this home ice advantage, considering that we looked through all the teams' records to find the best case? Do the Islanders really have a great home ice advantage or could this just have been by chance? If you toss a coin 26 times, can you get 17 (or more) heads just by chance? What *are* the chances?
d) Researchers perform thousands of studies. Considering the above, if researchers only publish the best, most statistically significant results, might we just be reading about Type I errors? If researchers alter their hypotheses and response variables in the course of a study (not uncommon), how does this affect any reported statistical significance?

38. Two-stage sample To study the problem of bullying, we sample 20 schools at random from Toronto, and then randomly select 60 students from each school to interview. We want to estimate the proportion of students who have been bullied in the past month. If we find that 300 out of the 1200 selected students have been bullied, can you determine the margin of error for this 25% estimate? If so, do it. If not, explain why not.

39. CADUMS survey The Canadian Alcohol and Drug Use Survey (CADUMS) from 2008 included random samples of about 1000 respondents from each Canadian province, except for Alberta and B.C., where larger samples of about 4600 and 4000, respectively, were taken, producing a total sample size of about 16 600. All were Canadian residents aged 15 years and over. The researchers collected information about alcohol and drug related behaviours. The survey included 1443 individuals aged 15–24. Researchers estimated that 52.9% percent of 15–24 year olds had used cannabis in their lifetime. The reported margin of error was 4.7% with 95% confidence.

a) If you try to confirm the margin of error, do you run into any problem? Explain. Why is the reported margin of error bigger or smaller than one might expect for a random sample of this size?
b) Using the given margin of error, calculate a 99% confidence interval for the true proportion and interpret your interval in context.

[7] www.nhl.com/ice/standings.htm?season=20082009&type=LEA

chapter

20 Inferences About Means

Where are we going?

We've learned how to generalize from the data at hand to the world at large for proportions. But not all data are as simple as "yes" or "no." In this chapter, we'll learn how to make confidence intervals and test hypotheses for the mean of a quantitative variable.

Who	Secondary school students
What	Time to travel to school
Units	Minutes
When	2007–2008
Where	Ontario
Why	To learn about the time spent by Ontario students travelling to school
How	Taking a random sample from the Census At School data base

Time				
30	10	8	30	5
8	7	15	10	35
15	10	25	22	20
25	30	10	25	8
15	18	25	15	10
25	5	2	5	25
20	15	47	20	20
13	20	5	15	12

A curve has been found representing the frequency distribution of values of the means of such samples (from a normal population), when these values are measured from the mean of the population in terms of the standard deviation of the sample . . .

—William Gosset

Travelling back and forth to work or school can be a real pain (though a good seat on the bus or subway can provide a chance to read, study, maybe snooze . . .). Since 2000, the International CensusAtSchool Project has surveyed over a million school students from Canada, the U.K., Australia, New Zealand, and South Africa using educational activities conducted in class. School participation is on a voluntary basis. Over 30 000 Canadian students participated in 2007–08. One question commonly asked in the survey is, "How long does it usually take you to travel to school?"

So just how long does it take Ontario students to get to school? Times vary from student to student, but knowing the average would be helpful. As we've learned, a single number or estimate that is almost surely wrong is not as useful as a range of values (or confidence interval) that is almost surely correct. Using the random data selector from the CensusAtSchool project, the responses (in minutes) were obtained for a random sample of 40 participating Ontario secondary school students from 2007–2008.[1]

These data differ from data on proportions in one important way. Proportions are summaries of individual responses, which have two possible values, such as "yes" and "no," "male" and "female," or "1" and "0." Quantitative data, however, usually report a quantitative value for each individual. Now, recall the three rules of data analysis and *plot the data,* as we have done here.

With quantitative data, we summarize with measures of centre and spread, such as the mean and standard deviation. Because we want to make inferences, we'll need to think about sampling distributions, which will lead us to a new sampling model. But first, some review.

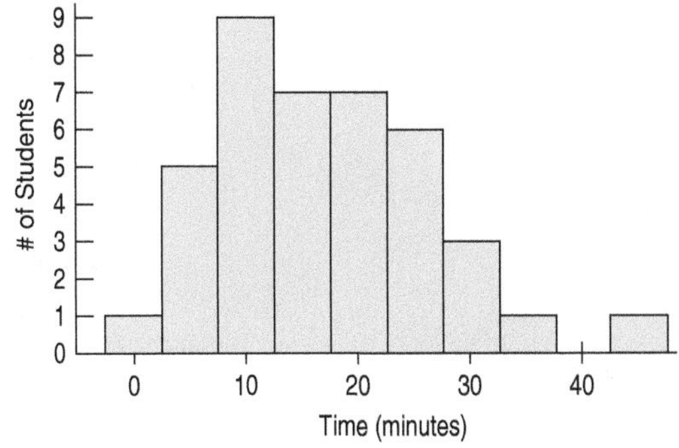

Figure 20.1
The travel times (to school) of Ontario secondary students appear to be unimodal and perhaps slightly right-skewed.

[1] www.censusatschool.ca.

20.1 The Central Limit Theorem Revisited

You've learned how to create confidence intervals and test hypotheses about proportions. We always centre confidence intervals at our best guess of the unknown parameter. Then, we add and subtract a margin of error. For proportions, that means $\hat{p} \pm ME$.

We found the margin of error as the product of the standard error, $SE(\hat{p})$, and a critical value, z^*, from the Normal table. So we had $\hat{p} \pm z^* SE(\hat{p})$.

We knew we could use the Normal distribution because the Central Limit Theorem told us (in Chapter 15) that the sampling distribution model for proportions is Normal.

Now we want to do exactly the same thing for means, and the Central Limit Theorem (still in Chapter 15) told us that the same Normal model works as the sampling distribution for means. Here again is this fundamental theorem:

THE CENTRAL LIMIT THEOREM (ABBREVIATED CLT)

When a random sample is drawn from any population with mean μ and standard deviation σ, its sample mean, $\bar{y}$, has a sampling distribution with the same *mean* μ but whose *standard deviation* is $\frac{\sigma}{\sqrt{n}}$ (and we write $\sigma(\bar{y})$ or $SD(\bar{y}) = \frac{\sigma}{\sqrt{n}}$).

No matter what population the random sample comes from, the *shape* of the sampling distribution is approximately Normal as long as the sample size is large enough. The larger the sample used, the more closely the Normal approximates the sampling distribution of the sample mean.

For Example USING THE CLT (AS IF WE KNEW σ)

Based on weighing thousands of animals, the American Angus Association reports that mature Angus cows have a mean weight of 1309 pounds (1 pound = 0.4536 kg) with a standard deviation of 157 pounds. This result was based on a very large sample of animals from many herds over a period of 15 years, so let's assume that these summaries are the population parameters and that the distribution of the weights was unimodal and not very severely skewed.

QUESTION: What does the CLT predict about the mean weight seen in random samples of 100 mature Angus cows?

ANSWER: It's given that weights of all mature Angus cows have $\mu = 1309$ and $\sigma = 157$ pounds. Because $n = 100$ animals is a fairly large sample, I can apply the Central Limit Theorem. I expect the resulting sample means $\bar{y}$ will average 1309 pounds and have a standard deviation of $SD(\bar{y}) = \frac{\sigma}{\sqrt{n}} = \frac{157}{\sqrt{100}} = 15.7$ pounds.

The CLT also says that the distribution of sample means follows a Normal model, so the 68–95–99.7 Rule applies. I'd expect that

- in 68% of random samples of 100 mature Angus cows, the mean weight will be between 1309 − 15.7 = 1293.3 and 1309 + 15.7 = 1324.7 pounds;
- in 95% of such samples, $1277.6 \le \bar{y} \le 1340.4$ pounds;
- in 99.7% of such samples, $1261.9 \le \bar{y} \le 1356.1$ pounds.

The CLT says that all we need to model the sampling distribution of $\bar{y}$ is a random sample of quantitative data.

And the true population standard deviation, σ.

Uh oh. That could be a problem. How are we supposed to know σ? With proportions, we had a link between the proportion value and the standard deviation of the sample proportion: $SD(\hat{p}) = \sqrt{\frac{pq}{n}}$. And there was an obvious way to estimate the standard deviation from the data: $SE(\hat{p}) = \sqrt{\frac{\hat{p}\hat{q}}{n}}$. But for means, $SD(\bar{y}) = \frac{\sigma}{\sqrt{n}}$, so knowing $\bar{y}$

STANDARD ERROR

Because we estimate the standard deviation of the sampling distribution model from the data, we'll call it a *standard error*. So we use the $SE(\bar{y})$ notation. Remember, though, that it's just the estimated standard deviation of the sampling distribution model for means.

doesn't tell us anything about $SD(\bar{y})$. We know n, the sample size, but the population standard deviation, σ, could be *anything*. So what should we do? We do what any sensible person would do: We estimate the population parameter σ with s, the sample standard deviation based on the data. The resulting standard error is $SE(\bar{y}) = \frac{s}{\sqrt{n}}$.

A century ago, people used this standard error with the Normal model, assuming it would work. And for large sample sizes it *did* work pretty well (as mentioned earlier in optional section 16.6). But they began to notice problems with smaller samples. The sample standard deviation, s, like any other statistic, varies from sample to sample. And this extra variation in the standard error was messing up the P-values and margins of error.

William S. Gosset is the man who first investigated this problem. He realized that not only do we need to allow for the extra variation with larger margins of error and P-values, but we even need a new sampling distribution model. In fact, we need a whole *family* of models, depending on the sample size, n. These models are unimodal, symmetric, bell-shaped models, but the smaller our sample, the more we must stretch out the tails. Gosset's work transformed Statistics, but most people who use his work don't even know his name.

20.2 Gosset's *t*

International Statistical Institute (ISI)

To find the sampling distribution of $\frac{\bar{y} - \mu}{s/\sqrt{n}}$, Gosset simulated it *by hand*. He drew 750 samples of size 4 by shuffling 3000 cards on which he'd written the heights of some prisoners and computed the means and standard deviations with a mechanically cranked calculator. (He knew μ because he was simulating and knew the population from which his samples were drawn.) Today, you could repeat in seconds on a computer the experiment that took him over a year. Gosset's work was so meticulous that not only did he get the shape of the new histogram approximately right, but he even figured out the exact *formula* for it from his sample. The formula was not confirmed mathematically until years later by Sir R. A. Fisher.

Gosset had a job that made him the envy of many. He was the chief experimental brewer for the Guinness Brewery in Dublin, Ireland. The brewery was a pioneer in scientific brewing and Gosset's job was to meet the demands of the brewery's many discerning customers by developing the best stout (a thick, dark beer) possible.

Gosset's experiments often required as much as a day to make the necessary chemical measurements or a full year to grow a new crop of hops. For these reasons, not to mention his health, his sample sizes were small—often as small as three or four.

When he calculated means of these small samples, Gosset wanted to compare them to a target mean to judge the quality of the batch. To do so, he followed common statistical practice of the day, which was to calculate z-scores and compare them to the Normal model. But Gosset noticed that with samples of this size, his tests weren't quite right. He knew this because when the batches that he rejected were sent back to the laboratory for more extensive testing, too often they turned out to be OK. In fact, about three times more often than he expected. Gosset knew something was wrong, and it bugged him.

Guinness granted Gosset time off to earn a graduate degree in the emerging field of Statistics, and naturally he chose this problem to work on. He figured out that when he used the standard error, $\frac{s}{\sqrt{n}}$, as an estimate of the standard deviation of the mean, the shape of the sampling model changed. He even figured out what the new model should be.

The Guinness Company may have been ahead of its time in using statistical methods to manage quality, but they also had a policy that forbade their employees to publish. Gosset pleaded that his results were of no specific value to brewers and was allowed to publish under the pseudonym "Student," chosen for him by Guinness's managing director. Accounts differ about whether the use of a pseudonym was to avoid ill feelings within the company or to hide from competing brewers the fact that Guinness was using statistical methods. In fact, Guinness was alone in applying Gosset's results in their quality assurance operations.

It was a number of years before the true value of "Student's" results was recognized. By then, statisticians knew Gosset well, as he continued to contribute to the young field of Statistics. But this important result is still widely known as **Student's *t*.**

Gosset's sampling distribution model is always bell-shaped, but the details change with different sample sizes. When the sample size is very large, the model is nearly Normal, but when it's small, the tails of the distribution are much heavier than the Normal. That means that values far from the mean are more common and that can be important for small samples (see Figure 20.2). So the Student's t-models form a whole *family* of related distributions that depend on a parameter known as **degrees of freedom**. The degrees of freedom of a distribution represent the number of independent quantities that are left after we've estimated the parameters. Here it's simply the number of data values, n, minus the number of estimated parameters. When we estimate one mean, that's just $n - 1$. We often denote degrees of freedom as df and the model as t_{df} with the degrees of freedom as a subscript.

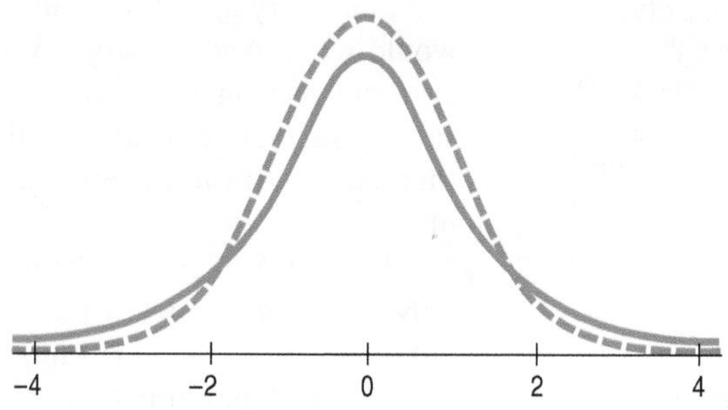

Figure 20.2

The t-model (solid curve) on 2 degrees of freedom has fatter tails than the Normal model (dashed curve). So the 68–95–99.7 Rule doesn't work for t-models with only a few degrees of freedom. It may not look like a big difference, but a t with 2 df is more than four times as likely to have a value greater than 2 compared to a standard Normal.

What Did Gosset See?

We can reproduce the simulation experiment that Gosset performed to get an idea of what he saw and how he reached some of his conclusions. Gosset drew 750 samples of size 4 from data on the heights of 3000 convicts. That population looks like this:[2]

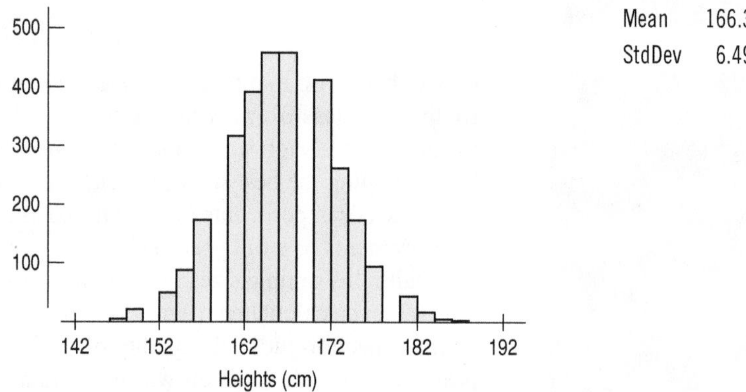

Following Gosset's example, we drew 750 independent random samples of size 4 and found their means and standard deviations.[3] As we (and Gosset) expected, the distribution of the means was even more Normal.[4]

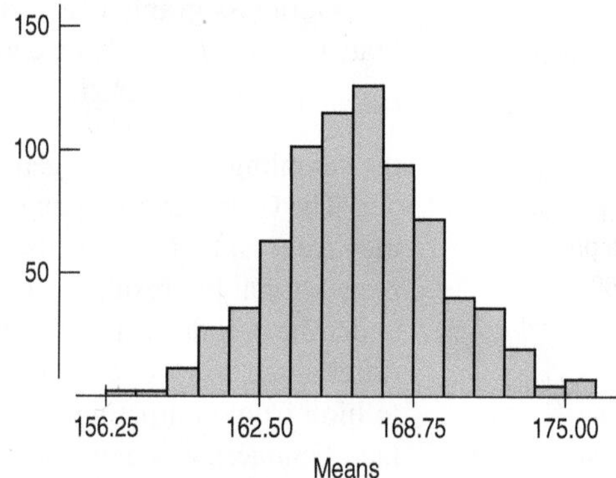

[2]If you have sharp eyes, you might have noticed some gaps in the histogram. Gosset's height data were rounded to the nearest 1/8 inch, which made for some gaps. Gosset noted that flaw in his paper.

[3]In fact, Gosset shuffled 3000 cards with the numbers on them and then dealt them into 750 piles of four each. That's not quite the same thing as drawing independent samples, but it was quite good enough for his purpose. We've drawn these samples in the same way.

[4]Of course, we don't know the means that Gosset actually got because we randomized using a computer and he shuffled 3000 cards, but this is one of the distributions he might have gotten, and we're pretty sure most of the others look like this as well.

Gosset's concern was for the distribution of $\frac{\bar{y} - \mu}{s/\sqrt{n}}$. We know $\mu = 166.301$ cm because we know the population, and $n = 4$. The values of $\bar{y}$ and s we find for each sample. Here's what the distribution looks like:

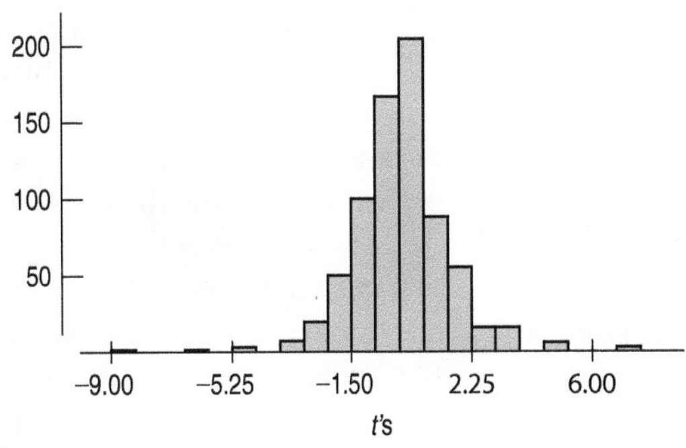

It's easy to see that this distribution is much thinner in the middle and longer in the tails than a Normal model we saw for the means themselves. This was Gosset's principal result.

20.3 A *t*-Interval for the Mean

To make confidence intervals or test hypotheses for means, we need to use Gosset's model. Which one? Well, for means, it turns out the right value for degrees of freedom is $df = n - 1$.

NOTATION ALERT

Ever since Gosset, *t* has been reserved in Statistics for his distribution.

A S *Activity:* **Estimating the Standard Error.** What's the average age at which people have heart attacks? A confidence interval gives a good answer, but we must estimate the standard deviation from the data to construct the interval.

A PRACTICAL SAMPLING DISTRIBUTION MODEL FOR MEANS

When certain assumptions and conditions[5] are met, the standardized sample mean,

$$t = \frac{\bar{y} - \mu}{SE(\bar{y})}$$

follows a Student's *t*-model with $n - 1$ degrees of freedom. We estimate the standard deviation of $\bar{y}$ with

$$SE(\bar{y}) = \frac{s}{\sqrt{n}}$$

When Gosset corrected the model for the extra uncertainty, the margin of error got bigger, as you might have guessed. When you use Gosset's model instead of the Normal model, your confidence intervals will be just a bit wider and your P-values just a bit larger. That's the correction you need. By using the *t*-model, you've compensated for the extra variability in precisely the right way.[6]

NOTATION ALERT

When we found critical values from a Normal model, we called them z^*. When we use a Student's *t*-model, we'll denote the critical values t^*.

ONE-SAMPLE *t*-INTERVAL FOR THE MEAN

When the assumptions and conditions[7] are met, we are ready to find the confidence interval for the population mean, μ. The confidence interval is

$$\bar{y} \pm t^*_{n-1} \times SE(\bar{y})$$

where the standard error of the mean $SE(\bar{y}) = \frac{s}{\sqrt{n}}$.

The critical value t^*_{n-1} depends on the particular confidence level, C, that you specify and on the number of degrees of freedom, $n - 1$, which we get from the sample size.

[5]You can probably guess what they are. We'll see them in the next section.

[6]Gosset, as the first to recognize the consequence of using s rather than σ, was also the first to give the sample standard deviation, s, a different letter than the population standard deviation σ,

[7]Yes, the same ones, and they're still coming in the next section.

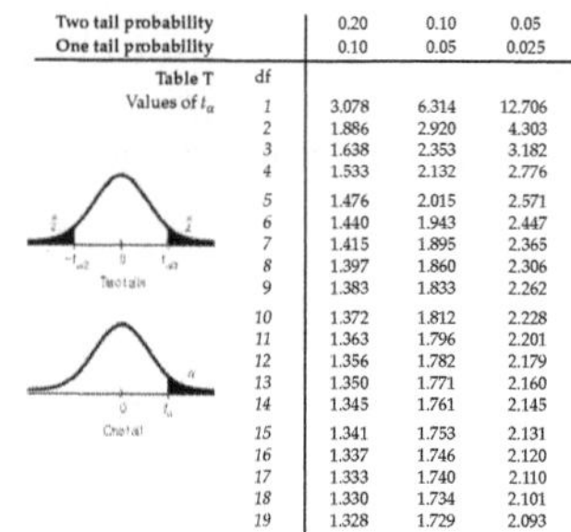
Part of Table T

A S ***Activity:* Building *t*-Intervals with the *t*-Table.** Interact with an animated version of Table T.

A S ***Activity:* Student's *t* in Practice.** Use a statistics package to find a *t*-based confidence interval; that's how it's almost always done.

As degrees of freedom increase, the shape of Student's *t*-model changes more and more slowly. Table T at the back of the book includes degrees of freedom between 100 and 1000 selected so that you can pin down critical values for just about any df. If your df's aren't listed, take the cautious approach by using the next lower df value, or use technology.

Using the *t*-Table to Find Critical Values

The Student's *t*-model is different for each value of degrees of freedom. Usually we find critical values and margins of error for Student's *t*-based intervals with technology. Calculators or statistics programs can give critical values for a *t*-model for any number of degrees of freedom and for any confidence level you choose.

But you can also use tables, such as Table T at the back of this book. The tables run down the page for as many degrees of freedom as can fit. For *enough* degrees of freedom, the *t*-model gets closer and closer to the Normal, so the tables give a final row with the critical values from the Normal model and label it "∞ *df*."

These tables are only a portion of the full tables, such as the one we used for the Normal model. We could have printed a table like Table Z for every df, but that's a lot of pages and not likely to be a bestseller. One way to shorten the book is to limit ourselves to only a few values. Although it might be nice to be able to get a critical value for a 93.4% confidence interval with 179 df, in practice we usually limit ourselves to 90%, 95%, 99%, and 99.9% and selected degrees of freedom. So, Table T fits on a single page with columns for selected confidence levels and rows for selected df's.[8]

For confidence intervals, the values in the table are usually enough to cover most cases of interest. If you can't find a row for the df you need, just be conservative and use the next smaller df in the table.

For Example A ONE-SAMPLE *t*-INTERVAL FOR THE MEAN

In 2004, a team of researchers published a study of contaminants in farmed salmon.[9] Fish from many sources were analyzed for 14 organic contaminants. The study expressed concerns about the level of contaminants found. One of those was the insecticide mirex, which has been shown to be carcinogenic and is suspected to be toxic to the liver, kidneys, and endocrine system. One farm in particular produced salmon with very high levels of mirex. After those outliers are removed, summaries for the mirex concentrations (in parts per million) in the rest of the farmed salmon are:

$$n = 150 \quad \bar{y} = 0.0913 \text{ ppm} \quad s = 0.0495 \text{ ppm}.$$

QUESTION: What does a 95% confidence interval say about mirex?

$$df = 150 - 1 = 149$$

$$SE(\bar{y}) = \frac{s}{\sqrt{n}} = \frac{0.0495}{\sqrt{150}} = 0.0040$$

$$t^*_{149} \approx 1.977 \text{ (from table } T\text{, using 140 df)}$$
$$\text{(actually, } t^*_{149} \approx 1.976 \text{ from technology)}$$

ANSWER: So the confidence interval for μ is

$$\bar{y} \pm t^*_{149} \times SE(\bar{y})$$
$$0.0913 \pm 1.977\,(0.0040)$$
$$0.0913 \pm 0.0079$$
$$(0.0834,\ 0.0992)$$

I'm 95% confident that the mean level of mirex concentration in farm-raised salmon is between 0.0834 and 0.0992 parts per million.

Student's *t*-models are all unimodal, symmetric, and bell-shaped, just like the Normal. But *t*-models with only a few degrees of freedom have noticeably longer tails

[8]You can also find tables and interactive tools on the Internet.

[9]Ronald A. Hites, Jeffery A. Foran, David O. Carpenter, M. Coreen Hamilton, Barbara A. Knuth, and Steven J. Schwager, 2004, "Global assessment of organic contaminants in farmed salmon," *Science* 303: 5655, pp. 226–229.

and larger standard deviation than the Normal. (That's what makes the margin of error bigger.) As the degrees of freedom increase, the t-models look more and more like the standard Normal. In fact, the t-model with infinite degrees of freedom is exactly the standard Normal.[10] This is great news if you happen to have an infinite number of data values, but that's not likely. *However, above about 60 degrees of freedom, it's very hard to tell the difference.* Of course, in the rare situation that we *know* σ, it would be foolish not to use that information, and if we don't have to estimate σ, we can use the Normal model.

z or t?

If you know σ, use the standard Normal model. (That's rare!) Whenever you use s to estimate σ, use t (though for large *df*, the t is well approximated by the standard Normal)

When σ is known. Administrators of a hospital were concerned about the prenatal care given to mothers in their part of the city. To study this, they examined the gestation times of babies born there. They drew a sample of 25 babies born in their hospital in the previous six months. Human gestation times for healthy pregnancies are thought to be well modelled by a Normal curve with a mean of 280 days and a standard deviation of 14 days. The hospital administrators wanted to test the mean gestation time of their sample of babies against the known standard. For this test, they used the established value for the standard deviation, 14 days, rather than estimating the standard deviation from their sample. Because they used the model parameter value for σ, they based their test on the standard Normal model rather than Student's t.

Assumptions and Conditions

Gosset found the t-model by simulation. Years later, Sir Ronald A. Fisher showed mathematically that Gosset was right and confirmed the assumptions needed by Gosset in discovering the t curve—that we are making *repeated independent draws from a Normally distributed population*. Now for our practical advice:

WHEN THE ASSUMPTIONS FAIL

When you check conditions, you usually hope to make a meaningful analysis of your data. The conditions serve as *disqualifiers*—you keep going unless there's a serious problem. If you find minor issues, note them and express caution about your results. If the sample is not an SRS, but you believe it's representative of some population, limit your conclusions accordingly. If there are outliers, perform the analysis both with and without them. If the sample looks bimodal, try to analyze subgroups separately. Only when there's major trouble—like a strongly skewed small sample or an obviously non-representative sample—are you unable to proceed at all.

Independence Assumption

Independence Assumption: The data values should be mutually independent. There's really no way to check independence of the data by looking at the sample, but you should think about whether the assumption is reasonable.

Randomization Condition: This condition is satisfied if the data arise from a random sample or suitably randomized experiment. Randomly sampled data, especially data from a simple random sample, are ideal—almost surely independent, with well-defined target population. If the data don't satisfy the **Randomization Condition**, then you should think about whether the values are likely to be independent for the variables you are concerned with and whether the sample you have is likely to be representative of the population you wish to learn about. Cluster and multistage samples, though, may have bigger SEs than suggested by our formula.

In the rare case that you have a sample that is a more than 10% of the population, you may want to consider using special formulas that adjust for that. But that's not a common concern for means. Without the correction, your SE will just err on the conservative side (be too high). This is actually a violation of the independence assumption, but a good one, since the effects are known and beneficial.

Normal Population Assumption

Student's t-models won't work for data that are badly skewed. How skewed is too skewed? Formally, we assume that the data are from a population that follows a Normal model. Practically speaking, there's no way to be sure this is true.

[10] Formally, in the limit as n goes to infinity.

And it's almost certainly *not* true. Models are idealized; real data are, well, real. The good news, however, is that even for small samples, it's sufficient to check the . . .

Nearly Normal Condition: The data come from a distribution that is unimodal and symmetric.

Check this condition by making a histogram or Normal probability plot. The importance of Normality for Student's t depends on the sample size. Just our luck: It matters most when it's hardest to check.[11]

For very small samples ($n < 15$ or so), the data should follow a Normal model fairly closely. Of course, with so little data, it's rather hard to tell. But if you do find outliers or clear skewness, don't use these methods.

For moderate sample sizes (n between 15 and about 40), the t methods will work reasonably well for mildly to moderately skewed unimodal data, but would perform badly in the presence of strong skewness or outliers. Make a histogram.

When the sample size is larger than 40, the t methods are generally quite safe to use, though very severe skewness can require much larger sample sizes (in which case a better approach might be to apply a non-linear transformation, like a logarithm). Be sure to make a histogram. If you find outliers in the data, it's always a good idea to perform the analysis twice, once with and once without the outliers, even for large samples. Outliers may well hold additional information about the data, but you may decide to give them individual attention and then summarize the rest of the data. If you find multiple modes, you may have different groups that should be analyzed and understood separately.

Guinness stout may be hearty, but the *t*-procedure is robust!

The one-sample t-test is an example of a *robust statistical test*. We say that it is robust with respect to the assumption of Normality, or against violations of Normality. This means that although the procedure is derived mathematically from an assumption of Normality, it can still often produce accurate results even when this assumption is violated. How well does the procedure tolerate violations of assumptions? How greatly do violations perturb the accuracy of P-values and confidence levels? These are questions about the robustness of the procedure.

The bigger the violations that can be tolerated, the greater is the robustness of the procedure. Robustness for most procedures will increase with the size of the sample. The robustness of the one-sample t-procedure is described above, where we see moderate robustness for sample sizes as small as 15 and remarkable robustness for samples over size 40. Pretty impressive! And the two-sample t-procedure to be discussed in the next chapter is even more robust. The usefulness of these t-procedures would be greatly compromised were it not for their high level of robustness. Similarly, many other common statistical procedures have good robustness, increasing their utility and value.

For Example CHECKING ASSUMPTIONS AND CONDITIONS FOR STUDENT'S *t*

RECAP: Researchers purchased whole farmed salmon from 51 farms in eight regions in six countries. The histogram below shows the concentrations of the insecticide mirex in 150 farmed salmon (after removing some outliers, mentioned earlier).

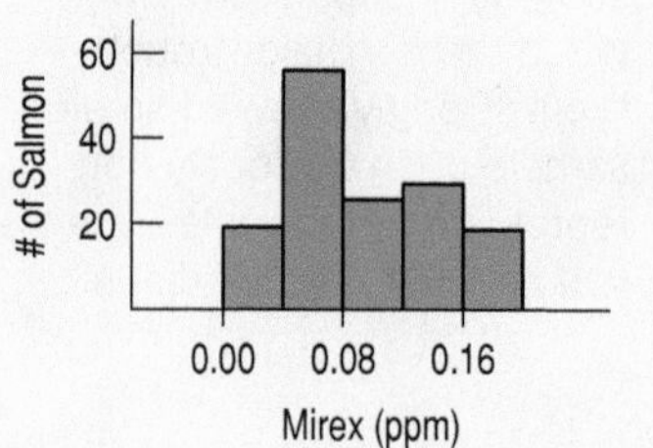

QUESTION: Are the assumptions and conditions for inference about the mean satisfied?

[11]There are formal tests of Normality, but they don't really help. When we have a small sample—just when we really care about checking Normality—these tests have very little power. So it doesn't make much sense to use them in deciding whether to perform a t-test. We don't recommend that you use them.

ANSWER:

- *Independence/Randomization:* The fish were not a random sample because no simple population existed to sample from. But they were raised in many different places, and samples were purchased independently from several sources, so they were likely to be nearly independent and to reasonably represent the population of farmed salmon worldwide.
- *Nearly Normal Condition:* The histogram of the data is unimodal. Although it may be somewhat skewed to the right, this is not a concern with a sample size of 150.

It's okay to use these data for inference about farm-raised salmon. Whew, now we know that we can actually trust that mechanical confidence interval calculation done earlier! Anyone can plug into a formula; the hard part is determining whether your procedure gives trustworthy results and answers.

Just Checking

Every five years, Statistics Canada conducts a census in order to compile a statistical portrait of Canada and its people. Prior to 2011, there were two forms: the short questionnaire, distributed to 80% of households, and the long questionnaire[12] (short-form questions plus additional questions), slogged through by the remaining one in five households, chosen at random. For estimates resulting from the additional questions appearing only on the long form, Statistics Canada would calculate a standard error.

1. Why did Statistics Canada need to calculate a standard error for long-form information, but not for the questions that appear on both the long and short forms?
2. The standard errors are calculated after re-weighting the individual results, to correct for differences between the sample proportion who are male, aged 15–24, etc., and the known (from the long form) population proportions who are male, aged 15–24, etc., so that the resulting estimates will be more precise (so, for example, if we know that 50% of residents in a region are male, and 52% of the 20% sample are male, each male is given a slightly lower weight or multiplier than each female to "correct" for the overrepresentation of males). Hence, a simple average (for quantitative variables) or simple proportion is not used. Can Statistics Canada use the *t*-model for standard errors and associated confidence intervals (for quantitative variables)? If simple (unweighted) averages were used instead, could we employ the *t*-model?
3. The standard errors calculated by Statistics Canada are bigger for geographic areas with smaller populations and for characteristics of small sub-groups in the area examined (such as people living in poverty in a middle-income neighbourhood). Why is this so? For example, why should a confidence interval for mean family income be wider for a sparsely populated area of farms in the Prairies than for a densely populated area in an urban centre? How does the *t*-model formula show that this will happen?

   [To deal with this problem, Statistics Canada classifies estimates based on "data quality" (the size of the associated standard error relative to the estimate), warns of low-quality (high standard error) estimates, and omits low-quality estimates with excessively high standard errors, since the latter are essentially noninformative and also have the potential to compromise confidentiality, due to the small number of cases]

[12]In June 2010, the minority Conservative government decided to do away with the mandatory long form and to replace it with the voluntary National Household Survey, in spite of significant opposition, citing privacy concerns.

A S *Activity:* **Intuition for *t*-based Intervals.** A narrated review of Student's *t*.

4. Suppose that in one census tract, there were 200 Aboriginal individuals in the 20% sample, and we estimate their mean annual earnings, along with the standard error and a 95% confidence interval, using the simple *t*-model. In another census tract, we would like to calculate a similar confidence interval, but there were only 50 Aboriginal people in the sample. What effect would the smaller number of Aboriginals in the second tract have on the 95% confidence interval? Specifically, which values used in the formula for the margin of error would change a lot and which would change only slightly? Approximately how much wider would the confidence interval based on 50 individuals be than the one based on 200 individuals?

Step-by-Step Example A ONE-SAMPLE *t*-INTERVAL FOR THE MEAN

Let's build a 90% confidence interval for the mean travel time to school for Ontario secondary school students. The interval that we'll make is called the **one-sample *t*-interval**.

Question: What can we say about the mean travel time to school for secondary school students in Ontario?

THINK ➡ Plan State what we want to know. Identify the parameter of interest.

Identify the variables and review the *W*'s.

I want to find a 90% confidence interval for the mean travel time to school for Ontario secondary school students. I have data on the travel time of 40 students in 2007–08.

Make a picture. Check the distribution shape and look for skewness, multiple modes, and outliers.

Here's a histogram of the 40 travel times.

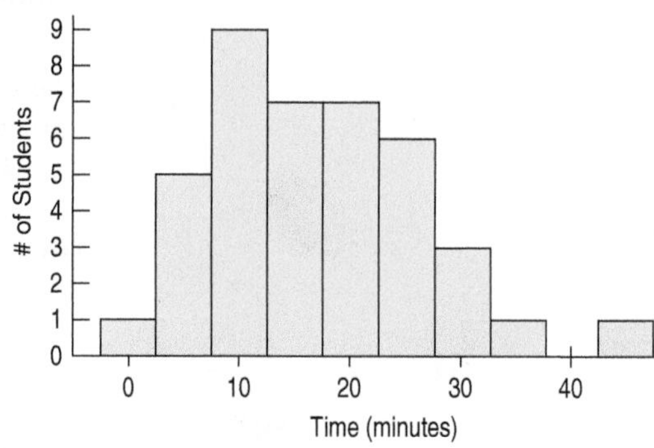

REALITY CHECK The histogram centres around 15–20 minutes, and the data lie between 0 and 50 minutes. We'd expect a confidence interval to place the population mean close to 15 or 20 minutes.

Model Think about the assumptions and check the conditions.

✓ **Independence Assumption:** These are independent selections from the stored data.

✓ **Randomization Condition:** Participation was voluntary but very broad-based, so I believe the students we randomly selected from the database should be reasonably representative of Ontario.

✓ **Nearly Normal Condition:** The histogram of the speeds is unimodal and slightly right-skewed, but not enough to be a concern.

State the sampling distribution model for the statistic.

Choose your method.

The conditions are satisfied, so I will use a Student's *t*-model with

$$(n - 1) = 39 \text{ degrees of freedom}$$

and find a **one-sample *t*-interval for the mean.**

SHOW ➡ Mechanics Construct the confidence interval. Be sure to include the units along with the statistics.	*Calculating from the data given at the beginning of this chapter:* $n = 40$ *students* $\bar{y} = 17.00$ *minutes* $s = 9.66$ *minutes* *The standard error of* $\bar{y}$ *is* $SE(\bar{y}) = \frac{s}{\sqrt{n}} = \frac{9.66}{\sqrt{40}} = 1.53$ *minutes*
The critical value we need to make a 90% interval comes from a Student's *t* table, a computer program, or a calculator. We have 40 – 1 = 39 degrees of freedom. The selected confidence level says that we want 90% of the probability to be caught in the middle, so we exclude 5% in each tail, for a total of 10%. The degrees of freedom and 5% tail probability are all we need to know to find the critical value.	*The 90% critical value is* $t^*_{39} = 1.685$ *(using software), so the margin of error is* $ME = t^*_{39} \times SE(\bar{y})$ $= 1.685(1.53)$ $= 2.58$ *minutes* *The 90% confidence interval for the mean travel time is* 17.0 ± 2.6 *minutes.*
REALITY CHECK The result looks plausible and in line with what we thought.	
TELL ➡ Conclusion Interpret the confidence interval in the proper context. When we construct confidence intervals in this way, we expect 90% of them to cover the true mean and 10% to miss the true value. That's what "90% confident" means.	*I am 90% confident that the interval from 14.4 to 19.6 minutes contains the true mean travel time to school for Ontario secondary school students.*

A S ***Activity:* The Real Effect of Small Sample Size.** We know that smaller sample sizes lead to wider confidence intervals, but is that just because they have fewer degrees of freedom?

Here's the part of the Student's *t* table that gives the critical value we needed. (See Table T in the back of the book.) To find a critical value, locate the row of the table corresponding to the degrees of freedom and the column corresponding to the probability you want. Our 90% confidence interval leaves 5% of the values on either side, so look for a one-tail probability of 0.05 at the top of the column or 90% confidence level at the bottom. The value in the table at that intersection is the critical value we need, but unfortunately, this concise table omits 39 *df*. The correct value lies between 1.684 and 1.690. Either be conservative and go with the bigger value, 1.690, or use software.

Using Table T to look up the critical value t for a 90% confidence level with 39 degrees of freedom.*

Two-tail One-tail	0.20 0.10	0.10 0.05	0.05 0.025	0.02 0.01	0.01 0.005
28	1.313	1.701	2.048	2.467	2.763
29	1.311	1.699	2.045	2.462	2.756
30	1.310	1.697	2.042	2.457	2.750
32	1.309	1.694	2.037	2.449	2.738
35	1.306	1.690	2.030	2.438	2.725
40	1.303	1.684	2.021	2.423	2.704
45	1.301	1.679	2.014	2.412	2.690
50	1.299	1.676	2.009	2.403	2.678
60	1.296	1.671	2.000	2.390	2.660

Of course, you can also create the entire confidence interval with the right computer software or calculator.

Make a Picture, Make a Picture, Make a Picture

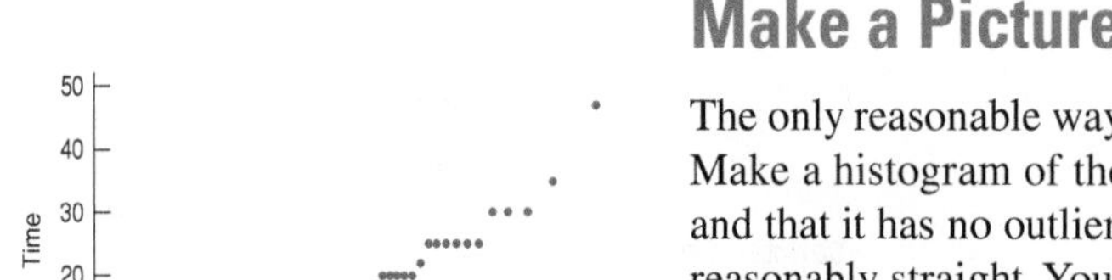

Figure 20.3

A Normal probability plot of travel times looks a bit curved but close enough to straight.

The only reasonable way to check the Nearly Normal Condition is with graphs of the data. Make a histogram of the data and verify that its distribution is unimodal and symmetric and that it has no outliers. You should also make a Normal probability plot to see that it's reasonably straight. You'll be able to spot deviations from the Normal model more easily with a Normal probability plot, but it's easier to understand the particular nature of the deviations from a histogram.

If you have a computer or graphing calculator doing the work, there's no excuse not to look at *both* displays as part of checking the Nearly Normal Condition.

Interpreting Confidence Intervals

Confidence intervals for means offer new, tempting, and wrong interpretations. Here are some things you *shouldn't* say:

- ***Don't say,*** "*90% of all Ontario secondary students* take between 14.4 and 19.6 minutes to get to school." The confidence interval is about the *mean* travel time, not about the times of *individual* students.
- ***Don't say,*** "We are 90% confident that *a randomly selected student* will take between 14.4 and 19.6 minutes to get to school." This false interpretation is also about individual students rather than about the *mean* of their times. We are 90% confident that the *mean* travel time of all secondary students is between 14.4 and 19.6 minutes.
- ***Don't say,*** "The mean student travel time is 17.0 minutes, 90% *of the time.*" That's about means, but still wrong. It implies that the true mean varies, when in fact it is the confidence interval that would have been different had we gotten a different sample.
- Finally, ***don't say,*** "90% *of all samples* will have mean travel times between 14.4 and 19.6 minutes." That statement suggests that this interval somehow sets a standard for every other interval. In fact, this interval is no more (or less) likely to be correct than any other. You could say that 90% of all possible samples will produce intervals that actually do contain the true mean time. (The problem is that, because we'll never know where the true mean time really is, we can't know if our sample was one of those 90%.)
- ***Do say,*** "90% of intervals found in this way cover the true value." Or make it more personal and say, "I am 90% confident that the true mean travel time is between 14.4 and 19.6 minutes."

SO WHAT *SHOULD* WE SAY?

Since 90% of random samples yield an interval that captures the true mean, we *should* say, "I am 90% confident that the interval from 14.4 to 19.6 minutes contains the mean travel time for all Ontario secondary students." It's also okay to say something less formal: "I am 90% confident that the average travel time for all secondary students is between 14.4 and 19.6 minutes." Remember: *Our uncertainty is about the interval, not the true mean.* The interval varies randomly. The true mean travel time is neither variable nor random—just unknown.

20.4 Hypothesis Test for the Mean

Students and their parents are naturally concerned about how long the commute to school takes. Suppose the Ministry of Education claims that the average commute time for secondary students is no greater than 15 minutes. But you're not so sure, particularly after collecting some data and finding a sample mean higher than 15 minutes. Maybe this is just chance variation or maybe we've found real evidence of excessive commute times. How can we tell the difference? This calls for a hypothesis test called the **one-sample *t*-test for the mean**.

You already know enough to construct this test. The test statistic looks just like the others we've seen. It compares the difference between the observed statistic and a hypothesized value to the standard error of the observed statistic. We've seen that, for means, the appropriate probability model to use for P-values is Student's t with $n - 1$ degrees of freedom.

A S *Activity:* **A *t*-Test for Wind Speed.** Watch the video in the preceding activity, and then use the interactive tool to test whether there's enough wind for electricity generation at a site under investigation.

ONE-SAMPLE *t*-TEST FOR THE MEAN

The assumptions and conditions for the one-sample *t*-test for the mean are the same as for the one-sample *t*-interval. We test the hypothesis $H_0: \mu = \mu_0$ using the statistic

$$t = \frac{\bar{y} - \mu_0}{SE(\bar{y})}$$

The standard error of $\bar{y}$ is $SE(\bar{y}) = \frac{s}{\sqrt{n}}$

When the conditions are met and the null hypothesis is true, this statistic follows a Student's *t*-model with $n - 1$ degrees of freedom. We use that model to obtain a P-value.

If you have to make a decision, set your α-level *a priori*, and reject H_0 if $P < \alpha$.

For Example A ONE-SAMPLE *t*-TEST FOR THE MEAN

RECAP: Researchers tested 150 farm-raised salmon for organic contaminants. They found the mean concentration of the carcinogenic insecticide mirex to be 0.0913 parts per million, with standard deviation 0.0495 ppm. As a safety recommendation to recreational fishers, the Environmental Protection Agency's (EPA) recommended "screening value" for mirex is 0.08 ppm.

QUESTION: Are farmed salmon contaminated beyond the level permitted by the EPA?

ANSWER: (We've already checked the conditions; see p. 571.)

$$H_0: \mu = 0.08$$ [13]

$$H_A: \mu > 0.08$$

These data satisfy the conditions for inference; I'll do a one-sample *t*-test for the mean:

$$n = 150,\ df = 149$$

$$\bar{y} = 0.0913,\ s = 0.0495$$

$$SE(\bar{y}) = \frac{0.0495}{\sqrt{150}} = 0.0040$$

$$t = \frac{0.0913 - 0.08}{0.0040} = 2.825$$

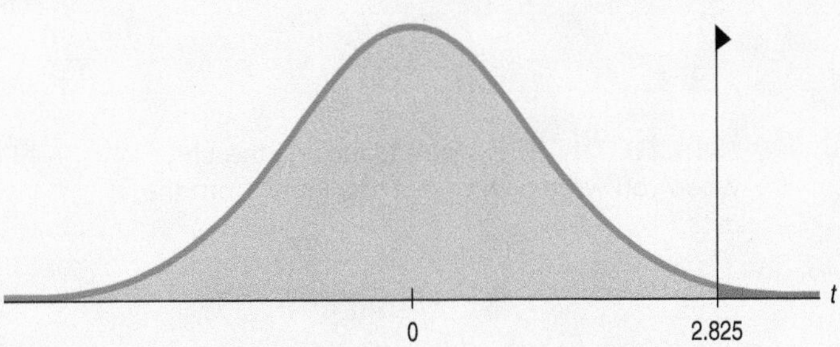

$$P(t_{149} > 2.825) = 0.0027 \text{ (from technology)}$$

Such a low P-value provides overwhelming evidence that, in farm-raised salmon, the mirex contamination level does exceed the EPA screening value.

[13] The *true* null hypothesis is $H_0: \mu \le 0.08$, but we can only test one null value for μ. $\mu = 0.08$ is the *conservative* choice, since if we can reject $\mu = 0.08$ in favour of a larger μ, we can even more convincingly reject any μ smaller than 0.08. Just plug in something smaller than 0.08 for μ_0 and you can see the *t*-statistic get bigger and more statistically significant.

What if, in the example above about farm-raised salmon, you had used the standard Normal distribution instead of the t distribution? You would get essentially the same P-value. This is sometimes referred to as the *large sample z-test*, since the Normal distribution will work just fine as the sampling model when you plug $SE(\bar{y}) = \frac{s}{\sqrt{n}}$ in place of $SD(\bar{y}) = \frac{\sigma}{\sqrt{n}}$ in the denominator of the standardized statistic—*when n is large. Only, only, only* when n is large.

Step-by-Step Example A ONE-SAMPLE *t*-TEST FOR THE MEAN

The Ministry of Transportation claims that secondary students can get to their schools in 15 minutes or less, on average (okay, we confess, we made up this claim).

Question: Do the data convincingly refute this claim?

THINK ➡ Plan State what we want to know. Make clear what the population and parameter are.

Identify the variables and review the *W*s.

I want to know whether the mean travel time for students exceeds the Ministry's claim. I have a sample 40 travel times from 2007–08.

Hypotheses The null hypothesis is that the true mean travel time is equal to the claim. Because we're interested in whether travel times are excessive, the alternative is one-sided.

H_0: Mean travel time $\mu = 15$ minutes

H_A: Mean travel time $\mu > 15$ minutes

Make a picture. Check the distribution for major skewness, multiple modes, and outliers.

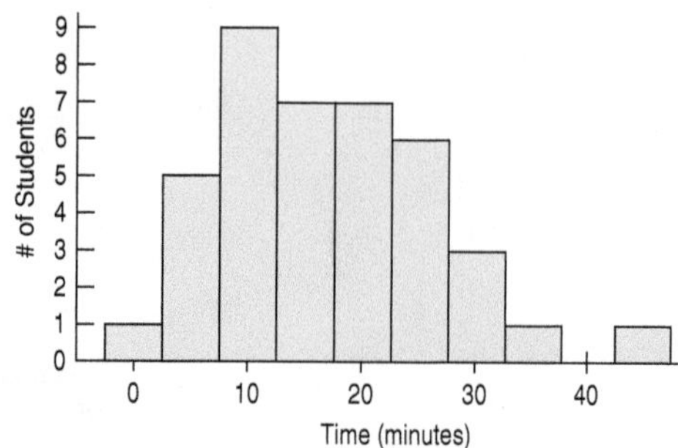

REALITY CHECK The histogram is clustered around 10–20 minutes, so we'd be surprised to find that the true mean was much higher than that. (The fact that 15 is within the confidence interval we've just found confirms this suspicion.)

Model Think about the assumptions and check the conditions.

State the sampling distribution model. (Be sure to include the degrees of freedom.)

Choose your method.

✓ **Independence Assumption:** Discussed earlier.

✓ **Randomization Condition:** Discussed earlier.

✓ **Nearly Normal Condition:** Discussed earlier.

The conditions are satisfied, so I'll use a Student's t-model with $(n - 1) = 39$ degrees of freedom to do a **one-sample *t*-test for the mean.**

SHOW ➡ Mechanics Be sure to include the units when you write down what you know from the data.

From the data,

$n = 40$ students

$\bar{y} = 17.0$ minutes

$s = 9.66$ minutes

$$SE(\bar{y}) = \frac{s}{\sqrt{n}} = \frac{9.66}{\sqrt{40}} = 1.53 \text{ minutes.}$$

The t-statistic calculation is just a standardized value, like z. We subtract the hypothesized mean and divide by the standard error.

$$t = \frac{\bar{y} - \mu_0}{SE(\bar{y})} = \frac{17.0 - 15.0}{1.53} = 1.31$$

We use the null model to find the P-value. Make a picture of the t-model centred at zero. Since this is an upper-tail test, shade the region to the right of the observed t-value.

(The observed mean is 1.31 standard errors above the hypothesized value.)

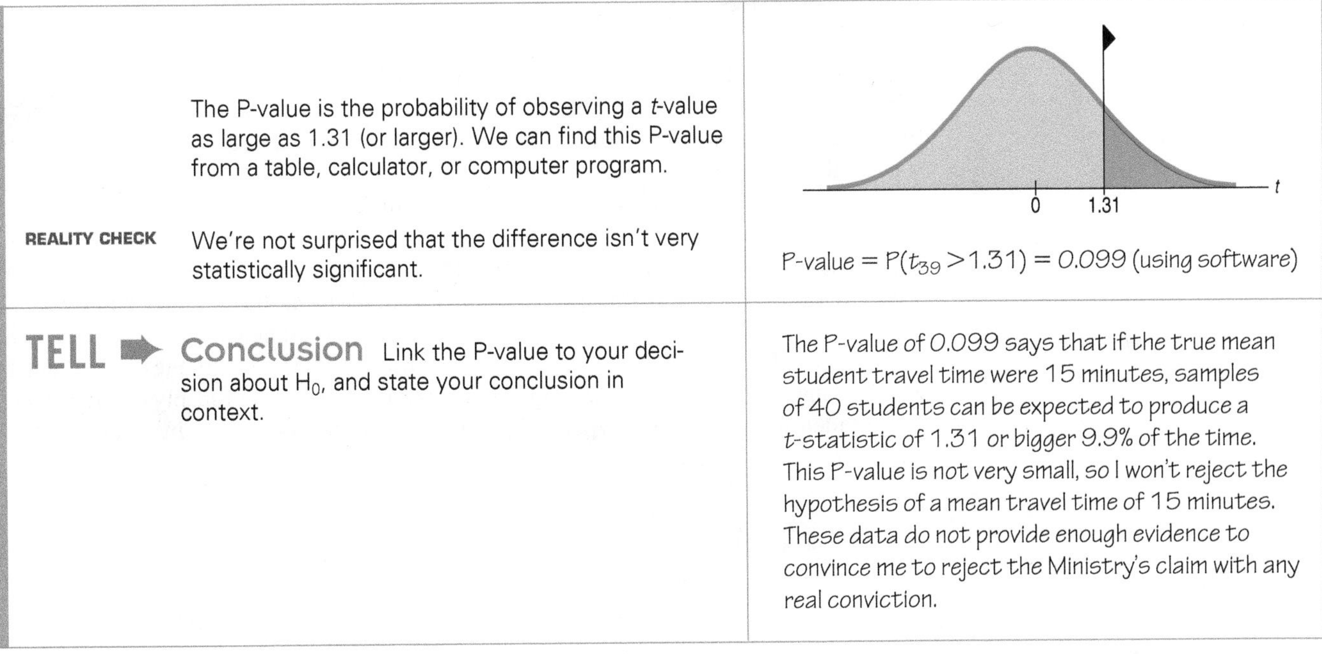

The P-value is the probability of observing a t-value as large as 1.31 (or larger). We can find this P-value from a table, calculator, or computer program.

REALITY CHECK We're not surprised that the difference isn't very statistically significant.

P-value = $P(t_{39} > 1.31) = 0.099$ (using software)

TELL ➡ Conclusion Link the P-value to your decision about H_0, and state your conclusion in context.

The P-value of 0.099 says that if the true mean student travel time were 15 minutes, samples of 40 students can be expected to produce a t-statistic of 1.31 or bigger 9.9% of the time. This P-value is not very small, so I won't reject the hypothesis of a mean travel time of 15 minutes. These data do not provide enough evidence to convince me to reject the Ministry's claim with any real conviction.

For hypothesis tests, the computed t-statistic can take on any value, so the value you get is not likely to be one found in the table. The best we can do is to trap a calculated t-value between two columns. Just look across the row with the appropriate degrees of freedom to find where the t-statistic falls. The P-value will be between the two values at the heads of the columns. Report that the P-value falls between these two values. Usually that's good enough.

For Example FINDING P-VALUES FROM TABLE T

RECAP: We've computed a one-sample t-test for the mean mirex contamination in farmed salmon, finding $t = 2.825$ with 149 df. In the earlier example, we found the P-value with technology.

QUESTION: How can we estimate the P-value for this upper-tail test using Table T?

ANSWER: I seek $P(t_{149} > 2.825)$. Table T has neither a row for 149 df nor an entry that is exactly 2.825. Here's the part of Table T where I'll need to work; roughly the right degrees of freedom and t-values:

Two-tail probability One-tail probability		0.20 0.10	0.10 0.05	0.05 0.025	0.02 0.01	0.01 0.005
Values of t_α (One tail)	df					
	140	1.288	1.656	1.977	2.353	2.611
	180	1.286	1.653	1.973	2.347	2.603

Since 149 df doesn't appear in the table, I'll be conservative and use the next lower df value that does appear. In this table, that's 140 df. Looking across the row for 140 df, I see that the largest t-value in the table is 2.611. According to the column heading, a t-value this large or larger will occur with probability 0.005. My t-value of 2.825 is larger than this, so I know that the probability of a t-value that large must be even smaller. I can report $P < 0.005$.[14]

[14]If the alternative was instead H_A: $\mu \neq 0.08$, we would report $p < 2(0.005) = 0.01$, since values in both tails would now support H_A.

Statistical Significance and Importance

Recall that "statistically significant" does not mean "actually important" or "meaningful," even though it sort of sounds that way. In this example, it does seem possible that travel times may average to a bit above 15 minutes. If so, perhaps a larger sample would show statistical significance.

So, should we try for a bigger sample? The difference between 17 minutes and 15 minutes doesn't seem very meaningful, and even if statistically significant, it would be hard to convince the government of a need to build more schools or the public to spend more money on improving transportation modes. Looking at the confidence interval, we can say with 90% confidence that the mean travel time is somewhere between 14.4 and 19.6 minutes. Even in the worst case, if the mean travel time is 19.6 minutes, would this be a bad enough situation to convince anyone to spend more money? Probably not. It's always a good idea when we test a hypothesis to also check the confidence interval and think about the likely values for the mean.

Just Checking

One disadvantage of using both long and short census forms is that estimates of characteristics that are reported on the short form will not exactly match the long-form estimates.

Short form summary measures are computed from a complete census, so they are the "true" values—something we don't usually have when we do inference.

5. Suppose we use long-form data to make 95% confidence intervals for the mean age of residents for each of 100 census tracts. How many of these 100 intervals should we expect will fail to include the true mean age (as determined from the complete census data)?

6. Based only on a long-form sample, we might test a null hypothesis about the mean household income in a region. Would the power of the test increase or decrease if a region returns more long forms?

Intervals and Tests

Confidence intervals and hypothesis tests look at the data from different perspectives. A hypothesis test starts with a *proposed parameter value* and asks if the *data* are consistent with that value. If the observed statistic is too far from the proposed parameter value, it is less plausible that the proposed value is the truth. So we reject the null hypothesis. By contrast, a confidence interval starts with the *data* and finds an interval of plausible values for where the parameter may lie. The 90% confidence interval for the mean school travel time was 17.0 $\pm$ 2.6 minutes, or (14.4 minutes, 19.6 minutes). If someone hypothesized that the mean time was really 15 minutes, how would you feel about it? How about 25 minutes?

Because the confidence interval included the time of 15.0 minutes, it certainly looks like 15 minutes might be a plausible value for the true mean school travel time. "Plausible" sounds rather like "acceptable" as a null hypothesis, and indeed this is the case. If we wanted to test the null hypothesis that the true mean is 15 minutes, and we find that 15 lies within some confidence interval, it follows that 15 minutes is a plausible null hypothesis—at some alpha level—but what alpha level? This depends on the confidence level of the confidence interval.

Confidence intervals and significance tests are built from the same calculations. Here's the connection: The confidence interval contains all possible values for the parameter that would not be rejected, as null hypotheses, in a test (after matching up test alpha level and confidence level, as discussed below).

More precisely, a level C confidence interval contains *all* of the plausible null hypothesis values that would *not* be rejected by a two-sided hypothesis test at alpha level $1 - C$. So a 95% confidence interval matches up with a $1 - 0.95 = 0.05$, or 5% significance level test for these data.

Confidence intervals are naturally two-sided, so they match up exactly with two-sided hypothesis tests. When the hypothesis is one-sided, as in our example, it matches up exactly with a one-sided confidence interval (which we are not covering in this text).

To relate a one-sided hypothesis test to a two-sided confidence interval, proceed as follows: Check to see if the level C confidence interval misses the null value and supports the alternative hypothesis—that is, lies entirely within the range of values of the alternative hypothesis. If so, you can reject the null hypothesis at the $(1 - C)/2$ level of significance (or $P < (1 - C)/2$). If not, the test will fail to reject the null hypothesis at the $(1 - C)/2$ level (or $P > (1 - C)/2$).

So if we were to use our 90% confidence interval of (14.4, 19.6) to test H_0: $\mu = \mu_0$ vs H_A: $\mu > \mu_0$, then any value for μ_0 smaller than 14.4 would have to be rejected as a null hypothesis, not at the 10% level, but rather at the 5% level of significance ($P < 0.05$), since $(1 - 0.90)/2 = 0.05$.

Degrees of Freedom

Don't divide by *n*.
Some calculators offer an alternative button for standard deviation that divides by n instead of $n - 1$. Try sticking a wad of gum over the "n" button so you won't be tempted to use it. Use $n - 1$.

The parameter of the t curve, its $df = n - 1$, might have reminded you of the value we divide by to find the standard deviation of the data (since, in fact, it's the same number). When we introduced that formula, we promised to later say more about why we divide by $n - 1$ rather than by n.

If only we knew the true population mean, μ, we would use it to calculate the sample standard deviation, giving us:[15]

$$s = \sqrt{\frac{\Sigma(y - \mu)^2}{n}} \qquad \text{(Equation 20.1)}$$

But we don't know μ, so we naturally use $\bar{y}$ instead, and that causes a problem. For any sample, the data values will generally be closer to their own sample mean than to the true population mean, μ. Why is that? Imagine that we take a simple random sample of 10 students who just wrote the final exam in your very large introductory Statistics course. Suppose that the mean test score (for all students) was 70. The sample mean, $\bar{y}$, for *these* 10 students won't be exactly 70. Are the 10 students' scores closer to 70 or $\bar{y}$? They will tend to be closer to their own average $\bar{y}$. So, when we calculate s using $\Sigma(y - \bar{y})^2$ instead of $\Sigma(y - \mu)^2$ in Equation 20.1, our standard deviation estimate is too small. How can we fix it? The amazing mathematical fact is that we can fix it by dividing by $n - 1$ instead of by n. This difference is much more important when n is small than when it's big. The t-distribution inherits this same number and we call $n - 1$ the degrees of freedom.[16]

20.5 Determining the Sample Size

How large a sample do we need? The simple answer is "more." But more data cost money, effort, and time, so how much is enough? Suppose your computer just took half an hour to download a movie you want to watch. You're not happy. You hear about a program that claims to download movies in less than 15 minutes. You're interested enough to spend $29.95 for it, but only if it really delivers. So you get the free evaluation copy and test it by downloading that movie 10 different times. Of course, the mean download time is not exactly 15 minutes as

[15]Statistics textbooks often use equation numbers so they can talk about equations by name. We haven't needed equation numbers yet, but we admit it's useful here, so this is our first.

[16]Here is another way to think about *df.* If the data are say: 4, 5, 9, the mean is 6, and the deviations are -2, -1, $+3$. The sum of deviations from the sample mean must equal zero, so since the first two deviations sum to -3, the last one *must* be $+3$. Only $n - 1$ deviations are truly *free to vary* (unlike the deviations about μ, all n of which are free to vary). Dividing a sum of squared deviations by its *df* is generally the best way to convert such a sum to an average.

claimed. Observations vary. If the margin of error were 4 minutes, though, you'd probably be able to decide whether the software is worth the money. Doubling the sample size would require several more hours of testing and would reduce your margin of error to a bit under 3 minutes. You'll need to decide whether that's worth the effort.

As we make plans to collect data, we should have some idea of how small a margin of error we need to be able to draw a useful conclusion. Armed with the target *ME* and confidence level, we can find the sample size we'll need. Almost.

We know that for a mean, $ME = t^*_{n-1} \times SE(\bar{y})$ and that $SE(\bar{y}) = \frac{s}{\sqrt{n}}$, so we can determine the sample size by solving this equation for n:

$$ME = t^*_{n-1}\frac{s}{\sqrt{n}}$$

The good news is that we have an equation; the bad news is that we won't know most of the values we need to solve it. When we thought about sample size for proportions in Chapter 16, we ran into a similar problem. There we had to guess a working value for p to compute a sample size. Here, we need to know s. We don't know s until we get some data, but we want to calculate the sample size *before* collecting the data. A guess is often good enough, but if you have no idea what the standard deviation might be, or if the sample size really matters (for example, because each additional individual is very expensive to sample or experiment on), a small *pilot study* can provide you with a rough estimate of the standard deviation.

That's not all. Without knowing n, we don't know the degrees of freedom and we can't find the critical value, t^*_{n-1}. One common approach is to use the corresponding z^* value from the Normal model. If you've chosen a 95% confidence level, then just use 2, following the 68–95–99.7 Rule. If your estimated sample size is, say, 60 or more, it's probably okay—z^* was a good guess. If it's smaller than that, you may want to add a step, using z^* at first, finding n, and then replacing z^* with the corresponding t^*_{n-1} and calculating the sample size once more.

For Example FINDING SAMPLE SIZE

A company claims its program will allow your computer to download movies quickly. We'll test the free evaluation copy by downloading a movie several times, hoping to estimate the mean download time with a margin of error of only 4 minutes. We think the standard deviation of download times is about 5 minutes.

QUESTION: How many trial downloads must we run if we want 95% confidence in our estimate with a margin of error of only 4 minutes?

ANSWER: Using $z^* = 1.96$, solve

$$4 = 1.96\frac{5}{\sqrt{n}}$$

$$\sqrt{n} = \frac{1.96 \times 5}{4} = 2.45$$

$$n = (2.45)^2 = 6.0025$$

That's a small sample size, so I'll use $(6 - 1) = 5$ degrees of freedom[17] to substitute an appropriate t^* value. At 95%, $t^*_5 = 2.571$. Solving the equation one more time:

$$4 = 2.571\frac{5}{\sqrt{n}}$$

$$\sqrt{n} = \frac{2.571 \times 5}{4} \approx 3.214$$

$$n = (3.214)^2 \approx 10.33$$

To make sure the *ME* is no larger, I'll round up, which gives $n = 11$ runs. So, to get an *ME* of 4 minutes, I'll find the downloading times for 11 movies.

[17] Ordinarily we'd round the sample size up. But at this stage of the calculation, rounding *down* is the safer choice. Can you see why?

Sample size calculations are *never* exact. But, it's always a good idea to know whether the sample size is large enough to give you a good chance of being able to tell you what you want to know, before you collect any data.

On the other hand, when we are testing a null hypothesis, we will be concerned with our ability to detect departures from the null that might be of considerable practical importance, so our focus shifts from the margin of error to *the power of the test.* Power calculations for the t-test are more complicated than those for the single proportion test (as illustrated in Chapter 17), but the basic idea is the same. Specify the difference from the null that you believe is big enough to be of practical importance, then determine the sample size (using software) that achieves the desired power (such as 0.8 or 0.9).

Make sure your test is adequately powered for important alternatives, or you risk letting those big important effects (departures from the null)—just what you were looking for—slip by undiscovered. Turning up the power of a test is like turning up the power of a microscope, allowing you to see and discern even small things more clearly—in the case of tests, it allows you to see genuine effects or differences more clearly. With low power, you may end up seeing nothing clearly, so you fall back on the status quo of the null.

Let's return to the chapter example where we tested the null hypothesis of a mean mirex content in farm-raised salmon of 0.08 ppm, the recommended screening level for this contaminant. True levels slightly above 0.08 ppm might not matter all that much, but suppose that a level as high as 0.10 ppm was considered dangerously high, meriting major remedial action. We would want to ensure that our test will lead us to correctly reject the null hypothesis when such a high contamination level is actually present. Tell your software the following:

- *Statistical test to be used.* Here we need the one-sample t-test.
- *Alpha level of your test.* Let's choose the very common $\alpha = 0.05$.
- *The null hypothesis.* In this example, it is the screening level of $\mu = 0.08$ ppm.
- *Directionality of alternative.* Let's make it one-sided: $\mu > 0.08$, since we only care to detect high levels of mirex.
- *Particular alternative (effect size) considered to be of practical importance.* This is the alternative (to the null) that we want to have a good chance to detect, should it be true. We decided that $\mu = 0.10$ ppm was a dangerously high level, so that is the alternative that we enter. For purposes of comparison, we'll also consider $\mu = 0.09$ ppm.
- *Your guess at the standard deviation of the measurements.* Let's guess that approximately $\sigma = 0.05$, perhaps from some available data or pilot study, or just by making an educated guess. Why can't we use the s from our study? Well, remember that this is usually a planning exercise, so the study hasn't been run yet!
- *Desired power.* Let's aim for a rather high power of 0.95. This means that 95 times in 100 when we have a situation as bad as 0.10 ppm, we will correctly reject the null hypothesis and conclude that mirex levels are too high.

Okay! Ready . . . aim . . . fire (up your software). Below is some typical output:

Power and Sample Size

One-Sample t Test
Testing mean = null (versus > null)
Calculating power for mean = null + difference
Alpha = 0.05 Assumed standard deviation = 0.05

Difference	Sample Size	Target Power	Actual Power
0.01	272	0.95	0.950054
0.02	70	0.95	0.952411

The sample size is for each group.

It appears that the researchers didn't need to test 150 salmon; 70 would have sufficed, if 0.02 ppm above the screening value is where serious problems occur. But if they felt they needed to detect a lower mirex level, like 0.09 ppm (0.01 above the screening level), 272 salmon would be required for testing. Note we have demanded a rather high power of 0.95. If you reduce this target power, smaller sample sizes will result.

Actually, this power calculation is quite doable without using the computer if the sample size is not too small—at least 40 or 50. Check the last two exercises at the end of this chapter if you'd like to go through the actual calculations without using software, for moderately big samples.

*20.6 The Sign Test

Another and perhaps more simple way to test the Ministry's claim of 15 minutes average school travel time would be to ignore the actual travel time data and just ask each student, "Does it take longer than 15 minutes to get to school?" So rather than record the numerical times in minutes, we could just record a "yes" (or "1") for students who take longer than 15 minutes and a "no" (or "0") for students who take less than 15 minutes (and we'll ignore those who say it takes them exactly 15 minutes).

But what is the actual null hypothesis that could be tested from such 0–1 data? Well, 15 minutes would be some sort of centre if roughly equal numbers of students took more than 15 minutes and less than 15 minutes. Aha! That would then make 15 minutes not the mean, but rather the median travel time, and so our null hypothesis would say that the median is 15 minutes. If this null hypothesis were true, we'd expect the proportion of students who take longer than 15 minutes to be 50%. On the other hand, if the true median time were greater than 15 minutes, we'd expect to have more than 50% of students with travel times exceeding 15 minutes.

What we've done is turn the quantitative data about travel times into a set of yes-or-no values (Bernoulli trials from Chapter 14). And we've turned a question about the median time into a test of a proportion (Is the proportion of students who take more than 15 minutes to get to school greater than 0.50?). We already know how to conduct a test of proportions, so this isn't a new situation. (Can you see why we had to throw out the data points exactly equal to 15?)

When we test a hypothesized median by counting the number of values above and below that value, it's called a **sign test**. The sign test is a ***distribution-free method*** (or *non-parametric* method), so called because there are no *distributional* assumptions or conditions on the data. Specifically, because we are no longer working with the original quantitative data, we aren't requiring the Nearly Normal Condition.

We already know all we need for the sign test Step-by-Step:

Step-by-Step Example *A SIGN TEST

THINK ➡ Plan State what we want to know.	I want to know whether there is evidence that the median travel time to school for secondary school students exceeds 15 minutes.
Identify the parameter of interest. Here, it is the population median. Identify the variables and review the *W*s.	I have 34 students for the test (six students with travel times of 15 minutes were omitted) and have recorded whether or not their travel times exceeded 15 minutes.
Hypotheses Write the null and alternative hypotheses. There is not a great need to plot the data. Medians are resistant to the effects of skewness or outliers.	H_0: The median travel time to school for Ontario secondary students is 15 minutes. Equivalently, the proportion of student travel times exceeding 15 minutes is 50%: H_0: $p = 0.50$.

	H_A: The true proportion of students taking more than 15 minutes is more than 0.50, or $p > 0.50$.
Model Think about the assumptions and check the conditions. The sign test doesn't require the Nearly Normal Condition.	✓ **Independence Assumption:** Previously checked. ✓ **Randomization Condition:** Previously checked. ✓ **10% Condition:** The data are from a large number of students (so no special adjustment is needed to our SD formula).
If the **Success/Failure Condition** fails, we can still calculate a P-value using the Binomial model for the observed count of Successes.	✓ **Success/Failure Condition:** Both $np_0 = 34(0.5) = 17$ and $nq_0 = 34(0.5) = 17$ are greater than 10, showing that I expect at least 10 successes and at least 10 failures. Hence the Normal model for proportions may be used.
Choose your method.	Because the conditions are satisfied, I'll do a **sign test.** This is just a test of $p_0 = 0.5$.
SHOW ➡ **Mechanics** We use the null model to find the P-value—the probability of observing a proportion as far from the hypothesized proportion as the one we observed, or even farther. The P-value is the probability of observing a sample proportion as large as 0.529 (or larger) when the null hypothesis is true: $P = P(\hat{p} \geq 0.529 \mid p = 0.50)$	$SD(\hat{p}) = \sqrt{\frac{0.5 \times 0.5}{34}} = 0.0857$ Of the 34 students, 18 had times over 15 minutes (six indicated exactly 15 minutes and were dropped), so the observed proportion, $\hat{p}$, is 0.529. 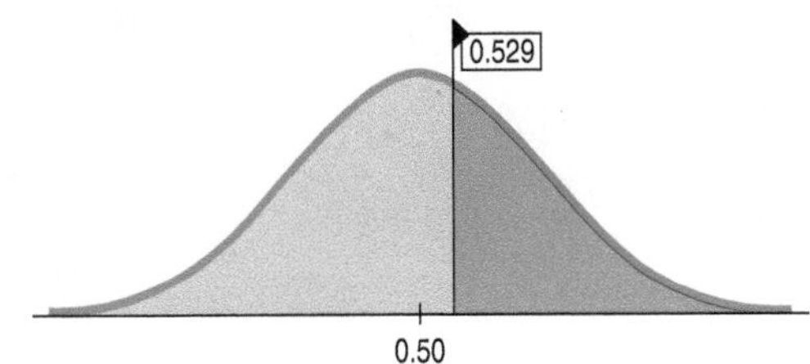
The probability of observing a value 0.34 standard deviations or more above the mean of a Normal model can be found by computer, calculator, or table.	$z = \frac{0.529 - 0.5}{0.0857} = 0.34$, so it is 0.34 standard deviations above the hypothesized proportion. The P-value is $P(z > 0.34) = 0.367$.
TELL ➡ **Conclusion** Link the P-value to your decision, then state your conclusion in the proper context.	The P-value of 0.367 is not very small, so I fail to reject the null hypothesis. There is insufficient evidence to suggest that the median travel time is greater than 15 minutes.

The sign test is simpler than the *t*-test, and it requires fewer assumptions. We need only yes/no data. We still should check for Independence and the Randomization Condition, but we no longer need the Nearly Normal Condition. When the data satisfy all the assumptions and conditions for a *t*-test on the mean, we usually prefer the *t*-test because it is more powerful than the sign test; for the same data, the P-value from the

t-test would be smaller than the P-value from the sign test. (In fact, the P-value for the t-test here was 0.099.) That's because the t-test uses the actual quantitative data values, which contain much more information than just knowing whether those same values are over 15. The more information we use, the more potential for statistical significance.

On the other hand, the sign test works even when the data have outliers or a skewed distribution—problems that can distort the results of the t-test and reduce its power. When we have doubts whether the conditions for the t-test are satisfied, it's a good idea to perform a sign test.[18]

WHAT CAN GO WRONG?

The most fundamental issue you face is knowing when to use Student's t methods.

- **Don't confuse proportions and means.** When you treat your data as categorical, counting successes and summarizing with a sample proportion, make inferences using the (usually Normal model based) methods you learned about in Chapters 16 through 19. When you treat your data as quantitative, summarizing with a sample mean, make your inferences using Student's t methods.

Student's t methods work well when the Normality Assumption is roughly true. Naturally, many of the ways things can go wrong turn out to be different ways that the Normality Assumption can fail. It's always a good idea to look for the most common kinds of failure. It turns out that you can even fix some of them.

- **Beware of multimodality.** The Nearly Normal Condition clearly fails if a histogram of the data has two or more modes. When you see this, look for the possibility that your data come from two groups. If so, your best bet is to try to separate the data into different groups. (Use the variables to help distinguish the modes, if possible. For example, if the modes seem to be composed mostly of men in one and women in the other, split the data according to sex.) Then you could analyze each group separately.
- **Beware of severely skewed data.** Make a Normal probability plot and a histogram of the data. If the data are very skewed, you might try re-expressing the variable. Re-expressing may yield a distribution that is more nearly unimodal and symmetric, more appropriate for Student's t inference methods for means. Re-expression cannot help if the sample distribution is not unimodal. Some people may object to re-expressing the data, but unless your sample is very large, you just can't use the methods of this chapter on data that are severely skewed.
- **Set outliers aside – respectfully.** Student's t methods are built on the mean and standard deviation, so we should beware of outliers when using them. When you make a histogram to check the Nearly Normal Condition, be sure to check for outliers as well. If you find some, consider doing the analysis twice, both with the outliers excluded and with them included in the data, to get a sense of how much they affect the results.

  The suggestion that you can perform an analysis with outliers removed may be controversial in some disciplines. Setting aside outliers is seen by some as "cheating." But an analysis of data with outliers left in place is *always* wrong. The outliers violate the Nearly Normal Condition and also the implicit assumption of a homogeneous population, so they invalidate inference procedures. An analysis of the nonoutlying points, along with a separate discussion of the outliers, is often much more informative and can reveal important aspects of the data.

  How can you tell whether there are outliers in your data? The "outlier nomination rule" of boxplots can offer some guidance, but it's just a very rough rule of thumb and not an absolute definition. The best practical definition is that a value is an outlier if removing it substantially changes your conclusions about the data. You won't want a single value to

[18] It's probably a good idea to routinely compute both. If they agree, then the inference is clear. If they differ, it may be interesting and important to see why.

DON'T IGNORE OUTLIERS

As tempting as it is to get rid of annoying values, you can't just throw away outliers and not discuss them. It isn't appropriate to lop off the highest or lowest values just to improve your results.

determine your understanding of the world unless you are very, very sure that it is absolutely correct and truly "belongs" to your target population. Of course, when the outliers affect your conclusion, this can lead to the uncomfortable state of not really knowing what to conclude. Such situations call for you to use your knowledge of the real world and your understanding of the data you are working with.[19]

Of course, Normality issues aren't the only risks you face when doing inferences about means. Remember to *Think* about the usual suspects.

- **Watch out for bias.** Measurements of all kinds can be biased. If your observations differ from the true mean in a systematic way, your confidence interval may not capture the true mean. And there is no sample size that will save you. A bathroom scale that's five pounds off will be five pounds off even if you weigh yourself 100 times and take the average. We've seen several sources of bias in surveys, and measurements can be biased, too. Be sure to think about possible sources of bias in your measurements.
- **Make sure cases are independent.** Student's t methods also require the sampled values to be mutually independent. Think hard about whether there are likely violations of independence in the data collection method. If there are, be very cautious about using these methods.
- **Make sure that data are from an appropriately randomized sample.** Ideally, all data that we analyze are drawn from a simple random sample or are generated by a completely randomized experimental design. When they're not, be careful about making inferences from them. You may still compute a confidence interval or get the mechanics of the P-value right, but this might not save you from making a serious mistake in inference. For other types of random samples, more complicated SE formulas apply. Cluster sampling in particular may have a much bigger SE than given by our formula.
- **Interpret your confidence interval correctly.** Many statements that sound tempting are, in fact, misinterpretations of a confidence interval for a mean. You might want to have another look at some of the common mistakes (as explained on p. 574). Keep in mind that a confidence interval is about the mean of the population, not about the means of samples, individuals in samples, or individuals in the population.
- **Choose your alternative hypothesis based only on what you are trying to prove.** Never choose a one-sided alternative after seeing which way the data are pointing, or you will incorrectly report a P-value half its true size. If you have any doubt about the nature of the alternative, go with the conservative choice of a two-sided alternative.

CONNECTIONS

The steps for finding a confidence interval or hypothesis test for means are just like the corresponding steps for proportions. Even the form of the calculations is similar. As the z-statistic did for proportions, the t-statistic tells us how many standard errors our sample mean is from the hypothesized mean. For means, though, we have to estimate the standard error separately. This added uncertainty changes the model for the sampling distribution from standard Normal to t.

As with all of our inference methods, the randomization applied in drawing a random sample or in randomizing a comparative experiment is what generates the sampling distribution. Randomization is what makes inference in this way possible at all.

The new concept of degrees of freedom connects back to the denominator of the sample standard deviation calculation, as shown earlier.

There's just no escaping histograms and Normal probability plots. The Nearly Normal Condition required to use Student's t can be checked best by making appropriate displays of the data. When we first used histograms, we looked at their shape and, in particular, checked whether they were unimodal and symmetric, and whether they showed any outliers. Those are just the features we check for here. The Normal probability plot zeros in on the Normal model a little more precisely.

[19] An important reason for you to know Statistics rather than let someone else analyze your data.

What Have We Learned?

Learning Objectives

Know the sampling distribution of the mean.

- To make inferences using the sample mean, we typically will need to estimate its standard deviation. This *standard error* is given by:

$$SE(\bar{y}) = \frac{s}{\sqrt{n}}.$$

- When we use the *SE* instead of the *SD*, the sampling distribution model that allows for the additional uncertainty is Student's *t*.

Construct confidence intervals for the true mean, μ.

- A confidence interval for the mean has the form $\bar{y} \pm ME$.
- The Margin of Error is $ME = t^*_{df} SE(\bar{y})$.
- Find t^* values by technology or from tables.
- When constructing confidence intervals for means, the correct degrees of freedom is $n - 1$.
- Check the Assumptions and Conditions before using any sampling distribution for inference.

Perform hypothesis tests for the mean using the standard error of $\bar{y}$ as a ruler and then finding the *P*-value from Student's t^* model on $n - 1$ degrees of freedom.

Write clear summaries to interpret a confidence interval or state a hypothesis test's conclusion.

Find the sample size needed to produce a given margin of error or to produce desired power in a test of hypothesis.

Review of Terms

Student's *t* A family of distributions indexed by its degrees of freedom. The *t*-models are unimodal, symmetric, and bell-shaped, but generally have fatter tails and a narrower centre than the Normal model. As the degrees of freedom increase, *t*-distributions approach the standard Normal (p. 565).

Degrees of freedom for Student's *t*-distribution For the application of the *t*-distribution in this chapter, the degrees of freedom are equal to $n - 1$, where n is the sample size (p. 566).

One-sample *t*-interval for the mean A one-sample *t*-interval for the population mean is

$$\bar{y} \pm t^*_{n-1} \times SE(\bar{y}), \text{ where } SE(\bar{y}) = \frac{s}{\sqrt{n}}$$

The critical value t^*_{n-1} depends on the particular confidence level, *C*, that you specify and on the number of degrees of freedom, $n - 1$ (p. 567).

One-sample *t*-test for the mean The one-sample *t*-test for the mean tests the hypothesis H_0: $\mu = \mu_0$ using the statistic $t = \frac{\bar{y} - \mu_0}{SE(\bar{y})}$ where the standard error of $\bar{y}$ is $SE(\bar{y}) = \frac{s}{\sqrt{n}}$ (p. 574).

Sign test A distribution-free test of a hypothesized median (p. 582).

On the Computer INFERENCE FOR MEANS

Statistics packages offer convenient ways to make histograms of the data. Even better for assessing near-Normality is a Normal probability plot. When you work on a computer, there is simply no excuse for skipping the step of plotting the data to check that it is nearly Normal. Beware: Statistics packages don't agree on whether to place the Normal scores on the *x*-axis (as we have done) or the *y*-axis. Read the axis labels.

Any standard statistics package can compute a hypothesis test. Here's what the package output might look like in general (although no package we know gives the results in exactly this form):[20]

A S *Activity:* **Student's *t* in Practice.** We almost always use technology to do inference with Student's *t*. Here's a chance to do that as you investigate several questions.

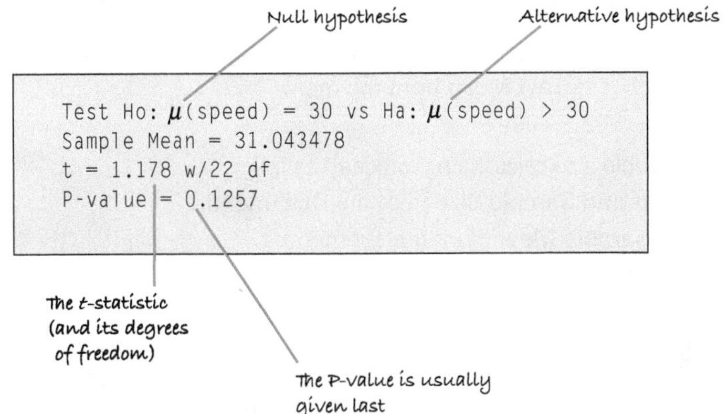

The package computes the sample mean and sample standard deviation of the variable and finds the P-value from the *t*-distribution based on the appropriate number of degrees of freedom. All modern statistics packages report P-values. The package may also provide additional information, such as the sample mean, sample standard deviation, *t*-statistic value, and degrees of freedom. These are useful for interpreting the resulting P-value and telling the difference between a meaningful result and one that is merely statistically significant. Statistics packages that report the estimated standard deviation of the sampling distribution usually label it "standard error" or "SE."

Inference results are also sometimes reported in a table. You may have to read carefully to find the values you need. Often, test results and the corresponding confidence interval bounds are given together. And often you must read carefully to find the alternative hypotheses. Here's an example of that kind of output:

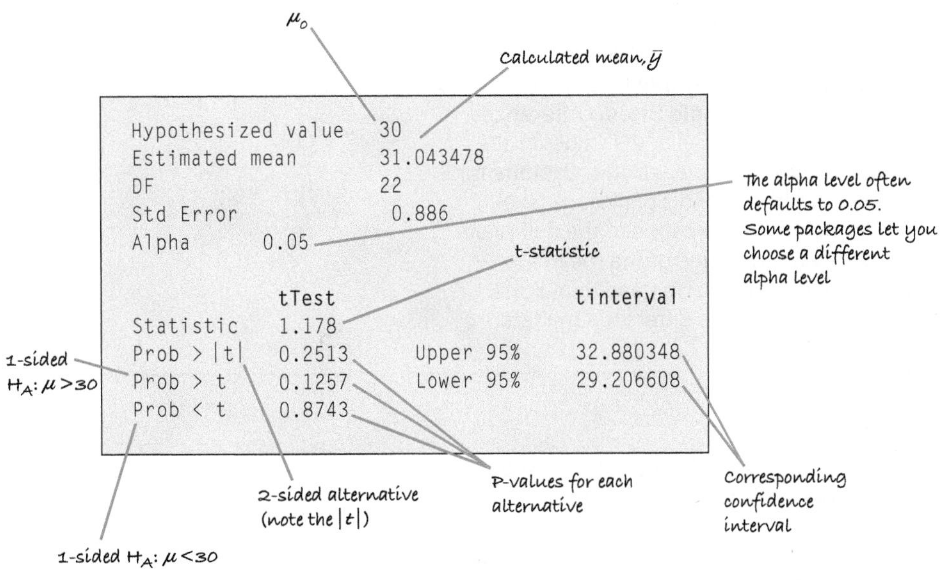

[20]Many statistics packages keep as many as 16 digits for all intermediate calculations. If we had kept as many, our results in the Step-By-Step section would have been closer to these.

EXCEL

Specify formulas. Find t^* with the TINV(alpha, df) function.

COMMENTS

Not really automatic. There's no easy way to find P-values or to perform power and sample size calculations in Excel.

JMP

From the **Analyze** menu, select **Distribution**.
For a confidence interval, scroll down to the "Moments" section to find the interval limits.
For a hypothesis test, click the red triangle next to the variable's name, and choose **Test Mean** from the menu.
Then, fill in the resulting dialogue.

For power and sample size calculations, proceed as follows:

- Choose **Power and Sample Size** from the **DOE** menu.
- Choose **One Sample Mean** from the submenu.
- Indicate the **Difference** in the means that you are hoping to detect, the **Alpha value**, and choose a one-or two-sided alternative.
- Guess at the **Std Dev**.
- Fill in either your desired **Sample size** or **Power**. The one you leave blank will be calculated.
- Click **Continue**.

COMMENTS

"Moment" is a fancy statistical term for means, standard deviations, and other related statistics.

MINITAB

From the **Stat** menu, choose the **Basic Statistics** submenu.
From that menu, choose **1-sample t. . . .**
Then, fill in the dialogue.

For power and sample size calculations:
From the **Stat** menu, choose the **Basic Statistics**, then **Power and Sample Size**, then **1-Sample t. . . .** In the dialogue box, fill in any two of **Sample Sizes, Differences, Power values**. Make your best guess at the value for the **Standard deviation**. And be sure to check the **Options** for the correct alternative hypothesis and significance level. For **Difference**, fill in the difference between the null value for the mean and the alternative value of the mean at which you are doing the calculation. No need to indicate the null value anywhere, as only the difference matters.

COMMENTS

The dialogue offers a clear choice between confidence interval and test.

R

To test the hypothesis that $\mu =$ mu (default is mu = 0) against an alternative (default is two-sided) and to produce a confidence interval (default is 95%), create a vector of data in x and then:

- **t.test**(x, alternative = c("two.sided", "less", "greater"), mu = 0, conf.level = 0.95)

provides the t-statistic, P-value, degrees of freedom, and the confidence interval for a specified alternative.

In package {pwr} (not installed by default), to perform a sample size or power calculation for a one-sample *t*-test,

- **pwr.t.test(n= , d= , sig.level= , power= , type= "one.sample")** returns **one** argument that is not specified in the function.

For example, for fixed $\alpha = 5\%$, 80% power, and effect size, *d*, of 0.5,

pwr.t.test(d=0.5, sig.level=0.05, power=0.8, type="one.sample")

will return a sample size of 33.36713 (i.e., 34).

Use the **alternative= "two.sided," "less," or "greater"** attribute to perform one-sided tests. In the case of using "less," your effect size should be negative.

COMMENTS

The effect size, *d*, required by *R* is equal to the difference between the alternative and null means *divided by the population standard deviation*. Since you often won't have data when doing this calculation, the SD needs to be guessed. A pilot study can help.

SPSS

From the **Analyze** menu, choose the **Compare Means** submenu.

From that, choose the **One-Sample *t*-test** command.

You need the IBM SPSS SamplePower add-on for power and sample size calculations.

COMMENTS

The commands suggest neither a single mean nor an interval. But the results provide both a test and an interval.

STATCRUNCH

To do inference for a mean using summaries:

- Click on **Stat**.
- Choose **T Statistics » One sample » with summary**.
- Enter the **Sample mean**, **Sample std dev**, and **Sample size**.
- Click on **Next**.
- Indicate **HypothesisTest**, then enter the hypothesized **Null mean**, and choose the **Alternative** hypothesis.

OR

- Indicate **Confidence Interval**, and then enter the **Level** of confidence.
- Click on **Calculate**.

To do inference for a mean using data:

- Click on **Stat**.
- Choose **T Statistics » One sample » with data**.
- Choose the variable **Column**.
- Click on **Next**.
- Indicate **HypothesisTest**, then enter the hypothesized **Null mean**, and choose the **Alternative** hypothesis.

OR

- Indicate **Confidence Interval**, then entre the **Level** of confidence.
- Click on **Calculate**.

Power & Sample size calculations are readily available, **using Stat » T Statistics » One Sample » Power/Sample size**.

- Click on **HypothesisTest Power**.
- Fill in all the boxes except for the one that you want to determine, either **Power** or **Sample Size**.
- Make a guess at **the Standard deviation**.

TI-83/84 PLUS

Finding a confidence interval:

In the **STAT TESTS** menu, choose **8:TInterval**. You may specify that you are using data stored in a list, or you may enter the mean, standard deviation, and sample size. You must also specify the desired level of confidence.

Power and sample size calculations not provided.

Testing a hypothesis:

In the **STAT TESTS** menu, choose **2:T-Test**. You may specify that you are using data stored in a list, or you may enter the mean, standard deviation, and size of your sample. You must also specify the hypothesized model mean and whether the test is to be two-tail, lower-tail, or upper-tail.

Exercises

1. *t*-models, part I Using the *t* tables, software, or a calculator, estimate
   a) the critical value of *t* for a 90% confidence interval with df = 17.
   b) the critical value of *t* for a 98% confidence interval with df = 88.
   c) $P(t \geq 2.09$ if 4 df)
   d) $P(|t| > 1.78$ if 22 df)

2. *t*-models, part II Using the *t* tables, software, or a calculator, estimate
   a) the critical value of *t* for a 95% confidence interval with df = 7.
   b) the critical value of *t* for a 99% confidence interval with df = 102.
   c) $P(t \geq 2.19$ if 41 df)
   d) $P(|t| > 2.33$ if 12 df)

3. *t*-models, part III Describe how the shape, centre, and spread of *t*-models change as the number of degrees of freedom increases.

4. *t*-models, part IV (last one!) Describe how the critical value of *t* for a 95% confidence interval changes as the number of degrees of freedom increases.

5. Cattle Researchers give livestock a special feed supplement to see if it will promote weight gain. They report that the 77 cows studied gained an average of 56 pounds, and that a 95% confidence interval for the mean weight gain this supplement produces has a margin of error of ±11 pounds. Some students wrote the following conclusions. Did anyone interpret the interval correctly? Explain any misinterpretations.
   a) 95% of the cows studied gained between 45 and 67 pounds.
   b) We're 95% sure that a cow fed this supplement will gain between 45 and 67 pounds.
   c) We're 95% sure that the average weight gain among the cows in this study was between 45 and 67 pounds.
   d) The average weight gain of cows fed this supplement will be between 45 and 67 pounds 95% of the time.
   e) If this supplement is tested on another sample of cows, there is a 95% chance that their average weight gain will be between 45 and 67 pounds.

6. Viewing hours Software analysis of the weekly hours spent by Canadian secondary school students viewing television, videos, or movies from a random sample of 200 students produced the *t*-interval shown below. Which conclusion, from the choices below, is correct? What's wrong with the others?

   With 90% Confidence,
   $8.6 < \mu$ (weekly viewing hours) < 10.8

   a) If we took many random samples of Canadian secondary students, about 9 out of 10 of them would produce this confidence interval.
   b) If we took many random samples of Canadian secondary students, about 9 out of 10 of them would produce a confidence interval that contained the mean weekly television, video, or movie viewing time of all Canadian secondary students.
   c) About 9 out of 10 Canadian secondary students spend between 8.6 and 10.8 hours per week on television, videos, or movies.
   d) About 9 out of 10 of the students surveyed spend between 8.6 and 10.8 hours per week on television, video, or movie viewing.
   e) We are 90% confident that the average time spent viewing television, videos, or movies by secondary students in Canada is between 8.6 and 10.8 hours per week.

7. Meal plan After surveying students at Dartmouth College, a campus organization calculated that a 95% confidence interval for the mean cost of food for one term (of three in the Dartmouth trimester calendar) is ($1372, $1562). Now the organization is trying to write its report and is considering the following interpretations. Comment on each.
   a) 95% of all students pay between $1372 and $1562 for food.
   b) 95% of the sampled students paid between $1372 and $1562.
   c) We're 95% sure that students in this sample averaged between $1372 and $1562 for food.
   d) 95% of all samples of students will have average food costs between $1372 and $1562.
   e) We're 95% sure that the average amount all students pay is between $1372 and $1562.

8. Snow Based on meteorological data for the past century, a local television weather forecaster estimates that the region's average winter snowfall is 58 cm, with a margin of error of 5 cm. Assuming he used a 95% confidence interval, how should viewers interpret this news? Comment on each of these statements (assuming a lack of systematic climate change):
   a) During 95 of the past 100 winters, the region got between 53 cm and 63 cm of snow.
   b) There's a 95% chance that the region will get between 53 cm and 63 cm of snow this winter.
   c) There will be between 53 cm and 63 cm of snow on the ground for 95% of winter days.
   d) Residents can be 95% sure that the area's average snowfall is between 53 cm and 63 cm.
   e) Residents can be 95% confident that the average snowfall during the past century was between 53 cm and 63 cm per winter.

T **9. Pulse rates** A medical researcher measured the pulse rates (beats per minute) of a sample of randomly selected adults and found the following Student's t-based confidence interval:

With 95.00% Confidence,
$70.887604 < \mu(\text{Pulse}) < 74.497011$

a) Explain carefully what the software output means.
b) What is the margin of error for this interval?
c) If the researcher had calculated a 99% confidence interval, would the margin of error be larger or smaller? Explain.

10. Crawling Data collected by child development scientists produced this confidence interval for the average age (in weeks) at which babies begin to crawl:

t-Interval for μ (95.00% Confidence): $29.202 < \mu(\text{age}) < 31.844$

a) Explain carefully what the software output means.
b) What is the margin of error for this interval?
c) If the researcher had calculated a 90% confidence interval, would the margin of error be larger or smaller? Explain.

11. CEO compensation A sample of 20 CEOs from the *Forbes* 500 shows total annual compensations ranging from a minimum of \$0.1 million to \$62.24 million. The average for these 20 CEOs is \$7.946 million. Here's a histogram:

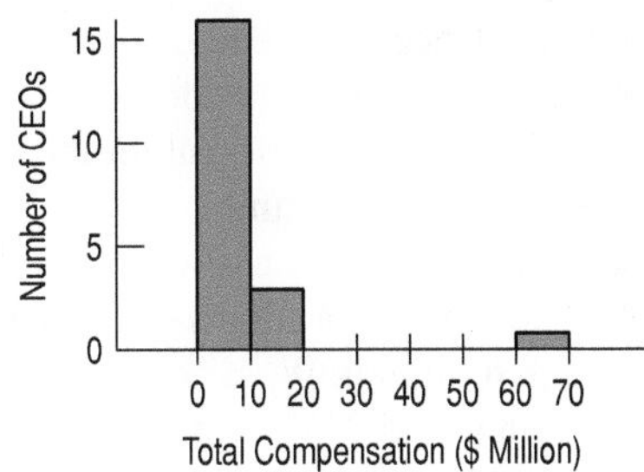

Based on these data, a computer program found that a 95% confidence interval for the mean annual compensation of all *Forbes* 500 CEOs is (1.69, 14.20) \$ million. Why should you be hesitant to trust this confidence interval?

12. Credit card charges A credit card company takes a random sample of 100 cardholders to see how much they charged on their card last month. Here's a histogram:

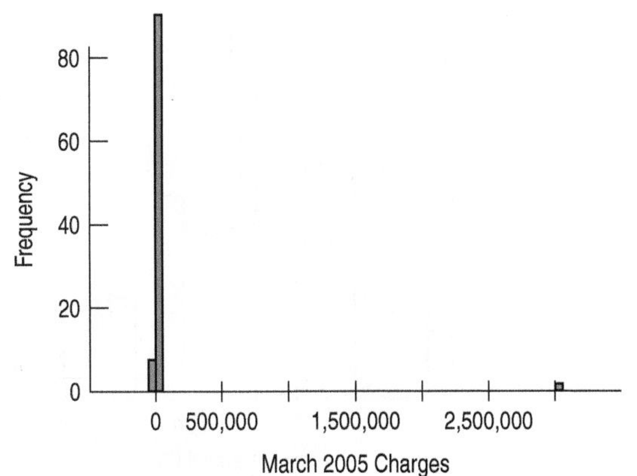

A computer program found that the resulting 95% confidence interval for the mean amount spent in March 2013 is (−\$28366.84, \$90691.49). Explain why the analysts didn't find the confidence interval useful, and explain what went wrong.

13. Normal temperature The researcher described in Exercise 9 also measured the body temperatures of that randomly selected group of adults. The data he collected are summarized below. We wish to estimate the average (or "normal") temperature among the adult population.

Summary	
Count	52
Mean	36.83°C
Median	36.78°C
MidRange	37.00°C
StdDev	0.38
Range	1.55
IntQRange	0.58

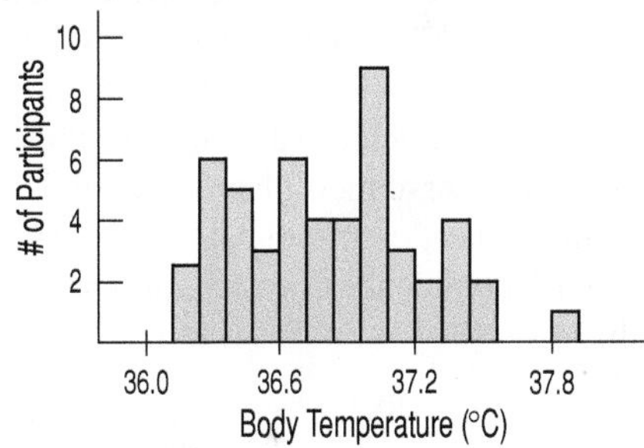

a) Are the necessary conditions for a t-interval satisfied? Explain.
b) Find a 98% confidence interval for mean body temperature.
c) Explain the meaning of that interval.
d) Explain what "98% confidence" means in this context.
e) 37°C is commonly assumed to be "normal." Do these data suggest otherwise? Explain.

14. Parking Hoping to lure more shoppers downtown, a city builds a new public parking garage in the central business district. The city plans to pay for the structure through parking fees. During a two-month period (44 weekdays), daily fees collected averaged \$126, with a standard deviation of \$15.

a) What assumptions must you make in order to use these statistics for inference?
b) Write a 90% confidence interval for the mean daily income this parking garage will generate.
c) Explain in context what this confidence interval means.
d) Explain what "90% confidence" means in this context.
e) The consultant who advised the city on this project predicted that parking revenues would average \$130 per day. Based on your confidence interval, do you think the consultant was correct? Why?

15. Normal temperatures, part II Consider again the statistics about human body temperature in Exercise 13.

a) Would a 90% confidence interval be wider or narrower than the 98% confidence interval you calculated before? Explain. (You should not need to compute the new interval.)

b) What are the advantages and disadvantages of the 98% confidence interval?

c) If we conduct further research, this time using a sample of 500 adults, how would you expect the 98% confidence interval to change? Explain.

d) How large a sample would you need to estimate the mean body temperature to within 0.05 degrees with 98% confidence?

16. Parking II Suppose that, for budget planning purposes, the city in Exercise 14 needs a better estimate of the mean daily income from parking fees.

a) Someone suggests that the city use its data to create a 95% confidence interval instead of the 90% interval first created. How would this interval be better for the city? (You need not actually create the new interval.)

b) How would the 95% interval be worse for the planners?

c) How could they achieve an interval estimate that would better serve their planning needs?

d) How many days' worth of data must they collect to have 95% confidence of estimating the true mean to within $3?

17. Speed of light In 1882, Michelson measured the speed of light (usually denoted c as in Einstein's famous equation $E = mc^2$). His values are in km/sec and have 299 000 subtracted from them. He reported the results of 23 trials with a mean of 756.22 and a standard deviation of 107.12.

a) Find a 95% confidence interval for the true speed of light from these statistics.

b) State in words what this interval means. Keep in mind that the speed of light is a physical constant that, as far as we know, has a value that is true throughout the universe.

c) What assumptions must you make in order to use your method?

T 18. Better light After his first attempt to determine the speed of light (described in Exercise 17), Michelson conducted an "improved" experiment. In 1897, he reported results of 100 trials with a mean of 852.4 km/sec and a standard deviation of 79.0.

a) What is the standard error of the mean for these data?

b) Without computing it, how would you expect a 95% confidence interval for the second experiment to differ from the confidence interval for the first? Note at least three specific reasons why they might differ, and indicate the ways in which these differences would change the interval.

c) According to Stigler (who reports these values), the true speed of light is 299 710.5 km/sec, corresponding to a value of 710.5 for Michelson's 1897 measurements. What does this indicate about Michelson's two experiments? Explain, using your confidence interval.

T 19. Departures 2011 What are the chances your flight will leave on time? The U.S. Bureau of Transportation Statistics of the Department of Transportation publishes information about airline performance. Here are a histogram and summary statistics for the percentage of flights departing on time each month from 1995 thru September 2011. (www.transtats.bts.gov/HomeDrillChart.asp)

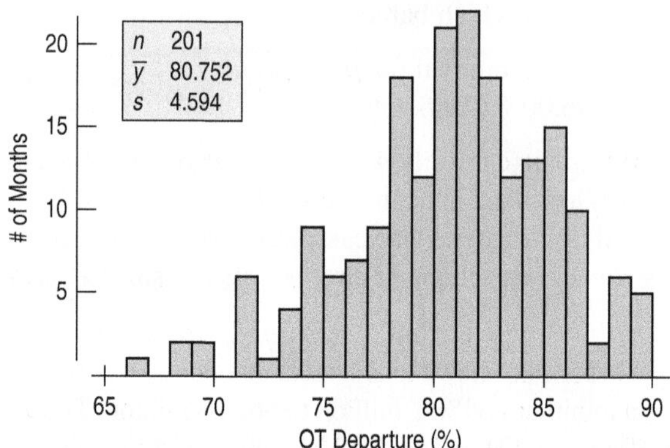

There is no evidence of a trend over time.

a) Check the assumptions and conditions for inference.

b) Find a 90% confidence interval for the true percentage of flights that depart on time.

c) Interpret this interval for a traveller planning to fly.

d) Suppose the number of flights differs considerably from month to month. What are you actually estimating in part b)? What might you recommend doing instead?

T 20. Arrivals 2011 Will your flight get you to your destination on time? The U.S. Bureau of Transportation Statistics reported the percentage of flights that were late each month from 1995 through September of 2011. Here's a histogram, along with some summary statistics:

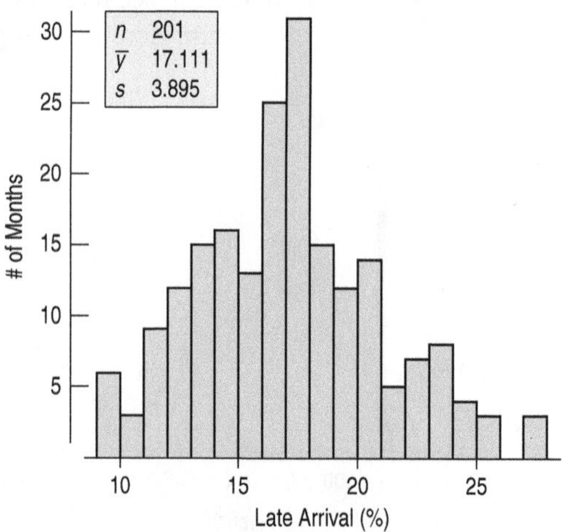

We can consider these data to be a representative sample of all months. There is no evidence of a time trend.

a) Check the assumptions and conditions for inference about the mean.

b) Find a 99% confidence interval for the true percentage of flights that arrive late.

c) Interpret this interval for a traveller planning to fly.

d) The *t* test (or confidence interval) is sometimes referred to as a "small sample" procedure. Why would it be okay to use a z^* value instead of a t^* value in constructing your confidence interval in part b)?

T **21. Farmed salmon, second look** This chapter's For Examples looked at mirex contamination in farmed salmon. We first found a 95% confidence interval for the mean concentration to be 0.0834 to 0.0992 parts per million. Later, we rejected the null hypothesis that the mean did not exceed the EPA's recommended safe level of 0.08 ppm based on a P-value of 0.0027. Explain how these two results are consistent. Your explanation should discuss the confidence level, the P-value, and the decision.

22. Hot dogs A nutrition lab tested 40 hot dogs to see if their mean sodium content was less than the 325 mg upper limit set by regulations for "reduced sodium" franks. The lab failed to reject the null hypothesis that the hot dogs did not meet this requirement, with a P-value of 0.142. A 90% confidence interval estimated the mean sodium content for this kind of hot dog at 317.2 to 326.8 mg. Explain how these two results are consistent. Your explanation should discuss the confidence level, the P-value, and the decision.

23. Pizza A researcher tests whether the mean cholesterol level among those who eat frozen pizza exceeds the value considered to indicate a health risk. She gets a P-value of 0.07. Explain in this context what the "7%" represents.

24. Golf balls The United States Golf Association (USGA) sets performance standards for golf balls. For example, the initial velocity of the ball may not exceed 250 feet per second when measured by an apparatus approved by the USGA. Suppose a manufacturer introduces a new kind of ball and provides a sample for testing. Based on the mean speed in the test, the USGA comes up with a P-value of 0.34. Explain in this context what the "34%" represents.

25. TV safety The manufacturer of a metal stand for home television sets must be sure that its product will not fail under the weight of the television. Since some larger sets weigh nearly 300 pounds (about 136 kg), the company's safety inspectors have set a standard of ensuring that the stands can support an average of over 500 pounds. Their inspectors regularly subject a random sample of the stands to increasing weight until they fail. They test the hypothesis H_0: $\mu = 500$ against H_A: $\mu > 500$, using the level of significance $\alpha = 0.01$. If the stands in the sample fail to pass this safety test, the inspectors will not certify the product for sale to the general public.

a) Is this an upper-tail or lower-tail test? In the context of the problem, why do you think this is important?

b) Explain what will happen if the inspectors commit a Type I error.

c) Explain what will happen if the inspectors commit a Type II error.

26. Catheters During an angiogram, heart problems can be examined via a small tube (a catheter) threaded into the heart from a vein in the patient's leg. It's important that the company that manufactures the catheter maintain a diameter of 2.00 mm. (The standard deviation is quite small.) Each day, quality control personnel make several measurements to test H_0: $\mu = 2.00$ against H_A: $\mu \neq 2.00$ at a significance level of $\alpha = 0.05$. If they discover a problem, they will stop the manufacturing process until it is corrected.

a) Is this a one-sided or two-sided test? In the context of the problem, why do you think this is important?

b) Explain in this context what happens if the quality control people commit a Type I error.

c) Explain in this context what happens if the quality control people commit a Type II error.

27. TV safety revisited The manufacturer of the metal television stands in Exercise 25 is thinking of revising its safety test.

a) If the company's lawyers are worried about being sued for selling an unsafe product, should they increase or decrease the value of α? Explain.

b) In this context, what is meant by the power of the test?

c) If the company wants to increase the power of the test, what options does it have? Explain the advantages and disadvantages of each option.

28. Catheters again The catheter company in Exercise 26 is reviewing its testing procedure.

a) Suppose the significance level is changed to $\alpha = 0.01$. Will the probability of a Type II error increase, decrease, or remain the same?

b) What is meant by the power of the test the company conducts?

c) Suppose the manufacturing process is slipping out of proper adjustment. As the actual mean diameter of the catheters produced gets farther and farther above the desired 2.00 mm, will the power of the quality control test increase, decrease, or remain the same?

d) What could they do to improve the power of the test?

29. Marriage In 1960, census results indicated that the age at which Canadian women first married had a mean of 22.6 years. It is widely suspected that young people today are waiting longer to get married. We want to find out if the mean age of first marriage has increased during the past 40 years.

a) Write appropriate hypotheses.
b) We plan to test our hypotheses by selecting a random sample of 40 women who married for the first time last year. Do you think the necessary assumptions for inference are satisfied? Explain.
c) Describe the approximate sampling distribution model for the mean age in such samples.
d) The women in our sample married at an average age of 27.2 years, with a standard deviation of 5.3 years. What is the P-value for this result?
e) Explain (in context) what this P-value means.
f) What is your conclusion?

30. Fuel economy A company with a large fleet of cars hopes to keep gasoline costs down and sets a goal of attaining a fleet average of at most 9 litres per 100 km. To see if the goal is being met, they check the gasoline usage for 50 company trips chosen at random, finding a mean of 9.40 L/100 km and a standard deviation of 1.81 L/100 km. Is this strong evidence that they have failed to attain their fuel economy goal?

a) Write appropriate hypotheses.
b) Are the necessary assumptions to make inferences satisfied?
c) Describe the sampling distribution model of mean fuel economy for samples like this.
d) Find the P-value.
e) Explain what the P-value means in this context.
f) State an appropriate conclusion.

T **31. Ruffles** Students investigating the packaging of potato chips purchased six bags of Lay's Ruffles marked with a net weight of 28.3 grams. They carefully weighed the contents of each bag, recording the following weights (in grams): 29.3, 28.2, 29.1, 28.7, 28.9, 28.5.

a) Do these data satisfy the assumptions for inference? Explain.
b) Find the mean and standard deviation of the observed weights.
c) Create a 95% confidence interval for the mean weight of such bags of chips.
d) Explain in context what your interval means.
e) Comment on the company's stated net weight of 28.3 grams.

T **32. Doritos** Some students checked six bags of Doritos marked with a net weight of 28.3 grams. They carefully weighed the contents of each bag, recording the following weights (in grams): 29.2, 28.5, 28.7, 28.9, 29.1, 29.5.

a) Do these data satisfy the assumptions for inference? Explain.
b) Find the mean and standard deviation of the observed weights.
c) Create a 95% confidence interval for the mean weight of such bags of chips.
d) Explain in context what your interval means.
e) Comment on the company's stated net weight of 28.3 grams.

T **33. Popcorn** Yvon Hopps ran an experiment to test optimum power and time settings for microwave popcorn. His goal was to find a combination of power and time that would deliver high-quality popcorn with only 10% of the kernels left unpopped, on average. After experimenting with several bags, he determined that power 9 at four minutes was the best combination.

a) He concluded that this popping method achieved the 10% goal. If it really does not work that well, what kind of error did Hopps make?
b) To be sure that the method was successful, he popped eight more bags of popcorn (selected at random) at this setting. All were of high quality, with the following percentages of unpopped popcorn: 7, 13.2, 10, 6, 7.8, 2.8, 2.2, 5.2. Does this provide evidence that he met his goal of an average of no more than 10% unpopped kernels? Explain.

T **34. Ski wax** Bjork Larsen was trying to decide whether to use a new racing wax for cross-country skis. He decided that the wax would be worth the price if he could average less than 55 seconds on a course he knew well, so he planned to test the wax by racing on the course eight times.

a) Suppose that he eventually decides not to buy the wax, but it really would lower his average time to below 55 seconds. What kind of error would he have made?
b) His eight race times were 56.3, 65.9, 50.5, 52.4, 46.5, 57.8, 52.2, and 43.2 seconds. Should he buy the wax? Explain.

T **35. Chips Ahoy** In 1998, as an advertising campaign, the Nabisco Company announced a "1000 Chips Challenge," claiming that every 18-ounce (about 625 grams) bag of their Chips Ahoy cookies contained at least 1000 chocolate chips. Dedicated Statistics students at the Air Force Academy (no kidding) purchased some randomly selected bags of cookies and counted the chocolate chips. Some of their data are given below. (*Chance*, 12, no. 1[1999])

1219	1214	1087	1200	1419	1121	1325	1345
1244	1258	1356	1132	1191	1270	1295	1135

a) Check the assumptions and conditions for inference. Comment on any concerns you have.
b) Create a 95% confidence interval for the average number of chips in bags of Chips Ahoy cookies.
c) What does this evidence say about Nabisco's claim? Use your confidence interval to test an appropriate hypothesis and state your conclusion.

T 36. Yogourt *Consumer Reports* tested 14 brands of vanilla yogourt and found the following numbers of calories per serving:

160	200	220	230	120	180	140
130	170	190	80	120	100	170

a) Check the assumptions and conditions for inference.
b) Create a 95% confidence interval for the average calorie content of vanilla yogourt.
c) A diet guide claims that you will get an average of 120 calories from a serving of vanilla yogourt. What does this evidence indicate? Use your confidence interval to test an appropriate hypothesis and state your conclusion.
d) *Perform a sign-test to test the hypothesis that the median number of calories is 120. Is your conclusion similar to what you found in part c)?

T 37. Maze Psychology experiments sometimes involve testing the ability of rats to navigate mazes. The mazes are classified according to difficulty, as measured by the mean length of time it takes rats to find the food at the end. One researcher needs a maze that will take rats an average of about one minute to solve. He tests one maze on several rats, collecting the data shown.

Time (sec)	
38.4	57.6
46.2	55.5
62.5	49.5
38.0	40.9
62.8	44.3
33.9	93.8
50.4	47.9
35.0	69.2
52.8	46.2
60.1	56.3
55.1	

a) Plot the data. Do you think the conditions for inference are satisfied? Explain.
b) Test the hypothesis that the mean completion time for this maze is 60 seconds. What is your conclusion?
c) Eliminate the outlier, and test the hypothesis again. What is your conclusion?
d) Do you think this maze meets the "one-minute average" requirement? Explain.
e) *Perform a sign-test to see if the median time is one minute or less, keeping the outlier in the data set. Does your conclusion change from the one you arrived at in part d)?

38. Braking A tire manufacturer is considering a newly designed tread pattern for its all-weather tires. Tests have indicated that these tires will provide better gas mileage and longer tread life. The last remaining test is for braking effectiveness. The company hopes the tire will allow a car travelling at 100 km/h to come to a complete stop within an average of 38 metres after the brakes are applied. They will adopt the new tread pattern unless there is strong evidence that the tires do not meet this objective. The distances (in metres) for 10 stops on a test track were 39.3, 39.0. 39.6, 40.2, 41.1, 37.5, 31.1, 38.1, 39.0, and 39.6. Should the company adopt the new tread pattern? Test an appropriate hypothesis and state your conclusion. Explain how you dealt with the outlier and why you made the recommendation you did.

T 39. Driving distance 2011 How far do professional golfers drive a ball? (For non-golfers, the drive is the shot hit from a tee at the start of a hole and is typically the longest shot.) Here's a histogram of the average driving distances of the 186 leading professional golfers by end of November 2011 along with summary statistics (www.pgatour.com).

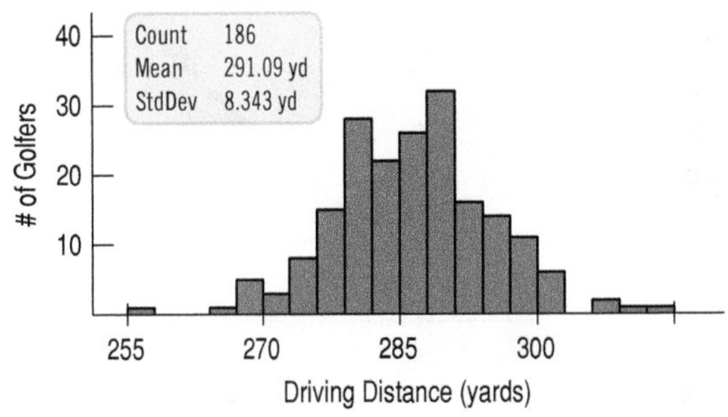

a) Find a 95% confidence interval for the mean drive distance.
b) Interpreting this interval raises some problems. Discuss.
c) The data are the mean driving distance for each golfer. Is that a concern in interpreting the interval? (*Hint:* Review the What Can Go Wrong warnings of Chapter 8. Chapter 8?! Yes, Chapter 8.)
d) If instead we used these golfers' individual drive distances, what problem would this create for our inferential procedures?

T 40. Wind power Should you generate electricity with your own personal wind turbine? That depends on whether you have enough wind on your site. To produce enough energy, your site should have an annual average wind speed of at least eight miles per hour (mph), according to the Wind Energy Association. One candidate site was monitored for a year, with wind speeds recorded every six hours. A total of 1114 readings of wind speed averaged 8.019 mph with a standard deviation of 3.813 mph. You've been asked to make a statistical report to help the landowner decide whether to place a wind turbine at this site.

a) Discuss the assumptions and conditions for using Student's t inference methods with these data. Here are some plots that may help you decide whether the methods can be used:

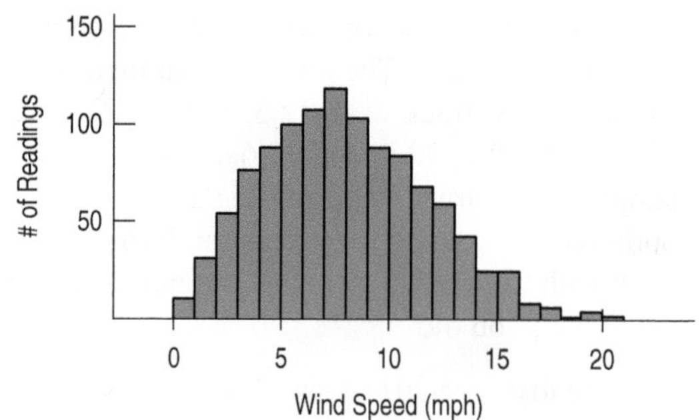

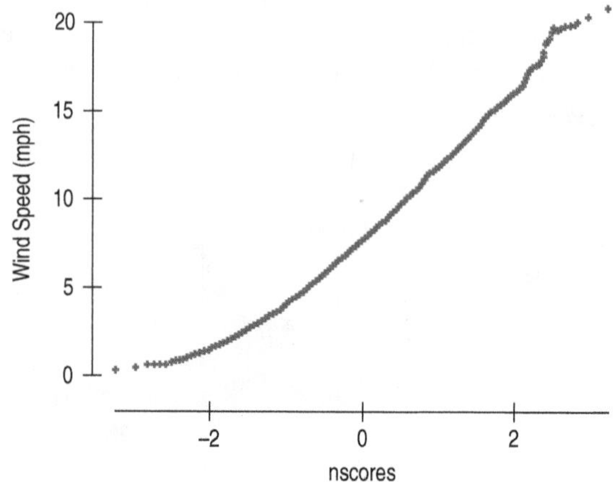

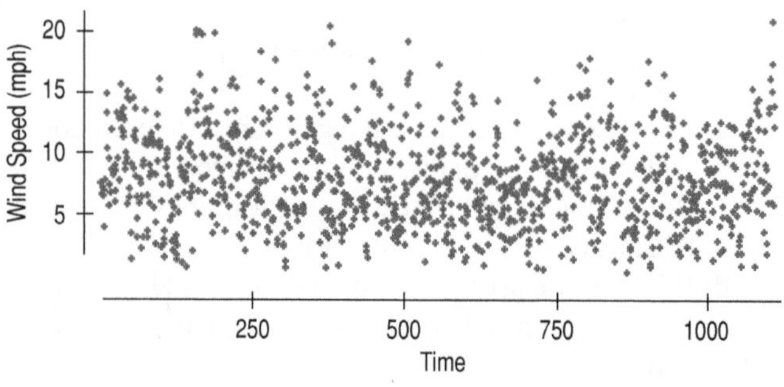

b) What would you tell the landowner about whether this site is suitable for a small wind turbine? Explain.

c) Why could we easily analyze data like this even before Gosset's discovery of the t distribution?

41. Worst of times Below is a sample randomly selected from all the Census Metropolitan Areas (CMAs) and Census agglomerations (CAs) in Canada showing the percentage change in the number of persons unemployed between May 2008 and May 2009 (during the deep 2008–2009 recession) for each area.

Area	% Change
Victoria	182.8
Alma	10.4
Salaberry-de-Valleyfield	64.0
Penticton	180.0
Campbell River	139.6
Woodstock	127.1
Baie-Comeau	6.6
Whitehorse	62.2
Hawkesbury, Ont. part	60.0
London	96.5
Prince Albert	62.2
Red Deer	304.3
Swift Current	200.0
Port Hope	122.2
Port Alberni	102.7
Ottawa-Gatineau, Gatineau part	53.7
Norfolk	129.3
Trois-Rivières	29.3
Labrador City	110.3
Nanaimo	136.6

Source: Adapted from Statistics Canada, Employment Insurance Statistics Maps, 73-002-XWE2009002 June 2009, Released August 25, 2009.

a) Estimate with a 95% confidence interval the true mean percentage increase in the number of unemployed persons per CMA/CA in Canada over this period.

b) Is your 95% confidence level quoted in part a) trustworthy? Check what you should.

c) The overall Canadian change in unemployment numbers was an increase of 71.5%. If we took similar repeated random samples and calculated such 95% confidence intervals over and over, would you expect them to catch this 71.5% figure 95% of the time? Why or why not?

42. Mercury sushi Torontonians (including one of your authors) seem to love their sushi, but is it always safe? The *New York Times* bought pieces of tuna sushi from a number of restaurants and stores in New York City in October 2007 and tested them for mercury levels. The results were not good. At most, consuming just six pieces per week would put you beyond an acceptable consumption level of mercury (49 micrograms of mercury per week for a person of average weight of 70 kg). Let's hope Toronto would fare better—but then again, the article states that experts believe similar results would be observed elsewhere, particularly for bluefin tuna sushi (the most common type in the survey). Analysts examined at least two pieces from each place and calculated the methylmercury level in parts per million. Results below are for the piece of sushi with the highest mercury level for the restaurants surveyed. The pieces vary in size, so also shown is how many pieces per week it would take to exceed the acceptable mercury intake of 49 micrograms per week.

Restaurants	Methylmercury (parts per million)	Number of pieces to reach Rfd
Bar Masa	0.49	8.6
Blue Ribbon Sushi	1.40	2.6
Japonica	0.86	1.6
Jewel Bako	0.83	5.2
Megu	0.87	7.7
Monster Sushi (22 West 46th Street)	0.56	4.7
New York Times cafeteria	0.50	6.4
Nobu Next Door	1.00	6.2
Sushi of Gari	1.04	3.6
Sushi Seki	1.04	4.9
Sushi Yasuda	0.79	9.9
Yuka	0.61	3.3
Yuki Sushi	0.86	4.1

Source: From the *New York Times*, January 23, 2008, © 2008 The New York Times. All rights reserved. Used by permission and protected by the copyright laws of the United States. The printing, copying, redistribution, or retransmission of this content without express written permission is prohibited. www.newyorktimes.com.

a) Give a 95% confidence interval for the mean mercury concentration level (per *worst* piece) if we can consider this to be a representative sample of New York City restaurants. Now check to see if that figure of 95% confidence is really trustworthy (that is, check and comment on any necessary conditions).

b) Give a 95% confidence interval for the mean number of (worst) pieces required to exceed health guidelines if we can assume this to be a representative sample of New York City restaurants. Now check to see if that 95% confidence level figure is really trustworthy (that is, check and comment on any necessary conditions).

43. Simulations Use your computer software to generate a sample of size 20 from a Normal distribution with a mean of 50 and a standard deviation of 10.

a) From the sample, calculate a 90% confidence interval for the population mean. Does it contain the number 50?

b) Repeat part a) for 99 fresh samples. How many confidence intervals out of 100 contained the number 50? What percent of confidence intervals would you expect to contain the number 50 if you repeated these simulations many times? If $X =$ the number of confidence intervals out of 100 that contain 50, what is the distribution of X?

44. More simulations Use your computer software to generate a sample of size 30 from a (continuous) uniform distribution on the interval 0 to 1.

a) From the sample, calculate an 80% confidence interval for the population mean. Does it contain the true mean?

b) Repeat part a) for 49 fresh samples. How many confidence intervals out of 50 contained the true mean? What percent of confidence intervals would you expect to contain the true mean if you repeated these simulations many times? If $X =$ the number of confidence intervals out of 50 that contain the true mean of this uniform distribution, what is the distribution of X?

45. Still more simulations Use your computer software to generate a sample of size 15 from an exponential distribution with a mean of 1 (if a parameter is requested, set it equal to 1.0). Plot the data and describe the shape of this distribution.

a) From the sample, calculate a 90% confidence interval for the population mean. Does it contain the true mean?

b) Repeat part a) for 99 fresh samples. How many confidence intervals out of 100 contained the true mean of 1.0? What percent of confidence intervals would you expect to contain 1.0 if you repeated these simulations many times? If there are difficulties answering this question, explain. If we changed the sample size to 100, would that affect your answer?

46. Yet more simulation. Use your computer software to generate a sample of size 100 from an exponential distribution with a mean of 1 (if requested, set scale $= 1.0$ and threshold $= 0.0$). Plot the data and describe the shape of this distribution.

a) From the sample, calculate a 90% confidence interval for the population mean. Does it contain the true mean?

b) Repeat part a) for 99 fresh samples. How many confidence intervals out of 100 contained the true mean of 1.0? What percent of confidence intervals would you expect to contain 1.0 if you repeated these simulations many times? Justify your answer.

47. Even more simulations Use your computer software to generate a sample of size 5 from a Normal distribution with mean of 50 and a standard deviation of 5. (For example, these might be Guinness stout measurements for a batch that you are checking for adequate quality.)

a) From the sample, test the null hypothesis that the population mean is 50 (our requirement for passing the batch) at the 10% significance level versus a two-sided alternative. Did you reject the null hypothesis? Did you make an incorrect decision? Did you pass or fail a good or bad batch of stout? An incorrect decision here would constitute what type of error?

b) Repeat part a) for 99 fresh samples. In how many tests did you reject the null hypothesis? In what percent of tests would you expect to reject this null hypothesis at the 10% level if you repeated these simulations many times? If X = the number of tests out of 100 in which you reject the null hypothesis at the 10% significance level, what is the distribution of X?

48. Final simulations Use your computer software to generate a sample of size 5 from a Normal distribution with a mean of 50 and a standard deviation of 5. (For example, these might be Guinness stout measurements for a batch that you are checking for adequate quality.)

a) From the sample, test the null hypothesis that the population mean is 60 (our requirement for passing the batch) at the 10% significance level versus a two-sided alternative. Did you reject the null hypothesis? Was this a correct decision or an error? Did you pass or fail a good or bad batch of stout? What type of error would a wrong decision here constitute?

b) Repeat part a) for 99 fresh samples. In how many tests did you reject the null hypothesis? This number is an estimate of something—what, exactly? If you have the appropriate software, use it to determine what would be the long-run percentage of such rejections.

49. Calculating power For the chapter example about salmon mirex levels, let's do an approximate calculation of the power of the test for an alternative of 0.09 ppm. For samples of size n $\geq$ 30, we can approximate the t distribution by the standard Normal distribution in these rough calculations. We also need to make a guess at the value of the unknown parameter σ, but here a study has already been run, so let's just take that $\sigma = 0.0495$ as the guess for σ and round it to 0.05. Since we are guessing at σ, this calculation is just an approximation, but usually that's all we need.

a) Setting alpha at 0.05, find the critical value for a standard Normal z-statistic. Write the criterion for rejection of the null in terms of the t-statistic. You should have a criterion like $\frac{\bar{y} - 0.08}{s/\sqrt{150}} > z^*$, with a specific number for the critical value.

b) Now another approximation. Pretend that the sample standard deviation s will equal the true population standard deviation σ. Funny thing to do, since s is random, not a constant, but this works well enough as an approximation when n is not too small (and otherwise we'd be stuck!). Rewrite the criterion above with just the sample mean on the left side, that is, find out just how big the sample mean must be for you to reject the null hypothesis (after setting $s = \sigma$). You should have a criterion like $\bar{y} > \bar{y}^*$ with a specific number for the $\bar{y}^*$ critical value.

c) Calculate the probability that $\bar{y}$ is bigger than $\bar{y}^*$, assuming the true mean is equal to 0.09 (standardize properly and use the standard Normal table). What you've now got is the power—the probability of making the right decision (to reject $\mu = 0.08$ ppm) should the true mean $\mu = 0.09$ ppm.

d) For a small n, though, this does not work well, since you have to take into account more properly the random variation present in the sample standard deviation s, in which case using statistical software is recommended. If your software does power calculations for the one-sample t-test, use it to confirm your calculation above. Also using your software, determine how low the power drops:
 i. if you halve your sample size ($n = 75$).
 ii. if you halve the sample size yet again (to $n = 38$).

50. More power to you For the chapter example about school travel times, let's do an approximate calculation of the power of the test for an alternative of 20 minutes. For samples of size $n \geq 30$, we can approximate the t distribution by the standard Normal distribution in these rough calculations. We also need to make a guess at the value of the unknown parameter σ, but here, a study has already been run so let's just make that $s = 9.66$ minutes as the guess for σ and round it to 10 minutes. Since we are guessing at σ, this calculation is just an approximation, but usually that's all we need.

a) Setting alpha at 0.05, find the critical value for a standard Normal z-statistic. Write the criterion for rejection of the null in terms of the t-statistic. You should have a criterion like $\frac{\bar{y} - 15}{s/\sqrt{40}} > z^*$, with a specific number for the critical value.

b) Now another approximation. Pretend that the sample standard deviation s will equal the true population standard deviation σ. Funny thing to do, since s is random, not a constant, but this works well enough as an approximation when n is not too small (and otherwise we'd be stuck!). Rewrite the criterion above with just the sample mean on the left side; that is, find out just how big the sample mean must be for you to reject the null hypothesis (after setting $s = \sigma$). You should have a criterion like $\bar{y} > \bar{y}^*$ with a specific number for the $\bar{y}^*$ critical value.

c) Calculate the probability that $\bar{y}$ is bigger than $\bar{y}^*$, assuming the true mean is equal to 20 minutes (standardize properly and use the standard Normal table). What you've now got is the power—the probability of making the right decision (to reject $\mu = 15$ minutes) should the true mean $\mu = 20$ minutes.

d) For small n, this does not work well, since you have to take into account more properly the random variation present in the sample standard deviation s, in which case using statistical software is recommended. If your software does power calculations for the one-sample t-test, use it to confirm your calculation above. Also using your software, determine how low the power drops:
 i. if you halve the sample size ($n = 20$).
 ii. if you double the sample size (to $n = 80$).

Just Checking ANSWERS

1. Questions on the short form are answered by everyone in the population. This is a census, so means or proportions *are* the true population values. The long forms are given to just a sample of the population. When we estimate parameters from a sample, we use a confidence interval to take sample-to-sample variability into account.
2. They don't know the population standard deviation, so they must use the sample standard deviation as an estimate. The additional uncertainty is taken into account by *t*-models if we are using an unweighted average.[21] We don't know what model to use for a weighted average (perhaps a *t* model but with a different SE formula).
3. The margin of error for a confidence interval for a mean depends, in part, on the standard error, $SE(\bar{y}) = \frac{s}{\sqrt{n}}$.

   Since n is in the denominator, smaller sample sizes lead to larger SEs and correspondingly wider intervals. Long forms returned by one in every five households in a less populous area will produce a smaller sample.
4. The t^* value would change a little, while n and its square root change a lot, making the interval much narrower for the larger sample. The smaller sample is one fourth as large, so the confidence interval would be roughly twice as wide.
5. We expect 95% of such intervals to cover the true value, so five of the 100 intervals might be expected to miss.
6. The power would increase if we have a larger sample size.

MathXL

MyStatLab

Go to MathXL at www.mathxl.com or MyStatLab at www.mystatlab.com. You can practise exercises for this chapter as often as you want. The guided solutions will help you find answers step by step. You'll find a personalized study plan available to you too!

[21]Though ideally a finite population correction factor should be applied to the SE formula, as discussed in Chapter 15, since the sample is more than 10% of the population.

chapter

21

Comparing Means

Nothing is good or bad, but by comparison.

—*Thomas Fuller*

Where are we going?

Does taking echinacea help you get over a cold faster? Does playing Mozart to babies in the womb make them smarter? Do students who use Minitab software learn Statistics better than those who do not? Many of the decisions made in business, medicine, and science compare the mean of two groups. Comparing two groups is very much like testing one. The standard errors are different, but the basic concepts are very similar.

Who	AA alkaline batteries
What	Length of battery life while playing a CD continuously
Units	Minutes
Why	Class project
When	1998

A S *Video:* **Can Diet Prolong Life?** Watch a video that tells the story of an experiment. We'll analyze the data later in this chapter.

Should you buy generic rather than brand-name batteries? A Statistics student designed a study to test battery life. He wanted to know whether there was any real difference between brand-name batteries and a generic brand. To estimate the difference in mean lifetimes, he kept a battery-powered CD player[1] continuously playing the same CD, with the volume control fixed at five, and measured the time until no more music was heard through the headphones. (He ran an initial trial to find out approximately how long that would take, so that he didn't have to spend the first three hours of each run listening to the same CD.) For his trials, he used six sets of AA alkaline batteries from two major battery manufacturers: a well-known brand name and a generic brand. He measured the time in minutes until the sound stopped. To account for changes in the CD player's performance over time, he randomized the run order by choosing sets of batteries at random.

Here are his data (times in minutes):

Brand Name	Generic
194.0	190.7
205.5	203.5
199.2	203.5
172.4	206.5
184.0	222.5
169.5	209.4

Studies that compare two groups are common throughout both science and industry. We might want to compare the effects of a new drug with the traditional therapy, the fuel efficiency of two car engine designs, or the sales of new products in two different test cities. In fact, battery manufacturers do research like this on their products and competitors' products.

21.1 Comparing Means of Independent Samples

Plot the Data

The natural display for comparing two groups is boxplots of the data for the two groups, placed side by side. Although we can't make a confidence interval or test a hypothesis from the boxplots themselves, you should always start with boxplots when comparing groups. Let's look at the boxplots of the battery test data, in Figure 21.1.

It sure looks like the generic batteries lasted longer. And we can see that they were also more consistent. But is the difference large enough to change our battery-buying

[1]Once upon a time, not so very long ago, there were no iPods. At the turn of the century, people actually carried CDs around, and devices to play them. We bet you can find one on eBay.

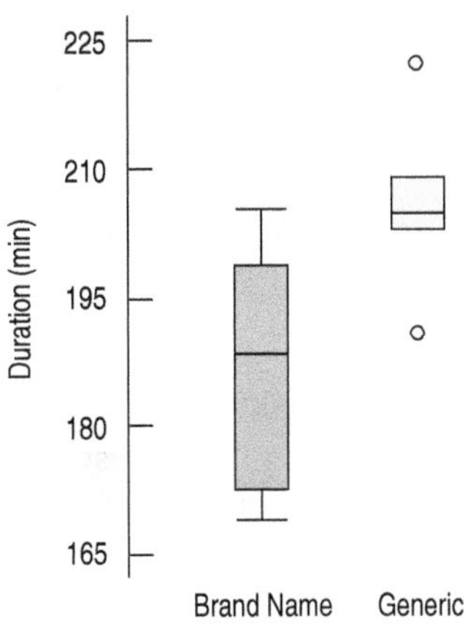

Figure 21.1

Boxplots comparing the brand-name and generic batteries suggest a difference in duration.

behaviour? Can we be confident that the difference is more than just random fluctuation? That's why we need statistical inference.

The boxplot for the generic data identifies two possible outliers. That's interesting, but with only six measurements in each group, the outlier nomination rule is not very reliable. Both of the extreme values are plausible results, and the range of the generic values is smaller than the range of the brand-name values, even with the outliers. So we're probably better off just leaving these values in the data.

Standard Error and Confidence Interval

The population model parameter of interest is the difference between the *mean* battery lifetimes of the two brands, $\mu_1 - \mu_2$. The statistic of interest is the difference in the two observed means, $\bar{y}_1 - \bar{y}_2$. We'll start with this statistic to build our confidence interval, but we'll need to know its standard deviation and its sampling model. Then we can build confidence intervals and find P-values for hypothesis tests.

We know that, for independent random quantities, the variance of their *difference* is the *sum* of their variances, $Var(Y - X) = Var(Y) + Var(X)$. To find the standard deviation of the difference between the two independent sample means, we add their variances and then take a square root:

$$\begin{aligned} SD(\bar{y}_1 - \bar{y}_2) &= \sqrt{Var(\bar{y}_1) + Var(\bar{y}_2)} \\ &= \sqrt{\left(\frac{\sigma_1}{\sqrt{n_1}}\right)^2 + \left(\frac{\sigma_2}{\sqrt{n_2}}\right)^2} \\ &= \sqrt{\frac{\sigma_1^2}{n_1} + \frac{\sigma_2^2}{n_2}} \end{aligned}$$

Of course, we still don't know the true standard deviations of the two groups, σ_1 and σ_2, so as usual, we'll use the estimates, s_1 and s_2. Using the estimates gives us the *standard error*:

$$SE(\bar{y}_1 - \bar{y}_2) = \sqrt{\frac{s_1^2}{n_1} + \frac{s_2^2}{n_2}}$$

We'll use the standard error to see how big the difference really is. Because we are working with means and estimating the standard error of their difference using the data, we shouldn't be surprised that the sampling model is a Student's *t*.

For Example FINDING THE STANDARD ERROR OF THE DIFFERENCE IN SAMPLE MEANS

Can you tell how much you are eating from how full you are? Or do you need visual cues? Researchers[2] constructed a table with two ordinary 18 fluid oz.[3] soup bowls and two identical-looking bowls that had been modified to slowly, imperceptibly, refill as they were emptied. They randomly assigned experiment participants to the bowls, and served them tomato soup. Those eating from the ordinary bowls had their bowls refilled by ladle whenever they were one-quarter full. If people judge their portions by internal cues, they should eat about the same amount. How big a difference was there in the amount of soup consumed? The table summarizes their results.

	Ordinary bowl	Refilling bowl
n	27	27
$\bar{y}$	8.5 oz	14.7 oz
s	6.1 oz	8.4 oz

[2]Brian Wansink, James E. Painter, and Jill North, 2005, "Bottomless bowls: Why visual cues of portion size may influence intake," *Obesity Research*, Vol. 13, No. 1.

[3]1 fluid oz. = 29.6 mL and 18 fluid oz. = 532 mL.

QUESTION: How much variability do we expect in the difference between the two means? Find the standard error.

ANSWER: $SE(\bar{y}_{refill} - \bar{y}_{ordinary}) = \sqrt{\frac{s_r^2}{n_r} + \frac{s_o^2}{n_o}} = \sqrt{\frac{8.4^2}{27} + \frac{6.1^2}{27}} = 2.0 \text{ oz.}$

The confidence interval we build is called a **two-sample *t*-interval** (for the difference in means). The corresponding hypothesis test is called a **two-sample *t*-test**. The interval looks just like all the others we've seen—the statistic plus or minus an estimated margin of error:

$$(\bar{y}_1 - \bar{y}_2) \pm ME$$
$$\text{where } ME = t^* \times SE(\bar{y}_1 - \bar{y}_2)$$

The ME is just a critical value times the standard error. Here we use a Student's *t*-model to find the appropriate critical t^*-value corresponding to our chosen confidence level, just as with the one-sample *t* procedure. It is essential to use the *t* model here, and not the standard normal (unless the *df* are quite large, in which case z^* and t^* values are essentially the same).

What are we missing? Only the degrees of freedom for the Student's *t*-model. Unfortunately, *that* formula is strange.

The deep, dark secret is that the sampling model isn't *really* Student's *t*, but only something close. The trick is that by using a special, adjusted degrees-of-freedom value, we can make it so close to a Student's *t*-model that nobody can tell the difference. The adjustment formula is straightforward but doesn't help our understanding much, so we leave it to the computer or calculator. (If you are curious and really want to see the formula, look in the footnote.[4])

A SAMPLING DISTRIBUTION FOR THE DIFFERENCE BETWEEN TWO MEANS

When the assumptions and conditions (described below) are met, we can model the sampling distribution of the standardized sample difference between the means of two independent groups

$$t = \frac{(\bar{y}_1 - \bar{y}_2) - (\mu_1 - \mu_2)}{SE(\bar{y}_1 - \bar{y}_2)}$$

by a Student's *t*-model with a number of degrees of freedom found with a special formula. We estimate the standard error with

$$SE(\bar{y}_1 - \bar{y}_2) = \sqrt{\frac{s_1^2}{n_1} + \frac{s_2^2}{n_2}}$$

Assumptions and Conditions

Now we've got everything we need. Before we can make a two-sample *t*-interval or perform a two-sample *t*-test, though, we have to check the assumptions and conditions.

[4] $$df = \frac{\left(\frac{s_1^2}{n_1} + \frac{s_2^2}{n_2}\right)^2}{\frac{1}{n_1 - 1}\left(\frac{s_1^2}{n_1}\right)^2 + \frac{1}{n_2 - 1}\left(\frac{s_2^2}{n_2}\right)^2}$$

Are you sorry you looked? This formula usually doesn't even give a whole number. If you are using a table, you'll need a whole number, so round down to be safe. If you are using technology, it's even easier. The approximation formulas that computers and calculators use for the Student's *t*-distribution can deal with fractional degrees of freedom, so it's okay to pass the fractional degrees of freedom from this formula through to those formulas.

Independence Assumption

Independence Assumption: Within each group, we need independent responses from the individuals. We can't check that for certain, but we see randomization as evidence of independence, so we *can* check the following:

Randomization Condition: For surveys, are they random samples (or if not, selected independently, and reasonably representative of the target populations)?[5] For experiments, was the experiment properly randomized?

If the samples are bigger than 10% of the populations in a survey, random draws are no longer approximately independent. No problem, just note that your results are conservative (and overestimate the SE).

Normal Population Assumption

As we did before with Student's t-models, we should check the assumption that the underlying populations are *each* Normally distributed. We check the:

Nearly Normal Condition: We must check this for *both* groups; a violation by either one violates the condition. As we saw for single sample means, the Normality Assumption matters most when sample sizes are very small. For samples of $n < 10$ in either group, you should not use these methods if the histogram or Normal probability plot shows clear skewness.[6] For n's of 10 or so, a moderately skewed histogram is okay, but you should not work with strongly skewed data or data containing clear outliers. For larger samples (n's ≥ 20), data skewness is even less of an issue, but you should still be on the lookout for outliers, extreme skewness, and multiple modes.

Robustness.
The two-sample t procedure is even more *robust* against violations of Normality than the one-sample t, particularly for equal n's and similarly shaped distributions, in which case even $n_1 = n_2 = 5$ may be sufficient to give reliable conclusions.

Independent Groups Assumption

Independent Groups Assumption: To use the **two-sample t methods**, the two groups we are comparing must be independent of each other. In fact, this test is sometimes called the *independent samples* t-test. No statistical test can verify this assumption. You have to think about how the data were collected. The assumption would be violated, for example, if one group consisted of husbands and the other group of their wives. Whatever we measure on couples might naturally be related. Similarly, if we compared subjects' performances before some treatment with their performances afterward, we'd expect a relationship of each "before" measurement with its corresponding "after" measurement. In cases such as these, where the observational units in the two groups are related or matched, *the two-sample methods of this chapter can't be applied*. When this happens, we need to use a different (actually quite simple) procedure, which we'll see in the next chapter.

A S *Activity:* **Does Restricting Diet Prolong Life?** This activity lets you construct a confidence interval to compare life spans of rats fed two different diets.

For Example CHECKING ASSUMPTIONS AND CONDITIONS

RECAP: Researchers randomly assigned people to eat soup from one of two bowls: 27 got ordinary bowls that were refilled by ladle, and 27 others ate from bowls that secretly refilled slowly.

QUESTION: Can the researchers use their data to make inferences about the role of visual cues in determining how much people eat?

ANSWER:

- **Independence/Randomization:** Since subjects were randomly assigned to treatments, independence between subjects is plausible. And the researchers should also conduct the experiment in such a way that individuals do not influence each other's consumption.

[5] Unfortunately, for cluster or multistage sampling, the SE formula given above will *underestimate* the true variability present when the clusters are somewhat homogeneous (as is often the case). A more complicated formula for the SE should be used.

[6] Consider a transformation (like logs) if both groups exhibit similar skewness.

- **Nearly Normal Condition:** The histograms for both groups look unimodal but somewhat skewed to the right. However, both groups are large enough (27) to allow use of t-methods.

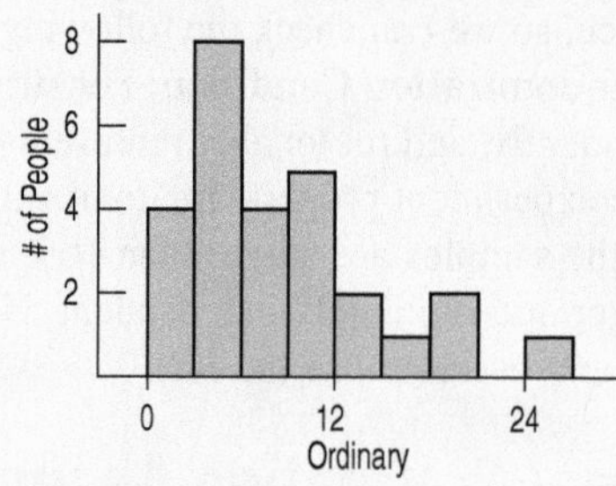

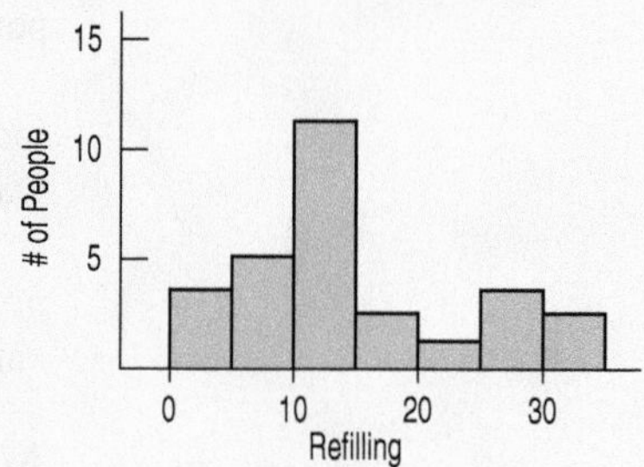

- **Independent Groups Assumption:** Randomization to treatment groups guarantees this.

It's okay to construct a two-sample t-interval for the difference in means.

TWO-SAMPLE t-INTERVAL FOR THE DIFFERENCE BETWEEN MEANS

When the conditions are met, we are ready to find the confidence interval for the difference between means of two independent groups, $\mu_1 - \mu_2$. The confidence interval is

$$(\bar{y}_1 - \bar{y}_2) \pm t^*_{df} \times SE(\bar{y}_1 - \bar{y}_2)$$

where the standard error of the difference of the means is

$$SE(\bar{y}_1 - \bar{y}_2) = \sqrt{\frac{s_1^2}{n_1} + \frac{s_2^2}{n_2}}$$

The critical value t^*_{df} depends on the particular confidence level, C, that you specify and on the number of degrees of freedom, which we get from the sample sizes and a special formula.

AN EASIER RULE?

The formula for the degrees of freedom of the sampling distribution of the difference between two means is long, but the number of degrees of freedom is always at least the smaller of the two n's minus 1. So wouldn't it be easier to just use that value? You could do that, and use the lower value. But *that* approximation can be a poor choice because it can give fewer than *half* the degrees of freedom you're entitled to from the correct formula. Underestimating the df gives you a *conservative* result (wider confidence interval than necessary).

For Example FINDING A CONFIDENCE INTERVAL FOR THE DIFFERENCE IN SAMPLE MEANS

RECAP: Researchers studying the role of internal and visual cues in determining how much people eat conducted an experiment in which some people ate soup from bowls that secretly refilled. The results are summarized in the table.

	Ordinary bowl	Refilling bowl
n	27	27
$\bar{y}$	8.5 oz.	14.7 oz.
s	6.1 oz.	8.4 oz.

We've already checked the assumptions and conditions, and have found the standard error for the difference in means to be $SE(\bar{y}_{refill} - \bar{y}_{ordinary}) = 2.0$ oz.

QUESTION: What does a 95% confidence interval say about the difference in mean amounts eaten?

ANSWER: The observed difference in means is

$$\bar{y}_{refill} - \bar{y}_{ordinary} = (14.7 - 8.5) = 6.2 \text{ oz.}$$

$$df = 47.46 \quad t^*_{47.46} = 2.011 \text{ (Table gives } t^*_{45} = 2.014)$$

$$ME = t^* \times SE(\bar{y}_{refill} - \bar{y}_{ordinary}) = 2.011(2.0) = 4.02 \text{ oz.}$$

The 95% confidence interval for $\mu_{\text{refill}} - \mu_{\text{ordinary}}$ is

$$6.2 \pm 4.02, \text{ or } (2.18, 10.22) \text{ oz.}$$

I am 95% confident that people eating from a subtly refilling bowl will eat an average of between 2.18 and 10.22 more ounces of soup than those eating from an ordinary bowl.

Step-by-Step Example A TWO-SAMPLE *t*-INTERVAL

Judging from the boxplot, the generic batteries seem to have lasted about 20 minutes longer than the brand-name batteries. Before we change our buying habits, what should we expect to happen with the next batteries we buy?

Question: How much longer might the generic batteries last?

THINK ➡ **Plan** State what we want to know.

Identify the *parameter* you wish to estimate. Here our parameter is the difference in the means, not the individual group means.

Identify the *population(s)* about which you wish to make statements. We hope to make decisions about purchasing batteries, so we're interested in all the AA batteries of these two brands.

Identify the variables and review the *W*'s.

Make a picture. Boxplots are the display of choice for comparing groups. We'll also want to check the distribution of each group. Histograms or Normal probability plots do a better job there.

I have measurements of the lifetimes (in minutes) of six sets of generic and six sets of brand-name AA batteries from a randomized experiment. I want to find an interval that is likely, with 95% confidence, to contain the true difference $\mu_G - \mu_B$ between the mean lifetime of the generic AA batteries and the mean lifetime of the brand-name batteries.

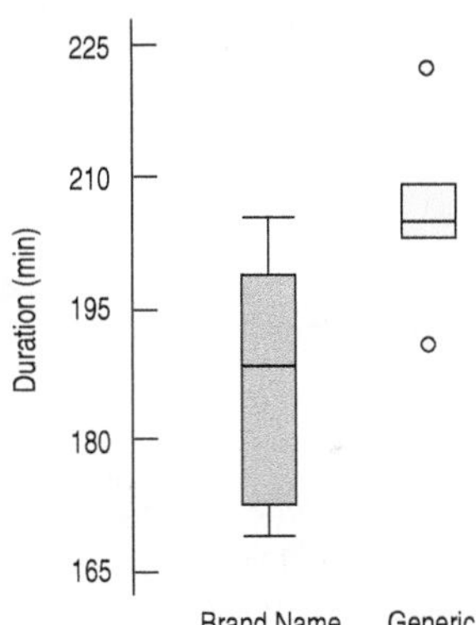

REALITY CHECK From the boxplots, it appears our confidence interval should be centred near a difference of 20 minutes. We don't have a lot of intuition about how far the interval should extend on either side of 20.

Model Think about the appropriate assumptions and check the conditions to be sure that a Student's *t*-model for the sampling distribution is appropriate.

✓ **Randomization Condition:** The sets of batteries were selected at random from those available for sale. Not exactly an SRS, but a reasonably representative random sample. And randomizing the run order helps ensure independence of the observations.

✓ **Independent Groups Assumption:** Batteries manufactured by two different companies and purchased in separate packages should be independent.

✓ **Nearly Normal Condition:** Considering how small the sample sizes, the histograms look reasonably unimodal and symmetric:

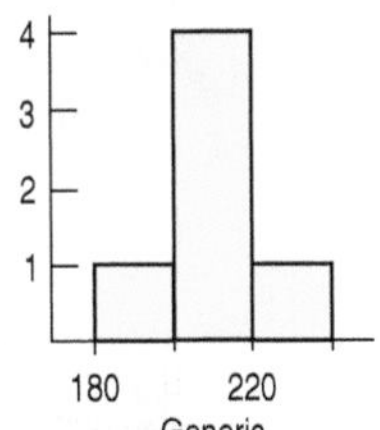

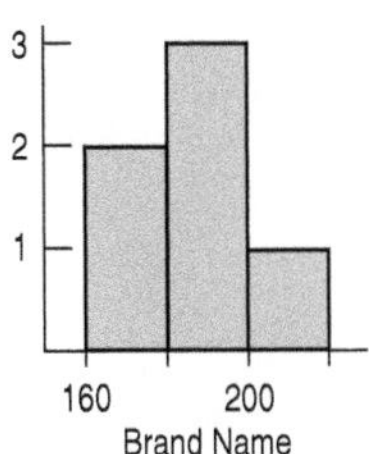

State the sampling distribution model for the statistic. Here the degrees of freedom will come from that messy approximation formula.

Specify your method.

Under these conditions, it's okay to use a Student's t-model.

I'll use a two-sample t-interval.

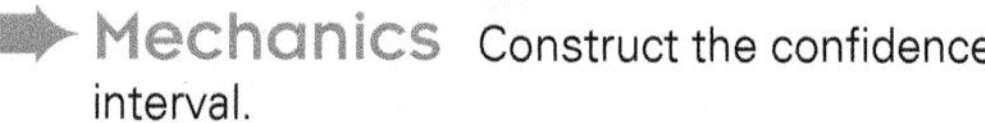

SHOW ➡ Mechanics Construct the confidence interval.

Be sure to include the units along with the statistics. Use meaningful subscripts to identify the groups.

Use the sample standard deviations to find the standard error of the sampling distribution.

We have three choices for degrees of freedom. The best alternative is to let the computer or calculator use the aforementioned special formula for *df*. This gives a fractional degree of freedom (here $df = 8.98$), and technology can find a corresponding critical value. In this case, it is $t^* = 2.263$.

I know

$$n_G = 6 \qquad n_B = 6$$
$$\bar{y}_G = 206.0 \text{ min} \qquad \bar{y}_B = 187.4 \text{ min}$$
$$s_G = 10.3 \text{ min} \qquad s_B = 14.6 \text{ min}$$

The groups are independent, so

$$\begin{aligned} SE(\bar{y}_G - \bar{y}_B) &= \sqrt{SE^2(\bar{y}_G) + SE^2(\bar{y}_B)} \\ &= \sqrt{\frac{s_G^2}{n_G} + \frac{s_B^2}{n_B}} \\ &= \sqrt{\frac{10.3^2}{6} + \frac{14.6^2}{6}} \\ &= \sqrt{\frac{106.09}{6} + \frac{213.16}{6}} \\ &= \sqrt{53.208} \\ &= 7.29 \text{ min} \end{aligned}$$

Or we could round the formula's *df* value down to an integer so we can use a *t* table. That gives 8 *df* and a critical value $t^* = 2.306$.

The *df* formula calls for 8.98 degrees of freedom.[7]

The corresponding critical value for a 95% confidence level from a Student's t-model with 8.98 *df* is $t^* = 2.263$.

The very conservative *easy rule* says to use only $6 - 1 = 5$ *df*. That gives a critical value $t^* = 2.571$. The corresponding confidence interval is about 14% wider—a high price to pay for a small savings in effort.

So the margin of error is

$$\begin{aligned} ME &= t^* \times SE(\bar{y}_G - \bar{y}_B) \\ &= 2.263(7.29) \\ &= 16.50 \text{ min} \end{aligned}$$

The 95% confidence interval is

$$(206.0 - 187.4) \pm 16.5 \text{ min}$$
$$\text{or } 18.6 \pm 16.5 \text{ min}$$
$$= (2.1, 35.1) \text{ min.}$$

TELL ➡ Conclusion Interpret the confidence interval in the proper context.

Less formally, you could say, "I'm 95% confident that generic batteries last an average of 2.1 to 35.1 minutes longer than brand-name batteries."

I am 95% confident that the interval from 2.1 minutes to 35.1 minutes captures the average amount of time by which generic batteries outlast brand-name batteries for this task (a rather wide interval, due to the small sample sizes, but note that it is completely above zero, giving good evidence that the generic do last longer, on average). If generic batteries are cheaper, there seems little reason not to use them.

[7]If you try to find the degrees of freedom with that messy special formula (We dare you! It's in the footnote on page 602) using the values above, you'll get 8.99. The minor discrepancy is because we rounded the standard deviations to make the exposition clearer.

A S **Activity: Find Two-Sample t-Intervals.** Who wants to deal with that ugly df formula? We usually find these intervals with a statistics package. Learn how here.

Just Checking

Carpal tunnel syndrome (CTS) causes pain and tingling in the hand. It can be bad enough to keep sufferers awake at night and restrict their daily activities. Researchers studied the effectiveness of two alternative surgical treatments for CTS (Mackenzie, Hainer, and Wheatley, *Annals of Plastic Surgery*, 2000). Patients were randomly assigned to have endoscopic or open-incision surgery. Four weeks later the endoscopic surgery patients demonstrated a mean pinch strength of 9.1 kg compared to 7.6 kg for the open-incision patients.

1. Why is the randomization of the patients into the two treatments important?
2. A 95% confidence interval for the difference in mean strength is about (0.04 kg, 2.96 kg). Explain what this interval means.
3. Why might we want to examine such a confidence interval in deciding between these two surgical procedures?
4. Why might you want to see the data before trusting the confidence interval?

21.2 The Two-Sample *t*-Test

If you bought a used camera in good condition from a friend, would you pay the same as you would if you bought the same item from a stranger? A researcher at Cornell University[8] wanted to know how friendship might affect simple sales such as this. She randomly divided subjects into two groups and gave each group descriptions of items they might want to buy. One group was told to imagine buying from a friend they expected to see again. The other group was told to imagine buying from a stranger.

Here are the prices they offered for a used camera in good condition:

Who	University students
What	Prices offered for a used camera
Units	$
Why	Study of the effects of friendship on transactions
When	1990s
Where	Cornell University

Price Offered for a Used Camera ($)	
Buying from a Friend	**Buying from a Stranger**
275	260
300	250
260	175
300	130
255	200
275	225
290	240
300	

The researcher who designed this study had a specific concern. Previous theories had doubted that friendship had a measurable effect on pricing. She hoped to find an effect on friendship. This calls for a hypothesis test—in this case, a **two-sample *t*-test for the difference between means.**[9]

A Test for the Difference between Two Means

You already know enough to construct this test. The test statistic looks just like the others we've seen. First, we find the difference between the observed group means and compare

[8]J. J. Halpern, "The Transaction Index: A method for standardizing comparisons of transaction characteristics across different contexts," *Group Decision and Negotiation*, 6: 557–572.

[9]Because it is performed so often, this test is usually just called a "two-sample *t*-test."

A S *Activity:* **The Two-Sample *t*-Test.** How different are beef hot dogs and chicken dogs? Test whether measured differences are statistically significant.

this with a hypothesized value for that difference. When we write the null hypothesis, we could write: H_0: $\mu_1 = \mu_2$. But sometimes we'll want to see if the difference is something other than 0, so we usually write it slightly differently. We'll focus on the hypothesized *difference* between the two means and we'll call that hypothesized difference Δ_0 ("delta naught"). It's so common for that hypothesized difference to be zero that we often just assume $\Delta_0 = 0$. In that case, we'll write: H_0: $\mu_1 - \mu_2 = 0$, but in general we'll write: H_0: $\mu_1 - \mu_2 = \Delta_0$.

Then, as usual, we compare the observed difference with its standard error (the same standard error that we used for the confidence interval). We already know that, for a difference between independent means, we can find P-values from a Student's *t*-model on that same special number of degrees of freedom.

NOTATION ALERT

Δ_0—delta naught—isn't so standard that you can assume everyone will understand it. We use it because it's the Greek letter (good for a parameter) "D" for "difference."

TWO-SAMPLE *t*-TEST FOR THE DIFFERENCE BETWEEN MEANS

The conditions for the two-sample *t*-test for the difference between the means of two independent groups are the same as for the two-sample *t*-interval. We test the hypothesis

$$H_0: \mu_1 - \mu_2 = \Delta_0$$

where the hypothesized difference is almost always 0, using the statistic

$$t = \frac{(\bar{y}_1 - \bar{y}_2) - \Delta_0}{SE(\bar{y}_1 - \bar{y}_2)}$$

The standard error of $\bar{y}_1 - \bar{y}_2$ is

$$SE(\bar{y}_1 - \bar{y}_2) = \sqrt{\frac{s_1^2}{n_1} + \frac{s_2^2}{n_2}}$$

When the conditions are met and the null hypothesis is true, this statistic can be closely modelled by a Student's *t*-model with a number of degrees of freedom given by the special formula. We use that model to obtain a P-value.

Step-by-Step Example A TWO-SAMPLE *t*-TEST FOR THE DIFFERENCE BETWEEN TWO MEANS

Recall: We want to know if people will pay the same amount, on average, when buying from a friend or a stranger. The usual null hypothesis is that there's no difference in means. That's just the right null hypothesis for the camera purchase prices.

Question: Is there a difference in the price people would offer a friend rather than a stranger?

THINK ➡ **Plan** State what we want to know. Identify the *parameter* you wish to estimate. Here our parameter is the difference in the means, not the individual group means. Identify the variables and check the *W*'s.	I want to know whether people are likely to offer a different amount for a used camera when buying from a friend than when buying from a stranger. I wonder whether the difference between mean amounts is zero. I have bid prices from eight subjects buying from a friend and seven buying from a stranger, found in a randomized experiment.
Hypotheses State the null and alternative hypotheses. The research claim is that friendship changes what people are willing to pay.[10] The natural null hypothesis is that friendship makes no difference.	H_0: The difference in mean price offered to friends and mean price offered to strangers is zero: $\mu_F - \mu_S = 0.$

[10]This claim is a good example of what is called a "research hypothesis" in many social sciences. The only way to check it is to deny that it's true and see where the resulting null hypothesis leads us.

We didn't start with any knowledge of whether friendship might increase or decrease the price, so we choose a two-sided alternative.

H_A: The difference in mean prices is not zero:

$$\mu_F - \mu_S \neq 0.$$

Make a picture. Boxplots are the display of choice for comparing groups. We'll also want to check the distribution of each group. Histograms or Normal probability plots do a better job there.

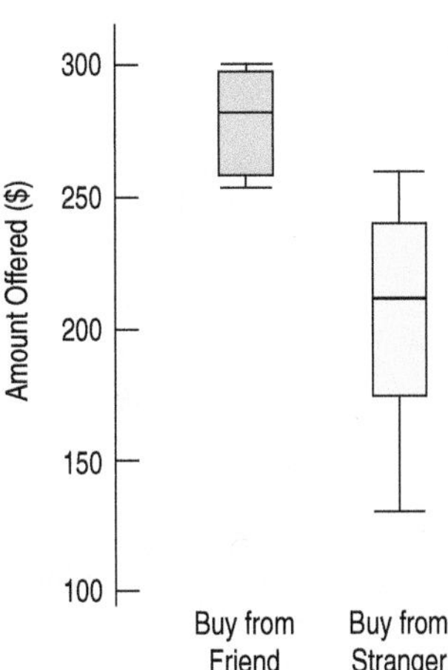

REALITY CHECK Looks like the prices are higher if you buy from a friend, but we can't be sure. You can't tell from looking at the boxplots whether the difference is statistically significant—the plot shows spreads on the same scale as the data, and we know those don't add. You'll need to add the *variances* to get a suitable ruler for comparing the difference.

Model Think about the assumptions and check the conditions.

✓ **Randomization Condition:** The experiment was randomized. Subjects were assigned to treatment groups at random and so the prices paid should be independent.

✓ **Independent Groups Assumption:** Randomizing the subjects gives independent groups.

✓ **Nearly Normal Condition:** Histograms of the two sets of prices are reasonably unimodal and close to symmetric, considering the small samples sizes:

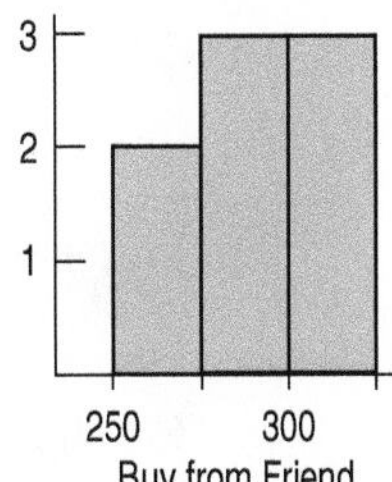

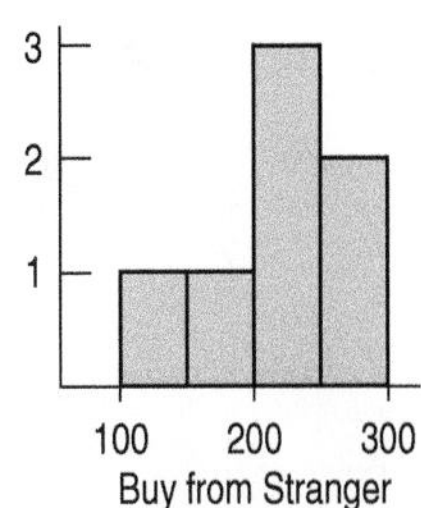

State the sampling distribution model.

Specify your method.

The assumptions are reasonable and the conditions are okay, so I'll use a Student's *t*-model to perform a two-sample t-test.

SHOW ➡ Mechanics List the summary statistics. Be sure to use proper notation.

From the data:

$n_F = 8 \qquad n_S = 7$

$\bar{y}_F = \$281.88 \qquad \bar{y}_S = \$211.43$

$s_F = \$18.31 \qquad s_S = \$46.43$

Use the null model to find the P-value. First determine the standard error of the difference between sample means.	For independent groups, $SE(\bar{y}_F - \bar{y}_S) = \sqrt{SE^2(\bar{y}_F) + SE^2(\bar{y}_S)} = \sqrt{\frac{s_F^2}{n_F} + \frac{s_S^2}{n_S}} = \sqrt{\frac{18.31^2}{8} + \frac{46.43^2}{7}} = 18.70$
	The observed difference is $(\bar{y}_F - \bar{y}_S) = 281.88 - 211.43 = \$70.45$
Find the t-value. Make a picture. Sketch the t-model centred at the hypothesized difference of zero. Because this is a two-tailed test, shade the region to the right of the observed t value and the corresponding region in the other tail. A statistics program or graphing calculator can find the P-value using the fractional degrees of freedom from the special formula. If you are doing a test like this without technology, you could use the smaller sample size to quickly determine conservative degrees of freedom, which would equal $n_2 - 1 = 6$.[11]	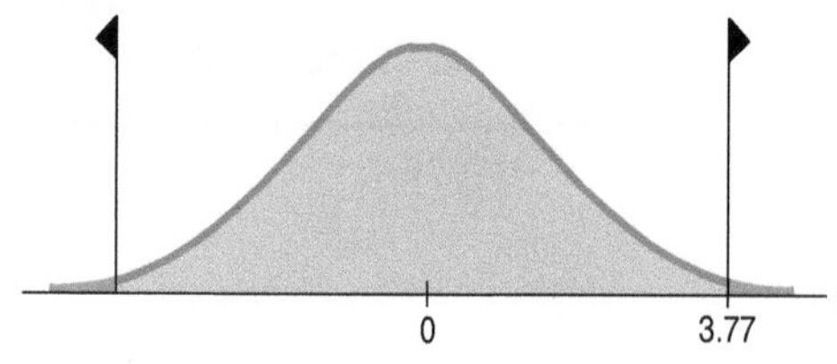 $t = \frac{(\bar{y}_F - \bar{y}_S) - (0)}{SE(\bar{y}_F - \bar{y}_S)} = \frac{70.45}{18.70} = 3.77$ The approximation formula gives 7.62 degrees of freedom.[12] P-value $= 2P(t_{7.62} > 3.77) = 0.006$
TELL ➡ Conclusion Link the P-value to your decision about the null hypothesis, and state the conclusion in context. Be cautious about generalizing to items whose prices are outside the range of those in this study.	If there were no difference in the mean prices, a difference this large would occur only six times in 1000. That's too rare to believe, so I reject the null hypothesis and conclude that people are likely to offer a friend more than they'd offer a stranger for a used camera (and possibly for other, similar items).

Just Checking

Recall the experiment comparing patients four weeks after surgery for carpal tunnel syndrome. The patients who had endoscopic surgery demonstrated a mean pinch strength of 9.1 kg compared to 7.6 kg for the open-incision patients.

5. What hypotheses would you test?
6. The P-value of the test was less than 0.05. State a brief conclusion.
7. The study reports work on 36 "hands," but there were only 26 patients. In fact, seven of the endoscopic surgery patients had both hands operated on, as did three of the open-incision group. Does this alter your thinking about any of the assumptions? Explain.

[11] Keep in mind that you err conservatively with this formula; your reported P-value would be inflated. With 6 degrees of freedom, P would equal 0.009 here. If that is small enough for you to reject the null hypothesis, there is little need for a more accurate calculation!

[12] If you were daring enough to calculate that messy degrees-of-freedom formula by hand with the values given here, you'd get about 7.74. Without a computer, you would likely round this down to 7 df and use the tables. Computers also work with more precision for the standard deviations than we gave in our example. All give about the same result for the P-value, so it doesn't really matter—the conclusion would be the same.

For Example A TWO-SAMPLE *t*-TEST

Pearson Education

Many office "coffee stations" collect voluntary payments for the food consumed. Researchers at the University of Newcastle upon Tyne performed an experiment to see whether the image of eyes watching would change employee behaviour.[13] They alternated pictures (seen here) of eyes looking at the viewer with pictures of flowers each week on the cupboard behind the "honesty box." They measured the consumption of milk to approximate the amount of food consumed and recorded the contributions (in £) each week per litre of milk. The table summarizes their results.

	Eyes	Flowers
n (# weeks)	5	5
$\bar{y}$	0.417 £/L	0.151 £/L
s	0.1811 £/L	0.067 £/L

QUESTION: Do these results provide evidence that there really is a difference in honesty even when it's only photographs of eyes that are "watching"?

ANSWER:

$$H_0: \mu_{eyes} - \mu_{flowers} = 0$$
$$H_A: \mu_{eyes} - \mu_{flowers} \neq 0$$

- **Independence Assumption**: This study is an experiment, though they alternated assignment of treatments (to weeks), rather than randomized. But average contributions in alternating weeks should be fairly unrelated or independent.
- **Nearly Normal Condition:** I don't have the data to check, but it seems unlikely there would be outliers in either group. I could be more certain if I could see histograms for both groups.
- **Independent Groups Assumption:** Though there is pairing of an 'eyes' week with a 'flowers' week (due to alternation) the results from weeks with eyes should be pretty much independent of the results from weeks with flowers.

It's okay to do a two-sample *t*-test for the difference in means:

$$SE(\bar{y}_{eyes} - \bar{y}_{flowers}) = \sqrt{\frac{s^2_{eyes}}{n_{eyes}} + \frac{s^2_{flowers}}{n_{flowers}}} = \sqrt{\frac{0.1811^2}{5} + \frac{0.067^2}{5}} = 0.0864$$

$$\text{df} = 5.07$$

$$t = \frac{\bar{y}_{eyes} - \bar{y}_{flowers} - 0}{SE(\bar{y}_{eyes} - \bar{y}_{flowers})} = \frac{0.417 - 0.151}{0.0864} = 3.08$$

$$P\text{-value} = 2 \times P(t_5 > 3.08) = 2 \times 0.0137 = 0.027$$

Assuming the data were free of outliers, the low P-value leads me to reject the null hypothesis. This study provides evidence that people will leave higher average voluntary payments for food if pictures of eyes are "watching."

(Note: In Table T we can see that at 5 df, $t = 3.08$ lies between the critical values for $P = 0.02$ and $P = 0.05$, so using that table, we could report $P < 0.05$.)

[13] Melissa Bateson, Daniel Nettle, and Gilbert Roberts, "Cues of being watched enhance cooperation in a real-world setting," *Biol. Lett. doi*:10.1098/rsbl.2006.0509.

21.3 The Pooled t-Test

You may encounter another way to test whether the means of two groups are different called the **pooled t-test**. We present it both for historical reasons and because it's important in the analysis of designed experiments (Chapter 25). Before 1949, it wasn't known that the statistic we saw on page 602 could be modelled by a Student's t distribution—even one with that crazy number of degrees of freedom. So, another, simpler test was used. Although simpler, this test has a big assumption—it assumes that the variances of the two groups are the same.

That might not seem like a big deal, but it's a very restrictive assumption. Just because the means of two groups are assumed equal in a test doesn't mean that their variances are as well. Generally, knowing the mean doesn't tell you *anything* about the variance. But there are cases where the assumption makes some sense—especially in the analysis of designed experiments. If you're trying to decide whether the new computer code you designed actually makes downloading files faster, the null hypothesis is that the mean of your method is the same as the standard code. But, if it has no effect on the mean, it might be safe to assume that it has no effect on the variance either. In such a case, the pooled t-test may be useful.

Pooled t-tests have a couple of advantages. They often have a few more degrees of freedom than the corresponding two-sample test and a much simpler degrees of freedom formula. But these advantages come at a price: You have to pool the variances and think about another assumption. The assumption of equal variances is a strong one, is often not true, and is difficult to check. For these reasons, we recommend that you generally use the (unpooled) two-sample t-test instead.

The pooled t-test is the theoretically correct method only when we have good reason to believe that the variances are equal. And (as we will see shortly) there are times when this makes sense. Keep in mind, however, that it's never wrong *not* to pool.

Details of the Pooled t-Test

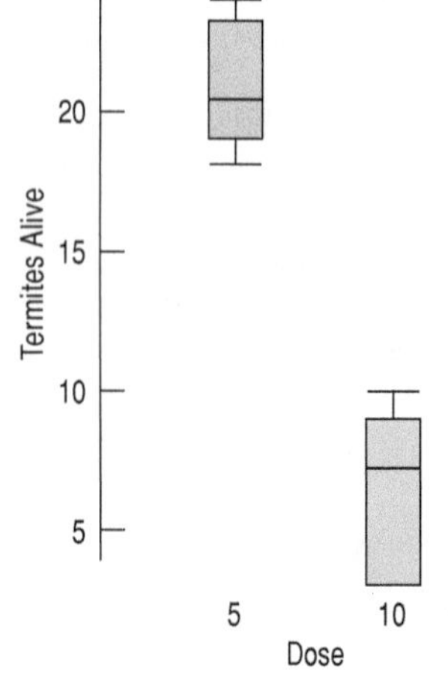

Termites cause billions of dollars of damage each year to homes and other buildings, but some tropical trees seem to be able to resist termite attack. A researcher extracted a compound from the sap of one such tree and tested it by feeding it at two different concentrations to randomly assigned groups of 25 termites.[14] After five days, eight groups fed the lower dose had an average of 20.875 termites alive, with a standard deviation of 2.23. But six groups fed the higher dose had an average of only 6.667 termites alive, with a standard deviation of 3.14. Is this a large enough difference to declare the sap compound effective in killing termites? In order to use the pooled t-test, we must make the **Equal Variance Assumption** that the variances of the two populations from which the samples have been drawn are equal. That is, $\sigma_1^2 = \sigma_2^2$. (Of course, we could think about the standard deviations being equal instead.) The corresponding **Similar Spreads Condition** really just consists of looking at the boxplots to check that the spreads are not very different. We were going to make boxplots anyway, so there's really nothing new here.

$$s_{\text{pooled}}^2 = \frac{(8-1)2.23^2 + (6-1)3.14^2}{(8-1)+(6-1)} = 7.01$$

Once we decide to pool, we estimate the common variance by combining numbers we already have:

$$s_{\text{pooled}}^2 = \frac{(n_1-1)s_1^2 + (n_2-1)s_2^2}{(n_1-1)+(n_2-1)}$$

(If the two sample sizes are equal, this is just the average of the two variances.)

Now, we just substitute this pooled variance in place of each of the variances in the standard error formula. Remember, the standard error formula for the difference of two independent means is

$$SE(\bar{y}_1 - \bar{y}_2) = \sqrt{\frac{s_1^2}{n_1} + \frac{s_2^2}{n_2}}$$

[14]Adam Messer, Kevin McCormick, H. H. Sunjaya, Ferny Tumbel Hagedorm, and J. Meinwald, "Defensive role of tropical tree resins: antitermitic sesquiterpenes from Southeast Asian Dipterocarpaceae," *J Chem Ecology*, 16:122, pp. 3333–3352.

We just substitute the common pooled variance for each of the two variances in this formula and simplify. That makes the pooled standard error formula much simpler:

$$SE_{pooled}(\bar{y}_1 - \bar{y}_2) = \sqrt{\frac{7.01}{8} + \frac{7.01}{6}} = 1.43$$

$$SE_{\text{pooled}}(\bar{y}_1 - \bar{y}_2) = \sqrt{\frac{s^2_{\text{pooled}}}{n_1} + \frac{s^2_{\text{pooled}}}{n_2}} = s_{\text{pooled}}\sqrt{\frac{1}{n_1} + \frac{1}{n_2}}$$

The formula for degrees of freedom for the Student's t-model is simpler, too. It was so complicated for the two-sample t that we stuck it in a footnote.[15] Now it's just $df = n_1 + n_2 - 2$.

$$t = \frac{20.875 - 6.667}{1.43} = 9.936$$

Substitute the pooled-t estimate of the standard error and its degrees of freedom into the steps of the confidence interval or hypothesis test, and you'll be using the pooled-t method. For the termites, $\bar{y}_1 - \bar{y}_2 = 14.208$, giving a t-value $= 9.936$ with 12 df and a P-value ≤ 0.0001. Of course, if you decide to use a pooled-t method, you must defend your assumption that the variances of the two groups are equal.

POOLED *t*-TEST AND CONFIDENCE INTERVAL FOR MEANS

The conditions for the pooled t-test for the difference between the means of two independent groups (commonly called a "pooled t-test") are the same as for the two-sample t-test discussed earlier with the additional assumption that the variances of the two groups are the same. We test the hypothesis

$$H_0: \mu_1 - \mu_2 = \Delta_0$$

where the hypothesized difference, Δ_0, is almost always 0, using the statistic

$$t = \frac{(\bar{y}_1 - \bar{y}_2) - \Delta_0}{SE_{pooled}(\bar{y}_1 - \bar{y}_2)}$$

The standard error of $\bar{y}_1 - \bar{y}_2$ is

$$SE_{pooled}(\bar{y}_1 - \bar{y}_2) = \sqrt{\frac{s^2_{pooled}}{n_1} + \frac{s^2_{pooled}}{n_2}} = s_{pooled}\sqrt{\frac{1}{n_1} + \frac{1}{n_2}}$$

where the pooled variance is

$$s^2_{pooled} = \frac{(n_1 - 1)s_1^2 + (n_2 - 1)s_2^2}{(n_1 - 1) + (n_2 - 1)}$$

When the conditions are met and the null hypothesis is true, we can model this statistic's sampling distribution with a Student's t-model with $(n_1 - 1) + (n_2 - 1)$ degrees of freedom. We use that model to obtain a P-value for a test or a margin of error for a confidence interval.

The corresponding confidence interval is

$$(\bar{y}_1 - \bar{y}_2) \pm t^*_{df} \times SE_{pooled}(\bar{y}_1 - \bar{y}_2)$$

where the critical value t^* depends on the confidence level and is found with $(n_1 - 1) + (n_2 - 1)$ degrees of freedom.

A S ***Activity:* The Pooled *t*-Test.** It's those hot dogs again. The same interactive tool can handle a pooled t-test, too. Take it for a spin here.

Everybody Out of the Pool!

We're testing whether the means are equal, so we admit that we don't *know* whether they are equal. Doesn't it seem a bit much to just *assume* that the variances are equal? Well, yes—but there are some special cases to consider. So when *should* you use the pooled method rather than the unpooled method?

Never.

What, never?

Well, hardly ever.

You see, when the variances of the two groups are in fact equal, the two methods give pretty much the same result. (For the termites, the two-sample t statistic is not much

[15] But not this one. See page 602.

different—9.44 with 8 df—and the P-value is still < 0.001.) Pooled methods have a small advantage (slightly narrower confidence intervals, slightly more powerful tests) mostly because they usually have a few more degrees of freedom, but the advantage is slight.

When the variances are *not* equal, the pooled methods are just not valid and can give poor results. You have to use the unpooled two-sample method instead.

WHEN SHOULD I USE THE POOLED-*t* TEST?

Because the advantages of pooling are small, and you are allowed to pool only rarely (when the Equal Variances Assumption is met), *don't. It's never wrong not to pool.*

As the sample sizes get bigger, the advantages that come from a few more degrees of freedom make less and less difference. So the advantage (such as it is) of the pooled method is greatest when the samples are small—just when it's hardest to check the conditions. And the difference in the degrees of freedom is greatest when the variances are not equal—just when you can't use the pooled method anyway. Our advice is to use the unpooled two-sample *t* methods to compare means.

Why did we devote a whole section to a method that we don't recommend using? The answer is that pooled methods are actually very important in Statistics. It's just that the simplest of the pooled methods—those for comparing two means—have good alternatives in the unpooled two-sample methods that don't require the extra assumption. Lacking the burden of the Equal Variances Assumption, the unpooled two-sample methods apply to more situations and are safer to use.

Why not test whether the variances are equal?

There are tests of whether two variances are equal, but they are very sensitive to their assumptions and work poorly for small sample sizes. So they don't help in choosing between the pooled and the unpooled tests.

Pooling

Pooled methods show up in several places in Statistics, and often in more complex situations than a two-sample test. As you've seen, the most sensitive part of most inference procedures is how we estimate the standard deviation of the test statistic so we can use it as a ruler to judge significance or construct a confidence interval. Anything we can do to improve our standard error estimate pays us back with improved power, shorter confidence intervals, and sometimes, simply with a valid method. So if you think the groups you are comparing have the same standard deviation, you should look for a pooled method that takes advantage of that.

Here's a common situation. In a randomized experiment, we start by assigning our experimental units to treatments at random, as the experimenter did with the termites. Because of this random assignment, each treatment group is a random sample from the same population,[16] so before they were treated, each group had not just the same mean, but also the same variance. The pooled *t*-test assumes that the treatment doesn't change the variance. Our null hypothesis says that the treatment didn't change the means, so it's not much of a stretch to think that the treatments made no difference *at all.* For example, we might suspect that the treatment is no different from the placebo. So, it might make sense to assume that the variances have remained equal. It's still an assumption, and there are conditions that need to be checked (make the boxplots, make the boxplots, make the boxplots), but at least it's a plausible assumption.

This line of reasoning is important. The methods used to analyze comparative experiments *do* pool variances in exactly this way and defend the pooling with a version of this argument. Chapter 25 (Analysis of Variance) introduces these methods.

For Example A POOLED *t*-TEST

RECAP: Remember the experiment showing that people would consume more soup if their bowls were secretly being refilled as they ate? Could it be that those subjects with the refilling bowl just enjoyed eating soup more? The experimenters asked the participants to estimate how many ounces of soup they had eaten. Their responses are summarized in the table.

	Ordinary bowl	Refilling bowl
n	27	27
$\bar{y}$	8.2	9.8
s	6.9	9.2

[16]That is, the population of experimental subjects. Remember that to be valid, experiments do not need a representative sample drawn from a population because we are not trying to estimate a population model parameter.

QUESTION: Is there evidence that those who ate more knew what they had done?

ANSWER:

$$H_0: \mu_{refill} - \mu_{ordinary} = 0$$
$$H_A: \mu_{refill} - \mu_{ordinary} \neq 0$$

The Independence Assumptions and Randomization Condition checked out earlier. Also,

Nearly Normal Condition: The histograms for both groups look unimodal and skewed to the right. The similar shapes and lack of clear outliers (the value around 30 in the left histogram appears to be just part of the long tail extending to the right) in groups *as large as these* suggest that *t*-methods should be very safe to use.

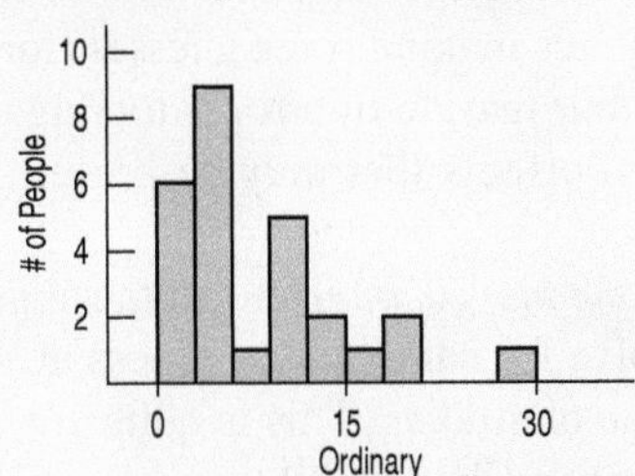

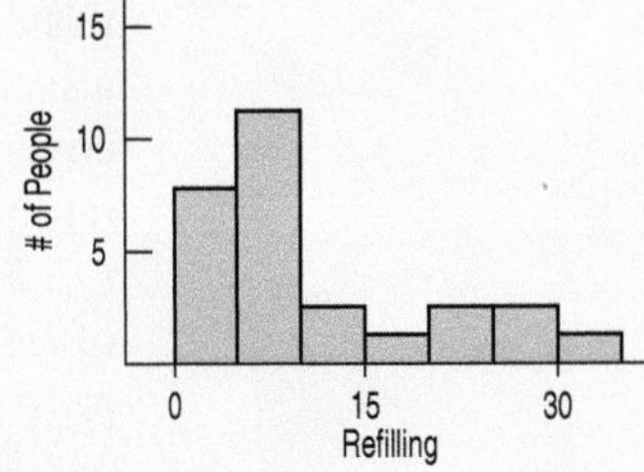

- **Equal Variances Assumption:** Subjects were assigned randomly to the two treatments, so I can assume that before they ate any soup, both groups' volume-estimating abilities had equal variances.
- **Similar Spreads Condition:** The boxplots show comparable spreads for the two groups.

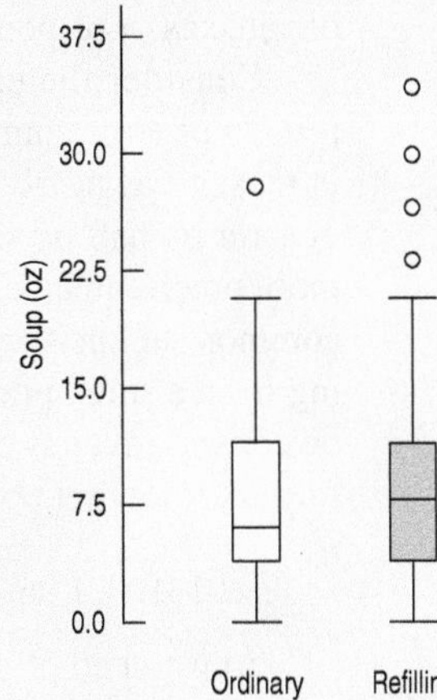

A pooled t-test should be quite reliable here.

$$s^2_{pooled} = \frac{(n_r - 1)s_r^2 + (n_o - 1)s_o^2}{(n_r - 1) + (n_o - 1)} = \frac{26 \times 9.2^2 + 26 \times 6.9^2}{(26) + (26)} = 66.125$$

$$SE_{pooled}(\bar{y}_o - \bar{y}_r) = \sqrt{\frac{s^2_{pooled}}{n_r} + \frac{s^2_{pooled}}{n_o}} = s_{pooled}\sqrt{\frac{1}{n_r} + \frac{1}{n_o}} = 8.13$$

$$\sqrt{\frac{1}{27} + \frac{1}{27}} = 2.213$$

$$df = (27 - 1) + (27 - 1) = 52$$

$$t = \frac{(9.8 - 8.2) - 0}{2.213} = 0.723$$

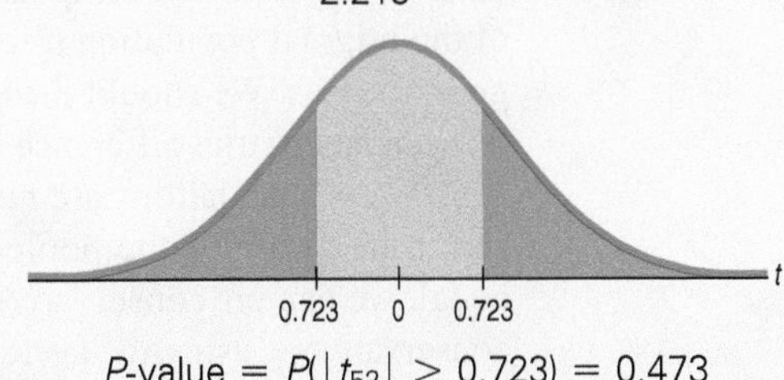

$$P\text{-value} = P(|t_{52}| > 0.723) = 0.473$$

Nonparametrics.
Tired of making and checking so many assumptions and conditions? What if there are severe violations? If a transformation doesn't fix things, or you just don't feel right about changing the measurement scale (by nonlinear transformation), you can track down and try an alternative *nonparametric* (distribution-free) procedure. The proper one here is called the Wilcoxon rank sum test, which is discussed in Chapter 29.

An event that happens nearly half of the time by chance isn't very extraordinary, so I would not reject the null hypothesis. There's no evidence that the subjects who ate more soup from the refilling bowl could tell that they did so.

21.4 Determining the Sample Size

When planning a study, it's important to try to determine an optimal sample size, though this usually requires making some guesses (for example, at standard deviations). Too small a sample, and your margin of error is too big or the power of the test is inadequate to discover important effects that may be present. Too big a sample, and time, money, and resources are needlessly wasted.

To control the margin of error, set the appropriate formula equal to the desired margin of error and solve for sample size. Guess at what you do not know, perhaps using some prior knowledge or studies. This is quite similar to the discussion of *sample size* for the one-sample case in Chapter 20 (see page 579), but now there are two sample sizes to determine, so let's make the simplifying assumption that the two sample sizes are equal. And there are good reasons for aiming for equal sample sizes; for example, it improves the robustness of the procedure. You may work with unequal sample sizes if you wish: just set $n_1 = 2n_2$ if you want sample 1 to be, say, twice as big as sample 2. But henceforth we will assume $n_1 = n_2$, and let $n =$ the common sample size (so that $2n$ will be the total number of subjects or responses actually required).

Consider the earlier example comparing brand-name and generic batteries. The margin of error computed was 16.5 minutes—rather large, considering that the difference in means was nearly as large at 18.6 minutes. We know that we could decrease the margin of error by half by simply multiplying both sample sizes by 2^2 or 4, but let's proceed in a more precise manner. Suppose we want the margin of error to equal five minutes. Let the common sample size equal n. Choose 95% confidence. We'll view the earlier data as coming from a small pilot study, and take the two sample standard deviations (rounded up, to be conservative) as estimates of the true standard deviations needed in the margin of error formula. Now set up and solve the following:

Set $5.0 = 1.96\sqrt{\frac{10.3^2}{n} + \frac{14.6^2}{n}}$ and then square both sides and solve. We get $n = 50$.

So we need 50 batteries of each type (of course, you'd better weigh your desire for precision against the costs incurred and time involved in draining so many batteries). Note that we use a z-value of 1.96 to start with and only need to change this if our sample size turns out to be fairly small (as in the Chapter 20 example). $n = 50$ is big enough to justify use of the standard Normal distribution instead of the t-distribution, particularly since we know that the degrees of freedom cannot be less than $n - 1 = 49$, and likely should be considerably more.

On the other hand, when performing a *test of significance*, our concern will be about having *adequate power* to detect important differences. Consider the example of killing termites with a sap compound, where we wanted to see if a higher concentration would kill more termites. The investigators might feel that a difference of less than two or three termites killed (out of the 25 termites at the start), on average, would not be a big enough effect to consider this compound a success, but that a difference of, say, five termites (20% of the original population size) would be a large enough effect to make this compound a great success. We should then choose a sample size that will give us a high power when the (unknown) true difference in the means is 5.0.

Power calculations are not possible unless we assume equal variances (another reason to be aware of the pooled test), but even if we believe that the variances are not equal, we can err conservatively by guessing at both, then using the higher one (or less conservatively, using the average of the two standard deviations). The standard deviations from our (assumed) pilot study were 2.23 and 3.14, so let's take 3.0 as a rough guess at the common or average standard deviation. Now *fire up your statistics software,* look for *power and sample size* calculations for the *two-sample t-test*, and input the following:

- *Alpha level of the test.* Let's go with 5%.
- *Nature of the alternative hypothesis* (the null is a difference of zero). Choose the two-sided option.
- *Size of the difference you want to detect, if present.* Set this equal to 5.0 killed termites.
- *Your guess at the common sigma.* We are going with a guess of 3.0 killed termites.
- *Desired power.* Let's aim for 0.90.

Here is some typical output:

Power and Sample Size

Two-Sample *t*-Test
Testing mean 1 = mean 2 (versus not =)
Calculating power for mean 1 = mean 2 + difference
Alpha = 0.05 Assumed standard deviation = 3

Difference	Sample Size	Target Power	Actual Power
5	9	0.9	0.912548

The sample size is for each group.

So, we only need nine groups fed the lower dose and nine groups fed the higher dose to achieve the desired power for a difference of five dead termites, on average. If the true mean difference in dead termites between the two dosages is five dead termites (out of 25), then 9 times out of 10 (actually 91 times out of 100) we will make the correct decision and correctly detect a dosage effect.

WHAT CAN GO WRONG?

- **Don't use two-sample methods when the samples aren't independent of each other.** These methods give wrong answers when this assumption of independence is violated. Good randomization is usually the best assurance of independent groups. Make sure there is no relationship between the two groups. Matched-pairs designs in which the observations are deliberately related arise often and are important, but can't be analyzed by the methods of this chapter. The next chapter deals with them.
- **Look at the plots.** The usual (by now) cautions about checking for outliers and non-Normal distributions apply, of course. The simple defense is to make and examine boxplots. You may be surprised how often this step saves you from the wrong or even absurd conclusions that can be generated by a single undetected outlier. You don't want to conclude that two methods have very different means just because one observation is atypical.
- **Be cautious in drawing conclusions if you apply inference methods where there was no randomization.** If the data do not come from representative random samples or a properly randomized experiment, any conclusions about the differences between the groups may be completely unjustified. In a prospective or retrospective study, there is always the danger that some lurking variables are the real reason for an observed difference. Conclusions of causality require well-designed controlled randomized experiments.

Do what we say, not what we do . . . Precision machines used in industry often have a bewildering number of parameters that have to be set, so experiments are performed in an attempt to try to find the best settings. Such was the case for a hole-punching machine used by a well-known computer manufacturer to make printed circuit boards. The data were analyzed by one of the (U.S.) authors, but because he was in a hurry, he didn't look at the boxplots first and just performed *t*-tests on the experimental factors. When he found extremely small P-values even for factors that made no sense, he plotted the data. Sure enough, there was one observation 1 000 000 times bigger than the others. It turns out that it had been recorded in microinches (millionths of an inch), while all the rest were in inches.

The inference methods we are learning follow a consistent pattern. Although standard errors may differ, the step-by-step procedures are almost identical. For example, we decide whether a statistic is large by dividing it by its standard error.

We first examined side-by-side boxplots in Chapter 4. There, we made general statements about the shape, centre, and spread of each group. When we compared groups, we asked whether their centres looked different compared to how spread out the distributions were. Here we've made that kind of thinking precise, with confidence intervals for the difference and tests of whether the means are the same.

We use Student's *t* as we did for single sample means, and for the same reasons: We are using standard errors from the data to estimate the standard deviation of the sample statistic. As before, to work with Student's *t*-models, we need to check the Nearly Normal Condition. Histograms and Normal probability plots are the best methods for such checks.

What Have We Learned?

Learning Objectives

Know how to construct and interpret the two-sample *t*-interval for the difference between means of two independent groups.

- Know the requisite Assumptions and Conditions regarding independence and Normality.

Be able to perform and interpret a two-sample *t*-test of the difference between the means of two independent groups.

- Understand the relationship between testing and providing a confidence interval.
- The most common null hypothesis is that the means are equal.

Recognize that in special cases in which it is reasonable to assume equal variances between the groups, we can also use a pooled *t*-test for testing the difference between means.

- This may make sense particularly for randomized experiments in which the randomization has produced groups with equal variance to start with and the null hypothesis is that a treatment under study has had no effect.

Keep in mind that these two sample methods require both independence of observations within each group and independence between the groups themselves.

- These methods are not appropriate for paired or matched data.

Review of Terms

Two-sample *t* methods Two-sample *t* methods allow us to draw conclusions about the difference between the means of two quantitative populations, based on independent samples. The unpooled two-sample methods make relatively few assumptions about the underlying populations, so they are usually the method of choice for comparing two sample means. However, the

Student's *t*-models are only approximations for their true sampling distribution, and there is a special rule for estimating degrees of freedom (p. 603).

Two-sample *t*-interval for the difference between means A confidence interval for the difference between the means of two independent groups found as

$$(\bar{y}_1 - \bar{y}_2) \pm t^*_{df} \times SE(\bar{y}_1 - \bar{y}_2)$$

where

$$SE(\bar{y}_1 - \bar{y}_2) = \sqrt{\frac{s_1^2}{n_1} + \frac{s_2^2}{n_2}}$$

and the number of degrees of freedom is given by a special formula (see footnote 4 on page 602) (p. 604).

Two-sample *t*-test for the difference between means A hypothesis test for the difference between the means of two independent groups. It tests the null hypothesis

$$H_0: \mu_1 - \mu_2 = \Delta_0$$

where the hypothesized difference, Δ_0, is almost always 0, using the statistic

$$t = \frac{(\bar{y}_1 - \bar{y}_2) - \Delta_0}{SE(\bar{y}_1 - \bar{y}_2)}$$

with the number of degrees of freedom given by the special formula (p. 607).

Pooled-*t* methods Pooled-*t* (also called *pooled two-sample t*) methods provide inferences about the difference between the means of two independent populations under the assumption that both populations have the same standard deviation. When the assumption is justified, pooled-*t* methods generally produce slightly narrower confidence intervals and more powerful significance tests than unpooled two-sample *t* methods. When the assumption is not justified, they generally produce worse results, giving inaccurate or wrong conclusions.

We recommend that you generally use the unpooled two-sample *t* methods (p. 612).

Pooled *t*-test A hypothesis test for the difference in the means of two independent groups when we are willing and able to assume that the variances of the groups are equal. It tests the null hypothesis

$$H_0: \mu_1 - \mu_2 = \Delta_0$$

where the hypothesized difference, Δ_0, is almost always 0, using the statistic

$$t = \frac{(\bar{y}_1 - \bar{y}_2) - \Delta_0}{SE_{pooled}(\bar{y}_1 - \bar{y}_2)}$$

where the pooled standard error is defined as for the pooled interval (see below) and the degrees of freedom is $(n_1 - 1) + (n_2 - 1)$ (p. 612).

Pooled *t*-interval A confidence interval for the difference between the means of two independent groups used when we are willing and able to make the additional assumption that the variances of the groups are equal. It is found as

$$(\bar{y}_1 - \bar{y}_2) \pm t^*_{df} \times SE_{pooled}(\bar{y}_1 - \bar{y}_2)$$

where

$$SE_{pooled}(\bar{y}_1 - \bar{y}_2) = \sqrt{\frac{s^2_{pooled}}{n_1} + \frac{s^2_{pooled}}{n_2}} = s_{pooled}\sqrt{\frac{1}{n_1} + \frac{1}{n_2}}$$

the pooled variance is

$$s^2_{pooled} = \frac{(n_1 - 1)s_1^2 + (n_2 - 1)s_2^2}{(n_1 - 1) + (n_2 - 1)}$$

and the number of degrees of freedom is $(n_1 - 1) + (n_2 - 1)$ (p. 613).

Pooling We may sometimes combine, or *pool*, data from two or more populations to estimate a parameter (such as a common variance) when we are willing to assume that the estimated parameter is the same in both populations. Using more data may lead to a more reliable estimate. However, pooled estimates are appropriate only when the required assumptions are believed to be (nearly) true (p. 614).

On the Computer TWO-SAMPLE METHODS FOR MEANS

Here's some typical computer package output with comments:

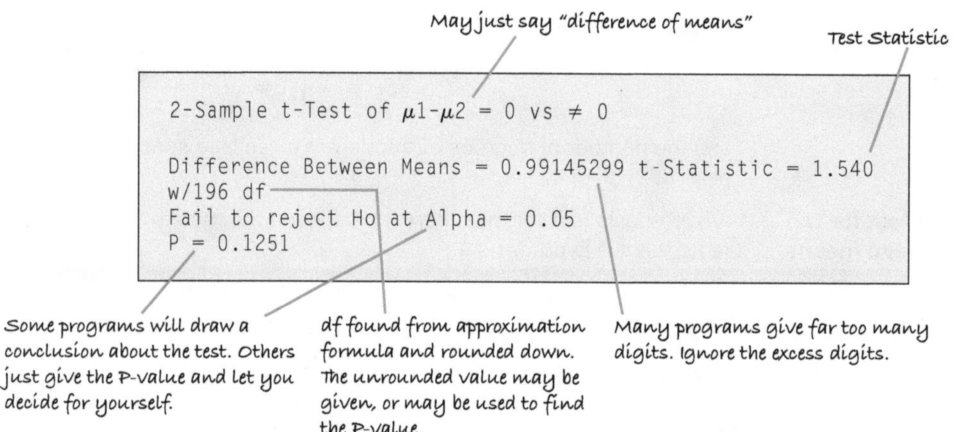
May just say "difference of means"
Test Statistic
2-Sample t-Test of μ1-μ2 = 0 vs ≠ 0
Difference Between Means = 0.99145299 t-Statistic = 1.540
w/196 df
Fail to reject Ho at Alpha = 0.05
P = 0.1251
Some programs will draw a conclusion about the test. Others just give the P-value and let you decide for yourself.
df found from approximation formula and rounded down. The unrounded value may be given, or may be used to find the P-value.
Many programs give far too many digits. Ignore the excess digits.

Most statistics packages compute the test statistic for you and report a P-value corresponding to that statistic. And, of course, statistics packages make it easy to examine the boxplots and histograms of the two groups, so you have no excuse for skipping this important check.

Some statistics software automatically tries to test whether the variances of the two groups are equal. Some automatically offer both the unpooled two-sample-*t* and pooled-*t* results. Ignore the test for the variances; it has little power in any situation in which its results could matter. If the pooled and unpooled two-sample methods differ in any important way, you should stick with the unpooled two-sample method. Most likely, the equal variance assumption needed for the pooled method has failed.

The degrees of freedom approximation usually gives a fractional value. Most packages seem to round the approximate value down to the next smallest integer (although they may actually compute the P-value with the fractional value, gaining a tiny amount of power).

There are two ways to organize data when we want to compare two independent groups. The data can be in two lists, as in the table at the start of this chapter. Each list can be thought of as a variable. In this method, the variables in the batteries example would be *brand-name* and *generic*. Graphing calculators usually prefer this form, and some computer programs can use it as well.

There's another way to think about the data. What is the response variable for the battery life experiment? It's the *time* until the music stopped. But the values of this variable are in both columns, and actually there's an experiment factor here, too—namely, the *brand* of the battery. So, we could put the data into two different columns, one with the *time* in it and one with the *brand*. Then the data would look like this:

Time	Brand
194.0	Brand name
205.5	Brand name
199.2	Brand name
172.4	Brand name
184.0	Brand name
169.5	Brand name
190.7	Generic
203.5	Generic
203.5	Generic
206.5	Generic
222.5	Generic
209.4	Generic

This way of organizing the data makes sense as well. Now the factor and the response variables are clearly visible. You'll have to see which method your program requires. Some packages even allow you to structure the data either way.

EXCEL

From the **Analyze** menu, select **Fit y by x.** Select variables.
From the **Data Tab, Analysis Group**, choose **Data Analysis**.
Alternatively (if the Data Analysis Tool Pack is not installed), in the Formulas tab, choose **More functions.** > **Statistical** > TTEST, and specify Type = 3 in the resulting dialogue.
Fill in the cell ranges for the two groups, the hypothesized difference, and the alpha level.

Power and sample size calculations are not included in the standard data analysis toolkit.

COMMENTS

Excel expects the two groups to be in separate cell ranges. Notice that, contrary to Excel's wording, we do not need to assume that the variances are *not* equal; we simply choose not to assume that they *are* equal.

JMP

From the **Analyze** menu, select **Fit y by x**.
Select variables: a **Y, Response** variable that holds the data and an **X, Factor** variable that holds the group names. JMP will make a dotplot.
Click the **red triangle** in the dotplot title, and choose **Unequal variances.** The *t*-test is at the bottom of the resulting table. Find the P-value from the Prob>F section of the table (they are the same).

For power and sample size calculations, select **Power and Sample Size** from the **DOE** menu. Choose **Two Sample Means** from the submenu. Indicate the **Difference** in the means that you are hoping to detect, the **Alpha** value, and choose a one- or two-sided alternative. Guess at the common (or larger) **Std Dev.** Fill in either your desired **Sample size or Power**. The one you leave blank will be calculated. Click **Continue.**

COMMENTS

JMP expects data in one variable and category names in the other. Don't be misled: There is no need for the variances to be unequal to use two-sample *t* methods.

MINITAB

From the **Stat** menu, choose the **Basic Statistics** submenu.
From that menu, choose **2-sample t . . .** Then, fill in the dialogue.

For power and sample size calculations, choose **Basic Statistics** from the **Stat** menu, then **Power and Sample Size**, then **2-Sample t**. In the dialogue box, fill in two of **Sample sizes, Differences, Power values**, and Minitab will calculate the remaining one.

COMMENTS

The dialogue offers a choice of data in two variables, or data in one variable and category names in the other.

The power and sample size calculation is based on the pooled test, so you have to give a guess for only one standard deviation. If you believe that the two standard deviations differ substantially, guess at the average of the two, or be conservative and guess at the bigger one.

R

To test the hypothesis that $\mu_1 = \mu_2$ against an alternative (default is two-sided):

- Create a vector for the data of each group, say *X* and *Y*, and produce the confidence interval (default is 95%):
- t.test(X,Y, alternative = c("two.sided", "less", "greater"), conf.level = 0.95)

will produce the *t*-statistic, degrees of freedom, P-value, and confidence interval for a specified alternative.

In package {pwr}, to perform a sample size or power calculation for a two-sample t-test:

- **pwr.t.test(n= , d= , sig.level= , power= , type= "two.sample")** returns **one** argument that is not specified in the function.

For unequal groups sizes, use:

- **pwr.t2n.test(n1= , n2= , d= , sig.level= , power=)** instead.

The effect size (d) is equal to the difference between the two group means, *divided by the pooled standard deviation*—which usually needs to be guessed (perhaps from a pilot study). The calculation is based on the pooled *t*-test, hence you can only specify one SD. A pilot study can help.

COMMENTS

This is the same function as for the one-sample *t*-test, but with two vectors. In R, the default *t*-test does not assume equal variances in the two groups (default is **var.equal = FALSE**). To get the "pooled" *t*-test, add **var.equal = TRUE** to the function call.

SPSS

From the **Analyze** menu, choose the **Compare Means** submenu.
From that, choose the **Independent-Samples t-test** command. Specify the data variable and "group variable." Then, type in the labels used in the group variable. SPSS offers both the two-sample and pooled-*t* results in the same table.

For power and sample size calculations, you will need the IBM SPSS SamplePower add-on.

COMMENTS

SPSS expects the data in one variable and group names in the other. If there are more than two group names in the group variable, only the two that are named in the dialogue box will be compared.

STATCRUNCH

To do inference for the difference between two means using summaries:

- Click on **Stat**.
- Choose **T Statistics » Two sample » with summary**.
- Enter the **Sample mean**, **Standard deviation**, and sample **Size** for each group.
- De-select **Pool variances**.
- Click on **Next**.
- Indicate **Hypothesis Test**, then enter the hypothesized **Null mean** difference (usually 0), and choose the **Alternative** hypothesis.

  OR

  Indicate **Confidence Interval**, and then enter the **Level** of confidence.
- Click on **Calculate.**

To do inference for the difference between two means using data:

- Click on **Stat**.
- Choose **T Statistics » Two sample » with data**.
- Choose the variable Column for each group.
- De-select **Pool variances**.
- Click on **Next**.
- Indicate **Hypothesis Test**, then enter the hypothesized **Null mean** difference (usually 0), and choose the **Alternative** hypothesis.

  OR

  Indicate **Confidence Interval**, and then enter the **Level** of confidence.
- Click on **Calculate**

Power and Sample size calculations are readily available using **Stat > T Statistics > Two Sample > Power/Sample size.**

- Click on **Hypothesis Test Power**.
- Fill in all the boxes except for the one that you want to determine, either **Power** or **Sample Size**.
- Make a guess at the **Pooled std. dev.**

TI-83/84 PLUS

For a confidence interval:

- In the **STAT TESTS** menu, choose **0:2-SampTInt.**
- Specify if you are using data stored in two lists, or enter the means, standard deviations, and sizes of both samples.
- Indicate whether to pool the variances (when in doubt, say no), and specify the desired level of confidence.

To test a hypothesis:

- In the **STAT TESTS** menu, choose **4:2-SampTTest.**
- Specify if you are using data stored in two lists, or enter the means, standard deviations, and sizes of both samples.
- Indicate whether to pool the variances (when in doubt, say no) and specify whether the test is to be two-tail, lower-tail, or upper-tail.

Power and sample size calculations are not available.

Exercises

1. Hot dogs In its July 2007 issue, *Consumer Reports* examined the calorie content of two kinds of hot dogs: meat (usually a mixture of pork, turkey, and chicken) and all-beef. The researchers purchased samples of several different brands. The meat hot dogs averaged 111.7 calories, compared to 135.4 for the beef hot dogs. A test of the null hypothesis that there's no difference in mean calorie content yields a P-value of 0.124. Would a 95% confidence interval for $\mu_{meat} - \mu_{beef}$ include 0? Explain.

2. Hot dogs and sodium The *Consumer Reports* article described in Exercise 1 also listed the sodium content (in mg) for the various hot dogs tested. A test of the null hypothesis that beef hot dogs and meat hot dogs don't differ in the mean amounts of sodium yields a P-value of 0.11. Would a 95% confidence interval for $\mu_{meat} - \mu_{beef}$ include 0? Explain.

3. Hot dogs and fat The *Consumer Reports* article described in Exercise 1 also listed the fat content (in grams) for samples of beef and meat hot dogs. The resulting 90% confidence interval for $\mu_{meat} - \mu_{beef}$ is $(-6.5, -1.4)$.

a) The endpoints of this confidence interval are negative numbers. What does that indicate?

b) What does the fact that the confidence interval does not contain 0 indicate?

c) If we use this confidence interval to test the hypothesis that $\mu_{meat} - \mu_{beef} = 0$, what is the corresponding alpha level?

4. Washers In its June 2007 issue, *Consumer Reports* examined the relative merits of top-loading and front-loading washing machines, testing samples of several different brands of each type. One of the variables the article reported was "cycle time,"—the number of minutes it took each machine to wash a load of clothes. Among the machines rated good to excellent, the 98% confidence interval for the difference in mean cycle time $(\mu_{top} - \mu_{front})$ is $(-40, -22)$.

a) The endpoints of this confidence interval are negative numbers. What does that indicate?

b) What does the fact that the confidence interval does not contain 0 indicate?

c) If we use this confidence interval to test the hypothesis that $\mu_{top} - \mu_{front} = 0$ what is the corresponding alpha level?

5. Hot dogs, 2nd helping In Exercise 3, we saw a 90% confidence interval of $(-6.5, -1.4)$ grams for $\mu_{meat} - \mu_{beef}$, the difference in mean fat content for meat versus all-beef hot dogs. Explain why you think each of the following statements is true or false:

a) If I eat a meat hot dog instead of a beef dog, there's a 90% chance I'll consume less fat.

b) 90% of meat hot dogs have between 1.4 and 6.5 grams less fat than a beef hot dog.

c) I'm 90% confident that meat hot dogs average 1.4–6.5 grams less fat than the beef hot dogs.

d) If I were to get more samples of both kinds of hot dogs, 90% of the time the meat hot dogs would average 1.4–6.5 grams less fat than the beef hot dogs.

e) If I tested more samples, I'd expect about 90% of the resulting confidence intervals to include the true difference in mean fat content between the two kinds of hot dogs.

6. Second load of wash In Exercise 4, we saw a 98% confidence interval of $(-40, -22)$ minutes for $\mu_{top} - \mu_{front}$, the difference in time it takes top-loading and front-loading washers to do a load of clothes. Explain why you think each of the following statements is true or false:

a) 98% of top loaders are 22 to 40 minutes faster than front loaders.

b) If I choose the laundromat's top loader, there's a 98% chance that my clothes will be done faster than if I had chosen the front loader.

c) If I tried more samples of both kinds of washing machines, in about 98% of these samples I'd expect the top loaders to be an average of 22 to 40 minutes faster.
d) If I tried more samples, I'd expect about 98% of the resulting confidence intervals to include the true difference in mean cycle time for the two types of washing machines.
e) I'm 98% confident that top loaders wash clothes an average of 22 to 40 minutes faster than front-loading machines.

7. **Learning math** The Core Plus Mathematics Project (CPMP) is an innovative approach to teaching Mathematics that engages students in group investigations and mathematical modelling. After field tests in 36 high schools over a three-year period, researchers compared the performances of CPMP students with those taught using a traditional curriculum. In one test, students had to solve applied algebra problems using calculators. Scores for 320 CPMP students were compared to those of a control group of 273 students in a traditional Math program. Computer software was used to create a confidence interval for the difference in mean scores.[17]

Conf level: 95% Parameter: μ(CPMP) – μ(Ctrl)
Interval: (5.573, 11.427)

a) What is the margin of error for this confidence interval?
b) If we had created a 98% CI, would the margin of error be larger or smaller?
c) Explain what the calculated interval means in this context.
d) Does this result suggest that students who learn Mathematics with CPMP will have significantly higher mean scores in Algebra than those in traditional programs? Explain.

T 8. **Stereograms** Stereograms appear to be composed entirely of random dots. However, they contain separate images that a viewer can "fuse" into a three-dimensional (3D) image by staring at the dots while defocusing the eyes. An experiment was performed to determine whether knowledge of the form of the embedded image affected the time required for subjects to fuse the images. One group of subjects (group NV) received no information or just verbal information about the shape of the embedded object. A second group (group VV) received both verbal information and visual information (specifically, a drawing of the object). The experimenters measured how many seconds it took for the subject to report that he or she saw the 3D image.

2-Sample t-Interval
Conf level = 90% df = 70
μ(NV) $-$ μ(VV) interval: (0.55, 5.47)

a) Interpret your interval in context.
b) Does it appear that viewing a picture of the image helps people "see" the 3D image in a stereogram?
c) What is the margin of error for this interval?
d) Explain carefully what the 90% confidence level means.
e) Would you expect a 99% confidence level to be wider or narrower? Explain.
f) Might that change your conclusion in part b)? Explain.

9. **CPMP, again** During the study described in Exercise 7, students in both CPMP and traditional classes took another Algebra test that did not allow them to use calculators. The table below shows the results. Are the mean scores of the two groups significantly different?

Performance on Algebraic Symbolic Manipulation Without Use of Calculators

Math Program	n	Mean	SD
CPMP	312	29.0	18.8
Traditional	265	38.4	16.2

a) Write an appropriate hypothesis.
b) Do you think the assumptions for inference are satisfied? Explain.
c) Here is computer output for this hypothesis test. Explain what the P-value means in this context.

2-Sample t-Test of $\mu_1 - \mu_2 \neq 0$
t-Statistic $= -6.451$ w/574.8761 df
$P < 0.0001$

d) State a conclusion about the CPMP program.

T 10. **Memory** Does ginkgo biloba enhance memory? In an experiment to find out, subjects were assigned randomly to take ginkgo biloba supplements or a placebo. Their memory was tested to see whether it improved. Here are boxplots comparing the two groups and some computer output from a two-sample t-test computed for the data.

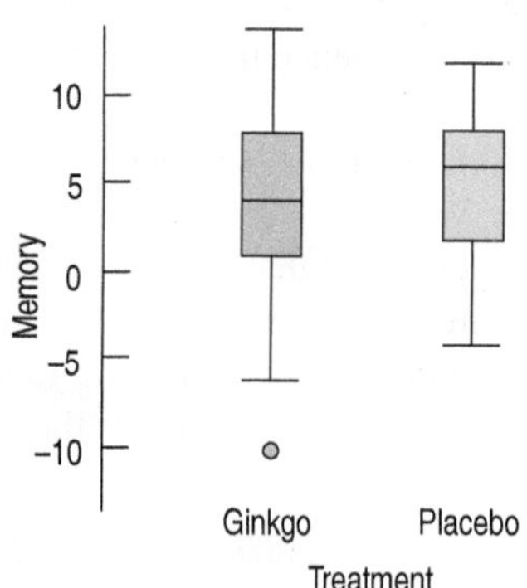

2-Sample t-Test of $\mu_G - \mu_P > 0$
Difference Between Means $= -0.9914$
t-Statistic $= -1.540$ w/196 df
$P = 0.9374$

[17] *Journal for Research in Mathematics Education*, 31, no. 3 [2000].

a) Explain what the P-value means in this context.
b) State your conclusion about the effectiveness of ginkgo biloba.
c) Proponents of ginkgo biloba continue to insist that it works. What type of error do they claim your conclusion makes? Explain.

11. **Commuting** A man who moves to a new city sees that there are two routes he could take to work. A neighbour who has lived there a long time tells him Route A will average five minutes faster than Route B. The man decides to experiment. Each day, he flips a coin to determine which way to go, driving each route for 20 days. He finds that Route A takes an average of 40 minutes, with standard deviation 3 minutes, and Route B takes an average of 43 minutes, with standard deviation 2 minutes. Histograms of travel times for the routes are roughly symmetric and show no outliers.
a) Find a 95% confidence interval for the difference in average commuting time for the two routes. (From technology, df = 33.1.)
b) Should the man believe the neighbour's claim that he can save an average of five minutes a day by always driving Route A? Explain.

12. **Pulse rates** A researcher wanted to see whether there is a significant difference in resting pulse rates for men and women. The data she collected are displayed in the boxplots and summarized below.

	Sex	
	Male	Female
Count	28	24
Mean	72.75	72.625
Median	73	73
StdDev	5.37225	7.69987
Range	20	29
IQR	9	12.5

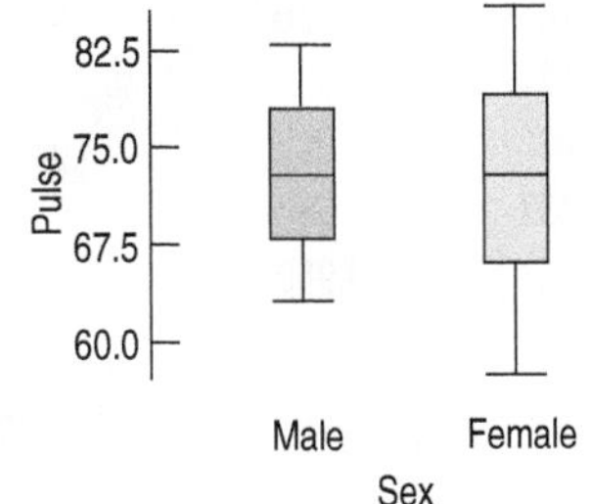

a) What do the boxplots suggest about differences between male and female pulse rates?
b) Is it appropriate to analyze these data using the methods of inference discussed in this chapter? Explain.
c) Create a 90% confidence interval for the difference in mean pulse rates.
d) Does the confidence interval confirm your answer to part a)? Explain.

T 13. **Cereal** The data below show the sugar content (as a percentage of weight) of several national brands of children's and adults' cereals. Create and interpret a 95% confidence interval for the difference in mean sugar content. Be sure to check the necessary assumptions and conditions.

Children's cereals: 40.3, 55, 45.7, 43.3, 50.3, 45.9, 53.5, 43, 44.2, 44, 47.4, 44, 33.6, 55.1, 48.8, 50.4, 37.8, 60.3, 46.6

Adults' cereals: 20, 30.2, 2.2, 7.5, 4.4, 22.2, 16.6, 14.5, 21.4, 3.3, 6.6, 7.8, 10.6, 16.2, 14.5, 4.1, 15.8, 4.1, 2.4, 3.5, 8.5, 10, 1, 4.4, 1.3, 8.1, 4.7, 18.4

T 14. **Egyptians** Some archaeologists theorize that ancient Egyptians interbred with several different immigrant populations over thousands of years. To see if there is any indication of changes in body structure that might have resulted, they measured 30 skulls of male Egyptians dated from 4000 B.C.E. and 30 others dated from 200 B.C.E. (Source: A. Thomson and R. Randall-Maciver, *Ancient Races of the Thebaid,* Oxford: Oxford University Press, 1905.)

Maximum Skull Breadth (mm)			
4000 B.C.E.	4000 B.C.E.	200 B.C.E.	200 B.C.E.
131	131	141	131
125	135	141	129
131	132	135	136
119	139	133	131
136	132	131	139
138	126	140	144
139	135	139	141
125	134	140	130
131	128	138	133
134	130	132	138
129	138	134	131
134	128	135	136
126	127	133	132
132	131	136	135
141	124	134	141

a) Are these data appropriate for inference? Explain.
b) Create a 95% confidence interval for the difference in mean skull breadth between these two eras.
c) Do these data provide evidence that the mean breadth of males' skulls changed over this period? Explain.
d) The standard deviations are very similar, so pooled methods should work well here. Calculate a 95% pooled t confidence interval for the difference and compare with the interval in part b.

T 15. **Hurricanes** The data below show the number of hurricanes (category 3 or higher) recorded annually before and after 1970. Create an appropriate visual display, and determine whether these data are appropriate for testing whether there has been a change in the frequency of hurricanes.

1944–1969	1970–2010
3, 2, 1, 2, 4, 3, 7, 2, 3, 3, 2, 5, 2, 2, 4, 2, 2, 6, 0, 2, 5, 1, 3, 1, 0, 3	2, 1, 0, 1, 2, 3, 2, 1, 2, 2, 2, 3, 1, 1, 1, 3, 0, 1, 3, 2, 1, 2, 1, 1, 0, 5, 6, 1, 3, 5, 3, 3, 2, 3, 6, 7, 2, 2, 5, 2, 5

16. Streams Researchers collected samples of water from streams in the Adirondack Mountains to investigate the effects of acid rain. They measured the pH (acidity) of the water and classified the streams with respect to the kind of substrate (type of rock over which they flow). A lower pH means the water is more acidic. Here is a plot of the pH of the streams by substrate (limestone, mixed, or shale):

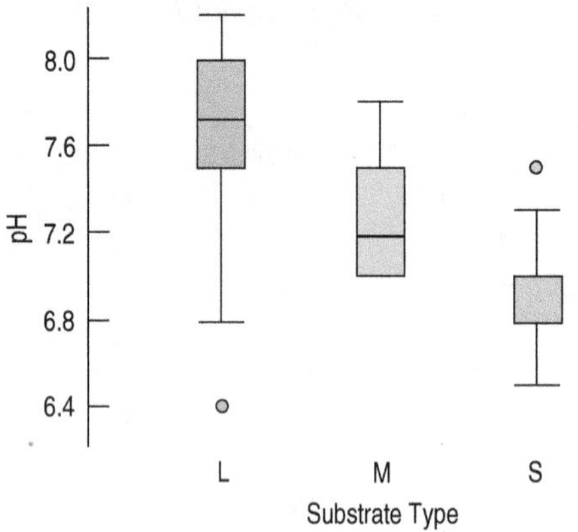

Here are selected parts of a software analysis comparing the pH of streams with limestone and shale substrates:

2-Sample t-Test of $\mu_1 - \mu_2$

Difference Between Means = 0.735

t-Statistic = 16.30 w/133 df

$p < 0.0001$

a) State the null and alternative hypotheses for this test.
b) From the information you have, do the assumptions and conditions appear to be met?
c) What conclusion would you draw?

17. Reading An educator believes that new reading activities for elementary schoolchildren will improve reading comprehension scores. She randomly assigns Grade 3 students to an eight-week program in which some will use these activities and others will experience traditional teaching methods. At the end of the experiment, both groups take a reading comprehension exam. The back-to-back stem-and-leaf display shows their scores. Do these results suggest that the new activities are better? Test an appropriate hypothesis and state your conclusion.

New Activities		Control
	1	07
4	2	068
3	3	377
96333	4	12222238
9876432	5	355
721	6	02
1	7	
	8	5

18. CPMP and word problems The study of the new CPMP Mathematics methodology described in Exercise 7 also tested students' abilities to solve word problems. This table shows how the CPMP and traditional groups performed. What do you conclude?

Math Program	n	Mean	SD
CPMP	320	57.4	32.1
Traditional	273	53.9	28.5

19. Baseball American League baseball teams play their games with the designated hitter rule, meaning that pitchers do not bat. The league believes that replacing the pitcher, traditionally a weak hitter, with another player in the batting order produces more runs and generates more interest among fans. Below are the average numbers of home runs hit per game in American League and National League stadiums for the 2011 season.

American	National	American	National
1.500	1.354	0.913	0.948
1.267	1.314	0.903	0.941
1.230	1.160	0.880	0.919
1.186	1.110	0.789	0.862
1.144	1.095	0.786	0.799
1.060	1.062	0.708	0.774
1.037	0.987		0.735
0.987	0.950		0.596

a) Create an appropriate display of these data. What do you see?
b) With a 95% confidence interval, estimate the mean number of home runs hit in American League games.
c) Coors Field, in Denver, stands a mile above sea level, an altitude far greater than that of any other major league ball park. Some believe that the thinner air makes it harder for pitchers to throw curve balls and easier for batters to hit the ball a long way. Do you think the 1.354 home runs hit per game at Coors is unusual? Explain.

d) Explain why you should not use two separate confidence intervals to decide whether the two leagues differ in average number of runs scored.
e) Using a 95% confidence interval, estimate the difference between the mean number of home runs hit in American and National League games.
f) Interpret your interval.
g) Does this interval suggest that the two leagues may differ in average number of home runs hit per game?

20. Hard water In an investigation of environmental causes of disease, data were collected on the annual mortality rate (deaths per 100 000) for males in 61 large towns in England and Wales. In addition, the water hardness was recorded as the calcium concentration (parts per million, or ppm) in the drinking water. The data set also notes, for each town, whether it was south or north of Derby. Is there a significant difference in mortality rates in the two regions? Here are the summary statistics.

Summary of:		Mortality		
For categories in:		Derby		
Group	**Count**	**Mean**	**Median**	**StdDev**
North	34	1631.59	1631	138.470
South	27	1388.85	1369	151.114

a) Test appropriate hypotheses, and state your conclusion.
b) The boxplots of the two distributions show an outlier among the data north of Derby. What effect might that have had on your test?

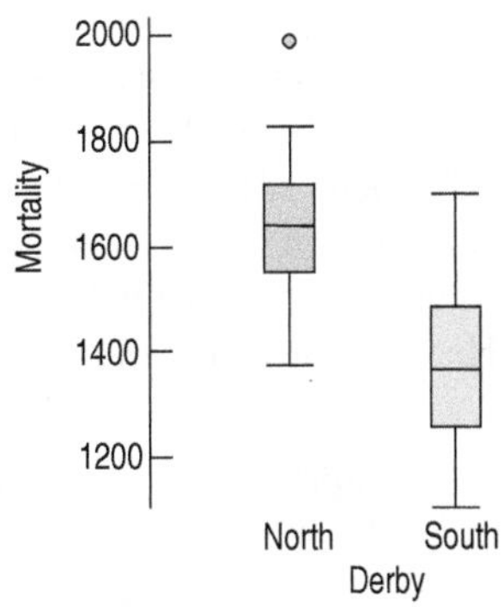

c) Repeat part a) using the pooled t-test, and compare results. Since the standard deviations are so similar, we would expect the pooled test to work well here.

21. Job satisfaction A company institutes an exercise break for its workers to see if it will improve job satisfaction, as measured by a questionnaire that assesses workers' satisfaction. Scores for 10 randomly selected workers before and after implementation of the exercise program are shown below. The company wants to assess the effectiveness of the exercise program. Explain why you can't use the methods discussed in this chapter to do that. (Don't worry, we'll give you another chance to do this the right way.)

Worker Number	Job Satisfaction Index	
	Before	**After**
1	34	33
2	28	36
3	29	50
4	45	41
5	26	37
6	27	41
7	24	39
8	15	21
9	15	20
10	27	37

22. Summer school Having done poorly on their Math final exams in June, six students repeat the course in summer school, then take another exam in August. Both results are shown below for each student. If we consider these students as representative of all students who might attend this summer school in other years, do these results provide evidence that the program is worthwhile? Can you analyze the data with methods discussed in this chapter?

June exam	54	49	68	66	62	62
Aug. exam	50	65	74	64	68	72

23. Sex and violence In June 2002, the *Journal of Applied Psychology* reported on a study that examined whether the content of television shows influenced the ability of viewers to recall brand names of items featured in the commercials. The researchers randomly assigned volunteers to watch one of three programs, each containing the same nine commercials. One of the programs had violent content, another sexual content, and the third neutral content. After the shows ended, the subjects were asked to recall the brands of products that were advertised. The table below summarizes the results.

	Program Type		
	Violent	**Sexual**	**Neutral**
No. of subjects	108	108	108
Brands recalled			
Mean	2.08	1.71	3.17
SD	1.87	1.76	1.77

a) Do these results indicate that viewer memory for ads may differ depending on program content? A test of the hypothesis that there is no difference in ad memory between programs with sexual content and those with violent content has a P-value of 0.136. State your conclusion.

b) Is there evidence that viewer memory for ads may differ between programs with sexual content and those with neutral content? Test an appropriate hypothesis, and state your conclusion.
c) Repeat part a) using the pooled t-test, and compare the P-values. Why might the pooled t-test be justifiable here?

24. Ad campaign You are a consultant to the marketing department of a business preparing to launch an ad campaign for a new product. The company can afford to run ads during one television show, and has decided not to sponsor a show with sexual content. You read the study described in Exercise 23, then use a computer to create a confidence interval for the difference in mean number of brand names remembered between the groups watching violent shows and those watching neutral shows.

TWO-SAMPLE T

95% CI FOR $\mu_{viol} - \mu_{neut}$: $(-1.578, -0.602)$

a) At the meeting of the marketing staff, you have to explain what this output means. What will you say?
b) What advice would you give the company about the upcoming ad campaign?

25. Sex and violence II In the study described in Exercise 23, the researchers also contacted the subjects again, 24 hours later, and asked them to recall the brands advertised. Results are summarized below.

	Program Type		
	Violent	Sexual	Neutral
No. of subjects	101	106	103
Brands recalled			
Mean	3.02	2.72	4.65
SD	1.61	1.85	1.62

a) Is there a significant difference in viewers' abilities to remember brands advertised in shows with violent versus neutral content?
b) Find a 95% confidence interval for the difference in mean number of brand names remembered between the groups watching shows with sexual content and those watching neutral shows. Interpret your interval in this context.

26. Ad recall In Exercises 23 and 25, we see the number of advertised brand names people recalled immediately after watching television shows and 24 hours later. Strangely enough, it appears that they remembered more about the ads the next day. Should we conclude this is true in general about people's memory of television ads?

a) Suppose one analyst conducts a two-sample hypothesis test, using the two means displayed earlier for violent show viewers, to see if memory of brands advertised during violent television shows is higher 24 hours later. The P-value of his test is 0.00013. What might he conclude?
b) Explain why his procedure was inappropriate. Which of the assumptions for inference was violated?
c) Suggest a design that could compare immediate brand-name recall with recall one day later, if we are restricted to using methods discussed in this chapter.

27. Hungry? Researchers investigated how the size of a bowl affects how much ice cream people tend to scoop when serving themselves.[18] At an "ice cream social," people were randomly given either a 17 fluid oz (503 mL) or a 34 fluid oz (1006 mL) bowl (both large enough that they would not be filled to capacity). They were then invited to scoop as much ice cream as they liked. Did the bowl size change the selected portion size? Here are the summaries:

Small Bowl		Large Bowl	
n	26	n	22
$\bar{y}$	5.07 oz	$\bar{y}$	6.58 oz
s	1.84 oz	s	2.91 oz

Test an appropriate hypothesis and state your conclusions. For assumptions and conditions that you cannot test, you may assume that they are sufficiently satisfied to proceed.

28. Thirsty? Researchers randomly assigned participants either a tall, thin "highball" glass or a short, wide "tumbler," each of which held 355 mL. Participants were asked to pour a shot (44.3 mL) into their glass. Did the glass shape make a difference in how much liquid they poured?[19] Here are the summaries:

highball		tumbler	
n	99	n	99
$\bar{y}$	42.2 mL	$\bar{y}$	60.9 mL
s	16.2 mL	s	17.9 mL

Test an appropriate hypothesis and state your conclusions. For assumptions and conditions that you cannot test, you may assume that they are sufficiently satisfied to proceed.

[18] Brian Wansink, Koert van Ittersum, and James E. Painter, 2006, "Ice cream illusions: Bowls, spoons, and self-served portion sizes," *Am J Prev Med.*

[19] Brian Wansink and Koert van Ittersum, 2005, "Shape of glass and amount of alcohol poured: Comparative study of effect of practice and concentration," *BMJ*:331; 1512–1514.

29. **Battle of the sexes** A June 2009 *Toronto Star* study of 100 downtown Toronto parkers attempted "to resolve scientifically the question of who is most competent at pulling into a parking space . . . men or women?"[20] It reported that women are better at parallel parking, though they take longer. Researchers counted the distance from the curb, in inches, and reported an average distance of 9.3 inches for 47 women and 11.1 inches for 46 men. Below are the data:

Distances for males:

0.5 1.0 1.5 1.5 1.5 3.0 3.5 5.0 6.0 6.0 7.0
7.0 7.0 7.0 7.0 8.0 8.0 8.5 8.5 9.0 9.0 9.5
10.0 10.0 11.0 12.0 12.0 12.0 13.0 14.0 14.0 14.0
14.0 14.0 14.0 14.5 14.5 14.5 17.0 18.0 18.0 19.0
19.0 20.0 21.0 48.0

Distances for females:

2.0 2.5 3.0 3.0 3.0 3.0 3.0 4.0 4.5 4.5 4.5
5.0 5.0 5.0 6.0 6.0 7.0 7.0 7.0 7.0 7.5 8.0
8.5 8.5 9.0 9.0 9.0 9.0 10.0 10.0 12.0 12.0 12.0
12.0 12.5 13.0 13.0 13.0 14.0 14.0 14.0 15.5 16.0
18.0 20.0 21.0 25.0

Respond to the *Star*'s claim. Do you accept it? Make your case carefully and properly.

30. **Ontario students, U.K. students** Below are usual travel times to school, in minutes, for 40 British secondary school students selected using the CensusAtSchool random data selector (http://rds.censusatschool.org.uk/). Participation was voluntary but there was broad participation from schools, teachers, and students in the U.K in this exercise in statistical literacy.

45 5 4 15 50 20 20 20 20 20 25 35 15 30 20
10 45 10 3 60 25 20 5 15 5 15 17 30 40 20
10 30 10 15 20 10 15 17 10 25

In Chapter 20, we selected 40 Ontario students in a similar fashion (www.censusatschool.ca), with the following travel times:

30 10 8 30 5 8 7 15 10 35 15 10 25 22 20
25 30 10 25 8 15 18 25 15 10 25 5 2 5 25
20 15 47 20 20 13 20 5 15 12

a) Test for a difference in mean travel time between Ontario students and British students. Also give a 95% confidence interval for the difference.
b) Are your P-value and confidence level, in part a) trustworthy?

31. **Crossing Ontario** Between 1954 and 2003, swimmers have crossed Lake Ontario 43 times. Both women and men have made the crossing. Here are some plots (we've omitted a crossing by Vikki Keith, who swam a round trip—north to south to north—in 3390 minutes):

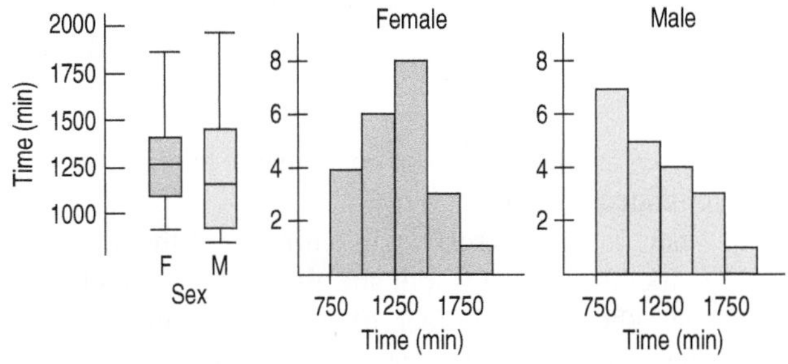

Summary statistics are as follows:

Summary of Time (min)			
Group	Count	Mean	StdDev
F	22	1271.59	261.111
M	20	1196.75	304.369

How much difference is there between the mean amount of time (in minutes) it would take female and male swimmers to swim the lake?

a) Construct and interpret a 95% confidence interval for the difference between female and male crossing times.
b) Comment on the assumptions and conditions.
c) Repeat part a) using the pooled t method, and compare results. Why might we expect the pooled t method to work well here?

32. **Still swimming** Here's some additional information about the Ontario crossing times presented in the previous exercise. It is generally thought to be harder to swim across the lake from north to south. Indeed, this has been done only five times. Every one of those crossings was by a woman. If we omit those five crossings, the boxplots look like this:

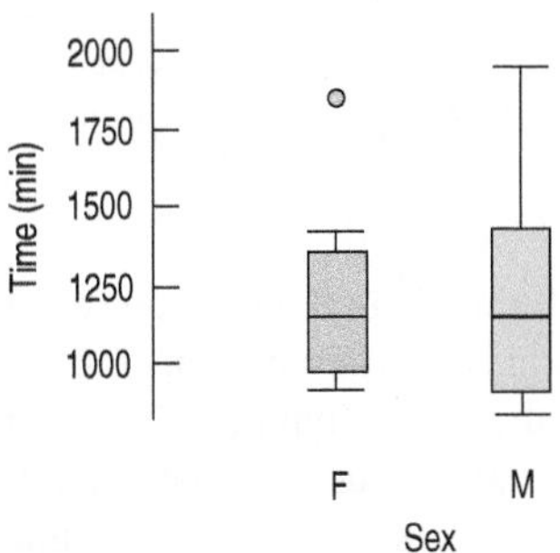

The female outlier is Vikki Keith, who crossed the lake swimming only butterfly stroke. Omitting that extraordinary swim gives the following summary statistics:

[20]Diana Zlomislic, "A new angle on the battle of the sexes." *Toronto Star*, June 6, 2009: A23. Data supplied by Andrew Bailey.

Summary of Time (min)			
Group	Count	Mean	StdDev
F	16	1175.25	177.794
M	20	1196.75	304.369

a) Do women or men appear to be faster at swimming across the lake? Support your answer by interpreting a confidence interval.

b) Vikki Keith was responsible for two of the more remarkable crossings, but she also swam Lake Ontario two other times. In fact, of the 36 crossings in this analysis, seven were repeat crossings by a swimmer who'd crossed the lake before. How does this fact affect your thoughts about the confidence interval?

33. Running heats In Olympic running events, preliminary heats are determined by random draw, so we should expect that the abilities of runners in the various heats to be about the same, on average. Here are the times (in seconds) for the 400–m women's run in the 2004 Olympics in Athens for preliminary heats 2 and 5. Is there any evidence that the mean time to finish is different for randomized heats? Explain. Be sure to include a discussion of assumptions and conditions for your analysis.

Country	Name	Heat	Time
USA	HENNAGAN Monique	2	51.02
BUL	DIMITROVA Mariyana	2	51.29
CHA	NADJINA Kaltouma	2	51.50
JAM	DAVY Nadia	2	52.04
BRA	ALMIRAO Maria Laura	2	52.10
FIN	MYKKANEN Kirsi	2	52.53
CHN	BO Fanfang	2	56.01
BAH	WILLIAMS-DARLING Tonique	5	51.20
BLR	USOVICH Svetlana	5	51.37
UKR	YEFREMOVA Antonina	5	51.53
CMR	NGUIMGO Mireille	5	51.90
JAM	BECKFORD Allison	5	52.85
TOG	THIEBAUD-KANGNI Sandrine	5	52.87
SRI	DHARSHA K V Damayanthi	5	54.58

34. Swimming heats In the previous exercise, we looked at the times in two different heats for the 400-m women's run from the 2004 Olympics. Unlike track events, swimming heats are *not* determined at random. Instead, swimmers are seeded so that better swimmers are placed in later heats. Here are the times (in seconds) for the women's 400-m freestyle from heats 2 and 5. Do these results suggest that the mean times of seeded heats are not equal? Explain. Include a discussion of assumptions and conditions for your analysis.

Country	Name	Heat	Time
ARG	BIAGIOLI Cecilia Elizabeth	2	256.42
SLO	CARMAN Anja	2	257.79
CHI	KOBRICH Kristel	2	258.68
MKD	STOJANOVSKA Vesna	2	259.39
JAM	ATKINSON Janelle	2	260.00
NZL	LINTON Rebecca	2	261.58
KOR	HA Eun-Ju	2	261.65
UKR	BERESNYEVA Olga	2	266.30
FRA	MANAUDOU Laure	5	246.76
JPN	YAMADA Sachiko	5	249.10
ROM	PADURARU Simona	5	250.39
GER	STOCKBAUER Hannah	5	250.46
AUS	GRAHAM Elka	5	251.67
CHN	PANG Jiaying	5	251.81
CAN	REIMER Brittany	5	252.33
BRA	FERREIRA Monique	5	253.75

35. Tees Does it matter what kind of tee a golfer places the ball on? The company that manufactures Stinger tees claims that the thinner shaft and smaller head will lessen resistance and drag, reducing spin and allowing the ball to travel farther. In August 2003, Golf Laboratories, Inc. compared the distance travelled by golf balls hit off regular wooden tees to those hit off Stinger tees. All the balls were struck by the same golf club using a robotic device set to swing the club head at approximately 95 miles per hour. Below are some summary statistics from the test. Assume that six balls were hit off each tee and that the data were suitable for inference.

		Total Distance (yards)	Ball Velocity (mph)	Club Velocity (mph)
Regular tee	Avg.	227.17	127.00	96.17
	SD	2.14	0.89	0.41
Stinger tee	Avg.	241.00	128.83	96.17
	SD	2.76	0.41	0.52

Is there evidence that balls hit off the Stinger tees would on average have a higher initial velocity?

36. Golf again Given the test results on golf tees described in the previous exercise, is there evidence that balls hit off Stinger tees would travel farther? Again, assume that six balls were hit off each tee and that the data were suitable for inference.

37. Statistics journals When a professional statistician has information to share with colleagues, he or she will submit an article to one of several Statistics journals for publication. This can be a lengthy process; typically, the article must be circulated for peer review and perhaps edited before being accepted for publication. Then the article must wait in line with other articles before actually appearing in print. In *Chance* magazine, Eric Bradlow and Howard Wainer reported on this delay for several journals. For 288 articles published in *The American Statistician*, the mean length of time between initial submission and publication was 21 months, with a standard deviation of 8 months. For 209 *Applied Statistics* articles, the mean time to publication was 31 months, with a standard deviation of 12 months. Create and interpret a 90% confidence interval for the difference in mean delay, and comment on the assumptions that underlie your analysis.

38. Music and memory Is it a good idea to listen to music when studying for a big test? In a study conducted by some Statistics students, 62 people were randomly assigned to listen to rap music, music by Mozart, or no music while attempting to memorize objects pictured on a page. They were then asked to list all the objects they could remember. Here are summary statistics for each group:

	Rap	Mozart	No Music
Count	29	20	13
Mean	10.72	10.00	12.77
SD	3.99	3.19	4.73

a) Does it appear that it is better to study while listening to Mozart than to rap music? Test an appropriate hypothesis, and state your conclusion.
b) Create a 90% confidence interval for the mean difference in memory score between students who study to Mozart and those who listen to no music at all. Interpret your interval.

39. Rap Using the results of the experiment described in the previous exercise, does it matter whether one listens to rap music while studying, or is it better to study without music at all?
a) Test an appropriate hypothesis and state your conclusion.
b) If you concluded that there is a difference, estimate the size of that difference with a confidence interval and explain what your interval means.

T **40.** Cuckoos Cuckoos lay their eggs in the nests of other (host) birds. The eggs are then adopted and hatched by the host birds. But the potential host birds lay eggs of different sizes. Does the cuckoo change the size of her eggs for different foster species? The numbers in the table are lengths (in mm) of cuckoo eggs found in nests of three different species of other birds. The data are drawn from the work of O.M. Latter in 1902 and were used in a fundamental textbook on statistical quality control by L.H.C. Tippett (1902–1985), one of the pioneers in that field.

Cuckoo Egg Length (mm)		
Foster Parent Species		
Sparrow	**Robin**	**Wagtail**
20.85	21.05	21.05
21.65	21.85	21.85
22.05	22.05	21.85
22.85	22.05	21.85
23.05	22.05	22.05
23.05	22.25	22.45
23.05	22.45	22.65
23.05	22.45	23.05
23.45	22.65	23.05
23.85	23.05	23.25
23.85	23.05	23.45
23.85	23.05	24.05
24.05	23.05	24.05
25.05	23.05	24.05
	23.25	24.85
	23.85	

Investigate the question of whether the mean length of cuckoo eggs is the same for different species, and state your conclusion.

41. Tee it up again Consider again Exercise 35 about the Stinger golf tees. Six balls were used on each type of tee. If we wanted the margin of error in the estimate of the mean difference to be just 1.0 yard, with 95% confidence, how many balls should have been struck off each type of tee? Be conservative and set each standard deviation to 3.0 yards.

T **42.** Harder water Consider again Exercise 20 about hard water in England and Wales. If we want the margin of error for our estimate of the difference in mortality rates to be only 50 (per 100 000), with 95% confidence, how big should the sample sizes be (assume two equal sample sizes)? You can use the standard deviations given previously as your estimates.

43. Fore! Consider again Exercise 35 about the Stinger golf tees (by the way, "fore" is shouted in golf to warn of a ball flying your way). Perhaps a difference of as little as one yard, on average, is important enough to justify using a new tee type. Using software, determine how many balls should be driven off each type of tee so that the test will have a power of 0.95 when the true mean difference is three yards. Assume, conservatively, that both standard deviations equal about 3.0 yards.

T **44.** Hard water again Consider again Exercise 20 about hard water in England and Wales. If we want to successfully detect a difference of 100 (per 100 000), in mortality rates, 9 times in 10, how many towns should actually be sampled (assume two equal sample sizes)? You can use 150.0 as your estimate of both standard deviations.

Just Checking ANSWERS

1. Randomization should balance unknown sources of variability in the two groups of patients and helps us believe the two groups are independent.
2. We can be 95% confident that after four weeks, endoscopic surgery patients will have a mean pinch strength between 0.04 kg and 2.96 kg higher than open-incision patients.
3. The lower bound of this interval is close to 0, so the difference may not be great enough that patients could actually notice the difference. We may want to consider other issues such as cost or risk in making a recommendation about the two surgical procedures.
4. Without data, we can't check the Nearly Normal Condition.
5. H_0: Mean pinch strength is the same after both surgeries.
   H_A: Mean pinch strength is different after the two surgeries.
6. With a P-value this low, we reject the null hypothesis. We can conclude that mean pinch strength differs after four weeks in patients who undergo endoscopic surgery versus patients who have open-incision surgery. Results suggest that the endoscopic surgery patients may be stronger, on average.
7. If some patients contributed two hands to the study, then the observations within the groups may not be mutually independent. It is reasonable to assume that two hands from the same patient might respond in similar ways to similar treatments.

MathXL
MyStatLab

Go to MathXL at www.mathxl.com or MyStatLab at www.mystatlab.com. You can practise exercises for this chapter as often as you want. The guided solutions will help you find answers step by step. You'll find a personalized study plan available to you too!

chapter

22 Paired Samples and Blocks

To call in the statistician after the experiment is done may be no more than asking him to perform a post-mortem examination: he may be able to say what the experiment died of.

—Ronald Fisher

Where are we going?

How much will an LSAT prep course raise students' scores? Do married women do more housework than their husbands? (but seriously, just how much more do they do per week?) Questions like these look at paired variables. When pairs of quantitative observations go together naturally, they can't be independent, so the methods we used in the last chapter won't work. In this chapter, you'll see what to do with paired data.

Who	Olympic speed-skaters
What	Time for women's 1500 m
Units	Seconds
When	2006
Where	Torino, Italy
Why	To see whether one lane is faster than the other

Yuri Kadobnov/AFP/Getty Images

Speed-skating races are run in pairs. Two skaters start at the same time, one on the inner lane and one on the outer lane. Halfway through the race, they cross over, switching lanes so that each will skate the same distance in each lane. Even though this seems fair, at the 2006 Olympics some fans thought there might have been an advantage to starting on the outside. After all, the winner, Cindy Klassen,[1] started on the outside and skated a remarkable 1.47 seconds faster than the silver medalist.

Here are the data for the women's 1500-m race:

Inner Lane		Outer Lane	
Name	**Time**	**Name**	**Time**
OLTEAN Daniela	129.24	(no competitor)	
ZHANG Xiaolei	125.75	NEMOTO Nami	122.34
ABRAMOVA Yekaterina	121.63	LAMB Maria	122.12
REMPEL Shannon	122.24	NOH Seon Yeong	123.35
LEE Ju-Youn	120.85	TIMMER Marianne	120.45
ROKITA Anna Natalia	122.19	MARRA Adelia	123.07
YAKSHINA Valentina	122.15	OPITZ Lucille	122.75
BJELKEVIK Hedvig	122.16	HAUGLI Maren	121.22
ISHINO Eriko	121.85	WOJCICKA Katarzyna	119.96
RANEY Catherine	121.17	BJELKEVIK Annette	121.03
OTSU Hiromi	124.77	LOBYSHEVA Yekaterina	118.87
SIMIONATO Chiara	118.76	JI Jia	121.85
ANSCHUETZ THOMS Daniela	119.74	WANG Fei	120.13
BARYSHEVA Varvara	121.60	VAN DEUTEKOM Paulien	120.15
GROENEWOLD Renate	119.33	GROVES Kristina	116.74
RODRIGUEZ Jennifer	119.30	NESBITT Christine	119.15
FRIESINGER Anni	117.31	KLASSEN Cindy	115.27
WUST Ireen	116.90	TABATA Maki	120.77

We can view this skating event as an experiment testing whether the lanes were equally fast. Skaters were randomly assigned to lanes. The boxplots of times recorded in the inner and outer lanes (Figure 22.1 on next page) don't show much difference. But

[1] Winnipeg native and tied with Clara Hughes as Canada's all-time most decorated Olympian.

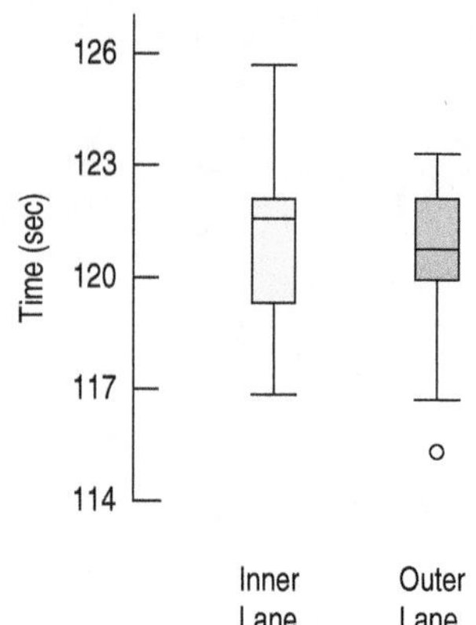

Figure 22.1

Using boxplots to compare times in the inner and outer lanes shows little because it ignores the fact that the skaters raced in pairs (with one exception, Daniela Oltean, who raced alone and whose time of 129.24 is not shown above).

that's not the right way to compare these times. Conditions can change during the day. The data are recorded for races run in pairs (with the exception of Daniela Oltean, who raced alone), so the two groups are not independent.

22.1 Paired *t*-Test

Paired Data

Data such as these are called **paired**. We have the times for skaters in each lane for each race. The races are run in pairs, so they can't be independent. And since they're not independent, we can't use the two-sample *t* methods. Instead, we can focus on the *differences* in times for each racing pair.

Paired data arise in a number of ways. Perhaps the most common way is to compare subjects with themselves before and after a treatment. When pairs arise from an experiment, the pairing is a type of *blocking*. When they arise from an observational study, it is a form of *matching*. Paired data may be quantitative or categorical, but we'll only discuss quantitative data in this text (for categorical data, look up McNemar's test).

For Example IDENTIFYING PAIRED DATA

Do flexible schedules reduce the demand for resources? The Lake County, Illinois, Health Department experimented with a flexible four-day work week. For a year, the department recorded the mileage driven by 11 field workers on an ordinary five-day work week. Then, the department changed to a flexible four-day work week and recorded mileage for another year.[2] Here are the data in miles (1 mile = 1.6 km):

Name	5-Day Mileage	4-Day Mileage
Jeff	2798	2914
Betty	7724	6112
Roger	7505	6177
Tom	838	1102
Aimee	4592	3281
Greg	8107	4997
Larry G.	1228	1695
Tad	8718	6606
Larry M.	1097	1063
Leslie	8089	6392
Lee	3807	3362

QUESTION: Why are these data paired?

ANSWER: The mileage data are paired because each driver's mileage is measured before and after the change in schedule. I'd expect drivers who drove more than others before the schedule change to continue to drive more afterward, so the two sets of mileages can't be considered independent.

Pairing isn't a problem; it's an opportunity. If you know the data are paired, you can take advantage of that fact—in fact, you *must* take advantage of it. You *may not* use the two-sample methods of the previous chapter when the data are paired. Remember: Those methods rely on the Pythagorean Theorem of Statistics, and that *requires the two samples be independent.* Paired data aren't. You must decide whether the data are paired from understanding how they were collected and what they mean (check the *W*'s). There is no test to determine whether the data are paired.

[2]Charles S. Catlin, "Four-day work week improves environment," *Journal of Environmental Health*, Denver, 59:7.

Once we recognize that the speed-skating data are matched pairs, it makes sense to consider the difference in times for each two-skater race. So we look at the *pairwise* differences:

Skating Pair	Inner Time	Outer Time	Inner – Outer
1	129.24		
2	125.75	122.34	3.41
3	121.63	122.12	−0.49
4	122.24	123.35	−1.11
5	120.85	120.45	0.40
6	122.19	123.07	−0.88
7	122.15	122.75	−0.60
8	122.16	121.22	0.94
9	121.85	119.96	1.89
10	121.17	121.03	0.14
11	124.77	118.87	5.90
12	118.76	121.85	−3.09
13	119.74	120.13	−0.39
14	121.60	120.15	1.45
15	119.33	116.74	2.59
16	119.30	119.15	0.15
17	117.31	115.27	2.04
18	116.90	120.77	−3.87

A S *Activity:* **Differences in Means of Paired Groups.** Are married couples typically the same age, or do wives tend to be younger than their husbands, on average?

One skater raced alone, so we'll omit that race. Because it is the differences we care about, we'll treat them as if they were the data, ignoring the original two columns. Now that we have only one column of values to consider, we can use a simple one-sample t-test. Mechanically, **a paired t-test** is just a one-sample t-test for the means of these pairwise differences. The sample size is the number of pairs.

22.2 Assumptions and Conditions

Paired Data Condition

Paired Data Condition: The data must be quantitative and paired. You can't just decide to pair data when in fact the samples are independent. When you have two groups with the same number of observations, it may be tempting to match them up. Don't, unless you are prepared to justify your claim that the data are paired.

On the other hand, be sure to recognize paired data when you have them. Remember, two-sample t methods aren't valid without independent groups, and paired groups aren't independent. Although this is a strict requirement, it is one that can be easy to check if you understand how the data were collected.

Independence Assumption

Independence Assumption: If the data are paired, the *groups* are not independent. For these methods, it's the *differences* that must be independent of each other. There's no reason to believe that the difference in speeds of one pair of races could affect the difference in speeds for another pair.

Suitable randomization can both prevent bias *and* produce independence where needed in a design. With paired data, randomness can arise in many ways. The pairs may be a random sample. In an experiment, the order of the two treatments may be randomly assigned, or the treatments may be randomly assigned to one member of each pair. In a before-and-after study, we may believe that the observed differences are a representative sample from a population of interest. If we have any doubts, we'll need to include a control group to be able to draw conclusions. What we want to know usually focuses our attention on where the randomness should be. In our example, skaters were assigned to the lanes at random.

If the study involves sampling, and your sample is bigger than 10% of the population (which is rare), be sure to acknowledge that your standard error and P-value are in fact conservative (too big) estimates.

Normal Population Assumption

We need to assume that the population of *differences* follows a Normal model. We don't need to check the individual groups.

Nearly Normal Condition: This condition can be checked with a histogram or Normal probability plot of the *differences*—but not of the individual groups. As with the one-sample *t*-methods, *robustness* against departures from Normality increases with the sample size; in other words, the Normality assumption matters less the more pairs we have to consider. You may be pleasantly surprised when you check this condition. Even if your original measurements are skewed or bimodal, the *differences* may be nearly Normal. After all, the individual who was way out in the tail on an initial measurement is likely to still be out there on the second one, giving a perfectly ordinary difference.

For Example CHECKING ASSUMPTIONS AND CONDITIONS

RECAP: Field workers for a health department compared driving mileage on a five-day work schedule with mileage on a new four-day schedule. To see if the new schedule changes the amount of driving they do, we'll look at paired differences in mileages before and after.

Name	5-Day mileage	4-Day mileage	Difference
Jeff	2798	2914	−116
Betty	7724	6112	1612
Roger	7505	6177	1328
Tom	838	1102	−264
Aimee	4592	3281	1311
Greg	8107	4997	3110
Larry G.	1228	1695	−467
Tad	8718	6606	2112
Larry M.	1097	1063	34
Leslie	8089	6392	1697
Lee	3807	3362	445

QUESTION: Is it okay to use these data to test whether the new schedule changes the amount of driving?

ANSWER:

✓ **Paired Data Condition:** The data are paired because each value is the mileage driven by the same person before and after a change in work schedule.

✓ **Independence Assumption:** The driving behaviour of any individual worker is independent of the others, so the differences are mutually independent. And the results may well be reasonably representative of what might happen for field workers in other years.

✓ **Nearly Normal Condition:** The histogram of the mileage differences is unimodal and symmetric:

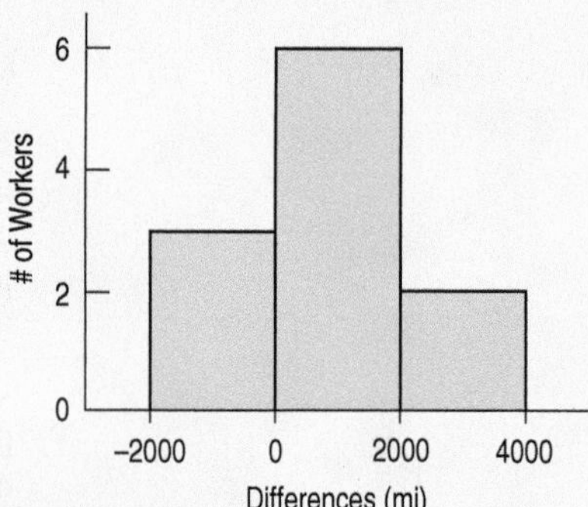

Since the assumptions and conditions are satisfied, it's okay to use paired-*t* methods for these data.

The steps of testing a hypothesis for paired differences are very much like the steps for a one-sample t-test for a mean. After all, it's just the one-sample t-test applied to a single sample of *differences*.

THE PAIRED t-TEST

When the assumptions and conditions are met, we are ready to test whether the mean of paired differences is significantly different from Δ_0. We test the hypothesis

$$H_0: \mu_d = \Delta_0$$

where the d's are the pairwise differences and Δ_0 is almost always 0.

We use the statistic

$$t = \frac{\bar{d} - \Delta_0}{SE(\bar{d})}$$

where $\bar{d}$ is the mean of the pairwise differences and

$$SE(\bar{d}) = \frac{s_d}{\sqrt{n}}$$

is the ordinary standard error of the mean, applied to the differences, with $n =$ the number of pairs.

When the conditions are met and the null hypothesis is true, we can model the sampling distribution of this statistic with a Student's t-model with $n - 1$ degrees of freedom, and use that model to obtain a P-value.

Step-by-Step Example A PAIRED t-TEST

Vladislav Gajic/Shutterstock

Question: Was there a difference in speeds between the inner and outer speed-skating lanes at the 2006 Winter Olympics?

THINK ➡ Plan State what we want to know. Identify the *parameter* we wish to estimate. Here our parameter is the mean difference in race times. Identify the variables and check the *W*'s.	I want to know whether there really was a difference in the speeds of the two lanes for speed skating at the 2006 Olympics. I have data for 17 pairs of racers at the women's 1500-m race.
Hypotheses State the null and alternative hypotheses. Although fans suspected one lane was faster, we can't use the data we have to specify the direction of a test. We (and Olympic officials) would be interested in a difference in either direction, so we'd better test a two-sided alternative.	H_0: Neither lane offered an advantage: $\mu_d = 0$ H_A: The mean difference is different from zero: $\mu_d \neq 0$
REALITY CHECK The individual differences are all in seconds. We should expect the mean difference to be comparable in magnitude.	

Model Think about the assumptions and check the conditions.

State why you think the data are paired.

✓ **Paired Data Condition:** The data are paired because racers compete in pairs.

Think about what we hope to learn and where the randomization comes from.

✓ **Independence/Randomization:** Lane assignments are random, and each race should be independent of the others, so the differences are mutually independent and free of bias.

Make a picture—just one. There is no need to plot separate distributions of the two groups[3]—that entirely misses the pairing. For paired data, it's the Normality of the *differences* that we care about. Treat those paired differences as you would a single variable, and check the Nearly Normal Condition with a histogram or a Normal probability plot.

✓ **Nearly Normal Condition:** The histogram of the differences is unimodal and fairly symmetric:

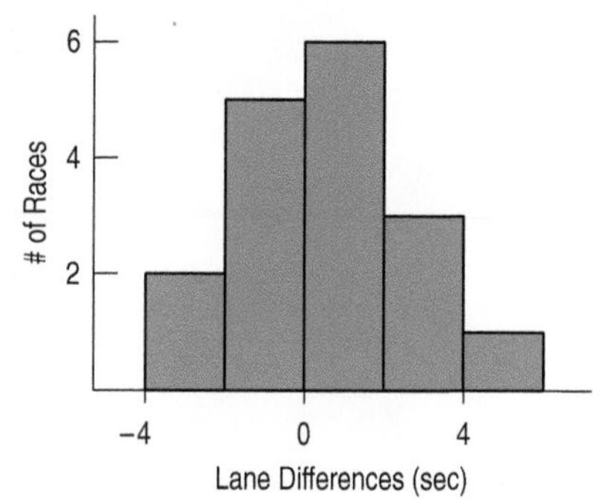

Specify the sampling distribution model.

Choose the method.

The conditions are met, so I'll use a Student's t-model with $(n - 1)$ degrees of freedom, and perform a **paired t-test.**

SHOW ➡ Mechanics

n is the number of *pairs*—in this case, the number of races.

$\bar{d}$ is the mean difference.

s_d is the standard deviation of the differences.

The data give

$$n = 17 \text{ pairs}$$

$$\bar{d} = 0.499 \text{ seconds}$$

$$s_d = 2.333 \text{ seconds}$$

Find the standard error and the t-score of the observed mean difference. There is nothing new in the mechanics of the paired-t methods. These are the mechanics of the t-test for a mean applied to the differences.

I estimate the standard deviation of $\bar{d}$ using

$$SE(\bar{d}) = \frac{s_d}{\sqrt{n}} = \frac{2.333}{\sqrt{17}} = 0.5658$$

So $$t = \frac{\bar{d} - 0}{SE(\bar{d})} = \frac{0.499}{0.5658} = 0.882$$

Make a picture..Sketch a t-model centred at the hypothesized mean of 0. Because this is a two-tail test, shade both the region to the right of the calculated t value and the corresponding region in the lower tail.

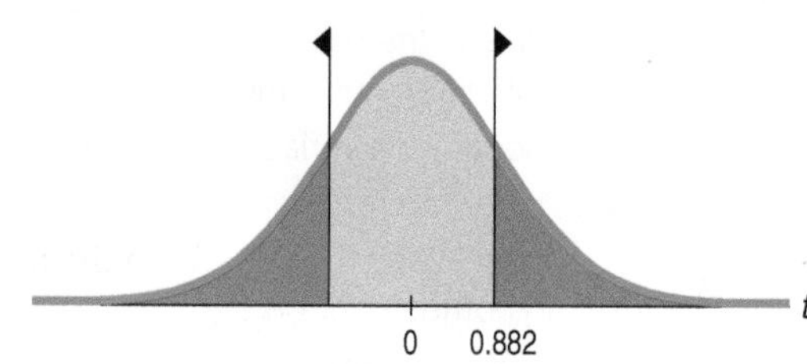

Find the P-value, using technology.

$$\text{P-value} = 2P(t_{16} > 0.882) = 0.39$$

REALITY CHECK The mean difference is 0.499 seconds. That may not seem like much, but a smaller difference determined the silver and bronze medals. The standard error is about this big, so a t-value less than 1.0 isn't surprising. Nor is a large P-value.

[3]Though often done to provide a more complete visual impression of both the absolute and *relative* effect size.

TELL ➡ Conclusion Link the P-value to your decision about H_0, and state your conclusion in context.	The P-value is large. Events that happen more than a third of the time are not remarkable. So, even though there is an observed difference between the lanes, I can't conclude that it isn't due simply to random chance. The fans may have interpreted a random fluctuation in the data as one lane's advantage. There's insufficient evidence to declare any lack of fairness.

For Example DOING A PAIRED *t*-TEST

RECAP: We want to test whether a change from a five-day work week to a four-day work week could change the amount driven by field workers of a health department. We've already confirmed that the assumptions and conditions for a paired *t*-test are met.

QUESTION: Is there evidence that a four-day work week would change how many miles workers drive?

ANSWERS: H_0: The change in the health department workers' schedules didn't change the mean mileage driven; the mean difference is zero:

$$\mu_d = 0$$

H_A: The mean difference is different from zero:

$$\mu_d \neq 0$$

The conditions are met, so I'll use a Student's *t*-model with $(n - 1) = 10$ degrees of freedom and perform a paired *t*-test.

The data give

$$n = 11 \text{ pairs}$$
$$\bar{d} = 982 \text{ miles}$$
$$s_d = 1139.6 \text{ miles}$$
$$SE(\bar{d}) = \frac{s_d}{\sqrt{n}} = \frac{1139.6}{\sqrt{11}} = 343.6$$
$$t = \frac{\bar{d} - 0}{SE(\bar{d})} = \frac{982.0}{343.6} = 2.86$$

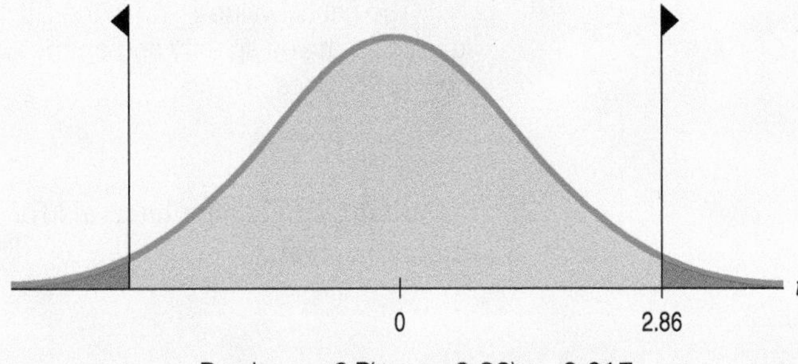

$$P\text{-value} = 2P(t_{10} > 2.86) = 0.017$$

The P-value is small, so I reject the null hypothesis and conclude that the change in work week did lead to a change in average driving mileage.

It appears that changing the work schedule may reduce the mileage driven by workers. We should propose a course of action, but it's hard to tell from the hypothesis test whether the reduction matters. Is the difference in mileage important in the sense of reducing air pollution or costs, or is it merely statistically significant? To help make that decision, we should look at a confidence interval. If the difference in mileage proves to be large in a practical sense, then we might recommend a change in schedule for the rest of the department.

22.3 Paired *t* Confidence Interval

Who	170 randomly sampled couples
What	Ages
Units	Years
When	Recently
Where	Britain

In developed countries, the average age of women is generally higher than that of men. After all, women tend to live longer. But if we look at *married couples*, husbands tend to be slightly older than wives. How much older, on average, are husbands? We have data from a random sample of 200 British couples, the first seven of which are shown below. Only 170 couples provided ages for both husband and wife, so we can work only with that many pairs. Let's form a confidence interval for the mean difference of husband's and wife's age for these 170 couples. Here are the first seven pairs:

Wife's Age	Husband's Age	Difference (husband – wife)
43	49	6
28	25	–3
30	40	10
57	52	–5
52	58	6
27	32	5
52	43	–9
⋮	⋮	⋮

Clearly, these data are paired. The survey selected *couples* at random, not individuals. We're interested in the mean age difference within couples. How would we construct a confidence interval for the true mean difference in ages?

PAIRED *t*-INTERVAL

When the assumptions and conditions are met, we are ready to find the confidence interval for the mean of the paired differences. The confidence interval is

$$\bar{d} \pm t^*_{n-1} \times SE(\bar{d})$$

where the standard error of the mean difference is $SE(\bar{d}) = \frac{s_d}{\sqrt{n}}$.

The critical value t^* from the Student's *t*-model depends on the particular confidence level, C, that you specify and on the degrees of freedom, $n - 1$, which is based on the number of pairs, n.

Making confidence intervals for matched pairs follows exactly the steps for a one-sample *t*-interval.

Step-by-Step Example A PAIRED *t*-INTERVAL

Question: How big a difference is there, on average, between the ages of husbands and wives?

THINK ➡ Plan State what we want to know. Identify the variables and check the *W*'s. Identify the parameter you wish to estimate. For a paired analysis, the parameter of interest is the mean of the differences in the entire population.	I want to estimate the mean difference in age between husbands and wives. I have a random sample of 200 British husbands and wives, 170 of whom provided both ages.

Model Think about the assumptions and check the conditions.

✓ **Paired Data Assumption:** The data are paired because they are on husbands and wives.

✓ **Independence Assumption:** The data are from a randomized survey, so husbands and wives should be independent of each other.

Make a picture. We focus on the differences, so a histogram or Normal probability plot is best here.

✓ **Nearly Normal Condition:** The histogram of the husband–wife differences is unimodal and symmetric:

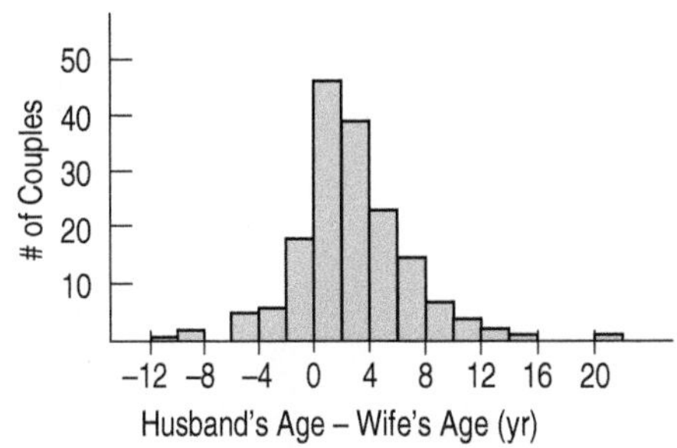

REALITY CHECK The histogram shows husbands are often older than wives (because most of the differences are greater than 0). The mean difference seen here of around two years is reasonable.

State the sampling distribution model.

Choose your method.

The conditions are met, so I can use a Student's *t*-model with $(n - 1) = 169$ degrees of freedom and find a **paired *t*-interval.**

SHOW ➡ Mechanics

n is the number of *pairs*, here the number of couples.

$\bar{d}$ is the mean difference.

s_d is the standard deviation of the differences.

Be sure to include the units along with the statistics.

The critical value we need to make a 95% interval comes from a Student's *t* table, a computer program, or a calculator.

$n = 170$ couples

$\bar{d} = 2.2$ years

$s_d = 4.1$ years

I calculate the standard error of $\bar{d}$ as

$$SE(\bar{d}) = \frac{s_d}{\sqrt{n}} = \frac{4.1}{\sqrt{170}} = 0.31 \text{ years.}$$

The df for the *t*-model is $n - 1 = 169$.

The 95% critical value for t_{169} (from the table) is conservatively estimated to be 1.977.

The margin of error is

$$ME = t^*_{169} \times SE(\bar{d}) = 1.977(0.31) = 0.61$$

So the 95% confidence interval is

$$2.2 \pm 0.6 \text{ years,}$$

or an interval of (1.6, 2.8) years.

REALITY CHECK This result makes sense. Our everyday experience confirms that an average age difference of about two years is reasonable.

TELL ➡ Conclusion Interpret the confidence interval in context.

I am 95% confident that British husbands are, on average, 1.6 to 2.8 years older than their wives.

22.4 Effect Size and Sample Size

When we examined the speed-skating times, we failed to reject the null hypothesis, so we couldn't be certain whether there really was a difference between the lanes. Maybe there wasn't any difference, or maybe whatever difference there might have been was just too small to matter at all. Were the fans right to be concerned?

We can't tell from the hypothesis test, but using the same summary statistics, we can find that the corresponding 95% confidence interval for the mean difference is $(-0.70 < \mu_d < 1.70)$ seconds.

A confidence interval is a good way to get a sense for the size of the effect we're trying to understand. That gives us a plausible range of values for the true mean difference in lane times. If differences of 1.7 seconds were too small to matter in 1500-m Olympic speed skating, we'd be pretty sure there was no need for concern.

But in fact, except for the gold–silver gap, the successive gaps between each skater and the next-fastest one were *all* less than the high end of this interval, and most were right around the middle of the interval:

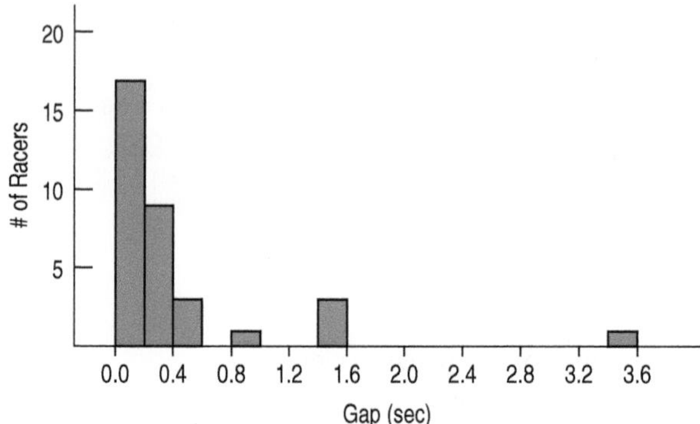

So even though we were unable to discern a real difference, the confidence interval shows that the effects we're considering *just may be* big enough to be important. We may want to continue this investigation by checking out other races on this ice and being alert for possible differences at other venues.

Effect size issues should make you think about *sample size* issues. Here, the sample size may have been in some sense fixed, but since it is possible to continue this investigation by monitoring other races on this ice, we can ask the usual questions:

- How small a margin of error or confidence interval should we plan for in order to have sufficiently precise and useful information for decision-making purposes?

Or thinking in terms of hypothesis testing, we might need to ask:

- How big an effect size (difference) is sufficiently important so that its discovery should be assured with high probability by our test procedure?

We can then determine the sample size accordingly. And since this is just a one-sample *t*-procedure (based on differences), the issues are exactly the same as when we discussed sample size determination in Chapter 20. And we attack it the same way.

If we wanted the margin of error to equal, say, 0.50 seconds, with 95% confidence, we would solve the following equation for n (using the sample standard deviation of 2.333 from this study as our estimated standard deviation):

$$0.50 = \frac{1.96(2.333)}{\sqrt{n}}$$

Solving for n gives us $n = 83.6$, or 84 paired skater races that should be observed to attain the desired margin of error of one-half second.

On the other hand, suppose we decided that a difference of one second was large enough to make an important practical difference to the racers. In that case, we might decide to set the power of a test of hypothesis equal to some high probability, such as 0.90, when the true difference equals one second. This would give us a good chance of detecting a problem of this magnitude (and correctly rejecting the null hypothesis). Using software, we see below that 60 races would need to be observed to attain the desired power when the *true difference* is 1.0 seconds. But if a half-second difference was considered too important to miss, the required sample size, shown below, jumps to 231.

Power and Sample Size

1-Sample *t*-Test

Testing mean = null (versus not = null)

Calculating power for mean = null + difference

Alpha = 0.05 Assumed standard deviation = 2.333

Difference	Sample Size	Target Power	Actual Power
1.0	60	0.9	0.904173
0.5	231	0.9	0.900382

For Example LOOKING AT EFFECT SIZE WITH A PAIRED-*t* CONFIDENCE INTERVAL

RECAP: We know that, on average, the switch from a five-day work week to a four-day work week reduced the amount driven by field workers in that Illinois health department. However, finding that there is a significant difference doesn't necessarily mean that difference is meaningful or worthwhile. To assess the size of the effect, we need a confidence interval. We already know the assumptions and conditions are met.

QUESTION: By how much, on average, might a change in work week schedule reduce the amount driven by workers?

ANSWER: $\bar{d} = 982\ mi$ $\quad SE(\bar{d}) = 343.6$ $\quad t^*_{10} = 2.228\ (for\ 95\%)$

$$ME = t^*_{10} \times SE(\bar{d}) = 2.228(343.6) = 765.54$$

So the 95% confidence interval for μ_d is

$$982 \pm 765.54 \text{ or } (216.46, 1747.54) \text{ fewer miles.}$$

With 95% confidence, I estimate that by switching to a four-day work week, employees would drive on average 216 to 1748 fewer miles per year. With high gas prices, this could save a lot of money.

This is a very wide confidence interval; we were fortunate to find a significant difference with such a small sample. Planning the sample size (number of workers) more carefully could have narrowed the interval, making it more informative. Or, we could have set the sample size to ensure that a genuine mean reduction of, say, 500 miles would likely be discovered in a test of hypothesis (by running a *power & sample size* analysis with appropriate software).

22.5 Blocking

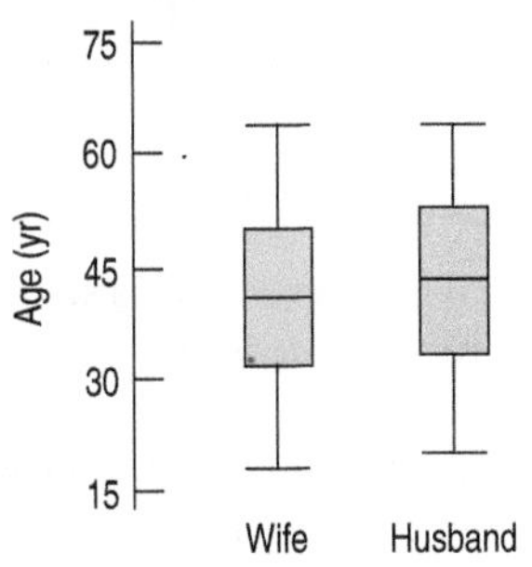

Figure 22.2

This display is interesting but lacks important information. It doesn't do much good to compare all the wives as a group with all the husbands. We care about the paired differences.

Because the sample of British husbands and wives includes both older and younger couples, there's a lot of variation in the ages of the men and in the ages of the women. In fact, that variation is so great that a boxplot of the two groups (Fig. 22.2) would show little difference. But that would not be the most appropriate plot. It's the *difference* we care about. Pairing isolates the extra variation and allows us to focus on the individual differences. In Chapter 11, we saw how we could design an experiment with blocking to filter out the variability between identifiable groups of subjects, allowing us to better see variability among treatment groups due to their response to the treatment. A paired design is an example of blocking.

When we pair, we have roughly half the degrees of freedom of a two-sample test. You may see discussions that suggest that in "choosing" a paired analysis we "give up" these degrees of freedom. This isn't really true, though. If the data are inherently paired, then there never were additional degrees of freedom, and we have no "choice."

Matching pairs generally removes so much extra variation that it more than compensates for having only half the degrees of freedom, so it is usually a good design choice when you plan a study. Of course, inappropriate matching when the groups are in fact independent (say, by matching on the first letter of the last name of subjects) would cost degrees of freedom without the benefit of reducing the variance.

What's Independent?

The methods of this chapter require the pairs to be independent of each other. They make no assumptions about the individuals—only about the pairs. By contrast, many other inference methods require independence among all the individuals.

You will sometimes see the paired t methods of this chapter referred to as methods for "dependent" samples to differentiate them from the independence we required in the previous chapter. But in fact, the only "dependence" in paired data is the pairing itself. That's not a failure of statistical independence; it's a feature of the design for the data collection.

Just Checking

Think about each of the situations described below.

- Would you use a two-sample t or paired-t method (or neither)? Why?
- Would you perform a hypothesis test or find a confidence interval?

1. Random samples of 50 men and 50 women are asked to imagine buying a birthday present for their best friend. We want to estimate the difference in how much they are willing to spend.
2. Mothers of twins were surveyed and asked how often in the past month strangers had asked whether the twins were identical.
3. Are parents equally strict with boys and girls? In a random sample of families, researchers asked a brother and sister from each family to rate how strict their parents were.
4. Forty-eight overweight subjects are randomly assigned to either aerobic or stretching exercise programs. They are weighed at the beginning and at the end of the experiment to see how much weight they lost.
   a. We want to estimate the mean amount of weight lost by those doing aerobic exercise.
   b. We want to know which program is more effective at reducing weight.
5. Couples at a dance club were separated and each person was asked to rate the band. Do men or women like this band more?

*22.6 The Sign Test for Paired Data

With paired data, we've been using a simple t-test for the paired differences. This suggests that if we want a distribution-free method, it would be natural to compute a sign test on the paired differences and test whether the *median* of the differences is 0. That's exactly what we do. The test is very simple. We record a 0 for every paired difference that's negative and a 1 for each positive difference, ignoring pairs for which the difference is exactly 0. We test the associated proportion $p = 0.5$ using a z-test if the number of pairs is at least 20 (so that we expect at least 10 successes and failures), or compute the exact Binomial probabilities if the number of pairs is less than 20 (as discussed in Chapter 14).

Let's try it on the married couples data. Of the 170 couples, there are 119 with the husband older, 32 with the wife older, and 19 in which the husband and wife are the same age. So, we test $p = 0.5$, with a sample proportion of $119/151 = 0.788$. The Success/Failure

Condition is easily satisfied. Applying the one-proportion z-test to the differences gives a z-score of 7.08, with a P-value < 0.00001. We can be pretty confident that the median is not 0. (Just for comparison, the t-statistic for testing $\mu_d = 0$ is 7.097 with 169 df—almost an identical result.)

As with other distribution-free tests, the advantage of the ***sign test for matched pairs*** is that we don't require the Nearly Normal Condition for the paired differences. Because it looks only at the direction of the difference, the sign test isn't affected by outliers—extraordinarily large differences—which can be an advantage. On the other hand, when the assumptions of the *paired t-test* are met, the paired t-test is more powerful than the sign test.

WHAT CAN GO WRONG?

- **Don't use a two-sample *t*-test when you have paired data.** See the What Can Go Wrong? discussion in Chapter 21.
- **Don't use a paired-*t* method when the samples aren't paired.** Just because two groups have the same number of observations doesn't mean they can be paired, even if they are shown side-by-side in a table. We might have 25 men and 25 women in our study, but they might be completely independent of one another. If they were siblings or spouses, we might consider them paired. Remember that you cannot *choose* which method to use based on your preferences. If the data are from two independent samples, use two-sample t methods. If the data are from an experiment in which observations were paired, you must use a paired method. If the data are from an observational study, you must be able to defend your decision to use matched pairs or independent groups.
- **Don't forget outliers.** The outliers we care about now are in the differences. A subject who is extraordinary both before and after a treatment may still have a perfectly typical difference. But one outlying difference can completely distort your conclusions. Be sure to plot the differences (even if you also plot the data).
- **Don't look for the difference between the means of paired groups with side-by-side boxplots.** The point of the paired analysis is to remove extra variation. The boxplots of each group still contain that variation. Comparing them is often done, but such a graph gives a very incomplete and possibly misleading picture.

CONNECTIONS

The most important connection is to the concept of blocking that we first discussed when we considered designed experiments in Chapter 11. Pairing is a basic and very effective form of blocking.

Of course, the details of the mechanics for paired t-tests and intervals are identical to those for the one-sample t-methods. Everything we know about those methods applies here.

The connection to the two-sample methods of the previous chapter is that when the data are naturally paired, those methods are not appropriate because paired data fail the required condition of independence.

What Have We Learned?

Learning Objectives

Recognize when data are paired or matched.

Know how to construct a confidence interval for the mean difference in paired data.

Know how to perform a hypothesis test about the mean difference (usually with a null of zero representing no difference between the groups.)

Review of Terms

Paired data Data are paired when the observations are collected in pairs or the observations in one group are naturally related to observations in the other. The simplest form of pairing is when we measure each subject twice—often before and after a treatment is applied. More sophisticated forms of pairing in experiments are a form of blocking and arise in other contexts. Pairing in observational and survey data is a form of matching (p. 634).

Paired *t*-test A hypothesis test for the mean of the pairwise differences of two groups. It tests the null hypothesis

$$H_0: \mu_d = \Delta_0$$

where the hypothesized difference is almost always 0, using the statistic

$$t = \frac{\bar{d} - \Delta_0}{SE(\bar{d})}$$

which has a t_{n-1} sampling model under the null hypothesis, where $SE(\bar{d}) = \frac{s_d}{\sqrt{n}}$, and n is the number of pairs (p. 635).

Paired-*t* confidence interval A confidence interval for the mean of the pairwise differences between independent groups found as

$$\bar{d} \pm t^*_{n-1} \times SE(\bar{d}), \text{ where } SE(\bar{d}) = \frac{s_d}{\sqrt{n}} \text{ and } n \text{ is the number of pairs (p. 640).}$$

On the Computer PAIRED *t*

Most statistics programs can compute paired-*t* analyses. Some may want you to find the differences yourself and use the one-sample *t* methods. Those that perform the entire procedure will need to know the two variables to compare. The computer, of course, cannot verify that the variables are naturally paired. Most programs will check whether the two variables have the same number of observations, but some stop there, and that can cause trouble. Most programs will automatically omit any pair that is missing a value for either variable (as we did with the British couples). You must look carefully to see whether that has happened.

As we've seen with other inference results, some packages pack a lot of information into a simple table, but you must locate what you want for yourself. Here's a generic example with comments:

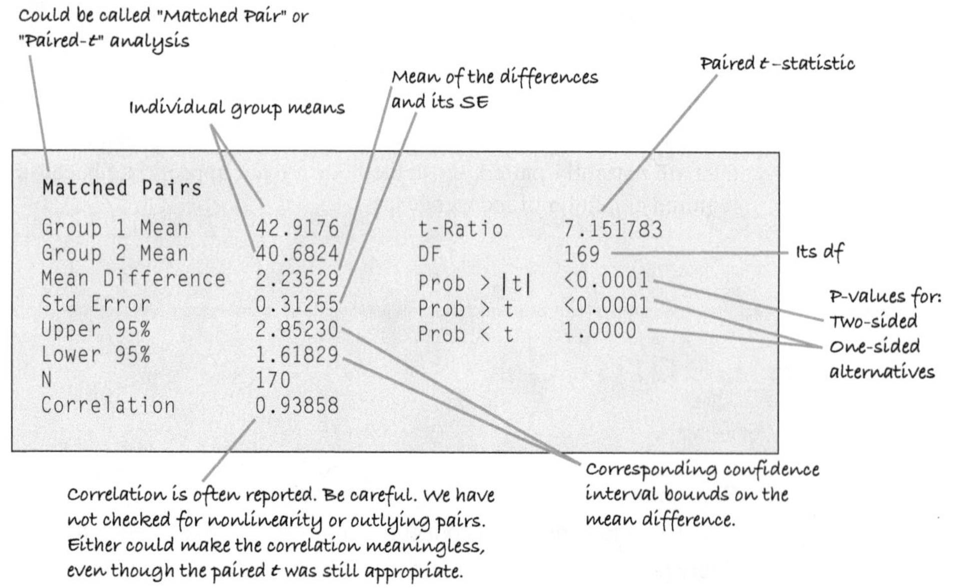
```
Matched Pairs
Group 1 Mean      42.9176     t-Ratio    7.151783
Group 2 Mean      40.6824     DF         169
Mean Difference   2.23529     Prob > |t| <0.0001
Std Error         0.31255     Prob > t   <0.0001
Upper 95%         2.85230     Prob < t   1.0000
Lower 95%         1.61829
N                 170
Correlation       0.93858
```

Other packages try to be more descriptive. It may be easier to find the results, but you may get less information from the output table.

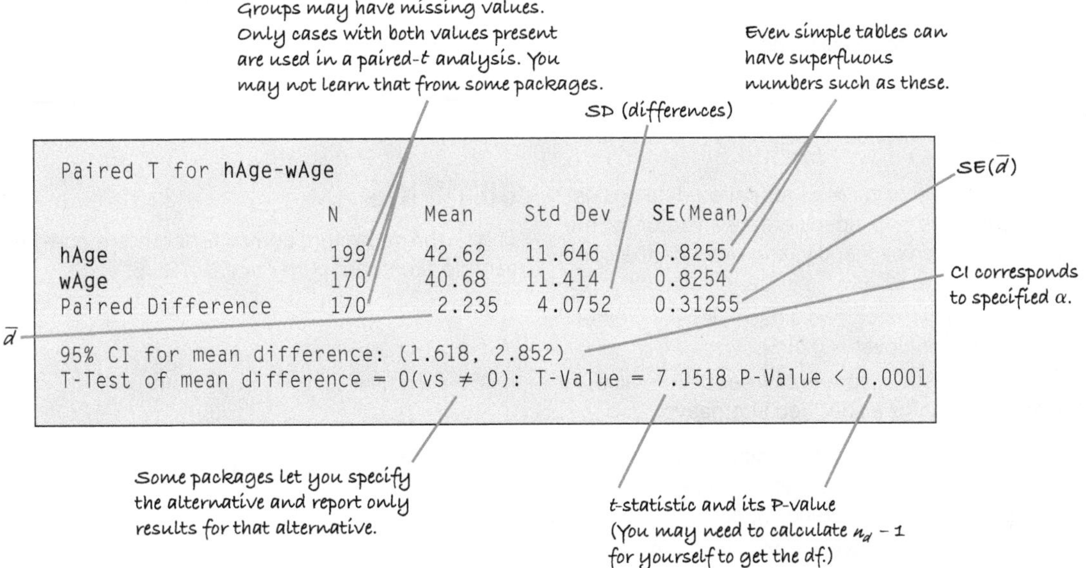

Groups may have missing values. Only cases with both values present are used in a paired-t analysis. You may not learn that from some packages.
Even simple tables can have superfluous numbers such as these.
SD (differences)
SE($\bar{d}$)
Paired T for hAge-wAge
N Mean Std Dev SE(Mean)
hAge 199 42.62 11.646 0.8255
wAge 170 40.68 11.414 0.8254
Paired Difference 170 2.235 4.0752 0.31255
CI corresponds to specified α.
$\bar{d}$
95% CI for mean difference: (1.618, 2.852)
T-Test of mean difference = 0(vs ≠ 0): T-Value = 7.1518 P-Value < 0.0001
Some packages let you specify the alternative and report only results for that alternative.
t-statistic and its P-value (You may need to calculate $n_d - 1$ for yourself to get the df.)

Computers make it easy to examine the boxplots of the two groups and the histogram of the differences—both important steps. Some programs offer a scatterplot of the two variables. That can be helpful. In terms of the scatterplot, a paired *t*-test is about whether the points tend to be above or below the 45° line $y - x$. (Note, though, that pairing says nothing about whether the scatterplot should be straight. That doesn't matter for our *t*-methods.)

EXCEL

- In Excel 2003 and earlier, select **Data Analysis** from the **Tools** menu.
- In Excel 2007, select **Data Analysis** from the **Analysis** group on the **Data** tab.
- From the **Data Analysis** menu, choose ***t*-test: paired two-sample for Means**. Fill in the cell ranges for the two groups, the hypothesized difference, and the alpha level.

COMMENTS

Excel expects the two groups to be in separate cell ranges.

Warning: Do not compute this test in Excel without checking for missing values. If there are any missing values (empty cells), Excel will usually give a wrong answer. Excel compacts each list, pushing values up to cover the missing cells, and then checks only that it has the same number of values in each list. The result is mismatched pairs and an entirely wrong analysis.

JMP

- From the **Analyze** menu, select **Matched Pairs**.
- Specify the columns holding the two groups in the **Y Paired Response** dialogue.
- Click **OK**.

For power and sample size calculations, see Chapter 20.

MINITAB

- From the **Stat** menu, choose the **Basic Statistics** submenu. From that menu, choose **Paired t. . .**
- Then, fill in the dialogue.

For power and sample size calculations, see Chapter 20.

COMMENTS

Minitab takes "First sample" minus "Second sample."

R

To test the hypothesis that $\mu_1 = \mu_2$ for paired data against an alternative (default is two-sided), create a vector for the data of each group, such as x and y, and produce the confidence interval (default is 95%):

- **t.test**(x,y, alternative = c("two.sided", "less", "greater", paired = TRUE, conf.level = 0.095)

will produce the *t*-statistic, degrees of freedom, P-value, and confidence interval for a specified alternative.

In package {pwr}, to perform a sample size or power calculation for a matched pairs *t*-test, either recognize this as just a one-sample *t* test (on differences), or use:

- pwr.t.test(n= , d= , sig.level= , power= , type= "paired") returns **one** argument that is not specified in the function.

COMMENTS

This is the same function as for both the one- and two-sample *t*-test, but with paired=TRUE.

SPSS

- From the **Analyze** menu, choose the **Compare Means** submenu.
- From that, choose the **Paired-Samples *t*-test** command.
- Select pairs of variables to compare, and click the arrow to add them to the selection box.

For power and sample size calculations, you will need the IBM SPSS SamplePower add-on.

COMMENTS

You can compare several pairs of variables at once. Options include the choice to exclude cases missing in any pair from all tests.

STATCRUNCH

To do inference for the mean of paired differences:

- Click on **Stat**.
- Choose **T Statistics » Paired**.
- Choose the **Column** for each variable.
- Check **Save differences** so you can look at a histogram to be sure the Nearly Normal condition is satisfied.
- Click on **Next**.
- Indicate **Hypothesis Test,** then enter the hypothesized **Null mean** difference (usually 0), and choose the **Alternative** hypothesis.

OR

- Indicate **Confidence Interval**, and then enter the **Level** of confidence.
- Click on **Calculate**.

TI-83/84 PLUS

If the data are stored in two lists, say, L1 and L2, create a list of the differences:

L1 − L2 → L3. (The arrow is the STO button.)

- Since inference for paired differences uses one-sample *t*-procedures, select **2:T-Test** or **8:TInterval** from the **STAT TESTS** menu.
- Specify as your data the list of differences you just created in L3, and apply the procedure.

Exercises

1. More eggs? Can a food additive increase egg production? Agricultural researchers want to design an experiment to find out. They have 100 hens available. They have two kinds of feed: the regular feed and the new feed with the additive. They plan to run their experiment for a month, recording the number of eggs each hen produces.
   a) Design an experiment that will require a two-sample *t*-procedure to analyze the results.
   b) Design an experiment that will require a matched-pairs *t*-procedure to analyze the results.
   c) Which experiment would you consider the stronger design? Why?

2. MTV Some students do homework with the television on. (Anyone come to mind?) Some researchers want to see if people can work as effectively with distraction as they can without it. The researchers will time some volunteers to see how long it takes them to complete some relatively easy crossword puzzles. During some of the trials, the room will be quiet; during other trials in the same room, a television will be on, tuned to MTV.
   a) Design an experiment that will require a two-sample *t*-procedure to analyze the results.
   b) Design an experiment that will require a matched-pairs *t*-procedure to analyze the results.
   c) Which experiment would you consider the stronger design? Why?

3. Sex sells? Ads for many products use sexual images to try to attract attention to the product. But do these ads bring people's attention to the item that was being advertised? We want to design an experiment to see if the presence of sexual images in an advertisement affects people's ability to remember the product.
   a) Describe an experimental design that would require a matched-pairs *t*-procedure to analyze the results.
   b) Describe an experimental design that would require an independent sample procedure to analyze the results.

4. Freshman 15 Many people believe that students tend to gain weight during first year at university (allegedly 15 pounds, on average). Suppose we plan to conduct a study to see if this is true.
   a) Describe a study design that would require a matched-pairs *t*-procedure to analyze the results.
   b) Describe a study design that would require a two-sample *t*-procedure to analyze the results.

5. Diabetes and weight loss Researchers at the University Obesity Research Centre in Australia were interested in the effect on Type II diabetes of surgically induced weight loss. One group of 30 obese patients received conventional therapy with a focus on weight loss, while laparoscopic adjustable gastric banding (a surgical intervention) with conventional diabetes care was used for another group of 30 obese patients. One concern was weight loss, as researchers did find a strong association between diabetes remission and weight loss. Weights of all subjects were recorded at the start (baseline) and after two years.
   a) What procedure would you use to test if subjects given the surgical treatment weigh less on average after two years as compared with baseline (at entry)? State your null hypothesis. What would be the distribution of the test statistic? The P-value was < 0.001. What can you conclude?
   b) What procedure would you use to test if surgical treatment produces greater weight loss, on average, than conventional treatment? State your null hypothesis. What would be the distribution of the test statistic? The P-value was < 0.001. What can you conclude?

T 6. Rain Simpson, Alsen, and Eden (*Technometrics* 1975) reported the results of trials in which clouds were seeded and the amount of rainfall recorded. The authors reported on 26 seeded and 26 unseeded clouds in order of the amount of rainfall, largest amount first. Here are two possible tests to study the question of whether cloud seeding works. Which test is appropriate for these data? Explain your choice. Using the test you select, state your conclusion.

   Paired *t*-Test of $\mu(1 - 2)$
   Mean of Paired Differences $= -277.39615$
   t-Statistic $= -3.641$ w/25 df
   $P = 0.0012$

   2-Sample *t*-Test of $\mu 1 - \mu 2$
   Difference Between Means $= -277.4$
   t-Statistic $= -1.998$ w/33 df
   $P = 0.0538$

   a) Which of these tests is appropriate for these data? Explain.
   b) Using the test you selected, state your conclusion.

T 7. Friday the 13th, I The *British Medical Journal* published an article titled, "Is Friday the 13th Bad for Your Health?" Researchers in Britain examined how Friday the 13th affects human behaviour. One question was whether people tend to stay at home more on that day. The data below are the number of cars passing Junctions 9 and 10 on the M25 motorway for consecutive Fridays (the 6th and 13th) for five different periods.

Year	Month	6th	13th
1990	July	134 012	132 908
1991	September	133 732	131 843
1991	December	121 139	118 723
1992	March	124 631	120 249
1992	November	117 584	117 263

Here are summaries of two possible analyses:

Paired t-Test of mu(1 − 2) = 0 vs. mu(1 − 2) > 0

Mean of Paired Differences: 2022.4

t-Statistic = 2.9377 w/4 df

P = 0.0212

2-Sample t-Test of mu1 = mu2 vs. mu1 > mu2

Difference Between Means: 2022.4

t-Statistic = 0.4273 w/7.998 df

P = 0.3402

a) Which of the tests is appropriate for these data? Explain.
b) Using the test you selected, state your conclusion.
c) Are the assumptions and conditions for inference met?

8. Friday the 13th, II The researchers in Exercise 7 also examined the number of people admitted to emergency rooms for vehicular accidents on 12 Friday evenings (six each on the 6th and 13th).

Year	Month	6th	13th
1989	October	9	13
1990	July	6	12
1991	September	11	14
1991	December	11	10
1992	March	3	4
1992	November	5	12

Based on these data, is there evidence that more people are admitted, on average, on Friday the 13th? Here are two possible analyses of the data:

Paired t-Test of mu(1 − 2) = 0 vs. mu(1 − 2) < 0

Mean of Paired Differences = 3.333

t-Statistic = 2.7116 w/5 df

P = 0.0211

2-Sample t-Test of mu1 = mu2 vs. mu1 < mu2

Difference Between Means = 3.333

t-Statistic = 1.6644 w/9.940 df

P = 0.0636

a) Which of these tests is appropriate for these data? Explain.
b) Using the test you selected, state your conclusion.
c) Are the assumptions and conditions for inference met?

9. Online insurance I After seeing countless commercials claiming people can get cheaper car insurance from an online company, a local insurance agent was concerned that he might lose some customers. To investigate, he randomly selected profiles (type of car, coverage, driving record, etc.) for 10 of his clients and checked online price quotes for their policies. The comparisons are shown in the table. His statistical software produced the following summaries (where *PriceDiff* = *Local* – *Online*):

Variable	Count	Mean	StdDev
Local	10	799.200	229.281
Online	10	753.300	256.267
PriceDiff	10	45.9000	175.663

Local	Online	PriceDiff
568	391	177
872	602	270
451	488	−37
1229	903	326
605	677	−72
1021	1270	−249
783	703	80
844	789	55
907	1008	−101
712	702	10

At first, the insurance agent wondered whether there was some kind of mistake in this output. He thought the Pythagorean Theorem of Statistics should work for finding the standard deviation of the price differences—in other words, that $SD(Local - Online) = \sqrt{SD^2(Local) + SD^2(Online)}$.

But when he checked, he found that $\sqrt{(229.281)^2 + (256.267)^2} = 343.864$, not 175.663 as given by the software. Tell him where his mistake is.

10. Windy, part I To select the site for a wind turbine, wind speeds were recorded at several potential sites evtery six hours for a year. Two sites not far from each other looked good. Each had a mean wind speed high enough to qualify, but we should choose the site with a higher average daily wind speed. Because the sites are near each other and the wind speeds were recorded at the same times, we should view the speeds as paired. Here are the summaries of the speeds (in kilometres per hour):

Variable	Count	Mean	StdDev
site2	1114	11.993	5.771
site4	1114	11.664	5.505
site2 – site4	1114	0.328	4.105

Is there a mistake in this output? Why doesn't the Pythagorean Theorem of Statistics work here? In other words, shouldn't

$$SD(site2 - site4) = \sqrt{SD^2(site2) + SD^2(site4)}?$$

But $\sqrt{(5.771)^2 + (5.505)^2} = 7.976$, not 4.105 as given by the software. Explain why this happened.

11. Online insurance II In Exercise 9, we saw summary statistics for 10 drivers' car insurance premiums quoted by a local agent and an online company. Here are displays for each company's quotes and for the difference (*Local* – *Online*):

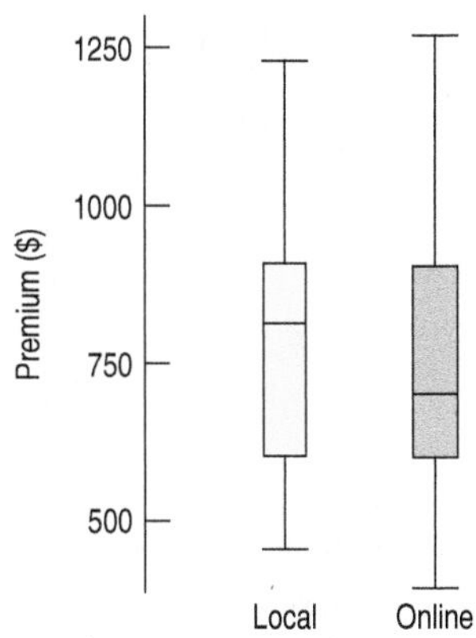

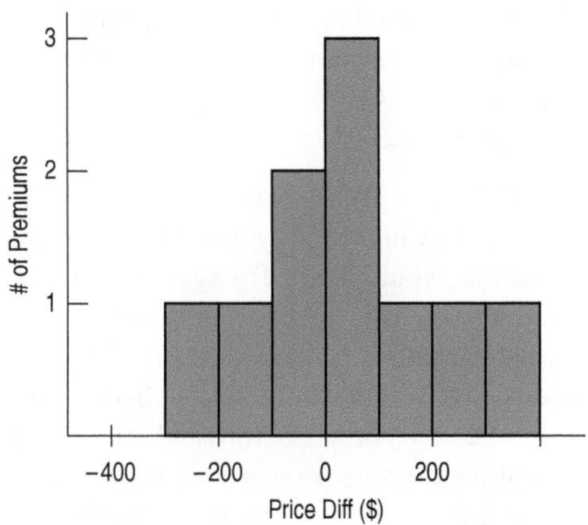

a) Which of the summaries would help you decide whether the online company offers cheaper insurance? Why?

b) The standard deviation of *PriceDiff* is quite a bit smaller than the standard deviation of prices quoted by either the local or online companies. Discuss why.

c) Using the information you have, discuss the assumptions and conditions for inference with these data.

12. Windy, part II In Exercise 10, we saw summary statistics for wind speeds at two sites near each other, both being considered as locations for a wind turbine. The data, recorded every six hours for a year, showed each of the sites had a mean wind speed high enough to qualify, but how can we tell which site is best? Here are some displays:

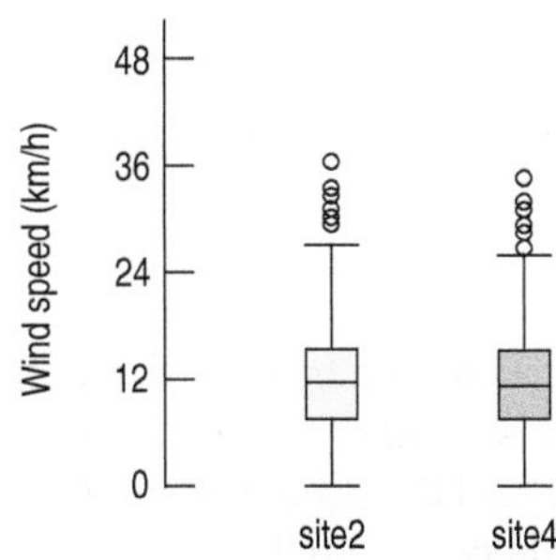

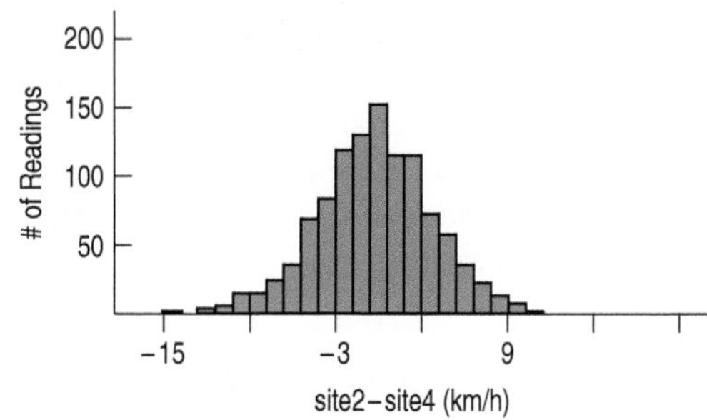

a) The boxplots show outliers for each site, yet the histogram shows none. Discuss why.

b) Which of the summaries would you use to select between these sites? Why?

c) Using the information you have, discuss the assumptions and conditions for paired *t* inference for these data. (*Hint:* Think hard about the independence assumption in particular.)

13. Online insurance III Exercises 9 and 11 give summaries and displays for car insurance premiums quoted by a local agent and an online company. Test an appropriate hypothesis to see if there is evidence that drivers might save money by switching to the online company.

14. Windy, part III Exercises 10 and 12 give summaries and displays for two potential sites for a wind turbine. Test an appropriate hypothesis to see if there is evidence that either of these sites has a higher average wind speed.

15. Temperatures The table below gives the average high temperatures in January and July for several European cities. Write a 90% confidence interval for the mean temperature difference between summer and winter in Europe. Be sure to check conditions for inference, and clearly explain what your interval means.

	Mean High Temperatures (°C)	
City	Jan.	July
Vienna	1	24
Copenhagen	2	22
Paris	6	24
Berlin	2	23
Athens	12	32
Rome	12	31
Amsterdam	4	21
Madrid	8	31
London	7	23
Edinburgh	6	18
Moscow	–6	24
Belgrade	3	29

16. NY Marathon 2011 The table below shows the winning times (in minutes) for men and women in the New York City Marathon between 1978 and 2011. Assuming that performances in the Big Apple resemble performances elsewhere, we can think of these data as a sample of performance in marathon competitions. Create a 90% confidence interval for the mean difference in winning times for male and female marathon competitors. (www.tcsnycmarathon.org)

Year	Men	Women	Year	Men	Women
1978	132.2	152.5	1995	131.0	148.1
1979	131.7	147.6	1996	129.9	148.3
1980	129.7	145.7	1997	128.2	148.7
1981	128.2	145.5	1998	128.8	145.3
1982	129.5	147.2	1999	129.2	145.1
1983	129.0	147.0	2000	130.2	145.8
1984	134.9	149.5	2001	127.7	144.4
1985	131.6	148.6	2002	128.1	145.9
1986	131.1	148.1	2003	130.5	142.5
1987	131.0	150.3	2004	129.5	143.2
1988	128.3	148.1	2005	129.5	144.7
1989	128.0	145.5	2006	130.0	145.1
1990	132.7	150.8	2007	129.1	143.2
1991	129.5	147.5	2008	128.7	143.9
1992	129.5	144.7	2009	129.3	148.9
1993	130.1	146.4	2010	128.3	148.3
1994	131.4	147.6	2011	125.1	143.3

T **17. Push-ups** Every year the students at Central Secondary School take a physical fitness test during their gym classes. One component of the test asks them to do as many push-ups as they can. Results for one class are shown below, separately for boys and girls. Assuming that students at Central are assigned to gym classes at random, create a 90% confidence interval for how many more push-ups boys can do than girls, on average, at that school.

Boys	17	27	31	17	25	32	28	23	25	16	11	34
Girls	24	7	14	16	2	15	19	25	10	27	31	8

T **18. Brain waves** An experiment was performed to see whether sensory deprivation over an extended period of time has any effect on the alpha-wave patterns produced by the brain. To determine this, 20 subjects, inmates in a Canadian prison, were randomly split into two groups. Members of one group were placed in solitary confinement. Those in the other group were allowed to remain in their own cells. Seven days later, alpha-wave frequencies were measured for all subjects, as shown in the following table.[4]

[4]P. Gendreau et al., 1972, "Changes in EEG alpha frequency and evoked response latency during solitary confinement," *Journal of Abnormal Psychology* 79: 54–59.

Nonconfined	Confined
10.7	9.6
10.7	10.4
10.4	9.7
10.9	10.3
10.5	9.2
10.3	9.3
9.6	9.9
11.1	9.5
11.2	9.0
10.4	10.9

a) What are the null and alternative hypotheses? Be sure to define all the terms and symbols you use.
b) Are the assumptions necessary for inference met?
c) Perform the appropriate test, indicating the formula you used, the calculated value of the test statistic, and the P-value.
d) State your conclusion.

T **19. Job satisfaction** (When you first read about this exercise break plan in Chapter 21, you did not have an inference method that would work. Try again now.) A company institutes an exercise break for its workers to see if it will improve job satisfaction, as measured by a questionnaire that assesses workers' satisfaction. Scores for 10 randomly selected workers before and after the implementation of the exercise program are shown in the table.

a) Identify the procedure you would use to assess the effectiveness of the exercise program, and check to see if the conditions allow the use of that procedure.
b) Test an appropriate hypothesis and state your conclusion.
c) If your conclusion turns out to be incorrect, what kind of error would you be making?
*d) Use the sign test to test the appropriate hypothesis. Do your conclusions change from those in part b)?

Worker Number	Job Satisfaction Index Before	After
1	34	33
2	28	36
3	29	50
4	45	41
5	26	37
6	27	41
7	24	39
8	15	21
9	15	20
10	27	37

T **20. Summer school** (When you first read about the summer school issue in Chapter 21, you did not have an inference method that would work. Try again now.) Having done poorly on their Math final exams in June, six students repeat the course in summer school and take another exam in August.

June	54	49	68	66	62	62
Aug.	50	65	74	64	68	72

a) If we consider these students to be representative of all students who might attend this summer school in other years, do these results provide evidence that the program is worthwhile?
b) Your conclusion, of course, may be incorrect. If so, which type of error would you be making?

T 21. **Yogourt** Do these data suggest that there is a significant difference in calories between servings of strawberry and vanilla yogourt? Test an appropriate hypothesis and state your conclusion. Don't forget to check assumptions and conditions!

	Calories per Serving	
Brand	**Strawberry**	**Vanilla**
Astro	210	200
Breyer's Lowfat	220	220
Columbo	220	180
Dannon Light'n Fit	120	120
Dannon Lowfat	210	230
Dannon la Crème	140	140
Great Value	180	80
La Yogurt	170	160
Mountain High	200	170
Stonyfield Farm	100	120
Yoplait Custard	190	190
Yoplait Light	100	100

T 22. **Gasoline** Many drivers of cars that can run on regular gas actually buy premium in the belief that they will get better gas mileage. To test that belief, we use 10 cars from a company fleet in which all the cars run on regular gas. Each car is filled first with either regular or premium gasoline, decided by a coin toss, and the mileage for that tank is recorded. Then, the mileage is recorded again for the same cars for a tank of the other kind of gasoline. We don't let the drivers know about this experiment.

Here are the results (litres per 100 km):

Car #	1	2	3	4	5	6	7	8	9	10
Regular	14.7	11.8	11.5	10.7	10.2	10.7	8.7	9.4	8.7	8.4
Premium	12.4	10.7	9.8	9.8	9.4	9.4	9.0	9.0	8.4	7.4

a) Is there evidence that cars get significantly better fuel economy with premium gasoline?
b) How big might that difference be? Check a 90% confidence interval.
c) Even if the difference is significant, why might the company choose to stick with regular gasoline?

d) Suppose you had done a "bad thing." (We're sure you didn't.) Suppose you had mistakenly treated these data as two independent samples instead of matched pairs. What would the significance test have found? Carefully explain why the results are so different.
e) *Use the sign test to test the appropriate hypothesis. Do your conclusions change from those in part a)?

T 23. **Braking** In a test of braking performance, a tire manufacturer measured the stopping distance for one of its tire models. On a test track, a car made repeated stops from a speed of 100 km per hour. The test was run on both dry and wet pavement, with results as shown in the table. (Note that actual *braking distance,* which takes into account the driver's reaction time, is much longer, typically nearly 300 feet, or 91.5 metres, at 100 km/h!)
a) Write a 95% confidence interval for the mean dry pavement stopping distance. Be sure to check the appropriate assumptions and conditions, and explain what your interval means.
b) Write a 95% confidence interval for the mean increase in stopping distance due to wet pavement. Be sure to check the appropriate assumptions and conditions, and explain what your interval means.

Stopping Distance (ft)	
Dry Pavement	**Wet Pavement**
145	211
152	191
141	220
143	207
131	198
148	208
126	206
140	177
135	183
133	223

T 24. **Braking, test 2** For another test of the tires in Exercise 23, the company tried them on 10 different cars, recording the stopping distance for each car on both wet and dry pavement. The following table shows the results.

	Stopping Distance (ft)	
Car	**Dry Pavement**	**Wet Pavement**
1	150	201
2	147	220
3	136	192
4	134	146
5	130	182
6	134	173
7	134	202
8	128	180
9	136	192
10	158	206

a) Write a 95% confidence interval for the mean dry pavement stopping distance. Be sure to check the appropriate assumptions and conditions, and explain what your interval means.
b) Write a 95% confidence interval for the mean increase in stopping distance due to wet pavement. Be sure to check the appropriate assumptions and conditions, and explain what your interval means.

T **25. Math and gender 2009** Below are mean PISA (Programme for International Student Assessment) math scores for very large national samples of 15-year-old male and female students in a random sample of the Organisation for Economic Co-operation and Development (OECD) and various other industrialized nations.[5]

Country	Male Average	Female Average
Singapore	565	559
Norway	500	495
Korea	548	544
Switzerland	544	524
Sweden	493	495
Australia	519	509
Brazil	394	379
Argentina	394	383
Liechtenstein	547	523
Trinidad and Tobago	410	418
Italy	490	475
Tunisia	378	366
Dubai (UAE)	454	451
Hungary	496	484
Colombia	398	366
Romania	429	425

a) Create a 90% confidence interval for the mean difference in average math scores between genders. Be sure to justify the procedure that you use.
b) What does your confidence interval say about the claim that there is no difference between genders? About the claim that the mean difference is at least 10 points, with males scoring higher? (Actually, the biggest gender gap was in reading scores, with females scoring higher on average in all OECD countries, and more than 50 points higher in some countries.)
c) Is your confidence interval an estimate of the average difference between males and female scores for students aged 15 in all the participating countries? Or is it an estimate of something else? Explain.

[5]Based on data from OECD PISA 2009 online database, http://pisa2009.acer.edu.au, accessed on Feb 22, 2013.

T **26. Sex sells, part II** In Exercise 3, you considered the question of whether sexual images in ads affected people's abilities to remember the item being advertised. To investigate, a group of Statistics students cut ads out of magazines. They were careful to find two ads for each of 10 similar items, one with a sexual image and one without. They arranged the ads in random order and had 39 subjects look at them for one minute. Then, they asked the subjects to list as many of the products as they could remember. The following table shows their data. Is there evidence that the sexual images mattered?

	Ads Remembered			Ads Remembered	
Subject Number	Sexual Image	No Sex	Subject Number	Sexual Image	No Sex
1	2	2	21	2	3
2	6	7	22	4	2
3	3	1	23	3	3
4	6	5	24	5	3
5	1	0	25	4	5
6	3	3	26	2	4
7	3	5	27	2	2
8	7	4	28	2	4
9	3	7	29	7	6
10	5	4	30	6	7
11	1	3	31	4	3
12	3	2	32	4	5
13	6	3	33	3	0
14	7	4	34	4	3
15	3	2	35	2	3
16	7	4	36	3	3
17	4	4	37	5	5
18	1	3	38	3	4
19	5	5	39	4	3
20	2	2			

T **27. Strike three** Advertisements for an instructional video claim that the techniques will improve the ability of young amateur baseball pitchers to throw strikes and that, after undergoing the training, players will be able to throw strikes on at least 60% of their pitches. To test this claim, we have 20 players throw 50 pitches each, and we record the number of strikes. After the players participate in the training program, we repeat the test. The table on the next page shows the number of strikes each player threw before and after the training.
a) Is there evidence that after training players can throw strikes more than 60% of the time?
b) Is there evidence that the training is effective in improving a player's ability to throw strikes?
*c) Use the sign test to test the appropriate hypothesis. Do your conclusions change from those in part b)?

Number of Strikes (out of 50)		Number of Strikes (out of 50)	
Before	**After**	**Before**	**After**
28	35	33	33
29	36	33	35
30	32	34	32
32	28	34	30
32	30	34	33
32	31	35	34
32	32	36	37
32	34	36	33
32	35	37	35
33	36	37	32

T **28. Freshman 15, revisited** In Exercise 4, you thought about how to design a study to see if it's true that students tend to gain weight during their first year at university. Well, Cornell Professor of Nutrition David Levitsky did just that. He recruited students from two large sections of an introductory health course. Although they were volunteers, they appeared to match the other first-year students in terms of demographic variables, such as sex and ethnicity. The students were weighed during the first week of the semester, then again 12 weeks later. Based on Professor Levitsky's data, estimate the mean weight gain in first-year students with a confidence interval. What can you conclude? Is there a serious problem? (Weights are in pounds.)

Subject Number	Initial Weight	Terminal Weight	Subject Number	Initial Weight	Terminal Weight
1	171	168	35	148	150
2	110	111	36	164	165
3	134	136	37	137	138
4	115	119	38	198	201
5	150	155	39	122	124
6	104	106	40	146	146
7	142	148	41	150	151
8	120	124	42	187	192
9	144	148	43	94	96
10	156	154	44	105	105
11	114	114	45	127	130
12	121	123	46	142	144
13	122	126	47	140	143
14	120	115	48	107	107
15	115	118	49	104	105
16	110	113	50	111	112
17	142	146	51	160	162
18	127	127	52	134	134
19	102	105	53	151	151
20	125	125	54	127	130
21	157	158	55	106	108
22	119	126	56	185	188
23	113	114	57	125	128
24	120	128	58	125	126
25	135	139	59	155	158
26	148	150	60	118	120
27	110	112	61	149	150
28	160	163	62	149	149
29	220	224	63	122	121
30	132	133	64	155	158
31	145	147	65	160	161
32	141	141	66	115	119
33	158	160	67	167	170
34	135	134	68	131	131

T **29. Wheelchair marathon 2010** The Boston Marathon has had a wheelchair division since 1977. Who do you think is typically faster, the men's marathon winner on foot or the women's wheelchair marathon winner? Because the conditions differ from year to year, and speeds have improved over the years, it seems best to treat these as paired measurements. Here are summary statistics for the pairwise differences in finishing time (in minutes): (www.boston.com/sports/marathon/history/champions/)

Andy Crawford/Dorling Kindersley, Ltd.

Summary of wheelchrF – runM

N = 34

Mean = –3.57

SD = 35.083

a) Comment on the assumptions and conditions.

b) Assuming that these times are representative of such races and the differences appeared acceptable for inference, construct and interpret a 95% confidence interval for the mean difference in finishing times.

c) Would a hypothesis test at $\alpha = 0.05$ reject the null hypothesis of no difference? What conclusion would you draw?

30. Marathon start-up years 2010 When we considered the Boston Marathon in the previous exercise, we were unable to check the Nearly Normal Condition. Here's a histogram of the differences:

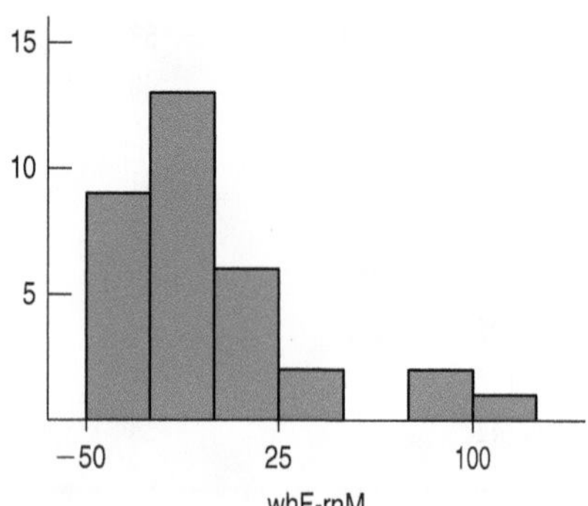

Those three large differences are the first three years of wheelchair competition: 1977, 1978, and 1979. Often the start-up years of new events are different; later on, more athletes train and compete. If we omit those three years, the summary statistics change as follows:

Summary of wheelchrF – runM

N = 31

Mean = –12.794

SD = 18.64

a) Comment on the assumptions and conditions.
b) Assuming that these times are representative of such races, construct and interpret a 95% confidence interval for the mean difference in finishing time.
c) Would a hypothesis test at $\alpha = 0.05$ reject the null hypothesis of no difference? What conclusion would you draw?

31. BST In the U.S., some dairy cows receive injections of BST, a hormone intended to spur greater milk production. After the first injection, a test herd of 60 Ayrshire cows increased their mean daily production from 47 pounds to 61 pounds of milk. The standard deviation of the increases was 5.2 pounds. We want to estimate the mean increase a farmer could expect in his own cows.

a) Check the assumptions and conditions for inference.
b) Write a 95% confidence interval.
c) Explain what your interval means in this context.
d) Given the cost of BST, a farmer believes he cannot afford to use it unless he is sure of attaining at least a 25% increase in milk production. Based on your confidence interval, what advice would you give him?

32. BST II In the experiment about hormone injections in cows described in Exercise 31, a group of 52 Jersey cows increased average milk production from 43 pounds to 52 pounds per day, with a standard deviation of 4.8 pounds. Is this evidence that the hormone may be more effective in one breed than the other? Test an appropriate hypothesis and state your conclusion. Be sure to discuss any assumptions you make.

33. Sample size & margin of error For the chapter example about switching from a five-day to a four-day work week, how big should the sample have been to narrow the width of the reported confidence interval to 500 miles? Use the sample standard deviation given in the example as your standard deviation guess.

34. BST III Review Exercise 31 about BST injections for Ayrshire cows. Approximately how big would the sample have to be in order to narrow the width of the reported confidence interval to about 1.00 pounds?

35. Sample size & power For the chapter example about switching from a five-day to a four-day work week, *use software* to determine how big the sample size should have been in order to attain a power of 0.90 for the test of hypothesis (at the 5% level) when there is an actual annual mean mileage reduction of 500 miles. Use the sample standard deviation given in the example as your estimate of the true standard deviation. Explain in words what the probability of 0.90 represents here.

36. BST IV Review Exercise 31 about BST injections for Ayrshire cows. *Use software* to estimate how big the sample should have been if we wanted to be 90% certain of correctly detecting an increase in average milk production when in fact the true mean increase is 2.0 pounds. Assume this to be a one-sided hypothesis test, at the 5% level, and use the sample standard deviation calculated in Exercise 31 as an estimate of the true standard deviation.

Just Checking ANSWERS

1. These are independent groups sampled at random, so use a two-sample *t* confidence interval to estimate the size of the difference.
2. There is only one sample. Use a one-sample *t*-interval.
3. A brother and sister from the same family represent a matched pair. The question calls for a paired *t*-test.
4. **a.** A before-and-after study calls for paired *t*-methods. To estimate the loss, find a confidence interval for the before–after differences.
   b. The two treatment groups were assigned randomly, so they are independent. Use a two-sample *t*-test to assess whether the mean weight losses differ.
5. Most likely, couples would discuss the band or even decide to go to the club because they both like a particular band. While we may not know to what degree their responses may be related, the sampling design has us selecting the individuals in linked pairs, so use paired-*t*.

chapter

23 Comparing Counts

Life is the art of drawing sufficient conclusions from insufficient premises.

—*Samuel Butler*

Where are we going?

Is your favourite colour related to how much education you've had? A survey found a higher percentage of adults with only a high school education named blue and a lower percentage said red compared to adults with more education. Could this be just random fluctuation, or is the distribution of colour preference different for different education levels? We saw tables of counts and percentages in Chapter 2. In this chapter, we'll see how to test the strength of the patterns we saw in those tables.

Who	Executives of Fortune 400 companies
What	Zodiac birth sign
Why	Maybe the researcher was a Gemini and naturally curious?

A S *Activity:* **Children at Risk.** See how a contingency table helps us understand the different risks to which an incident exposed children.

Does your zodiac sign predict how successful you will be later in life? *Fortune* magazine collected the zodiac signs of 256 heads of the largest 400 companies. Here are the number of births for each sign:

Births	Sign	Births	Sign
23	Aries	18	Libra
20	Taurus	21	Scorpio
18	Gemini	19	Sagittarius
23	Cancer	22	Capricorn
20	Leo	24	Aquarius
19	Virgo	29	Pisces

Birth totals by sign for 256 *Fortune* 400 executives.

We can see some variation in the number of births per sign, and there *are* more Pisces, but is that enough to claim that successful people are more likely to be born under some signs than others?

23.1 Goodness-of-Fit

If births were distributed uniformly across the year, we would expect about 1/12 of them to occur under each sign of the zodiac. That suggests 256/12, or about 21.3 births per sign. How closely do the observed proportions of births per sign fit this simple "null" model?

A hypothesis test to address this question is called a test of "**goodness-of-fit**." The name suggests a certain badness-of-grammar, but it is quite standard. After all, we are asking whether the model that births are uniformly distributed over the signs fits the data good . . . er, well. Goodness-of-fit involves testing a hypothesis. We have specified a model for the distribution and want to know whether it fits. A confidence interval wouldn't make much sense because we are not estimating a single parameter.

If the question were about only one astrological sign (for example, "Are executives more likely to be Pisces?"[1]), we could use a one-proportion z-test and ask if the true proportion of executives with that sign is equal to 1/12. However, here we have 12 hypothesized proportions, one for each sign. We need a test that considers all of them simultaneously and gives an overall idea of whether the observed distribution differs from the hypothesized one.

[1]A question actually asked us by someone who was undoubtedly a Pisces.

"All creatures have their determined time for giving birth and carrying fetus, only a man is born all year long, not in determined time, one in the seventh month, the other in the eighth, and so on till the beginning of the eleventh month."

—*Aristotle*

For Example FINDING EXPECTED COUNTS

Birth month may not be related to success as a CEO, but what about on the rink? It has been proposed by some researchers that children who are the older ones in their class at school naturally perform better in sports, and that these children then get more coaching and encouragement. Could that make a difference in who makes it to the professional level in sports?

Canada has contributed 71.7% of the NHL players from the start of the league until the end of the 2009 season (4594 players). We also have demographic statistics for the population birth percentages for the years 1970–2005, excepting '72, '91 and '98.[2] We'll consider the birth months of the players born in those years, and test whether the observed distribution of hockey players' birth months shows just random fluctuations or whether it represents a real deviation from the national pattern.

Month	Hockey Player Count	National Birth %	Month	Hockey Player Count	National Birth %
1	127	8.0	7	88	8.7
2	116	7.6	8	74	8.5
3	117	8.7	9	84	8.5
4	127	8.6	10	79	8.2
5	118	8.9	11	75	7.8
6	110	8.5	12	69	7.9
			Total	1184	

QUESTION: We need expected counts. How can we find them?

ANSWER: There are 1184 players in this set of data. I'd expect 8% of them to have been born in January, and 1184(0.08) = 94.72. I won't round off, because expected "counts" needn't be integers. Multiplying 1184 by each of the birth percentages gives the expected counts shown in the table in the margin.

Month	Expected	Month	Expected
1	94.72	7	103.01
2	89.98	8	100.64
3	103.01	9	100.64
4	101.82	10	97.09
5	105.38	11	92.35
6	100.64	12	93.54

Assumptions and Conditions

These data are organized in tables, as we saw in Chapter 2, and the assumptions and conditions reflect that. Rather than displaying each individual's categorical response, we often work with just the summary counts for the various categories. In our example, we didn't display the birth signs of each of the 256 executives, only the totals for each sign.

Counted Data Condition: The data must be *counts* for the categories of a categorical variable. This might seem a simplistic—even silly—condition. But many kinds of values can be assigned to categories, and it is unfortunately common to find the methods of this chapter applied incorrectly to proportions, percentages, or measurements just because they happen to be organized in a contingency table. So, check to be sure you really have counts.

Independence Assumption

Independence Condition: The *individuals* who make up the *counts* in the cells should be independent of each other. The easiest case is when the individuals who are counted in the cells are sampled independently from some population. That's what we'd like to have if we want to draw conclusions about that population. Randomness can arise in other ways, though. For example, these *Fortune* 400 executives are not a random sample, but we might still think that their birth dates are randomly distributed throughout the year. If we want to generalize to a large population, we should check the randomization condition.

[2]Addona, Vittorio and Yates, Philip A. (2010) "A Closer Look at the Relative Age Effect in the National Hockey League." *Journal of Quantitative Analysis in Sports*: Vol. 6: Iss. 4, Article 9. Available at: www.bepress.com/jqas/vol6/iss4/9

Randomization Condition: The individuals who have been counted and whose counts are available for analysis should be a random sample from some population.

Sample Size Assumption

We must have enough data for the methods to work. We usually check the following:

Expected Cell Frequency Condition: We should expect to see at least five individuals in each cell.

The expected cell frequency condition sounds like—and is, in fact, quite similar to—the condition that np and nq be at least 10 when we tested proportions. In our astrology example, assuming equal births in each month leads us to expect 21.3 births per month, so the condition is easily met here.

For Example CHECKING ASSUMPTIONS AND CONDITIONS

RECAP: Are professional hockey players more likely to be born in some months than in others? We have observed and expected counts for the 1184 players born in select years from 1970–2005.

QUESTION: Are the assumptions and conditions met for performing a goodness-of-fit test?

ANSWER:

- ✓ **Counted Data Condition:** I have month-by-month counts of hockey player births.
- ✓ **Independence Assumption:** These births were independent.
- ✓ **Randomization Condition:** Although they are not a random sample, we can take these players to be representative of players past and future.
- ✓ **Expected Cell Frequency Condition:** The expected counts range from 89.98 to 105.38, all much greater than five.

It's okay to use these data for a goodness-of-fit test.

NOTATION ALERT

We compare the counts *observed* in each cell with the counts we *expect* to find. The method for finding the expected counts depends on the model.

Calculations

We have observed a count in each category from the data, and have an expected count for each category from the hypothesized proportions. Are the *differences* just natural sampling variability, or are they so large that they indicate something important? It's natural to look at the *differences* between these observed and expected counts, denoted $(Obs - Exp)$. We'd like to think about the total of the differences, but just adding them won't work because some differences are positive and others negative. We've been in this predicament before—once when we looked at deviations from the mean, and again when we dealt with residuals. In fact, these *are* residuals. They're just the differences between the observed data and the expected counts given by the (null) model. We handle these residuals in essentially the same way we did in regression: We square them. That gives us positive values and focuses attention on any cells with large differences from what we expected. Because the differences between observed and expected counts generally get larger the more data we have, we also need to get an idea of the *relative* sizes of the differences. To do that, we divide each squared difference by the expected count for that cell.

The test statistic, called the **chi-square** (or chi-squared) **statistic**, is found by adding up the sum of the squares of the deviations between the observed and expected counts divided by the expected counts:

$$\chi^2 = \sum_{all\ cells} \frac{(Obs - Exp)^2}{Exp}$$

NOTATION ALERT

The only use of the Greek letter χ in Statistics is to represent this statistic and the associated sampling distribution. This is another violation of our "rule" that Greek letters represent population parameters. Here we are using a Greek letter simply to name a family of distribution models and a statistic.

The **chi-square statistic** is denoted χ^2, where χ is the Greek letter chi (pronounced "ky" as in "sky"). It refers to a family of sampling distribution models we have not seen before called (remarkably enough) the *chi-square models*.

This family of models, like the Student's *t*-models, differs only in the number of degrees of freedom. The number of degrees of freedom for a goodness-of-fit test is *the number of categories* -1. For the zodiac example, we have 12 signs, so our χ^2 statistic has 11 degrees of freedom.

Chi-Square P-Values

The chi-square statistic is used only for testing hypotheses, not for constructing confidence intervals. If the observed counts don't match the expected, the statistic will be large. It can't be "too small." That would just mean that our model *really* fit the data well. So the chi-square test is always one-sided. If the calculated statistic value is large enough, we'll reject the null hypothesis. What could be simpler?

If you don't have technology handy, it's easy to read the χ^2 table (Table X in Appendix C).

A portion of Table X.

Right-Tail Probability		0.10	0.05	0.025	0.01	0.005
	df					
	1	2.706	3.841	5.024	6.635	7.879
Values of χ^2_α	2	4.605	5.991	7.378	9.210	10.597
	3	6.251	7.815	9.348	11.345	12.838
	4	7.779	9.488	11.143	13.277	14.860
	5	9.236	11.070	12.833	15.086	16.750
	6	10.645	12.592	14.449	16.812	18.548
	7	12.017	14.067	16.013	18.475	20.278
	8	13.362	15.507	17.535	20.090	21.955
	9	14.684	16.919	19.023	21.666	23.589
	10	15.987	18.307	20.483	23.209	25.188
	11	17.275	19.675	21.920	24.725	26.757
	12	18.549	21.026	23.337	26.217	28.300
	13	19.812	22.362	24.736	27.688	29.819
	14	21.064	23.685	26.119	29.141	31.319

The usual selected probability values are at the top of the columns. As with the *t*-tables, we have only selected probabilities, so the best we can do is to trap a P-value between two of the values in the table. Just find the row for the correct number of degrees of freedom, and read across to find where your calculated χ^2 value falls. Of course, technology can find an exact P-value, and that's usually what we'll see.

Even though its mechanics make use of only one side of the sampling model, the interpretation of a chi-square test is in some sense *many*-sided. With more than two proportions, there are many ways the null hypothesis can be wrong. By squaring the differences, we made all the deviations positive, whether our observed counts were higher or lower than expected. There's no direction to the rejection of the null model. All we know is that it doesn't fit.

For Example DOING A GOODNESS-OF-FIT TEST

RECAP: We're looking at data on the birth months of Canadian NHL players. We've checked the assumptions and conditions for performing a χ^2 test.

QUESTIONS: What are the hypotheses, and what does the test show?

ANSWER: H_0: The distribution of birth months for Canadian NHL players is the same as that for the general population.

H_A: The distribution of birth months for Canadian NHL players differs from that of the rest of the country.

Using technology, $\chi^2 = \sum \frac{(Obs - Exp)^2}{Exp} = 54.09$ with $df = 12 - 1 = 11$

$P(\chi^2_{11} \geq 54.09) < 0.005$.

Because of the small P-value, I reject H_0; there's evidence that birth months of Canadian NHL players have a different distribution from the rest of us.

Step-by-Step Example A CHI-SQUARE TEST FOR GOODNESS-OF-FIT

We have counts of 256 executives in 12 zodiac sign categories. The natural null hypothesis is that birth dates of executives are divided equally among all the zodiac signs. The test statistic looks at how closely the observed data match this idealized situation.

Question: Are zodiac signs of CEOs distributed uniformly?

THINK ➡ Plan State what you want to know. Identify the variables and check the *W*'s.

I want to know whether births of successful people are uniformly distributed across the signs of the zodiac. I have counts of 256 *Fortune* 400 executives, categorized by their birth sign.

Hypotheses State the null and alternative hypotheses. For χ^2 tests, it's usually easier to do that in words than in symbols.

H_0: Births are uniformly distributed over zodiac signs.[3]

H_A: Births are not uniformly distributed over zodiac signs.

Model Make a picture. The null hypothesis is that the frequencies are equal, so a bar chart (with a line at the hypothesized "equal" value) is a good display.

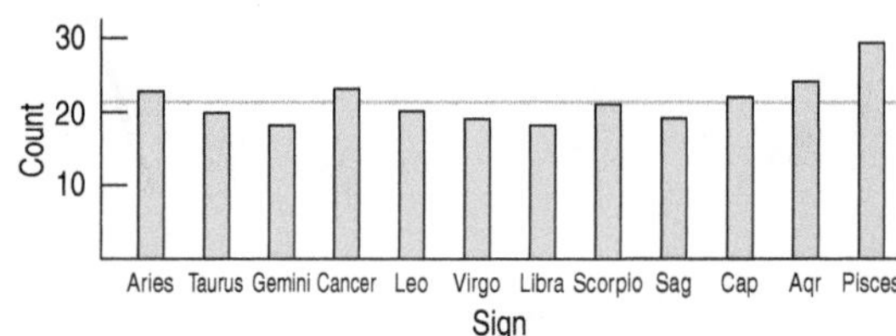

The bar chart shows some variation from sign to sign, and Pisces is the most frequent. But it is hard to tell whether the variation is more than I'd expect from random variation.

Think about the assumptions and check the conditions.

- ✓ **Counted Data Condition:** I have counts of the number of executives in 12 categories.
- ✓ **Independence Assumption:** The birth dates of executives should be independent of each other.
- ✓ **Randomization Condition:** This is a convenience sample of executives, but there's no reason to suspect bias.
- ✓ **Expected Cell Frequency Condition:** The null hypothesis expects that 1/12 of the 256 births, or 21.333, should occur in each sign. These expected values are all at least five, so the condition is satisfied.

[3]It may seem that we have broken our rule of thumb that null hypotheses should specify parameter values. If you want to get formal about it, the null hypothesis is that

$$p_{\text{Aries}} = p_{\text{Taurus}} = \cdots = p_{\text{Pisces}} = 1/12.$$

That is, we hypothesize that the true proportions of births of CEOs under each sign are equal. The role of the null hypothesis is to specify the model so that we can compute the test statistic. That's what this one does.

Specify the sampling distribution model.
Name the test you will use.

The conditions are satisfied, so I'll use a χ^2 model with $12 - 1 = 11$ degrees of freedom and do a chi-square goodness-of-fit test.

SHOW ➡ Mechanics Each cell contributes an $\frac{(Obs - Exp)^2}{Exp}$ value to the chi-square sum.

We add up these components for each zodiac sign. If you do it by hand, it can be helpful to arrange the calculation in a table. We show that after this Step-By-Step.

The P-value is the area in the upper tail of the χ^2 model above the computed χ^2 value.

The χ^2 models are skewed to the high end, and change shape depending on the degrees of freedom. The P-value considers only the right tail. Large χ^2 statistic values correspond to small P-values, which lead us to reject the null hypothesis.

The expected value for each zodiac sign is 21.333.

$$\chi = \sum \frac{(Obs - Exp)^2}{Exp} = \frac{(23 - 21.333)^2}{21.333} + \frac{(20 - 21.333)^2}{21.333} + \cdots$$

$$= 5.094 \text{ for all 12 signs.}$$

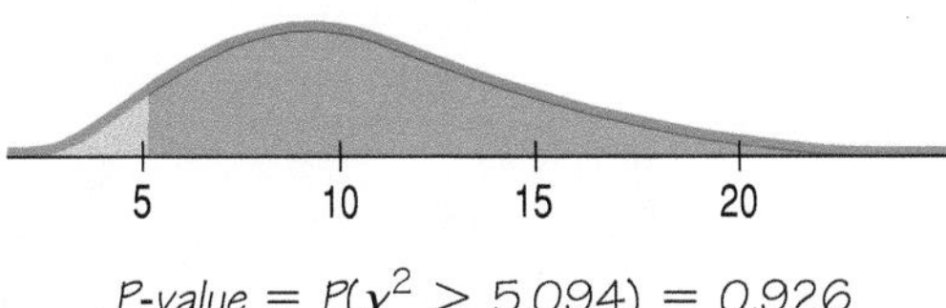

$$P\text{-value} = P(\chi^2 > 5.094) = 0.926$$

TELL ➡ Conclusion Link the P-value to your decision. Remember to state your conclusion in terms of what the data mean, rather than just making a statement about the distribution of counts.

The P-value of 0.926 says that if the zodiac signs of executives were in fact distributed uniformly, an observed chi-square value of 5.09 or higher would occur about 93% of the time. This certainly isn't unusual, so I fail to reject the null hypothesis, and conclude that these data show virtually no evidence of nonuniform distribution of zodiac signs among executives.

The Chi-Square Calculation

Let's make the chi-square procedure very clear. After clearly stating H_0 and H_A, the steps for calculations are:

1. **Find the expected values.** These come from the null hypothesis model. Every model gives a hypothesized proportion for each cell. The expected value is the product of the *total number of observations times this proportion.*

   For our example, the null model hypothesizes *equal* proportions. With 12 signs, 1/12 of the 256 executives should be in each category. The expected number for each sign is 21.333.
2. **Compute the residuals.** Once you have expected values for each cell, find the residuals, *Observed – Expected.*
3. **Square the residuals.**
4. **Compute the components.** Now find the **component,** $\frac{(Observed - Expected)^2}{Expected}$, for each cell.
5. **Find the sum of the components.** That's the chi-square statistic.
6. **Find the degrees of freedom.** It's equal to *the number of cells minus one*. For the zodiac signs, that's $12 - 1 = 11$ degrees of freedom.
7. **Test the hypothesis.** Large chi-square values mean lots of deviation from the hypothesized model, so they give small P-values. Look up the critical value from a table of chi-square values, or use technology to find the P-value directly.

A S *Activity:* **The Chi-Square Test.** This animation completes the calculation of the chi-square statistic and the hypothesis test based on it.

The steps of the chi-square calculations are often laid out in tables. Use one row for each category, and columns for observed counts, expected counts, residuals, squared residuals, and the contributions to the chi-square total, like this:

Sign	Observed	Expected	Residual = (*Obs* − *Exp*)	$(Obs - Exp)^2$	Component = $\frac{(Obs - Exp)^2}{Exp}$
Aries	23	21.333	1.667	2.778889	0.130262
Taurus	20	21.333	−1.333	1.776889	0.083293
Gemini	18	21.333	−3.333	11.108889	0.520737
Cancer	23	21.333	1.667	2.778889	0.130262
Leo	20	21.333	−1.333	1.776889	0.083293
Virgo	19	21.333	−2.333	5.442889	0.255139
Libra	18	21.333	−3.333	11.108889	0.520737
Scorpio	21	21.333	−0.333	0.110889	0.005198
Sagittarius	19	21.333	−2.333	5.442889	0.255139
Capricorn	22	21.333	0.667	0.444889	0.020854
Aquarius	24	21.333	2.667	7.112889	0.333422
Pisces	29	21.333	7.667	58.782889	2.755491
					$\Sigma = 5.094$

How big is big? When we calculated χ^2 for the zodiac sign example, we got 5.094. That value would have been big for z or t, leading us to reject the null hypothesis. Not here. Were you surprised that $\chi^2 = 5.094$ had a huge P-value of 0.926? What *is* big for a χ^2 statistic, anyway?

Think about how χ^2 is calculated. In every cell, any deviation from the expected count contributes to the sum. Large deviations generally contribute more, but if there are a lot of cells, even small deviations can add up, making the χ^2 value larger. So the more cells there are, the higher the value of χ^2 has to get before it becomes noteworthy. For χ^2, then, the decision about how big is big depends on the number of degrees of freedom.

Unlike the Normal and t families, χ^2 models are skewed. Curves in the χ^2 family change both shape and centre as the number of degrees of freedom grows. Here, for example, are the χ^2 curves for 5 and 9 degrees of freedom.

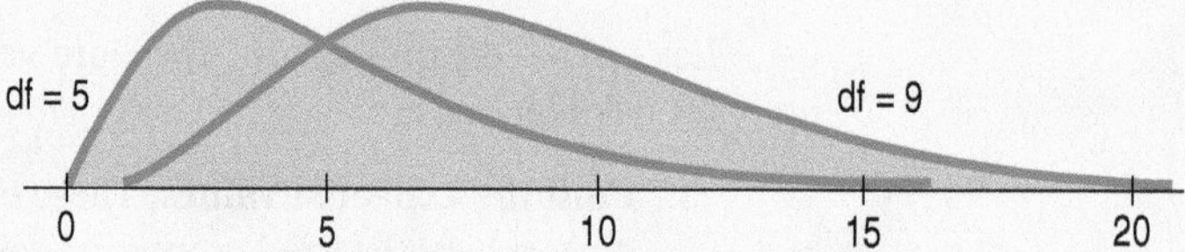

Notice that the value $\chi^2 = 10$ might seem somewhat extreme when there are 5 degrees of freedom, but appears to be rather ordinary for 9 degrees of freedom. Here are two simple facts to help you think about χ^2 models:

- The mode is at $\chi^2 = df - 2$. (Look at the curves; their peaks are at 3 and 7, see?)
- The expected value (mean) of a χ^2 model is its number of degrees of freedom. That's a bit to the right of the mode—as we would expect for a skewed distribution.

Our test for zodiac birthdays had 11 df, so the relevant χ^2 curve peaks at 9 and has a mean of 11. Knowing that, we might have easily guessed that the calculated χ^2 value of 5.094 wasn't going to be significant.

But I Believe the Model . . .

Goodness-of-fit tests are likely to be performed by people who have a theory of what the proportions *should* be in each category and who believe their theory to be true. Unfortunately, the only *null* hypothesis available for a goodness-of-fit test is that the theory is true. And as we know, the hypothesis-testing procedure allows us only to *reject* the null or *fail to reject* it. We can never confirm that a theory is, in fact, true, which is often what people want to do.

Unfortunately, they're stuck. At best, we can point out that the data are consistent with the proposed theory. But this doesn't *prove* the theory. The data *could* be consistent with the model even if the theory were wrong. In that case, we fail to reject the null hypothesis, but can't conclude anything for sure about whether the theory is true.

And we can't fix the problem by turning things around. Suppose we try to make our favoured hypothesis the alternative. Then, it is impossible to pick a single null. For example, suppose, as a doubter of astrology, you want to prove that the distribution of executive births is uniform. If you choose uniform as the null hypothesis, you can only *fail* to reject it. So you'd like uniformity to be your alternative hypothesis. Which particular violation of equally distributed births would you choose as your null? The problem is that the model can be wrong in many, many ways. There's no way to frame a null hypothesis the other way around. There's just no way to prove that a favoured model is true.

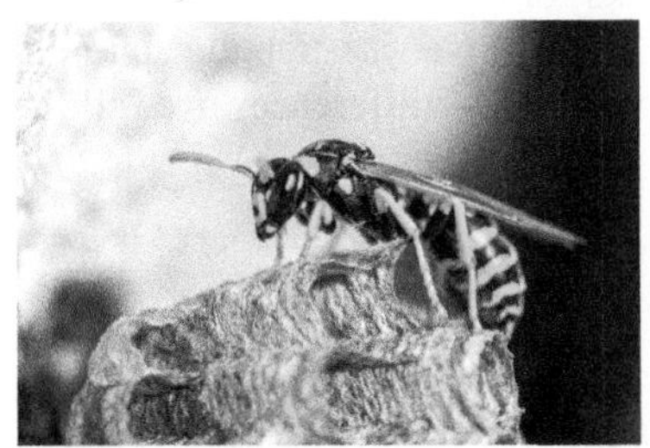
McCarthys_PhotoWorks/Fotolia

WHY CAN'T WE PROVE THE NULL?

A biologist wanted to show that her inheritance theory about fruit flies was valid. It says that 10% of the flies should be type 1, 70% type 2, and 20% type 3. After her students collected data on 100 flies, she did a goodness-of-fit test and found a P-value of 0.07. She started celebrating, since her null hypothesis wasn't rejected—that is, until her students collected data on 100 more flies. With 200 flies, the P-value dropped to 0.02. Although she knew the answer was probably no, she asked the statistician, somewhat hopefully, if she could just ignore half the data and stick with the original 100. By this reasoning we could always "prove the null" just by not collecting much data. With only a little data, the chances are good that they'll be consistent with almost anything. But they also have little chance of disproving anything, either. In this case, the test has no power. Don't let yourself be lured into this scientist's reasoning. With data, more is always better. But you can't ever prove that your null hypothesis is true.

23.2 Comparing Observed Distributions

Who	Graduates from four areas of study at a university
What	Post-graduation activities
When	2006
Why	Survey for general information

Many universities survey graduating classes to determine the plans of the graduates. We might wonder whether the plans of students are the same at different areas of study. Here's a summary table for class of 2006 graduates from one university. Each **cell** of the table shows how many students from a particular area of study made a certain choice.

Table 23.1

Post-graduation activities of the class of 2006 from a large university for several areas of study.

	Agriculture	Arts & Sciences	Engineering	Social Science	Total
Employed	379	305	243	125	**1052**
Grad School	186	238	202	96	**722**
Other	104	123	37	58	**322**
Total	**669**	**666**	**482**	**279**	**2096**

Because class sizes are so different for each area of study, we see differences better by examining the proportions for each class rather than the counts:

Table 23.2

Activities of graduates as a percentage of respondents from each area of study.

	Agriculture	Arts & Sciences	Engineering	Social Science	Total
Employed	56.7%	45.8	50.4	44.8	**50.2**
Grad School	27.8	35.7	41.9	34.4	**34.4**
Other	15.5	18.5	7.7	20.8	**15.4**
Total	**100**	**100**	**100**	**100**	**100**

We already know how to test whether *two* proportions are the same. For example, we could use a two-proportion z-test to see whether the proportion of students choosing graduate school is the same for Agriculture students as for Engineering students. But now we have more than two groups. We want to test whether the students' choices are the same across all four disciplines. The z-test for two proportions generalizes to a **chi-square test of homogeneity**.

Chi-square again? It turns out that the mechanics of this test are *identical* to the chi-square test for goodness-of-fit that we just saw. (How similar can you get?) Why a different name, then? The goodness-of-fit test compared counts with a theoretical model. But here we're asking whether choices are the same among different groups, so we find the expected counts for each category directly from the data. As a result, we count the degrees of freedom slightly differently as well.

A S *Video:* **The Incident.** You may have guessed which famous incident put women and children at risk. Here you can view the story complete with rare film footage.

The term "homogeneity" means that things are the same. Here, we ask whether the post-graduation choices made by students are the *same* for these four areas of study. The homogeneity test comes with a built-in null hypothesis: We hypothesize that the distribution does not change from group to group. The test looks for differences large enough to step beyond what we might expect from random sample-to-sample variation. It can reveal a large deviation in a single category, or small—but persistent—differences over all the categories—or anything in between.

Assumptions and Conditions

The assumptions and conditions are the same as for the chi-square test for goodness-of-fit. The **Counted Data Condition** says that these data must be counts. You can't do a test of homogeneity on proportions, so we have to work with the counts of graduates given in the first table. Also, you can't do a chi-square test on measurements. For example, if we had recorded GPAs for these same groups, we wouldn't be able to determine whether the mean GPAs were different using this test.[4]

When sampling from larger populations, we would like the cases within each group to be selected randomly. We need to know whether the **Independence Assumption** both within and across groups is reasonable. As usual, proper randomization can make the assumption plausible.

Often though when we test for homogeneity, we aren't interested in some larger population, in which case we don't really need a random sample. (We would need one if we wanted to draw a more general conclusion—say, about the choices made by all members of the class of 2006.) Don't we need *some* randomness, though? Fortunately, we can think of the null hypothesis as a model in which the counts in the table are distributed as if each student chose a plan randomly according to the overall proportions of the choices, regardless of the student's class. As long as we don't want to generalize, we don't have to check the **Randomization Condition**.

We still must be sure we have enough data for this method to work. The **Expected Cell Frequency Condition** says that the expected count in each cell must be at least five. We'll confirm that as we do the calculations.

Calculations

The null hypothesis says that the proportions of graduates choosing each alternative should be the same for all four areas of study. So for each choice, the expected area of study proportion is just the overall proportion of students making that choice. The expected counts are those proportions applied to the number of students in each graduating class.

For example, overall, 1052, or about 50.2%, of the 2096 students who responded to the survey were employed. If the distributions are homogeneous (as the null hypothesis asserts), then 50.2% of the 669 Agriculture graduates (or about 335.8 students) should be employed. Similarly, 50.2% of the 482 Engineering grads (or about 241.96) should be employed.

[4]To do that, you'd use a method called analysis of variance, discussed in Chapter 25.

Working in this way, we (or, more likely, the computer) can fill in expected values for each cell. Because these are theoretical values, they don't have to be integers. The expected values look like this:

Table 23.3

Expected values for the 2006 graduates.

	Agriculture	Arts & Sciences	Engineering	Social Science	Total
Employed	335.777	334.271	241.290	140.032	**1052**
Grad School	230.448	229.414	166.032	96.106	**722**
Other	102.776	102.315	74.048	42.862	**322**
Total	**669**	**666**	**482**	**279**	**2096**

Now check the **Expected Cell Frequency Condition**. Indeed, there are at least five individuals expected in each cell.

Following the pattern of the goodness-of-fit test, we compute the component for each cell of the table. For the highlighted cell, employed students graduating from the Agriculture, that's

$$\frac{(Obs - Exp)^2}{Exp} = \frac{(379 - 335.777)^2}{335.777} = 5.564$$

Summing these components across all cells gives

$$\chi^2 = \sum_{all\ cells} \frac{(Obs - Exp)^2}{Exp} = 54.5$$

How about the degrees of freedom? We don't really need to calculate all the expected values in the table. We know there is a total of 1052 employed students, so once we find the expected values for three of the areas of study, we can determine the expected number for the fourth by just subtracting. Similarly, we know how many students graduated from each area of study, so after filling in two rows, we can find the expected values for the remaining row by subtracting. To fill out the table, we need to know the counts in only $R - 1$ rows and $C - 1$ columns. So the table has $(R - 1)(C - 1)$ degrees of freedom.

NOTATION ALERT

For a contingency table, R represents the number of rows and C the number of columns.

In our example, we need to calculate only two choices in each column and counts for three of the four areas of study, for a total of $2 \times 3 = 6$ degrees of freedom. We'll need the degrees of freedom to find a P-value for the chi-square statistic.

Step-by-Step Example A CHI-SQUARE TEST FOR HOMOGENEITY

We have reports from four areas of study on the post-graduation activities of the university's 2006 graduating classes.

Question: Are students' choices of post-graduation activities the same across all the areas of study?

THINK ➡ Plan State what you want to know.

Identify the variables and check the *W*'s.

I want to know whether post-graduation choices are the same for students from each of four areas of study. I have a table of counts classifying each area of study's class of 2006 respondents according to their activities.

Hypotheses State the null and alternative hypotheses.

H_0: Students' post-graduation activities are distributed in the same way for all four areas of study.

H_A: Students' plans do not have the same distribution.

Model Make a picture: A side-by-side bar chart shows the four distributions of post-graduation activities. Plot column percents to remove the effect of class size differences. A segmented bar chart would also be an appropriate choice.

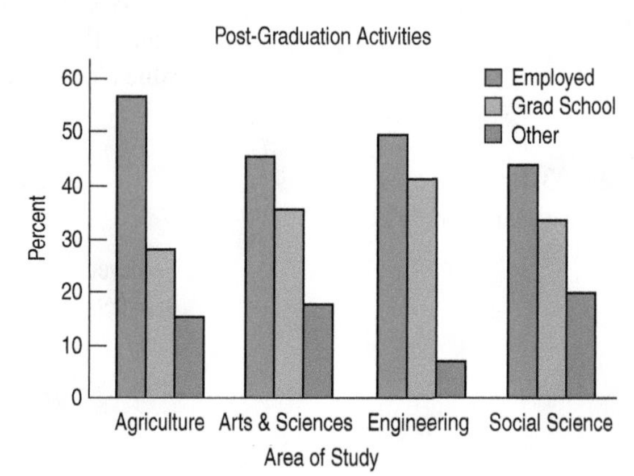

A side-by-side bar chart shows how the distributions of choices differ across the four areas of study.

Think about the assumptions and check the conditions.

✓ **Counted Data Condition:** I have counts of the numbers of students in categories.

✓ **Independence Assumption:** Student plans should be largely independent of each other. The occasional friends who decide to join the Canadian Teaching Agency together or couples who make grad school decisions together are too rare to affect this analysis.

✓ **Randomization Condition:** I don't want to draw inferences to other universities or other classes, so there is no need to check for a random sample.

✓ **Expected Cell Frequency Condition:** The expected values (shown below) are all at least five.

State the sampling distribution model.

Name the test you will use.

The conditions seem to be met, so I can use a χ^2 model with $(3 - 1) \times (4 - 1) = 6$ degrees of freedom and do a chi-square test of homogeneity.

SHOW ➡ Mechanics Show the expected counts for each cell of the data table. You could make separate tables for the observed and expected counts, or put both counts in each cell as shown here. While observed counts must be whole numbers, expected counts rarely are—don't be tempted to round those off.

Calculate χ^2.

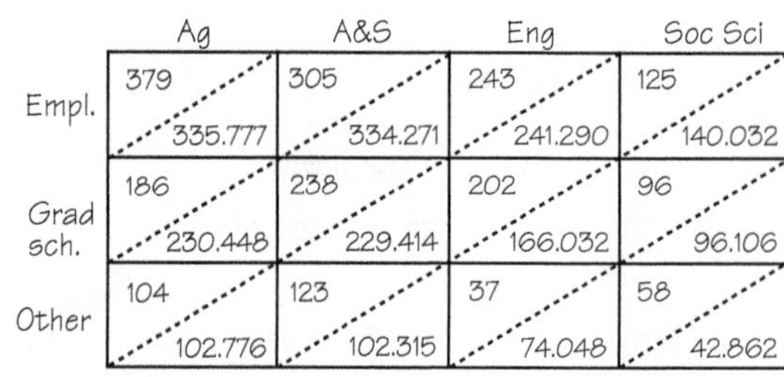

	Ag	A&S	Eng	Soc Sci
Empl.	379 / 335.777	305 / 334.271	243 / 241.290	125 / 140.032
Grad sch.	186 / 230.448	238 / 229.414	202 / 166.032	96 / 96.106
Other	104 / 102.776	123 / 102.315	37 / 74.048	58 / 42.862

$$\chi^2 = \sum_{all\ cells} \frac{(Obs - Exp)^2}{Exp} = 54.5$$

The shape of a χ^2 model depends on the degrees of freedom. A χ^2 model with 6 df is skewed to the high end.

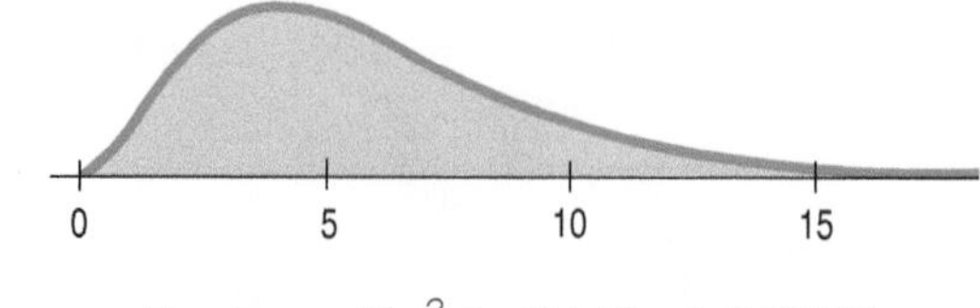

The P-value considers only the right tail. Here, the calculated value of the χ^2 statistic is off the scale, so the P-value is quite small.

P-value $= P(\chi^2 > 54.5) < 0.0001$

TELL ➡ Conclusion State your conclusion in the context of the data. You should specifically talk about whether the distributions for the groups appear to be different.	The P-value is very small, so I reject the null hypothesis and conclude that there's evidence that the post-graduation activities of students from these four areas of study don't have the same distribution.

If you find that simply rejecting the hypothesis of homogeneity is a bit unsatisfying, you're in good company. Okay, so the post-graduation plans are different. What we'd really like to know is what the differences are, where they're the greatest, and where they're the smallest. The test for homogeneity doesn't answer these interesting questions, but it does provide some evidence that can help us.

23.3 Examining the Residuals

Whenever we reject the null hypothesis, it's a good idea to examine residuals. (We don't need to do that when we fail to reject because when the χ^2 value is small, all of its components must have been small.) For chi-square tests, we want to compare residuals for cells that may have very different counts. So we're better off standardizing the residuals. We know the mean residual is zero,[5] but we need to know each residual's standard deviation. When we tested proportions, we saw a link between the expected proportion and its standard deviation. For counts, there's a similar link. To standardize a cell's residual, we just divide by the square root of its expected value:

$$c = \frac{(Obs - Exp)}{\sqrt{Exp}}$$

Notice that these **standardized residuals** are just the square roots of the **components** we calculated for each cell, and their sign indicates whether we observed more cases than we expected or fewer.

The standardized residuals give us a chance to think about the underlying patterns and to consider the ways in which the distribution of post-graduation plans may differ from area of study to area of study. Now that we've subtracted the mean (zero) and divided by their standard deviations, these are z-scores. If the null hypothesis were true, we could even appeal to the Central Limit Theorem, think of the Normal model, and use the 68–95–99.7 Rule to judge how extraordinary the large ones are.

Here are the standardized residuals for the class of 2006 data:

Table 23.4

Standardized residuals can help show how the table differs from the null hypothesis pattern.

	Ag	A&S	Eng	Soc Sci
Employed	2.359	−1.601	0.069	−1.270
Grad School	−2.928	0.567	2.791	−0.011
Other	0.121	2.045	−4.305	2.312

A S *Activity:* **Calculating Standardized Residuals.** Women were at risk, too. Standardized residuals help us understand the relative risks.

The column for Engineering students immediately attracts our attention. It holds both the largest positive and the largest negative standardized residuals. It looks like Engineering graduates are more likely to go on to graduate work and are very unlikely to take time off for "volunteering and travel, among other activities" (as the "Other" category is explained). By contrast, Agriculture graduates seem to be readily employed and less likely to pursue graduate work immediately after graduation.

[5]Residual = observed − expected. Because the total of the expected values is set to be the same as the observed total, the residuals must sum to zero.

For Example LOOKING AT χ^2 RESIDUALS

RECAP: Some people suggest that school children who are the older ones in their class naturally perform better in sports and therefore get more coaching and encouragement. To see if there's any evidence for this, we looked at Canadian NHL players born since 1970. A goodness-of-fit test found their birth months to have a distribution that's significantly different from the rest of us. The table shows the standardized residuals.

Month	Residual	Month	Residual
1	3.32	7	−1.48
2	2.74	8	−2.66
3	1.38	9	−1.66
4	2.49	10	−1.84
5	1.23	11	−1.81
6	0.93	12	−2.54

QUESTION: What is different about the distribution of birth months among Canadian NHL players?

ANSWER: It appears that, compared to the general population, fewer hockey players than expected were born in August and December, and more than expected were born in January and February. The clear pattern of positive residuals in the first half of the year and negative residuals in the second half supports the conjecture that being older is an advantage in terms of a career as a pro athlete.

Just Checking

Tiny black potato flea beetles can damage potato plants in a vegetable garden. These pests chew holes in the leaves, causing the plants to wither or die. They can be killed with an insecticide, but a canola oil spray has been suggested as a nonchemical "natural" method of controlling the beetles. To conduct an experiment to test the effectiveness of the natural spray, we gather 500 beetles and place them in three Plexiglas® containers. Two hundred beetles go in the first container, where we spray them with the canola oil mixture. Another 200 beetles go in the second container, and we spray them with the insecticide. The remaining 100 beetles in the last container serve as a control group; we simply spray them with water. Then, we wait six hours and count the number of surviving beetles in each container.

1. Why do we need the control group?
2. What would our null hypothesis be?
3. After the experiment is over, we could summarize the results in a table like this. How many degrees of freedom does our χ^2 test have?

	Canola oil	Insecticide	Water	Total
Survived				
Died				
Total	200	200	100	500

4. Suppose that, all together, 125 beetles survived. (That's the first-row total.) What's the expected count in the first cell—survivors among those sprayed with the canola oil?

5. If it turns out that only 40 of the beetles in the first container survived, what's the calculated component of χ^2 for that cell?
6. If the total calculated value of χ^2 for this table turns out to be around 10, would you expect the P-value of our test to be large or small? Explain.

23.4 Independence

A study from the University of Texas Southwestern Medical Center examined whether the risk of hepatitis C was related to whether people had tattoos and to where they got their tattoos. Hepatitis C causes about 10 000 deaths each year in the United States, but often lies undetected for years after infection.

The data from this study can be summarized in a two-way table, as follows:

Table 23.5

Counts of patients classified by their hepatitis C test status according to whether they had a tattoo from a tattoo parlour or from another source, or had no tattoo.

	Hepatitis C	No Hepatitis C	Total
Tattoo, parlour	17	35	52
Tattoo, elsewhere	8	53	61
None	22	491	513
Total	47	579	626

Who	Patients being treated for non–blood-related disorders
What	Tattoo status and hepatitis C status
When	1991, 1992
Where	Texas

These data differ from the kinds of data we've considered before in this chapter because they categorize subjects from a single group on two categorical variables rather than on only one. The categorical variables here are *hepatitis C status* ("Hepatitis C" or "No Hepatitis C") and *tattoo status* ("Parlour," "Elsewhere," "None"). We've seen counts classified by two categorical variables displayed like this in Chapter 2, so we know such tables are called contingency tables. **Contingency tables** categorize counts on two (or more) variables so that we can see whether the distribution of counts on one variable is contingent on the other.

The natural question to ask of these data is whether the chance of having hepatitis C is *independent* of tattoo status. Recall that for events **A** and **B** to be independent, $P(\mathbf{A})$ must equal $P(\mathbf{A}|\mathbf{B})$. Here, this means the probability that a randomly selected patient has hepatitis C should not change when we learn the patient's tattoo status. We examined the question of independence in just this way in Chapter 13, but we lacked a way to test it. The rules for independent events are much too precise and absolute to work well with real data. **A chi-square test for independence** is the test called for here.

A S *Activity:* **Independence and Chi-Square.** This unusual simulation shows how independence arises (and fails) in contingency tables.

If *hepatitis status* is independent of tattoos, we'd expect the proportion of people testing positive for hepatitis to be the same for the three levels of *tattoo status*. This sounds a lot like the test of homogeneity. In fact, the mechanics of the calculation are identical.

The difference is that now we have two categorical variables measured on a single population. For the homogeneity test, we had a single categorical variable measured independently on two or more populations. But now we ask a different question: "Are the variables independent?" rather than "Are the groups homogeneous?" These are subtle differences, but they are important when we draw conclusions.

The only difference between the test for homogeneity and the test for independence is in what you . . .

For Example WHICH χ^2 TEST?

Data from a travel agent in Cobourg, Ontario, describe the vacation locations (France, Bali, or Australia) and gender of 77 travellers.

		Location			
		France	Bali	Australia	Total
Gender	**Male**	3	16	22	**41**
	Female	12	15	9	**36**
	Total	**15**	**31**	**31**	**77**

QUESTIONS: Which test would be appropriate to examine whether gender is a factor in the choice of the vacation location? What are the hypotheses?

ANSWER: These data represent one travel agent in Cobourg, categorized on two variables, *Location* and *Gender*. I'll do a chi-square test of independence.

H_0: The choice of vacation location is independent of the gender of the traveller.

H_A: Decisions about the location of the vacation are not independent of the traveller's gender.

Assumptions and Conditions

A S ***Activity:* Chi-Square Tables.** Work with MyStatLab's interactive chi-square table to perform a hypothesis test.

Of course, we still need counts and enough data so that the expected values are at least five in each cell.

If we're interested in the independence of variables, we usually want to generalize from the data to some population. In that case, we'll need to check that the data are a representative random sample from that population.

Step-by-Step Example A CHI-SQUARE TEST FOR INDEPENDENCE

We have counts of 626 individuals categorized according to their "tattoo status" and their "hepatitis status."

Question: Are tattoo status and hepatitis status independent?

THINK ➡ Plan State what you want to know. Identify the variables and check the *W*'s.

I want to know whether the categorical variables tattoo status and hepatitis status are statistically independent. I have a contingency table of 626 Texas patients with an unrelated disease.

Hypotheses State the null and alternative hypotheses.

We perform a test of independence when we suspect the variables may not be independent. We are on the familiar ground of making a claim (in this case, that knowing *tattoo status* will change probabilities for *hepatitis C status*) and testing the null hypothesis that it is *not* true.

H_0: Tattoo status and hepatitis status are independent.[6]

H_A: Tattoo status and hepatitis status are not independent.

Model Make a picture. Because these are only two categories—Hepatitis C and No Hepatitis C—a simple bar chart of the distribution of tattoo sources for hep C patients shows all the information.

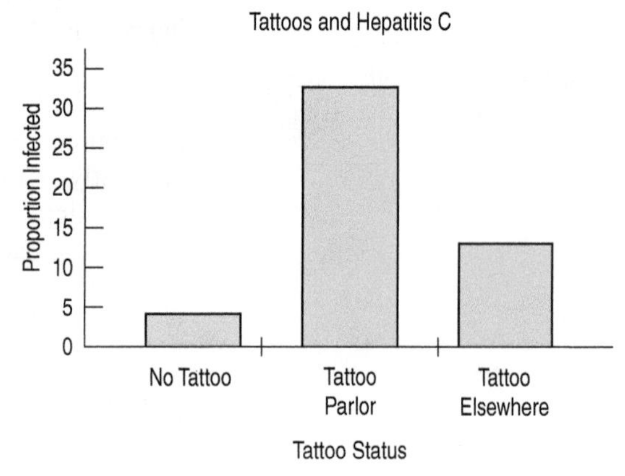

[6]Once again, parameters are hard to express. The hypothesis of independence itself tells us how to find expected values for each cell of the contingency table. That's all we need.

The bar chart suggests strong differences in the relative frequency of tattoo sources among hepatitis C patients.

Think about the assumptions and check the χ^2 conditions.

✓ **Counted Data Condition:** I have counts of individuals categorized on two categorical variables.

✓ **Independence Assumption:** The people in this study are likely to be independent of each other.

✓ **Randomization Condition:** These data are from a retrospective study of patients being treated for something unrelated to hepatitis. Although they are not an SRS, they were selected to avoid biases.

✓ **Expected Cell Frequency Condition:** The expected values do not meet the condition that all are at least five.

This table shows both the observed and expected counts for each cell. The expected counts are calculated exactly as they were for a test of homogeneity; in the first cell, for example, we expect $\frac{52}{626}$ (that's 8.3%) of 47.

Warning: Be wary of proceeding when there are small expected counts. If we see expected counts that fall far short of five or if many cells violate the condition, we should not use χ^2. (We will soon discuss ways you can fix the problem.) If you do continue, always check the residuals to be sure those cells did not have a major influence on your result.

	Hepatitis C	No Hepatitis C	Total
Tattoo, parlour	17 / 3.904	35 / 48.096	52
Tattoo, elsewhere	8 / 4.580	53 / 56.420	61
None	22 / 38.516	491 / 474.484	513
Total	47	579	626

Specify the model.

Name the test you will use.

Although the Expected Cell Frequency Condition is not satisfied, the values are close to five. I'll go ahead, but I'll check the residuals carefully. I'll use a χ^2 model with $(3 - 1) \times (2 - 1) = 2$ df and do a **chi-square test of independence.**

SHOW ➡ Mechanics Calculate χ^2.

The shape of a chi-square model depends on its degrees of freedom. With 2 df, the model looks quite different, as you can see here. We still care only about the right tail.

$$\chi^2 = \sum_{all\ cells} \frac{(Obs - Exp)^2}{Exp} = 57.91$$

P-Value $= P(\chi^2 > 57.91) < 0.0001$

TELL ➡ Conclusion Link the P-value to your decision. State your conclusion about the independence of the two variables.

(We should be wary of this conclusion because of the small expected counts. A complete solution must include the additional analysis, recalculation, and final conclusion discussed in the following section.)

The P-value is very small, so I reject the null hypothesis and conclude that *hepatitis status* is not independent of *tattoo status*. Because the Expected Cell Frequency Condition was violated, I need to check that the two cells with small expected counts did not influence this result too greatly.

For Example CHI-SQUARE MECHANICS

RECAP: We have data that allow us to investigate whether the locations of vacations are independent of travellers' gender.

Gender		Location			
		France	Bali	Australia	Total
	Male	3	16	22	41
	Female	12	15	9	36
	Total	**15**	**31**	**31**	**77**

QUESTIONS: What are the degrees of freedom for this test? What is the expected frequency of going to France for a vacation for a female traveller? What is that cell's component in the χ^2 computation? And how is the standardized residual for that cell computed?

ANSWER: This is a 2×3 contingency table, so df $= (2-1)(3-1) = 2$. Overall, 15 of 77 travellers went to France. If vacation destinations are independent of gender, then I'd expect $\frac{(36 \times 15)}{77} = 7.01$ of the 36 females to have vacationed in France.

That cell's term in the χ^2 calculation is

$$\frac{(12 - 7.01)^2}{7.013} = 3.55$$

The standardized residual for that cell is

$$\sqrt{3.55} = 1.88$$

Examine the Residuals

Each cell of the contingency table contributes a term to the chi-square sum. As we did earlier, we should examine the residuals because we have rejected the null hypothesis. In this instance, we have an additional concern that the cells with small expected frequencies not be the ones that make the chi-square statistic large.

Our interest in the data arises from the potential for improving public health. If patients with tattoos are more likely to test positive for hepatitis C, perhaps physicians should be advised to suggest blood tests for such patients.

The standardized residuals look like this:

Table 23.6

Standardized residuals for the hepatitis and tattoos data. Are any of them particularly large in magnitude?

	Hepatitis C	No Hepatitis C
Tattoo, parlour	6.628	−1.888
Tattoo, elsewhere	1.598	−0.455
None	−2.661	0.758

The chi-square value of 57.91 is the sum of the squares of these six values. The cell for people with tattoos obtained in a tattoo parlour who have hepatitis C is large and positive, indicating there are more people in that cell than the null hypothesis of independence would predict. Maybe tattoo parlours are a source of infection or maybe those who go to tattoo parlours also engage in risky behaviour.

The second-largest component is a negative value for those with no tattoos who test positive for hepatitis C. A negative value says that there are fewer people in this cell than independence would expect. That is, those who have no tattoos are less likely to be infected with hepatitis C than we might expect if the two variables were independent.

What about the cells with small expected counts? The formula for the chi-square standardized residuals divides each residual by the square root of the expected frequency. Too small an expected frequency can arbitrarily inflate the residual and lead to an inflated chi-square statistic. Any expected count close to the arbitrary minimum of five calls for checking that cell's standardized residual to be sure it is not particularly large. In this case, the standardized residual for the "Hepatitis C and Tattoo, elsewhere" cell is not particularly large, but the standardized residual for the "Hepatitis C and Tattoo, parlour" cell is large.

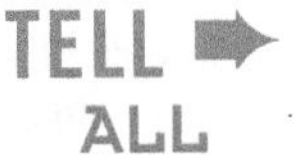

We might choose not to report the results because of concern with the small expected frequency. Alternatively, we could include a warning along with our report of the results. Yet another approach is to combine categories to get a larger sample size and correspondingly larger expected frequencies, if there are some categories that can be appropriately combined. Here, we might naturally combine the two rows for tattoos, obtaining a 2 × 2 table:

Table 23.7

Combining the two tattoo categories gives a table with all expected counts greater than 5.

	Hepatitis C	No Hepatitis C	Total
Tattoo	25	88	113
None	22	491	513
Total	47	579	626

TELL ➡ ALL

This table has expected values of at least five in every cell, and a chi-square value of 42.42 on 1 degree of freedom. The corresponding P-value is < 0.0001.

We conclude that *tattoo status* and *hepatitis C status* are not independent. The data *suggest* that tattoo parlours may be a particular problem, but we don't have enough data to draw that conclusion.

For Example WRITING CONCLUSIONS FOR χ^2 TESTS

RECAP: We're looking at vacation destination data to see if they show evidence of gender preference. With 2 df, technology calculates $\chi^2 = 10.60$, a P-value of approximately 0.0050, and these standardized residuals:

Gender	Location: France	Bali	Australia
Male	−1.76	−0.12	1.35
Female	1.88	0.13	1.44

QUESTIONS: What is your conclusion?

ANSWER: The very low P-value leads me to reject the null hypothesis. There is strong evidence that vacation destinations are associated with traveller gender.

From the standardized residuals, it appears that female travellers are more likely to go to France for vacation.

Just Checking

Which of the three chi-square tests—goodness-of-fit, homogeneity, or independence—would you use in each of the following situations?

7. A restaurant manager wonders whether customers who dine on Friday nights have the same preferences among the four "chef's special" entrées as those who dine on Saturday nights. One weekend he has the wait staff record which entrées were ordered each night. Assuming these customers to be typical of all weekend diners, he'll compare the distributions of meals chosen Friday and Saturday.

8. Company policy calls for parking spaces to be assigned to everyone at random, but you suspect that may not be so. There are three lots of equal size: lot A, next to the building; lot B, a bit farther away; and lot C, on the other side of the highway. You gather data about employees at middle management level and above to see how many were assigned parking in each lot.

9. Is a student's social life affected by where the student lives? A campus survey asked a random sample of students whether they lived in a dormitory, in off-campus housing, or at home, and whether they had been out on a date 0, 1–2, 3–4, or 5 or more times in the past two weeks.

23.5 Chi-Square and Causation

Chi-square tests are common. Tests for independence are especially widespread. Unfortunately, many people interpret a small P-value as proof of causation. We know better. Just as correlation between quantitative variables does not demonstrate causation, a failure of independence between two categorical variables does not show a cause-and-effect relationship between them, nor should we say that one variable *depends* on the other.

The chi-square test for independence treats the two variables symmetrically. There is no way to differentiate the direction of any possible causation from one variable to the other. In our example, it is unlikely that having hepatitis causes one to crave a tattoo, but other examples are not so clear.

In this case, it's easy to imagine that lurking variables are responsible for the observed lack of independence. Perhaps the lifestyles of some people include both tattoos and behaviours that put them at increased risk of hepatitis C, such as body piercings or even drug use. Even a small subpopulation of people with such a lifestyle among those with tattoos might be enough to create the observed result. After all, we observed only 25 patients with both tattoos and hepatitis.

In some sense, a failure of independence between two categorical variables is less impressive than a strong, consistent, linear association between quantitative variables. Two categorical variables can fail the test of independence in many ways, including ways that show no consistent pattern of failure. Examination of the chi-square standardized residuals can help you think about the underlying patterns.

WHAT CAN GO WRONG?

- **Don't use chi-square methods unless you have counts.** All three of the chi-square tests apply only to counts. Other kinds of data can be arrayed in two-way tables. Just because numbers are in a two-way table doesn't make them suitable for chi-square analysis. Data reported as proportions or percentages can be suitable for chi-square procedures, *but only after they are converted to counts.* If you try to do the calculations without first finding the counts, your results will be wrong.

A S *Simulation:* **Sample Size and Chi-Square.** Chi-square statistics have a peculiar problem. They don't respond to increasing the sample size in quite the way you might expect.

- **Beware of large samples.** Beware of *large* samples?! That's not the advice you're used to hearing. The chi-square tests, however, are unusual. You should be wary of chi-square tests performed on very large samples. No hypothesized distribution fits perfectly, no two groups are exactly homogeneous, and two variables are rarely perfectly independent. The degrees of freedom for chi-square tests don't grow with the sample size. With a sufficiently large sample size, a chi-square test can always reject the null hypothesis. Confidence intervals for proportions and differences between proportions, along with bar charts or pie charts, will help you assess the practical importance of your findings.
- **Don't say that one variable "depends" on the other just because they're not independent.** Dependence suggests a model or a pattern, but variables can fail to be independent in many different ways. When variables fail the test for independence, you might just say they are "associated."

CONNECTIONS

Chi-square methods relate naturally to inference methods for proportions. We can think of a test of homogeneity as stepping from a comparison of two proportions to a question of whether three or more proportions are equal. The standard deviations of the residuals in each cell are linked to the expected counts much like the standard deviations we found for proportions.

Independence is, of course, a fundamental concept in Statistics. But chi-square tests do not offer a general way to check on independence for all those times we have had to assume it.

Stacked bar charts or side-by-side pie charts can help us think about patterns in two-way tables. A histogram or boxplot of the standardized residuals can help locate extraordinary values.

What Have We Learned?

Learning Objectives

Recognize hypotheses about categorical variables. We use one of three related methods. All look at counts of data in categories, and all rely on chi-square models, a new family indexed by its degrees of freedom.

- Goodness-of-fit tests compare the observed distribution of a single categorical variable to an expected distribution based on a theory or model.
- Tests of homogeneity compare the distribution of several groups for the same categorical variable.
- Tests of independence examine counts from a single group for evidence of an association between two categorical variables.

For each test, find the expected cell frequencies. Evaluate the chi-square statistic, and know the correct *df*. Find the *P*-value using the upper tail only.

For each test, check the assumptions and corresponding conditions:

- Counted data condition.
- Independence assumption; randomization makes independence more plausible.
- Sample size assumption with the expected cell frequency condition; expect at least 5 observations in each cell.

Mechanically, these tests are almost identical. Although the tests use a one-sided upper tail critical region when looking for evidence against the null hypothesis, the alternative hypothesis is actually many-sided, because there are many ways that a table of counts can deviate significantly from what we hypothesized. When the null hypothesis is rejected, examine the standardized residuals in order to better understand patterns in the table.

Review of Terms

Chi-square statistic Can be used to test whether the observed counts in a frequency distribution or contingency table match the counts we would expect according to some model. It is calculated as

$$\chi^2 = \sum_{all\ cells} \frac{(Obs - Exp)^2}{Exp}$$

Chi-square statistics differ in how expected counts are found, depending on the question asked (p. 660).

Chi-square model Chi-square models are skewed to the right. They are parameterized by their degrees of freedom and become less skewed with increasing degrees of freedom (p. 661).

Chi-square test of goodness-of-fit A test of whether the distribution of counts in one categorical variable matches the distribution predicted by a model is called a test of goodness-of-fit. In a chi-square goodness-of-fit test, the expected counts come from the predicting model. The test finds a P-value from a chi-square model with the number of categories in the categorical variable −1 degrees of freedom (p. 662).

Chi-square component The components of a chi-square calculation are

$$\frac{(Observed - Expected)^2}{Expected}$$

found for each cell of the table (p. 663).

Cell One element of a two-way table corresponding to a specific row and a specific column. Table cells can hold counts, percentages, or measurements on other variables, or they can hold several values (p. 665).

Chi-square test of homogeneity A test comparing the distribution of counts for two or more groups on the same categorical variable is called a test of homogeneity. A chi-square test of homogeneity finds expected counts based on the overall frequencies, adjusted for the totals in each group under the (null hypothesis) assumption that the distributions are the same for each group. We find a P-value from a chi-square distribution with $(\#Rows - 1) \times (\#Cols - 1)$ degrees of freedom, where *#Rows* gives the number of categories and *#Cols* gives the number of independent groups (or vice versa) (p. 666).

Standardized residual In each cell of a two-way table, a standardized residual is the square root of the chi-square component for that cell with the sign of the *Observed – Expected* difference:

$$\frac{(Obs - Exp)}{\sqrt{Exp}}$$

When we reject a chi-square test, an examination of the standardized residuals can sometimes reveal more about how the data deviate from the null model (p. 669).

Chi-square test of independence A test of whether two categorical variables are independent examines the distribution of counts for one group of individuals classified according to both variables. A chi-square test of independence finds expected counts by assuming that knowing the marginal totals tells us the cell frequencies, assuming that there is no association between the variables. This turns out to be the same calculation as a test of homogeneity. We find a P-value from a chi-square distribution with $(\#Rows - 1) \times (\#Cols - 1)$ degrees of freedom, where *#Rows* gives the number of categories in one variable and *#Cols* gives the number of categories in the other (p. 671).

Contingency table A two-way table that classifies individuals according to two categorical variables (p. 671).

Two-way table Each cell of a two-way table shows counts of individuals. One way classifies a sample according to a categorical variable. The other way can classify different groups of individuals according to the same variable or classify the same individuals according to a different categorical variable (p. 671).

On the Computer CHI-SQUARE

Most statistics packages associate chi-square tests with contingency tables. Often chi-square is available as an option only when you make a contingency table. This organization can make it hard to locate the chi-square test and may confuse the three different roles that the chi-square test can take. In particular, chi-square tests for goodness-of-fit may be hard to find or missing entirely. Chi-square tests for homogeneity are computationally the same as chi-square tests for independence, so you may have to perform the mechanics as if they were tests of independence and interpret them afterward as tests of homogeneity.

Most statistics packages work with data on individuals rather than with the summary counts. If the only information you have is the table of counts, you may find it more difficult to get a statistics package to compute chi-square. Some packages offer a way to reconstruct the data from the summary counts so they can then be passed back through the chi-square calculation, finding the cell counts again. Many packages offer chi-square standardized residuals (although they may be called something else)..

EXCEL

Excel offers the function **CHITEST (actual_range, expected_range)**, which computes a chi-square value for homogeneity. Both ranges are of the form UpperleftCell:LowerRightCell, specifying two rectangular tables that must hold counts (although Excel will not check for integer values). The two tables must be of the same size and shape.

COMMENTS

Excel's documentation claims this is a test for independence and labels the input ranges accordingly, but Excel offers no way to find expected counts, so the function is not particularly useful for testing independence. You can use this function only if you already know both tables of counts or are willing to program additional calculations.

JMP

From the **Analyze** menu, select **Fit Y by X**.

- Select variables: a Y, Response variable that holds responses for one variable, and an X, Factor variable that holds responses for the other. Both selected variables must be nominal or ordinal.

JMP will make a plot and a contingency table. Below the contingency table, JMP offers a **Tests** panel. In that panel, the chi-square for independence is called a **Pearson ChiSquare**. The table also offers the P-value.

- Click on the Contingency Table title bar to drop down a menu that offers to include a **Deviation** and **Cell Chi square** in each cell of the table.

COMMENTS

JMP will choose a chi-square analysis for a **Fit Y by X** if both variables are nominal or ordinal (marked with an N or O), but not otherwise. Be sure the variables have the right type.

Deviations are the 'observed – expected' differences in counts. Cell chi-squares are the squares of the standardized residuals. Refer to the deviations for the sign of the difference.

Look under **Distributions** in the **Analyze** menu to find a chi-square test for goodness-of-fit.

MINITAB

- From the **Stat** menu, choose the **Tables** submenu.
- From that menu, choose **Chi Square Test. . . .**
- In the dialogue, identify the columns that make up the table.

Minitab will display the table and print the chi-square value and its P-value.

COMMENTS

Alternatively, select the **Cross Tabulation . . .** command to see more options for the table, including expected counts and standardized residuals.

R

In library(stats)

- **chisq.test(x, p = p_0)** will perform a goodness-of-fit test for a distribution stored in the vector x. The test will check if the distribution of x follows that of vector p_0, or if the distribution is uniform if p is missing.
- **chisq.test(x)** will perform an independence test on a matrix x, which must have at least two rows and columns.

SPSS

- From the **Analyze** menu, choose the **Descriptive Statistics** submenu.
- From that submenu, choose **Crosstabs. . . .**
- In the Crosstabs dialogue, assign the row and column variables from the variable list. Both variables must be categorical.
- Click the **Cells** button to specify that standardized residuals should be displayed.
- Click the **Statistics** button to specify a chi-square test.

COMMENTS

SPSS offers only variables that it knows to be categorical in the variable list for the Crosstabs dialogue. If the variables you want are missing, check that they have the right type.

STATCRUNCH

To perform a goodness-of-fit test:

- Enter the counts in one column and the expected counts in a second.
- Click on **Stat**.
- Choose **Goodness-of-fit » Chi-Square test**.
- Select the columns you filled in previously and click on **Calculate**.

To perform a test for independence:

- Enter the contingency table starting in column 2.
- Label the rows with the variable name in column 1.
- Click on Stat.
- Choose **Tables » Contingency » with summary**.
- Select the columns making up the table, starting from var2.
- Choose var1 under "Row labels in".
- Select any options you'd like, and click on **Calculate**.

TI-83/84 PLUS

The TI-83 does not have a routine for the chi-square goodness-of-fit test.

To test a hypothesis of homogeneity or independence, enter the data as a matrix:

- Push the **MATRIX** button, and choose to **EDIT** matrix.
- Specify the dimensions of the table, rows × columns, then enter the appropriate counts.
- To do the test, choose **C: χ^2-Test** from the **STAT TESTS** menu. Note that the calculator automatically stores the expected counts in a matrix you specify.

Exercises

1. Which test? For each of the following situations, state whether you'd use a chi-square goodness-of-fit test, a chi-square test of homogeneity, a chi-square test of independence, or some other statistical test:

a) A brokerage firm wants to see whether the type of account a customer has (silver, gold, or platinum) affects the type of trades that customer makes (in person, by phone, or on the Internet). It collects a random sample of trades made for its customers over the past year and performs a test.

b) That brokerage firm also wants to know if the type of account affects the size of the account (in dollars). It performs a test to see if the mean size of the account is the same for the three account types.

c) The academic research office at a large university wants to see whether the distribution of courses chosen (Humanities, Social Science, or Science) is different for its residential and nonresidential students. It assembles last semester's data and performs a test.

2. Which test again? For each of the following situations, state whether you'd use a chi-square goodness-of-fit test, a chi-square test of homogeneity, a chi-square test of independence, or some other statistical test:

a) Is the quality of a car affected by what day it was built? A car manufacturer examines a random sample of the warranty claims filed over the past two years to test whether defects are randomly distributed across days of the work week.

b) A medical researcher wants to know if blood cholesterol level is related to heart disease. She examines a database of 10 000 patients, testing whether the cholesterol level (in milligrams) is related to whether or not a person has heart disease.

c) A student wants to find out whether political leaning (liberal, moderate, or conservative) is related to choice of major. He surveys 500 randomly chosen students and performs a test.

3. Dice After getting trounced by your little brother in a children's game, you suspect the die he gave you to roll may be unfair. To check, you roll it 60 times, recording the number of times each face appears. Do these results cast doubt on the die's fairness?

Face	Count
1	11
2	7
3	9
4	15
5	12
6	6

a) If the die is fair, how many times would you expect each face to show?
b) To see if these results are unusual, will you test goodness-of-fit, homogeneity, or independence?
c) State your hypotheses.
d) Check the conditions.
e) How many degrees of freedom are there?
f) Find χ^2 and the P-value.
g) State your conclusion.

4. **M&M's** As noted in an earlier chapter, the Masterfoods Company says that until very recently yellow candies made up 20% of its milk chocolate M&M's, red another 20%, and orange, blue, and green 10% each. The rest are brown. On his way home from work the day he was writing these exercises, one of the authors bought a bag of plain M&M's. He got 29 yellow ones, 23 red, 12 orange, 14 blue, 8 green, and 20 brown. Is this sample consistent with the company's stated proportions? Test an appropriate hypothesis and state your conclusion.
a) If the M&M's are packaged in the stated proportions, how many of each colour should the author have expected to get in his bag?
b) To see if his bag was unusual, should he test goodness-of-fit, homogeneity, or independence?
c) State the hypotheses.
d) Check the conditions.
e) How many degrees of freedom are there?
f) Find χ^2 and the P-value.
g) State a conclusion.

5. **Nuts** A company says its premium mixture of nuts contains 10% Brazil nuts, 20% cashews, 20% almonds, and 10% hazelnuts, and the rest are peanuts. You buy a large can and separate the various kinds of nuts. Upon weighing them, you find there are 112 grams of Brazil nuts, 183 grams of cashews, 207 grams of almonds, 71 grams of hazelnuts, and 446 grams of peanuts. You wonder whether your mix is significantly different from what the company advertises.
a) Explain why the chi-square goodness-of-fit test is not an appropriate way to find out.
b) What might you do instead of weighing the nuts in order to use a χ^2 test?

6. **Mileage** A salesman who is on the road visiting clients thinks that, on average, he drives the same distance each day of the week. He keeps track of his mileage for several weeks and discovers that he averages 122 kilometres on Mondays, 203 kilometres on Tuesdays, 176 kilometres on Wednesdays, 181 kilometres on Thursdays, and 108 kilometres on Fridays. He wonders if this evidence contradicts his belief in a uniform distribution of kilometres across the days of the week. Explain why it is not appropriate to test his hypothesis using the chi-square goodness-of-fit test.

7. **Blood type** According to geneticists, in a general population, 42% of people are type A, 10% are type B, 3% are type AB, and 45% are type O. In 1919, a study was conducted for variance populations. In particular, a sample of 116 residents on Bongainville Island was tested. It was observed that 74 were type A, 12 were type B, 11 were type AB, and 19 were type O. Do the blood types of residents on Bongainville Island reflect the general population? Test an appropriate hypothesis and state your conclusion.

8. **Violence against women 2006** In its study *When Men Murder Women: An Analysis of 2006 Homicide Data*, the Violence Policy Center reported that 1836 women were murdered by men in 2006.[7] Of these victims, a weapon could be identified for 1670 of them. Of those for whom a weapon could be identified, 909 were killed by guns, 334 by knives or other cutting instruments, 227 by other weapons, and 200 by personal attack (battery, strangulation, etc.). The FBI's Uniform Crime Report says that, among all murders in the United States, the weapon use rates were as follows: guns 63.4%, knives 13.1%, other weapons 16.8%, personal attack 6.7%. Is there evidence that violence against women involves different weapons than other violent attacks in the United States?

9. **Fruit flies** Offspring of certain fruit flies may have yellow or ebony bodies and normal wings or short wings. Genetic theory predicts that these traits will appear in the ratio 9:3:3:1 (9 yellow, normal: 3 yellow, short: 3 ebony, normal: 1 ebony, short). A researcher checks 100 such flies and finds the distribution of the traits to be 59, 20, 11, and 10, respectively.
a) Are the results this researcher observed consistent with the theoretical distribution predicted by the genetic model?
b) If the researcher had examined 200 flies and counted exactly twice as many in each category—118, 40, 22, 20—what conclusion would he have reached?
c) Why is there a discrepancy between the two conclusions?

T 10. **Pi** Many people know the mathematical constant π is approximately 3.14. But that's not exact. To be more precise, here are 20 decimal places: 3.14159265358979323846. Still not exact, though. In fact, the actual value is irrational, a decimal that goes on forever without any repeating pattern. But notice that there are no 0s and only one 7 in the 20 decimal places above. Does that pattern persist, or do all the digits show up with equal frequency? The table shows the number of times each digit appears in the first million digits. Test the hypothesis that the digits 0 through 9 are uniformly distributed in the decimal representation of π.

[7] www.vpc.org

The first million digits of π	
Digit	Count
0	99 959
1	99 758
2	100 026
3	100 229
4	100 230
5	100 359
6	99 548
7	99 800
8	99 985
9	100 106

11. Hurricane frequencies The U.S. National Hurricane Center provides data that list the numbers of large (category 3, 4, or 5) hurricanes that have struck the United States by decade since 1851.[8] Here are the data:

Decade	Count	Decade	Count
1851–1860	6	1931–1940	8
1861–1870	1	1941–1950	10
1871–1880	7	1951–1960	9
1881–1890	5	1961–1970	6
1891–1900	8	1971–1980	4
1901–1910	4	1981–1990	4
1911–1920	7	1991–2000	5
1921–1930	5	2001–2010	7

Recently, there's been some concern that perhaps the number of large hurricanes has been increasing. The natural null hypothesis would be that the frequency of such hurricanes has remained constant.

a) With 96 large hurricanes observed over the 16 periods, what are the expected value(s) for each cell?
b) What kind of chi-square test would be appropriate?
c) State the null and alternative hypotheses.
d) How many degrees of freedom are there?
e) The value of χ^2 is 12.67. What is the P-value of the test?
f) State your conclusion.
g) Look again at the last decade. Suppose we only had data up to 2006 (where in fact the hurricane count was already at 7). Would this alter our conclusion at all?

12. Lottery numbers The fairness of Lotto 6/49 was recently called into question. The lottery publishes historical statistics on its Website.[9] The table shows the number of times each of the 49 numbers was drawn in the main lottery and as the "bonus" number as of July 2009:

Number	Count	Bonus	Number	Count	Bonus
1	343	57	26	344	59
2	338	39	27	367	57
3	339	61	28	314	51
4	342	48	29	350	37
5	333	60	30	354	53
6	333	39	31	386	56
7	362	55	32	360	52
8	337	59	33	338	54
9	338	56	34	391	63
10	319	51	35	334	47
11	322	74	36	343	64
12	334	54	37	348	59
13	327	58	38	348	55
14	317	49	39	349	66
15	309	30	40	376	46
16	338	52	41	341	65
17	340	50	42	347	60
18	329	62	43	376	58
19	352	51	44	339	41
20	374	47	45	369	55
21	345	59	46	371	55
22	329	56	47	374	59
23	339	59	48	328	45
24	317	59	49	338	51
25	325	69			

a) What kind of test should we perform?
b) What are the appropriate expected value(s) for the test?
c) State the null and alternative hypotheses.
d) How many degrees of freedom are there?
e) What is the P-value of the test?
f) State your conclusion.

13. Childbirth, part 1 There is some concern that if a woman has an epidural to reduce pain during childbirth, the drug can get into the baby's bloodstream, making the baby sleepier and less willing to breastfeed. In December 2006, the *International Breastfeeding Journal* published results of a study conducted at Sydney University. Researchers followed up on 1178 births, noting whether the mother had an epidural and whether the baby was still nursing after six months. The results are summarized in the following table.

		Epidural?		
		Yes	No	Total
Breastfeeding @ 6 months?	Yes	206	498	**704**
	No	190	284	**474**
	Total	**396**	**782**	**1178**

a) What kind of test would be appropriate?
b) State the null and alternative hypotheses.

[8]www.nhc.noaa.gov
[9]www.olg.ca/lotteries/viewFrequentWinningNumbers.do

14. Does your doctor know? A survey[10] of articles from the *New England Journal of Medicine* (NEJM) classified them according to the principal statistics methods used. The articles recorded were all noneditorial pieces appearing during the indicated years. Let's look at whether these articles used statistics or not.

	Publication Year			
	1978–79	1989	2004–05	Total
No stats	90	14	40	144
Stats	242	101	271	614
Total	332	115	311	758

Has there been a change in the use of statistics?

a) What kind of test would be appropriate?

b) State the null and alternative hypotheses.

15. Childbirth, part 2 In Exercise 13, the table shows results of a study investigating whether the aftereffects of epidurals administered during childbirth might interfere with successful breastfeeding. We're planning to do a chi-square test.

a) How many degrees of freedom are there?

b) The smallest expected count will be in the epidural/no breastfeeding cell. What is it?

c) Check the assumptions and conditions for inference.

16. Does your doctor know?, part 2 The table in Exercise 14 shows whether NEJM medical articles during various time periods included statistics or not. We're planning to do a chi-square test.

a) How many degrees of freedom are there?

b) The smallest expected count will be in the 1989/no stats cell. What is it?

c) Check the assumptions and conditions for inference.

17. Childbirth, part 3 In Exercises 13 and 15, we began examining the possible impact of epidurals on successful breastfeeding.

a) Calculate the component of chi-square for the epidural/no breastfeeding cell.

b) For this test, $\chi^2 = 14.87$. What is the P-value?

c) State your conclusion.

18. Does your doctor know?, part 3 In Exercises 14 and 16, we began examining whether the use of statistics in NEJM medical articles has changed over time.

a) Calculate the component of chi-square for the 1989/no stats cell.

b) For this test, $\chi^2 = 25.28$. What is the P-value?

c) State your conclusion.

19. Childbirth, part 4 In Exercises 13, 15, and 17, we tested a hypothesis about the impact of epidurals on successful breastfeeding. The table shows the test's residuals.

		Epidural?	
		Yes	No
Breastfeeding at 6 months?	Yes	−1.99	1.42
	No	2.43	−1.73

a) Show how the residual for the epidural/no breastfeeding cell was calculated.

b) What can you conclude from the standardized residuals?

20. Does your doctor know?, part 4 In Exercises 14, 16, and 18, we tested a hypothesis about whether the use of statistics in NEJM medical articles has changed over time. The table shows the test's residuals.

	1978–79	1989	2004–05
No stats	3.39	−1.68	−2.48
Stats	−1.64	0.81	1.20

a) Show how the residual for the 1989/no stats cell was calculated.

b) What can you conclude from the patterns in the standardized residuals?

21. Childbirth, part 5 In Exercises 13, 15, 17, and 19, we looked at a study examining epidurals as one factor that might inhibit successful breastfeeding of newborn babies. Suppose a broader study included several additional issues, including whether the mother drank alcohol, whether this was a first child, and whether the parents occasionally supplemented breastfeeding with bottled formula. Why would it not be appropriate to use chi-square methods on the 2×8 table with yes/no columns for each potential factor?

22. Does your doctor know?, part 5 In Exercises 14, 16, 18, and 20, we considered data in articles in the NEJM. The original study listed 23 different statistics methods. (The list read: *t*-tests, contingency tables, linear regression. . . .) Why would it not be appropriate to use chi-square methods on the 23×3 table with a row for each method?

T **23. *Titanic*** Here is a table we first saw in Chapter 2 showing who survived the sinking of the *Titanic* based on whether they were crew members or passengers booked in first-, second-, or third-class staterooms:

	Crew	First	Second	Third	Total
Alive	212	202	118	178	710
Dead	673	123	167	528	1491
Total	885	325	285	706	2201

a) If we draw an individual at random from this table, what is the probability that we will draw a member of the crew?

[10]Suzanne S. Switzer and Nicholas J. Horton, 2007, "What your doctor should know about statistics (but perhaps doesn't)." *Chance*, 20:1.

b) What is the probability of randomly selecting a third-class passenger who survived?
c) What is the probability of a randomly selected passenger surviving, given that the passenger was a first-class passenger?
d) If someone's chances of surviving were the same regardless of their status on the ship, how many members of the crew would you expect to have lived?
e) State the null and alternative hypotheses we would test here.
f) Give the degrees of freedom for the test.
g) The chi-square value for the table is 187.8, and the corresponding P-value is barely greater than 0. State your conclusions about the hypotheses.

T **24. NYPD and sex discrimination** The table shows the rank attained by male and female officers in the New York City Police Department (NYPD). Do these data indicate that men and women are equitably represented at all levels of the department?

		Male	Female
Rank	Officer	21 900	4 281
	Detective	4 058	806
	Sergeant	3 898	415
	Lieutenant	1 333	89
	Captain	359	12
	Higher ranks	218	10

a) What is the probability that a person selected at random from the NYPD is a female?
b) What is the probability that a person selected at random from the NYPD is a detective?
c) Assuming no bias in promotions, how many female detectives would you expect the NYPD to have?
d) To see if there is evidence of differences in ranks attained by males and females, will you test goodness-of-fit, homogeneity, or independence?
e) State the hypotheses.
f) Test the conditions.
g) How many degrees of freedom are there?
h) Find χ^2 and the P-value.
i) State your conclusion.
j) If you concluded that the distributions are not the same, analyze the differences using the standardized residuals of your calculations.

25. *Titanic*, again Examine and comment on this table of the standardized residuals for the chi-square test you looked at in Exercise 23.

	Crew	First	Second	Third
Alive	−4.35	9.49	2.72	−3.30
Dead	3.00	−6.55	−1.88	2.27

26. NYPD again Examine and comment on this table of the standardized residuals for the chi-square test you looked at in Exercise 24.

	Male	Female
Officer	−2.34	5.57
Detective	−1.18	2.80
Sergeant	3.84	−9.14
Lieutenant	3.58	−8.52
Captain	2.46	−5.86
Higher ranks	1.74	−4.14

27. Birth order and study choice Students in an Introductory Statistics class at a large university were classified by birth order and by discipline they study.

		Birth Order (1 = oldest or only child)				
		1	2	3	4 or more	Total
Discipline	Arts and Sciences	34	14	6	3	57
	Agriculture	52	27	5	9	93
	Social Science	15	17	8	3	43
	Professional	13	11	1	6	31
	Total	114	69	20	21	224

		Expected Values Birth Order (1 = oldest or only child)			
		1	2	3	4 or more
Discipline	Arts and Sciences	29.0089	17.5580	5.0893	5.3438
	Agriculture	47.3304	28.6473	8.3036	8.7188
	Social Science	21.8839	13.2455	3.8393	4.0313
	Professional	15.7768	9.5491	2.7679	2.9063

a) What kind of chi-square test is appropriate—goodness-of-fit, homogeneity, or independence?
b) State your hypotheses.
c) State and check the conditions.
d) How many degrees of freedom are there?
e) The calculation yields $\chi^2 = 17.78$ with P = 0.0378. State your conclusion.
f) Examine and comment on the standardized residuals below. Do they challenge your conclusion? Explain.

		Standardized Residuals Birth Order (1 = oldest or only child)			
		1	2	3	4 or more
Discipline	Arts and Sciences	0.92667	20.84913	0.40370	21.01388
	Agriculture	0.67876	20.30778	21.14640	0.09525
	Social Science	−1.47155	1.03160	2.12350	20.51362
	Professional	−0.69909	0.46952	21.06261	1.81476

28. Bingeing and politics Students in a large Statistics class at a university were asked to describe their political position and whether they had engaged in binge drinking (five drinks at a sitting for a man; four for a woman).

	Political Position			
Binge Drinking	**Conservative**	**Moderate**	**Liberal**	**Total**
In the past 3 days	7	19	14	**40**
In the past week	9	19	28	**56**
In the past month	3	8	23	**34**
Some other time	8	15	22	**45**
Never	9	22	20	**51**
Total	**36**	**83**	**107**	**226**

Expected Values			
	Conservative	**Moderate**	**Liberal**
In the past 3 days	6.37168	14.6903	18.9381
In the past week	8.92035	20.5664	26.5133
In the past month	5.41593	12.4867	16.0973
Some other time	7.16814	16.5265	21.3053
Never	8.12389	18.7301	24.1460

a) What kind of chi-square test is appropriate—goodness-of-fit, homogeneity, or independence?
b) State your hypotheses.
c) State and check the conditions.
d) How many degrees of freedom are there?
e) The calculation yields $\chi^2 = 10.10$, with $P = 0.2578$. State your conclusion.
f) Would you expect that a larger sample might find statistical significance? Explain.

T **29. Cranberry juice** It's common folk wisdom that drinking cranberry juice can help prevent urinary tract infections in women. In 2001, the *British Medical Journal* reported the results of a Finnish study in which three groups of 50 women were monitored for these infections over six months. One group drank cranberry juice daily, another group drank a lactobacillus drink, and the third drank neither of those beverages, serving as a control group. In the control group, 18 women developed at least one infection, compared to 20 of those who consumed the lactobacillus drink and only 8 of those who drank cranberry juice. Does this study provide supporting evidence for the value of cranberry juice in warding off urinary tract infections?
a) Is this a survey, a retrospective study, a prospective study, or an experiment? Explain.
b) Will you test goodness-of-fit, homogeneity, or independence?
c) State the hypotheses.
d) Test the conditions.
e) How many degrees of freedom are there?
f) Find χ^2 and the P-value.
g) State your conclusion.
h) If you concluded that the groups are not the same, analyze the differences using the standardized residuals of your calculations.

T **30. Cars** A random survey of autos parked in the student lot and in the staff lot at a large university classified the brands by country of origin, as seen in the table. Are there differences in the national origins of cars driven by students and staff?

	Driver	
Origin	**Student**	**Staff**
American	107	105
European	33	12
Asian	55	47

a) Is this a test of independence or homogeneity?
b) Write appropriate hypotheses.
c) Check the necessary assumptions and conditions.
d) Find the P-value of your test.
e) State your conclusion and analysis.

31. Computer-assisted ordering In the book *Essentials of Marketing Research*, Dillon, Madden, and Firtle discuss the relationship between delivery time and computer-assisted ordering. The table shows data from a sample of 40 firms. Is there evidence of an association between computer-assisted ordering and delivery time? Do a test.

	Below Industry Average	**Equal to Industry Average**	**Above Industry Average**
Not computer-assisted	4	12	8
Computer-assisted	10	4	2

a) Is this a test of homogeneity or independence?
b) Write an appropriate hypothesis.
c) Are the conditions for inference satisfied?
d) Find the P-value for your test.
e) State a complete conclusion.

T **32. Fish diet** Medical researchers followed 6272 Swedish men for 30 years to see if there was any association between the amount of fish in their diet and prostate cancer.[11]

Fish Consumption	**Total Subjects**	**Prostate Cancers**
Never/seldom	124	14
Small part of diet	2621	201
Moderate part	2978	209
Large part	549	42

a) Is this a survey, a retrospective study, a prospective study, or an experiment? Explain.
b) Is this a test of homogeneity or independence?
c) Do you see evidence of an association between the amount of fish in a man's diet and his risk of developing prostate cancer?
d) Does this study prove that eating fish does not prevent prostate cancer? Explain.

[11]"Fatty fish consumption and risk of prostate cancer," *Lancet*, June 2001.

33. Grades Two professors teach an introductory Statistics course. The table shows the distribution of final grades they awarded. Is one of these professors an "easier" grader?

	Prof. Alpha	Prof. Beta
A	3	9
B	11	12
C	14	8
D	9	2
F	3	1

a) Will you test goodness-of-fit, homogeneity, or independence?
b) Write appropriate null hypotheses.
c) Find the expected counts for each cell, and explain why the chi-square procedures are not appropriate for this table.

T **34. Full moon** Some people believe that a full moon elicits unusual behaviour in people. The table shows the number of arrests made in a small town during weeks of six full moons and six other randomly selected weeks in the same year. We wonder if there is evidence of a difference in the types of illegal activity that take place.

	Full Moon	Not Full
Violent (murder, assault, rape, etc.)	2	3
Property (burglary, vandalism, etc.)	17	21
Drugs/Alcohol	27	19
Domestic abuse	11	14
Other offenses	9	6

a) Will you test goodness-of-fit, homogeneity, or independence?
b) Write appropriate null hypotheses.
c) Find the expected counts for each cell, and explain why the chi-square procedures are not appropriate for this table.

35. Grades, again In some situations where the expected cell counts are too small, as in the case of the grades given by Professors Alpha and Beta in Exercise 31, we can complete an analysis anyway. We can often proceed after combining cells in some way that makes sense and also produces a table in which the conditions are satisfied. Here we create a new table displaying the same data, but calling D and F grades "Below C":

	Prof. Alpha	Prof. Beta
A	3	9
B	11	12
C	14	8
Below C	12	3

a) Find the expected counts for each cell in this new table, and explain why a chi-square procedure is now appropriate.
b) With this change in the table, what has happened to the number of degrees of freedom?
c) Test your hypothesis about the two professors, and state an appropriate conclusion.

T **36. Full moon, next phase** In Exercise 32, you found that the expected cell counts failed to satisfy the conditions for inference.
a) Find a sensible way to combine some cells that will make the expected counts acceptable.
b) Test a hypothesis about the full moon and state your conclusion.

37. Eating in front of the TV Roper Reports asked a random sample of people in 30 countries whether they agreed with the statement "I like to nibble while reading or watching TV." Allowable responses were "Agree completely," "Agree somewhat, "Neither disagree nor agree," "Disagree somewhat," "Disagree completely," and "I Don't Know/No Response." Does a person's age influence their response? Here are data from 3792 respondents in the 2006 sample of five countries (China, India, France, United Kingdom, and United States) for three age groups (teens, 30s (30–39), and over 60):

	Agree Completely	Agree Somewhat	Neither Disagree nor Agree	Disagree Somewhat	Disagree Completely
Teen	369	540	299	175	106
30s	272	522	325	229	170
60+	93	207	153	154	178

a) Make an appropriate display of these data.
b) Does a person's age seem to affect their response to the question about nibbling?

38. Eating in front of the TV, II In Exercise 35, we saw a survey of people who were asked if they agreed with the statement "I like to nibble while reading or watching TV." Does a person's culture tend to influence the response? In the table are random samples from each of the five countries that responded:

	Agree Completely	Agree Somewhat	Neither Disagree nor Agree	Disagree Somewhat	Disagree Completely
China	253	479	382	244	139
France	188	407	328	296	308
India	315	395	246	227	332
U.K.	307	578	330	218	114
U.S.	308	632	331	201	73

a) Make an appropriate display of the data.

b) Does what country a person lives in seem to affect their response to the question about nibbling? Explain.

T **39. Racial steering** A subtle form of racial discrimination in housing is "racial steering." Racial steering occurs when real estate agents show prospective buyers only homes in neighbourhoods already dominated by those buyers' race. This violates the *Canadian Human Rights Act*. According to an article in *Chance* magazine,[12] tenants at a large apartment complex recently filed a lawsuit alleging racial steering. The complex is divided into two parts: Section A and Section B. The plaintiffs claimed that white potential renters were steered to Section A, while African-Americans were steered to Section B. The table displays the data that were presented in court to show the locations of recently rented apartments. Do you think there is evidence of racial steering?

	New Renters		
	White	Black	Total
Section A	87	8	95
Section B	83	34	117
Total	170	42	212

T **40. *Titanic*, redux** Newspaper headlines at the time, and traditional wisdom in the succeeding decades, have held that women and children escaped the *Titanic* in greater proportions than men. Here's a table with the relevant data. Do you think that survival was independent of whether the person was male or female? Defend your conclusion.

	Female	Male	Total
Alive	343	367	710
Dead	127	1364	1491
Total	470	1731	2201

41. Steering, revisited You could have checked the data in Exercise 39 for evidence of racial steering using two-proportion z procedures.

a) Find the z-value for this approach, and show that when you square your z-value, you get the value of χ^2 you calculated in Exercise 39.

b) Show that the resulting P-values are the same.

T **42. Survival on the *Titanic*, one more time** In Exercise 40, you could have checked for a difference in the chances of survival for men and women using two-proportion z procedures.

a) Find the z-value for this approach.

b) Show that the square of your calculated value of z is the value of χ^2 you calculated in Exercise 40.

c) Show that the resulting P-values are the same.

T **43. Race and education** Data from the U.S. Census Bureau show levels of education attained by age 30 for a sample of U.S. residents.

	Not HS Grad	HS Diploma	College Grad	Adv. Degree
White	810	6429	4725	1127
Black	263	1598	549	117
Hispanic	1031	1269	412	99
Other	66	341	305	197

Do these data highlight significant differences in education levels attained by these groups?

T **44. Pregnancies** Most pregnancies result in live births, but some end in miscarriages or stillbirths. A 2005 Statistics Canada report examined those outcomes, broken down by the age of the mother. The table shows counts consistent with that report. Is there evidence that the distribution of outcomes is not the same for these age groups? Do you have any concerns about your analysis?

Age of Mother	Live Births	Fetal Losses
Under 20	14 013	586
20–24	55 318	1422
25–29	105 566	2121
30–34	107 524	2161
35 or over	59 672	2158

T **45. Race and education, part 2** Consider only the people who have graduated from high school from Exercise 43. Do these data suggest there are significant differences in opportunities for black and Hispanic Americans who have completed high school to pursue university or advanced degrees?

	HS Diploma	College Grad	Adv. Degree
Black	1598	549	117
Hispanic	1269	412	99

T **46. Education by age** Use the survey results in the table to investigate differences in education level attained among different age groups in Canada.

Education Level	Age Group				
	25–34	35–44	45–54	55–64	≥65
Not HS grad	27	50	52	71	101
HS	82	19	88	83	59
1–3 years college	43	56	26	20	20
≥ 4 years college	48	75	34	26	20

47. Ranking universities In 2004, the Institute of Higher Education at Shanghai's Jiao Tong University evaluated the world's universities. Among their criteria were the size of the institution, the number of Nobel Prizes and Fields Medals won by faculty and alumni, and the faculty's research output. This ranking of the top 502 universities included 200 in North or Latin America, 209 in Europe, and 93 in the rest of the world (Asia, Pacific, and Africa). A closer examination of the top 100 showed

[12]Vol. 14, no. 2 (2001).

55 in the Americas, 37 in Europe, and eight elsewhere. Is there anything unusual about the geographical distribution of the world's top 100 universities?

48. Ranking universities, redux In Exercise 45, you read about the world's top universities, as ranked by the Institute of Higher Education. The table shows the geographical distribution of these universities by groups of 100. (Not all groups have exactly 100 because of ties in the rankings.) Do these institutions appear to be randomly distributed around the world, or does there appear to be an association between ranking and location?

		Ranking				
		1–100	101–200	201–300	301–400	401–500
Location	North/Latin America	55	46	37	26	36
	Europe	37	42	46	46	38
	Asia/Africa/Pacific	8	13	17	30	25

Just Checking ANSWERS

1. We need to know how well beetles can survive six hours in a Plexiglas® box so that we have a baseline to compare the treatments.
2. There's no difference in survival rate in the three groups.
3. $(2 - 1)(3 - 1) = 2$ df
4. 50
5. 2
6. The mean value for a χ^2 with 2 df is 2, so 10 seems pretty large. The P-value is probably small.
7. This is a test of homogeneity. The clue is that the question asks whether the distributions are alike.
8. This is a test of goodness-of-fit. We want to test the model of equal assignment to all lots against what actually happened.
9. This is a test of independence. We have responses on two variables for the same individuals.

part VI Review

Learning About the World

Quick Review

We continue to explore how to answer questions about the statistics we get from samples and experiments. In this Part, those questions have been about means—means of one sample, two independent samples, or matched pairs—and about proportions in several categories and in relationships between categorical variables. Here's a brief summary of the key concepts and skills:

- A confidence interval uses a sample statistic to estimate a range of possible values for the parameter of a population model.
- A hypothesis test proposes a model for the population, then examines the observed statistics to see if the model is plausible.
- Statistical inference procedures for proportions are based on the Central Limit Theorem. We can use Normal models to make inferences about a single proportion or the difference of two proportions.
- Statistical inference procedures for means require a new sampling model, since we usually don't know the population standard deviation. Student's t-models take the additional uncertainty of independently estimating the standard deviation into account.
  - We can use t-models to make inferences about one mean, the difference of two independent means, or the mean of paired differences.
  - No inference procedure is valid unless the underlying assumptions are true. Always think about the assumptions and check the conditions before proceeding.
  - Because t-models assume that samples are drawn from Normal populations, data in the sample should appear to be nearly Normal for small samples. Skewness and outliers are particularly problematic. For larger samples, the *robustness* of the t procedures allows us to relax considerably the Normality assumption.
  - When there are two variables, you must think carefully about how the data were collected. You may use two-sample t procedures only if the groups are independent.
  - Unless there is some obvious reason to suspect that two independent populations have the same standard deviation, you should not pool the variances. It is never wrong to use unpooled t procedures.
  - If two groups are somehow paired, the data are *not* from independent groups. You must use matched-pairs t procedures and test the mean difference rather than the difference in the means.
- Not all sampling distributions are unimodal, symmetric, or bell-shaped. Inferences about distributions of counts use chi-square models, which are unimodal but skewed to the high end. Nevertheless, the sampling distribution plays the same role in inference, helping us to translate between statistics from data and probabilities.
  - To see if an observed distribution is consistent with a proposed model, use a chi-square goodness-of-fit test.
  - To see if two or more observed distributions could have arisen from populations with the same model, use a test of homogeneity.
  - To see if two random categorical variables are independent, perform a chi-square test of independence.
- You can now use statistical inference to answer questions about means, proportions, distributions, and associations.
  - No inference procedure is valid unless the underlying assumptions are true. Always check the conditions before proceeding.
  - You can make inferences about a single proportion or about the difference between two proportions using Normal models.
  - You can use t-models to make inferences about one mean, about the difference between two independent means, or about the mean of paired differences.
  - You can make inferences about distributions using chi-square models.
  - You can make inferences about association between categorical variables using chi-square models.

Now for some opportunities to review these concepts. Be careful. You have a lot of thinking to do. These review exercises mix questions about proportions and means. You have to determine which of our inference procedures is appropriate in each situation. Then, you have to check the proper assumptions and conditions. Keeping track of those can be difficult, so first we summarize the many procedures with their corresponding assumptions and conditions. Look them over carefully . . . then, on to the Exercises!

Assumptions for Inference	And the Conditions that Support or Override Them
Proportions (z)	
• **One sample**	
1. Individuals are independent.	1. SRS and $<$ 10%* of the population.
2. Sample is sufficiently large.	2. Successes and failures $\geq$ 10.
• **Two sample**	
1. Samples are independent.	1. (Think about how the data were collected.)
2. Data in each sample are independent.	2. Both are SRSs and $<$ 10%* of populations OR random allocation.
3. Both samples are sufficiently large.	3. Successes and failures $\geq$ 10 for both.
Means (t)	
• **One sample** (df $= n - 1$)	
1. Individuals are independent.	1. SRS
2. Population has a Normal model.	2. Histogram is unimodal and symmetric.*
• **Matched pairs** (df $= n - 1$)	
1. Data are paired.	1. (Think about the design.)
2. Individuals are independent.	2. SRS OR random allocation.
3. Population of differences is Normal.	3. Histogram of differences is unimodal and symmetric.**
• **Two independent groups** (df from technology)	
1. Groups are independent.	1. (Think about the design.)
2. Data in each group are independent.	2. SRSs OR random allocation.
3. Both populations are Normal.	3. Both histograms are unimodal and symmetric.*
Distributions/Association (χ^2)	
• **Goodness of fit** (df = # of cells − 1; one variable, one sample compared with population model)	
1. Data are counts.	1. (Are they?)
2. Data in sample are independent.	2. SRS.
3. Group is sufficiently large.	3. All expected counts $\geq$ 5.
• **Homogeneity** [df $= (r-1)(c-1)$; samples from many populations compared on one variable]	
1. Data are counts.	1. (Are they?)
2. Data in samples are independent.	2. SRSs OR random allocation.
3. Groups are sufficiently large.	3. All expected counts $\geq$ 5.
• **Independence** [df $= (r-1)(c-1)$; sample from one population classified on two variables]	
1. Data are counts.	1. (Are they?)
2. Data are independent.	2. SRS.
3. Group is sufficiently large.	3. All expected counts $\geq$ 5.

*not critical though, since $>$10% makes our reported confidence levels and P-values conservative
**less critical as n increases

Review Exercises

1. **Crawling** A study published in 1993 found that babies born at different times of the year may develop the ability to crawl at different ages! The author of the study suggested that these differences may be related to the temperature at the time the infant is six months old.[1]
   a) The study found that 32 babies born in January crawled at an average age of 29.84 weeks, with a standard deviation of 7.08 weeks. Among 21 July babies, crawling ages averaged 33.64 weeks with a standard deviation of 6.91 weeks. Is this difference significant?
   b) For 26 babies born in April, the mean and standard deviation were 31.84 and 6.21 weeks, while for 44 October babies, the mean and standard deviation of crawling ages were 33.35 and 7.29 weeks. Is this difference significant?
   c) Are these results consistent with the researchers' claim? (We'll examine these data in more detail in a later chapter.)

T 2. **Mazes and smells** Can pleasant smells improve learning? Researchers timed 21 subjects as they tried to complete paper-and-pencil mazes. Each subject attempted a maze both with and without the presence of a floral aroma. Subjects were randomized with respect to whether they did the scented trial first or second. The table shows some of the data collected. Is there any evidence that the floral scent improved the subjects' ability to complete the mazes?[2]

Time to Complete the Maze (sec)			
Unscented	Scented	Unscented	Scented
25.7	30.2	61.5	48.4
41.9	56.7	44.6	32.0
51.9	42.4	35.3	48.1
32.2	34.4	37.2	33.7
64.7	44.8	39.4	42.6
31.4	42.9	77.4	54.9
40.1	42.7	52.8	64.5
43.2	24.8	63.6	43.1
33.9	25.1	56.6	52.8
40.4	59.2	58.9	44.3
58.0	42.2		

3. **BC birth weights** Below are the means and standard deviations of birth weights of newborns born in BC during 1981–2000 to some different groups, for births following a 40-week gestation period (the modal gestation period).[3] For the entire BC population, the mean birth weight was 3558 grams.

Group	# Births	Mean Weight	Standard Deviation
First Nations people	20 578	3645 g	466 g
Canadians of Chinese origin	14 586	3393 g	391 g

   a) Find a 95% confidence interval for the average birth weight for each group above.
   b) Does either group differ significantly from the general BC population?
   c) Do the means for these two groups differ significantly from each other? Include the P-value. If so, find a confidence interval for the mean difference.

T 4. **Drugs** In a full-page ad that ran in many U.S. newspapers in August 2002, a Canadian discount pharmacy listed costs of drugs that could be ordered from a Web site in Canada. The table compares prices (in US$) for commonly prescribed drugs.

Drug Name	Cost per 100 Pills: United States	Canada	Percent Savings
Cardizem	131	83	37
Celebrex	136	72	47
Cipro	374	219	41
Pravachol	370	166	55
Premarin	61	17	72
Prevacid	252	214	15
Prozac	263	112	57
Tamoxifen	349	50	86
Vioxx	243	134	45
Zantac	166	42	75
Zocor	365	200	45
Zoloft	216	105	51

   a) Give a 95% confidence interval for the average savings in dollars.
   b) Give a 95% confidence interval for the average savings in percent.
   c) Which analysis do you think is more appropriate? Why?
   d) In small print the newspaper ad says, "Complete list of all 1500 drugs available on request." How does this comment affect your conclusions above?

T 5. **Pottery** Archaeologists can use the chemical composition of clay found in pottery artifacts to determine whether different sites were populated by the same

[1]Benson and Janette, *Infant Behavior and Development*, 1993.
[2]A. R. Hirsch and L. H. Johnston, "Odors and Learning." Chicago: Smell and Taste Treatment and Research Foundation.
[3]British Columbia Vital Statistics Agency

ancient people. They collected five samples of Romano-British pottery from each of two sites in Great Britain—the Ashley Rails site and the New Forest site—and measured the percentage of aluminum oxide in each. Based on these data, do you think the same people used these two kiln sites? Base your conclusion on a 95% confidence interval for the difference in aluminum oxide content of pottery made at the sites.[4]

Ashley Rails	19.1	14.8	16.7	18.3	17.7
New Forest	20.8	18.0	18.0	15.8	18.3

T **6. Diet** Thirteen overweight women volunteered for a study to determine whether eating specially prepared crackers before a meal could help them lose weight. The subjects were randomly assigned to eat crackers with different types of fibre (bran fibre, gum fibre, both, and a control cracker). Unfortunately, some of the women developed uncomfortable bloating and upset stomachs. Researchers suspected that some of the crackers might be at fault. The contingency table of cracker versus bloat shows the relationship between the four different types of crackers and the reported bloating. The study was paid for by the manufacturers of the gum fibre. What would you recommend to them about the prospects for marketing their new diet cracker?

		Bloat	
		Little/None	Moderate/Severe
Cracker	Bran	11	2
	Gum	4	9
	Combo	7	6
	Control	8	4

7. Gehrig Ever since baseball legend Lou Gehrig developed amyotrophic lateral sclerosis (ALS), this deadly condition has been commonly known as Lou Gehrig's disease. Some believe that ALS is more likely to strike athletes or the very fit. Columbia University neurologist Lewis P. Rowland recorded personal histories of 431 patients he examined between 1992 and 2002. He diagnosed 280 as having ALS: 38% of them had been varsity athletes; the other 151 had other neurological disorders, and only 26% of them had been varsity athletes. (*Science News*, Sept. 28, 2002)

a) Is there evidence that ALS is more common among athletes? Attack this two ways: using a *t*-test and a chi-square test: How are these two tests related?

b) What kind of study is this? How does that affect the inference you made in part a)?

T **8. Teen drinking** A study of the health behaviour of school-aged children asked a sample of 15-year-olds in several different countries if they had been drunk at least twice. The table shows the results. Give a 95% confidence interval for the difference in the rates for males and females. Be sure to check the assumptions that support your chosen procedure, and explain what your interval means.[5]

		Percent of 15-Year-Olds Drunk at Least Twice	
		Female	Male
Country	Denmark	63	71
	Wales	63	72
	Greenland	59	58
	England	62	51
	Finland	58	52
	Scotland	56	53
	No. Ireland	44	53
	Slovakia	31	49
	Austria	36	49
	Canada	42	42
	Sweden	40	40
	Norway	41	37
	Ireland	29	42
	Germany	31	36
	Latvia	23	47
	Estonia	23	44
	Hungary	22	43
	Poland	21	39
	USA	29	34
	Czech Rep.	22	36
	Belgium	22	36
	Russia	25	32
	Lithuania	20	32
	France	20	29
	Greece	21	24
	Switzerland	16	25
	Israel	10	18

9. Genetics Two human traits controlled by a single gene are the ability to roll one's tongue and whether one's ear lobes are free or attached to the neck. Genetic theory says that people will have neither, one, or both of these traits in the ratio 1:3:3:9 (1—attached, noncurling; 3—attached, curling; 3—free, noncurling; 9—free, curling). An Introductory Biology class of 122 students collected the data shown. Are they consistent with the genetic theory? Test an appropriate hypothesis, and state your conclusion.

	Trait			
	Attached, noncurling	Attached, curling	Free, noncurling	Free, curling
Count	10	22	31	59

[4]A. Tubb, A. J. Parker, and G. Nickless, 1980, "The analysis of Romano-British pottery by atomic absorption spectrophotometry." *Archaeometry*, 22: 153–171.

[5]*Health and Health Behavior Among Young People*. Copenhagen: World Health Organization, 2000.

10. Waiting for homicide In London, England, the number of days between homicides was recorded for the period April 2004–March 2007, and is displayed below.[6] Days is on the x-axis, with counts on the y-axis.

a) Estimate with a 95% confidence interval the average time between homicides in London (estimate as best you can what you need from the graph).

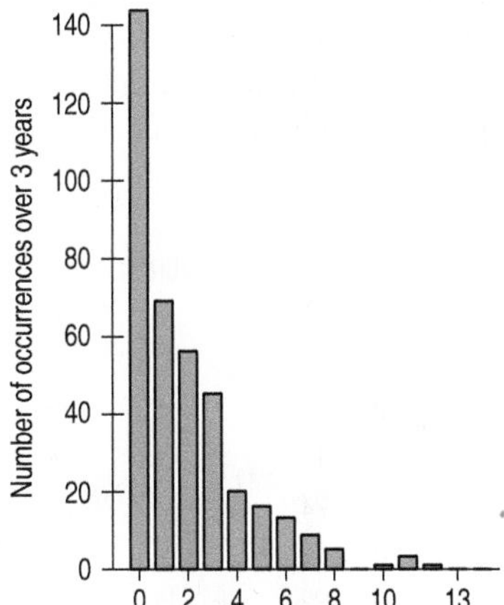

b) Can you trust the reported confidence level of 95% for your interval?

11. Babies The National Perinatal Statistics Unit of the Sydney Children's Hospital reports that the mean birth weight of all babies born in Australia in 1999 was 3360 grams. A British Columbia hospital reports that the average weight of 112 babies born there in 1999 was 3464 g, with a standard deviation of 595 g. If we believe the BC babies born at this hospital fairly represent newborn babies in BC, is there any evidence that BC babies and Australian babies do not weigh the same amount at birth?

12. Renters weigh less On June 16, 2009, the *Windsor Star* described a study that found female homeowners tend to weigh more than women who are leasing a residence.[7] In this survey of 600 women, homeowners outweighed renters by 12.5 pounds (5.7 kg), on average.

a) Are you impressed by the difference? What else might you like to know?

b) If 300 women were in each group, what would be the P-value for the observed difference of 12.5 pounds if the pooled standard deviation was 30 pounds?

c) If 300 women were in each group, what would the pooled standard deviation have to be in order to make a difference of 12.5 pounds statistically significant at the 1% level?

13. Feeding fish Large mouth bass is one type of fish raised by the aquaculture industry. Researchers wanted to know whether the fish would grow better if fed a natural diet of fathead minnows or an artificial diet of food pellets. They stocked six ponds with bass fingerlings weighing about 8 grams. For one year, the fish in three of the ponds were fed minnows, and the others were fed the commercially prepared pellets. The fish were then weighed and measured. The bass fed a natural food source had a higher average length (19.6 cm) and weight (95.9 g) than those fed the commercial fish food (17.3 cm and 72.0 g, respectively). The researchers reported P-values for both measurements to be less than 0.001.

a) Explain to someone who has not studied Statistics what the P-values mean here.

b) What advice should the researchers give the people who raise largemouth bass?

c) If that advice turns out to be incorrect, what type of error occurred?

T **14. Risk** A study of auto safety determined the number of driver deaths per million vehicle sales for the model years 2001–2004, classified by type of vehicle. The data below are for six midsize models and six SUVs. We wonder if there is evidence that drivers of SUVs are safer, hoping to create a 95% confidence interval for the difference in driver death rates for the two types of vehicles. Are these data appropriate for this inference? Explain.[8]

Midsize	11	14	30	55	85	118
SUV	13	21	44	51	115	127

T **15. Twins** In 2000, *The Journal of the American Medical Association* published a study that examined a sample of pregnancies that resulted in the birth of twins. Births were classified as preterm with intervention (induced labour or Caesarean), preterm without such procedures, or term or post-term. Researchers also classified the pregnancies by the level of prenatal medical care the mother received (inadequate, adequate, or intensive). The table below summarizes the data from the years 1995–1997. Figures are in thousands of births.[9]

Twin Births 1995–1997 (in thousands)

Level of Prenatal Care		Preterm (induced or Caesarean)	Preterm (without procedures)	Term or postterm	Total
	Intensive	18	15	28	61
	Adequate	46	43	65	154
	Inadequate	12	13	38	63
	Total	76	71	131	278

Is there evidence of an association between the duration of the pregnancy and the level of care received by the mother?

[6] D. Spiegelhalter & A. Barnett, 2009, "London murders: A predictable pattern?" *Significance*. Reprinted with permission of John Wiley & Sons Inc.
[7] M. Harris, 2009, "Women who rent weigh less, survey says." *Windsor Star*, June 16, 2009, p. B8.

[8] *Insurance Institute for Highway Safety*, Status Report, Vol. 42, No. 4, April 19, 2007.
[9] *JAMA*, 284[2002]:335–341.

16. Twins, again After reading of the *JAMA* study in Exercise 15, a large city hospital examined their records of twin births for several years, and found the data summarized in the table below. Is there evidence that the way the hospital deals with pregnancies involving twins may have changed?

Outcome of Pregnancy		1990	1995	2000
	Preterm (induced or Caesarean)	11	13	19
	Preterm (without procedures)	13	14	18
	Term or postterm	27	26	32

17. Age In a study of how depression may impact one's ability to survive a heart attack, researchers reported the ages of the two groups they examined. The mean age of 2397 patients without cardiac disease was 69.8 years (SD = 8.7 years), while for the 450 patients with cardiac disease the mean and standard deviation of the ages were 74.0 and 7.9, respectively.

a) Create a 95% confidence interval for the difference in mean ages of the two groups.

b) How might an age difference confound these research findings about the relationship between depression and ability to survive a heart attack?

18. Smoking In the depression and heart attack research described in Exercise 17, 32% of the diseased group were smokers, compared with only 23.7% of those free of heart disease.

a) Create a 95% confidence interval for the difference in the proportions of smokers in the two groups.

b) Is this evidence that the two groups in the study were different? Explain.

c) Could this be a problem in analyzing the results of the study? Explain.

19. Computer use A Gallup telephone poll of 1240 teens conducted in 2001 found that boys were more likely than girls to play computer games, by a margin of 77% to 65%. An equal number of boys and girls were surveyed.

a) What kind of sampling design was used?

b) Give a 95% confidence interval for the difference in game playing between boys and girls.

c) Does your confidence interval suggest that among all teens a higher percentage of boys than girls play computer games?

20. Hearing Fitting someone for a hearing aid requires assessing the patient's hearing ability. In one method of assessment, the patient listens to a tape of 50 English words. The tape is played at low volume, and the patient is asked to repeat the words. The patient's hearing ability score is the number of words perceived correctly. Four tapes of equivalent difficulty are available so that each ear can be tested with more than one hearing aid. These lists were created to be equally difficult to perceive in silence, but hearing aids must work in the presence of background noise. Researchers had 24 subjects with normal hearing compare two of the tapes when a background noise was present, with the order of the tapes randomized. Is it reasonable to assume that the two lists are still equivalent for purposes of the hearing test when there is background noise? Base your decision on a confidence interval for the mean difference in the number of words people might misunderstand.[10]

	List			List	
Subject	A	B	Subject	A	B
1	24	26	13	36	32
2	32	24	14	32	34
3	20	22	15	38	32
4	14	18	16	14	18
5	32	24	17	26	20
6	22	30	18	14	20
7	20	22	19	38	40
8	26	28	20	20	26
9	26	30	21	14	14
10	38	16	22	18	14
11	30	18	23	22	30
12	16	34	24	34	42

21. Forest fires Below is a random sample of fire sizes in hectares (ha) from the Large Fire Database at Natural Resources Canada for fires that covered over 200 ha that occurred in Canada in the late 1990s.[11]

900.0	2 000.0	804.2	1 682.4	559.0	5 000.0	309.0
3 790.0	2 450.0	1 380.0	12 630.0	4 243.6	557.9	1 560.0
485.6	3 215.0	12 492.0	2 300.0	389.9	1 366.0	2 612.0
1 200.0	208.0	55 000.0	40 759.7	250.0	6 607.2	774.1
6 284.0	349.8	4 381.0	7 400.0	255.3	304.0	1 200.0
545.9	627.0	1 777.0	400.0	1 657.2	2 443.4	547.8
14 200.0	631.0	1 280.0	221.0	450.0	1 550.7	17 400.0
459.0	2 513.0	34 209.0	4 274.1	356.0	4 4261.0	1 400.0
347.9	2 800.0	5 650.0	12 557.0	19 320.8	30 070.0	3 261.2
321.3	19 735.6	435.0	6 465.5	1 817.0	13 433.0	339.0
3 580.7	739.0	799.5	4 170.0	3 000.0	414.9	14 072.5
500.0	27 500.0	225.4	750.0	611.0	300.0	1 642.4
10 240.0	447.6	48 451.1	1 500.0	297.0	825.0	850.0
333.2	17 597.0	431.1	300.0	3267.1	604.6	10 600.0
17 531.0	520.0					

a) Estimate with a 90% confidence interval the true mean fire size for all fires in Canada over this period.

[10]Faith Loven, *A Study of the Interlist Equivalency of the CID W-22 Word List Presented in Quiet and in Noise.* University of Iowa, 1981.

[11]Reproduced with permission from Natural Resources Canada, Canadian Forest Service, 2013. http://cwfis.cfs.nrcan.gc.ca/home.

b) The actual mean of all fires over 200 ha during this period was 6398. Did your interval in a) catch this value? If you were to repeat the sampling procedure over and over, obtaining independent random samples from the population of fires, and then calculate confidence intervals from each sample, how often would your confidence interval contain the value 6398? Suppose you obtained 200 independent random samples from the entire population and counted the number of samples that contained the value 6398; call it X. What would be the distribution of the random variable X?

c) Examine the shape of the distribution for the sample. Do you have any qualms about the validity of your analysis in part a)?

d) What measure of centre do you consider most appropriate here? If you preferred to work with a different measure of centre, what inferential procedure is available to you for testing hypotheses? (One way to form a 95% confidence interval would be to place in it all null values that could not be rejected in a two-sided test at the 5% level.)

e) We might also wonder if we are using the most appropriate measurement scale. Should we change it? If so, how would you suggest proceeding?

T **22. Newspapers** Who reads the newspaper more, men or women? Eurostat, an agency of the European Union (EU), conducts surveys on several aspects of daily life in EU countries. Recently, the agency asked samples of 1000 respondents in each of 14 European countries whether they read the newspaper on a daily basis. Below are the data by country and sex.

	% Reading a Newspaper Daily	
Country	**Men**	**Women**
Belgium	56.3	45.5
Denmark	76.8	70.3
Germany	79.9	76.8
Greece	22.5	17.2
Spain	46.2	24.8
Ireland	58.0	54.0
Italy	50.2	29.8
Luxembourg	71.0	67.0
Netherlands	71.3	63.0
Austria	78.2	74.1
Portugal	58.3	24.1
Finland	93.0	90.0
Sweden	89.0	88.0
U.K.	32.6	30.4

a) Examine the differences in the percentages for each country. Which of these countries seem to be outliers? What do they have in common?

b) After eliminating the outliers, is there evidence that in Europe men are more likely than women to read the newspaper?

T **23. Meals** A university student is on a "meal program." His budget allows him to spend an average of $10 per day for the term. He keeps track of his daily food expenses for two weeks; the data are given in the table. Is there strong evidence that he will overspend his food allowance? Explain.

Date	Cost ($)	Date	Cost ($)
7/29	15.20	8/5	8.55
7/30	23.20	8/6	20.05
7/31	3.20	8/7	14.95
8/1	9.80	8/8	23.45
8/2	19.53	8/9	6.75
8/3	6.25	8/10	0
8/4	0	8/11	9.01

T **24. Gender and school** Below are primary school enrolment percentages (enrolment of the official age group for primary education expressed as a percentage of the corresponding population) by gender for a sample of countries. Compute a 95% confidence interval for the difference between genders, interpret it, and be sure to examine any important assumptions. Is there evidence of a systemic gender effect on primary education accessibility?

Location	Net Primary School Enrolment Ratio Female (%), 2006	Net Primary School Enrolment Ratio Male (%), 2006
Algeria	94	96
Benin	73	87
Burkina Faso	42	52
Burundi	73	76
Cape Verde	87	88
Central African Republic	38	53
Congo	52	58
Eritrea	43	50
Ethiopia	62	68
Gambia	64	59
Ghana	64	63
Guinea	66	77
Kenya	76	75
Lesotho	74	71
Liberia	39	40
Madagascar	96	96
Malawi	94	88
Mali	54	67
Mauritania	82	78
Mauritius	96	94
Mozambique	73	79
Namibia	79	74
Niger	37	50
Sao Tome and Principe	95	97

Location	Net Primary School Enrolment Ratio Female (%), 2006	Net Primary School Enrolment Ratio Male (%), 2006
Senegal	70	71
Togo	75	86
United Republic of Tanzania	97	98
Zambia	94	90
Zimbabwe	88	87
Bahamas	89	87
Barbados	96	97
Belize	97	97
Bolivia	95	94
Colombia	88	89
Cuba	97	96
Dominica	79	75
Dominican Republic	78	76
Ecuador	97	96
El Salvador	94	94
Guatemala	92	96
Honduras	97	96
Mexico	97	98
Nicaragua	90	90
Panama	98	99
Peru	97	96
Saint Lucia	97	99
Suriname	98	95
United States	93	91
Uruguay	100	100
Venezuela	91	91
Djibouti	34	42
Egypt	92	96
Jordan	90	89
Kuwait	83	84
Lebanon	82	82
Morocco	85	91
Oman	75	73
Pakistan	57	73
Qatar	94	93
Tunisia	97	96
United Arab Emirates	88	88
Andorra	83	83
Armenia	84	80
Austria	98	97
Azerbaijan	83	86
Belarus	88	90
Belgium	97	97
Bulgaria	92	93
Croatia	90	91
Cyprus	99	99
Denmark	96	95
Estonia	94	95

Location	Net Primary School Enrolment Ratio Female (%), 2006	Net Primary School Enrolment Ratio Male (%), 2006
Finland	97	97
France	99	98
Georgia	91	88
Germany	98	98
Greece	99	100
Hungary	88	89
Iceland	97	98
Ireland	95	94
Israel	97	96
Italy	98	99
Kazakhstan	90	90
Kyrgyzstan	85	86
Lithuania	89	90
Luxembourg	98	96
Netherlands	97	99
Norway	98	98
Poland	96	96
Portugal	98	98
Republic of Moldova	88	88
Romania	93	93
Russian Federation	91	91
Slovenia	95	96
Spain	99	100
Sweden	95	95
Switzerland	89	89
Tajikistan	95	99
Turkey	89	93
Ukraine	90	90
United Kingdom	99	98
Bhutan	79	79
India	87	90
Indonesia	94	97
Maldives	97	97
Myanmar	100	99
Thailand	94	94
Australia	97	96
Brunei Darussalam	94	94
Cambodia	89	91
Fiji	91	91
Japan	100	100
Lao People's Democratic Republic	81	86
Mongolia	93	90
New Zealand	99	99
Philippines	92	90
Vanuatu	87	88

Source: Reproduced, with the permission of the publisher, from World Health Statistics-socioeconomics and demographics-net primary school enrollment for 2006.

25. Teach for America Several programs attempt to address the shortage of qualified teachers in the U.S. by placing uncertified instructors in schools with acute needs—often in inner cities. A 1999–2000 study compared students taught by certified teachers with others taught by undercertified teachers in the same schools. Reading scores of the students of certified teachers averaged 35.62 points with standard deviation 9.31. The scores of students instructed by undercertified teachers had mean 32.48 points with standard deviation 9.43 points on the same test. There were 44 students in each group. The appropriate *t* procedure has 86 degrees of freedom. Is there evidence of lower scores with uncertified teachers? Discuss.[12]

26. Labour force The Labour Force Survey conducted by Statistics Canada provided employment figures for men and women in 2006. The following table shows the values reported for men and women who were employed full time and part time separately. Is there evidence that employment and sex are dependent?

	Male	Female
Full time	6972	5157
Part time	483	1351

a) Test an appropriate hypothesis.
b) State your conclusion, including an analysis of differences you find (if any).

27. Streams Researchers in the Adirondack Mountains collect data on a random sample of streams each year. One of the variables recorded is the substrate of the stream—the type of soil and rock over which they flow. The researchers found that 69 of the 172 sampled streams had a substrate of shale. Construct a 95% confidence interval for the proportion of Adirondack streams with a shale substrate. Clearly interpret your interval in this context.

T **28. Legionnaires' disease** In 1974, the Bellevue-Stratford Hotel in Philadelphia was the scene of an outbreak of what later became known as Legionnaires' disease. The cause of the disease was finally discovered to be bacteria that thrived in the air-conditioning units of the hotel. Owners of the Rip Van Winkle Motel, hearing about the Bellevue-Stratford, replaced their air-conditioning system. The following data are the bacteria counts in the air of eight rooms, before and after a new air-conditioning system was installed (measured in colonies per cubic foot of air). The objective is to find out whether the new system has succeeded in lowering the bacterial count. You are the statistician assigned to report to the hotel whether the strategy has worked. Base your analysis on a confidence interval. Be sure to list all your assumptions, methods, and conclusions.

Room Number	Before	After
121	11.8	10.1
163	8.2	7.2
125	7.1	3.8
264	14	12
233	10.8	8.3
218	10.1	10.5
324	14.6	12.1
325	14	13.7

29. PISA reading scale We looked at PISA math scores earlier in the text. What about reading scores? The means and SEs of PISA reading scale scores for Canadian 15-year-old students are shown below, by gender.[13]

Gender	Mean	SE
All	527	2.4
Males	511	2.8
Females	543	2.5

a) The study involved about 22 000 15-year-old students from nearly 1000 schools across Canada. The standard deviation for all the scores was reported as 97. Why do you think we can't divide the standard deviation by the square root of the sample size to get the reported SE of 2.4? Give a confidence interval for the Canadian population mean, stating any assumptions you are making.
b) The difference between average scores for females and males in the study is 32. It was also reported that the SE of the difference in means is 2.3. Why do you think this does not agree with what you would get using the Pythagorean rule? Give a confidence interval for the population difference between males and females, and state any assumptions you are making.

30. Bipolar kids The June 2002 issue of *American Journal of Psychiatry* reported that researchers used medication and psychotherapy to treat children aged 7 to 16 who exhibit bipolar symptoms. After two years, symptoms had cleared up in only 26 of the 89 children involved in the study.
a) Write a 95% confidence interval, and interpret it in this context.
b) If researchers subsequently hope to produce an estimate of treatment effectiveness for bipolar disorder that has a margin of error of only 6%, how many patients should they study?

T **31. Online testing** The Educational Testing Service is now administering several of its standardized tests online, such as the TOEFL exam. Since taking a test on a computer is different from taking a test with pencil and paper,

[12] *The Effectiveness of "Teach for America" and Other Under-certified Teachers on Student Academic Achievement: A Case of Harmful Public Policy*. Education Policy Analysis Archives, 2002.

[13] www.pisa.oecd.org

researchers wondered if the scores will be the same. To investigate this question, they created two versions of a TOEFL-type test and got 20 volunteers to participate in an experiment. Each volunteer took both versions of the test, one with pencil and paper and the other online. Subjects were randomized with respect to the order in which they sat for the tests (online/paper) and which form they took (Test A, Test B) in which environment. The table summarizes the scores (out of a possible 20).

Subject	Paper	Online
	Test A	Test B
1	14	13
2	10	13
3	16	8
4	15	14
5	17	16
6	14	11
7	9	12
8	12	12
9	16	16
10	7	14
11	8	13
12	11	13
13	15	17
14	11	13
15	13	14
16	9	9
17	15	9
18	14	15
19	16	12
20	8	10

a) Were the two forms (A/B) of the test equivalent in terms of difficulty? Test an appropriate hypothesis and state your conclusion.
b) Is there evidence that testing environment (paper/online) matters? Test an appropriate hypothesis and state your conclusion.

32. Bread Tobermory Bakery is trying to predict how many loaves to bake. In the past 100 days, the bakery has sold between 95 and 140 loaves per day. Here are a histogram and the summary statistics for the number of loaves sold for the past 100 days.

Summary of sales	
Mean	103
Median	100
StdDev	9.000
Min	95
Max	140
Lower 25th %tile	97
Upper 25th %tile	105.5

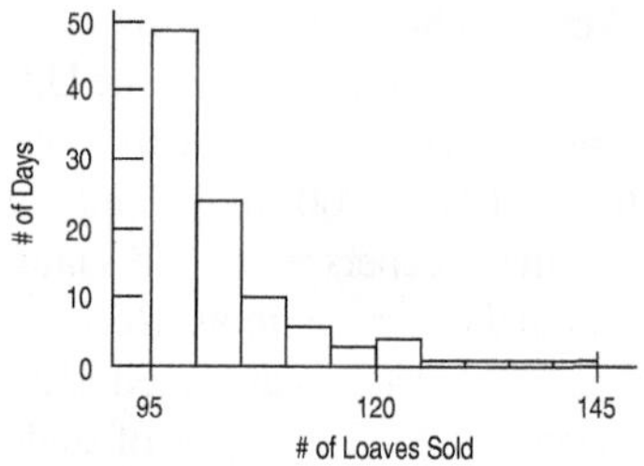

a) Can you use these data to estimate the number of loaves sold on the busiest 10% of all days? Explain.
b) Explain why you can use these data to construct a 95% confidence interval for the mean number of loaves sold per day.
c) Calculate a 95% confidence interval and carefully interpret what that confidence interval means.
d) If the bakery would have been satisfied with a confidence interval whose margin of error was twice as wide, how many days' data could they have used?
e) When the bakery opened, the owners estimated that they would sell an average of 100 loaves per day. Does your confidence interval provide strong evidence that this estimate was incorrect? Explain.

33. Irises Can measurements of the petal length of flowers be of value when you need to determine the species of a certain flower? Here are the summary statistics from measurements of the petals of two species of irises.[14]

	Species	
	Versicolor	Virginica
Count	50	50
Mean	55.52	43.22
Median	55.50	44
StdDev	5.519	5.362
Min	45	30
Max	69	56
Lower quartile	51	40
Upper quartile	59	47

a) Make parallel boxplots of petal lengths for the two species.
b) Describe the differences seen in the boxplots.
c) Write a 95% confidence interval for this difference.
d) Explain what your interval means.
e) Based on your confidence interval, is there evidence of a difference in petal length? Explain.

34. Insulin and diet A study published in the *Journal of the American Medical Association* examined people to see if they showed any signs of IRS (insulin resistance syndrome) involving major risk factors for Type 2 diabetes and heart disease. Among 102 subjects who consumed

[14]R. A. Fisher, 1936, "The use of multiple measurements in axonomic problems." *Annals of Eugenics*, 7: 179–188.

dairy products more than 35 times per week, 24 were identified with IRS. In comparison, IRS was identified in 85 of 190 individuals with the lowest dairy consumption, fewer than 10 times per week.

a) Is this strong evidence that IRS risk is different in people who frequently consume dairy products than in those who do not? Attack this two ways: use a z-test and a chi-square test. How are these tests related?

b) Does this prove that dairy consumption influences the development of IRS? Explain.

35. World Series If the two teams playing in the World Series are evenly matched, the probability that each team wins any game is 0.5. Then, the probability that the Series ends with one of the teams sweeping four straight games would be $2(0.5)^4 = 0.125$. Further probability calculations indicate that 25% of all World Series should last five games, 31.25% should last six games, and the other 31.25% should last the full seven games. The table shows the number of games it took to decide all the World Series from 1922 (when the seven-game format was set) through 2012. Do these results indicate that the teams are usually equally matched? Give statistical evidence to support your conclusion.

Length of series	4 games	5 games	6 games	7 games
Number of times	19	18	19	33

T **36. Rainmakers?** In an experiment to determine whether seeding clouds with silver iodide increases rainfall, 52 clouds were randomly assigned to be seeded or not. The amount of rain they generated was then measured (in acre-feet). Create a 95% confidence interval for the average amount of additional rain created by seeding clouds. Explain what your interval means.

	Unseeded Clouds	Seeded Clouds
Count	26	26
Mean	164.588	441.985
Median	44.200	221.600
SD	278.426	650.787
IntQRange	138.600	337.600
25 %ile	24.400	92.400
75 %ile	163	430

37. Lay's As a project for an Introductory Statistics course, students checked six bags of Lay's potato chips marked with a net weight of 35.4 grams. They carefully weighed the contents of each bag, recording the following weights (in grams): 35.5, 35.3, 35.1, 36.4, 35.4, 35.5. Is there evidence that the mean weight of bags of Lay's is less than advertised?

a) Write appropriate hypotheses.

b) Do these data satisfy the assumptions for inference? Explain.

c) Test your hypothesis using all six weights.

d) Retest your hypothesis with the one unusually high weight removed.

e) What would you conclude about the stated net weight?

T **38. Colour or text?** In an experiment, 32 volunteer subjects are briefly shown seven cards, each displaying the name of a colour printed in a different colour (example: red, blue, and so on). The subject is asked to perform one of two tasks: memorize the order of the words or memorize the order of the colours. Researchers record the number of cards remembered correctly. Then, the cards are shuffled and the subject is asked to perform the other task. The tables display the results for each subject. Is there any evidence that either the colour or the written word dominates perception?

a) What role does randomization play in this experiment?

b) State appropriate hypotheses.

c) Are the assumptions necessary for inference reasonable here?

d) Perform the test.

e) State your conclusion.

Subject	Colour	Word	Subject	Colour	Word
1	4	7	17	4	3
2	1	4	18	7	4
3	5	6	19	4	3
4	1	6	20	0	6
5	6	4	21	3	3
6	4	5	22	3	5
7	7	3	23	7	3
8	2	5	24	3	7
9	7	5	25	5	6
10	4	3	26	3	4
11	2	0	27	3	5
12	5	4	28	1	4
13	6	7	29	2	3
14	3	6	30	5	3
15	4	6	31	3	4
16	4	7	32	6	7

39. And it means? Every statement about a confidence interval contains two parts—the level of confidence and the interval. Suppose that an insurance agent estimating the mean loss claimed by clients after home burglaries created the 95% confidence interval ($1644, $2391).

a) What's the margin of error for this estimate?

b) Carefully explain what the interval means.

c) Carefully explain what the 95% confidence level means.

40. Batteries We work for the "Watchdog for the Consumer" consumer advocacy group. We've been asked to look at a battery company that claims its batteries last an average of 100 hours under normal use. There have been several complaints that the batteries don't last that long, so we decide to test them. To do this, we select 16 batteries and run them until they die. They lasted a mean of 97 hours, with a standard deviation of 12 hours.

a) One of the editors of our newsletter (who does not know statistics) says that 97 hours is a lot less than the advertised 100 hours, so we should reject the company's claim. Explain to him the problem with doing that.
b) What are the null and alternative hypotheses?
c) What assumptions must we make in order to proceed with inference?
d) At a 5% level of significance, what do you conclude?
e) Suppose that, in fact, the average life of the company's batteries is only 98 hours. Has an error been made in part d)? If so, what kind?

41. Hamsters How large are hamster litters? Among 47 golden hamster litters recorded, there were an average of 7.72 baby hamsters, with a standard deviation of 2.5 hamsters per litter.

a) Create and interpret a 90% confidence interval.
b) Would a 98% confidence interval have a larger or smaller margin of error? Explain.
c) How many litters must be used to estimate the average litter size to within 1.0 baby hamsters with 95% confidence?

42. Family planning Before the introduction of birth control pills, many young women experienced unplanned pregnancies. A 1954 study of 1438 pregnant women examined the association with the women's education levels, producing these data:

	Education Level		
	< 3 yr HS	3 + yr HS	Some college
Number of pregnancies	591	608	239
% unplanned	66.2%	55.4%	42.7%

Do these data provide evidence of an association between family planning and education level?[15]

43. Recruiting In September 2002, CNN reported on a method of grad student recruiting by the Haas School of Business at U.C.-Berkeley. The school notifies applicants by formal letter that they have been admitted, and also e-mails the accepted students a link to a Web site that greets them with personalized balloons, cheering, and applause. The director of admissions says this extra effort at recruiting has really worked well. The school accepts 500 applicants each year, and the percentage who actually choose to enrol at Berkeley has increased from 52% the year before the Web greeting to 54% this year.

a) Create a 95% confidence interval for the change in enrolment rates.
b) Based on your confidence interval, are you convinced that this new form of recruiting has been effective? Explain.

44. Splitting hairs Samples of five hairs are taken from each of four cancer patients' heads, and the zinc content is measured for each of the 20 specimens. We want to see if the zinc levels are lower in the hair of cancer patients than in the general population (as is the case for blood zinc levels). From the data, would you be able to construct a 95% confidence interval for the mean hair zinc concentration in the patient population sampled, by computing the mean and standard deviation of the 20 numbers, plugging into the usual one-sample t-statistic, and then obtaining the P-value from the t-distribution with 19 degrees of freedom? Explain. If you did proceed in this way, would your margin of error be predictably off the mark in some way? (Too large? Too small?)

[15] *Fertility Planning and Fertility Rates by Socio-Economic Status*, Social and Psychological Factors Affecting Fertility, 1954.

chapter

24 Inferences for Regression

Truth will emerge more readily from error than from confusion.

—Francis Bacon

Where are we going?

A scatterplot of IQ versus brain size shows a mildly positive association. Could this just be due to chance? A hypothesis test is clearly needed. We can estimate the slope, but how reliable is our estimate? In this chapter, we'll build on what we know about tests and confidence intervals for means, and learn to make inferences about regression models.

Who	250 male subjects
What	Body fat and waist size
Units	%Body fat and inches
When	1990s
Where	United States
Why	Scientific research

TERMINOLOGY ALERT

The variable we are trying to predict is called the *y-variable* or *response* or *dependent variable*, while the variable whose observed value will be used to enable the prediction is called the *x-variable* or *explanatory variable* or *predictor*.

Three percent of a man's body is essential fat. (For a woman, the percentage is closer to 12.5%.) As the name implies, essential fat is necessary for a normal, healthy body. Fat is stored in small amounts throughout your body. Too much body fat, however, can be dangerous to your health. For men between 18 and 39 years old, a healthy percent body fat ranges from 8% to 19%. (For women of the same age, it's 21% to 32%.)

Measuring body fat can be tedious and expensive. The "standard reference" measurement is by dual-energy X-ray absorptiometry (DEXA), which involves two low-dose X-ray generators and takes from 10 to 20 minutes.

How close can we get to a useable prediction of body fat from easily measurable variables such as *height*, *weight*, or *waist size*? Figure 24.1 shows a scatterplot of *%Body Fat* plotted against *Waist size* for a sample of 250 males of various ages.

In Chapter 7, we modelled relationships like this by fitting a least squares line. The plot is clearly straight, so we can find that line. The equation of the least squares line for these data is

$$\widehat{\%Body\ Fat} = -42.7 + 1.7\ Waist$$

The slope says that, on average, *%Body Fat* is greater by 1.7% for each additional inch around the waist.

Figure 24.1

Percent *Body Fat* versus *Waist size* for 250 men of various ages. The scatterplot shows a strong, positive, linear relationship.

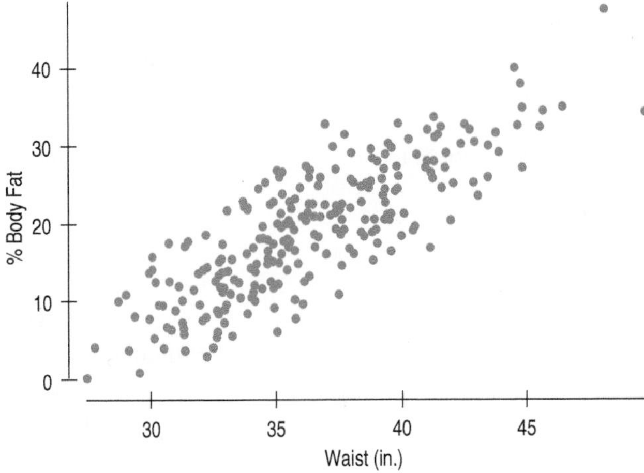

How useful is this model? When we fit linear models before, we used them to describe the relationship between the variables and we interpreted the slope and intercept as descriptions of the data. Now, we'd like to know what the regression model can tell us beyond the 250 men in this study. To do that, we'll want to make confidence intervals and test hypotheses about the slope and intercept of the regression line.

24.1 A Regression Model

The Population and the Sample

When we found a confidence interval for a mean response, we could imagine a single, true underlying value for the mean. When we tested whether two means or two proportions were equal, we imagined a true underlying difference. But what does it mean to do inference for regression? We know better than to think that even if we knew every population value, the data would line up perfectly on a straight line. After all, even in our sample, not all men who have 38-inch waists have the same *%Body Fat*. In fact, there's a whole distribution of *%Body Fat* for these men shown in Figure 24.2.

This is true at each *Waist* size. In fact, we could depict the distribution of *%Body Fat* at different *Waist sizes* as shown in Figure 24.3.

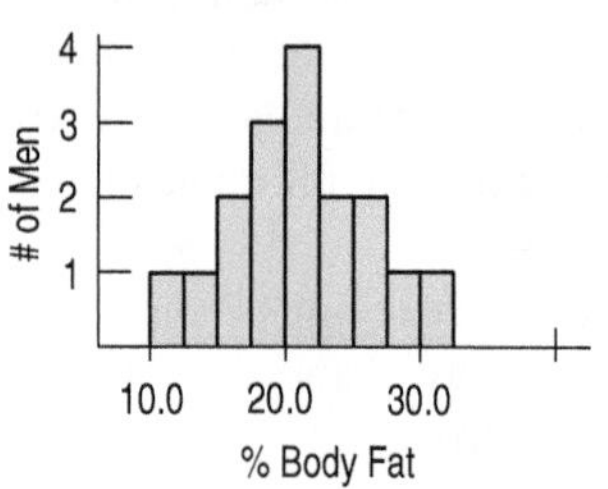

Figure 24.2
The distribution of *%Body Fat* for men with a *Waist size* of 38 inches is unimodal and symmetric.

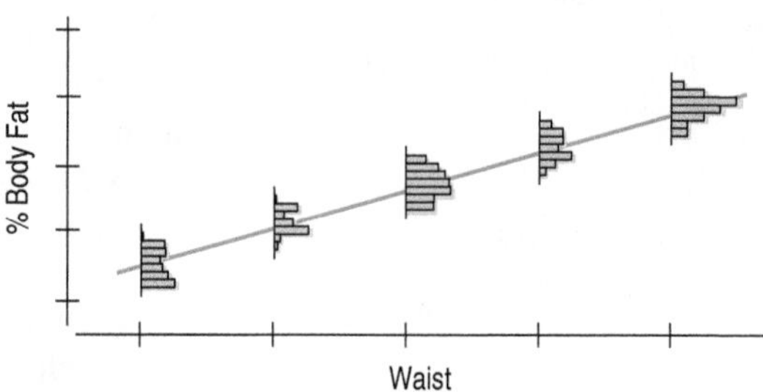

Figure 24.3
There's a distribution of *%Body Fat* for each value of *Waist size*. We'd like the means of these distributions to line up.

But we want to *model* the relationship between *%Body Fat* and *Waist size* for all men. To do that, we imagine an idealized regression line. The model assumes that the *means* of the distributions of *%Body Fat* for each *Waist size* fall along the line, even though the individuals are scattered around it. We know that this model is not a perfect description of how the variables are associated, but it may be useful for predicting *%Body Fat* and for understanding how it's related to *Waist size*.

NOTATION ALERT

This time we used up only one Greek letter for two things. Lower-case Greek β (beta) is the natural choice to correspond to the *b*'s in the regression equation. We used β before for the probability of a Type II error, but there's little chance of confusion here.

If only we had all the values in the population, we could find the slope and intercept of this *idealized regression line* explicitly by using least squares. Following our usual conventions, we write the idealized line with Greek letters and consider the coefficients (the slope and intercept) to be *parameters:* β_0 is the intercept and β_1 is the slope. Corresponding to our fitted line of $\hat{y} = b_0 + b_1x$, we write

$$\mu_y = \beta_0 + \beta_1 x$$

Why μ_y instead of $\hat{y}$? Because this is a model. There is a distribution of *%Body Fat* for each *Waist size*. The model places the *means* of the distributions of *%Body Fat* for each *Waist size* on the same straight line.

Of course, not all the individual y's are at these means. In fact, most of the y's lie either above or below the line. So, for each data point, the model makes an error. These errors are random and, of course, can be positive, zero, or negative. They are model errors, so we use a Greek letter ε to represent them.

When we put the errors into the equation, we can account for each individual y:

$$y = \beta_0 + \beta_1 x + \varepsilon$$

This equation is now true for each data point (since there is an ε to soak up the deviation), so the model gives a value of y for any value of x.

For the body fat data, an idealized model like this provides a summary of the relationship between *%Body Fat* and *Waist size*. Like all models, it simplifies the real situation. We know there is more to predicting body fat than *Waist size* alone. But the advantage of a model is that the simplification might help us to think about the situation and assess how well *%Body Fat* can be predicted from simpler measurements.

We estimate the β's by finding a regression line, $\hat{y} = b_0 + b_1x$, as we did in Chapter 7. The residuals, $e = y - \hat{y}$, are the sample-based versions of the errors, ε. We'll use them to help us assess the regression model.

We know that least squares regression will give reasonable estimates of the parameters of this model from a random sample of data. Our challenge is to account for our uncertainty in how well they do. For that, we need to make some assumptions about the model and the errors.

Assumptions and Conditions

A S ***Activity:* Conditions for Regression Inference.** View an illustrated discussion of the conditions for regression inference.

In Chapter 7, when we fit lines to data, we needed to check only the Straight Enough Condition. Now, when we want to make inferences about the coefficients of the line, we'll have to make more assumptions. Fortunately, we can check conditions to help us judge whether these assumptions are reasonable for our data. And as we've done before, we'll make some checks *after* we find the regression equation.

Also, we need to be careful about the order in which we check conditions. If our initial assumptions are not true, it makes no sense to check the later ones. So now we number the assumptions to keep them in order.

1. Linearity Assumption

CHECK THE SCATTERPLOT

The shape must be linear, or we can't use linear regression at all.

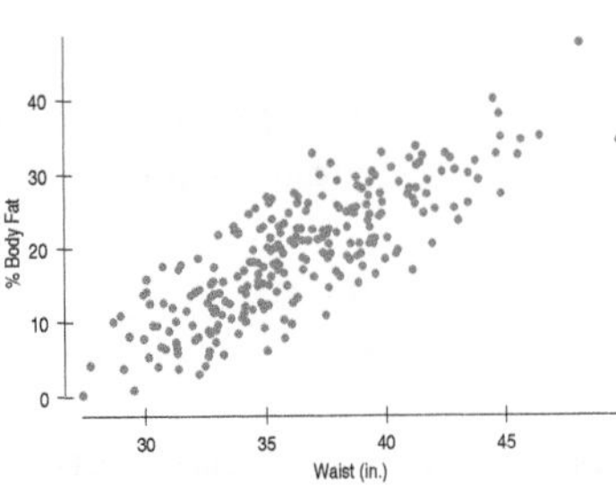

CHECK THE RESIDUAL PLOT (1)

The residuals should appear to be randomly scattered.

If the true relationship is far from linear and we use a straight line to fit the data, our entire analysis will be useless, so we always check this first.

The **Straight Enough Condition** is satisfied if a scatterplot looks straight. It's generally not a good idea to draw a line through the scatterplot when checking. That can fool your eye into seeing the plot as more straight. Sometimes it's easier to see violations of the Straight Enough Condition by looking at a scatterplot of the residuals against the predicted values, $\hat{y}$. That plot will have a horizontal direction and should have no pattern if the condition is satisfied. For a model with only one explanatory variable x, a scatterplot of the residuals against x will give the same information as the scatterplot of the residuals against $\hat{y}$. However, for a multiple linear regression model (a linear model with two or more explanatory variables), as discussed in Chapter 27, scatterplots of residuals against individual explanatory variables and against $\hat{y}$ give different information. The former shows only the relationship between each explanatory variable and y; the latter shows the linearity of the entire model.

If the scatterplot is straight enough, we can go on to some assumptions about the errors. If not, stop here, or consider re-expressing the data to make the scatterplot more nearly linear. For the *%Body Fat* data, the scatterplot is beautifully linear. Of course, the data must be quantitative for this to make sense. Check the **Quantitative Data Condition**.

2. Independence Assumption

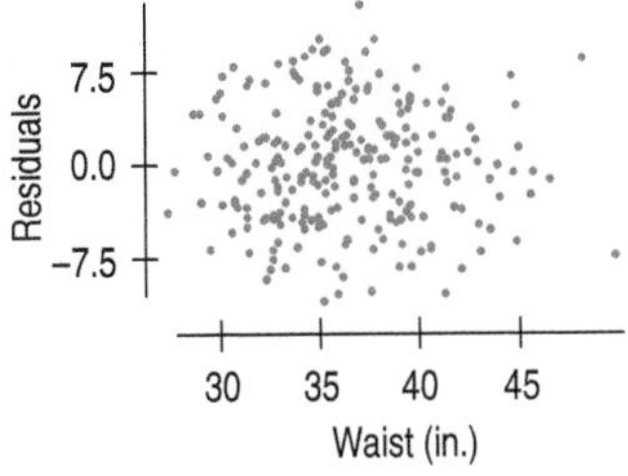

Figure 24.4

The residuals show only random scatter when plotted against *Waist size.*

The errors in the true underlying regression model (the ε's) must be mutually independent. As usual, there's no way to be sure that the Independence Assumption is true.

Usually, when we care about inference for the regression parameters, it's because we think our regression model might apply to a larger population. In such cases, we can check a **Randomization Condition** that the individuals are a representative sample from that population.

We can also check displays of the regression residuals for evidence of patterns, trends, or clumping, any of which would suggest a failure of independence. In the special case when the response y is related to time, a common violation of the Independence Assumption

CHECK THE RESIDUAL PLOT (2)

The vertical spread of the residuals should be roughly the same everywhere.

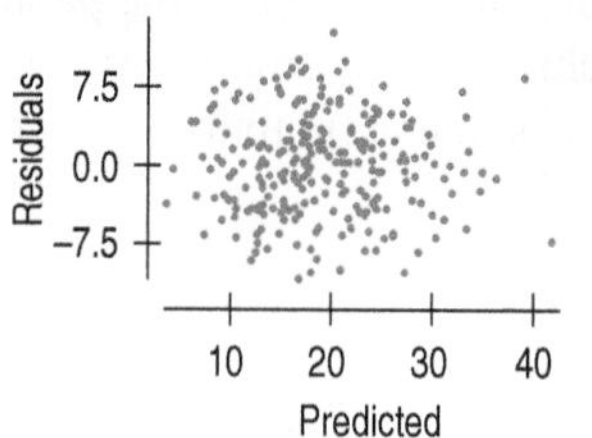

Figure 24.5

A scatterplot of residuals against predicted values can help check for plot thickening. Note that this plot looks identical to the plot of residuals against *Waist size*. For a regression of one response variable on one predictor, these plots differ only in the labels on the x-axis.

CHECK A HISTOGRAM OF THE RESIDUALS

The distribution of the residuals should be unimodal and symmetric.

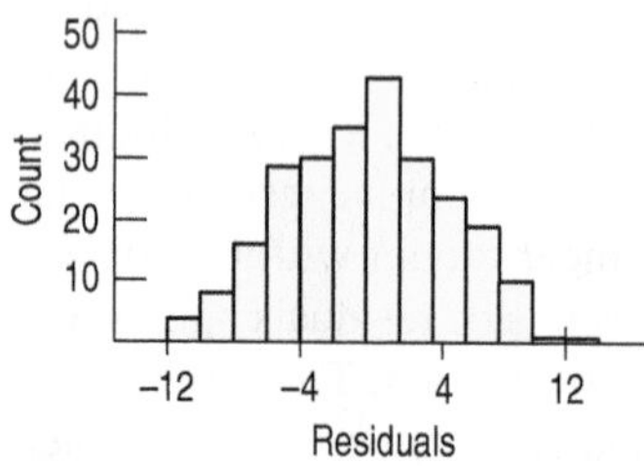

Figure 24.6

A histogram of the residuals is one way to check whether they are nearly Normal. Alternatively, we can look at a Normal probability plot.

is for the errors to be correlated. (The error our model makes today may be similar to the one it made yesterday.) This violation can be checked by plotting the residuals against the order of occurence (in other words, the time plot of the residuals and looking for patterns).

The *%Body Fat* data were collected on a sample of men taken to be representative. The subjects were not related in any way, so we can be pretty sure that their measurements are independent. The residual plot shows no pattern.

3. Equal Variance Assumption

The variability of y should be about the same for all values of x. In Chapter 7, we looked at the standard deviation of the residuals (s_e) to measure the size of the scatter. Now, we'll need this standard deviation to build confidence intervals and test hypotheses. The standard deviation of the residuals is the building block for the standard errors of all the regression parameters. But it makes sense only if the scatter of the residuals is the same everywhere. In effect, the standard deviation of the residuals "pools" information across all of the individual distributions at each x-value, and pooled estimates are appropriate only when they combine information for groups with the same variance.

Practically, what we can check is the **Does the Plot Thicken? Condition**. A scatterplot of y against x offers a visual check. Fortunately, we've already made one. Make sure the spread around the line is nearly constant. Be alert for a "fan" shape or some other tendency for the variation to grow or shrink in one part of the scatterplot. Often, it is better to look at the residuals plotted against the predicted values, $\hat{y}$, as shown in Figure 24.5. With the slope of the line removed, it's easier to see patterns left behind. For the body fat data, the spread of *%Body Fat* around the line is remarkably constant across waist sizes from 30 inches to about 45 inches.

If the plot is straight enough, the data are independent, and the plot doesn't thicken, you can move on to the final assumption.

4. Normal Population Assumption

We assume the errors around the idealized regression line at each value of x follow a Normal model. We need this assumption so that we can use a Student's t-model for inference.

As we have at other times when we've used Student's t, we'll settle for the residuals satisfying the Nearly Normal Condition and the Outlier Condition. Look at a histogram or Normal probability plot of the residuals.[1]

The histogram of residuals in the *%Body Fat* regression certainly looks nearly Normal (Figure 24.6). As we have noted before, the Normality Assumption becomes less important as the sample size grows, because the model is about means and the Central Limit Theorem takes over.

If all four assumptions were true, the idealized regression model would look like this:

Figure 24.7

The regression model has a distribution of y-values for each x-value. These distributions follow a Normal model with means lined up along the line and with the same standard deviations.

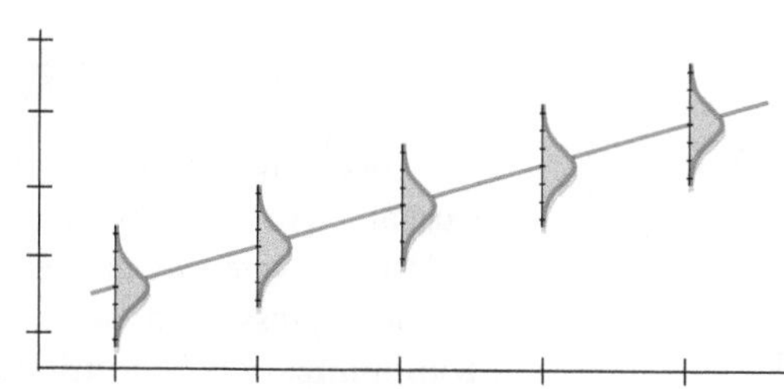

[1] *This* is why we have to check the conditions in order. We have to check that the residuals are independent and that the variation is the same for all x's so that we can lump all the residuals together for a single check of the Nearly Normal Condition.

At each value of x, there is a distribution of y-values that follows a Normal model, and each of these Normal models is centred on the line and has the same standard deviation. Of course, we don't expect the assumptions to be exactly true, and we know that all models are wrong, but the linear model is often close enough to be very useful.

For Example CHECKING ASSUMPTIONS AND CONDITIONS

Stocktrek/Photodisc/Getty Images

Look at the Moon with binoculars or a telescope, and you'll see craters formed by thousands of impacts. Earth, being larger, has been hit even more often. Meteor Crater in Arizona was the first recognized impact crater and was identified as such only in the 1920s. With the help of satellite images, more and more craters have been identified; now more than 180 are known. These are only a small sample of all the impacts Earth has experienced: Only 29% of Earth's surface is land, and many craters have been covered or eroded away. Astronomers have recognized a roughly 35-million-year cycle in the frequency of cratering, although the cause of this cycle is not fully understood. Below is a scatterplot of the known impact craters from the most recent 35 million years.[2] We've taken logs of both age (in millions of years ago) and diameter (km) to make the relationship simpler.

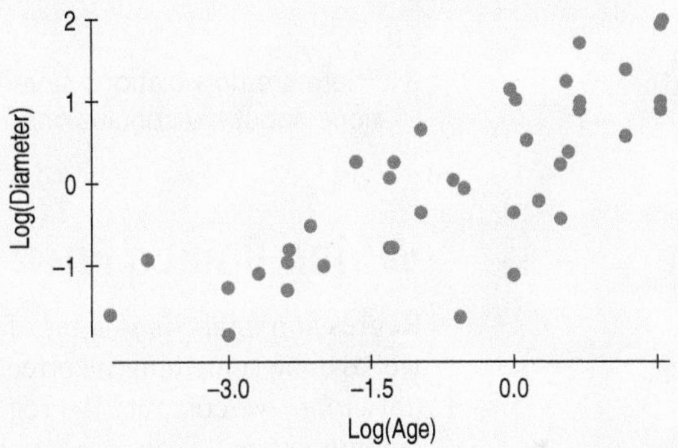

Who	39 impact craters
What	Diameter and Age
Units	km and millions of years ago
When	Past 35 million years
Where	Worldwide
Why	Scientific research

QUESTION: Are the assumptions and conditions satisfied for fitting a linear regression model to these data?

ANSWER:

✓ **Linearity Assumption:** The scatterplot satisfies the Straight Enough Condition.

✓ **Independence Assumption:** Impacts are likely to be generally independent.

✓ **Randomization Condition:** These are the only known craters, and may differ from others that have disappeared and have not yet been found. I'll need to be careful not to generalize my conclusions too broadly.

✓ **Does the Plot Thicken? Condition:** After fitting a linear model, I find these residuals:

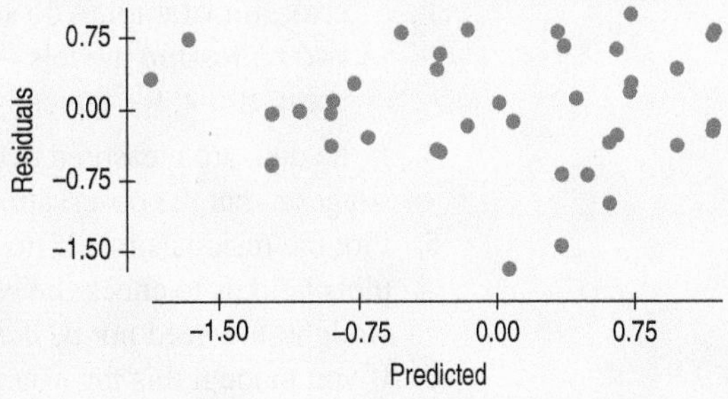

[2]You can find data, pictures, and much more information about the Earth Impact Database at www.unb.ca/passc/ImpactDatabase.

Two points seem to give the impression that the residuals may be more variable for higher predicted values than for lower ones, but this doesn't seem to be a serious violation of the Equal Variance Assumption.

✓ **Nearly Normal Condition:** A Normal probability plot suggests a bit of skewness in the distribution of residuals, and the histogram confirms that.

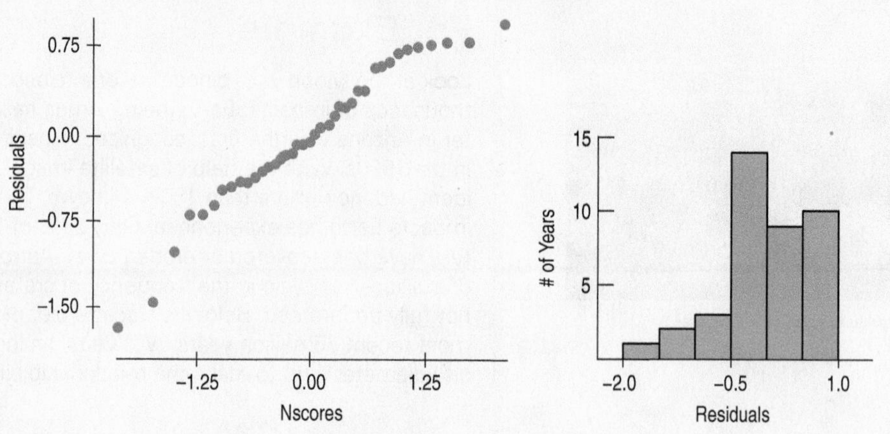

There are no violations severe enough to stop my regression analysis, but I'll be cautious about my conclusions.

Which Comes First: The Conditions or the Residuals?

Regression analysis is a bit of a "chicken-and-the-egg" problem. In order to decide if we've done something incorrect, we need to check the residuals. But we get the residuals only *after* we compute the regression line! So, where do we start? Here is a suggested method that will keep you "straight":

1. Make a scatterplot of the data and check the Straight Enough Condition. If the relationship looks curved, you can try re-expressing the data. If that doesn't work, you may need polynomial terms or a non-linear model.[3]
2. If the data appear straight enough, fit a regression line and find the residuals, e, and the predicted values, $\hat{y}$.
3. Make a scatterplot of the residuals vs. the predicted values (residuals go on the y-axis). Many software programs will do this for you.[4] The plot should have no obvious patterns, bends, curvature, or outliers.
   a. If you see curvature, this is an indication that the relationship may not have been linear after all.
   b. Outliers are trickier. If you can justify removing them (because of a data entry error, for example), do so, and go back to Step 1. Otherwise you may need to fit two regression models—one with the outliers and one without. If they agree, great. If not, we typically mention both scenarios.
4. If the data are measured over time, plot the residuals vs. time. If you see a pattern, it suggests that the observations may not have been independent.
5. Plot the residuals in a histogram and a Normal Probability Plot (NPP). Both of these plots help us to check the Nearly Normal Condition. The NPP should be pretty straight, but need not be *perfect*.
6. If you made it this far, you are ready to go ahead with inference. Typically, we are interested in the *slope* of the line, but occasionally the *intercept* as well.

[3]Which means you'll have to continue studying Statistics!
[4]We did it once by hand. It was really tedious . . .

Step-by-Step Example REGRESSION INFERENCE

Let's see how much more we can learn about body fat and waist size from a regression model.

Questions: What is the relationship between *%Body Fat* and *Waist size* in men? How does our linear model predict body fat from waist size, and how well does it do the job?

THINK ➡ **Plan** Specify the question of interest.

Name the variables and report the *W*'s.

Identify the parameters you want to estimate.

I have quantitative body measurements on 250 adult males from the BYU Human Performance Research Center. I want to understand the relationship between *%Body Fat* and *Waist size.*

Model Think about the assumptions and check the conditions.

Make pictures. For regression inference, you'll need a scatterplot, a residual plot, and either a histogram or a Normal probability plot of the residuals.

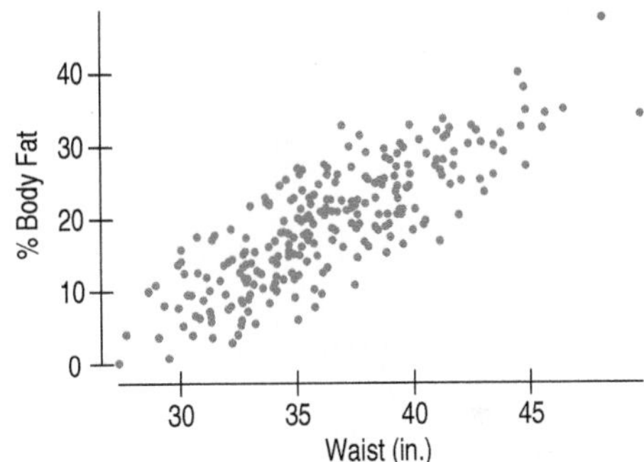

✓ **Straight Enough Condition:** There's no obvious bend in the original scatterplot of the data or in the plot of residuals against predicted values.

✓ **Independence Assumption:** These data are not collected over time, and there's no reason to think that the *%Body Fat* of one man influences the *%Body Fat* of another.

(We've seen plots of the residuals already. See Figures 24.5 and 24.6.)

✓ **Does the Plot Thicken? Condition:** Neither the original scatterplot nor the residual scatterplot shows any changes in the spread about the line.

✓ **Nearly Normal Condition, Outlier Condition:** A histogram of the residuals is unimodal and symmetric. The Normal probability plot of the residuals is quite straight, indicating that the Normal model is reasonable for the errors.

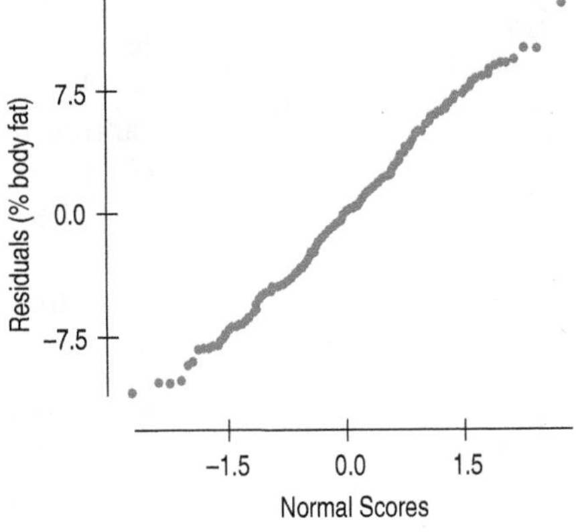

Choose your method.	Under these conditions a regression model is appropriate.
SHOW ➡ **Mechanics** Let's just "push the button" and see what the regression looks like. You can find the formula for the regression equation in Chapter 7, and the standard error formulas will be shown a bit later, but regressions are almost always computed with a computer program or calculator. Write the regression equation.	Here's the computer output for this regression: Dependent variable is: %BF R-squared = 67.8% s = 4.713 with 250 − 2 = 248 degrees of freedom **Variable** / **Coeff** / **SE(Coeff)** / ***t*-ratio** / ***P*-value** Intercept / −42.734 / 2.717 / −15.7 / <0.0001 Waist / 1.70 / 0.0743 / 22.9 / <0.0001 The estimated regression equation is $\widehat{\%Body\ Fat} = -42.73 + 1.70\ Waist$
TELL ➡ **Conclusion** Interpret your results in context.	The R^2 for the regression is 67.8%. *Waist size* seems to account for about 2/3 of the *%Body Fat* variation in men. The slope of the regression says that *%Body Fat* increases by about 1.7 percentage points per additional inch of *Waist size*, on average.
More Interpretation We haven't worked it out in detail yet, but the output gives us numbers labelled as *t*-statistics and corresponding P-values, and we have a general idea of what those mean. (Now it's time to learn more about regression inference so we can figure out what the rest of the output means.)	The standard error of 0.07 for the slope is much smaller than the slope itself, so it looks like the estimate is reasonably precise. And there are a couple of t-ratios and P-values given. Because the P-values are small, it appears that some null hypotheses can be rejected.

Intuition About Regression Inference

A S *Simulation:* **Simulate the Sampling Distribution of a Regression Slope.** Draw samples repeatedly to see for yourself how slope can vary from sample to sample. This simulation experiment lets you build up a histogram to see the sampling distribution.

Wait a minute! We've just pulled a fast one. We've pushed the "regression button" on our computer or calculator but haven't discussed where the standard errors for the slope or intercept come from. We know that if we had collected similar data on a different random sample of men, the slope and intercept would be different. Each sample would have produced its own regression line, with slightly different b_0 and b_1. This sample-to-sample variation is what generates the sampling distributions for the coefficients.

There's only one regression model; each sample regression is trying to estimate the same parameters, β_0 and β_1. We expect any sample to produce a b_1 whose expected value is the true slope, β_1. What about its standard deviation? What aspects of the data affect how much the slope (and intercept) vary from sample to sample?

- **Spread around the line.** Below are two situations in which we might do regression. Which situation would yield the more consistent slope? That is, if we were to sample over and over from the two underlying populations that these samples come from and compute all the slopes, which group of slopes would vary less?

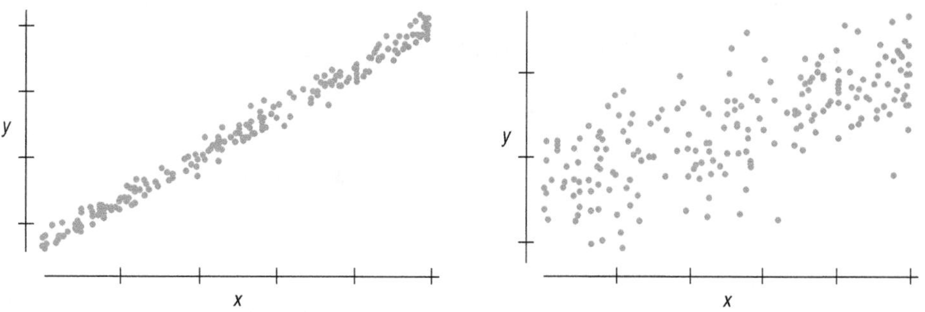

Figure 24.8
Which of these scatterplots shows a situation that would give the more consistent regression slope estimate if we were to sample repeatedly from its underlying population?

n − 2?

For standard deviation (in Chapter 3), we divided by $n - 1$ because we didn't know the true mean and had to estimate it. Now, it's later in the course and there's even more we don't know. Here we don't know *two* things: the slope and the intercept. If we knew them both, we'd divide by n and have n degrees of freedom. When we estimate both, however, we adjust by subtracting two so we divide by $n - 2$ and (as we will soon see) have two fewer degrees of freedom.

Clearly, data like those in the left plot give more consistent slopes.

Less scatter around the line means the slope will be more consistent from sample to sample. The spread around the line is measured with the **residual standard deviation,** s_e. You can always find s_e in the regression output, often just labelled s. You're probably not going to calculate the residual standard deviation by hand. As we noted when we first saw this formula in Chapter 7, it looks a lot like the standard deviation of y, only now subtracting the predicted values rather than the mean and dividing by $n - 2$ instead of $n - 1$:

$$s_e = \sqrt{\frac{\sum(y - \hat{y})^2}{n - 2}}$$

The less scatter around the line, the smaller the residual standard deviation and the stronger the relationship between x and y.

Some people prefer to assess the strength of a regression by looking at s_e rather than R^2. After all, s_e has the same units as y, and because it's the standard deviation of the errors around the line, it tells you how close the data are to our model. By contrast, R^2 is the proportion of the variation of y accounted for by x. Why not look at both?

- **Spread of the x values:** Here are two more situations. Which of these would yield more consistent slopes?

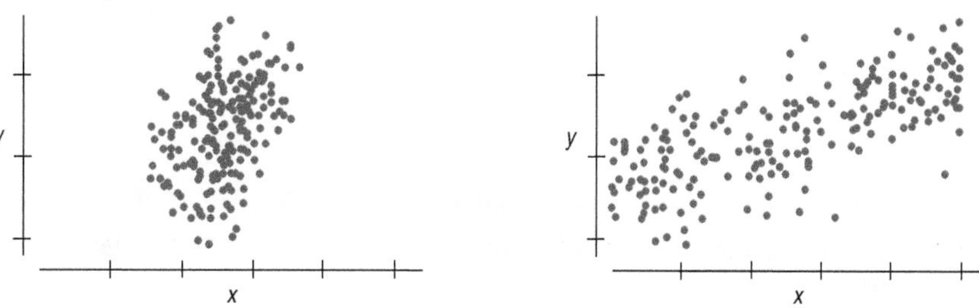

Figure 24.9
Which of these scatterplots shows a situation that would give the more consistent regression slope estimate if we were to sample repeatedly from the underlying population?

A plot like the one on the right has a broader range of x-values, so it gives a more stable base for the slope. We'd expect the slopes of samples from situations like that to vary less from sample to sample. A large standard deviation of x, s_x, provides a more stable regression.

- **Sample size.** Here we go again. What about these two?

It shouldn't be a surprise that having a larger sample size, n, gives more consistent estimates from sample to sample.

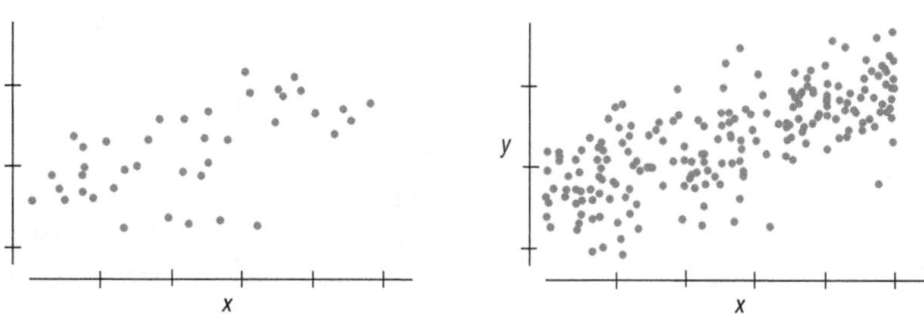

Figure 24.10
Which of these scatterplots shows a situation that would give the more consistent regression slope estimate if we were to sample repeatedly from the underlying population?

24.2 Standard Errors for the Parameters

First, and Most Importantly, *the Slope*

Three aspects of the scatterplot, then, affect the standard error of the regression slope:

- Spread around the line: s_e
- Spread of x values: s_x
- Sample size: n

A S *Activity:* **Regression Slope Standard Error.** See how $SE(b_1)$ is constructed and where the values used in the formula are found in the regression output table.

These are in fact the *only* things that affect the standard error of the slope. An alternative formula often used for the SE of the slope is:

$$SE(b_1) = \frac{s_e}{\sqrt{n-1}\,s_x}$$

The error standard deviation, s_e, is in the *numerator*, since spread around the line *increases* the slope's standard error. The denominator has both a sample size term $\sqrt{n-1}$ and s_x, because increasing either of these *decreases* the slope's standard error. Although you'll probably never have to calculate it by hand, the formula for the standard error is

A S *Simulation: x*-**Variance and Slope Variance.** You don't have to just imagine how the variability of the slope depends on the spread of the *x*'s.

$$SE(b_1) = \frac{s_e}{\sqrt{\sum(x_i - \bar{x})^2}}$$

NOTATION ALERT:

Don't confuse the standard deviation of the residuals, s_e, with the standard error of the slope, $SE(b_1)$. The first measures the scatter around the line, and the second tells us how reliably we can estimate the slope.

We know the b_1's vary from sample to sample. As you'd expect, their sampling distribution model is centred at β_1, the slope of the idealized regression line. Now we can estimate its standard deviation with $SE(b_1)$. What about its shape? Here William Gosset (Chapter 20) comes to the rescue again. When we standardize the slopes by subtracting the model mean and dividing by their standard error, we get a Student's t-model, this time with $n-2$ degrees of freedom:

$$\frac{b_1 - \beta_1}{SE(b_1)} \sim t_{n-2}$$

NOTATION ALERT:

In Statistics, the '~' symbol is a shortcut for saying "Is distributed as." For the case on this page, it means the standardized slope is distributed as a t random variable with $n-2$ degrees of freedom.

A SAMPLING DISTRIBUTION FOR REGRESSION SLOPES

When the conditions are met, the standardized estimated regression slope,

$$t = \frac{b_1 - \beta_1}{SE(b_1)}$$

follows a Student's t-model with $n-2$ degrees of freedom. We estimate the standard error with

$$SE(b_1) = \frac{s_e}{\sqrt{n-1}\,s_x}, \text{ where } s_e = \sqrt{\frac{\sum(y-\hat{y})^2}{n-2}}$$

n is the number of data values, and s_x is the ordinary standard deviation of the x-values.

For Example FINDING STANDARD ERRORS

RECAP: Recent terrestrial impact craters seem to show a relationship between age and size that is linear when re-expressed using logarithms.

Here are summary statistics and regression output from a statistics package:

Variable	Count	Mean	StdDev
logAge	39	−0.656310	1.57682
logDiam	39	0.012600	1.04104

Dependent variable is: logDiam
R squared = 63.6%
s = 0.6362 with 39 − 2 = 37 degrees of freedom

Variable	Coefficient	SE(coeff)	*t*-ratio	P-value
Intercept	0.358262	0.1106	3.24	0.0025
logAge	0.526674	0.0655	8.05	< 0.0001

QUESTIONS: How are the standard error of the slope and the *t*-ratio for the slope calculated? (And aren't you glad the software does this for you?)

ANSWER: $SE(b_1) = \frac{s_e}{\sqrt{n-1} \times s_x} = \frac{0.6362}{\sqrt{39-1} \times 1.57682} = 0.0655$

Assuming no association $(\beta_1 = 0), t_{37} = \frac{b_1 - \beta_1}{SE(b_1)} = \frac{0.526674 - 0}{0.0655} = 8.05$

What About the Intercept?

The same reasoning applies for the intercept. We could write

$$\frac{b_0 - \beta_0}{SE(b_0)} \sim t_{n-2}$$

and use it to construct confidence intervals and test hypotheses, but often the value of the intercept isn't something we care about. The intercept usually isn't interesting. Most hypothesis tests and confidence intervals for regression are about the slope. But in case you really want to see the formula for the standard error of the intercept, we've put it in a footnote.[5]

24.3 Regression Inference

WHAT IF THE SLOPE WERE 0?

If $b_1 = 0$, our prediction is $\hat{y} = b_0 + 0x$. The equation collapses to just $\hat{y} = b_0$. Now, x is nowhere in sight, so y doesn't depend on x at all.

And b_0 would turn out to be $\bar{y}$. Why? We know that $b_0 = \bar{y} - b_1\bar{x}$, but when $b_1 = 0$, that becomes simply $b_1 = \bar{y}$. It turns out, then, that when the slope is 0, the equation is just $\hat{y} = \bar{y}$; at every value of x, we always predict the mean value for y.

Now that we have the standard error of the slope and its sampling distribution, we can test a hypothesis about it and make confidence intervals. The usual null hypothesis about the slope is that it's equal to 0. Why? Well, a slope of zero would say that the response y doesn't tend to change linearly when x changes—in other words, there is no linear association between the two variables. If the slope were zero, there wouldn't be much left of our regression equation.

So a null hypothesis of a zero slope questions the entire claim of a linear relationship between the two variables—and often that's just what we want to know. In fact, every software package or calculator that does regression simply assumes that you want to test the null hypothesis that the slope is really zero.

To test H_0: $\beta_1 = 0$, we find

$$t = \frac{b_1 - 0}{SE(b_1)}$$

which has a t_{n-2} distribution under the null hypothesis. This is just like every t-test we've seen: a difference between the statistic and its hypothesized value, divided by its standard error.

---

[5]$SE(b_0) = s_e\sqrt{\frac{1}{n} + \frac{\bar{x}^2}{\sum(x - \bar{x})^2}}$

For our body fat data, the computer found the slope (1.7), its standard error (0.0743), and the ratio of the two: $(1.7 - 0)/0.0743 = 22.9$ (see p. 708). Nearly 23 standard errors from the hypothesized value certainly seems big. The P-value (< 0.0001) confirms that a *t*-ratio this large would be very unlikely to occur if the true slope were zero.

Maybe the standard null hypothesis isn't all that interesting here. Did you have any doubts that *%Body Fat* is related to *Waist size*? A more sensible use of these same values might be to make a confidence interval for the slope, instead.

We can build a confidence interval in the usual way, as an estimate plus or minus a margin of error. As always, the margin of error is just the product of the standard error and a critical value. Here the critical value comes from the *t*-distribution with $n - 2$ degrees of freedom, so a 95% confidence interval for β is

$$b_1 \pm t^*_{n-2} \times SE(b_1)$$

For the body fat data, $t^*_{248} = 1.970$, so that comes to $1.7 \pm 1.97 \times 0.074$, or an interval from 1.55 to 1.85 *%Body Fat* per additional inch of *Waist size*.

For Example INTERPRETING A REGRESSION MODEL

RECAP: On a log scale, there seems to be a linear relationship between the diameter and the age of recent terrestrial impact craters. We have regression output from statistics software:

Dependent variable is: logDiam
R squared = 63.6%
s = 0.6362 with 39 − 2 = 37 degrees of freedom

Variable	Coefficient	SE(coeff)	*t*-ratio	P-value
Intercept	0.358262	0.1106	3.24	0.0025
logAge	0.526674	0.0655	8.05	≤ 0.0001

QUESTIONS: What is the regression model, and what can it tell us?

ANSWER: For terrestrial impact craters younger than 35 million years, the logarithm of diameter grows linearly with the logarithm of *Age*: $\widehat{log\ Diam} = 0.358 + 0.527\ logAge$. The P-value for each coefficient's *t*-statistic is very small, so I'm quite confident that neither is zero. Based on my model, I conclude that, on average, the older a crater is, the larger it tends to be. This model accounts for 63.6% of the variation in logDiam.

Although it is possible that impacts (and their craters) are getting smaller, it is more likely that I'm seeing the effects of age on craters. Small craters are probably more likely to erode or become buried, or otherwise be difficult to find as they age. Larger craters may survive the huge expanses of geologic time more successfully.

Just Checking

Researchers in food science studied how big people's mouths tend to be. They measured mouth volume by pouring water into the mouths of subjects who lay on their backs. Unless this is your idea of a good time, it would be helpful to have a model to estimate mouth volume more simply. Fortunately, mouth volume is related to height. (Mouth volume is measured in cubic centimetres and height in metres.)

The data were checked and deemed suitable for regression. Here is some computer output:

Summary of	Mouth Volume
Mean	60.2704
StdDev	16.8777

Dependent variable is: Mouth Volume
R squared = 15.3%
s = 15.66 with 61 − 2 = 59 degrees of freedom

Variable	Coefficient	SE(coeff)	t-ratio	P-value
Intercept	−44.7113	32.16	−1.39	0.1697
Height	61.3787	18.77	3.27	0.0018

1. What does the *t*-ratio of 3.27 tell us about this relationship? How does the P-value help our understanding?
2. Would you say that measuring a person's height could reliably be used as a substitute for the wetter method of determining how big a person's mouth is? What numbers in the output helped you reach that conclusion?
3. What does the value of s_e add to this discussion?

Another Example

Pearson Education

Every spring, Nenana, Alaska, hosts a contest in which participants try to guess the exact minute that a wooden tripod placed on the frozen Tanana River will fall through the breaking ice. The contest started in 1917 as a diversion for railroad engineers, with a jackpot of $800 for the closest guess. It has grown into an event in which hundreds of thousands of entrants enter their guesses on the Internet[6] and vie for as much as $300 000.

Because so much money and interest depends on the time of breakup, it has been recorded to the nearest minute with great accuracy ever since 1917. And because a standard measure of breakup has been used throughout this time, the data are consistent. An article in *Science*[7] used the data to investigate global warming—whether greenhouse gasses and other human actions have been making the planet warmer. Others might just want to make a good prediction of next year's breakup time.

Of course, we can't use regression to tell the *causes* of any change. But we can estimate the *rate* of change (if any) and use it to make better predictions.

Here are some of the data; a scatterplot follows in Figure 24.11.

Who	Years
What	Year, day, and hour of ice breakup
Units	*x* is in years since 1900. *y* is in days after midnight Dec. 31.
When	1917–present
Where	Nenana, Alaska
Why	Wagering, but proposed to look at global warming

Year (since 1900)	Breakup Date (days since midnight Dec. 31)	Year (since 1900)	Breakup Date (days since midnight Dec. 31)
17	119.4792	30	127.7938
18	130.3979	31	129.3910
19	122.6063	32	121.4271
20	131.4479	33	127.8125
21	130.2792	34	119.5882
22	131.5556	35	134.5639
23	128.0833	36	120.5403
24	131.6319	37	131.8361
25	126.7722	38	125.8431
26	115.6688	39	118.5597
27	131.2375	40	110.6437
28	126.6840	41	122.0764
29	124.6535	⋮	⋮

[6]www.nenanaakiceclassic.com
[7]"Climate change in nontraditional data sets." *Science* 294 [26 October 2001]: 811

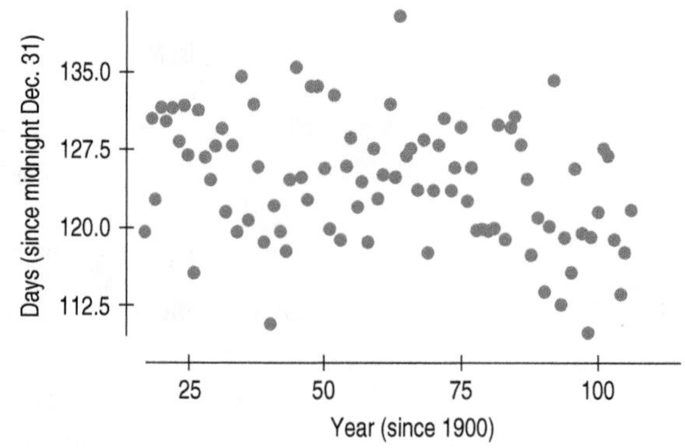

Figure 24.11
Has the date of ice breakup on the Tanana River changed since 1917, when record keeping began? Earlier breakup dates might be a sign of global warming.

A S ***Activity:* A Hypothesis Test for the Regression Slope.** View an animated discussion of testing the standard null hypothesis for slope.

Step-by-Step Example A REGRESSION SLOPE *t*-TEST

The slope of the regression gives the change in Nenana ice breakup date per year.

Questions: Is there sufficient evidence to claim that ice breakup times are changing? If so, how rapid is the change?

THINK ➡ **Plan** State what you want to know.

Identify the *parameter* you wish to estimate. Here our parameter is the slope.

Identify the variables and review the *W*'s.

I wonder whether the date of ice breakup in Nenana has changed over time. The slope of that change might indicate climate change. I have the date of ice breakup annually since 1917, recorded as the number of days and fractions of a day until the ice breakup.

Hypotheses Write your null and alternative hypotheses.

H_0: There is no change in the date of ice breakup: $\beta_1 = 0$

H_A: Yes, there is: $\beta_1 \neq 0$

Model Think about the assumptions and check the conditions.

✓ **Straight Enough Condition:** I have quantitative data with no obvious bend in the scatterplot.

✓ **Independence Assumption:** These data are a time series, which raises my suspicions that they may not be independent. To check, here's a plot of the residuals against time, the *x*-variable of the regression:

Make pictures. We saw a scatterplot of the data earlier. Because it seems straight enough, we can find and plot the residuals.

Usually, we check for suggestions that the Independence Assumption fails by plotting the residuals against the predicted values. Patterns and clusters in that plot raise our suspicions. But when the data are measured over time, it is always a good idea to plot residuals against time to look for trends and oscillations.

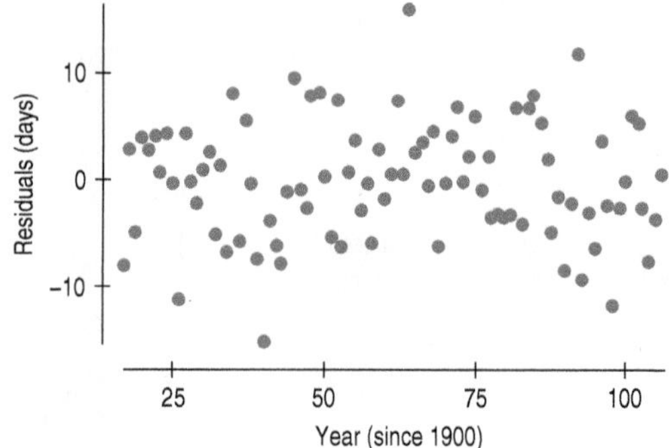

I see a hint that the data oscillate up and down, which suggests some failure of independence, but not so strongly that I can't proceed with the analysis. These data are not a random sample, so I'm reluctant to extend my conclusions beyond this river and these years.

✓ **Does the Plot Thicken? Condition:** The residuals plot shows no obvious trends in the spread.

✓ **Nearly Normal Condition, Outlier Condition:** A histogram of the residuals is unimodal and symmetric.

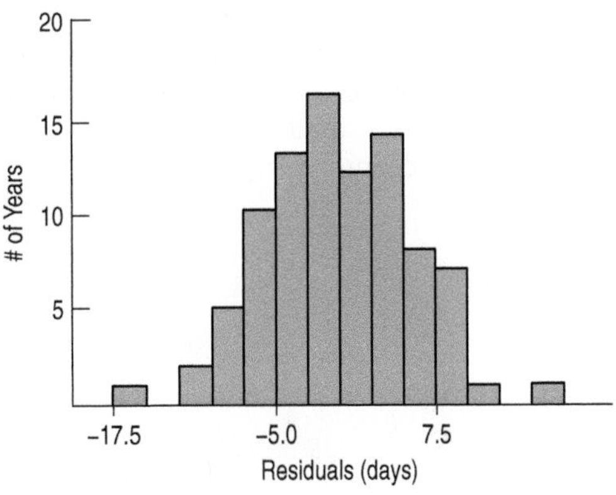

State the sampling distribution model.

Under these conditions, the sampling distribution of the regression slope can be modelled by a Student's t-model with $(n - 2) = 89$ degrees of freedom.

Choose your method.

I'll do a **regression slope t-test.**

SHOW ➡ Mechanics The regression equation can be found from the formulas in Chapter 7, but regressions are almost always found from a computer program or calculator.

The P-values given in the regression output table are from the Student's t-distribution on $(n - 2) =$ 88 degrees of freedom. They are appropriate for two-sided alternatives.

Here's the computer output for this regression:

Dependent variable is: Breakup Date

R-squared = 11.3%
s = 5.673 with 91 − 2 = 89 degrees of freedom

Variable	Coeff	SE(Coeff)	t-ratio	P-value
Intercept	128.950	1.525	84.6	<0.0001
Years Since 1900	−0.07606	0.0226	−3.36	0.0012

The estimated regression equation is

$$\widehat{Date} = 128.95 - 0.076 \text{ Years Since } 1900.$$

TELL ➡ Conclusion Link the P-value to your decision and state your conclusion in the proper context.

The P-value of 0.0012 means that the association we see in the data is unlikely to have occurred by chance. I reject the null hypothesis, and conclude that there is strong evidence that, on average, the ice breakup is occurring earlier each year. But the oscillation pattern in the residuals raises concerns.

SHOW ➡ **Create a confidence interval for the true slope** More	A 95% confidence interval for β_1 is $b_1 \pm t^*_{89} \times SE(b_1)$ $-0.076 \pm (1.987)(0.0226)$ or $(-0.12, -0.03)$ *days per year.*
TELL ➡ **Interpret the interval** Simply rejecting the standard null hypothesis doesn't guarantee that the size of the effect is large enough to be important. Whether we want to know the breakup time to the nearest minute or are interested in global warming, a change measured in hours each year is big enough to be interesting. More	I am 95% confident that the ice has been breaking up, on average, between 0.03 days (about 40 minutes) and 0.12 days (about three hours) earlier each year since 1917.

But is it global warming? So the ice is breaking up earlier. Temperatures are higher. Must be global warming, right?

Maybe.

An article challenging the original analysis of the Nenana data proposed a possible confounding variable. It noted that the city of Fairbanks is upstream from Nenana and suggested that the growth of Fairbanks could have warmed the river. So maybe it's not global warming.

Or maybe global warming is a lurking variable, leading more people to move to a now balmier Fairbanks and also leading to generally earlier ice breakup in Nenana.

Or maybe there's some other variable or combination of variables at work. We can't set up an experiment, so we may never really know.

Only one thing is for sure. When you try to explain an association by claiming cause and effect, you're bound to be on thin ice.[8]

24.4 Inference for the Mean of *y* and for a Future Value of *y*

Once we have a useful regression, how can we indulge our natural desire to predict, without being irresponsible? We know how to compute predicted values of y for any value of x. We first did that in Chapter 7. This predicted value would be our best estimate, but it's still just an informed guess.

Now, however, we can find standard errors. We can use those to construct a confidence interval for the predictions and to report our uncertainty honestly.

Standard Errors for Predicted Values

From our model of *%Body Fat* and *Waist size*, we might want to use *Waist size* to get a reasonable estimate of *%Body Fat*. A confidence interval can tell us how precise that prediction will be. The precision depends on the question we ask, however, and there are two questions: Do we want to know the mean *%Body Fat* for *all* men with a *Waist size* of, say, 38 inches? Or do we want to predict the *%Body Fat* for a particular man with a 38-inch waist before he climbs onto the X-ray table?

What's the difference between the two questions? The predicted *%Body Fat* is the same, but one question leads to an answer much more precise than the other. We can estimate

[8]How *do* scientists sort out such messy situations? Even though they can't conduct an experiment, they *can* look for replications elsewhere. A number of studies of ice on other bodies of water have also shown earlier ice breakup times in recent years. That suggests they need an explanation that's more comprehensive than just Fairbanks and Nenana.

the *mean %Body Fat* for *all* men whose *Waist size* is 38 inches with a lot more precision than we can predict the *%Body Fat* of a *particular individual* whose *Waist size* happens to be 38 inches. Both are interesting questions.

We start with the same prediction in both cases. We are predicting the response value for a new individual, one that was not part of the original data set. To emphasize this, we'll call his x-value "x sub new" and write it x_ν.[9] Here, x_ν is 38 inches. The regression equation predicts *%Body Fat* as $\hat{y}_\nu = b_0 + b_1 x_\nu$.

Now that we have the predicted value, we construct both intervals around this same number. Both intervals take the form

$$\hat{y}_\nu \pm t^*_{n-2} \times SE$$

Even the t^* value is the same for both. It's the critical value (from Table T or technology) for $n - 2$ degrees of freedom and the specified confidence level. The intervals differ because they have different standard errors. Our choice of ruler depends on which interval we want.

The standard errors for prediction depend on the same kinds of things as the coefficients' standard errors. If there is more spread around the line, we'll be less certain when we try to predict the response. Of course, if we're less certain of the slope, we'll be less certain of our prediction. If we have more data, our estimate will be more precise. And there's one more piece: If we're farther from the centre of our data, our prediction will be less precise. This last factor is new, but makes intuitive sense: It's a lot easier to predict a data point near the middle of the data set than to extrapolate far from the centre.

Each of these factors contributes uncertainty—that is, variability—to the estimate. Because the factors are independent of each other, we can add their variances to find the total variability. The resulting formula for the standard error of the estimated *mean* value explicitly takes into account each of the factors:

$$SE(\hat{\mu}_\nu) = \sqrt{SE^2(b_1) \cdot (x_\nu - \bar{x})^2 + \frac{s_e^2}{n}}$$

Individual values vary more than means, so the standard error for a single predicted value has to be larger than the standard error for the estimated mean. In fact, the standard error of a single predicted value has an *extra* source of variability: the variation of individuals' responses around their own mean. That appears as the extra variance term, s_e^2, at the end under the square root:

$$SE(\hat{y}_\nu) = \sqrt{SE^2(b_1) \cdot (x_\nu - \bar{x})^2 + \frac{s_e^2}{n} + s_e^2}$$

Mean vs. Individual Predictions For the Nenana Ice Classic, someone who planned to place a bet would want to predict this year's breakup time. By contrast, scientists studying global warming are likely to be more interested in the mean breakup time. If you want to gamble, be sure to take into account that the variability is greater when predicting for a single year.

Remember to keep this distinction between the two kinds of standard errors when looking at computer output. The smaller one is for the estimated *mean* value, and the larger one is for a predicted *individual* value.[10]

Confidence Intervals for Predicted Values

Now that we have standard errors, we can ask how well our analysis can estimate the mean *%Body Fat* for men with 38-inch waists. The regression output table provides most of the numbers we need:

$$s_e = 4.713$$
$$n = 250$$

[9]Yes, this is a bilingual pun. The Greek letter ν is called "nu."

[10]You may see the standard error expressions written in other, equivalent ways. The most common alternatives are $SE(\hat{\mu}_\nu) = s_e\sqrt{\frac{1}{n} + \frac{(x_\nu - \bar{x})^2}{\sum(x - \bar{x})^2}}$ and $SE(\hat{y}_\nu) = s_e\sqrt{1 + \frac{1}{n} + \frac{(x_\nu - \bar{x})^2}{\sum(x - \bar{x})^2}}$

$$SE(b_1) = 0.074, \text{ and from the data we need to know that}$$
$$\bar{x} = 36.3$$

The regression model gives a predicted value at $x_\nu = 38$ of

$$\hat{y}_\nu = -42.7 + 1.7(38) = 21.9\%$$

Let's find the 95% **confidence interval for the mean** *%Body Fat* for all men with 38-inch waists. We find the standard error from the formula:

$$SE(\hat{\mu}_\nu) = \sqrt{0.074^2(38 - 36.3)^2 + \frac{4.713^2}{250}} = 0.32\%$$

The t^* value that excludes 2.5% in either tail with $250 - 2 = 248$ df is (according to the tables) 1.97.

Putting it all together, we find the margin of error as

$$ME = 1.97(0.32) = 0.63\%$$

So, we are 95% confident that the mean *%Body Fat* for men with 38-inch waists is

$$21.9\% \pm 0.63\%$$

Suppose, instead, we want to predict the *%Body Fat* for an individual man with a 38-inch waist. We need the larger standard error:

$$SE(\hat{y}_\nu) = \sqrt{0.074^2(38 - 36.3)^2 + \frac{4.713^2}{250} + 4.713^2} = 4.72\%$$

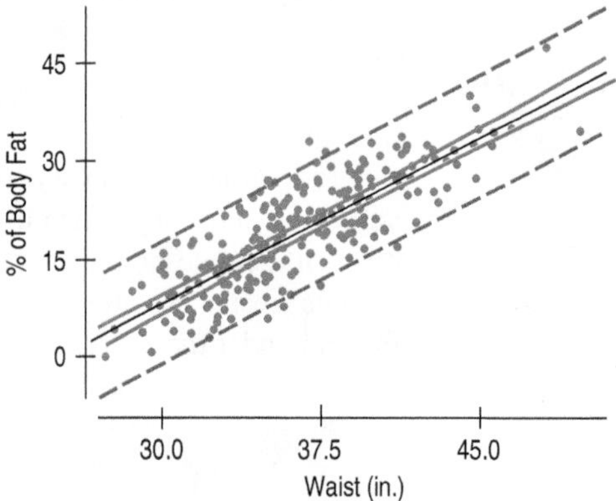

Figure 24.12

A scatterplot of *%Body Fat* versus *Waist size* with a least squares regression line. The solid green lines near the regression line show the extent of the 95% confidence intervals for mean *%Body Fat* at each *Waist size.* The dashed red lines show the prediction intervals. Most of the points are contained within the prediction intervals, but not within the confidence intervals. As the sample gets bigger, the confidence intervals get narrower and narrower, unlike the prediction intervals which get narrower, but in the limit just get closer and closer to $\mu_\nu \pm 1.96\sigma_\varepsilon$.

The corresponding margin of error is

$$ME = 1.97(4.72) = 9.30\%$$

so the **prediction interval** is

$$21.9\% \pm 9.30\%$$

We can think of this interval as having a 95% chance of capturing the true *%Body Fat* of a randomly selected man whose waist is 38 inches.[11] Notice how much wider this interval is than the previous one. As we've known since Chapter 15, the sample mean is much less variable than a randomly selected individual value, and our regression estimate of a population mean is a very similar creature.

Keep in mind this distinction between the two kinds of confidence intervals: The narrower interval is a confidence interval for the mean response at x_ν, and the wider interval is a prediction interval for an individual response with that x-value.

[11]Technically, it's a little more complicated, but it's very close to this.

For Example INTERVALS FOR PREDICTIONS

RECAP: We've found a linear model for the relationship between the logDiam and logAge for terrestrial impact craters:

$$\widehat{\log Diam} = 0.358 + 0.527\log Age.$$

Let's look at some confidence intervals, keeping in mind that the model is in terms of the logs of these variables, so we must work with logs and then transform back to original units when we are done. (Many of the values substituted into the formulas are from the software outputs shown in the For Example boxes on pages 711 and 712.)

QUESTION #1: If we wish to hunt for new craters that are 5 million years old, what's a 95% confidence interval for the mean size we'd expect them to be?

ANSWER #1:

$$\widehat{\log Diam} = 0.358 + 0.527\log Age$$
$$= 0.358 + 0.527\log(5)$$
$$= 0.726$$
$$SE(\hat{\mu}) = \sqrt{SE^2(b_1)\cdot(x - \bar{x})^2 + \frac{s_e^2}{n}}$$
$$= \sqrt{0.0655^2(\log(5) - (-0.65631))^2 + \frac{0.6362^2}{39}}$$
$$= 0.1351$$

From Table T, t^* 2.030, so a 95% CI is $0.726 \pm 2.030(0.1351)$, or (0.452, 1.00).

Since these represent logs of the diameters, I'll find the original units:

$$10^{0.452}\text{km and } 10^{1.00}\text{ km}.$$

I'm 95% confident that terrestrial craters 5 million years old average between 2.8 and 10 kilometres in diameter. Actually, means do not simply transform back, unlike individual observations, so consider this a rough approximation.

QUESTION #2: There's just been a news report announcing the discovery of a new crater that is 10 million years old. How large do you expect it to be? What is a 95% prediction interval for your estimate?

ANSWER #1: For a crater 10 million years old, $\widehat{\log Diam} = 0.358 + 0.527\log(10) = 0.885$ *and* $10^{0.885} = 7.674$. I expect the new crater to be about 7.7 km in diameter.

$$SE(\hat{y}) = \sqrt{SE^2(b_1)\cdot(x - \bar{x})^2 + \frac{s_e^2}{n} + s_e^2}$$
$$= \sqrt{0.0655^2(\log(10) - (-0.65631))^2 + \frac{0.6362^2}{39} + 0.6362^2}$$
$$= 0.653$$

A 95% prediction interval is $0.885 \pm 2.030(0.653) = (-0.441, 2.211)$; in original units, that's (0.36, 162.55).

I'm 95% confident that the newly discovered 10 million-year-old crater is between 0.36 and 162.55 kilometres in diameter. (Without a lot of data and a very strong association, prediction intervals often aren't very precise!)

MATH BOX

So where do those messy formulas for standard errors of predicted values come from? They're based on many of the ideas we've studied so far. Start with regression, add random variables, then throw in the Pythagorean Theorem, the Central Limit Theorem, and a dose of algebra. Mix well. . . .

We begin our quest with an equation of the regression line. Usually we write the line in the form $\hat{y} = b_0 + b_1x$. We found the intercept, b_0, as $b_0 = \bar{y} - b_1\bar{x}$. Substituting in the formula for the line and rearranging terms yields $\hat{y} = b_1(x - \bar{x}) + \bar{y}$.

To predict a y-value for x_ν, we just substitute into the equation:

$$\hat{y}_\nu = b_1(x_\nu - \bar{x}) + \bar{y}$$

Remember that a point on the line is the mean of y-values at that x-value, and the predicted value from the regression equation is our best estimate of this mean (see Figure 24.7), so we can write

$$\hat{\mu}_y = b_1(x_\nu - \bar{x}) + \bar{y}$$

To create a confidence interval for the mean value, we need to measure the variability in this prediction:

$$Var(\hat{\mu}_y) = Var(b_1(x_\nu - \bar{x}) + \bar{y})$$

We now call on the Pythagorean Theorem of Statistics once more: The slope, b_1, and mean, $\bar{y}$, are independent,[12] so their variances add:

$$Var(\hat{\mu}_y) = Var(b_1(x_\nu - \bar{x}) + Var(\bar{y})$$

The horizontal distance from our specific x-value to the mean, $x_\nu - \bar{x}$, is a constant:

$$Var(\hat{\mu}_y) = (Var(b_1))(x_\nu - \bar{x})^2 + Var(\bar{y})$$

Let's write that equation in terms of standard deviations:

$$SD(\hat{\mu}_y) = \sqrt{(SD^2(b_1))(x_\nu - \bar{x})^2 + SD^2(\bar{y})}$$

Because we'll need to estimate these standard deviations using sample statistics, we're really dealing with standard errors:

$$SE(\hat{\mu}_y) = \sqrt{(SE^2(b_1))(x_\nu - \bar{x})^2 + SE^2(\bar{y})}$$

As shown in Chapter 14, the standard deviation of $\bar{y}$ is $\frac{\sigma}{\sqrt{n}}$.

Here, we'll estimate σ using s_e the sample standard deviation of the residuals. Plugging this in, we get:

$$SE(\hat{\mu}_y) = \sqrt{(SE^2(b_1))(x_\nu - \bar{x})^2 + \left(\frac{s_e}{\sqrt{n}}\right)^2}$$

$$= \sqrt{(SE^2(b_1))(x_\nu - \bar{x})^2 - \frac{s_e^2}{n}}$$

And there it is—the standard error we need to create a confidence interval for an estimated mean value.

When we try to predict an individual value of y, we also must worry about how far the true point may lie above or below the regression line. We represent that uncertainty by adding another term, e, to the original equation:

$$y = b_1(x_\nu - \bar{x}) + \bar{y} + e$$

To make a long story short (and the equation a wee bit longer), that additional term simply adds one more standard error to the sum of the variances:

$$SE(\hat{y}) = \sqrt{(SE^2(b_1))(x_\nu - \bar{x})^2 + \frac{s_e^2}{n} + s_e^2}$$

[12]This takes a bit of work to prove, but it makes sense. Each can take on any value without affecting the other. For example, imagine adding 10 to every y-value. That would change the mean, but not the slope. Or we could add 10 to the y-value for the highest x and subtract 10 from the y-value for the lowest x. That would change the slope, but not the mean.

24.5 Correlation Test

In Chapter 2, we learned that the estimated slope can be calculated by

$$b_1 = r\, s_y / s_x$$

with r being the correlation coefficient of x and y. The slope inherits the sign of the correlation coefficient.

Fisher showed that when the *population correlation coefficient* denoted as ρ is 0, then,

$$\frac{r\sqrt{n-2}}{\sqrt{1-r^2}} \sim t_{n-2}$$

Notice that when $\rho = 0$ (when there is no correlation), the slope must be zero ($\beta_1 = 0$). In other words, the above equation can be used for testing H_0: $\beta_1 = 0$.

For our body fat data, the correlation coefficient is 0.8234. Hence,

$$t = 0.8234\sqrt{250-2}/\sqrt{1-(0.8234)^2} = 22.9$$

which is the same as before. The *P*-value is less than 0.0001, which confirms that a *t*-ratio this large (with $250 - 2 = 248$ df) would be very unlikely to occur if the true slope were zero.

For Example CORRELATION TEST

RECAP: On a log scale, the correlation coefficient of age and size is 0.7975, with a sample size of 39.

QUESTION: Are the two variables linearly related?

ANSWER: Assuming no linear relationship, $\rho = \beta_1 = 0$:

$$t = r\sqrt{n-2}/\sqrt{1-r^2} = 0.7975\sqrt{39-2}/\sqrt{1-(0.7975)^2} = 8.05$$

The P-value is less than 0.0001 (using the t_{37} distribution), so I am quite confident that a linear relationship exists between age and size on a log scale.

(Note that this *t* value is the same as the *t*-ratio for the slope on page 712.)

Equivalently, we can test H_0: $\rho = 0$ simply by testing H_0: $\beta_1 = 0$. We might also wish to construct a confidence interval for ρ. However, the strong conditions required and the details involved make it a topic we elect to omit in this text.

*24.6 Logistic Regression

The Pima Indians of southern Arizona are a unique community. Their ancestors were among the first people to cross over into the Americas some 30 000 years ago. For at least two millennia, they have lived in the Sonoran Desert near the Gila River. Known throughout history as a generous people, they have given of themselves for the past 30 years by helping researchers at the National Institutes of Health study certain diseases like diabetes and obesity. Young Pima Indians often marry other Pimas, making them an ideal group for genetic researchers to study. Pimas also have an extremely high incidence of diabetes.

Researchers investigating factors for increased risk of diabetes examined data on 768 adult women of Pima Indian heritage. One possible predictor is the body mass index, BMI, calculated as weight/height^2, where weight is measured in kilograms and height in metres.

We are interested in the relationship between *BMI* and the incidence of *Diabetes*. We might start by looking at boxplots of *BMI* for each group:

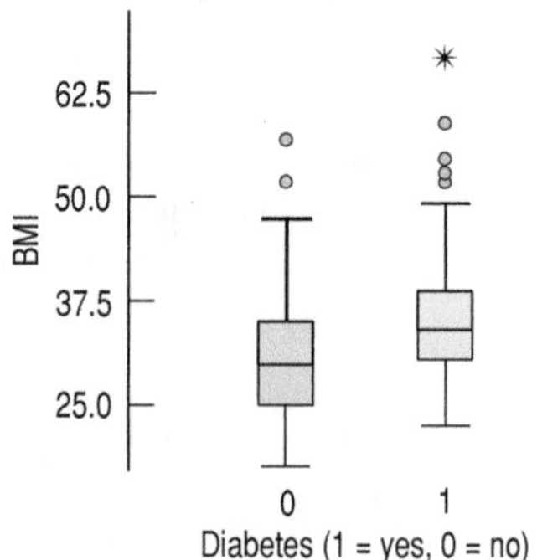

Figure 24.13
Side-by-side boxplots for the two *Diabetes* groups (1 = has diabetes; 0 = doesn't have diabetes) show significantly elevated body mass index (BMI) for the women who have (BMI) diabetes.

From the boxplots, we see that the group with diabetes has a higher mean *BMI*. (A *t*-test would show the difference to be more than 9 SEs from 0 with a P-value < 0.0001.) There is clearly a relationship. Here we've displayed *BMI* as the response and *Diabetes* as the predictor. But the researchers are interested in predicting the increased risk of *Diabetes* due to increased *BMI*, not the other way around.

Reversing the roles, we could code having *Diabetes* as "1" and *No Diabetes* as "0" and make a scatterplot:

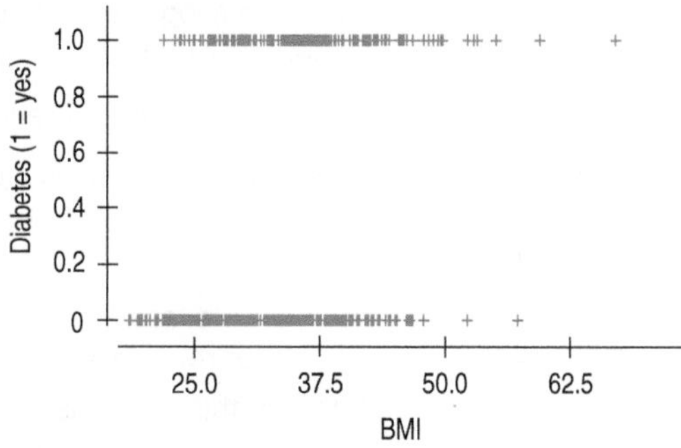

Figure 24.14
A scatterplot of *Diabetes* by *BMI* shows a shift in *BMI* for the two groups, but is not easy to interpret because *Diabetes* is a dichotomous (two-valued) variable.

Diabetes is usually treated as a categorical variable, but what if we treat it as a quantitative variable and fit a linear regression to these data? We would get the following regression line:

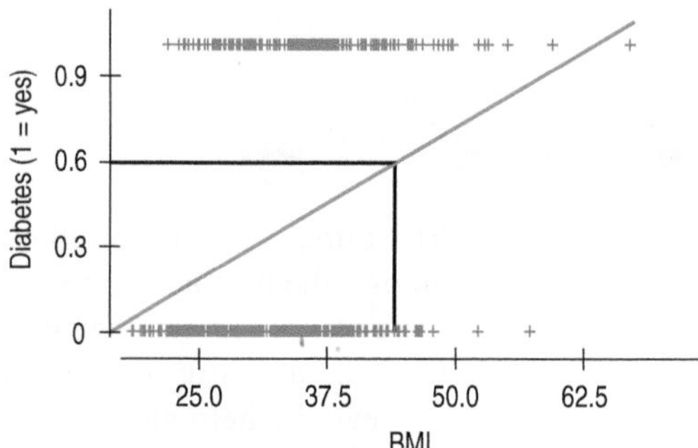

Figure 24.15
A regression of *Diabetes* on *BMI* shows an increasing likelihood of having diabetes with increasing *BMI*. Of course, a linear regression is not strictly appropriate unless *Diabetes* is treated as a quantitative variable.

The equation says: Predicted *Diabetes* = –0.351 + 0.022 *BMI*. Does this make *any* sense? Suppose someone had a *BMI* of 44. The equation predicts 0.60. What would you guess is the chance that she will have diabetes? If you said about 60%, then you're using the line to model the *probability* of having *Diabetes*.

There are some obvious problems with this model, though. What's the probability that someone with a *BMI* of 10 has diabetes? It's low, but the equation predicts –0.13, obviously an impossible probability. And if we imagined someone with a *BMI* of 70, we might suspect that the probability of her having diabetes is pretty high, but not the predicted value of 1.16.

We can fix this problem just by setting all negative probabilities to 0 and all probabilities greater than 1 to 1. That would give a model that looked like this:

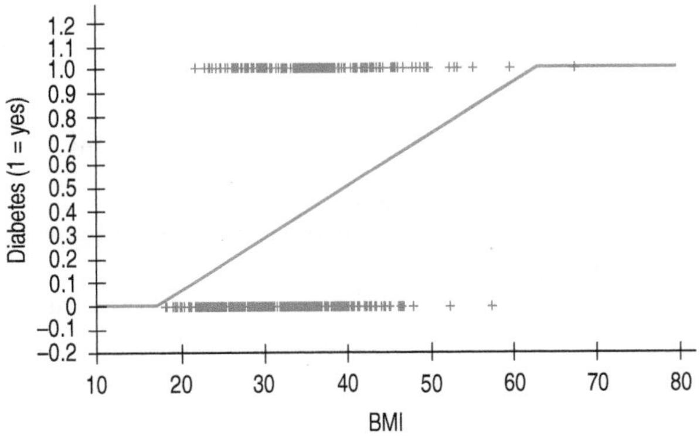

Figure 24.16

This model eliminates the problem of probability values that are negative or that exceed 1, but the corners are not aesthetically pleasing, nor do they make scientific sense.

This avoids one problem, but it can't really be correct. The occurrence of *Diabetes* can't be either certain ($p = 1$) or impossible ($p = 0$) based only on *BMI*. And it's not aesthetically pleasing. Real-world changes are likely to be smooth, so we prefer models with smooth transitions rather than corners. That makes good predictive sense, too. There's no reason to expect sharp changes at certain *BMI* values. So instead, we can use a smooth curve to model the probability of having *Diabetes*, like this:

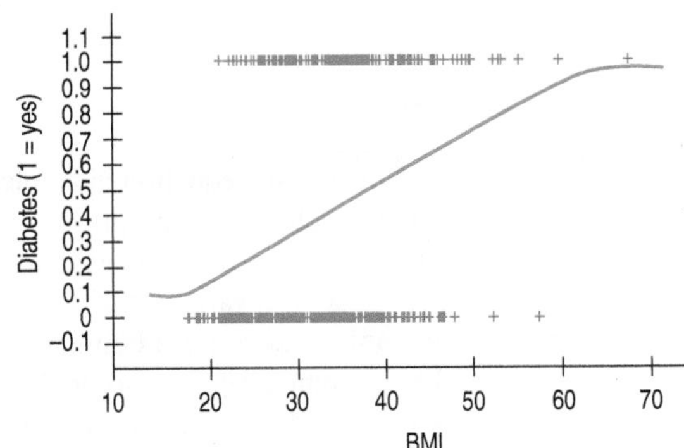

Figure 24.17

The smooth curve models the probability of having *Diabetes* as a function of *BMI* in a sensible way. The logistic curve shown here is just one of a number of choices for the form of the curve, but all are fairly similar in shape and in the resulting predicted probabilities.

This smoother version is a sensible way to model the probability of having *Diabetes* as a function of *BMI*. There are many curves in mathematics with shapes like this that we might use for our model. One of the most common is the *logistic curve*. The regression based on this curve is called **logistic regression.**

The equation for logistic regression can be written like an ordinary regression on a transformed *y*-variable:

$$In\left(\frac{\hat{p}}{1-\hat{p}}\right) = b_0 + b_1 x.$$

The expression on the left-hand side gives the *logistic* curve. The logarithm used is the natural, or base *e*, log, although that doesn't really matter to the shape of the model.

It turns out that the logistic curve has a particularly useful interpretation. Racetrack enthusiasts know that when *p* is a probability, $\frac{p}{1-p}$ is the *odds* in favour of a success. For example, when the probability of success, *p*, = 1/3, we'd get the ratio 1/3/2/3 =1/2. We'd say that the odds in favour of success are 1:2 (or we'd probably say the odds *against* it are 2:1).

Logistic regression models the *logarithm* of the odds as a linear function of x. In fact, nobody really thinks in terms of the log of the odds ratio. But it's the combination of that ratio and the logarithm that gets us the nice S-curved shape. What is important is that we can work backward from a log odds ratio to get the probability—which is often easier to think about.

Because we're not fitting a straight line to the data directly, we can't use ordinary least squares to estimate the parameters b_0 and b_1.[13] Instead, we use special *nonlinear* methods that require a good deal more computation—but the computers take care of that for us, and can work backward through the fitted equation to give predictions in terms of probabilities.

For the Pima Indians data set, the equation is

$$\text{In}\left(\frac{\hat{p}}{1-\hat{p}}\right) = -4.0 + 0.102\,BMI$$

A computer output will typically provide a table like the following:

Term	Estimate	Std Error	ChiSquare	*P*-value
Intercept	−3.9967	0.4288	86.8576	< 0.00001
BMI	0.1025	0.0126	66.0872	< 0.00001

We usually don't interpret the slope itself (unless you happen to think naturally in log odds), but we can perform a test on whether the slope is 0, similar to the t-test we did for linear regression. Unlike linear regression, the ratio of the estimate to its standard error does not have a t distribution, but the square of that ratio has a χ^2 distribution.[14] The P-values for both the slope and the intercept clearly indicate that neither is 0.

Once we have decided that the slope is not 0, we can use the model to predict the probabilities. Solving the equation $\text{ln}\left(\frac{\hat{p}}{1-\hat{p}}\right) = b_0 + b_1x$ for $\hat{p}$ gives

$$\hat{p} = \frac{1}{1 + e^{-(b_0+b_1x)}}$$

and the logistic equation guarantees that the estimate of $\hat{p}$ will be between 0 and 1. Fortunately, technology can provide these probability estimates, producing the curve shown previously and giving an estimate for the probability at any *BMI* value.

Response variables that are dichotomous (having only two possible values) like the variable *Diabetes* are common, so there's a widespread need to model data like these. Phone companies want to predict who will switch carriers, credit card companies want to know which transactions are likely to be fraudulent, and loan companies want to know who is most likely to declare bankruptcy. All of these are potential application areas for logistic regression. It's not surprising that logistic regression has become an important modelling tool in the toolbox of analysts in science and industry in the past decade. By understanding the basics of what logistic regression can do, you can expand your ability to apply regression to many other important applications.

WHAT CAN GO WRONG?

In this chapter, we've added inference to the regression explorations we looked at in Chapters 7 and 8. Everything covered in those chapters that could go wrong with regression can still go wrong. It's probably a good time to review Chapter 8. We'll let you do that.

[13]This is a tricky point. We've fit regressions with transformed y's before without special methods. If our data consisted of observed proportions of people, we could transform the data using $\log(p/(1-p))$ and use linear regression to fit the equation. But all we have are individual 0's and 1's. To be able to fit this equation with our raw data, we need special nonlinear methods.

[14] Here's that nonlinear fitting popping up again. Don't let it bother you. You've seen χ^2 models before, and all you really need to look at for a test is the P-value.

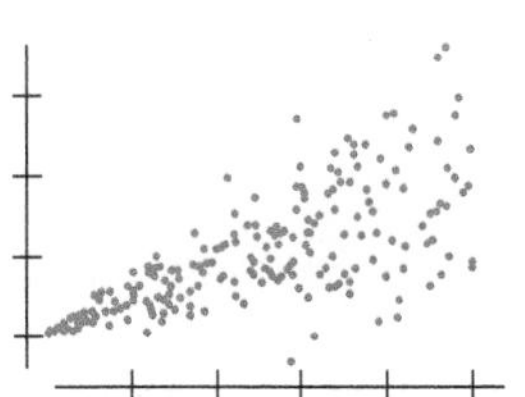

With inference, we've put numbers on our estimates and predictions, but these numbers are only as good as the model. Here are the main things to watch out for:

- **Don't fit a linear regression to data that aren't straight.** This is the most fundamental assumption. If the relationship between x and y isn't approximately linear, there's no sense in fitting a straight line to it.
- **Watch out for the plot thickening.** The common part of confidence and prediction intervals is the estimate of the error standard deviation, the spread around the line. If it changes with x, the estimate won't make sense. Imagine making a prediction interval for these data.

  When x is small, we can predict y precisely, but as x gets larger, it's much harder to pin y down. Unfortunately, if the spread changes, the single value of s_e won't pick that up. The prediction interval will use the average spread around the line, with the result that we'll be too pessimistic about our precision for low x-values and too optimistic for high x-values. A re-expression of y is often a good fix for changing spread.
- **Make sure the errors are Normal.** When we make a prediction interval for an individual, the Central Limit Theorem can't come to our rescue. For us to believe the prediction interval, the errors must be from the Normal model. Check the histogram and Normal probability plot of the residuals to see if this assumption looks reasonable.
- **Watch out for extrapolation.** It's tempting to think that because we have prediction *intervals*, they'll take care of all of our uncertainty so we don't have to worry about extrapolating. Wrong. The interval is only as good as the model. The uncertainty our intervals predict is correct only if our model is true. There's no way to adjust for wrong models. That's why it's always dangerous to predict for x-values that lie far from the centre of the data.
- **Watch out for high-influence points and outliers.** We always have to be on the lookout for a few points that have undue influence on our estimated model—and regression is certainly no exception.
- **Watch out for one-tailed tests.** Because tests of hypotheses about regression coefficients are usually two-tailed, software packages report two-tailed P-values. If you are using software to conduct a one-tailed test about slope, you'll need to divide the reported P-value in half.

Regression inference is connected to almost everything we've done so far. Scatterplots are essential for checking linearity and whether the plot thickens. Histograms and Normal probability plots come into play to check the Nearly Normal Condition. And we're still thinking about the same attributes of the data in these plots as we were in the first part of the book.

Regression inference also is connected to just about every inference method we have seen for measured data. The assumption that the spread of data about the line is constant is essentially the same as the assumption of equal variances required for the pooled-t methods. Our use of all the residuals together to estimate their standard deviation is a form of pooling.

Inference for regression is closely related to inference for means, so your understanding of means transfers directly to your understanding of regression. Here's a table that displays the similarities:

	Means	Regression Slope
Parameter	μ	β_1
Statistic	$\bar{y}$	b_1
Population spread estimate	$s_y = \sqrt{\frac{\Sigma(y-\bar{y})^2}{n-1}}$	$s_e = \sqrt{\frac{\Sigma(y-\hat{y})^2}{n-2}}$
Standard error of the statistic	$SE(\bar{y}) = \frac{s_y}{\sqrt{n}}$	$SE(b_1) = \frac{s_e}{s_x\sqrt{n-1}}$
Test statistic	$\frac{\bar{y}-\mu_0}{SE(\bar{y})} \sim t_{n-1}$	$\frac{b_1-\beta_1}{SE(b_1)} \sim t_{n-2}$
Margin of error	$ME = t^*_{n-1} \times SE(\bar{y})$	$ME = t^*_{n-2} \times SE(b_1)$

What Have We Learned?

Learning Objectives

Understand that the "true" population regression line does not fit the population data perfectly, but is an idealized summary or model of the relationship.

- Know how to examine your data and a scatterplot of y vs. x for violations of assumptions that would make inference for regression unwise or invalid.

Know how to examine displays of the residuals from a regression to double-check that the conditions required for regression have been met. In particular:

- Know how to judge linearity and constant variance from a scatterplot of residuals against predicted values.
- Know how to judge Normality from a histogram and Normal probability plot.

Remember to be especially careful to check for failures of the Independence Assumption when working with data recorded over time.

- To search for patterns, examine scatterplots both of x against time and of the residuals against time.

Know how to test the standard hypothesis that the true regression slope is zero.

- Be able to state the null and alternative hypotheses. Know where to find the relevant numbers in standard computer regression output.

Be able to find a confidence interval for the slope of a regression based on the values reported in a standard regression output table. And be able to interpret it, in practical terms.

Construct and interpret a confidence interval for the predicted mean value corresponding to a specified value, x_ν.

- $\hat{y} \pm t^*_{n-2} \times SE(\hat{\mu}_\nu)$, where $SE(\hat{\mu}_\nu) = \sqrt{SE^2(b_1) \times (x_\nu - \bar{x})^2 + \frac{s_e^2}{n}}$

Construct and interpret a confidence interval for an individual predicted value corresponding to a specified value, x_ν.

- $\hat{y}_\nu \pm t^*_{n-2} \times SE(\hat{y}_\nu)$, where $SE(\hat{y}_\nu) = \sqrt{SE^2(b_1) \times (x_\nu - \bar{x})^2 + \frac{s_e^2}{n} + s_e^2}$

Be able to summarize a regression, and interpret the P-value of the t-statistic for the slope to test the standard null hypothesis.

Review of Terms

Conditions for inference in regression (and checks for some of them)

- Straight Enough Condition for linearity—check that the scatterplot of y against x has linear form and that the scatterplot of residuals against predicted values has no obvious pattern (p. 703).
- Independence Assumption—think about the nature of the data; check a residuals plot (p. 703).
- Does the Plot Thicken? Condition for constant variance—check that the scatterplot shows consistent spread across the range of the x-variable, and that the residuals plot has constant variance, too; a common problem is increasing spread with increasing predicted values—the plot thickens! (p. 704).
- Nearly Normal Condition for Normality of the residuals—check a histogram or NPP of the residuals (p. 704).

Residual standard deviation The spread of the data around the regression line is measured with the residual standard deviation, s_e (p. 709):

$$s_e = \sqrt{\frac{\sum(y - \hat{y})^2}{n - 2}} = \sqrt{\frac{\sum e^2}{n - 2}}$$

***t*-test for the regression slope** When the assumptions are satisfied, we can perform a test for the slope coefficient. We usually test the null hypothesis that the true value of the slope is zero against the alternative that it is not. A zero slope would indicate a complete absence of linear relationship between y and x.

To test H_0: $\beta_1 = 0$ we find

$$t = \frac{b_1 - 0}{SE(b_1)}$$

where

$$SE(b_1) = \frac{s_e}{\sqrt{n-1}s_x}, \quad s_e = \sqrt{\frac{\Sigma(y-\hat{y})^2}{n-2}}$$

n is the number of cases, and s_x is the standard deviation of the x-values. We find the P-value from the Student's t-model with $n - 2$ degrees of freedom.

Alternatively, to test H_0: $\beta_1 = 0$ (or $\rho = 0$), we find

$$t = \frac{r\sqrt{n-2}}{\sqrt{1-r^2}}$$

where r is the correlation coefficient. We find the P-value from the student's t-model with $n - 2$ degrees of freedom (p. 710).

Confidence interval for the regression slope

When the assumptions are satisfied, we can find a confidence interval for the slope parameter from $b_1 \pm t^*_{n-2} \times SE(b_1)$. The critical value, t^*_{n-2}, depends on the confidence level specified and on a Student's t-model with $n - 2$ degrees of freedom (p. 712).

Confidence interval for a mean response

At any x-value, the fitted y-value estimates the mean of the entire (Normally distributed) population of y-values or responses, corresponding to that particular value of x. We can form a confidence interval for the true y population mean at a specified x-value, x_ν, as $\hat{y}_\nu \pm t^*_{n-2} \times SE(\hat{\mu}_\nu)$, where

$$SE(\hat{\mu}_\nu) = \sqrt{SE^2(b_1) \cdot (x_\nu - \bar{x})^2 \pm \frac{s_e^2}{n}}$$

The critical value, t^*_{n-2}, depends on the specified confidence level and the Student's t-model with $n - 2$ degrees of freedom (p. 718).

Prediction interval for an individual response

We can also form a (wider) confidence interval to capture a single future observation on the response y at some particular x-value. The interval takes the form $\hat{y}_\nu \pm t^*_{n-2} \times SE(\hat{y}_\nu)$, where

$$SE(\hat{y}_\nu) = \sqrt{SE^2(b_1) \cdot (x_\nu - \bar{x})^2 + \frac{s_e^2}{n} + s_e^2}$$

The critical value, t^*_{n-2}, depends on the specified confidence level and the Student's t-model with $n - 2$ degrees of freedom (p. 718).

On the Computer REGRESSION ANALYSIS

All statistics packages make a table of results for a regression. These tables differ slightly from one package to another, but all are essentially the same. We've seen two examples of such tables already.

All packages offer analyses of the residuals. With some, you must request plots of the residuals as you request the regression. Others let you find the regression first and then analyze the residuals afterward. Either way, your analysis is not complete if you don't check the residuals with a histogram or Normal probability plot and a scatterplot of the residuals against x or the predicted values.

You should, of course, always look at the scatterplot of your two variables before computing a regression.

Regressions are almost always found with a computer or calculator. The calculations are too long to do conveniently by hand for data sets of any reasonable size. No matter how the regression is computed, the results are usually presented in a table that has a standard form. Here's a portion of a typical regression results table, along with annotations showing where the numbers come from:

A S ***Activity:* Regression on the Computer.** How fast is the universe expanding? And how old is it? A prominent astronomer used regression to astound the scientific community. Read the story, analyze the data, and interactively learn about each of the numbers in a typical computer regression output table.

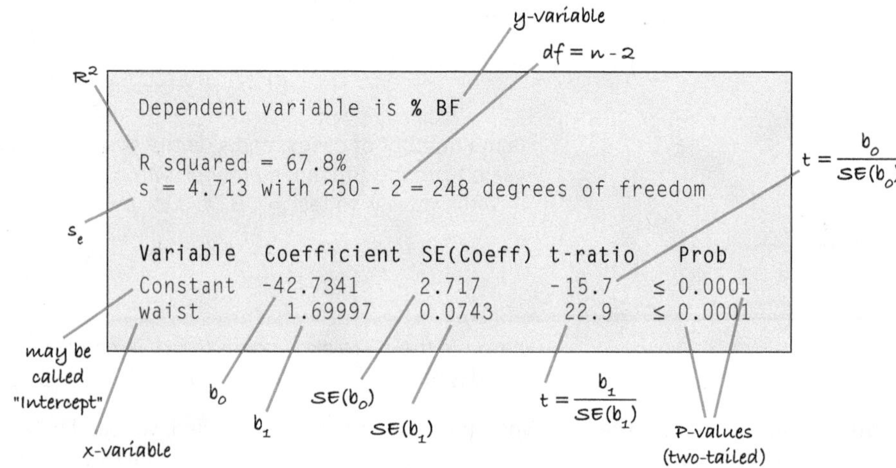

The regression table gives the coefficients (once you find them in the middle of all this other information), so we can see that the regression equation is

$$\widehat{\%BF} = -42.73 + 1.7\ Waist$$

and that the R^2 for the regression is 67.8%.

The column of t-ratios gives the test statistics for the respective null hypotheses that the true values of the coefficients are zero. The corresponding P-values are also usually reported.

EXCEL

- In Excel 2003 and earlier, select Data Analysis from the **Tools** menu. In Excel 2007, select Data Analysis from the **Analysis Group** on the Data tab.
- Select Regression from the **Analysis Tools** list.
- Click the **OK** button.
- Enter the data range holding the *y*-variable in the box labelled "Y-range."
- Enter the range of cells holding the *x*-variable in the box labelled "X-range."
- Select the **New Worksheet Ply** option.
- Select **Residuals** options. Click the **OK** button.

COMMENTS

The *y* and *x* ranges do not need to be in the same rows of the spreadsheet, although they must cover the same number of cells. But it is a good idea to arrange your data in parallel columns as in a data table.

Although the dialogue offers a Normal probability plot of the residuals, the data analysis add-in does not make a correct probability plot, so don't use this option.

JMP

- From the **Analyze** menu, select **Fit Y by X**.
- Select variables: a Y, Response variable, and an X, Factor variable. Both must be continuous (quantitative).
- JMP makes a scatterplot.
- Click on the red triangle beside the heading labelled **Bivariate Fit . . .** and choose **Fit Line**. JMP draws the least squares regression line on the scatterplot and displays the results of the regression in tables below the plot.

- The portion of the table labelled "Parameter Estimates" gives the coefficients and their standard errors, *t*-ratios, and P-values.

COMMENTS

JMP chooses a regression analysis when both variables are "continuous." If you get a different analysis, check the variable types.

The parameter table does not include the residual standard deviation s_e. You can find that as Root Mean Square Error in the Summary of Fit panel of the output.

MINITAB

- Choose **Regression** from the **Stat** menu.
- Choose **Regression . . .** from the **Regression** submenu.
- In the Regression dialogue, assign the *y*-variable to the Response box and assign the *x*-variable to the Predictors box.
- Click the **Graphs** button.
- In the Regression-Graphs dialogue, select **Standardized residuals**, and check **Normal plot of residuals** and **Residuals versus fits**.
- Click the **OK** button to return to the Regression dialogue.
- Click the **OK** button to compute the regression.

COMMENTS

You can also start by choosing a **fitted line plot** from the Regression submenu to see the scatterplot first—usually good practice.

R

- **confint(fit, level=0.95)** will give a 95% CI for the slope of each variable in the model saved to "fit".
- **predict(fit, newdata, interval = c("none", "confidence", "prediction")**, level = 0.95) will give a 95% interval of your choice using the model "fit" and a data frame "newdata", which should contain the values of the predictor variable at which you'd like to make predictions.

SPSS

- Choose **Regression** from the **Analyze** menu.
- Choose **Linear** from the **Regression** submenu.
- In the Linear Regression dialogue that appears, select the *y*-variable and move it to the dependent target. Then move the *x*-variable to the independent target.
- Click the **Plots** button.
- In the Linear Regression Plots dialogue, choose to plot the *SRESIDs against the *ZPRED values.
- Click the **Continue** button to return to the Linear Regression dialogue.
- Click the **OK** button to compute the regression.

STATCRUNCH

To compute a regression with confidence and prediction intervals:

- Click on Stat.
- Choose Regression » Simple Linear.
- Choose X and Y variable names from the list of columns.
- Choose Confidence Intervals.
- Choose Predict Y for X = _____ and enter X_ν.
- Click on Calculate.

TI-83/84 PLUS

Under **STAT TESTS** choose **LinRegTTest**. Specify the two lists where the data are stored, and (usually) choose the two-tail option. In addition to reporting the calculated value of *t* and the P-value, the calculator will tell you the coefficients of the regression equation (*a* and *b*), the values of r^2 and *r*, and the value of *s* you need for confidence or prediction intervals.

Exercises

T **1. Hurricane predictions 2012** Let's look at some data from the National Oceanic and Atmospheric Administration about their success in predicting hurricane tracks. Here is a scatterplot of the error (in nautical miles) for predicting hurricane locations 72 hours in the future versus the year in which the prediction (and the hurricane) occurred:

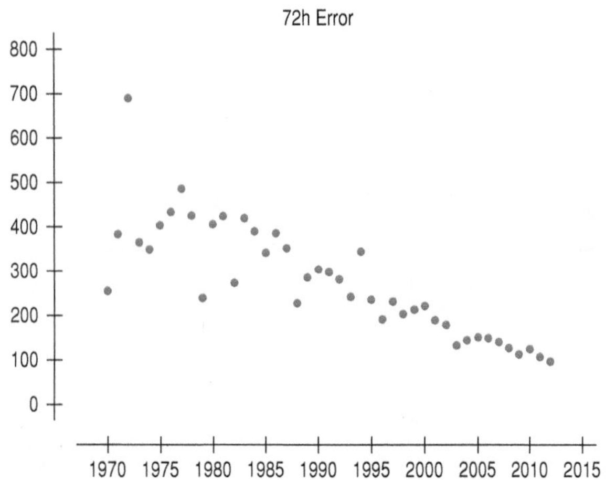

Here is the regression analysis:

Dependent variable is: 72Error
R squared = 71.1%
s = 68.7 with 43 − 2 = 41 degrees of freedom

Variable	Coefficient	SE(Coeff)	*t*-ratio	P-value
Intercept	454.558	20.5942	22.07	< 0.0001
Year since 1970	−8.4790	0.8443	−10.04	< 0.0001

a) Explain in words and numbers what the regression says.

b) State the hypothesis about the slope (both numerically and in words) that describes how hurricane prediction quality has changed.

c) Assuming that the assumptions for inference are satisfied, perform the hypothesis test and state your conclusion. Be sure to state it in terms of prediction errors and years.

d) Explain what the R-squared means in terms of this regression.

T **2. Drug use** The *2011 World Drug Report* investigated the prevalence of drug use as a percentage of the population aged 15 to 64. Data from 22 European countries are shown in the following scatterplot and regression analysis. (*Source: World Drug Report*, 2011. www.unodc.org/unodc/en/data-and-analysis/WDR-2011.html)

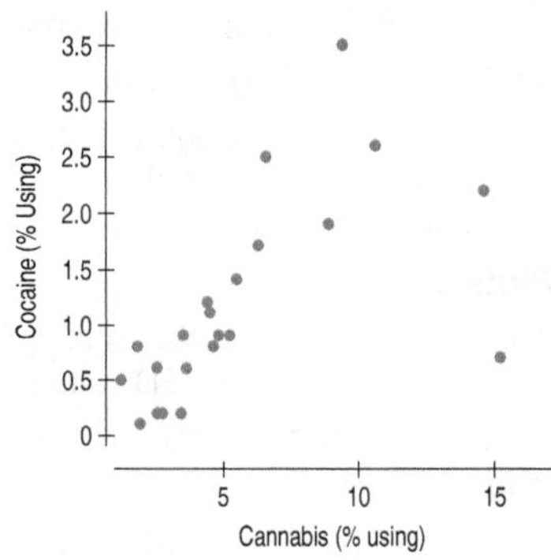

Dependent variable is Cocaine
R-squared = 38.1%
s = 0.724 with 22 – 2 = 20 degrees of freedom

Variable	Coefficient	SE(Coeff)	t-ratio	P-value
Intercept	0.35707	0.2757	1.295	0.210
Cannabis%	0.14264	0.0406	3.512	0.002

a) Explain what the regression says in context.
b) State the hypothesis about the slope (both numerically and in words) that describes how use of marijuana is associated with other drugs.
c) Assuming that the assumptions for inference are satisfied, perform the hypothesis test and state your conclusion in context.
d) Explain what R-squared means in context.
e) Do these results indicate that marijuana use leads to the use of harder drugs? Explain.

T **3. Movie budgets** How does the cost of a movie depend on its length? Data on the cost (millions of dollars) and the running time (minutes) for major release films of 2005 are summarized in these plots and computer output:

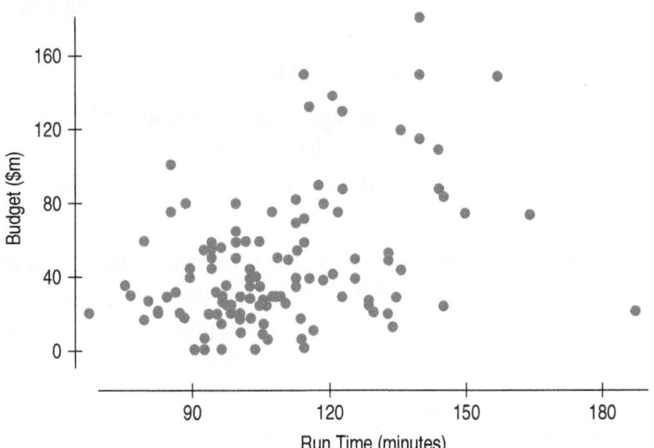

Dependent variable is: Budget($M)
R squared = 15.4%
s = 32.95 with 120 – 2 = 118 degrees of freedom

Variable	Coefficient	SE(Coeff)	t-ratio	P-value
Intercept	–31.3869	17.12	–1.83	0.0693
Run Time	0.714400	0.1541	4.64	≤0.0001

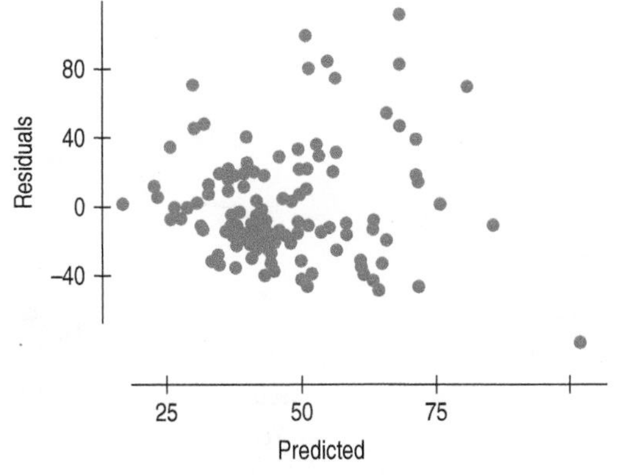

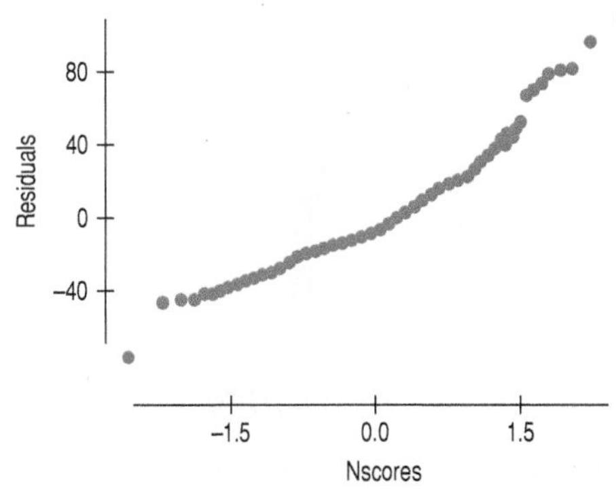

a) Explain in words and numbers what the regression says.
b) The intercept is negative. Discuss its value, taking note of the P-value.
c) The output reports $s = 32.95$. Explain what that means in this context.
d) What is the value of the standard error of the slope of the regression line?
e) Explain what that means in this context.

T **4. House prices** How does the price of a house depend on its size? Data from Saratoga, New York, on 1064 randomly selected sold houses include data on price ($1000s) and size (1000 ft^2), producing the following graphs and computer output:

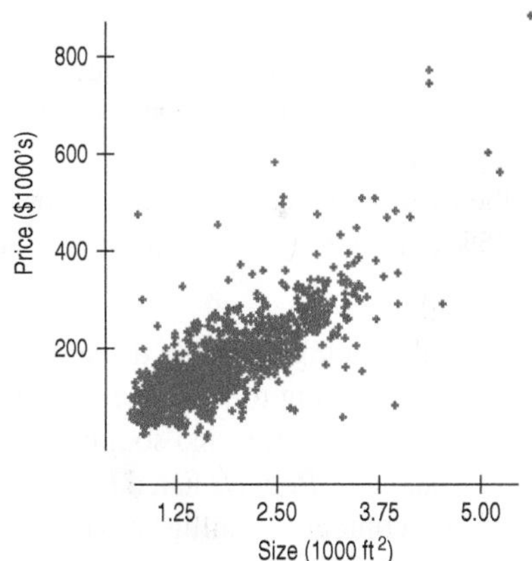

Dependent variable is: Price
R squared = 59.5%
s = 53.79 with 1064 – 2 = 1062 degrees of freedom

Variable	Coefficient	SE(Coeff)	t-ratio	P-value
Intercept	–3.11686	4.688	–0.665	0.5063
Size	94.4539	2.393	39.5	≤0.0001

Residuals
300
150
0
–150
125
250
375
500
Predicted

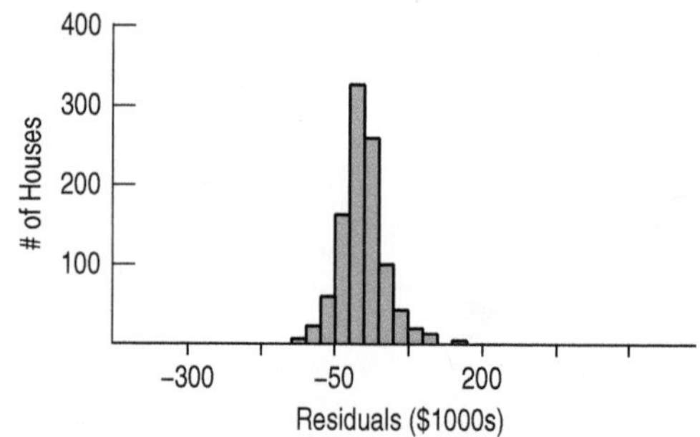

a) Explain in words and numbers what the regression says.
b) The intercept is negative. Discuss its value, taking note of its P-value.
c) The output reports $s = 53.79$. Explain what that means in this context.
d) What is the value of the standard error of the slope of the regression line?
e) Explain what that means in this context.

5. Movie budgets: the sequel Exercise 3 shows computer output examining the association between the length of a movie and its cost.
a) Check the assumptions and conditions for inference.
b) Find a 95% confidence interval for the slope and interpret it.

6. Second home Exercise 4 shows computer output examining the association between the size of houses and their sale prices.
a) Check the assumptions and conditions for inference.
b) Find a 95% confidence interval for the slope and interpret it.

7. Hot dogs Healthy eating probably doesn't include hot dogs, but if you are going to have one, you'd probably hope it's low in both calories and sodium. In its July 2007 issue, *Consumer Reports* listed the number of calories and sodium content (in milligrams) for 13 brands of all-beef hot dogs it tested. Examine the association, assuming that the data satisfy the conditions for inference.

Dependent variable is: Sodium
R squared = 60.5%
s = 59.66 with 13 – 2 = 11 degrees of freedom

Variable	Coefficient	SE(Coeff)	*t*-ratio	P-value
Constant	90.9783	77.69	1.17	0.2663
Calories	2.29959	0.5607	4.10	0.0018

a) State the appropriate hypotheses about the slope.
b) Test your hypotheses and state your conclusion in the proper context.

8. Cholesterol 2007 Does a person's cholesterol level tend to change with age? Data collected from 1406 adults aged 45 to 62 produced the regression analysis shown. Assuming that the data satisfy the conditions for inference, examine the association between age and cholesterol level.

Dependent variable is: Chol
s = 46.16

Variable	Coefficient	SE(Coeff)	*t*-ratio	P-value
Intercept	194.232	13.55	14.3	≤0.0001
Age	0.771639	0.2574	3.00	0.0028

a) State the appropriate hypothesis for the slope.
b) Test your hypothesis and state your conclusion in the proper context.

9. Second frank Look again at Exercise 7's regression output for the calorie and sodium content of hot dogs.
a) The output reports $s = 59.66$. Explain what that means in this context.
b) What is the value of the standard error of the slope of the regression line?
c) Explain what that means in this context.

10. More cholesterol Look again at Exercise 8's regression output for age and cholesterol level.
a) The output reports $s = 46.16$. Explain what that means in this context.
b) What is the value of the standard error of the slope of the regression line?
c) Explain what that means in this context.

11. Last dog Based on the regression output seen in Exercise 7, create a 95% confidence interval for the slope of the regression line and interpret your interval in the proper context.

12. Cholesterol, finis Based on the regression output seen in Exercise 8, create a 95% confidence interval for the slope of the regression line and interpret your interval in the proper context.

13. Marriage age 2008 The scatterplot suggests a decrease in the difference in ages at first marriage for Canadian men and women since 1921. We want to examine the regression to see if this decrease is significant.

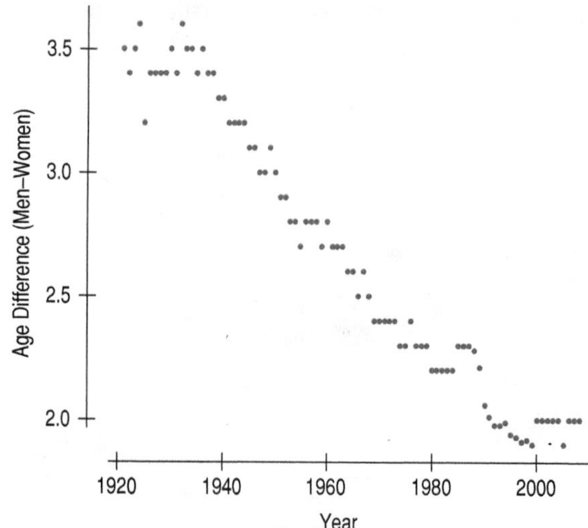

Dependent variable is: Men–Women
R squared = 95.75%
s = 0.1139 with 88 – 2 = 86 degrees of freedom

Variable	Coefficient	SE(Coeff)	*t*-ratio	P-value
Intercept	43.9879	0.9387888	46.86	< 0.0001
Year	–0.021036	0.0004778	–44.02	< 0.0001

a) Write appropriate hypotheses.

b) Here are the residuals plot and a histogram of the residuals. Do you think the conditions for inference are satisfied? Explain.

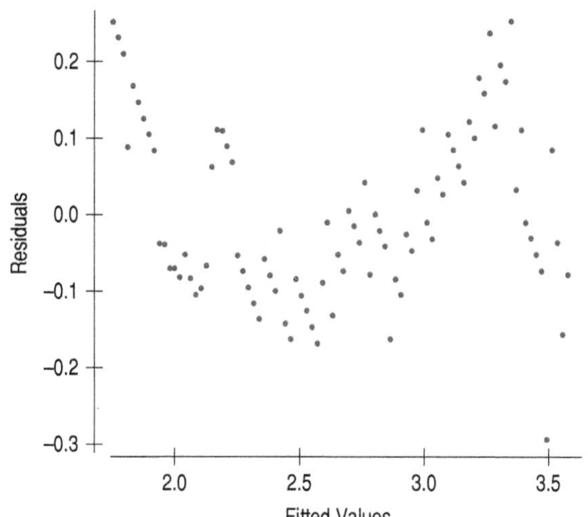

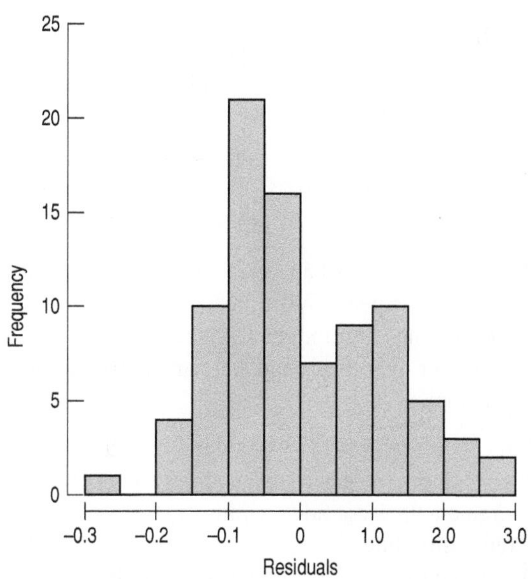

c) Test the hypothesis and state your conclusion about the trend in age at first marriage.
Data: www4.hrsdc.gc.ca/.3ndic.1t.4r@-eng.jsp?iid=78

14. Used cars 2013 Classified ads in a newspaper offered several used Toyota Corollas for sale. The following lists the ages of the cars and the advertised prices.

Age	Price
1	13 900
1	14 610
2	13 998
2	16 887
3	10 900
3	12 995
3	14 995
4	10 999
4	11 988
4	11 995
4	12 495
5	8 988
5	9 488
5	10 488
5	10 989
7	11 400
8	6 995
9	6 988
10	10 000
12	2 700

a) Make a scatterplot for these data.

b) Do you think a linear model is appropriate? Explain.

c) Find the equation of the regression line.

d) Check the residuals to see if the conditions for inference are met.

15. Marriage age 2008, again Based on the analysis of differences in marriage ages given in Exercise 13, give a 95% confidence interval for the rate at which the age gap is closing. Clearly explain what your confidence interval means.

16. Used cars 2013, again Based on the analysis of used car prices you did for Exercise 14, create a 95% confidence interval for the slope of the regression line and explain what your interval means in context.

17. Fuel economy A consumer organization has reported test data for 50 car models. We will examine the association between the weight of the car (in thousands of pounds) and the fuel efficiency (in miles per gallon). Shown are the scatterplot, summary statistics, and regression analysis:

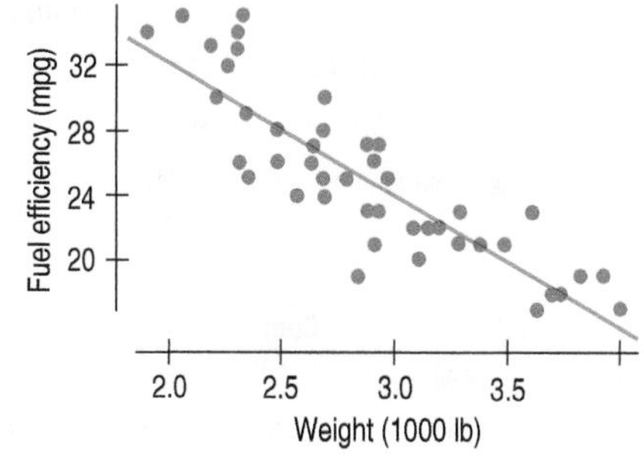

Variable	Count	Mean	StdDev
MPG	50	25.0200	4.83394
wt/1000	50	2.88780	0.511656

Dependent variable is: MPG
R-squared = 75.6%
s = 2.413 with 50 − 2 = 48 df

Variable	Coefficient	SE(Coeff)	*t*-ratio	P-value
Intercept	48.7393	1.976	24.7	≤0.0001
Weight	−8.21362	0.6738	−12.2	≤0.0001

a) Is there strong evidence of an association between the weight of a car and its gas mileage? Write an appropriate hypothesis.
b) Are the assumptions for regression satisfied? Check the following plots.
c) Test your hypothesis and state your conclusion.

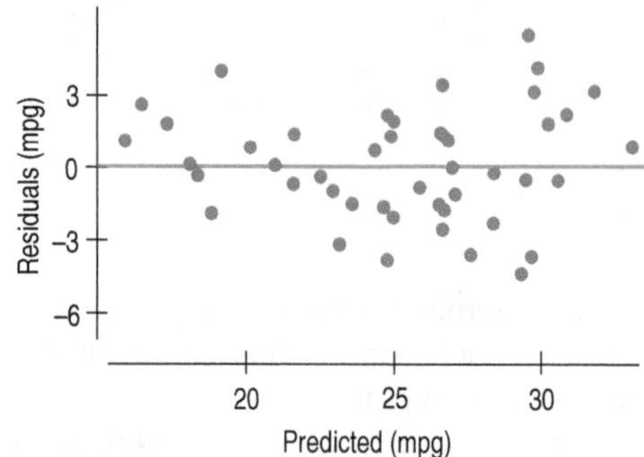

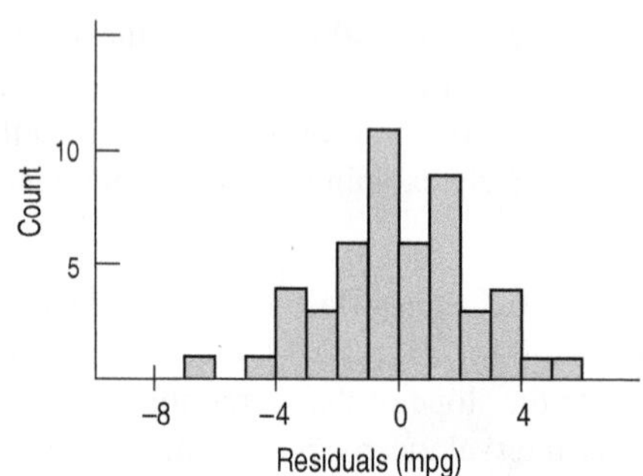

18. Cockpit noise How strong is the association between cockpit noise level (in decibels) and airspeed (in knots indicated air speed, KIAS) for a Boeing 727, an older type of aircraft lacking the design improvements of newer planes? Here are summaries and plots of the data:

Variable	Count	Mean	Median	StdDev	Range
Speed	61	332.5082	335	48.7957	190
Noise	61	89.6803	90	3.9487	15

Dependent variable is: Noise
R-squared = 89.47%
s = 1.2923 with 61 − 2 = 59 df

Variable	Coefficient	SE(Coeff)	*t*-ratio	P-value
Intercept	64.2294	1.1484	55.9073	≤0.0001
Speed	0.0765	0.0034	22.3867	≤0.0001

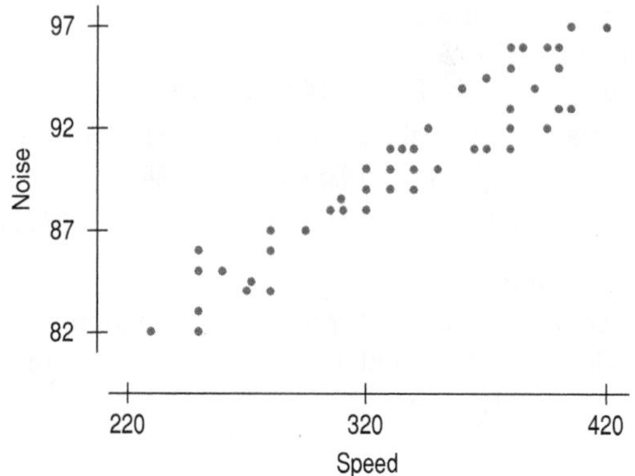

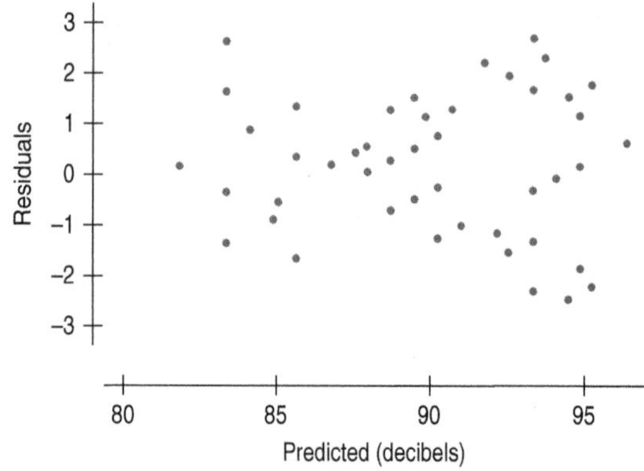

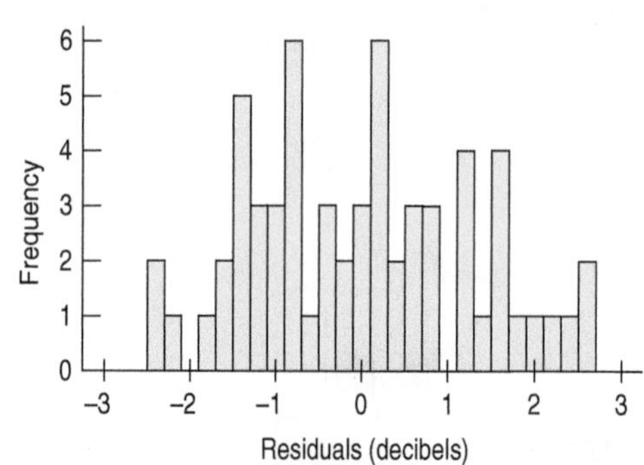

a) Is there evidence of an association between cockpit noise level and airspeed? Write an appropriate hypothesis.
b) Discuss the assumptions for inference.
c) Test your hypothesis and state an appropriate conclusion.

T **19. Fuel economy, part II** Consider again the data in Exercise 17 about the gas mileage and weights of cars.
a) Create a 95% confidence interval for the slope of the regression line.
b) Explain in this context what your confidence interval means.

20. Cockpit noise, part II Consider the cockpit noise data from Exercise 18.
a) Find a 90% confidence interval for the slope of the true line describing the association between cockpit noise level and airspeed.
b) Explain in this context what your confidence interval means.

21. Fuel economy, part III Consider again the data in Exercise 17 about the gas mileage and weights of cars.

a) Create a 95% confidence interval for the average fuel efficiency among cars weighing 2500 pounds, and explain what your interval means.
b) Create a 95% prediction interval for the gas mileage you might get driving your new 3450-pound SUV, and explain what that interval means.

22. Cockpit noise again Consider the cockpit noise data from Exercise 18 (Challenging exercise).

a) Find a 90% confidence interval for the mean speed of all Boeing 727s with a noise level of 90 decibels.
b) Find a 90% prediction interval for the speed of a Boeing 727 if you know that its noise level is 95 decibels.

23. Cereal People are often concerned about the amount of calories and sodium in their cereal. Data for 77 cereals were examined and judged acceptable for inference. The 77 cereals had between 50 and 160 calories per serving and between 0 and 320 mg of sodium per serving. The regression analysis is shown:

Dependent variable is: Sodium
R-squared = 9.0%
s = 80.49 with 77 − 2 = 75 degrees of freedom

Variable	Coefficient	SE(Coeff)	*t*-ratio	P-value
Intercept	21.4143	51.47	0.416	0.6786
Calories	1.29357	0.4738	2.73	0.0079

a) Is there an association between the number of calories and the sodium content of cereals? Explain.
b) Do you think this association is strong enough to be useful? Explain.

24. Brain size Does your IQ depend on the size of your brain? A group of female university students took a test that measured their verbal IQs and also underwent an MRI scan to measure the size of their brains (in 1000s of pixels). The scatterplot and regression analysis are shown, and the assumptions for inference were satisfied.

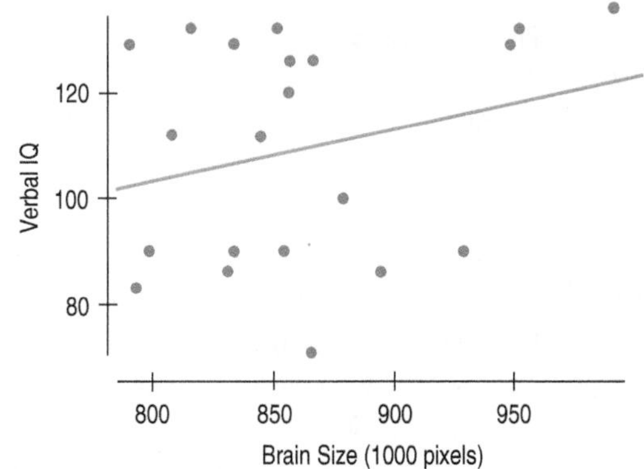

Dependent variable is: IQ_Verbal
R-squared = 6.5%

Variable	Coefficient	SE(Coeff)
Intercept	24.1835	76.38
Size	0.098842	0.0884

a) Test an appropriate hypothesis about the association between brain size and IQ.
b) State your conclusion about the strength of this association.

25. Another bowl Further analysis of the data for the breakfast cereals in Exercise 23 looked for an association between fibre content and calories by attempting to construct a linear model. Several graphs are shown. Which of the assumptions for inference are violated? Explain.

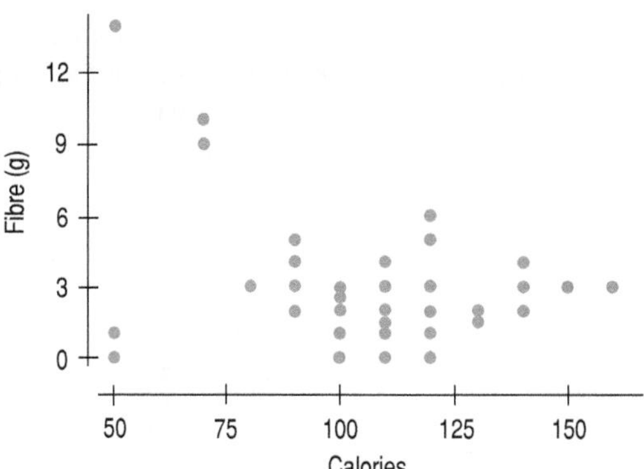

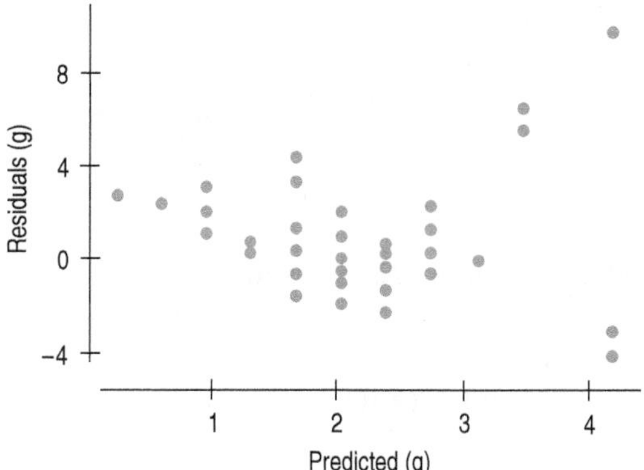

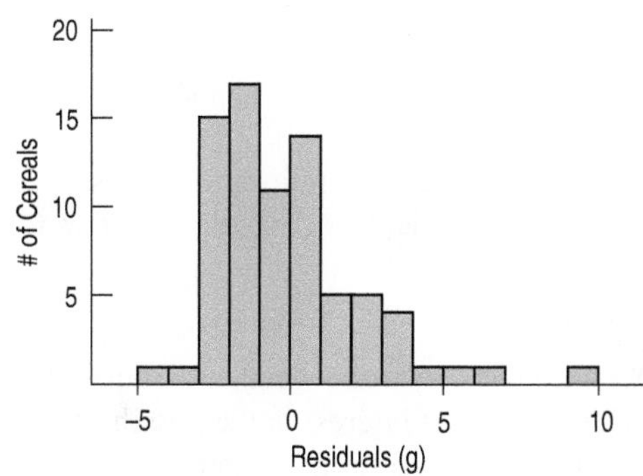

26. Heptathlon, round 2 Consider again the 2012 Women's heptathlon data from the Chapter 7 exercises . If we regress the High Jump result on the Long Jump (avg 6.06 m) for the top 29 women, we get the following results:

Dependent variable is: High Jump
R-squared: 38.58%

Variable	Coefficient	SE(Coeff)	*t*-ratio	P-value
Intercept	0.64931	0.27610	2.352	0.026237*
LongJump	0.18759	0.04555	4.119	0.000323***

Residual standard error: 0.06024 on 27 degrees of freedom

Assuming the conditions for inference are met,

a) Estimate the average high jump result for a long jump of 5.90m, and give a 95% interval for this estimate.

b) Predict the high jump result for a heptathlete who jumps 6.25m in the long jump, and give a 95% interval for this prediction.

c) Test the hypothesis that the correlation coefficient between high jump and long jump is zero.

27. Acid rain Biologists studying the effects of acid rain on wildlife collected data from 163 streams in the Adirondack Mountains. They recorded the pH (acidity) of the water and the BCI, a measure of biological diversity. Following is a scatterplot of BCI against pH:

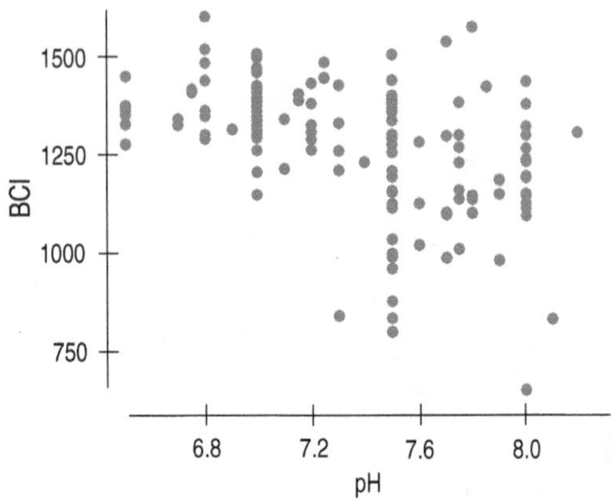

And here is part of the regression analysis:

Dependent variable is: BCI
R-squared = 27.1%
s = 140.4 with 163 − 2 = 161 degrees of freedom

Variable	Coefficient	SE(Coeff)
Intercept	2733.37	187.9
pH	−197.694	25.57

a) State the null and alternative hypotheses under investigation.

b) Assuming that the assumptions for regression inference are reasonable, find the t- and P-values for the test.

c) State your conclusion.

28. El Niño Concern over the weather associated with El Niño has increased interest in the possibility that the climate on Earth is getting warmer. The most common theory relates an increase in atmospheric levels of carbon dioxide (CO_2), a greenhouse gas, to increases in temperature. Here is part of a regression analysis of the mean annual CO_2 concentration in the atmosphere, measured in parts per million (ppm), at the top of Mauna Loa in Hawaii and the mean annual air temperature over both land and sea across the globe, measured in degrees Celsius. The scatterplots and residuals plots indicated that the data were appropriate for inference.

Dependent variable is: Temp
R-squared = 33.4%
s = 0.0809 with 37 − 2 = 35 degrees of freedom

Variable	Coefficient	SE(Coeff)
Intercept	15.3066	0.3139
CO_2	0.004	0.0009

a) Write the equation of the regression line.

b) Is there evidence of an association between CO_2 level and global temperature?

c) Do you think predictions made by this regression will be very accurate? Explain.

29. Ozone The Environmental Protection Agency is examining the relationship between the ozone level (in parts per million) and the population (in millions) of U.S. cities. Part of the regression analysis is shown.

Dependent variable is: Ozone
R-squared = 84.4%
s = 5.454 with 16 − 2 = 14 df

Variable	Coefficient	SE(Coeff)
Intercept	18.892	2.395
Pop	6.650	1.910

a) We suspect that the greater the population of a city, the higher its ozone level. Is the relationship significant? Assuming the conditions for inference are satisfied, test an appropriate hypothesis and state your conclusion in context.

b) Do you think that the population of a city is a useful predictor of ozone level? Use the values of both R^2 and s in your explanation.

30. Sales and profits A business analyst was interested in the relationship between a company's sales and its profits. She collected data (in millions of dollars) from a random sample of *Fortune* 500 companies, and created the regression analysis and summary statistics shown. The assumptions for regression inference appeared to be satisfied.

	Profits	Sales
Count	79	79
Mean	209.839	4178.29
Variance	635 172	49 163 000
Std Dev	796.977	7011.63

Dependent variable is: Profits
R-squared = 66.2% s = 466.2

Variable	Coefficient	SE(Coeff)
Intercept	−176.644	61.16
Sales	0.092498	0.0075

a) Is there a significant association between sales and profits? Test an appropriate hypothesis and state your conclusion in context.

b) Do you think that a company's sales serve as a useful predictor of its profits? Use the values of both R^2 and s in your explanation.

31. Ozone, again Consider again the relationship between the population and ozone level of U.S. cities that you analyzed in Exercise 29.

a) Give a 90% confidence interval for the approximate increase in ozone level associated with each additional million city inhabitants.

b) For the cities studied, the mean population was 1.7 million people. The population of Boston is approximately 0.6 million people. Predict the mean ozone level for cities of that size with an interval in which you have 90% confidence.

T 32. More sales and profits Consider again the relationship between the sales and profits of *Fortune* 500 companies that you analyzed in Exercise 30.

a) Find a 95% confidence interval for the slope of the regression line. Interpret your interval in context.

b) Last year the drug manufacturer Eli Lilly, Inc., reported gross sales of \$9 billion (that's \$9000 million). Create a 95% prediction interval for the company's profits and interpret your interval in context.

33. Start the car! In October 2002, *Consumer Reports* listed the price (in dollars) and power (in cold cranking amps) of auto batteries. We want to know if more expensive batteries are generally better in terms of starting power. Here are several software displays:

Dependent variable is: Power
R-squared = 25.2%
s = 116.0 with 33 − 2 = 31 degrees of freedom

Variable	Coefficient	SE(Coeff)	*t*-ratio	P-value
Intercept	384.594	93.55	4.11	0.0003
Cost	4.14649	1.282	3.23	0.0029

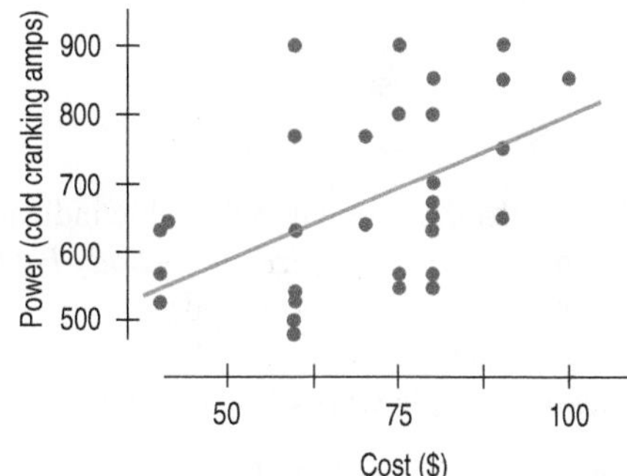

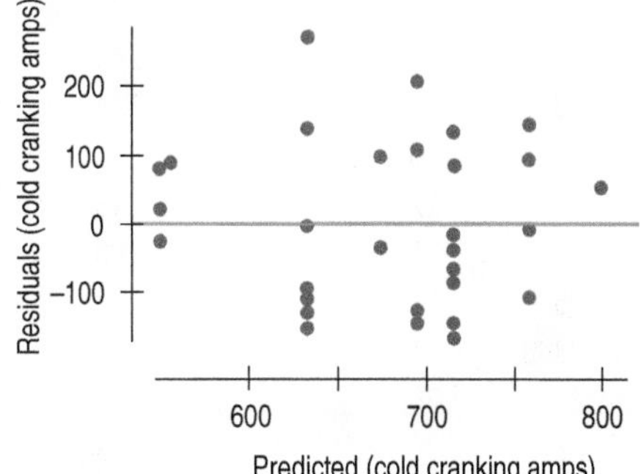

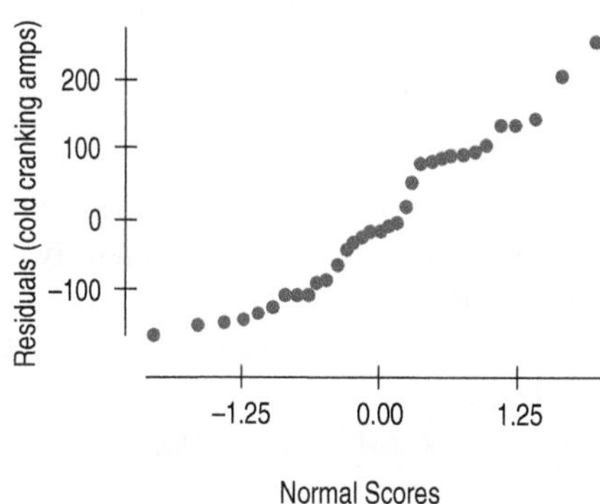

a) How many batteries were tested?

b) Are the conditions for inference satisfied? Explain.

c) Is there evidence of an association between the cost and cranking power of auto batteries? Test an appropriate hypothesis and state your conclusion.

d) Is the association strong? Explain.

e) What is the equation of the regression line?

f) Create a 90% confidence interval for the slope of the true line.

g) Interpret your interval in this context.

T 34. Crawling Researchers at the University of Denver Infant Study Center wondered whether temperature might influence the age at which babies learn to crawl. Perhaps the extra clothing that babies wear in cold weather would restrict movement and delay the age at which they started crawling. Data were collected on 208 boys and 206 girls. Parents reported the month of the baby's birth and the age (in weeks) at which their child first crawled. The table gives the average outside temperature (°F) when the babies were six months old and average crawling age (in weeks) for each month of the year. Make the plots and compute the analyses necessary to answer the following questions.

Birth Month	6-Month Temperature	Average Crawling Age
Jan.	66	29.84
Feb.	73	30.52
Mar.	72	29.70
April	63	31.84
May	52	28.58
June	39	31.44
July	33	33.64
Aug.	30	32.82
Sept.	33	33.83
Oct.	37	33.35
Nov.	48	33.38
Dec.	57	32.32

a) Would this association appear to be weaker, stronger, or the same if data had been plotted for individual babies instead of using monthly averages? Explain.

b) Is there evidence of an association between temperature and crawling age? Test an appropriate hypothesis and state your conclusion. Don't forget to check the assumptions.

c) Create and interpret a 95% confidence interval for the slope of the true relationship.

35. Tablet computers In October 2011, cnet.com listed the battery life (in hours) and luminous intensity (i.e., screen brightness, in cd/m²) for a sample of tablet computers. We want to know if brighter screens drain the battery more quickly.[15]

Dependent variable is Video battery life (in hours)
R-squared = 27.9%
s = 1.913 with 23 − 2 = 21 degrees of freedom

Variable	Coefficient	SE(Coeff)	t-Ratio	P-Value
Intercept	2.8467073	1.6628386	1.7119566	0.10163
Brightness	0.014080549	0.0049373503	2.851843	0.00955

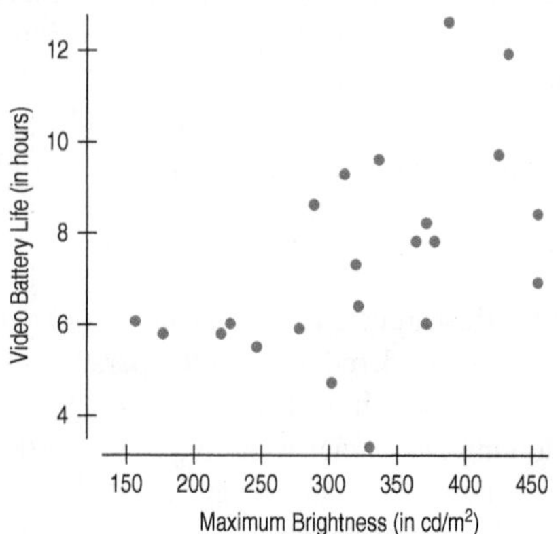

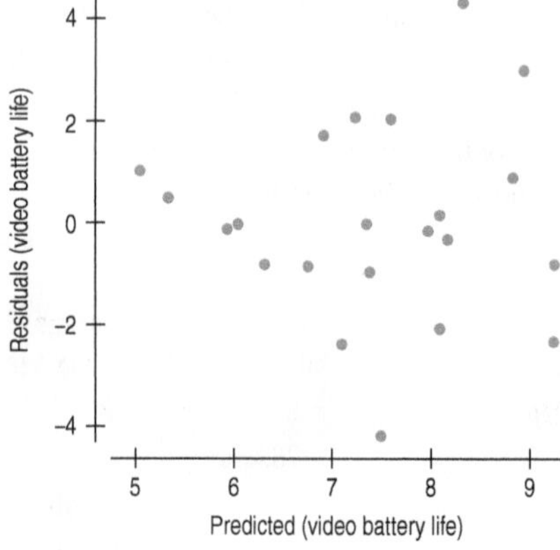

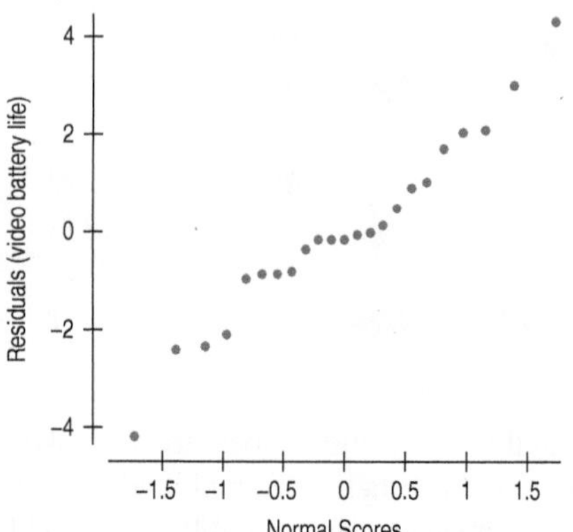

a) How many tablet computers were tested?
b) Are the conditions for inference satisfied? Explain.
c) Is there evidence of an association between maximum brightness of the screen and battery life? Test an appropriate hypothesis and state your conclusion.
d) Is the association strong? Explain.
e) What is the equation of the regression line?
f) Create a 90% confidence interval for the slope of the true line.
g) Interpret your interval in this context.

36. Strike two Remember the Little League instructional video discussed in Chapter 22? Ads claimed that the techniques would improve the performances of Little League pitchers. To test this claim, 20 Little Leaguers threw 50 pitches each, and we recorded the number of strikes. After the players participated in the training program, we repeated the test. The table shows the number of strikes each player threw before and after the training. A test of paired differences failed to show that this training was effective in improving a player's ability to throw strikes. Is there any evidence that the effectiveness of the video depends on the player's initial ability to throw strikes? Test an appropriate hypothesis and state your conclusion.

Number of Strikes (out of 50)

Before	After	Before	After
28	35	33	33
29	36	33	35
30	32	34	32
32	28	34	30
32	30	34	33
32	31	35	34
32	32	36	37
32	34	36	33
32	35	37	35
33	36	37	32

37. Body fat Do the data shown in the table indicate an association between *Waist size* and *%Body Fat*?
a) Test an appropriate hypothesis and state your conclusion.
b) Give a 95% confidence interval for the mean *%Body Fat* found in people with 40-inch waists.

Waist (in.)	Weight (lb)	Body Fat (%)	Waist (in.)	Weight (lb)	Body Fat (%)
32	175	6	33	188	10
36	181	21	40	240	20
38	200	15	36	175	22
33	159	6	32	168	9
39	196	22	44	246	38
40	192	31	33	160	10
41	205	32	41	215	27
35	173	21	34	159	12
38	187	25	34	146	10
38	188	30	44	219	28

[15] http://www.cnet.com/news/cnet-tablet-battery-life-results/

38. Body fat, again Use the data from Exercise 37 to examine the association between weight and *%Body Fat*.

a) Find a 90% confidence interval for the slope of the regression line of *%Body Fat* on weight.
b) Interpret your interval in context.
c) Give a 95% prediction interval for the *%Body Fat* of an individual who weighs 165 pounds.

39. Grades The data set below shows midterm scores from an Introductory Statistics course.

First Name	Midterm 1	Midterm 2	Homework
Timothy	82	30	61
Karen	96	68	72
Verena	57	82	69
Jonathan	89	92	84
Elizabeth	88	86	84
Patrick	93	81	71
Julia	90	83	79
Thomas	83	21	51
Marshall	59	62	58
Justin	89	57	79
Alexandra	83	86	78
Christopher	95	75	77
Justin	81	66	66
Miguel	86	63	74
Brian	81	86	76
Gregory	81	87	75
Kristina	98	96	84
Timothy	50	27	20
Jason	91	83	71
Whitney	87	89	85
Alexis	90	91	68
Nicholas	95	82	68
Amandeep	91	37	54
Irena	93	81	82
Yvon	88	66	82
Sara	99	90	77
Annie	89	92	68
Benjamin	87	62	72
David	92	66	78
Josef	62	43	56
Rebecca	93	87	80
Joshua	95	93	87
Ian	93	65	66
Katharine	92	98	77
Emily	91	95	83
Brian	92	80	82
Shad	61	58	65
Michael	55	65	51
Israel	76	88	67
Iris	63	62	67
Mark	89	66	72
Peter	91	42	66
Catherine	90	85	78
Christina	75	62	72
Enrique	75	46	72
Sarah	91	65	77
Thomas	84	70	70
Sonya	94	92	81
Michael	93	78	72
Wesley	91	58	66
Mark	91	61	79
Adam	89	86	62
Jared	98	92	83
Michael	96	51	83
Kathryn	95	95	87
Nicole	98	89	77
Wayne	89	79	44
Elizabeth	93	89	73
John	74	64	72
Valentin	97	96	80
David	94	90	88
Marc	81	89	62
Samuel	94	85	76
Brooke	92	90	86

a) Fit a model predicting the second midterm score from the first.
b) Comment on the model you found, including a discussion of the assumptions and conditions for regression. Is the coefficient for the slope statistically significant?
c) A student comments that because the P-value for the slope is very small, Midterm 2 is very well predicted from Midterm 1. So, he reasons, next term the professor can give just one midterm. What do you think?

40. Grades? The professor teaching the Introductory Statistics class discussed in Exercise 39 wonders whether performance on homework can accurately predict midterm scores.

a) To investigate it, she fits a regression of the sum of the two midterms scores on homework scores. Fit the regression model.
b) Comment on the model, including a discussion of the assumptions and conditions for regression. Is the coefficient for the slope "statistically significant"?
c) Do you think she can accurately judge a student's performance without giving the midterms? Explain.

41. Swimming Ontario Since Marilyn Bell first swam across Lake Ontario in 1954, swimmers have crossed the lake 58 times (as of August, 2013). Have swimmers been getting faster or slower, or has there been no change in swimming speed? A regression predicting time for the swim (in minutes) against year of the swim (for the years 1974–2010) looks like this:

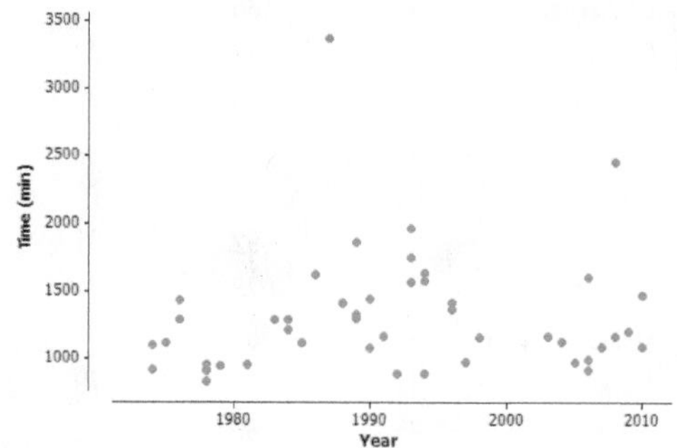

Regression Analysis: Time (min) versus Year
The regression equation is
Time (min) = −9931 + 5.64 Year

Predictor	Coef	SE Coef	T	P
Constant	−99.31	11694	−0.85	0.400
Year	5.636	5.873	0.96	0.342

S = 449.925 R-Sq = 2.0% R-Sq(adj) = 0.0%

a) What do you hope to learn from these data? Define the variables, and state the null and alternative hypotheses about the slope.
b) Check assumptions and conditions.
c) If they are satisfied, complete the analysis.
d) The longest time is for a swim in which Vicki Keith swam round-trip, crossing the lake twice. If we remove it from the data, the new analysis is shown. Check the conditions again.

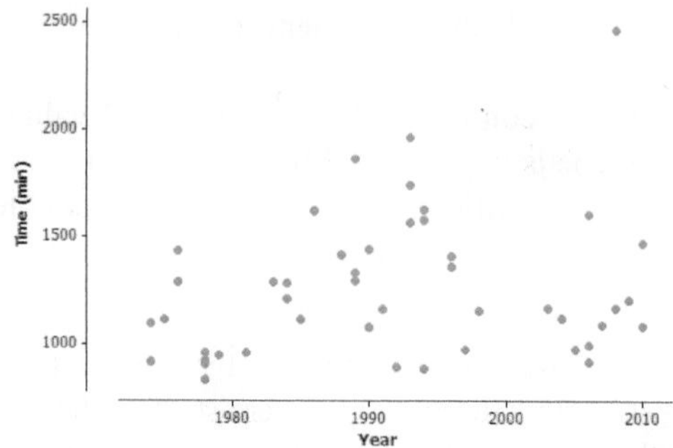

The regression equation is
Time (min) = −13111 + 7.21 Year

Predictor	Coef	SE Coef	T	P
Constant	−13111	8490	−1.54	0.130
Year	7.211	4.263	1.69	0.098

S = 326.099 R-Sq = 6.0% R-Sq(adj) = 3.9%

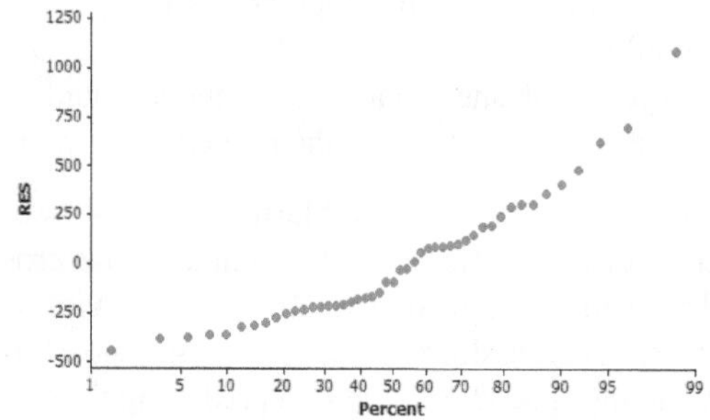

a) If the conditions are satisfied, complete the analysis.
b) The slope itself didn't change very much when we removed the outlier, but the P-value did. Consider the formula for the t-ratio and explain what changed.

42. All the efficiency money can buy 2011 A sample of 84 model-2011 cars from an online information service was examined to see how fuel efficiency (as highway mpg) relates to the cost (Manufacturer's Suggested Retail Price in dollars) of cars. Here are displays and computer output:

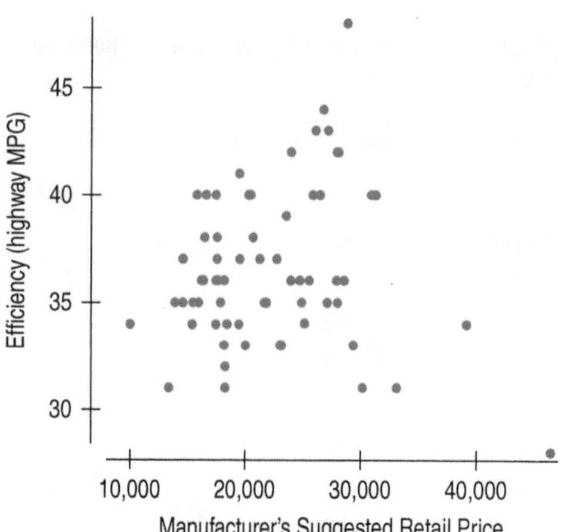

Dependent variable is MPG
R-squared = 0.0216%
s = 3.54

Variable	Coefficient	SE(Coeff)	t-Ratio	P-Value
Intercept	36.514	1.496	24.406	<0.0001
Slope	−8.089E−6	6.439E−5	−0.1256	0.900

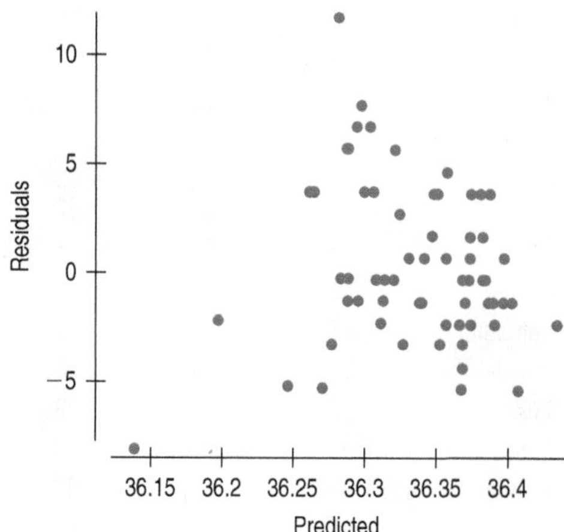

a) State what you want to know, identify the variables, and give the appropriate hypotheses.
b) Check the assumptions and conditions.
c) If the conditions are met, complete the analysis.

43. Education and mortality The software output below is based on the mortality rate (deaths per 100 000 people) and the education level (average number of years in school) for 58 U.S. cities.

Variable	Count	Mean	StdDev
Mortality	58	942.501	61.8490
Education	58	11.0328	0.793480

Dependent variable is: Mortality
R-squared = 41.0%
s = 47.92 with 58 − 2 = 56 degrees of freedom

Variable	Coefficient	SE(Coeff)
Intercept	1493.26	88.48
Education	249.9202	8.000

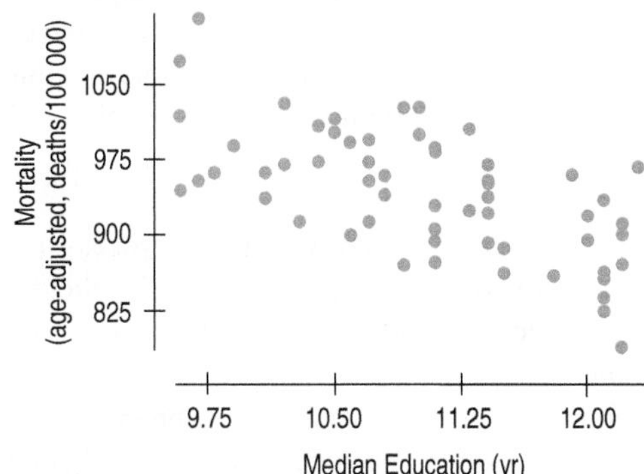

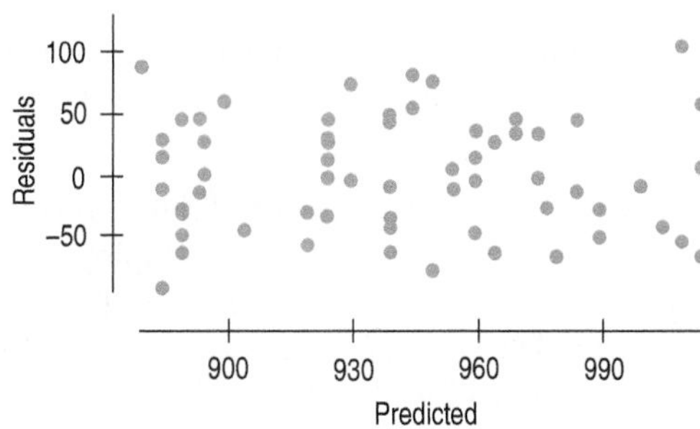

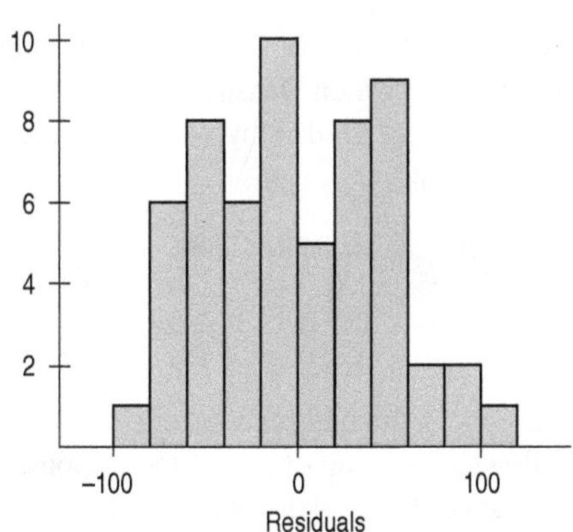

a) Comment on the assumptions for inference.
b) Is there evidence of a strong association between the level of education in a city and the mortality rate? Test an appropriate hypothesis and state your conclusion.
c) Can we conclude that getting more education is likely (on average) to prolong your life? Why or why not?
d) Find a 95% confidence interval for the slope of the true relationship.
e) Explain what your interval means.
f) Find a 95% confidence interval for the average mortality rate in cities where the adult population completed an average of 12 years of school.

T **44. Property assessments** The software outputs below provide information about the size (in square feet) of 18 homes in Ithaca, New York, and the city's assessed value of those homes.

Variable	Count	Mean	StdDev	Range
Size	18	2003.39	264.727	890
Value	18	60946.7	5527.62	19710

Dependent variable is: Value
R-squared = 32.5%
s = 4682 with 18 − 2 = 16 degrees of freedom

Variable	Coefficient	SE(Coeff)
Intercept	37108.8	8664
Size	11.8987	4.290

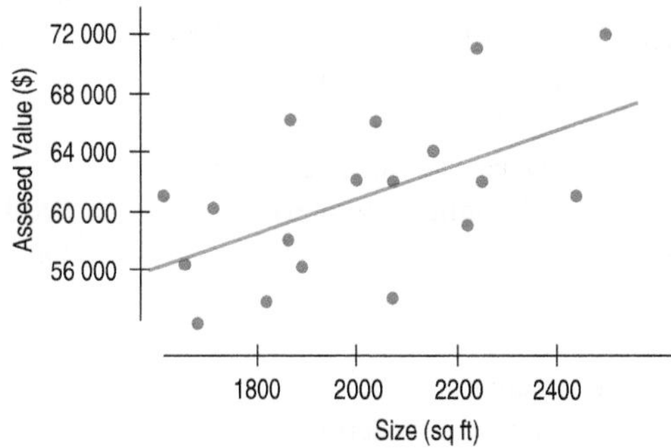

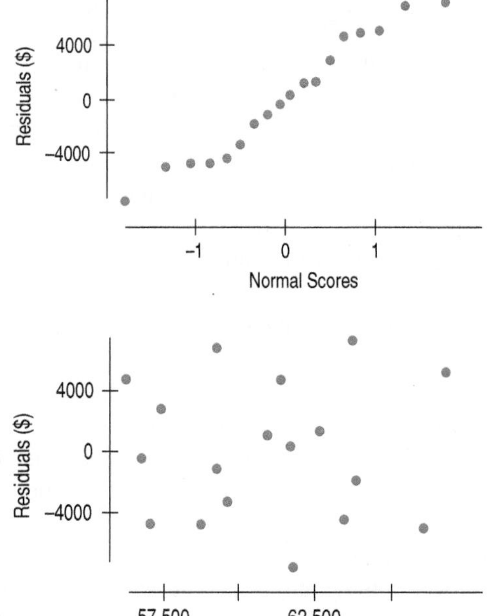

a) Explain why inference for linear regression is appropriate with these data.
b) Is there a significant association between the size of a home and its assessed value? Test an appropriate hypothesis and state your conclusion.

c) What percentage of the variability in assessed value is explained by this regression?
d) Give a 90% confidence interval for the slope of the true regression line, and explain its meaning in the proper context.
e) From this analysis, can we conclude that adding a room to your house will increase its assessed value? Why or why not?
f) The owner of a home measuring 2100 square feet files an appeal, claiming that the $70 200 assessed value is too high. Do you agree? Explain your reasoning.

45. Seasonal spending The authors have noticed that spending on credit cards decreases after the Christmas spending season (as measured by amount charged on a credit card in December). The data set online at www.pearsoncanada.ca/deveaux contains the monthly credit card charges of a random sample of 99 cardholders.
a) Build a regression model to predict January spending from December's spending.
b) How much, on average, will cardholders who charged $2000 in December charge in January?
c) Give a 95% confidence interval for the average January charges of cardholders who charged $2000 in December.
d) From part c), give a 95% confidence interval for the average decrease in the charges of cardholders who charged $2000 in December.
e) What reservations, if any, do you have about the confidence intervals you made in parts c) and d)?

46. Seasonal spending II. As we saw in Exercise 45, financial analysts know that January credit card charges will generally be much lower than those of the month before. What about the difference between January and the next month? Does the trend continue? The data set online at www.pearsoncanada.ca/deveaux contains the monthly credit card charges of a random sample of 99 cardholders.
a) Build a regression model to predict February charges from January's charges.
b) How much, on average, will cardholders who charged $2000 in January charge in February?
c) Give a 95% confidence interval for the average February charges of cardholders who charged $2000 in January.
d) From part c), give a 95% confidence interval for the average decrease in the charges of cardholders who charged $2000 in January.
e) What reservations, if any, do you have about the confidence intervals you made in parts c) and d)?

47. CPU Performance Below is a sample of computer processor performance and price data, obtained from CPU-Benchmark.net in July 2013. The full dataset is online at www.pearsoncanada.ca/deveaux.

CPU	Performance	Price ($)
AMD FX-6100 Six-Core	5380	104.99
Intel Celeron G530 @ 2.40GHz	2274	48.99
AMD FX-6200 Six-Core	6231	134.99
AMD FX-6300 Six-Core	6459	139.99
Intel Celeron G540 @ 2.50GHz	2297	49.99
AMD FX-8320 Eight-Core	8229	179.99
AMD FX-8350 Eight-Core	9239	209.99

a) Make a scatterplot of the data. Do you see any points of high leverage, high influence, or with large residuals? If so, remove the data points that are causing the problem (and do not draw any conclusions in this domain).
b) Fit a linear regression model to the remaining data. Check the residuals plot and the Normal probability plot. Do you see any problems?
c) AMD has just come out with a new processor that has scored 8000 on the performance test. Predict the price of this processor and give a 95% interval around this prediction.
d) Intel has just come out with a new processor that has scored 2000 on the performance test. Predict the price of this processor and give a 95% interval around this prediction.
e) You might have expected the interval widths in the previous two parts to be quite different, yet they're not. Comment on why they are so close.

***48. Right-to-work laws** Are so-called "right-to-work" laws related to the percent of public sector employees in unions and the percent of private sector employees in unions? This data set looks at these percentages for the states in the United States in 1982. The dependent variable is whether the state had a right-to-work law or not. The computer output for the logistic regression is given here. (Source: N. M. Meltz, "Interstate and Interprovincial Differences in Union Density," *Industrial Relations*, 28:2 [Spring 1989], 142–158 by way of DASL.)

Logistic Regression Table

Predictor	Coeff	SE(Coeff)	Chisq	P
Intercept	6.1995	1.78724	12.04	0.001
publ	−0.1062	0.0475	5.02	0.025
pvt	−0.2230	0.0811	7.56	0.006

a) Write out the estimated regression equation.
b) The following are scatterplots of the response variable against each of the explanatory variables. Does logistic regression seem appropriate here? Explain.

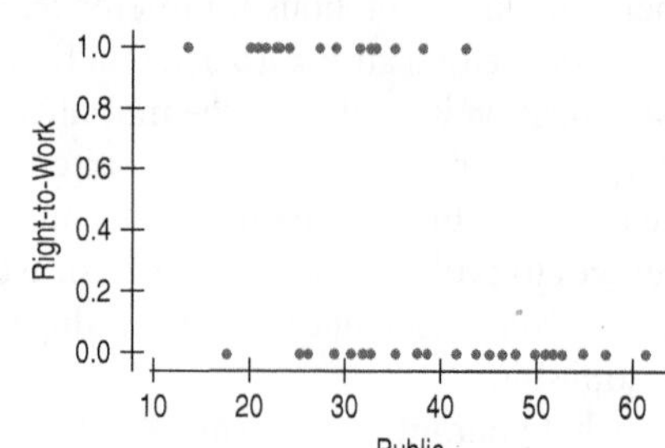

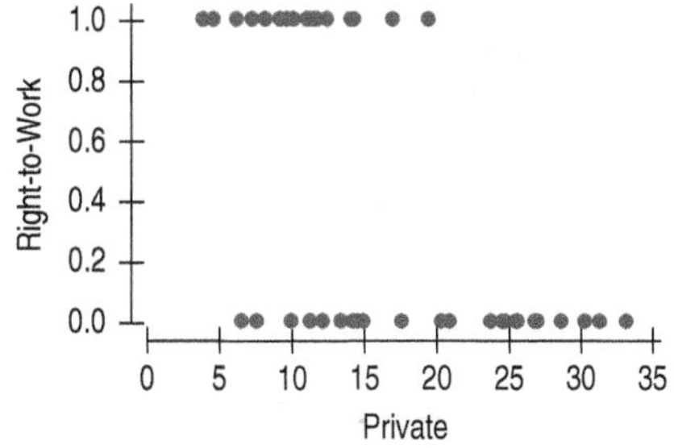

***49. Cost of higher education** Are there fundamental differences between liberal arts colleges and universities? In this case, we have information on the top 25 liberal arts colleges and the top 25 universities in the Unites States. We will consider the type of school as our response variable and will use the percent of students who were in the top 10% of their high school class and the amount of money spent per student by the college or university as our explanatory variables. The output from this logistic regression is given on the following page.

Logistic Regression Table

Predictor	Coeff	SE(Coeff)	Chisq	P
Intercept	−13.1461	3.98629	10.89	0.001
Top 10%	0.0845	0.0396345	4.54	0.033
$/Student	0.0003	0.0000860	9.12	0.003

a) Write out the estimated regression equation.

b) Is percent of students in the top 10% of their high school class statistically significant in predicting whether or not the school is a university? Explain.

c) Is the amount of money spent per student statistically significant in predicting whether or not the school is a university? Explain.

Just Checking ANSWERS

1. A high t-ratio of 3.27 indicates that the slope is different from zero—that is, that there is a linear relationship between height and mouth size. The small P-value says that a slope this large would be very unlikely to occur by chance if, in fact, there was no linear relationship between the variables.
2. Not really. The R^2 for this regression is only 15.3%, so height doesn't account for very much of the variability in mouth size.
3. The value of s tells the standard deviation of the residuals. Mouth sizes have a mean of 60.3 cubic centimetres. A standard deviation of 15.7 in the residuals indicates that the errors made by this regression model can be quite large relative to what we are estimating. Errors of 15 to 30 cubic centimetres would be common.

chapter

25

Analysis of Variance

If you torture the data long enough, it will confess.

—Ronald Coase

Where are we going?

In Chapter 21, we compared the mean lifetimes of generic and brand name batteries. But our supermarket carries four different "name" brands of batteries and two cheaper generic brands. Are all these brands equally good? How can we compare them? We could run a *t*-test for each of the 15 head-to-head comparisons, but in this chapter, we'll learn a better way to compare more than two groups.

Who	Hand washings by four different methods, assigned randomly and replicated eight times each
What	Number of bacteria colonies
How	Sterile media plates incubated at 36°C for two days

25.1 Inference for a Difference in Means

Did you wash your hands with soap before eating? You've undoubtedly been asked that question a few times in your life. Mom knows that washing with soap eliminates most of the germs you've managed to collect on your hands. Or does it? A student decided to investigate just how effective washing with soap is in eliminating bacteria. To do this, she tested four different methods—washing with water only, washing with regular soap, washing with antibacterial soap (ABS), and spraying hands with antibacterial spray (AS) (containing 65% ethanol as an active ingredient). Her experiment consisted of one experimental factor, the washing *method*, at four levels.

She suspected that the number of bacteria on her hands before washing might vary considerably from day to day. To help even out the effects of those changes, she generated random numbers to determine the order of the four treatments. Each morning, she washed her hands according to the treatment randomly chosen. Then, she placed her right hand on a sterile media plate designed to encourage bacteria growth. She incubated each plate for two days at 36°C, after which she counted the bacteria colonies. She replicated this procedure eight times for each of the four treatments.

A side-by-side boxplot of the numbers of colonies seems to show some differences among the treatments:

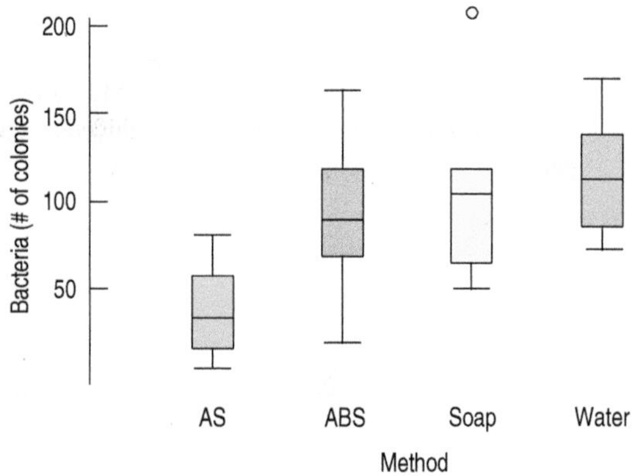

Figure 25.1

Boxplots of the bacteria colony counts for the four different washing methods suggest some differences between treatments.

When we first looked at a quantitative variable measured for each of several groups in Chapter 4, we displayed the data this way with side-by-side boxplots. And, when we compared the boxes, we asked whether the centres seemed to differ, using the spreads of the boxes to judge the size of the differences. Now, we want to make this more formal by testing a hypothesis. We'll make the same kind of comparison, comparing the variability among the means with the spreads of the boxes. It looks like the alcohol spray has lower bacteria counts, but as always, we're skeptical. Could it be that the four methods really have the same mean counts and we just *happened* to get a difference like this because of natural sampling variability?

What is the null hypothesis here? It seems natural to start with the hypothesis that *all the group means are equal.* That would say that it doesn't matter what method you use to wash your hands because the mean bacteria count will be the same. We know that even if there were no differences at all in the *means* (for example, if someone replaced all the solutions with water), there would still be sample-to-sample differences. We want to see, statistically, whether differences as large as those observed in the experiment could naturally occur by chance in groups that have equal means. If we find that the differences in washing *methods* are so large that they would occur only very infrequently in groups that actually have the same mean, then, as we've done with other hypothesis tests, we'll reject the null hypothesis and conclude that the washing *methods* really have different means.[1]

For Example A ONE-FACTOR EXPERIMENT

Hand clinics commonly use contrast baths to reduce swelling and stiffness after surgery. Patients' hands are immersed alternately in warm and cool water. (That's the *contrast* in the name.) Sometimes, the treatment is combined with mild exercise. Although the treatment is widely used, it had never been verified that it accomplished the stated outcome goal of reducing swelling.

Researchers[2] randomly assigned 59 patients who had received carpal tunnel release surgery to one of three treatments: contrast bath, contrast bath with exercise, and (as a control) exercise alone. Hand therapists, who did not know how the subjects had been treated, measured hand volumes in millilitres before and after treatments by measuring how much water the hand displaced when submerged. The change in hand volume (after treatment minus before) was reported as the outcome.

QUESTION: Specify the details of the experiment's design. Identify the subjects, the sample size, the experimental factor, the treatment levels, and the response. What is the null hypothesis? Was randomization employed? Was the experiment blinded? Was it double-blinded?

ANSWER: Subjects were patients who received carpal tunnel release surgery. Sample size is 59 patients. The factor was contrast bath treatment with three levels: contrast baths alone, contrast baths with exercise, and exercise alone. The response variable is the change in hand volume. The null hypothesis is that the mean changes in hand volume will be the same for the three treatment levels. Patients were randomly assigned to treatments. The study was single-blind because the evaluators were blind to the treatments. It was not (and could not be) double-blind because the patients had to be aware of their treatments.

Are the Means of Several Groups Equal?

We saw in Chapter 21 how to use a t-test to see whether two groups have equal means. We compared the difference in the means to a standard error estimated from all the data. And when we were willing to assume that the underlying group variances were equal, we pooled the data from the two groups to find the standard error.

[1]The alternative hypothesis is that "the means are not *all* equal." Be careful not to confuse that with "all the means are different." With 11 groups, we could have 10 means equal to each other and 1 different. The null hypothesis would still be false.

[2]Janssen, Robert G., Schwartz, Deborah A., and Velleman, Paul F., "A Randomized Controlled Study of Contrast Baths on Patients with Carpal Tunnel Syndrome," *Journal of Hand Therapy*, **22**:3, pp. 200–207. The data reported here differ slightly from those in the original paper because they include some additional subjects and exclude some outliers.

Now, we have more groups, so we can't just look at differences in the means.[3] Even if the null hypothesis were true, and the means of the populations underlying the groups were equal, we'd still expect the sample means to vary a bit. We could measure that variation by finding the variance of the means. How much should they vary? Well, if we look at how much the data themselves vary, we can get a good idea of how much the means should vary. And if the underlying means are actually different, we'd expect that variation to be larger.

To get an idea of how it works, let's start by looking at the following two sets of boxplots:

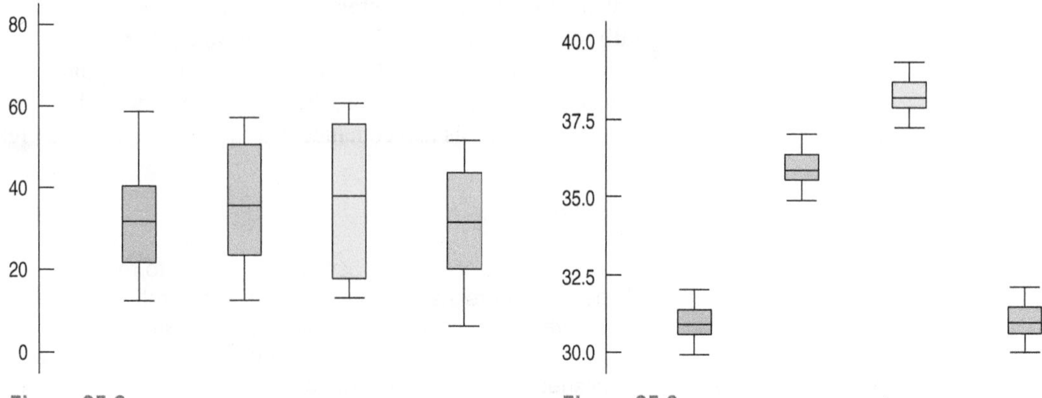

Figure 25.2
It's hard to see the difference in the means in these boxplots because the spreads are large relative to the differences in the means.

Figure 25.3
In contrast with Figure 25.2, the smaller variation makes it much easier to see the differences among the group means.

We're trying to decide if the means are different enough for us to reject the null hypothesis. If they're close, we'll attribute the differences to natural sampling variability. It's easy to see that the means in the second set differ. It's hard to imagine that the means could be that far apart just from natural sampling variability alone. How about the first set? It looks as though these observations *could* have occurred from treatments with the same means.[4] This much variation among groups does seem consistent with equal group means.

Believe it or not, the two sets of treatment means in both figures are the same. (They are 31, 36, 38, and 31, respectively.) So why do the figures look so different? In the second figure, the variation *within* each group is so small that the differences *between* the means stand out. This is what we looked for when we compared boxplots by eye back in Chapter 4. And it's the central idea of the F-test. We compare the differences *between* the means of the groups with the variation *within* the groups. When the differences between means are large compared with the variation within the groups, we reject the null hypothesis and conclude that the means are not equal. In the first figure, the differences among the means look as though they could have arisen just from natural sampling variability from groups with equal means, so there appears to be little evidence to reject H_0.

How can we make this comparison more precise statistically? All the tests we've seen have compared differences of some kind with a ruler based on an estimate of variation. And we've always done that by looking at the ratio of the statistic to that variation estimate. Here, the differences among the means will show up in the numerator, and the ruler we compare them with will be based on the underlying standard deviation—that is, on the variability *within* the treatment groups.

[3] You might think of testing all pairs, but that method generates too many Type I errors. We'll see more about this later in the chapter. It turns out that we can build a hypothesis test to check whether the variation in the means is bigger than we'd expect it to be just from random fluctuations. We'll need a new sampling distribution model, called the F-model, but that's just a different table to look at in the back of the book.

[4] Of course, with a large enough sample, we can detect any small differences. For experiments with the same sample size, it's easier to detect differences when the variation *within* each box is smaller.

For Example BOXPLOTS FOR COMPARING TREATMENTS

RECAP: Fifty-nine post-surgery patients were randomly assigned to one of three treatment levels. Changes in hand volume were measured. Here are the boxplots of the data. The recorded values are volume after treatment, volume before treatment, so positive values indicate swelling. Some swelling is to be expected.

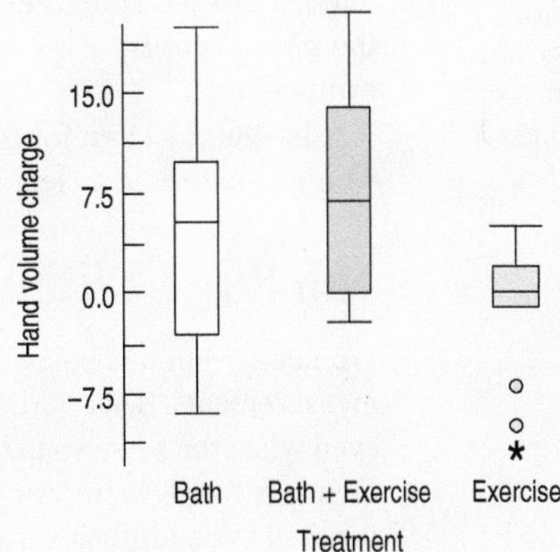

QUESTION: What do the boxplots say about the results?

ANSWER: There doesn't seem to be much difference between the two contrast bath treatments. The exercise-only treatment may result in less swelling.

How Different Are They?

The challenge here is that we can't take a simple difference as we did when comparing two groups. In the hand-washing experiment, we have differences in mean bacteria counts across *four* treatments. How should we measure how different the four group means are? With only two groups, we naturally took the difference between their means as the numerator for the *t*-test. It's hard to imagine what else we could have done. How can we generalize that to more than two groups? When we've wanted to know how different many observations were, we measured how much they vary, and that's what we do here.

How much natural variation should we expect among the means if the null hypothesis were true? If the null hypothesis were *true*, then each of the treatment means would estimate the *same* underlying mean. If the washing methods are all the same, it's as if we're just estimating the mean bacteria count on hands that have been washed with plain water. And we have several (in our experiment, four) different, independent estimates of this mean. Here comes the clever part. We can treat these estimated means as if they were observations, and simply calculate their (sample) variance. This variance is the measure we'll use to assess how different the group means are from each other. It's the generalization of the difference between means for only two groups.

The more the group means resemble each other, the smaller this variance will be. The more they differ (perhaps because the treatments actually have an effect), the larger this variance will be.

Level	n	Mean
Alcohol spray	8	37.5
Antibacterial soap	8	92.5
Soap	8	106.0
Water	8	117.0

For the bacteria counts, the four means are listed in the table to the left. If you took those four values, treated them as observations, and found their sample variance, you'd get 1245.08. That's fine, but how can we tell whether it is a big value? Now, we need a model, and the model is based on our null hypothesis that all the group means are equal. Here, the null hypothesis is that it doesn't matter what washing method you use, the mean bacteria count will be about the same:

$$H_0: \mu_1 = \mu_2 = \mu_3 = \mu_4 = \mu.$$

Why variances? We've usually measured variability with standard deviations. Standard deviations have the advantage that they're in the same units as the data. Variances have the advantage that for independent variables, the variances add. Because we're talking about sums of variables, we'll stay with variances before we get back to standard deviations.

As always when testing a null hypothesis, we'll start by assuming that it is true. And if the group means are equal, then there's an overall mean, μ—the bacteria count you'd expect all the time after washing your hands in the morning. And each of the observed group means is just a sample-based estimate of that underlying mean.

We know how sample means vary. The variance of a sample mean is σ^2/n. With eight observations in a group, that would be $\sigma^2/8$. The estimate that we've just calculated, 1245.08, should estimate this quantity. If we want to get back to the variance of the *observations*, σ^2, we need to multiply it by 8. So $8 \times 1245.08 = 9960.64$ should estimate σ^2.

Is 9960.64 large for this variance? How can we tell? We'll need a hypothesis test. You won't be surprised to learn that there is just such a test, thanks to Sir Ronald Fisher.

The Ruler Within

We need a suitable ruler for comparison—one based on the underlying variability in our measurements. That variability is due to the day-to-day differences in the bacteria count even when the same soap is used. Why would those counts be different? Maybe the experimenter's hands were not equally dirty, or she washed less well some days, or the plate incubation conditions varied. We randomized just so we could see past such things.

We need an independent estimate of σ^2, one that doesn't depend on the null hypothesis being true, and one that won't change if the groups have different means. As in many quests, the secret is to look "within." We could look in *any* of the treatment groups and find its variance. But which one should we use? The answer is *all* of them!

At the start of the experiment (when we randomly assigned experimental units to treatment groups), the units were drawn randomly from the same pool, so each treatment group had a sample variance that estimated the same σ^2. If the null hypothesis is true, then not much has happened to the experimental units—or at least, their means have not moved apart. It's not much of a stretch to believe that their variances haven't moved apart much either. (If the washing methods are equivalent, then the choice of method would not affect the mean *or* the variability.) So each group variance still estimates a common σ^2.

We're assuming that the null hypothesis is true. If the group variances are equal, then the common variance they all estimate is just what we've been looking for. Since all the group variances estimate the same σ^2, we can pool them to get an overall estimate of σ^2. Recall that we pooled data to estimate variances when we tested the null hypothesis that two proportions were equal—and for the same reason. It's also exactly what we did in a pooled *t*-test. We'll denote the variance estimate we get by pooling, as before, by s_p^2.

Level	*n*	Mean	Std Dev	Variance
Alcohol spray	8	37.5	26.56	705.43
Antibacterial soap	8	92.5	41.96	1760.64
Soap	8	106.0	46.96	2205.24
Water	8	117.0	31.13	969.08

For the bacteria counts, the standard deviations and variances are listed above. If we pool the four variances (here we can just average them because all the sample sizes are equal), we'd get $s_p^2 = 1410.10$. In the pooled variance, each variance is taken around its *own* treatment mean, so the pooled estimate doesn't depend on the treatment means being equal. But the estimate in which we took the four means as observations and took their variance does. That estimate gave 9960.64. That seems a lot bigger than 1410.10. Might this be evidence that the four means are not equal?

Let's see what we've got. We have an estimate of σ^2 from the variation *within* groups of 1410.10. That's just the variance of the *residuals* pooled across all groups. Because it's a pooled variance, we could write it as s_p^2. Traditionally, this quantity is also called the **mean square error**, or sometimes the ***mean square within***, and is denoted by $\mathbf{MS_E}$. These names date back to the early twentieth century when the methods were developed. But we also have a *separate* estimate of σ^2 from the variation *between* the groups because we know how much means ought to vary. For the hand-washing data, when we took the variance of the four means and multiplied it by n, we got 9960.64. We expect this to estimate σ^2, too, *as long as we assume the null hypothesis is true.* We call this quantity the **mean square treatment** (or sometimes the ***mean square between***) and denote it by $\mathbf{MS_T}$.

The *F*-statistic

Now, we have two different estimates of the underlying variance. The first one, the MS_T, is based on the differences *between* the group means. If the group means are equal, as the null hypothesis asserts, it will estimate σ^2. But, if they are not, it will give some bigger value. The other estimate, the MS_E, is based only on the variation *within* the groups around each of their own means, and doesn't depend at all on the null hypothesis being true.

So how do we test the null hypothesis? When the null hypothesis is true, the treatment means are equal, and both MS_E and MS_T estimate σ^2. Their ratio, then, should be close to 1.0. But, when the null hypothesis is false, the MS_T will be *larger* because the treatment means are not equal. The MS_E is a pooled estimate in which the variation within each group is found around its own group mean, so differing means won't inflate it. That makes the ratio MS_T/MS_E perfect for testing the null hypothesis. When the null hypothesis is true, the ratio should be near 1. If the treatment means really are different, the numerator will tend to be larger than the denominator, and the ratio will tend to be bigger than 1.

NOTATION ALERT

Capital *F* is used only for this distribution model and statistic in honour of Sir Ronald Fisher.

Of course, even when the null hypothesis *is* true, the ratio will vary around 1 just due to natural sampling variability. How can we tell when it's big enough to reject the null hypothesis? To be able to tell, we need a sampling distribution model for the ratio. We call the ratio MST/MSE the F-statistic. By comparing this statistic with the appropriate ***F*-distribution**, we (or the computer) can get a P-value.

The ***F*-test** is simple. It is one-tailed because any differences in the means make the F-statistic larger. Larger differences in the treatment effects lead to the means being more variable, making the MS_T bigger. That makes the F-ratio grow. So the test is significant if the F-ratio is big enough. In practice, we find a P-value, and big F-statistic values go with small P-values.

You might think we should call such an analysis the *analysis of means*, since it's the equality of means we're testing. But we use variances within and between groups to do this, so the entire analysis is called the analysis of variance, commonly abbreviated **ANOVA**. Like Student's t-models, the F-models are a family. F-models depend on not one, but two, degrees of freedom parameters. The degrees of freedom come from the two variance estimates and are sometimes called the *numerator df* and the *denominator df*. The *mean square treatment*, MS_T, is the sample variance of the observed treatment means. If you think of them as observations, then since there are k groups, this variance has $k - 1$ degrees of freedom. The *mean square error*, MS_E, is the pooled estimate of the variance within the groups. If there are n observations in each group, then we get $n - 1$ degrees of freedom from each, for a total of $k(n - 1)$ degrees of freedom.

NOTATION ALERT

What, first little *n* and now big *N*? In an experiment, it's standard to use *N* as our count of *all* the cases and *n* for the number in each treatment group.

A simpler way of tracking the degrees of freedom is to start with all the cases. We'll call that N. Each group has its own mean, costing us a degree of freedom, or k in total. So we have $N - k$ degrees of freedom for the error. When the groups all have equal sample size, that's the same as $k(n - 1)$, but this way works even if the group sizes differ.

We say that the F-statistic, MS_T/MS_E, has the F-distribution with $k - 1$ and $N - k$ degrees of freedom. We usually write these parameters as subscripts, numerator df first, then denominator df. On page 750, you can see what the density curve actually looks like.

Back to Bacteria!

For the hand-washing experiment, the $MS_T = 9960.64$. The $MS_E = 1410.14$. If the treatment means were equal, the *mean square treatment* should be about the same size as the *mean square error*, about 1410. But it's 9960.64, which is 7.06 times bigger. In other words, $F = 7.06$. This F-statistic has $4 - 1 = 3$ and $32 - 4 = 28$ degrees of freedom.

An F-value of 7.06 is bigger than 1, but we can't tell for sure whether it's big enough to reject the null hypothesis until we check the $F_{3,28}$ model to find its P-value. (Usually, that's most easily done with technology, but we can use printed tables.) It turns out the P-value is 0.0011. In other words, if the treatment means were actually equal, we would expect the ratio MS_T/MS_E to be 7.06 or larger about 11 times out of 10 000, just from natural sampling variability. That's not very likely, so we reject the null hypothesis and conclude that the means are different. We have strong evidence that the four different methods of hand washing are not equally effective at eliminating germs.

The ANOVA Table

You'll often see the mean squares and other information put into a table called the ANOVA table. Here's the table for the hand-washing experiment:

Calculating the ANOVA table This table has a long tradition stretching back to when ANOVA calculations were done by hand. Major research labs had rooms full of mechanical calculators operated by women. (Yes, always women; women were thought—by the men in charge, at least—to be more careful at such an exacting task.) Three women would perform each calculation, and if any two of them agreed on the answer, it was taken as the correct value.

Analysis of Variance Table

Source	DF	Sum of Squares	Mean Square	F-ratio	P-value
Method	3	29882	9960.64	7.0636	0.0011
Error	28	39484	1410.14		
Total	31	69366			

The ANOVA table was originally designed to organize the calculations. With technology, we have much less use for that. We'll show how to calculate the sums of squares later in the chapter, but the most important quantities in the table are the F-statistic and its associated P-value. When the F-statistic is large, the mean square treatment is large compared to the mean square error, and provides evidence that in fact the means of the groups are not all equal.

You'll almost always see ANOVA results presented in a table like this. After nearly a century of writing the table this way, statisticians (and their technology) aren't going to change. Even though the table was designed to facilitate hand calculation, computer programs that compute ANOVAs still present the results in this form. Usually the P-value is found next to the F-ratio. The P-value column may be labelled with a title such as "Prob > F," "sig," or "Prob." Don't let that confuse you; it's just the P-value.

You'll sometimes see the two mean squares referred to as the *mean square between* and the *mean square within*—especially when we test data from observational studies rather than experiments. ANOVA is often used for such observational data, and as long as certain conditions are satisfied, there's no problem with using it in that context.

For Example *F*-TEST

RECAP: An experiment to determine the effect of contrast bath treatments on swelling in post-surgical patients recorded hand volume changes for patients who had been randomly assigned to one of three treatments. Here is the analysis of variance for these data:

Analysis of variance for hand volume change

Source	DF	Sum of Squares	Mean Square	*F*-ratio	P-value
Treatment	2	716.159	358.080	7.4148	0.0014
Error	56	2704.38	48.2926		
Total	58	3420.54			

QUESTION: What does the ANOVA say about the results of the experiment? Specifically, what does it say about the null hypothesis?

ANSWER: The F-ratio of 7.4148 has a P-value that is quite small. We can reject the null hypothesis that the mean change in hand volume is the same for all three treatments.

The *F*-table

Usually, you'll get the P-value for the F-statistic from technology. Any software program performing an ANOVA will automatically "look up" the appropriate one-sided P-value for the F-statistic. If you want to do it yourself, you'll need an F-table. (There's one in Appendix C called **Table F**.) F-tables are usually printed only for a few values of α, often 0.05, 0.01, and 0.001. They give the critical value of the F-statistic with the appropriate number of degrees of freedom determined by your data, for the α level that you select. If your F-statistic is greater than that value, you know that its P-value is less than that α level. So, you'll be able to tell whether the P-value is greater or less than 0.05, 0.01, or 0.001, but to be more precise, you'll need technology (or an interactive table like the one in MyStatLab).

Here's an excerpt from an F-table for $\alpha = 0.05$:

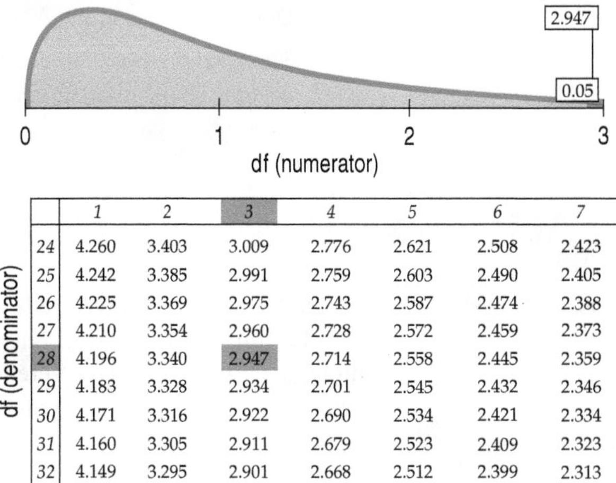

	1	2	3	4	5	6	7
24	4.260	3.403	3.009	2.776	2.621	2.508	2.423
25	4.242	3.385	2.991	2.759	2.603	2.490	2.405
26	4.225	3.369	2.975	2.743	2.587	2.474	2.388
27	4.210	3.354	2.960	2.728	2.572	2.459	2.373
28	4.196	3.340	2.947	2.714	2.558	2.445	2.359
29	4.183	3.328	2.934	2.701	2.545	2.432	2.346
30	4.171	3.316	2.922	2.690	2.534	2.421	2.334
31	4.160	3.305	2.911	2.679	2.523	2.409	2.323
32	4.149	3.295	2.901	2.668	2.512	2.399	2.313

Figure 25.4

Part of an F-table showing critical values for $\alpha = 0.05$, and highlighting the critical value, 2.947, for 3 and 28 degrees of freedom. We can see that only 5% of the values will be greater than 2.947 with this combination of degrees of freedom.

Notice that the critical value for 3 and 28 degrees of freedom at $\alpha = 0.05$ is 2.947. Since our F-statistic of 7.06 is larger than this critical value, we know that the P-value is less than 0.05. We could also look up the critical value for $\alpha = 0.01$ and find that it's 4.568 and the critical value for $\alpha = 0.001$ is 7.193. So, our F-statistic sits between the two critical values 0.01 and 0.001, and our P-value is slightly *greater* than 0.001. Technology can find the value precisely. It turns out to be 0.0011.

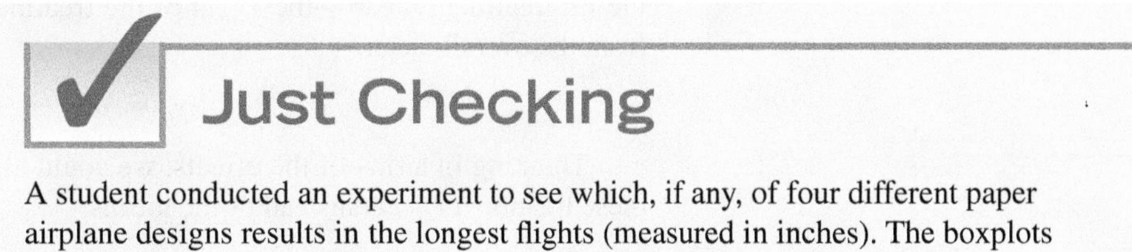

Just Checking

A student conducted an experiment to see which, if any, of four different paper airplane designs results in the longest flights (measured in inches). The boxplots look like this:

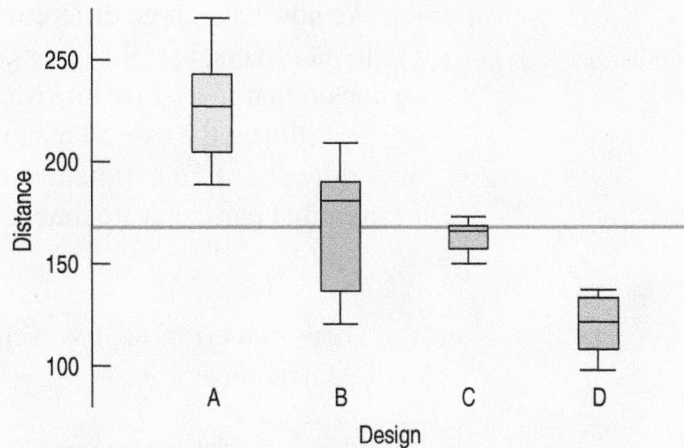

The ANOVA table shows:

Analysis of Variance					
Source	DF	Sum of Squares	Mean Square	F Ratio	Prob > F
Design	3	51991.778	17330.6	37.4255	< .0001
Error	32	14818.222	463.1		
Total	35	66810.000			

1. What is the null hypothesis?
2. From the boxplots, do you think that there is evidence that the mean flight distances of the four designs differ?
3. Does the F-test in the ANOVA table support your preliminary conclusion in 2?
4. The researcher concluded that "there is substantial evidence that all four of the designs result in different mean flight distances." Do you agree?

25.2 The ANOVA Model

To understand the ANOVA, let's start by writing a model for what we observe. We start with the simplest interesting model: one that says that the only differences of interest among the groups are the differences in their means. So, we'll characterize each group in terms of its mean and assume that any variation around that mean is just random error:

$$y_{ij} = \mu_j + \varepsilon_{ij}$$

That is, each observation is the sum of the mean for the treatment it received plus a random error. Our null hypothesis is that the treatments made no difference—that is, that the means are all equal:

$$H_0\text{: } \mu_1 = \mu_2 = \cdots = \mu_k$$

It will help our discussion if we think of the overall mean of the experiment and consider the treatments as adding or subtracting an effect to this overall mean. Thinking in this way, we could write μ for the overall mean and τ_j for the deviation from this mean to get to the jth treatment mean—the *effect* of the treatment (if any) in moving that group away from the overall mean:

$$y_{ij} = \mu + \tau_j + \varepsilon_{ij}$$

Thinking in terms of the effects, we could also write the null hypothesis in terms of these treatment *effects* instead of the means:

$$H_0\text{: } \tau_1 = \tau_2 = \cdots = \tau_k = 0$$

We now have three different kinds of components in our model: the overall mean, the treatment effects, and the errors. We'll want to estimate them from the data. Fortunately, we can do that in a straightforward way.

To estimate the overall mean, μ, we use the mean of all the observations: $\bar{\bar{y}}$ (called the "grand mean."[5]) To estimate each treatment effect, we find the difference between the mean of that particular treatment and the grand mean:

$$\hat{\tau}_j = \bar{y}_j - \bar{\bar{y}}$$

There's an error, ε_{ij}, for each observation. We estimate those with the residuals from the treatment means: $e_{ij} = y_{ij} - \bar{y}_j$.

[5]The father of your father is your grandfather. The mean of the group means should probably be the grandmean, but we usually spell it as two words.

Now, we can write each observation as the sum of three quantities that correspond to our model:

$$y_{ij} = \bar{\bar{y}} + (\bar{y}_j - \bar{\bar{y}}) + (y_{ij} - \bar{y}_j)$$

What this says is simply that we can write each observation as the sum of

- the grand mean,
- the effect of the treatment it received, and
- the residual.

Or:

$$Observation = Grand\ mean + Treatment\ effect + Residual$$

If we look closely at the equivalent equation, $y_{ij} = \bar{\bar{y}} + (\bar{y}_j - \bar{\bar{y}}) + (y_{ij} - \bar{y}_j)$, it doesn't really seem like we've done anything. In fact, collecting terms on the right-hand side will give back just the observation, y_{ij} again. But this decomposition is actually the secret of the analysis of variance. We've split each observation into "sources"—the grand mean, the treatment effect, and error.

Where does the residual term come from? Think of the annual report from any *Fortune* 500 company. The company spends billions of dollars each year and at the end of the year, the accountants show where each penny goes. How do they do it? After accounting for salaries, bonuses, supplies, taxes, etc., etc., etc., what's the last line? It's always labelled "other" or "miscellaneous." Using "other" as the difference between all the sources they know and the total they start with, they can always make it add up perfectly. The residual is just the statisticians' "other." It takes care of all the other sources we didn't think of or don't want to consider, and makes the decomposition work by adding (or subtracting) back in just what we need.

Let's see what this looks like for our hand-washing data. Here are the data again, displayed a little differently:

	Alcohol	AB Soap	Soap	Water
	51	70	84	74
	5	164	51	135
	19	88	110	102
	18	111	67	124
	58	73	119	105
	50	119	108	139
	82	20	207	170
	17	95	102	87
Treatment Means	37.5	92.5	106	117

The grand mean of all observations is 88.25. Let's put that into a similar table:

Alcohol	AB Soap	Soap	Water
88.25	88.25	88.25	88.25
88.25	88.25	88.25	88.25
88.25	88.25	88.25	88.25
88.25	88.25	88.25	88.25
88.25	88.25	88.25	88.25
88.25	88.25	88.25	88.25
88.25	88.25	88.25	88.25
88.25	88.25	88.25	88.25

The treatment *means* are 37.5, 92.5, 106, and 117, respectively, so the treatment *effects* are those minus the grand mean (88.25). Let's put the treatment effects into their table:

Alcohol	AB Soap	Soap	Water
−50.75	4.25	17.75	28.75
−50.75	4.25	17.75	28.75
−50.75	4.25	17.75	28.75
−50.75	4.25	17.75	28.75
−50.75	4.25	17.75	28.75
−50.75	4.25	17.75	28.75
−50.75	4.25	17.75	28.75
−50.75	4.25	17.75	28.75

Finally, we compute the residuals as the differences between each observation and its treatment mean:

Alcohol	AB Soap	Soap	Water
13.5	−22.5	−22	−43
−32.5	71.5	−55	18
−18.5	−4.5	4	−15
−19.5	18.5	−39	7
20.5	−19.5	13	−12
12.5	26.5	2	22
44.5	−72.5	101	53
−20.5	2.5	−4	−30

Now, we have four tables for which each

$$Observation = Grand\ Mean + Treatment\ Effect + Residual$$

(You can verify, for example, that the first observation, $51 = 88.25 + (-50.75) + 13.5$).

Why do we want to think in this way? Think back to the boxplots in Figures 25.2 and 25.3. To *test* the hypothesis that the treatment effects are zero, we want to see whether the treatment effects are large *compared* to the errors. Our eye looks at the variation *between* the treatment means and compares it to the variation *within* each group.

The ANOVA separates those two quantities into the *treatment effects* and the *residuals*. Sir Ronald Fisher's insight was how to turn those quantities into a statistical test. We want to see if the treatment effects are large compared with the residuals. To do that, we first compute the *sums of squares* of each table. Fisher's insight was that dividing these sums of squares by their respective degrees of freedom lets us test their ratio by a distribution that he found (which was later named the F in his honour). When we divide a sum of squares by its degrees of freedom we get the associated *mean square.*

When the mean square treatment is *large* compared to the mean square error, this provides evidence that the treatment means are different. And we can use the F-distribution to see how large "large" is.

The sums of squares for each table are easy to calculate. Just take every value in the table, square it, and add them all up. For the *methods,* the treatment sum of squares, $SS_T = (-50.75)^2 + (-50.75)^2 + \cdots + (28.75)^2 = 29\,882$. There are four treatments, and so there are 3 degrees of freedom. So,

$$\mathrm{MS_T} = SS_T/3 = 29\,882/3 = 9960.64$$

In general, we could write the treatment sum of squares as

$$SS_T = \sum\sum(\bar{y}_j - \bar{\bar{y}})^2$$

Be careful to note that the summation is over the *whole* table, rows and columns. That's why there are two summation signs.

And,

$$\mathrm{MS_T} = SS_T/(k - 1)$$

The table of residuals shows the variation that remains after we remove the overall mean and the treatment effects. These are what's left over after we account for what we're interested in—in this case, the treatments. Their variance is the variance *within* each group that we see in the boxplots of the four groups. To find its value, we first compute the error sum of squares, SS_E, by summing up the squares of every element in the residuals table. To get the mean square (the variance), we have to divide it by $N - k$ rather than by $N - 1$ because we found them by subtracting each of the k treatment means.

So,

$$SS_E = (13.5)^2 + (-32.5)^2 + \ldots + (-30)^2 = 39\,484$$

and

$$\mathrm{MS_E} = SS_E/(32 - 4) = 1410.14$$

As equations:

$$SS_E = \sum\sum(y_{ij} - \bar{y}_i)^2$$

and

$$\mathrm{MS_E} = SS_E/(N - k)$$

Now where are we? To test the null hypothesis that the treatment means are all equal, we find the F-statistic:

$$F = \mathrm{MS_T}/\mathrm{MS_E}$$

and compare that value to the F-distribution with $k - 1$ and $N - k$ degrees of freedom. When the F-statistic is large enough (and its associated P-value small), we reject the null hypothesis and conclude that at least one mean is different.

There's another amazing result hiding in these tables. If we take each of these tables, square every observation, and add them up, the sums add as well!

$$SS_{Observations} = SS_{Grand\ Mean} + SS_T + SS_E$$

The $SS_{\text{Observations}}$ is usually very large compared to $\mathrm{SS_T}$ and $\mathrm{SS_E}$, so when ANOVA was originally done by hand, or even by calculator, it was hard to check the calculations using this fact. The first sum of squares was just too big. So, usually you'll see the ANOVA table use the "Corrected Total" sum of squares. If we write

$$Observation = Grand\ Mean + Treatment\ Effect + Residual$$

we can naturally write

$$Observation - Grand\ Mean = Treatment\ Effect + Residual$$

Mathematically, this is the same statement, but numerically it is more stable. What's amazing is that the sums of the squares still add up. That is, if you make the first table of observations with the grand mean subtracted from each, square those, and add them up, you'll have the SS_{Total} and

$$SS_{Total} = SS_T + SS_E$$

That's what the ANOVA table shows. If you find this surprising, you must be following along. The tables add up, so sums of their elements must add up. But it is not at all obvious that the sums of the *squares* of their elements should add up, and this is another great insight of the Analysis of Variance.

MATH BOX

In a general set up, the data can be displayed as

Group	1	2	...	k
Observation	y_{11}	y_{12}	...	y_{1k}
	y_{21}	y_{22}	...	y_{2k}
	...	...	...	...
	$y_{n_1 1}$	$y_{n_2 2}$	...	$y_{n_2 k}$
Total	T_1	T_2	...	T_k
Sample size	n_1	n_2	...	n_k
Average	$\bar{y}_1$	$\bar{y}_2$	...	$\bar{y}_k$

where $y_{ij} = i^{th}$ observation of the jth group, $T_j = y_{1j} + y_{2j} + \ldots + y_{nj}$, and $\bar{y}_j = \frac{T_j}{n_j}$. Then the total # of observations and the grand mean are $N = n_1 + n_2 + \cdots + n_k$ and $\bar{\bar{y}} = \frac{\Sigma\Sigma y_{ij}}{N}$. Since we have

$$y_{ij} - \bar{\bar{y}} = (\bar{y}_j - \bar{\bar{y}}) + (y_{ij} - \bar{y}_j)$$

we can square both sides of the equation:

$$(y_{ij} - \bar{\bar{y}})^2 = [(\bar{y}_j - \bar{\bar{y}}) + (y_{ij} - \bar{y}_j)]^2$$

By summing over the observations within each group and then over the groups, we get

$$\Sigma\Sigma(y_{ij} - \bar{\bar{y}})^2 = \Sigma\Sigma[(\bar{y}_j - \bar{\bar{y}}) + (y_{ij} - \bar{y}_j)]^2$$

Notice that the sum of squares equation can be simplified to

$$\Sigma\Sigma(y_{ij} - \bar{\bar{y}})^2 = \Sigma\Sigma(\bar{y}_j - \bar{\bar{y}})^2 + \Sigma\Sigma(y_{ij} - \bar{y}_j)^2$$

because $\Sigma\Sigma(\bar{y}_j - \bar{\bar{y}})(y_{ij} - \bar{y}_j) = 0$. So we get

$$SS_{Total} = SS_T + SS_E$$

where $SS_T = \Sigma\Sigma(\bar{y}_j - \bar{\bar{y}})^2 = \Sigma n_j(\bar{y}_j - \bar{\bar{y}})^2$ with degrees of freedom $= k - 1$, $SS_E = \Sigma\Sigma(y_{ij} - \bar{y}_j)^2$ with degrees of freedom $= N - k$, and $SS_{Total} = \Sigma\Sigma(\bar{y}_{ij} - \bar{\bar{y}})^2$ with degrees of freedom $= N - 1$.

From these formulas, the ANOVA table can be obtained (though using software is recommended.)

Back to Standard Deviations

We've been using the variances because they're easier to work with. But when it's time to think about the data, we'd really rather have a standard deviation because it's in the units of the response variable. The natural standard deviation to think about is the standard deviation of the residuals.

The variance of the residuals is staring us in the face. It's the MS_E. All we have to do to get the **residual standard deviation** is take the square root of MS_E:

$$s_p = \sqrt{MS_E} = \sqrt{\frac{\Sigma e^2}{N - k}}$$

The *p* subscript is to remind us that this is a *pooled* standard deviation, combining residuals across all *k* groups. The denominator in the fraction shows that finding a mean for each of the *k* groups cost us one degree of freedom for each.

This standard deviation should "feel" right. That is, it should reflect the kind of variation you expect to find in any of the experimental groups. For the hand-washing data, $s_p = \sqrt{1410.14} = 37.6$ bacteria colonies. Looking back at the boxplots of the groups, we see that 37.6 seems to be a reasonable compromise standard deviation for all four groups.

Plot the Data . . .

Just as you would never perform a linear regression without looking at the scatterplot of y versus x, you should never embark on an ANOVA without first examining side-by-side boxplots of the data comparing the responses for all of the groups. You already know what to look for—we talked about that in Chapter 4. Check for outliers within any of the groups and correct them if there are errors in the data. Get an idea of whether the groups have similar spreads (as we'll need) and whether the centres seem to be alike (as the null hypothesis claims) or different. If the spreads of the groups are very different—and especially if they seem to grow consistently as the means grow—you should consider re-expressing the response variable to make the spreads more nearly equal. Doing so is likely to make the analysis more powerful and more correct. Likewise, if the boxplots are skewed in the same direction, you may be able to make the distributions more symmetric with a re-expression.

Don't ever carry out an analysis of variance without looking at the side-by-side boxplots first. The chance of missing an important pattern or violation is just too great.

25.3 Assumptions and Conditions

When we checked assumptions and conditions for regression, we had to take care to perform our checks in order. Here we have a similar concern. For regression, we found that displays of the residuals were often a good way to check the corresponding conditions. That's true for ANOVA as well.

Independence Assumptions

The groups must be independent of each other. No test can verify this assumption. You have to think about how the data were collected. The assumption would be violated, for example, if we measured subjects' performance before some treatment, again in the middle of the treatment period, and then again at the end.[6]

The data *within* each treatment group must be independent as well. The data must be drawn independently and at random from the target population, or generated by a randomized comparative experiment.

We check the **Randomization Condition:** Were the data collected with suitable randomization? For surveys, are the data drawn from each group a representative random sample of that group? For experiments, were the treatments assigned to the experimental units at random?

We were told that the hand-washing experiment was randomized.

Equal Variance Assumption

The ANOVA requires that the variances of the treatment groups be equal. After all, we need to find a pooled variance for the MS_E. To check this assumption, we can check that the groups have similar variances:

[6]There is a modification of ANOVA, called *repeated measures* ANOVA, that deals with such data. (If the design reminds you of a paired-t situation, you're on the right track, and the lack of independence is the same kind of issue we discussed in Chapter 22.)

Similar Variance Condition: There are some ways to see whether the variation in the treatment groups seems roughly equal:

- Look at side-by-side boxplots of the groups to see whether they have roughly the same spread. It can be easier to compare spreads across groups when they have the same centre, so consider making side-by-side boxplots of the residuals. If the groups have different spreads, it can make the pooled variance—the MS_E—larger or smaller, depending on how imbalanced the groups are. Consequently, unequal spread can be an especially bad problem because we don't always know whether the test is more conservative or less. With balanced groups though the F test is quite robust against small to moderate differences in spread, causing only a very minor inflation of the Type I error rate.
- Look at the original boxplots of the response values again. In general, do the spreads seem to change *systematically* with the centres? One common pattern is for the boxes with bigger centres to have bigger spreads. This kind of systematic trend in the variances is more of a problem than random differences in spread among the groups and should not be ignored. Fortunately, such systematic violations are often helped by re-expressing the data. (If, in addition to spreads that grow with the centres, the boxplots are skewed with the longer tail stretching off to the high end, then the data are pleading for a re-expression. Try taking logs of the dependent variable for a start. You'll likely end up with a much cleaner analysis.)
- Look at the residuals plotted against the predicted values. Often, larger predicted values lead to larger magnitude residuals. This is another sign that the condition is violated. (This may remind you of the **Does the Plot Thicken? Condition** of regression. And it should.) When the plot thickens (to one side or the other), it's usually a good idea to consider re-expressing the response variable. Such a systematic change in the spread is a more serious violation of the Equal Variance Assumption than slight variations of the spreads across groups.

Let's check the conditions for the hand-washing data. Here's a boxplot of residuals by group and residuals by predicted value:

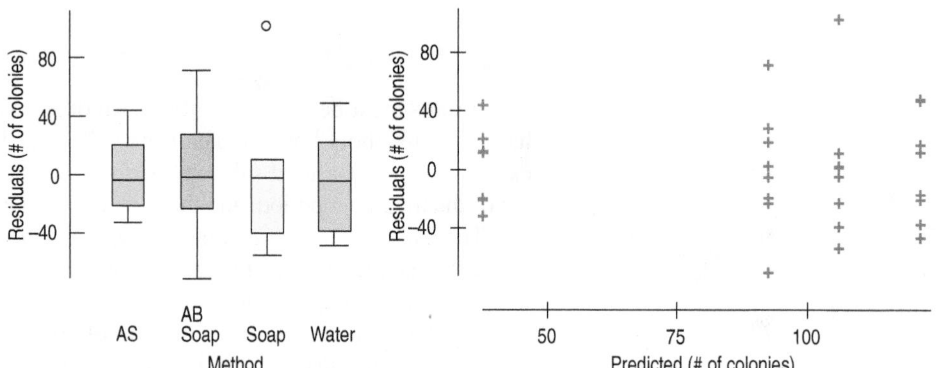

Figure 25.5

Boxplots of residuals for the four hand-washing methods and a plot of residuals versus predicted values. There's no evidence of a systematic change in variance from one group to the other or by predicted value. Neither plot shows a violation of the condition. The IQRs (the box heights) are quite similar and the plot of residuals versus predicted values does not show a pronounced widening to one end. The pooled estimate of 37.6 colonies for the error standard deviation seems reasonable for all four groups.

Normal Population Assumption

Like Student's *t*-tests, the *F*-test requires the underlying errors to follow a Normal model. As before when we've faced this assumption, we'll check a corresponding **Nearly Normal Condition**.

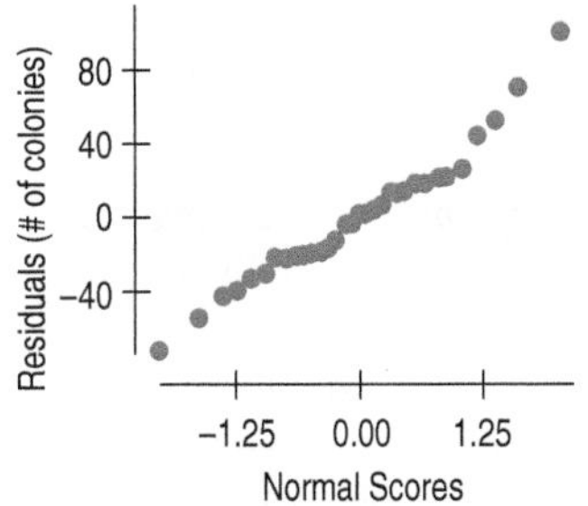

Figure 25.6

The hand-washing residuals look nearly Normal in this Normal probability plot. Because we really care about the Normal model within each group, the Normal population assumption is violated if there are outliers in any of the groups. Check for outliers in the boxplots of the values for each treatment group. The soap group of the hand-washing data shows a possible outlier, so we might want to compute the analysis again without that observation. (For these data, it turns out to make little difference.)

Technically, we need to assume that the Normal model is reasonable for the populations underlying *each* treatment group. We can (and should) look at the side-by-side boxplots for indications of skewness. Certainly, if they are all (or mostly) skewed in the same direction, the Nearly Normal Condition fails (and re-expression is likely to help).

In experiments, we often work with fairly small groups for each treatment, and it's nearly impossible to assess whether the distribution of only six or eight numbers is Normal (though sometimes it's so skewed or has such an extreme outlier that we can see that it's not). Here we are saved by the Equal Variance Assumption, which we've already checked. The residuals have their group means subtracted, so the mean residual for each group is 0. If their variances are equal, we can group all the residuals together for the purpose of checking the Nearly Normal Condition.

Check Normality with a histogram or a Normal probability plot of all the residuals together. The hand-washing residuals look nearly Normal in the Normal probability plot (Figure 25.6), although, as the boxplots showed, there's a possible outlier in the soap group (actually, just barely above the upper fence).

ONE-WAY ANOVA *F*-TEST

We test the null hypothesis $H_0: \mu_1 = \mu_2 = \ldots = \mu_k$ against the alternative that the group means are not all equal. We test the hypothesis with the *F*-statistic, $F = \frac{MS_T}{MS_E}$, where MS_T is the Treatment Mean Square, reflecting variability among the means of the treatment groups, and MS_E is the Error Mean Square, found by pooling the variances within each of the treatment groups. If the *F*-statistic is large enough, we reject the null hypothesis.

Step-by-Step Example ANALYSIS OF VARIANCE

In Chapter 4, we looked at side-by-side boxplots of four different containers for holding hot beverages. The experimenter wanted to know which type of container would keep his beverages hot longest. To test it, he heated water to a temperature of 180°F, placed it in the container, and then measured the temperature of the water again 30 minutes later. He randomized the order of the trials and tested each container eight times. His response variable was the difference in temperature (in °F) between the initial water temperature and the temperature after 30 minutes. Let's test whether these containers really perform differently.

THINK ➡ Plot Plot the side-by-side boxplots of the data.

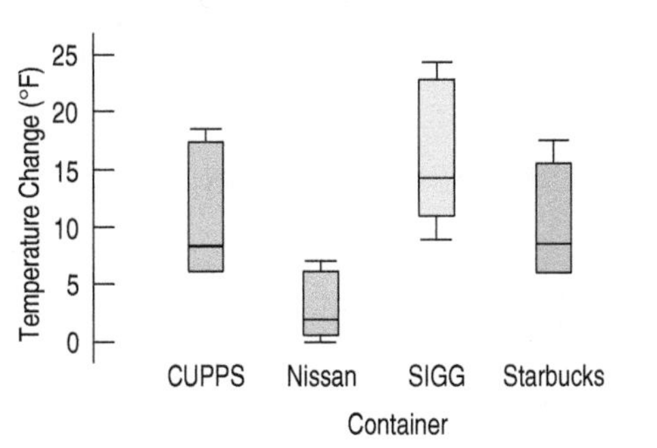

Plan State what you want to know and the null hypothesis you wish to test. For ANOVA, the null hypothesis is that all the treatment groups have the same mean. The alternative is that at least one mean is different.

I want to know whether there is any difference among the four containers in their ability to maintain the temperature of a hot liquid for 30 minutes. I'll write μ_k for the mean temperature difference for container k, so the null hypothesis is that these means are all the same:

$$H_0: \mu_1 = \mu_2 = \mu_3 = \mu_4.$$

Think about the assumptions and check the conditions.

The alternative is that the group means are not all equal.

✓ **Independence Assumption:** The "experimental units" in this experiment are cups of heated water. It's easy to believe that one cup of water is independent of another. It also seems reasonable that the performance of one tested cup should be independent of other cups.

✓ **Randomization Condition:** The experimenter performed the trials in random order.

✓ **Similar Variance Condition:** The Nissan mug variation seems to be a bit smaller than the others. I'll look later at the plot of residuals versus predicted values to see if the plot thickens.

SHOW ➡ **Mechanics** Fit the ANOVA model.

Analysis of Variance

Source	DF	Sum of Squares	Mean Square	F-ratio	P-value
Container	3	714.1875	238.063	10.713	<0.0001
Error	28	622.1875	22.221		
Total	31	1336.3750			

✓ **Nearly Normal Condition, Outlier Condition:** The Normal probability plot is not very straight, but there are no outliers.

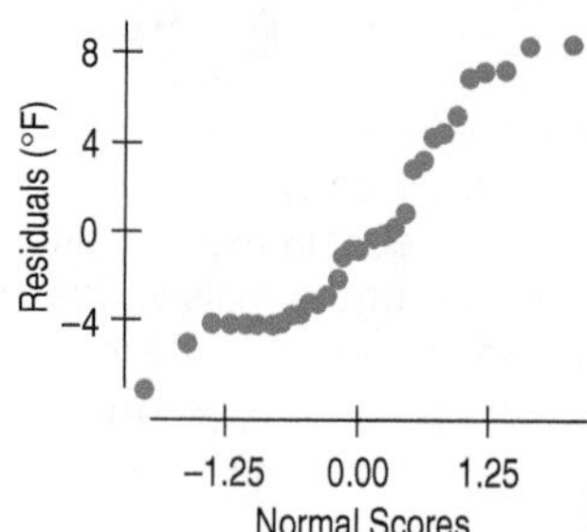

The histogram shows that the distribution of the residuals is skewed to the right:

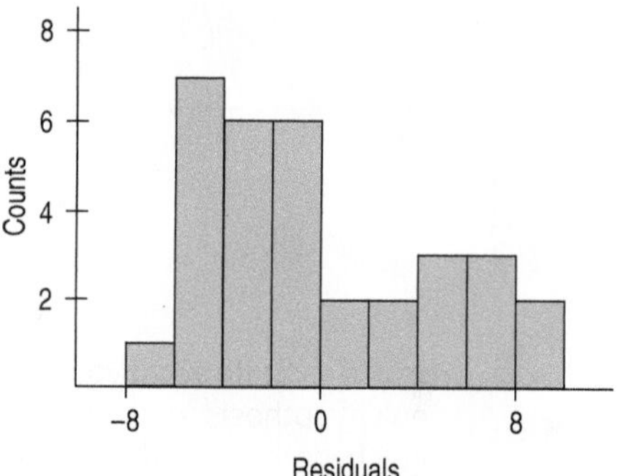

The table of means and SDs shows that the standard deviations grow along with the means. Possibly a re-expression of the data would improve matters.

	Under these circumstances, I cautiously find the P-value for the *F*-statistic from the *F*-model with 3 and 28 degrees of freedom.
SHOW ➡ Show the table of means.	From the ANOVA table, the mean square error, MS_E, is 22.22, which means that the standard deviation of all the errors is estimated to be $\sqrt{22.22} = 4.71$ degrees F. This seems like a reasonable value for the error standard deviation in the four treatments (with the possible exception of the Nissan mug).

Level	*n*	Mean	Std Dev
CUPPS	8	10.1875	5.20259
Nissan	8	2.7500	2.50713
SIGG	8	16.0625	5.90059
Starbucks	8	10.2500	4.55129

The ratio of the mean squares gives an *F*-ratio of 10.7134 with a P-value of < 0.0001.

TELL ➡ **Interpretation** Tell what the *F*-test means.	An *F*-ratio this large would be very unlikely if the containers all had the same mean temperature difference.
THINK ➡ State your conclusions. More precise conclusions might require a re-expression of the data.	**Conclusions:** Even though some of the conditions are mildly violated, the *P*-value is so extremely small that even if it is a bit off the mark, it would still be quite small. I conclude that the means are not all equal.

For Example ANALYSIS OF VARIANCE FOR HAND VOLUME CHANGE

RECAP: An ANOVA for the contrast baths experiment had a statistically significant *F*-value. Here are summary statistics for the three treatment groups:

Group	Count	Mean	StdDev
Bath	22	4.54545	7.76271
Bath+Exercise	23	8	7.03885
Exercise	14	−1.07143	5.18080

QUESTION: What can you conclude about these results?

ANSWER: We can be confident that there is a difference. However, it is the exercise treatment that appears to reduce swelling the most. We might conclude (as the researchers did) that contrast bath treatments are of limited value.

The Balancing Act

The two examples we've looked at so far share a special feature. Each treatment group has the same number of experimental units. For the hand-washing experiment, each washing method was tested eight times. For the cups, there were also eight trials for each cup. This feature (the equal numbers of cases in each group, not the number eight) is called **balance**, and experiments that have equal numbers of experimental units in each treatment are said to be balanced or to have balanced designs.

Balanced designs are a bit easier to analyze because the calculations are simpler, so we usually try for balance. But in the real world we often encounter unbalanced data. Participants drop out or become unsuitable, plants die, or maybe we just can't find enough experimental units to fit a particular criterion.

Everything we've done so far works just fine for unbalanced designs except that the calculations get a bit more complicated. Where once we could write n for the number of experimental units in a treatment, now we have to write n_k and sum more carefully. Where once we could pool variances with a simple average, now we have to adjust for the different n's. Technology clears these hurdles easily, so you're safe thinking about the analysis in terms of the simpler balanced formulas and trusting that the technology will make the necessary adjustments.

25.4 Comparing Means

When we reject H_0, it's natural to ask which means are different. No one would be happy with an experiment to test 10 cancer treatments that concluded only with "We can reject H_0—the treatments are different!" We'd like to know more, but the F-statistic doesn't offer that information.

What can we do? If we can't reject the null, we've got to stop. There's no point in further testing. If we've rejected the simple null hypothesis, however, we *can* do more. In particular, we can test whether any pairs or combinations of group means differ. For example, we might want to compare treatments against a control or a placebo, or against the current standard treatment.

In the hand-washing experiment, we could consider plain water to be a control. Nobody would be impressed with (or want to pay for) a soap that did no better than water alone. A test of whether the antibacterial soap (for example) was different from plain water would be a simple test of the difference between two group means. To be able to perform an ANOVA, we first check the Similar Variance Condition. If things look okay, we assume that the variances are equal. If the variances *are* equal, then a pooled t-test is appropriate. Even better (this is the special part), we already have a pooled estimate of the standard deviation based on *all* of the tested washing methods. That's s_p, which, for the hand-washing experiment, was equal to 37.55 bacteria colonies.

The null hypothesis is that there is no difference between water and the antibacterial soap. As we did in Chapter 21, we'll write that as a hypothesis about the difference in the means:

$H_0: \mu_W - \mu_{ABS} = 0$. The alternative is

$H_A: \mu_W - \mu_{ABS} \neq 0$.

The natural test statistic is $\bar{y}_W - \bar{y}_{ABS}$ and the (pooled) standard error is

$$SE(\bar{y}_W - \bar{y}_{ABS}) = s_p\sqrt{\frac{1}{n_W} + \frac{1}{n_{ABS}}}$$

The difference in the observed means is $117.0 - 92.5 = 24.5$ colonies. The standard error comes out to 18.775. The t-statistic, then, is $t = \frac{24.5}{18.775} = 1.31$. To find the P-value, we consult the Student's t-distribution on $N - k = 32 - 4 = 28$ degrees of freedom. The P-value is about 0.1—not small enough to impress us. So, we can't discern a significant difference between washing with the antibacterial soap and just using water.

Level	n	Mean	Std Dev
Alcohol spray	8	37.5	26.56
Antibacterial soap	8	92.5	41.96
Soap	8	106.0	46.96
Water	8	117.0	31.13

Our t-test asks about a simple difference. We could also ask a more complicated question about groups of differences. Does the average of the two soaps differ from the alcohol spray, for example? Complex combinations like these are called *contrasts*. Finding the standard errors for contrasts is straightforward but beyond the scope of this book. We'll restrict our attention to the common question of comparing simple pairs of treatments after H_0 has been rejected.

*Bonferroni Multiple Comparisons

Our hand-washing experimenter *was* pretty sure that alcohol would kill the germs even before she started the experiment. But alcohol dries the skin and leaves an unpleasant smell. She was hoping that one of the antibacterial soaps would work as well as alcohol so she could use that instead. That means she really wanted to compare *each* of the other treatments against the alcohol spray. We know how to compare two of the means with a t-test. But now we want to do several tests, and each test poses the risk of a Type I error. As we do more and more tests, the risk that we might make a Type I error grows bigger than the α level of each individual test. If we do enough tests, we're almost sure to reject one of the null hypotheses by mistake—and we'll never know which one.

There is a defense against this problem. In fact, there are several defenses. As a class, they are called **methods for multiple comparisons**. All multiple comparison methods require that we first be able to reject the overall (or global) null hypothesis with the ANOVA's F-test. Once we've rejected the overall null, then we can think about comparing several—or even all—pairs of group means.

Let's look again at our test of the water treatment against the antibacterial soap treatment. This time we'll look at a confidence interval instead of the pooled t-test. We did a test at significance level $\alpha = 0.05$. The corresponding confidence level is $1 - \alpha = 95\%$. For *any* pair of means, a confidence interval for their difference is $(\bar{y}_1 - \bar{y}_2) \pm ME$, where the margin of error is

$$ME = t^* \times s_p\sqrt{\frac{1}{n_1} + \frac{1}{n_2}}$$

As we did in the previous section, we get s_p as the pooled standard deviation found from *all* the groups in our analysis. Because s_p uses the information about the standard deviation from *all* the groups, it's a better estimate than we would get by combining the standard deviation of just two of the groups. This uses the **Equal Variance Assumption** and "borrows strength" in estimating the common standard deviation of all the groups. We find the critical value t^*- from the Student's t-model corresponding to the specified confidence level found with $N - k$ degrees of freedom, and the n_k's are the number of experimental units in each of the treatments.

To reject the null hypothesis that the two group means are equal, the difference between them must be larger than the ME. That way, 0 won't be in the confidence interval for the difference. When we use it in this way, we call the margin of error the **least significant difference (LSD** for short). If two group means differ by more than this amount, then they are significantly different at level α *for each individual test*.

For our hand-washing experiment, each group has $n = 8$, $s_p = 37.55$, and df $= 32 - 4 = 28$. From technology or Table T, we can find that t^* with 28 df (for a 95% confidence interval) is 2.048. So

$$\text{LSD} = 2.048 \times 37.55 \times \sqrt{\frac{1}{8} + \frac{1}{8}} = 38.45 \text{ colonies,}$$

and we could use this margin of error to make a 95% confidence interval for any difference between group means. Any two washing methods whose means differ by more than 38.45 colonies could be said to differ at $\alpha = 0.05$ by this method.

Of course, we're still just examining individual pairs. If we want to examine *many* pairs simultaneously, there are several methods that adjust the ME or LSD so that the resulting confidence intervals provide appropriate tests for all the pairs. And, in spite of making *many* such intervals, the overall Type I error rate stays at (or below) α.

One such method is called the **Bonferroni method**. This method adjusts the LSD to allow for making many comparisons. The result is a wider margin of error called the **minimum significant difference**, or **MSD**. The MSD is found by replacing t^* with a slightly larger number. That makes the confidence intervals wider for each contrast and the corresponding Type I error rates lower for *each* test. And it keeps the *overall* Type I error rate at or below α.

The Bonferroni method distributes the error rate equally among all the tests. It divides the error rate among J tests by calculating each confidence interval at confidence level $1 - \frac{\alpha}{J}$ instead of the original $1 - \alpha$. To signal this adjustment, we label the critical value t^{**} rather than t^*. For example, to make the three confidence intervals comparing the alcohol spray with the other three washing methods, and preserve our overall α risk below 5%, we'd construct each with a confidence level of

$$1 - \frac{0.05}{3} = 1 - 0.01667 = 0.98333$$

The only problem with this is that t-tables don't have a column for 98.33% confidence (or, correspondingly, for $\alpha = 0.01667$). Fortunately, technology has no such constraints.[7] For the hand-washing data, if we want to examine the three confidence intervals comparing each of the other methods with the alcohol spray, the t^{**}-value (on 28 degrees of freedom) turns out to be 2.546. That's somewhat larger than the individual t^*-value of 2.048 that we would have used for a single confidence interval. And the corresponding ME is 42.02 colonies (rather than 38.45 for a single comparison). The larger critical value along with correspondingly wider intervals is the price we pay for making multiple comparisons.

Many statistics packages assume that you'd like to compare all pairs of means. Some will display the result of these comparisons in a table like this:

Level	*n*	Mean	Groups
Alcohol spray	8	37.5	A
Antibacterial soap	8	92.5	B
Soap	8	106.0	B
Water	8	117.0	B

This table shows that the alcohol spray is in a class by itself and that the other three hand-washing methods are statistically indistinguishable from one another.

25.5 ANOVA on Observational Data

So far we've applied ANOVA only to data from designed experiments. That's natural for several reasons. The primary one is that, as we saw in Chapter 11, randomized comparative experiments are specifically designed to compare the results for different treatments. The overall null hypothesis, and the subsequent tests on pairs of treatments in

[7]The electronic t-tables provided in MyStatLab let you add new columns to the t-table at any alpha level, so you can do the Bonferroni calculation easily.

ANOVA, address such comparisons directly. In addition, as we discussed earlier, the **Equal Variance Assumption** (which we need for all of the ANOVA analyses) is often plausible in a randomized experiment because the treatment groups start out with sample variances that all estimate the same underlying variance of the collection of experimental units.

Sometimes, though, we just can't perform an experiment. When ANOVA is used to test equality of group means from observational data, there's no *a priori* reason to think the group variances might be equal at all. Even if the null hypothesis of equal means were true, the groups might easily have different variances. But if the side-by-side boxplots of responses for each group show roughly equal spreads and symmetric, outlier-free distributions, you can use ANOVA on observational data.

Observational data tend to be messier than experimental data. They are much more likely to be unbalanced. If you aren't assigning subjects to treatment groups, it's harder to guarantee the same number of subjects in each group. And because you are not controlling conditions as you would in an experiment, things tend to be, well, less controlled. The only way we know to avoid the effects of possible lurking variables is with control and randomized assignment to treatment groups, and for observational data, we have neither.

ANOVA is often applied to observational data when an experiment would be impossible or unethical. (We can't randomly break some subjects' legs, but we *can* compare pain perception among those with broken legs, those with sprained ankles, and those with stubbed toes by collecting data on subjects who have already suffered those injuries.) In such data, subjects are already in groups, but not by random assignment.

Be careful; if you have not assigned subjects to treatments randomly, you can't draw *causal* conclusions even when the F-test is significant. You have no way to control for lurking variables or confounding, so you can't be sure whether any differences you see among groups are due to the grouping variable or to some other unobserved variable that may be related to the grouping variable.

Because observational studies often are intended to estimate parameters, there is a temptation to construct pooled confidence intervals for the group means for this purpose. Although these confidence intervals are statistically correct, be sure to think carefully about the population that the inference is about. The relatively few subjects that you happen to have in a group may not be a simple random sample of any interesting population, so their "true" mean may have only limited meaning.

Step-by-Step Example ONE MORE EXAMPLE

Here's an example that exhibits many of the features we've been discussing. It gives a fair idea of the kinds of challenges often raised by real data.

A study in a liberal arts department attempted to find out who watches more television in university: men or women? Varsity athletes or nonathletes? Student researchers asked 197 randomly selected students questions about their backgrounds and about their television-viewing habits. The researchers found that men watch, on average, about 2.5 hours per week more television than women, and that varsity athletes watch about 3.5 hours per week more than those who are not varsity athletes. But is this the whole story? To investigate further, they divided the students into four groups: male athletes (MA), male nonathletes (MNA), female athletes (FA), and female nonathletes (FNA). Let's do the ANOVA step-by-step.

THINK ➡ Variables Name the variables, report the *W*'s, and specify the questions of interest.

I have the number of hours spent watching television in a week for 197 randomly selected students. We know their sex and whether they are varsity athletes or not. I wonder whether television watching differs according to sex and athletic status.

Plot Always start an ANOVA with side-by-side boxplots of the responses in each of the groups. Always.

Here are the side-by-side boxplots of the data:

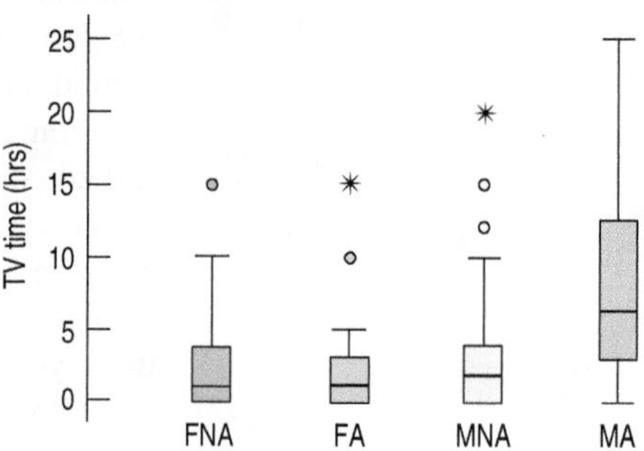

This plot suggests problems with the data. Each box shows a distribution skewed to the high end, and outliers pepper the display, including some extreme outliers. The box with the highest centre (MA) also has the largest spread. These data just don't pass our first screening for suitability. This sort of pattern calls for a re-expression.

Here are the boxplots for the square root of television hours.

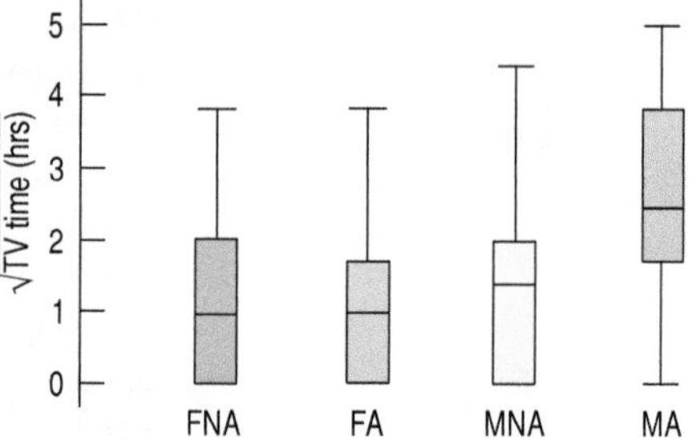

The spreads in the four groups are now very similar and the individual distributions closer to symmetric. And now there are no outliers.

✓ **Independence Assumption:** Because this is a random sample, the assumption of independence is reasonable.

✓ **Randomization Condition:** The data come from a random sample of students.

✓ **Similar Variance Condition:** The boxplots show similar spreads. I may want to check the residuals later.

The ANOVA table looks like this:

Source	DF	Sum of Squares	Mean Square	F-ratio	P-value
Group	3	47.24733	15.7491	12.8111	<0.0001
Error	193	237.26114	1.2293		
Total	196	284.50847			

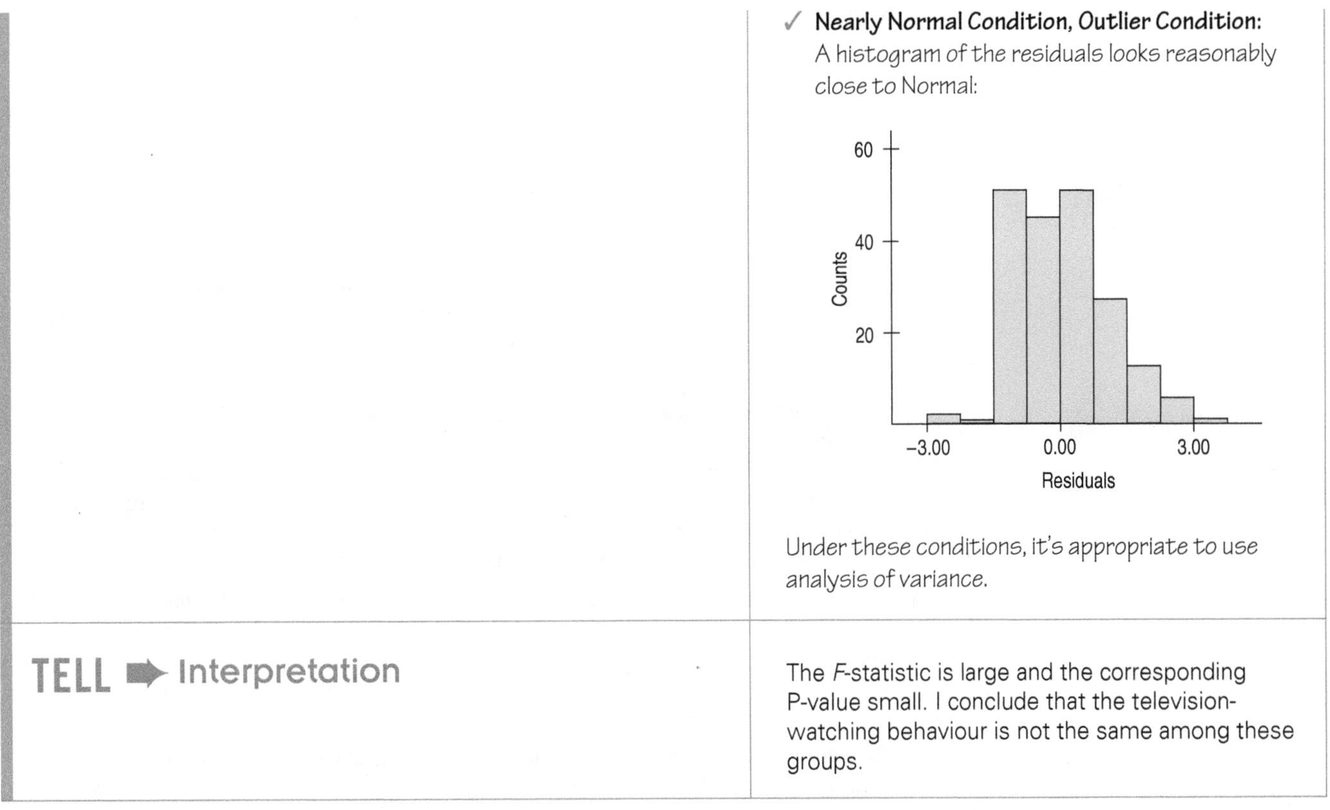

✓ Nearly Normal Condition, Outlier Condition:
A histogram of the residuals looks reasonably close to Normal:
60
40
20
Counts
−3.00
0.00
3.00
Residuals
Under these conditions, it's appropriate to use analysis of variance.

TELL ➡ Interpretation

The F-statistic is large and the corresponding P-value small. I conclude that the television-watching behaviour is not the same among these groups.

*So Do Male Athletes Watch More TV?

Here's a Bonferroni comparison of all pairs of groups:

	Difference	Std. Err.	P-Value
FA–FNA	0.049	0.270	0.9999
MNA–FNA	0.205	0.182	0.8383
MNA–FA	0.156	0.268	0.9929
MA–FNA	1.497	0.250	<0.0001
MA–FA	1.449	0.318	<0.0001
MA–MNA	1.292	0.248	<0.0001

Three of the differences are very significant. It seems that among women, there's little difference in television watching between varsity athletes and others. Among men, though, the corresponding difference is large. And among varsity athletes, men watch significantly more television than women.

But wait. How far can we extend the inference that male athletes watch more television than other groups? The data came from a random sample of students made during the week of March 21. If the students carried out the survey correctly using a simple random sample, we should be able to make the inference that the generalization is true for the entire student body during that week.

Is it true for other universities? Is it true throughout the year? The students conducting the survey followed it up by collecting anecdotal information about television watching of male athletes. It turned out that during the week of the survey, the NCAA men's basketball tournament was televised. This could explain the increase in television watching for the

Differing Standard Errors? In case you were wondering . . . The standard errors are different because this isn't a balanced design. Differing numbers of experimental units in the groups generate differing standard errors.

male athletes. It could be that the increase extends to other students at other times, but we don't know that. Always be cautious in drawing conclusions too broadly. Don't generalize from one population to another.

WHAT CAN GO WRONG?

- **Watch out for outliers.** One outlier in a group can change both the mean and the spread of that group. It will also inflate the mean square error, which can influence the F-test. The good news is that ANOVA fails on the safe side by losing power when there are outliers. That is, you are less likely to reject the overall null hypothesis if you have (and leave) outliers in your data. But they are not likely to cause you to make a Type I error.
- **Watch out for changing variances.** The conclusions of the ANOVA depend crucially on the assumptions of independence and constant variance, and (somewhat less seriously as n increases) on Normality. If the conditions on the residuals are violated, it may be necessary to re-express the response variable to approximate these conditions more closely. ANOVA benefits so greatly from a judiciously chosen re-expression that the choice of a re-expression might be considered a standard part of the analysis.
- **Be wary of drawing conclusions about causality from observational studies.** ANOVA is often applied to data from randomized experiments for which causal conclusions are appropriate. If the data are not from a designed experiment, however, the analysis of variance provides no more evidence for causality than any other method we have studied. Don't get into the habit of assuming that ANOVA results have causal interpretations.
- **Be wary of generalizing to situations other than the one at hand.** Think hard about how the data were generated to understand the breadth of conclusions you are entitled to draw.
- **Watch for multiple comparisons.** When rejecting the null hypothesis, you can conclude that the means are not *all* equal. But you can't start comparing every pair of treatments in your study with a t-test. You'll run the risk of inflating your Type I error rate. Use a multiple comparisons method when you want to test many pairs.

CONNECTIONS

We first learned about side-by-side boxplots in Chapter 4. There we made general statements about the shape, centre, and spread of each group. When we compared groups, we asked whether their centres looked different compared with how spread out the distributions were. Now we've made that kind of thinking precise. We've added confidence intervals for the difference and tests of whether the means are the same.

We pooled data to find a standard deviation when we tested the hypothesis of equal proportions. For that test, the assumption of equal variances was a consequence of the null hypothesis that the proportions were equal, so it didn't require an extra assumption. Means don't have a linkage with their corresponding variances, so to use pooled methods we must make the additional assumption of equal variances. But in a randomized experiment, that's a plausible assumption.

Chapter 11 offered a variety of designs for randomized comparative experiments. Each of those designs can be analyzed with a variant or extension of the ANOVA methods discussed in this chapter. Entire books and courses deal with these extensions, but all follow the same fundamental ideas presented here.

ANOVA is closely related to the regression analyses we saw in Chapter 24. (In fact, most statistics packages offer an ANOVA table as part of their regression output.) The assumptions are similar—and for good reason. The analyses are, in fact, related at a deep conceptual (and computational) level, but those details are beyond the scope of this book.

The pooled two-sample t-test for means is a special case of the ANOVA F-test. If you perform an ANOVA comparing only two groups, you'll find that the P-value of the F-statistic is exactly the same as the P-value of the corresponding pooled t-statistic. That's because, in this special case, the F-statistic is just the square of the t-statistic. The F-test is more general. It can test the hypothesis that several group means are equal.

What Have We Learned?

Learning Objectives

We learned, in Chapter 21, how to test whether the means of two groups are equal. Now in this chapter, we've extended that to testing whether the means of *several* groups are equal.

- Recognize situations for which ANOVA is the appropriate analysis.
- Know how to check conditions to verify the assumptions before we proceed with inference.
- Be able to perform an ANOVA using a statistics package or calculator for one response variable and one factor with any number of levels.

We learned in Chapter 4 that a good first step in looking at the relationship between a quantitative response and a categorical grouping variable is to look at side-by-side boxplots. We've seen that it's still a good first step before formally testing the null hypothesis.

- Know how to examine your data for violations of conditions that would make ANOVA unwise or invalid.
- Be able to explain the contents of an ANOVA table, in particular the roles of the MST, MSE, and the pooled standard deviation.

We've learned that the *F*-test is a generalization of the *t*-test that we used for testing two groups. Under certain assumptions, the statistic used to test whether the means of k groups are equal is distributed as an *F*-statistic with $k - 1$ and $N - k$ degrees of freedom.

- If the *F*-statistic is large enough, we reject the null hypothesis that all the means are equal.

We've also learned that when the null hypothesis is rejected and we conclude that there are differences, we need to adjust the confidence intervals for the pair-wise differences between means.

- Recognize when a further analysis of differences between group means would be appropriate.
- Be able to perform several subsequent tests using a multiple comparisons procedure.

Review of Terms

Mean square treatment (or between) (MS_T) Another estimate of the error variance—under the assumption that the treatment means are all equal. If the (null) assumption is not true, the MS_T will tend to be larger than the error variance (p. 748).

***F*-distribution** The sampling distribution of the *F*-statistic when the null hypothesis that the treatment means are equal is true. It has two degrees of freedom parameters, one for the numerator, $k - 1$, and one for the denominator, $N - k$, where N is the total number of observations and k is the number of groups (p. 748).

ANOVA An analysis method for testing equality of means across treatment groups (p. 748).

Mean square error (or within) (MS_E) The estimate of the error variance obtained by *pooling* the variances of each treatment group. The square root of the MS_E is the estimate of the error standard deviation, s_p (p. 749).

ANOVA table A table that shows the degrees of freedom, the mean square treatment, the mean square error, their ratio, the *F*-statistic, and its P-value. Other quantities of lesser interest are usually included as well (p. 749).

***F*-statistic** The ratio MS_T/MS_E. When the *F*-statistic is sufficiently large, we reject the null hypothesis that the group means are equal (p. 750).

ANOVA model The model for a one-way (one response, one factor) ANOVA is

$$y_{ij} = \mu_j + \varepsilon_{ij}$$

Estimating above with $y_{ij} = \bar{y}_j + e_{ij}$ gives predicted values $\hat{y}_{ij} = \bar{y}_j$ and residuals $e_{ij} = y_{ij} - \bar{y}_j$ (p. 751).

Residual standard deviation Gives an idea of the underlying variability of the response values:

$$s_p = \sqrt{MS_E} = \sqrt{\frac{\Sigma e^2}{N - k}}$$

(p. 755).

Assumptions for ANOVA (and conditions to check)

- Equal Variance Assumption (Similar Variance Condition). Look at side-by-side boxplots to check for similar spreads, or look at residuals versus predicted to see if the plot thickens (p. 756).
- Normal Population Assumption (Nearly Normal Condition). Check a histogram or Normal probability plot of the residuals (p. 756).
- Independence Assumption. Think about the design of the experiment or, if an observational study, how the data were collected (p. 759).

***F*-test** Tests the null hypothesis that all the group means are equal against the alternative that they are not all equal. We reject the hypothesis of equal means if the *F*-statistic exceeds the critical value from the *F*-distribution corresponding to the specified significance level and degrees of freedom (p. 757).

Balance An experiment's design is balanced if each treatment level has the same number of experimental units. Balanced designs make calculations simpler and are generally more powerful (p. 761).

Multiple comparisons If we reject the null hypothesis of equal means, we often then want to investigate further and compare pairs of treatment group means to see if they differ. If we want to test several such pairs, we must adjust for performing several tests to keep the overall risk of a Type I error from growing too large. Such adjustments are called methods for multiple comparisons (p. 762).

Least significant difference (LSD) The standard margin of error in the confidence interval for the difference of two means. It has the correct Type I error rate for a single test, but not when performing more than one comparison (p. 762).

***Bonferroni method** One of many methods for adjusting the length of the margin of error when testing the differences between several group means (p. 763).

***Minimum significant difference (MSD)** The Bonferroni method's margin of error for the confidence interval for the difference of two group means. The MSD can be used to test differences of several pairs of group means. If their difference exceeds the MSD, they are different at the overall α rate (p. 763).

On the Computer ANOVA ON THE COMPUTER

Most analyses of variance are found with computers, and all statistics packages present the results in an ANOVA table much like the one we discussed. Technology also makes it easy to examine the side-by-side boxplots and check the residuals for violations of the assumptions and conditions.

Statistics packages offer different choices among possible multiple comparisons methods (although Bonferroni is quite common). This is a specialized area. Get advice or read further if you need to choose a multiple comparisons method.

As we saw in Chapter 4, there are two ways to organize data recorded for several groups. We can put all the response values in a single variable and use a second, "factor," variable to hold the group identities. Most statistics packages expect the data to be in stacked format because this form also works for more complicated experimental designs. Some packages can work with either form, and some use one form for some things and the other for others. (Be careful, for example, when you make side-by-side boxplots; be sure to give the appropriate version of the command to correspond to the structure of your data.)

Most packages offer to save residuals and predicted values and make them available for further tests of conditions. In some packages you may have to request them specifically.

EXCEL

- From the tools menu, select **Data Analysis**.
- Select **Anova Single Factor** from the list of analysis tools.
- Click the **OK** button.
- Enter the data range in the box provided.
- Check the **Labels in First Row** box, if applicable.
- Enter an alpha level for the *F*-test in the box provided.
- Click the **OK** button.

COMMENTS

The data range should include two or more columns of data to compare. Unlike all other statistics packages, Excel expects each column of the data to represent a different level of the factor. However, it offers no way to label these levels. The columns need not have the same number of data values, but the selected cells must make up a rectangle large enough to hold the column with the most data values.

JMP

- From the **Analyze** menu select **FitY by X**.
- Select variables: a quantitative Y, Response variable, and a categorical X, Factor variable.
- JMP opens the **Oneway** window.
- Click on the red triangle beside the heading, select **Display Options**, and choose **Boxplots**.
- From the same menu, choose the **Means/ANOVA.t-test** command.
- JMP opens the one-way ANOVA output.

COMMENTS

JMP expects data in "stacked" format with one response and one factor variable.

MINITAB

- Choose **ANOVA** from the Stat menu.
- Choose **One-way . . .** from the **ANOVA** submenu.
- In the One-way Anova dialogue, assign a quantitative Y variable to the Response box and assign a categorical X variable to the Factor box.
- Check the **Store Residuals** check box.
- Click the **Graphs** button.
- In the ANOVA-Graphs dialogue, select **Standardized residuals**, and check No**rmal plot of residuals** and **Residuals versus fits**.
- Click the **OK** button to return to the Regression dialogue.
- Click the **OK** button to compute the regression.

COMMENTS

If your data are in unstacked format, with separate columns for each treatment level, choose **One-way (unstacked)** from the **ANOVA** submenu.

R

- **aov(formula, data=dataset)** will perform an ANOVA.

COMMENTS

If the dataset has a column Y with the response variable, and a column X of factors, then **formula** will be $Y \sim X$.

SPSS

- Choose **Compare Means** from the **Analyze** menu.
- Choose **One-way ANOVA** from the **Compare Means** submenu.
- In the One-Way ANOVA dialogue, select the Y-variable and move it to the dependent target. Then move the X-variable to the independent target.
- Click the **OK** button.

COMMENTS

SPSS expects data in stacked format. The **Contrasts** and **Post Hoc** buttons offer ways to test contrasts and perform multiple comparisons. See your SPSS manual for details.

STATCRUNCH

To perform a one-way ANOVA:

- Enter the factor in one column and the response variable in a second
- Click on **Stat**
- Choose **ANOVA** » **One Way**
- Under **Compare values in a single column**, select the columns you filled in previously and click on **Calculate**

COMMENTS

In addition to the ANOVA output, STATCRUNCH also displays group means.

Exercises

1. Popcorn A student runs an experiment to test four different popcorn brands, recording the number of kernels left unpopped. She pops measured batches of each brand four times, using the same popcorn popper and randomizing the order of the brands. After collecting her data and analyzing the results, she reports that the F-ratio is 13.56.

a) What are the null and alternative hypotheses?

b) How many degrees of freedom does the treatment sum of squares have? How about the error sum of squares?

c) Assuming that the conditions required for ANOVA are satisfied, what is the P-value? What would you conclude?

d) What else about the data would you like to see in order to check the assumptions and conditions?

2. Skating A figure skater tried various approaches to her Salchow jump in a designed experiment using five different places for her focus (arms, free leg, midsection, take-off leg, and free). She tried each jump six times in random order, using two of her skating partners to judge the jumps on a scale from 0 to 6. After collecting the data and analyzing the results, she reports that the F-ratio is 7.43.

a) What are the null and alternative hypotheses?

b) How many degrees of freedom does the treatment sum of squares have? How about the error sum of squares?

c) Assuming that the conditions are satisfied, what is the P-value? What would you conclude?

d) What else about the data would you like to see in order to check the assumptions and conditions?

3. Gas mileage A student runs an experiment to study the effect of three different mufflers on gas mileage. He devises a system so that his Jeep Wagoneer uses gasoline from a one-litre container. He tests each muffler eight times, carefully recording the number of kilometres he can go in his Jeep Wagoneer on one litre of gas. After analyzing his data, he reports that the F-ratio is 2.35 with a P-value of 0.1199.

a) What are the null and alternative hypotheses?

b) How many degrees of freedom does the treatment sum of squares have? How about the error sum of squares?

c) What would you conclude?
d) What else about the data would you like to see in order to check the assumptions and conditions?
e) If your conclusion in part c) is wrong, what type of error have you made?

4. Darts A student interested in improving her dart-throwing technique designs an experiment to test four different stances to see whether they affect her accuracy. After warming up for several minutes, she randomizes the order of the four stances, throws a dart at a target using each stance, and measures the distance of the dart in centimetres from the centre of the bull's-eye. She replicates this procedure 10 times. After analyzing the data, she reports that the *F*-ratio is 1.41.
a) What are the null and alternative hypotheses?
b) How many degrees of freedom does the treatment sum of squares have? How about the error sum of squares?
c) What would you conclude?
d) What else about the data would you like to see in order to check the assumptions and conditions?
e) If your conclusion in part c) is wrong, what type of error have you made?

5. Cat hair A Toronto veterinarian is interested to see if the amount of hair a cat will shed is related to the type of food consumed by the cat. The vet randomly divides 60 cats from a kennel into three groups and gives one group wet food and another group dry food. The third group acts as a control and receives no food that day. She measures the weight of cat hair remaining in the cage at the end of the day, and performs an ANOVA, obtaining an F-ratio of 2.58.
a) What are the null and alternative hypotheses?
b) How many degrees of freedom does the treatment sum of squares have? How about the error sum of squares?
c) Assuming that the conditions are satisfied, what is the P-value? What would you conclude?
d) Do you have any concerns about this experiment design?

6. Consulting A researcher from another department wants to know if he used the correct analysis. He collected resting heart rates and diastolic blood pressure for 100 students from five different majors, and performed an ANOVA. Since heart rate and diastolic blood pressure have a similar mean and variance, he decided to use a Bonferroni test to compare the heart rate of Biology majors to the blood pressure of physical education majors, and found no significant difference. Do you see any problems with this technique?

T **7. Activating baking yeast** To shorten the time it takes him to make his favourite pizza, a student designed an experiment to test the effect of sugar and milk on the activation times for baking yeast. Specifically, he tested four different recipes and measured how many seconds it took for the same amount of dough to rise to the top of a bowl. He randomized the order of the recipes and replicated each treatment four times.

Here are the boxplots of activation times from the four recipes:

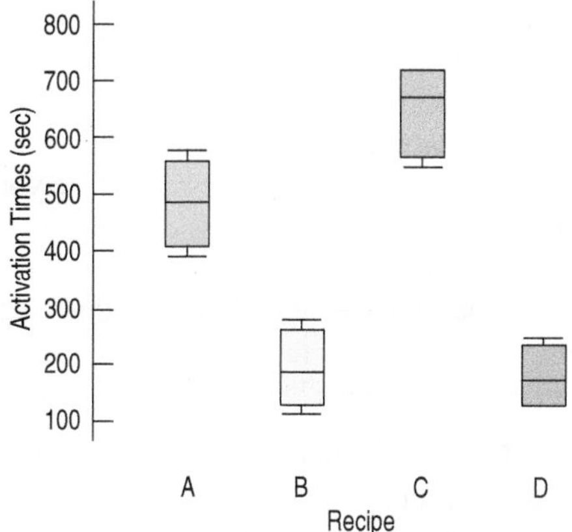

The ANOVA table is shown below.

Analysis of Variance

Source	DF	Sum of Squares	Mean Square	F-ratio	P-value
Recipe	3	638967.69	212989	44.7392	<0.0001
Error	12	57128.25	4761		
Total	15	696095.94			

a) State the hypotheses about the recipes (both numerically and in words).
b) Assuming that the assumptions for inference are satisfied, perform the hypothesis test and state your conclusion. Be sure to state it in terms of activation times and recipes.
c) Would it be appropriate to follow up this study with multiple comparisons to see which recipes differ in their mean activation times? Explain.

T **8. Frisbee throws** A student performed an experiment with three different grips to see what effect it might have on the distance of a backhanded Frisbee throw. She tried it with her normal grip, with one finger out, and with the Frisbee inverted. She measured in paces how far her throw went. The boxplots and the ANOVA table for the three grips are shown below:

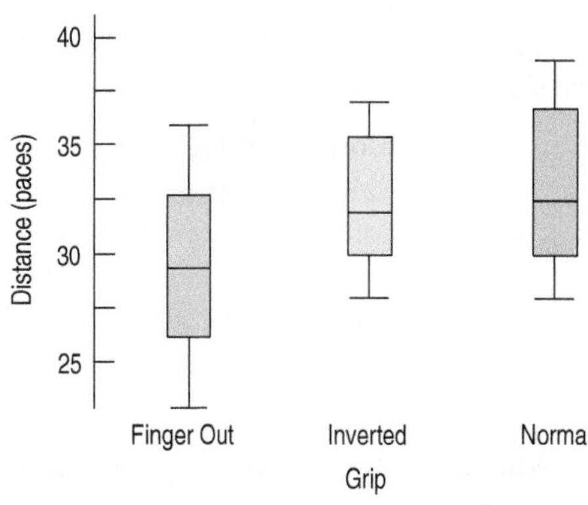

Analysis of Variance

Source	DF	Sum of Squares	Mean Square	F-ratio	P-value
Grip	2	58.58333	29.2917	2.0453	0.1543
Error	21	300.75000	14.3214		
Total	23	359.33333			

a) State the hypotheses about the grips.
b) Assuming that the assumptions for inference are satisfied, perform the hypothesis test and state your conclusion. Be sure to state it in terms of Frisbee grips and distance thrown.
c) Would it be appropriate to follow up this study with multiple comparisons to see which grips differ in their mean distance thrown? Explain.

9. Eye and hair colour In Chapter 4, Exercise 30, we saw a survey of 1021 school-age children conducted by randomly selecting children from several large urban elementary schools. Two of the questions concerned eye and hair colour. In the survey, the following codes were used:

Hair colour:	Eye colour:
1 = Blonde	1 = Blue
2 = Brown	2 = Green
3 = Black	3 = Brown
4 = Red	4 = Grey
5 = Other	5 = Other

The Statistics students analyzing the data were asked to study the relationship between eye and hair colour. They produced this plot:

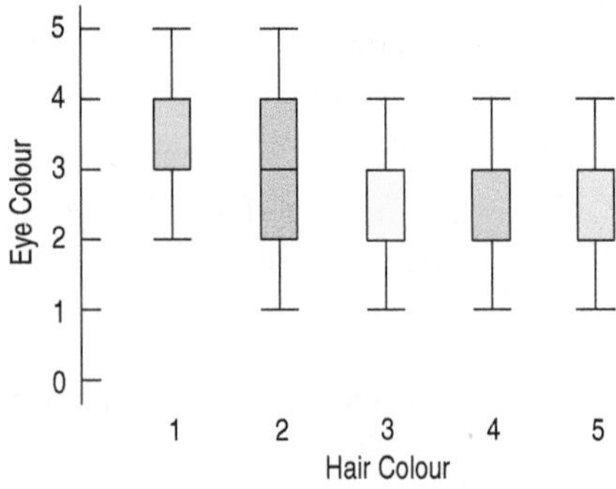

They ran an analysis of variance with eye colour as the response and hair colour as the factor. The ANOVA table they produced is shown below.

Analysis of Variance

Source	DF	Sum of Squares	Mean Square	F-ratio	P-value
Hair colour	4	1.46946	0.367365	0.4024	0.8070
Error	1016	927.45317	0.912848		
Total	1020	928.92263			

What suggestions do you have for the Statistics students? What alternative analysis might you suggest?

10. U.S. zip codes revisited The intern from the marketing department at the Holes-R-Us online piercing salon (Chapter 3, Exercise 37) has recently finished a study of the company's 500 customers. He wanted to know whether people's zip codes vary by the last product they bought. They have 16 different products, and the ANOVA table of zip code by product shows the following:

ANOVA table

Source	DF	Sum of Squares	Mean Square	F-ratio	P-value
Product	15	3.836e10	2.55734e9	4.9422	<0.0001
Error	475	2.45787e11	517445573		
Total	490	2.84147e11			

(Nine customers were not included because of missing zip code or product information.)

What criticisms of the analysis might you make? What alternative analysis might you suggest?

T 11. Wines revisited The boxplots we saw in Chapter 4, Exercise 10, display case prices (in dollars) of wines produced by wineries along three of the Finger Lakes.

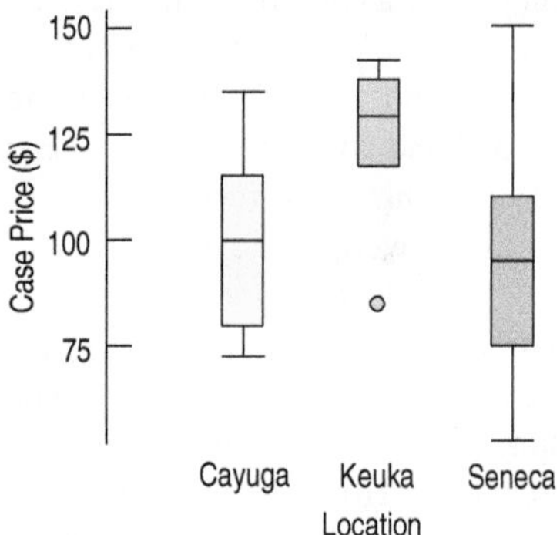

a) What are the null and alternative hypotheses? Talk about prices and location, not symbols.
b) Do the conditions for an ANOVA seem to be met here? Why or why not?

T 12. Tellers A bank is studying the time that it takes six of its tellers to serve an average customer. Customers line up in the queue and then go to the next available teller. Here is a boxplot of the last 140 customers and the times it took each teller:

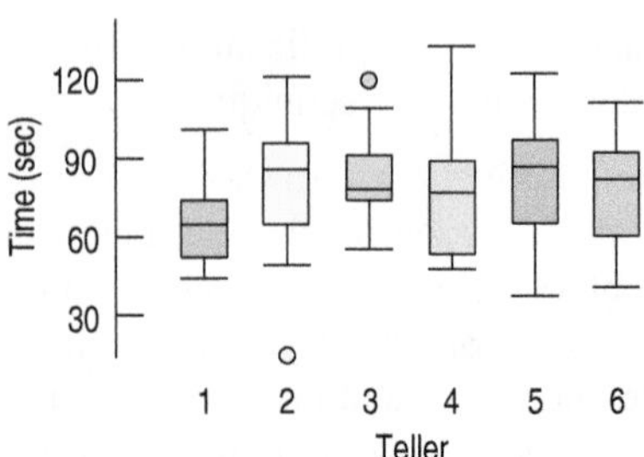

Analysis of Variance

Source	DF	Sum of Squares	Mean Square	F-ratio	P-value
Teller	5	3315.32	663.064	1.508	0.1914
Error	134	58919.1	439.695		
Total	139	62234.4			

a) What are the null and alternative hypotheses?
b) What do you conclude?
c) Would it be appropriate to run a multiple comparisons test (for example, a Bonferroni test) to see which tellers differ from each other? Explain.

T **13. Hearing** A researcher investigated four different word lists for use in hearing assessment (Loven, 1981). She wanted to know whether the lists were equally difficult to understand in the presence of a noisy background. To find out, she tested 96 subjects with normal hearing, randomly assigning 24 to each of the four word lists, and measured the number of words perceived correctly in the presence of background noise. Here are the boxplots of the four lists:

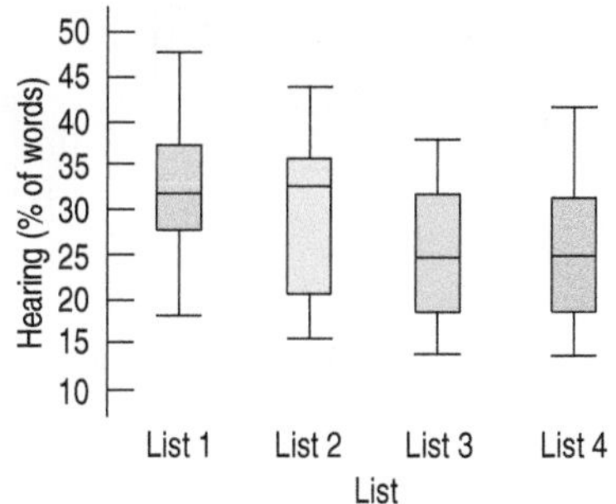

Analysis of Variance

Source	DF	Sum of Squares	Mean Square	F-ratio	P-value
List	3	920.4583	306.819	4.9192	0.0033
Error	92	5738.1667	62.371		
Total	95	6658.6250			

a) What are the null and alternative hypotheses?
b) What do you conclude?
c) Would it be appropriate to run a multiple comparisons test (for example, a Bonferroni test) to see which lists differ from each other in terms of mean percent correct? Explain.

14. Yogourt An experiment to determine the effect of several methods of preparing cultures for use in commercial yogourt was conducted by a food science research group. Three batches of yogourt were prepared using each of three methods: traditional, ultrafiltration, and reverse osmosis. A trained expert then tasted each of the nine samples, presented in random order, and judged them on a scale from 1 to 10. A partially complete analysis of variance table of the data is shown below.

An incomplete ANOVA Table for the Yogourt Data

Source	DF	Sum of Squares	Mean Square	F-ratio	P-value
Treatment		17.300			
Residual		0.460			
Total		17.769			

a) Calculate the mean square of the treatments and the mean square of the error.
b) Form the F-statistic by dividing the two mean squares.
c) The P-value of this F-statistic turns out to be 0.000017. What does this say about the null hypothesis of equal means?
d) What assumptions have you made in order to answer part c)?
e) What would you like to see in order to justify the conclusions of the F-test?
f) What is the average size of the error standard deviation in the judge's assessment?

15. Smokestack scrubbers Particulate matter is a serious form of air pollution often arising from industrial production. One way to reduce the pollution is to put a filter, or scrubber, at the end of the smokestack to trap the particulates. An experiment to determine which smokestack scrubber design is best was run by placing four scrubbers of different designs on an industrial stack in random order. Each scrubber was tested five times. For each run, the same material was produced, and the particulate emissions coming out of the scrubber were measured (in parts per billion). A partially complete analysis of variance table of the data is shown below.

An incomplete ANOVA Table for the Smokestack Data

Source	DF	Sum of Squares	Mean Square	F-ratio	P-value
Treatment		81.2			
Residual		30.8			
Total		112.0			

a) Calculate the mean square of the treatments and the mean square of the error.
b) Form the F-statistic by dividing the two mean squares.
c) The P-value of this F-statistic turns out to be 0.00000949. What does this say about the null hypothesis of equal means?
d) What assumptions have you made in order to answer part c)?
e) What would you like to see in order to justify the conclusions of the F-test?
f) What is the average size of the error standard deviation in particulate emissions?

T **16. Eggs** A student wants to investigate the effects of real versus substitute eggs on his favourite brownie recipe. He enlists the help of 10 friends and asks them to rank each of eight batches on a scale from 1 to 10. Four of the batches were made with real eggs, four with substitute eggs. The judges tasted the brownies in random order. Here is a boxplot of the data:

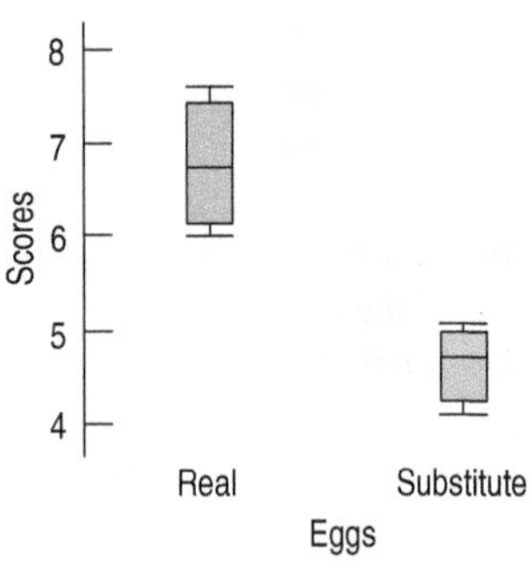

Analysis of Variance

Source	DF	Sum of Squares	Mean Square	F-ratio	P-value
Eggs	1	9.010013	9.01001	31.0712	0.0014
Error	6	1.739875	0.28998		
Total	7	10.749883			

The mean score for the real eggs was 6.78 with a standard deviation of 0.651. The mean score for the substitute eggs was 4.66 with a standard deviation of 0.395.

a) What are the null and alternative hypotheses?
b) What do you conclude from the ANOVA table?
c) Do the assumptions for the test seem to be reasonable?
d) Perform a two-sample pooled t-test of the difference. What P-value do you get? Show that the square of the t-statistic is the same (to rounding error) as the F-ratio.

T **17. Auto noise filters** In a statement to a U.S. Senate Public Works Committee, a senior executive of Texaco, Inc., cited a study on the effectiveness of auto filters on reducing noise. Because of concerns about performance, two types of filters were studied, a standard silencer and a new device developed by the Associated Octel Company. Here are the boxplots from the data on noise reduction (in decibels) of the two filters. Type 1 = standard; Type 2 = Octel.

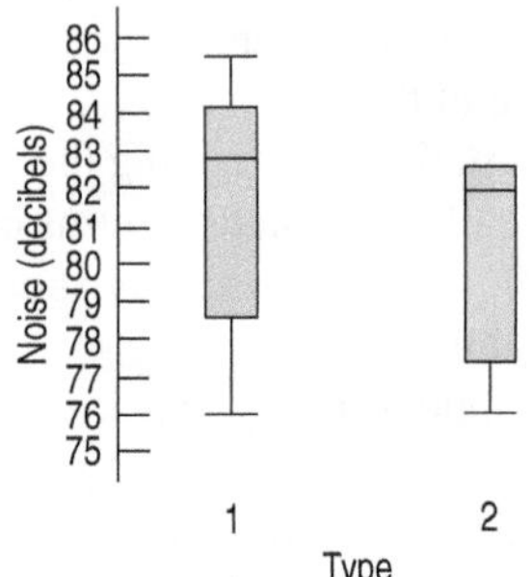

Analysis of Variance

Source	DF	Sum of Squares	Mean Square	F-ratio	P-value
Type	1	6.31	6.31	0.7673	0.3874
Error	33	271.47	8.22		
Total	34	2.77			

Level	n	Mean	StdDev
Standard	18	81.5556	3.22166
Octel	17	80.7059	2.43708

a) What are the null and alternative hypotheses?
b) What do you conclude from the ANOVA table?
c) Do the assumptions for the test seem to be reasonable?
d) Perform a two-sample pooled t-test of the difference. What P-value do you get? Show that the square of the t-statistic is the same (to rounding error) as the F-ratio.

T **18. School system** A school district superintendent wants to test a new method of teaching arithmetic in Grade 4 at his 15 schools. He plans to select eight students from each school to take part in the experiment, but to make sure they are roughly of the same ability, he first gives a test to all 120 students. Here are the scores of the test by school:

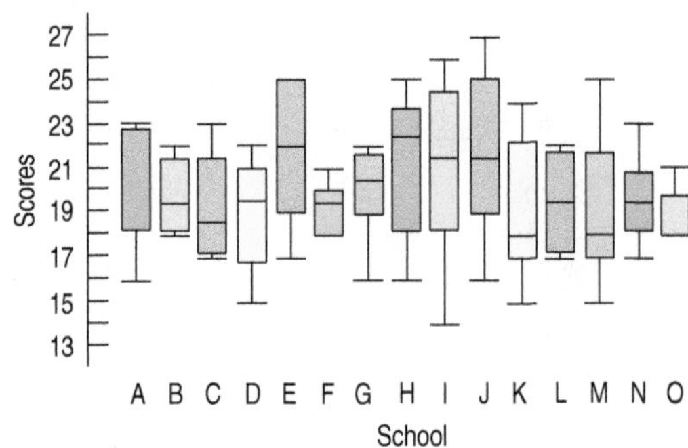

The ANOVA table shows:

Analysis of Variance

Source	DF	Sum of Squares	Mean Square	F-ratio	P-value
School	14	108.800	7.7714	1.0735	0.3899
Error	105	760.125	7.2392		
Total	119	868.925			

a) What are the null and alternative hypotheses?
b) What does the ANOVA table say about the null hypothesis? (Be sure to report this in terms of scores and schools.)
c) An intern reports that he has done t-tests of every school against every other school and finds that several of the schools seem to differ in mean score. Does this match your finding in part b)? Give an explanation for the difference, if any, of the two results.

T **19. Fertilizers** A biology student is studying the effect of 10 different fertilizers on the growth of mung bean sprouts. She sprouts 12 beans in each of 10 different petri dishes, and adds the same amount of fertilizer to each dish. After one week she measures the heights of the

120 sprouts in millimetres. Here are boxplots and an ANOVA table of the data:

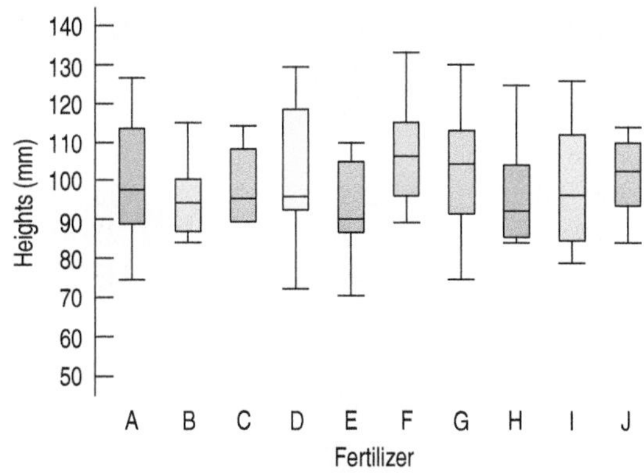

Analysis of Variance

Source	DF	Sum of Squares	Mean Square	F-ratio	P-value
Fertilizer	9	2073.708	230.412	1.1882	0.3097
Error	110	21331.083	193.919		
Total	119	23404.791			

a) What are the null and alternative hypotheses?

b) What does the ANOVA table say about the null hypothesis? (Be sure to report this in terms of heights and fertilizers).

c) Her lab partner looks at the same data. He says that he did *t*-tests of every fertilizer against every other fertilizer and finds that several of the fertilizers seem to have significantly higher mean heights. Does this match your finding in part b)? Give an explanation for the difference, if any, between the two results.

20. Cereals Grocery stores often place similar types of cereal on the same shelf. The same data set we met in the Step-By-Step of Chapter 7 keeps track of the shelf as well as the sugar, sodium, and calorie content of 77 cereals. Does sugar content vary by shelf? Here's a boxplot and an ANOVA table for the 77 cereals:

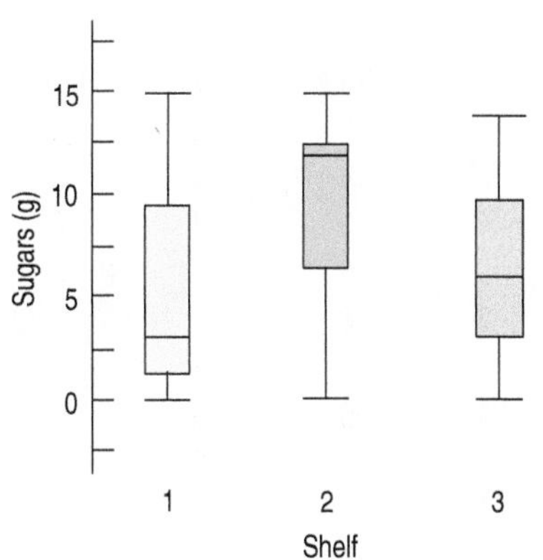

Analysis of Variance

Source	DF	Sum of Squares	Mean Square	F-ratio	P-value
Shelf	2	248.4079	124.204	7.3345	0.0012
Error	74	1253.1246	16.934		
Total	76	1501.5325			

Level	n	Mean	StdDev
1	20	4.80000	4.57223
2	21	9.61905	4.12888
3	36	6.52778	3.83582

a) What are the null and alternative hypotheses?

b) What does the ANOVA table say about the null hypothesis? (Be sure to report this in terms of sugars and shelves.)

c) Can we conclude that cereals on shelf 2 have a higher mean sugar content than cereals on shelf 3? Can we conclude that cereals on shelf 2 have a higher mean sugar content than cereals on shelf 1? What *can* we conclude?

d) To check for significant differences between the shelf means, we can use a Bonferroni test, whose results are shown below. For each pair of shelves, the difference is shown along with its standard error and significance level. What does it say about the questions in part c)?

Dependent Variable: SUGARS

	(I) SHELF	(J) SHELF	Mean Difference (I-J)	Std. Error	P-value	95% Confidence Interval Lower Bound	95% Confidence Interval Upper Bound
Bonferroni	1	2	−4.819(*)	1.2857	0.001	−7.969	−1.670
		3	−1.728	1.1476	0.409	−4.539	1.084
	2	1	4.819(*)	1.2857	0.001	1.670	7.969
		3	3.091(*)	1.1299	0.023	0.323	5.859
	3	1	1.728	1.1476	0.409	−1.084	4.539
		2	−3.091(*)	1.1299	0.023	−5.859	−0.323

*The mean difference is significant at the 0.05 level.

21. Cereals redux We also have data on the protein content of cereals by their shelf number. Here are the boxplot and ANOVA table:

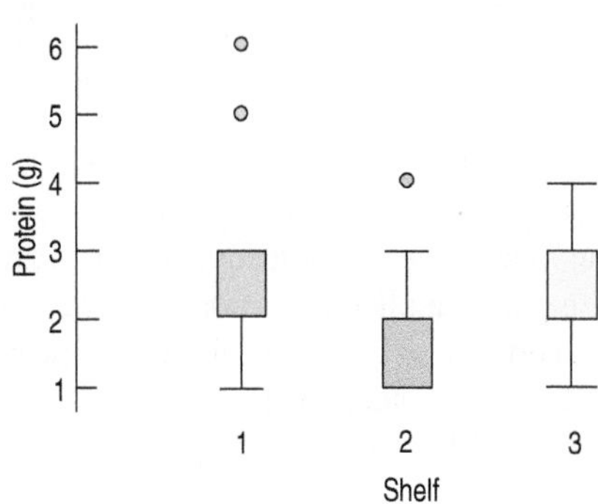

Analysis of Variance

Source	DF	Sum of Squares	Mean Square	F-ratio	P-value
Shelf	2	12.4258	6.2129	5.8445	0.0044
Error	74	78.6650	1.0630		
Total	76	91.0909			

Means and Std Deviations

Level	n	Mean	StdDev
1	20	2.65000	1.46089
2	21	1.90476	0.99523
3	36	2.86111	0.72320

a) What are the null and alternative hypotheses?

b) What does the ANOVA table say about the null hypothesis? (Be sure to report this in terms of protein content and shelves.)

c) Can we conclude that cereals on shelf 2 have a lower mean protein content than cereals on shelf 3? Can we conclude that cereals on shelf 2 have a lower mean protein content than cereals on shelf 1? What *can* we conclude?

d) To check for significant differences between the shelf means, we can use a Bonferroni test, whose results are shown below. For each pair of shelves, the difference is shown along with its standard error and significance level. What does it say about the questions in part c)?

Dependent Variable: PROTEIN

	(I) SHELF	(J) SHELF	Mean Difference (I-J)	Std. Error	P-value	95% Confidence Interval	
						Lower Bound	Upper Bound
Bonferroni	1	2	0.75	0.322	0.070	−0.04	1.53
		3	−0.21	0.288	1.000	−0.92	0.49
	2	1	−0.75	0.322	0.070	−1.53	0.04
		3	−0.96(*)	0.283	0.004	−1.65	−0.26
	3	1	0.21	0.288	1.000	−0.49	0.92
		2	0.96(*)	0.283	0.004	0.26	1.65

*The mean difference is significant at the 0.05 level.

22. Downloading To see how much of a difference time of day made on the speed at which he could download files, a second-year university student performed an experiment. He placed a file on a remote server and then proceeded to download it at three different time periods of the day. He downloaded the file 48 times in all, 16 times at each time of day, and recorded the time in seconds that the download took.

Early (7 a.m.) Time (sec)	Evening (5 p.m.) Time (sec)	Late night (12 a.m.) Time (sec)
68	299	216
138	367	175
75	331	274
186	257	171
68	260	187
217	269	213
93	252	221
90	200	139
71	296	226
154	204	128
166	190	236
130	240	128
72	350	217
81	256	196
76	282	201
129	320	161

a) State the null and alternative hypotheses, being careful to talk about download time and time of day as well as parameters.

b) Perform an ANOVA on these data. What can you conclude?

c) Check the assumptions and conditions for an ANOVA. Do you have any concerns about the experimental design or the analysis?

d) (Optional) Perform a multiple comparisons test to determine which times of day differ in terms of mean download time.

23. Analgesics A pharmaceutical company tested three formulations of a pain relief medicine for migraine headache sufferers. For the experiment, 27 volunteers were selected and 9 were randomly assigned to one of three drug formulations. The subjects were instructed to take the drug during their next migraine headache episode and to report their pain on a scale of 1 = no pain to 10 = extreme pain 30 minutes after taking the drug.

Drug	Pain	Drug	Pain	Drug	Pain
A	4	B	6	C	6
A	5	B	8	C	7
A	4	B	4	C	6
A	3	B	5	C	6
A	2	B	4	C	7
A	4	B	6	C	5
A	3	B	5	C	6
A	4	B	8	C	5
A	4	B	6	C	5

a) State the null and alternative hypotheses, being careful to talk about drug and drug pain as well as parameters.
b) Perform an ANOVA on these data. What can you conclude?
c) Check the assumptions and conditions for an ANOVA. Do you have any concerns about the experimental design or the analysis?
d) (Optional) Perform a multiple comparisons test to determine which drugs differ in terms of mean pain level reported.

24. Popsicle sticks In a bridge-building competition, popsicle sticks are often used to construct joints by gluing two or more sticks together. A high school class performed an experiment to test several different factors, including: four types of *Glue* (hot, white, carpenter's, Gorilla); two types of *Clamping* (clamped, not clamped); three types of *Overlap* distance (S, M, L); two types of *Colour* (C, B); and three numbers of *Sticks* in the joint (2, 3, 4).

Twelve batches of roughly 30 sticks were tested at various combinations of the factors. One research question of interest was whether coloured sticks broke at a different stress than regular beige sticks. Some analysis is shown below.

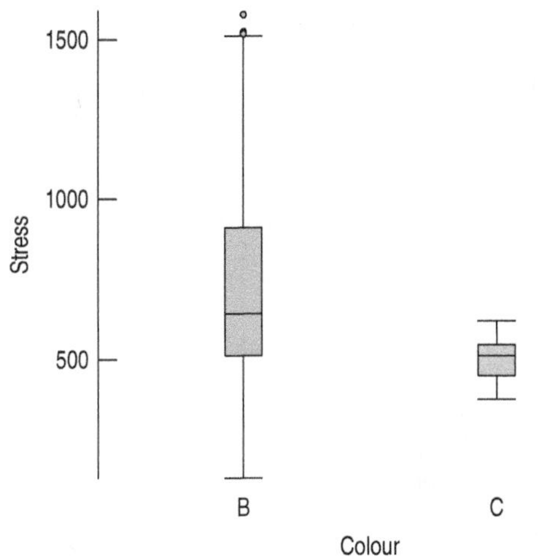

Analysis of Variance

Source	DF	Sum of Squares	Mean Square	F-ratio	P-value
Colour	1	1303181	1303181	18.01	< 0.0001
Error	350	25320104	72343		
Total	351	26623285			

Level	n	Mean	StdDev
B	323	722.3	279.72
C	29	501.0	66.79

a) How many factors are present in this experiment? How many possible treatments are there if the design were to be completely balanced?
b) Does it appear from the boxplots that coloured sticks break at a lower stress than beige sticks? Does the ANOVA output support this conclusion?
c) Do you have any concerns about this analysis? What are they?
d) Do you see any problems with the study design? What could you have done differently when designing the study to best answer the research question?

25. Popsicle sticks, redux Consider again the popsicle stick data from the previous question. Suppose we want to see which glue type is strongest (the original purpose of the experiment, as we had no idea). Some boxplots are shown below, as well as the ANOVA output.

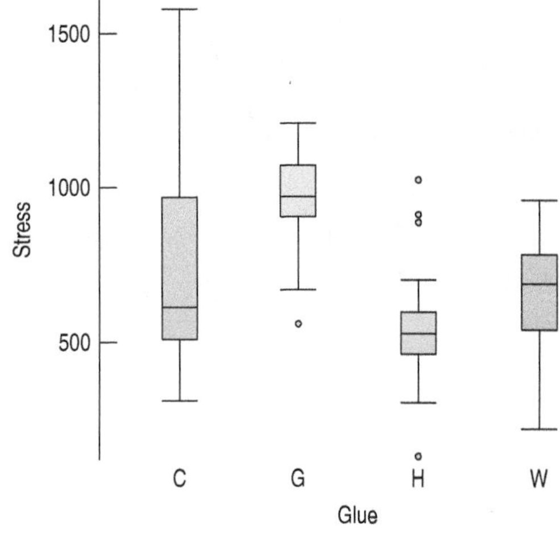

Analysis of Variance

Source	DF	Sum of Squares	Mean Square	F-ratio	P-value
glue	3	3686845	1228948	18.65	< 0.0001
Error	348	22936439	65909		
Total	351	26623284			

Level	n	Mean	StdDev
C	209	737.2	301.53
G	25	957.0	146.53
H	59	535.7	146.85
W	59	648.1	197.35

The global test is very significant, so there are some differences among the glues. It looks like Gorilla may be the best, and Hot the worst. Carpenter's has the second highest mean, but is it significantly worse than Gorilla? Why not do a follow-up test to see if this difference is statistically significant. Perform a two-sample t-test. What do you conclude? Do you have any concerns about this method?

26. Popsicle sticks, another taste The students decide to compare the breaking force for 28 joints made with each glue type. One joint could not be tested and was removed. An incomplete Minitab ANOVA table is shown below.

An incomplete ANOVA table for the Popsicle stick data

Source	DF	SS	MS	F	P
glue	3	724.32	241.44		
Error					
Total	110	1237.41			

a) What are the null and alternative hypotheses?
b) How many error degrees of freedom does this experiment have?
c) Fill in the sum of squares and the mean square for the error row.
d) What do you conclude, in the context of this study? What else would you like to see?

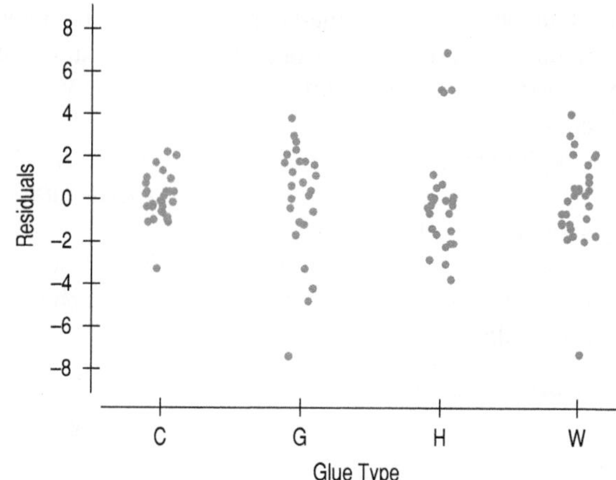

27. Popsicle sticks, dessert A residual plot and normal probability plot for the popsicle stick experiment is shown below. Comment on the following aspects of the graphs:
a) Equal variance assumption
b) Outliers
c) Normality of residuals
d) Do you feel justified performing inference in this case?

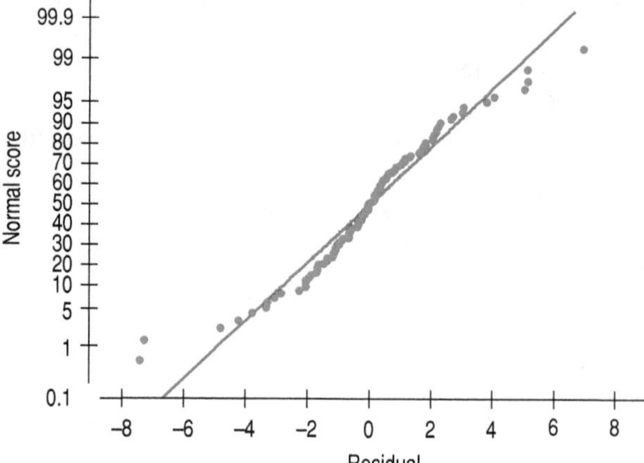

Just Checking ANSWERS

1. The null hypothesis is that the mean flight distance for all four designs is the same.
2. Yes, it looks as if the variation between the means is greater than the variation within each boxplot.
3. Yes, the F-test rejects the null hypothesis with a P-value < 0.0001.
4. No. The alternative hypothesis is that *at least* one mean is different from the other three. Rejecting the null hypothesis does not imply that all four means are different.

chapter

26 Multifactor Analysis of Variance

Where are we going?

When production problems forced a computer manufacturer to halt operations, teams of engineers, scientists, production workers, and managers were put together to solve the problem. They found themselves overwhelmed with many potential causes. But experiments that test one factor at a time would have been far too inefficient. And they might have missed important combinations of factors. In this chapter, we'll see how to look at two factors at a time, and we'll discover why two-factor designs are so much more interesting than one-factor experiments.

Who	Attempts at hitting a bull's-eye on a dartboard
What	Distance from the centre of the target
Units	Inches
How	Two-factor experiment, randomized order of runs

Without randomization, your inferences haven't got a chance.

—*George Cobb*

How accurately can you throw a dart? It probably depends on which hand you use and on how far from the target you're standing. A student designed an experiment to test the effects of both factors. He used three levels of the factor *Distance*: near, middle, and far—and the two natural levels left and right for *Hand.* (He was right-handed.) In random order, he threw six darts at a dartboard under each of the six treatment conditions, and measured the distance each dart landed from the bull's-eye in inches.[1]

It makes sense that dart throws would be more accurate from nearer distances, especially when using his regular throwing hand. Side-by-side boxplots of the distance from the bull's-eye by each factor seem to support our intuition:

Figure 26.1

Boxplots of accuracy (distance in inches from the bull's-eye) plotted for each experimental factor, *Distance* and *Hand.*

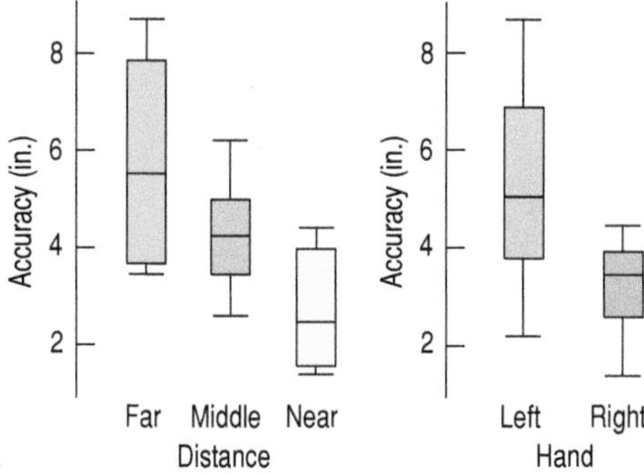

It appears that both the *Distance* and the *Hand* he used affected the accuracy of his throw, and by about the same amount. How can we measure the effect of each variable? And how can we test whether the observed differences are larger than we'd expect just from sampling variation?

Two Factors at Once?!

Unlike the experiments we've seen so far, this experiment has not one but two factors. You might expect this to confuse things, but the surprising fact is that it actually improves the experiment and its analysis. It also lets us see something we'd never be able to see if we ran two different one-factor experiments.

True, with two factors, we have two hypothesis tests. Each of those hypotheses asks whether the mean of the response variable is the same for each of the treatment levels. In comparing means, we know that any excess variation just gets in the way. Back in Chapter 11, we considered ways to remove or avoid extra variation in designing experiments.

[1]He called the response variable "accuracy," although it measures inches away from the target. That means that bigger values correspond to less accuracy, not more.

A two-factor experiment does just that. If both factors do in fact have an effect on the response (as we'd expect *Hand* choice and *Distance* to do for dart accuracy), then removing the effects of one factor should make it easier to see and assess the effects of the other. We'll have removed variability due to an identifiable cause.

For Example HOW FAR CAN A PAPER AIRPLANE FLY?

A group of students set up an experiment to investigate factors that determine how far a paper airplane will fly. Among the factors are whether the plane is constructed of copier paper or heavier construction paper, whether it is a "dart" or a glider, and whether the ends of the wings are folded up into "flaps." We'll combine the paper and shape factors into a single design factor with four levels: *dart/copier, dart/construction, glider/copier,* and *glider/construction.* The second factor, *flaps,* has two levels: *flaps* and *none.*

The students constructed four planes of each of the four designs. For each design, two planes were given flaps and two were not. They flew each plane four times and carefully measured the distance of the flight.

QUESTION: What are the details of the experiment design?

ANSWER: We have two factors: *design* and *flaps.* There are thus 4 × 2 = 8 treatment groups. Each treatment group has two planes and each plane was flown four times for a total of eight observations. The response variable is the distance flown.

26.1 An ANOVA Model

The model for one-way ANOVA represented each observation as the sum of simple components. It broke each observation into the sum of three effects: the grand mean, the treatment effect, and an error:

$$y_{ij} = \mu + \tau_j + \varepsilon_{ij}$$

Now we have two factors, *Hand* and *Distance.* So each treatment is a combination of a hand assignment and a distance. For example, throwing with the left hand from a near distance would be one treatment. Our model should reflect the effects of both factors. We've already used τ for the treatment effects for the first factor, so for the second factor, we'll denote the effects by γ. Now, for each observation, we write:

$$y_{ijk} = \mu + \tau_j + \gamma_k + \varepsilon_{ijk}$$

Here we see the i-th observation at level j of factor A and level k of factor B. Why the i subscript? We're told that the student recorded six trials at each treatment, so the subscript i denotes the different observations at each combination of the two factors. The error has three subscripts because we associate a possibly different error with each observation. (If there were only one observation at each treatment, we wouldn't need the extra subscript.)

When you write out a model, you'll usually find it clearer to name the factors rather than using Greek letters. So the model will look like this:

$$y_{ijk} = \mu + \textit{Hand effect}_j + \textit{Distance effect}_k + \textit{Error}_{ijk}$$

To estimate the effect due to each level of a factor, we take the *difference* between the mean response at that level of the factor and the grand mean, $\bar{\bar{y}}$. So, to find the effect of throwing with the left hand, we'd find the mean accuracy of *all* runs using the left hand and subtract the grand average from that. This effect is also referred to as a *main effect* since we are averaging over all levels of the other factor.

In general, we'd like to know whether the mean dart accuracy changes with the levels of either factor. The model helps us to think about this question. Our null hypothesis on each factor is that the effects of that factor are all zero. So we can write two null hypotheses we'd like to test as

$$H_0: \tau_1 = \tau_2 \quad \text{and} \quad H_0: \gamma_1 = \gamma_2 = \gamma_3$$

where the τ's (taus) denote the (main) effects due to the two levels of *Hand* and the γ's (gammas) denote the (main) effects due to the three levels of *Distance*.

The alternative hypotheses are that the factor's effects are not all equal. The details of the analysis are pretty much the same as for a one-factor ANOVA. We want to compare the differences in mean across the levels of the factor with the underlying variability within the treatments. But now with two factors, that underlying variability has the effect of both factors removed from it. So the error term holds the variability that's left after removing the effects of both factors. Here's the ANOVA table for the dart experiment:

Analysis of Variance Table

Source	DF	Sum of Squares	Mean Square	*F*-Ratio	P-value
Distance	2	51.041	25.521	28.561	<0.0001
Hand	1	36.693	36.693	44.415	<0.0001
Error	32	28.595	0.894		
Total	35	116.329			

This analysis splits up the total variation into three sources: the variation due to changes in *Distance*, the variation due to changes in *Hand*, and the error variation. The square root of the error mean square, 0.946 inches, estimates the standard deviation we'd expect if we made repeated observations at any treatment condition. Both F-ratios are large. Looking back at the boxplots, we shouldn't be surprised. It was clear that there were differences in accuracy among *near*, *middle*, and *far* even without accounting for the fact that each level contained both hands. The F-ratio for *Distance* is 28.56. The probability of an F-ratio that big occurring by chance is very small (0.0001). So we can reject the hypothesis that the mean effects of *near*, *middle*, and *far* distances are equal.

The means of *left* and *right* don't look the same either. The F-ratio of 44.42 with a P-value of 0.0001 leads us to reject the null hypothesis about hand effects as well.

Of course, there are assumptions and conditions to check. The good news is that most are the same in the **two-way ANOVA** as for one-way ANOVA.

26.2 Assumptions and Conditions

Even before checking any assumptions, you should have already plotted the data. Boxplots give a first look that draws our attention to the aspects tested by the ANOVA.

Plot the Data . . .

You could start by examining side-by-side boxplots of the data across levels of *each factor* as we did above, looking for outliers within any of the groups and correcting them if they are errors, or consider omitting them if we can't correct them.[2] (There don't seem to be any.) We'd like to get an idea of whether the groups have similar spreads (as we'll want) and whether the centres seem to be alike (as the null hypotheses claim) or different.

The problem of just looking at the boxplots is that the responses at each level of one factor contain *all levels of the other factor*. For example, the responses for the *left Hand* were measured at all three *Distances*. As a result, the boxplots show more variability than they need to. Our model will deal with both factors so it will take care of this extra variability, but the simple side-by-side boxplots can't do that.

[2]Unlike regression coefficients, which can be changed drastically by outliers, ANOVA tests tend to fail on the "safe side." That is, outliers generally inflate the error mean square, which reduces the F-statistics and increases the P-values. That makes it harder to reject the null hypotheses, so we are more likely to make Type II errors. That's generally thought to be safer than increasing the Type I errors. For example, we are not likely to be mistakenly led to approve a drug that isn't really effective. But it also means that we might fail to approve a drug that really is effective. Generally, it is best to remove outliers and report them separately.

On the other hand, if you do see a difference here, that's a difference *in spite of this extra variation*, so it's quite likely that the F-test will reject the null hypothesis for that factor. Just be careful not to prematurely rule out the possibility of differences that aren't apparent.

A better alternative might be to make boxplots for each factor level *after removing the effects of the other factor.* How would we do that? We could compute a one-way ANOVA on one factor and find the residuals. Then make boxplots of those *residuals* for each level of the other factor. We might call such a display a **partial boxplot.**[3] By removing the variability due to the other factor, you should see differences in the responses among the levels of the factor you are graphing more readily. For example, here are the boxplots of dart toss accuracy plotted for the two levels of *Hand*, both in the original data and after removing the effects of *Distance*:

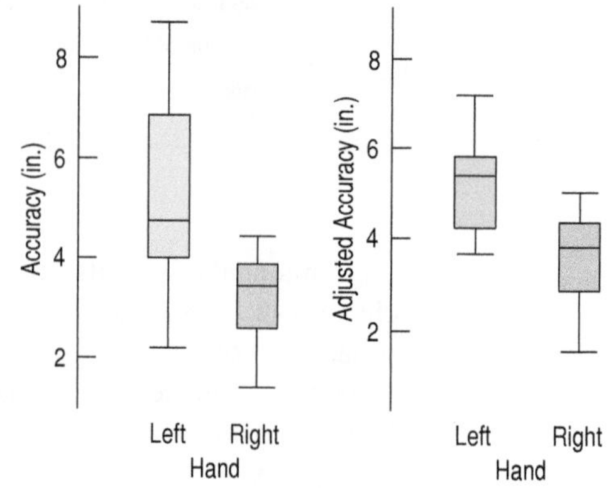

Figure 26.2

Ordinary boxplots of accuracy by *Hand*, on the left, show the effect of changing *Hand* less clearly than the corresponding partial boxplots on the right, which show the effects of *Hand* after the effect of *Distance* has been removed. The effect of changing the factor *Hand* is much easier to see without the unwanted variation caused by changing the other factor, *Distance*.

Removing the variation due to *Distance* makes the differences between the two *Hands* in the partial boxplots much clearer. We'll want to do this for each factor, removing the effects of the other factor by computing the one-way ANOVA and finding the residuals. Here are the usual boxplots for *Distance* and the partial boxplots for *Distance*, removing the variability due to *Hand:*

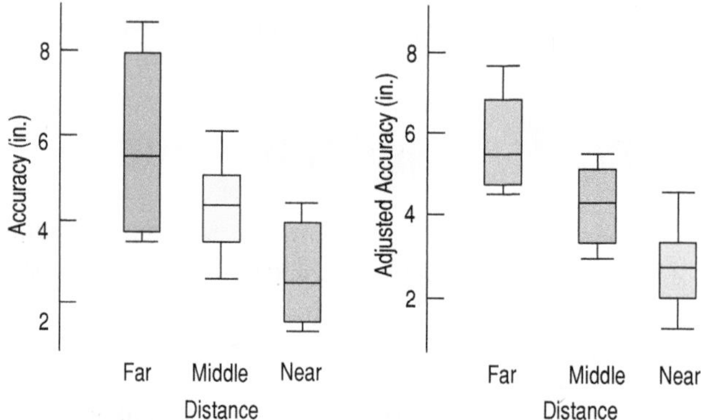

Figure 26.3

Partial boxplots after removing the effect of *Hand* (on the right) show the differences in accuracy among the three *Distances* more clearly than the boxplots of accuracy by *Distance* (on the left).

We'll use the partial boxplots for diagnostic purposes here, just as we used regular boxplots in a one-way analysis. If the partial boxplots were all skewed in the same direction, or if their spreads changed systematically from level to level across a factor, we would have considered re-expressing the response variable to make them more symmetric. Re-expression would be likely to make the analysis more powerful and more correct.

Additivity

Each observation is assigned a treatment that combines two factor levels, one from each factor. Our model assumes that we can just add the effects of these two factor levels together:

$$y_{ijk} = \mu + Hand\ effect_j + Distance\ effect_k + Error_{ijk}$$

[3]This is not a standard term. You are not likely to find it in a statistics package. But it is a natural term because we've removed *part* of the variability.

That's a nice simple model, but it may not be a good one for our data. Just like linearity, **additivity** is an assumption. Like all assumptions, it helps us by making the model simpler, but it may be wrong for our data. Of course, we can't know for sure, but we can check the **Additive Enough Condition**. You won't be surprised that we check it with displays.

For the effects of *Hand* and *Distance* to be additive, changing hands must make the *same* difference in accuracy no matter what distance you throw from (and *vice versa*). Is this reasonable for this experiment? Alternatively, we could ask whether moving away from the target makes a greater difference to accuracy when throwing with the weaker hand. That's a *conditional* question, so we look at how accuracy changes by *Distance* for each *Hand* separately.

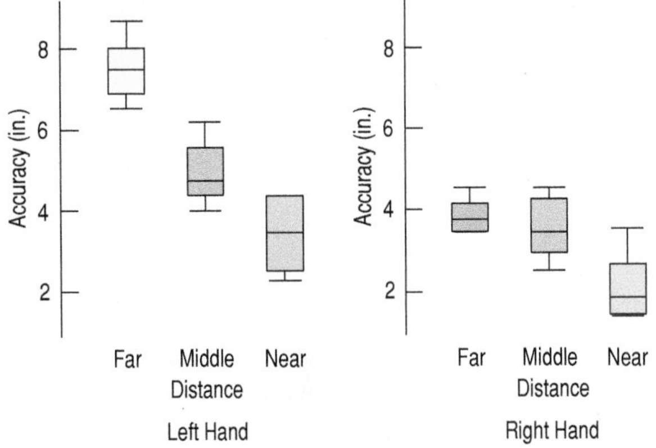

Figure 26.4
Boxplots of accuracy by *Distance* conditional on each *Hand*. Changing the *Distance* seems to have a greater effect on left *Hand* accuracies.

The changes in accuracy due to *Distance* do look larger for the left *Hand*, especially at the far distance.

When the effects of one factor *change* for different levels of another factor, we say that there's an **interaction**. To show the interaction, we often plot the treatment means in a single display, called an **interaction plot** (also known as the means plot). This plot shows the average of the observations at *each* level of one factor broken up by the levels of the other factor.

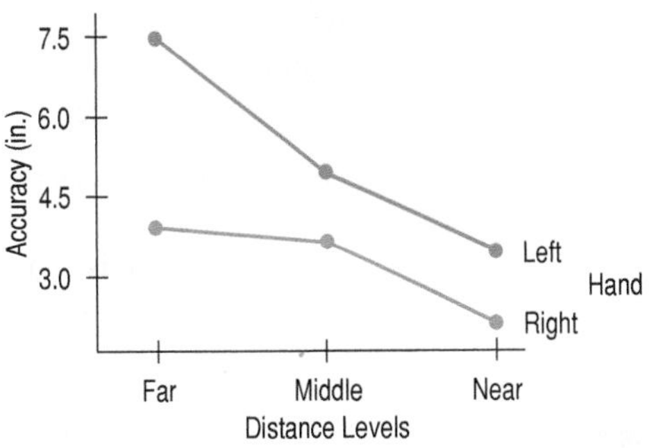

Figure 26.5
An interaction plot showing the mean accuracy at the six combinations of *Distance* and *Hand*. Connecting lines for right and left hands shows the greater effect of the far *Distance* on left-*Hand* accuracy than on right-*Hand* accuracy.

Here it's easier to see that throwing from the far *Distance* has more of an effect for the left *Hand*. If the effect of *Distance* were constant, the lines in this plot would be parallel. If the lines aren't parallel, that's evidence that the effects are not additive. But how parallel is enough? Are these lines sufficiently far from parallel for us to doubt whether our data satisfy the **Additive Enough Condition**? A graphical check like this is usually good enough to proceed. Later in the chapter, we'll see a way to include a term for nonadditivity in our model and test for it.

Independence Assumptions

The **Independence Assumption** for the two-factor model is the same as in one-way ANOVA. That is, the observations within each treatment group must be independent of each other. As usual, no test can verify this assumption. Ideally, the data should be generated by a randomized comparative experiment or be drawn independently and at random from your target population. Without randomization of some sort, our P-values are meaningless. We have no sampling distribution models and no valid inference.

Check the **Randomization Condition**. Were the data collected with suitable randomization? For surveys, are the data for each group a representative random sample of that group? For experiments, was the experiment properly randomized?

We were told that the order of treatment conditions in the dart experiment was randomized. This is especially crucial for an experiment like this because of the possibility of either learning or tiring. Your accuracy might increase after a few throws as you warm up, but after an hour or two, your concentration—and your aim—might start to wander.

If we are satisfied with the Additivity and Independence Assumptions, then we can look at the ANOVA table. For the dart data, we might have some doubts about the additivity, but we'll go ahead for now.

Source	DF	Sum of Squares	Mean Square	*F*-ratio	P-value
Distance	2	51.041	25.521	28.561	≤ 0.0001
Hand	1	36.693	36.693	44.415	≤ 0.0001
Error	32	28.595	0.894		
Total	35	116.329			

These may look like small P-values, but inference requires additional assumptions. Before we believe the *F*-tests and use the P-values to make decisions, we have other assumptions to think about.

Equal Variance Assumption

Like the one-way ANOVA, the two-factor ANOVA requires that the variances of all treatment groups be equal. It's the residuals after fitting both effects that we'll pool for the mean square error, so it's their variance whose stability we care about most. We check this assumption by checking the **Similar Variance Condition.**

In one-way ANOVA, we looked at the side-by-side boxplots of the groups to see whether they had roughly the same spread. But here, we need to check for equal spread across all treatment groups. The easiest way to check this is to plot the residuals from the two-way ANOVA model.

- Look at the residuals plotted against the predicted values. If the variance is constant, the plot should be patternless. If the plot thickens (to one side or the other), it's a sign that the variance is changing systematically. Then it's usually a good idea to consider re-expressing the response variable.

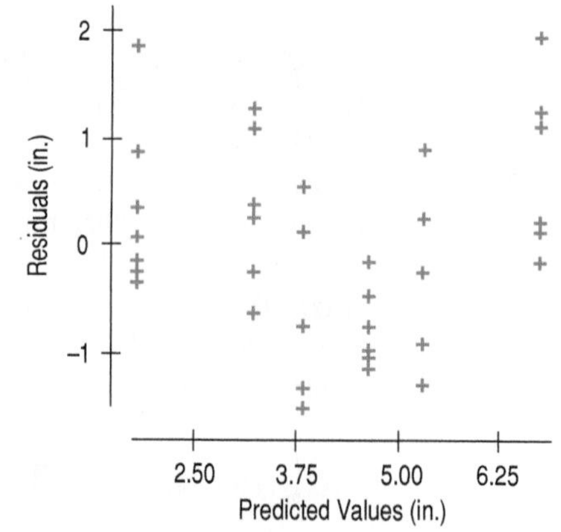

Figure 26.6

A scatterplot of residuals versus predicted values from the two-way ANOVA model for accuracy shows a U-shaped pattern. This suggests that a condition is violated.

This plot doesn't seem to show changing variance, but it's certainly not patternless. Residuals at both ends of the predicted values are higher than those in the middle. This could mean that something is missing from our model. Could that nagging suspicion we had about the additivity be resurfacing? Stay tuned.

- You can also plot the residuals grouped by each factor. Here's a boxplot of residuals by factor level for each factor in the dart experiment:

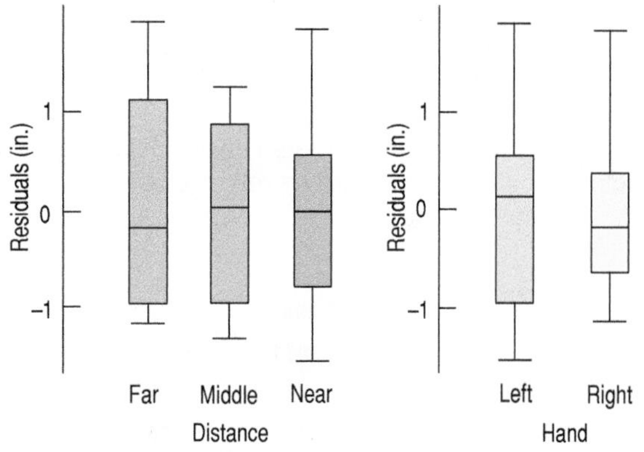

Figure 26.7

Boxplots of the residuals from the two-way ANOVA of the dart accuracy experiment show roughly equal variability when plotted for each factor.

Remember, we can't use the original boxplots to check changing spread. From those, it seemed that the spread of the near level was smaller than that of the far level. But much of that difference was due to changes in the way the factor *Hand* behaves at different *Distances* rather than a change in the underlying variance. By contrast, these boxplots of residuals look fine.

Normal Error Assumption

As with one-way ANOVA, the *F*-tests require that the underlying errors follow a Normal model. We'll check a corresponding **Nearly Normal Condition** with a Normal probability plot or histogram of the residuals.

Figure 26.8

A Normal probability plot of the residuals from the two-way ANOVA model for dart accuracy seems reasonably straight.

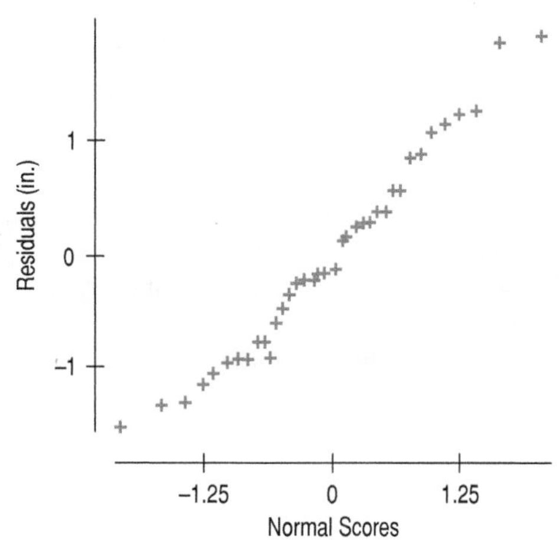

This condition appears to be satisfied.

What Have We Learned so Far?

The ANOVA table confirmed what we saw in the boxplots. Both *Distance* and *Hand* affect accuracy of throwing a dart, at least at the levels chosen in this experiment. The small P-values give us confidence that the differences we see are due to the factors and not to chance. We did worry that the Additive Enough Condition might not be satisfied, and there seems to be something amiss in the plot of residuals versus predicted values. Before we examine that more, let's look at a different example with two factors.

Step-by-Step Example TWO-FACTOR ANALYSIS OF VARIANCE

Another student, who prefers the great outdoors to damp pub basements, wonders whether leaving her tennis balls in the trunk of her car for several days after the can has been opened affects their performance, especially in the winter when it can get quite cold. She also wonders if the more expensive brand might retain its bounce better. To investigate, she performed a two-factor experiment on *Brand* and *Temperature*, using two *Brands* and three levels of *Temperature*. She bounced three balls under each of the six treatment conditions by first randomly selecting a brand and, for that ball, randomly selecting whether to leave it at room temperature or to put it in the refrigerator or the freezer. After eight hours, she dropped the balls from a height of 1 metre to a concrete floor, recording the *Height* of the bounce in centimetres.

This is a completely randomized replicated design in two factors. The factor *Brand* has two levels: premium and standard. The factor *Temperature* has three levels: room, fridge, and freezer. The null hypotheses are that *Brand* has no effect and that *Temperature* has no effect.

Who Tennis balls of two *Brands* allocated to three *Temperatures*.

What *Height* of the ball after it bounces from an initial height of 100 centimetres

Units Centimetres

How Two-factor experiment, randomized order of runs

Brand	Temperature	Bounce Height (cm)
Standard	Freezer	37
Standard	Fridge	59
Standard	Room	59
Premium	Freezer	45
Premium	Fridge	60
Premium	Room	63
Standard	Freezer	37
Standard	Fridge	58
Standard	Room	60
Premium	Freezer	39
Premium	Fridge	64
Premium	Room	62
Standard	Freezer	41
Standard	Fridge	60
Standard	Room	61
Premium	Freezer	37
Premium	Fridge	63
Premium	Room	61

THINK ➡ State what we want to know and the null hypotheses we wish to test. For two-factor ANOVA, the null hypotheses are that each factor has the same mean at each of its levels. The alternatives are that the means are not all equal.

I want to know whether the storage *Temperature* affects the mean bounce *Height* of tennis balls. I also want to know whether the two *Brands* have different mean bounce *Heights*. Writing τ_j for the effect of *Brand*$_j$, the null hypothesis is that the two *Brand* effects are the same:

$$H_0: \tau_P = \tau_S$$

Writing γ_k for the effect of *Temperature* level k, then the other null hypothesis is

$$H_0: \gamma_{room} = \gamma_{fridge} = \gamma_{freezer}$$

The alternative for the first hypothesis is that the effects of the two *Brands* on the bounce *Height* are different. The alternative for the second is that the effects of the *Temperature* levels are not all equal.

Plot Examine the side-by-side partial boxplots of the data.

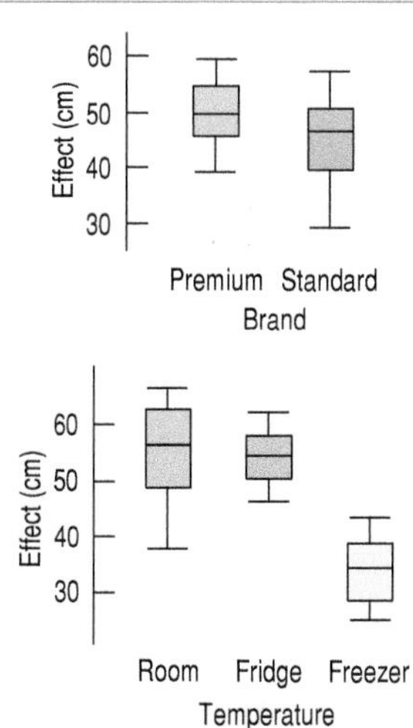

It's pretty clear that *Temperature* has an effect, but we can't be sure about the effect of *Brand* just from the plots.

Plan Think about the assumptions and check the appropriate conditions.

✓ **Independence Assumption:** Because the order was randomized, I can assume the observations are independent. However, I might want to check to see if the balls of each *Brand* came from the same box. It might be better to have selected balls of the same *Brand* from different boxes.

✓ **Randomization Condition:** The experimenter performed the trials in random order.

✓ **Additive Enough Condition:** Here is the plot of the mean bounce *Heights* at each treatment condition:

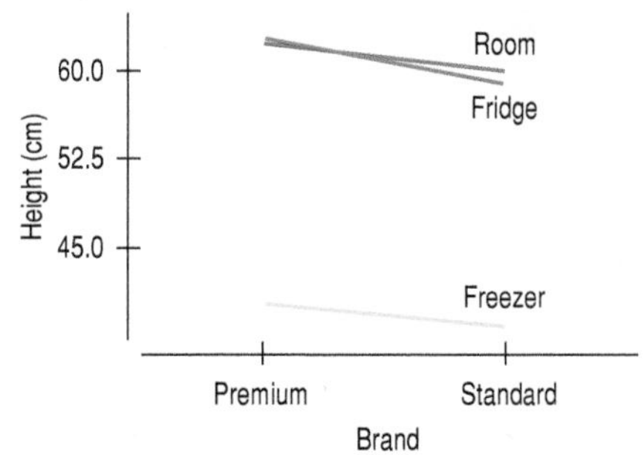

The lines look reasonably parallel. Now, I can fit the additive model in two factors:

$y_{ijk} = \mu + \textit{Brand effect}_j + \textit{Temp effect}_k + \textit{Error}_{ijk}.$

Under these conditions, it's appropriate to fit the two-way analysis of variance model.

Show the ANOVA table.

Analysis of Variance for Height

Source	DF	Sum of Squares	Mean Square	F-ratio	P-value
Temp	2	1849.33	924.665	209.55	<0.0001
Brand	1	26.8889	26.8889	6.0935	0.0271
Error	14	61.7778	4.41270		
Total	17	1938.00			

The error mean square is 4.41, which means that the common error standard deviation is estimated to be $\sqrt{4.41}$, a bit more than 2 centimetres. This seems like a reasonable accuracy for measuring bounce *Heights* of tennis balls dropped from a height of 1 metre.

Before testing the hypotheses, I need to check some more conditions:

✓ **Similar Variance Condition:** A plot of residuals versus predicted values shows some increased spread on the left, but not a huge difference and with only five points it's hard to say.

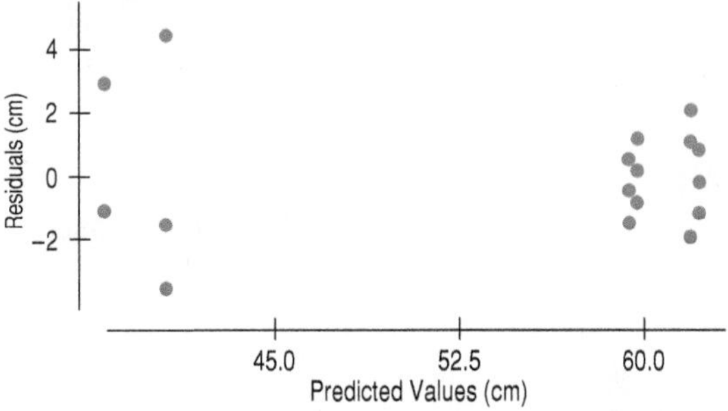

✓ **Nearly Normal Condition, Outlier Condition:** A Normal probability plot of the residuals looks okay, too:

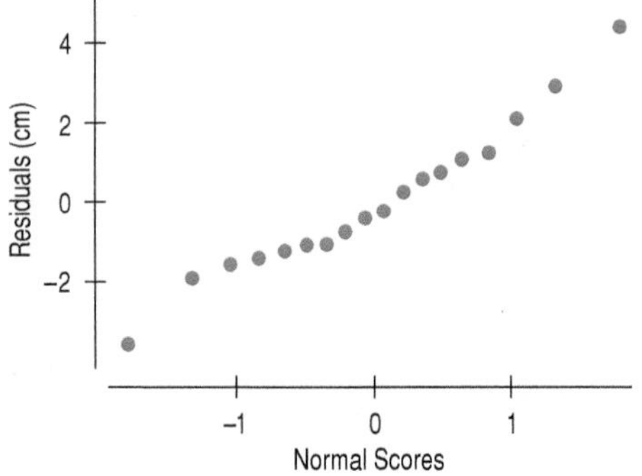

Under these conditions, it is appropriate to perform and interpret the *F*-test.

SHOW ➡ Mechanics Discuss the tests.

The *F*-ratio for *Brand* is 6.09, with a P-value of 0.0271. The *F*-ratio for *Temperature* is 209.55, with a P-value less than 0.0001.

Display the main effect estimates for each level of the significant factors. Each represents the average difference in response between that level of the factor and the overall mean for the data. They always sum to zero for a factor, since they are deviations from the mean. Remember, significance does not guarantee that the differences are meaningful from a practical standpoint.

The overall mean bounce *Height* is 53.67 cm.

The *Brand* effects are

Premium	1.22
Standard	−1.22

The *Temperature* effects are

Room	7.33
Fridge	7.00
Freezer	−14.33

TELL ➡ Interpretation Tell what the *F*-tests mean.

P-values this small would be very unlikely if the factors had no effect. I conclude that the means of the two *Brand* levels are not equal. I also conclude that the means of the three *Temperature* levels are not all equal.

Overall, the premium *Brand* bounces about 2.44 cm (1.22 − (−1.22)) higher than the standard *Brand*, averaging across the *Temperature* ranges. Generally, there seems to be little difference between the average bounce *Height* for room and fridge *Temperatures*, but tennis balls from the freezer bounce almost 22 cm less, on average.

Conclusions: It appears that the premium *Brand* outperforms the standard *Brand* overall. It looks from the plots like our student needn't be too worried about playing tennis on a cold day, but if the *Temperature* drops below freezing, she should try to warm the tennis balls first.[4]

For Example LEARNING FROM BOXPLOTS

RECAP: Two factors were tested in an experiment to investigate the effects of paper airplane design on flight distance.

Below are boxplots of distance by each of the factors:

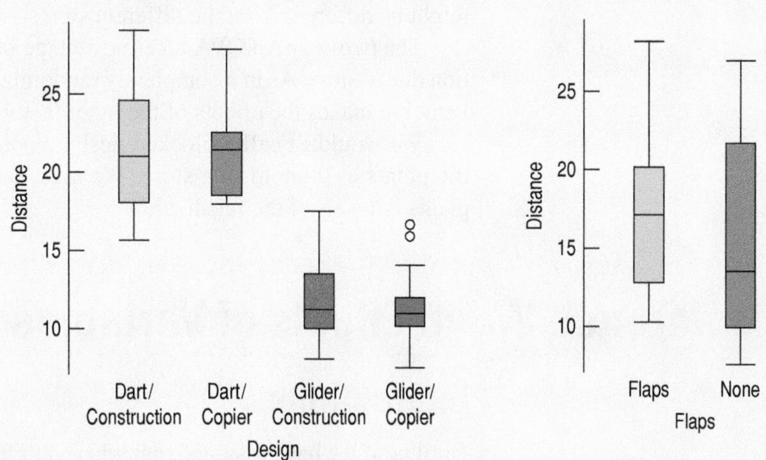

QUESTION: What can we learn from the boxplots?

ANSWER: It appears that the darts travel farther than the gliders. There does not appear to be an effect for flaps. The spreads of all the groups are comparable. There are two observations plotted as outliers for the glider, copier design, but they don't seem to be very extreme and are probably not a problem.[5]

[4]It's quite obvious from the plots and the tables of effects that there's a difference between freezer and the other temperatures, but little difference between room and fridge. A Bonferroni test would confirm this.

[5]But is independence a problem? Note that two planes were used for each treatment, with four flights per plane. The planes are not identical so there may be some differences on average between them. This violates the independence assumption. It is possible to enter the 'plane' effect into our model, but this is a bit more complicated than we can deal with here. If this effect was small, we're good.

For Example REJECTING THE NULL HYPOTHESIS

RECAP: An experiment on paper airplanes considered two factors: design and flaps. Here is the two-factor ANOVA table:

Source	DF	Sum of Squares	Mean Square	F-ratio	P-value
Design	3	1 549.50	516.499	56.610	0.0001
Flap	1	27.0140	27.0140	2.9608	0.0905
Error	59	538.303	9.12378		
Total	63	2 114.81			

QUESTION: State and test the null hypotheses. What can you conclude?

ANSWER: The first null hypothesis is that the mean distance flown will be the same for each design, after allowing for any effects of flaps. That hypothesis is rejected. As we saw in the boxplots, the darts fly farther.

The second null hypothesis is that the mean distance will be the same whether or not flaps are present, after allowing for the differences due to design. The P-value for that factor appears to be too large to allow us to reject the null hypothesis.

Blocks

Back in Chapter 11, we thought about a *randomized block design*.[6] We imagined obtaining tomato plants from two stores and randomly assigning plants from each store to one of three fertilizer treatments. We can view the store as a factor in this design. We now have a two-factor model, with store and fertilizer as factors. The ANOVA looks just the same, but now we're not likely to be interested in testing the store effect. We *assumed* that the plants might be different from the different stores—that's why we used a blocked design.

The two-way ANOVA takes advantage of the blocking to remove the unwanted variation due to store. As in a completely randomized two-factor design, removing the unwanted variation makes the effects of the other factor easier to see.

We wouldn't call a blocked design *completely randomized* because we can't assign the plants at random to a store. We must randomize *within* the blocks when we assign plants to levels of the fertilizer factor.

26.3 Inference When Effects of Variables Are Related

Interactions

Until now, we have assumed that whatever effects the two factors have on the response can be modelled by simply adding the separate (main) effects together. What if that's not good enough? In the dart-throwing experiment, we worried that the experimenter's accuracy when throwing from farther away may deteriorate *even more* when he uses his nondominant hand. Looking at the interaction plot added evidence to that suspicion.

It doesn't look like the effect of *Distance* is constant for the two *Hands*. In particular, moving from middle to far has a much larger effect when using the left-hand (see Figure 26.9 on the next page). It's not unusual for the effect of a factor's level to depend on the level of the other factor. Unfortunately, our model doesn't take this into account. So, we need a new model.

[6] You may have noticed in the tennis ball example that balls couldn't be randomly assigned to a brand. Brand is actually a blocking factor, and the experiment is a randomized block design. But it doesn't change the analysis at all.

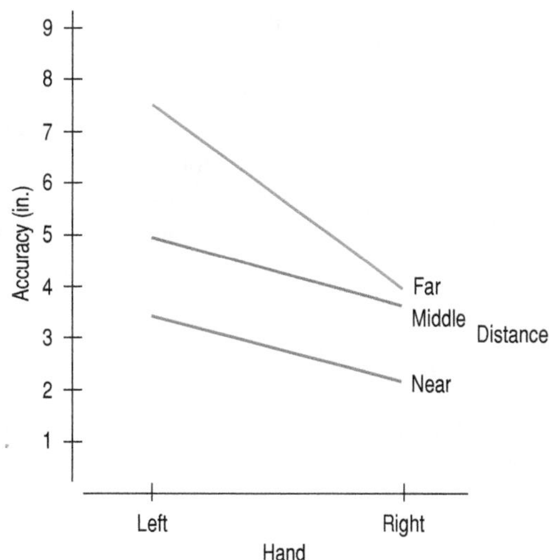

Figure 26.9

Interaction plot of accuracy by *Hand* and *Distance*. The effect of distance appears to be greater for the left hand. We can also say that the effect of changing from right to left hand is bigger at the far distance.

When the effect of one factor *changes* depending on the levels of the other factor, the factors are said to *interact*. We can model this interaction by adding a term to our model and testing whether the adjustment is worthwhile with an *F*-test.

We include a new term to our model, adding an interaction effect at every *combination* of the levels of the other two factors. The model can now be written:

$$y_{ijk} = \mu + \tau_j + \gamma_k + (\tau\gamma)_{jk} + \varepsilon_{ijk}$$

The new term, $(\tau\gamma)_{jk}$, represents the *interaction* effect at level *j* of factor 1 and level *k* of factor 2. This is the **Two-way ANOVA model with interaction**.

We add a corresponding line to the ANOVA table. Of course, we'll have to spend some degrees of freedom on this new term. We'll have to take them from the error term where we parked all the leftover degrees of freedom after fitting the two factors. We'll need the *product* of each factor's degrees of freedom. For the darts, this is only two degrees of freedom (one from *Hand* and two from *Distance*), but for factors with more levels, this could get expensive.

Here's the new ANOVA table for the dart experiment:

ANOVA Table for Accuracy

Source	DF	Sum of Squares	Mean Square	F-ratio	P-value
Distance	2	51.044	25.522	41.977	<0.0001
Hand	1	39.690	39.690	65.28	<0.0001
Distance × Hand	2	10.355	5.178	8.516	0.0012
Error	30	18.240	0.608		
Total	35	119.329			

What has changed? Almost everything. (Of course, the total sum of squares remains the same because it was calculated from the observations minus the overall mean, and the overall mean is still the same.) The interaction term is significant. Because we've now removed a significant source of variability, our error mean square is smaller in spite of the loss of two degrees of freedom.

The *F*-test for the interaction leads us to reject the null hypothesis of no interaction effect and indicates that the effect of *Distance* really does change depending on which *Hand* is used. (Or, equivalently, the effect of *Hand* changes depending on the *Distance*.)

What about the main effects? When a significant interaction term is present, it's difficult to *Tell* about the main effect of each factor. How much does switching from right hand to left hand matter? It depends! It depends on the distance. It no longer makes sense to talk about the main effects of each factor separately. The main effects are just averages of the effects over all the levels of the other factor. As we saw with Simpson's paradox (in Chapter 2), it can be misleading to average over very different situations.

When a significant interaction is present, the best way to display the results is with an interaction plot.

Now that we have the interaction term in our model, we recalculate the residuals. They're the values that are left over after accounting for the overall mean, the effect of each factor, and the interaction effect. Here's a plot of residuals versus predicted values for the two-factor model with interaction:

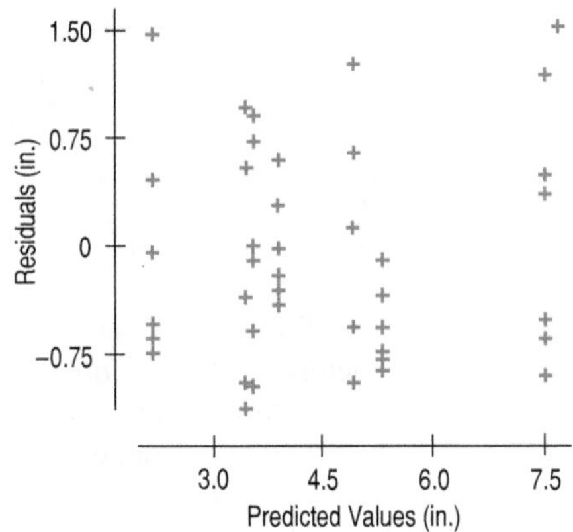

Figure 26.10
Residuals for the two-way ANOVA of dart accuracy with an interaction term included. Now, unlike in Figure 26.6, there is no U-shaped pattern.

How has the interaction term helped? These residuals have less structure. Fitting the interaction term succeeded in removing structure from the error. This new model seems to satisfy the assumptions more successfully, and so our inferences are likely to be closer to the truth.

If the interaction is significant, the interpretation of the analysis depends crucially on the interaction term, and the interaction plot shows most of what to *Tell* about the data. When there's a significant interaction term, the effect of each factor depends on the level of the other factor, so it may not make sense to talk about the effects of a factor by itself. However, we may still be able to talk about the main effects. For example, we can see that the right *Hand* is better *on average* than the left at any *Distance*, and the closer you are to the target the better the accuracy, no matter which hand you use. So both main effects are significant, and we can discuss them. A significant interaction effect tells us more. It says that *how much* better the performance is at near distances *depends* on which hand you use.

If the lines in the interaction plot cross, you need to be careful. Here's an interaction plot from an experiment on fuels and alternative engine designs:

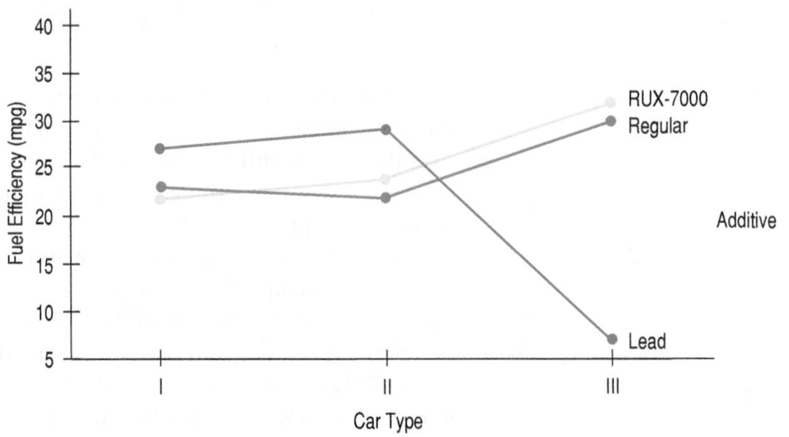

Figure 26.11
An interaction plot of gas mileage by car type and gas additive. If the lines cross, as they do here, be careful of interpreting the main effects.

It looks like the fuel with lead is great for two of the engine designs, but a disaster for the third. Suppose someone asked you to summarize the effect of adding lead. Nothing simpler than saying it *depends* on the engine type makes sense. *On average*, it lowers gas mileage, but is that relevant? The real message is that sometimes it improves mileage and sometimes it hurts. You won't want to *tell anything* about the main effect. Its F-test is irrelevant.

Whether the lines cross or not, including an interaction term in our model is always a good way to start. But sometimes we just can't. Suppose we had run the dart experiment with only one dart at each treatment. A design like that is called an **unreplicated two-factor design**. Imagine that the throw at the combination of left *Hand* and far *Distance* missed the bull's-eye by a mile. Would this be evidence of an interaction between *Distance* and *Hand*, or was it just a bad throw? How could we tell? The obvious thing to do would be to *repeat* the treatment. If he consistently did poorly at this combination, we'd want to model the interaction. If the other throws were no worse than we'd expect from the two main effects, then we might set the outlier aside. We can't distinguish an interaction effect from error unless we replicate the experiment. In fact, the design tells us that. Without replication, if we try to fit an interaction term, there are exactly zero degrees of freedom left for the error. That makes any inference or testing impossible. If we are willing to *assume* that the additive model is adequate, then we can reduce the number of runs required and use an unreplicated design. In that case, although the residuals are indistinguishable from the interaction effects, the residual plot may help reveal whether the assumption made sense.

When you do replicate, it's best to replicate *all* treatment conditions equally. (Unbalanced and other more complicated designs can be analyzed, but they are beyond the scope of this book.)

Just Checking

The student conducting the paper airplane design decides to do a follow-up experiment to investigate two other factors in more detail.

Using his favourite design, he uses one of three different-sized weights and places the weight either on the nose or on the tail of the plane. At each combination, he performs four test flights (randomizing the order of the flights). An interaction plot looks like this:

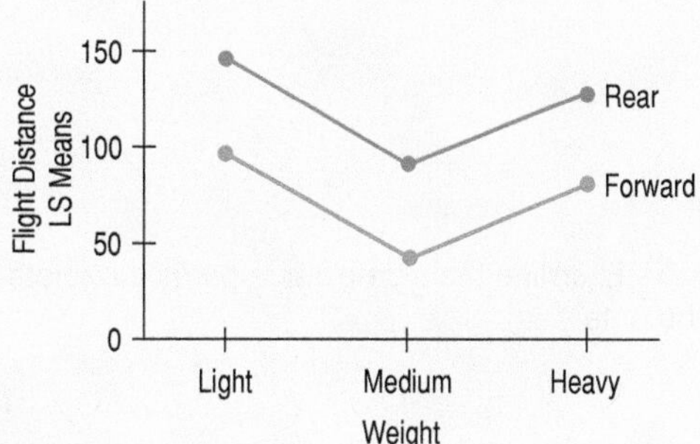

The ANOVA table shows:

Analysis of Variance

Source	DF	Sum of Squares	Mean Square	F-ratio	Prob > F
Position	1	13371.9314	13371.9314	53.386	≤0.0001
Weight	2	12625.5486	6312.7743	25.203	≤0.0001
Position × Weight	2	10.6208	5.3104	0.021	0.9790
Error	18	4508.501	250.47		
Total	23	30516.602			

1. What are the null hypotheses?
2. From the interaction plot, do the effects appear to be additive?
3. Does the *F*-test in the ANOVA table support your conclusion in part 2?
4. Does the position of the weight seem to matter?
5. Does the amount of weight appear to affect flight distance?
6. What would you recommend to the student to increase flight distance?

Step-by-Step Example TWO-FACTOR ANOVA WITH INTERACTION

In Chapter 25, we looked at how much television four groups of students watched on average. Let's look at their grade point averages (GPAs). Back in that chapter, we treated the four groups (male athletes, female athletes, male nonathletes, and female nonathletes) as four levels of the factor group. Now we recognize that there are really two factors: the factor sex with levels male and female and the factor varsity with levels yes and no. Let's analyze the GPA data with a two-factor ANOVA.

As with many Social Science studies, this isn't an experiment. (Some fans might object if we assigned students at random to varsity teams.[7]) Instead, it is an application of ANOVA to an observational study.

THINK ➡ State what we want to know and the null hypotheses we wish to test. For two-factor ANOVA with interaction, the null hypotheses are that all the levels of a factor have the same mean for each factor and that the interaction effect is 0. The alternatives are that at least one effect is not 0.

I want to know whether the mean *GPA* is the same for men and women, whether the mean *GPA* is the same for athletes and nonathletes, and whether the factors interact. Writing τ_j for the effect of sex level j, the null hypothesis is that the effects are the same:

$$H_0: \tau_M = \tau_F$$

Writing γ_k for the effect of varsity level k, the other null hypothesis is

$$H_0: \gamma_Y = \gamma_N$$

Finally, the null hypothesis for interaction is that the interaction terms are all 0.

The alternative for the first hypothesis is that the mean *GPA* is different for men and women. The alternative for the second is that the mean *GPA* is different for varsity athletes than for other students. The alternative for the last is that there is an interaction effect.

Plot Examine the side-by-side partial boxplots of the data.

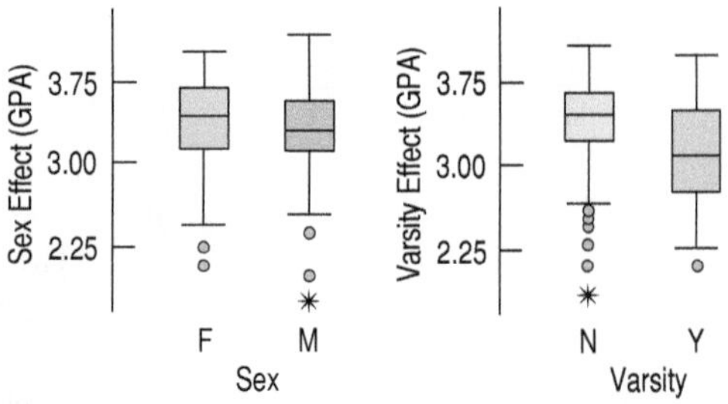

It appears that there is no difference in mean *GPA* between men and women (after accounting for varsity status), but there may be a difference between varsity and nonvarsity athletes.

[7] And randomly assigning the other factor would be problematic.

Plan Think about the assumptions and check the conditions.

The study is a random sample, so students' GPAs should be independent of each other.

✓ **Independence Assumption, Randomization Condition:** The study was based on a random sample of students.

✓ **Additive Enough Condition:** Here's the interaction plot of the means at each treatment condition:

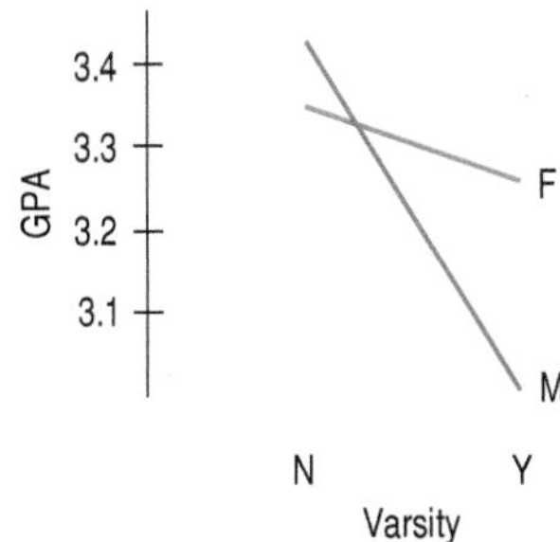

The lines cross and are not close to parallel, so I'll add an interaction term to the model and fit the two-way ANOVA with interaction:

$$y_{ijk} = \mu + Sex\ effect_j + Varsity\ effect_k + Interaction_{jk} + Error_{ijk}.$$

Show the ANOVA table.

Analysis of Variance for GPA

Source	DF	Sum of Squares	Mean Square	F-ratio	P-value
Sex	1	0.3040	0.3040	1.7681	0.1852
Varsity	1	2.4345	2.4345	14.1871	0.0002
Sex × Varsity	1	1.0678	1.0678	6.2226	0.0134
Error	196	33.6397	0.1716		
Total	199	37.4898			

✓ **Similar Variance Condition:** A plot of residuals versus predicted shows no thickening or other patterns:

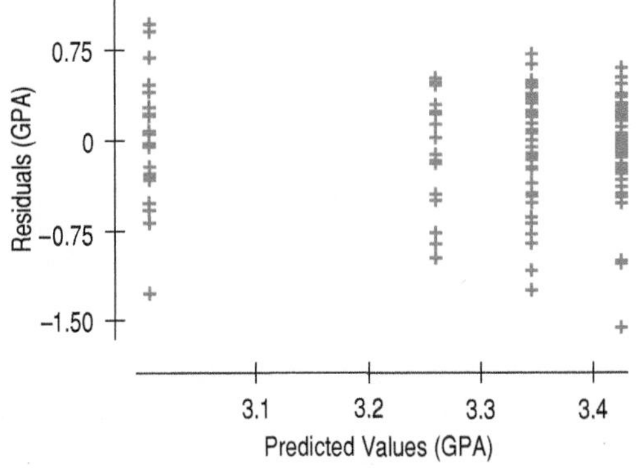

✓ **Nearly Normal Condition, Outlier Condition:** The Normal probability plot of the residuals (on the next page) is reasonable as well.

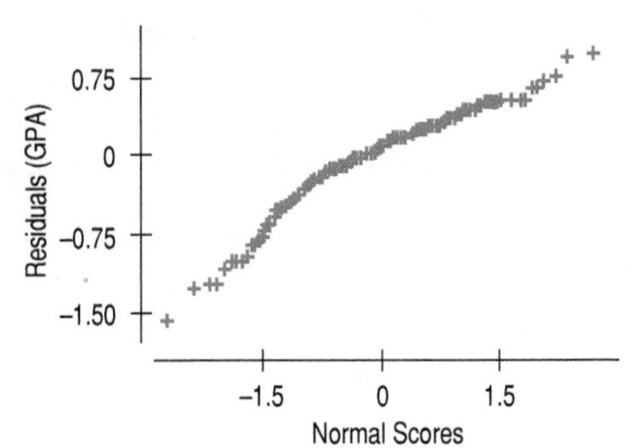

Under these conditions it is appropriate to interpret the *F*-tests and their P-values.

SHOW ➡ Mechanics

The error mean square is 0.172 points, which means that the common error standard deviation is estimated to be $\sqrt{0.172} = 0.415$ points (nearly half a letter grade).

The *F*-ratio for the interaction term is 6.223, with a P-value of 0.0134. Based on this, I reject the hypothesis of no interaction. Because there is interaction, I am cautious in testing the main effects, but will examine the means further to understand the factor effects.

Overall mean *GPA* = 3.32.

Level of Varsity	Effect
N	0.1281
Y	−0.1281

Show the table of means.

The main effect of *varsity* is significant, lowering *GPA* by about a quarter of a grade point averaged over men and women, but this effect is much greater for men than for women. Here are the four group mean *GPAs*:

Group	Mean GPA
Female, nonvarsity	3.35
Female, varsity	3.26
Male, nonvarsity	3.43
Male, varsity	3.00

TELL ➡ Interpretation

Tell what the *F*-test means.

The interaction term is significant, and it appears that most of the effect we see is due to a difference of 0.43 in mean *GPAs* between male athletes and nonathletes. It doesn't look like any of the other differences are large enough to be important. Probably none of them are statistically significant, either.

For Example INTERACTION BETWEEN VARIABLES

RECAP: The ANOVA for paper airplane distance on two factors, design and flaps, showed a significant effect for design, but not for flaps. But maybe there is an interaction. Here is the ANOVA with the interaction term, as well as an interaction plot.

Source	DF	Sum of Squares	Mean Square	F-ratio	P-value
Design	3	1549.50	516.499	81.065	0.0001
Flap	1	27.014	27.0140	4.2399	0.0441
Design × Flap	3	181.503	60.5009	9.4956	0.0001
Error	56	356.800	6.37144		
Total	63	2114.81			

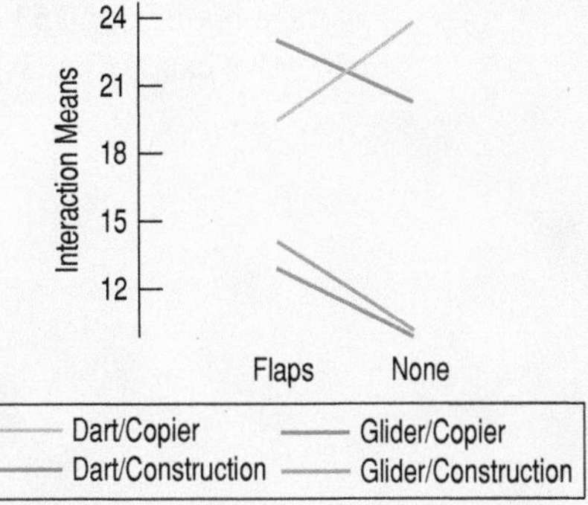

QUESTION: Interpret the interactions. What do they say about paper airplane design?

ANSWER: Including the interaction term reveals that the flaps factor is significant. But it matters in different ways for different designs. Flaps help improve the distance of both kinds of gliders, but they hinder the performance of darts made of lighter copier paper (although they help darts made of construction paper).

What Next?

This chapter introduced some of the issues that arise in advanced experimental design and analysis. We have scratched the surface of this large and important area by showing the analysis of a balanced two-way design with interaction.

Of course, there's no reason to stop with two factors. And if we have more than two, there are many ways in which they may interact. As you can imagine, real-life experiments and observational studies may pose many interesting challenges. Anyone who is serious about analyzing or designing multifactor studies should consult a professional statistician. You now know the vocabulary and principles, so you are well equipped to do that.

*How ANOVA Works—the Gory Details

You'll probably never carry out a two-way ANOVA by hand. But to understand the model it can help to see exactly how it decomposes each observation into components. Our tennis ball model (without an interaction term) says

$$y_{ijk} = \mu + Brand\ effect_j + Temp\ effect_k + Error_{ijk}.$$

To apply this model to data, we need to estimate each of these quantities:

$$y_{ijk} = \bar{\bar{y}} + \widehat{Brand\ effect_j} + \widehat{Temp\ effect_k} + Residual_{ijk}.$$

Here are the data again, displayed a little differently:

		Temperature		
		Room	Fridge	Freezer
Brand	Premium	63	60	45
		62	64	39
		61	63	37
	Standard	59	59	37
		60	58	37
		61	60	41

The first term in the model is μ, the overall mean. It's reasonable to estimate that with the grand mean, $\bar{\bar{y}}$, which is just the mean of all the observations. For the tennis balls, the average bounce was 53.67 cm. The grand mean has the same value for every observation, which we can lay out as a table like this:

		Temperature		
		Room	Fridge	Freezer
Brand	Premium	53.67	53.67	53.67
		53.67	53.67	53.67
		53.67	53.67	53.67
	Standard	53.67	53.67	53.67
		53.67	53.67	53.67
		53.67	53.67	53.67

Now, we turn to the *Brand* effects. To estimate the effect of, say, the *premium Brand*, we find the mean of all *premium* balls and subtract the grand mean. That's 55.89 cm − 53.67 cm = 1.22 cm. In other words, the *premium Brand* balls bounced an average of 1.22 cm higher than the overall average. The *standard Brand* balls bounced an average of 51.45 cm, so their effect is 51.45 cm − 53.67 cm = −1.22 cm. Of course, with only two levels, the fact that these two effects are equal and opposite is no coincidence. Effects *always sum to* 0.

We can put the *Brand* effects into a table like this:

		Temperature		
		Room	Fridge	Freezer
Brand	Premium	1.22	1.22	1.22
		1.22	1.22	1.22
		1.22	1.22	1.22
	Standard	−1.22	−1.22	−1.22
		−1.22	−1.22	−1.22
		−1.22	−1.22	−1.22

We can find a *Temperature* effect table in a similar way. We take the treatment average at each level of *Temperature* and subtract the grand mean. Here's the table:

		Temperature		
		Room	Fridge	Freezer
Brand	Premium	7.33	7.00	−14.33
		7.33	7.00	−14.33
		7.33	7.00	−14.33
	Standard	7.33	7.00	−14.33
		7.33	7.00	−14.33
		7.33	7.00	−14.33

We're missing only one table, the *Residuals*. Finding the residuals is easy, too. The additive model represents each observation as the sum of these three components. As always, the residuals are the differences between the observed values and those given by the model. Look at the first ball in the table. It was a *premium Brand* ball at *room Temperature*. From the effects, we'd predict it to bounce 53.67 (grand mean) +1.22 (from being premium) +7.33 (from being at room temperature) = 62.22 cm. Because it actually bounced 63 cm, its residual is 63 – 62.22 = 0.78 cm. If we compute all the residuals this way (or have the computer do it), we'd find the residual table to be

		Temperature		
		Room	Fridge	Freezer
Brand	Premium	0.78	−1.89	4.44
		−0.22	2.11	−1.56
		−1.22	1.11	−3.56
	Standard	−0.78	−0.44	−1.11
		0.22	−1.44	−1.11
		1.22	0.56	2.89

What we have done with all these tables is to decompose each original bounce height into the three estimated components plus a residual:

Bounce Height = *mean Bounce Height* + *Brand effect* + *Temperature effect* + *Residual.*

These tables actually display the workings of the ANOVA. Each term in the ANOVA model corresponds to one of these tables. Because the overall mean doesn't vary, we usually leave it out of the ANOVA table. Each Sum of Squares term in the ANOVA is simply the sum of the squares of all the values in its table.

The degrees of freedom are easy to find too. Look at the grand mean table. Because it has the same value for all observations, it has 1 degree of freedom. The *Brand* effect table seems to have 2 degrees of freedom, but we know that the two effects always sum to 0. So it really has only 1. In general, the degrees of freedom for a factor is one less than its number of levels. So *Temperature* has 2 degrees of freedom.

The original table has 18 observations and so 18 degrees of freedom. But what about the residuals? Remarkably enough, the degrees of freedom from the tables add. That means that we can find the degrees of freedom for the residuals by subtraction. They must have 18 – (1 + 1 + 2) = 14 degrees of freedom.

The mean squares are just the sum of squares for each effect divided by the degrees of freedom. The F-ratios are the ratios of each factor's Mean Square to the Error Mean Square. You can check the degrees of freedom and the mean squares in the step-by-step ANOVA table. The original table has 18 observations and so 18 degrees of freedom, but most programs choose not to display the overall mean and subtract out the degree of freedom associated with it to give what's sometimes called a "corrected" total or just "total." That's why our ANOVA table has only 17 degrees of freedom in "total."

WHAT CAN GO WRONG?

- **Beware of unreplicated designs unless you are sure there is no interaction.** Without replicating the experiment for each treatment combination, there is no way to distinguish the interaction terms from the residuals. If you are designing a two-factor experiment, you must be willing to assume that there is no interaction if you choose not to replicate. In such a case, you can fit an additive model only in the two factors. If there is an interaction, it will show up in the error term. You should examine a residual plot to help reveal a possible interaction effect. You must be prepared to defend the assumption of no interaction and the decision not to replicate.

- **Don't attempt to fit an interaction term to an unreplicated two-factor design.** If you have an unreplicated two-factor experiment or observational study, you'll find that if you try to fit an interaction term, you'll get a strange ANOVA table. The design exhausts the degrees of freedom for error, so fitting the interactions leaves no degrees of freedom for residuals. That wipes out the mean square errors, F-ratios, and P-values, which may appear in the computer output as dots, dashes, or some other indication of things gone wrong. Remove the interaction term from the model and try again.
- **Be sure to fit an interaction term when it exists.** When the design is replicated, it is always a good idea to fit an interaction term. If it turns out not to be statistically significant, you can then fit a simpler two-factor main effects model instead.
- **When the interaction effect is significant, don't interpret the main effects.** Main effects can be very misleading in the presence of interaction terms. Look at the interaction plot below.

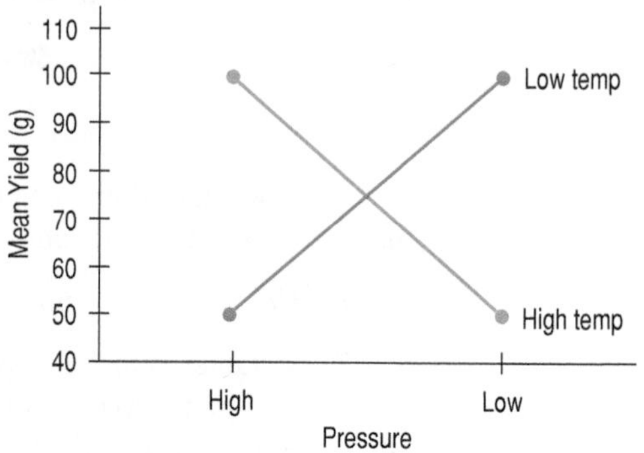

Figure 26.12
An interaction plot of yield by temperature and pressure. The main effects are misleading. There is no (main) effect of pressure because the average yield at the two pressures is the same. That doesn't mean that pressure has no effect on the yield. In the presence of an interaction effect, be careful when interpreting the main effects.

The experiment was run at two temperatures and two pressure levels. High amounts of material were produced at high pressure with high temperature and at low pressure with low temperature. What's the effect of temperature? Of pressure? Both main effects are 0, but it would be silly (and wrong) to say that neither temperature nor pressure was important. The real story is in the interaction.

- **Always check for outliers.** As in any analysis, outliers can distort your conclusions. An outlier can inflate the error mean square so much that it may be hard to discern any effect, whether it exists or not. Use the partial boxplots to search for outliers. Consider setting outliers aside and re-analyzing the results. An outlier can make an interaction term appear significant. For example, a single male varsity athlete with a very low GPA could account for the results we saw.
- **Check for skewness.** If the underlying data distributions are skewed, you should consider a transformation (re-expression) to make them more symmetric.
- **Beware of unbalanced designs and designs with empty cells.** We've been assuming that the data are *balanced* over the design—that we have equal numbers of observations in each cell. There are methods that will easily compensate for small amounts of imbalance, but empty cells and other more serious violations of balance require different methods and additional assumptions.

We first discussed designing experiments in Chapter 11. Now we know how to analyze a two-factor design. In Chapter 25, we saw that the one-way ANOVA generalized the pooled t-test to more than two levels of one factor. Now we've added a second factor.

Of course, we're still relying on boxplots, Normal probability plots, and scatterplots of residuals to help us check conditions and understand relationships. What we first said we'd look for in these displays back in Chapters 3, 4, 5, and 6 is still what we're concerned with here.

What Have We Learned?

Learning Objectives

In Chapter 25, we learned that the Analysis of Variance is a natural way to extend the *t*-test for testing the means of two groups to compare several groups. Often, those groups are different levels of a single factor. Now we've learned that we can extend the Analysis of Variance to designs with more than one factor.

- Understand the advantages of an experiment in two factors.
- Know how to set up an additive model in two factors.
- Be able to use a statistics package to compute a two-way ANOVA.

We've learned that partial boxplots are a good way to examine the effect of each factor on the response. We can carry out separate tests of hypotheses for each of the factors, testing whether the factor has any discernible effect on the response across levels of the other factor.

- Know how to make partial boxplots.
- Be able to interpret main effects in a two-way ANOVA.

We've also learned that sometimes factors can interact with each other. When we have at least two observations at each combination of factor levels, we can add an interaction term to our model to account for the possible interaction.

- Know how to examine the Additivity Assumption and when to consider an interaction term.
- Know how to make an interaction plot for replicated data.
- Be able to use an interaction plot to explain an interaction effect.
- Be able to distinguish when a discussion of main effects is appropriate in the presence of a significant interaction.

Of course, we still need to check the appropriate assumptions and conditions as we did for simple ANOVA.

All the methods we learned for displaying data in the early chapters are still important. And we've begun to see the deep connection between the design concepts we learned in Chapter 11, the exploratory methods we learned in the early chapters, and the inference methods of the latter part of the book.

Review of Terms

Partial boxplot Plots the response at each of the levels of one factor with the effects of the second factor removed. A simple way to generate partial boxplots of factor A is to first save the residuals from a one-way ANOVA of the response on factor B. Add the overall mean to the residuals and make boxplots of the resulting data against factor A (p. 784).

Interaction When the effects of the levels of one factor change depending on the level of the other factor, the two factors are said to interact. When interaction terms are present, it is misleading to talk about the main effect of one factor because how large it is *depends* on the level of the other factor (p. 785).

Interaction plot A plot that shows the means at each treatment combination, highlighting the factor effects and their behaviour at all the combinations (p. 785).

Two-way ANOVA model (main effects model) The model for a two-way (one response, two factors) ANOVA is

$$y_{ijk} = \mu + \tau_j + \gamma_k + \varepsilon_{ijk}$$

where τ_j represents the effect of (level j) of factor one and γ_k represents the effect of (level k) of factor two. The subscript i designates the ith replication of the treatment combination (p. 785).

Additivity, additive model A model in two factors is additive if it simply adds the effects of the two factors and doesn't include an interaction term (p. 785).

Assumptions for the two-way ANOVA model (and conditions to check)

- **Equal Variance Assumption:** Check the **Similar Variance Condition** by looking at side-by-side partial boxplots of the groups and a plot of residuals against predicted values. A common problem is *increasing* spread with increasing predicted values—the *plot thickens* (p. 786).
- **Independence Assumptions:** The observations within each treatment group must be independent of each other. Think about the nature of the data and the **Randomization Condition**. Check a residual plot (p. 786).
- **Normal Error Assumption:** Check the **Nearly Normal Condition** by *looking at a* normal probability plot of the residuals (p. 787).
- **Additivity Assumption:** We assume that we can *add* the effects of the two factors together. Check the **Additive Enough Condition** that the effects of one factor are about the same across all levels of the other factor. An interaction plot should show roughly parallel lines. *Of course, this assumption (and associated condition) is not required when an interaction term is included in the model* (p. 792).

Two-way ANOVA model with interaction For a replicated two-way design, an interaction term can also be fit. The resulting model is

$$y_{ijk} = \mu + \tau_j + \gamma_k + (\tau\gamma)_{jk} + \varepsilon_{ijk}$$

where τ_j represents the effect of (level j) of factor one and γ_k represents the effect of (level k) of factor two. Now $\tau\gamma_{jk}$ represents the effect of the interaction at levels j of factor one and k of factor two. The subscript i designates the ith replication of the treatment combination (p. 793).

On the Computer TWO-WAY ANOVA

Some statistics packages distinguish between models with one factor and those with two or more factors. You must be alert to these differences when analyzing a two-factor ANOVA. It's not unusual to find ANOVA models in several different places in the same package.

Usually, you must specify the interaction term yourself. That's because these features of statistics packages typically are designed for models with three or more factors where the number of interactions can explode unless the data analyst carefully selects the interactions to be included in the model.

EXCEL

- The Excel Data Analysis Add-in offers a two-way ANOVA "with and without replication."

COMMENTS

Excel requires that the data be in a special format and cannot deal with unbalanced data.

JMP

- From the **Analyze** menu, select **Fit Model**.
- Select variables: and **Add** them to the **Construct Model Effects** box.
- To specify an interaction, select both factors and press the **Cross** button.
- Click **Run Model**.
- JMP opens a **Fit Least Squares** window.
- Click on the red triangle beside each effect to see the means plots for that factor. For the interaction term, this is the interaction plot.
- Consult the JMP documentation for information about other features.

COMMENTS

JMP expects data in "stacked" format with one continuous response and two nominal factor variables.

MINITAB

- Choose **ANOVA** from the **Stat** menu.
- Choose **Two-way. . . .** from the **ANOVA** submenu.
- In the Two-Way ANOVA dialogue, assign a quantitative Y variable to the Response box and assign the categorical X factors to the Factor box.
- Click **Fit additive model** if you do not want interaction in your model.
- Check the **Store Residuals** box.
- Click the **Graphs** button.
- In the ANOVA-Graphs dialogue, select **Standardized residuals**, and check **Normal plot of residuals** and **Residuals versus** fits.

COMMENTS

Minitab expects "stacked" format data for two-way designs. You can also use the **Balanced Anova** choice from the **ANOVA** submenu.

R

- **aov(formula, data=dataset)** will perform a two-way ANOVA.
- **model.tables(aovObject, "mean")** will give the overall, group, and marginal means
- **interaction.plot(x1, x2, y)** will provide an interaction plot.

COMMENTS

If the dataset has a column Y with the response variable, and columns X_1 and X_2 of factors, then **formula** will be $Y \sim X_1 + X_2$ for an additive model, and $Y \sim X_1{*}X_2$ for a model with an interaction.

SPSS

- Choose **Analyze > General Linear Model > Univariate**.
- Assign the response variable to the **Dependent Variable** box.
- Assign the two factors to the **Fixed Factor(s)** box. This will fit the model with interactions by default.
- To omit interactions, click on **Model**. Select **Custom**. Highlight the factors. Select **Main Effects** under the **Build Terms** arrow and click the arrow.
- Click **Continue** and **OK** to compute the model.

STATCRUNCH

To perform a two-way ANOVA:

- Enter the factors in columns 1 and 2, and the response variable in the third.
- Click on **Stat**.
- Choose **ANOVA » Two Way**.
- Choose the corresponding columns from your table, as well as any options you'd like.
- Click on **Calculate**.

COMMENTS

In addition to the ANOVA output, STATCRUNCH also displays group means and the interaction plot.

TI-83/84 PLUS

COMMENTS

You need a special program to compute a two-way ANOVA on the TI-83.

Exercises

1. Popcorn revisited A student runs a two-factor experiment to test how microwave power and temperature affect popping. She chooses three levels of *Power* (low, medium, and high) and three *Times* (three minutes, four minutes, and five minutes), running one bag at each condition. She counts the number of uncooked kernels as the response variable.

a) What are the null and alternative hypotheses for the main effects?

b) How many degrees of freedom does each factor sum of squares have? How about the error sum of squares?

c) Should she consider fitting an interaction term to the model? Why or why not?

2. Gas mileage revisited A student runs an experiment to study the effect of *Tire Pressure* and *Acceleration* on gas mileage. He devises a system so that his Jeep Wagoneer uses gasoline from a one-litre container. He uses three levels of tire pressure (low, medium, and full) and two levels of acceleration, either holding the pedal steady or pumping it every few seconds. He randomizes the trials, performing four runs under each treatment condition, carefully recording the number of kilometres he can go in his Jeep Wagoneer on one litre of gas.

a) What are the null and alternative hypotheses for the main effects?

b) How many degrees of freedom does each treatment sum of squares have? How about the error sum of squares?

c) Should he consider fitting an interaction term to the model? Why might it be a good idea?

d) If he fits an interaction term, how many degrees of freedom would it have?

3. Popcorn again Refer to the experiment in Exercise 1. After collecting her data and analyzing the results, the student reports that the F-ratio for *Power* is 13.56 and the F-ratio for *Time* is 9.36.

a) What are the P-values?

b) What would you conclude?

c) What else about the data would you like to see in order to check the assumptions and conditions?

4. Gas mileage again Refer to the experiment in Exercise 2. After analyzing his data, the student reports that the F-ratio for *Tire Pressure* is 4.29 with a P-value of 0.030, the F-ratio for *Acceleration* is 2.35 with a P-value of 0.143, and the F-ratio for the interaction effect is 1.54 with a P-value of 0.241.

a) What would you conclude?

b) What else about the data would you like to see in order to check the assumptions and conditions?

c) If your conclusion about the *Acceleration* factor in part a) is wrong, what type of error have you made?

5. Crash analysis The U.S. National Highway Transportation Safety Administration runs crash tests in which stock automobiles are crashed into a wall at 56.33 km/hr with dummies in both the passenger's and the driver's seats. The THOR Alpha crash dummy is capable of recording 134 channels of data on the impact of the crash at various sites on the dummy. In this test, 335 cars are crashed. The response variable is a measure of head injury. Researchers want to know if which seat the dummy is in affects head injury severity, as well as whether the type of car affects severity.

Here are partial boxplots for the two different *Seat* (driver, passenger) and the six different *Size* classifications (compact, light, medium, mini, pickup, van):

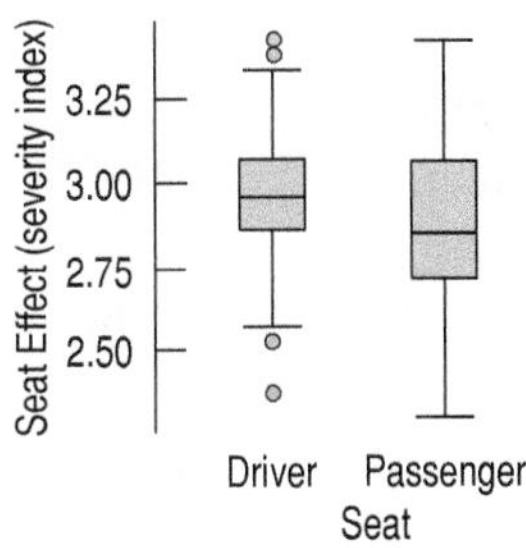

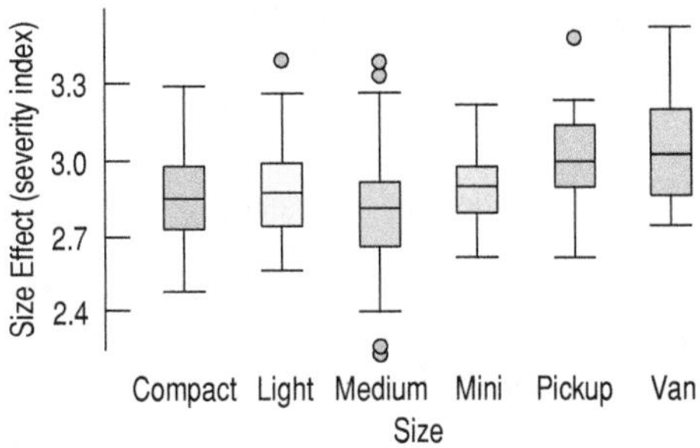

An interaction plot shows:

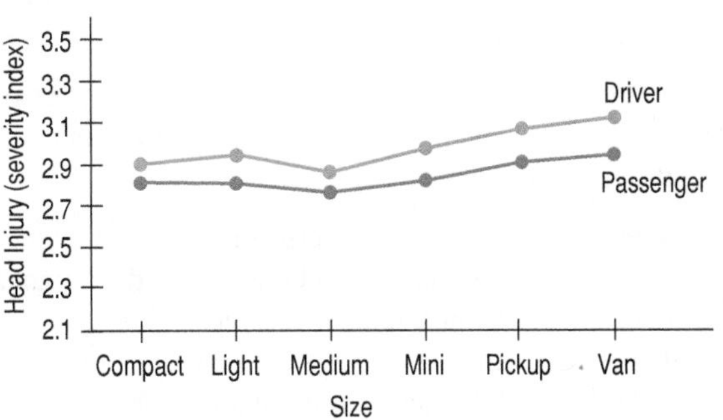

A scatterplot of residuals versus predicted values shows:

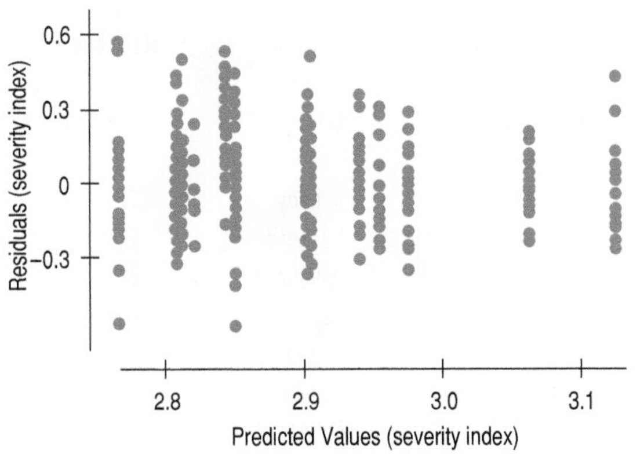

The ANOVA table follows:

Analysis of Variance for Head Severity: 293 cases

Source	DF	Sum of Squares	Mean Square	F-ratio	P-value
Seat	1	0.88713	0.88713	25.501	<0.0001
Size	5	1.49253	0.29851	8.581	<0.0001
Seat × Size	5	0.07224	0.01445	0.415	0.838
Error	282	9.8101	0.03479		
Total	293	12.3853			

a) State the hypotheses about the main effects (both numerically and in words).

b) Are the conditions for two-way ANOVA met?

c) If so, perform the hypothesis tests and state your conclusion. Be sure to state it in terms of head injury severity, seats, and vehicle types.

6. Sprouts An experiment on mung beans was performed to investigate the environmental effects of salinity and water temperature on sprouting. Forty beans were randomly allocated to each of 36 petri dishes that were subject to one of four levels of *Salinity* (0, 4, 8, and 12 ppm) and one of three *Temperatures* (32°, 34°, or 36°C). After 48 hours, the biomass of the sprouts was measured.

Here are partial boxplots of biomass on *Salinity* and *Temperature*:

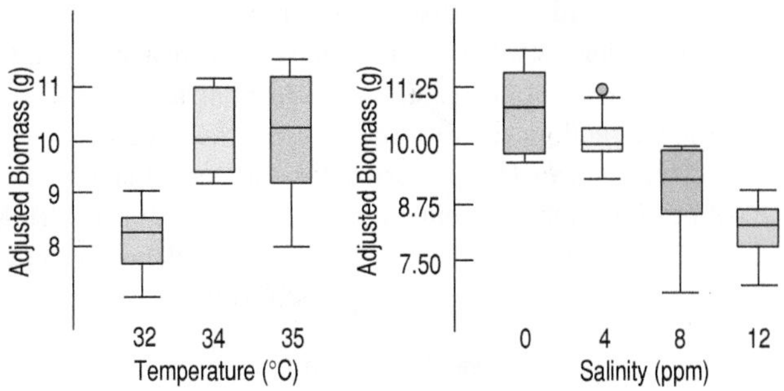

The interaction plot shows:

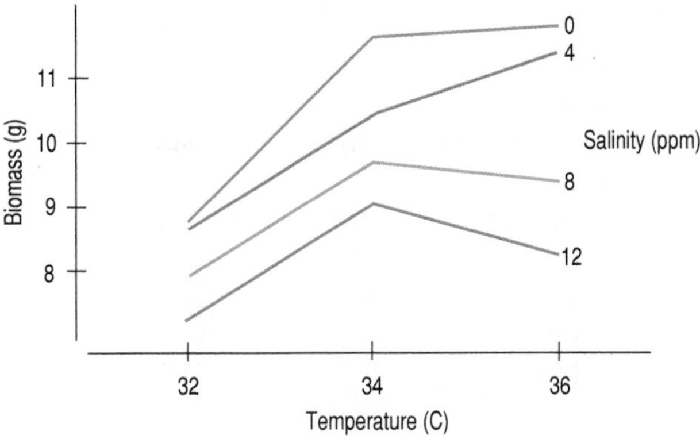

A two-way ANOVA model is fit, and the following ANOVA table results:

Analysis of Variance for Biomass (g)

Source	DF	Sum of Squares	Mean Square	F-ratio	P-value
Salinity	3	36.4701	12.1567	16.981	<0.0001
Temp	2	34.7168	17.3584	24.247	<0.0001
Salinity × Temp	6	5.2972	0.8829	1.233	0.3244
Error	24	17.1816	0.7159		
Total	35	93.6656			

The plot of residuals versus predicted values shows:

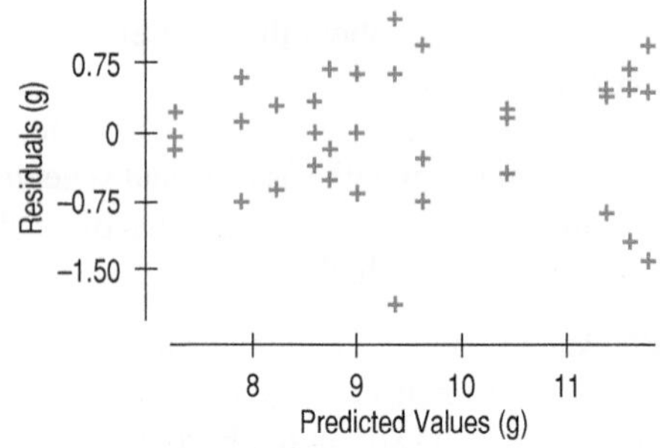

a) State the hypotheses about the factors (both numerically and in words).
b) Are the conditions for two-way ANOVA met?
c) Perform the hypothesis tests and state your conclusions. Be sure to state your conclusions in terms of biomass, salinity, and water temperature.

7. **Baldness and heart disease** A retrospective study examined the link between baldness and the incidence of heart disease. In the study, 1435 middle-aged men were selected at random and examined to see whether they showed signs of *Heart Disease* (or not) and what amount of *Baldness* they exhibited (none, little, some, or much). A student runs a two-factor ANOVA on these data (given below) and finds the following ANOVA table:

Source	DF	Sum of Squares	Mean Square	F-ratio	P-value
Baldness	3	62441.375	20813.792	17.956	0.020
Heart disease	1	1485.125	1485.125	1.281	0.340
Error	3	3477.375	1159.125		

a) Comment on her analysis. What problems, if any, do you find?
b) What sort of analysis might you do instead?

Baldness	Heart Disease	Number of Men
None	Yes	251
None	No	331
Little	Yes	165
Little	No	221
Some	Yes	195
Some	No	185
Much	Yes	52
Much	No	35

8. **Fish and prostate** In the Chapter 2 Step-By-Step, we looked at a Swedish study that asked 6272 men how much fish they ate and whether or not they had prostate cancer. Here are the data:

Eat Fish?	Prostate Cancer? No	Prostate Cancer? Yes
Never/seldom	110	14
Small part of diet	2420	201
Moderate part	2769	209
Large part	507	42

Armed with the methods of this chapter, a student performs a two-way ANOVA on the data. Here is her ANOVA table:

Source	DF	Sum of Squares	Mean Square	F-ratio	P-value
Fish	3	3110203.0	1036734.3	1.3599	0.4033
Prostate cancer	1	3564450.0	3564450.0	4.6756	0.1193
Error	3	2287051.0	762350.0		

a) Comment on her analysis. What problems, if any, do you find?
b) What sort of analysis might you do instead?

9. **Baldness and heart disease again** Refer back to Exercise 7. Perform your own analysis of the data to see if baldness and heart disease are related. Do your conclusions support the claim that baldness is a cause of heart disease? Explain.

10. **Prostate and fish** Refer back to Exercise 8. Perform your own analysis of the data to see if eating fish and contracting prostate cancer are related.

T 11. **Basketball shots** A student performed an experiment to see if her favourite sneakers and the time of day might affect her free-throw percentage. She tried shooting with and without her favourite sneakers and in the early morning and at night. For each treatment combination, she shot 50 baskets on four different occasions, recording the number of shots made each time. She randomized the treatment conditions by drawing numbers out of a hat. Here are her data:

Time of Day	Shoes	Shots Made
Morning	Others	25
Morning	Others	26
Night	Others	27
Night	Others	27
Morning	Favourite	32
Morning	Favourite	22
Night	Favourite	30
Night	Favourite	34
Morning	Others	35
Morning	Others	34

Time of Day	Shoes	Shots Made
Night	Others	33
Night	Others	30
Morning	Favourite	33
Morning	Favourite	37
Night	Favourite	36
Night	Favourite	38

a) What are the null and alternative hypotheses?
b) Write a short report on your findings, being sure to include diagnostic analysis as well as practical conclusions.

12. Washing For his final project, Jonathan examined the effects of two factors on how well stains are removed when washing clothes. On each of 16 new white handkerchiefs, he spread a teaspoon of dirty motor oil (obtained from a local garage). He chose four *Temperature* settings (each of which is a combination of wash and rinse: cold-cold, cold-warm, warm-hot, and hot-hot) and four *Cycle* lengths (short, med short, med long, and long). After its washing, each handkerchief was dried in a dryer for 20 minutes and hung up. He rounded up 10 family members to judge the handkerchiefs for cleanliness on a scale of 1 to 10 and used the average score as his response. Here are the data:

Temp	Cycle	Score
Cold-cold	Med long	3.7
Warm-hot	Med long	6.5
Cold-warm	Med long	4.9
Hot-hot	Med long	6.5
Cold-cold	Long	4.6
Warm-hot	Long	8.3
Cold-warm	Long	4.7
Hot-hot	Long	9.1
Cold-cold	Short	3.4
Warm-hot	Short	5.6
Cold-warm	Short	3.8
Hot-hot	Short	7.1
Cold-cold	Med short	3.1
Warm-hot	Med short	6.3
Cold-warm	Med short	5
Hot-hot	Med short	6.1

You may assume, as Jonathan did, that interactions between *Temperature* and *Cycle* are negligible. Write a report showing what you found about washing factors and stain removal.

13. Sprouts again The students running the sprouts experiment (Exercise 6) also kept track of the number of beans sprouted (out of 40) for each of the 36 dishes. The partial boxplots of *Sprouts* plotted against *Salinity* and *Temperature* are shown below.

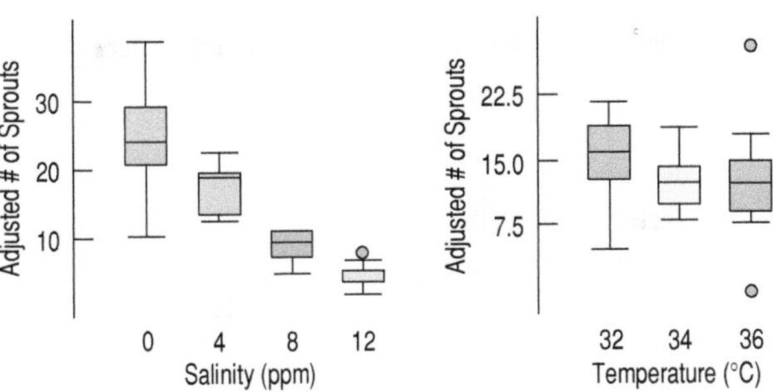

An ANOVA table shows:

Source	DF	Sum of Squares	Mean Square	F-ratio	P-value
Salinity	3	2014.750	671.583	23.657	<0.0001
Temp	2	57.556	28.778	1.014	0.3779
Salinity × Temp	6	96.000	16.000	0.564	0.7549
Error	24	681.333	28.389		
Total	35	2849.639			

An interaction plot shows:

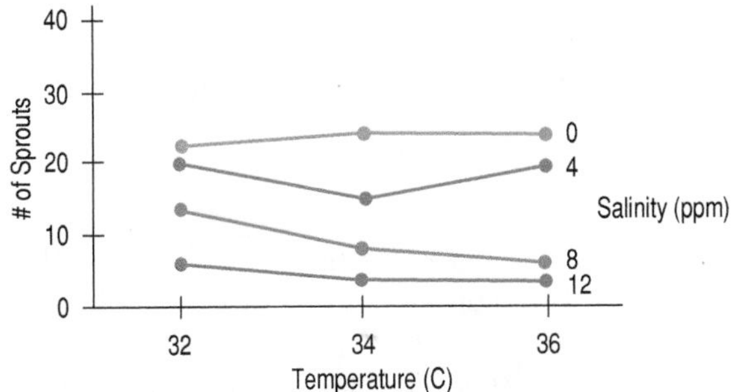

A plot of residuals vs. predicted values shows:

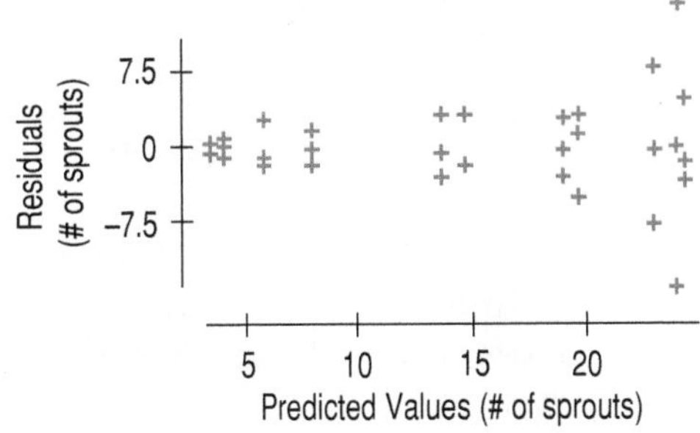

a) State the hypotheses about the factors (both numerically and in words).
b) Perform the hypothesis tests and state your conclusions. Be sure to check conditions.

14. Containers revisited Building on the cup experiment of the Chapter 3 Step-By-Step, a student selects one type of container and designs an experiment to see whether the type of *Liquid* stored and the outside *Environment* affect the ability of a cup to maintain temperature. He randomly chooses an experimental condition and runs each twice:

Liquid	Environment	Change in Temperature
Water	Room	13
Water	Room	14
Water	Outside	31
Water	Outside	31
Coffee	Room	11
Coffee	Room	11
Coffee	Outside	27
Coffee	Outside	29

After fitting a two-way ANOVA model, he obtains the following interaction plot, ANOVA table, and effects table:

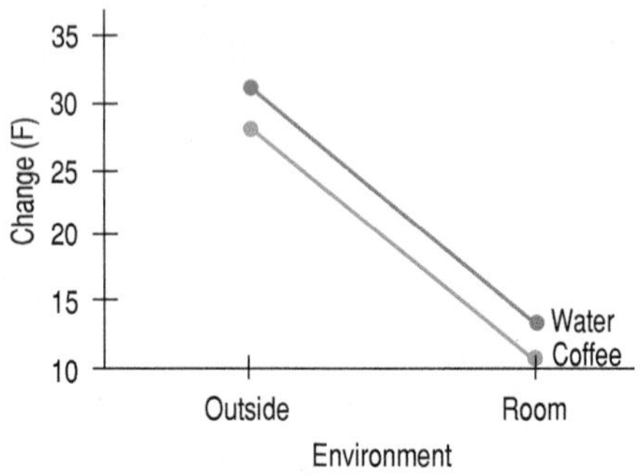

Source	DF	Sum of Squares	Mean Square	F-ratio	P-value
Liquid	1	15.125	15.125	24.2	0.0079
Envirn	1	595.125	595.125	952.2	<0.0001
Interaction	1	0.125	0.125	0.2	0.6779
Error	4	2.500	0.625		
Total	7	612.875			

Term	Estimate
Overall mean	20.875
Liquid II [Coffee]	−1.375
Liquid II [Water]	1.375
Environment II [Outside]	8.625
Environment II [Room]	−8.625

a) State the null and alternative hypotheses.
b) Test the hypotheses at $\alpha = 0.05$.
c) Perform a residual analysis.
d) Summarize your findings.

15. Gas additives An experiment to test a new gasoline *Additive*, Gasplus, was performed on three different car *Types*: a sports car, a minivan, and a hybrid. Each car was tested with both Gasplus and regular gas on 10 different occasions and their gas mileage was recorded. Here are the partial boxplots:

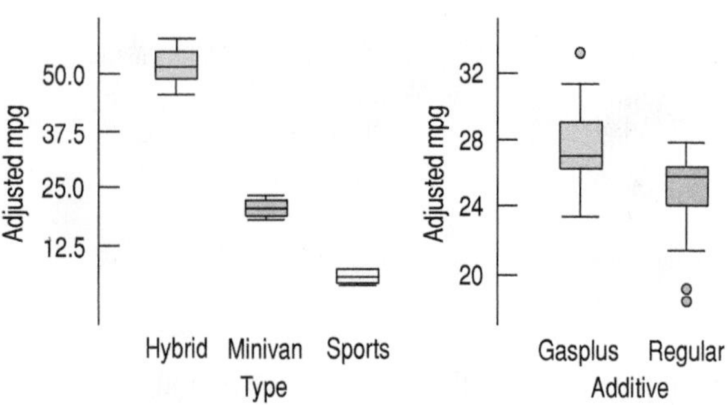

A two-way ANOVA with interaction model was run, and the following ANOVA table resulted:

Source	DF	Sum of Squares	Mean Square	F-ratio	P-value
Type	2	23175.4	11587.7	2712.2	<0.0001
Additive	1	92.1568	92.1568	21.57	<0.0001
Type × Additive	2	51.8976	25.9488	6.0736	0.0042
Error	54	230.711	4.27242		
Total	59	23550.2			

A plot of the residuals versus predicted values showed:

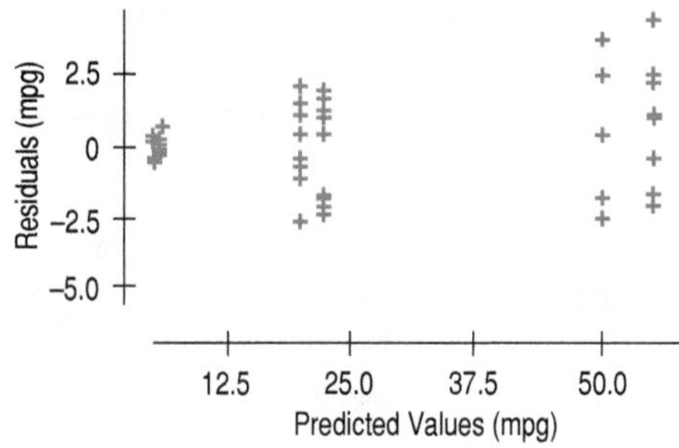

What conclusions about the *Additive* and car *Types* do you draw? Do you see any potential problems with the analysis?

16. Chromatography A gas chromatograph is an instrument that measures the amounts of various compounds contained in a sample by separating the constituents. Because different components are flushed through the system at different rates, chromatographers are able to both measure and distinguish the constituents of the sample. A counter is placed somewhere along the instrument that records how much material is passing at various times. By looking at the counts at various times, the chemist is able to reconstruct the amounts of various compounds present. The total count is proportional to the amount of the compound present.
An experiment was performed to see whether slowing down the flow rate would increase total counts. A mixture was produced with three different *Concentration* levels: low, medium, and high. The two *Flow Rates* used were slow and fast. Each mixture was run five times and the total count recorded each time. Partial boxplots for *Concentration* and *Flow Rates* show:

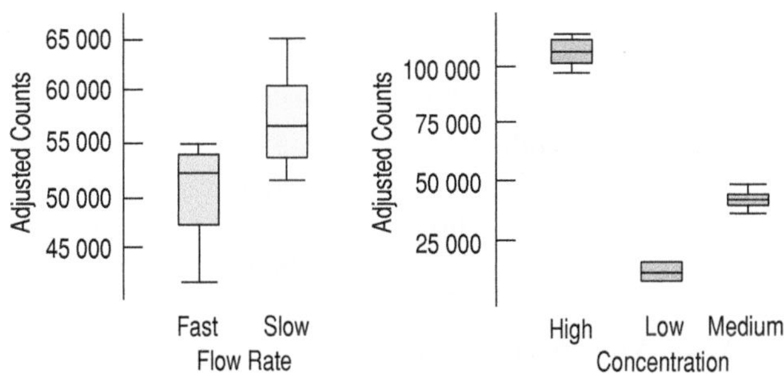

A two-way ANOVA with interaction was run, and the following ANOVA table resulted:

Source	DF	Sum of Squares	Mean Square	F-ratio	P-value
Conc	2	483655E5	241828E5	1969.44	<0.0001
Flow rate	1	364008E3	364008E3	29.65	<0.0001
Interaction	2	203032E3	101516E3	8.27	0.0019
Error	24	294698E3	122791E3		
Total	29	492272E5			

A plot of residuals versus predicted values showed:

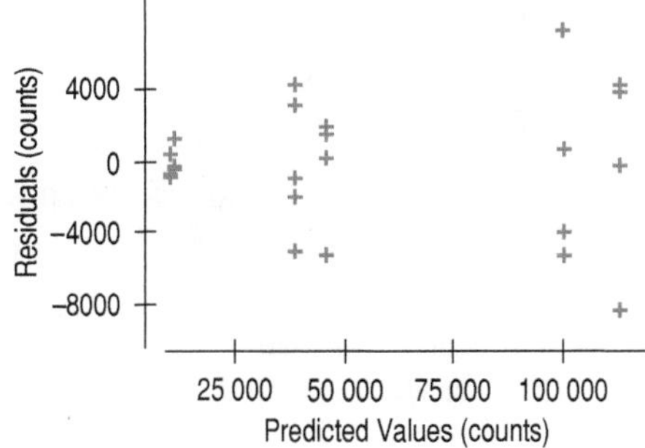

What conclusions about the effect of flow rate do you draw? Do you see any potential problems with the analysis?

17. Gas additives again Refer back to the experiment in Exercise 15. Instead of mpg, redo the analysis using log(mpg) as the response. Do your conclusions change? How? Are the assumptions of the model better satisfied?

18. Chromatography again Refer back to the experiment in Exercise 16. Instead of total counts, the analysis was redone using log(total counts) as the response. Do your conclusions change? How? Are the assumptions of the model better satisfied?

Analysis of Variance

Source	DF	Sum of Squares	Mean Square	F-ratio	P-Value
Type	2	10.1254	5.06268	5923.1	<0.0001
Additive	1	0.026092	0.0260915	30.526	<0.0001
Interaction	2	7.57E-05	3.78E-05	0.044265	0.9567
Error	54	0.046156	8.55E-04		
Total	59	10.1977			

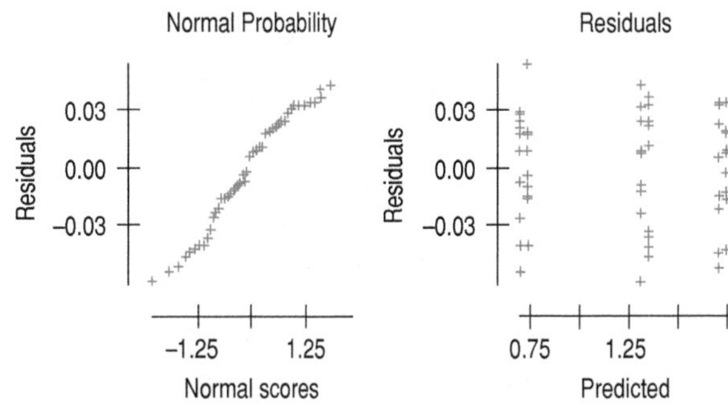

19. Batteries again A student experiment was run to test the performance of four *Brands* of batteries under two different *Environments* (room temperature and cold). For each of the eight treatments, two batteries of a particular brand were put into a flashlight. The flashlight was then turned on and allowed to run until the light went out. The number of minutes the flashlight stayed on was recorded. Each treatment condition was run twice.

Partial boxplots showed:

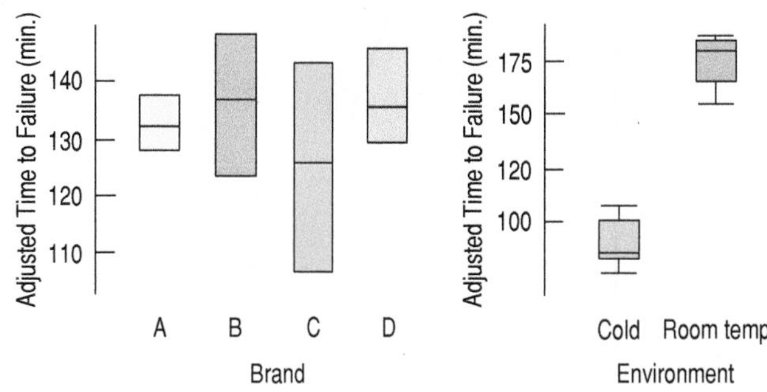

An interaction plot showed:

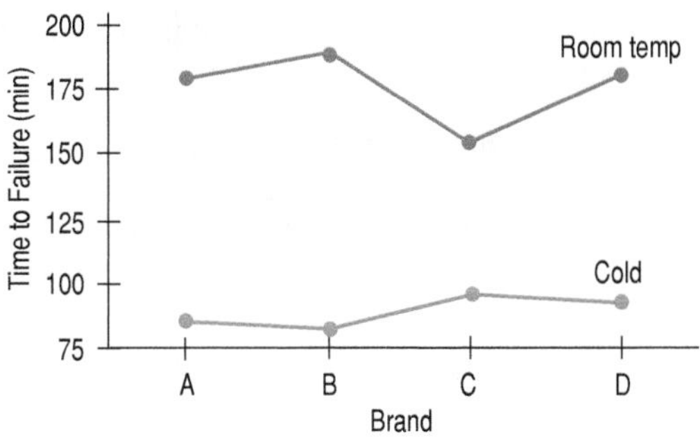

An ANOVA table showed:

Source	DF	Sum of Squares	Mean Square	F-ratio	P-value
Envir	1	30363.1	30363.1	789.93	<0.0001
Brand	3	338.187	112.729	2.9328	0.0994
Interaction	3	1278.19	426.063	11.085	0.0032
Error	8	307.5	38.4375		
Total	15	32286.9			

a) What are the main effect null and alternative hypotheses?
b) From the partial boxplots, do you think that the *Brand* has an effect on the time the batteries last? How about the condition?
c) Do the conclusions of the ANOVA table match your intuition?
d) What does the interaction plot say about the performance of the brands?
e) Why might you be uncomfortable with a recommendation to go with the cheapest battery (brand C)?

20. Peas In an experiment on growing sweet peas, a team of students selected two factors at four levels each and recorded *Weight*, *Stem Length*, and *Root Length* after six days of growth. They grew plants using various amounts of *Water* and *Quickgrow* solution, a fertilizer designed to help plants grow faster. Each factor was run at four levels: little, some, moderate, and full. They grew two plants under each of the 16 conditions. The response variable was *Weight* in mg.

An interaction plot shows:

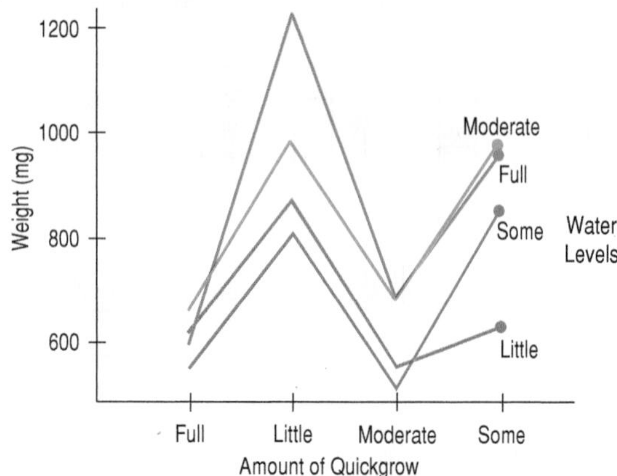

Because of this, a two-way ANOVA with interaction model was fit to the data, resulting in the following ANOVA table:

Analysis of Variance for Weight

Source	DF	Sum of Squares	Mean Square	F-ratio	P-value
Water	3	246255	82084.8	2.5195	0.0948
QS	3	827552	275851	8.4669	0.0013
Water × QS	8	176441	22055.1	0.6770	0.7130
Error	16	521275	32579.7		
Total	31	1771523			

Residuals plots showed no violations of the variance or Normality conditions. A table of effects for Quickgrow shows:

Level of Quickgrow	Effect
Little	213.8
Some	97.1
Moderate	−155.5
Full	−155.4

A table of effects for water shows:

Level of Water	Effect
Little	−91.1
Some	−81.9
Moderate	66.4
Full	106.6

Write a report summarizing what the students have learned about the effects of *Water* and *Quickgrow* solution on the early stages of sweet pea growth as measured by weight.

21. Batteries once more Another student analyzed the battery data from Exercise 19, using a one-way ANOVA. He considered the experimental factor to be an eight-level factor consisting of the eight possible combinations of *Brand* and *Environment*. Here are the boxplots for the eight treatments and a one-way ANOVA:

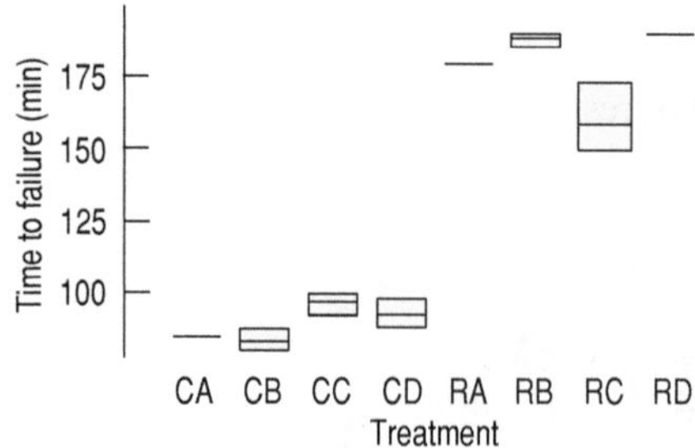

Source	DF	Sum of Squares	Mean Square	F-ratio	P-value
Treatment	7	31895.8	4556.54	93.189	<0.0001
Error	8	391.167	48.8958		
Total	15	32286.9			

Compare this analysis with the one performed in Exercise 19. Which one provides a better understanding of the data? Explain.

22. Containers, one more time Another student performs a one-way ANOVA on the container data of Exercise 14, using the four treatments water room, water outside, coffee room, and coffee outside. Perform this analysis and comment on the differences between this analysis and the one in Exercise 14.

Just Checking ANSWERS

1. The null hypotheses are that the mean flight distance for all three weights is the same, that the mean flight distance for both positions is the same, and that the interaction effects are all 0.
2. Yes, the effects appear to be additive. The lines are nearly parallel.
3. Yes, the F-test provides no evidence to reject the null hypothesis with a P-value of ≥ 0.9.
4. Yes, we reject the null hypothesis with a P-value ≤ 0.0001.
5. Yes, we reject the null hypothesis with a P-value ≤ 0.0001.
6. Because both factors are significant, it appears that using the light weight in the rear position may result in the longest mean flight distance.

chapter

27 Multiple Regression

Old stories are called teaching, new ones research.

—*Professor Edward Chamberlin*

Where are we going?

We've seen that the top wind speed in a hurricane depends on the central barometric pressure. But what about the sea surface temperature? Can we include other variables in our model? Linear models are often useful, but the world is usually not so simple that a two-variable model does the trick. For a more realistic understanding, we need models with several variables.

Who	250 male subjects
What	Body fat and waist size
Units	%body fat and inches
When	1990s
Where	United States
Why	Scientific research

In Chapter 24, we tried to predict the percent body fat of male subjects from their waist size, and we did pretty well. The R^2 of 67.8% says that we accounted for almost 68% of the variability in *%Body fat* by knowing only *Waist size*. We completed the analysis by performing hypothesis tests on the coefficients and looking at the residuals.

But that remaining 32% of the variance has been bugging us. Couldn't we do a better job of accounting for *%Body fat* if we weren't limited to a single predictor? In the full data set, there were 15 other measurements on the 250 men. We might be able to use other predictor variables to help us account for the leftover variation that wasn't accounted for by *Waist size*.

What about *Height*? Does *Height* help to predict *%Body fat*? Men with the same waist size can vary from short and corpulent to tall and emaciated. Knowing a man has a 50-inch waist tells us that he's likely to carry a lot of body fat. If we found out that he was seven feet tall, that might change our impression of his body type. Knowing his *Height* as well as his *Waist size* might help us make a more accurate prediction.

27.1 Multiple Regression Motivation

Does a regression with *two* predictors even make sense? It does—and that's fortunate, because the world is too complex a place for simple linear regression alone to model it. A regression with two or more predictor variables is called a **multiple regression**. We use a statistics program on a computer to do all the calculations for multiple regression.

For simple regression, we found the **least squares** solution, the one whose coefficients made the sum of the squared residuals as small as possible. For multiple regression, we'll do the same thing but this time with more coefficients. A statistics package can find the coefficients of the least squares model easily.

Here's a typical example of a multiple regression table:

Dependent variable is: %Body fat
R-squared = 71.3% R-squared (adjusted) = 71.1%
s = 4.460 with 250 − 3 = 247 degrees of freedom

Variable	Coefficient	SE(Coeff)	*t*-ratio	P-value
Intercept	−3.10088	7.686	−0.403	0.6870
Waist	1.77309	0.0716	24.8	≤ 0.0001
Height	−0.60154	0.1099	−5.47	≤ 0.0001

You should recognize most of the numbers in this table.

A NOTE ON TERMINOLOGY

When we have two or more predictors and fit a linear model by the least squares method, we are formally said to fit a least squares linear multiple regression. Most folks just call it "multiple regression." You may also see the abbreviation OLS used with this kind of analysis. It stands for "Ordinary Least Squares."

R^2 has the same interpretation in multiple regression as in simple regression (though it is now the square of the correlation between the observed responses and the predicted responses). Here, for example, R^2 gives the fraction of the variability of *%Body fat* accounted for by the *multiple* regression model. (With *Waist size* alone predicting *%Body fat*, the R^2 was 67.8%.) The multiple regression model accounts for 71.3% of the variability in *%Body fat*. We shouldn't be surprised that R^2 has gone up. It was the hope of accounting for some of that leftover variability that led us to try a second predictor ('R^2 adjusted' in the output will be discussed later in this chapter).

The standard deviation of the residuals is still denoted s (or sometimes s_e to distinguish it from the standard deviation of y).

The degrees of freedom calculation follows our rule of thumb: The degrees of freedom is the number of observations (250) minus one for each coefficient estimated—for this model, three.

For each predictor, we have a coefficient, its standard error, a t-ratio, and the corresponding P-value. As with simple regression, the t-ratio measures how many standard errors the coefficient is away from 0. So, using a Student's t-model, we can use its P-value for testing whether the true coefficient is different from 0.

Using the coefficients from this table, we can write the regression model:

$$\widehat{\%Body\ fat} = -3.10 + 1.77\ Waist\ size - 0.60\ Height.$$

As before, we define the residuals as

$$Residuals = \%Body\ fat - \widehat{\%\ Body\ fat}.$$

We've fit this model with the same least squares principle: The sum of the squared residuals is as small as possible for any choice of coefficients.

So, What's New?

With so much of the multiple regression looking just like simple regression, why devote an entire chapter (or two) to the subject?

There are several answers to this question. First—and most important—the *meaning* of the coefficients in the regression model has changed in a subtle but important way. Because that change is not obvious, multiple regression coefficients are often misinterpreted. We'll show some examples to help make the meaning clear.

Second, multiple regression is an extraordinarily versatile calculation, underlying many widely used statistics methods. A sound understanding of the multiple regression model will help you to understand these other applications.

Third, multiple regression offers our first glimpse into statistical models that use more than two quantitative variables. The real world is complex. Simple models of the kind we've seen so far are a great start, but often they're just not detailed enough to be useful for understanding, predicting, and decision making. Models that use several variables can be a big step toward realistic and useful modelling of complex phenomena and relationships.

For Example REAL ESTATE

As a class project, students in a large Statistics class collected publicly available information on recent home sales in their home towns. We have 894 properties. These are not a random sample, but they may be representative of home sales nationwide during a short period of time.

Variables available include the price paid, the size of the living area (sq. ft.), the number of bedrooms, the number of bathrooms, the year of construction, the lot size (acres), and a coding of the location as urban, suburban, or rural made by the student who collected the data.

Here's a regression to model the sale price from the living area (sq. ft.) and the number of bedrooms.

Dependent variable is: Price
R squared = 14.6% R squared (adjusted) = 14.4%
s = 266899 with 894 − 3 = 891 degrees of freedom

Variable	Coefficient	SE(Coeff)	t-ratio	P-value
Intercept	308100	41148	7.49	0.0001
Living area	135.089	11.48	11.8	0.0001
Bedrooms	−43346.8	12844	−3.37	0.0008

QUESTION: How should we interpret the regression output?

ANSWER: The model is

$$\widehat{Price} = 308\,100 + 135\text{ LivingArea} - 43\,347\text{ Bedrooms}$$

The R-squared says that this model accounts for only 14.6% of the variation in price. The value of *s* also leads us to doubt that this model would provide very good predictions because the standard deviation of the residuals is more than \$266 000. Nevertheless, we may be able to learn about home prices because the P-values of the coefficients are all very small, so we can be quite confident that none of them is really zero.

What Multiple Regression Coefficients Mean

We said that height might be important in predicting body fat in men. What is the relationship between *%Body fat* and *Height* in men? We know how to approach this question; we follow the three rules.

Look at the scatterplot in Figure 27.1. It doesn't look like *Height* tells us much about *%Body fat*. You just can't tell much about a man's *%Body fat* from his *Height*. Or can you? Remember, in the multiple regression model, the coefficient of *Height* was −0.60, had a *t*-ratio of −5.47, and had a very small P-value. So it *did* contribute to the *multiple* regression model. How could that be?

The answer is that the multiple regression coefficient of *Height* takes account of the other predictor, *Waist size*, in the regression model.

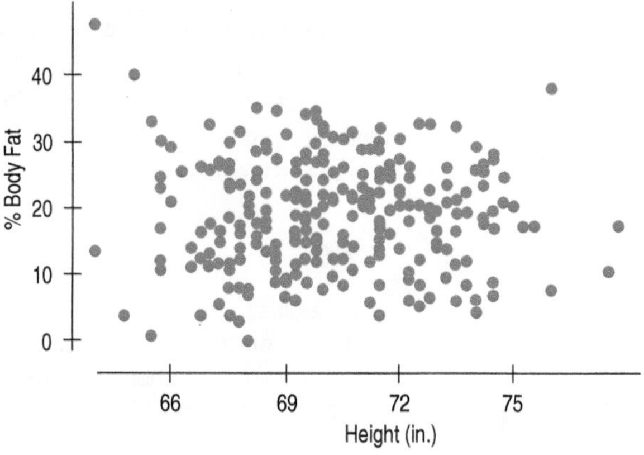

Figure 27.1
The scatterplot of *%Body fat* against *Height* seems to say that there is little relationship between these variables.

To understand the difference, let's think about all men whose waist size is about 37 inches—right in the middle of our sample. If we think only about *these* men, what do we expect the relationship between *Height* and *%Body fat* to be? Now a negative association makes sense because taller men probably have less body fat than shorter men *who have the same waist size*. Let's look at the plot:

Figure 27.2

When we restrict our attention to men with waist sizes between 36 and 38 inches (points in blue), we can see a relationship between *%Body fat* and *Height*.

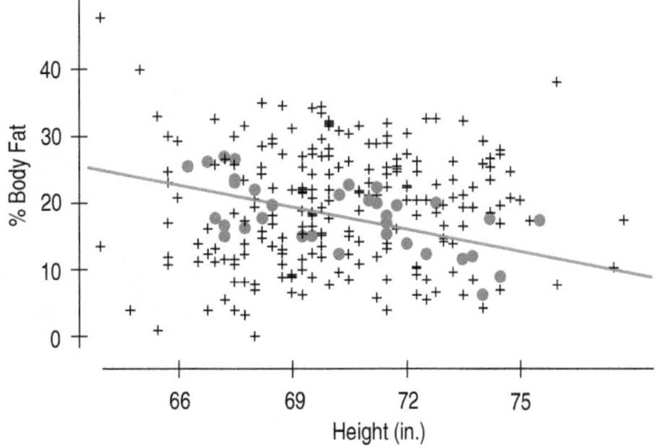

Here we've highlighted the men with waist sizes between 36 and 38 inches. Overall, there's little relationship between *%Body fat* and *Height*, as we can see from the full set of points. But when we focus on *particular* waist sizes, there *is* a relationship between body fat and height. This relationship is *conditional* because we've restricted our set to only those men within a certain range of waist sizes. For men with that waist size, an extra inch of height is associated with a decrease of about 0.60% in body fat. If that relationship is consistent for each *waist* size, then the multiple regression coefficient will estimate it. The simple regression coefficient simply couldn't see it.

Residuals

As their name reminds us, residuals are what's left over after we fit a model. That lets us remove the effects of some variables. The residuals are what's left.

We've picked one particular waist size to highlight. How could we look at the relationship between *%Body fat* and *Height* conditioned on *every individual Waist size at the same time*? Once again, residuals come to the rescue.

We plot the residuals of *%Body fat* after a regression on *Waist size* against the residuals of *Height* after regressing *it* on *Waist size*. This display is called a *partial regression plot*. It shows us just what we asked for: the relationship of *%Body fat* to *Height* after removing the linear effects of *Waist size*.

Figure 27.3

A partial regression plot for the coefficient of *Height* in the regression model has a slope equal to the coefficient value in the multiple regression model.

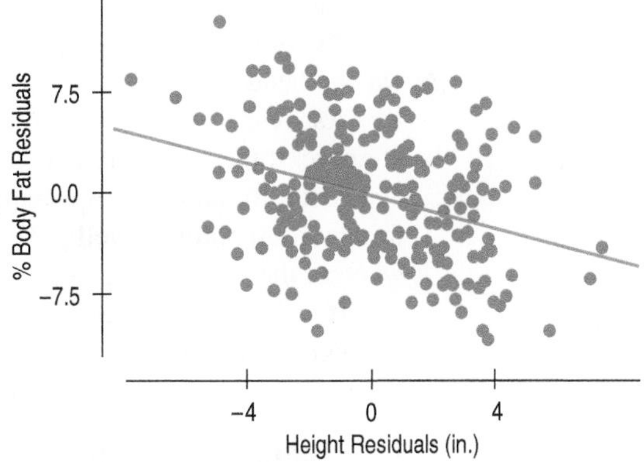

A **partial regression plot** (also known as added variable plot, adjusted variable plot, and individual coefficient plot) for a particular predictor has a slope that is the same as the *multiple* regression coefficient for that predictor. Here, it's –0.60. It also has the same residuals as the full multiple regression,[1] so you can spot any outliers or influential points and tell whether they've affected the estimation of this particular coefficient.

[1]If this reminds you of the partial boxplots you saw in Chapter 26, you're on the right track. Here, we've removed the effects of all the other predictors by plotting their residuals against the predictor of interest.

Many modern statistics packages offer partial regression plots as an option for any coefficient of a multiple regression. For the same reasons that we always look at a scatterplot before interpreting a simple regression coefficient, it's a good idea to make a partial regression plot for any multiple regression coefficient that you hope to understand or interpret.

27.2 The Multiple Regression Model

We can write a multiple regression model like this, numbering the predictors arbitrarily (we don't care which one is x_1), writing β's for the model coefficients (which we will estimate from the data), and including the errors in the model:

$$y = \beta_0 + \beta_1 x_1 + \beta_2 x_2 + \varepsilon$$

Of course, the multiple regression model is not limited to two predictor variables, and regression model equations are often written to indicate summing any number (a typical letter to use is k) of predictors. That doesn't really change anything, so we'll often stick with the two-predictor version just for simplicity. But don't forget that we can have many predictors.

The assumptions and conditions for the multiple regression model sound nearly the same as for simple regression, but with more variables in the model, we'll have to make a few changes.

Assumptions and Conditions

Linearity Assumption

We are fitting a linear model.[2] For that to be the right kind of model, we need an underlying linear relationship. But now we're thinking about several predictors. To see whether the assumption is reasonable, we'll check the Straight Enough Condition for *each* of the predictors.

Straight Enough Condition: Scatterplots of y against each of the predictors are reasonably straight. As we have seen with *Height* in the body fat example, the scatterplots need not show a strong (or any!) slope; we just check that there isn't a bend or other nonlinearity. For the body fat data, the scatterplot is beautifully linear in *Waist size*, as we saw in Chapter 24. For *Height*, we saw no relationship at all, but at least there was no bend.

CHECK THE RESIDUAL PLOT (PART 1A)

The residuals should appear to have no pattern with respect to the predicted values.

(PART 1B)

The residuals should appear to have no pattern with respect to each of the predictors.

As we did in simple regression, it's a good idea to check the residuals for linearity after we fit the model. It's good practice to plot the residuals against the predicted values and check for patterns, especially for bends or other nonlinearities. (We'll watch for other things in this plot as well.) We can also plot the residuals against each predictor to see whether the response, y, and the predictor, x_i, are reasonably linearly related.

If we're willing to assume that the multiple regression model is reasonable, we can fit the regression model by least squares. But we must check the other assumptions and conditions before we can interpret the model or test any hypotheses.

Independence Assumption

As with simple regression, the errors in the true underlying regression model must be independent of each other. As usual, there's no way to be sure that the Independence Assumption is true. Fortunately, even though there can be many predictor variables, there is only one response variable and only one set of errors. The Independence Assumption concerns the errors, so we check the corresponding conditions on the residuals.

[2]By *linear*, we mean that each x appears simply multiplied by its coefficient and added to the model. No x appears in an exponent or some other more complicated function. That means that as we move along any x-variable, our prediction for y will change at a constant rate (given by the coefficient) if nothing else changes.

Randomization Condition: The data should arise from a random sample or randomized experiment. Randomization assures us that the data are representative of some identifiable population. If you can't identify the population, you can interpret the regression model only as a description of the data you have, and you can't interpret the hypothesis tests at all because they are about a regression model for that population. Regression methods are often applied to data that were not collected with randomization. Regression models fit to such data may still do a good job of modelling the data at hand, but without some reason to believe that the data are representative of a particular population, you should be reluctant to believe that the model generalizes to other situations.

CHECK THE RESIDUAL PLOT (PART 2)

The residuals should appear to be randomly scattered and show no patterns or clumps when plotted against the predicted values or against *time* (if data taken at different time points).

We also check displays of the regression residuals for evidence of patterns, trends, or clumping, any of which would suggest a failure of independence. As discussed in Chapter 24, plotting the residuals against *time* will allow us to see whether the errors exhibit time dependence, a special form of violating the Independence Assumption.

The body fat data were collected on a sample of men. The men were not related in any way, so we can be pretty sure that their measurements are independent.

Equal Variance Assumption

CHECK THE RESIDUAL PLOT (PART 3)

The spread of the residuals should be fairly uniform when plotted against any of the x's or against the predicted values.

The variability of the errors should be about the same for all values of *each* predictor. To see if this is reasonable, we look at scatterplots.

Does the Plot Thicken? Condition: Scatterplots of the regression residuals against each x variable or against the predicted values, $\hat{y}$, offer a visual check. The spread around the line should be nearly constant. Be alert for a "fan" shape or other tendency for the variability to grow or shrink in one part of the scatterplot.

Here are the residuals plotted against *Waist size* and *Height*. Neither plot shows patterns that might indicate a problem.

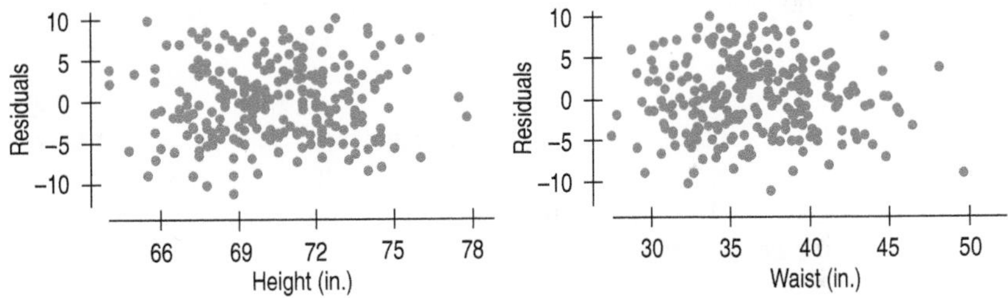

Figure 27.4

Residuals plotted against each predictor show no pattern. That's a good indication that the Straight Enough Condition and the Does the Plot Thicken? Condition are satisfied.

If residual plots show no pattern, if the residuals are plausibly independent, and if the plots don't thicken, we can feel good about interpreting the regression model. Before we test hypotheses, however, we must check one final assumption.

Normality Assumption

We assume that the errors around the idealized regression model at any specified values of the x-variables follow a Normal model. We need this assumption so that we can use a Student's t-model for inference. As with other times when we've used Student's t, we'll settle for the residuals satisfying the Nearly Normal Condition.

Nearly Normal Condition: Because we have only one set of residuals, this is the same set of conditions we had for simple regression. Look at a histogram or Normal probability plot of the residuals. The histogram of residuals in the body fat regression certainly looks nearly Normal, and the Normal probability plot is fairly straight. And, as we have said before, the Normality Assumption becomes less important as the sample size grows.

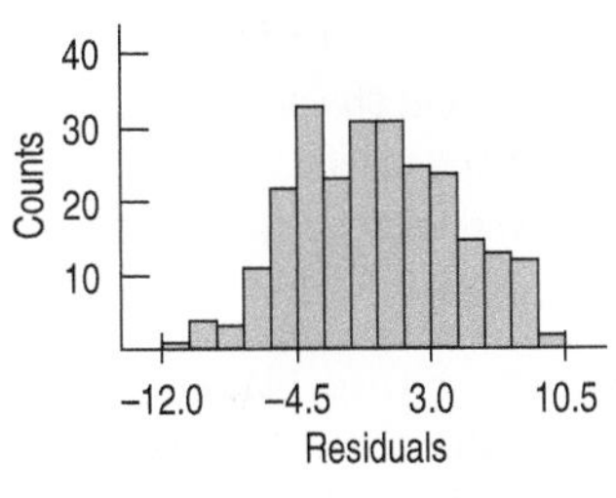

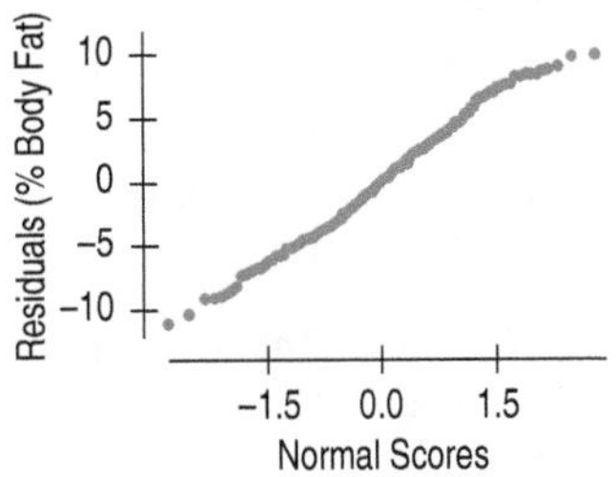

Figure 27.5

Check a histogram of the residuals. The distribution of the residuals should be unimodal and symmetric. Or, check a Normal probability plot to see whether it is straight.

Let's summarize all the checks of conditions that we've made and the order that we've made them:

1. Check the Straight Enough Condition with scatterplots of the y-variable against each x-variable.
2. If the scatterplots are straight enough (that is, if it looks like the regression model is plausible), fit a multiple regression model to the data. (Otherwise, either stop or consider re-expressing an x-variable or the y-variable.)

3. Find the residuals and predicted values.
4. Make a scatterplot of the residuals against the predicted values.[3] This plot should look patternless. Check in particular for any bend (which would suggest that the data weren't all that straight after all) and for any thickening. If there's a bend, and especially if the plot thickens, consider re-expressing the y-variable and starting over. You should also check for outliers here. If you find any, investigate them for errors, or consider doing the analysis with and without them.
5. Make scatterplots of the residuals against each predictor. The plots should look patternless. Check in particular for any bend, which would suggest the y-variable and x-variable do not have a straight line relationship. If this is the case, consider re-expressing the x-variable and starting over. You can also identify serious outliers and high leverage points in this step.
6. Think about how the data were collected. Was suitable randomization used? Are the data representative of some identifiable population? Make a timeplot of the residuals. This plot should look patternless. Any evidence of patterns suggests violation of the Independence Assumption.
7. If the conditions check out this far, feel free to interpret the regression model and use it for prediction. If you want to investigate a particular coefficient, make a partial regression plot for that coefficient.
8. If you wish to test hypotheses about the coefficients or about the overall regression, then make a histogram and Normal probability plot of the residuals to check the Nearly Normal Condition.

Step-by-Step Example MULTIPLE REGRESSION

Let's try finding and interpreting a multiple regression model for the body fat data.

THINK ➡ **Variables** Name the variables, report the *W*'s, and specify the questions of interest.	I have quantitative body measurements on 250 adult males from the BYU Human Performance Research Center. I want to understand the relationship between *%Body fat*, *Height*, and *Waist size*.
Plan Think about the assumptions and check the conditions.	✓ **Straight Enough Condition:** There is no obvious bend in the scatterplots of *%Body Fat* against either x-variable. And the scatterplot of residuals against predicted values below (on the next page) shows no pattern that would suggest nonlinearity. ✓ **Independence Assumption:** These data are not collected over time, and there's no reason to think that the *%Body fat* of one man influences that of another. I don't know whether the men measured were sampled randomly, but the data are presented as being representative of the male population of the United States.
Now we can find the regression and examine the residuals.	✓ **Does the Plot Thicken? Condition:** The scatterplot of residuals against predicted values below shows no obvious changes in the spread about the line.

[3]In Chapter 24, we noted that a scatterplot of residuals against the predicted values looked just like the plot of residuals against x. But for a multiple regression, there are several x's. Now the predicted or $\hat{y}$ values are a combination of the x's—in fact, they're the combination given by the regression equation we have computed. So they combine the effects of all the x's in a way that makes sense for our particular regression model. That makes them a good choice to plot against.

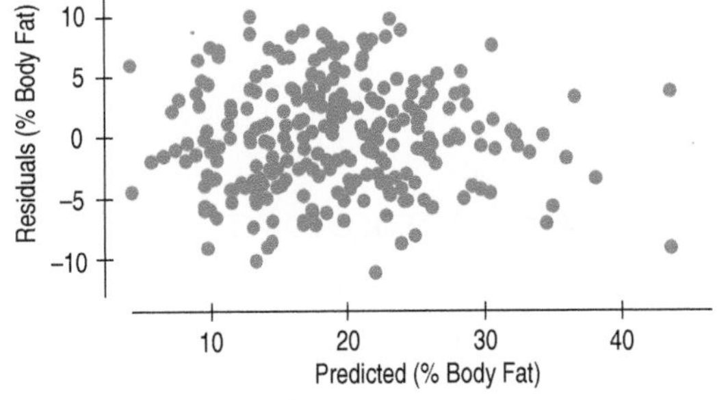

Actually, we need the Nearly Normal Condition only if we want to do inference.

✓ **Nearly Normal Condition, Outlier Condition:** A histogram of the residuals is unimodal and symmetric.

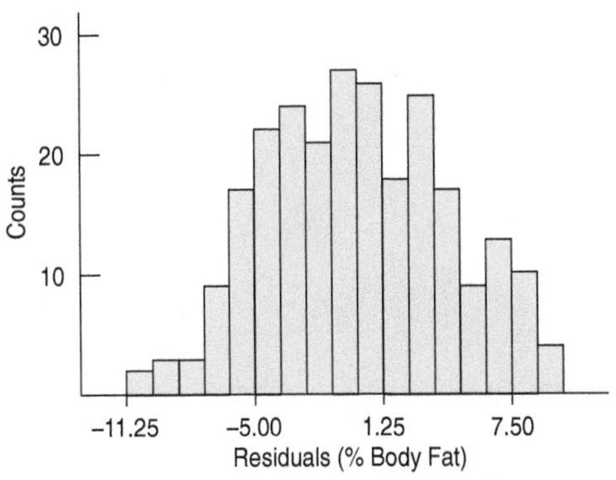

The Normal probability plot of the residuals is reasonably straight:

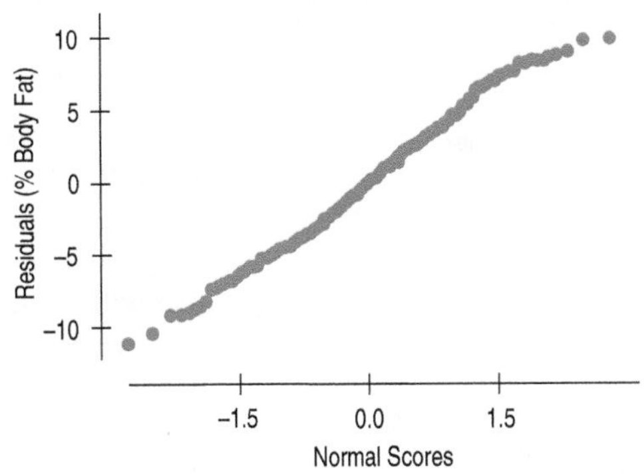

Choose your method.

Under these conditions a full multiple regression analysis is appropriate.

SHOW ➡ Mechanics

Here is the computer output for the regression:

Dependent variable is: %body fat
R-squared = 71.3% R-squared (adjusted) = 71.1%
s = 4.460 with 250 – 3 = 247 degrees of freedom

Source	Sum of Squares	DF	Mean Square	F-ratio	P-value
Regression	12216.6	2	6108.28	307	<0.0001
Residual	4912.26	247	19.8877		

Variable	Coefficient	SE(Coeff)	*t*-ratio	P-value
Intercept	–3.10088	7.686	–0.403	0.6870
Waist	1.77309	0.0716	24.8	<0.0001
Height	–0.60154	0.1099	–5.47	<0.0001

	The estimated regression equation is $\widehat{\%Body\ fat} = -23.10 + 1.77\ Waist\ size - 0.60\ Height$
TELL ➡ **Interpretation**	The R^2 for the regression is 71.3%. *Waist size* and *Height* together account for about 71% of the variation in *%Body fat* among men. The regression equation indicates that each inch in *Waist size* is associated with about a 1.77 increase in *%Body fat* among men who are of a particular height. Each inch of *Height* is associated with a decrease in *%Body fat* of about 0.60 among men with a particular waist size.
More Interpretation	The standard errors for the slopes of 0.07 (*Waist size*) and 0.11 (*Height*) are both small compared with the slopes themselves, so it looks like the coefficient estimates are fairly precise. The residuals have a standard deviation of 4.46%, which gives some indication of how precisely we can predict *%Body fat* with this model.

27.3 Multiple Regression Inference

I Thought I Saw an ANOVA Table . . .

There are several hypothesis tests in the multiple regression output, but all of them talk about the same thing. Each is concerned with whether the underlying model parameters are actually zero.

The first of these hypotheses is one we skipped over for simple regression (for reasons that will be clear in a minute). Now that we've looked at ANOVA (in Chapter 25),[4] we can recognize the ANOVA table sitting in the middle of the regression output. Where'd that come from?

The answer is that now that we have more than one predictor, there's an overall test we should consider before we do more inference on the coefficients. We ask the global question, "Is this multiple regression model any good at all?" That is, would we do as well using just $\bar{y}$ to model y? What would that mean in terms of the regression? Well, if all the coefficients (except the intercept) were zero, we'd have

$$\hat{y} = b_0 + 0x_1 + \ldots + 0x_k$$

and we'd just set $b_0 = \bar{y}$.

To address the overall question, we'll test

$$H_0: \beta_1 = \beta_2 = \ldots = \beta_k = 0.$$

(That null hypothesis looks very much like the null hypothesis we tested with an F-test in the analysis of variance in Chapter 25.) This equation is often referred to as the "overall F-test" or "global F-test" in the multiple regression context.

[4] If you skipped over Chapters 25 and 26, you can just take our word on this and read on.

We can test this hypothesis with a statistic that is labelled with the letter F (in honour of Sir Ronald Fisher, the developer of analysis of variance). In our example, the F-value is 307 on 2 and 247 degrees of freedom. The alternative hypothesis is just that the slope coefficients aren't all equal to zero, and bigger F-values mean smaller P-values. If the null hypothesis were true, the F-statistic would be near 1. The F-statistic here is quite large, so we can easily reject the null hypothesis and conclude that the multiple regression model is better than just using the mean.[5] (On page 750 you can see an F density curve and an F-table).

Why didn't we do this for simple regression? Because the null hypothesis would have just been that the lone model slope coefficient was zero, and we were already testing that with the t-statistic for the slope. Theoretically, for the simple linear regression, we can show that the *square* of that t-statistic is equal to the F-statistic in the ANOVA table, so it really was the identical test.

Testing the Coefficients

Once we check the F-test and reject the null hypothesis—and, if we are being careful, *only* if we reject that hypothesis—we can move on to checking the test statistics for the individual coefficients. Those tests look like what we did for the slope of a simple regression in Chapter 24. For each coefficient we test

$$H_0: \beta_j = 0$$

against the (typically two-sided) alternative that it isn't zero. The regression table gives a standard error for each coefficient and the ratio of the estimated coefficient to its standard error. If the assumptions and conditions are met (and now we need the Nearly Normal Condition), these ratios follow a Student's t-distribution.

$$t = \frac{b_j - 0}{SE(b_j)}$$

How many degrees of freedom? We have a rule of thumb that works here. The degrees of freedom is the number of data values minus the number of coefficients estimated (including the intercept). For our regression on two ($k = 2$) predictors, that's $n - 3$. You shouldn't have to look up the t-values. Almost every regression report includes the corresponding P-values.

We can build a confidence interval in the usual way, as an estimate $\pm$ a margin of error. As always, the margin of error is just the product of the standard error and a critical value. Here the critical value comes from the t-distribution on (number of data values)—(number of coefficients estimated) degrees of freedom. So a confidence interval for β_1 is

$$b_1 \pm t^*_{n-k-1} SE(b_1)$$

The tricky parts of these tests are that the standard errors of the coefficients now require harder calculations (which we leave to technology), and the meaning of a coefficient, as we have seen, depends on all the *other* predictors in the multiple regression model.

That last bit is important. If we fail to reject the null hypothesis for a multiple regression coefficient, it does **not** mean that the corresponding predictor variable has no linear relationship to y. It means that the corresponding predictor contributes nothing to modelling y *after allowing for all the other predictors.*

[5] There are F tables in the back of the book, and they work pretty much as you'd expect. Most regression tables include a P-value for the F-statistic, but often there's no need to perform this particular test in a multiple regression. Usually we just glance at the F-statistic to see that it's reasonably far from 1.0, the value it would have if the true coefficients were really all zero.

How's That, Again?

This last point bears repeating. The multiple regression model looks so simple and straightforward:

$$y = \beta_0 + \beta_1 x_1 + \cdots + \beta_k x_k + \varepsilon$$

It *looks* like each β_j tells us the effect of its associated predictor, x_j, on the response variable, y. But that is not so. This is, without a doubt, the most common error that people make with multiple regression:

- It is possible for there to be no simple relationship between y and x_j, and yet β_j in a *multiple* regression can be significantly different from 0. We saw this happen for the coefficient of height in our example.
- It is also possible for there to be a strong two-variable relationship between y and x_j, and yet β_j in a multiple regression can be almost 0 with a large P-value, so we cannot reject the null hypothesis that the true coefficient is zero. For example, if we're trying to model the horsepower of a car using both its weight and its engine size, it may turn out that the coefficient for engine size is nearly 0. That *doesn't* mean that engine size isn't important for understanding horsepower. It simply means that after allowing for the weight of the car, the engine size doesn't give much *additional* information.
- It is even possible for there to be a significant linear relationship between y and x_j in one direction, and yet β_j can be of the *opposite* sign and strongly significant in a multiple regression. More expensive cars tend to be bigger, and since bigger cars have worse fuel efficiency, the price of a car has a slightly negative association with fuel efficiency. But in a multiple regression of fuel efficiency on weight and price, the coefficient of price may be positive. If so, it means that *among cars of the same weight*, more expensive cars have better fuel efficiency. The simple regression on price, though, has the opposite direction, because, *overall*, more expensive cars are bigger. This switch in sign may seem a little strange at first, but it's not really a contradiction at all. It's due to the change in the *meaning* of the coefficient of price when it is in a multiple regression rather than in a simple regression.

So we'll say it once more: The coefficient of x_j in a multiple regression depends as much on the *other* predictors as it does on x_j. Remember that when you interpret a multiple regression model.

For Example INTERPRETING COEFFICIENTS

We looked at a multiple regression to predict the price of a house from its living area and the number of bedrooms. We found the model

$$\widehat{Price} = 308\,100 + 135\ LivingArea - 43\,347\ Bedrooms$$

However, common sense says that houses with more bedrooms are usually worth more. And, in fact, the simple regression of price on bedrooms finds the model

$$\widehat{Price} = 33\,897 + 40\,234\ Bedrooms$$

and the P-value for the slope coefficient is 0.0005.

QUESTION: How should we understand the coefficient of bedrooms in the multiple regression?

ANSWER: The coefficient of bedrooms in the multiple regression does not mean that houses with more bedrooms are generally worth less. It must be interpreted taking into account the other predictor (living area) in the regression. If we consider houses with a given amount of living area, those that devote more of that area to bedrooms either must have smaller bedrooms or less living area for other parts of the house. Those differences could result in reducing the home's value.

Just Checking

Recall the regression in Chapter 7 to predict hurricane maximum wind speed from central barometric pressure. Another researcher, interested in the possibility that global warming was causing hurricanes to become stronger, added the variable year as a predictor and obtained the following regression:
Dependent variable is: Max. Winds (kn)
275 total cases of which 113 are missing

R squared = 77.9% R squared (adjusted) = 77.6%
s = 7.727 with 162 − 3 = 159 degrees of freedom

Source	Sum of Squares	DF	Mean Square	F-ratio
Regression	33446.2	2	16723.1	280
Residual	9493.45	159	59.7072	

Variable	Coefficient	s.e. of Coeff	*t*-ratio	prob
Intercept	1009.99	46.53	21.7	≤0.0001
Central Pressure	−0.933491	0.0395	−23.6	≤0.0001
Year	−0.010084	0.0123	−0.821	0.4128

1. Interpret the R^2 of this regression.
2. Interpret the coefficient of central pressure.
3. The researcher concluded that "there has been no change over time in the strength of Atlantic hurricanes." Is this conclusion a sound interpretation of the regression model?

Another Example: Modelling Infant Mortality

Who	U.S. states
What	Various measures relating to children and teens
When	1999
Why	Research and policy

Infant mortality is often used as a general measure of the quality of healthcare for children and mothers. It is reported as the rate of deaths of newborns per 1000 live births. Data recorded for each of the 50 states of the United States may allow us to build regression models to help understand or predict infant mortality. The variables available for our model are *Child deaths* (deaths per 100 000 children aged 1–14), percent of teens (ages 16–19) who drop out of high school (*HS drop%*), percent of low-birth weight babies (*Low BW%*), *Teen births* (births per 100 000 females aged 15–17), and *Teen deaths* by accident, homicide, and suicide (deaths per 100 000 teens ages 15–19).[6]

All of these variables were displayed and found to have no outliers.[7] One useful way to check many of our conditions is with a **scatterplot matrix**. This is an array of scatterplots set up so that the plots in each row have the same variable on their *y*-axis and those in each column have the same variable on their *x*-axis. This way every pair of variables is graphed. On the diagonal, rather than plotting a variable against itself, you'll usually find either a Normal probability plot or a histogram of the variable.

The individual scatterplots show at a glance that each of the relationships is straight enough for regression. There are no obvious bends, clumping, or outliers. And the plots don't thicken. So it looks like we can examine some multiple regression models with inference.

[6]The data are available from the Kids Count section of the Annie E. Casey Foundation, and are all for 1999.
[7]In the interest of complete honesty, we should point out that the original data include the District of Columbia, but it proved to be an outlier on several of the variables, so we've restricted attention to the 50 states here.

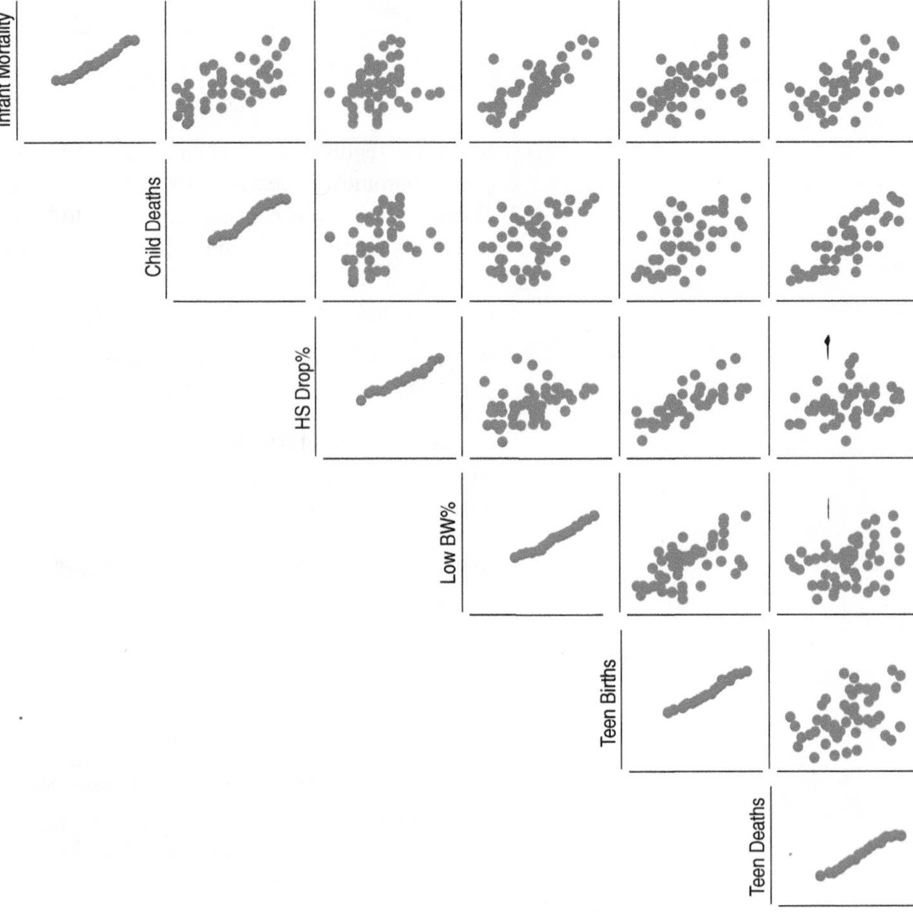

Figure 27.6
A scatterplot matrix shows a scatterplot of each pair of variables arrayed so that the vertical and horizontal axes are consistent across rows and down columns. The diagonal cells may hold Normal probability plots (as they do here), histograms, or just the names of the variables. These are a great way to check the Straight Enough Condition and to check for simple outliers. To see what is on the horizontal axis of any plot, simply rotate the closest y-axis label by 90 degrees clockwise.

Step-by-Step Example INFERENCE FOR MULTIPLE REGRESSION

Let's try to model infant mortality with all of the available predictors.

THINK ➡ Hypotheses State what we want to know.

I wonder whether all or some of these predictors contribute to a useful model for infant mortality.

First, I'll check the overall null hypothesis that asks whether the entire model is better than just modelling *y* with its mean:

(Hypotheses on the intercept are not particularly interesting for these data.)

H_0: The model itself contributes nothing useful, and all the slope coefficients are zero:

$$\beta_1 = \beta_2 = \ldots = \beta_k = 0$$

H_A: At least one of the β_j is not 0.

If I reject this hypothesis, then I'll test a null hypothesis for each of the coefficients of the form:

Plan State the null model.

H_0: The *j*-th variable contributes nothing useful, after allowing for the other predictors in the model: $\beta_j = 0$.

H_A: The *j*-th variable makes a useful contribution to the model: $\beta_j \neq 0$.

Think about the assumptions and check the conditions.

✓ **Straight Enough Condition, Outlier Condition:** The scatterplot matrix shows no bends, clumping, or outliers.

✓ **Independence Assumption:** These data are based on random samples and can be considered independent.

These conditions are enough to compute the regression model and find residuals.

✓ **Does the Plot Thicken? Condition:** The residual plot shows no obvious trends in the spread:

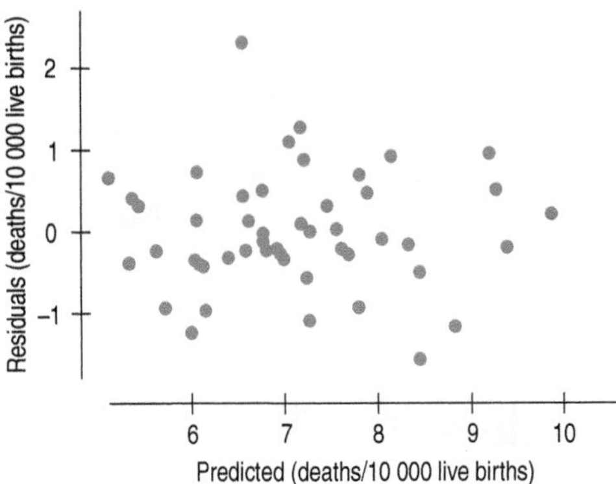

✓ **Nearly Normal Condition:** A histogram of the residuals is unimodal and symmetric.

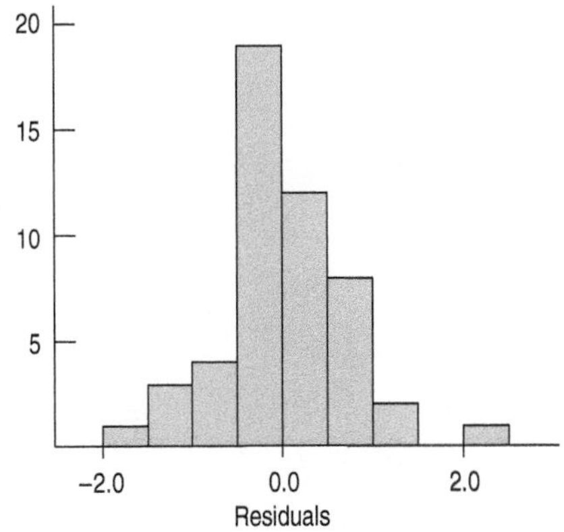

I don't see any significant outliers in the histogram.

Choose your method.

Under these conditions I can continue with the multiple regression analysis.

SHOW ➡ Mechanics Multiple regressions are always found from a computer program.

Computer output for this regression looks like this:

Dependent variable is: Infant mort
R-squared = 71.3 % R-squared (adjusted) 68.0 %
s = 0.7520 with 50 − 6 = 44 degrees of freedom

Source	Sum of Squares	DF	Mean Square	F-ratio
Regression	61.7319	5	12.3464	21.8
Residual	24.8843	44	0.565553	

The P-values given in the regression output table are from the Student's t-distribution on $(n-6) = 44$ degrees of freedom. They are appropriate for two-sided alternatives.

Variable	Coefficient	SE(Coeff)	t-ratio	P-value
Intercept	1.63168	0.9124	1.79	0.0806
Child Deaths	0.03123	0.0139	2.25	0.0292
HS Drop%	−0.09971	0.0610	−1.63	0.1096
Low BW%	0.66103	0.1189	5.56	<0.0001
Teen Births	0.01357	0.0238	0.57	0.5713
Teen Deaths	0.00556	0.0113	0.49	0.6245

Consider the hypothesis tests.

The F-ratio of 21.8 on 5 and 44 degrees of freedom is certainly large enough to reject the default null hypothesis that the regression model is no better than using the mean infant mortality rate. So I will examine the individual coefficients.

Under the assumptions we're willing to accept, and considering the conditions we've checked, the individual coefficients follow Student's t-distributions on 44 degrees of freedom.

Most of these coefficients have relatively small t-ratios, so I can't be sure that their underlying values are not zero. Two of the coefficients, *Child deaths* and *Low BW%*, have P-values less than 5%. So I can be confident that in this model both of these variables are unlikely to really have zero coefficients.

TELL ➡ Interpretation

Overall, the R^2 indicates that more than 71% of the variability in infant mortality can be accounted for with this regression model.

After allowing for the linear effects of the other variables in the model, an increase in child deaths of 1 death per 100 000 is associated with an increase of 0.03 deaths per 1000 live births in the infant mortality rate. And an increase of 1% in the percentage of live births that are low birth weight is associated with an increase of 0.66 deaths per 1000 live births.

*Testing a Set of Predictors in a Model

For the multiple regression model,

$$y = \beta_0 + \beta_1 x_1 + \ \ldots \ + \beta_g x_g + \beta_{g+1} x_{g+1} + \ \ldots \ + \beta_k x_k + \varepsilon$$

we use the F-test to test the null hypothesis that the overall regression is no improvement over just modelling y with its mean:

$$H_0: \beta_1 = \beta_2 = \ \ldots \ = \beta k = 0$$

And the t-ratios for the coefficients are used to test the significance of each regression coefficient. Besides these two tests, sometimes we might want to test the hypothesis that the overall regression is no improvement over modelling y with predictors $x_1, \ldots, x_g$ only:

$$y = \beta_0 + \beta_1 x_1 + \ \ldots \ + \beta_g x_g + \varepsilon$$

That is, we are testing

$$H_0: \beta_{g+1} = \beta_{g+2} = \ \ldots \ = \beta_k = 0$$

which is a statistical way of saying that the predictors $x_{g+1}, \ldots, x_k$ are not needed and may be removed from the original model.

Let us revisit the *Infant mortality* example. We have already determined that

$$InfantMort = \beta_0 + \beta_1 LowBW\% + \beta_2 ChildDeaths + \beta_3 HSDrop\% + \beta_4 TeenBirths + \beta_5 TeenDeaths + \varepsilon$$

is a significant model. Individually, *HSDrop%*, *TeenBirths*, and *TeenDeaths* may not make a useful contribution to the model. However, rather than examining them individually, we can examine if the above model is no improvement over the model

$$InfantMort = \beta_0 + \beta_1 LowBW\% + \beta_2 ChildDeaths\% + \varepsilon.$$

In other words, we are testing the hypothesis:

$$H_0: \beta_3 = \beta_4 = \beta_5 = 0.$$

How can we test this hypothesis?

Before we answer the question, we need to introduce some new terminology. The model with all the predictors in it is known as the *Complete Model*. If H_0 is true, the model that deletes the predictors we think may be worthless is known as the *Reduced Model*. For the *Infant Mortality* example, the Complete Model is:

$$InfantMort = \beta_0 + \beta_1 LowBW\% + \beta_2 ChildDeaths + \beta_1 HSDrop\% + \beta_4 TeenBirths + \beta_5 TeenDeaths + \varepsilon$$

and the Reduced Model

$$InfantMort = \beta_0 + \beta_1 LowBW\% + \beta_2 ChildDeaths\% + \varepsilon$$

If the null hypothesis is true, then the Reduced Model is as good a fit to the data as the Complete Model.

Recall that $SS_{Residuals}$ of a regression model measures the portion of the total variation not explained by the model (or equivalently, $SS_{Regression}$ measures the portion of the total variation explained by the model). Therefore by examining the difference in the $SS_{Residuals}$'s between the two models (equivalent to the difference in the $SS_{Regression}$, because SS_{Total} remains unchanged), we have the extra variation explained by the additional predictors in the model. The Complete Model must have a smaller $SS_{Residuals}$ (larger $SS_{Regression}$), unless you are extremely unlucky, in which case you might get equality. The difference is known as SS_{Drop}, which can be calculated by

$$\begin{aligned} SS_{Drop} &= SS_{Residuals}(Reduced) - SS_{Residuals}(Complete) \\ &= SS_{Regression}(Complete) - SS_{Regression}(Reduced) \end{aligned}$$

So, what we need to do is to take this difference and convert into an appropriate test statistic. As in Chapter 25, we cannot directly compare *SS*'s since they have different numbers of squared deviations, but mean squared deviations are comparable. By comparing the extra variation explained by the additional predictors with the mean squared error, we can see if this is a significant source of variation. More specifically, the test statistic is an *F*-statistic which takes the form

$$\frac{MS_{Drop}}{MS_{Residuals}(Complete)}$$

where

$$\begin{aligned} MS_{Drop} &= \frac{SS_{Drop}}{k - g} \\ MS_{Residuals}(Complete) &= \frac{SS_{Residuals}(Complete)}{n - (k + 1)}. \end{aligned}$$

Note that $(k - g)$ is just the number of predictors we are testing. The P-value of the test can then be obtained from the *F*-distribution with $(k - g)$ and $(n - (k + 1))$. Again the smaller the P-value, the more the evidence is against the null hypothesis.

Now back to our example. We have $n = 50$; $k = 5$; $g = 2$; $SS_{Residuals}(Complete) = 24.884$, $SS_{Residuals}(Reduced) = 26.938$. Therefore $SS_{Drop} = 26.938 = 24.884 - 2.054$ and the *F*-value is

$$\frac{2.054/(5 - 3)}{24.884/(50 - (5 + 1))} = 1.816.$$

Using a computer, the corresponding P-value from the F-distribution with 2 and 44 degrees of freedom is 0.1746. (If we are using the tables in this text, we can only report that P-value > 0.1.) With a large P-value, we have little evidence that the complete model is an improvement over a reduced model. That is, we have little evidence that these three "teen" variables (*HS drop%*, *Teen births*, and *Teen deaths*) would add anything useful to the model including just *Low BW%* and *Child deaths*.[8]

27.4 Comparing Multiple Regression Models

We have more variables available to us than we used when we modelled infant mortality. Moreover, several of those we tried don't seem to contribute to the model. How do we know that some other choice of predictors might not provide a better model? What exactly *would* make an alternative model better?

These are not easy questions. There is no simple measure of the success of a multiple regression model. Many people look at the R^2 value, and certainly we are not likely to be happy with a model that accounts for only a small fraction of the variability of y. But that's not enough. You can always drive the R^2 up by piling on more and more predictors, but models with many predictors are hard to understand. Keep in mind that the meaning of a regression coefficient depends on all the *other* predictors in the model, so it is best to keep the number of predictors as small as possible.

Regression models should make sense. Predictors that are easy to understand are usually better choices than obscure variables. Similarly, if there is a known mechanism by which a predictor has an effect on the response variable, that predictor is usually a good choice for the regression model.

How can we know whether we have the best possible model? The simple answer is that we can't. There's always the chance that some other predictors might bring an improvement (in higher R^2 or fewer predictors or simpler interpretation).

Adjusted R^2

You may have noticed that the full regression tables shown in this chapter include another statistic we haven't discussed. It is called adjusted R^2 and sometimes appears in computer output as R^2 (adjusted). The **adjusted R^2** statistic is a rough attempt to adjust for the simple fact that when we add another predictor to a multiple regression, the R^2 can't go down and will almost surely go up. Only if we were to add a predictor whose estimated coefficient turned out to be exactly zero would the R^2 remain the same. This fact makes it difficult to compare alternative regression models that have different numbers of predictors.

We can write a formula for R^2 using the sums of squares in the ANOVA table portion of the regression output table:

$$R^2 = \frac{SS_{Regression}}{SS_{Regression} + SS_{Residual}} = 1 - \frac{SS_{Residual}}{SS_{Total}}$$

Adjusted R^2 simply substitutes the corresponding *mean squares* for the SS's:[9]

$$R^2_{adj} = 1 - \frac{MS_{Residual}}{MS_{Total}}$$

Because the mean squares are sums of squares divided by degrees of freedom, they are adjusted for the number of predictors in the model. As a result, the adjusted R^2 value

[8] Note that looking at the three non-significant individual t-ratios in the earlier output does not assure us of a similar conclusion, since if you drop one predictor from the model, you have to refit the new model before assessing any other predictors.

[9] We learned about mean squares in Chapter 25. A mean square is just a sum of squares divided by its appropriate degrees of freedom. Mean squares are just special types of variances.

won't necessarily increase when a new predictor is added to the multiple regression model. That's fine. But adjusted R^2 no longer simply tells the fraction of variability accounted for by the model and it isn't even bounded by 0 and 100%, so it can be awkward to interpret.

As we'll see in the next chapter, comparing alternative regression models is a challenge, especially when they have different numbers of predictors. The search for a summary statistic to help us choose among models is the subject of much contemporary research in Statistics. Adjusted R^2 is one common—but not necessarily the best—choice often found in computer regression output tables. It's certainly not the sole decision criterion when you compare different regression models.

WHAT CAN GO WRONG?

Interpreting Coefficients

- **Don't interpret regression causally.** Regressions are usually applied to observational data. Without deliberately assigned treatments, randomization, and control, we can't draw conclusions about causes and effects. We can never be certain that there are no variables lurking in the background, causing everything we've seen. Don't interpret b_1, the coefficient of x_1 in the multiple regression, by saying, "If we were to change an individual's x_1 by 1 unit (holding the other x's constant), it would change his y by b_1 units." We have no way of knowing what applying a change to an individual would do.
- **Be cautious about interpreting a regression model as predictive.** Yes, we do call the x's predictors, and you can certainly plug in values for each of the x's and find a corresponding *predicted value* $\hat{y}$. But the term "prediction" suggests extrapolation into the future or beyond the data, and we know that we can get into trouble when we use models to estimate $\hat{y}$ values for x's not in the range of the data. Be careful not to extrapolate very far from the span of your data. In simple regression, it was easy to tell when you extrapolated. With many predictor variables, it's often harder to know when you are outside the bounds of your original data.[10] We usually think of fitting models to the data more as modelling than as prediction, so that's often a more appropriate term.
- **Don't claim to "hold everything else constant" for a single individual.** It's often meaningless to say that a regression coefficient says what we expect to happen if all variables but one were held constant for an individual and the predictor in question changed. While it's mathematically correct, it often just doesn't make any sense. We can't gain a year of experience or have another child without getting a year older. Instead, we *can* think about all those who fit given criteria on some predictors and ask about the conditional relationship between y and one x for those individuals. The coefficient –0.60 of height for predicting *%Body fat* says that among men of the same waist size, those who are one inch taller in height tend to be, on average, 0.60% lower in *%Body fat*. The multiple regression coefficient measures that average conditional relationship.
- **Don't think that the sign of a coefficient is special.** Sometimes our primary interest in a predictor is whether it has a positive or negative association with y. As we have seen, though, the sign of the coefficient also depends on the other predictors in the model. Don't look at the sign in isolation and conclude that "the direction of the relationship is positive (or negative)." Just like the value of the coefficient, the sign is about the relationship after allowing for the linear effects of the other predictors. The sign of a variable can change depending on which other predictors are in or out of the model. For example, in the regression model for infant mortality, the coefficient of *HS drop%* was negative and its P-value was fairly small, but the simple association between *Dropout rate* and *Infant mortality* is positive. (Check the matrix plot.)
- **If a coefficient's *t*-statistic is not significant, don't interpret it at all.** You can't be sure that the value of the corresponding parameter in the underlying regression model isn't really zero.

[10]With several predictors we can wander beyond the data because of the *combination* of values, even when individual values are not extraordinary. For example, both 28-inch waists and 76-inch heights can be found in men in the body fat study, but a single individual with both these measurements would not be at all typical. The model we fit is probably not appropriate for predicting the *%Body fat* for such a tall and skinny individual.

WHAT ELSE CAN GO WRONG?

- **Don't fit a linear regression to data that aren't straight.** This is the most fundamental regression assumption. If the relationship between the x's and y isn't approximately linear, there's no sense in fitting a linear model to it. What we mean by "linear" is a model of the form we have been writing for the regression. When we have two predictors, this is the equation of a plane, which is linear in the sense of being flat in all directions. With more predictors, the geometry is harder to visualize, but the simple structure of the model is consistent; the predicted values change consistently with equal size changes in any predictor.

  Usually we're satisfied when plots of y against each of the x's are straight enough. We'll also check a scatterplot of the residuals against the predicted values for signs of nonlinearity.
- **Watch out for the plot thickening.** The estimate of the error standard deviation shows up in all the inference formulas. But that estimate assumes that the error standard deviation is the same throughout the range of the x's so that we can combine (pool, actually) all the residuals when we estimate it. If s_e changes with any x, these estimates won't make sense. The most common check is a plot of the residuals against the predicted values. If plots of residuals against several of the predictors all show a thickening, and especially if they also show a bend, then consider re-expressing y. If the scatterplot against only one predictor shows thickening, consider re-expressing that predictor.
- **Make sure the errors are nearly Normal.** All of our inferences require that the true errors be modelled well by a Normal model. Check the histogram and Normal probability plot of the residuals to see whether this assumption looks reasonable.
- **Watch out for high-influence points and outliers.** We always have to be on the lookout for a few points that have undue influence on our model, and regression is certainly no exception. Partial regression plots are a good place to look for influential points and to understand how they affect each of the coefficients. Chapter 28 discusses this issue in greater depth.

CONNECTIONS

We would never consider a regression analysis without first making scatterplots. The aspects of scatterplots that we always look for—their direction, form, and strength—relate directly to regression.

Regression inference is connected to just about every inference method we have seen for measured data. The assumption that the spread of data about the line is constant is essentially the same as the assumption of equal variances required for the pooled-t methods. Our use of all the residuals together to estimate their standard deviation is a form of pooling.

Of course, the ANOVA table in the regression output connects to our consideration of ANOVA in Chapter 25. This, too, is not coincidental. Multiple regression, ANOVA, pooled t-tests, and inference for means are all part of a more general statistical model known as the General Linear Model (GLM).

What Have We Learned?

Learning Objectives

In Chapter 24, we learned to apply our inference methods to linear regression models. Now we've seen that much of what we know about those models is also true for multiple regression. Know how to perform a multiple regression, using the technology of your choice.

- Technologies differ, but most produce similar-looking tables to hold the regression results. Know how to find the values you need in the output generated by the technology you are using.

Understand how to interpret a multiple regression model.

- The meaning of a multiple regression coefficient depends on the other variables in the model. In particular, it is the relationship of *y* to the associated *x* after removing the linear effects of the other *x*'s.

Be sure to check the Assumptions and Conditions before interpreting a multiple regression model.

- The **Linearity Assumption** asserts that the form of the multiple regression model is appropriate. We check it by examining scatterplots. If the plots appear to be linear, we can fit a multiple regression model.
- The **Independence Assumption** requires that the errors made by the model in fitting the data be mutually independent. Data that arise from random samples or randomized experiments usually satisfy this assumption.
- The **Equal Variance Assumption** states that the variability around the multiple regression model should be the same everywhere. We usually check the **Equal Spread Condition** by plotting the residuals against the predicted values. This assumption is needed so that we can pool the residuals to estimate their standard deviation, which we will need for inferences about the regression coefficients.
- The **Normality Assumption** says that the model's errors should follow a Normal model. We check the **Nearly Normal Condition** with a histogram or normal probability plot of the residuals. We need this assumption to use Student's t models for inference, but for larger sample sizes, it is less important.

Know how to state and test hypotheses about the multiple regression coefficients.

- The standard hypothesis test for each coefficient is

$$H_0: \beta_j = 0 \text{ vs.}$$
$$H_A: \beta_j \neq 0$$

- We test these hypotheses by referring the test statistic

$$\frac{b_j - 0}{SE(b_j)}$$

to the Student's *t* distribution on $n - k - 1$ degrees of freedom, where *k* is the number of coefficients estimated in the multiple regression.

Interpret other associated statistics generated by a multiple regression

- R^2 is the fraction of the variation in *y* accounted for by the multiple regression model.
- Adjusted R^2 attempts to adjust for the number of coefficients estimated.
- The *F*-statistic tests the overall hypothesis that the regression model is of no more value than simply modeling *y* with its mean.
- The standard deviation of the residuals,

$$s_e = \sqrt{\frac{\sum e^2}{n - k - 1}}$$

provides an idea of how precisely the regression model fits the data.

Review of Terms

Multiple regression A linear regression with two or more predictors whose coefficients are found to minimize the sum of the squared residuals is a least squares linear multiple regression. But it is usually just called a multiple regression. When the distinction is needed, a least squares linear regression with a single predictor is called a simple regression. The multiple regression model is (p. 814)

$$y = \beta_0 + \beta_1 x_1 + \ldots + \beta_k x_k + \varepsilon$$

Least Squares We still fit multiple regression models by choosing the coefficients that make the sum of the squared residuals as small as possible. This is called the method of least squares (p. 815).

Partial regression plot The partial regression plot for a specified coefficient is a display that helps in understanding the meaning of that coefficient in a multiple regression. It has a slope equal to the coefficient value and shows the influences of each case on that value. Partial regression plots display the residuals when y is regressed on the *other* predictors against the residuals when the specified x is regressed on the other predictors (p. 817).

ANOVA table The Analysis of Variance table that is ordinarily part of the multiple regression results offers an *F*-test to test the null hypothesis that the overall regression is no improvement over just modeling y with its mean:

$$H_0: \beta_1 = \beta_2 = \ldots = \beta_k = 0$$

If this null hypothesis is not rejected, then you should not proceed to test the individual coefficients (p. 822).

***t*-ratios for the coefficients** The *t*-ratios for the coefficients can be used to test the null hypotheses that the true value of each coefficient is zero against the alternative that it is not (p. 828):

Scatterplot matrix A scatterplot matrix displays scatterplots for all pairs of a collection of variables, arranged so that all the plots in a row have the same variable displayed on their y-axis and all plots in a column have the same variable on their x-axis. Usually, the diagonal holds a display of a single variable such as a histogram or Normal probability plot, and identifies the variable in its row and column (p. 825).

Adjusted R^2 An adjustment to the R^2 statistic that attempts to allow for the number of predictors in the model. It is sometimes used when comparing regression models with different numbers of predictors (p. 830).

$$R^2_{adj} = 1 - \frac{MS_{Residual}}{MS_{Total}}$$

On the Computer REGRESSION ANALYSIS

All statistics packages make a table of results for a regression. If you can read a package's regression output table for simple regression, then you can read its table for a multiple regression. You'll want to look at the ANOVA table, and you'll see information for each of the coefficients, not just for a single slope.

Most packages offer to plot residuals against predicted values. Some will also plot residuals against the x's. With some packages you must request plots of the residuals when you request the regression. Others let you find the regression first and then analyze the residuals afterward. Either way, your analysis is not complete if you don't check the residuals with a histogram or Normal probability plot and a scatterplot of the residuals against the x's or the predicted values.

One good way to check assumptions before embarking on a multiple regression analysis is with a scatterplot matrix. This is sometimes abbreviated SPLOM in commands.

Always use a computer or programmable calculator to run multiple regressions. Before computers were available, a full multiple regression analysis could take months or even years of work.

EXCEL

- From the **Tools** menu, select Data Analysis.
- Select Regression from the **Analysis Tools** list.
- Click the **OK** button.
- Enter the data range holding the Y-variable in the box labelled "Y-range."
- Enter the range of cells holding the X-variables in the box labelled "X-range."
- Select the **New Worksheet Ply** option.
- Select **Residuals** options. Click the **OK** button.

COMMENTS

The Y and X ranges do not need to be in the same rows of the spreadsheet, although they must cover the same number of cells. But it is a good idea to arrange your data in parallel columns, as in a data table. The X-variables must be in adjacent columns. No cells in the data range may hold non-numeric values.

Although the dialogue offers a Normal probability plot of the residuals, the data analysis add-in does not make a correct probability plot, so don't use this option.

JMP

- From the **Analyze** menu select **Fit Model.**
- Specify the response, Y. Assign the predictors, X, in the **Construct Model Effects** dialogue box.
- Click on **Run Model.**

COMMENTS

JMP chooses a regression analysis when the response variable is "Continuous." The predictors can be any combination of quantitative or categorical. If you get a different analysis, check the variable types.

MINITAB

- Choose **Regression** from the **Stat** menu.
- Choose **Regression** . . . from the **Regression** submenu.
- In the Regression dialogue, assign the Y-variable to the Response box and assign the X-variables to the Predictors box.
- Click the **Graphs** button.
- In the Regression-Graphs dialogue, select **Standardized residuals,** and check **Normal plot of residuals** and **Residuals versus fits.**
- Click the **OK** button to return to the Regression dialogue.
- Click the **OK** button to compute the regression.

R

To fit a regression model of Y on X_1 and X_2:

- **lm(Y ~ X_1 + X_2, data=dataset)** produces the linear model.
- **summary(lm(Y ~ X_1 + X_2, data=dataset))** produces more information about the model.

COMMENTS

The above model assumes that your variables are in a data frame. If not, you may ignore the last option.

SPSS

- Choose **Regression** from the **Analyze** menu.
- Choose **Linear** from the **Regression** submenu.
- When the Linear Regression dialogue appears, select the Y-variable and move it to the dependent target. Then, move the X-variables to the independent target.
- Click the **Plots** button.
- In the Linear Regression Plots dialogue, choose to plot the *SRESIDs against the *ZPRED values.
- Click the **Continue** button to return to the Linear Regression dialogue.
- Click the **OK** button to compute the regression.

STATCRUNCH

To perform a multiple regression:

- Enter the response variables and all predictors in separate columns of the data table.
- Click on **Stat.**
- Choose **Regression » Multiple Linear.**
- Choose the corresponding columns from your table.
- Click on **Calculate.**

COMMENTS

You can also model interactions (not studied in this chapter, but similar to those discussed in Chapter 26).

TI-83/84

You need a special program to compute a multiple regression on the TI-83.

Exercises

1. Interpretations A regression performed to predict selling price of houses found the equation

$$\widehat{Price} = 169\,328 + 35.3\,Area + 0.718\,Lotsize - 6543\,Age$$

where price is in dollars, area is in square feet, lotsize is in square feet, and age is in years. The R^2 is 92%. One of the interpretations below is correct. Which is it? Explain what's wrong with the others.

a) Each year a house ages, it is worth $6543 less.
b) Every extra square foot of area is associated with an additional $35.30 in average price, for houses with a given lotsize and age.
c) Every dollar in price means lotsize increases 0.718 square feet.
d) This model fits 92% of the data points exactly.

2. More interpretations A household appliance manufacturer wants to analyze the relationship between total sales and the company's three primary means of advertising (television, magazines, and radio). All values were in millions of dollars. They found the regression equation

$$\widehat{Sales} = 250 + 6.75\,TV + 3.5\,Radio + 2.3\,Magazines.$$

One of the interpretations below is correct. Which is it? Explain what's wrong with the others.

a) If they did no advertising, their income would be $250 million.
b) Every million dollars spent on radio makes sales increase $3.5 million, all other things being equal.
c) Every million dollars spent on magazines increases television spending by $2.3 million.
d) Sales increase on average about $6.75 million for each million spent on television, after allowing for the effects of the other kinds of advertising.

3. Predicting final exams How well do tests given during the semester predict performance on the final? One class had three tests during the semester. Computer output of the regression gives

Dependent variable is: Final
s = 13.46 R-Sq = 77.7% R-Sq(adj) = 74.1%

Predictor	Coeff	SE(Coeff)	*t*	P-value
Intercept	−6.72	14.00	−0.48	0.636
Test1	0.2560	0.2274	1.13	0.274
Test2	0.3912	0.2198	1.78	0.091
Test3	0.9015	0.2086	4.32	<0.0001

Analysis of Variance

Source	DF	SS	MS	F	P-value
Regression	3	11961.8	3987.3	22.02	< 0.0001
Error	19	3440.8	181.1		
Total	22	15402.6			

a) Write the equation of the regression model.
b) How much of the variation in final exam scores is accounted for by the regression model?
c) Explain in context what the coefficient of Test3 scores means.
d) A student argues that clearly the first test doesn't help to predict final performance. She suggests that this exam not be given at all. Does Test1 have no effect on the final exam score? Can you tell from

this model? (*Hint:* Do you think test scores are related to each other?)

T **4. Scottish hill races 2008** Hill running—races up and down hills—has a written history in Scotland dating back to the year 1040. Races are held throughout the year at different locations around Scotland. A recent compilation of information for 91 races (for which full information was available and omitting two unusual races) includes the *Distance* (km), the *Climb* (m), and the *Record time* (minutes). A regression to predict the men's records as of 2008 looks like this:

Dependent variable is: Men's Time (mins)
R-squared 98.0% R-squared (adjusted) 98.0%
s = 6.623 with 90 − 3 = 87 degrees of freedom

Source	Sum of Squares	DF	Mean Square	F-ratio
Regression	189204	2	94602.1	2157
Residual	3815.92	87	43.8612	

Variable	Coefficient	SE(Coeff)	*t*-ratio	P-value
Intercept	−10.3723	1.245	−8.33	<0.0001
Climb (m)	0.034227	0.0022	15.7	<0.0001
Distance (km)	4.04204	0.1448	27.9	<0.0001

a) Write the regression equation. Give a brief report on what it says about men's record times in hill races.
b) Interpret the value of R^2 in this regression.
c) What does the coefficient of *Climb* mean in this regression?

5. Home prices Many variables have an impact on determining the price of a house. A few of these are size of the house (*Sq ft*), *Lotsize*, and number of *Bathrooms.* Information for a random sample of homes for sale in the Statesboro, GA, area was obtained from the Internet. Regression output modelling the asking price with *Square Footage* and number of *Bathrooms* gave the following result:

Dependent Variable is: Asking Price
s = 67013 R-Sq = 71.1% R-Sq (adj) = 64.6%

Predictor	Coeff	SE(Coeff)	*t*	P-value
Intercept	−152037	85619	−1.78	0.110
Baths	9530	40826	0.23	0.821
Sq ft	139.87	46.67	3.00	0.015

Analysis of Variance

Source	DF	SS	MS	F	P-value
Regression	2	99303550067	49651775033	11.06	0.004
Residual	9	40416679100	4490742122		
Total	11	1.39720E+11			

a) Write the regression equation.
b) How much of the variation in home asking prices is accounted for by the model?
c) Explain in context what the coefficient of *Square Footage* means.

d) The owner of a construction firm, upon seeing this model, objects because the model says that the number of bathrooms has no effect on the price of the home. He says that when *he* adds another bathroom, it increases the value. Is it true that the number of bathrooms is unrelated to house price? (*Hint:* Do you think bigger houses have more bathrooms?)

T **6. More hill races 2008** Here is the regression for the women's records for the same Scottish hill races we considered in Exercise 4:

Dependent variable is: Women's Time (mins)
R-squared 96.7% R-squared (adjusted) 96.7%
s = 10.06 with 90 − 3 = 87 degrees of freedom

Source	Sum of Squares	DF	Mean Square	F-ratio
Regression	261029	2	130515	1288
Residual	8813.02	87	101.299	

Variable	Coefficient	SE(Coeff)	*t*-ratio	P-value
Intercept	−11.6545	1.891	−1.16	≤0.0001
Climb (m)	0.045195	0.0033	13.7	≤0.0001
Distance (km)	4.43427	0.2200	20.2	≤0.0001

a) Compare the regression model for the women's records with that found for the men's records in Exercise 4.
Here's a scatterplot of the residuals for this regression:

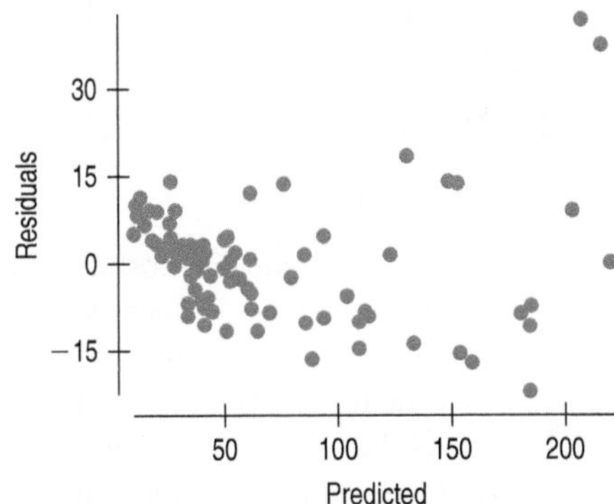

b) Discuss the residuals and what they say about the assumptions and conditions for this regression.

7. Predicting finals II Here are some diagnostic plots for the final exam data from Exercise 3. These were generated by a computer package and may look different from the plots generated by the packages you use. (In particular, note that the axes of the Normal probability plot are swapped relative to the plots we've made in the text. We only care about the pattern of this plot, so it shouldn't affect your interpretation.) Examine these plots and discuss whether the assumptions and conditions for the multiple regression seem reasonable.

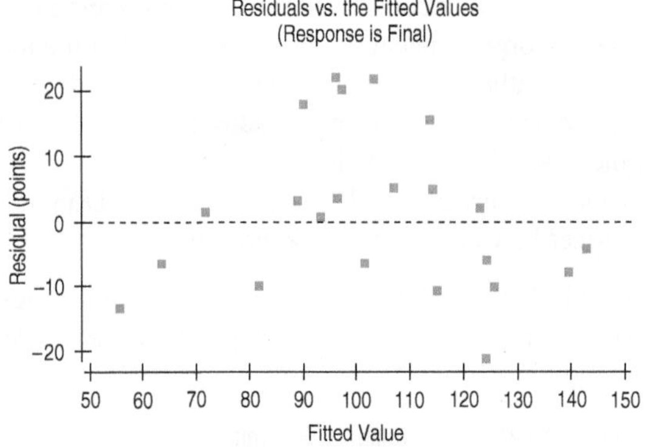

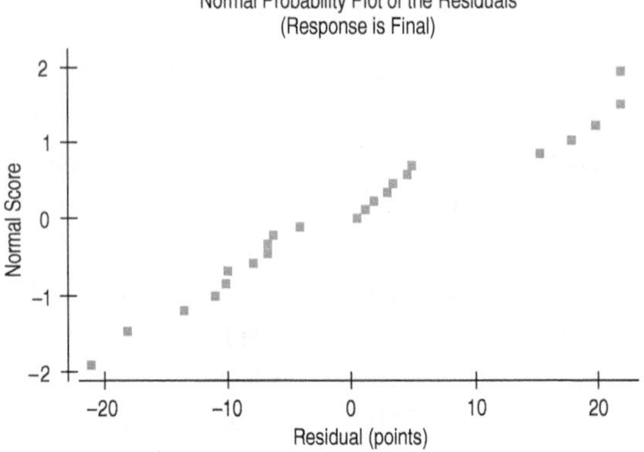

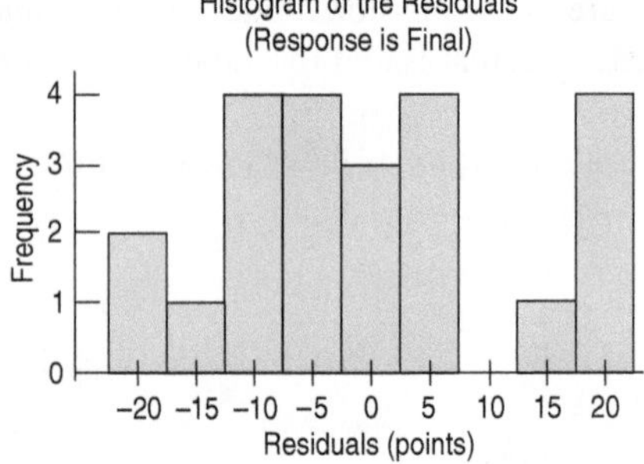

8. **Home prices II** Here are some diagnostic plots for the home prices data from Exercise 5. These were generated by a computer package and may look different from the plots generated by the packages you use. (In particular, note that the axes of the Normal probability plot are swapped relative to the plots we've made in the text. We only care about the pattern of this plot, so it shouldn't affect your interpretation.) Examine these plots and discuss whether the assumptions and conditions for the multiple regression seem reasonable.

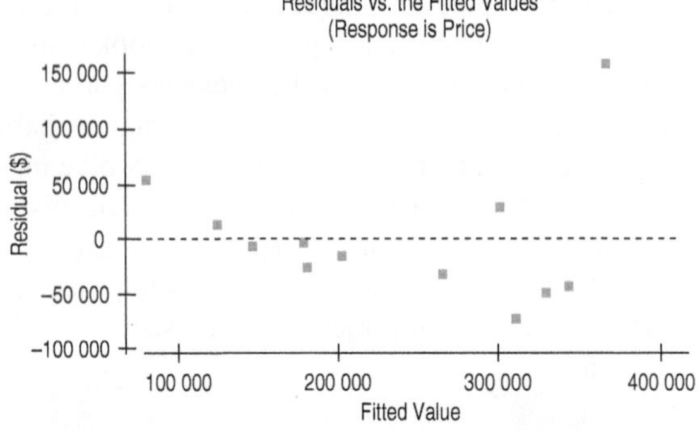

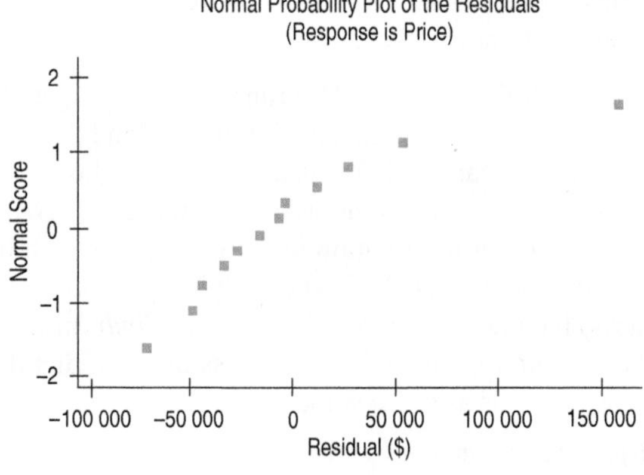

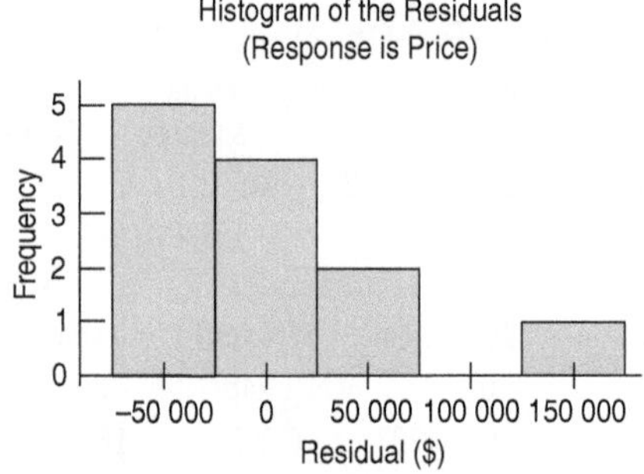

9. **Receptionist performance** The American Federation of Labor has undertaken a study of 30 receptionists' yearly salaries (in thousands of dollars). The organization wants to predict *Salaries* from several other variables.

The variables considered to be potential predictors of salary are:

X1 = months of service

X2 = years of education

X3 = score on standardized test

X4 = words per minute (wpm) typing speed

X5 = ability to take dictation in words per minute

A multiple regression model with all five variables was run on a computer package, resulting in the following output:

Variable	Coefficient	Std. Error	*t*-value
Intercept	9.788	0.377	25.960
X1	0.110	0.019	5.178
X2	0.053	0.038	1.369
X3	0.071	0.064	1.119
X4	0.004	0.307	0.013
X5	0.065	0.038	1.734

s = 0.430 $R^2 = 0.863$

Assume that the residual plots show no violations of the conditions for using a linear regression model.

a) What is the regression equation?

b) From this model, what is the predicted *Salary* (in thousands of dollars) of a receptionist with 10 years (120 months) of experience, Grade 9 education (nine years of education), a 50 on the standardized test, 60 wpm typing speed, and the ability to take 30 wpm dictation?

c) Test whether the coefficient for words per minute of typing speed (X4) is significantly different from zero at $\alpha = 0.05$.

d) How might this model be improved?

e) A correlation of *Age* with *Salary* finds $r = 0.682$, and the scatterplot shows a moderately strong positive linear association. However, if X6 = *Age* is added to the multiple regression, the estimated coefficient of *Age* turns out to be $b_6 = -0.154$. Explain some possible causes for this apparent change of direction in the relationship between *Age* and *Salary*.

10. GPA and SATs A large section of Stat 200 was asked to fill out a survey on grade point average and SAT scores. A regression was run to find out how well Math and Verbal SAT scores could predict academic performance as measured by GPA. The regression was run on a computer package with the following output:

Response: GPA

	Coefficient	Std Error	t-ratio	Prob > |t|
Intercept	0.574968	0.253874	2.26	0.0249
SAT Verbal	0.001394	0.000519	2.69	0.0080
SAT Math	0.001978	0.000526	3.76	0.0002

a) What is the regression equation?

b) From this model, what is the predicted GPA of a student with an SAT Verbal score of 500 and an SAT Math score of 550?

c) What else would you want to know about this regression before writing a report about the relationship between SAT scores and grade point averages? Why would these be important to know?

11. Body fat revisited The data set on body fat contains 15 body measurements on 250 men from 22 to 81 years old. Is average *%Body fat* related to *Weight*? Here's a scatterplot:

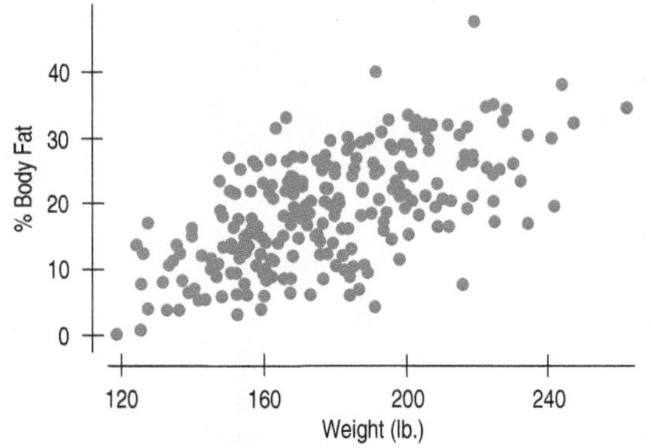

And here's the simple regression:

Dependent variable is: Pct BF
R-squared = 38.1% R-squared (adjusted) = 37.9%
s = 6.538 with 250 – 2 = 248 degrees of freedom

Variable	Coefficient	SE(Coeff)	t-ratio	P-value
Intercept	–14.6931	2.760	–5.32	<0.0001
Weigh t	0.18937	0.0153	12.4	<0.0001

a) Is the coefficient of *%Body fat* on *Weight* statistically distinguishable from 0? (Perform a hypothesis test.)

b) What does the slope coefficient mean in this regression?

We saw before that the slopes of both *Waist size* and *Height* are statistically significant when entered into a multiple regression equation. What happens if we add *Weight* to that regression? Recall that we've already checked the assumptions and conditions for regression on *Waist size* and *Height* in the chapter. Here is the output from a regression on all three variables:

Dependent variable is: Pct BF
R-squared = 72.5% R-squared (adjusted) = 72.2%
s = 4.376 with 250 – 4 = 246 degrees of freedom

Source	Sum of Squares	DF	Mean Square	F-ratio
Regression	12418.7	3	4139.57	216
Residual	4710.11	246	19.1468	

Variable	Coefficient	SE(Coeff)	t-ratio	P-value
Intercept	–31.4830	11.54	–2.73	0.0068
Waist	2.31848	0.1820	12.7	<0.0001
Height	–0.224932	0.1583	–1.42	0.1567
Weight	–0.100572	0.0310	–3.25	0.0013

c) Interpret the slope for *Weight*. How can the coefficient for *Weight* in this model be negative when its coefficient was positive in the simple regression model?

d) What does the P-value for *Height* mean in this regression? (Perform the hypothesis test.)

12. Breakfast cereals We saw in Chapter 7 that the calorie content of a breakfast cereal is linearly associated with its sugar content. Is that the whole story? Here's the output of a regression model that regresses *Calories* for each serving on its *Protein* (g), *Fat* (g), *Fibre* (g), *Carbohydrate* (g), and *Sugars* (g) content.

Dependent variable is: calories
R-squared = 84.5% R-squared (adjusted) = 83.4%
s = 7.947 with 77 – 6 = 71 degrees of freedom

Source	Sum of Squares	DF	Mean Square	F-ratio
Regression	24367.5	5	4873.50	77.2
Residual	4484.45	71	63.1613	

Variable	Coefficient	SE(Coeff)	t-ratio	P-value
Intercept	20.2454	5.984	3.38	0.0012
Protein	5.69540	1.072	5.32	<0.0001
Fat	8.35958	1.033	8.09	<0.0001
Fibre	–1.02018	0.4835	–2.11	0.0384
Carbo	2.93570	0.2601	11.3	<0.0001
Sugars	3.31849	0.2501	13.3	<0.0001

Assuming that the conditions for multiple regression are? met,

a) What is the regression equation?
b) Do you think this model would do a reasonably good job at predicting calories? Explain.
c) To check the conditions, what plots of the data might you want to examine?
d) What does the coefficient of *Fat* mean in this model?

13. Body fat again Chest size might be a good predictor of body fat. Here's a scatterplot of *%Body fat* versus *Chest size*.

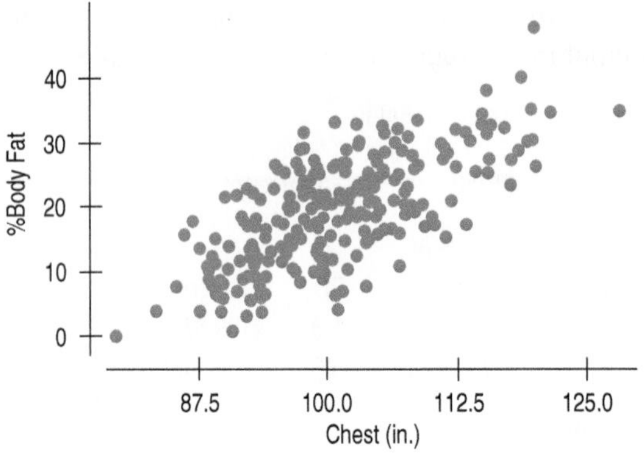

A regression of *%Body Fat* on *Chest Size* gives the following equation:

Dependent variable is: Pct BF
R-squared = 49.1% R-squared (adjusted) = 48.9%
S = 5.930 with 250 − 2 = 248 degrees of freedom

Variable	Coefficient	SE(Coeff)	t-ratio	P-value
Intercept	−52.7122	4.654	−11.3	< 0.0001
Chest Size	0.712720	0.0461	15.5	< 0.0001

a) Is the slope of *%Body fat* on *Chest size* statistically distinguishable from 0? (Perform a hypothesis test.)
b) What does the answer in part a) mean about the relationship between *%Body fat* and *Chest size*?

We saw before that the slopes of both *Waist size* and *Height* are statistically significant when entered into a multiple regression equation. What happens if we add *Chest size* to that regression? Here is the output from a regression on all three variables:

Dependent variable is: Pct BF
R-squared = 72.2% R-squared (adjusted) = 71.9%
s = 4.399 with 250 − 4 = 246 degrees of freedom

Source	Sum of Squares	DF	Mean Square	F-ratio	P-value
Regression	12368.9	3	4122.98	213	<0.0001
Residual	4759.87	246	19.3491		

Variable	Coefficient	SE(Coeff)	t-ratio	P-value
Intercept	2.07220	7.802	0.266	0.7908
Waist	2.19939	0.1675	13.1	<0.0001
Height	−0.561058	0.1094	−5.13	<0.0001
Chest Size	−0.233531	0.0832	−2.81	0.0054

c) Interpret the coefficient for *Chest size*.
d) Would you consider removing any of the variables from this regression model? Why or why not?

14. Grades The table below shows the five scores from an Introductory Statistics course. Find a model for predicting final exam score by trying all possible models with two predictor variables. Which model would you choose? Be sure to check the conditions for multiple regression.

Name	Final	Midterm 1	Midterm 2	Project	Homework
Timothy F.	117	82	30	10.5	61
Karen E.	183	96	68	11.3	72
Verena Z.	124	57	82	11.3	69
Jonathan A.	177	89	92	10.5	84
Elizabeth L.	169	88	86	10.6	84
Patrick M.	164	93	81	10	71
Julia E.	134	90	83	11.3	79
Thomas A.	98	83	21	11.2	51
Marshall K.	136	59	62	9.1	58
Justin E.	183	89	57	10.7	79
Alexandra E.	171	83	86	11.5	78
Christopher B.	173	95	75	8	77
Justin C.	164	81	66	10.7	66
Miguel A.	150	86	63	8	74
Brian J.	153	81	86	9.2	76
Gregory J.	149	81	87	9.2	75
Kristina G.	178	98	96	9.3	84
Timothy B.	75	50	27	10	20
Jason C.	159	91	83	10.6	71
Whitney E.	157	87	89	10.5	85
Alexis P.	158	90	91	11.3	68
Nicholas T.	171	95	82	10.5	68
Amandeep S.	173	91	37	10.6	54
Irena R.	165	93	81	9.3	82
Yvon T.	168	88	66	10.5	82
Sara M.	186	99	90	7.5	77
Annie P.	157	89	92	10.3	68
Benjamin S.	177	87	62	10	72
David W.	170	92	66	11.5	78
Josef H.	78	62	43	9.1	56
Rebecca S.	191	93	87	11.2	80
Joshua D.	169	95	93	9.1	87
Ian M.	170	93	65	9.5	66
Katharine A.	172	92	98	10	77
Emily R.	168	91	95	10.7	83
Brian M.	179	92	80	11.5	82
Shad M.	148	61	58	10.5	65
Michael R.	103	55	65	10.3	51
Israel M.	144	76	88	9.2	67
Iris J.	155	63	62	7.5	67
Mark G.	141	89	66	8	72
Peter H.	138	91	42	11.5	66
Catherine R.M.	180	90	85	11.2	78

Christina M.	120	75	62	9.1	72
Enrique J.	86	75	46	10.3	72
Sarah K.	151	91	65	9.3	77
Thomas J.	149	84	70	8	70
Sonya P.	163	94	92	10.5	81
Michael B.	153	93	78	10.3	72
Wesley M.	172	91	58	10.5	66
Mark R.	165	91	61	10.5	79
Adam J.	155	89	86	9.1	62
Jared A.	181	98	92	11.2	83
Michael T.	172	96	51	9.1	83
Kathryn D.	177	95	95	10	87
Nicole M.	189	98	89	7.5	77
Wayne E.	161	89	79	9.5	44
Elizabeth S.	146	93	89	10.7	73
John R.	147	74	64	9.1	72
Valentin A.	160	97	96	9.1	80
David T.O.	159	94	90	10.6	88
Marc I.	101	81	89	9.5	62
Samuel E.	154	94	85	10.5	76
Brooke S.	183	92	90	9.5	86

15. Fifty states 2009 Here is a data set on various measures of the 50 United States. The *Murder* rate is per 100 000, high school graduation rate is in % (*HS Grad*), *Income* is per capita income in dollars, *Illiteracy* rate is per 1000, and *Life expectancy* is in years. Find a regression model for *Life expectancy* with three predictor variables by trying all four of the possible models.

a) Which model appears to do the best?

b) Would you leave all three predictors in this model?

c) Does this model mean that by changing the levels of the predictors in this equation, we could affect *Life expectancy* in that state? Explain.

d) Be sure to check the conditions for multiple regression. What do you conclude?

State	Life expectancy	Murder rate08	Graduation Rate	Income/ cap07	Illiteracy
Alabama	74.4	7.6	62%	27557	0.15
Alaska	77.1	4.1	67	34316	0.09
Arizona	77.5	6.3	59	28088	0.13
Arkansas	75.2	5.7	72	25563	0.14
California	78.2	5.8	68	35352	0.23
Colorado	78.2	3.2	68	34902	0.10
Connecticut	78.7	3.5	75	46021	0.09
Delaware	76.8	6.5	73	34533	0.11
Florida	77.5	6.4	59	32693	0.20
Georgia	75.3	6.6	54	28452	0.17
Hawaii	80.0	1.9	69	33369	0.16
Idaho	77.9	1.5	78	26530	0.11
Illinois	76.4	6.1	78	34290	0.13
Indiana	76.1	5.1	74	28587	0.08
Iowa	78.3	2.5	93	29784	0.07
Kansas	77.3	4	76	31268	0.08
Kentucky	75.2	4.6	71	26457	0.12
Louisiana	74.2	11.9	69	29557	0.16
Maine	77.6	2.4	78	28677	0.07
Maryland	76.3	8.8	75	39136	0.11
Massachusetts	78.4	2.6	75	41740	0.10
Michigan	76.3	5.4	75	29837	0.08
Minnesota	78.8	2.1	82	34896	0.06
Mississippi	73.6	8.1	62	24530	0.16
Missouri	75.9	7.7	75	29245	0.07
Montana	77.2	2.4	83	27602	0.09
Nebraska	77.8	3.8	85	31015	0.07
Nevada	75.8	6.3	58	34424	0.16
New Hampshire	78.3	1	71	35302	0.06
New Jersey	77.5	4.3	75	41835	0.17
New Mexico	77.0	7.2	65	26766	0.16
New York	77.7	4.3	70	40296	0.22
North Carolina	75.8	6.5	63	28604	0.14
North Dakota	78.3	0.5	88	29633	0.06
Ohio	76.2	4.7	77	29657	0.09
Oklahoma	75.2	5.8	74	29044	0.12
Oregon	77.8	2.2	67	29580	0.10
Pennsylvania	76.7	5.6	82	32986	0.13
Rhode Island	78.3	2.8	72	33560	0.08
South Carolina	74.8	6.8	62	26374	0.15
South Dakota	77.7	3.2	80	28833	0.07
Tennessee	75.1	6.6	60	28301	0.13
Texas	76.7	5.6	67	31624	0.19
Utah	78.7	1.4	81	26523	0.09
Vermont	78.2	2.7	84	31184	0.07
Virginia	76.8	4.7	74	35162	0.12
Washington	78.2	2.9	70	34368	0.10
West Virginia	75.1	3.3	82	25118	0.13
Wisconsin	77.9	2.6	85	30655	0.07
Wyoming	76.7	1.9	81	36760	0.09

16. Breakfast cereals again We saw in Chapter 7 that the calorie count of a breakfast cereal is linearly associated with its sugar content. Can we predict the calories of a serving from its vitamin and mineral content? Here's a multiple regression model of *Calories* per serving on its *Sodium* (mg), *Potassium* (mg), and *Sugars* (g):

Dependent variable is: Calories

R-squared = 38.4% R-squared (adjusted) = 35.9%
s = 15.60 with 77 − 4 = 73 degrees of freedom

Source	Sum of Squares	DF	Mean Square	F-ratio	P-value
Regression	11091.8	3	3697.28	15.2	<0.0001
Residual	17760.1	73	243.289		

Variable	Coefficient	SE(Coeff)	*t*-ratio	P-value
Intercept	83.0469	5.198	16.0	<0.0001
Sodium	−0.05721	0.0215	2.67	0.0094
Potass	0.01933	0.0251	−0.769	0.4441
Sugars	2.38757	0.4066	5.87	<0.0001

Assuming that the conditions for multiple regression are met,

a) What is the regression equation?

b) Do you think this model would do a reasonably good job at predicting calories? Explain.

c) Would you consider removing any of these predictor variables from the model? Why or why not?

d) To check the conditions, what plots of the data might you want to examine?

17. Burger King revisited Recall the Burger King menu data from Chapter 7. BK's nutrition sheet lists many variables. Here's a multiple regression to predict calories for Burger King foods from *Protein* content (g), *Total fat* (g), *Carbohydrate* (g), and *Sodium* (mg) per serving:

Dependent variable is: Calories
R-squared = 100.0% R-squared (adjusted) = 100.0%
s = 3.140 with 31 − 5 = 26 degrees of freedom

Source	Sum of Squares	DF	Mean Square	F-ratio
Regression	1419311	4	354828	35994
Residual	256.307	26	9.85796	

Variable	Coefficient	SE(Coeff)	*t*-ratio	P-value
Intercept	6.53412	2.425	2.69	0.0122
Protein	3.83855	0.0859	44.7	<0.0001
Total fat	9.14121	0.0779	117	<0.0001
Carbs	3.94033	0.0336	117	<0.0001
Na/Serv.	−0.69155	0.2970	−2.33	0.0279

a) Do you think this model would do a good job of predicting calories for a new BK menu item? Why or why not?

b) The mean of *Calories* is 455.5 with a standard deviation of 217.5. Discuss what the value of *s* in the regression means about how well the model fits the data.

c) Does the R^2 value of 100.0% mean that the residuals are all actually equal to zero?

18. Popsicle sticks Recall the Popsicle stick data from Chapter 25. Suppose we want to fit a multiple regression model to predict the breaking stress from the overlap area and the breaking force, on the medium overlap joints only. Taking logs of all three variables, we get the following output:

Dependent variable is: log(stress)
R-squared = 100% R-squared (adjusted) = 100%
S = 0.0001746 on 287 degrees of freedom

Variable	Coefficient	SE(Coeff)	*t*-ratio	P-value
Intercept	9.191	0.0008452	10875	≤0.0001
log(force)	1.000	0.00002631	38002	≤0.0001
log(area)	−1.000	0.0001666	−6004	≤0.0001

This output looks too good to be true, with $R^2 = 100\%$ and MSE of virtually zero. Can you explain what happened?

Just Checking ANSWERS

1. 77.9% of the variation in maximum wind speed can be accounted for by multiple regression on central pressure and year.

2. In any given year, hurricanes with a central pressure that is 1 mb lower can be expected to have, on average, winds that are 0.933 kn faster.

3. First, the researcher is trying to prove his null hypothesis for this coefficient and, as we know, statistical inference won't permit that. Beyond that problem, we can't even be sure we understand the relationship of wind speed to year from this analysis. For example, both *Central pressure* and *Wind speed* might be changing over time, but their relationship might stay the same during any given year.

MathXL

MyStatLab

Go to MathXL at www.mathxl.com or MyStatLab at www.mystatlab.com. You can practise exercises for this chapter as often as you want. The guided solutions will help you find answers step by step. You'll find a personalized study plan available to you too!

chapter

28 Multiple Regression Wisdom

Where the world ceases to be the scene of our personal hopes and wishes, where we face it as free beings admiring, asking and observing, there we enter the realm of Art and Science.

—*Albert Einstein*

Where have we been?

We've looked ahead in each of the preceding chapters, but this is a good time to take stock. Wisdom in building and interpreting multiple regressions uses all that we've discussed throughout this book—even histograms and scatterplots. But most important is to keep in mind that we use models to help us understand the world with data. This chapter is about building that understanding even when we use powerful, complex methods. And that's been our purpose all along.

Who	Roller coasters
What	See Table 28.1. (For multiple regression, we have to know the "what" and the units for each variable.)
Where	Worldwide
When	All were in operation in 2003.
Source	The Roller Coaster DataBase, www.rcdb.com

Roller coasters are an old thrill that continue to grow in popularity. Engineers and designers compete to make them bigger and faster. Canada's fastest roller coaster (as of July 2013) is the *Leviathan* at Canada's Wonderland in Vaughan, Ontario. With a cost of $28 million to build, it reaches 306 feet tall (93 m) and maxes out at 92 mph (148 km/h). And at 3 minutes and 28 seconds, it's also one of the longest rides around! Can we learn what makes a roller coaster fast? Or how long the ride will last? Below are data on some of the fastest roller coasters in the world today. Here are the variables and their units:

- *Type* indicates what kind of track the roller coaster has. The possible values are "wooden" and "steel." (The frame usually is of the same construction as the track, but doesn't have to be.)
- *Duration* is the duration of the ride in seconds.
- *Speed* is the top speed in miles per hour.
- *Height* is the maximum height above ground level in feet.
- *Drop* is the greatest drop in feet.
- *Length* is the total length of the track in feet.
- *Inversions* reports whether riders are turned upside down during the ride. It has the values "yes" or "no."

Name	Park	Country	Type	Duration (sec.)	Speed (mph)	Height (ft.)	Drop (ft.)	Length (ft.)	Inversion?
New Mexico Rattler	Cliff's Amusement Park	USA	Wooden	75	47	80	75	2750	No
Fujiyama	Fuji-Q Highlands	Japan	Steel	216	80.8	259.2	229.7	6708.67	No
Goliath	Six Flags Magic Mountain	USA	Steel	180	85	235	255	4500	No
Great American Scream Machine	Six Flags Great Adventure	USA	Steel	140	68	173	155	3800	Yes
Hangman	Wild Adventures	USA	Steel	125	55	115	95	2170	Yes
Hayabusa	Tokyo SummerLand	Japan	Steel	108	60.3	137.8	124.67	2559.1	No
Hercules	Dorney Park	USA	Wooden	135	65	95	151	4000	No
Hurricane	Myrtle Beach Pavilion	USA	Wooden	120	55	101.5	100	3800	No

Table 28.1

A small selection of roller coasters from the larger data set available online at www.pearsoncanada.ca/deveaux.

It's always a good idea to explore the data before starting to build a model. Let's first consider the ride's duration. We have that information for only 63 of the 80 coasters in our data set, but there's no reason to believe that the data are missing in any patterned way so we'll look at those 63 coasters. The average duration for these coasters is 142 seconds, but one ride is as short as 28 seconds and another as long as 240 seconds. It makes sense that the duration of the ride should depend on the length of the track. Here's the scatterplot of duration against length and the regression (with residual plot):

Figure 28.1
Duration of the ride appears to be linearly related to the length of the track.

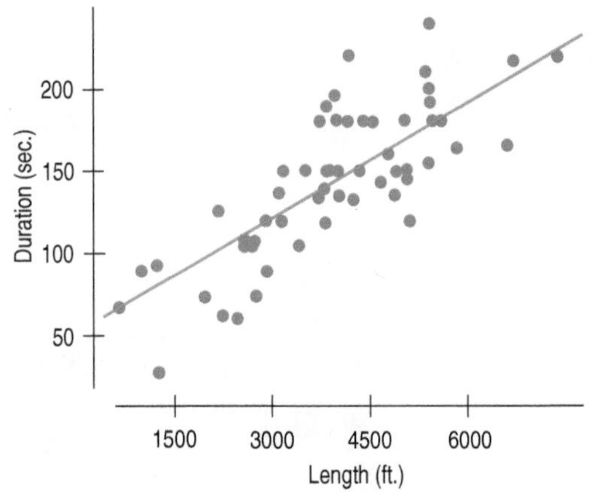

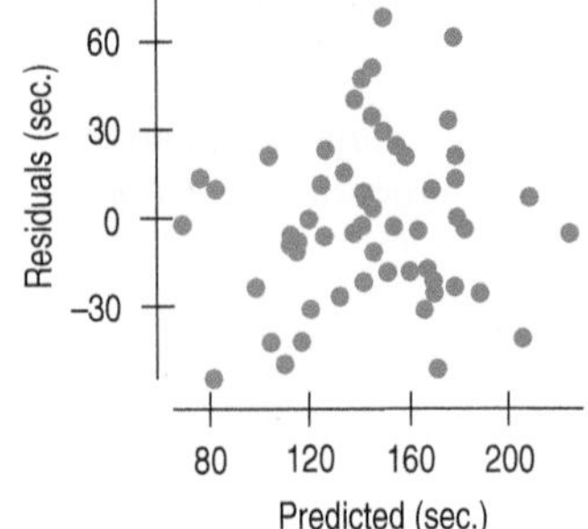

Dependent variable is: Duration
R-squared = 62.0% R-squared (adjusted) = 61.4%
s = 27.23 with 63 – 2 = 61 degrees of freedom

Source	Sum of Squares	DF	Mean Square	F-ratio
Regression	73901.7	1	73901.7	99.6
Residual	45243.7	61	741.700	

Variable	Coefficient	SE(Coeff)	t-ratio	P-value
Intercept	53.9348	9.488	5.68	≤0.0001
Length	0.0231	0.0023	9.98	≤0.0001

The regression conditions seem to be met, and the regression makes sense. We'd expect longer tracks to give longer rides, and we see that the duration of the ride increases by about 0.0231 seconds per foot of track—about 23 seconds more for each additional thousand feet of track.

28.1 Indicators

The *Leviathan* at Canada's Wonderland
robyelo357/Fotolia

Of course, there's more to these data. One interesting variable might not be one you'd naturally think of. Many modern coasters have "inversions." That's a nice way of saying that they turn riders upside down, with loops, corkscrews, or other devices. These inversions add excitement, but they must be carefully engineered, and that enforces some speed limits on that portion of the ride.

We'd like to add the information of whether the roller coaster has an inversion to our model. Until now, all our predictor variables have been quantitative. Whether or not a roller coaster has any inversions is a categorical variable ("yes" or "no"). Can we introduce the categorical variable *Inversions* as a predictor in our regression model? What would it mean if we did?

Let's start with a plot. Here's the same scatterplot of *Duration* against *Length*, but now with the roller coasters that have inversions shown as red x's and a separate regression line drawn for each type of roller coaster.

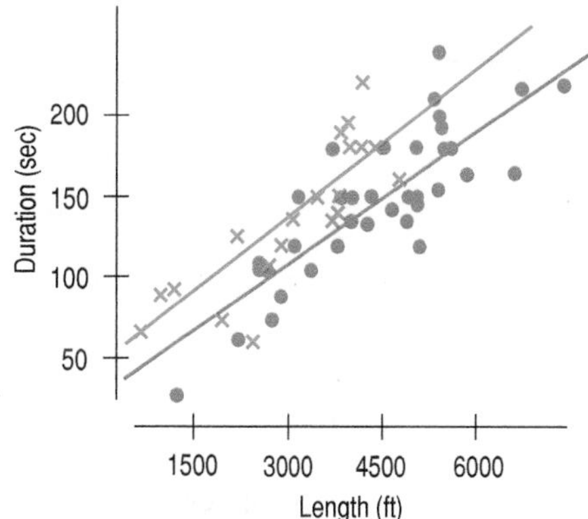

Figure 28.2
The two lines fit to coasters with inversions and without are roughly parallel.

It's easy to see that, for a given length, the roller coasters with inversions take a bit longer, and that for each type of roller coaster, the slopes of the relationship between *Duration* and *Length* are not quite equal but are similar.

We could split the data into two groups—coasters without inversions and those with inversions—and compute the regression for each group. That would look like this:

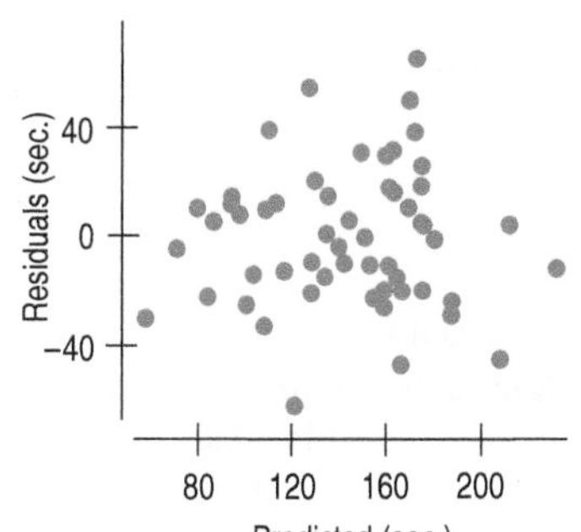

Dependent variable is: Duration
Cases selected according to: No inversions
R-squared = 69.4% R-squared (adjusted) = 68.5%
s = 25.12 with 38 − 2 = 36 degrees of freedom

Variable	Coefficient	SE(Coeff)	*t*-ratio	P-value
Intercept	25.9961	14.10	1.84	0.0734
Length	0.0274	0.003	9.03	≤0.0001

Dependent variable is: Duration
Cases selected according to: Inversions
R-squared = 70.5% R-squared (adjusted) = 69.2%
s = 23.20 with 25 − 2 = 23 degrees of freedom

Variable	Coefficient	SE(Coeff)	*t*-ratio	P-value
Intercept	47.6454	12.50	3.81	0.0009
Length	0.0299	0.004	7.41	≤0.0001

In other words, the regression model for roller coasters without inversion is

$$\widehat{Duration} = 26.00 + 0.027\,Length$$

and the regression model for roller coasters with inversion is

$$\widehat{Duration} = 47.65 + 0.030\,Length$$

The result is consistent with what the scatterplot showed: The slopes are very similar, but the intercepts are quite different.

When we have a situation like this with roughly parallel regressions for each group,[1] there's an easy way to add the group information to a single regression model. We make up a special variable that *indicates* what type of roller coaster we have, giving it the value 1 for roller coasters that have inversions and the value 0 for those that don't. (We could have reversed the coding; it's an arbitrary choice[2]). Such variables are called **indicator variables** (also known as indicators, dummies, or dummy variables) because they indicate which of two (or sometimes more than two) categories each case is in.

[1]The fact that the individual regression lines are nearly parallel is really part of the Linearity Assumption and Straight Enough Condition. You should check that the lines are nearly parallel before using this method, or read on to see what to do if they are not parallel enough.

[2]Some implementations of indicator variables use −1 and 1 for the levels of the categories.

When we add our new indicator, *Inversions*, to the regression model, the model looks like this:

Dependent variable is: Duration
R-squared = 70.4%   R-squared (adjusted) = 69.4%
s = 24.24 with 63 − 3 = 60 degrees of freedom

Variable	Coefficient	SE(Coeff)	t-ratio	P-value
Intercept	22.3909	11.39	1.97	0.0539
Length	0.028239	0.0024	11.7	< 0.0001
Inversions	30.0824	7.290	4.13	< 0.0001

The regression model is

$$\widehat{Duration} = 22.39 + 0.03\,Length + 30.08\,Inversions$$

This looks like a better model than the simple regression for all the data. The R^2 is larger, the t-ratios of both coefficients are large, and the residuals look reasonable. But, what does the coefficient for *Inversions* mean?

From the above regression model, for roller coasters without inversion (*Inversion* is 0), the model is

$$\begin{aligned}\widehat{Duration} &= 22.39 + 0.028\,Length + 30.08(0)\\ &= 22.39 + 0.028\,Length\end{aligned}$$

and for roller coasters with inversion (*Inversion* is 1), the model is

$$\begin{aligned}\widehat{Duration} &= 22.39 + 0.028\,Length + 30.08(1)\\ &= 52.47 + 0.028\,Length\end{aligned}$$

Notice how the indicator works in the model. When there is an inversion, the value 1 for the indicator causes the amount of the indicator's coefficient, 30.08, to be *added* to the prediction. When there is no inversion, the indicator is zero, so nothing is added. Looking back at the scatterplot, we can see that this is exactly what we need. The difference between the two lines is a vertical shift of about 30 seconds.

This may seem a bit confusing at first. We usually think of the coefficients in a multiple regression as slopes. For indicator variables, however, they act differently. They're vertical shifts that keep the slopes for different categories of a categorical predictor apart.

For Example USING INDICATOR VARIABLES

As a class project, students in a large Statistics class collected publicly available information on recent home sales in their hometowns. There are 894 properties. These are not a random sample, but they may be representative of home sales during a short period of time, nationwide. In Chapter 27, we looked at these data and constructed a multiple regression model. Let's look further. Among the variables available is an indication of whether the home was in an urban, suburban, or rural setting.

QUESTION: How can we incorporate information such as this in a multiple regression model?

ANSWER: We might suspect that homes in rural communities might differ in price from similar homes in urban or suburban settings. We can define an indicator (dummy) variable to be 1 for homes in rural communities and 0 otherwise. A scatterplot shows that rural homes have, on average, lower prices for a given living area:

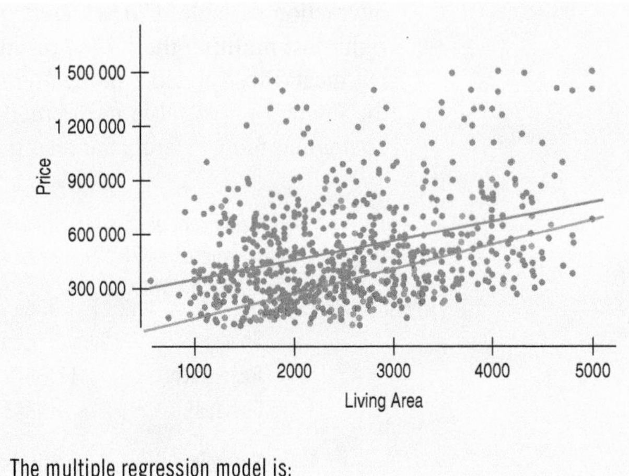

The multiple regression model is:
Dependent variable is: Price
R squared = 18.4% R squared (adjusted) = 18.2%
s = 260 996 with 894 − 3 = 891 degrees of freedom

Variable	Coefficient	SE(Coeff)	*t*-ratio	P-value
Intercept	230945	25706	8.98	0.0001
Living area	112.534	9.353	12.0	0.0001
Rural	−172359	23749	−7.26	0.0001

The coefficient of rural indicates that, for a given living area, rural homes sell for, on average, about $172 000 less.

Adjusting for Different Slopes

What if the lines aren't parallel? An indicator variable that is 0 or 1 can only shift the line up and down. It can't change the slope, so it works only when we have lines with the same slope and different intercepts.

Let's return to the Burger King data we looked at in Chapter 7, and look at how *Calories* are related to *Carbohydrates* (*Carbs* for short). Figure 28.3 in the margin shows the scatterplot.

Figure 28.3
Calories of Burger King foods plotted against *Carbohydrates* seems to fan out.

It's not surprising to see that more *Carbs* goes with more *Calories*, but the plot seems to thicken as we move from left to right. Could there be something else going on?[3]

Burger King foods can be divided into two groups: those with meat (including chicken and fish) and those without. When we colour the plot (red for meat, blue for nonmeat) and look at the regressions for each group, we see a different picture (in Figure 28.4).

Clearly, meat-based dishes contribute more calories from their carbohydrate content than do other Burger King foods. But we can't account for the kind of difference we see here by just including an indicator variable in a regression. It isn't just the height of the lines that is different; they have entirely different slopes. How can we deal with that in our regression model?

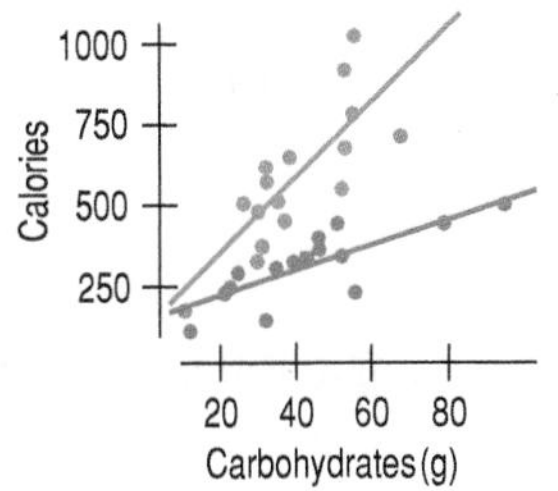

Figure 28.4
Plotting the meat-based and nonmeat items separately, we see two distinct linear patterns.

The trick is to adjust the slopes with another constructed variable. This one is the product of an indicator for one group and the predictor variable. The coefficient of this constructed **interaction term** in a multiple regression gives an adjustment to the slope, b_1, to be made for the individuals in the indicated group.[4] Here we have the indicator variable *Meat*, which is 1 for meat-containing foods and 0 for the others. We then construct an

[3] Would we even ask if there weren't?
[4] You've seen interaction effects in two-way ANOVA back in Chapter 26. Interaction terms, such as these, are exactly the same idea.

interaction variable, *Carbs*Meat,* which is just the product of those two variables. That's right; just multiply them. The resulting variable has the value of *Carbs* for foods containing meat (those coded 1 in the *Meat* indicator) and the value 0 for the others. By including the interaction variable in the model, we can adjust the slope of the line fit to the meat-containing foods. Here's the resulting analysis:

Dependent variable is: Calories
R-squared = 78.1% R-squared (adjusted) = 75.7%
s = 106.0 with 32 – 4 = 28 degrees of freedom

Source	Sum of Squares	df	Mean Square	F-ratio
Regression	1119979	3	373326	33.2
Residual	314843	28	11244.4	

Variable	Coefficient	SE(Coeff)	*t*-ratio	P-value
Intercept	137.395	58.72	2.34	0.0267
Carbs(g)	3.93317	1.113	3.53	0.0014
Meat	–26.1567	98.48	–0.266	0.7925
Carbs*Meat	7.87530	2.179	3.61	0.0012

The regression model is

$$\widehat{Calories} = 137.40 + 3.93\,Carbs - 26.16\,Meat + 7.88\,Carbs{*}Meat$$

What does the coefficient for the indicator *Meat* mean? It provides a different intercept to separate the meat and nonmeat items at the origin (where *Carbs* = 0). For these data, there is a different slope, but the two lines nearly meet at the origin, so there seems to be no need for an additional adjustment. The estimated difference of 26.16 calories is small; that's why the coefficient for *Meat* has a small *t*-statistic.

By contrast, the coefficient of the interaction term, *Carbs*Meat*, changes the slope of the lines. In particular, the slope relating calories to carbohydrates is steeper by 7.875 calories per carbohydrate gram for meat-containing foods than for meat-free foods. Its small P-value suggests that this difference is real.

Let's consider the regression model more carefully. For nonmeat items (*Meat* = 0), the regression model is

$$\begin{aligned}\widehat{Calories} &= 137.40 + 3.93\,Carbs - 26.16(0) + 7.88\,Carbs{*}0 \\ &= 137.40 + 3.93\,Carbs\end{aligned}$$

and for the meat items (*Meat* = 1), the regression model is

$$\begin{aligned}\widehat{Calories} &= 137.40 + 3.93\,Carbs - 26.16(1) + 7.88\,Carbs{*}1 \\ &= 111.24 + 11.81\,Carbs\end{aligned}$$

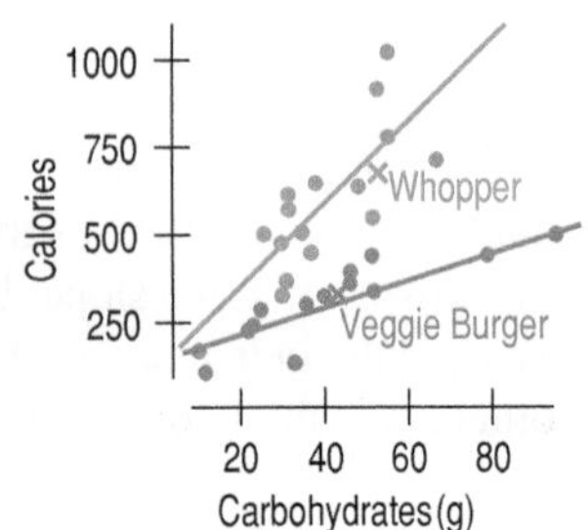

Figure 28.5
The Whopper and Veggie Burger belong to different groups.

Let's see how these adjustments work. A BK Whopper has 53 g of *Carbs* and is a meat dish. The model predicts its *Calories* as

$$137.395 + 3.93317 \times 53 - 26.1567 \times 1 + 7.8753 \times 53 \times 1 = 737.1,$$

not far from the measured calorie count of 680. By contrast, the Veggie Burger, with 43 g of *Carbs*, is predicted to have

$$137.395 + 3.93317 \times 43 - 26.1567 \times 0 + 7.87530 \times 0 = 306.5 \text{ calories},$$

not far from the 330 measured officially. The last two terms in the equation for the Veggie Burger are zero because the indicator for meat is 0 for the Veggie Burger.

28.2 Diagnosing Regression Models: Looking at the Cases

We often use regression analyses to try to understand the world. By working with the data and creating models, we can learn a great deal about the relationships among variables. As we saw with simple regression, sometimes we can learn as much from the cases that *don't* fit the model as from the bulk of cases that do. Extraordinary cases often tell us more about the world simply by the ways in which they fail to conform and the reasons we can discover for those deviations.

If a case doesn't conform to the others, we should identify it and, if possible, understand why it is different. As in simple regression, a case can be extraordinary by standing away from the model in the y direction or by having unusual values in an x-variable. In multiple regression it can also be extraordinary by having an unusual *combination* of values in the x-variables. Deviations in the y direction show up in the residuals. Deviations in the x's show up as *leverage.*

Leverage

Recent events have focused attention on airport screening of passengers, but screening has a longer history. The *Sourcebook of Criminal Justice Statistics Online* lists the numbers of various violations found by airport screeners in the U.S. for each of several types of violations in each year from 1977 to 1999. Here's a regression of the number of long guns (rifles and the like) found versus the number of times false information was discovered.

Dependent variable is: Long guns
R-squared = 7.8%   R-squared (adjusted) = 3.4%
s = 38.67 with 23 − 2 = 21 degrees of freedom

Variable	Coefficient	SE(Coeff)	*t*-ratio	P-value
Intercept	78.9069	13.60	5.80	≤0.0001
False info	0.242679	0.1823	1.33	0.1975

That summary doesn't look like it's a particularly successful regression. The R^2 is only 7.8%, and the P-value for *False Info* is large. But a look at the scatterplot tells us more:

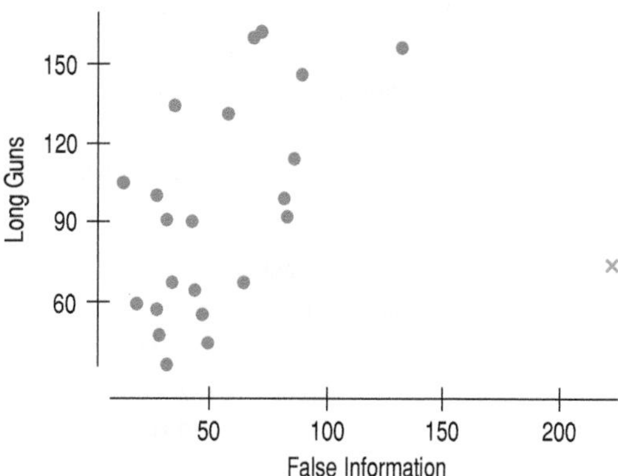

Figure 28.6

A high-leverage point can hide a strong relationship, so that you can't see it in the regression. Make a plot.

The unusual case is from 1988, when, for some reason, the number of false information reports jumped to over 200. The resulting case has high leverage because it is so far from the x-values of the other points. It's easy to see the influence of that one high-leverage case if we look at the regression lines with and without that case.

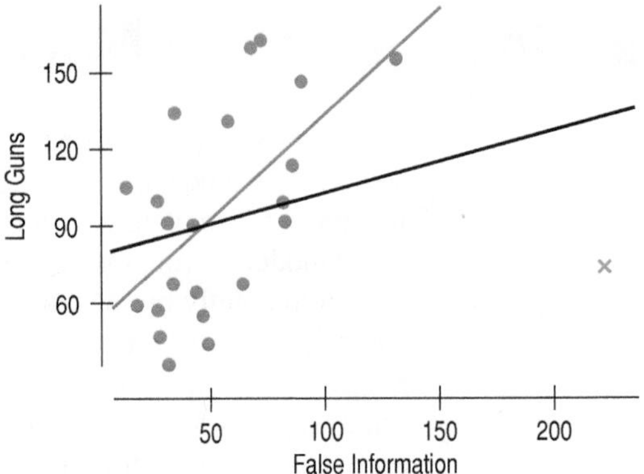

Figure 28.7

A single high-leverage point can change the regression slope quite a bit. The line omitting the point for 1988 is quite different from the line that includes the outlier.

The **leverage** of a case measures its ability to move the regression model all by itself by just moving the case in the *y* direction. In Chapter 8, when we had only one predictor variable, we could *see* high leverage points because they stood far from the mean of *x*. But now, with several predictors, we can't count on seeing them in our plots.

Fortunately, we can put a number on the leverage. If we keep everything else the same, change the *y*-value of a case by 1.0, and find a new regression, the leverage of that case is the amount by which the *predicted* value at that case would change. (The direction of the change, of course, would be the same as the direction in which we moved the case.) Leverage can never be greater than 1.0—we wouldn't expect the line to move *farther* than we move the case, only to try to keep up. Nor can it be less than 0.0—we'd hardly expect the line to move in the *opposite* direction. A point with minimal leverage has very little effect on the regression model, although it does participate in the calculations of R^2, *s*, and the *F*- and *t*-statistics.

For the airport inspections, the leverage of 1988 is 0.63. That's quite high. If there had been even one fewer long gun discovered that year (decreasing the *observed* *y*-value by 1), the *predicted* *y*-value for 1988 would have decreased by 0.63, dragging the regression line down still farther. For comparison, the next highest leverage of any other case is only 0.155.

The leverage of a case is a measure of how far that case is from the centre of the *x*'s. As always in statistics, we expect to measure that distance with a ruler based on a standard deviation—here, the standard deviation of the *x*'s. And that's really all the leverage is: an indication of how far each point is away from the centre of all the *x*-values, measured in standard deviations. Fortunately, there's a less tedious way to calculate leverage than moving each case in turn, but it's beyond the scope of this book and you'd never want to do it by hand anyway. So, just let the computer do the computing and think about what the result *means*. Most statistics programs calculate leverage values, and you should examine them.

A case can have large leverage in two different ways:

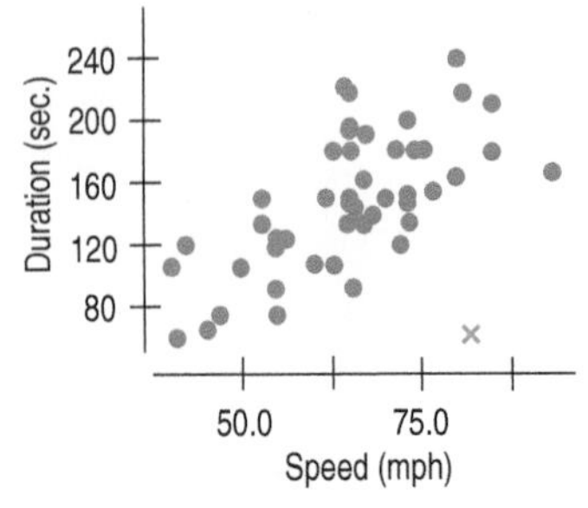

- It might be extraordinary in one or more individual variables. For example, the fastest or slowest roller coaster may stand out.
- It may be extraordinary in a *combination* of variables. For example, one roller coaster stands out in the scatterplot (in the margin) of *Duration* against *Speed*. It isn't extraordinarily fast and others have shorter duration, but the combination of high speed and short duration is unusual. Looking at leverage values can be a very effective way to discover cases that are extraordinary on a combination of *x*-variables.

There are no tests for whether the leverage of a case is too large. The average leverage value among all cases in a regression is (k + 1)/n, where k + 1 is the number of betas in the model, so sometimes we classify a point as high if it has leverage greater than 2 (k + 1)/n (twice the mean). Another common approach is just to make a histogram of the leverages. Any case whose leverage stands out in a histogram of leverages probably deserves special attention. You may decide to leave the case in the regression or to see how the regression model changes when you delete the case, but you should be aware of its potential to influence the regression.

For Example DIAGNOSING A REGRESSION

Here's another regression model for the real estate data we looked at earlier in the chapter.

Dependent variable is: Price
R squared = 23.1% R squared (adjusted) = 22.8%
s = 253 709 with 893 − 5 = 888 degrees of freedom

Variable	Coefficient	SE(Coeff)	*t*-ratio	P-value
Intercept	322470	40192	8.02	≤0.0001
Living area	92.6272	13.09	7.08	≤0.0001
Bedrooms	−69720.6	12764	−5.46	≤0.0001
Bathrooms	82577.6	13410	6.16	≤0.0001
Rural	−161575	23313	−6.93	≤0.0001

QUESTION: What do diagnostic statistics tell us about these data and this model?

ANSWER: A boxplot of the leverage values shows one rather large leverage point:

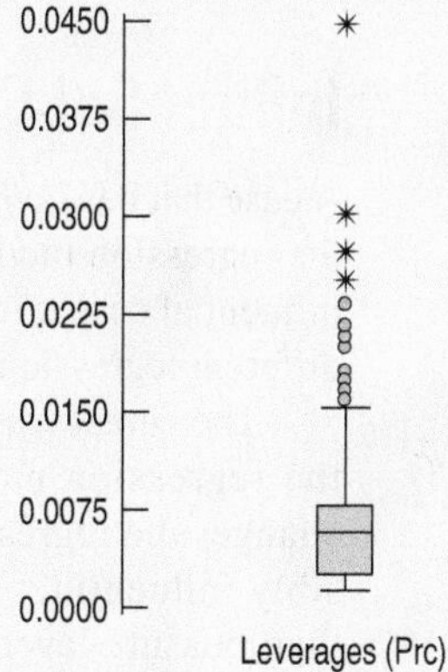

Investigation of that case reveals it to be a home with 8 bedrooms and only 2.5 bathrooms that sold for $489 900. This is a particularly unusual combination, especially for a home with that value. If we were pursuing this analysis further, we'd want to check the records for this house to be sure that the number of bedrooms and bathrooms were reported accurately.

Residuals and Standardized Residuals

Residuals are not all alike. Consider a point with leverage 1.0. That's the highest a leverage can be, and it means that the line follows the point perfectly. So, a point like that must have a zero residual. And, since we know the residual exactly, that residual has zero standard deviation. This tendency is true in general: The larger the leverage, the *smaller* the standard deviation of its residual.[5]

When we want to compare values that have differing standard deviations, it's a good idea to standardize them.[6] We can do that with the regression residuals, dividing

[5]Technically, $SD(e_i) = \sigma\sqrt{1 - h_i}$ where h_i is the leverage of the i-th case, e_i is its residual, and σ is the standard deviation of the regression model errors.

[6]Be cautious when you encounter the term "standardized residual." It is used in different books and by different statistics packages to mean quite different things. Be sure to check the meaning.

each one by an estimate of its own standard deviation. When we do that, the resulting values follow a Student's t-distribution. In fact, these standardized residuals are called **Studentized residuals**. It's a good idea to examine the Studentized residuals (rather than the simple residuals) to assess the Nearly Normal Condition and the **Does the Plot Thicken? Condition**. Any Studentized residual that stands out from the others deserves your attention.[7]

It may occur to you that we've always plotted the *unstandardized* residuals when we made regression models. And we've treated them as if they all had the same standard deviation when we checked the **Nearly Normal Condition**. It turns out that this was a simplification. It didn't matter much for simple regression, but for multiple regression models, it's a better idea to use the Studentized residuals when checking the Nearly Normal Condition. (Of course, Student's t isn't exactly Normal either—that's why we say "nearly" Normal.)

IT ALL FITS TOGETHER DEPARTMENT

Make an indicator variable for a single case—that is, construct a variable that is 0 everywhere except that it is 1 just for the case in question. When you include that indicator in the regression model, its t-ratio will be what that case's externally Studentized residual was in the original model without the indicator. That tells us that an externally Studentized residual can be used to perform a t-test of the null hypothesis that a case is *not* an outlier. If we reject that null hypothesis, we can call the point an outlier.[8]

Influential Cases

A case that has *both* high leverage and large Studentized residuals is likely to have changed the regression model substantially all by itself. Such a case is said to be **influential**. An influential case cries out for special attention because removing it is likely to give a very different regression model.

The surest way to tell whether a case is influential is to omit it[9] and see how much the regression model changes. You should call a case "influential" if omitting it changes the regression model by enough to matter for *your* purposes. To identify possibly influential cases, check the leverage and Studentized residuals. Two statistics that measure leverage—DFFITs and DFBETAs—are typically provided by many statistics packages. A statistic that combines leverage and Studentized residuals into a single measure of influence—Cook's distance—is also generally offered. If either measure is unusually large for a case, that case should be checked as a possibly influential point.

When a regression analysis has cases that have both high leverage and large Studentized residuals, it would be irresponsible to report only the regression on all the data. You should also compute and discuss the regression found with such cases removed, and discuss the extraordinary cases individually if they offer additional insight. If your interest is to understand the world, the extraordinary cases may well tell you more than the rest of the model. If your only interest is in the model (for example, because you hope to use it for prediction), then you'll want to be certain that the model wasn't determined by only a few influential cases, but instead was built on the broader base of the bulk of your data.

---

[7]There's more than one way to Studentize residuals according to how you estimate σ. You may find statistics packages referring to *externally Studentized residuals* and *internally Studentized residuals*. It is the *externally* Studentized version that follows a t-distribution, so those are the ones we recommend.

[8]Finally we have a test to decide whether a case is an outlier. Up until now, all we've had was our judgment based on how the plots looked. But you must still use your common sense and understanding of the data to decide *why* the case is extraordinary and whether it should be corrected or removed from the analysis. That important decision is *still* a judgment call.

[9]Or, equivalently, include an indicator variable that selects only for that case.

Step-by-Step Example DIAGNOSING A MULTIPLE REGRESSION

Let's consider what makes a roller coaster fast and then diagnose the model to understand more. Roller coasters get their speed from gravity (the "coaster" part), so we'd naturally look to such variables as the height and largest drop as predictors. Let's make and diagnose that multiple regression.

THINK ➡ Variables Name the variables, report the *W*'s, and specify the questions of interest.

I have data on 75 roller coasters that give their top *Speed* (mph), maximum *Height*, and largest *Drop* (both in feet).

Plot

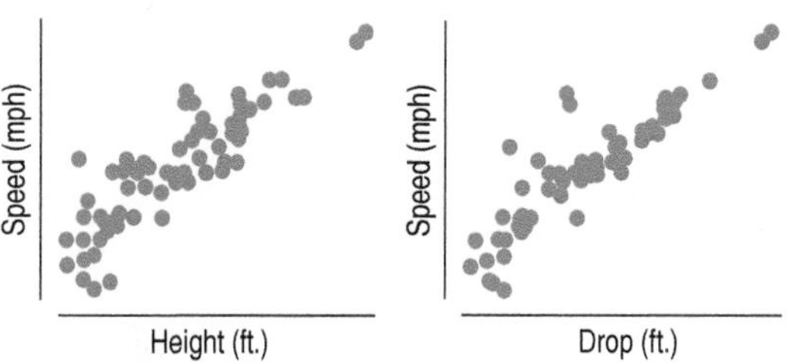

Plan Think about the assumptions and check the conditions.

✓ **Straight Enough Condition:** The plots look reasonably straight.

✓ **Independence Assumption:** There are only a few manufacturers of roller coasters worldwide. Coasters made by the same company may be similar in some respects, but each roller coaster in our data is individualized for its site, so the coasters are likely to be independent.

Because these conditions are met, I computed the regression model and found the Studentized residuals.

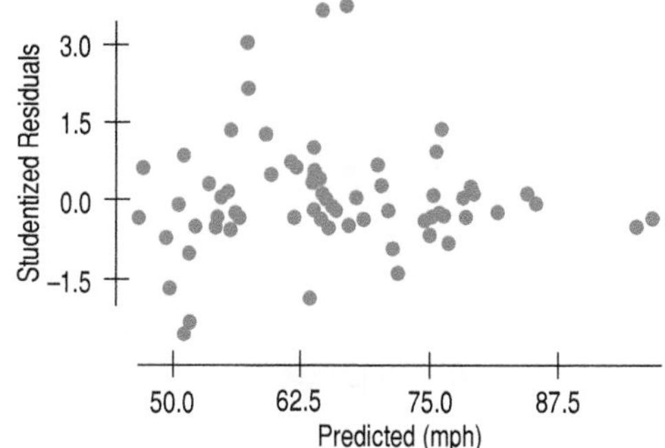

✓ **Straight Enough Condition (2):** The values for one roller coaster don't seem to affect the values for the others in any systematic fashion. This makes the Independence Assumption more plausible.

✓ **Does the Plot Thicken? Condition:** The scatterplot of Studentized residuals against predicted values shows a slight decrease in spread as the predicted values increase, but it's hard to be certain with so few observations at high speed. There do seem to be some large residuals that might be outliers. I will be cautious interpreting results from this regression.

Actually, we need the Nearly Normal Condition only if we want to do inference, but it's hard not to look at the P-values, so we usually check it out. In a multiple regression, it's best to check the Studentized residuals, although the difference is rarely large enough to change our assessment of the Normality.

✓ **Nearly Normal Condition:** A histogram of the Studentized residuals is unimodal and symmetric.

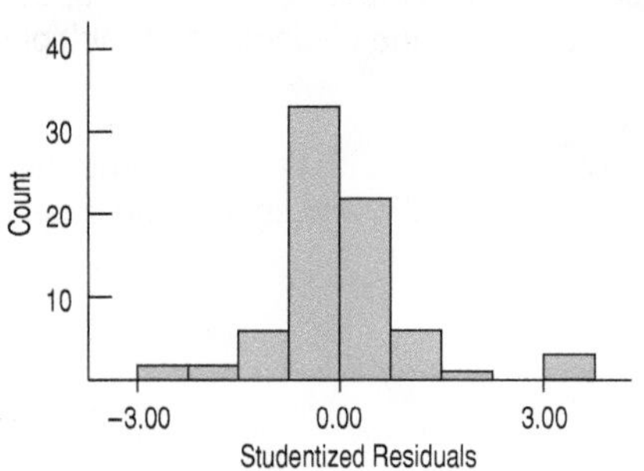

✓ **Outlier Condition:** The histogram shows three residuals separated from the others at the high end. I'll want to look at those.

Choose your method.

Under these conditions, the multiple regression model is appropriate.

SHOW ➡ Mechanics

Here is the computer output for the regression:

Dependent variable is: Speed
R-squared = 85.2% R-squared (adjusted) = 84.8%
s = 4.633 with 75 – 3 = 72 degrees of freedom

Source	Sum of Squares	DF	Mean Square	F-ratio
Regression	8902.58	2	4451.29	207
Residual	1545.50	72	21.4652	

Variable	Coefficient	SE(Coeff)	*t*-ratio	P-value
Intercept	36.9098	1.502	24.6	≤0.0001
Height	0.067218	0.0195	3.46	0.0009
Drop	0.124870	0.0190	6.57	≤0.0001

The estimated regression equation is

$$\widehat{Speed} = 36.91 + 0.067\ Height + 0.125\ Drop.$$

TELL ➡ Interpretation

The R^2 for the regression is 85.2%. *Height* and *Drop* account for 85% of the variation in *Speed* in roller coasters like these. Both *Height* and *Drop* contribute significantly to the *Speed* of a roller coaster.

Diagnosis

Leverage Most computer regression programs will calculate leverage. There is a leverage value for each case.

It may or may not be necessary to remove high leverage points from the model, but it's certainly wise to know where they are and, if possible, why they are unusual.

A histogram of the leverages shows one roller coaster with a rather high leverage of more than 0.24.

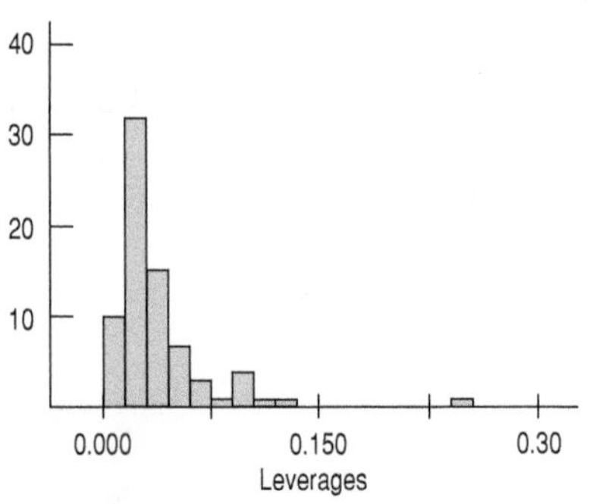

This high-leverage point is the Oblivion coaster in Alton, England. Neither the *Height* nor the *Drop* is extraordinary. To see what's going on, I made a scatterplot of *Drop* against *Height* with *Oblivion's* shown as a red x.

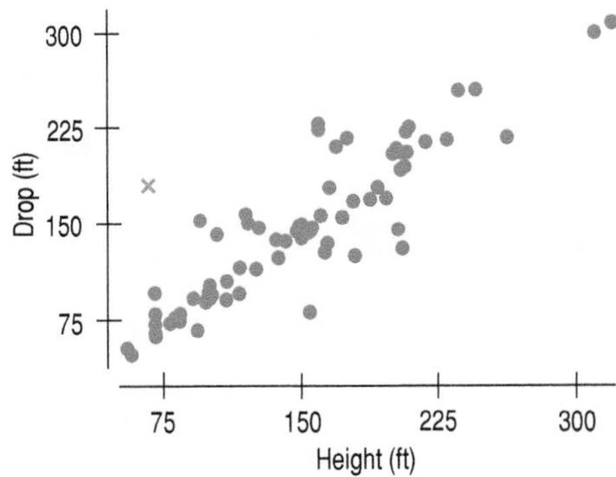

Although *Oblivion's* maximum height is a modest 65 feet, it has a surprisingly long drop of 180 feet. Looking it up, I discovered that the unusual feature of the *Oblivion's* coaster is that it plunges riders down a deep hole below the ground.

Big Residuals At this point, we might consider recomputing the regression model after removing these three coasters. That's what we do in the next section.

The histogram of the Studentized residuals (previous page) also nominates some cases for special attention. That bar on the right of the histogram holds three roller coasters with large positive residuals: the Xcelerator; Hypersonic XCL; and Volcano, the Blast Coaster. New technologies, such as hydraulics or compressed air, are used to launch all three roller coasters. These three coasters are different in that their speed doesn't depend only on gravity.

Diagnosis Wrapup

What have we learned from diagnosing the regression? We've discovered four roller coasters that may be influencing the model. And for each of them, we've been able to understand why and how they differed from the others. The oddness of *Oblivion* in plunging into a hole in the ground may cause us to value *Drop* as a predictor of *Speed* more than *Height*.

The three cases with large residuals turned out to be different from the other roller coasters because they are "blast coasters" that don't rely solely on gravity for their acceleration. Although we can't count on always discovering why unusual cases are special, diagnosing them raises the question of what about them might be different. Understanding unusual cases can help us understand our data better.

When there are unusual cases that may possibly be influential, we always want to consider the regression model without them:

Dependent variable is: Speed
R-squared = 92.7% R-squared (adjusted) = 92.1%
s = 3.331 with 72 − 3 = 69 degrees of freedom

Variable	Coefficient	SE(Coeff)	*t*-ratio	P-value
Intercept	36.4715	1.084	33.7	≤0.0001
Drop	0.175161	0.0151	11.6	≤0.0001
Height	0.016047	0.0154	1.04	0.3013

Without the three blast coasters, *Height* no longer appears to be important in the model, so we might try omitting it:

Dependent variable is: Speed
R-squared = 92.1% R-squared (adjusted) = 92.0%
s = 3.333 with 72 − 2 = 70 degrees of freedom

Variable	Coefficient	SE(Coeff)	t-ratio	P-value
Intercept	36.7620	1.048	35.1	≤0.0001
Drop	0.189301	0.0066	28.7	≤0.0001

That looks like a good model. It seems that our diagnosis has led us back to a simple regression.

INDICATORS FOR INFLUENCE

One good way to examine the effect of an extraordinary case on a regression is to construct a special indicator variable that is zero for all cases *except* the one we want to isolate. Including such an indicator in the regression model has the same effect as removing the case from the data, but it has two special advantages. First, it makes it clear to anyone looking at the regression model that we have treated that case specially. Second, the *t*-statistic for the indicator variable's coefficient can be used as a test of whether the case is influential. If the P-value is small, then that case really didn't fit well with the rest of the data. Typically, we name such an indicator with the identifier of the case we want to remove. Here's the last roller coaster model in which we have removed the influence of the three blast coasters by constructing indicators for them instead of by removing them from the data. Notice that the coefficients for the other predictors are just the same as the ones we found by omitting the cases.

Dependent variable is: Speed
R-squared = 92.1% R-squared (adjusted) = 92.0%
s = 3.333 with 74 − 5 = 69 degrees of freedom

Variable	Coefficient	SE(Coeff)	t-ratio	P-value
Intercept	36.7620	1.048	35.1	≤0.0001
Drop	0.189301	0.0066	28.7	≤0.0001
Xcelerator	19.4678	3.536	5.51	≤0.0001
HyperSonic	17.5842	3.386	5.19	≤0.0001
Volcano	17.0282	3.534	4.82	≤0.0001

The P-values for the three indicator variables confirm that each of these roller coasters doesn't fit with the others.

28.3 The Best Multiple Regression Model

"It is the mark of an educated mind to be able to entertain a thought without accepting it."

—*Aristotle*

When many possible predictors are available, we will naturally want to select only a few of them for a regression model. But which ones? The first and most important thing to realize is that often there is no such thing as the "best" regression model. (After all, all models are wrong.) Several alternative models may be useful or insightful. The "best" for one purpose may not be best for another. And the one with the highest R^2 may not be best for many purposes. There is nothing wrong with continuing to work with several models without choosing among them.

Multiple regressions are subtle. The coefficients often don't mean what they at first appear to mean. The choice of which predictors to use determines almost everything about the regression.

Predictors interact with each other, which complicates interpretation and understanding. So it is usually best to build a *parsimonious* model, using as few predictors as you can. On the other hand, we don't want to leave out predictors that are theoretically or practically important. Making this tradeoff is the heart of the challenge of selecting a good model. The best regression models, in addition to satisfying the assumptions and conditions of multiple regression, have:

- relatively few predictors to keep the model simple.
- a relatively high R^2 indicating that much of the variability in y is accounted for by the regression model.
- a relatively small value of s, the standard deviation of the residuals, indicating that the magnitude of the errors is small.
- relatively small P-values for their F- and t-statistics, showing that the overall model is better than a simple summary with the mean, and that the individual coefficients are reliably different from zero.
- no cases with extraordinarily high leverage that might dominate and alter the model.
- no cases with extraordinarily large residuals, and Studentized residuals that appear to be nearly Normal. Outliers can alter the model and certainly weaken the power of any test statistics. And the Nearly Normal Condition is required for inference.
- predictors that are reliably measured and relatively unrelated to each other.

The term "relatively" in this list is meant to suggest that you should favour models with these attributes over others that satisfy them less, but, of course, there are many trade-offs and no absolute rules.

You can deal with cases with high leverage or large residuals by introducing indicator variables.

In addition to favouring predictors that can be measured reliably, you may want to favour those that are less expensive to measure, especially if your model is intended for prediction with values not yet measured.

Seeking Multiple Regression Models Automatically

How can we find the best multiple regression model? The list of desirable features we just looked at should make it clear that there is no simple definition of the "best" model. A computer can try all combinations of the predictors to find the regression model with the highest R^2, or optimize some other criterion,[10] but models found that way are not best for all purposes and may not even be particularly good for many purposes.

Another alternative is to have the computer build a regression "stepwise." In a **stepwise** regression, at each step, a predictor is either added to or removed from the model. The predictor chosen to add is the one whose addition increases the R^2 the most (or similarly improves some other measure). The predictor chosen to remove is the one whose removal reduces the R^2 the least (or similarly loses the least on some other measure). The hope is that by following this path, the computer can settle on a good model. The model will gain or lose a predictor only if that change in the model makes a big enough change in the performance measure. The changes stop when no more changes pass the criterion.

Stepping in the wrong direction. Here's an example of how stepwise regression can go astray. We might want to find a regression to model *Horsepower* in a sample of cars from the cars' engine size *(Displacement)* and *Weight*. The simple correlations are as follows:

	HP	Disp	Wt
Horsepower	1.000		
Displacement	0.872	1.000	
Weight	0.917	0.951	1.000

[10] This is literally true. Even for many variables and a moderately large number of cases, it is computationally possible to find the "best subset" of predictors that maximizes R^2. Many statistics programs offer this impressive capability; note that if there are just 10 possible predictors under consideration, there are 1024 different subsets of predictors to compare!

Because *Weight* has a slightly higher correlation with *Horsepower*, stepwise regression will choose it first. Then, because *Weight* and engine size *(Displacement)* are so highly correlated, once *Weight* is in the model, *Displacement* won't be added to the model. But *Weight* is, at best, a lurking variable leading to both the need for more horsepower and a larger engine. Don't try to tell an engineer that the best way to increase horsepower is to add weight to the car and that the engine size isn't important! From an engineering standpoint, *Displacement* is a far more appropriate predictor of *Horsepower*, but stepwise regression can't find that model.

Stepwise methods can be valuable when there are hundreds or thousands of potential predictors, as can happen in data mining applications. They can build models that are useful for prediction or as starting points in the search for better models. Because they do each step *automatically*, however, stepwise methods are inevitably affected by influential points and nonlinear relationships. A better strategy might be to mimic the stepwise procedure yourself, but more carefully. You could consider adding or removing a variable yourself with a careful look at the assumptions and conditions each time a variable is considered. That kind of guided stepwise method is still not guaranteed to find a good model, but it may be a sensible way to search among the potential candidates.

Building Multiple Regression Models

You can build a regression model by adding variables to a growing regression. This is generally referred to as the *forward selection* method. Each time you add a predictor, you hope to account for a little more of the variation in the response. What's left over are the residuals. At each step, consider the predictors still available to you. Those that are most highly correlated with the current residuals are the ones that are most likely to improve the model.

If you see a variable with a high correlation at this stage and it is *not* among those that you thought were important, stop and think about it. Is it correlated with another predictor or with several other predictors? Don't let a variable that doesn't make sense enter the model just because it has a high correlation, but at the same time, don't exclude a predictor just because you didn't initially think it was important. (That would be a good way to make sure that you never learn anything new.) Finding the balance between these two choices underlies the art of successful model building.

Alternatively, you can start with all available predictors in the model and remove those with small t-ratios. This is generally referred to as the *backward elimination* method. At each step, make a plot of the residuals to check for outliers, and check the leverages (say, with a histogram of the leverage values) to be sure there are no high-leverage points. Influential cases can strongly affect which variables appear to be good or poor predictors in the model. It's also a good idea to check that a predictor doesn't appear to be unimportant in the model only because it's correlated with other predictors in the model. It may (as is true of *Displacement* in the example of predicting *Horsepower*) actually be a more useful or meaningful predictor than some of those in the model.

In either method, adding or removing a predictor will usually change *all* of the coefficients, sometimes by quite a bit.

Step-by-Step Example BUILDING MULTIPLE REGRESSION MODELS

Let's return to the Kids Count infant mortality data. In Chapter 27, we fit a large multiple regression model in which several of the t-ratios for coefficients were too small to be discernibly different from zero. Maybe we can build a more parsimonious model. Which model should we build?

The most important thing to do is to think about the data. Regression models can and should make sense. Many factors can influence your choice of a model, including the cost of measuring particular predictors, the reliability or possible biases in some predictors, and even the political costs or advantages to selecting predictors.

THINK ➡ **Variables** Name the available variables, report the *W*'s, and specify the question of interest or the purpose of finding the regression model.

I have data on the 50 states. The available variables are (all for 1999):

Infant mortality (deaths per 1000 live births)

Low birth weight (low BW%—%babies with low birth weight)

Child deaths (deaths per 100 000 children ages 1–14)

%Poverty (percent of children in poverty in the previous year)

HS Drop% (percent of teens who are high-school dropouts, ages 16–19)

Teen births (births per 100 000 females ages 15–17)

Teen deaths (by accident, homicide, and suicide; deaths per 100 000 teens ages 15–19)

We examined a scatterplot matrix and the regression with all potential predictors in Chapter 27.

I hope to gain a better understanding of factors that affect infant mortality.

Plan Think about the assumptions and check the conditions.

✓ **Straight Enough Condition:** The scatterplot matrix shows no bends, clumping, or outliers in any of the scatterplots.

✓ **Independence Assumption:** These data are based on random samples.

With this assumption and condition satisfied, I can compute the regression model and find residuals.

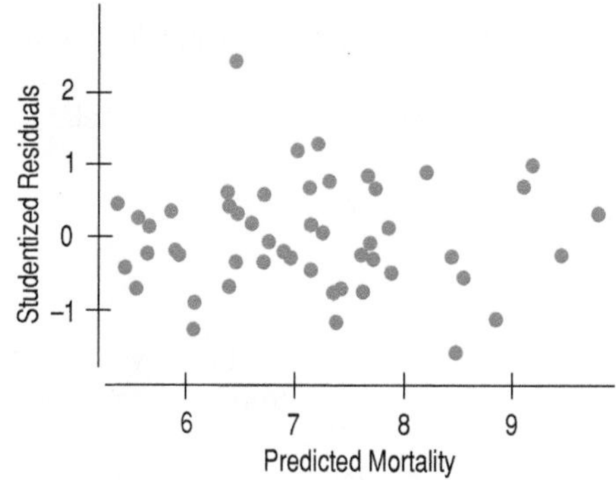

Remember that in a multiple regression, rather than plotting residuals against each of the predictors, we usually plot Studentized residuals against the predicted values.

✓ **Does the Plot Thicken? Condition:** This scatterplot of Studentized residuals versus predicted values for the full model (all predictors) shows no obvious trends in the spread.

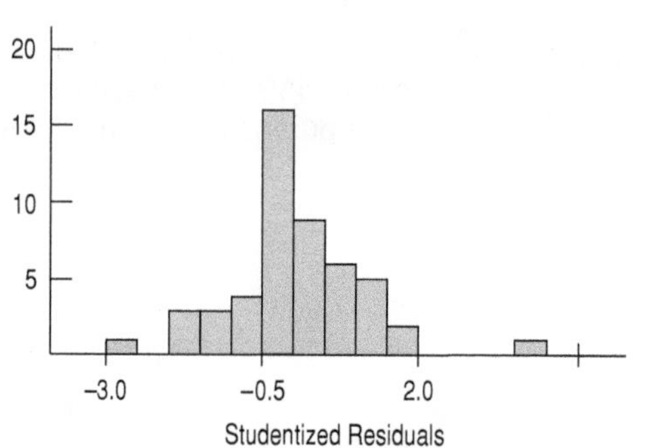

Nearly Normal Condition, Outlier Condition: A histogram of the Studentized residuals from the full model is unimodal and symmetric, but it seems to have an outlier. The unusual state is South Dakota. I'll test whether it really is an outlier by making an indicator variable for South Dakota and including it in the predictors.

SHOW ➡ Mechanics Multiple regressions are always found from a computer program.

I'll start with the full regression and work backward:

Dependent variable is: Infant mort
R-squared = 78.7% R-squared (adjusted) = 75.2%
s = 0.6627 with 50 − 8 = 42 degrees of freedom

Variable	Coefficient	SE(Coeff)	*t*-ratio	P-value
Intercept	1.31183	0.8639	1.52	0.1364
Low BW%	0.73272	0.1067	6.87	≤0.0001
Child Deaths	0.02857	0.0123	2.31	0.0256
%Poverty	−5.3026e-3	0.0332	−0.160	0.8737
HS Drop%	−0.10754	0.0540	−1.99	0.0531
Teen Births	0.02402	0.0234	1.03	0.3111
Teen Deaths	−1.5516e-4	0.0101	−0.015	0.9878
S. Dakota	2.74813	0.7175	3.83	0.0004

For model building, look at the P-values only as general indicators of how much a predictor contributes to the model.

The coefficient for the S. Dakota indicator variable has a very small P-value, so that case is an outlier in this regression model (and we can just carry along this indicator, which has the effect of removing any influence of the outlier). *Teen births*, *Teen deaths*, and *%Poverty* have large P-values and look like they are less successful predictors in this model.

You shouldn't remove more than one predictor at a time from the model because each predictor can influence how the others contribute to the model. If removing a predictor from the model doesn't change the remaining coefficients very much (or reduce the R^2 by very much), that predictor wasn't contributing very much to the model. When we carry an indicator for a data point, we don't need to check its actual influence as the indicator variable controls for this.

I'll remove *Teen deaths* first:

Dependent variable is: Infant mort
R-squared = 78.7% R-squared (adjusted) = 75.7%
s = 0.6549 with 50 − 7 = 43 degrees of freedom

Variable	Coefficient	SE(Coeff)	*t*-ratio	P-value
Intercept	1.30595	0.7652	1.71	0.0951
Low BW%	0.73283	0.1052	6.97	≤0.0001
Child Deaths	0.02844	0.0085	3.34	0.0018
%Poverty	−5.3548e-3	0.0326	−0.164	0.8703
HS Drop%	−0.10749	0.0533	−2.02	0.0501
Teen Births	0.02402	0.0231	1.04	0.3053
S. Dakota	2.74651	0.7014	3.92	0.0003

Removing *Teen births* and *%Poverty*, in turn, gives this model:

Dependent variable is: Infant mort
R-squared = 78.1% R-squared (adjusted) = 76.2%
s = 0.6489 with 50 − 5 = 45 degrees of freedom

Variable	**Coefficient**	**SE(Coeff)**	***t*-ratio**	**P-value**
Intercept	1.03782	0.6512	1.59	0.1180
Low BW%	0.78334	0.0934	8.38	≤0.0001
Child Deaths	0.03104	0.0075	4.12	0.0002
HS Drop%	−0.06732	0.0381	−1.77	0.0837
S. Dakota	2.66150	0.6899	3.86	0.0004

Adjusted R^2 can increase when you remove a predictor if that predictor wasn't contributing very much to the regression model.

Compared with the full model, the R^2 has come down only very slightly, and the adjusted R^2 has actually increased. The P-value for *HS drop%* is somewhat bigger than the standard .05 level, so if we have no particular practical reasons to keep it, let's take that variable out, and then the model looks like this:

Dependent variable is: Infant mort
R-squared = 76.6% R-squared (adjusted) = 75.1%
s = 0.6638 with 50 − 4 = 46 degrees of freedom

Variable	**Coefficient**	**SE(Coeff)**	***t*-ratio**	**P-value**
Intercept	0.760145	0.6465	1.18	0.2457
Child Deaths	0.026988	0.0073	3.67	0.0006
Low BW%	0.750461	0.0937	8.01	≤0.0001
S.Dakota	2.74057	0.7042	3.89	0.0003

This looks like a *good* model. It has a reasonably high R^2 and small P-values for each of the coefficients.

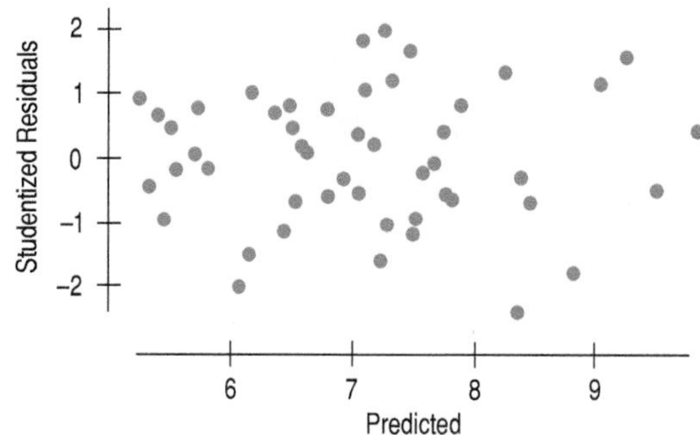

Before deciding that any regression model is a "keeper," remember to check the residuals.

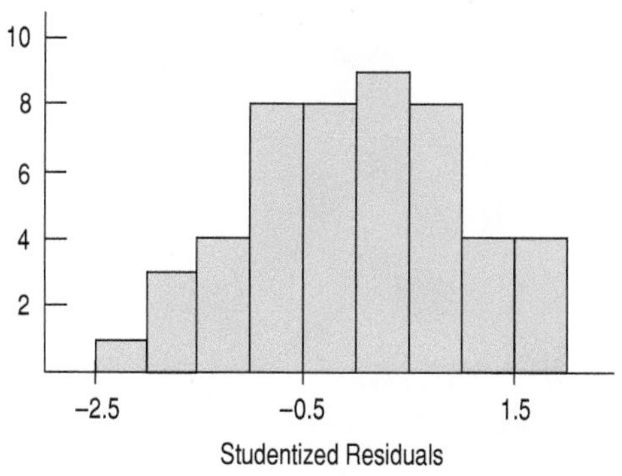

Summarize the features of this model.

Here's an example of an outlier that might help us learn something about the data or the world. Whatever makes South Dakota's infant mortality rate so much higher than the model predicts, it might be something we could address with new policies or interventions.

The scatterplot of Studentized residuals against predicted values shows no structure, and the histogram of Studentized residuals is nearly Normal. So this looks like a good model for *Infant mortality.* The coefficient for S. Dakota is still very significant, so I'd prefer to keep South Dakota separate and look into why its *Infant mortality rate* is so much higher (2.74 deaths per 1000 live births) than we would otherwise expect from its *Child death* rate and *Low birth weight* percent.

Let's try the other way and build a regression model using the forward selection method. If just building a model for prediction, we might start with the variable having the highest correlation with *Infant Mortality*. But let's assume we are mostly interested in studying how teen behaviour might be predictive of infant mortality.

The data include variables that concern young adults: *Teen births, Teen deaths*, and the *HS drop%.*

Both *Teen births* and *Teen deaths* are promising predictors, but births to teens seem more directly relevant. Here's the regression model:

Dependent variable is: Infant mort
R-squared = 29.3% R-squared (adjusted) = 27.9%
s = 1.129 with 50 − 2 = 48 degrees of freedom

Variable	Coefficient	SE(Coeff)	*t*-ratio	P-value
Intercept	4.96399	0.5098	9.74	≤0.0001
Teen Births	0.081217	0.0182	4.47	≤0.0001

One way to select variables to add to a growing regression model is to find the correlation of the residuals of the current state of the model with the potential new predictors. Predictors with higher correlations can be expected to account for more of the remaining residual variation if we include them in the regression model.

The correlations of the residuals with other (teen related) predictors look like this:

	Resids
HS Drop%	−0.188
Teen Deaths	0.333
%Poverty	0.105

Teen deaths looks like a good choice to add to the model:

Dependent variable is: Infant mort
R-squared = 39.1% R-squared (adjusted) = 36.5%
s = 1.059 with 50 − 3 = 47 degrees of freedom

Variable	Coefficient	SE(Coeff)	*t*-ratio	P-value
Intercept	3.98643	0.5960	6.69	≤0.0001
Teen Births	0.057880	0.0191	3.04	0.0039
Teen Deaths	0.028228	0.0103	2.75	0.0085

Finally, I'll try adding *HS drop%* to the model:

Dependent variable is: Infant mort
R-squared = 44.0% R-squared (adjusted) = 40.4%
s = 1.027 with 50 − 4 = 46 degrees of freedom

Variable	Coefficient	SE(Coeff)	*t*-ratio	P-value
Intercept	4.51922	0.6358	7.11	≤0.0001
Teen Births	0.097855	0.0272	3.60	0.0008
Teen Deaths	0.026844	0.0100	2.69	0.0099
HS Drop%	−0.164347	0.0819	−2.01	0.0506

Here is one more step, adding *%Poverty* to the model:

Notice that adding a predictor that does not contribute to the model can reduce the adjusted R^2.

Dependent variable is: Infant mort
R-squared = 44.0% R-squared (adjusted) = 39.1%
s = 1.038 with 50 − 5 = 45 degrees of freedom

Variable	Coefficient	SE(Coeff)	*t*-ratio	P-value
Intercept	4.49810	0.7314	6.15	≤0.0001
Teen Births	0.09690	0.0317	3.06	0.0038
Teen Deaths	0.02664	0.0106	2.50	0.0160
HS Drop%	−0.16397	0.0830	−1.98	0.0544
%Poverty	3.1053e-3	0.0513	0.061	0.9520

The P-value for *%Poverty* is quite high, so I prefer the previous model.

The regression that models *Infant mortality* on *Teen births, Teen deaths,* and *HS drop%* may be worth keeping as well. But, of course, we're not finished until we check the residuals:

Here are the residuals:

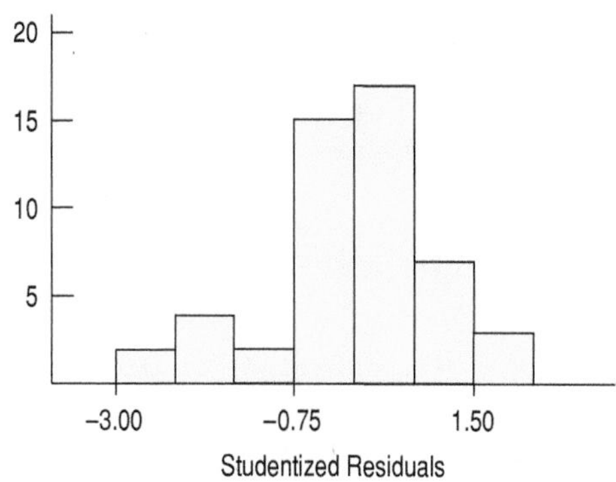

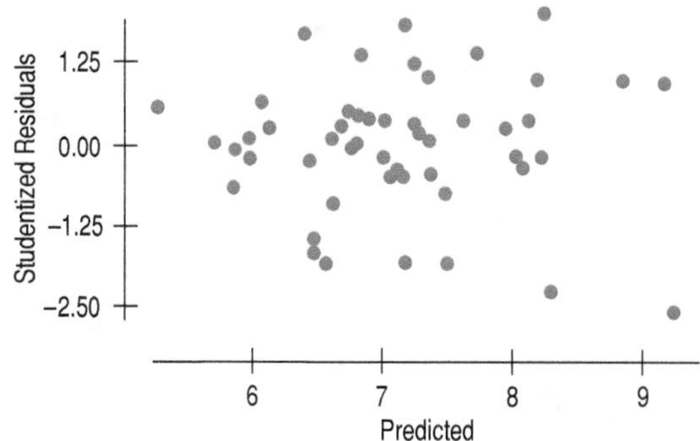

This histogram hints of a low mode holding some large negative residuals, and the scatterplot shows two in particular that trail off at the bottom right corner of the plot, Texas and New Mexico. These states are neighbours and may share some regional attributes. To be careful, I'll try removing them from the model. Or equivalently, I'll construct two indicator variables that are 1 for the named state and 0 for all others:

Dependent variable is: Infant mort
R-squared = 58.9% R-squared (adjusted) = 54.2%
s = 0.8997 with 50 − 6 = 44 degrees of freedom

Variable	Coefficient	SE(Coeff)	*t*-ratio	P-value
Intercept	4.15748	0.5673	7.33	≤0.0001
Teen Births	0.13823	0.0259	5.33	≤0.0001
Teen Deaths	0.02669	0.0090	2.97	0.0048
HS Drop%	−0.22808	0.0735	−3.10	0.0033
New Mexico	−3.01412	0.9755	−3.09	0.0035
Texas	−2.74363	0.9748	−2.81	0.0073

Removing the two outlying states has improved the model noticeably. The indicators for both states have small P-values, so I conclude that they were in fact outliers for this model. The R^2 has improved to 58.9%, and the P-values of all the other coefficients have been reduced.

A final check on the residuals from this model shows that they satisfy the regression conditions:

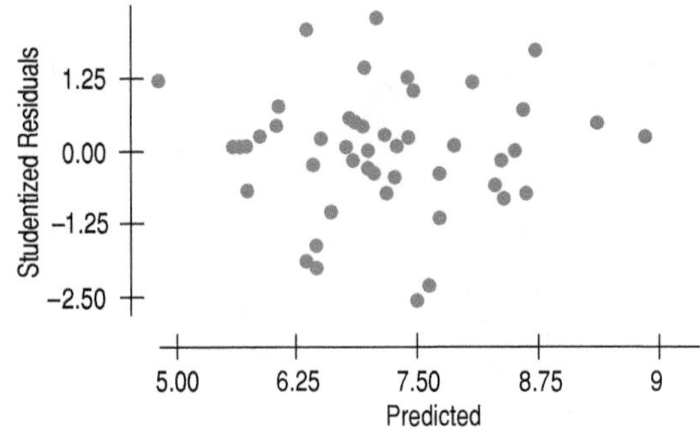

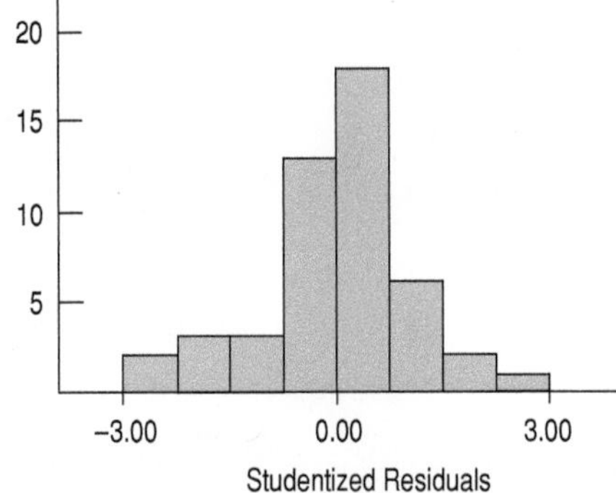

This model is an alternative to the first one I found. It has a smaller R^2 (58.9%) and larger *s* value, but it might be useful for understanding the relationships between these teen-related variables and *Infant* mortality.

TELL ➡ Compare and contrast the models.

I have found two reasonable regression models for *Infant mortality*. The first finds that *Infant mortality* can be modelled by *Child deaths* and *%Low birth weight*, removing the influence of South Dakota:

For a more complete understanding of infant mortality, we should look into South Dakota's early childhood variables and the teen-related variables in New Mexico and Texas. We might learn as much about infant mortality by understanding why these states stand out—and how they differ from each other—as we would from the regression models themselves.

$$\widehat{\text{Infant Mortality}} = 0.76 + 0.027\ \text{Child Deaths} + 0.75\ \text{LowBW\%}$$

It may be worthwhile to look into why South Dakota is so different from the other states.

The other model focused on teen behaviour, modelling *Infant mortality* by *Teen births*, *Teen deaths*, and *HS Drop%*, removing the influence of Texas and New Mexico:

$$\widehat{\text{Infant mortality}} = 4.16 + 0.138\ \text{Teen births} + 0.027\ \text{Teen deaths} - 0.228\ \text{HS Drop\%}$$

The coefficient of *HS Drop%* is the opposite sign of the simple relationship between *Infant deaths* and *HS Drop%*.

Each model has nominated different states as outliers. For a more complete understanding of infant mortality, it might be worthwhile to look into why these states are outliers in these models.

Which model is better? That depends on what you want to know. Remember—all models are wrong. But both may offer useful information and insights about infant mortality and its relationship with other variables, as well as the states that stood out and why they differ from the others.

Just Checking

1. Give two ways that we use histograms to support the construction, inference, and understanding of multiple regression models.
2. Give two ways that we use scatterplots to support the construction, inference, and understanding of multiple regression models.
3. What role does the Normal model play in the construction, inference, and understanding of multiple regression models?

28.4 Regression Roles

We build regression models for a number of reasons. One reason is to model how variables are related to each other in the hope of understanding the relationships. Another is to build a model that might be used to predict values for a response variable when given values for the predictor variables. A third reason is to identify and control values that seem to be too high or too low (think of salaries, sports agents, or the output of a production line).

When we hope to understand, we are often particularly interested in simple, straightforward models in which predictors are as unrelated to each other as possible. We are especially happy when the *t*-statistics are large, indicating that the predictors each contribute to the model. We are likely to want to look at partial regression plots to understand the coefficients and to check that no outliers or influential points are affecting them.

When prediction is our goal, we are more likely to care about the overall R^2 and MS_E. Good prediction occurs when much of the variability in y is accounted for by the model. We might be willing to keep variables in our model that have relatively small t-statistics simply for the stability that having several predictors can provide. We care less whether the predictors are related to each other because we don't intend to interpret the coefficients anyway.

When control is our goal, we are mainly interested in outlier detection. We examine plots to see if any points have extreme response values for a given level of the predictor variable. It may be the case that a particular hockey player is overpaid, given his contribution to the team (goals, assists, games played, etc... .) or it may be that his contribution simply can't be measured numerically! This is as much an art as a science.

In all roles, we may include some predictors to "get them out of the way." Regression offers a way to approximately control for factors when we have observational data because each coefficient measures effects *after removing the effects* of the other predictors. Of course, it would be better to control for factors in a randomized experiment, but often that's just not possible.

*28.5 Indicators for Three or More Levels

It's easy to construct indicators for a variable with two levels; we just assign 0 to one level and 1 to the other. But variables like *Month* or *Class* often have several levels. You can construct indicators for a categorical variable with several levels by constructing a separate indicator for each of these levels. There's just one trick: You have to choose one of the categories as a "baseline" and *leave out* its indicator. Then the coefficients of the other indicators can be interpreted as the amount by which their categories differ from the baseline, after allowing for the linear effects of the other variables in the model.[11]

Make sure your collection of indicators doesn't exhaust all the categories. One category must be left out to serve as a baseline or the regression model can't be found. For the two-category variable *Inversions*, we used "no inversion" as the baseline and coasters with an inversion got a 1. We needed only one variable for two levels. If we wished to represent *Month* with indicators, we would need 11 of them. We might, for example, define *January* as the baseline, and make indicators for *February*, *March*, ..., *November*, and *December*. Each of these indicators would be 0 for all cases except for the ones that had that value for the variable month. Why not just a single variable with "1" for *January*, "2" for *February*, and so on? That is quite unlikely to work. It would impose the pretty strict assumption that the responses to the months are equally spaced—that is, that the change in our response variable from January to February is the same in both direction and amount as the change from July to August. That's a pretty severe restriction and may not be true for many kinds of data. Using 11 indicators releases the model from that restriction, but, of course, at the expense of having 10 fewer degrees of freedom for all of our t-tests.

WHAT CAN GO WRONG?

Collinearity

Let's look at the infant mortality data one more time. One good predictor of *Infant mortality* is *Teen deaths.*

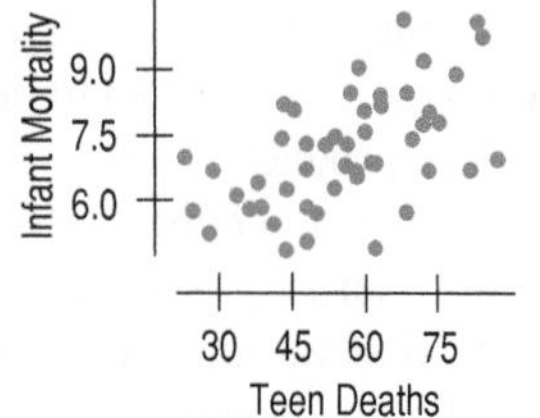

Dependent variable is: Infant mort
R-squared = 27.2% R-squared (adjusted) = 25.7%
s = 1.146 with 50 − 2 = 48 degrees of freedom

Variable	Coefficient	SE(Coeff)	t-ratio	P-value
Intercept	4.73979	0.5866	8.08	≤0.0001
Teen Deaths	0.042129	0.0100	4.23	0.0001

[11]There are alternative coding schemes that compare all the levels with the mean. Make sure you know how the indicators are coded.

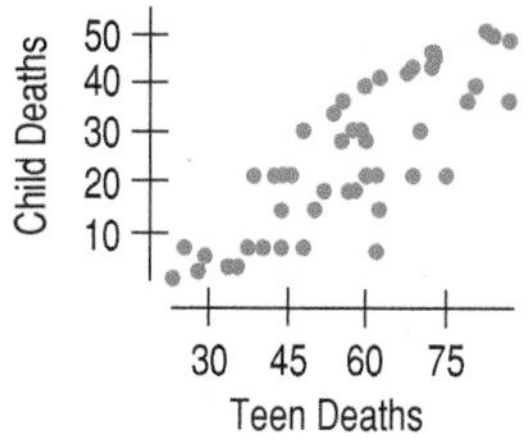

Figure 28.8
Child deaths and *Teen deaths* are linearly related.

Teen deaths has a positive coefficient (as we might expect) and a very small P-value. Suppose we now add *Child deaths rate* (CDR) to the regression model:

Dependent variable is: Infant mort
R-squared = 42.6% R-squared (adjusted) = 40.1%
s = 1.029 with 50 − 3 = 47 degrees of freedom

Variable	Coefficient	SE(Coeff)	*t*-ratio	P-value
Intercept	5.79561	0.6049	9.58	≤0.0001
Teen Deaths	−1.86877e–3	0.0153	−0.122	0.9032
Child Deaths	0.059398	0.0168	3.55	0.0009

Suddenly, *Teen deaths* has a small negative coefficient and a very large P-value. What happened? The problem is that *Teen deaths* and *Child deaths* are closely associated. The coefficient of *Teen deaths* now reports how *Infant mortality* is related to *Teen deaths after allowing for the linear effects* of *Child deaths* on both variables.

You may find this problem referred to as "multicollinearity." But there is no such thing as "unicollinearity"—we need at least two predictors for there to be a linear association between them—so there is no need for the extra two syllables.

When we have several predictors, we must think about how the predictors are related to each other. When predictors are unrelated to each other, each provides new information to help account for more of the variation in *y*. Just as we need a predictor to have a large enough variability to provide a stable base for simple regression, when we have several predictors, we need for them to vary in different directions for the multiple regression to have a stable base. If you wanted to build a deck on the back of your house, you wouldn't build it with supports placed just along one diagonal. Instead, you'd want the supports spread out in different directions as much as possible to make the deck stable. We're in a similar situation with multiple regression. When predictors are highly correlated, they line up together, which makes the regression they support balance precariously. Even small variations can rock it one way or the other. We can build a more stable model when predictors have low correlation and the points are spread out.

When two or more predictors are linearly related, they are said to be **collinear**. The general problem of predictors with close (but perhaps not perfect) linear relationships is called the problem of **collinearity**.

WHY NOT JUST LOOK AT THE CORRELATIONS?

It's sometimes suggested that we examine the table of correlations of all the predictors to search for collinearity. But this will find only associations among *pairs* of predictors. Collinearity can—and does—occur among *several* predictors working together. You won't find that more subtle collinearity with a correlation table.

Fortunately, there's an easy way to assess collinearity. To measure how much one predictor is linearly related to the others, just find the regression of that predictor on the others[12] and look at the R^2. That R^2 gives the fraction of the variability of the predictor in question that is accounted for by the other predictors. So $1 - R^2$ is the amount of the predictor's variance that is left after we allow for the effects of the other predictors. That's what the predictor has left to bring to the regression model. And we know that a predictor with little variance can't do a good job of predicting.[13]

Collinearity can hurt our analysis in yet another way. We've seen that the variance of a predictor plays a role in the standard error of its associated coefficient. Small variance leads to a larger SE. In fact, it's exactly this leftover variance that shows up in the formula for the SE of the coefficient. That's what happened in the infant mortality example.

As a final blow, when a predictor is collinear with the other predictors, it's often difficult to figure out what its coefficient means in the multiple regression. We've blithely talked about "removing the effects of the other predictors," but now when we do that, there may not be much left. What is left is not likely to be about the original predictor, but more about the fractional part of that predictor not associated with the others. In a regression of *Horsepower* on *Weight* and engine *Size*, once we've removed the effect of weight on horsepower, engine size doesn't tell us anything *more* about horsepower. That's certainly not the same as saying that engine size doesn't tell us anything about horsepower. It's just that most cars with big engines also weigh a lot.

When a predictor is collinear with the other predictors in the model, two things can happen:

1. Its coefficient can be surprising, taking on an unanticipated sign or being unexpectedly large or small.
2. The standard error of its coefficient can be large, leading to a smaller *t*-statistic and correspondingly large P-value.

[12]The residuals from this regression are plotted as the *x*-axis of the partial regression plot for this variable. So if they have a very small variance, you can see it by looking at the *x*-axis labels of the partial regression plot, and get a sense of how precarious a line fit to the partial regression plot—and its corresponding multiple regression coefficient—may be.

[13]The statistic $1/(1 - R^2)$ for the R^2 found from the regression of one predictor on the other predictors in the model is also called the *variance inflation factor*, or *VIF*, in some computer programs and books.

One telltale sign of collinearity is the paradoxical situation in which the overall *F*-test for the multiple regression model is significant, showing that at least one of the coefficients is discernably different from zero, and yet most or all of the individual coefficients have small *t*-values, each in effect, denying that *it* is the significant one.

What should you do about a collinear regression model? The simplest cure is to remove some of the predictors. That both simplifies the model and generally improves the *t*-statistics. And, if several predictors give pretty much the same information, removing some of them won't hurt the model. Which should you remove? Keep the predictors that are most reliably measured, least expensive to find, or even those that are politically important.

A Mathematics department at a large university built a regression model to help them predict success in graduate study. They were shocked when the coefficient for Mathematics GRE score was not significant. But the Math GRE was collinear with some of the other predictors, such as math course GPA and Verbal GRE, which made its slope not significant. They decided to omit some of the other predictors and retain Math GRE as a predictor because that model seemed more appropriate—even though it predicted no better (and no worse) than others without Math GRE.

- **Beware of collinearities.** When the predictors are linearly related to each other, they add little to the regression model after allowing for the contributions of the other predictors. Check the R^2 when each predictor is regressed on the others. If these are high, consider omitting some of the predictors.
- **Don't check for collinearity only by looking at pairwise correlations.** Collinearity is a relationship among any number of the predictors. Pairwise correlations can't always show that. (Of course, a high pairwise correlation between two predictors does indicate collinearity of a special kind.)
- **Don't be fooled when high-influence points and collinearity show up together.** A single high-influence point can be the difference between your predictors being collinear and seeming not to be collinear. (Picture that deck supported only along its diagonal and with a single additional post in another corner. Supported in this way, the deck is stable, but the height of that single post completely determines the tilt of the deck, so it's very influential.) Removing a high-influence point may surprise you with unexpected collinearity. Alternatively, a single value that is extreme on several predictors can make them appear to be collinear when in fact they would not be if you removed that point. Removing that point may make apparent collinearities disappear (and would probably result in a more useful regression model).

WHAT ELSE CAN GO WRONG?

Movie advice In the Oscar-winning movie *The Bridge on the River Kwai* and in the book on which it is based,[14] the character Colonel Green famously says, "As I've told you before, in a job like yours, even when it's finished, there's always one more thing to do." It is wise to keep Colonel Green's advice in mind when building, analyzing, and understanding multiple regression models.

- **Beware missing data.** Values may be missing or unavailable for any case in any variable. In simple regression, when the cases are missing for reasons that are unrelated to the variable we're trying to predict, that's not a problem. We just analyze the cases for which we have data. But when several variables participate in a multiple regression, any case with data missing on any of the variables will be omitted from the analysis. You can unexpectedly find yourself with a much smaller set of data than you started with. When comparing regression models with different predictors, be especially careful that the cases participating in the models are the same.
- **Remember linearity.** The Linearity Assumption (and the Straight Enough Condition) require linear relationships among the variables in a regression model. As you build and compare regression models, be sure to plot the data to check that it is straight. Violations of this assumption make everything else about a regression model invalid (In fact, it is quite simple to add, say, quadratic terms to your model and fit by least squares—similar to adding an interaction term. Though not linear as defined in this chapter, the model is still linear in a bigger sense—called a General Linear Model—and hence easy to fit and use for inference).

[14]The author of the book, Pierre Boulle, also wrote the book and script for *Planet of the Apes*. The director, David Lean, also directed *Lawrence of Arabia*.

- **Check for parallel regression lines.** When you introduce an indicator variable for a category, check the underlying assumption that the other coefficients in the model are essentially the same for both groups. If not, consider adding an interaction term.

CONNECTIONS

Now that we understand indicator variables, we can see that multiple regression and ANOVA are really the same analysis. If the only predictor in a regression is an indicator variable that is 1 for one group and 0 for the other, the t-test for its coefficient is just the pooled t-test for the difference in the means of those groups. In fact, most of the Student's t-based methods in this book can be seen as part of a more general statistical model known as the General Linear Model (GLM). That accounts for why they seem to be so connected, using the same general ideas and approaches.[15] We've generalized the concept of leverage that we first saw in Chapter 8. Everything we said about how to think about these ideas in Chapters 8 and 24 still applies to the multiple regression model.

Don't forget that the Straight Enough Condition is essential to all of regression. At any stage in developing a model, if the scatterplot that you check is not straight, consider re-expressing the variables (or adding higher order terms to your model) to make the relationship straighter.

What Have We Learned?

Learning Objectives

In Chapter 27, we learned that multiple regression is a natural way to extend what we know about linear regression models to include several predictors. As with other chapters in this book whose titles spoke of greater "wisdom," this chapter has drawn us deeper into the uses and cautions of multiple regression. Now we've learned that multiple regression is both more powerful and more complex than it may appear at first.

- Be able to use a statistics package to diagnose a multiple regression model.

We can incorporate categorical data by using indicator variables, modelling relationships that have parallel slopes but at different levels for different groups. With interaction terms, we can allow for different slopes as well. We can create indicator variables that isolate individual cases to remove their influence from the model while exhibiting how they differ from the other points and testing whether that difference is statistically significant.

- Know how to define and use indicator variables to introduce categorical variables as predictors in a multiple regression model.
- Be able to interpret the coefficients found for indicator variables in a multiple regression.

We've learned to beware of unusual cases. A single case can have high leverage, allowing it to influence the entire regression. Such cases should be treated specially, possibly by fitting the model both with and without them or by including indicator variables to isolate their influence.

- Understand how individual cases can influence a regression model.

[15]It has been wistfully observed that if only we could start the course by teaching multiple regression, everything else would just be simplifications of the general method. Now that you're here, you might try reading the book backward, contradicting the King's advice to the White Rabbit, which we quoted in Chapter 1.

- Know how to examine histograms of leverages and of Studentized residuals to identify extraordinary cases that deserve special attention.
- Know how to check for high-leverage cases by identifying cases whose leverage stands apart from the others.
- Know how to check for cases with large Studentized residuals.
- Be able to discuss the influence that a case with high leverage or a large Studentized residual may have in a regression.

We've learned that in complex models, we have to be careful in interpreting the coefficients. Associations among the predictors can change the coefficients to values that can be quite different from the coefficient in the simple regression of a predictor and the response, even changing the sign.

- Know how to recognize when a regression model may suffer from collinearity.
- Be able to check for collinearity and discuss its consequences.
- Be careful in interpreting regression coefficients when the predictors are collinear.
- Avoid the pitfalls of interpreting the sign of the coefficient as if it were different from the coefficient itself.

And we've learned that building multiple regression models is an art that speaks to the central goal of statistical analysis: understanding the world with data. We've learned that there is no right model. We've seen that the same response variable can be modelled with several alternative models, each showing us different aspects of the data and of the relationships among the variables and nominating different cases as special and deserving of our attention.

- Know how to build a multiple regression model, selecting predictors from a larger collection of potential predictors.

We've seen that everything we've discussed throughout this book fits together to help us understand the world. The graphical methods are the same ones we learned in the early chapters, and the inference methods are those we originally developed for means. In short, there's been a consistent tale of how we understand data to which we've added more and more detail and richness, but which has been consistent throughout.

Review of Terms

Indicator variable A variable constructed to indicate for each case whether it is in a designated group or not. A common way to assign values to indicator variables is to let them take on the values 0 and 1, where 1 indicates group membership (p. 845).

Interaction term A constructed variable found as the product of a predictor and an indicator variable. An interaction term adjusts the *slope* of the cases identified by the indicator against the predictor (p.847).

Leverage The leverage of a case measures how far its x-values are from the centre of the x's and, consequently, how much influence it can exert on the regression model. Points with high leverage can determine a regression model and should, therefore, be examined carefully (p. 849).

Studentized residual When a residual is divided by an independent estimate of its standard deviation, the result is a Studentized residual. The type of Studentized residual that has a t-distribution is an *externally* Studentized residual (p. 852).

Influential case A case is *influential* on a multiple regression model if, when it is omitted, the model changes by enough to matter for your purposes. (There is no specific amount of change defined to declare a case influential.) Cases with high leverage and large Studentized residual are likely to be influential (p. 852).

Stepwise regression An automated method of building regression models in which predictors are added to or removed from the model one at a time in an attempt to optimize a measure of the success of the regression. Stepwise methods rarely find the best model and are easily influenced by influential cases, but they can be valuable in winnowing down a large collection of candidate predictors (p. 857).

Collinearity When one (or more) of the predictors can be fit closely by a multiple regression on the other predictors, we have collinearity. When collinear predictors are in a regression model, they may have unexpected coefficients and often have inflated standard errors (and correspondingly small *t*-statistics) (p. 866).

On the Computer **REGRESSION ANALYSIS**

Statistics packages differ in how much information they provide to diagnose a multiple regression. Most packages provide leverage values. Many provide far more, including statistics that we have not discussed. But for all, the principle is the same. We hope to discover any cases that don't behave like the others in the context of the regression model and then to understand why they are special.

Many of the ideas in this chapter rely on the concept of examining a regression model and then finding a new one based on your growing understanding of the model and the data. Regression diagnosis is meant to provide steps along that road. A thorough regression analysis may involve finding and diagnosing several models.

EXCEL

Excel does not offer diagnostic statistics with its regression function.

COMMENTS

Although the dialogue offers a Normal probability plot of the residuals, the data analysis add-in does not make a correct probability plot, so don't use this option. The "standardized residuals" are just the residuals divided by their standard deviation (with the wrong df), so they, too, should be ignored.

JMP

- From the **Analyze** menu select **Fit Model.**
- Specify the response, Y. Assign the predictors, X, in the **Construct Model Effects** dialogue box.
- Click on **Run Model.**
- Click on the red triangle in the title of the Model output to find a variety of plots and diagnostics available.

COMMENTS

JMP chooses a regression analysis when the response variable is "Continuous."

MINITAB

- Choose **Regression** from the **Stat** menu.
- Choose **Regression** . . . from the **Regression** submenu.
- In the Regression dialogue, assign the Y variable to the Response box and assign the X-variables to the Predictors box.
- Click the **Storage** button.
- In the Regression Storage dialogue, you can select a variety of diagnostic statistics. They will be stored in columns of your worksheet.
- Click the **OK** button to return to the Regression dialogue.
- To specify displays, click **Graphs,** and check the displays you want.
- Click the **OK** button to return to the Regression dialogue.
- Click the **OK** button to compute the regression.

COMMENTS

You will probably want to make displays of the stored diagnostic statistics. Use the usual Minitab methods for creating displays.

R

To fit a regression model of Y on X_1 and X_2:

- **fit <- lm(Y ~ X_1 + X_2, data=dataset)** produces the linear model and stores it in an object called **'fit'**.
- **summary (fit)** produces four diagnostic plots including the residual plot and the NPP.

COMMENTS

The above model assumes that your variables are in a data frame. If not, you may ignore the last option.

SPSS

- Choose **Regression** from the **Analyze** menu.
- Choose **Linear** from the **Regression** submenu.
- When the Linear Regression dialogue appears, select the Y-variable and move it to the dependent target. Then move the X-variables to the independent target.
- Click the **Save** button.
- In the Linear Regression Save dialogue, choose diagnostic statistics. These will be saved in your worksheet along with your data.
- Click the **Continue** button to return to the Linear Regression dialogue.
- Click the **OK** button to compute the regression.

STATCRUNCH

To perform a multiple regression:

- Enter the response variables and all predictors in separate columns of the data table.
- Click on **Stat**.
- Choose **Regression** » **Multiple Linear**.
- Choose the corresponding columns from your table and click on **Next**.
- Choose a variable selection method if you wish, and click on **Next**.
- On the final screen you can save residuals, predicted values, etc., to a column, and plot them manually.
- Click on **Calculate**.

COMMENTS

You can also model interactions when entering predictor variables.

TI-83/84 PLUS

COMMENTS

You need a special program to compute a multiple regression on the TI-83.

Exercises

1. **Climate change 2009** Recent concern with the rise in global temperatures has focused attention on the level of carbon dioxide (CO_2) in the atmosphere. The National Oceanic and Atmospheric Administration (NOAA) records the CO_2 levels in the atmosphere atop the Mauna Loa volcano in Hawaii, far from any industrial contamination, and calculates the annual overall temperature of the atmosphere and the oceans using an established method.[16] Here is a regression predicting Mean Annual Temperature Anomaly (°C away from the 20th-century mean) from annual CO_2 levels (parts per million). We'll examine the data from 1959 to 2009.

Dependent variable is: Global Temperature Anomaly
R-squared = 85.3% R-squared (adjusted) = 84.7%
s = 0.0850 with 51 − 3 = 48 degrees of freedom

Source	Sum of Squares	df	Mean Square	F-ratio
Regression	2.00798	2	1.00399	3.36
Residual	0.346811	48	0.007225	

Variable	Coefficient	SE(Coeff)	*t*-ratio	P-value
Intercept	30.9595	12.52	2.47	0.0170
Year	−0.019404	0.0072	−2.71	0.0094
CO_2	0.022413	0.0049	4.55	≤0.0001

A histogram of the externally Studentized residuals looks like this:

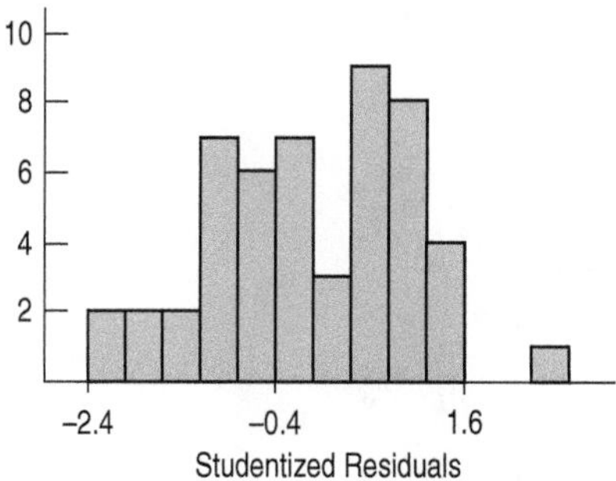

[16]See www.esrl.noaa.gov/gmd/ccgg/trends/ for the CO_2 levels and www.ncdc.noaa.gov/monitoring-references/faq/anomalies.php for the temperatures.

a) Comment on the distribution of the Studentized residuals.

b) It is widely understood that global temperatures have been rising consistently during this period. But the coefficient of *Year* is negative and its P-value is small. Does this contradict the common wisdom?

T 2. **Pizza** Consumers' Union rated frozen pizzas. Their report includes the number of *Calories*, *Fat* content, and *Type* (cheese or pepperoni, represented here as an indicator variable that is 1 for cheese and 0 for pepperoni). Here's a regression model to predict the *Score* awarded each pizza from these variables:

Dependent variable is: Score
R-squared = 28.7% R-squared (adjusted) = 20.2%
s = 19.79 with 29 − 4 = 25 degrees of freedom

Source	Sum of Squares	df	Mean Square	F-ratio
Regression	3947.34	3	1315.78	3.36
Residual	9791.35	25	391.654	

Variable	Coefficient	SE(Coeff)	*t*-ratio	P-value
Intercept	−148.817	77.99	−1.91	0.0679
Calories	0.743023	0.3066	2.42	0.0229
Fat	−3.89135	2.138	−1.82	0.0807
Type	15.6344	8.103	1.93	0.0651

a) What is the interpretation of the coefficient of cheese in this regression?

b) What displays would you like to see to check assumptions and conditions for this model?

3. **Healthy breakfast** In Chapter 27, we considered a regression model for the calories in breakfast cereals that looked like this:

Dependent variable is: Calories
R-squared = 84.5% R-squared (adjusted) = 83.4%
s = 7.947 with 77 − 6 = 71 degrees of freedom

Source	Sum of Squares	df	Mean Square	F-ratio
Regression	24367.5	5	4873.50	77.2
Residual	4484.45	71	63.1613	

Variable	Coefficient	SE(Coeff)	*t*-ratio	P-value
Intercept	20.2454	5.984	3.38	0.0012
Protein	5.69540	1.072	5.32	≤0.0001
Fat	8.35958	1.033	8.09	≤0.0001
Fibre	−1.02018	0.4835	−2.11	0.0384
Carbo	2.93570	0.2601	11.3	≤0.0001
Sugars	3.31849	0.2501	13.3	≤0.0001

Let's take a closer look at *Fibre*. Here's the partial regression plot for *Fibre* in that regression model:

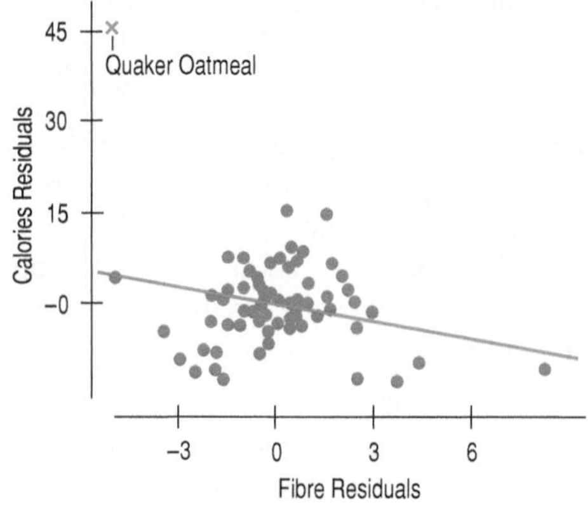

a) The line on the plot is the least squares line fit to this plot. What is its slope? (You may need to look back at the facts about partial regression plots in Chapter 27.)

b) One point is labelled as corresponding to Quaker Oatmeal. What effect does this point have on the slope of the line? (Does it make it larger, smaller, or have no effect at all?)

Here is the same regression with Quaker Oatmeal removed from the data:

Dependent variable is: Calories
77 total cases of which + is missing
R-squared = 93.9% R-squared (adjusted) = 93.5%
s = 5.002 with 76 − 6 = 70 degrees of freedom

Source	Sum of Squares	DF	Mean Square	F-ratio
Regression	27052.4	5	5410.49	216
Residual	1751.51	70	25.0216	

Variable	Coefficient	SE(Coeff)	*t*-ratio	P-value
Intercept	−1.25891	4.292	−0.293	0.7701
Protein	3.88601	0.6963	5.58	≤0.0001
Fat	8.69834	0.6512	13.4	≤0.0001
Fibre	0.250140	0.3277	0.763	0.4478
Carbo	4.14458	0.2005	20.7	≤0.0001
Sugars	3.96806	0.1692	23.4	≤0.0001

c) Compare this regression with the previous one. In particular, which model is likely to make the best predictions of calories? Which seems to fit the data better?

d) How would you interpret the coefficient of *Fibre* in this model? Does *Fibre* contribute significantly to modelling calories?

T **4. Fifty states** Let's look again at the data from the 50 United States given in Chapter 27. The *Murder* rate is per 100 000, *HS Graduation* rate is in %, *Income* is per capita income in dollars, *Illiteracy* rate is per 1000, and *Life expectancy* is in years. We are trying to find a regression model for *Life expectancy*.

Here's the result of a regression on all the available predictors:

Dependent variable is: Lifeexp
R-squared = 67.0% R-squared (adjusted) = 64.0%
s = 0.8049 with 50 − 5 = 45 degrees of freedom

Source	Sum of Squares	DF	Mean Square	F-ratio
Regression	59.1430	4	14.7858	22.8
Residual	29.1560	45	0.6479	

Variable	Coefficient	SE(Coeff)	*t*-ratio	P-value
Intercept	69.4833	1.325	52.4	≤0.0001
Murder	−0.261940	0.0445	−5.89	≤0.0001
HS grad	0.046144	0.0218	2.11	0.0403
Income	1.24948e−4	0.0002	0.516	0.6084
Illiteracy	0.276077	0.3105	0.889	0.3787

Here's a histogram of the leverages and a scatterplot of the externally Studentized residuals against the leverages:

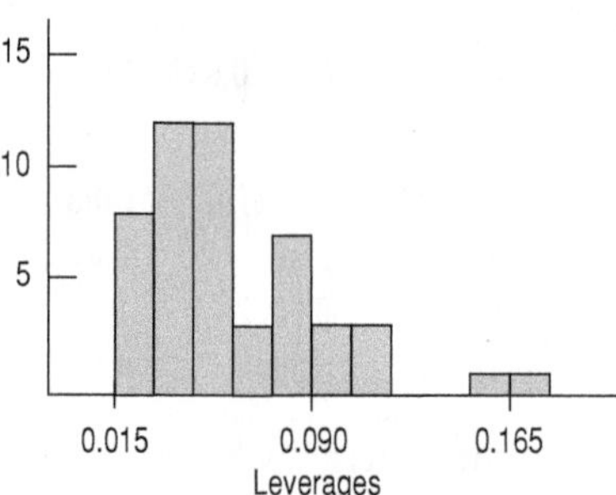

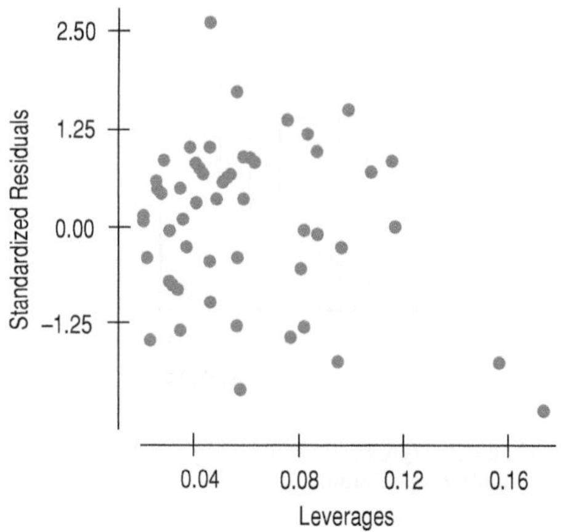

a) The two states with high leverages and large (negative) Studentized residuals are Nevada and Alaska. Do you think they are likely to be influential in the regression? From just the information you have here, why or why not?

Here's the regression with indicator variables for Alaska and Nevada added to the model to remove those states from affecting the model:

Dependent variable is: Lifeexp
R-squared = 74.1% R-squared (adjusted) = 70.4%
s = 0.7299 with 50 − 7 = 43 degrees of freedom

Variable	Coefficient	SE(Coeff)	t-ratio	P-value
Intercept	66.9280	1.442	46.4	≤0.0001
Murder	−0.207019	0.0446	−4.64	≤0.0001
HS grad	0.065474	0.0206	3.18	0.0027
Income	3.91600e−4	0.0002	1.63	0.1105
Illiteracy	0.302803	0.2984	1.01	0.3159
Alaska	−2.57295	0.9039	−2.85	0.0067
Nevada	−1.95392	0.8355	−2.34	0.0241

b) What evidence do you have that Nevada and Alaska are outliers with respect to this model? Do you think they should continue to be treated specially? Why?

c) Would you consider removing any of the predictors from this model? Why or why not?

5. A nutritious breakfast In Exercise 16 of Chapter 27, we considered a multiple regression model for predicting calories in breakfast cereals. The regression looked like this:

Dependent variable is: Calories
R-squared = 38.4% R-squared (adjusted) = 35.9%
s = 15.60 with 77 − 4 = 73 degrees of freedom

Source	Sum of Squares	DF	Mean Square	F-ratio	P-value
Regression	11091.8	3	3697.28	15.2	≤0.0001
Residual	17760.1	73	243.289		

Variable	Coefficient	SE(Coeff)	t-ratio	P-value
Intercept	83.0469	5.198	16.0	≤0.0001
Sodium	0.057211	0.0215	2.67	0.0094
Potassium	−0.019328	0.0251	−0.769	0.4441
Sugars	2.38757	0.4066	5.87	≤0.0001

Here's a histogram of the leverages and a partial regression plot for potassium in which the three high-leverage points are plotted with red x's. (They are *All-Bran, 100% Bran,* and *All-Bran with Extra Fibre.)*

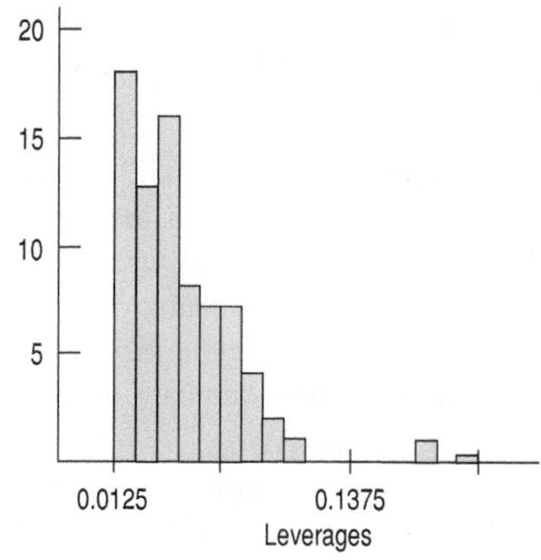

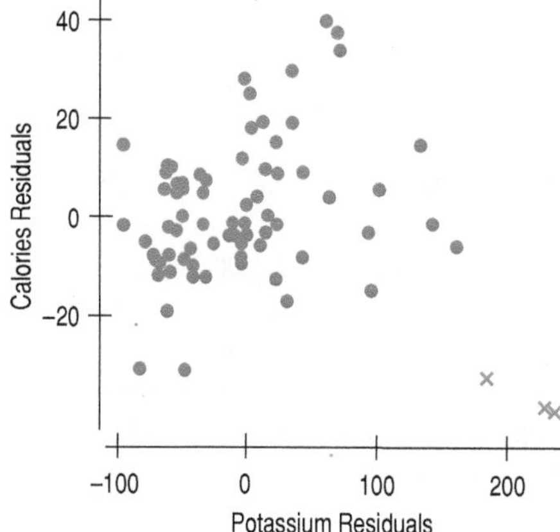

With this additional information, answer the following:

a) How would you interpret the coefficient of *Potassium* in the multiple regression?

b) *Without doing any calculating,* how would you expect the coefficient and t-statistic for *Potassium* to change if we were to omit the three high-leverage points?

Here's a histogram of the externally Studentized residuals. The selected bar, holding the two most negative residuals, holds the two bran cereals that had the largest leverages.

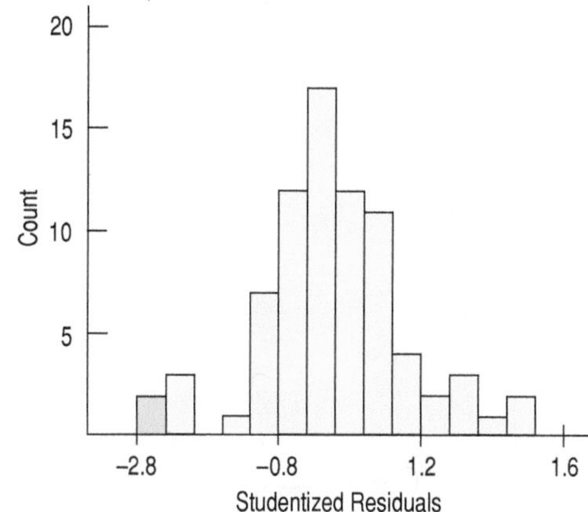

With this additional information, answer the following:

c) What term would you apply to these two cases? Why?

d) Do you think they should be omitted from this analysis? Why or why not? (*Note*: There is no correct choice. What matters is your reasons.)

6. Scottish hill races 2008 In Chapter 27, Exercises 4 and 6, we considered data on hill races in Scotland. These are overland races that climb and descend hills—sometimes several hills in the course of one race. Here is a regression analysis to predict the *Women's record* times from the *Distance* and total vertical *Climb* of the races:

Dependent variable is: Womens record
R-squared = 96.7% R-squared (adjusted) = 96.7%
s = 10.06 with 90 − 3 = 87 degrees of freedom

Source	Sum of Squares	DF	Mean Square	F-ratio
Regression	261029	2	130515	1288
Residual	8813.02	87	101.299	

Variable	Coefficient	SE(Coeff)	t-ratio	P-value
Intercept	−11.6545	1.891	−6.16	≤0.0001
Distance	4.43427	0.2200	20.2	≤0.0001
Climb	0.045195	0.0033	13.7	≤0.0001

Here is the scatterplot of externally Studentized residuals against predicted values, as well as a histogram of leverages for this regression:

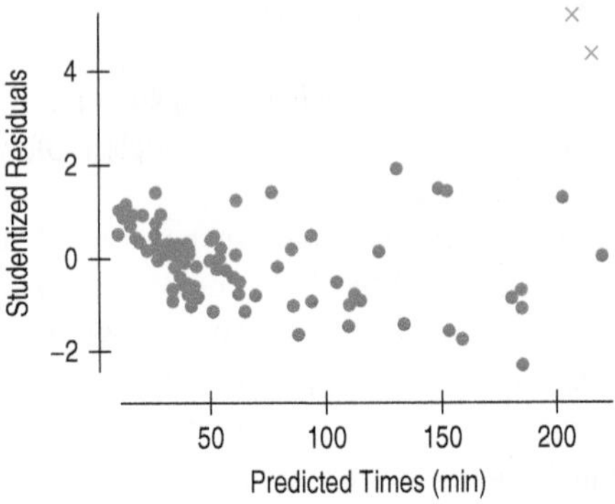

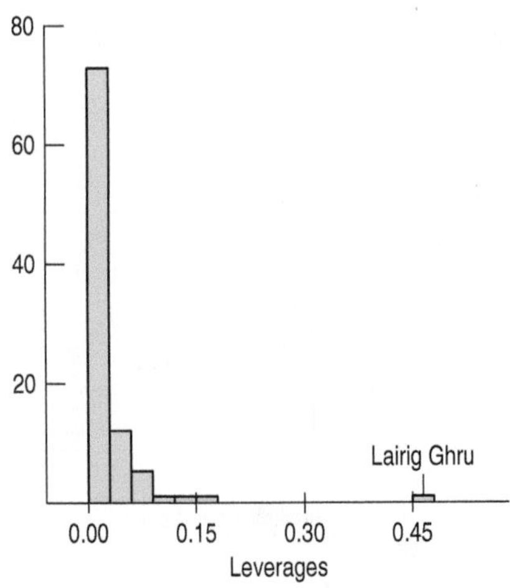

a) Comment on what these diagnostic displays indicate.

b) The two races with the largest Studentized residuals are the Arochar Alps race and the Glenshee 9. Both are relatively new races, having been run only one or two times with relatively few participants. What effects can you be reasonably sure they have had on the regression? What displays would you want to see to investigate other effects? Explain.

c) If you have access to a suitable statistics package, make the diagnostic plots you would like to see and discuss what you find.

7. Traffic delays The Texas Transportation Institute studies traffic delays. Data the institute published for the year 2001 include information on the *Total delay per person* (hours per year spent delayed by traffic), the *Average arterial road speed* (mph), the *Average highway road speed* (mph), and the *Size* of the city (small, medium, large, very large). The regression model based on these variables looks like this. The variables *Small*, *Large*, and *Very large* are indicators constructed to be 1 for cities of the named size and 0 otherwise.

Dependent variable is: Delay/person
R-squared = 79.1% R-squared (adjusted) = 77.4%
s = 6.474 with 68 − 6 = 62 degrees of freedom

Source	Sum of Squares	DF	Mean Square	F-ratio
Regression	9808.23	5	1961.65	46.8
Residual	2598.64	62	41.9135	

Variable	Coefficient	SE(Coeff)	t-ratio	P-value
Intercept	139.104	16.69	8.33	≤0.0001
Arterial mph	−2.04836	0.6672	−3.07	0.0032
HiWay mph	−1.07347	0.2474	−4.34	≤0.0001
Small	−3.58970	2.953	−1.22	0.2287
Large	5.00967	2.104	2.38	0.0203
Very large	3.41058	3.230	1.06	0.2951

a) Explain how the coefficients of *Small*, *Large*, and *Very large* account for the size of the city in the model. Why is there no coefficient for *Medium*?

b) What is the interpretation of the coefficient of *Large* in this regression model?

8. Gourmet pizza Here's a plot of the Studentized residuals against the predicted values for the regression model found in Exercise 2:

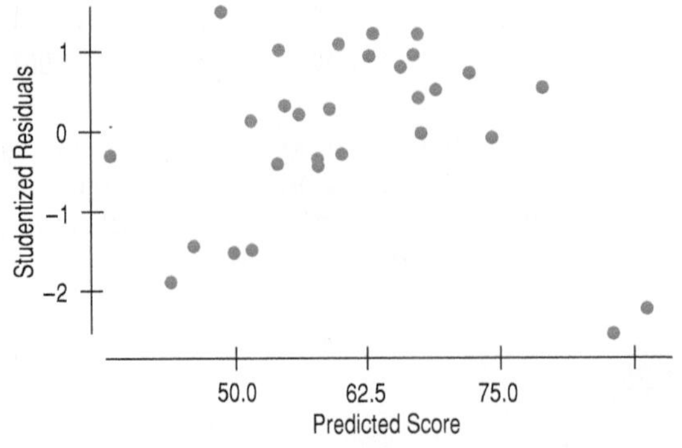

The extraordinary cases in the plot of residuals are Reggio's and Michelina's, two gourmet pizzas.

a) Interpret these residuals. What do they say about these two brands of frozen pizza? Be specific—that is, talk about the *Scores* they received and might have been expected to receive.

We can create indicator variables to isolate these cases. Adding them to the model results in the following:

Dependent variable is: Score
R-squared = 65.2% R-squared (adjusted) = 57.7%
s = 14.41 with 29 − 6 = 23 degrees of freedom

Source	Sum of Squares	df	Mean Square	F-ratio
Regression	8964.13	5	1792.83	8.64
Residual	4774.56	23	207.590	

Variable	Coefficient	SE(Coeff)	t-ratio	P-value
Intercept	−363.109	72.15	−5.03	≤0.0001
Calories	1.56772	0.2824	5.55	≤0.0001
Fat	28.82748	1.887	−4.68	0.0001
Cheese	25.1540	6.214	4.05	0.0005
Reggio's	−67.6401	17.86	−3.79	0.0010
Michelina's	−67.0036	16.62	−4.03	0.0005

b) What does the coefficient of *Michelina's* mean in this regression model? Do you think that Michelina's pizza is an outlier for this model for these data? Explain.

T 9. **More traffic** Here's a plot of Studentized residuals against *Arterial mph* for the model from Exercise 7. The plot is coloured according to *City size*, and regression lines are fit for each size.

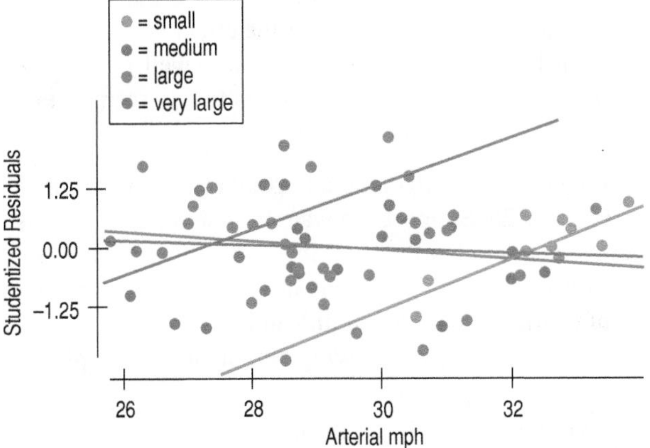

a) The model from Exercise 7 includes indicators for *City size*. Considering this display, have these indicator variables accomplished what is needed for the regression model? Explain.

We constructed additional indicators as the product of *Small* with *Arterial* mph and the product of *Very large* with *Arterial* mph. Here's the resulting model:

Dependent variable is: Delay/person
R-squared = 80.7% R-squared (adjusted) = 78.5%
s = 6.316 with 68 − 8 = 60 degrees of freedom

Source	Sum of Squares	DF	Mean Square	F-ratio
Regression	10013.0	7	1430.44	35.9
Residual	2393.82	60	39.8970	

Variable	Coefficient	SE(Coeff)	t-ratio	P-value
Intercept	153.110	17.42	8.79	≤0.0001
Arterial mph	−2.60848	0.6967	−3.74	0.0004
HiWay mph	−1.02104	0.2426	−4.21	≤0.0001
Small	−125.979	66.92	−1.88	0.0646
Large	4.89837	2.053	2.39	0.0202
Very large	−89.4993	63.25	−1.41	0.1623
AM*sml	3.81461	2.077	1.84	0.0712
AM*VLg	3.38139	2.314	1.46	0.1491

b) What does the predictor *AM*Sml* (*Arterial* mph by *Small*) do in this model? Interpret the coefficient.

c) Does this appear to be a good regression model? Would you consider removing any predictors? Why or why not?

T 10. **Another slice of pizza** A plot of Studentized residuals against predicted values for the regression model found in Exercise 8 now looks like this. It has been coloured according to *Type* of pizza, with separate regression lines fitted for each type:

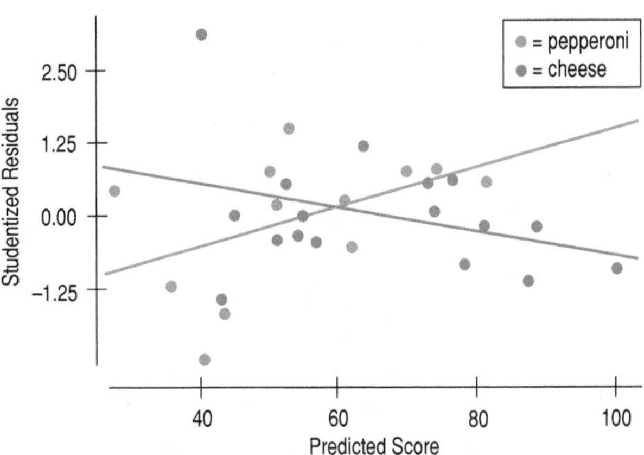

a) Comment on this diagnostic plot. What does it say about how the regression model deals with cheese and pepperoni pizzas?

Based on this plot, we constructed yet another variable consisting of the indicator *Cheese* multiplied by *Calories*:

Dependent variable is: Score
R-squared = 73.7% R-squared (adjusted) = 66.5%
s = 12.82 with 29 − 7 = 22 degrees of freedom

Source	Sum of Squares	DF	Mean Square	F-ratio
Regression	10121.4	6	1686.90	10.3
Residual	3617.32	22	164.424	

Variable	Coefficient	SE(Coeff)	t-ratio	P-Value
Intercept	−464.498	74.73	−6.22	≤0.0001
Calories	1.92005	0.2842	6.76	≤0.0001
Fat	−10.3847	1.779	−5.84	≤0.0001
Cheese	183.634	59.99	3.06	0.0057
Cheese*cals	−0.461496	0.1740	−2.65	0.0145
Reggio's	−64.4237	15.94	−4.04	0.0005
Michelina's	−51.4966	15.90	−3.24	0.0038

b) Interpret the coefficient of *Cheese*Cals* in this regression model.

c) Would you prefer this regression model to the model of Exercise 8? Explain.

11. Influential traffic? Here are histograms of the leverage and Studentized residuals for the regression model of Exercise 9.

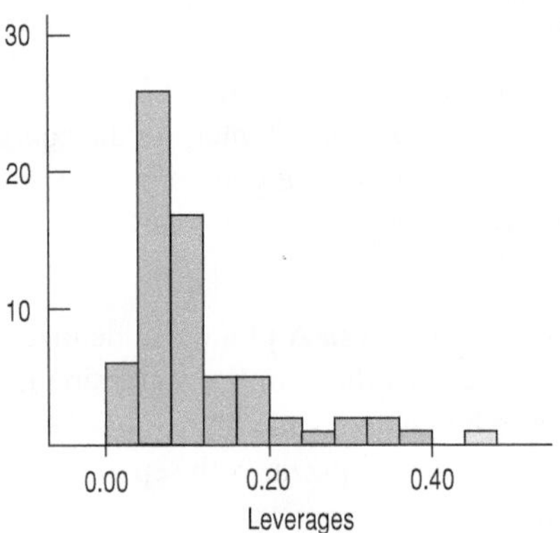

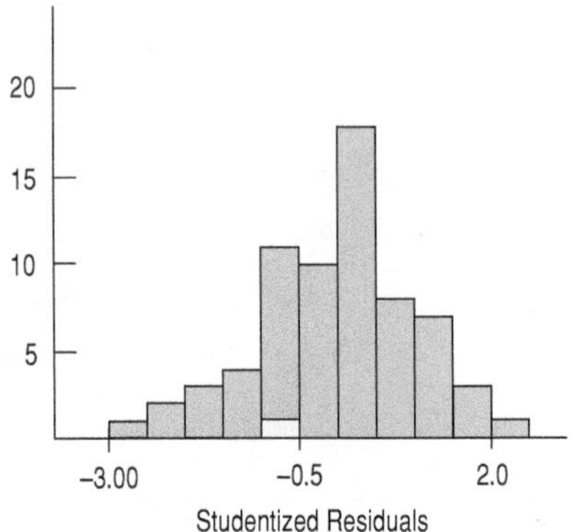

The city with the highest leverage is Colorado Springs, CO. It's highlighted in both displays.

Do you think Colorado Springs is an influential case? Explain your reasoning.

12. The final slice Here's the residual plot corresponding to the regression model of Exercise 10:

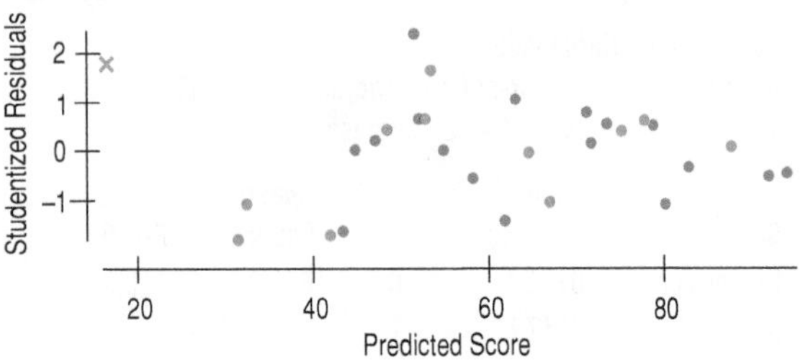

The extreme case this time is Weight Watchers Pepperoni (makes sense, doesn't it?). We can make one more indicator for *Weight Watchers*. Here's the model:

Dependent variable is: Score
R-squared = 77.1% R-squared (adjusted) = 69.4%
s = 12.25 with 29 − 8 = 21 degrees of freedom

Source	Sum of Squares	DF	Mean Square	F-ratio
Regression	10586.8	7	1512.41	10.1
Residual	3151.85	21	150.088	

Variable	Coefficient	SE(Coeff)	t-ratio	P-value
Intercept	2525.063	79.25	−6.63	≤0.0001
Calories	2.10223	0.2906	7.23	≤0.0001
Fat	210.8658	1.721	−6.31	≤0.0001
Cheese	231.335	63.40	3.65	0.0015
Cheese*cals	−0.586007	0.1806	−3.24	0.0039
Reggio's	−66.4706	15.27	−4.35	0.0003
Michelina's	−52.2137	15.20	−3.44	0.0025
Weight W ...	−8.3265	16.09	1.76	0.0928

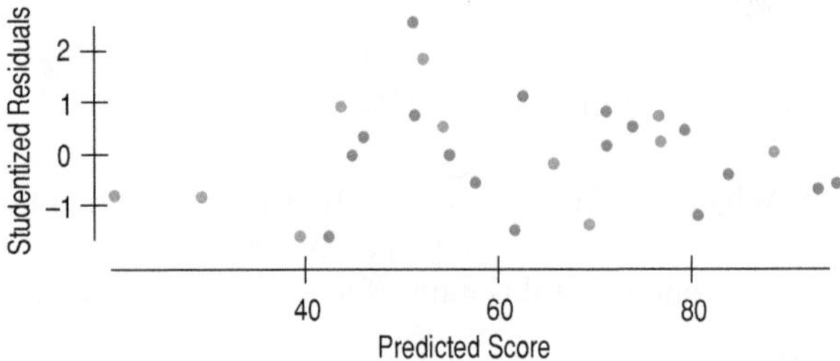

a) Compare this model with the others we've seen for these data. In what ways does this model seem better or worse than the others?

b) Do you think the indicator for *Weight Watchers* should be in the model? (Consider the effect that including it has had on the other coefficients as well.)

c) What do the Consumers' Union tasters seem to think makes for a really good pizza?

13. Popsicle sticks Recall the Popsicle stick data from Chapter 25. Suppose we want to fit a multiple regression model to predict the breaking stress from the overlap area, allowing for different slopes depending on the level of overlap (small, medium, and large). Fitting this model, we get the following output and scatter plot.

Dependent variable is: log(stress)
R-squared = 12.4% R-squared (adjusted) = 11.1%
S = 259.7 on 346 degrees of freedom

Variable	Coefficient	SE(Coeff)	t-ratio	P-value
Intercept	253.5732	1231.2640	0.206	0.837
area	0.7773	4.8330	0.161	0.872
overlapM	413.6458	1254.0276	0.330	0.742
overlapS	1646.5704	1314.0899	1.253	0.211
area*overlapM	−0.5074	5.0368	−0.101	0.920
area*overlapS	−10.6376	6.5436	−1.626	0.105

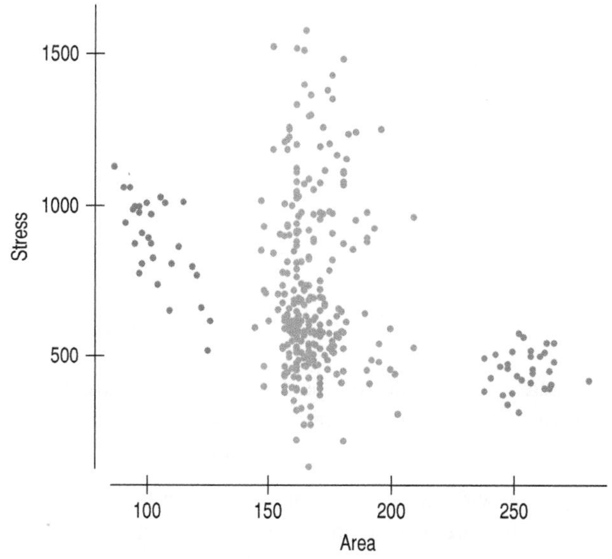

a) What does the coefficient for *Area* mean in this question?

b) It seems like there might be a linear relationship between stress and area for the small (green) overlap joints, but not the other two sizes, judging by the scatter plot. Yet, none of the slope P-values are significant. Explain why that came to be.

Assuming the residuals work out nicely, and holding everything else constant:

c) How much do we expect the stress to change on large overlap joints with a unit increase in area?

d) How much do we expect the stress to change on small overlap joints with an increase of 10 in area?

Just Checking ANSWERS

1. Histograms are used to examine the shapes of distributions of individual variables. We check especially for multiple modes, outliers, and skewness. They are also used to check the shape of the distribution of the residuals for the Nearly Normal Condition.
2. Scatterplots are used to check the Straight Enough Condition in plots of y versus any of the x's. They are used to check plots of the residuals or Studentized residuals against the predicted values, against any of the predictors, or against time to check for patterns. Scatterplots are also the display used in partial regression plots, where we check for influential points and unexpected subgroups.
3. The Normal model is needed only when we use inference; it isn't needed for computing a regression model. We check the Nearly Normal Condition on the residuals.

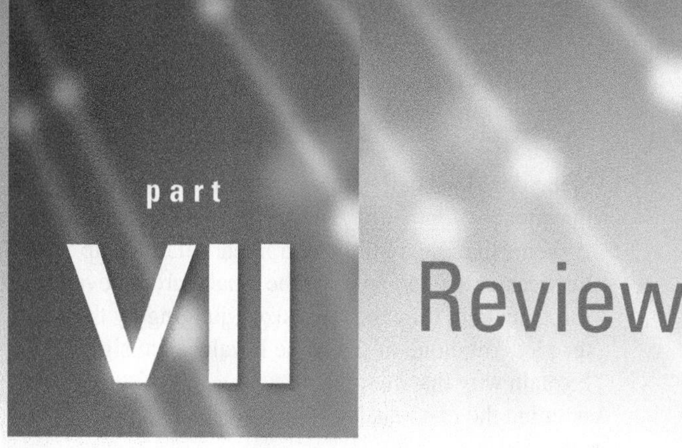

Review

Inference When Variables Are Related

Quick Review

This part introduces inference for the most widely used class of statistical models: linear models. You may have read one, two, or all of the chapters in this part. They have a consistent central theme, and it is one you'll likely see if you study more Statistics.

- Linear models predict a single quantitative variable from one or more other variables. The predictor variables can be quantitative or categorical. Linear models require only that the effects of the predictor variables be added together (rather than, for example, being multiplied or exponentiated).
- When the predictors are categorical, we are finding an analysis of variance (ANOVA) model.
- When the predictors are quantitative, we are finding a regression model. With two or more predictors, we have a *multiple* regression model.
- The overall fit of a linear model is tested with an F-test. The null hypothesis for this test is that we'd do just as well modelling y with its mean.
- If the overall F-test is significant, we can move on to consider the individual coefficients. For regression, this means considering the t-statistics. For ANOVA, it means first considering the F-statistics for individual factors. We may then move on to consider *contrasts* among different levels of significant factors.
- Models with two or more predictors can fail conditions in more ways than those with only one, and some of these can be quite subtle. These models require care in checking assumptions and conditions, and especially in checking for outliers and high-leverage points.
- If the response variable is categorical but dichotomous (has only two possible values) and the single predictor is quantitative, then you can use a logistic regression model.
- If the response variable is categorical and the predictor is categorical as well, then use the chi-squared methods from Chapter 23.

Remember (have we said this often enough yet?): Never use any inference procedure without first checking the assumptions and conditions. We summarized those in the last unit; have another look.

Here come more opportunities to practise using these concepts and skills, but we've also sprinkled in some exercises that call for methods learned in earlier chapters, so stay alert.

Quick Guide to Inference

Inference about?	One sample or two?	Procedure	Model	Parameter	Estimate	SE	Chapter
Proportions	One sample	1-Proportion z-interval	z	p	$\hat{p}$	$\sqrt{\frac{\hat{p}\hat{q}}{n}}$	16
		1-Proportion z-Test				$\sqrt{\frac{p_0 q_0}{n}}$	17, 18
	Two independent groups	2-Proportion z-Interval	z	$p_1 - p_2$	$\hat{p}_1 - \hat{p}_2$	$\sqrt{\frac{\hat{p}_1\hat{q}_1}{n_1} + \frac{\hat{p}_2\hat{q}_2}{n_2}}$	19
		2-Proportion z-Interval				$\sqrt{\frac{\hat{p}\hat{q}}{n_1} + \frac{\hat{p}\hat{q}}{n_2}}, \hat{p} = \frac{y_1 + y_2}{n_1 + n_2}$	19
Means	One sample	t-Interval t-Test	t df $= n - 1$	μ	$\bar{y}$	$\frac{s}{\sqrt{n}}$	20
	Two independent groups	2-sample t-Interval 2-Sample t-Test	t df from technology	$\mu_1 - \mu_2$	$\bar{y}_1 - \bar{y}_2$	$\sqrt{\frac{s_1^2}{n_1} + \frac{s_2^2}{n_2}}$	21
	n Matched pairs	Paried t-Interval Paried t-Test	t df $= n - 1$	μ_d	$\bar{d}$	$\frac{s_d}{\sqrt{n}}$	22
Distributions	One sample	Goodness of fit	χ^2 df $= cells - 1$			$\sum \frac{(obs - exp)^2}{exp}$	23
	Many independent samples	Homogeneity χ^2 Test	χ^2 $df = (r - 1)(c - 1)$				
Independence (two categorical variables)	One sample	Independence χ^2 Test					
Association (two quantitative variables)	One sample	Linear Regression t-Test or Confidence Interval for β	t $df = n - 2$	β_1	b_1	$\frac{s_e}{s_x\sqrt{n - 1}}$ (compute with technology)	24
		Confidence Interval for μ_ν		μ_ν	$\hat{y}_\nu$	$\sqrt{SE^2(b_1) \cdot (x_\nu - \bar{x})^2 + \frac{s_e^2}{n}}$	
		Prediction Interval for y_ν		y_ν	$\hat{y}_\nu$	$\sqrt{SE^2(b_1) \cdot (x_\nu - \bar{x})^2 + \frac{s_e^2}{n} + s_e^2}$	

Review Exercises

1. Tableware Nambe Mills manufactures plates, bowls, and other tableware made from an alloy of several metals. Each item must go through several steps, including polishing. To better understand the production process and its impact on pricing, the company checked the polishing *Time* (in minutes) and the retail *Price* (in \$) of these items. The regression analysis is shown below. The scatterplot showed a linear pattern, and residuals were deemed suitable for inference.

Dependent variable is: Price
R-squared = 84.5%
s = 20.50 with 59 − 2 = 57 degrees of freedom

Variable	Coefficient	SE(Coeff)
Intercept	−2.89054	5.730
Time	2.49244	0.1416

a) How many different products were included in this analysis?
b) What percentage of the variation in retail price is explained by the polishing time?
c) Create a 95% confidence interval for the slope of this relationship.
d) Interpret your interval in this context.

2. Hard water In an investigation of environmental causes of disease, data were collected on the annual *Mortality* rate (deaths per 100 000) for males in 61 large towns in England and Wales. In addition, the water hardness was recorded as the *Calcium* concentration (parts per million, or ppm) in the drinking water. Here are the scatterplot and regression analysis of the relationship between mortality and calcium concentration.

Dependent variable is: Mortality
R-squared = 43%
s = 143.0 with 61 − 2 = 59 degrees of freedom

Variable	Coefficient	SE(Coeff)
Intercept	1676	29.30
Calcium	−3.23	0.48

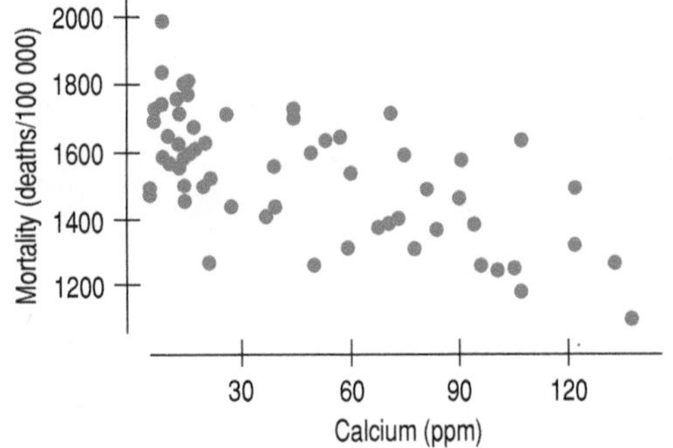

a) Is there an association between the hardness of the water and the mortality rate? Write the appropriate hypothesis.
b) Assuming the assumptions for regression inference are met, what do you conclude?
c) Create a 95% confidence interval for the slope of the true line relating calcium concentration and mortality.
d) Interpret your interval in context.

3. Mutual funds In March 2002, *Consumer Reports* listed the rate of return for several large cap mutual funds over the previous three-year and five-year periods. ("Large cap" refers to companies worth over \$10 billion.)
a) Create a 95% confidence interval for the difference in rate of return for the three- and five-year periods covered by these data. Clearly explain what your interval means.
b) It's common for advertisements to carry the disclaimer that "past returns may not be indicative of future performance," but do these data indicate that there was an association between three-year and five-year rates of return?

	Annualized Returns (%)	
Fund name	Three-year	Five-year
Ameristock	7.9	17.1
Clipper	14.1	18.2
Credit Suisse Strategic Value	5.5	11.5
Dodge & Cox Stock	15.2	15.7
Excelsior Value	13.1	16.4
Harbor Large Cap Value	6.3	11.5
ICAP Discretionary Equity	6.6	11.4
ICAP Equity	7.6	12.4
Neuberger Berman Focus	9.8	13.2
PBHG Large Cap Value	10.7	18.1
Pelican	7.7	12.1
Price Equity Income	6.1	10.9
USAA Cornerstone Strategy	2.5	4.9
Vanguard Equity Income	3.5	11.3
Vanguard Windsor	11.0	11.0

4. Football A student runs an experiment to test four different grips on his football throwing distance, recording the distance in yards that he can throw a bag of 20 footballs using each grip. He randomizes the grip used each time by drawing numbers out of a hat until each grip has been used five times. After collecting his data and analyzing the results, he reports that the P-value of his test is 0.0032.
a) What kind of test should he have performed?
b) What are the null and alternative hypotheses?
c) Assuming that the conditions required for the test are satisfied, what would you conclude?
d) What else about the data would you like to see to check the assumptions and conditions?
e) What might he want to test next?
f) Suggest one problem with this experiment design that can be fixed, and one that likely cannot.

5. **Golf** A student runs an experiment to test four different golf clubs on her putting accuracy, recording the distance in centimetres from a small target that she places on the green. She randomizes the club used each time by drawing numbers out of a hat until each club has been used six times. After collecting her data and analyzing the results, she reports that the P-value of her test is 0.0245.
   a) What kind of test should she have performed?
   b) What are the null and alternative hypotheses?
   c) Assuming that the conditions required for the test are satisfied, what would you conclude?
   d) What else about the data would you like to see to check the assumptions and conditions?
   e) What might she want to test next?

6. **Wild horses** Large herds of wild horses can become a problem on some U.S. federal lands in the West. Researchers hoping to improve the management of these herds collected data to see if they could predict the number of foals that would be born based on the size of the current herd. Their attempt to model this herd growth is summarized in the output shown.

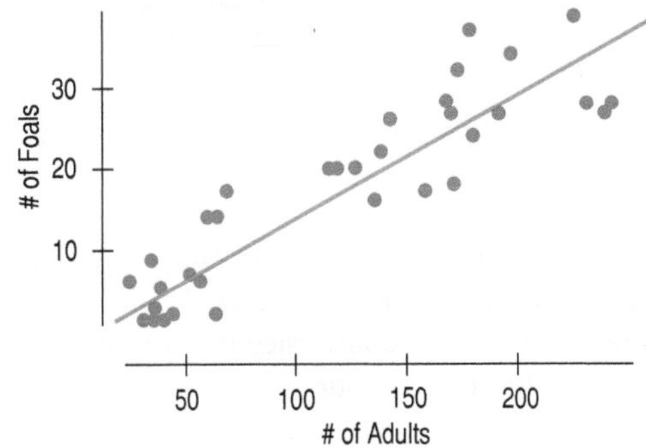

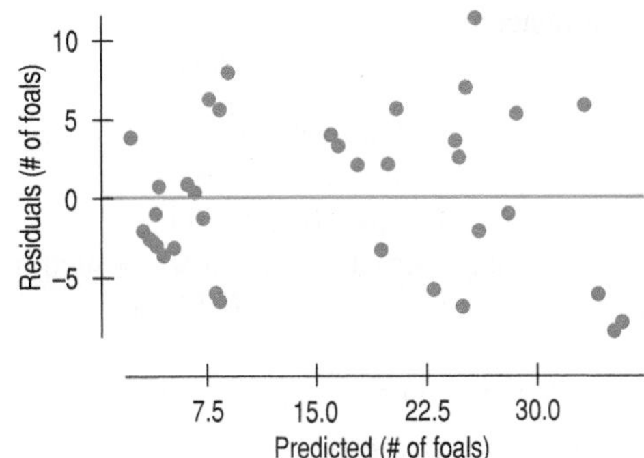

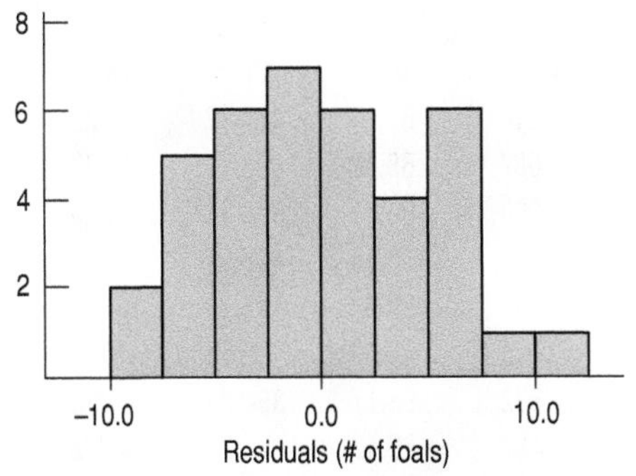

Variable	Count	Mean	StdDev
Adults	38	110.237	71.1809
Foals	38	15.3947	11.9945

Dependent variable is: Foals
R-squared = 83.5%
s = 4.941 with 38 − 2 = 36 degrees of freedom

Variable	Coefficient	SE(Coeff)	t-ratio	Prob
Intercept	−1.57835	1.492	−1.06	0.2970
Adults	0.153969	0.0114	13.5	≤0.0001

   a) How many herds of wild horses were studied?
   b) Are the conditions necessary for inference satisfied? Explain.
   c) Create a 95% confidence interval for the slope of this relationship.
   d) Explain in this context what that slope means.
   e) Suppose that a new herd with 80 adult horses is located. Estimate, with a 90% prediction interval, the number of foals that may be born.

7. **North and hard** The data on water hardness and mortality considered in Exercise 2 also included information on whether the town was north or south of Derby. When that information is coded in a variable as 1 for "north of Derby" and 0 for "south of Derby," and that variable is included in the regression, the resulting table looks like this:

Dependent variable is: Mortality
R-squared = 56.2% R-squared (adjusted) = 54.7%
s = 126.4 with 61 − 3 = 58 degrees of freedom

Variable	Coefficient	SE(Coeff)	t-ratio	P-value
Intercept	1537.50	42.02	36.6	≤0.0001
Calcium	−2.16011	0.4979	−4.34	≤0.0001
North of Derby	158.892	37.87	4.20	≤0.0001

   a) What is the name for the kind of variable that is being used for being north of Derby?
   b) What does the coefficient of that variable mean in this regression?
   c) Would you prefer this regression to the one of Exercise 2? Explain why or why not.

8. **Horses again, a bit less wild** In an attempt to control the growth of the herds of wild horses discussed in Exercise 6, managers sterilized some of the stallions. The variable *Sterilized* is coded 0 for herds and years in which no stallions were sterilized and 1 for herds and years in which some stallions were sterilized. The resulting regression looks like this:

Dependent variable is: Foals
R-squared = 84.8% R-squared (adjusted) = 83.9%
s = 4.814 with 38 − 3 = 35 degrees of freedom

Source	Sum of squares	DF	Mean square	F-ratio
Regression	4511.85	2	2255.92	97.3
Residual	811.229	35	23.1780	

Variable	Coefficient	SE(Coeff)	*t*-ratio	P-value
Intercept	6.23667	4.801	1.30	0.2024
Adults	0.112271	0.0268	4.18	0.0002
Sterilized	−6.43657	3.769	−1.71	0.0965

a) What is the name for the kind of variable used here to represent sterilizing?
b) What is the interpretation of the coefficient of *Sterilized*?
c) Does sterilizing appear to work? That is, do these data show a statistically significant effect of sterilizing stallions on the number of foals born?

9. Bowling A student runs an experiment to test how different factors affect her bowling performance. She uses three levels for the weight of the ball (low, medium, and high) and two approaches (standing and walking), throwing four balls at each condition and choosing the conditions in random order. She counts the number of pins knocked down as the response variable.
a) What are the null and alternative hypotheses for the main effects?
b) How many degrees of freedom does each factor sum of squares have? How about the error sum of squares?
c) Should she consider fitting an interaction term to the model? Why or why not? How many degrees of freedom would it take?

10. Video racing A student runs an experiment to test how different factors affect his reaction time while playing video games. He uses a specified race course with random hazards and times how long it takes him (in seconds) to finish four laps under a variety of experimental conditions. As factors, he uses three different types of *Mouse* (cordless ergonomic, cordless regular, or corded) and keeps the *Lights* on or off. He measures his time once at each condition. To avoid a learning effect, he runs the race five times before selecting the conditions in random order.
a) What are the null and alternative hypotheses for the main effects?
b) How many degrees of freedom does each factor sum of squares have? How about the error sum of squares?
c) Should he consider fitting an interaction term to the model? Why or why not? How many degrees of freedom would it take?

11. Resumé fraud In 2002, the Veritas Software company found out that its chief financial officer did not actually have the MBA he had listed on his resumé. They fired him, and the value of the company's stock dropped 19%. Kroll, Inc., a firm that specializes in investigating such matters, said that they believe as many as 25% of background checks might reveal false information. How many such random checks would they have to do to estimate the true percentage of people who misrepresent their backgrounds to within ±5% with 98% confidence?

T **12. Paper airplanes** In preparation for a regional paper airplane competition, a student tried out her latest design. The distances her plane travelled (in feet) in 11 trial flights are given here. (The world record is an astounding 193.01 feet!) The data were 62, 52, 68, 23, 34, 45, 27, 42, 83, 56, and 40 feet.

Here are some summaries:

Count	11
Mean	48.3636
Median	45
StdDev	18.0846
StdErr	5.45273
IntQRange	25
25th %tile	35.5000
75th %tile	60.5000

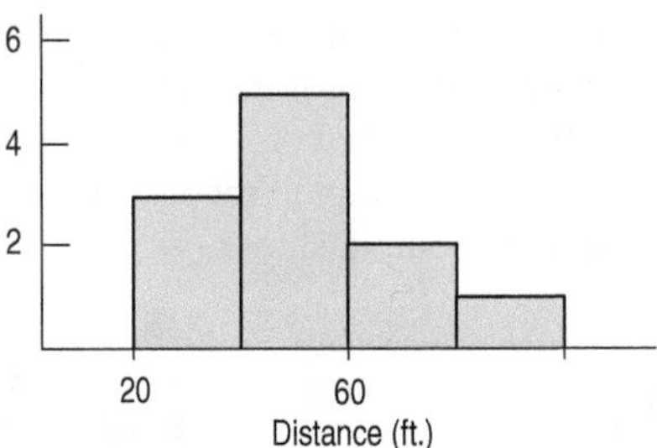

a) Construct a 95% confidence interval for the mean distance.
b) Based on your confidence interval, is it plausible that the true mean distance is 40 feet? Explain.
c) How would a 99% confidence interval for the mean distance differ from your answer in part a)? Explain briefly, without actually calculating a new interval.
d) How large a sample size would the student need to get a confidence interval half as wide as the one you got in part a), at the same confidence level?

T **13. Nuclear power** Here are data on 32 light water nuclear power plants. The variables are:
Cost: In \$100 000, adjusted to 1976 base.
Date: Date the construction permit was issued in years after 1900. Thus, 68.58 is roughly halfway through 1968.
MWatts: Power plant net capacity in megawatts.

We are interested in the cost of the plants as a function of *Date* and *MWatts*.

Cost	MWatts	Date	Cost	MWatts	Date
345.39	514	67.92	457.12	822	68.42
460.05	687	68.58	690.19	792	68.33
452.99	1065	67.33	350.63	560	68.58
443.22	1065	67.33	402.59	790	68.75
652.32	1065	68.00	412.18	530	68.42
642.23	1065	68.00	495.58	1050	68.92
272.37	822	68.17	394.36	850	68.92
317.21	457	68.42	423.32	778	68.42

Cost	MWatts	Date	Cost	MWatts	Date
712.27	845	69.50	473.64	538	70.42
289.66	530	68.42	697.14	1130	71.08
881.24	1090	69.17	207.51	745	67.25
490.88	1050	68.92	288.48	821	67.17
567.79	913	68.75	284.88	886	67.83
665.99	828	70.92	280.36	886	67.83
621.45	786	69.67	217.38	745	67.25
608.80	821	70.08	270.71	886	67.83

a) Examine the relationships between *Cost* and *MWatts* and between *Cost* and *Date*. Make appropriate displays and interpret them with a sentence or two.
b) Find the regression of *Cost* on *MWatts*. Write a sentence that explains the relationship as described by the regression.
c) Make a scatterplot of residuals versus predicted values and discuss what it shows. Make a Normal probability plot or histogram of the residuals. Discuss the four assumptions needed for regression analysis and indicate whether you think they are satisfied here. Give your reasons.
d) State the standard null hypothesis for the slope coefficient and complete the *t*-test at the 5% level. State your conclusion.
e) Estimate the cost of a 1000-MWatt plant. Show your work.
f) Compute the residuals for this regression. Discuss the meaning of the R^2 in this regression. Plot the residuals against *Date*. Does it appear that *Date* can account for some of the remaining variability?
g) Compute the multiple regression of *Cost* on both *MWatts* and *Date*. Compare the coefficient in this regression with those you have found for each of these predictors.
h) Would you expect *MWatts* and *Date* to be correlated? Why or why not? Examine the relationship between *MWatts* and *Date*. Make a scatterplot and find the correlation coefficient, for example. It's only because of the extraordinary nature of this relationship that the relationships you saw at earlier steps were this simple.

T **14. Barbershop music** At a barbershop music singing competition, choruses are judged on three scales: *Music* (quality of the arrangement, etc.), *Performance*, and *Singing*. The scales are supposed to be independent of each other, and each is scored by a different judge, but a friend claims that he can predict a chorus's singing score from the other two scores. He offers the following regression based on the scores of all 34 choruses in a recent competition:

Dependent variable is: Singing
R-squared = 90.9% R-squared (adjusted) = 90.3%
s = 6.483 with 34 − 3 = 31 degrees of freedom

Variable	Coefficient	SE(Coeff)	*t*-ratio	P-value
Intercept	2.08926	7.973	0.262	0.7950
Performance	0.793407	0.0976	8.13	≤0.0001
Music	0.219100	0.1196	1.83	0.0766

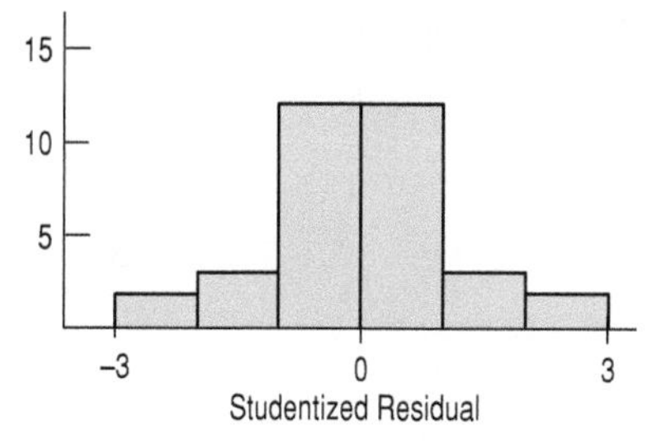

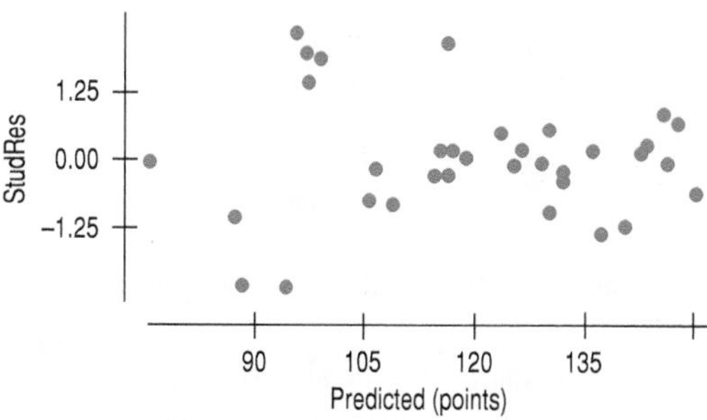

a) What do you think of your friend's claim? Can he predict singing scores? Explain.
b) State the standard null hypothesis for the coefficient of performance and complete the *t*-test at the 5% level. State your conclusion.
c) Complete the analysis. Check assumptions and conditions to the extent you can with the information provided.

15. Sleep Using a simple random sample, a student group asked 450 students about their sleep and study habits and received about 200 responses. The group wanted to know if the average amount of sleep varied by *Sex* (F or M) or by *Year* (first, second, third, or fourth) of the respondent. Partial boxplots of the amount of sleep last night by the two factors are shown below with the interaction plot:

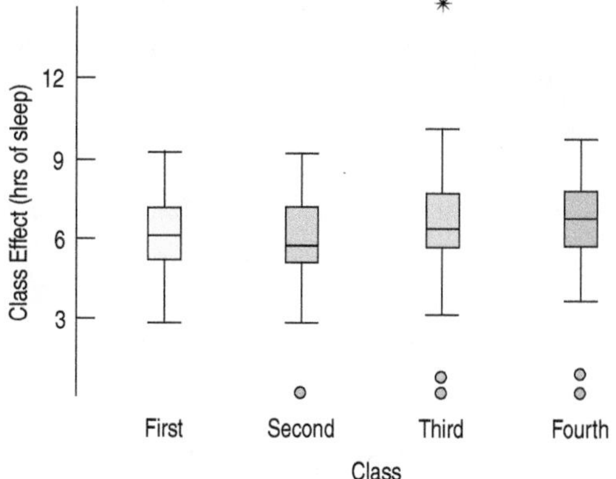

The ANOVA table shows:

Source	DF	Sum of squares	F-ratio	P-value
Sex	1	8.0658	2.2821	0.1325
Year	3	8.8990	0.8393	0.4739
Sex x Year	3	19.4075	1.8303	0.1431
Error	189	670.6457		
Total	196	707.0180		

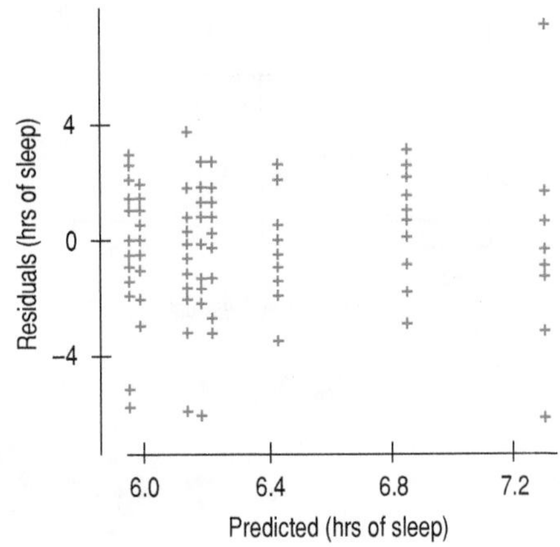

a) What are the null and alternative hypotheses for the main effects?
b) What effects appear to be significant?
c) What, if any, reservations do you have about the conclusions?

16. Study habits The survey in Exercise 15 also asked students about the number of hours they studied. Those doing the survey wanted to know if the average amount of studying varied by *Sex* or by *Class* of the respondent. Partial boxplots of the hours studied last night by the two factors are shown below:

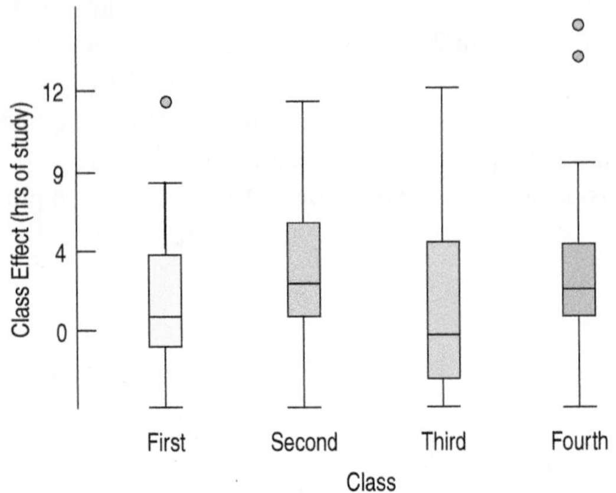

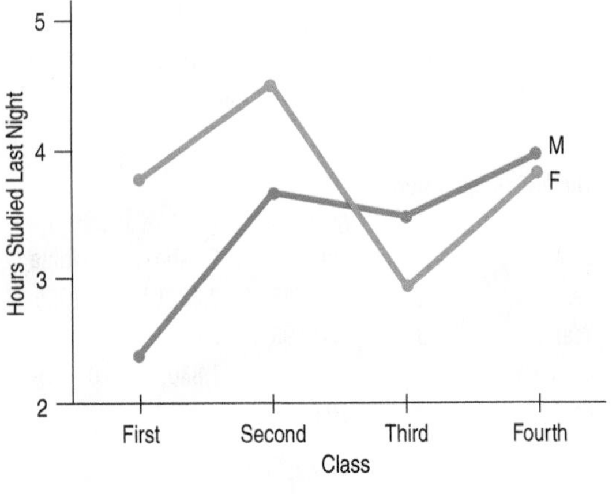

Source	DF	Sum of squares	F-ratio	P-value
Sex	1	7.1861	1.0744	0.3013
Class	3	34.9922	1.7440	0.1595
Sex × Class	3	27.8188	1.3865	0.2483
Error	189	1259.8049		
Total	196	1329.8020		

A plot of residuals versus predicted shows:

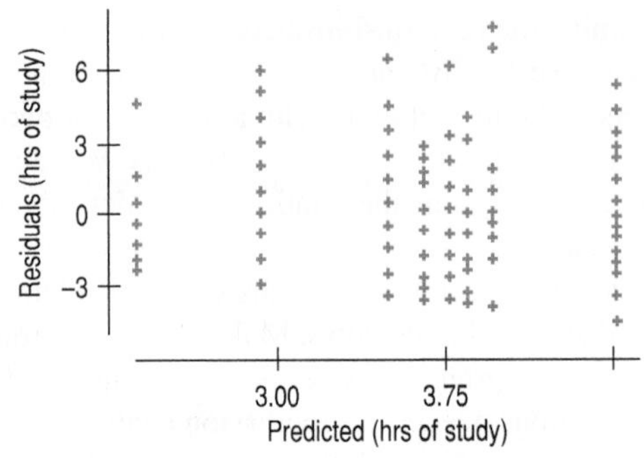

a) What are the null and alternative hypotheses for the main effects?
b) Should those doing the survey consider fitting an interaction term to the model? Why or why not?
c) What effects appear to be significant?
d) What, if any, reservations do you have about the conclusions?

17. Pregnancy In 1998, a San Diego reproductive clinic reported 42 live births to 157 women under the age of 38, but only seven successes for 89 clients aged 38 and older. Is this evidence of a difference in the effectiveness of the clinic's methods for older women?
a) Test the appropriate hypotheses using the two-proportion z-procedure.
b) Repeat the analysis using an appropriate chi-square procedure.
c) Explain how the two results are equivalent.

18. Lost baggage The Bureau of Transportation Statistics of the U.S. Department of Transportation reports statistics about airline performance. For July 2012, they reported the following number of bags lost per 1000 passengers.

Airline	Lost Bags
Airtran	1.73
Jetblue	2.18
Delta	2.43
US Airways	2.46
American	2.99
Southwest	3.33
United	4.84
Skywest	6.03
Expressjet	6.84

Are the large airlines roughly equal in their baggage performance? Perform a chi-square goodness-of-fit test, or explain why that would not be appropriate.

19. Old Faithful As you saw in an earlier chapter, Old Faithful isn't all that faithful. Eruptions do not occur at uniform intervals, and may vary greatly. Can we improve our chances of predicting the time *Interval* until the next eruption if we know the *Duration* of the previous eruption?

a) Describe what you see in this scatterplot.

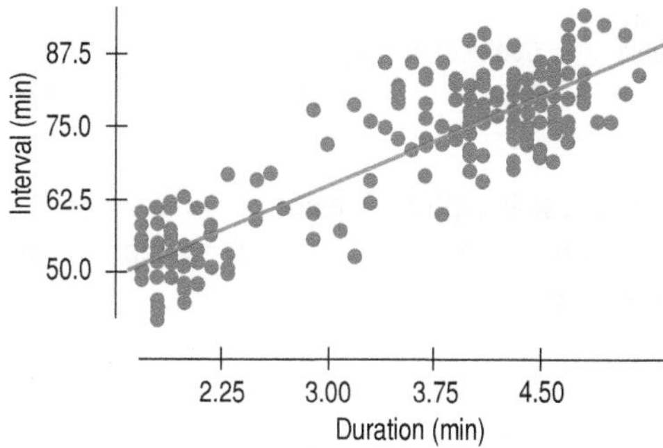

b) Write an appropriate hypothesis.

c) Here are a histogram of the residuals and the residuals plot. Do you think the assumptions for inference are met? Explain.

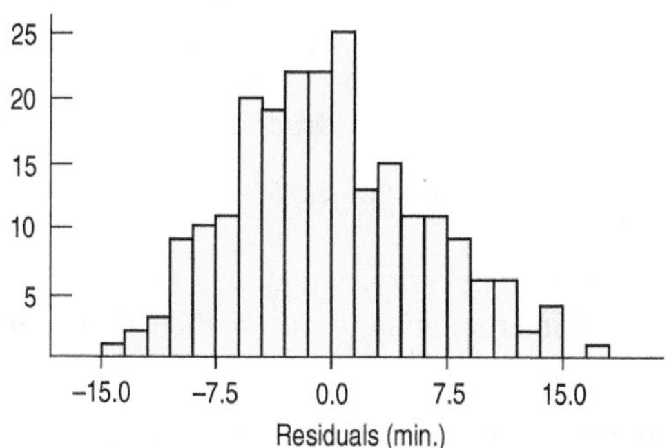

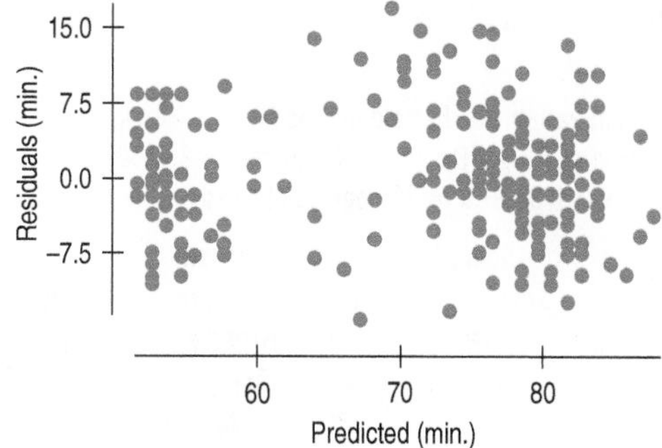

d) State a conclusion based on this regression analysis:

Dependent variable is: Interval
R-squared = 77.0%
s = 6.159 with 222 − 2 = 220 degrees of freedom

Variable	Coefficient	SE(Coeff)	*t*-ratio	P-value
Intercept	33.9668	1.428	23.8	≤0.0001
Duration	10.3582	0.3822	27.1	≤0.0001

Variable	Mean	StdDev
Duration	3.57613	1.08395
Interval	71.0090	12.7992

e) The second table shows the summary statistics for the two variables. Create a 95% confidence interval for the mean *Interval* following a two-minute eruption.

f) You arrive at Old Faithful just as an eruption ends. Witnesses say it lasted four minutes. Create a 95% prediction interval for the length of time you will wait to see the next eruption.

20. Togetherness Are good grades in high school associated with family togetherness? A simple random sample of 142 high school students was asked how many meals per week their families ate together. Their responses produced a mean of 3.78 meals per week, with a standard deviation of 2.2. Researchers then matched these responses against the students' grade-point averages. The scatterplot appeared to be reasonably linear, so they went ahead with the regression analysis, seen below. No apparent pattern emerged in the residuals plot.

Dependent variable: GPA
R-squared = 11.0%
s = 0.6682 with 142 − 2 = 140 df

Variable	Coefficient	SE(Coeff)
Intercept	2.7288	0.1148
Meals/wk	0.1093	0.0263

a) Is there evidence of an association? Test an appropriate hypothesis and state your conclusion.

b) Do you think this association would be useful in predicting a student's grade-point average? Explain.

c) Are your answers to parts a) and b) contradictory? Explain.

21. Is Old Faithful getting older? The data on Old Faithful eruptions we saw in Exercise 19 include another variable, the *Day* on which the eruption occurred (where 1 is the first day and each successive day just counts one more). The correlation of *Interval* (minutes until the next eruption) with *Day* is −0.004. But when we include *Day* in a multiple regression along with the *Duration* (in minutes) of the previous eruption, we get the following model:

Dependent variable is: Interval
R-squared = 77.6% R-squared (adjusted) = 77.3%
s = 6.092 with 222 − 3 = 219 degrees of freedom

Variable	Coefficient	SE(Coeff)	*t*-ratio	P-value
Intercept	35.2463	1.509	23.4	≤0.0001
Duration	10.4348	0.3794	27.5	≤0.0001
Day	−0.126316	0.0523	−2.42	0.0166

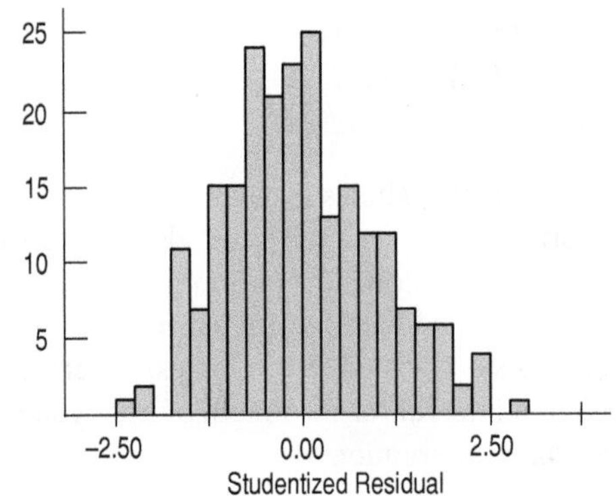

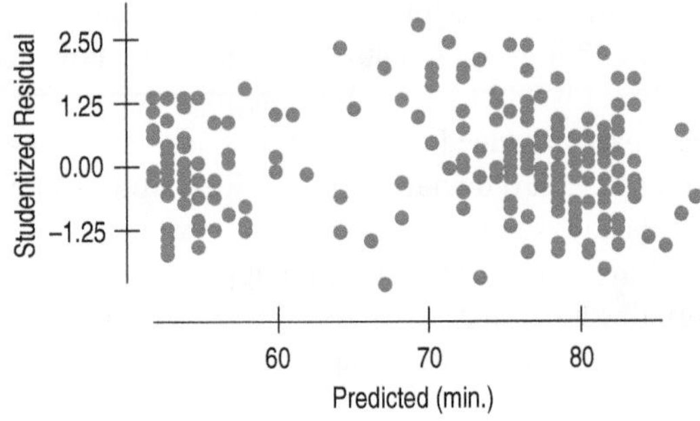

a) What is the model fit by this regression?
b) Is the *Interval* changing over time? Perform a formal test of the relevant hypothesis.
c) Doesn't the small P-value for *Day* contradict the correlation between *Interval* and *Day* being virtually zero? Explain.
d) Is the amount of change in *Interval* due to *Day* meaningful?

22. Lefties and music In an experiment to see if left- and right-handed people have different abilities in music, subjects heard a tone and were then asked to identify which of several other tones matched the first. Of 76 right-handed subjects, 38 were successful in completing this test, compared with 33 of 53 lefties. Is this strong evidence of a difference in musical abilities based on handedness?

23. Preemies Do the effects of being born prematurely linger into adulthood? Researchers examined 242 Cleveland area children born prematurely between 1977 and 1979, and compared them with 233 children of normal birth weight; 24 of the "preemies" and 12 of the other children were described as being of "subnormal height" as adults.[17] Is this evidence that babies born with a very low birth weight are more likely to be smaller than normal adults?

T **24. Teen traffic deaths 2007** The Insurance Institute for Highway Safety publishes data on a variety of traffic-related risks. One report gives the numbers of male and female teenagers killed in highway accidents during each year from 1975 to 2002. Here is a regression predicting *Female deaths* by *Year:*

Dependent variable is: Female deaths
R-squared = 57.4% R-squared (adjusted) = 56.1%
s = 182.6 with 33 = 2 = 31 degrees of freedom

Source	Sum of squares	DF	Mean square	F-ratio
Regression	1396025	1	1396025	41.9
Residual	1033964	31	33353.7	

Variable	Coefficient	SE(Coeff)	t-ratio	P-value
Intercept	45074.0	6648	6.78	≤0.0001
Year	−21.6006	3.339	−6.47	≤0.0001

a) Here's a scatterplot of residuals vs. predicted values for this regression. Discuss the assumptions and conditions of the regression.

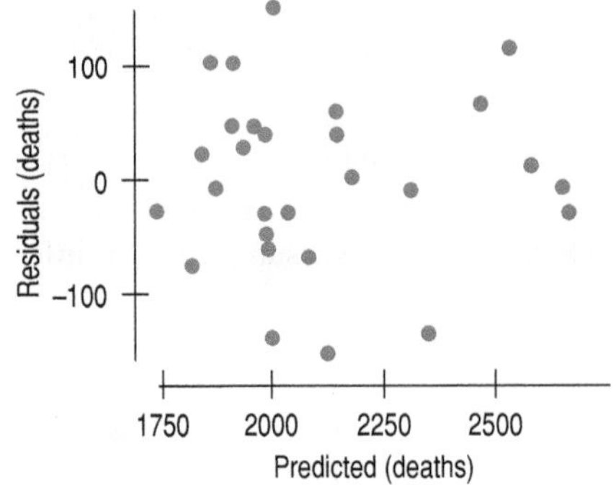

b) What is the meaning of the R^2 value of 52.3% for this regression?
c) What is the regression equation?
d) Give an interpretation of the coefficient of *Year* in this regression.

T **25. More teen traffic 2007** The data discussed in Exercise 25 included the numbers of male teen traffic deaths as well. We can add that as a predictor to obtain the following model:

Dependent variable is: Female deaths
R-squared = 89.5% R-squared (adjusted) = 88.8%
s = 92.5 with 33 − 3 = 30 degrees of freedom

Source	Sum of squares	DF	Mean square	F-ratio
Regression	2175799	2	1087899	128
Residual	254191	30	8473.02	

Variable	Coefficient	SE(Coeff)	t-ratio	P-value
Intercept	−20186.2	7583	−2.66	0.0124
Year	10.5383	3.749	2.81	0.0086
Male deaths	0.271321	0.0283	9.59	≤0.0001
Year	20.3628	4.326	4.71	≤0.0001

a) How does this regression compare with the regression of Exercise 25? Which would you prefer to use? Why?

[17] "Outcomes in young adulthood for very-low-birth-weight infants," *New England Journal of Medicine*, 346, no. 3 [January 2002].

b) What does the coefficient of *Year* mean in this regression?

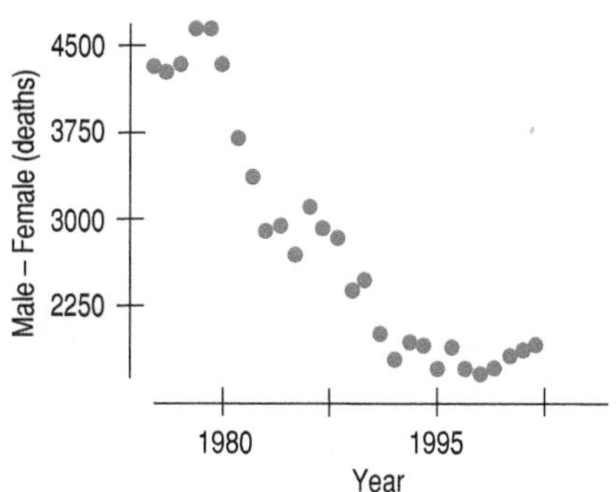

c) Considering both the regression model of Exercise 25 and this one, would you say that female teen traffic fatalities have been increasing or decreasing over time? How does the scatterplot above help to explain what is happening?

26. Typing For a class project, Nick designed and carried out an experiment to see if the room *Temperature* and the wearing of *Gloves* affected his typing speed. He ran each combination of hot and cold temperature and gloves on and off eight times, recording the net number of words typed (words typed minus mistakes). Partial boxplots and interaction plots show:

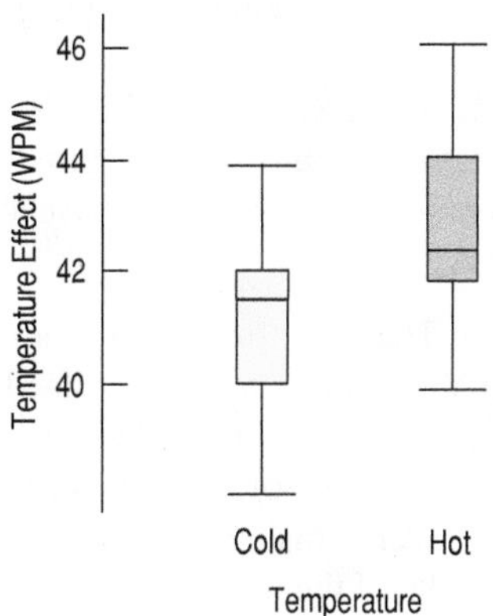

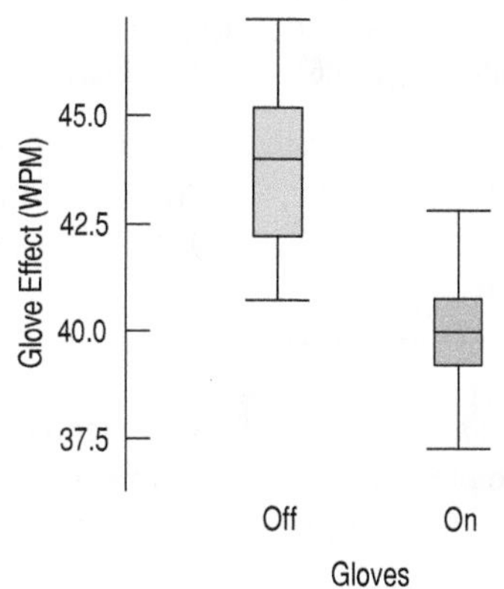

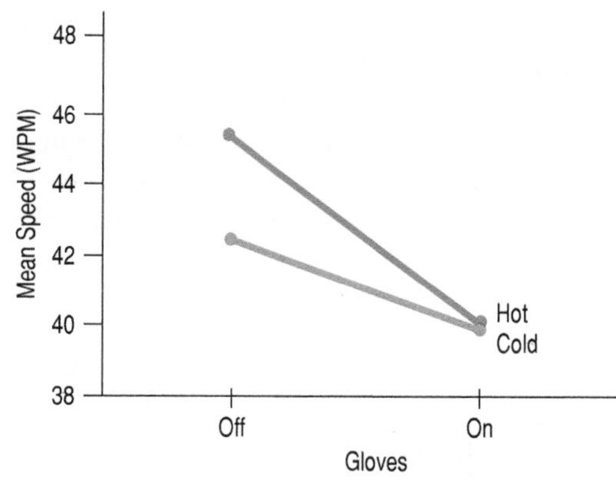

The ANOVA table shows:

Source	DF	Sum of squares	F-ratio	P-value
Gloves	1	120.1250	57.2511	<0.0001
Temperature	1	18.0000	8.5787	0.0067
Gloves*Temp	1	15.1250	7.2085	0.0121
Error	28	58.7496		
Total	31	212.0000		

A scatterplot of residuals versus predicted values shows:

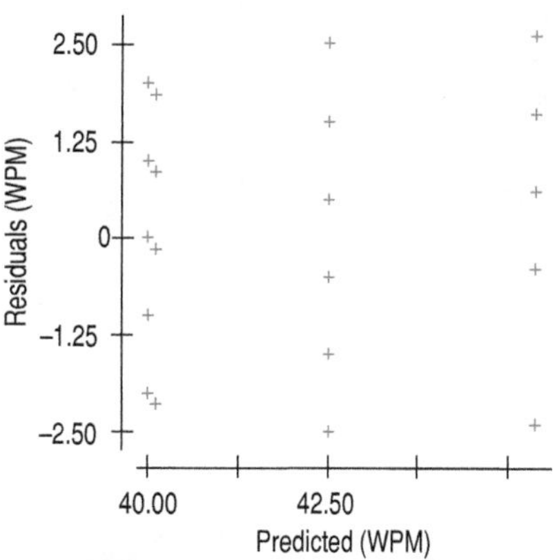

a) What are the null and alternative hypotheses for the main effects?

b) Given the partial boxplots, what do you suspect the *F*-test will say about the two main effects?

c) Should he consider fitting an interaction term to the model? Why or why not?

d) If he does fit an interaction term, do you suspect it will be significant on the basis of the interaction plot? Explain.

e) Which effects appear to be significant?

f) Describe the effects of the factors.

g) What is the size of the estimated standard deviation of the errors? Does this seem reasonable given the partial boxplots? Explain.

h) If Nick wants to increase his typing speed, what recommendations would you give him?

i) What reservations, if any, do you have about the conclusions?

27. Typing again Nick (see Exercise 26) designed a follow-up experiment to see if having *Music* or *Television* on would affect his typing speed. In particular, he'd like to know if he can type just as effectively with the music and/or the television on while he types. He ran each combination of *Music* and *Television* on or off eight times, recording the net number of words typed (words typed minus mistakes). Partial boxplots and interaction plots show:

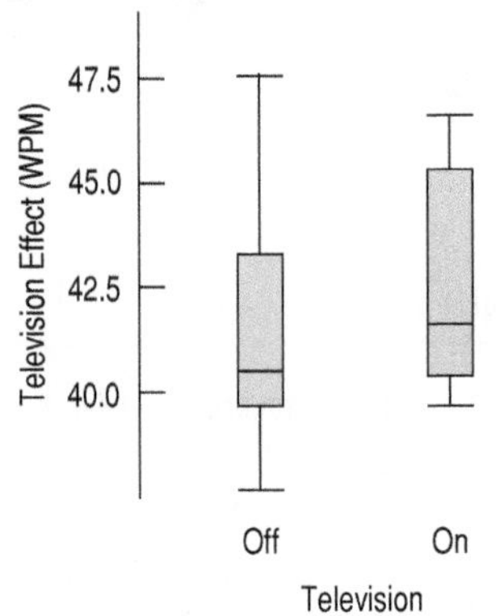

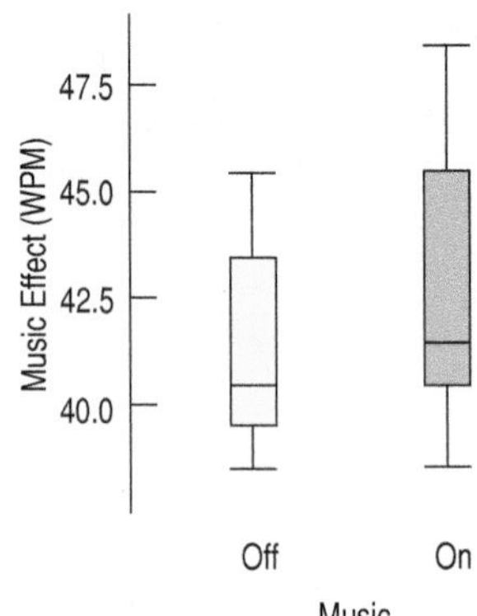

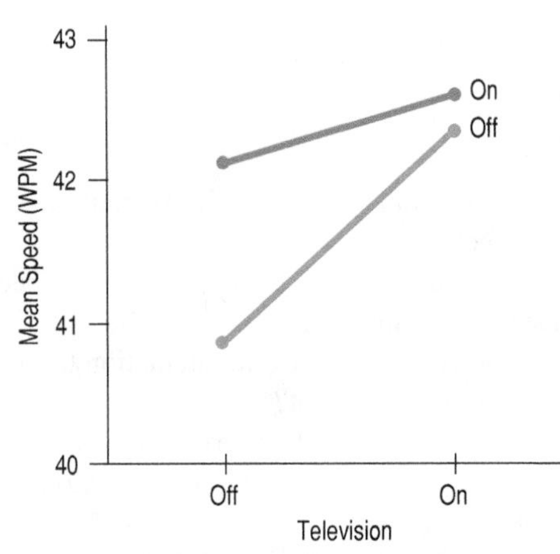

The ANOVA table shows:

Source	DF	Sum of squares	F-ratio	P-value
Music	1	4.5000	0.6380	0.4312
Television	1	8.0000	1.1342	0.2960
Interaction	1	2.0000	0.2835	0.5986
Error	28	197.5000		

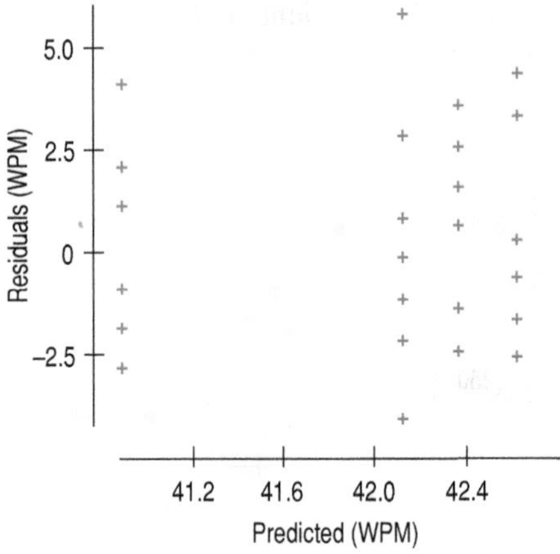

a) What are the null and alternative hypotheses for the main effects?
b) Given the partial boxplots, what do you suspect the F-test will say about the two main effects?
c) Should he consider fitting an interaction term to the model? Why or why not?
d) If he does fit an interaction term, do you suspect it will be significant on the basis of the interaction plot? Explain.
e) What effects appear to be significant?
f) Describe the effects of the factors.
g) What is the size of the estimated standard deviation of the errors? Does this seem reasonable given the partial boxplots? Explain.
h) If Nick wants to increase his typing speed, what recommendations would you give him?
i) What reservations, if any, do you have about the conclusions?

28. NY Marathon The *New York Times* reported the results of the 2003 NY Marathon by listing time brackets and the number of racers who finished within that bracket. Because the brackets are of different sizes, we look at the number of racers finishing per minute against the time at the middle of the time bracket. The resulting regression looks like this:

Dependent variable is: #/minute
R-squared = 10.9% R-squared (adjusted) = 9.5%
s = 78.45 with 63 − 2 = 61 degrees of freedom

Variable	Coefficient	SE(Coeff)	t-ratio	P-value
Intercept	45.1407	51.74	0.873	0.3863
Mid time ...	0.519037	0.1899	2.73	0.0082

a) How would you interpret the coefficient of *Mid time*?

Here's a scatterplot of the Studentized residuals against the predicted values for this regression:

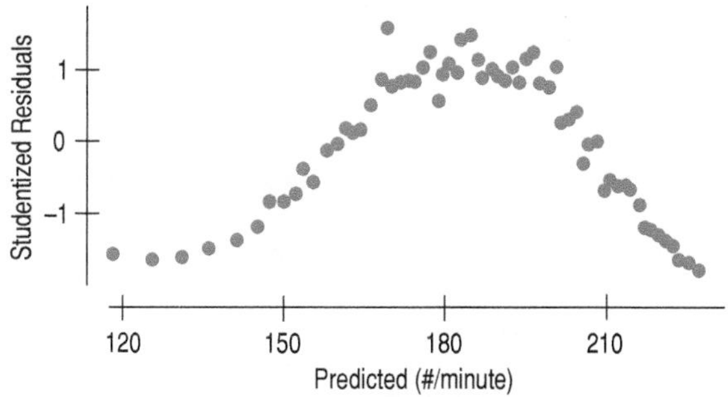

b) Comment on the regression in the light of this plot.

29. Births The U.S. National Vital Statistics Report provides information on live *Births* (per 1000 women), according to the age of the woman (in five-year brackets—*Age* used here is the midpoint of the bracket) and the *Year* from 1990 to 1999. The report isolates births to women younger than 20 as a separate category. Looking only at women 20 years old and older, we find the following regression:

Dependent variable is: Births
R-squared = 98.1% R-squared (adjusted) = 98.0%
s = 15.55 with 50 − 3 = 47 degrees of freedom

Variable	Coefficient	SE(Coeff)	t-ratio	P-value
Intercept	4422.98	1527	2.90	0.0057
Age	−15.1898	0.3110	−48.8	≤0.0001
Year	−1.89830	0.7656	−2.48	0.0168

a) Write out the regression model

b) How would you interpret the coefficient of *Year* in this regression? What happened to pregnancy rates during the decade of the 1990s?

Here's a scatterplot of the Studentized residuals against the predicted values.

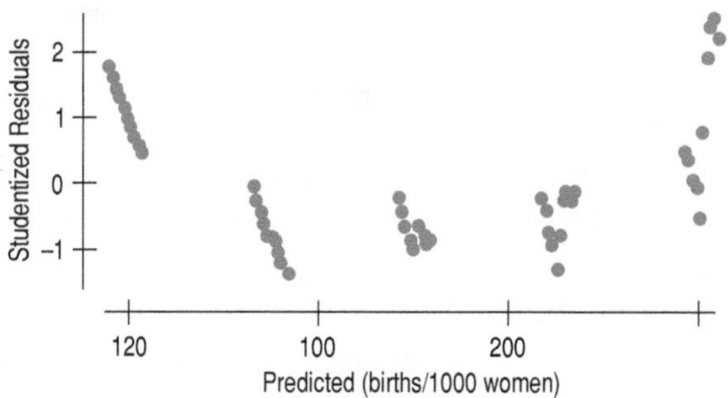

c) What might you do to improve the regression model?

30. Depression and the Internet The September 1998 issue of the *American Psychologist* published an article reporting on an experiment examining "the social and psychological impact of the Internet on 169 people in 73 households during their first 1 to 2 years on-line." In the experiment, a sample of households was offered free Internet access for one or two years in return for allowing their time and activity online to be tracked. The members of the households who participated in the study were also given a battery of tests at the beginning and again at the end of the study. One of the tests measured the subjects' levels of depression on a four-point scale, with higher numbers meaning the person was more depressed. Internet usage was measured in average number of hours per week. The regression analysis examines the association between the subjects' depression levels and the amounts of Internet use. The conditions for inference were satisfied.

Dependent variable is: Depression After
R-squared = 4.6%
s = 0.4563 with 162 − 2 = 160 degrees of freedom

Variable	Coefficient	SE(coeff)	t-ratio	Probe
Constant	0.565485	0.0399	14.2	≤0.0001
Intr_use	0.019948	0.0072	2.76	0.0064

a) Do these data indicate that there is an association between Internet use and depression? Test an appropriate hypothesis and state your conclusion clearly.

b) One conclusion of the study was that those who spent more time online tended to be more depressed at the end of the experiment. News headlines said that too much time on the Internet can lead to depression. Does the study support this conclusion? Explain.

c) As noted, the subjects' depression levels were tested at both the beginning and the end of this study; higher scores indicated the person was more depressed. Results are summarized in the table. Is there evidence that the depression level of the subjects changed during this study?

Depression Level
162 subjects

Variable	Mean	StdDev
DeprBfore	0.730370	0.487817
DeprAfter	0.611914	0.461932
Difference	−0.118457	0.552417

31. Learning math Developers of a new math curriculum called Accelerated Math compared performances of students taught by their system with control groups of students in the same schools who were taught using traditional instructional methods and materials. Statistics about pretest and posttest scores are shown in the table.[18]

a) Did the groups differ in average math score at the start of this study?

b) Did the group taught using the Accelerated Math program show a significant improvement in test scores?

c) Did the control group show significant improvement in test scores?

d) Were gains significantly higher for the Accelerated Math group than for the control group?

[18] J. Ysseldyke and S. Tardrew, *Differentiating Math Instruction*, Renaissance Learning, 2002.

		Instructional Method	
		Acc. math	Control
Number of students		231	245
Pretest	Mean	560.01	549.65
	St. Dev	84.29	74.68
Posttest	Mean	637.55	588.76
	St. Dev	82.9	83.24
Individual gain	Mean	77.53	39.11
	St. Dev.	78.01	66.25

32. Pesticides A study published in 2002 in the journal *Environmental Health Perspectives* examined the sex ratios of children born to workers exposed to dioxin in Russian pesticide factories. The data covered the years 1961 to 1988 in the city of Ufa, Bashkortostan, Russia. Of 227 children born to workers exposed to dioxin, only 40% were male. Overall in the city of Ufa, the proportion of males was 51.2%. Is this evidence that human exposure to dioxin may results in the birth of more girls? (An interesting note: It appeared that paternal exposure was most critical; 51% of babies born to mothers exposed to the chemical were boys.)

T **33. Dairy sales** Peninsula Creameries sells both cottage cheese and ice cream. The CEO recently noticed that in months when the company sells more cottage cheese, it seems to sell more ice cream as well. Two of his aides were assigned to test whether this is true or not. The first aide's plot and analysis of sales data for the past 12 months (in millions of pounds for *Cottage Cheese* and for *Ice Cream*) appears below.

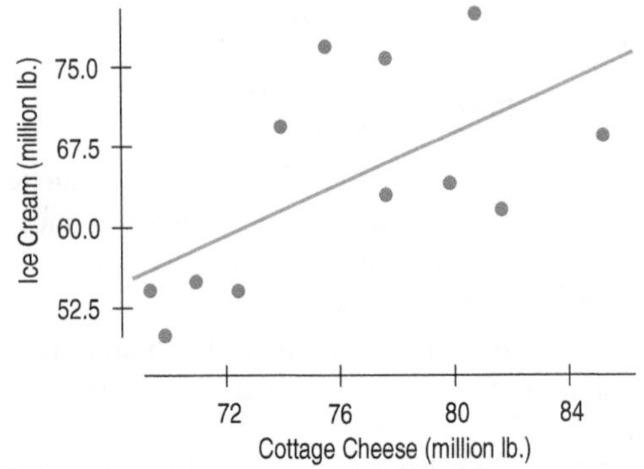

Dependent variable is: Ice cream
R-squared = 36.9%
s = 8.320 with 12 − 2 = 10 degrees of freedom

Variable	Coefficient	SE(Coeff)	*t*-ratio	Probe
Constant	−26.5306	37.68	−0.704	0.4975
Cottage C...	1.19334	0.4936	2.42	0.0362

The other aide looked at the differences in sales of ice cream and cottage cheese for each month, and created the following output:

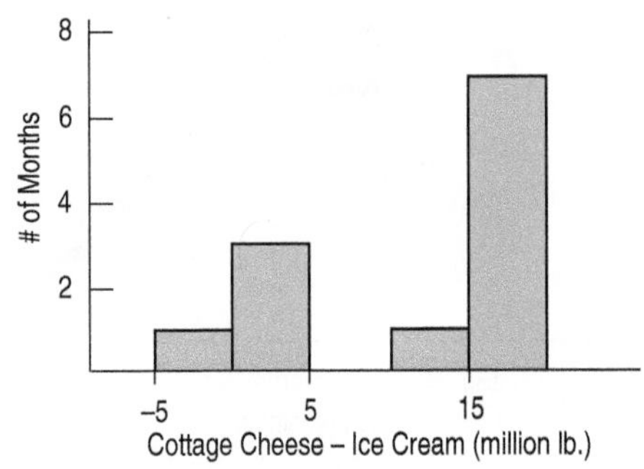

Cottage Cheese-Ice Cream

Count	12
Mean	11.8000
Median	15.3500
StdDev	7.99386
IntQRange	14.3000
25th %tile	3.20000
75th %tile	17.5000

Test H_0: μ(CC − IC) = 0 vs H_a: μ(CC − IC) ≠ 0
Sample Mean = 11.800000 *t*-Statistic = 5.113 w/11 df
Prob = 0.0003
Lower 95% bound = 6.7209429
Upper 95% bound = 16.879057

a) Which analysis would you use to answer the CEO's question? Why?
b) What would you tell the CEO?
c) Which analysis would you use to test whether the company sells more cottage cheese or ice cream in a typical year? Why?
d) What would you tell the CEO about this other result?
e) What assumptions are you making in the analysis you chose in part a)? What assumptions are you making in the analysis in part c)?
f) Next month's cottage cheese sales are 82 million pounds. Ice cream sales are not yet available. How much ice cream do you predict Peninsula Creameries will sell?
g) Give a 95% confidence interval for the true slope of the regression equation of ice cream sales by cottage cheese sales.
h) Explain what your interval means.

34. Video pinball A student runs an experiment to test how different factors affect his score while playing video pinball. Here are the results of 16 runs of an experiment performed in random order. Factor *Eyes* has two levels: both open and right eye closed. Factor *Tilt* has two levels: tilt on and tilt off. The response is the score of one ball at a combination of factors. Each combination was repeated four times in random order.

a) What are the null and alternative hypotheses for the main effects?
b) Analyze the data and write a short report on your findings. Include appropriate graphics and diagnostic plots.

Eyes	Tilt	Score
Both	On	67 059
Right eye	Off	21 036
Both	On	59 520
Right eye	Off	3 100
Both	Off	61 272
Right eye	On	55 957
Both	Off	72 472
Right eye	On	18 460
Right eye	On	16 556
Both	Off	89 553
Right eye	On	37 950
Both	Off	74 336
Right eye	Off	700
Both	On	79 037
Right eye	Off	36 591
Both	On	74 610

35. Javelin Brianna, a member of the track and field team, runs an experiment to test how different factors affect her javelin throw. She wants to know if the more expensive (premium) javelin is worth the extra price and is curious to know how much warming up helps her distance. She tries all four combinations of the two *Javelins* (standard and premium) and *Preparation* (no warm-up—cold—and warm-up), repeating each combination twice in random order. She measures the distance of her throw to the nearest metre. Here are the results of the eight runs.

Javelin	Preparation	Distance (metres)
Premium	Warm-up	46
Standard	Warm-up	39
Premium	Cold	37
Standard	Cold	30
Premium	Warm-up	45
Standard	Warm-up	40
Premium	Cold	35
Standard	Cold	32

a) What are the null and alternative hypotheses for the main effects?

b) Analyze the data and write a short report on your findings. Include appropriate graphics and diagnostic plots and a recommendation for Brianna to optimize her javelin distance.

36. Eye and hair colour A survey of 1021 school-age children was conducted by randomly selecting children from several large urban elementary schools. Two of the questions concerned eye and hair colour. In the survey, the following codes were used:

Hair Colour	Eye Colour
1 = Blond	1 = Blue
2 = Brown	2 = Green
3 = Black	3 = Brown
4 = Red	4 = Grey
5 = Other	5 = Other

The Statistics students analyzing the data were asked to study the relationship between eye and hair colour.

a) One group of students produced the output shown below. What kind of analysis is this? What are the null and alternative hypotheses? Is the analysis appropriate? If so, summarize the findings, being sure to include any assumptions you've made and/or limitations to the analysis. If it's not an appropriate analysis, explicitly state why not.

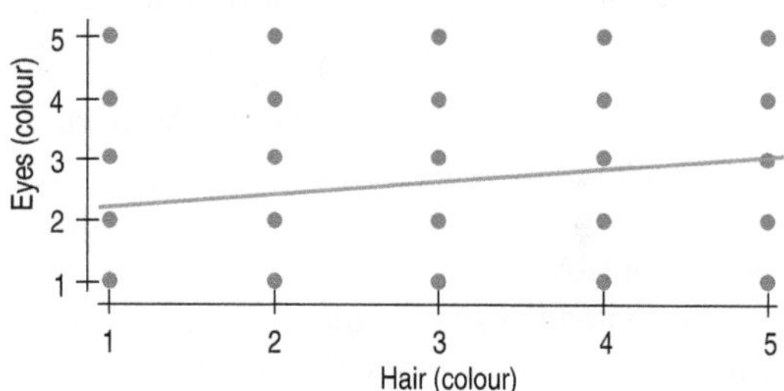

Dependent variable is: Eye Colour
R-squared = 3.7%
s = 1.112 with 1021 − 2 = 1019 degrees of freedom

Variable	Coefficient	SE(Coeff)	*t*-ratio	Probe
Constant	1.99541	0.08346	23.9	≤0.0001
Hair Colour	0.211809	0.03372	0.28	≤0.0001

b) A second group of students used the same data to produce the output shown below. What kind of analysis is this? What are the null and alternative hypotheses? Is the analysis appropriate? If so, summarize the findings, being sure to include any assumptions you've made and/or limitations to the analysis. If it's not an appropriate analysis, explicitly state why not.

Table contents: Counts
Standardized Residuals

Hair Colour	Eye Colour 1	2	3	4	5
1	143 7.6754	30 0.417988	58 −5.88169	15 −0.63925	12 −0.314506
2	90 −2.57141	45 0.290189	215 1.72235	30 0.491885	20 −0.0824592
3	28 −5.39425	15 22.3478	190 6.28154	10 −1.76376	10 −0.803818
4	30 2.06116	15 2.71589	10 −4.0554	10 2.37402	5 0.759931
5	10 −0.521945	5 0.332621	15 −0.941918	5 1.36326	5 −.07578

$$\sum \frac{(observed - Expected)^2}{Expected} = 223.6 \quad \text{P-value} = 0.00001$$

37. Infliximab In an article appearing in the journal *Lancet* in 2002, medical researchers reported on the experimental use of the arthritis drug infliximab in treating Crohn's disease. In a trial, 573 patients were given initial five-mg injections of the drug. Two weeks later, 335 had responded positively. These patients were then randomly assigned to three groups. Group I received continued injections of a placebo, Group II continued with five mg of infliximab, and Group III received 10 mg of the drug. After 30 weeks, 23 of 110 Group I patients were in remission, compared with 44 of 113 Group II and 50 of 112 Group III patients. Do these data indicate that continued treatment with infliximab is of value for Crohn's disease patients who exhibit a positive initial response to the drug?

T **38. LA rainfall** The Los Angeles Almanac Web site reports recent annual rainfall (in inches), as shown in the table.

a) Create a 90% confidence interval for the mean annual rainfall in L.A.

b) If you wanted to estimate the mean annual rainfall with a margin of error of only two inches, how many years' data would you need?

c) Do these data suggest any change in annual rainfall as time passes? Check for an association between rainfall and year.

Year	Rain (in.)	Year	Rain (in.)
1980	8.96	1991	21.00
1981	10.71	1992	27.36
1982	31.28	1993	8.14
1983	10.43	1994	24.35
1984	12.82	1995	12.46
1985	17.86	1996	12.40
1986	7.66	1997	31.01
1987	12.48	1998	9.09
1988	8.08	1999	11.57
1989	7.35	2000	17.94
1990	11.99	2001	4.42

39. TV and athletics Using a simple random sample, a student asked 200 students questions about their study and workout habits, and received 124 responses. One of the questions asked, "Do you participate in intramural (IM) athletics, varsity athletics, or no athletics?" while the second asked, "How many hours of television did you watch last week?" The student wants to see if participation in athletics is associated with amount of television watched. Here are the boxplots of *Television watching* by *Athletic participation*:

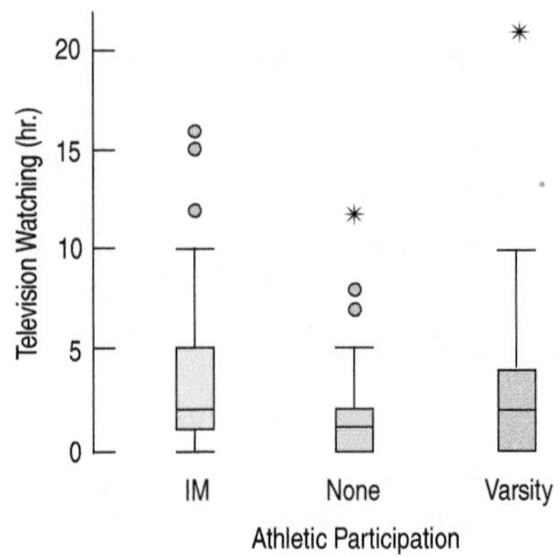

Analysis of Variance

Source	DF	Sum of squares	Mean square	F-ratio	P-value
Athl. Part.	2	128.271	64.1353	4.2343	0.0167
Error	121	1832.72	15.1465		
Total	123	1960.99			

a) State the hypothesis about the students (both numerically and in words).

b) Do the conditions for ANOVA appear to be satisfied? What concerns, if any, do you have?

c) Assuming that the assumptions for inference are satisfied, perform the hypothesis test and state your conclusion. Be sure to state it in terms of television watching and athletic participation.

d) An analysis of the data with all the points highlighted as outliers removed was performed. The *F*-test showed a P-value of 0.0049. How does this affect your answer to part b)?

40. Weight and athletics Using the same survey as in Exercise 39, the student examined the relationship between *Athletic Participation* and *Weight*. Here are the boxplots of *Weight* by *Athletic* participation:

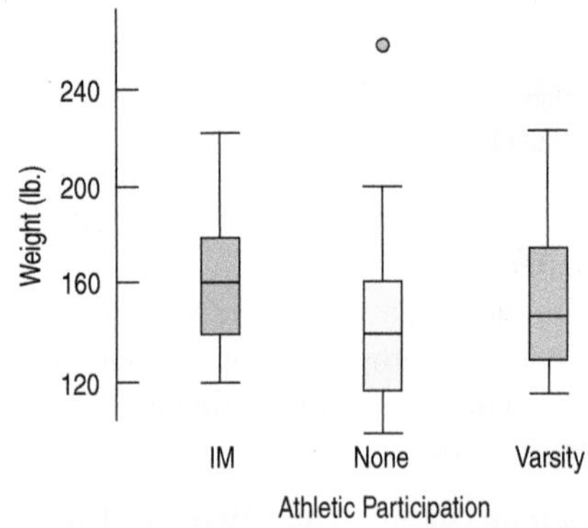

Analysis of Variance

Source	DF	Sum of squares	Mean square	F-ratio	P-value
Athl. Part.	2	9025.55	4512.78	5.7368	0.0042
Error	121	95183.2	786.638		
Total	123	104209			

a) State the null hypothesis about the students (both numerically and in words).
b) Do the conditions for ANOVA appear to be satisfied? What concerns, if any, do you have?
c) Assuming that the assumptions for inference are satisfied, perform the hypothesis test and state your conclusion. Be sure to state it in terms of *Weight* and *Athletic participation*. What might explain the apparent relationship?
d) An analysis of the data with the points highlighted as an outlier removed was performed. The *F*-test showed a P-value of 0.0030. How does this affect your answer to part b)?

41. Weight loss A weight loss clinic advertises that its program of diet and exercise will allow clients to lose 10 pounds in one month. A local reporter investigating weight reduction gets permission to interview a randomly selected sample of clients who report the given weight losses during their first month in this program. Create a confidence interval to test the clinic's claim that typical weight loss is 10 pounds.

Pounds	Lost
9.5	9.5
13	9
9	8
10	7.5
11	10
9	7
5	8
9	10.5
12.5	10.5
6	9

42. Cramming Students in two basic Spanish classes were required to learn 50 new vocabulary words. One group of 45 students received the list on Monday and studied the words all week. Statistics summarizing this group's scores on Friday's quiz are given. The other group of 25 students did not get the vocabulary list until Thursday. They also took the quiz on Friday after "cramming" Thursday night. Then, when they returned to class the following Monday they were retested—without advance warning. Both sets of test scores for these students are shown.

Group 1
Fri.
Number of students = 45
Mean = 43.2 (of 50)
StDev = 3.4
Students passing (score ≥ 40) = 33%

Group 2			
Fri.	Mon.	Fri.	Mon.
42	36	50	47
44	44	34	34
45	46	38	31
48	38	43	40
44	40	39	41
43	38	46	32
41	37	37	36
35	31	40	31
43	32	41	32
48	37	48	39
43	41	37	31
45	32	36	41
47	44		

a) Did the week-long study group have a mean score significantly higher than that of the overnight crammers?
b) Was there a significant difference in the percentages of students who passed the quiz on Friday?
c) Is there any evidence that when students cram for a test their "learning" does not last for three days?
d) Use a 95% confidence interval to estimate the mean number of words that might be forgotten by crammers.
e) Is there any evidence that how much students forget depends on how much they "learned" to begin with?

43. Education versus income The information below examines the *Median income* and *Median education* level (years in school) for several U.S. cities.

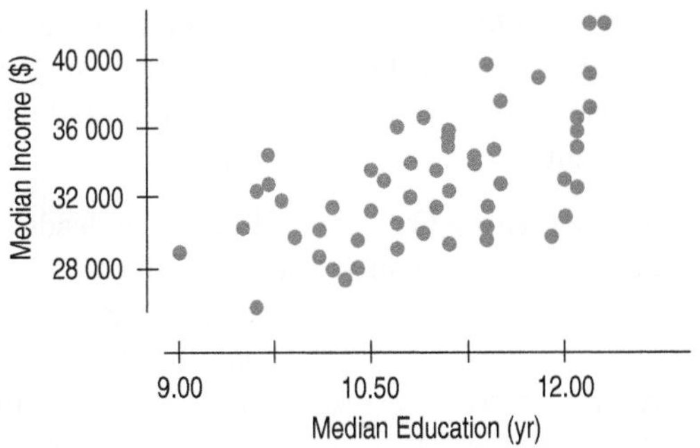

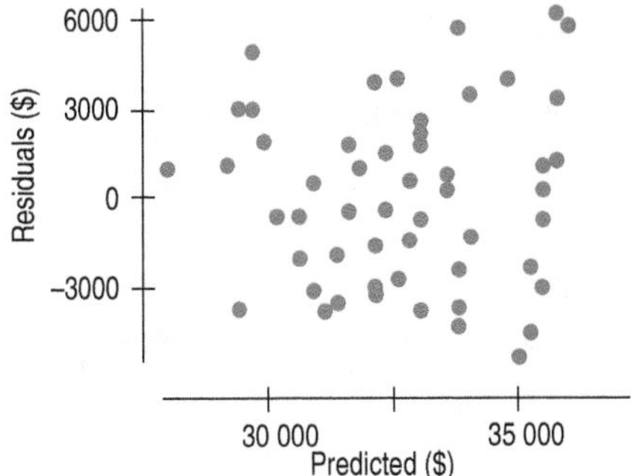

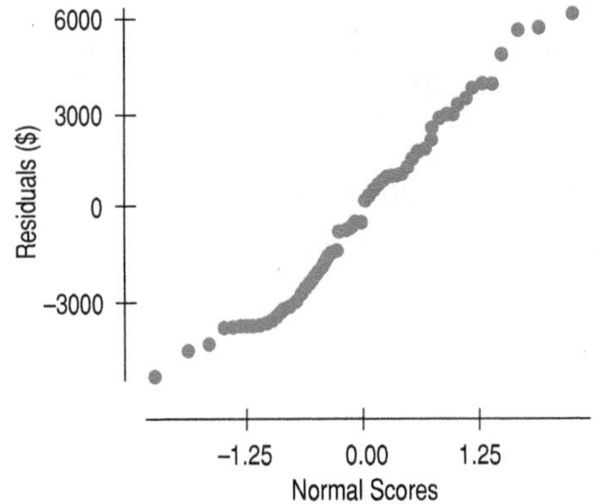

Variable	Count	Mean	StdDev
Education	57	10.9509	0.848344
Income	57	32742.6	3618.01

Dependent variable is: Income
R-squared = 32.9%
s = 2991 with 57 − 2 = 55 degrees of freedom

Variable	Coefficient	SE(Coeff)	*t*-ratio	Probe
Constant	5970.05	5175	1.15	0.2537
Education	2444.79	471.2	5.19	≤0.0001

a) Do you think the assumptions for inference are met? Explain.

b) Does there appear to be an association between education and income levels in these cities?

c) Would this association appear to be weaker, stronger, or the same if data were plotted for individual people rather than for cities in aggregate? Explain.

d) Create and interpret a 95% confidence interval for the slope of the true line that describes the association between income and education.

e) Predict the *Median income* for cities where residents spent an average of 11 years in school. Describe your estimate with a 90% confidence interval, and interpret that result.

44. Airport screening Concern with terrorism leads us to look at the records of airport screening for the years between 1977 and 1999 as provided by *the Sourcebook of Criminal Justice Statistics Online*. We find the following regression predicting the incidence of false information from other problems discovered while screening passengers:

Dependent variable is: False info
R-squared = 8.7% R-squared (adjusted) = −9.5%
s = 48.41 with 19 − 4 = 15 degrees of freedom

Source	Sum of Squares	DF	Mean square	F-ratio
Regression	3364.40	3	1121.47	0.479
Residual	35150.2	15	2343.35	

Variable	Coefficient	SE(Coeff)	*t*-ratio	P-value
Intercept	27.8181	65.24	0.426	0.6759
Long guns	0.545680	0.6010	0.908	0.3783
Handguns	−4.25650e-3	0.0357	−0.119	0.9066
Explosives	−0.093442	0.0952	−0.982	0.3417

a) Does this appear to be a successful model for incidence of false information? Why or why not?

Here's a scatterplot of the Studentized residuals versus the predicted values:

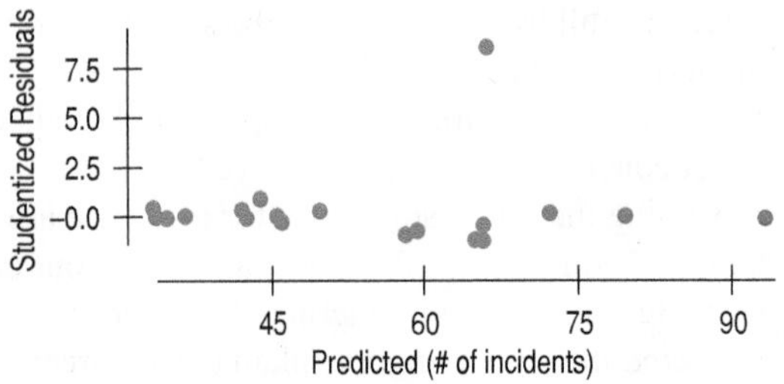

The outlying value is the year 1988. We created an indicator variable for 1988 and included it in the model with the following result:

Dependent variable is: False info
R-squared = 86.8% R-squared (adjusted) = 83.0%
s = 19.05 with 19 − 5 = 14 degrees of freedom

Source	Sum of Squares	DF	Mean square	F-ratio
Regression	33436.2	4	8359.05	23.0
Residual	5078.44	14	362.746	

Variable	Coefficient	SE(Coeff)	*t*-ratio	P-value
Intercept	47.2152	25.76	1.83	0.0882
Long guns	0.750819	0.2375	3.16	0.0069
Handguns	−0.023583	0.0142	−1.66	0.1190
Explosives	−0.089368	0.0374	−2.39	0.0316
1988	181.310	19.91	9.10	≤0.0001

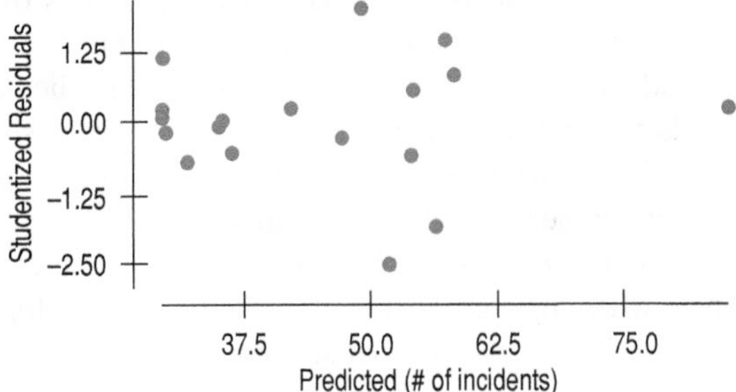

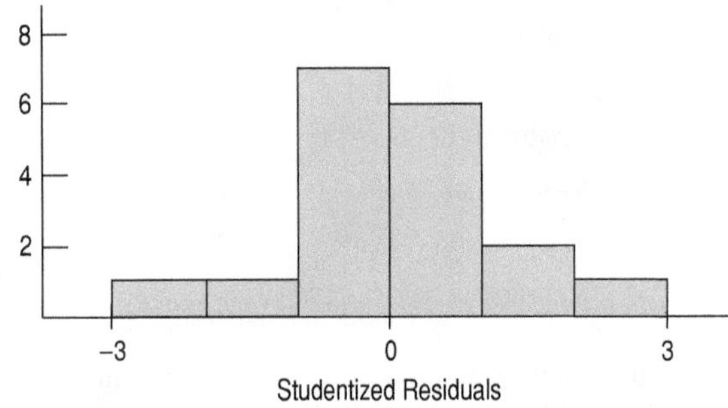

b) Complete the analysis. Check the assumptions and conditions, so far as you can with the information provided. Summarize the model. Discuss any concerns you may still have about the data or the model.

c) Would you remove any of the predictors from the model? If so, which one and why?

chapter

29 Rank-Based Nonparametric Tests

Where are we going?

The standard or "classical" approaches to statistical analysis of quantitative data discussed so far assume that the data are random samples from Normal distributions. We then estimate the Normal curve parameters, μ and σ, that will give a good fit to the data. Such an approach is said to be "parametric" in nature, since we choose a density curve model, and then estimate its parameters. If the Normal model doesn't fit the data, we can try a transformation, or try to find another density curve that will be a good fit. But, is there another way, a method that doesn't impose any particular curve or distribution on the data; a *distribution-free* method? If there's no density curve, there are no parameters to estimate, yet it is precisely those parameters that capture the information needed for inference. Is it possible to extract enough information from the data, *nonparametrically*, to enable inference? We'll see how in this chapter.

I have not failed. I have just found 10,000 ways that won't work.

—*Thomas Edison*

In the course Statistics 220, offered at the University of Toronto in the summer of 2006, there were six students who were Biology majors and five who were Psychology majors. Their final course grades were:

Biology Majors	Psychology Majors
C+	B
B−	C
A−	D−
D+	B+
A+	C+
B+	

We might ask if there is evidence of a difference in achievement levels between the two types of students. The grades appear higher for the Biology majors, but are they significantly higher? Though this was not a random sample, many students each year take Statistics 220, so it may be reasonable to view these students as a random sample of the many Biology and Psychology students who are required to take this course.

If the numerical grades were available to us, or if we were to convert the letter grades into corresponding numerical grades, we might be able to proceed using a t-statistic—after checking the data for possible violations of the assumption of Normality. Though the two-sample t-test is rather robust against violations of Normality, particularly for larger samples, the samples in this case are quite small, so noticeable departures from Normality would be of concern.

There are times when an alternative approach may be necessary or desirable, one that is less restrictive in its assumptions. Might there be some other way of attacking this problem? Surely, there must be more than one way to skin a cat (oops . . . [1])!

29.1 Wilcoxon Rank Sum Test

Let's start by ranking all 11 observations from low to high, so that the lowest grade receives a rank of 1, the next lowest a rank of 2, and so on, until we reach the highest grade, which receives a rank of 11. How should we deal with grades that are identical?

[1] Please forward to the authors suggestions for colourful but *politically correct* alternative expressions. Thank you.

While we might consider just flipping a coin to decide which student gets which rank, the usual approach is to assign the same rank value to each of the observations that are tied, where the rank value is the average of the ranks that the observations would have been assigned if not tied. So, if the third, fourth, and fifth biggest values in the data are identical, each one's rank will be recorded as $(3 + 4 + 5)/3 = 4$.

Below are the ranks of our 11 observations. Note that we rank them all together, as if this were one big sample of 11 numbers. Let's also compute the sum of ranks for each group, denoted T_1 and T_2. We will need them for making inferences.

Biology Majors	Rank	Psychology Majors	Rank
C+	4.5	B	7
B−	6	C	3
A−	10	D−	1
D+	2	B+	8.5
A+	11	C+	4.5
B+	8.5		
Rank Sum:	$42 = T_1$		$24 = T_2$

Who	Students taking STA220
What	Course (letter) grades in STA220
Why	To compare performance of students of different majors
When	Summer 2006
Where	University of Toronto

Alternatively, we could put all the grades in order, as below, while keeping track of which grades come from which sample (**bold** indicate Psychology majors), and then add up the ranks in bold (or not in bold), to compute the rank sum for Psychology (or Biology) majors:

D−	D+	**C**	**C+**	C+	B−	**B**	**B+**	B+	A−	A+
1	2	**3**	**4.5**	4.5	6	**7**	**8.5**	8.5	10	11

A common element of the methods in this chapter is the replacement of observations by their ranks (so, in some sense, we are employing a transformation approach, using the "rank-transformation"). This frees us from potentially risky parametric assumptions and leads to a rather simple form of analysis, since the rank values of 1, 2, 3, . . . never change (while raw data vary from sample to sample). And, surprisingly, little loss of information or power occurs when we replace the raw data with their ranks.

We want to know if the Biology majors generally tend to perform better or worse than the Psychology majors. Lacking a parametric representation, we think of this as a systematic shift higher or lower for the values in one distribution when compared with the values in the other distribution. (We will define *systematic shift* more precisely later in the chapter.) In a nonparametric setting, our null hypothesis simply states that the populations are identical. We test:

H_0: The two distributions of grades are identical.

vs.

H_A: The values in one distribution are systematically higher or lower than the values in the other distribution.

If one group is superior to the other, it should show up in generally higher ranks, so maybe we should compare the rank averages for the two groups. But we can do something even simpler. We have already calculated the two rank sums, denoted T_1 and T_2. The sum of all of the ranks (1 to 11) equals $1 + 2 + 3 + \ldots + 11 = 66$. And, in general, the sum of the first n integers 1, 2, 3, . . . and so on up to n is equal to $n(n + 1)/2$. So if we know one rank sum, the other is completely determined. Since the rank sum of the first sample is 42, we can predict, without even counting, that the other rank sum must equal the grand sum of 66 minus 42, or 24. Hence a test can be based on either of the two rank sums. We do not need to compare two rank sums or averages with each other (as in a two-sample t-test). We pick one rank sum, say $T_2 = 24$, and then determine whether it is unusually big or small.

What do we know about T_2? It could be as small as $1 + 2 + 3 + 4 + 5 = 15$, or as big as $11 + 10 + 9 + 8 + 7 = 45$. It appears that our expectation under H_0 might be right in the middle of these numbers, or equal to 30, and indeed this is true. Under the null hypothesis, the rankings in any sample are just a random selection of all rankings, so each sample should, on average, receive its fair share, which would be a rank sum proportionate to its size. Sample two has 5 of the 11 grades, so its expected rank sum is $5/11$ of the sum of all ranks, or $5/11 \times 66 = 30$.

Under the null, T_2 can range from 15 to 45, with a mean of 30. Here it equals 24, somewhat below its mean. But how unusual is this? In other words, what is the P-value?

If we had chosen to work with the other rank sum, T_1, a similar analysis would tell us that it could be anywhere between 21 and 51, with a mean of 36. The actual rank sum is 42, and comparing this high number with its expectation of 36 is exactly the same as comparing the second sample's low rank sum of 24 with its expectation of 30. So either rank sum will do. But when using tables of critical values, the convention is to use the rank sum corresponding to the smaller sample (with fewer observations). Here this is sample two.

We calculated $T_2 = 24$. We need to find the probability of getting such a small rank sum. A deck of cards and some time (well, lots of time) would suffice. Remove every card except the 1, 2, . . . , 10, and Jack, of say hearts (think of the Jack as 11). Under the null hypothesis, the ranks for sample two are a random draw of five cards from this deck. Deal out five cards, and count the rank sum. You now have one observation, by simulation, on T_2. Repeat, repeat, repeat . . . , thousands of times, to approximate the distribution of T_2. Find the proportion of times that the rank sum was 24 or below, and you'll have the P-value (but double it for a two-sided alternative). This is a good way to think about the P-value, but not exactly practical (though simulating by computer would be easy enough).

For the Normal curve parametric approach, our null model draws each sample from that well-shuffled deck of cards used by Gossett (in Chapter 20), where all the numbers in the deck, piled up, produce a bell-shaped population. Here, our null model shuffles 11 cards numbered 1 to 11, and draws five of them.

Some basic knowledge of combinations or permutations is really all that's needed. We have to count all the ways that any particular rank sum can occur (for example, $T_2 = 17$ only for rankings of 1, 2, 3, 4, 7 or 1, 2, 3, 5, 6), and divide that count by ${}_{11}C_5$ (which, as in Chapter 14, denotes the number of ways of choosing 5 items from 11 items, and equals 462) to compute the probability of obtaining this rank sum. If we did this for every possible rank sum, we could then find any P-value by summing the appropriate individual probabilities.

Fortunately, this dreary combinatorial work has already been done for you in tables of critical values for the **Wilcoxon Rank Sum test**. The corresponding table in Appendix C gives some tail probabilities for the rank sum corresponding to the smaller sized sample (for equal sample sizes, use either rank sum). In our example, the rank sum for the smaller sample is $T_2 = 24$. From the table, for $n_1/n_2 = 5/6$, the closest critical value is 20, and the probability of a rank sum ≤ 20 is approximately 0.05. So the P-value for our rank sum of 24 is bigger than 0.05×2, after doubling the lower tail area for our two-tailed test. At the 10% level of significance, we do not have evidence of a difference in course grades between the two majors in our study.

If a table is unavailable, or if your sample sizes do not appear in the table, what to do? Recall that the Central Limit Theorem tells us that averaging or summing many observations will often produce Normality,[2] and indeed it is the case here that if both sample sizes exceed 10, the rank sums will be approximately Normally distributed. Then all you need is the parameters of the Normal model. It can be shown that

$$E(T_i) = n_i(n_1 + n_2 + 1)/2$$

and

$$\text{Var}(T_i) = n_1 n_2(n_1 + n_2 + 1)/12$$

for $i = 1$ or 2, so either rank sum can be standardized using the above formula, and then the P-value can be found from the standard Normal distribution. It is possible to improve

[2]The Central Limit Theorem discussed in Chapter 15 has been generalized in various ways over the years. Under appropriate conditions, a sum or average of random variables—even when they do not all have the same distribution and are not independent—will have an approximately Normal distribution, as long as you sum enough of them. The conditions essentially ensure that one variable is not very dominant in its effects over the others. As long as they all make small contributions to the sum, Normality often results.

slightly on this formula using a continuity correction in the top and a correction for ties in the bottom, but we will omit these corrections here. Unless there are a great number of ties relative to the sample size, the correction for ties makes almost no difference.

Assumptions and Conditions

Since this is a distribution-free procedure, there are no distributional assumptions such as Normality to check, but pay attention to the other common design issues mentioned in Chapters 21 and 25, such as:

Independence Assumptions

Independence Groups Assumption: The groups must be independent of each other. No test can verify this, so think about how the survey or experiment was conducted.
Independence Assumption: The data within each group must be independent as well—in other words, the individuals should be drawn independently and at random from the target population, or properly randomized in an experiment.

Check the **Randomization Condition.** For surveys, were the data drawn using a proper random sampling scheme, or can the data be viewed as independent observations reasonably representative of some populations if collected in a less controlled observational setting? For an experiment, were the experimental units assigned at random to the treatments?

Ordinal or Quantitative Data Condition

Ordinal Data Condition: The data in our example were *ordinal* in nature—we knew which grades were higher or lower, but not the exact amount of the difference between two grades. If we had the students' actual numerical grades, which are on a precise quantitative scale, we still could have proceeded exactly as before, since numerical data can easily be converted to ranks. *Quantitative data can always be reduced to an ordinal scale.* So the Wilcoxon test can be used for quantitative data or for purely ordinal data.

As another example, suppose that we wanted to compare opinions in Ontario and Alberta about a proposed greenhouse emissions bill, where the possible responses in a survey were "strongly disagree," "mildly disagree," "neutral," "mildly agree," or "strongly agree." These responses are on an ordinal scale, so we can rank all the responses from each province, though we would have a large number of ties—which the Normal approximation with correction for ties can handle well enough. (Warning: Be wary of assigning numbers to ordinal data and treating the results as if they were actual quantitative data.)

WILCOXON RANK SUM TEST[3]

When the assumptions and conditions for the Wilcoxon Rank Sum test are met, we can test:

H_0: The two distributions are identical.

vs.

H_A: The values in one distribution are systematically higher or lower than the values in the other distribution.

Calculate the rank sum for the smaller sample. Then estimate the P-value using tables of critical values (or the Normal approximation, if either sample size is off the table).

Due to the nonparametric nature of the test, we cannot compare distributions using parameters of density curves. A general definition of *systematically higher* would say that the probability massed to the right of any particular value is always greater (or sometimes the same) in one of the distributions. But for two similarly shaped quantitative populations, this test may be viewed as comparing medians, and software will typically provide a confidence interval for the difference in the medians.

Computer software may report a confidence interval for the difference in population medians. This interval is only valid for two quantitative populations possessing similarly shaped distributions.

[3]There is an alternative but equivalent form of this test known as the Mann-Whitney test.

For Example THE WILCOXON RANK SUM TEST WITH QUANTITATIVE DATA

Consider the Chapter 21 opening example, comparing the lifetimes of generic versus brand-name batteries. The battery lifetime measurements in minutes are shown in two columns below. The 12 lifetimes were ordered, from low to high, with each assigned a rank, displayed to the right of the lifetime.

Brand Name	Rank	Generic	Rank
194.0	5	190.7	4
205.5	9	203.5	7.5
199.2	6	203.5	7.5
172.4	2	206.5	10
184.0	3	222.5	12
169.5	1	209.4	11

QUESTION: Does the Wilcoxon test provide evidence of a difference between the two brands?

ANSWER: The sum of the ranks for brand-name batteries, denoted T_1, is 26.0. We find from our tables ($n_1 = 6$, $n_2 = 6$) that $P(T_1 \le 26) \le 0.025$, so the P-value is double this, or ≤ 0.05, which provides some evidence of a difference. In Chapter 21, the 95% confidence interval barely missed zero, indicating that the P-value for a test of no difference would be slightly less than 0.05. In fact, a precise calculation with a two-sample *t*-test gives P = 0.032, so the parametric test sees the difference (assuming it exists) slightly more clearly. Of course, this is assuming that the parametric calculation is trustworthy, and with so little data, it's hard to tell. However, the similarity between the two P-values should be somewhat reassuring.

If the samples were larger, we could use the Normal approximation to estimate the P-value. Let's check it out here anyway:

$$E(T_1) = 6(13)/2 = 39$$

and

$$\text{Var}(T_1) = (6)(6)(13)/12 = 39 \quad \text{or} \quad \sigma = 6.245, \text{ so we get:}$$

$z = \dfrac{(26 - 39)}{6.245} = -2.08$, yielding a P-value = 0.038, which is a pretty decent approximation to the P-value from the exact Wilcoxon test, which, more accurately calculated (from software), would be 0.042.

This Wilcoxon test has the advantage of a greater range of applicability compared with the *t*-test, primarily due to less restrictive assumptions about the data. It may give you trustworthy P-values when those calculated from the *t*-test are deemed suspect.

This raises another interesting question: How does the Wilcoxon test stack up against the *t*-test when the data are in fact Normally distributed? The issue is one of power. We would expect the Wilcoxon test to be less powerful than the *t*-test since it does not use as much of the information present in the data. We have thrown out the numbers in favour of their ranks, so some loss of power should result (just as power would decrease if we lost some observations), but how much loss?

It turns out that the power of the two-sample *t*-test for samples of size 95 is approximately the same as the power of the Wilcoxon test for samples of size 100 (for Normally distributed data).

Quite surprising! The loss of power in using the Wilcoxon test rather than the *t*-test is equivalent to a loss of 5% of sample size, so those ranks do a pretty good job of holding on to the relevant information present in the data. And when severe skewness or extreme outliers are present, the Wilcoxon test will usually have greater power to discern a genuine difference than the *t*-test.

Just Checking

1. If we compare data in two samples, one of size three and one of size four,
   a. What will be the sum of all the ranks?
   b. What are the biggest and smallest rank sums possible for the larger sample?
   c. What would be the expected or average value for the rank sum of the larger sample if there were no actual difference between the two populations?

29.2 Kruskal-Wallis Test

Suppose now that we have more than two independent samples of measurements to compare. We need a generalization of the Wilcoxon Rank Sum test. Below are bacteria count data for four different hand-washing methods, as examined in a Chapter 25 example:

Method:	Alcohol	AB Soap	Soap	Water
	51	70	84	74
	5	164	51	135
	19	88	110	102
	18	111	67	124
	58	73	119	105
	50	119	108	139
	82	20	207	170
	17	95	102	87

Transform the data to ranks, ranking them as one big group from 1 to 32 (using tied ranks when necessary). Ranks are shown in parentheses below. As a check, note that the sum of all the ranks should equal $n(n + 1)/2 = 32(33)/2 = 528$.

	Alcohol	AB Soap	Soap	Water
	51(7.5)	70(11)	84(15)	74(13)
	5(1)	164(30)	51(7.5)	135(28)
	19(4)	88(17)	110(23)	102(19.5)
	18(3)	111(24)	67(10)	124(27)
	58(9)	73(12)	119(25.5)	105(21)
	50(6)	119(25.5)	108(22)	139(29)
	82(14)	20(5)	207(32)	170(31)
	17(2)	95(18)	102(19.5)	87(16)
Rank Sums:	**46.5**	**142.5**	**154.5**	**184.5**

Unlike the Wilcoxon test, we cannot simply focus on one rank sum for these data (again, any one of the rank sums would be redundant, but that still leaves three free to vary). We need a measure of the variation among all the rank sums (adjusted for differences in sample size), and this is provided by the **Kruskal-Wallis statistic**:

$$H = \left(\frac{12}{N(N + 1)}\sum \frac{T_i^2}{n_i}\right) - 3(N + 1)$$

where T_i are the rank sums, and N = total sample size.

The Kruskal-Wallis statistic measures variation among the rank sums and grows bigger as the differences among them increase. So, a very large value for H provides evidence of a genuine difference among the populations. To assess this, we need an appropriate sampling model. If the samples are not too small, the chi-square distribution does the job nicely.[4] The number of degrees of freedom in this distribution is the number of squared terms in the summation (or the number of samples) minus 1.

This chi-square model is actually a large-sample approximation, so for very small samples, it would be best to hunt down a table of exact critical values (or similar software). Again, we omit a correction for ties that could be inserted into the formula above.

Assumptions and Conditions

The assumptions and conditions for the Kruskal-Wallis test are the same as for the Wilcoxon Rank Sum test, but the Kruskal-Wallis test is suitable for *comparing any number of independent groups*, whereas the Wilcoxon test can only be used to compare two independent groups.

KRUSKAL-WALLIS TEST

When the assumptions and conditions for the Kruskal-Wallis test are met, we may assess:

H_0: The distributions are all identical.

vs.

H_A: At least two of the distributions differ (systematically higher or lower) from each other.

Calculate the Kruskal-Wallis test statistic. Then, estimate the P-value using the upper tail (only) of the chi-square model with $t - 1$ degrees of freedom, where $t =$ the number of samples.

For Example

Continuing with our example, let's test the null hypothesis that the hand-washing method has no effect on bacteria count. We take the rank sums calculated earlier and plug into the Kruskal-Wallis statistic:

$$H = \left(\frac{12}{32(32+1)} \times \frac{46.5^2 + 142.5^2 + 154.5^2 + 184.5^2}{8}\right) - 3(32+1) = 15.17$$

The P-value $= P(\chi^2_{3df} > 15.17) = 0.002$ (from software), so we have very strong evidence that the washing method does indeed affect the bacteria count.

Comparing with the parametric analysis in Chapter 25, we see that the P-value here is less impressive, though still quite small. This mostly reflects the fact that the ANOVA paid much more attention to the big difference between the alcohol spray and the other methods, whereas this test takes note that the alcohol spray values were lower, without paying attention to how much lower.

Just Checking

2. If there are four samples with three responses in each, what is the biggest value possible for the Kruskal-Wallis statistic? And what would be the estimated P-value in that case?

[4]The operation of squaring and summing often produces a quantity with approximately a chi-square distribution, after some doodling, whereas a ratio of two independent sums of squares, as in a one-way ANOVA, can often be worked into a random variable with an F-distribution, after some doodling (dividing each by df).

29.3 Wilcoxon Signed Rank Test for Paired Data

Suppose we want to compare two similarly priced pinot noir (red) wines newly marketed in Ontario: a Russian River (California) pinot noir and a Niagara region pinot noir. Some experienced wine tasters are available for the study. Should we proceed using a completely randomized design or a randomized block design? Randomly dividing tasters into two groups—one to try the Russian River and the other to try the Niagara wine—makes no sense, since we would be confounding differences in tasters with wine differences. Clearly, *blocking* or *pairing* to filter out (see Chapter 11) tasters' differences is the way to go. Each taster should taste each wine in a random order. Below are the results from a tasting, with the names changed to protect the innocent (wines abbreviated as RR and NG). The tasters rated both wines on a scale of 0 to 10, with 10 being best:

Taster#	1	2	3	4	5	6	7	8	9	10
RR	10	8	10	9	7	8	10	8	8	6
NG	8	8	7	5	6	8	4	5	3	7

To test the null hypothesis of equivalence of the wines, we might try the parametric approach, calculating the 10 differences, assuming them to be (hopefully) Normal, and performing a *t*-test. But how else might we proceed?

In Chapter 22, we discussed the *sign test* approach. Note that we have seven who prefer the Russian River, one who prefers the Niagara, and two ties. Ignoring the ties, we have a simple model under the hypothesis of equivalence:

If X = number who prefer the RR (number of "+" differences), then X is Binomially distributed with $n = 8, p = 0.5$.

The P-value equals $2 \times P(X \geq 7) = .07$ by direct calculation, using the Binomial distribution (or Binomial tables or software).

But the sign test is quite wasteful of information, since it throws away the numbers while only retaining their signs. Maybe, as with the first Wilcoxon test discussed in this chapter, we can gain by replacing the numerical differences with their ranks. First though, take the *absolute value* of each difference (denoted |d|), and then rank those absolute differences from low to high (ignoring zero differences):

Taster#	1	2	3	4	5	6	7	8	9	10
RR	10	8	10	9	7	8	10	8	8	6
NG	8	8	7	5	6	8	4	5	3	7
difference:	+2	*0*	*+3*	+4	+1	0	+6	+3	+5	−1
|d|	2		*3*	4	*1*		*6*	*3*	5	*1*
rank|d|	*3*		*4.5*	6	*1.5*		*8*	*4.5*	7	*1.5*

Let $T+$ = the sum of the ranks corresponding to the positive differences, which equals 34.5.

Let $T-$ = the sum of the ranks corresponding to the negative differences, which equals 1.5.

Consider the null hypothesis:

H_0: There is no true systematic difference in ratings between the two wines.

If this hypothesis is true, then we expect $T+$ and $T-$ to be close in value. But if the null is false, we expect them to be far apart; in other words, we expect a rather large or rather small $T+$ (or $T-$), keeping in mind that these two numbers always add up to the same thing: $n(n + 1)/2$, which equals $8(9)/2 = 36$, in this case.

Is $T+$ significantly large? This is the same as asking whether $T-$ is significantly small, since either of these rank sums is just 36 minus the other.

The standard tables for this test provide critical values for the (numerically) smaller of the two rank sums.

So, for convenience, when using tables, we'll likewise define our test statistic to be the smaller of ($T+$, $T-$), which we'll denote as T. Here, $T+ = 34.5$ and $T- = 1.5$, so $T = 1.5$.

For any specified value of n (the number of non-zero differences), we can refer to the table of critical values for the Wilcoxon Signed Rank test in Appendix C to find critical values for a two-sided or one-sided alternative.

In our wine-tasting example, we have assumed a two-sided alternative. In this case, either a small $T+$ or a small $T-$ would be evidence of a difference between the wines. From the table, we find the critical $T^* = 3$ at $\alpha = 0.05$ for a two-sided alternative. We calculated $T = 1.5$, which is < 3, and hence has a P-value < 0.05, so we may reject the null hypothesis of equivalence of the wines at the 5% level.[5]

If the alternative were one-sided, stating that the RR has higher ratings than the NG, we would have proof if $T+$ is big, or T$-$ is small, and from the table we see that a $T-$ smaller than 3 has a P-value below 0.025. So we would be able to reject H_0 at the 2.5% level.

For one-sided alternatives, be careful to check that the data actually point in the direction of and support the alternative, and not in the exact opposite direction, particularly when using tables of critical values. More generally, always plot and inspect the data right at the start, and then make sure your final conclusions bear some resemblance to the story that your plots tell.

On the other hand, if the alternative hypothesis stated that the NG has ratings higher than the RR, our conclusion would be immediate. Such a hypothesis could only be supported if there were many minuses and a large $T-$ (or small $T+$). But here $T-$ is quite small (T+ large), so the P-value must be very large indeed (in fact, $> 1 - 0.025$)!

If $n \geq 25$, then under H_0, $T+$ (or $T-$) has an approximately Normal distribution with

$$E(T + \text{ or } T-) = n(n+1)/4 \quad \text{and} \quad \text{Var}(T + \text{ or } T-) = n(n+1)(2n+1)/24$$

Even in our wine-tasting example, with much smaller n, the Normal approximation produces a result similar to the tables:

$$z = \frac{1.5 - 18}{\sqrt{8(9)(17)/24}} = -2.31$$

and from the standard Normal table, the P-value = 2(.01) = 0.02. More detailed Wilcoxon tables would give a P-value equal to 0.024 for $T = 2$. If instead we chose to use $T+$ in this calculation, we would get $z = +2.31$ and the same P-value.

Again, we omit corrections for ties and for continuity from the z-statistic formula. The correction for ties can be very useful, since throwing out and ignoring a substantial number of ties (as in the exact test) can bias the test against the null hypothesis (ties clearly support the null hypothesis). Software will always apply a correction for ties in the z-statistic.

Assumptions and Conditions

The responses must be quantitative. Other than that, there are no distributional assumptions to check, but pay attention to other common design issues for paired data studies (discussed in Chapter 22), such as:

Paired Data Assumption

Paired Data Assumption: There must be a clear linkage between responses in the two samples (for example, from the same person, or from the same pair of siblings, or from two individuals paired together *a priori* to improve the precision of the treatment comparison). Do not try to create some linkage yourself with the data values. It must be part of the study design.

Independence Assumptions

Independence Assumption: The groups are not independent due to pairing, but the *differences* must be independent of each other. Independence must be ensured by the study

[5]We are assuming that the tasters were well trained and able to apply the same rating scale consistently. A difference of three for one taster must be genuinely bigger than a difference of two for another taster, or the rankings used in our test would not make much sense.

design, as no test can verify this assumption (though plotting the differences versus any lurking variable, such as time order, may help point out possible problems), so think about how the data were collected.

Check the **Randomization Condition:** Were the data collected with suitable randomization? For surveys (or before–after studies), were the pairs drawn using a proper random sampling scheme, or, if collected in a less controlled observational setting, can they be viewed as reasonably representative of some population? In experiments, either the treatments should be randomly assigned to one member of each pair or the order of the treatments should be randomized, depending on the nature of the study.

Computer software may report a confidence interval for the median difference. This interval is only valid if we make the additional assumption that the numerical differences have a symmetrically shaped distribution. The Wilcoxon Signed Rank test can also be used with a single sample to test for a hypothesized median, but this likewise requires an assumption of symmetry.

WILCOXON SIGNED RANK TEST

When the appropriate assumptions and conditions are met, we may test:

H_0: There is no systematic difference within pairs.

vs.

H_A: The values are systematically higher or lower for one of the distributions or treatments.

Calculate $T+$ = the sum of ranks of the positive differences, and $T-$ = the sum of ranks of the negative differences. Then estimate the P-value using a table of critical values for $T = \text{minimum}\{T-, T+\}$, or, if the sample size is off the table, using the Normal approximation for either rank sum.

Step-by-Step Example ANALYZING PAIRED DATA WITH THE WILCOXON SIGNED RANK TEST

"Hormones may boost profits," stated the headline of science reporter Anne McIlroy's April 15, 2008, article in the *Globe and Mail.* To examine this theory, in June 2005, Canadian researcher John Coates took saliva samples twice a day for eight days from 17 men working on a London (England) mid-size trading floor, trading a wide range of assets, with largest exposure to German interest rate futures. He classified each trader according to whether his testosterone level was high or low on that day (compared with the trader's median over the period). High testosterone days differed from trader to trader, and high days differed from low days by 25%, on average, in testosterone level. He also recorded the daily profits or losses (P&L) in pounds sterling of each trader (from 11 a.m. to 4 p.m.). The *Globe and Mail* also published boxplots comparing P&L on high testosterone days with profits on low testosteroned days, similar to the plots shown here:[6]

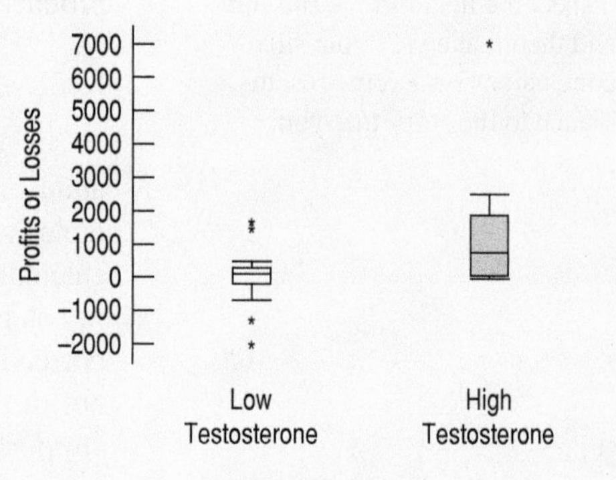

THINK ➡ **Plan** State what we want to know.

Make clear what we want to estimate or test.

Identify variables and check the *W*'s.

I want to determine if there is evidence of a testosterone effect on daily profits for traders similar to those examined in Coates's study. For each trader, we have recorded profit and loss averages for high testosterone days (testosterone level above each trader's own median) during an eight-day period in June 2005, and another profit average for low testosterone days during the same eight-day period.

Hypotheses State the null and alternative hypotheses.

We want to know if profits shift higher or lower according to testosterone level. Don't make assumptions; go with the two-sided alternative.

H_0: There is no difference in the distribution of profits between high testosterone days and low testosterone days.

H_A: Profits tend to be either generally higher or generally lower on a trader's high testosterone days than on his low testosterone days.

[6] "Hormones may boost profits," by Anne McIlroy. *Globe and Mail*, April 15, 2008, L1; J.M. Coates and J. Herbert, Proceedings of the National Academy of Sciences of the 0USA (*PNAS*) 2008; Vol. 105, pp. 6167–6172, available at www.pnas.org/content/105/15.toc. Copyright 2008 National Academy of Sciences. USA.

REALITY CHECK There appears to be a difference from the boxplots, but for a proper analysis we must recognize the paired structure of the data (each trader appears once in each of the two boxplots), which is not visible.

Model Think about the assumptions and check the conditions.

Are the data paired? How?

We took an average on both high and low testosterone days for the same trader, so each trader provided paired responses.

Was randomization used or implied?

This was not a random sample of traders, but we believe them to be representative, particularly of futures traders. There was nothing special about this time period or trading floor.

Are responses independent?

We'll assume that each trader acts independently of the others.

Check any distributional assumptions of the intended analysis.

All that we require for the Wilcoxon test is quantitative scale data. But let's check for Normality, too, since for Normal data, an appropriate t-test would be more powerful.

Make a picture of the *differences,* which form the actual data in our analysis—perhaps a histogram or Normal probability plot. One of the authors contacted Dr. Coates and asked for the data (even if we could read the points in the boxplots, we wouldn't know how to pair them!).

Dr. Coates kindly provided the following data from his study

Trader	P&L Low 11 a.m. Testosterone	P&L High 11 a.m. Testosterone
1	127	−63
2	113	725
3	−2	−5
4	−292	1780
5	−98	262
6	−1308	1401
7	88	2092
8	361	887
9	−700	2500
10	143	82
11	1448	1547
12	1625	7000
13	−127	38
14	−2066	1941
15	183	268
16	483	513
17	−115	−4

There is a high outlier in one of the boxplots, which would be a concern if we were performing, say, an independent samples t-test, but here, it's the differences that matter. They are, in order by trader

(High − Low): −190, 612, −3, 2072, 360, 2709, 2004, 526, 3200, −61, 99, 5375, 165, 4007, 85, 30, 111, and plotted below.

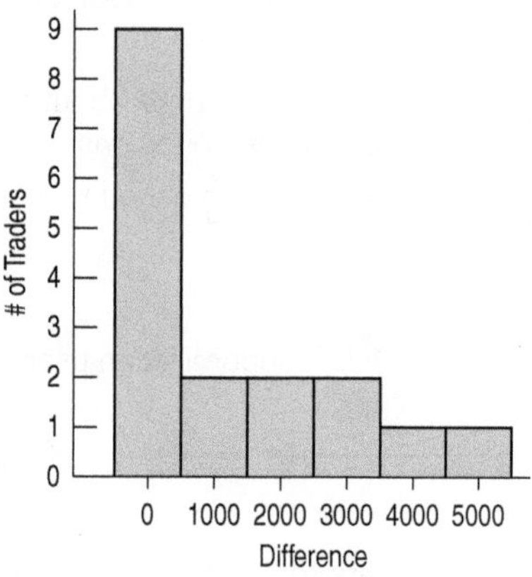

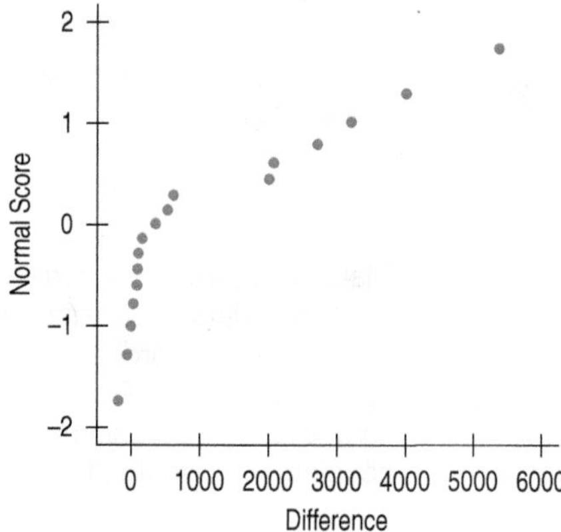

Choose the method of analysis.

The data are highly skewed. A paired (one-sample) t-test on the differences is risky. If we try it, we get $t = 3.30$ with $P = 0.008$. But it would be rather hard to trust this P-value. The sign test gives us a P-value of 0.0127 based on 14 positive differences out of 17 differences. But let's extract more information about the differences by applying the **Wilcoxon Signed Rank test.**

SHOW ➡ Mechanics

Compute the difference for each trader. Throw out any zero differences.

List the differences in order of magnitude. Sum the ranks for the positive differences and for the negative differences. Use the sum that is smaller as test statistic, as required by the table of critical values.

Test your rank sum for significance by using the table of critical values, software or the Normal approximation (if appropriate).

Here are the differences, in order by magnitude, with ranks shown below.

Diff:	−3	30	−61	85	99	111	165	−190	360
Rank/Diff/:	1	2	3	4	5	6	7	8	9

Diff:	526	612	2004	2072	2709	3200	4007	5375
Rank/Diff/:	10	11	12	13	14	15	16	17

The smaller rank sum is clearly for the negative differences: $T- = 1 + 3 + 8 = 12$. According to the table of critical values, we find that the P-value corresponding to this sum is < 0.002, since $P \leq 0.002$ for the bigger rank sum of 14 displayed in the table. The other rank sum must be $n(n + 1)/2 - 12 = 17(18)/2 - 12 = 141$, which could also

	be used if working with the Normal approximation. The Normal approximation yields a standard Normal statistic: $z = \frac{(141 - 76.5)}{21.12} = 3.05$, which gives a P-value of 0.002.
TELL ➡ **Conclusion** Link the P-value to your decision about the null hypothesis, and state your conclusion in context.	The P-value is tiny, so we have strong evidence of a difference, and we see that higher profits are associated with higher testosterone levels.

29.4 Friedman Test for a Randomized Block Design

In Chapter 26, we mentioned that a two-way layout of data may arise when one of the factors is a blocking variable (see Chapter 11 to review the concept of blocking). In this case, we often have just one observation at each combination of block and treatment, and then proceed after making the (often reasonable) assumption of an additive model. If we have a randomized block design with just two treatments, the Wilcoxon Signed Rank test can be used to compare the treatments. For more than two treatments, it turns out that a simple twist on the Kruskal-Wallis procedure does the trick quite nicely.

The table displays data showing weight gain, in grams, for nine young rats, three from each of three litters. Three different diets (the treatments) were assigned at random to the three siblings in litter one (block 1). This was repeated for the other two litters (blocks 2 and 3).

Treatment:	1	2	3
Block 1:	90	95	80
Block 2:	77	90	72
Block 3:	75	84	64

We need to convert to ranks, but now we *rank separately within each block*, which gives the following ranks (in parentheses):

Sample:	1	2	3
Block 1:	90(2)	95(3)	80(1)
Block 2:	77(2)	90(3)	72(1)
Block 3:	75(2)	84(3)	64(1)
Rank Sums:	6	9	3

To test for significant variation among the treatment rank sums, we can use the Friedman statistic:

$$S = \left(\frac{12}{bt(t+1)}\sum T_i^2\right) - 3b(t+1),$$

where T_i are the rank sums for the various treatments, b = the number of blocks, and t = the number of treatments.

The Friedman statistic measures variation among the three rank sums, and gets bigger as the sums get further apart from each other. It equals zero when the three sums are identical. So we should reject the null hypothesis that all treatments are identical if the Friedman statistic is *unusually big*. For moderate sample sizes, we can get a good approximation to the sampling distribution using the chi-square model with $t - 1$ degrees of freedom. Tables of exact critical values are recommended for very small sample sizes (like this one!).

Assumptions and Conditions

Our assumptions and conditions are generally the same as for the Wilcoxon Signed Rank test, except that we may compare any number of treatments where the responses are classified according to treatment and block (in a randomized block design). The data may be on either an ordinal or quantitative scale.

FRIEDMAN TEST

When the appropriate assumptions and conditions are met, we may test:

H_0: The treatments are all identical.

vs.

H_A: At least one of the treatments has responses systematically higher or lower than some other treatment.

Calculate the Friedman test statistic. Estimate the P-value using the upper tail of the chi-square model with $t - 1$ degrees of freedom, where t = number of treatments being compared. Or use a table of exact critical values for very small samples.

For Example

Continuing with our rat weight gain example, if we apply the Friedman statistic, we get:

$$S = [\frac{12}{3 \times 3(3 + 1)}(6^2 + 3^2 + 9^2)] - 3 \times 3(3 + 1) = 6.0$$

which gives a P-value = 0.05 using the χ^2 distribution with 2 df. We have some evidence that weight gain depends on diet, so it follows that superior weight gains are achievable by proper choice of diet.

Note that the evidence here was as strong as it could possibly be for treatment differences, since we had all the "1" rankings in one diet, all the "2" rankings in another diet, and all the "3" rankings in the remaining diet. This is the maximal spread among the rank totals. So for such small samples, with only three responses in each of three treatments, the best that we can ever hope to attain is 5% significance (and just barely). Considering the small sample size in this example, it would also be advisable to search for software or critical value tables for the exact test, in order to verify the accuracy of our *P*-value.

29.5 Rank Correlation

With bivariate data, we sometimes wish to define a measure of correlation that is valid and useful under less restrictive assumptions than the correlation coefficient defined in Chapter 6, where we first discussed bivariate quantitative data. Maybe converting observations to ranks will do the trick. The correlation coefficient defined in Chapter 6 is commonly referred to as *Pearson's product moment correlation.* One formula used to calculate it is:

$$r = \frac{\sum(x_i - \bar{x})(y_i - \bar{y})}{\sum(x_i - \bar{x})^2}$$

Recall that this correlation coefficient is a measure of the strength of linear (straight-line type) association between two variables X and Y. Let's generalize the notion of linear relation to the notion of **monotonic relation**. Consider the following graphs:

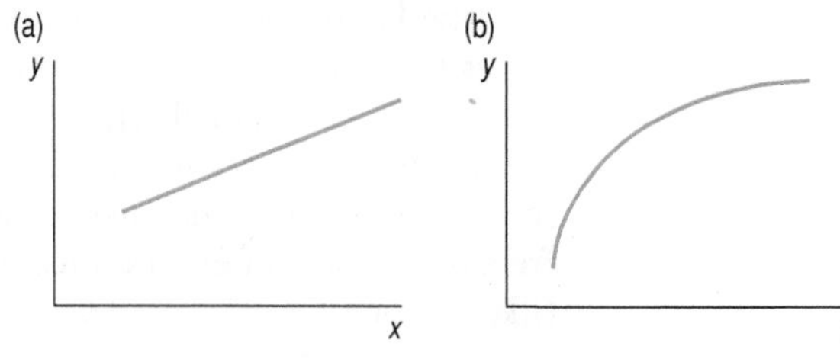

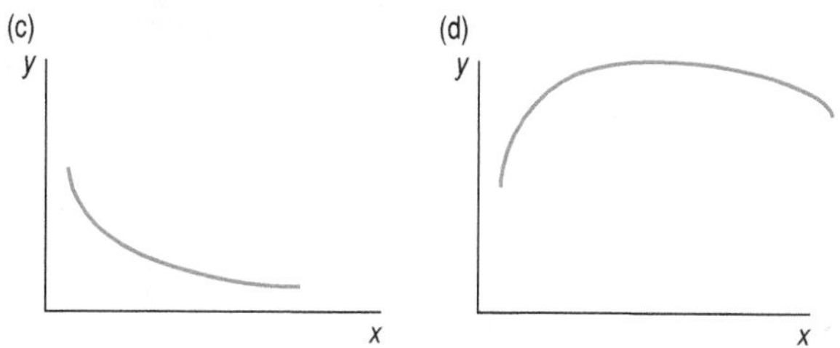

The relations in graphs (a), (b), and (c) are all perfect monotonic relations. In other words,

- as x increases, y increases (a positive monotonic relation); or
- as x increases, y decreases (a negative monotonic relation)

Only in graph (a) is this monotonic relation also a linear one. A linear relation is monotonic, but a monotonic relation need not be linear. In graph (d), the relation is not monotonic, since y neither goes up consistently nor goes down consistently with increases in x. About all we can say is that there is a positive monotonic relation over the first half of the range of x values, and over the second half there is a negative monotonic relation. However, the overall relationship is not monotonic.

Just as Pearson's r measures the strength of linear association, we can define a measure of the strength or extent of monotonic association. Replace the x values by their ranks, from 1 to n. Replace the y values by their ranks, from 1 to n. Compute Pearson's r for these two columns of rankings rather than for the original observations. This produces a measure of the linear association of the ranks rather than the linear association of the actual data. The resulting value is **Spearman's correlation,** or r_s.

Consider the following simple data:

x:	1	2	3	4	5
y:	1	2	4	8	16

Plotted, these data look like:

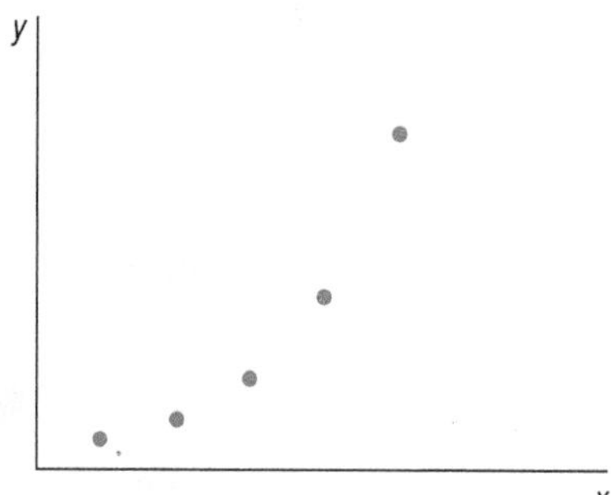

Pearson's correlation r is equal to 0.933 in this example. This number *partly* captures the strong rise in y with increases in x, but is less than 1 because we do not have a perfect linear relation. However, the data are a perfect monotonic relation, so r_s should capture this fact by producing a value of 1. Let's check:

Converting to ranks (x's and y's ranked separately), we find:

x rank:	1	2	3	4	5
y rank:	1	2	3	4	5

So, the ranks, unlike the actual data, exhibit a perfect linear association. Hence, r_s will equal 1, since Pearson's r applied to a perfect linear association must equal 1.

But there is a simpler way to calculate r_s. Instead of ranking the data and then plugging the rankings into the formula for Pearson's r, it can be shown that, in the absence of ties, we can reduce that formula to the following (applied to the ranks, not the raw data):

$$r_s = 1 - \frac{6\sum d_i^2}{n(n^2 - 1)}$$

where d_i is the difference between the ranks for the ith observation on x and on y.

Even if there are some ties, this simple formula will provide a good enough approximation to the more precise calculation applying Pearson's r to the ranks.

So r_s may be interpreted as a measure of the strength of the linear relation between the rankings of x and the rankings of y. Like r, this measure varies in value from -1 to 1, with the values of -1 or $+1$ indicating a perfect monotonic relationship between x and y and a value like $r_s = 0.7$, for example, indicating a moderate extent of positive monotonicity(in other words, a moderately strong increasing—but not necessarily linear—relation).

When compared with Pearson's r, there are a few possible advantages to using Spearman's nonparametric measure of correlation:

- greater resistance to the effects of a small number of observations (such as outliers) on the calculated correlation
- less restrictive assumptions regarding the form of the relationship (so associations other than linear ones may also be appropriate)
- less restrictive distributional assumptions when testing the significance of the observed association (in other words, Normality of X or Y is not required)

Sometimes, we may wish to test the null hypothesis that there is no relationship between the variables X and Y:

$$H_0\text{: } \rho_s = 0 \text{ where } \rho_s \text{ is the population equivalent of } r_s$$

For small samples, we can readily find tables of critical values to complete the test, but if n is not too small (say $n > 10$), we can plug r_s into the following t-statistic (which also works when testing the significance of the Pearson correlation, though under stronger parametric assumptions):

$$t = \frac{r\sqrt{n - 2}}{\sqrt{1 - r^2}}$$

and evaluate significance using the t_{n-2} distribution.

Assumptions and Conditions

Assumptions are rather minimal. Looking back at the assumptions for Pearson's correlation in Chapter 6, we see that the concerns expressed about nonlinearity and outlier effects have vanished. The **Quantitative Variables Condition** may be relaxed to some degree, as we could also apply Spearman's correlation in situations with ordinal responses—for example, if we wanted to correlate opinion in a sample of Canadian adults about two different House of Commons bills, where the possible responses were "strongly disagree," "mildly disagree," "neutral," "mildly agree," or "strongly agree." We could still rank all the ordinal responses for each bill, though with a large number of ties.

In Chapter 24, we considered a test of significance for correlation using the equivalent test on the slope. But now we only need to be concerned with the **Independence Assumption** (of the pairs), and we can ignore the other (distributional and linearity) assumptions mentioned in that chapter. Check the **Randomization Condition** that the individuals are or may be considered to be a representative sample from the population to which we would like our conclusions to apply.

SPEARMAN'S RANK CORRELATION

For two quantitative or ordinal variables, we can measure the strength of monotonic association between them using Spearman's correlation.

To compute Spearman's correlation, rank the responses separately for each variable, and then compute Pearson's correlation for the two columns of rankings, or use the shortcut formula:

$$r_s = 1 - \frac{6\sum d_i^2}{n(n^2 - 1)}$$

where each d_i is the different between the ranks for the ith observation on X and Y.

Assuming the data are a simple random sample from some population, we can test:

H_0: No monotonic association in the population ($\rho_s = 0$).

To estimate the P-value, use tables of critical values for rs or plug into the t-statistic discussed earlier (using $n - 2$ df):

$$t = \frac{r\sqrt{n-2}}{1 - r^2} \quad (\text{if } n > 10)$$

For Example

Below is a scatterplot for a sample of 12 Toronto Blue Jay baseball players, where X = the number of home runs scored in 2009 and Y = salary in 2009 (in \$10 000's):[7]

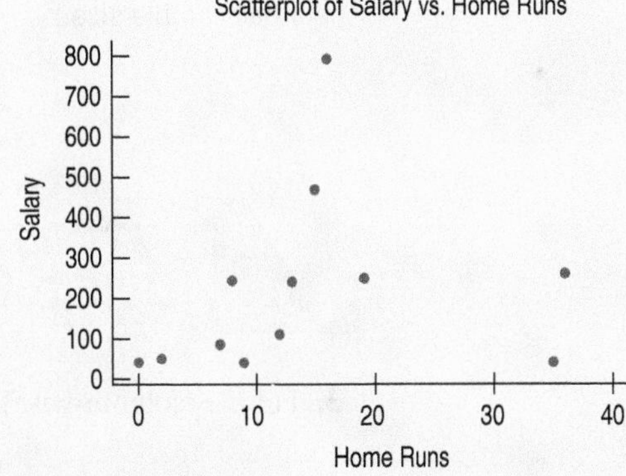

QUESTION: How would you characterize the relationship between home runs and salary? Below are the actual data.

Home Runs	Salary	Home Runs	Salary
36	259	13	240
15	469	7	85
35	41	9	40
12	110	2	50
19	250	8	243
16	795	0	41

[7]http://toronto.bluejays.mlb.com/stats, www.sportscity.com/mlb/salaries/toronto-blue-jays-salaries

Pearson's correlation for the data above is equal to 0.183. If we replace the data with their ranks, we get:

Home Run Rank	Salary Rank	Home Run Rank	Salary Rank
12	10	7	7
8	11	3	5
11	2.5	5	1
6	6	2	4
10	9	4	8
9	12	1	2.5

Using software to calculate Pearson's correlation for the two columns of ranks, we find Spearman's correlation = 0.518. Alternatively, one could plug into the alternative formula for r_s to get

$$r_s = 1 - \frac{6\sum d_i^2}{n(n^2 - 1)}$$

$$= 1 - \frac{6(2^2 + 3^2 + 8.5^2 + 0^2 + 1^2 + 3^2 + 0^2 + 2^2 + 4^2 + 2^2 + 4^2 + 1.5^2)}{12(12^2 - 1)}$$

$$= 1 - 0.481 = 0.519$$

We see that Pearson's $r = 0.18$, whereas Spearman's $r_S = 0.52$. There is some indication that the relation might not be very linear, and there are a few unusual observations. This limits the applicability of Pearson's r, which is noticeably smaller in value than Spearman's r. The latter is less affected by the unusual observations and pays no attention to any curvature present, but picks up on the moderately monotone positive relation (an increase in home runs is accompanied by an increase in salary value most of the time—what a shock, eh?). The bigger value for r_S relays this message more clearly. (However, a test of significance only produces a P-value of 0.08, due to the small sample size.)

Just Checking

3. For the (column-wise) paired data below

x:	1	2	3	4	5	6	7
y:	1	4	9	16	9	4	1

which correlation measure is more useful?

WHAT CAN GO WRONG?

- **Make sure that the data values are independent and from an appropriately randomized sample.** Though we avoid making unnecessary distributional assumptions in these tests, for valid inferences, we still need observations that are independent of each other (and from independent groups for the Wilcoxon Rank Sum or Kruskal-Wallis tests), with proper random sampling or randomization. Selecting survey data in clusters, for example, violates independence (though analysis is still possible, just more complicated), and convenience samples may be hard to generalize.

- **Beware of small samples with low power.** With distribution-free procedures, power and sample size calculations are not possible, but we should still be aware that, as usual, tests based on small samples may let important treatment effects slip by undetected due to lack of power. Power calculations using a similar parametric test may provide some rough guidance on sample size.
- **Be cautious about interpreting confidence intervals provided by statistical software for these procedures.** We have not discussed the confidence intervals for medians, or for differences in medians, that are often provided by statistical software in the output for these procedures. There are additional assumptions required for such confidence intervals to be valid (for example, similarly shaped distributions for the Wilcoxon Rank Sum test). Be sure to learn more about the particular procedure before attempting to interpret such an interval.

In Chapters 20, 21, 22, 25, and 26, we learned how to test hypotheses about the means of single samples and multiple samples for both independent sample and blocked designs. Those tests were based on parametric assumptions such as Normality and equality of variance. In this chapter, we removed these assumptions and learned how to proceed nonparametrically, continuing the discussion of nonparametric testing that began in Chapters 20 and 22 with the sign test. We also discussed a nonparametric version of the Pearson product-moment correlation, which was introduced in Chapter 6.

These procedures can provide valid conclusions where parametric tests become questionable due to violations of underlying assumptions, but they will have lower power if the parametric assumptions actually hold.

Basic concepts of hypothesis testing stay the same: Find a test statistic with a known distribution (under the null hypothesis), use this distribution to find the P-value for the observed sample, and interpret the P-value in the usual way.

What Have We Learned?

Learning Objectives

There is no need to restrict ourselves solely to the well-known and oft-used "classical" or Normal distribution-based methods when making inferences. There are other means for extracting relevant evidence about a hypothesis from the data. If concerned about being able to use one of the common parametric inferential methods after checking the necessary assumptions or conditions, seek out another approach, with different or fewer assumptions—a nonparametric procedure.

- Know how to use rank transformations to eliminate parametric modelling of data, and how to perform a variety of rank-based tests.
- A P-value is a P-value is a P-value. It is always interpreted the same way, as the weight of evidence against the null hypothesis, regardless of the test procedure.

Review of Terms

Rank sum Sum of the ranks in a sample after replacing the responses with their respective rankings (p. 897).

Wilcoxon Rank Sum test Nonparametric rank-based test for a systematic difference between two distributions in an independent samples study (p. 897).

Kruskal-Wallis statistic Used to test for systematic differences among any number of distributions in an independent samples study, after converting the data to ranks (p. 902).

Wilcoxon Signed Rank test Nonparametric rank-based test for a systematic difference between two distributions in a paired data study (p. 904).

Friedman statistic Used to test for systematic differences among the treatments in a randomized block design, after converting the data to ranks within each block (p. 909).

Spearman's correlation A measure of monotonic association between two quantitative or ordinal variables based on the ranks of the data values (p. 911).

Monotonic relationship An association where one variable increases whenever the other variable increases, or decreases whenever the other variable increases (p. 912)

On the Computer NONPARAMETRIC METHODS

Statistical packages generally provide an assortment of nonparametric procedures. Often, they are listed separately under the heading "Nonparametrics" or something similar, but they may pop up as available options when you execute more common parametric procedures. Computer output is similar to the output for parametric tests, showing P-values and often confidence intervals (for medians or differences in medians). However, if you want to use the confidence interval, be sure to read up on the procedure, as additional data assumptions are required.

EXCEL

Nonparametric procedures are not included in Excel's standard data analysis toolkit, though there is a RANK function (or RANK.AVG in Excel 2010) available.

JMP

JMP includes nonparametric procedures as additional options (under the **red triangle** or in the dialogue box) once you have run the corresponding parametric procedure:

- For two independent samples, run **Analyze>Fit y by x**, as in Chapter 21. Click the red triangle to find **Nonparametric** options, including the **Wilcoxon (Rank Sum) Test.**
- To compare several independent groups, run **Analyze>Fit y by x**, as in Chapter 25, then click the red triangle for **Nonparametric** options. Choose **Wilcoxon Test** (this produces the Kruskal-Wallis generalization of the Wilcoxon test).
- For a paired data comparison, select **Analyze>Matched Pairs,** as in Chapter 22. Click the red triangle at the top of the analysis and select **Wilcoxon Signed Rank.** The Friedman test for blocked data is not available.
- For correlation analysis, choose **Analyze>Multivariate Methods>Multivariate,** enter the variables, hit **OK**, then click the red triangle and choose **Nonparametric Correlations>Spearman's ρ.**

MINITAB

- From the **Stat** menu, choose the **Nonparametrics** submenu. The **Mann-Whitney** test is the same as the Wilcoxon Rank Sum test.
- For the Wilcoxon Signed Rank test, first calculate the differences and store them in a column, then input this column to **1-Sample Wilcoxon.**
- Minitab does not list Spearman's rank correlation under **Nonparametrics**, so to calculate this, use **Data>Rank** and convert each data column to corresponding ranks. Then use **Stat>Basic Statistics> Correlation** to correlate the ranks, which will also provide a P-value (valid if n > 10).

COMMENTS

Stat>Tables>Cross Tabulation and Chi-Square>Other Stats>Correlation coefficients for ordinal categories will also calculate Spearman's correlation, but no P-value is reported.

R

- For a Wilcoxon test, use wilcox.test(y1,y2,paired=FALSE); change to paired=TRUE for the signed rank test.
- To do the Kruskal-Wallis test, use kruskal.test(y~A) similar to an ANOVA model.

SPSS

- Select **Analyze>Nonparametric Tests** from the menu, then choose your option according to your data structure (either **Independent Samples** or **Related Samples**).

COMMENTS

For independent samples, make sure your data are stacked. Note that in the output, SPSS refers to the Wilcoxon Rank Sum test by its alternate name, the Mann-Whitney test.

STATCRUNCH

Several nonparametric procedures are available under **Stat > Nonparametrics**

TI-83/84 PLUS

These procedures are not automated, though they can be executed via a series of steps described in the TI-83/84 Statistics Handbook.

Exercises

1. **Course grades again** Though the samples in the chapter-opening Statistics course grades example were small, use the Normal approximation to estimate the P-value for these data, and compare with the previously reported P-value. If the grade of A– were changed to A+, how would this affect the Wilcoxon test? How would it affect the *t*-test (after replacing letter grades with numerical values)?

2. **Course grades amended** Add two additional Biology students, both with an A grade, to the chapter-opening Statistics grade data. How do you think this will affect the P-value? Redo the test and report the P-value again using the table of exact critical values. Though the samples were small, use the Normal approximation to estimate the P-value for these data, and compare with the exact distribution P-value.

3. **Music students** Fourteen music students from two different schools gave performances and were ranked as follows (best first, worst last). Students from Lawrence School of Music are in bold.
Joe Bini **Jack Bob Shuang** John Sally **Mary** Mo **Celina** Zheng **Aditi** Rodrigo Astrud
Test for a significant difference in performance between students from the two schools.

4. **Art students** Thirteen art students from two different classes painted portraits, which were ranked as follows (best first, worst last). Artists from Andrey's class are in bold.
John Ben **Jen Bo Sharon** Joe Sam **Matt** May Zee **Ashish** Rod Amy
Test for a significant difference in performance between students from the two classes.

T 5. **Downloading again** In Exercise 22 of Chapter 25, we examined how much of a difference time of day makes on the speed at which a university student can download files. Analyze the same data using an appropriate nonparametric procedure, and compare your results with the results from the one-way ANOVA.

Early (7 a.m.) Time (sec)	Evening (5 p.m.) Time (sec)	Late night (12 a.m.) Time (sec)
68	299	216
138	367	175
75	331	274
186	257	171
68	260	187
217	269	213
93	252	221
90	200	139
71	296	226
154	204	128
166	190	236
130	240	128
72	350	217
81	256	196
76	282	201
129	320	161

T 6. **Analgesics again** In Exercise 23 of Chapter 25, we compared the effect on migraine pain of three different drug formulations. Analyze the data using an appropriate nonparametric procedure, and compare your results with the results from the one-way ANOVA.

Drug	Pain	Drug	Pain	Drug	Pain
A	4	B	6	C	6
A	5	B	8	C	7
A	4	B	4	C	6
A	3	B	5	C	6
A	2	B	4	C	7
A	4	B	6	C	5
A	3	B	5	C	6
A	4	B	8	C	5
A	4	B	6	C	5

7. **Extreme values** Considering the information in Exercise 6, if drug A were even more effective—say, if all the data values there were reduced by two (to 2, 3, 2, . . .) —how would the one-way ANOVA and the Kruskal-Wallis test be affected? In which case would the effect be greater, and why?

T 8. **Extreme values again** Consider again the hand-washing data in this chapter and our Kruskal-Wallis test. Would doubling the values in the sample for hand-washing with water have a bigger effect on the Kruskal-Wallis test or on the ANOVA *F*-test? Explain. If we doubled the values in this sample once more, how would it affect the tests?

T 9. **Washing again** In Exercise 12 of Chapter 26, we examined the effect of temperature setting and cycle length on cleanliness. Assume that temperature is the factor of interest, and that cycle length is of minor interest, similar to a blocking variable. Analyze the temperature effect using Friedman's test, and compare with results from the ANOVA in Chapter 26.

Temp	Cycle	Score
Cold-cold	Med long	3.7
Warm-hot	Med long	6.5
Cold-warm	Med long	4.9
Hot-hot	Med long	6.5
Cold-cold	Long	4.6
Warm-hot	Long	8.3
Cold-warm	Long	4.7
Hot-hot	Long	9.1
Cold-cold	Short	3.4
Warm-hot	Short	5.6
Cold-warm	Short	3.8
Hot-hot	Short	7.1
Cold-cold	Med short	3.1
Warm-hot	Med short	6.3
Cold-warm	Med short	5
Hot-hot	Med short	6.1

10. Competition Suppose we want to compare four different competitive conditions on the success of the grass species *Agropyron*. These conditions are different ways of introducing a competing species into the plots. We use four plots in one region, four in another region, and four in a third region. Conditions are assigned randomly in each region, and the regions may vary slightly in fertility level. The table displays (coded) biomass measurements for the 12 plots. Analyze the results with a parametric procedure and with a nonparametric procedure. If the P-values differ greatly, explain why.

	Condition 1	Condition 2	Condition 3	Condition 4
Region 1	20	11	25	4
Region 2	42	25	39	16
Region 3	53	32	58	19

T **11. Women's 1500-m skate** Consider the speed skating example discussed in Chapter 22.

Inner Lane		Outer Lane	
Name	**Time**	**Name**	**Time**
OLTEAN Daniela	129.24	(no competitor)	
ZHANG Xiaolei	125.75	NEMOTO Nami	122.34
ABRAMOVA Yekaterina	121.63	LAMB Maria	122.12
REMPEL Shannon	122.24	NOH Seon Yeong	123.35
LEE Ju-Youn	120.85	TIMMER Marianne	120.45
ROKITA Anna Natalia	122.19	MARRA Adelia	123.07
YAKSHINA Valentina	122.15	OPITZ Lucille	122.75
BJELKEVIK Hedvig	122.16	HAUGLI Maren	121.22
ISHINO Eriko	121.85	WOJCICKA Katarzyna	119.96
RANEY Catherine	121.17	BJELKEVIK Annette	121.03
OTSU Hiromi	124.77	LOBYSHEVA Yekaterina	118.87
SIMIONATO Chiara	118.76	JI Jia	121.85
ANSCHUETZ THOMS Daniela	119.74	WANG Fei	120.13
BARYSHEVA Varvara	121.60	van DEUTEKOM Paulien	120.15
GROENEWOLD Renate	119.33	GROVES Kristina	116.74
RODRIGUEZ Jennifer	119.30	NESBITT Christine	119.15
FRIESINGER Anni	117.31	KLASSEN Cindy	115.27
WUST Ireen	116.90	TABATA Maki	120.77

a) Apply the sign test to these data.
b) Apply the Wilcoxon Signed Rank test to these data.
c) Compare results from a) and b) with the *t*-test results in Chapter 22. Compare the P-values among these three procedures and, if possible, explain some of the differences.

T **12. Wine tasting** For the data presented in this chapter about comparing two wines, add two additional tasters to the data, with one giving the Russian River and Niagara wines ratings of 8 and 6, respectively, and the other giving them ratings of 9 and 5, respectively.
a) Execute the *t*-test for these data. Is the P-value here trustworthy?
b) Execute the sign test and the Wilcoxon Signed Rank test.
c) Compare P-values among the three procedures, and explain the differences.

T **13. Gasoline again** Using an appropriate nonparametric procedure, tackle the hypothesis posed in Exercise 22a of Chapter 22, where we studied whether premium gas produces superior mileage over regular gas. Compare your results with the results from the parametric approach.

Here are the results (litres per 100 km):

Car #	1	2	3	4	5	6	7	8	9	10
Regular	14.7	11.8	11.5	10.7	10.2	10.7	8.7	9.4	8.7	8.4
Premium	12.4	10.7	9.8	9.8	9.4	9.4	9.0	9.0	8.4	7.4

T **14. Job satisfaction again** Using an appropriate nonparametric procedure, tackle the hypothesis posed in Exercise 19b of Chapter 22 about the effects of an exercise program on job satisfaction, and compare your results with the results from the parametric approach.

Worker	Job Satisfaction Index	
Number	**Before**	**After**
1	34	33
2	28	36
3	29	50
4	45	41
5	26	37
6	27	41

Worker Number	Job Satisfaction Index Before	After
7	24	39
8	15	21
9	15	20
10	27	37

15. Life expectancy 2010, again Consider again Exercise 31 in Chapter 8, where we displayed births per woman and life expectancy for 26 countries:

Country	Birth Rate (births/ 1000 population)	Life Expectancy
Argentina	18	77
Bahamas, The	16	70
Barbados	13	74
Belize	27	68
Bolivia	26	67
Canada	10	81
Chile	15	77
Colombia	18	74
Costa Rica	17	78
Dominican Republic	22	74
Ecuador	21	75
El Salvador	25	72
Guatemala	28	70
Honduras	26	70
Jamaica	20	74
Mexico	20	76
Nicaragua	23	72
Panama	20	77
Paraguay	28	76
Puerto Rico	12	79
United States	14	78
Uruguay	14	76
Venezuela	21	74
Virgin Islands	12	79

a) Calculate Spearman's correlation and Pearson's correlation for all 26 countries. Remove the obvious outlier, and recompute the two correlations.
b) Explain any differences observed between Pearson's correlation and Spearman's correlation in part a).

16. Gestation, again Look back at Exercise 25 in Chapter 8, which displays gestation periods and life expectancies for 18 mammal species.
a) Calculate Spearman's correlation and Pearson's correlation.
b) Remove the humans, and recompute the two correlations.
c) Compare the two correlations in a) and b), and explain any big differences between them.
d) Now remove the elephants too. If you were to recompute the correlations, which do you predict will change the most (and why)? Recompute both to check your prediction.

17. Antidepressants, again Consider our example in Exercise 16, Chapter 6, in which we discussed drug improvement scores versus placebo improvement scores. Would you expect Spearman's or Pearson's correlation to be bigger? Why? Pearson's correlation was given as 0.898. Compute Spearman's correlation.

18. Pace of course Survey researchers often ask people to rank their opinions or attitudes on a scale that lacks units. (Common scales run from 1 to 5, 1 to 7, or 1 to 10.) For example, we might ask:
"How would you assess the pace of your Statistics course so far?"
1 = Way too slow 2 = A little too slow 3 = About right 4 = A little too fast 5 = Way too fast
Scales of this sort that attempt to measure attitudes numerically are called Likert scales, after their developer, Rensis Likert. Likert scales have order. The higher the number, the faster you think the course has gone, but a 4 doesn't necessarily mean *twice* as fast as 2. Suppose we take a sample of students from this course and record the GPA, major field of study, and opinion about pace for each student.
a) What measure would you use to estimate the strength of the association between perceived pace and GPA? Explain briefly.
b) How would you test for an association between perceived pace and major field of study?

19. Buying from a friend In Chapter 21, we examined the effect of friendship on transactions where prices offered were compared for buying from a friend versus buying from a stranger. Use an appropriate nonparametric technique to compare the two conditions. Compare results with those from the parametric test executed in Chapter 21.[8] Here are the prices they offered for a used camera in good condition:

Price Offered for a Used Camera ($) Buying from a Friend	Buying from a Stranger
275	260
300	250
260	175
300	130
255	200
275	225
290	240
300	

[8] J. J. Halpern, "The Transaction Index: A method for standardizing comparisons of transaction characteristics across different contexts," *Group Decision and Negotiation*, 6: 557–572.

20. Ontario and U.K. students, again Look back at Exercise 30 in Chapter 21, where we tested for a difference in mean travel time to school between Ontario and U.K. secondary school students. Use an appropriate nonparametric procedure to test for a difference, and compare your results with the parametric approach.

U.K students travel times (minutes)

45 5 4 15 50 20 20 20 20 20 25 35 15 30 20
10 45 10 3 60 25 20 5 15 5 15 17 30 40 20
10 30 10 15 20 10 15 17 10 25

Ontario students travel times (minutes)

30 10 8 30 5 8 7 15 10 35 15 10 25 22 20
25 30 10 25 8 15 18 25 15 10 25 5 2 5 25
20 15 47 20 20 13 20 5 15 12

21. Egyptians again Consider the data from Exercise 14 in Chapter 21 comparing skull breadths of ancient Egyptians from two different eras. Use an appropriate nonparametric technique to test for a difference between the two eras. Compare your results with those from the appropriate parametric test.

Maximum Skull Breadth (mm)			
4000 B.C.E.	4000 B.C.E.	200 B.C.E.	200 B.C.E.
131	131	141	131
125	135	141	129
131	132	135	136
119	139	133	131
136	132	131	139
138	126	140	144
139	135	139	141
125	134	140	130
131	128	138	133
134	130	132	138
129	138	134	131
134	128	135	136
126	127	133	132
132	131	136	135
141	124	134	141

22. Cuckoos again In Exercise 40 of Chapter 21, we looked at cuckoo egg lengths for three species. Use an appropriate nonparametric approach to test for a difference in egg lengths between each pair of species. Compare your results with results from the appropriate parametric test for the sparrows versus robins comparison.

Cuckoo Egg Length (mm)		
Foster Parent Species		
Sparrow	Robin	Wagtail
20.85	21.05	21.05
21.65	21.85	21.85
22.05	22.05	21.85
22.85	22.05	21.85
23.05	22.05	22.05
23.05	22.25	22.45
23.05	22.45	22.65
23.05	22.45	23.05
23.45	22.65	23.05
23.85	23.05	23.25
23.85	23.05	23.45
23.85	23.05	24.05
24.05	23.05	24.05
25.05	23.05	24.05
	23.25	24.85
	23.85	

23. Using cards Using a deck of cards, how could you estimate the P-value for the Wilcoxon Rank Sum test where the two rank sums are 11 and 34 for two independent samples of size four and five, respectively?

24. Marooned If you were marooned on a desert island with no books but with plenty of time, how would you estimate the P-value for the results we obtained in our chapter-opening grades example, using no math (but pretend there are no ties)?

25. Exact P-values How many distinct hands are possible if you were to deal yourself five cards from a deck of 11 distinct cards (denoted ${}_{11}C_5$, or "11 choose 5" and defined in Chapter 14 for the Binomial model as the number of ways of choosing the trials in which to place the successes)? What would be the exact P-value if the rank sum for the smaller sample in this chapter's course grades example were equal to 15 (which could only happen if this sample had the ranks 1, 2, 3, 4, and 5)? What would be the P-value for a rank sum of 16? (How many ways can you get ranks adding up to 16?) For a rank sum of 17?

26. More P-values How many distinct hands are possible if you were to deal yourself four cards from a deck of nine distinct cards (denoted ${}_9C_4$ and defined in Chapter 14 for the Binomial model)? What would be the exact P-value if the rank sums for two independent samples of size four and size five were 10 and 35, respectively?

Just Checking ANSWERS

1. **a.** Sum of all ranks = 7(8)/2 = 28.
   b. Biggest = 7 + 6 + 5 + 4 = 22, smallest = 1 + 2 + 3 + 4 = 10
   c. Expected value = (4/7) × 28 = 16
2. We would have to have ranks 1, 2, and 3 in one sample, 4, 5, and 6 in another sample, etc., to produce rank sums of 6, 15, 24, and 33.

   So $H = \frac{12}{12 \times 13} \times (6^2/3 + 15^2/3 + 24^2/3 + 33^2/3)] - 3(13) = 10.38$ and $0.01 < P < .025$ from the chi-square distribution with 3 df.
3. Neither, since the graph goes up then down, which is nonlinear and not monotonic to any degree. Both correlations are zero. Neither picks up the strong (quadratic) relationship.

MathXL

MyStatLab

Go to MathXL at www.mathxl.com or MyStatLab at www.mystatlab.com. You can practise exercises for this chapter as often as you want. The guided solutions will help you find answers step by step. You'll find a personalized study plan available to you too!

chapter

30 The Bootstrap

Bootstrapping is like breathing your own carbon dioxide.

—Chairman George

"Pulling oneself up by the bootstraps" may be a physical impossibility in the real world, but in the statistical world we can pull ourselves up from the data itself to inference about the larger population without using mathematical power (though we do need to tie our bootstraps to a computer to help pull us up). Statistics Canada and many other researchers increasingly rely on *bootstrapping* to produce standard errors, confidence limits and P-values (the basic tools of inference), particularly for situations in which no textbook answer can be found. In the late 1980s, as cheap computer power was growing quickly, Bradley Efron and others[2] saw a way to harness the power of computer simulation to "bootstrap" answers. While there are various refinements possible, we will just introduce the basic idea, with application to estimating standard errors and confidence intervals. Unfortunately, "canned" bootstrapping capabilities of statistical software packages are still quite limited at this time.

Where are we going?

In previous chapters, we discussed inference for means and proportions arising from simple study designs such as simple random samples. But in real-world studies, things often get much more complicated. Complex surveys, estimators, and models cannot be handled by our earlier methods. So, contact a good mathematician, you may be thinking! But even super-mathematicians have limited superpowers. We often need to replace complex (or impossible) mathematical derivations by a conceptually simple, data-based, computer-intensive simulation method called *bootstrapping*. With bootstrapping, we can replace limited math power with nearly unlimited computer power. We avoid messy formulas (phew!), simulate sampling distributions, and emphasize basic concepts. And it really works![1]

The full chapter is available online at www.pearsoncanada.ca/deveaux and on the eText in MyStatLab.

[1]Well, most of the time. Statisticians have been busy studying when it works and what may cause it to fail.

[2]Including Efron's PhD student and former University of Toronto student and professor Rob Tibshirani.

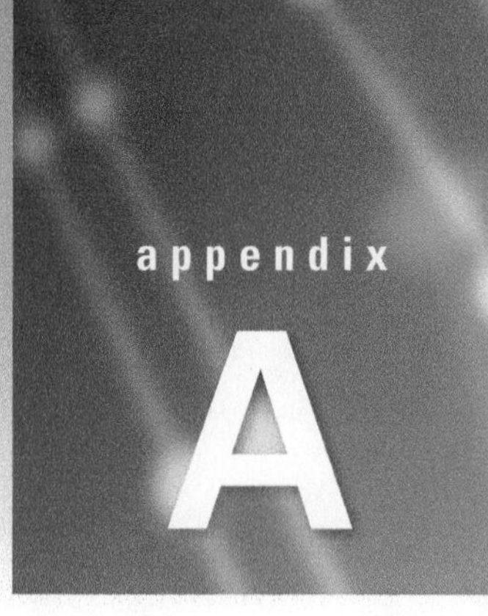

appendix A Answers

Here are the answers to the odd-numbered exercises for the chapters and the reviews. As we said in Chapter 1, the answers provide outlines of the complete solution. Your solution should follow the model of the Step-By-Step examples, where appropriate. You should explain the context, show your reasoning and calculations, and draw conclusions. For some problems, what you decide to include in an argument may differ somewhat from the answers here. However, make sure the numerical part of your answer matches the numbers in the answers shown, aside from small differences due to rounding errors or differences in the definitions of quartiles (which may also affect boxplots).

Chapter 1

1. Answers will vary.

3. *Who:* 40 undergraduate women.
What: Whether or not the women could identify the sexual orientation of men based on a picture.
Why: To see if ovulation affects a woman's ability to identify sexual orientation of a male.
How: Showing very similar photos to the women, with half gay.
Variables: Categorical variable: "He's gay" or "He's not gay."

5. *Who:* China/India/Chindia funds listed at globeinvestor.com.
What: 1 month, 1 year, and 5 year returns for each fund.
When: The most recent periods of time.
Where: globeinvestor.com Web site.
Why: To compare investment returns, for future investment decisions.
How: globeinvestor.com uses reports from the fund companies.
Variables (Q=quantitative, C=categorical): 1 month return (Q, percentage), 1 year return (Q, percentage), 5 year return, annualized (Q, percentage)

7. *Who:* All airline flights in Canada.
What: Type of aircraft, number of passengers, whether departures and arrivals were on schedule, and mechanical problems.
When: This information is currently reported.
Where: Canada.
Why: This information is required by Transport Canada and the Canadian Transportation Agency.
How: Data is collected from airline flight information.
Variables: Type of aircraft, departure and arrival timeliness, and mechanical problems are categorical variables, and number of passengers is a quantitative variable.

9. *Who*: Automobiles.
What: Make, country of origin, type of vehicle, and age of vehicle (probably in years).
When: Not specified.
Where: A large university.
Why: Not specified.
How: A survey was taken in campus parking lots.
Variables: Make, country of origin, and type of vehicle are categorical variables, and age of vehicle is a quantitative variable.

11. *Who:* Workers who buy coffee in an office.
What: Amount of money contributed to the collection tray.
Where: Newcastle.
Why: To see if people behave more honestly when feeling watched.
How: Counting money in the tray each week.
Variables: Amount contributed (pounds) is a quantitative variable.

13. *Who:* 54 bears.
What: Weight, neck size, length (no specified units), and sex.
When: Not specified.
Where: Not specified.
Why: Since bears are difficult to weigh, the researchers hope to use the relationships between weight, neck size, length, and sex of bears to estimate the weight of bears, given the other, more observable features of the bear.
How: Researchers collected data on 54 bears they were able to catch.
Variables: Weight, neck size and length are quantitative variables, and sex is a categorical variable. No units are specified for the quantitative variables.
Concerns: The researchers are (obviously!) only able to collect data from bears they were able to catch. This method is a good one, as long as the researchers believe the bears caught are representative of all bears, in regard to the relationships between weight, neck size, length, and sex.

15. *Who*: Doughnut types for sale at Tim Hortons.
What: Various nutritional characteristics (see variables below).
When: Not stated, but presumably the measurements were taken recently.
Where: Tim Hortons Web site.
Why: To help customers make good nutritional choices.
How: Not specified, but presumably at some specialized food analysis type lab.
Variables: All eight variables are quantitative: Number of calories (kca/s), amounts of trans fat (g), total fat (g), sodium (mg), sugar (g), protein (g), % daily value of iron (percentage), and % daily value of calcium (percentage). Units found by going to the Web site.

17. *Who:* 882 births.
What: Mother's age (in years), length of pregnancy (in weeks), type of birth (Caesarean, induced, or natural), level of prenatal care (none, minimal, or adequate), birth weight of baby (unit of measurement not specified, but probably pounds and ounces), gender of baby (male or female), and baby's health problems (none, minor, major).
When: 1998–2000.
Where: Large city hospital.
Why: Researchers were investigating the impact of prenatal care on newborn health.
How: It appears that they kept track of all births in the form of hospital records, although it is not specifically stated.
Variables: There are three quantitative variables: mother's age, length of pregnancy, and birth weight of baby. There are four categorical variables: type of birth, level of prenatal care, gender of baby, and baby's health problems.

19. *Who:* Experiment volunteers.
What: Herbal cold remedy or sugar solution, and cold severity.
When: Not specified.
Where: Major pharmaceutical firm.
Why: Scientists were testing the efficacy of an herbal compound on the severity of the common cold.

How: The scientists set up a controlled experiment.
Variables: Type of treatment (herbal or sugar solution) is categorical, and severity rating is quantitative.
Concerns: The severity of a cold seems subjective and difficult to quantify. Also, the scientists may feel pressure to report negative findings about the herbal product.

21. *Who:* Streams.
What: Name of stream, substrate of the stream (limestone, shale, or mixed), acidity of the water (measured in pH), temperature (in degrees Celsius), and BCI (unknown units).
When: Not specified.
Where: Upstate New York.
Why: Research is conducted for an Ecology class.
How: Not specified.
Variables: Name and substrate of the stream are categorical variables, and acidity, temperature, and BCI are quantitative variables.

23. *Who:* 41 refrigerators.
What: Brand, cost (probably in dollars), size (in cu. ft.), type, estimated annual energy cost (probably in dollars), overall rating, and repair history (in percent requiring repair over the past five years).
When: 2002.
Where: United States.
Why: The information was compiled to provide information to the readers of *Consumer Reports.*
How: Not specified.
Variables: Brand, type, and overall rating are categorical variables. Cost, size, estimated energy cost, and repair history are quantitative variables.

25. *Who:* Kentucky Derby races.
What: Year, winner, jockey, trainer, owner, and time (in minutes, seconds, and hundredths of a second.
When: 1875–2012.
Where: Churchill Downs, Louisville, Kentucky.
Why: It is interesting to examine the trends in the Kentucky Derby.
How: Official statistics are kept for the race each year.
Variables: Date, winner, jockey, trainer, and owner are categorical variables. Duration is a quantitative variable.

Chapter 2

1. Answers will vary.

3. Answers will vary.

5. The relative frequency distribution is shown below:

Cause of fire	Percentage
Lightning	46.94
Human activities	51.64
Unknown	1.42

Most forest fires are caused by human activities and lightning with only 1.42% of the forest fires due to unknown causes.

7. According to the Monitoring the Future study, teen smoking brand preferences differ somewhat by region. Although Marlboro is the most popular brand in each region, with about 58% of teen smokers preferring this brand in each region, teen smokers from the South prefer Newports at a higher percentage than teen smokers from the West, 22.5% to approximately 10%, respectively. Camels are more popular in the West, with 9.5% of teen smokers preferring this brand, compared to only 3.3% in the South. Teen smokers in the West are also more likely to have to particular brand than teen smokers in the South. 12.9% of teen smokers in the West have no particular brand, compared to only 6.7% in the South. Both regions have about 9% of teen smokers that prefer one of over 20 other brands.

9. a) Grounding, accounting for 160 spills, is the most frequent cause of oil spillage for these 460 spills. A substantial number of spills, 132, were caused by collision. Less prevalent causes of oil spillage in descending order of frequency were loading/discharging, other/unknown causes, fire/explosions, and hull failures.
b) If being able to differentiate between these close counts is required, use the bar chart. Since each spill only has one cause, the pie chart is also acceptable as a display, but it's difficult to tell whether, for example, there is a greater percentage of spills caused by fire/explosions or hull failure. If you want to showcase the causes of oil spills as a fraction of all 460 spills, use the pie chart.

11. There's no title, the percentages total only 93%, and the three-dimensional display distorts the sizes of the regions.

13. a) A bar chart would be appropriate. A pie chart would not because these are counts and not fractions of a whole.
b) The *Who* for these data is athletic trainers who used cryotherapy, which should be a cause for concern. A trainer who treated many patients with cryotherapy would be more likely to have seen complications than one who used cryotherapy rarely. We would prefer a study in which the *Who* referred to patients so we could assess the risks of each complication.

15. a) The relative frequency distribution of quadrant location is given below. Not all proportions are equal. In particular, the relative frequency for Quadrant 4 is approximately twice the other frequencies.

Quadrant	Quadrant 1	Quadrant 2	Quadrant 3	Quadrant 4
Relative Frequency	0.18	0.21	0.22	0.39

b) The relative frequency distribution of quadrant location is given below. There seems to have some similarity with that in part a. For example, Quadrant 4 has the highest relative frequency and Quadrant 1 has the lowest. However the values don't look very similar.

Quadrant	Quadrant 1	Quadrant 2	Quadrant 3	Quadrant 4
Relative Frequency	0.12	0.24	0.28	0.36

17. a) The females in this course were 45.5% Liberal, 46.8% Moderate, and 7.8% Conservative.
b) The males in this course were 43.5% Liberal, 38.3% Moderate, and 18.3% Conservative.
c) A segmented bar chart comparing the distributions is at the right.
d) Politics and sex do not appear to be independent in this course. Although the percentage of liberals was roughly the same for each sex, females had a greater percentage of moderates and a lower percentage of conservatives than males.

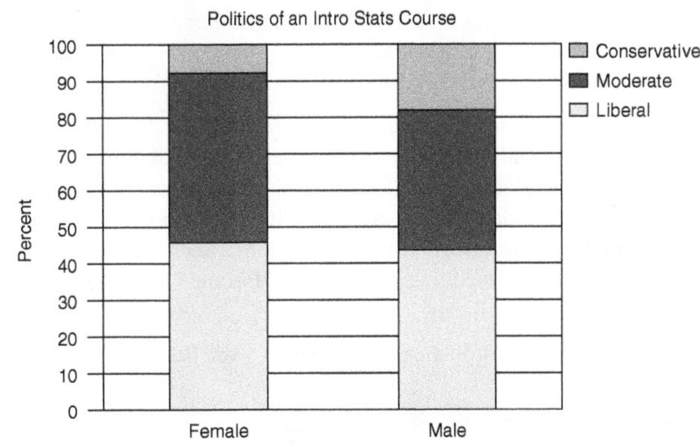

19. a) 68.1% b) 30.1% c) 94.4% d) 74.1%
e) If language knowledge were independent of Province, we would expect the percentage of French-speaking residents of Quebec to be the same as the overall percentage of Canadians who speak French. Since 30.1% of all Canadians speak French while 94.4% of residents

of Quebec speak French, there is evidence of an association between language knowledge and province.

21. a) 9.3% b) 24.7% c) 80.8%
d) No, there appears to be no association between weather and ability to forecast weather. On rainy days, his forecast was correct 79.4% of the time. When there was no rain, his forecast was correct 81.0% of the time.

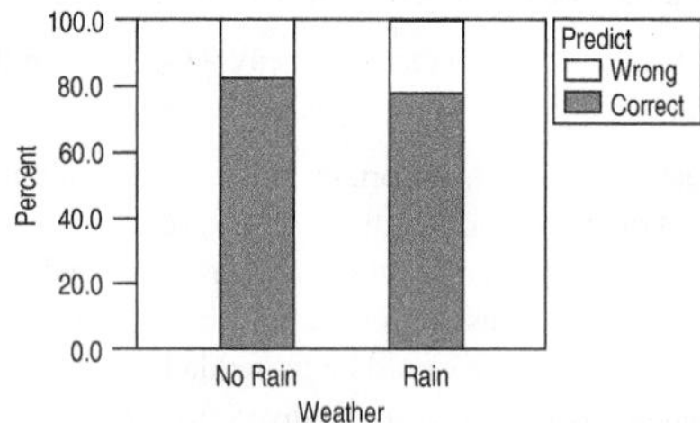

23. a) 4.8% b) 49.7%
c) Each age category appears to have about 50% male and 50% female drivers. The segmented bar chart shows a pattern in the deviations from 50%. At younger ages, males form the slight majority of drivers. This percentage shrinks until the percentages are 50% male and 50% for middle aged drivers. The percentage of male drivers continues to shrink until, at around age 45, female drivers hold a slight majority. This continues into the 85 and over category.

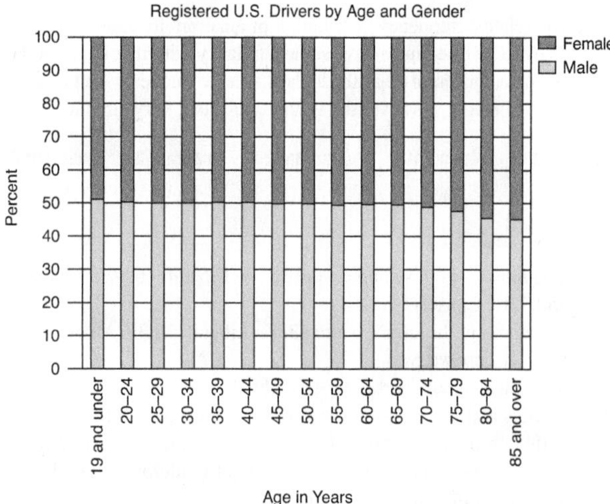

d) There appears to be a slight association between age and gender of U.S. drivers. Younger drivers are slightly more likely to be male, and older drivers are slightly more likely to be female.

25. These data provide no evidence that Prozac might be helpful in treating anorexia. About 71% of the patients who took Prozac were diagnosed as "Healthy," while about 73% of the patients who took a placebo were diagnosed as "Healthy." Even though the percentage was higher for the placebo patients, this does not mean that Prozac is hurting patients. The difference between 71% and 73% is not likely to be statistically significant.

27. a) The marginal distribution of genotype is given below:

Genotype	Marginal percentage
GG	42.71
GT	45.08
TT	12.21

b) The conditional distributions of genotype for the four categories of smokers are given in columns 2–5 of the table below.

	Cigarettes per day				
Genotype	1–10	11–20	21–30	31 and more	All
GG	48.06	42.60	38.32	36.75	42.71
GT	42.96	44.75	47.39	48.28	45.08
TT	8.99	12.65	14.29	14.98	12.21
All	100.00	100.00	100.00	100.00	100.00

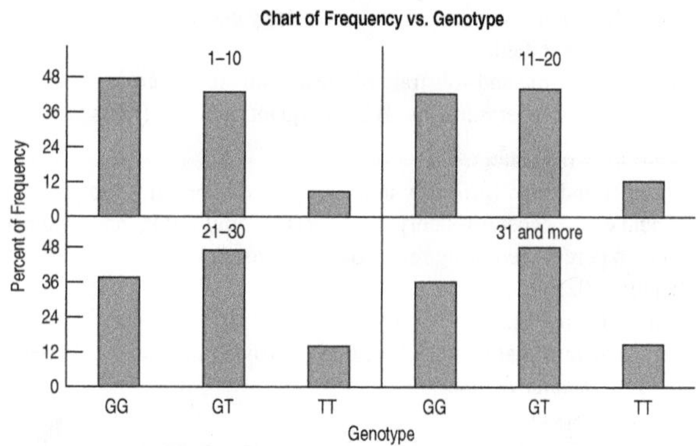

c) Though not a very noticeable difference, the percentages of smokers with genotype GT (also TT) are slightly higher among heavy smokers. However this is only an observed association. This does not prove that presence of T increases susceptibility to nicotine addiction. We cannot conclude that this crease was caused by the presence of T. There can be many factors associated with the presence of T and some of these factors might be the reason for increase in susceptibility to nicotine addiction.

29. Yes, the risk of bone fractures is about twice as high (10%) among the people who were taking SSRIs than among those who were not (5%).

31. The two-way table and the conditional distributions (percentages) of 'car accident' (crash or non-crash) for cell phone owners and non-cell phone owners are given below. The proportion of crashes is higher for cell phone owners than for non-cell phone owners.

	Cell phone owner	Non-cell phone owner	All
Crash	20	10	30
Non-crash	58	92	150
All	78	102	180

Cell Contents: Count

	Cell phone owner	Non-cell phone owner	All
Crash	25.64	9.80	16.67
Non-crash	74.36	90.20	83.33
All	100.00	100.00	100.00

Cell Contents: % of Column

b) On the basis of this study, cannot conclude that the use of a cell phone increases the risk of a car accident. This is only an observed association cell phone ownership and the risk of car accidents. We cannot conclude that the higher proportion of accidents was caused by the use of a cell phone. There can be lots of other factors common to cell phone owners and some of those factors can be the reason for the accidents.

33. a)

Blood Pressure	Total	Percent
Low	95	20.0%
Normal	232	48.9%
High	147	31.0%

b)

Blood Pressure	Under 30	Percent
Low	27	27.6%
Normal	48	49.0%
High	23	23.5%

Blood Pressure	30–49	Percent
Low	37	20.7%
Normal	91	50.8%
High	51	28.5%

Blood Pressure	Over 50	Percent
Low	31	15.7%
Normal	93	47.2%
High	73	37.1%

c)

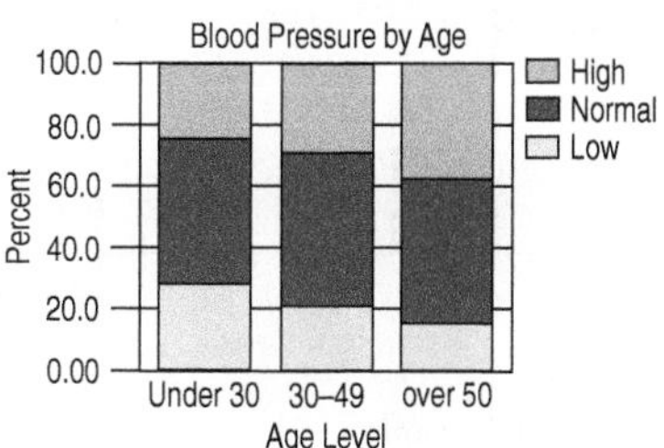

d) As age increases, the percent of adults with high blood pressure increases. On the other hand, the percent of adults with low blood pressure decreases.

e) No, but it gives an indication that it might. There might be additional reasons that explain the differences in blood pressures.

35. a) The second column includes some individuals in the 3rd, 4th, and the 5th columns, so it is not a standard contingency table.

b) Use "aboriginal population not included in columns 3, 4 and 5" (or call them "other aboriginals").

c) 0.08% (approx) d) 1.93% (approx) e) 4.24%

f) 1.95% g) 85.39% h) 98.94% i) 45.54% j) 3.63%

k) A table of percentages of total provincial population for each aboriginal identity group (Inuit, Metis, N.A. Indian) for NL, Ontario, Saskatchewan, and Alberta is given on the next page. The second table is a bit easier if using MINITAB. The side-by-side bar charts on the next page show that Saskatchewan has the highest proportion of N.A. Indian and Metis proportions. Ontario, Saskatchewan and Alberta have very small proportion of Inuit.

Region	Percent N.A. Indian	Percent Metis	Percent Inuit
Newfoundland and Labrador	3.80764%	1.51004%	1.23504%
Ontario	1.58954%	0.67986%	0.02656%
Saskatchewan	10.23137%	5.19945%	0.02875%
Alberta	3.26992%	2.71498%	0.05563%

Group	Newfoundland and Labrador	Ontario	Saskatchewan	Alberta
N.A. Indian	3.80764%	1.58954%	10.23137%	3.26992%
Metis	1.51004%	0.67986%	5.19945%	2.71498%
Inuit	1.23504%	0.02656%	0.02875%	0.05563%

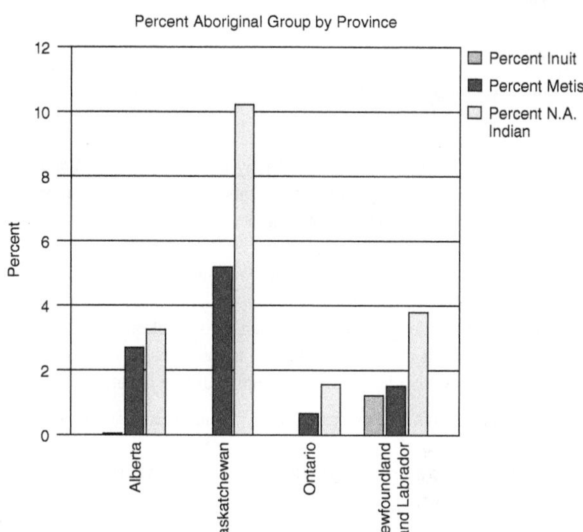

37. a) The marginal totals have been added to the table:

Procedure		Discharge delayed		
		Large Hospital	Small Hospital	Total
	Major surgery	120 of 800	10 of 50	130 of 850
	Minor surgery	10 of 200	20 of 250	30 of 450
	Total	130 of 1000	30 of 300	160 of 1300

160 of 1300, or about 12.3% of the patients had a delayed discharge.

b) Yes. Major surgery patients were delayed 130 of 850 times, or about 15.3% of the time. Minor Surgery patients were delayed 30 of 450 times, or about 6.7% of the time.

c) Large Hospital had a delay rate of 130 of 1000, or 13%. Small Hospital had a delay rate of 30 of 300, or 10%. The small hospital has the lower overall rate of delayed discharge.

d) Large Hospital: Major Surgery 15% delayed and Minor Surgery 5% delayed. Small Hospital: Major Surgery 20% delayed and Minor Surgery 8% delayed. Even though small hospital had the lower overall rate of delayed discharge, the large hospital had a lower rate of delayed discharge for each type of surgery.

e) No. While the overall rate of delayed discharge is lower for the small hospital, the large hospital did better with *both* major surgery and minor surgery.

f) The small hospital performs a higher percentage of minor surgeries than major surgeries. 250 of 300 surgeries at the small hospital were minor (83%). Only 200 of the large hospital's 1000 surgeries were minor (20%). Minor surgery had a lower delay rate than major surgery (6.7% to 15.3%), so the small hospital's overall rate was artificially inflated. Simply put, it is a mistake to look at the overall percentages. The real truth is found by looking at the rates after the information is broken down by type of surgery, since the delay rates for each type of surgery are so different. The larger hospital is the better hospital when comparing discharge delay rates.

39. a) 42.6%

b) A higher percentage of males than females were admitted: males: 47.2% to females: 30.9%.

c) Program 1: Males 61.9%, females 82.4%
Program 2: Males 62.9%, females 68.0%
Program 3: Males 33.7%, females 35.2%
Program 4: Males 5.9%, females 7.0%

d) The comparisons in part c show that males have a *lower* admittance rate in every program, even though the overall rate shows males with a higher rate of admittance. This is an example of Simpson's paradox.

Chapter 3

1. Answers will vary.

3. Answers will vary.

5. a) Unimodal (near 0) and skewed. Many seniors will have 0 or 1 speeding tickets. Some may have several, and a few have more than that.
b) Probably unimodal and slightly skewed to the right. It is easier to score 15 strokes over the mean than 15 strokes under the mean.
c) Probably unimodal and symmetric. Weights may be equally likely to be over or under the average.
d) Probably bimodal. Men's and women's distributions may have different modes. It may also be skewed to the right, since it is possible to have every long hair, but hair length can't be negative.

7. a) Bimodal. Looks like two groups. Modes are near 6% and 46%. No real outliers.
b) Looks like two groups of cereals, a low-sugar and a high-sugar group.

9. a) 22.7% b) 47.3%
c) The distribution is symmetric. The centre (the median) is around 32. The scores range from 0–60, but the few scores close to zero may be outliers (with a gap of just one we might not be able to conclude as a clear outlier, but they are somewhat unusual compared to the rest of the scores).

11. a) The distribution is right skewed and so it is logical to expect the mean to be greater than the median. The median is in the interval 0.45–0.55.
b) The disruption is right skewed and could be bimodal. The median is between 0.45 and 0.55. The values of the percentage of rejected ballots range from 0–2.5 (approx.). The largest values (after 1.8) could be outliers, so might warrant further investigation.

13. a) \$1001.50 b) \$1025, \$850, \$1200 c) \$835, \$350

15. a) Median will be unaffected. The mean will be larger.
b) The range and standard deviation will increase; IQR will be unaffected.

17. a) The standard deviation will be larger for set 2, since the values are more spread out. SD(set 1) = 2.2, SD(set 2) = 3.2.
b) The standard deviation will be larger for set 2, since 11 and 19 are farther from 15 than are 14 and 16. Other numbers are the same. SD(set 1) = 3.6, SD(set 2) = 4.5.
c) The standard deviation will be the same for both sets, since the values in the second data set are just the values in the first data set + 80. The spread has not changed. SD(set 1) = 4.2. SD(set 2) = 4.2.

19. a) Mean \$525, median \$450
b) Two employees earn more than the mean.
c) The median because of the outlier
d) The IQR will be least sensitive to the outlier of \$1200, so it would be the best to report. But really no one simple measure of spread here can provide as good a summary of spread as the 5-number summary.

21. The stem and leaf plot, a dotplot, and the five-number summary (plus mean) for these data are given below. The distribution looks slightly right-skewed (mean larger than median). The median number of slot machines is 428. The interquartile range is 731 − 299 = 432.

Stem and leaf plot:

1	2	3	4	5	6	7	8	9	10
7	0035	0033	0224	3	000	0678	66		0

(6|0 means 600 slot machines)

Dotplot:

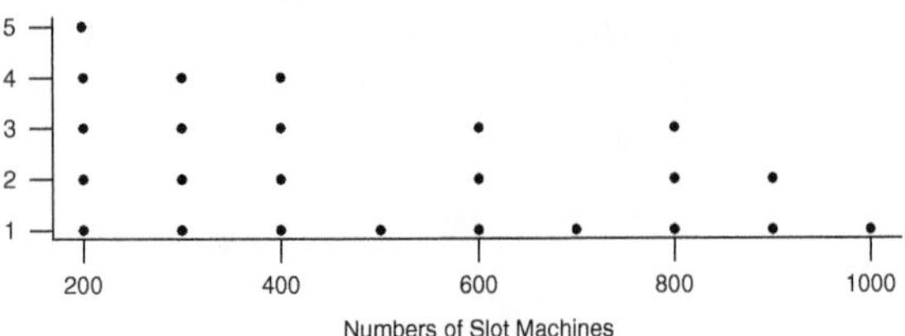

Descriptive Statistics: numbers of slot machines

Min.	1st Qu.	Median	3rd Qu.	Max.
170	299	428	731	1000

23. The histogram shows some low outliers in the distribution of height estimates. These are probably poor estimates and will pull the mean down. The median is likely to give a better estimate of the professor's true height.

25. The distribution of the number of homeruns hit by Joe Carter during the 1983–1998 seasons is skewed to the left, with a typical number of homeruns per season in the high 20s to low 30s. The season in which Joe hit no homeruns looks to be an outlier. With the exception of this no-homerun season, Joe's total number of homeruns per season was between 13 and 35. The median is 27 homeruns.

27. a) A dotplot of the number of hurricanes each year from 1944 through 2010 is displayed. Each dot represents a year in which there were that many hurricanes.

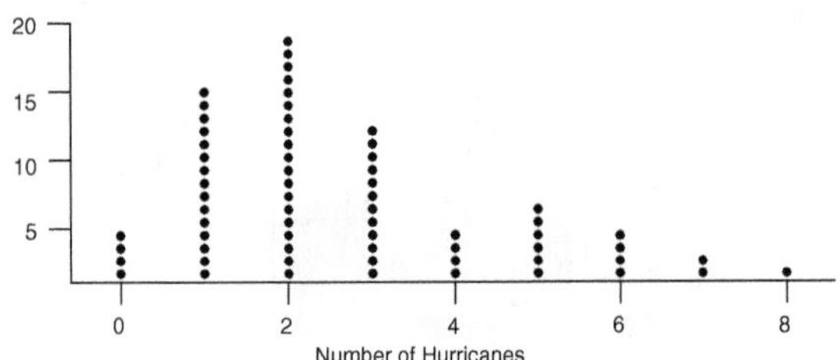

b) The distribution of the number of hurricanes per year is unimodal and skewed to the right, with centre around 2 hurricanes per year. The number of hurricanes per year ranges from 0 to 8. There are no outliers. There may be a second mode at 5 hurricanes per year, but since there were only 6 years in which 5 hurricanes occurred, this may simply be natural variability.

29. a) This is not a histogram. The horizontal axis should the number of home runs per year, split into bins of a convenient width. The vertical axis should show the frequency; that is, the number of years in which Carter hit a number of home runs within the interval of each bin. The display shown is a bar chart/time plot hybrid that simply displays the data table visually. It is of no use in describing the shape, centre, spread, or unusual features of the distribution of home runs hit per year by Carter.
b) The histogram is shown below.

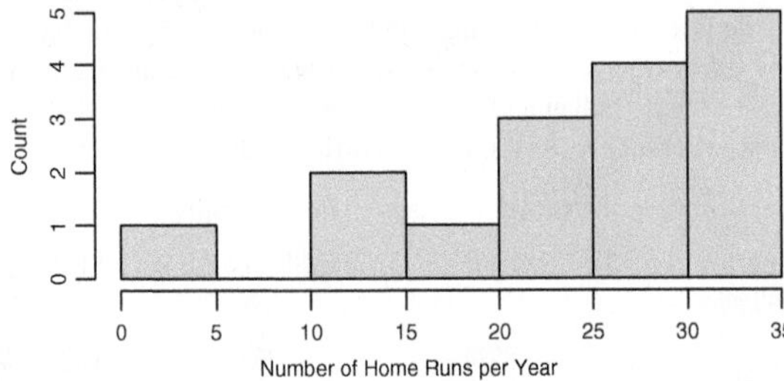

31. The distribution of the pH readings of water samples in Allegheny County, Pennsylvania, is bimodal. A roughly uniform cluster is centred on a pH of 4.4. This cluster ranges from pH of 4.1 to 4.9. Another smaller, tightly packed cluster is centred on a pH of 5.6. The two readings in the middle seem to belong to neither cluster.

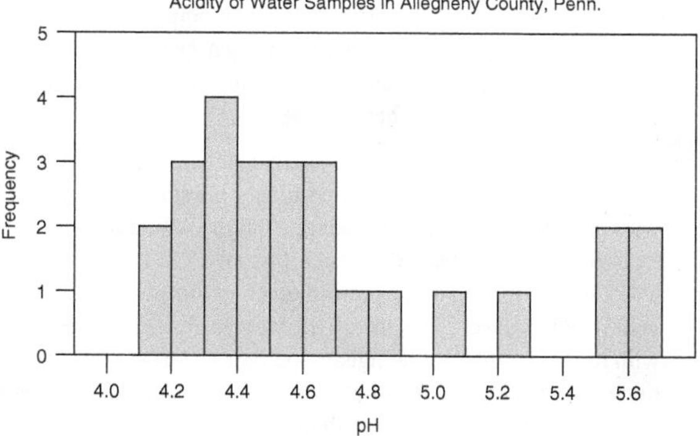

33. a)

−0	0	0	1	1	2	2	3	3	4	4
1	112334444	55566	234	5		8				6

Percent Increase (rounded)
(4|6 means 46 percent)

b) Minimum = −1.0, Q1 = 3.0, Median = 5.0, Q3 = 12.5, Maximum = 46.0

c) The mean is 8.62 (or 8.57 with the rounded data). Yes, the mean must be larger than the median because the distribution is right skewed.

d) The price increases range from −1% to 46%. The largest value 46 (possibly 28 also) is an outlier. The median increase is 5%. The distribution is right skewed.

e) The data sorted by the percentage increase is shown below. This shows large increases in the Prairie and Atlantic regions.

Data Display

Row	Metropolitan Area	Increase
1	Windsor	−0.6
2	Victoria	1.2
3	Charlottetown	1.4
4	Saint John, Fredericton and Moncton	2.4
5	Ottawa–Gatineau	3.1
6	Kitchener	3.4
7	Québec	3.9
8	Hamilton	3.9
9	London	4.0
10	St. Catharines–Niagara	4.3
11	Montréal	4.5
12	Toronto and Oshawa	4.5
13	Calgary	5.3
14	Vancouver	6.1
15	Greater Sudbury and Thunder Bay	6.3
16	St. John's	12.0
17	Halifax	12.8
18	Edmonton	13.5
19	Winnipeg	15.0
20	Regina	27.8
21	Saskatoon	46.2

35. a) mean = 6.4 median = 7 10% trimmed mean = 6.875

b) 5% trimmed mean = 4.5.

c) The data in part (a) are left skewed. The median would be the most appropriate measure for centre, followed by the trimmed mean, then the mean. The data in part (b) are right skewed. The median would be the most appropriate measure for centre, followed by the trimmed mean, then the mean.

37. Zip codes are categorical data, not quantitative. The summary statistics are meaningless. A bar chart would be more appropriate.

39. a)

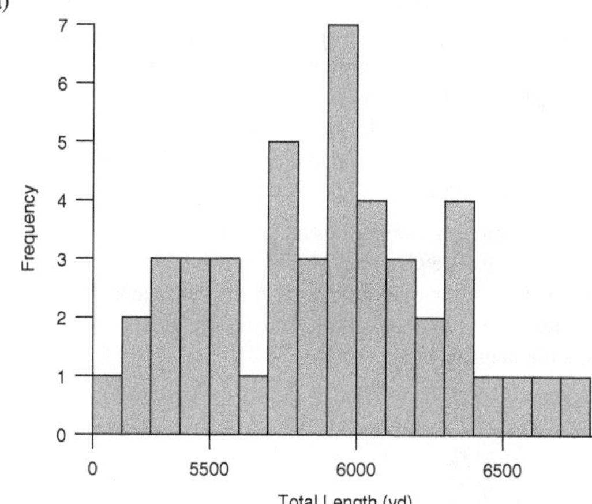

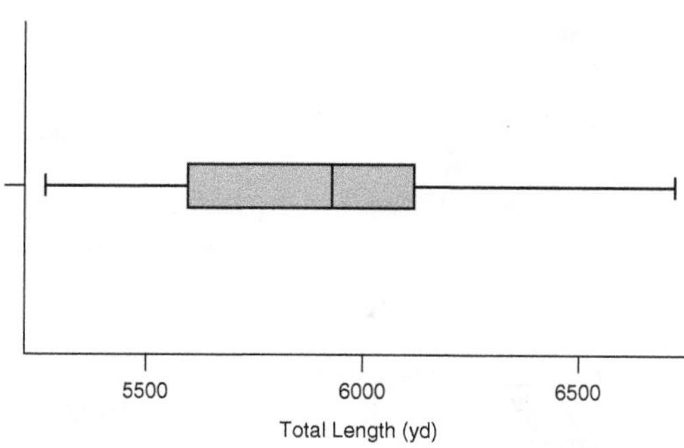

b) Between 5585.75 yd and 6131 yd.

c) Mean and SD, since this is roughly unimodal and symmetric.

d) The distribution of the lengths of all the golf courses in Vermont is roughly unimodal and symmetric. The mean length of the golf courses is approximately 5900 yards. Vermont has golf courses anywhere from 5185 yards to 6796 yards long. There are no outliers in the distribution.

e) Close to empirical rule for bell-shaped distributions.

41. a) The 5-number summary of the national averages is given below:

Descriptive Statistics: Ave Score

Min.	1st Qu.	Median	3rd Qu.	Max.
419	487	496.5	514	546

The IQR, mean, and the standard deviation of the national averages is given below:

Descriptive Statistics: Ave Score

Variable	Mean	StDev	IQR
Avg. Score	495.7	29.66	27

The distribution of average scores is left-skewed. The long left tail pulls the mean down toward the smaller values (in fact, three of these small values are smaller than Q1 − 1.5 × IQR, and are thus outliers). The outliers attract the mean toward them, whereas the median is resistant to outliers.

b) Since there are outliers in the data set, the 5-number summary is better than mean and standard deviation. The values in the 5-number summary are resistant to outliers. The mean and the standard deviation are not resistant measures.

c) Thirty-four countries participated in the program. The highest national average is 546 and the lowest is 418. The median national average is 496.3. The interquartile range is 27.2. Twenty-five percent (i.e., 9 countries) of the participating countries had a national average 487.1 or below, and at least 25% of the countries had a national average of 514.34 or above. Canada's national average (which is 526.8) is in the top 25% of all participating countries, more specifically, the 5th highest of all participating countries.

d) About two-thirds of students scored between 439 and 615. Only about 5% scored less than 351 or more than 703. Only a real math genius could have scored above 791.

43. a) A histogram (or a stemplot) is an appropriate graphical display. Both these displays are given below. The distribution is right skewed. This means the portion of large registry groups is relatively small. There are some (one or two) outliers. The median size is 717. The interquartile range is 1220.

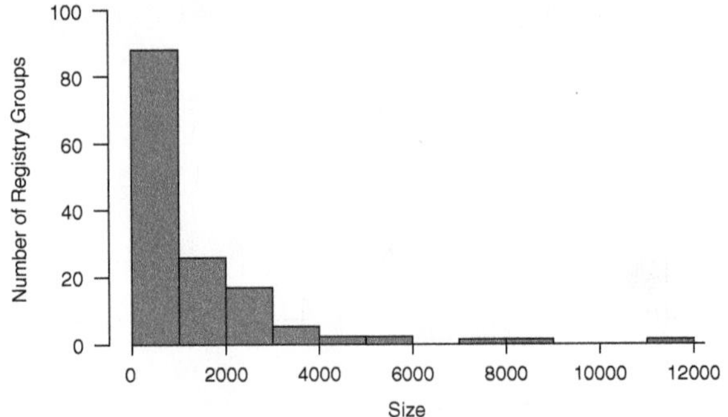

0	011122222222223333333444444444444444
0	5555555555555666666666666677777777777777788888899
1	000000111122222344
1	56677889999
2	01112233344
2	5567
3	0234
3	8
4	033
4	
5	
5	57
6	
6	
7	4
7	
8	0
8	
9	
9	
10	
10	
11	2

(7|4 means 7400)

The 5-number summary is:

Minimum	Q1	Median	Q3	Maximum
41	452	717	1672	11202

b) If we calculate the band sizes, most band sizes will be same as the registry group sizes since only Six Nations of the Grand River in Ontario consists of more than one registry group. This very large band, consisting of 13 registry groups, will increase considerably the mean and standard deviation, while having a much smaller effect on the more resistant median and IQR. The histogram will have a bigger gap due to this very large band.

45. a) The histogram of the proportion of city residents who are bilingual is shown below. The highest proportion of bilinguals is in Montreal with more than 50 percent bilinguals, and the lowest proportion is in Brantford with less than 5 percent bilinguals. The median is about 8%. The distribution of the proportion of city residents who are bilingual is right skewed. This means a relatively smaller number of cities with high proportion of bilinguals and more cities with a low proportion of bilinguals. The distribution of proportion of bilinguals appears bimodal—it looks like the distribution of a few cities with a large proportion of bilinguals has a distinct shape compared to the other cities. Summary statistics are given below. About 25% of the cities have less than 7% bilinguals.

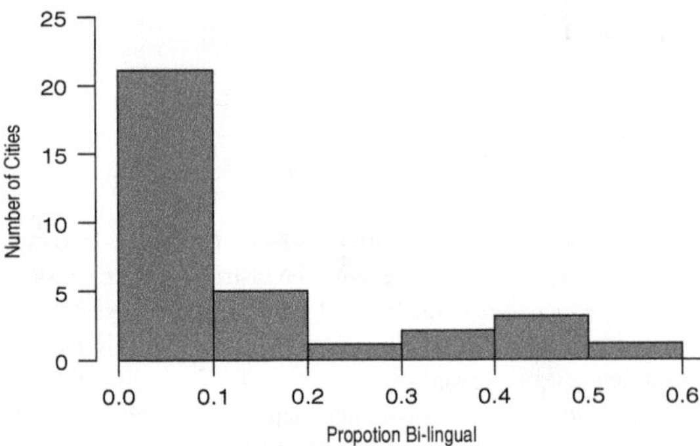

Descriptive Statistics: Proportion Bilingual

Mean	StDev	Min.	Q1	Median	Q3	Max.
0.153	0.145	0.042	0.066	0.077	0.148	0.539

b) The dotplot below shows the cities in Quebec and New Brunswick in red. They have relatively high proportions of bilinguals.

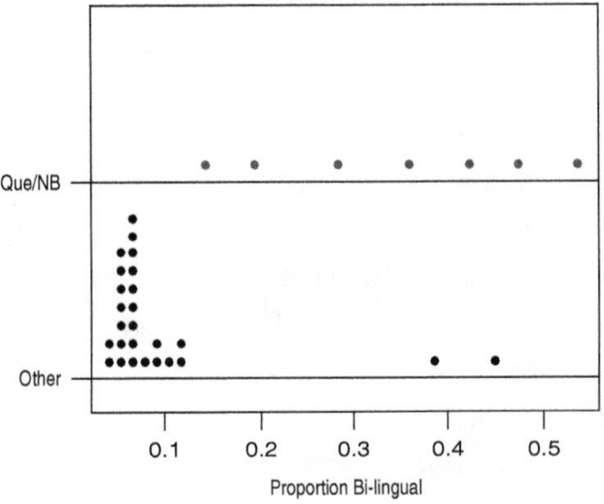

c) The histogram of the proportion of city residents who speak neither English nor French is shown below. The highest proportion of residents who speak neither English nor French is in Vancouver with more than 5%, and the lowest proportion is in Saguenay with about 0.03%. The median is about 0.69%. The distribution of the proportion of city residents who speak neither English nor French is right skewed. This means there is a relatively smaller number of cities with a high proportion of residents who speak neither English nor French and more cities with a low proportion. Three cities

(Vancouver, Abbotsford, and Toronto, the bars above 0.031 on the histogram) appear to be outliers.

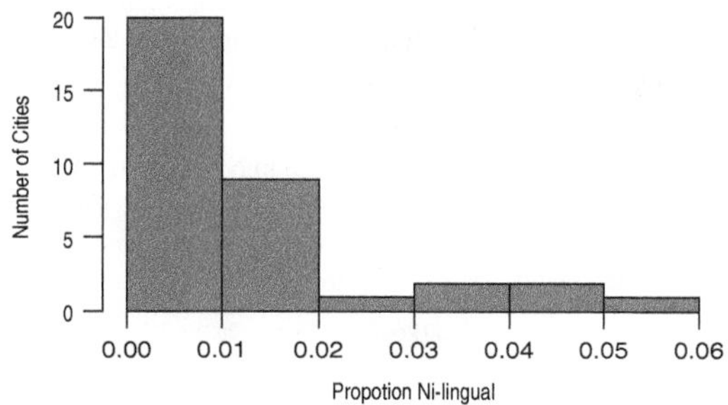

Descriptive Statistics: Proportion Ni-lingual

Mean	StDev	Minimum	Q1	Median	Q3	Maximum
0.0111	0.0135	0.0003	0.0031	0.0069	0.0144	0.0560

47. a) The closeness of the mean and median suggests that the distribution of the percentage of voter turnout in the 2011 election is approximately symmetric.
b) The distribution of the percentage of voter turnout in the 2011 election is approximately symmetric, with mean 61.068 and standard deviation 5.772.
c) 72.612 and 49.524
d) The overall percentage won't be exactly the same as the mean of the percentages as the number of eligible voters is not exactly the same in each electoral district. However, since the number of eligible voters in each riding is pretty close, we should expect the overall percentage to be fairly close to 61.07%.

49. a) The mean will be smaller because it depends on the sum of all of the values. The standard deviation will be smaller because we have truncated the distribution at 32. The range will be smaller for the same reason. The statistics based on quantiles of the data (median, Q1, Q3, IQR) won't change because the distribution was truncated to the right of Q3.
b) The spread of the distribution will be relatively smaller because we have added observations near the middle. Thus, we expect the standard deviation and IQR to be smaller. We should only see a small change to the mean and median because the added observations are near the middle of the distribution. The range will not change.
c) The spread of the distribution will be relatively larger because we have added observations at about 2 minutes left and right of the mean. Thus, we expect the standard deviation and IQR to be bigger. Because the additional observations were added symmetrically to either side of the mean, there should be little change to the mean or median, and the range will remain unchanged.
d) Subtracting one minute from each time shifts the entire distribution to the left by one minute. The mean and median should decrease by one minute. However, there has been no change to the shape of the distribution, so the standard deviation, IQR, and range will all remain the same.
e) The values added at 35 are further from the mean than the values at 29.5, so the mean will increase. The overall spread of the distribution has increased, so the standard deviation will increase. The added observations at 35 minutes lie above Q3, so the IQR will increase. Because the same number of observations was added on either side of the median, its value won't change. The minimum and maximum values have not changed, so the range remains unchanged as well.
f) We have moved some values outside of the Q1 to Q3 interval further away from the centre of the distribution. Q1 and Q3 remain unchanged, so the IQR remains unchanged. Because the adjustment is symmetric, the mean and median are unchanged. The range remains the same because the minimum and maximum are unchanged. The standard deviation increases because the spread of the distribution has increased.

51. a) **Descriptive Statistics: Height (class midpoint)**

Variable	N	Mean	StDev
Height (class mid)	130	67.069	3.996

Descriptive Statistics: Height (actual)

Variable	N	Mean	StDev
Height (actual)	130	67.115	3.792

The mean and the standard deviation of the actual data are very close to those calculated using the midpoints. For grouped data, we assume that all the values in a class are equal to the midpoint. This assumption is usually reasonable unless the class width is big.

b) Mean $= \dfrac{\sum_i f_i m_i}{\sum_i f_i}$, *Variance* $= \dfrac{\sum_i f_i(m_i - \overline{X})}{\sum_i f_i - 1}$

53. a) The last wide bin now has a rather big rectangle above it, with height equal to about 13. The number of run times exceeding 32.5 appears to have grown!

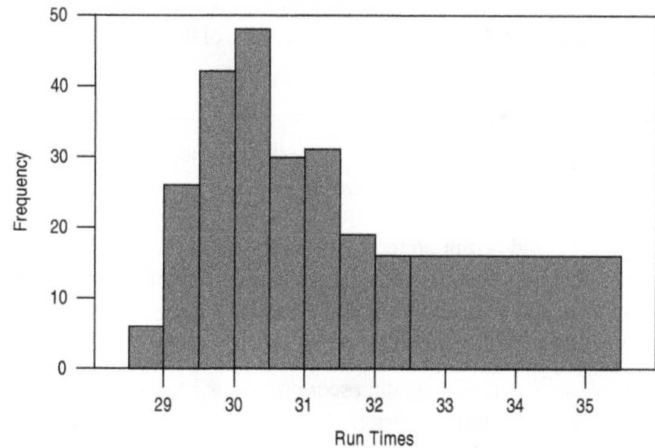

b) Now the rectangle height over the last bin equals the average of the 6 rectangle heights in the original histogram, i.e., the area of this rectangle equals the total area of the 6 rectangles it replaced. We have been true to the Area Principle, and kept area proportional to frequency. (Since area is height multiplied by width, the rectangle height must be proportional to frequency divided by bin width.)

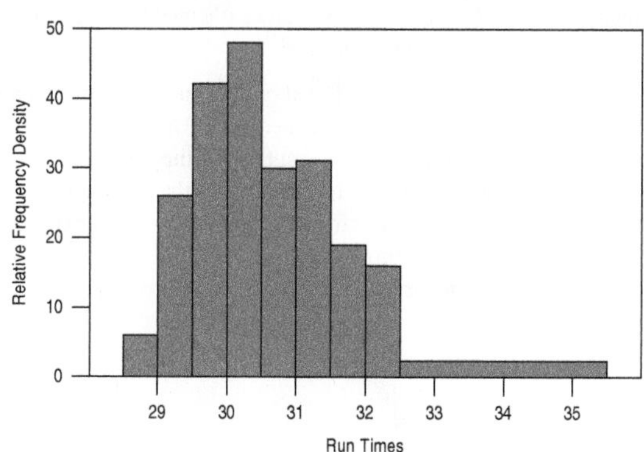

55. a) The histogram and boxplot of the distribution of ages both show that a typical crowd crush victim was approximately 18–20 years of age, that the range of ages is 36 years, and that there are two outliers, one at age 36–38 and another at age 46–48.
b) This histogram shows that there may have been two modes in the distribution of ages, one at 18–20 years of age and another at 22–24 years of age, but not shown in the boxplot.

c) Median is the better measure of centre, since the distribution of ages has outliers.

d) IQR is a better measure of spread, since the distribution of ages has outliers.

Chapter 4

1. Answers will vary.

3. a) Roughly symmetric. The mean and median are reasonably close.

b) No, all data are within the fences.

c)

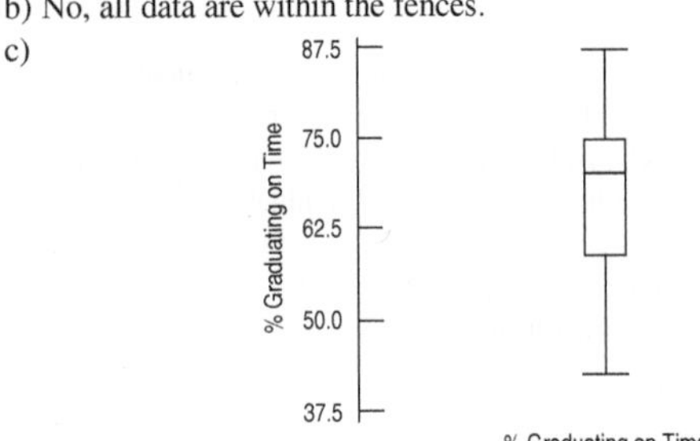

d) The 48 universities graduate, on average, about 68% of freshman "on time," with percents ranging from 43% to 87%. The middle 50% of these universities graduate between 59% and 75% of their freshman in 4 years.

5.

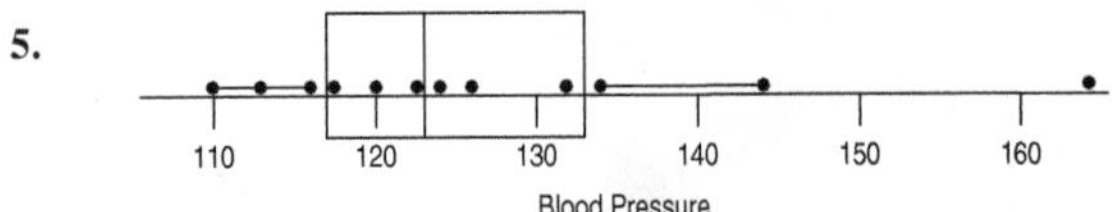

7. a) They should be put on the same scale, from 0 to 20 days.

b) Men have a mode at 1 day, then tapering off from there. Women have a mode near 5 days, with a sharp drop afterward.

c) A possible reason is childbirth.

9. a) Both women have a median score of about 17 points per game, but Yan is much more consistent. Her IQR is about 2 points, while Bini's is over 10.

b) If the coach wants a consistent performer, she should take Yan. She'll almost certainly deliver somewhere between 15 and 20 points. But, if she wants to take a chance and needs a "big game," she should take Bini. Bini scores over 24 points about a quarter of the time. (On the other hand, she scores under 11 points about as often.)

11. Class A is Class 1. The median is 60, but has less spread than Class B, which is Class 2. Class C is Class 3, since it's median is higher, which corresponds to the skew to the left.

13. a) *Who:* 45 volunteers. *What:* Level of caffeine consumption and memory test score. *When:* Not specified. *Where:* Not specified. *Why:* The student researchers want to see the possible effects of caffeine on memory. *How:* It appears that the researchers imposed the treatment of level of caffeine consumption in an experiment. However, this point is not clear.

b) Caffeine level: categorical; Test score: quantitative.

c)

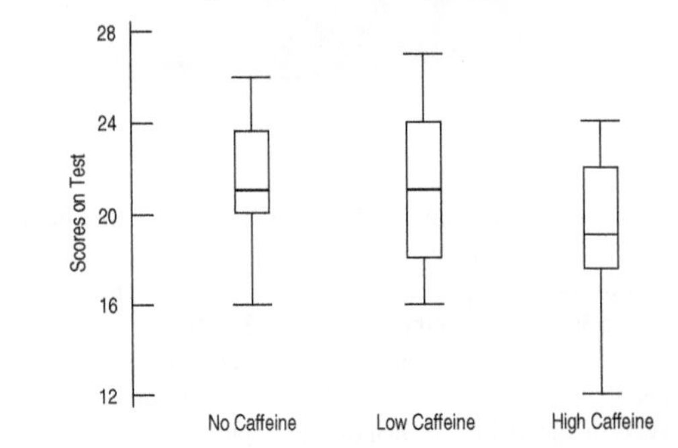

d) The participants scored about the same with no caffeine and low caffeine. The medians for both were 21 points, with slightly more variation for the low-caffeine group. The group consuming high caffeine had lower memory scores in general, with a median score of about 19. No one in the high caffeine group scored above 24, but 25% of each of the other groups scored above 24.

15. a) Comparative boxplots are at the right.

b) The distribution of population growth in NE/MW states is unimodal, symmetric and tightly clustered around 5% growth. The distribution of population growth in S/W states is much more spread out, with most states having population growth between 5% and 25%.

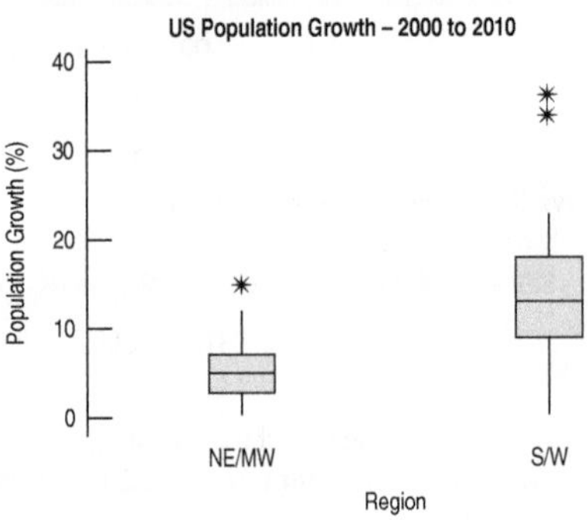

A typical state had about 15% growth. There were two outliers, with 34% and 35% growth, respectively. Generally, the growth rates in the S/W states were higher and more variable than the rates in the NE/MW states.

17. a) About 36 mph.

b) Q1 = 35 mph and Q3 = 37 mph, approximately.

c) Range = 7 mph and IQR = 2 mph

d) An approximate boxplot of winning Kentucky Derby Speeds is below.

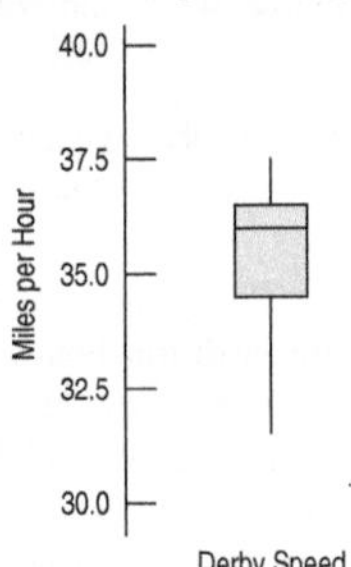

e) The distribution of winning speeds in the Kentucky Derby is skewed to the left. The lowest winning speed is just under 31 mph, and the fastest speed is about 38 mph. The median speed is approximately 36 mph, and 75% of winning speeds are above 35 mph. Only a few percent of winners have had speeds below 33 mph. The middle 50% of winning speeds are between 35 and 37 mph.

19. a) Boys b) Boys c) Girls

d) The boys appeared to have more skew, as their scores were less symmetric between quartiles. The girls' quartiles are the same distance from the median, although the left tail stretches a bit farther to the left.

e) Girls. Their median and upper quartiles are larger. The lower quartile is slightly lower, but close.

f) $14\,(4.2) + 11\,(4.6)/25 = 4.38$

21. a) The median for males is slightly higher than that for females. There are two low outliers in each boxplot. The median of the differences between the mean scores for males and females is about 10. All but one country has the mean for males greater than that for females. Only one country has mean score for females greater than that for males and that is identified as an outlier.

b) In the boxplots below, Canada, the U.S., and Mexico are indicated by the symbols x, U, and M, respectively. In the boxplots for males, Israel, Turkey, Chile, and Mexico are outliers because their mean scores are unusually low. In the boxplot for females, Mexico and Chile are outliers because their mean scores are unusually low.

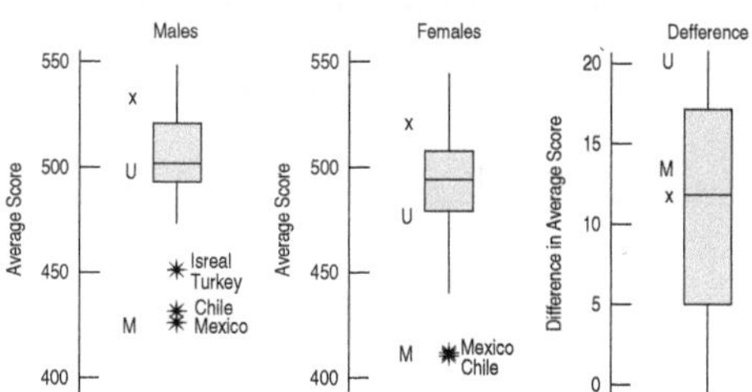

c) The 5-number summaries for the male scores, female scores, and differences are given below:

Column	Min	Q1	Median	Q3	Max
Male	425.4	492.9	501.3	520.4	547.8
Female	410.4	479.2	494.1	507.5	544.5
Difference	−1.75	5.00	11.8	17.1	21.8

The average female score for Mexico is 411.8, which is less than $Q1 - 1.5^{*}\ \mathrm{IQR} = 436.8$ and is thus an outlier.

d) The separate boxplots for males and females show the outliers (Turkey and Mexico) compared with the other scores. The boxplot for differences show the outlier (Iceland) compared with the differences (male – female) in other countries. The graph for differences is more useful for learning about gender differences in OECD countries.

23. There appears to be an outlier. That distance should be removed before looking at a plot of the drilling distances. With the outlier removed, we can determine that the slow drilling process is more accurate.

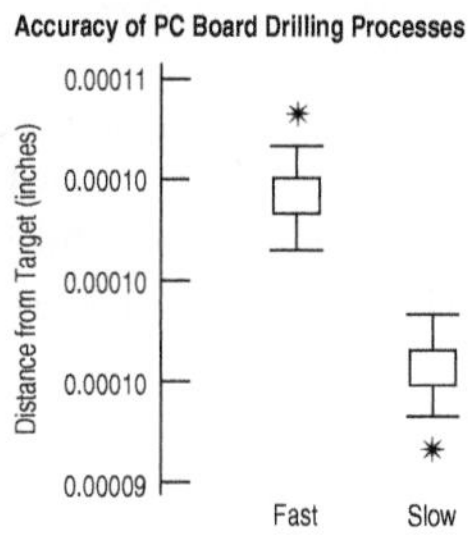

25. a) The side by side boxplots for Canada and the U.S. are given below. Homicide, aggravated assaults and robberies are higher in the U.S. Break and enter is higher in Canada. Theft and motor vehicle theft is about the same in Canada and the U.S. The pattern appears to be similar in the two countries, for example homicide is the least frequent among the six types in both countries and theft is the most frequent.

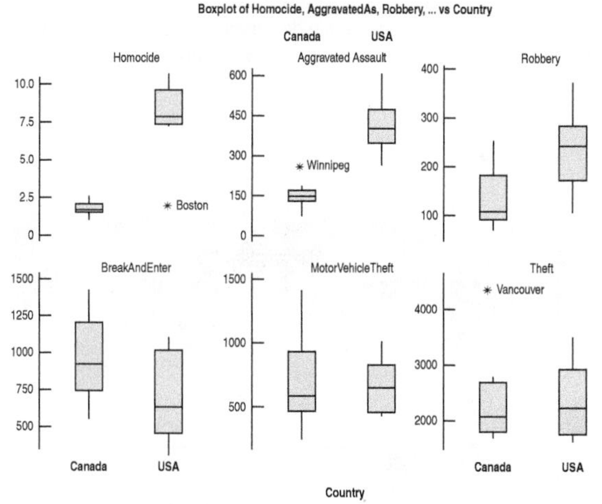

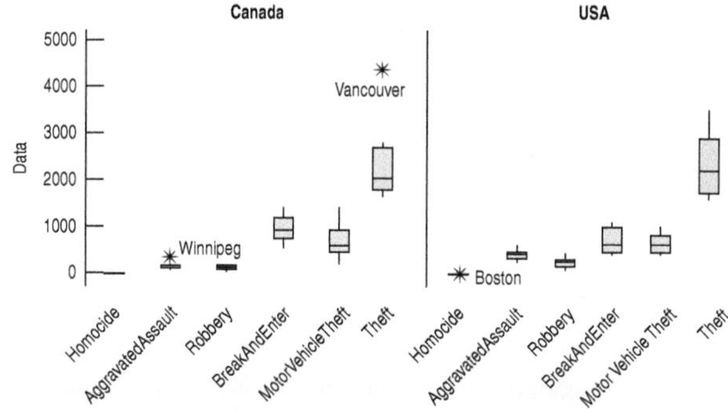

b) Winnipeg has aggravated assaults in the U.S. range and Boston has homicides in the Canadian range. Vancouver is a high outlier in Theft.

c) Population in the city and poverty can be lurking variables.

27. a) Side-by-side boxplots for % voter turnout are given below. PEI has the highest voter turnout (highest median, highest first quartile, and the highest third quartile). Newfoundland has the lowest % voter turnout (lowest median, first quartile, and the third quartile, other than the territories). There are whiskers for the Territories boxplot because there are only three observations (ridings) in the Territories.

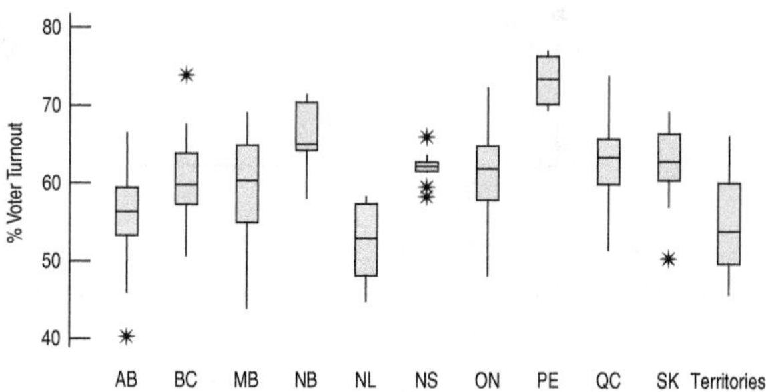

b) They are outliers compared to the observations (ridings) within those territories, but they are not outliers compared to all Canadian ridings.

c) The boxplot Canadian ridings (overall) is given below. There are outliers: Cardigan in PEI is the only positive outlier. The negative outliers are Calgary East, Nunavut, Random-Burin-St. George's, Bonavista-Gander-Grand Falls-Windsor, Churchill, and Fort McMurray-Athabasca. Fort McMurray-Athabasca is the only outlier in the combined boxplot that was also an outlier within its province/territory. Cardigan is a country-level outlier but is not an outlier within PEI because of the overall high voter turnout in PEI. With the exception of Calgary East, the negative outliers all seem to be in relatively remote areas of the country, which could account for the low turnout.

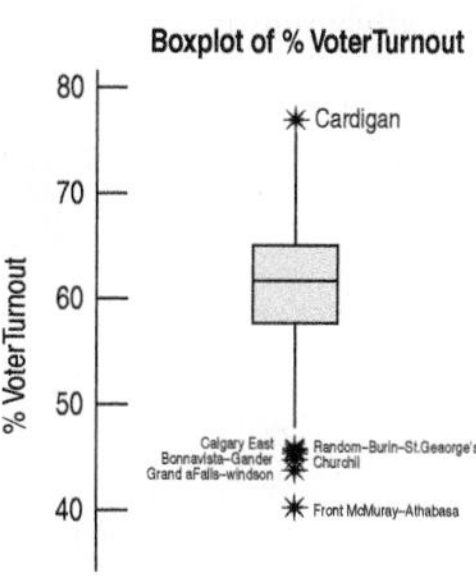

d) The mean must be smaller than the median in Manitoba, while in BC, the mean must be larger than the median.

29. a) The side-by-side boxplots of homicide rates are shown below. Large cities have the highest median homicide rate. Medium-sized cities have the least variable homicide rate. There are two large outliers.

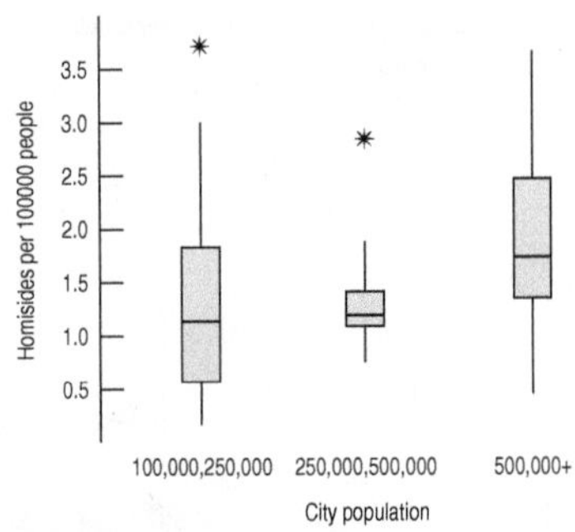

b) City size is a quantitative variable. In our analysis in (a) above we used the variable city size to create a categorical variable (with three categories). This helps us get some idea about how the homicide rates are related to the city size, but the actual city size values have more information about the city size and forming categories combining cities with sizes within certain ranges (even though the actual sizes are different) leads to some loss of information in the data. A plot of homicide rate versus population size (a scatterplot) will show the relation between the two variables.

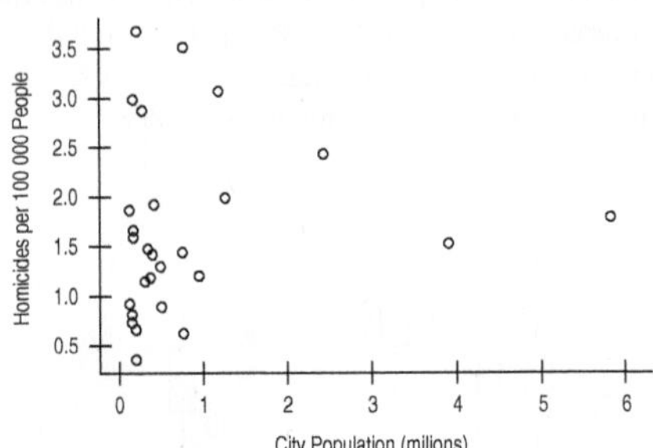

c) The side-by-side boxplots of homicide rates are shown below. Quebec has the lowest homicide rate and the West has the highest.

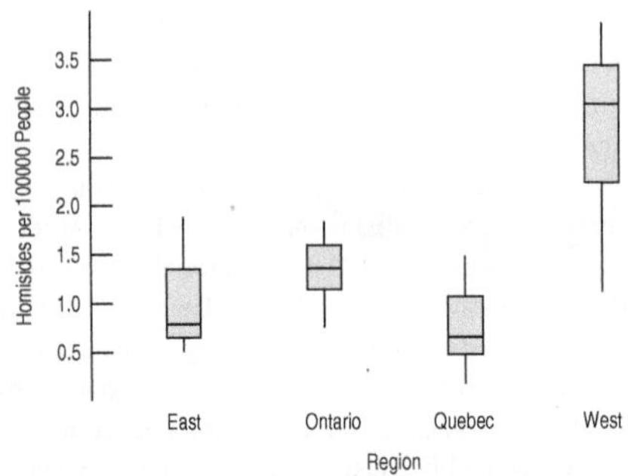

d) If we have the annual data for the 11-year period, we could examine the variation in homicide rate over time.

31. a) **Descriptive Statistics: points scored**

Variable	Minimum	Q1	Median	Q3	Maximum
points scored	48.0	98.3	145.5	192.8	215.0

b)

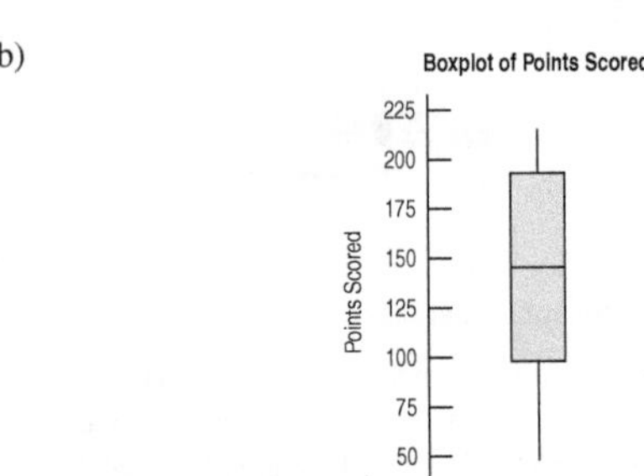

c) $48 > Q1 - 1.5\ IQR = -43.45$ and so 48 is not an outlier.

d) The plot of the total points versus year is shown below. The total points decreases with time. The boxplot does not show the relationship with time.

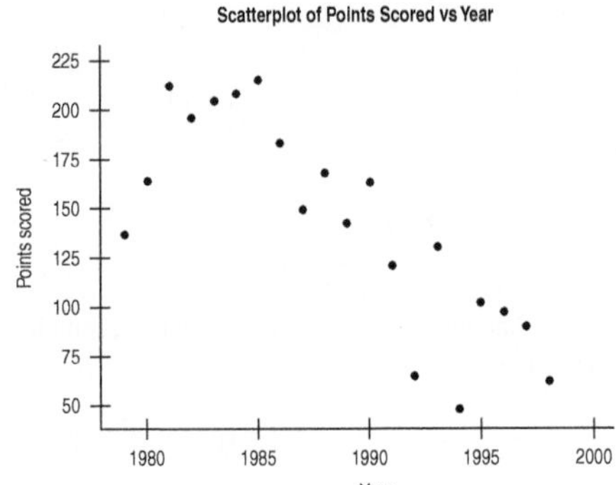

33.

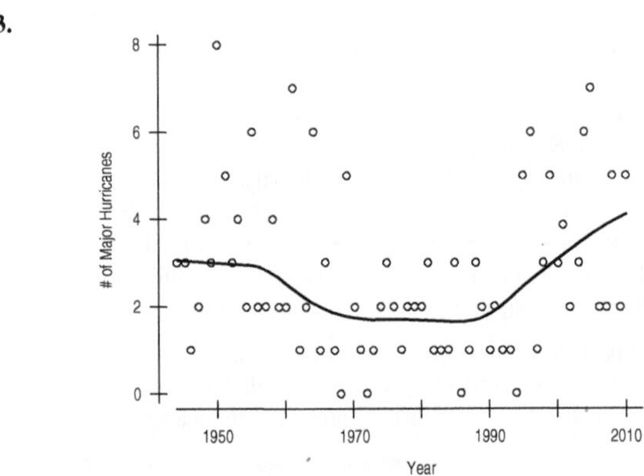

It appears that the number of hurricanes had a lull during the period from the mid-1960s to about 1990, and has increased fairly rapidly since. Their level is at least as high now as it was at the beginning of the time period.

35. a) The distribution is unimodal and slightly left skewed.

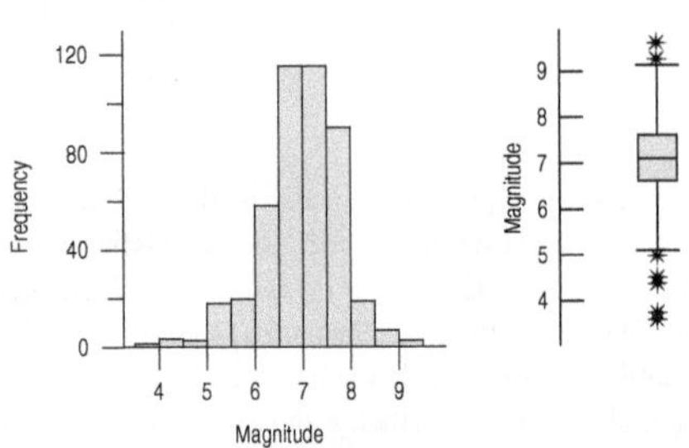

b) It doesn't appear that magnitude depends on time. That is, over time it doesn't appear that the distribution has really changed all that much.

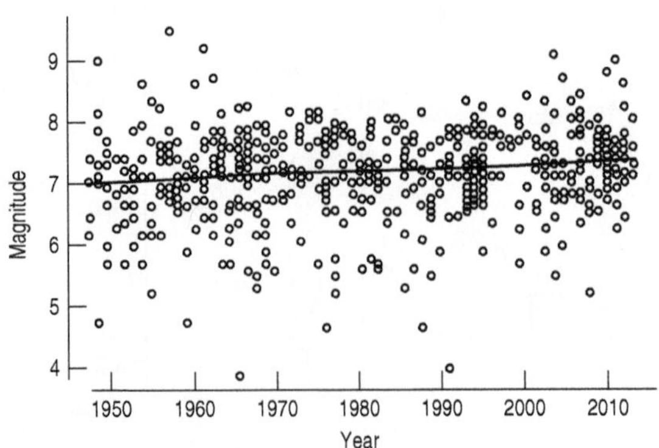

c) There are several countries with only 1 observation and thus only appear as a single point. There are also too many countries. We could group countries together in similar regions.

d) The quakes seemed to be the biggest in Peru. The smallest ones seemed to occur in Greece. The three countries with the biggest quakes were Chile (9.5 in 1960), the U.S. (9.2 in 1964), and Indonesia (9.1 in 2004). The smallest were in Greece (4.5 in 1962) and the U.S. (4.5 in 1952).

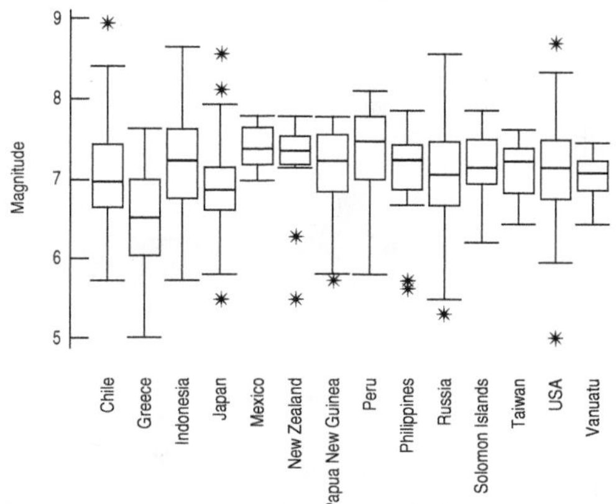

37. a) The logarithm makes the histogram more unimodal and symmetric. It is easy to see that the center is around 3.5 in log assets.
b) The company's assets are approximately 2,500 million dollars.
c) The company's assets are approximately 1,000 million dollars.

39. a) Fusion time and group.
b) Fusion time is quantitative (units = seconds). Group is categorical.
c) Both distributions are skewed to the right with high outliers. The boxplot indicates that visual information may reduce fusion time. The median for the Verbal/Visual group seems to be about the same as the lower quartile of the No/Verbal group.

41. a) 24 000, 7500. The median band size is slightly higher in Saskatchewan but it is hard to read the difference. Boxplots do not show the mean. Both of the means must be greater than the medians.
b) Add gridlines.
c) Right skewed. Many suspect outliers may just be part of the long upper tail of this skewed distribution, rather than genuine outliers.
d) Median and IQR do not change.
e) Because the distribution of band size is right skewed in both provinces.
f) At least 1500. Between 1500 and 2500 in size.
g) The distribution of the transformed sizes is much less skewed in both provinces.
h) Six Nations is still an outlier but it is much less extreme (less unusual) that it was before the transformation. There are additional outliers, most notably the Lucky Man band (size = 110) in North Central Saskatchewan.

Chapter 5

1. a) Skewed to the right; mean is higher than median.
b) \$350 and \$950
c) Minimum \$350. Mean \$750. Median \$550. Range \$1200. IQR \$600. Q1 \$400. SD \$400.
d) Minimum \$330. Mean \$770. Median \$550. Range \$1320. IQR \$660. Q1 \$385. SD \$440.

3. Lowest score = 910. Mean = 1230. SD = 120. Q3 = 1350. Median = 1270. IQR = 240.

5. In January, a high temperature of 13°C is less than 2 standard deviations above the mean. In July, a high temperature of 13° is more than two standard deviations below the mean. So it is less likely to happen in July.

7. a) Megan b) Anna

9. a) About 1.81 standard deviations below the mean
b) 1000 (z = 21.81) is more unusual than 1250 (z = 1.17).

11. a) Mean = 1152 1000 = 152 pounds; SD is unchanged at 84 pounds.
b) Mean = 0.40(1152) = \$460.80; SD = 0.40(84) = \$33.60.

13. Min = 0.40(980) 20 = \$372; median = 0.40(1140) 20 = \$436; SD = 0.40(84) = \$33.60; IQR = 0.40(102) = \$40.80.

15. College professors can have between 0 and maybe 40 (or possibly 50) years of experience. A standard deviation of 1/2 year is impossible, because many professors would be 10 or 20 SDs away from the mean, whatever it is. A SD of 16 years would mean that 2 SDs on either side of the mean is plus or minus 32, for a range of 64 years. That's too high. So, the SD must be 6 years.

17. a)

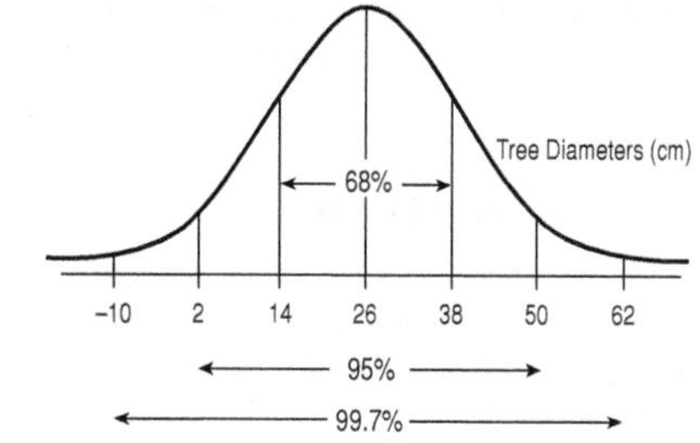

b) Between 2 cm and 50 cm. c) 2.5% d) 34% e) 16%

19. Since the histogram is not unimodal and symmetric, it is not wise to have faith in numbers from the Normal model.

21. a) 16% b) 13.56%
c) The percentages in parts a and b do not agree because the Normal model is not appropriate in this situation.
d) The histogram of 2010 Winter Olympic Downhill times is skewed to the right. The Normal model is not appropriate for the distribution of times, because the distribution is not symmetric.

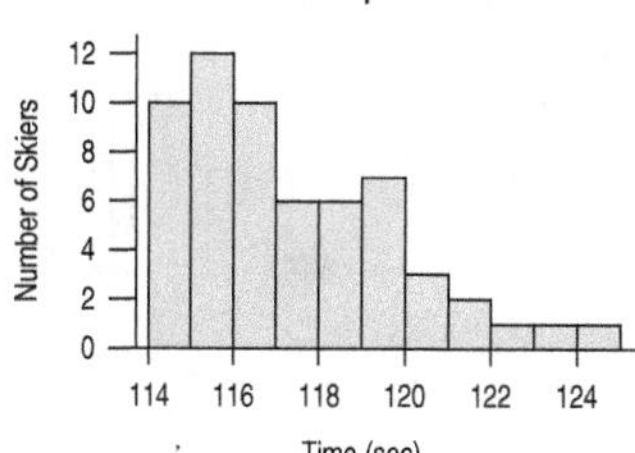

23. a) 16%
b) 16% of the students should be watching less than 1.27 hours a week, but no one can watch less than 0 hours, so the model doesn't fit well.
c) Data are strongly skewed to the right, not symmetric.

25. a) 6.68% b) 98.78% c) 71.63% d) 38.29%

27. a) 0.842 b) 0.675 c) 1.881 d) 1.645 to 1.645

29. a) 12.2% b) 71.6% c) 23.3%

31. a) 1130.7 lb b) 1347.4 lb c) 113.3 lb

33. a)

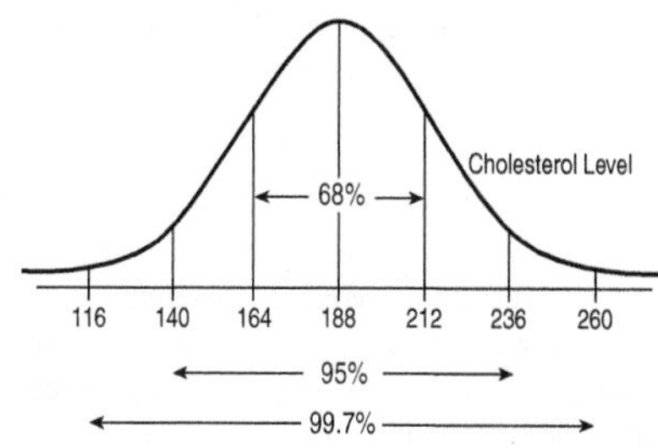

b) 30.85% c) 17.00% d) 32 points e) 212.9 points

35. a) 11.1 % b) (90.72, 103.28) cm c) 103.28 cm

37. About 1% of Greek men are expected to be taller than the average Dutch man. He wouldn't stand out radically, but he'd probably have a hard time keeping a "low profile"!

39. a) The smallest size is 50.672 cm and the biggest is 61.328 cm.
b) Heads in the range 54.488 cm and 57.512 cm.
c) z-values for 51 cm and 54 cm are 2.28 and 1.11 respectively. The proportion is 12.22%.
d) 57.872 cm e) Less than or equal to 52.31 cm.

41. a) 16 %
b) If the distribution is normal, the 10th and the 90th percentile are 463.2 and 616.8, which are not very different from the given values.
c) For a normal distribution with mean = 540 and standard deviation 60, IQR = 80.4. This is not that far from the given value.
d) 1.13 %

43. a) 5.3 grams b) 6.4 grams c) Younger because SD is smaller.
d) Mean 62.7 grams, SD 6.2 grams

45. a) The Normal probability plot is not straight, so there is evidence that the distribution of the lengths of songs in Corey's music library is not Normal.
b) The distribution of the lengths of songs in Corey's music library appears to be skewed to the right. The Normal probability plot shows that the longer songs in Corey's library are much longer than the lengths predicted by the Normal model. The song lengths are much longer than their quantile scores would predict for a Normal model.

47. The histogram, boxplot, and NPP are shown below. All three plots show several low outliers. Apart from these outliers, the histogram looks close to symmetric (maybe slightly left skewed). The boxplot has similar information. The normal probability plot shows a slight curvature (to the right). The many low outliers make the distribution look slightly left skewed.

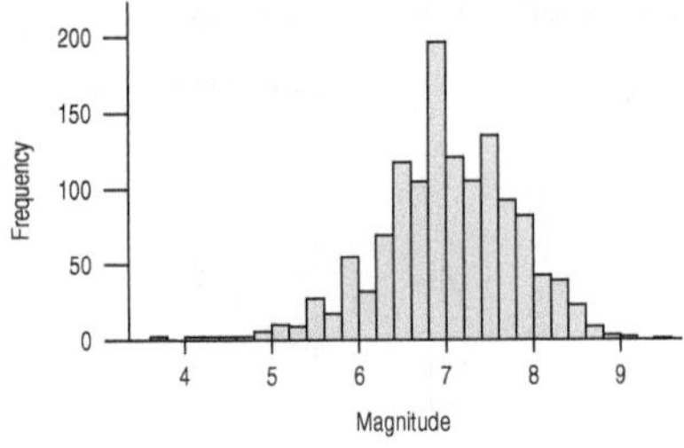

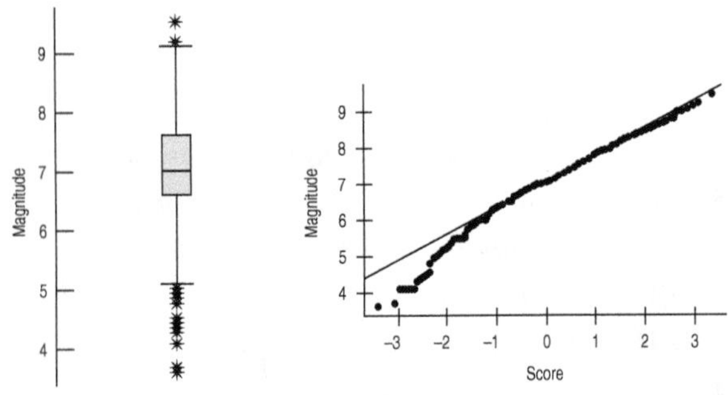

49. a) The histogram is close to a bell shape and the normal probability plot is close to a straight line. Usually it is easier to judge straightness. It can be difficult to judge *bell-shapedness* with small samples.
b) The histograms and the normal probability plots for the four simulated samples from the normal (0, 1) are given on the next page. The normal probability plots are close to straight lines. We expect this because the data were simulated from a normal distribution. The shapes of the histograms vary from sample to sample even though the data are from a normal distribution.

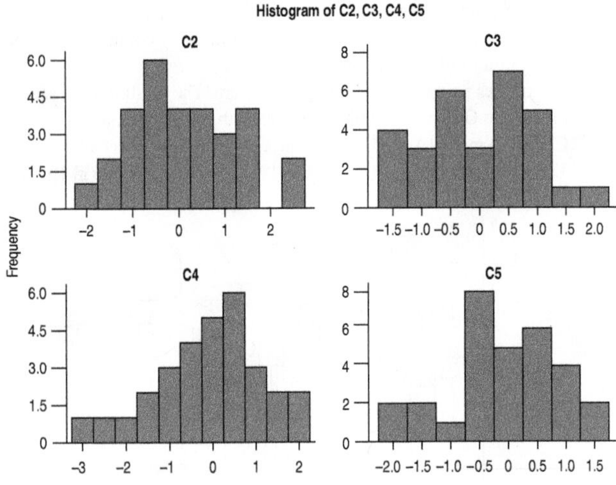

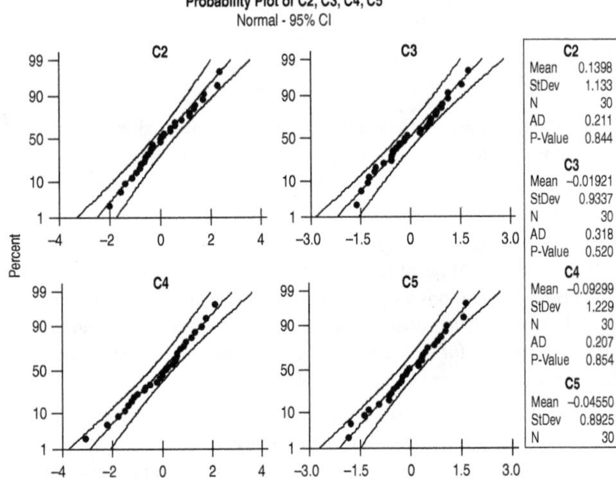

51. a) The histogram is close to a bell shape and the normal probability plot is close to a straight line. Usually is easier to judge straightness. It can be difficult to judge *bell-shapedness* in with small samples.
b) The normal probability plots are close to straight lines. We expect this because the data were simulated from a normal distribution. The shapes of the histograms vary from sample to sample even though the data are from a normal distribution.

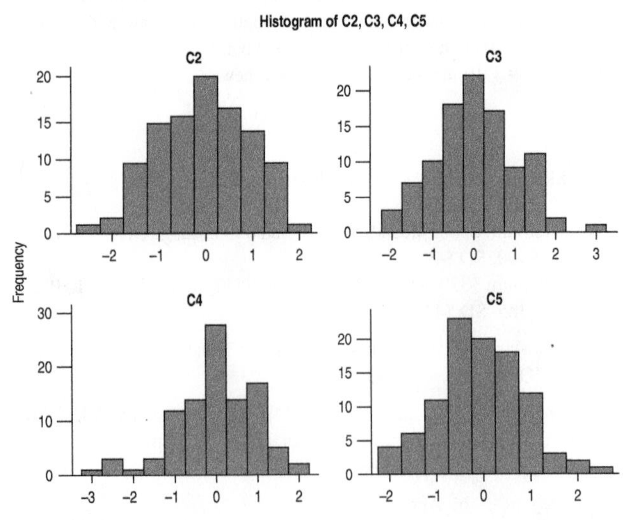

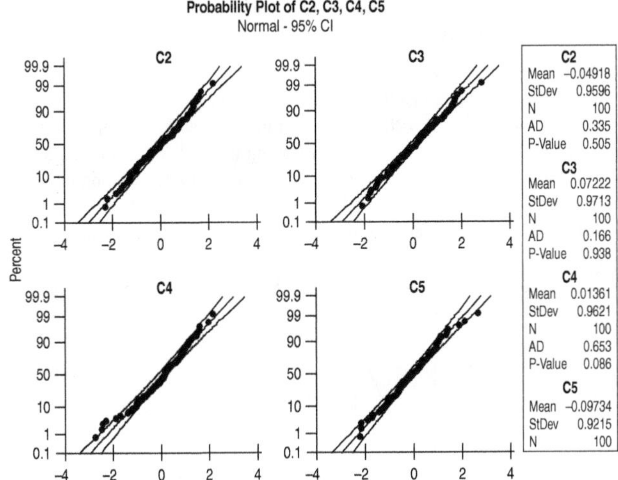

Part I Review

1. a) A histogram for the homicide rates is shown below:

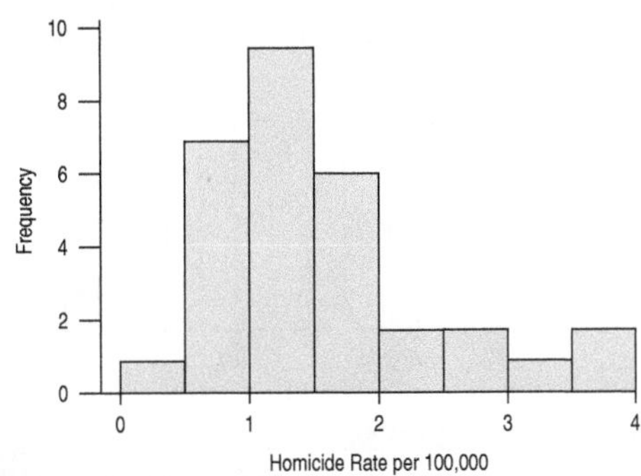

The homicide rates have median 1.36, quartiles Q1 = 0.92 and Q3 = 1.92, and interquartile range 1.00. The distribution is skewed to the right.

b) It appears that there are substantial differences between regions in Canada. The Prairies have the highest homicide rates, and Quebec has the lowest homicide rates.

3. a)

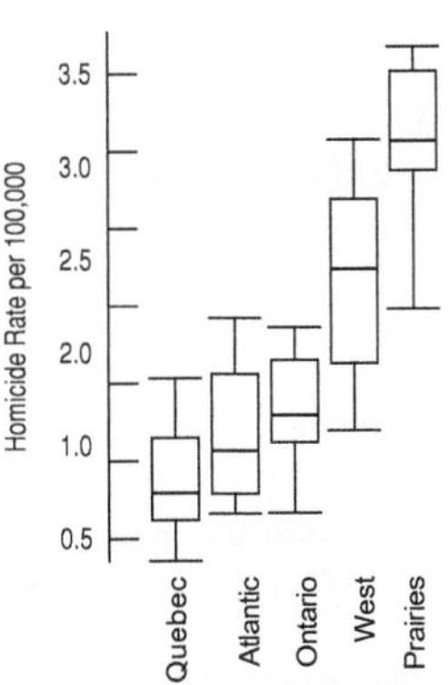

If enough sopranos have a height of 65 inches, this can happen.

b) The distribution of heights for each voice part is roughly symmetric. The basses are slightly taller than the tenors. The sopranos and altos have about the same median height. Heights of basses and sopranos are more consistent than those of altos and tenors.

5. a) It means their heights are also more variable.

b) The z-score for women to qualify is 2.40, compared with 1.75 for men, so it is harder for women to qualify.

7. a) *Who:* Local residents near a Canadian university.

What: Age, whether or not the respondent attended college, and whether or not the respondent had a favourable opinion of a Canadian university.

When: Not specified.

Where: Region around a Canadian university.

Why: The information will be included in a report to the university's directors.

How: 850 local residents were surveyed by phone.

b) There is one quantitative variable, age, probably measured in years. There are two categorical variables, college attendance (yes or no), and opinion of the university (favourable or unfavourable).

c) There are several problems with the design of the survey. No mention is made of a random selection of residents. Furthermore, there may be a non-response bias present. People with an unfavourable opinion of the university may hang up as soon as the staff member identifies himself or herself. Also, response bias may be introduced by the interviewer. The responses of the residents may be influenced by the fact that employees of the university are asking the questions. There may be a greater percentage of favourable responses to the survey than truly exist.

9. a) These are categorical data, so mean and standard deviation are meaningless.

b) Not appropriate. Even if it fits well, the Normal model is meaningless for categorical data.

11. a)

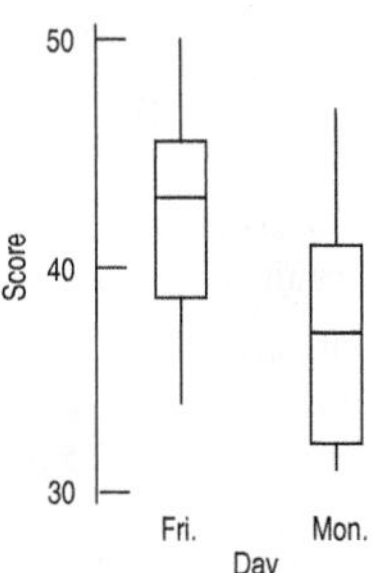

b) The scores on Friday were higher by about 5 points. This is a drop of more than 10% of the average score and shows that students fared worse on Monday after preparing for the test on Friday. The spreads are about the same, but the scores on Monday are a bit skewed to the right.

c)

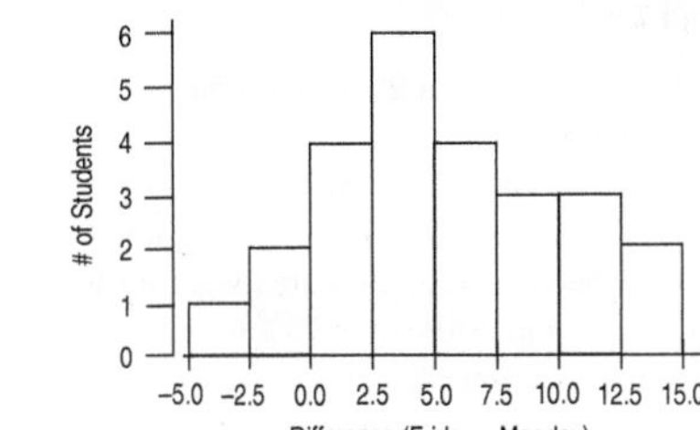

d) The changes (Friday – Monday) are unimodal and centered near 4 points, with a spread of about 5 (SD). They are fairly symmetric, but slightly skewed to the right. Only 3 students did better on Monday (had a negative difference).

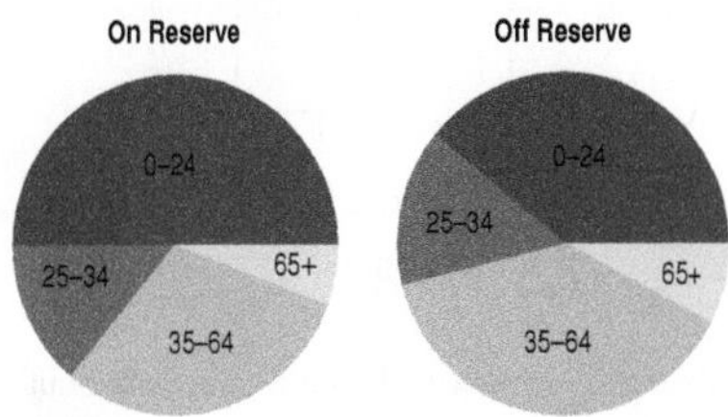

13. a) There is a higher proportion living on reserves in the 0–24 age group and a lower proportion living on reserves in the 35–64 age group.

b) Side by side barplots for the four conditional residence distributions are shown below. For the youngest age group, there are many more living on reserve. As the age groups increase in age, the percentage living on reserve declines.

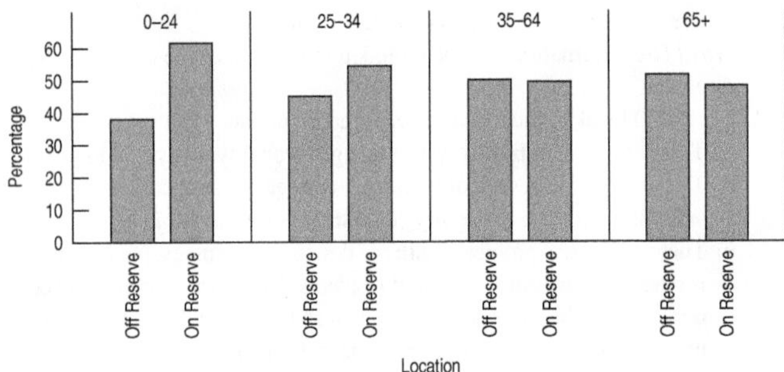

c) The residence variable is categorical. The age variable is quantitative, but has been made categorical for this display.

d) Yes. The following mean ages will occur if only 1/8 of the males live on reserves, but 3/4 of the females live on reserves.

Residency	Male	Female
On	30	40
Off	50	60
Overall	47.5	45

15. a) Annual mortality rate for males (quantitative) in deaths per 100 000 and water hardness (quantitative) in parts per million.

b) Calcium is skewed right, possibly bimodal. There looks to be a mode down near 12 ppm that is the centre of a fairly tight symmetric distribution and another mode near 62.5 ppm that is the centre of a much more spread out, symmetric (almost uniform) distribution. Mortality, however, appears unimodal and symmetric with the mode near 1500 deaths per 100 000.

17. a) They are on different scales.

b) January's values are lower and more spread out.

c) Roughly symmetric but slightly skewed to the left. There are more low outliers than high ones. Centre is around 40 degrees with an IQR of around 7.5 degrees.

19. a) Bimodal with modes near 2 and 4.5 minutes. Fairly symmetric around each mode.

b) Because there are two modes, which probably correspond to two different groups of eruptions, an average might not make sense.

c) The intervals between eruptions are longer for long eruptions. There is very little overlap. More than 75% of the short eruptions had intervals less than about an hour (62.5 minutes), while more than 75% of the long eruptions had intervals longer than about 75 minutes. Perhaps the interval could even be used to predict whether the next eruption will be long or short.

21. a)

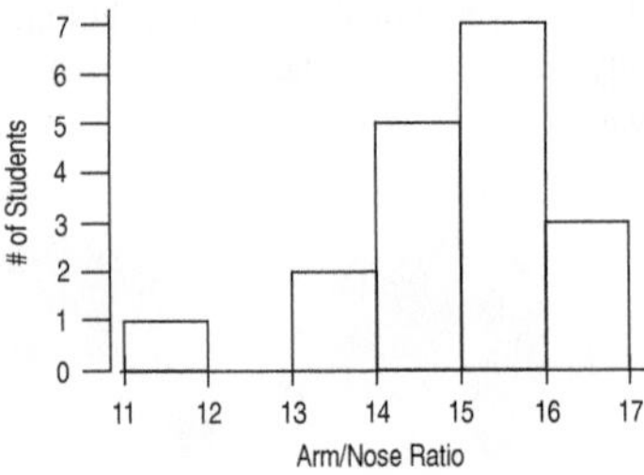

The distribution is left skewed with a centre of about 15. It has an outlier between 11 and 12.

b) Even though the distribution is somewhat skewed, the mean and median are close. The mean is 15.0 and the SD is 1.25.

c) Yes. 11.8 is already an outlier. 9.3 is more than 4.5 SDs below the mean. It is a very low outlier.

23. If we look only at the overall statistics, it appears that the follow-up group is insured at a much lower rate than those not traced (11.1% of the time as compared with 16.6%). But most of the follow-up group were black, who have a lower rate of being insured. When broken down by race, the follow-up group actually has a higher rate of being insured for both blacks and whites. So the overall statistic is misleading and is attributable to the difference in race makeup of the two groups.

25. a)

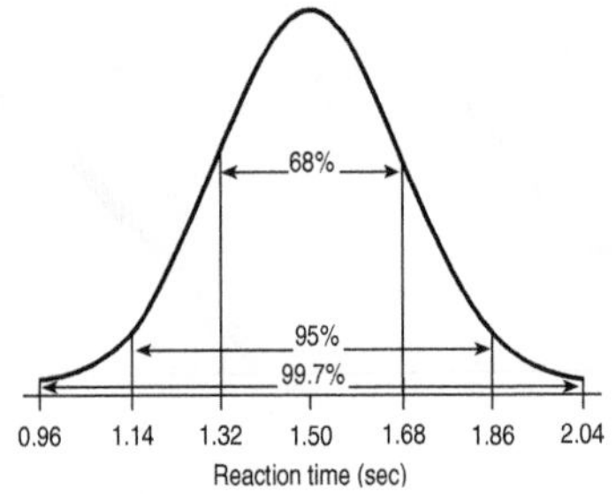

b) According to the model, reaction times are symmetric with centre at 1.5 seconds. About 95% of all reaction times are between 1.14 and 1.86 seconds.

c) 8.2% d) 24.1%

e) Quartiles are 1.38 and 1.62 seconds, so the IQR is 0.24 seconds.

f) The slowest 1/3 of all drivers have reaction times of 1.58 seconds or more.

27. a)

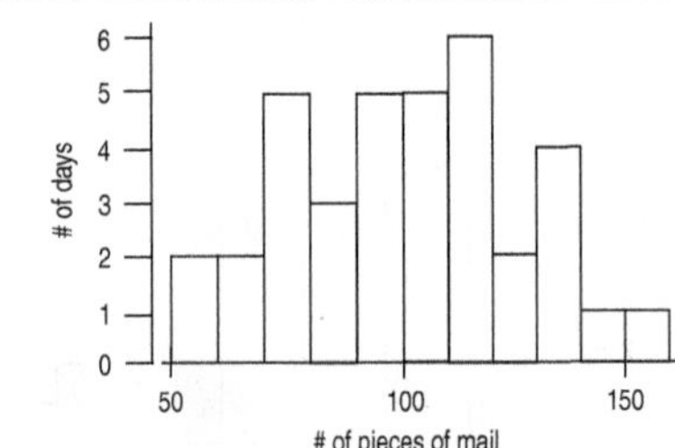

b) Mean \$100.25, SD \$25.54.

c) The distribution is somewhat symmetric and unimodal, but the centre is rather flat, almost uniform.

d) 64%. The Normal model seems to work reasonably well, since it predicts 68%.

29. a) *Who:* 100 health food store customers.
What: Have you taken a cold remedy?, and Effectiveness (scale 1 to 10).
When: Not stated.
Where: Not stated.
Why: Promotion of herbal medicine.
How: In-person interviews.

b) Have you taken a cold remedy?—categorical. Effectiveness—categorical or ordinal.

c) No. Customers are not necessarily representative, and the Council had an interest in promoting the herbal remedy.

31. a) 38 cars

b) Possibly because the distribution is skewed to the right.

c) Centre—median is 148.5 cubic inches. Spread—IQR is 126 cubic inches.

d) No. It's bigger than average, but smaller than more than 25% of cars. The upper quartile is at 231 inches.

e) No. 1.5 IQR is 189, and 105–189 is negative, so there can't be any low outliers. 231 + 189 = 420. There aren't any cars with engines bigger than this, since the maximum has to be at most Q1 + Range = 105 + 275 = 380 cubic inches.

f) Because the distribution is skewed to the right, this is probably not a good approximation.

g) Mean, median, range, quartiles, IQR, and SD all get multiplied by 16.4.

33.

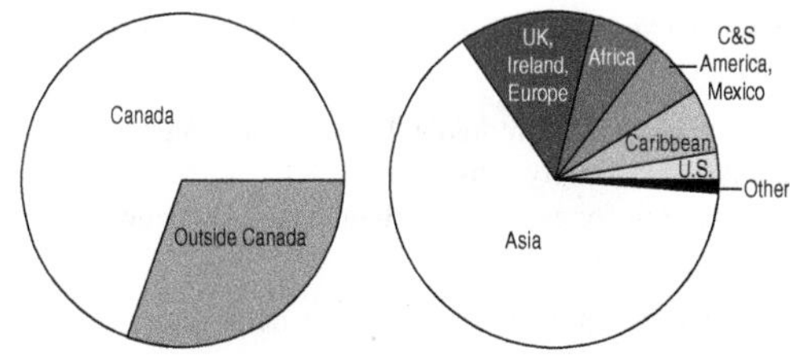

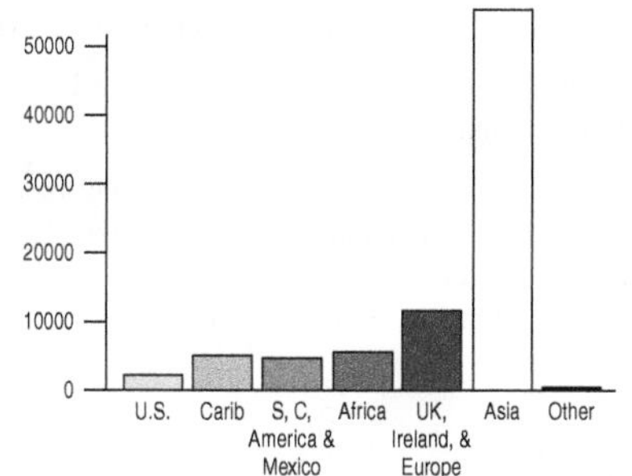

35. a) Side by side boxplots for the two teams are shown below.

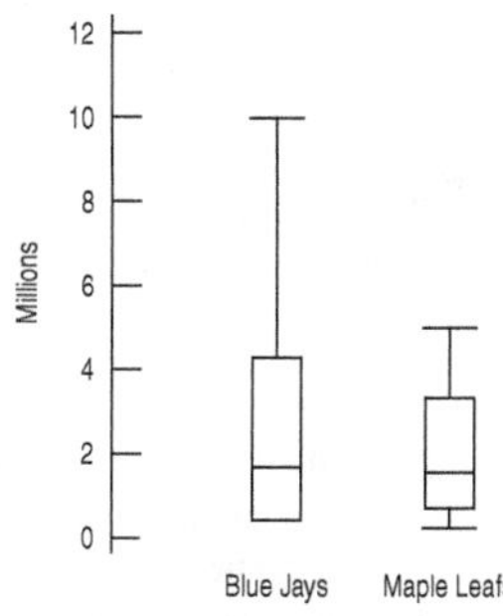

The median salaries for the two teams are similar, $1 675 000 for the Blue Jays and $1 725 000 for the Maple Leafs. The Blue Jay salaries are more spread out, with IQR $3 888 500 versus $2 500 000 for the Maple Leafs. There are many more very high salaries for the Blue Jays.

b) As a superstar I would prefer playing for the Blue Jays, where the very high salaries occur. As a new player I would prefer the Maple Leafs, as the minimum is higher ($500 000 versus $392 200). As an average player I would prefer the Blue Jays as the mean salaries are higher ($3 417 367 versus $2 099 545).

c) As an owner, I would prefer the salary structure of the Maple Leafs as the total salary budget is much smaller ($46 190 000 versus $82 016 800).

37. a) 0.43 hour b) 1.4 hours c) 0.89 hour (or 53.4 minutes)
d) Survey results vary, and the mean and the SD may have changed.

Chapter 6

1. a) Weight in ounces: explanatory; Weight in grams: response. (Could be other way around.) To predict the weight in grams based on ounces. Scatterplot: positive, straight, strong (perfectly linear relationship).

b) Circumference: explanatory. Weight: response. To predict the weight based on the circumference. Scatterplot: positive, linear, moderately strong.

c) Shoe size: explanatory; GPA: response. To try to predict GPA from shoe size. Scatterplot: no direction, no form, very weak.

d) Miles driven: explanatory; Gallons remaining: response. To predict the gallons remaining in the tank based on the miles driven since filling up. Scatterplot: negative, straight, moderate.

3. a) Altitude: explanatory; Temperature: response. (Other way around possible as well.) To predict the temperature based on the altitude. Scatterplot: negative, possibly straight, weak to moderate.

b) Ice cream cone sales: explanatory. Air conditioner sales: response—although the other direction would work as well. To predict one from the other. Scatterplot: positive, straight, moderate.

c) Age: explanatory; Grip strength: response. To predict the grip strength based on age. Scatterplot: curved down, moderate. Very young and elderly would have grip strength less than that of adults.

d) Reaction time: explanatory; Blood alcohol level: response. To predict blood alcohol level from reaction time test. (Other way around is possible.) Scatterplot: positive, nonlinear, moderately strong.

5. a) None b)3 and 4 c) 2, 3, and 4 d) 1 and 2
e) 3 and possibly 1

7. There seems to be a very weak—or possibly no—relation between brain size and performance IQ.

9. a)

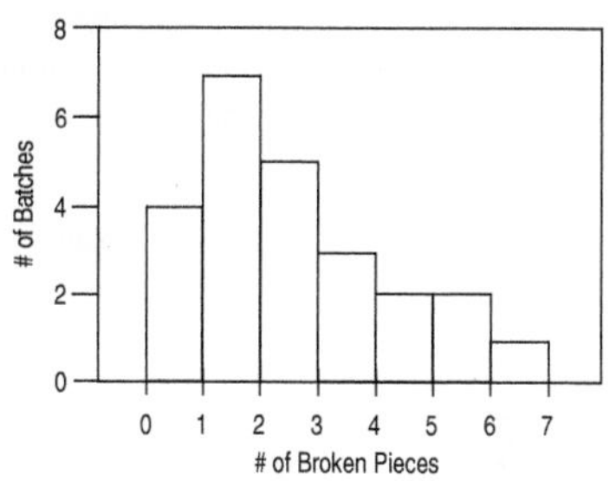

b) Unimodal, skewed to the right. The skew.

c) The positive, somewhat linear relation between batch number and broken pieces.

d) Answers may vary.

11. a) 0.006 b) 0.777 c) 0.923 d) 0.487

13. a) True.
b) False. The correlation will remain the same.
c) False. Correlation has no units.

15. a) Yes. It shows a linear form and no extreme outliers.
b) There is a strong, positive, linear association between drop and speed; the greater the coaster's initial drop, the higher the top speed.

17. a) −0.649
b) The correlation will not change. Correlation does not change with change of units.
c) There is a moderately strong negative linear association between number of calories consumed and time at the table for toddlers. Longer times are associated with fewer calories.
d) There appears to be a relation between time at the table and calories consumed, but there is no reason to believe increased time *causes* decreased calorie consumption. There may be a lurking variable.

19. a) A scatterplot of expected fuel economy vs. horsepower is shown below.

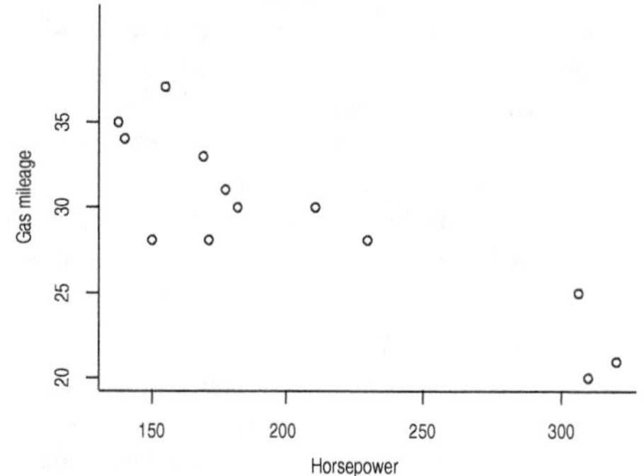

b) There is a strong, negative, straight association between horsepower and mileage of the selected vehicles. There don't appear to be any outliers. All of the cars seem to fit the same pattern. Cars with more horsepower tend to have lower mileage.

c) Since the relationship is linear with no outliers, correlation is an appropriate measure of strength. The correlation between horsepower and mileage of the selected vehicles is -0.909.

d) There is a strong linear relationship in the negative direction between horsepower and highway gas mileage. Lower fuel efficiency is associated with higher horsepower.

e) The scatterplot of fuel consumption versus horsepower is shown below. There is a strong, positive, straight association between fuel consumption and horsepower. The correlation between the fuel consumption and horsepower is stronger than the correlation between mileage and horsepower, and the direction of association is reversed (correlation = 0.921). The strength of correlation changed here because the transformation from mileage to consumption is non-linear. Both plots are roughly linear, but I might prefer the transformed scale, since the correlation is slightly higher.

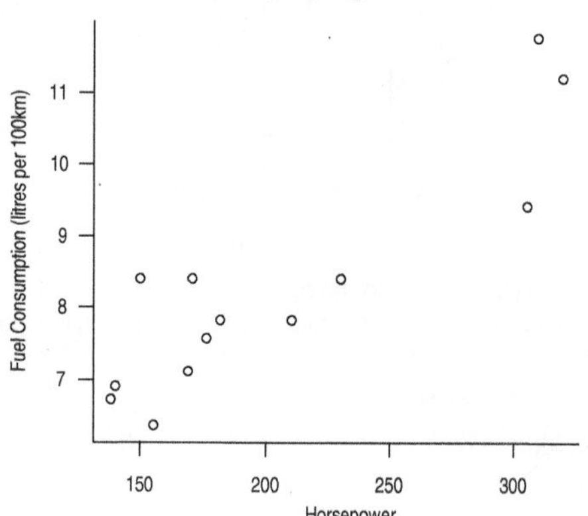

21.

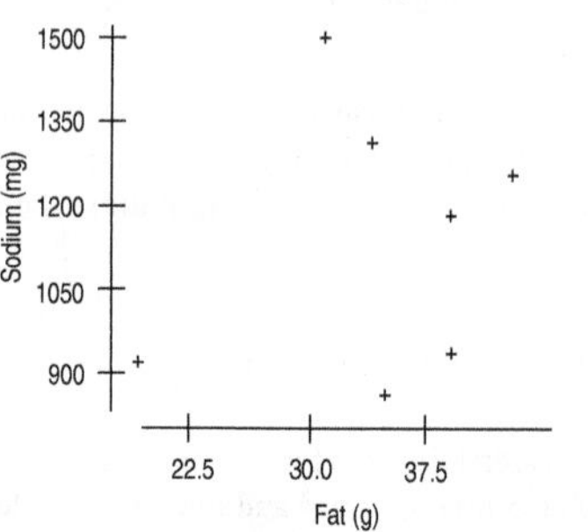

(Plot could have explanatory and predictor variables swapped.) Correlation is 0.199. There does not appear to be a relation between sodium and fat content in burgers, especially without the low-fat low-sodium item. The correlation of 0.199 shows a weak relationship, even with the outlier included.

23. a) Number of runs scored and attendance are quantitative variables, the relationship between them appears to be straight and there are no outliers, so calculating a correlation is appropriate.

b) The association between attendance and runs scored is positive, straight, and moderate in strength. Generally, as the number of runs scored increases, so does attendance.

c) There is evidence of an association between attendance and runs scored, but a cause-and-effect relationship between the two is not implied. There may be lurking variables that can account for the increases in each. For example, perhaps winning teams score more runs and also have higher attendance. We don't have any basis to make a claim of causation.

25. There may be an association but not a correlation unless the variables are quantitative. There could be a correlation between average number of hours of TV watched per week per person and the number of crimes committed per year. Even if there is a relationship, it doesn't mean one causes the other.

27. a) Actually, yes, taller children will tend to have higher reading scores, but this doesn't imply causation.

b) Older children are generally both taller and are better readers. Age is the lurking variable.

29. The scatterplot is not linear; correlation is not appropriate.

31. a) Assuming the relation is linear, a correlation of 20.772 shows a strong relation in a negative direction.

b) Continent is a categorical variable. Correlation does not apply.

33. This is categorical data even though it is represented by numbers. The correlation is meaningless.

35. The scatterplot below shows that the association between duration and length is straight, positive, and moderate, with no outliers. Generally, rides on coasters with a greater length tend to last longer. The correlation between length and duration is 0.698, indicating a moderate association.

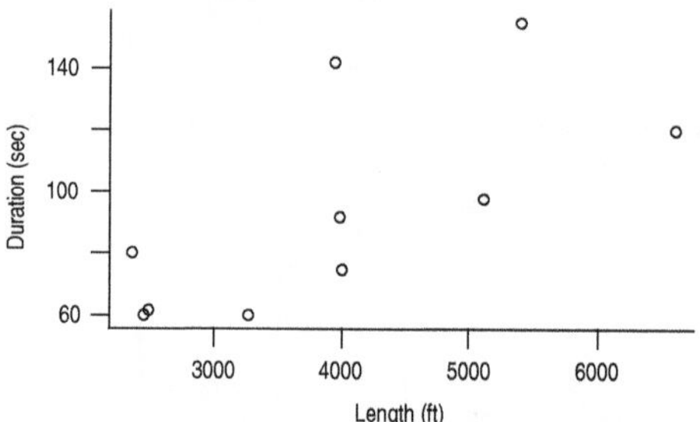

37. a) We would expect that as one variable (say length of ride) increases, the rank will improve, which means it will decrease.

b) Drop has the strongest correlation ($r = -0.193$), but even that correlation is very weak. The scatterplot shows no apparent association. The number one ranked coaster, Bizarro, has a fairly typical drop. There appear to be other factors that influence the rank of coaster more than any of the ones measured in this data set.

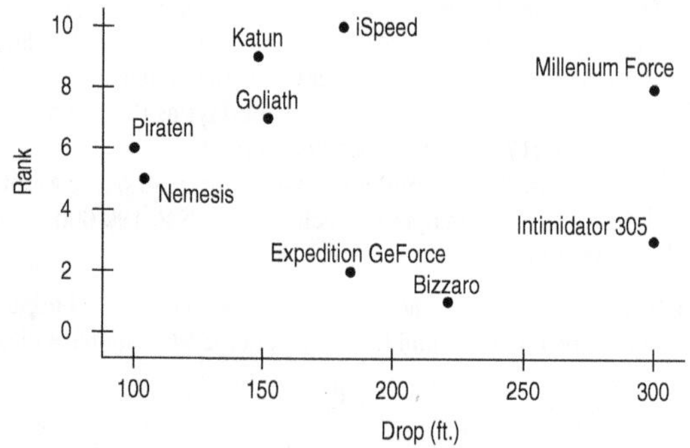

c) There may be other variables that account for the ranking. For example, other quantitative variables such as number of loops and number of corkscrews, categorical variables such as whether the coaster is made of wood or steel, and whether or not there are tunnels may all have an affect on the rank.

39. a) **Correlations: Year, Bust size (in), Waist size (in), Hips size (in)**

	Year	Bust size	Waist size
Bust size	1.000		
Waist size	0.992	0.992	
Hips size	0.936	0.936	0.929

b) All these relationships are linear, strong and positive in direction, with no clear outliers. The relationship between bust size and year is the strongest.

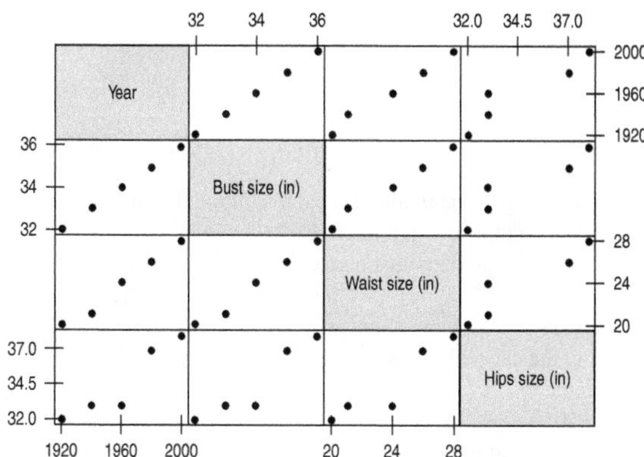

c) The correlations between body measures are not quite surprising as we expect some proportionality between them (unless otherwise women's body shape is changing over time.). The positive correlation with time can happen for example if we are getting bigger (increasing weight) over time due to various reasons (maybe lack of activity because most of the work we did in the good old days are taken over by machines).

41. a) Armspan and height have the highest correlation at 0.781.
b) Linear, positive, moderately strong relationship.
c) Relationship looks similar for both genders, but slightly stronger for females than for males.

43. a)

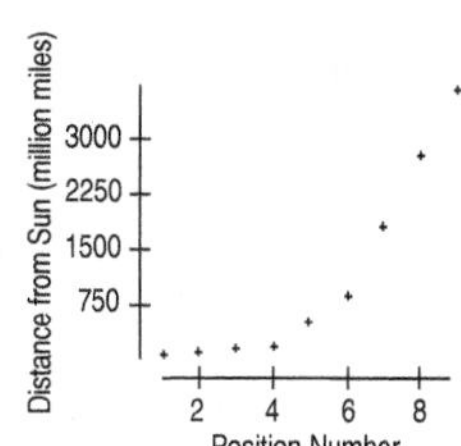

The relation between position and distance is very strong, positive, and curved. There is very little scatter from the trend.
b) The relation is not linear.

c)

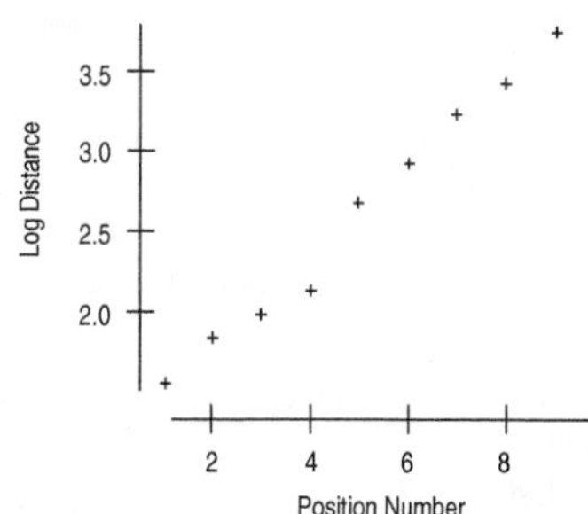

The relation between position number and log of distance appears to be roughly linear.

45. a) 0.995. This is very close to 1.
b) Most points are concentrated to small region and just two points far away from them but in the same direction. This can make the correlation very high and correlation is not an appropriate measure in situations like this.

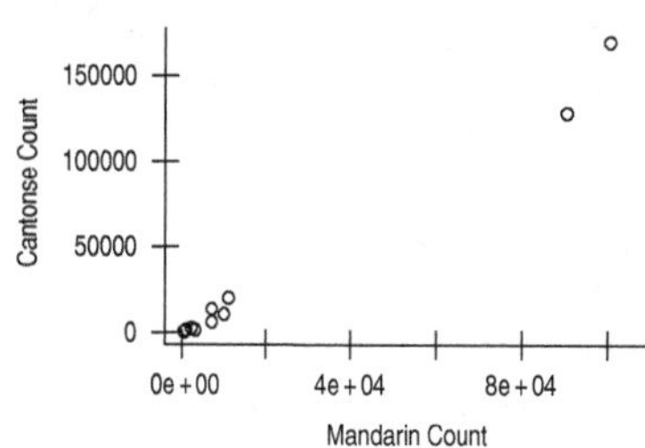

c) Yes, now there is concentration of points to smaller communities. The correlation is 0.985. The correlation is now an appropriate measure.

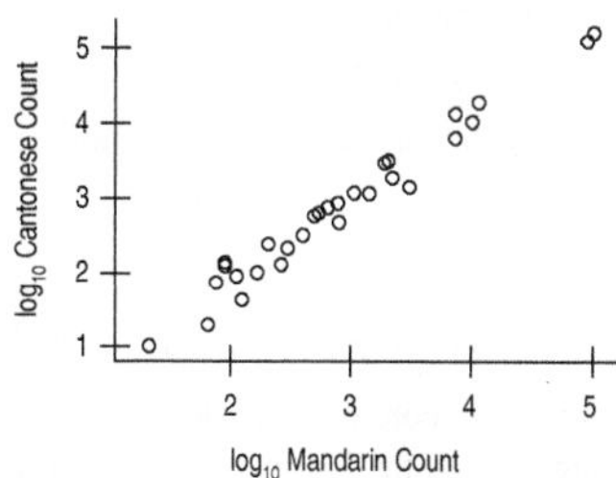

d) In larger cities it is reasonable to expect relatively larger counts for both these groups (as well as for any other group) and thinking this way, the larger Mandarin counts might not be because of the larger Cantonese counts. It is even possible that these two counts are not even related. In other words what we are suggesting is that the population in those cities is a possible lurking variable. If we think population is a possible lurking variable, we can use proportions of Cantonese and Mandarin speaker counts as the variables.

Chapter 7

1.

$\bar{x}$	s_x	$\bar{y}$	s_y	r	$\hat{y} = b_0 + b_1x$
a) 10	2	20	3	0.5	$\hat{y} = 12.5 + .75x$
b) 2	0.06	7.2	1.2	−0.4	$\hat{y} = 23.2 - 8x$
c) 12	6	152	30	−0.8	$\hat{y} = 200 - 4x$
d) 2.5	1.2	25	100	0.6	$\hat{y} = -100 + 50x$

3. a) Model is appropriate.
b) Model is not appropriate. Relationship is nonlinear.
c) Model may not be appropriate. Spread is changing.

5. The line $\hat{y} = 7 + 1.1x$ minimizes the sum of squared vertical distances from the points to the line.

7. a) The response variable is the asking price (in thousands of dollars) and the explanatory variable is the size (in square feet).
b) Millions of dollars per square feet.
c) The price usually increases as the size increases and so the slope must be positive.

9. 70.2% of the variability in Price is explained by the linear regression of Price on size.

11. a) 0.84
b) Price should be 0.84 standard deviations above the average price.
c) Price should be 1.68 standard deviations below the average price.

13. a) For every square foot increase in size, the price is expected to increase by 0.37 thousand dollars.
b) 419.3 dollars.
c) The predicted price is 493.3 thousand dollars and asking price = 487 300 dollars. The value 6000 is the residual for a 1200-square-foot condo with price 1200 dollars.

15. a) Probably. The residuals show some initially low points, but there is no clear curvature.
b) The linear model on Tar content accounts for 92.4% of the variability in Nicotine.

17. a) $r = \sqrt{R^2} = \sqrt{0.924} = 0.961$
b) Nicotine should be 1.922 SDs below average.
c) Tar should be 0.961 SDs above average.

19. a) $\widehat{Nicotine} = 0.15403 + 0.065052\ (Tar)$
b) 0.414 mg
c) Nicotine content increases by 0.065 mg of nicotine per milligram of tar.
d) We'd expect a cigarette with no tar to have 0.154 mg of nicotine.
e) 0.1094 mg

21. 300 pounds/foot. If a "typical" car is 15 feet long, all of 3, 30, and 3000 would give ridiculous weights.

23. a) R^2 does not tell whether the model is appropriate, but measures the strength of the linear relationship. High R^2 could also be due to an outlier.
b) Predictions based on a regression line are for average values of y for a given x. The actual wingspan will vary around the prediction.

25. a) Probably not. Your score is better than about 97.5% of people, assuming scores follow the Normal model. Your next score is likely to be closer to the mean.
b) The friend probably should retake the test. His score is better than only about 16% of people. His score is likely to be closer to the mean.

27. a) Moderately strong, fairly straight, and positive. Possibly some outliers (higher than expected math scores).
b) The student with 500 verbal and 800 math.
c) Positive, fairly strong linear relationship. 46.9% of variation in math scores is explained by verbal scores.
d) $\widehat{Math} = 217.692 + 0.662\ (Verbal)$
e) Every point of verbal score adds an average 0.662 points to the predicted math score.
f) 548.5 points g) 53.0 points

29. a) 0.685 b) $\widehat{Vetbal} = 162.106 + 0.709\ (Math)$
c) The observed verbal score is higher than predicted from the math score.
d) 516.7 points e) 559.6 points
f) Regression to the mean. Someone whose math score is below average is predicted to have a verbal score below average, but not as far (in SDs). So if we use that verbal score to predict math, they will be even closer to the mean in predicted math score than their observed math score. If we kept cycling back and forth, eventually we would predict the mean of each and stay there.

31. a)

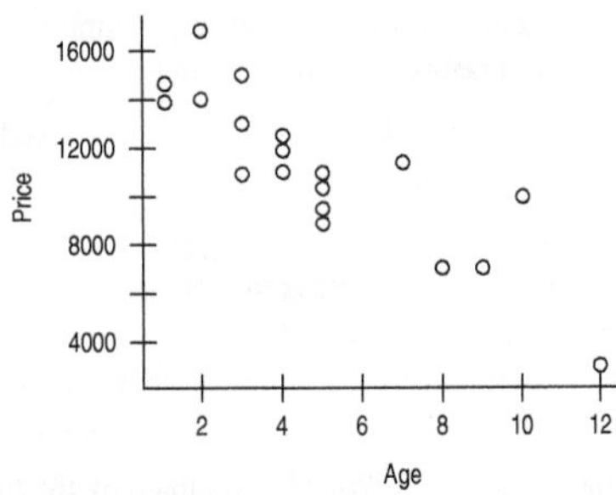

b) There is a strong, negative, linear association between price and age of used Toyota Corollas. A few observations are somewhat unusual, but not very unusual.
c) The scatterplot provides evidence that the relationship is Straight Enough. A linear model will likely be appropriate.
d) $r = -0.865$ e) 74.8%
f) The relationship is not perfect. Other factors, such as options, condition, and mileage, explain the rest of the variability in price.

33. a) The scatterplot from Exercise 31 shows that the relationship is straight, so the linear model is appropriate. The regression equation to predict the price of a used Toyota Corolla from its age is: Price = \$15,729 − \$935.9 × Age.

The regression output is below:

Coefficients:

	Estimate	Std. Error	t value	$Pr(>\|t\|)$
(Intercept)	15729.0	723.2	21.749	2.26e − 14 ***
Age	−935.9	127.9	−7.315	8.56e − 07 ***

b) According to the model, for each additional year in age, the car is expected to drop \$936 in price.
c) The model predicts that a new Toyota Corolla (0 years old) will cost \$15 729.
d) \$10 114
e) Buy the car with the negative residual. Its actual price is lower than predicted.
f) −\$2050
g) No, predicting a price after 17 years would be an extrapolation. In fact, the predicted price using the model would be negative.

35. a)

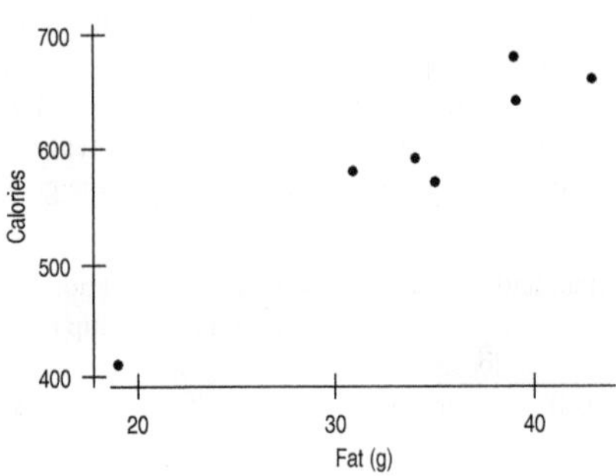

b) 92.3% of the variation in calories can be accounted for by the fat content.
c) $\widehat{Calories} = 210.954 + 11.0555\ (Fat)$
d)

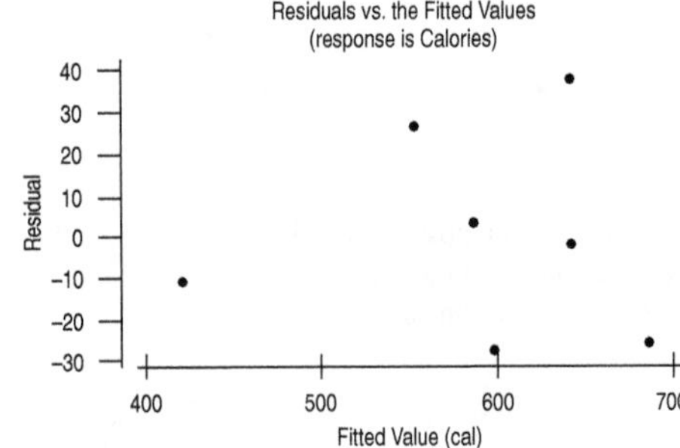

Residuals show no clear pattern, so the model seems appropriate.
e) Could say a fat-free burger still has 211.0 calories, but this is extrapolation (no data close to 0).
f) Every gram of fat adds 11.06 calories, on average.
g) 553.5 calories

37. a) The regression was for predicting calories from fat, not the other way around.
b) $\widehat{Fat} = -14.9622 + 0.083471\ (Calories)$
Predict 35.1 grams of fat.

39. a) This model predicts that every bridge will be deficient. The intercept is less than 5 and the slope is negative, so all predicted conditions are less than 5.
b) This model says bridges in New York City are decreasing in condition at only 0.004 per year—less rapidly than bridges in Tompkins County.
c) The R^2 for this model is only 2.6%. I don't think it has much predictive value.

41. a) $r = 0.92$
b) 84% of the variability in mean temperature can be accounted for by variability in CO_2 level.
c) $\widehat{MeanTemp} = 11.0276 + 0.0089CO_2$
d) The model predicts that an increase in CO_2 level of 1 ppm is associated with an increase of 0.0089°C in mean temperature.
e) According to the model, the mean temperature is predicted to be 11.0276°C when there is no CO_2 in the atmosphere. This is an extrapolation outside of the range of data, and isn't very meaningful in context, since there is always CO_2 in the atmosphere. We want to use this model to study the change in CO_2 level and how it relates to the change in temperature.
f) The residuals plot shows no apparent patterns. The linear model appears to be an appropriate one.
g) 14.59 °C

43. a) $\widehat{\%Fat} = -27.3763 + 0.249874\,(Weight)$
b) Residuals look randomly scattered around 0, so conditions are satisfied.
c) % Body Fat increases on average by 0.25 percent per pound of Weight.
d) Reliable is relative. R^2 is 48.5%, but residuals have a standard deviation of 7%, so variation around the line is large.
e) 0.9%

45. a) $\widehat{HighJump} = 2.42 - 0.004690(800\ m\ time)$. According to the model, the predicted high jump decreases by an average of 0.0047 metres for each additional second in 800 metre time.
b) 6.7%
c) Yes, good high jumpers tend to be fast runners. The slope of the association is negative. Faster runners tend to jump higher, as well.
d) The residuals plot is fairly patternless. The scatterplot shows a slight tendency for less variation in high jump height among the slower runners than the faster runners. Overall, the linear model is appropriate.

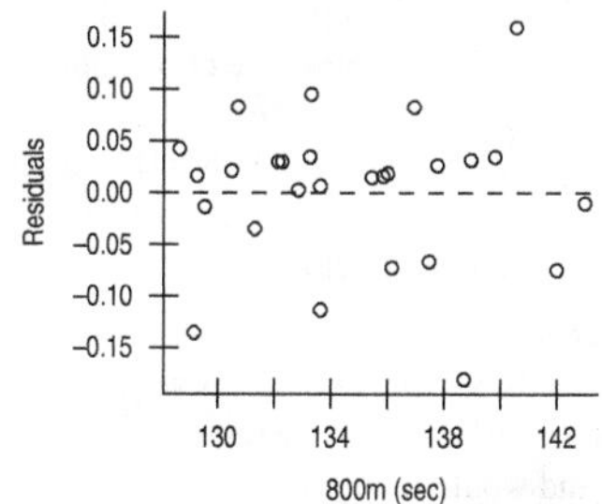

e) The linear model is not particularly useful for predicting high jump performance. First of all, 6.7% of the variability in high jump height is accounted for by the variability in 800 metre time, leaving 93.3% of the variability unaccounted for. Secondly, the residual standard deviation is 0.07288 metres, which is not much smaller than the standard deviation of all high jumps, 0.074 metres. Predictions are not likely to be accurate.

47. a) As calcium levels increase, mortality rate decreases. Relationship is fairly strong, negative, and linear.
b) $\widehat{Mortality} = 1676 - 3.23(Calcium)$
c) Mortality decreases 3.23 deaths per 100 000, on average, for each part per million of calcium. The intercept indicates a baseline mortality of 1676 deaths per 100 000 with no calcium, but this is extrapolation.
d) Exeter has 348.6 fewer deaths per 100 000 than the model predicts.
e) 1353 deaths per 100 000
f) Calcium concentration explains 43.0% of the variation in death rate per 100 000 people.

49. a) The association between speed and stopping distance is strong, positive, and appears straight. Higher speeds are generally associated with greater stopping distances. The linear regression model, with equation $\widehat{Stopping\ Distance} = -20.667 + 1.151\,(Speed)$. The model explains 96.7% of the variability in stopping distance. However, the residuals plot has a curved pattern. The linear model is not appropriate. A model using re-expressed variables should be used.

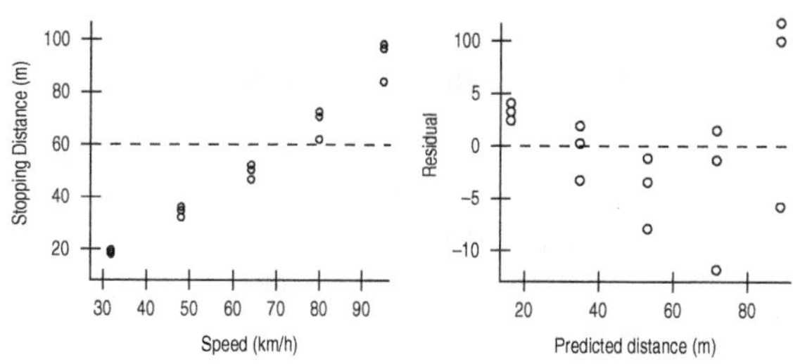

b) Stopping distances appear to be relatively higher for higher speeds. This increase in the rate of change might be able to be straightened by taking the square root of the response variable, stopping distance.
c) The model for the re-expressed data is $\widehat{\sqrt{Stopping\ distance}} = 1.777 + 0.08226\,(Speed)$. The residuals plot is better than the linear fit before re-expressing the data, and shows no pattern. 98.4% of the variability in the square root of the stopping distance can be explained by the model.

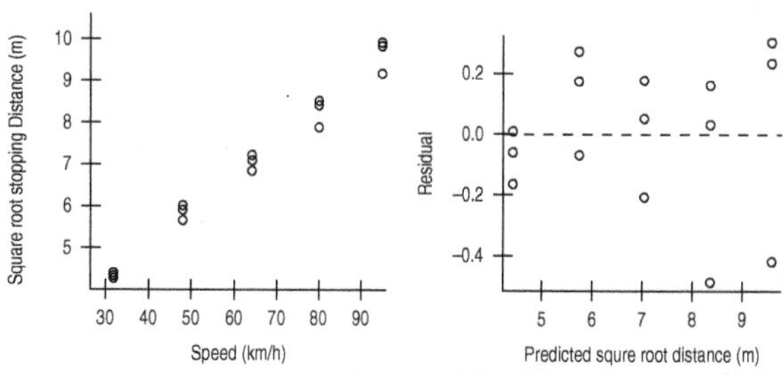

d) According to the model, a car traveling 90 km/h is expected to require approximately 84.3 metres to come to a stop.

51. a)

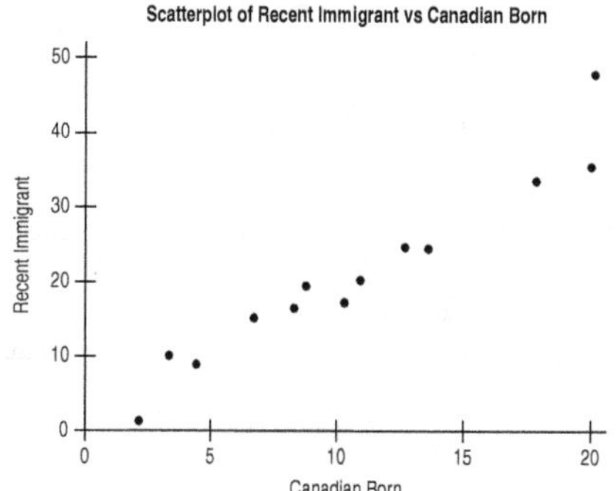

There is a strong positive linear relationship between the immigrant percentage and the Canadian born percentage. No outliers.
b) The fitted line is given below.

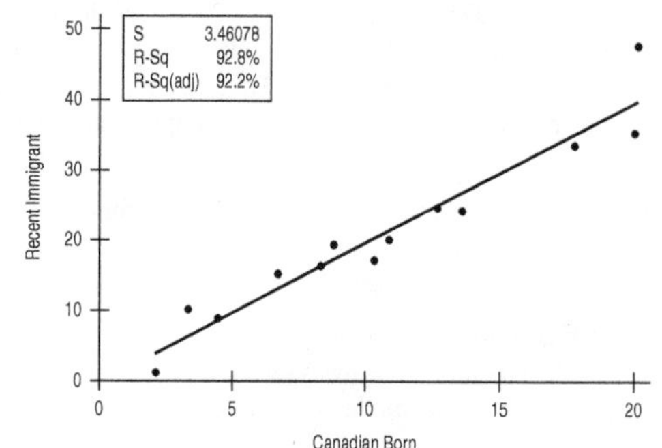

c) 1.99% (This is the slope.)
d) Montreal has a large residual compared to other observations. The predicted value for Montreal is 39.8 but the actual value is 47.8.
e) The plot of residuals vs. the Canadian born % is given below. It doesn't look quite random. For example, most of the points after Canadian born %, 10% are negative. The model overestimates the immigrant percentage for cities with a Canadian born % of 10 or more. The plot of residuals vs. Canadian born % without Montreal is given below. This problem is much less, but the variability of residuals seems to decrease with Canadian born %.

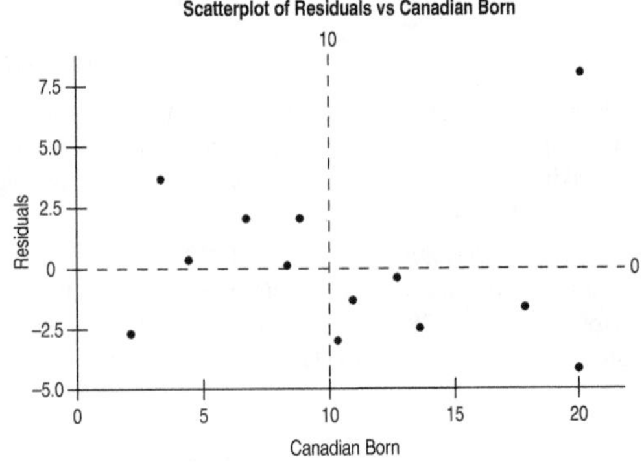

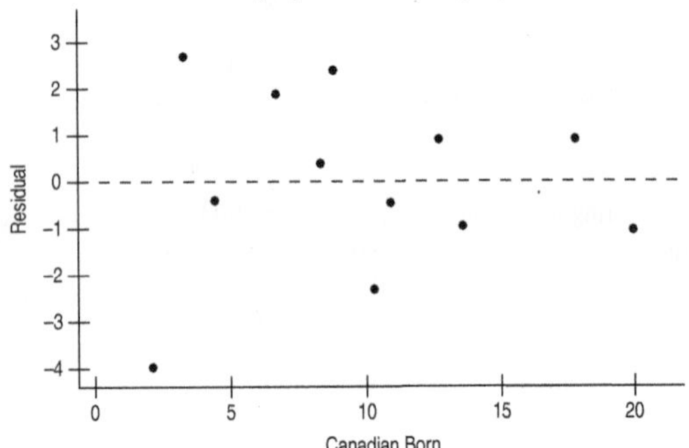

53. a) 79 b) 75.4
c) For a test score of 85, you predict an exam score of 79, but if the exam score is 79, you predict a test score of 75.4, not 85. Predictions are not reversible in direction. The regression model changes when the response does, and our prediction is always nearer the mean.

Chapter 8

1. a) The trend appears to be somewhat linear up to about 1940, but from 1940 to about 1970 the trend appears to be nonlinear and slightly positive. From 1975 to the present, the trend appears linear and positive.
b) Relatively strong for certain periods.
c) No, as a whole the graph is clearly nonlinear. However, certain time periods, like 1975 to present, have a high correlation.
d) Overall, no. You could fit a linear model to the period from 1975 to 1995, but this seems unnecessary. The ages for each year are reported, and, given the fluctuations in the past, extrapolation seems risky.

3. a) Fairly linear, negative, strong.
b) Gas mileage decreases an average 7.652 mpg for each thousand pounds of weight.
c) No. Residuals show a curved pattern.

5. a) No. We need to see the scatterplot first to see if the conditions are satisfied, and models are always wrong.
b) No, the linear model might not fit the data at all.

7. a) Millions of dollars per minute of run time.
b) Costs for longer movies increase at the same rate.
c) Regardless of run time, dramas cost less by about $20 million.

9. a) The use of Oakland airport has been growing at about 59 700 passengers/year, starting from about 282 000 in 1990.
b) 71% of the variation in passengers is accounted for by this model.
c) Errors in predictions based on this model have a standard deviation of 104 330 passengers.
d) No, that would extrapolate too far from the years we've observed.
e) The negative residual is September 2001. Air traffic was artificially low following the attacks on 9/11.

11. a) 1) The point has high leverage and a small residual.
2) The point is not influential. It has the *potential* to be influential, because its position far from the mean of the explanatory variable gives it high leverage. However, the point is not *exerting* much influence, because it reinforces the association.
3) If the point were removed, the correlation would become weaker. The point heavily reinforces the positive association. Removing it would weaken the association.
4) The slope would remain roughly the same, since the point is not influential.

b) 1) The point has high leverage and probably has a small residual.
2) The point is influential. The point alone gives the scatterplot the appearance of an overall negative direction, when the points are actually fairly scattered.
3) If the point were removed, the correlation would become weaker. Without the point, there would be very little evidence of linear association.
4) The slope would increase, from a negative slope to a slope near 0. Without the point, the slope of the regression line would be nearly flat.

c) 1) The point has moderate leverage and a large residual.
2) The point is somewhat influential. It is well away from the mean of the explanatory variable, and has enough leverage to change the slope of the regression line, but only slightly.
3) If the point were removed, the correlation would become stronger. Without the point, the positive association would be reinforced.
4) The slope would increase slightly, becoming steeper after the removal of the point. The regression line would follow the general cloud of points more closely.

d) 1) The point has little leverage and a large residual.
2) The point is not influential. It is very close to the mean of the explanatory variable, and the regression line is anchored at the point $(\bar{x},\bar{y})$, and would only pivot if it were possible to minimize the sum of the squared residuals. No amount of pivoting will reduce the residual for the stray point, so the slope would not change.
3) If the point were removed, the correlation would become slightly stronger, decreasing to become more negative. The point detracts from the overall pattern, and its removal would reinforce the association.
4) The slope would remain roughly the same. Since the point is not influential, its removal would not affect the slope.

13. 1) Point e is very influential. Its addition will give the appearance of a strong, negative correlation like $r = -0.90$.
2) Point d is influential (but not as influential as point e). Its addition will give the appearance of a weaker, negative correlation like $r = -0.40$.
3) Point c is directly below the middle of the group of points. Its position is directly below the mean of the explanatory variable. It has no influence. Its addition will leave the correlation the same, $r = 0.00$.
4) Point b is almost in the centre of the group of points, but not quite. Its addition will give the appearance of a very slight positive correlation like $r = 0.05$.
5) Point *a* is very influential. Its addition will give the appearance of a strong, positive correlation like $r = 0.75$.

15. Perhaps high blood pressure causes high body fat, high body fat causes high blood pressure, or both could be caused by a lurking variable such as a genetic or lifestyle issue.

17. a) The graph shows that, on average, students progress at about one reading level per year. This graph shows averages for each grade. The linear trend has been enhanced by using averages.
b) Very close to 1.
c) The individual data points would show much more scatter, and the correlation would be lower.
d) A slope of 1 would indicate that for each 1-year grade level increase, the average reading level is increasing by 1 year.

19. a) *Cost* decreases by \$2.13 per degree of average daily *Temp*. So warmer temperatures indicate lower costs.
b) For an avg. monthly temperature of 0°F, the cost is predicted to be \$133.
c) Too high; the residuals (observed – predicted) around 32°F are negative, showing that the model overestimates the costs.
d) \$111.70 e) About \$105.70
f) No, the residuals show a definite curved pattern. The data are probably not linear.
g) No, there would be no difference. The relationship does not depend on the units.

21. a) 0.88
b) Interest rates during this period grew at about 0.25% per year, starting from an interest rate of about 0.64%.
c) Substituting 50 in the model yields a predicted of about 13%.
d) Not really. Extrapolating 20 years beyond the end of these data would be dangerous and unlikely to be accurate.

23. a) The two models fit comparably well, but they have very different slopes.
b) This model predicts the interest rate in 2000 to be 3.24%, much lower than the other model predicts.
c) We can trust the new predicted value because it is in the middle of the data used for the regression.
d) The best answer is "I can't predict that."

25. a) Stronger. Both slope and correlation would increase.
b) Restricting the study to nonhuman animals would justify it.
c) Moderately strong.
d) For every year increase in life expectancy, the gestation period increases by about 15.5 days, on average.
e) About 270.5 days.

27. a) Removing hippos would make the association stronger, since hippos are more of a departure from the pattern.
b) Increase.
c) No, there must be a good reason for removing data points.
d) Yes, removing it lowered the slope from 15.5 to 11.6 days per year.

29. a) Modelling decisions may vary, but the important idea is using a subset of the data that allows us to make an accurate prediction for the year in which we are interested. We might model a subset to predict the marriage age in 2015, and model another subset to predict the marriage age in 1911. The linear model used to predict average female marriage age from year is: $Age = -210.827 + 0.127832(Year)$. According to the model, the average age at first marriage for women in 2020 will be 27.19 years old. Care should be taken with this prediction, however. It represents an extrapolation of 10 years beyond the highest year, and the residuals plot shows a pattern.
b) This prediction is for a year that is 10 years higher than the highest year for which we have an average female marriage age. Don't place too much faith in this extrapolation.
c) An extrapolation of more than 50 years into the future would be absurd. There is no reason to believe the trend would continue. In fact, given the situation, it is very unlikely that the pattern would continue in this fashion. The model given in part a) predicts that the average marriage age will be 32.5 years in 2065. Realistically, that seems quite high.

31. a) The scatterplot of birth rate and life expectancy is below. The association is moderate, linear, and negative. Countries with higher birth rates tend to have lower life expectancies. There is one outlier, Paraguay, with a birthrate of 28 births per 1000 people, and a life expectancy of 76 years.

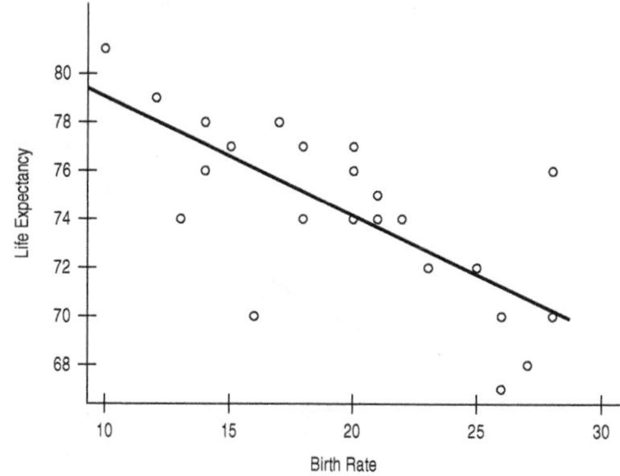

b) $LifeExpectancy = 83.957 - 0.487BirthRate$
c) 53.3% of the variability in life expectancy is explained by variability in the birthrate.
d) The residuals plot, below, is reasonably scattered. The linear model is appropriate.

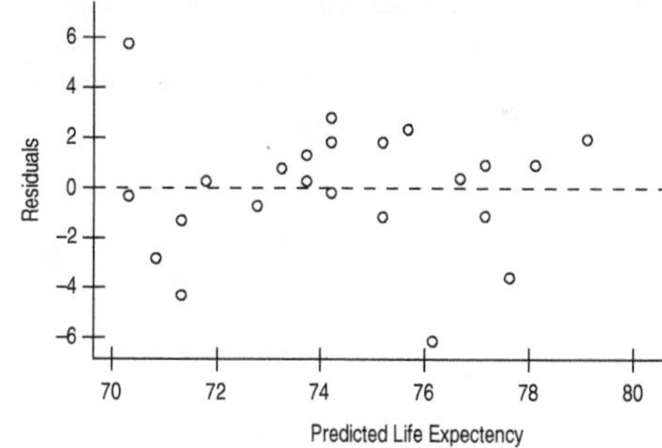

e) Paraguay has a large residual. Its higher than average life expectancy continues to stand out.
f) The data point for Paraguay is not extraordinarily unusual. You may have chosen not to set it aside. If you did set it aside, the recomputed regression is
$LifeExpectancy = 85.296 - 0.57BirthRate$. $R^2 = 65.4\%$.
g) The government leaders should not suggest that women have fewer children in order to raise the life expectancy. Although there is evidence of an association between the birth rate and life expectancy, this does not mean that one causes the other. There may be lurking variables involved, such as economic conditions, social factors, or level of health care.

33. a) The scatterplot of the percent earning over \$150 000 versus year is given below. The percent earning over \$150 000 increases over time, but not quite linearly. In the period 1986–1992, the relationship has a different form than that in the period 1993–2011. The relationship in the period 1993–2011 is strong, and roughly linear.

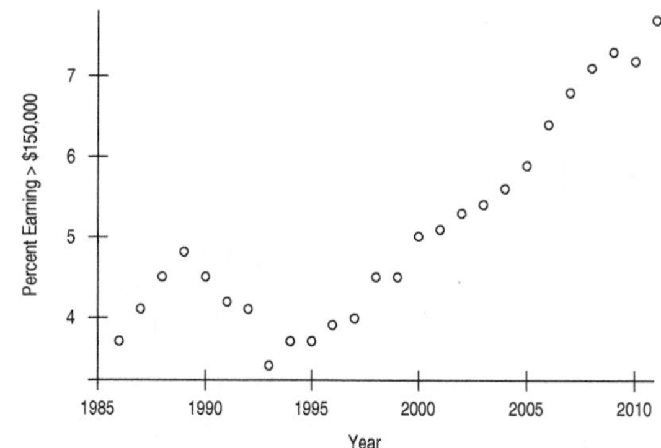

b) The regression equation is: $\widehat{Percent > \$150000} = -288 + 0.14667\,Year$. 76.03% of the variation in the percentage earning over \$150 000 is explained by the straight line relationship (this is R^2).

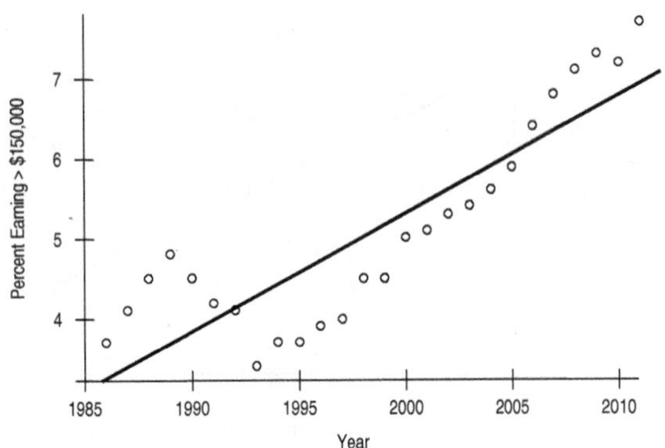

c) The predicted percent for 1997 is $-288 + 0.14667(1997) = 4.87\%$. The actual value (in the data set) is 4.0%. The predicted value is a bit too high—the residual is −0.87%.

d) Dropping the first seven years (from 1986 to 1992), we obtain the following model: $\widehat{Percent > \$150000} = -481 + 0.2428\,Year$. 98.4% of the variation in the percentage earning over \$150 000 is explained by the linear relationship (this is R^2). This model fits the later data better, and is more appropriate because the trend after 1992 appears to be close to linear.

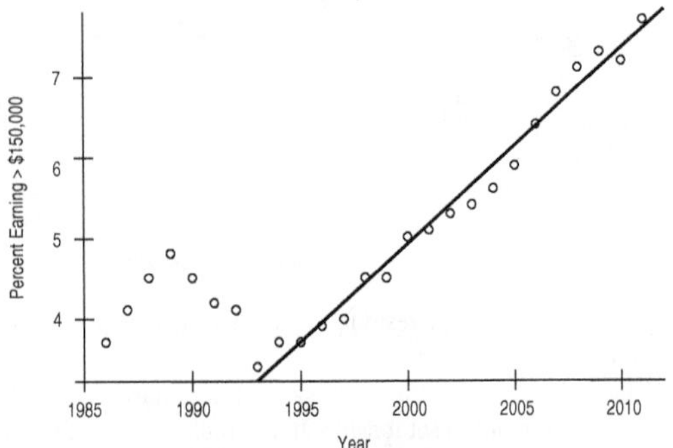

The predicted percent for 2013 is $-481 + 0.2428(2013) = 8.066\%$. The predicted percent for 2040 is $-481 + 0.2428(2040) = 14.62\%$. Both 2013 and 2040 are out of the period for which we had information (to estimate the model) and so both these are extrapolations. 2040 is way too far from this period (1986–2001) and so the predicted value is not very reliable. We should not be very confident that the linear increase will continue until 2040 given the earlier data from 1986-1992, but the prediction for 2013 might be closer to accurate.

35. a) The scatterplot of the difference in median earnings versus year is given below. The difference decreases with year, but not quite linearly. The relationship is strong with no outliers.

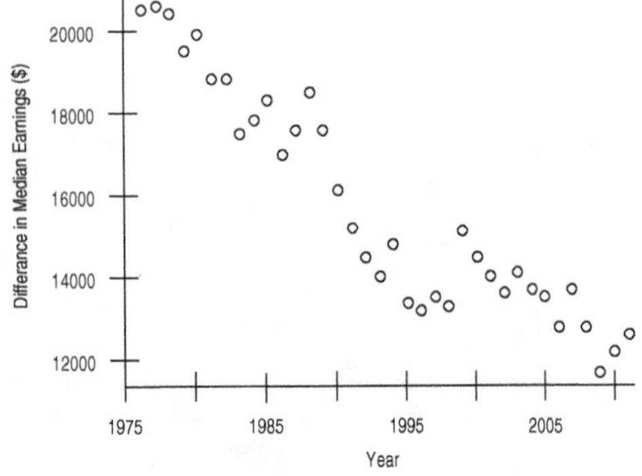

b) The linear regression model (the straight line fit) is: Difference in Median earnings = 497806 − 241.84(Year). The R^2 for the model is 87.9%, i.e. 87.9% of the variation in the differences in medians is explained by this model. The plot of residuals versus year is given below. The curving pattern on this plot indicates that a higher order model (e.g., a quadratic model) or some other curving pattern is needed to describe the relationship.

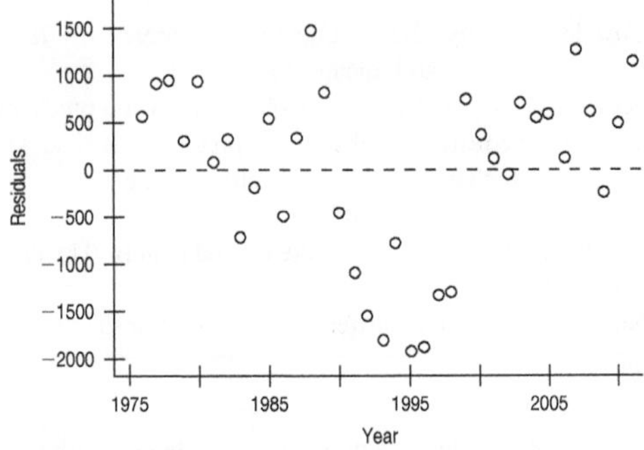

c) The predicted difference for 2070 (assuming that the form of the relationship will remain the same after 2011, and that the linear model is appropriate) is $497806 - 241.84 \times 2070 = -2804$. The difference is likely to be higher than this as we can see that the decrease in the difference reduces with time.

d) It looks like a quadratic curve fits the data better.

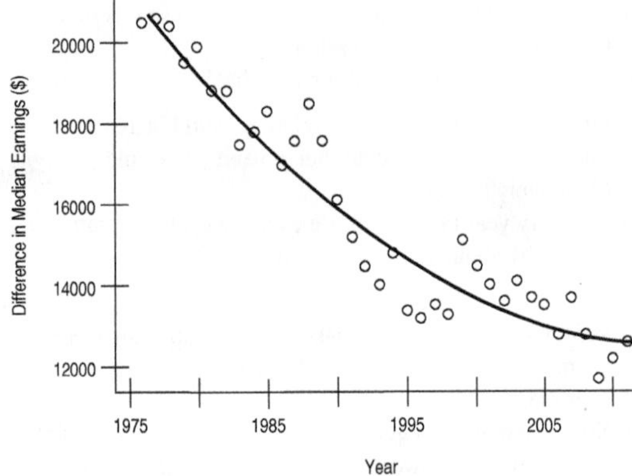

e) The plot of residuals for the quadratic model is much more random (free of patterns) than that for the linear model and so this is a more appropriate model than the linear model.

37. a) The plot of male mean score versus female mean score with the regression is shown below.

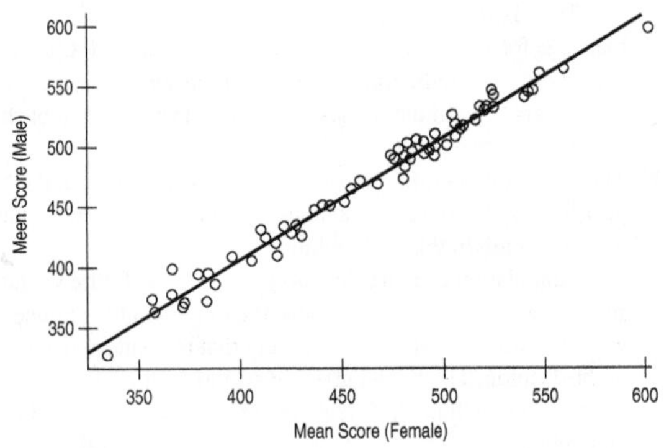

b) The plot of residuals versus the mean scores for females is shown below. The plot looks random (free of patterns) and so a straight line model seems appropriate. There is one large residual whose magnitude is greater than 20 points (Colombia).

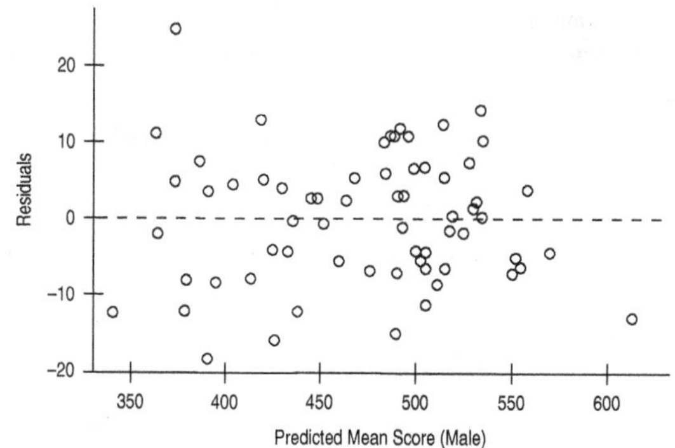

c) The histogram and the normal probability plot of residuals are shown below. The histogram looks roughly bell shaped and the normal probability plot is close to a straight, both suggesting that the residuals are approximately normally distributed. Colombia is noticeable in the right tail of the distribution.

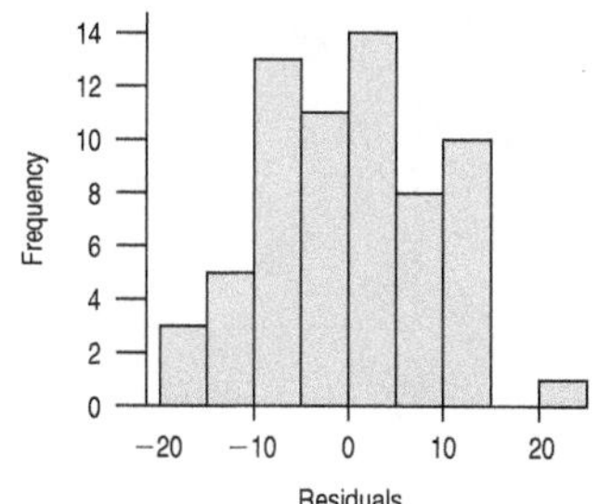

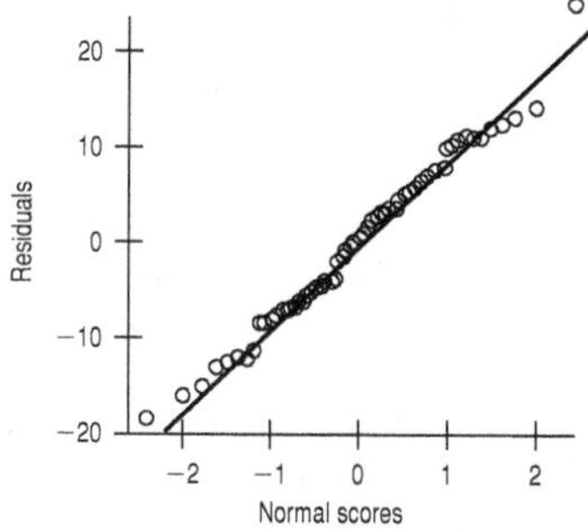

d) The plot of residuals versus row number is shown below. It seems that the OECD countries are less variable than the non-OECD countries.

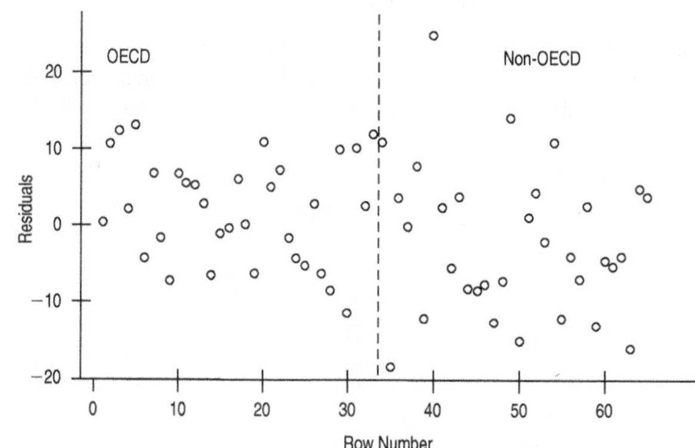

e) Shanghai-China and Kyrgyzstan have the greatest leverage on the fit. Their x-values (i.e., the means for females) are far from the average of all x-values. The average female scores in Shanghai-China is 601 and the average female score in Kyrgyzstan is 334.

f) Slope close to 1 implies when the a country's average for females increases by a certain value the country average for males also increases by approximately the same value.

When the slope is 1, the estimate of the Y-intercept is given by $\overline{Y} - \overline{X} =$ Average for males – Average for females = 471.98 – 463.18 = 8.80, so the estimated prediction equation is: Mean score (Male) = 8.80 + Mean score (Female).

g) The correlation will decrease as we can see from the scatter plot that the points with both male and female averages greater than 500 look less linear. The other points following the same pattern on the scatterplot make it look more linear and so add strength to the linear relationship.

39. a) Correlations: Year, Bust size (in), Waist size (in), Hips size (in)

	Year	Bust size	Waist size
Bust size	1.000		
Waist size	0.992	0.992	
Hips size	0.936	0.936	0.929

b) These values are averages. The correlations with averages are usually high since they do not show the variability in individual observations.

c) The prediction equation is Waist size (in) = –182 + 0.105 (Year). The predicted waist size for 1950 is 22.75 and that for 2080 is 36.40 inches. The year 2080 is way outside the period for which we have information and so the prediction for 2080 is not very reliable (extrapolation).

d) The prediction equation is Waist size (in) = –47.6 + 2.10 (Bust size (in)). The predicted waist size for UK women if their average bust size gets up to 44 is 44.8. The average bust size is outside the range of data on bust sizes and so this prediction is not very reliable.

41. a) *Arm span* is the best predictor of *Middle finger*. 32% of the variation in *Middle finger* measurements can be explained by a linear association *Arm span*.

b) *Height* is the best predictor of *Arm span*. 61% of the variation in *Arm span* measurements can be explained by a linear association *Height*.

c) The prediction equation is *Arm span* = 16.8 + 0.897 *Height*. The slope of this regression equation is 0.987. That means for every unit in increase in *Height*, the mean *Arm span* increases by 0.987 units.

d)

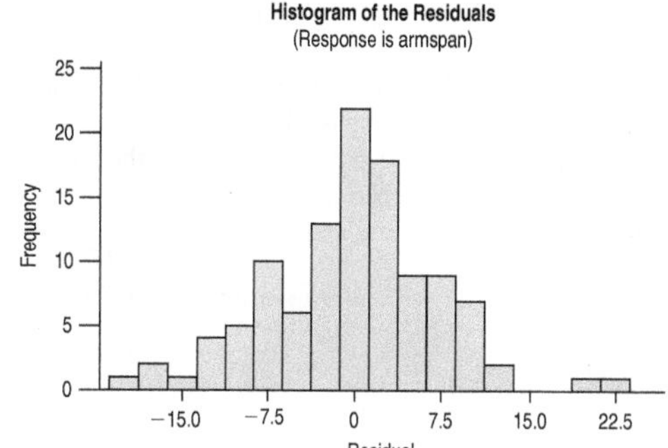

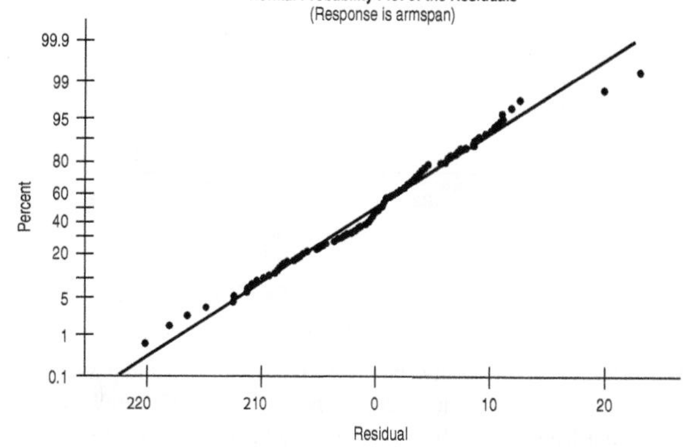

The histogram is close to a bell-shape and the normal scores plot is close to a straight, both indicating that there are no serious violations of the assumption of normality of residuals.

e) The plot of residuals against height and the plot of residuals against predicted values are shown below. They look pretty random, indicating no series violations in the assumption of the assumption of independents of error terms.

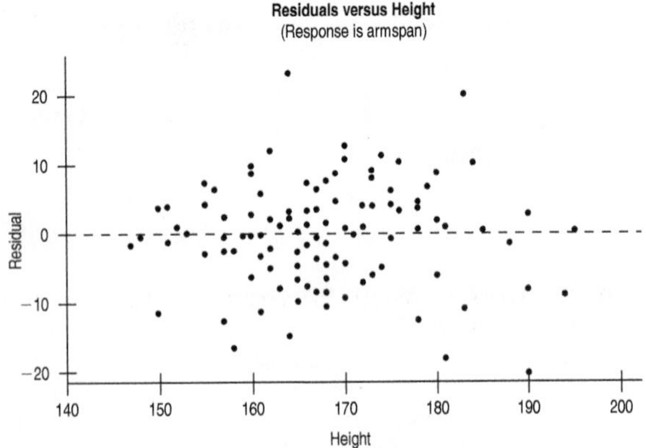

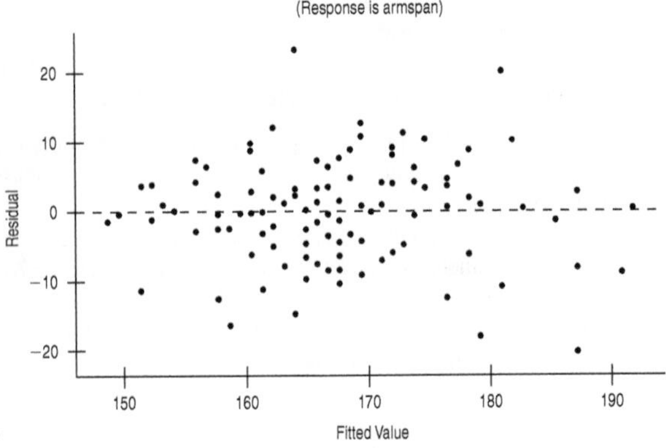

f) The individual with height 164 and arm span 187 (i.e., observation number 75) has the most unusual arm span (3.13 units longer than the average arm span for individuals with height = 164).

g) The plot of residuals versus gender (with genderCode=0 for girls and 1 for boys) is shown below.

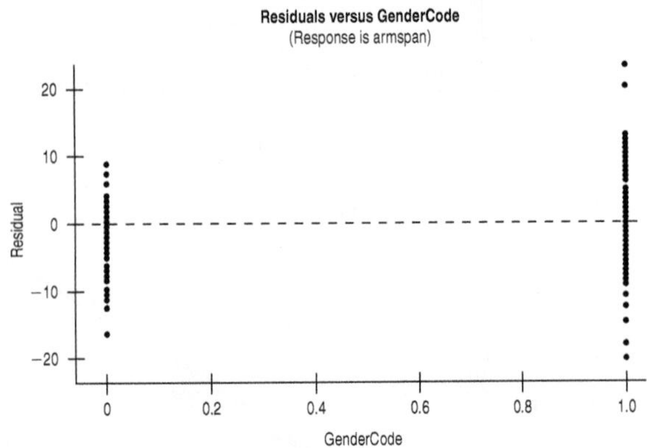

The residuals are mostly negative for girls in dilating that this regression model often overestimates the arm span of girls.

43. a) The association between distance from the sun and planet year is strong, positive, and curved concave upward. Generally, planets farther from the sun have longer years than closer planets.

b) The rate of change in length of year per unit distance appears to be increasing, but not exponentially. Re-expressing with the logarithm of each variable may straighten a plot such as this:

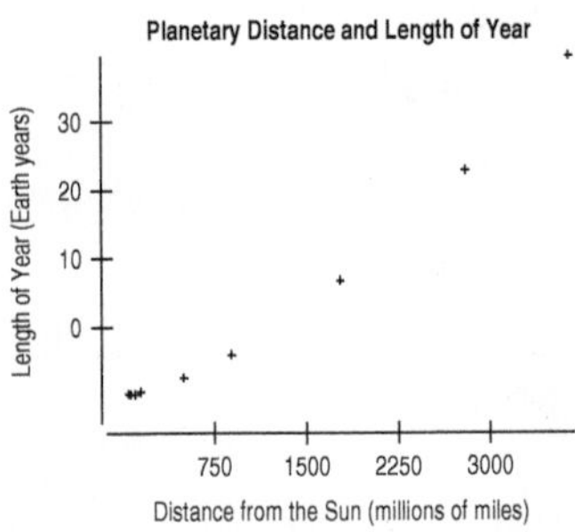

The scatterplot and residuals plot for the linear model relating log(Distance) and log(Length of Year) appear below.

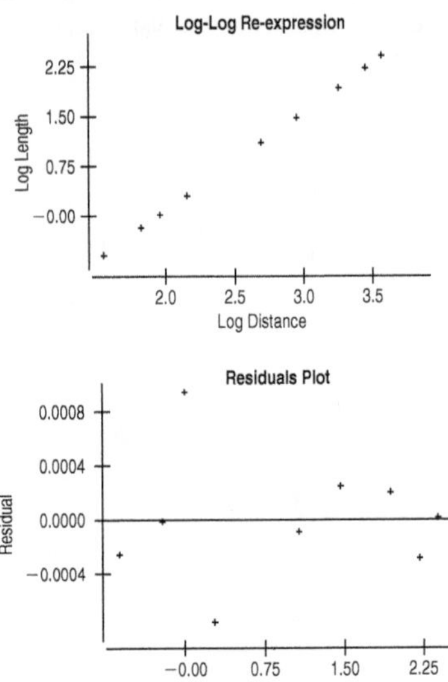

The regression model for the log-log re-expression is:

$\widehat{\log(Length)} = -2.95 + 1.5(\log(Distance))$.

c) The model explains 100% of the variability in the log of the length of the planetary year, at least according to the accuracy of the statistical software. The residuals plot is scattered, and the residuals are all extremely small. This is a very accurate model.

45. a) The value of R^2 is 60.6%. That is, 60.6% of the variation in the neonatal brain weight is explained by the linear regression model. The plot of residuals versus predicted values shows a curing pattern indicating that a straight line model is not appropriate. This can also be seen on the scatterplot of the neonatal brain weight versus gestation, but the curvature is more prominent in the plot of residual versus predicted values.

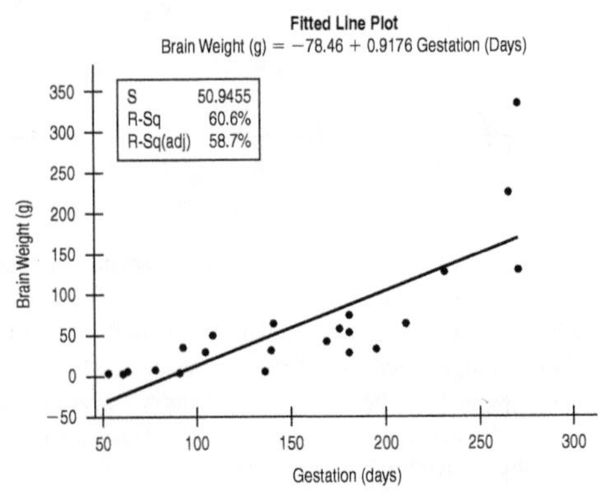

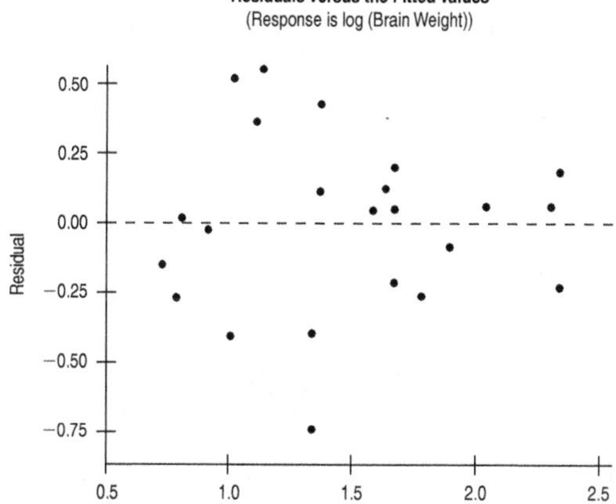

b) The value of R^2 is 71.6%. That is, 71.6% of the variation in the logarithms of the neonatal brain weight is explained by this regression model. The plot of residuals versus predicted values looks random (free of patterns), indicating that this model is appropriate for describing the relationship between the two variables.

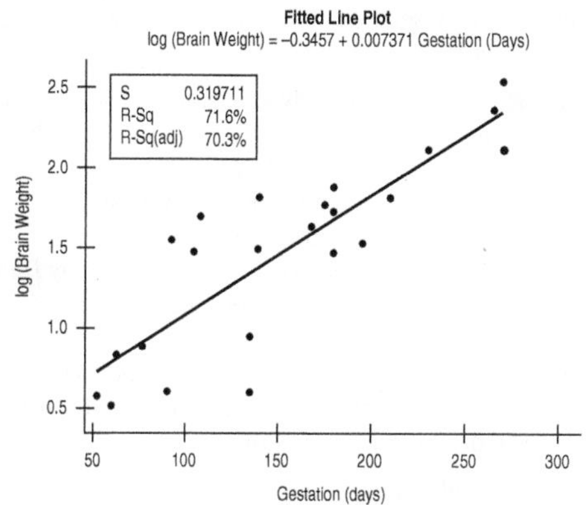

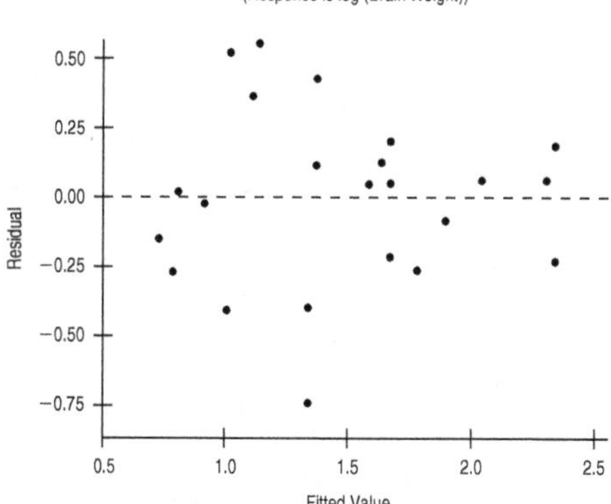

c) The plot of the square roots of the neonatal brain weight versus gestation with the regression line is shown below. The value of R^2 for this model is 74.09%. That is, 74.09% of the variation in the square roots of the neonatal brain weight is explained by this regression model. The plot of residuals versus predicted values is also given below. It looks like the variability of residuals increases with the fitted value (and also the plot of residuals versus predicted values for the log transformation looks a bit more random than that for the square root model) and so the log transformation is a bit better (even though the R^2 is a bit higher for the square root transformation).

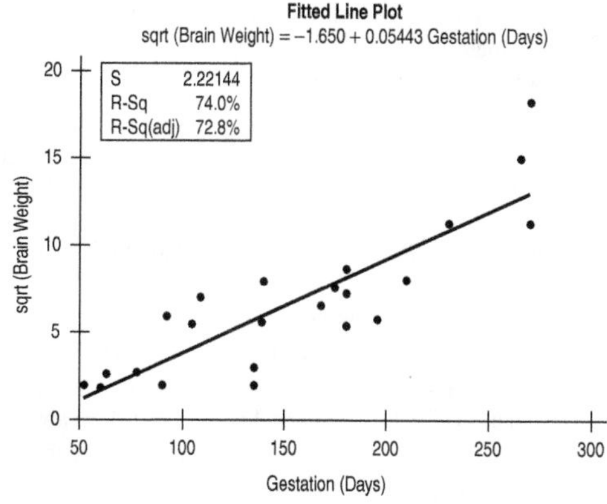

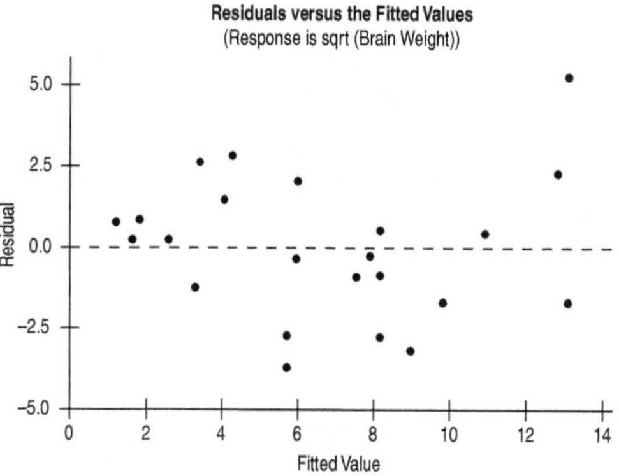

d) The predicted log brain weight for a 200-day gestation period is 1.82 and so the predicted brain weight is 66.1g.

e) The predicted square root brain weight for a 200-day gestation period is 9.23 and so the predicted brain weight is 85.19g.

Part II Review

1. % over 50: $r = 0.69$
% under 20: $r = -0.71$
% Full-time Fac.: $r = 0.09$
% Gr. on time: $r = -0.51$

3. a) There does not appear to be a linear relationship.
b) Nothing, there is no reason to believe that the results for the Finger Lakes region are representative of the vineyards of the world.
c) $\widehat{CasePrice} = 92.765 + 0.567284(Years)$
d) Only 2.7% of the variation in case price is accounted for by the ages of vineyards. Most of that is due to two outliers. We are better off using the mean price rather than this model.

5. a) (i) The correlation is negative, approximately $r = -0.7$.
(ii) The correlation is negative, approximately $r = 20.85$.
(iii) The correlation is essentially zero.
(iv) The correlation is negative, approximately $r = 20.75$.
b) (i) The correlation is positive, approximately $r = 1$.
(ii) The correlation is essentially zero.
(iii) The correlation is positive, approximately $r = 0.8$.
(iv) The correlation is essentially zero.

7. a) −0.52
b) Negative, moderate, linear, but with more variation as pH increases.
c) The BCI would also be average.
d) The predicted *BCI* will be 1.56 SDs of BCI below the mean *BCI*.

9. a) $\widehat{Kills} - 50.016 + 0.139245 ThousandBoats$
b) For every additional 10 000 powerboats registered, the model predicts that an additional 1.392 manatees will be killed on average.

c) The model predicts that if no powerboats were registered, the number of manatee deaths would be approximately –50.016. This is an extrapolation beyond the scope of the data, and doesn't have much contextual meaning.
d) The predicted number is 77.25 kills. The actual number of kills was 83. The model under-predicted the number of kills by 5.75.
e) A negative residual suggests that the actual number of kills was below the number of kills predicted by the model.
f) Over time, the number of powerboat registrations has increased and the number of manatees killed has increased. The trend may continue, resulting in a greater number of manatee deaths in the future. Extrapolation is risky, however. The very trend we are seeing may result in political or societal action to attempt to decrease the number of manatee deaths.

11. a) 20.984 b) 96.9% c) 32.95 mph d) 1.66 mph
e) Slope will increase.
f) Correlation will weaken (become less negative).
g) Correlation is the same, regardless of units.

13. a) A scatterplot of the 2001 versus 1996 census counts is shown below. The correlation appears to be very close to 1. In fact, $r = 0.997$.

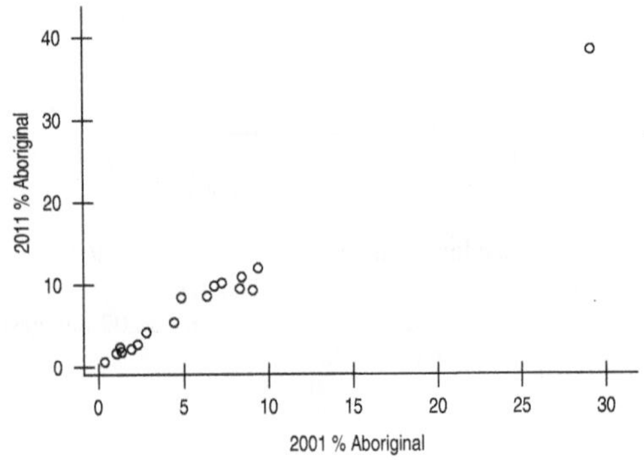

b) The fitted least squares line $\widehat{2011Percent} = 0.19439 + 1.29105(2001Percent)$.
c) The slope tells us that for each increase of 1% in the percentage of Aboriginals in 2001 there is a 1.29% increase in the percentage of Aboriginals in 2011.
d) Prince Albert has the highest percentage of Aboriginals. It also has the highest leverage of all census areas. However, if you set Prince Albert aside, there appear to be two large outliers (or perhaps even a pattern in the residuals) so that the linear model no longer seems appropriate.

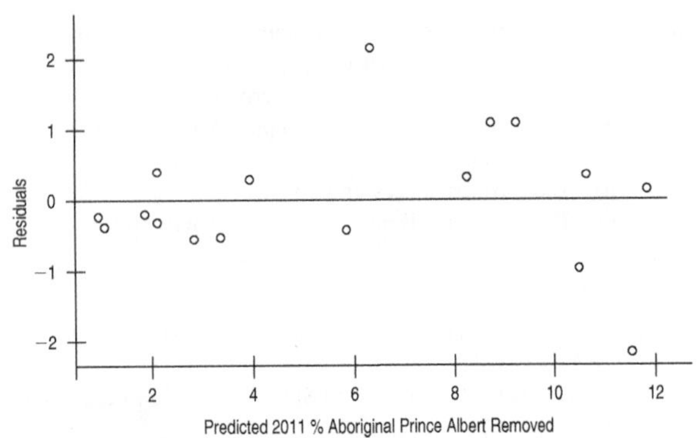

e) Taking logs seems to have resolved the issue. The new model is $\widehat{\log_{10}(2011Percent)} = 0.21015 + 0.89062 \log_{10}(2001Percent)$.

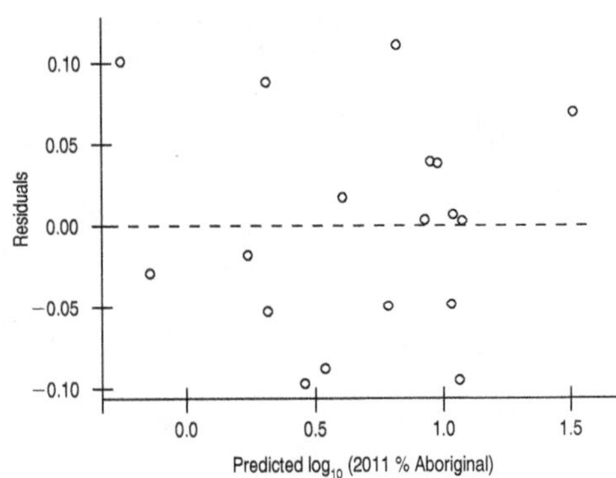

f) 12.6196

15. a) $\widehat{Horsepower} = 3.49834 + 34.3144(Weight)$
b) The weight is measured in thousands of pounds. The slope of the model predicts an increase of about 34.3 horsepower for each additional unit of weight. 34.3 horsepower for each additional thousand pounds makes more sense than 34.3 horsepower for each additional pound.
c) Since the residuals plot shows no pattern, the linear model is appropriate for predicting horsepower from weight.
d) According to the model, a car weighing 2595 pounds is expected to have 92.543 horsepower. The actual horsepower of the car is 115.0 horsepower.

17. a) The scatterplot shows a fairly strong, linear and positive relation. There seem to be two distinct clusters of data.
b) $\widehat{Interval} = 33.9668 + 10.3582(Duration)$
c) The time between eruptions increases by about 10.4 minutes per minute of *Duration* on average.
d) 77% of the variation in *Interval* is accounted for by *Duration*. The prediction is fairly accurate but not precise.
e) 75.4 minutes
f) A residual is the observed value minus the predicted value. So the residual = 79 – 75.4 = 3.6 minutes, indicating that the model under-estimated the interval in this case.

19. a) $r = 0.888$. Although r is high, you must look at the scatterplot and verify that the relation is linear in form.
b)

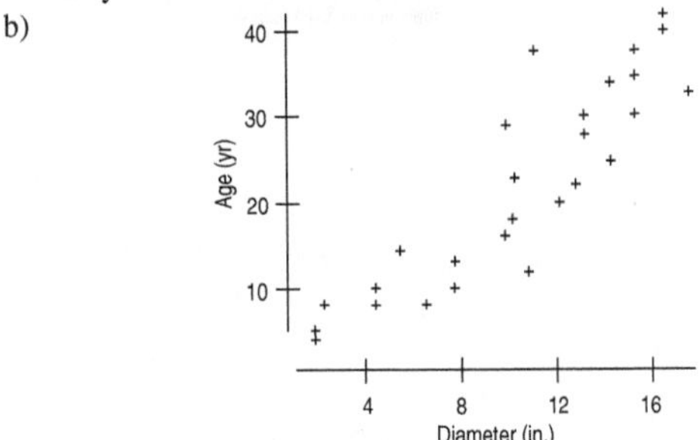

The association between diameter and age appears to be strong, somewhat linear, and positive.
c) $\widehat{Age} = -0.97 + 2.21 \times Diameter$
d)

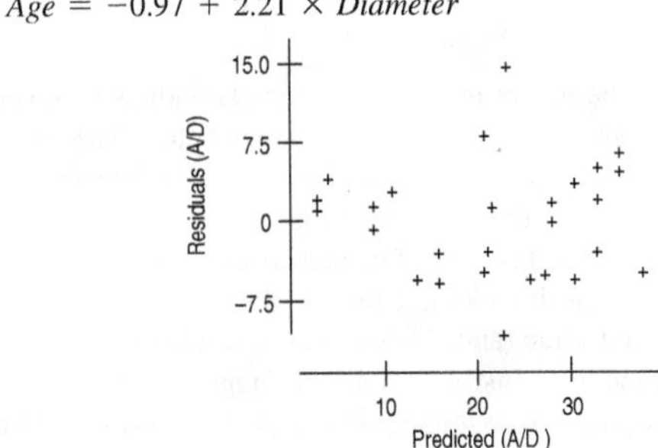

The residuals show a curved pattern (and two outliers).

e) The residuals for five of the seven largest trees (15 in. or larger) are positive, indicating that the predicted values underestimate the age.

21. Most houses have areas between 1000 and 5000 square feet. Increasing 1000 square feet would result in either 1000(.008) = 8 thousand dollars, 1000(.08) = 80 thousand dollars, 1000(.8) = 800 thousand dollars, or 1000(8) = 8000 thousand dollars. Only $80 000 is reasonable, so the slope must be 0.08.

23. The best one was the reciprocal square root re-expression, resulting in the equation $\frac{1}{\sqrt{DrainTime}} = 0.00243 + 0.219(Diameter)$.

25. There is a strong, roughly linear, negative association between mean January temperature and latitude. U.S. cities with higher latitudes generally have lower mean January temperatures. There are two outliers, cities with higher mean January temperatures than the pattern would suggest.

27. a) 71.9%
b) The negative correlation indicates that as the latitude increases, the average January temperature generally decreases.
c) $JanTemp = 108.796 - 2.111(Latitude)$
d) For each additional degree of latitude, the model predicts a decrease of approximately 2.1°F in average January temperature.
e) The model predicts that the mean January temperature will be approximately 108.8 °F when the latitude is 0°. This is an extrapolation, and may not be meaningful.
f) 24.4°F
g) In this context, a positive residual means that the actual average temperature in the city was higher than the temperature predicted by the model. In other words, the model underestimated the average January temperature.

29. a) The scatterplot shows a strong, linear, positive association.
b) There is an association, but it is likely that training and technique have increased over time and affected both jump performances.
c) Neither; the change in units does not affect the correlation.
d) The long jumper would jump 0.92 SDs above the mean long jump on average.

31. a) No relation; the correlation would probably be close to 0.
b) The relation would have a positive direction and the correlation would be strong, assuming students were studying French in each grade level. Otherwise, no correlation.
c) No relation; correlation close to 0.
d) The relation would have a positive direction and the correlation would be strong, since vocabulary would increase with each grade level.

33. $\widehat{Calories} = 560.7 - 3.08(Time)$
Each minute extra at the table results in 3.08 fewer calories being consumed on average. Perhaps the hungry children eat fast and eat more.

35. There seems to be a strong, positive, linear relationship with one high-leverage point (Northern Ireland) that makes the overall R^2 quite low. Without that point, the R^2 increases to 61.5%. Of course, these data are averaged across thousands of households, and so the correlation appears to be higher than individuals would be. Any conclusions about individuals would be suspect.

37. a) 30 818 pounds
b) 1302 pounds
c) 31 187.6 pounds
d) I would be concerned about using this relation if we needed accuracy closer than 1000 pounds or so, as the residuals are more than ±1000 pounds.
e) Negative residuals will be more of a problem, as the predicted weight would overestimate the weight of the truck; trucking companies might be inclined to take the ticket to court.

Chapter 9

1. Yes. You cannot predict the outcome beforehand.

3. a) Yes, who takes out the trash cannot be predicted before the flip of a coin.
b) No, it is not random, since you will probably name a favourite team.
c) Yes, your new roommate cannot be predicted before names are drawn.

5. Rolling the pair of dice is the component.

7. To simulate, you could roll dice and note whether or not "doubles" came up. A trial would be completed once "doubles" came up. You would count up the number of rolls until "doubles" for the response variable. Alternatively, you could use the digits 1, 2, 3, 4, 5, 6 on random digits table and disregard digits 7, 8, 9, and 0. Using the table, note the first legal digit for the first die, and then note the next legal digit for the second die. A double would indicate rolling doubles.

9. A machine pops up numbered balls. If it were truly random, the outcome could not be predicted and the outcomes would be equally likely. It is random only if the balls generate numbers in equal probabilities.

11. Use two-digit numbers 00–99; let 00–02 = defect, 03–99 = no defect

13. a) 06 Northwest Territories and 05, Newfoundland.
b) 08, Nunavut. We need more random digits to find the next province.

15. If the lottery is random, it doesn't matter which number you play; all are equally likely to win.

17. a) The outcomes are not equally likely; for example, tossing 5 heads does not have the same probability as tossing 0 or 9 heads, but the simulation assumes they are equally likely.
b) The even–odd assignment assumes that the player is equally likely to score or miss the shot. In reality, the likelihood of making the shot depends on the player's skill.
c) Suppose a hand has 4 aces. This might be represented by 1, 1, 1, 1, and any other number. The likelihood for the first ace in the hand is not the same as for the second or third or fourth. But with this simulation, the likelihood is the same for each.

19. The conclusion should indicate that the simulation *suggests* that the average length of the line would be 3.2 people. Future results might not match the simulated results exactly.

21. a) The component is one voter voting. An outcome is a vote for our candidate or not. Use two random digits, giving 00–54 a vote for your candidate and 55–99 for the underdog.
b) A trial is 100 votes. Examine 100 two-digit random numbers, and count how many people voted for each candidate. Whoever gets the majority of votes wins that trial.
c) The response variable is whether the underdog wins or not.

23. Answers will vary. According to the simulation, the probability of getting a complete set of pictures is expected to be about 51.5%.

25. Answers will vary. According to the simulation, the probability of getting all 6 multiple-choice questions correct is expected to be about 26%.

27. a) Answers will vary, but you should win about 10% of the time.
b) You should win at the same rate with any number.

29. Answers will vary, but you should win about 10% of the time.

31. Answers will vary. According to the simulation, the number of driving tests required to pass is expected to be about 1.9.

33. Answers will vary. According to the simulation, the player is expected to score about 1.24 points.

35. Do the simulation in two steps. First simulate the payoffs. Then count until $500 is reached. Answers will vary, but average should be near 10.2 customers.

37. Answers will vary. According to the simulation, the expected number of children in the family is about 3.

39. Answers will vary. According to the simulation, expect to roll the die about 7.5 times.

41. Answers may vary. According to simulation, the underdog is expected to win the World Series about 39% of the time.

43. Answers will vary. According to the simulation, all players are expected to be paired with someone other than the person with whom he or she came to the party about 37.5% of the time.

45. Answers will vary. Each driver is a component. One way to model this component is to generate random digits 00–99. Let 01–12 represent a driver that is talking on his or her cell phone, and let 13–99 and 00 represent a driver that is not talking on his or her cell phone. A trial consists of 10 pairs of digits. The response variable is whether or not at least 4 of the simulated drivers were talking on their cell phones. The simulated percentage of the time that 4 or more drivers of 10 are talking on their cell phones if the true rate of usage is 12% is the number of successes divided by the total number of trials. You should expect to find 4 or more drivers talking among 10 drivers only about 2% of the time. Based on what you saw at the bus stop, you'd suspect that the legislator's claim of 12% usage is probably too low.

47. Answers may vary. According to the simulation, all four tires are expected to last this long about 38.5% of the time.

Chapter 10

1. a) No. It would be nearly impossible to get exactly 500 males and 500 females from every country by random chance.
   b) A stratified sample, stratified by whether the respondent is male or female.

3. a) Voluntary response.
   b) We have no confidence at all in estimates from such studies.

5. a) The population of interest is all adults in the United States aged 18 and older.
   b) The sampling frame is U.S. adults with landline telephones.
   c) Some members of the population (e.g, many college students) don't have landline phones, which could create a bias.

7. a) Population—Human resources directors of Fortune 500 companies.
   b) Parameter—Proportion who don't feel surveys intruded on their work day.
   c) Sampling Frame—List of HR directors at Fortune 500 companies.
   d) Sample—23% who responded.
   e) Method—Questionnaire mailed to all (nonrandom).
   f) Bias—Nonresponse. Hard to generalize because who responds is related to the question itself.

9. a) Population—All U.S. adults.
   b) Parameter—Proportion who have used and benefited from alternative medicine.
   c) Sampling Frame—All Consumers Union subscribers.
   d) Sample—Those who responded (random).
   e) Method—Questionnaire to all (nonrandom).
   f) Bias—Nonresponse. Those who respond may have strong feelings one way or another.

11. a) Population—Adults.
   b) Parameter—Proportion who think drinking and driving is a serious problem.
   c) Sampling Frame—Bar patrons.
   d) Sample—Every 10th person leaving the bar.
   e) Method—Systematic sampling.
   f) Bias—Those interviewed had just left a bar. They probably think drinking and driving is less of a problem than do adults in general.

13. a) Population—Soil around a former waste dump.
   b) Parameter—Concentrations of toxic chemicals.
   c) Sampling Frame—Accessible soil around the dump.
   d) Sample—16 soil samples.
   e) Method—Not clear.
   f) Bias—Don't know if soil samples were randomly chosen. If not, may be biased toward more or less polluted soil.

15. a) Population—Snack food bags.
   b) Parameter—Weight of bags, proportion passing inspection.
   c) Sampling Frame—All bags produced each day.
   d) Sample—Bags in 10 randomly selected cases, 1 bag from each case for inspection.
   e) Method—Multistage sampling.
   f) Bias—Should be unbiased.

17. Bias. Only people watching the news will respond, and their preference may differ from that of other voters. The sampling method may systematically produce samples that don't represent the population of interest.

19. a) Voluntary response. Only those who both see the ad *and* feel strongly enough will respond.
   b) Cluster sampling. One school may not be typical of all.
   c) Attempted census. Will have nonresponse bias.
   d) Stratified sampling with follow-up. Should be unbiased.

21. a) This is a systematic sample.
   b) This sample is likely to be representative of those waiting in line for the roller coaster, especially if those people at the front of the line (after their long wait) respond differently from those at the end of the line.
   c) The sampling frame is patrons willing to wait for the roller coaster on that day at that time. It should be representative of the people in line, but not of all people at the amusement park.

23. a) This is a voluntary response sample.
   b) We have absolutely no confidence in estimates made from voluntary response samples. Voluntary response bias exists. Additionally, undercoverage limits the scope of any conclusions to the population of young adults, since only those that visited gamefaqs.com could have been chosen to be in the sample.

25. a) Answers will definitely differ. Question 1 will probably get many "No" answers, while Question 2 will get many "Yes" answers. This is wording bias.
   b) "Do you think standardized tests are appropriate for deciding whether a student should be promoted to the next grade?" (Other answers will vary.)

27. Only those who think it worth the wait are likely to be in line. Those who don't like roller coasters are unlikely to be in the sampling frame, so the poll won't get a fair picture of whether park patrons overall would favour still more roller coasters.

29. a) Biased toward yes because of "pollute." "Should companies be responsible for any costs of environmental cleanup?"
   b) Biased toward no because of "old enough to serve in the military." "Do you think the drinking age should be lowered from 21?"

31. a) Not everyone has an equal chance. People with unlisted numbers, people without phones, and those at work cannot be reached.
   b) Generate random numbers and call at random times.
   c) Under the original plan, those families in which one person stays home are more likely to be included. Under the second plan, many more are included. People without phones are still excluded.
   d) It improves the chance of selected households being included.
   e) This takes care of phone numbers. Time of day may be an issue. People without phones are still excluded.

33. a) Answers will vary.
   b) The parameter estimated by 10 measurements is the true length of your arm. The population is all possible measurements of your arm length.
   c) The population is now the arm lengths of your friends. The average now estimates the mean of the arm lengths of your friends.

d) These 10 arm lengths are unlikely to be representative of the community, or the country. Your friends are likely to be of the same age, and not very diverse.

35. a) Assign numbers 001 to 120 to each order. Use random numbers to select 10 transactions to examine.
b) Sample proportionately within each type. (Do a stratified random sample.)

37. a) Select three cases at random; then select one jar randomly from each case.
b) Use random numbers to choose 3 cases from numbers 61 through 80; then use random numbers between 1 and 12 to select the jar from each case.
c) No. Multistage sampling.

39. a) Depends on the Yellow Page listings used. If from regular (line) listings, this is fair if all doctors are listed. If from ads, probably not, as those doctors may not be typical.
b) Not appropriate. This cluster sample will probably contain listings for only one or two business types.

Chapter 11

1. Each of the 40 deliveries is an experimental unit. He has randomized the experiment by flipping a coin to decide whether or not to phone.

3. The factor is calling, and the levels are whether or not he calls the customer. The response variable is the tip percentage for each delivery.

5. By calling some customers but not others during the same run, the driver has controlled many variables, such as day of the week, season, and weather. The experiment was randomized because he flipped a coin to determine whether or not to phone and it was replicated because he did this for 40 deliveries.

7. Because customers don't know about the experiment, those that are called don't know that others are not, and vice versa. Thus, the customers are blind. That would make this a single-blind study. It can't be double-blind because the delivery driver must know whether or not he phones.

9. Yes. Driver is now a block. The experiment is randomized within each block. This is a good idea because some drivers might generally get higher tips than others, but the goal of the experiment is to study the effect of phone calls. Blocking on driver eliminates the variability in tips inherent to the driver.

11. Answers may vary. The cost or size of a delivery may confound his results. Larger orders may generally tip a higher or lower percentage of the bill.

13. a) No, this is not an experiment. There are no imposed treatments. This is anobservational study.
b) There can be other factors related to Facebook usage that are common among Facebook users and some of those factors might be the reason for the changes in the differences in average graduating grades.

15. a) This is a retrospective observational study.
b) That's appropriate because MS is a relatively rare disease.
c) The subjects were U.S. military personnel, some of whom had developed MS.
d) The variables were the vitamin D blood levels and whether or not the subject developed MS.

17. a) This was a randomized comparative, placebo-controlled experiment.
b) Yes, such an experiment is the right way to determine whether black cohosh has an effect.
c) 351 women aged 45 to 55 who reported at least two hot flashes a day.
d) The treatments were black cohosh, a multiherb supplement, a multiherb supplement plus advice, estrogen, and a placebo.
e) The response was the women's symptoms (presumably frequency of hot flashes.)

19. a) Experiment b) Bipolar disorder patients
c) Omega-3 fats from fish oil, two levels d) 2 treatments
e) Improvement, but there is no indication of how it is measured
f) Design not specified
g) Blind (due to placebo), unknown if double-blind
h) Individuals with bipolar disease improve with high-dose omega-3 fats from fish oil.

21. a) Observational study b) Prospective
c) Men and women with moderately high blood pressure and normal blood pressure, unknown selection process
d) Memory and reaction time
e) As there is no random assignment, there is no way to know that high blood pressure *caused* subjects to do worse on memory and reaction-time tests. A lurking variable may also be the cause.

23. a) Observational study b) Retrospective
c) Swedish men, unknown selection process
d) Occurrence of kidney cancer
e) As there is no random assignment, there is no way to know that the overweight or high blood pressure caused the higher risk for kidney cancer.

25. a) Experiment b) Postmenopausal women
c) Alcohol—2 levels; blocking variable—estrogen supplements (2 levels)
d) 1 factor (alcohol) at 2 levels = 2 treatments
e) Increase in estrogen levels f) Blocked g) Not blind
h) Indicates that alcohol consumption *for those taking estrogen supplements* may increase estrogen levels.

27. a) Experiment b) Locations in a garden
c) 1 factor: traps (2 levels) d) 2 treatments
e) Number of bugs in the trap f) Blocked by location
g) Not blind
h) One type of trap is more effective than the other

29. a) Observational study b) Retrospective
c) Women in Finland, unknown selection process with data from church records
d) Mothers' lifespans
e) As there is no random assignment, there is no way to know that having sons or daughters shortens or lengthens the lifespan of mothers.

31. a) Observational study b) Prospective
c) People with or without depression, unknown selection process
d) Frequency of crying in response to sad situations
e) There is no apparent difference in crying response (to sad movies) for depressed and nondepressed groups

33. a) Experiment b) Rats c) 1 factor: sleep deprivation; four levels
d) 4 treatments e) Glycogen content in the brain
f) No discussion of randomness g) Blinding is not discussed.
h) The conclusion could be that rats deprived of sleep have significantly lower glycogen levels and may need sleep to restore that brain energy fuel. Extrapolating to humans would be very speculative.

35. a) Experiment b) People experiencing migraines
c) 2 factors (pain reliever and water temperature), 2 levels each
d) 4 treatments e) Level of pain relief
f) Completely randomized over 2 factors
g) Blind, as subjects did not know if they received the pain medication or the placebo, but not blind, as the subjects will know if they are drinking regular or ice water.
h) It may indicate whether pain reliever alone or in combination with ice water gives pain relief, but patients are not blinded to ice water, so placebo effect may also be the cause of any relief seen caused by ice water.

37. a) Experiment b) Athletes with hamstring injuries
c) 1 factor: type of exercise program (2 levels)
d) 2 treatments
e) Time to return to sports
f) Completely randomized
g) No blinding—subjects must know what kind of exercise they do.
h) Can determine which of the two exercise programs is more effective.

39. They need to compare omega-3 results to something. Perhaps bipolarity is seasonal and would have improved during the experiment anyway.

41. a) Subjects' responses might be related to many other factors (diet, exercise, genetics, etc.). Randomization should equalize the two groups with respect to unknown factors.
b) More subjects would minimize the impact of individual variability in the responses, but the experiment would become more costly and time consuming.

43. People who engage in regular exercise might respond differently to the omega-3 fats, and that additional variability could obscure the effectiveness of this treatment.

45. Answers may vary. Use a random-number generator to randomly select 24 numbers from 01 to 24 without replication. Assign the first 8 numbers to the first group, the second 8 numbers to the second group, and the third 8 numbers to the third group.

47. a) First, they are using athletes who have a vested interest in the success of the shoe by virtue of their sponsorship. They should choose other athletes. Second, they should randomize the order of the runs, not run all the races with their shoes second. They should blind the athletes by disguising the shoes if possible, so they don't know which is which. The timers shouldn't know which athletes are running with which shoes, either. Finally, they should replicate several times, since times will vary under both shoe conditions.
b) Because of the problems in (a), the results they obtain may be biased in favour of their shoes. In addition, the results obtained for Olympic athletes maynot be the same as for runners in general.

49. a) Allowing athletes to self-select treatments could confound the results. Other issues such as severity of injury, diet, age, etc., could also affect time to heal, and randomization should equalize the two treatment groups with respect to any such variables.
b) A control group could have revealed whether either exercise program was better (or worse) than just letting the injury heal.
c) Doctors who evaluated the athletes to approve their return to sports should not know which treatment the subject had engaged in.
d) It's hard to tell. The difference of 15 days seems large, but the standard deviations indicate that there was a great deal of variability in the times.

51. a) The differences among the Mozart and quiet groups were more than would have been expected from ordinary sampling variation.
b)

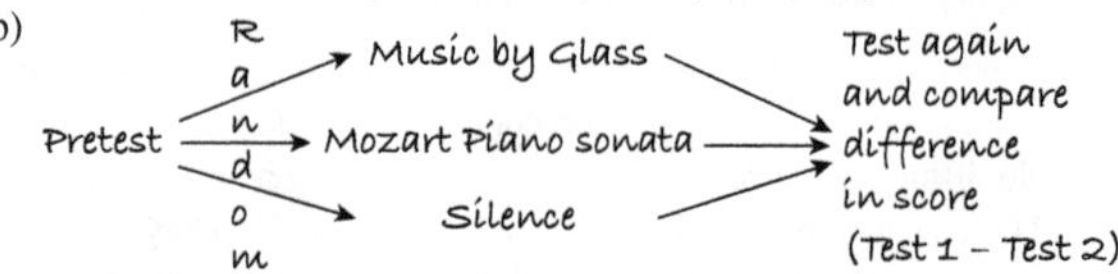

c) The Mozart group seems to have the smallest median difference and thus the *least* improvement, but there does not appear to be a significant difference.
d) No, if anything, there is less improvement, but the difference does not seem significant compared with the usual variation.

53. a) Observational. Randomly select a group of children, ages 10 to 13, have them taste the cereal, and ask if they like the cereal.
b) Answers may vary. Get volunteers ages 10 to 13. Each volunteer will taste both cereals, randomizing the order in which they taste them. Compare the percentage of favourable ratings for each cereal.

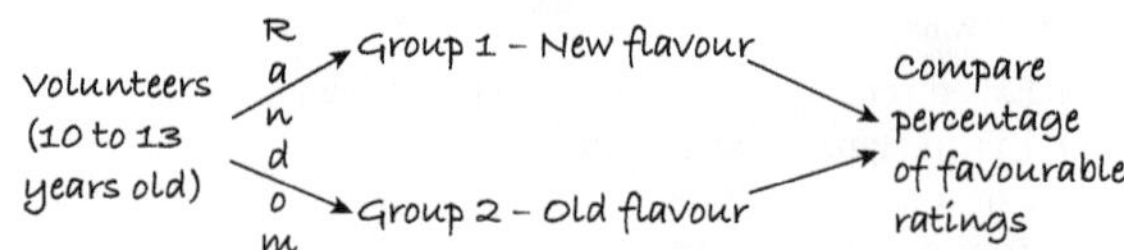

c) Answers may vary. From the volunteers, identify the children who watch Frump and identify the children who do not watch Frump. Use a blocked design to reduce variation in cereal preference that may be associated with watching the Frump cartoon.

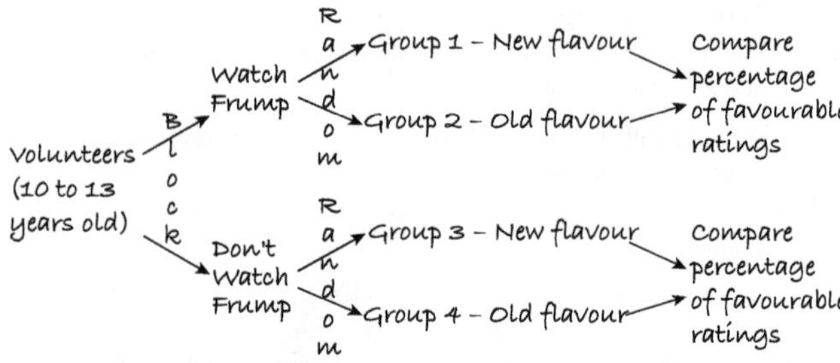

55. a) Observational, prospective study
b) The supposed relation between health and wine consumption might be explained by the confounding variables of income and education.
c) None of these. While the variables have a relation, there is no causality indicated for the relation.

57. a) Arrange the 20 containers in 20 separate locations. Use a random-number generator to identify the 10 containers that should be filled with water.
b) Guessing, the dowser should be correct about 50% of the time. A record of 60% (12 out of 20) does not appear to be significantly different.
c) Answers may vary. You would need to see a high level of success—say, 90% to 100%, that is, 18 to 20 correct.

59. Randomly assign half the reading teachers in the district to use each method. Students should be randomly assigned to teachers as well. Make sure to block both by school and grade (or control grade by using only one grade). Construct an appropriate reading test to be used at the end of the year, and compare scores.

61. a) They mean that the difference is higher than they would expect from normal sampling variability.
b) An observational study
c) No. Perhaps the differences are attributable to some confounding variable (e.g., people are more likely to engage in riskier behaviours on the weekend) rather than the day of admission.
d) Perhaps people have more serious accidents and traumas on weekends and are thus more likely to die as a result.

63. Answers may vary. This experiment has 1 factor (pesticide), at 3 levels (pesticide A, pesticide B, no pesticide), resulting in 3 treatments. The response variable is the number of beetle larvae found on each plant. Randomly select a third of the plots to be sprayed with pesticide A, a third with pesticide B, and a third with no pesticide (since the researcher also wants to know whether the pesticides even work at all). To control the experiment, the plots of land should be as similar as possible with regard to amount of sunlight, water, proximity to other plants, etc. If not, plots with similar characteristics should be blocked together. If possible, use some inert substance as a placebo pesticide on the control group, and do not tell the counters of the beetle larvae which plants have been treated with pesticides. After a given period of time, count the number of beetle larvae on each plant and compare the results.

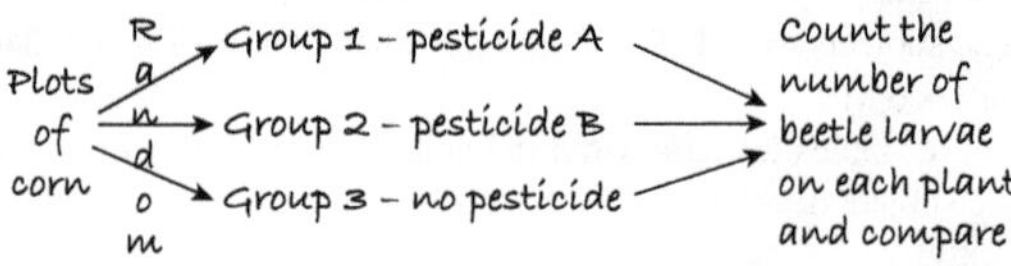

65. Answers may vary. Find a group of volunteers. Each volunteer will be required to shut off the machine with his or her left hand and right hand. Randomly assign the left or right hand to be used first. Complete the first attempt for the whole group. Now repeat the experiment with the alternate hand. Check the differences in time for the left and right hands.

67. a) Jumping with or without a parachute
b) Volunteer skydivers (the dimwitted ones)
c) A parachute that looks real but doesn't work
d) A good parachute and a placebo parachute

e) Whether parachutist survives the jump (or extent of injuries)
f) All should jump from the same altitude in similar weather conditions and land on similar surfaces.
g) Randomly assign people the parachutes.
h) The skydivers (and the people involved in distributing the parachute packs) shouldn't know who got a working chute. And the people evaluating the subjects after the jumps should not be told who had a real parachute either!

Part III Review

1. Observational prospective study. Indications of behaviour differences can be seen in the two groups. May show a link between premature birth and behaviour, but there may be lurking variables involved.

3. Experiment, matched by gender and weight. The experiment had one factor (diet), at two levels (allowing the dogs to eat as much as they want, or restricted diet), resulting in two treatments. The response variable was length of life. On average, dogs with a lower-calorie diet live longer.

5. Completely randomized experiment, with the treatment being receiving folic acid or not (one factor, two levels). The response variable is the number of precancerous growths, or simply the occurrence of additional precancerous growths. Neither blocking nor matching is mentioned, but in a study such as this one, it is likely that researchers and patients are blinded. Since treatments were randomized, it seems reasonable to generalize results to all people with precancerous polyps, though caution is warranted since these results contradict a previous study.

7. Sampling. No information is given about the sampling procedure, so hopefully the tested fireworks are selected randomly. It would probably be a good idea to test a few of each type of firework, so stratification by type seems likely. The population is all fireworks produced each day, and the parameter of interest is the proportion of duds. With a random sample, the manufacturers can make inferences about the proportion of duds in the entire day's production, and use this information to decide whether or not the day's production is suitable for sale.

9. Observational retrospective study. Researcher can conclude that for anyone's lunch, even when packed with ice, food temperatures are rising to unsafe levels.

11. Experiment, with a control group being the genetically engineered mice who received no antidepressant and the treatment group being the mice who received the drug. The response variable is the amount of plaque in their brains after one dose and after four months. There is no mention of blinding or matching. Conclusions can be drawn to the general population of mice and we should assume treatments were randomized. To conclude the same for humans would be risky, but researchers might propose an experiment on humans based on this study.

13. Experiment. Factor is gene therapy. Hamsters were randomized to treatments. Treatments were gene therapy or not. Response variable is heart muscle condition. Can conclude that gene therapy is beneficial (at least in hamsters).

15. Sampling. Population is all oranges on the truck. Parameter is proportion of unsuitable oranges. Procedure is probably simple random sampling. Can conclude whether or not to accept the truckload.

17. Observational prospective study. Physically fit men may have a lower risk of death from cancer.

19. Answers will vary. This is a simulation problem. Using a random digits table or software, call 0–4 a loss, and 5–9 a win for the gambler on a game. Use blocks of 5 digits to simulate a week's pick.

21. Answers will vary.

23. a) Experiment. Actively manipulated candy giving, diners were randomly assigned treatments, control group was those with no candy, lots of dining parties.
b) It depends on when the decision was made. If early in the meal, the server may give better treatment to those who will receive candy, which will bias the results.
c) A difference in response so large it cannot be attributed to natural sampling variability.

25. Everyone responding to the poll is online, and probably younger than the average voter. It may be the case that younger Canadians vote Liberal at a higher proportion than older Canadians. Older citizens also tend to show up to vote on election day in a higher proportion than young Canadians.

27. a) Simulation results will vary. Average will be around 5.8 points.
b) Simulation results will vary. Average will also be around 5.8 points.
c) Answers will vary.

29. a) Yes.
b) No. Residences without phones are excluded. Residences with more than one phone had a higher chance.
c) No. People who respond to the survey may be of age but not be registered voters.
d) No. Households who answered the phone may be more likely to have someone at home when the phone call was generated. These may not be representative of all households.

31. a) Does not prove it. There may be other confounding variables. Only way to prove this would be to do a controlled experiment.
b) Alzheimer's usually shows up late in life. Perhaps smokers have died of other causes before Alzheimer's can be seen.
c) An experiment would be unethical. One could design a prospective study in which groups of smokers and nonsmokers are followed for many years and the incidence of Alzheimer's is tracked.

33.

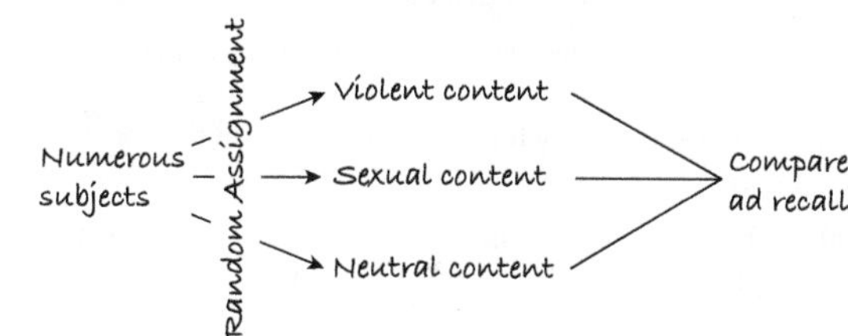

Numerous subjects will be randomly assigned to see shows with violent, sexual, or neutral content. They will see the same commercials. After the show, they will be interviewed for their recall of brand names in the commercials.

35. a) May have been a simple random sample, but given the relative equality in age groups, may have been stratified.
b) 38.2%
c) We don't know. If data were from precincts that are primarily Democratic or Republican, that would bias the results. Also, party preference and actual party registration are not exactly the same thing. Probably OK, though.
d) Do party affiliations differ for different age groups?

37. The factor in the experiment will be type of bird control. I will have three treatments: scarecrow, netting, and no control. I will randomly assign several different areas in the vineyard to one of the treatments, taking care that there is sufficient separation that the possible effect of the scarecrow will not be confounded. At the end of the season, the response variable will be the proportion of bird-damaged grapes.

39. a) We want all subjects treated as alike as possible. If there were no "placebo surgery," subjects would know this and perhaps behave differently.
b) The experiment was intended to see if there was a difference in the effectiveness of the two treatments. (If we wanted to generalize, we would need to assume that the results for these volunteers are the same as on all patients who might need this operation.)
c) Statistically significant means a difference in results so large it cannot be explained by natural sampling variability.

41. a) Use stratified sampling to select 2 first-class passengers and 12 from coach.
b) Number passengers alphabetically, 01 = Bergman to 20 = Testut. Read in blocks of two, ignoring any numbers more than 20. This gives 65, 43, 67, 11 (selects Fontana), 27, 04 (selects Castillo).
c) Number passengers alphabetically 001 to 120. Use the random number table to find three-digit numbers in this range until 12 have been selected.

43. Simulation results will vary.
(Use integers 00 to 99 as a basis. Use integers 0 to 69 to represent a tee shot on the fairway. If on the fairway, use digits 00 to 79 to represent on the green. If off the fairway, use 00 to 39 to represent getting on the green. If not on the green, use digits 00 to 89 to represent landing on the green. For the first putt, use digits 00 to 19 to represent making the shot. For subsequent putts, use digits 00 to 89 to represent making the shot.)

Chapter 12

1. a) S = {HH, HT, TH, TT}, equally likely
b) S = {0, 1, 2, 3}, not equally likely
c) S = {H, TH, TTH, TTT}, not equally likely
d) S = {1, 2, 3, 4, 5, 6}, not equally likely

3. In this context "truly random" should mean that every number is equally likely to occur.

5. There is no "Law of Averages." She would be wrong to think they are "due" for a harsh winter.

7. There is no "Law of Averages." If at bats are independent, his chance for a hit does not change based on recent successes or failures.

9. a) There is some chance you would have to pay out much more than the $300.
b) Many customers pay for insurance. The small risk for any one customer is spread among all.

11. a) Legitimate b) Legitimate c) Not legitimate (sum more than 1)
d) Legitimate e) Not legitimate (can't have negatives)

13. a) 0.72 b) 0.89 c) 0.28

15. a) 0.32 b) 0.408

17. a) 0.69 b) 0.03 c) 0.50 d) 0.31

19. a) 0.68 b) 0.32 c) 0.04

21. a) 0.09 b) 0.91 c) 0.09

23. a) 0.2125 b) 0.2875 c) 0.1208 d) 0.3792

25. 1/9

27. If the cards are selected at random (simple random sampling) from the deck of 52 cards, all sets of five cards have the same chance to be the selected set.

29. Answers will vary. Author simulations estimate the probability to be about 90%.

Chapter 13

1. $P(\text{dog or cat}) = P(\text{dog}) + P(\text{cat}) - P(\text{dog and cat})$
$= 0.25 + 0.29 - 0.12 = 0.42$

3. $$P(\text{football} \mid \text{no basketball}) = \frac{P(\text{football and no basketball})}{P(\text{no basketball})} = \frac{\frac{38}{100}}{\frac{60}{100}} \approx 0.633$$

(Or, use the table. Of the 60 people who don't like to watch basketball, 38 people like to watch football. 38/60 ≈ 0.633)

5. The overall survival rate, $P(S)$, was 0.323, yet the survival rate for first class passengers, $P(S \mid FC)$, was 0.625. Since, $P(S) \neq$, survival and ticket class are not independent. Rather, survival rate depended on class.

7. a) 0.0198 b) 0.328

9. a) 0.27 b) 0.128 c) 0.512 d) 0.271

11. a) Disjoint b) Independent
c) No. If two events are disjoint, whenever one event occurs, the one cannot occur and so they are not independent.

13. a) $(1/6)(1/6)(1/6) = 1/216 = .0046$
b) $(3/6)(3/6)(3/6) = 27/216 = 0.125$
c) $(4/6)(4/6)(4/6) = 64/216 = 0.2963$
d) $1 - (5/6)(5/6)(5/6) = 1 - 125/216 = 91/216 = 0.4213$
e) $1 - (1/6)(1/6)(1/6) = 1 - 1/216 = 215/216 = 0.9954$

15. a) 0.027 b) 0.063 c) 0.973 d) 0.013841287201

17. a) 0.438976 b) 0.729 c) 0.295031

19. 0.0776

21. a) 0.676 b) 0.0954 c) 0.324 d) 0.213

23. a) 1/2 b) 1 c) 2/26 d) 1/3

25. a) 0.11 b) 0.27 c) 0.407 d) 0.344

27. a) 15/38 b) 15/51 c) 21/69 d) 45/68

29. Having a fever and having a sore throat are not independent events, so: *P*(fever and sore throat) = 0.21

31. a) 0.145 b) 0.118 c) 0.414 d) 0.217

33. a) 0.318 b) 0.955 c) 0.071 d) 0.009

35. Yes. The overall probability of getting an ace is 1/13. If you consider just one suit, there is only 1 ace out of 13 cards, so the probability of getting an ace given that the card is a diamond, for instance, is 1/13.

37. a) Yes b) No. These events are disjoint.
c) No. some customers under 30 are buying comedies and so a customer selected at random can be both a customer buying a comedy and under 30 and so the events are not disjoint.
d) No. P(buying a comedy) = 0.4333 ≠ P(buying a comedy | under 30) = 0.33

39. High blood pressure and high cholesterol are not independent events. 28.8% of men with OK blood pressure have high cholesterol, while 40.7% of men with high blood pressure have high cholesterol.

41. a) 0.54 b) Not independent

43. a) No, the flight leaving on time and the luggage making the connection are not independent events. The probability is 0.95 if the flight is on time and only 0.65 if it is not on time.
b) 0.695

45. 0.975

47. a) No. The rate of absenteeism for the night shift is 2%, while the rate for the day shift is only 1%.
b) 1.4%

49. 57.1%

51. a) 0.2 b) 0.272 c) 0.353 d) 0.033

53. 0.563

Chapter 14

1. a) 19 b) 4.2

3. a) $50 b) $50

5. a)

Win	$0	$5	$10	$30
P(amount won)	$\frac{26}{52}$	$\frac{13}{52}$	$\frac{12}{52}$	$\frac{1}{52}$

b) $4.13 c) $4 or less (answers may vary)

7. a)

Children	1	2	3
P(Children)	0.5	0.25	0.25

b) 1.75 children c) 0.87 boys

Boys	0	1	2	3
P(Boys)	0.5	0.25	0.125	0.125

9. $27 000

11. a) 7 b) 1.89

13. $5.44

15. 0.83

17. a) 1.7 b) 0.9

19. $\mu = 0.64, \sigma = 0.93$

21. a) $50 b) $100

23. a) No. The probability of winning the second depends on the outcome of the first.
b) 0.42 c) 0.08

d)

Games won	0	1	2
P(Games won)	0.42	0.50	0.08

e) $\mu = 0.66, \sigma = 0.62$

25. a)

Number good	0	1	2
P(Number good)	0.067	0.467	0.467

b) 1.40 c) 0.61

27. a) 30; 6 b) 26; 5 c) 30; 5.39 d) 10; 5.39 e) 20; 2.83

29. a) 240; 12.8 b) 140; 24 c) 720; 34.18
d) 60; 39.40 e) 1200; 32

31. a) 1.8 b) 0.87 c) Cartons are independent of each other.

33. 13.6; 2.55 (assuming the hours are independent of each other)

35. a) 23.4; 2.97
b) We are assuming that trucks are ticketed independently.
c) An unusually bad month might be one in which the company got 30 or more tickets.

37. a) There will be many gains of $150 with a few large losses.
b) $\mu = \$300, \sigma = \$8485.28$
c) $\mu = \$1\,500\,000, \sigma = \$600\,000$
d) Yes. $0 is 2.5 SDs below the mean for 10 000 policies.
e) Losses are independent of each other. A major catastrophe with many policies in an area would violate the assumption.

39. a) No. More than two outcomes are possible.
b) Yes, assuming the people are unrelated to each other.
c) No. The chance of a heart changes as cards are dealt.
d) No, 500 is more than 10% of 3000.
e) If packages in a case are independent of each other, yes; otherwise, no.

41. a) Use single random digits. Let 0, 1 = Sidney Crosby. Examine random digits in groups of five, counting the number of 0's and 1's.

a) Results will vary.
b)

x	0	1	2	3	4	5
$P(x)$	0.33	0.41	0.20	0.05	0.01	0.0

43. Departures from the same airport during a 2-hour interval may not be independent. All could be delayed by weather, for example.

45. a) 50 b) $E(\text{heads}) = np = 100(0.5) = 50$ heads.

47. a) $\mu = 10.44, \sigma = 1.16$ b) 0.812 c) 0.475
d) 0.00193 e) 0.998

49. $\mu = 20.28, \sigma = 4.22$

51. a) 0.118 b) 0.324 c) 0.744 d) 0.580

53. a) $\mu = 56, \sigma = 4.10$
b) Yes, $np = 56 \geq 10, nq = 24 \geq 10$, serves are independent.
c) In a match with 80 serves, approximately 68% of the time she will have between 51.9 and 60.1 good serves, approximately 95% of the time she will have between 47.8 and 64.2 good serves, and approximately 99.7% of the time she will have between 43.7 and 68.3 good serves.
d) Normal, approx.: 0.014; Binomial, exact: 0.016

55. a) Assuming apples fall and become blemished independently of each other, Binom(300, 0.06) is appropriate. Since $np \geq 10$ and $nq \geq 10$, $N(18, 4.11)$ is also appropriate.
b) Normal, approx.: 0.072; Binomial, exact: 0.085
c) No, 50 is 7.8 SDs above the mean.

57. Normal, approx.: 0.053; Binomial, exact: 0.061

59. The mean number of sales should be 24 with SD 4.60. Ten sales is more than 3.0 SDs below the mean. He was probably misled.

61. a) 0.0869 b) 0.0364

63. a) 4 cases b) 0.9817

65. a) 0.107 b) $\mu = 24, \sigma = 2.19$
c) Normal, approx.: 0.819; Binomial, exact: 0.848

67. $\mu = 20, \sigma = 4$. I'd want *at least* 32 (3 SDs above the mean). (Answers will vary.)

69. Probably not. There's a more than 9% chance that he could hit 4 shots in a row, so he can expect this to happen nearly once in every 10 sets of 4 shots he takes. That does not seem unusual.

71. Yes. We'd expect him to make 22 shots, with a standard deviation of 3.15 shots. 32 shots is more than 3 standard deviations above the expected value, an unusually high rate of success.

73. a) $1/360$ b) 0.25

75. a) 1 oz b) 0.5 oz c) 0.023 d) $\mu = 4$ oz, $\mu = 0.5$ oz
e) 0.159 f) $\mu = 12.3$ oz, $\sigma = 0.54$ oz

77. a) 12.2 oz b) 0.51 oz c) 0.058

79. a) $\mu = 200.57$ sec, $\sigma = 0.46$ sec
b) No, $z = -2.36$. There is only 0.009 probability of swimming that fast or faster.

81. a) A = price of a pound of apples; P = price of a pound of potatoes; $Profit = 100A + 50P - 2$
b) $63.00 c) $20.62
d) Mean—no; SD—yes (independent sales prices)

83. a) Yes
b) No, since we are counting the number of events in a fixed number of trials.
c) 0.03125 d) 0.125

e) Very skewed since as you increase x, the power of q increases, and since q is less than 1, this part of the distribution formula decreases (geometrically), so the probabilities always decrease as x increases.

85. a) 2.0 b) 1.0 c) 3.0 d) 4.0
e) The smallest variance occurs when the investments are negatively correlated.

Part IV Review

1. a) 0.34 b) 0.27 c) 0.069
d) No, 2% of cars have both types of defects.
e) Of all cars with cosmetic defects, 6.9% have functional defects. Overall, 7.0% of cars have functional defects. The probabilities here are estimates, so these are probably close enough to say the defects are independent.

3. a) C = Price to China; F = Price to France; Total $= 3C + 5F$
b) $\mu = \$8100, \sigma = \$672.68$ c) $\mu = \$300, \sigma = \$180.28$
d) Means—no. Standard deviations—yes; ticket prices must be independent of each other for different countries, but all tickets to the same country are at the same price.

5. a) $\mu = \$0.20, \sigma = \$1.89$ b) $\mu = \$0.40, \sigma = \$2.67$

7. a) 0.999 b) 0.944 c) 0.993

9. a) 0.237 b) 0.015 c) 0.896

11. a) $\mu = 118.5, \sigma = 5.44$ b) Yes. $np \geq 10$ and $nq \geq 10$
c) Normal, approx.: 0.059; Binomial, exact: 0.073

13. Assuming that the first-year composition class consists of 25 randomly selected people, these may be considered Bernoulli trials. There are only two possible outcomes, having a specified language centre or not having the specified language centre. The probabilities of the specified language centers are constant at 80%, 10%, or 10%, for right, left, and two-sided language center, respectively. The trials are not independent, since the population of people is finite, but we will select fewer than 10% of all people.
a) Let L be the number of people with left-brain language control from $n = 25$ people. Use $Binom(25, 0.80)$.

$$P(\text{no more than } 15) = P(L \leq 15)$$
$$= P(L = 0) + \cdots + P(L = 15)$$
$$= {}_{25}C_0(0.80)^0(0.20)^{25} + \cdots + {}_{25}C_{15}(0.80)^{15}(0.20)^{10}$$
$$\approx 0.0173$$

According to the Binomial model, the probability that no more than 15 students in a class of 25 will have left-brain language centres is approximately 0.0173.
b) Let T = the number of people with two-sided language control from $n = 5$ people.
Use $Binom(5, 0.10)$.
$P(\text{none have two-sided language control}) = P(T = 0)$
$= {}_5C_0(0.10)^0(0.90)^5$
≈ 0.590
c) Using Binomial models:
$E(\text{left}) = 960$ people, $E(\text{right}) = 120$ people, $E(\text{two} - \text{sided}) = 120$ people
d) Let R = the number of people with right-brain language control. $E(R) = 120$ people and $SD(R) \approx 10.39$ people.
e) Since $npR = 120$ and $nqR = 1080$ are both greater than 10, the Normal model, $N(120, 10.39)$, may be used to approximate $Binom(1200, 0.10)$. According to the Normal model, about 68% of randomly selected groups of 1200 people could be expected to have between 109.61 and 130.39 people with right-brain language control. About 95% of randomly selected groups of 1200 people could be expected to have between 99.22 and 140.78 people with right-brain language control. About 99.7% of randomly selected groups of 1200 people could be expected to have between 88.83 and 151.17 people with right-brain language control.

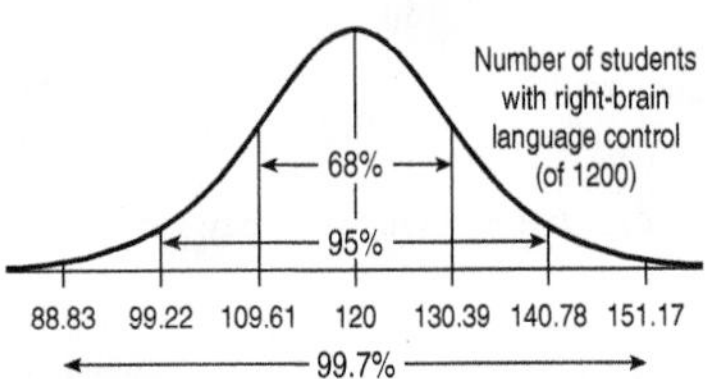

15. a) Men's heights are more variable than women's.
b) Men
c) M = Man's height; W = Woman's height; $M - W$ is how much taller the man is.
d) 5.1″ e) 3.75″ f) 0.913
g) If independent, it should be about 91.3%. We are told 92%. This difference seems small and may be due to natural sampling variability.

17. a) No. The chance is 1.6 3 10^{27}. b) 0.952 c) 0.063

19. −\$2080.00

21. a) 0.717 b) 0.588

23. a) $\mu = 100, \sigma = 8$ b) $\mu = 1000, \sigma = 60$
c) $\mu = 100, \sigma \approx 8.54$ d) $\mu = -50, \sigma = 10$
e) $\mu = 100, \sigma \approx 11.31$

25. a) Many do both, so the two categories can total more than 100%.
b) No. They can't be disjoint. If they were, the total would be 100% or less.
c) No. Probabilities are different for boys and girls. d) 0.0524

27. a) 21 days b) 1649.73 soms
c) Bring an extra 3300 soms. This gives you a cushion of about 157 soms for each days.

29. You'd expect 549.4 homeowners, with an SD of 13.46. 523 is about 2 SDs below the mean; somewhat unusual.

31. a) 0.005 b) 0.4746 c) 0.17

33. a) 15 b) 0.402

35. a) 38% b) 41% c) 19.6%
d) 19.6% of calculator users were computer users. In classes where calculatorswere not used, 22.4% of the classes used computers. Since the percentagesare different, there is evidence of an association between computer use andcalculator use. Calculator classes are less likely to use computers.

37. a) A Poisson model b) 0.1 failures c) 0.090 d) 0.095

39. a) 1/11 b) 7/22 c) 5/11 d) 0 e) 19/66

41. a) Expected number of stars with planets.
b) Expected number of planets with intelligent life.
c) Probability of a planet with a suitable environment having intelligent life.
d) f_l: If a planet has a suitable environment, the probability that life develops.
f_i: If a planet develops life, the probability that the life evolves intelligence.
f_c: If a planet has intelligent life, the probability that it develops radio communication.

43. 0.991

Chapter 15

1. a) Roughly Normal b) 0.36 c) 0.034

3. a) A Normal model is not appropriate since the histogram is strongly left skewed.
b) No. The 95% rule is based on the Normal model, and the Normal model is not appropriate here.

5. a) 0.0357 b) 400

7. All the histograms are centered near 0.05. As n gets larger, the histograms approach the Normal shape, and the variability in the sample proportions decreases.

9. a)

n	Observed mean	Theoretical mean	Observed st. dev.	Theoretical st. dev.
20	0.0497	0.05	0.0479	0.0487
50	0.0516	0.05	0.0309	0.0308
100	0.0497	0.05	0.0215	0.0218
200	0.0501	0.05	0.0152	0.0154

b) They are all quite close to what we expect from the theory.
c) The histogram is unimodal and symmetric for $n = 200$.
d) The success/failure condition says that np and nq should both be at least 10, which is not satisfied until $n = 200$ for $p = 0.05$. The theory predicted my choice.

11. a) Symmetric, because probability of heads and tails is equal.
b) 0.5
c) 0.125
d) $np = 8 < 10$

13. a) About 68% should have proportions between 0.4 and 0.6, about 95% between 0.3 and 0.7, and about 99.7% between 0.2 and 0.8.
b) $np = 12.5$, $nq = 12.5$; both are ≥ 10.
c)

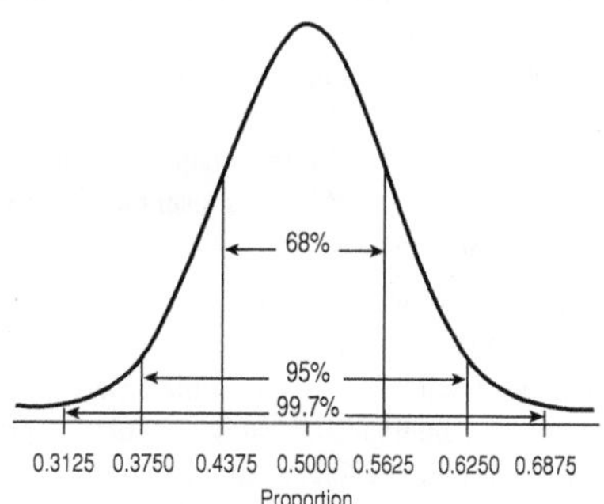

$np = nq = 32$; both are ≥ 10.
d) Becomes narrower (less spread around 0.5).

15. This is a fairly unusual result: about 2.26 SDs below the mean. The probability of that is about 0.012. So, in a class of 100 this is certainly a reasonable possibility.

17. a)

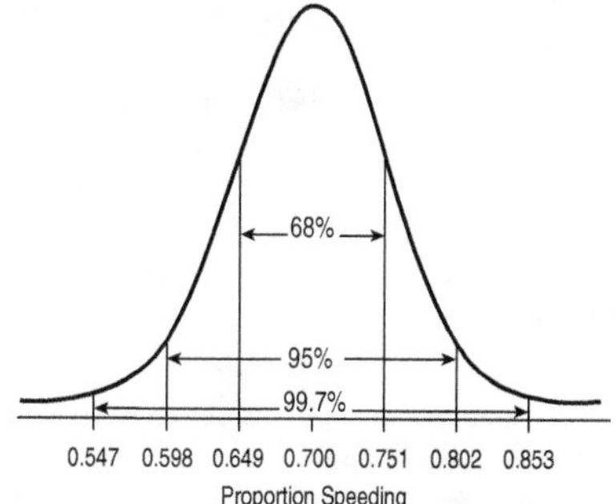

b) Both $np = 56$ and $nq = 24 \geq 10$. Drivers *may* be independent of each other, but if flow of traffic is very fast, they may not be. Or weather conditions may affect all drivers. In these cases they may get more or fewer speeders than they expect.

19. a) Assume that these children are typical of the population. They represent fewer than 10% of all children. We expect 20.4 nearsighted and 149.6 not; both are at least 10.

b)

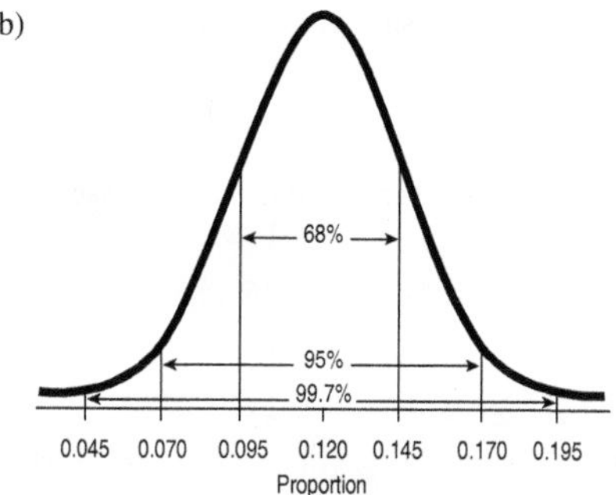

c) Probably between 12 and 29.

21. a) $\mu = 7\%$ $\sigma = 1.8\%$
b) Assume that clients pay independently of each other, that we have a random sample of all possible clients, and that these represent less than 10% of all possible clients. $np = 14$ and $nq = 186$ are both at least 10.
c) 0.048

23. 68% of the time within 0.014 (one SD), 95% of the time within 0.028 (two SD), and 99.7% of the time within 0.042 (three SD).

25. It is unlikely that rural families compose a SRS. However, proceeding as if they do, we find $\sigma(\hat{p}) = 1.10\%$ and so 16% would be a very unusual result, as it is outside the 3 $\sigma(\hat{p})$ range by the 68-95-99.7 rule.

27. Randomization condition: We must assume that the 400 voters were polled randomly 10% condition: 400 voters polled represent less than 10% of potential voters.

Success/Failure condition: $np = 192$ and $nq = 208$ are both greater than 10. Therefore, the sampling distribution model for q is Normal, with:

$$\mu_{\hat{p}} = p = 0.48$$

$$\sigma(\hat{p}) = \sqrt{\frac{pq}{n}} = \sqrt{\frac{(0.48)(0.52)}{400}} \approx 0.025$$

According to the Normal model, the probability that the newspaper's sample will lead them to predict victory (that is, predict referendum support above 50%) is approximately 0.212.

$$z = \frac{\hat{p} - \mu_{\hat{p}}}{\sqrt{\frac{pq}{n}}}$$

$$z = \frac{0.50 - 0.48}{\sqrt{\frac{(0.48)(0.52)}{400}}}$$

$$z = 0.801$$

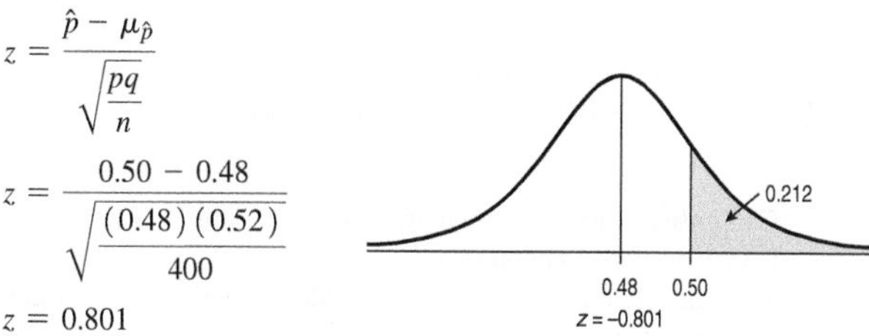

29. 0.088 using $N(0.08, 0.022)$ model

31. Answers will vary. Using $\mu + 3\sigma$ for "very sure," the restaurant should have 89 nonsmoking seats. Assumes customers at any time are independent of each other, a random sample, and represent less than 10% of all potential customers. $np = 72$, $nq = 48$, so Normal model is reasonable ($\mu = 0.60$, $\sigma = 0.045$).

33. a) Normal, centre at μ, standard deviation $\frac{\sigma}{\sqrt{n}}$
b) Standard deviation will be smaller. Centre will remain the same.

35. a) The histogram is unimodal and slightly skewed to the right, centred at 36 inches, with a standard deviation near 4 inches.
b) All the histograms are centred near 36 inches. As n gets larger, the histograms approach the Normal shape and the variability in the

sample means decreases. The histograms are fairly normal by the time the sample reaches size 5.

37. a)

n	Observed mean	Theoretical mean	Observed st. dev.	Theoretical st. dev.
2	36.314	36.33	2.855	2.842
5	36.314	36.33	1.805	1.797
10	36.341	36.33	1.276	1.271
20	36.339	36.33	0.895	0.899

b) They are all very close to what we would expect.
c) For samples as small as 5, the sampling distribution of sample means is unimodal and very symmetric.
d) The distribution of the original data is nearly unimodal and symmetric, so it doesn't take a very large sample size for the distribution of sample means to be approximately Normal.

39. Normal, $\mu = 73, \sigma = 1.4$. We assume that the students are randomly assigned to the seminars and represent less than 10% of all possible students, and that the individual's GPAs are independent of one another.

41. a) The mean of 100 coins varies less from sample to sample than the individual coins.

b) $SD(\bar{y}) = \frac{\sigma}{\sqrt{n}} = \frac{0.09}{\sqrt{100}} = 0.009$

43. a) 21.1% b) 276.8 days or more c) $N(266, 2.07)$ d) 0.002

45. a) There are more premature births than very long pregnancies. Modern practice of medicine stops pregnancies at about 2 weeks past normal due date.
b) Parts (a) and (b)—yes—we can't use Normal model if it's very skewed. Part (c)—no—CLT guarantees a Normal model for this large sample size.

47. a) $\mu = \$2.00, \sigma = \$3.61$ b) $\mu = \$4.00, \sigma = \$5.10$
c) 0.191. Model is $N(80, 22.83)$.

49. a) $\mu = 2.442, \sigma = 1.757$.
b) They would not be expected to follow a Normal Distribution. They would follow the frequency distribution given in the table.
c) Approximately $N(2.442,0.278)$

51. For $n = 36$, using $N(2.442, 0.293)$, $z = 1.91$ and $P(z > 1.91) = 0.0281$, so farm families may be larger on average.

53. a) $N(2.9, 0.045)$ b) 0.0131 c) 2.97 gm/mi

55. a) Can't use a Normal model to estimate probabilities. The distribution is skewed right—not Normal.
b) 4 is probably not a large enough sample to say the average follows the Normal model.
c) No. This is 3.16 SDs above the mean.

57. a) 0.0003. Model is $N(384, 34.15)$. b) \$427.77 or more

59. a) 0.3085. b) 0.3783 c) 0.0031 d) 0.0455

61. Answers may vary.

63. Answers may vary.

65. Answers may vary.

Chapter 16

1. a) This means that 49% of the 799 teens in the sample said they have misrepresented their age online. This is our best estimate of p, the proportion of all U.S. teens who would say they have done so.

b) $SE(\hat{p}) = \sqrt{\frac{(0.49)(0.51)}{799}} \approx 0.018$

c) Because we don't know p, we use $\hat{p}$ to estimate the standard deviation of the sampling distribution. So the standard error is our estimate of the amount of variation in the sample proportion we expect to see from sample to sample when we ask 799 teens whether they've misrepresented their age online.

3. a) We are 95% confident that, if we were to ask all U.S. teens whether they have misrepresented their age online, between 45.6% and 52.5% of them would say they have.
b) If we were to collect many random samples of 799 teens, about 95% of the confidence intervals we construct would contain the proportion of all U.S. teens who admit to misrepresenting their age online.

5. He believes the true proportion is within 4% of his estimate, with some (probably 95%) degree of confidence.

7. a) Population—all cars; sample—those actually stopped at the checkpoint; p—proportion of all cars with safety problems; $\hat{p}$—proportion actually seen with safety problems (10.4%); if sample (a cluster sample) is representative, then the methods of this chapter will apply.
b) Population—general public; sample—those who logged onto the Web site; p—population proportion of those who favour prayer in school; $\hat{p}$—proportion of those who voted in the poll who favoured prayer in school (81.1%); can't use methods of this chapter—sample is biased and nonrandom.
c) Population—parents at the school; sample—those who returned the questionnaire; p—proportion of all parents who favour uniforms; $\hat{p}$—proportion of respondents who favour uniforms (60%); should not use methods of this chapter, since not SRS (possible non-response bias).
d) Population—students at the college; sample—the 1632 students who entered that year; p—proportion of all students who will graduate on time; $\hat{p}$—proportion of that year's students who graduate on time (85.0%); can use methods of this chapter if that year's students (a cluster sample) are viewed as a representative sample of all possible students at the school.

9. a) Not correct. This implies certainty.
b) Not correct. Different samples will give different results. Many fewer than 95% will have 88% on-time orders.
c) Not correct. The interval is about the population proportion, not the sample proportion in different samples.
d) Not correct. In this sample, we *know* 88% arrived on time.
e) Not correct. The interval is about the parameter, not about the days.

11. a) False b) True c) True d) False

13. On the basis of this sample, we are 90% confident that the proportion of Japanese cars is between 29.9% and 47.0%.

15. a) $\hat{p} \pm z^*\sqrt{\frac{\hat{p}\hat{q}}{n}} = \left(\frac{42}{190}\right) \pm 1.960\sqrt{\frac{\left(\frac{42}{190}\right)\left(\frac{148}{190}\right)}{190}} = (0.162, 0.280)$
b) We are 95% confident that between 16.2% and 28.0% of all seafood packages sold in these three states are mislabelled.
c) The size of the population is irrelevant. If *Consumer Reports* had a random sample, 95% of intervals generated by studies like this are expected to capture the true proportion of seafood packages that are mislabelled.

17. a) 0.026
b) We are 90% sure that the true proportion of Canadians who are proud of health care is within ±2.6% of 58%.
c) Larger. To be more certain, we must be less precise.
d) 0.041 e) Less confidence
f) Our new interval (55.4%, 60.6%) does not include the old measurement of 50%. Even with the wider margins associated with the 99% CI, we still don't include 50%. Thus, we have strong evidence that the real proportion has changed.

19. a) (0.0465, 0.0491). The assumptions and conditions for constructing a confidence interval are satisfied.

b) The confidence interval gives the set of plausible values (with 95% confidence). Since 0.05 is outside the interval, that seems to be a bit too optimistic.

21. a) (12.7%, 18.6%)
b) We are 95% confident, based on this sample, that the proportion of all auto accidents that involve teenage drivers is between 12.7% and 18.6%.
c) About 95% of all random samples will produce confidence intervals that contain the true population proportion.
d) Contradicts. The interval is completely below 20%.

23. Probably nothing. Those who bothered to fill out the survey may be a biased sample.

25. a) Leading question bias b) (0.48, 0.54)
c) Smaller—the sample size is larger

27. a) (18.2%, 21.8%)
b) We are 98% confident, based on the sample, that between 18.2% and 21.8% of English children are deficient in vitamin D.
c) About 98% of all random samples will produce a confidence interval that contains the true proportion.

29. CI = (0.11, 0.19). The method used assumes independence of observations. This is unlikely to be true when observing individuals with a block. The method we used (assuming independence) underestimates the variability of the sample proportion and so the SE of the sample proportion.

31. a) (15.5%, 26.3%) b) 612
c) Sample may not be random or representative. Deer that may legally be hunted may not be representative of all sexes and ages.

33. a) 141 b) 318 c) 564

35. 1801

37. 384 total, using $\hat{p} = 0.15$

39. 90%

41. a) 99% m.e. = 0.03 and so a 99% CI = (0.49, 0.55).
b) The margin of increases as the sample size decreases. The sample size for BC must be smaller than the sample size for Canada and so, we should expect an increase in margin of error. The margin of error of the confidence interval for population proportions also depends on the sample proportion $\hat{p}$. The value of $\hat{p}(1 - \hat{p})$ takes its maximum possible value when $\hat{p} = 0.5$ and decreases as $\hat{p}$ moves away from 0.5 (either increases or decreases). It is most likely that the increase in the margin of error due to the decrease in sample size will dominate the decrease due to $\hat{p}$ moving away from 0.5 (since it only moves from 0.52 to 0.58, but the decrease in the sample size can be substantial).
c) 0.06 d) 8004

43. a) Answers may vary. (0.689751, 0.830249)
b) Answers may vary. 92% of the confidence intervals contain the true population proportion. For 100 simulated samples, if X = number of CIs that contain 0.8, the X should have a binomial distribution with $n = 100$ and $p = 0.9$.

45. a) $n\hat{p} = 5 < 10$ and so the sample size is not large enough for the distribution of the sample proportion to be approximately Normal.
b) (0.05, 0.25). We are 95% confident that between 5% to 25% of all puppies in this population has early hip dysplasia.

49. a) 7.9 b) (832.0, 872.8) km/sec
c) Michelson is 99% sure that this interval captures the true speed of light.
d) You would need to check that the data was distributed normally, as your use of the z^* critical values depends on this assumption. Also, the measurements would need to be independent and unbiased for our analysis to hold.
e) This value is well outside his 99% confidence interval. He was either very unlucky, or some of our assumptions about the normality and unbiasedness of his data are not correct. However, the result is still remarkably close for 1897!

51. To achieve a reduction of 50%: $n = 174$; for a ME of 10.0: $n = 415$

Chapter 17

1. a) The new drug is not more effective than aspirin, and reduces the risk of heart attack by 44%. ($p = 0.44$)
b) The new drug is more effective than aspirin, and reduces the risk of heart attack by more than 44%. ($p > 0.44$)

3. a) Since the P-value of 0.28 is big, there is insufficient evidence to conclude that the new drug is better than aspirin.
b) Since the P-value of 0.004 is very small, there is evidence that the new drug is more effective than aspirin.

5. The alternative hypothesis would be one-sided, because the only evidence that would support the friend's claim is guessing more than 25% of the suits correctly.

7. a) H_0: $p = 0.30$; H_A: $p < 0.30$ b) H_0: $p = 0.50$; H_A: $p \neq 0.50$
c) H_0: $p = 0.20$; H_A: $p > 0.20$

9. Statement d is correct.

11. No, we can say only that there is a 27% chance of seeing the observed effectiveness just from natural sampling variation. There is no *evidence* that the new formula is more effective, but we can't conclude that they are equally effective.

13. a) No. There's a 25% chance of losing twice in a row. That's not unusual.
b) 0.125 c) No, we expect that to happen 1 time in 8.
d) Maybe 5? The chance of 5 losses in a row is only 1 in 32, which seems unusual.

15. 1. Use p, not $\hat{p}$, in hypotheses.
2. The question was about failing to meet the goal, so H_A should be $p < 0.96$.
3. Did not check 0.04(200) = 8. Since $nq < 10$, the Success/Failure condition is violated. Didn't check 10% condition.
4. $188/200 = 0.94$; $SD(\hat{p}) = \sqrt{\frac{pq}{n}} = \sqrt{\frac{(0.96)(0.04)}{200}} \approx 0.014$.
5. z is incorrect; should be $z = \frac{0.94 - 0.96}{0.014} \approx -1.43$.
6. $P = P(z < -1.43) = 0.076$
7. There is only weak evidence that the new instructions do not work.

17. a) H_0: $p = 0.30$; H_A: $p > 0.30$
b) Possibly an SRS; we don't know if the sample is less than 10% of his customers, but it could be viewed as less than 10% of all possible customers; $(0.3)(80) \geq 10$ and $(0.7)(80) \geq 10$. Wells are independent only if customers don't have farms on the same underground springs.
c) $z = 0.73$; P-value $= 0.232$
d) If his dowsing is no different from standard methods, there is more than a 23% chance of seeing results as good as those of the dowser's, or better, by natural sampling variation.
e) These data provide no evidence that the dowser's chance of finding water is any better than normal drilling.

19. There is no need to conduct any hypothesis testing since the figures (i.e., 64.1% in 2013 and 60.7% in 2006) are already the parameters (recall that it is a census).

21. a) H_0: $p = 0.05$ vs. H_A: $p < 0.05$
b) We assume the whole mailing list has over 1 000 000 names. This is a random sample, and we expect 5000 successes and 95 000 failures.

c) $z = -3.178$; P-value $= 0.00074$, so we reject H_0; there is strong evidence that the donation rate would be below 5%.

23. a) $H_0: p = 0.296$, $H_A: p > 0.296$
b) The sample is representative. $240 < 10\%$ of all law school applicants. We expect $240(0.296) = 71.04$ to be admitted and $240(0.704) = 168.96$ not to be, both at least 10. $z = 1.69$; P-value $= 0.046$.
c) Although the evidence is weak, there is some indication that the program may be successful. Candidates should decide whether they can afford the time and expense.

25. $H_0: p = 0.20$; $H_A: p > 0.20$. SRS (not clear from information provided); 22 is more than 10% of the population of 150; $(0.20)(22) < 10$. Do not proceed with a test.

27. $H_0 = p = 0.03$; $H_A: p \neq 0.03$; $\hat{p} = 0.015$. One mother having twins will not affect another, so observations are independent; not an SRS; sample is less than 10% of all births. However, the mothers at this hospital may not be representative of all teenagers; $(0.03)(469) = 14.07 \geq 10$; $(0.97)(469) \geq 10$. $z = -1.91$; P-value $= 0.0556$. With a P-value this low, reject H_A. These data show some evidence that the rate of twins born to teenage girls at this hospital is less than the national rate of 3%. It is not clear whether this can be generalized to all teenagers.

29. $H_0: p = 0.25$; $H_A: p > 0.25$. SRS; sample is less than 10% of all potential subscribers; $(0.25)(500) \geq 10$; $(0.75)(500) \geq 10$. $z = 1.24$; P-value $= 0.1076$. The P-value is high, so do not reject H_0. These data do not show that more than 25% of current readers would subscribe; the company should not go ahead with the WebZine on the basis of these data.

31. $H_0: p = 0.40$; $H_A: p < 0.40$. Data are for all executives in this company and may not be able to be generalized to all companies; $(0.40)(43) \geq 10$; $(0.60)(43) \geq 10$. $z = -1.31$; P-value $= 0.0955$. Because the P-value is high, we fail to reject H_0. These data do not show that the proportion of women executives is less than the 40% of women in the company in general.

33. a) $H_0: p = 0.5$ vs $H_A: p \neq 0.5$. The test is two-sided.

35. $H_0: p = 0.90$; $H_A: p < 0.90$. $\hat{p} = 0.844$; $z = -2.05$; P-value $= 0.0201$. Because the P-value is so low, we reject H_0. There is strong evidence that the actual rate at which passengers with lost luggage are reunited with it within 24 hours is less than the 90% claimed by the airline.

37. a) Yes; assuming this sample to be a typical group of people, $P = 0.0008$. This cancer rate is very unusual.
b) No, this group of people may be atypical for reasons that have nothing to do with the radiation.

39. $H_0: p = 0.5$ vs $H_A: p \neq 0.5$. The test is two-sided (given). $\hat{p} = 0.4468$, $z = -2.31$, P-value $= 0.0208$. We have moderately strong evidence that more Canadians would choose Obama as their most admired figure.

Chapter 18

1. a) Two sided. Let p be the percentage of students who prefer Diet Pepsi. $H_0: p = 0.50$ vs. $H_A: p \neq 0.50$
b) One sided. Let p be the percentage of teenagers who prefer the new formulation. $H_0: p = 0.50$ vs. $H_A: p > 0.50$
c) One sided. Let p be the percentage of people who intend to vote for the override. $H_0: p = 2/3$ vs. $H_A: p > 2/3$
d) Two sided. Let p be the percentage of days that the market goes up. $H_0: p = 0.50$ vs. $H_A: p \neq 0.50$

3. If there is no difference in effectiveness, the chance of seeing an observed difference this large or larger is 4.7% by natural sampling variation.

5. $\alpha = 0.05$. Yes. The P-value is less than 0.05, so it's less than 0.10. But to reject H_0 at $\alpha = 0.01$, the P-value must be below 0.01, which isn't necessarily the case.

7. a) There is only a 1.1% chance of seeing a sample proportion as low as 89.4% vaccinated by natural sampling variation if 90% have really been vaccinated.
b) We conclude that p is below 0.9, but a 95% confidence interval would show that the true proportion is between (0.889, 0.899). Most likely, a decrease from 90% to 89.9% would not be considered important. On the other hand, with 1 000 000 children a year vaccinated, even 0.1% represents about 1000 kids—so this may very well be important.

9. a) (1.9%, 4.1%)
b) Because 5% is not in the interval, there is strong evidence that fewer than 5% of all men use work as their primary measure of success.
c) $\alpha = 0.01$; it's a lower-tail test based on a 98% confidence interval.

11. a) (0.065520, 0.079994)
b) Since 0.13 is not in the confidence interval, we reject the null hypothesis that $p = 0.13$. Since all values in the 95% CI lie below 0.13, we have some reasons to suspect a possible lack of rural applicants.

13. a) Type II error b) Type I error
c) By making it easier to get the loan, the bank has reduced the alpha level.
d) The risk of a Type I error is decreased and the risk of a Type II error is increased.

15. a) Power is the probability that the bank denies a loan that would not have been repaid.
b) Raise the cutoff score.
c) A larger number of trustworthy people would be denied credit, and the bank would miss the opportunity to collect interest on those loans.

17. a) The null hypothesis is that the obesity rate in First Nations youth has not decreased and the alternative hypothesis is that the obesity rate in First Nations youth has decreased.
b) Type I error in this situation is the error in concluding that the obesity rate in First Nations youth has decreased when in fact it has not decreased.
c) Type II error in this situation is the error in concluding that the obesity rate in First Nations youth has not decreased when in fact it has decreased.
d) If we make a type I error, this harms the federal and provincial governments. If we make a type II error, this harms the First Nations children as they are losing a program effective in reducing obesity rate.
e) The power of the test represents the probability of concluding that the obesity rate in First Nations youth has decreased when in fact it has decreased.

19. a) It is decided that the shop is not meeting standards when it is.
b) The shop is certified as meeting standards when it is not.
c) Type I d) Type II

21. a) The probability of detecting a shop that is not meeting standards.
b) 40 cars. Larger n.
c) 10%. More chance to reject H_0.
d) A lot. Larger differences are easier to detect.

23. a) One-tailed. The company wouldn't be sued if "too many" minorities were hired.
b) Deciding the company is discriminating when it is not.
c) Deciding the company is not discriminating when it is.
d) The probability of correctly detecting discrimination when it exists.
e) Increases power.
f) Lower, since n is smaller.

25. a) One-tailed. Software is supposed to decrease the dropout rate.
b) $H_0: p = 0.13$; $H_A: p < 0.13$
c) He buys the software when it doesn't help students.

d) He doesn't buy the software when it does help students.
e) The probability of correctly deciding the software is helpful.

27. a) We advise to buy a turbine, when in truth it will be a losing proposition financially.
b) We advise not to buy a turbine, when in truth it would have been a money-saving investment.
c) Decrease it so that there is less of a chance of recommending a bad investment in a turbine.
d) Power is the probability of advising to purchase a turbine when in fact it will be a money-saving investment.
e) Increase.
f) Make more measurements, say every 3 hours.
g) No, the small p-value indicates very strong and clear evidence that the wind speed exceeds 8 mph, but the wind speed might not exceed 8 mph by much. This can happen in this particular situation because of the large sample size, so it would be useful to look at a confidence interval as well to see how large the average wind speed might be.

29. a) $z = -3.21, p = 0.007$. The change is statistically significant. A 95% confidence interval is (2.3%, 8.5%). This is clearly lower than 13%. If the cost of the software justifies it, the professor should consider buying the software.
b) The chance of observing 11 or fewer dropouts in a class of 203 is only 0.07% if the dropout rate is really 13%.

31. a) $H_A: p = 0.30$, where p is the probability of heads.
b) Reject the null hypothesis if the coin comes up tails—otherwise fail to reject.
c) P(tails given the null hypothesis) $= 0.1 = \alpha$
d) P(tails given the alternative hypothesis) $=$ power $= 0.70$
e) Spin the coin more than once and base the decision on the sample proportion of heads.

33. a) 0.0464 b) Type I c) 37.6%
d) Increase the number of shots. Or keep the number of shots at 10, but increase alpha by declaring that 8, 9, or 10 will be deemed as having improved.

35. a) $\hat{p} > 0.6901003607$ b) 0.8925 c) $n = 147$

37. a) Answers may vary. $\hat{p} = 0.826667, n = 75, z = 0.58$, p-value $= 2 \times P(Z > 0.58) = 0.564 > 0.05$ and so there is no evidence to reject the null hypothesis $p = 0.8$ against the two-sided alternative.
b) Answers may vary. Ninety-four of the tests do not reject the null hypothesis at the 5% level and 6 of them reject the null hypothesis. We would expect 95% percent of tests to produce a correct decision if we repeated these simulations many times. If $X =$ number wrong decisions, then X has a binomial distribution with $n = 100$ and $p = 0.05$.

39. a) Answers may vary. $\hat{p} = 0.906667, n = 75, z = 3.91$, p-value $= 2 \times P(Z > 3.91) = 0.00 < 0.05$ and so we have evidence to reject the null hypothesis $p = 0.7$ against the two-sided alternative.
b) Answers may vary. Forty-seven of the tests reject the null hypothesis at the 5% percent level and 53 of them failed to reject the null hypothesis. The probability of a type II error $= 0.53$ and the power $= 0.47$. An estimate of power from a simulation based on 1500 samples (each of size 75) was 0.465333.
c) Exact power $= 0.467982$

41. Let $X =$ the number of significant tests. Then X is Binomial (10, 0.05), so $P(X \geq 1) = 0.40$. A single significant result in 10 trials is easily explained by chance, so we should not reject the null hypothesis.

Chapter 19

1. It is very unlikely that samples would show an observed difference this large if, in fact, there was no real difference in the proportions teens that have Facebook accounts and teens that have MySpace accounts.

3. Without knowing anything about how the polls were taken, we should be cautious about drawing conclusions from this study. For example, the second poll could have been taken at a support rally for this candidate, causing a large source of bias. If the same group was polled twice, we do not satisfy the independence assumption. Assuming proper polling techniques, a P-value this small would give us good evidence that the campaign ads are working.

5. The responses are not from two independent groups, but are from the same individuals.

7. a) Stratified b) 6% higher among males c) 4%
d)

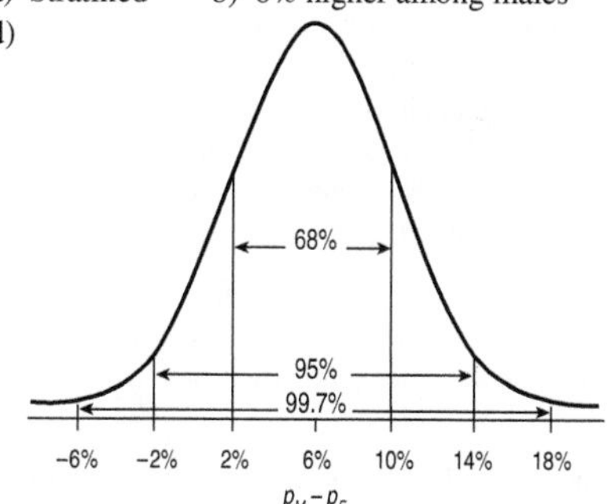

e) Yes. A poll result showing little difference is only 1–2 standard deviations below the expected outcome.

9. a) Yes. Random sample; less than 10% of the population; samples are independent; more than 10 successes and failures in each sample.
b) (0.055, 0.140)
c) We are 95% confident, based on these samples, that the proportion of American women age 65 and older who suffer from arthritis is between 5.5% and 14.0% more than the proportion of American men of the same age who suffer from arthritis.
d) Yes. The entire interval lies above 0.

11. a) 0.035 b) (0.356, 0.495)
c) We are 95% confident, based on these data, that the proportion of pets with a malignant lymphoma in homes where herbicides are used is between 35.6% and 49.5% higher than the proportion of pets with lymphoma in homes where no pesticides are used.

13. a) Yes. Subjects were randomly divided into independent groups, and more than 10 successes and failures were observed in each group.
b) (4.7%, 8.9%)
c) Yes, we're 95% confident that the rate of infection is 5–9 percentage points lower. That's a meaningful reduction, considering the 20% infection rate among the unvaccinated kids.

15. a) $H_0: p_V - p_{NV} = 0, H_A: p_V = p_{NV} < 0$
b) Because 0 is not in the confidence interval, reject the null. There's evidence that the vaccine reduces the rate of ear infections.
c) Type I
d) Babies would be given ineffective vaccinations.

17. a) Prospective study
b) $H_0: p_1 - p_2 = 0$; $H_A: p_1 - p_2 \neq 0$ where p_1 is the proportion of students whose parents disapproved of smoking who became smokers and p_2 is the proportion of students whose parents are lenient about smoking who became smokers.
c) Yes. We assume the students were randomly selected; they are less than 10% of the population; samples are independent; at least 10 successes and failures in each sample.
d) $z = -1.17$, P-value $= 0.2422$. These samples do not show evidence that parental attitudes influence teens' decisions to smoke.
e) If there is no difference in the proportions, there is about a 24% chance of seeing the observed difference or larger by natural sampling variation.
f) Type II

19. a) $(-0.065, 0.221)$

b) We are 95% confident that the proportion of teens whose parents disapprove of smoking who will eventually smoke is between 22.1% less and 6.5% more than for teens with parents who are lenient about smoking.
c) 95% of all random samples will produce intervals that contain the true difference.

21. a) No. This is observational data.
b) $H_0: p_1 - p_2 = 0$; $H_A: p_1 - p_2 \neq 0$. $z = 3.56$, P-value $= 0.0004$. With a P-value this low, we reject H_0. There is a significant difference in the clinic's effectiveness. Younger mothers have a higher birth rate than older mothers. Note that the Success/Failure Condition is met based on the pooled estimate of p.
c) We are 95% confident, based on these data, that the proportion of successful live births at the clinic is between 10.0% and 27.8% higher for mothers under 38 than in those 38 and older. However, the Success/Failure Condition is not met for the older women, so we should be cautious when using this interval.

23. a) $H_0: p_1 - p_2 = 0$; $H_A: p_1 - p_2 > 0$. $z = 1.18$, P-value $= 0.118$. With a P-value this high, we fail to reject H_0. These data do not show evidence of a decrease in the voter support for the candidate.
b) Type II error

25. a) We are 95% confident, based on this study, that between 67.0% and 83.0% of patients with joint pain will find medication A effective.
b) We are 95% confident, based on this study, that between 51.9% and 70.3% of patients with joint pain will find medication B effective.
c) Yes, they overlap. This might indicate no difference in the effectiveness of the medications, although this is not a proper test.
d) We are 95% confident that the proportion of patients with joint pain who will find medication A effective is between 1.7% and 26.1% higher than the proportion who will find medication B effective.
e) No. There is a difference in the effectiveness of the medications.
f) To estimate the variability in the difference of proportions, we must add variances. The two one-sample intervals do not. The two-sample method is the correct approach.

27. The conditions are satisfied to test $H_0: p_{young} = p_{old}$ against $H_A: p_{young} > p_{old}$. The one-sided P-value is 0.0619, so we may reject the null hypothesis. Although the evidence is not strong, *Time* may be justified in saying that younger men are more comfortable discussing personal problems.

29. Yes. With a low P-value of 0.003, reject the null hypothesis of no difference. There's evidence of an increase in the proportion of parents checking the Web sites visited by their teens.

31. Yes. With a low P-value of 0.0013, reject the null hypothesis of no difference. There is strong evidence to suggest that the percentages are different for the two groups: People from urban areas are more likely to agree with the statement than those from rural areas.

33. In order to use the methods discussed in this chapter, the two samples must be independent. Since we have the same 44 species-size combinations at two different times, they may not be independent. If this is the case, the methods we have discussed in this chapter cannot be used to carry out a test for comparing the two proportions.

35. a) $H_0: p_M - p_F = 0$ vs $H_A: p_M - p_F \neq 0$. $z = 1.41$, P-value $=$ 0.1586. The samples observed do not provide evidence against the null hypothesis at the 5% level of significance. That is, the samples do not provide sufficient evidence of a difference in the proportion of males and females who read newspapers on the subway.
b) $H_0: p_M - p_F = 0$ vs $H_A: p_M - p_F \neq 0$. $z = 1.96$, P-value $= 0.05$. The samples provide some evidence of a difference in the proportion of males and females who read novels on the subway.
c) Yes, the probability of one or more erroneous rejections of the null hypothesis (known as the family error rate) in the series increases as the number of tests in the series increases. For a 5% level test, the probability of type I error is 0.05. If we perform 40 tests, then the probability that all 40 tests will fail to reject when all of them are true is less than or equal to 0.95^{40} and the probability of rejecting null hypothesis in one or more of these 40 tests when all of them are true is greater than or equal to 0.8714878434, which is much higher than the acceptable level (0.05).

37. a) The table shows the conditional distributions as percentages.

Age	Male victim	Female victim
0–11	3.24	6.79
12–29	46.76	27.78
≥30	50.00	65.43
Total	100	100

b) $z = -0.3362$ and $P < .001$; very strong evidence against the null hypothesis.
c) $z = -1.9187$ and $P = 0.0550$; weak evidence against the null hypothesis.
d) $z = 4.1803$ and $P < .001$; very strong evidence against the null hypothesis.
e) There are proportionally fewer female than male victims in the 12–29 age group and more in the over 30 age group. An important problem with doing multiple tests is that the overall error rate, or probability of concluding at least one of the null hypotheses is incorrect, is inflated.
f) $z = 0.60$ and $P = 0.55$; no evidence against the null hypothesis of no gender effect.

39. a) 1538 b) 388

Part V Review

1. H_0: There is no difference in cancer rates, $p_1 - p_2 = 0$.
H_A: The cancer rate in those who use the herb is higher, $p_1 - p_2 > 0$.

3. a) 10.29
b) Not really. The z-score is -1.11. Not any evidence to suggest that the proportion for Monday is low.
c) Yes. The z-score is 2.26 with a P-value of 0.024 (two-sided).
d) Some births are scheduled for the convenience of the doctor and/or the mother.

5. a) $H_0: p = 0.40$; $H_A: p < 0.40$
b) Random sample; less than 10% of all California gas stations, $0.4(27) = 10.8$, $0.6(27) = 16.2$. Assumptions and conditions are met.
c) $z = -1.49$, P-value $= 0.0677$
d) With a P-value this high, we fail to reject H_0. These data do not provide evidence that the proportion of leaking gas tanks is less than 40% (or that the new program is effective in decreasing the proportion).
e) Yes, Type II.
f) Increase α, increase the sample size.
g) Increasing α—increases power, lowers chance of Type II error, but increases chance of Type I error. Increasing sample size—increases power, costs more time and money.

7. a) The researcher believes that the true proportion of "A's" is within 10% of the estimated 54%, namely, between 44% and 64%.
b) Small sample.
c) No, 63% is contained in the interval.

9. a) Pew uses a 95% confidence level. We can be 95% confident that the true proportion is within 2% of 13%—that is, that it is between 11% and 15%.
b) The cell phone group would have the larger ME because its sample size is smaller.
c) CI = (78.5%, 85.5%)
d) The ME is 0.035, which is larger than the 2% ME in part a, largely because of the smaller sample size. It is larger than the ME in part b, mostly because 0.82 is smaller than 0.87 and proportions closer to 0.50 have larger MEs.

11. a) Bimodal
b) μ, the population mean. Sample size does not matter.
c) $\frac{\sigma}{\sqrt{n}}$; sample size does matter.
d) It becomes closer to a Normal model and narrower as the sample size increases.

13. H_0: There is no difference, $p_1 - p_2 = 0$. H_A: Early births have increased, $p_1 - p_2 < 0$. $z = -0.729$, P-value $= 0.2329$. Because the P-value is so high, we do not reject H_0. These data do not show an increase in the incidence of early birth of twins.

15. a) H_0: There is no difference, $p_1 - p_2 = 0$. H_A: Treatment prevents deaths from eclampsia, $p_1 - p_2 < 0$.
b) Samples are random and independent; less than 10% of all pregnancies (or eclampsia cases); more than 10 successes and failures in each group.
c) 0.8008
d) There is insufficient evidence to conclude that magnesium sulfate is effective in preventing eclampsia deaths.
e) Type II
f) Increase the sample size or increase the level of significance.
g) Increasing sample size: decreases variation in the sampling distribution, is costly. Increasing level of significance: increases likelihood of rejecting the null hypothesis, but increases the chance of a Type I error.

17. a) The actual margin of error is $ME = 1.96\sqrt{\frac{\hat{p}(1-\hat{p})}{1000}}$ ≈ 0.0201, or about 2%.
b) We cannot do this without knowing the sample size for Quebec. If we are told Quebec's population is 23% of Canada's, we could base a calculation on an approximate sample size for Quebec of $0.23(1000) = 230$, which gives $ME \approx 0.0474$, or about 4.74%.
c) The margin of error is inversely proportional to the square root of the sample size. Increasing the sample size by a factor f reduces the ME for both all of Canada and just Quebec by $\frac{1}{\sqrt{f}}$, so we want $\frac{1}{\sqrt{f}} = 1/4$ or $f = 16$. The sample size for Canada would need to increase from 1000 to $1000 \times 16 = 16000$.

19. a) H_0: There is no difference, $p = 0.126$. H_A: The fatal accident rate is lower in teenage girls, $p < 0.126$. $z = -0.748$, P-value $= 0.227$. Because the P-value is high, we fail to reject H_0. There is little evidence that the fatal accident rate is lower for teenage girls than for teens in general.
b) If the proportion is really 12.6%, we will see the observed proportion (11.34%) or lower in about 22.7% of samples of size 388 due to sampling variation.

21. a) One would expect many small fish, with a few large ones.
b) We don't know the exact distribution, but we know it's not Normal.
c) Probably not. With a skewed distribution, a sample size of five is not a large enough sample to say the sampling model for the mean is approximately Normal.
d) 0.961

23. a) Yes. $(60)(0.80) = 48$, $(60)(0.20) = 12$. Both are ≥ 10.
b) 0.834
c) Higher. Bigger sample means smaller standard deviation for $\hat{p}$.
d) Answers will vary. For $n = 500$, the probability is 0.997.

25. a) 54.4% to 62.5%
b) Based on this study, with 95% confidence the proportion of Crohn's disease patients who will respond favourably to infliximab is between 54.4% and 62.5%.
c) 95% of all such random samples will produce confidence intervals that contain the true proportion of patients who respond favourably.

27. At least 423, assuming that p is near 50%.

29. a) Assume random sample; certainly less than 10% of all preemies and normal babies; more than 10 failures and successes in each group. 1.7% to 16.3% greater for children of normal birth weight.
b) Since 0 is not in the interval, there is evidence that preemies have a lower high school graduation rate than children of normal birth weight.
c) Type I, since we rejected the null hypothesis.

31. a) H_0: The computer is undamaged. H_A: The computer is damaged.
b) 20% of good PCs will be classified as damaged (bad), while all damaged PCs will be detected (good).
c) 3 or more.
d) 20%
e) By switching to two or more as the rejection criterion, 7% of the good PCs will be misclassified, but only 10% of the bad ones will, increasing the power from 20% to 90%.

33. a) The company is interested only in confirming that the athlete is well-known.
b) Type I: the company concludes that the athlete is well-known, but that's not true. It offers an endorsement contract to someone who lacks name recognition. Type II: the company overlooks a well-known athlete, missing the opportunity to sign a potentially effective spokesperson.
c) Type I would be more likely, but Type II less likely.

35. I am 95% confident that the proportion of U.S. adults who favour nuclear energy is between 7 and 19 percentage points higher than the proportion who would accept a nuclear plant near their area.

37. a) Using a two sample test for equality of proportions, with pooled $\hat{p} = 0.6220, \hat{p}_1 = 0.5854, \hat{p}_2 = 0.65858, SE_{\text{pooled}}(\hat{p}_1 - \hat{p}_2) = 0.107094$, we get $z \approx -0.6835$ which gives $P < .49$ and no evidence against the null hypothesis that the two rates are equal.
b) For the Canucks, $\hat{p} = 0.5488, SE_{\text{pooled}}(\hat{p}_1 - \hat{p}_2) \approx 0.1099$, $\hat{p}_1 = 0.5854, \hat{p}_2 = 0.5487, z = 0.6658$, and $P = 0.51$, so there is no evidence of a difference in home-away difference in winning proportion. For the Flames, $SE_{\text{pooled}}(\hat{p}_1 - \hat{p}_2) \approx 0.1096, \hat{p}_1 = 0.5610$, $\hat{p}_2 = 0.6585, z = 1.7802, P = 0.075$, so there is weak evidence against the null hypothesis of no difference in winning proportions for home and away games for the Flames.
c) When you look at all the teams in the NHL, you expect some large differences in home and away winning percentages. If the winning proportion is the same for home and away games, then the number won at home is binomial with probability .5 and n = total number of games won. For the Islanders, $n = 26$ and $P(X \geq 17) = .0843$ so it is not all that unusual to win such a large proportion at home.
d) Publication bias results when insignificant results are not reported. Because the number of insignificant studies not reported is large, many of the reported significant results could be type I errors. When researchers select their hypotheses on the basis of observed results, significance is also overstated, i.e., results appear more significant than they actually are.

39. a) Based on a random sample of 1443, the margin of error should be $ME = 0.0258$ or 2.6%. The published ME is larger at 4.7%. This is not a random sample, but a stratified sample where random samples are taken within each province. In addition, the results are post-stratified by age to focus on the 15–24 age group. The published ME takes these factors into account.
b) A 99% confidence interval requires adjustment of the confidence level from 0.95 to 0.99 in the margin of error, using $\hat{p} \pm 2.58\,ME/1.96$. Here $\hat{p} = 0.529$, and $ME = 0.047$, so the confidence interval is $0.529 \pm 2.58(0.047)/1.96$ or (0.4671, 0.5909). An interval constructed in this way will contain the true proportion 0.99 of the time.

Chapter 20

1. a) 1.74 b) 2.37 c) 0.0524 d) 0.0889

3. Shape becomes closer to Normal; centre does not change; spread becomes narrower.

5. a) The confidence interval is for the population mean, not the individual cows in the study.
b) The confidence interval is not for individual cows.
c) We *know* the average gain in this study was 56 pounds!
d) The average weight gain of all cows does not vary. It's what we're trying to estimate.
e) No. There is not a 95% chance for another sample to have an average weight gain between 45 and 67 pounds. There is a 95% chance that another sample will have its average weight gain within two standard errors of the true mean.

7. a) No. A confidence interval is not about individuals in the population.
b) No. It's not about individuals in the sample, either.
c) No. We know the mean cost for students in the sample was $1467.
d) No. A confidence interval is not about other sample means.
e) Yes. A confidence interval estimates a population parameter.

9. a) Based on this sample, we can say, with 95% confidence, that the mean pulse rate of adults is between 70.9 and 74.5 beats per minute.
b) 1.8 beats per minute
c) Larger

11. The assumptions and conditions for a t-interval are not met. The distribution is highly skewed to the right and there is a large outlier.

13. a) Yes. Randomly selected group; the histogram is not unimodal and symmetric, but it is not highly skewed and there are no outliers, so with a sample size of 52, there is no problem.
b) (36.70, 36.96)
c) We are 98% confident that the interval 36.7°C to 36.96°C contains the true mean body temperature for adults.
d) 98% of all random samples of size 52 will produce intervals that contain the true mean body temperature of adults.
e) Since the interval is completely below the body temperature of 37°C, there is strong evidence that the true mean body temperature of adults is lower than 37°C.

15. a) Narrower. A smaller margin of error, so less confident.
b) Advantage: more chance of including the true value. Disadvantage: wider interval.
c) Narrower; due to the larger sample, the SE will be smaller.
d) About 313

17. a) (709.9, 802.5)
b) We are 95% confident that the interval 299 709.9 to 299 802.5 km/sec contains the speed of light.
c) Assumed that the measurements are independent of each other and that the distribution of the population of all possible measurements is Normal.

19. a) There is no time trend. The histogram looks unimodal, and slightly skewed to the left.
b) (80.22, 81.29)
c) We are 90% confident that the interval from 80.22% to 81.29% contains the true mean monthly percentage of on-time flight departures.
d) If the number of flights per month is known, use all the data to construct an interval for the overall proportion.

21. The 95% confidence interval lies entirely above the 0.08 ppm limit, evidence that mirex contamination is too high and consistent with rejecting the null. We used an upper-tail test, so the P-value should therefore be smaller than $\frac{1}{2}(1 - 0.95) = 0.025$, and it was.

23. If in fact the mean cholesterol of pizza eaters does not indicate a health risk, then only 7 of every 100 samples would have mean cholesterol levels as high (or higher) as observed in this sample.

25. a) Upper-tail. We want to show it will hold 500 pounds (or more) easily.
b) They will decide the stands are safe when they're not.
c) They will decide the stands are unsafe when they are in fact safe.

27. a) Decrease α. This means a smaller chance of declaring the stands safe if they are not.
b) The probability of correctly detecting that the stands are capable of holding more than 500 pounds.
c) Decrease the standard deviation—probably costly. Increase the sample size—takes more time for testing and is costly. Increase α—more Type I errors. Increase the "design load" to be well above 500 pounds—again, costly.

29. a) H_0: $\mu = 22.6$; H_A: $\mu > 22.6$
b) We have a random sample of the population. Population may not be normally distributed, as it would be easier to have a few much older women at their first marriage than some very young women. However, with a sample size of 40, it should be safe to proceed.
c) The standard deviation of the population, $\sigma(\bar{y})$ will be estimated by $SE(\bar{y}) = \frac{s}{\sqrt{n}}$, and we will use a Student's t model, with $40 - 1 = 39$ degrees of freedom.
d) The P-value is <0.00001.
e) If the mean age at first marriage is still 22.6 years, there is a near-zero chance of getting a sample mean of 27.2 years or older simply from natural sampling variation.
f) Since the P-value is low, we reject the null hypothesis.

31. a) Probably a representative sample; the Nearly Normal Condition seems reasonable from a Normal probability plot. The histogram is nearly uniform, with no outliers or skewness.
b) $\bar{y} \approx 28.78$ grams, $s \approx 0.40$ grams
c) (28.36, 29.21) grams
d) Based on this sample, we are 95% confident the average weight of the content of Ruffles bags is between 28.36 and 29.21 grams.
e) The company is erring on the safe side, as it appears that, on average, it is putting in slightly more chips than stated.

33. a) Type I; he mistakenly rejected the null hypothesis that $p = 0.10$ (or worse).
b) Yes. These are a random sample of bags and the Nearly Normal Condition is met; $t = -2.51$ with 7 df for a one-sided P-value of 0.0203.

35. a) Random sample; the Nearly Normal Condition seems reasonable from a Normal probability plot. The histogram is roughly unimodal and symmetric with no outliers.
b) (1187.9, 1288.4) chips
c) Based on this sample, the mean number of chips in an 18-ounce bag is between 1187.9 and 1288.4, with 95% confidence. The *mean* number of chips is clearly greater than 1000. However, if the claim is about individual bags, then it's not necessarily true. If the mean is 1188 and the SD deviation is near 94, then 2.5% of the bags will have fewer than 1000 chips, using the Normal model. If in fact the mean is 1288, the proportion below 1000 will be less than 0.1%, but the claim is still false.

37. a) The Normal probability plot is relatively straight, with one outlier at 93.8 sec. Without the outlier, the conditions seem to be met. The histogram is roughly unimodal and symmetric with no other outliers.
b) $t = 22.63$, P-value $= 0.0160$. With the outlier included, we might conclude that the mean completion time for the maze is not 60 seconds; in fact, it is less.
c) $t = 24.46$, P-value $= 0.0003$. Because the P-value is so small, we reject H_0. Without the outlier, we see strong evidence that the average completion time for the maze is less than 60 seconds. The outlier here did not change the conclusion.
d) The maze does not meet the "one-minute average" requirement. Both tests rejected a null hypothesis of a mean of 60 seconds.

e) H_0: the median is 60 seconds, or $p = 0.5$ vs H_A: the median is not 60 seconds, or $p \neq 0.5$.
$\hat{p} = 0.7619, z = 2.40$, P-value = 0.0164 (two-sided).
We have good evidence to reject the null hypothesis that the median time for completion is 60 seconds, just as in part d).

39. a) (289.9, 292.3)
b) These data are not a random sample of golfers. The top professionals are not representative and were not selected at random. We might consider the 2009 data to represent the population of all professional golfers, past, present, and future.
c) The data are means for each golfer, so they are less variable than if we looked at all the separate drives.
d) Independence assumption violated.

41. a) (75.35, 142.63)
b) The plots suggest that the distribution of the data is close enough to Normal but there is one outlier (Red Deer). When there are outliers in the data set, the sample mean and the standard deviation are not good summary statistics of the data set and so the 95% confidence level quoted in (a) may not be trustworthy. The summary statistics and the confidence intervals calculated from the data set with and without the outlier are shown below. The mean is approximately 10 percentage points higher with the outlier and the standard deviation is about 15 percentage points higher.

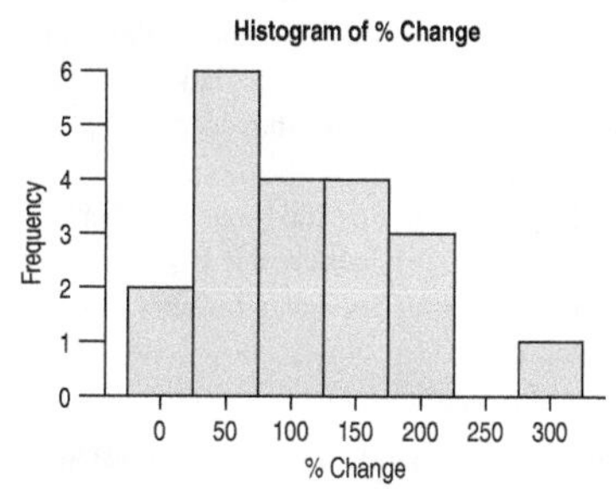

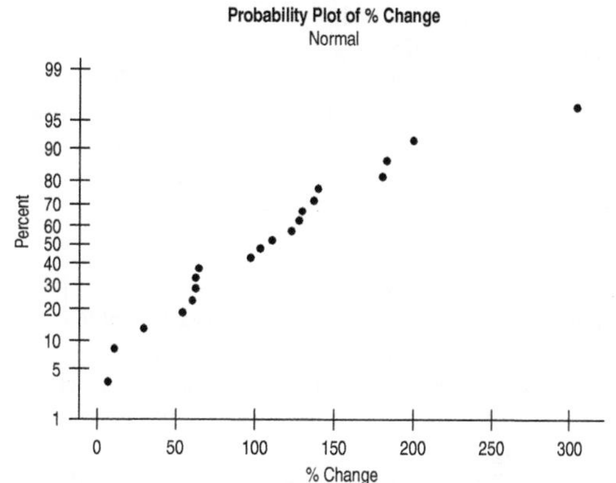

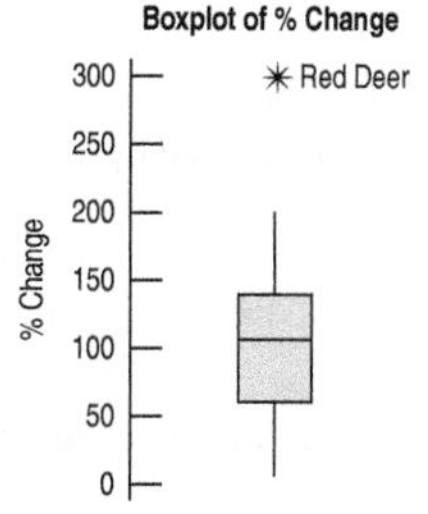

	n	Mean	StDev	SE Mean	95% CI
With the outlier	20	108.990	71.871	16.071	(75.353, 142.627)
Without the outlier	19	98.7105	56.7602	13.0217	(71.3530, 126.0681)

c) No, because our interval is for an average change per urban area, not an average per capita.

43. a) Answers may vary. 90% CI is (48.42, 55.74). This interval includes the mean (i.e., 50).
b) Answers may vary. 91 of intervals contain 50. We expect 90% percent of intervals to contain 50, if we repeated these simulations many times. *X* has a binomial distribution with $n = 100$ and $p = 0.95$.

45. a) Answers may vary. It is right skewed.

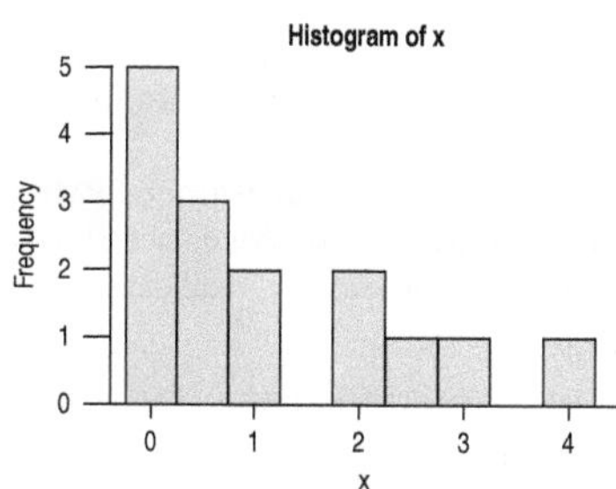

90% CI = (0.6408, 1.7987); This interval includes the mean (i.e., 1.0).
b) Answers may vary. 82 of the intervals contain the mean 1.0. The sample mean of a sample of size 15 from an exponential distribution will not have a Normal distribution and so this interval will not have the requested confidence level (i.e., 90%). If we change the sample size to 100, then the distribution of the sample mean will become approximately Normal and so approximately 90% of the CIs will contain 1.0.

47. a) Answers may vary. Mean = 51.6308, Standard deviation = 5.4848, Standard error of the mean = 2.4529, *t*-statistic = 0.66, P-value = 0.543 > 0.10, and so we do not reject the null hypothesis. Passed a good batch. Since the null hypothesis is true, the error of rejecting is the possible error here. That is type I error.
b) Answers may vary. We expect to reject 10% of the tests. Note that this is the type I error and the probability of a type I error is the level of the test. The distribution of *X* is binomial with ($n = 100, p = 0.10$).

49. a) $z^* = 1.645$ for $\alpha = 0.05$ and so the criterion is reject the null hypothesis if $\frac{\bar{y} - 0.08}{s/\sqrt{150}} > 1.645$.
b) Reject the null hypothesis if $\bar{y} > 0.08671568438$.
c) The probability of rejecting the null hypothesis is 0.7881.
d) Using software (MINITAB), the power at sample sizes 150, 75, and 38 with a difference 0.01 (i.e., 0.09–0.08) and $\alpha = 0.05$ is 0.786, 0.528, and 0.332 respectively. Note that the power decreases as the sample size decreases.

Chapter 21

1. Yes. The high P-value means that we lack evidence of a difference, so 0 is a possible value for $\mu_{Meat} - \mu_{Beef}$.

3. a) Plausible values of $\mu_{Meat} - \mu_{Beef}$ are all negative, so the mean fat content is probably higher for beef hot dogs.
b) The difference is significant. c) 10%

5. a) False; the confidence interval is about means, not about individual hot dogs.
b) False; the confidence interval is about means, not about individual hot dogs.

c) True
d) False; CIs based on other samples will also try to estimate the true difference in population means; there's no reason to expect other samples to conform to this result.
e) True

7. a) 2.927 b) Larger
c) Based on this sample, we are 95% confident that students who learn Math using the CPMP method will score, on average, between 5.57 and 11.43 points better on a test solving applied Algebra problems with a calculator than students who learn by traditional methods.
d) Yes; 0 is not in the interval.

9. a) H_0: $\mu_C - \mu_T = 0$ vs. H_A: $\mu_C - \mu_T \neq 0$
b) Yes. Groups are independent, though we don't know if students were randomly assigned to the programs. Sample sizes are large.
c) If the means for the two programs are really equal, there is less than a 1 in 10 000 chance of seeing a difference as large as or larger than the observed difference just from natural sampling variation.
d) On average, students who learn with the CPMP method do significantly worse on Algebra tests that do not allow them to use calculators than students who learn by traditional methods.

11. a) (1.36, 4.64)
b) No; 5 minutes is beyond the high end of the interval.

13.

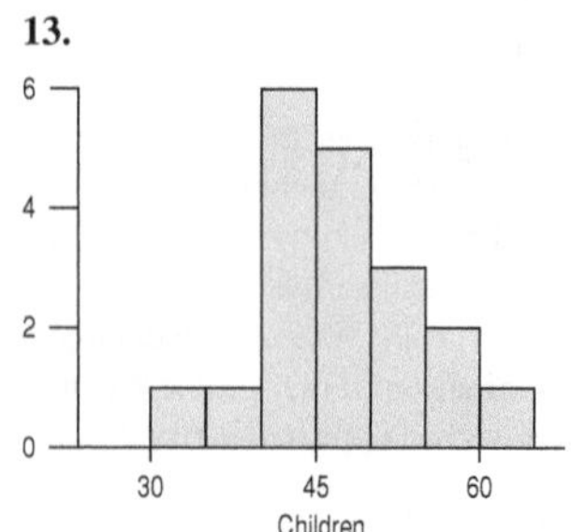

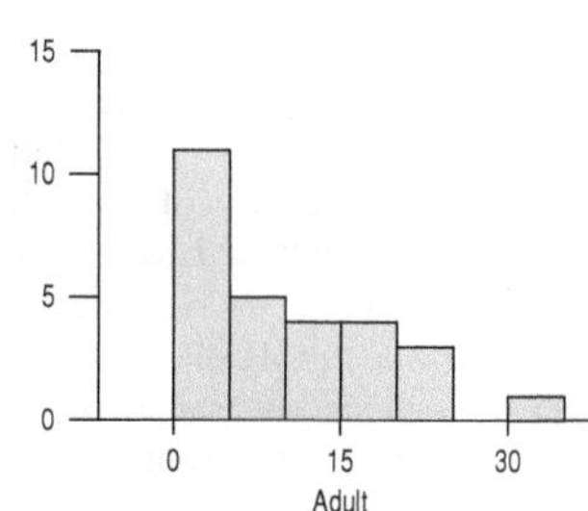

Independent group assumption—percentage of sugar in children's cereals is unrelated to the percentage of sugar in adults' cereals; Randomization condition—reasonable to assume that the cereals are representative of all adults and children's cereals in regard to sugar content; Nearly normal condition—histogram of adult cereal sugar content is skewed to the right, but the sample sizes are of reasonable size, allowing us to proceed. Based on these samples, with 95% confidence, children's cereals average between 32.49% and 40.80% more sugar content than adults' cereals.

15.

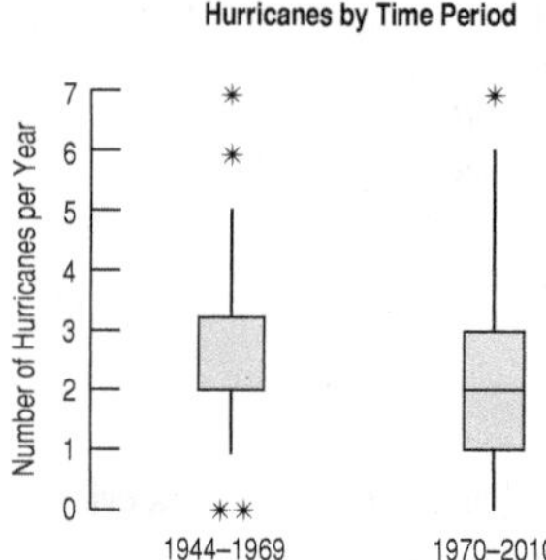

These data present some concerns: the sample is not from a randomized experiment; the data for 1944–1969 are not symmetric; and there are four outliers. Nevertheless, the data from both groups show the same degree of moderate right skewness, and their sample sizes are large enough ($n > 20$) and similar enough in shape to trust the robustness of the t-test.

17. The histograms of the scores are unimodal and symmetric; $t = 2.076$ and P-value = 0.0228. There is evidence that the students taught using the new activities have a higher mean score on the reading comprehension test than the students taught using traditional methods.

19. a) Both are roughly symmetric with roughly the same centre. The distribution of average number of home runs hit appears a bit more spread out for the American League.

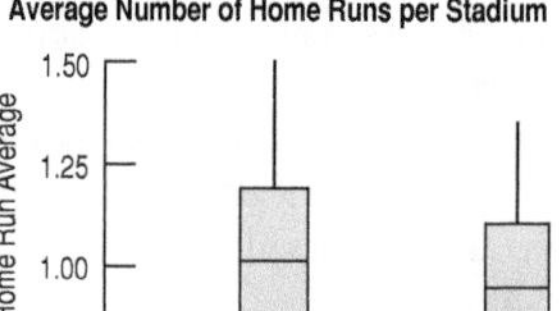

b) We are 95% confident that the mean number of home runs hit per game in American League stadiums is between 0.90 and 1.16.
c) The average of 1.354 home runs hit per game in Coors Field is not unusual. It is the highest average in the National League, but by no means an outlier.
d) If you attempt to use two confidence intervals to assess a difference in means, you are actually adding standard deviations. But it's the variances that add, not the standard deviations. The two-sample difference of means procedure takes this into account.
e) (−0.107, 0.212)
f) We are 95% confident that the mean number of home runs in American League stadiums is between 0.107 home runs lower and 0.212 home runs higher than the mean number of home runs in National League stadiums.
g) No; 0 is in the interval.

21. These are not two independent samples. These are before and after scores for the same individuals.

23. a) These data do not provide evidence of a difference in ad recall between shows with sexual content and violent content.
b) H_0: $\mu_S - \mu_N = 0$ vs. H_A: $\mu_S - \mu_N \neq 0$. $t = -6.08$, df = 213.99, P-value = 5.5×10^{-9}. Because the P-value is low, we reject H_0. These data suggest that ad recall between shows with sexual and neutral content is different; those who saw shows with neutral content had higher average recall.
c) P-value = 0.136. The pooled t-test makes sense since this is an experiment and the sample variances are similar.

25. a) H_0: $\mu_V - \mu_N = 0$ vs. H_A: $\mu_V - \mu_N \neq 0$. $t = -7.21$, df = 201.93, P-value = 1.1×10^{-11}. Because of the very small P-value, we reject H_A. There is a significant difference in mean ad recall between shows with violent content and neutral content; viewers of shows with neutral content remember more brand names, on average.
b) With 95% confidence, the average number of brand names remembered 24 hours later is between 1.45 and 2.41 higher for viewers of neutral content shows than for viewers of sexual content shows, based on these data.

27. H_0: $\mu_{big} - \mu_{small} = 0$ vs. H_A: $\mu_{big} - \mu_{small} \neq 0$; bowl size was assigned randomly; amount scooped by individuals and by the two groups should be independent. With 34 df, $t = 2.104$ and P-value = 0.0428. The low P-value leads us to reject the null hypothesis and conclude that there is a difference in the average amount of ice cream that people scoop when given a bigger bowl.

29. The data for males has an extreme outlier; we should remove it before using t-tests.

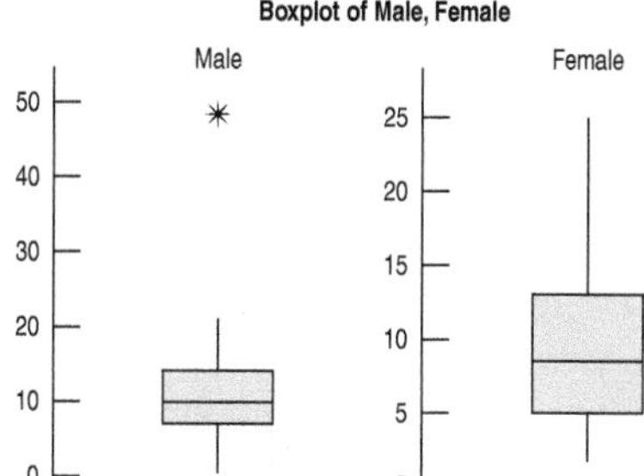

The boxplots and some summary statistics after removing the extreme outlier are shown below. We see that the spreads in the boxplots (and also the standard deviations in the summary statistics) are very similar and so we can use the pooled t-test.

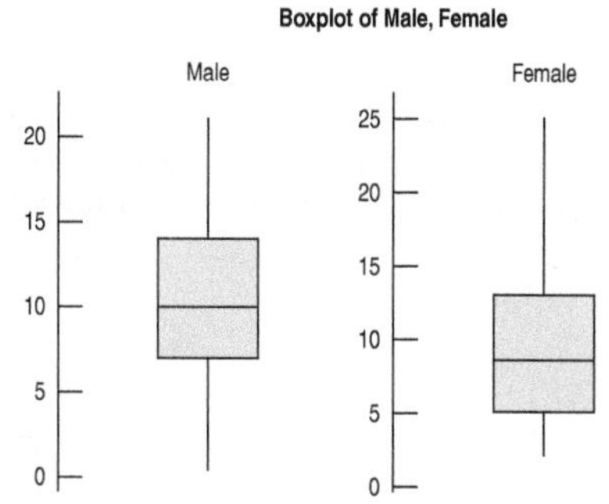

Descriptive Statistics: Male, Female

Variable	N	Mean	SE Mean	StDev
Male	45	10.322	0.820	5.498
Female	47	9.309	0.777	5.326

Test H_0: $\mu_M - \mu_F = 0$ vs. H_A: $\mu_M - \mu_F > 0$. $S_{pooled} = 5.4107$, $SE_{pooled} = 1.1285$. $t = 0.90$, P-value = 0.186 0.10. Even if the population means were equal, there is a high chance for $\bar{Y}_M - \bar{Y}_F$ to be greater than or equal to 1.013 and so the data do not provide evidence against the null hypothesis. We do not reject the null hypothesis. The data do not provide sufficient evidence to conclude that women are better at parallel parking.

31. a) We can be 95% confident that the interval -74.84 ± 178.05 minutes includes the true difference in mean crossing times between men and women. Because the interval includes zero, we cannot be confident that there is any difference at all.

b) **Independence Assumption**: There is no reason to believe that the swims are not independent or that the two groups are not independent of each other.

Randomization Condition: The swimmers are not a random sample from any identifiable population, but they may be representative of swimmers who tackle challenges such as this.

Nearly Normal Condition: The boxplots show no outliers. The histograms are unimodal; the histogram for men is somewhat skewed to the right.

c) $(-251.2, 101.6)$. Very similar to part a) but narrower due to more degrees of freedom. The pooled t-test might be appropriate because the sample variances of the two groups are similar.

33. Independent Groups Assumption: The runners are different women, so the groups are independent. The Randomization Condition is satisfied since the runners are selected at random for these heats.

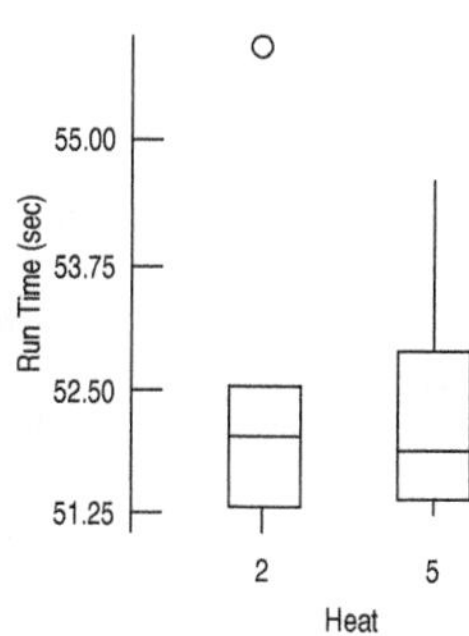

Nearly Normal Condition: The boxplots show an outlier, but we will proceed and then redo the analysis with the outlier deleted. When we include the outlier, $t \approx 0.035$ with a two-sided P-value of 0.97. With the outlier deleted, $t = -1.14$, with P = 0.2837. Either P-value is so large that we fail to reject the null hypothesis of equal means and conclude that there is no evidence of a difference in the mean times for runners in unseeded heats.

35. With $t \approx 4.57$ and a very low P-value of 0.0013, we reject the null hypothesis of equal mean velocities. There is strong evidence that golf balls hit off Stinger tees will have a higher mean initial velocity.

37. The 90% confidence interval for the difference in mean number of months to publication between *Applied Statistics* and *The American Statistician* is (8.43, 11.57) months. We assume that the articles are a random sample of the kinds of articles typically submitted to these two journals and that they represent fewer than 10% of all articles that could have been submitted. Both samples are large.

39. a) H_0: $\mu_R - \mu_N = 0$ vs. H_A: $\mu_R - \mu_N < 0$. $t = -1.36$, df = 20.00, P-value = 0.0945. Because the P-value is large, we fail to reject H_0. These data show no evidence of a difference in mean number of objects recalled between listening to rap or no music at all.

b) Didn't conclude any difference.

41. $n = 69$

43. Using MINITAB, the required sample size is 27 for each sample.

Power and Sample Size

2-Sample t Test

Testing mean 1 = mean 2 (versus not =)

Calculating power for mean 1 = mean 2 + difference

Alpha = 0.05 Assumed standard deviation = 3

	Sample Target		
Difference	Size	Power	Actual Power
3	27	0.95	0.950077

Chapter 22

1. a) Randomly assign 50 hens to each of the two kinds of feed. Compare production at the end of the month.

b) Give all 100 hens the new feed for 2 weeks and the old food for 2 weeks, randomly selecting which feed the hens get first. Analyze the differences in production for all 100 hens.

c) Matched pairs. Because hens vary in egg production, the matched-pairs design will control for that.

3. a) Show the same people ads with and without sexual images, and record how many products they remember in each group. Randomly decide which ads a person sees first. Examine the differences for each person.

b) Randomly divide volunteers into two groups. Show one group ads with sexual images and the other group ads without. Compare how many products each group remembers.

5. a) The paired t-test is appropriate here. The weight measurement before and after surgical treatment is paired by subject. H_0: There is no difference in weight due to the surgical treatment. Since the *P*-value is very low, we conclude that the mean weight is lower after surgical treatment than before.

b) The two-sample *t*-test is appropriate here. The subjects are not paired across the control and experimental groups. H_0: There is no difference in weight loss between the two treatment groups. Since the *P*-value is very low, we conclude that the mean weight loss of the group receiving surgical treatment is higher than the group receiving conventional treatment.

7. a) The paired *t*-test is appropriate since we have pairs of Fridays in 5 different months. Data from adjacent Fridays within a month may be more similar than data from randomly chosen Fridays.

b) We conclude that there is evidence (P-value of 0.0212) that the mean number of cars found on the M25 motorway on Friday the 13th is less than on the previous Friday.

c) We don't know if these Friday pairs were selected at random. Obviously, if these are the Fridays with the largest differences, this will affect our conclusion. The Nearly Normal Condition appears to be met by the differences, but the sample size is small.

9. Adding variances requires that the variables be independent. These price quotes are for the same cars, so they are paired. Drivers quoted high insurance premiums by the local company will be likely to get a high rate from the online company, too.

11. a) The histogram—we care about differences in price.

b) Insurance cost is based on risk, so drivers are likely to see similar quotes from each company, making the differences relatively smaller.

c) The price quotes are paired; they were for a random sample of fewer than 10% of the agent's customers; the histogram of differences looks approximately Normal.

13. H_0: $\mu(Local - Online) = 0$ vs. H_A: $\mu(Local - Online) > 0$; with 9 df, $t = 0.83$. With a high P-value of 0.215, we don't reject the null hypothesis. These data don't provide evidence that online premiums are lower, on average.

15.

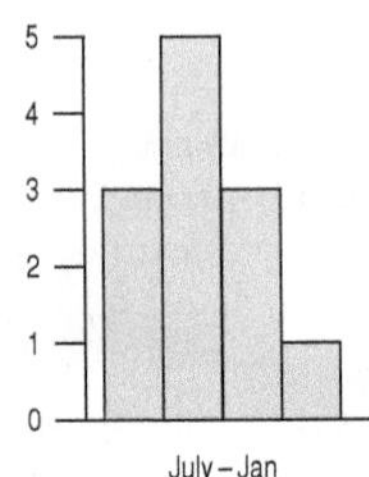

Data are paired for each city; cities are independent of each other; histogram shows the temperature differences are roughly unimodel and symmetric. This is probably not a random sample, so we might be wary of inferring that this difference applies to all European cities. Based on these data, we are 90% confident that the average temperature in European cities in July is between 32.3°F and 41.3°F higher than in January.

17. Based on these data, we are 90% confident that boys, on average, can do between 1.6 and 13.0 more push-ups than girls (independent samples—not paired).

19. a) Paired sample test. Data are before/after for the same workers; workers randomly selected; assume fewer than 10% of all this company's workers; histogram of differences shows them to be roughly unimodel and symmetric.

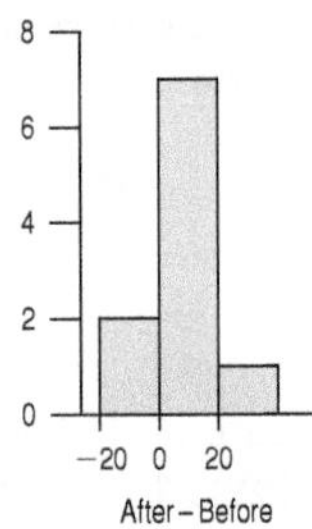

b) H_0: $\mu_d = 0$ vs. H_A: $\mu_d > 0$. $t \approx 3.60$, P-value = 0.0029. Because $P < 0.01$, reject H_0. These data show that average job satisfaction has increased after implementation of the exercise program.

c) Type I error

d) The p-value for 8 +'s and 2 −'s is $P(X = 0) + P(X = 1) + P(X = 2) = 0.0547$, where *X* is the number of −'s. Thus, we would not reject the null hypothesis, unlike in part b).

21. H_0: $\mu_D = 0$ *vs.* H_A: $\mu_D \neq 0$. Data are paired by brand; brands are independent of each other; fewer than 10% of all yogourts (questionable); histogram of differences shows an outlier for Great Value:

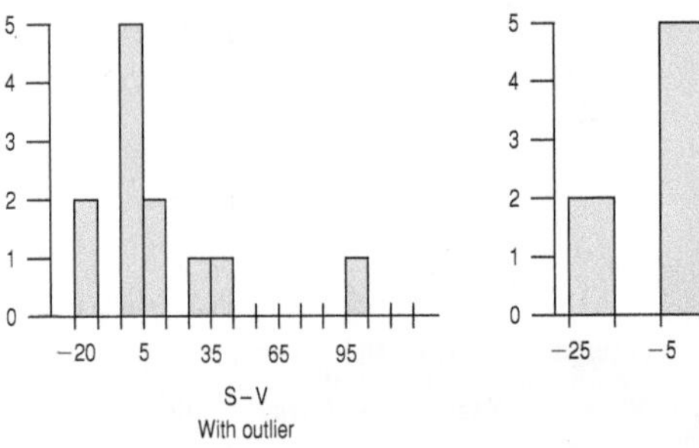

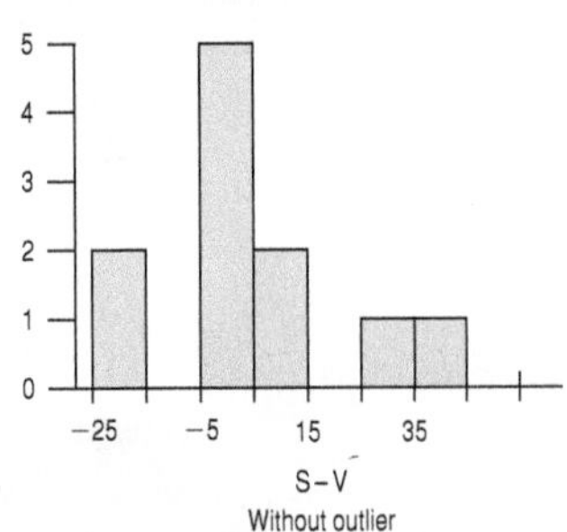

With the outlier included, the mean difference (Strawberry – Vanilla) is 12.5 calories with a *t*-stat of 1.332, with 11 df, for a P-value of 0.2098. Deleting the outlier, the difference is even smaller, 4.55 calories with a *t*-stat of only 0.833 and a P-value of 0.4241. With P-values so large, we do not reject H_0. We conclude that the data do not provide evidence of a difference in mean calories.

23. a) Not a simple random sample, but most likely representative; stops most likely independent of each other; histogram is roughly unimodel and symmetric.

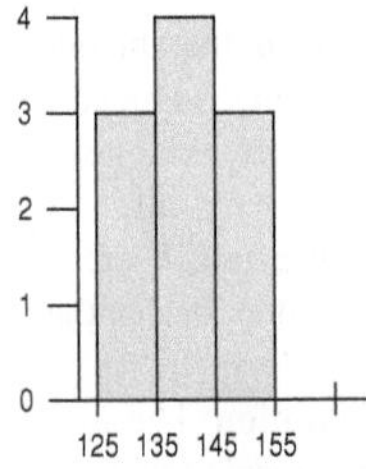

Based on these data, with 95% confidence, the average braking distance for these tires on dry pavement is between 133.6 and 145.2 feet.

b) Not simple random samples, but most likely representative; stops most likely independent of each other; less than 10% of all possible wet stops; Normal probability plots are relatively straight. Based on these data, with 95% confidence, the average increase in distance for these tires on wet pavement is between 51.4 and 74.6 feet.

25. a) These students are not random, having been selected from writers of the PISA, but they are probably representative of all students. They are less than 10% of all students. A histogram shows a roughly normal distribution with some left skew. Using paired *t*-test. 90% CI = (8.362, 14.971)

b) The confidence interval does not include zero, we have strong evidence that there is a non-zero difference between genders in math scores. The confidence interval includes 10, we cannot conclude that the mean difference is at least 10 points.
c) It is an estimate of average difference in MATH scores, not scores.

27. a) 60% is 30 strikes; H_0: $\mu_A = 30$ vs. H_A: $\mu_A > 30$. $t = 6.06$, P-value $= 3.92 \times 10^{-6}$. With a very small P-value, we reject H_0. There is very strong evidence that players can throw more than 60% strikes after training, based on this sample.
b) H_0: $\mu_d = 0$ vs. H_A: $\mu_d > 0$. $t \approx 0.135$, P-value $= 0.4472$. With such a high P-value, we do not reject H_0. These data provide no evidence that the program has improved pitching in these Little League players.
c) H_0: The median difference in number of strikes thrown before and after training is zero. H_A: The median difference in number of strikes thrown before and after training is greater than zero. Taking differences in number of strikes thrown as After – Before, we get the following:
+, +, +, −, −, −, 0, +, +, +, 0, +, −, −, −, −, + , −, −, −
(two ties). The p-value for 8 +'s and 10 −'s is
$P(X = 0) + P(X = 1) + \cdots + P(X = 10) = 0.760$, where X is the number of −'s. Thus, we would not reject the null hypothesis, just like in part b).

29. a) The data are clearly paired. Even if the individual times show a trend of improving speed over time, the differences may well be independent of each other. They are subject to random year-to-year fluctuations, and we may believe that these data are representative of similar races. We don't have any information with which to check the Nearly Normal condition.
b) With 95% confidence, the mean time difference is between −15.81 minutes and 8.67 minutes.
c) The interval contains 0, so we would not reject the hypothesis of no mean difference at $\alpha = 0.05$. We are unable to discern a difference between the female wheelchair times and the male running times.

31. a) Same cows before and after injection; the cows should be representative of others of their breed; cows are independent of each other; fewer than 10% of all cows; don't know about nearly Normal differences.
b) (12.66, 15.34)
c) Based on this sample, with 95% confidence, the average increase in milk production for Ayrshire cows given BST is between 12.66 and 15.34 pounds per day.
d) 0.25(47) = 11.75. The average increase is much more than this, so we would recommend he go to the extra expense.

33. $n = 80$. This sample size is large enough for the use of z^* in place of t^*_{n-1}.

35. Using MINITAB, we need a sample of size 57.
Power and Sample Size
1-Sample t Test
Testing mean = null (versus not = null)
Calculating power for mean = null + difference
Alpha = 0.05

Assumed standard deviation = 1139.6

	Sample Target		
Difference	Size	Power	Actual Power
500	57	0.9	0.902376

The probability that the test will conclude that the mean difference is non-zero when the true mean difference is 500 is 0.90.

Chapter 23

1. a) Chi-square test of independence. We have one sample and two variables. We want to see if the variable *Account Type* is independent of the variable *Trade Type*.
b) Other test. The variable *Account Size* is quantitative, not counts.
c) Chi-square test of homogeneity. We want to see if the distribution of one variable, *Courses*, is the same for two groups (resident and non-resident students).

3. a) 10 b) Goodness-of-fit
c) H_0: The die is fair (all faces have $p = 1/6$). H_A: The die is not fair.
d) Count data; rolls are random and independent; expected frequencies are all bigger than 5.
e) 5 f) $\chi^2 = 5.6$, P-value = 0.3471
g) Because the P-value is high, do not reject H_0. The data show no evidence that the die is unfair.

5. a) Weights are quantitative, not counts.
b) Count the number of each kind of nut, assuming the company's percentages are based on counts rather than weights.

7. H_0: The distribution of blood types of residents on Bongainville Island represents the distribution of blood types of the general population; H_A: The distribution of blood types of residents on Bongainville Island represents the distribution of blood types of the general population. One cell has expected count less than 5, the chi-square procedures are not appropriate. Cells would have to be combined in order to proceed. Combine Blood type B with blood type AB. $\chi^2 \approx 38.39$, df = 2, *P*-value = 0.0000. Because the *P*-value is so low, we reject H_0. There is strong evidence that the distribution of blood types of residents on Bongainville Island does not represent the distribution of blood types of the general population.

9. a) $\chi^2 \approx 5.671$, df = 3, *P*-value = 0.1288. With a *P*-value this high, we fail to reject H_0. Yes, these data are consistent with those predicted by genetic theory.
b) $\chi^2 \approx 11.342$, df = 3, *P*-value = 0.0100. Because of the low P-value, we reject H_0. These data provide evidence that the distribution is not as specified by genetic theory.
c) With small samples, many more data sets will be consistent with the null hypothesis. With larger samples, small discrepancies will show evidence against the null hypothesis.

11. a) 6 b) Goodness of Fit
c) H_0: The number of large hurricanes remains constant over decades. H_A: The number of large hurricanes has changed.
d) 15 e) 0.63
f) The very high P-value means these data offer no evidence that the numbers of large hurricanes has changed.
g) The final period is only 6 years rather than 10 and already 7 large hurricanes have been observed. Perhaps this decade will have an unusually large number of such hurricanes.

13. a) Independence
b) H_0: Breastfeeding success is independent of having an epidural. H_A: There's an association between breastfeeding success and having an epidural.

15. a) 1 b) 159.34
c) Breastfeeding behaviour should be independent for these babies. They are fewer than 10% of all babies; we assume they are representative. We have counts, and all the expected counts are at least 5.

17. a) 5.90 b) P-value = 0.000
c) The P-value is very low, so reject the null. There's strong evidence of an association between having an epidural and subsequent success in breastfeeding.

19. a) $c = \dfrac{Obs - Exp}{\sqrt{Exp}} = \dfrac{190 - 159.34}{\sqrt{159.34}} = 2.43$

b) It appears that babies whose mothers had epidurals during childbirth are much less likely to be breastfeeding 6 months later.

21. These factors would not be mutually exclusive. There would be yes or no responses for every baby for each.

23. a) 40.2% b) 8.1% c) 62.2% d) 285.48
e) H_0: Survival was independent of status on the ship. H_A: Survival depended on the status.
f) 3
g) We reject the null hypothesis. Survival depended on status. We can see that first-class passengers were more likely to survive than passengers of any other class.

25. First class passengers were most likely to survive, while third class passengers and crew were under-represented among the survivors.

27. a) Independence.
b) H_0: *College* is independent of *Birth Order*. H_A: There is an association between *Birth Order* and *College*.
c) Counted Data Condition—these are counts of students; Randomization Condition—a class is not a random sample, but there's little reason to think this particular group of students isn't representative; Expected Cell Frequency Condition—the expected frequencies are low for both the Social Science and Professional Colleges, for third and fourth or higher birth orders (check residuals later).
d) 9
e) With a P-value this low, we reject the null hypothesis. There is some evidence of an association between *Birth Order* and *College*.
f) Unfortunately, 3 of the 4 largest standardized residuals are in cells with expected counts less than 5. We should be very wary of drawing conclusions from this chi-square test.

29. a) Experiment—actively imposed treatments (different drinks)
b) Homogeneity
c) H_0: The rate of urinary tract infection is the same for all three groups. H_A: The rate of urinary tract infection is different among the groups.
d) Count data; random assignment to treatments; all expected frequencies larger than 5.
e) 2 f) $\chi^2 = 7.776$, P-value ≈ 0.020
g) With a P-value this low, we reject H_0. These data provide reasonably strong evidence that there is a difference in urinary tract infection rates between cranberry juice drinkers, lactobacillus drinkers, and the control group.
h) The standardized residuals are

	Cranberry	Lactobacillus	Control
Infection	−1.87276	1.191759	0.681005
No infection	1.245505	−0.79259	−0.45291

From the standardized residuals (and the sign of the residuals), it appears those who drank cranberry juice were less likely to develop urinary tract infections and those who drank lactobacillus were more likely to have infections.

31. a) Independence.
b) H_0: delivery time is independent from computer-assisted ordering. H_A: There is an association between delivery time and computer-assisted ordering.
c) Counted data, probably a random sample; one cell has expected count less than 5, the chi-square procedures are not appropriate. Cells would have to be combined in order to proceed. Combine last two columns.
d) $\chi^2 \approx 8.86$, P-value ≈ 0.003
e) Since the *P*-value is small, we reject H_0. There is evidence of an association between delivery time and computer-assisted ordering.

33. a) Homogeneity.
b) H_0: The distribution of grades is the same for the two professors. H_A: The distribution of grades is different for the two professors.
c) The expected counts are:

	Prof. Alpha	Prof. Beta
A	6.667	5.333
B	12.778	10.222
C	12.222	9.778
D	6.111	4.889
F	2.222	1.778

Since three cells have expected counts less than 5, the chi-square procedures are not appropriate. Cells would have to be combined in order to proceed.

35. a)

	Prof. Alpha	Prof. Beta
A	6.667	5.333
B	12.778	10.222
C	12.222	9.778
Below C	8.333	6.667

All expected frequencies are now larger than 5.
b) Decreased from 4 to 3.
c) $\chi^2 = 9.306$, P-value ≈ 0.0255. Because the P-value is so low, we reject H_0. The grade distributions for the two professors are different. Professor Alpha gives fewer A's and more grades below C than Professor Beta.

37. a) Answers will vary. Stacked bar charts or pie charts would work well.
b) Chi-square test of independence. One person's response should not influence another's. We have counts, an SRS of less than 10% of the population, and all expected cell frequencies are much larger than 5. With 8 df, $\chi^2 \approx 190.96$. Because the P-value < 0.001, reject the null. There is strong evidence to suggest that responses are not independent of age.

39. $\chi^2 \approx 14058$, df = 1, *P*-value ≈ 0.0002. With a P-value this low, we reject H_0. There is evidence of racial steering. Blacks are much less likely to rent in Section A than Section B.

41. a) $z \approx 3.74936$, $z^2 \approx 14.058$
b) The resulting P-values were both ≈ 0.0002. The two tests are equivalent.

43. $\chi^2 \approx 2815.968$, df = 9, P-value < 0.0001. Because the P-value is so low, we reject H_0. There are definite differences in education levels attained by the groups. The largest component indicates that Hispanics are more likely not to have high-school diplomas. The next-largest residual indicates that fewer whites than expected do not have a high-school diploma.

45. $\chi^2 \approx 0.870$, df = 2, P-value ≈ 0.6471. Because the P-value is so large, we do not reject H_0. These data show no evidence of a difference in higher education opportunities between blacks and Hispanics.

47. A goodness-of-fit test with 2 df yields $\chi^2 \approx 12.26$, with P-value = 0.002. Compared to the top 500 universities, the top 100 seem to be more heavily concentrated in the Americas and sparser in the Asia, Africa, and Pacific regions.

Part VI Review

1. a) H_0: $\mu_{Jan} - \mu_{Jul} = 0$; H_A: $\mu_{Jan} - \mu_{Jul} \neq 0$. $t = -1.94$, df = 43.0, P-value = 0.0590. Since P-value is fairly low, reject the null. These data show some evidence of a difference in mean *Age* to crawl between January and July babies.
b) H_0: $\mu_{Apr} - \mu_{Oct} = 0$; H_A: $\mu_{Apr} - \mu_{Oct} \neq 0$. $t = -0.92$; df = 59.40; P-value = 0.3610. Since P-value is high, do not reject the null; these data do not provide evidence of a significant difference

between April and October with regard to the mean age at which crawling begins.

c) These results are not consistent with the claim.

3. a) For First Nations, the confidence interval is (3638.6, 3651.4). For Immigrants of Chinese origin, the interval is (3386.7, 3399.3).

b) The value from the BC population, 3558, is not in either confidence interval, so both groups differ significantly from the BC population.

c) CI = (243.01, 260.99)

5. We are 95% confident that the difference in the mean percentage of aluminum oxide content of the pottery at the two sites is between –3.37% and 1.65%. Since 0 is in the interval, there is no evidence that the aluminum oxide content at the two sites is different. It would be reasonable for the archaeologists to think that the same ancient people inhabited the sites.

7. a) The observed difference between the proportions is $0.38 - 0.26 = 0.12$. Since the P-value $= 0.0060$ is very low, we reject the null hypothesis. There is strong evidence that the proportion of ALS patients who are athletes is greater than the proportion of patients with other disorders who are athletes.

b) This was a retrospective observational study. In order to make the inference, we must assume that the patients studied are representative of all patients with neurological disorders.

9. H_0: The proportions of traits are as specified by the ratio 1:3:3:9. H_A: The proportions of traits are not as specified. $\chi^2 = 5.01$, P-value $=$ 0.1711. Since the P-value is high, we fail to reject the null hypothesis. There is no evidence that the proportions of traits are anything other than 1:3:3:9.

11. We test the hypotheses: H_0: $\mu = 3360$; H_A: $\mu \neq 3360$, which gives $P = 0.067$, or $0.05 < P < 0.10$. There is only weak evidence against the null hypothesis that the means are the same.

13. a) If there is no difference in the average fish sizes, the chance of observing a difference this large, or larger, just by natural sampling variation is 0.1%.

b) There is evidence that largemouth bass that are fed a natural diet are larger. The researchers would advise people who raise largemouth bass to feed them a natural diet.

c) If the atdvice is incorrect, the researchers have committed a Type I error.

15. $\chi^2 = 6.14$; P-value $\approx$ 0.1887. Since the P-value $\approx$ 0.1887 is high, we fail to reject the null hypothesis. There is no evidence of an association between duration of pregnancy and level of prenatal care in twin births.

17. a) We are 95% confident that the mean age of patients with cardiac disease is between 3.39 and 5.01 years higher than the mean age of patients without cardiac disease.

b) Older patients are at greater risk for a variety of health problems, thus if an older patient does not survive a heart attack, the researchers will not know to what extent depression was involved because there will be a variety of other possible variables influencing the death rate. Additionally, older patients may be more (or less) likely to be depressed than younger ones.

19. a) It is unlikely that an equal number of boys and girls were contacted strictly by chance. It is likely that this was a stratified random sample, stratified by gender.

b) We are 95% confident that the proportion of computer gamers is between 7.0% and 17.0% higher for boys than for girls.

c) Since the interval lies entirely above 0, there is evidence that a greater percentage of boys play computer games than girls.

21. a) There are $n = 100$ fires, with mean $\bar{y} = 6056$ and standard deviation $s = 10811.98$. The 90% confidence interval is $6056 \pm 1.65(10811.98)/\sqrt{100} =$ or (4260.35, 7850.77).

b) The interval does contain the actual value of 6398. If we took 200 independent random samples from the population and counted the number of intervals which contained the actual mean 6398, this number would follow a binomial distribution with index $n = 200$ and probability $p = 0.10$.

c) The distribution is extremely skewed, as shown in the histogram below. However, the sample size is quite large, so the confidence interval should be valid.

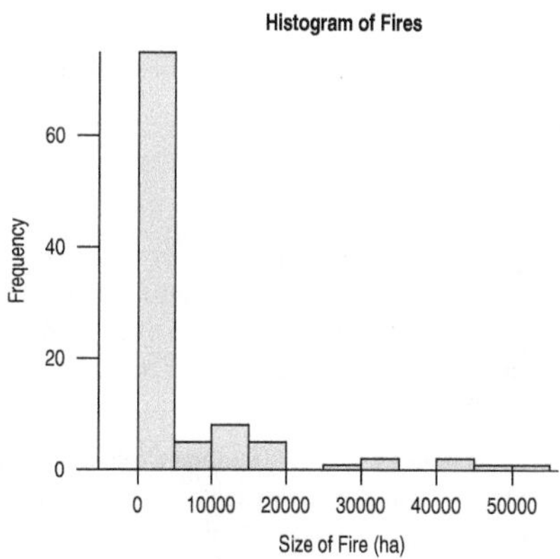

d) The median is an appropriate measure of center in this case. The sign test could be used to assess whether the true median has a particular value. A confidence interval would contain all values for the median for which the result of the sign test is not significant.

e) A transformation is appropriate here. The histogram for the logarithm transformation is shown below, and the distribution is much more symmetric than before.

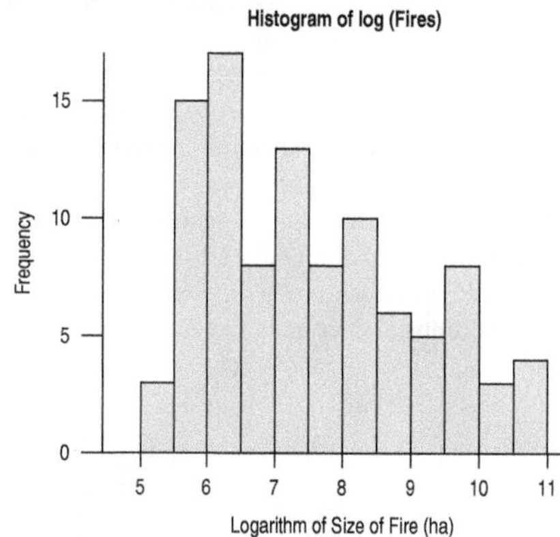

23. H_0: The university student's mean daily food expense is \$10. ($\mu = 10$); H_A: The university student's mean daily food expense is greater than \$10. ($\mu > 10$)

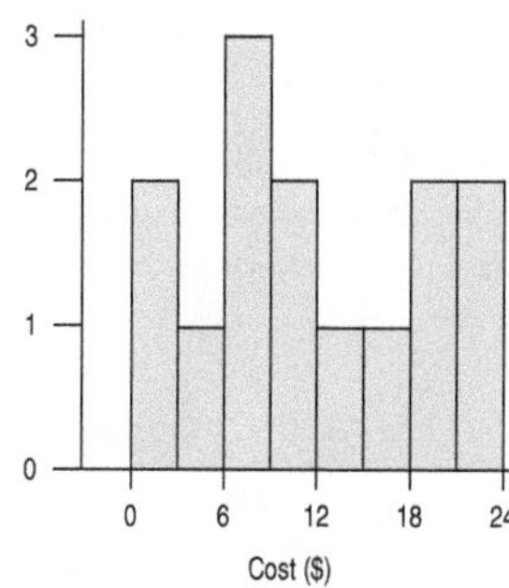

Randomization condition: Assume that these days are representative of all days.

Nearly Normal condition: The histogram of daily expenses is fairly unimodal and symmetric. It is reasonable to think that this sample came from a Normal population. The expenses in the sample had a mean of 11.4243 dollars and a standard deviation of 8.05794 dollars. Since the conditions for inference are satisfied, we can model the sampling distribution of the mean daily expense with a Student's t model, with $14 - 1 = 13$ degrees of freedom. We will perform a one-sample t-test. Since the P-value $= 0.2600$ is high, we fail to reject the null hypothesis. There is no evidence that the student's average spending is more than \$10 per day.

$$t = \frac{\bar{y} - \mu_0}{SE(\bar{y})}$$

$$t = \frac{11.4243 - 10}{\frac{8.05794}{\sqrt{14}}}$$

$$t \approx 0.66$$

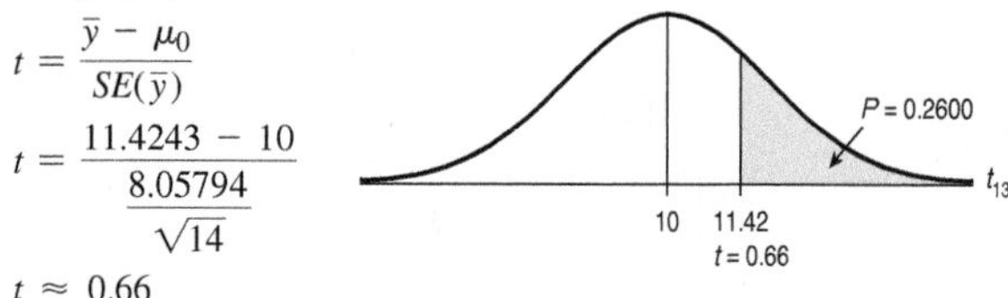

25. $H_0: \mu_C - \mu_U = 0$, $H_A: \mu_C - \mu_U > 0$, $t = 1.57$, df $= 86$, P-value $= 0.0598$. The P-value is just greater than 0.05. Although there may be some indication that students of certified teachers achieve higher mean reading scores than students of uncertified teachers, we cannot reject the null hypothesis at the 5% level.

27. Based on these data we are 95% confident that the proportions of streams in the Adirondacks with shale substrates is between 32.8% and 47.4%.

29. a) The students scores are dependent on what school they went to, so the normal standard deviation calculations do not apply. Assuming the standard errors provided in the table account for the dependence, the confidence interval for overall score is (522.296, 531.704).

b) Again, a dependence issue with school. The confidence interval is (27.49, 36.51), indicating that females score higher.

31. a) $H_0: \mu_d = 0$, $H_A: \mu_d \neq 0$, $t \approx 0.381$, df $= 19$, P-value $= 0.7078$. With a high P-value, we do not reject H_0. There is no evidence that the tests differ in mean difficulty.

b) $H_0: \mu_d = 0$, $H_A: \mu_d \neq 0$. The boxplot of the differences shows three outliers. With these outliers removed, the histogram is roughly unimodal and symmetric. A paired t-test with them included shows $t = -0.2531$ with 19 df and a P-value of 0.8029, showing no evidence of a difference in mean scores. With the three outliers removed, the t-test shows $t = -1.1793$ with 16 df and a P-value of 0.2555—still no evidence of a difference in means. We conclude the testing environment does not affect mean score.

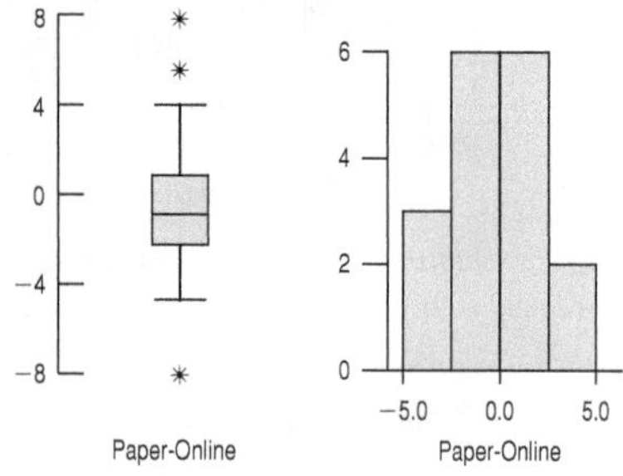

33. a) Parallel boxplots of the distributions of petal lengths for the two species of flower are shown below. No units are specified, but millimetres seems like a reasonable guess.

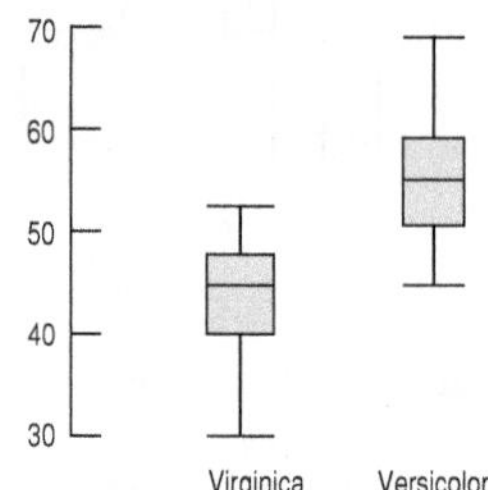

b) *Versicolor* generally has longer petals than *virginica.*

c) (10.14, 14.46)

d) We are 95% confident, based on the information given, that the average petal length for *versicolor* irises is between 10.14 and 14.46 mm longer than that of *virginica* irises.

e) Yes. The entire interval is above 0.

35. $\chi^2 = 10.146$, P-value $= 0.0174$. We reject the null hypothesis that teams are evenly matched at $\alpha = 0.05$. If they were, we would expect to see fewer 4-game series.

37. a) H_0: The mean weight of bags of Lays is 35.4 grams. ($\mu = 35.4$)
H_A: The mean weight of bags of Lays is less than 35.4 grams. ($\mu < 35.4$)

b) **Randomization condition:** It is reasonable to think that the 6 bags are representative of all bags of Lays.
Nearly Normal condition: The histogram of bags weights shows one unusually heavy bag. Although not technically an outlier, it probably should be excluded for the purposes of the test. (We will leave it in for the preliminary test, then remove it and test again.)

c) The bags in the sample had a mean weight of 35.5333 grams and a standard deviation in weight of 0.450185 grams. Since the conditions for inference are satisfied, we can model the sampling distribution of the mean weight of bags of Lays with a Student's t-model, with $6 - 1 = 5$ degrees of freedom. We will perform a one-sample t-test.

$$t = \frac{\bar{y} - \mu_0}{SE(\bar{y})}$$

$$t = \frac{35.5333 - 35.4}{\frac{0.450185}{\sqrt{6}}}$$

$$t \approx 0.726$$

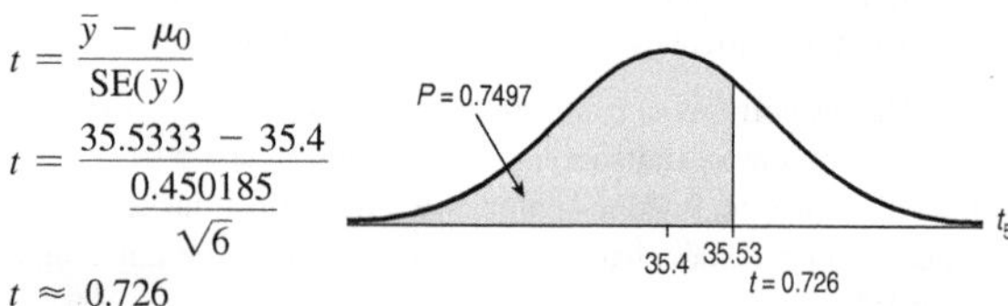

Since the P-value $= 0.7497$ is high, we fail to reject the null hypothesis. There is no evidence to suggest that the mean weight of bags of Lays is less than 35.4 grams.

d) With the one unusually high value removed, the mean weight of the 5 remaining bags is 35.36 grams, with a standard deviation in weight of 0.167332 grams. Since the conditions for inference are satisfied, we can model the sampling distribution of the mean weight of bags of Lays with a Student's t-model, with $5 - 1 = 4$ degrees of freedom. We will perform a one-sample t-test. Since the P-value $=$ 0.3107 is high, we fail to reject the null hypothesis. There is no evidence to suggest that the mean weight of bags of Lays is less than 35.4 grams.

$$t = \frac{\bar{y} - \mu_0}{SE(\bar{y})}$$

$$t = \frac{35.36 - 35.4}{\frac{0.167332}{\sqrt{5}}}$$

$$t = -0.53$$

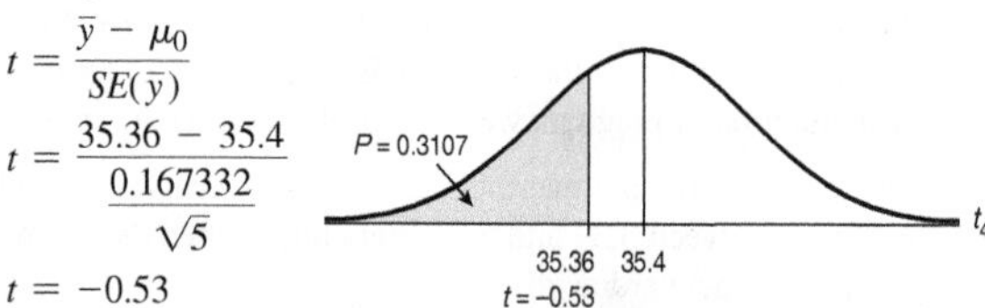

e) Neither test provides evidence that the mean weight of bags of Lays is less than 35.4 grams. It is reasonable to believe that the mean weight of the bags is the same as the stated weight. However, the sample sizes are very small, and the tests have very little power to detect lower mean weights. It would be a good idea to weigh more bags.

39. a) \$373.50

b) They are 95% confident that the average *Loss* in a home burglary is between \$1644 and \$2391, based on their sample.

c) 95% of all random samples will produce confidence intervals that contain the true mean *Loss.*

41. a) Based on these data, we are 90% confident that the average hamster litter will have between 7.11 and 8.33 babies.

b) Larger—to be more confident we need a wider interval.

c) About 27.

43. a) Based on the data, we are 95% confident that the difference in recruiting is between –4.2% and 8.2%.

b) No. The interval contains 0 and negative values—it may actually be the case that the new strategy lowers acceptance rates.

Chapter 24

1. a) $\widehat{Error} = -454.448 - 8.479\,(year)$; according to the model, the error made in predicting a hurricane's path was about 455 nautical miles, on average, in 1970. It has been declining at a rate of about 8.48 nautical miles per year.
b) $H_0: \beta_1 = 0$; there has been no change in prediction accuracy. $H_A: \beta_1 \neq 0$; there has been a change in prediction accuracy.
c) With a P-value < 0.001, I reject the null hypothesis and conclude that prediction accuracies have in fact been changing during this period.
d) 71.1% of the variation in hurricane prediction accuracy is accounted for by this linear model on time.

3. a) $\widehat{Budget} = -31.387 + 0.714\,(Runtime)$. The model suggests that movies cost about $714 000 per minute to make.
b) A negative starting value makes no sense, but the P-value of 0.07 indicates that we can't discern a difference between our estimated value and zero. The statement that a movie of zero length should cost $0 makes sense.
c) Amounts by which movie costs differ from predictions made by this model vary, with a standard deviation of about $33 million.
d) 0.154 $m/min
e) If we constructed other models based on different samples of movies, we'd expect the slopes of the regression lines to vary, with a standard deviation of about $154 000 per minute.

5. a) The scatterplot is straight enough, and the residuals plot looks unpatterned.
b) I'm 95% confident that the cost of making longer movies increases at a rate of between 0.41 and 1.02 million dollars per minute.

7. a) $H_0: \beta_1 = 0$; there's no association between calories and sodium content in all-beef hot dogs. $H_A: \beta_1 \neq 0$: there is an association.
b) Based on the low P-value (0.0018), I reject the null. There is evidence of an association between the number of calories in all-beef hot dogs and their sodium contents.

9. a) Among all-beef hot dogs with the same number of calories, the sodium content varies, with a standard deviation of about 60 mg.
b) 0.561 mg/cal
c) If we tested many other samples of all-beef hot dogs, the slopes of the resulting regression lines would be expected to vary, with a standard deviation of about 0.56 mg of sodium per calorie.

11. I'm 95% confident that for every additional calorie, all-beef hot dogs have, on average, between 1.03 and 3.57 mg more sodium.

13. a) H_0: Difference in age between men and women at first marriage has not been decreasing since 1921. ($\beta_1 = 0$)
H_A: Difference in age between men and women at first marriage has been decreasing since 1921. ($\beta_1 < 0$)
b) Residual plot shows a curved pattern; histogram is not particularly unimodal and symmetric, and it is skewed to the right.
c) $t = -44.02$, P-value < 0.0001. With such a low P-value, we reject H_0. These data show evidence that difference in age between men and women at first marriage appears to be decreasing over time.

15. Based on these data, we are 95% confident that the mean difference in age between men and women at first marriage decreases by between 0.020 and 0.022 years in age per year.

17. a) H_0: *Fuel Economy* and *Weight* are not (linearly) related, $\beta_1 = 0$. H_A: *Fuel Economy* changes with *Weight*, $\beta_1 \neq 0$. P-value < 0.0001, indicating strong evidence of an association.
b) Yes, the conditions seem satisfied. Histogram of residuals is unimodal and symmetric; residual plot looks OK, but some "thickening" of the plot with increasing values.
c) $t = -12.2$, P-value < 0.0001 These data show evidence that *Fuel Economy* decreases with the *Weight* of the car.

19. a) $(-9.57, -6.86)$ mpg per 1000 pounds.
b) Based on these data, we are 95% confident that *Fuel Efficiency* decreases between 6.86 and 9.57 miles per gallon, on average, for each additional 1000 pounds of *Weight*.

21. a) Based on this regression, we are 95% confident that 2500-pound cars will average between 27.34 and 29.07 miles per gallon.
b) Based on the regression, a 3450-pound car will get between 15.44 and 25.36 miles per gallon, with 95% confidence.

23. a) Yes. $t = 2.73$, P-value = 0.0079. With a P-value so low, we reject H_0. There is a positive relationship between *Calories* and *Sodium* content.
b) No. $R^2 = 9\%$ and s appears to be large, although without seeing the data, it is a bit hard to tell.

25. Plot of *Calories* against *Fibre* does not look linear; this is borne out in the residuals plot, which also shows increasing variance as predicted values get large. The histogram of residuals is right skewed.

27. a) H_0: No (linear) relationship between *BCI* and *pH*, $\beta_1 = 0$. H_A: There is a relationship, $\beta_1 \neq 0$.
b) $t = -7.73$ with 161 df; P-value < 0.0001
c) With such a small p-value, this means that the association we see in the data is unlikely to occur by chance. We reject the null hypothesis, and conclude that there is strong evidence of a linear relationship between *BCI* and *pH*. Streams with higher *pH* tend to have lower *BCI*.

29. a) H_0: No linear relationship between *Population* and *Ozone*, $\beta_1 = 0$. H_A: *Ozone* increases with *Population*, $\beta_1 > 0$. $t \approx 3.48$, P-value = 0.0018. With a P-value so low, we reject H_0. These data show evidence that *Ozone* does increase with *Population*.
b) Yes, *Population* accounts for 84% of the variability in *Ozone* level, and s is just over 5 parts per million.

31. a) Based on this regression, each additional million residents corresponds to an increase in average ozone level of between 3.29 and 10.01 ppm, with 90% confidence.
b) Based on the regression, the mean *Ozone* level for cities with 600 000 people is between 18.47 and 27.29 ppm, with 90% confidence.

33. a) 33 batteries
b) The scatterplot of the data is roughly linear with lots of scatter; plot of residuals vs. predicted values shows no overt patterns; Normal probability plot of residuals is reasonably straight.
c) H_0: No linear relationship between *Cost* and *Power*, $\beta_1 = 0$. H_A: *Power* increase with cost, $\beta_1 > 0$. $t \approx 3.23$; P-value $0.0029/2 = 0.0015$. With a P-value so low, we reject H_0. These data provide evidence that more expensive batteries do have more cranking amps.
d) No. $R^2 = 25.2\%$ and $s = 116$ amps. Considering that the range of amperage is only about 400 amps, an s of 116 is not very useful.
e) $\widehat{Power} = 384.594 + 4.14649\,(Cost)$
f) (1.97, 6.32) cold cranking amps per dollar
g) *Power* increase, on average, between 1.97 and 6.32 per dollar of battery *Cost* increase, with 90% confidence.

35. a) 23
b) **Straight enough condition:** The scatterplot is roughly straight, but scattered. **Independence assumption:** The residuals plot shows no pattern. **Does the plot thicken? condition:** The spread of the residuals is consistent. **Nearly Normal condition:** The Normal probability plot of residuals is reasonably straight.
c) H_0: There is no linear relationship between maximum brightness and battery life, $\beta_1 = 0$. H_A: There is a positive linear relationship between maximum brightness and battery life, $\beta_1 > 0$. $t \approx 2.85$, P-value = 0.00955, df = 21. We reject the null hypothesis, and conclude that there is strong evidence of a positive linear relationship between battery life and screen brightness. Tablets with greater screen brightness tend to have longer battery life.

d) Since $R^2 = 27.9\%$, only 27.9% of the variability in battery life can be accounted for by screen brightness. The residual standard deviation is 1.913 hours. That's pretty large, considering the range of battery life is only about 9 hours. Although there is strong evidence of a linear association, it is too weak to be of much use. Predictions would tend to be very imprecise.

e) The equation of the line of best fit for these data points is: $\widehat{Hours} = 2.8467 + 0.01408\,(Screen\ Brightness)$, battery life measure in hours and screen brightness measured in cd/m^2.

f) There are 21 degrees of freedom, so use $t^*_{21} = 1.721$. $b_1 \pm t^*_{n-2} \times SE(b_1) = 0.01408 \pm (1.721) \times 0.004937 \approx (0.0055, 0.0226)$

g) There are 21 degrees of freedom, so use $t^*_{21} = 1.721$.

37. a) H_0: No linear relationship between *Waist* size and *%Body Fat*, $\beta_1 = 0$. H_A: *%Body Fat* changes with *Waist* size, $\beta_1 \neq 0$. $t \approx 8.1$; P-value < 0.0001. *%Body Fat* seems to increase with *Waist* size.

b) With 95% confidence, mean *%Body Fat* for people with 40-inch waists is between 23.58 and 29.02, based on this regression.

39. a) The regression model is $\widehat{Midterm_2} = 12.005 + 0.72\,(Midterm_1)$ with output:

Dependent variable is: Midterm 2

No Selector

R squared = 19.9% R squared (adjusted) = 18.6%

s = 16.78 with 64 − 2 = 620 degrees of freedom

Source	Sum of Squares	Of	Mean Square	F-ratio
Regression	4337.14	1	4337.14	15.4
Residual	17459.5	62	281.604	

Variable	Coefficient	s.o. of Coeff	t-ratio	prob
Constant	12.0054	15.96	0.752	0.4546
Midterm 1	0.720990	0.1837	3.92	0.0002

b) The scatterplot shows a weak, somewhat linear, positive relationship. There are several outlying points, but removing them only makes the relationship slightly stronger. There is no obvious pattern in the residual plot. The regression model appears appropriate. The small P-value for the slope shows that the slope is statistically distinguishable from 0 even though the R^2 value of 0.199 suggests that the overall relationship is weak.

c) No. The R^2 value is only 0.199 and the value of s of 16.8 points indicates that she would not be able to predict performance on *Midterm*2 very accurately.

41. a) We want to know whether the times to swim across Lake Ontario have changed over time. We know the time of each swim in minutes and the year of the swim. H_0: $\beta_1 = 0$ (no linear relationship); H_A: $\beta_1 \neq 0$.

b) The Straight Enough Condition fails because there is an outlier.

c) We cannot continue this analysis.

d) With the outlier removed, the relationship looks straight enough. Swims are separate, but some swimmers performed the feat more than once, so swims may not be entirely independent. There is no thickening of the plot. The normal probability plot of the residuals looks straight. The conditions appear to be met.

e) The analysis shows that swim *Times* have been increasing at about 14.9 minutes per *Year.* The P-value of 0.0158 is small enough to reject the null hypothesis, so this looks like a real increase.

f) The SE depends on three things: the standard deviation of the residuals around the line, σ, the standard deviation of x, and the number of data values. Only the first of these was changed substantially by removing the outlier. Indeed, we can see that s_e was 447.3 with the outlier present and only 277.2 when it was removed. This accounts for most of the change in the SE and therefore in the P-value. Because the outlier was at a value of x near the middle of the x range, it didn't affect the slope very much.

43. a) Data plot looks linear; no overt pattern in residuals; histogram of residuals roughly symmetric and unimodal.

b) H_0: No linear relationship between *Education* and *Mortality*, $\beta_1 = 0$. H_A: $\beta_1 \neq 0$. $t \approx -6.24$; P-value < 0.001. There is evidence that cities in which the mean education level is higher also tend to have a lower mortality rate.

c) No. Data are on cities, not individuals. Also, these are observational data. We cannot predict causal consequences from them.

d) (−65.95, −33.89) deaths per 100 000 people.

e) *Mortality* decreases, on average, between 33.89 and 65.95 deaths per 100 000 for each year of average *Education,* based on the regression.

f) Based on the regression, the average *Mortality* for cities with an average of 12 years of *Education* will be between 874.239 and 914.196 deaths per 100 000 people.

45. a) $\widehat{January} = 120.73 + 0.6995(December)$ (We are told this is an SRS. One cardholder's spending should not affect another's. These are quantitative data with no apparent bend in the scatterplot. The residual plot shows some increased spread for larger values of January charges. A histogram of residuals is unimodal and slightly skewed to the right with several high outliers.)

b) \$1519.73 c) (\$1330.24, \$1709.24)

d) (\$290.76, \$669.76)

e) The residuals show increasing spread, so the confidence intervals may not be valid. I would be skeptical of interpreting them too literally.

47. a) Yes, three points that are outliers and have high leverage.

b) Residuals plot show some increased spread for larger values of fitted values.Normal probability plot of residuals is reasonably straight.

c) (\$170.05, \$300.60) d) (\$5.56, \$133.79)

e) The squared standard error of b1 is so small compared to s_e^2.

49. a) $\widehat{\text{Logit}(Type)} = -13.1461 + 0.08455\ Top\ 10\% + 0.000259\$/Student$

b) Yes, the percent of students in the top 10% is statistically significant, since the P-value of 0.033 is less than $\alpha = 0.05$.

c) Yes, the amount of money spent her student is statistically significant, since the P-value of 0.003 is less than $\alpha = 0.05$.

Chapter 25

1. a) H_0: $\mu_1 = \mu_2 = \mu_3 = \mu_4$ vs. the alternative that not all means are equal. Here μ_k refers to the mean number of popcorn kernels left unpopped for brand k.

b) MS_T has 3 df; MS_E has 12.

c) An $F_{3,12}$ value of 13.56 has a P-value of 0.00037. The data provide strong evidence that the means are not all equal. The brands do not produce the same mean number of unpopped kernels.

d) I would like to see side-by-side boxplots of the treatment groups, a Normal probability plot of the residuals, a histogram of the residuals, and a plot of the residuals vs. the predicted values.

3. a) The null hypothesis is that the mean mileage driven on one litre of gas using each muffler is the same. The alternative is that not all means are equal.

b) MS_T has 2 df; MS_E has 21 df.

c) We would not reject the null hypothesis of equal means with a P-value of 0.1199. The data from this experiment provide no evidence that the muffler type affects gas mileage.

d) I would like to see side-by-side boxplots of the treatment groups, a normal probability plot of the residuals, a histogram of the residuals, and a plot of the residuals vs. the predicted values.

e) Type II error

5. a) H_0: $\mu_1 = \mu_2 = \mu_3$ vs. the alternative that not all means are the same. Here μ_k refers to the mean weight of cat hair remaining in cage at the end of the day for cats on food type k.
b) MS_T has 2 df; MS_E has 57 df.
c) P-value= 0.0846. We fail to reject the null hypothesis and conclude that there is no evidence that the mean weights of cat hair for the three groups are not all the same.
d) The runs were randomized. We may need to be cautious in generalizing the results to all cats since each breed could react differently.

7. a) H_0: $\mu_1 = \mu_2 = \mu_3 = \mu_4$ vs. the alternative that not all means are equal. Here μ_k refers to the mean activation time using recipe k.
b) The data provide strong evidence with a P-value of 0.0001 to reject the null hypothesis and conclude that the means are not all equal. This experiment provides strong evidence that the mean activation times differ among the recipes.
c) Yes, because we have rejected the null hypothesis, we can proceed with a multiple comparisons method to compare all the groups.

9. An ANOVA is not appropriate because *Eye Colour* is a categorical variable. The students should consider analyzing the data with a χ^2 test of independence between eye and hair colour.

11. a) The null hypothesis is that the mean *Case Price* for each region is the same. The alternative is that not all means are equal.
b) The spreads of *Case Price* do not appear to be similar, violating the similar variance condition. There is also an outlier in the Keuka group. It would be a good idea to look at a plot of the residuals from an ANOVA vs. the predicted values to see if the variance increases with predicted value.

13. a) The null hypothesis is that the means of the four lists are all equal. The alternative hypothesis is that not all means are equal.
b) The data provide moderately strong evidence with a P-value of 0.0033 that the means of the four lists are not equal.
c) Yes, we reject the null hypothesis at a = .01. We can proceed to do multiple comparisons to compare groups.

15. a) $MS_T \approx 27.067, MS_E = 1.925$ b) $F \approx 14.0606$
c) The data provide very strong evidence that the means are not equal. The design does affect mean emissions.
d) We have assumed that the experimental runs were performed in random order, that the variances of the treatment groups are similar, and that the residuals are nearly Normal.
e) A boxplot of the scores by design, a plot of residuals vs. predicted values, a Normal probability plot, and a histogram of the residuals.
f) $s_p = \sqrt{MS_E} = \sqrt{1.925} \approx 1.387$ ppb

17. a) The null hypothesis is that the mean *Noise Levels* are the same for both types of filter. The alternative is that they are different.
b) The data provide no evidence that the means are different.
c) Yes, the conditions look reasonable. It would be good to examine the residuals as well.
d) The pooled estimate of the variance is equal to MS_E. The t-ratio approximately 0.87596 and $t^2 = (0.87596)^2 \approx 0.7673$ which is equal to the F-ratio. The two-sided P-value for the t with 33 df is 0.3874. The P-value agrees with the P-value for the F-statistic.

19. a) The null hypothesis is that the mean *Height* is the same for all fertilizer treatment groups. The alternative is that not all the means are equal.
b) The data provide no evidence that the mean *Heights* differ by fertilizer treatment.
c) The lab partner's Type I error rate is higher than that of the ANOVA test because he is performing multiple t-tests without compensating by using a multiple comparisons method.

21. a) The null hypothesis is that the mean *Protein Content* is the same for the cereals on each *Shelf*. The alternative is that not all the means are equal.
b) The P-value of 0.0044 provides strong evidence that the mean *Protein Content* is not the same for each *Shelf*. But the outliers evident in the boxplots should lead us to be cautious in interpreting this result.
c) We cannot conclude that cereals on *Shelf* 2 have a lower mean *Protein Content* than cereals on *Shelf* 3 or that cereals on *Shelf* 2 have a lower mean *Protein Content* than cereals on *Shelf* 1. We can conclude only that the means are not all equal.
d) The Bonferroni test shows that $\alpha = 0.05$. Now we can conclude that the mean *Protein Content* of cereals on *Shelf* 2 is *not* equal to the mean *Protein Content* on *Shelf* 3. The other pairwise comparisons are not significantly different at $\alpha = 0.05$.

23. a) The null hypothesis is that the mean *Pain Level* reported is equal for the three *Drugs*. The alternative is that not all the means are equal.

Analysis of Variance

Source	DF	Sum of Squares	Mean Square	F-ratio	P-Value
Drug	2	28.2222	14.1111	11.9062	0.0003
Error	24	28.4444	1.1852		
Total	26	56.6666			

b) The F-statistic is 11.9062 with 2 and 24 degrees of freedom, resulting in a P-value equal to 0.0003. We reject the null hypothesis and conclude that there is strong evidence that the mean pain level reported is different in at least one of the three drugs.
c) The boxplots show no evidence of violation of the similar variance condition (the IQRs are very close). The outliers in drug B are a slight concern, but there are two subjects reporting pain level 8. This value is just beyond the upper fence for drug B. The Normal probability plot is reasonably straight (considering that the data are reported as integers) and the residual vs. predicted plot shows no increase in variance for high predicted values.

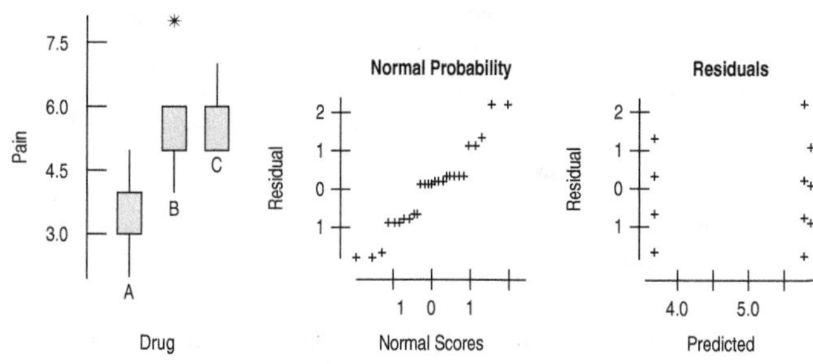

d) A Bonferroni test shows that *Drug* A's mean *Pain Level* as reported is significantly below the other two, but that *Drug* B's and *Drug* C's means are indistinguishable (at $\alpha = 0.05$).

25. This is unacceptable as we looked at the results before forming our hypothesis. This follow-up test would only be permissible if we had planned it before conducting the experiment.

27. a) The distributions of the residuals show similar spreads for the four glue types.
b) Some outliers are found in the hot and white glue type.
c) The normal probability plot of residuals shows some small deviation from the identity line from both ends.
d) Conditions seem satisfied and we could proceed with our usual inferences.

Chapter 26

1. a) H_0: $\gamma_1 = \gamma_2 = \gamma_3$ where γ represents the effect of the *Power* level vs. H_A: Not all of the *Power* levels have the same effect on the response. H_0: $\tau_1 = \tau_2 = \tau_3$ where τ represents the effect of *Time* level vs. H_A: Not all of the *Time* levels have the same effect on the response.

b) The *Power* sum of squares has 2 df; the *Time* sum of squares has 2 df; the error sum of squares has 4 df.
c) There are no degrees of freedom left for the interaction term. She must assume that the interaction effects are negligible.

3. a) The P-value for *Power* is 0.0165. The P-value for *Time* is 0.0310.
b) We reject the null hypothesis that *Power* has no effect and conclude that the mean number of uncooked kernels is not equal across all 3 *Power* levels. We also reject the null hypothesis that *Time* has no effect and conclude that the mean number of uncooked kernels is not the same across all 3 *Times.*
c) Partial boxplots, scatterplots of residuals vs. predicted values, and a Normal probability plot of the residuals to check the assumptions.

5. a) H_0: $\gamma_1 = \gamma_2$ where γ represents the effect of the *Seat* level vs. H_A: One of the *Seat* levels has a different effect. The null hypothesis states that *seat* choice (driver vs. passenger) has no effect on mean head injury sustained. The alternative states that it does. H_0: $\tau_1 = \tau_2 = \tau_3 = \tau_4 = \tau_5 = \tau_6$ where τ represents the effect of the *Size* level vs. H_A: The *Size* level effects are not all equal. The null hypothesis states that *Size* of vehicle (the 6 represented) has no effect on mean *Head Injury* sustained. The alternative states that it does.
b) The conditions appear to be met. The effects are additive enough, the data we assume were collected independently, the boxplots show that the variance is roughly constant, and there are no patterns in the scatterplot of residuals vs. predicted values.
c) There is no significant interaction. The P-values for both *Seat* and *Size* are 0.0001. Thus, we reject the null hypotheses and conclude that both the *Seat* and the *Size* of car affect the severity of head injury. From the partial boxplots we see that the mean *Head Injury* level is higher for the driver's side. The effect of driver's seat seems to be roughly the same for all 6 car sizes.

7. a) A two-factor ANOVA must have a quantitative response variable. Here the response is whether they exhibited baldness or not, which is a categorical variable. A two-factor ANOVA is not appropriate.
b) We could use a chi-square analysis to test whether *Baldness* and *Heart Disease* are independent.

9. a) A chi-square test of independence gives a chi-square statistic of 14.510 with a P-value of 0.0023. We reject the hypothesis that *Baldness* and *Heart Disease* are independent.
b) No, the fact that these are not independent does NOT mean that one causes the other. There could be a lurking variable (such as age) that influences both.

11. a) H_0: $\gamma_1 = \gamma_2$ where γ represents the effect of the *Time of Day* on *Shots Made*. H_A: The means of *Shots Made* differ across the two *Time of Day* levels. The null hypothesis states that *Time of Day* has no effect on *Shots Made*. The alternative states that it does. H_0: $\tau_1 = \tau_2$ where τ represents the kind of *Shoes* worn vs. H_A: The means of *Shots Made* differ across the two types of *Shoes*. The null hypothesis states that type of *Shoes* worn has no effect on *Shots Made*. The alternative states that it does.
b) Partial boxplots show little effect of either *Time of Day* or *Shoes* on *Shots Made.*

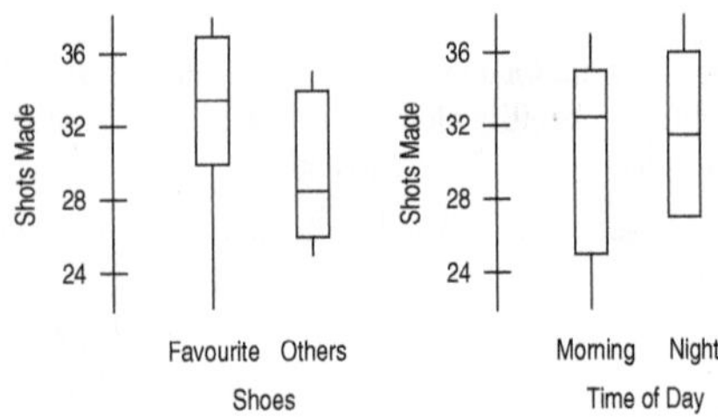

Neither the scatterplot of residuals vs. predicted values nor the Normal probability plots of residuals show any conditions that aren't met. We assume the number of shots made were independent from one treatment condition to the next.

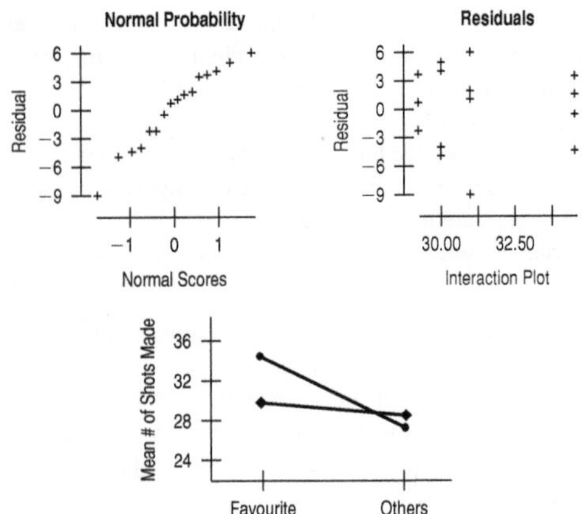

An interaction plot shows a possible interaction effect. It looks as though the favourite shoes may make more of a difference at night. However, the *F*-test shows that the null hypothesis of no interaction effect is not rejected. None of the effects appears to be significant. It looks as though she cannot conclude that either *Shoes* or *Time of Day* affect her mean *Shots Made.*

13. a) H_0: $\gamma_1 = \gamma_2 = \gamma_3$ where γ represents the effect of the *Temperature* level vs. H_A: Not all the γ are equal. The null hypothesis states that *Temperature* level (32°, 34°, and 36°C) has no effect on the *Number of Sprouts*. The alternative states that it does. H_0: $\tau_1 = \tau_2 = \tau_3 = \tau_4$ where τ represents the effect of the *Salinity* level vs. H_A: Not all the τ are equal. The null hypothesis states that *Salinity* level (0, 4, 8, and 12 ppm) has no effect on the *Number of Sprouts.* The alternative states that it does.
b) The partial boxplots show that *Salinity* appears to have an effect on the *Number of Sprouts,* but *Temperature* does not. The ANOVA supports this. The P-value for *Salinity* is < .0001, but the P-value for *Temperature* is 0.3779, providing no evidence of a temperature effect over this range of temperatures. There appears to be no interaction as well. There appears to be more spread in *Number of Sprouts* for the lower *Salinity* levels (where the response is higher). This is evident in both the partial boxplot and the residual vs. predicted values. This is cause for some concern, but most likely does not affect the conclusions.

15. The ANOVA table shows that both *Car Type* and *Additive* affect *Gas Mileage,* with P-values 0.0001. There is a significant interaction effect as well that makes interpretation of the main effects problematic. However, the residual plot shows a strong increase in variance, which makes the whole analysis suspect. The Similar Variance Condition appears to be violated.

17. The ANOVA table now shows only main effects to be significant:

Source	DF	Sum of Squares	Mean Square	F-ratio	P-value
Type	2	10.1254	5.06268	5923.1	< 0.0001
Additive	1	0.026092	0.0260915	30.526	< 0.0001
Interaction	2	7.57E-05	3.78E-05	0.044265	0.9567
Error	54	0.046156	8.55E-04		
Total	59	10.1977			

The residuals now show *no* violation of the conditions:

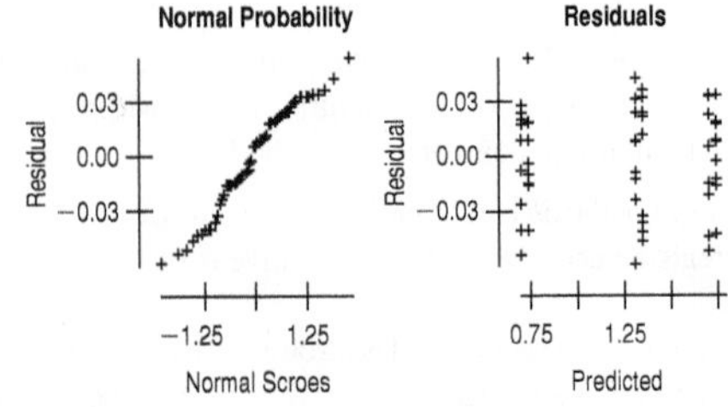

We can conclude that both the *Car Type* and *Additive* have an effect on mileage and that the effects are constant (in log10 *Mpg*) over the values of the various levels of the other factor.

19. a) H_0: $\gamma_1 = \gamma_2$ where γ represents the effect of the *Environment* level vs. H_A: The mean *Time* the battery lasted is not the same for the two conditions. H_0: $\tau_1 = \tau_2 = \tau_3 = \tau_4$ where τ represents the effect of the *Brand* vs. H_A: Not all the τ are equal.
b) From the partial boxplot, it is not clear that *Brand* has an effect. The condition clearly has an effect.
c) Yes, the *Brand* effect has a P-value of 0.099, while the *Environment* effect is clearly significant with a P-value 0.0001.
d) There is also an interaction, however, which makes the statement about *Brands* problematic. Not all brands are affected by the environment in the same way. Brand C, which works best in the cold, performs worst at room temperature.
e) Because it performs the worst of the four at room temperature.

21. In this one-way ANOVA, we can see that the means vary across treatments. (However, boxplots with only 2 observations are not appropriate.) By looking closely, it seems obvious that the four flashlights at room temperature lasted much longer than the ones in the cold. But it is much harder to see whether the means of the four brands are different, or whether they differ by the same amounts across both environmental conditions. The two-way ANOVA with interaction makes these distinctions clear.

Chapter 27

1. a) Doesn't mention the other predictors. b) This is correct.
c) Can't predict from *y* to *x*.
d) R^2 is about the fraction of variability accounted for by the regression model, not the fraction of data values.

3. a) $\widehat{Final} = -6.72 + 0.2560(Test1) + 0.3912(Test2) + 0.9015(Test3)$
b) R^2 is 77.7%, so 77.7% of the variation in final grade is accounted for by the regression model.
c) After allowing for the linear effects of the other predictors, each point on *Test3* is associated with an average increase of 0.9015 points on the final.
d) Test scores are probably collinear. If all we are concerned about is predicting final exam score, *Test1* may not add much to the regression. However, we'd expect it to be associated with final exam score.

5. a) $\widehat{Price} = -152037 + 9530(Baths) + 139.87(Sqft)$
b) 71.1% of the variation in asking price is accounted for by this regression model.
c) For homes with the same number of bathrooms, asking price increases, on average, by about \$139.87 per square foot.
d) The number of bathrooms is probably correlated with the size of the house (even after considering the square footage of the bathroom itself). That correlation may account for the coefficient of *Baths* not being discernibly different from zero. Moreover, the regression model does not predict what will happen when a house is modified (for example, by converting existing space into a bathroom).

7. The plot of residuals vs. predicted values looks bent, rising in the middle and falling at both ends. This violates the Straight Enough Condition. The Normal probability plot and the histogram of the residuals suggest that the highest five residuals (which we know are in the middle of the predicted value range) are extraordinarily high. These data may benefit from a re-expression.

9. a) $\widehat{Salary} = 9.788 + 0.11(Service) + 0.053(Education) + 0.071(Score) + 0.004(Speed) + 0.065(Dictation)$
b) 29.2 thousand dollars
c) H_0 has a *t*-statistic value of 0.013. That is not significant at any reasonable alpha level.
d) Omitting typing *Speed* would simplify the model and might result in one that was almost as good. Other predictors might also be omitted, but we can't tell that from what we know here.
e) *Age* may be collinear with other predictors in the model. In particular, it is likely to be highly associated with months of *Service*.

11. a) H_0: $\beta = 0$; H_A: $\beta \neq 0$; $t = 12.4$ on 248 df; $P < 0.0001$, so we reject the null hypothesis. The coefficient of *Weight* is statistically discernible from zero.
b) Each pound of weight is associated, on average, with an increase of 0.189 in *%Body Fat.*
c) After removing the linear effects of *Waist* and *Height,* each pound of *Weight* is associated on average with a decrease of 0.10 in *%Body Fat.* Alternatively, for men with the same *Waist* and *Height,* each pound of *Weight* is associated, on average, with a decrease of 0.10 in *%Body Fat.* The change in coefficient and sign is a result of including the other predictors. We expect *Weight* to be correlated with both *Waist* and *Height.* It may be collinear with them.
d) The P-value of 0.1567 says that if the coefficient of *Height* in this model is truly zero, we could expect to observe a sample regression coefficient as far from zero as the one we have here about 15.6% of the time.

13. a) P-value is less than 0.0001, so we can reject the null hypothesis that the true slope is zero.
b) It says that there is a clear relationship between *%Body Fat* and *Chest Size*. In fact, *%Body Fat* grows, on average, by 0.71272 % per inch of *Chest Size*.
c) After allowing for the linear effects of *Waist* and *Height*, each inch of *Chest Size* is associated, on average, with a decrease of 0.233531 in *%Body Fat*. This coefficient is statistically discernible from zero (P = 0.0054).
d) Each of the variables appears to contribute to the model. There appears to be no advantage in removing one.

15. a) The only model that seems to do poorly is the one that omits *Murder*. The other three are hard to choose among.
b) Each of the models has at least one coefficient with a large P-value. This predictor variable could be omitted to simplify the model without degrading it much.
c) No. Regression models can't be interpreted in this way.
d) Plots of the residuals highlight some states as possible outliers. You may want to consider setting them aside to see if the model changes.

17. a) With an R^2 of 100%, the model should make excellent predictions.
b) The value of *s*, 3.140 calories, is very small compared with the initial standard deviation of *Calories*. This indicates that the model fits the data quite well, leaving very little variation unaccounted for.
c) No, the residuals are not all zero. Indeed, we know that their standard deviation, *s*, is 3.140. But they are very small compared with the original values. The true value of R^2 was rounded up to 100%.

Chapter 28

1. a) The distribution of the Studentized residuals might be bimodal. We should check for evidence of two groups in the data.
b) Answers may vary. It is also known that carbon dioxide levels changed in character in the late 1970s, and began a more rapid increase than had been seen earlier in the 20th century. This is an example of Simpson's Paradox.

3. a) The slope of a partial regression plot is the coefficient of the corresponding predictor—in this case, −1.020.
b) Quaker oatmeal makes the slope more strongly negative. It appears to have substantial influence on this slope.
c) Not surprisingly, omitting Quaker oatmeal changes the coefficient of *Fibre*. It is now positive (although not significantly different from 0). This second regression model has a higher R^2, suggesting that it fits the data better. Without the influential point, the second regression is probably the better model.
d) The coefficient of *Fibre* is not discernibly different from zero. We have no evidence that it contributes significantly to *Calories* after allowing for the other predictors in the model.

5. a) After allowing for the effects of *Sodium* and *Sugars*, each gram of *Potassium* is associated with a 0.019-calorie decrease in *Calories*.
b) Those points pull the slope of the relationship down. Omitting them should increase the value of the coefficient of *Potassium*. It would likely become positive, since the remaining points show a positive slope in the partial regression plot.
c) These appear to be influential points. They have both high leverage and large residual, and the partial regression plot shows their influence.
d) If our goal is to understand the relationships among these variables, then it might be best to omit these cereals because they seem to behave as if they are from a separate subgroup.

7. a) This set of indicators uses *Medium* size as its base. The coefficients of *Small*, *Large*, and *Very Large* estimate the average change in amount of *Delay/Person* relative to the amount of *Delay/Person* for *Medium* size cities found for each of the other three sizes.
b) If there were an indicator variable for *Medium* as well, the four indicators would be collinear, so the coefficients could not be estimated.

9. a) An assumption required for indicator variables to be useful is that the regression models fit for the different groups identified by the indicators be parallel. These lines are not parallel.
b) The coefficient of *Am* × *Sml* adjusts the slope of the regression model fit for the small cities. We would say that the slope of *Delay/Person* on *Arterial Mph* for *Small* cities (after allowing for the linear effects of the other variables in the model) is 1.20613.
c) The regression model seems to do a good job. The R^2 shows that 80.7% of the variability in *Delay/Person* is accounted for by the model. Most of the P-values for the coefficients are small. The coefficients concerning the very large cities have larger P-values, but that may be due to having a relatively small number of such cities. It may still be wise to keep those predictors in the model.

11. Colorado Springs has high leverage, but it does not have a particularly high Studentized residual. It appears that the point has influence but doesn't exert it. Removing this case from the regression may not result in a large change in the model, so the case is probably not influential.

13. a) It is the expected change in stress for a unit change in area, for a large overlap joint, holding everything else constant.
b) There are way more medium joints than small joints. Since the relationship is very weak for medium joints, the P-value is "swamped" by them.
c) 0.7773 d) −98.6

Part VII Review

1. a) 59 products b) 84.5% c) (2.21, 2.78) dollars per minute
d) Based on this regression, average *Price* increases between \$2.21 and \$2.78 for each minute of polishing *Time*, with 95% confidence.

3. a) Based on these data, with 95% confidence 5-year yields are between 3.15% and 5.93% higher than 3-year yields on average (paired data).
b) Yes (at least for this data set). The regression line is 5-year = 6.93 + 0.719 3-year. H_0: $\beta_1 = 0$ against H_A: $\beta_1 \neq 0$ has $t \approx 4.27$; P-value = 0.0009. Since P is so small, we reject H_0. There is evidence of an association. (But we don't know that this was an SRS or even a representative sample of large cap funds.)

5. a) An ANOVA and an *F*-test.
b) The null hypothesis is that the mean *Distance* from the target is the same for all 4 *Clubs*. The alternative is that it is not.
c) Conclude that the mean *Distance* from the target is not the same for all the *Clubs*.
d) Boxplots of *Distance* by *Club*, residual vs. predicted value scatterplot, and a Normal probability plot of the residuals.
e) A multiple comparison test to see which club is best.

7. a) Indicator variable.
b) The *Mortality* rate north of Derby is, on average, 158.9 deaths per 100 000 higher than south of Derby, after allowing for the linear effects of the water hardness.
c) This appears to be a better regression. The indicator has a strongly significant coefficient, and the R^2 for the overall model has increased from 43% to 56.2%.

9. a) The null hypotheses are that the mean number of *Pins* is the same for all three *Weights* and for both *Approaches*. The alternatives are that the means are not all the same for the three *Weights* or for the two *Approaches*.
b) *Weight* has 2 df; *Approach* has 1. The error has 20 df.
c) If the interaction plot shows any evidence of not being parallel, she should fit an interaction term, using 2 df to fit it.

11. 404 checks

13. a) Both are straight enough. Larger plants are more expensive. Plants got more expensive over time.

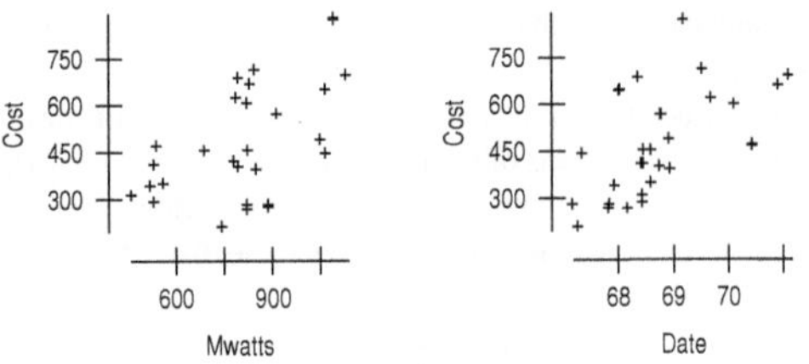

b) *Cost* of nuclear plants increased by \$42 383 on average for each *Megawatt* of power.
c) Relationship is straight enough; there is no reason to doubt independence. Residual plot shows reasonably constant variance. Residuals are nearly Normal.

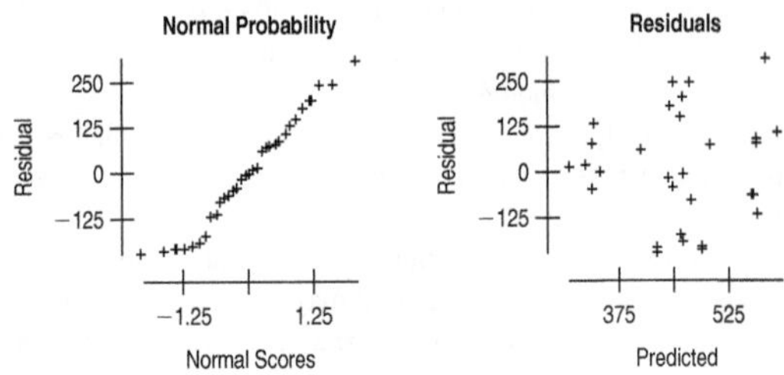

d) $t = 2.93$ on 30 df, P = 0.0064
e) Cost est = 111.7412 + 432.83 = 535.57, or \$535 570 000.
f) The R^2 means that 22.3% of the variation in *Cost* of nuclear plants can be accounted for by the linear relationship between *Cost* and the size of the plant, measured in megawatts (*Mwatts*). A scatterplot of residuals against *Date* shows a strong linear pattern, so it looks like *Date* could account for more of the variation.

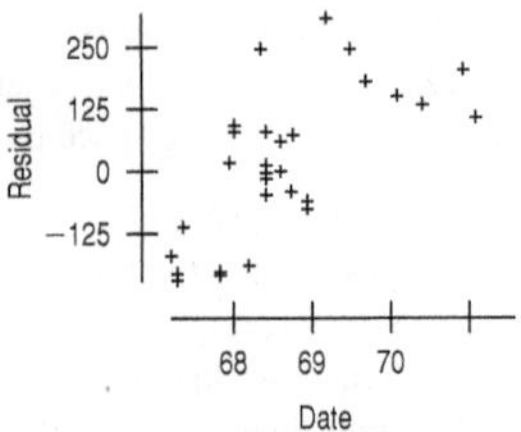

Dependent variable is: Cost
R-squared = 58.4% R-squared (adjusted) = 55.5%
s = 113.4 with 32 3 = 29 degrees of freedom

Variable	Coefficient	SE(Coeff)	t-ratio	P-value
Intercept	−6790.88	1378	−4.93	≤ 0.0001
Mwatts	0.413225	0.1076	3.84	0.0006
Date	100.776	20.07	5.02	≤ 0.0001

g) The coefficient of *Mwatts* changed very little.

h) Because the coefficient changed little when we added *Date* to the model, we can expect that *Date* and *Mwatts* are relatively uncorrelated. In fact, their correlation is 0.02.

15. a) Null hypotheses are that mean hours slept are the same for both sexes and that the mean hours slept are the same for all four classes. Alternatives are that the mean hours slept are not the same for both sexes and that not all four classes have the same mean hours slept.
b) Yes, an interaction term should be fit because the Additive Enough Condition does not seem to be met.
c) None appear to be significant.
d) There are a few outliers. The Constant Variance Condition appears to be met, but we do not have Normal probability plots of residuals. The main concern is that we may not have enough power to detect differences between groups.

17. a) $H_0: p_{<38} - p_{\geq 38} = 0$; $H_A: p_{<38} - p_{\geq 38} \neq 0$. $z = 3.56$; P-value = 0.0004. With such a small P-value, we reject H_0. We conclude there is evidence of a difference in effectiveness; it appears the methods are not as good for older women.
b) $\chi^2 = 12.70$; P-value = 0.0004. Same conclusion.
c) The P-values are the same; $z^2 = (3.563944)^2 = 12.70 = \chi^2$.

19. a) Positive direction, generally linear trend; moderate scatter.
b) H_0: There is no linear relationship between *Interval* and *Duration*. $\beta_1 = 0$. H_A: There is a linear relationship $\beta_1 \neq 0$.
c) Yes; histogram is unimodal and roughly symmetric; residuals plot shows random scatter.
d) $t \approx 27.1$; P-value = 0.001. With such a small P-value, we reject H_0. There is a significant positive linear relationship between *Duration* and time to next eruption of Old Faithful.
e) Based on this regression, the average time *Interval* to next eruption after a 2-minute eruption is between 53.24 and 56.12 minutes, with 95% confidence.
f) Based on this regression, we will have to wait between 63.23 and 87.57 minutes after a 4-minute eruption, with 95% confidence.

21. a) Predicted interval = 35.2463 + 10.4348 *Duration* 0.126316 *Day*. We predict the next eruption to be in 35 minutes plus 10 minutes for each minute of the *Duration* of the previous eruption minus 0.126 minutes (about 7.5 seconds) per *Day*.
b) The null hypothesis of no change (zero coefficient) is rejected at the .05 level because the P-value is 0.0166. It appears that there is change over time, after we account for the *Duration* of the previous eruption.
c) The coefficient for *Day* is not about the two-variable association, but about the relationship between *Interval* and *Day* after allowing for the linear effects of *Duration*.
d) In any event, the amount of change is only about 7.5 seconds per day—not enough to make much difference.

23. $H_0: p_P - p_N = 0$; $H_A: p_P - p_N > 0$. $z = 1.96$; P-value = 0.0249. With such a small P-value, we reject H_0. These data provide moderately strong evidence to suggest that preemies are more likely than babies of normal weight to be of "subnormal height" as adults.

25. a) This regression has a much higher R^2 and much smaller standard deviation of the residuals.
b) Female traffic *Deaths* are estimated to have increased at the rate of 20.4 per *Year* after allowing for the linear effects of male traffic deaths.
c) They've probably been decreasing over time. In the second regression, male and female deaths are likely to be collinear.

27. a) Null hypotheses are that the mean number of *Words Typed* is the same for both *Music* on and off and for both *Television* on and off. The alternatives are that the mean number of words typed are different for the two levels of music and for television on and off.
b) Based on the partial boxplots, both effects appear to be small and probably not statistically significant.
c) Yes, an interaction term should be fit because the Additive Enough Condition does not seem to be met.
d) The interaction term may be real. It appears that the effect of *Television* is stronger with *Music* off than on.
e) None of the three effects are significant.
f) None of the effects seem strong. He seems to type just as well with the music and/or television on.
g) $s = \sqrt{\frac{197.5}{28}} = 2.66$ words per minute. Yes, this seems consistent with the size of the variation shown in the partial boxplots.
h) Tell him that he can turn on the music and/or the TV. They do not seem to affect his typing speed.
i) Even though we haven't seen a histogram or Normal probability plot of residuals, the other conditions seem to be satisfied. For the levels of the factors that he used, it seems that to the level of experimental error present, neither the music nor the TV affects his typing speed.

29. a) Predicted *Births* = 4422.98 – 15.1898(*Age*) – 1.898(*Year*).
b) Births seem to be declining at the rate of 1.89 births per 1000 women each year, after allowing for differences in the age of the women.
c) The scatterplot shows both clumping and a curved relationship. We might want to re-express *Births* or add a quadratic term to the model. The clumping is due to having data for each year of the decade of the 90s for age bracket of women. It probably does not indicate a failure of the linearity assumption.

31. a) $t = 1.42$, P-value = 0.1574. Since $P > 0.05$, we do not reject H_0. The two groups did not differ in ability at the start of the study.
b) $t = 15.11$; P-value = 0.0001. The group taught using the Accelerated Math program showed a significant improvement.
c) $t = 9.24$; P-value = 0.0001. The control group showed a significant improvement in test scores.
d) $t = 5.78$; P-value < 0.0001. The Accelerated Math group had significantly higher gains than the control group.

33. a) The regression—he wanted to know about association.
b) There is a moderate relationship between *Cottage Cheese* and *Ice Cream* sales; for every million pounds increase in *Cottage Cheese,* 1.19 million pounds more of ice cream are sold on average.
c) Testing if the mean difference is 0 (matched pairs t-test). Regression won't answer this question.
d) The company sells more cottage cheese than ice cream, on average.
e) Part a—linear relationship; residuals have a Normal distribution; residuals are independent with equal variation about the line. c—Observations are independent; differences are approximately Normal; less than 10% of all possible months' data.
f) About 71.32 million pounds g) (0.09, 2.29)
h) From this regression, every million pounds increase in *Cottage Cheese* sold is associated with an increase in *Ice Cream* sales of between 0.09 and 2.29 million pounds.

35. a) H_0: Both javelin *Brands* give the same average *Distance*. H_A: There is a difference in average *Distance* between *Brands*. H_0: Both *Preparation* conditions lead to the same average *Distance*. H_A: There is a difference in average *Distance* between *Preparation* conditions.
b) Partial boxplots show that both factors seem to have effects:

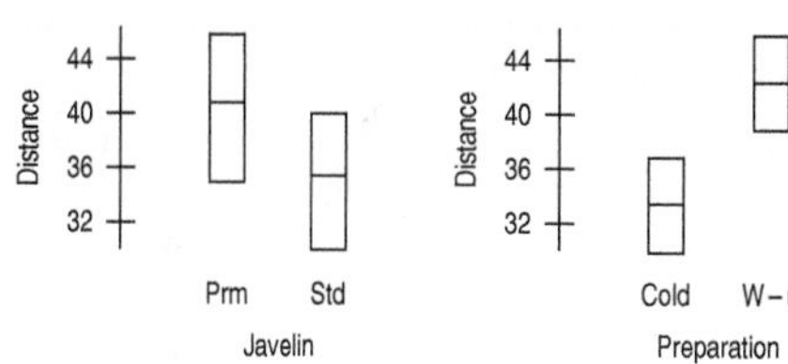

No interaction seems to be present. The Additive Enough Condition appears to be satisfied:

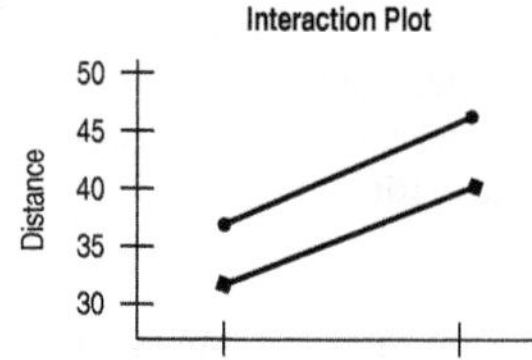

The conditions appear to be met to fit a two-way ANOVA without interaction:

Source	DF	Sum of squares	F-ratio	P-value
Javelin	1	60.58	55	0.0007
Preparation	1	162	147.27	< 0.0001
Error	5	5.5		

Both effects are significant. It appears that depending on the cost, the premium javelin may be worth it—it increases the distance about 5.5 metres on average. Warming up increases distance about 9 metres on average. She should always warm up and consider using the premium javelin.

37. $\chi^2 = 14.96$; P-value ≈ 0.0006. With such a small P-value, we reject H_0. These data provide evidence that continued treatment with infliximab is of value.

39. a) The null hypothesis is that mean *TV Watching* is the same for all three groups. The alternative is that it is not.

b) The variance for the None group appears to be slightly smaller, and there are outliers in all three groups. We don't have a Normal probability plot of the residuals, but we suspect that the data may not be Normal enough.

c) The *F*-test indicates that the mean amount of *TV Watching* is not the same for the three groups.

d) It seems that the differences are evident even when the outliers are removed. It now seems that the conclusion is valid.

41. Based on these data, the average *Weight Loss* for the clinic is between 8.24 and 10.06 pounds, with 95% confidence. The clinic's claim is plausible.

43. a) Yes. Data plot is linear; residuals plot shows random scatter; histogram of residuals is roughly Normal.

b) Yes; $t \approx 5.19$; P-value < 0.0001.

c) Weaker. Individuals are more variable than averages.

d) Based on this regression, every extra year of *Median Education* in a city is associated with an increase of between \$1500.48 and \$3389.10 in *Median Income*.

e) Based on the regression, *Median Income* for cities where residents spent an average of 11 years in school is between \$32 198.81 and \$33 526.67, with 90% confidence.

Chapter 29

1. If we assume that the rank sum has a normal distribution, its mean = 30 and the variance = 30. z = −1.10 and p-value = 0.2714 (for a non-directional test).

3. P-value for observed sum of 43 is greater than 0.05 × 2, so the difference is still not significant at the 5% level.

5. H = 31.64, $\chi^2_{2,0.05} = 5.991$, $H > \chi^2_{2,0.05}$ and so reject H_0: all populations have the same distributions. The results from the one-way ANOVA are copied below and based on that we reject the null hypothesis of equal population means.

One-way ANOVA: Time(sec) versus Time of day

Source	DF	SS	MS	F	P
Time of day	2	204641	102320	46.03	0.000
Error	45	100020	2223		
Total	47	304661			

S = 47.15 R-Sq = 67.17% R-Sq(adj) = 65.71%

```
                                  Individual 95% CIs For Mean Based on
                                  Pooled StDev
Level              N    Mean  StDev  -----+---------+---------+---------+----
Early(7a.m.)      16  113.38  47.65  (---*---)
Evening(5p.m.)    16  273.31  52.19                              (---*---)

Latenight(12a.m.) 16  193.06  40.90              (---*---)
                                     -----+---------+---------+---------+----
                                        120       180       240       300
```

Pooled StDev = 47.15

7. The one-way ANOVA shows a greater difference. F-value changes from 11.91 to 43.97 as the value of the Kruskal-Wallis statistics (H) only changes from 13.74 to 17.43.

Kruskal-Wallis Test: pain2 versus Drug

Kruskal-Wallis Test on pain2

Drug	N	Median	Ave	Rank	Z
A	9	2.000	5.0	4.17	
B	9	6.000	18.0	1.85	
C	9	6.000	19.0	2.31	
Overall	27	14.0			

H = 17.43 DF = 2 P = 0.000
H = 17.98 DF = 2 P = 0.000 (adjusted for ties)

One-way ANOVA: Pain 2 versus Drug

Source	DF	SS	MS	F	P
Drug	2	104.22	52.11	43.97	0.000
Error	24	28.44	1.19		
Total	26	132.67			

S = 1.089 R-Sq = 78.56% R-Sq(adj) = 76.77%

```
                          Individual 95% CIs For Mean Based on
                          Pooled StDev
Level  N   Mean  StDev  ----+---------+---------+---------+-----
A      9  1.667  0.866  (----*----)
B      9  5.778  1.481                          (----*----)
C      9  5.889  0.782                          (----*----)
                        ----+---------+---------+---------+-----
                          1.5       3.0       4.5       6.0
```

Pooled StDev = 1.089

9. S = 10.88, $\chi^2_{3,0.05} = 7.815$, $H > \chi^2_{2,0.05}$ and so reject H_0: all populations have the same distributions.
The results from the ANOVA are copied below and based on that we reject the null hypothesis of equal population means.

Two-way ANOVA: Score versus Temp, Cycle

Source	DF	SS	MS	F	P
Temp	3	33.2519	11.0840	23.47	0.000
Cycle	3	7.1969	2.3990	5.08	0.025
Error	9	4.2506	0.4723		
Total	15	44.6994			

S = 0.6872 R-Sq = 90.49% R-Sq (adj) = 84.15%

11. a) There are 7 negative differences and 10 positive differences, and the p-value of the sign test is 0.6291. The P-value is large so there's insufficient evidence to declare any difference.
b) W = 93.0 and the p-value is 0.449. The P-value is large so there's insufficient evidence to declare any difference.
c) The p-values for the paired *t*-test, Wilcoxon's signed rank test, and sign test are 0.391, 0.449 and 0.6291.

13. Wilcoxon's signed rank test shows a significant improvement (p-value = 0.005). Paired *t*-test similarly shows a significant improvement (p-value = 0.001).

15. a) **Correlations for data without outlier**
Pearson correlation of Woman and Exp. = −0.168
Spearman's correlation of RankWoman and RankExp = −0.651

Correlations for data without outlier
Pearson correlation of Woman and Exp. = −0.796
Spearman's correlation of RankWoman and RankExp = −0.838
b) The removal of the outlier caused a significant change (−0.168 to −0.796) but only a small change in Spearman's correlation (−0.651 to −0.838).

17. The curvature in the scatterplot will be less for the ranked data, so Spearman's correlation should be greater. The Spearman's correlation = 0.94.

19. Both Wilcoxon's rank sum test (Mann-Whitney test) and two sample *t*-test show a significant difference between the populations (*p*-values 0.0026 for the Wilcoxon's rank sum test and 0.007 for the two sample *t*-test).
Mann-Whitney Test and CI: Buying from a Friend, Buying from a Stranger

	N	Median
Buying from a Friend	8	282.50
Buying from a Stranger	7	225.00

Point estimate for ETA1-ETA2 is 60.00
95.7% CI for ETA1-ETA2 is (29.98, 125.02)
W = 90.5
Test of ETA1 = ETA2 vs ETA1 not = ETA2 is significant at 0.0026
The test is significant at 0.0025 (adjusted for ties)
Two-Sample T-Test and CI: Buying from a Friend, Buying from a Stranger
Two-sample T for Buying from a Friend vs Buying from a Stranger

	N	Mean	StDev	SE Mean
Buying from a Fr	8	281.9	18.3	6.5
Buying from a St	7	211.4	46.4	18

Difference = mu (Buying from a Friend) mu (Buying from a Stranger)
Estimate for difference: 70.4464
95% CI for difference: (26.2146, 114.6783)
T-Test of difference = 0 (vs not =): T-Value = 3.77 P-Value = 0.007 DF = 7

21. Both Wilcoxon's rank sum test (Mann-Whitney test) and two sample *t*-test show a significant difference between the populations (*p*-values 0.0016 for the Wilcoxon's rank sum test and 0.001 for the two sample *t*-test).
Mann-Whitney Test and CI: 4000BCE, 200BCE

	N	Median
4000BCE	30	131.00
200BCE	30	135.00

Point estimate for ETA1-ETA2 is −4.00
95.2 Percent CI for ETA1-ETA2 is (−7.00, −2.00)
W = 701.5
Test of ETA1 = ETA2 vs ETA1 not = ETA2 is significant at 0.0016
The test is significant at 0.0016 (adjusted for ties)
Two-Sample T-Test and CI: 4000BCE, 200BCE

Two-sample T for 4000BCE vs 200BCE				
	N	Mean	StDev	SE Mean
4000BCE	30	131.37	5.13	0.94
200BCE	30	135.63	4.04	0.74

Difference = mu (4000BCE) – mu (200BCE)
Estimate for difference: −4.26667
95% CI for difference: (−6.65627, −1.87706)
T-Test of difference = 0 (vs not =): T-Value = −3.58 P-Value = 0.001 DF = 54

23. Choose 1–10 and Jack from Spades. Mix them up well and then draw 4 of them. Note if the total is ≤ 11. Repeat thousands of times. Divide # time ≤ 11 by # repetitions. This estimated the probability for ≤ 11. Double for two sided test.

25. There are 462 distinct possible hands.
P-value for sum of 15 = $1/462 \times 2$
≤ 16: P-value = $2/462 \times 2$
≤ 17: P-value = $4/462 \times 2$

Chapter 30

1. Samples with all three identical measurements, like {2.0, 2.0, 2.0} are least likely since there is only one way such a sample can arise (with probability = $1/3 \times 1/3 \times 1/3$). Other samples have higher probabilities since they can arise in multiple ways.

3.

Sample	mean
1 (2 2 6 10)	5
2 (2 2 6 14)	6
3 (2 2 10 14)	7
4 (6 6 2 10)	6
5 (6 6 2 14)	7
6 (6 6 10 14)	9
7 (10 10 2 6)	7
8 (10 10 2 14)	9
9 (10 10 6 14)	10
10 (14 14 2 6)	9
11 (14 14 2 10)	10
12 (14 14 6 10)	11

The standard deviation of these sample means is the standard error and is equal to 1.907.

The standard deviation of the 4 values (2.0, 6.0, 10.0, 14.0) is

$s = \sqrt{\frac{\sum_i (x_i - \bar{x})^2}{n - 1}} = 5.16$ and the theoretical formula gives

$SE(\bar{X}) = \frac{s}{\sqrt{n}} = \frac{5.16}{\sqrt{4}} = 2.58.$

5. The answers will differ, depending on the resamples drawn. The means of the 200 resamples are shown below:

Mean								
7.825	4.700	7.075	6.125	5.775	6.050	7.175	5.725	3.550
7.850	7.675	4.650	7.400	5.200	6.475	7.475	9.075	5.975
6.400	7.650	6.150	5.650	7.500	4.875	7.650	6.475	6.275
6.750	7.075	6.150	6.600	4.450	6.150	8.150	5.775	5.800
4.775	6.225	5.675	6.550	6.825	5.550	6.800	6.775	5.875
7.300	6.675	4.375	4.850	7.350	4.900	5.350	5.775	6.575
6.675	5.625	7.325	5.700	6.650	5.925	6.625	6.250	4.800
8.325	8.925	3.875	7.575	7.000	6.150	7.175	4.625	8.375
5.950	6.500	6.400	7.400	6.425	5.475	3.900	6.650	8.725
7.475	7.150	8.675	8.350	6.300	5.725	5.875	5.850	9.950
6.575	7.775	7.225	5.500	6.900	6.575	6.600	6.625	7.825
8.325	6.900	6.450	7.975	5.400	5.800	6.600	7.025	5.250
8.925	7.725	6.750	8.150	3.750	7.000	6.325	6.650	6.600
7.625	6.875	8.325	6.150	7.425	7.750	5.800	7.725	8.075
5.700	6.525	6.075	7.075	5.450	6.800	6.125	4.700	6.300
7.050	5.700	6.250	6.125	7.150	5.675	5.650	5.825	5.275
6.450	5.000	6.575	5.800	6.600	7.200	4.650	4.950	5.800
7.450	6.500	6.550	6.725	8.675	5.650	5.975	6.150	6.600
7.075	7.450	7.200	7.975	7.000	8.825	8.050	5.875	6.775
5.750	5.825	6.375	4.375	8.025	5.800	7.675	6.750	8.500
5.625	7.350	7.250	5.000	8.000	5.400	7.775	6.500	9.275
6.700	7.550	4.875	5.700	5.175	6.675	6.450	7.250	5.775
6.350	7.425							

Mean and the standard deviation of the 200 sample means:

Descriptive Statistics: Mean

Variable	N	Mean	StDev
Mean	200	6.5449	1.1357

Bootstrap estimate of the standard error of the mean percentage of unpopped kernels = 1.1357

The standard deviation of the 8 values in the data

set is $s = \sqrt{\frac{\sum_i (x_i - \bar{x})^2}{n - 1}} = 3.6$ and the theoretical formula gives

$SE(\bar{X}) = \frac{s}{\sqrt{n}} = \frac{3.6}{\sqrt{8}} = 1.3.$

7. a) The answers will differ, depending on the resamples drawn. The medians of the 200 resamples are shown below:

Median							
1214.0	1207.0	1231.5	1244.0	1214.0	1258.0	1258.0	1258.0
1229.0	1216.5	1200.0	1291.5	1258.0	1219.0	1216.5	1214.0
1231.5	1244.0	1258.0	1244.0	1258.0	1270.0	1167.5	1195.5
1251.0	1195.5	1264.0	1244.0	1200.0	1244.0	1251.0	1195.5
1258.0	1251.0	1251.0	1214.0	1219.0	1258.0	1195.5	1200.0
1219.0	1214.0	1191.0	1257.0	1238.5	1244.0	1214.0	1244.0
1209.5	1219.0	1282.5	1161.5	1219.0	1258.0	1229.0	1219.0
1216.5	1251.0	1269.5	1163.0	1282.5	1214.0	1219.0	1216.5
1325.0	1200.0	1219.0	1214.0	1295.0	1269.5	1282.5	1135.0
1231.5	1219.0	1219.0	1195.5	1191.0	1244.0	1207.0	1219.0
1222.0	1195.5	1264.0	1307.5	1231.5	1264.0	1244.0	1195.5
1231.5	1257.0	1258.0	1257.0	1251.0	1207.0	1216.5	1209.5
1174.5	1276.5	1229.0	1163.0	1229.0	1161.5	1282.5	1297.5
1231.5	1231.5	1207.0	1191.0	1257.0	1251.0	1310.0	1222.0
1207.0	1214.0	1251.0	1236.0	1231.5	1231.5	1270.0	1219.0
1244.0	1219.0	1244.0	1229.0	1238.5	1238.5	1257.0	1270.0
1251.0	1251.0	1244.0	1214.0	1251.0	1264.0	1257.0	1200.0
1276.5	1251.0	1219.0	1195.5	1282.5	1238.5	1251.0	1207.0
1231.5	1244.0	1251.0	1195.5	1207.0	1200.0	1205.0	1251.0
1264.0	1219.0	1216.5	1251.0	1264.0	1191.0	1191.0	1200.0
1216.5	1251.0	1229.0	1264.0	1216.5	1264.0	1251.0	1214.0
1258.0	1251.0	1238.5	1251.0	1167.5	1251.0	1257.0	1264.0
1231.5	1244.0	1251.0	1216.5	1244.0	1214.0	1244.0	1229.0
1251.0	1238.5	1264.0	1251.0	1219.0	1269.5	1251.0	1258.0
1209.5	1207.0	1195.5	1200.0	1200.0	1257.0	1295.0	1229.0

Mean and the standard deviation of the 200 sample medians:

Descriptive Statistics: Median

Variable	N	Mean	StDev
Median	200	1233.4	30.4

Bootstrap estimate of the standard error of the median = 30.4

b) The answers will differ, depending on the resamples drawn. The means of the 200 resamples are shown below:

Mean						
1215.31	1245.31	1198.00	1229.88	1264.88	1253.13	1236.06
1239.94	1243.19	1246.25	1244.50	1224.19	1251.81	1217.88
1269.44	1262.88	1228.19	1240.31	1256.00	1222.69	1250.38
1260.75	1249.50	1263.31	1261.75	1227.00	1217.50	1242.63
1221.38	1245.06	1238.88	1276.19	1226.25	1274.19	1286.75
1252.75	1247.81	1246.63	1213.13	1207.50	1267.88	1187.00
1244.38	1270.56	1220.13	1223.06	1222.56	1256.69	1264.56
1258.75	1257.31	1269.94	1260.75	1249.19	1250.44	1236.38
1181.50	1227.69	1202.31	1228.06	1256.94	1254.88	1227.81
1213.06	1252.88	1274.25	1253.69	1217.81	1255.13	1257.44
1206.19	1234.25	1199.13	1217.81	1274.94	1284.94	1257.00
1271.31	1231.69	1268.50	1255.44	1238.19	1253.63	1246.25
1249.13	1241.94	1238.88	1260.19	1230.44	1224.19	1229.69
1252.13	1254.00	1278.38	1235.50	1228.19	1242.94	1274.50
1253.56	1251.94	1252.19	1262.50	1238.31	1229.56	1227.25
1260.63	1244.19	1227.69	1217.69	1223.31	1253.75	1271.75
1227.00	1285.00	1208.44	1263.13	1222.94	1240.94	1223.94
1213.25	1274.13	1206.75	1229.88	1240.38	1249.06	1247.94
1194.25	1223.81	1223.81	1224.31	1247.38	1272.94	1236.25

1234.88	1259.44	1229.75	1195.81	1238.13	1261.94	1252.50
1222.00	1231.13	1248.44	1197.50	1240.69	1262.81	1221.75
1230.31	1200.56	1252.56	1255.06	1234.75	1237.31	1294.00
1285.63	1231.31	1278.94	1229.25	1247.69	1240.75	1232.19
1235.00	1195.94	1273.19	1255.75	1237.38	1236.56	1265.13
1209.00	1221.44	1266.25	1270.31	1242.31	1255.75	1180.81
1251.31	1230.56	1260.88	1222.69	1225.13	1231.88	1210.31
1219.00	1208.19	1217.00	1239.88	1263.50	1183.56	1270.38
1240.38	1230.75	1184.13	1264.19	1266.63	1238.88	1236.31
1257.75	1242.00	1236.44	1279.00			

Mean and the standard deviation of the 200 sample means:

Descriptive Statistics: Mean

Variable	N	Mean	StDev
Mean	200	1241.0	22.7

Bootstrap estimate of the standard error of the mean = 22.7

Mean has a smaller standard error than the median and so the mean is a better estimator.

c) The formula for the standard error of the sample median is somewhat complicated and so bootstrap methods can be useful in estimating the standard error of the sample median. For the sample mean, a simple formula is available for the standard error and so it is not necessary to use bootstrap methods in part (b). It can be calculated as follows:

The standard deviation of the 8 values in the data set is $s = \sqrt{\dfrac{\sum_i (x_i - \bar{x})^2}{n - 1}} = 94.3$ and the theoretical formula gives $SE(\bar{X}) = \dfrac{s}{\sqrt{n}} = \dfrac{94.3}{\sqrt{16}} = 23.6$.

9. The answers will differ, depending on the resamples drawn. The trimmed means of the 200 resamples are shown below:

Trimmed Mean

1214.14	1235.71	1175.93	1222.64	1238.86	1225.93	1226.79
1192.00	1205.29	1236.43	1207.79	1238.71	1255.36	1288.07
1226.14	1212.14	1258.36	1239.50	1265.71	1234.86	1229.71
1266.29	1240.07	1222.86	1242.50	1304.57	1217.86	1190.00
1229.14	1230.71	1237.07	1239.93	1251.71	1264.93	1277.50
1245.86	1267.64	1231.64	1192.79	1234.79	1233.29	1240.71
1254.43	1242.00	1229.29	1226.71	1255.43	1207.71	1233.07
1266.07	1217.50	1193.93	1245.43	1258.57	1247.36	1158.79
1238.57	1250.21	1220.29	1211.14	1194.86	1232.71	1214.00
1249.43	1234.07	1224.86	1271.21	1225.14	1232.79	1249.71
1243.71	1199.79	1229.14	1213.57	1214.14	1255.86	1214.71
1226.79	1194.64	1232.64	1243.21	1218.29	1245.64	1253.21
1250.36	1218.21	1244.64	1229.57	1225.43	1251.64	1220.93
1248.57	1240.36	1255.71	1191.43	1254.29	1241.64	1219.36
1227.86	1260.57	1240.50	1246.00	1223.29	1261.93	1238.93
1272.36	1237.07	1211.57	1262.14	1201.14	1201.36	1203.93
1235.21	1259.43	1228.07	1236.43	1263.79	1198.07	1215.00
1251.43	1230.93	1241.29	1210.57	1208.43	1228.43	1213.21
1285.29	1231.14	1257.50	1286.14	1222.29	1281.36	1261.43

1221.36	1260.21	1248.21	1207.00	1220.36	1227.86	1262.36
1247.50	1230.50	1239.79	1248.43	1221.50	1212.21	1209.64
1256.86	1193.21	1224.93	1211.93	1241.29	1275.43	1225.86
1208.71	1255.50	1232.07	1224.43	1206.71	1232.64	1227.21
1261.14	1248.14	1240.43	1185.93	1243.50	1243.43	1206.36
1230.86	1247.36	1224.86	1221.36	1265.93	1296.93	1192.36
1242.71	1238.00	1236.07	1188.43	1216.29	1237.07	1241.79
1277.64	1244.79	1219.43	1192.07	1230.14	1197.21	1260.14
1225.14	1158.86	1257.64	1214.43	1229.00	1248.57	1271.00
1228.36	1242.21	1263.64	1234.21			

Mean and the standard deviation of the 200 sample trimmed means:

Descriptive Statistics: Trimmed Mean

Variable	N	Mean	StDev
Trimmed Mean	200	1233.8	24.1

Bootstrap estimate of the standard error of the mean = 24.1

Since there is no easy formula for the standard error, the Bootstrap methods are very useful in estimating it.

11. The standard errors with $B = 50$, 200 and 500 are shown below:

Descriptive Statistics: Mean

Variable	N	StDev
Mean	50	20.1

Descriptive Statistics: Mean

Variable	N	StDev
Mean	200	23.1

Descriptive Statistics: Mean

Variable	N	StDev
Mean	5 00	22.9

There are no big differences in the estimates of the standard error.

13. The answers will differ, depending on the resamples drawn. 200 resamples were drawn with replacement from each sample and their sample means and the differences between those sample means were calculated. These differences are shown below:

Difference in Means

22.7071	−12.5101	−6.2727	−3.6010	−12.9545	−10.2677	−13.8081
−14.2576	−13.2828	−11.4949	−19.3131	−15.3535	−5.7172	−14.3030
−17.3131	−7.5000	−5.8182	−9.9899	−2.8889	−16.9848	−8.0758
−11.4040	−7.8182	−9.2576	−8.3586	−8.1869	−2.6768	−9.3838
−13.9343	−10.6212	−0.7980	−12.6818	−7.5859	−15.0253	−6.7525
−9.0859	−4.4192	−10.9848	−13.4091	−14.4646	−10.7374	−9.2121
−6.1566	−12.2323	−14.1616	0.7071	−6.8030	−18.9747	−8.0909
−11.4697	−14.3081	−11.2121	−12.6364	−4.0051	−15.7121	−5.6919
−8.7576	−8.4798	−16.1263	−8.5303	−3.6162	−11.4444	−13.0606
−9.4343	−16.7475	−3.2071	−11.2273	−8.7424	−11.5303	−9.4646
−9.5101	−9.6970	−10.2626	−7.5202	−12.1010	−11.3081	−1.7475
−13.3434	−11.8838	−13.6919	−11.2424	−10.2273	−12.5808	−12.9343
−15.5051	−6.3081	−14.0758	−5.8889	−13.8990	−13.6768	−15.7374
−3.8434	−14.2980	−5.9545	−1.1313	−12.6970	210.3232	−10.8232
−3.4444	−10.8182	−13.5101	−3.4798	−16.4040	−5.6313	−16.3333

−11.8232	−8.4293	−20.1061	−17.5808	−6.0354	−11.2879	−13.2879
−11.3081	−9.8990	−12.8889	−6.7525	−16.0152	−13.7121	−7.3283
−11.2677	−13.5707	−7.9343	−11.8889	−4.5000	−8.2222	−17.2525
−10.3990	−10.4394	−15.1364	1.9899	−10.6111	−12.2828	−8.2677
−9.1313	−14.9444	−16.4596	−13.7879	−14.7424	−8.6263	−10.3636
−15.8081	−12.8737	−14.1869	−10.4495	−6.9899	−9.3485	−5.8384
−4.8636	−9.5000	−13.8889	−2.5101	−10.8838	−6.8687	−13.7929
−9.8232	−16.1667	−12.7172	−7.7828	−10.1364	−9.6818	−8.1919
−18.1212	−6.1970	−9.7778	−11.3990	−16.6616	−7.4899	−13.5505
−8.2374	−9.8737	−6.4091	−9.3788	−3.6111	−8.6212	−16.8535
−7.9495	−7.8788	−8.3232	−8.9192	−11.4040	−13.0354	−7.1010
−12.1313	−7.5101	−9.1970	−13.8333	−9.2374	−3.8687	−9.7778
−2.4242	−17.6667	−13.3485	−17.6970	−7.2980	−11.0859	−7.7626
−6.8586	−15.2929	−0.0556	−9.6010			

Mean and the standard deviation of these 200 differences:

Descriptive Statistics: Difference in Means

Variable	N	Mean	StDev
Difference in Me	200	10.281	4.229

Bootstrap estimate of the standard error of the difference of the two sample means = 4.229

The means and the standard deviations of the two original samples are:

	N	Mean	StDev	SE Mean
Control	22	41.8	16.6	3.5
NewActivities	18	51.7	11.7	2.8

The standard error of the difference between the two means calculated using the formula $= \sqrt{3.5^2 + 2.8^2} = 4.48$.

15. a) The answers will differ, depending on the resamples drawn. 200 resamples were drawn with replacement from each sample and their sample means and the differences between those sample means were calculated. These differences are shown below:

Difference in Means

278.875	−79.125	−84.750	−88.625	−78.250	−68.875	−80.750
−85.750	−97.625	−81.375	−74.875	−77.500	−68.750	−75.875
−84.625	−99.875	−89.875	−74.250	−78.375	−50.000	−78.375
−77.250	−89.875	−64.125	−77.125	−72.375	−66.875	−96.500
−74.500	−85.625	−55.250	−74.250	−89.000	−82.250	−58.375
−86.500	−107.000	−71.375	−58.500	−64.750	−80.875	−86.000
−68.500	−94.375	−61.375	−84.750	−97.125	−73.250	−75.125
−92.000	−65.875	−87.375	−65.625	−73.500	−71.375	−82.250
−99.750	−97.250	−105.875	−90.375	−68.000	−71.750	−66.250
−90.000	−90.500	−103.000	−66.875	−92.500	−71.500	−107.250
−67.375	−82.500	−56.875	−64.375	−69.750	−78.125	−80.125
−73.500	−89.500	−113.000	−76.500	−91.875	−87.000	−92.625
−72.125	−61.500	−53.750	−73.250	−71.750	−85.500	−67.125
−68.500	−53.375	−62.750	−68.500	−91.625	−57.500	−81.250
−60.750	−66.125	−71.625	−62.500	−100.875	−108.250	−85.000
−68.250	−84.625	−76.625	−74.750	−80.625	−80.750	−67.125
−88.750	−76.000	−78.250	−84.750	−41.375	−69.250	−79.750
−71.875	−78.750	−93.875	−97.750	−99.875	−70.000	−64.625
−78.125	−81.250	−92.000	−61.750	−76.250	−60.125	−82.250
−84.125	−95.500	−78.125	−92.125	−88.250	−80.125	−80.625
−94.125	−85.125	−65.750	−73.000	−103.625	−93.250	−66.375
−84.250	−64.500	−96.375	−69.000	−59.625	−61.250	−80.125
−94.375	−101.500	−62.625	−70.750	−72.375	−108.000	−62.750
−93.750	−90.125	−79.375	−83.875	−91.750	−83.750	−84.625
−65.250	−78.375	−60.625	−97.000	−90.500	−91.500	−88.000
−41.125	−70.000	−75.000	−87.375	−91.125	−96.500	−87.875
−83.750	−85.375	−84.000	−90.375	−75.625	−74.625	−60.750

Mean and the standard deviation of these 200 differences:

Descriptive Statistics: Difference in Means

Variable	N	Mean	StDev
Difference in Me	200	79.051	13.227

Bootstrap estimate of the standard error of the difference of the two sample mean = 13.227

b) The histogram of the above 200 differences is shown below. Based on this we see that it is unlikely to have a difference of zero between the two means.

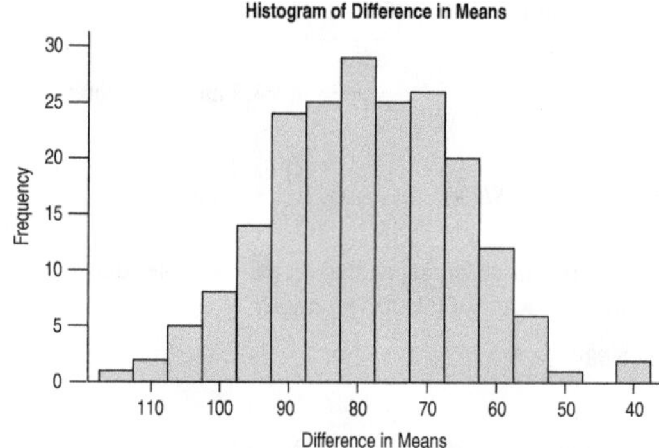

c) Assuming normality, we can test for the equality of the two means using a two-sample t-test as shown below. The test indicates a significant difference between the two means. The standard error the difference between the two sample means $= \sqrt{9.4^2 + 11.0^2} = 14.5$

Two-Sample T-Test and CI: Alcohol, Water

Two-sample T for Alcohol vs Water

	N	Mean	StDev	SE Mean
Alcohol	8	37.5	26.6	9.4
Water	8	117.0	31.1	11

Difference = mu (Alcohol) – mu (Water)
Estimate for difference: −79.5000
95% CI for difference: (−110.7561, −48.2439)
T-Test of difference = 0 (vs not =): T-Value = −5.49
P-Value = 0.000 DF = 13

17. The answers will differ, depending on the resamples drawn. The sample correlations of the 200 resamples are shown below (arranged in increasing order):

Correlation						
0.200598	0.203693	0.204524	0.207919	0.215304	0.235844	0.237436
0.238880	0.244068	0.244185	0.257506	0.257902	0.259889	0.260567
0.276175	0.276681	0.284213	0.288857	0.289554	0.290694	0.291995
0.293896	0.294071	0.299159	0.303086	0.308969	0.309029	0.310699
0.314050	0.314882	0.317267	0.318481	0.319164	0.320704	0.320732
0.321724	0.322511	0.326242	0.327452	0.333717	0.334874	0.335372
0.335381	0.335547	0.336336	0.336496	0.339413	0.340369	0.340912
0.341166	0.347025	0.347889	0.348145	0.355494	0.356217	0.356886
0.357040	0.357523	0.359749	0.360636	0.361713	0.362485	0.362552
0.364712	0.365204	0.365483	0.365646	0.366315	0.367589	0.368848
0.369505	0.369814	0.370334	0.370755	0.373089	0.373641	0.374363
0.375042	0.377443	0.378099	0.381677	0.383962	0.387595	0.388788
0.389049	0.390227	0.391184	0.392672	0.392751	0.392829	0.392989
0.393648	0.394665	0.396201	0.396239	0.397026	0.397457	0.398197
0.398434	0.399689	0.399905	0.400579	0.403271	0.403397	0.405644
0.406203	0.407440	0.408306	0.408397	0.413348	0.413432	0.414477
0.417260	0.417652	0.418136	0.418289	0.418745	0.419970	0.420130
0.421059	0.421371	0.421837	0.422472	0.425090	0.426623	0.426945
0.427333	0.427606	0.429286	0.430954	0.431234	0.434076	0.436650
0.437073	0.437815	0.441259	0.442450	0.444264	0.445058	0.445478
0.446602	0.446658	0.448464	0.448745	0.449599	0.453086	0.453126
0.453946	0.454934	0.455412	0.456402	0.457730	0.458857	0.459299
0.464396	0.464589	0.464644	0.465341	0.468707	0.471396	0.471593
0.472326	0.472346	0.474062	0.475949	0.476441	0.476713	0.476988
0.477399	0.478365	0.480248	0.484047	0.486055	0.488018	0.491710
0.496624	0.497230	0.505020	0.505194	0.505447	0.506256	0.510945
0.512534	0.517102	0.520673	0.527540	0.532018	0.534094	0.537388
0.543511	0.554632	0.555805	0.565348	0.567340	0.575615	0.584636
0.587042	0.594028	0.603128	0.630306			

Mean and the standard deviation of the 200 sample correlations:

Descriptive Statistics: Correlation

Variable	N	Mean	StDev
Correlation	200	0.40093	0.08501

Bootstrap estimate of the standard error of the sample correlation = 0.08501

For the 95% confidence interval, just pick the 5th values from the beginning and the end of the ordered correlations (i.e. approximate 2.5th and 97.5th percentiles). For this simulation it is approximately (0.215, 0.585). This interval does not include 0 and so based on this simulation, it looks unlikely for the correlation to be zero.

19. The answers will differ, depending on the resamples drawn. First, let's use the data for girls. The sample correlations of the 200 resamples are shown below (arranged in increasing order):

Correlation					
−0.321010	−0.275845	−0.271973	−0.261744	−0.260817	−0.259400
−0.259379	−0.258802	−0.255268	−0.248051	−0.243988	−0.238521
−0.235636	−0.222072	−0.215914	−0.211232	−0.210164	−0.208636
−0.199853	−0.198295	−0.198176	−0.197089	−0.190203	−0.188319
−0.187085	−0.184813	−0.174361	−0.160582	−0.153964	−0.146144
−0.139733	−0.137177	−0.136438	−0.133837	−0.127474	−0.126103
−0.125033	−0.119803	−0.113999	−0.110191	−0.108768	−0.106701
−0.098932	−0.097855	−0.095221	−0.093627	−0.093526	−0.089290
−0.089174	−0.088060	−0.087128	−0.086593	−0.082884	−0.080907
−0.079873	−0.079679	−0.078529	−0.075911	−0.072921	−0.071370
−0.069974	−0.067689	−0.067596	−0.066988	−0.066319	−0.064885
−0.063946	−0.061063	−0.059526	−0.053220	−0.049746	−0.046907
−0.043603	−0.040824	−0.037671	−0.036998	−0.035953	−0.035439
−0.030301	−0.029476	−0.027310	−0.026724	−0.017858	−0.017856
−0.017781	−0.016938	−0.014821	−0.013000	−0.012991	−0.012891
−0.001774	−0.000744	0.001490	0.004282	0.005490	0.011196
0.013219	0.014151	0.020654	0.024523	0.025532	0.025935
0.026738	0.028162	0.028313	0.030939	0.032546	0.032579
0.033360	0.033395	0.035334	0.039032	0.040978	0.041382
0.043915	0.044167	0.044317	0.046596	0.050111	0.050114
0.050543	0.050557	0.051210	0.053277	0.055379	0.055784
0.057610	0.058157	0.059672	0.060228	0.062768	0.063330
0.066720	0.068692	0.069586	0.071245	0.072224	0.074427
0.074434	0.076648	0.077000	0.084452	0.090321	0.095748
0.101561	0.105715	0.106341	0.107063	0.111600	0.122567
0.126200	0.127850	0.128500	0.130718	0.137376	0.137529
0.138116	0.144142	0.144433	0.150663	0.151233	0.151799
0.152964	0.153198	0.153364	0.153968	0.155570	0.158814
0.159701	0.160349	0.160804	0.161471	0.164167	0.164735
0.168804	0.172396	0.174599	0.178065	0.179041	0.184083
0.187504	0.187697	0.189127	0.189726	0.195518	0.207653
0.217265	0.232921	0.246649	0.264007	0.289946	0.292456
0.293703	0.321656	0.325494	0.336030	0.362482	0.381601
0.406137	0.455074				

Mean and the standard deviation of the 200 sample correlations:

Descriptive Statistics: Correlation

Variable	N	Mean	StDev
Correlation	200	0.0121	0.1501

Bootstrap estimate of the standard error of the sample correlation = 0.1501.

For the 95% confidence interval, just pick the 5th values from the beginning and the end of the ordered correlations (i.e., approximate 2.5th and 97.5th percentiles). For this simulation it is approximately (−0.261, 0.336). This interval includes 0, and this indicates that it is possible for the correlation to be zero. Now let's look at the data for boys. The sample correlations of the 200 resamples are shown below (arranged in increasing order):

Correlation						
0.076594	0.107678	0.125385	0.140372	0.155502	0.160641	0.172120
0.172132	0.174294	0.174588	0.177429	0.182077	0.214746	0.215070
0.218464	0.221345	0.225599	0.240627	0.244953	0.247703	0.250812
0.254627	0.255681	0.271429	0.273017	0.278207	0.278938	0.279310
0.282017	0.290109	0.290581	0.292909	0.301122	0.303799	0.306616

0.310674	0.312524	0.313305	0.316432	0.317332	0.320252	0.321984
0.322839	0.329484	0.331207	0.331209	0.332068	0.333575	0.334149
0.334986	0.336402	0.336782	0.337067	0.338324	0.339721	0.346060
0.346460	0.346617	0.348044	0.349744	0.350283	0.351098	0.355799
0.356485	0.356816	0.358150	0.359293	0.359542	0.359709	0.360302
0.367204	0.368202	0.368209	0.370324	0.371143	0.371183	0.373382
0.374312	0.376598	0.378019	0.381652	0.383867	0.384290	0.385205
0.385621	0.387691	0.387809	0.388980	0.389173	0.391049	0.393676
0.395479	0.397773	0.397908	0.399205	0.400247	0.401568	0.402527
0.402804	0.403806	0.405901	0.409505	0.409676	0.409910	0.419915
0.425402	0.427731	0.428168	0.429053	0.431885	0.432701	0.435306
0.435441	0.435753	0.437879	0.438817	0.439452	0.440544	0.440976
0.442923	0.442937	0.444322	0.444385	0.445133	0.445464	0.445654
0.448416	0.448727	0.449361	0.449573	0.450670	0.453238	0.454556
0.455494	0.457354	0.457592	0.459474	0.459728	0.460971	0.461082
0.462003	0.462496	0.463298	0.467316	0.469853	0.470442	0.478627
0.480062	0.482836	0.482921	0.483915	0.486714	0.489304	0.495848
0.496183	0.496862	0.499062	0.501704	0.501815	0.503066	0.504244
0.505002	0.506247	0.510371	0.511047	0.511141	0.519738	0.520178
0.524034	0.525570	0.526142	0.529668	0.530248	0.530294	0.531708
0.533834	0.542166	0.545207	0.545345	0.547910	0.549403	0.551635
0.555601	0.558277	0.564112	0.569155	0.574222	0.575517	0.584477
0.590003	0.590216	0.592314	0.604588	0.612498	0.637047	0.643558
0.655124	0.656766	0.669667	0.676449			

Mean and the standard deviation of the 200 sample correlations:

Descriptive Statistics: Correlation

Variable	N	Mean	StDev
Correlation	200	0.40543	0.11695

Bootstrap estimate of the standard error of the sample correlation = 0.11695.

For the 95% confidence interval, just pick the 5th values from the beginning and the end of the ordered correlations (i.e., approximate 2.5th and 97.5th percentiles). For this simulation it is approximately (0.156, 0.644). This interval does not include 0 and so based on this simulation, it looks unlikely for the correlation to be zero.

21. The answers will differ due to randomness.

The sample of 30 observations generated from a normal distribution with a known mean (of, say, $\mu = 65$) and a known standard deviation (of, say, $\sigma = 5$) and mean and the standard deviation of the sample are shown below:

59.1787	60.3044	67.0518	65.4541	66.6844	64.1374	53.5239
63.8788	63.1310	61.0043	61.5790	64.4940	69.0911	64.5024
65.0824	69.6807	56.6600	67.2456	55.8048	64.2704	58.5570
72.8720	70.2878	65.8018	65.4818	61.4000	59.0011	66.4736
70.4672	60.4118					

Variable	N	Mean	SE Mean	StDev
C1	30	63.784	0.842	4.612

The means of the 200 resamples from the above sample and the histogram of those 200 sample means are shown below:

Mean						
63.7447	63.8069	65.1318	64.1033	63.1662	63.7740	63.4635
62.7708	61.9800	63.1597	64.3513	64.0902	62.0807	64.5234
62.8716	63.6124	63.8453	62.2446	62.8600	63.1033	63.4459
63.7811	61.5864	65.5432	63.8388	64.7823	63.1369	61.8433
64.5491	65.5470	65.1019	62.6649	64.2730	63.1670	65.5454
62.9408	64.0521	64.8397	63.2670	64.0826	64.7394	64.7204
63.9626	62.8048	62.3764	64.5700	64.6598	63.3908	63.0051
62.6054	62.9576	63.7873	63.5749	64.4591	63.2810	62.1458
64.0168	62.9784	61.6540	62.9439	64.8909	63.3646	63.5682
63.2562	63.0667	63.0963	64.7465	63.4385	63.9771	64.4517
64.6042	64.2514	64.4759	64.0430	64.5107	63.6507	62.9070
62.6761	63.3080	63.2322	64.8306	63.2788	63.9941	63.7376
63.5327	64.2255	65.7598	62.9145	63.2147	63.4901	63.5470
62.5478	63.8546	63.6364	63.9743	63.8648	63.0994	63.8441
63.8297	63.5441	64.7062	64.2176	65.2195	64.6484	63.7047
64.1748	63.9021	62.4065	63.5739	63.1801	63.8946	62.6778
63.9002	64.9348	63.9331	64.0164	64.2924	64.1491	64.3211
64.8336	63.8866	62.7561	63.9456	63.7989	62.4128	62.2528
63.6070	63.5246	63.5601	62.1951	64.7309	63.7079	64.0152
63.1925	63.6502	64.0483	64.2712	63.8390	62.2928	61.6062
62.9311	62.8511	64.2809	62.0376	63.6406	64.2565	64.0274
63.6495	64.5283	63.0867	61.9325	63.5002	63.8589	63.4350
65.2124	63.2876	63.9733	64.7694	62.5573	65.2039	65.5147
62.9923	64.6568	64.0628	62.6455	62.6041	65.2800	64.6694
63.8130	63.5205	65.0073	64.9451	64.8911	63.3968	63.8602
63.5086	65.3508	63.4538	63.2068	63.8054	63.4675	63.3915
62.8429	63.5545	63.5763	64.1105	64.0116	63.7011	63.7861
62.7053	65.1161	63.2066	62.7636	63.7947	64.3211	64.4501
64.0046	62.7599	62.8489	63.1579			

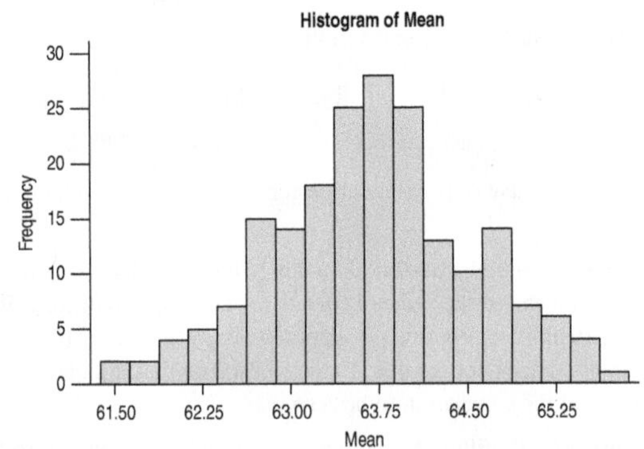

The mean and the standard deviation (the bootstrap standard error) of the 200 means:

Descriptive Statistics: Mean

Variable	N	Mean	StDev
Mean	200	63.696	0.857

The distribution is symmetric centred around 63.75 (close the mean of the sample from the normal distribution).

23. The answers will differ due to randomness.
The sample of 30 observations generated from a normal distribution with a known mean (of, say, $\mu = 65$) and a known standard deviation (of, say, $\sigma = 5$) and mean and the standard deviation of the sample are shown below:

C1

62.2614	64.2326	58.1349	67.8349	69.2504	63.4110	59.0211
63.5680	67.9223	74.6798	60.7009	66.4197	72.8893	50.5429
62.2437	70.6465	59.2842	68.9763	56.7706	65.4978	65.1057
66.2726	61.4787	64.9551	67.2118	60.5744	54.7737	68.4119
67.2500	62.5024					

Descriptive Statistics: C1

Variable	N	Mean	StDev
C1	30	64.094	5.295

The standard deviations of the 200 resamples from the above sample and the histogram of those 200 sample standard deviations are shown below:

Std deviation

4.66001	4.97557	5.01786	5.50359	5.66153	6.28903	3.99984
3.77663	5.00275	5.33451	5.42411	4.45115	6.66293	5.58117
4.54422	4.45296	5.17018	5.43343	5.67459	5.63363	4.88812
6.01936	5.38827	4.54415	5.91733	6.09181	5.22481	5.79918
5.57581	6.43161	5.14508	6.16960	4.83818	5.41626	3.96438
6.16723	4.51435	4.44443	4.80409	5.38541	5.31486	4.94219
5.32203	5.40650	5.39299	5.76170	4.56696	6.16282	4.96266
6.35409	5.25756	4.53397	5.18265	5.26029	5.53973	4.98305
4.97266	5.51348	5.25206	4.25202	4.79806	5.49566	4.48149
4.32959	4.26105	4.38888	4.84733	4.95296	4.16755	5.38639
5.18504	4.90396	5.41484	4.67684	5.54291	5.45316	4.89019
4.67559	5.42218	6.40685	5.81436	4.36872	4.73909	5.31081
5.97795	5.74172	5.04325	5.61432	5.50440	4.36547	5.30276
4.67350	5.86654	4.12740	5.25279	4.96729	4.90542	3.99175
5.95744	5.70628	5.45084	5.88647	4.56510	4.82318	5.63808
4.24854	6.26507	4.88949	6.06017	3.83108	5.42714	5.65611
3.88603	4.03755	4.50240	5.85583	4.61996	4.37661	5.08605
4.29964	6.30767	4.85154	5.53200	4.87651	6.12621	5.45537
4.49753	4.50189	4.73739	5.73770	4.94640	5.74848	6.36516
4.83004	4.95062	5.35772	4.84219	5.45511	5.83329	4.94890
4.95356	5.08019	4.44861	5.53432	6.38075	5.78856	4.95368
4.52841	4.29279	5.26769	5.65561	4.56361	4.78608	5.30467
5.69703	5.49891	4.46331	4.16347	4.45143	5.40321	4.70341
3.99124	5.09801	5.52631	4.17098	6.60630	5.40862	6.57939
7.22280	5.50610	4.12690	4.48439	5.01642	5.74896	5.73342
6.17881	4.02427	4.57888	5.19126	5.05119	5.01790	5.01334
4.54014	4.04914	5.23537	5.15926	5.00286	4.41312	4.44039
5.88761	5.03478	3.49237	5.69494	4.42845	5.26108	4.98231
4.79918	5.15061	5.81604	4.67808			

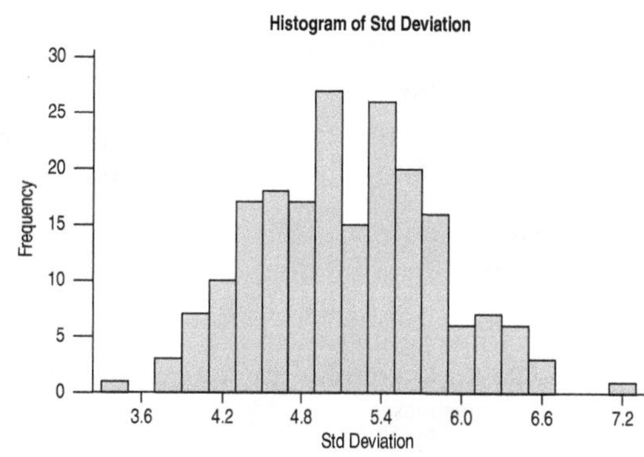

The mean and the standard deviation (the bootstrap standard error) of the 200 sample standard deviations:

Descriptive Statistics: Std deviation

Variable	N	Mean	StDev
Std deviation	200	5.1370	0.6634

The histogram is somewhat symmetric, with centre around 5.3 and ranging approximately from 3 to 7.

25. The answers will differ due to randomness.
The medians of the 200 resamples from the above sample and the histogram of those 200 sample medians are shown below:

Correlation

0.785698	0.613946	0.586843	0.501893	0.526401	0.396375
0.474427	0.241789	0.759066	0.644223	0.487315	0.519607
0.447036	0.471923	0.590963	0.522964	0.604221	0.546688
0.494286	0.774408	0.448096	0.678720	0.519220	0.486575
0.614990	0.309607	0.501002	0.647937	0.771385	0.596495
0.473838	0.557004	0.624819	0.640079	0.642788	0.458855
0.566706	0.267041	0.710863	0.591193	0.614932	0.654733
0.559902	0.530933	0.679616	0.648657	0.373612	0.348047
0.822826	0.605153	0.595275	0.657847	0.410380	0.666054
0.502975	0.626914	0.557472	0.527009	0.510862	0.534441
0.540771	0.576620	0.686379	0.560456	0.814902	0.747498
0.765250	0.315355	0.576484	0.584094	0.663106	0.448412
0.661316	0.686589	0.722054	0.141658	0.610311	0.706787
0.576382	0.513419	0.624309	0.580795	0.416723	0.471252
0.470901	0.586005	0.346633	0.620381	0.695170	0.641054
0.358568	0.524399	0.623996	0.364895	0.603395	0.642135
0.624157	0.601091	0.490992	0.663136	0.337762	0.561285
0.716795	0.709292	0.599795	0.664173	0.401727	0.758430
0.436710	0.569211	0.453085	0.373676	0.676526	0.610883
0.725133	0.431512	0.720192	0.669769	0.720809	0.431717
0.669493	0.413389	0.667052	0.711596	0.592035	0.636997
0.558879	0.780707	0.760007	0.588851	0.609481	0.605180
0.676924	0.382512	0.685123	0.611094	0.657267	0.435200
0.663943	0.628098	0.477997	0.568765	0.543731	0.540185
0.602348	0.719019	0.546733	0.745190	0.617606	0.686872
0.717183	0.022223	0.524841	0.690547	0.565033	0.369898
0.159915	0.541112	0.681576	0.543320	0.081134	0.702172

0.546986	0.676251	0.260466	0.559879	0.516077	0.613583
0.573403	0.697825	0.636131	0.593778	0.309336	0.615788
0.591174	0.731522	0.500375	0.633493	0.702300	0.584927
0.738964	0.436784	0.481917	0.557847	0.803252	0.807568
0.443769	0.491287	0.669460	0.560219	0.560573	0.703965
0.523546	0.567730	0.297033	0.691778	0.613286	0.541919
0.466034	0.467393				

The mean and the standard deviation (the bootstrap standard error) of the 200 sample correlations:

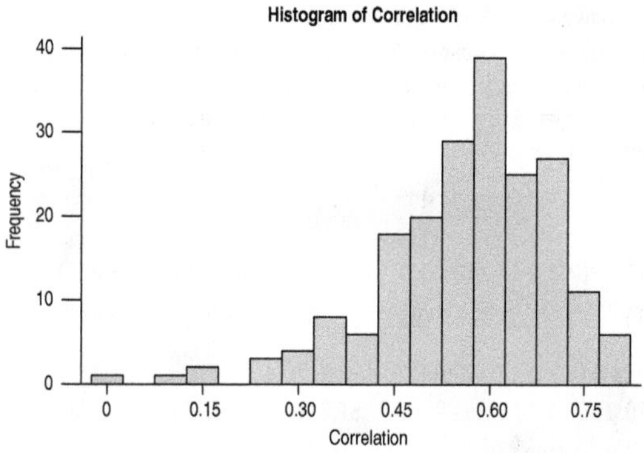

Descriptive Statistics: Correlation

Variable	N	Mean	StDev
Correlation	200	0.56823	0.13731

The histogram shows a left-skewed (negatively-skewed) shape with a mode close to 0.6.

appendix

B Index

Numbers

A

B

C

appendix

C Tables

Row	TABLE OF RANDOM DIGITS									
1	96299	07196	98642	20639	23185	56282	69929	14125	38872	94168
2	71622	35940	81807	59225	18192	08710	80777	84395	69563	86280
3	03272	41230	81739	74797	70406	18564	69273	72532	78340	36699
4	46376	58596	14365	63685	56555	42974	72944	96463	63533	24152
5	47352	42853	42903	97504	56655	70355	88606	61406	38757	70657
6	20064	04266	74017	79319	70170	96572	08523	56025	89077	57678
7	73184	95907	05179	51002	83374	52297	07769	99792	78365	93487
8	72753	36216	07230	35793	71907	65571	66784	25548	91861	15725
9	03939	30763	06138	80062	02537	23561	93136	61260	77935	93159
10	75998	37203	07959	38264	78120	77525	86481	54986	33042	70648
11	94435	97441	90998	25104	49761	14967	70724	67030	53887	81293
12	04362	40989	69167	38894	00172	02999	97377	33305	60782	29810
13	89059	43528	10547	40115	82234	86902	04121	83889	76208	31076
14	87736	04666	75145	49175	76754	07884	92564	80793	22573	67902
15	76488	88899	15860	07370	13431	84041	69202	18912	83173	11983
16	36460	53772	66634	25045	79007	78518	73580	14191	50353	32064
17	13205	69237	21820	20952	16635	58867	97650	82983	64865	93298
18	51242	12215	90739	36812	00436	31609	80333	96606	30430	31803
19	67819	00354	91439	91073	49258	15992	41277	75111	67496	68430
20	09875	08990	27656	15871	23637	00952	97818	64234	50199	05715
21	18192	95308	72975	01191	29958	09275	89141	19558	50524	32041
22	02763	33701	66188	50226	35813	72951	11638	01876	93664	37001
23	13349	46328	01856	29935	80563	03742	49470	67749	08578	21956
24	69238	92878	80067	80807	45096	22936	64325	19265	37755	69794
25	92207	63527	59398	29818	24789	94309	88380	57000	50171	17891
26	66679	99100	37072	30593	29665	84286	44458	60180	81451	58273
27	31087	42430	60322	34765	15757	53300	97392	98035	05228	68970
28	84432	04916	52949	78533	31666	62350	20584	56367	19701	60584
29	72042	12287	21081	48426	44321	58765	41760	43304	13399	02043
30	94534	73559	82135	70260	87936	85162	11937	18263	54138	69564
31	63971	97198	40974	45301	60177	35604	21580	68107	25184	42810
32	11227	58474	17272	37619	69517	62964	67962	34510	12607	52255
33	28541	02029	08068	96656	17795	21484	57722	76511	27849	61738
34	11282	43632	49531	78981	81980	08530	08629	32279	29478	50228
35	42907	15137	21918	13248	39129	49559	94540	24070	88151	36782
36	47119	76651	21732	32364	58545	50277	57558	30390	18771	72703
37	11232	99884	05087	76839	65142	19994	91397	29350	83852	04905
38	64725	06719	86262	53356	57999	50193	79936	97230	52073	94467
39	77007	26962	55466	12521	48125	12280	54985	26239	76044	54398
40	18375	19310	59796	89832	59417	18553	17238	05474	33259	50595

Table F

Denominator df / α = .01	Numerator df 1	2	3	4	5	6	7	8	9	10	11	12	13	14	15	16	17	18	19	20	21	22
1	4052.2	4999.3	5403.5	5624.3	5764.0	5859.0	5928.3	5981.0	6022.4	6055.9	6083.4	6106.7	6125.8	6143.0	6157.0	6170.0	6181.2	6191.4	6200.7	6208.7	6216.1	6223.1
2	98.50	99.00	99.16	99.25	99.30	99.33	99.36	99.38	99.39	99.40	99.41	99.42	99.42	99.43	99.43	99.44	99.44	99.44	99.45	99.45	99.45	99.46
3	34.12	30.82	29.46	28.71	28.24	27.91	27.67	27.49	27.34	27.23	27.13	27.05	26.98	26.92	26.87	26.83	26.79	26.75	26.72	26.69	26.66	26.64
4	21.20	18.00	16.69	15.98	15.52	15.21	14.98	14.80	14.66	14.55	14.45	14.37	14.31	14.25	14.20	14.15	14.11	14.08	14.05	14.02	13.99	13.97
5	16.26	13.27	12.06	11.39	10.97	10.67	10.46	10.29	10.16	10.05	9.96	9.89	9.82	9.77	9.72	9.68	9.64	9.61	9.58	9.55	9.53	9.51
6	13.75	10.92	9.78	9.15	8.75	8.47	8.26	8.10	7.98	7.87	7.79	7.72	7.66	7.60	7.56	7.52	7.48	7.45	7.42	7.40	7.37	7.35
7	12.25	9.55	8.45	7.85	7.46	7.19	6.99	6.84	6.72	6.62	6.54	6.47	6.41	6.36	6.31	6.28	6.24	6.21	6.18	6.16	6.13	6.11
8	11.26	8.65	7.59	7.01	6.63	6.37	6.18	6.03	5.91	5.81	5.73	5.67	5.61	5.56	5.52	5.48	5.44	5.41	5.38	5.36	5.34	5.32
9	10.56	8.02	6.99	6.42	6.06	5.80	5.61	5.47	5.35	5.26	5.18	5.11	5.05	5.01	4.96	4.92	4.89	4.86	4.83	4.81	4.79	4.77
10	10.04	7.56	6.55	5.99	5.64	5.39	5.20	5.06	4.94	4.85	4.77	4.71	4.65	4.60	4.56	4.52	4.49	4.46	4.43	4.41	4.38	4.36
11	9.65	7.21	6.22	5.67	5.32	5.07	4.89	4.74	4.63	4.54	4.46	4.40	4.34	4.29	4.25	4.21	4.18	4.15	4.12	4.10	4.08	4.06
12	9.33	6.93	5.95	5.41	5.06	4.82	4.64	4.50	4.39	4.30	4.22	4.16	4.10	4.05	4.01	3.97	3.94	3.91	3.88	3.86	3.84	3.82
13	9.07	6.70	5.74	5.21	4.86	4.62	4.44	4.30	4.19	4.10	4.02	3.96	3.91	3.86	3.82	3.78	3.75	3.72	3.69	3.66	3.64	3.62
14	8.86	6.51	5.56	5.04	4.69	4.46	4.28	4.14	4.03	3.94	3.86	3.80	3.75	3.70	3.66	3.62	3.59	3.56	3.53	3.51	3.48	3.46
15	8.68	6.36	5.42	4.89	4.56	4.32	4.14	4.00	3.89	3.80	3.73	3.67	3.61	3.56	3.52	3.49	3.45	3.42	3.40	3.37	3.35	3.33
16	8.53	6.23	5.29	4.77	4.44	4.20	4.03	3.89	3.78	3.69	3.62	3.55	3.50	3.45	3.41	3.37	3.34	3.31	3.28	3.26	3.24	3.22
17	8.40	6.11	5.19	4.67	4.34	4.10	3.93	3.79	3.68	3.59	3.52	3.46	3.40	3.35	3.31	3.27	3.24	3.21	3.19	3.16	3.14	3.12
18	8.29	6.01	5.09	4.58	4.25	4.01	3.84	3.71	3.60	3.51	3.43	3.37	3.32	3.27	3.23	3.19	3.16	3.13	3.10	3.08	3.05	3.03
19	8.18	5.93	5.01	4.50	4.17	3.94	3.77	3.63	3.52	3.43	3.36	3.30	3.24	3.19	3.15	3.12	3.08	3.05	3.03	3.00	2.98	2.96
20	8.10	5.85	4.94	4.43	4.10	3.87	3.70	3.56	3.46	3.37	3.29	3.23	3.18	3.13	3.09	3.05	3.02	2.99	2.96	2.94	2.92	2.90
21	8.02	5.78	4.87	4.37	4.04	3.81	3.64	3.51	3.40	3.31	3.24	3.17	3.12	3.07	3.03	2.99	2.96	2.93	2.90	2.88	2.86	2.84
22	7.95	5.72	4.82	4.31	3.99	3.76	3.59	3.45	3.35	3.26	3.18	3.12	3.07	3.02	2.98	2.94	2.91	2.88	2.85	2.83	2.81	2.78
23	7.88	5.66	4.76	4.26	3.94	3.71	3.54	3.41	3.30	3.21	3.14	3.07	3.02	2.97	2.93	2.89	2.86	2.83	2.80	2.78	2.76	2.74
24	7.82	5.61	4.72	4.22	3.90	3.67	3.50	3.36	3.26	3.17	3.09	3.03	2.98	2.93	2.89	2.85	2.82	2.79	2.76	2.74	2.72	2.70
25	7.77	5.57	4.68	4.18	3.85	3.63	3.46	3.32	3.22	3.13	3.06	2.99	2.94	2.89	2.85	2.81	2.78	2.75	2.72	2.70	2.68	2.66
26	7.72	5.53	4.64	4.14	3.82	3.59	3.42	3.29	3.18	3.09	3.02	2.96	2.90	2.86	2.81	2.78	2.75	2.72	2.69	2.66	2.64	2.62
27	7.68	5.49	4.60	4.11	3.78	3.56	3.39	3.26	3.15	3.06	2.99	2.93	2.87	2.82	2.78	2.75	2.71	2.68	2.66	2.63	2.61	2.59
28	7.64	5.45	4.57	4.07	3.75	3.53	3.36	3.23	3.12	3.03	2.96	2.90	2.84	2.79	2.75	2.72	2.68	2.65	2.63	2.60	2.58	2.56
29	7.60	5.42	4.54	4.04	3.73	3.50	3.33	3.20	3.09	3.00	2.93	2.87	2.81	2.77	2.73	2.69	2.66	2.63	2.60	2.57	2.55	2.53
30	7.56	5.39	4.51	4.02	3.70	3.47	3.30	3.17	3.07	2.98	2.91	2.84	2.79	2.74	2.70	2.66	2.63	2.60	2.57	2.55	2.53	2.51
32	7.50	5.34	4.46	3.97	3.65	3.43	3.26	3.13	3.02	2.93	2.86	2.80	2.74	2.70	2.65	2.62	2.58	2.55	2.53	2.50	2.48	2.46
35	7.42	5.27	4.40	3.91	3.59	3.37	3.20	3.07	2.96	2.88	2.80	2.74	2.69	2.64	2.60	2.56	2.53	2.50	2.47	2.44	2.42	2.40
40	7.31	5.18	4.31	3.83	3.51	3.29	3.12	2.99	2.89	2.80	2.73	2.66	2.61	2.56	2.52	2.48	2.45	2.42	2.39	2.37	2.35	2.33
45	7.23	5.11	4.25	3.77	3.45	3.23	3.07	2.94	2.83	2.74	2.67	2.61	2.55	2.51	2.46	2.43	2.39	2.36	2.34	2.31	2.29	2.27
50	7.17	5.06	4.20	3.72	3.41	3.19	3.02	2.89	2.78	2.70	2.63	2.56	2.51	2.46	2.42	2.38	2.35	2.32	2.29	2.27	2.24	2.22
60	7.08	4.98	4.13	3.65	3.34	3.12	2.95	2.82	2.72	2.63	2.56	2.50	2.44	2.39	2.35	2.31	2.28	2.25	2.22	2.20	2.17	2.15
75	6.99	4.90	4.05	3.58	3.27	3.05	2.89	2.76	2.65	2.57	2.49	2.43	2.38	2.33	2.29	2.25	2.22	2.18	2.16	2.13	2.11	2.09
100	6.90	4.82	3.98	3.51	3.21	2.99	2.82	2.69	2.59	2.50	2.43	2.37	2.31	2.27	2.22	2.19	2.15	2.12	2.09	2.07	2.04	2.02
120	6.85	4.79	3.95	3.48	3.17	2.96	2.79	2.66	2.56	2.47	2.40	2.34	2.28	2.23	2.19	2.15	2.12	2.09	2.06	2.03	2.01	1.99
140	6.82	4.76	3.92	3.46	3.15	2.93	2.77	2.64	2.54	2.45	2.38	2.31	2.26	2.21	2.17	2.13	2.10	2.07	2.04	2.01	1.99	1.97
180	6.78	4.73	3.89	3.43	3.12	2.90	2.74	2.61	2.51	2.42	2.35	2.28	2.23	2.18	2.14	2.10	2.07	2.04	2.01	1.98	1.96	1.94
250	6.74	4.69	3.86	3.40	3.09	2.87	2.71	2.58	2.48	2.39	2.32	2.26	2.20	2.15	2.11	2.07	2.04	2.01	1.98	1.95	1.93	1.91
400	6.70	4.66	3.83	3.37	3.06	2.85	2.68	2.56	2.45	2.37	2.29	2.23	2.17	2.13	2.08	2.05	2.01	1.98	1.95	1.92	1.90	1.88
1000	6.66	4.63	3.80	3.34	3.04	2.82	2.66	2.53	2.43	2.34	2.27	2.20	2.15	2.10	2.06	2.02	1.98	1.95	1.92	1.90	1.87	1.85

Table F (cont.)

Numerator df

α = .01	23	24	25	26	27	28	29	30	32	35	40	45	50	60	75	100	120	140	180	250	400	1000
1	6228.7	6234.3	6239.9	6244.5	6249.2	6252.9	6257.1	6260.4	6266.9	6275.3	6286.4	6295.7	6302.3	6313.0	6323.7	6333.9	6339.5	6343.2	6347.9	6353.5	6358.1	6362.8
2	99.46	99.46	99.46	99.46	99.46	99.46	99.46	99.47	99.47	99.47	99.48	99.48	99.48	99.48	99.48	99.49	99.49	99.49	99.49	99.50	99.50	99.50
3	26.62	26.60	26.58	26.56	26.55	26.53	26.52	26.50	26.48	26.45	26.41	26.38	26.35	26.32	26.28	26.24	26.22	26.21	26.19	26.17	26.15	26.14
4	13.95	13.93	13.91	13.89	13.88	13.86	13.85	13.84	13.81	13.79	13.75	13.71	13.69	13.65	13.61	13.58	13.56	13.54	13.53	13.51	13.49	13.47
5	9.49	9.47	9.45	9.43	9.42	9.40	9.39	9.38	9.36	9.33	9.29	9.26	9.24	9.20	9.17	9.13	9.11	9.10	9.08	9.06	9.05	9.03
6	7.33	7.31	7.30	7.28	7.27	7.25	7.24	7.23	7.21	7.18	7.14	7.11	7.09	7.06	7.02	6.99	6.97	6.96	6.94	6.92	6.91	6.89
7	6.09	6.07	6.06	6.04	6.03	6.02	6.00	5.99	5.97	5.94	5.91	5.88	5.86	5.82	5.79	5.75	5.74	5.72	5.71	5.69	5.68	5.66
8	5.30	5.28	5.26	5.25	5.23	5.22	5.21	5.20	5.18	5.15	5.12	5.09	5.07	5.03	5.00	4.96	4.95	4.93	4.92	4.90	4.89	4.87
9	4.75	4.73	4.71	4.70	4.68	4.67	4.66	4.65	4.63	4.60	4.57	4.54	4.52	4.48	4.45	4.41	4.40	4.39	4.37	4.35	4.34	4.32
10	4.34	4.33	4.31	4.30	4.28	4.27	4.26	4.25	4.23	4.20	4.17	4.14	4.12	4.08	4.05	4.01	4.00	3.98	3.97	3.95	3.94	3.92
11	4.04	4.02	4.01	3.99	3.98	3.96	3.95	3.94	3.92	3.89	3.86	3.83	3.81	3.78	3.74	3.71	3.69	3.68	3.66	3.64	3.63	3.61
12	3.80	3.78	3.76	3.75	3.74	3.72	3.71	3.70	3.68	3.65	3.62	3.59	3.57	3.54	3.50	3.47	3.45	3.44	3.42	3.40	3.39	3.37
13	3.60	3.59	3.57	3.56	3.54	3.53	3.52	3.51	3.49	3.46	3.43	3.40	3.38	3.34	3.31	3.27	3.25	3.24	3.23	3.21	3.19	3.18
14	3.44	3.43	3.41	3.40	3.38	3.37	3.36	3.35	3.33	3.30	3.27	3.24	3.22	3.18	3.15	3.11	3.09	3.08	3.06	3.05	3.03	3.02
15	3.31	3.29	3.28	3.26	3.25	3.24	3.23	3.21	3.19	3.17	3.13	3.10	3.08	3.05	3.01	2.98	2.96	2.95	2.93	2.91	2.90	2.88
16	3.20	3.18	3.16	3.15	3.14	3.12	3.11	3.10	3.08	3.05	3.02	2.99	2.97	2.93	2.90	2.86	2.84	2.83	2.81	2.80	2.78	2.76
17	3.10	3.08	3.07	3.05	3.04	3.03	3.01	3.00	2.98	2.96	2.92	2.89	2.87	2.83	2.80	2.76	2.75	2.73	2.72	2.70	2.68	2.66
18	3.02	3.00	2.98	2.97	2.95	2.94	2.93	2.92	2.90	2.87	2.84	2.81	2.78	2.75	2.71	2.68	2.66	2.65	2.63	2.61	2.59	2.58
19	2.94	2.92	2.91	2.89	2.88	2.87	2.86	2.84	2.82	2.80	2.76	2.73	2.71	2.67	2.64	2.60	2.58	2.57	2.55	2.54	2.52	2.50
20	2.88	2.86	2.84	2.83	2.81	2.80	2.79	2.78	2.76	2.73	2.69	2.67	2.64	2.61	2.57	2.54	2.52	2.50	2.49	2.47	2.45	2.43
21	2.82	2.80	2.79	2.77	2.76	2.74	2.73	2.72	2.70	2.67	2.64	2.61	2.58	2.55	2.51	2.48	2.46	2.44	2.43	2.41	2.39	2.37
22	2.77	2.75	2.73	2.72	2.70	2.69	2.68	2.67	2.65	2.62	2.58	2.55	2.53	2.50	2.46	2.42	2.40	2.39	2.37	2.35	2.34	2.32
23	2.72	2.70	2.69	2.67	2.66	2.64	2.63	2.62	2.60	2.57	2.54	2.51	2.48	2.45	2.41	2.37	2.35	2.34	2.32	2.30	2.29	2.27
24	2.68	2.66	2.64	2.63	2.61	2.60	2.59	2.58	2.56	2.53	2.49	2.46	2.44	2.40	2.37	2.33	2.31	2.30	2.28	2.26	2.24	2.22
25	2.64	2.62	2.60	2.59	2.58	2.56	2.55	2.54	2.52	2.49	2.45	2.42	2.40	2.36	2.33	2.29	2.27	2.26	2.24	2.22	2.20	2.18
26	2.60	2.58	2.57	2.55	2.54	2.53	2.51	2.50	2.48	2.45	2.42	2.39	2.36	2.33	2.29	2.25	2.23	2.22	2.20	2.18	2.16	2.14
27	2.57	2.55	2.54	2.52	2.51	2.49	2.48	2.47	2.45	2.42	2.38	2.35	2.33	2.29	2.26	2.22	2.20	2.18	2.17	2.15	2.13	2.11
28	2.54	2.52	2.51	2.49	2.48	2.46	2.45	2.44	2.42	2.39	2.35	2.32	2.30	2.26	2.23	2.19	2.17	2.15	2.13	2.11	2.10	2.08
29	2.51	2.49	2.48	2.46	2.45	2.44	2.42	2.41	2.39	2.36	2.33	2.30	2.27	2.23	2.20	2.16	2.14	2.12	2.10	2.08	2.07	2.05
30	2.49	2.47	2.45	2.44	2.42	2.41	2.40	2.39	2.36	2.34	2.30	2.27	2.25	2.21	2.17	2.13	2.11	2.10	2.08	2.06	2.04	2.02
32	2.44	2.42	2.41	2.39	2.38	2.36	2.35	2.34	2.32	2.29	2.25	2.22	2.20	2.16	2.12	2.08	2.06	2.05	2.03	2.01	1.99	1.97
35	2.38	2.36	2.35	2.33	2.32	2.30	2.29	2.28	2.26	2.23	2.19	2.16	2.14	2.10	2.06	2.02	2.00	1.98	1.96	1.94	1.92	1.90
40	2.31	2.29	2.27	2.26	2.24	2.23	2.22	2.20	2.18	2.15	2.11	2.08	2.06	2.02	1.98	1.94	1.92	1.90	1.88	1.86	1.84	1.82
45	2.25	2.23	2.21	2.20	2.18	2.17	2.16	2.14	2.12	2.09	2.05	2.02	2.00	1.96	1.92	1.88	1.85	1.84	1.82	1.79	1.77	1.75
50	2.20	2.18	2.17	2.15	2.14	2.12	2.11	2.10	2.08	2.05	2.01	1.97	1.95	1.91	1.87	1.82	1.80	1.79	1.76	1.74	1.72	1.70
60	2.13	2.12	2.10	2.08	2.07	2.05	2.04	2.03	2.01	1.98	1.94	1.90	1.88	1.84	1.79	1.75	1.73	1.71	1.69	1.66	1.64	1.62
75	2.07	2.05	2.03	2.02	2.00	1.99	1.97	1.96	1.94	1.91	1.87	1.83	1.81	1.76	1.72	1.67	1.65	1.63	1.61	1.58	1.56	1.53
100	2.00	1.98	1.97	1.95	1.93	1.92	1.91	1.89	1.87	1.84	1.80	1.76	1.74	1.69	1.65	1.60	1.57	1.55	1.53	1.50	1.47	1.45
120	1.97	1.95	1.93	1.92	1.90	1.89	1.87	1.86	1.84	1.81	1.76	1.73	1.70	1.66	1.61	1.56	1.53	1.51	1.49	1.46	1.43	1.40
140	1.95	1.93	1.91	1.89	1.88	1.86	1.85	1.84	1.81	1.78	1.74	1.70	1.67	1.63	1.58	1.53	1.50	1.48	1.46	1.43	1.40	1.37
180	1.92	1.90	1.88	1.86	1.85	1.83	1.82	1.81	1.78	1.75	1.71	1.67	1.64	1.60	1.55	1.49	1.47	1.45	1.42	1.39	1.35	1.32
250	1.89	1.87	1.85	1.83	1.82	1.80	1.79	1.77	1.75	1.72	1.67	1.64	1.61	1.56	1.51	1.46	1.43	1.41	1.38	1.34	1.31	1.27
400	1.86	1.84	1.82	1.80	1.79	1.77	1.76	1.75	1.72	1.69	1.64	1.61	1.58	1.53	1.48	1.42	1.39	1.37	1.33	1.30	1.26	1.22
1000	1.83	1.81	1.79	1.77	1.76	1.74	1.73	1.72	1.69	1.66	1.61	1.58	1.54	1.50	1.44	1.38	1.35	1.33	1.29	1.25	1.21	1.16

Denominator df

Table F (cont.)

α = .05	Numerator df 1	2	3	4	5	6	7	8	9	10	11	12	13	14	15	16	17	18	19	20	21	22
Denominator df 1	161.4	199.5	215.7	224.6	230.2	234.0	236.8	238.9	240.5	241.9	243.0	243.9	244.7	245.4	245.9	246.5	246.9	247.3	247.7	248.0	248.3	248.6
2	18.51	19.00	19.16	19.25	19.30	19.33	19.35	19.37	19.38	19.40	19.40	19.41	19.42	19.42	19.43	19.43	19.44	19.44	19.44	19.45	19.45	19.45
3	10.13	9.55	9.28	9.12	9.01	8.94	8.89	8.85	8.81	8.79	8.76	8.74	8.73	8.71	8.70	8.69	8.68	8.67	8.67	8.66	8.65	8.65
4	7.71	6.94	6.59	6.39	6.26	6.16	6.09	6.04	6.00	5.96	5.94	5.91	5.89	5.87	5.86	5.84	5.83	5.82	5.81	5.80	5.79	5.79
5	6.61	5.79	5.41	5.19	5.05	4.95	4.88	4.82	4.77	4.74	4.70	4.68	4.66	4.64	4.62	4.60	4.59	4.58	4.57	4.56	4.55	4.54
6	5.99	5.14	4.76	4.53	4.39	4.28	4.21	4.15	4.10	4.06	4.03	4.00	3.98	3.96	3.94	3.92	3.91	3.90	3.88	3.87	3.86	3.86
7	5.59	4.74	4.35	4.12	3.97	3.87	3.79	3.73	3.68	3.64	3.60	3.57	3.55	3.53	3.51	3.49	3.48	3.47	3.46	3.44	3.43	3.43
8	5.32	4.46	4.07	3.84	3.69	3.58	3.50	3.44	3.39	3.35	3.31	3.28	3.26	3.24	3.22	3.20	3.19	3.17	3.16	3.15	3.14	3.13
9	5.12	4.26	3.86	3.63	3.48	3.37	3.29	3.23	3.18	3.14	3.10	3.07	3.05	3.03	3.01	2.99	2.97	2.96	2.95	2.94	2.93	2.92
10	4.96	4.10	3.71	3.48	3.33	3.22	3.14	3.07	3.02	2.98	2.94	2.91	2.89	2.86	2.85	2.83	2.81	2.80	2.79	2.77	2.76	2.75
11	4.84	3.98	3.59	3.36	3.20	3.09	3.01	2.95	2.90	2.85	2.82	2.79	2.76	2.74	2.72	2.70	2.69	2.67	2.66	2.65	2.64	2.63
12	4.75	3.89	3.49	3.26	3.11	3.00	2.91	2.85	2.80	2.75	2.72	2.69	2.66	2.64	2.62	2.60	2.58	2.57	2.56	2.54	2.53	2.52
13	4.67	3.81	3.41	3.18	3.03	2.92	2.83	2.77	2.71	2.67	2.63	2.60	2.58	2.55	2.53	2.51	2.50	2.48	2.47	2.46	2.45	2.44
14	4.60	3.74	3.34	3.11	2.96	2.85	2.76	2.70	2.65	2.60	2.57	2.53	2.51	2.48	2.46	2.44	2.43	2.41	2.40	2.39	2.38	2.37
15	4.54	3.68	3.29	3.06	2.90	2.79	2.71	2.64	2.59	2.54	2.51	2.48	2.45	2.42	2.40	2.38	2.37	2.35	2.34	2.33	2.32	2.31
16	4.49	3.63	3.24	3.01	2.85	2.74	2.66	2.59	2.54	2.49	2.46	2.42	2.40	2.37	2.35	2.33	2.32	2.30	2.29	2.28	2.26	2.25
17	4.45	3.59	3.20	2.96	2.81	2.70	2.61	2.55	2.49	2.45	2.41	2.38	2.35	2.33	2.31	2.29	2.27	2.26	2.24	2.23	2.22	2.21
18	4.41	3.55	3.16	2.93	2.77	2.66	2.58	2.51	2.46	2.41	2.37	2.34	2.31	2.29	2.27	2.25	2.23	2.22	2.20	2.19	2.18	2.17
19	4.38	3.52	3.13	2.90	2.74	2.63	2.54	2.48	2.42	2.38	2.34	2.31	2.28	2.26	2.23	2.21	2.20	2.18	2.17	2.16	2.14	2.13
20	4.35	3.49	3.10	2.87	2.71	2.60	2.51	2.45	2.39	2.35	2.31	2.28	2.25	2.22	2.20	2.18	2.17	2.15	2.14	2.12	2.11	2.10
21	4.32	3.47	3.07	2.84	2.68	2.57	2.49	2.42	2.37	2.32	2.28	2.25	2.22	2.20	2.18	2.16	2.14	2.12	2.11	2.10	2.08	2.07
22	4.30	3.44	3.05	2.82	2.66	2.55	2.46	2.40	2.34	2.30	2.26	2.23	2.20	2.17	2.15	2.13	2.11	2.10	2.08	2.07	2.06	2.05
23	4.28	3.42	3.03	2.80	2.64	2.53	2.44	2.37	2.32	2.27	2.24	2.20	2.18	2.15	2.13	2.11	2.09	2.08	2.06	2.05	2.04	2.02
24	4.26	3.40	3.01	2.78	2.62	2.51	2.42	2.36	2.30	2.25	2.22	2.18	2.15	2.13	2.11	2.09	2.07	2.05	2.04	2.03	2.01	2.00
25	4.24	3.39	2.99	2.76	2.60	2.49	2.40	2.34	2.28	2.24	2.20	2.16	2.14	2.11	2.09	2.07	2.05	2.04	2.02	2.01	2.00	1.98
26	4.23	3.37	2.98	2.74	2.59	2.47	2.39	2.32	2.27	2.22	2.18	2.15	2.12	2.09	2.07	2.05	2.03	2.02	2.00	1.99	1.98	1.97
27	4.21	3.35	2.96	2.73	2.57	2.46	2.37	2.31	2.25	2.20	2.17	2.13	2.10	2.08	2.06	2.04	2.02	2.00	1.99	1.97	1.96	1.95
28	4.20	3.34	2.95	2.71	2.56	2.45	2.36	2.29	2.24	2.19	2.15	2.12	2.09	2.06	2.04	2.02	2.00	1.99	1.97	1.96	1.95	1.93
29	4.18	3.33	2.93	2.70	2.55	2.43	2.35	2.28	2.22	2.18	2.14	2.10	2.08	2.05	2.03	2.01	1.99	1.97	1.96	1.94	1.93	1.92
30	4.17	3.32	2.92	2.69	2.53	2.42	2.33	2.27	2.21	2.16	2.13	2.09	2.06	2.04	2.01	1.99	1.98	1.96	1.95	1.93	1.92	1.91
32	4.15	3.29	2.90	2.67	2.51	2.40	2.31	2.24	2.19	2.14	2.10	2.07	2.04	2.01	1.99	1.97	1.95	1.94	1.92	1.91	1.90	1.88
35	4.12	3.27	2.87	2.64	2.49	2.37	2.29	2.22	2.16	2.11	2.07	2.04	2.01	1.99	1.96	1.94	1.92	1.91	1.89	1.88	1.87	1.85
40	4.08	3.23	2.84	2.61	2.45	2.34	2.25	2.18	2.12	2.08	2.04	2.00	1.97	1.95	1.92	1.90	1.89	1.87	1.85	1.84	1.83	1.81
45	4.06	3.20	2.81	2.58	2.42	2.31	2.22	2.15	2.10	2.05	2.01	1.97	1.94	1.92	1.89	1.87	1.86	1.84	1.82	1.81	1.80	1.78
50	4.03	3.18	2.79	2.56	2.40	2.29	2.20	2.13	2.07	2.03	1.99	1.95	1.92	1.89	1.87	1.85	1.83	1.81	1.80	1.78	1.77	1.76
60	4.00	3.15	2.76	2.53	2.37	2.25	2.17	2.10	2.04	1.99	1.95	1.92	1.89	1.86	1.84	1.82	1.80	1.78	1.76	1.75	1.73	1.72
75	3.97	3.12	2.73	2.49	2.34	2.22	2.13	2.06	2.01	1.96	1.92	1.88	1.85	1.83	1.80	1.78	1.76	1.74	1.73	1.71	1.70	1.69
100	3.94	3.09	2.70	2.46	2.31	2.19	2.10	2.03	1.97	1.93	1.89	1.85	1.82	1.79	1.77	1.75	1.73	1.71	1.69	1.68	1.66	1.65
120	3.92	3.07	2.68	2.45	2.29	2.18	2.09	2.02	1.96	1.91	1.87	1.83	1.80	1.78	1.75	1.73	1.71	1.69	1.67	1.66	1.64	1.63
140	3.91	3.06	2.67	2.44	2.28	2.16	2.08	2.01	1.95	1.90	1.86	1.82	1.79	1.76	1.74	1.72	1.70	1.68	1.66	1.65	1.63	1.62
180	3.89	3.05	2.65	2.42	2.26	2.15	2.06	1.99	1.93	1.88	1.84	1.81	1.77	1.75	1.72	1.70	1.68	1.66	1.64	1.63	1.61	1.60
250	3.88	3.03	2.64	2.41	2.25	2.13	2.05	1.98	1.92	1.87	1.83	1.79	1.76	1.73	1.71	1.68	1.66	1.65	1.63	1.61	1.60	1.58
400	3.86	3.02	2.63	2.39	2.24	2.12	2.03	1.96	1.90	1.85	1.81	1.78	1.74	1.72	1.69	1.67	1.65	1.63	1.61	1.60	1.58	1.57
1000	3.85	3.00	2.61	2.38	2.22	2.11	2.02	1.95	1.89	1.84	1.80	1.76	1.73	1.70	1.68	1.65	1.63	1.61	1.60	1.58	1.57	1.55

Table F (cont.)

Denominator df	Numerator df																					
α = .05	23	24	25	26	27	28	29	30	32	35	40	45	50	60	75	100	120	140	180	250	400	1000
1	248.8	249.1	249.3	249.5	249.6	249.8	250.0	250.1	250.4	250.7	251.1	251.5	251.8	252.2	252.6	253.0	253.3	253.4	253.6	253.8	254.0	254.2
2	19.45	19.45	19.46	19.46	19.46	19.46	19.46	19.46	19.46	19.47	19.47	19.47	19.48	19.48	19.48	19.49	19.49	19.49	19.49	19.49	19.49	19.49
3	8.64	8.64	8.63	8.63	8.63	8.62	8.62	8.62	8.61	8.60	8.59	8.59	8.58	8.57	8.56	8.55	8.55	8.55	8.54	8.54	8.53	8.53
4	5.78	5.77	5.77	5.76	5.76	5.75	5.75	5.75	5.74	5.73	5.72	5.71	5.70	5.69	5.68	5.66	5.66	5.65	5.65	5.64	5.64	5.63
5	4.53	4.53	4.52	4.52	4.51	4.50	4.50	4.50	4.49	4.48	4.46	4.45	4.44	4.43	4.42	4.41	4.40	4.39	4.39	4.38	4.38	4.37
6	3.85	3.84	3.83	3.83	3.82	3.82	3.81	3.81	3.80	3.79	3.77	3.76	3.75	3.74	3.73	3.71	3.70	3.70	3.69	3.69	3.68	3.67
7	3.42	3.41	3.40	3.40	3.39	3.39	3.38	3.38	3.37	3.36	3.34	3.33	3.32	3.30	3.29	3.27	3.27	3.26	3.25	3.25	3.24	3.23
8	3.12	3.12	3.11	3.10	3.10	3.09	3.08	3.08	3.07	3.06	3.04	3.03	3.02	3.01	2.99	2.97	2.97	2.96	2.95	2.95	2.94	2.93
9	2.91	2.90	2.89	2.89	2.88	2.87	2.87	2.86	2.85	2.84	2.83	2.81	2.80	2.79	2.77	2.76	2.75	2.74	2.73	2.73	2.72	2.71
10	2.75	2.74	2.73	2.72	2.72	2.71	2.70	2.70	2.69	2.68	2.66	2.65	2.64	2.62	2.60	2.59	2.58	2.57	2.57	2.56	2.55	2.54
11	2.62	2.61	2.60	2.59	2.59	2.58	2.58	2.57	2.56	2.55	2.53	2.52	2.51	2.49	2.47	2.46	2.45	2.44	2.43	2.43	2.42	2.41
12	2.51	2.51	2.50	2.49	2.48	2.48	2.47	2.47	2.46	2.44	2.43	2.41	2.40	2.38	2.37	2.35	2.34	2.33	2.33	2.32	2.31	2.30
13	2.43	2.42	2.41	2.41	2.40	2.39	2.39	2.38	2.37	2.36	2.34	2.33	2.31	2.30	2.28	2.26	2.25	2.25	2.24	2.23	2.22	2.21
14	2.36	2.35	2.34	2.33	2.33	2.32	2.31	2.31	2.30	2.28	2.27	2.25	2.24	2.22	2.21	2.19	2.18	2.17	2.16	2.15	2.15	2.14
15	2.30	2.29	2.28	2.27	2.27	2.26	2.25	2.25	2.24	2.22	2.20	2.19	2.18	2.16	2.14	2.12	2.11	2.11	2.10	2.09	2.08	2.07
16	2.24	2.24	2.23	2.22	2.21	2.21	2.20	2.19	2.18	2.17	2.15	2.14	2.12	2.11	2.09	2.07	2.06	2.05	2.04	2.03	2.02	2.02
17	2.20	2.19	2.18	2.17	2.17	2.16	2.15	2.15	2.14	2.12	2.10	2.09	2.08	2.06	2.04	2.02	2.01	2.00	1.99	1.98	1.98	1.97
18	2.16	2.15	2.14	2.13	2.13	2.12	2.11	2.11	2.10	2.08	2.06	2.05	2.04	2.02	2.00	1.98	1.97	1.96	1.95	1.94	1.93	1.92
19	2.12	2.11	2.11	2.10	2.09	2.08	2.08	2.07	2.06	2.05	2.03	2.01	2.00	1.98	1.96	1.94	1.93	1.92	1.91	1.90	1.89	1.88
20	2.09	2.08	2.07	2.07	2.06	2.05	2.05	2.04	2.03	2.01	1.99	1.98	1.97	1.95	1.93	1.91	1.90	1.89	1.88	1.87	1.86	1.85
21	2.06	2.05	2.05	2.04	2.03	2.02	2.02	2.01	2.00	1.98	1.96	1.95	1.94	1.92	1.90	1.88	1.87	1.86	1.85	1.84	1.83	1.82
22	2.04	2.03	2.02	2.01	2.00	2.00	1.99	1.98	1.97	1.96	1.94	1.92	1.91	1.89	1.87	1.85	1.84	1.83	1.82	1.81	1.80	1.79
23	2.01	2.01	2.00	1.99	1.98	1.97	1.97	1.96	1.95	1.93	1.91	1.90	1.88	1.86	1.84	1.82	1.81	1.81	1.79	1.78	1.77	1.76
24	1.99	1.98	1.97	1.97	1.96	1.95	1.95	1.94	1.93	1.91	1.89	1.88	1.86	1.84	1.82	1.80	1.79	1.78	1.77	1.76	1.75	1.74
25	1.97	1.96	1.96	1.95	1.94	1.93	1.93	1.92	1.91	1.89	1.87	1.86	1.84	1.82	1.80	1.78	1.77	1.76	1.75	1.74	1.73	1.72
26	1.96	1.95	1.94	1.93	1.92	1.91	1.91	1.90	1.89	1.87	1.85	1.84	1.82	1.80	1.78	1.76	1.75	1.74	1.73	1.72	1.71	1.70
27	1.94	1.93	1.92	1.91	1.90	1.90	1.89	1.88	1.87	1.86	1.84	1.82	1.81	1.79	1.76	1.74	1.73	1.72	1.71	1.70	1.69	1.68
28	1.92	1.91	1.91	1.90	1.89	1.88	1.88	1.87	1.86	1.84	1.82	1.80	1.79	1.77	1.75	1.73	1.71	1.71	1.69	1.68	1.67	1.66
29	1.91	1.90	1.89	1.88	1.88	1.87	1.86	1.85	1.84	1.83	1.81	1.79	1.77	1.75	1.73	1.71	1.70	1.69	1.68	1.67	1.66	1.65
30	1.90	1.89	1.88	1.87	1.86	1.85	1.85	1.84	1.83	1.81	1.79	1.77	1.76	1.74	1.72	1.70	1.68	1.68	1.66	1.65	1.64	1.63
32	1.87	1.86	1.85	1.85	1.84	1.83	1.82	1.82	1.80	1.79	1.77	1.75	1.74	1.71	1.69	1.67	1.66	1.65	1.64	1.63	1.61	1.60
35	1.84	1.83	1.82	1.82	1.81	1.80	1.79	1.79	1.77	1.76	1.74	1.72	1.70	1.68	1.66	1.63	1.62	1.61	1.60	1.59	1.58	1.57
40	1.80	1.79	1.78	1.77	1.77	1.76	1.75	1.74	1.73	1.72	1.69	1.67	1.66	1.64	1.61	1.59	1.58	1.57	1.55	1.54	1.53	1.52
45	1.77	1.76	1.75	1.74	1.73	1.73	1.72	1.71	1.70	1.68	1.66	1.64	1.63	1.60	1.58	1.55	1.54	1.53	1.52	1.51	1.49	1.48
50	1.75	1.74	1.73	1.72	1.71	1.70	1.69	1.69	1.67	1.66	1.63	1.61	1.60	1.58	1.55	1.52	1.51	1.50	1.49	1.47	1.46	1.45
60	1.71	1.70	1.69	1.68	1.67	1.66	1.66	1.65	1.64	1.62	1.59	1.57	1.56	1.53	1.51	1.48	1.47	1.46	1.44	1.43	1.41	1.40
75	1.67	1.66	1.65	1.64	1.63	1.63	1.62	1.61	1.60	1.58	1.55	1.53	1.52	1.49	1.47	1.44	1.42	1.41	1.40	1.38	1.37	1.35
100	1.64	1.63	1.62	1.61	1.60	1.59	1.58	1.57	1.56	1.54	1.52	1.49	1.48	1.45	1.42	1.39	1.38	1.36	1.35	1.33	1.31	1.30
120	1.62	1.61	1.60	1.59	1.58	1.57	1.56	1.55	1.54	1.52	1.50	1.47	1.46	1.43	1.40	1.37	1.35	1.34	1.32	1.30	1.29	1.27
140	1.61	1.60	1.58	1.57	1.57	1.56	1.55	1.54	1.53	1.51	1.48	1.46	1.44	1.41	1.38	1.35	1.33	1.32	1.30	1.29	1.27	1.25
180	1.59	1.58	1.57	1.56	1.55	1.54	1.53	1.52	1.51	1.49	1.46	1.44	1.42	1.39	1.36	1.33	1.31	1.30	1.28	1.26	1.24	1.22
250	1.57	1.56	1.55	1.54	1.53	1.52	1.51	1.50	1.49	1.47	1.44	1.42	1.40	1.37	1.34	1.31	1.29	1.27	1.25	1.23	1.21	1.18
400	1.56	1.54	1.53	1.52	1.51	1.50	1.50	1.49	1.47	1.45	1.42	1.40	1.38	1.35	1.32	1.28	1.26	1.25	1.23	1.20	1.18	1.15
1000	1.54	1.53	1.52	1.51	1.50	1.49	1.48	1.47	1.46	1.43	1.41	1.38	1.36	1.33	1.30	1.26	1.24	1.22	1.20	1.17	1.14	1.11

Table F (cont.)

Denominator df / α = .1	1	2	3	4	5	6	7	8	9	10	11	12	13	14	15	16	17	18	19	20	21	22
	Numerator df																					
1	39.9	49.5	53.6	55.8	57.2	58.2	58.9	59.4	59.9	60.2	60.5	60.7	60.9	61.1	61.2	61.3	61.5	61.6	61.7	61.7	61.8	61.9
2	8.53	9.00	9.16	9.24	9.29	9.33	9.35	9.37	9.38	9.39	9.40	9.41	9.41	9.42	9.42	9.43	9.43	9.44	9.44	9.44	9.44	9.45
3	5.54	5.46	5.39	5.34	5.31	5.28	5.27	5.25	5.24	5.23	5.22	5.22	5.21	5.20	5.20	5.20	5.19	5.19	5.19	5.18	5.18	5.18
4	4.54	4.32	4.19	4.11	4.05	4.01	3.98	3.95	3.94	3.92	3.91	3.90	3.89	3.88	3.87	3.86	3.86	3.85	3.85	3.84	3.84	3.84
5	4.06	3.78	3.62	3.52	3.45	3.40	3.37	3.34	3.32	3.30	3.28	3.27	3.26	3.25	3.24	3.23	3.22	3.22	3.21	3.21	3.20	3.20
6	3.78	3.46	3.29	3.18	3.11	3.05	3.01	2.98	2.96	2.94	2.92	2.90	2.89	2.88	2.87	2.86	2.85	2.85	2.84	2.84	2.83	2.83
7	3.59	3.26	3.07	2.96	2.88	2.83	2.78	2.75	2.72	2.70	2.68	2.67	2.65	2.64	2.63	2.62	2.61	2.61	2.60	2.59	2.59	2.58
8	3.46	3.11	2.92	2.81	2.73	2.67	2.62	2.59	2.56	2.54	2.52	2.50	2.49	2.48	2.46	2.45	2.45	2.44	2.43	2.42	2.42	2.41
9	3.36	3.01	2.81	2.69	2.61	2.55	2.51	2.47	2.44	2.42	2.40	2.38	2.36	2.35	2.34	2.33	2.32	2.31	2.30	2.30	2.29	2.29
10	3.29	2.92	2.73	2.61	2.52	2.46	2.41	2.38	2.35	2.32	2.30	2.28	2.27	2.26	2.24	2.23	2.22	2.22	2.21	2.20	2.19	2.19
11	3.23	2.86	2.66	2.54	2.45	2.39	2.34	2.30	2.27	2.25	2.23	2.21	2.19	2.18	2.17	2.16	2.15	2.14	2.13	2.12	2.12	2.11
12	3.18	2.81	2.61	2.48	2.39	2.33	2.28	2.24	2.21	2.19	2.17	2.15	2.13	2.12	2.10	2.09	2.08	2.08	2.07	2.06	2.05	2.05
13	3.14	2.76	2.56	2.43	2.35	2.28	2.23	2.20	2.16	2.14	2.12	2.10	2.08	2.07	2.05	2.04	2.03	2.02	2.01	2.01	2.00	1.99
14	3.10	2.73	2.52	2.39	2.31	2.24	2.19	2.15	2.12	2.10	2.07	2.05	2.04	2.02	2.01	2.00	1.99	1.98	1.97	1.96	1.96	1.95
15	3.07	2.70	2.49	2.36	2.27	2.21	2.16	2.12	2.09	2.06	2.04	2.02	2.00	1.99	1.97	1.96	1.95	1.94	1.93	1.92	1.92	1.91
16	3.05	2.67	2.46	2.33	2.24	2.18	2.13	2.09	2.06	2.03	2.01	1.99	1.97	1.95	1.94	1.93	1.92	1.91	1.90	1.89	1.88	1.88
17	3.03	2.64	2.44	2.31	2.22	2.15	2.10	2.06	2.03	2.00	1.98	1.96	1.94	1.93	1.91	1.90	1.89	1.88	1.87	1.86	1.86	1.85
18	3.01	2.62	2.42	2.29	2.20	2.13	2.08	2.04	2.00	1.98	1.95	1.93	1.92	1.90	1.89	1.87	1.86	1.85	1.84	1.84	1.83	1.82
19	2.99	2.61	2.40	2.27	2.18	2.11	2.06	2.02	1.98	1.96	1.93	1.91	1.89	1.88	1.86	1.85	1.84	1.83	1.82	1.81	1.81	1.80
20	2.97	2.59	2.38	2.25	2.16	2.09	2.04	2.00	1.96	1.94	1.91	1.89	1.87	1.86	1.84	1.83	1.82	1.81	1.80	1.79	1.79	1.78
21	2.96	2.57	2.36	2.23	2.14	2.08	2.02	1.98	1.95	1.92	1.90	1.87	1.86	1.84	1.83	1.81	1.80	1.79	1.78	1.78	1.77	1.76
22	2.95	2.56	2.35	2.22	2.13	2.06	2.01	1.97	1.93	1.90	1.88	1.86	1.84	1.83	1.81	1.80	1.79	1.78	1.77	1.76	1.75	1.74
23	2.94	2.55	2.34	2.21	2.11	2.05	1.99	1.95	1.92	1.89	1.87	1.84	1.83	1.81	1.80	1.78	1.77	1.76	1.75	1.74	1.74	1.73
24	2.93	2.54	2.33	2.19	2.10	2.04	1.98	1.94	1.91	1.88	1.85	1.83	1.81	1.80	1.78	1.77	1.76	1.75	1.74	1.73	1.72	1.71
25	2.92	2.53	2.32	2.18	2.09	2.02	1.97	1.93	1.89	1.87	1.84	1.82	1.80	1.79	1.77	1.76	1.75	1.74	1.73	1.72	1.71	1.70
26	2.91	2.52	2.31	2.17	2.08	2.01	1.96	1.92	1.88	1.86	1.83	1.81	1.79	1.77	1.76	1.75	1.73	1.72	1.71	1.71	1.70	1.69
27	2.90	2.51	2.30	2.17	2.07	2.00	1.95	1.91	1.87	1.85	1.82	1.80	1.78	1.76	1.75	1.74	1.72	1.71	1.70	1.70	1.69	1.68
28	2.89	2.50	2.29	2.16	2.06	2.00	1.94	1.90	1.87	1.84	1.81	1.79	1.77	1.75	1.74	1.73	1.71	1.70	1.69	1.69	1.68	1.67
29	2.89	2.50	2.28	2.15	2.06	1.99	1.93	1.89	1.86	1.83	1.80	1.78	1.76	1.75	1.73	1.72	1.71	1.69	1.68	1.68	1.67	1.66
30	2.88	2.49	2.28	2.14	2.05	1.98	1.93	1.88	1.85	1.82	1.79	1.77	1.75	1.74	1.72	1.71	1.70	1.69	1.68	1.67	1.66	1.65
32	2.87	2.48	2.26	2.13	2.04	1.97	1.91	1.87	1.83	1.81	1.78	1.76	1.74	1.72	1.71	1.69	1.68	1.67	1.66	1.65	1.64	1.64
35	2.85	2.46	2.25	2.11	2.02	1.95	1.90	1.85	1.82	1.79	1.76	1.74	1.72	1.70	1.69	1.67	1.66	1.65	1.64	1.63	1.62	1.62
40	2.84	2.44	2.23	2.09	2.00	1.93	1.87	1.83	1.79	1.76	1.74	1.71	1.70	1.68	1.66	1.65	1.64	1.62	1.61	1.61	1.60	1.59
45	2.82	2.42	2.21	2.07	1.98	1.91	1.85	1.81	1.77	1.74	1.72	1.70	1.68	1.66	1.64	1.63	1.62	1.60	1.59	1.58	1.58	1.57
50	2.81	2.41	2.20	2.06	1.97	1.90	1.84	1.80	1.76	1.73	1.70	1.68	1.66	1.64	1.63	1.61	1.60	1.59	1.58	1.57	1.56	1.55
60	2.79	2.39	2.18	2.04	1.95	1.87	1.82	1.77	1.74	1.71	1.68	1.66	1.64	1.62	1.60	1.59	1.58	1.56	1.55	1.54	1.53	1.53
75	2.77	2.37	2.16	2.02	1.93	1.85	1.80	1.75	1.72	1.69	1.66	1.63	1.61	1.60	1.58	1.57	1.55	1.54	1.53	1.52	1.51	1.50
100	2.76	2.36	2.14	2.00	1.91	1.83	1.78	1.73	1.69	1.66	1.64	1.61	1.59	1.57	1.56	1.54	1.53	1.52	1.50	1.49	1.48	1.48
120	2.75	2.35	2.13	1.99	1.90	1.82	1.77	1.72	1.68	1.65	1.63	1.60	1.58	1.56	1.55	1.53	1.52	1.50	1.49	1.48	1.47	1.46
140	2.74	2.34	2.12	1.99	1.89	1.82	1.76	1.71	1.68	1.64	1.62	1.59	1.57	1.55	1.54	1.52	1.51	1.50	1.48	1.47	1.46	1.45
180	2.73	2.33	2.11	1.98	1.88	1.81	1.75	1.70	1.67	1.63	1.61	1.58	1.56	1.54	1.53	1.51	1.50	1.48	1.47	1.46	1.45	1.44
250	2.73	2.32	2.11	1.97	1.87	1.80	1.74	1.69	1.66	1.62	1.60	1.57	1.55	1.53	1.51	1.50	1.49	1.47	1.46	1.45	1.44	1.43
400	2.72	2.32	2.10	1.96	1.86	1.79	1.73	1.69	1.65	1.61	1.59	1.56	1.54	1.52	1.50	1.49	1.47	1.46	1.45	1.44	1.43	1.42
1000	2.71	2.31	2.09	1.95	1.85	1.78	1.72	1.68	1.64	1.61	1.58	1.55	1.53	1.51	1.49	1.48	1.46	1.45	1.44	1.43	1.42	1.41

Table F (cont.)

	Numerator df																					
$\alpha = .1$	23	24	25	26	27	28	29	30	32	35	40	45	50	60	75	100	120	140	180	250	400	1000
Denominator df																						
1	61.9	62.0	62.1	62.1	62.1	62.2	62.2	62.3	62.3	62.4	62.5	62.6	62.7	62.8	62.9	63.0	63.1	63.1	63.1	63.2	63.2	63.3
2	9.45	9.45	9.45	9.45	9.45	9.46	9.46	9.46	9.46	9.46	9.47	9.47	9.47	9.47	9.48	9.48	9.48	9.48	9.49	9.49	9.49	9.49
3	5.18	5.18	5.17	5.17	5.17	5.17	5.17	5.17	5.17	5.16	5.16	5.16	5.15	5.15	5.15	5.14	5.14	5.14	5.14	5.14	5.14	5.1
4	3.83	3.83	3.83	3.83	3.82	3.82	3.82	3.82	3.81	3.81	3.80	3.80	3.80	3.79	3.78	3.78	3.78	3.77	3.77	3.77	3.77	3.76
5	3.19	3.19	3.19	3.18	3.18	3.18	3.18	3.17	3.17	3.16	3.16	3.15	3.15	3.14	3.13	3.13	3.12	3.12	3.12	3.11	3.11	3.11
6	2.82	2.82	2.81	2.81	2.81	2.81	2.80	2.80	2.80	2.79	2.78	2.77	2.77	2.76	2.75	2.75	2.74	2.74	2.74	2.73	2.73	2.72
7	2.58	2.58	2.57	2.57	2.56	2.56	2.56	2.56	2.55	2.54	2.54	2.53	2.52	2.51	2.51	2.50	2.49	2.49	2.49	2.48	2.48	2.47
8	2.41	2.40	2.40	2.40	2.39	2.39	2.39	2.38	2.38	2.37	2.36	2.35	2.35	2.34	2.33	2.32	2.32	2.31	2.31	2.30	2.30	2.30
9	2.28	2.28	2.27	2.27	2.26	2.26	2.26	2.25	2.25	2.24	2.23	2.22	2.22	2.21	2.20	2.19	2.18	2.18	2.18	2.17	2.17	2.16
10	2.18	2.18	2.17	2.17	2.17	2.16	2.16	2.16	2.15	2.14	2.13	2.12	2.12	2.11	2.10	2.09	2.08	2.08	2.07	2.07	2.06	2.06
11	2.11	2.10	2.10	2.09	2.09	2.08	2.08	2.08	2.07	2.06	2.05	2.04	2.04	2.03	2.02	2.01	2.00	2.00	1.99	1.99	1.98	1.98
12	2.04	2.04	2.03	2.03	2.02	2.02	2.01	2.01	2.01	2.00	1.99	1.98	1.97	1.96	1.95	1.94	1.93	1.93	1.92	1.92	1.91	1.91
13	1.99	1.98	1.98	1.97	1.97	1.96	1.96	1.96	1.95	1.94	1.93	1.92	1.92	1.90	1.89	1.88	1.88	1.87	1.87	1.86	1.86	1.85
14	1.94	1.94	1.93	1.93	1.92	1.92	1.92	1.91	1.91	1.90	1.89	1.88	1.87	1.86	1.85	1.83	1.83	1.82	1.82	1.81	1.81	1.80
15	1.90	1.90	1.89	1.89	1.88	1.88	1.88	1.87	1.87	1.86	1.85	1.84	1.83	1.82	1.80	1.79	1.79	1.78	1.78	1.77	1.76	1.76
16	1.87	1.87	1.86	1.86	1.85	1.85	1.84	1.84	1.83	1.82	1.81	1.80	1.79	1.78	1.77	1.76	1.75	1.75	1.74	1.73	1.73	1.72
17	1.84	1.84	1.83	1.83	1.82	1.82	1.81	1.81	1.80	1.79	1.78	1.77	1.76	1.75	1.74	1.73	1.72	1.71	1.71	1.70	1.70	1.69
18	1.82	1.81	1.80	1.80	1.80	1.79	1.79	1.78	1.78	1.77	1.75	1.74	1.74	1.72	1.71	1.70	1.69	1.69	1.68	1.67	1.67	1.66
19	1.79	1.79	1.78	1.78	1.77	1.77	1.76	1.76	1.75	1.74	1.73	1.72	1.71	1.70	1.69	1.67	1.67	1.66	1.65	1.65	1.64	1.64
20	1.77	1.77	1.76	1.76	1.75	1.75	1.74	1.74	1.73	1.72	1.71	1.70	1.69	1.68	1.66	1.65	1.64	1.64	1.63	1.62	1.62	1.61
21	1.75	1.75	1.74	1.74	1.73	1.73	1.72	1.72	1.71	1.70	1.69	1.68	1.67	1.66	1.64	1.63	1.62	1.62	1.61	1.60	1.60	1.59
22	1.74	1.73	1.73	1.72	1.72	1.71	1.71	1.70	1.69	1.68	1.67	1.66	1.65	1.64	1.63	1.61	1.60	1.60	1.59	1.59	1.58	1.57
23	1.72	1.72	1.71	1.70	1.70	1.69	1.69	1.69	1.68	1.67	1.66	1.64	1.64	1.62	1.61	1.59	1.59	1.58	1.57	1.57	1.56	1.55
24	1.71	1.70	1.70	1.69	1.69	1.68	1.68	1.67	1.66	1.65	1.64	1.63	1.62	1.61	1.59	1.58	1.57	1.57	1.56	1.55	1.54	1.54
25	1.70	1.69	1.68	1.68	1.67	1.67	1.66	1.66	1.65	1.64	1.63	1.62	1.61	1.59	1.58	1.56	1.56	1.55	1.54	1.54	1.53	1.52
26	1.68	1.68	1.67	1.67	1.66	1.66	1.65	1.65	1.64	1.63	1.61	1.60	1.59	1.58	1.57	1.55	1.54	1.54	1.53	1.52	1.52	1.51
27	1.67	1.67	1.66	1.65	1.65	1.64	1.64	1.64	1.63	1.62	1.60	1.59	1.58	1.57	1.55	1.54	1.53	1.53	1.52	1.51	1.50	1.50
28	1.66	1.66	1.65	1.64	1.64	1.63	1.63	1.63	1.62	1.61	1.59	1.58	1.57	1.56	1.54	1.53	1.52	1.51	1.51	1.50	1.49	1.48
29	1.65	1.65	1.64	1.63	1.63	1.62	1.62	1.62	1.61	1.60	1.58	1.57	1.56	1.55	1.53	1.52	1.51	1.50	1.50	1.49	1.48	1.47
30	1.64	1.64	1.63	1.63	1.62	1.62	1.61	1.61	1.60	1.59	1.57	1.56	1.55	1.54	1.52	1.51	1.50	1.49	1.49	1.48	1.47	1.46
32	1.63	1.62	1.62	1.61	1.60	1.60	1.59	1.59	1.58	1.57	1.56	1.54	1.53	1.52	1.50	1.49	1.48	1.47	1.47	1.46	1.45	1.44
35	1.61	1.60	1.60	1.59	1.58	1.58	1.57	1.57	1.56	1.55	1.53	1.52	1.51	1.50	1.48	1.47	1.46	1.45	1.44	1.43	1.43	1.42
40	1.58	1.57	1.57	1.56	1.56	1.55	1.55	1.54	1.53	1.52	1.51	1.49	1.48	1.47	1.45	1.43	1.42	1.42	1.41	1.40	1.39	1.38
45	1.56	1.55	1.55	1.54	1.53	1.53	1.52	1.52	1.51	1.50	1.48	1.47	1.46	1.44	1.43	1.41	1.40	1.39	1.38	1.37	1.37	1.36
50	1.54	1.54	1.53	1.52	1.52	1.51	1.51	1.50	1.49	1.48	1.46	1.45	1.44	1.42	1.41	1.39	1.38	1.37	1.36	1.35	1.34	1.33
60	1.52	1.51	1.50	1.50	1.49	1.49	1.48	1.48	1.47	1.45	1.44	1.42	1.41	1.40	1.38	1.36	1.35	1.34	1.33	1.32	1.31	1.30
75	1.49	1.49	1.48	1.47	1.47	1.46	1.45	1.45	1.44	1.43	1.41	1.40	1.38	1.37	1.35	1.33	1.32	1.31	1.30	1.29	1.27	1.26
100	1.47	1.46	1.45	1.45	1.44	1.43	1.43	1.42	1.41	1.40	1.38	1.37	1.35	1.34	1.32	1.29	1.28	1.27	1.26	1.25	1.24	1.22
120	1.46	1.45	1.44	1.43	1.43	1.42	1.41	1.41	1.40	1.39	1.37	1.35	1.34	1.32	1.30	1.28	1.26	1.26	1.24	1.23	1.22	1.20
140	1.45	1.44	1.43	1.42	1.42	1.41	1.41	1.40	1.39	1.38	1.36	1.34	1.33	1.31	1.29	1.26	1.25	1.24	1.23	1.22	1.20	1.19
180	1.43	1.43	1.42	1.41	1.40	1.40	1.39	1.39	1.38	1.36	1.34	1.33	1.32	1.29	1.27	1.25	1.23	1.22	1.21	1.20	1.18	1.16
250	1.42	1.41	1.41	1.40	1.39	1.39	1.38	1.37	1.36	1.35	1.33	1.31	1.30	1.28	1.26	1.23	1.22	1.21	1.19	1.18	1.16	1.14
400	1.41	1.40	1.39	1.39	1.38	1.37	1.37	1.36	1.35	1.34	1.32	1.30	1.29	1.26	1.24	1.21	1.20	1.19	1.17	1.16	1.14	1.12
1000	1.40	1.39	1.38	1.38	1.37	1.36	1.36	1.35	1.34	1.32	1.30	1.29	1.27	1.25	1.23	1.20	1.18	1.17	1.15	1.13	1.11	1.08

Table X
Values of χ^2_α

Right-tail probability		0.10	0.05	0.025	0.01	0.005
	df					
	1	2.706	3.841	5.024	6.635	7.879
	2	4.605	5.991	7.378	9.210	10.597
	3	6.251	7.815	9.348	11.345	12.838
	4	7.779	9.488	11.143	13.277	14.860
	5	9.236	11.070	12.833	15.086	16.750
	6	10.645	12.592	14.449	16.812	18.548
	7	12.017	14.067	16.013	18.475	20.278
	8	13.362	15.507	17.535	20.090	21.955
	9	14.684	16.919	19.023	21.666	23.589
	10	15.987	18.307	20.483	23.209	25.188
	11	17.275	19.675	21.920	24.725	26.757
	12	18.549	21.026	23.337	26.217	28.300
	13	19.812	22.362	24.736	27.688	29.819
	14	21.064	23.685	26.119	29.141	31.319
	15	22.307	24.996	27.488	30.578	32.801
	16	23.542	26.296	28.845	32.000	34.267
	17	24.769	27.587	30.191	33.409	35.718
	18	25.989	28.869	31.526	34.805	37.156
	19	27.204	30.143	32.852	36.191	38.582
	20	28.412	31.410	34.170	37.566	39.997
	21	29.615	32.671	35.479	38.932	41.401
	22	30.813	33.924	36.781	40.290	42.796
	23	32.007	35.172	38.076	41.638	44.181
	24	33.196	36.415	39.364	42.980	45.559
	25	34.382	37.653	40.647	44.314	46.928
	26	35.563	38.885	41.923	45.642	48.290
	27	36.741	40.113	43.195	46.963	49.645
	28	37.916	41.337	44.461	48.278	50.994
	29	39.087	42.557	45.722	59.588	52.336
	30	40.256	43.773	46.979	50.892	53.672
	40	51.805	55.759	59.342	63.691	66.767
	50	63.167	67.505	71.420	76.154	79.490
	60	74.397	79.082	83.298	88.381	91.955
	70	85.527	90.531	95.023	100.424	104.213
	80	96.578	101.879	106.628	112.328	116.320
	90	107.565	113.145	118.135	124.115	128.296
	100	118.499	124.343	129.563	135.811	140.177

Table *Z*

Areas under the standard Normal curve

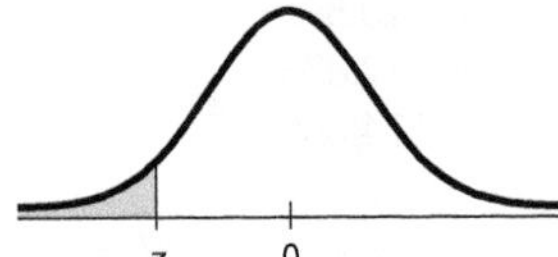

Second decimal place in z										
0.09	*0.08*	*0.07*	*0.06*	*0.05*	*0.04*	*0.03*	*0.02*	*0.01*	*0.00*	z
									0.0000†	−3.9
0.0001	0.0001	0.0001	0.0001	0.0001	0.0001	0.0001	0.0001	0.0001	0.0001	−3.8
0.0001	0.0001	0.0001	0.0001	0.0001	0.0001	0.0001	0.0001	0.0001	0.0001	−3.7
0.0001	0.0001	0.0001	0.0001	0.0001	0.0001	0.0001	0.0001	0.0002	0.0002	−3.6
0.0002	0.0002	0.0002	0.0002	0.0002	0.0002	0.0002	0.0002	0.0002	0.0002	−3.5
0.0002	0.0003	0.0003	0.0003	0.0003	0.0003	0.0003	0.0003	0.0003	0.0003	−3.4
0.0003	0.0004	0.0004	0.0004	0.0004	0.0004	0.0004	0.0005	0.0005	0.0005	−3.3
0.0005	0.0005	0.0005	0.0006	0.0006	0.0006	0.0006	0.0006	0.0007	0.0007	−3.2
0.0007	0.0007	0.0008	0.0008	0.0008	0.0008	0.0009	0.0009	0.0009	0.0010	−3.1
0.0010	0.0010	0.0011	0.0011	0.0011	0.0012	0.0012	0.0013	0.0013	0.0013	−3.0
0.0014	0.0014	0.0015	0.0015	0.0016	0.0016	0.0017	0.0018	0.0018	0.0019	−2.9
0.0019	0.0020	0.0021	0.0021	0.0022	0.0023	0.0023	0.0024	0.0025	0.0026	−2.8
0.0026	0.0027	0.0028	0.0029	0.0030	0.0031	0.0032	0.0033	0.0034	0.0035	−2.7
0.0036	0.0037	0.0038	0.0039	0.0040	0.0041	0.0043	0.0044	0.0045	0.0047	−2.6
0.0048	0.0049	0.0051	0.0052	0.0054	0.0055	0.0057	0.0059	0.0060	0.0062	−2.5
0.0064	0.0066	0.0068	0.0069	0.0071	0.0073	0.0075	0.0078	0.0080	0.0082	−2.4
0.0084	0.0087	0.0089	0.0091	0.0094	0.0096	0.0099	0.0102	0.0104	0.0107	−2.3
0.0110	0.0113	0.0116	0.0119	0.0122	0.0125	0.0129	0.0132	0.0136	0.0139	−2.2
0.0143	0.0146	0.0150	0.0154	0.0158	0.0162	0.0166	0.0170	0.0174	0.0179	−2.1
0.0183	0.0188	0.0192	0.0197	0.0202	0.0207	0.0212	0.0217	0.0222	0.0228	−2.0
0.0233	0.0239	0.0244	0.0250	0.0256	0.0262	0.0268	0.0274	0.0281	0.0287	−1.9
0.0294	0.0301	0.0307	0.0314	0.0322	0.0329	0.0336	0.0344	0.0351	0.0359	−1.8
0.0367	0.0375	0.0384	0.0392	0.0401	0.0409	0.0418	0.0427	0.0436	0.0446	−1.7
0.0455	0.0465	0.0475	0.0485	0.0495	0.0505	0.0516	0.0526	0.0537	0.0548	−1.6
0.0559	0.0571	0.0582	0.0594	0.0606	0.0618	0.0630	0.0643	0.0655	0.0668	−1.5
0.0681	0.0694	0.0708	0.0721	0.0735	0.0749	0.0764	0.0778	0.0793	0.0808	−1.4
0.0823	0.0838	0.0853	0.0869	0.0885	0.0901	0.0918	0.0934	0.0951	0.0968	−1.3
0.0985	0.1003	0.1020	0.1038	0.1056	0.1075	0.1093	0.1112	0.1131	0.1151	−1.2
0.1170	0.1190	0.1210	0.1230	0.1251	0.1271	0.1292	0.1314	0.1335	0.1357	−1.1
0.1379	0.1401	0.1423	0.1446	0.1469	0.1492	0.1515	0.1539	0.1562	0.1587	−1.0
0.1611	0.1635	0.1660	0.1685	0.1711	0.1736	0.1762	0.1788	0.1814	0.1841	−0.9
0.1867	0.1894	0.1922	0.1949	0.1977	0.2005	0.2033	0.2061	0.2090	0.2119	−0.8
0.2148	0.2177	0.2206	0.2236	0.2266	0.2296	0.2327	0.2358	0.2389	0.2420	−0.7
0.2451	0.2483	0.2514	0.2546	0.2578	0.2611	0.2643	0.2676	0.2709	0.2743	−0.6
0.2776	0.2810	0.2843	0.2877	0.2912	0.2946	0.2981	0.3015	0.3050	0.3085	−0.5
0.3121	0.3156	0.3192	0.3228	0.3264	0.3300	0.3336	0.3372	0.3409	0.3446	−0.4
0.3483	0.3520	0.3557	0.3594	0.3632	0.3669	0.3707	0.3745	0.3783	0.3821	−0.3
0.3859	0.3897	0.3936	0.3974	0.4013	0.4052	0.4090	0.4129	0.4168	0.4207	−0.2
0.4247	0.4286	0.4325	0.4364	0.4404	0.4443	0.4483	0.4522	0.4562	0.4602	−0.1
0.4641	0.4681	0.4721	0.4761	0.4801	0.4840	0.4880	0.4920	0.4960	0.5000	−0.0

†For $z \leq -3.90$, the areas are 0.0000 to four decimal places.

Table *Z* (cont.)

Areas under the standard Normal curve

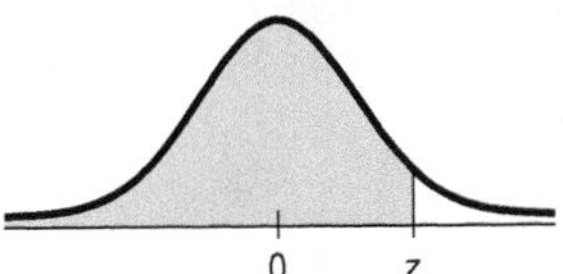

	Second decimal place in z									
z	0.00	0.01	0.02	0.03	0.04	0.05	0.06	0.07	0.08	0.09
0.0	0.5000	0.5040	0.5080	0.5120	0.5160	0.5199	0.5239	0.5279	0.5319	0.5359
0.1	0.5398	0.5438	0.5478	0.5517	0.5557	0.5596	0.5636	0.5675	0.5714	0.5753
0.2	0.5793	0.5832	0.5871	0.5910	0.5948	0.5987	0.6026	0.6064	0.6103	0.6141
0.3	0.6179	0.6217	0.6255	0.6293	0.6331	0.6368	0.6406	0.6443	0.6480	0.6517
0.4	0.6554	0.6591	0.6628	0.6664	0.6700	0.6736	0.6772	0.6808	0.6844	0.6879
0.5	0.6915	0.6950	0.6985	0.7019	0.7054	0.7088	0.7123	0.7157	0.7190	0.7224
0.6	0.7257	0.7291	0.7324	0.7357	0.7389	0.7422	0.7454	0.7486	0.7517	0.7549
0.7	0.7580	0.7611	0.7642	0.7673	0.7704	0.7734	0.7764	0.7794	0.7823	0.7852
0.8	0.7881	0.7910	0.7939	0.7967	0.7995	0.8023	0.8051	0.8078	0.8106	0.8133
0.9	0.8159	0.8186	0.8212	0.8238	0.8264	0.8289	0.8315	0.8340	0.8365	0.8389
1.0	0.8413	0.8438	0.8461	0.8485	0.8508	0.8531	0.8554	0.8577	0.8599	0.8621
1.1	0.8643	0.8665	0.8686	0.8708	0.8729	0.8749	0.8770	0.8790	0.8810	0.8830
1.2	0.8849	0.8869	0.8888	0.8907	0.8925	0.8944	0.8962	0.8980	0.8997	0.9015
1.3	0.9032	0.9049	0.9066	0.9082	0.9099	0.9115	0.9131	0.9147	0.9162	0.9177
1.4	0.9192	0.9207	0.9222	0.9236	0.9251	0.9265	0.9279	0.9292	0.9306	0.9319
1.5	0.9332	0.9345	0.9357	0.9370	0.9382	0.9394	0.9406	0.9418	0.9429	0.9441
1.6	0.9452	0.9463	0.9474	0.9484	0.9495	0.9505	0.9515	0.9525	0.9535	0.9545
1.7	0.9554	0.9564	0.9573	0.9582	0.9591	0.9599	0.9608	0.9616	0.9625	0.9633
1.8	0.9641	0.9649	0.9656	0.9664	0.9671	0.9678	0.9686	0.9693	0.9699	0.9706
1.9	0.9713	0.9719	0.9726	0.9732	0.9738	0.9744	0.9750	0.9756	0.9761	0.9767
2.0	0.9772	0.9778	0.9783	0.9788	0.9793	0.9798	0.9803	0.9808	0.9812	0.9817
2.1	0.9821	0.9826	0.9830	0.9834	0.9838	0.9842	0.9846	0.9850	0.9854	0.9857
2.2	0.9861	0.9864	0.9868	0.9871	0.9875	0.9878	0.9881	0.9884	0.9887	0.9890
2.3	0.9893	0.9896	0.9898	0.9901	0.9904	0.9906	0.9909	0.9911	0.9913	0.9916
2.4	0.9918	0.9920	0.9922	0.9925	0.9927	0.9929	0.9931	0.9932	0.9934	0.9936
2.5	0.9938	0.9940	0.9941	0.9943	0.9945	0.9946	0.9948	0.9949	0.9951	0.9952
2.6	0.9953	0.9955	0.9956	0.9957	0.9959	0.9960	0.9961	0.9962	0.9963	0.9964
2.7	0.9965	0.9966	0.9967	0.9968	0.9969	0.9970	0.9971	0.9972	0.9973	0.9974
2.8	0.9974	0.9975	0.9976	0.9977	0.9977	0.9978	0.9979	0.9979	0.9980	0.9981
2.9	0.9981	0.9982	0.9982	0.9983	0.9984	0.9984	0.9985	0.9985	0.9986	0.9986
3.0	0.9987	0.9987	0.9987	0.9988	0.9988	0.9989	0.9989	0.9989	0.9990	0.9990
3.1	0.9990	0.9991	0.9991	0.9991	0.9992	0.9992	0.9992	0.9992	0.9993	0.9993
3.2	0.9993	0.9993	0.9994	0.9994	0.9994	0.9994	0.9994	0.9995	0.9995	0.9995
3.3	0.9995	0.9995	0.9995	0.9996	0.9996	0.9996	0.9996	0.9996	0.9996	0.9997
3.4	0.9997	0.9997	0.9997	0.9997	0.9997	0.9997	0.9997	0.9997	0.9997	0.9998
3.5	0.9998	0.9998	0.9998	0.9998	0.9998	0.9998	0.9998	0.9998	0.9998	0.9998
3.6	0.9998	0.9998	0.9999	0.9999	0.9999	0.9999	0.9999	0.9999	0.9999	0.9999
3.7	0.9999	0.9999	0.9999	0.9999	0.9999	0.9999	0.9999	0.9999	0.9999	0.9999
3.8	0.9999	0.9999	0.9999	0.9999	0.9999	0.9999	0.9999	0.9999	0.9999	0.9999
3.9	1.0000[†]									

[†]For $z \geq 3.90$, the areas are 1.0000 to four decimal places.

Table T

Values of t_α

Two tails: $-t_{\alpha/2}$, 0, $t_{\alpha/2}$, $\frac{\alpha}{2}$ in each tail

One tail: 0, t_α, α

Two-tail probability	0.20	0.10	0.05	0.02	0.01	
One-tail probability	0.10	0.05	0.025	0.01	0.005	
df						df
1	3.078	6.314	12.706	31.821	63.657	1
2	1.886	2.920	4.303	6.965	9.925	2
3	1.638	2.353	3.182	4.541	5.841	3
4	1.533	2.132	2.776	3.747	4.604	4
5	1.476	2.015	2.571	3.365	4.032	5
6	1.440	1.943	2.447	3.143	3.707	6
7	1.415	1.895	2.365	2.998	3.499	7
8	1.397	1.860	2.306	2.896	3.355	8
9	1.383	1.833	2.262	2.821	3.250	9
10	1.372	1.812	2.228	2.764	3.169	10
11	1.363	1.796	2.201	2.718	3.106	11
12	1.356	1.782	2.179	2.681	3.055	12
13	1.350	1.771	2.160	2.650	3.012	13
14	1.345	1.761	2.145	2.624	2.977	14
15	1.341	1.753	2.131	2.602	2.947	15
16	1.337	1.746	2.120	2.583	2.921	16
17	1.333	1.740	2.110	2.567	2.898	17
18	1.330	1.734	2.101	2.552	2.878	18
19	1.328	1.729	2.093	2.539	2.861	19
20	1.325	1.725	2.086	2.528	2.845	20
21	1.323	1.721	2.080	2.518	2.831	21
22	1.321	1.717	2.074	2.508	2.819	22
23	1.319	1.714	2.069	2.500	2.807	23
24	1.318	1.711	2.064	2.492	2.797	24
25	1.316	1.708	2.060	2.485	2.787	25
26	1.315	1.706	2.056	2.479	2.779	26
27	1.314	1.703	2.052	2.473	2.771	27
28	1.313	1.701	2.048	2.467	2.763	28
29	1.311	1.699	2.045	2.462	2.756	29
30	1.310	1.697	2.042	2.457	2.750	30
32	1.309	1.694	2.037	2.449	2.738	32
35	1.306	1.690	2.030	2.438	2.725	35
40	1.303	1.684	2.021	2.423	2.704	40
45	1.301	1.679	2.014	2.412	2.690	45
50	1.299	1.676	2.009	2.403	2.678	50
60	1.296	1.671	2.000	2.390	2.660	60
75	1.293	1.665	1.992	2.377	2.643	75
100	1.290	1.660	1.984	2.364	2.626	100
120	1.289	1.658	1.980	2.358	2.617	120
140	1.288	1.656	1.977	2.353	2.611	140
180	1.286	1.653	1.973	2.347	2.603	180
250	1.285	1.651	1.969	2.341	2.596	250
400	1.284	1.649	1.966	2.336	2.588	400
1000	1.282	1.646	1.962	2.330	2.581	1000
∞	1.282	1.645	1.960	2.326	2.576	∞
Confidence levels	80%	90%	95%	98%	99%	

Wilcoxon Rank Sum Test

The table gives some lower and upper tail probabilities, under H_0, for T = sum of the ranks for the smaller of two independent samples (or for either sample if $n_1 = n_2$). Probabilities shown at the top are either $P(T \leq crit)$ or $P(T \geq crit)$.

For some sample sizes, for practical reasons related to the discreteness of the distribution, we have chosen to display tabled (critical) values that have a slightly higher probability (in parentheses) than what is written at the top of the table, but which are as close as possible. Otherwise, these probabilities are accurate (within a small rounding factor) or are conservative (high).

	Prob to left		//		Prob to right	
n_1/n_2	.01	.025	.05 //	.05	.025	.01
3/3			6	15		
3/4		6(.029)	7(.057)	17(.057)	18(.029)	
3/5	6(.018)	7(.036)	8(.07)	19(.07)	20(.036)	21(.018)
3/6	6	7	8	22	23	24
3/7	6	8(.033)	9(.058)	24(.058)	25(.033)	27
3/8	7	8	9	27	28	29
3/9	7	8	10	29	31	32
3/10	8(.014)	9	11(.056)	31(.056)	33	34(.014)
4/4	10(.014)	11(.029)	12(.057)	24(.057)	25(.029)	26(.014)
4/5	10	11	13(.056)	27(.056)	29	30
4/6	11	12	14(.057)	30(.057)	32	33
4/7	12	13	15(.055)	33(.055)	35	36
4/8	12	14	16(.055)	36(.055)	38	40
4/9	13	15	17(.053)	39(.053)	41	43
4/10	14	16(.027)	18(.053)	42(.053)	44(.027)	46
5/5	16	18(.028)	19	36	37(.028)	39
5/6	17	19	20	40	41	43
5/7	18	20	22(.053)	43(.053)	45	47
5/8	19	21	23	47	49	51
5/9	20	22	25(.056)	50(.056)	53	55
5/10	21	24(.028)	26	54	56(.028)	59
6/6	24	26	28	50	52	54
6/7	26	28	30	54	56	58
6/8	27	29	32(.054)	58(.054)	61	63
6/9	28	31	34(.057)	62(.057)	65	68
6/10	30	33(.028)	35	67	69(.028)	72
7/7	34	37(.027)	39	66	68(.027)	71
7/8	36	39(.027)	41	71	73(.027)	76
7/9	38	41(.027)	43	76	78(.027)	81
7/10	39	43(.028)	46(.054)	80(.054)	83(.028)	87
8/8	46	49	52	84	87	90
8/9	48	51	55(.057)	89(.057)	93	96
8/10	50	54(.027)	57	95	98(.027)	102
9/9	59	63	66	105	108	112
9/10	62	66(.027)	69	111	114(.027)	118
10/10	74	79	83(.053)	127(.053)	131	136

If sample sizes are off the table, use the normal approximation for the distribution of T = rank sum of sample i, where $E(T) = n_i(n_1 + n_2 + 1)/2$ and $V(T) = n_1 n_2(n_1 + n_2 + 1)/12$. (In the presence of many ties, this approach will be conservative.)

Adapted from Table A.6, Hollander & Wolf, *Nonparametric Statistical Methods*, Wiley (1973).

Tables for the Wilcoxon Signed Rank Test

Below are critical values for the two-sided signed rank test. Reject H_0 at indicated level if T = smaller of $\{T+, T-\} \leq$ critical value. Divide the given p-value by 2 to obtain p-values for one-sided alternatives, using either $T+$ or $T-$ as appropriate to your alternative (a small value for which of these will support your H_a?).

n	$p \leq .1$	$p \leq .05$	$p \leq .02$	$p \leq .01$	$p \leq .002$
5	0				
6	2	0			
7	3	2	0		
8	5	3	1	0	
9	8	5	3	1	
10	10	8	5	3	0
11	13	10	7	5	1
12	17	13	9	7	2
13	21	17	12	9	4
14	25	21	15	12	6
15	30	25	19	15	8
16	35	29	23	19	11
17	41	34	27	23	14
18	47	40	32	27	18
19	53	46	37	32	21
20	60	52	43	37	26
21	67	58	49	42	30
22	75	65	55	48	35
23	83	73	62	54	40
24	91	81	69	61	45
25	100	89	76	68	51

If off the table, use the normal approximation for the distribution of the rank sum T, with $E(T) = n(n+1)/4$ and $V(T) = n(n+1)(2n+1)/24$

Adapted from Table A.6, Hollander & Wolf, *Nonparametric Statistical Methods*, Wiley (1973).

Assumptions for Inference	And the Conditions that Support or Override them
Proportions (z)	
• **One sample**	
1. Individuals are independent.	1. SRS and $<$ 10% of the population.
2. Sample is sufficiently large.	2. Successes and failures $\geq$ 10.
• **Two sample**	
1. Samples are independent.	1. (Think about how the data were collected.)
2. Data in each sample are independent.	2. Both are SRSs and $<$ 10% of populations OR random allocation.
3. Both samples are sufficiently large.	3. Successes and failures $\geq$ 10 for both.
Means (t)	
• **One sample** (df $= n - 1$)	
1. Individuals are independent.	1. SRS and $<$ 10% of the population.
2. Population has a Normal model.	2. Histogram is unimodal and symmetric.*
• **Two independent Samples** (df from technology)	
1. Samples are independent.	1. (Think about the design.)
2. Data in each sample are independent.	2. SRSs and $<$ 10% OR random allocation.
3. Both populations are Normal.	3. Both histograms are unimodal and symmetric.*
• **Matched pairs** (df $= n - 1$)	
1. Data are matched; n pairs.	1. (Think about the design.)
2. Individuals are independent.	2. SRSs and $<$ 10% OR random allocation.
3. Population of differences is Normal.	3. Histogram of differences is unimodal and symmetric.
Distributions/Association (χ^2)	
• **Goodness of fit** (df = # of cells −1; one variable, one sample compared with population model)	
1. Data are counts.	1. (Are they?)
2. Data in sample are independent.	2. SRS and $<$ 10% of the population.
3. Sample is sufficiently large.	3. All expected counts $\geq$ 5.
• **Homogeneity** [df $= (r - 1)(c - 1)$; samples from many populations compared on one variable]	
1. Data are counts.	1. (Are they?)
2. Data in samples are independent.	2. SRSs and $<$ 10% OR random allocation.
3. Groups are sufficiently large.	3. All expected counts $\geq$ 5.
• **Independence** [df $= (r - 1)(c - 1)$; sample from one population classified on two variables]	
1. Data are counts.	1. (Are they?)
2. Data are independent.	2. SRSs and $<$ 10% of the population.
3. Group is sufficiently large.	3. All expected counts $\geq$ 5.
Regression with *R* predictors (t, $df = n - k - 1$)	
• **Association** of each quantitative predictor with the response variable	
1. Form of relationship is linear.	1. Scatterplots of y against each x are straight enough. Scatterplot of residuals against predicted values shows no special structure.
2. Errors are independent.	2. No apparent pattern in plot of residuals against predicted values.
3. Variability of errors is constant.	3. Plot of residuals against predicted values has constant spread, doesn't "thicken."
4. Error follow a Normal model.	4. Histogram of residuals is approximately unimodal and symmetric, or Normal probability plot is resonably straight.*
Analysis of Variance (F, df depends on number of factors and number of levels in each.)	
• **Equality** of the mean response across levels of categorical predictors	
1. Additive Model (if there are 2 factors with no interaction term).	1. Interaction plot shows parallel lines (otherwise include an interaction term if possible).
2. Independent errors.	2. Randomized experiment or other suitable randomization.
3. Equal variance across treatment levels.	3. Plot of residuals against predicted values has constant spread, Box plots (partial boxplots for 2 factors) show similar spreads.
4. Error follow a Normal model.	4. Histogram of residuals is unimodal and symmetric, or Normal probability plot is resonably straight.

(*Less critical as n increases)